Preface

The *Handbook of Structural Concrete* is intended as an international reference work on the current state of the art and science of structural concrete, to provide authoritative coverage of the principles, procedures and data relating to

(a) fundamental behaviour

(b) design, analysis, construction, maintenance and demolition

(c) design philosophies and their expression in codes of practice.

The *Handbook* has been designed to meet the needs of practising civil and structural engineers, consulting engineering and contracting firms, concrete materials producers and users, research institutes, universities and colleges. It is presented in six parts: Introduction; Materials; Design and Analysis; Construction; Structures; Practical Considerations. The 41 chapters have been contributed by an international group of authors, including some of the most distinguished engineers in the field of structural concrete.

Within the broad subject of civil and structural engineering, structural concrete is an area in which new and powerful concepts in design and analysis have emerged in recent years and in which changes have continually occurred in practice and procedure, as evidenced by the regular appearance of new codes of practice in various countries. The *Handbook of Structural Concrete* is a comprehensive reference work which reflects such emergence of new concepts and the changes in practice and procedure. A product of international efforts, the *Handbook* is a symbol of international cooperation and a recognition of the fact that the continual exchange of ideas and knowledge between engineers in different parts of our world has contributed significantly to the emergence of new concepts and the advances in structural concrete engineering practice.

The Editors wish to thank the chapter authors for their high quality contributions and for the thoroughness with which the manuscripts have been prepared. They also thank their publishers, and especially Mr Ian Pringle and Mr John Hindley of Pitmans who contributed significantly to the successful development of the *Handbook*. In the course of the work the Editors were greatly helped by the encouragement and advice from friends and colleagues, particularly Professor Edward G Nawy Chairman of the Department of Civil and Environmental Engineering at Rutgers University and Emeritus Professor R C Coates of the University of Nottingham and President of the Institution of Civil Engineers 1978–79.

F K Kong   R H Evans   Edward Cohen   Frederic Roll

v

Conversion Factors

SI Units to/from US Customary Units

Length

$$1 \text{ m} = 3.281 \text{ ft} = 1.094 \text{ yd}$$
$$1 \text{ mm} = 0.03937 \text{ in}$$
$$1 \text{ ft} = 0.3048 \text{ m}$$
$$1 \text{ in} = 25.4 \text{ mm} \ (\text{exactly})$$

Area

$$1 \text{ m}^2 = 10.76 \text{ ft}^2 = 1.196 \text{ yd}^2$$
$$1 \text{ mm}^2 = 0.001550 \text{ in}^2$$
$$1 \text{ ft}^2 = 0.0929 \text{ m}^2$$
$$1 \text{ in}^2 = 645.2 \text{ mm}^2$$

Volume

$$1 \text{ m}^3 = 35.31 \text{ ft}^3 = 1.308 \text{ yd}^3$$
$$1 \text{ mm}^3 = 61.02 \times 10^{-6} \text{ in}^3$$
$$1 \text{ litre} = 0.2642 \text{ US gallon}$$
$$1 \text{ ft}^3 = 0.02832 \text{ m}^3$$
$$1 \text{ in}^3 = 16.39 \times 10^3 \text{ mm}^3$$
$$1 \text{ yd}^3 = 0.7646 \text{ m}^3$$

Moment of inertia (Second Moment of Area)

$$1 \text{ mm}^4 = 2.403 \times 10^{-6} \text{ in}^4$$
$$1 \text{ in}^4 = 0.4162 \times 10^6 \text{ mm}^4$$

Mass

$$1 \text{ kg} = 2.205 \text{ lb}$$
$$1 \text{ lb} = 0.4536 \text{ kg}$$

Density

$$1 \text{ kg/m}^3 = 0.06243 \text{ lb/ft}^3$$
$$1 \text{ lb/ft}^3 = 16.02 \text{ kg/m}^3$$

Force (Loading)

$$1 \text{ N} = 0.2248 \text{ lb}$$
$$1 \text{ kN} = 0.2248 \text{ kip}$$
$$1 \text{ kN/m} = 0.06582 \text{ kip/ft}$$
$$1 \text{ kN/m}^2 = 0.02088 \text{ kip/ft}^2$$
$$1 \text{ lb} = 4.448 \text{ N}$$
$$1 \text{ kip} = 4.448 \text{ kN}$$
$$1 \text{ kip/ft} = 14.59 \text{ kN/m}$$
$$1 \text{ kip/ft}^2 = 47.88 \text{ kN/m}^2$$

Stress

$$1 \text{ MPa} \ (= 1 \text{ MN/m}^2) = 145.0 \text{ psi}$$
$$1 \text{ N/mm}^2 = 145.0 \text{ psi}$$
$$1 \text{ kN/mm}^2 = 145.0 \text{ ksi}$$
$$1 \text{ psi} = 6.895 \times 10^{-3} \text{ MPa}$$
$$= 6.895 \times 10^{-3} \text{ N/mm}^2$$

Moment (Torque)

$$1 \text{ Nm} = 0.7376 \text{ lb-ft}$$
$$1 \text{ kNm} = 0.7376 \text{ kip-ft}$$
$$1 \text{ lb-ft} = 1.356 \text{ Nm}$$
$$1 \text{ kip-ft} = 1.356 \text{ kNm}$$

Contents

Part I Introduction

1 Looking into the future

T Y Lin   Philip Y Chow

T. Y. Lin International, San Francisco, USA

Contents

Summary

This chapter ventures a glimpse into the future of structural concrete from the vantage points based on past achievements and experiences, as well as the future needs of the human society, including the dual problems of population growth and depletion of resources.

Three main components of structural engineering are examined: materials, designs, and construction.

Future improvements in materials include improvement of concrete itself, and of its reinforcement and additives. The development of plant and equipment will progress concurrently, probably at a faster rate, in the following areas: forming, precasting, large span structures, and offshore construction.

Design and analysis in the future will be dominated by computer technology, the considerations for optimization, the study into the behavior of materials, the establishment of a uniform international code and the effects on environmental protection and aesthetics.

A scenario of the future is presented including land structures, e.g. buildings, bridges, offshore, ocean and arctic structures, and energy structures such as prestressed concrete pressure vessels and ocean energy conversion vessels.

1 Introduction

Crystal-balling is a mystic art that defies logical explanation. Yet one could try to conjure up a picture of the future of structural concrete on a logical basis that is founded on past achievements and experiences, for the future of structural concrete is linked to the future of a society that is shaped by many factors, each representing a trend of its own. Thus, the future of human society has to do with its needs and aspirations which themselves change with the changes of social values and constraints, under societal pressures brought on by population growth and the depletion of known resources. These factors are mostly interrelated and continuously interacting one with another in a process of evolution to which mankind adapts itself. The future is also an extrapolation of the past. In looking into the future, therefore, one would do well to look first at the past.

2 The engineering past

Historically, engineering technology progresses in an evolutionary cycle that encompasses motivation, invention, development, experience and improvement. From simple beginnings, where inventions were mostly the product of chance encounters, e.g. the fabled Newton and the falling apple that begot the law of the equilibrium of forces, technological advances today are often the result of a long process of iterations, trials and errors, each but a building block among many that eventually makes up the magnificent edifice. And because of the multiplicity of modern technology, where development in one area can significantly impact other areas, a technical discipline today tends to stand less and less alone in the expanding technological family.

Structural concrete engineering is no exception. Already in some of the newer applications, structural concrete is considered not only for its strength and durability, but also for its other properties that are not normally associated with its traditional functions, such as its insulating properties against heat and radioactivity.

Let us take a look at the past and the present of structural concrete through its three main components: materials, design and construction. Since much of these will be described in other chapters of this book, we will focus our attention only on the major items.

2.1 Materials

The basic concrete materials—cement, aggregates and mild and high-strength steels—have been continuously improved to reach the present high level of quality, consistency and dependability. The availability of high-strength steel has led to the development of prestressed concrete, initiated by Freyssinet in France, and the opening of a whole new field of structural concrete. By transforming structural concrete from a state of passive combination of steel and concrete into an active state, prestressed concrete technique produces a much more efficient material combination that makes better use of the high tensile strength of steel and the compressive strength of concrete.

The development of light-weight concrete (LWC) represents another milestone in the advance of concrete technology. This and other new materials offer extraordinary qualities to meet special needs. Some of these are discussed below.

2.1.1 Light-weight concrete

LWC is not new. It is normally made of expanded shale, weighs around 1842 kg/m³ (115 lb/ft³) and has compressive strength of up to 33 MPa (5000 psi). What is new is a wide range of improved, if more costly, products that are even lighter and stronger than the conventional LWC. One such material is the polymer-filled concrete developed for marine application by the US Navy's Civil Engineering Laboratory (CEL) in Port Hueneme, California. The rationale for this invention was that, since the strength of LWC is limited by the voids in the concrete, it would be improved (by about 26%) if the voids are filled with polymeric materials having the same density as sea water. Another is a LWC that is made stronger by curing with a vacuum process. As yet, the cost of producing these special concretes is high. CEL's polymer-filled concrete, for example, was reported to cost about 30% more than normal LWC.

2.1.2 Fiber-reinforced concrete

FRC has been used commerically for more than a decade, and has been the subject of institutional studies (e.g. ACI Committee 544). Its use in the past has been irregular, and limited generally to the manufacture of products used in corrosive environments. The greatest advantage of FRC is its toughness and its resistance against corrosion, especially when non-metallic fibers are used. FRC can be pumped and can be advantageously used in restricted areas that cannot be reinforced in the normal manner.

2.1.3 Polymer concrete (PC) and polymer-impregnated concrete (PIC)

PC and PIC are included in the discussion of structural concrete in view of their potentials as structural materials for the future. They differ from the conventional structural concrete only in using a polymer as the cementing or impregnating agent.

Since its development by the Brookhaven National Laboratory and the Bureau of Reclamation in the US in 1965, PIC has opened up tremendous possibilities in expanding the use of structural concrete. Certain polymers have the effect of increasing the tensile and compressive strengths of ordinary concrete three to four times, improving its freeze–thaw behavior, and reducing its permeability to less than one-third the normal values.

PIC is made by impregnating the concrete that has been cleaned and dried, with a monomer, such as methyl methacrylate (MMA) or styrene (S). Impregnation is by pressure injection or, in the case of a flat, level surface, simply by soaking with monomer for several hours. The monomer is then polymerized. The classification and the properties of PIC are shown in Table 2–1.

The use of PC or PIC has been limited to special applications where its higher cost is justified. Another deterrent is the troublesome procedure and the special equipment required for its application. The workers have to be protected from the

Table 2-1   Characteristics of MMA and S-impregnated concrete[a]

Properties	Control	MMA-impregnated concrete		S-impregnated concrete	
		R process	T–C process	R process	T–C process
Compressive strength, psi	5267	20 255	18 161	14 710	9986
Modulus of elasticity, 10^6 psi	3.5	6.3	6.2	7.7	7.4
Tensile strength, psi	416	1 627	1 508	1 205	840
Water absorption, per cent	6.4	1.08	0.34	0.51	0.70
Abrasion, in	0.0497	0.0163	0.0147	0.040	0.0365
Water permeability 10^{-4} ft/yr	5.3	0.8	1.4	NA*	1.5
Freeze–thaw durability No. of cycles (% weight loss)	490(25.0)	750(4.0)	750(0.5)	620(6.5)	620(0.5)
Hardness impact (L) Hammer	32.0	55.3	52.0	48.2	50.1
Corrosion by sulphate solution 300-day exposure % expansion	0.144	0	NA[b]	0	NA[b]
Acid corrosion in 15% HCl. 84-day exposure, % weight loss	10.4	3.64	3.49	5.5	4.2

[a] Source: Frondiston-Yannas, S. A. and Dietz, G. L., Economic feasibility of polymer-impregnated concrete as a building material, J. Prestressed Concr. Inst., Vol. 22, No. 4, 1977.
[b] Data not available.

dangers of coming into contact with the chemical or inhaling the fume during casting operation.

2.1.4 Epoxy adhesives for concrete rehabilitation

The development of efficient concrete adhesives has contributed to the greater use of structural concrete. A number of epoxy adhesives are commercially available today for grouting joints and cracks in the concrete, which would make the grouted portion even stronger than the concrete that is uncracked. Construction methods have been developed in conjunction with the availability of these adhesives, e.g. the segmental construction method for long, precast and prestressed girders. Precast concrete sections or segments are assembled, then epoxied and prestressed together to form the complete structure. Epoxy injection is an important part of structural repairs today. Many damaged structures that might require total replacement at one time, can now be completely reinstated by epoxy injection, often without disrupting normal operations.

2.1.5 Flowing concrete

Flowing concrete is a term used in describing concrete that exhibits exceptional workability without strength loss by the use of superplasticizers. Superplasticizers made of sulphonated formaldehyde condensates, etc., offer many advantages to the users of structural concrete: low manpower requirement for concreting operation, ease of placement, better finishes, and accessibility to congested areas. The use of superplasticizers is reported to be popular in Japan, the USA and European countries, particularly Germany.

2.1.6 Fly ash

Fly ash has been used as concrete material for a long time. As early as the third century BC, the Romans combined it with lime and water to make cement for their concrete. Structures that were built with fly-ash concrete still stand as historic monuments today, e.g. the Coliseum in Rome.

The rediscovery and the use of fly ash in recent years have gained momentum with the increased use of coal and the production of fly ash by power plants. Much research and study into the properties of fly ash have been carried out and its proper use has been promoted by organizations, such as the National Ash Association of the USA. Researchers on fly ash have reported its advantages of higher strength, better workability and higher sulphate resistance.

2.2 Design

Engineering design practice, both in the method of analysis and in the tools with which the analysis is performed, has undergone remarkable changes in the last two decades. With powerful computers becoming commonly available, structures can now be analyzed to solve complex problems involving three-dimensional, time-dependent and non-linear conditions. What used to be laboriously computed by the slide-rule or calculator in days or weeks, can now be produced with better accuracy and reliability by computers in a matter of minutes. Already computerization is taking place in the engineering offices, performing many otherwise time-consuming tasks. The day will soon arrive when the entire engineering production process will be entrusted to an integrated computer system that receives instructions from one end and produces the finished product—design, analysis, drafting, specifications, cost estimation and contract documents—at the other.

2.3   Construction

Progress in engineering construction has been keeping pace with related technological advances. Construction methods are becoming more sophisticated, better controlled, and construction plants and equipment are becoming bigger, faster, more specialized, efficient and powerful. Computerization is also taking place in the construction field, in the form of electronic controls of construction plants and equipment, and increasing automation of construction activities, particularly in the production of building materials and components.

Major construction plant and equipment for structural concrete consist of lift-crane, jacks, concrete mixing and delivery equipment, formwork and scaffolding, and prestressing equipment. Lift-cranes and jacks for hoisting and pulling heavy loads that are available for the project, would determine the size and weight of pay loads, and the speed of construction. For this reason, construction plant and equipment tend to grow in size and power. Some of the heaviest equipment today include Manitowoc's crawler-mounted Model 6000W crane that has a maximum lifting capacity of 500 ton, and VSL's center-hole jack that pulls 580 ton. With this equipment, it is now possible to fabricate the entire roof structure of a large building on the ground, a hangar for instance, then lift it up to its final position several meters above the ground. This procedure has proven to be very economical and efficient in construction cost and time.

Concrete production and delivery systems have similarly been improved. Mixing plants are bigger, faster and more automatic. Concrete in major projects is commonly delivered by concrete pumps that are capable of pumping concrete to a height of over 160 m (530 ft) at pressures of the order of 10 MPa (1500 psi). Concrete may be delivered by truck-mounted pump and placing boom, or by a stationary placing boom that is separate from the pumping system. The latter can be placed in a central location at the construction site, and has larger coverage than a truck-mounted boom.

Formwork for concrete has likewise undergone much improvement to achieve economy and speedy construction. These are accomplished three ways: (1) by increasing the repetitive use of the forms, e.g. in precast construction; (2) by reducing the time taken to erect and remove formwork, e.g. fly-form or table-form; and (3) by slip-forming. Slip-forming is probably the most economical method to build tall, constant-section structures, such as a grain silo.

Concrete is now commonly used in large-span structures, particularly structures that are designed to support heavy roof loads, e.g. the Moscone Convention Center in San Francisco, which was built to support a 1 m (3 ft) layer of top soil. The Center (Fig. 2-1) houses a 25 500 m^2 (275 000 ft^2) exhibition hall with a column-free area of 84 m (275 ft) by 244 m (800 ft). The roof is supported by a series of 16 arches arranged in pairs. The thrust from each arch is carried by 4800 t (5300 short tons) of post-tensioning force in a tie located in the foundation mat (Fig. 2-2).

Figure 2-1   Inside of the exhibition hall of Moscone Convention Center, San Francisco   (Structural engineers: T. Y. Lin International; architects: Hellmuth, Obata & Kassabaum)

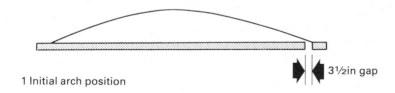

1 Initial arch position

3½in gap

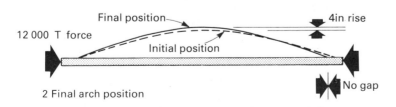

Final position

4in rise

12 000 T force

Initial position

2 Final arch position

No gap

Figure 2-2   Stressing of tied arches for exhibition hall

3   Societal needs and pressures

Human society is the product of human endeavors, which in turn reflect the needs and the pressures to which the society is subject. Many still believe that the most important factors in the shaping of the society are population growth, the depletion of resources and the human spirit of adventure. The three are really interrelated to some extent. Thus, the growth of population is one of the reasons for the depletion of resources on earth.

3.1   Population growth

Population growth gives rise to societal pressures that manifest themselves in economic, political, social and religious strife. It is perhaps the single most important factor behind the redistribution of population on earth. Many mass movements of people are recorded in history, a recent one being the exodus of refugees from Vietnam. It was Thomas R. Malthus who said that nature has a way of controlling population growth by providing limited subsistence, which causes wars and famines that decimate the population when it grows too fast. Perhaps. But he did not reckon with the ingenuity of Man in reducing natural death and improving food productivity by leaps and bounds. The combination of these two factors has brought the equilibrium between population and subsistence to a higher level, perhaps aggravating the problem should an imbalance occur. Even with intensive effort at family planning, the world population is growing at the rate of 1.8% each year, and will double the present figure of 3.3×10^9 people in 39 years. Increased population means increased demands for land and resources, forcing the expansion of established human frontiers into areas previously considered undesirable for human habitation. Today, supported by modern technology, and goaded by economic gains, human colonies have begun to appear in some of the most arid and frigid regions of the world.

Another relatively untouched area that may be ripe for human conquest is the ocean. With literally untold riches to be plucked from the bottom of the ocean, coupled with the rapid development of deep-sea tools and equipment that is now taking place, the day will not be far off when human colonization of the ocean floor becomes a matter of routine.

3.2   Depletion of resources

Resources on earth, particularly energy resources, have been depleting at a fast rate due to population growth, and a high per capita consumption rate needed to sustain an affluent society. According to statistics, worldwide consumption of energy has more than doubled itself every 15 years since 1925.

Energy shortage has spurred massive efforts by industrialized countries to find and develop alternative sources of energy. Some of these efforts have direct bearing on the use of structural concrete, e.g. as the floating platform for the ocean thermal energy conversion power plant (OTEC), or as pressure vessels for coal conversion plants.

Almost as serious as the depletion of oil and gas is the shortage of fresh water. Since water is continuously being replenished by nature, water shortage is only relative to the increasing demand due to population and industrial growths. Water is now reclaimed from the sea through desalination processes. It is envisaged that ice resources in Arctic regions may also be tapped in the world of tomorrow.

Some researchers have taken comfort in the ingenuity and ability of Man to survive and overcome whatever difficulties that may come his way. Has he not, in the short space of half a century, harnessed the atom, produced the computer, travelled in outer space and increased the productivity of his farmland ten-fold? Some believe the world is, in fact, on the verge of a technological revolution that will transform the present consumptive industrial society into an information-based society, where the depleting source of energy, oil, is replaced by the microchip as the new raw material. The advent of genetic engineering will bring about changes that will drastically change the way of life as we know it today. It will enable plants to flourish in hostile environments, and turn desert into farmland. Bacteria will be put to work to produce synthetic hydrocarbons for energy. The mastery of the genetic chain will do wonders to our resources. The world will, in fact, be heading toward plenty, instead of scarcity and want.

3.3 The effect on structural concrete

Notwithstanding the divergence of expert opinion on the effect of population growth and depletion of resources on the future of human society, it is clear that human society is poised to enter a vastly different world that can either be an exhausted and depleted place where conservation engineering, the desalination of sea water and population control are the orders of the day, or a thriving world of microchips, computers and an integrated system of production, communication and distribution. In either case, structural concrete will be there to play a substantial role.

3.4 Technical breakthroughs and constraints

The rate of technical progress can be accelerated or retarded by the occurrence of extraordinary circumstances that may or may not be foreseen. Accelerative circumstances are contributed by technological breakthroughs or new discoveries. Conversely, environmental protection and economic depression will slow down technological advance.

Some constraints are artificially inspired, e.g. the development of a deep-sea mining industry in international waters will require international agreement that can be drawn out for a long time with uncertain results.

Technical breakthroughs will occur at an increasing rate in technologically advanced countries, as new development in one area can be quickly capitalized, forcing the developmental pace of other areas. The whole process is fuelled by the manpower and money that are being poured into research efforts, and by the efficient communication system that has greatly narrowed the gap between development and application.

4 Development in the foreseeable future

Technological development must keep pace with the development and the needs of society. Because of the multiplicity and the breadth of technological frontiers today, technological advance in one area will impact another area, and vice versa. Technological spin-offs from space researches, e.g. the microchip, is a case in point. As a result, many industries are now enjoying the benefits of cheaper and better computers and space satellites in their operation.

In the foreseeable future, structural concrete can be expected to play a big part in various new areas that are now undergoing intensive development. In the energy field, structural concrete is already the candidate material for the construction of ocean energy conversion plants, such as the OTEC plant that converts solar energy into electrical energy (Fig. 4-1), and the DAM-ATOLL plant that converts wave energy and the large reactor pressure vessels for coal conversion. In the growing field of water recovery from the sea, initial efforts have been made to use structural concrete in an unprecedentedly large evaporator structure with production capacities of $250\,000\,\mathrm{m}^3/\mathrm{d}$ (66 Mgal/d). (Fig. 4-2.)

Figure 4-1 Spar-shaped ocean thermal energy conversion power plant by Lockheed Missiles Company (Structural consultants: T. Y. Lin International)

Figure 4-2 250 000 m³/d (66 Mgal/d) prestressed concrete desalination plant proposed for Saudi Arabia by Bechtel Corporation (Structural consultants: T. Y. Lin International)

Let us see what kind of materials and equipment will make these forthcoming developments possible.

4.1 Materials

Materials for structural concrete will undoubtedly continue to be maintained at high standards, or improved. Materials will be developed to meet special requirements created by the new industries mentioned before. These will be discussed under the various headings below.

4.1.1 Concrete

Concrete will be made stronger and lighter in a continuous effort to expand the use of structural concrete in long-span and high-rise structures. Although a concrete strength of 34 MPa (5000 psi) is commonly used in the conventional structures, ultimate strengths of 61 or 68 MPa (9000 or 10 000 psi) can be produced, at a price. Light-weight concrete of 1840 kg/m³ (115 pcf) density and 34 MPa (5000 psi) ultimate strength is already available, if more costly than the ordinary hard-rock concrete.

As concrete becomes lighter and stronger, concrete bridges will become still longer and concrete buildings taller. Concrete in new applications, such as the deep-sea, Arctic or pressure-vessel structures, may require special treatment to give it special properties to meet the new requirements. One of the more important properties is the impermeability of concrete to pressure penetration. Others include the enhanced resistance against fire, corrosion, fatigue and freeze–thaw cycles.

A number of new developments are already taking place that will impart to the concrete the required special properties. Fibrous concrete, Polymer concrete and polymer-impregnated concrete are a few that have been mentioned earlier.

4.1.2   Reinforcement

Steel reinforcement and prestressing steels have reached a high level of development and are therefore not expected to see much improvement in the foreseeable future. Steel could be made with special properties required by a project, e.g. enhanced resistance against corrosion, and against high and low temperatures.

The development of non-metallic reinforcement might be the next step in the evolution of reinforcement technology. Reinforcement for ferro-cement, i.e. layers of wire mesh impregnated with mortar, could also be another area of development.

4.1.3   Concrete additives

More and better purpose-specific additives will be developed to meet new demands in the forthcoming era that ushers in the high-pressure concrete vessels, the deep-sea habitats, and the Arctic concrete oil-production platforms. There will be additives that make concrete more resistant against pressure and heat, more water-tight and workable, and less susceptible to creep and shrinkage strains.

Additives, such as superplasticizer that greatly increases workability without loss of strength, are already available commercially. So are polymer-impregnated concrete and polymer concrete, which can be used to increase the toughness and the resistance against pressure penetration.

Structural concrete has also to improve its anti-corrosion properties in its service in chemical plants. Cladding with PIC panels could be an answer, if the panels can be protected from excessive heat.

4.2   Plant and equipment

Plant and equipment for the foreseeable future may be considered in the following categories:

   (a) Forming and concreting in situ
   (b) Precast construction
   (c) Large-span and high-rise construction
   (d) Offshore construction

4.2.1   Forming and concreting in situ

The technique will change with the advance of technology in the years ahead. One example is the greater use of computer techniques. New plant and equipment on site could be expected to be more automatically controlled, producing, handling and delivering concrete faster, and further from the mixing plant. Improvements will be concentrated on shortening construction time. Thus, vertical and horizontal slip-forming will find increased application. So will construction methods that eliminate rigid formwork altogether, e.g. ferro-cement that is 'formed' by the steel mesh itself, and the compressed-air form used to construct shell-roof structures over large areas. In bridges, traveling forms mounted on launching trusses will find wide application.

4.2.2 Precast construction

Precast work has its origin in an attempt to speed up construction. It will no doubt continue to expand in this role in the foreseeable future, particularly with the availability of larger plants and equipment. Precast sections tend to increase in variety, configuration and size. The future structure, for instance, could be simply an assembly of finished precast units which are bound together by stressing, welding or grouting to form the complete building.

For a central precasting facility, the size and shape of the precast pieces are restricted by public roads that provide access to the site. To obviate this difficulty, the 'mobile' precast plant will be developed and widely used.

A related area of precast construction, but on an international scale, is the fabrication of an entire industrial plant in steel or concrete in an industrially advanced country, and transporting it on barges to an offshore or coastal site for installation. There have been many instances of industrial plants constructed this way. Most were for the oil-rich countries in the Middle East, and for the remote, environmentally unfavorable regions, such as the petroleum production facilities in Prudhoe Bay in Alaska.

4.2.3 Large-span structures

Here again, improvement objectives will be centered on constructing these large, unsupported spans for indoor stadiums, convention centers, etc., more quickly and economically. Development based on prefabrication technique is envisaged. This may consist of constructing the structure on the ground, then lifting and installing it in its final position, as mentioned before.

This method has been practiced sporadically with make-shift equipment in the past. It was given a big boost by the development of large-capacity, center-hole lifting jacks in Switzerland that could lift roof structures weighing thousands of tons.

4.2.4 Offshore construction

Offshore structures are now entirely fabricated on shore. When completed, they are carried on barges or towed in a floating mode to the site for installation. For most steel production or other platforms, offshore construction consists simply of driving piles and assembling the prefabricated sections on site. Site work for the concrete gravity platforms is even simpler, i.e. securing the platform after instal-ling by ballasting, grouting the voids at the bearing surface, and depositing, if necessary, anti-scour rock armour around the periphery of the platform. These construction procedures are not expected to change much, but will find wider application in the foreseeable future.

However, further along in the future will be the development of a new set of construction techniques and equipment that enable Man to carry out actual construction work on the sea floor itself, in his advance on the 'inner space'.

5 Design and analysis

The design and analysis of concrete structures will enter a new era of sophistication with the help of (a) the common availability of computerized design aids; (b) better understanding of the behavior of materials; (c) the development of world standards; (d) the coming of age of the science of optimization; (e) better appreciation and enforcement of environmental protection and aesthetics requirements; and (f) the development of a global design concept. These are discussed separately in the following.

5.1 Computerized design

The next decade or two will see a proliferation of the uses of computers in all phases of design and analysis. Computerized design and analysis will become standard practice, especially for the more conventional types. Computer programs will become the common tool for most structural designs, replacing the design handbooks of today. At the same time, engineers and the public will come to understand better the limitations, shortcomings and the pitfalls of an over-reliance on computers. One direct result is the inability of the engineer to function independently of the computer, make rational decisions, and relate to the physical situation. He does not see the forest, only the trees. There is another shortcoming in over-emphasizing computer techniques in the training of engineers. The more time they spend with computers means less time in acquiring a basic understanding of the physical behavior of structures, particularly concrete structures.

Aided by computers, the analysis of concrete structures will become more sophisticated. It will take into consideration complex parameters involving multidimensional conditions and time-dependent behavior of composite materials, and the ability to seek out optimum designs. The development of sophisticated programs will be coupled with enlightened thinking and realistic concepts, such as partial prestressing methods, that can better cope with the analysis of cracked concrete acting in conjunction with its reinforcement in the plastic range.

5.2 Behavior of materials

The behavior of materials will have to be better understood in order to achieve a better design. Much has been accomplished in the recent past. The permeability of concrete, its resistance against fire and fatigue, its behavior under cryogenic conditions, the mechanics of concrete cracks, and the phenomenon of autogeneous healing in humid conditions, have all been subjects of intensive studies. However, much still remains to be accomplished. The engineer has, for instance, yet to learn of the behaviour of reinforcing bars in composite action with concrete, their bond to the concrete, the behavior across concrete cracks, and the changes in composition action between steel and concrete, with time. Better understanding of materials will make it possible to combine prestressed and reinforced concrete into one integrated design entity, and to better provide for the effects of the other dimensions in the design.

5.3 Uniform standards and codes

The evolution of a world code of practice on structural concrete, the process of which has begun with recommendations and standards set up by international organizations, such as the Fédération Internationale de la Précontrainte (FIP) and the International Standards Organization, will one day bring it to reality. The standardization of engineering practices is a matter of some urgency, in view of the international character of the modern engineering projects, e.g. the multinational activities in oil-producing areas in the Middle East, and the North Sea. The proliferation of nuclear installations is another case in point. The failure of a nuclear installation in one country can inflict upon another country the same damage as where the failure occurs.

The world code may specify performance rather than empirical values, allowing adaptations to local conditions, in line with an increasing awareness of the pitfalls of a code that is made up of fixed criteria. Even though criteria are quantitively set for a given structure, they should be flexible enough to permit a justifiable and judicious adjustment of the numbers specified in the code. Only the basic criteria—safety, economy, performance, serviceability and environmental fitness—should remain unchanged.

5.4 The science of optimization

The development of the science of optimization with the help of computers, both from physical and mathematical standpoints, will enable engineers to make better use of the expertise, materials, manpower and time that combine to make a better, more efficient and economic structure.

Optimization has evolved as an efficient method to cope with the modern project which involves many disciplines of expertise, each of which poses its own conditions on the project, and all require to be assessed from a global point of view. As pointed out before, in the growing complexity and magnitude of the technological family, no one discipline may stand apart from the other disciplines. All must work together in a trade-off process in the best total interest of the project.

The future engineer will have to accomplish more than structural designs. Like a band conductor, he will have to blend the basic purposes of the project with related societal and human considerations in a grand optimized design.

5.5 Environmental protection and aesthetics

Society will place increasing emphasis on the protection of the environment and the aesthetic aspect of the structures. More and more, engineers will realize that they must work with planners, architects, environmentalists and the public, to ensure that their structures will enhance environmental values. Rating methods will be devised to evaluate the aesthetic merits of various alternative schemes, their relative costs and cost–benefit ratios. It should be noted in this regard that the enhancement of environmental protection and aesthetics is not necessarily accompanied by an increase of cost. Indeed, some of the most aesthetic structures in the past had been achieved with savings in construction costs due to the rationality and simplicity of the design.

5.6 Global design

Finally, from the progress in the various areas described above will emerge the concept of the global design of structures that is based on the combination, collaboration and coordination of every discipline and interest that are involved in a project. The idea is akin to the optimization of a design by adjusting the requirements of the various participating disciplines, but on a grander scale. It sprang from the principle that since everybody involved with the project exerts an influence on its outcome, it would be best to bring everybody in at the developmental stage. Global design will include and require the early input and cooperation from the financiers, developers, approving authorities, builders, operators, as well as the various professional disciplines and the specialists associated with the project. Its implementation will require the establishment of a comprehensive control and coordination organization that operates with the help of computers and closed-circuit communication systems.

6 A scenario of the future

The future of structural concrete depends on the relative prominence of the influencing factors as described before. If the continuation of the present trends and progress can be assumed, the most significant influence will be exerted by an accelerated advance on all technological fronts that will bring society into a world of new materials, new techniques, automation, productivity, and a new way of life. Structural concrete will be fully integrated with other materials and disciplines in the construction of complex and massive projects, and in the development of new use areas. It will not be sufficient for the future structural engineer to solve a problem from the structural standpoint alone. He will have greater appreciation of the ramifications of a structural problem, and seeks an optimized solution with others in the project team.

In both the design office and in the field, the computers will take over all routine tasks, including: engineering analysis, design drawing, specification writing, cost estimating, construction scheduling, procuring, and plant, equipment and manpower control. The engineer will still exercise judgements and make decisions, but will rely on computer support to perform his role. Computer software will become so sophisticated that its development and maintenance will become a specialized field in itself.

By taking over manual tasks, the future computerized operation will be efficient, precise and better controlled. Major projects will be executed like a military operation, with little room for failures. Indeed, failure and risk analyses will themselves become an important part of project development. For the former, it is to avoid the serious consequences and costs of failures, and for the latter, to have a better evaluation of the project's success. Future projects will have much better chance of keeping to cost budgets, time schedules, and of success.

Let us see what the concrete structure in various fields will look like in the future.

6.1 Land structures

6.1.1 *Buildings*
Low-rise and medium-rise buildings will generally be built of concrete because other materials will be in short supply or too costly for these applications. Precast

construction will be in common use, and precast elements will come in a greater variety of size, shape and weight. Energy conservation will be an important design feature. Insulation against temperature and sound will be improved, and used in combination with thinner concrete sections to save material and weight. Connection in the field will be reduced as precast components are made in larger sections. Connection methods by bolting, welding, grouting and epoxying will be improved, with accent on construction speed.

The availability of stronger and lighter concrete will in effect raise the economic height of multi-story concrete office and residential buildings. With optimization and improved analytical techniques, high-rise concrete buildings will be better able to resist earthquakes. Improvement will be all-pervasive. The ductile design of concrete buildings, for instance, will strive to avoid steel congestion at the intersections of structural members; and shear wall construction, particularly those tall and slender ones in high-rise buildings, will be considered in conjunction with the other structural systems for earthquake resistance. The so-called tube-in-tube design (e.g. Kaiser Cement's 60-story building in the USA) will see further development and find new popularity as wind and earthquake-resisting structures.

Long-span and high-rise parking structures will continue to be built of prestressed concrete using precast construction. So will long-span thin-shell concrete roof structures. For extremely long roof spans, it will be more economic to use steel frames and fabric cladding, if maintenance, wind resistance and structural stiffness do not present any problem.

The future buildings have been conceptualized occasionally in forward-looking publications. An interesting concept of the future transportation terminal using unusual concrete structures is shown in Fig. 6-1, and a futuristic 65-story, column-free prestressed concrete highrise building, in Fig. 6-2.

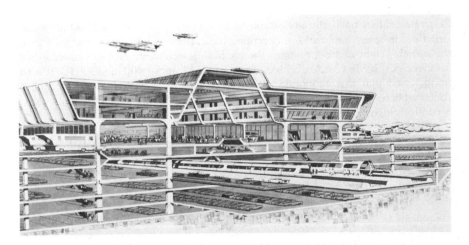

Figure 6-1 A concept of a transportation terminal in the future (Designer: Alfred Yee)

Figure 6-2 Futuristic 65-story, column-free prestressed concrete high-rise building (Design engineers: T. Y. Lin International)

6.1.2 *Bridges*

Concrete bridges have already established themselves as economic and aesthetic structures in many applications, excepting the extremely long-span bridges where the lighter weight of steel remains the controlling economic factor. Concrete has been competitive in cost for bridges up to a span of about 400 m (1300 ft). It is to be expected that this economic span will be increased as lighter and stronger concrete becomes available, and advanced design techniques of cable-stayed concrete bridges are developed.

Bridge construction will be intensified to keep pace with the growth of population and economy, and with the replacement of old and obsolete bridges. Other areas that will contribute to the expansion of bridge construction include the separation of grade crossings, the construction of mass transit systems in big cities, and the improvement of railways. This does not mean, however, that concrete materials and construction will one day completely replace steel. Quite the contrary, there will be considerable room for steel in bridge construction, remembering that steel construction techniques are also undergoing improvement at the same time. One improvement area could be the development of prestressed, post-tensioned steel structures, and their full composite action with concrete decks.

The longitudinal precast I, T and other sections will remain favorites for relatively short-span bridges. For medium spans of up to about 61 m (200 ft). T-shaped girders with stressed cables to take tension, combined with the concrete deck which takes compression, have proved to be an extremely economical solution. For the large cantilever and continuous spans, the box section will be more economical, particularly in the vicinity of the piers where the bottom flange is subjected to high compression. These structural forms can be expected to continue and improve in the foreseeable future.

Precast segmental construction will continue to develop, not only in box sections, but also in wing-shape and other sections. The joinery of these segments will be further improved with or without epoxies and match casting. Although precast concrete is commonly associated with speedy construction, in-place concrete construction by using special plant and equipment has proven to be economical and successful in a number of recent bridge projects. Best known, perhaps, is the traveler for segmental cantilever construction (Fig. 6-3), which is extended into cable-stayed bridges (Fig. 6-4).

A number of innovative and forthcoming construction techniques have been developed over the last decade, e.g. the incremental launching of concrete superstructure that is cast back of the abutments, the lifting and placing over water of entire prefabricated spans brought to the site by barges, the erection of upside-down suspension bridges over deep gorges (Fig. 6-5).

Certain bridge sections will tend to be standardized in the future, while special designs are required to meet unusual conditions. High on the list of developing bridge structures will include the cable-stayed concrete bridge. This has great

Figure 6-3   The 850 m (2790 ft) Rio Dulce Bridge in Guatemala with a main span of 189 m (620 ft) (Design engineers: T. Y. Lin International)

Figure 6-4   The 'hanging arc' design of Ruck-A-Chucky Bridge for California's American River. The curved span of 396 m (1300 ft) is suspended by cables anchored on the hillsides (Design engineers: T. Y. Lin International)

potential as it utilizes both steel and concrete to their best advantages. Present technology already permits these bridges to span generally between 152 m (500 ft) and 610 m (2000 ft). These span limits will no doubt be raised in time to come.

A number of basic layouts of cable-stayed bridges have been developed. These will be refined, improved and expanded. Other generic or unique designs will be conceived to meet new challenges and to match the development of new materials, equipment and related technologies.

6.1.3   *Other land structures*
Other land structures, some of which are new and under development, promise to be substantial users of structural concrete. These include storage facilities for hazardous materials that require special resistance against temperature and chemical attacks, prestressed-concrete pressure vessels capable of withstanding pressures of 27 MPa (4000 psi) and above, and underground installations and tunnels.

Figure 6-5 The Rio Colorado Bridge in Costa Rica, designed as an inverted segmental truss to span a 200 m (600 ft) canyon (Design engineers: T. Y. Lin International)

6.2 Offshore and ocean structures

Being the new frontier of human expansion, the ocean will provide new opportunities for the use of structural concrete. Structural concrete has established itself firmly as a suitable material for marine application some ten years ago, as crude-oil storage in the Ekofisk Field, North Sea, and as gravity platforms since. Among current efforts in developing alternative sources of energy, structural concrete will be the leading material for such construction as ocean energy conversion power plants, floating industrial plant, airports, barges and ships.

There are many advantages to a floating industrial plant. It can be built economically with established expertise and facilities in an industrially advanced country, then towed to the site and installed without having to contend with the problems associated with construction in a developing country. The following statistics will show the viability of this method of construction. By the end of 1980, 43 plants had been built on floating vessels. Of these, 28% was for power generation or water production, 15% for chemical and petroleum industries, and 14% for the accommodation of workers. One of the largest floating plants to date is a paper mill built by Japan's ICI Corporation, for operation in Brazil.

Figure 6-6   Prestressed concrete LPG storage barge for service in Indonesian waters (Structural engineers: ABAM Engineers)

Offshore floating plants are reported to cost 25 to 30% less than similar plants built on land.

Structural concrete has featured prominently in offshore vessels, where cruising speed is not an important criterion. A recent one is the LPG barge built in Tacoma, Washington, USA, for services in Indonesia. The barge measures 145 m (461 ft) long, 17 m (56 ft) deep, and 41.5 m (136 ft) wide. It has a displacement of 59 000 t (65 000 short tons), and a storage capacity of 375 000 barrels (Fig. 6-6).

There is considerable interests in the USA in exploring the possibility of using prestressed concrete in the construction of ships where cruising speed is important. The working stress, price and efficiency of a concrete ship have been compared with those of a steel ship, or simply, the stress ratio, the price ratio, and the efficiency ratio, with the following observations.

For the same volume of material, the stress ratio between steel and concrete is about 13 to 14; the price ratio between prestressed concrete and steel in the order of 16; and the efficiency ratio, between 1 and 1.2, is in favor of concrete. From a purely structural point of view, this comparision leads to the conclusion that a ship's hull constructed in prestressed concrete should cost about 75% of one fabricated in steel. However, the concrete ship is heavier and therefore not efficient in fuel consumption. There are other inefficiencies and inconveniences that will tip the scales against the concrete ship for the time being. The situation will change as stronger, lighter, and smoother concrete is developed, e.g. the use of PIC surfaces to reduce water friction. In the new development, the concrete ships may not necessarily follow the configuration of the conventional ship. The disadvantage of weight, for instance, could be reduced if the concrete ship takes the shape of a semi-submersible, like a submarine, with all but the bridge completely submerged when it is under way.

Included in this structural category are those structures that one day will be installed on ocean floor hundreds of feet below the water's surface. Beginning with the well-known Mohole project in the 1960s, the ocean floor has been the subject of intensive investigation, including feasibility studies for the installation of deep-sea human habitats. The operation depth of divers also been steadily increased with the development of saturation diving.

The next decade or two will certainly witness the accelerated advance by Man on the sea for economic and other gains.

6.3 Energy structures

This relatively new category of structures is being developed in connection with alternative sources of energy—solar heat, wave, wind, tide, and the conversion of coal. It is expected that structural concrete will play a substantial part in this new field. Some of these structures have already been mentioned, such as the large high-pressure reactor vessels for coal conversion and the ocean energy conversion power plants. These and others will now be further discussed.

6.3.1 *Prestressed concrete pressure vessels (PCPV)*

The PCPV was developed some 20 years ago for housing the nuclear reactor, in view of its many desirable properties in this application. It is rigid, durable and relatively free of maintenance, has good failure characteristics, and provides insulation against radioactivity. One would have expected the PCPV to go on from there and enter the non-nuclear pressure-vessel field. This had not happened, because (a) there was no compelling reason for the industry to change from using the well-developed, and well-tried steel vessels, to something that is relatively unknown; and (b) the deterrent effect of the existing PCPV codes on nuclear vessels.

Studies in the possible use of PCPVs for new coal conversion plants were conducted by the Oak Ridge National Laboratory, the University of Kentucky, the Ralph M. Parson/T. Y. Lin International team, and others in the USA and elsewhere in recent years. These had pointed to the feasibility and viability of the PCPV. Studies had shown it to be considerably cheaper than steel vessels. In one case in the Parson/Lin study, that of the combined dissolver–separator vessel (DSV), a single PCPV was designed with the same capacity as 18 steel vessels. This had resulted in a saving of about 70% of the steel vessel cost. A cross-section of the DSV is shown in Fig. 6-7. The comparison was made with steel vessels with solid plate shell. If multi-layer steel shell is used, the saving would be less, but would still be sizeable.

PCPVs for the non-nuclear industry will cater for a wide range of operating temperature and pressure that may far exceed those normally encountered by a nuclear PCPV. New designs, primarily based on cracked-concrete techniques, have been developed to increase the resistance of PCPV against high pressures. Like the nuclear PCPV, an insulating and cooling system is used on the inside

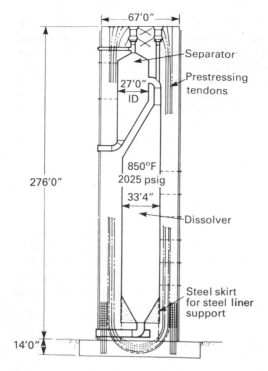

Figure 6-7 Dissolver–separator vessel by Ralph M Parsons (Structural engineers; T. Y. Lin International)

surface of the PCPV to limit the temperature of the concrete to the usual 93°C (200°F).

Interest in the non-nuclear PCPV is evidenced in the recent formation of technical groups and committees by engineering institutions (FIP Task Group, and ACI–ASME Joint Committee) to make recommendations or to prepare the ground for an eventual code of practice.

6.3.2 *Ocean energy conversion vessels*

Energy in the ocean is present in the forms of solar heat, wave and tide. Of these, the conversion of the kinetic and potential energy of the tides is the most straightforward.

The conversion of solar energy is more intricate. One method that has been actively pursued in the USA is the so-called ocean thermal energy conversion method (OTEC). An OTEC plant operates on the temperature difference between ocean surface and deep water, and uses this difference to evaporate and condense alternatively a working fluid, commonly ammonia, which is sent through the turbine-generators to produce electricity.

An early system proposed by Lockheed Missiles & Space Co. used a spar-shaped platform for greater stability, with a diameter of 76.3 m (250 ft) and a height of 85.4 m (280 ft). Cold water is drawn from a depth of about 457 m, (1500 ft) through a large-diameter cold-water pipe. A unique feature of this proposal is the four detachable power modules, each complete with its own evaporator, condensor and turbine-generators, which are mounted around the periphery of the vessel. Refer to Fig. 4-1.

Subsequent work sponsored by the US Department of Energy (DOE) included an investigation of other configurations: ship, semi-submersible, submersible, circular barge and tuned sphere. Each has its merits, but the ship configuration made of concrete was found by Lockheed to be the most feasible and viable solution (Fig. 6-8).

Another ocean energy conversion system, also developed by the Lockheed/T. Y. Lin International team, is the DAM-ATOLL system that converts wave energy into electricity. The DAM-ATOLL takes the shape of a large concrete dome with a vertical center core that houses the turbine. Waves riding up the dome are directed by a system of vanes to spiral down the core, driving the turbine. The turbine in turn drives the generator located above the water to produce electricity.

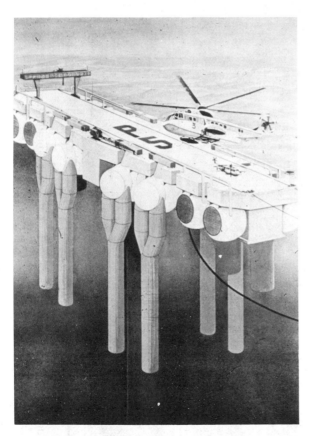

Figure 6-8 An OTEC plant in ship configuration by Lockheed Missiles and Space Co. (Structural engineers: T. Y. Lin International)

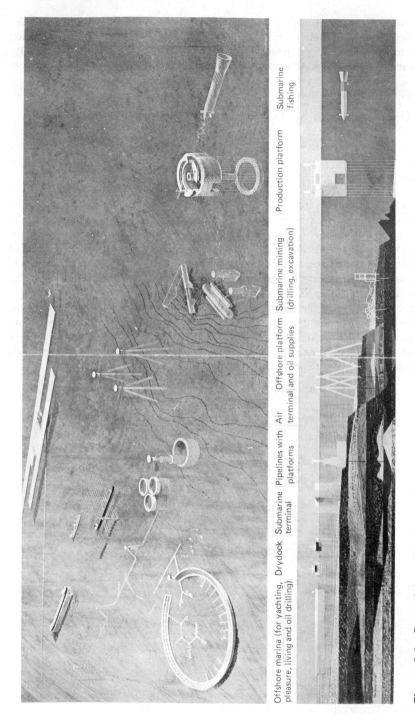

Offshore marina (for yachting,   Drydock   Submarine   Pipelines with   Air   Offshore platform   Submarine mining   Production platform   Submarine
pleasure, living and oil drilling)   terminal   platforms   terminal and oil supplies   (drilling, excavation)   fishing

Figure 6-9   Potential applications of structural concrete in the ocean by Hydro Concrete System

There are many innovations designed to produce energy from alternative sources. Some possibilities are shown in Fig. 6-9, taken from a promotion brochure. For a more thorough review of these possibilities, the readers may refer to the many publications that specifically deal with this subject.

6.3.3 Arctic structures

The Arctic regions may be due for extensive development to tap some of the riches they contain. A beginning has been made with Prudhoe Bay and the construction of the Alaskan pipeline. Much experience and knowledge have already been gathered in coping with the difficult hostile environment and the permanently frosted ground.

The extreme cold there would normally favor the use of structural concrete over steel. Low-temperature steel may be used, but it is more expensive. For massive application, such as the gravity platform for oil or gas production, there is really no substitute for concrete. In order for the platform to withstand severe ice loads, the Arctic platforms have been given special configurations that present minimum resistance against ice forces (the monopod), or assume the hourglass shape that has the effect of deflecting ice sheet upward as it impinges upon the platform, breaking it up in the process (Fig. 6-10).

In the final analysis, the growth of human colonies in the Arctic, or anywhere else for that matter, will depend on the extent to which its natural resources may be developed. The spirit of adventure and survival will take human beings anywhere. Concrete structures of all types will complement and enhance such efforts and help make their endeavors worthwhile.

Figure 6-10 Proposed drilling and production platform for Arctic Region by Ben C. Gerwick and Hans O. Jahns

References

Lin, T. Y., The future of prestressed concrete—a long look ahead, *J. Prestressed Concr. Inst.*, Vol. 21, No. 5, Sept. 1976, pp. 204–215.

Lin, T. Y., Role of prestressed concrete in energy development, *J. Prestressed Concr. Inst.*, Vol. 22, No. 3, May–June, 1977.

Lin, T. Y. and Chow, P. Y., Economics and problem areas of structural and constructional options for large prestressed concrete ships and vessels, *Continuing education in engineering*, University of California, on Concrete Ships and Vessels, Sept. 1975.

Lin, T. Y., Chow, P. Y. and Ngo, DE,. Innovations in prestressed concrete pressure vessel design, *Trans. 5th Int. Conf. Struct. Mech. Reactor Technol.*, Vol. H, pp. H3/3 1–8 August, 1979.

Second international congress on polymer in concrete, University of Texas at Austin, Oct. 1978.

FIP symposia on partial prestressing and practical construction in prestressed and reinforced concrete, Romania, Sept. 1980.

2 Lessons from the past— achievements and failures

S C C Bate

Consultant, Harry Stanger Ltd, England

Contents

Summary

All structural engineering activity is accompanied by the risk that the engineers' objectives in design and construction may not be achieved and a much smaller risk that failure may occur. It is important that the low level of the incidence of these events should be widely appreciated and that, when failures do occur, their causes should be thoroughly examined to ensure that the lessons learnt are of benefit to the whole community.

The widespread attention paid to some failures may detract from the effective determination of their causes and may inhibit the reporting of other failures and the investigation of their causes. It may also obscure the substantial achievements that have been and are being made in concrete construction.

The aim of this chapter is to review what has been achieved, set the occurrence of defects and failures in perspective, and give sufficient information on a number of failures for the appropriate lessons to be learnt. In this way, a background is provided for the other chapters.

1   Introduction

The history of progress in engineering, not only of the development of the technology of construction in concrete, shows that all technological development is founded on the practical trial and application of ideas and the modification of those ideas in the light of experience. Engineering development is seldom the result of the activity of an individual and, in the section which follows, it will be clearly seen that this has applied to the structural use of concrete. The progress made has been the result of the simultaneous or successive efforts of a large number of engineers in a continuously widening technical community. The rate at which it has progressed has depended on the rate of dissemination of information of experience gained in different parts of the world. It is this need for the widest communication of information that has led to the proliferation of technical publications that has occurred during recent years.

There are, however, factors associated with technical developments that tend to cause some delay in their application, since those engineers, who have the enterprise to seek the introduction of new ideas, usually wish to see some benefit from the risks that they may have incurred. If the resulting experience includes a number of problems and even possibly a failure, there is, not surprisingly, some reluctance to give the experience widespread publicity. It is unfortunately true that some of the most forward looking engineers have been amongst those that have suffered the anxiety of experiencing structural failures. It is, however, to the advantage of the profession as a whole that, where there is information on defects and failures as well as on achievements, it should be coordinated and brought together for publication, in order to widen the knowledge of other engineers. This is the primary purpose of this *Handbook*, which deals almost entirely with developments and achievements. The aim of this chapter is to examine the lessons that can be learnt, giving particular attention to those that may be gained from failures, which have been the subject of authoritative investigation.

When there is a structural failure which is of general concern, the non-technical press, probably reflecting public opinion, often asks why engineers have not learnt from previous experience, presumably assuming that the engineers are unaware of it. Such questions were asked about the use of high alumina cement in the UK following the collapse of part of the roof in a school in London (Section 5.4), when its use had been discontinued in France and Germany and it had been the subject of a comprehensive review in a paper by Neville [1] given to the Institution of Civil Engineers in 1963, which described a number of failures. It was not appreciated that an expert committee set up by the Institution of Structural Engineers [2] in 1964 had reviewed its characteristics in concrete and issued a report giving recommendations on how the cement should be used. The question that should have been asked was why this technical re-appraisal had not achieved its objective subsequently in practice. Other pertinent questions that should also be asked are how it came about that the structure at Ronan Point (London, England) was designed and built in the way it was after the experience of the Aldershot (England) collapse (Sections 5.2 and 5.3), and that the failure of the cooling tower at Ardeer (Scotland) occurred even though it had been redesigned to take the recommendations of the report on the collapse of the towers at

Ferrybridge (England) into account. These questions do not have ready answers but the aim in drawing them to the attention of engineers is that they may stimulate engineers to adopt a more objective and critical attitude to the design and construction of concrete structures.

A number of failures are described in this chapter with the object of identifying their causes and in sufficient detail to bring out any underlying factors, which may have contributed to the failures by influencing decisions made in design and construction. Since this requires some detailed knowledge of the background, a number of the failures included in the review are those in which the author was directly involved in the investigations. The chapter gives more attention to failures than to achievements because it is less difficult to draw lessons from the examination of structures that have failed than from successful structures where engineering judgement and intuition have been exercised in the application of new ideas.

Even the identification of a successful structure presents some difficulty because it implies that it is successful on all counts, which include: fitness for purpose, adequate strength and stiffness, competitive first cost and low maintenance costs, durability and adaptability. For failure, the structure needs only to fall short on one of these requirements, and can only be regarded as a success or otherwise on completion of its service life. The Congress Hall in Berlin [3] was judged a major achievement at its opening, but its merits as an example of an advanced structure were totally offset by its inadequate durability. When they were built, the blocks of flats of large panel construction were claimed as the solution of the housing shortage in densely built-up areas, but by the time of the collapse of Ronan Point, it was already becoming recognized that they were detrimental to community life and it was mainly for this reason that construction was discontinued thereafter. In this context, the prestressed concrete railway sleeper may be judged as one of the most successful forms of concrete construction since it is now nearly 40 years since manufacture was started [4] and they are still in use in large numbers. The most useful way of judging the achievements in the concrete industry and the lessons that may be derived from them is to review the stages of development in design concepts and methods, in materials and in construction techniques.

The review of the significance of developments in concrete construction will be considered in two parts: first, the evolutionary phase when theoretical design methods were either unknown or of very limited use, before research was introduced to supplement experience; and secondly, the era of modern technology when the rate of development has been accelerated by using sophisticated analytical techniques assisted by research in design. There is no clearly defined boundary in time between these approaches to construction, some engineers being much in advance of others in their thought and attitudes; it is a matter for some regret that what some practices tolerate today might well have been unacceptable yesterday, although the reverse is also certainly true. This review will be followed by consideration of what can be learnt from the various surveys that have been made to determine the performance of construction in service and by an analysis of each of a number of incidents of failure and the circumstances that led to their occurrence. The chapter will be completed by a summary of the main conclusions drawn from engineering experience in the past.

2 Early developments[1]

One indicator of the level of advancement of early civilizations was the extent to which they used calcareous cements. It was not, however, until the Roman period that a recognizable technology evolved, which was based initially on information obtained from the Greeks and which the Romans with their usual efficiency then proceeded to apply throughout their empire. Roman mortars were made with lime after slaking, which was mixed with sand, and concrete was produced by placing the mortar in formwork and adding pieces of rock or stone, placed in layers for better quality work. In the construction of vaults, the timber forms were sometimes supported by ribs of brick-work, which were left in position as a permanent part of the structure when the timber was removed. It was fully appreciated that, if certain volcanic deposits were finely ground and added to the lime, the resulting mortars were stronger and were more durable in both fresh and sea-water. These deposits were originally obtained from Pozzuoli on the bay of Naples, but in other parts of the Roman Empire, other suitable volcanic materials were sought and, where these were not available, ground tiles were used. Aggregates were carefully selected and the evidence shows that they were washed when necessary to remove dirt or silt and screened to sizes appropriate to their use. The durability of some of these mortars, such as that in the Roman breakwater at Naples, has sometimes been attributed to some unknown constituent, but investigation suggests that it is due to the denseness of the mortar obtained by thorough compaction. The endurance of their structures shows the value of enforcement of good specifications and effective oversight of workmanship.

The Romans embedded bronze cramps in stone masonry and in later centuries iron was used but the rusting and disruption of the stonework soon led to its abandonment, particularly when it was found that slender masonry construction could be built without the use of metal. It was not, therefore, until cements could be relied on to maintain an internal alkaline environment that it became feasible to incorporate iron in concrete. The first application appears to have been in 1848, when Lambot built a boat of reinforced concrete which was exhibited in Paris in 1855. This combination of cement and iron was patented in both France and Britain as a substitute for timber. Meanwhile, in 1849, Monier had started building tubs in reinforced concrete for planting trees such as orange trees, and Coignet, who had poured a concrete building *in-situ* in a timber framework in 1847, applied a similar procedure in 1852 to the encasement of an iron framework in concrete. In 1854, Wilkinson took out a British patent for the construction of concrete floors and beams using a mesh of flat iron bars and second-hand steel wire rope, which was raised over the supports and sagged near the bottom of the beam at mid-span; this system was used for the construction of a number of buildings in the Newcastle area. In 1855, Coignet patented the use of concrete for various purposes in buildings, which included embedding a square mesh of iron bars in concrete for floor construction. By 1861, he was advocating the use of concrete reinforced by iron bars for many of the functions for which reinforced concrete is used today.

1 This section has drawn on references [5–9].

After Wilkinson's early start in the development of reinforced concrete, there was little new activity in Britain for nearly half a century, apart from the introduction of expanded metal as reinforcement in floors from 1889 onwards. The developments in France, however, widened in scope and application. Monier's had grown from flower pots to portable containers, and in 1872 he constructed a small reservoir, having already taken out patents in both France and Britain for applications to pipes, tanks and bridges. In 1885, Wayss took over the exploitation of the Monier patents in Germany and Austria to such good effect that Wayss and Freitag built 320 arch bridges between 1887 to 1891. Until this stage, Monier's ideas had not been based on scientific principles, and Wayss therefore undertook a substantial program of investigation and commissioned Koenen to derive a method of computation. The results of this work were incorporated by Wayss in a textbook published in 1887. The method of calculation advocated assumed that the neutral axis was at mid-depth, that tensile stresses were carried by the steel alone and that the transfer of stress between steel and concrete was through adhesion between the materials. Later, Neumann at Brno introduced the ratio of the moduli of elasticity into the calculations in defining the depth of the neutral axis. At about this time, Melan, also in Brno, invented a form of construction for floors and vaults which was subsequently widely applied in bridges and introduced by von Emperger to the USA. By 1902, theoretical studies supported by experimental work had progressed to such an extent that the firm of Wayss and Freitag published a major review of their research results with recommendations for design by Mörsch, which formed the basis for regulations in a number of the German states.

Whilst these developments were taking place in the Austro–German sphere of influence, Coignet was extending the ideas developed by his father in France. In 1892, his proposals for using reinforced concrete instead of masonry for sewers were accepted for the new main drainage system for Paris with very substantial savings in materials. Two years later, in collaboration with de Tedesco, he applied the results of his experience and experimental work to drawing up a practical method for the design of reinforced concrete members. They assumed that plane sections remained plane, that Hooke's law applied with a constant modular ratio, and that the concrete in the region of the tensile reinforcement would be cracked and so without tensile resistance. There was wide disagreement on the validity of these assumptions. The outcome has been described by Hamilton [7] in the following.

> The practical common-sense view, however, was that the theoretical niceties were not justified in the face of the facts with which every experimenter was familiar. Cement was manufactured from natural material of variable composition by a process which was by no means always strictly controlled. Aggregates, also, varied widely in their physical properties. Even with the best selected materials, the quality of concrete still depended on the accuracy with which the ingredients were measured and on the thoroughness with which they were mixed, placed and consolidated. The influence of the water/cement ratio on strength was not then suspected, and the way in which reinforced concrete resisted shearing forces was but vaguely conceived in the nineties and in some quarters for long afterwards.

This was written in 1956 of the situation which existed some 60 years earlier. A similar attitude is developing today to modern theoretical concepts of structural concrete for reasons which are similar in nature if not in degree.

In France during this period, Considère was investigating the strength of reinforced concrete beams and columns in a series of systematically planned experiments, which included a study of the strength enhancement of columns by the use of helical binding, and at about the same time, Hennebique was developing his ideas for the practical application of reinforced concrete construction, which were taken to Britain by Mouchel, where they stimulated new interest in the use of concrete structures. By 1897, there was wide-spread agreement on design principles in France, which enabled a series of lectures to be given by Rabut to the students at the École des Ponts et Chaussées, and in 1902, a major review of reinforced concrete construction by Christophe was published in Paris and Liège. Official regulations were issued in France in 1906 largely as a result of the efforts of Considère and Mesnager, the City Engineer of Paris.

Until 1904, when Coignet opened an office there, the main developments in Britain in reinforced concrete were due to Mouchel in exploiting the Hennebique patents, although a number of British companies were building patented systems of fire-proof floors. Following considerable professional demand, the Royal Institute of British Architects set up a committee in 1905, which included a number of eminent engineers, to enquire into the proper use of reinforced concrete. They reported in 1907, giving authoritative guidance on design and construction. This was the forerunner to the London County Council byelaws, issued shortly afterwards, and the British Standard Codes of Practice which appeared much later.

In the USA, the development of reinforced concrete construction proceeded along more independent lines. Following experimental work carried out for him in New York and London, Hyatt patented his ideas in 1878, having previously published privately an account of his tests, which were aimed at the construction of economic forms of fireproof roofs and floors. His proposals do not, however, appear to have been used in practice. A few years later in 1884, Ransome patented the twisted square reinforcing bar, which he had found to be stronger than the plain bar, and, in 1888, he built his first major building with reinforced concrete beams and arches, and cast-iron columns. By 1904, he had introduced a system of composite-construction which consisted of precast concrete beams with helical shear reinforcement to link to the cast *in-situ* floor or roof slab. In the meanwhile, Norcross had patented a system of flat slab construction in 1902, which began to be widely used by 1910. Codification started in 1903 when the Borough of Manhattan introduced regulations for the design of reinforced concrete into their building code.

This brief summary of the early history of reinforced concrete has been described in order to show how progress was made in its development into a constructional form which is still familiar today. At the beginning of this century, design was almost exclusively in the hands of the specialist contractors, but the stage had been reached where there was a general desire for a better understanding of the material in order that it could be applied more widely by the engineering profession and that the adequacy of the construction could be checked more readily in the public interest. As a result, there was a start made to obtain agreement on codes of practice and regulations which eventually led to the regulatory documents now adopted in most countries and more recently to the concept of international codes. It was unlikely that the progress made in the inter-war period would have been made if these developments in codification had not occurred. In the early stages, however, it is quite clear that the rate of development and the way in which it went depended on the vigour and enthusiasm of individual entrepreneurs, who were often remarkably successful with no more than a rudimentary knowledge of theoretical principles. No real progress could be made in obtaining general acceptance of reinforced concrete, however, until engineers had established working principles and satisfied themselves by experimental investigation that these provided a sound basis for design.

No mention has yet been made of the evolution of prestressed concrete, mainly for the reason that it did not become a practical proposition until very much later, although the idea can be traced back to a German patent taken out by Doehring in 1886 and to a US patent by Jackson in 1888. The potential advantages of prestressing, mainly the freedom from cracks in the concrete, were reasonably well understood by 1910, but no satisfactory method had been found for dealing with the losses of stress in the steel resulting from creep and shrinkage of the concrete. So that, although the increased use of reinforced concrete could be advocated at that stage, even though the effect of water/cement ratio on strength was not established by Abrams until 1918, and the long-term research on the creep and shrinkage, which showed that limiting values were eventually reached at levels of stress in the concrete permissible in design, had not been commenced. Once this information was available, practical progress was made by a number of individual engineers in developing a number of proprietary systems in much the same way as reinforced concrete had developed about 40 years before. A method for prestressing circular tanks was introduced by Hewett in the USA in 1923. Prestressed concrete bridge construction started in Germany in 1928, when Dischinger's ideas were used in the construction of a bridge at Saale. As a result of his experience in building the major three-span arch bridge, the Plougastel Bridge, near Brest, Freyssinet worked out the principles for applying prestressing to a wide range of structures, which have since found world-wide application on a very substantial scale. There was only time for some of this work to start before the Second World War came 9 years later, which gave much impetus to different forms of specialized construction in a number of countries before normal peacetime development could be resumed. Wartime experience did, however, provide opportunities for gaining confidence in the form of construction much more quickly than would otherwise have been possible, since some increases in the risk of structural failure were then accepted.

By the early 1920s, the pattern had been set for the developments in building concrete structures that took place during the next 25 to 30 years. In this time, however, much of the research was done on which modern achievements have been based.

3 Modern achievements

3.1 Developments during the last three decades

In the period which immediately followed the war, there was an urgent need for new construction in the majority of the industrialized nations to make good the deficiencies resulting from the deferment of building during the war, as well as the wartime damage and destruction of all types of construction. The shortages of materials, in particular of steel, gave substantial impetus to innovation and the development of new constructional techniques. The end of hostilities also brought a more general desire for higher standards of living with a consequential need for the generation of more energy and for the adaptation of the technologies developed for warlike activities to the enhancement of peace-time living. The demands were, therefore, not only for housing, schools and hospitals, but also for industrial buildings, roads, bridges and power stations, including both coal-fired and hydroelectric as well as the embryonic needs for generation from nuclear sources. The stimulation to the construction industry in general and to the concrete industry in particular deriving from that time is reflected in much of the content of this *Handbook*, which is primarily concerned with modern achievements.

3.2   Materials

The materials used in structural concrete now are mostly little different from those that were in use or were available 30 years ago. There have, however, been major improvements in methods of production and their characteristics have been adjusted to make them more suitable for use in concrete in the light of the better understanding of what is needed for various forms of structural concrete.

There has been a substantial increase in the use of deformed bars for reinforcement of higher tensile strength than plain mild steel attained either by cold-working or the use of a naturally harder steel. Although there are instances where stainless steels or galvanized bars have been used for reinforcement, these are not common and reliance is normally placed on the protection against corrosion afforded by the alkalinity of the concrete. While cold-drawn wire is favoured for use in tendons in a number of countries, heat-treated wires are also in use, although the view is held that the latter are more susceptible to corrosion and its effects associated with high stresses; improvements in production procedures have, however, led to smaller losses of prestress due to the creep of steel. The most important change has been in the development of strand, which consists of a core with one or more layers of wires wound around it; it too is produced with low relaxation characteristics. High-tensile steel bars offer an alternative form of tendon when high prestressing forces need to be concentrated and anchored. Another material, developed within recent years for use in concrete, is glass fibre. Glass has the merit that it is much less sensitive to attack from the atmosphere than steel, but the demerit that glasses in general are attacked by the alkaline environment in concrete. Forms of glass fibre have, nevertheless, been developed, which are resistant, although, until it is proved otherwise, it seems wise to use these fibres for non-structural components where the tensile resistance of the concrete is required during handling and construction. See Chapter 6.

Portland cement remains the most commonly used cement in concrete, and it seems unlikely that this situation will change in the foreseeable future. Here, too, there have been important improvements in manufacture aimed at lower cost, fuel economy and avoiding atmospheric pollution. Improvements are seldom achieved, however, without some offsetting disadvantage and, in this case, there has been a tendency for the alkali content of the cement to rise, which leads to a greater risk of alkali–aggregate reaction (Section 5.7) although it may confer better protection to embedded steel. Currently, much attention is being given to the utilization of waste materials, but Portland blast-furnace cement has been in use for many years, in which the amount of slag may vary between 35 and 85% by weight, the remainder being Portland cement. A more recent development of slag cement has been super-sulphated cement comprising a mixture of Portland cement and calcium sulphate with at least 75% of granulated slag. These cements have particular application where high chemical durability is essential and also in massive structures such as dams where low rates of heat evolution are required. In a few countries, such as Britain and West Germany (Section 5.4), the high early strengths obtained with high alumina cement appeared attractive to manufacturers of precast prestressed concrete components—an abortive development, which proved unsatisfactory in each of those countries for somewhat different reasons. This cement is valuable for certain uses, however, as a cement in refractory concrete, where most of it is used, and in concrete in marine work or in sulphate soils.

A wide variety of materials is added to concrete for a variety of reasons. The best established, resulting from experience in airfield construction are the air-entraining agents giving enhanced resistance to frost action. Others are concerned with increasing the rate of hardening of the concrete or improving the workability of concrete, although any of these based on calcium chloride have become increasingly discredited as a result of experience with the corrosion of embedded steel (Section 5.5). Probably the most remarkable development has been the introduction of superplasticizers, which have been used extensively in West Germany and Japan. These materials are complex organic compounds, which enable concretes with low water/cement ratios to be placed with ease. They are claimed to have the advantages that concrete can be placed readily in areas of congested reinforcement, that more complex shuttering can be used and that pours can be larger, leading to a reduction in the number of construction joints. The use of fly ash from power stations burning pulverized fuel has been used for a considerable period as an aid to workability, for example in pumping concrete, but it is being used also as a partial replacement for cement since it has some cementitious value in concrete and a slower rate of generation of heat of hydration with benefit in placing concrete in massive structures.

Present concern with the avoidance of waste and with the conservation of natural resources is encouraging the greater use of waste materials for the manufacture of aggregates. Most of the development work was done 20 years ago and some experience has been gained in a number of countries in their application, but their use has not been extensive because of the extra cost that is thought to be involved. In most cases, the aggregates are lightweight and they produce structural grade concrete of about three quarters of the density of the corresponding grade of concrete made with natural gravel or crushed rock aggregates. The saving in weight of tall buildings and long-span bridges is therefore considerable and the resulting reduction in the cost of foundations can offset at least the additional cost of the aggregate with a further saving in the cost of handling since larger volumes of concrete can be lifted with the same lifting equipment.

The major change in Britain in the usage of aggregate within the last few

years has been the much greater supplies of natural aggregate that are now
dredged from the sea, which has largely come about as a result of the exhaustion
of conveniently located inland deposits of gravel and the high cost of transport by
road from less accessible pits. The marine aggregates introduce a small but not
negligible amount of chloride and shell into the concrete so that extra care is
needed in the selection of materials where special concrete is required.

3.3 Design

In some areas of construction, the use of concrete has succeeded the use of
masonry and in others it has succeeded the use of steel but, to start with, it was
regarded as a substitute for the longer established materials. It replaced masonry
in dams, brickwork in large underground pipe-lines and sewers, and steel in
bridges, single-story and multistory buildings. More recently, it has also supplanted steel in large pressure vessels for nuclear power stations and is being used
increasingly in offshore construction in deep water. At various times in these
applications, concrete was adopted as an innovation, but the stage has now been
reached where the choice of concrete for such structures is natural. This has come
about for a number of reasons; it offers more flexibility in choice of structural
form than other structural materials, and modern methods of analysis, such as
finite element methods, in combination with the use of computers allow the
engineer to analyze these complex structures with some confidence. As a result of
research on concrete, as a material, and on the behaviour of structural models,
comprehensive data on the performance of reinforced and prestressed concrete
structures have been established to back up this confidence.

Improvements in the understanding of the performance of concrete construction have taken place side-by-side with developments in the philosophy of design.
The earliest codes of practice for design of structural concrete provided for safety
by introducing safety factors on the strengths of materials to derive permissible
stresses which were used with elastic methods of analysis. At an early stage, this
approach was found to be unsatisfactory for the design of reinforced concrete
columns, since the effects of creep and shrinkage soon led to the permissible
stresses being exceeded, and so an ultimate strength method of calculation was
adopted for use with permissible stresses. Later, a similar approach was applied to
the design of beams. Although it is unwelcome to many designers on the grounds
of unnecessary complication, the limit state method of design, in which each of
the conditions of strength and serviceability required of the structure are considered separately, is being introduced in national and international codes of
practice. This method of design permits the most appropriate methods of calculation to be used with reliability methods based on statistical data to cover the
variability of loading and performance of materials, which can be accommodated
as the necessary information becomes available. Whilst short cuts are required for
straightforward routine design, the method should be particularly beneficial for
the design of special structures and for dealing with innovations.

3.4 Construction

Very few changes in methods of concrete construction have appeared during the
past 30 years. Ready-mix concrete was available; pumping was used for placing

concrete; concrete was compacted by vibration and pressure or vacuum processes could be applied; and a number of other constructional processes had been adopted in practice. The change that has taken place is in the extent to which mechanization and mechanical handling have become commonplace with a consequent reduction in the previous labour intensiveness in the industry. One example of the change is that there is now very little mixing of concrete on site, even on relatively small building sites; nearly all concrete cast *in-situ* in buildings is supplied ready mixed. For major civil engineering works, each site is likely to have its own batching plant. For many massive structures, such as dams, nuclear pressure vessels and offshore structure, concrete is pumped to the formwork and there is an increasing tendency for the size of pours to be increased; great care is then needed in the design of the concrete mix and the selection of the admixture to ensure that excessive temperature rise is avoided, which would result in cracking.

The advent of prestressing has led to the development of new techniques for bridge construction based on the cantilevering methods first evolved in West Germany for large-span bridges and, for bridges of shorter span, of the composite system of precast prestressed beams with a reinforced concrete deck. Special methods of post-tensioning have been produced for prestressing the large pressure vessels for nuclear power stations, which stress the concrete in three dimensions and enable containment structures to be built of a size not yet practicable with other materials of construction.

Where possible, precasting in the factory has replaced casting on site for components used in building, a very common form of floor and roof construction being the precast prestressed beam with cast *in-situ* screed. Apart from kerbs and paving slabs, for which pre-cast concrete has been used for many years, precasting has also been applied to the production of a substantial range of claddings and building components, including the precast segments for underground tunnels as a replacement for cast iron used previously. The period since the war has seen the disappearance of the clay roofing tile and its replacement by the concrete tile in a number of countries. The brick industry has also suffered at the expense of the concrete industry, since the brick or ceramic block inner leaf has disappeared and the lightweight aggregate or aerated concrete block with better thermal insulation has taken its place.

3.5 Research

In the earliest days of the development of structural concrete, progress was made by trial and error, and research, as it is now recognized, was largely confined to the natural sciences. At a very early stage, however, it was realized that this form of construction could only develop in the longer term if there was an understanding of its behaviour and of the principles necessary for a basis of design. It was not, however, until the 1920s and 1930s that much of the fundamental research was done that enabled the developments of the last 30 years to take place. The research of that earlier period also provided the foundation for that of the present day. During that inter-war period, the factors which governed the strength of concrete were established and the role of reinforcement in concrete construction was determined; a better understanding was obtained of the nature of creep and shrinkage of concrete and limiting values were estimated, without which the practical application of prestressing would not have become possible. It was during this period, too, that the necessity of research as an adjunct to future development was established in the progressive sections of the construction industry. The construction of prestressed concrete pressure vessels and, more recently, of offshore structures was preceded by comprehensive research programmes as a matter of course.

4   Assessments of structural performance

4.1   Information reviewed

Once a structure has been completed, the safety and stability of the construction may receive little further attention, except that certain structures, such as bridges and some buildings, are subject to periodic inspection. Since few are designed to facilitate access to essential features and, in modern buildings in particular, these may be hidden behind permanent or semi-permanent finishes, even inspection may reveal little of the real condition of the structure and whether it continues to conform with the designer's original intentions. Many buildings may only be examined to identify maintenance and decorative needs. Whatever is done by way of inspection and irrespective of whether it is useful from a structural point of view, the results are usually only reported to the owner and no coordinated reviews of the general condition of structures and their deterioration are available. Information on the performance of structures has come from a few surveys of the condition of some existing buildings, including tests on structures prior to demolition, and individual investigations or surveys of structural failures.

Most of the extensive surveys that have been made in recent years have been instituted as a result of the discovery of defects or failures, which could be widespread, in order to identify other defective structures. The resulting reports then concentrate on unsatisfactory construction and the remedial measures needed, giving scant attention to the condition of those structures found to be satisfactory. The general conclusions that are drawn, therefore, tend to be adversely biased with respect to construction as a whole. An example is provided by the experience in Britain with high alumina cement in precast, prestressed concrete components; the feedback from engineers responsible for structural appraisals indicates that more than 90% of the buildings with these components have been kept in service without any remedial measures with no further failures for more than seven years, in spite of the apparently high level of failure experienced initially. A small proportion of this construction, found to be unsatisfactory, was defective for reasons not associated with the cement.

Even less information on the incidence of inadequacy in structures is available from the investigations of individual failures or from the reviews of experience of failures, since these cannot be related to the total amount of construction. Nevertheless, such information is important in determining how the incidence can be reduced. Studies of existing construction independently of the occurrence of defects is much less common and can therefore only be indicative rather than representative of normal practice.

In this review of the performance of structures, reference is made to the various surveys of construction reported by the Building Research Establishment in England [10–12], to the tests of the Old Dental Hospital in Johannesburg [13] and of the New York World's Fair structures [14], before considering several reviews of defects and failures in concrete structures. These are considered in some detail with brief references to other experience.

4.2 Tests on the Old Dental Hospital, Johannesburg

Shortly after the Second World War, a major scheme was developed by the South African Railways for the extension of Johannesburg station, which entailed the demolition of the Old Dental Hospital built some ten years previously in 1942. This three-story building was therefore relatively modern, having been designed to comply with building regulations based on the recommendations of the Reinforced Concrete Structures Committee of DSIR, which preceded the British Standard Code of Practice for reinforced concrete issued in 1948.

The building was approximately 30 m (100 ft) square in plan with a central courtyard. It was of framed construction with columns, beams and floors of reinforced concrete, which was cast *in-situ*. Although of three-stories with part of a fourth story, it was designed for extension later to six stories and the columns and foundations were proportioned accordingly. Most of the floors were designed for a superimposed loading of 2.4 kN/m^2 (50 psf)*, some with an allowance for partitions, while for the remaining floors, the superimposed loading was taken as 3.8 kN/m^2 (80 psf). For most of the construction a 1:2:4 concrete was used with specified works cube strength of 15.5 MPa (2250 psi) at 28 days. Mild-steel reinforcement was used with a permissible tensile stress of 124 MPa (18 000 psi) and the permissible bending compressive stress for the concrete was 5.2 MPa (750 psi). The view was expressed at the time of the tests that the design, method of construction and the supervision exercised was representative of normal practice for a building of this kind.

By the time the tests [13] were made, the hospital had been in service for about ten years, and for this time it had been subjected to the effects of loading, temperature changes and earth tremors. Some cracking of the structure had occurred and considerable variation of both the workmanship and quality of the concrete was observed. Nevertheless, the loading tests did not disclose any unexpected weaknesses although unexpected strength was found in some of the tests. Within the range of working loads, little cracking developed and the strains and deflections recorded were nearly always less than these calculated ignoring the tensile strength of concrete, but agreed more closely if it was assumed that cracking had not occurred. For members subjected to bending in one direction only, ultimate strength methods of calculation generally underestimated strength by 15 to 25%. For slabs subjected to bending in two directions, the strength was much underestimated, in one instance by as much as three times. The tests on the floors of beams and slabs showed that the beams acted essentially as stiffening ribs and that the torsional stiffness of beams provided considerable end-restraint to other beams and slabs.

The main conclusion drawn from all the tests was that the then conventional

* 1 MPa = 1 MN/m^2 = 1 N/mm^2.

methods of design for reinforced-concrete framed buildings were very conservative and that, if advantage were taken of the interaction in structures, considerable economies could be made.

4.3 Tests of New York World's Fair structures

Following the 1964–1965 New York World's Fair, a number of the purpose-built structures were made available for test [14]. Many of the buildings were of unusual design and the results of any tests on these would not have wide application. The three selected for test were chosen as being representative of common forms of construction. One of these, the Rathskeller Structure, was of reinforced concrete, while the other two structures tested were of structural steel. The Rathskeller Structure was a box-like building of single story height, which supported temporary timber buildings with pedestrian access on its roof. The building was approximately 55×36.5 m (178×118 ft) with a waffle slab about 0.6 m (2 ft) thick supported on columns at about 9 m (29 ft) centres in each direction. The reinforcement was specified as intermediate grade billet steel and the concrete was required to have a compressive strength of 24 MPa (3500 psi) at 28 days. The design imposed load was 14.4 kN/m^2 (300 psf) and design conformed with the provisions of the 1956 ACI Building Code using the elastic analysis procedure for the calculation of moments and shears. The object of the test programme was to determine the shear strength of an interior panel, the strength of the column–slab connection at edge panels and the strengthening of a slab due to arch action.

The structure was inspected before starting the tests and some cracking was found which was attributed to settlement, amounting to about 180 mm (7 in) at one corner of the building. The average yield strength of the reinforcement ranged from 290 to 370 MPa ($42\,000$–$53\,700$ psi) and the equivalent cylinder strength of the concrete obtained from cores was between 24 and 45 MPa (3500–6525 psi). It was concluded that the behaviour of the structure conformed to design theory and possessed an adequate margin of safety. There was, however, some weakness in shear at the external column junctions with the roof slab and the flexural strength of the slabs was not attained in any of the three tests, its development being interrupted by shear failure at a column junction. It was thought that the initial cracking damage may have influenced both performance and strength. Under service loading, the performance of the structure was satisfactory; the steel stresses were low and the deflections were within acceptable limits; there were no undue increases in crack width.

4.4 Qualitative studies of buildings

Two series of studies of the condition of a number of buildings, including some built in reinforced concrete, were undertaken in Britain by the Building Research Station in 1954 [10] (in collaboration with the Cement and Concrete Association) and in the period between 1959 and 1963 [11, 12]. The object of the first survey was to collect evidence on the durability of reinforced concrete in buildings in order to establish the time likely to elapse before major repairs are needed and before further repairs or replacement may ultimately become necessary. The second survey was concerned with the performance of a number of buildings for which detailed information on their construction and past history was available, to determine how well they had fulfilled their designers' intentions.

Buildings between 20 and 50 years old were examined in a number of categories: office buildings, warehouses granaries, factories and buildings subject to chemical attack. The general conclusion drawn from the survey was that reinforced concrete was fundamentally a suitable material for permanent building since some of the buildings inspected had lasted for more than 50 years without requiring any special maintenance. Some buildings, however, which were only 20 years old, had developed serious defects requiring remedial measures. The defects were usually the result of the penetration of water, causing rusting of the steel reinforcement, due to too little concrete cover incompletely compacted. Other factors reducing durability were faulty design, use of unsuitable materials, bad construction and attack by airborne or water-borne chemicals. The commonest design defect was detailing the reinforcement in such a way that it interfered with the placing and compaction of the concrete and could be aggravated by misplacing the reinforcing bars so that the cover was both permeable and inadequate. The report recommended that cover of concrete should not be less than 37 mm ($1\frac{1}{2}$ in) to exposed surfaces. Another common design fault was the omission of provision for shedding rain water from projecting features. Several cases were reported of the inappropriate use of high alumina cement and it was suggested that its use should be confined to work underground or to situations where the temperature during curing and in service could always be kept low. Inadequate compaction, leakage from formwork and insufficient care at construction joints were identified as other faults in construction which reduced durability. The survey showed that well-mixed and thoroughly consolidated concrete, in which no reinforcement came within 25 mm (1 in) of any exposed face, would stand exposure for over 50 years to the atmosphere of an industrial town without serious deterioration. The recommendations of this report are as relevant in 1981 as they were 25 years ago.

The other studies of buildings mentioned dealt with individual structures in much more detail and were not concerned solely with durability but with overall success or otherwise. Of the 15 buildings examined, only two could be described as being primarily of reinforced concrete although the others had floors and sometimes other structural components of reinforced concrete. In two of these latter buildings, problems were experienced with corrosion of the reinforcement, in one case in precast concrete mullions with too little cover of concrete and in the other in a floor slab, cast in 1905, using a breeze aggregate with inadequate compaction, which was exposed to the weather without any protection for more than a year. The two structures constructed primarily of reinforced concrete were the Royal Horticultural Society's second hall built in 1928 and an office building of framed construction put up in 1936. The buildings were unique for different reasons; the main structure of the exhibition hall consisted of six arches spanning 21 m (69 ft) and rising 18 m (59 ft) above the floor, which were instrumented to obtain some of the earliest measurements on the creep and shrinkage of concrete; the office building comprised a basement, ground floor and six upper stories, and was supported at foundation level on a 100 mm (4 in) thickness of cork to provide insulation against traffic vibration and noise. When these structures were inspected in 1958 and 1962 respectively, no structural defects in the concrete were seen.

4.5 Commentary on assessments

None of the buildings considered in this section was selected for study as a result of being defective or possibly defective, and in this sense these buildings are, therefore, representative of concrete construction. They are, however, very few in number and, in so far as they are representative, they relate to structures built a considerable time ago. Since then, there has been a number of changes in design and construction, which have aimed at more economic construction. Nevertheless, it is clear that design was generally conservative (except for shear at column heads, which has been dealt with in the revisions of codes) and reduction in the overall factors of safety seems more than justified—in the appraisal of structures of high alumina cement concrete, it was clear that the real margins in more modern construction were still appreciably greater than those assumed in design. An area of much more concern is that of durability since the trends have been towards the use of less cement in the concrete and reduced requirements for cover until relatively recently. While the assessments suggested that durability was usually adequate, cases were noted which were not entirely satisfactory. More problems of durability now seem likely to be experienced, although it is being increasingly recognized that concrete is less durable than was sometimes supposed. In the future, a need can be seen for much more emphasis on supervision of workmanship, since durability is so dependent on proper compaction of the concrete.

5 Lessons from failures

5.1 Incidence of failure

It is difficult to obtain a full assessment of the number of structures that fail in relation to the total population of structures or, more particularly, for certain categories of structures which would enable sensitive areas to be identified. Some of the reports, tabulating failures, will be referred to later, but it is inevitable that not all failures become known and many may well be rectified before events become serious as a result of rigorous checking procedures. The available statistics may therefore be useful in deducing the lessons that should be learned, but are less so when attempts are made to determine whether or not the incidence of failure is acceptable. The evidence from the number of deaths among the general public due to structural collapses, which may be fully recorded in fact, number about six annually in the UK, and is so small that it has no statistical significance when compared with about 8000 deaths that occur on the roads annually and a similar number who die in accidents in the home.

A review of the building collapses investigated by the Building Research Establishment in the period from 1973 to 1977 inclusive [15] is summarized in Table 5-1. It shows that in all there were nine buildings of which six were owned by public authorities, two were in private ownership and one, although structurally complete, was still to be finished by the contractor. A few other failures were known to have occurred in this period but were not investigated and are not therefore included; some in private ownership may not have been publicized. The materials of construction included steel, concrete, masonry and timber and all were single-span roofs of prefabricated construction; that four were of concrete is probably not a disproportionate number. Since the spans in most of the cases were relatively long, the dead load was a high proportion of the total design load and hence this type of construction is likely to be more sensitive than, for example, floors. Floors are also likely to have a greater built-in continuity and composite action with a slab and its finishes, which are further reasons for the failure of roofs being more common. A part of this chapter will therefore deal with a number of failures of low-rise precast concrete construction, including most of those shown in Table 5-1.

Other parts of this chapter will deal with the failure of the multistory large-panel structure, Ronan Point, which led to revisions being adopted in the approach to design not only in Britain but in many other areas of the world, and with cooling towers which, because of their special character and construction appear to have presented special difficulties. The reasons why problems were encountered with high alumina cement concrete and with the use of calcium chloride will be examined, as well as the causes of failure due to the performance of certain natural aggregates in concrete.

5.2 Aldershot and low-rise construction

In the period from the end of the war to the early 1960s, there was an urgent need to renew and increase the stock of buildings at a time when skilled labour was in short supply. Innovative methods of construction were therefore adopted using a much wider range of factory produced components than hitherto; these included structural components of a variety of materials with a predominance in the use of precast concrete. Not all the developments were entirely satisfactory for some of the types of material and a few structural failures were experienced. One of these was a precast concrete barracks building at Aldershot (England), which collapsed, in 1963, during construction when almost structurally complete [16].

The design of the building, which was one of four similar officers mess buildings, had been adapted from a system of construction widely used in the reconstruction of barracks in Aldershot. They were, however, unique in design, comprising three stories and a penthouse with precast reinforced concrete beams, columns and floor slabs; only the floors of the stairwells were cast in place. The top story was partly offset from the other stories below as an architectural feature; the middle story was of open plan with external glazing only but the other stories had non-load-bearing partition walls. The buildings were about 19 m (62 ft) square in plan (Fig. 5-1).

At the time of collapse on 21 July 1963, two of the buildings, including the one that failed, were almost complete except for some of the cladding and the filling of some of the column joints; the other two buildings were still incomplete structurally. After the collapse, which had given some warning of its imminence by prior cracking of the concrete at some of the joints and which was almost total, the other structures were demolished (Fig. 5-2).

The report of the Aldershot failure [16] identified a number of defects in design, in the manufacture of the components and in the construction on site, which included the following:

(a) The beam-to-column connections had a nominal bearing of 38 mm ($1\frac{1}{2}$ in), which was half that recommended in the then current Code of Practice for a

Table 5-1 Structural collapses of buildings investigated during the period 1973 to 1977

Item	Location	Type of building and form of construction	Date of construction	Date and time of failure	Nature of failure	Cause
1	University of Leicester	Reading room. Precast prestressed roof beams with pre-cast edge beam and column units. Span 12.4 m	1964	Late morning 12 June 1973	Sudden collapse of part of the roof after failure of the edge beam	Inadequate bearing and low strength high alumina cement concrete in edge beam; poor control of cement content
2	Camden School	Assembly hall. Precast prestressed composite beams on precast edge beams and columns. Span 12 m	1954–5	About 22 00 h 13 June 1973	Sudden collapse of whole roof	Inadequate bearing of roof beams on edge beams aggravated by corrosion of steel and conversion of high alumina cement concrete
3	Sir John Cass's Foundation and Red Coat C of E Secondary School Stepney	Swimming pool. Precast prestressed concrete roof beams on concrete edge beams with brick walls. Span 10.1 m	1965–6	Late morning and again early evening 8 Feb 1974	Slight warning before sudden collapse of first beam; second beam collapsed later	Conversion of high alumina cement concrete subsequently attacked by sulphate
4		Warehouse roof with prestressed concrete and main and secondary beams. Main span 18 m	1962	About 07 00 h 16 May 1974	Sudden collapse of one main beam with post-tensioned steel	Corrosion of tendon due to 2–4% calcium chloride in concrete by wt of cement

No.	Name	Description	Year built	Date of failure	Type of failure	Cause
5	Ilford County High School for boys	Swimming pool. Timber roof supported by plywood box beams. Span 12.6 m	1963	During night 2–3 Oct 1974	Collapse of central third of roof	Failure of a glued joint with some deterioration of glue and timber due to condensation and to excess dead loading possibly with ponding
6		Sports hall. Welded steel lattice inter-connected portal frames with light-weight concrete deck. Span 17 m	1974 Construction effectively complete	During working hours. 12 Nov 1974	Sudden collapse of whole roof and part of walls	Inadequate welding of joints
7		Community centre. Finger jointed timber beams. Span 9 m	1968?	During night 26–27 Sept 1975	Collapse of one roof beam	Lack of quality control in the finger jointing process
8	Rock Ferry Comprehensive School Birkenhead	Sports hall. Timber trussed rafters supported on block masonry walls. Span 18.3 m	1973–4	During afternoon of Saturday 3 July 1976	Sudden collapse of whole roof and part of walls.	Inadequate lateral stability of trusses
9		Woodworking machine shop. Pitched roof of two lattice steel beams bolted at mid-span supported on brick walls. Span 9 m	1957	About 09 00 h 10 Jan 1977	Sudden collapse of whole roof	Under-design of connecting bolts and over-loading by ceiling insulation and ducting. Snow had fallen during the night

Figure 5-1 General view of one of the nearly completed buildings at Aldershot showing the absence of bracing in the first story and the offset second story

bearing of precast concrete on concrete. The actual length of bearing on site was often less than this, and reinforcement, which was intended to extend into the *in-situ* concrete in the column joint, was flush with the ends of the beams and therefore ineffective in providing continuity to hold the structure together.

(b) The overhang of the upper story introduced for aesthetic reasons required special beam connections, which were constricted in size to fit the beam geometry. These particular joints were made by forming a nib on the end of the secondary beams which was supported by a corbel on the side of the main beams, and located by a dowel bar; the nibs and corbels were intended to be reinforced around the dowel bars, but this reinforcement was omitted or displaced in some instances, and the concrete section was insufficient to resist the shear stress (Fig. 5-3).

(c) It was assumed in design that the columns were axially loaded and no provision was made to resist bending due to eccentric loading from beams or to withstand wind loading. No allowance was made for the reduced stiffness of the columns where these passed through the floors and the dimensions were reduced.

It was concluded that the most likely cause of failure was fracture of one of the nib-to-corbel joints in the floor beams at second-floor level, where cracking was in fact found in one of the other buildings. The collapse of the whole structure would

Figure 5-2   View of the Aldershot building after the collapse with part of the central core, which had cast *in-situ* floors, still standing

then have followed since none of the joints could withstand any form of accidental lateral loading.

The Aldershot report was widely disseminated and served to reinforce the misgivings felt by many engineers about some of the developments taking place in precast concrete construction as well as in other forms of construction. As a result, a symposium was organized by the Institution of Structural Engineers in London in 1966 [17], where the problems of industrialization and their implications for structural engineers were discussed. Although further misgivings were aired, no specific recommendations were made and it was not until the collapse of Ronan Point two years later (referred to later in this section) that specific provisions were made in British Codes of Practice. Before this time, however, a number of precast structures with inadequate bearings and with little or no arrangements for continuous ties had been built which were later to come to light as a result of a further small number of failures. Two of these occurred on consecutive days in June 1973 [18], the first in the Bennett Building at the University of Leicester and the second in the roof of the assembly hall of the Camden School for girls in London (Table 5-1, Items 1, 2).

At the Camden school [19], the whole of the roof over the seating area of the assembly hall collapsed with little prior warning—fortunately during the late

Figure 5-3 One of the fractured corbels of a second floor beam at Aldershot showing the position of the dowel bar and absence of the reinforcing links

evening when it was unoccupied (Fig. 5-4). The hall had been built during 1954 to 1955 as part of an urgent renewal programme necessitated by war damage. The roof, which had a clear span of about 13 m (42.6 ft), consisted of rectangular precast, prestressed concrete beams, 390 × 100 mm, (15.3 × 4 in), spaced about 0.6 m (2 ft) apart (Fig. 5-5). They were prestressed with 5 mm (0·2 in) diameter pretensioned wires and reinforced with two longitudinal bars of 5 mm diameter, to which were fixed vertical links of 5 mm diameter. The links projected into the cast *in-situ* concrete deck and shear connection was also provided by castellations along the top of the precast beams. At each end of the beams, there was a nib, 220 mm (8.7 in) in depth with a bearing length of 38 mm (1½ in), which was reinforced by a shear bar of 10 mm diameter in the form of a horizontal U. The structural screed varied in thickness between 38 and 90 mm (1½ and 3½ in), being cast against permanent formwork of corrugated asbestos sheeting; it was covered by a lightweight concrete screed from 50 to 100 mm (2–4 in) in thickness for thermal insulation and to provide a fall across the roof and finished with a waterproof membrane of 19 mm (0.75 in) of asphalt. The structural screed was linked to the main precast reinforced concrete beams with reinforcement consisting of bars of 5 mm diameter at a spacing of 300 mm (11.8 in). These main beams spanned between precast

Figure 5-4 Assembly hall, Camden School for Girls, London

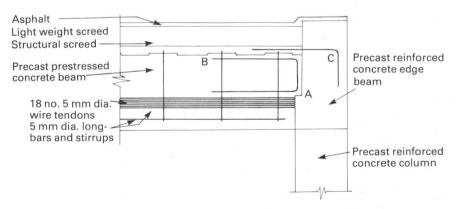

Asphalt
Light weight screed
Structural screed

Precast prestressed
concrete beam

18 no. 5 mm dia.
wire tendons
5 mm dia. long-
bars and stirrups

B

C

A

Precast reinforced
concrete edge
beam

Precast reinforced
concrete column

Figure 5-5 The joint between the roof beam and the edge beam at the
Camden school, showing the reinforcement and the narrow bearing
A **38** mm wide bearing
B **10** mm diameter shear bar
C **5** mm diameter continuity bar

columns, which were fixed to the adjacent classroom block along one side of the hall and free-standing along the other side, and supported the roof beams on a continuous corbel 38 mm ($1\frac{1}{2}$ in) wide. In the failure, the whole of the concrete roof collapsed to the floor below.

It was concluded in the report [19] that the failure had been initiated by the fracture of the nib at the end of the roof beam nearest to the stage, which was followed by the successive failure of each of the supports to the roof beams along this side of the hall, either due to the fracture of the nibs or of the corbel. The cause of the collapse was attributed to

(a) Insufficient bearing of the roof beams on the edge beams to allow effective reinforcement of the nibs (Fig. 5-6).
(b) Termination of the prestressing wires within the span of the roof beams without provision for the continuation of the longitudinal steel into the nibs.
(c) Insufficient structural cross-tying of the building.
(d) Reduction in the strength of the concrete in the prestressed roof beams, which had been made of high alumina cement.
(e) Deterioration and corrosion of the continuity reinforcement between the edge beams and the roof—(Fig. 5-7).

Figure 5-6   One cause of failure at the Camden school: the insufficient bearing of a roof beam on an edge beam

Figure 5-7 Another cause of failure at the Camden school: deterioration and corrosion of the continuity reinforcement between the edge beams and roof

In the later comments made on this particular failure, there has been a tendency to concentrate attention on the fact that the roof beams were cast in high alumina cement concrete and to ignore other important aspects of the failure, such as the nature of the bearing and the effects of corrosion. Although the bearing was appreciably narrower than the recommended width, tests on the ends of beams salvaged from the debris showed that, even when the width of bearing was reduced to 25 mm, a factor of safety against failure of 1.9 was obtained, which at first sight would appear to justify the design, but only serves to emphasize that laboratory conditions do not reproduce those experienced in real buildings. The importance of detailing reinforcement in such a way that it can be effective in the manner intended is also important. In this case, the shear bar in the nib at the end of the roof beams could not provide satisfactory reinforcement in shear and, if positioned wrongly, could introduce a plane of weakness in shear, as it may have done in the first beam to fail. Damage to the connecting steel between the roof beam during manufacture and subsequent corrosion resulted in this steel being ineffective well within the expected life of the structure, when it should have prevented the collapse or at least given warning of its imminence—bars of larger diameter would have been much more durable.

The collapse at the University of Leicester was also due to failure of an inadequate bearing, which might have been avoided by providing a wider bearing and improved detailing of the reinforcement at the joint between the edge beam and the roof beams. In this instance too, the use of high alumina cement concrete has masked some of the conclusions that should be drawn. The edge beams were of precast reinforced concrete, for which high alumina cement was chosen in order that the concrete should be dark in colour to satisfy an architectural requirement. High alumina cement was also used for the pretensioned roof beams to speed up their manufacture. The investigation after the collapse of part of the roof showed that one of the bearings on the edge beam had failed and that other edge beams exhibited signs of distress at similar bearings. Although the bearing area was inadequate, the primary cause of failure was the low strength of the concrete resulting from the omission of part of the cement content from the mix, which was further seriously reduced by the effects of the conversion of a high alumina cement concrete with an excessive water content. The prestressed beams were subjected to loading tests and retained as part of the building since their strength was found to be adequate after the conversion of the concrete. Thus, although certain aspects of the design were unsatisfactory and unexpected deterioration in strength of the concrete occurred, the failure originated in the lack of supervision and control in the precast works and the initial choice of the cement for use in the edge beams for a reason which appears to be trivial in retrospect.

Immediately following the failures at Camden and Leicester, where the buildings had been designed by the same consultant, an inspection was instituted of some 16 educational buildings with bearings of a similar character, most of which were strengthened. Some of these buildings also contained beams made with high alumina cement concrete, which did not appear to present any special problems due to loss of strength. In these cases, the use of this cement was intermittent as a substitute for Portland cement when it was necessary to increase the rate of output from the precast works to avoid hold-ups on site. Although there was some examination of the recommendations dealing with the use of high alumina cement at that stage, it was not until the collapse at Stepney in the following year that the evidence on performance suggested a need for major reappraisal (Section 5.4).

Three other failures of low-rise construction are also considered here, since they occurred with common forms of structure and their collapse had widespread implications; these were (Table 5-1 Items 3, 4, 8)

(a) The roof of a swimming pool building in a school at Stepney, London, which was built in 1965–6 and had two of its roof beams fail in February 1974 [20]. The failure was attributed to the low strength and poor quality of high alumina cement concrete, which resulted in serious chemical attack. The problems that this raised for prestressed concrete made with this cement are examined in Section 5.4.

(b) A prestressed concrete roof beam in a warehouse, erected in 1962, failed in May 1974 as a result of the fracture of the prestressing wires. These were severely corroded due to excessive amounts of calcium chloride in the concrete. The use of this additive was not permitted by the specification for the beams and, at the time, it appeared to be an isolated occurrence, but subsequent experience showed that it was not uncommon for calcium chloride to be incorporated in prestressed concrete units. The consequences are dealt with in more detail in Section 5.5.

(c) The third of the failures was that of the complete roof of a school gymnasium built in 1973–4 at Birkenhead (England) which collapsed in July 1976 [21]. The roof structure was of timber trussed rafters supported on masonry walls of concrete blockwork. The primary cause of failure was due to overall buckling of the slender rafters as the result of the omission of diagonal bracing. The situation was, however, aggravated by the slenderness of the walls and the provision of movement joints at each corner of the building, which detracted from the overall robustness of the structure. This particular failure raises the whole question of the stability of buildings with structures comprised of two or more different materials, which has not so far been adequately treated in codes of practice.

Reference has already been made in this chapter to the occurrence of failures in structures of materials other than concrete and that a number of these were of low-rise construction of a type often used for the construction of educational and other public buildings. These often have roofs of relatively long span, which are particularly vulnerable, since they are directly exposed to extremes of the environment, have a high value for the ratio of dead to live load and hence are permanently loaded to a high proportion of their service load; they are commonly prefabricated, so having little built-in robustness. Floor construction is not inherently so vulnerable since each of these factors is less adverse. There appears, therefore, to be a case for the introduction of greater conservatism in the future in the design of roofs than in the design of floors.

The stability of low-rise buildings [22] of what has been described as hybrid construction presents major problems, since the requirement, which also applies to high-rise construction, is that the structure must not only be capable of sustaining the clearly defined dead, imposed and wind loads in safety without loss of serviceability but also survive abnormal situations, such as extremes of loading or excessive weakness. The abnormal situations are difficult to quantify and to take into account in design. For concrete construction, these difficulties are most serious with concrete blockwork and where precast components are used in conjunction with brickwork, where there are inherent tensile weaknesses. Elsewhere, much can be achieved by the incorporation of steel reinforcing ties.

A number of the low-rise buildings which have failed were built for educational use. At the time, they were needed urgently at minimum cost in labour and materials and so they favoured the introduction of innovatory ideas and the development of building systems and standardized structural components. The ideas that were successful have become traditional and those that were not tend to be remembered because a single failure led to the requirement that all buildings using that system or those components should be appraised. Sometimes, the failure was compounded by adapting the system to an unfamiliar use without sufficient critical thought, as at Aldershot. A new design of connection might be introduced with insufficient preparation to ensure that it could be built, would not deteriorate during the life of the structure and was insensitive to inadequate workmanship, as at Aldershot, Camden and Leicester. Lack of clarity in the definition of responsibility for design and construction has contributed to failures, as at Aldershot where the designer worked for the contractor and not for the building owner and at the school in Birkenhead where the design of the trusses and of the walls were carried out in different organizations. Insufficient supervision and control of workmanship have also played their part, at Leicester where too little cement was used in the mix, at Stepney where too much water was used in the concrete and in the warehouse where calcium chloride was added in spite of the specification, which required otherwise.

5.3   Ronan Point and multistory buildings

The collapse of part of a multistory block of flats, known as Ronan Point, following a gas explosion on 16 May 1968 [23] received world-wide attention and directed engineers towards a more positive approach to structural safety in their designs (Fig. 5-8). It had, however, even wider implications since it drew the

Figure 5-8   Ronan Point: progressive collapse

attention of the public to the problems encountered by engineers, which had hitherto been regarded as none of their concern.

Inevitably, most of the technical reports on this failure concentrated on the inadequacies in the design of this and many other similar structures rather than on the circumstances which led to the acceptance, almost without question, of unsatisfactory and deficient concepts for their design. If similar origins of failure are to be avoided in the future, some thought must be given to the climate of engineering opinion, in addition to the technical aspect, which led experienced engineers to design, approve and build structures later found to require costly structural repairs.

The dramatic nature of the failure has also resulted in an exaggerated view of the magnitude of the failure. While it was clearly too serious to be accepted as a consequence of a gas explosion in one flat, it should be noted that the total floor area destroyed in the collapse was not more than 6% of the total floor area. Fortuitously, the number of casualties were relatively few; four people were killed, 17 were admitted to hospital but only three of these were detained, of whom one died later due to other causes. The list of casualties would have been much more severe had three of the most severely damaged flats been occupied at the time and had the explosion occurred when the occupants were in the sitting rooms in the SE corner of the block, which was almost completely destroyed.

The London Borough of Newham, which owned this block of flats, like the other Boroughs of East London, had suffered very severe damage to housing as a result of bombing during the war. By 1968, nearly 17 000 new dwellings had been built by the Borough for rehousing, but it still had 8 000 names on its waiting list and additionally needed to clear 9 000 slums. Construction of new dwellings was therefore extremely urgent. Until the mid-1950s, most of the new building had been of two-story houses and three-story blocks of flats, which gave a density of about 70 people to the acre. There had then been a change in central government policy, which thenceforward encouraged the building of blocks of flats of from eight to 23 stories with a density of between 140 and 150 persons per acre. A major limitation to the expansion of the rate of building was the shortage of skilled labour, and consequently construction from factory-made components had to be adopted even though it tended to be more expensive than more conventional construction. The Borough Council examined a number of different industrialized building systems before selecting that used for Ronan Point and a number of similar blocks. The design was prepared by a firm of consulting engineers, which was a subsidiary of a large firm of contractors with responsibility for the work on site and a major interest in the precasting firm supplying the components. Byelaw approval was obtained and the Borough's engineer gave a certificate of compliance, which was required before an Exchequer subsidy could be paid. It was in this climate that the construction was planned.

The construction of Ronan Point started in July 1966 and was completed about two months before the gas explosion occurred shortly before 0548 hours on 16 May 1968. The building, which was about 24×18 m (79×59 ft) in plan, consisted of 22 floors of precast concrete construction erected on a single story podium of reinforced concrete built on site. It was a load-bearing wall structure with stairwell and lift shaft. Spine walls were located on either side of the longitudinal central corridor with transverse walls including the flank walls, which were all

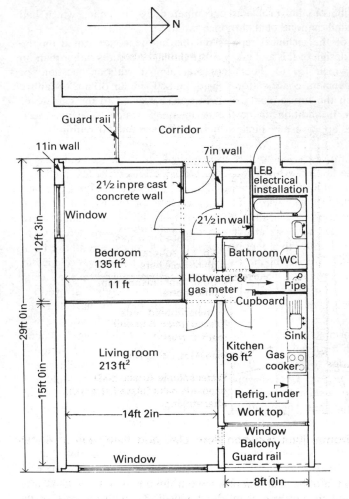

Figure 5-9 Layout of Ronan Point flat

load-bearing. The other external walls and a few of the partition walls were non-load-bearing (Fig. 5-9). The loadbearing walls were reinforced, about 2.4 m (7.8 ft) in height and 2.7 m (8.8 ft) long. The floor units, which had a span of about 4.4 m (14.4 ft) and a depth of about 0.25 m (10 in), were also about 2.7 m in width so that joints in many of the walls and floors were almost in the same vertical plane. Continuous reinforcement was provided in the transverse direction across the building only in the joint between the floor and flank wall units; short galvanized steel straps were also used to connect the lifting bolts at the top of the flank wall units and bolts in the floor units; other reinforcement in the joints between floor units comprised only short bars (Fig. 5-10).

The explosion occurred in a corner flat due to leakage of town gas at a cooker connection in the kitchen on the 18th floor, and propagated into the sitting room

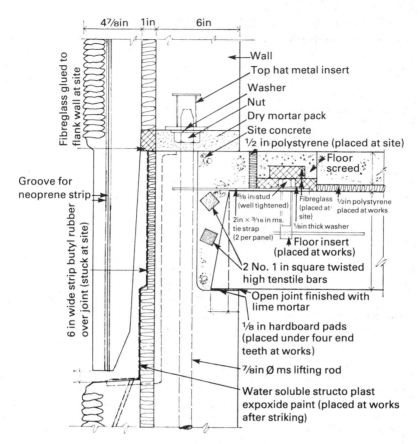

Figure 5-10 Horizontal joint between floor slab and flank wall in Ronan point

in the SE corner and into the bedroom blowing out three panels of the flank wall. Their removal caused the collapse of all the panels above with the ends of the floors which they were supporting. This collapse then caused the collapse of two bays of wall and one bay of floor units progressively downwards in all the lower stories. Only superficial damage was caused, however, to the reinforced concrete podium. Damage to other flats was also slight.

The tribunal set up to enquire into the cause of the failure concluded that the maximum explosion pressure in the sitting room on the 18th floor had a mean value of 21 kN/m² (3 psi) and that it had probably been two to three times as high in the kitchen and more than four times as high in the entrance hall. This explosion was regarded as not being of exceptional violence. Other conclusions reached by the tribunal [23] included the following:

(a) The risk of a gas explosion occurring in a flat in a high block is no greater than in any other form of dwelling but in a block the size of Ronan Point, with 110 flats and a life of 60 years, there is a 2% risk of a gas explosion

causing structural damage in the lifetime of the block. In other words, one block in 50 may suffer in this way some time during its lifetime.
(b) Progressive collapse is not an inevitable feature of system-built blocks. It can be avoided by the introduction of sufficient steel reinforcement to give continuity at the joints, and the adoption of a plan-form which provides for the arrangement of the load-bearing walls in such a way that the load is carried by alternative paths if part of the structure fails.

These conclusions were drawn partly from the evidence of experience of other construction and partly from experimental data obtained in the laboratory from tests on assemblies of panel and floor units supplied by the precast concrete manufacturer.

Experience of the results of an explosion which destroyed wall panels at ground-floor and first-floor levels in a large panel block of flats in North Africa was quoted. In this instance, the damage was confined to the immediate area of the explosion because effective tying reinforcement had been placed in the floors to hold the structure together.

Seven years after the collapse at Ronan Point, another gas explosion was experienced in a 16-story block of flats, Mersey House in Bootle (England) [24]. It occurred in the caretaker's flat at ground-floor level at 0245 hours on 28 August 1975. The block was designed and built at about the same time as Ronan Point and the designer did not, therefore, give any special attention to the possible effect of gas explosions. The form of construction was, however, different since it was composed of a cast *in-situ* reinforced concrete framework with in-fill walls of no-fines concrete cast with the framework. The explosion resulted from a leak of natural gas in the kitchen, which built up into a gas layer under the ceiling, as is thought to have occurred at Ronan Point. It was estimated that the peak pressure reached was about 35 kN/m^2 (5 psi) on the basis of the damage to Mersey House and the adjacent blocks. Extensive damage was caused to the windows of two nearby blocks and to some other dwellings nearby. Severe damage was caused within the caretaker's flat where parts of the external walls and the internal partitions were destroyed and two of the reinforced concrete columns suffered considerable displacement. The floor of the flat above was deflected upwards causing damage to the partitions and there was also some deformation of the structure at this level. There was some damage in the entrance hall and at higher levels through the release of pressure into the lift shaft. The caretaker, who was believed to have been in the kitchen at the time of the explosion, suffered extensive burns and other injuries and was removed to hospital with some of the other inhabitants with shock or less serious injuries.

Although the building was not designed to comply with the requirements for tying reinforcement introduced into the Code of Practice, CP 110, after the Ronan Point failure, the incident showed that

(a) accidental explosions of the severity envisaged in the amendments to the *Building Regulations* made in 1968 could occur with sufficient frequency to require attention in design;
(b) had the provisions made in the Code of Practice after 1968 been followed in the design, the resulting damage at Mersey House would have probably

been somewhat less than was actually experienced, although this was not such to cause undue concern.

Study of the incident at Mersey House provided valuable confirmation for the recommendations made by the tribunal on Ronan Point with regard to the incidence and magnitude of accidental explosion resulting from leakage of gas from main supplies.

The benefit gained from *in-situ* construction at Mersey House in resisting excessive damage when compared with Ronan Point is not just due to the fact that one was of *in-situ* concrete and the other of precast concrete. It was the result of the former building having continuity of reinforcement in three dimensions as a consequence of being a framed structure of traditional design, whereas precast construction inevitably has a large number of discontinuities at joints between components and there is some difficulty in providing continuity of reinforcement. To avoid progressive collapse in any form of construction, there must be sufficient ductility at all sections in the structure to ensure that the debris from a local failure does not directly or indirectly destroy a disproportionately large part of the structure. That progressive collapse may in certain circumstances occur with cast *in-situ* reinforced concrete in conventional construction was illustrated by the collapse during construction of a tower block at Bailey's Crossroads in Virginia, USA, in March 1973 when 14 workers were killed and 34 injured [25].

The tower block in question at Bailey's Crossroads was similar in design and construction to two others at the Skyline Centre which had already been completed and occupied. The building comprised cast *in-situ* concrete columns of normal weight concrete and cast *in-situ* flat slab floors of lightweight aggregate concrete. Its structure was of reinforced concrete with brick masonry cladding, and it was one of two high-rise apartment buildings, which were intended to be of 26 stories with a four-story garage between them. At the time of the incident, the block which was damaged was almost complete structurally to the 24th story with the formwork in position for casting part of the next floor; the second block was about 6 m (20 ft) above the foundations and the garage, which had reinforced columns and post-tensioned and unbonded flat slabs for its floors, was complete to the whole of the first floor and part of the second floor. Each of the floors in the apartment block were cast in four sections, the overall dimensions of each floor being 118×23 m (387×75 ft). Shortly before the accident occurred, concrete had been placed in the third section of floor at the 24th story level; then the form-work supporting the 23rd floor from the 22nd floor was removed, precipitating the collapse of this third section of floor at each floor level throughout the height of the building as well as the floors of the garage. It was concluded subsequently that props were removed before the concrete had gained sufficient maturity and therefore a floor failure for some other reason at a later age might not have led to progressive collapse; nevertheless, it suggests that some conventional cast *in-situ* reinforced concrete construction could be vulnerable to this form of failure.

The investigation of the Ronan Point failure included loading tests in the laboratory of assemblies of components made to determine the strength and behaviour of the joint between the floor slabs and the flank wall. When there was no vertical load on the flank wall, the horizontal force required to disrupt the joint was equivalent to a uniform pressure of 23 kN/m^2 (3.3 psi); when a vertical load equal to the dead load on the flank wall at the level where the explosion occurred was applied in a test, the strength at failure was in excess of 28 kN/m^2 (4 psi). In each case, the joint failed in horizontal shear with little deformation. When a similar joint with the additional tie steel required by the revised Code of Practice, CP 110, was tested, however, it sustained a force equivalent to a uniform pressure of 40 kN/m^2 (5.8 psi) without failure, but both floors and walls were very severely deformed in flexure. The extra cost of providing this amount of tie steel was small, but when the behaviour in this test is considered in relation to structures, such as Ronan Point, it must be concluded that, had Ronan Point been reinforced in this way, the floor and wall units above the explosion would not have separated at the joints and so produce the debris causing the successive collapse of the floors below. The damage done to the structure in the immediate region of the explosion might have been less, but this would have depended on the extent to which the increase in resistance to lateral loading would have allowed venting of the explosion through the external non-load-bearing walls and so released the build-up of pressure.

Although the report of the tribunal [23] on the Ronan Point failure was critical of the design of the block for wind loading, subsequent research on wind loading on tall blocks suggested that the deficiencies were not serious in this respect. It could therefore have been said that the designers complied with the requirements of the *Building Regulations* and current Codes of Practice in their design, since they were not required to consider the effects of accidental loadings, such as gas explosions, although subsequent investigation showed that their incidence was significant in causing structural damage and casualties. Even if the designers had been required to consider the effects of gas explosions, they would probably have regarded the results of the first two tests, referred to above, as satisfactory in showing adequate resistance to lateral loading, since the paramount need to tie the structure together might not have been obvious in the absence of first-hand experience. The engineering lessons from Ronan Point, therefore, are the avoidance of rigidity in the interpretation of design requirements and the application of imagination and intuition at the outset when making the decision on the structural design concept to adopt. A more general observation may also be appropriate, that social necessity or other adminstrative pressures, such as the need to meet specific dates in a financial programme, should not be allowed to interfere with the processes of effective design checks.

A matter which must be of some concern is that the results of tests, as in the case of the tests on the bearings following the Camden failure, could have been misleading and suggested that the joints were of adequate strength when what was required was ductility. Particular care is therefore necessary to ensure that the differences between real conditions in the structure and those simulated in the laboratory are given proper weight.

5.4 High alumina cement concrete

Although there had been a number of occasions when the structural use of high alumina cement concrete had been brought into question in the UK and other

European countries, its use in structures was not prohibited in the UK, although some of the other countries took a different view. Following the collapse of the roof at the Camden school and of the roof beams at Leicester, described in Section 5.2, it was concluded that in each case the quality of the concrete did not conform to the recommendations in the Code of Practice, CP 116, for the use of the cement [26], and that it would be unreasonable, therefore, to propose that its use should be discontinued in the precast concrete industry where many factories based their production cycle for prestressed components on its very rapid hardening characteristics. The collapse of the two beams at Stepney, in February 1974, completely changed the perspective because these had been made to a specification which conformed to the recommendations of the current British Standard Codes of Practice, although it was ascertained on investigation that the temperatures in the building should have led to a modification in the design stresses.

The Stepney swimming pool building [20] was built as one of a number of buildings in a school development programme in which the walls were of brickwork with precast prestressed concrete beams for the roof structures. The roof over the swimming pool, which was almost identical with that of the adjacent gymnasium, (Fig. 5-11) had a clear span of 10.1 m (33 ft), and consisted of precast joists, 0.30 m (12 in) deep, at 1.5 m (5 ft) spacing; these supported woodwool slabs 50 mm (2 in) in thickness, which were overlaid with a Portland cement screed and waterproof membrane of asphalt. The construction was completed in September 1966. It was inspected in 1973 following the Camden failure and its structural condition was reported as being satisfactory.

The pool was in use immediately before the collapse of the first beam, but some warning was given of its imminence when fragments fell into the pool, which was

Figure 5-11   Stepney swimming pool building

Figure 5-12 Stepney swimming pool after collapse of the second beam

immediately cleared of people. The beam failed shortly afterwards and a second beam collapsed several hours later (Fig. 5-12). It was clear from the condition of the debris that the concrete was chocolate brown in colour, very friable and weak, showing the presence of a white deposit, which was later identified as ettringite resulting from sulphate attack. Since the roof was over a heated swimming pool, the temperature and humidity conditions were of first concern in view of the advice against using high alumina cement concrete in these conditions given in the Codes of Practice. It had been reported that roof leakage and considerable condensation had been experienced and estimates of temperature suggested that it would be between 28 and 30°C for long periods, sometimes rising to 35°C at roof level. Thus, acceleration of conversion with loss of strength due to the formation of the more stable but weaker calcium aluminate hydrate was expected, which was confirmed by analysis. The source of the sulphate attack was identified as the plaster, applied as a finish to the under-side of the woodwool slabs. As a check on these findings, tests were made on the similar beams in the adjacent gymnasium, where the temperature was about 10°C lower, the condensation less and the plaster finish absent. The tests showed that the concrete here was also highly converted but to a lesser extent, and that it had lost strength by a serious amount. It therefore was necessary to initiate extensive investigations on the performance of high alumina cement concrete and to carry out selective checks on res, since it was estimated that there were between 30 000 and 50 000 ɟs containing prestressed concrete beams made to the same specification.

The causes of the loss of strength by the concrete and the other factors affecting the structural collapse were analyzed in some detail. The defects could be traced to procedures adopted both in design and in the manufacture of the prestressed concrete components. Since these have general implications for engineers, they will be considered in some detail.

At the time the buildings were designed, the effect of temperature on the strength of high alumina cement concrete had been established for storage at 18°C and at 38°C and it was also known that, if the temperature rose above 27°C, the loss of strength became much more severe. This information appeared very shortly afterwards in the Code of Practice for precast concrete, which was first issued in 1965. Since the temperature in the swimming pool might be as high as 24°C, there was little margin to allow for the effects of solar gain at roof level. It might therefore have been appropriate to assume lower stresses for design, such as those later given in the Code for conditions where higher temperatures might be experienced. Subsequent examination of other structures showed that it was unusual to allow for the effects of the loss of strength in prestressed units of high alumina cement concrete, even when they were used in even warmer environments, e.g., over boiler rooms for major heating installations. The reasons for the wise-spread adoption of what was shown later to be an unduly optimistic interpretation of available data are not clear. It is possible that it was the result of the strongly expressed views that the conversion of high alumina cement concrete was not common in temperate climates and that, even if it did occur, the consequences with respect to loss of strength were not unduly serious provided that the water/cement ratio of the concrete was kept below 0.40 [27].

Even in general terms, the nature of the conversion of high alumina cement in concrete was not then understood by many engineers and, because of the complexity of the hydration processes, it tends to remain something of a mystery. When high alumina cement is hydrated [6] at normal temperatures, the products include monocalcium aluminate decahydrate and dicalcium aluminate octahydrate, which give the concrete its high early strength [28]. At the temperatures experienced in buildings and most other structures, the aluminates convert to the more stable tricalcium aluminate hexahydrate, which is of appreciably greater density, and hence there is a tendency for the matrix to become porous. The rate at which conversion takes place is critically dependent on temperature and to some extent on water/cement ratio. The effects are shown in Figs 5-13 and 5-14 [29]. The first of these figures shows that, at 18–20°C, concrete with a free water/cement ratio of 0.4 is about 50% converted after 10 years and that, at 38°C, the level of conversion of similar concrete is in excess of 80% after only 3 months. Figure 5–14 indicates that water/cement ratio has little effect at the higher temperature of 38°C, but is more significant at 18–20°C, taking place almost twice as quickly when the free water/cement ratio is 0.7 as when it is 0.3. The effects of conversion on strength are more difficult to determine since the minimum strength may not be reached for 5 to 10 years at the lower temperature, although it is reached at the higher temperature in a much shorter period, depending on the time elapsing after casting before heating takes place. If the concrete is heated immediately after casting, the minimum strength occurs after about 5 days but, if a period of 24 h elapses, the minimum is only reached after about 3 months.

The effects of temperature and free water/cement ratio on minimum strength

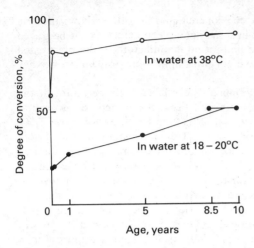

Figure 5-13 Relationships between degree of conversion and age for high alumina cement concrete with a free water/cement ratio of 0.4 for storage in water at 18–20°C and at 38°C

are shown diagrammatically in Fig. 5-15, which has been inferred from experimental data. The relationships between free water/cement ratio and minimum strength at the highest and at the lowest temperatures shown are reasonably well established, but the precise temperature at which there is a substantial drop in strength is not so clearly defined, although for concretes with low water/cement ratios it has been shown to occur at between 25 and 30°C. From these relationships, it may be seen that, provided the free water/cement ratio does not exceed

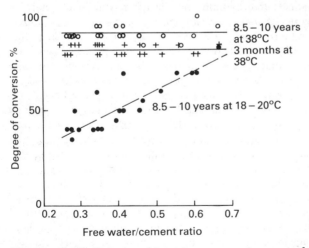

Figure 5-14 Relationships between degree of conversion and free water/cement ratio for high alumina cement concretes at different ages and at different temperatures of water storage

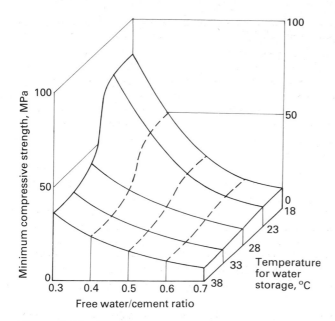

Figure 5-15 Diagrammatic representation of the interrelationship between minimum crushing strength, free water/cement ratio and temperature for water storage for high alumina cement concrete

0.4 and the temperature does not rise above 25°C, the converted strength is still that of a high-strength concrete; if the water/cement ratio or the temperature exceed these respective values, the additional loss of strength is serious, but if both are exceeded the effect is much more serious. The effect of water/cement ratio on strength appears to be due to the fact that the amount of water required for the complete hydration of the unconverted calcium aluminate hydrates is about twice that required by the tricalcium aluminate formed by conversion, and there is therefore release of water leading to increased porosity and reduced strength. When low water/cement ratios are used, there is insufficient water for complete hydration initially, but as conversion takes place, further hydration occurs, reducing the tendency for increased porosity, and thus assisting in the retention of strength. The rate of conversion is greater at higher temperatures and this results in increased crystal size of the tricalcium aluminate hydrate and lower strength. Chemical environment may also accelerate the rate of conversion and certain aggregates with freeable alkalis may also cause a more rapid loss of strength. Whilst the calcium aluminate content of the cement plays the predominant role in defining the strength of the concrete during the earlier part of its life, other hydration processes occur which lead to a gradual increase in strength with time and hence the strength of the concrete after reaching a minimum subsequently shows a small recovery.

The recommendations for the use of high alumina cement at the time when the Stepney school was being built have already been mentioned. While they were not based on as clear an understanding of the behaviour of the cement in concrete as that summarized in the preceding paragraph, they were reasonably sound, except that they implied that the loss of strength due to conversion did not necessarily occur in temperate climates and so too little emphasis was placed on the effect of temperature at any stage in the life of the concrete. Apart from the use of high alumina cement concrete in warm conditions without any limitation on design stresses, there was also a tendency to use the concrete in prestressed joists of standardized sections, some of which were sensitive to loss of compressive strength in the concrete. The beams used in the roof of the Stepney school were in this category; they were originally produced for use in composite construction with a concrete roof deck to be cast in place later in Portland cement concrete, and the dimensions of the top flange were therefore relatively smaller than those for other similar joists of I-section intended for non-composite construction. Tests on prototypes suggested that they might also be used for long-span lightweight roofs and a number were used in this way. The cross-section of one of these joists is shown in Fig. 5-16, and the effect of the strength of the concrete on their factor

Figure 5-16 Cross-section of one of the beams from the Stepney swimming pool roof—in contrast to this beam, the beams which failed showed an absence of large sized gravel from the top flange

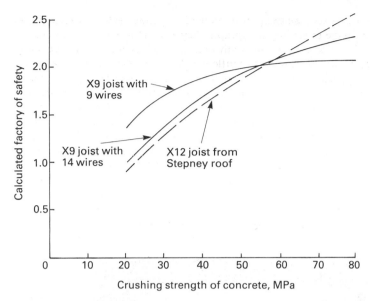

Figure 5-17 Calculated factor of safety for the Stepney beams as a function of the crushing strength of the concrete. The comparison is shown with under and over-reinforced X9 prestressed concrete joists

of safety is given in Fig. 5-17. The specified strength of the concrete in these joists was 52 MPa (7500 psi) at one day and experimental evidence showed that on conversion this could drop to about 20 MPa (2900 psi) for the specified water/cement ratio; Fig. 5-17 shows that this corresponds to a reduction in the factor of safety from about 1.8 to less than 0.9. Some of the cores cut from the beams gave mean equivalent cube strengths as low as 14 MPa (2000 psi) and the strength at the point of failure could have been even lower.

At the time of the manufacture of the beams, particular attention was paid to the control of concrete production to ensure that a low water/cement ratio was obtained with the lean mix proportions used of $1:8\frac{1}{2}$ by weight of cement to total aggregate. A siphon-can test, originally developed for determining specific gravity, was therefore used to determine water content. Critical examination of the results of this test, made in the investigation of the failure, showed, however, that an error in the test measurement of $\frac{1}{2}$% correspond to an error of 10% in the estimation of water/cement ratio. The test procedure adopted therefore gave the appearance of a rigorousness of control that did not exist in reality. Examination of the data on strength suggested that the free water/cement ratio for the Stepney beams may have been as high as 0.5. Another factor in production which may have influenced the quality of the joists was that, when a smooth top surface to the flange was required to support woodwool or similar cladding units, a finish of high alumina cement mortar was placed in the top of the moulds. This layer was intended to be a finish only, but in the joists that failed it extended through a considerable depth of the top flange in some places, and a mortar of unknown quality may therefore have been required to sustain the maximum compressive stresses in the joists. Once the structure had been completed, a decision was made to render the underside of the woodwool slabs with gypsum plaster and it is very doubtful whether the significance of this decision was appreciated, although it might have been relatively unimportant had the high alumina cement concrete conformed to the specification.

Overall, the main question to ask about the manufacture of the Stepney beams is whether it was feasible, as a routine production process, to specify a free water/cement ratio not greater than 0.4 for concrete mixes whose mix proportions by weight of cement to aggregate were 1 to 8 or more, in view of the consequences of exceeding this water content, which are much more serious for high alumina cement concrete than for Portland cement concrete.

The collapse at Stepney led to the structural appraisal of a large number of buildings and other structures where high alumina cement had been used [30, 31]. Nearly all of these were built with mass-produced prestressed concrete components and the majority were found to be structurally adequate, usually because they were incorporated compositely in the construction. In all, ten structures were reported as having failed or being close to collapse [32]; in the majority of these cases, deterioration of the high alumina cement concrete was the primary cause of failure but, in a few cases, the primary cause was some other structural defect, such as inadequate bearings, and then the loss of strength in the high alumina cement concrete was only a contributory factor. More than 100 structures were reported to have been repaired or replaced. Evidence from the appraisals showed, however, that the general level of conversion in buildings was appreciably higher than had been expected from laboratory studies and that the mean equivalent cube strength of the concrete obtained from tests on cores taken mainly from prestressed components was 37 MPa (5370 psi) with a range from 8.5 to 84 MPa (1230–12 000 psi), which may be compared with a Code recommendation for a design mean strength of 57 MPa (8270 psi). It is therefore clear that the Code, which was based on laboratory data, substantially underestimated the effects of temperature in real conditions and the variations in concrete quality that would actually occur.

5.5   Calcium chloride in structural concrete (See also Ch. 9, Section 1.3)

In May 1977, the British Standards Institution issued an amendment to the Code of Practice for the structural use of concrete, CP 110 [33]. It read as follows:

> Experience shows that corrosion of prestressing tendons, reinforcement and embedded metal usually results from a combination of factors including excess addition of calcium chloride caused by failure to maintain specified dosage, departure from specified mix proportions, poor compaction, inadequate cover and poor design.
>
> It is therefore strongly recommended that calcium chloride should never be added in prestressed concrete, reinforced concrete and concrete containing embedded metal.

Prior to that amendment, the Code had permitted the use of calcium chloride in reinforced concrete provided that the total anhydrous calcium chloride, plus the equivalent value of anhydrous calcium chloride calculated from the chloride in the aggregate, did not exceed 1.5% by weight of the cement. It went on to say that it should never be used in pretensioned, prestressed concrete, or in the main concrete of post-tensioned, prestressed concrete unless there was a durable barrier between the main concrete and the tendons. From the first introduction of BS Codes for structural concrete in 1948, clauses had been included permitting its use, although more recently they had tended to reduce the level permissible. The recommendations had been based on the results of research as well as engineering experience. It had been reported, for example, after an experimental investigation on reinforced concrete [34] that the use of 2% flake calcium chloride (i.e., 1.5% anhydrous calcium chloride) by weight of the cement in a dense, well-compacted concrete would have little effect on the degree of corrosion of the reinforcement, but it was suggested bars smaller than about 10 mm (0.4 in) in diameter might suffer a considerable loss of strength in a porous or badly compacted concrete. A report on the corrosion of prestressing steel published at about the same time [35] concluded that this quantity of calcium chloride in dense, normally cured, ordinary Portland cement concrete would have no structural significance; it warned that excessive use of calcium chloride would result in serious corrosion and appreciable reduction in the strength of the prestressing wire and that the use of other cements or steam curing increased the likelihood of corrosion. The reasons for the change of view expressed in the 1977 amendment requires examination therefore, since the original attitude had been held for 30 years and appeared to be supported by research.

The factors which led to the virtual banning in the UK of the addition of calcium chloride to structural concrete containing embedded metal arose mainly from accumulated experience and field investigations made as a result of failures in many parts of the world. One particular field study undertaken by the Building Research Establishment [36] serves to illustrate the lessons that may be learnt from this problem in concrete technology. Reference has already been made to the failure of a prestressed concrete roof beam in a warehouse in May 1974 as the result of the excessive use of calcium chloride. This building was one form of the Intergrid system of construction, which had been developed during the period immediately following the war, mainly for the construction of schools and similar types of building. As a result of the failure, inspections were made of a few other structures to ascertain whether calcium chloride had been used in the components, but the results were negative. Since the specification for the production of the components was specific in prohibiting the use of the additive, no further investigations were made. In October 1976, however, there was a report of deterioration of tendons in pretensioned Intergrid columns caused by the presence of excessive amounts of chloride in a school, which had been built about 20 years before. The local authorities owning similar school buildings with columns thought to have been made by the same precast concrete manufacturer were therefore asked to inspect their buildings. A number of these columns were found to have some rusting of tendons, and all owners of buildings were then advised to carry out inspections with the full cooperation of the proprietors of the system and their consulting engineers. The Building Research Establishment extended its research programme on the corrosion of steel in concrete to include the analysis of the reports of inspections by building owners, to provide advice on the general condition of the buildings, and to find the effects of calcium chloride in prestressed concrete from actual construction.

The Intergrid system, as first developed in 1952, used light modular precast concrete elements of small cross-section, which were prestressed together with post-tensioned wire tendons to produce the floor and roof construction, together with precast, pretensioned columns to provide support. Since then, a wide variety of developments had taken place; larger pretensioned beams were used in floors and roofs, and a number of special individual structures were built. In all, over a period of 24 years, more than 200 contracts had been completed at 171 sites with components from at least 18 manufacturers from 24 works. So far as the Intergrid system of construction was concerned, it was found that the majority of the buildings appeared to be in good condition and the incidence of visible defects associated with corrosion of embedded steel was low. One 20-year-old pretensioned column, which was seriously cracked, had a break in one tendon. Visible cracking, spalling or rust staining was found in about a quarter of the columns at two sites, in a few columns at 17 sites and in a few beams at 16 sites. At a total of 9 sites, external tendons to post-tensioned beams were found to have sustained some surface corrosion and, at one site, these tendons in the area of the kitchen and showers had lost a significant amount of their cross-sectional area as a result of rusting. For more than two thirds of the buildings for which chemical analyses of the concrete were available, i.e., from 109 sites, the chloride content was low, being less than 0.6% for the equivalent of anhydrous calcium chloride by weight of the cement. For the majority of the sites, therefore, it is unlikely that chloride was added. It was concluded, however, that it had been added in the manufacture of the beams used in nine contracts and the columns used in five contracts. It was not possible to identify a level of chloride below which corrosion of the tendons would not occur. The likelihood of corrosion of embedded steel and the rate at which it occurred appeared to be dependent on the combined action of a number of factors. The most important of these were the chloride content of the concrete, the age of the concrete, its alkalinity in the proximity of the steel and the humidity of the environment. For the risk of progressive corrosion to become significant in the expected life of a structure where the steel remains in alkaline surroundings and the concrete is directly exposed to moisture, it was estimated that the chloride content should not exceed the low level given above. In buildings where the concrete may be introduced in the aggregates, the conclusion meant effectively that the deliberate addition of chloride could not be recommended.

When the general results of this field investigation are compared with those of the laboratory study, referred to earlier, it becomes clear that the conflict in the conclusions arises mainly as a result of differences in the quality of concrete produced in the laboratory and that manufactured in the precasting works. The BRE investigation revealed a considerable amount of information on the degree of variation experienced in production of concrete, which was not previously available since there have been few opportunities for wide-spread sampling of concrete from a relatively homogenous group of structures.

Generally, the quality of the concrete on visual inspection appeared to be good, including the regions where the corrosion of tendons occurred. The cement contents of the concrete samples were mostly within the range from 14 to about 28%, i.e., ranging over a ratio of 1 to 2. It should, however, be remembered that the samples taken were small and, while being relevant to considerations of the corrosion of tendons which is very localized initially, probably have rather less application to considerations of strength. The chloride content of the concrete in components made after 1964 was generally low and suggested that very little, if any, was added subsequently. For some of the buildings in the early period, the content was high. On one contract, the amount, expressed as the equivalent anhydrous calcium chloride by weight of the cement, ranged from 0.6 to about 10%. More commonly, the range was in the proportion of about 1 to 2. Thus, in practice, when the amount used on average was, for example, $1\frac{1}{2}$%, it would be almost inevitable that the amount in any particular region could vary from $\frac{3}{4}$ to $2\frac{1}{4}$%. This was not apparently taken into account when the Code recommendations for reinforced concrete were set at $1\frac{1}{2}$%, which represented laboratory concrete where variations of this nature are small, and much of the conflict between the results of the laboratory studies and field investigations is then resolved.

These comments on calcium chloride in structural concrete would not be complete without a few details on the prestressed concrete beam which collapsed, as mentioned previously, in 1974. The building, which was a non-standard Intergrid structure, had a roof of 18 m (59 ft) span with main beams at a spacing of 7.6 m (25 ft), which were about 12 years old when one collapsed without warning. The beam was formed from three precast units and prestressed by three post-tensioned cables, each of 12 wires 7 mm (0.28 in) in diameter, contained in metal ducts and grouted. Examination of the failure revealed fractured tendons resulting from severe corrosion at one of the joints between units. Adjacent to this joint, the chloride content in the concrete was equivalent on average to about 2% anhydrous calcium chloride by weight of the cement and was about 1.3% on average in the grout. The grout had not completely filled the ducts and it was found to have been saturated with water. The duct formers were of mild steel and were not continuous across the joints, so that chloride ions could come in contact with the tendons; the ducts had also corroded and, for this reason as well, did not provide an effective barrier—elsewhere the chloride content in the grout was appreciably lower. The cause of the failure was therefore clear and, in the circumstances, not inexplicable. It was observed that, although there was no apparent external sign of the corrosion of the tendons, the corrosion of secondary reinforcement at the ends of the beams was sufficient to crack the concrete and cause some surface rust staining.

5.6 Cooling towers at Ferrybridge

During a severe westerly gale in November 1965, three out of eight cooling towers, all of which were structurally complete, collapsed at the Ferrybridge 'C' Power Station in Yorkshire, England [37]. A committee was set up immediately by the Central Electricity Generating Board to examine the collapse of the towers and to make recommendations. Their report, published in the following year, concluded that the failures were due primarily to a serious underestimation of wind loading in the design, which led to tensile failure of the vertical reinforcement in the lower part of the shells. Constructional defects were not regarded as significant. The report drew attention to a number of aspects of design and construction of this type of structure, which can provide future guidance to all structural engineers, and therefore merits further study.

Each of the towers, which were of conventional hyperbolic form, was 114 m

(374 ft) in height and 91 m (298 ft) in diameter at the base. They were staggered in two parallel rows 107 m (351 ft) apart. The thickness of the shells was 127 mm (5 in), which had been used for other cooling towers, although these were the largest that had been built at this thickness up to that time in the UK. It was normal practice for similar groups of towers to be spaced apart by a diagonal distance of 1.5 times the base diameter but, in this case, the ratio was 1.42 because of the need to provide a pillar of support in the coal measures below. There were, therefore, several departures from experience in their design and construction.

The wind conditions at the time of the failures were reported as being very severe, sufficiently severe for workmen to be withdrawn from the towers. Estimates of wind speed at the site from meteorological data recorded in the locality showed that the maximum mean hourly speed at 10 m (33 ft) height was 18–20 m/s (45 mile/h) and, therefore, that the gale was not exceptional but could be expected to recur at least once in five years. The estimated maximum wind speeds at 10 m for Ferrybridge, likely to occur on average once in 50 years, were a mean hourly wind speed of 24 m/s (54 mile/h) and a maximum gust of 43 m/s (96 mile/h). More recent meteorological data have led to an upward revision of this gust speed to 46 m/s. Estimates of gust speed from damage to the roof of the nearby power station suggested that it may have reached as much as 45 m/s during the gale causing the collapses. It is therefore evident that there was some defect in the design of the towers.

The design was based on conventional membrane theory and detailed scrutiny of the method, in the course of the inquiry, did not suggest that this introduced any significant error in the calculation of the stresses in the structures. Little information on the mode of failure could be obtained from examination of the debris since this was very fragmented and was mainly confined within the area of each base. Examination of the towers which were left standing showed that vertical cracks had developed, extending downwards from the top of the shells through the full thickness of the walls and distributed around the circumference. One tower was found to have horizontal cracking in the main body of the shell and three towers, including this tower, were observed to be cracked vertically from the ring beam upward to near the top. The latter cracking was attributed to minor differential settlement of the foundations. Eye witness accounts were reported for the failure of the second and third towers to collapse, the first having occurred without warning; these described ovalling and rippling just below the throat, followed by the flexing in and out of a substantial area and then by horizontal rippling across this area until a hole appeared there and the top toppled forward and the wall caved inwards with substantial collapse following, which seemed to be similar for each of the three towers. Some rippling movements were seen later in the towers that remained standing. After the preliminary examination of information available, the inquiry examined five modes of failure, which were considered as possible: foundation failure; shear failure at the top of the shell; instability of the shell; vibrational failure; and tensile failure within the shell. The results of a number of wind tunnel studies on model hyperbolic shells were available before the collapse and at the committee's request these were supplemented by further wind tunnel tests and analyses.

The committee found from their investigations that the layout of the station buildings and cooling towers had an important effect on the wind loading. As a result, the estimated vertical tensile stresses in the towers which collapsed were significantly greater than in the towers to windward which remained standing, or would have been in an isolated tower. Calculations showed that these stresses corresponding to the estimates made of the wind speeds experienced were sufficient to cause failure. Various aspects in the approach to design were also regarded as contributing to the cause of failure. The specification for the design did not give the duration to be assumed for the maximum mean wind speed or the return period to be used in design, guidance was not clear on the interpretation of wind tunnel data, and the figures used were lower than had been used previously in the design of towers of smaller size. The design procedure adopted relied on safety factors applied to the strengths of the materials to provide the margin against failure without any estimate being made of the conditions that would cause failure. Since the tensile reinforcement required was then calculated from the difference between tensile uplift stresses due to wind loads and compressive stresses due to dead loads, the amount of reinforcement provided was very sensitive to the errors made in estimating the wind loads. Had safety factors on the loads been introduced with values, which depended on the reliability of the assumptions made in determining the magnitude of each source of load, the significance of these errors in the estimates of wind loading would have been very much reduced. This improved approach to design is obtained by the inclusion of various combinations of partial safety factors for loads in limit state design, and the resulting additional complication would seem to be completely justified for this type of construction.

The main finding of the committee, after examining all the possible causes of failure, was that the failure was the result of a serious underestimate of wind loading in design leading to tensile failure of the vertical reinforcement in the lower part of the shells. The examination ruled out the other possible causes of the Ferrybridge failures although it was felt that forced vibration resulting from vortex shedding might have been a secondary contributory factor.

The dimensional accuracy of the cooling towers at Ferrybridge was carefully supervised during construction as these were the largest towers to have been built at that time. The departures from circularity on the first of the towers were found to be small, but they were greater on those built later. For one of the towers, the deviations were as much as 200 mm (8 in) and the rate of deviation or correction with height was greater than for the first tower. The thickness of the shells had an average value of 127 mm (5 in), which was the design dimension, with a standard deviation of 8 mm (0.3 in). The reinforcement, which should have been located in the centre of the wall, was an average 8 mm from the centre towards the inner face with a standard deviation of 11 mm. These dimensional inaccuracies were not regarded in the inquiry as being of importance, but constructional inaccuracy was found to be the likely cause of the collapse of a large reinforced concrete hyperbolic cooling tower, which failed in 1973 in Scotland.

The cooling tower at Ardeer [38] had already been designed when the report on the Ferrybridge failures was published and its design was therefore reappraised in the light of the recommendations of that report. Changes were then made in the design of the tower, which was 107 m (351 ft) high above the soffit with a maximum shell diameter of 79 m (259 ft); the thickness of the shell was increased from 127 to 152 mm (6 in) and the maximum gust speed was increased; the load factor against failure due to uplift was then calculated to be 1.3. During construction, substantial deviations from the specified shape of the tower were discovered and, following a reappraisal of the design, the thickness of the upper portion was increased to reduce meridional and uplift stresses. At the time of the collapse of the tower in September 1973, it was estimated that the maximum 3s gust speed was about 25 m/s (56 mile/h) at 10 m height. This was appreciably less than that assumed in design, and the form of the failure was unexpected since the tower collapsed in a direction approximately at right angles to the direction of the wind. The largest values for the imperfections in shape had been recorded as being mainly in the middle of the tower at heights between 30 and 75 m (98–246 ft). The maximum inward deviation of the shell was about 530 mm 21 in, but in one region an inward deviation of about 230 mm (9 in) was adjacent to an outward deviation of the same amount. The average radial error between the elevations of 48 and 55 m (157 and 180 ft) and of between 55 and 61 m (200 ft) was, in each case, about 200 mm (7.8 in). In addition to the records of geometric imperfections, records were maintained of the growth of meridional cracking throughout the life of the structure, and it was reported that shortly before the failure one of these cracks, a dominant double crack, had penetrated the major area of dimensional deviation. Analysis showed that the presence of substantial deviations in the shell caused significant tensile stresses in an uncracked shell at about 70° on each side of the windward meridian, which could be sustained by the concrete. Once cracking had occurred through the zone of maximum tensile stress, however, the situation became much more serious. Failure would then occur at a much lower wind speed due to the fracture of the tensile reinforcement. Such a mode of failure was consistent with what actually occurred and the growth of the critical crack explained why previously gales of similar severity had been sustained without failure. This particular collapse illustrates the importance of the supervision of construction for structures that are particularly sensitive to dimensional accuracies and that such sensitivity may become greater with time.

A later failure in the USA [39] illustrates a very different aspect of the design and construction of cooling towers and other structures requiring specialized methods of construction. In April 1978, one of two hyperbolic cooling towers, which were under construction for a power station at Willow Island, West Virginia, collapsed causing the death of the 51 men working on the scaffold at the time. After the collapse, the National Bureau of Standards was called in by the Occupational Safety and Health Administration to investigate the cause of failure. The first of the two towers, which were 131 m (430 ft) in height with a base diameter of 109 m (358 ft), had been completed, and the shell of the second was being concreted when the failure occurred. The scaffolding and the lift of concrete cast on the previous day collapsed. The inquiry into the failure included a field investigation at the site, a study of the system of construction, laboratory tests on components used in construction and theoretical analysis of the partially completed shell.

A patented system of lift form was used for construction, which had been employed in building a number of other towers without any fatal accidents. It was designed to enable a $1\frac{1}{2}$ m (5 ft) lift of concrete to be placed daily and, when the failure occurred, the 29th lift was being placed. Construction of shells of this kind with thin walls of large diameter, which varied with height, have led to the development of special construction techniques, and this lift-form system, like others, relied for its support on the structure as it was being built. The conclusion of the inquiry into the failure of the tower at Willow Island was that there was no inadequacy in the construction equipment, but that the failure was initiated by failure of the anchorage of the concrete hoist in the concrete cast on the previous day; the cylinder crushing strength of this concrete was estimated to have been $1\frac{1}{2}$ MPa (200 psi), which was less than the strength required to resist the local stresses induced in construction. It was therefore recommended for the future that the key operation of form removal and movement should be controlled by an engineer and that specific provisions should be made for testing field-cured concrete specimens before the removal of forms to ensure that the strength of the concrete is sufficient to bear the construction loads likely to be imposed.

These three cases of failure of a highly specialized form of construction illustrate very clearly the need to review the 'traditional' approach to structural safety in design and construction when innovations introduce a greater degree of sensitivity to inadequacy.

5.7 Aggregate anomalies

The increasing use of marine dredged sands and gravels in concrete and the experience of corrosion of embedded steel when calcium chloride has been added as a concrete admixture have resulted in the UK in a revision of the recommendations governing permissible levels of chloride in concrete. It has been set at 0.1% for prestressed concrete for the percentage of chloride ion by weight of the cement. This limit was chosen in the light of general experience with chloride in concrete, bearing in mind that the cross-sectional area of tendons is small in relation to surface area and hence that they are particularly sensitive to corrosion and loss of strength. A consequence is, however, that, when marine aggregates are used for prestressed work, they require efficient washing. For sulphate-resisting Portland cement with embedded metal, the limit on chloride content permitted is 0.2% and, for ordinary Portland cement with embedded metal, the limit is 0.35% for 95% of results and 0.5% for all results. It should be noted that these figures were adopted as a reasonable precaution and not as a result of failure of concrete made with marine aggregate.

In a number of parts of the world, concrete failures have been experienced as a result of alkali–aggregate reaction—a form of failure which, for many years, was completely unknown in the UK. However, within the last few years, there have been reports of a few cases of damage, which were not sufficiently severe to require immediate attention. They were identified as being the result of alkali–silica reactivity [40, 41] resulting from the presence of silica in the form of opal or chert, which reacts with the sodium or potassium hydroxide solution released by the hydrating cement to produce an alkali–silica gel; this draws in water and can produce sufficient pressure to cause expansion and disruption of the concrete. It has been found that there is a pessimum value for the proportion of reactive material in the aggregate and that the amount of expansion is dependent on the mix proportions and the alkali content of the cement. It seems possible that the slight increase in the occurrence of this form of damage in the UK may be linked

Table 6-1 Causes of defects in structures

Origin of defect	Details of defect	Examples in the failures described
Design fault	Incomplete understanding of structural behaviour including sensitivity to accidents	At Ronan Point, the possibility of a gas explosion had been ignored since there was no specific requirement for its consideration in design In the school at Birkenhead, the roof trusses were not braced; it appeared to be assumed that the walls braced the timber trusses and the trusses braced the concrete block masonry walls At Aldershot, the instability of unbraced slender precast concrete columns in a three-story building was not checked
	Incorrect assessment of loading including dead, imposed and wind loads, and dynamic effects	Ignoring the effect of the temperatures in flat roofs boiler rooms, etc., on the rate of conversion of high alumina cement concrete At Ferrybridge Power Station, incorrect interpretation of wind tunnel data and insufficient allowance for the close spacing of the cooling towers. Failure to assess the margins against failure under wind loading Failure to assess or to assess correctly the loads due to placing concrete at Willows Island cooling tower and in building at Bailey's Cross—the most common cause of collapse of concrete bridges
	Choice of unsuitable materials	Use of high alumina cement in prestressed concrete where the components were sensitive to failure due to loss of strength Use of aggregates which react with cement and cause deterioration

Errors in detailing	Inattention to the need for practical limits for tolerances in precast construction at Aldershot and the Camden school
	Inadequate bearing areas for the supports in precast work at Aldershot, in the Camden school and at the University of Leicester
Construction faults	
Poor control of workmanship	Insufficient cement in the concrete at Leicester and too much water at Stepney
	Excessive use of calcium chloride even though its use was prohibited in the Intergrid system of prestressing
	Inadequate prestress in some high alumina cement concrete components
Inadequate control of progress	Premature removal of falsework at Bailey's Cross
	Insufficient strength of concrete to sustain construction stresses at Willows Island cooling tower
Departures from working drawings	Structurally significant errors in the form of the Ardeer cooling tower
	Omission and misplacement of reinforcement in precast components at Aldershot
Unauthorized changes of use	There have been a number of failures due to changes in loading with a change in occupancy and failures have also been experienced as a result of modification to the structure—none was included in the earlier section
Inadequate maintenance	Failure to maintain weather-proofness, particularly of flat roofs
	The problems associated with the use of high alumina cement and calcium chloride with prestressed concrete were aggravated by water penetration causing dampness in buildings

with the tendency for the alkali content of cements to increase with more efficient methods of cement manufacture and the winning of aggregate from the sea where its real origin cannot be traced.

Problems of the shrinkage of natural aggregates in concrete, which have been experienced in Scotland [42], have been sufficiently serious to require the replacement of some buildings. The majority of aggregates available in the UK, whether gravel or crushed rock, do not shrink or expand but certain aggregates have been found, which exhibit large volume changes on wetting and drying. The shrinkage of concretes made with these aggregates can be as much as four times that of concrete made with non-shrinking aggregates. A survey of aggregates has shown that some igneous rocks of the basalt and dolerite types and some sedimentary rocks, especially greywacke and mudstone, can be shrinkable; other types, such as granite, limestone, quartzite and fresh felsite, are not. The problem is most serious in reinforced concrete buildings, particularly those with under-floor heating, where the concrete is able to dry out completely, and is much less serious in massive civil engineering structures where the concrete has little opportunity to dry at all. Damage due to the use of shrinkable aggregates has occurred in South Africa, Australia and New Zealand, and laboratory data suggest that it could occur in other parts of the UK in addition to Scotland. For exposed concretes, air entrainment reduces the risk of damage and the introduction of compression steel in floors and beams helps to reduce excessive deflection. Guidance on the use of these aggregates and on a method of test have been developed for general application.

The nature of the aggregate can, as already described, contribute to the loss of strength of high alumina cement concrete and, although it is not of great significance, it serves to illustrate further that the use of natural aggregates in concrete can present some difficulties. Aggregates, such as granite, shales, micaceous sandstones and schists, may contain releaseable alkalis, which result in more rapid conversion and may contribute to chemical attack on high alumina cement concrete. It appeared that an aggregate of this type aggravated the deterioration of the concrete in the beams which collapsed at the Stepney school in 1974 (Section 5.4).

It should be appreciated that the frequency of problems of concrete damage resulting from the use of an unsuitable aggregate are rare in most countries. Where they do exist, local experience of the durability of concrete will usually provide a guide to suitability, and the likelihood of needing to carry out extensive testing on an aggregate, once its rock type has been identified, is likely to be small.

6   Lessons—concluding commentary

The descriptions of the incidents given in the previous section have been brief and have dealt with a relatively small number of failures. A much wider range have been described in a number of other publications [43–52], but the direct causes can be assigned to a small number of technical possibilities under the following headings, which provide a means of identifying the engineering lessons: design faults; construction faults; unauthorized changes of use; and inadequate maintenance. This has been done in Table 6-1 for the failures described earlier and, even though these incidents tend to be the most spectacular since they have called for investigations and published reports, they illustrate the causes as clearly as the commoner and more prosaic forms of structural unfitness for use, such as unwanted cracking and deflection.

From the summary given in Table 6-1 and from the collated data from other reviews, the following conclusions of a general technical character can be drawn:

(a) It is relatively easy to establish the technical causes of structural failures.
(b) Failures are very seldom due to any lack of basic scientific or technical knowledge, but rather to a failure to ensure that the knowledge was properly applied or to an absence of engineering judgement.
(c) Failures are usually due to human error, which are likely to be the result of mismanagement—the acceptance of the wrong priorities, misunderstanding or lack of communication between those with different responsibilities within the project.

Since human error is the primary cause of failure, the most effective way of avoiding failures in the future is to ensure that the responsibility for structural safety is in the hands of competent engineers, well educated and trained, and capable of seeing beyond their own immediate responsibilities. Clearly, the quality of the education and training of engineers and their supporting staff, the aptness and validity of codes of practice, standards and regulations, and the calibre of the individuals who choose to be engineers and technicians have a direct relevance to the quality of the finished product. These can, however, only attain the standards required if there is a coherent and articulate profession giving active encouragement to their development.

In this chapter and in making these comments, much more attention has been given to the failures experienced than to the achievements that have led to the highly sophisticated forms of concrete construction that have become essential to the foundation of modern industralized society. These achievements far outweigh the failures and need to be remembered when the incidence of failures is over-emphasized. It is easier to recognize human error than the enterprise of the individuals over a period of more than a century, who have contributed to the development of the use of structural concrete. While engineers responsible for different aspects of these developments can be identified by name, the evidence suggests that, had the innovation not taken place where it did when it did, it would have happened elsewhere soon afterwards. To avoid concluding too complacently, the final comment on lessons from the past might be left to a mathematician:

> That's the reason they're called lessons,' the Gryphon remarked, 'because they lessen from day to day.
>
> Lewis Carroll

Acknowledgments

Figures 5-1 to 5-12 and 5-16 are Crown Copyright and appear by courtesy of the Director of the Building Research Establishment and by permission of the Controller of HM Stationery Office.

References

1. Neville, A. M., A study of the deterioration of structural concrete made with high alumina cement, *Proc. Inst. Civ. Eng.* Vol. 25, July 1963, pp. 287–324.
2. Institution of Structural Engineers, *Report on the use of high alumina cement*, London, 1964.
3. Rusty tendons felled Berlin Hall, *New Civ. Eng.*, 10 July 1980, p. 5. Berlin collapse report confirms tendon overload theory, *New Civ. Eng.*, 16 Oct. 1980, p. 8.
4. British Standards Institution, War emergency BS 968: 1941, *Concrete sleepers*, London, 1941.
5. Davey, N., *A history of building materials*, Phoenix House, London, 1961. Davey, N., Roman concrete and mortar, *Struct. Eng.*, Vol. 52, No. 6, June 1974, pp. 193–195.
6. Lea, F. M., *The chemistry of cement and concrete*, 3rd Edn, Arnold, London, 1973.
7. Hamilton, S. B., *A note on the history of reinforced concrete in buildings*, National Building Studies Special Report, No. 24, DSIR, HM Stationery Office, London, 1956, 30 pp.
8. Thomas, F. G., Prestressed concrete, *Proc. conf.*, Feb., 1949, Institution of Civil Engineers, London, 1949, 132 pp.
9. Matthews, D. D., The background to CP 110: 1972 The structural use of concrete, *Unified code symposium*, Institution of Structural Engineers and the Concrete Society, London, 1973.
10. *The durability of reinforced concrete in buildings: a report on a survey undertaken in 1954, jointly by the Building Research Station and the Cement and Concrete Association*, National Building Studies Special Report, No. 25, DSIR, HM Stationery Office, London, 1956, 32 pp.
11. Hamilton, S. B., Bagenal, H. and White, R. B., *A qualitative study of some buildings in the London area*, National Building Studies Special Report, No. 33. DSIR, HM Stationery Office, London, 1964, 163 pp.
12. White R. B., *Qualitative studies of buildings*, National Building Studies Special Report, No. 39, DSIR, HM Stationery Office, London, 1966, 59 pp.
13. Ockleston, A. J., *Tests on the Old Dental Hospital, Johannesburg*, Concrete Association, Papers Nos 1–6, Johannesburg, 1956.
14. National Academy of Sciences, *Full-scale testing of New York World's Fair structures Vol. 2: The Rathskeller Structure*, 1969.
15. Bate, S. C. C., Practical problems of the structural assessment of buildings, *Colloq. inspect. mainten. struct.*, Cambridge, 1978, International Association for Bridge and Structural Engineering, British Group, London, 1978.
16. Building Research Station, *Technical statement by the Building Research Station on the collapse of a precast building under construction for the Ministry of Public Building and Works*, HM Stationery Office, 1963, 18 pp.
17. Institution of Structural Engineers, *Proc. symp. industrialized building and the structural engineer*, London, 17–19 May 1966, 332 pp.
18. Scott, G., *Building disasters and failures*, Construction Press, Lancaster, 1976, 168 pp.
19. Bate, S. C. C., *Report of the collapse of the assembly hall of the Camden*

School for Girls, Report of Special Investigation by the Building Research Station for the Department of Education and Science, HM Stationery Office, 1973, 16 pp.

20. Bate, S. C. C., *Report on the failure of roof beams at Sir John Cass's Foundation and Red Coat Church of England Secondary School, Stepney*, Building Research Establishment Current Paper, CP 58/74, 1974, 18 pp.

21. Menzies, J. B. and Grainger, G. D., *Report on the Collapse of the Sports Hall at Rock Ferry Comprehensive School, Birkenhead*, Building Research Establishment Current Paper, CP 69/76, 1976, 13 pp.

22. Institution of Structural Engineers, *Proc. symp. low-rise bldgs hybrid const.*, 5 July 1978, London, 53 pp.

23. *Report of the inquiry into the collapse of flats at Ronan Point, Canning Town*, HM Stationery Office, London, 1968, 71 pp.

24. Mainstone, R. J. and Butlin, R. N., *Report of an explosion at Mersey House, Bootle, Lancs*, Building Research Establishment Current Paper, CP 34/76, 1976.

25. Bailey's Crossroads synopsis, *Civ. Eng. (New York)* Vol. 45, No. 11, Nov. 1975, pp. 59–61.

26. British Standards Institution, CP 116: 1965, *The structural use of precast concrete*, London, 1965.

27. Masterman, O. J., High alumina cement concrete with data concerning conversion, *Civ. Eng. Publ. Wks Rev.*, Vol. 56, No. 657, 1961, pp. 483–486.

28. *High alumina cement concrete in buildings*, Building Research Establishment Current Paper, CP 34/75, 1975, 50 pp.

29. Bate, S. C. C., High alumina cement concrete in existing building superstructures. To be published 1982.

30. Building Regulations Advisory Committee, *Report by Sub-Committee P (High alumina cement concrete)*, Department of the Environment, London, 1975., 12 pp.

31. Report of Working Party on high alumina cement concrete, *Struct. Eng.*, Vol. 54, No. 9, Sept. 1976, pp. 352–361.

32. Bate, S. C. C., High alumina cement concrete—an assessment from laboratory and field studies, *Struct. Eng.*, Vol. 58A, No. 12, Dec. 1980.

33. British Standards Institution, CP 110: 1972 (as amended 1977) *The structural use of concrete*, London, 1980, Parts 1–3, 156 pp., 104 pp., and 93 pp.

34. Blenkinsop, J. C., Effect on normal $\frac{3}{8}$ in. reinforcement of calcium chloride additions to dense and porous concretes, *Mag. Concr. Res.*, Vol. 15, No. 43, March 1963, pp. 33–38.

35. Roberts, M. H., Effect of calcium chloride on the durability of pre-tensioned wire in prestressed concrete, *Mag. Concr. Res.*, Vol. 14, No. 42, Nov. 1962, pp. 143–154.

36. Building Research Establishment, *The structural condition of Intergrid buildings of prestressed concrete*, HM Stationery Office, London, 1978, 62 pp.

37. Central Electricity Generating Board, *Report of the Committee of Inquiry into collapse of cooling towers at Ferrybridge, Monday 1 November 1965*, London, 1966, 16 pp.

38. Kemp, K. O. and Croll, J. G. A., The role of geometric imperfections in the collapse of a cooling tower, *Struct. Eng.* Vol. 54, No. 1, Jan. 1976, pp. 33–37.

39. Lew, H. S., West Virginia cooling tower collapse caused by inadequate concrete strength, *Civ. Eng. (New York)*, Vol. 50, No. 2, Feb. 1980, pp. 62–67.

40. Hobbs, D. W., Expansion of concrete due to alkali–silica reaction—an explanation, *Mag. Concr. Res.*, Vol. 30, No. 105, Dec. 1978, pp. 215–220.

41. Hobbs, D. W., *Influence of mix proportions and cement alkali content upon expansion due to the alkali–silica reaction*, Cement and Concrete Association Technical Report 534, 1980, 31 pp.

42. *Shrinkage of natural aggregates in concrete*, Building Research Station Digest 35 (2nd Series), HM Stationery Office, London, 1963.

43. McKaig, T. H., *Building failures—case studies in construction and design*, McGraw-Hill, New York, Toronto and London, 1963. 261 pp.

44. Feld, J., *Lessons from failures of concrete structures*, American Concrete Institute, Monograph No. 1, 1964, 179 pp.

45. Smolira, M., *Analysis of defects in concrete and brick structures during construction and in service*, Department of the Environment, London, 1969, 142 pp.

46. Smith, W. D., Bridge failures, *Proc. Inst. Civ. Eng.*, Vol. 60, Aug. 1976, pp. 367–382.

47. Melchers, R. E., *Studies of civil engineering failures—a review and classification*, Monash University, Civil Engineering Research Reports, No. 6/1976, 1976, 34 pp.

48. Blockley, D. I., Analysis of structural failures, *Proc. Inst. Civ. Eng.*, Vol. 62, Feb. 1977, pp. 51–74.

49. Sibly, P. G. and Walker, A. C., Structural accidents and their causes, *Proc. Inst. Civ. Eng.*, Vol. 62, May 1977, pp. 191–208.

50. Wearne, S. H., A review of reports of failures, *Proc. Inst. Mech. Eng.*, Vol. 193, No. 20, 1979, pp. 125–136.

51. Hauser, R., Lessons from European failures, *Concr. Int.*, Vol. 1, No. 12, Dec. 1979, pp. 21–25.

52. Walker, A. C., Study and analysis of the first 120 failure cases, *Proc. symp. struct. failures in buildings*, Institution of Structural Engineers, London, 30 April 1980, 78 pp.

3 Design philosophy and structural decisions

Sir Alan Harris

Harris and Sutherland, London, England

Contents

Summary

The building of concrete structures is seen as an art that is an action of the practical intelligence devoted to making things—useful things, which may achieve beauty. Much knowledge is needed; not for its own sake but in so far only as it is useful. Design is the designation of what is to be built and is central to building. The designer aims to satisfy the intended function at minimum cost and safely. The process of design is examined and is found to consist in a sequence of appreciation, conception, critical appraisal, decision, checking, elaboration and instructions.

The economics of the use of the material are governed by the very low cost of the basic materials. High costs of forming and shaping are thus acceptable only where low weight or refinement of shape are of direct economic value. The behaviour of the material under load is examined both by itself, when reinforced and when prestressed.

1 Introduction

The building of concrete structures is an art, by which is meant that it is an activity of the practical intelligence devoted to making something. It is thus distinct from science, which is an activity of the speculative intelligence seeking knowledge for its own sake and further distinct from prudence, which is an activity of the practical intelligence seeking the most effective course of action in the pursuit of some end.

As we shall see, the art with which we are concerned has need of science and prudence in the course of its practice. It is an art which seeks first to satisfy some useful purpose, though it may hope to achieve beauty.

For its execution, much labour, plant and money are needed; heavy responsibilities must be accepted. The sheer ponderability of concrete construction profoundly affects the nature of the art; it cannot well be seen as an expressive art, still less as ecstatic.

It is an art which requires much knowledge for its execution—knowledge which clarifies its purpose and facilitates its task. The basic and essential requirement is knowledge of its materials—how they are made, shaped, erected and assembled; how they stand up to use; how they weather; and how they fail. The intention with which this knowledge is sought, however, is quite different from that of the scientist; it is not sought for its own sake but only in so far as it is useful—and that at a given time and place.

The more knowledge about his task possessed by the engineer, the greater his freedom of action. On the other hand, he may not build beyond his knowledge.

2   Design

By design is meant the determination of what is to be built and the preparation of the instructions needed to get it built.

Design dominates structural engineering; construction is subservient to the intentions of the designer. There must be nothing in the structure not foreseen by the designer, although excellence in construction can perfect the beauty of a good design and even redeem a mediocre one.

There are three major purposes which the designer seeks to achieve in his structure. They are that the structure shall fulfil its function, that it shall be economic and that it shall be safe—in that order. There is no point in ensuring the safety of a structure which is too expensive to build, nor in constructing economically one which does not fulfill the required function. The detailed consideration of the safety of a structure is made only at a late stage of the design process.

In addition to these purposes, the very magnitude of the responsibilities of the designer imposes on him the pursuit of excellence, an excellence which, he may hope, will sometimes achieve beauty in the finished work.

These purposes will be examined.

2.1   Function

The structural designer is concerned with the *total* function of his structure, not merely the support of loads as expressed in codes of practice, etc. There may be many and complex actions on the structure to be resisted: fire, chemical attack, weathering, ageing, fatigue and other causes of deterioration. The designer will seek such robustness in his structure as is needed to counter not only use, but likely ill-use as well as probable or possible accident. Water-tightness, thermal and sound insulation are often capital features of the function. The reality of the 'load', meaning the forces which are to be borne by the structure, is likely to be remote from a simply expressed force per unit area and will most often consist of a spectrum of loads and their combinations occurring with varying frequency, the evaluation of which may pose considerable analytical problems.

Above all, the designer will seek the reflection of the purposes and uses of the structure in its material, form and strength.

2.2 Economy

In general, everyone seeks satisfaction at least cost. Money is the convenient measure of cost, but in the last resort cost is measured by the human effort needed to attain that satisfaction.

Structural engineering is no exception. Even if the structure is to give an impression of lavishness, of richness, of magnificence, its achievement at minimum cost will always be sought. All design decisions, right down to the choice of the smallest detail, are thus economic decisions; no such decisions can be made, either implicitly or explicitly, without at least an opinion on resulting cost.

There is a difficulty here. The determination of first cost is straightforward (though not always easy) but the evaluation of cost in use, maintenance costs and their relation to first cost is much more difficult. Each can be reduced at the expense of the other; somewhere there must be an optimum. The realistic search for least total cost would often lead to much higher expenditure on first cost.

Few owners of structures seem able to take account of final cost. The two components, of course, are commonly paid for from different funds and the distribution of cost between the two has financial and fiscal consequences of some magnitude. In practice, it is rare that anything but least first cost is sought and the designer using his judgment in seeking to avoid excessive maintenance may have to struggle to get an extra capital expenditure to achieve it.

2.3 Safety

To the lay public, safety means freedom from danger; but the engineer knows that total freedom from danger in his structures is unattainable, no matter how much money and effort is spent on design and construction. Communication on matters concerning safety thus has its problems. 'Is it safe?' asks the owner. 'The odds against failure are a million to one' [or whatever the figure be] replies the engineer. 'You mean it isn't really safe?' replies the owner. Etymologically speaking, he has a strong case.

Clearly, the consequences of failure of any likely structure are grave; there are few structures to which someone does not regularly entrust his life.

The public, indeed, is more sensitive to deaths resulting from the collapse of buildings than from many other causes. It has been said that we are ten times as sensitive to risks which we cannot avoid, such as that of a roof collapsing, as risks which we accept voluntarily, such as driving a motor car. If the avoidance of loss of life be the purpose, it can be argued that there are ways in which money can be better spent than on avoiding structural failure, but the argument excites little popular response. The designer must accept that his structures should not fall down.

Safety is a particular aspect of a more general desire—reliability. The owner of a structure requires that he might use it without interruption, whether the interruption be caused by maintenance, repair or collapse, and we shall see that reliability, the probability that an undesirable eventuality should not occur, is the key concept. The designer's estimate of reliability will be partly analytical but also partly a matter of judgment; the accident-prone structure may be difficult to represent in numerical terms but is recognizable to the experienced designer.

2.4 Excellence

The satisfaction of function at an acceptable cost by a reliable structure is the least that is expected of the designer but, as we have suggested, more is required of him. The typical product of the structural designer occupies much of the field of vision of the man in street, who rightly requires that it be not too displeasing to the eye.

Beauty may be desired, can be sought by the designer but is beyond his conscious grasp. Excellence in design, however, is directly his concern.

When the designer has tried hard to satisfy the needs of function, economy and safety and looks at the result of his labours, he will know that he has excelled when simplicity, unity and clarity are to be seen in it.

Simplicity is that impression of ease common to excellence in all the arts, in much the same way as lightness of touch is the perfection of strength; it is an absence of needless intricacies and complexities. It is attained only by effort. Note that the simple does not mean the elementary—it does not imply the use solely of simply supported beams. Indeed, the attainment of simplicity in the structure may well imply complexity in the analysis.

Unity means that the work is seen as a single thing; an organic unity, not an accretion of disparate parts. It is not to be obtained by hiding diverse parts behind an eye-catching screen—it is surprising how many ways they have of showing through.

Clarity is intelligibility: the structure is seen for what it is and what it does, without confusion. The clear is not necessarily the obvious; clarity is consistent with mystery, that sense of a richness yet to be revealed.

Necessity is perhaps the key. Paradoxically, if there be nothing in a structure but what is needed, the effect is not of privation but of wealth.

These are the signs which show that the design has been well done; how splendid the effect when they have been achieved! Emerson said, 'We ascribe beauty to that which is simple, which has no superfluous parts and which exactly answers its end'. If excellence in design be achieved, beauty will sometimes follow.

3 The design process

The stages in the design process are:

(a) the appreciation of a need;
(b) the conception of a project to satisfy the need;

(c) the appraisal of the project;
(d) decision;
(e) checking and elaboration;
(f) preparation of instructions;
(g) modifications during construction.

These stages will be considered in detail.

3.1 Appreciation of a need

The first stage is to meet the client. It is not impossible for a designer to be his own client but in civil engineering works of any magnitude, it is highly unlikely.

The client is as various as mankind—indeed, he is mankind. Sometimes, he is well-informed concerning what he needs, sometimes he is ignorant. Whoever he be, the encounter of client and designer is not to be seen as merely the briefing of an agent. The designer may be expected to know of possibilities unknown to the client, whose first statement of his needs may well be based on inadequate ideas of what can be done. What he thinks he wants and what he really wants may be very different; the first and perhaps the most significant task of the designer is to discover the client's real needs. This task is creative; great works of art rarely spring from anything but a creative sympathy between client and designer, a sympathy which the latter must do all he can to foster. But he should not be more clever than is necessary; he must not avoid the simple silly solution if it suits.

The designer will now seek to find out all he can about the task. To clarify it, he may have extensive recourse to science and to mathematical techniques, whose role in works of any complexity may be as valuable at this stage of design as at any other. He will also seek to know every possible practical detail, from the nature of the site and its climate and the source of materials and labour to the likely state of the market when construction is to begin.

The designer will be avid for fact, remembering that he will never have enough and that his information will seldom be wholly reliable.

3.2 Conception of a project

From the search for information, the designer now moves to the invention of a project which will satisfy the needs as he has come to understand them.

The conception of a project is an act of the creative imagination, an act which cannot be forced, cannot indeed be favoured except by removing obstacles and inhibitions. The starting point is the mass of fact which the designer has built up. He will focus his mind on this, meditate on it, brood over it, contemplate it. He will rearrange it, try to forget it to let the subconscious play its part. The hope is that ideas will spring to the mind; often they do, sometimes surprisingly complete, at other times in profusion, sometimes not at all. The incidence of ideas is independent of conscious volition and may happen at the most unlikely, indeed embarrassing, moments. While he cannot rely upon whether, much less when, an idea will occur, the designer may be certain that if he has not worked hard at understanding his task and if he has not concentrated fiercely upon such facts as he has assembled, it never will. Not ever.

This inventive faculty (and every new design is in some sense an invention) is a gift of the gods. There is no accounting for it, no analyzing it, no planning for it. It is far from uncommon, particularly among the young; it responds to cultivation and exercise—and can be stifled by discouragement. It is only a tool in the hands of a sentient being and the determination to use it is more important than the sharpness of its edge. It is important that at this stage of design, when the designer is examining what he knows of the work and is trying out tentative ideas of form, material and methods of construction, the critical faculty be left in abeyance— imagination must be uninhibited. The man who knows all the reasons why any idea will not work has no place at this time.

Some say that deliberate techniques can favour the conception of a project— 'wild thinking', 'brainstorming' sessions, relaxed weekend parties for the design team. So they can, but they have the danger that invention will get submerged in a prolixity of bright ideas.

There is no substitute for hard work and inspiration. Try the first; pray for the second.

3.3 Appraisal of projects

It is likely that, after the previous stage, the designer has ended up with a number of very various ideas of what the work might be. He must now examine them critically. Do they satisfy the needs? Are they feasible structurally? Are they easy to construct? Are they economic in first and final cost? What are the hidden snags? If there are shortcomings, can they be overcome?

If innocence and inexperience, with the freedom of imagination which goes with them, have their value at the concept stage, now comes the time for experience, for the man who knows all the snags. Indeed, good designers are known almost as much for the ruthlessness with which they criticize their own schemes as for their fecundity in producing them—and designers learn to sacrifice their darlings immediately they are found wanting. A confident designer knows he can soon think up a new and a better scheme stimulated by the very problems he has met; only a man doubtful of his powers feels that this one, the only one, must be made to succeed at all costs.

Appraisal requires some approximate development of the concept, but an experienced designer will probably need to carry out little structural analysis—no more, perhaps, than is enough to establish orders of magnitude of sizes of members. The matters on which he concentrates are more likely to be functional suitability, economy and, above all, ease of erection. If he does not give early thought to this last, he may find that his structure cannot be built at all. Economy, moreover, depends more on establishing a straightforward clear-cut method of erection with successive operations following each other without interference than it does on quantities of material.

There is a dilemma here: the designer should normally allow the contractor to choose the method of construction in the light of the plant and labour which he has available, but the designer must assume a method on which to base his design. In practice, if a designer can think of one way of building a structure, he can think of six and he may rely on a contractor to think of a seventh; if he can think of none, then none there probably is. Moreover, a coherent erection scheme with no inconsistencies will usually permit of a variant.

It is not unusual for reflection on erection methods to reveal some detail on the practicability of which the whole scheme depends and for the designer, at this early stage, to be obliged to develop it fully before he can appraise its practicability.

The designer may return to the client during the process of appraisal. New possibilities may have appeared or inadequacies been revealed; the finance may need rethinking. It is rare that the development and appraisal of ideas for the structure do not affect the original requirements. This spiral process of formulation of needs, collection of fact, conception, appraisal and re-examination of needs may be repeated several times with sometimes a new leap. Often it is only after much concentrated thought and laboured sketching and calculating that the simple, the obvious scheme occurs.

3.4 Decision

These intriguing intellectual exercises can continue indefinitely, the design doubtless approaching asymptotically some platonic ideal. But perfection is not for this world and in the meantime the client wants his structure. Decision is needed.

Decision is an act of the practical intellect concerned with prudence, with the conduct of affairs, the attainment of a practical end. It is a moral act and there is as much in it of the will as of the intelligence; not only must the designer know what should be done but must decide to do it and make the decision effective. It is now that practical consequences become real; now that the ponderous wheels are set in motion which will result in construction; now that responsibility is accepted for good or ill.

Like any other act of prudence, decision involves the balancing of contradictory good. To decide soon is to please the client and reduce the drawing office costs but the design may be the poorer for it. How often has one seen designs of infernal complexity produced in a frantic hurry when a few weeks longer would have permitted a fresh start and the achievement of simplicity? To leave it longer may perhaps produce a better design (though this is not certain—there is a point in the process beyond which a good design becomes over elaborate) but often a good design on time is useful while an exquisite masterpiece a few months late is simply too late.

Many design concepts at the stage of decision have in them features which, while good in themselves, are incompatible; to decide is to choose one and reject the other and to fumble and half choose is to invite confusion. Behind design decision lies a sense of the responsibilities involved; the designer knows that if he decides wrongly, dissatisfaction, trouble on site, disaster, death await.

The character of the designer is nowhere more manifest than now; nowhere are the virtues of the man of action more valuable—singlemindedness, courage, probity, devotion to the task. Here again, a keen and fertile mind is only an implement in the hands of a being possessed of volition; more important than its quality are the determination and decision with which it is used.

When the material of the structure, its form and method of erection have been settled, the major decisions have been made; keen analysis and detailing in the remaining stages of design will not have a large effect on either the function or the economy of the structure.

3.5 Checking and elaboration

The structure must now be fully drawn out and, as it is elaborated, it is checked for safety and economy.

Checking for safety has three steps:

(a) the definition of an action on the structure;
(b) the analysis of the effects of this action;
(c) the comparison of these effects with a criterion of adequacy.

We have seen above that there are many actions on a structure in addition to loads defined in codes of practice and regulations. It is just as important to estimate their magnitude as it is to analyze correctly their effects. They may be the subject of careful investigation on large structures, but even small structures must be designed in such a manner that their effect, however approximately evaluated, can be supported. Many conventions for proportions and minimum thicknesses of sections, minimum reinforcement, secondary reinforcement, etc. have been based, even if only intuitively, upon the need to support these secondary effects; such conventions may well be ignored if the designer so chooses, but only if he has satisfied himself fully that the alternatives are equally effective.

The definition of an action in an official document, is a useful guide to a designer but does not relieve him from the responsibility for correctly assessing the actions on his structure to be taken into account in design. Change of use may have to be allowed for; many standard loads are based on an unspoken assumption as to structural form and for an unusual form may be substantially in error, giving either an excessive or an inadequate load.

The power of the techniques of structural analysis has hugely increased these last two decades and no great difficulty exists in analyzing the most complex structures behaving in the most complicated manner. The traditional warning is worth repeating, however: concentration on an elaborate technique can obscure the purposes of the analysis. The designer needs a clear view of the process and the ability to judge the credibility of the answers at a glance.

Criteria of adequacy are in principle related to two limiting states of the structure, rupture and serviceability, i.e., that state at which the structure ceases to be a structure and collapses and that state at which it ceases to fulfil its function for reasons other than collapse. The rupture state is easy to define however difficult it may be to calculate, especially when combinations of different actions with different probabilities of occurrence have to be considered; collapse is at least recognizable.

Serviceability is more difficult. Deflection of a beam may well cause cracking of ceilings or rupture of partitions; excessive vibration may render a structure unusable; if a structure is cracked it may be unable to fulfil its function of retaining water. None of these conditions, however, can be generalized—deflection and vibration acceptable in one structure may not be so in another. Even with a water-retaining structure, a crack in bending tension may be permissible, particularly if occurring on the outside of the structure, although one in overall tension would not. The cracking load in prestressed concrete is often taken as a serviceability limit (at least, it is well defined) but it is illogical since the adoption of this criterion would render reinforced concrete unusable; moreover, in practice, prestressed concrete structures are often serviceable for an indefinite period under actions which cause cracking.

Hence, while serviceability is always the concern of the designer, the concept of a limit state of serviceability is not yet sufficiently developed to be of much use to him and his calculations will relate to rupture with occasional checks on deflection, etc.

It is rare that the maximum value of an action can be defined. It can happen, as with an open-topped water-retaining structure, but nearly all actions show a broad spectrum in which increasing values occur with decreasing frequency and it is usually impossible to define a value so large that it is certain that it will never be exceeded. There are also rare and extreme events, such as vehicle collision, explosion, as well as the natural phenomena of wind, wave, flood and earthquake, all of which can act at 'freak' intensities. An aircraft might fall on the structure; so might a meteorite.

In addition, the worst effects are often caused by combinations of several actions; to employ extreme values for all of them as a basis for design may lead to a prohibitively expensive structure. On the other hand, the strength of the materials of which the structure is made vary considerably, a variation further compounded by that of workmanship.

Probability provides a common measure whereby all these variables may be assessed and rationally combined.

Unfortunately, design based consistently on probability is not yet possible due to difficulties in quantifying. In the first place, the statistics are inadequate, particularly those relating to actions on the structure. There is difficulty, moreover, in defining an acceptable risk of failure, particularly since many, if not most, failures are the result of human failings, sometimes bizarre, which do not lend themselves to statistical treatment. In the second place, as we have seen, the public does not like structures which collapse and is made uneasy by talk of probabilities of failure.

We have in consequence what may be called a semi-probabilistic method of ensuring structural safety, in which probability is used where possible (e.g., in estimating strength of materials) and values chosen where it is not to provide a general level of strength similar to that produced by classic methods of design. To some extent, this gives the worst of two worlds: no significant structural economy is obtained and the design process is complicated. In practice, it is not as bad as that since the design of many concrete structures was already implicitly based on rupture loads (e.g., two-way slabs, flat slabs) and probability provides the designer with a logical and coherent treatment using which many awkward features of design are found to fall into place.

There are three types of action: long-term or steady actions, oft-repeated actions and dynamic high-energy actions.

For long-term or steady actions, the critical type of rupture is that which occurs under static load, and the criterion will be an adequate margin between rupture and the conditions during service.

For oft-repeated actions, rupture caused by fatigue is critical and the criterion will be the value of elastic stress caused by the action.

For dynamic actions, such as those of wave or earthquake, the energy characteristics of the structure, e.g., the energy required to cause rupture, the energy absorbed resiliently during loading, the energy dissipated, resonant frequencies of oscillation, the response of the structure to the action, etc., will need investigation to ensure that there exists an adequate margin against rupture.

The elaboration of the structure will proceed in step with its checking for safety. It is important that all details be consistent with the chosen method of erection. Difficulties of detailing will be encountered which can often be easily solved by some technique not hitherto incorporated in the works—welding or precasting in an *in-situ* reinforced concrete structure, *in-situ* concrete in a precast structure. Such intrusions will be extremely expensive and consistency must be maintained in the interests of practicability and economy.

3.6   Preparation of instructions

The preparation of instructions for the builder include drawings, descriptive documents and specifications. These last are to be seen not only as contractually defensive documents but also as part of the description of the works.

This stage constitutes the major work load of the design office; ease of comprehension by the man on site and the avoidance of confusion are the signs by which excellence may be recognized.

3.7   Construction

The designer's task has not finished when construction has started. Few jobs are free of incident, incidents whose gravity only the designer may be able to appraise. Modifications may be called for either to counter unexpected circumstances or to cope with changes of intention on the part of the client. Such modifications must be dealt with by the designer who alone understands the basis of his design.

4   Concrete as seen by the designer

Concrete attracts designers, perhaps because so much in a concrete structure is at his choice—material strength, method of construction, structural form, shape of elements and type of behaviour under load. It is a sort of *materia prima*, matter without form, of which the designer can make what he wishes. He will need to know the basic economics of the material, how it behaves under steady, repeated and dynamic load and the nature of its final rupture.

We will consider first the material itself and then the forms in which it can be used structurally—mass, reinforced and prestressed.

4.1   Economics of the material itself

Concrete used as blinding under roads or foundations may well have a strength of little more that 4 MPa (580 psi); concrete used as a structural material will be taken as having a strength of 40 MPa (it will often be stronger); small cylinders have been made experimentally, compacted by pressure before setting, with a strength of 400 MPa. The materials (though not the proportions) in all of these are the same, and all are among the least expensive and least energy-consuming of commonly used engineering materials. The difference between the three lies in the compaction of the aggregate and, more important, of the cement.

With increased compaction goes an increase not only in strength but in all the desirable properties of concrete—impermeability, durability, dimensional stability, elastic stiffness. Even appearance is improved; with good compaction, that powdery, chalky surface becomes dense-looking and slate-grey and, if prepared with care, resembles a dark coloured marble.

So compaction is good for concrete, but there is a reason why it is not sought as an end in itself.

The raw materials of concrete (cement, fine and coarse aggregate) being so inexpensive, a concrete structure of some complication will have a materials cost which is small compared with the costs of formwork and of mixing, placing and compacting the concrete. Thus, to use high strength and complicated shapes can add hugely to the cost of the finished structure, unlike with expensive materials, such as light alloys or stainless steel, where it is often worth while saving material at the cost of considerable structural complication. There are, however, exceptions.

The classic use of concrete is for static, land-based structures of modest span. In such structures, it is unusual for the saving of weight to be of economic value *in itself*; the only overall economy lies in the reduction of quantities of concrete, which, as we have seen, is of small cost. It is for this reason that there has been little progress in taking advantage of higher strengths than 40 MPa, even at a time when 80 MPa, even 100 MPa (14 500 psi) is not difficult of achievement either in a precasting factory or on site.

It has been found worth while to benefit from the better quality of high-strength concrete but not to seek the reduction in quantities which would result from higher stresses.

There is, however, a future for concrete in a different category of structures— what one might call 'mobile' structures. The exemplar of such is the aeroplane. Every kilogram of weight in an aircraft structure has to be subtracted from the payload and thus from the profits of operation; it is worth paying large sums in first cost to save weight. There is a limit, of course, but it is high; within that limit, saving weight is economic.

Ships are similar. Concrete ships have a long history and a reputation for low first cost, low maintenance costs and robustness. Unfortunately, their large structural weight has impaired their viability in use. By relating first cost, running costs, structural weight and payload, however, it may be shown that readily attainable increases in concrete strength make concrete ships an economic proposition even if the finished material costs four or five times the cost of normal structural concrete. This margin offers considerable scope for expenditure in placing and forming the concrete.

Furthermore, running costs are sensitive to hull shape and a complex shape may well be economic despite its expense. In this sort of structure, therefore, maximum overall economy may well be obtained by using much higher strengths and much more complex shapes than would be justifiable in classic static structures.

Offshore platforms are another instance; the economic penalty on extra weight is heavy. Weight aloft impairs stability, thus needing a bigger structure below and, in turn, greater bouyancy, all the additions multiplying one another.

One may expect the development of such structures using higher-strength concrete but on one condition. Present design as expressed in codes has attitudes to weight and economy appropriate to static structures deeply rooted in it. These attitudes are wrong for mobile structures and the application of building and bridge codes to such structures imposes a grave handicap on the use of concrete.

4.2 Behaviour of the material itself

Figure 4-1 shows the typical stress-strain curve of a test piece under compressive stress increased monotonically to rupture. The following zones may be distinguished.

Zone A: stable microcracking; stress 0–$0.55f_{cu}$ The approximate proportion 0.55 is not sensitive to the value of f_{cu}.

Strain is partly elastic and partly caused by non-reversible rupture of crystalline bonds, known as microcracking, hence the bend of the curve. The rupture is mostly by sliding between unhydrated cement particles and is accompanied by reduction in volume. On condition that adequate moisture be present, the crystalline bonds will be re-established when the load is released or maintained steadily. The release of the load will be as shown in Fig. 4-2, with a hysteresis loop and a residual deflection. Successive load cycles tend to a stable residual deflection with a hysteresis loop, the energy represented by the area of the loop being dissipated into heat.

The propagation of microcracks under loads within this range is *stable* and the concrete can support an unlimited number of applications of such loads.

Zone B: unstable micro cracking; stress $0.55f_{cu}$–$0.80f_{cu}$ Microcracking is stable under a single application of load but increases with repeated applications until the microcracks link together to form macrocracks which in turn progress in

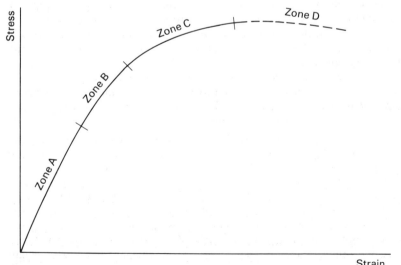

Figure 4-1   Typical stress–strain curve for concrete in compression

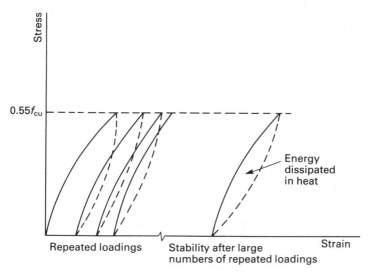

Figure 4-2   Typical stress–strain curves for repeated loadings in compression

extent until rupture ensues. Strains increase with repetition of load. Final rupture is similar to rupture caused by a load monotonically increased and is accompanied by large strains; it is not brittle. As in zone A, after the release of the load or under a steady load, cracks, both micro and macro, will tend to seal up by the re-establishment of crystalline bonds accompanied by some increase in strength.

Zone C: unstable macrocracking; stress $0.8f_{cu}$–$1.0f_{cu}$ Under a single application of load in this zone, microcracks are formed as the load increases, which then link together to form macrocracks. These macrocracks will extend and lead to rupture with either the repetition or the maintenance of the load.

The material is disaggregated and its apparent volume increases.

After the release of the load (but not under steady load), cracks will seal up as in zones A and B.

Zone D: post-rupture stress   With a load possessing sufficient energy—potential (piled-up weights), strain (an extensible test rig) or dynamic (an impact)—as soon as the concrete reaches the maximum stress which it can support, it collapses.

Where, however, the strain is controlled, as by material elsewhere in a hyperstatic structure or by a testing machine applying a constant rate of strain, or where the load is more properly seen as an imposed strain, or, again, where the load is responsive to strain, it is not unknown for the concrete to continue to support a reducing value of stress with very large strains—as much as three or four times the strain at f_{cu}. This behaviour is not easily accounted for and, with present knowledge, is not to be relied upon but its existence in special circumstances is well accredited.

Behaviour in tension
The rupture stress and strain in tension are much smaller than in compression but the properties of the curve are similar and the zones of behaviour described above are broadly reproduced, including the zone of post-rupture stress behaviour.

Stress concentrations
The behaviour of concrete under load represents the continual rupture and reconstitution of crystalline bonds and the existence of complex microcracks and elastic discontinuities; notches and discontinuities on a structural scale are in comparison of no significance. A concrete structure is thus effectively not notch-sensitive.

4.3 Mass concrete

Mass concrete is used for some of the largest structures of modern times—indeed, for size, only the biggest earthfill structures exceed them.

For such structures, the form will be conceived in the light of the behaviour of the material as sketched out above, but also taking account of the following features related to construction—which, indeed, is usually the major problem.

Heat of hydration   In large masses, this can cause high temperatures in the mass during setting, causing loss of strength and extensive cracking. The designer will give thought early on to the size of pour and problems of cooling.

Shrinkage   This is related to the drying out of the mix. In an equable climate and with well compacted concrete, this drying out will not penetrate farther than a few feet at most. A large, mass-concrete block can thus under ideal conditions have a very small drying-out shrinkage.

Climatic variations will play a similar role in causing both shrinkage and contraction but will also be attenuated by the size of the mass.

In a concrete member in which drying out and climatic variations can affect the whole of the volume, e.g., thin-walled structures and paving slabs, the resultant strains are of the same order as the tensile strains at rupture and can cause extensive cracking if the structure is not free to deform. Moreover, where, for instance, a paving slab is laid continuously without joints, freak weather (e.g., a hot sun following a downpour of rain at the end of a warm, wet season) can cause large transient compressive strains, capable of inducing high stresses and creep. If this condition be followed by cold, dry weather, the probability of cracking is high.

4.4 Reinforced concrete

Two forms of reinforced concrete can fail in a brittle manner.

The first is either ties or beams with reinforcement such that the tensile strength of the concrete is greater than that of the steel. When the concrete fails in tension, there is a sudden release of tensile force which the steel cannot support; the cracking load of the concrete is higher than the rupture load of the steel. The failure load of the concrete is little affected by the presence of the steel.

The second is a beam with reinforcement such that the tensile strength of the steel is greater than that of the compressive strength of the concrete above the neutral axis; when the latter reaches its rupture strength, it fails suddenly.

These forms are generally to be avoided.

With intermediate proportions of reinforcement, whether in ties or beams, rupture occurs as a result of failure of the steel. With a beam, the concrete will indeed fail in compression but it is a failure caused by the extension of the steel beyond its yield point causing a raising of the neutral axis and consequent inadequacy of the compressive strength of the concrete.

The behaviour of a reinforced concrete beam of this sort will be as indicated diagrammatically in Fig. 4-3. The full curve shows behaviour under a single load up to rupture, the dashed lines show the return curves after the release of loads of different magnitudes.

The following zones may be distinguished though they are in no way clear cut.

Zone A: from zero load to cracking load   Cracking often occurs under that load which produces a tensile stress equal to the tensile strength of the concrete as established by bending tests on concrete prisms without reinforcement, but not invariably—the phenomenon is often more complex. The occurrence of tensile strain beyond the point of maximum tensile stress has a large effect on cracking but is not fully understood save that it is sensitive to the rate of loading, the bond of steel, the diameter of the bars relative to the dimensions of the beam section, the shrinkage of concrete, etc. Even when cracking occurs, it may be little greater than that microcracking which we have seen to be a normal accompaniment of loading—depending again on a number of factors similar to those affecting tensile strain. The effect of cracking on the corrosion of the steel is itself complex. By and large, the cracking of reinforced concrete has as much of mystery and paradox as of certainty and the folk-lore of the material, which would have that it is excellent

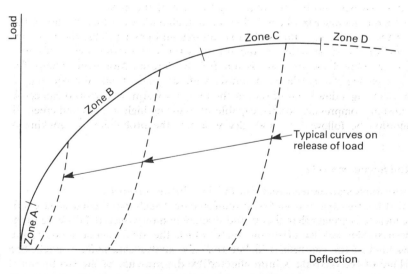

Figure 4-3   Typical load–deflection curve of a reinforced concrete beam

under steady, long-maintained loads but less good under repeated and high-energy loads, is perhaps as good a summary of what we know as any.

Sufficient repetition of load in the upper half of this zone will cause cracking.

Zone B: quasi-elastic behaviour of the cracked section   The steel behaves elastically and the concrete is generally within zone A of material behaviour (stable microcracking) but overlapping into zone B (unstable microcracking) at the upper end. The non-linearity of the concrete behaviour causes the neutral axis to rise, resulting in a larger departure from linearity in the behaviour of the beam.

Repeated loading causes increased deflection and crack widths, but the beam is probably able to support large numbers of repetitions. Release of the load leaves an appreciable residual deflection.

Zone C: non-elastic behaviour of the cracked section terminating in rupture   The stress in the steel exceeds the elastic limit and the stress in the concrete enters zone C, the zone of unstable microcracking. Deflections are large; the repetition or maintenance of the load will lead to collapse. Release of the load leaves a large residual deflection.

Zone D: Post-rupture load behaviour   The ability of the beam to support large deflections but with decreasing load is primarily caused by the large non-elastic strain of the steel prior to rupture but the concrete makes a contribution. If the compressive flange is transversely reinforced with closed stirrups or links, its ability to support large compressive strains is greatly increased. The importance of this zone lies in the large area beneath the deflection curve and, in consequence, in the large energy at rupture, energy of which a small fraction only is resilient but nearly all of which is dissipated. This renders reinforced concrete ideal for high-energy loads (earthquake, explosion) of rare occurrence.

4.5   Prestressed concrete

Given a concrete section and a cross-sectional area of high-tensile steel at a defined location in the section, the ultimate strength of a prestressed concrete beam has been determined subject only to such variations as are described below. Brittle failures similar to those in reinforced concrete can occur with extreme proportions of steel, but it will be assumed that such proportions have been avoided. There then exists an extra variable at the control of the designer—the initial tension in the steel; the behaviour of the beam will differ profoundly according to the value chosen for this initial tension. Figure 4-4 shows the range of behaviour.

At one extreme, where the steel has not been tensioned, behaviour is similar to that of reinforced concrete. The difference is that, since the steel is of high tensile strength, the steel strains are large and the neutral axis rises rapidly to near the upper surface and the full strength of the steel cannot be utilized. The cracking load is small and well defined.

At the other extreme, the steel is stressed to near its elastic limit. The cracking point is little lower than the rupture load. Since the steel has been given an initial extension, less strain is needed for it to reach its ultimate tensile strength but

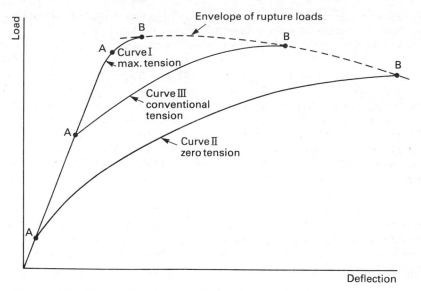

Figure 4-4 Effect of initial steel tension on load–deflection curve of a prestressed concrete beam
A Cracking load
B Rupture load

there is adequate strain for the neutral axis to rise to the highest level consistent with concrete strength. The rupture strength of the given arrangement of concrete and steel is at its maximum.

With intermediate values of initial tension in the steel (that shown represents a typical design according to current practice whereby the tension in the steel after relaxation is approximately $0.6f_{su}$), the cracking load will be much lower, the ultimate slightly lower than that with maximum initial tension.

What is the significance of these differences of behaviour?

Note first the significance of the cracking load in prestressed concrete. From zero load to cracking load there is very little difference in the value of the force in the steel, the increased bending moment being supported by a displacement of the centre of thrust in the concrete, thus increasing the moment arm. In the concrete, stresses vary between limits fixed by the designer as being capable of being supported indefinitely. As a result, until loads appreciably exceed the cracking load, there is no danger of failure under repeated loads.

In curve I (maximum initial tension in steel), the cracking load is a high proportion of the static rupture load; so too is the fatigue limit—indeed, it is doubtful whether any known structural material can withstand indefinite repetitions of so large a proportion of its static rupture load. On the other hand, deflection at rupture is small as is the energy to cause rupture; behaviour tends towards the brittle. This is the extreme of resistance to repeated loads.

In curve II (zero initial tension), the behaviour is the reverse. The cracking load is small as is the fatigue limit; deflections at rupture are large as is the energy at

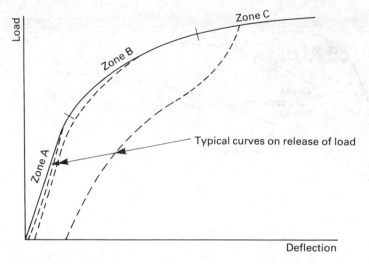

Figure 4-5 Typical load–deflection curve of a prestressed concrete beam

rupture; intermediate loads cause large irreversible deflections representing large dissipation of energy. This is the extreme of resistance to a single-high-energy load.

Curve III shows the behaviour which is typical of beams designed in accordance with current codes. Figure 4-5 shows the load–deflection curve of a typical prestressed concrete beam with characteristics similar to those shown in curve III of Fig. 4-4. The dashed lines show the strains on release of loads of various magnitudes.

In zone A, residual deflections on release of the load are small; as we have seen, repetition of the load in this zone will not produce failure.

In zone B, residual deflections on release of the load are larger but still small. The area under the curve after loading is large but this energy is not dissipated but for the most part stored resiliently and the beam is still quasi-elastic. The inflection in the curve of deflection is of value in redistributing the load in redundant structures, especially since this redistribution is *reversible* and residual deflections on release of the load remain small. The beam in this zone can support steady loads indefinitely but repeated loads can cause failure.

In zone C, residual deflections on release of the load are larger but still much smaller than those in curve II. Dissipated energy is significant but a large part of the energy absorbed is restored resiliently until rupture is near. Steady loads in this zone will cause failure as will repeated loads.

Post rupture–load behaviour has been observed in beams of the type shown in curves II and III of Fig. 4-4; it seems not to have been investigated in those shown in curve I.

Part II Materials

4 Selection of materials to improve performance of concrete in service

D Campbell-Allen   H Roper

University of Sydney, Australia

Contents

Summary

The chapter concentrates on the influence of cements, aggregates, pozzolans and admixtures on the durability of concrete. Although other cements are mentioned, the main attention is paid to Portland cements and blended Portland cements.

Concrete performance is considered under normal atmospheric conditions (including freezing and thawing conditions), under aggressive environments (including acid and sulphate-bearing groundwaters), under industrial use, in hydraulic and marine structures and finally under extremes of temperature. Possible sources of internal disruption include alkali-aggregate reaction and unsoundness of cement. The mechanism of corrosion of steel in concrete is outlined and suggestions are made for improving the resistance of steel to corrosion, after noting that, in the great majority of structures, Portland cement concrete properly designed and constructed provides complete protection against corrosion for the design life.

Under many conditions, the durability of concrete is dependent on the production of dense concrete having an adequate cement content. This chapter should therefore be read in conjunction with the chapters concerned with concrete production and construction procedures.

1   Properties of concrete and material selection

The properties of concrete required in any application are adequate strength, long-term durability and appropriate economy. The concrete must also provide physical and chemical conditions under which steel reinforcement either remains passive or corrodes sufficiently slowly not to cause distress during the design-life period. These necessary properties are achieved in any particular set of circumstances by the selection of suitable materials, the choice of mix proportions and the use of proper methods of placement and curing. All these three aspects interact, as mix proportions and placement methods depend on material selection.

In this chapter, the influence of cements, aggregates, pozzolans and admixtures on the resulting durability of concrete are discussed. Throughout the chapter, it will be noticed that it is not only the inherent nature of the constituent materials that affects properties. The combination of the materials into concrete has also a vital influence. Reference is therefore necessary to Chapters 5 to 9 and to Chapter 27 in conjunction with this chapter.

The great bulk of structural concrete is made from Portland cements or from Portland cement blended with slags, fly ash or other pozzolanic materials, and most of the discussion is therefore related to these types of cement.

For some applications, special binders such as supersulphated slag cements, which form no hydroxides on setting, expansive and shrinkage compensating cements, or high alumina cement (ciment fondu) may be considered. It should be appreciated that the application of experience gained in the use of Portland cement concretes to these special materials can be misleading. As an example, consider briefly high alumina cement. This cement gains strength very rapidly and may at 24 h reach strengths comparable to those of 28-day moist-cured Portland cement. It was first developed as a cement having special resistance to sulphate groundwater attack and may be considered for use under those conditions. Particularly under warm, humid conditions, the hydrated cement may undergo a slow conversion from metastable calcium aluminate hydrate to more stable compounds: this change is accompanied by an increase in porosity and a marked reduction in strength. A number of structural failures resulting from this conversion have been widely reported, and anyone considering the use of high alumina cement should refer to the literature and note in particular the very stringent requirements of a 0.4 maximum water/cement ratio for use in structural concretes [1, 2, 3].

Although many of the properties of Portland cement concrete, and particularly its resistance to atmospheric and other attack, are determined by the nature of the hydrated cement paste, aggregates can also contribute to durability or lack of it. Most specifications for concrete aggregates call for material which passes a number of somewhat arbitrary tests. Tests which are commonly regarded as relating to durability are those concerned with resistance to abrasion and crushing, such as the Los Angeles abrasion test, the test for aggregate crushing value and for 10% fines value; those concerned with the presence of weak and friable particles, and with soundness, as for example determined by the sodium sulphate soundness test; those concerned with possible chemical reactivity with cements, such as the presence of organic impurities (including sugars) and the potential for reactivity with cement alkalis. These tests often do not provide satisfactory discrimination between aggregates that will, and those that will not, perform satisfactorily in practice under any specified set of ambient conditions. Although some guidance can be obtained from conducting such tests, the final judgement must be based on the probable performance of concrete containing the aggregates under the particular conditions proposed for use. Generally much more useful advice on the likely behaviour of aggregates can be obtained from a study carried out by a petrographer skilled in the examination of concrete aggregates, than from reliance on the results of specification tests.

Finely divided particulate material (with either pozzolanic or latent hydraulic properties) have become common additives to concrete mixes, and with the demand for greater energy savings their use will continue to expand. As a rule, their compositions and physical nature vary widely, and even for slag cements derived solely from iron blast furnaces, no general agreement exists on the best hydraulic factors to use in the prediction of strength development [4]. The task of the materials engineer becomes even more daunting when considering the quantitative determination of pozzolanic activity, which Sersale describes as a complex and unresolved problem [5]. Choice of materials of this sort therefore must depend to a great extent on experience derived from tests on single sources, rather than on a class of such additives.

Many chemical admixtures act to change the nature of the plastic concrete so that it may be more readily placed and compacted. Superplasticizers are the most dramatic examples. Admixtures should only be used when they are required for a specific purpose and in these cases reference should be made to literature on the particular subject [6]. Frequently, the effectiveness of an admixture is altered by changing the content of fine material in the mix, and the effects of the use of an admixture in conjunction with pozzolans should therefore always be investigated. Chemical admixtures, particularly those used for air entrainment, may also be much changed in their effects if two admixtures are combined, a matter that becomes particularly important when the properties of the hardened concrete are thereby changed. For example, bubble size and distribution required for frost resistance may be totally changed if a superplasticizer is used in conjunction with an air-entraining agent [7]. Potentially serious long-term effects arise if admixtures containing chlorides are used. This situation is common when lignosulphonates, which operate as water-reducing and retarding agents, have added to them calcium chloride as an accelerator in order to provide a water-reducing agent which does not change set characteristics. The first effect is to increase drying shrinkage and the second is to reduce the capacity of such concrete to prevent steel corrosion. Although the potential corrosion problem may be reduced by the substitution of triethanolamine as the set regulator, Rixom shows that shrinkage is still increased with respect to plain concrete [6]. A careful examination of the description of such admixtures is therefore essential, and some background should be acquired as to which properties are changed by their use, either individually, or when combined [8].

2   Concrete in atmospheric conditions

Well-made concrete with an adequate cement content has been found to be durable in normal atmospheric conditions, provided that conditions leading to internal disruption, discussed in Section 5, are avoided. The choice of materials, apart from those leading to such disruption, does not therefore have a marked effect on such durability. Disintegration by weathering, if it does occur, is caused mainly by freezing and thawing and, to a lesser extent, by expansion and contraction, both differential and under external restraint, resulting from wetting and drying and temperature cycles. Concrete which is dried and kept dry is less susceptible to damage by freezing and thawing. If concrete is liable to become wet and then subjected to freeze–thaw cycles, rapid deterioration can occur unless special precautions are taken.

It has long been established that the purposeful entrainment of air, by the use of an air-entraining agent, can lead to a very marked increase in the resistance of concrete to freeze–thaw cycles. Figure 2-1 shows this effect in terms of a durability factor. The air, to be effective, should constitute about 8% by volume of the mortar fraction, and the optimum proportion in concrete is therefore about 4% when aggregate of 20 mm size is used. The bubbles must be stable and non-coalescing and range in size from about 20 to 300 μm. For optimum durability they should be dispersed throughout the mortar at an average spacing of 200 μm. Recognized air-entraining agents achieve this desirable bubble structure, provided the dose is appropriately adjusted to take account of cement content, quantity of fine material, temperature and slump of the mix. There is no guarantee that air entrained, when recognized AEAs are used in conjunction with other chemical admixtures, will have the desirable structure.

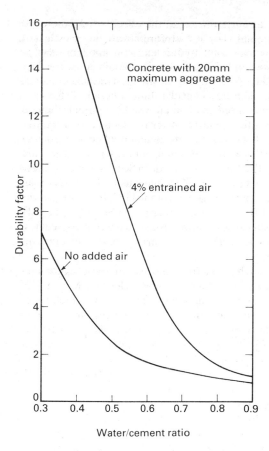

Figure 2-1   Relation between durability and water/cement ratio for air-entrained and non-air-entrained concrete

The influence of aggregates on resistance to freezing and thawing has usually been studied by checking the performance of aggregates in concrete under test freeze–thaw conditions. The most important characteristics affecting resistance to freeze–thaw and physical soundness generally are the size, abundance and continuity of pores and channelways within the particles. Particles containing an abundance of channelways less than 4 μm in diameter are the most prone to reduce durability. Kaneuji *et al.* [9] have shown that durability of concretes (expressed as expected durability factor *EDF*) made with different aggregates can be related to the total intruded pore volume (*PV*) (determined by mercury intrusion) and the median pore diameter (*MD*) in the aggregates. They propose a relationship of the form

$$EDF = (K_1/PV) + (K_2 MD) + K_3$$

They suggest that this approach is superior to a selection based solely on absorption, which is the commonly used criterion at the present time.

The drying shrinkage of concrete made in accordance with good practice is usually considered to be of the order of 450 to 600 micro-strain after 90 days drying at 50% relative humidity (RH). Apart from certain prestressed structures, shrinkage of concrete is usually not taken into consideration in design. From time to time, aggregate materials have been used in concrete which, although they do not appreciably affect strength properties, lead to shrinkage up to three times as high as the assumed maximum. Such aggregates are not limited to a particular rock type nor location. In South Africa, the rock types and derived sands are sandstones, mudstones and marls of the Karoo area [10]. In Scotland, where there are many sources of such aggregate, the rock types are igneous rocks of the basalt and dolerite groups, together with greywacke and mudstone [11]. In Australia, the worst materials are breccias derived from volcanic pipes in the Sydney area [12]. Many of these materials pass the usual specification tests for concrete aggregates; their deleterious nature can be determined by shrinkage tests or petrographic studies.

Long-term durability problems which stem from the use of these aggregates include cracking, crazing and unacceptable deflections. Polivka and Mehta [13] suggest that the use of expansive cement can greatly reduce the cracking of concrete which could be expected if such marginal aggregates are used in combination with Portland cement. A more realistic approach seems to be the imposition of limits of shrinkage for particular uses of concrete, as is recommended in Scotland [14].

3 Concrete in aggressive conditions

3.1 Groundwater

3.1.1 Acid groundwater

Portland cement is a highly alkaline material. Acid solutions react with it so as to cause complete disintegration if a low pH is maintained. Acid levels of, say, pH < 5 occur in certain soils, at times in concrete sewers, and in land fills where acidic material has been dumped. Soft waters, which are not saturated with calcium or magnesium salts, generally cause only superficial attack unless they are fast flowing, in which case mortar may be removed and consequent deeper attack can follow. Such soft waters may also cause carbonation of the concrete, a factor disadvantageous from the viewpoint of corrosion of reinforcement.

In a study of conditions related to failed buried prestressed pipelines, Phillips [15] noted that the ground waters were generally acid with pH values ranging from 4.5 to 6.8. Humic acid was present in some of the water. Concrete cannot be expected to stand up to such conditions without developing problems.

Tuthill [16] warns that snow water in mountain streams and reservoirs is often particularly aggressive because it is cold, pure, and consequently lime hungry, and contains carbon dioxide which exists as a mild carbonic acid solution.

Under these conditions, a number of counter measures can be considered.

(a) Portland blast-furnace slag cement or Portland pozzolan cement should be used in place of normal Portland cement. If normal Portland cement is used, it should be of a 'low lime' type, in which the proportion of C_2S is greater than that of C_3S.

(b) The water should be treated before it is admitted to the concrete structure by passing it over beds of calcium carbonate rock.

(c) Detailed attention should be given to design and construction to reduce the permeability and capillarity of the concrete and to reduce cracking in the concrete.

3.1.2 *Sulphate-bearing groundwater*

In the presence of moisture, an expansive reaction takes place between the tricalcium aluminate phase of Portland cement and the sulphate ions to produce calcium sulpho-aluminate hydrates. The precise mechanism has been studied for many years, but it is still subject to debate. Nevertheless, the outcome of such reactions have resulted in spectacular degradation of concrete structures in various parts of the world. Sulphate-resistant cements were specifically developed to increase resistance under this type of attack. The special requirement for sulphate-resistant cement as given in many cement specifications is that the C_3A shall not exceed 5.0% (BS 4027 : 1980 sets this limit at 3.5%). Sulphate-resistant cement should be used if soluble sulphates are present in concentrations which exceed 0.1% in the surrounding soil or 150 ppm in the groundwater. The limit of 5% has been in use since 1936 and was supported by subsequent work in North America [17, 18]. Analysis of more recent test results gathered over some years show that protection against sulphate attack is not provided in all cases when the C_3A content is low [19]. In the extreme case when C_3A content is zero, some increase in susceptibility to sulphate attack has been observed [20]. Mehta [21] pointed out that, although the formation of ettringite from the reaction of sulphates with calcium aluminate hydrates produces expansive disruptions, attack by acidic sulphate solutions may involve the conversion of $Ca(OH)_2$ and calcium silicate hydrate gel to gypsum (calcium sulphate), leading to surface softening and strength loss. He concluded that the sulphate-resistant behaviour of both Portland and blended-Portland cements seems to depend on the total reactive aluminate and calcium hydroxide present in the cement paste at the time of sulphate exposure. Whatever the nature of the cement, it is clear from all investigations that if concrete is to perform satisfactorily in sulphate-bearing soils or water it must be highly impermeable [22]. Air entrainment has been shown to improve resistance to sulphate attack, but this is largely due to the associated reduction in water/cement ratio rather than to the provision of space for the deposition of otherwise expansive reaction products [23]. Calcium chloride in concrete greatly reduces its resistance. Sulphate-resistant Portland cement produces concrete which is more resistant to sulphate attack than does Portland blast-furnace slag cements with slag in the range 15–65% or supersulphated cement.

Certain pozzolans produce a very marked improvement on the resistance of concrete to sulphate attack, sometimes increasing life expectancy from below 50 years to above 100 years. Certain bituminous coal fly ashes have been found to be particularly successful [24]. Indeed, Kalousek *et al.* [19] even suggested that rather than ASTM Type V cement, pozzolanic fly-ash cements made with clinker which contains less than 6.5% C_3A and less than 12% C_4AF should be considered as a sulphate resistant cement.

3.2   Industrial use

3.2.1   *Chemical attack*

Many industrial processes expose concrete to attack by corrosive substances, such as oils, sugars, organic and inorganic acids. In some cases, the process combines such attack with severe abrasion, as for example in stock yards and slaughter houses. The rate of deterioration of concrete under such attack depends on the particular attacking substance and can in all cases be slowed down by the use of a dense concrete surface of low permeability. However, under severe attack, even the best made Portland cement concrete will not be satisfactory for long. A change of binder to supersulphated cement will improve performance under animal and vegetable oils and sugars.

Solutions to deterioration problems may have to be sought by polymer impregnation or by the provision of liners using stainless steel or plastics. A requirement of any liner is that it should have adequate chemical resistance and material durability and should be able to be applied without the formation of pinholes and in such a way that external groundwater pressures do not remove it.

The materials for polymer impregnation are expensive and their price may be up to 20 times the cost of Portland cement. Nevertheless, this expense may well be justified under conditions of severe attack, such as in sulphuric acid production plants. At such a plant in Japan, cast iron or coated stainless-steel beds were expected to last less than one year. Pump beds made from polymer impregnated concrete were in good condition after two years service [25].

Particular conditions of attack arise in sewers and sewage treatment works. Typical corrosion of concrete sewers occurs only above the water line, severe attack being immediately above the average daily level. Along the sides, above this level some corrosion may be evident, increasing in severity until the crown is reached. In severe cases, the crown may be perforated. The first evidence is the appearance of white powdery efflorescence. Initial deterioration is slow but, as it progresses, it becomes more rapid, the initial deposits flaking off the concrete. As its final development, the surface concrete is reduced to a soft putty-like material from which the aggregate falls.

Sulphur compounds present in sewage are reduced to sulphides by microorganisms in those parts of the system where the dissolved oxygen in the sewage becomes depleted as a result of biological action. Some of the sulphides thus formed escape to the sewer atmosphere in the form of hydrogen sulphide gas. This dissolves in the condensed moisture on the pipe walls above the sewage flowline and on the walls of manholes and sumps, where it is converted by sulphur-oxidizing bacteria, in the presence of oxygen, to sulphuric acid. The acid causes deterioration of the pipe material, the severity of the corrosion depending on conditions in the sewer and on the degree of resistance of the pipe material to acid attack.

Various actions to slow down attack have been adopted. These include the provision of a sacrificial layer of concrete which can be corroded away without affecting the structural performance; the use of calcareous alkaline aggregate to neutralize the acid; the use of high alumina cement which has a somewhat greater resistance to acid attack. These are regarded by some as only palliatives to be used only if mildly aggressive conditions are expected [26]. Protective plastic linings, such as plasticized PVC linings mechanically locked to the concrete, have a long history of successful use in severely aggressive conditions.

3.2.2 Wearing resistance

Floors and slabs can be produced to have a high degree of resistance to traffic and other wear. The quality of a concrete slab or floor is very dependent on achieving a hard and durable surface which is plane and free of cracks. The properties of the surface are determined by the quality of the concreting operations. Furthermore, the timing of these operations and of the finishing techniques is critical. Otherwise, undesirable conditions occur at the wearing surface, which may lead to soft or dusting surfaces, permeable concrete, cracking and poor durability.

ACI Committee 302 [27] describes desirable materials and practices for the successful production of floors to withstand different levels of wear. The importance of early effective curing is stressed. The wear resistance of pavements made with Portland cement, Portland cement and fly ash, and Portland blast-furnace slag cement is in all three cases increased by early curing [28]. It should, however, be noted that much damage to floors, particularly from hard-tyred traffic such as fork-lifts, occurs at cracks in the floor and these cracks may be the result of deficiencies in structural design rather than in surface quality.

The influence of concrete properties on wear resistance is affected by the type of wear involved. Many test procedures have been developed over the last 50 years, but many are also outmoded as the type of traffic to which industrial floors are subjected has changed. Tests do not therefore necessarily mirror the actual behaviour of concrete in a floor, where, in use, abrasion and impact occur in conjunction with flexural and shearing loading.

The common defects of scaling (loss of top surface at an early age when exposed to traffic) and dusting (the presence of a loose powdery surface on hardened concrete) are believed to be primarily due to premature finishing operations undertaken while bleed water is still present. The other common defect of crazing, formed by surface shrinkage during the early stages of hardening, is frequently caused by the application of a dry cement shake to the surface with the object of obtaining an early and easy finish.

Under conditions of very severe wear, it is often desirable to introduce special wear-resisting aggregates. The common materials used for this purpose are

(a) Metallic types:
    (1) pearlitic iron turnings—specially selected and processed;
    (2) crushed cast iron chilled grit.
(b) Non-metallic types:
    (1) silicon carbide grains (carborundum);
    (2) fused alumina grains (corundum);
    (3) natural emery grains (alumina and magnetite).

Both groups of materials, if carefully used, provide a considerable increase in wear resistance. Of the metallic types, the pearlitic turnings have been shown to be superior to the chunky cast iron grains, as they are less prone to loosening from the cement matrix under impact. Where slip resistance is of prime importance or where rusting of the iron would be objectionable, the non-metallic types are commonly used.

There is some indication that silicon carbide and alumina grains are more resistant to wear than finishes containing magnetite, but this is probably a function of the test method and may not apply in actual operational conditions.

It has been suggested that fibre-reinforced concrete is superior in wear resistance to conventional concrete, but, although superior performance in pavements has been demonstrated, this improved performance is largely the result of superior crack resistance rather than resistance to abrasion [29].

Liquid hardeners have been proposed but there is not much technical support for their use except possibly where very poor and porous concrete requires remedial action.

3.3   Hydraulic structures

Hydraulic structures may be subject to physical erosion, arising from particles carried by flowing water, from the collapse of cavities formed in water in which pressure is varying, and from fluctuating pressures on and within the concrete under conditions of unsteady flow. The first two sources of erosion have been discussed in some detail by ACI Committee 210 [30] but little attention has been paid to the third aspect.

3.3.1   *Cavitation erosion*
Vapour bubbles will form in running water wherever the pressure in the liquid is reduced to its vapour pressure. These vapour bubbles flow downstream and on entering a zone of higher pressure collapse with great impact. Repeated collapse of such cavities near the surface of concrete will cause pitting. Cavitation pitting is readily recognized from the holes or pits formed, which are distinguished from the smoother worn surface produced by abrasion from solids in flowing water. Damage may arise in locations where boundary irregularities and changes of profile cause local reductions of pressure, but has not been found likely to occur in open channels when stream velocities are less than 12 m/s. In closed conduits where the air pressure is reduced by the water flow, damage due to cavitation has been found to occur at stream velocities as low as 8 m/s, and at higher velocities the forces are sufficient to erode away large quantities of high-quality concrete and even to penetrate steel plates in a comparatively short time.

Some things can be done to defer damage by cavitation. It has been conclusively demonstrated that the resistance of concrete to forces of cavitation (and abrasion) increases with increased compressive strength of the concrete. ACI Committee 210 recommends that under cavitation conditions the mix should be proportioned for a strength of at least 40 MPa (6000 psi). There is no observed influence of cement type on resistance to cavitation. In the selection of aggregates, good bond is more important than hardness, as the action of cavitation is generally one of plucking out large particles. Larger particles are more easily removed than smaller ones and therefore there is an advantage in limiting the maximum size of aggregate to say 20 mm. Work by Watkins and Samarin [31] confirms that resistance to cavitational erosion depends primarily on the quality of the concrete matrix. The nature of the coarse aggregate was found to be of significance when it became exposed after the erosion of the matrix.

Abrupt irregularities should be avoided if the water velocity is likely to be greater than 12 m/s. Limits of irregularities suggested are: parallel to the direction of flow, 6 mm; at right angles to the direction of flow, 3 mm. For higher velocities over 30 m/s, irregularities should be reduced to half these limits. Unformed surfaces should be finished with the minimum of working to prevent the formation of a poor quality surface layer. Tests on cavitation resistance of concrete containing steel fibres and concrete impregnated with methylmethacrylate monomer subsequently polymerized have shown a marked increase in resistance (from 47 h for conventional high-strength concrete to 130 h for fibre-reinforced and for polymer-impregnated concretes). Still further increased resistance was noted for fibre-reinforced polymer-impregnated concrete [32]. In spite of the improved performance that may be obtained by the introduction of fibres and polymers, if cavitation forces exist it is unlikely that concrete will remain undamaged indefinitely. The best concrete that can be made will not withstand the forces of cavitation for a prolonged period and the only solution is to design and construct to eliminate as far as possible the forces involved.

3.3.2 *Abrasion*
The erosion of concrete by water-borne solids can be as severe as that caused by cavitation. Concrete in the invert of the diversion tunnel at Anderson Park Dam, in the USA, was worn away to a depth of 80 mm while it was in use for 43 months carrying water containing large quantities of silt, sand and gravel. Some dry-packed mortar patches were completely eroded away. Large particles can be moved at comparatively low velocities and therefore concrete resistant to abrasion should be used if there is any likelihood of this condition arising. The rate of erosion is dependent on the quantity, shape, size and hardness of the transported solids, the velocity of the water and the quality of the concrete. As with cavitation resistance, resistance to abrasion is increased with increasing concrete strength, but in contrast, the abrasion resistance is apparently increased with larger sizes of aggregate [33].

3.3.3 *Fluctuating pressures*
As noted above, fluctuating water pressures in unsteady flow conditions may cause the disruption of concrete, particularly if the concrete is permeable or is cracked and internal fluid pressures can be built up. There is very little test information relating to resistance to this form of attack, but it can be concluded in general terms that a dense impermeable surface is desired both for resistance to fluctuating pressures and for resistance to high velocity.

3.4 Sea-water

The behaviour of concrete in sea-water is significantly different in three different zones:

(a) the zone above high-tide level, where the build up of salts can occur in evaporative drying cycles and where an abundance of atmospheric oxygen is available;

(b) the inter-tidal zone, where the concrete is kept saturated but is subject to cyclic exposure to air and is subject to abrasion and attachment of some organisms;

(c) the totally submerged zone, where oxygen availability is limited, and as depth increases, more rapid penetration of the sea-water into the concrete is brought about by hydrostatic pressure [34].

In zones (a) and (b), world-wide surveys show that the greatest problems of durability arise from the corrosion of the reinforcing steel rather than from chemical or physical disintegration of the concrete [35, 36, 37]. Disintegration of concrete *per se*, if it does occur, is much influenced by the ambient temperatures. As for zone (c), there is no evidence of degradation in permanently immersed concrete. However, there always exists the possibility that, in this zone, embedded steel may be losing metal to form either low-oxygen compounds, which are less liable to disrupt the concrete, or mobile products which appear elsewhere.

The most potent source of the physical disintegration of concrete in the zone above high tide is freezing and thawing. The resistance can be markedly improved by the introduction of air entrainment. The performance of concrete under these conditions has also been shown to be affected by choice of aggregates, for which the best guidance that can be given is that the coarse aggregates most contributing to frost degradation show the lowest bond with the cement paste [38]. In the inter-tidal zone, physical attack occurs as a result of water movement and cavitation, and abrasion from suspended solids, such as silt, sand and ice. High-quality concrete, as discussed in Sections 3.2 and 3.3, must be provided to resist these conditions.

Chemical attack by sea-water arises especially because it contains significant amounts of magnesium sulphate. Biczok [39] quotes a figure of 2800 mg/l as the average concentration of SO_4^{2-} ions in Atlantic ocean water, a figure which, according to the US Bureau of Reclamation, puts such water into the 'severe' range as far as sulphate attack is concerned [40]. Both mechanisms of sulphate attack, as discussed in Section 3.1.2, can occur with this level of sulphate concentration. In the splash zone, the salt concentrations can be higher than in the surrounding sea-water. The use of sulphate-resistant cement, low in C_3A, does not necessarily therefore produce structures free of disintegration. The increased protection to steel observed when higher C_3A contents are used should also be noted. It is difficult to generalize on the resistance of different cements to sea-water attack. The resistance is related to the mineralogical composition, which may differ from the potential composition calculated by the Bogue method. The differences are due to the presence of sulphates other than $CaSO_4$ and to the presence of foreign ions in the basic compounds [41]. Two forms of C_2S exist (α, β) with the α- form containing alkali in solid solution and two forms of C_3A (cubic and orthorhombic) again with the orthorhombic containing alkali. The cubic form has been found to have greater resistance to attack by magnesium sulphates. Cement fineness can also affect resistance. In coarsely ground cement, encased C_3A in grains can react with sea-water to produce ettringite at a stage when the cement paste has already hardened. Local stress and cracking then results.

Blast-furnace slag cements have been found generally to improve the resistance of concrete to sea-water attack. To achieve this improvement, a considerable proportion of slag is required. In France, cements containing more than 65% slag are admitted by the Permanent Commission on Hydraulic Binders and Admixtures (COPLA) as marine cements. Alternatively, pozzolans may be used to improve resistance, but not all pozzolans are equally effective.

The effects of aggregate types on resistance has not been widely reported, but there are indications that feldspar rocks start to be attacked sooner. Alkali–aggregate reaction may also be enhanced in sea-water [42].

It should once again be stressed that if sound impermeable concrete with adequate cement content is properly mixed and placed, durable structures can be produced. A water/cement ratio of less than 0.45 and preferably less than 0.40 and a cement content of 390 kg/m^3 are recommended by Gerwick [43].

4 Temperature extremes

4.1 High-temperature operating conditions

Hardened concrete may be subjected deliberately or accidently to heat. The response of concrete is, of course, dependent on the temperature involved and can be considered in three ranges.

(a) Extreme climatic conditions involve temperatures up to 50°C. Such extreme temperatures are likely to be associated with low humidity and may also be accompanied by thermal and hygral cycles.

(b) For some industrial applications, as for example aluminium plants or the surrounds of industrial kilns, temperatures well above ambient may be encountered and these temperatures are likely to be applied cyclically, generally with a rather long period. In the event of an accident, in for example a nuclear power plant, high temperatures may be reached rapidly and be accompanied by unusual loads to produce a serious design requirement. It has often been held that operating concrete temperatures should not exceed 65°C. Above this temperature, the changing properties of concrete must be taken into account, but provided this is done a feasible design can be developed for temperatures up to about 300°C [44]. For even higher temperatures, refractory concretes become necessary.

(c) Furnaces and accidental fires may involve temperatures up to 1200°C. The response to such high temperatures is a function not only of the temperature reached but also of the time gradient. Fire resistance is discussed in Chapter 14.

4.2 The effect of high temperature on hardened concrete

The major influences of heat are due to:

(a) removal of evaporable water;
(b) removal of combined water;
(c) disruption from the disparity of the expansion of the concrete components and thermal stress;
(d) alteration of aggregate;
(e) alteration of cement paste;
(f) change of aggregate–paste bond.

As a result of the interaction of these influences, the reported results of the effect of heat on properties are conflicting. In particular, rates of heating and consequently temperature and moisture gradients are particularly significant.

4.2.1   Behaviour of mortar

Dry heat applied to mortar removes moisture and may, at temperatures up to 100°C, slightly increase strength. At about 400°C, dehydration of $Ca(OH)_2$ occurs and involves a major reduction of mortar strength. There is evidence that some changes also occur at quite low temperature and that these changes are independent of loss of water. Hansen and Eriksson [45] have shown that the physical properties of cement mortars deteriorate at temperatures as low as 30°C even though loss of water is eliminated. X-ray diffraction analysis has shown changes in cement pastes at temperatures of 65°C [46].

4.2.2   Compressive strength of concrete

The effects of high temperature as found by several investigators are summarized in Fig. 4-1. The curves probably represent the full spread of data and other reported findings generally fit within the extremes. Notwithstanding the wide scatter, the curves fall into three groups: the aluminous cement concrete is most sensitive to temperature change; the normal Portland cement concretes fill the central portion of the figure; and finally the specially prepared 'heat resistant concrete' apparently improves in character even up to 650°C. Above 700°C, the curves for the various materials tend to converge and the strengths are generally below 50% of the unheated concretes.

The results reported are strengths obtained at room temperature after the concrete has been fired at the high temperature. Some loss of strength appears to take place in the cooling process. Work by Abrams [47] has confirmed this, as unstressed residual strengths (specimens heated, cooled and then tested) were found to be somewhat lower than the strengths of comparison specimens tested at high temperatures.

Cycles of temperature change can have a progressive effect on the reduction of strength. In an investigation using dolerite aggregate, one cycle between 20 and 250°C resulted in no strength loss. One cycle between 20 and 300°C produced a 15% strength loss. After five cycles, a 35% loss had occurred and a further 4% occurred at the end of ten cycles. Cycles having 200°C maximum had similar cumulative effects [48].

Different aggregates affect the strength (and other properties) very differently. Concrete made with aggregates which have been heat treated in the process of production, such as fire-clay, brick and expanded shale, show generally less reduction of strength at temperatures up to 300°C than do natural aggregates such as limestone [46]. Concretes containing carbonate aggregates when tested maintained more than 75% of their original strengths at temperatures up to 650°C. For siliceous aggregates, the corresponding temperature was 425°C for an equivalent strength decrease. At 575°C, quartz undergoes a phase transformation, accompanied by a 0.4% increase in volume. This phase change is often associated with the spalling of heated concrete.

The effects of moisture content on the strength and other structural properties of concrete have been discussed in detail by Lankard et al. [49]. They conclude

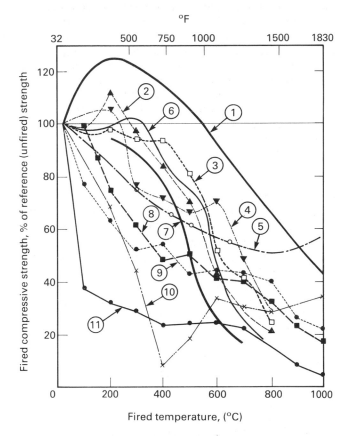

Figure 4-1 **Effect of temperature on the compressive strength of concrete as studied by various investigators:**

1 Heat resistant concrete; fire clay aggregate
2 Sandstone aggregate
3 Limestone aggregate
4 Expanded-slag aggregate
5 British aluminous cement; expanded shale
6 Gravel aggregate
7 Portland cement; river sand and gravel
8 Aluminous cement; anorthosite aggregate
9 Aluminous cement; expanded shale
10 Aluminous cement; phonolite aggregate
11 Aluminous cement; ilmenite aggregate

that, for situations in which the free moisture can evaporate as heating commences, little or no change will take place in the compressive strength on heating to 260°C. If free moisture is retained in the concrete during heating, deterioration in the structural properties can be expected in increasing degrees of severity at all levels of temperature up to 260°C. In this case, siliceous aggregates are recommended as diminishing the degree of deterioration.

4.2.3 Tensile strength of concrete
Tensile strength is more susceptible to the influence of heat than is compressive strength [46]. With dolerite aggregate, one cycle between 20 and 300°C produced about 15% loss and ten cycles over 50% loss. Noticeable macroscopic cracking took place on the concrete surface at about four cycles. A similar loss of tensile strength was observed with limestone aggregate, but with fireclay and, to a less extent, with expanded shale aggregates, the reduction of tensile strength was nearer to that of mortar. It was apparent from the tests that the bond between aggregate and mortar was affected by the heat cycles in different ways. In the case of the dolerite and limestone for example, the oxidation of iron compounds forming parts of the aggregate surfaces appeared to produce a reduction of bond.

4.2.4 Elastic modulus of concrete
In the range of temperature up to 300°C, the modulus of elasticity is more adversely affected than the compressive strength and is further reduced by cycling. Values as low as 24% of the unheated modulus have been obtained after ten cycles between 20 and 300°C. This reduction may be helpful in reducing thermal stresses.

4.2.5 Concrete under transient temperatures
During rapid rise and fall of temperature, the response of concrete is affected by the interaction of thermal expansion, drying (and hence shrinkage), cracking due to thermal incompatibility and enhanced creep at high temperatures. Under fast rates of heating, the influence of creep in reducing stresses is lessened. Thus, high stresses can be induced in members if the heating is sufficiently rapid, and the failure of material or instability may result. Subsequent cooling can induce tensile stresses if creep has been sufficiently large during heating, and hence cracking, and even failure, may occur after cooling [50].

4.3 The effect of high temperature on steel

The influence of temperature on steel appears as a change in yield stress, ultimate strength and modulus of elasticity both at the elevated temperature and after cooling. The effects depend on the type of steel and are more marked with work-hardened high-yield steel than with mild steel.

4.3.1 Reinforcing steel
The strength of hot-rolled high-tensile reinforcement is not reduced if the temperature does not reach 300°C, but at temperatures of 500–600°C, the yield strength is reduced to the order of the working stress and the elastic modulus is reduced by one-third [51, 52].

The strength of a loaded member is directly affected by the reduction of yield stress and the reduction in Young's modulus may also be significant. If the reinforcement is exposed to the direct effects of heating following spalling and the removal of cover, this reduction in stiffness can lead to the buckling of the reinforcement between ties in compression members.

After cooling from temperatures in the 500–600°C range, the strength of hot-rolled and cold-worked reinforcement is not significantly affected, but after heating to 800°C the strength of hot-rolled reinforcement is reduced by about 5%, while for cold-worked steel the yield is reduced by about 30% and the ultimate strength by 20%. Cold-worked steel may gain strength, through age hardening, after being heated to 200–300°C. Rapid cooling from an elevated temperature may in some steels, particularly those with high carbon content, cause embrittlement [51].

4.3.2 *Prestressing steel*

A significant property for prestressing steel is its tensile strength, since the stress in a tendon under service load is in general related to this value. Most of the data available have been obtained on wires rather than strands. A loss of strength occurs in prestressing wires at lower temperatures than for reinforcing steel [53]. After cooling, there is a permanent loss of strength for wires which have been at temperatures above 350°C.

Just as significant in the performance of prestressed members is the enhanced relaxation which occurs at high temperature, leading to increased deflection and cracking.

4.4 Low-temperature effects

Concrete which is subjected to low temperatures and particularly freezing conditions during the curing period suffers delayed hydration and the strength is reduced.

After satisfactory curing, concrete has successfully been used at extremely low temperatures (−160°C) for the storage of liquified natural gas. Such use will induce a cycle of cooling every time a container is filled. In addition, lack of control, as for example a leak, may produce a sudden temperature change. Some knowledge of concrete properties under these conditions is therefore necessary for successful design and operation. Concrete which is cooled once and tested cold shows no major change in strength unless the concrete is initially saturated with water to a critical level. For such saturated concrete tested at below freezing temperatures, an increase of compressive strength from 80 to 150% has been observed and an increase in indirect tensile (splitting) strength from 60 to 70% [54]. These improvements are unfortunately of little value, as when the concrete is allowed to thaw, a reduction in strength below the initial value is to be expected and further cycles of freezing and thawing have a further marked effect. The damage has been related to the thermal strain, which itself depends largely on the moisture content and the pore structure. This structure depends on the cement type, and the responses of Portland cement and Portland blast-furnace slag cement have been shown to be considerably different. A typical thermal strain–temperature relation is shown in Fig. 4-2. With concrete possessing this sort of

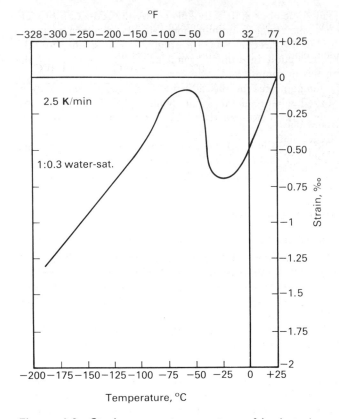

Figure 4-2   Strain versus temperature of hydrated cement paste

response, damage from temperature cycles increases with temperature drop down to $-70°C$. At this temperature, expansion ceases and damage is not increased by cycles to much lower temperatures, such as $-170°C$. Portland cement mortar, with a siliceous aggregate and a water/cement ratio of 0.5, gives the results shown in Table 4-1 when saturated and thawed in water.

As pointed out by Rostasy et al. [55], these great losses of strength only occur in the specimens initially saturated with water. Cycles of freezing using liquid nitrogen $(-192°C)$ on air-dry concrete cause some loss of modulus of elasticity and presumably therefore of strength. Air entrainment gives some benefit under these conditions [56]. Under operating conditions, it becomes important to ensure that concrete will not become critically saturated either by liquid water ingress or prolonged condensation from the surroundings. Investigations by Rostasy et al. [55] into the pore-size distribution suggested that pores in the size range from 40 to 80×10^{-9} m were irreversibly enlarged by the action of ice crystallization and that this may have resulted from the breakdown of partition walls between neighbouring pores.

Table 4-1   Strength after 12 cycles as a percentage of the initial strength

Strength	Cement type	Minimum temperature		
		−30°C	−70°C	−170°C
Compressive	Portland	50	40	45
Tensile	Portland	40	30	30
Compressive	Portland blast- furnace slag	70	50	40
Tensile	Portland blast- furnace slag	50	20	40

5 Internal disruption of concrete

5.1 Alkali–aggregate reaction

The chemical reactions, nature and characteristics of aggregates associated with alkali attack and consequent internal disruption of concrete have been reviewed in recent publications [57, 58]. Despite 40 years of research, the chemical reactions between cement and aggregates causing expansion are by no means fully understood, and no clear and generally accepted theory has yet been proposed which explains all the observed chemical and physical changes.

Three types of alkali–aggregate reaction are however known to occur:

(a) Alkali–silica reaction, which is the classic type and involves cherts, flints, chalcedony, opal, naturally occurring acid-volcanic glass and most manufac-tured glasses. The alkali–silica reaction involves hydroxide attack on sus-ceptible aggregates, resulting in alkali-silicate products which can take up water. The consequent expansion will produce mechanical or osmotic pressure such as to crack the concrete. As pointed out by Diamond [59], reactivity is not a direct consequence of the presence of alkalis, which, however, strongly influence the content of hydroxide ions in the pore liquor.

(b) Alkali-carbonate rock reaction, for which the texture of the aggregate must be such that larger crystals of dolomite are surrounded by, and scattered in, a fine-grained matrix of calcite and clay. Two reactions have been reported to occur with rocks of this sort: dedolomitization with the forma-tion of calcite accompanied by brucite $(Mg(OH)_2)$ and a rim silicification reaction. The rim zones become less porous and enriched in silica. Accord-ing to Mather [60], the rim silicification reactions may not be harmful.

(c) Alkali–silicate reaction, ascribed to some phyllosilicates which react to form swelling gel. These reactions are not yet well documented, but rocks of the Malmesbury Group in South Africa are described by Oberholster and Brand [61] to be expansive and to be different in nature to the former described rock types. In New Zealand, expansions have been reported for concretes made with Matahina greywacke-argillite by St John and Smith [62]. In both the last two cases, the reactions occur even when low alkali cement is used.

Leaving aside the alkali–silicate reaction, it may be confidently stated that, for the other two types of reaction, three conditions must be satisfied:

(a) Sufficient alkali must be present in the concrete.
(b) The presence of a reactive constituent in the aggregate is necessary.
(c) The right environmental conditions must exist.

Cement is the commonest source of alkali, but alkalis may be introduced by aggregates containing alkali salts, or by the use of certain fly ashes or other pozzolanic materials. Although sodium and potassium are both important, to simplify the nomenclature when describing the quantity of these alkalis present in cement, it is usual to calculate the total alkali as equivalent sodium oxide, where

$$\% Na_2O \text{ equivalent} = \% Na_2O + (0.658 \times \% K_2O).$$

In the USA and Germany, an upper limit of 0.6% Na_2O equivalent is accepted for cement used in combination with reactive aggregates (ASTM C150–77 and West German DIN 1164). However, expansions have been observed with an Na_2O equivalent as low as 0.4%.

A series of standard test methods have been reputed to be of assistance in evaluating potential alkali reactive aggregate. ASTM provides such a series in ASTM C289–76, *Test for potential reactivity of aggregates (chemical method)*; ASTM C227–76, *Test for potential alkali activity of cement–aggregate combinations (mortar-bar method)*; and ASTM C586–75, *Test for potential alkali reactivity of carbonate rocks for concrete aggregates (rock cylinder method)*. Recently, criticism has been levelled at these tests, and in particular C289 is suspect. Some of the test methods are specific to certain types of aggregate, and hence it is essential to apply petrographic methods to determine which type of reactivity is to be expected in order that a suitable test method may be selected. This may be done in accordance with ASTM C295–73, *Petrographic examination of aggregates for concrete.*

Despite the guidance of standard tests, the application of engineering judgement is required once a rock has been classified as potentially reactive. As a first step, avoid the use of such aggregate by appropriate procedures such as selective quarrying. If it is not feasible to do this, then specify the use of low-alkali cement, the minimum aggregate size that is economic, and consider possibly blending to minimize expansion by moving away from the pessimum proportion of reactive aggregate. Consider also the use of admixtures such as blast-furnace slag, calcined shale or pulverized fuel ash, certain of which have been shown to inhibit the expansive effects by pozzolanic or other reactions with the hydroxide ions.

Concrete is most susceptible to expansion when subject either to continual dampness or to cycles of wetting and drying, especially when such cycles are combined with cycles of heating and cooling. Concrete maintained in a continuously dry condition has a reduced potential for expansion even though it may contain reactive aggregates in conjunction with a high-alkali cement.

In structures, the time taken for the reaction to manifest itself in phenomena such as cracking and gel exudation varies considerably, but signs of disruption are generally observed after four to seven years. For all the reactive aggregates the deterioration commonly takes the form of pattern cracking where the concrete is free to expand. After the initial crack pattern is formed, the principal development is further opening of the cracks, probably due to the expansion of the concrete inside the outer shell, without an equivalent expansion in the outer layers. Frequently, the pattern is accentuated by a discolouration of the concrete adjacent to the crack. In cases where concrete is restrained against expansion in one direction, the patterns of cracks tend to be parallel to the axis of restraint.

In wet locations, exudations of resinous gel may be observed on concrete surfaces; these may be accompanied by popouts caused by such gel. Expansion and warping may either be visually obvious or be measurable on some structural members. Reaction rims may be observable around certain of the aggregate particles in the concrete. More detailed descriptions of field observations are given in the case of alkali–silica reactions by Idorn [63]. Mather's paper [60] appears in a *Transportation Research Record* entirely devoted to alkali–carbonate reactions, in which structures affected by expansion are described, and the paper by Oberholster and Brand [61] on alkali–silicate reaction describes the field problems using the third type of aggregate.

In recent literature, there has been an increase in reports of actual damage being directly ascribed to alkali reactivity, but even more importantly changing technological factors suggest increased probability of future damage. These include progressively increasing alkali contents of cement produced in conventional dry-process kilns with energy-efficient pre-heater systems, and wide-spread use of aggregates previously considered marginal or unsuitable. The design and site engineer will therefore have to be more aware of such problems in the future, since the only remedial procedures which are available once the problems are observed to exist are those which reduce or restrict water ingress to the deeper layers of concrete.

5.2 Unsoundness of cement

Bogue [64] uses the term soundness of cement 'to indicate freedom from, or degree of, volume change, warping or crack formation of a cement paste as determined by standard tests involving the boiling or steaming of neat pats of the paste for stated periods at stated ages'. The object of soundness tests is to accelerate the hydration of free CaO and crystalline MgO contained in a cement, these oxides being the primary cause of unsoundness and expansion of cement paste in concrete exposed to moist conditions over a period of years. Uncontrolled reactions between C_3A and gypsum may also lead to unsoundness; consequently, most cement standards limit the maximum SO_3 content of cement, as well as the free CaO and MgO contents. Several common test procedures, namely, the Le Chatelier test (see BS 12), the autoclave test (see ASTM C151–77) and the pat-crack test (see DIN 1164) are used at cement plants as part of their regular quality control operation. The Le Chatelier test is concerned mainly with unsoundness caused by the presence of free lime and also detects unsoundness due to reactions between gypsum and C_3A, whereas the autoclave test reflects expansions due to free lime, crystalline magnesium oxide and the C_3A reactions. Mehta [65], discussing the status of the autoclave test in the United States, has been critical of its retention.

Although there was a considerable amount of anxiety expressed during the

1930s about the problem of both MgO and CaO unsoundness of field concrete, no recent problems directly attributable to these materials have been encountered. This is probably due to the chemical limits set on these compounds, the greater control of raw feed composition, and the better firing conditions of modern kilns. At the present time there is considerable evidence to suggest that in slag-cement production even high MgO slags do not lead to unsoundness [4], but if periclase were permitted to crystallize, then this need not always be the case.

6   Corrosion of reinforcement in concrete

At the outset, it must be said that, in a structure properly designed and constructed for the appropriate conditions of use, there should be no problem of steel corrosion in concrete. Unfortunately, these highly desirable conditions are not always achieved. In a review such as this, one must therefore dwell to a great extent on the unusual or negative factors which cause corrosion, rather than discuss the many structures which throughout their life span show no problems of this sort. Concrete provides a high degree of protection against corrosion because of its alkalinity, and well-made concrete having low permeability minimizes the penetration of oxygen, chloride ions and carbon dioxide, all of which influence corrosion activity. Despite this, the cost of corrosion of steel in concrete is high; in the US, the annual costs of bridge-deck corrosion repair have been estimated at between \$160 m and \$500 m [66].

6.1   Conditions for corrosion

For corrosion of any metal to occur, the following conditions must be fulfilled.

(a) An electron sink area (anode) must be present at which the de-electronation reaction (i.e. metal dissolution, oxidation) occurs.

(b) An electronic conductor (in most cases the metal itself) is required to carry the electrons to the electron source area (cathode) where an electronation (reduction) reaction occurs.

(c) An ionic conductor must be present to keep the ion current flowing, and to function as a medium for the electronic reaction.

In the case of reinforcement corrosion in concrete, anodes and cathodes develop either as small, closely spaced areas on the bar itself, or as areas of reinforcement often up to 3 m apart. The electronic conductor is the reinforcement itself, and the ionic conductor or electrolyte is the liquid held within the concrete pores.

At the anode, iron is oxidized to ferrous ions:

$$Fe \rightleftarrows Fe^{+2} + 2e^- \qquad E^0 = -0.440 \text{ V (standard electrode potential)} \qquad (6\text{-}1)$$

The following subsequent reactions may occur:

$$Fe^{2+} + 2OH^- \xrightarrow{\;H_2O\;} FeO \cdot (H_2O)_x \qquad (6\text{-}2)$$

$$FeO \cdot (H_2O)_x + \tfrac{1}{2}H_2O + \tfrac{1}{4}O_2 \rightarrow Fe(OH)_3 + (x-1)H_2O \qquad (6\text{-}3)$$

When the iron reacts to form ferric oxide, the change in volume is four-fold.

Concrete is highly basic and usually an adequate supply of oxygen is available so the cathodic reaction is generally considered to be

$$\tfrac{1}{2}H_2O + \tfrac{1}{4}O_2 + e^- \rightleftarrows OH^- \qquad E^0 = 0.401 \text{ V} \qquad (6\text{-}4)$$

In concrete, the stoppage of an electric current is brought about most commonly by polarization at the cathode, i.e. the slowing down of reaction (6-4). Thus, when corrosion of reinforcement can occur due to some interference with the passive film as discussed later, the condition leading to its inhibition is primarily a very low degree of permeability of the concrete.

6.2 Helpful Pourbaix diagrams

A useful start in understanding the corrosion of steel in concrete is the examination of two experimental Pourbaix diagrams for iron. In these, two parameters E_H, (the standard hydrogen potential) and pH (the hydrogen ion concentration) have been plotted on Cartesian coordinates for iron immersed in solutions in the absence of chlorides (Fig. 6-1) and in the presence of 10^{-3} g ion/l (355 ppm)

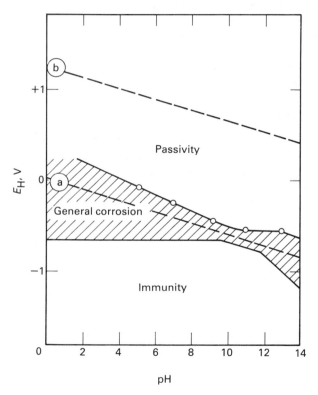

Figure 6-1   Behaviour of iron in chloride-free solutions: experimental conditions of immunity, general corrosion and passivity

    a Equilibrium hydrogen reaction
    b Equilibrium oxygen reaction

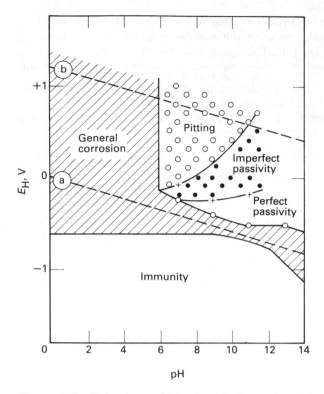

Figure 6-2 Behaviour of iron in solutions containing chloride (10^{-3} g ion/l, i.e. 355 ppm Cl^-): experimental conditions of immunity, general corrosion, perfect passivity, imperfect passivity and pitting

Cl^{-2} (Fig. 6-2). The limitations of these types of diagram have been expressed by Pourbaix [67]. An important limitation is that data of this type are often given for pure iron rather than for steel and alloy influences are absent. For the purpose of understanding the general corrosion processes, pure iron and mild steel may be considered to act in somewhat similar ways. Accepting this, it may be concluded from Fig. 6-1 that a pH above about 11, which would be the condition for steel in contact with uncarbonated concrete, and E_H greater than about -0.5 V, a passive condition would prevail. The steel would be protected by a continuous strongly adherent layer of Fe_2O_3, Fe_3O_4, or an equivalent hydrate, and would not undergo general corrosion. As the pH falls, the tendency for general corrosion to occur increases, hence the significance of the high pH of the pore liquor in concrete may be appreciated. In Fig. 6-2 it is observed that the increase in the field of 'general corrosion' and the development of fields of 'pitting' and 'imperfect passivity' now reduce 'perfect passivity' to a relatively small area of the diagram. It can be concluded that, in the presence of chloride ions, corrosion may occur despite the high pH of the concrete pore solution. That such a high pH will exist in water extracted from cement paste, even when it contains up to 5% sea salt, has been demonstrated by Gjorv *et al.* [68]. This is not to say that the pH at an anodic area of a bar in such a paste will also remain high, since reactions at localized cells will dictate such conditions.

6.3 Local cell development

The development of very local cells, leading to pitting, and hence reduction of area, is particularly important in the case of prestressed tendons. The development of pits may be due to the unfavourable storage conditions of tendons prior to installation, as described by Fountain *et al.* [69], or to the corrosive conditions active in service, as described by Phillips [15] for wires from the Geehi aqueduct in Australia. In both cases, the mechanism can be described as in Fig. 6-3. It should be noted that, at the pit, the site of corrosion, the area is depleted of oxygen and the process is autocatalytic in nature. The pH at the pit tip may be about 4 compared to 11 or higher elsewhere on the corroding surface.

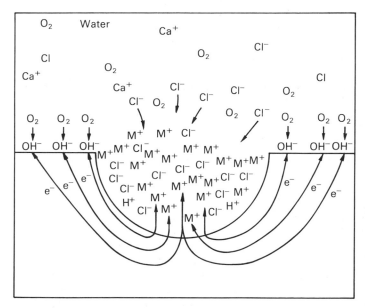

Figure 6-3 Pitting corrosion of a metal in a salt solution

6.4 Differential cell development

Let it be argued that one is able to block the access of O_2 to the steel-concrete interface all over a structure; reaction 6-4 would stop and corrosion would cease. What would be the effect of stopping its access to only one half of the area of the bar? One might argue that half the corrosion would occur. This is not so, however, for corrosion may actually be enhanced due to the development of a differential cell between the oxygen-rich and oxygen-poor sections of a structure. Thus, since reinforcement in concrete structures generally consists of systems of electrically interconnected bars or wires, the entire system may develop large-scale corrosion cells of considerable size, if conditions are right for their formation and continued operation. Anodic and cathodic areas, separated from each other by distances of 0.5 to 4 m, under potential differences of 0.5 V, have been measured [70]. Under such conditions, serious corrosion leading to the cracking of concrete, arising from the pressure of accumulated corrosion products, has been observed. In advanced cases, spalling of the concrete from the plane of the reinforcement has occurred.

Other factors may also lead to the existence of differential cells; these include differences in alkalinity from one area to another and differences in the concentration of Cl^- ions. Bleeding, segregation and poor consolidation of concrete may also cause differences in environment between the upper and lower surfaces of individual reinforcing bars. Temperature differences within the concrete may also create differences in the electrochemical potential. Repaired areas, where, for example, one section of steel is embedded in epoxy and the other in Portland cement, may also produce a differential cell. Differences in the steel are caused, for example, by welds, and ordinary bars connected to zinc-dipped bars may also lead to differential currents. A recent claim by the Danish Corrosion Institute [71] is that offshore concrete platforms could be in serious danger of intense corrosion caused by this phenomenon being overlooked by designers. Huge corrrosion cells develop where external steel, often provided in the form of pipe supports, is in direct contact with internal reinforcement.

6.5 Water in concrete and its influence on electrical resistance

Returning to reaction 6-4, one must ask what would occur if water were to be excluded from the concrete? Such exclusion is much harder to achieve in practice than one would first believe. There is a great affinity between concrete and water, and, even well below 100% RH, the micro-pores of the hydrated cement paste are filled with hydrous solutions. The water in these pores, which vary from perhaps 1×10^{-9} to 8×10^{-9} m in diameter, will have properties such as vapour pressure, freezing characteristics and mobility different from those of free water in bulk. Because of the relatively larger size of capillary spaces, proportionately less of the capillary water is strongly adsorbed on their walls, and it is therefore more volatile than is gel water. Nevertheless, the water in the capillary spaces is almost completely evaporable only at humidities below 40%. Above 40% RH, 'capillary condensation' occurs, that is, the capillaries become filled with water by precipitation from the vapour phase. Thus, to ensure that relatively freely mobile water is absent from the vicinity of steel, one would have to hold the local RH to below 40%.

The electrical resistance of concretes is increased by air drying and therefore the drying of concrete is advantageous in combating corrosion. Monfore [72] reports data for a series of dry, 100 mm cubes, with internal electrodes, which were given two applications of various coatings, and after the coatings had hardened, the cubes were immersed in water. Resistance changes were measured after several periods of immersion. A coating of epoxy was most effective, but after 28 days of immersion in water, the resistance of the coated concrete was only moderately greater than that of the uncoated specimen. Monfore notes that these effects are in significant contrast to epoxy coatings on steel. Some coating materials actually resulted in decreased electrical resistance of the concrete after immersion.

6.6 Carbonation effects

Since carbonation is known to influence the rate of corrosion of steel, it is important to have some background knowledge of this phenomenon. Considerable amounts of data are available from papers by Hamada [73] and Meyer [74]. Verbeck [75] showed that the carbonation rate was at a maximum at about 50% RH, and was considerably reduced for both wet and very dry concrete. If, therefore, concrete were kept for the most part at about 50% RH with occasional wetting, the corrosion rate of an embedded bar may be greater than if the concrete remained wet continuously, an outcome different to that expected for steel exposed to the atmosphere.

It is accepted that concrete, on exposure to atmospheric conditions, will carbonate, and the engineer knowingly, or unknowingly, places a sacrificial layer of calcium hydroxide between his reinforcing steel and the atmosphere to react with the carbon dioxide of the air and hopefully prevent its ingress to steel level. He also decreases the permeability to carbon dioxide by making the concrete as dense as he can within the constraints of construction procedure and cost. The concrete cover to the reinforcement thus plays a two-part role, as a barrier to carbon dioxide gas ingress and as a chemical absorber for that gas. The amount of calcium hydroxide available for reaction and the effect of cracks are both important in this regard.

Well-hydrated Portland cements have contents of $Ca(OH)_2$ varying from 15 to 30% by weight of unhydrated cement [76]. In contrast to normal Portland cements, however, other types of cement, such as high alumina cements, Portland blast-furnace cements and Portland pozzolan cements all have various amounts of $Ca(OH)_2$ bound to reactive siliceous materials, and hence the reserve basicity is correspondingly lower.

6.7 Cracking and permeability

The permeability of well-compacted concrete is held, in the absence of cracking, to be mainly determined by the permeability of the cement paste. However, it has been shown that many rocks have greater permeabilities than mature pastes [77], and some corrosion problems can be in part attributed to the use of very porous aggregates such as sandstone and inferior light-weight aggregates [78]. Penetration of calcium chloride in solution into concrete and paste was measured by Ost et al. [79]. They found that for both concrete and paste the penetration was strongly influenced by water/cement ratio, and further, that the rate of penetration into concrete was greater than into the paste. Verbeck [80] recalculated some of the data to obtain an almost linear relationship between permeability of concrete to chloride ions and water/cement ratio.

A paper by Schiessel [81] is a memorable work dealing with concrete cover, and in particular the effects of cracking on that cover. In the first part of the paper, laws governing the spread of carbonation of concrete and the speed of corrosion of steel in carbonated concrete are discussed. According to Schiessel, the value of the effective diffusion constant of oxygen, D_B, varies between 2 and 4 mm^2/s for 20 to 30 year-old concrete which has carbonated to a depth of about 15–20 mm. It is suggested by him that the diffusion rate is of lesser importance during the carbonation stage than it is during the corrosion stage, i.e. for oxidation-reduction to occur, O_2 must be available. It is shown that under the conditions outlined, namely, for concrete carbonated to the depth of the steel, the influence of the thickness of cover is not very significant, but that there is a relationship between concrete strength (as reflecting cement content) and corrosion rate.

In the second part of the paper, theoretical results are compared with experimental results. The test specimens (reinforced concrete beams with constantly open cracks and non-reinforced concrete specimens) were exposed to the air under different climatic conditions. The comparison shows good agreement, and data are presented which relate the increase in depth of carbonation at a crack with crack width. For both uncracked concrete and crack-free locations, the thickness of the carbonated layer tends to a limit. It is concluded the corrosion or the corrosion protection of the reinforcement depends only on the carbonation of concrete, as influenced by crack width and concrete cover.

It is pointed out that in considering cracks in concrete a division between types of crack must be made. For example, the influence of the ribs on a deformed bar on carbon dioxide ingress is to increase permeability in some regions of the crack while decreasing it in others. Furthermore, the conclusions reached in the paper are only for cracks transverse to the steel, and the author notes that following the development of wide transverse cracks, longitudinal cracks develop. This latter type shows greater potential to cause corrosion and can lead to spalling of the cover, particularly where they are associated with thicker rod diameters and in corner regions of beams. Schiessel suggests that one should use thicker beams when thicker diameter bars are used. It should be noted that cracks along the bar lengths may also be caused by adverse conditions, such as high plastic shrinkage and settlement cracking, factors not mentioned by the author.

Taking as an example a flexural member with transverse cracks, continually open, and having an average width of 0.30 mm and concrete cover of 25 mm, Schiessel derives a probability of about 55% for a corrosion layer of a certain thickness, $t_m > 0.01$ mm, to form. Were the average crack widths to be limited to 0.15 mm, then the probability of the development of the same corrosion layer thickness, $t_m > 0.01$ mm, decreases to about 35%. Now, however, the number of cracks compared to the first case is nearly doubled, so the decrease in crack width does not greatly help in practice.

He suggests that an advantageous step may be taken by using thinner reinforcing rods, but admits that thin rods suffer a greater proportional loss of cross-section than thicker rods for the same corrosion layer depth. It is indicated that some division of reinforcement between that required for crack control and that necessary for load carrying capacity may be advisable, a view shared by Leonhardt [82].

Tremper [83] commenting earlier on the crack size debate states:

> Tremper [84] found that cracks of considerable width in blocks exposed for ten years in a moist but substantially salt-free environment did not cause significant corrosion of embedded steel.

Hausmann [85] states:

> Experience has shown that minor cracks penetrating to the steel are normally not damaging and frequently heal in a moist environment. Protection is also provided at the base of the crack by alkaline material from the adjacent concrete.

Gewertz [86] discussing the conditions found in the San Mateo-Hayward Bridge in California states:

> The deterioration taking place was independent of surface cracking which normally might have been considered to afford easy ingress for the corrosive effects of a marine environment. As a practical matter, therefore, it does not appear that cracks in concrete are of decisive effect with respect to the development of corrosion.

It would appear, therefore, from a general survey of literature that, apart from lean concrete mixes and under very severe corrosion conditions, transverse flexural cracks do not play a major role in increasing corrosion. In many cases, however, the cracks associated with corroding bars are not of this type. Many cracks examined in the field tend to run parallel with bars and may either be due to the expansive and hence disruptive forces caused by the volume increase of the steel as it oxidizes, or are due to shrinkage, thermal movements or problems with concrete in the plastic state. Elaborate calculations of crack widths due to flexural loads would appear to be valueless for corrosion protection if the other types of crack are ignored. One line of action which might be effective would be to use thinner reinforcing rods, following the German work [81, 82].

7   Methods used to improve corrosion resistance

Several approaches have been used in attempts to increase the durability of reinforcement in concrete. They include the following:

(a) corrosion inhibitor use;
(b) Portland cement concrete modification;
(c) membrane system installation;
(d) steel coating;
(e) cathodic protection.

7.1   Corrosion inhibitors

A corrosion inhibitor is an admixture that is used in concrete to keep the metal embedded in concrete from progressive corrosion.

7.1.1 Types of inhibitors

Anodic inhibitors (alkalis, phosphates, chromates, nitrites, benzoates) function by decreasing the reaction at the anode. They react with the existing corrosion products to form an extremely insoluble adherent coating on the metal surface. For example, nitrite inhibition is due to a rapid oxidation reaction leading to the precipitation of a relatively impervious ferrous iron film. Organic types replace water at sites on the water–steel interface, thus decreasing corrosion. Cathodic inhibitors (calcium carbonate, aluminium oxide and magnesium oxide) act to stifle the cathodic reaction. They are generally less effective since they do not form films on the anode. A mixed inhibitor may affect both anodic and cathodic processes.

A 'safe' inhibitor is defined as one which reduces the total corrosion without increasing the intensity of attack on unprotected areas. 'Dangerous' inhibitors produce increased rates of attack on those areas. Such increased rates can be due to lack of sufficient inhibitor, presence of enough chlorides to prevent complete protection or the presence of crevices into which the inhibitor does not rapidly diffuse. Anodic inhibitors are generally considered 'dangerous'. An exception is sodium benzoate, which if used in insufficient quantity causes general rather than local attack. Rosenberg et al. [87] discussing calcium nitrite for use in reinforced concrete have pointed out the difference between localized corrosion in, say, a steel tank and failure in reinforced concrete due to disruption of concrete following corrosion product build up. They do not discuss the possible problem which may arise due to accentuated pitting corrosion of prestressing tendons in the presence of a deficient proportion of inhibitor. They do, however, suggest that low concentrations of calcium nitrite may not effectively close off the anode to further reaction, but add that this condition will not accelerate corrosion even for a short time in the reinforced concrete system. Cathodic inhibitors are generally 'safe', but zinc sulphate has been reported as an exception.

7.1.2 Use in concrete

Griffin [88] suggests that very little evidence has been given as to the effectiveness of inhibitors in concrete. He suggests that some may reduce strength (sodium benzoate and sodium nitrite) whereas stannous chloride, which may act as an effective inhibitor, increases it. Rosenberg et al. [87], who have worked on calcium nitrite as a corrosion inhibitor, suggest that, at 2% addition, the setting time is shortened but the 28-day strength is increased by about 15%. They suggest that set acceleration can be offset by the use of a small amount of conventional retarder. Problems may, however, arise in the use of such chemicals, as they may interact with other admixtures such as air-entraining agents and hence lead to changes in the bubble structure of the paste and reduced durability.

Most of the research on inhibitors has been done under laboratory conditions. Little has appeared in print of actual use under site conditions. This is unfortunate since the use of a chemical in the concrete mix would be an ideal method of counteracting corrosion, particularly under marine exposure conditions.

7.2 Modifications to Portland cement concrete

One way in which corrosion protection may be increased is to modify the concrete so as to decrease its permeability. The method which has met with greatest success so far has been impregnation with various polymers and in particular methyl methacrylate. In a study by Clear [89] it was indicated in an interim report that polymer impregnation with methyl methacrylate was effective in preventing substantial chloride migration into a series of concrete slabs.

In the same study, a latex-modified concrete (water–styrene butadiene emulsion) and an epoxy-modified concrete (water-dispersable epoxy binder and amino-amine curing agent) were also tested. In both cases, the organics were added directly to the concrete mix. It was found that the latex-modified concrete overlay was effective in preventing ingress of chlorides but the epoxy-modified concrete overlay was not, due to the development of large, localized channels. Probably the most significant fact was that a low water/cement ratio (0.32) Portland cement concrete overlay was itself effective, up to that time, in preventing chloride ingress except where the overlay was not properly consolidated. Review papers of concrete polymer use are presented in an ACI publication [90]. The reader is also referred to Chapter 8.

It is feasible to mix wax beads into concrete to protect reinforcing steel in bridge decks from corrosion. Such a procedure was followed on a railway overpass near Seattle, USA. Electric blankets produced a temperature of 80°C below the surface in 5 h. The heat melted the wax, forming an interior barrier to protect the reinforcing [91]. Jenkins and Butler [92] have described a method of internally sealing concrete using inert thermoplastic particles in the mix which were warmed to fuse and flow into capillaries in the hardened mass. Such special procedures are obviously costly and can be justified only rarely and then for extreme exposure conditions. Personal observation of some bridge decks on which such procedures have been used has indicated that considerable disruption of concrete occurs during heating, and that durability problems may eventually arise in the concrete due to such cracking rather than corrosion *per se*.

One problem which still has not been fully studied is the effect of partial embedment of steel in, say, methyl methacrylate impregnated concrete, with the remainder left in ordinary concrete. At the junction of these two materials, conditions for a cell are probably developed. For ordinary steel, corrosion at a point is probably acceptable, but in the case of a prestressed tendon such a condition could be much more dangerous.

7.3 Membrane systems

In certain structures, such as bridge decks, water-proof membranes may be used between the concrete and asphaltic wearing courses. Such membranes were used during the early 1960s on over 30 bridges in New York and consisted of polyester resin or epoxy-modified coal tar. Cady [93] suggests that there is evidence that such measures do retard corrosion. More recently, pre-formed membranes have been employed. Frascoia [94] describes the evaluation of 29 different membrane systems on Vermont bridges between 1971 and 1977, and indicates that although some membrane systems are capable of providing the desired protection, the most important factors in their success are correct pavement design and paving procedures.

7.4 Steel coating

Coatings are sometimes considered for steel that is to be embedded in concrete exposed to adverse corrosive conditions. There are both benefits and disadvantages to their use, and any benefit can only be optimized by carefully considering the specific job. The more obvious of these considerations are

(a) Do the expected service life and structure exposure warrant coating of the steel?

(b) If coating is desirable, is a field job required or may the coating be applied prior to fabrication of the reinforcing for the structure?

(c) Do transportation and subsequent fabrication pose a significant danger to the coating?

(d) In view of the exposure conditions, is the choice of coating dictated by these conditions rather than the other considerations?

Backstrom [95] considers each of these factors in some detail and indicates that at this stage the data available on the use of coatings on steel in concrete are far from adequate to permit unqualified use of such coatings.

7.4.1   Organic coatings

Organic coatings on reinforcement include coal-tar enamels, epoxies, chlorinated and other rubbers, vinyls, phenolics, neoprenes and urethanes. In considering the literature, most of these have significant disadvantages, but the epoxy group seems to have the best potential for use. Despite the fact that epoxy coatings provided excellent corrosion protection of prestressing steel in tests by Moore *et al.* [96], the authors state that they were lacking in wear resistance and ease of application. Furthermore, they were also relatively high in cost.

Work was undertaken by Clifton *et al.* [97] to ascertain the feasibility of using organic coatings, especially epoxies, to protect the steel reinforcing bars embedded in concrete of bridge decks. Corrosion in decks is often caused by the chloride ions from de-icing salts. Altogether, 47 different coating materials were evaluated to some extent, consisting of 21 liquid and 15 powder epoxies; five polyvinyl chlorides; three polyurethanes, one polypropylene; one phenolic nitrile, and one zinc-rich coating. The chemical and physical durabilities, chloride permeabilities, and protective qualities of the coatings were assessed. The bond between bars and concrete was measured by both pullout and creep tests.

Only the epoxy-coated bars had acceptable bond and creep characteristics when embedded in concrete and provided adequate protection of steel from corrosion. The powder-epoxy coatings overall performed better than the liquid epoxies, and four powder-epoxy coatings were identified as promising materials to be used on reinforcing bars embedded in concrete decks of experimental bridges.

Since that time *Specification for epoxy coated reinforcing steel* has been distributed by the US Federal Highway Administration, and considerable work has been undertaken to assess performance both in laboratories and in the bridge structures. The first such bridge deck was constructed in 1973 and is described by Kilareski [98]. The reinforcing bar, having been cleaned to a near-white finish by shot or grit, is heated to about 230°C, passed through a set of electrostatic spray guns, which apply the charged dry epoxy powder to the bar so that a coating of about 0.15 mm is applied. The bar is then either water quenched or air cooled, by which time the epoxy has melted, flowed and cured on the steel and is ready for touch-up repairs on holidays. Added to the expense of manufacture, the great care required in transporting and placing epoxy-coated reinforcement would appear to be a significant disadvantage to its general use.

7.4.2   Metallic coatings

Metallic coatings are generally limited to zinc and nickel, although Bird and Strauss [99] have suggested that cadmium may be a good choice of coat under marine exposure conditions.

In considering zinc as a coating for reinforcement, it should be noted that it has the longest in-service history of any coating material. Extensive laboratory testing by, for example, Bresler and his co-workers [100, 101, 102] has indicated that galvanized coatings should provide improved resistance to the corrosion of steel by serving as physical barriers to chloride attack on the steel and, when disrupted, will provide sacrificial protection. Rehm and Lammke [103] have discussed the reaction products of zinc in cement mortars and concretes and have discussed the importance of chromate contents necessary for the prevention of hydrogen evolution problems. Chromate as CrO_3 amounting to 0.004% by weight of cement should be present or must be added to avoid hydrogen evolution.

Two significant surveys of structures undertaken by the Construction Technology Laboratories of the Portland Cement Association indicate that galvanized reinforcing bars have performed very effectively in concrete where chloride concentrations are well above the level at which corrosion of untreated steel will occur [104, 105]. The first of these dealt with the performance of galvanized reinforcement in concrete bridge decks subject to de-icing salts. The results of this investigation showed that galvanized steel clearly out-performed untreated steel where a corrosive environment existed. The second investigation was the evaluation of the corrosion resistance of galvanized reinforcing steel in concrete exposed to severe sea-water environments in Bermuda. Backstrom [95] also indicates satisfactory performance of galvanized materials.

It would appear, however, from the literature that in the presence of Cl^- ions zinc coating does not always provide increased protection [106]. A further problem, always present when two metals are coupled in an electrolyte, should be considered prior to the use of galvanized reinforcement. Hahin et al. [107] discussed the effects of coupling hot-dip galvanized bars to bare carbon steel bars. Galvanic current output became stable after 20 days for $CaCl_2$ concentration in the concrete of less than 1%. For concentrations of 1% or greater, current fluctuations were cyclic, increasing as the percentage of $CaCl_2$ increased. At 4% $CaCl_2$, current output rose sharply. This suggests that, above 1% concentration, inhibition was not achieved under the test conditions. It is important that if galvanized reinforcement is used, it be used throughout the structure and not coupled to other metal inserts, bar chairs, ties or ungalvanized reinforcement.

The results of an extensive testing of nickel-coated bar has recently been presented by Baker et al. [108]. The purpose of their study was to determine if nickel-coated steel had any advantage over bare or zinc-coated steel when used for concrete reinforcement in marine environments. The nickel-coated bars were produced from commercially nickel-plated 100 mm steel billets. Sulphur-free pure nickel was electro-deposited from a Watts-type nickel bath on to the descaled and cleaned billets. They were then hot-rolled to size after heating to about 1000°C in a commercial rebar mill. It was demonstrated that the nickel coating resisted corrosion during the course of the evaluation program which spanned a period of 11 years. It was concluded that metallic coated rebars (Ni and Zn) were definitely better than carbon or low-alloy steels in their overall performance in reinforced structures subject to marine environments. Furthermore, the results indicated that a 0.025 mm nickel coating was sufficient to achieve improved reinforced concrete performance. The cost factor will no doubt be important in any possible future use of nickel-clad bar, and further work will be necessary to assess couple effects in real structures.

7.5   Cathodic protection

Cathodic protection current may be applied in one of two ways. First, by a direct-current power source such as a battery, dc generator or transformer/rectifier system. Such a system is termed 'power impressed' as opposed to the second or 'sacrificial anode system', in which the anode provides the necessary protection current by virtue of being more base than the material to be protected. As a rule, the sacrificial anode system appears to be preferred in the protection of steel in concrete structures.

Robinson [109] discusses the use of cathodic protection to prevent steel corrosion of reinforced and prestressed concrete structures that have been damaged, subjected to stray current interference or installed in severely corrosive environments. He states that there are no widely accepted cathodic protection criteria for steel in concrete. Structures are sometimes protected as if they were bare or organically coated steel structures and on this basis the assumption is made that the steel is protected only if sufficient current is applied to maintain a steel potential of -0.85 V to a copper sulphate electrode (CSE). This generally results in the over-protection of steel in concrete.

Uncorroded steel in concrete normally has a potential ranging from 0 to -0.30 V to a CSE, which is 0.35 to 0.50 V more noble than bare steel in soil. It has been demonstrated by Hausmann [110, 111] that the corrosion of steel in concrete exposed to a high chloride environment can be prevented if sufficient cathodic protection current is applied to shift the steel polarization potential to -0.50 V to a CSE. For a structure subject to stray current discharge Hausmann [85] suggests that the polarization potential should be shifted to -0.40 V to a CSE. Many investigators recommend that the maximum potential value with concrete-coated steel should be limited to about -1.1 V. This limitation is necessary to avoid loss of bond between the concrete and the steel as a result of hydrogen evolution at the cathodic steel surface.

Most practical experience on cathodic protection of reinforced and prestressed concrete structures is related to buried pipeline systems. The effectiveness of controlling steel corrosion in this way has been demonstrated for many years by results both from field and laboratory studies. The first attempt to apply cathodic protection to a corroding bridge deck was described by Stratfull [112] in 1959, but not until 1974 was a technique described which could efficiently control this type of steel corrosion [113]. Since then, cathodic protection has been applied to a number of bridge decks [114, 115].

Concrete structures are becoming increasingly important in offshore gas and oil production, concrete gravity platforms being in use in the North Sea for both production and storage. Gjorv and Vennesland [116] noted that in 1979 cathodic protection had not yet been used to protect embedded steel in such structures, but that in many the reinforcing system is electrically connected to structural steel components being cathodically protected, and thus large portions of the reinforcement are themselves being subjected to cathodic protection. The possibility of unconnected parts of the reinforcement corroding anodically should not be overlooked. Heuze [117] also discusses the application of cathodic protection to concrete offshore structures.

Installation and operation costs of such protection systems are generally high and the conditions under which they may easily be installed are not often encountered in civil engineering practice; hence at the present time their use must be considered limited.

8 Cracking and the influence of materials

Cracks may affect the durability, function and appearance of a concrete structure. As has been indicated in Section 6, the protection afforded by concrete against the corrosion of reinforcement is also much influenced by the extent and width of certain types of crack.

Leonhardt [82] has pointed out that many builders harbour the illusion that freedom from cracks can be assured if concrete tension stresses due to loading can be limited. In fact, cracks result from a wide variety of causes and are certainly not confined to situations where tensile stresses are calculated to be present. Surveys have shown that much cracking is caused by defects in design and construction procedures [118]. In addition, the choice of materials or faulty mix design may encourage cracking as a result of plastic shrinkage and settlement, of alkali–aggregate reactivity, or even cement unsoundness. Some cracks can be directly traced to the use of materials which induce high-drying shrinkage, as shrinkage when associated with restraint will inevitably result in cracking.

It has to be recognized that, under most conditions other than continuous immersion under water at a uniform and constant temperature, the potential volume decrease in any concrete will be greater than its tensile strain capacity. Some reduction of cracking may be achieved by the selection of materials which provide minimum temperature generation during hydration and have the lowest potential drying shrinkage. As there is no way in which cracking can be totally avoided, other than by triaxial prestressing, the only route open is to provide sufficient reinforcing steel, appropriately arranged in suitable sizes, to control cracks to an acceptably small width.

Most codes of practice give little guidance on the way in which the quantity and distribution of steel should be determined. They generally content themselves with proposing a minimum area of steel expressed as a fraction of the concrete area to be reinforced, to be selected in accordance with the yield strength of the steel. For the effective control of cracking much more is needed. In particular, the area of steel to be used has to be related to the likely temperature drop. The early rise of temperature above ambient and the consequent temperature drop after hardening depends on the heat of hydration of the cement, the cement content of the mix, the curing procedures and the size of the member. Subsequent temperature changes depend on the time of the year at which casting occurs and seasonal climatic changes. The influences of materials and ambient conditions on potential shrinkages are discussed in Chapter 5, from which it can be seen that values of shrinkage may range from close to zero to figures in excess of 1500×10^{-6} strain. Clearly, such large ranges of movement cannot be properly accommodated by only one fixed amount of steel.

A useful approach to the selection of steel areas has been presented by Falkner [119] and a more simplified version is suggested by Campbell-Allen and Hughes [120].

References

1. Midgley, H. G., Conversion of high alumina cement, *Mag. Concr. Res.*, Vol. 27, No. 91, June 1975, pp. 59–77.
2. Teychenne, D. C., Long-term research into the characteristics of high alumina cement concrete, *Mag. Concr. Res.*, Vol. 27, No. 91, June 1975, pp. 78–102.
3. Neville, A. M., *High alumina cement concrete*, Halsted Press, New York, 1975, 201 pp.
4. Smolczyk, H. G., Slag structure and identification of slags, *Princ. Rep. 7th int. congr. chem. cem.*, Paris, 1980, Theme III-I, pp. 3–17.
5. Sersale, R. Structure and characterization of pozzolans and of fly ashes, *Princ. rep. 7th int. congr. chem. cem.*, Paris, 1980, Theme IV-I, pp. 3-18.
6. Rixom, M. R., *Chemical admixtures for concrete*, Spon, London, 1978, 234 pp.
7. Mielenz, R. C. and Sprouse, J. H. High range water-reducing admixtures: effect on the air-void system in air-entrained and non-air-entrained concrete, *Superplasticizers in concrete*, American Concrete Institute, SP-62, 1979, pp. 167–192.
8. ACI Committee 212, Guide for use of admixtures in concrete, *Concr. Int.* Vol. 3, No. 5, May 1981, pp. 53–65.
9. Kaneuji, M., Winslow, D. M. and Dolch, W. L., The relationship between an aggregate's pore size distribution and its freeze thaw durability in concrete, *Cem. & Concr. Res.*, Vol. 10, No. 3, May 1980, pp. 433–441.
10. Roper, H., Volume changes of concrete affected by aggregate type, *J. Res. Dev. Labs Portld Cem. Ass.*, Vol. 2, No. 3, Sept. 1960, pp. 13–19.
11. Snowdon, L. C. and Edwards, A. G., The moisture movement of natural aggregate and its effect on concrete, *Mag. Concr. Res.*, Vol. 14, No. 41, July 1962, pp. 109–116.
12. Roper, H. and Ryan, W. G., A preliminary assessment of the contribution to shrinkage of some cement and aggregates in Australia, *RILEM int. colloq. shrinkage hydraul. concr.*, Madrid, 1968, Section IV-D, 18 pp.
13. Polivka, M. and Mehta, P. K., Use of aggregates producing high shrinkage with shrinkage-compensating cements, *ASTM Spec. Tech. Publ.*, No. 597, June 1975, pp. 36–44.
14. Building Research Station, *Shrinkage of natural aggregates in concrete*, Digest 35 (2nd Ser.), MTP Construction, Lancaster, 1974, pp. 51–55.
15. Phillips, E., *Survey of corrosion of prestressing steel in concrete water-retaining structures*, Australian Water Resources Council, Technical Paper No. 9, 1975, 143 pp.

16. Tuthill, L. H., Resistance to chemical attack, *ASTM Spec. Tech. Publ.*, No. 169-A, April 1966, pp. 275–289.
17. McMillan, F. R., Stanton, T. E., Tyler, I. L. and Hansen, W. C., Long time study of cement performance in concrete, Chapter 5, Concrete exposed to sulfate soils, *Bull. Res. Dept. Portld Cem. Ass.*, No. 30, Dec. 1949, 64 pp.
18. Lerch, W. A., Performance test for potential sulfate resistance of Portland cement, *Bull. ASTM*, No. 212, Feb. 1956, pp. 37–44.
19. Kalousek, G. L., Porter, L. C. and Benton, E. J., Concrete for long-time service in sulfate environment, *Cem. & Concr. Res.*, Vol. 2, No. 1, Jan. 1972, pp. 79–89.
20. Bellport, B. P., Combating sulphate attack on Bureau of Reclamation projects, Swenson, E. G. (Ed.), *Performance of concrete—resistance of concrete to sulphate and other environmental conditions: symposium in honour of Thorbergur Thorvaldson*, University of Toronto Press, 1968, pp. 77–92.
21. Mehta, P. K., Evaluation of sulfate-resisting cements by a new test method, *J. Am. Concr. Inst.*, Proc. Vol. 72, No. 10, Oct. 1975, pp. 573–575.
22. Hansen, W. C., The chemistry of sulphate-resisting Portland cements, Swenson, E. G. (Ed.), *Performance of concrete—resistance of concrete to sulphate and other environmental conditions: symposium in honour of Thorbergur Thorvaldson*, University of Toronto Press, 1968, pp. 18–55.
23. Verbeck, G. J., Field and laboratory studies of the sulphate resistance of concrete, Swenson, E. G. (Ed.), *Performance of concrete—resistance of concrete to sulphate and other environmental conditions: symposium in honour of Thorbergur Thorvaldson*, University of Toronto Press, 1968, pp. 113–124.
24. Kalousek, G. C., Porter, L. C. and Harbse, E. M., Past, present and potential developments of sulfate-resisting concretes, *J. Test. & Eval.*, Vol. 4, No. 5, Sept. 1976, pp. 347–354.
25. Dikeou, J. T., Review of worldwide developments and use of polymers in concrete, *Polymers in concrete*, Concrete Society, London, 1976, pp. 2–8.
26. Baker, C. A., PVC linings for corrosion protection of concrete in sewerage systems, *Aust. Corros. Eng.*, Vol. 19, No. 4, April 1975, pp. 15–24.
27. ACI Committee 302, Proposed ACI standard recommended practice for concrete floor and slab construction, *J. Am. Concr. Inst.*, Proc. Vol. 65, No. 8, Aug. 1968, pp. 577–610.
28. Ryan, W. G., Williams, R. T. and Munn, R. L., Durability of concrete containing various cementitious materials in an experimental concrete road pavement, *Symp. concrete for engineering, engineering for concrete*, Institution of Engineers, Australia, Brisbane, Aug. 1977, pp. 29–34.
29. Johnston, C. D., Steel fibre reinforced concrete pavement—second interim performance report, Neville, A. M. (Ed.), *RILEM symp., fibre reinforced cement and concrete*, 1975, pp. 409–418.
30. ACI Committee 210, Erosion resistance of concrete in hydraulic structures, *J. Am. Concr. Inst.*, Proc. Vol. 52, No. 3, Nov. 1955, pp. 259–271.
31. Watkins, R. D. and Samarin, A., Cavitation erodibility of concrete, *Proc. conf. serviceability concr.*, Institution of Engineers, Australia, Melbourne, Aug. 1975, pp. 54–59.
32. Houghton, D. L., Borge, O. E. and Paxton, J. A., Cavitation resistance of some special concretes, *J. Am. Concr. Inst.*, Proc. Vol. 75, No. 12, Dec. 1978, pp. 664–667.

33. Neville, A. M., *Hardened concrete: physical and mechanical aspects*, American Concrete Institute, Monograph No. 6, 1971, p. 94.

34. Bury, M. R. C. and Domone, P. L. J., The role of research in the design of concrete offshore structures, *6th. annu. offshore technol. conf.*, Houston, Texas, May 1974, Paper No. OTC 1949, pp. 155–168.

35. Browne, R. D. and Domone, P. L. J., The long term performance of concrete in the marine environment, *Symp. offshore struct.*, Institution of Civil Engineers, London, Oct. 1974, pp. 49–59.

36. Gjorv, O. E., Control of steel corrosion in concrete sea structures, *Corrosion of metals in concrete*, American Concrete Institute, SP-49, 1975, pp. 1–9.

37. Seki, H. Deterioration of coastal structures in Japan, *Durability of concrete*, American Concrete Institute, SP-47, 1975, pp. 293–315.

38. Mather, K., Concrete weathering at Treat Island, Maine, *Performance of concrete in marine environment*, American Concrete Institute, SP-65, 1980, pp. 101–11.

39. Biczok, I., *Concrete corrosion, concrete protection*, Chemical Publishing Co, New York, 1967, pp. 178 and 353.

40. Bureau of Reclamation, *Concrete manual*, 8th edn, US Dept Interior, 1975, p. 11.

41. Regourd, M., Physico-chemical studies of cement pastes, mortars and concretes exposed to sea water, *Performance of concrete in marine environment*, American Concrete Institute, SP-65, 1980, pp. 63–82.

42. Regourd, M., Bisserry, P., Evers, G., Hornain, H. and Mortureux, B., Ettringite et thaumasite dans le mortier de la digue du large du Port de Cherbourg, *Annls Inst. Tech. Bâtim.*, No. 358, 1978, pp. 1–14.

43. Gerwick, B., Practical method of ensuring durability of prestressed concrete ocean structures, *Durability of concrete*, American Concrete Institute, SP-47, 1975, pp. 317–324.

44. David, J. N., Heat resistant Portland cement concrete to withstand combined high temperatures and structural loads, *J. Am. Concr. Inst.*, *Proc.* Vol. 69, No. 2, Feb. 1972, pp. 118–124.

45. Hansen, T. C. and Eriksson, L., Temperature change effect on behaviour of cement paste and concrete under load, *J. Am. Concr. Inst.*, *Proc.* Vol. 63, No. 4, Apr. 1966, pp. 489–504.

46. Campbell-Allen, D. and Desai, P. M., The influence of aggregate on the behaviour of concrete at elevated temperatures, *Nucl. Eng. & Des.*, Vol. 6, No. 1, Aug. 1967, pp. 65–77.

47. Abrams, M. S., Compressive strength of concrete at temperatures to 1600°F, *Temperature and concrete*, American Concrete Institute, SP-25, 1971, pp. 33–58.

48. Campbell-Allen, D., Low, E. W. E. and Roper, H., An investigation of the effect of elevated temperatures on concrete for reactor vessels, *Nucl. Struct. Eng.*, Vol. 2, No. 4, Oct. 1965, pp. 382–388.

49. Lankard, D. R., Birkimer, D. L., Fondriest, F. F. and Snyder, M. J., Effects of moisture content on the structural properties of Portland cement concrete exposed to temperature up to 500°F, *Temperature and concrete*, American Concrete Institute, SP-25, 1971, pp. 59–102.

50. Fisher, R. *Über das Verhalten von Zementmörtel und Beton bei höheren*

Temperaturen, Deutscher Ausschuss für Stahlbeton, Heft 214, 1970, pp. 61–28.

51. Bannister, J. L., Steel reinforcement and tendons for structural concrete, *Concrete*, Vol. 2, No. 7, July 1968, pp. 295–306.

52. Kordina, K., Fire rating in buildings, *Symp. prestressed concrete in building*, FIP-CIA, Sydney, Sept. 1976, 22 pp.

53. Cahill, T., The behaviour of prestressing wire at elevated temperatures, *Symp. fire resistance of prestressed concrete*, Braunschweig, June 1965, Bauverlag, Wiesbaden, Berlin, pp. 101–107.

54. Monfore, G. E. and Lentz, A. E., Physical properties of concrete at very low temperatures, *Bull. Res. Dept. Portld Cem. Ass.*, No. 145, May 1962, pp. 33–39.

55. Rostasy, F. S., Schneider, U. and Wiedemann, G., Behaviour of mortar and concrete at extremely low temperatures, *Cem. & Concr. Res.*, Vol. 9, March 1979, pp. 365–376.

56. Goto, Y. and Miura, T., Mechanical properties of concrete at very low temperatures, *Proc. 21st Jpn congr. mat. res.*, Tokyo, Oct. 1977, pp. 157–159.

57. Calleja, J., Durability, *Princ. rep. 7th int. congr. chem. cem.*, Paris, 1980, Theme VII-2, pp. 1–47.

58. Poole, A. B. (Ed.), *Symposium on the effect of alkalies on the properties of concrete*, Cement and Concrete Association, London, 1976, 373 pp.

59. Diamond, S., A review of alkali–silica reaction and expansion mechanisms of reactive aggregates, *Cem. & Concr. Res.*, Vol. 6, No. 4, April 1976, pp. 549–560.

60. Mather, B., Developments in specification and control, *Transpn. Res. Rec.*, No. 525, 1974, pp. 38–42.

61. Oberholster, R. E. and Brand, M. P., Report on reactive concrete aggregate from the Cape Peninsular, South Africa, Poole, A. B. (Ed.), *Symp. effect alkalies prop. concr.*, Cement and Concrete Association, London, 1976, pp. 291–304.

62. St John, D. A. and Smith, L. M., Expansion of concrete containing New Zealand argillite aggregate, Poole, A. B. (Ed.), *Symp. effect alkalies prop. concr.*, Cement and Concrete Association, London, 1976, pp. 319–352.

63. Idorn, G. M., *Durability of concrete structures in Denmark: a study of field behaviour and microscopic features*, Danmarks Tekniske Hojskole, København, 1967, 208 pp.

64. Bogue, R. H., *The chemistry of Portland cement*, Reinhold, New York, 1955, p. 674.

65. Mehta, P. K., History and status of performance tests for evaluation of soundness of cements, *ASTM Spec. Tech. Publ.*, No. 663, Dec. 1977, pp. 35–60.

66. Cook, A. R., Deicing salts and the longevity of reinforced concrete, *Corrosion 80*, NACE, Chicago, March 1980, Paper No. 132, pp. 132/1–10.

67. Pourbaix, M., *Atlas of electrochemical equilibria in aqueous solutions*, Pergamon Press, Brussels, 1966, 644 pp.

68. Gjorv, O. E. and Vennesland, O., Sea salts and alkalinity of concrete, *J. Am. Concr. Inst., Proc.* Vol. 73, No. 9, Sept. 1976, pp. 512–516.

69. Fountain, M. J., Blackie, D. and Mortimer, D., Corrosion protection of prestressing tendons, *Proc. int. conference experience in the design, construction and operation of prestressed concrete pressure vessels and containments for nuclear reactors*, Institution of Mechanical Engineers, University of York, England, 1975, pp. 237–244.

70. Stratfull, R. F., Half-cell potentials and the corrosion of steel in concrete, *Highway Res. Rec.*, No. 433, 1973, pp. 12–22.

71. Danish Corrosion Institute, Huge corrosion cells—a hidden danger, *Offshore Eng.*, Oct. 1976, p. 73.

72. Monfore, G. E., The electrical resistivity of concrete, *J. Res. Dev. Labs Portld Cem. Ass.*, Vol. 10, No. 2, May 1968, pp. 35–48.

73. Hamada, M., Neutralization (carbonation) of concrete and corrosion of reinforcing steel, *Proc. 5th int. symp. chem. cem.*, Tokyo, 1968, Vol. III, pp. 343–384.

74. Meyer, A., Investigations on the carbonation of concrete. *Proc. 5th int. symp. chem cem.*, Tokyo, 1968, Vol. III, pp. 394–401.

75. Verbeck, G., Carbonation of hydrated Portland cement, *ASTM Spec. Tech. Publ.*, No. 205, 1958, pp. 17–36.

76. Pressler, E. E., Brunauer, S., Kantro, D. L. and Weiss, G. H., Determination of the free calcium hydroxide contents of hydrated Portland cements and calcium silicates, *Analyt. Chem.* Vol. 33, No. 7, June 1961, pp. 877–882.

77. Powers, T. C., Copeland, L. E., Hayes, J. C. and Mann, H. M., Permeability of Portland cement pastes, *J. Am. Concr. Inst., Proc.* Vol. 51, 1954, pp. 285–298.

78. Campbell-Allen, D. and Roper, H., Durability of precast facades, *Symp. concr. 1981—Towards better concrete structures*, Institution of Engineers, Australia, Adelaide, June 1981, pp. 67–72.

79. Ost, B. and Monfore, G. E., Penetration of chloride into concrete, *Bull. Res. Dept. Portld Cem. Ass.*, No. 192, Jan. 1966, pp. 46–52.

80. Verbeck, G., Mechanisms of corrosion of steel in concrete, *Corrosion of metals in concrete*, American Concrete Institute, SP-49, 1975, pp. 21–38.

81. Schiessel, P., *Zur Frage der zulässigen Rissbreite und der erforderlichen Betondeckung im Stahlbetonbau unter besonderer Berücksichtigung der Karbonatisierung des Betons*, Deutscher Ausschuss für Stahlbeton, Heft 255, 1976, 175 pp.

82. Leonhardt, F., Rissebeschränkung, *Beton Stahlbetonbau*, Vol. 71, No. 1, Jan. 1976, pp. 14–20.

83. Tremper, B., Corrosion of reinforcing steel, *ASTM Spec. Tech. Publ.*, No. 169-A, April 1966, pp. 220–229.

84. Tremper, B., The corrosion of reinforcing steel in cracked concrete, *J. Am. Concr. Inst., Proc.* Vol. 43, June 1947, pp. 1137–1144.

85. Hausmann, D. A., Electrochemical behaviour of steel in concrete, *J. Am. Concr. Inst., Proc.* Vol. 61, No. 2, Feb. 1964, pp. 171–187.

86. Gewertz, M. W., Causes and repair of deterioration to a California bridge due to corrosion of reinforcing steel in a marine environment, Pt. 1, Methods of repair, *Bull. Highw. Res. Bd*, No. 182, Jan. 1958, pp. 1–17.

87. Rosenberg, A. M., Gaides, J. M., Kossivas, T. G. and Previte, R. W., A

corrosion inhibitor formulated with calcium nitrite for use in reinforced concrete, *ASTM Spec. Techn. Publ.*, No. 629, 1976, pp. 89–99.

88. Griffin, D. F., Corrosion inhibitors for reinforced concrete, *Corrosion of metals in concrete*, American Concrete Institute, SP-49, 1975, pp. 95–102.

89. Clear, K. C., Rebar corrosion in concrete: effect of special treatments, *Corrosion of metals in concrete*, American Concrete Institute, SP-49, 1975, pp. 71–82.

90. *Polymers in concrete*, American Concrete Institute, SP-58, 1978, 420 pp.

91. Clear, K. C., Internally sealed concrete for bridge decks, *Corrosion 76*, NACE, Houston, Texas, March 1976, Paper No. 23, 10 pp.

92. Jenkins, G. H. and Butler, J. M., *Internally sealed concrete*, Federal Highway Administration, Washington, DC, Report No. FHWA-RD-75-20, 1975.

93. Cady, P. D., Corrosion of reinforcing steel in concrete—a general overview of the problem, *ASTM Spec. Tech. Publ.*, No. 629, 1976, pp. 3–11.

94. Frascoia, R. I., Vermont's experience with bridge deck protective systems, *ASTM Spec. Tech. Publ.*, No. 629, 1976, pp. 69–81.

95. Backstrom, T. E., Use of coatings on steel embedded in concrete, *Corrosion of metals in concrete*, American Concrete Institute, SP-49, 1975, pp. 103–114.

96. Moore, D. G., Klodt, D. T. and Hansen, R. J., *Protection of steel in prestressed concrete bridges*, National Co-operative Highway Research Program, Report No. 90, 1970, 86 pp.

97. Clifton, J. R., Beeghly, H. F. and Mathley, R. G., *Non-metallic coatings for concrete reinforcing bars*, National Bureau of Standards, Building Science Series 65, Aug. 1975, 34 pp.

98. Kilareski, W. P., Epoxy coatings for corrosion protection of reinforcement steel, *ASTM Spec. Tech. Publ.*, No. 629, 1976, pp. 82–88.

99. Bird, C. E. and Strauss, P. K., Metallic coating for reinforcing steel, *Mater. Prot.*, Vol. 6, No. 7, July 1967, pp. 48–52.

100. Bresler, B. and Cornet, I., Corrosion of steel and galvanized steel in concrete, *Mater. Prot.*, Vol. 5, No. 4, April 1966, pp. 69–72.

101. Bresler, B. and Cornet, I., Mechanism of steel corrosion in concrete structures, *Mater. Prot.*, Vol. 7, No. 3, March 1968, pp. 45–47.

102. Bresler, B., Cornet, I. and Ishikawa, T., Electrochemical study of the corrosion behaviour of galvanized steel in concrete, *Proc. 4th int. congr. metallic corros.*, Amsterdam, Sept. 1969, pp. 556–559.

103. Rehm, G. and Lammke, A., Korrosionsverhalten verzinkter Stahle in Zementmortel und Beton, *Betonstein Z.*, Vol. 36, 1970, pp. 46–60.

104. Perenchio, W. and Stark, D., *The performance of galvanized reinforcement in concrete bridge decks*, Portland Cement Association, ILZRO Project ZE-206, July 1974–Oct. 1975, 80 pp.

105. Stark, D., *Galvanized reinforcement in concrete containing chlorides*, Portland Cement Association, ILZRO Project ZE-247, Oct. 1976–April 1978, 35 pp.

106. Griffin, D. F., *Effectiveness of zinc coating on reinforcing steel in concrete exposed to a marine environment*, Technical Note N-1032, Naval Civil Engineering Laboratory, Port Hueneme, California, 1969, 23 pp.

107. Hahin, C., Cato, S. and Matthersen, W., *Chloride sensitivity of the corrosion*

rate of zinc coated reinforcing bars, Construction Engineering Research Laboratory, Champaign, Illinois, Interim Report No. m-191, 1976, 44 pp.

108. Baker, E. A., Money, K. L. and Sanborn, C. B., Marine corrosion of bare and metallic coated steel reinforcing rods in concrete, *ASTM Spec. Tech. Publ.*, No. 629, 1976, pp. 30–50.

109. Robinson, R., Cathodic protection of steel in concrete, *Corrosion of metals in concrete*, American Concrete Institute, SP-49, 1975, pp. 83–93.

110. Hausmann, D. A., Steel corrosion in concrete, *Mater. Prot.*, Vol. 6, No. 11, Nov. 1967, pp. 19–23.

111. Hausmann, D. A., Criteria for cathodic protection of steel in concrete, *Mater. Prot.*, Vol. 8, No. 10, Oct. 1969, pp. 23–25.

112. Stratfull, R. F., Progress report on inhibiting the corrosion of steel in a reinforced concrete bridge, *Corrosion*, Vol. 15, No. 6, June 1959, pp. 331t–334t.

113. Stratfull, R. F., *Experimental cathodic protection of a bridge deck*, Paper 53rd Annual Meeting of Highway Research Board, Jan. 1974, 55 pp.

114. Fromm, H. J., Protection of rebar in concrete bridge decks, *Mater. Perform.*, Vol. 16, Nov. 1977, pp. 21–29.

115. Schutt, W. R., Practical experience with bridge deck cathodic protection, *Corrosion 78*, NACE, Houston, Texas, March 1978, Paper No. 7, 4 pp.

116. Gjorv, O. E. and Vennesland, O., Cathodic protection of steel in offshore platforms, *Corrosion 79*, NACE, Atlanta, Georgia, March 1979, Paper No. 139, 10 pp.

117. Heuze, B., Cathodic protection on concrete offshore platforms, *Corrosion 79*, NACE, Atlanta, Georgia, March 1979, Paper No. 138, 26 pp.

118. Campbell-Allen, D., *The reduction of cracking in concrete*, University of Sydney/Cement and Concrete Association of Australia, 1979, 165 pp.

119. Falkner, H., *Zur Frage der Rissbildung durch Eigen-und Zwängspannungen infolge Temperatur in Stahlbetonbauteilen*, Deutscher Ausschuss für Stahlbeton, Heft 208, 1969, 99 pp.

120. Campbell-Allen, D. and Hughes, G. W., Reinforcement to control thermal and shrinkage cracking, *Civ. Eng. Trans. Inst. Eng. Aust.*, Vol. CE23, No. 3, Aug. 1981, pp. 158–165.

5 High-strength concrete

Edward G Nawy   P N Balaguru

Rutgers University, New Brunswick, USA

Contents

Notation

A_1, A_2, A_3	stress–strain relationship constants
A_{sp}	cross-sectional area of spiral wire
b	width of the section
D_c	diameter of the core, out-to-out of spiral
E_c	Young's modulus of concrete
f^c	confinement pressure
f_c	compressive stress in concrete
f'_c	cylinder compressive strength of concrete
f_{ccmax}	maximum compressive strength of confined concrete
f_{ccu}	usable compressive strength of confined concrete
f_{cr}	compressive strength of trial mix
f_r	modulus of rupture
f_s	hoop stress in the spiral wire

f'_t	tensile strength of concrete
f_y	yield strength of steel
J	invariant
l_1	half initial length of microcrack
l_2	length of the branching microcrack
q	external compressive load intensity
s	spacing (pitch) of spiral or standard deviation
z	distance between points of maximum and zero moment; distance between critical section and point of contraflexure
α, β	crack angle measured from the loading direction
ε	strain
ε_c	strain in confined concrete
ε_u	fracture strain of concrete
κ_0, κ_a	confining stress constants

κ, η	CEB constants for stress–strain relationships	ρ_s	volume of spiral reinforcement
K_{Ic}, K_{IIc}	macrocracking strength factors	σ	stress
ν	Poisson's ratio	$\sigma_1, \sigma_2, \sigma_3$	principal stresses
		τ_o	octahedral stress

Summary

Concrete having compressive cylinder strength, f_c', exceeding 6000 psi (41.4 MPa) is designated as high-strength concrete. Several methods are available to obtain such concretes. In general, plain, high-strength concrete contains strong aggregates, a higher Portland cement content, and a low water/cement ratio. The addition of water-reducing admixtures, fly ash and pozzolith are not uncommon in high-strength concretes. For special applications, high strengths can also be attained by using polymers.

To provide the most relevant information to designers in the limited space available, this chapter will deal with factors affecting concrete strength, the micro and macromechanics of high-strength concrete, stress–strain relationships under uniaxial and multiaxial loading, behavior under sustained loading, inelastic behavior and ductility of structural elements, modified concretes, and design considerations in the use of high-strength concrete in buildings.

Several charts, plots, and equations are presented to give an overall view of this topic, together with an extensive list of references. It will not be uncommon in the near future to use concretes of high strength values of the order of 20 000 psi (138 MPa). To achieve such high strengths special techniques and control processes are required. The condensed material presented is intended to give the basic information needed for achievement of such high strengths.

1 Introduction

High-strength concrete is a relative term. Normal-strength concrete in one practice can be considered as high-strength concrete in another. In this chapter, concrete having compressive cylinder strength f_c' exceeding 6000 psi (41.4 MPa) is designated as high strength concrete:

(a) To produce concrete above 6000 psi (41.4 MPa), more stringent quality control on ingredients, mixing and placing are essential.
(b) Most code recommendations at present cover concretes within this range.

Several methods are possible to achieve high strength. In general, plain high-strength concrete contains strong aggregates, a higher Portland cement content and a low water/cement ratio. The addition of water-reducing admixtures, fly ash and pozzolith are not uncommon in high-strength concretes. For special applications, high strengths can also be attained by using polymers. High strength can be advantageously used in many structural components such as columns and shear walls in high-rise buildings, precast and prestressed structural elements and in systems where durability is an important factor.

In order to provide the most relevant information to designers in the limited space available, the following topics are presented in this chapter:

(a) Factors affecting concrete strength.
(b) Micro- and macromechanics of high-strength concrete.
(c) Stress–strain relationships under uniaxial and multiaxial loading.
(d) Behavior under sustained loading.
(e) Inelastic behavior and ductility of structural elements.
(f) Modified concretes.
(g) Design considerations.

1.1 Factors affecting concrete strength

The major factors affecting concrete strength are

(a) Water/cement ratio.
(b) Cement/aggregate ratio.
(c) Properties of the aggregates, namely strength, hardness, grading, shape and surface texture.
(d) Air content.

These factors control mortar strength and the bond between the mortar and the aggregate, and hence the strength of the finished product. Grading of the aggregate is of major significance in determining the strength of the product and studies have shown that gap grading provides better results than continuous grading [1]. The aggregate compressive strength should be comparable to that of the concrete and particular attention has to be given to cleanliness of both the fine and the coarse aggregate.

High-strength concrete normally has a higher cement/aggregate ratio and a lower water/cement ratio. The ability to work with lower water/cement ratio and the expanding need for higher strength in compression have contributed immensely to the development of high-strength concrete. Improvements in the placing, vibration, compacting and finishing techniques have resulted in lower water/cement ratios. Moreover, a variety of admixtures have been developed to increase considerably the workability of the concrete as measured in terms of the slump without the need for excess mixing water. Use of other water reducers, such as superplasticizers, is a relatively recent development in this direction. Additionally, the use of polymers to reduce the percentage of voids in the concrete has resulted in higher strengths. Polymer-impregnated concretes and polymer modified concretes are such examples.

2 Micro and macromechanics of high-strength concrete

2.1 Microscopic and macroscopic behavior

The behavior of concrete under compressive and tensile loads is a function of the level of developing cracks. Volume changes in the cement paste prior to loading generate interfacial microscopic bond cracks between the aggregate–mortar interface. With the addition of compressive short-term loading, no additional significant cracks form until the load reaches approximately 30% of the concrete compressive strength [2], whereby additional cracks initiate throughout the matrix. As loading increases, fracture lines and fracture surfaces develop with a rapid propagation of macrocracks through the medium.

Micromechanics studies of cracking mechanisms in concrete and concrete mortar are intended to predict the fundamental behavior of concretes through analysis of the microscopic cracking behavior of the constituent materials, particularly those of the cement-paste matrix. The variability within the materials and the interaction of the components make it difficult to deduce a workable theoretical model for the prediction of the mechanical properties of the medium such as stress–strain relationships. However, such studies can give valuable insight into the behavior of the cement-paste matrix that can result in high-strength concrete.

Macromechanics studies cover the planar behavior of elements and interpret the development of fracture planes and curvilinear surfaces through such elements and bodies. They also deal with crack propagation and fracture as in the case of micromechanics evaluations.

2.2 Micro and macromechanics theories

The theories most widely used in micro and macromechanics are (a) crack propagation theory and (b) fracture theory. The gist of these theories and their utilization to attain high strength are presented herein.

As to crack formation, cement-paste behavior needs to be initially considered at the microscopic level. This is due to the fact that the characteristic hydraulic radius of the pore system of the paste is of the order of magnitude of 20 Å, whch is negligible compared to the pores present in concrete. The behavior of mortar is dealt with at the mesolevel, where pores are at least ten times larger than the gel pores. The characteristics of pores in the mortar depend mainly on the water/cement ratio. Concrete which has pores and cracks formed by bleeding, capillary shrinkage and plastic shrinkage is dealt with at the macrolevel.

Wittman presented a model for crack propagation using the combination of linear elastic fracture mechanics and Monte-Carlo computer simulation techniques to simulate the randomness of the pores and cracks [3]. The development of the model can be summarized as follows:

(1) Development of an equation to predict the initiation and growth of a crack in a single pore of an infinite matrix. A single crack is found to develop and stabilize along the compressive axis, with the crack length increasing with the increase in load.

(2) The single pore is then replaced by a number of equivalent-size pores located at distances along a straight line. It was shown in this case that, when the combined length of crack reaches a critical value, unstable propagation of cracking occurs, leading to failure.

(3) An arbitrary inclined crack in the homogeneous plate is then considered under uniaxial compression. This arbitrary crack is found to extend at both ends along the compressive axis in response to increasing load. The crack propagation becomes unstable after the crack length exceeds the critical

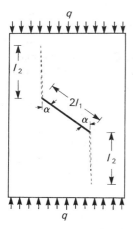

Figure 2-1   Branching cracks development from a random crack

length, causing failure. The relation between the load and the crack lengths, as in Fig. 2-1 and [3], can be expressed as

$$q = \frac{\sqrt{l_2}}{l_1} \frac{K_{Ic}}{2A(\alpha, \rho)} \frac{\sqrt{\pi}}{l_1}$$ ((2-1))

where

q = external compressive load intensity
l_1 = half initial length of inclined crack
l_2 = length of the branching crack
$A(\alpha, \rho) = \sin^2 \alpha \cos \alpha - \rho \sin^3 \alpha$
K_{Ic} = strength factor

(4) Next, coarse aggregate is introduced into the matrix. First, crack propagation around a single aggregate is examined. It has been shown in [3] that a partial linear bond crack which exists along the interface between the aggregate and mortar will grow in an unstable manner until its length exceeds the projected length of the aggregate. Then the crack grows parallel to the load as in step 3 and propagates in a stable manner under increasing load. This crack thereafter combines with the mortar cracks and eventually leads to failure.

(5) Then, the complete model consisting of randomly oriented aggregates in a mortar matrix is considered. In this case, the crack which is propagating from one aggregate piece meets the next aggregate piece. Depending on the fracture mechanics criteria and the angle of inclination of the second interface with respect to the compressive axis, the crack may stop, go around the second aggregate piece, or penetrate through it without deviation (Fig. 2-2). The appropriate equations which govern the above possibilities were developed as follows.

The crack which grows parallel to the load and meets the second aggregate piece grows around the interface or within the mortar medium depending on the

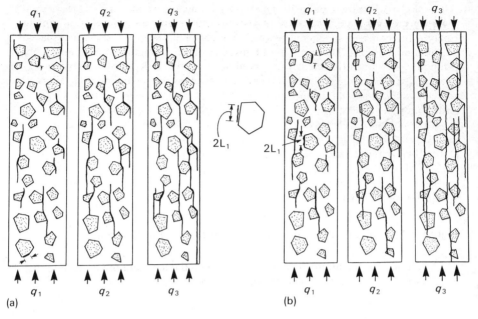

Figure 2-2   Crack propagation comparison
(a) Normal-strength concrete
(b) High-strength concrete

strength factors K_{Ic} for the mortar and K_{IIc} for the interface. The corresponding equations can be written as

$$q_I = \frac{2K_{Ic}\sqrt{\pi l_2/L_1}}{A(\alpha,\rho)[3\cos\frac{1}{2}\beta + \cos\frac{3}{2}\beta] - 3C(\alpha,\rho)[\sin\frac{1}{2}\beta + \sin\frac{3}{2}\beta]} \tag{2-2}$$

and

$$q_{II} = \frac{2K_{IIc}\sqrt{\pi l_2/L_1}}{A(\alpha,\rho)[\sin\frac{1}{2}\beta + \sin\frac{3}{2}\beta] + C(\alpha,\rho)[\cos\frac{1}{2}\beta + 3\cos\frac{3}{2}\beta]} \tag{2-3}$$

where

l_1 = half length of the bond crack
β = angle similar to α in the second aggregate piece
$C(\alpha,\rho) = (\sin\alpha\cos\alpha - \rho\sin^2\alpha)\cos\alpha$

It should be noted that further crack growth depends on both α and β. The crack which meets the second aggregate piece will pass through the aggregate if the aggregate strength is low in comparision with the resistance of the matrix and interface. This can be represented by the equation

$$q_I^A = \frac{K_{Ic}^A}{A(\alpha,\rho)}\frac{1}{2L_1}\sqrt{\pi l_2} \tag{2-4}$$

where K_{Ic}^{A} is the stress intensity factor for the aggregate and q_i^{A} is the normal stress in the aggregate. The above equations can be utilized using a computer to analyze the concrete specimen in which aggregates of arbitrary shape are placed at random positions. The distinction between normal-strength concrete and high-strength concrete can be clearly observed at this stage. If K_{Ic} of the matrix is less than K_{Ic}^{A} of the aggregate, then the cracks go around the aggregate and failure is developed as a result of inclined or shear cracks. If the two factors are of the same order, then the cracks propagate directly through the aggregates without deviating from the compressive stress axis and split tension failure occurs where the cracks run parallel to the loading axis. The computer simulation of crack growth in normal and high-strength concrete, which supports the above hypothesis, is presented in Fig. 2-2. Figure 2-3 shows pictures of two contrasting fracture surfaces of test specimens. It is also possible to use the crack propagation theory for sustained and high loading rates [3], but continued research is necessary to solve realistic problems.

(a)

(b)

Figure 2-3 Fracture surfaces in tensile splitting testing as a measure of tensile strength
        (a) Mortar failure in tensile splitting $f_t' = 450$ psi (3.10 MPa)
        (b) Aggregate failure in tensile splitting $f_t' = 1550$ psi (10.69 MPa)

Some of the other methods used in conjunction with fracture mechanics are (a) stochastic theory, (b) blunt crack band propagation theory, and (c) evaluation using numerical or finite element methods [4–9].

In the stochastic theory [4], concrete is assumed to consist of a number of elements which crack independently. The tensile strengths of the various elements are simulated using a probability density function. The influence of rate of loading temperature and size effect on the mean value of strength can be studied using this model.

The blunt crack theory was proposed by Bazant and Cedolin [5], based on the fact that in concrete many microcracks are present at loads far below the failure load. It was assumed that the crack grows as a smeared crack band rather than a single sharp crack. This approach is particularly useful for finite element modelling because the crack can be smeared over a finite element or elements, assuming that the crack zone can transmit loads only parallel to the crack direction.

Hilleborg et al. [6] presented a method which combines fracture mechanics and finite element analysis. They assumed that a plastic zone is created near a crack tip capable of transmitting stresses. The material behavior is characterized by choosing the appropriate functions for stress variations.

As a general summary of the foregoing discussion one can conclude [7–9] that

(a) Bond cracks develop around the coarse aggregate even before external loading.
(b) Under external loading, bond cracking is more likely to occur around larger coarse aggregates.
(c) At the same level of failure load, high-strength concrete develops significantly less cracking than normal-strength concrete. This is true for both mortar cracks and combined bond–mortar cracks.
(d) The number of potential failure planes decreases with the increase in strength.
(e) Micro and macromechanical behaviour of high-strength concrete approaches the behavior of homogeneous materials.

3 Stress–strain relationship under uniaxial and multiaxial loading

3.1 Uniaxial behavior

The uniaxial response of any structural material provides the information necessary for the design of most structural elements and systems. While extensive work exists for defining the ascending branch of the stress–strain diagram, more limited information exists on defining the descending branch up to fracture, particularly for high-strength concretes. Some of the general observations on the strength, f'_c, and the strain, ε_c, of concrete are [8–14]

(a) Stiffness increases as the compressive strength, f'_c, increases. The rate of increase decreases with higher strengths and length of the initial portion of the f_c–ε ascending branch increases with the increase in strength.
(b) Strain at ultimate strength increases with f'_c.
(c) The fracture strain, ε_u, seems to decrease with the increase in strength. This could also be a function of the stiffness of the testing machine, with the

possibility that much higher strain could be recorded using special testing techniques.

(d) The energy-absorption capacity per unit volume does not increase in proportion to strength.

3.1.1 *Stress–Strain relationship in uniaxial compression*

The stress–strain relationship in compression consists of a curved ascending branch and a descending branch of a usually flatter slope, as seen in Fig. 3-1. The modulus of concrete, E_c, depends mainly on the concrete strength and the rate of loading, and the slopes of both the ascending and the descending branches are steeper for higher strength concretes. A mathematical model for the stress–strain diagram of concrete in compression must at least satisfy the following criteria [10]:

(a) $f_c = 0$ at $\varepsilon = 0$ where the curve passes through the origin.
(b) $d\sigma/d\varepsilon = E_c$ at $\varepsilon = 0$ where the slope of the stress–strain curve at the origin is equal to the modulus of elasticity.
(c) $d\sigma/d\varepsilon = 0$ at $\varepsilon = \varepsilon_0$ where the stress–strain curve has a maximum ordinate.
(d) The descending branch of an analytical curve must be in agreement with experimental results, i.e. it must pass through at least one experimentally defined point which is beyond the peak.

A general expression developed in [11] by Balaguru can give the stress–strain relationship up to the experimentally observed fracture strain, such as 0.004 in Fig. 3.1, as applied to the experimental results of [12] for concretes of cylinder compressive strengths $f_c' = 7500$ psi (51.7 MPa) to $f_c' = 14\,000$ psi (96.5 MPa).

$$f_c = A_1 \sin\left(\frac{\pi\varepsilon}{0.0070}\right) + A_2 \sin\left(\frac{\pi\varepsilon}{0.0035}\right) + A_3 \sin\left(\frac{\pi\varepsilon}{0.00233}\right) \tag{3-1}$$

where

$$A_1 = 7.15 \sin\left[\pi(f_c' - 3.1)/13.1\right]$$
$$A_2 = 12 - 10.3 \sin\left[\pi(14.5 - f_c')/17\right]$$
$$A_3 = 5.3 \sin\left[\pi(f_c' - 1.85)/16\right] - 5$$

An idealized stress–strain diagram primarily applicable to normal concrete is shown in Fig. 3-2. It is recommended by CEB in [15] for short-term loading up to fracture strain ε_{cu} and gives

$$\frac{f_c}{f_c'} = \frac{\kappa\eta - \eta^2}{1 + (\kappa - 2)\eta} \tag{3-2}$$

where

$$\eta = \varepsilon_c/\varepsilon_{c_1}$$
$$\varepsilon_{c_1} = 0.0022 \text{ (maximum strain is axial compression)}$$
$$\kappa = (1.1E_c)\varepsilon_{c_1}/f_c'$$

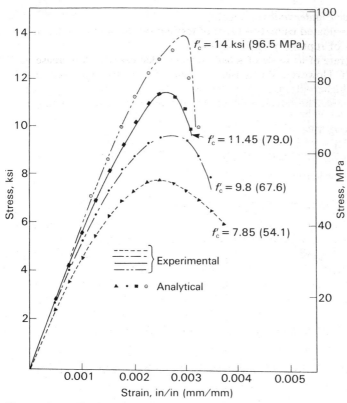

Figure 3-1 Typical stress–strain diagrams of high-strength concrete: f'_c is the compressive strength. Note 1 ksi = 1000 psi

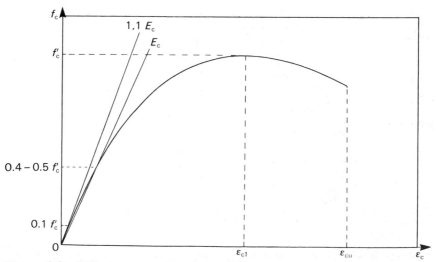

Figure 3-2 CEB stress–strain relationship

3.1.2 Tensile behavior under uniaxial loading

Tensile behavior is evaluated using two types of testing: (a) the split cylinder test, and (b) the modulus of rupture or bending test.

In both cases, the rate of increase of tensile strength decreases with increase in compressive strength. However, the tensile strength, f'_t, can be expressed as a function of $\sqrt{f'_c}$ for higher-strength concretes as in the case of normal strength concrete, as in Fig. 3-3a. The modulus of rupture as a measure of tensile strength

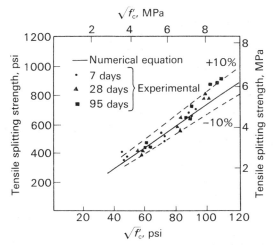

Figure 3-3a Tensile splitting strength versus the square root of compressive strength

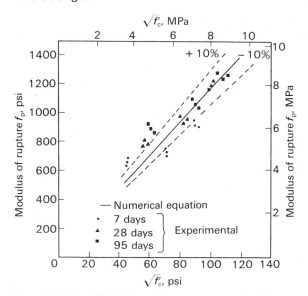

Figure 3-3b Modulus of rupture versus the square root of compressive strength

for an f'_c range of 3000–12 000 psi (21–83 MPa) can be expressed as (Fig. 3-3b)

$$f_r = 11.7\sqrt{f'_c}\,\text{psi} \quad \text{or} \quad f_r = 0.94\sqrt{f'_c}\,\text{MPa} \tag{3-3}$$

The tensile splitting strength, f'_t, for the same compressive strength range can be expressed as

$$f'_t = 6.8\sqrt{f'_c}\,\text{psi} \quad \text{or} \quad f'_t = 0.54\sqrt{f'_c}\,\text{MPa} \tag{3-4}$$

3.1.3   Modulus of elasticity and Poisson's ratio

The modulus of elasticity, E_c, of concrete increases with compressive strength but at a lower rate than the strength. The following equation seems to satisfactorily express the relation between E_c and f'_c:

$$E_c = (40\,000\sqrt{f'_c} + 10^6)\,\text{psi} \quad \text{or} \quad E_c = (3322\sqrt{f'_c} + 6900)\,\text{MPa} \tag{3-5}$$

for 3000 psi (21 MPa) $\leq f'_c \leq$ 12 000 psi (83 MPa).

Poisson's ratio, ν, was found to be close to 0.2, as in the case of normal-strength concrete. The variation of ν with increasing stress was also found to be similar to that of normal-strength concrete [8, 16].

3.2   Multiaxial behavior

Knowledge of multiaxial behavior is necessary for the design of special structural components such as beam–column junctions, anchorage zones in prestressed concrete elements, reactor vessels and shell-type offshore structures. The triaxial behavior of concrete in the elastic range is simpler than its behavior in the inelastic range. In the elastic range, the material behavior is defined by Young's modulus and Poisson's ratio. In the inelastic range, microcracking plays an important role, particularly for high-strength concrete where maintenance of ductility requires higher confining stresses. The effect of multiaxial pressure on ductility and plasticity of the material is of major importance, particularly where safer failure behavior of columns in high-rise buildings is a major factor.

The total-strain theory, sometimes referred to as the deformation theory, is applicable to the interpretation of triaxial behavior. It is basically more applicable to proportional or near proportional loading, as it does not provide for the relationship between shear strains and normal stresses necessary in considering the inelastic behavior of concrete. But the material can be idealized as isotropic for a relationship to be obtained between (a) normal stress and strain, (b) octahedral shear stress and strain.

Octahedral stresses are the stresses that develop on the sides of an octahedral element formed by planes whose normals form equal angles with the principal stress axis. The octahedral stress, τ_0, can be related to the normal principal stresses σ_1, σ_2, and σ_3 in the form:

$$\tau_0 = \tfrac{1}{3}[(\sigma_1 - \sigma_2)^2 + (\sigma_2 - \sigma_3)^2 + (\sigma_3 - \sigma_1)^2]^{\frac{1}{2}} = \tfrac{1}{3}[2J]^{\frac{1}{2}} \tag{3-6}$$

where J is an invariant.

Equation 3-6 postulates that failure takes place when the octahedral shear stress is equal to $(\sqrt{2})/3$ times the yield stress in pure shear or $(\sqrt{2})/3$ times the yield stress in uniaxial tension. But the expression requires more rigorous effort for experimental application to concrete. Other work on concrete under high triaxial stress applies the Mohr theory in such a manner that the concrete cylinders are treated as though they are subjected to biaxial combined stress. Mohr's assumption that the intermediate principal stress has no influence on failure makes the expressions easier to apply [17] such as

$$\sigma_1^{1.33} = 135(\sigma_3 + 480) \tag{3-7}$$

where σ_1 is the principal axial stress and σ_3 is the principal confining lateral compressive stress of failure in psi. Or, from [18–20],

$$\sigma_1 = f'_c + 4.5\sigma_3 \tag{3-8a}$$

where σ_3 is in compression, and

$$\sigma_1 = f'_c + 30\sigma_3 \tag{3-8b}$$

where σ_3 is in tension. If a material safety factor of 1.5 is considered, Eqn 3-8a becomes

$$\sigma_1 = 0.67f'_c + 3.0\sigma_3 \tag{3-8c}$$

3.2.1 Stress–strain relationships under multiaxial loading

The stress–strain relationships under (a) biaxial and (b) triaxial loading can be summarized as follows:

(a) *Biaxial loading.* No appreciable differences seems to exist between the relative performance of normal-strength and high-strength concretes. The

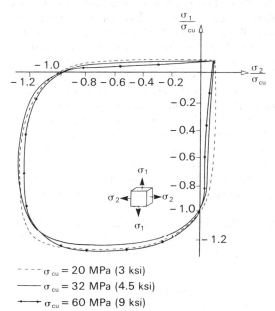

--- $\sigma_{cu} = 20$ MPa (3 ksi)
— $\sigma_{cu} = 32$ MPa (4.5 ksi)
—•→ $\sigma_{cu} = 60$ MPa (9 ksi)

Figure 3-4 Comparison of the biaxial behavior of normal-strength and high-strength concrete: σ_{cu} is the compressive strength

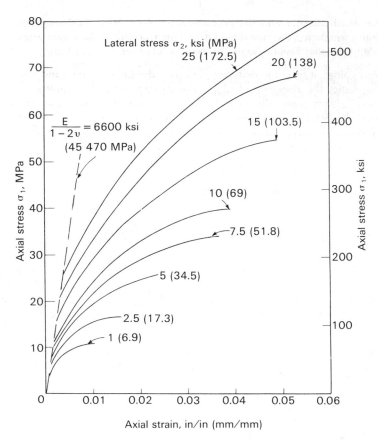

Figure 3-5   Stress–strain diagrams for concrete under triaxial loading. Note 1 ksi = 1000 psi

non-dimensional biaxial failure surface for concrete cylinder strength ranging from 3000 to 9000 psi (20–60 MPa) shown in Fig. 3-4 illustrates this as in [18, 20]. It shows that in the case of compression–compression, normal and high-strength concretes appear to behave the same way. For tension–tension, a decrease of the ratio between tension and compression strength seems to occur for higher compressive strengths, which is a behavior consistent with the behavior in uniaxial tension.

(b) *Triaxial loading.* Figure 3-5, based on experimental work in [17], shows plots of various combinations of axial and lateral triaxial compressive stresses up to an axial principal level σ_1 at a failure stress of 86 470 psi (597 MPa) for concretes of cylinder strength $f'_c = 4000$ psi (27.6 MPa). The relationship between the axial stress, σ_1, and the confining stress, σ_3, at failure can be seen in Fig. 3-6a based on Eqn 3-7 and Fig. 3-6b based on

Eqn 3-8a, which also presents the experimental values. Figure 3-7 gives the octahedral normal stress–strain relationships for triaxially loaded concretes [16, 20, 21] having f'_c ranging from 2200 to 9000 psi (15.3–62.1 MPa).

From the foregoing, it can be deduced that the design engineer and the researcher can predict the response of high-strength concrete under multiaxial loading conditions in structures which can be subjected to such loads. Once the principal stresses are determined at the critical sections, the failure strength at those sections based on a particular compressive strength, f'_c, can be determined.

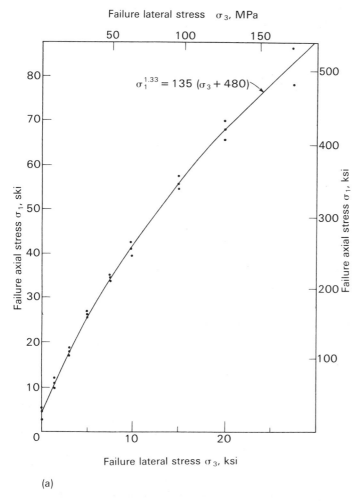

(a)

Figure 3-6a **Axial to lateral stress relationship at failure: 6×12 in (150×300 mm) cylinders, water/cement ratio 0.58, fog-cured, 90-day strength. Note 1 ksi = 1000 psi**

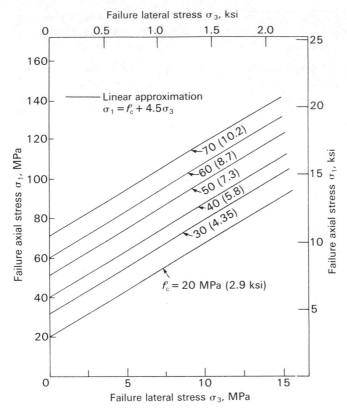

Figure 3-6b   Axial to lateral stress relationships at failure for various concrete strengths. Note 1 ksi = 1000 psi

4   Rate and duration of loading effects

4.1   Creep and time dependence

Creep can be represented as the sum of basic creep and drying creep (which is similar to shrinkage). Mix composition, specimen age and strength of the constituents and the resulting concrete have a profound influence on the creep behavior of the material.

One of the important earlier works on creep and creep recovery by Roll [24, 25] presented generalized models for creep prediction of highly stressed cylinders, including the effects of sustained uniform stress. Neville [26] reported extensive work on the creep of concrete as a function of the age of loading, and Branson [27] developed creep and shrinkage expressions taking into account the effects of most major parameters. Meyers *et al.* [28] presented empirical equations for predicting long-term creep and shrinkage coefficients and the error coefficient involved, and Bazant [29, 30] discussed refinements in the evaluations of the effective modulus and long-term deformations.

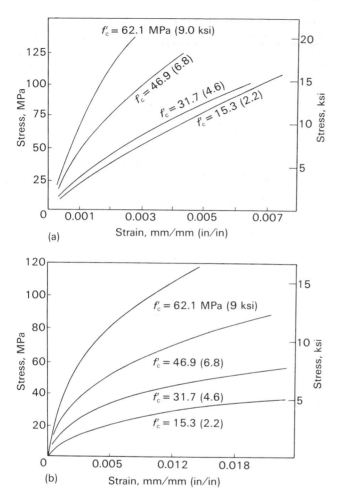

Figure 3-7 Octahedral stress–strain relationships: (a) normal and (b) shear.
Note 1 ksi = 1000 psi

Chapter 11 of this Handbook dwells in detail on this subject. It is important to recognize that existing data have been generally accumulated on concretes of strengths not higher than 6000 to 8000 psi (41.37–54.92 MPa). However, for higher strength concretes, the principles and trends hold except for possible differences in the value of the coefficient of variation.

As in the case of compressive strength, lower water/cement ratio results in better creep performance, namely, both basic creep and total creep decrease with the decrease in water/cement ratio. This seems logical if we consider that part of the creep mechanism involves the solid particles migrating from loaded weakly-bonded areas in the cement gel to load-free areas. Concrete with lower water/cement ratio has less pores. Hence, the microstructure is more extensively bonded and therefore particle migration decreases.

Since most high-strength concretes have a low water/cement ratio, it can be expected that the creep strains will be less for high-strength concrete than normal-strength concrete. The other factors which have subtle effects on concrete strength also have similar effects on creep strain. These factors can be summarized as follows:

(a) age at loading;
(b) relative humidity;
(c) size of the member;
(d) cement content.

The time dependence for high-strength concrete can be expected to be basically the same as for normal concretes. Similarly, the acceleration of creep due to cyclic loading can be expected to have the same effect in both concretes.

4.2 Shrinkage strains and the effect of temperature

Moisture transfer and shrinkage mainly depend on porosity, which in turn depends on the water/cement ratio. Hence, the lower water/cement ratio used in high-strength concrete results in advantageous shrinkage behavior. However, the aggregate/cement ratio, which decreases for high-strength concrete, will increase the shrinkage of high-strength concrete.

The effects of temperature on time-dependent strains are twofold:
(a) The reduction of viscosity at high temperature increases the flow of particles.
(b) The acceleration of hydration which increases the strength reduces the time-dependent strains.

The first factor cannot have a different effect on high-strength concrete than on normal-strength concrete because the molecular structure is not changed. The second factor should decrease the creep strains in high-strength concrete.

4.3 Experimental investigations

Experimental investigations on the time-dependent behavior of high-strength concrete are so far limited [31–35]. However, the following general observations can be made:

(a) As compared with normal-strength concrete, the compressive strength, f'_c, of high-strength concrete under sustained loading is closer to the f'_c value under short-term loading.
(b) Sustained high levels of stress produce detrimental effects on the material, as in the case of normal-strength concrete. The effect is not as severe as that on normal-strength concrete.
(c) Specific creep of high-strength concrete is less than that of normal-strength concrete. Creep strains for the same stress/strength ratio appears to be the same for high-strength and normal-strength concretes.
(d) High-strength concrete has a smaller drying creep to basic creep ratio.
(e) In the case of high-strength concrete, specific creep (creep strain per unit stress) remains constant for much higher stress levels than in normal-strength concrete.

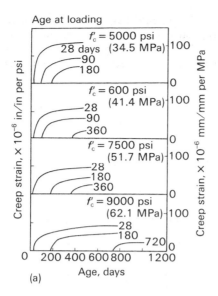

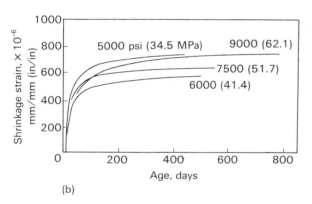

Figure 4-1 (a) Creep and (b) shrinkage data for high-strength concrete

(f) Drying shrinkage seems to be independent of the strength of concrete. Figure 4-1 relates age to creep and shrinkage determined from field observations [33] of columns 48 × 48 in (1220 × 1220 mm) in cross-section.

4.4 Behavior under other types of loading

The triaxial properties of non-linear creep and strain-rate effects are not sufficiently known even for normal concrete. Knowledge of this behavior of high-strength concrete under impact, dynamic or fatigue loading is still limited. It can be expected, however, that the fatigue behavior of high-strength concrete would not be significantly different from that of normal-strength concrete. The stress–strain diagram for high-strength concrete under dynamic or low-cycle fatigue loading is expected to be basically similar in trend to that of normal-strength concrete, though high-strength concrete could be more vulnerable under impact loading because of its brittleness.

5 Confined concretes

The confinement of concrete can considerably improve its strength and deformation capacity. The increase in deformation capacity is useful for normally loaded structures but essential in the case of structures subjected to seismic loads. The basic hypotheses on the effect of confinement [10] can be summarized as follows:

(a) Laterally confined reinforced concrete sections comprise a core enclosed in ties or hoops enveloped by thin concrete shell cover. Both the core and the shell undergo the same deformation magnitude, but follow different stress paths because of the presence of lateral stress on the core.

(b) The stress–strain behavior of the core and the cover follows the discussion presented in Section 3.

(c) As the normal principal compressive stress increases, the concrete expands, straining the lateral reinforcement, which in turn applies confining pressure. In most cases, the confining pressure is zero in the unstressed specimen. Such pressure can be produced on the externally unstressed specimen by prestressing the hoop reinforcement.

(d) Because of the complexity of interaction of the concrete and the reinforcement, it is complicated to develop and use a reliable constitutive equation. Simpler empirical expressions are used to evaluate the increase in strength and deformation capacity as a function of confinement.

(e) The most effective lateral reinforcement is the one which can provide a confining pressure analogous to hydrostatic pressure, since such pressure can considerably increase the compressive resistance of concrete (Fig. 3-5).

Several investigations exist on the influence of confinement on normal-strength concrete. This influence on high-strength concrete is basically similar. Figure 5–1 presents the stress–strain relationship for unconfined and confined concretes having a compressive cylinder strength $f'_c = 6000$ psi (41.3 MPa) [36]. The concrete specimens used were 6×18 in (150×450 mm) cylinders confined with wire spirals spaced at 0.5 in (13 mm) to 0.75 in (20 mm). The yield strength of the confining wires ranged from 40 to 100 ksi (276–689 MPa). Similarly, Fig. 5-2 from [37] shows the effect of confinement on concrete of 9500 psi (65.5 MPa) compressive strength, where the hoop stress value in the confining hoops exceeds 60 000 psi (413.5 MPa) in spirals at a pitch of 1/2 in (12.7 mm) and 1 in (25.4 mm). It can be observed that confinement is very effective in developing large deformability, hence ductility. The ultimate fracture strains, ε_u, are about 8–10 times the strain at peak loading.

The effectiveness of the confinement can be characterized by two material constants κ_0 and κ_u which determine the increase in compressive strength [38]. The maximum compressive strength of confined concrete, f_{ccmax}, can be related to the unconfined compression strength, f'_c:

$$f_{ccmax} = f'_c + \kappa_0 f^c \tag{5-1}$$

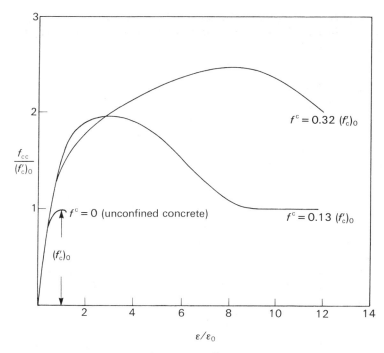

Figure 5-1   Lateral confinement effect

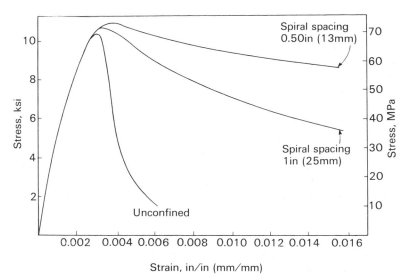

Figure 5-2   Stress–strain relationships for high-strength confined concretes. Note 1 ksi = 1000 psi

where f^c is the confinement pressure developed by the spiral and depends on the geometric and material characteristics of the spiral. The value of f^c can be determined using the expression.

$$f^c = 2A_{sp}f_s/SD_c \tag{5-2}$$

where

$\quad A_{sp}$ = cross-sectional area of spiral wire
$\quad D_c$ = diameter of core, out-to-out of spiral
$\quad s$ = spacing (pitch) of spiral
$\quad f_s$ = hoop stress in the spiral wire

For practical purposes, the magnitude of f_{ccmax} alone is not sufficient if the shell cover to the confining reinforcement spalls, thereby transferring additional load to the core in the case of columns while sustaining large deformations. A reduced compressive strength, f_{ccu}, will have to be adopted, using a material constant κ_u:

$$f_{ccu} = f'_c + \kappa_u f^c \tag{5-3}$$

An average value of $\kappa_u = 4$ can be used, as in [38].

Extrapolating the results available for normal concrete [38], the effectiveness of confinement decreases with the increase in concrete strength. Therefore, for high-strength concrete, a hoop reinforcement with a higher yield strength should be used in order to achieve satisfactory results.

For compression members, the ACI code [23] allows a maximum usable strain value of 0.004 at the extreme concrete compression fibers for f'_c not greater than 8000 psi (55.12 MPa), the ratio of end moments $M_1/M_2 \leq 0$, and the hoop reinforcement $\rho_s \geq 460/f_y$, where ρ_s is the ratio of the volume of the confining hoop including compression steel to the total volume of the confined concrete core, and f_y is the yield strength (in psi) of the confining steel. This value of strain is reasonable although considerably higher strain can be attained through appropriate confinement [39].

The maximum concrete strain, ε_c, of the confined concrete can be evaluated from the following simplified expression [40]:

$$\varepsilon_c = 0.003 + 0.02b/z + 0.2\rho_s \tag{5-4}$$

or from the following expression [41] which more accurately defines the contribution of ρ_s:

$$\varepsilon_c = 0.003 + 0.02b/z + (\rho_s f_y/14.5)^2 \tag{5-5}$$

where

$\quad z$ = distance between points of maximum and zero moment or the distance between the critical section and the point of contraflexure in a member
$\quad b$ = width of the section
$\quad f_y$ = yield strength (in ksi) of the confining steel

and the term $(\rho_s f_y/14.5)^2 \geq 0.001$.

6   Modified high-strength concretes

These concretes are generally produced using polymerizing material. They can be broadly classified as (a) polymer-modified concrete (PMC) and (b) polymer-impregnated concrete (PIC). Both PMC and PIC provide considerable increase in compressive strength and other improved mechanical properties. A combination of polymerization and autoclaving can be used to produce very high strengths. In polymer-modified concretes, polymers in the form of emulsions of natural or synthetic latexes or resins are added during the process of mixing the concrete components and thus are more adaptable for field applications. Polymer impregnation for PIC using various monomers is achieved using sophisticated and expensive equipment to saturate precast high-strength concrete elements in chambers and subject them to elevated temperature or radiation to achieve the finished high-strength product.

6.1   Polymer-modified concrete

Latex systems based on copolymers of styrene–baladiene, vinylidene chloride, and acrylics, natural and nitrite rubber have been successfully used to obtain PMC [42]. Recent research at Rutgers University by Nawy *et al.* provides comprehensive results on PMC obtained using a liquid epoxy resin (EPON 828) and hardener (CA 640). They obtained a compressive strength of 13 531 psi (93.2 MPa) and a tensile strength of 1538 psi (10.6 MPa) [43–48]. The following are pertinent information regarding the mix design, the behavior of PMC under compression, shear and bending forces, the performance of PMC and normal concrete–PMC composite beams, and the freeze–thaw resistance characteristics of the concrete.

6.1.1   *Mix design*
The PMC can be designed using the nomogram presented in Fig. 6-1 [46]. The nomogram relates the compressive strength and slump to the liquid/cement ratio, water/cement ratio, and polymer/cement ratio. Therefore, the most important variable, namely, the liquid/cement ratio for the specified compressive strength, and the slump can be readily obtained. It should be noted that it is possible to obtain the same compressive strength for two different slump variables. For example, from Fig. 6-1, 10 000 psi (68.9 MPa) can be obtained using the liquid/cement ratio of either 0.65 or 0.55. The first choice which has a water/cement ratio of 0.2 and a polymer/cement ratio of 0.45 would produce a slump of 3 in (76 mm). The second choice with a water/cement ratio of 0.2 and a polymer/cement ratio of 0.35 would produce a slump of 1.5 in (38 mm). For reasons of economy, it is better to use as low a polymer/cement ratio as possible.

The optimum mix proportions of other ingredients were found to be 1 : 1.62 : 2.44 in the order of cement : sand : coarse aggregate [45]. Normal Portland cement, graded natural sand and $\frac{3}{4}$ in size (20 mm) crushed stone aggregate are recommended. The aggregates should be clean and of good quality.

The following procedure is recommended to obtain the best results. First, mix properly washed and dried coarse aggregate with the blended resin and hardener. Then, add the sand to the mixture. Finally, add the cement paste to obtain the PMC.

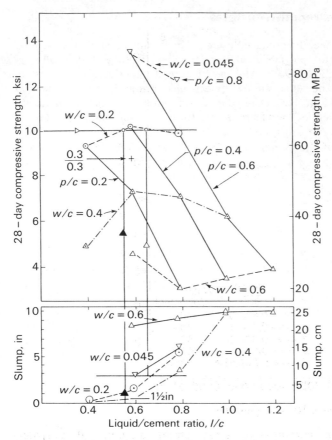

Figure 6-1 Nomogram for mix design of polymer-modified concrete

The following additional observations can be effectively used by designers [43]:

(a) The optimum polymer/cement ratio lies betwen 0.3 and 0.45 for water/cement ratios less than 0.6.
(b) Liquid polymers of the type used in the investigation [43, 50], such as epoxy resin EPON 828 and hardener CA 460, give better results if their viscosities are low.
(c) The optimum resin/hardener ratio was found to be 100:68.
(d) Curing under moderate heat has a negative effect.

6.1.2 Mechanical properties and applications of PMC

Typical stress–strain curves of specimens subjected to uniaxial compression are presented in Fig. 6-2 and in [44]. It can be observed from the curves that PMC has better toughness properties than regular high-strength concrete (Fig. 3-1). It can also be noted that the increase in Young's modulus is small compared to the

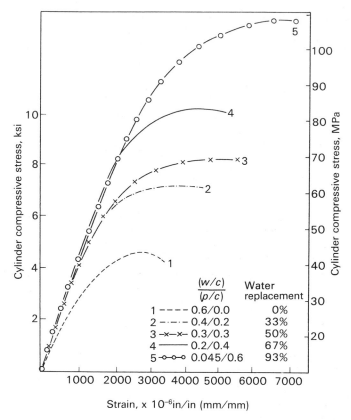

Figure 6-2 General compressive stress–strain diagram of polymer-modified concrete

increase in compressive strength. The following other conclusions can be drawn using [43–50] regarding the mechanical properties and application of PMC.

(a) Tensile strength can be increased more than $3-3\frac{1}{2}$ times by adding polymer to the concrete at the mixing stage.

(b) The bond shear transfer capacity of PMC is about twice that of the normal concrete.

(c) Two-layer composite systems with PMC as top layer can be effectively used for structures exposed to adverse loading such as stress, thermal and chemical conditions.

(d) The strength and stiffness behavior of PMC–normal concrete two-layer composite beams are superior to those of total normal concrete beams. Such composite beams can be designed using the simple principles of mechanics.

(e) Permeability of PMC is negligible as compared to normal concrete and its resistance to freeze-thaw cycles and thus durability is much superior to that of normal concrete.

6.2 Polymer-impregnated concrete

PIC is a high-strength concrete. It is produced by the impregnation and polymerization of suitable monomers through concrete elements, usually through the use of liquid vinyl monomer systems. Such systems include styrene, methyl methacrylate, butyl acrylate or other easily polymerizable materials. The monomer viscosity can be altered by blending higher or lower-viscosity monomers, which normally contain an inhibitor for preventing premature polymerization of the monomer. Precast concrete elements of dense concrete, such as plates or shallow beams, are used. They are dried to constant weight, placed in vacuum chambers and saturated with a monomer prior to polymerization [49]. The optimum drying temperature is approximately 300°F (150°C) for a period of 4 to 12 h.

There are basically three methods of polymerization of monomers: (a) thermal catalytic, (b) catalyst, and (c) radiation. The thermal-catalytic process is the most practical since it is the fastest. It involves adding to the monomer small quantities of compounds which generate free radicals on heating and are consumed during the reaction. Such radicals in the third method are achieved by the use of ionizing radiation, such as the gamma-rays emitted by cobalt 60.

The strength in compression, tensile splitting and flexure of the PIC can be as high as four times the strengths of the control specimens, attaining cylinder compressive strengths in excess of 20 000 psi (137.9 MPa) as in [50, 51]. The improvement in strength is mainly a function of the degree of impregnation of the cement matrix, the higher the water/cement ratio the higher the polymer content necessary.

In general, the use of PIC is restricted by the size of the chambers which determine the size of the precast elements that can be produced. Field production of PIC is relatively difficult and expensive as compared to PMC, which can be produced using the same techniques in field mixing as normal concrete. PIC also exhibits gradual reduction of strength with increased temperature. However, higher strengths can be achieved through impregation than through field mixing.

7 Design considerations

7.1 Mix design

The production of high-strength concrete requires special attention to a number of factors including water/cement ratio, gradation and mineralogy of the aggregates, composition and fineness of cement, and curing.

The most fundamental factor in the production of high-strength concrete is the use of high-quality cement paste. The compressive strength of cement paste increases with fineness of the cement [52]. Also, the rate of increase of compressive strength increases if the water/cement ratio is reduced to less than 0.4. Hence, it is theoretically possible to achieve a concrete strength equal to the cement clinker strength, which could be considerably higher than 45 000 psi (310 MPa).

For low water/cement ratios, smaller size coarse aggregates give better results. For water/cement ratios less than 0.4, the maximum size of the coarse aggregate should be limited to 3/8 in (10 mm) [52, 53]. Three characteristics of the coarse aggregate, namely, compressive strength, bonding potential with cement paste, and low water absorption capacity are important in the production of high-strength concrete. Gravel and trap rock with compressive strengths of about 40 000 psi (275 MPa), excellent bonding properties and low water absorption equalities, have been successfully used to produce high-strength concrete.

If a low water/cement ratio is used, the importance of curing increases exponentially. In such a concrete, one must not only avoid the escape of moisture from the specimen but also provide extra water for hydration [54]. The members should be cured using water ponding or moist air surroundings. Saturated aggregates can be used to provide water within the specimen for continuous hydration.

The following guideline can be used for the trial mix design [23]. If f'_c is the specified compressive strength, the mix should be designed for a compressive strength f_{cr}, where

$$f_{cr} = f'_c + \Delta f'_c \tag{7-1}$$

$$\Delta f'_c = 1.3435s \geq (2.326s - 500)\,\text{psi} \tag{7-2a}$$

or

$$\Delta f'_c = (2.326s - 3.5)\,\text{MPa} \tag{7-2b}$$

where s is the standard deviation calculated using at least 30 consecutive tests. If sufficient test results are not available, for high-strength concretes of compressive strength $f'_c > 5000$ psi (35 MPa), f_{cr} should be taken as $(f'_c + 1400)$ psi or $(f'_c + 10)$ MPa. This relation between the specified strength, f'_c, and the strength, f_{cr}, for which the mix has to be designed provides a probability of less than 1-in-100 for a cylinder compression strength, f'_c, below the design-specified f'_c.

The mix for high strength can be designed as in normal-strength concrete [55] except that a low water/cement ratio should be used. Neville, in [56], discusses in detail and with charts the various parameters affecting the achievement of high strength in concrete. Tables 7-1 [33] and 7-2 [52] give mix proportions used to obtain high-strength concretes.

Table 7-1   Mix proportions used in Chicago Water Tower Place [33]

| Concrete design strength psi | Quantities per yd³ | | | | | |
| | Cement lb | Aggregate | | Water lb | Fly ash lb | Pozzolith fl oz |
		Fine lb	Coarse lb			
9000	846	1025	1800	330	100	25.4
7500	729	1250	1695	315	100	21.9
6000	705	1190	1825	303	100	21.1
5000	541	1385	1850	268	100	16.2
4000	447	1485	1860	257	100	13.4

Table 7-2 Mix proportions for high-strength concrete [52]

Ref. no.	Cement	Aggregate	W/C ratio by wt	Admixture	Remarks	Sand %	Compressive strength, psi					
							1 day	3 days	7 days	28 days	90 days	1 year
1	A	Elgin sand and gravel	0.25	None	Unworkable	35	6 620	7 260	7 800	7 000	7 430	8 440
2	A	Elgin sand and gravel	0.30	None		35	6 120	7 660	8 190	8 940	10 430	11 270
3	A	Elgin sand and gravel	0.35	None		35	4 620	6 500	7 030	8 230	9 530	10 440
4	A	Elgin sand and gravel	0.35	2% CaCl$_2$		35	7 300	8 360	8 680	9 240	10 740	11 080
5	A	Elgin sand and gravel	0.30	2% CaCl$_2$	Unworkable	35	8 210	7 750	8 430	7 600	9 060	11 110
6	A	Elgin sand and gravel	0.35	None	Dry aggregate	35	3 560	5 590	5 630	6 800	7 870	8 460
7	A	Elgin sand and gravel	0.30	None	Dry aggregate	35	4 640	6 510	6 650	7 240	8 540	9 350
8	A	Elgin sand and gravel	0.25	None	Dry aggregate, unworkable	35	4 530	3 500	4 150	8 540	5 000	3 170
9	A	Elgin sand, dresser traprock	0.32	None		35	6 910	8 010	8 600	9 260	9 810	10 450
10	A	Elgin sand and gravel	0.30	None	Ponded 1st 24 h.	31	6 020	8 070	9 020	9 500	10 650	11 470
11	A	Elgin sand and gravel	0.28	None	No 30 mesh sand	31	6 110	8 140	8 870	9 450	10 550	11 350
12	A	Elgin sand and gravel	0.27	None	No 16 mesh sand	31	6 020	7 440	8 300	9 360	10 210	10 840
13	A	Elgin sand, Eau Claire gravel	0.29	None		31	5 530	7 330	7 840	8 690	9 050	9 850
14	A	Elgin sand and gravel	0.30	None	$\frac{3}{4}$ in max. size	47	5 670	9 500	10 880	11 170	12 120	13 320
15	A	Elgin sand and gravel	0.30	None	$\frac{3}{8}$ in max. size	63	5 710	9 820	10 510	11 870	12 730	14 000
16	A	Elgin sand and gravel	0.35	None		63	4 330	8 400	10 200	10 770	11 470	12 690
17	A	Elgin sand, dresser traprock	0.30	None		63	6 320	10 400	12 700	13 150	14 680	15 150
18	A	Elgin sand and gravel	0.30	None	Gap-graded sand	40	5 290	8 480	8 880	10 000	10 760	10 750
19	A	Elgin sand and gravel	0.30	None	Gap-graded sand	60	5 900	8 880	9 200	10 200	10 660	11 390
20	A	Elgin sand and gravel	0.35	2% CaCl$_2$		63	7 730	9 450	10 320	11 500	11 970	12 610

No.	Type	Aggregate	w/c	Admixture	Notes							
21	A	Elgin sand, dresser traprock	0.35	2% $CaCl_2$		63	7 540	9 850	10 900	12 590	12 800	13 850
22	A	Ottawa silica sand, Dresser traprock	0.30	None	Gap-graded sand	63	4 140	8 450	10 730	12 280	14 160	13 260
23	A	Elgin sand and gravel	0.30	0.25% Na_2Co_3, 0.25 lignosulfonate		63	6 840	9 440	10 060	10 660	11 570	9 800
24	A	Elgin sand and gravel	0.30	0.1% sucrose, 2% $CaCl_2$		63	8 420	12 050	12 710	13 970	14 190	14 000
25	B	Elgin sand and gravel	0.30	None	Grading as received	63	4 670	8 980	10 600	11 730	12 240	13 150
26	A	Clinker 23 K	0.30	0.1% sucrose, 2% $CaCl_2$		80	8 870	12 370	14 670	15 240	17 400	15 520
27	A	Elgin sand, dresser traprock	0.30	0.1% sucrose, 2% $CaCl_2$		63	8 030	11 320	12 190	13 240	14 620	13 680
28	C	Elgin sand and gravel	0.24	None	Admixtures added 1 min late	63	1 910	5 220	7 040	9 000	10 950	12 020
29	C	Elgin sand and gravel	0.24	0.1% sucrose, 2% $CaCl_2$	Admixtures added 1 min late	63	2 660	6 560	7 540	9 550	10 950	12 220
30	D	Elgin sand and gravel	0.27	None	Admixtures added 1 min late	63	1 250	2 840	4 190	7 510	8 560	9 940
31	D	Elgin sand and gravel	0.27	0.1% sucrose, 2% $CaCl_2$	Admixtures added 1 min late	63	1 020	4 250	5 480	7 950	8 210	11 040
32	C	Elgin sand and gravel	0.22	0.1% sucrose, 2% $CaCl_2$	Admixtures added 1 min late	63	3 430	8 090	9 680	11 400	11 470	13 390
33	D	Elgin sand and gravel	0.23	0.1% sucrose, 2% $CaCl_2$	Admixtures added 1 min late	63	1 780	5 750	7 430	9 700	12 040	14 280

Reference numbers 1 through 13 represent concrete which contained coarse aggregate up to $1\frac{1}{2}$ in (38 mm) maximum size.
All others except reference number 14 contained nothing larger than $\frac{3}{8}$ in (9.5 mm).
All specimens from reference number 10 through 33 were ponded for the first 24 h then placed in the moist room.
Note 1000 psi = 6.89 MPa; 1 inch = 25.4 mm.

7.2   Design of structural components using high-strength concrete

Design engineers increasingly find extensive economic advantages in using high-strength concrete in the range of 6000–12 000 psi (42–84 MPa). However, thus far the primary application of such concrete has been limited to columns in high-rise building and in prestressed elements. Lack of other general applications can be attributed to insufficient information available on the engineering properties of high-strength concrete and the difficulties in field quality control. The following are general recommendations for the design engineer:

(a) High-strength concrete elements can be designed using the same provisions used for normal-strength concrete, including the assumption of an equivalent rectangular compressive stress block.
(b) The ductility factor of concrete beams does not decrease with increase in concrete strength if the same reinforcement percentage is used.
(c) The tensile splitting strength and modulus of rupture of high-strength concrete is proportional to the square root of the compressive strength of concrete, f_c', as in the case of normal-strength concrete. Consequently, the serviceability behavior of beams, beam–columns, and slabs as determined by deflection and crack control can be evaluated using the same methodology as in normal-strength concrete design.
(d) Behavior under sustained loading also appears to be similar to that of normal-strength concrete, namely, creep and shrinkage strains will be essentially the same if the stress/strength ratio is similar.
(e) Multiaxial loading seems to produce the same effects as in normal-strength concrete. However, for confined concrete, more confining pressure is needed in the case of high-strength concrete to produce an equivalent increase in compressive strength. High-strength concrete columns become more slender, hence more attention has to be given to cases of spalling of the cover.

References

1. Ramakrishnan, V., Contribution of Gap-Grading to the Development of Low Cost and High Strength Concrete, *Proc. int. conf. mater. constr. developing countries*, Bangkok, Aug. 1978, pp. 395–405.
2. Hsu, Thomas T. C., Slate, Floyd O., Sturman, G. M. and Winter, G., Microcracking of plain concrete and the shape of the stress–strain curve, *J. Am. Concr. Inst., Proc.*, Vol 60, No. 2, Feb. 1963, pp. 209–224.
3. Whittman, F. H., Micromechanics of achieving high strength and other superior properties, *Proc. high strength concr. conf.*, University of Illinois, Chicago, Nov. 1979, pp. 8–30.
4. Mihashi, H. and Izumi, M., A stochastic theory for concrete fracture, *Cem. & Concr. Res.*, Vol. 7, 1977, pp. 411–422.
5. Bazant, Z. P., and Cedolin, L., Blunt crack band propagation in finite element analysis, *J. Eng. Mech. Div., Am. Soc. Civ. Eng.*, Vol. 105, 1979, pp. 297–315.
6. Hillerborg, A., Modeer, M. and Petersson, M., Analysis of crack formation

and crack growth in concrete by means of fracture mechanics and finite elements, *Cem. & Concr. Res.*, Vol. 6, 176, pp. 773–782.

7. ACI Committee 224, Control of cracking in concrete structures, *J. Am. Concr. Inst., Proc.*, Vol. 69, No. 12, Dec. 1972, pp. 717–753; updated *Concr. Int.*, Vol. 2, No. 10, Oct. 1980, pp. 35–76.

8. Carrasquillo, R. L., Nilson, A. H. and Slate, F. O., *Microcracking and engineering properties of high strength concrete*, Department of Structural Engineering, Cornell University, Report No. 80–1, 1980, 254 pp.

9. Nilson, A. H. and Slate, F. O., *Structural properties of very high strength concrete*, Second Progress Report, Department of Structural Engineering, Cornell University, Jan. 1979, 62 pp.

10. Sargin, M., *Stress–strain relationships for concrete and the analysis of structural concrete sections*, Publication No. 4, Solid Mechanics Division, University of Waterloo, 1971, 167 pp.

11. Balaguru, P. N., A numerical equation for the stress–strain curve of concrete and nonlinear analysis of RC beams, *J. Civ. Eng. Des.*, Vol. 2, No. 2, 1980, pp. 347–360.

12. Kaar, P. H., Hanson, N. W. and Copell, H. T., Stress–strain characteristics of high-strength concrete, *Douglas McHenry international symposium on concrete and concrete structures*, American Concrete Institute, SP-55, 1978, pp. 161–186.

13. Wang, P. T., Shah, S. P. and Naaman, A. C., Stress-strain curves of normal and lightweight concrete in compression, *J. Am. Concr. Inst., Proc.*, Vol. 75, No. 11, Nov. 1978, pp. 603–611.

14. Hognestad, E., Hanson, N. W. and McHenry, D., Concrete stress distribution in ultimate strength design, *J. Am. Concr. Inst., Proc.*, Vol. 52, No. 4, Dec. 1955, pp. 455–479.

15. CEB-FIP, *Model code for concrete structures*, 3rd Edn, Vols 1 and 2, Comité Euro-International du Béton, Paris, 1978, 348 pp.

16. Gerstle, K. H., Material behavior under various types of loading, *Proc. high strength concr. conf.*, University of Illinois, Chicago, Nov. 1979, pp. 43–78.

17. Balmer, G. G., Jones, V. and McHenry, D., *Shearing strength of concrete under high triaxial stress—computation of Mohr's envelope as a curve*, Structural Research Laboratory Report No. SP-23, US Department of Interior, Bureau of Reclamation, 1949, 26 pp.

18. Kupfer, H., The behavior of concrete under multiaxial short-term loading, Deutscher Ausschluss für Stahlbeton, Heft 229, 1973.

19. Hobbs, D. W., Pomeroy, C. D. and Newman, J. B., Design stresses for concrete structures subjected to multiaxial stresses, *Struct. Eng.*, Vol. 55, No. 4, April 1977, pp. 151–164.

20. Gerstle, K. H. *et al.*, Strength of concrete under multiaxial stress state, *Douglas McHenry international symposium on concrete and concrete structures*, American Concrete Institute, SP-55, 1978, pp. 103–131.

21. Kotosovos, M. D. and Newman, J. B., A mathematical description of the deformation behavior of concrete, *Mag. Concr. Res.*, Vol. 31, No. 107, June, 1979, pp. 77–90.

22. Leslie, K. E., Rajagopalan, K. S. and Everard, N. J., Flexural behavior of high strength concrete beams, *J. Am. Concr. Inst., Proc.*, Vol. 73, No. 7, Sept.

1976, pp. 517–521; discussion, *J. Am. Concr. Inst.*, Vol. 74, No. 3, March 1977, pp. 140–145.

23. ACI Committee 318, *Building code requirements for reinforced concrete*, (ACI Standard 318–77) and *Commentary to building code requirements for reinforced concrete*, American Concrete Institute, Detroit, 1977 and 1983.

24. Freudenthal, A. M. and Roll, F., Creep and creep recovery of concrete under high compressive stress *J. Am. Concr. Inst., Proc.*, Vol. 54, No. 12, June 1951, pp. 1118–1142.

25. Roll, F., Long-time creep recovery of highly stressed concrete cylinders, *Creep of concrete* American Concrete Institute, SP-9, 1964, pp. 95–114.

26. Neville, A. M., *Creep of concrete: plain, reinforced and prestressed*, North-Holland, Amsterdam, 1970, 622 pp.

27. Branson, D. E., *Deformation of concrete structures*, McGraw-Hill, New York, 1977, 545 pp.

28. Meyers, B. L., Branson, D. E. and Schumann, C. G., Prediction of creep and shrinkage behavior for design from short-time tests, *J. Prestressed Concr. Inst., Proc.*, Vol. 17, No. 3, May 1972, pp. 29–45.

29. Bazant, Z., Prediction of creep effects using age-adjusted effective modulus method, *J. Am. Concr. Inst., Proc.*, Vol. 69, No. 4, April 1972, pp. 212–217.

30. Bazant, Z. and Panula, L., Practical prediction of time dependent deformation of concrete, *Mater. & Struct.*, Vol. 11, 1978, pp. 307–328, 415–434; Vol. 12, 1979, pp. 169–183.

31. Carrasquillo, R. L., Nilson, A. H. and Slate, F. O., High strength concrete: an annotated bibliography 1930–1979, *Cement, Concrete and Aggregate. J.*, Vol. 2, No. 1, July 1980, pp. 1–19.

32. Swamy, R. N. and Anand, K. L., Shrinkage and creep of high strength concrete, *Civ. Eng. and Pub. Works Rev.*, Vol. 68, No. 807, Oct. 1973, pp. 859–868.

33. Russell, H. G. and Corley, W. G. G., Time-dependent behavior of columns in Water Tower Place, *Douglas McHenry international symposium on concrete and concrete structures*, American Concrete Institute, SP-55, 1978, pp. 347–373; also Portland Cement Association, Research and Development Bulletin, RD052.01B, 1977, 10 pp.

34. Parenchio, W. F. and Klieger, P., *Some physical properties of high strength concrete*, Portland Cement Association, Research and Development Bulletin, RD056, 1978, 6 pp.

35. Ngab, A. S., Nilson, A. H. and Slate, F. O., *Behavior of high strength concrete under sustained compressive stress*, Cornell University, Report No. 80–2, New York, Feb. 1980, 201 pp.

36. Bresler, B. and Bertero, V. V., Influence of high strain rate and cyclic loading behavior of unconfined and confined concrete in compression, *Proc. 2nd Can. conf. earthquake eng.*, Hamilton, 5–6 June 1975, 25 pp.

37. Ahmad, S. H. and Shah, S. P. Stress–strain curves of confined concrete, *J. Am. Concr. Inst., Proc.*, 1982. (In press).

38. Bertero, V. V., Inelastic behavior of structural elements and structures, *Proc. high strength concr. conf.*, University of Illinois, Chicago, 1979, pp. 96–167.

39. Nawy, E. G., Danesi, R. F. and Grosko, J. J., Rectangular spiral binders

effects on the plastic hinge rotation capacity in reinforced concrete beams, *J. Am. Concr. Inst., Proc.*, Vol. 65, No. 12, Dec. 1968, pp. 1001–1010.

40. Corley, W. G., Rotational capacity of reinforced concrete beams, *J. Struct. Div. Am. Soc. Civ. Eng.*, Vol. 92, No. 5 ST, Oct. 1966, pp. 121–146.

41. Kaar, P. E. and Corley, W. G., Properties of confined concrete for design of earthquake resistant structures, *Proc. 6th world conf. earthquake eng.*, New Delhi, Jan. 1977.

42. Manson, J. A., Overview of current research on polymer concrete: materials and future needs, *Applications of polymer concrete*, American Concrete Institute, SP-69, 1981, pp. 1–19.

43. Sun, P. F., Nawy, E. G. and Sauer, J. A., Properties of epoxy cement concrete system, *J. Am. Concr. Inst., Proc.*, Vol. 72, No. 11, Nov. 1975, pp. 608–613.

44. Nawy, E. G., Ukadike, M. M. and Sauer, J. A., High strength field modified concretes, *J. Struct. Div. Am. Soc. Civ. Eng.*, Vol. 103, No. ST12, Dec. 1977, pp. 2307–2322.

45. Nawy, E. G., Ukadike, M. M. and Sauer, J. A., Optimum polymer content in concrete modified by liquid epoxy resins, *Polymers in concrete*, American Concrete Institute, SP-58, 1978, pp. 329–355.

46. Nawy, E. G., Ukadike, M. M. and Balaguru, P. N., Investigation of concrete—PMC composite, *J. Struct. Div. Am. Soc. Civ. Eng.*, Vol. 108, No. ST5, May 1982, pp. 1049–1063.

47. Nawy, E. G. and Ukadike, M. M., Shear transfer in concrete and polymer modified concrete members subjected to shearing loads, *Am. Soc. Test. Mater., J. Testing and Evaluation* Vol. 11, No. 2, March 1983, pp. 89–99.

48. Balaguru, P. N., Ukadike, M. M. and Nawy, E. G., Freeze–thaw resistance of polymer-modified concrete, to be published.

49. ACI Committee 548 *Polymers in concrete*, American Concrete Institute, Detroit, 1977, 92 pp.

50. ACI Committee 548, *Applications of polymer concrete*, American Concrete Institute, SP-69, 1981, 228 pp.

51. ACI Committee 548, *Polymers in concrete*, American Concrete Institute, SP-58, 1978, 420 pp.

52. Perenchio, W. F., *An evaluation of some of the factors involved in producing very high strength concrete*, Portland Cement Association, Research and Development, Bulletin RD014, Skokie, 1973, 7 pp.

53. Walker, S. and Bloem, D., Effect of aggregate size on properties of concrete, *J. Am. Concr. Inst. Proc.*, Vol. 57, No. 8, Sept. 1960, pp. 283–298.

54. Klieger, P., *Early high strength concrete for prestressing*, Portland Cement Association, Bulletin 91, Skokie, March 1958.

55. ACI Committee 211, *Recommended practice for selecting proportions of normal and heavy-weight concrete*, American Concrete Institute, Detroit, 1977, 20 pp.

56. Neville, A. M., *Properties of concrete*, 3rd Edn Pitman, London, 1981, 779 pp.

6 Fiber reinforced concrete

Surendra P Shah

Northwestern University, Evanston, Illinois, USA

Contents

Notation

d	diameter of fibers
E	modules of elasticity
E_{f*}	equivalent modules of elasticity of fibers in direction of tensile load
E_m	modules of elasticity of matrix
l	length of fibers
V_f	volume fraction of fibers
x	crack spacing
ε_{mu}	strain in matrix corresponding to σ_{cc}
σ_{cc}	stress in composite just prior to first cracking
σ_{mu}	ultimate stress in the matrix
τ	bandstrength between fibers and matrix

Summary

Fiber reinforced concrete is made of hydraulic cements and aggregates, reinforced with discontinuous and discrete fibers. Fibers suitable for reinforcing concrete have been produced from steel, glass, and organic polymers. Naturally occurring asbestos fibers and vegetable fibers (sisal, jute, etc.) are also used for reinforcement.

Addition of fibers has made conventional, plain concrete more versatile, more flexible in methods of production and more competitive as a construction material. Fiber reinforced concrete is increasingly being used for precast products, airport runways, blast and impact resistant structures, tunnel linings, among others. Many types of applications, methods of fabrication and new types of fibers are currently being explored.

This chapter provides a summary of some recent developments on fiber reinforced concrete. Short-term and long-term properties of concrete reinforced with different types of fibers are presented. Basic concepts behind fiber reinforcement are emphasized so that a reader can keep abreast of this still-evolving field.

1   Introduction

1.1   Definition

Fiber reinforced concrete is concrete made of hydraulic cements with aggregates of various sizes and amounts and reinforced with discontinuous and discrete fibers. Fibers suitable for reinforcing concrete have been produced from steel, glass and organic polymers. Naturally occurring asbestos fibers and vegetable fibers, such as sisal and jute, are also used for reinforcement. Fibers are available in various sizes and shapes. Generally, the length and the diameter of the fibers used for concrete do not exceed, respectively, 50 mm and 500 μm.

1.2   Historical perspective

Fibers have been used to reinforce brittle materials since ancient times; straw was used to reinforced sunbaked bricks and horse hair was used to reinforce masonry mortar and plaster. In more recent times, large-scale commercial use of asbestos fibers in a cement paste matrix began with the invention of the Hatschek process in 1900. Asbestos cement construction products are widely used throughout the world today. However, partly because of the health hazards associated with asbestos fibers, different types of fiber have been developed in the last twenty to thirty years.

Utilization of glass fibers in concrete was first attempted in Russia in the late 1950s [1]. It was quickly established that ordinary glass fibers, including the borosilicate E-glass fibers, are prone to attack by the alkaline environment of cement paste. Considerable development work was directed towards producing a form of alkali-resistant glass fibers containing zirconium by Majumdar and his colleagues at the Building Research Establishment in England [2]. This work has stimulated considerable commercial activity in glass-fiber reinforced concrete building products.

Experimental trials and patents involving the use of discontinuous steel reinforcing elements, i.e. nails, wire segments and metal chips, to improve the properties of concrete date from 1910. In recent times, the first major attempt of evaluating the potential of steel fibers as a reinforcement for concrete was the investigation by Romualdi and Batson [3] in the United States in the early 1960s. Since then, a substantial amount of research, development, experimentation and industrial application of steel-fiber reinforced concrete have occurred.

Initial attempts at using synthetic fibers (Nylon, polypropylene) were not as successful as those using glass or steel fibers [4, 5]. However, better understanding of the concepts behind fiber reinforcement, new methods of fabrication and new types of organic fiber have led researchers to conclude that both natural and synthetic fibers can successfully reinforce concrete [6, 7].

Considerable research, development and applications are still taking place all over the world. The enthusiasm and interest shown in the new fiber-reinforced concrete construction materials are evidenced from the numerous research papers and several international symposia and the state-of-the-art reports issued by professional societies. The American Concrete Institute's technical committee on fiber-reinforced concrete wrote a report in 1973 [8] and is currently (1981) revising the report. RILEM's committee on fiber-reinforced cement composites has also published a report [9]. A recommended practice for glass-fiber reinforced panels has been published by the Prestressed Concrete Institute [10].

1.3   Fiber-reinforced concrete and reinforced concrete

Plain, unreinforced concrete has a low tensile strength and a low tensile strain before fracture. These shortcomings are traditionally overcome by reinforcing concrete. Reinforcing bars are continuous and are carefully placed in the structures to optimize their performance. Fibers are discontinuous and generally are randomly distributed throughout the concrete matrix. As a result, the reinforcing performance of steel fibers, for example, is inferior to that of steel reinforcing rods. In addition, fibers are likely to be considerably more expensive than conventional steel rods. Thus, fibers are not likely to replace conventional reinforced concrete. However, because of the flexibility in methods of fabrication, fiber reinforced concrete can be an economic and useful construction material. For example, thin (10–20 mm) precast fiber-reinforced concrete panels appear to be economically viable in the US and UK for curtain walls of buildings and other applications involving thin sheets. In this and other uses of fiber reinforced concrete, fibers are not replacing conventional reinforced concrete, but making plain concrete more versatile, competitive and useful.

2   Basic concepts and mechanical properties

2.1   Glass-fiber reinforced polymers

The current interest in microreinforced composite materials stems largely from research by Griffith on the strength of glass fibers. In a paper published in 1920,

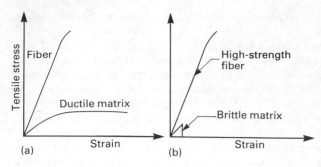

Figure 2-1 Tensile stress–strain curves for fibers with ductile and brittle matrices

he reported that the strength of glass fibers increased many fold as their diameter decreased. Although these small-diameter glass fibers are very strong in tension, they are brittle, notch-sensitive and cannot be used by themselves in larger lengths. When glass fibers are embedded in a polymeric matrix such as polyester resin, it is possible to utilize the high-tensile strength of glass and at the same time provide a notch-insensitive composite. The highly successful composites are known as 'Fiberglas'. The volume fraction of glass in such composites is up to 40% of the total volume. The relative stress–strain curves in tension for glass fibers and an epoxy matrix is shown in Fig. 2-1a. The incorporation of glass fibers increases both the stiffness and the strength of the composite. Since the failure strain of the matrix is higher than that of the glass fibers, it is possible to realize the full potential strength of fibers.

2.2 Strong fibers and brittle matrices

The relative stress–strain curves in tension for a strong fiber (steel, glass, asbestos, polypropylene, sisal, kevlar, etc.) and a brittle matrix (Portland cement, gypsum, etc.) are shown in Fig. 2-1b. The strain at fracture for the brittle matrix is considerably smaller (less than 1/50) than that for the fiber. As a result, when a fiber-reinforced brittle matrix composite is loaded, the matrix will crack long before the fiber can be fractured. Once the matrix is cracked, one of the following types of failure of the composite occurs:

(a) The composite fractures immediately after the matrix cracking. This type of behavior is shown in Fig. 2-2a.
(b) Although the maximum load on the composite is essentially the same as that of the matrix alone, the composite continues to carry decreasing load after the peak, as shown in Fig. 2-2b. The post-cracking resistance is primarily provided by the pulling out of fibers from the cracked surfaces. Although no significant increase in the tensile strength of the composite is observed, a considerable increase in the toughness of the composite (area under the complete stress-strain curve) can be obtained.
(c) Even after the cracking of the matrix, the composite continues to carry increasing tensile stresses; the peak stress and the peak strain of the composite are greater than those of the matrix alone; and during the

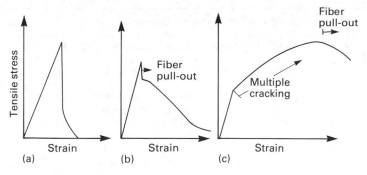

Figure 2-2 Three possible composite stress–strain curves for fiber-reinforced brittle matrices

inelastic range (between the first cracking and the peak) multiple cracking of the matrix occurs (Fig. 2-2c). It is clear that this mode of failure results in the optimum performance of both the matrix and the fibers. Various aspects of this type of behavior are discussed in what follows (for composites subjected to uniaxial tension).

2.3 Elastic range

Until the initial cracking of the matrix, it is reasonable to assume that both the fibers and the matrix behave elastically and that there is no slippage between the fibers and the matrix. In that case, the stress in the composite just prior to the first cracking in the matrix, σ_{cc}, and the corresponding strain in the matrix, ε_{mu}, are related as follows:

$$\sigma_{cc} = E_m(1 - V_f)\varepsilon_{mu} + E_f^* V_f \varepsilon_{mu} \tag{2-1}$$

Table 2-1 Some typical fiber properties

Fiber	Diameter μm	Density 10^3 kg/m^3	Young's modulus GPa	Tensile strength GPa	Elongation at break (%)
Asbestos					
chrysotile	0.02–20	2.55	164	3.1	2–3
crocidolite	0.1–20	3.37	196	3.5	2–3
Polypropylene	20–200	0.09	5–10	0.5	10–20
Nylon (high	>4	1.14	4	0.9	~15
tenacity)					
Kevlar					
PRD 49	~10	1.45	133	2.9	2.6
PRD 29	12	1.44	69	2.9	4.0
Sisal	10–50	1.5	5–10	0.8	~3
Glass	9–15	~2.6	~80	2–4	2–3.5
Steel	5–500	7.8	200	1–3	3–4

Table 2-2   Typical properties of fiber reinforced concrete

Fiber type	Length mm	Diameter μm	Volume %	Matrix proportions by weight	Flexural strength MPa
Asbestos	5–10	0.02–20	8–16	Only Portland cement, no coarse aggregates	15–25[a]
Glass	20–50	500–500[b]	4–6	Sand: cement ~0.5 Water: cement ~0.3	20–30[c]
Steel	12–25	~500	1–2	Sand: cement ~2.0 Water: cement ~0.45 Admixtures	8–12[e]
Polypropylene	12–25[d]	20–200	4–8	Sand: cement ~1.0 Water: cement ~0.35 Admixtures	6–15[f]

[a] May vary with the direction of testing
[b] Dimensions of fiber bundle made up of about 200 filaments
[c] Properties of about 10 mm thick precast sheets
[d] Continuous sheets are also used
[e] Conventionally mixing procedure
[f] Depends upon method of manufacturing

where E_m is the modulus of elasticity of matrix, V_f the volume fraction of fibers and E_f^* the equivalent modulus of elasticity of the fibers in the direction of tensile loading. Only if the fibers are sufficiently long and are aligned in the direction of loading, then $E_f^* = E_f$, the modulus of elasticity of the fibers. Otherwise, E_f^* can be as little as one third of the actual modulus. The properties of various fibers (including their elastic moduli) used for reinforcing concrete are shown in Table 2-1. The typical properties in which fibers are utilized in concrete construction are shown in Table 2-2. Combining the information given in Tables 2-1 and 2-2 and Eqn 2-1, it can be seen that, in general, the composite stress at first cracking will

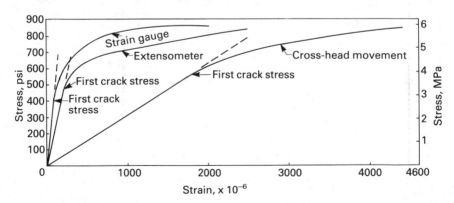

Figure 2-3   Stress–strain curves in tension for a steel-fiber reinforced specimen: $V_f = 3\%$, $l = 25.4$ mm (1 in), $d = 0.25$ mm (0.01 in)

not be substantially different from the tensile strength of the matrix. The experimental evidence (Figs 2-3 and 2-4) confirms that the presence of fibers does not significantly alter the properties of the unreinforced matrix until the first cracking [11, 12]. Note that the determination of the stress at first cracking is dependent on the definition of first cracking and the methods of measurement (Fig. 2-3).

2.4   Inelastic range

The composite will carry increasing loads after the first cracking of the matrix only if the pull-out resistance of the fibers at the first crack is greater than the load at first cracking. The pull-out resistance of the fibers will depend on the average bond strength, τ, between the fibers and the matrix, the number of fibers crossing the crack and the length and diameter of the fibers. Thus, it is not surprising that the product of the quantity V_f and the aspect ratio l/d (where l is the length and d the diameter of the fibers) has been found to greatly influence the post-cracking behavior of fiber reinforced concrete (Fig. 2-4). If either the bond strength, τ, or the quantity $V_f l/d$ is too low, then the behavior shown in Fig. 2-2b is observed.

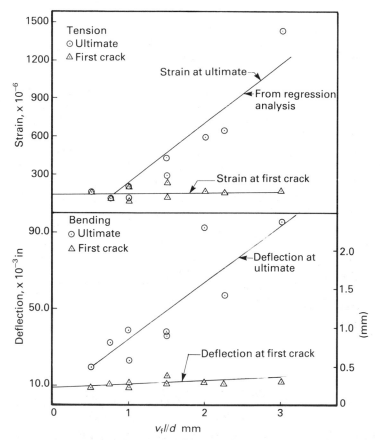

Figure 2-4   Strains and deflections for steel-fiber reinforced specimens

At the cracked section, the matrix does not resist any tension and the fibers carry the entire load taken by the composite. With an increasing load on the composite, the fibers will tend to transfer the additional stress to the matrix through bond stresses. If these bond stresses do not exceed the bond strength, then there may be additional cracking in the matrix. This process of multiple cracking will continue until either the fibers fail or the accumulated local debonding will lead to fiber pull-out. Naaman and Shah have derived a theoretical relationship for predicting the number of cracks at the end of multiple cracking for continuous fibers [13]:

$$x = 1.5 \frac{V_m}{V_f} \frac{\sigma_{mu}d}{2\tau} \tag{2-2}$$

where x is the crack spacing. Note that if x is very large then only one or two very wide cracks are observed before the peak composite stress is reached (Fig. 2-2b). If x is very small, a large number of very fine cracks are observed (Fig. 2-2c).

The process of multiple cracking is a key element in assuring the maximum performance of strong fibers in a brittle matrix. With multiple cracking, it is possible to achieve substantially larger composite peak stress and especially peak strain as compared to those at first matrix cracking. Multiple cracking reduces crack widths, which is an important design consideration [14]. The extent of multiple cracking can be increased by increasing the volume fraction of fibers, by improving the bond strength, as well as by increasing the aspect ratio of fibers.

2.5 Pull-out behavior of fibers

It is clear from the above discussion that the processes of debonding and subsequent pulling out are critical in understanding and predicting the behavior of fiber reinforced concrete. Our understanding of the basic interfacial debonding and stress-transfer phenomena and of the interaction among matrix properties (modulus of elasticity, tensile strength, etc.), fiber orientation and shape of fibers is not sufficient to enable satisfactory predictions of the entire stress–strain curve of the composite [15–18].

2.6 Phenomena of cracking

Unlike glass-fiber reinforced plastics, the primary contribution of the addition of steel, glass or organic fibers is evident only after the cracking of the matrix. An unreinforced matrix ruptures catastrophically; properly designed fiber reinforced concrete fails gradually, accompanied by multiple cracking and the subsequent pulling out of fibers. As a result, fiber reinforced concrete has at least an order of magnitude higher toughness (ductility, impact resistance). Although most of the applications of fiber reinforced concrete rely on this increased toughness rather than on increased strength, there are no experimental or analytical methods to quantify the increased resistance to cracking provided by fibers. Research in applying fracture mechanics concepts to quantify toughness and experimentally to evaluate impact resistance is continuing [19–22].

3   Fabrication

Fiber reinforced concrete can be manufactured in a variety of methods. Fibers can be combined with cement, aggregate and water in a concrete mixer or a ready-mix truck. This method has an advantage that one does not have to alter substantially the normal and most common method of manufacturing ordinary concrete. To achieve the uniform distribution of fibers within the matrix with the conventional method of mixing, greater workability is necessary. The required workability is obtained by limiting the volume fraction and the aspect ratio of the fibers. Thus, although for mechanical performance it is desirable to use as high a value of V_f and l/d as possible, the workability requirement limits the maximum values of these two parameters. The required increased workability is generally also obtained by increasing the cement content (Table 2-2) and by using fly ash [23].

For glass-fiber reinforced concrete, it has been possible to use higher aspect ratios and volume fractions by using the spray technique. Fibers are fed to a compressed-air gun which cuts the fibers to the required length at the spray head and sprays them simultaneously with the cement slurry onto a forming surface. This process is used to manufacture thin sheet-like glass-fiber reinforced concrete products. Asbestos cement sheets are commonly produced by combining fibers, cement and water in a liquid-like slurry suspension and by filtering and removing thin layers on a moving belt [6]. Since a higher volume fraction of fibers can be used with these two methods, a much higher value of flexural strength is obtained for asbestos cement and glass-fiber reinforced concrete sheets (Table 2-2).

Steel fibers have been extensively used in shotcrete for mine and tunnel linings and for rock slope stabilization [24]. Although in most cases steel fibers were placed by the dry process, the wet process has also been used.

4   Durability

Glass fibers are attacked by the moist alkaline environment of cement paste. Despite the improvement in durability and strength retention achieved by the introduction of alkali-resistant glass composition, there is still some attack on glass. Glass-fiber reinforced composites which were stored for 20,years in the natural weather conditions in the United Kingdom had essentially the same flexural strength as that of the unreinforced matrix, although after 28 days the flexural strength of the composite specimens was about three times that of plain matrix. This means that the contribution of fibers has practically disappeared after a long-term exposure to wet conditions. As a result of this, the Prestressed Concrete Institute's guideline recommends the use of 28 days first crack strength (proportional limit) as the design strength for glass-fiber reinforced concrete products [10]. Research is being conducted to improve the durability of glass fibers by incorporating polymer latex in the cement matrix [25] and coating the glass fibers [26].

Steel fibers, asbestos and polypropylene fibers are chemically stable in the cement-paste matrix. The high alkalinity of cement paste protects steel from being corroded. However, the corrosion of steel fibers can become a problem when the matrix becomes cracked. The severity of the problem will depend on the severity of exposure and cracking.

5 Applications

5.1 Asbestos fibers

The naturally available mineral fibers, asbestos, have been successfully combined with Portland cement paste to form the product called asbestos cement. Asbestos cement has been the most widely used fiber reinforced concrete composite. The world consumption of asbestos fibers for making building products such as sheets, shingles, pipes, tiles and corrugated roofing elements is about 3×10^6 t/year. The reasons for the wide use of asbestos fibers are many. They are naturally available and as a result are relatively inexpensive. They have a desirable thermal, mechanical and chemical resistance. The methods of manufacturing asbestos cement products, such as the Hatschek or Magnani process, are well developed and suited for precast factory products.

Asbestos cement products contain about 8 to 16% by volume of asbestos fibers. The flexural strength of asbestos cement board is roughly two to four times that of the unreinforced matrix (Table 2-2). However, since the fibers are relatively short (10 mm), the composite has a brittle behavior (Fig. 2-2a) and has a low impact strength.

Only a few countries are the major producers of asbestos fibers and in these countries the better grades are being depleted. Because of the depletion and a lung cancer, called asbestosis, associated with asbestos, a great demand exists for a suitable substitute.

5.2 Glass fibers

Glass fibers are used in large quantities with a resin matrix in the production of Fiberglas composites. The common glass fibers, called E-glass, are believed to be chemically attacked by the highly alkaline environment of cement paste, and as a result the glass-fiber reinforced cement composites lose their strength with time. The strength reduction will depend on the chemical nature of the glass and the cement paste as well as on the environment. The rate of strength reduction can be reduced by using the so-called alkali-resistant glass fibers, by using an organic coating over the fibers, or by reducing the relative humidity of the environment. Considerable research is being done to establish the long-term properties of glass-fiber reinforced cement products [25, 26].

Glass fibers are made from up to 200 to 400 individual filaments, 5 to 15 μm in diameter, which are lightly bonded to make up a strand. These strands can be chopped into various lengths or combined to make cloth, mat or tape. Commercial applications of glass fibers have involved the addition of fibers either by the conventional mixing procedure or by the spray-up process. When glass fibers (or any other fibers) are incorporated using the fabrication techniques which are standard for normal concrete, then it is not possible to mix more than about 2% by volume of fibers of up to about 25 mm long. With this amount and length of fibers, the flexural strength is at most about doubled and the ductility, toughness of impact resistance is increased about tenfold.

With the spray-up process, it is possible to incorporate higher volume (6%) and length (50 mm) of fibers with a low water/cement ratio matrix (0.4) resulting in a superior composite (Table 2-2). In this process, both the chopped glass fibers and the cement slurry are sprayed with pressurized air onto a surface. If the process is done in a factory, then the excess water can be removed by applying suction.

Applications of glass-fiber reinforced concrete (mortar) include a variety of precast products, curtain wall and surface bonded masonry construction.

5.3   Steel fibers

Steel wire is produced by a series of hot and cold-working methods. Round steel fibers are produced by cutting or chopping the wire, typically having diameters in the range 0.25–0.76 mm. Flat-sheet fibers having typical cross-sections in the range 0.15–0.41 mm in thickness and 0.25–0.90 mm in width are produced by shearing sheets. Fibers have also been produced from hot melt extract. Deformed fibers which are loosely bonded with a water soluble glue in the form of a bundle are also available. Since individual fibers often cluster together, their uniform distribution in the matrix is often difficult. This may be avoided by adding fiber bundles which separate during the mixing process. Since the failure of fiber reinforced concrete is by fibers pulling out, the use of deformed fibers increases the pull-out strength and, consequently, the mechanical properties of the composite.

Normal mix proportions of steel-fiber reinforced concrete consist of about 1–2% by volume of up to 25 mm-long fibers (aspect ratio of 75–100) and a higher than usual cement content (Table 2-2). The larger volume of long fibers would

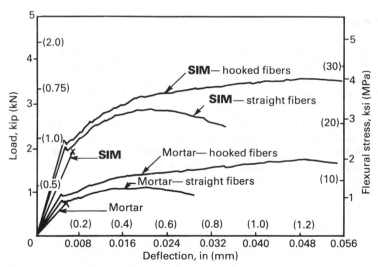

Figure 5-1   Load–deflection curves for steel-fiber reinforced mortar: $V_f =$ 2%, straight fibers 25 × 0.25 mm, hooked fibers 37.5 × 0.4 mm. SIM—sulphur impregnated mortar

give higher mechanical performance, if they could be uniformly distributed. But problems of workability and uniform distribution increase with increasing volume and length of fibers. The improvement with steel fibers include about a twofold to threefold increase in flexural strength and, more important, up to a twentyfold increase in crack resistance or toughness. The corrosion of steel fibers has not been observed to be a problem, perhaps because of the high cement content and the large number of fibers. However, the long-term aspects of the corrosion of fibers at a cracked surface have not been fully investigated.

The efficiency of steel fibers can be increased by increasing the matrix strength. This can be seen in Fig. 5-1, where some results of steel-fiber reinforced sulphur-impregnated mortar are shown [27].

Applications of steel fibers include highway pavements, airport runways, refractory concrete and tunnel linings, with sprayed fiber reinforced concrete.

5.4 Organic fibers

Organic fibers include synthetic fibers, such as polypropylene, or natural fibers, such as sisal. Organic fibers are lighter in weight and may be more chemically inert than either steel or glass fibers. They also may be cheaper, especially if they are natural. However, since these fibers have poorer bond and a lower modulus of elasticity than mineral or metallic fibers, when they are used in similar proportions as the other types of fibers, they do not show similar improvements. However, the recent work with which the author was involved [7] has shown that it is possible to obtain multiple cracking and the stress–strain curve of the type shown

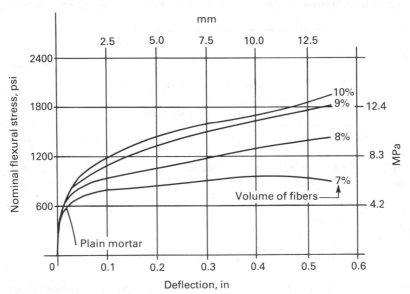

Figure 5-2   Results of flexural tests on mortar reinforced with agave fibers: average values of series of three specimens

in Fig. 2-2c with vegetable fibers provided a large enough volume fraction (7%, 50 mm length, agave type of fibers) is used (Fig. 5-2). The problems of mixing and uniform dispersion were solved by adding a superplasticizer. Although additional research is needed, it appears that organic fibers can be substituted for mineral or metallic fibers in many applications.

References

1. Biryukovich, K. L. and Yu, D. L., *Glass fiber reinforced cement*, translated by G. L. Cairns, CERA Translation, No. 12, Civil Engineering Research Association, London, 1965, 41 pp.
2. Majumdar, A. J., Properties of fiber cement composites, *Proc. RILEM symp.*, London, 1975, Construction Press, Lancaster, 1976, pp. 279–314.
3. Romualdi, J. P. and Batson, G. B., Mechanics of crack arrest in concrete, *J. Eng. Mech. Div. Am. Soc. Civ. Eng.*, Vol. 89, No. EM3, June 1963, pp. 147–168.
4. Monfore, G. E., A review of fiber reinforced of Portland cement paste, mortar and concrete, *J. Res. Dev. Labs Portl. Cem. Ass.*, Vol. 10, No. 3, Sept. 1968, pp. 36–42.
5. Goldfein, S., *Plastic fibrous reinforcement for Portland cement concrete*, US Army Corps of Engineers Ohio River Division Laboratories, Cincinnati, Technical Report No. 1757-TR, 1965, 56 pp.
6. Hannan, D. J., *Fiber cements and fiber concretes*, Wiley, New York, 1978, 219 pp.
7. Castro, J. and Naaman, A. E., Cement mortar reinforced with natural fibers, *J. Am. Concr. Inst., Proc.*, Vol. 78, No. 2, Jan. 1981, pp. 69–78.
8. ACI Committee 544, Revision of state-of-the-art report (ACI 544 TR-73) on fiber reinforced concrete, *J. Am. Concr. Inst., Proc.*, Nov. 1973, Vol. 70, No. 11, pp. 729–744.
9. RILEM Technical Committee 19-FRC, Fibre concrete materials, *Mater. & Struct., Test Res.*, Vol. 10, No. 56, 1977, pp. 103–120.
10. Prestressed Concrete Institute, Recommended practice for glass fiber reinforced concrete panels, *J. Prestressed Concr. Inst.*, Vol. 76, No. 1, Jan. 1981, pp. 25–93.
11. Shah, S. P. and Rangan, B. V., Fibre reinforced concrete properties, *J. Am. Concr. Inst., Proc.*, Vol. 68, No. 2, Feb. 1971, pp. 126–135.
12. Shah, S. P. and Naaman, A. E., Mechanical properties of glass and steel fiber reinforced mortar, *J. Am. Concr. Inst., Proc.*, Vol. 73, No. 5, Jan. 1976, pp. 50–53.
13. Naaman, A. E. and Shah, S. P., Tensile tests on ferrocement, *J. Am. Concr. Inst., Proc.*, Vol. 68, No. 9, Sept. 1971, pp. 693–698.
14. Kormeling, H. A., Reinhardt, H. W. and Shah, S. P., Static and fatigue properties of concrete beams reinforced with continuous bars and with fibers, *J. Am. Concr. Inst., Proc.*, Vol. 77, No. 1, Jan–Feb. 1980, pp. 36–43.
15. Naaman, A. E. and Shah, S. P., Pull-out mechanism of steel fiber reinforced concrete, *J. Struct. Div. Am. Soc. Civ. Eng.*, Vol. 102, No. ST8, Aug. 1976, pp. 1537–1548.

16. Stroeven, P., de Haan, Y. M., Bouter, C. and Shah, S. P., Pull-out tests of steel fibers, *RILEM symp. testing and test methods of fiber cement composites*, 1978, Construction Press, Lancaster, pp. 345–354.
17. Shah, S. P., Stroeven, P., Dalhuisen, D. and van Stekelenburg, P., Complete stress–strain curves for steel fibre reinforced concrete in uniaxial tension and compression, *RILEM symp. testing and test methods of fiber cement composites*, 1978, Construction Press, Lancaster, pp. 399–408.
18. Stroeven, P. and Shah, S. P., Use of radiography-image analysis for steel fiber reinforced concrete, *RILEM symp. testing and test methods of fiber cement composites*, 1978, Construction Press, Lancaster, pp. 275–288.
19. Velazco, G., Visalvanich, K. and Shah, S. P., Fracture behavior and analysis of fiber reinforced concrete beams, *Cem. & Concr. Res.*, Vol. 10, 1980, pp. 41–51.
20. Wecharatana, M. and Shah, S. P., Resistance to crack growth in Portland cement composites, *Proc. fracture concr.*, American Society of Civil Engineers National Convention, Florida, 1980. (See also *Journal, ASCE*, Vol. 108, No. ST6, June 1982, pp. 1400–1413).
21. Suaris, W. and Shah, S. P., Inertial effects in the instrumental impact testing of cementitious composites, *ASTM Journal of Cement, Concrete and Aggregates*, Vol. 3, No. 2, 1981, pp. 77–83.
22. Shah, S. P., Materials characterization for fragmentation, *Proc. NATO–ARI symp., Paris, 1980*, Plenum Press, New York, 1981, pp. 245–262.
23. Kesler, C. E., Mix design considerations, *Proc. conf. M-28, fibrous concr.—constr. mater. for seventies*, Construction Engineering Research Laboratory, Champaign, Ill., 1972, pp. 29–37.
24. Henager, C. H., Properties of steel fibre reinforced mortar and concrete, *Proc. symp. fibrous concr.*, Concrete International 1980, Concrete Society, London Construction Press, Lancaster, pp. 29–47.
25. Bijen, J. and Geurts, E., Sheets and pipes incorporating polymer film material in a cement matrix, *Proc. symp. fibrous concr.*, Concrete International, 1980, Concrete Society, London/Construction Press, Lancaster, pp. 194–202.
26. Majumdar, A. J., Properties and performance of GRC, *Proc. symp. fibrous concr.*, Concrete International, 1980, Concrete Society, London/Construction Press, Lancaster, pp. 69–86.
27. Shah, S. P., Naaman, A. E. and Smith, R. H., Investigation of concrete impregnated with sulfur at atmospheric pressure, *Polymers in concrete*, American Concrete Institute, SP-58, 1978, pp. 399–416.

7 Structural lightweight concrete

Thomas A Holm

Solite Corporation, Mt Marion, NY, USA

Contents

Notation

E_c modulus of deformation of the concrete

E_{ca} modulus of deformation of the coarse aggregate particle

E_{cd} dynamic modulus of deformation

E_m modulus of deformation of the 'mortar' fraction

f_c' compressive strength of a 150×300 mm cylinder

f_{ct} tensile splitting strength of a 150×300 mm cylinder

k thermal conductivity (Imperial units)

λ thermal conductivity (SI units)

ρ density

Summary

The present state of the technology and design of structural lightweight aggregate concrete are summarized. The chapter presents and interprets the data on structural lightweight concretes from many laboratory studies and the accumulated experience resulting from increased and successful in-service field applications. It includes, in a condensed form, the inherent characteristics and methods of production of expanded aggregates, the physical properties of concretes incorporating structural lightweight aggregates and recommendations for design procedures. Recommendations are included for specifications, proportioning, mixing and placing methods as well as pumping and field control procedures. Also included are other physical properties, such as durability and thermal insulation, that provide the designer with comprehensive data relating to the overall performance of structures produced from structural lightweight concrete.

1   Introduction

There are numerous techniques by which the density of concrete can be reduced, but this chapter, in keeping with the overall concept of a handbook devoted to structural concrete, will focus attention only on fully compacted, structural lightweight concretes produced from high-quality structural grade, expanded aggregates. No coverage is given to extremely low-density, strictly insulating concretes composed of perlites, vermiculites, or expanded plastic aggregates. Marginal strength concretes produced from naturally occurring low-density volcanic aggregates, cellular concretes, highly aerated and no-fines concretes having large interstitial voids will also be omitted.

1.1   General considerations

Applications of structural lightweight aggregate concretes may be arbitrarily separated into three categories: structural (S), structural–insulating (SI) and insulating (I). Definitions of what constitutes 'structural' concrete may vary throughout the world with ranges shaped by historical factors and geographical conditions. For example, the principal early North American application of lightweight aggregate was in the production of concrete masonry units. This is still true today, and of the nearly four thousand million concrete masonry units produced annually in the USA more than half are produced from expanded aggregates to load-bearing strength requirements that are identical to those required for units composed of natural aggregates. Similarly, when North American concrete applications first used lightweight concrete to reduce the self-weight of building frames, it was approached on the basis that the structural design properties would be relatively unchanged.

Equilibrium densities of hardened lightweight concretes for structural and structural–insulating applications vary throughout the world with ranges comparable to those mentioned for strength. Broadly speaking, these densities can vary from 1300 to 1900 kg/m^3 (80–120 lb/ft^3). To attempt a presentation of all-inclusive data is a futile exercise and hardly justified as each geographical area contains producers of expanded aggregates who provide comprehensive technical property and design data appropriate for their particular expanded aggregate and tailored towards acceptance by governing buildings codes and regulations.

The transition of Système International notations may cause some departures from time tested, well known expressions. In this chapter the words 'lightweight', 'low density' and 'expanded' are considered equivalent expressions. Similarly, 'structural lightweight aggregate concrete', 'structural low-density aggregate concrete', and 'structural low-density concrete' will also be considered interchangeable, but in general referred to simply as 'structural lightweight concrete'. The term 'all light-weight' indicates concretes in which both the coarse and fine fractions are expanded aggregates; the term 'sand lightweight' indicates concretes with coarse expanded aggregate and the fine fraction composed of natural sand. Conversions between SI and older notational systems was conducted on a 'soft' basis, rounding off numbers to appropriate orders of accuracy.

2   Structural low-density aggregates

2.1   Definitions

Fine lightweight aggregates are composed of expanded minerals suitable for the production of structural lightweight concrete, properly graded with 85–100% passing the 5 mm ($\frac{3}{16}$ in) screen, having a dry loose bulk density less than 1120 kg/m^3 (70 lb/ft^3) and complying with the requirements of relevant national material standards. Coarse lightweight aggregates are composed of expanded minerals which are suitable for the production of structural lightweight concrete, having a dry loose bulk density less than 880 kg/m^3 (55 lb/ft^3) and complying with the relevant national material standards.

2.2   Internal structure of expanded aggregates

Expanded aggregates used in structural concrete have a low particle density due to the cellular structure of the individual aggregate particles. This cellular structure within the particles is formed at high temperature, generally 1100°C (2000°F) or higher, by the formation of gases due to the reaction of heat on certain constituents in the raw materials, coincident with incipient fusion of the mineral so that gases causing expansion are trapped in the viscous, pyroplastic mass. High-quality expanded aggregates are produced when small-size, well-distributed, non-interconnected pores are enveloped in a strong continuous vitreous phase (Fig. 2-1).

2.3   Production of expanded aggregates

Uniform high-quality expanded aggregates are produced in modern automated manufacturing plants from raw materials that are generally:

(a) Suitable natural deposits of shales, clays or slates.
(b) By-products of other industries, such as iron blast furnace slags or fly ash from the burning of powdered coal in thermoelectric power plants.

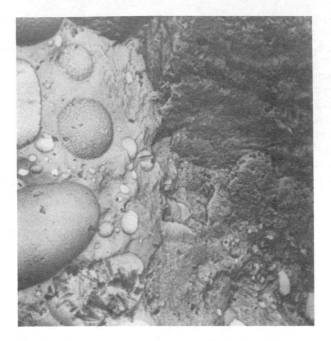

Figure 2-1 Scanning electron micrograph of 60-year-old structural light-weight concrete. Note that the contact zone between aggregate and hydrated paste is almost indistinguishable

Usage of these materials stems from environmentally sound reasoning, as they utilize materials that have limited applications in their natural state, thus minimizing the demands of the construction industry on the finite resources of natural sands, stones and gravels.

Methods of production of expanded aggregates include the rotary kiln process (a long, slowly rotating, nearly horizontal cylinder lined with refractory materials similar to cement plants); the sintering process wherein a bed of raw materials is carried by a traveling grate under ignition hoods; and the rapid agitation of molten slag with controlled amounts of water. No single description will apply to any raw material or process and the reader is urged to consult with local producers for physical and mechanical properties of expanded aggregates and low density concrete.

2.4 Aggregate properties

2.4.1 Specific density

Specific density is the mass of the dry material per unit volume of poreless solid material. It is useful in determining the particle porosity. For materials of a clay, shale or slate origin, this value is approximately 2500 kg/m^3; for slags the value is considerably higher.

2.4.2 Particle density
The particle density of an aggregate is the ratio between the mass of a quantity of the material and the volume occupied by the individual particles contained in that quantity. This volume includes the pores *within* the particles but does *not* include the voids *between* the particles. In general, the volume of the particles is determined from the volume they displace when submerged in water, with the penetration of water into the particles during the test limited by previous saturation. The particle density depends both on the specific density of the vitreous material and on the pore content within the particles and generally increases when particle sizes decreases.

2.4.3 Bulk density
The bulk density of an aggregate is the ratio between the mass of a given quantity of material and the total volume occupied by it. This volume includes the voids between as well as within the particles. Bulk density is a function of the density, size, gradation, moisture content and shape of the particle, as well as the method of packing the material (loose, vibrated, rodded) and thus varies for different materials as well as for different sizes and gradations of a particular material.

2.4.4 Voids content
The voids content (within particle pores and between particles' voids) can be determined from measured values of particle and bulk density. If, for example, measurements on a sample of expanded coarse aggregate are

Bulk density, ρ_{bulk} 770 kg/m^3 (48 lb/ft^3)
Particle density, ρ_{part} 1400 kg/m^3
Specific density, ρ_{spec} 2500 kg/m^3

then the pores content of an individual particle is

$$\frac{2500-1400}{2500} \times 100 = 44\%$$

and the bulk interstitial voids (between particles) are

$$\frac{1400-770}{1400} \times 100 = 45\%$$

2.4.5 Particle shape and surface texture
Expanded aggregates have differences in particle shape and texture depending on the source and the method of production. The shape may be cubical, rounded, angular or irregular. Surface textures may range from relatively smooth with small exposed pores to irregular with small to large exposed pores. Particle shape and surface texture influence mix proportioning factors, such as workability, coarse-to-fine aggregate ratio, cement content, water requirements as well as other physical properties.

2.4.6 Gradation
Usual theories of gradation of aggregate apply to expanded aggregates with the exception that, for most expanded aggregates, the density increases with decreasing particle size. This being the case, usual normal-weight gradation standards

should not be applied to expanded aggregates as these recommendations are established by weight of each sieve fraction, where in fact ideal behavior is developed through volumetric considerations.

2.4.7   *Moisture content*

Due to their cellular structure, expanded aggregates are capable of absorbing more water than normal-weight aggregates. Based upon a 24 h absorption test, expanded aggregates generally absorb from 5 to 25% by weight of dry aggregate. By contrast, normal-weight aggregates usually will absorb less than 2% of moisture. The moisture contents of stockpiled aggregate may be as high as 5 to 10% or more. The important distinction is that the moisture of expanded aggregates is largely absorbed into the interior of the particles whereas in normal-weight aggregates it is largely on the surface. Recognition of this differ-ence is crucial in mix proportioning, batching and control. Rate of absorption is dependent on the size and distribution characteristics of the pores of an expanded aggregate, particularly those close to the surface [3]. Internally absorbed water within the particle is not immediately available to the cement as mixing water.

2.4.8   *Modulus of deformation of aggregate particles*

The modulus of deformation of concrete is a function of the moduli of its constituents. Concrete may be considered to be a two-phase material consisting of

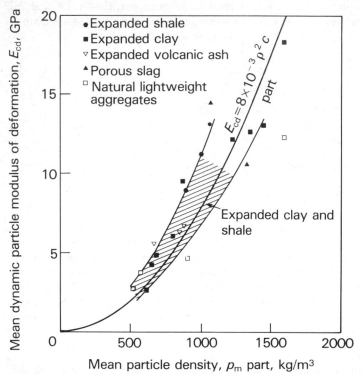

Figure 2-2   Relationship between the mean particle density (12–16 mm diameter) and the mean dynamic modulus of deformation, E_{cd}, for the particles of artificial and natural lightweight aggregates [2]

coarse aggregate inclusion within a continuous 'mortar' fraction that includes cement, water, entrained air and fine aggregate. Dynamic measurements [2, 5] made on aggregates alone have revealed a considerable scatter (Fig. 2-2) about a parabola corresponding to the function: $E_{cd} = 0.008 \, \rho^2$, where E_{cd} is the dynamic modulus of deformation of the particle in MPa and ρ is the dry mean particle density in kg/m^3. Dynamic moduli for usual expanded aggregates have a range of 5000–20 000 MPa, whereas the range for strong ordinary aggregates is approximately 30 000 MPa for quartz to 100 000 MPa for basaltic rock.

3 Proportioning, mixing and placing structural lightweight concrete

3.1 Proportioning

In general, the proportioning rules and techniques used for ordinary concrete mixes apply to lightweight concrete with the added considerations of special attention to concrete density and the influence of the water absorption of the expanded aggregate. The majority of both ordinary and lightweight concretes are proportioned by absolute volume methods in which the fresh concrete produced by a combination of materials is considered equal to the sum of the absolute volumes of cement, aggregates, net water and entrained air. Proportioning by this method requires the determination of water absorption and particle density of the separate sizes of aggregates [6].

In proportioning a lightweight concrete mix, the engineer is concerned with obtaining desirable, predictable values of properties from the concrete proposed for a particular application. Specifications for lightweight concrete usually require minimum values for compressive and tensile strength, maximum values for workability, both minimum and maximum values for air content and, finally, a maximum limitation on fresh bulk density. From the load-resisting considerations of structural members, reduced density can lead to improved economy of structures and is therefore an important consideration in the proportioning of lightweight concrete mixtures. While this property depends primarily on the particle density of the expanded or normal-weight aggregates, it is also influenced by the cement, water and air contents, and to a small degree by the proportions of coarse to fine aggregate.

The water/cement ratio of lightweight concretes is generally avoided due to the uncertainty of calculating that portion of the total water in the mix which is part of the water/cement ratio. Water absorbed in the aggregate prior to mixing is not included as part of the cement paste, and complications are introduced by the absorption of some indeterminate part of the water added at the mixer. This absorbed water is available, however, for continued hydration of the cement after normal curing has ceased. The general practice with lightweight aggregates is to proportion the mix on the basis of a cement content at a given workability for a particular aggregate.

Air entrainment in lightweight concrete, as in normal-weight concrete, improves durability and resistance to scaling. In concretes made with some angular lightweight aggregates, it is an effective means of improving workability of otherwise harsh mixtures. With moderate air contents, mixing-water requirements are lowered while maintaining the same workability as well as reducing bleeding and segregation. Recommended ranges of total air content of usual structural lightweight concretes are

Maximum size of aggregate *Air content per cent by volume*
20 mm ($\frac{3}{4}$ in)   4 to 8
10 mm ($\frac{3}{8}$ in)   5 to 9

Higher air contents to develop greater thermal resistance, or for lowering densities on semi-structural concrete fills, are frequently used with reduction of the strength-making characteristics as a natural consequence. Additions of water reducers, retarders, superplasticizers, etc. result in improved concrete characteristics for lightweight concretes in a manner similar to the experience with normal-weight concretes.

3.2   Placing, finishing and curing

With properly proportioned structural lightweight concrete, there is little or no difference in delivering and placing, when compared with ordinary concretes. The most important consideration in handling concrete is to avoid separation of the coarse aggregate from the mortar fraction. Basic principles required to secure a well placed concrete include:

(a) well-proportioned workable mixes that use a minimum quantity of added water;
(b) equipment capable of expeditiously moving the concrete;
(c) proper consolidation in the forms;
(d) quality workmanship in finishing.

Generally speaking, well proportioned structural lightweight concretes can be placed, screeded and floated with less effort than that required for ordinary concrete. Excessive vibrating should be avoided, as this practice serves to drive the heavier mortar down from the surface where it is required for finishing. On completion of the final finishing, curing operations similar to ordinary concrete should begin as soon as possible. Lightweight concretes batched with aggregates containing high moisture contents carry their own internal water supply for curing within the aggregate and are therefore less sensitive to poor curing practice [7] or unfavorable ambient conditions. This 'internal curing' water is transferred from the expanded aggregate to the mortar phase as evaporation takes place on the surface of the concrete, thus continuously maintaining moisture balance while replacing moisture essential for continued hydration through a period dependent on ambient conditions and the as-batched moisture content of the expanded aggregate.

3.3   Pumping

The capacity of expanded aggregates to continuously absorb relatively large amounts of water with respect to time or increased pressure is the main difference in pumping characteristics. For this reason, it is essential to presoak the expanded aggregate before its introduction into the mix prior to pumping. Presoaking can be accomplished by applying water directly to the aggregate on the belts and to the stockpiles for at least 24 h but preferably 72 h, through hose or sprinkler system. Presoaking is best accomplished at aggregate production plants where uniformity of moisture content is achieved by applying water through spray bars on moving belts. Other techniques, such as thermal and vacuum saturation methods, may be feasible when extended drying periods are possible prior to experiencing freezing temperatures.

Presoaking reduces the rate of an aggregate to absorb water, minimizing the transfer of water out of the mortar fraction which, in turn, causes the loss of consistency during the pumping operation. The higher moisture content developed during presoaking results in an increased moist-particle density, which in turn develops higher fresh-concrete densities. This increased water content due to presoaking will eventually diffuse out of the concrete, developing a somewhat longer period of internal curing as well as a larger decrease from fresh to equilibrium density than that usually associated with lightweight concretes placed with lower moisture contents. Concrete and aggregate suppliers should be consulted about pumping aids and mix adjustments that may be necessary for consistent pumpability.

3.4 Field control

Changes in absorbed moisture, gradation or density of the expanded aggregates or variations in the volume of entrained air suggest frequent checks of the fresh content at the job site in order to allow adjustments to develop consistent concrete characteristics. Standardized field tests of consistency, fresh density and entrained air content should be employed to verify that field concretes conform to the approved mix as well as the specification. In general, when variations in fresh density exceed ±2%, an adjustment in batch weights may be required to restore the specified concrete properties. Air content should not vary more than ±1.5% from the specified value, thus avoiding adverse effects on strength, workability and durability.

4 Physical and mechanical properties of structural lightweight concrete

4.1 Compressive strength

The usual compressive strengths necessary for acceptable performance in concrete frame applications can be obtained by the presently available structural expanded aggregates. Strength levels of 20–35 MPa (3000–5000 psi) when measured on 150×300 mm (6×12 in) cylinders are common practice in North America [8], while in other areas cube strengths of 10–20 MPa (1500–3000 psi) are considered structurally functional. Perception of what constitutes 'structural lightweight concrete' varies from locations where the motivation is to improve the strength/density structural frame efficiency, to other areas where the primary concern is to develop thermally resistant exterior panels with moderate load-carrying capacity.

All expanded aggregates have a maximum achievable strength 'ceiling' above which further additions of cementitious materials do not appreciably increase the concrete strength [9]. This strength 'ceiling' is a function of the vitreous material and the quantity, size, shape and distribution of the enveloped pores, with the decisive factor being the strength of the largest individual particle. Strength ceilings of some aggregates may be significantly improved by reducing the maximum size which, usually, has the lowest particle density and initiates the failure mechanism by the crushing of the aggregate.

When aggregate particles are less rigid than the surrounding mortar, they do not participate fully in the transmission of internal forces. Lightweight concrete cannot reach the strength of the 'mortar' fraction as long as there is a significant elastic mismatch between the aggregate and the mortar. As the elastic rigidity of the lightweight aggregate particle is increased, a point of 'elastic compatibility' will be reached, wherein the coarse aggregate and the surrounding 'mortar' fraction have similar moduli of deformation: $E_{ca} \approx E_m \approx E_c$, where E_{ca} is the modulus of deformation of the coarse aggregate, E_m the modulus of deformation of the mortar fraction (fine aggregate, cement paste and air), and E_c is the modulus of deformation of the concrete.

The elastic mismatch that occurs between some extremely rigid natural aggregates and the mortar fraction causes stress concentrations that precipitate failure in the mortar fraction long before the aggregate's strength capacity is developed. It is for this reason that several investigators report the unexpected results of expanded aggregates with densities of approximately 1300–1400 kg/m^3 reaching equal or higher strength levels than companion normal-density concretes, at equal cement contents [4, 5]. The interaction between the coarse aggregate and an enveloping mortar phase of relatively equal elastic rigidity has been reported by several investigators [10, 11] and satisfactorily explains the development of high-strength lightweight concretes. The continuous transition from the crushing of extremely low-density aggregates, to that state wherein the strength and rigidity of the expanded aggregate match that of the mortar fraction, has been documented in one report that extended the comparisons to the usual elastic mismatch occurring in normal density concretes [3].

Improved long-term strength gain ratios are possible when expanded aggregates having a sufficiently high particle strength are batched at a moisture content equal to, or exceeding, that achieved by soaking for 24 h. This long-term strength gain is due to the continuous hydration of the binder with the moisture available from the slowly released reservoir of water absorbed within the pores of the expanded aggregates. This process of 'internal curing' is further enhanced if a pozzolan (fly ash or suitable expanded aggregate fine fraction) is introduced into the mix. It is well known that the pozzolanic reaction of a finely divided alumino-silicate material with the calcium hydroxide liberated as cement hydrates is contingent upon the availability of moisture. The pozzolanic activity of certain expanded fine aggregates is a dependable advantage when curing takes place at high temperature but with normal ambient conditions should not be used as a cement replacement but rather as a valuable additional virtue.

With some materials, compressive strength can be increased by the partial replacement of expanded fine aggregates with high-quality natural sands. Local expanded aggregate producers should be consulted for this trade-off between structural efficiency and material costs.

4.2 Density

High-quality expanded aggregates can produce structural concretes with densities 20–40% less than normal-weight concretes without compromising structural strength. Different geographical areas have developed wide variation of the strength/density relationships, as is evident in Fig. 4-1. With almost 70% of the absolute volume of a concrete mix taken up by aggregates, the resulting density of the concrete is chiefly a function of the particle density of the aggregate fraction. Expanded aggregate producers are, in general, the best source of information regarding the numerous combinations of cement aggregates, air-entraining admixtures, etc., that may be custom-tailored to specific strength, rigidity and thermal requirements.

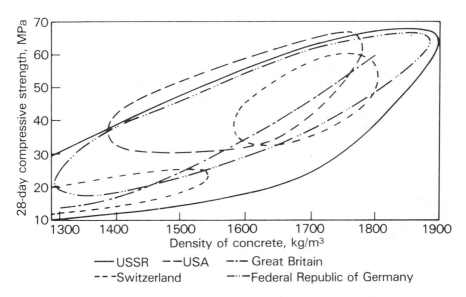

Figure 4-1   Relationship between the compressive strength and density of structural lightweight concrete [2]

4.3 Modulus of deformation

The modulus of deformation of concrete depends on the modulus of each constituent (hydrated paste, expanded and normal-density aggregates) and their relative mix proportion. Normal-density concretes have a higher modulus because the moduli of the natural aggregate particles (and parent rock formations) are greater than the moduli of expanded aggregate particles. For practical design conditions, the modulus of deformation of concretes with densities between 1400 and 2500 kg/m^3 (90–155 lb/ft^3) and within normal strength ranges [8] may be assumed to follow the formula:

$$E_c = 0.043\sqrt{\rho^3 f_c'}$$

where E_c denotes the secant modulus in MPa, ρ the density in kg/m^3, and f'_c the compressive strength, in MPa, of a 150×300 mm (6×12 in) cylinder.

This or any other formula should only be considered as a first approximation, as the modulus is significantly affected ($\pm 25\%$) by moisture, aggregate type and other variables. The formula clearly overestimates the modulus for high-strength, low-density concretes, where limiting values are determined by the modulus of the expanded aggregate [9]. When design conditions require an accurate evaluation of E_c, as for example in the One-Shell Plaza Tower in Houston, Texas, in which the entire frame of floors, columns and foundations was constructed with moderate to high-strength lightweight concretes [12], laboratory tests should be conducted on specific concretes proposed for the project (Fig. 4–2).

Figure 4-2 One-Shell Plaza Tower, Houston, Texas [2]

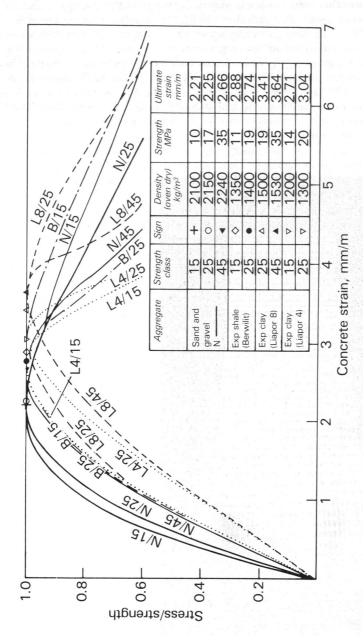

Aggregate	Strength class	Sign	Density (oven dry) kg/m³	Strength MPa	Ultimate strain mm/m
Sand and gravel N ——	15	+	2100	10	2.21
	25	○	2150	17	2.25
	45	▼	2240	35	2.66
Exp shale (Berwilit)	15	◇	1350	11	2.88
	25	●	1400	19	2.74
Exp clay (Liapor 8)	25	◁	1500	19	3.41
	45	◀	1530	35	3.64
Exp clay (Liapor 4)	15	▷	1200	14	2.71
	25	▷	1300	20	3.04

Figure 4-3   Stress–strain curves of lightweight concretes made with expanded shale, expanded clay and sand-and-gravel at a constant rate of deformation varying between 0.25 and 1.0 mm m⁻¹ min⁻¹ [2]

Stress–strain relationships for lightweight and ordinary concretes are similar, and the usual stress block recommendations apply to both concrete types [2, 8]. Figure 4-3 shows the stress–strain diagrams for lightweight concretes composed of expanded shales and clays, as well as for ordinary concretes, obtained from tests [13] where a constant strain rate was maintained during loadings so that the descending section of the curve could also be plotted. The compressive strain reached at peak loads ranged between 2.74 and 3.64% for lightweight concretes and between 2.21 and 2.66% for ordinary concretes. Other tests [9] on high-strength lightweight concretes, loaded through a steel beam, showed similar results (Fig. 4-4).

Poisson's ratio, when tested by resonance methods [14], indicated variations influenced by age, strength and other factors averaging out at 0.21. Static methods also determined similar variations with results averaging 0.20. Tests conducted on expanded and normal-weight aggregate particles alone by ultrasonic pulse velocity measurement techniques produced variations between 0.16 and 0.28 [11]. For practical design conditions, a value of 0.2 for usual service stresses is generally assumed but, should special conditions warrant, further testing by static or dynamic methods on specially proportioned concretes may be conducted.

At high stress/strength levels, the value of Poisson's ratio increases rapidly, signaling the onset of microcracking that in time leads to volumetric expansion of the concrete and ultimately failure. The stress/strength ratio at which the initiation of failure may be considered to have started is higher for lightweight concretes than for ordinary concretes due to the better elastic matching and superior bond between the coarse aggregates and mortar fractions [15, 16].

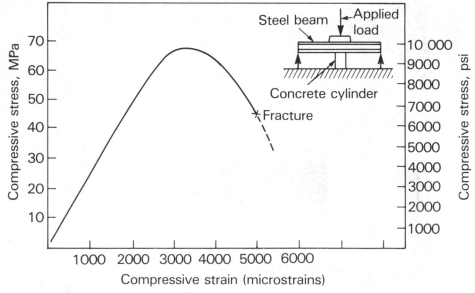

Figure 4-4   Stress versus strain—uniaxial compression. Solite lightweight concrete: 100×200 mm (4×8 in) cylinder, density 1870 kg/m³ (117 lb/ft³), static modulus 24 000 MPa (3.5 × 10⁶ psi) [9]

4.4 Tensile strength

Shear and torsion resistance, anchorage, bond strength and resistance to cracking
are related to the tensile strength of concrete, which, in turn, is dependent upon
the tensile strength of coarse aggregate and mortar phase and the degree to which
the two phases are securely bonded. Traditionally, tensile strength has been
defined as a function of compressive strength, but this is known to be only a first
approximation that does not reflect aggregate particle strength, surface characteristics and the concrete's moisture content and distribution. The splitting tensile
strength (Fig. 4-5) is widely used throughout North America as a simple, practical
method of determining design criteria and is generally considered a more reliable
indicator than beam flexural tests, which are extremely sensitive to moisture
content and distribution at the time of tests [17].

Tests have shown that the diagonal tensile strengths of beams and slabs closely
correlate with the splitting strengths of concrete [18]. These tests demonstrated
the wide dispersion of tensile splitting results for different combinations of

Figure 4-5 Indirect tensile splitting test on 100×200 mm (4×8 in) high-
strength lightweight concrete cylinder [9]

materials, so, in general, the designer should consult with the available supplier for laboratory test data on the splitting strength. Special test data should be developed for unusual structures or for projects where the development of early-age tensile strengths in precast or tilt-up members is essential.

Attention to the moisture condition of the concrete is important as it has been found that laboratory test data on the tensile strength of low-density concretes which undergo drying is more relevant in respect to the behavior of concrete in actual structures. During the drying of the concrete, moisture loss progresses at a slow rate into the interior of concrete members, resulting in the development of tensile stresses on the outer envelope that balance the compressive stresses in the still moist interior zones. The tensile resistance to external loading of lightweight concretes undergoing drying will be less than that of similar, but continously moist-cured concretes [19].

In general, the splitting tensile strength of all-lightweight concrete varies from approximately 75 to 100% of normal-weight concretes of equal compressive strength. The blending of normal-weight fine aggregates normally increases the tensile strength of these concretes, especially those subject to early drying.

4.5 Bond strength

Field performance has indicated satisfactory behavior of lightweight concrete with respect to bond. Pull-out tests are normally used to measure bond strength and while there is significant data developed from extensive tests [20], there is a consensus that the conditions of most laboratory bond tests are not comparable to those existing in a full-sized structure. These tests results cannot therefore be directly applied to practice but rather allow comparisons between concretes of differing strengths, densities and mix composition. Because of the lower particle strength, lighweight concretes have generally lower bond capacities than normal-weight concrete. North American practice is to dispense with the some-what academic conception of bond and require slightly longer embedment lengths for reinforcement in lightweight concretes that are related to the lower tensile strength ratios—the reciprocal of 0.85 for sanded lightweight and 0.75 for all lightweight concretes [8].

4.6   Shrinkage

As with ordinary concretes, the shrinkage of structural lightweight concretes is principally determined by

(a) The quality (inherent shrinkage characteristics) of the cement paste frac-
    tion.
(b) The restraining influence provided by the aggregate fraction.
(c) The relative proportions of absolute volume occupied by the shrinkage
    medium (cement paste fraction) and the restraining skeletal structure
    (aggregate fraction)
(d) Ambient humidity and temperature.

Aggregate characteristics determine the amount of cementitious binder fraction (the shrinking fraction) necessary to produce a required strength at a given workability. Gradation and particle shape have a direct influence on water demand and individual particle strength (particularly the larger sizes) directly determines the volume of cement paste necessary to meet specified strength levels. Once that interaction has been established, it is the rigidity of the aggregate fraction that restrains the shrinkage of the cement paste. In general, where the expanded aggregate does not require the inclusion of significantly larger proportions of cement paste than normal aggregates, the shrinkage of the low-density concrete is generally, but not always, somewhat greater than that of ordinary concrete due to the lower rigidity of the expanded aggregate [20]. The development of shrinkage is slower, and the time required to reach a plateau of equilibrium is longer when the expanded aggregate is batched at a high moisture content. The shrinkage of high-strength lightweight concrete is closely similar to that of high-strength ordinary concrete of the same binder content [9].

4.7 Creep

Numerous long-time creep tests on structural low-density concrete have demonstrated the general similarity in performance to that of normal density concretes [14, 20, 21]. The factors known to have a significant influence on the magnitude of creep of normal concrete, i.e. aggregate type, cement type, mix-water content, entrained-air content, age at initial loading, magnitude of applied stress, method of curing, size of specimen, ambient relative humidity and period of loading, have a similar effect on lightweight concrete.

The creep of concrete is largely the result of the deformation of the cement paste under continuous loading with creep increasing directly with the content of the hardened cement paste of the concrete. Figure 4-6 shows the creep of low and ordinary density concretes from several European and North American investigations, demonstrating the wide range of results for both concretes [20–22]. For the same cement paste content, the creep of lightweight concretes is similar to that of normal-weight concretes. If the particle density of the expanded aggregate is sufficiently high (1400 kg/m^3), lightweight concretes can achieve strengths equal to normal-weight concretes at equal cement contents. With good particle shape and gradation, the mixing water demands for workability will also be similar.

4.8 Durability

Horizontal structures composed of structural lightweight concrete must be resistant to freezing and thawing and to the disruptive effects caused by deicing salts. Generally, deterioration is not likely to occur in vertically exposed columns and walls, except in areas which are continuously exposed to water.

Despite difficulties in correlating laboratory accelerated freeze–thaw testing methods with field observations, numerous programs in North America [23, 24] and in Europe [25] have researched the influence of air content, cement content, moisture content of aggregates, drying times and testing environment with generally similar conclusions; 'air entrained lightweight concretes proportioned with high quality binder provide satisfactory durability results when tested under usual laboratory freeze–thaw program' [26]. Limited data on the resistance to weight loss in freeze–thaw tests in the presence of deicing salts indicate similar performance between structural lightweight and normal-weight concretes. Comprehensive investigations into the long-term weathering performance of bridge decks

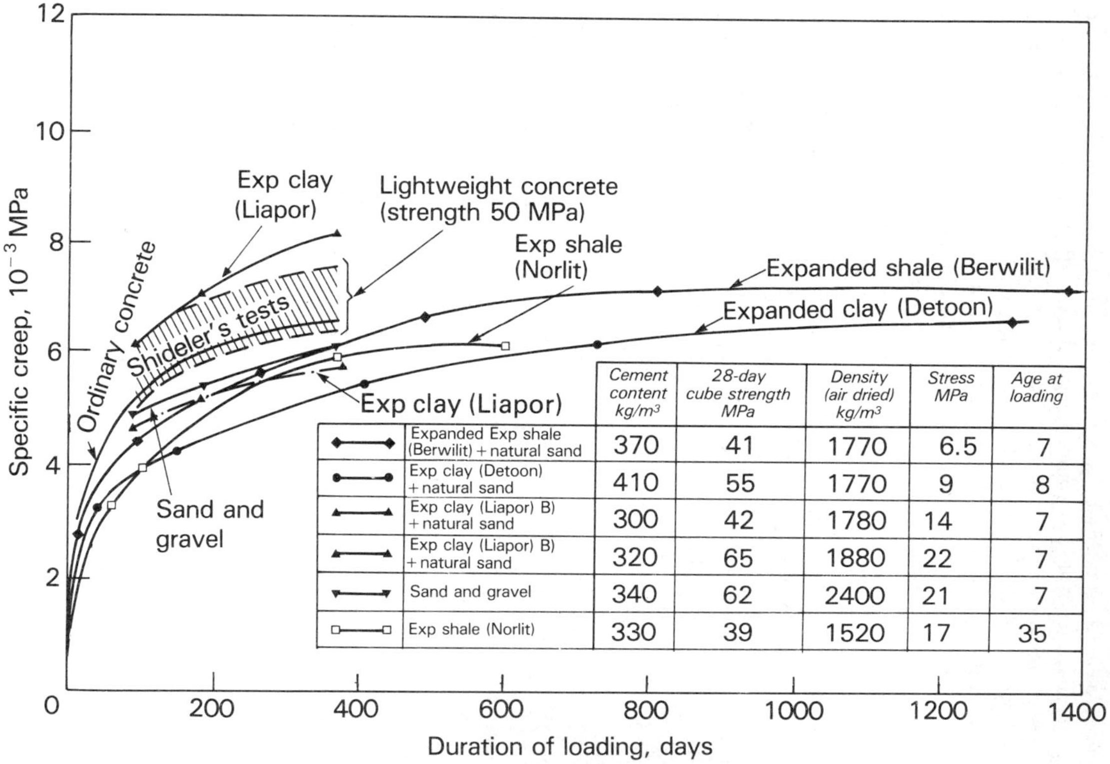

		Cement content kg/m³	28-day cube strength MPa	Density (air dried) kg/m³	Stress MPa	Age at loading
◆—◆	Expanded Exp shale (Berwilit) + natural sand	370	41	1770	6.5	7
●—●	Exp clay (Detoon) + natural sand	410	55	1770	9	8
▲—▲	Exp clay (Liapor) B) + natural sand	300	42	1780	14	7
▲—▲	Exp clay (Liapor) B) + natural sand	320	65	1880	22	7
▼—▼	Sand and gravel	340	62	2400	21	7
□—□	Exp shale (Norlit)	330	39	1520	17	35

Figure 4-6 Relationship between the creep and the time under loading for lightweight concrete compared with ordinary concrete [2, 20, 21, 22]

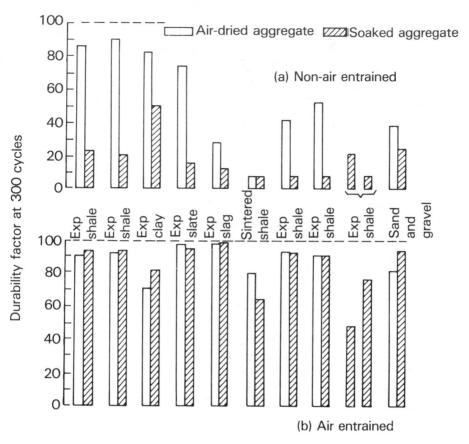

Figure 4-7   Influence of air entrainment and of moisture content of aggregate on the durability of concrete (strength about 35 MPa) [32]. The durability factor is the ratio of the dynamic modulus of deformation after a specified number of freezing and thawing cycles to the modulus before exposure [23]

[27] and marine structures [26] support the findings of laboratory investigations and suggest that properly proportioned and placed lightweight concretes perform equal to, or better than, normal-weight concretes.

The USS Selma, a 7500 ton reinforced expanded shale concrete tanker launched in 1919, is perhaps the most vivid testament to the durability and corrosion resistance of structural lightweight concrete. After several years of service, this vessel was purposely sunk in 1922 in Galveston Bay, Texas, where it has remained partially submerged ever since.

The hull has been in sea water for over 60 years with a bandwidth of approximately 1 m exposed to wetting and drying in salt air caused by wind and tides. The deck and upper section of the hull have been exposed to salt air and occasionally awash with sea water due to wave action in storms (Fig. 4-8).

Figure 4-8   USS Selma, a lightweight concrete ship launched in 1919. This photograph was taken in 1953, 31 years after it was purposely sunk. [26]

A study of the concrete was sponsored by the Expanded Shale Clay and Slate Institute in 1953 and reported in *Story of the Selma—expanded shale concrete endures the ravages of time*. Surprisingly little corrosion was noted despite a cover of only 16 mm ($\frac{5}{8}$ in).

In 1980, the Selma's performance was reviewed again, with cores (Fig. 4-9) taken for an update on the extensive tests conducted in 1953 [26]. Examination of both the submerged and waterline area cores of the 60-year-old lightweight concrete revealed little deterioration. Form marks on the surface and no evidence of microcracking were observed when viewed under a magnifying glass. The concrete was extremely well compacted (the first known application of internal vibration techniques) with only a small number of macroscopic air voids. The coarse aggregate (minus $\frac{1}{2}$ in) was well bonded to the mortar fraction, uniformly distributed and without any indication of water gain under large aggregate particles, indicating a fluid, easily placed, but not wet concrete.

Core samples cut through the steel reinforcing bars showed little evidence of rusting. There was almost no sign of corrosion at the reinforcing bars–concrete interface that could be observed when cut sections of steel were inadvertently separated from broken cores.

Compressive strength measured on the concrete cores, normalized to 150×300 mm (6×12 in) cylinders, was 70 MPa (10.2×10^3 psi) and the compressive modulus of deformation was 25 000 MPa (3.59×10^6 psi). These values are

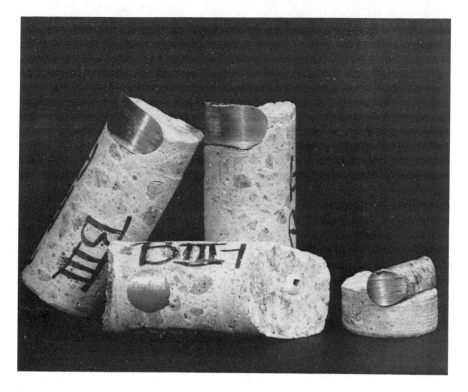

Figure 4-9   Cores taken from USS Selma in 1980

comparable to the results of the cores and cubes test results from the 1953 investigation.

Explanation of this unusual resistance to weathering and corrosion is difficult to quantify and may include some or all of the following physical and chemical mechanisms:

(a) Superior resistance to microcracking due to the high bond of the aggregate-to-mortar fraction combined with the reduction of the inner stresses due to elastic matching of the coarse aggregate and the mortar phase.

(b) High ultimate strain capacity provided by a concrete with a high strength-to-modulus ratio, as well as a high stress-to-strength ratio at which the dilation of the concrete commences.

(c) A well dispersed void system provided by the fine expanded aggregate fraction that may serve an absorption function in weathering resistance as well as reducing salt concentration levels in the mortar phase.

(d) Long-term pozzolanic action provided by expanded fine aggregate with the proper physical and chemical characteristics that could combine with the calcium hydroxide liberated during cement hydration. This could minimize the leaching of soluble compounds and in addition may reduce the possibility of sulphate salt disruptive behavior.

(e) Low mortar permeability provided by the combination of high cement contents and the water available in the expanded aggregate for internal curing.

The water permeabilities of ordinary and lightweight aggregate concretes are of the same order of magnitude, as the resistance of concrete to the penetration of water is principally a function of the capillary porosity of the mortar fraction [28]. Continued hydration that is possible when water is contained within the aggregate combined with the higher binder contents develops higher quality, lower porosity, and lower permeable mortar fractions. Reported experience of the concrete ship program of the United States in both world wars indicated dry bulkheads and almost no moisture penetration problems. All other factors being equal, resistance to corrosion is primarily a function of the permeability of the mortar fraction (cement, air, water and fine aggregate).

When stored under water, fully compacted, lightweight concretes will generally absorb from 12 to 22% by volume as compared to less than 12% for normal-weight concretes. This increase is due to the pores of the expanded aggregate particle being filled by diffusion while continuously submerged. Concretes that are not submerged ultimately achieve equilibrium at a moisture content of about 5% by volume wherein most of the water is contained in the paste fraction with little difference between concretes of different aggregate types.

The abrasion resistance of concrete depends on the strength, hardness and toughness of the cement paste and the aggregates, as well as on the bond between the constituents. Most lightweight aggregates suitable for structural concretes are composed of solidified glassy material that is comparable to quartz on the Moh scale of hardness. However, because of its vesicular, porous structure, the net resistance to wearing forces is less than that of a solid particle of similar composition. While laboratory wear tests indicate greater losses for bare lightweight concretes, garage floors and bridge decks subjected to truck traffic show wearing performance similar to that of normal-weight concretes. Limitations would only be necessary in commercial applications where steel-wheeled industrial vehicles are present, but these surfaces generally receive specially prepared surface treatments in any event. If a surface is likely to be exposed to extreme wearing conditions, an abrasion resistant top surface containing high abrasion resistance aggregates may be cast on top of the fresh lightweight concrete.

4.9 Heat flow properties

4.9.1 Thermal conductivity

Thermal conductivity is a specific property of a material and is a measure of the rate at which energy (heat) passes perpendicularly through a unit area of homogeneous material of unit thickness for a temperature gradient of one degree. Thermal conductivity of concrete depends mainly on its density and moisture content but is also influenced by the size and distribution of the pores, the chemical composition of the solid components, their internal structure (crystalline, amorphous) and the test temperature. Crystalline materials (e.g. quartz) conduct heat better than amorphous materials (e.g. burned clays, ceramics etc.).

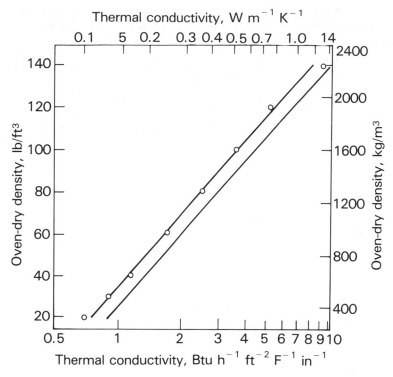

Figure 4-10   Relation of average thermal conductivity values of concrete to its dry density [1]

Thermal conductivity is generally measured on oven-dry samples in a guarded hot-plate assemblage. Figure 4-10 shows the results of the analysis of over 400 published thermal conductivity test results on concretes with densities from 320 to 3200 kg/m³ (20–200 lb/ft³) which suggest the equation

$$\lambda = 0.072 \exp (0.001\,25\rho)$$

where λ is the thermal conductivity in $W\,m^{-1}\,K^{-1}$ and ρ the density in kg/m³. The corresponding equation in Imperial units is

$$k = 0.5 \exp (0.02\rho)$$

where k is the thermal conductivity in $Btu\,h^{-1}\,ft^{-2}\,{}^\circ F^{-1}\,in^{-1}$ and ρ the density in lb/ft³.

For a given concrete, an accurate value of the thermal conductivity based upon tests in a guarded hot-plate (for oven-dry specimens) or a heat flowmeter (for rapid testing when specimens contain moisture) is preferable to an estimated value, but the formula provides guidance for estimating the thermal conductivity in an oven-dry condition and, in addition, may readily be revised for air-dry conditions. When thermal resistance values are part of the project specifications, the addition of crystalline natural aggregates should be avoided, as the resulting thermal conductivity of the mixture will increase at a rate faster than that predicted by density alone [30, 31].

Increasing the free-moisture content of hardened concrete causes an increase of thermal conductivity. As most conductivity data are reported for oven-dry concrete, it is essential to know the moisture contents of the concrete in equilibrium with its in-service environment, and then apply a modification factor for estimating the conductivity under service conditions. Various investigations [29, 32] have reported long-term moisture contents for concretes with the average for lightweight concretes being 4% by volume. As a practical matter, considering the many variations of density, mix composition, in-service ambient conditions, etc., a reasonable approximation would be to increase the in-place thermal conductivity by 20% over test oven-dry values [29].

4.9.2 Thermal expansion

The coefficient of thermal expansion of concretes is principally determined by the expansion characteristic of the aggregates, the volumetric proportions and the moisture conditions of the concretes. This is hardly surprising, as aggregates compose approximately 70% of the total volume of concrete. Reports indicate

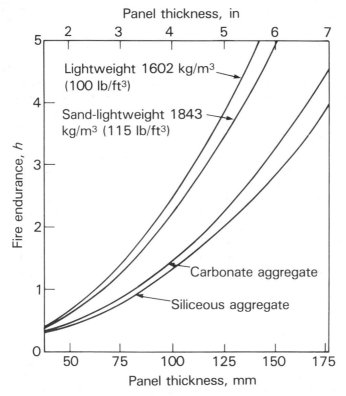

Figure 4-11 Fire endurance (heat transmission) of concrete slabs as a function of thickness for naturally dried specimens [1]

approximate values of 8×10^{-6} per °C for lightweight concretes with relatively small differences between lightweight aggregate types, and average values of 11×10^{-6} per °C for normal-weight concretes with a considerably wider range of values dependent on the mineralogical characteristics of the aggregates (33–35).

4.9.3 Fire resistance

Structural lightweight concretes have greater fire-resistance ratings than normal-weight concretes because of their lower thermal conductivity, lower coefficient of thermal expansion and the inherent fire stability of an aggregate already heated to over 1100°C. An illustration of recent results of fire tests of various types of concrete (at equal strengths of approximately 30 MPa) is given in Fig. 4-11 [36].

Similarly, the greater insulating resistance of lightweight concretes impedes the temperature rise of embedded reinforcing steel thus allowing slightly lower cover requirements than those required for normal-weight concretes. The strength retention characteristics of lightweight concretes are similar to those of carbonate aggregate concretes with approximately 75% of their original strength being maintained at 650°C [37].

5 Design of structural lightweight concrete

5.1 Compressive strength

Engineers have designed structurally efficient, economic projects with structural lightweight concrete since the advent of high-quality manufactured expanded aggregate over 60 years ago. During most of this period, designs were based upon the fundamental properties of concrete properly modified for the incorporation of an expanded aggregate, without the guidance of design codes or recommended practices specifically directed toward structural lightweight concretes. With the adoption of the 1963 ACI Building Code [38], structural lightweight concrete gained full recognition as an acceptable structural material. Now the building codes and standards of most countries provide design requirements and materials classifications appropriate to structural lightweight concrete, with the approach somewhat varying from country to country. In most codes, structural lightweight concrete is accommodated by the insertion of additional articles or factors that modify the design techniques and formulas by the effect brought about by the differing physical properties (e.g. reduced modulus and tensile strength). In others, a separate chapter covering different design factors is added to amend the basic concrete code, while in at least one instance a complete separate code is issued [39].

In general, structural engineers make their designs for a specific project based upon detailed information on the properties of concrete made with lightweight aggregates geographically available. With the world-wide increase in research exclusively devoted to the properties of expanded aggregates and the concretes made with them, the original situation of limited design recommendations has now reached the state where there is more data available for the design of structural lightweight concrete than for ordinary concretes. Lightweight concretes have now been shown by extensive laboratory tests and by long-term field performance to behave structurally in the same manner as ordinary concretes, while providing some significantly improved properties, the chief being reduced self-weight and increased thermal resistance.

The design of columns using structural lightweight concrete is essentially the same as for normal-weight concrete. The reduced modulus should be used in the code sections in which slenderness effects are considered. Large-scale laboratory test programs [40, 41] comparing the time-dependent behavior of structural lightweight and normal-weight columns produced the following facts:

(a) Instantaneous shortening caused by initial loading can be accurately pre-dicted by elastic theory, with greater shortening of lightweight concrete columns due to the lower modulus of deformation of lightweight concrete. This reduced modulus should be used in long-column formulas where slenderness effects are important.

(b) Time-dependent shortenings of lightweight concretes are somewhat greater than those of normal-weight concretes when small unreinforced specimens are compared. However, these differences are minimized when large reinforced concrete columns are tested. Measured time-dependent shorten-ings were compared with those theoretical predictions with a satisfactory correlation.

(c) Test results of ultimate strengths were compared with theory with good correlations. Concrete type (lightweight and ordinary) and previous loading history had no effect on this correlation.

(d) The lightweight concrete columns generally had slightly greater ultimate strain capacity than normal weight concrete columns when unreinforced. When reinforced, the strain capacities were similar.

5.2 Deflection

At equal strength levels, the modulus of deformation of structural lightweight concrete is less than that of ordinary concretes. All design characteristics that incorporate this property (e.g. deflection characteristics, buckling effects in long columns or slender beams, and elastic shortening in compression zones) should reflect the code recommendations or data specifically developed for the local materials. In some codes, a lower limit on span/depth ratio is provided [8]. When calculating flexural deflections, the influence of the lower modulus is partially offset by the movement of the neutral axis in the cross-section towards the maximum tensile fibre so that the moment of inertia of the cracked section under load is increased. The increased modular ratio adds to this effect by increasing the equivalent cross-section corresponding to the reinforcing steel area.

The shrinkage and creep of structural lightweight concrete is generally, but not always, greater than ordinary concretes and takes longer to reach equilibrium. These volume changes cause additional time-dependent movement above the immediate elastic deflection for all concretes. In general, these differences are small and North American practice leads to the same factor for structural lightweight concrete as that of ordinary concretes. The multiplier of the elastic deformation for additional long-time deflections for flexural members (lightweight and ordinary) is obtained by multiplying the immediate deflection caused by the sustained load considered, by the factor

$$\left(2 - 1.2 \times \frac{\text{area of compression reinforcement}}{\text{area of non-prestressed tension reinforcement}} \geq 0.6\right)$$

More refined approaches to estimating deflection are, in general, not warranted.

Structural lightweight concrete is particularly adaptable to seismic design because of the significant reduction in dead weight. Many multistory buildings and bridge structures have effectively utilized lightweight concrete in areas subject to earthquakes, especially in high seismic activity areas of the United States and Japan. The lateral forces activity upon a structure during earthquake motions is directly proportional to the inertia of that structure which in turn is related to the mass. These lower lateral forces may be calculated by recognized formulas and are applied with the other load factors.

5.3   Tensile strength and shear resistance

The air-dry tensile strength of structural lightweight concrete is lower than that of normal-density concrete of the same strength. In uniaxial tension, cracks will form at lower stresses than in ordinary concrete, but in flexural members the lower modulus of deformation will partly compensate for this by lowering the neutral axis and hence reducing the tensile strain. In restrained members subject to volume changes, the lower modulus and coefficient of expansion of lightweight concrete will cause the development of lower stresses. Lightweight concretes, subject to shear and diagonal tension, behave in the same manner as normal-weight concrete members. The lower tensile strengths are accommodated in some codes by reduced diagonal tensile capacity factors [8] (0.75 for all lightweight aggregate and 0.85 for sanded lightweight aggregate concretes). Similar reduction factors are promulgated for torsional strength and modulus of rupture. Some codes have provided for the opportunity of determining specific modification factors [8] that are developed by the insertion of a laboratory-tested indirect tensile splitting strength in the formula: $f_{ct}/6.7\sqrt{f_c'}$, where f_{ct} is the tensile splitting strength, and f_c' the cylinder compressive strength.

Since reductions in self-weight lead to a significant reduction in total load on lightweight concrete members, shear capacity reduction by as much as 25% in ordinary concretes does not necessarily lead to an increase in web reinforcement, or, for that matter, to a decrease in relative structural efficiency.

5.4   Development of reinforcement

North American practice provides for a modification factor for an increased development length of reinforcement on all lightweight aggregate and sanded lightweight aggregate concretes that parallels the tensile strength portions of the code.

5.5   Thermal design considerations

In concrete elements exposed to environmental conditions, the use of expanded aggregates will provide several distinct advantages over natural aggregate concrete [42], including:

(a) The lower diffusivity provides a greater thermal inertia that lengthens the time for exposed members to reach any steady-state temperature.

(b) Due to this resistance, the effective interior temperature change will be smaller under transient temperature conditions. This greater time lag will moderate the transient temperature effects.

(c) The lower coefficient of linear thermal expansion that is developed in the concrete due to the contribution of the lower expansion characteristics of the expanded aggregate itself is a fundamental design consideration in exposed members. The expansion and contraction of exposed columns of tall buildings induce shearing forces and bending moments into floor frames connected to interior members that are subject to unchanging interior structural members. The architectural decision to locate glass window lines must of necessity take into account the conductivity and the expansion coefficients of the exposed concretes.

(d) The lower modulus of deformation will develop lower stress changes in members exposed to thermal strains.

A comprehensive thermal investigation [43] studying the shortening developed by the average temperature of an exposed column restrained by the interior frame concluded that the axial shortening effects were about 30% smaller and the stresses due to restrained bowing were about 35% less when using structural lightweight concrete. The analysis, conducted on a 20-story concrete frame, used the following assumptions:

	Ordinary concrete	Lightweight concrete
Thermal conductivity	$1.73 \text{ W m}^{-1} \text{ K}^{-1}$ ($12 \text{ Btu h}^{-1} \text{ ft}^{-2} \, {}^\circ\text{F}^{-1} \text{ in}^{-1}$)	$0.72 \text{ Wm}^{-1} \text{ K}^{-1}$ ($5.0 \text{ Btu h}^{-1} \text{ ft}^{-2} \, {}^\circ\text{F}^{-1} \text{ in}^{-1}$)
Coefficient of linear thermal expansion	9.9×10^{-6} per $^\circ$C (5.5×10^{-6} per $^\circ$F)	8.10×10^{-6} per $^\circ$C (4.5×10^{-6} per $^\circ$F)
Modulus of deformation for $f'_c = 28$ MPa (4000 psi)	25.000 MPa (3.6×10^6 psi)	17.000 MPa (2.5×10^6 psi)

Exact analysis requires the knowledge of physical property data on local aggregate obtained from expanded and ordinary aggregate suppliers.

Numerous practical examples of the isotherms and average temperatures developed in exposed columns made in lightweight as well as ordinary concretes are included, demonstrating the practical considerations of how the thermal inertia of structural lightweight concrete also serves to minimize condensation. With concrete frame buildings reaching for greater heights, the structural approach to controlling the temperature movements requires an exact understanding of the contribution of the superior thermal response of structural lightweight concrete.

5.6 Design of prestressed lightweight concrete

Prestressed lightweight concrete has been used extensively throughout the world in roofs, walls, and floors of buildings, in bridge decks and in nearly every application in which prestressed ordinary-weight concrete has been employed. For these applications, the reduced dead weight with its lower structural, seismic and foundation loads, the better thermal insulation and better fire resistance have usually been the determining factors in the selection of prestressed lightweight concrete [44].

Prestressed lightweight concrete has been used in several variations of composite action with ordinary-weight concrete, including

(a) Prestressed lightweight concrete beams with ordinary weight concrete cast-in-place slabs.
(b) Prestressed lightweight concrete beams with lightweight concrete cast-in-place slabs.
(c) Prestressed ordinary weight concrete beams with cast-in-place lightweight concrete slabs.

Prestressed lightweight concrete must be composed of aggregates of high quality (particle strength, rigidity and gradation) and the resulting concrete should have high strength. The following is a summary of the usual properties of prestressed lightweight concrete:

(a) Density. The range is usually 1600–1920 kg/m^3 (100–120 lb/ft^3).
(b) Compressive strength. In general, high-strength concrete is used in prestressed applications with commercial strength ranging from 28 to 40 MPa (4000–6000 psi)
(c) Modulus of deformation. When accurate values of the modulus are required, it is suggested that laboratory tests be conducted on the mix proposed for the application. In general, most code formulas for the modulus of deformation overestimate the modulus of high-strength structural lightweight concretes. This overestimation is caused by the increasing influence of the elastic mismatch of high-modulus mortar fractions compared to the modulus of deformation of the expanded aggregate particles.
(d) Combined loss of prestress. This is usually about 110 to 115% of the total losses for normal-weight concrete when both are subjected to normal curing; 124% of the total losses when both are subjected to steam curing.
(e) Thermal properties. The greater thermal insulation of lightweight aggregate concrete has a significant effect on prestressing applications, because of the following factors: greater temperature differential in service between the side exposed to sun and the inside may cause greater camber; better response to steam curing; greater suitability for winter concreting; better fire resistance resulting in thinner slabs.
(f) Dynamic considerations. Prestressed lightweight concrete appears to behave similarly to normal-weight concrete and may be better in some applications due to its greater resilience and lower modulus of elasticity.
(g) Cover requirements. Concrete cover as protection of reinforcement against weather and other effects is essentially the same for lightweight concretes as that for ordinary concretes. Where fire requirements dictate the cover requirements, the insulating effects developed by the lower density, as well as the fire stability offered by a pre-burned aggregate, may be used to considerable advantage.

5.7   General specifications

Lightweight concretes may be specified and proportioned on the basis of laboratory trial batches or on the statistical analysis of field experience. Most expanded
aggregate suppliers have mix proportioning information available for their material, and many producers provide field control and technical service to assure the
quality of field concrete. The relationship between targeted and specified
minimum strength requirements for lightweight concrete is the same as that for
ordinary concretes with the same degree of field control. The analysis of the
load-carrying capacity of a lightweight concrete structure either by cores or load
tests is the same as that for ordinary concrete. Splitting tensile strength tests
should not be used as a basis for field acceptance of lightweight concrete.

   The fresh density of expanded aggregate concretes is a function of mix
proportioning, field air contents, water demands and the particle density and
moisture content of the expanded aggregate as batched. As the concrete matures,
the density will decrease due to moisture loss, which in turn will be a function of
ambient conditions and the surface area/volume ratio of the member. Specifications should be in terms of fresh density, as limits of acceptability should be
controlled at placement. Design loads should be based upon equilibrium density
which for most conditions and members may be assumed to be reached at 2
months (Fig. 5-1). Extensive tests [24] conducted during a North American
durability study demonstrated that, despite wide initial variations of aggregate
moisture content, equilibrium density may be determined by adding 50 kg/m^3
(3 lb/ft^3) to the density obtained after complete drying out at a temperature of
105°C (221°F). European recommendations [2] are similar.

   Where cement and dry aggregate batch weights are known, the design density

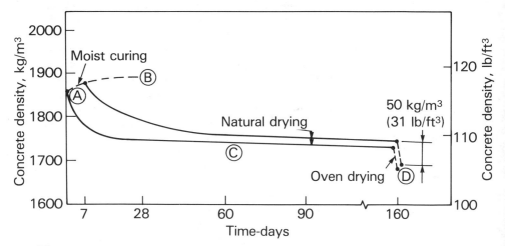

Figure 5-1   Density versus time of drying for sanded lightweight concrete
A Fresh density-field and specification control
B Density gain due to moist curing
C Equilibrium density (approximately 60 days)
D Oven-dry density

may be calculated by

$$\rho = A + 1.2C + 50 \text{ kg/m}^3$$

where A is the dry weight of aggregate content, and C the cement content in kg/m^3 of compacted concrete.

For severe exposure conditions, use high-quality structural expanded aggregates that have a proven record of durable performance in severe weathering exposure. Use a minimum cement content of 360 kg/m^3 (611 lb/yd^3) for structures with moderate exposure (building facades, garages) and 440 kg/m^3 (748 lb/yd^3) for concretes with severe exposure (bridge decks, marine environments). Use minimum air contents of 6% for 20 mm ($\frac{3}{4}$ in), and 7.5% for 10 mm ($\frac{3}{8}$ in) aggregate. Place concretes at reasonable slumps of 75 mm (3 in), using the best recommended practice for compaction, finishing and curing.

Acknowledgements

The two most authoritative sources of information, the ACI *Guide for structural lightweight concrete* [1] and the CEB–FIP *Manual of design and technology: lightweight aggregate concrete* [2], form the basis of this chapter and are heavily referred to. In general, the selection of methodology was determined with the view of providing the reader with optimum guidelines for the use of a more structurally efficient alternative to ordinary concrete. References to specific building codes or national standards are incorporated where these recommendations have resulted in the satisfactory field service of numerous structural lightweight concrete structures.

Specifically the author and publishers gratefully acknowledge permission to use copyright material as follows: from the American Concrete Institute to reproduce two figures from *Guide for structural lightweight aggregate concrete* (ACI 213 R-79) and two photographs from *Performance of concrete in a marine environment*, SP65 (1980); from Longman Group Limited for the reproduction of one photograph and one diagram in the *Proceedings of the Second International Congress on Lightweight Concrete* published by Construction Press (1980); and from Longman Group Limited and the Comité Eurointernational du Béton for the use of six figures appearing in the *CEB/FIB Manual of design and technology: lightweight aggregate concrete* published by Construction Press (1977).

References

1. ACI Committee 213, *Guide for structural lightweight aggregate concrete*, ACI 213 R-79, American Concrete Institute, Detroit, 1979, pp. 33–62.
2. CEB–FIP, *Manual of design and technology: lightweight aggregate concrete*, 1977, 168 pp.
3. Bremner T. W., *Influence of aggregate structure on low density concrete*, Doctoral Thesis, Imperial College of Science, London, June 1981, 225 pp.
4. New York City Housing Authority, *Lightweight concrete testing program*, Unpublished report, May 1964, 368 pp.

5. Muller-Rochholz, J., Determination of the elastic properties of lightweight aggregate by ultrasonic pulse velocity measurement, *Int. J. Lightweight Concr.*, Vol. 1, No. 2, Dec. 1979, pp. 87–90.

6. ACI Committee 211, *Recommended practice for selecting proportions for structural lightweight concrete*, ACI 211.2–69, Revised 1977, American Concrete Institute, Detroit, 1977, pp. 21–39.

7. Campbell, R. H. and Tobin, R. E., Core and cylinder strengths of natural and lightweight concrete, *J. Am. Concr. Inst., Proc.*, Apr. 1967, pp. 190–195.

8. ACI Committee 318, *Building code requirements for reinforced concrete*, ACI Standard 318–77, American Concrete Institute, Detroit, 1977, pp. 1–103.

9. Holm, T. A., Physical properties of high strength lightweight aggregate concrete, *2nd Int. Congr. Lightweight Concr.*, London, April 1980; pp. 187–204.

10. Schulz, B., Entwurf mittel und hochfester Leichtbeton Mischungen, *Beton*, No. 3, 1978, pp. 95–97.

11. Muller-Rochholz, J., *Influence of lightweight aggregate characteristics on the strength of lightweight concrete*, Doctoral Dissertation, Technische Hochschule, Aachen, 1978, 322 pp.

12. Khan, F. R., Stockbridge, G. and Brown, E. J., Quality control of high strength light-weight concrete for One Shell Plaza, *Lightweight concrete*, American Concrete Institute, SP-29, 1971, pp. 15–34.

13. Oleszkiewicz, S., The analysis of lightweight concrete deformation, *CEB Symposium*, Cracow, Special Publication No. 3, 1973, pp. 60–70.

14. Reichard, T. W., Creep and drying shrinkage of lightweight and normal weight concretes, *Monogr. Natn Bur. Stand.*, No. 74, Mar. 1965, 30 pp.

15. Berg, O., Several features of investigations of the strength and deformation properties of lightweight concretes, *Proc. conf. agloporite and agloporite concrete*, Minsk, 1964, pp. 191–198 (In Russian).

16. Holm, T. A. and Bremner, T., *Homogeneous concrete—elastic matching of aggregates and mortar*, Unpublished report.

17. Grieb, W. E. and Werner, G., Comparison of splitting tensile strength of concrete with flexural and compressive strengths, *Proc. Am. Soc. Test. Mater.*, Vol. 62, 1962, pp. 972–995.

18. Hanson, J. A., Tensile strength and diagonal tension resistance of structural light-weight concrete, *J. Am. Concr. Inst., Proc.*, Vol. 58, No. 1, July 1961, pp. 1–39.

19. Hanson, J. A., Effects of curing and drying environments on splitting tensile strength of concrete, *J. Am. Concr. Inst., Proc.*, Vol. 65, No. 40, July 1962, pp. 535–543.

20. Shideler, J. J., Lightweight aggregate concrete for structural use, *J. Am. Concr. Inst., Proc.*, Vol. 54, No. 4, Oct. 1957, pp. 299–328.

21. Weigler, H. and Karl, S., *Reinforced lightweight concrete–manufacture, properties and design*, Bauverlag, Wiesbaden, 1972, 260 pp.

22. Forschungsinstitut der Zementindustrie, *Shrinkage and creep tests on lightweight concretes made with Liapor lightweight aggregate*, Düsseldorf, 1970.

23. Klieger, P., and Hanson, J. A., Freezing and thawing tests of lightweight aggregate concrete, *J. Am. Concr. Inst., Proc.*, Vol. 57, No. 38, Jan. 1961, pp. 779–796.

24. Expanded Shale Institute, *Freeze–thaw durability of structural lightweight concrete*, Lightweight Concrete Information Sheet No. 13, 1970, 4 pp.
25. Weigler, H. and Karl, S., Salt scaling resistance and abrasion resistance of lightweight concretes, *Betonstein-Zeitung*, Vol. 34, 1968, No. 5, pp. 225–240, No. 11, pp. 581–583.
26. Holm, T. A., Performance of structural lightweight concrete in a marine environment, *Performance of concrete in marine environment*, American Concrete Institute, SP-65, 1980, pp. 589–608.
27. Expanded Shale Institute, *Bridge deck survey*, Bethesda, 1960, 23 pp.
28. Haynes, H. H., Permeability of concrete in sea water, *Performance of concrete in marine environment*, American Concrete Institute, SP-65, 1980, pp. 21–38.
29. Valore, R. C., Jr, Calculation of U-values of hollow concrete masonry, *J. Am. Concr. Inst., Proc.*, Vol. 2, No. 2 Feb. 1980, pp. 40–63.
30. Schüle, W. and Kupke, *Thermal conductivity of expanded clay concretes without and with addition of quartz sand*, Institut für Bauphvsik, Stuttgart, 1972, pp. 17–24.
31. Valore, R. C., Jr, and Holm, T. A., *Thermal conductivity of concrete as a function of its constituents*, Unpublished report.
32. Valore, R. C., Jr, Insulating concretes, *J. Am. Concr. Inst., Proc.*, Vol. 53, No. 5, Nov. 1956, pp. 509–532.
33. Price, W. H. and Cordon, W. A., Tests of lightweight aggregate concrete designed for monolithic construction, *J. Am. Concr. Inst., Proc.*, Vol. 45, No. 8, Apr. 1949, pp. 581–600.
34. Monfore, G. E. and Lentz, A. E., Physical properties of concrete at very low temperatures, *J. Portld Cem. Ass. Dev. Labs.*, Vol. 4, No. 2, May 1962, 24 pp.
35. Petersen, D. H., Burned shale and expanded slag concretes with and without air entraining mixtures, *J. Am. Concr. Inst., Proc.*, Vol. 45, No. 1, Oct. 1948, pp. 165–174.
36. Abrams, M. S. and Gustaferro, A. H., Fire endurance of concrete slabs as influenced by thickness, aggregate type and moisture, *J. Res. Dev. Labs*, Vol. 10, No. 2, May 1968, 24 pp.
37. Abrams, M. S., Compressive strength of concrete at temperatures to 1600°F, *Temperature and concrete*, American Concrete Institute, SP-25, Detroit, 1971, pp. 33–58.
38. ACI Committee 318, *Building code requirements for reinforced concrete*, ACI Standard 318-63, American Concrete Institute, Detroit, 1963, 137 pp.
39. Gerritse, A., Design considerations for reinforced lightweight concrete, *Concrete International, London*, Construction Press, Lancaster, 1980, pp. 57–69.
40. Pfeifer, D. W., Reinforced lightweight concrete columns, *J. Struct. Div. Am. Soc. Civ. Eng.*, Vol. 95, No. ST1, Jan. 1969, pp. 57–82.
41. Holm, T. A. and Pistang, J., Time-dependent load transfer in reinforced lightweight concrete columns, *J. Am. Concr. Inst., Proc.*, Vol. 63, No. 11, Nov. 1966, pp. 1231–1246.
42. Fintel, M. and Khan, F. R., Effects of column exposure in tall structures, Temperature variations and their effects, *J. Am. Concr. Inst., Proc.*, Dec. 1965, pp. 1533–1556; Analysis for length changes of exposed columns, Aug. 1966

pp. 843–864; Design considerations and field observations of buildings, Feb. 1968, pp. 99–100.

43. Fintel, M., *Thermal investigation*, Unpublished communication, Portland Cement Association Skokie.

44. Report of the FIP commission on prestressed lightweight concrete, *J. Prestressed Concr. Inst.*, Vol. 12, No. 3, June 1967, 27 pp.

8 Polymers in concrete

David W Fowler

University of Texas at Austin, USA

Contents

Notation

C	creep coefficient
c	distance from compression face to neutral axis
E	modulus of elasticity
f'_{cp}	ultimate compressive strength
f_{min}	minimum flexural stress
f_{max}	maximum flexural stress
f_r	modulus of rupture
f_t	splitting tensile strength
K	wobble coefficient
N	number of cycles
p	polymer loading
R	relaxation of prestressing
T	time (days)
μ	friction coefficient

Summary

Three categories of concrete-polymer material are generally recognized: polymer-impregnated concrete (PIC), polymer concrete (PC), and polymer Portland cement concrete (PPCC). PIC is made by impregnating cured Portland cement concrete with a low viscosity monomer which is subsequently polymerized. Fully-impregnated components have excellent compressive strength and durability. Partial-depth impregnation has been used on bridges and hydraulic structures to improve durability. PC is produced by using a polymer as a binder for aggregate. PC has been widely used to repair Portland cement concrete in pavements, bridges, and hydraulic structures and to produce precast components such as building panels, flooring, insulation, drains, facing panels, and manholes. PC has very good strength and durability properties. PPCC is made by adding a monomer or polymer to fresh concrete. These materials generally have improved strengths and very good durability properties. PPCC is widely used for floor and bridge overlays and for repairing Portland cement concrete. The chemicals used to produce concrete polymer materials are often volatile, flammable, and toxic. Care must be exercised in the use of these materials in construction.

1   Introduction

Concrete–polymer materials *are relatively new* in the construction industry. There are several composites that offer very good mechanical and durability properties. There are three basic types of concrete-polymer materials.

(a) *Polymer-impregnated concrete* (PIC) is 'a hydrated Portland cement concrete which has been impregnated with a monomer and is subsequently polymerized *in situ*' [1].

(b) *Polymer concrete* (PC) is 'a composite material formed by polymerizing a monomer and aggregate mixture. The polymerized monomer acts as the binder for the aggregate' [1].

(c) *Polymer–Portland cement concrete* (PPCC) is 'a premixed material in which either a monomer or polymer is added to a fresh concrete mixture in a liquid, powdery or dispersed phase, and subsequently allowed to cure, and if needed, polymerized in place' [1]. The terms polymer cement concrete (PCC) and resin concrete have also been used in the literature to refer to this same material.

Monomers and polymers are used for the purpose of changing the properties of the final material. Polymers used in small amounts to modify the properties of fresh concrete are defined as admixtures and are not considered in this chapter.

Concrete–polymer materials were not widely known until the early 1970s. However, PC was used in the early 1950s to produce floor tile and prefabricated wall panels. PPCC in the form of a latex which could be added to mortar was also developed in the 1950s. These mortars were used to produce prefabricated brick panels and bridge deck overlays. A patent for PIC was issued in the Soviet Union in 1954, but apparently little work was performed on PIC until after research performed in the United States was published [2].

Concrete–polymer materials received much wider recognition after the results of research begun in 1966 and sponsored by the United States government were published [3–7]. The thrust of the research was directed toward PIC after the initial tests indicated that the structural and durability properties of PIC were much greater than the concrete from which it was produced. For several years, most of the research in polymers in concrete was performed in PIC.

In recent years, PC has received considerable attention from researchers. Applications include the repair of concrete, overlays, pipe, and utility components. PPCC has been used most widely in overlays, but has not received as much publicity, perhaps, as the other two materials since the addition of many monomers to the fresh concrete has not been successful in improving the properties of the hardened concrete. However, the concept of PPCC continues to have considerable appeal because of the similarity to process technology for conventional Portland cement concrete. Reference [65] is a good summary of current research in concrete polymer materials.

Sulfur concrete and sulfur-infiltrated concrete are generally included in the broad definition of concrete–polymer materials. However, these materials will not be discussed in this chapter.

2 Definitions

A *monomer* is a small, simple molecule which can be chemically linked together into a long, repeated chain-like structure with a high molecular weight which is known as a *polymer*. The chemical process of linking the molecules together is known as *polymerization*. If more than one kind of monomer is used, the process is called *copolymerization* and the resulting material a *copolymer*.

Inhibitors are materials added to monomers to prevent premature polymerization during shipping and storage. *Initiators*, also referred to as *catalysts*, are chemicals which, when added to monomers, start the polymerization process. *Promoters*, or *accelerators*, are chemicals used to accelerate the polymerization process.

Polymerization can be achieved by several methods. *Thermal-catalytic* polymerization is accomplished by heating a monomer to which small amounts of initiator have been added; free radicals are generated. This method is widely used for PIC, especially for field applications.

Promoted-catalytic polymerization utilizes a promoter in the monomer to increase the decomposition rate of the initiator. This method permits polymerization to occur at ambient temperatures, without the addition of external heat, at temperatures of 0°C or higher. This method is used for producing PC. Polymerization can also be accomplished by the use of ionizing *radiation* such as gamma radiation emitted by cobalt-60. Initiators and promoters are not required. This method is used in the production of PIC, especially in laboratory or manufacturing facilities [1].

3   Polymer-impregnated concrete

PIC is produced by drying the precast Portland cement concrete to remove moisture, full or partial saturation with monomer, and *in situ* polymerization. The precast Portland cement concretes used to produce PIC can be made from practically all types of aggregate, cement and admixture that are used in current concrete practice. Any method of curing the concrete may be used although high-pressure steam-cured concrete usually produces the highest strength PIC. Dense concretes require less monomer than lower quality, more porous concretes or concrete made with lightweight aggregate.

3.1   Monomers

Many monomers have been used to produce PIC, although vinyl monomers, e.g. methyl methacrylate, styrene, and acrylonitrile, have been the most widely investigated. Methyl methacrylate (MMA) has been the most widely used monomer because of its low viscosity (less than 1 cP), relatively low cost, and good mechanical properties. It is commonly used to produce transparent covers for lighting fixtures, acrylic sheets, and for encapsulating objects such as coins. For thermal-catalytic polymerization a commonly-used formulation is 95 wt% MMA and 5 wt% trimethylol propane trimethacrylate (TMPTMA), a multi-functional monomer known as a cross-linking agent. TMPTMA is used to increase the polymerization rate, especially for field applications. A small amount (0.5 to 1 wt%) of initiator, e.g. benzoyl peroxide (BzP) or azobis (isobutyronitrile), is used to generate free radicals when heat is supplied.

3.2   Impregnation processes

3.2.1   *Full impregnation*
Full impregnation of a concrete specimen is usually feasible only in laboratory or plant facilities since the entire specimen must be dried and fully saturated with monomer. Drying temperatures must be in excess of 100°C, and 150°C is near optimum. Temperatures in excess of 150°C may result in decreased strengths. Drying should be uniform and the drying time will be dependent on the thickness of the concrete. Large specimens may require careful heating and cooling techniques to prevent cracking.

Monomer saturation may begin when the concrete is cooled to a temperature that will not cause premature polymerization, usually about 38°C or less for MMA.

Evacuation of air in the concrete by use of a vacuum chamber, immersion in monomer, and application of a modest overpressure (10 psi or 70 kPa) reduce the time required for complete saturation. Without evacuation, overpressures approximately ten times higher must be applied to achieve saturation in about 1 h.

The following process has been shown to provide high quality PIC using dense concrete specimens with a cross-sectional dimension of up to 300 mm [1]:

(1) Dry to constant weight at 105°C, which requires about 24 h.
(2) Maintain vacuum of ~30 inHg (100 kPa) for about 30 min. Figure 3–1 shows a precast concrete bridge slab being loaded into a vacuum chamber which also serves as the impregnation chamber.

Figure 3-1   Precast concrete bridge slab being loaded into vacuum/impregnation Chamber (*Courtesy* US Bureau of Reclamation)

(3) Apply monomer under vacuum and then pressurize to 10 psi (70 kPa). If vacuum is not used, pressure-soak for 60 min at 100 psi (700 kPa).
(4) Remove section from monomer and place under water, or drain monomer from impregnation chamber and then fill with water.
(5) Polymerize monomer *in situ* with radiation, hot water or steam.
(6) Remove water and clean specimen.

Several encapsulation techniques have been used to minimize monomer evaporation losses during polymerization. Wrapping the specimen in polyethylene and/or immersing the specimens in water has been found to be effective [1].

3.2.2   Partial-depth impregnation

Extensive research has been conducted to develop processes for impregnating conventional Portland cement concrete to depths of 0.5–1.5 in (10–40 mm) [9–18]. Partial-depth impregnation (sometimes called surface impregnation) is desirable when durability rather than strength is of primary concern. Partial-depth impregnation can be performed under many field conditions whereas full-depth impregnation may not be feasible because of the difficulty of obtaining full-depth drying, evacuation, and applying overpressure. It should be noted that partial-depth impregnation will not result in polymer loadings within the pores as high as obtained when the full-impregnation process is used, and the mechanical properties of the impregnated zone will not be as high.

Based on the experience gained in the bridges impregnated prior to November 1977, recommendations for bridge deck impregnation include [17]

(a) *Surface preparation.* The surface should be clean of contaminants, such as oil, grease, or dirt, that would retard penetration of the monomer into the concrete. Solvents or sandblasting are effective in removing some contaminants.

(b) *Drying.* Clean concrete sand should be placed to a depth of $\frac{3}{8}-\frac{1}{2}$ in (10–13 mm) prior to drying. The slab should be dried full-width or curb-to-curb for 6–8 h at a surface temperature of 250–300°F (121–139°C). The temperature rise at the surface should not exceed 100°F h (55°C h) with the variation on the surface not to exceed ±25°F (±14°C). Figure 3-2 shows infrared heaters in place for drying concrete.

(c) *Cooling.* The slab should be cooled to a rate not exceeding 100°F h (38°C h) to a surface temperature of 100°F (38°C) or less.

(d) *Monomer system.* The monomer system should consist of 95 wt% MMA, 5 wt% TMPTMA, and 0.5 wt% AIBN or AMVN initiator.

(e) *Monomer application.* Monomer should be applied at a rate of 0.6–1.0 psf (30–50 N/m^2) depending upon polymer depth required. The monomer should be permitted to soak for 4–6 h prior to polymerization. Figure 3-3 shows monomer being applied to a sand-covered slab. A membrane should be placed over the area during soaking.

(f) *Polymerization.* A surface temperature of 150–190°F (66–88°C) should be applied for 2–4 h for polymerization. No open-flame heat sources or

Figure 3-2   Infrared heaters in place for drying concrete

Figure 3-3 Monomer applied to sand-covered slab (*Courtesy* US Army Corps of Engineers)

high-temperature elements which could cause combustion of the monomer vapor should be used. Steam heat is preferable.

Inadequate depth of impregnation may result if any of these requirements are not met.

The determination of polymer depth and quality of impregnation have been made by different techniques. Generally, the depth of impregnation can be made visually, as shown in the top of the impregnated concrete in Fig. 3-4 [17]. Other techniques that have proven successful are acid etching [3, 19]; color enhancement by use of phenolphthalein in conjunction with a microscopic examination [19]; petrographic examination of polished sections viewed with reflected polarized light under the microscope [19]; nondestructive resistivity measurements [20]; thermal analysis [3]; and pyrolysis coupled with infrared spectroscopy [21].

The improvement in durability of partial-depth polymer-impregnated concrete has been found to be very significant. Freeze–thaw resistance has been found to be significantly improved because of the increased resistance to water penetration [13, 22]. Skid resistance has been shown to be equal or better than that for unimpregnated controls [10, 13]. Resistance to chloride intrusion is also increased significantly [21].

Surface cracks were reported in some bridges after the impregnation process, apparently due to the absence of the thin layer of sand on the surface during drying [16]. The sand seems to minimize thermal shock during the application of

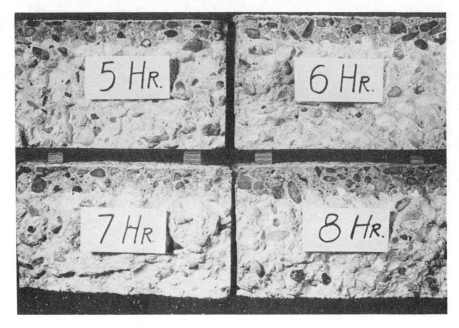

Figure 3-4   Depth of impregnation as function of soak time (University of Texas)

heat and the thermal gradient during the cool down. A crack survey made before and after the impregnation on one large bridge, in which a sand cover was used during drying, indicated 6% more cracks after impregnation. However, the cracks, which were hairline in width, could not be conclusively attributed to the impregnation process [17]. Recent unpublished studies indicate that some partial-depth impregnated bridge slabs may permit chloride intrusion through the hairline cracks. Currently, there is a moratorium on partial-depth impregnation of bridge decks on which deicing salt will be used.

3.3   Properties

The strength and durability properties of PIC, particularly PIC produced by the full-impregnation process, are generally much greater than for the unimpregnated concrete (Table 3-1). It should be noted that the values shown in Table 3-1 are primarily for fully-impregnated PIC made with MMA. The stress–strain relationship for PIC produced with MMA has been shown to be very linear, with practically no ductility. The addition of monomers with lower glass transition temperatures (T_g) result in much more ductile behavior. Figure 3-5 shows the stress–strain curves for PIC made with varying combinations of MMA and butyl acrylate (BA) [8]. The stress–strain curves were generated from repeated high-intensity loading of PIC cylinders. It can be seen that the ultimate stress reduces and the ultimate strain increases with increasing percentages of butyl acrylate.

Table 3-1   Typical properties of concrete–polymer materials

Property	PPCC	Ordinary concrete	PIC	PC
Compressive strength, psi	4000–8000	4000–5000	15–22 000	6000–20 000
Tensile strength, psi	600–900	300–350	1200–1600	1000–2000
Modulus of rupture, psi	1400–1800	400–600	2000–2700	1200–4000
Modulus of elasticity, psi	$1.5–2.2 \times 10^6$	$3–4 \times 10^6$	$4–6.5 \times 10^6$	$2–4 \times 10^6$
Abrasion resistance, factor of improvement	10	—	—	—
Water absorption	—	5–6	0.3–0.6	—
Freeze–thaw resistance, no. of cycles/ % wt loss	—	700/25	2000–4000/0–2	1500/0–1

Most of the increases in durability properties can be explained by the fact the pores are partially filled with polymer, which greatly restricts the intrusion of water or acids into the concrete. Both strength and durability are strongly dependent upon the fraction of the porosity of the cement phase which is filled with polymer [4].

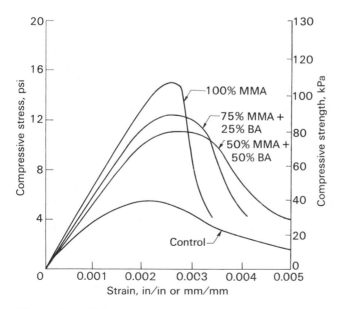

Figure 3-5 PIC stress–strain curves made with varying combination of MMA and BA [8]

3.4   Structural behavior and design

The high compressive and tensile strengths, high modulus of elasticity, and low creep make PIC a potentially promising material for post-tentioned beams. Tests on I-shaped post-tensioned MMA—PIC beams (Fig. 3–6) with a variable number of wires in each tendon have been reported [8, 23]. Figure 3–7 shows the load–deflection response for an unimpregnated control (PBC-2) and four PIC beams. The number in each beam designation indicates the number of 0.25 in (6.4 mm) high-strength wires in each tendon. The compressive strength of the control was 5500 psi (37.9 MPa) and for the PIC was 15 100 psi (104 MPa). An assumed triangular stress distribution used in the deflection prediction produced good agreement with experimental results. A rectangular stress block with an equivalent stress of $0.90f'_{cp}$, where f'_{cp} is the ultimate compressive stress, and a depth of $0.70c$, where c is the distance from the top of the beam to the neutral axis, were found to yield good prediction of ultimate moment. Other tentative recommended design criteria from reference [8] are given in Table 3-2. These values apply to the parameters used in this study.

A theoretical investigation was reported on the ultimate strength and ductility of PIC beams and columns with both brittle and ductile stress–strain behavior. The strength and ductility of the members were found to be improved [24]). Reference [25] gives the results of experimental testing on a PIC specimen with varying ductility. Beams, columns, and cones were reinforced with steel fibers and mesh and a combination of polymer and steel. It was concluded that the impregnation of concrete members reinforced with mesh or fibers with a ductile copolymer results in high strengths, high moduli and high ductility.

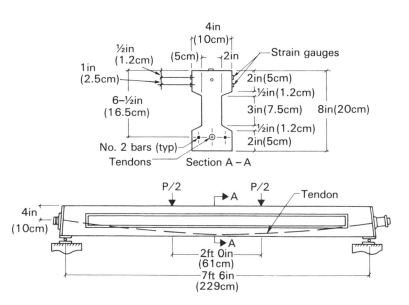

Figure 3-6   Dimensions and loading for post-tensioned PIC beams [8]

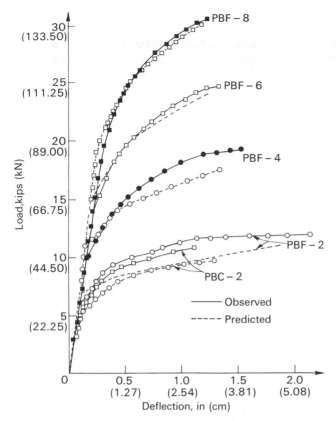

Figure 3-7 Load–deflection response of unimpregnated control (PBC-2) and PIC post-tensioned beams. The number alongside each set of curves indicates number of wires) [8]

3.5 PIC applications

Bridges throughout the United States have been surface-impregnated, and on several the contractors were selected by competitive bidding for a price of $31/m^2$ or less. Costs have varied widely as a result of differences in bridge size, working conditions, location and process. References [14, 16–18] describe the impregnation of several bridges. However, for the reasons discussed in Section 3.2, bridge-deck impregnation is not currently being used in areas where deicing salts are applied.

Impregnation techniques were developed for the walls of a dam outlet which were repaired after being seriously damaged by cavitation and erosion. Infrared heaters were used to dry 10×10 ft (3×3 m) areas on each wall. After cool-down, stainless steel panels were pressed tightly against the walls to provide a space of about 3 mm between panels and wall to hold the monomer during soaking (Fig. 3–8). The monomer was permitted to soak for 4–6 h and then drained. The

Table 3-2   Recommended design criteria for PIC beams[a] [8]

Criterion	Recommended value
Compressive strength, psi	$f'_{cp} = 1050 + 2450p$, where $p = \%$ polymer loading
Modulus of rupture, psi	$f_{rp} = 6\sqrt{f'_{cp}}$
Tensile strength, psi	$f_{tp} = 10\sqrt{f'_{cp}}$
Modulus of elasticity, psi	$E = 2\,150\,000 + 0.0003f'_{cp}$
Wobble coefficient	
Waterproof paper wrapping	$K = 0.0050$
Sheathed metal conduits	$K = 0.50$
Friction coefficient	
Waterproof paper wrapping	$\mu = 0.0005$
Sheathed metal conduits	$\mu = 0.10$
Creep coefficient	
$0.15f'_{cp} < f_{cp} \leq 0.3f'_{cp}$	$C = 0.36T/(24 + T)$, where $T =$ time in days
$f'_{cp} \leq 0.15f'_{cp}$	$C = 0.30T/(24 + T)$
Relaxation of prestressing Steel[b], per cent	$R = 3.15T/(1.70 + T)$

[a] Limited to PIC made from MMA
[b] 0.25 in (6.4 mm) grade 240 prestressing steel

Figure 3-8 Monomer impregnation chamber for vertical walls [40]
(*Courtesy* US Army Corps of Engineers)

panels were filled with hot water, which was kept hot by electrical resistance heaters attached to the backs of the panels. The hot water was used to polymerize the monomer to a depth of $\frac{3}{4}$–$1\frac{1}{4}$ in (19–32 mm). Later inspections have revealed that the walls have performed very well, as has the floor of the stilling basin, which was impregnated to a depth of about $\frac{1}{2}$ in (13 mm) [40, 41].

The use of polymer impregnation to restore the structural integrity of a building has been demonstrated [49]. A badly-deteriorated ceiling slab of a jail was impregnated with a MMA system to provide acceptable strengths at a cost of about $97/m^2.

Current applications of fully-impregnated PIC are very limited, although many applications have been proposed. Precast PIC bridge decking has been investigated [42] in the United States. Initially, 16 ft × 4 ft × 6 in (4.88 m × 1.22 × 150 mm) panels with tongue-and-groove joints were pretensioned and then fully impregnated. Current work is underway to construct post-tensioned bridge decking. A PIC tunnel support and lining system made of seven segments which formed a ring section with a diameter of 8 ft (2.4 m) has been developed. Tests on a PIC system which had a segment thickness of 2 in (51 mm) and $3\frac{1}{2}$ in (89 mm) of concrete back-fill could support 60% more load than a similar system composed entirely of conventional concrete [43].

PIC reinforced beams have been tested to determine their potential as a replacement for wood supports in mines [44]. Beams made with lightweight aggregate concrete and normal-weight concrete were made and subjected to flexural tests. A system of PIC crib supports has been developed for undergound mine supports.

Tests on pipe made of PIC, conventionally-reinforced and steel fiber-reinforced pipe, were conducted. Three-point and hydrostatic burst tests were conducted on full-scale sections with a diameter of 40 in (914 mm). Durability tests were performed on cores from the pipe subjected to attack by sulfate ions and sulfuric acid. PIC pipe had the greatest strength and durability of the three materials tested [45].

The use of PIC for concrete sea structures has been studied. Concrete ships, offshore structures, underwater oil-storage vessels, and ocean thermal-energy plants were included in the study [46]. Spherical model hulls have been tested, and a cost comparison indicated that PIC was competitive with regular concrete and steel [47].

PIC has been evaluated for use in multistage flash distillation vessels for desalination plants [6, 7]. It was concluded that prestressed PIC vessels would be cost-competitive with more conventional types.

Other potential applications include curbstones [45], plumbing fixtures [1], pump beds [48], flooring [48], and cover plates for cable pits [48]. But, in spite of the many possible applications, fully-impregnated PIC products have not yet been produced commercially to any significant degree. The most likely reason is the high initial equipment costs required for an efficient, large-volume manufacturing facility.

4 Polymer concrete

Polymer concrete (PC) is produced by using polymer as a binder for the aggregate. A wide range of aggregates and monomers may be used, although the cost and properties of the PC may be strongly influenced by the gradation and monomer. A well-graded aggregate may require as little as 5 to 8% monomer by weight, while more than 15% may be required for gap-graded aggregate.

Polymer concrete is used for field applications, such as the repair of concrete, and for manufactured products. For field applications, either user-formulated PC systems or commercially-available prepackaged systems are used. Several mixing techniques can be used. For manufactured products, continuous processing has been used for years in Europe and is now being used in other countries to produce a wide variety of components.

4.1 Monomer systems

Many monomer systems have been investigated for use in PC. The formulations usually have more components than those used for PIC because of the desire for polymerization at ambient conditions. Most PC monomer systems contain one or more monomers, one or more promoters, initiators, and, in some cases, a multifunctional monomer (cross-linking agent).

Additives are sometimes included to modify the properties. Plasticizers, flame retardants, silane coupling agents (to improve the bond between polymer and aggregates), and colorants are commonly-used additives.

Initially, most of the work with PC utilized polyester–styrene resin systems and, to a lesser extent, furan, epoxy, and vinyl ester systems [1]. Polyurethane has been developed and marketed commercially in the UK [1] and is now being used in other countries.

MMA has received the attention of most of the work in PC development in the United States in the last several years, especially for the repair of concrete. Benzoyl peroxide (BzP) is perhaps the most commonly-used initiator, and dimethylparatoluidine (DMPT) and dimethyl aniline (DMA) are widely used promoters with user-formulated MMA systems. TMPTMA, a multi-functional monomer, is used in some user-formulated systems to increase the rate of polymerization [26–29]. Polyester–styrene and vinyl ester systems generally use cobalt naphthenate and/or DMA as the promoters and methyl ethyl ketone (MEK) peroxide as the initiator [1, 29, 30].

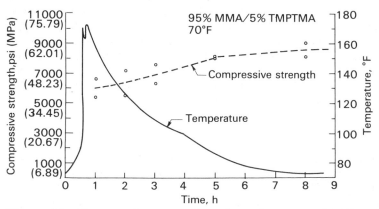

Figure 4-1 Compressive strength and exotherm as a function of time

There are several commercially-available polymer concrete systems that use MMA as the basic monomer. One system consists of two components—a liquid monomer and a package of premixed fine aggregate, polymers, initiators, and pigments [31]—and was originally developed in Europe; other similar systems are now available in many countries.

Other commercially-available systems include epoxy and furan resin. One furan system requires that the initiator first be mixed with the aggregate and then the resin be added.

Most of the monomer systems are formulated to provide a work time of 15–60 min. The peak exotherm, which is typically 20–50°C greater than ambient temperatures, is usually reached in 45–90 min after the PC is placed. Figure 4-1 indicates the strength and exotherm of MMA PC as a function of time [67]. For highway repairs, traffic can usually be restored in $1\frac{1}{2}$–2 h after the PC is placed.

4.2 Aggregate

Aggregate should be sound, and free of dirt, asphalt, and other organic materials. The aggregate should be dry, preferably with a moisture content of 0.5% or less to insure adequate bond to the polymer. The aggregate gradation varies according to the application, but generally the gradation should provide a relatively low-void volume and have sufficient fines to provide workability [26, 28]. Coarse aggregate gradation for extending commercially-available PC is specified by the manufacturer.

4.3 Properties

The mechanical properties of PC are strongly influenced by the viscoelastic behavior of the polymer and, to some extent, the properties of the aggregate. The response of most polymers used in PC is a function of time, temperature, and the molecular structure. Creep is much more pronounced in PC than in PIC, especially at elevated temperatures. Shrinkage can vary widely depending upon the polymer type and polymer loading. PC is usually more ductile than Portland cement concrete or PIC produced with MMA (Fig. 4-2), except when it is cast at low temperatures. Typical properties for PC are shown in Table 3-1. The properties can be expected to have a greater range than those for PIC or PPCC since PC, even when made from the same monomer, can be produced from many types and gradations of aggregate under a wide range of environmental and working conditions. Even when PC is used as an overlay or repair material for Portland cement concrete, however, the compressive and tensile strengths of PC are nearly always greater than for the concrete. Of more importance, perhaps, are the shrinkage, creep, coefficient of thermal expansion, ductility, stiffness, and bond strength to concrete. The importance and contribution of each of these properties are not yet fully understood.

The results of creep studies for MMA PC are shown in Fig. 4-3. The ultimate compressive strength, f'_{cp}, modulus of elasticity, E, and the modulus of rupture, f_r, are shown on the graph. No failure occurred in the specimens at stress levels of 0.3 and $0.4f'_{cp}$; however, failure occurred in about 50 days at a stress of $0.5f'_{cp}$ [66].

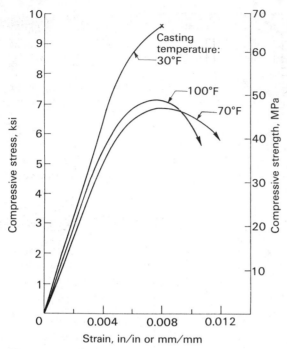

Figure 4-2   Stress–strain curves for MMA PC cast at different temperatures: specimens 3×6 in (76×152 mm) cylinders; monomers 95% MMA/5% TMPTMA; loading rate 47.16 psi/s (325 kPa/s)

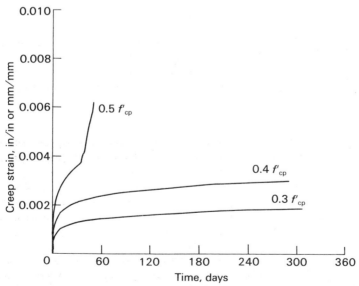

Figure 4-3   Creep strain of MMA PC for different stress levels at room temperature (elastic strain not included) [66]: f'_{cp} = 9000 psi (62 MPa); E = 2.8×10⁶ psi (19.3 GPa) f_r = 2000 psi (13.8 MPa)

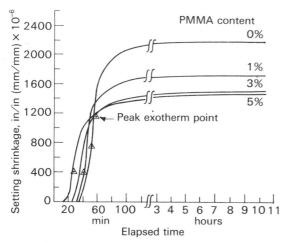

Figure 4-4   Shrinkage of MMA PC as a function of time and PMMA level [67]

The shrinkage of PC is also relatively high for some formulations. Figure 4-4 indicates shrinkage for user-formulated MMA PC made with varying percentages (by weight of aggregate) of polymethyl methacrylate (PMMA) powder. PMMA is added in some formulation to serve as a thickener which results in a non-workable material. Without the PMMA, the shrinkage was about 0.0022 in/in (mm/mm) [67].

The coefficient of thermal expansion of PC is usually about two to three times greater than that for Portland cement concrete.

The flexural fatigue strength of MMA–PC is shown in Fig. 4-5. Three ratios of

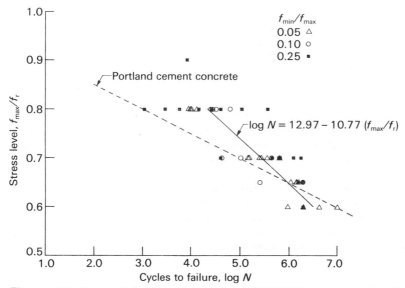

Figure 4-5   Flexural fatigue strength for MMA PC compared to Portland cement concrete [66].

minimum to maximum stress (f_{min}/f_{max}) were used. The stress level (f_{max}/f_r) is shown as a function of log N, where N is the number of cycles to failure. A typical fatigue strength for Portland cement concrete is shown for comparison. In excess of 10^6 cycles, Portland cement concrete is seen to exhibit a higher stress level than PC; however, since the stress level is a function of f_r, the ultimate strength, and f_r for PC (about 2000 psi or 13.8 MPa) is considerably higher than that for Portland cement concrete, PC performs quite well in fatigue loadings [66].

4.4   Repair applications of Portland cement concrete

PC has been used very successfully as a repair material for Portland cement concrete, especially for bridge decks [26–28]. The rapid curing, excellent mechanical and durability properties, and excellent bond to Portland cement have made PC, especially those using MMA, a very desirable patching material for bridges, pavements, and other concrete structures [26–32, 71, 72].

All unsound concrete should be removed and all surfaces to which PC will bond should be cleaned from asphalt or other organic materials and should be dry. Reinforcing steel should have corrosion scale removed. Formwork, if required, should be monomer-tight and should be treated with release agents or oil so that forms can be removed.

In user-formulated MMA systems, the monomer system is mixed just prior to placement. The pot life is measured from the time the initiator is added. Adequate safety precautions, discussed in Section 6, should be taken during the storage of chemicals, handling, and mixing.

Figure 4-6   Monomer poured over pre-placed aggregate in pavement repair

Figure 4-7   Nozzle of in-line mixing system for injecting monomer into pre-placed aggregate

The PC may be placed in several ways, including:

(a) Pre-mixed aggregate is placed in the repair, and the monomer solution is sprinkled or poured over the aggregate (Fig. 4-6) or injected with the nozzle of an in-line mixing system. (Fig. 4-7). This is the simplest method, but may result in more segregation.

(b) Pre-mixing the aggregate and monomer either by hand or with a mixer. The wetted aggregate tends to segregate less when placed.

The PC should be vibrated to improve consolidation. Placing the PC in layers not exceeding 3 in (75 mm) facilitates vibration. The surface can be finished by screeding and/or troweling. If excess monomer accumulates on the surface, additional fine aggregate should be placed to avoid 'slick' spots. The repair should be covered with polyethylene to minimize evaporation of the monomer.

For emergency repairs when only wet aggregate is available, it has been shown that the use of steel fibers improves the strength of PC [70].

If one of the commercially-produced PC systems is used, the recommended procedures is specified by the manufacturer.

With commercially-available MMA and furan PC systems, the concrete surface to which the PC is placed should be primed to improve the bond. Care should be taken when a MMA PC is placed adjacent to asphalt since the MMA acts as a solvent for the asphalt, and the presence of the asphalt tends to inhibit polymerization of the MMA.

Figure 4-8   Commercially-available PC mixed in conventional mixer

The commercially-available PC systems can be mixed in a conventional con-crete mixer (Fig. 4-8) or, in some cases, mixed in bags provided by the manufac-turer. In furan PC systems, the catalyst is first mixed with the aggregate and then the resin is added.

Many PC repairs ranging from 0.4 in (10 mm) thick to full depth have been documented for highway structures and pavements [26, 27, 29, 31, 32]. PC for the repair of structural beams, concrete cladding, floors, dams, and tanks has also been shown to be feasible.

4.5   PC overlays

PC overlays have been developed to provide (a) a highly impermeable membrane to prevent intrusion of water and chlorides; (b) a surfacing material to repair spalled and deteriorated concrete surfaces; and (c) restore skid resistance. The advantages of PC overlays are their good bond to Portland cement concrete, their relatively shallow depth, and their fast cure times.

One overlay utilized a polyester–styrene resin as a binder for well-graded aggregate. The top $\frac{1}{4}$ in (6 mm) of surface mortar is removed by scarification. A tack coat of resin is applied to the concrete surface. The polymer concrete is mixed in a mortar mixer and placed on the deck. The PC is screeded to a thickness of about $1\frac{3}{4}$ in (45 mm) and then consolidated by a roller or asphalt compactor to a thickness of about $1\frac{1}{2}$ in (38 mm). Traffic can be resumed approxi-mately $2\frac{1}{2}$ h after the last PC batch has been placed [30].

Another PC overlay used either a polyester resin or vinyl ester resin as a binder. The catalyzed and promoted monomer system is applied to the deck, which must be clean, dry and repaired if deteriorated areas exist. A thin layer of fine aggregate is broadcast over the monomer and compacted. After polymerization of the monomer, the process is repeated until four layers have been applied. The thickness of the overlay is in the range of 0.5–0.6 in (13–16 mm) [50, 51].

A third PC overlay utilized MMA as the basic momomer. After screed rails, which have a thickness of 0.6 in (16 mm), have been attached to the clean, dry concrete surface, a layer of well-graded fine aggregate is applied to a uniform thickness of about 0.2 in (5 mm), and a layer of about 0.4 in (10 mm) aggregate is broadcast uniformly and compacted with light pressure. An application of monomer similar to that used for impregnation is applied with a spray bar. After about 30 min to permit the monomer to soak into the concrete, a more viscous monomer system, which contains about 30 wt% of MMA syrup, is applied to the aggregate and permitted to polymerize. The thickness of the overlay is from 0.5 to 0.6 in (13–16 mm) [52].

The commercially-available MMA PC systems are used for overlays and floor toppings in thicknesses of approximately $\frac{3}{8}$ in. (10 mm).

The PC overlays have the advantage over polymer impregnation of requiring much less drying, much less cooling and monomer soaking time, and no externally-applied heat. As a result the process time is much less.

4.6 Other PC applications

PC for use in high-temperature geothermal environments has been developed. One monomer system containing styrene, acrylonitrile, acrylamide, and divinyl

Figure 4-9 Utility manhole made of polyester PC [34]

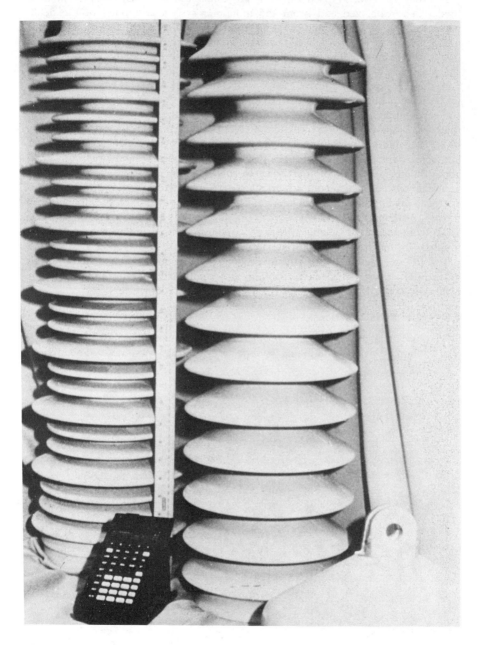

Figure 4-10   Electric insulators made of polyester-styrene PC [69]

Figure 4-11   Manufacture of PC building panels [39]

benzene used with an aggregate consisting of silica sand and Portland cement produces a compressive stress of 23 000 to 30 000 psi (159–207 MPa) at 20°C and is thermally stable to about 240°C. Field tests have been successful [33]. The use of siloxanes in conjunction with silica flour and Portland cement to produce PC for the completion of geothermal wells has been shown. Tests indicate that compressive strength is essentially constant (10 450 psi or 72 MPa) over a temperature range of 25–350°C with little weight loss of polymer [68].

Manholes made of polyester PC have been mass-produced in Japan (Fig 4-9). Over 40 000 of the large manholes have been produced [34].

A plant in West Germany has been manufacturing wall panels, cable channels, facades, drains, and curb stones. The PC is made with MMA [36]. Another manufacturer makes continuous casting machines which are capable of producing a wide range of products including sewer channels, sanitary ware, facing panels, floor tile, and window sills. Several monomers can be used [37].

Polyester–styrene is being used to make PC for electric insulators (Fig. 4-10) which have been shown to provide more efficient designs than were possible with porcelain [35, 69]. In Czechoslovakia, furan resin concrete has been used to make pipe for transporting chemical wastes [38].

Since 1958, precast polymer concrete building panels made with polyesters have been produced (Figs 4-11 and 4-12). Exposed aggregate facings, fiberglass cloth reinforcing and insulation have been incorporated [39].

Figure 4-12   Building with PC building panels [39]

5   Polymer Portland cement concrete

PPCC has been of considerable interest to researchers and users because of the similarity to current concrete technology. However, PPCC has not received as much attention as PIC or PC since most of the successful polymers that have been identified for PPCC have been proprietary.

5.1   Monomers and polymers

Most monomers used successfully with PIC and PC have not worked well when added to the fresh concrete. Most of these organic monomers, including MMA, are practically insoluble in water, generally interfere with the hydration of cement, and are subject to slow alkaline degradation. Strength is often found to be reduced in the hardened concrete. Acrylonitrile and styrene are exceptions in that they appear to be alkaline resistant and provide improvements in properties compared to unmodified controls [1].

Polymer latexes, which are very small ($0.05–1.0$ μm diameter) spherical particles held in suspension, have been used very successfully in PPPC, usually referred to as latex-modified concrete (LMC). Polymer latexes are usually copolymer systems of two or more monomers including vinyl acetate, styrene, vinyl chloride, and butadiene. Reference [1] summarizes the properties of some of the polymer latexes. LMC is the most widely used PPCC in the United States.

Several thermoplastics have been evaluated for use in PPCC. Polyvinyl esters were found to be subject to rapid alkaline degradation. Polyacrylics are superior to polyvinyl esters but require dry curing and limited exposure to wet conditions. Polyvinyldiene chloride–polyvinyl chloride copolymer and styrene–butadiene formulations are commercially available for use in PPCC [1].

Several elastomeric latexes are available for PPCC, including acrylonitrile–butadiene [53], Neoprene [53], and styrene–butadiene [53]. Water-soluble polymers have been added to fresh concrete. Thermosetting polymers include epoxies, amino resins, formaldehyde derivatives, and polyesters [53]. Polyorgano-siloxanes have been used in the USSR to improve the strength and watertightness of concrete [54].

Thermosetting resins have been used to produce PPCC. A polyester formulation has been used in the United Kingdom to produce 'Estercrete' [55]. When the system is mixed with water, the catalyst is activated.

Epoxies have been developed which can be added to fresh concrete and will produce improvements in properties of the hardened concrete [19, 56, 58, 73].

5.2   Properties

A range of typical properties of PPCC is presented in Table 3-1 There are several references in the literature which review the properties of PPCC [1, 57]. Polymer latexes alter the properties of the fresh and hardened concrete. In fresh concrete, the latexes will usually increase workability and allow reductions in water content. They will often delay the setting time. Anti-foaming agents are often added to control the excessive amounts of air that are entrained by the latexes. LMC generally has improved durability due to reduced porosity because of the lower water-cement ratio and the partial filling of pores with polymer. The polymer also tends to seal existing pores and microcracks, resulting in reduced water permeability. The frost resistance of LMC is generally excellent.

The chemical resistance depends upon the type and amount of polymer used. Abrasion and impact resistance are increased. LMC exhibits excellent adhesive properties which makes it attractive for repairing and overlaying concrete [1].

Reference [1] provides a general summary of properties of many types of PPCC. Properties of LMC are described in [60, 63]. Several investigations have reported on the properties of epoxy modified concretes (EMC) [58, 59, 61, 62]. EMC develops compressive strengths in the range of 8000 psi (55 MPa) and tensile strengths of about 700 psi (4.8 MPa). Resistance to freeze–thaw exposure is very good. Protection against chloride-ion intrusion has been found to be very effective.

5.3   Process technology

The process by which latex-modified concrete (LMC) is used is very similar to that of Portland cement concrete. The latex is added to the Portland cement concrete. Normally, LMC uses a cement-rich mix design [1]. Air-entraining agents are not used because polymer latexes entrain large amounts of air. Since the latexes act as water-reducing agents, the water content of the mix is reduced by an amount equal to the polymer volume or water in the polymer.

LMC is prepared and placed in the way that conventional concrete is placed. Excessive vibration should be avoided to prevent excess water from collecting on the surface.

Curing for LMC is much shorter than for Portland cement concrete. LMC usually moist-cured for no more than 3 days, after which it is air-cured. Steam heat is not recommended.

EMC is also produced by conventional processes. After the dry ingredients are added to the mixer or transit-mix truck, the resin, curing agent and water, in no particular order, are added. The mix has a relatively high cement content. Air-entraining agents are not recommended, but the EMC usually has about 3% entrapped air. The recommended water/cement ratio is 0.27 to 0.32 and the optimum epoxy/cement ratio is about 0.2.

5.4   Applications of PPCC

Latex-modified concrete (LMC) has been developed and used for many years for patching and overlaying deteriorated bridge decks [74]. The excellent bond strength, freeze–thaw resistance, resistance to penetration of chlorides, and its ease of application have made it a widely used material [63]. Latex-modified mortar has also been used for producing prefabricated brick panels which have sufficient strength for handling. Panels as large as 10×15 ft (3×4.6 m) have been prefabricated [63].

Epoxy modified concrete has been developed for the deep-patching and resurfacing of bridge decks and the resurfacing of food-packing plant floors. It is also recommended for industrial floors, sewer pipe and piling. A repair of an 28×280 ft (8.5×85 m) bridge deck is described which required 46 yd^3 (35 m^3) of epoxy modified concrete [58].

Increasing the water/cement ratio to 0.4 may result in cracking during curing. The concrete surface should not have excessive water, although small amounts of water can be added to improve trowelling. No special curing is required [73].

The overlay of the roof of a large water storage tank is described and material and labor costs are provided in [73].

6 Safety considerations

The chemicals and techniques used in the production of concrete-polymer materials require careful consideration to handling. The chemicals create the most obvious safety hazards, but the use of high-temperature drying and curing systems poses the need for adequate planning. A thorough discussion of safety has been published which addresses chemicals and construction practice [64]. Manufacturers of chemicals are good sources of safety and handling literature.

6.1 Chemicals

Generally, monomers are volatile, combustible and toxic liquids. Practice has shown, however, that prolonged stability and safety can be achieved by following recommended storage and handling practices. Manufacturers provide safety instructions which should be carefully followed.

To prevent premature polymerization, monomers contain inhibitors. Monomers should be stored in cool, shaded areas free of ignition sources. Most common monomers are toxic, and some handling precautions are necessary. For example, a safe limit of 100 ppm of MMA on an 8 h time-weighted average concentration, to which employees may be exposed, has been established. Normally, irritation to the eyes or respiratory membranes will become noticeable long before explosive or flammable concentrations are reached. Very small concentrations of MMA can be smelled, and most people would find concentrations much less than 100 ppm objectionable. Some people may be allergic to MMA, but no case of ill health has been reported.

The polymer latexes used to produce latex-modified concrete are safe to handle and store. Repeated exposure of the skin to latexes may cause irritation, and eyes should be washed thoroughly after contact with latex.

The initiators generally used are either organic peroxides or azo compounds. They should be stored in a cool place away from sources of ignition. They should be stored separately from monomers and, in particular, away from promoters or other oxidizing materials. Benzoyl peroxide in dry, granular solid form must be handled very carefully since it is decomposed by heat, shock, and reaction with many chemicals. However, it is now available in paste and dispersion forms which make it safer to store and use. The azo initiators are not shock sensitive, but some types should be stored at lower temperatures to prevent decomposition. Manufacturers' instructions for handling and storage of chemicals should be carefully followed. The initiator used with furan resin is first mixed with aggregate before adding the resin. The initiator should not be mixed directly with the resin to avoid a potentially dangerous reaction.

Promoters should be stored and handled similar to monomers. Extreme care should be taken that promoters not be mixed or come into contact with initiators in concentrated form because the mixture can react explosively.

6.2 Construction practices

Workmen should receive safety training in the handling of chemicals, but care should be exercised that employees not be made to fear the use of the chemicals. Respirators should be available even if concentrations stay below acceptable safe levels. Goggles, gloves, and boots should be worn to prevent the contact of workmen's skin and clothing with chemicals. Smoking must be prohibited in areas where chemicals are stored and used. Chemical fire extinguishers should be placed in strategic locations. Water is ineffective as a fire extinguisher and, in fact, may tend to spread a fire involving monomers which are less dense than water.

The storage and handling of chemicals should be in strict accordance with manufacturers' instructions. All chemicals and mixing vessels should be labeled and kept clean. Mixing should be done with wooden paddles or other non-sparking devices. If monomers are stored for long periods of time, the inhibitor level should be checked to avoid the danger of bulk polymerization.

When a catalyzed monomer is left over, it should be disposed of before bulk polymerization occurs. The monomer can be placed in an open-top container, and promoter and initiator can be added so that polymerization will occur. The polymerized monomer can then be disposed in conventional landfills.

References

1. ACI Committee 548, *Polymers in concrete*, American Concrete Institute, Detroit, 1977, 92 pp.
2. Dikeou, J. T., Polymers in concrete: new construction achievements on the horizon, *Proc. 2nd int. congr. polymers in concr.*, 1978, University of Texas at Austin, pp. 1–8.
3. Dikeou, J. T. *et al. Concrete–polymer materials*, Third Topical Report, USBR REC-ERC-71-6 and BNL 50275 (T-602), Jan. 1971,
4. Dikeou, J. T., Kukacka, L. E., Backstrom, J. E. and Steinberg, M., Polymerization makes tougher concrete, *J. Am. Concr. Inst., Proc.*, Vol. 66, No. 10, Oct. 1969, pp. 829–839.
5. Kukacka, L., Colombo, P., Steinberg, M. and Monowitz, B., *Concrete–polymer composites for underwater applications*, BNL 14267.
6. Steinberg, M., Kukacka, L. E., Colombo, P., Auskern, A., Reich, M. and Pike, R., *Concrete–polymer materials for highway applications*, Progress Report No. 1, BNL 15395, Sept. 1970.
7. Steinberg, M., Colombo, P. and Kukacka, L. E. (to US Atomic Energy Commission), *US Patent* 3567, March 2, 1971.
8. Limsuwan, E. *et al.*, *Behavior of post-tensioned polymer-impregnated concrete beams*, Research Report No. 114–6, Center for Highway Research, University of Texas at Austin, June 1978, 287 pp.
9. Fowler, D. W., Houston, J. T. and Paul, D. R., Polymer-impregnated concrete surface treatments for bridge decks, *Polymers in concrete*, American Concrete Institute, SP-40, 1973, pp. 93–118.

10. Fowler, D. W. and Paul, D. R., Partial polymer impregnation of highway bridge decks, *Transp. Res. Rec.*, No. 542, 1975, pp. 9–19.

11. Mehta, H. C. *et al.*, Innovations in impregnation techniques for highway concrete, *Transp. Res. Rec.*, No. 542, Polymer Concrete, 1975, pp. 29–40.

12. Paul, D. R. and Fowler, D. W., Surface impregnation of concrete bridge decks with polymer, *J. Appl. Polymer Sci.*, Vol. 19, 1971, pp. 281–301.

13. Yimprasert, P., Fowler, D. W. and Paul, D. R., *Durability, strength, and method of application of polymer-impregnated concrete for slabs*, Research Report No. 114–4, Center for Highway Research, University of Texas at Austin, Jan. 1976, 222 pp.

14. Webster, R. P., Fowler, D. W. and Paul, D. R., Bridge deck impregnation in texas, *Polymers in concrete*, American Concrete Institute, SP-58, 1978, pp. 249–266.

15. Smoak, W. G., *Polymer impregnation of new concrete bridge deck surfaces*, Report No. FHWA-RD-78-5, Federal Highway Administration, Washington, DC, Jan. 1978, 37 pp.

16. Hallin, J. P., Field evaluation of the polymer impregnation of new bridge deck surfaces, *Polymers in concrete*, American Concrete Institute, SP-58 1978, pp. 267–280.

17. Bartholomew, J., Fowler, D. W. and Paul, D. R., Current status of bridge deck impregnation, *Proc. 2nd int. congr. polymers in concr.*, 1978, University of Texas at Austin, pp. 399–412.

18. Smoak, W. G., *Development and field evaluation of a technique for polymer impregnation of new concrete bridge deck surfaces*, Report No. FHWA-RD-76-95, Federal Highway Administration, Washington, DC, Sept. 1976, 161 pp.

19. Patty, T., Petrographic identification of polymer in surface-treated bridge decks, *Proc. 2nd int. congr. polymers in concr.*, 1978, University of Texas at Austin, pp. 245–253.

20. Locke, C. and Hsu, C. M., Measurement of the depth of partial polymer impregnation of concrete, *Proc. 2nd int. congr. polymers in concr.*, 1978, University of Texas at Austin, pp. 25–35.

21. Heller, F. C., *Determination of polymer impregnation depth in concrete*, Report GR-4-77, US Bureau of Reclamation, Denver, July 1977.

22. Fowler, D. W., Paul, D. R. and Yimprasert, P., Corrosion protection of reinforcing provided by polymer-impregnated concrete, *J. Am. Concr. Inst., Proc.* Vol. 75, No. 10, Oct. 1978, pp. 520–525.

23. Limsuwan, E., Fowler, D. W., Paul, D. R. and Burns, N. H., Flexural behavior of post-tensioned, polymer-impregnated concrete beams, *Proc. 2nd int. congr. polymers in concr.*, 1978, University of Texas at Austin, pp. 361–380.

24. Subrahmanyam, B. V., Babu, K. G., Rajamane, N. P. and Neelamegan, M., Strength and plastic ductility of polymer impregnated concrete beams and columns. *Proc. 2nd int. congr. polymers in concr.* 1978, University of Texas at Austin, pp. 229–244.

25. Mehta, H. C., Chen, W. F., Manson, J. A. and Vanderhoff, J. W., Stress–strain behavior of polymer-impregnated concrete beams, columns and

shells, *Polymers in concrete*, American Concrete Institute, SP-58, 1978, pp. 161–186.

26. Fowler, D. W. and Paul, D. R., Polymer-concrete for repair of bridge decks, *Proc. 2nd int. congr. polymers in concr.*, 1978, University of Texas at Austin, pp. 337–350.

27. Fontana, J., Webster, R. and Kukacka, L., Rapid patching of deteriorated concrete using polymer concrete, *Proc. 2nd int. congr. polymers in concr.*, 1978, University of Texas at Austin, pp. 105–119.

28. Kukacka, L. E. and Fontana, J., Polymer concrete patching materials, *Implementation Package* 77–11, Vols 1 and 2, Federal Highway Administration, Washington, DC, April 1977, 27 pp.

29. DePuy, G. W. and Selander, C. E., Polymer–concrete trials and tribulations, *Proc. 2nd int. congr. polymers in concr.*, 1978, University of Texas at Austin, pp. 461–482.

30. Jenkins, J. C., Beecroft, G. W. and Quinn, W. J., *Polymer concrete overlay test program users' manual*, Federal Highway Administration, Washington, DC, Dec. 1977.

31. Peschke, H. J., Eighteen years experience with polymers in concrete in Europe, *Proc. 2nd int. congr. polymers in concr.*, 1978, University of Texas at Austin, pp. 447–460.

32. McNerney, M. T., An investigation into the use of polymer–concrete for rapid repair of airfield pavements, *Proc. 2nd int. congr. polymers in concr.*, 1978, University of Texas at Austin, pp. 431–446.

33. Kukacka, L. E., Polymer concrete materials for use in geothermal energy processes, *Proc. 2nd int. congr. polymers in concr.*, 1978, University of Texas at Austin, pp. 157–172.

34. Imamura, K., Toyokawa, K. and Murai, N., Precast polymer concrete for utilities applications, *Proc. 2nd int. congr. polymers in concr.*, 1978, University of Texas at Austin, pp. 173–186.

35. Gunasekaran, M. and Perry, E. R., Polymer concrete for high voltage electrical insulation, *Proc. 2nd int. congr. polymers in concr.*, 1978, University of Texas at Austin, pp. 187–191.

36. Koblischek, P. J., Acryl–concrete, *Proc. 2nd int. congr. polymers in concr.*, 1978, University of Texas at Austin, pp. 413–430.

37. Kreis, R., World-wide application and industrial manufacturing of polymer concrete, *Proc. 2nd int. congr. polymers in concr.*, 1978, University of Texas at Austin, pp. 519–575.

38. Bares, R. A., Furane resin concrete and its application to large diameter sewer pipes, *Polymers in concrete*, American Concrete Institute, SP-58, 1978, pp. 41–74.

39. Prusinski, R. C., Study of commercial development in precast polymer concrete, *Polymers in concrete*, American Concrete Institute, Detroit, SP-58, 1978, pp. 75–101.

40. Schrader, E. K., Fowler, D. W., Kaden, R. A. and Stebbins, R. J., Polymer impregnation: used in concrete repairs of cavitation/erosion damage, *Polymers in concrete*, American Concrete Institute, Detroit, SP-58, 1978, pp. 225–248.

41. Schrader, E. K., The use of polymers in concrete to resist cavitation/erosion

damage, *Proc. 2nd int. congr. polymers in concr.*, 1978, University of Texas at Austin, pp. 283–309.

42. Lockman, W. T. and Cowan, W. C., *Polymer-impregnated precast structural concrete bridge deck panels*, Report No. FHWA-RD-75-121, Federal Highway Administration, Washington, DC, Oct. 1975, 181 pp.

43. Carpenter, L. R., Cowan, W. C., and Spencer, R. W., *Polymer-impregnated tunnel support and lining*, Report No. REC-ERC-73-23, US Bureau of Reclamation, Dec. 1973, 162 pp.

44. Goris, J. M., unpublished data, US Department of Interior, Bureau of Mines, Spokane, Washington, Dec. 1973.

45. Sopler, B., Fiorato, A. E. and Lenschow, R., A study of partially impregnated polymerized concrete specimens, *Polymers in concrete*, American Concrete Institute, Detroit, Michigan, SP-40, 1973, pp. 149–151.

46. Gerwick, B. C., Application of polymers to concrete sea structures, *Proc. 2nd int. congr. polymers in concr.*, 1978, University of Texas at Austin, pp. 37–43.

47. Albertson, M. D. and Haynes, N. H., Polymer-impregnated concrete spherical hulls for seafloor structures, *Polymers in concrete*, American Concrete Institute, SP-40, 1973, pp. 119–148.

48. Tazawa, E., and Kobayashi, E., Properties and applications of polymer impregnated cementious materials, *Polymers in concrete*, American Concrete Institute, SP-40, 1973, pp. 57–92.

49. Kaeding, A. O., Building structural restoration using monomer impregnation, *Polymers in concrete*, American Concrete Institute, SP-58, 1978, pp. 281–298.

50. Webster, R., Fontana, J. and Kukacka, L. E., *Thin polymer concrete overlays*, Report No. FHWA-TS-78-225, Federal Highway Administration, Washington, DC, Feb. 1970.

51. Fontana, J. and Webster, R., Thin, sand-filled, resin overlays, *Proc. 2nd int. congr. polymers in concr.*, 1978, University of Texas at Austin, pp. 89–103.

52. Hsu, H. T., Fowler, D. W., Paul, D. R. and Miller, M., An *Investigation of the use of polymer–concrete overlays for bridge decks*, Research Report No. 114–7, Center for Highway Research, University of Texas at Austin, March 1979, 141 pp.

53. Synthetic resins in building construction, *Proc. RILEM symp.*, Paris, 1967,

54. Antonova, I. T. *et al.*, Use of polymer cement concrete on a base of furfural resin, *Hydrotech. Constr.* Vol. 8, 1970.

55. Nutt, W. O., An evaluation of a polymer–cement composite, *Composites*, Vol. 1, No. 4, June 1970, pp. 234–238.

56. *Epoxy Portland cement concretes*, Technical Bulletin PC-10, Celanese Chemical Company, Louisville, March 1972.

57. Riley, V. R. and Razl, I., Polymer additives for cement composite: a review *Composites*, Vol. 5, No. 1, June 1974, pp. 27–33.

58. McClain, R. R., Epoxy modified cement admixtures, *Proc. 2nd int. cong. polymers in concr.*, 1978, University of Texas at Austin, pp. 483–501.

59. Nawy, E. G., Ukadike, M. M. and Sauer, J. A., Optimum polymer content in concrete modified by liquid epoxy resins, *Polymers in concrete*, American Concrete Institute, SP-58, 1978, pp. 329–355.

60. Akihama, S., Morita, H., Watanabe, S. and Chida, H., Improvement of

mechanical properties of concrete through the addition of polymer latex, *Polymers in concrete*, American Concrete Institute, SP-40, 1973, pp. 319–338.

61. Raff, R., and Austin, H., Epoxy polymer modified concretes, *Polymers in concrete*, American Concrete Institute, SP-40, 1973, pp. 339–346.

62. Popovics, S. and Tamas, S., Investigation on Portland cement pastes and mortars modified by the addition of epoxy, *Polymers in concrete*, American Concrete Institute, SP-58, 1978, pp. 357–366.

63. Eash, R. and Shafer, H., Reactions of polymer latexes with Portland cement concrete, *Transp. Res. Rec.*, No. 542, 1975, pp. 1–8.

64. Fowler, D. W. *et al.*, Safety aspects of concrete–polymer materials, *Polymers in concrete*, American Concrete Institute, SP-58, 1978, pp. 123–138.

65. Manson, J. A., Overview of current research on polymer concrete, *Applications of polymer concrete*, American Concrete Institute, SP-69, 1981, pp. 1–19.

66. Hsu, H., unpublished research, University of Texas at Austin.

67. Fowler, D. W., Meyer, A. H., McCullough, B. F. and Paul, D. R., Methyl methacrylate polymer-concrete for bomb damage repair, Final Report, to be published.

68. Zeldin, A. N., Kukacker, L. E. and Carciello, N., New, novel well-cementing polymer concrete composite, *Applications of polymer concrete*, American Concrete Institute, SP-69, 1981, pp. 73–92.

69. Perry, E. R., Polymer concretes and the electric power industry, *Applications of polymer concrete*, American Concrete Institute, SP-69, 1981, pp. 63–72.

70. Fowler, D. W., Meyer, A. H. and Paul, D. R., Techniques to improve strength of polymer concrete made with wet aggregate, *Applications of polymer concrete*, American Concrete Institute, SP-69, 1981, pp. 107–122.

71. Fontana, J. J. and Bartholomew, J., Use of concrete polymer materials in the transportation industry, *Applications of polymer concrete*, American Concrete Institute, SP-69, 1981, pp. 21–44.

72. McNerney, M. T., Research in progress: rapid all-weather pavement repair with polymer concrete, *Applications of polymer concrete*, American Concrete Institute, SP-69, 1981, pp. 93–106.

73. Christie, S. H., McClain, R. R. and Melloan, J. H., Epoxy-modified Portland cement concrete, *Applications of polymer concrete*, American Concrete Institute, SP-69, 1981, pp. 155–168.

74. Kuhlman, L. A., Performance history of latex-modified concrete overlays, *Applications of polymer concrete*, American Concrete Institute, SP-69, 1981, pp. 123–144.

9 Admixtures for concrete

Raymond J Schutz

Protex Industries, Inc. Denver, Colorado, USA

Contents

Summary

This chapter describes concrete admixtures, their uses and the modifications they impart to plastic and hardened concrete. The reaction mechanisms and economic aspects are also covered.

The chapter is divided into accelerating admixtures, air entraining admixtures, water reducing and set controlling admixtures, including the high range water reducers (superplasticizers), finely divided mineral admixtures including cementitious materials, pozzolans, pozzolanic and cementitious materials and relatively inert mineral admixtures and miscellaneous admixtures including gas forming, coloring and damp-proofing admixtures.

1 Introduction

An admixture for concrete is defined by the American Concrete Institute as 'a material other than water, aggregates and hydraulic cement used as an ingredient of concrete or mortar, and added to the concrete immediately before or during its mixing'. The compounds covered in this chapter conform to that definition. Admixtures have been used almost since the inception of the art of concreting. It is reported that the ancient Roman builders used ox blood, pig tallow and milk as admixtures in their concrete and masonry structures. Research has shown that ox blood is an excellent proteinaceous air-entraining agent. During the early part of the century it was common practice to add common brown laundry soap to concrete as a waterproofing agent. These soaps were rich in stearates and acted as a combination of air-entraining and damp-proofing agents. It is doubtful whether these early uses of admixtures were carried out on a scientific basis; however, the users did realize their benefits. From these crude beginnings the use of admixtures has progressed to a scientifically based technology.

2 Modification of properties

Admixtures may be used to modify the properties of concrete in such a way as to make it more suitable for the work at hand or for economy, or for other purposes such as saving energy or increasing durability. In some instances an admixture may be the only means of achieving the desired results.

2.1 Modification of fresh concrete

Admixtures may be used to modify one or more of the following properties:

(a) To increase workability without increasing water content or to decrease the water content at the same workability.
(b) To retard or accelerate both initial and final setting times.
(c) To reduce or prevent settlement.
(d) To create slight expansion in concrete and mortar.
(e) To modify the rate or capacity for bleeding or both.
(f) To reduce segregation of concrete, mortars and grouts.
(g) To improve penetration and or pumpability of concretes, mortars and grouts.
(h) To reduce rate of slump loss.

2.2 Modification of hardened concrete

Admixtures may be used to modify one or more of the following properties:

(a) To retard or reduce heat generation during early hardening.
(b) To accelerate the rate of strength development.
(c) To increase the strength of concrete or mortar (compressive, tensile or flexural).
(d) To increase the durability or resistance to severe conditions of exposure, including the application of de-icing salts.

(e) To decrease the capillary flow of water.
(f) To decrease the permeability to liquids.
(g) To control the expansion caused by the reaction of alkalis with certain aggregate constituents.
(h) To produce cellular concrete.
(i) To increase the bond of concrete to steel reinforcement.
(j) To increase the bond between old and new concrete.
(k) To improve impact resistance and abrasion resistance.
(l) To inhibit the corrosion of embedded metal.
(m) To produce colored concrete or mortar.

3 Economic aspects of the use of admixtures

Admixtures may increase or decrease the cost of concrete by lowering the cement requirements for a given strength, changing the volume of the mixture, or lowering the cost of the placing and handling operations. Control of the setting time of concrete may result in economy, such as decreasing the waiting time for floor finishing or extending the time in which the concrete remains plastic, thereby eliminating bulkheads and construction joints. Accelerating strength development can speed job progress and thus allow the quicker re-use of formwork.

4 Accelerating admixtures

There are many chemicals that will accelerate the hardening of Portland cement. Among these are the soluble chlorides, carbonates, silicates, fluorides, hydroxides, and organic compounds such as triethanolamine and calcium formate. For specialized purposes, such as quick-setting mortars for sealing leaks, there are proprietary compounds available with setting times as short as 15 s; however, the most universally employed accelerator for Portland cement is calcium chloride. Most proprietary compounds sold as accelerators are based on calcium chloride, although many of these compounds may contain other admixtures as well. Triethanolamine in combination with other admixtures is also used to a very limited extent.

4.1 Soluble inorganic salts

4.1.1 *Quick setting admixtures*
Proprietary quick-setting admixtures for the hand-sealing of leaks or the setting of anchor bolts or other inserts, are available with setting times as short as 15 s. These are generally based on soluble carbonates, aluminates or silicates. Setting times will be dependent on the salt and its concentration. These admixtures are generally marketed as aqueous solutions ready to use undiluted with cement alone or with cement and sand mixtures. Setting times can generally be prolonged by dilution with water.

4.1.2   *Shotcrete accelerators*

Dry-process shotcrete accelerators are available as liquids or powders. Powdered admixtures are generally preferred as their addition to the dry mixture can be controlled by automatic feeders synchronized with the feed rate of the shotcrete machine. Liquid accelerators must be injected at the nozzle and dosage rate cannot be controlled to match the output of the machine.

Powdered accelerators for dry-process shotcrete will be based on blends of soluble carbonates and aluminates. The blend may be adjusted to suit the chemical composition of the cement. The proposed ASTM specification for the compatibility of shotcrete accelerators with Portland cement using Gilmore needles requires an initial set in less than 3 min and a final set in less than 9 min when used at a dosage of 3% by weight of cement. Setting times can be controlled within limits by varying dosage (Fig. 4-1).

Liquid accelerators for dry-process and wet-process shotcrete are either solutions of aluminates or sodium or potassium silicates. The performance of these liquids is generally inferior to that of the powdered compounds in regard to both setting time and ultimate strength.

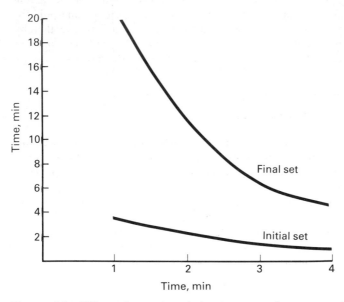

Figure 4-1   Effect of powdered shotcrete accelerator on time of set: Type I cement, water/cement ratio 0.3, Gilmore needles (ASTM C266–77)

4.1.3   *Calcium chloride*

Calcium chloride is the most commonly used accelerator for concrete. Most proprietary accelerators are based on calcium chloride, although many of these compounds may contain other admixtures as well. Calcium chloride was used as an admixture as early as 1873 and a patent was issued to Miltar and Nichols in 1885 on the use of this chemical in concrete. It is widely available and economic.

Calcium chloride can be used safely in amounts up to 2% by weight of the

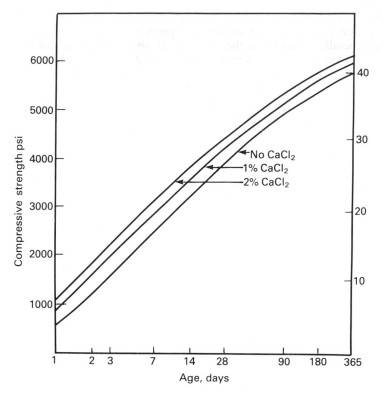

Figure 4-2   Rate of strength development of concrete with and without calcium chloride: cement content 286 kg/m³ (486 lb/yd³), water/cement ratio 0.53

cement, i.e. up to 2 lb per US bag (94 lb) or 1 kg per UK bag (50 kg). Larger amounts may be detrimental and will provide little additional advantages. The addition of up to 2% calcium chloride by weight of the cement will increase compressive and flexural strengths at early ages of all types of Portland cement. This increase will be in the range 2.8–6.9 MPa (400–1000 psi) at both 21 and 4.5°C (70 and 40°F) curing temperatures (Fig. 4-2; [1]). This increase in strength reaches a maximum in 1 to 3 days and thereafter will generally decrease. At 1 year, some increase may still be evident. Calcium chloride will not increase the flexural strength of concrete to the same degree as the compressive strength, and decreases in the flexural strength are generally obtained at or after 28 days.

Calcium chloride should not be used as an antifreeze agent for concrete. In the quantities normally used, the freezing point of the concrete is lowered only a negligible amount, less than 2°C (3.6°F). No materials are known which will substantially lower the freezing point of the water in concrete without being harmful to the concrete in other respects.

4.1.3.1 *Reaction mechanisms of calcium chloride*   The addition of calcium chloride changes the complex reaction of Portland cement and water. These changes are not fully understood, but calcium chloride can be considered as a catalyst which triggers the hydration of Portland cement. There does not appear to be any chemical reaction between the calcium chloride and the di- or tricalcium silicates, although their rate of reaction is increased. The calcium chloride is partially consumed during hydration, forming new reaction products between tricalcium aluminate and gypsum. The calcium chloride reacting with the tricalcium aluminate forms calcium chloro-aluminate. It is believed the new reaction product is predominately monochloro-aluminate, although it is possible that trichloro-aluminate may also be present as a reaction product.

4.1.3.2 *Effect on properties* Calcium chloride will also affect the following characteristics of concrete:

(a) Drying shrinkage and creep. It has been reported [1] that calcium chloride will increase drying shrinkage and creep when measured conventionally. However, an alternative hypothesis offered by Mather [2] and Bruere *et al.* [3] indicates that the influence of calcium chloride on these volume changes is dependent on the length of curing prior to drying or loading. The longer the concrete is allowed to cure, the less will be the effects on shrinkage and creep.

(b) Durability. Calcium chloride will lower the resistance of concrete to freezing and thawing and to attack by sulphates and other injurious solutions. The effect on freeze–thaw resistance may be offset by proper air entrainment, and on sulphate resistance by the use of Type V (sulphate resistant) cement [4].

(c) Rate of temperature rise. Calcium chloride will increase the rate of temperature rise due to the heat of hydration and in large sections will therefore increase the stresses caused by thermal contraction. (Fig. 4-3; [5]).

(d) Effect on embedded metals. Data are available which indicate that calcium chloride may cause stress corrosion of prestressing steel. On the other hand, calcium chloride will not accelerate the corrosion of full embedded reinforcing steel with adequate cover. Where large concentrations of stray currents are present, such as in concrete used in structures for electric railroads, powerhouses, or electrolytic-reduction plants, calcium chloride in the concrete can cause corrosion of adequately embedded reinforcing steel.

(e) Galvanized metal. Galvanized metal embedded in concrete containing calcium chloride may be expected to corrode at an accelerated rate.

(f) Combinations of metals. Combinations of metals, such as aluminum alloy conduit and steel reinforcing, should not be used in concrete containing calcium chloride as electrolytic corrosion may take place.

(g) Allowed limits of calcium chloride. The American Concrete Institute Committee 201, Durability of concrete, suggests the following limits for the chloride ion (Cl^-) in concrete:

(1) Prestressed concrete   0.06% by weight of cement

(2) Conventionally reinforced concrete in a moist environment and exposed to chloride   0.10% by weight of cement

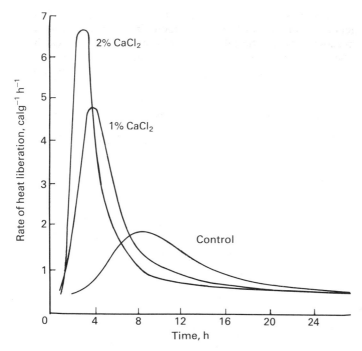

Figure 4-3   Effect of calcium chloride upon the rate of hydration at 23°C (73°F)

(3)	Conventionally reinforced concrete in a moist environment but not exposed to chloride	0.15% by weight of cement
(4)	Above-ground building construction where the concrete will stay dry (does not include locations where the concrete will be occasionally wetted, such as kitchens, parking garages and water-front structures)	No limit for corrosion

4.2   Soluble organic compounds

The most common organic accelerators are triethanolamine and calcium formate. Neither of these compounds is as efficient as calcium chloride in accelerating set or strength gain but both are non-corrosive. Both products will affect drying shrinkage and creep in a similar manner to calcium chloride.

Triethanolamine is an effective set accelerator but is rarely used alone since it reduces the strength of concrete. It is an ingredient in many normal setting or accelerating water reducing admixtures.

Calcium formate will accelerate both set and strength but, on an equal-weight basis, is approximately one half as efficient as calcium chloride. Since it costs more than calcium chloride, its use is confined to areas where the chloride ion is unacceptable and where its high cost per unit of acceleration is tolerable.

4.3 Solid admixtures

Occasionally, calcium aluminate cement is blended with Portland cement to accelerate setting. Although setting time can be shortened to varying degrees, even to minutes, by such blends, strength at setting or one day and later will be reduced significantly. Shrinkage and swelling on immersion will be increased drastically and durability will be poor [6].

5 Air-entraining agents

Air-entraining agents are the most widely used admixtures. Their use probably can be traced to the Roman Empire, although there is no evidence that these early builders understood the mechanisms involved. In the late 1930s, it was observed that cement from a certain mill in Rosendale, New York, USA, produced concrete with excellent durability. Investigations showed that this durability could be traced to the beef tallow used as a grinding aid. In 1941, Myron Swazy published one of the first papers indicating that the addition of organic materials, such as certain mineral oils, animal and vegetable fats and natural resins, produced concrete with improved durability and workability. However, Swazy did not credit this increase in durability to entrained air.

The first symposium on the effects of entrained air on durability was presented by the American Concrete Institute in 1944. Since that time, extensive research and field experience has provided conclusive proof of the role of entrained air on durability. Air entrainment should be required in all concrete subject to freezing and thawing or to the application of de-icing chemicals.

5.1 Effect of air entrainment on durability (resistance to cycles of freezing and thawing)

The mechanism by which freezing and thawing damages concrete is well understood. The action is physical rather than chemical. It involves the development of disruptive osmotic and hydraulic pressures during freezing, principally in the paste. The action of de-icing chemicals is similar. De-icing chemicals cause a high degree of saturation in the concrete and, in addition, solutions of de-icing chemicals have a lower vapor pressure than water, therefore little or no drying takes place between the cycles of melting and drying or freezing and thawing.

If the concrete contains voids of sufficient size to accommodate the volume change taking place as capillary water freezes and if these were spaced close enough to the capillaries to accommodate the expansion before disruptive forces develop, freezing would have little effect on concrete. Research and experience have shown that to accomplish this, it is necessary to have an empty space within 0.10 m (0.004 in) of every point in the hardened cement paste or a spacing factor of 0.20 mm (0.008 in) or less. Additional requirements are that the surface of the air voids be greater than 23.6 mm²/mm³/(600 in²/in³) of air volume [7, 8]. Maximum durability will be obtained if the mortar phase of the concrete contains 9–13% air

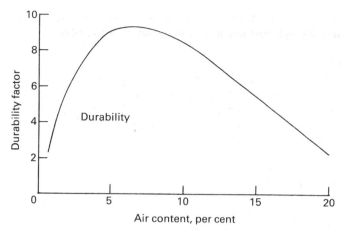

Figure 5-1   Effect of air content on durability: 38 mm (1.5 in) maximum size aggregate

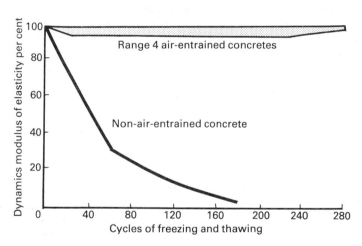

Figure 5-2   Effect of freezing and thawing on dynamic modulus (ASTM C666–77)

by volume. The effect of entrained air on durability is illustrated in Figs 5-1 and 5-2.

5.2   Materials for air entrainment

Air entraining admixtures are powerful organic surfactants which have the ability to stabilize foam with a bubble size of 0.076–1.27 mm (0.003–0.05 in) in diameter. Agents used for this purpose are salts of wood resins, synthetic detergents, salts of lignosulphonic acid, salts of petroleum acids, salts of proteinaceous materials, fatty and resinous acids and their salts, and salts of hydrocarbons. Materials meeting the above description may or may not produce a desirable air void system and therefore materials proposed for use as an air-entraining agent should be tested for conformance with ASTM C260–77, *Air entraining admixtures for concrete.*

5.3 Effect of entrained air on plastic concrete

Air entrainment alters the properties of plastic concrete. The presence of the minute air bubbles dispersed uniformily through the cement paste renders the concrete more workable and cohesive than concrete of equal water/cement ratio which is non-air-entrained.

Segregation and bleeding are also reduced, resulting in a reduction in the formation of pockets of water underneath aggregates and embedded items, and the accumulation of laitance on the surface of the concrete.

Concrete mixture proportions must be changed to accommodate for the volume of air entrained and the increase in workability usually accompanying air entrainment.

Figure 5-3 illustrates the typical water reductions possible due to entrained air.

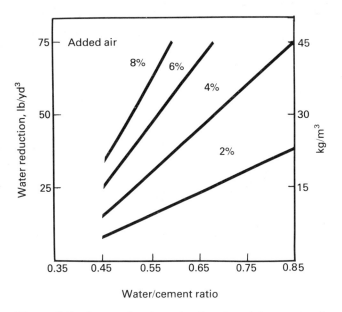

Figure 5-3 Approximate reduction in mixing water for air entrainment

The volume of this water reduction is less than the volume of entrained air, therefore, to compensate for the volume of entrained air, the fine aggregate volume must also be reduced. Typical reductions in fine aggregate required to maintain yield are illustrated in Fig. 5-4.

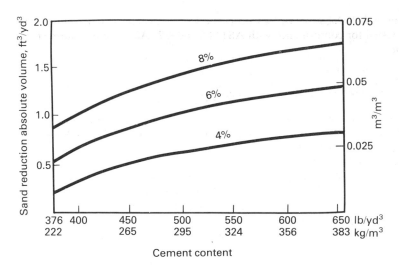

Figure 5-4 Approximate sand reduction required to compensate for entrained air

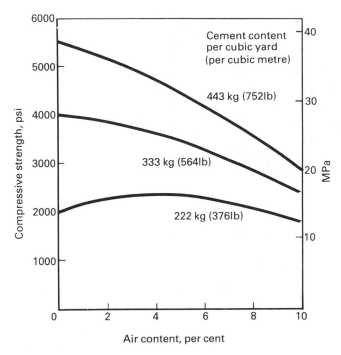

Figure 5-5 Effect of air content on compressive strength: 76–100 mm (3–4 in) slump, 25.4 mm (1 in) gravel

5.4 Effect of entrained air on hardened concrete

Despite the reduction in the water/cement ratio usually obtained by its use, air entrainment usually reduces strength, particularly in concretes of high or moderate cement content (Fig. 5-5). This strength reduction is generally proportional to the amount of air entrained, the range being 4–6% reduction in strength for each 1% by volume (STP) of air entrained. However, in lean concrete—concrete deficient in paste volume—entrained air may increase strength.

The fine, entrained bubbles break up the capillaries in the mortar phase, usually resulting in reduced absorption.

At a given water content, entrained air will increase the drying shrinkage; however, it will also increase the workability. At constant workability as measured by slump, the water content is reduced. Therefore, at constant workability in the practical ranges of air contents, the drying shrinkage is not changed significantly by air entrainment (Fig. 5-6; [9]).

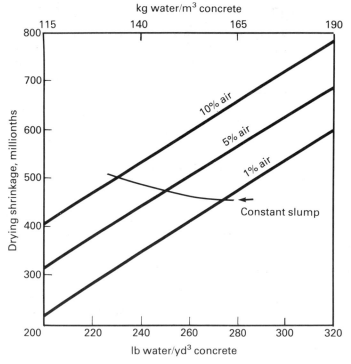

Figure 5-6 Effect of air content on drying shrinkage

5.5 Factors effecting air entrainment

The amount of air entrained for a given dosage of air entraining admixture will depend on many factors. As an example, an increase in fineness of the cement or pozzolan, in the fines aggregates, temperature of the concrete and mixing time will tend to decrease the air content of the concrete. Decreases in these and an increase in workability will tend to increase the air content. Since durability, strength and workability will be affected by the amount of air entrained, the concrete should be monitored and the dosage of air entraining admixture varied to obtain the desired air content.

6 Water-reducing and set-controlling admixtures

Certain organic compounds or mixtures of organic and inorganic compounds are used as admixtures to reduce the water requirement of the concrete. Water reduction results in a desirable reduction in the water/cement ratio for a given consistency (slump) and cement content, or in an increased consistency for the same water/cement ratio and cement content.

Generally, the effects of the use of these materials on the hardened concrete are increased compressive strength and some reduction in permeability and, in combination with adequate air entrainment, improved resistance to freezing and thawing. A reduction in the water/cement ratio increases the strength of the concrete but the gain in compressive strength frequently is greater than that which is indicated by that relationship alone.

Such admixtures may also modify the setting properties of concrete or mortars, usually resulting in retardation for those commonly used. Combinations of the water-reducing and retarding materials with accelerators may produce admixtures which retain the water-reducing property but are less retarding, non-retarding (neutral setting) or accelerating, the degree of these conditions depending upon the formulation (i.e. the relative amounts of each ingredient used in the formulation.) Such formulations may contain other materials to produce or to modify certain other effects, such as the inclusion of an air-entraining admixture to produce air-entrained concrete.

The specific effects of water-reducing and set-controlling admixtures will vary depending on cement composition, water/cement ratio, cement content, mixing time and ambient temperature, and other factors.

The American Society for Testing and Materials specification C494–80, *Chemical admixtures for concrete,* covers seven types of water-reducing and set-controlling admixtures based on performance not chemical composition. These are: Type A, Water-reducing; Type B, Retarding; Type C, Accelerating; Type D, Water-reducing and retarding; Type E, Water-reducing and accelerating; Type F, Water-reducing, high-range; Type G, Water-reducing, high-range and retarding.

6.1 Reaction mechanisms

6.1.1 *ASTM C494 Types A, B, C, D and E*
Several theories have been proposed to explain the action of organic water-reducing and set-controlling admixtures. Hansen [10] proposed that the admixtures were adsorbed onto the anhydrous cement particles, protecting them from initial reaction with water. Watanabe *et al.* [11] proposed that insoluble calcium salts precipitated onto the surfaces and similarly retarded the initial cement–water reaction. Young [12] proposed that admixture behavior seems best described by

complexing nucleations, the interference with the growth surface of hydration products rather than reaction with the anhydrous compounds.

All these admixtures contain hydroxyl, carboxyl or sulphonate groups or a combination of these groups and therefore can combine with calcium or aluminum ions. It is therefore probable that all these theories can be correct to a degree.

6.1.2 *ASTM C494 Types E and F: high-range water reducers (superplasticizers)*
The action of these admixtures is best described as that of powerful dispersing agents; they are adsorbed onto the cement particles and being anionic provide a negative charge to the particles rendering them mutually repulsive.

6.1.3 *ASTM C494 Type C: accelerating admixture*
See Section 4, Accelerating admixtures.

6.2 Chemical composition

The chemicals commonly used for water-reducing and set-controlling admixtures fall into five general classes:

(a) Lignosulphonic acids and their salts.
(b) Modifications and derivatives of lignosulphonic acids and their salts.
(c) Hydroxylated carboxylic acids and their salts.
(d) Modifications of carboxylic acids and their salts.
(e) Other materials:
   (1) Inorganic materials such as zinc salts, borates and phosphates.
   (2) Amines and their derivatives.
   (3) Carbohydrates, polysaccharides and sugar acids.
   (4) Certain polymeric compounds such as sulphonated naphthalene and melamine formaldehyde condensates, cellulose ethers, silicones and sulphonated hydrocarbons.

These admixtures can be used alone or in combination with those listed and with other organic or inorganic active or essentially inert substances.

6.3 Effect on plastic concrete

6.3.1 *Water reduction*

6.3.1.1 *ASTM C494 Types A, D and E.* These water-reducing admixtures reduce the water required for the same consistency (slump) of concrete up to 10% or more. Concrete containing lignosulphonate admixtures generally require 5–10% less water than comparable non-admixtured concrete. Hydroxylated carboxylic acid salts reduce the water requirements by 5–8%; carbohydrates, polysaccharides and sugar acids generally yield lower water reductions.

Water reduction with a given concrete will depend on the dosage of the admixture (Fig. 6-1). The dosage of a retarding admixture can be varied to offset the effect of high concrete temperatures on water requirements (Fig. 6-2).

For concrete of a given workability, water reduction will reduce the

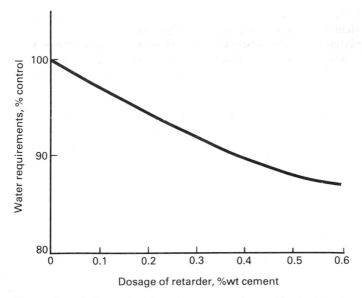

Figure 6-1   Effect of a Type D water reducer (lignin base) on water requirements

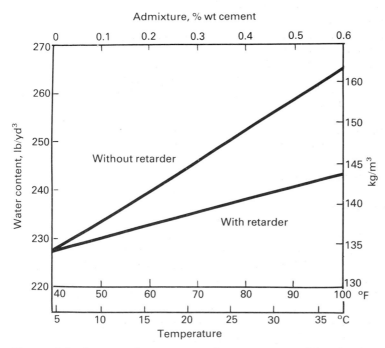

Figure 6-2   Increase in water requirements caused by rise in temperature can be offset by use of water-reducing retarders, ASTM C494, Type D

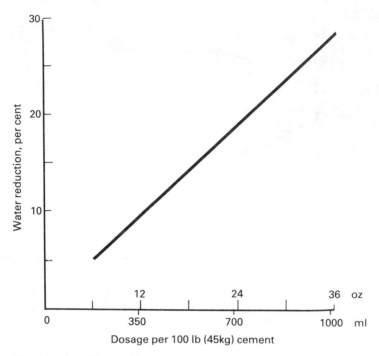

Figure 6-3   Effect of high-range water reducer on water reduction at constant slump

water/cement ratio for a given cement content or permit a reduction in cement content while maintaining the same water/cement ratio.

6.3.1.2   *ASTM C494 Types F and G.* These high-range water reducers (superplasticizers) permit water reductions as high as 30% (Fig. 6-3) or may be used at constant water content or slightly reduced water content to produce very workable or even self-levelling concrete (Fig. 6-4).

6.3.2   *Bleeding*
Admixtures of the hydroxylated carboxylic acid type tend to increase the bleeding rate and capacity. This may be of advantage when placing flat slab work under high-drying conditions. The supply of bleed water may equal or surpass the rate of evaporation of water from the surface, tending to prevent plastic shrinkage cracking.

Admixtures based on lignosulphonates tend to reduce bleeding, which is generally desirable as bleeding may cause the entrapment of water under aggregates and embedded items, and can cause an increase in laitance on the surface.

The high-range water reducers, Types F and G, can cause severe bleeding and segregation in concretes deficient in fines (aggregate fines or cement).

6.3.3   *Setting time*
When used at recommended dosages, the retarding admixtures Types B, D, C and G will retard setting time between 1 and 3 h. Increased dosages will increase

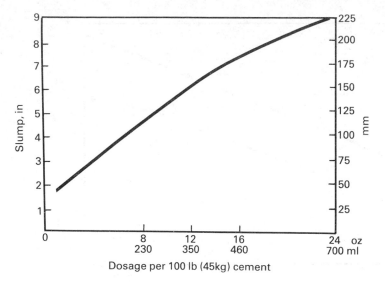

Figure 6-4 Effect of dosage of high-range water reducer on slump at constant water/cement ratio

setting time (Fig. 6-5). Prolonged retardation may be of advantage in eliminating construction joints or cracks due to form deflection. The acceleration of setting time by high temperatures can also be offset by varying the dosage of the retarder.

Concrete has been purposefully retarded as long as 19 days with no adverse effects on its ultimate strength or other properties. Where prolonged retardation is employed, care must be exercised to prevent the drying of the concrete.

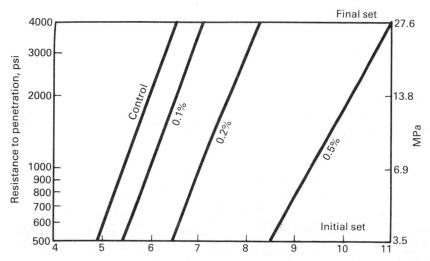

Figure 6-5 Effect of a lignin retarder (ASTM Type D) on time of set: Type II cement, 23°C (73°F), test method ASTM C403–77

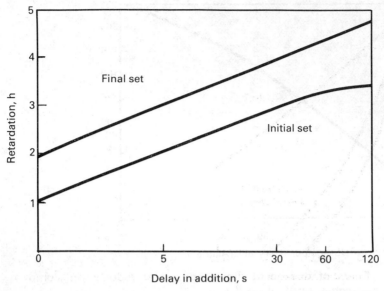

Figure 6-6 Effect of time of addition of a set-retarding admixture on time of set: cement content 333 kg/m³ (564 lb/yd³)

At recommended dosages, Type A and F admixtures will not affect the setting time significantly less than 1 h acceleration and $1\frac{1}{2}$ h retardation.

The time of addition of an admixture will affect the setting time. The setting times listed are for the immediate addition of the admixture to the concrete. That is, the admixture is added to a portion of the mixing water. If the mixing water is added to the concrete, any delay in adding the admixture will cause a significant increase in the setting time. The water-reducing admixture should be added at the same time in the mixing cycle, in order to obtain a uniform setting time among batches (Fig. 6-6; [13]).

For Type E admixtures, see Section 4.

6.3.4 Slump loss

In general, water-reducing and set-controlling admixtures will not decrease the rate of slump loss and in many instances may increase it.

The high-range water reducers Types F and G (superplasticizers) will increase slump loss, the degree depending on their type, dosage, cement composition and other factors (Fig. 6-7; [14]). However, Type F high-range water reducers may be used to redose the concrete to regain lost slump (Fig. 6-8; [15]).

6.3.5 Air entrainment

Admixtures of the lignosulphonate class may entrain air to various degrees, 2–6% or more. The air-entrainment properties may be controlled by modifying the formulation.

All the water-reducing admixtures enhance the air-entraining capabilities of

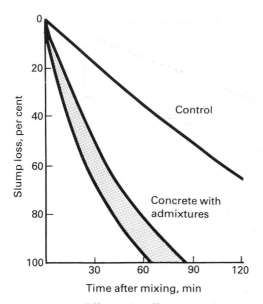

Figure 6-7   Effect of different high-range water reducers on slump loss: cement content 390 kg/m³ (658 lb/yd³)

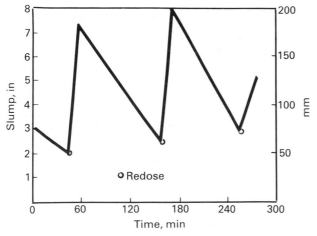

Figure 6-8   Effect of redosing a high-range water reducer on slump

air-entraining agents. Thus, lower dosages of air-entraining agents are required for a given air content when water reducers are used.

6.4   Effect on hardened concrete

6.4.1   Strength
At constant workability (slump), water-reducing admixtures Types A, D and E will reduce the water requirements and therefore increase the compressive

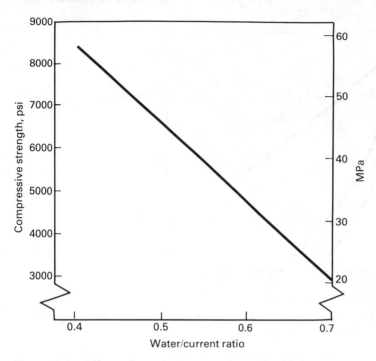

Figure 6-9   Effect of varying proportions of a high-range water reducer on the 28-day strength and water/cement ratio at constant slump 75–100 mm (3–4 in): cement content 370 kg/m³ (658 lb/yd³)

strength, due to the modification of the cement–water reaction. This increase in strength will be greater than that indicated by water reduction alone (Fig. 6-9; [16]). At the same cement content, strength increases of 25% are not uncommon. The high-range water reducers Types F and G, as their name implies, will reduce water to a greater degree, and workable concrete with very low water requirements can be produced, resulting in extremely high strengths. Generally, Type F and G admixtures will produce concrete strengths following the water/cement ratio relationship (Fig. 6-10). Increases in the flexural strength of concrete containing water-reducing admixtures are usually attained but they are not proportionally as great as the increase in the compressive strength.

6.4.2 Rate of strength gain

The normal setting water-reducing admixtures Types A, E and F will increase the strength of concrete at all ages. The retarding Types D and G may decrease the strength at early ages, but unless used at unusually high dosage rates they will produce an increase in the strength at 24 h.

The very low water/cement ratios which can be obtained with the high-range water reducers will result in very high early strengths. If advantage is taken of the water reduction possible, heat curing may be reduced or eliminated (Fig. 6-11; [17]).

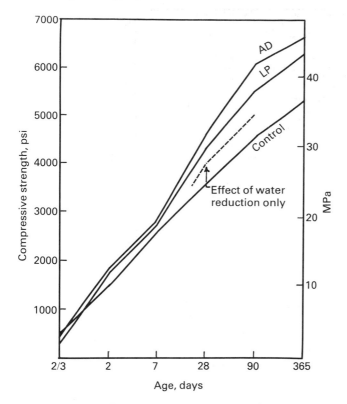

Figure 6-10 Effect of Type D water reducers on compressive strength: cement content 305 kg/m³ (517 lb/yd³)
AD   255 g (9 oz) lignin type
LP   85 g (3 oz) hydroxylated carboxylic acid type

6.4.3 *Economy*

Since water-reducing admixtures increase strength significantly, cement contents can be reduced to attain a given strength (Fig. 6-12). The cement thus saved exceeds the cost of the admixture, resulting in more economic concrete.

6.4.4 *Heat of hydration*

The use of a water-reducing admixture will not change the heat of hydration of a given concrete significantly. The rate of heat generation will change depending on whether an accelerating or retarding admixture is used. With thin concrete sections, retardation may allow sufficient time for the dissipation of the heat to the atmosphere, formwork or grade, which will lower the maximum concrete temperature. In mass concrete, there is insufficient surface-to-volume ratio for the dissipation of heat and the maximum temperature rise will not be significantly affected by the use of water reducers.

The total heat of hydration of concrete can be reduced by taking advantage of the cement reductions made possible by the use of water-reducing admixtures.

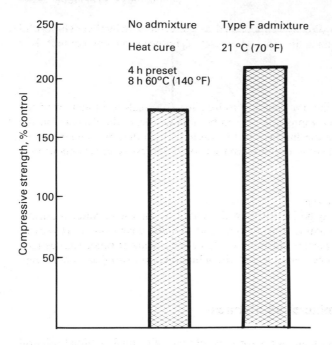

Figure 6-11   Relative 18-hour strength: Type I cement content 415 kg/m³ (705 lb/yd³); control 18 h at 21°C (70°F)

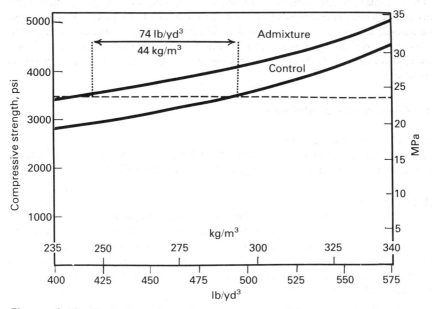

Figure 6-12   Typical savings in cement possible by use of a Type A water-reducing admixture: 127 mm (5 in) slump

Heat reduction will be directly proportional to the cement reduction. The use of water-reducing admixtures in mass concrete of reduced cement content is a standard practice.

6.4.5   *Durability*
Resistance to freezing and thawing and to de-icing chemicals is a function of the air-void system. Further improvements can be obtained from the decrease in the water/cement ratio and the increase in the strength imparted by water-reducing admixtures. A slight increase in the resistance to an aggressive environment may be realized.

6.4.6   *Shrinkage and creep*
Shrinkage and creep may be increased or decreased by the use of water reducers, depending on the admixture formulation and cement composition. However, changes in mixture proportions (cement content, water/cement ratio, etc.) usually act to reduce any detrimental effects of the admixtures on shrinkage and creep.

7   Finely divided mineral admixtures

The ACI Committee on admixtures for concrete classifies finely divided mineral admixtures into four types: (1) those which are cementitious; (2) those which are pozzolanic; (3) those which are both cementitious and pozzolanic; and (4) others.

These materials are usually available as powders, as fine as or finer than Portland cement. They are inexpensive to produce or are by-products. Compared to chemical admixtures, they are used in relatively high proportions—up to 30% or more by weight of cement. Being added at the concrete mixer and being materials other than Portland cement, aggregates and water, they are defined as admixtures.

7.1   Types of mineral admixture

7.1.1   *Cementitious materials*
This classification includes natural cements, hydraulic lime, slag cements (mixtures of blast-furnace slag and lime) and granulated blast-furnace slag.

7.1.2   *Pozzolans*
A pozzolan is defined 'as a siliceous or siliceous and aluminous material which in itself possesses little or no cementitious value, but will in finely divided form and in the presence of moisture, chemically react with calcium hydroxide at ordinary temperatures to form compounds possessing cementitious properties'. In this class would be certain fly ashes, volcanic glass, diatomaceous earths, and certain shales and clays, either heat treated or natural.

7.1.3   *Pozzolanic and cementitious materials*
Certain fly ashes produced from sub-bituminous coal or lignite have limited cementitious properties and also combine with lime, as do pozzolans.

7.1.4 *Other finely divided mineral admixtures*
These admixtures include finely divided quartz, silica sands, dolomitic and calcitic limestone, marble, granite and other rocks, asbestos waste, hydrated lime and other materials.

Until recently, these admixtures were considered relatively inert. However, many carbonate rocks are not chemically inert, and considerable differences in performance can result from their use. The pre-testing of these mineral admixtures is a prudent procedure.

7.2 Effects on plastic concrete

Being extremely fine powders, these mineral admixtures will influence the fresh paste in a manner similar to cement. They can be used to augment the cement in mixtures deficient in fines. Many concretes contain a larger amount of Portland cement than necessary for strength requirements to provide workability or pumpability. A portion or all of this excess cement may be replaced with a suitable mineral admixture.

The relatively chemically inert mineral admixtures have no direct effect on the required amount of cement for a given strength other than they may increase or decrease the total water requirements.

The cementitious and pozzolanic mineral admixtures may not only affect water requirements but, due to their contribution to strength development, are often considered part of the cementing material.

7.2.1 *Proportioning*
Techniques for proportioning concrete containing finely divided mineral aggregates are basically the same as for non-admixtured concrete. For optimum properties and maximum economy, the volume of admixture employed should be greater than the volume of cement replaced by its use. Since these admixtures have a fineness similar to Portland cement, they should be regarded as part of the paste volume when determining the percentage of coarse and fine aggregate.

The relatively inert mineral admixtures generally will have no effect on the amount of cement required other than their effect on water requirements.

The pozzolanic, pozzolanic–cementitious, and cementitious mineral admixtures affect water requirements and therefore cement content, and in addition react with the free lime generated by the cement–water reaction, developing cementing properties. They are usually used in the proportion of 15–35% by weight of the cement and in proportioning the concrete should be considered as part of the cementing medium.

7.2.2 *Workability and bleeding*
Where the available aggregate is deficient in fines (material passing a No. 200 sieve, 75 μm), the use of a finely divided mineral admixture of favorable particle size will reduce segregation and bleeding and increase workability. In all concretes, the use of finely divided mineral admixtures will increase the ratio of the surface area of the solids to the volume of water in the paste, generally increasing workability, placeability and pumpability, and in addition reducing bleeding and segregation. The placeability of concrete containing blast-furnace slag is generally greater than that indicated by static tests (slump) or water/cement ratio [18].

7.2.3 Setting time

Mineral admixtures generally do not affect the setting time of concrete. When they are used as cement replacements, the concrete so produced will have the setting characteristics of a concrete of similar reduced cement content. Leaner concretes will set slower than richer concretes.

This slower set caused by the reduced cement content can quite often be offset by use of a normal-setting Type A or accelerating Type E water-reducing admixture.

7.3 Effect on hardened concrete

7.3.1 Relatively inert mineral admixtures

In lean concretes, these admixtures may effect a strength increase by increasing paste volume. In medium to rich concretes, the increase in water requirements caused by their use may reduce strength.

7.3.2 Pozzolanic, pozzolanic–cementitious, and cementitious admixtures

The reactivity and water requirements of these materials will vary depending on their chemical composition, fineness and particle shape. Therefore, the characteristics cited are of necessity general in nature.

The initiation time for the lime-pozzolanic reaction will vary from 24 h to 4–6 weeks [19, 20]. Water is required for this reaction and therefore curing may be

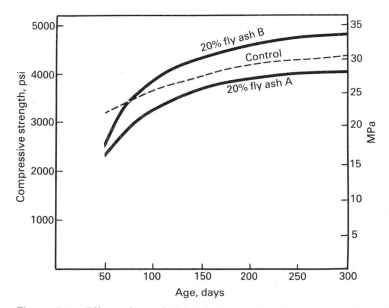

Figure 7-1   Effect of two fly ashes on compressive strength: total cementitious material 300 kg/m³ (510 lb/yd³)
   Fly ash A   Specific surface 3270 cm²/g
   Fly ash B   Specific surface 2410 cm²/g

more critical than with non-admixtured concrete. With pavement concrete, moisture will be supplied by the subgrade and rain, except in a few arid regions.

7.3.2.1 *Strength* Pozzolanic admixtures usually will increase the strength of concrete, especially at later ages. The strength increase at early ages will be dependent on the characteristics of the individual pozzolan. Figure 7-1 illustrates the effect on strength of two different fly ashes used as a replacement for 20% by weight of the Portland cement [21]. The fly ash A produced an increase in strength at all ages after 80 days while the coarser fly ash B exhibited a slightly lower strength than the control concrete. This emphasizes the need for pre-testing mineral admixtures.

Figure 7-2 shows the effect of a ground pumicite, a natural pozzolan used to replace 21 kg (47 lb) of cement in mass concrete 150 mm (6 in) maximum size

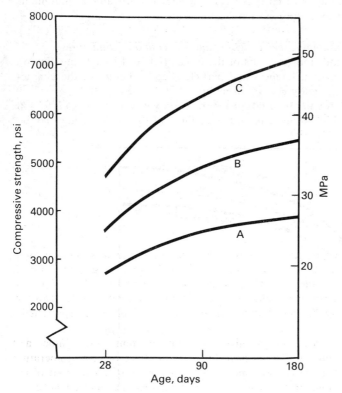

Figure 7-2   Effect of pumicite and pumicite and lignin-type water reducers on compressive strength of mass concrete for 150 mm (6 in) maximum size aggregate

A   139 kg cement per cubic metre (235 lb/yd³)
B   111 kg cement plus 28 kg pumicite per cubic metre
     (188 lb cement plus 47 lb pumicite per cubic yard)
C   Concrete B plus 0.37% pumicite-lignin-type admixture

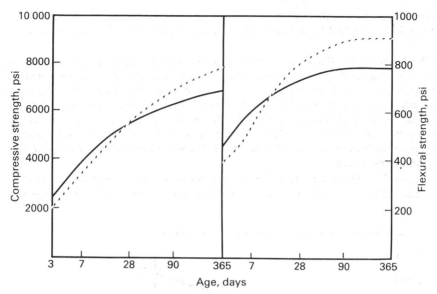

Figure 7-3   Effect of cement type on compressive and flexural strengths: average of five sources, water/cement ratio 0.46
. . . . . . Slag cements
- - - - - Portland cements

aggregate [22]. The effect of a lignin-type water reducer on this concrete is also shown.

Figure 7-3 shows the general effect of granulated blast-furnace slag on the flexural and compressive strengths of concrete. These results are an average of slag cements from five sources [23].

7.3.2.2   *Rate of strength gain*   When these admixtures are used as a cement replacement, the strengths at early ages are generally lower than non-admixtured, non-cement replaced concrete. Certain fly ashes produced from lignite or semi-bituminous coal may have sufficient cementitious properties to offset this low early strength. Water-reducing admixtures may be used to increase early strength.

7.3.2.3   *Temperature rise*   At a given cement content, pozzolanic and pozzolanic–cementitious admixtures have little or no effect on the temperature rise of concrete in place. However, since they can be used to replace part of the Portland cement, the temperature rise of the reduced-cement concrete will be less than that of the unreplaced concrete (Fig. 7-4). The use of pozzolans in lean cement-replaced concrete is very desirable for mass concrete.

7.3.2.4   *Durability*   The resistance of concrete to freezing and thawing is mainly dependent on its air-void system (see Section 9.5, air-entraining agents). Concrete containing finely divided mineral fillers with properly entrained air will have excellent resistance to freezing and thawing.

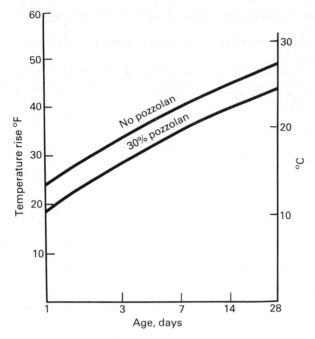

Figure 7-4   Effect of cement replacement with a pozzolan on adiabatic temperature rise: cementitious material content 166 kg/m³ (283 lb/yd³) mass concrete

The addition of fly ash, natural pozzolans and granulated slag in sufficient quantities will increase the sulphate resistance of concrete.

The addition of granulated slag in sufficient quantity can minimize or eliminate expansion in Portland cement concrete containing reactive aggregates. The effectiveness of fly ash in this role will be dependent on the chemical and mineralogical characteristics of the individual fly ash.

7.4   Economy

Most finely divided mineral admixtures are inexpensive to produce or are by-products. They can be effectively used as partial cement replacements, reducing the unit cost of concrete. Further economies can quite often be realized due to the increase in workability and placeability imparted to concrete by their use.

Various blends of granulated blast-furnace slag and Type I Portland cement may be used as substitutes for Type II (moderate sulphate-resistant), Type IV (low heat of hydration) and Type V (high-sulphate-resistant) cements. The use of these blends will increase the economy and versatility of concrete production.

8   Miscellaneous admixtures

8.1   Gas-forming admixtures

The gas-void content of concrete can be increased by the use of admixtures which generate or liberate bubbles of gas in the fresh concrete. These materials will tend to counteract settlement and bleeding and will cause gaseous expansion of the concrete in the plastic state.

The materials are not intended to form a void system which is considered a replacement for entrained air.

The expansion induced by gas-forming agents is in the plastic state and will not offset the volume changes of shrinkage and carbonation occurring after hardening.

8.1.1 *Materials*

Admixtures which produce gases are hydrogen peroxide, which releases oxygen, certain activated carbons, from which air is liberated, and metallic aluminum, which generates hydrogen. Of these, only aluminum is used extensively. Generally, unpolished aluminum powder is used, the addition rate being from 0.005 to 0.02% by weight of cement. Larger percentages may be used to produce light-weight cellular concrete.

To obtain a given expansion, the quantity of aluminum added will have to be increased as the temperature falls. Usually, twice as much is required at 20°F (−6.5°C) 40°F (4.5°C).

8.1.2 *Effect on concrete*

The generation or release of gas will cause a slight expansion of the fresh concrete. Where the concrete is restrained, this expansion will tend to increase the bond to horizontal steel with little decrease in strength. If the concrete is unrestrained, the voids formed will weaken the concrete. It is important when using these admixtures that the forms be tight and adequately sealed.

8.2 Bonding admixtures

Admixtures of this class are organic polymer emulsions (sometimes called latexes). They usually increase the air content of mortars or concretes.

8.2.1 *Materials*

The principal organic polymer emulsions used in these admixtures are polyvinyl acetate, acrylic or styrene-butadiene. The polyvinyl acetate types may or may not be re-emulsifiable. The re-emulsifiable polyvinyl acetate types should be restricted to interior or other areas not subject to immersion.

8.2.2 *Effect on concrete*

Bonding admixtures tend to render plastic concretes sticky and more difficult to place and finish. Part of this characteristic is due to their air entrainment and part to the nature of the latex itself.

These admixtures are used in the proportion of 5–20% polymer solids based on the cement content, a minimum of 12–15% being optimum. Concretes containing these admixtures exhibit excellent resistance to freezing and thawing, and to de-icing salts and chloride-ion penetration.

The particles of the polymer emulsion tend to coalesce rapidly. Once the emulsion has been added to the concrete, mixing, placing and finishing must be quickly performed to prevent the breaking of the coalesced film and the formation of cracks on drying.

The full benefit of the polymer can only be realized when the polymer film dries. This is partially accomplished by the water in the emulsion system being consumed by the hydration process. Moist curing is not generally necessary and if used should be limited to 24 h.

The compressive strength is not changed significantly by bonding admixtures but the bond, flexural and tensile strengths may be increased by a factor of two.

For the optimum bonding of patches and toppings of concretes or mortars containing these admixtures, a thin grout coat of the emulsion and Portland cement is applied immediately prior to placing the mortar or concrete. The bond obtained is often stronger than the overlay or the patch or the base concrete.

8.3 Coloring admixtures

Pigments specifically prepared for use in concrete and mortar are available both as natural and synthetic materials. They are resistant to fading on outdoor exposure and stable in the alkaline involvement inherent in cement paste.

8.3.1 *Materials*
The pigments listed below may be used to obtain a variety of colors:

Color	Pigment
Grey to black	Black iron oxide (red black), Mineral black
	Carbon black (blue black)
Blues	Phthalocyanine blue, Ultramarine
Reds	Red iron oxide
Ivory, cream, buff, yellow	Yellow iron oxide
Green	Chromium oxide, Phthalocyanine green
White	Titanium oxide

8.3.2 *Effect on concrete*
When used at an addition rate of 6% or less by weight of cement, these pigments generally have little effect on the properties of the plastic or hardened concrete. Addition rates up to 10% may be used. Being inert fines, they may increase water requirements at the higher addition rates and the effect on the properties of the concrete would be consistent with this increased water demand.

Unmodified carbon black may increase the quantity of air-entraining agent required for durability.

To ensure the color uniformity of concrete containing coloring admixtures, the water/cement ratio and the weight of all ingredients must be carefully controlled.

8.4 Damp-proofiing admixtures

These admixtures, by reducing the penetration of moisture into the visible pores of concrete, may retard the penetration of rain into a dry concrete wall but they do not affect the transmission of moisture through concrete, except where such concrete contains a paste having high porosity. Such a paste would be one of high water/cement ratio (over 0.6) and poorly cured. If the concrete has sufficiently low porosity and is well cured, damp-proofing agents have little merit.

Damp-proofing agents may have beneficial secondary effects, such as air entrainment or acceleration.

8.4.1 *Materials*
Soaps, salts of fatty acids, such as ammonium or calcium stearate or oleate—are commonly used; sometimes with the addition of calcium chloride, these soaps

entrain air. The soap content of the concrete should be a maximum of 0.2% based on the weight of the cement.

Butyl stearate, added as an emulsion at a dosage rate of 1% stearate based on the weight of the cement, imparts repellency without entraining air.

8.5 Other types of admixture

There are many other types of admixture available for specialized applications, such as grouting and oil-well cementing. A good reference is the American Concrete Institute's report and guide on the use of admixtures for concrete [24].

9 Conclusion

Considering the long history of structural concrete, the use of admixtures on a scientific basis is relatively new. The American Society for Testing and Materials published its first standard specification for air-entraining admixtures, C260, in 1950 and its specification on water-reducing and set-controlling admixtures, C494, in 1962. In this short time the use of admixtures in concrete has grown to be almost a standard practice. On the North American continent, all concrete for highway use contains air-entraining admixtures and the majority of the states and provinces use in addition a water-reducing admixture. All mass concrete for large dams contains air-entraining admixtures and water-reducing and retarding admixtures, plus pozzolanic admixtures where available. Over 90% of the ready-mix concrete produced in the United States and Canada is admixtured.

The judicious use of admixtures allows the engineer or producer to change the characteristics of concrete in the plastic or hardened state to be more suitable for its intended use. The setting time and rate of strength gain can be increased to speed job progress, retarded to allow for form deflection or to offset the effects of high or low ambient temperatures.

Economies can be realized in labor requirements for placing, by quicker re-use of formwork or savings in costly and energy-intensive cement. Even characteristics such as color or insulating value can be controlled by the use of admixtures.

Acknowledgement

Figures 6-6, 6-9 and 7-3 are copyright, American Society for Testing and Materials, 1916 Race Street, Philadelphia, PA 19103 and are reprinted with permission.

References

1. Shideler, J. J., Calcium chloride in concrete, *J. Am. Concr. Inst., Proc.*, Vol. 48, No. 7, March 1952 pp. 537–560.
2. Mather, B., Drying shrinkage—second report, *Highw. Res. News*, No. 15, Nov. 1964.
3. Bruere, G. M., Newbegin, J. D. and Wilson, L. M., *A laboratory investigation of the drying shrinkage of concrete containing various types of chemical admixtures*, Division of Applied Mineralogy, CSIRO, Technical Paper, No. 1, 1971.

4. Bureau of Reclamation, *Concrete manual*, 8th Edn, US Dept. Interior, 1975.
5. Lerch, W. A., *Res. Rep. Portld Cem. Ass.*, April 1944.
6. Feret, L. and Venaut, M., Effect on shrinkage and swelling of mixing different cements to obtain rapid set, *Revue Mater. Constr. Trav. Publ.*, No. 496, 1957.
7. Cordon, W. A., *Freezing and thawing of concrete mechanisms and control*, American Concrete Institute, Monograph, No. 3, 1966, 100 pp.
8. Macinnis, C. and Beaudoin, J. J., Mechanisms of frost damage in hardened cement paste, *Cem. & Concr. Res.*, Vol. 4, No. 2, March 1974.
9. Bureau of Reclamation, *Concrete manual*, 5th Edn US Dept. Interior, 1949.
10. Hansen, W. C., Action of calcium sulfate and admixtures in Portland cement pastes, *ASTM Spec. Tech. Publ.*, No. 266, 1960.
11. Watanabe, Y., Suzuki, S. and Nishi, S., Influence of saccharides and other organic compounds on the hydration of Portland cement, *J. Res. Onoda Cem. Co.*, Vol. 11, 1969.
12. Young, J. F., Reaction mechanisms of organic admixtures with hydrating cement compounds, concrete admixtures *Transp. Res. Rec.*, No. 564, 1976.
13. Dodson, J. H. and Farkas, E., Delayed addition of set retarding admixtures to Portland cement concrete, *67th Ann. Meet. Am. Soc. Test. Mater.*, June 1964.
14. Perenchio, W. F., Whiting, D. A. and Kantro, D. L., Water reduction, slump loss and entrained air void systems as influenced by superplasticizers, *Proc. int. symp. superplasticizers concr.*, Ottawa, May 1977.
15. Seabrook, R. T. and Malhotra, V. M., Accelerated strength testing of super-plasticized concrete and effect of repeated doses of superplasticizers on properties of concrete, *Proc. int. symp. superplasticizers concr.*, Ottawa, May 1977.
16. Tuthill, L. H., Adams, R. F. and Hemme, J. M., Observations in testing and use of water-reducing retarders, *ASTM Spec. Tech. Publ.*, No. 266, Oct. 1959.
17. Lafraugh, R. W., The use of superplasticizers in the precast industry, *Proc. int. symp. superplasticizers concr.*, Ottawa, May 1978.
18. Lankard, D. P., *The performance of blended cements*, Interim Technical Report to Federal Energy Administration by Battelle, Columbus, Ohio, Sept. 1977.
19. Jarrige, A., Fly ashes in concrete, *Review of the materials of construction*, 1970, p. 665.
20. Venaut, M., Investigation of fly ashes, *Review of the materials of construction*, 1962, pp. 565–567.
21. Kokubu, M. and Pamada, J., Fly ash cements, *Proc. 6th int. congr. chem. cem.*, Moscow, Sept. 1974.
22. Wallace, G. B., and Ove, E. L., Structural and lean concrete as affected by water reducing and set retarding admixtures, *Proc. Am. Soc. Test. Mater. 3rd Pacific Area Nat. Meet.*, Oct. 1939.
23. Grieb, W. E. and Werner, V., Final report on tests of concrete containing blast furnace slag cement, *Public Roads*, Aug. 1961.
24. ACI Committee 212, Admixtures for concrete, *Concr. Int.*, Vol. 3, No. 5, May 1981, pp. 24–79.

10 Failure criteria for concrete

D W Hobbs

Cement and Concrete Association, Slough, England

Contents

Notation

A	area; area over which pore pressure acts
A_u	true area of contact
C, C_s	material constants
d_s, d_w	shoulder and waist diameters of dumb-bell shaped specimen
E	Young's modulus; elastic modulus
E_a	Young's modulus of aggregate
E_{ct}, E_{c28}	Young's modulus of concrete at t and 28 days
E_{pt}	Young's modulus of cement paste at t days
F	shear or frictional force
f	compressive strength of a cube or plate
f_{cu}	characteristic cube strength
f_{cyl}	compressive strength of a cylinder
f_t, f_{28}	compressive strength of a cube at t and 28 days
K_0	factor related to E_a

$k, k_1, \ldots k_{11}$	constants	σ_n	normal compressive stress
p	pore pressure	σ_0	octahedral normal stress
Q	load	$\sigma_1, \sigma_2, \sigma_3$	major, intermediate and
T_0	uniaxial tensile strength		minor principal stresses
U_t	estimated tensile strength		(compressive stresses are
V_a	aggregate volume concent-		taken to be positive)
	ration	$\sigma_{1f}, \sigma_{2f}, \sigma_{3f}$	principal stresses at failure
w/c	water/cement ratio	σ_{1v}, σ_{3v}	major and minor principal
γ_m	material safety factor		stresses at which volumetric
$\epsilon_1, \epsilon_2, \epsilon_3$	strains in principal stress di-		strain starts to increase
	rections	σ_{1s}, σ_{3s}	serviceability limit state
ϵ_f	strain at failure		stresses
ϵ_t	tensile failure strain	σ_{1u}, σ_{3u}	ultimate limit state stresses
2θ	angle between normal to	τ	shear stress
	Mohr envelope and σ-axis	τ_b	material shear strength
μ	coefficient of internal fric-	τ_c	cohesive strength
	tion	τ_o	octahedral shear stress
ν	Poisson's ratio	ϕ	creep factor
σ	stress	ψ	a function

Summary

In this chapter the literature dealing with the strength, failure criteria and deformation of plain concrete under multiaxial stress is examined. Several stress states are considered including multiaxial compression, biaxial compression, tension plus biaxial compression, and compression plus tension. It is shown that under certain stress combinations concrete can support stresses and loads that are markedly higher than the cube or cylinder crushing strength, while under other stress combinations it is possible that the concrete will crack even though there is no stress acting as high as the uniaxial compressive strength. Thus under some stress combinations there are benefits which can be exploited in design. The lower bounds to the published strength data are used to obtain simple expressions which show whether structural concrete under prescribed stresses is safe according to ultimate and serviceability limit states.

1 Introduction

The strength and deformation properties of concrete are conventionally determined by loading cubes or cylinders in uniaxial compression between steel platens or by loading beams in bending. The strength is defined as the maximum stress supported by the concrete specimen. Within structural concrete, the stress does not always act in a single direction; frequently the stresses act in more than one direction [1]. Such a stress state is termed a complex, multiaxial or combined stress state. At any point in a loaded solid, it is possible to find three planes which are mutually orthogonal, on which no shearing forces act. The three stress components orthogonal to these planes define the stress system at the point and are known as principal stresses, classified as major, σ_1, intermediate, σ_2, and minor, σ_3, in decreasing order of magnitude. As concrete is normally loaded in compression, compressive stresses are taken to be positive.

Under certain combined stress combinations, concrete can support stresses and loads that are markedly higher than the cube or cylinder crushing strength, while under other stress combinations it is possible that the concrete will crack or fail even though there is no stress acting which is as high as the uniaxial compressive strength. Thus, under some stress combinations, there are benefits which can be exploited in design, but for less favourable combinations of stress, there may be an increased risk of cracking or failure if stress limitations are based solely on the uniaxial compressive strength of concrete.

If plain and reinforced concretes are subject to the same combined stress state then it is reasonable to assume that the behaviour of the concrete will be the same. Deformation and failure criteria for plain concrete can therefore be used to predict the deformation and failure of reinforced or prestressed structural elements of any shape of section [2]. In the present chapter, the literature dealing with the strength, failure criteria and deformation of plain concrete under multiaxial stress is examined. It is shown that several of the techniques used to determine the properties of concrete under multiaxial stresses are unsatisfactory. The published data are used to formulate simple expressions that show whether structural concrete under prescribed stresses is safe according to ultimate and serviceability limit states.

2 Combined stress in concrete: six simple examples

A combined stress state in concrete may result from the loading and the shape of the concrete section or from the loading and prestressing forces. Six simple examples of situations in which concrete is subject to combined stress are given below [1].

(a) If a concrete section, such as a column or pile, is restrained from expanding laterally by spiral reinforcement or by a steel tube, the compressive load that it can carry may exceed the uniaxial crushing strength by a substantial margin (Fig. 2-1a).

(b) If a localized compressive load is applied to a large concrete member, the concrete adjacent to the loaded area provides restraint to lateral displacement of the concrete beneath the loaded area. It is possible for the bearing stress to exceed the uniaxial compressive strength, particularly if the reinforcing steel is favourably placed (Fig. 2-1b).

(c) In concrete hinges, the concrete provides its own restraint to loading and compressive stresses several times the concrete cube strength can be supported without causing cracking of the concrete (Fig. 2-1c).

(d) The first three examples depend upon passive restraint for strength enhancement. Greater strength enhancement can be achieved by applying lateral compression to the concrete—prestressed-concrete machine-tool frames which have been used in Russia are an example (Fig. 2-1d).

(e) In some instances, the strength enhancement which is produced from physical restraint to lateral displacement of the concrete is already being

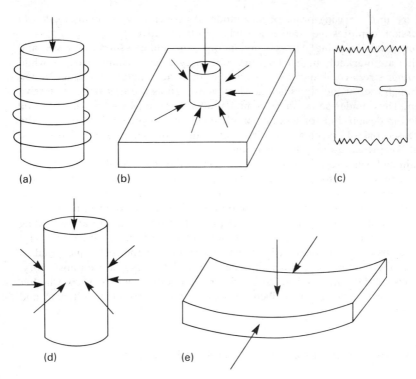

(a)   (b)   (c)

(d)   (e)

Figure 2-1   Some simple examples of combined stress in concrete

exploited. An example is the permissible stresses under bends in reinforcement where the concrete is subject to lateral restraint from the steel. The
British Code of Practice, CP 110, Clause 3.11.6.8, allows the stresses under
bars at working loads to be several times the usual working stress [3].

(f) If a concrete slab is loaded in flexure, the tensile strength may be reduced
by compressive loads in the plane of the slab. This compression could result
from thermal gradients or from loads transmitted from the structure. In the
example shown in Fig. 2-1e, the slab will crack below the load predicted
from a simple flexure test.

The above examples illustrate that concrete does not have a unique strength; its
strength is affected by the stresses which act simultaneously.

3   Behaviour of concrete in uniaxial compression

3.1   Stress–strain behaviour

For most concretes, the aggregate volume concentration ranges from between
0.55 and 0.80 and as a result the elastic properties of the aggregate have a very
significant effect upon the deformation. Stress–strain curves for rock, paste and

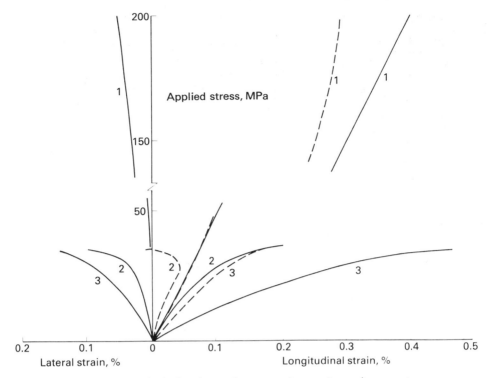

Figure 3-1   Stress–strain behaviour of aggregate, paste and concrete
1   Rock
2   Concrete, water/cement ratio 0.47
3   Paste, water/cement ratio 0.47
   − − − Volumetric strain

concrete loaded in uniaxial compression are illustrated in Fig. 3-1 [4]. Rocks that are used as the aggregate in concrete generally exhibit an approximately linear relationship between stress and longitudinal strain up to failure, whereas both paste and concrete exhibit markedly non-linear behaviour. At low applied stress, the 28-day elastic modulus of the paste fraction ranges from about 6 GPa $(0.87 \times 10^6 \text{ psi})$, when the water/cement ratio is 0.71, to about 15 GPa $(2.18 \times 10^6 \text{ psi})$, when the water/cement ratio is 0.35. The natural aggregates which are often used in concrete have Young's moduli which range from 40 to 100 GPa $(5.80–14.50 \times 10^6 \text{ psi})$.

At working stress levels, the Young's modulus of concrete may be determined from the theoretical expression [4, 5]

$$E_{ct} = E_{pt}\left[1 + \frac{2V_a(E_a - E_{pt})}{E_a + E_{pt} - V_a(E_a - E_{pt})}\right] \tag{3-1}$$

or alternatively from the empirical expression [6–8]

$$E_{ct} = E_{c28}(0.4 + 0.6f_t/f_{28}) \tag{3-2}$$

where E_{c28}, the Young's modulus of the concrete at 28 days, is given by

$$E_{c28} = K_0 + 0.2f_{28} \qquad\qquad (3\text{-}3)$$

E_{ct} and E_{pt} are the Young's moduli of the concrete and its cement paste at time t, E_a is the Young's modulus of the aggregate, V_a is the aggregate volume concentration, and f_t and f_{28} are the cube strengths at t days and 28 days. K_0 is a factor closely related to the Young's modulus of the aggregate and for most aggregates will lie in the range 14–26 GPa (2.03–3.77 × 10^6 psi). Where deflections are of great importance, K_0 should be determined by experiment.

From these equations, it can be noted that the stiffness of the concrete increases as the water/cement ratio is reduced, as the aggregate stiffness is increased and as the aggregate volume is increased. At higher loads, the longitudinal stiffness of concrete decreases continuously (Fig. 3-1), the departure from linearity being more marked for intermediate water/cement ratio concretes than low water/cement ratio concretes (Fig. 3-2). The approximate longitudinal strains at failure for aggregate, paste and concrete are 0.4–1%, 0.6–0.8% and 0.1–0.25%, respectively.

The stress–lateral strain curves for aggregate, paste and concrete are all non-linear (Fig. 3-1), the lateral stiffness decreasing continuously as the applied stress increases. The decrease is generally very marked as the peak stress is

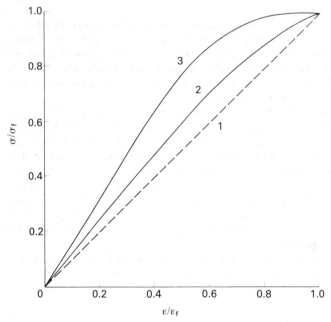

Figure 3-2   Relationship between normalized stress, σ/σ_f, and normalized strain, ϵ/ϵ_f
1   Linear stress–strain curve
2   Water/cement ratio 0.35, $V_a = 0.66$
3   Water/cement ratio 0.59, $V_a = 0.75$

approached. The approximate lateral strains when failure occurs are, for aggregates, 0.1 to 0.5 of their longitudinal failure strains, and, for pastes and concretes 0.3 to 0.5 of their longitudinal failure strains.

The stress–volumetric strain curves for aggregate, paste and concrete are generally non-linear (Fig. 3-1). For pastes there is a continuous reduction in volume as the applied stress is increased, and as the stress increases the slope of the stress–volumetric strain curve decreases. The behaviour of concrete is of similar form except that, at a stress of about 70 to 90% of the peak stress sustained, the volumetric strain increases. The approximate volumetric strain at this stress lies between 0.04 and 0.08%.

3.2 Compressive strength

The strength of concrete in compression may be governed by one or more of the following: (a) the strength of the bond between the cement paste and the aggregate, (b) the strength of the cement paste, and (c) the strength of the aggregate. It is generally considered that if the aggregate is stronger than the paste, then the paste–aggregate bond strength governs failure in compression [9, 10–12]. However, the strength of well-compacted concretes can be expressed as a function of the water/cement ratio, w/c, of the paste, namely,

$$f = k_1 k_2^{-(w/c)} \tag{3-4}$$

where k_1 and k_2 are constants, which implies that the compressive strength of concrete is governed by the strength of its paste fraction. Aggregate volume concentration [13] and aggregate grading [14] have little effect upon strength but there is evidence that changes in aggregate type can significantly affect strength [4]. For example, the use of crushed aggregate can give strengths up to 40% higher than those when a gravel is used.

It is apparent from Fig. 3-1 that damage in concrete occurs at applied stresses well below the maximum stress supported by the concrete. Some investigators [9] have deduced that small cracks are formed in concrete before loading and that upon loading these cracks propagate and grow. There is, however, not complete agreement regarding the mechanism of failure. Two possible modes of failure are illustrated in Fig. 3-3. In Fig. 3-3a, progressive damage is assumed to be caused by the formation of tensile cracks at the ends of a number of sliding shear cracks together with some crack opening along the cracks, whilst in Fig. 3-3b progressive damage is caused by the growth of the relatively long narrow cracks orientated in the direction of the applied stress.

A number of investigators [15–17] take the view that the fracture process can be divided into a number of discrete stages and have suggested that critical or discontinuity stress levels exist. This is not supported by the work of Spooner and Dougill [18]. To assess damage, concrete specimens were subjected to a series of cycles of loading and unloading and it was assumed that the reduction in the initial slope of the reloading curve is related to structural degradation of the material. Spooner and Dougill found that damage during the first loading cycle occurs in concrete at very low strains, about 200 micro-strains, and that once started the process is continuous. For the concretes tested, no evidence to support the existence of any 'critical' or 'discontinuity' stress was found.

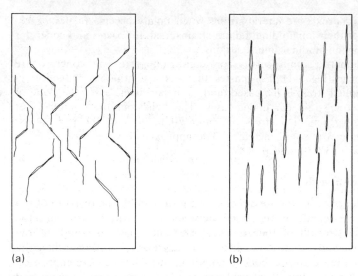

(a)   (b)

Figure 3-3   Two possible modes of failure

The stress at which the volumetric strain commences to increase is believed by some concrete technologists [15, 16] to be of particular significance since at this stress the rate at which the volumetric strain is increased by crack opening exceeds the rate at which it is reduced by elastic contraction and by compaction. Certainly some evidence does exist to support this view [19]. It can be noted from Fig. 3-4 that the long-term strength is approximately 80% of the peak stress

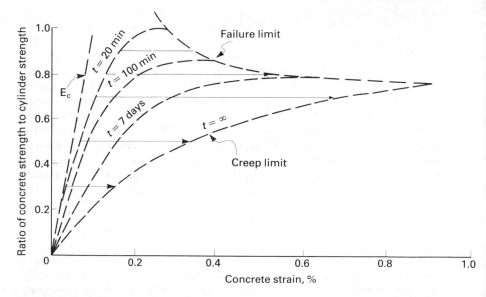

Figure 3-4   Influence of applied stress and time on concrete strain [19]

supported under short-term loading which is similar to the stress at which the volumetric strain increases.

3.3 Problems associated with the uniaxial compression test

In a uniaxial compression test, a cube, a prism or a cylinder with plane and parallel end faces is loaded at a constant rate between steel platens. Due to the rigidity of the steel platens, the concrete is subject to approximately uniform displacement at the steel–concrete interface and the lateral expansion of the concrete is reduced by the frictional restraint which develops at the interface. Concrete is therefore subject to a non-uniform distribution of stress which influences the strength and stress–strain behaviour of the specimens. The stress variations close to the loaded boundary are probably greater than those in the interior of the concrete where a condition of plain strain does not exist.

To eliminate frictional restraint when an elastic specimen is loaded, it is necessary to ensure that the lateral movements of the platen and specimen are the same. This can be achieved by using platens with the same ratio of Poisson's ratio to Young's modulus and with the same cross-section as the specimen. However, concrete is neither elastic nor homogeneous and this results in a non-uniform distribution of stress in the specimen regardless of the elastic properties of the platens.

The frictional restraint can be reduced by the use of lubricants or deformable layers at the loaded interface, but such techniques may set up radial tensile stresses which induce longitudinal splitting of the concrete. This is because the lubricant or deformable layer may be forced into the ends of the specimen or alternatively the deformable layer may extrude.

4   Behaviour of concrete in uniaxial tension

4.1 Stress–strain behaviour

No standard method exists for determining the uniaxial tensile strength of rocks, paste or aggregate. This is due to their low tensile failure strains which make it very difficult to induce an approximately uniform tensile strain and as a result little work on the stress–strain behaviour in uniaxial tension is reported in the literature. The work reported [4] shows that the stress–longitudinal strain curve for concrete deviates from proportionality at a stress well below the maximum stress sustained. Data on the complete stress–strain curves for concrete subject to direct tension have been reported by Evans and Marathe [20]. The elastic modulus in tension may be determined either from Eqn 3-1 or 3-2. The longitudinal tensile strain at failure ranges from about 0.005 to 0.01% and the lateral strain when failure occurs is about 0.002%.

4.2 Tensile strength

Changes in the water/cement ratio, aggregate grading and aggregate type affect tensile strength in a different way to compressive strength. The ratio of uniaxial tensile strength to compressive strength ranges from about 1/20 for a concrete of high strength to about 1/10 for a concrete of medium strength [21]. Changes in aggregate grading have been found to have a greater influence on tensile strength than compressive strength and it has been shown that tensile strength is dependent upon the type of aggregate used [21, 22]. The use of calcareous aggregates instead of siliceous aggregates increases tensile strength more than compressive strength and this has been attributed to the stronger paste–aggregate bond [23].

The phenomenon of bleeding in concrete, which results in water being trapped between aggregate particles, has been found by Hughes and Ash [24] to give differences of more than 2/1 between the tensile strength parallel and perpendicular to the direction of casting.

5 Behaviour of concrete under biaxial stress

5.1 Loading through solid platens

A number of investigators have attempted to determine the strength behaviour of concrete under biaxial compression [4, 25–27] and tension–compression [25, 28] by loading specimens in two orthogonal directions between pairs of dry or 'lubricated' solid platens. The latter platens were used in order to reduce frictional restraint. The technique used for biaxial compression together with some of the results obtained are shown in Fig. 5-1. The various investigators have obtained widely different results; for example, the ratio of strength under equal biaxial compression to the uniaxial compressive strength ranges from 3.6 [29] to 1.0 [25, 30], the highest strengths being obtained when concrete is loaded between dry platens. When concrete is loaded between lubricated platens, it is generally found that the strength in equal biaxial compression is between 10 and 30% higher than the strength in uniaxial compression.

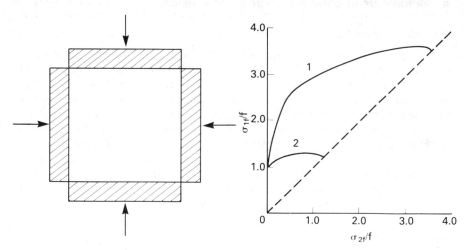

Figure 5-1 Biaxial compression through solid platens
1 Sundara Raja Iyengar *et al.* [29]
2 Vile [25]

With the above loading technique, it is assumed that the applied loads are distributed solely in the concrete specimen. This assumption is invalid since each load is carried partly by the concrete specimen and partly by the loading platens (and rams) acting in the transverse direction. Consequently, this technique overestimates strength [31].

5.2 Loading through brush platens

The frictional restraint which occurs when concrete is loaded through solid steel platens can be reduced by loading through brush platens. Several investigators [26, 27, 32, 33] have used this technique to study the behaviour of concrete under biaxial compression, compression–tension and biaxial tension. The individual filaments of the platens were 0.64 mm in diameter or 4×4 mm^2 in cross-section and were designed to load surface areas ranging from 50×50 mm^2 to 177×126 mm^2. Strength results obtained by Kupfer *et al.* [32] on concrete plates with a uniaxial compressive strength of 31 MPa (4496 psi), under biaxial stress, are given in Fig. 5-2. Similar results were obtained by Nelissen [33], Linse and Taylor [26]. When brush platens are used it is found that the strength in biaxial compression can be about 20% higher than the strength in uniaxial compression, that the tensile stress at failure decreases as a simultaneously acting compressive stress is increased, and that the strength in biaxial tension is approximately equal to the uniaxial tensile strength.

The stress–strain behaviour of concrete under biaxial stress is shown in Fig. 5-3. In biaxial tension, the concretes tested show approximately Hookean behaviour, whilst in biaxial compression the behaviour is similar to that in uniaxial compression. In tension–compression, when the compressive stress is at least five times the tensile stress, and in biaxial compression, the stress at minimum volume is approximately 80% of the maximum stress sustained.

Although the use of brush platens largely eliminates frictional restraint, the local stress variations at the loaded boundary due to the inhomogeneity of the

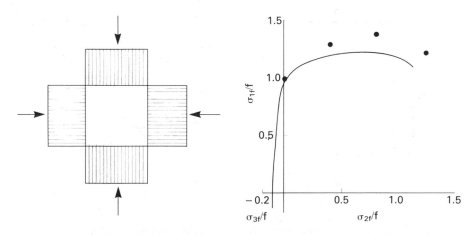

Figure 5-2 Biaxial stress through brush platens [32]
• Specimens loaded by fluid pressure [26]

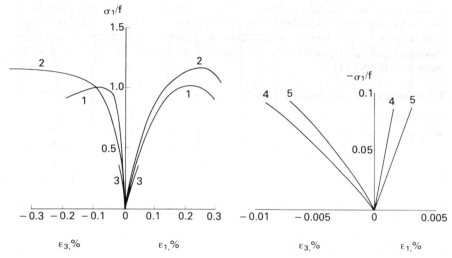

Figure 5-3   Stress–strain behaviour of concrete under biaxial stress
1   $\sigma_1/\sigma_2 = 1/0$
2   $\sigma_1/\sigma_2 = 1/1$
3   $\sigma_1/\sigma_3 = 1/-0.2$
4   $\sigma_1/\sigma_3 = 0/-1$
5   $\sigma_1/\sigma_3 = -1/-1$

concrete are still present. In addition, an indeterminate stress distribution is produced in the concrete around the circumference of each filament. Thus, loading through brush platens may initiate failure in concrete close to the platen rather than in the general body of the concrete and it is therefore not strictly correct to assume that the failure stress is given by the average applied stress.

5.3   Loading by fluid pressure

An alternative method of loading concrete in biaxial compression is by fluid pressure acting through a deformable cushion or membrane in contact with the concrete. The three techniques which have been used are shown in Fig. 5-4. Techniques (b) and (c) are limited to equal biaxial compression in which the radial stress equals the hoop stress. Parrott [34] has shown that the fluid loading type of platen applies a uniform stress and that the transverse stiffness of the fluid platen cannot be detected. Results obtained by Gerstle and Hon-yim Ko [26] using technique (a) are plotted in Fig. 5-2. The observed strength is similar to that observed with brush platens. In the case of techniques (b) and (c), the strength under equal biaxial compression is generally found to be between 0 and 15% [4, 31, 35–37] higher than the uniaxial compressive strength.

The stress–strain results obtained using fluid platens are similar to those obtained using brush platens (Fig. 5-3).

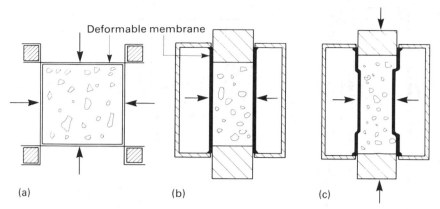

Figure 5-4   Loading concrete by fluid pressure acting through a flexible membrane

5.4   Loading and stressing hollow cylinders

To determine the strength of concrete in tension–compression, several investigators [2, 4] have subjected hollow cylinders of concrete either to torsion simultaneously with axial compression or alternatively to internal pressure and axial compression [4, 38]. Bellamy [39] determined the strength of concrete under biaxial compression by subjecting hollow cylinders to axial compression and external pressure. The stresses at failure were calculated from both thick-walled and thin-walled elastic theory. This latter theory assumes a uniform stress distribution across the wall. These investigators found that the tensile stress at failure decreased as a simultaneously acting compressive stress was increased and that under equal biaxial compression a stress three times as high as the uniaxial strength could be supported.

There is a number of serious objections which can be raised against the use of the above techniques. First, the hollow cylinders of concrete tested cannot be regarded as being homogeneous and isotropic since the maximum particle sizes used ranged from 0.18 to 0.63 of the wall thickness, consequently neither thin or thick-walled theory is justified. Secondly, a large stress gradient is induced across the wall by the applied couple, the internal pressure or the external pressure. So, when the specimens are broken only by an axial load, failure is equally likely to initiate anywhere in the concrete sample, but under biaxial stress failure is localized either at the external or internal circumference of the hollow cylinder where the stresses have their maximum values. This a serious objection since materials can survive greater stresses when they are localized than when they are approximately uniformly distributed.

6   Behaviour of concrete under triaxial stress

6.1   The conventional triaxial compression test: strength

The most widely used method of loading concrete under combined stress is the standard triaxial compression test. In this test, a concrete cylinder is subjected to a

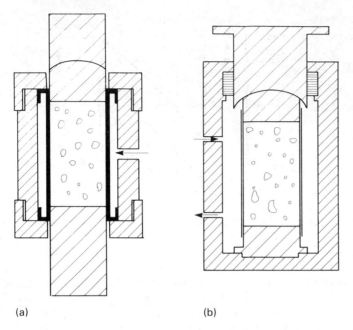

(a)   (b)

Figure 6-1   Apparatus used in the conventional triaxial compression test

fluid pressure along its length, $\sigma_2 = \sigma_3$, and is loaded axially, generally through steel platens by a larger compressive stress, σ_1. The length-to-diameter ratio of the specimens is generally between 2/1 and 3/1. Ingress of the pressure fluid into the concrete specimen is prevented by an impermeable membrane placed along the length of the cylinder. The use of the membrane restricts the development of pore-water pressure in the sample. Two typical triaxial pressure cells are shown in Fig. 6-1. In the triaxial compression test, it is preferable to use pistons and end pieces of the same cross-section as the concrete specimens, otherwise the confining pressure produces an axial component of stress upon the piston and also, due to volume changes of the cell, the control of the confining pressure is more difficult.

The first triaxial tests on jacketed samples of concrete were reported by Richart et al. [35] in 1928. Earlier tests on jacketed samples of mortar crushed in water under pressure were reported by Considere in 1903 and 1906 [40]. Since the late 1940s a considerable amount of work has been published on the strength of concretes, mortars and pastes using the conventional triaxial compression test [4]. The specimens tested were mainly moist-cured or water-cured for 27 days or more and in many instances were subsequently dried for a day or more prior to testing. Some workers dried the concretes at temperatures in excess of 50°C [41–44] and in some experiments the specimens were also dried in a vacuum dessicator [44].

In all the reported work on concretes, it has been found that the application of a confining pressure produces large increases in strength (Figs 6-2 and 6-3). Paste specimens, when subjected to a confining pressure, show a lower gain in strength

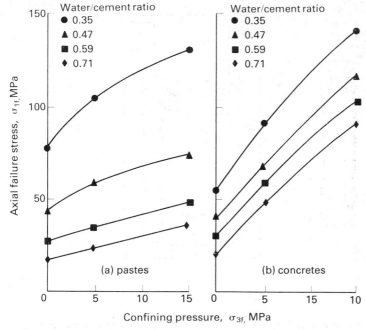

Figure 6-2   **Influence of water/cement ratio upon the strength of pastes and concretes in triaxial compression: age 56 days**

than concretes of a similar water/cement ratio and the resulting relationship between the principal stresses at failure has a more pronounced curvature [44, 45] (Fig. 6-2).

The axial failure stress for concrete, σ_{1f}, may be expressed approximately in terms of the confining pressure, σ_{3f}, and the uniaxial cylinder compressive strength by

$$\sigma_{1f} = k_1 \sigma_{3f} + f_{cyl} \tag{6-1}$$

where $\sigma_{3f} < f_{cyl}$ and k_1 is about 4 or 5. Mix proportions, aggregate type and grading, and cement properties only have a small influence on the magnitude of k_1. This equation only approximately fits the results because the slope of the curve for the axial failure stress plotted against confining pressure decreases with increasing confining pressure. A more accurate empirical expression is

$$\sigma_{1f} = k_2 \sigma_{3f}^{k_3} + f_{cyl} \tag{6-2}$$

where k_2 and k_3 are constants and where $k_3 < 1$. When normalized, the two previous relationships become

$$\sigma_{1f}/f_{cyl} = (k\sigma_{3f}/f_{cyl}) + 1 \tag{6-3}$$

and

$$\sigma_{1f}/f_{cyl} = k_4 (\sigma_{3f}/f_{cyl})^{k_5} + 1 \tag{6-4}$$

where $k_4 \simeq 4$ and $k_5 \simeq 0.9$.

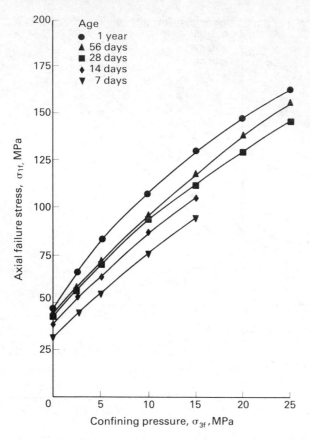

Figure 6-3   Influence of specimen age on the strength of concrete under triaxial compression: water/cement ratio 0.47

Equation 6-3, when $k_1 = 4$, is shown plotted in Fig. 6-4 together with the experimental strength results obtained on wet and air-dried concrete by a number of investigators [35, 36, 42, 43, 45–47]. In the case of oven-dried concrete, a greater increase in strength is produced by restraint [44]. A lower bound to the triaxial strength data is given by

$$\sigma_{1f}/f_{cyl} = 3(\sigma_{3f}/f_{cyl}) + 1 \qquad (6\text{-}5)$$

An alternative measure of concrete strength is the stress at minimum volume, σ_{1v}; this is the stress at which the volumetric strain starts to increase. Hobbs [45] found that the ratio of the minimum volume stress to the ultimate strength ranged from about 0.7 to 0.95 and was approximately independent of confining pressure, mix proportions or concrete age. The results obtained by Hobbs are shown plotted in Fig. 6-5. The equation of the line plotted is

$$\sigma_{1v}/f_{cyl} = 0.86 + 4.0\sigma_{3v}/f_{cyl} \qquad (6\text{-}6)$$

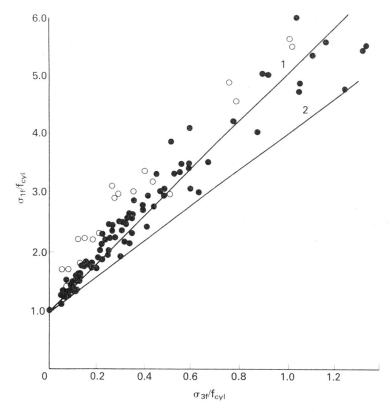

Figure 6-4 Strength of concrete under triaxial compression: results obtained by a number of investigators
○   Oven-dried concrete
1   Equation 6-3, $k_1 = 4$
2   Lower bound

The comparable ultimate strength equation obtained by Hobbs was

$$\sigma_{1f}/f_{cyl} = 1 + 4.8\sigma_{3f}/f_{cyl} \tag{6-7}$$

6.2   The conventional triaxial compression test: stress–strain behaviour

The effect of restraint upon the mode of failure of concrete is illustrated in Fig. 6-6. At confining pressures of less than about 5 MPa (725 psi), an apparently intact cone is formed at one or both ends of the specimen, the remaining parts of the specimen having a number of cracks running roughly parallel to the axis of the specimen. Concrete specimens tested at confining pressures greater than 5 MPa often exhibit gross failure and barrelling in the manner illustrated in the figure without cracking visible on the surface. Some specimens fail along a single plane inclined at an angle of between 20 and 30° to the specimen axis.

The stress–strain behaviour of concrete in triaxial compression is influenced by mix proportions [45] (Fig. 6-7). At low values of the differential stress, $\sigma_1 - \sigma_3$,

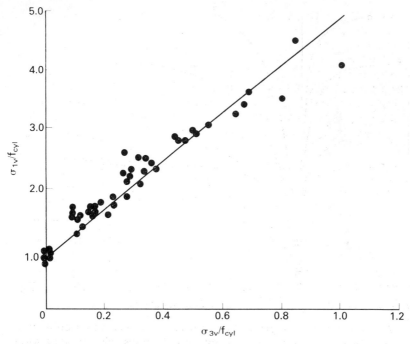

Figure 6-5 Relationship between normalized principal stresses at minimum volume
$$\sigma_{1v}/f_{cyl} = 0.86 + 4.0\ (\sigma_{3v}/f_{cyl})$$

the Young's modulus and Poisson's ratio of concrete are essentially independent of the applied stress system provided the confining pressure is less than $0.3f_{cyl}$. However, when the confining pressure is greater than $0.3f_{cyl}$, the Young's modulus and Poisson's ratio decrease with increasing σ_3 [45, 47], falling to roughly 50% of their value in uniaxial compression. Poisson's ratio falls from between 0.1 and 0.15 at $0 < \sigma_3 < 0.3f_{cyl}$ to about 0.05 at $\sigma_3 = f_{cu}$. The magnitude of the

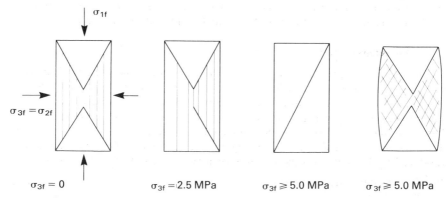

Figure 6-6   The effect of restraint upon the mode of failure of concrete

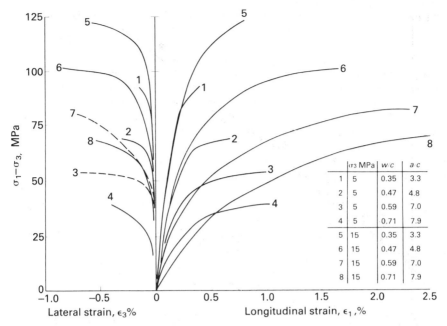

Figure 6-7   Influence of mix proportions upon the stress–strain behaviour of concrete subject to triaxial compression: age 56 days, flint gravel aggregate

measured Poisson's ratio is therefore lower than the figure, namely 0.2, quoted in CP 110 [3]. As the applied stress is increased, it can be seen from Fig. 6-7 that there is a continuous reduction in stiffness and further as the confining pressure is increased the strains which concrete can sustain prior to failure increase markedly. Newman [48] has reported stress–strain data for concrete up to confining pressures as high as 138 MPa (20 000 psi).

It is possible to express the stress–strain behaviour of concrete in triaxial compression by empirical equations. This has been done by Hobbs [45], Credolin *et al.* [49], Kotsovos and Newman [17] and Bazant and Tsubaki [50].

6.3   The extension test

In the conventional triaxial compression test, the intermediate principal stress is the same as the minor principal stress. It is also possible, using the triaxial compression cell, to carry out tests in which the intermediate principal stress is the same as the major principal stress. Such a test is known as an extension test. Extension tests can also be carried out under biaxial compression plus tension. These tests are carried out by stressing cylinders or dumbbells. The axial stress produced in the waist of a dumb-bell specimen is given by the expression

$$\sigma_3 = (4Q/d_w^2) - \sigma_1(d_s^2 - d_w^2)/d_w^2 \tag{6-8}$$

where σ_1 is the confining pressure and d_s and d_w are the diameters of the shoulder

and waist of the dumb-bell specimen, respectively. Thus, by varying the axial load
Q or by varying the waist diameter d_w it is possible to vary the ratio σ_3/σ_1 at
failure.

In the extension test, the concrete specimens crack along a single plane aligned
at right angles to the minimum principal stress. The ultimate strength results
obtained by Hobbs [31, 36, 45] and Kotsovos [37] are summarized in Fig. 6-8, the
experimental results being normalized with respect to the uniaxial compressive
strength. As the tensile stress is reduced, e.g., as the minimum principal stress is
increased, the compressive stress which the concrete can sustain increases, or
conversely as the major (compressive) principal stress is increased the tensile
stress which the concrete can sustain decreases. The shape of the failure envelope
in the tension–biaxial compression quadrant is dependent upon the water/cement

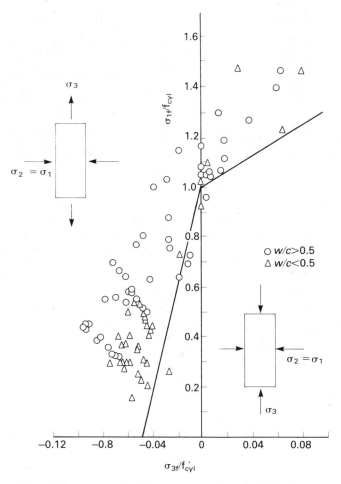

Figure 6-8   Strength of concrete under triaxial stress: extension test results,
$\sigma_{1f} = \sigma_{2f} > \sigma_{3f}$

ratio. This is because variations in the water/cement ratio have less effect on concrete's tensile strength than on its compressive strength.

The lower bounds to the strength data shown plotted in Fig. 6-8 may be taken to be

$$\sigma_{1f}/f_{cyl} = 3(\sigma_{3f}/f_{cyl}) + 1 \qquad \text{when } \sigma_{3f} > 0 \qquad\qquad (6\text{-}9)$$

and

$$\sigma_{1f}/f_{cyl} = 1 + 20(\sigma_{3f}/f_{cyl}) \qquad \text{when } \sigma_{3f} < 0 \qquad\qquad (6\text{-}10)$$

7 Influence of the intermediate principal stress

The most satisfactory techniques which have been used for loading concrete under multiaxial stress are

(a) Loading plates or cubes of concrete through brush platens—biaxial compression, tension–compression and biaxial tension.
(b) Loading plates, cubes or cylinders of concrete by fluid pressure acting through a membrane or cushion—biaxial compression.
(c) Combined loading of cylinders and dumb-bells of concrete by fluid pressure acting through a membrane plus loading through solid platens—triaxial compression and tension plus biaxial compression.

With these tests, the average stresses acting in the concrete are known and they are approximately uniform.

The results obtained using the above techniques are summarized in Figs 7-1

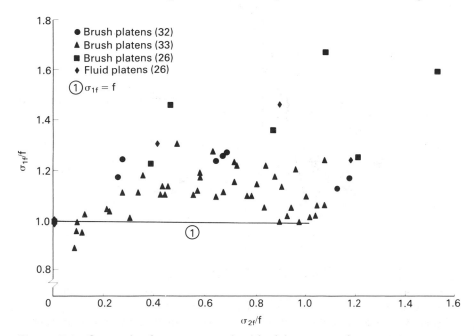

Figure 7-1 Strength of concrete under biaxial compression

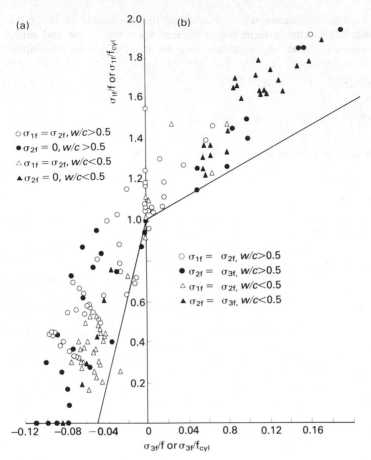

Figure 7-2 Strength of concrete under multiaxial stress when σ_{3f}/f and σ_{3f}/f_{cyl} are less than 0.20 [31–33, 36, 37, 45, 51]

and 7-2. Figure 7-1 shows strength results obtained in biaxial compression [26, 32, 33]. Figure 7-2a compares strength results obtained in tension–compression with results obtained in tension–biaxial compression. In the former case, the intermediate principal stress is zero; in the latter case, it is equal to the maximum compressive stress. Figure 7-2b compares the ultimate strength results obtained using the conventional triaxial compression test with these obtained using the 'extension' test. In the former case, the intermediate principal stress equals the smaller compressive stress; in the latter case, it equals the larger compressive stress.

Examination of Fig. 7-1 shows that the strength of concrete under biaxial compression is higher than the strength under uniaxial compressive stress, but the increase is small and for engineering purposes is of little significance. A reasonable lower bound to the biaxial compressive strength data is

$$\sigma_{1f}/f_{cyl} = 1 \qquad \text{when } \sigma_1 > \sigma_2 > \sigma_3 = 0 \qquad (7\text{-}1)$$

Examination of Fig. 7-2 shows no evidence that the magnitude of the intermediate principal stress has a significant influence upon the major and minor principal stresses at failure. It may therefore be concluded that the major principal stress which can be supported by concrete is dependent primarily upon the minor principal stress acting on the concrete, thus

$$\sigma_{1f} = \psi(\sigma_{3f}) \tag{7-2}$$

This conclusion is not accepted by many concrete technologists but is compatible with the experimental results. The lower bounds to the strength data shown plotted in Figs 7-1 and 7-2 are

$$\sigma_{1f}/f_{cyl} = 3(\sigma_{3f}/f_{cyl}) + 1 \qquad \text{when } \sigma_1 > \sigma_2 > \sigma_3 \geqslant 0 \tag{7-3}$$

and

$$\sigma_{1f}/f_{cyl} = 1 + 20(\sigma_{3f}/f_{cyl}) \qquad \text{when } \sigma_1 > \sigma_2 > 0 > \sigma_3 \tag{7-4}$$

8 Influence of pore pressure

Hobbs [36] has carried out both drained and undrained tests on concrete and found that the strength of concrete in drained tests was only about 3% higher than in the undrained tests. The increase in strength produced by restraint is therefore essentially independent of the degree of saturation of the concrete and Hobbs concluded that only low pore water pressures were developed in concrete in the undrained tests and therefore that the specimens were not water saturated. However, if concrete is subject to a hydrostatic pressure and the pressurizing fluid is free to enter all the pores in the concrete or alternatively enters the concrete accidentally through a punctured membrane then a sharp reduction in strength is observed. The pore pressure itself produces no damage in the concrete and failure only occurs if an external load is applied.

If a pore pressure p acts over an internal area A per unit area of the concrete, then the reduction in confining pressure by the pore pressure is Ap. The effective confining pressure is $\sigma_3 - Ap$, and if $\sigma_3 = p$ the effective confining pressure becomes $\sigma_3(1 - A)$. Thus, the strength when a confining pressure p exists in the concrete may be taken to be

$$\sigma_{1f} - Ap = k_1(\sigma_{3f} - Ap) + f_{cyl} \tag{8-1}$$

where $\sigma_{1f} - Ap$ and $\sigma_{3f} - Ap$ are known as the effective stresses. Terzaghi and Rendulic [52] suggested that for very porous materials $A \to 1$. Several workers [43, 44, 52–54] have in fact shown that the effective porosity of concrete, close to failure, is unity. Thus, the strength of concrete when the pore pressure is equal to p is given by

$$\sigma_{1f} - p = k_1(\sigma_{3f} - p) + f_{cyl} \tag{8-2}$$

So, if σ_{3f} is produced by fluid pressure and the fluid is free to enter the concrete pores the strength equation becomes

$$\sigma_{1f} - \sigma_{3f} = f_{cyl} \tag{8-3}$$

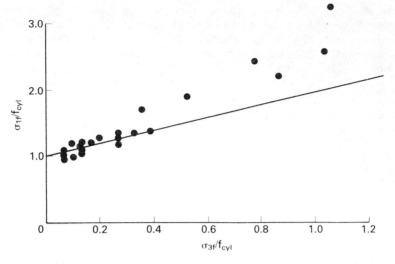

Figure 8-1 Triaxial compression data obtained on unsealed concrete specimens [41, 44]

This equation gives a reasonable lower bound to triaxial compression data obtained on unsealed concrete specimens (Fig. 8-1).

9 Strength of lightweight concretes under multiaxial stress

Little experimental work has been published on lightweight concretes. Some results recently obtained by the author are given in Table 9-1, and are plotted in Fig. 9-1 together with results obtained by Hansen [46]. The influence of the minor principal stress is not as great as for normal-weight concretes. The lower bound to the strength results shown plotted in Fig. 9-1 is

$$\sigma_{1f}/f_{cyl} = 2(\sigma_{3f}/f_{cyl}) + 1 \qquad \text{when } \sigma_3 \geqslant 0 \qquad (9\text{-}1)$$

Table 9-1 Triaxial compression results obtained on two lightweight concretes[a]

Confining pressure, σ_{3f}(MPa)	0	5	10	15	20
Lightweight A, σ_{1f}(MPa)	35.3	49.7	59.8	76.6	81.6
Lightweight B, σ_{1f}(MPa)	24.1	41.7	53.3	58.1	67.8

[a] Free water/cement ratio 0.59; age 4 months; lightweight concrete A contained lightweight coarse aggregate and flint sand; lightweight concrete B contained lightweight coarse and fine aggregate. The lightweight aggregate was pelletized and sintered pulverized fuel ash.

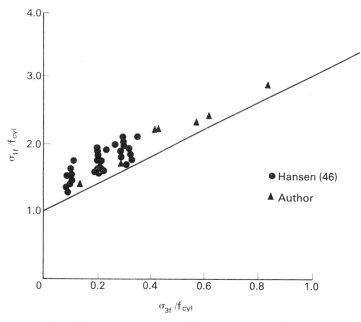

Figure 9-1   Strength of lightweight concrete under multiaxial compression

This equation may only be taken to be applicable to the lightweight concretes tested. It is unlikely, for example, to be applicable to concretes containing polystyrene.

10   Criteria for failure

On the basis of the work reported in [26], Gerstle *et al.* concluded that the strength of concrete under multiaxial stress is independent of the load history through which this failure state is attained. If, therefore, the assumption is made that failure occurs at the peak stress then the failure criterion for concrete subject to multiaxial stress may be expressed in terms of the principal stresses by

$$\sigma_{1f} = \psi(\sigma_{2f}, \sigma_{3f}) \qquad\qquad (10\text{-}1)$$

To obtain a failure criterion for concrete which adequately describes failure, various simple empirical assumptions and various simple physical assumptions have been made. A number of these criteria are discussed in this section.

10.1   Empirical failure criteria

10.1.1   *The maximum tensile strain criterion*
This criterion has recently been applied to concrete by Lowe [55]. It assumes that concrete is a linear elastic material up to a limiting value of principal tensile strain ϵ_t. At this strain, failure is assumed to occur. Thus, in uniaxial compression,

failure occurs when

$$f_{cyl} = E_c \epsilon_t / \nu \tag{10-2}$$

where ν is Poisson's ratio. Under multiaxial stress, the strain in the direction of the minimum principal stress is given by

$$\epsilon_3 = [\sigma_3 - \nu(\sigma_1 + \sigma_2)]/E_c \tag{10-3}$$

Substituting Eqn 10.2 into Eqn 10.3 gives

$$\sigma_{1f} = (\sigma_{3f}/\nu) - \sigma_{2u} + f_{cyl} \tag{10-4}$$

This simple criterion involves all three principal stresses. For the conventional triaxial compression test, $\sigma_2 = \sigma_3$, it becomes

$$\sigma_{1f} = 4\sigma_{3f} + f_{cyl} \qquad \text{when } \nu = 0.2 \tag{10-5}$$

For the extension test, Eqn 10-4 becomes

$$\sigma_{1f} = 2.5\sigma_{3f} + f_{cyl}/2 \quad \text{when } \nu = 0.2 \tag{10-6}$$

and for biaxial stress Eqn 10-4 becomes

$$\sigma_{1f} + \sigma_{2f} = f_{cyl} \tag{10-7}$$

It is apparent from these equations that the maximum tensile strain criterion correctly predicts the strength behaviour in the conventional triaxial compression test but incorrectly predicts that the strength goes down as the intermediate stress is increased. Also, the mode of failure is not in agreement with this criterion since concrete, when subject to multiaxial compression, can support very large 'tensile' strains.

10.1.2 *Maximum octahedral shear stress criterion*
This criterion assumes that failure occurs when the octahedral shear stress, τ_0, reaches a value C which is characteristic of the material:

$$[(\sigma_{1f} - \sigma_{2f})^2 + (\sigma_{2f} - \sigma_{3f})^2 + (\sigma_{3f} - \sigma_{1f})^2]^{\frac{1}{2}}/3 = C \tag{10-8}$$

For the conventional triaxial compression test and the extension test this criterion predicts

$$\sigma_{1f} = \sigma_{3f} + 3C/2^{\frac{1}{2}} \tag{10-9}$$

and for biaxial stress predicts

$$\sigma_{1f}^2 + \sigma_{2f}^2 - \sigma_{1f}\sigma_{2f} = 9C^2/2 \tag{10-10}$$

Thus, from Eqn 10-10, $\sigma_{1f} = f_{cyl}$ when $\sigma_{1f} = \sigma_{2f}$ and $\sigma_{1f} = 1.15f_{cyl}$ when $\sigma_{2f} = \sigma_{1f}/2$. The maximum octahedral shear stress criterion is in reasonable agreement with results obtained under biaxial compression but incorrectly predicts the behaviour of concrete under multiaxial compression and also when one or more of the stresses is tensile.

Bresler and Pister [2] have suggested that at failure the octahedral shear stress is a function of the octahedral normal stress σ_0, thus

$$\tau_0 = \psi(\sigma_0) \tag{10-11}$$

where $\sigma_0 = (\sigma_{1f} + \sigma_{2f} + \sigma_{3f})/3$. The form of the function $\psi(\sigma_0)$ is left to experiment. Bresler and Pister found that the experimental data could be closely approximated by a quadratic equation but several investigators [49, 56–58] have concluded that the octahedral shear stress cannot be expressed as a single simple function of the octahedral normal stress. This can be shown to be largely a consequence of assuming that the magnitude of the intermediate principal stress has a significant effect upon the major and minor principal stresses at failure.

10.1.3 Additional empirical failure criterion

In Section 6-1 it was shown that the strength of concrete under triaxial compression, $\sigma_2 = \sigma_3$, could be represented by the empirical Eqns 6-3 and 6-4. A more general equation applicable to all combined stress states has been proposed by Hobbs [45], namely,

$$\sigma_{1f} = f_{cyl}[(\sigma_{3f}/u_t) + 1]^{k_6} \tag{10-12}$$

where U_t, the estimated tensile strength, and k_6 are given by

$$U_t = (0.018 \pm 0.004)f_{cyl} + 2.3 \pm 0.1 \tag{10-13}$$

$$k_6 = (7.7 \pm 0.6)/f_{cyl} + 0.40 \pm 0.02 \tag{10-14}$$

Equations 10-12 to 10-14 give a close fit to the experimental results obtained on a range of concretes tested under triaxial compression and extension (Fig. 10-1). In the tension-compression or tension-biaxial compression quadrants

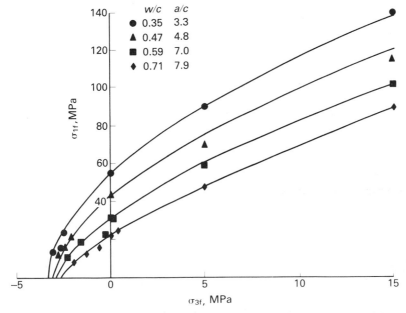

Figure 10-1 Comparison between the strengths calculated from Eqns 10–12 to 10–14 and some results obtained by Hobbs [45]: age 56 days, flint gravel aggregate

the equations are strictly only applicable to the concretes tested since the uniaxial tensile strength is more affected by aggregate type, aggregate grading and casting direction than is compressive strength.

10.2   Tensile failure criteria

10.2.1   *Griffith's criterion*
Griffith [59] assumed that brittle materials contained minute flaws and showed that, even when the applied stresses were compressive, tensile stresses could be set up close to the ends of suitably orientated flaws of sufficient magnitude to cause tensile failure. In the case of a material containing elliptically shaped cracks subject to a two–dimensional stress system, Griffith's criterion predicts that failure occurs when

$$(\sigma_{1f} - \sigma_{3f})^2 = 8T_0(\sigma_{1f} + \sigma_{3f}) \qquad \text{if } \sigma_1 + 3\sigma_3 > 0 \tag{10-15}$$

and

$$\sigma_{3f} = -T_0 \qquad \text{if } \sigma_1 + 3\sigma_3 < 0 \tag{10-16}$$

where T_0 is the uniaxial tensile strength. The equation of the Mohr envelope is

$$\tau^2 = 4T_0(\sigma + T_0) \tag{10-17}$$

When $\sigma_{3f} = 0$, $\sigma_{1f} = 8T_0 = f_{cyl}$; when $\sigma_{3f} = f_{cyl}$, $\sigma_{1f} = 3f_{cyl}$; and when $\sigma_{3f} = \frac{1}{2}f_{cyl}$, $\sigma_{1f} = 2.2f_{cyl}$. Thus, the predictions from Griffith's theory are in approximate agreement with the observed behaviour of concrete. The observed increase in concrete strength, under multiaxial compression is, however, greater than that predicted possibly because the flaws and cracks close under compression. To take this factor into account McClintock and Walsh [60] modified Griffith's theory and showed that, when the compressive stresses are sufficiently high to close all the cracks, the Griffith's criterion becomes identical with Coulomb's criterion.

10.2.2   *Extended Griffith's criterion*
Griffith's theory is only applicable to two dimensions. A possible extension to three dimensions proposed by Murrell [61, 62] is

$$(\sigma_{1f} - \sigma_{2f})^2 + (\sigma_{2f} - \sigma_{3f})^2 + (\sigma_{3f} - \sigma_{1f})^2 = 24T_0(\sigma_{1f} + \sigma_{2f} + \sigma_{3f}) \tag{10-18}$$

so $f_{cyl} = 12T_0$. For the triaxial test, $\sigma_2 = \sigma_3$

$$(\sigma_{1f} - \sigma_{3f})^2 = 12T_0(\sigma_{1f} + \sigma_{3f}) \tag{10-19}$$

and $d\sigma_{1f}/d\sigma_{3f} = 4$ when $\sigma_3 = 0$. For the extension test, $\sigma_1 = \sigma_2$

$$(\sigma_{1f} - \sigma_{3f})^2 = 12T_0(2\sigma_{1f} + \sigma_{3f}) \tag{10-20}$$

and $d\sigma_{1f}/d\sigma_{3f} = 2.5$ when $\sigma_{3f} = 0$. For biaxial compression, $\sigma_3 = 0$

$$\sigma_{1f}^2 - \sigma_{1f}\sigma_{2f} + \sigma_{2f}^2 = 12T_0(\sigma_{1f} + \sigma_{2f}) \tag{10-21}$$

so $d\sigma_{1f}/d\sigma_{2f} = 2$ when $\sigma_2 = 0$, and $\sigma_{1f} = 24T_0 = 2f_{cyl}$ when $\sigma_{1f} = \sigma_{2f}$. Equation 10-18 can also be written in the form

$$\tau_0 = 8T_0\sigma_0^{\frac{1}{2}} \tag{10-22}$$

It is clear from the above equations that the extended Griffith's criterion adequately predicts the strength behaviour of concrete in the triaxial test but incorrectly predicts the behaviour under extension and biaxial compression. This is because the extended Griffith's criterion overestimates the influence of the intermediate principal stress upon failure.

10.3 Shear failure criteria

Several shear failure criteria have been proposed, each one assumes that failure occurs by flow or fracture when the shear stress along the plane of failure has increased to a certain value determined by some function of the normal stress on the plane, thus

$$\tau = \psi(\sigma_n) \tag{10-23}$$

where σ_n and τ are the normal and shear stresses across the plane.

10.3.1 Coulomb's criterion
Coulomb in 1773 suggested that the failure criterion for a rock could be represented by

$$\tau = \tau_c + \mu\sigma_n \qquad \text{for } \sigma_n > 0 \tag{10-24}$$

where τ_c is the cohesive strength of the material and μ is the coefficient of internal friction. This criterion assumes that failure is not influenced by the magnitude of the intermediate stress.

In terms of the principal stresses, Coulomb's criterion may be written [61]

$$\sigma_{1f}[(\mu^2+1)^{\frac{1}{2}}-\mu]-\sigma_{3f}[(\mu^2+1)^{\frac{1}{2}}+\mu]=2\tau_c \tag{10-25}$$

which is a straight line intercepting the σ_1 axis at

$$f_{cyl} = 2\tau_c[(\mu^2+1)^{\frac{1}{2}}-\mu]^{-1} \tag{10-26}$$

The intercept on the σ_3 axis is not the tensile strength of the material. Substituting for f_{cyl} in Eqn 10-25 gives

$$\sigma_{1f} = f_{cyl}+[(\mu^2+1)^{\frac{1}{2}}+\mu]^2\sigma_{3f} \tag{10-27}$$

Coulomb's criterion predicts two possible directions for the failure planes inclined at equal angles $\frac{1}{2}\tan^{-1}(1/\mu)$ to the direction of maximum principal stress.

It was shown in Section 6 that the observed relationship between σ_{1f} and σ_{3f} for concrete under triaxial compression was approximately linear, as Coulomb's criterion predicts, with a slope of between 4 and 5. Thus, $0.7 < \mu < 0.8$ and therefore the predicted angle of fracture is approximately 30°.

10.3.2 Mohr's hypothesis
The Mohr failure curve is obtained experimentally and is the envelope to the Mohr failure circles of diameter $\sigma_{1f} - \sigma_{3f}$ obtained at various magnitudes of σ_{3f}. Each Mohr circle of diameter $\sigma_1 - \sigma_3$ gives the normal and shear stresses along all planes in the material and if the circle lies below the envelope, then failure does not occur, but if the circle touches the envelope, then Mohr proposed that failure occurs along a plane parallel to the direction of the intermediate principal stress

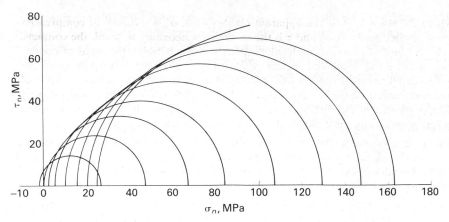

Figure 10-2 Mohr's envelope: age of concrete 1 year, water/cement ratio 0.47

and inclined at an angle θ to the maximum compressive stress, where 2θ is the angle between the normal to the envelope and the σ-axis.

Coulomb's criterion is equivalent to the assumption of a linear Mohr envelope. In the case of concrete, the envelope is concave downwards (Fig. 10-2). Leon [63] used a parabolic form for the Mohr envelope. This approach has the advantage that both compressive and tensile stress states can be treated by the same criterion.

10.3.3 Cohesion theory of friction

A commonly accepted theory which accounts for the frictional behaviour of sliding surfaces of material [64] is the cohesion theory of friction. This theory, in the author's view, is the one which most closely accords with the behaviour of concrete. In the theory, contact between two nominally flat surfaces is assumed to occur at the tips of widely spaced irregularities or asperities, even when the two surfaces are apparently smooth. It is further assumed that at the areas of contact there is such intimate contact between the two surface materials that molecular adhesion occurs. The shear or frictional force necessary to break the contact areas is

$$F = \tau_b A_u \tag{10-28}$$

where τ_b is the shear strength of the material and A_u is the true area of contact.

A number of models of increasing complexity, all of which were assumed to be composed of spherical granules deforming elastically under stress, have been considered by Archard [65]. In all but the simplest model, the number of contact areas is a function of the normal load and Archard showed that the true contact area ranges from $k_7 Q^{2/3}$ to almost $k_7 Q$ for the most complex model considered. The shear or frictional force necessary to break the contact areas is

$$F = \tau_b k_7 Q^k = \tau_b k_7 A^k \sigma_n^k \tag{10-29}$$

thus

$$\tau = k_8 \sigma_n^k \tag{10-30}$$

where $2/3 \leqslant k \leqslant 1$, A is the apparent contact area, σ_n is the normal compressive stress applied to the body and τ is the shear stress necessary to break the contacts.

In the case of concrete, molecular adhesion exists between the areas of contact before any stress is applied and the application of a compressive stress gives rise to increased areas of contact and additional areas of surface contact over which molecular adhesion occurs. If concrete behaves elastically, the additional adhesion disappears when the stresses are removed [66]. Thus, for a concrete which deforms elastically under load, the failure condition is given by

$$\tau = \tau_c + k_8 \sigma_n^k \qquad (10\text{-}31)$$

where τ_c is the shear strength of the concrete when the normal stress is zero. If failure occurs along the plane of maximum shear,

$$\sigma_{1f} - \sigma_{3f} = 2\tau_c + k_9(\sigma_{1f} + \sigma_{3f})^k \qquad (10\text{-}32)$$

and normalizing, the failure criterion becomes

$$(\sigma_{1f} - \sigma_{3f})/f_{cyl} = k_{10} + k_{11}[(\sigma_{1f} + \sigma_{3f})/f_{cyl}]^k \qquad (10\text{-}33)$$

In Fig. 10-3, $(\sigma_{1f} - \sigma_{3f})/f_{cyl}$ is shown plotted against $(\sigma_{1f} + \sigma_{3f})/f_{cyl}$ for a range of concretes tested under triaxial compression and extension. It can be seen from the figure that when k equals 1 Eqn 10-33 gives a close fit to the experimental data.

If the actual areas of contact within concrete are low compared to the apparent area of contact and a pore pressure, p, is acting in the concrete,

$$(\sigma_{1f} - \sigma_{3f})/f_{cyl} = k_{10} + k_{11}[(\sigma_{1f} + \sigma_{3f} - 2p)/f_{cyl}] \qquad (10\text{-}34)$$

and if $p = \sigma_{3f}$, Eqn 10-34 leads to

$$\sigma_{1f} - \sigma_{3f} = f_{cyl} \qquad (10\text{-}35)$$

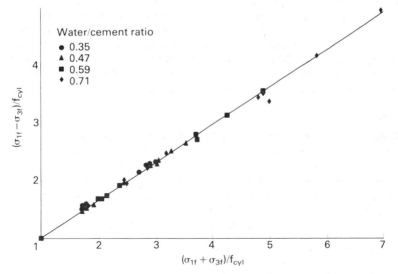

Figure 10-3   Relationship between $(\sigma_{1f} - \sigma_{3f})/f_{cyl}$ and $(\sigma_{1f} + \sigma_{3f})/f_{cyl}$: age of concrete 56 days

This equation closely fits triaxial compression test data obtained on concrete when the pressurizing fluid is free to enter the pores in the concrete (Fig. 8-1).

11   Design stresses for concrete structures subject to multiaxial stresses

The recommendations given in the Sections 11-1 to 11-3 have been proposed by Hobbs *et al.* [1] and are based upon much of the experimental data reported in the earlier sections. In the experimental work, measurements have been made on concretes with water/cement ratios greater than 0.35 and aggregate volume concentrations above 0.5. It is possible that their recommendations will not apply to concretes of lower water/cement ratio. Concretes with lower aggregate contents are unlikely to be used in structures. The strength measurements were made at ambient laboratory temperatures so the application of the data may not be valid outside the temperature range 5–30°C.

11.1   Criteria used in the prediction of safe working stresses

Judgement of the performance of a structure is based upon a number of interacting assumptions regarding the dead and live loads, the performance and variability of the component and the analysis of stresses in the structure, particularly in critical regions. As neither the loads nor the material properties are precisely known and because the actual stresses in a structure will not be exactly those predicted by analysis, it is necessary to introduce partial safety factors to ensure safety and serviceability of the structure. In the UK, the performance of concrete in a structure is currently based upon the characteristic cube strength, f_{cu}, and a factor of 0.67 is introduced to take account of the ratio between the characteristic cube strength and the strength of concrete in a flexural member. A material partial safety factor γ_m is then introduced to allow for possible differences in quality between the concrete in the structure and that in the test cubes. The value of γ_m depends upon the limit state and for ultimate strength is taken to be 1.5.

Concrete cannot carry stresses that are close to the ultimate strength for sustained periods of time nor can it withstand oscillating stresses that reach peak values close to the ultimate strength. This fact is taken into account by defining serviceability stress limits. The available evidence suggests that if the magnitude of the major principal stress is kept below 0.7 of the ultimate strength for the prevailing level of the minor principal stress, creep failure or fatigue are unlikely to occur.

When the designer calculates the acceptable values of the major principal stress, σ_1, for the ultimate limit states and for serviceability, calculations may be necessary for different situations that occur during the construction and life of the structure. In the calculation, the value of the minor principal stress, σ_3, is taken to be the lowest credible value that can be justified by the designer, throughout the stress history of the structure.

11.2 Design stresses for various stress combinations

11.2.1 Multiaxial compression: normal-weight concretes

The maximum stress that concrete in a structure can support is given by

$$\sigma_{1f} = 0.67f_{cu} + 3\sigma_3 \qquad \text{when } \sigma_3 < f_{cu} \tag{11-1}$$

Applying a material safety factor $\gamma_m = 1.5$ gives for the ultimate limit state of collapse

$$\sigma_{1u} = 0.45f_{cu} + 3\sigma_3 \tag{11-2}$$

and the serviceability limit state stresses are obtained from

$$\sigma_{1s} = 0.7\sigma_{1u} = 0.3f_{cu} + 2\sigma_3 \tag{11-3}$$

11.2.2 Multiaxial compression: lightweight aggregate concretes

The maximum stress that structural lightweight concrete can support is given by

$$\sigma_{1f} = 0.67f_{cu} + 2\sigma_3 \tag{11-4}$$

and the ultimate limit state of collapse and the serviceability limit state are

$$\sigma_{1u} = 0.45f_{cu} + 2\sigma_3 \qquad \text{if } \gamma_m = 1.5 \tag{11-5}$$

and

$$\sigma_{1s} = 0.3f_{cu} + 1.4\sigma_3 \tag{11-6}$$

11.2.3 Multiaxial compression: damaged concrete

Concrete that has been damaged by the loading can still carry load provided the concrete is constrained by reinforcement or by surrounding, undamaged material. The ultimate limit state, the state at which the broken material will collapse, is given by

$$\sigma_{1u} = 3\sigma_3 \qquad \text{for natural aggregates} \tag{11-7}$$

$$\sigma_{1u} = 2\sigma_3 \qquad \text{for lightweight aggregates} \tag{11-8}$$

The minor principal stress σ_3 here depends on the magnitude of σ_1 and σ_3 and is therefore defined as its minimum credible value for any active applied stress, σ_1.

11.2.4 Multiaxial compression: pore pressure

If concrete is subjected to a pressure from a fluid that can permeate the pores, the ultimate limit state for collapse becomes

$$\sigma_{1u} = 0.45f_{cu} + 3\sigma_3 - 2p \qquad \text{for normal-weight concrete} \tag{11-9}$$

$$\sigma_{1u} = 0.45f_{cu} + 2\sigma_3 - p \qquad \text{for lightweight concrete} \tag{11-10}$$

σ_3 is the lowest credible value and p the highest possible pore pressure that can be foreseen in the particular application. If the minor principal stress is applied hydrostatically by a fluid, such as sea-water at depth acting on unsealed concrete, then $\sigma_3 \to p$ within a short period of time and the ultimate limit state of collapse and the serviceability limit state are

$$\sigma_{1u} = 0.45f_{cu} + \sigma_3 \qquad\qquad (11\text{-}11)$$

$$\sigma_{1s} = 0.3f_{cu} + \sigma_3 \qquad\qquad (11\text{-}12)$$

11.2.5   Compression and tension

If one or more of the principal stresses is tensile, then the maximum compressive stress which can be supported by concrete in a structure without cracking occurring is

$$\sigma_{1f} = 0.67f_{cu} + 20\sigma_3 \qquad\qquad (11\text{-}13)$$

Since σ_3 is negative, the magnitude of the compressive principal stress that can be carried decreases as the tensile stress rises, and in determining σ_{1f} the maximum credible tensile value for σ_3 is used. If cracking must be avoided, the ultimate limit state is given by

$$\sigma_{1u} = 0.45f_{cu} + 20\sigma_3 \qquad\qquad (11\text{-}14)$$

The tensile stress, σ_{3f}, at which cracking may occur is given by

$$0 > \sigma_{3f} = (\sigma_1 - 0.67f_{cu})/20 \qquad\qquad (11\text{-}15)$$

the maximum credible value being assigned to σ_1. The ultimate limit state for cracking is

$$\sigma_{3u} = (\sigma_1 - 0.45f_{cu})/20 \qquad\qquad (11\text{-}16)$$

and the serviceability limit state for sustained or variable loads is

$$\sigma_{3s} = (\sigma_1 - 0.3f_{cu})/20 \qquad\qquad (11\text{-}17)$$

In most structures when concrete is subjected to tension plus compression, the concrete will crack. Under this stress combination, the crack spacing will be less than that for simple tension and hence the crack widths will also be reduced.

11.3   Deformation of concrete

The short-term strains at the serviceability limit state are compared with the failure strains in Fig. 11-1, where the strains are shown plotted against the normalized minor principal stress. The strains plotted are approximate and were deduced from experimental data obtained by Hobbs [45] and Newman [48] on normal weight concretes.

At stresses below the ultimate limit state, the Young's modulus of concrete, E_c, and its Poisson's ratio may be taken to be independent of the applied stress system provided the minor principal stress is less than $0.3f_{cu}$. However, when σ_3 is greater than $0.3f_{cu}$, there is evidence that the concrete Young's modulus and Poisson's ratio decrease with increasing σ_3 [45, 47], falling to roughly 50% of their value in uniaxial compression. Poisson's ratio falls from between 0.1 and 0.15 at $0 < \sigma_3 < 0.3f_{cu}$ to approximately 0.05 at $\sigma_3 = f_{cu}$. These values for Poisson's ratio are lower than the value, namely 0.2, suggested in CP 110 [3]. The short-term strains in concrete at working stress levels may be obtained using the principle of

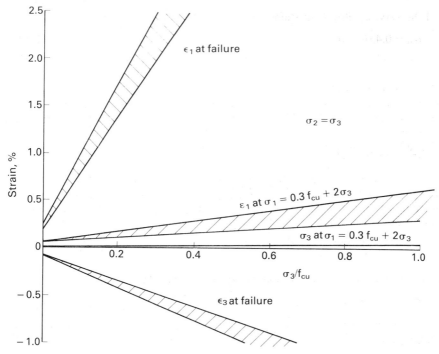

Figure 11-1   Variation of short-term failure strain and strain at the serviceability limit state with σ_3/f_{cu}

superposition:

$$\epsilon_1 = [\sigma_1 - (\sigma_2 + \sigma_3)\nu]/E_{ct} \tag{11-18}$$

$$\epsilon_2 = [\sigma_2 - (\sigma_3 + \sigma_1)\nu]/E_{ct} \tag{11-19}$$

$$\epsilon_3 = [\sigma_3 - (\sigma_1 + \sigma_2)\nu]/E_{ct} \tag{11-20}$$

Thus, if $\sigma_2 = \sigma_3 = f_{cu} = 40$ MPa (5800 psi), $E_{ct} = 36$ GPa (5.2×10^6 psi) and $\nu = 0.15$, then at the serviceability limit state the short-term concrete strains are

$$\epsilon_1 = 2[92 - 0.05(80)]/(36 \times 10^3) = 4900 \times 10^{-6}$$

and

$$\epsilon_3 = 2[40 - 0.05(132)]/(36 \times 10^3) = 1900 \times 10^{-6}$$

If $\sigma_2 = \sigma_3 = f_{cu}/3 = 13$, $E_{ct} = 36$ GPa and $\nu = 0.15$, then at the serviceability limit state

$$\epsilon_1 = [30 - 0.15(26)]/(36 \times 10^3) = 750 \times 10^{-6}$$

and

$$\epsilon_3 = [13 - 0.15(51)]/(36 \times 10^3) = 150 \times 10^{-6}$$

The calculated strains given above are in approximate agreement with the strains determined experimentally.

The time-dependent deformations under biaxial stress may be estimated by replacing E_{ct} in Eqns 11-18 to 11-20 by E_{ct}/φ, where φ is the creep factor [7, 8, 67] and by replacing ν by the instantaneous creep Poisson's ratio [67]. In the case of triaxial compression, no clear recommendations can be made regarding the time-dependent deformations. Until data are available, it is suggested that an estimate of the possible creep strains be made simply by replacing E_{ct} in Eqns 11-18 to 11-20 by E_{ct}/φ, due allowance being made for the reduction in E_{ct} and ν when σ_3 is greater than $0.3f_{cu}$.

12 Main conclusions

There are four main conclusions:

(a) The major principal stress which can be supported by concrete may be assumed, for engineering purposes, to be independent of the magnitude of the intermediate principal stress.

(b) The lower bounds to the strength data obtained on normal-weight concrete are

$$\sigma_{1f}/f_{cyl} = 3(\sigma_{3f}/f_{cyl}) + 1 \qquad \text{when } \sigma_1 > \sigma_2 > \sigma_3 \geq 0$$

and

$$\sigma_{1f}/f_{cyl} = 1 + 20(\sigma_{3f}/f_{cyl}) \qquad \text{when } \sigma_1 > \sigma_2 > 0 > \sigma_3$$

(c) The ultimate limit state of collapse for normal-weight concrete is

$$\sigma_{1u} = 0.45f_{cu} + 3\sigma_3 \qquad \text{when } \sigma_1 > \sigma_2 > \sigma_3 \geq 0$$

(d) The serviceability limit state for normal-weight concrete is

$$\sigma_{1s} = 0.3f_{cu} + 2\sigma_3 \qquad \text{when } \sigma_1 > \sigma_2 > \sigma_3 \geq 0$$

where σ_3, the minor principal stress, is the lowest credible value that can be justified by the designer throughout the stress history of the structure.

Acknowledgements

The author and publishers gratefully acknowledge the following permission to use copyright material: from the American Concrete Institute to reproduce one figure from the *ACI Journal*, Vol. 57 (1960–61) and use information given in two others in Vol. 66 (1969); from the Cement and Concrete Association to reproduce two figures from Technical Report 42.497, July 1974 (*Strength and deformation properties of plain concrete*, Part 3), and also to quote from Technical Report 42.484, September 1973 (*Strength and deformation of concrete under short-term loading: a review*); and from the Institution of Structural Engineers to reproduce two figures and some text from *The Structural Engineer*, Vol. 55, No. 4, April 1977.

References

1. Hobbs, D. W., Pomeroy, C. D. and Newman, J. B., Design stresses for concrete structures subject to multi-axial stresses, *Struct. Eng.*, Vol. 55, No. 4, April 1977, pp. 151–164.

2. Bresler, B. and Pister, K. S., Strength of concrete under combined stress, *J. Am. Concr. Inst.*, Vol. 55, No. 3, Sept. 1958, pp. 321–345.

3. British Standards Institution, CP110, Part 1: 1972 (as amended 1980). *The structural use of concrete*, London, 1980, 156 pp.+amend.

4. Hobbs, D. W., *The strength and deformation of concrete under short-term loading: a review*, Cement and Concrete Association Technical Report 42.484, 1973, 27 pp.

5. Hobbs, D. W. The dependence of the bulk modulus, Young's modulus, creep, shrinkage and thermal expansion of concrete upon aggregate volume concentration, *Mater. & Struct. Test. & Res.*, Vol. 4, No. 20, March-April 1971, pp. 107–114.

6. Teychenne, D. C., Parrott, L. J. and Pomeroy, C. D., *The estimation of the elastic modulus of concrete for the design of structures*, Building Research Establishment Current Paper, CP 23/78, 1978, 12 pp.

7. Parrott, L. J., *Simplified methods of predicting the deformation of structural concrete*, Cement and Concrete Association Development Report 3, 1979, 11 pp.

8. British Standards Institution, CP110, Part 1, *The structural use of concrete*, Draft for comment, 1982.

9. Hsu, T. T. C., Slate, P. O., Sturman, G. M. and Winter, G., Micro-cracking of plain concrete and the shape of the stress–strain curve, *J. Am. Concr. Inst.*, *Proc.*, Vol. 60, No. 2, Feb. 1963, pp. 209–224.

10. Shah, S. P. and Winter, G., Inelastic behaviour and fracture of concrete, *Causes, mechanism and control of cracking in concrete*, American Concrete Institute, SP-20, 1968, pp. 5–28.

11. Jones, R. and Kaplan, M. F., The effect of coarse aggregate on the mode of failure of concrete in compression and flexure, *Mag. Concr. Res.*, Vol. 9, No. 26, Aug. 1957, pp. 89–94.

12. Swamy, N., Aggregate–matrix interaction in concrete systems, *Structure, solid mechanics and engineering design*, Proc. civ. eng. conf., Southampton, 1969, Wiley–Interscience, New York and London, 1971, Paper 25, pp. 301–315.

13. Stock, A. F., Hannant, D. J. and Williams, R. I. T., The effect of aggregate concentration upon the strength and modulus of elasticity of concrete, *Mag. Concr. Res.*, Vol. 31, No. 109, Dec. 1979, pp. 225–234.

14. Hobbs, D. W., The compressive strength of concrete: a statistical approach to failure, *Mag. Concr. Res.*, Vol. 24, No. 80, Sept. 1972, pp. 127–138.

15. Newman, K., Criteria for the behaviour of plain concrete under complex states of stress, *The structure of concrete and its behaviour under load*, Proc. int. conf., London, Sept. 1965, Cement and Concrete Association, London, 1968, pp. 255–274.

16. Newman, K. and Newman, J. B., Failure theories and design criteria for plain concrete, *Structure, solid mechanics and engineering design*, Proc. civ. eng. conf., Southampton, 1969, Wiley–Interscience, 1971, Paper 77, pp. 963–995.

17. Kotsovos, M. D. and Newman, J. B., A mathematical description of the deformational behaviour of concrete under complex loading, *Mag. Concr. Res.*, Vol. 31, No. 107, June 1979, pp. 77–90.

18. Spooner, D. C. and Dougill, J. W., A quantitative assessment of damage sustained in concrete during compressive loading, *Mag. Concr. Res.*, Vol. 27, No. 92, Sept. 1975, pp. 151–160.

19. Rüsch, H., Researches toward a general flexural theory for structural concrete, *J. Am. Concr. Inst., Proc.*, Vol. 57, No. 1, July 1960, pp. 1–28.
20. Evans, R. H. and Marathe, M. S., Micro-cracking and stress–strain curves for concrete in tension, *Mater. & Struct. Test & Res.*, Vol. 1, No 1, Jan–Feb. 1968. pp. 61–64.
21. Johnston, C. D. and Sidwell, E. H., Testing concrete in tension and compression, *Mag. Concr. Res.*, Vol. 20, No. 65, Dec. 1968, pp. 221–228.
22. Johnston, C. D., Strength and deformation of concrete in uniaxial tension and compression, *Mag. Concr. Res.* Vol. 22, No. 70, Mar. 1970, pp. 5–16.
23. Gonnerman, H. F. and Shuman, E. C., Compression, flexure and tension tests of plain concrete, *Proc. Am. Soc. Test. Mater.*, Vol. 28, Part 2, 1928, pp. 527–552.
24. Hughes, B. P. and Ash, J. E., Anisotropy and failure criteria for concrete, *Mater. & Struct. Test. & Res.*, Vol. 3, No. 18, Nov–Dec. 1970, pp. 371–374.
25. Vile, G. W. D., The strength of concrete under short-term static biaxial stress, *The structure of concrete and its behaviour under load*, Proc. int. conf., London, Sept. 1965, Cement and Concrete Association, London, 1968, pp. 275–288.
26. Gerstle, K. H. *et al.*, Strength of concrete under multiaxial stress states, *Proc. McHenry symp.*, Mexico City, Oct. 1976, American Concrete Institute, SP 55–5.
27. Gerstle, K. H. *et al.*, Behaviour of concrete under multiaxial stress states, *J. Eng. Mech. Div. Am. Soc. Civ. Eng.*, Vol. 106, No. EM6, Dec. 1980, pp. 1383–1403.
28. Smith, G. M., Failure of concrete under combined tensile and compressive stresses, *J. Am. Concr. Inst.*, Vol. 50, No. 2, Oct. 1953, pp. 137–140.
29. Sundara Raja Iyengar, K. T., Chandrashekhara, K. and Krishnaswamy, K. T., Strength of concrete under biaxial compression, *J. Am. Conc. Inst., Proc.*, Vol. 62, No. 2, Feb. 1965, pp. 239–250.
30. Millard, F. H., MSc Thesis, University of Illinois, 1912.
31. Hobbs, D. W., *Strength and deformation properties of plain concrete subject to combined stress. Part 2: strength in multiaxial compression*, Cement and Concrete Association Technical Report 42.463, 1972, 7 pp.
32. Kupfer, J., Hilsdorf, H. K. and Rüsch, H., Behaviour of concrete under biaxial stresses, *J. Am. Conc. Inst., Proc.*, Vol. 66, No. 8, Aug. 1969, pp. 656–666.
33. Nelissen, L. J. M., Biaxial testing of normal concrete, *Heron*, Vol. 18, No. 1, 1972, 90 pp.
34. Parrott, L. J., An improved apparatus for biaxial loading of concrete specimens, *J. Strain Anal.*, Vol. 5, No. 3, 1970, pp. 169–176.
35. Richart, F. E., Brandtzaeg, A. and Brown, R. L., A study of the failure of concrete under combined stresses, *Bull. Ill. Univ. Eng. Exp. Stn*, No. 185, 1928, 104 pp.
36. Hobbs, D. W., *Strength and deformation properties of plain concrete subject to combined stress. Part 1: strength results obtained on one concrete*, Cement and Concrete Association Technical Report 42.451, 1970, 12 pp.
37. Kotsovos, M., *Failure criteria for concrete under generalised stress states*, PhD Thesis, Imperial College of Science and Technology, London, 1974, 284 pp.

38. McHenry, D. and Karni, J., Strength of concrete under combined tensile and compressive stress, *J. Am. Concr. Inst., Proc.*, Vol. 54, No. 10, April 1958, pp. 829–839.

39. Bellamy, C. J., Strength of concrete under combined stress. *J. Am. Concr. Inst., Proc.*, Vol. 58, No. 4, Oct. 1961, pp. 367–381.

40. Considere, A., *Experimental researches on reinforced concrete*, Translation and introduction by L. L. Moisseiff, 2nd Edn, McGraw-Hill, New York, 1906.

41. Gardner, N. J., Triaxial behaviour of concrete, *J. Am. Concr. Inst.*, Vol. 66, No. 2, Feb. 1969, pp. 136–146.

42. Balmer, G. G., *Determination of boundary porosity by triaxial compression tests of concrete*, US Bureau of Reclamation, Structural Research Laboratory Report, SP-15, 1947, 29 pp.

43. Balmer, G. G., *Shearing strength of concrete under high triaxial stress— computation of Mohr's envelope as a curve*, US Bureau of Reclamation, Structural Research Laboratory Report, SP-23, 1949, 26 pp.

44. Sims, J. R., Krahl, N. W. and Victory, S. P., Triaxial tests of mortar and neat cement cylinders, *Anniversary Volume*, International Association for Bridge and Structural Engineering, 1966, pp. 481–495.

45. Hobbs, D. W., *Strength and deformation properties of plain concrete subject to combined stress. Part 3: results obtained on a range of flint gravel aggregate concretes*, Cement and Concrete Association Technical Report 42.497, 1974, 20 pp.

46. Hansen, J. A., Strength of structural lightweight concrete under combined stress, *J. Res. Div. Labs Portld Cem. Ass.*, Vol. 5, No. 1, 1963, pp. 39–46.

47. Newman, J. B. and Newman, K., *Design criteria for concrete under combined states of stress*, CIRIA Contract: Criteria of Concrete Strength, Report 2, Vol. 1, Imperial College, London, 1973, 281 pp.

48. Newman, J. B., Concrete under complex stress; Chapter 5 in *Developments in concrete technology* I, edited by F. D. Lydon, Applied Science Publishers, London, 1979, pp. 151–219.

49. Cedolin, L., Crutzen, Y. R. J. and Dei Poli, S. D., *Stress-strain relationship and ultimate strength of concrete under triaxial loading conditions*, Research Centre for Reinforced and Prestressed Concrete Structures, Politecnico di Milano, 1976, pp. 123–137.

50. Bazant, Z. P. and Tsubaki, T., Total strain theory and path-dependence of concrete. *J. Eng. Mech. Div. Am. Soc. Civ. Eng.*, Vol. 106, No. EM6, Dec. 1980. pp. 1151–1173.

51. Saucier, K. L., *Equipment and test procedure for determining multiaxial tensile and combined tensile–compressive strength of concrete*, US Army Engineer, Waterways Experiment Station, Technical Report C-74-1, 1974, 15 pp. + figs and tables.

52. Terzaghi, K. and Rendulic, L., Die wirksame Flachenporosität des Betons, *Z. Öst. Ing. u. Archit. Ver.*, Vol. 86, No. 1/2, Jan. 1934, pp. 1–9.

53. McHenry, D., The effect of uplift pressure on the shearing strength of concrete, *Proc. 3rd int. conf. large dams*, Stockholm, 1948, Commisson Internationale des Grands Barrages, Paris, pp. 329–349, R. 48, Ques. 8.

54. Butler, J. E., The influence of pore pressure upon concrete, *Mag. Concr. Res.*, Vol. 33, No. 114, March 1981, pp. 3–17.

55. Lowe, P. G., Deformation and fracture of plain concrete, *Mag. Concr. Res.*, Vol. 30, No. 105, Dec. 1978, pp. 200–204.

56. William, K. J. and Warnke, E. P., Constitutive model for the triaxial behaviour of concrete, *Seminar on concrete structures subjected to multiaxial stresses*, Instituto Sperimentale Modeli e Strutture (ISMES), Bergamo, 17–19 May 1974, Paper III-I.

57. Mills, L. L. and Zimmerman, R. M., Compressive strength of plain concrete under multiaxial loading conditions, *J. Am. Concr. Inst., Proc.*, Vol. 67, No. 10, Oct. 1970, pp. 802–807.

58. Kotsovos, M. D., A mathematical description of the strength properties of concrete under generalised stress, *Mag. Concr. Res.*, Vol. 31, No. 108, Sept. 1979, pp. 151–158.

59. Griffith, A. A. The phenomena of rupture and flow in solids, *Phil. Trans. R. Soc.*, Vol. A221, 1921, pp. 163–198.

60. McClintock, F. A. and Walsh, J. B., Friction of Griffith cracks under pressure, *Proc. 4th US int. cong. appl. mech.*, 1962, pp. 1015–1021.

61. Jaeger, J. C. and Cook, N. G. W., *Fundamentals of rock mechanics*, Methuen London, 1969, 513 pp.

62. Murrell, S. A. F., A criterion for brittle fracture of rocks and concrete under triaxial stress and the effect of pore pressure on the criterion, Fairhurst, C. (Ed.), *Rock mechanics*, Proc. 5th rock mech. symp. Pergamon Press, Oxford, 1963, pp. 563–577.

63. Leon, A., Über die Rolle der Trennungsbrüche im Rahmen des Mohrschen Anstrengungshypothese, *Bauingenieur*, Vol. 31–32, 1934, pp 318–321.

64. Hobbs, D. W., A study of the behaviour of a broken rock under triaxial compression, and its application to mine roadways, *Int. J. Rock Mech. & Mining Sci.*, Vol. 3, 1966, pp. 11–43.

65. Archard, J. F., Elastic deformation and the laws of friction, *Proc. R. Soc.*, Vol. A243, 1958, pp. 190–205.

66. Johnson, K. L., A note on the adhesion of elastic solids, *Brit. J. Appl. Phys.*, Vol. 9, 1958, pp. 199–200.

67. Illston, J. M., Time-dependent deformations of concrete; Chapter 2 in *Developments in concrete technology* I, edited by F. D. Lydon, Applied Science Publishers, London, 1979, pp. 51–81.

11 Elasticity, shrinkage, creep and thermal movement of concrete

Bernard L Meyers   Eugene W Thomas

Bechtel Power Corporation, Gaithersburg, Maryland, USA

Contents

Notation

A	Cross-sectional area; coefficient (see Eqn 4-6)
A_s, A_s'	Tension and compression reinforcing steel cross-sectional area, respectively
B	Coefficient (see Eqn 4-7)
C_F	Carry-over factor
D	Flexural rigidity $Eh^3/12(1-\nu)$
D_c	Differential creep and shrinkage coefficient between the top of a beam and the bottom of the topping slab
E	Modulus of elasticity (may be a function of time); coefficient (see Eqn 4-10)
E'	Effective modulus of elasticity, accounting for ultimate creep effects
F	Prestress force after losses; coefficient (see Eqn 4-11)
F_i	Initial prestressing force
F_0	Prestress force at stress transfer
I	Cross-section moment of inertia
K^c, K^s	Correction factors for creep and shrinkage, respectively
K_C^c	Creep coefficient given as a function of relative humidity (see Eqn 2-5)
K_B, K_{sh}, K_s	Deflection coefficient for beams due to load application (K_B), due to shrinkage (K_{sh}) and due to load application on slabs (K_s)
$K_r, K_r^c, \bar{K}_r^c$	Reduction factor to relate creep strain to curvature for prestressed beams (K_r) and for non-prestressed beams (K_r^c), $(\bar{K}_r^c)$ is a reduction factor that accounts for both shrinkage and creep of reinforced members
K_T	Stiffness coefficient
L	Span length
L_T	Portion of beam span assumed to be cracked
M	Moment

M_{cr}	Moment at which concrete cracking initiates	n	Modular ratio (may be a function of time)
M_0	Maximum moment of a simply supported beam	n'	Effective modular ratio which accounts for the effect of creep (may be a function of time)
M_{FE}^A, M_{FE}^G	Fixed end moments due to axial thermal growth and due to a thermal gradient across the cross-section, respectively	p	Ratio of tension reinforcing steel to concrete cross-section
M_{max}	Maximum moment	p'	Ratio of compression reinforcing steel to concrete cross-section
R	Restraint factor		
T_∞	Creep and shrinkage factor	q	Uniformly distributed loading on a slab
a	Distance from a neutral axis of an uncracked section to the extreme tension fiber; portion of beam span from one end assumed to be cracked (see Figure 4-2)	r	Radius of gyration of a beam cross-section
		t	Time
		t_0	Time of initial loading
		t_1	Time at which composite action first occurs (see also, discussion following Eqn 3-23)
b	Width of a beam		
d	Distance from centroid of tension steel to extreme compression fiber		
e	Eccentricity of the prestressing steel from the beam centroid	y	Beam or slab deflection (may be a function of time)
e_{cs}	Distance between the centroid of the composite section and the centroid of the topping slab	Δ	Transverse displacement between the ends of a member
		Δ_A	Axial elongation of a member
		ΔF	Total prestress loss after elastic losses
e_{comp}	Eccentricity of the prestressing steel with respect to the composite section	Δf	Loss of prestress force
		ΔT_A	Change in temperature from a base or stress free condition
f_c	Concrete compression stress	ΔT_ϕ	Thermal gradient across cross-section of a member
f'_c	Ultimate concrete compression stress	α	Creep coefficient (ratio of creep strain to initial elastic strain-may be a function of time)
f_r	Modulus of rupture (see Eqn 3-7)		
f_s	Steel stress	$\alpha 28$	Creep coefficient for concrete initially loaded at 28 days (may be a function of time)
h	Cross-sectional height (or depth) of a member		
j	Distance between centroid of reinforcing steel and centroid of concrete compression force (working stress method) non-dimensionalized to (d)	α_T	Coefficient of expansion for concrete
		β	Coefficient used in Eqn 3-10
		ε_c	Creep function (time-dependent strain/unit stress)
k	Depth of concrete compression stress area in bending (working stress method) non-dimensionalized to (d)	ε_{sh}	Shrinkage strain (may be a function of time)
		ε_0	Initial strain
		ν	Poisson's ratio
m	Ratio of cracked to gross moment of inertia (see Eqn 4-12)	σ_0	Initial stress
		ϕ	Curvature

Subscripts (in addition to those included with the symbols):

AC	Air content	H	Relative humidity
B	Beam	L	Live load
comp	Pertaining to composite action	$\underline{L}$	Sustained live load
D	Dead load	$\underline{T}$	Topping slab; refers to transformed section when used with moment of inertia (I)
D_c	Pertaining to D_c		
F	Total composite section; fine aggregate content	b	Concrete composition

c	Concrete	p	Percent reinforcement; camber
cr	Cracked section	s	Concrete consistency; steel
d	Minimum member thickness	sh	Shrinkage
	(Eqn 2-3); age of concrete at	t	Theoretical thickness
	load application (Eqn 2-5)	t_0	Age of concrete at load appli-
e	Effective value; cross-section		cation
	dimension	w	Concrete density
g	Gross section	∞	Ultimate
m, (m)	Midspan	1, 2	Beam ends
i	Initial		

Superscripts (in addition to those included with list of symbols):

D_c	Pertains to D_c as defined in	c	Creep
	list of symbols	cs	Creep and shrinkage
M_s	Gain in prestress due to slab	s	Shrinkage
	dead load moment	'	Pertaining to compression
R	Relaxation of steel stress		steel, ultimate stress, or effec-
T	Thermal		tive properties

Summary

The time-dependent behavior of concrete structures is considered, including those movements which do not result from a change in applied load, i.e. creep, shrinkage and thermal movement. This behavior is evaluated in three distinct steps. First, the material behavior is estimated, i.e. how much creep, shrinkage and thermal movement is expected of the concrete used in the environment, and subjected to the loading history of the prototype structure. Secondly, the effect of these movements on structural stress, deflection and general structural behavior is considered. Finally, construction methods that assure the predicted behavior takes place are discussed. Each of these aspects of design are considered: materials behavior; design for creep and shrinkage; design for thermal movement; and construction methods. In addition, observations of time-dependent behavior on structures are presented in order to give the designer an appreciation for actual structural response.

1 Introduction

This chapter will consider the time-dependent behavior of concrete structures including those movements which do not result from a change in applied load, i.e. creep, shrinkage and thermal movement. Such movements, however, do result in changes in internal stress, if the structure is restrained from moving. In general, there are three steps required to consider adequately these time-dependent deformations in the design of structural systems. First, the material behavior must be estimated, i.e. how much creep, shrinkage and thermal movement can be expected of the concrete used in the environment, and subjected to the loading history of the prototype structure. Secondly, how these movements affect structural stress, cracking, deflection and general structural behavior must be considered. Finally, the structure must be constructed to assure that the predicted behavior takes place. This chapter considers each of these aspects of design: materal behavior; design for creep and shrinkage; design for thermal movement; and construction methods. In addition, a survey of observations of time-dependent behavior on actual structures is presented in order to give the designer a feel for actual structural response.

A great deal has been published on the subject of the material and structural behavior of systems subjected to time-dependent deformation. Methods are available ranging from simple, effective modulus methods using gross creep coefficients, to non-linear design methods using complex non-linear creep laws. Obviously, it is important to decide what level of complexity is required for the design in question. This chapter will discuss various levels of predicting both material and structural behavior. The section on construction methods suggests methods and material requirements that will assure that the structure will behave in a manner consistent with the design assumptions. Finally, the section on observation of structures will indicate where available data can be found, so that the designer can evaluate the potential behavior of similar structures.

2 Material behavior

A great deal of information is in the literature [1] dealing with hypotheses concerning the reasons for and explanations of the phenomena of creep and shrinkage of concrete. The sections that follow will not review or make judgments on any of these hypotheses, but will attempt to describe the phenomena and their effects in engineering terms.

2.1 Factors affecting uniaxial creep

The uniaxial time-dependent behavior of concrete is generalized in Fig. 2-1. A number of observations can be made:

(a) The ratio of creep strain plus initial elastic strain to initial elastic strain is, in general, between 2 and 3.

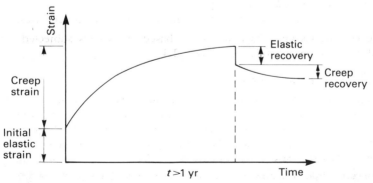

Figure 2-1 Uniaxial time-dependent behavior of concrete: generalized representation

(b) The ratio of elastic recovery plus creep recovery to initial elastic strain is approximately one.
(c) Creep deformation is a function of loading history, the environment of the system and the material constituents of the concrete.

In this section, a brief discussion of the above factors will be given.

It can be assumed that aggregates usually used in concrete do not creep, while paste and/or mortar exhibit large amounts of creep. Thus, the bulk of the time-dependent deformation that occurs in concrete can be assumed to be confined to the paste and the aggregate–paste interface. It can be further assumed that this deformation of concrete under load is made up of strain resulting from progressive microcracking at the aggregate–paste interface, from moisture exchange with the atmosphere, and from moisture movement within the system. The deformation associated with microcracking accounts for 10–25% of the total load-induced time-dependent deformation, and the deformation associated with moisture exchange with the atmosphere some 5–40%. Both are *not* recoverable. Migration of water within the concrete system is the cause of a large portion of both recoverable and non-recoverable creep accounting for 50–80% of the load-induced time-dependent deformation [2].

The concrete material properties that seem to have the greatest effect on the creep of concrete are

(a) The water/cement ratio. As the water/cement ratio is increased, the magnitude of creep is increased.
(b) The aggregate/paste ratio. As the aggregate/paste ratio is increased, the magnitude of creep is increased, for a given volume of material.

A number of environmental and loading factors also affect the creep of concrete. The major factors are

(a) For mature concrete, creep strains decrease as age at time of loading is increased, for concrete exposed to the atmosphere.
(b) Creep is approximately proportional to the applied stress for levels not exceeding about 40–60% of the ultimate strength of the concrete.
(c) As the relative humidity of the surrounding atmosphere is increased, the amount of creep decreases in a drying system.
(d) Creep increases with increased temperature.
(e) As the size of the member is increased, the amount of creep decreases.

The above conclusions and observations are based on systems subjected to uniaxial load, and will be quantified in Section 2.4.

2.2 Effect of loading history on creep

Although the time-dependent behavior of a large majority of structures is not affected by loading history, because of small variations in applied load, or because the structure is not creep sensitive, some structures can be significantly affected by loading history. It is important to note that other than a constant sustained load can either increase or decrease creep. For example [3]:

(a) If a concrete system is subjected to unloading and reloading to the same sustained load, creep response is increased if the time of unloading is short, and decreased if the time of unloading is long.

(b) If a concrete system is unloaded to a sustained load less than that initially applied, or completely unloaded and then reloaded to a level less than that originally applied, the system will exhibit some recovery, prior to developing positive creep. The creep after the stress level is reduced will be less than that of a system initially loaded to the reduced stress.

(c) Fluctuations in temperature and/or humidity will result in a greater creep than a sustained load applied at the average values of temperature and/or humidity.

(d) Creep often results in redistribution of stress in a complex structure, which will in turn result in a change in sustained stress even if the loads applied to the structure remain constant.

It is extremely difficult to predict accurately the loading and environmental history of most structures. However, it is possible to estimate their effects on structural response. This will be touched upon in Section 3.2.

2.3 Factors affecting drying shrinkage

The drying shrinkage behavior of concrete is generalized in Fig. 2-2. Although the rate of shrinkage is dependent on the environment and the rate of moisture loss to the surrounding environment, it can usually be assumed that shrinkage is complete after about 3 months. Because shrinkage and creep are both related to moisture migration, they are affected by many of the same variables. The major parameters affecting shrinkage are [4]

(a) The water content. As the water content is increased, drying shrinkage is increased. Thus, anything that increases the water requirements of a mix, (i.e. high slump, high temperatures of the fresh concrete, or smaller size coarse aggregate) will increase the shrinkage potential.

(b) Chemical admixtures. Most chemical admixtures do not substantially affect shrinkage. However, the use of calcium chloride results in a significant increase in drying shrinkage.

(c) Pozzolans. The use of pozzolans as a cement substitute often increases the water requirements and, therefore, drying shrinkage, and in some cases increases shrinkage without a substantial change in water requirements.

(d) Member size. Both the rate and final value of shrinkage decrease as member size increases.

Some of these observations will be quantified in Section 2.4.

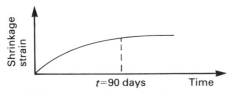

Figure 2-2 Drying shrinkage behavior of concrete: generalized representation

2.4 Methods for prediction of creep and shrinkage properties

2.4.1 Creep prediction

Hilsdorf and Muller [5] made a study of a number of the methods available to predict time-dependent strains of plain concrete. They concluded that the 'average coefficient of variation as a measure of the differences between predicted and measured creep functions ranges from 26 to 45 percent,' while 'the average error for the prediction of final creep strain ranges from 31 to 35 percent.' They also concluded that different methods of creep prediction should be used depending on the creep sensitivity of the structure in question. In this section, a number of prediction methods will be described. The engineer should use the method most suitable for the structure being designed and should account for the expected variability. (Variability is discussed in Section 6.)

The methods that will be discussed are based on the following general creep function:

$$\varepsilon_c(t, t_0) = \frac{1}{E_c(t_0)} + \frac{\alpha(t, t_0)}{E_c(t_0)} \tag{2-1}$$

where

$\varepsilon_c(t, t_0)$ = creep function (ratio of time-dependent strain to unit stress)
$E_c(t_0)$ = modulus of elasticity for concrete loaded at time t_0
$\alpha(t, t_0)$ = creep coefficient (ratio of creep strain to initial elastic strain) for concrete at time t, loaded at time t_0

Branson et al. [6] and ACI 209 [7] have suggested that $\alpha(t, t_0)$ be evaluated from

$$\alpha(t, t_0) = \frac{(t - t_0)^{0.6}}{10 + (t - t_0)^{0.6}} \alpha_\infty \tag{2-2}$$

and

$$\alpha_\infty = 2.35 K_H^c K_d^c K_s^c K_F^c K_{AC}^c K_{t_0}^c \tag{2-3}$$

where $2.35 = \alpha_\infty$ for standard conditions, and each coefficient K^c is a correction factor for conditions other than standard as follows:

K_H^c = relative humidity
K_d^c = minimum member thickness
K_s^c = concrete consistency
K_F^c = fine aggregate content
K_{AC}^c = air content
$K_{t_0}^c$ = age of concrete at load application

Graphical representations and general equations, as well as standard conditions for these correction factors are given in Fig. 2-3

CEB–FIP 70 [8] suggests that the creep function be written in terms of initial load application at 28 days. Therefore,

$$\varepsilon_c(t, t_0) = \frac{1}{E_c(t_0)} + \frac{\alpha_{28}(t, t_0)}{E_c(t_{28})} \tag{2-4}$$

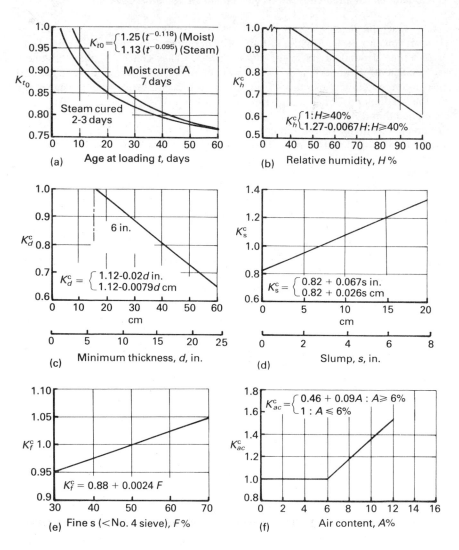

Figure 2-3 Creep correction factors for the ACI 209 method of creep prediction. Note that at standard conditions $K^c = 1.0$

(a) Age of concrete at load application
(b) Ambient relative humidity
(c) Minimum member thickness
(d) Concrete consistency
(e) Fine aggregate content
(f) Air content

where $\alpha_{28}(t, t_0)$ is the creep coefficient at any time, t, for concrete initially loaded at 28 days, and can be evaluated from

$$\alpha_{28}(t, t_0) = K^c_c K^c_d K^c_b K^c_e K^c_t \tag{2-5}$$

where K^c_c is a creep coefficient given as a function of relative humidity and each additional coefficient K^c are correction factors for conditions other than standard

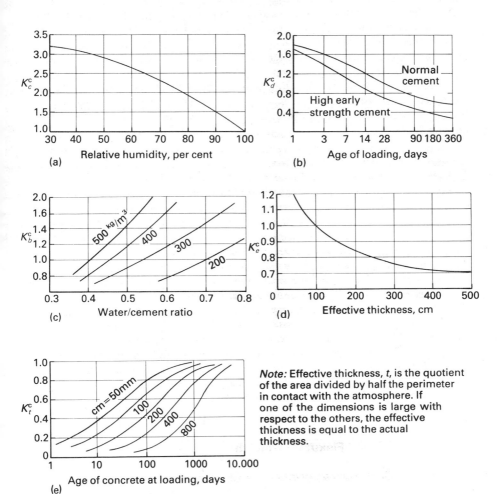

Figure 2-4   Creep correction factors for the CEB–FIP 70 method of creep prediction

(a) Ambient relative humidity
(b) Age of concrete at load application
(c) Concrete composition
(d) Cross-sectional dimension
(e) Theoretical thickness coefficient

as follows:

K_d^c = age of concrete at load application
K_b^c = concrete composition
K_e^c = cross-sectional dimension
K_t^c = theoretical thickness

Curves for these coefficients are given in Fig. 2-4.

It can be seen that these two methods are very similar. Both are empirical and were developed from available data. However, the initial data bases used to develop these relationships were not the same, and, therefore, they do not result in the same creep functions for the same conditions. At present, a joint ACI–CEB committee is attempting to reconcile these differences.

It should also be noted that the accuracy of both methods (ACI 209 and CEB–FIP 70) can be significantly increased when actual creep data can be used in the evaluation of the creep function. This will be discussed more in detail in the ensuing pages.

2.4.2 Simplified creep prediction

The ACI 209 and the CEB–FIP 70 methods can be simplified. Meyers and Branson [9] have proposed a simple graph to evaluate creep coefficients which can

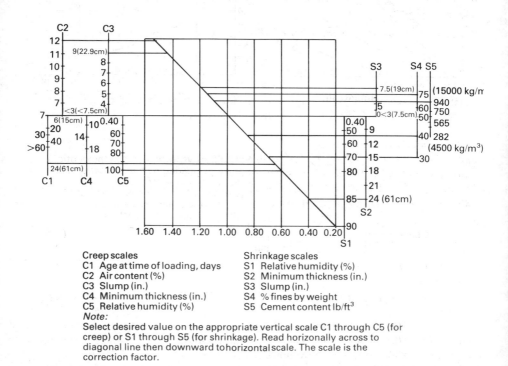

Creep scales
C1 Age at time of loading, days
C2 Air content (%)
C3 Slump (in.)
C4 Minimum thickness (in.)
C5 Relative humidity (%)

Shrinkage scales
S1 Relative humidity (%)
S2 Minimum thickness (in.)
S3 Slump (in.)
S4 % fines by weight
S5 Cement content lb/ft³

Note:
Select desired value on the appropriate vertical scale C1 through C5 (for creep) or S1 through S5 (for shrinkage). Read horizontally across to diagonal line then downward to horizontal scale. The scale is the correction factor.

Figure 2-5 Design aid for creep and shrinkage correction factors

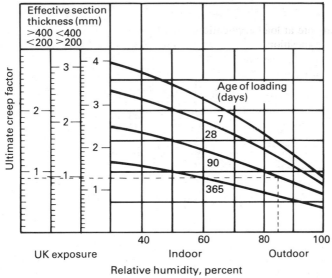

Note:
Read vertically from desired relative humidity to correct age of loading curve. Then horizontally to appropriate effective thickness scale to obtain ultimate creep factor.

Figure 2-6 Ultimate creep factor from the Concrete Society method 1977

be used with a modified form of Eqn 2-3 of the ACI 209 method. The curve is given in Fig. 2-5.

The Concrete Society [10] has also proposed a simple graph for predicting creep based on the CEB–FIP 70 method. This curve is given in Fig. 2-6.

2.4.3 Shrinkage prediction

Both ACI 209 and CEB–FIP 70 have also proposed empirical methods to predict shrinkage behavior. In the ACI 209 (Branson) method, the form of the shrinkage equation is very similar to the creep equation and shrinkage, ε_{sh}, at any time for standard conditions is predicted from

$$\varepsilon_{\text{sh}}(t, t_0) = \frac{(t - t_0)}{35 + (t - t_0)} \varepsilon_{\text{sh}\infty} \tag{2-6}$$

The ultimate shrinkage value $\varepsilon_{\text{sh}\infty}$ can be predicted for conditions other than standard using

$$\varepsilon_{\text{sh}\infty} = 800^{-6} K_H^s K_d^s K_s^s K_F^s K_B^s \dot{K}_{\text{AC}}^s \tag{2-7}$$

where $800 = \varepsilon_{\text{sh}\infty}$ for standard conditions and each coefficient K^s are correction factors for other than standard conditions. All coefficients are for the parameters as defined for creep except K_B^s, which is a coefficient for cement content.

Graphical representations and general equations, as well as standard conditions for these factors are given in Fig. 2-7.

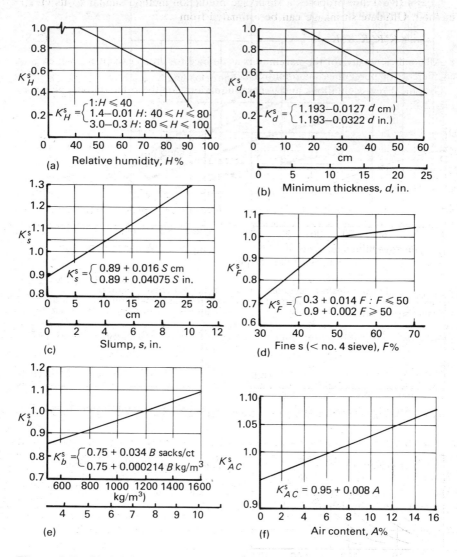

Figure 2-7 Shrinkage correction factors for the ACI 209 method of shrinkage prediction. Note that at standard conditions $K^s = 0$

(a) Ambient relative humidity
(b) Member minimum thickness
(c) Concrete consistency
(d) Fine aggregate content
(e) Cement content
(f) Air content

CEB–FIP 70 also proposes a shrinkage prediction method similar to its creep method. Ultimate shrinkage can be estimated from

$$\varepsilon_{sh\infty} = K_c^s K_d^s K_e^s K_p^s K_t^s \qquad (2\text{-}8)$$

All coefficients are for the parameters as defined for creep except K_p^s, which is a coefficient for percentage of reinforcement.

Graphical representations and general equations for these factors are given in Fig. 2-8.

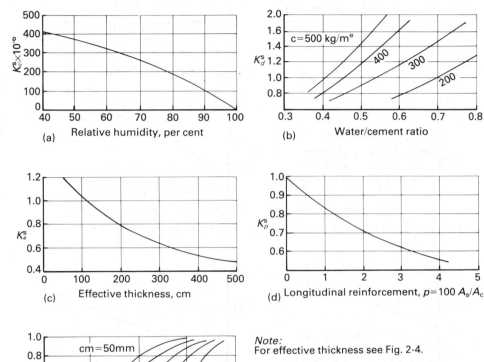

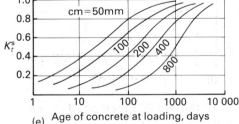

Figure 2-8   Shrinkage correction factors for the CEB–FIP 70 method of shrinkage prediction. Note that at standard conditions $K^s = 0$

        (a) Ambient relative humidity coefficient
        (b) Concrete composition coefficient
        (c) Cross-sectional dimensions coefficient
        (d) Longitudinal reinforcement coefficient
        (e) Theoretical thickness coefficient

2.4.4 *Simplified shrinkage prediction*
Figure 2-5 also includes a simplified shrinkage prediction method, as suggested by Meyers and Branson [9].

2.4.5 *Creep and shrinkage prediction for creep-sensitive structures*
The literature is replete with methods significantly more complicated than those described in the preceding pages. These methods, as suggested by Bazant [11], CEB–FIP 78 [12], Daye [13], etc., are often proposed for creep sensitive structures. It is suggested that the engineer select the appropriate method from his experience with the structure and the analysis method used.

Alternatively, it is suggested that when creep and shrinkage prediction is important, creep tests be performed and used with either the ACI 209 method or the CEB–FIP 70 method. Lengthy tests are not required. Meyers *et al.* [14] have shown that a creep prediction error coefficient of less than 20% can be obtained from 28-day tests. Actual 28-day data are easily substituted into the CEB–FIP 70 method, and in the ACI 209 method, it can be shown that

$$\alpha_\infty = \alpha(t_{28}, t_0)/0.425 \qquad\qquad (2\text{-}9)$$

Since all measurable shrinkage takes place in about 3 months, complete shrinkage tests can be easily performed.

3 Design for creep and shrinkage

3.1 Design for creep and shrinkage in general structures

Time-dependent strain, as discussed in the previous sections, substantially increases the initial or short-time deflection of concrete structural components. These increased deflections may affect the serviceability of the structure and, therefore, become an important design consideration. Excessive deflection can result in improperly fitting architectural components, and misalignment of machinery. These deflections may also cause increased cracking and, therefore, affect structural serviceability. In addition to serviceability considerations, time-dependent deformations may affect structural integrity. For example, prestressed beams are subject to losses in prestressing force caused by creep and shrinkage, which may reduce applied load carrying capacity, particularly if the prestressing tendons are unbonded. Load-carrying capacity is also reduced in structural components in which deflection affects stability, such as the lateral deflection of a column.

Various semi-empirical methods have been developed to determine time-dependent deflections of basic structural components, such as beams, slabs, and columns. Some of these methods which show favorable comparison with experimental data are presented below.

3.1.1 *Time-dependent deflections of reinforced concrete beams and one-way slabs*
Total time-dependent deflection of a reinforced concrete beam in flexure at any

time, t, is given by

$$y(t, t_0) = \overbrace{y_D(0) + y_L(0)}^{\text{short-time deflection}} + \overbrace{y_D^c(t, t_0) + y_L^c(t, t_0) + y^s(t, t_0)}^{\text{time-dependent deflection}} \qquad (3\text{-}1)$$

where $\underbrace{\qquad}_{\substack{\text{dead}\\\text{load}}}$ $\underbrace{\qquad}_{\substack{\text{live}\\\text{load}}}$ $\underbrace{\qquad}_{\substack{\text{creep deflection due}\\\text{to dead load, } y_D \text{ and}\\\text{sustained live load, } y_L}}$ $\underbrace{\qquad}_{\substack{\text{warpage due}\\\text{to shrinkage}}}$

The ultimate deflection with respect to time is obtained by substituting ultimate values of creep and shrinkage into the time-dependent terms.

3.1.2 Determination of short-time deflection

The short-time (initial) response of a reinforced concrete beam, assuming an uncracked section, may be determined from

$$y(0) = K_B \frac{M}{E_c I_g} L^2 \qquad (3\text{-}2)$$

where M is the moment, L the free span, E_c the modulus of elasticity, and I_g is the transformed, uncracked moment of inertia. K_B is a deflection coefficient that depends on the load distribution and support conditions, and may be determined by standard structural analysis techniques, such as moment-area, conjugate beam, or superposition methods. Examples of K_B for mid-span deflections of uniformly loaded and center-span concentrated loaded beams which may resist end moments are given in Eqns 3-3 and 3-4, respectively.

$$K_B = \frac{5}{48}\left(1 + \frac{M_1 + M_2}{10M_m}\right) \qquad (3\text{-}3)$$

$$K_B = \frac{4}{48}\left(1 + \frac{M_1 + M_2}{4M_m}\right) \qquad (3\text{-}4)$$

where M_1 and M_2 are end moments ($M_1 = M_2 = 0$ for simply supported beams) and M_m is the mid-span moment.

Accurate prediction of reinforced concrete deflection is significantly more difficult than for most structural materials, because of the variation in material properties, a non-linear stress versus strain relationship that varies with the age of the concrete, and the inability of concrete to resist substantial tension stresses. However, stress is proportional to strain up to about 0.4 to $0.5f_c'$ and an age-adjusted modulus of elasticity [15] $E_c(t)$ such as that given by Eqn 3-5 can be used for normal-weight concretes.

$$E_c(t) = 57\,600\left(\frac{f_{c,28}'}{0.75 + 7/t}\right)^{\frac{1}{2}} \text{psi} \quad \text{or} \quad 4780\left(\frac{f_{c,28}'}{0.75 + 7/t}\right)^{\frac{1}{2}} \text{kPa} \qquad (3\text{-}5)$$

where $f_{c,28}'$ is the concrete strength (in consistent units) at 28 days, and t is time in days.

Concrete's weakness in tension affects the moment–curvature relationship which, in turn, alters the member deflection. As a beam in flexure is subjected to increasing curvature, both tension and compression concrete stresses increase on opposite sides of the beam's neutral axis. The limited tension capacity may result in cracking on the tension side of the neutral axis. For small moments, little cracking occurs and the stiffness, EI/L, approximates that of the gross uncracked section. As the applied moment increases, cracking increases, reducing the contribution of the concrete on the tension side of the neutral axis to the beam stiffness. This causes a continuing reduction in stiffness, until cracking reaches the neutral axis. Thereafter, the stiffness, substantially reduced from that of the uncracked section, remains fairly constant resulting in a bi-linear moment and curvature relationship.

Several methods have been developed to approximate the variation in stiffness caused by cracking along the span. If a high degree of accuracy is desired, the beam may be subdivided into several elements, each properly accounting for the variation in stiffness as a function of the section properties and anticipated concrete cracking. However, such an approach is cumbersome and for most problems is not required. Semi-empirical methods are available which alter one or more of the parameters in Eqn 2-2 to approximate the variation in stiffness. These methods make adjustments to the curvature [16], alter M/I [17], provide correction factors for the flexural rigidity, EI [18], or identify an effective moment of inertia, such as that given in Eqn 3-6 [7, 19]:

$$I_e = \left(\frac{M_{cr}}{M_{max}}\right)^3 I_g + \left[1 - \left(\frac{M_{cr}}{M_{max}}\right)^3\right] I_T \tag{3-6}$$

where

$$M_{cr} = f_r I_g / a \tag{3-7}$$

and $I_e = I_g$ if $M_{max} \leq M_{cr}$, the moment at which concrete cracking initiates; I_T is the moment of inertia of the cracked transformed section; a is the distance from the neutral axis of the uncracked section to the extreme tension edge; and

$$f_r = 0.65(wf_c')^{\frac{1}{2}} \text{ psi} \quad \text{or} \quad 0.0135\,(wf_c')^{\frac{1}{2}} \text{ kPa} \tag{3-8}$$

where w is the concrete density in lb/ft^3 (or kg/m^3) and f_c' is the ultimate compressive stress in psi (or kPa).

An approximate method for estimating I_e, which eliminates some of the laborious computations is given in [20].

Equation 3-6 adopted by the ACI 318–77 code [21] as an acceptable method to compute short-time deflections, was initially developed to evaluate mid-span deflection for simply supported beams. The value of the effective moment of inertia, I_e, may be altered to reflect other beam end conditions. For continuous spans, ACI 318–77 [21] suggests an average of the mid-span (positive moment) and end-span (negative moment) values given by

$$I_e' = \tfrac{1}{2}[I_{e(m)} + \tfrac{1}{2}(I_{e(1)} + I_{e(2)})] \tag{3-9}$$

where the subscripts m, 1, and 2 refer to mid-span, and each beam end, respectively.

ACI Committee 435 [22] indicates that reasonable accuracy is obtained with the use of the midspan I_e for many cases involving continuous beams, but recommends Eqn 3–10 if the ratio of the end region to mid-span flexural rigidity,

EI, is 2 or larger and the support moment exceeds that of a fixed end.

$$I'_e = I_{e(m)}\left[1 - \left(\frac{M_1 + M_2}{2M_0}\right)^\beta\right] + \left(\frac{I_{e(1)} + I_{e(2)}}{2}\right)\left(\frac{M_1 + M_2}{2M_0}\right)^\beta \tag{3-10}$$

where β is at least 2 and if $0.1 < M_m/M_0 < 0.6$, $\beta = 4$ (M_0 is the mid-span moment of a simply supported beam of the same dimensions). For cantilever beams, the effective moment of inertia at the support as determined by Eqn 3-6 may be used.

3.1.3 Deflection due to shrinkage

Deflection of beams due to shrinkage may be predicted from

$$y^s(t, t_0) = K_{sh}\phi_{sh}(t, t_0)L^2 \tag{3-11}$$

where $\phi_{sh}(t, t_0)$ is the curvature (M/EI). The general practice is to assume that the reinforcing steel along the span is constant when determining K_{sh}. For continuous beams, the curvature is assumed to be the same value in the mid-span region (positive curvature for normal gravity type loading) and in the negative-curvature region near the end of the spans, with the inflection points arbitrarily taken at the span quarter points. Values of K_{sh} based on these conditions and using a constant value of $\phi_{sh}(t, t_0)$ (with appropriate signs for negative curvature) are 0.125, 0.625, and 0.5 for a simply supported beam, fixed end beam, and cantilever beam, respectively.

Several empirical methods have been developed for the determination of shrinkage curvature [23, 24]. The expression developed by Miller, and modified by Branson [25, 26], is shown in Eqn 3–12 and has been demonstrated to provide good correlation [25, 26] with experimental results.

$$\phi_{sh}(t, t_0) = 0.7\frac{\varepsilon_{sh}}{h}(t, t_0)(p - p')^{\frac{1}{3}}\left(\frac{p - p'}{p}\right)^{\frac{1}{2}} \quad \text{for} \quad (p - p') \leq 3\% \tag{3-12}$$

where $p = 100\,A_s/bh$ and $p' = 100\,A'_s/bh$ (b and h are the width and height of the beam cross-section).

3.1.4 Deflection due to creep

Two approaches are widely used for the determination of flexural deflections due to concrete creep. Both are based on a constant stress history, and both utilize short-time deflection as a measure of initial concrete strain.

One approach utilizes the increased strain due to creep in combination with the initial strain, ε_0, to define an effective modulus of elasticity, $E'_c(t)$

$$E'_c(t) = \frac{\sigma_0}{\varepsilon_0[1 + \alpha(t, t_0)]} \tag{3-13}$$

where $\sigma(0)$ is the initial stress. Then, the effective modular ratio becomes

$$n'_c(t) = \frac{E_s}{E'_c(t)} = \frac{E_s}{E_c(t)}[1 + \alpha(t, t_0)] = n[1 + \alpha(t, t_0)] \tag{3-14}$$

The effective modulus and effective modular ratio may be used in lieu of $E_c(t)$ and $n(t)$ to predict the effects of creep.

An approach [22] which has gained wide acceptance, both because of its ease of application and excellent correlation with experimental results, relates the creep effect to the short-time deflection as given in Eqn 3-15:

$$y^c(t, t_0) = K_r^c \alpha(t, t_0) y(0) \tag{3-15}$$

where K_r^c is a reduction factor empirically determined to relate creep strain to curvature. This factor accounts for shifts in the neutral axis caused by creep, and the effect reinforcing steel has on limiting creep strain. Several expressions have been developed to define K_r^c [26]. Equation 3-16, recommended by ACI Committee 209 [7] is acceptable for beams with a steel percentage greater than 1%.

$$K_r^c = 0.85 - 0.45 A_s'/A_s \qquad K_r^c \geq 0.40 \tag{3-16}$$

For ultimate deflection, it is convenient to combine both the creep and shrinkage effects. This is accomplished in Eqn 3-17:

$$y^{cs}(\infty) = \bar{K}_r^c T_\infty y(0) \tag{3-17}$$

where

$$\bar{K}_r^c = 1 - 0.6 A_s'/A_s \qquad \bar{K}_r^c \geq 0.40 \tag{3-18}$$

ACI 318 code recommends a value of 2.0 for the factor T_∞. See [26] for further discussion of K_r^c and $\bar{K}_r^c$.

3.1.5 Deflection of two-way slabs

The ACI Committee 435 [27] identifies several approaches for estimating the deflection of slabs, including classical methods [28], crossing-beam analogies, beam-strip methods, equivalent frames, and finite element methods [29]. However, the uncertainties associated with the prediction of beam deflections are increased by slab action and it can be shown that, for normal design, the additional effort required to perform such analyses does not result in additional accuracy. Therefore, total deflection of a two-way slab, or flat plate, including creep and shrinkage considerations may be estimated using Eqn 3-1. Short-time deflection of such a slab may be approximated by Eqn 3-19:

$$y(0) = K_s q L^4 / D \tag{3-19}$$

where q is a uniformly distributed loading, L the longer span, $D = Eh^3/12(1-\nu)$, the flexural rigidity, and K_s the slab deflection coefficient. Branson [26] and ACI 209 [7] provide numerical values of K_s for various edge conditions.

Neglecting Poisson's ratio and substituting I for $h^3/12$, Eqn 3-19 becomes

$$y(0) = \frac{K_s q L^4}{E_c(t) I} \tag{3-20}$$

The moment of inertia may be determined from Eqn 3-6. Since the effect of cracking is less for a slab than a beam, the CRSI suggests [30] that the average gross moment of inertia be used for two-way joist and waffle slabs. For solid slabs, the moment of inertia can be computed from Eqn 3-6, using an average value of M_{cr}/M_{max}. Long-time deflections may be calculated from Eqn 3-15.

3.1.6 *Loss of prestress and camber in prestressed beams*

Time-dependent creep and shrinkage in prestressed concrete members result in a substantial portion of the total losses in both prestress force and camber. These and other losses (i.e. anchorage losses, elastic shortening, relaxation of the steel, and frictional losses) must be considered to assure adequate serviceability.

The purpose of this section is not to provide analytical methods for the design of prestressed members, or even to discuss the determination of all such losses. Rather, the discussion will be limited to the prediction of creep and shrinkage effects [7, 26, 31].

The general expression for the loss of prestress force in beams is given by Eqn 3-21:

$$\Delta f_{\mathrm{s}}(t,t_0) = \underbrace{n(t)f_{\mathrm{c}}(0)}_{\substack{\text{elastic}\\\text{shortening}\\\text{losses (not}\\\text{applicable}\\\text{for post}\\\text{tensioning)}}} + \overbrace{\underbrace{\Delta f_{\mathrm{s}}^{\mathrm{c}}(t,t_0) + \Delta f_{\mathrm{s}}^{\mathrm{s}}(t,t_0) + \Delta f_{\mathrm{s}}^{\mathrm{R}}(t,t_0)}_{\substack{\text{losses due to creep, }\Delta f_{\mathrm{s}}^{\mathrm{c}},\\\text{and shrinkage, }\Delta f_{\mathrm{s}}^{\mathrm{s}}, \text{ in the}\\\text{concrete and relaxation,}\\\Delta f_{\mathrm{s}}^{\mathrm{R}}, \text{ in the steel}}} - \underbrace{(\Delta f_{\mathrm{s}}^{M_{\mathrm{s}}})(t,t_0) - (\Delta f_{\mathrm{s}}^{D_{\mathrm{c}}})(t,t_0)}_{\substack{\text{losses due to}\\\text{composite action}}}}^{\text{time-dependent losses}}$$

$$(3\text{-}21)$$

Axial shortening losses may be calculated from Eqn 3-22:

$$n(t)f_{\mathrm{c}}(0) = n\!\left(\frac{F_{\mathrm{i}}}{A_{\mathrm{B}}} + \frac{F_{\mathrm{i}}e^2}{I_{\mathrm{B}}} - \frac{M_{\mathrm{D}}e}{I_{\mathrm{B}}}\right) \tag{3-22}$$

where $n(t)$ is the modular ratio, F_{i} is the initial tensioning force, A_{B} and I_{B} are the area and moment of inertia of the beam section, M_{D} is the beam dead-load moment and e is the eccentricity of the prestressing steel from the beam centroid.

Creep losses are given by Eqn 3-23:

$$\Delta f_{\mathrm{s}}^{\mathrm{c}}(t,t_0) = nf_{\mathrm{c}}K_{\mathrm{r}}\underbrace{\left[\alpha(t_1,t_0)\left(1 - \frac{\Delta F(t_1)}{2F_0}\right)\right.}_{\substack{\text{creep of non-composite section}\\\text{or creep before}\\\text{composite action}}}$$

$$\underbrace{\left. + (\alpha(t,t_1) - \alpha(t_1,t_0))\left(1 - \frac{\Delta F(t) + \Delta F(t_1)}{2F_0}\right)\frac{I_{\mathrm{B}}}{I_{\mathrm{F}}}\right]}_{\text{creep losses after composite action}} \tag{3-23}$$

where t_0 and t_1 are, respectively, the time at initial prestressing and the time when composite action first occurs. (Note that for a non-composite beam, t_1 is the time at which loss of prestress is to be evaluated.) $\Delta F(t)/F_0$ is the total loss after elastic losses for normal-weight concrete and can be taken as 0.10, if the time between prestressing and load application is between 3 weeks and a month, and 0.14 between 2 and 3 months, with an ultimate value of 0.18. F_0 is the prestress force at stress transfer and I_{F} is the moment of inertia of the composite section. K_{r} is defined in Eqn 3-31.

The effect of shrinkage is given by Eqn 3-24:

$$\Delta f_c^s(t, t_0) = \frac{\varepsilon_{\mathrm{sh}}(t, t_0)E_s}{1 + np(1 + e^2/r^2)K_r}$$ (3-24)

where E_s is the modulus of elasticity of the steel, p is the ratio of prestressing steel to concrete area, and r is the ratio of gyration of the section $(I_B/A_B)^{\frac{1}{2}}$.

Several expressions [26], including Eqn 3-25 [31], are available to predict the relaxation effects, Δf_s^R, of stress-relieved steel:

$$\Delta f_s^R(t, t_0) = \frac{f_s(0)}{100} 1.5 \log_{10} t$$ (3-25)

where $f_s(0)$ is the initial stress in the prestressing wire, and t is given in hours. The ultimate loss for such wire can be taken as 7.5%.

The final two terms in Eqn 3-21 represent the gain in prestress due to the slab dead-load moment, M_T, given in Eqn 3-26 and differential creep and shrinkage, D_c, between the beam and topping slab (Eqn 3-27):

$$(\Delta f_s^{M_s})_T = \frac{n(t_1)M_T e}{I_B} \left(1 + \alpha(t, t_0)\frac{I_B}{I_F}\right)$$ (3-26)

where $n(t_1)$ is the modular ratio of the beam at initial composite action.

$$(\Delta f_s^{D_c}) = \frac{n(t_1)D_c A_T E_T e_{cs} e_{comp}}{3I_F}$$ (3-27)

where e_{cs} is the distance between the centroid of the composite section and the centroid of the topping slab, and e_{comp} is the eccentricity of the steel with respect to the composite section. An average strain value of 415×10^{-6} may be assumed for unshored construction. A_T and E_T are the slab area and modulus of elasticity (at occurrence of composite action), respectively.

The effective camber, $y(t, t_0)$, of a prestressed beam subject to composite action is given in Eqn 3-28 (the camber of a non-composite beam may be obtained by eliminating the composite effects):

short-time deflection   time-dependent deflection

$$y(t, t_0) = \underbrace{y_P(0)}_{\substack{\text{initial} \\ \text{camber} \\ \text{due to} \\ \text{pre-} \\ \text{stressing}}} - \underbrace{y_D(0) - y_L(0)}_{\substack{\text{dead, } y_D, \text{ and} \\ \text{live load, } y_L, \\ \text{deflection}}} - \underbrace{y_D^c(t, t_0) - y_L^c(t, t_0) + y_P^c(t, t_0)}_{\substack{\text{creep effects caused by} \\ \text{dead load, } y_D^c, \text{ sustained} \\ \text{live load, } y_L^c \text{ and camber,} \\ y_P^c}} - \underbrace{y_{comp}(t, t_1)}_{\substack{\text{differential} \\ \text{shrinkage be-} \\ \text{tween beam and} \\ \text{topping (com-} \\ \text{posite sections} \\ \text{only)}}}$$ (3-28)

where the short-time deflection is computed from Eqn 3-2. The uncracked moment of inertia of either the beam or the composite section may be used, depending upon the condition at load application. For live loads, the effective moment of inertia as determined by Eqn 3-6 may be used, except that special consideration may be necessary to determine the cracking moment. A detailed discussion is presented in [26]. For non-composite beams, the cracking moment can be determined from Eqn 3-29:

$$M_{cr} = Fe + \frac{FI_B}{A_B h} + \frac{f_r I_B}{h} - M_D \tag{3-29}$$

where h is the distance from the centroid of the uncracked section to the extreme tension fiber, and F is the prestress force after losses. The creep effect due to dead and sustained live load is given by Eqn 3-30.

$$y_D^c(t, t_0) + y_L^c(t, t_2) = K_r y_D(0) \left\{ \overbrace{\alpha(t_1, t_0)}^{\substack{\text{effect before} \\ \text{composite} \\ \text{action}}} + \overbrace{[\alpha(t, t_1) - (\alpha(t_1, t_0)]}^{\substack{\text{effect after composite} \\ \text{action}}} \frac{I_B}{I_F} \right\}$$

creep deflection caused by
beam dead load

$$+ K_r (y_D(0))_T \overbrace{\alpha(t, t_1) \frac{I_B}{I_F}}^{\substack{\text{creep deflection} \\ \text{caused by topping}}} + K_r y_L(0) \overbrace{\alpha(t, t_2) \frac{I_B}{I_F}}^{\substack{\text{creep deflection caused} \\ \text{by sustained live load}}} \tag{3-30}$$

where K_r is calculated from Eqn 3-31 and t_2 is estimate of sustained live load application.

$$K_r = \frac{1}{1 + A_s'/A_s} \tag{3-31}$$

where A_s and A_s' are the area of non-prestressed tension and compression steel in the member, respectively.

Equation 3-32 gives the creep effects caused by the initial camber:

$$y_p^c(t, t_0) = y_p(0) \left\{ \overbrace{\left[\frac{-\Delta F(t_1)}{F_0} + \left(1 - \frac{\Delta F(t_1)}{2F_0}\right) \alpha(t_1, t_0) \right]}^{\substack{\text{before composite action} \\ \text{(non-composite beam)}}} \right.$$

$$\left. + \overbrace{\left[-\frac{\Delta F(t) - \Delta F(t_1)}{F_0} + \left(1 - \frac{\Delta F(t_1) + \Delta F(t)}{2F_0}\right)(\alpha(t, t_1)) - \alpha(t_1, t_0)) \right] \frac{I_B}{I_F}}^{\text{after composite action}} \right\} \tag{3-32}$$

The final term in Eqn 3-21 accounts for deflection caused by differential creep and shrinkage between the beam and topping, if composite action occurs. This is given by Eqn 3-33 for a simply supported beam:

$$\dot{y}_{comp}(t, t_0) = D_c \frac{A_T E_T e_{cs}}{24 E_B(t) I_F} L^2 \tag{3-33}$$

3.2 Design for structures sensitive to time-dependent deformation

It is extremely difficult to define a structure that is sensitive to time-dependent deformation. A structure can be sensitive because of mix proportions, environmental conditions, and loading history. In addition, the structural configuration plays an important role. In general, light structures are more sensitive than massive ones. However, if the other parameters that affect time-dependent deformation result in small change in strain, many light structures may have negligible time-dependent deformation effects.

All of the methods described in previous sections consider uniform average values for mix proportions, environmental condition, loading history and structural size. Thus, the engineer must decide if these values represent the conditions of the prototype with sufficient accuracy and conservatism. In order to make such an evaluation, the parameters discussed in Sections 2.1, 2.2 and 2.3 must be reviewed and ranges estimated.

The determination of the importance of time-dependent deformations must consider two aspects of the analysis:

(a) The required accuracy of prediction of material properties.
(b) The effect of structural configuration and loading history.

If accuracy greater than that which can be obtained using the various prediction methods discussed in Section 2.4 for general structures is required, then testing as discussed in the same section should be considered.

Further, there are a number of analysis methods described in the literature that can be used to consider some of the loading history parameters discussed in Section 2.2. Some of these are: rate of creep [32], rate of flow [33], superposition [34], and age adjusted effective modulus method [11]. A detailed treatment of these methods is presented in [15].

4 Design for thermal movement

4.1 Axially loaded structures

As with most materials, hardened concrete undergoes volume changes with change in temperature. The magnitude of the change is primarily influenced by the type of aggregate. Troxell and Davis [35] report that the coefficient of expansion, α_T, ranges from 2.2 to 3.9×10^{-6} per °C with an average of 3.06×10^{-6} per °C. Concrete made with quartz has the highest value, and concretes made with sandstone, granite, basalt and some limestones have decreasingly lower values.

If concrete is restrained, thermal strain is accompanied by a change in stress. Equation 4-1 gives the compression stress in an axial concrete member caused by restrained ends.

$$f_c^T = R(\alpha_T \Delta T_A) E_c \tag{4-1}$$

where R is a restraint factor varying from 0 (for no end restaint) to 1 (for complete restraint, i.e. no movement) and ΔT_A is the change in temperature from a base, or stress free (often taken as 70°F). Since concrete has limited capacity to resist tension, little resistance is offered if the concrete is restrained against volume reductions caused by a drop in temperature. In fact, after the concrete reaches f_r, cracking occurs (see Eqn 3-8) eliminating the stress path necessary to resist tension.

4.2 Frame structures

Thermal loads on structures are seldom limited to axial deformation. Structural response generally includes axial, bending and shear stresses. The effect of thermal change on these stresses is influenced by the make up of the thermal loads and rigidity of the section.

Thermal loading in a structure generally occurs from a temperature differential across the structure. For analysis purposes, these differentials are resolved into a uniform axial temperature change, ΔT_A, and a gradient, ΔT_ϕ. This gradient creates a curvature $\alpha_T \Delta T_\phi / h$ in the section. Increasing curvature causes cracking on the tension side of the neutral axis. This reduces the stiffness, increases rotation, and, therefore, relieves stress.

The commentary to ACI 349, Appendix A [36] identifies three general approaches to evaluate thermal loads:

(a) Impose thermal loads on a monolithic section, evaluating the rigidity of the section and the stiffness of the structure based on an uncracked section analysis. Such an analysis is fairly simple but may be overly conservative since it does not consider the self-relieving nature of the thermal stress due to cracking and deformation.

(b) Consider cracking of concrete for all loads, both thermal and non-thermal (mechanical). Although this approach leads to the most accurate analysis, it is extremely complex. An iterative solution is required, sequentially applying loads, reevaluating the degree of cracking along the structure and revising member properties accordingly. These revised properties cause a redistribution of stresses. The process is continued until an equilibrium configuration is obtained.

(c) An approximate method as suggested by ACI Committee 349 [37], which considers the structure uncracked for the mechanical loads and cracked for the thermal loads. In this method, thermal loads may consist of a uniform axial temperature change and a gradient across the section. The structure is then analyzed for mechanical loads using conventional methods and un-cracked section properties to determine frame moments. It is then assumed that the addition of a thermal moment which has the same sign as that obtained for the mechanical loads increases cracking. However, adding a thermal moment of opposite sign to the mechanical moments may result in a final section that is uncracked. Therefore, in those portions of the span where these signs are the same, a cracked section is assumed, and if the signs are opposite, the section is assumed to be uncracked. This results in the general condition, shown in Fig. 4-1. The thermal moments from the uncracked analysis are added to the mechanical moments using moment distribution. The analysis requires the following assumptions:

(1) The uncracked moment of inertia, I_g, is calculated using the gross section, without reinforcing steel.

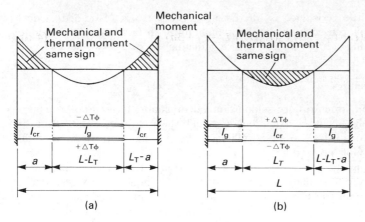

Figure 4-1 (a) End-cracked member
(b) Center-cracked member

(2) The cracked section moment of inertia, I_{cr}, neglects compression steel and assumes the tension reinforcement is at the tension face. Therefore, $I_{cr} = 6jk^2I_g$

(3) Resulting axial forces at any section are no greater than $2M/h$, where h is the depth of the section. This limitation is necessary to maintain the ratio of the actual cracked moment of inertia within 10% of that used in the analysis.

Expressions for fixed-end moments at the a-end of the beam (see Fig. 4-1) are given in Eqns 4-2 and 4-3 for the thermal gradient across the section, $\alpha_T\Delta T_\phi/h$, and the uniform axial temperature change respectively.

$$M_{FE}^G = \frac{E_c\alpha_T\Delta T_\phi bh^2}{12}\left[\frac{K_T(1-C_F)}{2}\right] \tag{4-2}$$

$$M_{FE}^A = \frac{E_cI_g\Delta}{L^2}[K_T(1+C_F)] \tag{4-3}$$

where K_T and C_F are the stiffness coefficient and carry-over factor, respectively, determined by Eqns 4-4 through 4-14 or from [37]).

For an end-cracked beam (see Fig. 4-1a)

$$K_T = \frac{2m}{1-(1-m)\left(\dfrac{L-L_T}{L}\right)\left(\dfrac{L+L_T-2a}{L}\right)-C_F\left[1-(1-m)\left(\dfrac{L-L_T}{L}\right)\left(\dfrac{L-L_T-2a}{L}\right)\right]} \tag{4-4}$$

$$C_F = \frac{\dfrac{L^2}{6}-(1-m)\left(\dfrac{L-L_T}{2L}\right)A}{\dfrac{L^2}{3}-(1-m)\left(\dfrac{L-L_T}{2L}\right)B} \tag{4-5}$$

where

$$A = (L_T - a)(L - L_T + 2a) + \tfrac{1}{3}(L - L_T)(L - L_T + 3a) \tag{4-6}$$

$$B = a(L - L_T + 2a) + \tfrac{2}{3}(L - L_T)(L - L_T + \tfrac{3}{2}a) \tag{4-7}$$

For a center-cracked beam (see Fig. 4-1b):

$$K_T = 2 \left/ \left\{ \frac{1}{2} + \left(\frac{1}{m} - 1\right)\left(\frac{L_T}{L}\right)\left(\frac{L - a - \frac{1}{2}L_T}{L}\right) - C_F\left[\frac{1}{2} + \left(\frac{1}{m} - 1\right)\left(\frac{L_T}{L}\right)\left(\frac{a + \frac{1}{2}L_T}{L}\right)\right] \right\} \right. \tag{4-8}$$

$$C_F = \frac{\frac{1}{6}L^2 + \left(\frac{1}{m} - 1\right)\dfrac{L_T}{L} E}{\frac{1}{3}L^2 + \left(\frac{1}{m} - 1\right)\dfrac{L_T}{L} F} \tag{4-9}$$

where

$$E = (L - L_T - a)\left(\frac{L_T + 2a}{2}\right) + L_T\left(\frac{L_T + 3a}{6}\right) \tag{4-10}$$

$$F = a\left(\frac{L_T + 2a}{2}\right) + L_T\left(\frac{2L_T + 3a}{6}\right) \tag{4-11}$$

and

$$m = 6jk^2 \tag{4-12}$$

$$k = [(pn)^2 + 2pn]^{\frac{1}{2}} - pn \tag{4-13}$$

$$j = 1 - \tfrac{1}{3}k \tag{4-14}$$

Δ is the transverse displacement between the end of the member, resulting from differential axial elongation of adjacent members according to Eqn 4-15.

$$\Delta_A = L(\alpha_T \Delta T_A) \tag{4-15}$$

4.2.1 Steps in conducting the analysis
These are

(1) Determine the moments on the structure caused by the mechanical loads.
(2) For each member compare the signs of the expected thermal moments to the moments obtained from Step 1. From the moments in Step 1 also determine a and L_T for each end of each member.
(3) Knowing a, L_T and the assumed cracking pattern, determine K_T and C_F either from Eqns 4–4 through 4-14, or from [37].
(4) For axially restrained members, determine the unstrained elongation from Eqn 4-15, where ΔT_A is the uniform difference in temperature from the stress-free temperature of the member.
(5) Distribute the unrestrained axial elongation at each end of the member in inverse proportion to the resistance (stiffness) of the attaching structure.
(6) Given K_T, C_F and the distribution of the axial elongation, determine the fixed-end moments for each end of each member from Eqns 4-3 and 4-4.

(7) If necessary, correct for side-sway using standard moment distribution techniques.

(8) Add the corrected end moments to the moments determined in Step 1.

(9) Confirm that resulting axial forces are no greater than $2M/h$.

4.3 Axisymmetric structures

ACI Committee 349 [37] has also developed an approximate method to evaluate thermal gradients on shells of revolution by adding the results of a cracked section analysis to the mechanical moments, determined by conventional uncracked methods.

For axisymmetric structures, a uniform temperature change is not restrained, except at external boundaries and discontinuities. This effect must be analyzed separately. Therefore, any forces created by the uniform axial temperature change must be considered as part of the mechanical loading on the axisymmetric section. The thermal gradient, however, is restrained by the axisymmetric geometry. This restraint creates section stresses that may be relieved by the cracked section analysis.

If the resulting normal force is no greater than $1.4M/h$ (this level of force does not substantially affect the centroid of the cracked section) the thermal moment is given by

$$M = E_c(t)\, \frac{\alpha_T \Delta T_\phi bh^2}{1-\nu} \{-0.152k^3 + 1.818\rho' n[(d'/d) - k](d'/d) + 0.909\rho n(1-k)\}$$

(4-16)

where d and d' are the distances from the extreme tension fiber to the center of the tension and the compression reinforcing steel, respectively, and

$$k = \{(2\rho'n + \rho n)^2 + 2[2\rho'n(d'/d) + \rho n]\}^{\frac{1}{2}} - (2\rho'n + \rho n)$$

(4-17)

5 Construction methods

A major structural problem associated with time and temperature-dependent deformation is cracking. There are numerous examples that immediately come to mind, ranging from simple shrinkage cracking to the failure of building cladding, caused by inattention to construction details necessary to accommodate creep, shrinkage and temperature deformation.

In this section, it is assumed that the engineer has taken full advantage of the design information presented in this chapter and has minimized the potential creep and shrinkage of the system as much as practicable. It is further assumed that materials and material properties that minimize time-dependent deformation have been specified.

Therefore, the only additional step the designer must take is to minimize the effects of restraint and settlement to further control cracking. The American Concrete Institute has prepared an excellent report *Control of cracking in concrete structures* [4]. Some of the advice given there will be summarized here.

Cracking due to restraint and settlement can be reduced if

(a) Construction joints of a depth not less than 10% of the wall thickness on both sides should be provided at intervals of one (for high walls) to three (for low walls) times the height of the wall.

(b) New concrete should be protected from temperature change or moisture loss until the concrete is strong enough to withstand the induced strain without cracking.

(c) Contraction joints should be used to relieve restraint at sharp changes in cross-section.

(d) Contraction joints should be used to allow slabs to change volume from all sides towards the center.

6 Observations on structures

It can be seen from the preceding sections that most of the prediction, design and analysis methods available for evaluating creep and shrinkage are empirical. This is necessitated by the fact that, although significant progress has been made toward understanding the creep and shrinkage behavior of concrete, neither an all-inclusive empirical relationship nor a complete theoretical formulation has been developed to describe these phenomena.

It is, therefore, important that measurements on structures be carefully performed and reported. Russell *et al.* [38] have developed a simple set of guidelines for taking such measurements and have reviewed some of the published data. In their work, they point out that significant variability can occur when taking creep and shrinkage measurements. This variability is equally valid for structural calculations, and is reviewed below.

Variability of the prediction method   Presently available prediction methods for creep properties have a coefficient of variation of approximately 25%. This variability can be reduced if creep properties are estimated from experimental data obtained from the concrete used in the structure.

Variability of structural analysis method   Analysis methods can range from simple elastic beam analysis to complex finite element procedures. The accuracy to which such methods predict stress and deformation vary significantly.

Variability of material properties   It can be assumed that the general variability (coefficient of variation) of material properties, i.e. strength and elastic modulus, are approximately 10% or less, for most structural grade concrete.

Variability of loads   Prior to evaluating such variability, the investigator should realize that, although structures are designed to withstand highly variable loads (i.e. wind, earthquake, etc.), the loads actually experienced by a structure usually are well defined (i.e. dead load, permanent live load, etc.). Although these latter loads are well defined, they are often significantly less than full design loads, and, therefore, result in lower stresses, deformations and cracking than those assumed in design.

Variability of environment The effect of seasonal and daily temperature and humidity variations on measured response should be considered, since creep and shrinkage behavior is significantly affected by such variation.

Therefore, even for carefully planned investigations, a variation of as much as ±50% from predicted deformations can be expected.

Actual field data are available for many structures, and can be studied to determine general behavior. References [39–61] present results for high-rise concrete buildings, [62–75] for concrete bridge structures, and [76–79] for prestressed concrete nuclear reactor containment structures. These references are typical and represent a small percentage of those available.

7 Conclusions

The evaluation of the effect of creep, shrinkage and temperature on structural systems is a complicated and inexact science. It can, however, be very important in insuring the structural integrity of the prototype. A checklist of important steps may help.

(a) Estimate the expected loading history and environment of the structure. The use of historical weather data may be helpful.

(b) Design the concrete mix as early as possible, using those materials available in the area that will produce the lowest creep and shrinkage consistent with economic considerations.

(c) Predict the creep and shrinkage response of the mix selected, using methods outlined in Section 2.

(d) Estimate the sensitivity of the structure to effects of time-dependent deformation.

(e) Design the structure using normal design practice and methods suggested in Sections 3 and 4.

(f) Estimate the range of variation that can be expected between the structure as designed and the structure as built, and insure that structural integrity is maintained throughout the estimated range of variation.

(g) Specify those construction methods that will insure the structure is constructed as designed.

The above steps are not difficult to perform and can result in significant economies over the life of the structure.

The information presented in this chapter just skims the surface of the subject. It is hoped that sufficient information has been presented to deal with the general subject of shrinkage, creep and temperature changes in concrete systems. However, if additional information is required, two rather significant texts are available. They are *Deformation of concrete structures* [26] by Branson and *Creep of concrete: plain, reinforced, and prestressed* [15] by Neville. The major emphasis of Branson's book is on the behavior of structural systems, while Neville's concentrates on material behavior. Together, they represent the state-of-the-art of the subjects considered in this chapter.

Acknowledgements

The authors and publishers thank *Concrete*, London, for permission to include one figure from *A simple design method for predicting the elastic modules and creep of structural concrete* (1977); the Comité Euro-international du Béton for permission to reproduce two figures from the *International recommendation for design and construction of structures* (1970), and the American Concrete Institute for permission to reproduce one figure from the *ACI Journal, Proceedings*, Vol. 69, No. 9, Sep. 1972 p. 553 (Ref. [9]).

References

1. ACI Committee 209, *Shrinkage and creep in concrete*, American Concrete Institute, Bibliography No. 7 and No. 10, 1967, 1972.
2. ACI Committee 209, Subcommittee 1, Effects of concrete constituents, environment, and stress on creep and shrinkage of concrete, *Designing for effects of creep, shrinkage, and temperature in concrete structures*, American Concrete Institute, SP-27, 1971, pp. 1–42.
3. Daye, M. A., and Meyers, B. L., Creep of Structural Systems, *6th int. conf. struct. mech. reactor technol.*, Vol. J(a), Paris, 1981, *pp.* J3-10.
4. ACI Committee 224, *Control of cracking in concrete structures*, American Concrete Institute, 224 R-80, July 1981.
5. Hilsdorf, H. K. and Muller, H. S., *Comparison of methods to predict time-dependent strains of concrete*, Institut für Baustofftechnologie Universität Karlsruhe, Oct. 1979.
6. Branson, D. E. and Christianson, M. L., Time-dependent concrete properties related to design, *Designing for effects of creep, shrinkage, and temperature in concrete structures*, American Concrete Institute, SP-27, 1971, pp. 257–277.
7. ACI Committee 209, Prediction of creep, shrinkage and temperature effects in concrete structures, *Designing for effects of creep, shrinkage, and temperature in concrete structures*, American Concrete Institute, SP-27, 1971, pp. 51–93.
8. CEB–FIP, *International recommendations for the design and construction of structures*, Paris and London, 1970.
9. Meyers, B. L. and Branson, D. E., Design Aid for Predicting Creep and Shrinkage Properties of Concrete, *J. Am. Concr. Inst, Proc.*, Vol. 69, No. 9, Sept. 1972, pp. 551–555.
10. Concrete Society, *A simple design method for predicting the elastic modulus and creep of structural concrete*, London, 1977.
11. Bazant, Z., Prediction of concrete creep effects using age-adjusted effective modulus method, *J. Am. Concr. Inst. Proc.*, Vol. 69, No. 4, 1972, pp. 212–216.
12. CEB–FIP, *Model code*, Paris, London and Berlin, 1978.
13. Daye, M. A., *Creep of concrete for reactor containment building*, unpublished report.
14. Meyers, B. L., Branson, D. E. and Schumann, C. G., Prediction of creep and shrinkage behavior for design from short-time tests, *J. Prestressed Concr. Inst.*, 17, May-June 1972, pp. 29–45.

15. Neville, A. M., *Creep of concrete: plain, reinforced and prestressed,* North Holland, Amsterdam, 1970.
16. Beeby, A. W., *Short-term deformations of reinforced concrete members,* Cement and Concrete Association Technical Report, TRA 408, London, March 1968.
17. CEB, Commission IVb, *Deformations,* Portland Cement Association, Foreign Literature Study 547, 1968.
18. Yu, W. W. and Winter, G., Instantaneous and long-time deflections of reinforced concrete beams under working loads, *J. Am. Concr. Inst. Proc.,* Vol. 57, No. 1, 1960, pp. 29–50.
19. Branson, D. E., Instantaneous and time-dependent deflections of simple and continuous reinforced concrete beams, Report No. 7, Part 1, Alabama Highway Research Dept, Bureau of Public Roads, Aug. 1963, pp. 1–78.
20. Grossman, J. S., Simplified computations for effective moment of inertia I_e and minimum thickness to avoid deflection computations, *J. Am. Concr. Inst. Proc.,* Vol. 78, No. 6, Nov. Dec. 1981, pp. 423–439.
21. ACI Committee 318 *Building code requirements for reinforced concrete,* ACI Standard 318–77, American Concrete Institute, Detroit, 1977.
22. ACI Committee 435, *Deflections of continuous concrete beams,* ACI 435. 5R-73 Reaffirmed 1979, ACI *manual of concrete practice,* Part 4, American Concrete Institute, Detroit, 1981, pp. 435.5R-1 to 435.5R-7.
23. Pauw, A. and Meyers, B. L., Effect of creep and shrinkage on the behavior of reinforced concrete members, *Symposium on creep of concrete,* American Concrete Institute, SP-9, 1964, pp. 129–156.
24. Miller, A., Warping of reinforced concrete due to shrinkage, *J. Am. Concr. Inst. Proc.,* Vol. 54, No. 11, May 1958, pp. 939–950.
25. ACI Committee 435, Deflections of reinforced concrete flexural members, *J. Am. Concr. Inst.* Vol. 63, No. 6, June 1966. ACI 435.2R-66, Reaffirmed 1979, *ACI manual of concrete practice,* Part 3, American Concrete Institute, Detroit, 1981, pp. 435.2R-1 to 435.2R-29.
26. Branson, D. E., *Deformations of concrete structures,* McGraw-Hill, New York, 1977. pp. 1–447.
27. ACI Committee 435, *Deflections of two-way reinforced concrete floor systems: state-of-the-art report,* ACI 435.6R-74, Reaffirmed 1979, *ACI manual of concrete practice,* Part 3, American Concrete Institute, Detroit, 1981, pp. 435. 6R-1 to 435.6R-24.
28. Timoshenko, S. and Woinowsky-Krieger, S., *Theory of plates and shells,* McGraw-Hill, New York, 1959.
29. Scordelis, A. C., Finite element analysis of reinforced concrete structures, *Proc. conf. finite element methods civ. eng.* McGill University, 1972.
30. Concrete Reinforcing Steel Institute, Serviceability checks, *CRSI Handbook,* Chicago, 1975, 65 pp.
31. Branson, D. E., Meyers, B. L. and Kripanarayanan, K. M., *Loss of prestress, camber, and deflection of non-composite and composite structures using different weight concrete,* Iowa State Highway Commission, Report No. 70–6 Aug. 1970.
32. Ross, A. D., Creep of concrete under variable stress, *J. Am. Concr. Inst. Proc.,* Vol. 54, No. 9, March 1965, pp. 739–758.

33. England, G. L. and Illstron, J. M., Methods of computing stress in concrete from history of measured strain, *Civ. Eng. Public Wks Rev.* Vol. 60, April, May and June 1965, pp. 513–517, 692–694, 846–847.

34. McHenry, D., A new aspect of creep in concrete and its application to design, *Proc. Am. Soc. Test. Mat.*, Vol. 43, 1943, pp. 1069–1084.

35. Troxell, G. C. and Davis, H. E., *Composition and properties of concrete*, McGraw-Hill, New York, 1956, 244 pp.

36. ACI Committee 349, *Code requirements for nuclear safety related concrete structures*, ACI Standard 349–76, American Concrete Institute, Detroit, 1976.

37. ACI Committee 349 (Cannon, R. and Fulton, J., prime authors), Reinforced concrete design for thermal effects on nuclear power plant structures, Report No. ACI 349-1R-80, *J. Am. Conc. Inst. Proc.*, Vol. 77, No. 6, Nov. Dec 1980, pp. 399–428.

38. Russell, H., Meyers, B. L. and Daye, M. A., *Creep and shrinkage in concrete structures*, Martinus Nijhoff, The Hague, 1982.

39. Beresford, F. D., Measurement of time-dependent behavior in concrete buildings, *4th Australian bldg res. congr.* Sydney, Aug. 1970; also Commonwealth Scientific and Industrial Research Organization Reprint No. 554.

40. Experimental lightweight flat plate structure, Part I—Measurements and observations during construction, *Construct. Rev.*, Vol. 34, No. 1, Jan. 1961.

41. Experimental lightweight flat plate structure, Part II—Deformation due to self weight, *Construct. Rev.*, Vol. 34, No. 3. March 1961.

42. Experimental lightweight flat plate structure, Part III—Long term deformations, *Construct. Rev.*, Vol. 34, No. 4, April 1961.

43. Experimental lightweight flat plate structure, Part IV—Design and erection of structure with concrete columns, *Construct. Rev.* Vol. 35, No. 2, Feb. 1962.

44. Beresford, F. D., Experimental lightweight flat plate structure, Part V—Deformation under lateral load, *Construct. Rev.* Vol. 35, No. 12, Dec. 1962.

45. Lewis, R. K., Experimental lightweight flat plate structure, Part VI—Design and erection of post-tensioning flat plate, *Construct. Rev.*, Vol. 36, No. 3, March 1963.

46. Beresford, F. D. and Blakey, F. A., Experimental lightweight flat plate structure, Part VII—A test to destruction, *Construct. Rev.*, Vol. 36, No. 6, June 1963.

47. Gamble, W. L., Experimental lightweight flat plate structure, Part VIII—Test to failure of prestressed slab, *Construct. Rev.*, Vol. 37, No. 10, Oct. 1964, pp.

48. Blakey, F. A., Australian experiments with flat plates, *J. Am. Concr. Inst.* Vol. 60, No. 4, April 1963, pp. 515–525.

49. Brotchie, J. F. and Beresford, F. D., Experimental study of prestressed concrete flat plate structure, *Civ. Eng. Trans. Inst. Eng., Austral.*, Vol. CE9, No. 2, Oct. 1967.

50. Taylor, P. J. and Heiman, J. L., Long-term deflection of reinforced concrete flat slabs and plates, *J. Am. Concr. Inst. Proc.*, Vol. 74, No. 11, Nov. 1977, pp. 556–561.

51. Parrott, L. J., Long-term deformation of concrete in a prestressed concrete floor, *Proc. conf. performance bldg. struct.*, Glasgow, April 1976.

52. Parrott, L. J., A study of some long-term strains measured in two concrete structures, *Int. symp. testing in situ concr. struct.*, Budapest, Sept. 1977.

53. Swamy, R. N. and Arumugasaamy, P., Deformations in service of reinforced concrete columns, *Douglas McHenry international symposium on concrete and concrete structures.* American Concrete Institute, SP-55, pp. 375–407.

54. Swamy, R. N. and Potter, M. M. A., Long-term movements in an eight-story reinforced concrete structure, *Proc. conf. performance bldg. struct.*, Glasgow, April 1976.

55. Pfeifer, D. W., Magura, D. D., Russell, H. G. and Corley, W. G., Time-dependent deformations in 70-story structure, *Designing for effects of creep, shrinkage, and temperature in concrete structures*, American Concrete Institute, SP-27, 1971, pp. 159–185.

56. Russell, H. G. and Corley, W. G., Time-dependent behavior of columns in Water Tower Place, *Douglas McHenry international symposium on concrete and concrete structures*, American Concrete Institute, SP-55, pp. 347–373; also Portland Cement Association, *Research and Development Bulletin*, RD052-01B.

57. Reith, R. D., Kokn, R. and Watson, P., Full-scale measurements of a reinforced concrete chimmney, *Proc. conf. performance bldg struct.*, Glasgow, April 1976.

58. Russell, H. G., Field instrumentation of concrete structures, *Full-scale load testing of structures*, ASTM Spec. Tech. Publ., No. 702, 1980.

59. Russell, H. G., and Shiu, K. N., Creep and shrinkage behavior of tall building and long span bridges, *Symp. fundam. res. creep and shrinkage of concrete*, Lausanne, Sept. 1980.

60. *High-strength concrete in Chicago high-rise buildings*, Chicago Committee on High-Rise Buildings, Task Force Report No. 5, Chicago, Feb. 1977.

61. Fintel, M. and Khan, F. R., Effects of column creep and shrinkage in tall structures—Analysis for differential shortening of columns and field observation of structures, *Designing for effects of creep, shrinkage, and temperature in concrete structures*, American Concrete Institute, SP-27, 1971, pp. 95–120.

62. Tyler, R. G., Creep shrinkage and elastic strain in concrete bridges in the United Kingdom, 1963–71, *Mag. Concr. Res.*, Vol. 28, No. 95, June 1976, pp. 55–84

63. Finsterwalder, U., Die Ergebnisse von Kriech- und Schwindmessungen an Spanbetonbauten [Results of creep and shrinkage measurements on prestressed concrete structures] *Beton- und Stahlbetonbau*, Vol. 53, No. 5, May 1958.

64. Roelfstra, P. E., Computerized structural analyses applied to large span bridges, *Symp. on Fundam. Res. Creep and Shrinkage of Concrete*, Lausanne, Sept. 1980.

65. Haas, W., Numerical analysis of creep and shrinkage in concrete structures, *Symp. Fundam. Res. Creep and Shrinkage of Concrete*, Lausanne, Sept. 1980.

66. Pauw, A., Time-dependent deflections of a box girder bridge, *Designing for effects of creep, shrinkage and temperature in concrete structures*, American Concrete Institute, SP-27, pp 141–158.

67. Pauw., A., and Breen, J. E., Field testing of two prestressed concrete girders,

Highway Research Board Bulletin, No. 307; *Prestressed Concrete structures— creep, shrinkage, deflection studies*, National Academy of Sciences/National Research Council, Publication No. 937, 1961.

68. Gamble, W. L., *Field investigation of a continuous composite prestressed I-beam highway bridge located in Jefferson County, Illinois*, Civil Engineering Studies, Structural Research Series No. 360, University of Illinois at Urbana-Champaign, Urbana, 1970.

69. Houdeshell, D. M., Anderson, T. C. and Gamble, W. L., *Field investigation of a prestressed concrete highway bridge located in Douglas County, Illinois*, Civil Engineering Studies, Structural Research Series No. 375, University of Illinois at Urbana-Champaign, Urbana, 1972.

70. Gamble, W. L. *Long-term behavior of a prestressed I-girder highway bridge in Champaign County, Illinois*, Civil Engineering Studies, Structural Research Series No. 470, University of Illinois at Urbana-Champaign, Urbana, 1979.

71. Mossiossian, V. and Gamble, W. L., *Time-dependent behavior of non-composite and composite prestressed concrete structures under field and laboratory conditions*, Civil Engineering Studies, Structural Research Series No. 384, University of Illinois at Urbana-Champaign, Urbana, 1972.

72. Hernandez, H. D. and Gamble, W. L., *Time-dependent prestress losses in pretensioned concrete construction*, Civil Engineering Studies, Structural Research Series No. 417, University of Illinois at Urbana-Champaign, Urbana, 1972.

73. Gamble, W. L., *Final summary report: field investigation of prestressed reinforced concrete highway bridges*, Civil Engineering Studies, Structural Research Series No. 479, University of Illinois at Urbana-Champaign, Urbana, 1980.

74. Corley, W. G., Instrumentation of Denny Creek Bridge, *Concr. Int.* Vol. 2, No. 11, Nov. 1980, 25 pp.

75. Holman, R. J., *Development of an instrumentation program for studying behavior of a segmental concrete box girder bridge*, Joint Highway Research Project, JHRP-77-4, Purdue University, Indiana, March 1977.

76. Baltimore Gas & Electric Corporation, Calvert Cliffs Nuclear Power Plant Unit No. 1, *Containment structural post-tensioning system, 1 and 3 years surveillance reports*, Bechtel Power Corp., Gaithersburg, July 1975 and May 1977.

77. Turkey Point Nuclear Power Plant Unit No. 3, *Containment structural post-tensioning system, six months, 1, 2, and 5 years surveillance reports*, Bechtel Power Corp., Gaithersburg, 1972, 1973, 1974, 1977.

78. *Concrete properties of Calvert Cliffs nuclear containment vessel—progress report No. 1*, University of California, Berkeley, 1971.

79. Jávor, T., *Creep observations of prestressed concrete bridges*, Research Institute of Civil Engineering, Bratislava, 1980.

Part III Design and Analysis

13 Structural elements: strength, serviceability and ductility

Edward G Nawy

Contents

12 Structural elements: strength, serviceability and ductility

Edward G Nawy

Rutgers University, New Brunswick, USA

Contents

Notation

a	depth of equivalent rectangular stress block as defined in Section 2.7
a_{cs}	stabilized mean crack spacing
A_c	area of core of spirally reinforced compression member measured to outside diameter of spiral
A_g	gross area of concrete section
A_i	area of each segment
A_s	area of steel reinforcement; area of steel per unit width of slab or plate
A_s'	area of compression reinforcement
A_{st}	total area of longitudinal reinforcement (bars or steel shapes)
A_t	concrete stretched area, namely, concrete area in tension. In

two-way slabs, it is for a 12 in (300 mm) width of slab strips

b length of long span

C_m a factor relating actual moment diagram to an equivalent uniform moment diagram

C_1–C_6 moment coefficient for two-way slabs (Table 4-1)

c distance from extreme compression fiber to neutral axis

d distance from extreme compression fiber to centroid of tension reinforcement

d' distance from extreme tension or compression fiber to centroid of reinforcement

d_c thickness of concrete cover measured from extreme tension fiber to center of bar or wire located closest thereto

E_c modulus of concrete

E_s Young's modulus of steel

EI flexural stiffness of compression member

F prestressing force in tendon

f_c compressive stress in concrete

f_c' cylinder compressive strength of concrete

$\sqrt{f_c'}$ square root of specified compressive strength of concrete

f_c'' concrete strength in the structure

f_d stress in the prestressing steel corresponding to the decompression load

f_{nt} stress in the prestressing steel at any load level beyond the decompression load

f_r modulus of rupture of concrete

f_s calculated stress in reinforcement at service loads

f_t' tensile splitting strength of concrete

f_y specified yield strength of non-prestressed reinforcement

h overall thickness of member

I_{cr} moment of inertia of cracked section transformed to concrete

I_e effective moment of inertia for computation of deflection

I_g gross moment of inertia

I_1 grid index in direction 1 $(=\phi_1 s_2 p_{t_1})$ closest to outer concrete fibers

I_2 grid index in direction 2 $(=\phi_2 s_1/p_{t_2})$ in direction perpendicular to 1

i number of segments into which slab is divided by idealized yield lines

j number of yield lines

K fracture coefficient dependent on load or reaction type and on the support condition

k effective length factor for compression members

k_{ec} flexural stiffness of equivalent column

L clear long span length

L_p plasticity length

l effective beam span

l_d development length

l_e equivalent embedment length of a hook

l_u unsupported length of compression member or height of column

M unit positive yield moment

M' unit negative yield moment

M_a maximum service load moment in span

M_{cr} cracking moment

m ratio of short span to long span for two-way action slabs

m' $m-1$ to account for area of concrete displaced by bars

M_n nominal moment strength at section

M_1–M_6 unit ultimate resisting moment of slab in Eqns 4-7 and 4-8

N_m axial force due to net compressive membrane action

P_b nominal axial load strength at balanced strain conditions

P_c critical load

P_n nominal axial load strength at given eccentricity

P_0 nominal axial load strength at zero eccentricity

P_u factored axial load at given eccentricity $\leq \phi P_n$, where ϕ is a reduction factor

q $A_s f_y/f_c' d$ reinforcement index per 1 in (1 mm) unit width of slab

R, R_i ratio of distance from neutral axis of the beam to the concrete outside tension face, h_2 to the distance from the neutral axis to the steel reinforcement centroid, h_1; it ranges from 1.25 to 2.56, with an average value of 1.25

S' effective short span length, $S + 2t$

S'/t thickness aspect ratio

s	spacing of stirrups or ties		closest to outer concrete fibers
T_{sn}	nominal compressive resisting		for which crack control check is
	force		to be made
t	total thickness of slab (in)	β_1	factor varies from 0.85 for $f'_y =$
u_m	maximum bond stress as a func-		4000 psi to 0.65 minimum; it
	tion of $\sqrt{f'_c}$		decreases at a rate of 0.05 per
V_c	nominal shear strength provided		1000 psi strength above
	by concrete		4000 psi [3]
V_s	nominal shear strength provided	δ	moment magnification factor
	by shear reinforcement	$\lambda_1, \lambda_2, \lambda_3$	parameters defining the yield
w_{lim}	limit load		line geometry
w_{max}	maximum flexural crack width	Δ	deflection at limit load, namely,
	at steel level		limit load deflection
w'_{max}	maximum flexural crack width	Δf_s	$(f_{nt} - f_d)$, net stress in the pre-
	at tensile face of concrete		stressing steel, or the magnitude
W	total uniform load per unit area		of the tensile stress in the non-
W_E	external energy imposed on		prestressing steel at any load
	each segment		level
W_I	internal energy dissipated at	ε_s	unit strain in the reinforcement
	yield sections in a unit length of	ν	Poisson's ratio
	yield line	ρ	ratio of non-prestressed tension
x	shorter overall dimension of rec-		reinforcement: A_s/bd for beams
	tangular part of cross-section		and $A_s/12d$ for slab
x_1	shorter center-to-center dimen-	$\bar{\rho}_b$	reinforcement ratio producing
	sion of closed rectangular		balanced strain conditions
	stirrup	ρ_s	ratio of volume of spiral rein-
y	longer overall dimension of rec-		forcement to total volume of
	tangular part of cross-section		core (out-to-out of spirals) of a
y_t	distance from centroidal axis of		spirally reinforced compression
	cross section, neglecting rein-		member
	forcement, to extreme fiber in	ρ_t	A_s/A_t, active steel ratio
	tension	ω	$\rho f_y/f'_c$, reinforcement index
y_1	longer center-to-center dimen-	$\sum 0$	sum of reinforcing element cir-
	sion of closed rectangular		cumferences
	stirrup	$\sum x^2 y$	torsional section properties
1	direction of reinforcing elements		

Summary

Analysis, hence design, is covered from essentially basic principles in the areas of flexure, shear, torsion, deflection, cracking and ductility behavior and improvement. It is a compact for quick understanding of these parameters while at the same time it introduces the engineer, as well as the researcher, to the state of the art through the selected list of 116 references presented. An attempt has been made to treat the subject from more than one approach, since the CEB philosophy differs in detail from that of the ACI. However, the laws of statics and physics are a common denominator to all approaches. Readers should be able without difficulty to extend their horizons in the art and science of proportioning concrete structural systems once a thorough understanding of the material presented is achieved.

1 Introduction: design versus analysis

Concrete cannot be considered homogeneous, isotropic or a linearly elastic material. Its strength in tension is about one-tenth of its compressive strength. Hence, the basic and classical equations and methods of analysis and design in compression, bending, shear and torsion are not directly applicable.

The expressions used for the analysis and design of concrete structural elements are the result of (a) exact mathematical formulation and deduction of equations, or (b) empirical formulation of the parameters governing the analysis equations, or (c) both (a) and (b). The types of stress or force involved determine the research background of the equations, such as whether flexure or shear is the subject of the research.

A large number of parameters has to be dealt with in proportioning a reinforced or prestressed concrete structural element, such as geometrical width, depth, area of reinforcement, steel strain, concrete strain, steel stress, etc. As a result, trial adjustment is necessary in the choice of the concrete sections, with assumptions based on conditions at the site, availability of the constituent materials, architectural and headroom requirements, the applicable codes, and environmental conditions. Such an array of parameters has to be considered because reinforced concrete is generally a site-constructed composite in contrast to the standard mill-fabricated beam and column sections in steel structures.

The trial and adjustment procedures for the choice of a concrete section lead to the convergence of analysis and design. Hence, it is proposed in this chapter that every design is an analysis once a trial section is chosen. The availability of handbooks, charts, computers, and programmable calculator programs support this approach as a more efficient, compact, and speedy instructional method in comparison with the traditional approach of treating the analysis of reinforced and prestressed concrete sections separately from pure design.

Since this chapter deals with all facets of analysis (hence design) in such a limited space, only basic principles of analysis will be presented. The material should serve as an introduction for the understanding of the strength, serviceability, and ductility behavior of reinforced concrete structural elements. The references listed at the end of this and other chapters will enable the designer to locate the source for as much additional detail as necessary.

2 Flexural strength

2.1 The compressive block and stress–strain relationships

The form of the distribution of stress along the depth of a reinforced concrete cross-section changes with load. The tensile zone of the section is neglected in the analysis after cracking and all the tensile force is assumed to be transferred to the reinforcement (Fig. 2-1). Several forms of stress distribution have been proposed [1], from the linear triangular form by Koenen in 1886 to the cubic rising parabola by Chambaud in 1949, to the equivalent rectangular block by Whitney in 1942. To date, the simplified equivalent rectangular block in Fig. 2-2 is accepted as the standard in analysis and design of reinforced concrete [2–4].

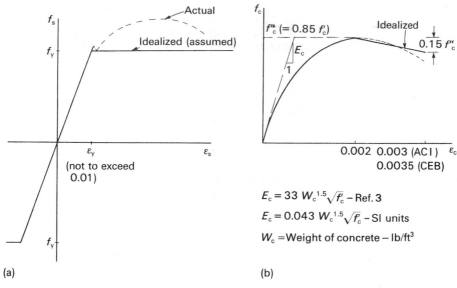

$E_c = 33 \, W_c^{1.5}\sqrt{f_c'}$ – Ref. 3

$E_c = 0.043 \, W_c^{1.5}\sqrt{f_c'}$ – SI units

W_c =Weight of concrete – lb/ft³

(a)   (b)

Figure 2-1   Idealized stress–strain diagrams: (a) steel; (b) concrete

The basic assumptions used are

(a) Plane sections before bending remain plane after bending.

(b) The tensile strength of concrete is neglected in the analysis.

(c) Identical strain variation (compatibility of strain) exists between the reinforcement and the surrounding concrete prior to yielding of the reinforcement.

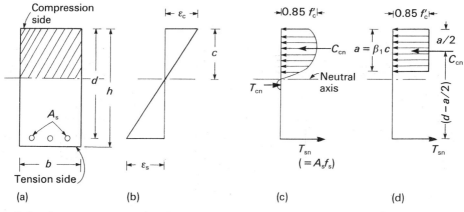

Subscript c = concrete; subscript s = steel

Figure 2-2   Stress and strain distribution across beam depth
(a) Beam cross-section
(b) Strain distribution
(c) Theoretical stress distribution
(d) Equivalent stress distribution

(d) Distribution of strain is linear across depth of the section.
(e) The maximum allowable compressive strain in the concrete in bending is 0.003 (ACI) or 0.0035 (CEB) and the maximum allowable compressive strain in axial compression is 0.002 for normal concrete.
(f) The depth, a, of the equivalent concrete block is β_1 times the depth from the extreme compression fibers. $\beta_1 = 0.85$ maximum for concrete strengths $f'_c = 4000$ psi (27.580 MPa) or less and decreases at the rate of 0.05 for every 1000 psi (6.895 MPa) up to a minimum of $\beta_1 = 0.65$.
(g) To ensure ductile behavior, all sections shall be under-reinforced, namely, that the area of tension steel should not exceed 75% of that needed for the balanced condition. (The balanced condition denotes simultaneous failure by yielding of the tension reinforcement and crushing of the concrete at the expression fibers.)

2.2 Singly reinforced sections

For the analysis (design) of beams reinforced at the tension side only, the following expressions for the internal forces and moment of resistance are derived from first principles of mechanics. The external design moment, M_u, should not exceed the internal moment, M_n. (ACI requires that $M_u = \phi M_n$, where ϕ is a reduction factor less than 1.0.)

From horizontal equilibrium (Fig. 2-2d),

$$C_{cn} = T_{sn} \qquad (2\text{-}1)$$

where C_{cn}, the nominal compressive resisting force (volume of the compressive block), equals $0.85f'_c ba$, and T_{sn}, the nominal tensile resisting force, equals $A_s f_y$ (on the assumption of yield of tension reinforcement). Hence,

$$A_s f_y = 0.85f'_c ba$$

giving

$$a = \frac{A_s f_y}{0.85f'_c b} \qquad (2\text{-}2)$$

ρ (reinforcement percentage) $= A_s/bd$, and ω (reinforcement index) $= \rho f_y/f'_c$, the nominal moment of resistance of the section from equilibrium of moments is

$$M_n = A_s f_y (d - \tfrac{1}{2}a) \qquad (2\text{-}3)$$

or

$$M_n = bd^2 f'_c \omega (1 - 0.59\omega) \qquad (2\text{-}4)$$

The balanced reinforcement percentage is (in Imperial units)

$$\rho_b = \frac{0.85\beta_1 f'_c}{f_y} \times \frac{87\,000}{87\,000 + f_y} \qquad (2\text{-}5)$$

such that ρ should not exceed $0.75\rho_b$.

2.3 Doubly reinforced sections

Doubly reinforced sections contain reinforcement both at the tension and the compression faces. They become necessary when architectural considerations require limitation on the depth of a section. Basically, these sections can only be treated as such when the mid-span tension reinforcement at the bottom fibers of a continuous beam is extended adequately through the adjacent spans so that full development length is reached. For a beam with compression reinforcement A'_s, the solution is to theoretically split the beam into two parts (Fig. 2-3): Case 1 for the singly reinforced part of the solution involving the equivalent rectangular block, as previously discussed in Section 2.2, with the area of tension reinforcement being $(A_s - A'_s)$; Case 2 for the two areas of equivalent steel, A'_s, at both the tension and compression faces to form the couple T_{n2} and C_{n2}, where $\rho = A_s/bd$ and $\rho' = A'_s/bd$.

(1) *Case* $(\rho - \rho')$

$$T_{n1} = C_{n1} = A_{s1}f_y$$

$$M_{n1} = A_{s1}f_y(d - \tfrac{1}{2}a)$$

(2) *Case* ρ'

$$T_{n2} = C_{n2} = M_{n2}/(d - d') \tag{2-6}$$

$$M_n - M_{n1} = A'_s f_y(d - d') \tag{2-7}$$

The total area

$$A_s = A_{s1} + A_{s2} \tag{2-8}$$

where $A_{s1} = (A_s - A'_s)$ and $A_{s2} = A'_s = T_{n2}/f_y$.
    The total moment of resistance $M_n = M_{n1} + M_{n2}$, hence

$$M_n = (A_s - A'_s)f_y(d - \tfrac{1}{2}a) + A'_s f_y(d - d') \tag{2-9}$$

This equation is valid *only* if A'_s reaches its yield strength f_y. Otherwise, treat as singly reinforced, or find actual stress in compression steel A'_s and use its value in Eqn 2-9.
    For A'_s to yield

$$(\rho - \rho') \geq 0.85\beta_1 \frac{f'_c d'}{f_y d} \times \frac{0.003 E_s}{0.003 E_s - f_y} \tag{2-10}$$

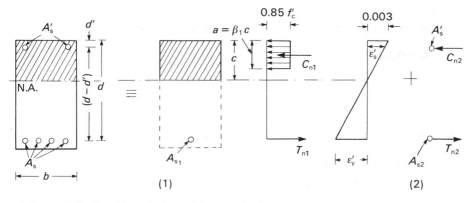

Figure 2-3 Doubly-reinforced beam design

Also f'_s, the stress in compression steel $(f'_s \le f_y)$ is given by

$$f'_s = \varepsilon'_s E_s = \varepsilon_c E_s \left(\frac{c-d'}{c}\right)$$

$$\rho_b = \bar{\rho}_b + \rho' f'_s / f_y \qquad \rho = \le 0.75 \bar{\rho}_b + \rho' f'_s / f_y \qquad (2\text{-}11)$$

In all of this discussion, adjustment for the concrete area replaced by the compression reinforcement is neglected, for practical design purposes, as insignificant.

It is to be noted that in cases where the compression reinforcement, A'_s, did not yield, the depth of the rectangular compressive block becomes

$$a = \frac{A_s f_y - A'_s f'_s}{0.85 f'_c b}$$

where $f'_s < f_y$ and

$$M_n = (A_s f_y - A'_s f'_s)(d - \tfrac{1}{2}a) + A'_s f'_s (d - d') \qquad (2\text{-}12)$$

Example 2-1
Design a doubly reinforced concrete beam with maximum effective depth $d = 25$ in (635 mm) and a resisting moment of 10.756×10^6 lbf in (1216 kN m). Given $f_y = 6 \times 10^4$ psi (413.7 MPa); $f'_c = 4000$ psi (27.6 MPa), minimum cover $d' = 2.5$ in (63.5 mm).

Solution   From trial and adjustment or experience, try section width $b = 14$ in (355.6 mm). To ensure that the compression steel A'_s has yielded, namely, $\varepsilon'_s \ge \varepsilon_y$, use Eqn 2-10:

$$(\rho - \rho') \ge 0.85 \beta_1 \frac{f'_c}{f_y} \frac{d'}{d} \frac{0.003 E_s}{(0.003 E_s - f_y)}$$

where $E_s = 29 \times 10^6$ psi

$$\ge 0.85 \times 0.85 \times \frac{4000}{60\,000} \times \frac{2.5}{25} \times \frac{87\,000}{(87\,000 - 60\,000)} \ge 0.0154$$

Now, $0.75 \bar{\rho}_b = 0.0216$, therefore $0.0154 < (\rho - \rho') < 0.0216$, where $\bar{\rho}_b$ stands for singly reinforced. Assume $(\rho - \rho') \approx 0.016$, then

$$A_{s1} = 0.016 \times 14 \times 25 = 5.60 \text{ in}^2 \ (3613.12 \text{ mm}^2)$$

Depth of rectangular compressive block

$$a = \frac{(A_s - A'_s) f_y}{0.85 f'_c b_w} = \frac{5.6 \times 6 \times 10^4}{0.85 \times 4000 \times 14} = 7.05 \text{ in } (179.07 \text{ mm})$$

$$M_{n1} = (A_s - A'_s) f_y (d - \tfrac{1}{2}a) = 5.60 \times 6 \times 10^4 (25 - \tfrac{1}{2} \times 7.05)$$
$$= 7.208 \times 10^6 \text{ lbf in } (814 \text{ kN m})$$

Alternatively, the reinforcement index is

$$\omega = \frac{5.60}{14 \times 25} \times \frac{6 \times 10^4}{4000} = 0.24$$

$$M_{n1} = b_w d^2 f'_c \omega (1 - 0.59\omega) = 14(25)^2 \times 4000(0.24)(1 - 0.59 \times 24)$$
$$= 7.208 \times 10^6 \text{ lbf in } (814 \text{ kN m})$$

$$M_{n2} = M_n - M_{n1} = (10.756 - 7.208) \times 10^6 = 3.548 \times 10^6 \text{ lbf in } (402 \text{ kN m})$$

But $M_{n2} = A'_s f_y (d - d')$. Hence,

$$A'_s = \frac{M_{n2}}{f_y(d - d')} = \frac{3.548 \times 10^6}{6 \times 10^4 (25 - 2.5)} = 2.64 \text{ in}^2 \ (1702.8 \text{ mm}^2)$$

Total A_s at tension side is given by

$$A_s = A_{s1} + A'_{s2} = 5.60 + 2.64$$
$$= 8.24 \text{ in}^2 \ (5316 \text{ mm}^2)$$

A check to validate that the beam is under-reinforced is additionally made as follows:

$$0.75\bar{\rho}_b = 0.0216 \qquad \rho' = \frac{2.64}{14 \times 25} = 0.0075$$

$$\text{Actual } \rho = \frac{8.24}{14 \times 25} = 0.0235$$

$$\text{Maximum } \rho \text{ for under-reinforced} = 0.75\bar{\rho}_b + \rho' f'_s / f_y$$
$$= 0.0216 + 0.0075 = 0.0291 > 0.0235$$

Hence, the beam is under-reinforced, namely, failure would occur by yielding of the tension steel. To ensure that a beam does not behave as non-reinforced, a check has to be made that the designed reinforcement ratio, ρ, always exceeds a minimum value of $200/f_y$ [3].

The external factored design movement, M_u, at the limit state of failure would be $M_n = \phi M_n$, where ϕ is a statistical reduction factor having a value less than 1.0 (ACI ϕ factor for failure is 0.90).

2.4 Flanged sections

In the non-rectangular section groups, T and L-beams are the most common cross-sections. Because slabs and beams are ordinarily monolithically cast as shown in Fig. 2-4, a beam automatically gains an extra width at the top, forming a flange in the section.

If the beams are cast monolithically with the floor, the effective flange width b (compression side) for T-beams is determined from the least of the following three conditions (Fig. 2-4a).

    (a) Effective overhang $r \leq 8h_f$
    (b) Effective overhang $r \leq$ half clear distance to next web
    (c) $b \leq \frac{1}{4}$ span of web

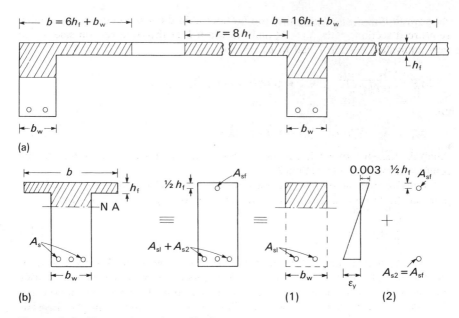

Figure 2-4 Flanged sections design
        (a) Flanged sections in monolithic floor system
        (b) T-beam transfer

For spandrel or end beams, the overhang length, r, should be the lesser of

$$r \leq 6h_f \quad \text{or} \quad r \leq \tfrac{1}{12} \text{ span of web}$$

or $r \geq \tfrac{1}{2}$ clear distance to next web

Once the flanged section's geometrical dimensions are determined, a test has to be made to determine whether the neutral axis is outside the flange or within it. If the neutral axis falls outside the flange, the section is as shown in Fig. 2-4. If the neutral axis falls inside the flange, the section is treated as a singly reinforced section having a width at the compression side equivalent to the flange width b and analyzed (designed) as discussed in Section 2.2.

A check for the position of the neutral axis can be made from first principles, where C_n has to equal T_n. Alternatively, if the flange thickness $h_f \leq 1.18\bar{\omega}d/\beta_1$, the section is a *real T*-beam where $\bar{\omega} = (A_s/bd)(f_y/f_c')$.

In the case of a flanged section such as a real T-beam, a theoretical reinforcement area, A_{sf}, can be deduced which would perform the same function as the flange overhang, such that

$$A_{sf}f_y = 0.85f_c'h_f(b - b_w) \tag{2-13a}$$

so that

$$A_{sf} = \frac{0.85f_c'h_f(b - b_w)}{f_y} \tag{2-13b}$$

As seen in Figs 2-4b, the section can be treated in the same manner as a doubly reinforced section, with $A_{sf} = A'_s$ and the width b at the compression side of the new equivalent section equal to b_w. Hence,

$$a = \frac{(A_s - A_{sf})f_y}{0.85f'_c b_w}$$

and

$$M_n = (A_s - A_{sf})f_y(d - \tfrac{1}{2}a) + A_{sf}f_y(d - \tfrac{1}{2}h_f) \tag{2-14}$$

Analysis (design) of such sections follows the same steps as in the doubly reinforced example of Section 2.3.1

To ensure that the flanged section is ductile by initial yielding of the reinforcement at the tension side, the maximum steel percentage $\rho \leq 0.75\rho_b$, where

$$\rho_b = \frac{b_w}{b}(\bar{\rho}_b + \rho_f) \tag{2-15}$$

and $\rho_f = A_{sf}/b_w d$. It should be noted that no compatibility check for the yield of the theoretical compression steel area is necessary as was the case in doubly reinforced beams. In flanged sections, A_{sf} is determined on the assumption that the concrete in the flange reaches the failure stress level $0.85f'_c$, and the theoretical reinforcement, A_{sf}, simultaneously yields. The external factored design moment $M_u = \phi M_n$, where ϕ has a similar value less than 1.0 as in the other types of beam.

2.5 Compression and uniaxial bending: columns

2.5.1 *Introduction*
Columns are compression members normally subjected to combined bending moment and axial load. Their strength is evaluated on the basis of the following principles:

(a) Linear strain distribution.
(b) No slippage between concrete and steel.
(c) Maximum concrete strain $\varepsilon_c = 0.003$ (0.0035 by CEB–FIP).
(d) Tensile resistance of the concrete is neglected.
(e) Determination whether failure is the result of material failure by initial yielding of the steel at the tension face, or initial crushing of the concrete at the compression face, or failure by loss of lateral structural stability, namely, buckling.

Initial material failure is applicable to short columns, namely, those with slenderness ratio not exceeding $kl_u/r = 22$ for non-braced columns. Initial buckling is applicable to columns whose slenderness ratio exceeds the limits given for short columns (see also Section 2.5.8).

2.5.2 *Limit state at material failure*
The same principles applied to beams are applicable to columns concerning the stress distribution and the equivalent rectangular block. Figure 2-5 differs from

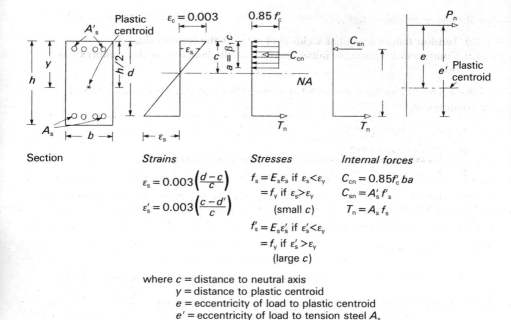

Figure 2-5   Stresses and forces in columns

Fig. 2-3 in the introduction of the additional longitudinal nominal force, P_n, at limit failure state acting at an eccentricity, e, from the geometric plastic centroid of the section. If c is the depth from the compressive face to the neutral axis, what determines the strength of the column would be the magnitude of the depth c. If the member is subjected to pure axial load, it would shorten uniformly without bending, namely, $c \approx \infty$

$$\varepsilon_s = \varepsilon'_s \text{ in compression } > \varepsilon_y$$

$$f_s = f'_s = f_y$$

and the nominal strength at zero eccentricity is

$$P_0 = 0.85 f'_c A_g + f_y (A_s + A'_s) \tag{2-16a}$$

or

$$P_0 = 0.85 f'_c A_g (1 + m\rho_g) \tag{2-16b}$$

where $m = f_y/(0.85 f'_c)$ and $\rho_g = (A_s + A'_s)/A_g$ with P_0 assumed to act at the plastic centroid. For practical design considerations, no adjustment is made for the area of concrete replaced by the reinforcing bars.

The distribution of stress and strain and the sets of equilibrium forces acting on the column section are as given in Fig. 2-5 with the appropriate values of the force acting on the section. Based on the magnitude of the strain in the steel reinforcement at the tension side, the section is subjected to one of two initial conditions of failure as follows:

(a) Tension failure by initial yielding of steel at the tension side.
(b) Compression failure by initial crushing of the concrete at the compression side.

The balanced condition occurs when failure develops simultaneously in tension or compression.

$P_n < P_b$ = tension failure

$P_n = P_b$ = balanced failure

$P_n > P_b$ = compression failure

In all these cases, the strain-compatibility relationship has to be maintained. The equilibrium expressions for forces and moments from Figs. 2-5 can be expressed as follows for non-slender columns.

Nominal failure force

$$P_n = C_{cn} + C_{sn} - T_n \tag{2-17}$$

Nominal resisting moment $M_n = P_n e$ about the plastic centroid

$$M_n = P_n e = C_{cn}(\bar{y} - \tfrac{1}{2}a) + C_{sn}(\bar{y} - d') + T_n(d - \bar{y}) \tag{2-18}$$

Equations 2-17 and 2-18 can be rewritten in terms of internal stresses:

$$P_n = (0.85 f_c' ba + A_s' f_y - A_s f_s) \tag{2-19}$$

$$M_n = P_n e' = [0.85 f_c' ba(d - \tfrac{1}{2}a) + A_s' f_y(d - d')] \tag{2-20}$$

Both Eqns 2-19 and 2-20 assume the compression steel to have yielded. Here, e' is measured from the center of the tension steel and the depth, a, of the equivalent rectangular block is less than the total depth, h, of the section. The stress f_s in Eqn 2-19 would become f_y in those columns where the initial limit state of failure is by yielding of the tension steel:

$$f_s = E_s \varepsilon_s = E_s \varepsilon_c (d - c)/c \tag{2-21}$$

2.5.3   Types of column
Columns are classified according to the form and arrangement of the reinforcement. Basically, three types of column are constructed (Fig. 2-6):

(a) Tied columns where the reinforcement consists of vertical bars transversely tied with separate closed ties.
(b) Spirally reinforced columns where closely spaced spirals enclose a circular concrete core containing vertical bars.
(c) Composite columns comprising structural steel shapes encased in concrete. The concrete is also reinforced with vertical bars with either separate ties or continuous spirals.

The tied columns are the most commonly used because of lower construction costs although the spirally reinforced columns are generally analytically more

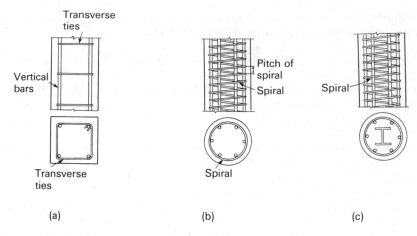

Figure 2-6   Types of column: (a) tied, (b) spiral, (c) composite

efficient due to the confinement of the concrete core. The percentage of the reinforcement in terms of the gross cross-sectional area can vary between 1 and 8%. A normal percentage which does not result in congestion of the reinforcement is approximately $2\frac{1}{2}$ to 4%.

2.5.4   Rectangular column sections

2.5.4.1   *Tension failure*   The initial limit state of failure occurs by the yielding of the steel at the tension face of the column. It results in cases of large eccentricity, e, when the load P_u is outside the kern section. Equations 2-19 and 2-20 are applicable in analysis (design) using yield strength f_y in Eqn 2-19 for the stress in the tension reinforcement. In this state, $\varepsilon_s = \varepsilon_y$ and $f_s = f_y$.

For symmetric reinforcement parallel to the axis of bending (i.e. $\rho = \rho'$), Eqns 2-19 and 2-20 can be combined as follows, where the area of concrete displaced by the compression steel area, A'_s, is taken into account:

$$P_n = 0.85f'_c bd\left\{-\rho+1-\frac{e'}{d}+\sqrt{\left(1-\frac{e'}{d}\right)^2+2\rho\left[m'\left(1-\frac{d'}{d}\right)+\frac{e'}{d}\right]}\right\} \qquad (2\text{-}22)$$

where $m = f_y/0.85f'_c$, $\rho = A_s/bd$ and $m' = m - 1$.

2.5.4.2   *Balanced condition*   This condition occurs when the tension steel yields at the same time that the concrete crushes at the face.

$$\varepsilon_s = \varepsilon_y \quad \text{and} \quad c_b = d\left(\frac{0.003}{0.003+\varepsilon_y}\right) = d\left(\frac{0.003E_s}{0.003E_s+f_y}\right)$$

From Eqns 2-19 and 2-20,

$$P_{bn} = (0.85f'_c ba_b + A'_s f_y - A_s f_y) \qquad (2\text{-}23)$$

$$M_{bn} = P_{bn}e_b = [0.85f'_c ba_b(d-d''-\tfrac{1}{2}a_b)+A'_s f_y(d-d'-d'')+A_s f_y d''] \qquad (2\text{-}24)$$

$$e_b = M_{bn}/P_{bn} \qquad (2\text{-}25)$$

where d' is the cover to center of reinforcement layer, and d'' the distance from plastic centroid to the centroid of the tension steel.

2.5.4.3 *Compression failure* For initial crushing of the concrete whereby the load eccentricity is small, the basic equilibrium Eqns 2-19 and 2-20 are applicable when the stress in the tensile reinforcement is below yield, namely, $f_s < f_y$. Their use in analysis (design) has to be made applying trial and adjustment procedures and ensuring strain-compatibility checks at all stages [6]. The procedure can be summarized as follows.

For a given section, geometry and eccentricity value e, assume a value of the depth, a, of the compressive block in the moment Eqn 2-20 to calculate P_n, hence assuming a value e for the neutral axis depth. Then, substitute the assumed value of c in Eqn 2-21 to obtain f_s. Apply the calculated value of P_n and f_s to the force Eqn 2-19 to calculate c. Repeat the process until the assumed and final values of c are sufficiently close. This trial and adjustment process rapidly iterates, particularly if a computer program or programmable calculator program is used.

In lieu of this method and for columns reinforced with symmetric reinforcement in single layers parallel to the axis of bending, an empirical expression based on Whitney's simplified interaction curve [5, 7] gives the following conservative solution:

$$P_n = \frac{A_s' f_y}{[e/(d-d')]+0.5} + \frac{bhf_c'}{(3he/d^2)+1.18} \qquad (2\text{-}26)$$

The nominal load, P_n, at the limit state of failure is assumed to decrease linearly from the axial load, P_0, of Eqn 2-16a for zero moment to the balanced P_{bn} as the moment increases from zero to M_{bn}. The maximum allowable P_n for material failure conditions, namely, for short columns (no buckling) is, according to [3],

$$P_u = \phi P_n = 0.80\phi P_0 \quad \text{for tied columns} \qquad (2\text{-}27a)$$

$$P_u = \phi P_n = 0.85\phi P_0 \quad \text{for spiral columns} \qquad (2\text{-}27b)$$

where ϕ is the strength reduction factor discussed later.

2.5.5 *Circular concrete columns*

2.5.5.1 *Tension failure* In lieu of the trial and adjustment procedure outlined for rectangular columns and applicable as detailed in [6] to circular sections, the Whitney simplified empirical but conservative approach is presented:

$$P_n = 0.85f_c'h^2\left[\sqrt{\left(\frac{0.85e}{h}-0.38\right)^2 + \frac{\rho_g m D_s}{2.5h}} - \left(\frac{0.85e}{h}-0.38\right)\right] \qquad (2\text{-}28)$$

where

h = diameter of section
D_s = diameter of the reinforcement cage center-to-center of the outer vertical bars
e = eccentricity to the plastic centroid of section
ρ_g = (gross steel area)/(gross concrete area)

2.5.5.2 Compression failure

$$P_n = \frac{3A_{st}+f_y}{\frac{3e}{D_s}+1.00} + \frac{A_y f_c'}{\frac{9.6he}{(0.8h+0.67D_s)^2}+1.18} \qquad (2.29)$$

2.5.6 General case
When columns are reinforced with reinforcement on all faces and those where the reinforcement in the parallel faces is non-symmetric, solutions have to be based on conditions of compatibility of strain where the basic equilibrium equations of forces and moments from such equations as 2-19 and 2-20 have to be developed.

2.5.7 Tied and spiral reinforcement
Tied reinforcement spacing is restricted to the least of the following three values:

(a) Least lateral column dimension
(b) 16 times the diameter of the longitudinal bars
(c) 48 times the diameter of the tie

Spiral reinforcement size and pitch are determined taking into consideration the triaxial confining effect of the spiral on the enclosed concrete core [8, 9]:

$$\rho_s = \frac{\text{volume of spiral in one loop}}{\text{volume of concrete core for pitch } s} = \frac{A_{sp}}{A_c}$$

$$= \frac{a_s \pi (D_c - d_b)}{\frac{1}{4}\pi D_c^2 s}$$

where a_s is the cross-section of spiral, and D_c the diameter of concrete core.
The spiral reinforcement can be determined from

$$\rho_s = 0.45\left(\frac{A_g}{A_c}-1\right)\frac{f_c'}{f_{sy}} \qquad (2\text{-}30)$$

where f_{sy} is the yield strength of the spiral reinforcement.

2.5.6 Column strength reduction factor ϕ
For members subject to flexure and relatively small axial loads, failure is initiated by the yielding of the tension reinforcement and takes place in a more ductile manner. Hence, for small axial loads, it is reasonable to permit an increase in the ϕ-factor from that required for pure compression members. When the axial load vanishes, the member is subjected to pure flexure, and the strength reduction factor, ϕ, becomes 0.90 [3, 4]. As the compression load increases, the ϕ-factor is reduced to 0.7 for tied columns and 0.75 for spiral columns.

The following expressions give variations in the value of ϕ for symmetrically reinforced compression members with columns whose effective depth is not less than 70% of the total depth and for steel reinforcement having a yield strength,

f_y, not exceeding 6×10^4 psi (431.7 MPa), so that for tied columns

$$\phi = 0.90 - \frac{0.20P_u}{0.1f'_cA_g} \geq 0.70 \qquad (2\text{-}31a)$$

and for spiral columns

$$\phi = 0.90 - \frac{0.15P_u}{0.1f'_cA_g} \geq 0.75 \qquad (2\text{-}31b)$$

where $P_u = \phi P_n$ is the external design column load.

These formulations are based on the load factors and reduction factors in use in the USA both for the SI and Imperial units with the concrete compressive strength, f'_c, used in them evaluated from cylinder tests. Although the strength calculations presented are made from adjusted basic principles of mechanics, the load and reduction factors in the CEB–FIP and other codes are not similar and require more rigorous evaluations. A comparison is made of the various codes in another chapter in this handbook. Also, [10] gives a detailed comparison of various analysis (design) methods.

2.5.7 Load–moment interaction diagrams (P–M diagrams)

The capacity of a concrete section to resist combined axial and bending loads is most conveniently expressed by the P–M interaction diagram, as shown in Fig. 2-7. Each point on the diagram represents one combination of nominal ultimate resisting load, P_n, and nominal ultimate resisting moment, M_n, corresponding to one certain neutral axis location. The interaction diagram is separated into the

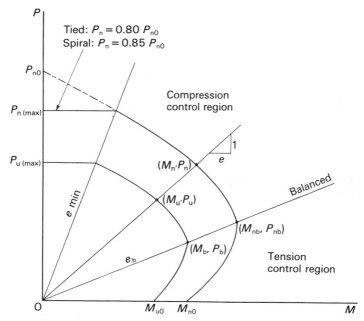

Figure 2-7   P–M typical column interaction diagram

tension control region and the compression control region by the balanced condition in points M_b, P_b. Interaction non-dimensional diagram sets of charts are available for various codes and units to facilitate speedy analysis and design.

2.5.8 *Limit state at buckling failure*

Considerable literature exists on the behavior of columns subjected to stability considerations [9–16]. If the column slenderness ratio exceeds the limits for short columns, the compression member will buckle prior to reaching material failure. The strain in the compression face of the concrete at buckling load will be less than $\varepsilon_c = 0.002$ (Fig. 2-1). Such a column would be a slender member subjected to combined axial and bending loads, deforming laterally and developing additional moment due to the $P\Delta$ effect, where P is the axial load and Δ the deflection of the column's buckled shape at its mid-height. Point B on Fig. 2-8 shows the buckled load caused by the magnification of moment due to $P\Delta$. Point A on the diagram gives the portion of critical load without the magnification effect.

The lower limits for neglecting the slenderness ratio, namely, disregarding stability analysis [3] are

Braced frames: $kl_u/r < [34 - 12(M_1/M_2)]$ (2-32a)

Unbraced frames: $kl_u/r < 22$ (2-32b)

where M_1 and M_2 are the moments at the opposite ends of the compression member.

If the value of kl_u/r is larger than that obtained from Eqn 2-32, two methods of

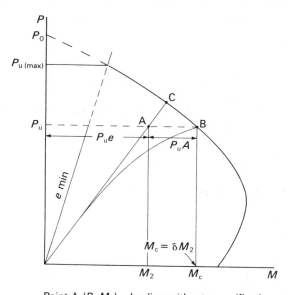

Point A (P_u, M_2)	loading without magnification
Point B (P_u, M_c)	actual loading
Point C	loading for short column

Figure 2-8 P–M load magnification diagram

stability analysis having general acceptance are recommended:

(a) The 'moment magnification method', where the member design is based on a magnified moment $M_c = \delta M_2$, where M_2 is the nominal moment based on simple analysis and $\delta \geq 1.0$.

(b) A 'second order' analysis, taking into consideration the effect of deflections. It must be used in cases where $kl_u/r > 100$.

It is to be noted that all columns have to be designed for a minimum value of eccentricity such as $e = 0.6 + 0.03h$, or more conservatively, $e - 0.1h$ for tied columns.

2.5.9 Moment magnification method

The moment magnifier, δ, is dependent upon the slenderness of the member, the restraint or applied moment at its ends, and the design cross-section such that [11, 12]

$$\delta = \frac{C_m}{1 - (P_u/\phi P_c)} \geq 1 \tag{2-33}$$

where

$$P_c = \text{Euler buckling load} = \pi^2 EI/(kl_u)^2 \tag{2-34}$$

and kl_u is the effective length (between points of inflection), l_u the length of compression member or height of column, and C_m a factor relating the actual moment diagram to an equivalent uniform moment diagram. For braced members subject to end loads only,

$$C_m = 0.6 + 0.4 \frac{M_1}{M_2} \geq 0.4$$

where $M_1 \leq M_2$ and $M_1/M_2 > 0$ if there is no point of inflection between ends. For other conditions, $C_m = 1.0$.

An estimate of EI must include the effects of cracking and creep under long term loading. For a heavily reinforced member,

$$EI = \frac{\frac{1}{5}E_c I_g + E_s I_s}{(1 + \beta_d)} \tag{2-35a}$$

For lightly reinforced members,

$$EI = \frac{E_c I_g/2.5}{(1 + \beta_d)} \tag{2-35b}$$

where

$$\beta_d = \frac{\text{design dead-load moment}}{\text{design total moment}}$$

The effective length, kl_u, in Eqn 2-34 is used to modify the basic moment magnification formula to account for end restraint other than pinned. kl_u represents the length of an auxiliary pin-ended column which has an Euler buckling load

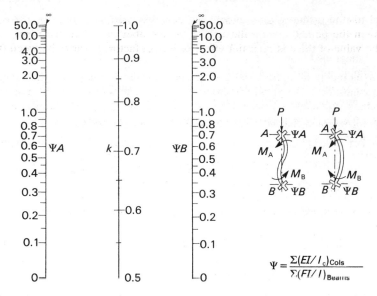

Figure 2-9   Effective length factor, *k*, for braced frames

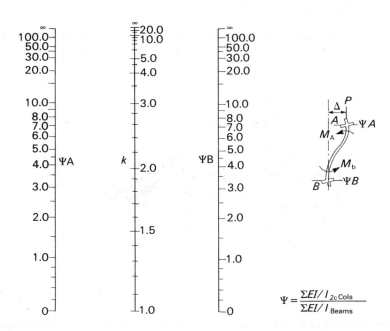

Figure 2-10   Effective length factor, *k*, for unbraced frames

equal to that of the column under consideration. Alternatively, it is the distance between the points of contraflexure of the member in its buckled form.

The value of the end-restraint effective length factor, k, varies between 0.5 and 2.0.

For members in a structural frame, end restraint lies between the hinged and fixed conditions. The value of actual k can be determined from the Jackson and Moreland alignment charts in Figs 2-9 and 2-10 [4, 17, 18]. Alternatively, the value of k can be calculated from the following expressions.

(a) For braced columns in frames, the smaller of the following two values:

$$k = 0.7 + 0.005(\psi_A + \psi_B) \leq 1.0 \tag{2-36a}$$

$$k = 0.85\psi_{min} \leq 1.0 \tag{2-36b}$$

(b) For unbraced columns in frames restrained at both ends,

$$k = \frac{20 - \psi_m}{20}\sqrt{1 + \psi_m} \quad \text{for } \psi < 2 \tag{2-36c}$$

$$k = 0.9\sqrt{1 + \psi_m} \quad \text{for } \psi \geq 2 \tag{2-36d}$$

(c) For unbraced compression members hinged at one end,

$$k = 2.0 + 0.3\psi \tag{2-36e}$$

where ψ_A and ψ_B are degrees of restraint at ends A and B.

$$\psi_A \text{ or } \psi_B = \frac{\Sigma(EI/l_u) \text{ for all columns}}{\Sigma(EI/l) \text{ for all beams}} \tag{2-36f}$$

A chart to determine, with minimum complications, when the slenderness effect needs to be considered is given in Fig. 2-11 [17] for columns braced against side-sway.

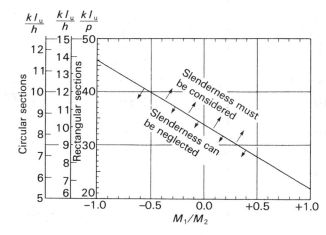

Figure 2-11   Slenderness ratio kl_u/r, below which the slenderness effect can be neglected for braced columns

2.5.10 *Second-order analysis*
This rigorous mathematical approach is needed if the slenderness ratio kl_u/r exceeds 100. The effect of deflection has to be taken into account and an appropriate reduced tangent modulus for concrete has to be used [19]. The designer, with the aid of computers, can solve with limited effort the set of simultaneous equations needed to determine the size of the reinforced concrete slender column, as detailed in [14, 19]. Charts can also be developed for the various eccentricity ratio and reinforcement and concrete strength combinations. It should be noted that the majority of columns do not necessitate such an analysis since the slenderness ratio kl_u/r is, in most of the usual cases, below 100.

Example 2-2
Design a rectangular tied column subjected to a uniaxial bending in a frame not braced against lateral sway. Given: $l_u = 18$ ft 0 in (5.55 m); factored external design load $P_u = P_n/\phi = 7.26 \times 10^5$ lbf (3229 kN); end moments $M_1 = 5.50 \times 10^5$ lbf in (203.88 kN m), $M_2 = 14.20 \times 10^5$ lbf in (160.43 kN m); $\beta_d = 0.5$, $\psi_A = 2.0$, $\psi_B = 3.0$; $f'_c = 5000$ psi (34.48 MPa); $f'_c = 6 \times 10^4$ psi (413.7 MPa).

Solution By trial and adjustment, a column section is assumed and analyzed. Try a section 21×21 in (533.4×533.4 mm). Actual eccentricity is given by

$$\frac{M_2}{P_u} = \frac{14.20 \times 10^5}{7.26 \times 10^5} = 1.96 \text{ in (49.8 mm)}$$

Assumed 10% eccentricity $= 0.1 \times 21$ in $= 2.10$ in (5.33 mm) Use $M_2 = 2.1 \times 7.26 \times 10^5 = 1.5246 \times 10^6$ lbf in (172.26 kN m). From chart in Fig. 2-10, $k = 1.7$, thus

$$\text{slenderness ratio, } kl_u/r = \frac{1.7 \times 18 \times 12}{0.3 \times 21} = 58.29$$

which is >22 and <100. Hence, use moment magnification method.

$$P_c = \pi^2 EI/(kl_u)^2$$

$$E_c = 3.3 w^{1.5}\sqrt{f'_c} = 3.3 \times 150^{1.5}\sqrt{5000}$$

$$= 4.29 \times 10^6 \text{ psi (29.58 GPa)}$$

$$I_g = \frac{21(21)^3}{12} = 16\,206.8 \text{ in}^4$$

$$EI = \frac{(E_c I_g/2.5)}{(1+\beta_d)} = \frac{4.29 \times 10^6 \times 16\,206.8}{2.5} \times \frac{1}{1+0.5}$$

$$= 18.54 \times 10^9 \text{ lbf in}^4$$

$$(kl_u)^2 = (1.7 \times 18 \times 12)^2 = 134.8 \times 10^3 \text{ in}^2$$

Hence,

$$P_c = \frac{\pi^2 \times 18.54 \times 10^9}{134.8 \times 10^3} = 1.356 \times 10^6 \text{ lb (6032 kN)}$$

The moment magnifier is

$$\delta = \frac{C_m}{1 - \dfrac{P_u}{\phi P_c}} = \frac{C_m}{1 - \dfrac{7.26 \times 10^5}{0.7 \times 1.356 \times 10^6}} = 4.255$$

because $C_m = 1.0$ for non-braced.

Design moment $M_c = \delta M_2 = 4.255 \times 14.20 \times 10^5 = 6.042 \times 10^6$ lbf in (682.66 kN m)

$$P_n = \frac{P_u}{\phi} = \frac{7.26 \times 10^5}{0.7} = 1.037 \times 10^6 \text{ lbf (4612.8 kN)}$$

$$M_n = \frac{6.042 \times 10^6}{0.7} = 8.632 \times 10^6 \text{ lbf in (1957.22 kN m)}$$

Hence, design a non-slender column section subjected to a nominal ultimate resisting force $P_n = 1.037 \times 10^6$ lbf and a nominal ultimate moment of resistance $M_n = 8.632 \times 10^6$ lbf in.

$$e_n = \frac{8.632 \times 10^6}{1.037 \times 10^6} = 8.32 \text{ in (211.3 mm)}$$

Analyze the assumed 21×21 in square section. Assume $\rho = \rho' \approx 1.25\%$.

$$A_s = A_s' = 0.0015 \ (21 \times 18.5) = 4.86 \text{ in}^2$$

Five No. 9 bars (5–28 mm diameter) on each face giving $A_s = A_s' = 5.00$ in² (3226 mm²)

$$c_b = d \times \frac{0.003 E_s}{0.003 E_s + f_y} = 18.5 \times \frac{87\,000}{87\,000 + 60\,000}$$

$$= 10.94 \text{ in (277.9 mm)}$$

Therefore $a_b = \beta_1 c_b = (0.85 - 0.05) \times 10.94 = 8.75$ in (222.3 mm)
From Eqn 2-23,

$$P_{bn} = 0.85 f_c' ba = 0.85 \times 5000 \times 21 \times 8.75$$

$$= 7.809 \times 10^5 \text{ lbf (3473.8 kN)}$$

From Eqn 2-24,

$$M_{bn} = 0.85 f_c' b a_b (d - d'' - \tfrac{1}{2} a_b) + A_s' f_y (d - d' - d'') + A_s f_y d''$$

$$d'' = \tfrac{1}{2}(21 - 2.5 - 2.5) = 8 \text{ in (203 mm)}$$

$$d - d' - d'' = 18.5 - 2.5 - 8.0 = 8 \text{ in (203 mm)}$$

Or

$$M_{bn} = 7.809 \times 10^5 \ (18.5 - 8.0 - \tfrac{1}{2} \times 8.75) + 2(5.0 \times 60 \times 8)$$

$$= (4.783 \times 4.800) \times 10^6 = 9.583 \times 10^6 \text{ lbf in (1082.72 kN m)}$$

$$e_b = \frac{9583}{780.94} = 12.27 \text{ in (311.66 mm)} > \text{actual } e_n = 8.32 \text{ in}$$

Hence, this column is controlled by compression failure and it is assumed that $\phi = 0.7$ is correct. Use Eqn 2-26 for solution:

$$P_n = \frac{A'_s f_y}{\dfrac{e}{(d-d')}+0.5} + \frac{bhf'_c}{\dfrac{3he}{d}+1.18}$$

$$= \frac{5.0 \times 60\,000}{\dfrac{8.32}{(18.5-2.5)}+0.5} + \frac{21 \times 21 \times 5000}{\dfrac{3 \times 21 \times 8.32}{(18.5)^2}+1.18} = 1.107 \times 10^6 \text{ lbf (4924 kN)}$$

which is greater than required $P_n = 1.037 \times 10^6$ lbf. So, for this design, it is possible to reduce the area of reinforcement slightly to lower the column resistance to $P \sim 1.040 \times 10^6$ lbf.

To design the ties:

Try No. 3 ties (9.52 mm diameter). Spacing least of $h = 21$ in (533.4 mm)

16 diameter No. 9 bar = 18 in (457.2 mm)

48 diameter No. 3 tie = 18 in (457.2 mm)

Hence, use $\frac{3}{8}$ in diameter (952 mm) closed ties spaced at 18 in (455 mm) center to center.

It is to be noted that ample tables and charts [15, 17, 19] exist for the analysis (design) of short and slender reinforced concrete columns. The discussion given here illustrates the basis for which such charts have been constructed, and the illustrative example demonstrates the basic steps of the calculation from the theory.

2.6 Columns in biaxial bending

The Bresler load contour method [20] of cutting the three-dimensional failure surface S in Fig. 2-12 at a constant value P_n to give an interaction plane relating M_{nx} and M_{ny}. In other words, the contour surface S can be viewed as a curvilinear surface which includes a family of curves, termed as the load contours.

The general non-dimensional equation for the load contour at a constant P_n may be expressed as follows:

$$\left(\frac{M_{nx}}{M_{0x}}\right)^{\alpha_1} + \left(\frac{M_{ny}}{M_{0y}}\right)^{\alpha_2} = 1.0 \tag{2-37}$$

where $M_{nx} = P_n e_y$ and $M_{ny} = P_n e_x$, and

$M_{0x} = M_{nx}$ at that axial load P_n for which M_{ny} or $e_x = 0$

$M_{0y} = M_{ny}$ at that axial load P_n for which M_{nx} or $e_y = 0$

α_1, α_2 = exponents depending on the cross section geometry, steel percentage and its location, and material stresses f'_c and f_y.

Equation 2-37 can be simplified using a common α exponent and introducing a

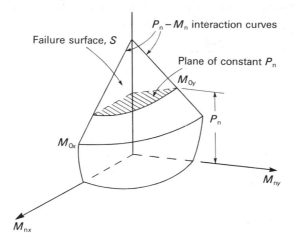

Figure 2-12 Failure interaction surface for biaxial column bending

factor β such that the M_{nx}/M_{ny} ratio would have the same value as the M_{0x}/M_{0y} ratio, as detailed by Parme et al. [22]. Such simplification leads to

$$\left(\frac{\beta M_{nx}}{M_{0x}}\right)^{\alpha} + \left(\frac{\beta M_{ny}}{M_{0y}}\right)^{\alpha} = 1.0 \tag{2-38}$$

where α would have a value of $(\log 0.5)/\log \beta$. Figure 2-13 gives a contour plot ABC from Eqn 2-38. For design purposes, the contour is approximated by two

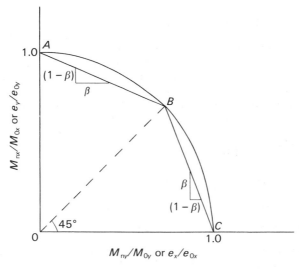

Figure 2-13 Modified interaction contour plot of constant P_n for biaxially loaded column

straight lines BA and BC, and Eqn 2-38 can be simplified to two conditions [23]:

(a) For AB, when $M_{ny}/M_{0y} < M_{nx}/M_{0x}$:

$$\frac{M_{nx}}{M_{0x}} + \frac{M_{ny}}{M_{0y}}\left(\frac{1-\beta}{\beta}\right) = 1.0 \qquad (2\text{-}39a)$$

(b) For BC, when $M_{ny}/M_{0y} > M_{nx}/M_{ny}$:

$$\frac{M_{ny}}{M_{0y}} + \frac{M_{nx}}{M_{0x}}\left(\frac{1-\beta}{\beta}\right) = 1.0 \qquad (2\text{-}39b)$$

For rectangular sections where the reinforcement is evenly distributed along all the column faces, the relationship

$$M_{0y}/M_{0x} \simeq b/h$$

gives the following revised expressions:

(a) For $M_{ny}/M_{nx} \geq b/h$:

$$M_{ny} + M_{nx}\left(\frac{b}{h}\right)\left(\frac{1-\beta}{\beta}\right) \simeq M_{0y} \qquad (2\text{-}40a)$$

(b) For $M_{ny}/M_{0y} \leq b/h$:

$$M_{nx} + M_{ny}\left(\frac{h}{b}\right)\left(\frac{1-\beta}{\beta}\right) \simeq M_{0x} \qquad (2\text{-}40b)$$

A plot of β for the analysis and design of such columns is given in Fig. 2-14.

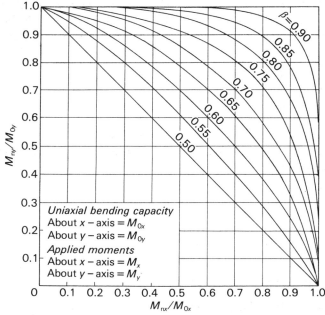

Figure 2-14 β-factor chart for rectangular columns in biaxial bending

In effect, the modified load-contour method can be summarized in Eqn 2-40 as a method for finding an equivalent moment M_{0x} and M_{0y} for designing columns as if they were uniaxially loaded.

Example 2-3: biaxial bending in columns
Select a rectangular cross-section for a biaxially loaded non-slender column which can have an ultimate axial resisting load of $P_n = 2.97 \times 10^5$ lbf (1321 kN) and an ultimate moment of resistance $M_{nx} = 2.46 \times 10^5$ lbf ft (333.5 kN m) and $M_{ny} = 9.85 \times 10^4$ lbf ft (133.6 kN m). It is given that $f'_c = 4000$ psi (27.5 MPa) and $f_y = 60\,000$ psi (413.7 MPa)

Solution Find an equivalent uniaxial bending moment about the x-axis assuming $\beta = 0.65$.

$$\frac{M_{nx}}{M_{ny}} = \frac{2.46 \times 10^5}{9.85 \times 10^4} = 2.5$$

Assume $h/b \simeq 2.5$, then

$$\text{equivalent } M_{0x} \simeq M_{nx} + M_{ny}\left(\frac{h}{b}\right)\left(\frac{1-\beta}{\beta}\right)$$

$$\simeq 2.46 \times 10^5 + 9.85 \times 10^4 \times 2.5\left(\frac{1-0.65}{0.65}\right)$$

$$= 3.78 \times 10^5 \text{ lbf ft } (512.5 \text{ kN m})$$

Assume a cross-section 12×20 in (304.8×508 mm) with a steel percentage of 2% on each of the two faces parallel to the x-axis. Hence, $\rho = \rho' = 0.02$, and $A_s = 0.02 \times 12(20 - 2.5) = 4.20$ m². Try four No. 9 bars on each face $= 4.0$ in² (2580.8 mm²). Also assume the cover d' to the center of the vertical bars $2\frac{1}{2}$ in (63.5 mm).
    The equivalent eccentricity is

$$e_y = \frac{3.78 \times 10^5 \times 12}{2.97 \times 10^5} = 15.3 \text{ in. } (387.9 \text{ mm})$$

This is a large eccentricity and an assumption can be made at this stage that tension failure controls, to be subsequently verified. From Eqn 2-22, substitute all appropriate values in

$$P_n = 0.85 f'_c b d \left\{ -\rho + 1 - \frac{e'}{d} \right.$$

$$\left. + \sqrt{\left(1 - \frac{e'}{d}\right)^2 + 2\rho(m-1)\left(1 - \frac{d'}{d}\right) + \frac{e'}{d}} \right\}$$

to get $P_n = 3.948 \times 10^5$ lbf $\gg$ actual 2.97×10^5 lbf. Hence, reduce the section to 12×18 in (304.8×457.2 mm). Assume that $\beta = 0.57$ and use $\rho = \rho' = 0.02$. Then,

$$M_{0x} = 2.46 \times 10^5 + 9.85 \times 10^4 \left(\frac{18}{12}\right)\left(\frac{1-0.57}{0.57}\right) = 3.573 \times 10^5 \text{ lbf ft } (484.7 \text{ kN m}).$$

The equivalent eccentricity is

$$e_y = \frac{357.5 \times 12}{297} = 14.44 \text{ in}$$

$$e' = d - \tfrac{1}{2}h + e = 15.5 - \tfrac{1}{2} \times 18 + 14.44 = 20.94 \text{ in}$$

$$\frac{e'}{d} = \frac{20.94}{15.5} = 1.35 \qquad \frac{d'}{d} = \frac{2.5}{15.5} = 0.16 \qquad (m-1) = 16.65$$

$$P_n = 0.85 \times 4000\ bd\{0.02 + 1 - 1.35 + \sqrt{(-0.35)^2 + 0.04 \times 16.65(1 - 0.16) + 1.35}\}$$

$$= 3400bd(-0.37 + 0.858) = 3400 \times 12 \times 15.5 \times 0.488$$

$$= 3.086 \times 10^5 \text{ lbf} \approx \text{actual } P_n = 2.97 \times 10^5 \text{ lbf}$$

This is acceptable.

A check has to be made whether the proper equation was used, namely, for tension failure where equivalent $e > e_b$. From Eqn 2-23, $P_{bn} = 0.85 f'_c ba_b$ and

$$c_b = \left(\frac{0.003E_s}{0.003E_s + f_y}\right)d = 0.592d = 9.176 \text{ in } (233.1 \text{ mm})$$

$$a_b = \beta_1 c_b = 0.85 \times 9.176 = 7.80 \text{ in } (198.1 \text{ mm})$$

$$P_{bn} = 0.85 \times 4000 \times 12 \times 7.80 = 3.182 \times 10^5 \text{ lbf } (1415.5 \text{ kN})$$

M_{bn} about plastic centroid of section from Eqn 2-24 for

$$A_s = A'_s = 0.02(12 \times 15.5) = 3.72 \text{ in}^2 (2400 \text{ mm}^2)$$

gives

$$M_{bn} = 4.525 \times 10^6 \text{ lbf in } (511 \text{ kN m}).$$

Thus, equivalent e_b is given by

$$e_b = \frac{M_{bn}}{P_{bn}} = \frac{4524.5}{318.22} = 14.2 \text{ in } (360.7 \text{ mm})$$

which is less than actual $e_y = 14.44$ in. Hence, initial failure is in tension by the yielding of steel at the tension side.

Adopt a section 12×18 in with four No. 9 bars (four 28.6 mm diameter bars) on each of the two faces parallel to the x-axis and add two No. 9 bars on each of the two other faces such that 12 No. 9 bars are used and design the transverse ties as previously described.

3   Shear, diagonal tension and torsional strength

3.1   Shear stress and diagonal tension

The classical Mohr theory in mechanics for principal stresses (Fig. 3-1) defines the shear stress for a homogeneous material. Modification becomes necessary to account for the fact that reinforced concrete is a non-homogeneous composite material whose strength in tension is about one-tenth of its strength in compression. The stress trajectories, namely, planes of principal stress, would cause cracks

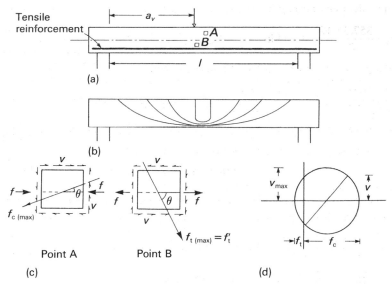

(a)

(b)

Point A   Point B

(c)   (d)

Figure 3-1   Principal stresses $f_{c(max)}$ at $f_{t(max)}$ in a beam
(a) Simply supported reinforced concrete beam
(b) Idealized cracked planes perpendicular to the stress trajectories
(c) Stresses at points A and B
(d) Mohr circle

to open in directions perpendicular to their planes (Fig. 3-1b). The basic Eqn 3-1 for principal stress,

$$f_{t(max)} = f'_t + \tfrac{1}{2}f_t + \sqrt{(\tfrac{1}{2}f_t)^2 + v^2} \tag{3-1}$$

has undergone considerable study for application to reinforced concrete. References [24, 25] give a comprehensive discussion of the adaptation of Eqn 3-1 to today's state-of-the-art format, where shear can more appropriately be called diagonal tension. The geometrical proportions of the beam, the ratio of the 'shear' span to depth, and the reinforcement percentage are the basic parameters controlling its diagonal tension behavior. Figure 3-2a gives a typical beam action failure for ratios $a_v/d = 2.5$ or higher [26, pp. 126–139]. Dowel action by horizontal splitting of the concrete around the tension bars would account for sustaining 15–25% of the shearing force while the compression zone would carry about 20–40% and the aggregate interlock 33–50%. Figure 3-2a gives an arch action type of sudden failure as in a deep beam. As a_v/d approaches a value of 4.5 or more, a flexure failure type would be expected. The shear reinforcement, whether it comprises inclined or vertical stirrups, would arrest the inclined cracks shown in Fig. 3-2a such that at least one stirrup would act on each possible crack.

Modification of the basic Eqn 3-1 by methods based on exact analytical physical models has been extremely difficult. Hence, a semi-empirical approach based on the physical interpretation of the various parameters involved has emerged over

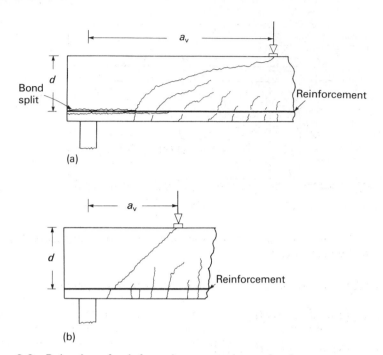

(a)

(b)

Figure 3-2 Behavior of reinforced concrete beam in shear
(a) Beam action diagonal tension failure for large a_v/d
(b) Arch action diagonal tension failure for small a_v/d

the past 40 years, involving such parameters as the steel percentage, the shear span to depth ratio, the inclined compressive chords of the truss analogy, the dowel effect of the main reinforcement, the aggregate interlock and the strength of the concrete, etc. Justifiably, different codes assign somewhat different weights to these various parameters, hence the difference in the formats of the equations. The approach concisely presented in this section follows the ACI design practice [3, 4] while [2, 23, 26, 27] give ample discussion of the variations in the other practices.

The model Eqn 3-2 gives the nominal shear capacity at failure of normal-weight concrete in the web of a reinforced concrete beam subjected to flexure and shear only:

$$V_c = [1.9\sqrt{f_c'} + 2500\rho_w V_u d/M_u]b_w d \qquad (3\text{-}2)$$

but not greater than $3.5\sqrt{f_c'}\,b_w d$. $V_u d/M_u$, which is a measure of the shear span to depth ratio, cannot exceed 1.0. Figure 3-3 is a transformed best-fit regression plot of Eqn 3-2.

A simpler, more conservative expression for V_c for normal weight concrete in beams subjected to flexure and shear only is

$$V_c = 2\sqrt{f_c'}\,b_w d \qquad (3\text{-}3)$$

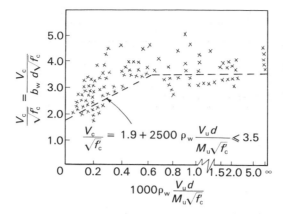

Figure 3-3 Plot of Eqn 3-2 for the shear resisting capacity, V_c, of a reinforced concrete web

When axial compression also exists, adjust V_c in Eqn 3-3 such that

$$V_c = 2\left(1 + \frac{N_u}{2000A_g}\right)\sqrt{f'_c}\, b_w d \qquad (3\text{-}4a)$$

When significant axial tension exists,

$$V_c = 2\left(1 + \frac{N_u}{500A_g}\right)\sqrt{f'_c}\, b_w d \qquad (3\text{-}4b)$$

N_u/A_g is expressed in psi.

If lightweight concrete is used, the values of $\sqrt{f'_c}$ in Eqns 3-2 through 3-5 shall be multiplied by a reduction factor $\lambda = 0.75$ for all lightweight concrete and $\lambda = 0.85$ for sanded lightweight concrete.

If $V_n = V_u/\phi$ is the nominal resisting capacity of the section and its web steel reinforcement, and V_s the nominal resisting force of the web stirrups, where $V_s = A_v f_y d/s$, then

$$V_n = V_c + V_s \qquad (3\text{-}5)$$

For vertical stirrups spacing s:

(a) s at a maximum $\frac{1}{2}d$ if $V_s \leq 4\sqrt{f'_c}\, b_w d$.
(b) s at a maximum $\frac{1}{4}d$ if $V_s > 4\sqrt{f'_c}\, b_w d$ but $< 8\sqrt{f'_c}\, b_w d$.
(c) If $V_s > 8\sqrt{f'_c}\, b_w d$, the section designed for flexure has to be enlarged.

3.2 Combined bending, shear and torsion

As in the case of design for shear (diagonal tension), the exact mathematical formulation of equations based on physical models is not possible because of the numerous parameters involved in concrete as a composite non-homogeneous material.

For over 60 years, the torsional analysis of concrete members has been based on either (a) the classical theory of elasticity developed through mathematical study coupled with either the soap film analogy method or the membrane analogy method [28]), or (b) the plastic theory represented by the sand heap analogy method [29]. In both, it was determined that

(1) Torsional moment = 2 × volume under blown membrane or sand heap
(2) Shear stress due to torsion = slope of the sand heap or the slope of the membrane

Also in both, the evaluation is essentially with respect to pure torsion. However, in the first approach, purely elastic behavior is entailed, while in the second, the behavior is purely plastic. The behavior of reinforced concrete, however, is better represented by the second approach, namely, the non-plastic hypothesis of behavior.

Since torsion rarely occurs in concrete structures without being accompanied by bending and shear, shear adjustments have to be made to account for combined torsion and shear stresses, represented by the transverse and longitudinal reinforcement.

The design for torsion must be based on the state of behavior of the beam close to failure. The stress resultants due to torsion in statically determinate beams can be determined from equilibrium conditions alone. In statically indeterminate systems, stiffness assumptions, redistribution and compatibility affect the stress resultants (compatibility torsion). Neglect of the full effect of the latter does not necessarily lead to failure but to possible excessive cracking.

Basically two approaches are being used: (a) the skew bending theory, and (b) the space truss analogy theory. While the former follows the plane deformation approach to plane sections subjected to bending, the latter follows the truss analogy approach used in evaluating the compression, tension, and shear forces in the stirrup reinforced web.

3.2.1 Skew bending theory

The failure plane or surface of the normal beam cross-section subjected to bending moments M_u only remains normal to the beam axis, as seen in Fig. 3-4. When torsional moments, T_u, are also present, the moment vector, M_u, is skewed, producing a skewing of the failure surface, as in Fig. 3-4. The compression zone of the new cross-section top fibers would no longer be straight and would subtend an angle θ to the original cross-section. The stirrups which are not subjected to stresses before cracking would start to carry a large portion of the torsional moment. The skew bending theory originally proposed by Lessig [30] and further extensively developed by others [27, 31, 32] idealizes the compression zone by considering it to be of uniform depth. It assumes the cracks on the remaining three faces to be uniformly spread, the steel hoops at those faces carrying the tensile forces at the cracks, and the longitudinal bars resisting shear through dowel action. Figure 3–5 shows all the forces acting on the skewly bent plane. The polygon in Fig. 3-5b gives the shear strength, F_c, attributable to the concrete and T_l, the longitudinal bar forces in the concrete compression zone. From [9], finding the equivalent to these vectors in terms of the internal stress in concrete, $k_1\sqrt{f_c'}$, and the geometrical constants of the section, $k_2 x^2 y$, leads to the expression for the

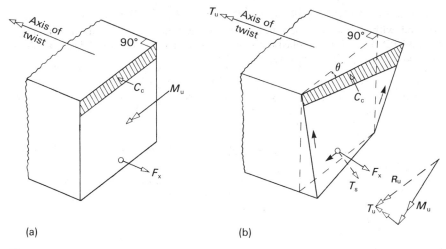

Figure 3-4   Skew bending due to torsion
(a) Bending before twist
(b) Bending and torsion

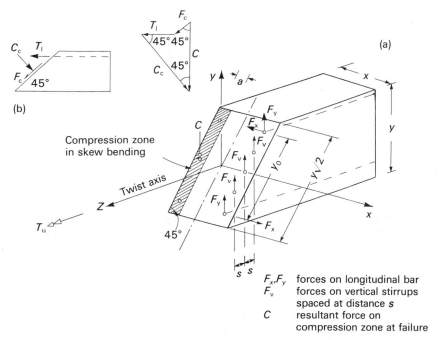

F_x, F_y	forces on longitudinal bar
F_v	forces on vertical stirrups spaced at distance s
C	resultant force on compression zone at failure

Figure 3-5   Forces on the skewly bent plane
(a) All forces acting on skew plane at failure
(b) Vector forces on compression zone

strength, T_c, attributed to the concrete alone in a rectangular section:

$$T_c = \left(\frac{2.4}{\sqrt{x}}\right)x^2 y \sqrt{f_c'} \qquad (3\text{-}6)$$

In Fig. 3-5b, the dowel forces F_x and F_y are assumed to be proportional to the cross-sectional areas of these bars. If a ratio is established between the proportion of torsional resistance given by the dowel forces F_x and F_y and the torsional resistance of the hoop forces F_v, torsional moments would be the summations

$$\sum f_v(\tfrac{1}{2}x_1) \quad \sum F_x(\tfrac{1}{2}y_0) \quad \sum F_y(\tfrac{1}{2}x_0) \quad \sum T_1(\tfrac{1}{2}x_0)$$

The dimensions x_1, y_1 are respectively the shorter and the longer center-to-center dimensions of the closed rectangular stirrups, and the dimensions x_0, y_0 are the corresponding center-to-center dimensions of the longitudinal bars at the corners of the stirrups. The resulting expression for the torsional strength, T_s, provided by the hoops and the longitudinal steel in the rectangular section is

$$T_s = \alpha_t \frac{x_1 y_1 A_t f_y}{s} \qquad (3\text{-}7)$$

where $\alpha_t = 0.66 + 0.33 y_1/x_1$, so that the total nominal ultimate torsional moment of resistance is $T_n = T_c + T_s$ or

$$T_n = \left(\frac{2.4}{\sqrt{x}}\right)x^2 y \sqrt{f_c'} + \left(0.66 + \frac{y_1}{x_1}\right)\left(\frac{x_1 y_1 A_t f_y}{2}\right) \qquad (3\text{-}8)$$

3.22 Space truss analogy theory

The space truss analogy theory is an extension of the model used in the design of the shear-resisting stirrups, in which the diagonal tension cracks, once they start to develop, are resisted by the stirrups. Because of the non-planar shape of the cross-sections due to the twisting moment, a space truss composed of the stirrups is used as the diagonal tension members, and idealized concrete strips at 45° between the cracks are used as the compression members (Fig. 3-6). This

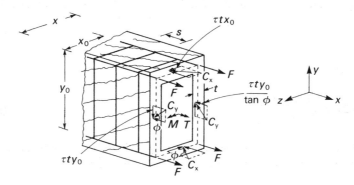

F	tensile force in each longitudinal bar
C_x	inclined compressive force on horizontal sides
C_y	inclined impressive force on vertical side
τt	shear flow force per unit length of wall

Figure 3-6 Forces on box concrete surface by space truss analogy

approach was developed by Lampert [33]. It is assumed that the reinforced concrete beam behaves in torsion similar to a thin-walled box with constant shear flow in the wall cross-section producing a constant torsional moment. The use of hollow-walled sections as compared to solid sections proved to give essentially the same ultimate torsional moment, provided the walls are not too thin [34, 35]. Such a conclusion is borne out by tests which have shown that the torsional strength of the solid sections is composed of the resistance of the closed stirrup cage, consisting of the longitudinal bars and transverse stirrups, and the idealized concrete inclined compression struts. The effective wall thickness of the hollow beam is taken as $\frac{1}{6}D_0$, where D_0 is the diameter of the circle inscribed in the rectangle connecting the corner longitudinal bars, namely, $D_0 = x_0$ in Fig. 3-6. In summary, the absence of the core does not affect the strength of such members in torsion, hence the acceptability of the space truss analogy approach based on hollow sections.

If the shear flow in the walls of the box section is τt, where τ is the shear stress, and F is the tensile force in each longitudinal bar at the corner, the force equilibrium equation would be

$$4F = 2\,\frac{\tau t x_0}{\tan \phi} + 2\,\frac{\tau t y_0}{\tan \phi} \tag{3-9a}$$

and the moments due to the shear flow forces would be

$$T_n = \tau t y_0 x_0 + \tau t x_0 y_0 \tag{3-9b}$$

If A_t is the area of the stirrup cross-section, and f_y is the yield strength of the stirrup spaced at distance s, then

$$A_t f_y = \tau t s \tan \phi \tag{3-10a}$$

Also, if A_l is the total area of the four longitudinal bars at the corners, then

$$F = \tfrac{1}{4} A_l f_y \tag{3-10b}$$

This will lead to

$$T_n = 2 x_0 y_0 \sqrt{\frac{A_l f_y A_t f_y}{2 s (x_0 + y_0)}} \tag{3-10c}$$

Substituting the appropriate terms into their equivalents in Eqn 3-10b, c (see [9]), the torsional moment of resistance, T_n, at failure would be

$$T_n = 2 \left(\frac{A_t f_y}{s} \right) x_0 y_0 \tag{3-11}$$

Note the similarity of Eqn 3-7 developed by the skew bending theory and Eqn 3-11 developed by the space truss analogy theory.

3.2.3 Sections subjected to combined loading

Both the skew bending theory and the space truss analogy theory concur on the interaction behavior in combined loading. Three basic combinations are considered here:

- (a) combined shear and torsion
- (b) combined bending and torsion
- (c) combined bending, shear and torsion

(a) *Combined shear and torsion* Figure 3-7a is the interaction curve for combined shear and torsion. It is a quarter of an arc for beams without web reinforcement, and flatter for beams with web steel [38]. T_n and V_n are the nominal ultimate torsional moment and shear acting simultaneously. T_{n0} is the ultimate value when torsion acts alone, and V_{n0} is the ultimate value when it acts alone.

$$\left(\frac{T_n}{T_{n0}}\right)^2 + \left(\frac{V_n}{V_{n0}}\right)^2 = 1 \tag{3-12}$$

(b) *Combined bending and torsion* The interaction diagram for the strength of beams subjected to combined bending and torsion is given in Fig. 3-7b [23, 36, 38, 39]. The interaction expression is given also for cases where the areas of the compression and tension steels are not equal.

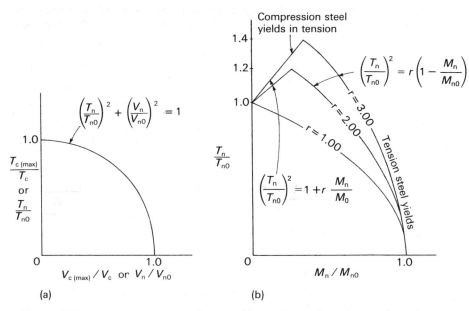

Figure 3-7 Interaction curves for combined shear, bending and torsion
(a) Shear and torsion interaction diagram
(b) Bending and torsion interaction diagram

(1) When tension steel yields in the tension zone:

$$\left(\frac{T_n}{T_{n0}}\right)^2 = r\left(1 - \frac{M_n}{M_{n0}}\right) \tag{3-13a}$$

(2) When tension yielding occurs in the flexural compression zone:

$$\left(\frac{T_n}{T_{n0}}\right)^2 = 1 + r\left(\frac{M_n}{M_{n0}}\right) \tag{3-13b}$$

where T_n is the Ultimate torsional moment, T_{n0} the Capacity in torsion only (Eqn 3-10), M_n the Ultimate bending moment, M_{n0} the Basic ultimate bending amount and $r = A_s f_y / A_s' f_y'$

(c) *Combined shear, bending and torsion*   Due to the combination of actions of these three parameters, an interaction surface applies. The CEB and the ACI approaches are quite similar [2, 3, 35]. The effect of shear and torsion is superimposed on the effect of bending and torsion from the two interaction cases of Eqns 3-12 and 3-13 to give

$$V_n = \frac{V_{n0}}{\sqrt{1 + \left(\frac{T_{n0}}{V_{n0}}\right)^2 \left(\frac{V_n}{T_n}\right)^2}} \tag{3-14a}$$

and

$$T_n = \frac{T_{n0}}{\sqrt{1 + \left(\frac{T_{n0}}{V_{n0}}\right)^2 \left(\frac{V_n}{T_n}\right)^2}} \tag{3-14b}$$

The basic ACI values for V_{n0} and T_{n0} for normal-weight curvature are

$$V_{n0} = 2\sqrt{f_c'}\, b_w d \quad \text{and} \quad T_{n0} = 0.8\sqrt{f_c'} \sum x^2 y$$

giving respectively the shear and torsion strengths provided by the concrete alone as follows:

$$V_c = \frac{2\sqrt{f_c'}\, b_w d}{\sqrt{1 + (2.5 C_t T_u / V_u)^2}} \tag{3-15a}$$

and

$$T_c = \frac{0.8\sqrt{f_c'} \sum x^2 y}{\sqrt{1 + (0.4 V_u / C_t T_u)^2}} \tag{3-15b}$$

where $C_t = b_w d / \sum x^2 y$ is a factor relating the shear and torsional stress properties. When lightweight concrete is used, the reduction factor, λ, is applied to $\sqrt{f_c'}$ as given in Section 3.1.

The ACI code of practice also stipulates:

(a) The limit for the strength provided by the closed stirrups is $V_{s(max)} \leq 8\sqrt{f_c'}\, b_w d$, beyond which the section has to be enlarged.
(b) $T_{s(max)} \leq 4 T_c$
(c) Torsional effects can be disregarded if $T_n < 0.5 f_c' \sum x^2 y$

(d) The length of the flange of a flanged section contributing to torsional stiffness is three times the flange thickness.

(e) For uniform torsional moment or shear, the critical section in a normal beam is at a distance d from the face of the support.

Example 3-1
An L-shaped end beam has a web width $b_w = 12$ in (304.8 mm) and a total depth $h = 20$ in (508.0 mm) including a 6 in (152.4 mm) flange as required for the bending moment. It is subjected to combined bending shear and torsion such that the torsional moment of resistance required at a distance d from the face of the support is $T_n = 6.74 \times 10^5$ lbf in (76.15 kN m) and the required ultimate shear resistance at d from the face of support is $V_n = 25.5 \times 10^3$ lbf (113.43 kN). The concrete strength $f_c' = 4000$ psi (27.58 MPa) and the steel yield strength $f_y = 60\,000$ psi (413.7 MPa). Design the reinforcement needed for the combined shear and torsion forces. The beam is made from normal-weight concrete.

Solution The effective depth $d = (20.0 - 2\frac{1}{2}) = 17.5$ in (444.5 mm). The width of the effective flange for torsion $= 3 \times 6$ in $= 18$ in (457.2 mm)

$$\sum x^2 y = x_1^2 y_1 + x_2^2 y_2 = (12)^2 (20) + (6)^2 (3 \times 6)$$
$$= 3528 \text{ in}^3 \ (57.8 \times 10^6 \text{ mm}^3)$$

$$C_t = b_w d / \sum x^2 y = 12 \times 17.5 / 3528 = 0.0595$$

$$V_c = \frac{2\sqrt{f_c'}\, b_w d}{\sqrt{1 + \left(2.5 C_t \dfrac{T_u}{V_u}\right)^2}} = \frac{2\sqrt{4000} \times 12 \times 17.5}{\sqrt{1 + \left(2.5 \times 0.059 \dfrac{6.74 \times 10^5}{25.5 \times 10^3}\right)}}$$

(The ratio T_u/V_u is the same as T_n/V_n here.)

$$V_c = 6544 \text{ lbf (29.11 kN)}$$

Hence,

$$V_s = (25.5 - 6.544) \times 10^3 = 18.956 \times 10^3 \text{ lbf (84.32 kN)}$$

$$T_c = \frac{0.8\sqrt{f_c'}\sum x^2 y}{\sqrt{1 + \left(\dfrac{0.4 V_u}{C_t T_u}\right)^2}} = \frac{0.8\sqrt{4000} \times 3528}{\sqrt{1 + \left(\dfrac{0.4 \times 25.5 \times 10^3}{0.0595 \times 6.74 \times 10^5}\right)^2}}$$

$$= 1.73 \times 10^5 \text{ lbf in (19.54 kN m)}$$

$$T_s = T_n - T_c = 674 - 173 = 5.01 \times 10^5 \text{ lbf in (56.60 kN m)}$$

which is less than $4T_c$, hence section is acceptable.
For two legs of the stirrup

$$\left|\frac{A_v}{s}\right| = \frac{V_s}{f_y d} = \frac{18.956 \times 10^3}{60\,000 \times 17.5}$$

$$= 0.0180 \text{ in}^2 \text{ per two legs per 1 in spacing}$$

$$x_1 \simeq 8 \text{ in} \qquad y_1 \simeq 16 \text{ in (406.5 mm)}$$

The maximum allowable spacing is 12 in or

$$s = \frac{x_1 + y_1}{4} = \frac{8 + 16}{4} = 6 \text{ in}$$

Hence, a spacing of 6 in (152.5 mm) controls.
Now

$$T_s = \frac{A_t \alpha_t x_1 y_1 f_y}{s}$$

where

$$\alpha_t = 0.66 + 0.33 y_1 / x_1 = 1.32 < 1.50$$

Therefore,

$$T_s = 5.01 \times 10^5 = \frac{A_t \times 1.32 \times 8 \times 16 \times 60\,000}{s}$$

or $A_t/s = 0.0494$ in^2 per *one* leg per 1 in spacing.
The minimum area of the closed stirrups is

$$A_v + 2A_t = 50 b_w s / f_y$$

$$= \frac{50 \times 12 \times s}{60\,000} = 0.005 s$$

Hence,

$$\text{minimum} \left| \frac{A_v + 2A_t}{s} \right| = 0.005$$

But

$$\text{required} \left| \frac{A_v}{s} + \frac{A_t}{s} \right| = 0.0180 + 2 \times 0.0494 = 0.1168 \text{ in}^2 \text{ per two legs}$$

per 1 in spacing

which is greater than 0.005, hence controls.
Try No. 5 stirrups (15.88 mm diameter), $A_t = 0.305$ m^2. The spacing is given by

$$s = \frac{2 \times 0.305}{0.1168} = 5.22 \text{ in } (132.65 \text{ mm}) < 6 \text{ in center-to-center}$$

Use $\frac{5}{8}$ in stirrups spaced at $5\frac{1}{4}$ in center-to-center.
Longitudinal steel to resist torsion is

$$A_l = 2 A_t \left(\frac{x_1 + y_1}{s} \right) = \frac{2(0.305)(8 + 16)}{5.22} = 2.80 \text{ in}^2 \text{ (1809.5 mm}^2)$$

Also,

$$A_l = \left[\frac{400 s}{f_y} \frac{T_u}{T_u + (V_u/3C_t)} - 2A_t \right] \left(\frac{x_1 + y_1}{s} \right)$$

gives a value less than 2.80 in^2.

Use ten No. 5 longitudinal bars = 3.05 in^2 (1967.9 mm^2) in addition to the bending moment steel to resist torsion and transverse closed stirrups No. 5 at $5\frac{1}{4}$ in (133 mm) center-to-center.

3.3 Brackets and corbels

These elements are short cantilevers projecting from girders, or columns, and are used to sustain concentrated loads transferred from crane girders or precast concrete beams. The following recommendations are based on experimental studies of the distribution of the principal stresses and stress trajectories in the brackets and the adjoining supporting member:

(a) The depth of a corbel measured at the outer edge of the bearing area should not be less than one-half of the required total depth of the corbel.
(b) The tension reinforcement should be anchored as close to the outer face as cover requirements permit. Welding the main bars to special devices, such as a cross bar equal in size to the main bar, is one method of accomplishing this end.
(c) The outer edge of the bearing area should not be closer than 2 in to the outer edge of the corbel.
(d) When corbels are designed to resist horizontal forces, the bearing plate should be welded to the tension reinforcement.

The designs have to be based on the shear span to depth ratio, a/d, where a is the moment arm from the point of application of the load to the face of the wall or column holding the bracket, and d is the effective depth of the bracket from the center of the top tensile steel to the bottom tip of the bracket. Deep-beam theory applies to brackets and corbels since the span to depth ratio is generally less than 1.0 [40]. This ratio is often less than 1/2, which would require the application of shear friction theory. Comprehensive details and analysis examples of brackets and corbels are given in [41].

The section adjacent to the support beams has to be designed to resist simultaneously the design exterior shear force, V_u, the design moment, $[V_u \times a + N_u(h - d)]$, and the design horizontal force, N_u. Additionally, the reinforcement at the front corbel face at the support has to be anchored either by welding to a transverse bar of equal strength or by bending back the bar to form a loop (McGregor in [42]). In both cases, the bearing area of the load should not project beyond the straight portion of the bars forming the tension steel.

3.4 Anchorage and length development of bars: bond stress

The normal methods of anchorage are: straight anchorage, curved anchorages, such as hooks, bends, and loops, and anchorages by mechanical devices. Their purpose is to develop enough bond length between the bonded reinforcement in reinforced concrete and the surrounding concrete, such that both materials act together as a composite. Such a condition becomes a requirement in order for the bars to be able to resist the flexural, shear, and torsional forces imposed by the external loads on the structural element or system.

Basically two types of bond stress occur: anchorage bond stresses and flexural bond stresses. In both, bond strength is based on the following behavioral parameters to varying degrees:

(a) The adhesion between the steel reinforcement and the concrete.
(b) The frictional resistance to sliding when the steel is being pulled out of the concrete.
(c) The gripping effect resulting from the shrinkage of the concrete.
(d) The quality of the concrete in the structural element and the shape and arrangement of the reinforcing bars, such as the use of deformed bars and the types of deformation.
(e) The mechanical anchorage effect of the ends of the bars due to the use of bends and hooks.

3.4.1 Anchorage bond

Assume L_d in Fig. 3-8a to be the length of the bar embedded in concrete being subjected to a net pulling force dT. If d_b is the diameter of the bar, u the bond stress, and f_s the stress in the reinforcing bar due to direct pull or bending stresses in a beam, the anchorage pulling force, V_u, would be $u\pi d_b L_d$ and equal to the

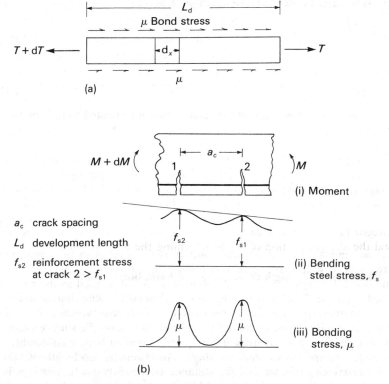

Figure 3-8 Bond stresses in reinforced concrete elements
(a) Pull-out anchorage bond in a bar
(b) Flexural bond

tensile force, dT, on the bar cross-section, namely, $dT = \frac{1}{4}\pi d_b^2 f_s$. Hence,

$$u\pi d_b L_d = \frac{1}{4}\pi d_b^2 f_s$$

from which the bond stress is given by

$$u = \frac{f_s d_b}{4L_d} \tag{3-16a}$$

and the development length

$$L_d = \frac{f_s}{4u} d_b \tag{3-16b}$$

3.4.2   Flexural bond

The change in stress along the length of a bar in a beam due to the variation of moment along the span is schematically shown in Fig. 3-8b. If jd is the lever arm of the couple T due to moment M, then $T = M/jd$ or $dT = dM/jd$. In terms of the moment difference between cracked sections 1 and 2,

$$dT = \frac{dM}{jd} \tag{3-17a}$$

also,

$$dT = udx \sum O \tag{3-17b}$$

where $\sum O$ is the circumference of the reinforcement subjected to the bond stress pull, to get

$$\frac{dM}{dx} = ujd \sum O$$

since $dM/dx = $ shear V. Hence,

$$u = \frac{V}{jd \sum O} \tag{3-18}$$

It is to be emphasized that flexural bond stresses develop in the compression reinforcement as well as in the tension reinforcement.

If the bond development length, L_d, from Eqn 3-16b is used as the basis, the ACI [3, 4] and the CEB [2] differ in rigor and details, taking into account the difference in the type of reinforcement used in European versus US practice. Generally, because reliance in the US practice is mostly on the surface deformations (corrugations) of the individual bars or the bundled bars, considerably less use is made of bent hooks than elsewhere. Additionally, the length of laps to effect the continuity of bars and the distance between two adjacent laps have different values in the two approaches although, in both, the same basic principles presented in Sections 3.4.1 and 3.4.2 were applied.

4 Structural concrete two-way slab and plate systems

The analysis and design of framed floor slab systems covers more than one aspect. The present state of knowledge permits reasonable evaluation of the following:

(a) Moment capacity
(b) Slab–column shear capacity
(c) Serviceability: crack control and deflection control

A brief historical survey of slab design development is necessary to put in proper perspective the evolution of the state of knowledge in this field in the last 50 years. The analysis of slab behavior in flexure up to the 1940s and early 1950s followed the classical theory of elasticity, particularly in the United States. The small deflection theory of plates, assuming the material to be homogeneous and isotropic, formed the basis of ACI Code recommendations with moment coefficient tables. The work principally by Westergaard and empirically allowing limited moment redistribution, guided the thinking of the code writers. Hence, the elastic solutions, complicated even for simple shapes and boundary conditions when no computers were available, made it mandatory to idealize and empiricize conditions beyond economic bounds.

In 1943, Johansen [43] presented his yield line theory for evaluating the collapse capacity of slabs. Since that time, extensive research into the ultimate behavior of reinforced concrete slabs has been undertaken. Studies by many investigations, such as those of Ockleston, Mansfield, Rzhanitsyn, Powell, Wood, Sawczuk, Gamble–Sozen–Siess, and Park [44–52] contributed immensely to further understanding of the limit state behavior of slabs and plates.

4.1 Moment capacity

The investigations to date in the area of collapse behavior have established that

(a) The ultimate strength of a slab is considerably higher than the current design methods predict.
(b) Membrane stresses, which were neglected in Johansen's yield line theory, have a considerable influence on the strength of the slab. The smaller the steel percentage, the greater the influence.
(c) The ultimate behavior of small-scale slab models does not basically differ from that of the full scale.
(d) Since the distortion of a slab at failure is concentrated in yield lines and hinge fields, it can be assumed that the energy absorbed by the slab is essentially confined to the yield-line fields.

As a result of the interest generated by the yield line theory, a limit approach was developed based on the general theory of plasticity and incorporating Johansen's theory, while assuming a square yield criterion. Upper bound solutions were deduced requiring a *valid* mechanism in the equivalence of internal dissipation of energy and external work, and lower bound solutions used requiring also that the stress fields satisfy equilibrium *everywhere*, and nowhere violating the yield criterion. Then the strip method, due to Hillerborg, was developed giving lower bound values. In this approach, a slab is assumed as sets of narrow strips intersecting in orthogonal directions, hence simulating simple beam design.

The following is a more detailed discussion of the methods previously enumerated.

4.1.1 The semi-elastic ACI Code approach

The ACI approach [3, 4] gives two alternatives for the design of a framed two-way action slab system: the direct design method and the equivalent frame method.

4.1.1.1 The direct design method In this method, updating the old empirical method of flat slab design, the basic premise is that the absolute sum of the positive and average negative bending moments in each direction shall not be less than $M_0 = \omega l_2 l_c^2/8$. It also stipulates that the successive span lengths in each direction shall not differ by more than one-third of the longer span, and if the slab panel is supported on beams on all sides, the relative stiffness of the beams in the two perpendicular directions shall not be less than 0.2 nor greater than 5.0, and that all loads shall be due to gravity only, with the lead uniformly distributed over the entire panel. These limitations and others render the direct design method more restrictive in use than the other ACI approach, namely, the equivalent frame method.

4.1.1.2 The equivalent frame method In this method, the structure is considered to be made up of frames on column lines taken longitudinally and transversely through the building. Each frame is considered to consist of a row of equivalent columns or supports and slab–beam strips. The method is applicable to all two-way action slabs, including flat slabs and flat plates.

Since slab strips with their supports are idealized into frames, relative stiffness of the interacting components has to be determined from the onset. The flexibility of the column is considered in this method as the sum of the flexibilities of the column in flexure and the beam–slab combination in the transverse direction in torsion, namely,

$$\frac{1}{k_{ec}} = \frac{1}{k_c} + \frac{1}{k_t}.$$

The torsional stiffness, k_t, is introduced to account for the moment 'leak' around the column, while k_c is independent of the torsional distribution along the beam, or of the beam torsional stiffness. A good detailed discussion is given in [15] of the evaluation of the stiffness coefficients k_{ec}, k_c, and k_t.

Once the bending moments at the critical sections are determined, they can be distributed to the slab column strip as in the direct design method previously discussed.

4.1.2 The yield line theory

Whereas the semi-elastic code approach applies to standard cases and shapes and has an inherent, excessively large safety factor with respect to capacity, the yield line theory is a plastic theory easy to apply to irregular shapes and boundary conditions. Provided that serviceability constraints are applied, it is the simplest

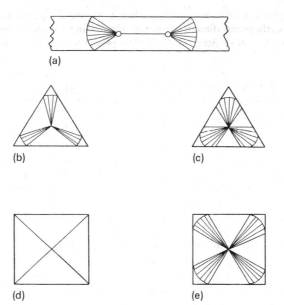

Figure 4-1 Typical collapse mechanisms in slabs
(a) Strip loaded at two points
(b) Triangular slab simply supported, uniformly loaded
(c) Triangular slab restrained, uniformly loaded
(d) Square slab simply supported, uniformly loaded
(e) Square slab restrained, uniformly loaded

method that the designer can use, representing the true behavior of reinforced concrete slabs. Figure 4-1 shows typical collapse mechanisms or yield lines which, once predicted by the design engineer, enable the solution to be readily obtained.

A simple stepped yield criterion was proposed by Jones and Wood [56], assuming that in an orthotropically reinforced slab, the yield moments precipitate yield lines at right angles to each reinforcement band. For each band of reinforcement taken on its own, the yield line is essentially considered to be divided into small steps parallel to and at right angles to the reinforcement (Fig. 4-2). It is also assumed that all the reinforcement crossing the yield lines has yielded. If M and μM are the moments in the two perpendicular directions, then the normal yield moments are

$$M_n = M \cos^2 \theta + \mu M \sin^2 \theta \qquad (4\text{-}1)$$

and the twisting moments are

$$M_T = (M - \mu M) \sin \theta \cos \theta \qquad (4\text{-}2)$$

If internal dissipation of energy is to be equal to the external work, and if all regions of the slab other than the yield lines are considered rigid, then only the normal moments are of importance, and the twisting moments are neglected in the basic theory.

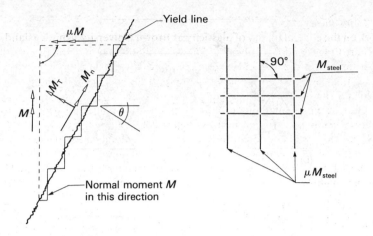

Figure 4-2 Stepped yield lines in orthogonally reinforced plate

In summary, equating internal energy to external work, Johansen's theory can be summed up by the simple work equation:

$$\sum\left[\iint p\Delta\,dx\,dy\right]=\sum\left[\theta_n\int M_n\,ds\right] \tag{4-3}$$

$$\underset{\text{segment}}{\text{each}}\qquad\qquad\underset{\text{line}}{\text{each}}$$

where M_n is defined in Eqn 4-1. As statical equivalents to correct for twists and shears on the yield lines, nodal forces have to be imposed as shown in Fig. 4-3.

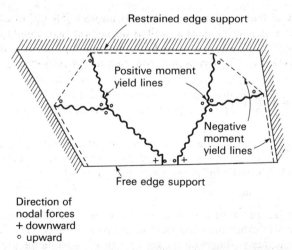

Figure 4-3 Nodal point forces in rectangular panel at failure

4.1.3 *Limit theory*

The work on the general theory of plasticity at Brown University, Rhode Island, such as Drucker, Prager and Greenberg's [57] was utilized later and Johansen's theory was incorporated. A square yield criterion is applied, where $M_n = M$ in all directions when $\mu = 1$ in Eqn 4–1, postulating a prefectly plastic, fully isotropic material from which a slab is assumed made.

The interest in developing a limit solution became necessary due to the possibility of finding a variation in the collapse field which can give a lower failure load. Hence, an upper bound solution requiring a *valid* mechanism [58] when supplying the work equation was sought, as well as a lower bound solution requiring that the stress field satisfies everywhere the differential equation of equilibrium, namely,

$$\frac{\partial^2 M_x}{\partial x^2} + \frac{\partial^2 M_y}{\partial y^2} - 2\frac{\partial^2 M_{xy}}{\partial x\,\partial y} = -w \tag{4-4}$$

Variable reinforcement permits the lower bound solution to be still valid. Mansfield's work [45] using equi-angular spiral shapes of failure mechanisms, and applying the calculus of variations for the first time in the solutions, Wood's work [48] using circular shapes, and other researchers, could therefore give these more accurate semi-exact predictions of the collapse load. It must be emphasized, however, that Johansen's yield line theory is an upper bound solution as long as a *valid* failure mechanism is used in predicting the collapse load.

Further amplification on the true behavior of the slabs was introduced by Wood and Park [52–54] taking into account the effects of membrane action, both compressive and tensile, on enhancing the value of the predicted collapse load.

In all the developments discussed, compatibility of deflections at all fields of a slab is assumed to be valid. The assumptions are admissible since Johansen's 'stepped' yield line and the limit square yield both assume a slab to be completely rigid until collapse, hence a zero deflection value everywhere, hence deflection compatibility.

Further work at Rutgers University, New Brunswick [59, 60] incorporated the deflection effect at high level loads as well as the compressive membrane effects in predicting the collapse load. In this work, the relationship between limit load and limit-load deflection can be developed from the energy principle:

$$\int_A E_E \, dA = \int_A E_1 \, dA \tag{4-5}$$

where subscripts E and I denote external and internal energies, respectively. Rewriting Eqn 4-5 to show the total internal energy as the summation of the internal energies dissipated in all the yield lines gives

$$\sum_1^{n=1} E_E A_{I_i} = \sum_1^{m=j} \int_l^j (E_1)_j \, dl \tag{4-6}$$

If a typical idealized yield-line failure mechanism for a rectangular slab restrained on all four boundaries is that shown in Figs 4-4 and 4-5, Eqn 4-6 can be

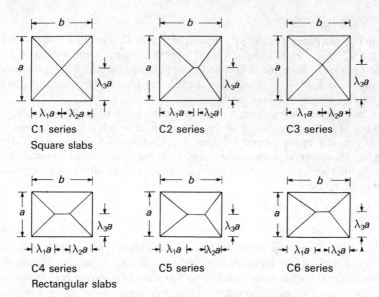

Figure 4-4   Idealized segmenting of slab panels at failure. $\lambda_1, \lambda_2, \lambda_3$ define the idealized yield line geometry

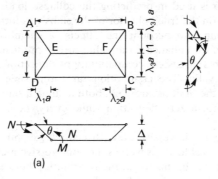

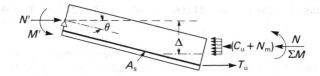

$N_m = C_u - T_u$, compressive membrane force
$M_1 = (M + M_m)$, where M_m is due to induced membrane force N_m

(b)

Figure 4-5   Idealized limit mechanism
(a) Typical idealized failure mechanism
(b) Detail of slab mechanism element

rewritten as follows:

$$\sum_{1l}^{i} \int (E_l)_i \, dl = \sum_{1l}^{i} \int [M' + M + (M_m - N_m \Delta)] \tag{4-7}$$

where $N_m = (C_u - T_u)$ and Δ is the deflection at the limit strength load (Fig. 4-5).

The energy dissipation approach, accounting for the compressive membrane action at large deflections, yields the following from Eqn 4-7 for the limit load, w_{lim}, per unit area, as detailed in [60] in terms of the ultimate resisting moment M_1 through M_6 and limit load deflection Δ:

$$w_{lim} = \frac{1}{a^2} [C_1 M_1 + C_2 M_2 (\Delta/t) + C_3 M_3 (\Delta/t)^2 - R_i] \tag{4-8}$$

where

$$R_i = C_4 M_4 + C_5 M_5 (\Delta/t) + C_6 M_6 (\Delta/t)^2 \tag{4-9}$$

R_i is the strength reduction for hinged boundaries. For restrained boundaries, $R_i = 0$. C_1 through C_6 are constants as shown in Table 4-1. The values of resisting moments M_1 through M_6 are as follows:

$$M_1 = f'_c d^2 [0.24(1 + d'/d) + q(1 - d'/d) - 0.07]$$

$$M_2 = f'_c d^2 [q(1 + d'/d) - 0.35(1 + d'/d) + 0.07]$$

$$M_3 = f'_c d^2 [0.12(1 + d'/d) - 0.02]$$

$$M_4 = f'_c d^2 [0.59(1 - d'/d) + 0.07]$$

$$M_5 = f'_c d^2 [0.24(1 + d'/d) - 0.07]$$

$$M_6 = f'_c d^2 [-0.12(1 + d'/d) + 0.02]$$

where $q = A_s f_y / f'_c d$ is the reinforcement index per 1 in unit width of slab.

Table 4-1   C-coefficients for moments in Eqns 4-8 and 4-9

$m - a/b$	Case 1			Case 2			Case 3		
	C_1	C_2	C_3	C_1/C_4	C_2/C_5	C_3/C_6	C_1/C_4	C_2/C_5	C_3/C_6
1.0	24.000	12.000	8.000	24.000	12.462	8.615	24.000	12.000	8.000
				5.769	*2.885*	*1.923*	*12.000*	*6.000*	*4.000*
0.9	21.766	11.713	8.362	21.830	12.063	8.807	21.815	11.758	8.406
				4.883	*2.442*	*1.628*	*11.054*	*6.025*	*4.349*
0.8	19.810	11.430	8.637	19.896	11.688	8.952	19.903	11.523	8.730
				4.104	*2.052*	*1.368*	*10.222*	*6.032*	*4.635*
0.7	18.071	11.154	8.848	18.154	11.333	9.059	18.200	11.283	8.978
				3.410	*1.705*	*1.137*	*9.476*	*6.017*	*4.864*
0.6	16.492	10.861	8.984	16.562	10.983	9.123	16.653	11.023	9.147
				2.789	*1.395*	*0.930*	*8.790*	*5.975*	*5.037*
0.5	15.035	10.552	9.058	15.081	10.622	9.136	15.222	10.740	9.246
				2.230	*1.115*	*0.743*	*8.149*	*5.908*	*5.161*

The values in italic type are moment coefficients for strength reduction

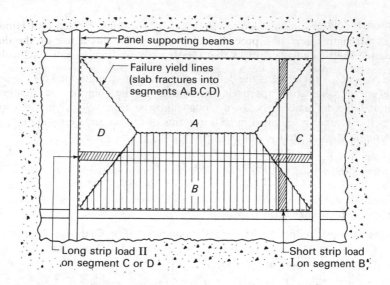

Figure 4-6 Strip method loading

Charts for values of moments M_1 through M_6 as well as the expressions developed for them for various steel and concrete strength combinations are easily developed, as in [60].

4.1.4 *The strip method*

This method was proposed by Hillerborg [61] as a lower bound solution, trying to fit the reinforcement to the stress fields. Since practical considerations require the reinforcement placed in orthogonal directions, Hillerborg set twisting moment $M_{xy} = 0$ in Eqn 4-4 and split the load intensity w (Fig. 4-6) into components αw and βw, where $\alpha + \beta = 1$, making the moments in the strips governed by

$$\frac{\partial^2 M_x}{\partial x^2} = -\alpha w \tag{4-10}$$

$$\frac{\partial^2 M_y}{\partial y^2} = -\beta w \tag{4-11}$$

By this proposal, the slab is transformed into intersecting beam strips, hence the name strip method.

4.2 Slab-column shear capacity

Particularly in flat plates, strength is frequently controlled by the shear and torsional stresses in the slab–column junction along the periphery at the face of the column. While some designs have assumed that part of the unbalanced bending moment results in non-uniform distribution of shear stresses, it is found that torsional stresses of appreciable magnitude also exist and have to be accounted for.

In work by Hawkins and Corley [62] and by Hawkins, Fallsen and Hinojosa [63], a strength calculation procedure is presented for direct use by the design engineer. The procedure assumes that the slab section framing into each column face can be idealized as part of a beam. Compatibility restrictions are ignored and each section is assumed to deform sufficiently for the development of the governing ultimate torque, bending moment, or shear for the beam. Two modes of failure are proposed: moment–torsion failure and shear–torsion failure. Their combined effect determines the total shear capacity of the slab–column junction. It is assumed that 60% of the moment is considered to be transferred by flexure ($d/2$ from face for regular sections) and 40% by eccentricity of the shear about the centroid of the critical section. The stress distribution is assumed to vary according to Fig. 4-7 for both interior and exterior columns. The shear V and the moment M are determined at the centroidal axis, c–c, of the diagram. The maximum stress is assumed to be empirically represented by

$$v_{AB} = \frac{V}{A_c} + \frac{KMC_{AB}}{J_c} \tag{4-12}$$

$$v_{CD} = \frac{V}{A_c} - \frac{KMC_{AB}}{J_c} \tag{4-13}$$

where $A_c = 2d(c_1 + c_2 + 2d)$, J_c is the torsional constant, and K is the stiffness transfer fraction (fraction of bending moment M to be transferred by eccentricity about the critical section at $d/2$).

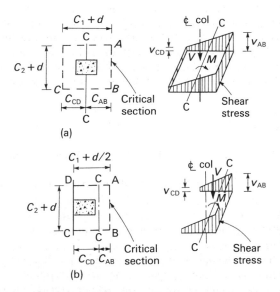

(a)

(b)

Figure 4-7   Assumed shear stress distribution at slab–column junction
(a) Interior column
(b) Edge column

5 Serviceability as determined by deflection control

Deflection as a measure of the serviceability of structures is one of the important parameters in the design of reinforced concrete structural systems. Long-term effects such as creep and shrinkage increase deflection with time and can lead to reduced serviceability, increased cracking, and a lowering of the reserve strength of many structures.

There are two approaches to the design of elements for controlled deflection:

(a) Empirical, namely the use of permissible limits of span-to-depth ratios.
(b) Computation of expected deflection.

The first approach is expectedly more conservative in many instances, while the latter requires more effort on the part of the designing engineer.

Considerable work is available on the evaluation of the deflection and camber of beams and other one-dimensional members. On the other hand, only limited information is available at present on the evaluation of the deflection and camber of two-way systems, such as slab and plate floor systems.

5.1 Deflection control in beams

Deflection is a time-dependent phenomenon. The final long-term deflection after years of sustained loading can be more than twice the initial deflection. Since deflection is not only a function of stress in the member but also of its stiffness, EI, and the quality of the component materials, deflection calculations at service-load levels are less accurate than calculations for bending and shear. Additionally, flexure theory neglects the contribution of the cracked concrete in the tensile zone. The cracked concrete contributes to the moment of inertia, I. Such a contribution is more difficult to calculate precisely, and, taken with the empirical value of the modulus, E_c, makes the deflection computations not very accurate.

The deflection at the mid-span of a beam is the result of the applied bending moment and is obtained by double integration of the curvature, $1/\rho$, given by

$$\frac{1}{\rho} = \frac{M}{EI} = \frac{d^2y}{dx^2}$$

over the beam span. On that basis, the deflection, Δ, can be expressed as

$$\Delta = kMl^2/E_cI \tag{5-1}$$

where k is a factor depending on the loading pattern and the boundary conditions. Values of k are available in most handbooks and texts on mechanics. A typical value of k is $\frac{5}{48}$ for uniformly loaded supply supported beams, and $\frac{1}{12}$ for centrally loaded beams.

Since the value of the moment of inertia is dependent on the degree of cracking, the gross moment of inertia, I_g, has to be reduced to an equivalent value, I_e, taking into account the transformed section cracked moment of inertia, I_{cr}. Branson's work [64–67] as adopted by the ACI [3–4] expresses the effective moment of inertia, I_e, as follows:

$$I_e = (M_{cr}/M_a)^3 I_g + [1 - (M_{cr}/M_a)^3] I_{cr} \tag{5-2}$$

where

$M_{cr} = f_r I_g / y_t$, the cracking moment

f_r = modulus of rupture of concrete, namely, the unit tensile strength of the concrete at incipient cracking

$= 7.5\sqrt{f_c'}$ for normal-weight concrete

$= 0.65\sqrt{wf_c'}$ for all concretes, w being the weight in lb/ft^3

y_t = distance from the neutral axis to the extreme concrete section fibers in tension

The value of I_e is intended to represent a single value for the entire span in simply supported beams or a single value between the inflection points in a continuous beam. Hence, the deflection, Δ, for short-term loading will be calculated from a modified Eqn 5-1 so that

$$\Delta = kML^2 / E_c I_e \tag{5-3}$$

Equation 5-2 was verified as equally applicable to both reinforced and prestressed concrete beams by Nawy *et al.* [68–70]. The idealized ACI bilinear moment–deflection relationship is given in Fig. 5-1a [3, 4, 66, 67].

The CEB approach [2] is also a bilinear approximation (Fig. 5-1b) where the

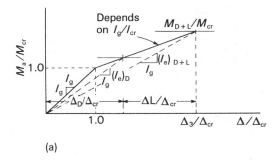

(a)

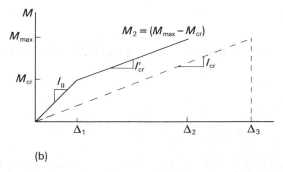

(b)

Figure 5-1　ACI and CEB moment–deflection relationships
(a) Idealized short-term moment–deflection diagram (ACI approach)
(b) Idealized short-term moment–deflection diagram (CEB approach)

deflection is given by

$$\Delta = \Delta_1 + \Delta_2 \leq \Delta_3 \tag{5-4}$$

where

Δ_1 = deflection at first cracking load calculated on the basis of $E_c I_g$ in Eqn 5-1
Δ_2 = deflection based on a section stiffness $(E_c I_{cr})' = 0.75 E_c I_{cr}$ for a moment
$M_2 = M_{max} - M_{cr}$
Δ_3 = deflection due to total moment, M_{max}, and section stiffness, $E_c I_{cr}$.

5.2 Long-term deflections

Long-term deflections are due to sustained loading resulting in increased deflection due to creep and shrinkage. If Δ_{iD} is the initial elastic deflection computed from Eqn 5-2 due to total sustained load, Δ_{cp} the creep deflection and Δ_{sh} the shrinkage deflection, then

$$\Delta_{total} = \Delta_{iD} + \Delta_{cp} + \Delta_{sh} \tag{5-5}$$

where, [66],

$$\Delta_{cp} = k_r C_t \Delta_{iD} \qquad \Delta_{sh} = \alpha_1 \phi_{sh} L^2$$

in which

α_1 = constant = 0.50 for cantilever beams, 0.125 for simply supported beams, 0.086 for beams continuous on one end only, and 0.063 for beams continuous on both ends
ϕ_{sh} = shrinkage curvature = $(\varepsilon_{sh}/d)(1 - \varepsilon_s/\varepsilon_{sh})$ with ε_{sh} being the free shrinkage strain k (average ultimate value at 40% humidity is 800×10^{-6})
     = ε_{sh}/h for $(\rho - \rho') > 3\%$
L = span length
$k_r = 0.85 - 0.45(A_s'/A_s) \geq 0.40$
C_t = creep coefficient depending on loading duration time t after curing
     = $\left(\dfrac{t^{0.6}}{10 + t^{0.6}}\right) C_u$ (average value of C_u at 40% humidity is 2.35)

An empirical, simplified expression for a deflection multiplier, λ, of the initial deflection, Δ_{iD}, for the combined shrinkage and creep deformation [3, 4, 66] is

$$\lambda = 2 - 1.2(A_s'/A_s) \geq 0.6 \tag{5-6a}$$

such that the total deflection

$$\Delta_{total} = \{1 + [2 - 1.2(A_s'/A_s)]\} \Delta_{iD} \tag{5-6b}$$

Another more refined expression for the long-term deflection multiplier, λ, can be used to account for the load duration:

$$\lambda = T\left(\frac{1}{1 + 50\rho'}\right) \tag{5-6c}$$

where

$T = 2.0$ for a 5 year load duration
$= 1.4$ for a 12 month load duration
$= 1.2$ for a 6 month load duration
$= 1.0$ for a 3 month period of sustained load duration

and $\rho' = A_s'/bd$ is the compressive reinforcement at the support for a cantilever and at mid-span for simple and continuous spans. (In most cases, ρ' at mid-span would be negligible unless closed transverse ties are used throughout span.)

It can be recognized from Eqns 5-6a and 5-6c that if no compression steel A_s' exist, the shrinkage and creep multiplier $\lambda = 2.0$. It is also to be noted that the maximum permissible deflection, Δ_{total}, for floors which do not support fragile partitions should not exceed a span of $L/360$.

5.3 Deflection of two-way slabs and plates

While adequate simple methods are available for deflection evaluation in beams and one-way members as presented in Sections 5.1 and 5.2, the same cannot be said for two-way slabs and plates. The complications arise due to many factors, including the influence of the side ratios of the floor panels, their aspect ratios, the flexibility and rotation of the monolithic supporting elements for these panels, the effect of cracking, the inhomogeneity of concrete, and the anisotropy of the reinforcement from one region to another.

Two approaches can be applied to the problem of deflection control:
(a) Direct control method through the prediction of deflection by mathematical computations. The slab or plate relative stiffness can be so proportioned that the predicted deflection falls within acceptable limit in terms of span-to-deflection ratios.
(b) Indirect control method through the limitation of the geometry to qualitatively accepted values of the span-to-depth ratios.

5.3.1 *Direct method of deflection evaluation*
The classical theory of elasticity—assuming thin plates of constant thickness, small deflections and low stress levels—has expressed the deflection relationship as a Lagrangian equation:

$$\frac{q}{D} = \frac{\partial^4 w}{\partial x^4} + \frac{2\partial^4 w}{\partial x^2 \partial y^2} + \frac{\partial^4 w}{\partial y^4}$$

where q is the transverse load, w the deflection, and the flexural rigidity $D = Eh^3/12(1 - \mu^2)$

Solutions are available for plates subjected to various loading and boundary conditions but of normal shapes such as circular, square and rectangular [71]. Timoshenko's simplified equation for a flat plate supported on columns without drop panels, where the cross-sectional dimensions of columns are negligible compared with the column spacing, gives deflection $w = \alpha q L^4/D$, where b is the longer span and α the coefficient varying with the side ratio.

However, such simplification is too crude for reliable prediction and various other analytical methods have been proposed in the last two decades. The following are brief summaries of these methods, given in [72].

5.3.1.1 *Cross-beam analogy* This method due to Marsh (1904) proposed replacing a uniformly loaded continuous slab by a gridwork of cross beams. The applied load is assumed to be divided between the beams spanning in the short direction and the long direction, assuming *equal* vertical deflection at the center of the panel. Marsh neglected the presence of the torsional moment in the slab. Marcus (1924) introduced modifying factors to account for the torsional moments. The method, however, was not of much value since it applied only to uniformly loaded rectangular panels on unyielding supports.

5.3.1.2 *Analogous gridwork method* The slab is first divided into strips in each direction. Then these strips are replaced by equivalent beams whose flexural and torsional properties are the same as those of the corresponding strips. At each point, a moment with associated rotation is applied. The total stiffness of a joint 0, for instance, is the sum of the flexural and torsional stiffness of all the members meeting at joint 0. Sets of simultaneous equations in terms of arbitrary joint displacements are solved for the unknown displacements.

5.3.1.3 *Finite element method* In this method, the slab is divided into a number of subregions or finite elements which are generally triangular, rectangular, or quadrilateral in shape. They are considered interconnected only at discrete points, called nodes, at the corners of the individual elements.

The continuous displacement quantities are expressed in terms of a finite number of displacements $w(x, y)$, called degrees of freedom at the nodes.

Therefore, for a rectangular plate bending element having 12 degrees of freedom, for instance (see [72], Fig. 8),

$$\{A\} = \{K_m\}\{D\}$$

where

$\{D\} = 12 \times 1$ column vector consisting of vertical displacements and rotations about each horizontal axis of each of the four corners.

$\{K\} = 12 \times 12$ element stiffness matrix at each of the four corners.

$\{A\} = 12 \times 1$ column vector consisting of transverse forces and bending moments at the nodes.

A difficulty in the application of the finite element method is obtaining a suitable force–displacement relationship between the nodal forces and the corresponding nodal displacements. This is particularly true in reinforced concrete because of its non-elastic behavior and time-dependent effects of cracking and deflection. The high expense of computer use can also sometimes be difficult to justify in a lengthy analysis.

5.3.1.4 *Equivalent frame method* The structure is divided into continuous frames centered on the column lines in each direction (Fig. 5-2). Each column is composed of a row of columns and a *broad* band of slab together with column line beams, *if any*, between panel center-lines.

By the requirements of statics, the entire applied load must be accounted for in each of the two orthogonal directions. In order to account for the torsional deformations of the support beams, an *equivalent column* is used whose flexibility

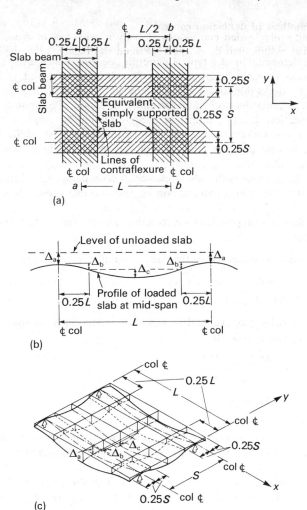

Figure 5-2   Equivalent frame method for deflection analysis
(a) Plate panel transferred into equivalent frames
(b) Profile of deflected shape at centre line
(c) Deflected shape of panel

is the *sum* of the flexibilities of the actual column and the torsional flexibility of the transverse beam or slab–beam strips. The design moments obtained from the equivalent frame analysis discussed in Section 4.1 are proportioned between the column strips and middle strips according to coefficients found from tests.

The slab–beam strips are not supported directly on the columns but by *transverse* beam–slab strips on the column lines.

5.4 Direct method of deflection control

The equivalent frame method for deflection, herein called the direct method, considers the deformation of a typical panel in *one direction at a time*. Therefore, the contribution in each direction of the two directions, x and y, is added to obtain the total deflection at any point in the slab or plate [72, 73].

First, the deflection due to bending in the x-direction is computed (Fig. 5-2a). Then the deflection due to bending in the y-direction is found. The mid-panel deflection can now be obtained as the sum of the center-span deflections of the column strip in one direction and that of the middle strip in the orthogonal direction (Fig. 5-2c).

One has to keep in mind that the presence of drop panels or column capitals in the column strips require consideration of the variable moments of inertia of those strips.

The deflection of each panel can be considered as the sum of three components: (a), (b) and (c):

(a) Basic mid-span deflection of the panel, assumed fixed at both ends, given by

$$\delta' = \frac{wl^4}{384E_c I_{\text{frame}}}$$

This has to be proportioned to separate deflection δ_c of the column strip and δ_s of the middle strip, such that

$$\delta_c = \delta' \times \frac{M_{\text{col strip}}}{M_{\text{frame}}} \times \frac{E_c I_{cs}}{E_c I_c}$$

$$\delta_s = \delta' \times \frac{M_{\text{slab strip}}}{M_{\text{frame}}} \times \frac{E_c I_{cs}}{E_c I_s}$$

where I_{cs} is the moment of inertia of the total frame, I_c the moment of inertia of the column strip, and I_s the moment of inertia of the middle slab strip.

(b) Center deflection, $\delta''_{\theta L} = \frac{1}{8}\theta L$, due to rotation at the left end while the right end is considered fixed, where $\theta_L = \text{Left } M_{\text{Net}}/K_{ec}$, and K_{ec} is the flexural stiffness of equivalent column (moment per unit rotation).

(c) Center deflection, $\delta''_{\theta R} = \frac{1}{8}\theta L$, due to rotation at the right end while the left end is considered fixed, where $\theta_R = \text{Right } M_{\text{Net}}/K_{ec}$.

Therefore,

$$\delta_{cx} \text{ or } \delta_{cy} = \delta_c + \delta''_{\theta L} + \delta''_{\theta R} \tag{5-7a}$$

$$\delta_{sx} \text{ or } \delta_{sy} = \delta_s + \delta''_{\theta L} + \delta''_{\theta R} \tag{5-7b}$$

(Use in Eqns 5-7a,b the values of $\delta_c, \delta''_{\theta L}$, and $\delta''_{\theta R}$ which correspond to the applicable span directions.)

From Figs 5-2b and 5-2c, the total central deflection is

$$\Delta = \delta_{sx} + \delta_{cy} = \delta_{sy} + \delta_{cx} \tag{5-8}$$

This approach has been found accurately applicable to both reinforced and prestressed concrete two-way multi-panel elements [74, 75]. Adjustment for long-term deflections would be made in accordance with the discussion in Sections 5.1 and 5.2.

Example 5-1
An interior panel of a multi-panel floor system subjected to a long-term intensity of load of 440 psf (21.067 kPa) has panel dimensions of 25 ft in the E–W direction (7.62 m) and 20 ft in the N–S direction (6.10 m) [9]. The slab panel is supported on beams whose size is 14×28 in for the 25 ft span beam (355.6×711.2 mm), and 12×24 in for the 20 ft span beam (304.8×604.6 mm). $E_c = 3.12 \times 10^6$ psi (21.51×10^6 kPa)

Net moment M from adjacent spans

E–W support 1	18×10^3 lbf ft (24.4 kN m)
E–W support 2	5.5×10^3 lbf ft (7.46 kN m)
N–S support 1	41.5×10^3 lbf ft (56.27 kN m)
N–S support 2	19.5×10^3 lbf ft (26.44 kN m)

The stiffness factor, k_{ec}, and the moment distribution factors for the equivalent frame are as follows:

	k_{ec}	*Moment distribution factors*	
		Column strip	*Middle strip*
		$(+ \ and \ -)$	$(+ \ and \ -)$
E–W	$404E_c$	81%	19%
N–S	$391E_c$	67.5%	32.5%

Find the maximum central deflection of the panel due to long-term loading.

Solution Gross moment of inertia properties can be easily calculated for the *T*-section at the column strip and the rectangular slab section at the middle strip. The numerical values of I_g from [9] for the sections used in this example are as follows:

	E–W	*N–S*
I_{cs}—total frame section (Fig. 5-3b)	66 845 in⁴	39 813 in⁴
I_c—column strip beam (Fig. 5-3c)	56 087 in⁴	33 869 in⁴
I_s—middle strip slab (Fig. 5-3d)	2 750 in⁴	3 433 in⁴

Deflections E–W direction (span = 25 ft)

Long-term $W_w = 440$ psf

$$\delta'_{25} = \frac{440 \times 20(25)^4 \times 1728}{384 \times 3.12 \times 10^6 \times 66\ 845} = 0.0742$$

$$\delta_c = 0.0742 \times 0.81 \times \frac{66\ 845}{56\ 087} = 0.0716 \text{ in}$$

$$\delta_s = 0.0742 \times 0.19 \times \frac{66\ 845}{2750} = 0.3426 \text{ in}$$

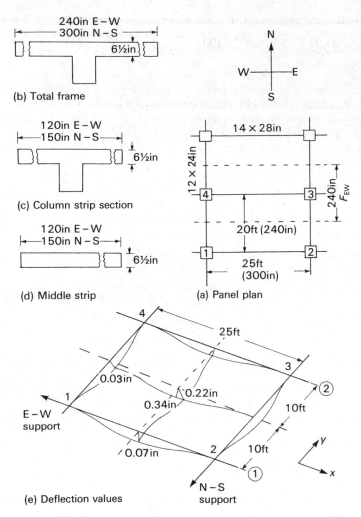

(b) Total frame

(c) Column strip section

(d) Middle strip

(a) Panel plan

(e) Deflection values

Figure 5-3 Example on equivalent frame deflection computations

Rotation at end 1 is

$$\theta_1 = \frac{M_1}{k_{ec}} = \frac{18 \times 12\,000}{404 \times 3.12 \times 10^6} = 1.71 \times 10^{-4}\ \text{rad}$$

and rotation at end 2 is

$$\theta_2 = \frac{M_2}{k_{ec}} = \frac{5.5 \times 12\,000}{404 \times 3.12 \times 10^6} = 0.053 \times 10^{-4}\ \text{rad}$$

where θ is the rotation at one end if the other is fixed.

The deflection adjustment, δ'', due to rotation at supports 1 and 2 is

$$\delta'' = -\frac{(1.71+0.53)\times 10^{-4}\times 300}{8} = -0.0084 \text{ in}$$

Therefore,

net $\delta_{cx} = 0.0716 - 0.0084 = 0.0632$ in, say 0.07 in

net $\delta_{sx} = 0.3426 - 0.0084 = 0.3342$ in, say 0.34 in

Deflections N–S direction (span = 20 ft)

$$\delta'_{20} = \frac{440\times 25(20)^4\times 1728}{384\times 3.12\times 10^6\times 39\,813} = 0.0638 \text{ in}$$

$$\delta_c = 0.0638\times 0.675\times \frac{39\,813}{33\,869} = 0.0506 \text{ in}$$

$$\delta_s = 0.0638\times 0.325\times \frac{39\,813}{3433} = 0.2405 \text{ in}$$

Rotation at end 1 is

$$\theta_1 = \frac{M_1}{k_{ec}} = \frac{41.5\times 12\,000}{391\times 3.12\times 10^6} = 4.08\times 10^{-4} \text{ rad}$$

and rotation at end 4 is

$$\theta_4 = \frac{M_2}{k_{ec}} = \frac{19.5\times 12\,000}{391\times 3.12\times 10^6} = 1.90\times 10^{-4} \text{ rad}$$

The deflection adjustment is

$$\delta'' = \frac{\theta L}{8} = -\frac{(4.08+1.90)\times 10^{-4}\times 240}{8} = -0.0179 \text{ in}$$

Therefore,

net $\delta_{cy} = 0.0506 - 0.0179 = 0.0327$ in, say 0.03 in

net $\delta_{sy} = 0.2405 - 0.0179 = 0.2226$ in, say 0.22 in

Total central deflection is

$$\Delta = \delta_{sx} + \delta_{cy} = \delta_{cy} + \delta_{cx}$$

$$\Delta_{E-W} = \delta_{sx} + \delta_{cy} = 0.34 + 0.03 = 0.37 \text{ in} \ (9.40 \text{ mm})$$

$$\Delta_{N-S} = \delta_{sy} + \delta_{cx} = 0.22 + 0.07 = 0.29 \text{ in} \ (7.36 \text{ mm})$$

Hence, average Δ at center of interior panel $= \frac{1}{2}(\Delta_{E-W}+\Delta_{N-S}) = 0.33$ in (8.38 mm)

Adjustment for cracked section
Using Branson's I_e (Eqn 5-2) and given

$$f_r = 7.5\sqrt{f'_c} = 410.8 \text{ psi} \ (2.83 \text{ MPa})$$

E–W:

y_t for the 240 in total frame (Fig. 5-3b) is 22.5 in (571.5 mm)

$I_{cr} = 45\ 333\ in^4$

Mid-span long-term $M_a = 3.20 \times 10^5$ lbf ft (433.9 kN m)

N–S:

y_t for the 300 in total frame (Fig. 5-3b) is 19.6 in (497.4 mm)

$I_{cr} = 27\ 000\ in^4$

Mid-span long-term $M_a = 2.58 \times 10^5$ lbf ft (349.8 kN m)

E–W

$$M_{cr} = \frac{f_r I_g}{y_t} = \frac{410.8 \times 66\ 845}{22.5 \times 12\ 000} = 1.016 \times 10^5 \text{ lbf ft (137.7 kN m)}$$

$$\frac{M_{cr}}{M_a} = \frac{1.016 \times 10^5}{3.20 \times 10^5} = 0.317 \quad \left(\frac{M_{cr}}{M_a}\right)^3 = 0.032 \quad I_g = 66\ 845\ in^4$$

$$I_e(\text{E–W}) = 0.032 \times 66\ 844 + (1 - 0.032)45\ 333 = 47\ 021\ in^4$$

N–S

$$M_{cr} = \frac{410.8 \times 39\ 813}{19.6 \times 12\ 000} = 9.271 \times 10^4 \text{ lbf ft (125.7 kN m)}$$

$$\frac{M_{cr}}{M_a} = \frac{9.271 \times 10^{-4}}{2.58 \times 10^{-5}} = 0.36 \quad \left(\frac{M_{cr}}{M_a}\right)^3 = 0.047 \quad I_g = 39\ 813\ in^4$$

$$I_e(\text{N–S}) = 0.047 \times 39\ 813 + (1 - 0.047) \times 27\ 000 = 27\ 602\ in^4$$

Deflection multiplier to adjust for cracked section is

$$\tfrac{1}{2}(\text{E–W}I_g/I_e + \text{N–S}I_g/I_e) = \tfrac{1}{2}(66\ 844/47\ 021 + 39\ 813/27\ 602) = 1.43$$

Therefore,

adjusted central deflection is 1.43 in $\times$ 0.33 in = 0.5 in

$$\frac{\Delta}{l} = \frac{0.5}{20 \text{ ft} \times 12 \text{ in}} = \frac{1}{480} < \frac{1}{360} \text{ normally allowed}$$

Therefore, deflection is acceptable.

5.5 Indirect method of deflection evaluation and control

Empirical guidelines for controlling deflections by limiting the ratio of member depth to its span are often used unless more accurate analyses, as given in Section 5.4, are necessary. Such control is necessary to prevent damage due to cracks appearing in walls, doors jamming, damage to plaster, ceiling and wall finishes, horizontal cracks above and below ceiling supports due to the excessive rotation of edge beams resulting from excessive deflection, and others. Hence, as an alternative to calculating deflection in every case, members can be proportioned in standard cases, and particularly in beams, so that their deflection could be within acceptable limits.

Different countries use different span/deflection ratios. Such ratios, usually from $h = L/16$ for a simply supported beam, to $h = L/28$ for a solid one-way slab, to $h = L/8$ for a cantilever beam, are generally somewhat conservative and the designer can use them as a preliminary guide.

More logical approaches for serviceability deflection determination would be to limit the span-to-depth ratio.

A typical span/deflection ratio, α_{max}, permitted by ACI [3, 4] for solid reinforced concrete beams and one-way slabs is given in Table 5-1 where L_c is the service live load included in load D_c, and D_c the service dead load including the sustained portion of live load $\alpha_{max} = L/\Delta$.

Deflection computations would have to be made in order to utilize Table 5-1 by using such expressions as Eqn 5-1 and 5-2 or other expressions specified by the codifying agency.

For two-way solid reinforced concrete slabs and plates, deflection control as a serviceability requirement is even more complicated. The guidelines given in the various codes differ more extensively than in the case of the beams. For example, the permissible deflection for a slab panel simply supported on all sides and

Table 5-1   Permissible span–deflection limitations (beams)

Type of member	Deflection to be considered	α_{max}
Flat roofs not supporting and not attached to non-structural elements likely to be damaged by large deflections	Immediate deflection due to live load, L_c	180
Floors not supporting and not attached to non-structural elements likely to be damaged by large deflections	Immediate deflection due to live load, L_c	360
Roof or floor construction supporting or attached to nonstructural elements likely to be damaged by large deflections	That part of the total deflection occurring after attachment of non-structural elements (sum of the long-time deflection due to all sustained loads, D_c, and the immediate deflection due to any additional live load, L_c	480
Roof or floor construction supporting or attached to nonstructural elements not likely to be damaged by large deflections		240

carrying partitions [77]), the minimum slab thickness, h, as specified by the different codes would be

France	$L/500$ or 4 in (100 mm)
Norway	$L/200$
Romania	$L/200$ to $L/750$
USA	$L/480$
USSR	$L/200$ to $L/400$

Unless rigorous computations are made, as presented in Section 5.4, which is advisable and often necessary, the designer has to follow an empirical preliminary process of thickness selection. A recommended guide is given in Table 5-2 for the preliminary selection of the total thickness of two-way slabs and plates supported on walls or beams based on ACI recommendations, and with a minimum allowable thickness of

(a) slabs or plates without beams or drop panels—5 in (127 mm);
(b) slabs with beams on all four edges and an average flexural stiffness ratio of edge beam to slab of 2.0 minimum $-3\frac{1}{2}$ in (88.9 mm).

In this table, taken from [78], the side ratio is

$$\beta = \frac{\text{clear long span}}{\text{clear short span}}$$

and

$$\beta_s = \frac{\text{total length of continuous edges}}{\text{total perimeter of slab panel}}$$

Table 5-2   Maximum shorter span/depth ratios (slabs on beams or walls)

Side ratio		Over 2	2.0	1.8	1.6	1.4	1.2	1.0
Type of slab	β_s	Span/depth ratio						
Simple support	0	20.0	23.2	24.0	25.0	28.0	31.9	37.3
Semi-continuous	$\frac{1}{2}$	24.0	27.8	28.8	30.0	31.7	34.2	39.6
Continuous	1	28.0	32.5	33.6	35.0	36.0	36.4	41.8

In cantilevers and panels with free edges, multiply the length normal to the free edge by 2.8 before entering the table.

Example 5-2: use of Table 5-2
For preliminary choice of slab thickness, take a panel 12×24 ft (3.66×7.32 m).

(a) Cantilever case of one long edge continuous, others free: Equivalent size $= 12$ ft $\times 2.8 = 33.60$ ft

$$\beta = \text{over } 2 \qquad \beta_s = 24/24 = 1.0$$

$$\text{Minimum } h = \frac{33.6 \times 12}{28.0} = 14.3 \text{ in say, } 14\frac{1}{2} \text{ in (368.3 mm)}$$

(b) One long edge free, one edge continuous; short edges simply supported:

$$\text{Equivalent size} = 12 \times 2.8 \times 24 \text{ ft}$$

$$= 33.60 \times 24 \text{ ft} (10.25 \times 7.32 \text{ m})$$

$$\beta = \frac{33.60}{24.0} = 1.40 \qquad \beta_s = \frac{24}{2 \times 33.6 + 24} = 0.26$$

Span/depth ratio (interpolated between 31.7 and 36.9) = 32.6

$$\text{Minimum } h = \frac{24 \times 12}{32.6} = 8.83 \text{ in, say 9 in (228.6 mm)}$$

(c) One long and one short edge free, other edges continuous:

$$\text{Equivalent size} = 12 \times 2.8 \times 24 \times 2.8 \text{ ft}$$

$$= 33.6 \times 67.2 (10.25 \times 20.50 \text{ m})$$

$$\beta = \frac{67.2}{33.6} = 2.0 \qquad \beta_s = \frac{12 + 24}{12 + 24} = 1.0$$

$$\text{Minimum } h = \frac{33.6 \times 12}{32.5} = 12.41 \text{ in, say } 12\tfrac{1}{2} \text{ in (315 mm)}$$

More refined analysis can then be used to determine the thickness needed for other considerations.

In the case of prestressed concrete structural elements, guidelines for approximate limits of span-to-depth ratios also differ in various practices. Generally, in a prestressed concrete element, a rule of thumb for estimating preliminary depth is to use about 70% of the depth of the conventional reinforced concrete member.

Accumulated experience reported by Lin and Burns [79] gives (Table 5-3) recommended limits for span-to-depth ratios as a general guide.

Table 5-3   Maximum span/depth ratios (prestressed concrete)

	Continuous spans		Simple spans	
	Roof	Floor	Roof	Floor
One-way solid slabs	52	48	48	44
Two-way solid slabs (supported on columns only)	48	44	44	40
Two-way waffle slabs (3 ft waffles)	40	36	36	32
Two-way waffle slabs (12 ft waffles)	36	32	32	28
One-way slabs with small cores	50	46	46	42
One-way slabs with large cores	48	44	44	40
Double tees and single tees (side by side)	40	36	36	32
Single tees (spaced 20 ft centers)	36	32	32	28

6   Serviceability as determined by crack control

The increasing demand on the national resources available, the advanced knowledge of the behavior of structural materials, and the quest for efficiency and economy in design have resulted in more refined codes of practice. Higher strength steels and concretes are being specified. More slender members and systems are being constructed.

Consequently, assurance of strength adequacy of structural elements is no longer sufficient for aesthetic or safe performance acceptance. Serviceability conditions have to be met involving cracking performance at normal load conditions, since concrete is weak in tension, thereby susceptible to cracking under adverse conditions of environment or high stress.

The magnitude of cracks and the extent of the cracked zones determine whether the structure can endure the corrosive or other adverse conditions continuously existing in its environment. Maximum utilization of the strength of the material leads to the acceptance of higher stress levels in the structure under normal conditions both in tension and compression. As a result, the cracking behavior is more critical today than in the past.

The majority of cracks are a result of the following actions to which concrete can be subjected [82]:

(a) Volumetric change caused by drying shrinkage, creep under sustained load, thermal stresses (including elevated temperatures) and chemical incompatibility of the concrete components.
(b) Direct stress due to applied loads, reactions, or continuity effects, reversible fatigue load, long-term deflection, camber in prestressed systems, or environmental effects, including differential movement in structural systems.
(c) Flexural stress due to bending.

While the net result of these three actions is the formulation of cracks, the mechanisms of their development cannot be considered identical. Volumetric change generates internal microcracking which may develop into full cracking, while direct (internal or external) stress or applied loads and reactions could either generate internal microcracking (as in the case of fatigue due to reversible load), or flexural microcracking leading to fully developed cracking.

6.1   Microcracking and drying shrinkage cracking

Microcracking can be mainly classified into two categories: (a) bond cracks at the aggregate—mortar interface, and (b) paste cracks within the mortar matrix. Interfacial bond cracks are caused by interfacial shear and tensile stresses due to early volumetric change without the presence of external load. Volume change caused by hydration and shrinkage could create tensile and bond stresses of sufficient magnitude to cause failure at the aggregate—mortar interface [80].

Shrinkage is a basic characteristic of hydraulic cement concrete. It is a result of loss of absorbed water and interlayer water from the calcium silicate hydrate gel formed during the hydration of cement. Thus, on exposure to dry environmental conditions, moisture diffuses towards the outer surface from the inner concrete layers in order to replace the moisture which is lost by evaporation of the surface layer's water.

Drying results in dimensional decrease. Since the magnitude of drying at the surface of the specimens is greater than that inside the concrete element, a differential change in dimensions between the outer surface and the inner layers leads to crack development as a result of developed tensile stress. Also, since the element as a whole body also shrinks, cracks develop resulting from the build-up of tensile stress in the element if it is restrained from shrinkage. Such restraint is usually the case, since concrete elements or members usually form part of a total interconnected system.

Shrinkage is a function of the drying time to which an element or a structure is subjected. Drying is a very slow process, which can take several years: it can take, for instance, 10 years under normal conditions to get rid of 75% of the moisture at 4 in (100 mm) below a concrete slab surface.

Shrinkage reinforcement properly designed in the required percentages and properly placed, together with the use of proper joints, would control to a large degree the magnitude and extent of shrinkage cracking. The minimum percentage recommended [3] is 0.2% of the concrete section for steels with yield strengths less than 60 000 psi (431.7 MPa) and 0.18% for steel with yield strengths higher than 60 000 psi or for welded wire fabric.

In addition, the shrinkage of a member which is part of a structural frame could impose additional forces, hence additional bending moments in the parts to which it is interconnected. The designer can provide for those stresses through the multipliers proposed in Section 5.2.

Also, depending on the characteristics of the aggregate, and the property of the resulting concrete and its age, the coefficient of shrinkage of concrete can vary between 0.0011 for zero humidity and 0.0005 for 85% humidity. On this basis, the axial forces caused by shrinkage or swelling can be accurately evaluated and added to the normal forces from external live load. In summary, by evaluating and allowing for the expected shrinkage forces in the total design, it is possible to control or prevent increased increased cracking with time.

6.2 Load-induced flexural cracking

External load results in direct and bending stresses causing flexural, bond, and diagonal tension cracks. Immediately after the tensile stress in the concrete exceeds its tensile strength, internal microcracks form. These develop into macrocracks that propagate to the external fiber zones of the element.

After the full development of the first crack in a reinforced concrete element, the stress in the concrete at the cracking zone is reduced to zero and is redistributed to the reinforcement. The distribution of ultimate bond stress, longutudinal tensile stress in the concrete and longitudinal tensile stress in the steel can be schematically represented in Fig. 6-1. When the ultimate bond stress in the concrete surrounding the reinforcement reaches the rupture strength of the concrete, cracks 1 and 2 open. The longitudinal tensile stress in the concrete at 1 and 2 is dynamically transferred to the steel at those two cleavages zones in the concrete,

Crack width is a primary function of the deformation of reinforcement between

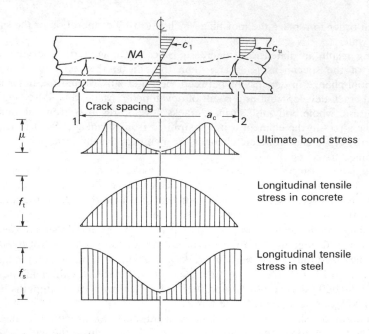

Figure 6-1   Schematic stress distributions between adjacent flexural cracks

the two adjacent cracks 1 and 2, assuming that the small concrete strain along the crack interval, a_c, is negligible. Hence, the crack width is a function of the crack spacing up to the level of stabilization of the crack spacing [81].

The major parameters affecting the development and characteristics of the cracks are: percentage of reinforcement, bond characteristics and size of bar, concrete cover, and concrete stretched area. On this basis, the following mathematical model may be proposed [81, 82] for the maximum crack width.

$$w = \alpha a_c^\beta \varepsilon_s^\gamma \tag{6-1}$$

where α, β and γ are non-linearity constants. The crack spacing, a_c, is a function of the factors enumerated previously and is inversely proportional to the bond strength and the active steel ratio, i.e. the steel percentage in terms of the concrete area in tension, and ε_s is the strain in the reinforcement induced by external load.

6.3   Crack control in reinforced concrete beams

The basic mathematical model presented in [81] and subsequent additional works by the present author and others [81–90] gave equations in varying format. Ferry-Borges [83, 84] led the European work which is now embodied in modified form in the CEB recommendations [2]. The 1972 report by the ACI Committee 224 on cracking [88] gives an extensive review of the cracking problem.

The CEB recommendations for crack control in beams [2] gives the mean crack width, w_m in terms of the mean crack spacing, s_{rm}, such that

$$w_m = \varepsilon_{sm} s_{rm} \tag{6-2a}$$

where ε_{sm} represents the average strain in the steel and is given by

$$\varepsilon_{sm} = \frac{f_s}{E_s}\left[1 - \kappa\left(\frac{f_{sr}}{f_s}\right)^2\right] \leq 0.4\frac{f_s}{E_s} \tag{6-2b}$$

and where

  $f_s =$ steel stress at the crack
  $f_{sr} =$ steel stress at the crack due to forces causing cracking at the tensile strength of the concrete
  $\kappa =$ bond coefficient, 1.0 for ribbed bars, reflecting influence of load repetitions and load duration

The mean cracking spacing is

$$s_{rm} = 2\left(c + \frac{s}{10}\right) + \kappa_2\kappa_3\frac{d_b}{Q_r} \tag{6-2c}$$

where

  $c =$ clear concrete cover
  $s =$ bar spacing, limited to $15d_b$
  $\kappa_2 = 0.4$ for ribbed bars.
  $\kappa_3 =$ depends on the shape of the stress diagram, 0.125 for bending
  $Q_r = A_s/A_t$
  $A_t =$ effective area in tension, depending on the arrangement of bars and type of external forces; it is limited by a line $c + 7d_b$ from the tension face for beams; in the case of slabs, not more than halfway to the neutral axis

A simplified formula can be derived for the mean crack width in beams with ribbed bars [88]:

$$w_m = 0.7\frac{f_s}{E_s}\left(3c + 0.05\frac{d_b}{Q_r}\right) \tag{6-3}$$

A characteristic value of the crack width, presumably equivalent to the probable maximum value, is given as $1.7w_m$.

The ACI recommendations [88] for crack control in beams and thick one-way slabs are based on statistical analysis of crack width data [86] from a large number of sources, resulting in the following expression:

$$w = k\beta f_s\sqrt[3]{d_c A}\times 10^{-3} \tag{6-4}$$

where

      f_s = steel stress at working load level (ksi) and = $0.60f_y$
      w = maximum crack width (in)
      d_c = thickness of cover from the tension fiber to the center of the bar closest
         thereto (in)
      k = 0.076
      f_y = yield strength of the reinforcement
      A = area of concrete symmetric with the reinforcing steel divided by the
         number of bars (in^2)
      β = ratio of the distance between the neutral axis and the tension face to the
         distance between the neutral axis and the centroid of the reinforcing
         steel ≈ 1.20 in beams

For beams reinforced with bundled bars, a modified term to account for the reduced bonded area of bars is given in [89].

6.4 Crack control in two-way slabs and plates

Flexural cracking behavior in concrete structural floors under two-way action is significantly different from that in one-way members. Crack control equations for beams underestimate the crack widths developed in two-way slabs and plates, and do not tell the designer how to space the reinforcement. Cracking in two-way slabs and plates is controlled primarily by the steel stress level and the spacing of the reinforcement in the two perpendicular directions. In addition, the clear concrete cover in two-way slabs and plates is nearly constant ($\frac{3}{4}$ in (19 mm) for interior exposure), whereas it is a major variable in the crack control equations for beams. The results from extensive tests on slabs and plates [81, 90–93] demonstrate this difference in behavior.

A fracture hypothesis developed by the present author in [90] is outlined herewith. Stress concentration develops initially at the points of intersection of the reinforcement in the reinforcing bars and at the welded joints of the wire mesh, namely, at grid nodal points A_1, B_1, A_2, and B_2 in Fig. 6-2. This stress concentration causes plastic deformation of the concrete at these locations as a result of the energy imposed by the external load per unit area of slab. The bond between the bar or wire and the concrete at these locations is destroyed and active cleavages start to generate fracture lines towards the paths of least resistance. Planes of discontinuity, which are paths of least resistance, are the interaction surfaces between the reinforcement grid lines and the surrounding concrete gel, namely, A_1B_1, A_1B_2, A_2B_2, and B_2B_1. The resulting fracture pattern is a total repetitive cracking grid, provided the spacing of the nodal points A_1, B_1, A_2 and B_2 is close enough to generate this preferred initial fracture mechanism of orthogonal cracks narrow in width.

If the spacing of the reinforcing grid intersections is too large, the magnitude of the stress concentration and the energy absorbed per unit grid is too low to generate cracks along the reinforcing wires or bars. As a result, the principal cracks follow diagonal yield-line cracking in the plain concrete field away from the reinforcing bars early in the loading history. These cracks are wide and few.

This hypothesis also leads to the conclusion that surface deformations of the

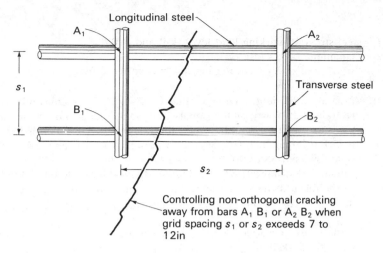

Figure 6-2   Grid unit in two-way action reinforcement

individual reinforcing elements have little effect in arresting the generation of the cracks or controlling their type or width in a two-way action slab or plate. In a similar manner, one may conclude that the scale effect on two-action cracking behavior is insignificant, since the cracking grid would be a reflection of the reinforcement grid if the preferred orthogonal narrow cracking widths develop. Therefore, to control cracking in two-way action floors, the major parameter to be considered is the reinforcement spacing in two perpendicular directions. Concrete cover has only a minor effect, since it is usually a small, constant value of 0.75 in (20 mm).

For a constant area of steel determined for bending in one direction, namely, for energy absorption per unit slab area, the smaller the spacing of the transverse bars or wires, the smaller should be the diameter of the longitudinal bars. The reason is that less energy has to be absorbed by the individual longitudinal bars. If one considers that the magnitude of fracture is determined by the energy imposed per specific volume of reinforcement acting on a finite element of the slab, a proper choice of the reinforcement grid size and bar size can control cracking into preferred orthogonal grids.

It must be emphasized that this hypothesis is important for serviceability and reasonable overload conditions. In relating orthogonal cracks to yield-line cracks, the failure of a slab ultimately follows the generally accepted rigid-plastic yield-line criteria.

As a result of this fracture hypothesis, the mathematical model in Eqn 6-1 and the statistical analysis of the data of 90 slabs tested to failure [90, 91], the following crack control equation emerged:

$$w = K\beta f_s \sqrt{\frac{d_{b_1} s_2}{Q_{t_1}}} \tag{6-5}$$

where the radical $G_1 = d_{b_1} s_2 / Q_{t_1}$ is termed the grid index, and can be transformed into

$$G_I = \left(\frac{s_1 s_2 d_c}{d_{b_1}} \frac{8}{\pi} \right)$$

and where

K = fracture coefficient, having a value $k = 2.8 \times 10^{-5}$ for uniformly loaded restrained two-way action square slabs and plates. For concentrated loads or reactions, or when the ratio of short to long span is less than 0.75 but larger than 0.5, a value of $k = 2.1 \times 10^{-5}$ is applicable. For a span aspect ratio of 0.5, $k = 1.6 \times 10^{-5}$

β = ratio of the distance from the neutral axis to the tensile face of the slab to the distance from the neutral axis to the centroid of the reinforcement grid (to simplify the calculations, though it varies between 1.20 and 1.35)

f_s = actual average service load stress level, or 40% of the design yield strength, f_y (ksi)

d_{b_1} = diameter of the reinforcement in direction 1 closest to the concrete outer fibers (in)

s_1 = spacing of the reinforcement in direction 1 (in)

s_2 = spacing of the reinforcement in perpendicular direction 2 (in)

1 = direction of the reinforcement closest to the outer concrete fibers; this is the direction for which crack control check is to be made

Table 6-1   Fracture coefficients for slabs and plates

Loading[a] type	Slab shape	Boundary condition[b]	Span ratio[c] S/L	Fracture coeff $10^5 K$
A	Square	4 edges r	1.0	2.1
A	Square	4 edges s	1.0	2.1
B	Rectangular	4 edges r	0.5	1.6
B	Rectangular	4 edges r	0.7	2.2
B	Rectangular	3 edges r 1 edge h	0.7	2.3
B	Rectangular	2 edges r 2 edges h	0.7	2.7
B	Square	4 edges r	1.0	2.8
B	Square	3 edges r 1 edge h	1.0	2.9
B	Square	2 edges r 2 edges h	1.0	4.2

[a] Loading type: A = concentrated; B = uniformly distributed.
[b] Boundary condition: r = restrained; s = simply supported; h = hinged.
[c] Span ratio: S = clear short span; L = clear long span.

Q_{t_1} = active steel ratio

$$= \frac{\text{area of steel } A_s \text{ per ft width}}{12(d_{b_1} + 2C_1)}$$

where C_1 is clear concrete cover measured from the tensile face of concrete to the nearest edge of the reinforcing bar in direction 1

w = crack width at face of concrete caused by flexural load (in)

Subscripts 1 and 2 pertain to the directions of reinforcement. Detailed values of the fraction coefficients for various boundary conditions are given in Table 6-1 and a graphical solution of Eqn 6-5 is given in Fig. 6-3 for $f_y = 60\,000$ psi (413.7 MPa) and $f_s = 40\%$. $f_y = 24\,000$ psi (165.5 MPa) for rapid determination of the reinforcement size and spacing method for crack control.

The grid index, G_I, specifies the size and spacing of the bars in the two perpendicular directions of any concrete floor system, and w_{max} is the maximum allowable crack width.

Charts for use of Eqn 6-5 are available in [94, 95].

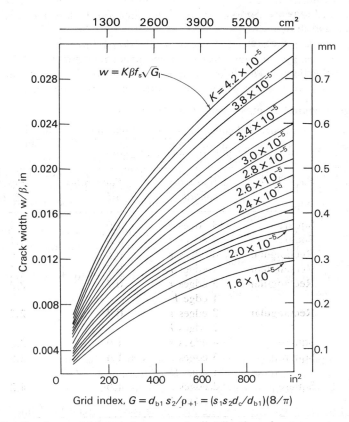

Figure 6-3 Crack control reinforcement distribution in two-way action slabs and plates for all exposure conditions: $f_y = 60\,000$ psi, $f_s = 24\,000$ psi ($= 0.40 f_y$)

6.5 Permissible crack width for serviceability

Different codes have varied limitations on the permissible crack width as a serviceability requirement. Based on a study [82] of the various conditions and experiences and the environmental conditions expected, Table 6-2 of the ACI Committee 224 on cracking gives a reasonable guide for permissible crack widths [88].

Table 6-2 Permissible crack widths

Exposure condition	Crack width	
	in	mm
Dry air or protective membrane	0.016	0.41
Humidity, moist air, soil	0.012	0.30
Deicing chemicals	0.007	0.18
Seawater and seawater spray;		
wetting and drying	0.006	0.15
Water retaining structures[a]	0.004	0.10

[a] Excluding non-pressure pipes.

6.6 Crack control in prestressed concrete beams

Crack control in prestressed concrete beams has become important because of the increased use of partial prestressing and also because of frequent overloading. Primary cracks form in the region of maximum bending moment when the external load reaches the cracking load. As loading is increased, additional cracks will form and the number of cracks will be stabilized when the stress in the concrete no longer exceeds its tensile strength at further locations regardless of load increase. This condition is important as it essentially produces the absolute minimum crack spacing which can occur at high steel stresses, to be termed the stabilized maximum crack spacing. Hence, the stabilized mean crack spacing, a_{cs}, is deduced as the mean value of the two extremes.

The total tensile force, T, transferred from the steel to the concrete over the stabilized mean crack spacing can be defined as

$$T = f a_{cs} u_m \sum O \qquad (6\text{-}6a)$$

where

f = a factor reflecting the distribution of bond stresses
u_m = maximum bond stress which is a function of $\sqrt{f_c'}$
$\sum O$ = sum of reinforcing elements' circumferences

The resistance, R, of the concrete area in tension, A_t, can be defined as

$$R = A_t f_t' \qquad (6\text{-}6b)$$

By equating Eqns 6-6a and 6-6b, the following expression for a_{cs} is obtained for the mean stabilized crack spacing [98, 99]:

$$a_{cs} = 1.20 A_t \sum O \qquad (6\text{-}7)$$

If Δf_s is the net stress in the prestressed tendon or the magnitude of the tensile stress in the normal steel at any crack width load level in which the decompression load (decompression here means $f_c = 0$ at the level of the reinforcing steel) is taken as the reference point, then for the prestressed tendon

$$\Delta f_s = f_{nt} - f_d \qquad (6\text{-}8a)$$

where f_{nt} is the stress in the prestressing steel at any load level beyond the decompression load, and f_d the stress in the prestressing steel corresponding to the decompression load. From the basic mathematical model in Eqn 6-1,

$$w_{max} = k' a_{cs} (\Delta f_s)^\alpha \qquad (6\text{-}8b)$$

where k' and α are constants.

6.6.1 Pretensioned beams

Equations 6-8a and 6-8b lead to an expression for the maximum crack width at the reinforcement level, as detailed in [99, 100]

$$w_{max} = 1.4 \times 10^{-5} a_{cs} (\Delta f_s)^{1.31} \qquad (6\text{-}9)$$

or the simpler term

$$w_{max} = 5.85 \times 10^{-5} \frac{A_t}{\sum O} (\Delta f_s) \qquad (6\text{-}10a)$$

and the maximum crack width (in) at the tensile face of the concrete is

$$w'_{max} = 5.85 \times 10^{-5} R_i \frac{A_t}{\sum O} (\Delta f_s) \qquad (6\text{-}10b)$$

where R_i is the distance ratio as defined in Notation, p. 12-2.

6.6.2 Post-tensioned beams

Similarly, for post-tensioned non-bonded beams [101]

$$w_{max} = 2.35 \times 10^{-5} a_{cs} (\Delta f_s)^{1.15} \qquad (6\text{-}11)$$

or the simpler term

$$w_{max} = 6.51 \times 10^{-5} \frac{A_t}{\sum O} (\Delta f_s) \qquad (6\text{-}12a)$$

The expression for w_{max} at the tensile face of the concrete for the post-tensioned beams becomes

$$w_{max} = 6.51 \times 10^{-5} R_i \frac{A_t}{\sum O} (\Delta f_s) \qquad (6\text{-}12b)$$

Detailed verification of the validity and applicability of Eqns 6-10 and 6-12 and comparisons with approaches in other codes are given in [102, 103].

It is recommended to locate the non-prestressed steel below the prestressed tendons. This is due to the fact that mild steel has larger diameters than the prestressed reinforcement, hence larger bond area with the surrounding concrete. Also, by placing the mild steel close to the tensile concrete face, cracks will be more evenly distributed, hence crack spacing and, consequently, width will be smaller.

The effect of the spacing of the stirrups on the crack spacing is not pronounced. The final crack spacing pattern does not necessarily follow the vertical shear reinforcement. Once the allowable crack width is established for the prevailing environmental conditions, the proper percentage of non-prestressed reinforcement can be determined to ensure serviceability behavior for the expressions presented. For long-term effects, modifiers to allow for the increase in crack width, such as given in Section 5.2, have to be used.

6.7 Crack control in prestressed tank walls

Because of the plate and shell action of large circular tanks, two-way action affects their cracking behavior. Vessay and Preston [104] applied the criteria for crack control in two-way action in slabs and plates modifying the term in Eqn 6-5 from [88, 90] for use in prestressed concrete tanks and reservoirs. The maximum crack width in terms of the tensile surface strain, ε_{ct}, is

$$w = 4.1 \times 10^{-6} \varepsilon_{ct} E_s / G_1 \qquad\qquad (6\text{-}13)$$

where G_1 is the grid index [90], $\varepsilon_{ct} = P_{ct} f_{si} / E_s$, f_{si} is the tendon pre-release stress, and P_{ct} a dimensionless parameter.

Here also, long-term effects have to be considered for controlling cracking in the walls and the slabs.

7 Ductility of structural elements

Reinforced concrete, unless under-reinforced, behaves in a brittle manner at high loading levels. In order for a concrete structural element to undergo large deformations and rotations at loads close to the limit state at failure, limitations on the percentage of reinforcement and the provision of special distribution and geometry of the reinforcing elements become necessary.

Most structural systems are continuous. Consequently, the redistribution of stresses caused by the bending moments from regions of higher stress at the supports to the less stressed regions along the span is possible. The extent of the redistribution depends upon the relative stiffness and ductility of the sections in the two regions. If plastic rotation continues at the support of a continuous structural element, for instance, while the collapse moment level is maintained, additional moments are carried at other sections along the span, leading to a higher collapse load of the total structure. The increase in the plastic rotation capability, hence higher ductile capacity, can thus not only lead to more economic systems, but also considerably increases the reserve safety margin in structures subjected to earthquake and dynamic loadings.

7.1 Rotational capacity and ductility

Considerable work has been accomplished in the past three decades for increasing the rotational capacity of concrete plastic hinges [105–115]. Most of this work, particularly [111, 112, 115], centered on research to increase to the limit the degree of confinement of the critical sections, leading to a large increase in the rotational capacity of the plastic hinges, hence a considerable increase in ductility. Figure 7-1a shows the load–deflection, load–strain or moment–rotation schematic relationship, whereby an increase in confinement can immeasurably increase the ductility (ε_c increased from 0.004 to 0.140 from tests in [111]). The moment–rotation relationship can be idealized either as a simplified bilinear diagram or a trilinear diagram (Fig. 7-1a) as recommended by the CEB Commission on Hyperstatic Structures [2]. State I in the trilinear relationship denotes the un-cracked linear elastic condition; state II denotes the cracked condition and state III denotes the plastic condition with full rotation of the developed hinges.

If R is the radius of curvature of the section in Fig. 7-1b, ϕ is the angle of curvature, namely, the rotation per unit length of the structural element, and dx is

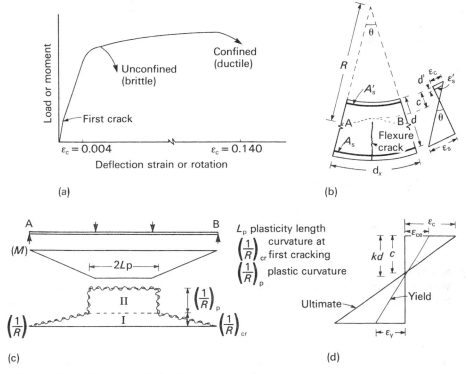

Figure 7-1   Elastic and plastic curvature in concrete sections
(a) Load–deflection or moment–rotation relationship
(b) Curvature of a cracked element
(c) Variation of curvature along a beam
(d) Change in neutral-axis position

an infinitesimal length of element along the span, then

$$\frac{dx}{R} = \frac{\varepsilon_c \, dx}{c} = \frac{\varepsilon_s \, dx}{d-c} \qquad (7\text{-}1a)$$

or curvature

$$\phi = \frac{1}{R} = \frac{M}{EI} = \left(\frac{\varepsilon_c + \varepsilon_s}{d}\right) \qquad (7\text{-}1b)$$

Figure 7-1c gives an idealized variation of curvature along the span of a beam. The rotation is given by

$$\theta_{AB} = \int_A^B \phi \, dx = \int_A^B \left(\frac{M}{EI}\right) dx \qquad (7\text{-}1c)$$

The plasticity length, L_p, defines the plastic hinging region where area II under the curvature diagram over the length $2L_p$ gives the plastic rotation of that hinge, hence the measure of ductility of that section.

Ductility is the ratio of the ultimate deformation to the deformation at first yield. Hence, the ductility factor can be defined as the ratio between the curvature at failure to the curvature at yield of the reinforcement, or the ratio between the ultimate rotation and the yield rotation. As a practical matter in design, the stress–strain diagram of concrete can be assumed to be linear up to 70% of the concrete cylinder strength, f_c'. Up to that limit, it is therefore possible to treat the section as elastic and determine the value of curvature ϕ_y at yield as detailed in [23], where

$$\phi_y = \frac{\varepsilon_s}{d(1-kd)} = \frac{f_y/E_s}{d(1-kd)} \qquad (7\text{-}2a)$$

The ultimate curvature

$$\phi_u = \varepsilon_c/c \qquad (7\text{-}2b)$$

Hence, the ductility factor

$$f = \frac{\phi_u}{\phi_y} = \frac{\varepsilon_c(d-c)}{(f_y/E_s)(a/\beta_1)} \qquad (7\text{-}2c)$$

where $c = a/\beta_1$, assuming in Fig. 7-1d that the change in the neutral axis position from kd at yield to c is negligible. From Eqn 7-2c, it can be recognized that increases in area A_s of the tension steel decreases ductility, since the depth c increases. An increase in the compression steel A_s' increases ductility as the depth c decreases, an increase in yield strength, f_y, decreases ductility, and an increase in concrete cylinder f_c' increases ductility.

From [23], if the modular ratio $n = E_s/E_c$, the ductility factor is

$$f = \frac{\phi_u}{\phi_y} = \frac{0.85\beta_1 E_s \varepsilon_c f_c'}{f_y^2(\rho-\rho')} \left\{1+(\rho+\rho')n-(\rho-\rho')^2 n^2 + 2\left(\rho+\frac{\rho'd''}{d}\right)n^{\frac{1}{2}}\right\} \qquad (7\text{-}3)$$

A ductility factor, f, can exceed a value of 3.0 to 4.0 depending on how low a steel percentage, ρ, is used.

7.2   Ultimate plastic rotation

From basic Eqn 7-1c, using the effective moment of inertia I_e [3, 66], the plastic rotation portion θ_p (radians) would be

$$\theta_p = \int_A^B \frac{M_p}{E_c I_e}\,dx = (\phi_u - \phi_y)L_p \tag{7-4}$$

where the total rotation, θ_{AB}, in Fig. 7-1 is

$$\theta_{AB} = \theta_e + \theta_p$$

and the elastic rotation, θ_e, is

$$\theta_e = \phi_y(\tfrac{1}{2}\,\text{span } L)$$

The basic equation for plastic rotation would thus become

$$\theta_p = \left(\frac{\varepsilon_c}{c} - \frac{\varepsilon_{ce}}{kd}\right)L_p \tag{7-5}$$

where the plasticity length, L_p, of the plastic hinge has to be evaluated in accordance with the applicable code, and the concrete strains ε_{ce} at yield and ε_c at failure (Fig. 7-1d) are calculated in the usual manner.

Several expressions are available for the plasticity length L_p, namely, the equivalent plastic hinge length on each side of the center of the critical section. The various empirical expressions of [106, 113, 114] give L_p values that can differ by about 20%. A reasonable expression [114] which is simple to apply for sufficiently increasing the ductility of concrete elements gives the plasticity length, L_p, on one side of the critical section as

$$L_p = 0.5d + 0.05z \tag{7-6}$$

and a maximum concrete strain ε_c as a function of the confining transverse steel

$$\varepsilon_c = 0.003 + 0.02b/z + 0.2\rho_s \tag{7-7}$$

where

$z =$ distance between the critical section and the point of contraflexure

$\rho_s = \dfrac{\text{volume of confining steel including compression steel}}{\text{volume of confined concrete core}}$

$b =$ width of beam section

$d =$ effective depth section

The term $0.2\rho_s$ in Eqn 7-7 can be substituted by $(\rho_s/f_y/14.5)^2$, as in [3, 4], as a more accurate contribution of ρ_s, where $(\rho_s f_y/14.5)^2 \geq 0.01$.

With the use of the confining binders, the resulting value of the ductility factor, $f = \phi_u/\phi_y$, becomes considerably higher than 3.0–5.0, which is expected when the concrete is confined.

The permissible local plastic rotation should be less than the value obtained from Eqn 7-5. The larger the difference between what Eqn 7-7 provides within practicable economic limits and the permissible limiting value in Fig. 7-2, the larger is the reserve ductility. Figure 7-2b is recommended by CEB (Ref. 2)

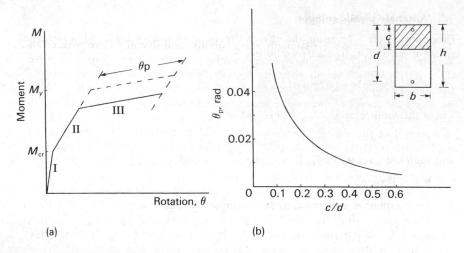

Figure 7-2   Plastic rotation in the idealized trilinear M–θ relationship
(a) Idealized trilinear diagram
(b) Permissible local plastic rotation disregarding confinement

assuming that the plastic rotations are to be concentrated at the critical sections. This diagram proposed by the CEB Commission on Hyperstatic Structures relates the angle of plastic rotation θ_p to the c/d ratio (where c is the depth of the compression zone) and does not include the favorable influence of the confining transverse binders.

Since the angle of rotation is a function of the concrete strain ε_c, the values of ε_c used in the plastic design of ductile structures have to be within reasonable limits in order to prevent discontinuity and premature loss of structural stability due to localized premature rupture of a hinge.

For ductile frames, particularly those subjected to seismic loading, practice according to the ACI recommendations [3] limits the steel percentage, $\rho = A_s/bd$, of the main longitudinal bars to 50% of the balanced percentage ρ_b as an upper limit. Such limitation is expected to produce a balanced strain condition $\varepsilon_c = 0.003$ in the extreme compression fibers of the *unconfined* concrete section when the steel tension bars reach the yield state f_y.

The limit on the allowable concrete strain ε_c differs in different practices and codes. According to the ACI recommendations, a reasonable limiting value for maximum permissible concrete strain, ε_c, at ultimate for non-confined concrete is $\varepsilon_c = 0.003$. The value recommended by CEB for rectangular sections without confinement ranges from $\varepsilon_c = 0.003$ for concrete having $f'_c = 7250$ psi (50 MPa) to $\varepsilon_c = 0.0036$ for concrete having $f'_c = 3000$ psi (20 MPa) as mean values in rectangular sections.

For concrete sections confined with transverse closed ties or spirals designed in accordance with Eqns 7-4 and 7-5, a reasonable limiting value for the permissible ultimate concrete strain, ε_c, would be 0.010.

8   Synthesis of design

The preceding sections on strength, serviceability, and ductility have covered the various phases that a design (analysis) undergoes. A trial and adjustment procedure is the initial step in choosing sections for flexural analysis to carry the positive moment along the span and the negative moment at the support, including the choice of the required reinforcement and its anchorage and development length.

A check of the shear capacity is made, and of the torsional capacity where applicable, to determine whether the section size determined in bending is sufficient to carry the external shearing and/or torsional forces. If it is, the stirrups are designed; if not, the section is enlarged.

It is then determined whether the sections chosen are adequately geometrically proportioned to control cracking and deflection to within permissible levels. Such a check would ensure long-term acceptable serviceability behavior, which is as important as the strength design itself. It is important to apply engineering judgement in choosing the values of the magnifying factors that affect the long-term, time-dependent cracking and deflection performance, namely, shrinkage, temperature, environment, and creep.

If the structural system is indeterminate and is to be designed for full redistribution of moments, ductility of the joints becomes most important. Increases in ductility can thus be produced through design of confining transverse ties or spirals at the critical zones, such as the supports in beams and in column ends. Such provisions for increases in ductility are of utmost importance in systems susceptible to seismic, dynamic, or cyclic loadings, and also whenever the plastic limit theory is used.

Equally important is the attention needed for reinforcing details. Reinforcement geometry and distribution determine whether the design can be appropriately implemented.

In summary, a synthesis of all the factors presented in this chapter, and attention to optimum combinations of sizes of the concrete sections and the percentage, distribution, and detailing of the reinforcement in those elements consequently result in the desired synthesized safe and economic design of the total system.

References

1. ASCE-ACI, Report of ASCE–ACI joint committee on ultimate strength design, *Proc. Struct. Div., Am. Soc. Civ. Eng.*, Vol. 81, No. 809, Oct. 1955, 68 pp.
2. CEB–FIP, *Model code for concrete structures*, Vols 1 and 2, 3rd Edn, Paris, 1978, 348 pp.
3. ACI Committee 318, *Building code requirements for reinforced concrete*, ACI Standard 318–77, American Concrete Institute, Detroit, 1977 and 1983.
4. ACI Committee 318C, *Commentary on building code requirements for reinforced concrete*, ACI 318 C, American Concrete Institute, Detroit, 1977 and 1983.

5. Whitney, C. S., Plastic theory of reinforced concrete design, *Proc. Am. Soc. Civ. Eng.*, Dec. 1940; and *Trans. Am. Soc. Civ. Eng.*, Vol. 107, 1942, pp. 251–326.
6. Mattock, A. H., Kriz, L. B. and Hognestad, E., Rectangular stress distribution in ultimate strength design, *J. Am. Concr. Inst., Proc.*, Vol. 57, No. 2, Feb. 1961, pp. 875–928.
7. Hognestad, E., *A study of combined bending and axial load in reinforced concrete members*, University of Illinois, Bulletin No. 399, Nov. 1951, 128 pp.
8. Huang, T., On the formula for spiral reinforcement, *J. Am. Concr. Inst., Proc.*, Vol. 61, No. 3 March 1964, pp. 351–353.
9. Wang, C. K. and Salmon, C. G., *Reinforced concrete design*, 3rd Edn, Harper Row, New York, 918 pp.
10. CEB–FIP, *Trial and comparison calculations based on the CEB–FIP model code for concrete structures*, Bulletin d'Information No. 129, Comité Euro-International du Béton, Paris, Oct. 1978, 406 pp.
11. American Institute of Steel Construction; *Specifications for the design, fabrication and erection of structural steel for building* and *Commentary on the specifications*, New York, 1969 and 1978, pp. 166.
12. Johnston, B G. (ed), *Structural Stability Research Council Guide to stability design criteria for metal structures*, Wiley, New York, 1966, 217 pp.
13. MacGregor, J. G. and Breen, J. E., Design of slender concrete columns, *J. Am. Concr. Inst., Proc.*, Vol. 67, No. 1, Jan. 1970, pp. 6–28.
14. Timoshenko, S. P. and Gere, J. M., *Theory of elastic stability*, 2nd edn, McGraw-Hill, New York, 1961, 541 pp.
15. Furlong, R. W. and Ferguson, P. M., Test of frames with columns in simple curvature, *Symposium on reinforced concrete*, American Concrete Institute, SP-13, 1966, pp. 55–73.
16. Massonnet, C., Stability considerations in the design of steel columns, *J. Struct. Div. Am. Soc. Civ. Eng.*, Vol. 85, No. ST7, Sept. 1959, pp. 75–111.
17. American Concrete Institute, *Design handbook in accordance with the strength design method*, Vol. 2: Columns, S-17A (78), 1978, 191 pp.
18. Canadian Portland Cement Association, *Metric design handbook for reinforced concrete elements*, Ottawa, 1980, 328 pp.
19. Broms, B. and Viest, I. M., Long reinforced concrete columns, *Trans. Am. Soc. Civ. Eng.*, Jan. 1958, pp. 309–400.
20. Bresler, B., Design criteria for reinforced concrete columns under axial load and biaxial bending, *J. Am. Concr. Inst., Proc.*, Vol. 57, No. 3, Nov. 1960, pp. 481–490.
21. Furlong, R. W., Ultimate Strength of Square Columns Under Biaxially Eccentric Load, *J. Am. Concr. Inst., Proc.*, Vol. 57, No. 9, March 1961, pp. 1129–1140.
22. Parme, A. L., Nieves, J. M. and Gouens, A., Capacity of reinforced rectangular columns subjected to biaxial bending, *J. Am. Concr. Inst., Proc.*, Vol. 63, No. 9, Sept. 1966, pp. 911–921.
23. Park R. and Paulay, T., *Reinforced concrete structures*, Wiley, New York, 1975, 768 pp.
24. Moody, K. G., Viest, I. M., Elstner, R. C. and Hognestad, E., Shear

strength of reinforced concrete beams, *Bull. Reinforced Concr. Res. Council Eng. Foundation*, No. 6, 1955, ACI papers No. 51–15, pp. 317–434.

25. ACI–ASCE Committee 426, The shear strength of reinforced concrete members, *J. Struc. Div. Am. Soc. Civ. Eng.*, Vol. 99, No. ST6, 1973, pp. 1091–1187.
26. CEB–FIP, *Shear and torsion, discussion of model code*, Bulletin d'Information No. 126, Comité Euro-International du Béton, Paris, June 1978, 378 pp.
27. Warner, R. F., Rangan, B. V. and Hall, A. S., *Reinforced concrete*, Pitman, 1976, 475 pp.
28. Timoshenko, S. and Goodier, J. N., *Theory of elasticity*, McGraw-Hill, New York, 1970, 567 pp.
29. Nadai, A., *Plasticity: a mechanics of the plastic state of matter*, McGraw-Hill, New York, 1931, 349 pp.
30. Lessig, N. N., Determination of large carrying capacity of reinforced concrete elements with rectangular cross-section subjected to flexure with torsion, *Zhelezonbetona*, 1959, pp. 5–28.
31. Collins, M. P., Walsh, P. F., Archer, F. E. and Hall, A. S., Ultimate strength of reinforced concrete beams subjected to combined torsion and bending, *Torsion of Structural Concrete*, American Concrete Institute, SP-18, 1968, pp. 379–402.
32. Below, K. D., Rangan, B. V. and Hall, A. S., Theory for Concrete Beams in Torsion and Bending, *J. Struct. Div.*, *Am. Soc. Civ. Eng.*, Vol. 101, No. ST8, Aug. 1975, pp. 1645–1660.
33. Lampert, P., *Torsion und Biegung von Stahlbetonbalken*, Institut für Baustatik, Zurich, Bericht No. 26, Jan 1970, 189 pp.
34. Hsu, T. C., Ultimate torque of reinforced concrete members, *J. Struct. Div. Am. Soc. Civ. Eng.*, Vol. 94, No. ST2, Feb. 1968, pp. 485–510.
35. Thurliman, B., Torsional strength of reinforced and prestressed concrete beams—CEB approach, *Concrete design—US and European practices*, American Concrete Institute, 1979, pp. 117–143.
36. Collins, M. P. and Mitchell, D., Shear and torsion design of prestressed and non-prestressed concrete beams, *J. Prestressed Concr. Inst.*, Sept.–Oct. 1980, pp. 32–100.
37. Ersoy, U. and Ferguson, P. M., Behavior and strength of concrete L-beams under combined torsion and shear, *J. Am. Concr. Inst., Proc.*, Vol. 64, No. 12, Dec. 1967, pp. 797–801.
38. Lampert, P. and Collins, M. P., Torsion, bending and confusion—an attempt to establish the facts, *J. Am. Concr. Inst., Proc.*, Vol. 69, No. 8, Aug. 1972, pp. 500–504.
39. Zia, P., What do we know about torsion in concrete members? *J. Struct. Div. Am. Soc. Civ. Eng.*, Vol. 96, No. ST6, June 1970, pp. 1185–1199.
40. Kriz, L. B. and Raths, C. H., Connections in precast concrete structures—strength of corbels, *J. Prestressed Concr. Inst.*, Vol. 10, No. 1, Feb. 1965, pp. 14–47.
41. Mattock, A. H., Design proposals for reinforced concrete corbels, *J. Prestressed Concr. Inst.*, Vol. 21, No. 3, May–June 1976, pp. 2–26.

42. CEB–FIP, *Concrete design—US and European practices*, Joint ACI–CEB symposium, Bulletin d'Information No. 113, Comité Euro-International du Béton, Paris, Feb. 1979, 345 pp.
43. Johansen, K. W., *Yield line theory*, Cement and Concrete Association, London, 1962, 181 pp.
44. Ockleston, A. J., *Loading tests on reinforced concrete slabs spanning in two directions*, Portland Cement Institute, Johannesburg, Paper No. 6, Oct. 1958, 26 pp.
45. Mansfield, E. H., Studies in collapse analysis of rigid-plastic plates with a square yield diagram, *Proc. R. Soc.*, Vol. 241, Aug. 1957, pp. 311–338.
46. Rzhanitsyn, A. R., *The shape at collapse of elastic plastic plates simply supported along the edges*, Translation of Chapter 8, from Rzhanitsyn's *Structural analysis taking account of plastic properties*, Moscow, 1954.
47. Powell, D. S., *Ultimate strength of concrete panels subjected to uniformly distributed loads*, MSc Thesis, University of Cambridge, 1956.
48. Wood, R. H., *Plastic and elastic design of slabs and plates*, Thames and Hudson, London, 1961, pp. 225–261.
49. Sawczuk, A., Membrane action in flexure of rectangular plates with re-strained edges, *Flexural mechanics of reinforced concrete*, American Concrete Institute, SP-12, 1964, pp. 347–358.
50. Sawczuk, A. and Winnicki, L., Plastic behavior of simply supported plates at moderately large deflections, *Int. J. Solids & Struct.*, Vol. 1, 1965, pp. 97–111.
51. Gamble, W. L., Sozen, M. A. and Siess, C. P., *An experimental study of a reinforced concrete two-way floor slab*, Civil Engineering Studies, Structural Res. Ser. No. 211, University of Illinois, 1961, 304 pp.
52. Park, R., Ultimate strength of rectangular concrete slabs under short-term uniform loading with edges restrained against lateral movement, *Proc. Inst. Civ. Eng.*, Vol. 28, June 1964, pp. 125–150.
53. Park, R., The ultimate strength and long-term behavior of uniformly loaded, two-way concrete slabs with partial lateral restraint at all edges, *Mag. Concr. Res.*, Vol. 16, No. 48, Sept. 1964, pp. 139–152.
54. Park, R., Tensile membrane behavior of uniformly loaded rectangular reinforced concrete slabs with fully restrained edges, *Mag. Concr. Res.*, Vol. 16, No. 46, March 1964, pp. 39–44.
55. Corley, G. W. and Jirsa, J. O., Equivalent frame analysis for slab design, *J. Am. Concr. Inst.*, *Proc.*, Vol. 67, No. 11, Nov. 1970, pp. 875–884.
56. Jones, L. L. and Wood, R. H., *Yield line analysis of slabs*, Thames and Hudson, London, 405 pp.
57. Drucker, D. C., Prager, W. and Greenberg, H. J. Extended limit design theorems for continuous media, *Q. Appl. Math.*, Vol. 9, 1952, pp. 381–384.
58. Wood, R. H., How slab design has developed in the past and what the indications are for future development, *Cracking, deflection and ultimate load of concrete slab systems*, American Concrete Institute, SP-30, 1972, pp. 203–221.
59. Igbal, M. and Derecho, T. A., *Design criteria for deflection capacity of conventionally reinforced concrete slabs*, Portland Cement Association, Skokie, Aug. 1979, 145 pp.

60. Hung, T. Y. and Nawy, E. G., Limit strength and serviceability factors in uniformly loaded, isotropically reinforced two-way slabs, *Cracking, deflection and ultimate load of concrete slab systems*, American Concrete Institute, SP-30, 1972, pp. 301–324.

61. Hillerborg, A., A plastic theory for the design of reinforced concrete slabs, *6th Congr. Ass. Bridge and Struct. Eng.*, Stockholm, 1960, 26 pp.

62. Hawkins, N. M. and Corley, W. G., Transfer of unbalanced moment and shear from plates to columns, *Cracking, deflection and ultimate load of concrete slab systems*, American Concrete Institute, SP-30, 1972, pp. 147–176.

63. Hawkins, N. M., Fallsen, H. B. and Hinojosa, R. C., Influence of column rectangularity on the behavior of flat plate structures, *Cracking, deflection and ultimate load of concrete slab systems*, American Concrete Institute, SP-30, 1972, pp. 127–146.

64. Branson, D. E., *Instantaneous and time dependent deflections of simple and continuous reinforced concrete beams*, Alabama Highway Research Report, No. 7, 1965, 78 pp.

65. Branson, D. E., Design procedures for computing deflections, *J. Am. Concr. Inst., Proc.*, Vol. 65, No. 9, Sept. 1968, pp. 730–742.

66. Branson, D. E., *Deformation of concrete structures*, McGraw-Hill, New York, 1977, 546 pp.

67. ACI Committee 435, *Revised ACI code and commentary*, American Concrete Institute, Nov. 1975 (Committee internal report).

68. Nawy, E. G. and Potyondy, J. G., Moment rotation, cracking and deflection of spirally bound pretensioned prestressed beams, *Eng. Res. Bull. Rutgers Univ.*, No. 51, 1970, 124 pp.

69. Nawy, E. G., and Potyondy, G. J., Flexural cracking behavior of pretensioned prestressed I and T beams, *J. Am. Concr. Inst., Proc.*, Vol. 68, No. 5, May 1971, pp. 355–360; Discussion and closure, Nov. 1971, pp. 873–877.

70. Nawy, E. G. and Huang, P. T., Crack and deflection control of pretensioned, prestressed beams, *J. Prestressed Concr. Inst.*, Vol. 22, No. 3, May–June 1977, pp. 30–47.

71. Timoshenko, S. and Woinowsky-Kreiger, S., *Theory of plates and shells*, McGraw-Hill, New York, 1959, 492 pp.

72. ACI Committee 435, State of the art report on deflection of two-way reinforced concrete floor systems, *Deflection of concrete structures*, American Concrete Institute, SP-43, 1974, pp. 55–81.

73. Nilson, A. H. and Walters, D. B., Deflection of two-way floor systems by the equivalent frame method, *J. Am. Concr. Inst., Proc.*, Vol. 72, No. 5, May 1975, pp. 210–218.

74. Nawy, E. G., Deflection control in two-way slab and plate systems, *Proc. joint conf. PCA–ACI–ASCE*, University of Miami, Jan. 1975, 30 pp.

75. Nawy, E. G. and Chakrabarti, P., Deflection of prestressed concrete flat plates, *J. Prestressed Concr. Inst.*, Vol. 21, No. 2, March–April 1976, pp. 86–102.

76. Nawy, E. G., Crack and deflection control in two-way slab and plate systems, *Proc. conf. Serviceability Concr. Struct.* (Control del agrietamiento y

las deflexiones de estructuras de concreto), Instituto Mexicano del Cemento y del Concreto, Mexico City, April, 1976.

77. Clarke, C. V., Neville, A. M. and Houghton-Evans, W., Deflection problems and treatment in various countries, *Deflection of concrete structures*, American Concrete Institute, SP-43, 1974, pp. 129–178.

78. Abolitz, A. L., Preliminary selection of thickness of two-way slabs supported by beams or walls, *J. Am. Concr. Inst., Proc.*, Nov. 1972, Vol. 69, No. 11, pp. 698–699.

79. Lin, T. Y. and Burns, N. H., *Design of prestressed concrete structures*, 3rd Edn, Wiley, 1981, 646 pp.

80. Hsu, T. C., Slate, F. O., Sturman, G. M. and Winter, G., Microcracking of plain concrete and the shape of the stress–strain curve, *J. Am. Concr. Inst., Proc.*, Vol. 60, No. 2, Feb. 1963, pp. 209–224.

81. Nawy, E. G., Crack width control in two-way concrete slabs centrally loaded, *Eng. Res. Bull. Rutgers Univ.*, No. 46, 1967, pp. 1–87.

82. Nawy, E. G., Crack control in reinforced concrete structures, *J. Am. Concr. Inst., Proc.*, Vol. 65, No. 2, Feb. 1968, pp. 825–836.

83. Ferry-Borges, J., Cracking and deformability of reinforced concrete beams, *Mem. Int. Ass. Bridge Struct. Eng.*, Vol. 26, 1966, pp. 75–95.

84. Ferry-Borges, J. and Lima, J. A., Crack and deformation similitude in reinforced concrete, *RILEM Bull.*, No. 7, June 1960, 26 pp.

85. Holmberg, A. and Lindgren, S., *Crack spacing and crack width due to normal force or bending moment*, Document D2, National Swedish Council for Building Research, Stockholm, 1970, 57 pp.

86. Gergely, P. and Lutz, L. A., Maximum crack width in reinforced concrete members, *Cause, mechanism and control of cracking in concrete*, American Concrete Institute, SP-20, 1968, pp. 87–117.

87. Beeby, A. W., *The prediction of cracking in reinforced concrete members*, PhD Thesis, University of London, 1971, 252 pp.

88. ACI Committee 224, Control of cracking in concrete structures, *J. Am. Concr. Inst., Proc.*, Dec. 1972, pp. 717–753. Updated report, *Conc. Int.*, Vol. 2, No. 10, Oct. 1980, pp. 35–76.

89. Nawy, E. G., Crack control in beams reinforced with bundled bars, *J. Am. Concr. Inst., Proc.*, Vol. 69, No. 10, Oct. 1972, pp. 637–639.

90. Nawy, E. G. and Blair, K. W., Further studies on flexural crack control in structural slab systems, *Cracking, deflection and ultimate load of concrete slab systems*, American Concrete Institute SP-30, 1971, pp. 1–41.

91. Nawy, E. G., Blair, K. W. and Hung, T. Y., Crack control, serviceability, and limit design of two-way action slabs and plates, *Eng. Res. Bull. Rutgers Univ.*, No. 53, 1972, 135 pp.

92. Nawy, E. G., Crack control through reinforcement distribution in two-way acting slabs and plates, *J. Am. Concr. Inst., Proc.*, Vol. 69, No. 4, Apr. 1972, pp. 217–219.

93. Nawy, E. G. and Blair, K. W., Discussion by Committee ACI 318 and author's closure to 'Further studies on flexural crack control in structural slab systems', *J. Am. Concr. Inst., Proc.*, Vol. 70, No. 1, Jan. 1973, pp. 61–63.

94. ACI Committee 340, *Ultimate strength design handbook*, American Concrete Institute, SP-17 1981, 508 pp.

95. US Steel Corporation, *Distribution of flexural reinforcing in two-way slabs*, Handbook, 1972, pp. 1–34.
96. Abeles, P. W., Fully and partly prestressed reinforced concrete, *J. Am. Concr. Inst., Proc.*, Vol. 31, No. 1, Jan. 1945, pp. 181–216.
97. Abeles, P. W., Static and fatigue tests on partially prestressed concrete construction, *J. Am. Concr. Inst., Proc.*, Vol. 51, No. 12, Dec. 1954, pp. 361–376.
98. Nawy, E. G. and Potyondy, J. G., Moment rotation, cracking and deflection of spirally bound, pretensioned prestressed beams, *Eng. Res. Bull. Rutgers Univ.*, No. 51, Nov. 1970, p. 96.
99. Nawy, E. G. and Potyondy, J. G., Flexural cracking behavior of pretensioned prestressed concrete J- and T-beams, *J. Am. Concr. Inst., Proc.*, Vol. 68, No. 5, May 1971, pp. 355–360.
100 Nawy, E. G. and Huang, P. T., Crack and deflection control of pretensioned prestressed beams, *J. Prestressed Concr. Inst.*, Vol. 22, No. 3, May–June, 1977, pp. 30–47.
101. Nawy, E. G. and Chiang, J. Y., Serviceability behavior and crack control of post-tensioned and prestressed beams, *J. Prestressed Concr. Inst.*, Vol. 25, No. 1, Jan–Feb 1980, pp. 74–95; Discussion and authors' closure, pp. 144–149.
102. Naaman, A. E. and Siriaksorn, A., Serviceability based design of partially prestressed beams, Part I: Analytic formulation, *J. Prestressed Concr. Inst.*, Vol. 24, No. 2, March–April 1979, pp. 65–89.
103. Siriaksorn, A. and Naaman, A. E., Serviceability based design of partially prestressed beams, Part II: Computerized design and parametric evaluation, *J. Prestressed Concr. Inst.*, Vol. 24, No. 3, May 1979, pp. 40–60.
104. Vessey, J. V. and Preston, R. L., Critical review of code requirements for circular prestressed concrete reservoirs, *Proc. 8th FIP Congress*, May 1978, Fédération Internationale de la Précontrainte, Paris, 1978, 11 pp.
105. Baker, A. L. L., *The ultimate load theory applied to the design of reinforced and prestressed concrete frames*, Concrete Publications, London, 1956, 91 pp.
106. Baker, A. L. L. and Amarakone, A. M. N., Inelastic hyperstatic frame analysis, *Flexural mechanics of reinforced concrete*, American Concrete Institute, SP-12, 1964, pp. 85–142.
107. Baker, A. L. L., *Limit state design of reinforced concrete*, Cement and Concrete Association, London, 1970, 345 pp.
108. Base, G. D. and Read, J. B., Effectiveness of helical binders in the compression zone of concrete beams, *J. Am. Concr. Inst., Proc.*, Vol. 62, No. 7, July 1965, pp. 763–779.
109. Mattock, A. H., Rotational Capacity of Hinging Regions in Reinforced concrete Beams. *Flexural mechanics of reinforced concrete*, American Concrete Institute, SP-12, 1964, pp. 143–181.
110. Cohn, M. Z. and Ghosh, S. K., Ductility of reinforced sections in bending, *Proc. int. symp. inelasticity and non-linearity in struct. concr.*,, University of Waterloo Press, June 1972, pp. 111–146.
111. Nawy, E. G., Danesi, R. F. and Gnosko, J. J., Rectangular spiral binders effects on the plastic hinge rotation capacity in reinforced concrete beams, *J. Am. Concr. Inst., Proc.*, Vol. 65, No. 12, Dec. 1968, pp. 1001–1010.

112. Nawy, E. G. and Salek, F., Moment–rotation relationships of non-bonded post-tensioned I- and T-beams, *J. Prestressed Concr. Inst.*, Vol. 13, No. 4, Aug. 1968, pp. 40–55.

113. Sawyer, H. A., Design of concrete frames for two failure states, *Flexural mechanics of reinforced concrete*, American Concrete Institute, SP-12, 1964, pp. 405–431.

114. Corley, W. G., Rotational capacity of reinforced concrete beams, *J. Struct. Div., Am. Soc. Civ. Eng.*, Vol. 92, No. ST5, Oct. 1966, pp. 121–146; Discussion by Mattock, A. H., No. ST2, April 1967, pp. 519–522.

115. Nawy, E. G. and Potyondy, J. G., *Moment–rotation, deflection, and cracking of spirally bound, pretensioned, prestressed concrete flanged beams*, Bureau of Engineering Research, Rutgers University, Bulletin No. 51, Sept. 1976, 124 pp.

116. CEB, *Structures hyperstatiques*, Bulletin d'Information No. 105, Comité Euro-International du Béton, Paris, Feb. 1976, pp. 1–259.

13 Structural performance as influenced by detailing

H P J Taylor

Dow Mac Concrete Limited, Stamford
England

Contents

Summary

This chapter shows how the choice of a detail can significantly influence the overall behaviour of a structure. The intention is to describe and illustrate the general principles of good practice in such a way that a designer will be able to develop and improve his own details from a sound basis.

1 Background to design

The design of a structure is conveniently and traditionally carried out in stages, each becoming progressively less generally and more particularly related to small areas. In the early stages, the overall stability of the structure is considered and a rough layout of members is proposed—a layout which, in the experience of the designer, will promise a good skeleton for a robust and economic structure. Later stages of the design will then be concerned with the assessment of loadings and the calculation of force resultants in the structural members. A detailed design of the members will then take place, followed by the detailing phase. In each phase, the process becomes less general, the tributary area for consideration of interaction with other parts of the structure becomes smaller but the level of intricacy of interaction becomes more intense.

At each stage, the process may be viewed as the combination of three basic activities, Analysis, Design and Appraisal. These same activities are carried out

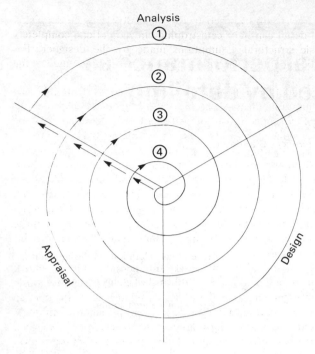

Analysis

① ② ③ ④

Appraisal

Design

Figure 1-1 The design process

over and over again progressively putting the flesh and features on the basic structural skeleton (Fig. 1-1).

In the figure, the four stages of the process would typically be

(a) Determine basic stability and structural skeleton.
(b) Assess loads and load effects on members.
(c) Proportion members including steel content.
(d) Detailed design of members and intersections.

Each turn of the spiral requires a repeat of the three basic activities:

(a) *Analysis* of the problem so far and definition of the next stage.
(b) *Design* of the next stage to satisfy the previous analysis.
(c) *Appraisal* of the results of the design stage to iron out incompatibilities with other parts of the scheme.

At each stage, it is necessary, if there are any remaining incompatibilities at the appraisal stage, to repeat the relevant turn of the spiral until a satisfactory fit is obtained. Design is, therefore, a process of constant refinement, repeated until all details are defined and no incompatibilities remain.

The word 'detail' has a number of meanings and usages, some of which are unfortunate as far as its use in structural design is concerned. The expression 'minor detail' or 'mere detail' seem to indicate that details are trivial and are insignificant when considering the whole. This is not the case. Very many structural failures are a direct result of unsatisfactory or badly designed details. A badly designed structural detail can have catastrophic effects and can completely negate some of the basic structural assumptions made by the designer. For example, joints may not be capable of carrying moments or forces assumed in the design.

The lack of attention to detailing originates from several sources. Detailing may be entrusted to a technician draughtsman who lacks the academic background to understand the structural concepts involved. Alternatively, detailing may be carried out by a young engineer with the correct academic background but without the practical grasp of what can be achieved or what has been used successfully in the past. A number of the 'rules of thumb' and standard methods used in a design office have developed from thought-out and rational ideas, the bases of which may have been lost over the years. These came from both satisfactory and bitter experience. The change of emphasis away from long-serving employees training and passing on their knowledge to juniors to more emphasis on training within academic establishments has made it easier for this feedback from practice to be lost.

This chapter is intended to be of use in recognition of this state of affairs. Detailing is too important to be left to chance; it must be carried out consciously and conscientiously and must be checked by the design team leader. In order to provide a basis for the rational consideration of details, the principal methods of force transfer from element to element within a concrete structure are described in Section 2 and a unifying way of looking at detailing problems is proposed in Section 3. Examples of some of the common details are given in Section 4 with good and bad solutions. The intention is to promote a satisfactory method of approach to the problem rather than to attempt to lay down hard and fast rules to the very large number of possible detailing situations formed in practice.

2   Mechanisms of force transfer

The usual methods of describing and calculating the force transfer between steel reinforcement and concrete are by bond, anchorage and bearing. In the former two cases, classical theories exist which are now seen to have limited validity and, in each case, boundary effects have a significant effect on failure mode and failure strength. In this section, the basic mechanisms are described in order to show how they may be sensitive to unexpected modes of failure in particular circumstances and also to show how, with other boundary conditions, the simple theories give unreasonably low predictions of ultimate strength.

2.1   Bond

In a linear bending member, bond at its most straightforward can be related to the rate of change of tensile force in the reinforcement, $\partial T/\partial x$. Bond stresses do not theoretically exist in members in regions of constant force or moment (they do, of course, exist to a limited extent between cracks in both tension and flexural members).

The traditional model used in describing bond is that of slip (Fig. 2-1a). In this

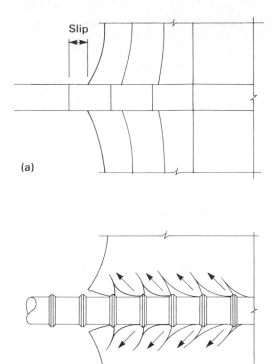

Figure 2-1   Bond mechanisms: (a) friction and (b) mechanical grip

model, the force transfer between the bar and its surrounding concrete is by friction, with the bar pulling out at an interface and the slip reducing progressively to a point of zero slip where the bond bar force has been passed totally to the surrounding concrete. This theory holds true to a certain extent for plain bars and is used almost universally in codes of practice as the basis for the design calculation method for all bar types.

Recent research [1–4] has shown that, for deformed bars, the bond forces are carried by a mechanical interlocking of a bar into the concrete (Fig. 2-1b). The displacements generated by the strains from the force transfer take place by a system of microcracks opening between inclined compressions 'struts' from the points of interlock between bar and concrete. This form of internal damage has been observed [1] and may be used not only to predict observed forms of bond failure but also to derive the most comprehensive and refined theory for cracking in reinforced concrete members [5].

The inclined compressive forces shown in Fig. 2-1b radiate from a bar in all directions, not just those shown in the plane of the paper. This radial compressive force field (Fig. 2-2) gives circumferential tensile stresses which produce cracks when a bar fails in bond and causes it to split out of the parent concrete (Fig. 2-3). This bond-force carrying mechanism may be enhanced by transverse reinforcement, circumferential binding reinforcement or by transverse compressive forces,

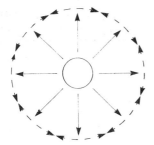

Figure 2-2   **Radial stresses around a bar**

all of which will restrain the occurrence and development of splitting cracks. Equally, tensile stress systems, either from bars bonded and acting in tension at right angles to the bar in question, or from transverse tensile or flexural tensile stress fields, may tend to assist in the development of bond splitting cracks. In details of joints, these additional transverse forces are a common occurrence and can be responsible for both enhanced and diminished bond performance of critical bars.

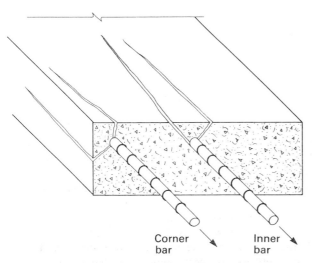

Corner   Inner
bar   bar

Figure 2-3   Failure mechanism: pull-out of deformed bar

2.2 Bearing

Forces are often transferred from one structural member, or component of a member, to another by bearing. In many situations, the force transfer is on a limited area and has high local stresses. A typical example is that of the force produced by an anchor block at the end of a prestressed post-tensioned concrete beam.

The direction of the elastic compressive stress is shown in Fig. 2-4. Just under the bearing plate, a transverse compressive stress is induced. Further away from

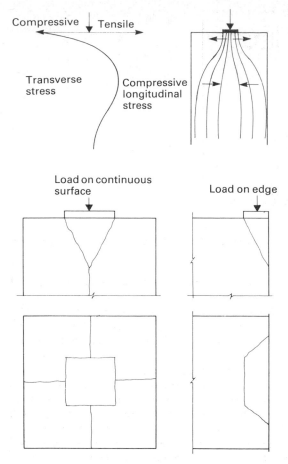

Figure 2-4 Splitting mechanism in bearing

the point of application of the load, as the compressive stress turns to become parallel with the longitudinal axis of the member, a transverse tensile stress is produced. Bearing failure, therefore, is often characterized by the splitting of the member in the line of action of the bearing force (Fig. 2-4) and also by a wedge of sound material shearing through the compressive zone, further accentuating the splitting action. This mode of failure may be observed beneath a loaded plate on the surface of a concrete member and may even be found beneath the plane end of a bar carrying compression.

In design codes, two aspects of bearing are often considered separately. The design of members in which there is no transverse reinforcement to carry the transverse forces shown in Fig. 2-4 is treated as a problem of purely limiting the compressive bearing stress applied to the surface of the member.

The CEB and ACI codes have equations which allow progressively larger direct bearing stresses as the ratio of loaded-to-unloaded area of the member surface

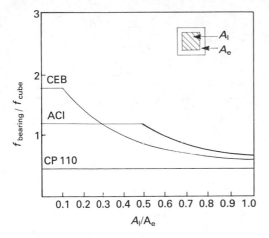

Figure 2-5   Permissible bearing stresses from three codes (square bearing area)

decreases. CP 110 [20] is much more conservative and has a constant value of allowable bearing stress regardless of the loaded-to-unloaded area ratio. This aspect of the subject has been reviewed by Williams [6], who presents a comparison of the code methods and a comprehensive list of all readily available test data. Figure 2-5 shows the comparison of the design codes for a prism with a square loaded area.

The other way in which codes treat bearing is usually written as a design method for end blocks in prestressed concrete beams. In this case, the design method provides for reinforcement to carry the bursting forces. The work of Zelinski and Rowe [7] provides a summary of the problems and the research which provides the basis of the design rules. A more recent report [8] complements this work and covers a number of other different situations and related problems. Although the end block work may seem removed from the problems of detailing joints, the design methods are of more general use than is often realized. Particular situations are beneath inserts and billets in joints in precast concrete construction and as an ancilliary consideration in anchorage problems. These and other similar situations will be pointed out later in this chapter.

2.3   Anchorage

The anchorage of bars may be with a straight bar of suitable bond length or with a hook. In the case of the straight bars, the force mechanisms described previously apply and failure modes are also as previously described. A bar anchored around a bend, or around another bar, acts by a combination of the previous two mechanisms and significantly each of these produces a failure mode which involves splitting. In a sense, a hooked bar produces a doubly dangerous situation, as the rate of change of bar force, considered as a vector, has a $\partial T/\partial x$ and a $\partial T/\partial \theta$ component at the same time. Thus the rate of loss (or gain) of bar force around a bend is greater than that in linear bond and the splitting forces are therefore increased.

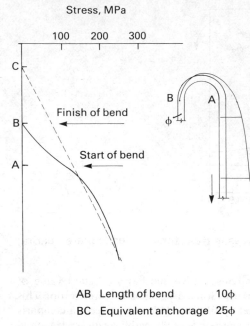

Stress, MPa

AB Length of bend   10φ
BC Equivalent anchorage 25φ

Figure 2-6 Anchorage around a bend

An interesting instrumented example of the advantage of a bend in anchorage has been reported by Franz [9]. In this case, a bar was instrumented by forming a hood and sawing it down the middle with a 1.5 mm wide cut. Strain gauges were glued to one of the semicircular section loops and the second loop was then glued back. Thus, the strain gauges did not mar the surface of the bars and therefore did not affect their bond performance. The test results, plotted in a slightly different way from that used by Franz, are illustrated in Fig. 2-6. The figure shows the bar force plotted linearly as it changes down the bar. The rate of change of bar force up to point A from the straight lead out from the bend is linear, showing the effect of the linear bond mechanism. From A to B, the rate of change of bar force is much greater due to the combined bond and bearing mechanism. Projecting the original line backwards to the point where a straight bar would be expected to have zero stress, it is interesting to note that the 10φ of bar length around the bend provides 25φ worth of straight bar bond. In fact, 25φ would be the anchorage value of such a bend according to CP 110.

The splitting effect of a hooked bar is something that should be guarded against in practical details. Where the structure is relatively continuous in the direction perpendicular to the plane of the beams or where reinforcement is passing through the hook, transverse restraining forces may build up, increasing the effectiveness of the hook. At the other extreme, it is dangerous on a frame joint to have a hook outside all other reinforcement as it can then easily split out of the member (Fig. 2-7). Although this example may seem obvious or trivial, such details have been used with disastrous results.

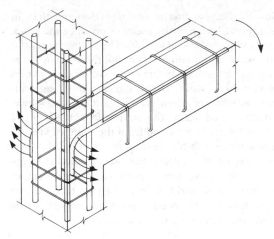

Figure 2-7   Splitting of hooked-bar anchorages

2.4 Shear

For the purpose of this section, the term 'shear' designates the transfer of forces across pre-cracked planes in an element. In uncracked concrete, complex stress conditions can be resolved by elastic analysis and the resultant principal stress states may be compared with the results of tests on concrete under states of complex stress. However, it is often the case that, in situations of reversed loading, it may be necessary for a detail loaded in one direction to require compressive and/or shear stresses to be transmitted across pre-cracked concrete (Fig. 2-8). The figure shows the top of a frame which is carrying transverse

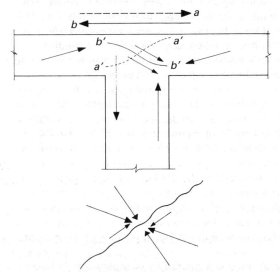

Forces carried across cracks

Figure 2-8 Shear carried across cracked zones

loading—from wind, for example—by swaying first in one direction and then in the other. In the a-direction, the internal forces in the member produce the crack $a'-a'$, and in the reverse direction, produce the crack $b'-b'$. The figure shows the internal forces with the b-direction sway, and it can be seen that the crack $a'-a'$ is carrying transverse compressive forces and stress. In this case, the crack is closed because the compressive stress is very high. But it is also the case in beams with inclined cracks in shear, that open cracks are often called upon to carry shear forces. A great deal of work has been carried out on this subject from the point of view of understanding shear as a basic phenomenon and from the point of view of the behaviour of concrete structures under seismic loading.

The force transfer mechanism by aggregate interlock has been studied in stress states of shear with the indirect compressive stress from fixed displacement regimes and in stress states of constant crack width held by varying compressive stress. For the seismic problem, tests have also been carried out on specimens with transverse reinforcement at varying angles to the cracked plane.

3 General considerations in design details

In this section, a general framework is laid down in which a basic approach to detailing problems is derived, which can be used in common and novel situations to produce basic, safe designs. As stated earlier, it is not the intention of this chapter to catalogue all details but to develop the correct method of approach.

One of the earliest and still most successful ways of understanding the behaviour of a beam in bending and shear is to consider an internal statically determinate system of forces in equilibrium with the applied load. Figure 3-1a gives a simple explanation of how stirrups work in a beam and shows how the tension 'tie' forces in the stirrups and longitudinal steel, interact with the compressive 'strut' forces in the compressive zone of the beam and diagonally in the web. This method gives, in its most straightforward form, a conservative prediction of the ultimate strength of a beam, and in its refined versions, as accurate a prediction as may yet be obtained. This method is not entirely general. Even in beams, if the strut angle is progressively reduced, thereby reducing the number of links needed until none is required at all (Fig. 3-1b), the basic internal strut and tie structure may still be found. In this case, however, when the span-to-depth ratio is very large ($L/d > 6$), diagonal cracks may form and make the mechanism unstable to the point at which it no longer gives safe ultimate strength predictions. On the other hand, Kani [10] has shown that quite novel ways of reinforcing beams in shear do give safe results (Fig. 3-1c).

Many detailing problems may be clarified by looking at them in this way. Internal statically determinate strut-and-tie systems may be sought and reinforcement provided accordingly. Particular attention should be focused on the points of intersection of the internal forces to see if they interact properly, considering the previously described force transfer mechanisms of bond, anchorage, bearing and aggregate interlock.

This approach is not sufficient to provide a guarantee of satisfactory behaviour of a detail. The method in itself may fail if the included angle between some of the forces is too large, as in beam shear, just as a lattice truss with a similar layout

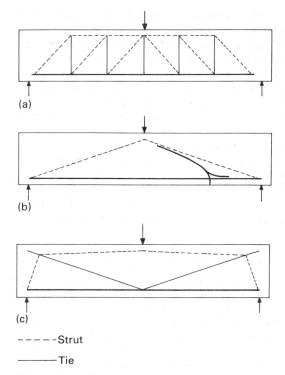

— — — —Strut

————————Tie

Figure 3-1   Truss approaches to shear

would be very inefficient. A further problem is that the approach does not consider the behaviour of the member under the service loads; in fact, the member would be required to have a number of cracks before the internal force system is developed. The satisfactory design of a detail therefore also requires the visualization of elastic stress fields on the uncracked elements, the position of likely cracks and the provision of steel to control their width. This may be done by experience, by elastic analysis (finite elements are useful in this respect) and, on important details, by test.

A simple method of visualizing the elastic behaviour of a detail is to make a readily deformable model, mark it with a regular grid and observe the change of shape of the grid when the model is deformed. A thin sheet of foam rubber has been found to be very useful for this work (Fig. 3-2).

In cases where repeated loading is expected, these same general principles apply. Details which work well in static loading usually work well if the loading is of a constant repeated nature, although in the latter case it is usually beneficial to have additional links or binding steel in joint zones to control the width of cracks and to limit shear deformations along diagonal cracks which may, for large loads, become progressive.

Seismic loading calls for special detailing techniques to cope with the special requirements for a structure with a high chance of survival of a few cycles of large

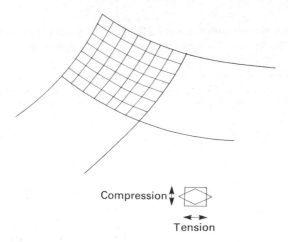

Compression
Tension

Figure 3-2   Deformation of a rubber model of a corner

imposed displacements. The same principles of design apply but it must be remembered that, for example, an opening joint detail on one cycle will be required to sustain a closing deformation on the return. This results in the characteristic X-shaped crack patterns at intersections. These coupled with the random application of vertical accelerations require that considerable amounts of binding reinforcement are provided to hold the cracked concrete together. Many details have been tested and general design methods are becoming available.

4   Practical illustrations

4.1   Corbels, half joints and nibs

Corbels have been the subject of a considerable amount of research, probably because of the catastrophic nature of corbel failure and probably also because the problem is well constrained and is fairly easily researched. Many design models have been used but this section considers the most successful approach, which is based on the detailing principles outlined earlier. Half joints and nibs have received less attention.

4.1.1   *Corbels*

Figure 4-1 shows the elastic state of stress on a corbel from the work of Franz and Niedenhoff [11]. Elastic stress fields, like the more simple deformable rubber models suggested earlier, give a good indication on the possibility of a 'truss' solution. In this case, the following conclusions can be reached:

(a) The compressive stress down the inclined face of the corbel is almost constant, as is the tensile stress parallel to the top face of the corbel.
(b) Stress concentration can be observed at the top and bottom of the corbel where it joins the main body of the column.

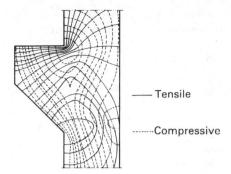

Figure 4-1 Corbel stress patterns from elastic models by Franz and Niedenhoff [11]

(c) If the corbel were made of a rectangular section, then a 'dead area' would exist in the triangular part of the corbel (missing in the figure).

This and other similar work by Mehmel and Becker [12] led to tests of reinforced concrete corbels and a design method based on the principles given earlier. Franz and Niedenhoff, from such work, proposed the truss analogy in Fig. 4-2. The failure modes of reinforced concrete corbels are consistent with the design approach.

(a) The tension member, T, may yield, producing very wide flexural cracks at the junction with the column.
(b) The compression strut, C, may crush.
(c) The tensile stresses acting transversely in the strut can cause vertical splitting and instability in the strut.
(d) If the corbel is heavily reinforced or is very short, a shear failure may occur in which a number of diagonal cracks form at the corbel to column interface.

The shear strength of a very short corbel in which $a_v/d < 2$ is greater than that of a beam because the geometry of the corbel forces the cracks to be more vertical than they would naturally want to be. In other words, the principal tensile stresses is not at 45° to the longitudinal axis of the corbel. Because corbels are not constrained in shape and they occupy the area $0.5 < a_v/d < 2$, shear failures can

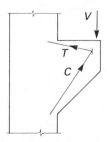

Figure 4-2 Simple truss analogy

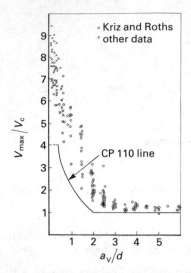

Figure 4-3 Basis for CP 110 shear enhancement method

occur on rare occasions and a shear criterion is often added to the simple truss methods of design. The method of enhancing a basic nominal shear stress, usually arrived at empirically by the use of a simple multiplying factor kd/a_V, is used in many design codes. The experimental basis of this approach is shown in Fig. 4-3.

The information is now assembled to produce a comprehensive design method. The CP 110 method is illustrated in Fig. 4-4. The method is based on the truss principles but as a corbel becomes long $(a_V/d > 2)$ and beam behaviour predominates, the beam design method of ensuring strain compatibility is also used. The method also requires the provision of an apparently arbitrary amount of horizontal tie steel below the main tie, 50% of the main tie steel. This acts as secondary

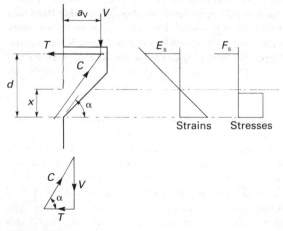

Figure 4-4 CP 110 approach to corbel design

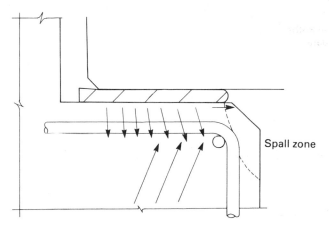

Figure 4-5   Bearing detail

tie steel and is distributed over the area of the strut to control any splitting cracks which may otherwise enable the concrete of the strut to fall away. A danger always exists of the detail of the intersection of the tie and inclined strut not working well and is a typical example of how the design should consider not only the main members of the strut but also their intersection. Figure 4-5 shows a detail under a beam supported on a corbel. It is essential, particularly if the bearing forces are large and the members are major, to ensure that the bearing force, by rotation of the supported member, cannot bear upon the corbel in front of the line of intersection of the strut and tie. This may be achieved by the use of suitable packing material and by providing large chamfers on the front of the corbel. Even so, an unsuitable bearing material (too soft a rubber) may produce horizontal creep stresses which accentuate the problem. This is one of the most difficult parts of a corbel design and it is often worth while to draw the detail out full size, including all the correct bends on the steel to see it it works correctly. Even if the front part of the corbel does spall away, a dangerous state does not necessarily exist; the load is merely forced back to the correct point of intersection of the strut and tie forces. Bearing stresses may be very high on the remaining concrete, however, and an untidy repair problem at the front of the corbel remains.

4.1.2   Half joints

Half joints are very similar to corbels in that they are limited projections from a longer member carrying large loads by cantilever action. A number of half joints was tested by Reynolds [13], who also developed a design method. Two possible truss methods exist for a half joint, the one proposed earlier for corbels and a further one using diagonal bars. Because a half joint does not have the well-bound vertical column steel found in a corbel, the failure crack pattern was found to be better modelled by the method using diagonal steel (Fig. 4-6). The detailing points of the intersection of the line of action of the bearing force and the tie still apply and for this reason it was found necessary to focus attention on a chamfer at

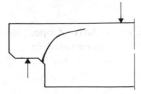

Typical cracking

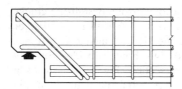

Design solution

Figure 4-6   Forces on a half joint

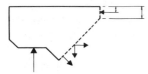

Figure 4-7   Reinforcement of a half joint

the end of the beam (packing material) and on the provision of horizontal steel to restrain splitting above the bearing point (Fig. 4-7).

4.1.3 *Nibs*

Nibs and boots are similar in action to corbels and half joints, as Clarke [14, 15] pointed out. In these cases, the loads are usually small and easy to satisfy with reinforcement. The detailing problem is in the fact that these members are usually required to be physically very small and normal methods of detailing the steel are difficult to apply. Clarke tested a number of practical details and reached the following conclusions. Both the truss arrangements shown in Fig. 4-8 are effective, although for practical reasons of handling and fixing bars, the half-joint type is more difficult to apply. Reinforcement proportioned in that way is effective and the shear span used in the calculation of shear enhancement, in accordance with Fig. 4-8, is a_V/d.

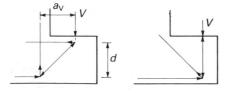

Figure 4-8   Nib truss systems

4.2 Corner joints

A number of failures of corner joints has been reported, particularly of opening corners [16]. Figure 4-9 shows the typical trajectories of the concentrated forces in a flexural corner, both for the opening case. In the closing case, the tensile forces are on the outside and the compressive forces on the inside. The equilibrating diagonal force across the corner is in compression. The most likely method of failure is for the bars round the outside of the corner to split the concrete in bearing. In a continuous slab-to-slab connection, bars perpendicular to the corner, inside the bend, stop this from occurring. On a beam-to-beam corner, it is possible for the main steel to split the concrete and come out of the corner sideways. To stop this, short steel bars under the bend or links across the diagonal of the corner are very effective.

Opening corners have caused most of the problems in the past. Figure 4-9 shows the line of action of the concentrated forces. In this case, the tensile forces are on the inside of the corner and the compressive forces on the outside. The diagonal resultant is tensile and it is this force that designers often neglect in their detailing. Four common methods of detailing corner reinforcement are shown in Fig. 4-10, all of which have been tested by a number of workers. Figures 4-11 and

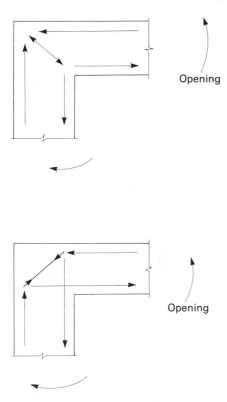

Figure 4-9   Forces in a corner joint

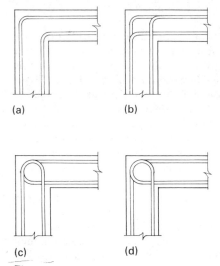

(a)   (b)

(c)   (d)

Figure 4-10   Methods of reinforcing corners

4-12 show the results brought together in a straightforward way. In the figures, the efficiency of the corner is shown plotted against the percentage of tensile steel in the members making up the corner. Efficiency is defined as the ratio of the flexural strength of the corner at test to the flexural strength of the elements making up the corner, and is expressed as a percentage. An efficiency of a detail should be 100% if the design requires it to be so. It may be seen that the efficiencies of all the four details are very low. At low steel percentages, the corners are under-reinforced and their strength is equal to that of the plain

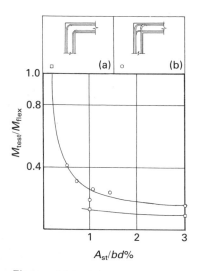

Figure 4-11   Corner tests

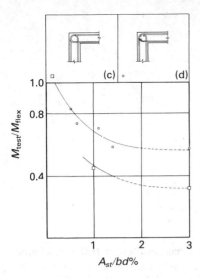

Figure 4-12 Corner tests

concrete, giving an efficiency of 100%. The most popular detail in practice, detail (b), is very weak at all practical efficiencies and only detail (d) shows a reasonable strength. Clearly, all these details are not suitable where a design has relied upon them in actual locations to have 100% efficiency or in statically determinate situations.

Reinforcement may be laid out in two truss configurations, as shown in Fig. 4-9. These two layouts have been tested and give good results (Fig. 4-13) although even these do not automatically guarantee 100% efficiency. Tests detail (f), in

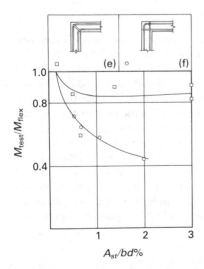

Figure 4-13 Efficient corner details

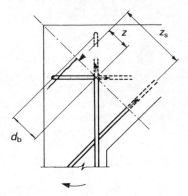

Figure 4–14 Splayed corner detail

which intersecting hairpins are used with a splay has resulted in 100% efficiency. The design method is shown in Fig. 4-14. In the figure, the line of action of all the forces are as shown; z is assumed to be equal to $0.75d_b$ and the hairpin steel is assumed to be 80% efficient. Provided z_s is not more than four times z, to guarantee ductility when all the steel is yielding, very high efficiencies are obtained for all steel percentages.

This section shows how the simple truss approach provides good design results to a difficult problem. It should be realized, however, that the use of an efficient detail may not be necessary in all situations. The use of some of the more common, less efficient details with splays and low-tensile steel percentages will result in designs in which diagonal cracks will not occur in the joint zone and satisfactory design behaviour will result.

4.3 Beam–column joints

Beam–column joints may be considered as two superimposed corner joints, one opening and one closing, and therefore show the weaknesses of both details. The internal concentrated forces in a joint zone are shown in Fig. 4-15. In such a beam–column joint, diagonal cracking parallel to the line of action of the internal diagonal strut will occur, as will the failure of the concrete beneath the bend of the beam steel. In beam–column joints, the column steel which passes through the joint is usually well bound with links and adds much to the strength of the detail.

Tests were carried out by Taylor [17] on the details illustrated in Fig. 4-16 and these showed that efficiencies of 100% could be obtained with detail (a) for tensile beam steel percentages of 1.0. A simple design method was developed which limited the beam steel percentage to obtain 100% efficiency, based on the geometry of the section and the enhanced shear strength of the short sections shown in Fig. 4-3. Using the notation in Fig. 4-15 the expression below was derived:

$$100\rho b = 100\left(3 + \frac{2d_c}{z_b}\right)\frac{b_c d_c}{b_d d_b}\frac{\beta V_f}{0.87 f_y}$$

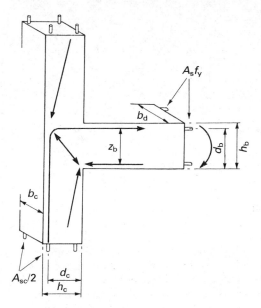

Figure 4-15   Forces in a beam–column joint

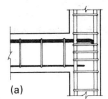

(a)

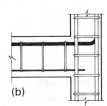

(b)

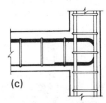

(c)

Figure 4-16   Beam–column joint test-layout by Taylor [17]

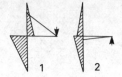

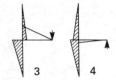

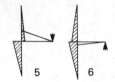

Figure 4-17   Moments in a column during a repeated reverse moment test

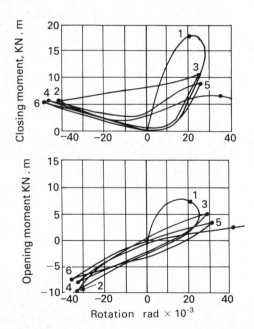

Figure 4-18   Moment–rotation plots/reverse moments

where

$100\rho b$ = limiting beam steel percentage

$\beta = 1 - \text{redistribution}/100$.

V_f = nominal beam shear strength of the column using $100A_{sc}/2b_c d_c$ as its steel percentage.

f_y = characteristic strength of the beam steel.

Use of this expression will not significantly restrict design but should ensure that beam–column joints have suitable efficiencies.

An interesting finding in Taylor's work was that the moment was not shared equally by the closing and opening parts of the corner. The closing corner was stiff and in a statically indeterminate situation attracted up to three times more moment than the opening part. Figure 4-17 shows how the beam moment is shared by the closing and opening parts of the corner during a repeated loading test. On the initial application of load, the closing corner took 70 per cent of the applied moment. On the reverse cycles of loading, because the beam tended to lengthen as its cracks did not completely close on the reverse cycle, the opening corner tended to close and closing corner did not open. The moment rotation plots for the two corner sections are shown in Fig. 4-18. This finding is important in considering earthquake loading, where cycles of such extreme deformations are possible, and shows how a test result—in this case only possible because a prop was provided at the beam end to simulate continuity—can be critically dependent upon test method and can be missed if too simple a test is used.

4.4   Precast concrete

A detailer of precast concrete faces a number of problems not encountered with *in situ* concrete, although the solutions and method of approach are often identical to those mentioned earlier. This section does not cover the problems of precast concrete joints in detail and restricts itself to general principles. Many manuals are available and treat the subject more exhaustively than is possible here. In particular, the Prestressed Concrete Institute and Institution of Structural Engineers manuals [18, 19] give valuable advice and should be consulted for particular cases.

The precast structure almost invariably consists of a number of elements, prefabricated and assembled *in situ* and at each connection some form of joint detail is required. The need for a rapid assembly of the structure, at least to free the crane at the early stages, leads often to a two-part approach. A joint should be designed to be safe and stable as soon as the crane is released even if more work, grouting, welding, etc., is to be carried out at a later stage.

The structural concept of many precast buildings is one in which continuity is primarily of vertical loads only and lateral stability is provided by shear walls or by cantilever columns. Common detailing problems are those of bearing in vertical column-to-column or wall-to-wall joints, detail problems in beam-to-column joints and in floor-to-beam joints.

4.4.1   *Column-to-column and wall-to-wall joints*

These joints are common, and have similar design features. Often, the joint is made in a temporary manner onto packers or levelling strips and is grouted or mortar packed at a later stage. The joint should therefore be designed so that temporary packers do not cause local over-stress and bursting, and so that the final grouting or packing can be carried out effectively and can be properly checked. Potential problems in these joints are

(a) In mortar joints, the best results are obtained if the mortar strength is at least 75 per cent of the strength of the column. Weak mortar causes the sides of the column to spall.
(b) Hard packers should not be used if more than 4 per cent of steel is required in the jointed members.
(c) The area around a hard packer in the temporary condition should be designed using end-block theory.
(d) Grouting should always be carried out in such a way that its effectiveness can be checked. The volume of grout should be measured and inspection should be thorough.
(e) Welding can be carried out but should be used in situations where working conditions are good and a good weld can be achieved.

4.4.2 Beam-to-column joints

A wide variety of beam-to-column joints are used and these often bring detailing problems of the half joint or corbel type, even if they do not specifically involve these layouts and have embedded inserts, etc. Attention should be given to the following aspects:

(a) Half joints and corbels should be designed using the methods described earlier.
(b) A check should be made for bursting under temporary packers, and these areas, as in column-to-column joints, should be designed specifically.
(c) Joints in which embedded steel sections or billets are used and are effective. The truss design approval is very useful in providing safe designs for the interaction of the steel member with the rest of the structure.
(d) In many beam-to-column joints, large concentrated forces are transmitted from member to member. If a new detail is established, it is prudent to test it, as the result of an unexpected failure in service would be disasterous.

Acknowledgements

The author acknowledges and thanks the Cement and Concrete Association and the Institution of Structural Engineers for permission to reproduce figures from their publications.

References

1. Goto, Y., Cracks formed in concrete around deformed tension bars, *J. Am. Concr. Inst., Proc.*, Vol. 68, No. 4, April 1971, pp. 244–251.
2. Tepfers, R., *A theory of bond applied to overlapped tension reinforcement splices for deformed bars*, Publication 73.2, Division of Concrete Structures, Chalmers Tekniska Högskola, Göteborg, 1973, pp. 58–72.

3. Tepfers, R., Cracking of concrete cover along anchored deformed reinforcing bars, *Mag. Concr. Res.*, March 1979, pp. 3–12.

4. Fergusson, P. M., *Reinforced concrete fundamentals*, Wiley, New York, 1973, pp. 184–191.

5. Beeby, A. W., The prediction of crack widths in hardened concrete, *Struct. Eng.*, Vol. 57A, No. 1, Jan. 1979, pp. 7–32.

6. Williams, A., *The bearing capacity of concrete over a limited area*, Cement and Concrete Association Technical Report No. 526, London, Aug. 1979, 70 pp.

7. Zclinski, G. and Rowe, R. E., *An investigation of the stress distribution in the anchorage zone of post-tensioned concrete members*, Cement and Concrete Association Research Report No. 9, London, 1960.

8. Clarke, J. L., *A guide to the design of anchor blocks for post-tensioned prestressed concrete members*, Construction Industry Research and Information Association Design Guide 1, London, June 1976.

9. Franz, G., Discussion: the connection of precast elements with loops, *Design philosophy and its application to precast concrete structures*, Cement and Concrete Association, London, 1967, pp. 63–66.

10. Kani, G. N. J., A rational theory for the function of web reinforcement. *J. Am. Concr. Inst., Proc.*, Vol. 66, No. 3, March 1969, pp. 185–197.

11. Franz, G. and Niedenhoff, H., *The reinforcement of brackets and short deep beams*, Cement and Concrete Association, London, Publication 61.114, 1964, pp. 17. [Translation of article in *Beton- und Stahlbetonbau*, Vol. 58, No. 5. 1963. pp. 112–120].

12. Mehmel, A., and Becker, G., Zur Schubbemessung des kurzen Kragarmes, *Bauingenieur*, Vol. 40, No. 6, 1965, pp. 224–231.

13. Reynolds, G. L., *The strength of half joints in reinforced concrete beams*, Cement and Concrete Association Report No. 42. 415, London, June 1969.

14. Clarke, J. L., *Behaviour and design of small nibs*, Cement and Concrete Association Report No. 42. 512, London, March 1976, 8 pp.

15. Clarke, J. L., *Behaviour and design of small continuous corbels*, Cement and Concrete Association Report No. 42. 513, London, March 1976, 11 pp.

16. Somerville, G. and Taylor, H. P. J., The influence of reinforcement detailing on the strength of concrete structures, *Struct. Eng.*, Vol. 50, No. 1, Jan. 1972, pp. 7–19.

17. Taylor, H. P. J., *The behaviour of insitu concrete beam–column joints*, Cement and Concrete Association Report No. 42. 492, London, May 1974, 32 pp.

18. *Connection details for precast–prestressed concrete buildings*, Precast Concrete Institute, Chicago, 1973, pp. 13–27.

19. *Structural joints and precast concrete*, Institution of Structural Engineers, London, 1978.

20. British Standards Institution, CP 110:1972 (as amended 1980), *The structural use of concrete*, London, 1980, Parts 1–3, 156 pp., 104 pp. and 93 pp.

14 Fire resistance–design and detailing

B K Bardhan-Roy

Jan Bobrowski and Partners, Twickenham, England

Contents

Notation

A_s	total area of tensile steel
A_{sv}	cross-sectional area of links
A_{sv}	total area of vertical steel within shear span
A_{sH}	area of hanger bars
B	effective bearing area
b	width of beam; minimum width in cross-section within shear span
b_1	effective width for I & T sections
d	effective depth
d_n	depth of compression block
$F_{suT°}$	tensile force of steel at $T°$
f_c	average uniform stress in concrete at failure
f_{cu}	characteristic concrete cube strength at 20°C
$f_{cT°}$	residual concrete strength at $T°$
f_{ps}	stress in prestressing tendon at ultimate condition
f_{su}	characteristic strength of steel at 20°C
f_y	characteristic yield stress of steel at 20°C
G_k	characteristic dead load
H	$(V_a \cos \theta)/\sin \theta$
h	total depth of section
l_d	distance of the point of contra-flexure from the support
l_s	shear span
M_a	bending movement due to applied load at section
$M_{RT°}$	moment of resistance at $T°$C
$M_{T°}$	moment capacity of section at $T°$C
Q_k	characteristic live load
V_a	shear due to applied load at section
$V_{crT°}$	critical shear capacity at $T°$C
z	lever arm
γ_g	dead load partial safety factor
γ_{mc}	concrete partial safety factor
γ_{ms}	steel partial safety factor
γ_q	live load partial safety factor
θ	angle as shown in Fig. 5.3
ϕ	(area of tensile reinforcement)/(area of concrete to effective depth)
$\psi_{cT°}$	fraction of strength of concrete at 20°C
$\psi_{sT°}$	fraction of strength of steel at 20°C

Other symbols as explained in the text

Summary

This chapter surveys the statutory requirements for fire resistance and examines the behaviour of structures under fire. Several methods of determining fire resistance are explained with comments on their relative merits: tests on representative models, the use of tabulated data for minimum dimensions and cover, and the comparatively recent analytical approach. The design recommendations of the following organizations are explained:

(a) Building Research Establishment, U.K.
(b) FIP–CEB
(c) American Concrete Institute
(d) Institution of Structural Engineers and The Concrete Society, U.K.

The chapter concludes with a section on detailing.

1　Introduction and general considerations

1.1　Introduction

The incidence of fire is unfortunately quite frequent in buildings. In spite of the growing awareness amongst the public and various efforts by the authorities concerned with safety, the statistical surveys still indicate an upward trend in the number of incidences of fire in almost all of the industrialized countries of the world.

The effects of fire are dreadful. Apart from the high risk of human casualties, loss of materials, property and wealth, there are other consequences—direct or indirect—such as disruption of services, loss of business and job, interruption to trade, etc., and when all these effects are expressed in economic terms, the actual fire losses run into astronomical figures. In the United Kingdom alone the direct and consequential losses due to fire are estimated to be in the region of £1000 m per year based on the figures of 1973 and 1974 [1]. By 1982 the figure had risen to £1500 m.

With so much at stake, some form of statutory control for the fire protection of buildings is almost inevitable and in most countries regulations, codes or by-laws are now in existence to this effect. In the City of London, fire regulations date back to 1667—the year following the Great Fire of London. The objectives of regulations or by-laws are principally two-fold: (a) safety of occupants, and (b) minimizing damage to property, materials and contents. The safety of occupants is sought by (a) the use of such materials and design parameters in the construction as would delay the growth of fire and prevent it from spreading and developing into conflagration; (b) the provision of adequate escape routes; and (c) ensuring easy access and working of fire-fighting equipment and rescue operations.

Minimizing the damaging effects of fire to property and contents and reducing the economic loss for the owners are usually covered by insurance activities and the requirements for achieving this objective are not mandatory unless the property or its contents happens to be of vital interest to the state or nation in some respect [2].

Overall fire-fighting measures include not only in-built permanent fire protection of the building components and materials, but also other devices such as detection, sprinkler installations and similar 'active' measures for fire-fighting. Detailed discussions of the 'active' measures are, of course, beyond the scope of this chapter.

1.2 Statutory requirements

Regulations, by-laws or codes normally specify the degree of fire resistance required for a construction based mainly on considerations of (a) the type of occupancy, i.e. the purpose for which the building will be used; (b) its position in the property; and (c) its floor area, height, cubic capacity and the contents. Other considerations, such as the distance of the building from the water mains or the nearest fire station, the installation of sprinklers or similar 'active' measures, may

Table 1-1   Fire resistance requirements by British *Building Regulations* for single-story buildings

Purpose group		Maximum floor area m^2	Minimum period of fire resistance for elements of structure h
I	Small residential	No limit	$\frac{1}{2}$
II	Institutional	3000	$\frac{1}{2}$
III	Other residential	3000	$\frac{1}{2}$
IV	Office	3000	$\frac{1}{2}$
		No limit	1
V	Shop	2000	$\frac{1}{2}$
		3000	1
		No limit	2
VI	Factory	2000	$\frac{1}{2}$
		3000	1
		No limit	2
VII	Assembly	3000	$\frac{1}{2}$
		No limit	1
VIII	Storage and general	500	$\frac{1}{2}$
		1000	1
		3000	2
		No limit	4

Note to Tables 1-1 *and* 1-2

For the purpose of the *Building Regulations*, the period of fire resistance to be taken as being relevant to an element of structure is the period included in the appropriate column in the line of entries, which specifies the floor area with which there is conformity or, if there are two or more such lines, in the topmost of those lines.

Table 1-2   Fire resistance requirements by British *Building Regulations* for buildings over one story

Purpose group	Maximum dimensions			Minimum period of fire resistance (in hours) for elements of structure forming part of:	
	Height m	Floor area m²	Cubic capacity m³	Ground story or upper story	Basement story
I Small residential:					
House having not more than 3 storys	No limit	No limit	No limit	$\frac{1}{2}$	1
House having 4 storys	No limit	250	No limit	1	1
House having any number of storys	No limit	No limit	No limit	1	$1\frac{1}{2}$
II Institutional	28	2000	No limit	1	$1\frac{1}{2}$
	Over 28	2000	No limit	$1\frac{1}{2}$	2
III Other residential:					
Building or part having not more than 2 storys	No limit	500	No limit	$\frac{1}{2}$	1
Building or part having 3 storys	No limit	250	No limit	1	1
Building having any number of storys	28	3000	8500	1	$1\frac{1}{2}$
Building having any number of storys	No limit	2000	5500	$1\frac{1}{2}$	2

IV Office	7.5	250	No limit	$\frac{1}{2}$	1
	7.5	500	No limit	$\frac{1}{2}$	1
	15	No limit	3500	1	1
	28	5000	14000	1	$1\frac{1}{2}$
	No limit	No limit	No limit	$1\frac{1}{2}$	2
V Shop	7.5	150	No limit	$\frac{1}{2}$	1
	7.5	500	No limit	$\frac{1}{2}$	1
	15	No limit	3500	1	1
	28	1000	7000	1	2
	No limit	2000	7000	2	4
VI Factory	7.5	250	No limit	$\frac{1}{2}$	1
	7.5	No limit	1700	$\frac{1}{2}$	1
	15	No limit	4250	1	1
	28	No limit	8500	1	2
	28	No limit	28000	2	4
	over 28	2000	5500	2	4
VII Assembly	7.5	250	No limit	$\frac{1}{2}$	1
	7.5	500	No limit	$\frac{1}{2}$	1
	15	No limit	3500	1	1
	28	1000	7000	1	$1\frac{1}{2}$
	No limit	No limit	7000	$1\frac{1}{2}$	2
VIII Storage and general	7.5	150	No limit	$\frac{1}{2}$	1
	7.5	300	No limit	$\frac{1}{2}$	1
	15	No limit	1700	1	1
	15	No limit	3500·	1	2
	28	No limit	7000	2	4
	28	No limit	21000	4	4
	over 28	1000	No limit	4	4

also play an important part sometimes. Requirements, of course, vary from country to country.

In the USA, the Model By-laws specify a minimum of $\frac{1}{2}$ h resistance to fire for domestic buildings if the height of the building is within 15 m (50 ft) and the floor area less than 230 m^2 (2500 ft^2). If these limits are exceeded, 1 h fire resistance is required. For public buildings and warehouses which are not used predominantly for storage, generally three different periods are specified: $\frac{1}{2}$ h if the building does not exceed 15 m in height and/or 3500 m^3 in capacity; 1 h for a building between 15 and 23 m in height and between 3500 and 7500 m^3 in capacity but not exceeding 700 m^2 in area; and 2 h if these limits are exceeded. For warehouses which are used predominantly for storage, periods of resistance of $\frac{1}{2}$, 1, 2 and 4 h are required depending on the height and capacity of the building [3].

In England and Wales, the Building Regulations [4] specify the minimum period of fire resistance required for different buildings, which varies from $\frac{1}{2}$ to 4 h, depending on the purpose group of the building, its floor area and height. Tables 1-1 and 1-2 show in detail the requirements of the British Building Regulations for single-story buildings and for buildings over one story, respectively [4].

Higher fire-resistance ratings often result in lower fire insurance premiums and for that reason it is possible that the client may demand higher than the mandatory minimum required for fire resistance in his building.

1.3   Fire resistance: definition, limiting criteria and tests

The fire resistance of a structure (frequently mentioned in the foregoing) is in fact a measure of its ability to withstand the effects of fire without reaching any of the limit states discussed later, and is expressed in terms of time as established by standard fire tests to BS 476, Part 8:1972 [5] in Great Britain and ASTM E119-79 [6] in the USA. In the standard fire tests, the structure or the element is exposed to a fire controlled to follow a standard time–temperature curve. Standard time–temperature curve to BS 476, Part 8 is illustrated by a firm line in Fig. 1-1 and the relationship can be represented by the equation $T = 345 \log_{10}(8t + 1)$, where T is the temperature rise in °C and t the time in minutes.

The time–temperature curves used in the USA and West Germany are also shown in Fig. 1-1 by a broken line and chain line, respectively. As can be seen, the difference between various national standards is very small and up to 30 min duration (corresponding to an increase in temperature of up to 821°C) the curves are identical [3].

The relationship between the fire severity in actual fire and that of a standard fire is not fully known. Recent research in a number of countries has shown that fire severity depends on three main factors: fire load or fuel for fire, ventilation and the geometry of the compartment. Some experiments with full-sized fires have been made at the Building Research Station, England, and a comparison of the results of fire severity in the experiments and in the standard fire is shown in Fig. 1-2 [7]. In that diagram, the individual numerals 60, 30, 15 and 7.5 indicate the fire load in terms of wood in kg/m^2 to be burnt, whereas the figures in brackets, (1/4) and (1/2), represent the proportion of openings in the wall. Thus, 60 kg/m^2 wood for a compartment in which only a quarter of the wall area consists of openings represents the worst condition of fire severity. It can be

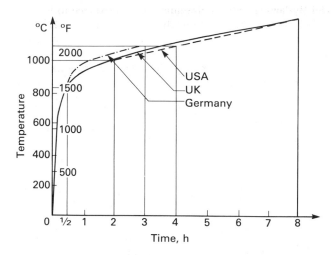

Figure 1-1   Standard time–temperature curve

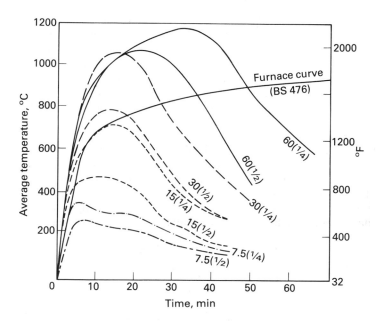

Figure 1-2   Average computed temperature as a function of time

observed that in a full-sized fire the rise of temperature is very rapid at the initial stages as in the standard curve, and within half an hour or so the peak temperature is reached, which could be much higher than the temperature in a standard fire at the corresponding period. After reaching the peak value, however, the higher temperature is not sustained (unlike the standard curve) but gradually decreases.

When a structure or its components is exposed to a fire environment, it is assumed to have reached its limit state if:

(a) The service load can no longer be supported. This criterion is known as the 'limit state of stability' or structural collapse. A stability limit is deemed to have been reached if the actual collapse occurs or if the deflection exceeds 1/30th of the clear span.

(b) It ceases to prevent the passage of flames and/or hot gases. This condition is termed the 'limit state of integrity' or 'flame barrier limit'. An integrity failure is deemed to occur when cracks and openings appear in the specimen through which flames or hot gases can pass which would cause flaming of a cotton-wool pad.

(c) It is unable to restrict excessive transfer of heat to the unexposed surface. This is known as the 'limit state of insulation' or 'thermal limit'. An insulation failure is deemed to occur when the *mean temperature* of the unexposed surface of the specimen increases by more than 140°C above the initial temperature or the temperature at *any point* on the unexposed surface rises 180°C above the initial temperature.

The flame barrier and thermal limit in items (b) and (c) are applicable only to floor and/or walls which perform a separating function. Although these are criteria for failure in fire, in a structural sense these can be compared to the serviceability limits as the structure (or the component) may still be repaired without undue cost provided the deformations are not excessive [3].

As indicated earlier, the fire-resistance rating of a structure or its component is derived from standard fire tests which are carried out in or above a furnace. The standard requires that the loading which will be applied to the specimen to be tested should correspond to the maximum critical service load taken in the design at normal temperature conditions, and that the loading will remain constant in the course of the test. The standard also requires that the specimens have similar end bearing and side conditions as are likely to be achieved in service. Where the actual end or side conditions cannot be reproduced or specified, the following condition would be applied.

Floors and beams	Simply supported at bearing and vertical loading
Columns	Restrained in position and direction at ends and vertical loading
Walls (load bearing)	Vertical edges free from restraint and vertical loading

In addition to the criterion of three limit states, the load-bearing elements must also be capable of supporting the load 24 h after the end of the heating period.

Should collapse occur during the heating period or during a re-load test, the notional maximum period for stability for the element is to be 80% of the time to collapse or the duration of heating if failure occurs during the re-load test [5].

2 Behaviour of structures under fire

When a flexural member—reinforced or prestressed—is exposed to fire (exposure to underside assumed), the bottom fibres of the member expand more than the top owing to the temperature differential and consequently a downward deflection occurs. During the initial stages of heating, the behaviour of an element is dominated by the material properties, which are still broadly similar to those at normal temperatures. This may tend to cause a brittle mode of failure in a structure. However, after the initial period of exposure, the properties are much affected by heating and high temperature in the sense that the strain capacity is increased, more ductility and redistribution of stresses occur with the result that large deformations can be accommodated before collapse, and the possibility of brittle failure diminishes [1].

If the member is freely supported, expansion takes place without any resistance and no reactive forces due to restraint develop. The temperature in the concrete, and also in the steel embedded in the concrete, continues to rise with consequent reduction in strength. The extent of the increase in steel temperature depends on its position in the concrete, the severity of fire, the number of exposed surfaces and the heat-penetration properties of the type of concrete used. When the temperature in tensile steel reaches the critical limit which reduces its strength to the actual stress due to loading, flexural collapse occurs.

If the heated member is restrained by the surrounding structures from free expansion, a compressive force or thrust is in fact induced in the heated member [8, 9]. In actual fire conditions, this will be possible if the fire is confined to a compartmented area containing the flexural member. The line of action of the thrust is near the bottom of the member initially, but moves upwards as the fire progresses. The thrust acts like an external prestressing force which increases the moment capacity of the flexural member and hence its resistance [8]. If, however, due to undercut or similar detailing at bearing, the line of action of the force goes above the neutral axis of the section at the critical point, the moment capacity may indeed be reduced.

The magnitude of this thrust depends on many factors and is very difficult, if not impossible, to assess accurately. Owing to a sharp decrease of stiffness and E-values of concrete at elevated temperatures and accelerated creep and consequent relaxation of stresses, the magnitude of thrust, in actual fact, is considerably less than that calculated by the use of the elastic properties of concrete and steel together with appropriate coefficients of expansion [8]. When the member contains some top reinforcement with sufficient development length, a substantial reserve of strength can be mobilized from the continuity effect, or ultimately even from the 'membrane tension state', provided the anchorage is effective and detailing of the member as well as the supporting structure is adequate [10]. The top steel being at the remote end from the exposed surface remains relatively cool, even after prolonged exposure, and retains a high percentage of its normal temperature strength and thus increases the resistance of the member.

In a continuous flexural member, downward deflection due to differential heating causes a rapid redistribution of moment until a hinge forms over the support by the yielding of the top steel. As stated earlier, the top steel being remote from the exposed surface, virtually retains its original strength. A considerable moment of resistance can therefore be developed at supports to counteract the load effect. The increase in the support moment, however, causes the point of contraflexure to move further away from the support, and if the continuity reinforcement at the top is not extended sufficiently beyond the support to accommodate the complete redistributed moment and the change in the position of inflection points, cracks will develop leading to an early failure in fire [9]. If effectively uninterrupted top reinforcement runs from support to support and the detailing at anchorage is adequate, the membrane tension state or suspension effect may also develop, considerably increasing its resistance to fire attack.

Expansion of vertical members may increase the load in these members (and therefore of members below and above them) and in a localized fire the horizontal members near the top of the affected column may be deflected upwards. The expansion of the roof or floor at the top of the column may cause excessive lateral movement of the column and unless it is made relatively flexible so that a kind of hinge is formed, such movement is likely to induce very high stresses in the column and result in appreciable shear distortion and ultimately in failure [9].

The study of fire tests, as well as actual cases, shows that for simply supported members the stability failure is governed by the time at which the temperature of steel reaches its critical value. This in turn is influenced by the thickness of concrete cover to reinforcement, the shape of the member and the type of concrete. In the case of a slab, the heat enters through one surface only, and so the shape is of little consequence. In the case of a column, heat penetrates from all four directions and for normal T-beams from three directions (for I-beams it is more complex). The width of the section in these cases greatly influences the temperature distribution in concrete. If a wide beam is used and the main reinforcement in it is placed further inside the section, increased fire resistance results from the increased protection.

In the continuous or restrained structure, the major part of the resistance is developed by the top steel and the reactive forces due to restraint of expansion/rotation, and as a result the importance of the bottom steel is diminished and hence, in most cases, the concrete cover becomes relatively unimportant.

The insulation properties of concrete, the heat transfer and conduction, vary enormously with the type of aggregate used. Calcareous (limestone) or lightweight aggregate has much lower thermal conductivity and diffusivity than gravel or similar siliceous aggregate. They are much less susceptible to physical and chemical changes at high temperatures which may cause spalling and cracking. In addition, the loss of compressive strength at elevated temperatures is also reduced in the case of concrete with carbonaceous or light-weight aggregates. Figure 2-1, taken from [11], shows the comparative reduction rate of strength of different types of concrete with temperature.

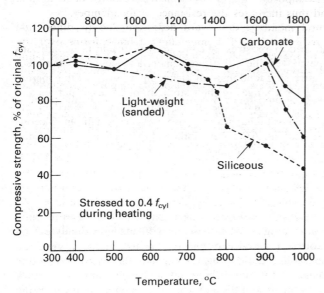

Figure 2-1   Variation of concrete strength with temperature

Tests in the USA and elsewhere confirm that the performance of light-weight aggregate concrete is much superior to that of dense gravel concrete in fire conditions.

The fire resistance is generally assessed by tests and the interpretation of test results conducted on the representative individual elements. It should be borne in mind that the performance of the whole structure is often quite different from that of an individual element or member because of the possible beneficial interaction of various members and also the possibility of an alternative path of load transfer. Detailing therefore plays a very important part in the enhancement of the fire rating of a structure.

There is basically no difference between a test to destruction of an element at normal temperature and a fire test. Broadly speaking, in the former the strength of the materials (concrete and steel) remains constant, and the load is gradually increased until failure occurs, whereas in a fire test the load remains constant and the collapse is caused by reducing the strength of the materials by an increase of temperature. The modes of failure, however, remain similar in either case. For example, if a flexural member (beam or slab) is load tested at normal temperature, the failure mode will most probably be pure flexure or the combined effects of shear and flexure or pure shear, depending on the detailing, geometry of cross-section and pattern of loading. For a similar element in the fire test, the mode of failure is not likely to be different with a similar pattern of loading. However, in past investigations of the fire resistance of beams and slabs, not a great deal of attention was paid to the role of shear as a critical factor of the limit state of stability [10].

3 Properties of materials

3.1 General

The behaviour of a reinforced or prestressed concrete structure in fire conditions is governed by the properties of the constituent materials—concrete and steel (reinforcement and/or prestressing tendons)—at high temperatures. Both concrete and steel undergo considerable changes in their strength, physical properties and stiffness by the effects of heating and some of the changes are not recoverable after subsequent cooling. Chemical changes may also occur, especially in siliceous aggregate concrete, due to heating. An understanding of these changes is essential in predicting or assessing the performance of the structure during fire and subsequent cooling.

3.2 Concrete

Concrete is a composite material consisting of aggregates and matrix as its basic components. The effects of heating on both these components individually as well as their interaction control the behaviour of concrete at high temperatures.

With the rise in temperature, the aggregates expand; the expansion of the matrix, on the other hand, is substantially offset and sometimes completely negated by shrinkage due to the evaporation of water [1]. The resultant expansion differential causes internal cracking in the concrete and reduction in its stiffness. The extent of this phenomenon differs considerably with the types of aggregate and is most pronounced in the case of concrete with siliceous aggregates, which at very high temperatures (575°C or above) also undergoes physical changes accompanied by a sudden expansion in volume, thus sometimes causing aggregate splitting and/or spalling.

Light-weight aggregates normally undergo various heating processes during manufacture and possess superior insulation properties. The physical compatibility between the matrix and the aggregate with regard to deformability and expansion characteristics is also much better in light-weight concrete than in dense concrete and as a result much less damage and internal stresses are expected in light-weight concrete during heating. For the same reason, it can also withstand cooling shock much better than gravel concrete. In a series of tests at Chalmers Technical University, Göteborg, Sweden, specimens 40 × 150 × 150 mm made of gravel concrete and of very light-weight aerated concrete of density 1100 kg/m^3 (called 3-L hydrophobe concrete) were heated up to 1000°C and then immersed in water at 20°C, followed by another cycle of heating to 1000°C and re-immersion in water. The dense concrete specimen exploded as soon as it was thrown into the water for the first time. The 3-L specimen was undamaged after 15 cycles [12].

Calcareous aggregates do not normally undergo physical changes during heating and are usually free from cracks and local damage. But at exceptionally high temperatures some chemical changes take place when free lime is liberated from the calcium carbonate. This chemical process is in fact beneficial during heating as it retards the temperature rise. During cooling, however, free lime chemically combines with atmospheric moisture and expands in volume causing cracks and damage [1].

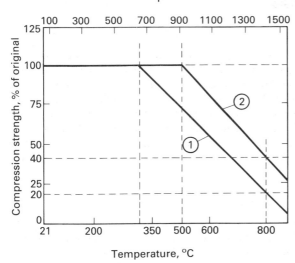

Figure 3-1   Variation of concrete strength with temperature (idealized)
1   Dense concrete
2   Light-weight concrete

Both light-weight and calcareous aggregate concrete possess low thermal conductivity (and hence low thermal diffusivity), which results in less temperature rise in light-weight or calcareous aggregate concrete than in one with siliceous aggregate, after equal exposure to fire.

All concrete loses strength at elevated temperatures, but the rate of reduction differs with the type of aggregate used. In Fig. 2-1, a comparison of the reduction of compressive strength of different concrete with temperature has been shown.

On the basis of these, as well as the British and European tests, a Joint Committee of the Institution of Structural Engineers and the Concrete Society (of Great Britain) produced, in its Interim Guidance [9], idealized strength reduction curves, which are presented in Fig. 3-1. The curves are useful for design purposes and, being lower bounds, give results on the safe side. It can be seen that light-weight aggregate concrete retains its original strength (i.e. the strength at 20°C) up to 500°C. Thereafter, reduction takes place and only 40% of the original strength is retained at 800°C. For dense concrete (with siliceous aggregate) the reduction generally commences at 350°C and only 20% of the original strength is retained at 800°C. For calcareous or carbonaceous aggregate concrete, reduction can be taken as similar to that of light-weight concrete.

The Young's modulus or E-value of concrete also drastically reduces with temperature, as can be seen from Fig. 3-2 taken from [1]. The reduction is more for concrete with higher water/cement ratio. Also, the moist-cured specimen generally shows more reduction than the air-cured specimen. On these curves, a suggested mean graph for the reduction of the E-value has been added, which may be used for design purposes.

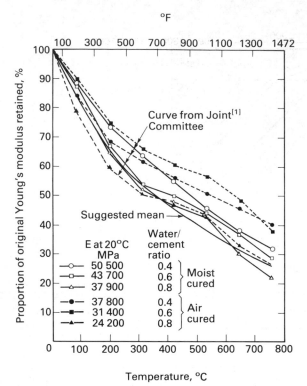

°F

Figure 3-2   Variation of *E*-value of concrete with temperature

The creep of concrete increases at an accelerated rate at high temperatures and is affected by the age, moisture content, type and strength of concrete and the stress/strength ratio [8, 13]. It considerably affects the deformation of the element in fire, especially when the exposure is for a long period. On the other hand, thermal and other stresses are relaxed and the magnitude of thrust which may develop due to thermal restraint is appreciably modified. It should perhaps be mentioned that neither the reduction in *E*-value nor the accelerated creep has any significant effect on the limit state of collapse condition [3].

Increase in the strain capacity and ductility of concrete at high temperatures has already been mentioned in Section 2. It has been established that the stress–strain curve of concrete in uniaxial compression retains the same general form at elevated temperatures, but for the peak stresses the range of strain is extended. In the guidelines [9] a semi-parabolic stress–strain diagram similar to one in the British Code of Practice CP 110 [14] for normal temperature design, has been suggested for design for fire resistance with the variation that, for temperatures above 500°C, the maximum strain can be taken as 0.006 (instead of 0.0035 at normal temperatures) with the peak stress at 0.67 times the actual cube strength

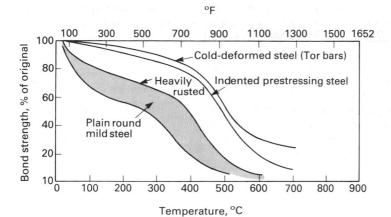

Figure 3-3   Relative bond strength of various bars as a function of temperature

corresponding to the level of temperature in the concrete, divided, of course, by the partial safety factors [9].

Information on the bond characteristics of concrete at elevated temperatures is very limited. In the last few years, some research has been carried out at the Technical University of Braunschweig, West Germany, on the subject. In a specially developed testing rig, pull-out tests were carried out for plain round mild-steel bars, indented prestressing wires and cold-worked deformed bars. The apparatus was so designed that a uniform semi-steady heating of the specimen was possible, and also the maximum load, 200 kN, could be applied at 800°C.

The results of the tests have been recently published by Diederichs and Schneider [15]. The reduction and rate of reduction of bond strength (at 20°C) with temperature for various types of steel are reproduced in Fig. 3-3.

The paper also concludes that the bond strength depends not only on the temperature level but also on the test procedure and the shape of the bar. The loss of bond strength for deformed bars at constant elevated temperature is of the same order as the loss of compressive strength of concrete. At the same strength, the plain round bars show a sharper decrease in bond strength.

3.3   Steel

Like concrete, steel also loses strength with temperature. The magnitude of the loss and the rate of reduction depends upon the type of steel and its manufacturing process. Mild steel and hot-rolled high-tensile steel retains only 50% of its normal temperature (20°C) yield stress at about 600°C, but up to a temperature of about 300°C the strength (yield stress) appears to be slightly higher than that at room temperature. For cold-worked high-tensile reinforcement, 50% reduction of the yield stress usually occurs at a temperature around 550°C. The rise in strength up to a temperature of 300°C is also less.

If reinforcing bars are heated up to 600°C, they virtually recover their full normal temperature strength when cooled to room temperature again. If, however, steel is heated to 800°C, then on cooling to room temperature full strength will not be regained. In the case of hot-rolled bars, this reduction is minimal (within 5%), but for cold-worked high-tensile bars this may be in the region of 30% [1].

The modulus of elasticity of reinforcement is also affected by temperature. Up to 600°C, E-values gradually reduce to 80% of their normal temperature (20°C) value but thereafter the rate of decrease is much higher [8]. The change in the E-value of reinforcement is practically the same for all types.

Prestressing steel loses its strength in larger proportion with temperature and the reduction usually starts from 150°C. Only 50% of its original tensile strength (i.e. at 20°C) is retained when the temperature is about 400 to 450°C. The reduction of the E value is also higher than in reinforcement and the relaxation of

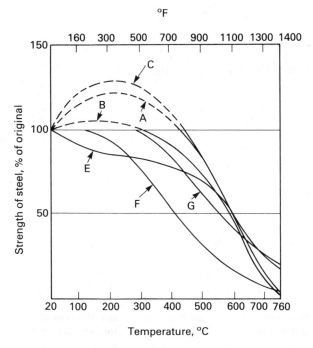

Figure 3-4   Variation of steel strength/yield stress with temperature
A   Hot-rolled high-yield bars (yield) ⎫
B   Cold-worked bar (yield)   ⎬ British tests
C   Mild-steel bar (yield)   ⎭
E   Hot-rolled steel (yield strength) ⎫
F   Cold-drawn prestressing steel
    (tensile strength)   ⎬ US tests
G   High-strength alloy-steel bars
    (tensile strength)   ⎭

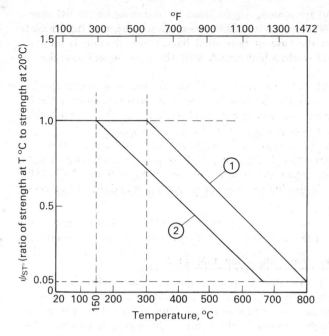

Figure 3-5 Variation of steel strength-yield stress with temperature (idealized)
 1   High-yield reinforcement bars
   Mild steel
   High-strength alloy-steel bars (e.g.
   Macalloy bars)
 2 Prestressing wires or strands

the prestressing forces in prestressed wire, bar or strands is greatly accelerated at high temperatures.

Figure 3-4 shows the US and British test results on the strength reduction of various types of steel with temperature, adapted from references [1] and [8].

In the Interim Guidance—*Design and detailing of concrete structures for fire resistance* [9], idealized design curves have been given to assess reduction in the strength of reinforcement and prestressing steel with temperature. The curves are reproduced in Fig. 3-5. The curves are simple and easy to use and at the same time, being lower bounds, give results on the safe side.

4   Temperature distribution in concrete elements

The knowledge of temperature distribution in the cross-section is essential for a fair assessment of fire resistance of a concrete element. As discussed in Section 2, temperature distribution depends on the shape of the element exposed to fire, the severity of fire and the type of aggregate used in the concrete. This is also influenced by moisture content and moisture–vapour movement in the concrete [2].

The classical equation for a transient heat flow in a concrete mass is given by

$$\frac{K_1}{C\gamma}\left[\frac{\partial^2 T}{\partial^2 x}+\frac{\partial^2 T}{\partial^2 y}\right]=\frac{\partial T}{\partial t}$$

where T is the temperature, t the time, γ the density of concrete, K_1 the thermal conductivity and C the specific heat of concrete.

The solution of this differential equation is rather involved and even with some simplified assumptions is difficult because of the complicated boundary conditions. Besides the equation does not take into account the influence of vapour movement inside the concrete.

Through international research and experiments, a good deal of information is now available about the insulating properties of concrete, the rate of penetration of heat and the temperature distribution in concrete exposed to a standard fire, which forms the basis of rational design for fire resistance.

Figure 4-1 shows the temperature in a slab exposed to standard fire for $\frac{1}{2}$, 1, 2, 3 and 4 h periods. The firm lines represent the temperature distribution for dense gravel or siliceous aggregate concrete (density over 2300 kg/m³). The broken lines indicate the temperature distribution in the light-weight aggregate concrete slab, where density will not exceed 2000 kg/m³. The figure has been adapted from [9]. These curves are also applicable to walls, because, like slabs, only one side of a wall is normally exposed to fire. There may, of course, be circumstances in

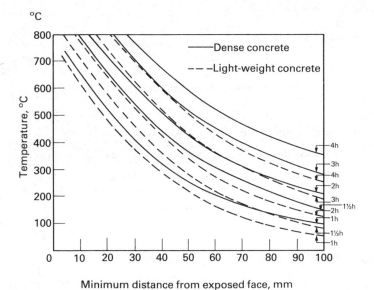

Figure 4-1 Temperature distribution in slab

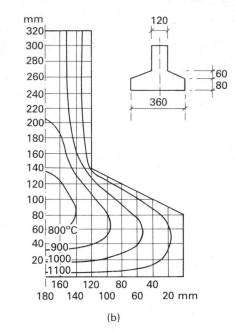

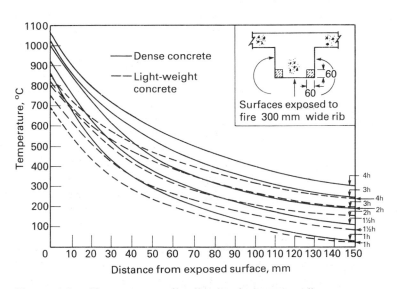

(a) (b)

Figure 4-2 Typical temperature profile after 2h in (a) T-beam and (b) I-beam

Figure 4-3a Temperature distribution in beam or rib.
Note that the temperature in the hatched area may be 10% higher than that obtained from the curves

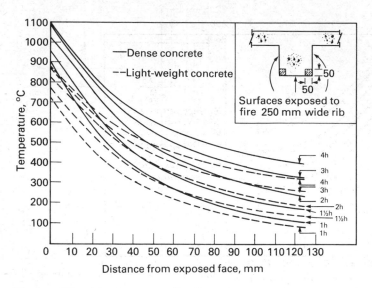

Figure 4-3b Temperature distribution in beam or rib.
Note that the temperature in the hatched area may be 10% higher than that obtained from the curves

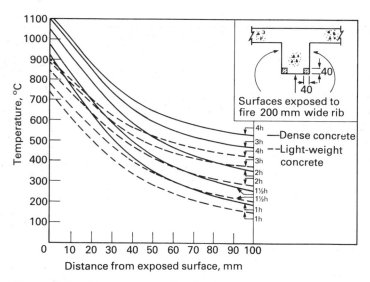

Figure 4-3c Temperature distribution in beam or rib.
Note that the temperature in the hatched area may be 10% higher than that obtained from the curves

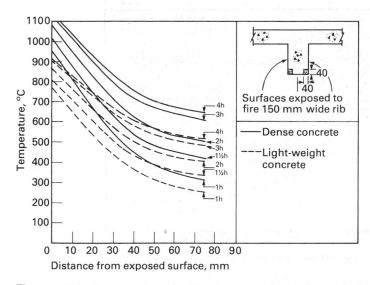

Figure 4-3d   Temperature distribution in beam or rib.
Note that the temperature in the hatched area may be 10% higher than that obtained from the curves

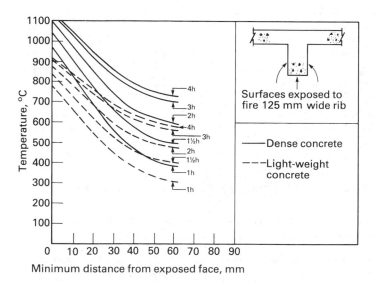

Figure 4-3e   Temperature distribution in beam or rib

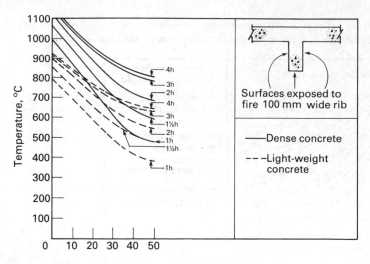

Figure 4-3f   Temperature distribution in beam or rib

practice when a wall may be heated on both sides if, for example, a fire spreads from room to room, or for external walls flames project from the windows.

Temperature distribution in beams is more complex as three or more surfaces could be exposed to fire. The width of the exposed surface influences the distribution, and a typical temperature profile in T-beams after 2 h exposure to standard fire, and in an I-beam for a similar period of exposure, is given in Fig. 4-2a, b [2]. The profiles are true for the particular widths shown in the figure and will vary with dimension and duration. For facility of design, some simplified, easy to use, design curves, similar to those for slabs, have been prepared for the temperature distribution in T-beams of varying width in [9] for $\frac{1}{2}$ to 4 h duration of exposure to standard fire. Figures 4-3a–f are reproduced from that document. For each width, the firm lines represent the distribution in gravel concrete and the broken lines the distribution in light-weight concrete of density below 2000 kg/m³.

5   Methods of determining the fire resistance of an element

The methods currently used for determining the fire resistance of a structural element are as follows:

(a) Tests of a full-sized member representative of the actual element in construction and the direct application of such tests results.

(b) Compliance with tabulated data for the minimum dimensions, sizes and cover to reinforcement compiled from various tests.

(c) Use of empirical rules based on the interpolation or extrapolation of fire test results having taken account of the variations from the test conditions.

(d) Analytical techniques based on tests and research studies, as well as the knowledge of the behaviour of the structure under ultimate conditions.

5.1 The design based on tests

Tests on representative members obviously are the most reliable way of establishing the fire resistance of a structural member and should be preferred whenever possible.

In general, the fire resistance of concrete elements is influenced by the following parameters:

(a) Size and shape of the element.
(b) Detailing, type and quality of reinforcement or prestressing tendon.
(c) The level of load supported and the pattern of loading.
(d) Type of concrete and aggregate.
(e) Condition at ends and bearing.
(f) Protective cover to reinforcement.

An accurate assessment of all the above factors is not feasible and only in representative tests is their influence duly reflected. In Section 1.3, the method of test and the limiting criteria of failure have been discussed. Fire tests are, however, expensive, time consuming and, sometimes, may not even be practical.

5.1.1 *Actual fire tests of structural members*

The effect of the bearing condition on the fire resistance of an element is clearly shown in the three tests carried out by the Fire Research Station, England. In all cases, 300 mm deep precast prestressed double-tee units of the same size and span (6.8 m) were used with 75 mm *in-situ* structural topping. In two tests, the ends were freely supported (equivalent to support over brickwork), whereas in the other case the ends were cast into concrete edge beams with properly anchored anti-crack reinforcement at the top. The width of the ribs, the concrete quality and the prestressing tendons were similar in all cases. The bottom cover (clear) to steel in the ribs was the same for all the units but the side cover in the unit for the third test (restrained case) was 12 mm less. The minimum width of the ribs was 127 mm in all specimens.

In both simply supported cases, failure occurred after about 1 h and 20 min (actually 1 h 18 min and 1 h 30 min). But in the third test, fire was continued for 4 h and no structural collapse occurred. During cooling, the slab part at the junction of two double-tee units failed [1, 16, 17].

Time-deflection curves show that, after 1 h fire exposure, the deflection in the case of the end-restrained beam was negligible and even after 2 h it was under 50 mm [16].

Recently, a fire test was performed at the Warrington Research Centre, England, on a specimen of Lytag light-weight aggregate concrete beam-and-slab construction (reinforced concrete). The overall depth of the beams was 350 mm inclusive of the 150 mm thick slab. The overall width of the construction was 2900 mm, the ribs being 1800 mm apart. The width of the rib was 180 mm. The beams were designed for partial continuity at normal temperature conditions. The span was 4.46 m, and the continuity steel was provided for superload (i.e. for live load and finishes) only. The minimum clear cover to the bottom reinforcement was 45 mm and to top reinforcement 25 mm. The test specimen was 6.070 m long, thus projecting 805 mm on each end of the furnace and the continuity effect was achieved by applying sufficient cantilever load at the projecting ends such that no rotation could take place. The specified characteristic strength of the concrete was 30 MPa. In actual fact, the specimen concrete cube showed an average strength of 48.25 MPa at the time of testing. The moisture content was measured on the day of test and was found to be 5.7% by weight.

The test was discontinued after 4 h 3 min. The deflection measured after 1, 2, 3 and 4 h was 23, 38, 50 and 56 mm, respectively. The test satisfied all the requirements of the stability test, integrity test and insulation test. After 24 h cooling, the specimen was successfully tested for re-loading also. According to the tabulated data, the construction could have only about 2 h fire resistance.

5.2 Design based on tabulated data

National codes of practice of most of the countries specify certain cover and/or minimum dimensions for structural members for a specific period of fire rating. The use of such tabulated data offers the quickest method of checking the fire resistance and its simplicity no doubt appeals to the designer. Data relating to the minimum dimension and cover have been compiled from a series of fire tests, but recommended values vary a great deal in different countries. The drawback of this method is that it is rather inflexible and fails to recognize any individual variations. In certain cases, the values may even prove unsafe.

5.2.1 BRE recommendations

In the UK, the Building Research Establishment has published a report [18] in which detailed recommendations have been made with respect to the fire-resistance rating and requirements of all building elements and materials. The tables were prepared by the Fire Research Station and are based on national information as well as consultations with the FIP Commission on Fire Resistance. The recommendations, inasmuch as they are applicable to concrete structures, have already been incorporated in the Interim Guidance [9] and are expected to be included in the British Code of Practice CP 110 when the revised edition comes into practice. It is also very likely that these recommendations will have 'deemed to satisfy' status of the *Building Regulations*. Tabulated data, of necessity, have to be on the conservative side as they tend to cover in a simple form the whole range of variations in the construction. The BRE data are presented in Tables 5-1 and 5-2 for dense gravel concrete and light-weight aggregate concrete, respectively. The following 'guidance notes' form an integral part of the tables and hence should be read in conjunction with them.

(a) *Cover*

  (1) For solid slabs, cover is the average distance from the heated surface to the reinforcement. With single-layer reinforcement, the actual distance is used. For multi-layer reinforcement, the average distance is calculated as

$$a_{av} = \sum A_s a_1 / \sum A_s$$

where A_s is the area of tensile steel and a_1 is the distance between the nearest exposed surface and the main steel. If a_1 for a particular reinforcement is less than half of a_{av}, than that reinforcement should be disregarded in the calculation of the ultimate resistance at high temperature.

(2) For rectangular beams, cover a_1 is the distance from the nearest exposed face to the steel. If the corner bars have a_1 less than half of a_{av}, these should be ignored in the calculation of the ultimate resistance at high temperature.

(3) For I-beams, the effective cover, having been determined as in (2), should be multiplied by 0.6 to allow for the additional heat transfer through the upper face of the flange. The web thickness of fully exposed I-section beams should not be less than 0.5 of the minimum width specified in the tables for beams for various fire-resistance periods.

(b) *Width* For all beams the width should be determined at the level of the lowest reinforcement.

(c) *Thickness* In the case of solid slabs, the actual thickness together with any directly applied non-combustible finish should be considered. For hollow slabs or ribs with filler blocks, the actual thickness should be converted into effective thickness as follows:

$$d = h\sqrt{S}$$

where d is the effective thickness, h the actual thickness and S the proportion of solid material per unit width.

(d) *End condition* The member can be regarded as continuous if the reinforcement provided at the top can develop restraining forces capable of resisting end moments present in normal loading conditions as well as those induced because of the heating of the elements.

(e) *Supplementary reinforcement* Supplementary reinforcement is required to reduce the possibility of the concrete spalling when the actual cover to the outermost steel exceeds 40 mm for dense concrete and 50 mm for light-weight concrete. Such reinforcement can consist of (i) expanded metal lath or wire fabric not less than 0.5 kg/m^2 and wire centres not exceeding 100 mm, or (ii) links at not more than 200 mm centres. The reinforcement should be located so that is not more than 20 mm from the exposed face.

(f) *Additional protection* If the size of the construction is found to be inadequate and it is not possible or desirable to increase it to meet the requirements of the table, the fire resistance may be enhanced by the application of a protective coating or construction. It is important in such a case to ensure proper adhesion of the coating.

Table 5-1 Minimum dimensions and cover for structural members in dense siliceous aggregate concrete for various periods of fire resistance (BRE)

	Member	End condition	Minimum dimension mm	For fire resistance of 4 h	3 h	2 h	1½ h	1 h
Reinforced concrete	Columns fully exposed	NA	Width	450	410	300	250	200
			Cover	35	35	35	30	25
	Columns 50% exposed	NA	Width	350	300	200	200	160
			Cover	35	30	25	25	25
	Walls/cols with one face exposed	NA	Thickness	240	200	160	140	120
			Cover	25	25	25	25	25
	Beams	Simply supported	Width	280	240	200	150	120
			Cover	80	70	60	40	30
		Continuous	Width	240	200	150	120	80
			Cover	70	60	50	35	20
	Floors with plain soffit	Simply supported	Thickness	170	150	125	110	95
			Cover	55	45	35	25	20
		Continuous	Thickness	170	150	125	110	95
			Cover	45	35	25	20	20
	Floors with ribbed open soffit	Simply supported	Thickness	150	135	115	105	90
			Width	175	150	125	110	90
			Cover	65	55	45	35	25
		Continuous	Thickness	150	135	115	105	90
			Width	150	125	110	90	80
			Cover	55	45	35	25	20
Prestressed concrete	Beams	Simply supported	Width	280	240	200	150	120
			Cover	90	80	70	55	40
		Continuous	Width	240	200	150	120	100
			Cover	80	70	55	40	30
	Floors with plain soffit	Simply supported	Thickness	170	150	125	110	95
			Cover	65	55	40	30	25
		Continuous	Thickness	170	150	125	110	95
			Cover	55	45	35	25	20
	Floors with ribbed open soffit	Simply supported	Thickness	150	135	115	105	90
			Width	200	175	150	135	110
			Cover	75	65	55	45	35
		Continuous	Thickness	150	135	115	105	90
			Width	175	150	125	110	75
			Cover	65	55	45	35	25

Table 5-2 Minimum dimensions and cover for structural members in light-weight concrete (density not greater than 2000 kg/m³) for various periods of fire resistance (BRE).

	Member	End condition	Minimum dimension mm	For fire resistance of				
				4 h	3 h	2 h	1½ h	1 h
Reinforced concrete	Columns fully exposed	NA	Width	260	320	240	200	160
			Cover	35	35	35	25	20
	Columns 50% exposed	NA	Width	275	250	185	160	130
			Cover	30	30	25	25	20
	Walls/cols with one face exposed	NA	Thickness	190	160	130	115	100
			Cover	25	25	25	20	20
	Beams	Simply supported	Width	250	200	160	130	100
			Cover	65	55	45	35	20
		Continuous	Width	200	150	110	90	80
			Cover	55	45	35	25	20
	Floors with plain soffit	Simply supported	Thickness	150	135	115	105	90
			Cover	45	35	25	20	15
		Continuous	Thickness	150	135	115	105	90
			Cover	35	25	20	20	15
	Floors with ribbed open soffit	Simply supported	Thickness	130	115	100	95	85
			Width	150	135	100	85	75
			Cover	55	45	35	30	25
		Continuous	Thickness	130	115	100	95	85
			Width	125	100	90	80	75
			Cover	45	35	30	25	20
Prestressed concrete	Beams	Simply supported	Width	250	200	160	130	110
			Cover	75	65	55	45	30
		Continuous	Width	200	150	125	100	90
			Cover	65	55	45	35	25
	Floors with plain soffit	Simply supported	Thickness	150	135	115	105	90
			Cover	60	45	35	30	20
		Continuous	Thickness	150	135	115	105	90
			Cover	45	35	30	25	20
	Floors with ribbed open soffit	Simply supported	Thickness	130	115	100	95	85
			Width	175	150	125	110	90
			Cover	65	55	45	35	30
		Continuous	Thickness	130	115	100	95	85
			Width	150	125	110	90	75
			Cover	55	45	35	30	25

5.2.2 FIP–CEB report

The FIP–CEB report, *Methods of assessment of the fire resistance of concrete structural members* [19], was published in 1978 and in the absence of 'national recommendations' the values in this document are usually accepted. One basic difference between the FIP–CEB tables and the BRE tables is that in the former no distinction has been made between a reinforced concrete and a prestressed concrete member with respect to the minimum cover and width, although it is known that the proportion of loss of strength of prestressing wires or strands at elevated temperatures is significantly higher than the loss of strength in reinforcement bars. The FIP–CEB recommendation for reinforced and prestressed concrete beams are given in Tables 5-3 and 5-4 for gravel concrete and lightweight aggregate concrete respectively. The tables are for simply supported end conditions only and without any end restraint. For continuous beams, if no test results are available, the fire resistance given in Tables 5-3 and 5-4 may be increased, as

Table 5-3 Minimum dimension and cover for reinforced and prestressed concrete beams in dense siliceous concrete for various periods of fire resistance (FIP–CEB)

Fire resistance period h	Beam width (b) and corresponding minimum axis distance (a) (See Fig. 5-1) mm					Minimum web thickness mm
$\frac{1}{2}$	b	80	120	160	200	80
	a	25	15	10	10	
1	b	120	160	200	200	100
	a	40	35	30	30	
$1\frac{1}{2}$	b	150	200	280	400	100
	a	55	45	40	35	
2	b	200	240	300	500	120
	a	65	55	50	45	
3	b	240	300	400	600	140
	a	80	70	65	60	
4	b	280	350	500	700	160
	a	90	80	75	70	

$$a_{st} = a + 10 \text{ mm} \qquad a_{st} = a$$

(1) Values are valid for moisture content not exceeding 2–3% by weight.
(2) For carbonaceous aggregate either the minimum dimension of the cross-section or the minimum value of the axis distance may be reduced by between 5 and 10%
(3) a_{st} is measured from the side and an a from soffit (see Fig. 5-1).

Table 5-4 Minimum dimensions and cover for reinforced or prestressed concrete beams in light-weight concrete (density up to 1200 kg/m³)
for various periods of fire resistance (FIP–CEB)

Fire resistance h	Beam width (b) and corresponding minimum axis distance (a) (See Fig. 5-1) mm					Minimum web thickness mm
$\frac{1}{2}$	b	80	120	160	200	80
	a	20	15	10	10	
1	b	100	160	200	300	80
	a	40	30	25	20	
$1\frac{1}{2}$	b	120	200	280	400	80
	a	55	40	35	30	
2	b	160	240	300	500	100
	a	65	50	40	35	
3	b	190	300	400	600	115
	a	80	65	55	50	
4	b	225	350	500	700	130
	a	90	75	65	55	

$$a_{st} = a + 10 \text{ mm} \qquad a_{st} = a$$

If the density of concrete is 1850 kg/m³, a 10% increase in either the minimum
dimension of cross-section or the minimum axis distance will be required. For
density between 1200 and 1850 kg/m³, the values may be interpolated.
a_{st} is measured from the side and an a from soffit (see Fig. 5-1).

a guide, by 30 min for beams having a width less than 180 mm and by 60 min for
beams with width equal to or over 200 mm. Table 5-5 show the FIP–CEB
proposal for slabs with freely supported end conditions. For continuous slabs,
similar increases as in the beams may be permitted except for two-way slabs. In
the case of light-weight concrete particularly, the FIP–CEB values appear to be
very conservative. This is, of course, expected, as the variation in the
properties of light-weight aggregate concrete used internationally are
considerable.

An important point in the FIP–CEB tables is that the reduction of cover (or
axis distance) with the increase in the width of the section, or vice versa for the
same period of fire resistance, has been suggested.

5.2.3 *PCI proposal*
Because of its inherent shortcomings, the tabulated data, except for columns or
walls, appears to be less popular in the USA than anywhere else. The American

Table 5-5 Minimum thickness and axis distance for reinforced and prestressed concrete slabs (FIP–CEB)

| Design details | Fire resistance rating | | | | | |
	$\frac{1}{2}$ h	1 h	$1\frac{1}{2}$ h	2 h	3 h	4 h
Dense aggregate concrete Slab thickness, h (mm)	60	80	100	120	150	175
One-way Axis distance, a (mm)	10	25	35	45	60	70
Two-way $l_y/l_x \leqslant 1.5^a$ Axis distance, a (mm)	10	10	15	20	30	40
Two-way $l_y/l_x \geqslant 2.0$ Axis distance, a (mm)	10	25	35	45	60	70
Light-weight concrete Slab thickness, h (mm)	60	65	80	95	120	140
One-way Axis distance, a (mm)	10	20	30	40	50	55
Two-way $l_y/l_x \leqslant 1.5$ Axis distance, a (mm)	10	10	10	15	25	30
Two-way $l_y/l_x \geqslant 2.0$ Axis distance, a (mm)	10	20	30	40	50	55

[a] l_x and l_y are the spans of the slab in two directions at right angles.
(1) Moisture content assumed for dense concrete is 2–3% by weight
(2) Values for light-weight concrete are valid for density not greater 1200 kg/m³. For density of 1850 kg/m³, the values (either thickness or axial distance) should be increased by 10%. For density between 1200 and 1850 kg/m³, the values may be interpolated.

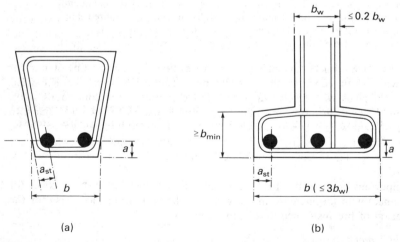

(a)   (b)

Figure 5-1 Parameters for Table 5-3 and 5-4

Table 5-6   Minimum dimensions for prestressed concrete members in dense concrete (PCI)

Member	End condition	Minimum dimension mm	For fire resistance of				
			4 h	3 h	2 h	$1\frac{1}{2}$ h	1 h
Beams	Simply supported	Width	305	255	195	155	115
		Cover	95	80	65	50	30
Beams or ribs	Continuous 1.2 m or less on centres	Width	110	98	79	65	65
		Cover	45	38	33	33	33
Beams	Continuous over 1.2 m on centres	Width	195	155	115	115	115
		Cover	65	50	45	45	45
Slabs (floors with plain soffit)	Simply supported	Thickness	178	158	127	110	90
		Cover	71	50	46	38	28
	Continuous	Thickness	178	158	127	110	90
		Cover	25	25	20	20	20

Insurance Association's [20] rating for fire resistance is virtually followed in all states of the USA. In 1971–72, the Prestressed Concrete Institute's Committee on Fire Resistance Rating prepared tabulated data for prestressed concrete members. These, however, were not officially published on the consideration that some of the values were rather conservative. In Table 5-6, the tentative PCI proposal for prestresssed concrete members (in dense concrete) is reproduced mainly to compare the values with those in the BRE report for similar members. Subsequently, the PCI published a report [11] on the fire resistance of post-tensioned structures. The recommendations in that report with respect to cover, etc., are similar to those in the original proposal. A comparison of Table 5-6 and the prestressed concrete part of Table 5-1 shows that while for simply supported end conditions the values in Table 5-6 are more severe, for continuous structures the PCI proposals are much less stringent than in the British recommendations. The continuous beam or slab in the PCI proposal is defined as that which has been designed for continuity for full superimposed load at least, and has a minimum of 20% of the total reinforcement required for maximum negative moment made continuous throughout the span. Beams and slabs which are built into surrounding structures in such a way that restraints to thermal expansion will occur, are also regarded equivalent to continuous, although for design in normal temperatures, the support condition may have been taken as simply supported.

An interesting point in the PCI proposal is that, for continuous beams spaced at close centres (less than 1.2 m or 4 ft), the requirements with regard to cover

Table 5-7   Fire endurances proposed by Hull and Ingberg for columns

Aggregate type	Minimum area of round or square cross-section cm^2	Concrete cover mm	Fire endurance classification h
Siliceous	710	38	$1\frac{1}{2}$
Siliceous	1290	38	$2\frac{1}{2}$
Siliceous (mesh in cover)	1290	38	$3\frac{1}{2}$
Siliceous	1613	64	3
Siliceous (mesh in cover)	1613	64	6
Traprock and slag	1290	38	4
Carbonate	1290	38	6

and width have been significantly reduced. In other words, the fire resistance of closely spaced continuous beams is considered to be appreciably higher than that of widely spaced continuous beams under similar conditions. Such a distinction is perhaps justifiable from the consideration that closely spaced beams tend to act like an anisotropic slab, ensuring better transverse distribution of load. The capacity of load distribution to adjacent ribs undoubtedly improves the fire performance, since it may reduce the risk (generally present in primary beams) of sudden collapse due to spalling. This fact has been recognized in the BRE report also, as a separate column has been provided for open ribbed slabs which need less stringent requirements with regard to dimensions, cover, etc. than beams, although not to the same extent as in the PCI proposal for beams spaced on 1.2 m centres or less [3].

The demarcation line for the spacing of ribs or beams (1.2 m or less) apparently has been drawn for practical considerations and from experience of standard double-tee units which have rib centres at 1.2 m centres usually. In the British proposal, there is no specific recommendation about the maximum spacing of ribs in an open ribbed slab, beyond which the ribs will be regarded as beams.

The BRE recommendations for column dimensions for different periods of fire resistance have been included in Tables 5-1 and 5-2. In most US building codes, columns larger than 305 mm in diameter or 305 mm square are assigned, respectively, 3 and 4 h fire rating classifications. ACI Committee 216 [8] suggests that for columns the fire endurance proposed by Hull and Ingberg, which is reproduced in Table 5-7, should be accepted.

5.3   Use of empirical rules

This method is virtually an extension of the previous two methods (i.e. tests and tabulated data). It requires an estimation of the effect of the alteration of load, boundary and support conditions and materials in the case of the actual element, from those used in the tests, and making a judgement about the modification needed in the application of the 'test results' or tabulated data based on test results to that particular element. For example, if a column is over-designed (i.e. if the column does not carry its capacity load at any time), then the relaxation of the requirement of the minimum dimension for a specific period of fire resistance should be possible. Studies at the Fire Research Station, England, suggest that it is possible to produce a correlation chart which will help to quantify the amount of relaxation in the minimum-dimension requirement of a column with the reduction in the load level.

5.4   Analytical approach

When all conditions are known and/or can reasonably be assumed, the fire resistance of an element can be reliably computed by an analytical technique from the 'stability' considerations. The knowledge of the temperature distribution in a concrete section when exposed to fire and the effects of high temperatures on the strength and other structural properties of the constituent materials (concrete and steel) of the element under consideration are essential for any computational approach. Full appreciation of the ultimate limit states of structures at normal temperatures is also essential to understand failure conditions/mechanism in fire and to formulate design principles. The advantage of the analytical method is that the limitations of tabulated data can be overcome and individual conditions, detailing and interaction can be taken into consideration and thus make the assessment more realistic.

As discussed in Sections 3 and 4, through international research and tests, enough information is now available about the behaviour and properties of concrete and steel when exposed to fire, and on the basis of such information design rules and recommendations can be established consistent with the limit-state philosophy of design. However, it should be stated that the application of design principles or the analytical technique is still confined mainly to flexural members (beams and slabs) and where failure is governed by the yielding of tensile steel. As yet, it has not been possible to extend the computational approach reliably to those structural members which are predominantly under compression, such as columns and walls. Although recently some analytical procedures have been suggested in the technical papers published by the National Research Council of Canada, Division of Building Research [21, 22], tabulated data or experience from fire tests still form the basis of assessment of fire resistance for columns, with the emphasis on good detailing.

5.4.1   *Recommendations for design of flexural members in* [9]

The Interim Guidance given by a Joint Committee of the Institution of Structural Engineers and the Concrete Society suggests a design method for fire resistance of flexural members which is comprehensive and covers all modes of failure likely to occur in a flexural member. As the concept of 'nominal shear stress' is inappropriate in cracked sections, the problem of shear has been dealt with by resolving, from first principles, the magnitudes and direction of actual forces and relying on the concrete taking compression and the steel taking tension.

Observations of tests in a flexural member confirm that there are basically three modes of failure:

(a) A flexural mode of failure characterized by vertical cracks (pure flexure failure, practically no influence of shear).
(b) A failure mode characterized by inclined cracks initiated by flexural tensile crack (failure due to the combined action of shear and flexure).
(c) A failure mode characterized by diagonal splitting under a biaxial stress field in the web (shear predominant and influence of flexure negligible).

The Interim Guidance incorporates design procedure for each mode of failure under fire conditions.

The member or element concerned is first designed for normal gravity and/or lateral loads under normal temperature conditions in accordance with the requirements of the national codes of practice and then checked for the appropriate period of fire resistance. The limit state of collapse condition needs only to be considered for fire design.

The following partial safety factors have been suggested:

Materials	Concrete	$\gamma_{mc} = 1.30$
	Steel	$\gamma_{ms} = 1.00$
Loads	Dead load	$\gamma_g = 1.05$
	Live load	$\gamma_q = 1.00$

As the requirements of fire resistance are intended for a completed building (when concrete is sufficiently old and mature) an age factor of 1.2 may be applied to concrete strength related to the 28-day characteristic strength, for design.

5.4.1.1 *Simply supported member—design for pure flexural mode of failure* This mode of failure is usually the governing consideration in the design when the shear span/effective depth ratio exceeds 6. Shear span is defined as the ratio of bending moment to shear at the section considered.

For a simply supported member without any end restraint or fixity, the mechanics of design are to determine, by using the design graphs in Fig. 4-1 for slabs or Figs 4-3a–f as appropriate for beams (or by calculation if possible), the temperature of the concrete at the level of tensile reinforcement corresponding to the period of exposure to the standard fire. From that, the residual strength of steel is determined by using the idealized curves given in Fig. 3-5 or any other reliable way. As in the case of normal temperature design, for prestressing tendons the residual strength will be related to characteristic tensile strength (f_{su}) whereas for reinforcement bars this will be related to characteristic yield stress (f_y).

The average temperature of concrete in the compression region is similarly determined using the appropriate curves, and the residual strength of concrete in compression is assessed by using the idealized graph in Fig. 3-1.

Having thus obtained the actual strength level of concrete and steel, the moment of resistance is computed at the critical section in a similar manner to normal temperature conditions (i.e. by equating the tensile force in the steel to the compression in concrete and multiplying the balanced force by the lever arm between the tensile force and the compressive force). If the simplified rectangular stress block is used in the calculation of the moment of resistance, the uniform concrete stress is taken as $0.67f_{cT^\circ}/\gamma_{mc}$, where f_{cT° is the residual concrete strength at temperature T°. Alternatively, a semi-parabolic stress–strain diagram, as described in Section 3.2 with the maximum strain up to 0.006 (at temperatures above 500°C) and the maximum stress of $0.67f_{cT^\circ}/\gamma_{mc}$, may be assumed.

If ψ_{sT° and ψ_{cT° respectively represent the fraction of normal temperature (20°C) strength of steel and concrete at T°, then for a rectangular beam

$$F_{suT^\circ} = A_s(f_y\psi_{sT^\circ})/\gamma_{ms}$$

where A_s is the total area of tensile steel, F_{suT° the tensile force at temperature T° and f_y the characteristic yield stress of steel at 20°C (for prestressing tendons f_y should be replaced by f_{su} or f_{ps} as appropriate).

The concrete strength at temperature T° is

$$f_{cT^\circ} = \psi_{cT^\circ}f_{cu} \times 1.2$$

where 1.2 is the age factor and f_{cu} the characteristic concrete cube strength at normal temperature (20°C).

The average uniform concrete stress at failure is given by

$$f_c = 0.67f_{cT^\circ}/\gamma_{mc} = 0.67\psi_{cT^\circ} \times f_{cu} \times 1.2/1.3$$

taking $\gamma_{mc} = 1.3$.

Therefore, the depth of the compression block is given by

$$d_n = F_{suT^\circ}/f_c b$$

where b is the width of the beam. Therefore,

$$M_{RT^\circ} = F_{suT^\circ}z$$

where $z = $ lever arm $= (d - \frac{1}{2}d_n)$, d being the effective depth, and M_{RT° is the moment of resistance at T°. If $M_{RT^\circ} \geqslant M_{aT^\circ}$ due to loading, i.e. due to $(1.05G_k + 1.00Q_k)$, then the section is satisfactory for that period of exposure.

5.4.1.2 *Design for combined bending and shear for simply supported members* If the shear span/effective depth ratio is between 2 and 6, this mode of failure may be critical. For a uniformly distributed load, the critical section is between $0.15L$ and $0.2L$, and for a concentrated load it is usually under the load.

The design procedure is as follows:

(a) Determine M_a and V_a (bending moment and shear) due to the applied load at the section.
(b) Determine the shear span $l_s = M_a/V_a$.
(c) Determine M_{T° (moment capacity of the section at T°) in N mm from the formula

$$M_{T^\circ} = 0.875dl_s\left(0.342b_1 + \frac{0.3M_{RT^\circ}}{d^2}\right)\sqrt{\frac{z}{l_s}}\sqrt[4]{\frac{16.66}{\Phi f_{su}\psi_{sT^\circ}}} \ngtr M_{RT^\circ}$$

where M_{RT° is as given in Section 5.4.1.1 (in N mm), b_1—shown for various sections in Fig. 5-2—is the effective width of section (in mm), and

$$\Phi = \frac{\text{area of tensile reinforcement}}{\text{area of concrete to effective depth}}$$

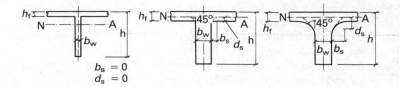

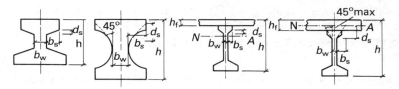

Figure 5-2 Effective width b_1 for various cross-sections

If $b_s < d_s$, $b_1 = b_w + 2b_s$

If $b_s > d_s$, $b_1 = b_w + 2d_s$

If $M_{T°} > M_a$, then only nominal stirrups will suffice. If $M_a > M_{T°}$, then stirrups should be provided such that

$$\frac{A_{sv}}{s} \geq \frac{V_a - (M_{T°}/l_s)}{(f_y \psi_{sT°})d}$$

where s is the pitch and $\gamma_{ms} = 1$. A_s = cross-sectional area of links.

5.4.1.3 *Design procedure for diagonal-splitting phenomenon for simply supported members* This mode of failure is possible when the shear span/effective depth ratio is 2 or less. The basic considerations of design for this mode are the splitting strength of concrete and the tensile strength of the horizontal and vertical steel made fully effective through adequate anchorage.

The critical shear capacity at temperature $T°$, $V_{crT°}$, is given by

$$V_{crT°} = B[(1.2 \times 0.7 f_{cu} \psi_{cT°})/\gamma_{mc}] \sin \theta$$

where B is the effective bearing area to be taken as the lesser of (a) $\frac{1}{3}bh$ and (b) $b \times$ width of bearing, b being the minimum width in the cross-section within the shear span and h the total depth of the section; θ is the angle as shown in Fig. 5-3; and $V_{crT°} \geq V_a$ (i.e. the shear due to the applied load $(1.05G_k + 1.0Q_k)$). In addition, the following requirements should also be satisfied:

$$A_s f_y \psi_{sT°} \geq H, \quad \text{where} \quad H = V_a \cos \theta / \sin \theta$$

and

$$A'_{sv} f_y \psi_{sT°} \cos \theta + A'_s f_y \psi_{sT°} \sin \theta \geq V_a$$

where A'_{sv} is the total area of the vertical steel within the length l_s and A'_s is the area of the horizontal steel in excess of what is required to resist H.

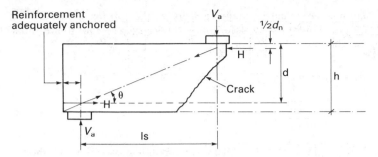

Reinforcement
adequately anchored

Figure 5-3 Force diagram for end blocks and brackets

5.4.1.4 Design for flexural members continuous, fixed or end restrained
Examples of truly freely supported concrete beams or slabs are rare in practice. In most cases, even though the member may be designed as simply supported at bearings it is monolithically built into the surrounding supporting structure with at least adequate anti-crack reinforcement at the top. The continuity reinforcement, or even the anti-crack bars mentioned above, improve the fire resistance of a member or element considerably, provided the detailing is satisfactory.

The design for such flexural members is similar to that for a simply supported member discussed above and should cover three modes of failure. The beneficial effect of thrust due to thermal restraint has been ignored.

The moments of resistance both at the supports and in the span should be calculated, and as long as the sum of these resistance moments is greater than the maximum free bending moment due to the applied load on the member the structural collapse cannot occur.

The calculation procedure can be summarized as follows:

(a) Determine the top steel, A_{st}, at the face of the supporting structure—only the steel having full anchorage on either side of the section should be considered—and compute the flexural moment of resistance at the top, $M_{RtT°}$, in a similar manner as in Section 5.4.1.1, taking into account the appropriate temperature of the steel and concrete and the corresponding strength reduction. The top steel, being remote from the exposed surface, usually retains its original strength. It is advisable to ignore 5 to 10 mm thickness of concrete in the compression zone, near the exposed surfaces, to allow for temperature shock or similar effect.

(b) With this $M_{RtT°}$ as the negative moment at support and the free moment due to the applied load in the span, determine the point of contraflexure, which is, say, a distance l_d away from the support.

(c) If l_d is less than or equal to $2d$ (d is the effective depth), then the adequacy of the section should be checked in accordance with Section 5.4.1.3. If this is adequate, then the maximum moment of resistance at the top should be taken as

$$M_{RtT°} = V_a \times 2d$$

(d) If $l_d > 2d$, then the actual moment capacity at the top (as influenced by shear), $M_{T°(top)}$, should first be determined (in accordance with the formula given in Section 5.4.1.2):

$$M_{T°(top)} = 0.875 dl_s \left[0.342 b_1 + \frac{0.3 M_{RtT°}}{d^2} \sqrt{\frac{z}{l_s}} \right]$$

$$\times \sqrt[4]{\frac{16.66}{\Phi f_y \psi_{sT°}}} \ngtr M_{RtT°}$$

assuming a certain value of l_s. The corresponding value of V should be found from the relationship

$$M_{T°(top)}/l_s = V$$

V should be equal to V_a. By trial and error, a value of l_s should be found such that V becomes equal to V_a. Then, the corresponding value of $M_{T°(top)}$ should be taken as the maximum negative moment of resistance at the support. However, if adequate shear reinforcement is available in accordance with Section 5.4.1.2, then $M_{RT°}$ can be taken as the maximum negative moment of resistance at the support.

(e) Determine the point of contraflexure and use suitable hanger bars. The area of such steel should be obtained from the relationship

$$A_{sH} = \frac{V_a}{f_y \psi_{sT°}}$$

where A_{sH} is the area of the hanger bars. Other notations as already explained.

(f) In the case of end restrained or fixed beams, the supporting structures should be designed to take the torsion, the magnitude of which should be taken as equal to the maximum negative moment of resistance in the member. The area of steel (both longitudinal and transverse) to resist torsion should be calculated in accordance with the provision in the code of practice for normal temperature design with the modification of steel stresses due to temperature rise and ignoring concrete resistance.

5.4.2 Design guidance by ACI Committee 216

The method suggested by Committee 216 [8] is based on the ultimate design principles of the *Building code requirements for reinforced concrete* (ACI 318) [23]. The suggestion relates only to flexural strength calculations—the problem of shear has not been included. In the ACI *Building code*, there is no partial safety factor for material, but the calculated 'capacity' is multiplied by a capacity reduction factor.

For fire design, the loading assumed is the actual service load without any factor either to the dead load or the live load.

A set of curves based on actual tests have been presented in the ACI guidelines, which gives the relationship between cover and temperature in embedded steel after different periods of exposure. For slabs, these curves are similar to

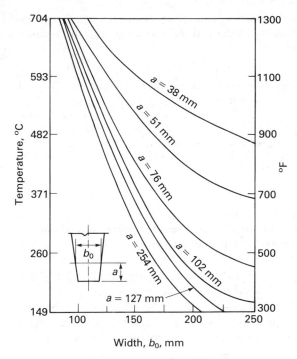

Figure 5-4 Temperature after 2 h along the vertical axis of beams made with sanded light-weight concrete

the ones shown in Fig. 4-1. For beams, for different periods of exposure, the temperature profile in the concrete section of various widths has been plotted, from which the temperature at any particular point after a specific period of exposure can be assessed. Graphs are also presented in the document for beams or ribs having reinforcement in the central vertical axis, correlating the width of the beam or rib and the temperature rise after a specific period of exposure. One such typical graph for light-weight concrete (sanded) ribs exposed for 2 h is shown in Fig. 5-4. From these figures, the temperature distribution in the concrete beam or slab at the level of steel can be obtained and then, from Fig. 3-4, the residual strength of steel. Similarly, the reduction in concrete strength at the compression zone can be obtained from Fig. 2-1, the average temperature distribution in the compression zone having been ascertained.

5.4.2.1 *Simply supported members* For simply supported rectangular beams or slabs, the moment of resistance, $M_{RT°}$, is then worked out from the equation

$$M_{RT°} = A_s f_y \psi_{sT°}(d - \tfrac{1}{2}d_n)$$

d_n being the depth of the rectangular stress block of compression and for a rectangular beam equal to

$$\frac{A_s f_y \psi_{sT°}}{b \times (0.85 f_{cyl} \psi_{cT°})}$$

where f_{cyl} is the cylinder strength of concrete at normal temperature (20°C) and other notations are as explained beforehand.

It should be noted that $\psi_{sT°}$ and $\psi_{cT°}$ are not exactly the same as in Section 5.4.1. In the case of prestressed concrete, f_{ps} should replace f_y, where f_{ps} indicates the stress in prestressing tendons at ultimate condition. In lieu of an analysis based on strain compatibility, the value of f_{ps} may be taken as

$$f_{ps} = f_{su}\left(1 - \frac{0.5 A_s f_{su}}{bd f_{cyl}}\right)$$

where f_{su} is the ultimate tensile strength of a prestressing tendon. $M_{RT°}$ should be equal to or greater than the applied moment due to service load.

5.4.2.2 *Continuous members* For continuous beams and slabs, first the temperature of the top steel at the support and the concrete compression zone (at the bottom half of the beam) should be determined, from which the negative moment is estimated. Then, the maximum positive moment should be estimated, taking due account of the redistribution. A trial-and-error procedure may be necessary to get the accurate value.

5.4.2.3 *Slabs or beams where thermal restraint occurs* As discussed before, if the surrounding structures restrict the thermal expansion of a member, a thrust acts on the member like a prestressed force, which enhances the fire resistance of the

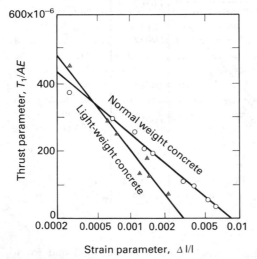

Figure 5-5 Relationship between thrust parameter and strain parameter

member. An estimate of the actual magnitude of thrust is virtually impossible. However, based on extensive fire tests covering various types of floor construction, such as individual beams, cored slabs and double-tee units, a relationship has been established between longitudinal expansion, expressed as a strain parameter $\Delta l/l$, and the thrust parameter T_1/AE, as shown in Fig. 5-5, where Δl is the allowed thermal expansion, l the length of the specimen exposed to fire, T_1 the thrust developed due to restraint, A the cross-sectional area of the member resisting thrust, and E the modulus of deformation (prior to fire exposure).

As a further step, from that relationship nomograms correlating thrust parame-

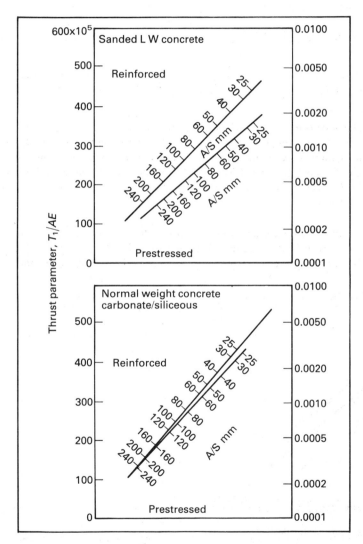

Figure 5-6   Nomogram relating thrust, strain and A/s ratio

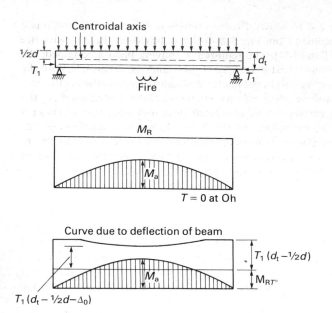

Figure 5-7 Moment diagrams for axially restrained beam during fire exposure

ter, strain parameter and the ratio of A/S (S being the heated perimeter) have been prepared which are reproduced in Fig. 5-6 from references [8] and [11].

The procedure for estimating thrust requirements is as follows.

The temperature distribution in the concrete section and the strength level of reinforcement and concrete having been ascertained, the moment of resistance, $M_{RT°}$, of the section is computed. If this is less than the bending moment, M_a, due to the applied load, then the additional resistance moment which occurs due to T_1 has to be considered, as indicated in Fig. 5-7. The bending moment due to the thrust T_1 depends on the lever arm $(d_t - \frac{1}{2}d)$, where d_t is the distance from the top face to the line of action of T_1 (which has to be assumed). From this, the

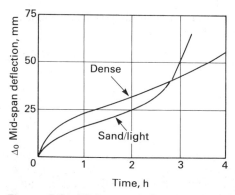

Figure 5-8 Mid-span deflection, Δ_0, of reference specimen

deflection at the centre has to be deducted, as shown in Fig. 5-7. The magnitude of deflection can be obtained from Fig. 5-8, which is based on the extensive fire tests considered in [11] and [24]. The required T_1 having been ascertained, the next step is to compute the thrust parameter T_1/AE and the ratio A/S. With the help of these two values, the strain parameter $\Delta l/l$ can be obtained directly from the nomogram. By multiplying the strain parameter by l (the heated length of the member), Δl or the expansion can be calculated. It is now necessary to check if the surrounding structure can withstand the thrust T_1 with the displacement Δl calculated. If the surrounding structure cannot withstand T_1 with the displacement Δl, then it will be necessary either to increase $M_{RT\circ}$ by increasing the cover and/or the reinforcement, or alternatively to stiffen the surrounding structure.

6 Spalling

Spalling is the general term for the dislodgement of small or large pieces of surface material from heated concrete. The phenomenon is unpredictable and leads to premature failure in tests or actual fires, as the loss of protective cover may expose steel to fire temperatures and accelerate strength reduction to a critical level.

Spalling is more common with siliceous aggregate; by comparison limestone and most of the manufactured light-weight aggregates generally show low susceptibility to spalling.

Spalling may broadly be classified into three types:

(a) *Explosive spalling* As the name suggests, this is violent in nature and normally occurs at an early stage of heating and mostly within the first half hour of the fire. It may cause the complete destruction of the member and occur suddenly without warning.

(b) *Local spalling* This kind of spalling includes the removal of materials from the surface locally and from the external corners. This can occasionally be violent and initiate extensive damages.

(c) *Sloughing-off* This is a gradual process of reduction in the cross-section, usually occurring at a very high temperature when the surface layers tend to separate from the main body. The section thus loses its strength and stiffness.

The exact causes of spalling are not known but the contributory factors appear to be the following, either singly or in combination.

(a) Physical and/or chemical changes in certain types of aggregate when heated, accompanied by a sudden expansion in volume.

(b) The pore pressure due to the expansion of water vapour or steam in conjunction with the compressive stresses in the concrete through prestress or applied loading. (The more impermeable the concrete is, the more susceptible it is to spalling as water present in the concrete cannot easily escape.)

(c) Secondary stresses developed as a result of non-uniform temperature distribution in the cross-section, and/or incompatibility between the aggregates and the matrix with respect to expansion and/or restraint of thermal expansion in the longitudinal direction.

It has been established that spalling becomes more severe with higher moisture content, as the intensity of pore pressure is likely to increase. A moisture content less than 5% by volume for siliceous aggregate concrete and 3% for light-weight aggregate concrete, has been suggested to prevent explosive spalling [9]. Tests in West Germany and the Netherlands, however, show that the probability of spalling is very low even if the moisture content limit is raised to 7% by volume both for dense and light-weight concrete [24]. By controlling the moisture content in the concrete and by putting supplementary reinforcement in the cover (Section 5.2.1), the risk of spalling can be diminished.

7　Detailing

The performance of a concrete structure—whether in fire or in a normal temperature situation—is actually governed by the way it has been detailed. The importance of detailing, therefore, cannot be over-emphasized. In a fire, the whole structure is tested to its ultimate condition—which otherwise very rarely occurs. so, any deficiency or weakness in the construction that may have been unnoticed in normal circumstances, becomes conspicuous in fire conditions.

Essentially, detailing should be such that the assumptions in the design remain valid and the intended method of load transfer is ensured. It has been observed that if 'traditional good detailing practices' are followed, and the workmanship is of an adequate standard, then concrete structures can endure fire very well and rarely present any hazard up to 2 h period of standard rating [1]. 'Traditional good detailing practices' normally should include

(a) Provision of adequate bearing area for the load bearing elements, especially for precast ones.
(b) Continuing sufficient bottom reinforcement (at least 50% of the maximum requirements in the span in the case of a simply supported member or 30% for a continuous member) onto the support with proper anchorage.
(c) The top reinforcement having adequate embedment length on either side of the support and extended at least 12 times the bar diameter or the effective depth of the section, whichever is greater, beyond the point where it is no longer required for calculation.
(d) Provision of at least the minimum percentage (0.12% for high-tensile and 0.2% for mild steel bars, of the plan area of the member) of properly anchored transverse reinforcement.
(e) Provision of adequate tie reinforcement in the transverse and longitudinal direction interconnecting all elements such that the whole structure can act together and the possibility of progressive collapse is eliminated.

Practical experience shows that almost anything that increases the load bearing capacity and overall stability of a structure under normal temperature, invariably enhances its resistance to fire damage; the reverse is also true. A good example of this is the Avianca Building in Bogota, Columbia, which withstood a heavy fire lasting over 12 h without collapse and could subsequently be in operation after repairs.

Although good detailing practices at normal temperature serve concrete structures admirably in fire conditions, some additional considerations may be necessary for fire. As stated earlier, the bond stress of concrete reduces appreciably at elevated temperatures. Some special attention therefore needs to be given to anchorage of reinforcement, and in special cases it may be prudent to provide positive end anchorage rather than to rely on an embedment length. Attention is also required at laps and splicing, especially where spalling may occur. In highly sensitive areas, it is recommended that spirals are added around the laps so that, even after local spalling, the structure may transfer the load satisfactorily.

The top bars should extend far into the span to cater for negative moment, when the point of contraflexure moves as a result of the redistribution of moments, already discussed in Section 2. Also, at the point of contraflexure, sufficient hanger bars in the shape of fully anchored stirrups need to be provided. Some top steel (20% or so of the requirement over the support) should be effectively continued through the span so that, in the ultimate state, after a reasonable deflection, provided the structural integrity is still retained, the suspension effect may develop.

For prestressed I-beams with a thin web, it is essential to provide end block where the steel should be properly anchored with spirals, mats or similar devices. The longitudinal steel in flanges should be protected, giving sufficient cover, and in addition it is desirable to provide well-protected stiffeners at suitable intermediate points. In this case, even if the web disintegrates, the beam can behave as a tied arch or a Vierendeel girder.

In Section 5, the beneficial effects of the restraint of thermal expansion by the surrounding structures was discussed. Care should be taken in detailing so that the members do not become very stiff, as otherwise collapse may occur due to excessive compression and/or shear. Similarly, in detailing columns, attempts should be made to make them relatively flexible to cope with the expansion of flexural members without appreciable shear distortion and the development of high stresses.

Acknowledgements

Figure 2-1 is reproduced from *Concrete manual*, 8th edition, by courtesy of the US Department of the Interior, Bureau of Reclamation Engineering and Research Center, Denver, Colorado, and Fig. 4-1 is reprinted from a paper by Campbell-Allen, Low and Roper, pp. 382–388 in *Nuclear Structural Engineering* (now *Nuclear Engineering and Design*) Vol. 2 (1965) No. 4 by courtesy of North Holland Publishing Company, Amsterdam.

References

1. Joint Committee of Institution of Structural Engineers and Concrete Society, *Fire resistance of concrete structures*, London, 1975.
2. FIP–CEB, *Recommendations for the design of reinforced and prestressed concrete structural members for fire resistance*, 1975.
3. Abeles P. W. and Bardhan-Roy, B. K., *Prestressed concrete designer's handbook*, 3rd Edn, Viewpoint Publication, Eyre & Spottiswoode Publications Ltd, Leatherhead, 1981, pp. 456–571.
4. *The Building Regulations 1976*, No. 1676, HM Stationery Office, London,

1976, pp. 58–65.

5. British Standards Institution, BS 476, Part 8: 1972—*Test methods and criteria for fire resistance of elements of building construction*, London, 1972.

6. American Society for Testing and Materials. *Standard methods of fire tests of building constructions and materials*, ANSI/ASTM E119-79, Book of ASTM Standards, Pt 18, Philadelphia, 1979.

7. Butcher, E. G., Chitty T. R. and Ashton L. A., *The temperature attained by steel in building fires*, Fire Research Technical Paper 15, HM Stationery Office, London, 1976.

8. ACI Committee 216, Guide for determining the fire endurance of concrete elements, *Concr. Int.*, Vol. 3, No. 2, Feb. 1981, pp. 13–47.

9. Joint Committee of Institution of Structural Engineers and Concrete Society, *Design and detailing of concrete structures for fire resistance*, Interim Guidance, London, 1978.

10. Bobrowski, J., Fire resistance, rational approach—basic philosophy, *Concrete design: US and European practices*, American Concrete Institute, SP-59, 1976, pp. 231–255.

11. Gustaferro, A. H., Fire resistance of post-tensioned structures, *J. Prestressed Concr. Inst.*, Vol. 18, No. 2, March–April 1973, pp. 38–63.

12. Berge, O., Hydrofob 3-L—betong ett attraktivt nytt byggnadsmaterial, *Nordisk Betong*, Vol. 6, No. 3, pp. 1–11, 1978.

13. Nasser, K. W. and Neville, A. M., Creep of old concrete at normal and elevated temperatures, *J. Am. Concr. Inst., Proc.* Vol. 64, No. 2, Feb. 1967, pp. 97–103.

14. British Standards Institution, CP 110, Part 1: 1972 (as amended 1980), *The structural use of concrete*, London, 1980, 156 pp. + amend.

15. Diederichs, U. and Schneider, U., Bond strength at high temperatures, *Mag. Concr. Res.*, Vol. 33, No. 115, June 1981 pp. 75–84.

16. Abeles, P. W. and Bobrowski, J., Fire resistance and limit state design for concrete, *Concrete*, Vol. 6, No. 4, pp. 1–4, April 1972.

17. Fire Research Station, *Report FROSI 5070*.

18. Read, R. E. H., Adams, F. C. and Cooke, G. M. E., *Guidelines for the construction of fire resisting structural elements*, Building Research Establishment Report, HM Stationery Office, London, 1980.

19. FIP–CEB, *Methods of assessment of the fire resistance of concrete structural members*, 1978.

20. American Insurance Association, *Fire resistance ratings*.

21. Lie, T. T. and Allen, D. E., *Calculation of the fire resistance of reinforced concrete columns*, Technical Paper 378, Division of Building Research, National Research Council of Canada, Ottawa.

22. Lie, T. T. and Allen, D. E., *Further studies in fire resistance of reinforced concrete columns*, Technical Paper 416, Division of Building Research, National Research Council of Canada, Ottawa.

23. ACI Committee 318, *Building code requirements for reinforced concrete*, ACI Standard 318-77, American Concrete Institute, Detroit, 1977.

24. Salvaggio, S. L. and Carlson, C. C., Restraint in fire tests of concrete floors and roofs, *ASTM Spec. Tech. Publ.*, No. 422, 1967, pp. 21–39.

25. Copier, W. J., The spalling of normal weight and lightweight concrete on exposure to fire, *Heron*, Vol. 24, No. 2, 1979, pp. 11–76.

15 Earthquake-resistant structures

Mark Fintel   S K Ghosh

Portland Cement Association
Skokie USA

Contents

Notation

Symbols are explained where they occur in the text. *See also* Glossary on pages 74–78.

Summary

This chapter gives an overview of the earthquake resistant design of concrete structures. Starting from the earthquake phenomenon, the concepts of the response spectrum and the design spectrum are introduced. The choice of design ground motion or design spectrum at a specific site is discussed. Methods of analysis of the dynamic response of structures to earthquake ground motions are described. The need for consideration of inelastic member behavior in such analysis is pointed out. The current empirical approach to earthquake resistant design of structures is explained in depth, starting from fundamentals. Deficiencies of the empirical approach are pointed out, and an alternative approach using inelastic response history analysis is developed and illustrated through a design example. Lessons learned from structural performance in recent earthquakes are enumerated. A few guides to detailing of structural members and joints for earthquake resistance are given. Finally, the evaluation and repair of structural damage caused by earthquakes is briefly discussed.

1 Background: earthquakes

1.1 Causes of earthquakes

The cause of earthquakes has been the subject of much speculation and conjecture over the centuries. Geologists now agree that earthquakes can be divided into two categories, depending upon their origin: *tectonic* and *volcanic*. Tectonic earthquakes are those associated with faulting or other structural processes. Volcanic earthquakes are those associated with volcanic eruptions, or subterranean movements of magma. The major portion of observed seismic activity is tectonic in origin.

It was not until the 1890s that the fracture of crustal rock was recognized as the cause of earthquakes. Ground breakage in earthquakes had been observed earlier, but it was always viewed as a consequence rather than as the cause. H. F. Reid carried the notion further, following the San Francisco earthquake of 1906. Reid theorized that the crustal rock in the vicinity of the fault was being strained gradually with time until finally the stress overtook the strength of the rock at the fault, and the rock fractured and slipped back toward a stress-free state. The

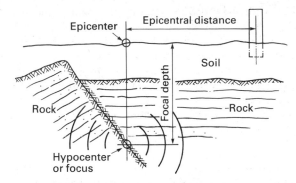

Figure 1-1   Definition of common earthquake-related terms

tremendous amount of strain energy that had accumulated in the rock during distortion was thus suddenly released, and propagated out in all directions from the fault break as shock waves, with accompanying reflections from the earth's surface as well as reflections and refractions as they traversed the earth's interior. This is briefly the Reid elastic rebound theory [1, 2].

The point of fracture within the earth's interior where the initial slip occurs, triggering a large-scale slippage sometimes extending over hundreds of miles, is called the earthquake *focus* or *hypocenter*. The point on the surface vertically above the focus is called the *epicenter* (Fig. 1-1).

Earthquakes may originate anywhere from near the surface of the earth to a depth of 700 km or more. Two discontinuities divide the earth internally into an outer shell—the *crust*; an intermediate shell—the *mantle*; and a central core (Fig. 1-2). Within the core there is a further division: the inner core may differ considerably from its surroundings in physical characteristics. The Mohorovicic discontinuity, separating the crust from the mantle, is 30 to 60 km deep in

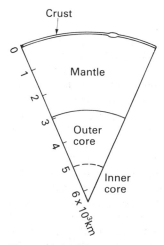

Figure 1-2 Major discontinuities in the earth

continental areas; in oceanic areas, it rises to within 10 or even 5 km of the sea surface. Shocks with focal depths of about 60 km or less are called shallow-focus earthquakes, while those with greater focal depths are broadly termed deep-focus earthquakes. The latter include two subclasses: intermediate, with foci between 60 and 300 km; and deep, with focal depths exceeding 300 km. Most earthquakes have shallow foci; they originate in the crustal layer, above the Mohorovicic discontinuity.

The exact process by which forces are generated within the earth's interior ultimately causing crustal upheavals (mountain building and earthquakes) is not well understood as yet [3]. In Richter's [1] words, 'The days are gone when all was cheerfully explained in terms of a cooling and contracting earth. Oceanographic and geological studies indicate that the earth's crust comprises several crustal plates, bounded by rifts and trenches mainly in the oceans, along which the plates are in continuous slow motion relative to one another. It is along the plate boundaries that most earthquakes occur. Along the west coast of North America, the Pacific plate is thrusting under the American plate, thus producing the California earthquakes. This is still an incomplete explanation because major earthquakes have occurred far from the plate boundaries. A more recent theory postulates currents of plastic flow, probably thermal convection currents, in the mantle of the earth. Such currents, fixed like the established currents of circulation in the oceans, would produce continuous and coherent stresses in a region of the earth, and could also account for earthquakes [3].

1.2 Nature of earthquake motion: seismic waves

The complex ground motion characterizing an earthquake is caused by the passage of waves of distortion through the affected area. Such motions are juxtapositions of a number of different wave types, among the most significant of which are P-waves, S-waves, Rayleigh or R-waves, and Love or Q-waves. Both P- and S-waves are body waves, traversing the interior of the earth and making up the high-frequency components of strong ground motion near the epicenter. In a P-wave, the particles vibrate in the direction of propagation, while in an S-wave the particles oscillate in a perpendicular plane. P-waves travel faster than S-waves. R- and Q-waves are surface waves which travel slightly slower than S-waves. In the R-wave, the vibration is in the direction of propagation, while in the Q-wave the particle displacement is transverse to that direction, with no vertical or longitudinal component. Surface waves form a significant portion of the strong ground motion in epicentral regions of shallow-focus earthquakes, and also make up the dominant long-period phases of shallow-focus earthquakes recorded at distances over 100 km from the epicenter [1, 4].

Generally, the amplitude of ground motion diminishes with distance from the epicenter. Under certain geologic conditions, particularly where a deep layer of saturated alluvium overlies bedrock, a substantial amplification of wave motion transmitted from the bedrock may occur. Also, the short-period, high-energy components of seismic waves are damped out much faster than longer period waves, so that at some distance from the epicenter, the ground motion consists predominantly of long-period oscillations.

1.3 Earthquake magnitude and intensity

Magnitude and intensity scales describe earthquakes in quantitative terms. Richter [1] has used analogy with radio transmissions to explain the difference between magnitude and intensity. Seismographs record waves of disturbance radiated from the earthquake source, just as receiving units catch radio waves transmitted by broadcasting stations. Magnitude can be compared to the power output in kilowatts of a broadcasting station. Local intensity is comparable to the signal strength on a receiver at a given locality. Intensity, like signal strength, falls off with distance from the source, although it also depends on local conditions and the pathway from the source to the point of recording.

1.3.1 *Magnitude*

The magnitude of an earthquake is a measure of its size at the source, independent of the place of observation. The local magnitude, defined by Richter in 1935, is the most widely used magnitude measure. It is calculated from measurements on seismographs, and is expressed in ordinary numbers and decimals. Physically, the magnitude has been correlated with the energy released by an earthquake, as well as with the fault rupture length and the maximum fault displacement [1]. The energy associated with an earthquake increases by more than a factor of 10 with each unit increase in Richter magnitude. The largest earthquakes recorded have had a Richter magnitude of 8.9. This is believed to be the largest possible magnitude on the Richter scale, the limit being set by the maximum amount of strain energy that can be stored by crustal rocks without rupture.

Earthquakes of magnitude less than about 5 are believed to be incapable of producing damaging effects. However, the destructiveness of an earthquake, although dependent partly on its magnitude, is a function of other equally important factors, such as focal depth, epicentral distance, local geology, and structural characteristics including resistance to damaging oscillations.

1.3.2 *Intensity*

The word 'intensity', when used to describe earthquakes, denotes the potential destructiveness of an earthquake at a given location. The most common intensity scale is the Modified Mercalli (MM) scale, given in Table 1-1. Also shown in the table are approximate values of the ground acceleration, $\ddot{x}_g$, corresponding to the different intensity ranges, as obtained from a relationship proposed by Gutenberg and Richter [5].

Following an earthquake, the assignment of an intensity to a given location is based on interviews with inhabitants and on observations of damage in the area. Assigned intensity values at different locations can be mapped as a series of isoseismals that separate regions of successive intensity ratings. These isoseismal or intensity maps reflect the attenuation of damage with distance from the source, and the extent of the felt area of the earthquake. The shapes and extent of the isoseismals may be influenced by the tectonic features of the area, indicating predominant directions of transmission of seismic waves and the manner in which the earthquake originates. An intensity map for the 1971 San Fernando earthquake is shown in Fig. 1-3 [6].

Table 1-1   The Modified Mercalli Intensity Scale and corresponding approximate ground accelerations [4]

MM scale	Ground acceleration, $\ddot{x}_g$ cm/s^2	
I Not felt except by a very few under especially favorable circumstances		
II Felt only by a few persons at rest, especially on upper floors of buildings. Delicately suspended objects may swing	2 3	
III Felt quite noticeably indoors, especially on upper floors of buildings, but many people do not recognize it as an earthquake. Standing motor cars may rock slightly. Vibrations like passing truck. Duration estimated	4 5 6	0.005 g
IV During the day felt indoors by many, outdoors by few. At night some awakened. Dishes, windows, doors disturbed; walls make creaking sound. Sensation like heavy truck striking building. Standing motor cars rocked noticeably.	7 8 9 10	0.01 g
V Felt by nearly everyone; many awakened. Some dishes, windows, etc., broken; a few instances of cracked plaster; unstable objects overturned. Disturbances of trees, poles, and other tall objects sometimes noticed. Pendulum clocks may stop	20 30	
VI Felt by all; many frightened and run outdoors. Some heavy furniture moved; a few instances of fallen plaster or damaged chimneys. Damage slight	40 50 60 70	0.05 g
VII Everybody runs outdoors. Damage negligible in buildings of good design and construction; slight to moderate in well-built ordinary structures; considerable in poorly built or badly designed structures; some chimneys broken. Noticed by persons driving motor cars	80 90 100	0.1 g
VIII Damage slight in specially designed structures; considerable in ordinary substantial buildings, with partial collapse; great in poorly built structures. Panel walls thrown out of frame structures. Fall of chimneys, factory stacks, columns, monuments, walls. Heavy furniture overturned. Sand and mud ejected in small amounts. Changes in well water. Disturbs persons driving motor cars	200 300	
IX Damage considerable in specially designed structures; well-designed frame structures thrown out of plumb; great in substantial buildings, with partial collapse. Buildings shifted off foundations. Ground cracked conspicuously. Underground pipes broken	400 500 600	0.5 g
X Some well-built, wooden structures destroyed; most masonry and frame structures destroyed with foundations; ground badly cracked. Rails bent. Landslides considerable from river banks and steep slopes. Shifted sand and mud. Water splashed over banks	700 800 900 1000	1 g
XI Few, if any, (masonry) structures remain standing. Bridges destroyed. Broad fissures in ground. Underground pipelines completely out of service. Earth slumps and land slips in soft ground. Rails bent greatly	2000 3000	
XII Damage total. Waves seen on ground surfaces. Lines of sight and level distorted. Objects thrown upward into the air	4000 5000 6000	5 g

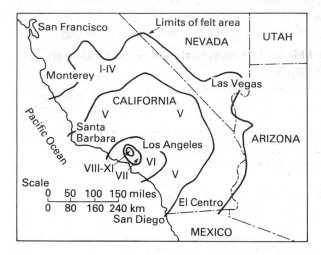

Figure 1-3   Modified Mercalli Intensity Map; San Fernando earthquake

Reported Modified Mercalli intensities are only a crude measure of the destructiveness of an earthquake. Other measures, more suitable for structural engineering applications, are discussed later in this chapter.

1.4   Characterization of strong-motion earthquakes

The best representation of earthquake ground motion is the acceleration time history which provides a complete record of ground shaking at the instrument site. However, for quantitatively comparing different records and for relating measured ground motion to potential structural damage, suitable parameters defining a given record are needed. The principal ground motion characteristics that affect dynamic structural response are duration, frequency content and intensity [7]. Duration refers to the length of a record during which relatively large amplitude pulses occur, with due allowance made for a reasonable build-up time. Frequency characteristics indicate the relative energy content of the different component waves (having different frequencies) that make up a motion. Intensity provides a characteristic measure of the amplitude of acceleration pulses in a record.

1.4.1   *Duration*

In evaluating various assessments of the duration of strong shaking, possible differences in the definitions of this parameter should be kept in mind. Housner [8] pointed out that the time history of an earthquake ground motion consists of (a) an initial segment in which the energy and vibration levels rapidly increase to high values, (b) a segment of uniformly strong shaking at these high values, and (c) a period of gradually attenuating vibration. He defined duration of strong shaking to correspond to item (b) above. Duration has been defined in much the same manner by others [9, 10].

1.4.2 *Frequency content*

A typical strong-motion accelerogram shows a complex series of oscillations that may be thought of as a superposition of simple, constant-amplitude sinusoidal waves, each with a different frequency, amplitude and phase. The frequency characteristics of a given input motion are important because of the phenomenon of resonance. Near maximum response to earthquake excitation can be expected if the dominant frequency components of the exciting motion nearly coincide with the dominant frequencies of a structure.

A convenient way of studying the frequency characteristics of an accelerogram is provided by *response spectra* [11, 12]. If the simple linear oscillator of mass m and stiffness k (Fig. 1-4) has viscous damping, its equation of motion is given by

$$m\ddot{x} + c\dot{x} + kx = -m\ddot{x}_g \tag{1-1}$$

where x is the relative displacement of the mass with respect to the base, $\dot{x}$ is the velocity of the mass relative to that of the ground, $\ddot{x}_g$ is the acceleration of the base,

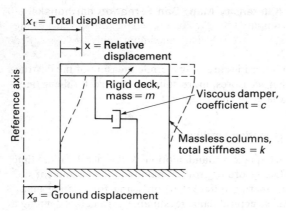

Figure 1-4   Lumped SDOF system subjected to base translation

and c is the damping coefficient. The vibratory response is given by

$$x(t, \omega, \beta) = \frac{1}{\omega_d} \int_0^t \ddot{x}_g(\tau) \exp\left[-\beta\omega_d(t-\tau)\right] \sin \omega_d(t-\tau)\, d\tau$$

$$= \frac{1}{\omega_d} \bar{X}(t, \omega, \beta) \tag{1-2}$$

where $\omega^2 = k/m = (2\pi/T)^2$, T is the natural period of vibration, $\beta = c/c_{cr} = c/2m\omega$ is the fraction of critical damping[1] (Fig. 1-5) and $\omega_d = \omega\sqrt{1-\beta^2}$. For $\beta < 0.2$, ω_d is practically equal to ω.

---

1 Critical damping is defined as the least damping coefficient for which the free response of a system is nonvibratory, i.e. for which it returns to the static position without oscillation after any excitation.

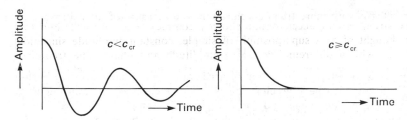

Figure 1-5   Non-periodic dying out of vibration

While Eqn 1-2 expresses deflection response, the effective earthquake force is

$$kx = m\omega^2 x = m\omega \bar{X} \tag{1-3}$$

where $\omega^2 x = \omega \bar{X}$ may be thought of as an effective acceleration.

To obtain the entire history of seismic displacements and forces, Eqns 1-2 and 1-3, may be unnecessary in most practical situations; it may be sufficient to determine only the maximum response quantities. The maximum force as well as displacement response can be computed by introducing the maximum value of the response function $\bar{X}$ into Eqns 1-2 and 1-3. This maximum value of $\bar{X}$ is called the *spectral pseudo-velocity*:

$$S_v = \bar{X}_{max} = \left[\int_0^t \ddot{x}_g(\tau)\exp\left[-\beta\omega(t-\tau)\right] \sin \omega(t-\tau)\, d\tau \right]_{max} \tag{1-4}$$

Maximum displacement equals the spectral pseudo-velocity divided by the circular frequency (Eqn 1-2). This ratio is called the *spectral displacement*:

$$S_d = S_v/\omega \tag{1-5}$$

Similarly, the maximum earthquake forces are seen from Eqn 1-3 to equal the product of the mass, the circular frequency and the spectral pseudo-velocity, leading to the following definition of the *spectral pseudo-acceleration*

$$S_a = \omega S_v = \omega^2 S_d \tag{1-6}$$

The physical significance of the spectral pseudo-velocity, S_v, can be explained [12] as follows. The maximum displacement corresponds to a condition of zero kinetic energy and maximum strain energy $\frac{1}{2}kS_d^2$. If this energy were in the form of kinetic energy $\frac{1}{2}m(\dot{x})^2 = \frac{1}{2}kS_d^2$, the maximum relative velocity would be

$$\dot{x} = \sqrt{k/m}\,S_d = \omega S_d = S_v \tag{1-5a}$$

Response spectra are discussed further in Section 3.1.

1.4.3   *Intensity*

The peak ground acceleration is presently the most widely used measure of the strength of ground shaking. The peak velocity and the peak displacement are also used at times. Housner [13] has proposed that 'spectrum intensity' be used as a measure of the severity of ground motion. Spectrum intensity is the area under the relative velocity response spectrum, between bounding values of the period range of interest (0.1–2.5 s in [13], 0.1–3.0 s in [7]). Others [14] have proposed using the root-mean-square (rms) acceleration as a measure of intensity:

$$\ddot{x}_{rms} = \left[\frac{1}{t}\int_0^t \ddot{x}^2 \, dt\right]^{\frac{1}{2}} \qquad (1\text{-}7)$$

If intensity is to reflect the variation of acceleration amplitudes over the period range of interest, then peak acceleration is a poor measure. Spectrum intensity, taken over the period range and for a reasonable damping value, is a more representative measure of intensity. A close correlation between the stationary rms acceleration and Housner's spectrum intensity has been shown to exist [7, 15].

2 Evaluation of seismic risk

Knowledge of the earthquake ground motions that can be expected at a site is crucial in earthquake engineering. It is worthwhile to examine how the severity, frequency and nature of such ground motions can be anticipated.

2.1 Geological evidence of seismic activity

Geological evidence of the seismic activity of a region, as available from sources such as local officials, engineers, seismologists, local building regulations, published papers or reference books, is a valuable qualitative tool in the evaluation of seismic risk at a site. It is helpful in estimating the likely locations, magnitude and frequency of seismic events. Also, on the basis of an examination of factors such as the types of fault movement prevalent on a given fault (Section 2.2), it may be possible to anticipate some of the characteristics of ground motions at a site.

2.2 Geological factors affecting ground shaking at a site

These can be related to the earthquake source mechanism, the source to site transmission path or the local site conditions [6, 16].

2.2.1 *Source mechanism*
(A) Nature of slippage:

(a) Fault type. When an earthquake occurs, its effect can depend on the type of causative faulting. Faults may be of the following types:

(1) Strike-slip faults (Fig. 2-1a) where relative horizontal displacement across the fault takes place along a nearly vertical fault plane.
(2) Reverse or compressive faults (Fig. 2-1b) in which the block above an inclined fault surface moves upward relative to the block below. A reverse fault may be low-angle or underthrust, or high-angle or overthrust.
(3) Normal or extensional faults (Fig. 2-1c) in which the block above an inclined fault surface moves downward relative to the block below.

Few pure examples of the above occur; most earthquake fault movements have components parallel and normal to the fault trace. Also, fault slip may take place

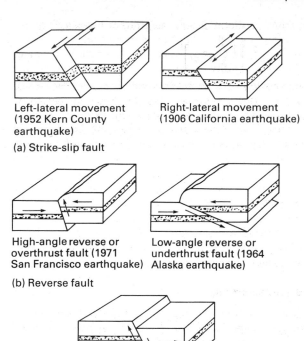

Left-lateral movement
(1952 Kern County
earthquake)

Right-lateral movement
(1906 California earthquake)

(a) Strike-slip fault

High-angle reverse or
overthrust fault (1971
San Francisco earthquake)

Low-angle reverse or
underthrust fault (1964
Alaska earthquake)

(b) Reverse fault

(c) Normal or extensional fault
(1954 Dixie Valley – Fairview Peak, Nevada earthquake)

Figure 2-1   Fault types [6]

along an irregular surface rather than along a plane, and there may be branch faulting.

  (b) Single or multiple-fault break. Complex seismic waves generated by multiple fault breaks are naturally different from simpler wave forms emanating from single-fault breaks.

(B) Source parameters:

  (a) Stress drop. As a result of fault movement, the rock stress across the fault changes from a high initial to a lower final value. The size of the stress drop influences the amplitude of seismic waves generated.
  (b) Total fault displacement. The larger the total displacement along the causative fault, the larger the energy released and the greater the earthquake magnitude. A larger displacement also results in a longer duration of fault movement and in larger amplitudes of long-period waves.
  (c) Length of fault break. A fault rupture of extensive length generates an earthquake of large magnitude. The length of the fault break can also significantly affect the duration of strong shaking.
  (d) Fault shape. A large-magnitude earthquake generally occurs on an elongated fault area whose length greatly exceeds its width. The dimensions of

the slipped area also influence the nature of the ground shaking, including frequency content, duration, and felt area of the earthquake.
(e) Proximity of fault plane to ground surface. The ground acceleration amplitudes depend more on the proximity of the faulting process to the ground surface than on the earthquake magnitude, especially in the near-field.

2.2.2 Transmission path from source to site

As seismic waves radiate from a fault, they undergo certain changes. Some of these result from geometric spreading, while there is also attenuation because of energy loss caused by travel through rocks. This energy loss is less important than the geometric attenuation in the near field. More important is the effect of dispersion of the waves resulting from changes in modulus of rigidity or of density of the rock. Dispersion seems to be a dominant factor, so that at a distance the ground motion exhibits the influence of travel-path geology [16].

2.2.3 Local site conditions

The seismic wave energy arriving at a site is carried primarily in the P- and S-waves transmitted through subsurface rock materials. These waves can reach the ground surface only by travelling through the underlying soils. Unfortunately, little is known about wave propagation in the softer deposits at the earth's surface. Some soft soil deposits vibrate with one predominant mode, such as in Mexico City in 1962, in which case the spectrum curves (Section 1.4.2) show a pronounced peak corresponding to the natural period of vibration of this mode. In other situations, the spectrum curves of ground motions recorded on alluvial deposits do not show pronouncedly resonant peaks.

Among the more important variables that determine the effects of local soil conditions on the ground shaking at a site are [6]:

(a) Depth to rock. Correlations between structural damage and the depth to rock at sites with similar soil conditions have indicated that high-rise buildings with long natural periods have sustained the greatest damage when sited on deep soil deposits, and vice versa. This suggests that the depth to rock affects the frequency content of the seismic waves transmitted to the ground surface.
(b) Nonlinear stress–strain properties of soil materials. The effective modulus of soils decreases and their material damping increases with increasing soil strain levels. Thus an increased amplitude of the subsurface input motions causes a reduced amplification, or even an attenuation, of these motions as they traverse the overlying soil deposits; the characteristic frequency of the surface motions also decreases.
(c) Geometry of layers and ground surface. The degrees of inclination of soil layers, or the presence of significant topographic features, can greatly influence the reflection and refraction of P- and S-waves, and thus the wave motion transmitted to the ground surface.

2.2.4 Need for quantitative information

Although studies of geological factors provide useful information on possible seismic activity at a site, this qualitative information must be supplemented with quantitative estimates of earthquake magnitude and frequency as outlined below.

2.3 Earthquake motion data base

Data presently available to serve as a basis for estimating earthquake-induced ground motions at a site consist of observational and instrumental records of past earthquakes, artificial earthquakes, and empirical scaling relationships based on past records.

2.3.1 *Records of past earthquakes*

Instrumental records of earthquake ground motions close to the epicenter are valuable in structural engineering. These usually consist of acceleration traces of motion along two perpendicular horizontal directions and in the vertical direction (the rotational components are usually unimportant). The records are obtained using strong-motion accelerographs (SMACs).

Although ground motions recorded at a site may not be repeated, strong-motion records, if available over a long period, reveal the general character of the ground motion and the effect of geologic conditions at a particular location. The current library of strong motion records from earthquakes in the United States (in corrected, digitized form) is available from the California Institute of Technology [17]. Where a number of accelerograms are available for a region, a set chosen by careful sampling can be used in dynamic response studies of proposed structures.

A set of acceleration traces that has often been used in dynamic response studies is that of the Imperial Valley (California) earthquake of 1940. The set was recorded at the El Centro instrument site, which rests on some 5000 ft (1524 m) of alluvium about 4 miles (6.4 km) away from the causative fault break. The set represents one of the strongest earthquakes ever recorded, and exhibits high-frequency, large-amplitude pulses lasting over a long duration. A plot of the north–south component of horizontal ground accelerations during the first 30 s of the above earthquake is shown in Fig. 2-2. Also shown are plots of the ground

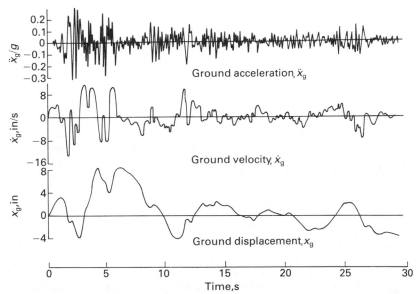

Figure 2-2   1940 El Centro (California) earthquake accelerogram: N–S component and corresponding derived velocity and displacement plots

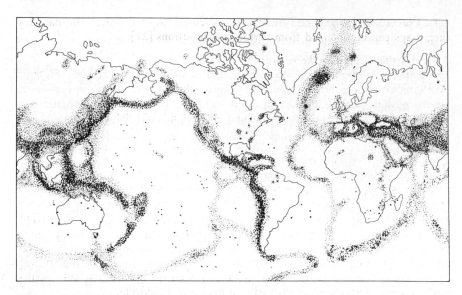

Figure 2-3 Map showing global seismicity during the present century. *Reprinted with permission from:* A. Holmes, *Principles of Physical Geology,* 3rd Edn, Van Nostrand Reinhold, 1978.

velocity and displacement, as obtained by successive integration. The maximum recorded ground acceleration in the NS direction was about 0.33 g.

There are important gaps in the currently available library of strong motion records [6]. In view of the recency and scarcity of such records, the so-called maps of seismic events are usually the best source of information on the seismicity of a site. A plot of earthquake epicenters all over the world, as calculated by the US Coast and Geodetic Survey for events with Richter magnitudes above 4 occurring in 1966, is shown in Reference [18]. Plots for other years show a similar distribution, and the cumulative picture is shown in Fig. 2-3.

Almost all major earthquakes, and about 95% of all recorded earthquakes, have occurred in two distinct zones. The principal one, which accounts for about 80% of all earthquakes, is the *Circum-Pacific seismic zone* bordering the Pacific ocean. The second major zone, called the *Alpide zone,* accounts for about 15% of all earthquakes and runs from east to west across Europe and northern India, turning southeastward to join the Circum-Pacific belt in New Guinea. A third zone, accounting for most of the remaining five or so per cent of all recorded earthquakes, is a narrow band along the center of mid-oceanic ridges, particularly in the Atlantic and Indian oceans.

For design purposes, more detailed information than is shown on Fig. 2-3 is required. Detailed information on the seismicity of the US is now available from a map that divides the country into a 0.5° × 0.5° grid [19]. For some active areas on the west coast, more detailed microzoning maps are available. A set of large maps covering the whole world, showing the position of all earthquakes with magnitude higher than 4 recorded between 1961 and 1969, has been published [20]. For a particular site, it is possible to construct a map indicating the locations, focal depths and magnitudes of all earthquakes (with magnitude higher than a certain value) recorded in this century within a certain radius of the site. The data for such maps can be obtained from various organizations [21].

2.3.2 *Artificial earthquake records*
In an attempt to remedy the present scarcity of strong-motion records, artificial accelerograms have been generated. These embody the basic properties of strong-motion records, as judged from available data and from engineering judgment concerning subsequent events. The artificial earthquake records developed by Jennings *et al.* [22] are widely used.

2.3.3 *Empirical scaling relationships*
Empirical scaling curves developed from prior strong-motion records can be useful in estimating peak accelerations and durations of potential ground shaking at a site. Many empirical expressions have been developed relating MM intensities and the peak horizontal acceleration at a given location [6, 23]. The current library of strong-motion records has also been used to relate peak ground motions to the magnitude of the earthquake and to the distance from the earthquake source [8, 24]. Considerable judgment must be exercised in using such empirical curves, since they are generally based on a limited number of available earthquake records corresponding primarily to events in California.

3 Design ground motion at a site

3.1 Response spectra

The design ground motion(s) at a site is often specified in terms of a response spectrum. The relative (pseudo-)velocity response spectra for the NS component of the 1940 El Centro record, for different amounts of damping, are shown in Fig. 3-1. The sharp peaks, conspicuous when damping is absent, reflect the resonant behavior of the system of Fig. 1-4 in certain frequency ranges. Significantly, even a moderate amount of damping not only reduces the response but also smooths out the jagged character of the spectral plot. Thus, the sharp peaks are not important in practice. Also, for design purposes, earthquake motions of varying frequency characteristics are customarily accounted for by using smoothed or averaged spectra based on a number of earthquake records, all reduced or normalized to a reference intensity.

Because of the relationships indicated by Eqn 1-6, a single plot can be constructed showing the variations of spectral acceleration, velocity and displacement with frequency. Figure 3-1 [25] shows such a plot, with log scales on all axes, the spectral displacement and acceleration being read along diagonal scales. The plot is typical of earthquake response spectra and confirms certain intuitively obvious aspects of the dynamic response of simple systems. For low-frequency systems—corresponding to a heavy mass supported by a light spring—the mass remains practically motionless when the base is seismically excited, the relative displacement of the mass with respect to the base being essentially equal to the base displacement. For high-frequency systems, exemplified by a light mass supported by a very stiff spring (on the right side of the plot), the mass simply moves with the base, so that it acquires essentially the same absolute acceleration as the base. In the intermediate frequency range, some modification of the

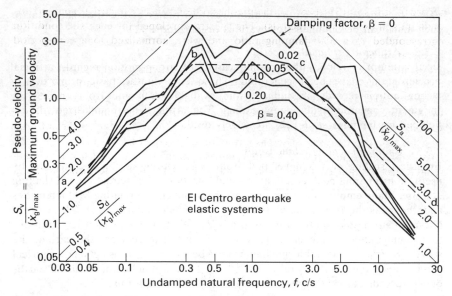

Figure 3-1 Deformation spectra for elastic SDOF systems subjected to 1940 El Centro earthquake: N–S component [25]

response parameters characterizing the motion of the mass relative to that of the base occurs. For linear systems with damping ratios of 5 to 10% subjected to the 1940 El Centro motion, the maximum amplification factors for displacement, velocity and acceleration are about 1.5, 2.0, and 2.5, respectively. A typical response spectrum curve in Fig. 3-1 can be approximately represented by three line segments shown as the dashed line a–b–c–d in the figure.

3.2 Establishing design ground motion(s) at a site

A number of procedures for selecting design (earthquake) ground motions at a site are currently available. Werner [26] has classified these as (a) site-independent, or (b) site-dependent using site-matched records, or site-response analysis.

3.2.1 *Site-independent procedures*
Site-independent procedures use standardized spectrum shapes developed from accelerograms that represent a variety of seismologic, geologic and local soil conditions. The use of site-independent spectra was first introduced by Housner [13]. Since then, other shapes have been developed, including those of Newmark and Hall [27] and Newmark, Blume and Kapur [28].

3.2.2 *Site-dependent procedures*

Site-matched records or spectra   Seed, Ugas and Lysmer [29] have developed site-dependent spectra based specifically on local site conditions. Ensemble average and mean-plus-one standard deviation spectra were developed from 104 site-matched records corresponding to four broad site classifications: rock (28 records), stiff soil (31 records), deep cohesionless soil (30 records), and soft-to-medium soil deposits (15 records). The spectra developed for each site condition corresponded to a 5% damping ratio and were normalized to a zero-period acceleration of 0.1 g.

A state-of-the-art recommendation for specifying earthquake ground shaking at a site in the United States has been drawn up by the Applied Technology Council [30]. The ATC has chosen to represent the intensity of design ground shaking by two parameters (Fig. 3-2a): effective peak acceleration (EPA) and effective peak velocity (EPV). EPA is expressed in terms of a dimensionless coefficient, A_a, which is equal to EPA expressed as a fraction of g (e.g. if EPA = 0.2 g, A_a = 0.2). EPV is expressed in terms of a dimensionless parameter, A_v, which is a velocity related acceleration coefficient (Fig. 3-2a). The ATC report furnishes detailed

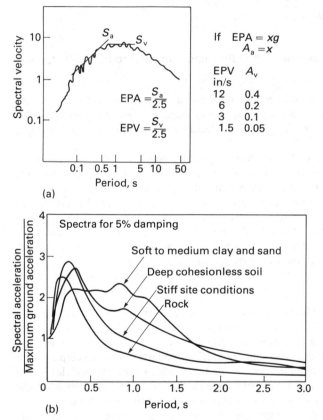

(a)

(b)

Figure 3-2 Applied Technology Council recommendations for specifying ground shaking at a site [30]
(a) Schematic representation showing how effective peak acceleration (EPA) and effective peak velocity (EPV) are obtained from a response spectrum.
(b) Average acceleration spectra for different site conditions (After Seed *et al.* [29])

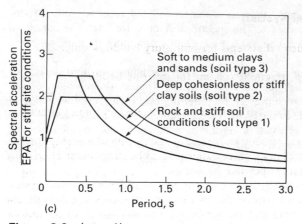

(c)

Figure 3-2   (contd.)
   (c) Spectral curves for use in building code, normalized for EPA for stiff
       site conditions

maps which divide the United States into seven map areas. The A_a and A_v coefficients for each map area are given. The probability is estimated at 90% that the recommended EPA and EPV at a given location will not be exceeded during a 50-year period.

The Seed–Ugas–Lysmer [29] spectral shapes for different soil conditions (Fig. 3-2b) were simplified to a family of three curves by combining the spectra for rock and stiff soil conditions (Fig. 3-2c). Recommended ground motion spectra (for 5% damping) for the different map zones are obtained by multiplying the normalized spectral values shown in Fig. 3-2c by the values of effective peak ground acceleration (and a correction factor of 0.8 for soft-to-medium stiff clay and sand type soils). Where the A_a and A_v values for a map area differ, the portion of the response spectrum controlled by velocity (the descending parts in Fig. 3-2c) should be increased in proportion to the EPV value, and the remainder of the response spectrum extended to maintain the same overall spectral form.

Site-response analyses   The use of site-matched records defines ground motions more nearly representative of the seismologic, geologic, and local soil conditions at a site than do the use of site-independent spectra. However, the local soil characteristics or the magnitudes and causative fault distances associated with earthquakes at a particular site may not be fully reflected by any motion in the currently available library of strong motion records. Also, the subsurface conditions at the site of many strong motion records are poorly defined, so that such records cannot be applied with complete confidence to a particular site. For these reasons, analytical evaluation of site response may at times become necessary [26].

Site-response analyses usually consist of three steps: the selection of an ensemble of representative subsurface input motions; the development of a mathematical model of the site soil deposits; and the evaluation of the various types of output from the analyses. These steps have been discussed in detail in [26], which should be consulted for further information.

4 Aseismic structural system

This section discusses structural systems for multistory buildings only.

4.1 Lateral load-resisting structural systems for concrete buildings

The three basic framing systems to resist lateral loads in concrete buildings are: (a) frames, (b) shearwalls coupled or acting individually, and (c) frames interacting with shearwalls.

Reinforced concrete frame structures depend on the rigidity of member connections for their resistance to lateral forces, and tend to be uneconomical (require high cost for lateral resistance) beyond 20 stories.

The introduction of shearwalls is an efficient way of stiffening a frame system (Fig. 4-1a). The interaction between the two elements reduces significantly the lateral deflection of the frame and the structural wall through a set of internal forces in the slabs. A judicious disposition of walls in plan allows them to function efficiently as vertical and lateral load resisting elements without interfering much with architectural requirements (Fig. 4-1b). Reinforced concrete buildings up to 70 stories high have been built using shearwall–frame interactive systems [31].

A modification of the conventional frame arrangement, suitable for buildings up to about 60 stories high, is the so-called 'framed tube' (Fig. 4-1c). The exterior

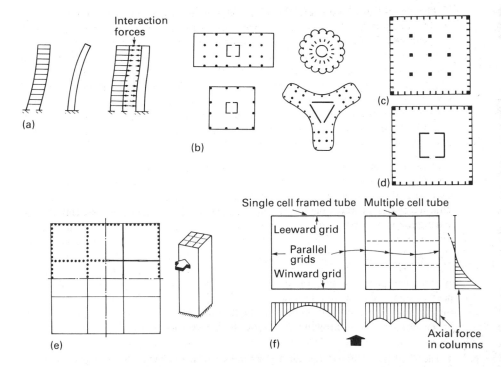

Figure 4-1 Lateral load resisting structural systems

columns are spaced closely together and are connected by spandrel beams to form an exterior grid that is usually designed to resist the bulk of the lateral load. The system represents a logical evolution of the conventional frame structure, combining high lateral stiffness with excellent torsional resistance, while retaining the planning flexibility allowed by isolated interior columns [31, 32].

For taller structures, an arrangement suitable for office buildings is the so-called 'tube-in-tube' system [31, 32] (Fig. 4-1d). This system provides a tall, stiff structure with wide column-free spaces between a central core and an external peripheral grid of closely spaced elements.

For still taller structures, especially where a large plan area is involved, intersecting planes of interior walls or closely spaced column-beam grids traversing the entire width of the building may be used (Fig. 4-1e). Connecting the exterior peripheral grids with interior walls reduces the shear lag[1] across the windward and leeward grids and allows the latter to participate to a greater extent in resisting the lateral load (Fig. 4-1f). The use of such interior vertical diaphragms essentially produces vertical multi-cell cantilever box beams or bundled tubes [31, 32].

Table 4-1, adapted from [33], is presented as a guide in the choice of an appropriate structural system for a new building. The ranges of suitability shown may vary somewhat depending upon the use of the building, the story heights, and the design live and wind loads.

Table 4-1 Guide to the selection of structural systems [33]

| Structural system | Number of stories[a] | | Seismic behavior |
	Office buildings	Apartment buildings hotels, etc.	
Frame	Up to 15	Up to 20	Very good
Shear wall		Up to 150	Good
Shear walls acting with frames	Up to 40	Up to 70	Very good
Single-framed tube	Up to 40	Up to 60	Very good
Tube-in-tube	Up to 80	Up to 100	Good

[a] The values given here are based on present day practice as well as trends indicated by current practice.

4.2 Planning and design considerations

A structure proportioned for gravity and wind loads represents a good preliminary design of an earthquake-resistant building. However, modifications in configuration and proportions to anticipate earthquake requirements should be made at an early stage. The following are some design considerations.

1 The decrease in the vertical forces transmitted to the columns as one moves from the corner toward the center of a frame subjected to lateral loads.

(a) *Drift limitation* For severe earthquake motion, an important concern is the stability of a structure under gravity loads while undergoing large lateral displacements. The SEAOC recommendations [34] mention an allowable drift due to the specified earthquake forces twice that allowed for wind. In applying such a limit, a distinction should be made between the drift produced by the code-specified static forces and the dynamic lateral displacements corresponding to a particular earthquake. The latter could be several times larger than the former [35]. The need for damage control may require a limit on the interstory drift as well (Section 6.6).

(b) *Avoidance of unnecessary torsion and force concentrations* Basically, a structure should (a) be simple, (b) be symmetrical, (c) be not too elongated in plan or elevation, and (d) have uniform and continuous distribution of strength. An earthquake will unerringly seek out and exploit every weakness in a structure. Proper attention to the above principles constitutes the basis of sound earthquake engineering.

Re-entrant angles are generally undesirable in plan layout. External elevator shafts and stairwells are undesirable because they introduce eccentricities and tend to act on their own in earthquakes.

A structure has more or less uniform and continuous distribution of strength if (a) the load-bearing members are uniformly distributed, (b) all columns and walls are continuous and without offset from roof to foundation, (c) all beams are free of offsets, (d) columns and beams are coaxial, (e) columns and beams are nearly the same width, (f) no principal members change section suddenly, and (g) the structure is as continuous and monolithic as possible.

(c) *Building multiple lines of defense* A high degree of static indeterminacy is desirable for earthquake resistance. Further, an advantageous sequence in the propagation of yielding should be established, so that damage in repairable and less critical areas will occur first. The principal gravity load-carrying units will then receive the greatest degree of protection.

(d) *Tying together of elements* The need to adequately tie together all the structural elements making up a building, or that portion of it which is intended to act as a unit, cannot be overemphasized. This applies to superstructure as well as foundation elements, particularly in buildings founded on relatively soft soil. Adequate connections should be provided across construction joints if they are required between parts of a building or between the main portion of a structure and an appendage.

(e) *Prevention of hammering* The different portions of a building, if not properly tied together, should be separated by a sufficient distance to prevent their hammering against each other. Any required passageway, corridor or bridge linking structurally separated parts should be so detailed as to allow free, unhindered movement during an earthquake.

(f) *Infilled frames* The presence of rigid infill totally changes the behavior of relatively flexible frames. If the infilling material is intended to act in combination

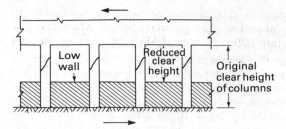

Figure 4-2 Effect of introducing low walls between columns originally designed for longer clearance height

with the enclosing frame, then proper reinforcement and connection to the enclosing frame are essential. The increased stiffness and modified behavior of the infilled frames should be considered in analysis. If the infill is made of fairly brittle material, such as glass or hollow brick masonry, and is not expected to contribute significantly to lateral resistance, then it should be effectively isolated from the surrounding frame by gaps or readily crushable or yielding material.

(g) *Reduction in the clear height of columns* The effect of introducing low walls between columns (Fig. 4-2) should be noted. The reduction in column height increases column stiffness with respect to bending in the plane of the wall, causing greater horizontal shears than would develop if the walls were absent. This is in addition to the effect of a decrease in the structural period (caused by increased column stiffness). The reduction in height also reduces the lateral deformation capacity of the columns in the plane of the wall. The use of such walls without allowing for their effects can cause severe distress in portions of the columns above the wall.

4.3 Possible control of seismic input

Structural features such as symmetry, absence of major discontinuities, etc., reduce sharp peak concentrations of earthquake-induced forces. The generally desirable objective of reducing seismic forces can logically be pursued further by introducing special devices or mechanisms into the structure. This approach has so far been limited to a very few applications. It is hoped that further research will develop the full potential of this possible course. The schemes proposed in the past to reduce the effect of ground motion on the structure have been reviewed in [36]. References [37, 38] contain up-to-date information on the subject.

4.4 Soil–structure interaction

Fundamental to aseismic design, as discussed in this chapter, is the assumption that the motion experienced by the base of a structure during an earthquake is the same as the free-field ground motion, i.e. the motion that would have occurred at the level of the foundation if no structure were present. This assumption applies in a strict sense only to structures founded on essentially rigid ground. For structures supported on soft soil, the foundation motion is generally different from the free-field motion, and may include, in addition to a lateral or translational component, a racking component that may be important for tall structures.

The flexibly supported structure differs from its rigidly supported counterpart also in that a substantial part of its vibrational energy may be dissipated into the supporting medium by radiation of waves and by hysteretic action in the soil. The importance of soil hysteresis increases with increasing intensity of ground shaking.

Two different approaches may be used to assess the effects of soil–structure interaction. The first involves modifying the free-field design ground motion and evaluating the response of the given structure to the modified motion of the foundation. The second approach involves modifying the dynamic properties of the structure and evaluating the response of the modified structure to the free-field design ground motion. The two approaches, properly implemented, give comparable results. The second approach, involving the use of free-field ground motion, is more convenient for design purposes.

The topic of soil–structure interaction is too broad to permit a meaningful discussion here. The reader is referred to the commentary on Chapter 6 of [30] for a good overview of the subject.

5 Response of structures to earthquakes

This section generally follows Clough's excellent treatment of the topic, as available in more detail in [39, 40].

5.1 Dynamic versus static structural analysis

In a structural dynamics problem, the loading and all aspects of the structural response vary with time, so that a solution must be obtained for each instant during the history of response.

There is a more important distinction between a static and a dynamic problem. When the simple column of Fig. 5-1a is subjected to a static lateral load, the

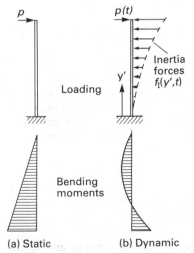

Loading

p $p(t)$

y' Inertia forces $f_1(y', t)$

Bending moments

(a) Static   (b) Dynamic

Figure 5-1 Essential difference between static and dynamic loading

internal forces may be evaluated by simple statics. If the same load is applied dynamically, the time-varying deflections involve accelerations which in turn generate inertia forces resisting the motion (Fig. 5-1b). The external loading, $p(t)$, that causes the motion and the inertia forces, $f_I(t)$, that resist its acceleration act simultaneously. The internal forces in the column must equilibrate this combined load system, so that it is necessary to know the inertia forces before the internal forces can be determined. The inertia forces depend upon the rate of loading and on the flexibility and mass characteristics of the structure. The basic difficulty of dynamic analysis lies in that the deflections which lead to the development of inertia forces are themselves influenced by the inertia forces.

5.2 Degrees of freedom

The complete system of inertia forces acting in a structure can be determined only by evaluating the acceleration of every mass particle. The analysis can be greatly simplified if the deflections of the structure can be specified adequately by a limited number of displacement components or coordinates. This can be achieved through the *lumped mass* or the *generalized coordinate* approach. In either case, the number of displacement components required to specify the positions of all significant mass particles in a structure is called the number of degrees of freedom of the structure. In the lumped-mass idealization, the mass of the structure is assumed concentrated at a number of discrete locations. An idea of the generalized coordinate approach is obtainable from Section 5.3.3.

5.3 Dynamics of multi-degree-of-freedom systems

5.3.1 *Equations of motion*
The building of Fig. 5-2 is used to illustrate multi-degree-of-freedom analysis. The mass of the structure is assumed concentrated at the floor levels (lumped-mass idealization) and subject to lateral displacements only. The equations of dynamic equilibrium of the system may be written as

$$[m]\{\ddot{x}\}+[c]\{\dot{x}\}+[k]\{x\}=\{p(t)\} \tag{5-1}$$

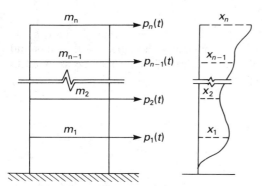

Figure 5-2 Lumped MDOF system with applied lateral loading and resulting displacements

where $\{\ddot{x}\}, \{\dot{x}\}$, and $\{x\}$ are the acceleration, velocity, and displacement vectors, respectively; $\{p(t)\}$ is the load vector; and $[m], [c]$, and $[k]$ are the mass, damping and stiffness matrices, respectively. The mass matrix for a lumped mass system is diagonal, i.e. the inertia force corresponding to any degree of freedom depends only on the acceleration in that degree of freedom. In general, it is not practical to evaluate the damping coefficients in matrix $[c]$, and damping is expressed in fractions of critical damping. The stiffness coefficient k_{ij} in matrix $[k]$ is the force corresponding to displacement coordinate i resulting from a unit displacement of coordinate j.

5.3.2 Vibration mode shapes and frequencies

The free vibration behavior of a structure corresponds to no damping ($[c] = [0]$) and no applied loading ($\{p\} = \{0\}$), so that Eqn 5-1 becomes

$$[m]\{\ddot{x}\} + [k]\{x\} = \{0\} \tag{5-2}$$

The motions of a system in free vibration are simple harmonic. Thus,

$$\{x\} = \{\bar{x}\} \sin \omega t \tag{5-3}$$

where $\{\bar{x}\}$ represents the amplitude of motion and ω is the circular frequency. Introducing Eqn 5-3 and its second derivative into Eqn 5-2

$$-\omega^2 [m]\{\bar{x}\} + [k]\{\bar{x}\} = \{0\} \tag{5-4}$$

Equation 5-4 is a form of eigenvalue equation [41]. Computer programs are available for the solution of very large eigenvalue equation systems. The solution to Eqn 5-4 for an n-degree-of-freedom system consists of a frequency ω_m and a mode shape $\{\varphi_m\}$ for each of its n modes of vibration (Fig. 5-3). The distinguishing feature of a mode of vibration is that a dynamic system can, under certain circumstances, vibrate in that mode alone; during such vibration the ratio of the displacements of any two masses remains constant with time. These ratios define the characteristic shape of the mode, the absolute amplitude of motion is arbitrary.

5.3.3 Modal equations of motion

An important simplification in Eqn 5-1 is possible because the vibration mode shapes of any multi-degree system are orthogonal with respect to the mass and stiffness matrices. The same type of orthogonality condition may be assumed to apply to the damping matrix as well:

$$\{\varphi_m\}^{\mathrm{T}}[m]\{\varphi_n\} = 0, \{\varphi_m\}^{\mathrm{T}}[c]\{\varphi_n\} = 0, \{\varphi_m\}^{\mathrm{T}}[k]\{\varphi_n\} = 0 \text{ for } m \neq n \tag{5-5}$$

Figure 5-3 shows that any arbitrary displaced shape of the structure may be expressed in terms of the amplitudes of mode shapes, treating them as generalized displacement coordinates:

$$\{x\} = [\varphi]\{X\} \tag{5-6}$$

in which $\{X\}$, the vector of the so-called normal coordinates of the system, represents the vibration mode amplitudes.

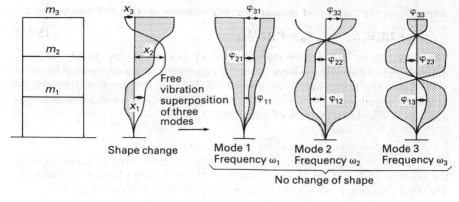

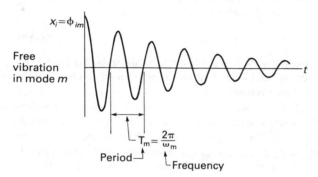

Figure 5-3   Mode superposition analysis of earthquake response

Substituting Eqn 5-6 and its derivatives into Eqn 5-1 and multiplying the resulting equation by the transpose of any mode shape vector yield

$$\{\varphi_m\}^{\mathrm{T}}[m]\{\varphi_m\}\ddot{X}_m + \{\varphi_m\}^{\mathrm{T}}[c]\{\varphi_m\}\dot{X}_m + \{\varphi_m\}^{\mathrm{T}}[k]\{\varphi_m\}X_m = \{\varphi_m\}^{\mathrm{T}}\{p(t)\} \qquad (5\text{-}7)$$

by virtue of the orthogonality properties of Eqn 5-5. Introducing

Generalized mass $M_m = \{\varphi_m\}^{\mathrm{T}}[m]\{\varphi_m\}$
Generalized damping $C_m = \{\varphi_m\}^{\mathrm{T}}[c]\{\varphi_m\}$
Generalized stiffness $K_m = \{\varphi_m\}^{\mathrm{T}}[k]\{\varphi_m\}$
Generalized loading $P_m(t) = \{\varphi_m\}^{\mathrm{T}}\{p(t)\}$ (5-8)

and recognizing that the generalized damping, stiffness and mass are related,

$$C_m = 2\beta_m\omega_m M_m \quad \text{and} \quad K_m = \omega_m^2 M_m \qquad (5\text{-}9)$$

where β_m is the fraction of critical damping in mode m, Eqn 5-7 becomes

$$\ddot{X}_m + 2\beta_m\omega_m\dot{X}_m + \omega_m^2 X_m = P_m(t)/M_m \tag{5-10}$$

Equation 5-10 shows that the equation of motion of any mode, m, of a multi-degree system is equivalent to the equation for a single-degree system (Eqn 1-1). Thus, the mode shapes or normal coordinates of a multi-degree system reduce its equations of motion to a set of independent or decoupled equations (as against the coupled equations of motion, Eqn 5-1).

5.3.4 Modal superposition analysis of earthquake response

The dynamic analysis of a multi-degree system by modal superposition requires the solution of Eqn 5-10 for each mode to obtain its contribution to response. The total response is obtained by superposing the modal effects (Eqn 5-6).

In the case of earthquake excitation (Fig. 5-4), the effective load is

$$\{p_{\text{eff}}(t)\} = -[m]\{1\}\ddot{x}_g(t) \tag{5-11}$$

where $\{1\}$ represents a unit vector of dimension n (the total number of degrees of freedom). Substituting Eqn 5-11 into Eqn 5-8,

$$P_{m_{\text{eff}}}(t) = -\{\varphi_m\}^{\text{T}}[m]\{1\}\ddot{x}_g(t) = -L_m\ddot{x}_g(t) \tag{5-12}$$

where

$$L_m = \{\varphi_m\}^{\text{T}}[m]\{1\} \tag{5-13}$$

represents the *earthquake participation factor* for mode m. Introducing Eqn 5-13 into Eqn 5-10, the equation of motion for mode m of a multi-degree system subject to earthquake excitation becomes

$$\ddot{X}_m + 2\beta_m\omega_m\dot{X}_m + \omega_m^2 X_m = -\frac{L_m}{M_m}\ddot{x}_g(t) \tag{5-14}$$

The response of the mth mode at any time t may be obtained, by analogy with Eqn 1-2, from

$$X_m(t) = -\frac{L_m}{M_m}\frac{1}{\omega_m}\int_0^t \ddot{x}_g(\tau)\exp\left[-\beta_m\omega_m(t-\tau)\right]\sin\omega_m(t-\tau)\,\mathrm{d}\tau \tag{5-15}$$

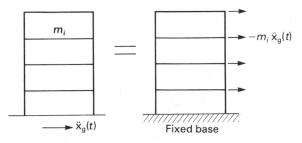

Figure 5-4 **Effective load caused by earthquake excitation**

or, using the symbol $\bar{X}_m(t)$ to represent the value of the integral at time t,

$$X_m(t) = -\frac{L_m}{M_m}\frac{\bar{X}_m(t)}{\omega_m} \tag{5-16}$$

The total response of the n-degree system may be obtained as follows:

Displacements   From Eqn 5-6

$$\{x\} = [\varphi]\{X\} = -[\varphi]\left\{\frac{L_m}{M_m}\frac{\bar{X}_m(t)}{\omega_m}\right\} \tag{5-17}$$

Elastic forces

$$\{f_s\} = [k]\{x\} = [k][\varphi]\{X\} = [m][\varphi][\Omega^2]\{X\} = -[m][\varphi]\left\{\frac{L_m}{M_m}\omega_m\bar{X}_m(t)\right\} \tag{5-18}$$

where $[\Omega^2]$ is a diagonal matrix of the squared modal frequencies ω_m^2. It should be noted that the elastic force vector associated with the mth mode is

$$\{f_{sm}\} = -[m]\{\varphi_m\}\frac{L_m}{M_m}\omega_m\bar{X}_m(t) \tag{5-18a}$$

Internal forces   These can be found from $\{f_s\}$ by statics.

Base shear   The base shear associated with the mth mode is the summation, over n stories, of the elastic forces associated with that mode:

$$V_m(t) = \{1\}^T\{f_{sm}(t)\} = -\{1\}^T[m]\{\varphi_m\}\frac{L_m}{M_m}\omega_m\bar{X}_m(t) = -\frac{L_m^2}{M_m}\omega_m\bar{X}_m(t) \tag{5-19}$$

from Eqn 5-13. The total base shear is

$$V(t) = -\sum_{m=1}^{n}\frac{L_m^2}{M_m}\omega_m\bar{X}_m(t) = -\sum_{m=1}^{n}\frac{W_m}{g}\omega_m\bar{X}_m(t) \tag{5-19a}$$

where

$$W_m = \frac{L_m^2}{M_m}g \tag{5-20}$$

W_m is the effective weight for mode m, and represents the portion of the total structural weight that is effective in developing base shear in the mth mode.

An important advantage of the mode superposition procedure is that an approximate solution can frequently be obtained by considering only the first few modes (sometimes, just the first or fundamental mode) in analysis.

An important limitation is that the mode superposition procedure is not applicable to any structure that is stressed beyond the elastic limit.

5.3.5   Response spectrum analysis

The entire response history of a multi-degree system is defined by Eqns 5-17 and 5-18, once the modal response amplitudes are determined (Eqn 5-16). The maximum response of any mode can be obtained from the earthquake response spectra, by procedures used earlier for single-degree systems.

On this basis, introducing S_{vm}, the spectral pseudo-velocity for mode m, into Eqn 5-16 leads to an expression for the maximum response of mode m:

$$|X_{m_{max}}| = \frac{L_m}{M_m} \frac{S_{vm}}{\omega_m} = \frac{L_m}{M_m} S_{dm} \qquad (5\text{-}21)$$

where S_{dm} is the spectral displacement for mode m. The distribution of maximum displacement in mode m is given (from Eqn 5-17) by

$$\{|x_{m_{max}}|\} = \{\varphi_m\} |X_{m_{max}}| = \{\varphi_m\} \frac{L_m}{M_m} S_{dm} \qquad (5\text{-}22)$$

Similarly, the distribution of maximum effective earthquake forces in the mth mode (from Eqn 5-18a) becomes

$$\{|f_{sm_{max}}|\} = [m]\{\varphi_m\} \frac{L_m}{M_m} \omega_m S_{vm} = [m]\{\varphi_m\} \frac{L_m}{M_m} S_{am} \qquad (5\text{-}23)$$

where S_{am} is the spectral pseudo-acceleration for mode m. From Eqn 5-19a, the maximum base shear in mode m is

$$V_{m_{max}} = \frac{W_m S_{am}}{g} = \frac{L_m^2}{M_m} S_{am} \qquad (5\text{-}24)$$

Because, in general, the modal maxima (e.g. $V_{m_{max}}$) do not occur simultaneously, they cannot be directly superimposed to obtain the maximum of the total response (e.g. V_{max}). The direct superposition of modal maxima provides an upper bound to the maximum of total response. A satisfactory estimate of the total response can usually be obtained from

$$V_{max} \doteq \sqrt{\sum V_{m_{max}}^2} \qquad (5\text{-}25)$$

As discussed, the summation needs to include only the lower few modes.

5.4   First-mode analysis of multi-degree-of-freedom systems

This is illustrated through an example adapted from [39].

A typical five-story building is shown in Fig. 5-5a; it is assumed to have a period $T_1 = 0.5$ s, and a damping ratio $\beta_1 = 10\%$ in the first mode. For these values, Fig. 5-5b gives the following spectral values: $S_{d1} = 0.48$ in (12 mm), $S_{v1} = 6.0$ in/s (152 mm/s), $S_{a1} = 76.0$ in/s^2 (=0.2 g, 1.96 m/s^2).

As is customary, the mass is assumed concentrated in the floor slabs. In a complete analysis, the lateral motion of each floor slab would be an independent degree of freedom. An approximate single-degree-of-freedom analysis for the building can be made by assuming that the lateral displacements are of a specified form. A reasonable assumption is that the displacements corresponding to the first mode increase linearly with height, i.e. $\varphi_{i1} = h_i/h_n$. The generalized mass and

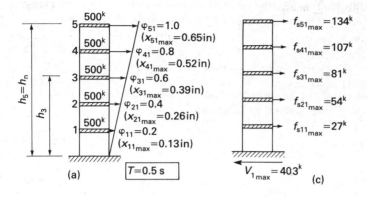

(a) $T=0.5\,\text{s}$ $V_{1\,\text{max}}=403^k$ (c)

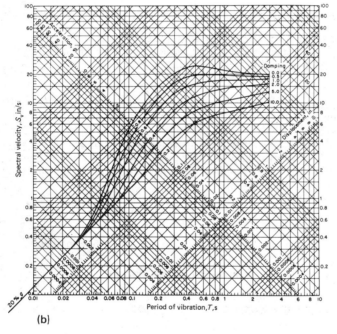

(b)

Figure 5-5   First-mode analysis of MDOF systems
(a) System analyzed and maximum displacement response
(b) Response spectra used in analysis
(c) Maximum earthquake forces: first subscript refers to floor level, second subscript to mode number

earthquake participation factors corresponding to the first mode are then given by

$$M_1 = \{\varphi_1\}^T[m]\{\varphi_1\} = \sum_i m_i \varphi_{i1}^2 \qquad (5\text{-}8a)$$

$$= \frac{500}{g}(0.2^2 + 0.4^2 + 0.6^2 + 0.8^2 + 1.0^2) = \frac{1100 \text{ kips}}{g} \qquad (5\text{-}13a)$$

$$L_1 = \{\varphi_1\}^T[m]\{1\} = \sum_i m_i \varphi_{i1} \qquad (5\text{-}13a)$$

$$= \frac{500}{g}(0.2 + 0.4 + 0.6 + 0.8 + 1.0) = \frac{1500 \text{ kips}}{g}$$

The maximum earthquake deflection is given by Eqn 5-22

$$\{x_{1max}\} = \{\varphi_1\}\frac{L_1}{M_1}S_{d1} = \{\varphi_1\}\frac{1500}{1100} \times 0.48 \text{ in} = \{\varphi_1\} \times 0.65 \text{ in} \qquad (5\text{-}22a)$$

For example,

$$x_{31max} = 0.6 \times 0.65 = 0.39 \text{ in, etc.} \quad \text{(Fig. 5-5a)}$$

The maximum base shear force is given by Eqn 5-24:

$$V_{1max} = \frac{W_1 S_{a1}}{g} = \frac{L_1^2}{M_1}S_{a1} = \frac{(1500)^2}{1100}\frac{76.0}{386} = 403 \text{ kips} \qquad (5\text{-}24a)$$

The forces at the various story levels may be obtained by distributing the base shear force according to Eqns 5-23 and 5-24:

$$\{f_{s1max}\} = [m]\{\varphi_1\}\frac{V_{1max}}{L_1} \qquad (5\text{-}23a)$$

or

$$f_{si1max} = \frac{m_i \varphi_{i1}}{L_1}V_{1max} = \frac{500}{1500}\varphi_{i1} \times 403 \text{ kips} = \varphi_{i1} \times 134.3 \text{ kips}$$

For example,

$$f_{s31max} = 0.6 \times 134.3 = 80.6 \text{ kips, etc.} \quad \text{(Fig. 5-5c)}$$

6 Earthquake-resistant design by building codes

The general pattern of seismic building code development in the United States is for code provisions to be adopted first by the Structural Engineers Association of California (SEAOC), then by the *Uniform building code*, and following that, by other regional codes. SEAOC recommendations, as embodied in the *Uniform building code*, also influence most seismic building codes of the world. The latest edition of the *Uniform building code* [42] will be used as the basis of discussion in this section.

Code provisions must be viewed as minimum requirements covering a broad class of structures of conventional configuration. Unusual structures may require design procedures beyond normal building code provisions.

The principal steps involved in the aseismic design of a typical concrete structure according to building code provisions are as follows.

(1) Determination of design earthquake forces:

    (a) calculation of base shear corresponding to computed or estimated fundamental period of vibration of the structure (a preliminary design of the structure is assumed here);

    (b) distribution of the base shear over the height of the building.

(2) Analysis of the structure under the (static) lateral forces calculated in step 1, as well as under gravity and wind loads, to obtain member design forces.

(3) Designing members and joints for the most unfavorable combination of gravity and lateral loads, and detailing them for ductile behavior.

The design base shear represents the total horizontal seismic force (service load level) that may be assumed acting parallel to the axis of the structure considered. The force in the other horizontal direction is assumed to act non-concurrently. Vertical seismic forces are not considered because the building is already designed for the vertical acceleration of gravity; the additional vertical acceleration due to an earthquake would normally be only a fraction of g. It can be accommodated by the factor of safety on gravity loads.

6.1 Design base shear

The total lateral force or base shear, V, in UBC-82 [42] is given by

$$V = ZIKCSW \tag{6-1}$$

Z is a numerical coefficient that depends on the seismic zone in which the structure is located (Fig. 6-1). Z has a value of 1 in zone 4, $\frac{3}{4}$ in zone 3, $\frac{3}{8}$ in zone 2 and $\frac{3}{16}$ in zone 1.

I is an occupancy importance factor, as given in Table 6-1.

K is a factor depending on the type of structural system used. It is intended to account for differences in the ductility or energy absorption capacity of various structural systems. Values of K (Table 6-2) range from 0.67 for a ductile moment-resisting space frame designed to resist the entire lateral load to 2.5 for elevated tanks. Note that frame–shearwall structures, which are designed for a K factor of 0.8, have to satisfy three independent loading requirements that call for three separate analyses.

C is a coefficient related to the stiffness of the structure:

$$C = 1/(15\sqrt{T_1}) \leqslant 0.12 \tag{6-2}$$

The fundamental period of vibration in seconds, T_1, may be determined from the formula:

$$T_1 = 0.05 h_n/\sqrt{D} \tag{6-3}$$

where h_n is the height (ft) of the structure above the base and D the dimension (ft) of the building in the direction parallel to the applied force. Alternatively, in

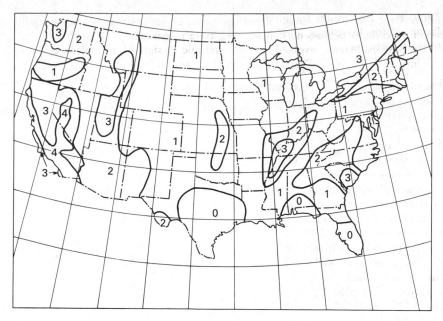

Figure 6-1 Seismic zone map of the USA [42]

       Zone 0 No damage

       Zone 1 Minor damage; distant earthquakes may cause damage to structures with fundamental periods greater than 1.0 s; corresponds to intensities V and VI of the MM Scale

       Zone 2 Moderate damage; corresponds to intensity VII of the MM Scale

       Zone 3 Major damage; corresponds to intensity VII and higher of the MM Scale

       Zone 4 Those areas with zone 3 determined by the proximity of certain major fault systems

Table 6-1 Values of occupancy importance factor I

Occupancy	I
Essential facilities	1.5
Assembly buildings	1.25
Other occupancy	1.0

Table 6-2 Horizontal force factor K for buildings and other structures

Type of structure	K
Ductile frames	0.67
Dual bracing systems[a]	0.80
Box systems	1.33
Other buildings	1.00
Elevated tanks	3.00
Other structures	2.00

[a] Shearwall-frame interactive systems: the frames and the shearwalls in interaction shall resist the total lateral force in accordance with their relative rigidities; the shearwalls and the frames shall have the capacity to independently resist 100% and 25% of the total lateral force, respectively.

buildings in which the lateral force-resisting system consists entirely of ductile moment-resisting space frames,

$$T_1 = 0.1n \tag{6-4}$$

where n is the number of stories.

The fundamental period of exceptionally stiff or light buildings may be significantly shorter than the estimate provided by Eqn 6-3 or 6-4. Especially for such buildings, and as a general check, the period should be recomputed by established methods of mechanics. An approximate formula based on Rayleigh's method (Eqn 12-3 of [42]) is convenient.

S in Eqn 6-1 is a coefficient for site–structure resonance. It is a function of the ratio of building and site periods, attaining its maximum value of 1.5 when the two periods coincide, and dropping off to a minimum value of 1 or slightly higher when the site and structure periods widely differ (Fig. 6-2). A supplemental publication designates ways of calculating the site period. The code stipulates limiting values of T_s, regardless of what a geotechnical site investigation may indicate. When T_s is not properly established, the value of S is to be taken equal to 1.5. The product of C and S need not exceed 0.14. Figure 6-3 shows minimum and maximum values of this product as related to the building period.

W in Eqn 6-1 is the total dead load of the structure and, for warehouse occupancy, 25% of the design storage load.

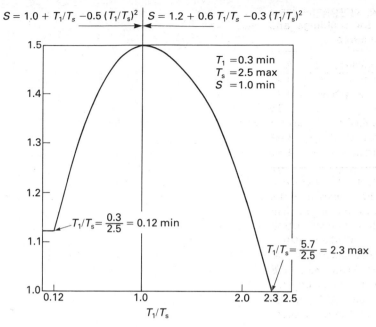

$$S = 1.0 + T_1/T_s - 0.5\,(T_1/T_s)^2 \quad | \quad S = 1.2 + 0.6\,T_1/T_s - 0.3\,(T_1/T_s)^2$$

$T_1 = 0.3$ min
$T_s = 2.5$ max
$S = 1.0$ min

$T_1/T_s = \dfrac{0.3}{2.5} = 0.12$ min

$T_1/T_s = \dfrac{5.7}{2.5} = 2.3$ max

T_1/T_s

Figure 6-2   Site–structure resonance factor, S, related to the fundamental building period, T_1, and the fundamental site period, T_s

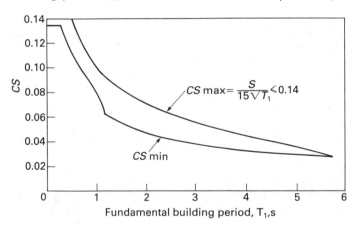

CS max $= \dfrac{S}{15\sqrt{T_1}} \leqslant 0.14$

CS min

Fundamental building period, T_1,s

Figure 6-3   Product CS, related to building period

6.2   Distribution of base shear along height of building

The total lateral force or base shear, V, is divided into two parts: a concentrated load, F_t, applied at the top of the structure, and the balance, $(V - F_t)$, distributed over the height of the building as concentrated loads at the floor levels. The distribution of $(V - F_t)$ is triangular, increasing from zero at the base to a maximum at the top.

The top load F_t is given by

$$F_t = 0$$

if $T_1 < 0.7$ sec, otherwise, $\qquad\qquad$ (6-5)

$$F_t = 0.07 T_1 V \leq 0.25 V$$

The magnitude of the distributed forces, F_x, making up the balance of the total lateral force, $(V - F_t)$, is given by

$$F_x = (V - F_t) w_x h_x \Big/ \sum_{i=1}^{n} w_i h_i \qquad\qquad (6\text{-}6)$$

where

> w_x, w_i = the portion of the total weight, W, located at or assigned to level x or i, respectively
> h_x, h_i = the height (ft) above the base to level x or i, respectively
> level n = the uppermost level in the main portion of the structure

At each level x, the force F_x is to be applied over the area of the building in accordance with the mass distribution at that level.

6.3 Comparison with first-mode analysis of multi-degree systems

The design base shear and the distribution of that shear along the height of a building, as prescribed in UBC-82 [42] and as obtained from the first-mode analysis of a multi-degree system (Section 5.4), are compared in Table 6-3. It can be seen that the code-prescribed distribution of base shear corresponds essentially to the fundamental mode of vibration response: the higher modes are accounted for through the force F_t applied at the top of flexible structures. As to magnitude, the base shear coefficient C of the code is compared with the base shear coefficient S_a/g from first mode analysis in Fig. 6-4. It is clear that the code forces, which are assumed to be elastically resisted by a structure, are substantially smaller than those which would develop if the structure were to respond elastically to an earthquake of 1940 El Centro intensity. Thus, code-designed buildings would be expected to undergo fairly large inelastic deformations when subjected to an earthquake of such intensity. The realization that it is economically unwarranted to design buildings to resist major earthquakes elastically, and the recognition of the capacity of structures possessing adequate strength and deformability to withstand major earthquakes inelastically, lie behind the relatively low forces specified by the codes, coupled with the special requirements for ductility particularly at and near member connections.

6.4 Balance between strength and ductility

The ability of a structure to deform past the elastic limit is usually measured in terms of ductility. Ductility in reinforced concrete structures in general is defined as the ratio of a specified distortion at a particular stage of the loading to that at the onset of yielding.

Table 6-3 First mode analysis of MDOF systems compared with UBC [42] seismic design provisions

	UBC-82	First-mode analysis	Comparison
Base shear	$V = ZIKCSW$ with $Z = 1.0$, highest value $I = 1.0$, usual value $K = 1.0$, typical value $S = 1.0$, typical value $V = CW$	$V_{1\max} = \dfrac{S_{a1}}{g} W_1$ from Eqn 5.24a $W_1 = \dfrac{L_1^2 g}{M_1} = \dfrac{(\sum_{i=1}^n m_i \varphi_{i1})^2}{\sum_{i=1}^n m_i \varphi_{i1}^2}\, g$ from 5-20 5-13a Eqns 5-8a $= \dfrac{(\sum_{i=1}^n w_i \varphi_{i1})^2}{\sum_{i=1}^n w_i \varphi_{i1}^2}$	$W_1 < W = (0.6-0.8)W$, typically C versus S_{a1}/g: see Figs 6-4, 6-6.
Distribution along height	$F_x = \dfrac{w_x h_x}{\sum_{i=1}^n w_i h_i}(V - F_t)$	$f_{sj1\max} = \dfrac{m_j \varphi_{j1}}{L_1} V_{1\max}$ Eqn 5-23a $= \dfrac{m_j \varphi_{j1}}{\sum_{i=1}^n m_i \varphi_{i1}} V_{1\max}$ $= \dfrac{w_j \varphi_{j1}}{\sum_{i=1}^n w_i \varphi_{i1}} V_{1\max}$ Eqn 5-13a	For $\varphi_{j1} = h_j/h_n$, as assumed in Section 5 (Fig. 5-5), $f_{sj1\max} = \dfrac{w_j h_j}{\sum_{i=1}^n w_i h_i} V_{1\max}$ Thus, UBC-82 distribution of base shear along height corresponds essentially to the fundamental mode of vibration response. F_t represents an attempt to account for the higher modes of vibration

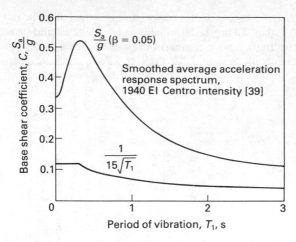

Figure 6-4 Comparison of base shear coefficients from UBC [42] and first-mode analysis of MDOF systems

Figure 6-5 shows the lateral force versus lateral displacement relationship for two structures with identical stiffnesses but different strengths, responding to the same earthquake. Structure A is able to resist the given earthquake elastically, and suffers the maximum deflection Δ_A. In structure B, when the lateral force reaches F_B, the structure reaches its elastic limit at a lateral displacement Δ_B.

It has been shown that the displacements which a regular medium-to-high-rise structure undergoes when subjected to a typical earthquake motion are of the same order of magnitude, whether the response is elastic or inelastic. Thus, structure B must be able to deform plastically from B to B', if it is to survive the earthquake.

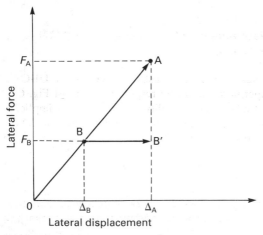

Figure 6-5 Force–displacement relationships of elastic and elasto-plastic systems

It can be seen that the designer may select a lower strength than the elastic response force, F_A, provided that inelastic deformability is available and the resulting damage is acceptable. In the above comparison, Δ_A is the same as Δ_{max},[1] the maximum deflection of structure B, and Δ_B is the same as Δ_y, deflection at yielding of structure B. The ratio $\Delta_A/\Delta_B = \Delta_{max}/\Delta_y = \mu_\Delta$ is the displacement or system ductility factor for the structure.

6.5 Magnitude of available ductility assumed in code

Response spectra for single-degree-of-freedom elasto-plastic systems having 10% critical damping are shown in Fig. 6-6 for displacement ductility factors of 1 (corresponding to an elastic system), 1.25, 2 and 4 [43]. It can be seen that even a relatively small amount of yielding, corresponding to values of μ_Δ of the order of 1.25 or 1.5, produces appreciable reductions in the base shear coefficient. The reductions from values for the elastic systems are roughly 20% for $\mu_\Delta = 1.25$, 50% for $\mu_\Delta = 2$, and 75% for $\mu_\Delta = 4$.

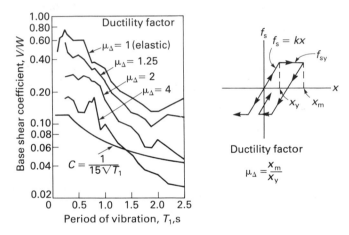

Figure 6-6 Magnitude of displacement ductility assumed in UBC [42]: S_a/g for $\beta = 0.10$ (1940 El Centro earthquake) [43]

The base shear coefficient versus period relationship prescribed in UBC-82 [42], $C = 1/(15\sqrt{T_1})$, has been superimposed on the response spectra of Fig. 6-6. It can be seen that an available displacement ductility of four to six is implicitly assumed in code-designed structures. The code contains requirements for detailing of members and their connections to ensure sufficient deformation capacity in the inelastic range.

6.6 Stability considerations

Because of the large inelastic deformations which a code-designed building may undergo during a strong earthquake, provisions must be made to ensure that the

---

1 In general, the resistance corresponding to Δ_{max} must be no less than a specified proportion (say, 80%) of the maximum lateral load carrying capacity of the structure.

structure maintain its integrity under the vertical loads while undergoing large lateral displacements. This is done in two ways:

(a) *Relative strengths of beams and columns at a joint*   The intent of the code is to have yielding initiated in the beams while the columns remain elastic long into their seismic response. This strong column–weak beam concept of design is implemented by requiring that the sum of the moment strengths of the columns meeting at a joint, under the design axial load, be greater than the sum of the moment strengths of the beams framing into the joint in the same plane. This provision reduces the risk of instability due to large inelastic deflections of the columns, and has been briefly discussed earlier (Section 4.2, item (c)).

(b) *Drift limitation*   This has been discussed under item (a), Section 4.2. UBC-82 [42] requires that

> Lateral deflections or drift of a story relative to its adjacent stories shall not exceed 0.005 times the story height unless it can be demonstrated that greater drift can be tolerated. The displacement calculated from the application of the required lateral forces shall be multiplied by $(1.0/K)$ to obtain the drift. The ratio $(1.0/K)$ shall not be less than 1.0.

It should be noted that the structural displacements computed by elastic analysis under the code-specified static loads are only a fraction of the actual displacements that may be caused by earthquake excitation. The code-imposed limit on interstory drift recognizes this difference.

6.7   Overturning moments

The distribution of story shear over height computed from the lateral forces of Eqns 6-5, 6-6 is intended to provide an envelope. Because of the contributions of several modes, shears in all stories do not attain their maxima simultaneously during an earthquake. Thus, the story moments statically consistent with the envelope of story shears are usually overestimates.

Results of dynamic analyses suggest that up to 20% reduction is, in general, reasonable for the story moments computed statically from the envelope of story shears, but that no reduction should be permitted in the upper stories of buildings [12, 44].

Formerly, many codes allowed large reduction in overturning moments relative to their values statically consistent with the story shears. These reductions appeared to be excessive in the light of the damage to buildings during the 1967 Caracas, Venezuela, earthquake, where a number of column failures could be attributed to overturning moments. In the current UBC, no reduction is allowed, although this may not be entirely justified.

6.8   Torsion

The inertia force acts through the center of mass. If there is a lack of symmetry, the center of rigidity, i.e. the location where the applied horizontal force would not produce any torsion, may not coincide with the center of mass. The product of the distance between these two points and the shear in the story gives a horizontal

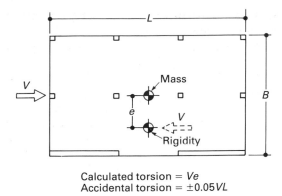

Calculated torsion = Ve
Accidental torsion = $\pm 0.05 VL$

Figure 6-7   Horizontal torsion in buildings

torsion that must be accounted for in design. In addition, there is a possibility of the seismic waves not acting uniformly across the building. Therefore, an accidental torsion is prescribed by the code; this torsion (even in fully symmetrical structures) is 5% of the product of the story shear and the maximum plan dimension at the level being considered (Fig. 6-7). Provisions need to be made for the increase in shear in vertical resisting elements due to the horizontal torsion; negative torsional shears should be neglected.

6.9   Distribution of shears and moments

Shear in any horizontal plane is to be distributed to the various lateral force-resisting elements in proportion to their rigidities, considering the rigidity of the horizontal bracing diaphragm. Concrete floor diaphragms can be considered infinitely stiff; only the stiffness of the vertical resisting elements are then of importance. However, if the floor diaphragm is flexible, the distribution of the forces becomes more uniform than determined using the rigid diaphragm assumption. A simplified approach to distributing horizontal shear among various resisting elements, considering flexibility of the floor diaphragm, has been suggested in Ref. [44].

6.10   Load factors and loading combinations for design

UBC-82 specifies that the strength of a structure or its components be not less than that required by any of the following load combinations:

$$U \geqslant \begin{cases} 1.4DL + 1.7LL & (6\text{-}7a) \\ 0.75[1.4DL + 1.7LL \pm (1.7WL \text{ or } 1.87EL)] & (6\text{-}7b) \\ 0.9DL \pm (1.3WL \text{ or } 1.43EL) & (6\text{-}7c) \\ 0.75[1.4(DL + TL) + 1.7LL] & (6\text{-}7d) \end{cases}$$

where

U = required strength

DL = dead loads or their related internal forces

LL = live loads or their related internal forces

WL = wind loads or their related internal forces

EL = load effects of earthquakes, or their relative internal forces

TL = load effects of contraction and expansion caused by temperature changes, creep, shrinkage, or differential settlement.

For ductile moment-resisting space frames in seismic zones 3 and 4, and for earthquake resisting concrete structural walls and braced frames, Eqns 6-7b,c for earthquake loading are to be modified to

$$U \geq \begin{cases} 1.4(DL + LL \pm EL) & \text{(6-8a)} \\ 0.9DL \pm 1.4EL & \text{(6-8b)} \end{cases}$$

provided, further, that $2.0E$ shall be used in Eqns 6-8a,b in calculating shear and diagonal tension stresses in shearwalls of buildings other than those complying with requirements for buildings with $K = 0.67$.

7 Alternatives to aseismic design by building codes

7.1 Inelastic analysis

A major earthquake can cause significant inelasticity in any standard multistory building designed by normal code procedures. The true dynamic behavior of such a structure can be determined only if allowance is made for nonlinearity in the analysis procedure.

For analysis considering nonlinear response, the equations of motion developed in Section 5.3.1 need to be used directly. The response of a nonlinear system may be considered as being essentially linear over small intervals of time. As yielding occurs in members and the stiffness of a structure changes, the response of the nonlinear system is obtained as the response of successively different linear systems.

Computer programs developed to undertake noniinear dynamic analysis of multistory frames [45–47] can generate a variety of useful information, such as maximum deformations and forces at all significant locations, as well as time history records of deformations and forces at particular points in the structure. However, there are uncertainties in the values of stiffness and damping to be used in conjunction with the structural model.

The volume of numerical work and the computer storage capacity required for inelastic dynamic analysis make it necessary to assume fairly simple force–deformation characteristics for structural members. Allowance is usually made only for yield hinges forming at the ends of members, the hinging being governed by the magnitude of the bending moment [48–50]. More elaborate force–deformation relationships for structural members incorporating axial load–moment interaction effects and inelastic shear behavior have recently become available, and have been used to a limited extent [51–54].

7.2 Design considering inelastic response

Three approaches to design that accounts for inelastic structural response are currently available, and are discussed below.

7.2.1 *Design based on inelastic response spectra*

Newmark [55, 56] has demonstrated use of the design spectrum to approximate inelastic behavior. The UBC-82 [42], ATC 3-06 [30], and other code approaches to seismic design are in principle based on the concept of the inelastic design spectrum. In ATC 3-06, the response modification factor, R, is used, in effect, to derive inelastic design spectra from the corresponding elastic spectra. The earthquake force derived from the inelastic design spectrum is the design base shear of the codes.

Shibata and Sozen [57] have proposed a 'substitute structure' method, where, instead of deriving an inelastic response spectrum from an elastic spectrum, a substitute structure is derived from the actual structure with properties adjusted to account for inelastic behavior, and the elastic design spectrum is directly used.

7.2.2 *Shortcomings of design approach based on inelastic design spectrum (including code-recommended design)*

These are

(a) Internal forces determined from elastic analysis under static loads specified in codes or derived from inelastic design spectra differ from forces caused by actual inelastic seismic structural response.

(b) The distribution and magnitude of inelastic deformations in structural members cannot be determined through elastic analysis under static forces given in codes or derived from inelastic spectra. Thus, special ductility details must be provided throughout the structure, although inelasticity may actually occur only in certain locations. Also, there is no certainty that the ductility provided through conformance with code-prescribed detailing requirements will always suffice.

(c) Elastic story drifts under static forces given in codes or derived from inelastic response spectra, even when multiplied with a ductility factor or a response modification factor, may differ from actual inelastic story drifts. Thus, the intended damage control and safety against instability may not be achieved.

(d) Manipulating the relationship between column and beam strengths at a joint, on the basis of results of gravity load and static lateral load analyses, may also be ineffective [58]. Actual earthquakes can cause axial forces in columns and other internal forces that may substantially alter the strength relationship intended by the designer.

7.2.3 *Inelastic design based on energy considerations*

Housner [59] proposed a simplified design method based on equating the energy input from an earthquake with the energy absorbed in a structure. Blume *et al.* [60] proposed a more elaborate and formalized 'reserve energy technique' and applied it to at least one 24-story reinforced concrete frame building. More recently, Larson [61] has discussed the need for considering the earthquake energy input, and its absorption, in design. Work toward a simple energy method for seismic design of structures is in progress at the University of British Columbia [62]. A great deal of judgment and intuition is needed to successfully apply the available energy methods.

7.2.4 Design utilizing inelastic response history analysis

Information on the amount and distribution of internal forces and deformations in yielding structures can be obtained only through inelastic response history analyses of structures subjected to earthquake input motions. Such analysis has often been used in research studies of seismic structural response [35, 53, 63].

The Muto Institute has developed a Dynamic Design Procedure (DDP) [64, 65] that utilizes dynamic analysis. The linear and nonlinear properties of equivalent 'stick' models are determined from tests on segments of the actual frames. In applications of DDP, peak ground accelerations of 0.1 g, 0.3 g, and 0.5 g have been taken as corresponding to 'minor or moderate,' 'severe,' and 'worst' earthquakes, respectively. Dynamic (elastic or inelastic, as appropriate) analyses of structures under actual recorded ground motions (normalized to the above peak ground accelerations) have been carried out, and compliance with several postulated design criteria has been checked.

Fintel and Ghosh [66, 67] have proposed an Explicit Inelastic Dynamic Design Procedure (EIDDP) that uses earthquake accelerograms as loading, dynamic inelastic response history analysis to determine member forces and deformations, and resistance from tests for proportioning the members.

7.2.4.1 Features of EIDDP These are

(a) The systematic selection of input motions that are likely to excite critical response in a particular structure forms an integral part of EIDDP.
(b) Nonlinear models of member hysteretic response, generalized from test results, are used in response history analysis.
(c) The designer exerts control over where in the structure the bulk of seismic energy dissipation for the selected input motion(s) takes place.
(d) Two reference stages are usually considered: a 'design earthquake' and an 'hypothetical maximum credible earthquake,' which are comparable with the serviceability and ultimate strength limit states. Just one or more than two reference stages can also be considered. Performance criteria for each reference stage may be chosen to suit the designer's particular requirements.

EIDPP has become feasible because of the development of computer programs with which two-dimensional inelastic response history analyses can be carried out relatively inexpensively. To date, EIDDP has been applied to the design of several multistory reinforced concrete buildings [68–72].

7.2.4.2 Principal elements of EIDDP The suggested design procedure entails:

(1) Preliminary layout and design of the structural system for gravity loads and for code-specified wind and earthquake forces.
(2) Modeling of the structure for dynamic analysis in two orthogonal directions

using multibay–multistory models, and determination of the natural undamped periods of the structure in each direction.

(3) Selection of design earthquake accelerograms with frequencies that may
critically excite the structure. By subjecting the model to these records
(normalized for intensity according to local seismicity), the one or two
producing the maximum response can be chosen. Quite often, the maximum
displacements and the largest bending moments or shear forces are caused
by different accelerograms. In such cases, more than one design earthquake
accelerogram must be used in further computation.

(4) Specifying performance criteria under the design and the hypothetical
maximum credible earthquakes.

(5) Determination of internal forces and deformations caused by the design
earthquake using inelastic response history analysis. Repeat of such analysis
with altered strength levels for the structural members until all performance
criteria are satisfied and an optimum relationship between strength and
deformability demands is obtained.

(6) Proportioning of members for strength and deformability.

(7) Checking that the structure has the needed margin of ductility to meet the
chosen safety criteria corresponding to the hypothetical maximum credible
earthquake. Checking for stability may become necessary if there is yielding
in the columns or walls.

7.2.4.3 *Input motion* For response history analysis, input motion accelerograms, rather than design response spectra, are needed. A procedure for the
selection of input motions to critically excite a given structure has been developed at the Portland Cement Association [7]. The procedure attempts to match
the frequency content of input motions, as indicated by their velocity response
spectra, with the frequency characteristics of a stucture. The procedure is
virtually site-independent and may, therefore, be somewhat conservative. The
hypothetical maximum credible earthquake for a site is considered to differ from
the design earthquake only in intensity.

7.2.4.4 *Dynamic analysis* The computer program DRAIN-2D [46] was
selected for inelastic response history analysis, based on a comparative evaluation
of several available programs. DRAIN-2D determines the seismic response of
plane inelastic structures, with masses lumped at nodal points, through step-by-
step integration of the coupled equations of motion, based on the assumption of a
constant response acceleration during each time step. Viscous damping having
both mass-proportional and stiffness-proportional components is used.

Program DRAIN-2D accounts for inelastic effects by allowing the formation of
'point hinges' at the ends of elements where the moments exceed the specified
yield moments. The moment versus end rotation characteristics of elements are
defined in terms of basic bilinear relationships. These develop into hysteretic
loops following the modified Takeda model [49] which incorporates post-yield
decreases in unloading and reloading stiffnesses (Fig. 7-1). P-delta effects can be
considered in DRAIN-2D analysis, but are seldom significant in concrete structures containing shearwalls.

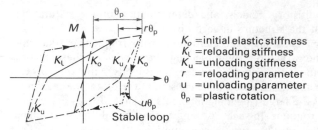

K_o = initial elastic stiffness
K_l = reloading stiffness
K_u = unloading stiffness
r = reloading parameter
u = unloading parameter
θ_p = plastic rotation

Figure 7-1 Modified Takeda model of the moment-rotation relationship of a reinforced concrete member segment

7.2.4.5 *Control of energy dissipation* In a building structure, the beams are not crucial to structural stability and are easy to repair. A structure can be designed such that, under severe earthquake-induced excitation, all inelastic deformations will be confined to designated beams, while the more vital elements such as columns and walls will remain elastic throughout their seismic response. This is best achieved by intentionally making the designated members weaker in relation to others. Intentionally weakened beams act as structural fuses which (Fig. 7.2): (a) preset the commencement of yielding during an earthquake, thus ensuring controlled energy dissipation in selected members; (b) reduce the moments, M_y, transferred from beams into columns, thus protecting the columns from hinging; (c) reduce the seismic axial forces $(M_y^- + M_y^+)/l$ (tension and compression) in the columns or walls; and (d) decrease the congestion of reinforcement, as well as shear stresses, in the joints.

Beam strength reductions, as discussed above, can be made only if the corresponding deformations and internal forces do not violate the chosen performance criteria. If the initially chosen strength levels result in unacceptable deformations, shears or axial forces, then adjustments must be made and the structure reanalyzed, until all performance criteria are satisfied. This iteration process can usually be accomplished with three to five analyses.

7.2.4.6 *Definition of ductility* Rotational ductility for wall and beam segments is defined in EIDPP as shown in Fig. 7-3.

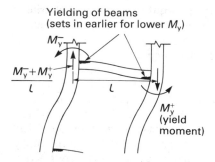

Figure 7-2 Intentionally weakened beams acting as structural fuses

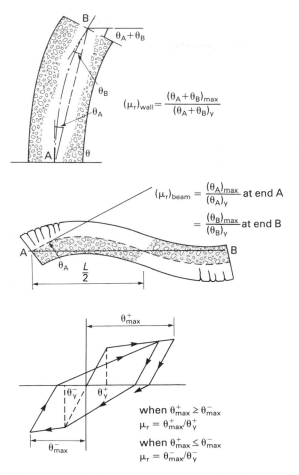

$$(\mu_r)_{wall} = \frac{(\theta_A + \theta_B)_{max}}{(\theta_A + \theta_B)_y}$$

$$(\mu_r)_{beam} = \frac{(\theta_A)_{max}}{(\theta_A)_y} \text{ at end A}$$

$$= \frac{(\theta_B)_{max}}{(\theta_B)_y} \text{ at end B}$$

when $\theta^+_{max} \geq \theta^-_{max}$
$\mu_r = \theta^+_{max}/\theta^+_y$
when $\theta^+_{max} \leq \theta^-_{max}$
$\mu_r = \theta^-_{max}/\theta^-_y$

Figure 7-3 Definition of rotational ductility. *Note:* In general, the resistance corresponding to θ_{max} must be no less than a specified proportion (say, 80%) of the moment capacity

7.2.4.7 Performance criteria For the design earthquake, serviceability criteria need to be formulated; for the maximum credible earthquake, safety against collapse must be assured.

Under the design earthquake, inelasticity may be disallowed, or may be confined only to the beams, with limits put on the inelastic deformations. The interstory drifts may be limited to assure damage control. Shear stresses in structural members, axial forces in vertical members, etc. may also be limited.

Under the maximum credible earthquake, some inelasticity may be allowed in columns and walls, although this should be kept to a minimum because of stability considerations. The maximum deformations in all members must be below the available deformation capacities. The total lateral deflection should be limited to assure stability.

Suitable limits on deformations, shear stresses, etc., constituting performance criteria, can be established only through correlation with observed deformations, cracking and damage in laboratory tests of structural members.

7.2.4.8 *Proportioning of members—combination of load effects*: Member proportioning has been discussed in [67].

7.2.4.9 *Advantages of EIDDP* These are

(a) The magnitude and location of inelastic deformations are determined, so that ductility details can be provided only where required and usable.
(b) A desired balance between strength and ductility can be designed into the various structural members.
(c) A predetermined sequence of plastification can be designed into a structure, thus regulating the onset and progress of energy dissipation.
(d) Progressively more relaxed performance criteria may be specified for increasing intensities of seismic excitation.
(e) The designer can creatively devise various structural configurations and distribute the energy dissipating elements where they will be the most effective without endangering structural stability.

7.2.4.10 *Shortcomings of EIDPP* These are

(a) Two-dimensional rather than three-dimensional analysis is used, so that torsion has to be considered separately. The approach is limited to reasonably symmetrical structures for which two-dimensional modeling is realistic. This shortcoming is common to all current design approaches.
(b) Shear forces obtained from dynamic analysis fluctuate a lot due to higher mode effects. When these are compared with shear capacities based on tests carried out under essentially static (monotonic or cyclic) loading, an unintended conservatism results.
(c) The use of an inelastic approach, although relatively inexpensive in terms of computer cost, requires special knowledge on the part of the designer. It is not intended at this time for routine designs.

7.3 Design example using explicit inelastic dynamic design procedure

7.3.1 *Building studied and its structural system*
A typical floor layout of the 16-story apartment building chosen for this design example is shown in Fig. 7-4a. The structure is a composite precast and cast-in-place system. The vertical hollow cores of the 8 in (200 mm) thick precast walls carry vertical continuity reinforcement, and are filled with concrete. Most of the slabs consist of 4 in (100 mm) thick prestressed precast units plus a $3\frac{1}{2}$ in (90 mm) thick cast-in-place topping which is composite with the precast units. The cast-in-place connections in the structural system provide for good continuity between walls and slabs, between adjacent slabs, and between successive lifts of walls. However, the precast walls have no vertical connections at their junctions. Two pairs of walls, in each orthogonal direction, separated by doorways, are coupled by the door lintels having a span-to-depth ratio of 1.43.

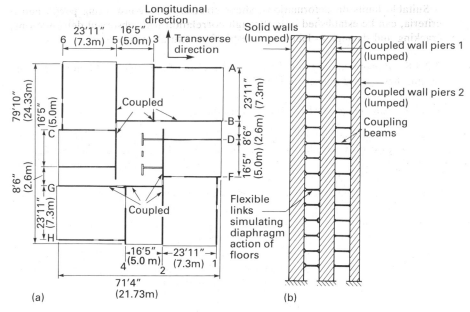

Figure 7-4 (a) Floor plan; (b) schematic section of building studied

7.3.2 Partial design problem considered

The coupling beams or door lintels were chosen to act as structural fuses. It was decided that under the design earthquake, only the coupling beams should yield and dissipate energy, while the walls should remain fully elastic. The coupling beams themselves were required to remain elastic up to at least $0.75 \times 1.7 = 1.3$ times the design wind loads.

Of major concern in the design process was the determination of appropriate strengths to be built into the walls and the coupling beams, in order to limit the deformation demands in critical regions to reasonable levels, while keeping the maximum moments, shear forces and axial forces in these regions acceptable. The determination of design strength levels for the coupling beams in the longitudinal direction only is discussed below.

7.3.3 Modeling of the structure

For purposes of analysis, the isolated walls in each orthogonal direction were lumped into a single solid wall, the area and moment of inertia of the lumped wall being equal to the sum of those of the individual walls. The coupled walls in each direction were similarly lumped into a single coupled wall. The lumped solid wall and the lumped coupled wall were connected through flexible links as shown in Fig. 7-4b. These links simulated the function of slabs which transmit little bending.

7.3.4 Input motion

For the analyses reported in this chapter, the input motion selected was the first 10 s of the E–W component of the 1940 El Centro record. The acceleration ordinates were multiplied with a factor to give the resulting accelerogram a spectrum intensity equal to 0.67 times that of the N–S component of the 1940 El Centro record for the design earthquake. The 1940 El Centro N–S record has a peak ground acceleration of 0.33 g.

7.3.5 Results of dynamic analyses

Seven inelastic dynamic analyses of the structure in the longitudinal direction under the design earthquake were carried out for different yield levels of the coupling beams. The lowest level of yield moments was about equal to the maximum elastic moments caused in these beams by twice the UBC [42] 25 psf zone wind forces. At the other extreme, the coupling beams were elastic. In all seven analyses, the walls were kept elastic throughout their seismic response. Five per cent of critical damping was assumed.

The results obtained from the seven analyses are presented in Fig. 7-5. Fig. 7-5a shows that the ductility requirements in the coupling beams increase drastically as the coupling beam strength levels decrease. The elastic coupling beams obviously require no ductility. Figure 7-5b shows that there is a trade-off between the shear capacity and the ductility requirements. The shear capacity requirements increase as the coupling beam strength levels increase, the shear capacity demands being the highest for the elastic coupling beams. Figure 7-5c shows the axial forces due to coupling in the coupled wall piers. These forces also increase with increasing coupling beam strength and may cause unduly high tensile and compressive forces in the piers of strongly coupled walls.

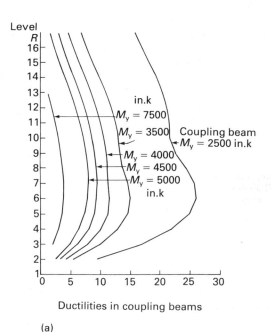

(a)

Figure 7-5   Choice of coupling beam strength: results of dynamic analysis

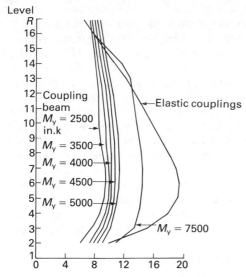

Nominal shear stresses on coupling
beam sections

(b)

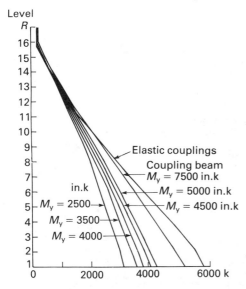

Max. axial forces in piers of coupled walls

(c)

Figure 7-5 (*contd.*)

7.3.6 *Choice of coupling beam strength*

Figure 7-5 clearly shows that it is advantageous in design to choose the lowest level of coupling beam strength, provided the corresponding required ductilities can be supplied. In this study, in view of the high ductility and shear capacity requirements in the coupling beams, a decision was made to reinforce them diagonally, as suggested by Paulay and Binney [73]. Any shear force such beams may be subjected to is carried in truss action directly by the diagonal reinforcement, so that relatively high shear force levels cease to be a problem. Also, the limit of available ductility in diagonally reinforced coupling beams is relatively high, and may be set at 15, based on available test results [73].

Based on a consideration of the actual diagonal reinforcement that can conveniently be protruded from the precast walls into the coupling beams, a yield level of 4000 in k (453 kN m) was selected for the lumped coupling beams. This appears to be a good choice in view of the results presented in Fig. 7-5. The corresponding ductilities required in the coupling beams are reasonable (the highest value is 11.6). The corresponding axial forces in the coupled wall piers

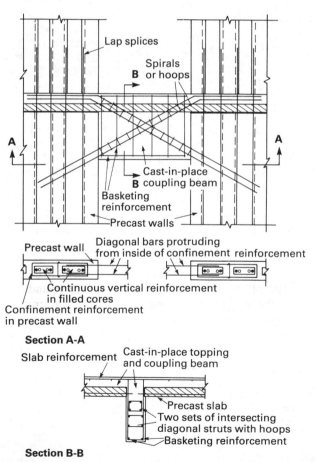

Figure 7-6 Coupling beam–precast wall joint details

are also at a reasonable level. Further increases in coupling-beam strength would substantially increase these forces, causing higher tension and compression.

7.3.7 *Proportioning of coupling beams*

The diagonally reinforced coupling beams were proportioned as suggested in [73]. The 4000 in k (453 kN m) yield level of the lumped coupling beams (2000 in k or 227 kN m for an individual beam) required the provision of two No. 6 (20 mm) Grade 60 bars on each face of each beam in each direction (totalling four bars in each direction or eight to a beam, Fig. 7-6). The four bars in each direction form a diagonal strut. It is important to confine the concrete within each diagonal strut by ties or spirals embracing all bars of the strut. Such ties should be spaced no more than 4 in (100 mm) apart.

8 Detailing for earthquake resistance

8.1 Providing against non-ductile failure

In buildings subjected to earthquake motion, both beams and columns are loaded in transverse shear in addition to bending, axial load (significant only in columns) and possibly torsion. Because of the abrupt nature of a failure due to shear (i.e., diagonal tension) or loss of anchorage of reinforcement, members should be proportioned so that their strength and ductility are controlled by flexure. Code provisions are intended to provide members with an inclined cracking load greater than their flexural capacity. Flexural failure is ensured to be ductile by limiting the maximum amount of longitudinal tensile reinforcement, so that failure is initiated by the yielding of such reinforcement. Codes also specify a minimum amount of flexural reinforcement, to prevent a sudden brittle-type flexural failure in lightly loaded members when tension cracking occurs.

Effective interaction between compression-resisting concrete and tension-carrying steel in reinforced concrete members requires the provision of adequate anchorage to develop the strength of the reinforcement. The abrupt character of bond or anchorage failure requires careful attention in design to avoid such failure before the flexural capacity of a member is reached.

The inclined cracking load of a beam without shear reinforcement is determined by the tensile strength of the concrete, the amount of longitudinal reinforcement and the ratio of the applied shear to moment acting at the critical section. The presence of a compressive load increases the apparent tensile strength of the concrete and hence increases the inclined cracking load. Once inclined cracking starts, the load on a beam is transferred across a crack by a combination of shear on the compression zone, aggregate interlock along the crack and dowel action of the longitudinal reinforcement. At this stage, the integrity of a beam and its capacity to carry more loads has to be maintained through the use of shear reinforcement.

8.2 Confinement reinforcement

Apart from the transverse reinforcement required by shear (diagonal tension), sufficient transverse confinement reinforcement needs to be provided at critical regions of a structure where inelastic action might occur. This is particularly true for structures that may experience moderate-to-strong earthquakes; the critical regions may be subjected to repeated, reversed cycles of deformation well into the inelastic range.

The use of sufficient confinement reinforcement in critical regions ensures their ductile performance by increasing the strength and strain capacity of the confined concrete in compression and by reducing the disruptive effect of large amplitudes of reversed loading on the concrete core. In beam–column connections, such confinement assures the maintenance of the vertical load-carrying capacity of the joint as well as adequate anchorage of any longitudinal beam steel embedded within the core. The increase in compressive strength of the confined concrete also helps to compensate for strength loss due to the spalled outer shell.

Transverse reinforcement that confines also acts as shear reinforcement, and vice versa. Shear reinforcement lying in planes transverse to the longitudinal axis of a member is the most effective since it resists shear in either direction and can also serve as hoops to confine the concrete and as ties to restrain the longitudinal reinforcement. Such reinforcement may be in the form of circular or rectangular spirals or individual stirrups. Tests of reinforced concrete beams under cyclic loading have clearly demonstrated the considerable energy-absorbing capacity of members well provided with transverse reinforcement.

8.3 Limitations on material strength

UBC-82 [42] requires a minimum specified 28-day strength of concrete, f'_c, of 3000 psi (21 MPa). Reinforcing is to be billet steel, Grade 40 or 60 bars. The actual yield stress, based on mill tests, must not exceed the minimum specified yield stress, f_y, by more than 18 000 psi (124 MPa). The ultimate tensile stress must be not less than 1.53 times the actual yield stress, based on mill tests.

A decrease in concrete strength or an increase in the yield strength of the tensile reinforcement decreases the rotational ductility of a member segment in flexure. The limitation on the difference between actual and specified yield stresses limits shear forces in flexural members, estimates of which must be based on the specified yield stress. The limitation on the ratio between the actual ultimate and yield stresses of the reinforcement ensures a minimum of post-yield strength and ductility in reinforced concrete sections and members.

8.4 Flexural members

The significant code [42] provisions relating to flexural members can be divided into five categories.

(a) *Dimensional restrictions* 'Flexural members shall not have a width–depth ratio of less than 0.3, nor shall the width be less than 10 inches nor more than the supporting column width plus a distance on each side of the column of three-fourths the depth of the flexural member.' [42] These restrictions are intended to ensure that beam dimensions generally lie within a range where experimental information is available.

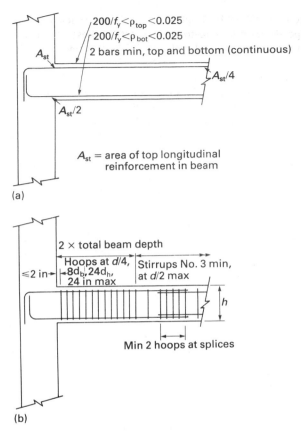

Figure 8-1   Reinforcement details for flexural members in UBC [42] seismic
zones 3,4

(a) Longitudinal reinforcement
(b) Transverse reinforcement

(b) *Limitations on flexural reinforcement* (Fig. 8-1a)   All flexural members must have a minimum reinforcement ratio, for top and bottom reinforcement, of $200/f_y$. The tensile reinforcement ratio must not exceed 2.5%.

To allow for the possibility of the positive moment at a beam end due to earthquake-induced lateral displacements exceeding the negative moment due to gravity loads, a minimum positive moment capacity at beam ends equal to 50% of the negative moment capacity is required. Bertero and Popov [74] observed that a considerable increase in the energy absorption and dissipation capacities of beams occurs if equal amounts of longitudinal steel are used near both the top and bottom surfaces of critical regions. The behavior of specimens having less reinforcement in the bottom than in the top was poorer. Therefore, it may be desirable to use positive reinforcement in an amount greater than that required by the code.

At least one-fourth the larger amount of tension reinforcement required for negative moment at either end of a flexural member is to be continuous through-out top of member. At least two bars must be provided at both top and bottom throughout the length of flexural members.

(c) *Splices* Tensile steel must not be spliced by lapping in a region of tension or reversing stress unless the region is confined by hoops or stirrup-ties. A stirrup-tie is continuous reinforcement of not less than a No. 3 bar bent to form a closed hoop which encloses the longitudinal steel and the ends of which have a standard 135° bend with a 10-bar-diameter extension or equivalent. Splices must not be located within the column or within a distance of twice the member depth from the column face. At least two stirrup-ties must be provided at all splices.

(d) *Anchorage* Flexural members framing into opposite sides of a column should have top and bottom reinforcement provided at ends of members continu-ous through the column if possible. Where this is impractical, the reinforcement must be extended, without horizontal offsets, to the far face of the confined concrete region within the column connection, terminating in standard 90° hooks. The length of required anchorage should be computed beginning at the near face of the column. Length of anchorage in confined regions, including hook and vertical extension, must be not less than 56% of development length computed by Section 2612(f) of [42], nor less than 24 in (610 mm).

(e) *Web reinforcement* (Fig. 8-1b) Web reinforcement is to be provided against shears resulting from factored gravity loads on members and from moment strengths of plastic hinges at member ends caused by lateral displacement. The moment strengths are to be computed without the φ-factor reduction and assuming the reinforcing yield strength to be 1.25 times the specified yield strength. Ultimate shear capacities should be computed with the φ-factor reduc-tion.

Web reinforcement perpendicular to longitudinal reinforcement is to be pro-vided throughout the length of flexural members. Minimum stirrup size should be No. 3, and the maximum spacing, $d/2$.

Stirrup-ties, at a maximum spacing of not over $d/4$, eight bar diameters, 24 stirrup-tie diameters or 12 in (305 mm), whichever is least, must be provided:

(a) At each end of all flexural members. The first stirrup-tie should be located no more than 2 in (50 mm) from the face of the column and the last, at least twice the member depth away from the column face.
(b) Wherever ultimate moment capacities or plastic hinges may develop in flexural members due to lateral frame displacement.
(c) Wherever *required* compression reinforcement occurs in beams.

In regions where stirrup-ties are required, they should be so arranged that every corner and alternate longitudinal bar has lateral support provided by the corner of a tie having an included angle of no more than 135°; no bar should be farther than 6 in (150 mm) clear on either side along the tie from such a laterally supported bar. Single or overlapping stirrup-ties and supplementary cross-ties may be used.

8.5 Columns

The significant code provisions relating to columns also fall under five categories.

(a) *Dimensional restrictions* 'The ratio of minimum to maximum column thickness shall be not less than 0.4 nor shall any dimension be less than 12 in' [42].

(b) *Limitations on longitudinal reinforcement* (Fig. 8-2a)   Area of longitudinal reinforcement must not be less than 0.01 nor more than 0.06 times gross area A_g of column section.

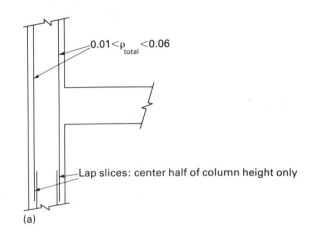

$0.01 < \rho_{total} < 0.06$

Lap slices: center half of column height only

(a)

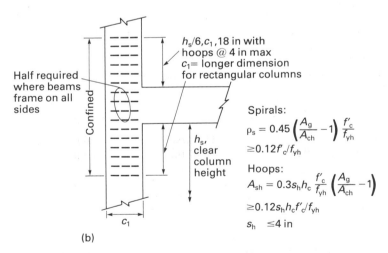

Half required where beams frame on all sides

Confined

$h_s/6, c_1, 18$ in with hoops @ 4 in max
c_1 = longer dimension for rectangular columns

h_s, clear column height

c_1

(b)

Spirals:

$$\rho_s = 0.45 \left(\frac{A_g}{A_{ch}} - 1 \right) \frac{f'_c}{f_{yh}}$$

$$\geq 0.12 f'_c/f_{yh}$$

Hoops:

$$A_{sh} = 0.3 s_h h_c \frac{f'_c}{f_{yh}} \left(\frac{A_g}{A_{ch}} - 1 \right)$$

$$\geq 0.12 s_h h_c f'_c/f_{yh}$$

$$s_h \leq 4 \text{ in}$$

Figure 8-2   Reinforcement details for columns of frames in UBC [42] seismic zones 3,4
      (a) Longitudinal reinforcement
      (b) Transverse reinforcement

(c) *Splices* Lap splices should be made within the center half of column height, and the splice length should be not less than 30 bar diameters. Continuity may also be effected by welding or by appropriate mechanical devices, provided not more than alternate bars are welded or mechanically spliced at any level and the vertical distance between these welds or splices of adjacent bars is not less than 24 in (610 mm).

(d) *Transverse reinforcement* The cores of columns need to be confined by transverse reinforcement, as specified below, or as required for shear.

(1) Special transverse reinforcement consisting of spiral or hoop reinforcement should be provided above and below beam–column connections, as shown in Fig. 8-2b, where

ρ_s = ratio of volume of spiral reinforcement to total volume of core (out-to-out of spirals)
A_g = gross area of section, in^2
A_{ch} = area of core measured out-to-out of hoop or spiral, in^2
f_{yh} = specified yield strength of spiral or hoop, psi
s_h = center-to-center spacing of hoops, in
h_c = larger core dimension of rectangular tied columns, in

Single or overlapping hoops may be provided to meet the total cross-sectional area requirement. Supplementary cross ties of the same size and spacing as hoops, using 135° minimum hooks engaging the peripheral hoop and secured to a longitudinal bar, may be used. Supplementary cross ties or legs of overlapping hoops should be spaced not more than 14 in (356 mm) on center transversely.

(2) Special transverse reinforcement shall be provided at any section where the ultimate capacity of the column is less than the sum of the shears, ($\sum V_u$), computed by Eqn 8-1 for all the beams framing into the column above the level under consideration:

$$V_u = (M_a + M_b)/l_{ab} + 1.1(V_d + V_l) \qquad (8\text{-}1)$$

where M_a and M_b are the ultimate moment capacities (computed without the φ-factor reduction and assuming the reinforcing yield strength to be 1.25 times the specified yield strength) of opposite sense at each hinge location of the member, $V_d + V_l$ is the simple span shear, and l_{ab} is the distance between M_a and M_b. For beams framing into opposite sides of the column, the moment components of Eqn 8-1 may be assumed to be of opposite sign.

(3) Where walls or stiff partitions do not continue from story to story, columns supporting such wall or partition loads must have special transverse reinforcement throughout their full height.

(4) Portions of column height not requiring special transverse reinforcement should have all vertical reinforcement enclosed by lateral ties at least No. 3 in diameter for smaller than No. 11 longitudinal bars, and at least No. 4 in diameter for No. 11 or larger bars. The ties should be spaced apart not over 16 bar diameters, 48 tie diameters or the least lateral column dimension.

The lateral support requirement for longitudinal bars, as given in Section 8.4e, also applies to all tied columns.

(e) *Shear reinforcement* The transverse reinforcement in columns subjected to bending and axial compression must satisfy the following requirement:

$$A_{sv}f_{yh}d_c/s = V_u/\varphi - V_c \qquad (8\text{-}2)$$

where

$A_{sv} =$ total cross-sectional area of special transverse reinforcement in tension within a distance s; two-thirds of such area is to be used in the case of circular spirals.

$d_c =$ dimension of the column core in the direction of load.

$s =$ spacing $\leqslant$ half minimum column dimension.

$V_c = v_c A_{ch}$. The shear stress carried by concrete, v_c, shall not exceed $2\sqrt{f_c'}$ unless a more detailed analysis is made in accordance with Section 2611(e) of [42]. v_c must be considered zero when $P_e/A_g < 0.12f_c'$, P_e being the maximum design axial load on a column during an earthquake.

V_u can be computed from the free body of a column shown in Fig. 8-3a. The moments M_c both act clockwise or both act counterclockwise under seismic loading. When a plastic hinge develops in a column end, M_c is the ultimate moment capacity of the column section. When plastic hinges develop in the beams, column moments are shown on the modified freebodies of beam-column joints in Fig. 8-3b. Under no condition does the transverse reinforcement in a column need to be heavier than that corresponding to the formation of plastic hinges at the top and bottom of the column. Ultimate moment capacities are to be computed without φ-factor reduction and under all possible vertical loading conditions and assuming the maximum reinforcing yield strength to be 1.25 times the specified yield strength.

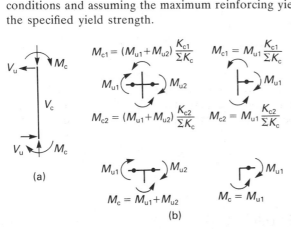

$$M_{c1} = (M_{u1} + M_{u2})\frac{K_{c1}}{\Sigma K_c} \qquad M_{c1} = M_{u1}\frac{K_{c1}}{\Sigma K_c}$$

$$M_{c2} = (M_{u1} + M_{u2})\frac{K_{c2}}{\Sigma K_c} \qquad M_{c2} = M_{u1}\frac{K_{c2}}{\Sigma K_c}$$

(a)

$$M_c = M_{u1} + M_{u2} \qquad M_c = M_{u1}$$

(b)

Figure 8-3   Column under seismic loading [75]
(a) Maximum total column shear
(b) Column moments in joints where plastic hinges develop in beams

8.6 Beam–column joints

Transverse reinforcement through beam–column connections should be proportioned as indicated in Section 8.5d(1). It should be checked for compliance with
the provisions of Section 8.5e, except that the V_u acting on the connection shall
be taken equal to the maximum shear in the connection, the computation of
which should take into account the column shear and the concentrated shears
developed from the forces in the beam reinforcement assumed stressed to f_y (Fig.
8-4).

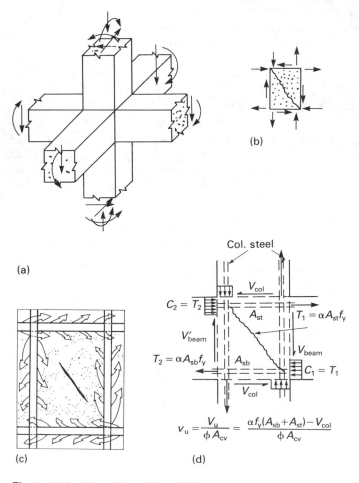

Figure 8-4   Shear in a joint [75]
(a) Forces in members at joint
(b) Resultant forces in joint
(c) Transfer of forces within joint
(d) Shear stress in joint

The amount of transverse reinforcement determined above may be reduced by one-half within a depth of connection taken equal to the depth of the shallowest framing member, where such members frame into all four sides of a column and where each framing member has a width equal to at least three-fourths the column width. When the distance from the side of a beam framing into a column face to the nearest corner of that column exceeds 4 in (100 mm) no reduction in the amount of transverse reinforcement is to be allowed.

References [75–77] should be consulted for more information on and insight into the seismic design of reinforced concrete beam–column joints.

8.7 Detailing of frame members and connections in zones of low seismicity

The requirements of Sections 8.3 through 8.6 are meant to apply to members and connections of ductile moment-resisting space frames in UBC [42] seismic zones 3 and 4. Ductile moment-resisting space frames in seismic zones 1 and 2 can be exempt from the above requirements provided the horizontal force factor, K, is not taken less than unity in computing the seismic base shear (Section 6.1.1); Section 2626 of UBC-82 [42] contains some minor detailing requirements for such frames.

8.8 Shearwalls

Buildings with lateral force resisting systems as described in Table 6-2 for a K of 0.80 must be provided with special vertical boundary elements. These must be reinforced as required for the columns in Section 8.5, with special transverse reinforcement extending over full height. The special transverse reinforcement may be omitted in seismic zones 1 and 2 when the combined dead load, live load and seismic stresses are not over one-half of those allowed. The boundary elements must be designed to carry all the vertical stresses from the wall loads, the tributary dead and live loads, and the horizontal seismic forces prescribed in the code [42]. In zones 1 and 2, the boundary elements may be designed to carry only the vertical stresses from tributary dead and live loads not supported by the shearwalls.

Wall reinforcement required to resist wall shear should be terminated with not less than a 90° bend plus a six-bar-diameter extension beyond the boundary reinforcing at vertical and horizontal end faces of wall sections. Wall reinforcement terminating in boundary columns or beams shall be fully anchored into the boundary elements.

Reference [78] contains detailing recommendations, including some modifications to code requirements, for isolated cantilever shearwalls. References [79, 80] should be consulted for behavioral and design information on shearwalls responding to seismic loading individually or in interaction with frames or other shearwalls. The detailing of coupling beams is also discussed in the above literature. Reference [81] discusses slab-coupling of earthquake-resisting shearwalls.

8.9 Slab–column and slab–wall joints

It is suggested that flat plate buildings be not used as seismic-resistant structures without the presence of some frames or walls to stiffen the structure.

Even with such stiffening elements present, substantial unbalanced moment may need to be transferred at the slab–column connection, and these should be made adequately ductile. The reversals of unbalanced moment which occur during an earthquake can lead to a shear failure in the slab around the column due to degradation of the shear strength. Hawkins [82] and Park [76] have suggested reinforcing details that enhance the seismic performance of column–slab joints.

A series of tests of reinforced concrete slab–wall joints has recently been completed in New Zealand [76]. The tests have not been fully reported as yet. Various arrangements of shear reinforcement were used in the slabs. It was found difficult to prevent shear failure at the critical toes of the wall; nevertheless, yield lines developed across the slab and allowed the full flexural strength of the slab to be developed.

9 Experience in recent earthquakes

Summarizing the experience in earthquakes of the last 20 years, as related to performance of reinforced concrete multistory structures, it can be stated that in each earthquake there were examples of satisfactory, sometimes excellent behavior (Fig. 9-1). In each earthquake there were also examples of poor behavior, sometimes even failure, of concrete structures (Fig. 9-2). The same, of course, can be said of steel structures. It is basically not the structural material which is good or bad—it is how the building is designed and constructed and how the materials are utilized that, in the final test, determine its earthquake performance.

(a)

Figure 9-1 Examples of good performance
(a) In the Managua (Nicaragua) earthquake of 1972
(b) In the Bucharest (Romania) earthquake of 1977

(b)

Figure 9-1   (*contd.*)

Figure 9-2   Collapsed three-story computer center, Bucharest, Romania, 1977

Among the lessons from the earthquakes of the last 20 years are a number of both positive and negative observations:

(a) Well-designed and well-constructed concrete structures demonstrated satisfactory performance in earthquakes in different areas of the world (Figs. 9-3).

(b) Concrete structures containing shearwalls performed considerably better, as compared to those consisting of moment-resisting frames only, showing superior behavior with respect to damage as well as safety against collapse [83]. This was due to the smaller interstory distortions effected by the shearwalls.

(c) Most of the collapses of multistory concrete structures resulted from shear failures of columns in ground stories, leading to large distortions of that story, and subsequent collapse (Fig. 9-2).

(d) In almost every earthquake, shear failures of short columns, restrained by masonry (captive columns), could be seen (Fig. 9-4).

(e) Brittle low-strength masonry filled into flexible skeletons failed in an explosive way, sometimes even damaging the frame (Fig. 9-5).

(f) Buildings of irregular plan layouts, with large eccentricities between the center of mass and the center of resistance to lateral forces, usually suffered much higher damage than buildings with regular plan layouts in which torsional effects were minimized.

Figure 9-3 The 17-story Plaza I Building that went through the Caracas, Venezuela, earthquake of 1967 without a scratch, while some of its ten-story neighbors collapsed or were severely damaged

Figure 9-4 Typical distress of columns unintentionally restrained by window parapets

Figure 9-5 Failure of brittle masonry filled into flexible skeleton

Figure 9-6   The flexible frame of this five-story building went through the Managua, Nicaragua, earthquake of 1972 without damage; however, the large interstory deformations severely damaged the brittle masonry

(g) Weaknesses or defects in materials, construction or design became obvious in an earthquake.

(h) Buildings having flexible frames, although often riding out an earthquake without structural distress, usually experienced a high degree of non-structural damage (to partitions, finishes, mechanical components, etc.) because of the large interstory distortions they underwent during the earthquake. (Fig. 9-6).

10   Repair of damage caused by earthquakes

10.1   Objectives

The objective to the repair of damage caused by earthquakes is the restoration or augmentation of the quake-resisting capabilities of a structure. Limits to the degree of necessary earthquake resistance should be formulated considering accepted seismic design criteria and current code regulations. It should be recognized that current seismic design provisions accept the possibility of damage in the case of high-intensity earthquakes. Repair of seismic damage must not affect the original and/or necessary overload capabilities of a structure which must remain in keeping with its use [84]. Original fire resistance ratings must also not be diminished by repair or strengthening procedures.

10.2 Economic considerations

The decision to repair, strengthen, or demolish a building must be based on economic considerations. The purpose of rehabilitation is to save or recover the investment which a building represented before sustaining damage. Thus, the cost of rehabilitation should bear a reasonable relationship to the original investment, or should be below the cost of a new and equivalent building. Rehabilitation costs must be justified by increases in market value, anticipated lifetime, expected revenue, and/or possible tax or depreciation benefits [30]. Earthquake insurance may at times be a factor to consider. Another possible factor may be government disaster assistance or other incentives. In specific cases of landmark buildings of special significance, rehabilitation may be justified on other than economic grounds.

10.3 Evaluation

A thorough examination of quake-inflicted damage, of seismic hazard in the damaged building, and of possible options toward abatement of such hazard, is absolutely crucial. A proper evaluation, which must be based on examination of available pertinent geotechnical, design, construction, and installation documentation and on-site inspection plus analytical studies, if necessary, should explore characteristics of the site, the structure, the design, the nonstructural elements, the damage, and the failure.

The strength, stiffness and ductility of the structure and of each of its parts need to be evaluated and the results compared with quake-inflicted damage. The quality of the construction and the load acting on the structure at the time of the quake should be considered in that comparison. If the conclusions from the evaluation are consistent with the damages, then the evaluation is correct. If not, additional research should be conducted until a reasonable cause and effect relationship is established [84].

If a structure has not collapsed under an earthquake, certain factors constituting its earthquake resistance, prevented it from doing so. The present characteristics of these resistances must be determined. This evaluation must consider that most structural elements have suffered from stresses, strains, and in some cases damages, which have partially exhausted their original capabilities for withstanding similar actions. The repair or strengthening project should satisfy the following relationship for the structure as well as for its components:

$$\begin{array}{c} \text{WEAKENED} \\ \text{RESISTANCE} \end{array} + \begin{array}{c} \text{REPAIR OR} \\ \text{STRENGTHENING} \end{array} = \begin{array}{c} \text{REQUIRED} \\ \text{RESISTANCE} \end{array}$$

10.4 Constraints

A repair or strengthening program must be executed within constraints that may be [84]

(a) Economic. Applicable budget limitations must not be exceeded.
(b) Architectural. Repairs can change the appearance or operation of a building; these must be kept within acceptable limits.
(c) Construction. There may be restrictions imposed by the physical location of

the building; the availability of material, labor, construction equipment, experienced contractors, etc.

(d) Continued use and availability of time. The partial or near-full use of a structure may have to continue through its rehabilitation. The time available for repair may also limit the process of rehabilitation.

10.5 Rehabilitation schemes

The rehabilitation of a structure may be effected in several possible ways:

(a) Structures may be repaired by replacing damaged material.

(b) Individual structural components and/or their connections may be strengthened by methods such as increasing element thickness or size, adding reinforcement, and/or increasing the strength of connections.

(c) Structural systems may be strengthened by providing additional shear walls or vertical bracing. Additional columns may sometimes be added to reduce spans. The designer must ensure that a consistent load path or possible alternate load paths are provided within the modified structural system.

(d) Upper stories may be removed to reduce the mass of a building.

(e) The period of a modified structure may be shortened, and the response characteristics consequently increased. Therefore, the earthquake resistance of a modified structure needs re-evaluation.

10.6 Repair techniques

10.6.1 *Materials*

The more common materials used to repair or strengthen reinforced concrete construction are [30] (a) shotcrete or 'gunite', which is pneumatically applied concrete; (b) epoxy resins or adhesives manufactured from petroleum products, usually consisting of two or more component chemicals mixed just before application; (c) epoxy mortar obtained by combining epoxy resins with sand aggregates; (d) fiber-reinforced concrete obtained by adding steel, glass or plastic fibers to normal or Type III Portland cement concrete; (e) gypsum cement concrete; (f) Portland cement concrete, usually using Type III cement for high early strength; (g) quick-setting cement mortar which is a nonhydrous, phosphate-magnesium cement with a liquid and a dry component mixed similarly to Portland cement concrete; (h) pre-placed aggregate concrete obtained by first filling the forms with clean, well-graded, coarse aggregate and then pumping mortar or grout into the void spaces; (i) reinforcing steel anchored in existing concrete—a common technique for providing anchorage is to drill a hole larger than the bar in existing concrete, fill it with epoxy, expansive cement, grout, sulfur, or other high-strength grouting material, push the bar into place and hold it until the grout is cured; (j) mechanical anchors which use wedging action to provide anchorage—some anchors resist both shear and tension, while others resist only tension.

10.6.2 *Repair*

Repair techniques must be selected according to the degree of damage and the level of repair to be accomplished.

If cracks are reasonably small (less than $\frac{1}{4}$ in wide), the simplest method of repair is by pressure-injection of epoxy. For cracks larger than $\frac{1}{4}$ in or for regions with crushed concrete, other treatment is required. Loose concrete should be removed and replaced with shotcrete, any one of special cement mortars, or new concrete. Selection of replacement material depends on the desired material characteristics.

In a severely damaged reinforced concrete member, reinforcement may have buckled, elongated with excessive yielding, or fractured in extreme cases. Reinforcement can be replaced with new steel, using butt welding, lap welding, or in some cases a splice. If practical, the repair should be made without removing the existing steel. The best approach depends on the amount of space available in the original member. Additional confinement steel should be added to inhibit future buckling of bars in this region.

When damage is severe, the need for additional shear or flexural reinforcement should be considered. If, however, development of the additional reinforcement into adjacent solid concrete is required, the repaired parts will be stronger than the adjacent existing material and failure in this adjacent portion would be probable during a subsequent earthquake.

In the case of damage to wall and floor diaphragms, it may be more economical to add new steel on the outside and to cover it with concrete rather than to repair the damaged material. The increased weight must be considered in re-analyzing building forces and in checking the foundations.

References

1. Richter, C. F., *Elementary seismology*, Freeman, San Francisco, 1958, 768 pp.
2. Berg, G. V., *Designing structures to resist earthquakes*, PS 532-1277-5M-BO, American Iron and Steel Institute, Washington, DC, 14 pp.
3. Eardley, A. J., The cause of mountain building—an enigma, *Am. Sci.*, June 1957, pp. 189–217.
4. Derecho, A. T. and Fintel, M., Earthquake-resistant structures, Fintel, M. (Ed.), *Handbook of concrete engineering*, Van Nostrand Reinhold, New York, 1974, pp. 356–432.
5. Gutenberg, B. and Richter, C. F., Earthquake magnitude, intensity, energy, and acceleration, *Bull. Seis. Soc. Am.*, Vol. 32, No. 3, July 1942, pp. 163–191; Vol. 46, No. 2, April 1956, pp. 105–145.
6. Werner, S. D., Engineering characteristics of earthquake ground motions, *Nucl. Eng. & Des.*, Vol. 36, 1976, pp. 367–395.
7. Derecho, A. T., Fugelso, L. E. and Fintel, M., *Structural walls in earthquake-resistant buildings—dynamic analysis of isolated structural walls—input motions*, Report on research sponsored by National Science Foundation (RANN), Grant No. ENV74-14766, Portland Cement Association, Skokie, 1977, 54 pp.
8. Housner, G. W., Intensity of earthquake ground shaking near the causative fault, *Proc. 3rd world conf. earthq. eng.*, Auckland and Wellington, New Zealand, 1965, Vol. 3, pp. III-94–III-111.

9. Bolt, B. A., Duration of strong ground motion, *Proc., 5th world conf. earthq. eng.*, Rome, 1973, Vol. 1, pp. 1304–1313.
10. Page, R. A. *et al.*, Ground motion values for use in the seismic design of the Trans-Alaska pipeline system, *Circ. US Geol. Surv.*, No. 672, 1972.
11. Housner, G. W., Strong ground motion, Weigel, R. L. (Ed.), *Earthquake engineering*, Prentice-Hall, Englewood Cliffs, NJ, 1979, pp. 75–91.
12. Newmark, N. M. and Rosenblueth, E., *Fundamentals of earthquake engineering*, Prentice-Hall, Englewood Cliffs, NJ, 1971, 640 pp.
13. Housner, G. W., Behavior of structures during earthquakes, *J. Eng. Mech. Div. Am. Soc. Civ. Eng.*, Vol. 85, No. EM4, Oct. 1959, pp. 109–129.
14. Arias, A., A measure of earthquake intensity, Hansen, R. J. (Ed.), *Seismic design for nuclear power plants*, MIT Press, Cambridge, Mass., 1970, pp. 438–483.
15. Liu, S. C., Statistical analysis and stochastic simulation of ground-motion data, *Bell. Syst. Tech. J.*, Vol. 47, Dec. 1968, pp. 2273–2298.
16. Housner, G. W., Important features of earthquake ground motion, *Proc. 5th world conf. earthq. eng.*, Rome, 1973, Vol. 1, pp. CLIX–CLXVIII.
17. Earthquake Engineering Research Laboratory, Pasadena, Analysis of strong motion earthquake accelerograms, Vols I–IV, Parts A–Y, California Institute of Technology, 1969–1975.
18. *Bull. Seism. Soc. Am.*, Vol. 57, No. 3, June 1967.
19. Culver, C. G., Lew, H. S., Hart, G. C. and Pinkham, C. W., *Natural hazards evaluation of existing buildings*, Building Science Series 61, National Bureau of Standards, Washington, DC, 1975, 958 pp.
20. National Earthquake Information Center, *World seismicity 1961–1969*, US Department of Commerce, Washington, DC, 1970, 15 maps.
21. Dowrick, D. J., *Earthquake resistant design*, Wiley, London, 1977, 374 pp.
22. Jennings, P. C., Housner, G. W. and Tsai, N. C., Simulated earthquake motions for design purposes, *Proc., 4th world conf. earthq. eng.*, Santiago, Chile, 1969, Vol. 1, pp. A-1 145–160.
23. Trifunac, M. D. and Brady, A. G., On the correlation of seismic intensity scales with the peaks of recorded strong ground motion, *Bull. Seism. Soc. Am.*, Vol. 65, No. 1, Feb. 1975, pp. 139–162.
24. Trifunac, M. D. and Brady, A. G., Correlation of peak acceleration, velocity and displacement with earthquake magnitude, distance and site conditions, *Earthq. Eng. & Struct. Dyn.*, Vol. 4, 1976, pp. 455–471.
25. Newmark, N. M., Current trends in the seismic analysis and design of high-rise structures, Wiegel, R. L. (Ed.), *Earthquake engineering*, Prentice-Hall, Englewood Cliffs, NJ, 1970, pp. 403–424.
26. Werner, S. D., Procedures for developing vibratory ground motion criteria at nuclear plant sites, *Nucl. Eng. & Des.*, Vol. 36, 1976, pp. 411–441.
27. Newmark, N. M. and Hall, W. J., Seismic Design Criteria for Nuclear Reactor Facilities, *Proc. 4th world conf. earthq. eng.*, Santiago, Chile, 1969, Vol. 2, pp. B-4 37–50.
28. Newmark, N. M., Blume, J. A. and Kapur, K. K., Seismic Design Spectra for Nuclear Power Plants, *J. Power Div. Am. Soc. Civ. Eng.*, Vol. 99, No. PO2, Nov. 1973, pp. 287–303.

29. Seed, H. B., Ugas, C. and Lysmer, J., *Site-dependent spectra for earthquake-resistant design*, Report EERC 74-12, University of California, Berkeley, 1974, 17 pp.

30. Applied Technology Council, *Tentative provisions for the development of seismic regulations for buildings*, ATC Publication ATC 3-06, US Government Printing Office, Washington, DC, 1978, 505 pp.

31. Derecho, A. T., Frames and frame–shear wall systems, *Response of multistory concrete structures to lateral forces*, American Concrete Institute, SP-36, 1973, pp. 13–37.

32. Khan, F. R., Influence of design criteria on selection of structural systems for tall buildings, Preprint, *Canad. struct. eng. conf.*, Toronto, 1972, 15 pp.

33. ACI Committee 442, Response of buildings to lateral force, *J. Am. Concr. Inst., Proc.*, Vol. 68, No. 2, Feb. 1971, pp. 81–106.

34. Seismology Committee of the Structural Engineers Association of California, *Recommended lateral force requirements and commentary*, 4th Edn Revised, San Francisco, 1980, 160 pp.

35. Clough, R. W. and Benuska, K. L., *FHA study of seismic design criteria of high-rise buildings*, Report HUD TS-3, Federal Housing Administration, Washington, DC, Aug. 1966, 354 pp.

36. Fintel, M. and Ghosh, S. K., Structural systems for earthquake resistant concrete buildings, *Proc., workshop on earthquake-resistant reinforced concrete building construction*, University of California, Berkeley, July 1977, Vol. 2, pp. 707–741.

37. Kelly, J. M., Aseismic base isolation: its history and prospects, *Joint sealing and bearing systems for concrete structures*, Vol. 1, American Concrete Institute, SP-70, 1981, pp. 549–586.

38. Kelly, J. M., Beucke, K. E. and Skinner, M. S., *Experimental testing of a friction-damped aseismic base isolation system with fail-safe characteristics*. Report EERC 80-18, University of California, Berkeley, 1980, 58 pp.

39. Clough, R. W., Earthquake response of structures, Wiegel, R. L. (Ed.) *Earthquake engineering*, Prentice-Hall, Englewood Cliffs, NJ, 1970, pp. 307–347.

40. Clough, R. W. and Penzien, J., *Dynamics of structures*, McGraw-Hill, New York, 1975, 634 pp.

41. Gere, J. M. and Weaver, J. W., *Matrix algebra for engineers*, Van Nostrand Reinhold, New York, 1965, 168 pp.

42. International Conference of Building Officials, *Uniform building code*, Whittier, California, 1982, 780 pp.

43. Veletsos, A. S. and Newmark, N. M., Effect of inelastic behavior on the response of simple systems to earthquake motions, *Proc., 2nd world conf. earthq. eng.*, Tokyo, 1960, Vol. 2, pp. 895–912.

44. Chopra, A. K. and Newmark, N. M., Analysis, Rosenblueth, E. (Ed.), *Design of earthquake resistant structures*, Wiley, New York, 1980, pp. 27–53.

45. Anderson, J. C. and Bertero, V. V., *Seismic behavior of multistory frames designed by different philosophies*, Report EERC 69-11, University of California, Berkeley, 1969.

46. Kanaan, A. E. and Powell, G. H., *DRAIN-2D: a general purpose computer*

program for dynamic analysis of inelastic plane structures, Reports No. EERC 73-6 and EERC 73-22, University of California, Berkeley, April 1973 (revised Sept. 1973) and Aug. 1975, 273 pp.

47. Otani, S., SAKE—a computer program for inelastic response of R/C frames to earthquakes, *Civ. Eng. Stud. Univ. Ill. Struct. Res. Ser.*, No. 413, Nov. 1975.

48. Clough, R. W., *Effect of stiffness degradation on earthquake ductility requirements*, Report No. EERC 66-16, University of California, Berkeley, 1966, 196 pp.

49. Takeda, T., Sozen, M. A. and nielsen, N. N., Reinforced concrete response to simulated earthquake, *J. Struct. Div. Am. Soc. Civ. Eng.*, Vol. 96, No. ST12, Dec. 1970, pp. 2557–2573.

50. Riddell, R. and Newmark, N. M., Statistical analysis of the response of nonlinear systems subjected to earthquakes, *Civ. Eng. Stud. Univ. Ill. Struct. Res. Ser.*, No. 468, Aug. 1979.

51. Takayanagi, T. and Schnobrich, W. C., Non-linear analysis of coupled wall systems, *Earthq. Eng. & Struct. Dyn.*, Vol. 7, No. 1, Jan.-Feb. 1979, pp. 1–22.

52. Saatcioglu, M. and Derecho, A. T., Dynamic inelastic response of coupled walls as affected by axial forces, *Nonlinear design of concrete structures*, Study No. 14, Solid Mechanics Division, University of Waterloo, Ontario, 1980, pp. 639–670.

53. Saatcioglu, M., *Inelastic behavior and design of earthquake resistant coupled walls*, PhD Thesis, Northwestern University, Evanston, Ill., 1981, 272 pp.

54. Takayanagi, T., Derecho, A. T. and Corley, W. G., Analysis of inenelastic shear deformation effects in reinforced concrete structural wall systems, *Nonlinear design of concrete structures*, Study No. 14, Solid Mechanics Division, University of Waterloo, Ontario, 1980, pp. 545–579.

55. Newmark, N. M. and Hall, W. J., A rational approach to seismic design standards for structures, *Proc. 5th world conf. earthq. eng.*, Rome, 1974, Vol. 2, pp. 2266–2275.

56. Newmark, N. M., Design of structures to resist seismic motions, *Proc. earthq. eng. conf.*, University of South Carolina, Jan. 1975, pp. 235–275.

57. Shibata, A. and Sozen, M. A., Substitute-structure method for seismic design in R/C, *J. Struct. Div. Am. Soc. Civ. Eng.*, Vol. 102, No. ST1, Jan. 1976, pp. 1–18.

58. Biggs, J. M., Lau, W. K. and Persinko, D., *Aseismic design procedures for reinforced concrete frames*, Publication No. R-79-21, Department of Civil Engineering, Massachusetts Institute of Technology, Cambridge, Mass., 1979, 79 pp.

59. Housner, G. W., Limit design of structures to resist earthquakes, *Proc. 1st world conf. earthq. eng.*, Berkeley, California, 1956, pp. 5.1–5.13.

60. Blume, J. A., Newmark, N. M. and Corning, L. H., *Design of multistory reinforced concrete buildings for earthquake motions*, Publication EB032.01D, Portland Cement Association, Skokie, 1961, 318 pp.

61. Larson, M. A., Needed improvements in aseismic design, *Struct. moments*, No. 3, Structural Engineers Association of Northern California, San Francisco, Feb. 1980, 4 pp.

62. McKevitt, W. E., Anderson, D. L., Nathan, N. D. and Cherry, S., Towards a simple energy method for seismic design of structures, *Proc. 2nd US nat. conf. earthq. eng.*, Stanford, California, Aug. 1979, pp. 383–392.

63. Derecho, A. T., Ghosh, S. K., Iqbal, M., Freskakis, G. N. and Fintel, M., *Structural walls in earthquake-resistant buildings—analytical investigation, dynamic analysis of isolated structural walls—parametric study*, Report on research sponsored by National Science Foundation (RANN), Grant no. ENV74–14766, Portland Cement Association, Skokie, 1978, 233 pp.

64. Muto Institute of Structural Mechanics, *Structural design of Shinjuku Mitsui Building (SMB)*, Tokyo, 1972, 5 pp.

65. Muto Institute of Structural Mechanics, *Structural design of International Telecommunications Center Building*, Tokyo, 1973, 7 pp.

66. Fintel, M. and Ghosh, S. K., The structural fuse: an inelastic approach to seismic design of buildings, *Civ. Eng.—ASCE*, Jan. 1981, pp. 48–51.

67. Fintel, M. and Ghosh, S. K., Explicit inelastic dynamic design procedure for aseismic structures, *J. Am. Concr. Inst., Proc.*, Vol. 79, No. 2, March–April 1982, pp. 110–118.

68. Fintel, M. and Ghosh, S. K., *Seismic resistance of a 16-story coupled wall structure: a case study using inelastic dynamic analysis*, Publication EB82.01D, Portland Cement Association, Skokie, 1980, 31 pp.

69. Fintel, M. and Ghosh, S. K., Case study of aseismic design of a 16-story coupled wall structure using inelastic dynamic analysis, *J. Am. Concr. Inst., Proc.*, Vol. 79, No. 3, May–June 1982, pp. 171–179.

70. Fintel, M. and Ghosh, S. K., Seismic resistance of a 31-story shear wall–frame building using dynamic inelastic response history analysis, *Proc., 7th world conf. earthq. eng.*, Istanbul, 1980, Vol. 4, pp. 379–386.

71. Fintel, M. and Ghosh, S. K., Application of inelastic response history analysis in the aseismic design of a 31-story frame–wall building, *Earthq. Eng. & Struct. Dyn.*, Vol. 9., No. 6, Nov.–Dec. 1981, pp. 543–556.

72. Inelastic seismic design idea simplifies concrete high-rise: novel technique relieves rebar congestion, *Eng. News Rec.* 28 Aug. 1980, pp. 58–59.

73. Paulay, T. and Binney, J. R., Diagonally reinforced coupling beams of shear walls, *Shear in reinforced concrete*, American Concrete Institute, SP-42, 1974, pp. 579–598.

74. Bertero, V. V. and Popov, E. P., Seismic behavior of ductile moment-resisting reinforced concrete frames, *Reinforced concrete structures in seismic zones*, American Concrete Institute, SP-53, 1977, pp. 247–291.

75. ACI–ASCE Committee 352, Recommendations for design of beam–column joints in monolithic reinforced concrete structures, *J. Am. Concr. Inst., Proc.*, Vol. 73, No. 7, July 1976, pp. 375–393.

76. Park, R., Accomplishments and research and development needs in New Zealand, *Proc. workshop on earthquake-resistant reinforced concrete building construction*, Vol. 2, University of California, Berkeley, June 1978, pp. 255–295.

77. Jirsa, J. O., Seismic behavior of R.C. connections (beam–column joints), *State-of-the-art in earthquake engineering 1981*, Turkish National Committee on Earthquake Engineering, Ankara, Oct. 1981, pp. 365–374.

78. Oesterle, R. G., Fiorato, A. E. and Corley, W. G., Reinforcement details for earthquake-resistant structural walls, *Concr. Int.*, Vol. 2, No. 12, Dec. 1980, pp. 55–66.
79. Paulay, T., Earthquake resistant structural walls, *Proc. workshop on earthquake-resistant reinforced concrete building construction*, Vol. 3, University of California, Berkeley, June 1978, pp. 1339–1365.
80. Paulay, T., Ductility of reinforced concrete shearwalls for seismic areas, *Reinforced concrete structures in seismic zones*, American Concrete Institute, SP-53, 1977, pp. 127–147.
81. Paulay, T. and Taylor, R. G., Slab coupling of earthquake-resisting shearwalls, *J. Am. Concr. Inst., Proc.*, Vol. 78, No. 2, March–April 1981, pp. 130–140.
82. Hawkins, N., Lateral load design considerations for flat-plate structures, *Nonlinear design of concrete structures*, Study No. 14, Solid Mechanics Division, University of Waterloo, Ontario, 1980, pp. 581–614.
83. Fintel, M., Ductile shear walls in earthquake resistant multistory buildings, *J. Am. Concr. Inst., Proc.*, Vol. 71, No. 6, June 1974, pp. 296–305.
84. Gallegos, H. and Rios, R., Earthquake—repair—earthquake, *Reinforced concrete structures in seismic zones*, American Concrete Institute, SP-53, 1977, pp. 463–477.

Glossary

Accelerogram: Record of ground acceleration caused by an earthquake. Components of motion along two perpendicular horizontal directions and in the vertical direction are usually recorded.

Alluvium: Soil, sand, gravel or similar material deposited by running water.

Aseismic: Earthquake resistant.

ATC: Applied Technology Council.

Base shear: Total horizontal seismic shear force at the base of a structure; is a function of horizontal acceleration of each of the masses of the structure relative to the base, the effects added algebraically at any instant.

Bedrock: The solid rock underlying surface formations.

Confinement reinforcement: Transverse reinforcement in a structural member that confines the concrete, increasing its strain capacity.

Core: The innermost portion of the earth underneath the mantle.

Coupled wall: Shearwalls connected through beams or slabs, usually at every floor level, with the longitudinal axes of the connecting elements lying in the plane of the shearwalls.

Crust: The upper layer or outer shell of the earth, from the surface to the Mohorovicic discontinuity. The thickness of the earth's crust varies from about 5 to 60 kilometers, being thicker in the continental areas and thinner under the sea.

Damping: The dissipation of energy during vibration due to internal and external friction. *Viscous damping*: A type of energy dissipation that occurs in a vibrating system; is represented by a force which resists motion and is proportional to velocity. Often expressed as a fraction of critical damping, 5% damping meaning damping equal to 5% of critical. *Critical damping*: The level of viscous damping that will cause a displaced system to return to its initial position without oscillation.

DDP: Dynamic Design Procedure

Deep-focus earthquakes: Shocks with focal depths exceeding 300 km.

Degree-of-freedom: The number of independent displacement components describing the motion of a system.

Design earthquake: The ground motion which an architect or engineer should have in mind when designing a building to provide protection for life safety. Presently specified as having a 90% probability of not being exceeded during the life of the structure (50 years), this probability being equivalent to a return period of 475 years. May be specified in terms of a design response spectrum, or a design earthquake accelerogram with duration, frequency content and intensity specified.

Drift: Ratio of lateral deflection to height of structure over which the deflection is measured.

Ductility: The ability of a member or structure to deform beyond its elastic limit without any significant loss of strength. The ductility discussed here is based either on lateral displacements of the structure (displacement ductility) or on rotations over the hinging region (rotational ductility).

Duration (of an earthquake): The length of an earthquake record during which pulses of reasonably large amplitude occur, with due allowance made for a build-up time.

Dynamic: Varying with time.

Earthquake input motion: Time-history of the ground acceleration resulting from earthquake excitation, used as input for response history analysis.

Effective peak acceleration (EPA): Proportional to spectral ordinates (accelerations) for periods in the range of 0.0 to 0.5 seconds. The constant of proportionality (for a smoothed 5%-damped spectrum) is set at a standard value of 2.5. See Ref. [30].

Effective peak velocity (EPV): Proportional to spectral ordinates (velocities) at a period of about 0 second. The constant of proportionality is again 2.5.

EIDDP: Explicit Inelastic Dynamic Design Procedure.

Elastic: Indicates a return to initial state upon unloading, without residual deformations.

Elastic static analysis: Analysis of a structure, assuming linear elastic material behavior, under time-independent loads.

Energy dissipation: Dispersion of the earthquake energy input into a structure by hysteresis, damping and other mechanisms. Energy dissipated by hysteresis equals the area enclosed within a complete loop of the nonlinear force-displacement curve of a member segment under cyclic loading conditions.

Energy dissipating mechanism: Specific phenomenon, e.g., inelastic deformation of a particular member, by which earthquake energy is dissipated.

Epicenter: The point of the earth's surface vertically above the focus.

Epicentral distance: Direct distance along the earth's surface from point of observation to the epicenter.

Fault: A surface of weakness in the earth's crust, with displacement of one side of the surface with respect to the other in a direction parallel to the surface. The surface, when not notably curved, is called the *fault plane*.

Flexural (bending) strength (capacity): Maximum bending moment that can be carried by a section in the absence of any significant shear force. It is a function of axial load on the section.

Focal depth: The vertical distance between the focus and the epicenter.

Focus (hypocenter): The point of fracture within the earth's interior where the initial slip of rock occurs in an earthquake.

Frame: An assemblage of linear structural elements capable of resisting gravity as well as lateral loads.

Free-field ground motion: The ground motion that would have occurred at the level of the foundation of a structure if no structure were present.

Frequency: Number of cycles of a periodic oscillation occurring in a unit of time is cyclical frequency. It is the reciprocal of the time taken for one complete cycle of free vibration of a particular mode for a given system.

Generalized coordinate approach: A means of approximately specifying the deflections of a structure by a limited set of displacement coordinates: $y(x) = \sum_{m=0}^{n} \Psi_m(x) Y_m$ where Y_m is the amplitude of the mth shape function $\Psi_m(x)$.

Harmonic motion (force): Repetitive motion (or force), with the amplitude of motion (or magnitude of force) varying sinusoidally with time.

Hinging region: The length of a number over which yielding occurs due to the bending moments exceeding the specified yield moments.

Hysteresis: The nonlinear force-displacement response of a material or section or member segment under cyclic reversible loading conditions.

Inelastic: Indicates an incomplete return to initial state on unloading, with resulting residual deformations.

Intensity (of an earthquake): A measure of the severity of ground shaking at a site.

Isoseismals: Curves that separate regions of successive intensity ratings.

Leeward: The side to which the wind blows (the face of a structure directly opposite the windward face).

Linear (response, behaviour, analysis, model, etc.): Indicates proportionality between force and displacement.

Lumped mass: An idealization of the distribution of mass in a structural system in which concentrated masses are assigned to the node points.

Magnitude (of an earthquake): A measure of the size of an earthquake at its source, independent of the place of observation. It has been correlated with the energy released by an earthquake.

Mantle: The intermediate layer or shell of the earth from the Mohorovicic discontinuity down to a second discontinuity (at an approximate depth of 2900 kilometers) that separates it from the core.

(Mathematical) model: A mathematical representation of a structure including its schematic depiction. Used in the static or dynamic analysis of structures.

Maximum credible earthquake: The most severe ground shaking that can ever be expected to occur at the site of a structure. A precise definition of this term is not currently available in the literature.

MDOF: Multi degree of freedom.

Mode of vibration: The distinguishing feature of a (normal or natural) mode of vibration is that a dynamic system can, under certain circumstances, vibrate in that mode alone, and during such vibration the ratio of the displacements of any two masses remains constant with time. These ratios define the characteristic shape of the mode. The term normal implies mathematical orthogonality between mode shapes. A system has exactly the same number of normal modes as degrees of freedom. *Fundamental mode*: The mode of vibration with the longest period. *Higher modes*: Modes of vibration excluding the fundamental mode.

Modal (superposition) analysis: A method of dynamic analysis wherein the response of a number of participating modes of a given elastic structure are superimposed to obtain the total system response. Each mode is treated as a single degree-of-freedom system. The response in each mode is obtained from a participation factor and the spectral response at the modal period. An approximation of the total response is commonly obtained by taking the square root of the sum of the squares of the individual modal responses.

Mohorovicic discontinuity: A discontinuity that separates the crust of the earth from its mantle.

MM: Modified Mercalli.

Nodes: Nodes are the locations of degrees-of-freedom in a structural model and often depict joints of the real structure as modeled.

Nonlinear (response, behaviour, analysis, model, etc.): Indicates lack of proportionality between force and displacement. The two primary sources of non-linearity usually are material properties (*material nonlinearity*) and large deformations (*geometric nonlinearity*).

Resonance: The phenomenon exhibited by a vibrating system which responds with maximum amplitude under the action of a harmonic excitation; this occurs when the frequency of the exciting force or motion is close to a natural frequency of the vibrating body.

Response spectrum: A plot with respect to frequency or period whose ordinate represents the maximum acceleration, velocity, or displacement for all single degree-of-freedom systems when excited by a specific ground motion. All three responses can be represented simultaneously on a tripartite or three-axis graph, usually using logarithmic scales.

SEAOC: Structural Engineers Association of California.

Seismic: Caused by or subject to ground motion, such as from an earthquake. Also, in the context of structural design procedures, commonly used to mean *aseismic*.

Seismic waves: Vibration of rock or soil particles that propagates out in all directions from the fault break in an earthquake.

Seismicity: The degree to which a location is subject to earthquake risk. Usually refers to the frequency of occurrence and magnitude of past earthquakes. Also, the expectation of future seismic activity.

Sequence of plastification: The sequence in which inelasticity or yielding spreads to the various structural members.

Shallow-focus earthquakes: Shocks with focal depths not exceeding 60 km.

Shear lag: The decrease in the vertical forces transmitted to the columns as one moves from the corner toward the center of a frame subjected to lateral loads.

Shear strength (capacity): Maximum shear force that can be carried by a member segment if prior failure due to axial load or flexure is precluded. It is a function of axial loads and bending moments acting on the member segment.

Shear (*structural*) *wall*: A vertical element that primarily resists in-plane horizontal shear forces and overturning moments.

SDOF: Single degree of freedom.

SMAC: Strong-motion accelerograph.

Spectral acceleration: The maximum absolute acceleration response of a single degree-of-freedom system having a given period.

Spectral displacement: The maximum relative displacement response of a single degree-of-freedom system having a given period.

Spectral velocity: The maximum relative velocity response of a single degree-of-freedom system having a given period.

Spectrum intensity: Area under the 5%-damped relative velocity response spectrum between periods of 0.1 and 3.0 seconds.

Stiffness: Force per unit displacement.

Strain (*tensile or compressive*): Unit charge in length.

Strain capacity: The largest strain that does not cause unacceptable damage or loss of strength.

Strong-motion accelerographs: Simple, damped oscillators in which the relative displacement between a moveable mass and a fixed base is proportional to the input acceleration. Strong-motion accelerographs do not usually start recording until triggered by an initial motion.

Tectonic: Pertaining to structures resulting from deformation of the earth's crust, especially faulting.

Time-history (*response history*) *analysis*: A step-by-step tracing, in small time increments, of the response of a structure to an earthquake accelerogram. At each time interval, a new analysis is carried out for the structure in its deformed state, as determined at the end of the previous time segment.

UBC: Uniform Building Code.

Volcanic: Associated with eruptions of volcanoes, or subterranean movements of magma.

Windward: The side from which the wind blows (the side of a structure facing the wind).

Yielding: A stage of response corresponding to the intersection between the initial elastic and the post-elastic branches of the force-displacement diagram of a section, hinging region, member or structure.

Yield moment (*level*): The bending moment at a section corresponding to the stage of yielding.

16 Design for fatigue

John M Hanson

Wiss Janney Elstner Associates, Northbrook, Ill, USA

Contents

Notation

A_s area of steel bar, in^2 or mm^2
A_w area of stirrup, in^2 or mm^2
d diameter of steel bar, in. or mm
D diameter of bend, in. or mm
e eccentricity, in. or mm
f'_c compressive strength of concrete, psi or MPa
f_{lim} fatigue limit, psi or MPa
f_{max} maximum stress of cycle, psi or MPa
f_{min} minimum stress of cycle, psi or MPa
f_{pu} ultimate strength of high strength steel, psi or MPa
f_r stress range, psi or MPa
freq   frequency, cycles per second
f_u ultimate strength of steel bar, psi or MPa
f_y steel yield strength, psi or MPa
h lug height of steel bar, in. or mm
Hz   hertz—frequency, cycles per second
k coefficient to consider the difference in concrete strength—0.85

N number of cycles
N_f total life in cycles
N_{fi} number of cycles that will cause failure at stress level i
N_i number of constant amplitude cycles at stress level i
p probability of failure
R ratio $= f_{min}/f_{max}$ for concrete, or $(f_{max} - f_{lim}) \times 100/f_{pu}$ for steel
r radius at base of transverse lug of steel bar, in. or mm
s standard deviation
S characteristic stress: stress range or a function of maximum and minimum stresses
V shear force, kips or N
α flexural gradient coefficient
β constant determined experimentally
γ material factor
ε_{max} maximum strain at a cycle
ε_{min} minimum strain at a cycle
ε_0 maximum strain in the first load cycle

Summary

Concern about fatigue has increased because of new uses of concrete, such as in offshore structures and railroad ties. In these types of structures, occasional overloadings are expected and the full design loading may be repeated for a very large number of cycles.

Substantial research has been conducted on fatigue of structural concrete members in the last two decades. This research is summarized here. Information about the strength of the component materials under cyclic loading may be used to design members in which flexure is predominant. Information about the effects of cyclic shear, torsion and bond is limited. However, recent experience emphasizes the importance of recognizing these effects.

Many countries now include design provisions for fatigue in their codes. Selected provisions are included in the chapter.

1   Introduction

Fatigue of a structural member occurs when cracking develops under repetitive loads that are less than the static load capacity. Substantial research has been conducted in recent years on the fatigue of concrete systems. Fortunately, this research has shown that the static load criteria under which most existing concrete structures have been designed have virtually precluded the possibility of fatigue failure in the primary load-carrying elements of the main members. Reports of fatigue failures in these primary elements are, apparently, nonexistent. However, a fatigue failure of anchor bolts embedded in a concrete pier caused a collapse of a large crane [1]. Also, a recent laboratory study in Japan [2] indicated that the fatigue resistance of bridge deck slabs to moving concentrated loads is much less than to repeated fixed-point loads. This suggests that the progressive failure of slabs may, in part, be attributed to fatigue where repeated shear and torsional forces are applied. Fatigue is also regarded as a cause of cracking and failure in concrete pavements.

Even though there are no reports of failures in primary elements of structural concrete members due to fatigue, there are a number of reasons for increasing concern. Adoption of strength design procedures and use of higher strength materials require these members to perform under higher stress levels. New uses are being made of concrete, such as in offshore structures and railroad ties, where the full design loading may be repeated for a very large number of cycles, or occasional overloadings are expected.

Research on fatigue dates back to the early 1900s. Most of the early research dealt with fatigue of plain concrete. In the 1960s, very extensive programs on the fatigue of reinforcing elements were carried out in Europe, Japan, and the United States. Committee 215 of the American Concrete Institute published a report [3] in 1974 that included an extensive list of references. More recent work has been directed to specialized applications. The proceedings [4] of a colloquium held in

1982 by the International Association for Bridge and Structural Engineering contain a substantial number of current and relevant papers.

In discussing fatigue, the differences between static, dynamic, fatigue, and impact loadings should be recognized. Static loading, or sustained loading, remains constant with time. However, a load which increases slowly is often called static loading; the maximum load capacity under such conditions is referred to as static strength. Dynamic loading varies with time in any arbitrary manner. Fatigue and impact loadings are special cases of dynamic loading.

A fatigue loading is a sequence of load repetitions. A distinction is made between high-cycle, low-amplitude and low-cycle, high-amplitude fatigue loading. The distinction depends on whether the repeated loading causes a failure at less than or more than an arbitrary number of cycles, usually considered in the range of 100–1000 cycles.

Most of the previous research has been directed toward high-cycle, low-amplitude fatigue loading, in the range of $1000–10^7$ cycles. The information in this chapter deals primarily with this type of loading. Higher cycle loadings may be expected, however, for bridges and elevated structures on expressways or rapid transit systems. Concrete structures supporting dynamic machines may be subjected to hundreds of millions of load cycles. Structures built in oceans may also be subjected to hundreds of millions of cycles of loading due to wave action. Low-cycle, high-amplitude fatigue loading of less than 1000 cycles may occur as a result of earthquakes or other events that cause a loading of the structure beyond its normal service level. Gerwick [5] has discussed the effect of this type of fatigue on sea structures.

This chapter is intended to provide information that will be useful in the design of concrete structures subjected to fatigue loading. If fatigue is a controlling factor, the reinforcing elements will be more likely to limit the life of the member than the concrete, particularly in members that are subjected to predominantly flexural high-cycle, low-amplitude loadings between 1000 and 10^7 cycles. Results of investigations of the fatigue strength of concrete and of reinforcing materials are presented in Section 2. Most of the data have been obtained from reviews of experimental investigations reported in the technical literature. Information on the behavior of members under repeated loading is presented in Section 3. In flexural members, the fatigue life is closely related to the fatigue strength of the component materials. However, environmental effects such as corrosion may limit the life of the members. Other factors are also important in those situations where shear, torsion or bond are significant. Finally, in Section 4 some of the design provisions in codes of various countries are summarized.

2 Fatigue strength of component materials

Test data on the fatigue strength of concrete or reinforcing steels are usually presented in the form of $S–N$, or Wohler, diagrams. S is a characteristic stress of the loading cycle, often either the stress range or a function of the maximum and minimum stress, and N is the number of cycles to failure. Fatigue data on concrete or reinforcing steels have commonly been shown in semi-log plots, where S is plotted linearly as a dependent variable and life is plotted as an independent variable on a log scale. In this form, the mean of the data can often be represented by a straight regression line. In contrast, fatigue data on the performance of structural steel members and welded details are generally shown in log–log plots.

Considerable effort has been directed in recent years to the evaluation of fatigue data on structural steel members and welded details with the aid of fracture mechanics techniques. These techniques are also now being applied to concrete and reinforcing steels in an exploratory manner, and they show promise of being useful in predicting life and in understanding the behavior of the materials.

Fatigue life is considered to occur in three stages: initiation of cracking, propagation of cracking, and fracture. In the propagation stage, microcracking is occurring in the concrete or cracking is growing in a steel element. During most of this stage, the cracking is growing slowly, followed by a short period of more rapid growth leading to fracture.

In designing for fatigue, recognition must be given to the variability that is inherent in the phenomenon. The mean regression line of a set of fatigue test data [6] is shown in Fig. 2-1. The assumption of a normal distribution of the fatigue lives about the mean is usually acceptable. With this assumption, for example, it may be stated that the probability of failure, p, at the lower dashed line is 5%. This dashed line is located 1.96 times the standard deviation, s, of the data below the mean. Fatigue data on concrete have often been represented in this manner.

It is more common and more useful for design, particularly with fatigue data on steel, to utilize a lower tolerance limit on the data. A tolerance limit associates a confidence level with a probability of survival $(1 - p)$. This confidence level is dependent on the number of test results in the sample. For the data shown in Fig. 2-1, 95% of the test results are expected to exceed, at a 95% confidence level, the lower tolerance limit. This limit is about twice the standard error of estimate of N below the mean.

Confidence intervals are also commonly shown on fatigue test data. These intervals, for example at the 95% level, indicate the range in which 95% percent of the means of the population represented by the data are expected to be included. However, it is never known if the S–N curve of the population actually lies within the confidence interval.

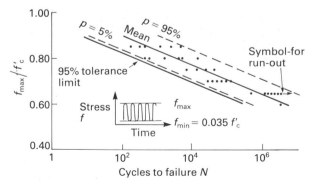

Figure 2-1 Variability in fatigue life of concrete in compression

2.1 Plain concrete

When concrete is subjected to increasing compressive load, Shah *et al.* [7] observed that the volume of the concrete ceased to decrease, and instead began to increase, at a stress level as low as approximately 50% of the compressive strength. The initial deviation from linearity of the relationship between stress and decrease in volume was related to a significant increase in microcracking of the aggregate–paste interface, while the stress at which the volume began to increase was related to a noticeable increase in microcracking through the matrix. Repetitive loading appeared to have a significant effect on the growth of microcracking [8]. As a result, the fatigue of plain concrete may be considered to be a process of progressive, permanent changes occurring within the concrete matrix under repetitive loading.

Under cyclic compressive loading of concrete specimens, a number of investigators have observed a decrease in the measured value of pulse velocity and an increase in acoustic emissions. These changes are related to the growth of the microcracking. More important to the designer is the increase in deformation that occurs during repeated loading, as illustrated by the data obtained by Holmen [9] and shown in Fig. 2-2. A rapid increase in strain occurs between 0 and about 10% of the total life, N_f. The increase in strain between 10 and about 80% is slow and uniform, after which the strain increases at an increasing rate until failure occurs. Furthermore, it appears that the strain consists of two components, one related to the microcracking and the other which is time-dependent and related to creep deformation. Expressions for estimating the strain have been developed [6, 9].

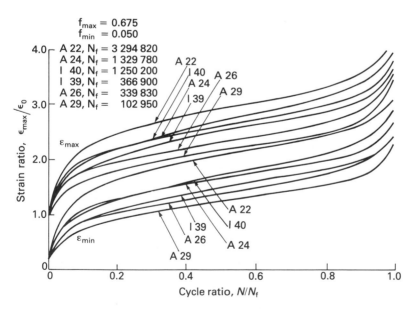

Figure 2-2 Variation in measured strain: ε_0 is the maximum strain in the first cycle; the frequency is 5 Hz

2.1.1   *Cyclic compression*

Numerous investigators have found that the fatigue life of a common group of concrete specimens tested in compression, each under a constant amplitude loading with the stress, f, varying from the same minimum to different maximums, may generally be presented by a straight line, as illustrated in Fig. 2-1. Furthermore, there is no apparent endurance limit below which the concrete will sustain an unlimited number of repetitions, although it should be noted that very little data are available for loadings greater than 10^7 cycles.

An expression for the fatigue of concrete in compression was developed by Aas-Jacobsen [10] based on linearity of the relationship between f_{max}, f_{min} and log N, as follows:

$$\log N = \frac{1-(f_{max}/f_c')}{\beta(1-(f_{min}/f_{max}))} \tag{2-1}$$

in which a value of β, based on a review of available data, equal to 0.064 was given. A more recent study by Tepfers and Kutti [11], which was based on 475 tests, including normal as well as lightweight aggregate concrete, has recommended using β equal to 0.0685 for estimating fatigue life, although it was noted that uncertainty in the value of β increased when f_{max} was greater than $0.8f_c'$. Equation 2-1 with $\beta = 0.0685$ is plotted in Fig. 2-3 for various values of $R = f_{min}/f_{max}$.

Another expression for the fatigue of concrete was reported by Kakuta *et al.* [12] at the 1982 IABSE colloquium in Lausanne, as follows:

$$\log N = 17\left[1 - \frac{(f_{max}-f_{min})/f_c'}{1-(f_{min}/f_c')}\right] \tag{2-2}$$

A comparison of test data with the line represented by Eqn 2-2, as reported in [12], is shown in Fig. 2-4.

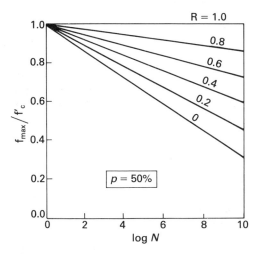

Figure 2-3   Representation of Eqn 2-1: $R = f_{min}/f_{max}$, $\beta = 0.0685$, $p = 50\%$

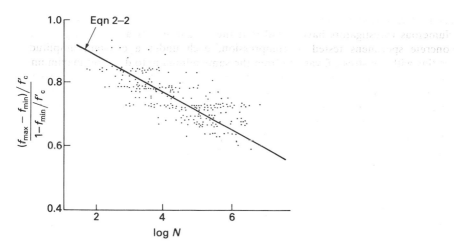

Figure 2-4 Comparison of Eqn 2-2 with test data

Very little information is available in a useful form about the dispersion of the fatigue life of concrete in compression. In a recent investigation, however, Siemes [13] concluded that the dispersion was not larger than should be expected on the basis of the dispersion in the static compressive strength of the concrete, f'_c. In this investigation, values of standard deviation, s, of $\log N$ were reported for 17 controlled groups of constant amplitude tests in which the primary variables were f_{max}, R and the frequency of loading. The value of s varied from 0.062 to 0.569, although it should be noted that the second highest value of s was 0.349. Discarding the high value, the average was 0.162.

In [12], a design equation for the fatigue of concrete in compression was presented which has been included in the *Tentative recommendations for the limit state design of concrete structures* by the Japan Society of Civil Engineers, as follows:

$$f_{max} - f_{min} = (0.9kf'_c - f_{min})\left(1 - \frac{\log N}{15}\right)$$ (2-3)

where k is a coefficient taken equal to 0.85 to consider the difference in concrete strength measured using standard cylinders and the in-place strength. This equation was reported to have been derived from Eq 2-2, whereby the fatigue life corresponds to a probability of failure of about 5%.

A comparison is made in Fig. 2-5 between values of $\log N$ computed from Eqns 2-2 and 2-3, assuming $k = 1$, for a value of $f_{min} = 0.1f'_c$. It may be seen that the separation of the lines, $\Delta \log N$, is nearly uniform and approximately equal to 1.92. Similar comparisons with f_{min} equal to 0 and $0.2f'_c$ show differences of approximately 1.74 and 2.10, respectively. If a shift of 1.9 is introduced into Eqn 2-1 for design purposes, on the assumption that the dispersions are comparable, the result is

$$\log N = \frac{1 - (f_{max}/f'_c)}{0.0685(1 - (f_{min}/f_{max}))} - 1.9$$ (2-4)

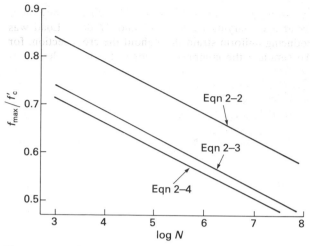

Figure 2-5   Comparison of Eqns 2-2, 2-3 and 2-4 for $f_{min} = 0.1f'_c$

Equation 2-4 is also shown in Fig. 2-5, and it appears that both are in reasonable agreement and quite suitable for design purposes.

On the basis that Eqn 2-3 is an approximation of a 5% probability of failure of the data for which Eqn 2-2 represents a 50% probability of failure, it is evident that the dispersion of a combination of test data representing various conditions is much greater than for controlled groups of data as reported in [13]. Consequently, a designer should realize that an improvement in fatigue strength at a 5% probability of failure may be obtained over that given by Eqns 2-3 or 2-4 if fatigue tests are conducted on representative specimens of the concrete.

Stress gradient has been shown [14] to influence the fatigue strength of concrete. Results of tests on $4 \times 6 \times 12$ in $(102 \times 152 \times 305$ mm$)$ concrete prisms under three different compressive strain gradients are shown in Fig. 2-6. The

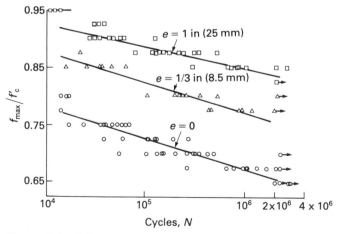

Figure 2-6   Effect of compressive strain gradient for $f_{min} = 0.1f'_c$ (Reproduced from [14] with permission)

prisms had a compressive strength of about 6000 psi (41.4 MPa). They were tested at a rate of 8.3 Hz at ages varying between 47 and 77 days. Load was applied concentrically, producing uniform strain throughout the cross-section, for the case marked $e = 0$. To simulate the compression zone of a beam, load was applied eccentrically in the other two cases, marked $e = \frac{1}{3}$ in (8.5 mm) and $e = 1$ in (25.4 mm). The loads were applied such that, during the first cycle of fatigue loading, the maximum strain at the extreme fiber was the same for all three sets of specimens. For the two eccentrically loaded cases, the minimum strain was half the maximum strain and zero, respectively.

From the mean curves shown in Fig. 2-6, it can be seen that the fatigue strength of the eccentrically loaded specimens is 15 to 18% higher than for the uniformly stressed specimens for a fatigue life of 40 000 to 10^6 cycles. For design, it has been pointed out [12] that the assumption of a rectangular stress distribution about the resultant compressive force will correct for the eccentricity and correlate closely with axial compressive fatigue strength.

2.1.2 Cyclic tension
A number of investigations have shown that the fatigue strength of concrete under loadings producing axial, splitting or flexural tension is about the same as for compression. Cyclic loading from compression to tension has been reported [15] to cause more damage than zero-to-tension loadings.

2.1.3 Effect of material properties
Numerous investigations have studied the effect of such factors as cement content, water/cement ratio, curing conditions, age at loading, amount of entrained air, and type of aggregate [16, 17]. Except for the latter factor, there is general agreement that these factors affect fatigue strength in a proportionate manner to the static strength of the concrete, as directly given by the previous expressions for fatigue strength. Furthermore, the fatigue strength of mortar is also comparable to concrete when expressed as a function of compressive strength.

The fatigue strength of lightweight aggregate concrete has also generally been considered to be related to static strength in the same manner as normal-weight concrete. Tepfers and Kutti [11] found no difference in susceptibility to fatigue between these types of concrete, and concluded that Eqn 2-1 with $\beta = 0.0685$ was applicable. However, Kakuta et al. [12] reported that the fatigue life of lightweight aggregate should be taken as 70% of that given by Eqn 2-2.

2.1.4 Rate of loading and load history
Most fatigue data have been obtained from tests in which the loads were alternated between constant minimum and maximum values and the rate of loading was between 1 and 15 Hz. Within this range, the rate of loading has not been regarded [3] to have a significant effect on fatigue strength, except when the maximum stress level is greater than 75% of the static strength. Under higher stresses, creep effects become more important, and lead to a reduction in fatigue strength with decreasing rate of loading. A recent investigation [13] also showed a reduction in fatigue life roughly equivalent to $\log N = 1$ when the rate of loading was decreased from 6 to 0.06 Hz.

In practice, structural concrete members are normally subjected to randomly varying loads with periods of rest. In laboratory tests on concrete specimens subjected to varying flexural stresses, rest periods were beneficial [18], increasing the fatigue strength at 10^7 cycles, expressed as f_{max}/f_c', by about 10%. Low-amplitude cyclic loading interspersed in higher-amplitude loading also has a beneficial effect [9].

Miner's hypothesis has been commonly used [3] to determine the accumulation of fatigue damage under varying stresses:

$$\sum_{i=1}^{k} \frac{N_i}{N_{fi}} = 1 \qquad (2\text{-}5)$$

where N_i equals the number of constant amplitude cycles at stress level i, N_{fi} equals the number of cycles that will cause failure at that stress level i, and k equals the number of stress levels. Recent investigations have led to differing views of the accuracy of Eqn 2-5. In one program [9], it was found that variable-amplitude loading seemed to be more damaging than predicted by Eqn 2-5, and that the number of cycles to failure was dependent on the loading sequence. Waagaard [20] has summarized the requirements for fatigue design of the *Veritas rules for the design, construction, and inspection of offshore structures*, which use a Miner's number of 0.2 rather than one. In another program [20], it was concluded that Miner's rule was remarkably good, and that the dispersion in the test results was fully explainable by the dispersion in static strength.

2.1.5   *Concrete submerged in water*

Several investigations [12, 20] have reported that the fatigue strength of concrete is reduced in ocean water. Water is trapped in opening and closing cracks, causing hydraulic fracturing and propagation of microcracks. According to Hawkins and Shah [22], crack propagation is more severe for higher stresses, saturated concrete, high temperatures and long time periods. Thus, for offshore structures, large-amplitude, low-frequency load cycles can be more critical than high-frequency, small-amplitude cycles.

An equation for design for the fatigue of concrete in compression in offshore structures was presented in [20] as follows:

$$\log N = 10 \left[\frac{1.0 - \dfrac{f_{max}}{\alpha\left(f_c'/\gamma\right)}}{1.0 - \dfrac{f_{min}}{\alpha\left(f_c'/\gamma\right)}} \right] \qquad (2\text{-}6)$$

where α is intended to take into account the effect of flexural gradient and γ is a material factor that was specified equal to 1.25. Equation 2-6 is more conservative than Eqn 2-3 or 2-4, apparently reflecting the anticipated reduction in fatigue strength when the concrete is in sea water.

2.2   Reinforcing bars

Fatigue of reinforcing (nonprestressed) bars occurs as the result of the initiation and propagation of a crack under cyclic loading. The bar will fracture when the stress intensity factor at the crack front reaches a critical value. A

Figure 2-7 Fatigue fracture of a reinforcing bar. (By courtesy of Portland Cement Association)

typical fatigue fracture of a reinforcing bar is shown in Fig. 2-7. This particular No. 11 (35 mm diameter) bar with a yield strength, f_y, equal to 60 000 psi (414 MPa) was at one time embedded in a concrete beam and subjected to repeated loads until the bar failed, as part of a major investigation of fatigue of reinforcing bars by Helgason et al. [23].

In Fig. 2-7, the orientation of the bar is the same as it was in the beam; the bottom of the bar was adjacent to the extreme tensile fibers in the beam. The smoother zone, with the dull, rubbed appearance, is the fatigue crack. The remaining zone of more jagged surface texture is the part that finally fractured in tension after the growing fatigue crack weakened the bar. It is noteworthy that the fatigue crack did not start from the bottom of the bar. Rather, it started along the side of the bar, at the base of one of the transverse lugs. This is a common characteristic of fatigue fractures of bars in concrete beams.

The lowest stress range known to have caused a fatigue failure of a straight hot-rolled deformed bar embedded in a concrete test beam and tested in air is 21 000 psi (144.8 MPa). This failure occurred after 1 250 000 cycles of loading on a beam containing a No. 11 (35 mm diamter) Grade 60 bar, when the minimum stress level was 17 500 psi (120.7 MPa). This bar was from the same lot as the bar shown in Fig. 2-7.

Laboratory data have been obtained from tests on bars in air and on bars embedded in concrete beams. Tests on bars in air are easier to conduct, but tests on bars in concrete are considered more representative of actual conditions. When a test is conducted on a bar embedded in a concrete beam, the fatigue

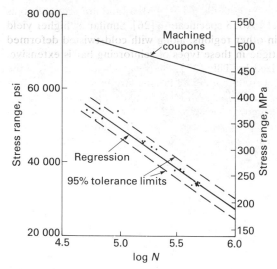

Figure 2-8   Results of 25 fatigue tests on bars in which stress range was the only variable

fracture invariably occurs in close proximity to a flexural crack initiating from the base of a lug or a manufacturer's mark. On a test in air, it has been found that the crack may initiate at irregularities along the surface of the bar [24]. That is probably the main reason that most test results on bars in air are generally a little lower than on bars from the same lot embedded in a concrete beam.

The results of a group of 25 fatigue tests [23] on T-shaped concrete beams, each containing a single No. 8 (25 mm diameter) Grade 60 bar from a common lot, are shown in Fig. 2-8. Thus, except for the applied stress range, all of the test conditions in this group are nominally identical. Lines representing the mean of the data and tolerance limits which have a 95% probability of including 95% of all test results from this population are shown in this figure. From these and other tests in this investigation, which incorporated variation of the effect of stress range, minimum stress, bar diameter, size of beam and grade of bar, it was found that stress range was the predominant factor determining fatigue life. Minimum stress level was also significant. For design purposes, it was concluded that there is a limiting stress range which may be considered to be a fatigue limit. At stress ranges above this limit, a reinforcing bar will have a finite life. Below this limit, the bar will have a long life, and may be able to sustain a virtually unlimited number of cycles. The transition from the finite life to long life region occurred in the range of one to two million cycles.

The fatigue strength of coupons machined from the center of the No. 8 Grade 60 bars is also shown in Fig. 2-8. It is evident that the fatigue strength of the bars is about one-half of the fatigue strength of the coupons. The main reasons for the reduction are the deformations and decarburization of the surface. These effects have been found to be cumulative [25].

Several types of bar are used for reinforcing a concrete beam. In North America, hot-rolled deformed bars are used almost exclusively. Most of these bars have a yield strength, f_y, equal to 60 000 psi (414 MPa) and are referred to as Grade 60 in the governing ASTM A615 specification [26]. Similar or higher yield bars are also used extensively in other regions, along with cold-twisted deformed or square bars. Research on fatigue in these types of reinforcing bar is extensive, and has recently been summarized by Tilly [27]. The research indicates that the type of bar is not an important variable affecting fatigue strength. Variables that are important from a design viewpoint, in addition to stress range and minimum stress, are the geometry of deformations on a bar, radius of bends, welding, and corrosion. The relatively minor effects of other factors, including bar size and orientation, yield strength, chemical composition, etc., are discussed in technical literature [3].

2.2.1 *Geometry of deformations*

Deformations on reinforcing bars are needed for good bond between the steel and the concrete. However, the base of a transverse lug or the intersection of a lug and a longitudinal rib may be the source of a stress concentration. These points of stress concentrations are where the fatigue fractures are usually observed to initiate in tests on bars embedded in concrete beams. It should be especially noted that brand marks on the surface of the bar may have stress concentrations that are similar to those of bar deformations.

Analytical studies [25, 28] have shown that the stress concentration of an external notch on an axially loaded bar may be appreciable. These studies indicated that the width, height, angle of rise, and base radius of a protruding deformation affect the magnitude of the stress concentration. Many reinforcing bar lugs have stress concentration factors of 1.5 to 2.0.

Tests on bars having a base radius varying from about 0.1 to 10 times the height of deformation have been reported [23, 29–31]. These tests indicate that, when the base radius is increased from 0.1 to about one to two times the height of the deformation, fatigue strength is increased appreciably. An increase in base radius beyond one to two times the height of the deformation does not show much effect on fatigue strength. However, Japanese tests [31] have shown that lugs with radii larger than two to five times the height of the deformation have reduced bond capacity.

Several studies have also included the effect of wear in the rolls at the steel mills on fatigue strength. In general, the difference in the fatigue strength of bars made with new or old rolls has been found to be small [32, 33].

It has been demonstrated [34] that the fatigue strength of a bar may be influenced by the orientation of the longitudinal ribs. In that study, an increased fatigue life was obtained when the longitudinal ribs were oriented in a horizontal plane rather than a vertical plane. This phenomenon is apparently associated with the location where the fatigue crack initiates. In other words, if there is a particular location on the surface of a bar which is more critical for fatigue than other locations, then the positioning of that location nearer to the bottom of the beam will tend to reduce the fatigue strength.

In the study of bars made by four different US manufacturers that was described in [23], the following equation was developed for the limiting stress range in the long-life region, below which fatigue damage is unlikely to occur in straight, hot-rolled bars with no welds:

$$f_r = 21\,000 - 0.33f_{min} + 8000(r/h) \quad \text{in psi} \qquad (2\text{-}7)$$

or

$$f_r = 145 - 0.33f_{min} + 55(r/h) \quad \text{in MPa}$$

where f_r is equal to the stress range, i.e., $f_{max} - f_{min}$, f_{max} and f_{min} are the maximum and minimum stress level (tensile stresses are positive), respectively, and r/h is the ratio of the radius at the base of a transverse lug to the lug height. This equation gives a value for stress range which includes an adjustment to represent approximately a 95% probability that 95% of the test results on one of the manufacturers' bars will exceed.

Although Eqn 2-7 is a lower limit for bars made by US manufacturers conforming to ASTM designation A615 [26], it is believed to be reasonably applicable to other reinforcing bars whose surface geometry has not been controlled in a manner that assures a higher fatigue strength. As discussed in Section 4, this equation has been adopted in US design specifications, and it gives a lower fatigue strength than values in European and Japanese specifications. A 1981 draft of Swiss Standard SIA 162 cites a fatigue limit of 28 700 psi (198 MPa) for high yield strength bars made in Switzerland.

An equation was also developed in [23] for a safe fatigue life for all stress ranges above the fatigue limit represented by Eqn 2-7. This equation is as follows:

$$\log N = 6.1044 - 4.07(10)^{-5}f_r - 1.38(10)^{-5}f_{min}$$
$$+ 0.71(10)^{-5}f_u - 0.0566A_s + 0.3233\,d(r/h) \qquad (2\text{-}8)$$

where f_r and f_{min} are in psi, A_s is the area of the bar in in^2 and d is the diameter of the bar in in. Other terms have been previously defined. Or

$$\log N = 6.1044 - 591(10)^{-5}f_r - 200(10)^{-5}f_{min}$$
$$+ 103(10)^{-5}f_u - 8.77(10)^{-5}A_s + 0.0127\,d(r/h)$$

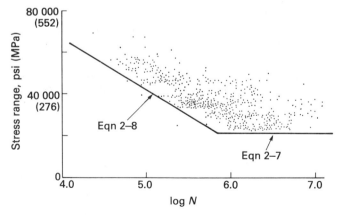

Figure 2-9 Published test results compared with Eqns 2-7 and 2-8: No. 6 (19 mm) bar, $f_{min} = 6000$ psi (41 MPa), $f_u = 90\,000$ psi (620 MPa), $r/h = 0.3$

where f_r and f_{min} are in MPa, A_s is in mm^2, and d is in mm. A comparison is made in Fig. 2-9 of $\log N$ computed from Eqns 2-7 and 2-8 for the given conditions with data from sources in both North America and Europe.

2.2.2 Bends

The fatigue strength of a bar will be reduced by a bend. In a North American investigation [35], fatigue tests were carried out on both straight and bent No. 8 (25 mm diameter) deformed bars embedded in concrete beams. The bends were through an angle of 45° around a pin of 6 in (152 mm) diameter. The fatigue strength of the bent bars was a little more than 50% below the fatigue strength of the straight bars. In one test, a bent bar embedded in a reinforced concrete beam failed in fatigue after sustaining 900 000 cycles of a stress range of 18 000 psi (124 MPa) imposed on a minimum stress of 5900 psi (40.6 MPa). In another test, application of 1 025 000 cycles produced a failure when the stress range and minimum stress were 16 400 and 19 100 psi (113 and 132 MPa), respectively.

A survey [36] of the fatigue behavior of reinforcing bars cited tests in West Germany showing the influence of the diameter of the bend, D, in relation to the diameter of the bar, d. When $D/d = 25$, there was no reduction in strength compared to straight bars. However, for smaller D/d ratios, reductions in fatigue strength within the ranges shown in Table 2-1 were found. Another review [12] of tests in Japan cited reductions in fatigue strengths of $\frac{1}{2}$ to $\frac{2}{3}$ for bent bars.

Table 2–1

D/d	Percentage reduction in fatigue strength
15	16–22
10	22–41
5	52–68

2.2.3 Welding

Tack welding of reinforcing bars is a practice that facilitates the placement of the steel. However, in an investigation [34] using two grades of bars with the same deformation pattern, it was found that the fatigue strength of bars with stirrups attached by tack welding was about one-third less than that of bars with stirrups attached by wire ties. For both grades of steel, the fatigue strength of bars with tack welding was about 20 000 psi (138 MPa) at 5 million cycles. All of the fatigue cracks were initiated at the weld locations.

Investigations have also been carried out to evaluate the behavior of butt-welded reinforcing bars in reinforced concrete beams [28, 37]. In tests conducted at a minimum stress level of 2000 psi (14 MPa) tension, the least stress range that produced a fatigue failure was 24 000 psi (165 MPa). It was observed that minimum stress level in the butt-welded joint was not a significant factor affecting the fatigue strength of the beams.

Work on the fatigue strength of welded wire fabric and welded bar mats has also been reported in the United States and West Germany. Most of this work is summarized in the ACI Committee 215 report [3]. The fatigue strength of these materials appears to be more variable than for straight unwelded bars, with reported values as low as 13 000 psi (90 MPa) at 5 million cycles.

In a recent investigation by Nurnberger [36], an *S–N* curve for a welded mat is shown with a long-life region above 2 million cycles where the fatigue limit was 18 000 psi (124 MPa). This investigation also reports an *S–N* curve for welded overlap splices of hot-rolled bars which exhibited a fatigue limit of 10 000 psi (69 MPa) above 2 million cycles.

In consideration of these limited findings, and also in view of the general lack of knowledge about the conditions that will cause brittle fracture in reinforcing bars, it would appear to be good practice to avoid welding in construction that will be subjected to repetitive loads. Where welding cannot be avoided, reference should be made to fatigue design criteria for comparable welded details in structural steel members, such as the AASHTO [38] or AISC [39] specifications in the United States. For example, treating the butt welded bar as a Category B condition, the allowable stress range is 16 000 psi (110 Pa) when over 2 million cycles of loading are expected. Similarly, treating a welded overlap splice as a Category E condition, the allowable stress range is 5000 psi (34.5 MPa) for over 2 million cycles.

2.2.4 *Mechanical splices*
Bennett [40] has reported tests on beams with mechanical splices of the main reinforcement in the maximum moment region. These tests were on No. 8 (25 mm diameter) cold-worked deformed bars spliced by lapping and by lapping and cranking in addition to the use of cold-forged swage sleeves and screw couplers. The bars had a yield strength of 61 600 psi (425 MPa). A beam with straight bars withstood 4 million cycles at a stress range of 18 900 psi (130 MPa) without

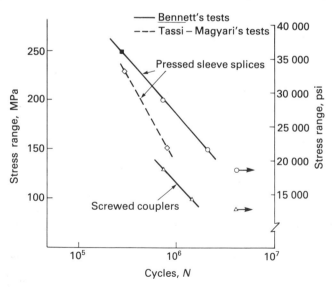

Figure 2-10 Fatigue strength of bars with mechanical splices

failure, whereas a beam with lapped and cranked bars failed at the bend point of the crank after only 100 000 cycles of loading at the same stress range. For the bars with swaged splices, fatigue fractures occurred where the bar entered the sleeve. For bars spliced with screwed couplers, failures occurred in the coupler at a high stress range and in the bar at the edge of the coupler for a lower stress range. Tassi and Magyari [41] have also reported tests on pressed sleeve splices conducted in air and also embedded in concrete. Failures occurred in the bar, sometimes where the bar entered the splice and in other cases away from the splice. The results of both Bennett's and Tassi and Magyari's tests are shown in Fig. 2-10. Only tests where the failures occurred at the splice are included. From these tests, it appears that the fatigue performance of the pressed sleeve splices is superior to the screwed couplers. However, it is evident that all splices must be carefully located in structures subject to repeated load.

2.2.5 *Effect of corrosion*

Information on the effect of corrosion on the fatigue strength of bars is limited and contradictory. Fatigue cracks initiate at the surface of a bar, commonly at the base of deformations. If in time the stress concentration at a surface pit were more severe than at a lug, a fatigue crack could initiate at that location. However, in an investigation [30] which included 'ordinary' rusted bars, the fatigue strength was equal to that of bars that had been cleaned prior to testing.

The corrosion of bars embedded with adequate cover in concrete is usually not a serious problem unless the concrete contains chlorides, or unless the concrete is subject to the application of chlorides, as in a bridge deck or parking garage. In these instances, delaminations of the concrete at the level of the steel are caused by the expansion of the corrosion by-products, leading to severe deterioration of the structure. However, fatigue of bars has not been reported even in heavily travelled bridge decks with severely corroded bars.

Several investigations have observed reductions in fatigue strength of bars that are exposed to sea water or chloride solutions. Submersion of beams may cause 'crack-blocking', increasing fatigue strength. Thus, a reduction is apparently due to an electrochemical effect on the reinforcing steel. Tilly [42] has reviewed a number of these investigations. In one case, a reduction in fatigue strength of 35% at 10^7 cycles was observed for bars tested at 145 Hz axially in air and in sea water. In another case, a reduction in fatigue strength of 22% at 10^7 cycles was observed in bars tested at 3 Hz embedded in concrete beams submerged in sea water. Also, an investigation is cited where tests on bars with 0.44% carbon in salt water exhibited a 24% reduction in fatigue strength at 2 million cycles when the rate of loading was reduced from 43 to 4 Hz.

A variety of conditions were included in another recent Australian investigation [43] on 24 mm (0.9 in) hot-rolled bars of 230 MPa (33 400 psi) grade and cold-twisted 410 MPa (59 500 psi) grade bars made by twisting, without tensioning, comparable 230 MPa (33 400 psi) grade bars. These bars were embedded in concrete beams and tested at 6.7 Hz in air, sea water or a 3% NaCl solution. Some of the bars were galvanized. Lines fitted to the test data are shown in Fig. 2-11.

The hot-rolled bars embedded in concrete and tested in air exhibited a finite-life and long-life region, with a fatigue limit of 218 MPa (32 000 psi). These

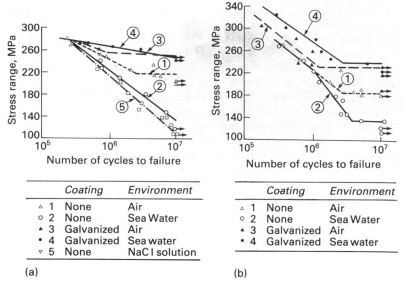

		Coating	Environment
△	1	None	Air
○	2	None	Sea Water
▲	3	Galvanized	Air
•	4	Galvanized	Sea water
▽	5	None	NaCl solution

(a)

		Coating	Environment
△	1	None	Air
○	2	None	Sea Water
▲	3	Galvanized	Air
•	4	Galvanized	Sea water

(b)

Figure 2-11   Effect of immersion in sea water or NaCl solution on fatigue strength
(a) Hot-rolled bars (Grade 230 MPa)
(b) Similar hot-rolled cold-twisted bars (Grade 410 MPa)

bars were reported to have an r/h ratio of 1.4 and were tested at a minimum stress level of 13 MPa. Under these test conditions, the fatigue strength computed from Eqn 2-7 is 218 MPa. Immersion of the lower half of the beam in sea water substantially reduced the fatigue strength and no fatigue limit was apparent out to 10^7 cycles. Galvanizing improved the fatigue strength of the bars in the beams in air, and also fully protected the bars in sea water.

The cold-twisted bars embedded in concrete and tested in air also exhibited a finite-life and long-life region. However, the cold-twisting reduced the fatigue strength in air. The fatigue strength was further reduced above 10^6 cycles when the beams were immersed in sea water. Galvanizing restored the fatigue strength to their pre-twisting level.

It was concluded in this investigation that the zinc was acting as an anode to protect the steel at a crack. However, the action was considered to be time dependent, since it depended on sacrificial wastage of the zinc. It was also pointed out that beams with galvanized bars tested in sea water have failed due to fatigue of the concrete in compression.

2.3   Prestressing tendons

Fatigue of prestressing tendons is similar to that of bars in that it occurs as the result of initiation and propagation of a crack under cyclic loading. In Fig. 2-12, fractures of three seven-wire, $\frac{7}{16}$ in (11 mm) diameter strand are shown [45]. All of the wires in Fig. 2-12a have failed in tension, exhibiting the commonly referred to

(a)

(b)

Fig. 2-12 Tensile and fatigue fractures of seven-wire strand
  (a) All tensile fractures
  (b) Both tensile and fatigue fractures
  (c) All fatigue fractures
(By courtesy of Portland Cement Association)

Figure 2-12 (*contd.*)   (c)

'cup and cone' fracture surfaces. These wires also show a significant reduction in cross-sectional area at the fracture. All of the wires in Fig. 2-12c have failed in fatigue. Each has a small area of relatively flat fatigue crack growth. The remaining area that fractured abruptly is inclined to the axis of the wire. Figure 2-12b includes both tensile and fatigue fractures.

The ACI Committee 215 report [3] contains a thorough review of work on fatigue of prestressing tendons and anchorages up to 1974. Since that time, work has been quite limited, apparently because of the prevailing view that fatigue should not be a controlling factor in the design of tendons. In a pretensioned or bonded post-tensioned member, a sufficient level of prestressing to either prevent or limit the extent of flexural cracking has been considered as an adequate safeguard against fatigue. Many bridge members in the United States have been designed allowing a nominal tensile stress in the bottom fibers of $6\sqrt{f'_c}$ psi $(0.5\sqrt{f'_c}$ MPa). In an unbonded post-tensioned member, it is nearly impossible to have sufficient stress variation in the tendon to cause a fatigue failure.

Recently, full-scale tests were carried out on a series of bridge girders by Rabbat *et al.* [45] in which fatigue fractures occurred in the strands at about 3 million cycles of loading producing a nominal tension of $6\sqrt{f'_c}$ psi $(0.5\sqrt{f'_c}$ MPa) in the bottom fibers. The calculated minimum and maximum stresses in the strands at a crack located at the midspan of the girders was 142 000 and 151 000 psi (980 and 1040 MPa), respectively. As a result of these unexpected failures, substantial additional work on fatigue of strand is presently underway in the United States. These tests were conducted for the purpose of assessing the bond fatigue resistance of girders with 'blanketed' strand, as discussed in Section 3.

Prestressing steels are classified as wire, strand, and bars. Wires are drawn steels and strands are manufactured from wires. Drawing increases the tensile strength of the wire, and produces a grain structure which inhibits crack nucleation and provides a smooth surface which reduces stress concentrations. Bars are usually hot-rolled alloy steels. Typically, hot-rolled steels have a tensile strength of 160 000 psi (1102 MPa) while drawn wires have strengths ranging between about 250 000 and 280 000 psi (1722 and 1929 MPa).

Most fatigue tests on prestressing steels are conducted in air, without embedment in concrete, because of the difficulty in calculating or measuring the stress in steels in prestressed beams. However, tests on wire and strand in air require special end gripping arrangements.

2.3.1   Wire and strand

Warner [46] found good correlation of constant cycle fatigue data from tests in air in both prestressing wire and strand manufactured in Australia as well as on strand manufactured in the United States, as shown in Fig. 2-13. The fatigue limit corresponding to the minimum stress level for the data shown in Fig. 2-13 is given in Table 2-2.

The equation of the curve shown in Fig. 2-13 is

$$\log N = \frac{1.169}{R} + 5.227 - 0.031R \qquad (2-9)$$

where R equals $(f_{max} - f_{lim}) \times 100/f_{pu}$, f_{max} and f_{min} are the maximum and minimum stress levels, and f_{pu} is the tensile strength of the strand. In this investigation, it was noted that ambient temperature was a secondary but nevertheless significant factor affecting observed fatigue life in both the wire and strand tests. Currently, cumulative damage tests were being conducted, and Warner also noted that the

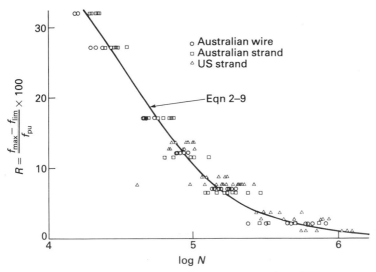

Figure 2-13   Comparison of wire and strand data [46]

Table 2-2

Source	Specimen	Minimum stress level $\dfrac{f_{min}}{f_{pu}} \times 100$	Fatigue limit $\dfrac{f_{lim}}{f_{pu}} \times 100$
Australia	Wire	40	53
Australia	Strand	40	53
United States	Strand	40	56.5
Australia	Wire	60	73
Australia	Strand	60	73
United States	Strand	60	73

results obtained were 'not seriously in conflict' with a linear damage hypothesis. It should also be noted that an estimate of the fatigue limit is needed before Eqn 2-9 can be used to calculate mean fatigue life. A further reduction in log N is needed for design to account for the dispersion of the data.

Naaman [47] has also evaluated European and United States test data on strand and presented the following equation for stress range, f_r, as a function of fatigue life:

$$f_r/f_{pu} = -0.123 \log N + 0.87 \qquad\qquad (2\text{-}10)$$

which apparently is intended to apply for values of f_r greater than approximately $0.05f_{pu}$ and f_{min} less than $0.6f_{pu}$. However, the only data below $0.13f_{pu}$ that were included in this evaluation were either from tests where f_{min} was zero or the strand failed in the grips. Although Naaman indicated that this equation can be considered a lower bound design line, further reduction would appear desirable.

All of the data included in the evaluation of wire and strand in the ACI Committee 215 report show that the fatigue strength at 2 million cycles, for minimum stress levels between 0.4 and $0.7f_{pu}$, exceeds $0.075f_{pu}$. Excluding some data from the USSR, Belgium and Japan, which was obtained from tests prior to 1969, the lowest fatigue strength at 2 million cycles was $0.13f_{pu}$.

2.3.2 Bars

The use of bars for prestressing is limited. Currently, only one system is available in the United States. This system has rolled-in deformations which make it possible to cut the bar at any point and screw on either an anchor or a coupler. The bars come in sizes ranging from $\frac{5}{8}$ in (16 mm) to $1\frac{3}{8}$ in (35 mm), with tensile strengths of 150 000 psi (1030 MPa) or 160 000 psi (1100 MPa).

Work previously reviewed in the ACI Committee 215 report [3] indicates that the fatigue strength of smooth bars is similar to those of strand. Tests on bars ranging between 1 and $1\frac{3}{8}$ in (25 and 35 mm) in diameter have shown that the fatigue limits of these bars are in excess of 0.1 times the tensile strength of the bar for 1 million cycles of loading at a minimum stress of 0.6 times the tensile strength.

2.3.3 Anchorages

The fatigue strength of anchorages is generally of concern only for unbonded construction, where stress changes in the steel are transmitted directly to the anchorage. There is wide variation in the fatigue strength of wire and strand in an anchorage. Test data as low as $f_r = 0.027f_{pu}$ are noted in the ACI Committee 215 report [3], although most data are in excess of $0.04f_{pu}$.

It is common practice in the United States to require verification of the fatigue behavior of a tendon assembly by test, as described in Section 4.1.

2.4 Anchor bolts

Anchor bolts are frequently used to transmit sizable loads into concrete members. In some cases, such as that reported in [1], the anchor bolts may be subjected to cyclic loading. As pointed out by Elfgren et al. [48], the most important factor governing the life of an anchor bolt is the stress range in the bolt. Prestressing reduces the stress range, and reductions of 50% or more can be achieved depending on the length and embedment. In critical usages, it is advisable periodically to check the level of prestress.

Frank [49] has recently completed an experimental investigation and review of the fatigue strength of anchor bolts. He concluded that the use of a Category E design condition in the AISC [39] or AASHTO [38] specifications was suitable for design of bolts without prestress; these specifications impose a limit for stress ranges not exceeding 5000 psi (34.5 MPa) for not more than 2 million cycles.

3 Behavior of concrete members under repeated loading

When a structural concrete element is subjected to a repetitive loading, the behavior will be dependent upon one or more of the predominant conditions, i.e. flexure, shear, torsion or bond. One of the earlier findings of research on fatigue was that the tensile cracking of the concrete occurs at lower levels of sustained or repeated load than under a static loading condition. To determine whether cracking will occur under repeated loading, a reduced value of tensile strength of the concrete should be assumed, as low as one-half of the value expected under a static condition. From a practical viewpoint, however, variations in loading or an occasional overloading may hasten the development of a full cracking pattern.

The initial response of an element will be inelastic to some degree, depending mainly on the development of cracking as discussed in the preceding paragraph, the level of the repeated loading, and on time-dependent deformation due to creep and shrinkage. Under a high-cycle, low-amplitude loading, the response of the element usually becomes both stable and elastic after a short period when flexure is predominant. However, when shear, torsion or bond stresses are significant, the response is likely to remain inelastic for a longer, substantial period. If in this period the element is submerged in sea water or other adverse environmental conditions exist, the response of the element may be affected. As discussed earlier, concrete has a shorter fatigue life in water than in air. Likewise, the fatigue strength of the reinforcement may be reduced in a corrosive environment.

Under a low-cycle, high-amplitude loading, the influence of time-dependent conditions along with the degradation of load-resisting mechanisms, such as crack interface shear transfer, the response should be expected to remain inelastic with increasing residual deformation. Gerwick [5] has discussed these effects on offshore structures.

3.1 Flexural fatigue

In tests conducted on beams with nonprestressed reinforcement, a single reinforcing element has commonly been used. Fatigue failures of these beams occur suddenly, with little sign of distress. Very few tests have been conducted on beams with multiple non-prestressed reinforcing elements. However, fatigue fractures of reinforcing bars were induced in bridges in the AASHTO road test [50] during special testing after the completion of the vehicular traffic tests. These beams may not exhibit significant distress before failure occurs, because of the high bond between the steel and concrete, unless there is sufficient redundancy in the overall structural system to permit redistribution of the loading.

Multiple reinforcing elements have generally been used in the tests on beams containing prestressed steel. Fatigue failures of these beams have occurred only after significant signs of distress in the form of the development of a very wide crack and increasing deflections. As an example [51], the load–deflection curve

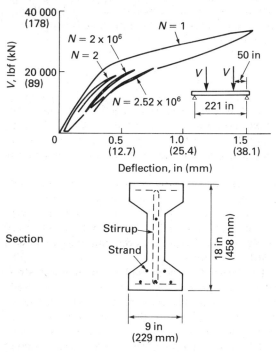

Figure 3-1 Deflection of a prestressed beam under static and cyclic loading (Reproduced from Hanson, J. M., Halsbos, C. L. and Vanhorn, D. A., 'Fatigue tests of prestressed concrete I-beams', *J. Struct. Div. ASCE*, Vol. 96, pp. 2443–2464, 1970).

for a pretensioned prestressed concrete I-beam in which a flexural fatigue failure occurred is shown in Fig. 3-1, along with a cross-section through the beam showing the location of the six $\frac{7}{16}$ in (11 mm) diameter seven-wire 270 000 psi (1860 MPa) grade strand. Compressive strength of the concrete was 7120 psi (49 MPa) at the time of test. This beam was subjected to an initial static loading of approximately 80% of the ultimate flexural capacity of the specimen, which was sufficient to fully develop flexural cracking and also cause significant inclined cracking in both shear spans. Next, the beam was subjected to 2 million cycles of a normal design loading with induced moments in the center of the span ranging between 19 and 45% of the flexural capacity of the specimens. Under this loading, the stress in the bottom fibers of the beam ranged between 890 psi (6.14 MPa) compression and 440 psi (3.0 MPa) tension, computed on the basis of an uncracked section. The tensile stress was approximately equal to $5.2\sqrt{f_c'}$ psi ($0.44\sqrt{f_c'}$ MPa). No damage was observed. The range of the cyclic loading was subsequently increased to between 19 and 50% of the flexural capacity. At the maximum load, the nominal tensile stress in the bottom fibers was 660 psi (4.5 MPa) or $7.8\sqrt{f_c'}$ psi ($0.65\sqrt{f_c'}$ MPa). The beam sustained 570 000 cycles of this increased above-design loading before the test was stopped because of extensive fatigue damage. The first indication of increased deflection corresponding to fatigue damage was evident at 455 000 cycles of the above design-loading. After the test was concluded, the prestressing strand was exposed and a total of 21 wire fractures were found in the three lower level strands. Four of these fractures were at locations other than at the major crack.

3.2 Shear

The effects of shear, torsion and bond are, in many cases, quite closely related. For example, the development of inclined cracking will, in the absence of sufficient shear reinforcement, lead to increased stress in the flexural steel, which may affect the bond strength. On the other hand, slip of reinforcement in an otherwise adequately reinforced member may induce a shear failure, particularly in prestressed construction.

A number of investigations of the shear fatigue behavior of both non-prestressed and prestressed beams have been reported. Chang and Kesler [52] observed that the shear fatigue strength of non-prestressed beams without web reinforcement was approximately 60% of their static shear strength. Fractures of stirrup reinforcement have been reported in fatigue tests on both non-prestressed and prestressed beams [51, 53–56].

Designers should recognize that inclined cracking will occur at lower stress under cyclic loading than under static loading. This, of course, is generally regarded as the limit state in a beam without web reinforcement. The number of cycles will be related to the tensile fatigue strength of the concrete. Non-prestressed beams without web reinforcement will therefore have a lower shear fatigue strength than prestressed beams. For design, the shear fatigue strength should probably not be taken greater than one-half of the static design shear strength. Prestressing contributes to the shear strength of a beam without web reinforcement by imposing a compressive stress in regions where inclined cracks may originate. The shear fatigue strength of these members may be estimated by reducing the portion of the contribution which is dependent on the tension in the concrete to one-half of the usual value.

Most structural beams will contain at least a minimum amount of shear reinforcement. However, this reinforcement does not delay the formation of the inclined cracks. As the cracks form, the stresses in the stirrups increase rapidly, until nearly all of the shear is carried by the stirrups. Tests have shown that fractures of the stirrup reinforcement occur where the inclined cracks cross the stirrups. The stress ranges in these stirrups under the cyclic loading, estimated by the usual design procedures, are substantially less than the stress ranges that cause fractures in tests which are reviewed in Section 2.2. Thus significant doweling forces are apparently induced in the stirrups, causing secondary bending stresses that lower the fatigue strength. Fractures have also been observed in tests near bends in stirrups if an inclined crack was developed at that location.

If inclined cracking occurs, the shear fatigue resistance will be related to the crack widths. As long as the crack widths are not large, the integrity of the shear resistance mechanisms will be maintained. Increasing crack widths will undoubtedly lead to failure.

For design in those situations where shear fatigue is of concern and a large number of cycles (more than one million) is expected, the recommendation of ACI Committee 357 [57] is appropriate, as follows:

> Where maximum shear exceeds the allowable shear on the concrete alone, and where the cyclic range is more than half the maximum allowable shear in the concrete alone, then all shear should be taken by the stirrups.

3.3 Torsion

Information on the torsional fatigue resistance of concrete members is limited. A recent investigation by Okada and Kojima [58] confirmed that the fatigue properties of plain and prestressed concrete under torsion were about the same as those of concrete under compression or flexure.

Unintended or unrecognized structural interactions may induce torsion in slabs and beams. These torsional forces may be surprisingly large in uncracked members, where the resistance is dependent on the tensile strength of the concrete. Under cyclic loading, the torsion may cause cracking that will subsequently contribute to the deterioration of the member.

In torsion, as in shear, it was found [58] that the primary stresses are redistributed to the bars after cracking, and that the torsional stiffness is reduced. However, failures occurred due to fatigue of the concrete rather than the steel.

3.4 Bond

Hawkins [55] has reviewed information and conducted tests on bond fatigue, leading to the conclusion that in the absence of cracks in the anchorage zone the bond strength for 1 million cycles will be about 60% of the static strength. When cracking occurs, the bond fatigue strength will be strongly dependent on shear effects.

Two areas are of particular importance for bond fatigue—railroad ties and prestressed beams with 'blanketed' strands.

Kaar and Hanson [59] have investigated the effect of subjecting rectangular beams simulating pretensioned concrete cross-ties to high bending moments near their ends. In this investigation, it was found that the repeated loading of a member with a crack in or near the transfer length can cause an early bond failure of the pretensioned strand. To insure long life, a tie should be designed so that any existing crack will not be opened by the expected repeated loading. In their tests, if an existing crack was allowed to open more than 0.001 in, the load could not be applied nearer the end than 2.2 times the strand transfer length for smooth $\frac{3}{8}$ in (9.5 mm) seven-wire strand in order to obtain a bond fatigue life of 3 million cycles. However, that distance could be reduced to the transfer length for strand that was lightly rusted or roughened by sandblasting.

Fatigue tests were conducted on full-size pretensioned girders with blanketed and draped strands by Rabbat *et al.* [45]. Blanketing means to unbond the strands for a specified length at the ends of the member. This procedure is frequently less costly than the practice of draping strand to reduce the tensile stress in the top of the member. However, in this case, the strands may be required to develop their bond strength in a region of flexural tension of the girder. Strand slip was observed in some of these tests, and it was concluded that the usual development length should be doubled when strand is debonded at the end of a member and tension at service load is allowed in the precompressed tensile zone.

4 Design provisions

Structural concrete members are normally proportioned to carry factored service loads. Fatigue is a serviceability condition that must be checked in the design process if the member is subjected to cyclic loading. This check is made by computing minimum and maximum stress levels under the anticipated cyclic loading at any potentially critical location and comparing the stress variation to the fatigue strength of the material. If there is variability in the cyclic loading, some method of accumulating the effect of the different load levels must be included. Miner's hypothesis, as expressed in Eqn 2-5, is frequently used for both the concrete and steel components of the member.

A satisfactory estimate of the stress levels in both the steel and the concrete can usually be made using the ordinary principles of flexural mechanics. However, these procedures become quite complicated for prestressed beams subjected to cyclic loading that induces cracking. These members are often referred to as partially prestressed. Aids have been developed [60, 61] for the analysis of these members.

While simple in concept, a check of a member for fatigue can become quite complicated when, for example, different load patterns are required to obtain maximum or minimum stress levels at a selected location. However, in many cases, the question will be whether or not the concrete member is on the threshold of fatigue distress. Hence it will be mainly important to project the number of cycles of maximum repeated loading which the member may conceivably resist during its design life.

Most building codes governing the design of concrete structures subjected to repeated loads now contain provisions for assessing fatigue. Some of these provisions are presented and briefly discussed in the following sections.

4.1   United States of America

Both the American Association of State Highway and Transportation Officials (AASHTO) and the American Railway Engineering Association (AREA) have adopted fatigue stress limits for design. Committees 215, 357 and 447 of the American Concrete Institute have also presented design recommendations.

The provisions in Section 1.5.38 of the 1977 AASHTO *Standard specifications for highway bridges* [38] require limitation of stresses at service loads in *non-prestressed* construction to the following:

(A) *Concrete*
The maximum compressive stress in the concrete shall not exceed $0.5f'_c$ at sections where stress reversals occur caused by live load plus impact at service load. This stress limit shall not apply to concrete deck slabs.

(B) *Reinforcement*
The range between a maximum tension stress and minimum stress in straight reinforcement caused by live load plus impact at service load shall not exceed:

$$f_f = 21.000 - 0.33f_{min} + 8000(r/h) \text{ in psi}$$

or

$$= [145 - 0.33f_{min} + 55(r/h)] \text{ in MPa}$$

where

f_f = stress range
f_{min} = algebraic minimum stress level, tension positive, compression negative
r/h = ratio of base radius to height of rolled on transverse deformation; when actual value is not known, use 0.3

Bends in primary reinforcement shall be avoided in regions of high stress range.

It should be noted that these provisions apply only to straight hot-rolled bars with no welds and with no stress raisers (including manufacturers' marks) more severe than deformations meeting the requirements of the American Society for Testing and Materials (ASTM) Designation A615.

Consideration of fatigue for prestressed construction in the AASHTO specifications is limited to anchorages and couplers of unbonded tendons, again apparently because adherence to static limitations is believed to preclude the possibility of fatigue failure. The requirement for unbonded tendons in Section 1.6.17 is as follows:

For unbonded tendons, a dynamic test shall be performed on a representative specimen and the tendon shall withstand, without failure, 500 000 cycles from 60 to 66% of its minimum specified ultimate strength, and also 50 cycles from 40 to 80% of its minimum specified ultimate strength. The period of each cycle involves the change from the lower stress level to the upper stress level and the back to the lower. The specimen used for the second dynamic test need not be the same used for the first dynamic test. Systems utilizing multiple strands, wires, or bars may be tested utilizing a test tendon of smaller capacity than the full size tendon. The test tendon shall duplicate the behavior of the full size tendon and generally shall not have less than 10% of the capacity of the full size tendon. Dynamic tests are not required on bonded tendons, unless the anchorage is located or used in such manner that repeated load applications can be expected on the anchorage.

4.2 Japan

The Japan Society of Civil Engineers introduced fatigue provisions for the design of reinforced concrete structures in 1967. New procedures were adopted in 1981 in the *Tentative recommendations for the limit states design of concrete structures.* These procedures have been summarized by Kakuta *et al.* [12] as follows:

(a) Fatigue limit states should be considered in the design of structures when the effects of variable loads are dominant.
(b) Design fatigue strength, which is obtained by dividing the characteristic strength by the appropriate partial safety factor for strength, should be larger than applied design variable stress.
(c) Characteristic fatigue strength shall be calculated from permanent stress and equivalent cycles, N, of the applied design variable stress. The equivalent cycles may be evaluated on the assumption of a linear cumulative damage theory, such as Miner's.
(d) Characteristic fatigue strength of concrete, f'_{rck}, may be calculated as follows:

$$f'_{rck} = (0.9kf'_{ck} - \sigma'_{cp})(1 - \log N/15)$$

where f'_{ck} denotes the characteristic static strength of concrete, σ'_{cp} the permanent compressive stress, and $k(=0.85)$ is a coefficient considering the difference of the concrete strengths between cylinder specimens and the structural members. Principally, the fatigue strength of reinforcing bars shall be determined on the basis of test results. The following values may be used for deformed bars with a diameter not larger than 32 mm:

$$f_{rsk} = (160 \text{ MPa} - \sigma_{sp}/3)10^{-0.2(\log N - 6)} \text{ for } \log N < 6$$
$$= (160 \text{ MPa} - \sigma_{sp}/3)10^{-0.1(\log N - 6)} \text{ for } \log N > 6$$

where σ_{sp} is the permanent tensile stress. However, the fatigue strength of bars with bends or welded connections must not be taken greater than half of the above value unless determined by test results.
(e) Applied design stresses due to flexure may be calculated by the elastic theory of a cracked section. However, the fatigue limit state for concrete in compression may be examined only for the stress at the location of compressive resultant. Applied design variable stress and permanent stress

in shear reinforcement can be calculated as follows:

$$\sigma_{wrk} = \frac{1.15(V_{md} - 0.5 V_{cd})s}{A_w d (\sin \alpha + \cos \alpha)(V_{rd}/V_{md})}$$

$$\sigma_{wrp} = \sigma_{wrk}(V_{pd}/V_{rd})$$

where V_{md} is the design maximum shear force, V_{cd} is the design ultimate shear force resisted by concrete, V_{pd} is the applied design permanent shear force, V_{rd} is the applied design variable shear force, A_w is the area of shear reinforcement within a distance s, d is the effective depth and α is the angle between shear reinforcement and the longitudinal axis of the member.

(f) Fatigue limit states for reinforced concrete beams may generally be examined only for longitudinal tensile reinforcement and shear reinforcement. Fatigue limit states for reinforced concrete slabs may generally be examined only for tensile reinforcement. The examination of fatigue limit state for reinforced concrete columns may generally be omitted.

4.3   Europe

Several European countries have codes containing fatigue design provisions. The following provisions are included in Appendix f of the 1978 CEB–FIP *Model code for concrete structures.*

f.1   *General*

Fatigue failure of a material is failure due to frequent repetition of stresses lower than its strength under static loading.

The stresses that are comparable with the fatigue strength should be determined by means of elastic methods, taking into account the dynamic effects, the effects of creep, the losses of prestress, etc. These stresses are defined as follows:

σ_{max} corresponding to a frequent action repeated $2 \cdot 10^6$ times at its maximum value;

σ_{min} corresponding either to a quasi-permanent action or to a frequent action repeated $2 \cdot 10^6$ times at its minimum value, as the case may be.

The condition to be checked is

$$\Delta\sigma = \sigma_{max} - \sigma_{min} \leqslant \Delta f_{rep}/\gamma_{fat} \tag{f.1}$$

Δf_{rep} denoting the strength under repeated load effects.

f.2   *Fatigue strength of the concrete*

The fatigue strength is defined as the 50% fractile deduced from test results.

f.3   *The fatigue strength of the steel reinforcement*

For the steel the characteristic strength is the 10% fractile and for the anchorage devices the 50% fractile, deduced from tests in which σ_{max} is repeated $2 \cdot 10^6$ times and where:

$\sigma_{max} = 0.7 f_{yk}$ (or $f_{0.2k}$) for reinforcing steel,

$\sigma_{max} = 0.85 f_{0.2}$ for prestressing tendons.

The reduction of the fatigue strength owing to curvature, welding, mechanical connections, end anchorages, etc. should be taken into account in the calculations, preferably on the basis of test results.

f.4 *Special considerations*

The task is to take into account all the stress concentrations which are liable to affect fatigue behavior but which are beyond the usual objectives of stress checks.

According to the notes which accompany Appendix f, the provisions represent a simplified approach of practical character which is appropriate for most conventional structures. Only in special cases is it necessary to check the cumulative effect of the repetitions at different stress levels, using the Palmgren–Miner rule for example, with appropriate limitations and a defined load spectrum, related to the expected life term of the structure. The random nature of the load repetitions is not taken into consideration.

Normally, the following numerical values are introduced:

For steel, γ_{fat} is applied to the characteristic value and is taken at 1.15, i.e. $\Delta f_{rep}/\gamma_{fat} = \Delta f_{sk}/1.15$.

For concrete, and for the anchorage devices, γ_{fat} is taken equal to 1.25 and applied to the mean value, i.e. $\Delta f_{rep}/\gamma_{fat} = \Delta f_{cm}/1.25$ or $\Delta f_{sm}/1.25$.

In the absence of test results or practical experience, the following lower limits are accepted ($\sigma_{min} = 0$) for Δf_{cm}:

0.6 for the stresses in the concrete and for bond stresses in high bond bars.
0.4 for the bond stresses in smooth round bars.

In the absence of test results, the following values can be adopted for Δf_{sk} ($\sigma_{min} = 0$):

Smooth bars	250 MPa
Prestressing tendons (without bond due to deformed shape)	200 MPa
Prestressing tendons (with bond due to deformed shape)	150 MPa
High-bond bars	150 MPa

In the absence of test results, Δf_{sk} may be reduced by the following coefficients:

Curvature: $(1-1.5 \varnothing r)$, where r denotes the radius of curvature
Spot welding: 0.4
Continuous seam welding: 0.4
Butt welding: 0.7

Bars of small diameter are to be preferred
The spacing between bars should not exceed
   $10\varnothing$ for the longitudinal reinforcement
   $5\varnothing$ for the transverse reinforcement

In general, checking tendons for fatigue in fully prestressed elements is not necessary. It should be noted however that the cracking moment can be reduced as a result of fatigue in tension of the concrete.

Acknowledgements

The assistance of Dr Luis F. Estenssoro in the preparation of this chapter is gratefully acknowledged. The author and publishers are also indebted to the Comité Euro-International du Béton for quotations from Appendix f—Fatigue from The *CEB-FIP International Recommendations*, 3rd Edn, 1978; to the International Association for Bridge and Structural Engineers for four figures from the *Proceedings of the Lausanne Colloquium*, 1982; and to Professor Y. Kakuta and colleagues for their procedural summary at that Colloquium. Articles 1.5.38 and the third paragraph of Article 1.6.17 on pages **16**–28/29 are from *Standard Specifications for Highway Bridges*, Twelfth Edition, Washington, D.C.: The American Association of State Highway and Transportation Officials, copyright 1977. Used with permission.

References

1. Hanson, J. M., Collapse of a cantilevered truss supporting a heavy crane, *Proc. Int. Ass. Bridge Struct. Eng. Colloq.*, Lausanne, 1982, *IABSE Rep.*, Vol. 37, pp. 569–576.
2. Okada, K., Okamura, H. and Sonoda. K., Fatigue failure mechanism of reinforced concrete bridge deck slabs, *Transp. Res. Rec.*, No. 664, 1978, pp. 136–144.
3. ACI Committee 215, Considerations for design of concrete structure subjected to fatigue loading, *J. Am. Conc. Inst., Proc.*, Vol. 71, No. 3, March 1974, pp. 97–121.
4. International Association for Bridge and Structural Engineering, Proceedings of Colloquium, Lausanne, 1982, *IABSE Reports*, Vol. 37, 895 pp.
5. Gerwick, B. G., High-amplitude low-cycle fatigue in concrete sea structures, *J. Prestressed Concr. Inst.*, Vol., 26 No. 5, Sept.–Oct. 1981, pp. 82–96.
6. Lenschow, R., Fatigue of concrete structures, *Proc. Int. Ass. Bridge Struct. Eng. Colloq.*, Lausanne, 1982, *IABSE Rep.*, Vol. 37, pp. 15–24.
7. Shah, S. P. and Chandra, S., Critical stress, volume change and microcracking of concrete, *J. Am. Concr. Inst., Proc.*, Vol. 65, No. 9, Sept. 1968, pp. 770–781.
8. Shah, S. P. and Chandra, S., Fracture of concrete subjected to cyclic and sustained loading, *J. Am. Concr. Inst., Proc.*, Vol. 67, No. 10, Oct. 1970, pp. 816–824.
9. Holmen, J. O., Fatigue of concrete by constant and variable amplitude loading, *Recent research in fatigue of concrete structures*, American Concrete Institute, SP-75, 1982, pp. 72–110.
10. Aas-Jacobsen, K., *Fatigue of concrete beams and columns*, Norwegian Institute of Technology, Bulletin No. 70–1, Trondheim, 1979, p. 148.
11. Tepfers, R. and Kutti, T., Fatigue strength of plain, ordinary, and lightweight concrete, *J. Am. Concr. Inst., Proc.*, Vol. 76, No. 5, May 1979, pp. 635–652.
12. Kakuta, Y., Okamura, H. and Kohno, M., New concepts for concrete fatigue design procedures in Japan, *Proc. Int. Ass. Bridge Struct. Eng. Colloq.*, Lausanne, 1982, *IABSE Rep.*, Vol. 37, pp. 51–58.
13. Siemes, A. J. M., Fatigue of plain concrete in uniaxial compression, *Proc. Int.*

Ass. Bridge Struct Eng. Colloq., Lausanne, 1982, *IABSE Rep.*, Vol. 37, pp. 283–292.

14. Ople, F. S. and Hulsbos, C. L., Probable fatigue life of plain concrete with stress gradient, *J. Am. Concr. Inst., Proc.*, Vol. 63, No. 1, Jan. 1966, pp. 59–81.

15. Cornelissen, H. A. W. and Reinhardt, H. W., Fatigue of plain concrete and uniaxial tension and in alternating tension–compression loading, *Proc. Int. Ass. Bridge Struct. Eng. Colloq.*, Lausanne, 1982, *IABSE Rep.*, Vol. 37, pp. 273–282.

16. Nordby, G. M., *A review of research—fatigue of concrete, J. Am. Concr. Inst., Proc.*, Vol. 55, No. 2, Aug. 1958, pp. 191–220.

17. Raithby, K. D., Flexural fatigue behavior of plain concrete, *Fatigue of engineering materials and structures*, Vol. 2, Pergamon Press, Oxford, 1979, pp. 269–278.

18. Hilsdorf, H. K., and Kesler, C. E., Fatigue strength of concrete under varying flexural stresses, *J. Am. Concr. Inst., Proc.*, Vol. 63, No. 10, Oct. 1966, pp. 1069–1076.

19. Miner, M. A., Cumulative damage in fatigue, *Trans. Am. Soc. Mech. Eng.*, Vol. 67, 1945, pp. A159–164.

20. Waagaard, K., Design recommendations for offshore concrete structures, *Proc. Int. Ass. Bridge Struct. Eng. Colloq.*, Lausanne, 1982. *IABSE Rep.*, Vol. 37, pp. 59–67.

21. Siemes, A. J. M., Miner's rule with respect to plain concrete variable-amplitude tests, *Recent research in fatigue of concrete structures*, American Concrete Institute, SP-75, 1982, pp. 343–372.

22. Hawkins, N. M. and Shah, S. P., American Concrete Institute considerations for fatigue, *Proc. Int. Ass. Bridge Struct. Eng. Colloq.*, Lausanne, 1982, *IABSE Rep.*, Vol. 37, pp. 41–49.

23. Helgason, T., Hanson, J. M., Somes, N. F., Corley, W. G. and Hognestad, E., *Fatigue strength of high-yield reinforcing bars*, National Cooperative Highway Research Program Report, No. 164, Transportation Research Board, Washinton, DC 1976, pp. 1–90.

24. Tilly, G. P. and Moss, D. S., Long endurance fatigue of steel reinforcement, *Proc. Int. Ass. Bridge Struct. Eng. Colloq.*, Lausanne, 1982, *IABSE Rep.*, Vol. 37, pp. 229–238.

25. Jhamb, I. C. and MacGregor, J. G. Stress concentrations caused by reinforcing bar deformations, *Abeles symposium: fatigue of concrete*, American Concrete Institute, SP-41, 1974, pp. 168–182.

26. American Society for Testing and Materials, *Standard specification for deformed billet-steel bars for concrete reinforcement*, A 615-81, Vol. 4, Philadelphia, 1981, pp. 597–602.

27. Tilly, G. P., Fatigue of steel reinforcement bars in concrete: a review, *Fatigue of engineering materials and structures.*, Vol. 2, Pergamon Press, Oxford, pp. 251–268.

28. Derecho, A. T. and Munse, W. H., Stress concentration at external notches in members subjected to axial loadings, *Bull. Ill. Univ. Eng. Exp. Stn*, No. 494, Jan. 1968, 51 pp.

29. Hanson, J. M., Burton, K. T. and Hognestad, Eivind. Fatigue tests of

reinforcing bars—effect of deformation pattern, *J. Res. Dev. Labs Portld Cem. Ass.*, Vol. 10, No. 3, Sept. 1968, pp. 2–13.

30. MacGregor, J. G., Jhamb, I. C. and Nuttall, N., Fatigue strength of hot-rolled reinforcing bars, *J. Am. Concr. Inst., Proc.*, Vol. 68, No. 3, March 1971, pp. 169–179.

31. Kokubu, Masatane and Okamura, Hajime, Fundamental study on fatigue behavior of reinforced concrete beams using high strength deformed bars, *Trans. Jap. Soc. Civ. Eng., Tokyo*, No. 122, Oct. 1965, pp. 1–28. [In Japanese with English summary].

32. Burton, K. T., Fatigue tests of reinforcing bars, *J. Res. Dev. Labs*, Vol. 7, No. 3, Sept. 1965, pp. 13–23.

33. Gronqvist, N.-O., Fatigue strength of reinforcing bars, *Second international symposium on concrete bridge design*, American Concrete Institute, SP-26, 1971, pp. 1011–1059.

34. Burton, K. T. and Hognestad, E., Fatigue rest of reinforcing bars—tack welding of stirrups., *J. Am. Concr. Inst., Proc.*, Vol. 64, No. 5, May 1967, pp. 244–252.

35. Pfister, J. F. and Hognestad, E., High strength bars as concrete reinforcement, Part 6: Fatigue tests, *J. Res. Dev. Labs.* Vol. 6, No. 1, Jan. 1964, pp. 65–84.

36. Nurnberger, U., Fatigue resistance of reinforcing steel, *Proc. Int. Ass. Bridge Struct. Eng. Colloq.*, Lausanne, 1982, *IABSE Rep.*, Vol. 37, pp. 213–220.

37. Sanders, W. W., Hoadley, P. G. and Munse, W. H., Fatigue behaviour of welded joints in reinforcing bars for concrete., *Welding J., Res. Suppl.*, Vol. 40, No. 12, 1961, pp. 529-s to 535-s.

38. American Association of State Highway and Transportation Officials, *Standard specifications for highway bridges*, 12th Edn., 1977, Washington, DC, 496 pp.

39. American Institute of Steel Construction, *Specification for the design, fabrication, and erection of structural steel for buildings*, Manual of steel construction, 8th Edn, Chicago, pp. 5–1 to 5–104.

40. Bennett, E. W. Fatigue tests of spliced reinforcement in concrete beams, *Recent research in fatigue of concrete structures*, American Concrete Institute, SP-75, 1982, pp. 177–193.

41. Tassi, G. and Magyari, B., Fatigue of reinforcements with pressed sleeve splices. *Proc. Int. Ass. Bridge Struct. Eng. Colloq.*, Lausanne, 1982, *IABSE Rep.*, Vol. 37, pp. 265–271.

42. Tilly, G. P., Fatigue of steel reinforcement bars in concrete: a review, *Fatigue of engineering materials and structures*, Vol. 2, Pergamon Press, Oxford, pp. 251–268.

43. Roper, H., Reinforcement for concrete substructures subject to fatigue, *Proc. Int. Ass. Bridge Struct. Eng. Colloq.*, Lausanne, 1982, *IABSE Rep.*, Vol. 37, pp. 239–245.

44. Hawkins, N. M. and Heaton, L. W., Fatigue characteristics of welded wire fabric, *Abeles Symposium: fatigue of concrete*, American Concrete Institute, SP-41, 1974, pp. 183–202.

45. Rabbat, B. G., Kaar, P. H., Russell, H. G. and Bruce, R. N., Jr., Fatigue tests of pretensioned girders with blanketed and draped strands, *J. Prestressed*

Concr. Inst., July–Aug. 1979, pp. 88–114.

46. Warner, R. F., Fatigue of partially prestressed concrete beams, *Proc. Int. Ass. Bridge Struct. Eng. Colloq.*, Lausanne, 1982, *IABSE Rep.*, Vol. 37, pp. 431–438.

47. Naaman, A. E. Fatigue in partially prestressed concrete beams, *Recent research in fatigue of concrete structures*, American Concrete Institute, SP-75, 1982, 25–46.

48. Elfgren, L., Cederwall, K., Guilltoft, K. and Broms, C. E., Fatigue of anchor bolts in reinforced concrete foundations, *Proc. Int. Ass. Bridge Struct. Eng. Colloq.*, Lausanne, 1982, *IABSE Rep.*, Vol. 37, pp. 463–470.

49. Frank, K. H. Fatigue strength of anchor bolts, *J. Struct. Div. Am. Soc. Civ. Eng.*, Vol. 106, No. ST6, June, 1980, pp. 1279–1243.

50. Highway Research Board, *AASHTO road test, Report* 4, *bridge research*, Special Report 61D, Publication No. 953, National Academy of Sciences, National Research Council, Washington, DC, 1962, 217 pp.

51. Hanson, J. M., Hulsbos, C. L. and Vanhorn, D. A., Fatigue tests of prestressed concrete I-beams, *J. Struct. Div. Am. Soc. Civ. Eng.*, Vol. 96, No. ST-11, Nov. 1970, pp. 2443–2464.

52. Chang, T. S. and Kesler, C. E., Static and fatigue strength in shear of beams with tensile reinforcement, *J. Am. Concr. Inst., Proc.*, Vol. 54, No. 12, June 1958, pp. 1033–1057.

53. Chang, T. S. and Kesler, C. E., Fatigue behavior of reinforced concrete beams, *J. Am. Concr. Inst., Proc.*, Vol. 55, No. 2, Aug. 1958, pp. 245–254.

54. Price, K. M. and Edwards, A. D., Fatigue strength in shear of prestressed concrete I-beams, *J. Am. Concr. Inst., Proc.*, Vol. 68, No. 4, April 1971, pp. 282–292.

55. Hawkins, N. M., Fatigue characteristics in bond and shear of reinforced concrete beams, *Abeles symposium: fatigue of concrete*, American Concrete Institute, SP-41, 1974, pp. 203–236.

56. Okamura, H. and Ueda, T. Fatigue behavior of reinforced concrete beams under shear force, *Proc. Int. Ass. Bridge Struct. Eng. Colloq.*, Lausanne, 1982, *IABSE Rep.*, Vol. 37, pp. 415–422.

57. ACI Committee 357, Guide for the design and construction of fixed offshore concrete structures, *J. Am. Concr. Inst., Proc.*, Vol. 75, No. 12, Dec. 1978, pp. 658–709.

58. Okada, K. and Kojima, T., Fatigue properties of concrete members subjected to torsion, *Proc. Int. Ass. Bridge Struct. Eng. Colloq.*, Lausanne, 1982, *IABSE Rep.*, Vol. 37, pp. 423–430.

59. Kaar, P. H. and Hanson, N. W., Bond fatigue tests of beams simulating pretensioned concrete crossties, *J. Prestressed Concr. Inst.*, Vol. 20, No. 5, Sept. 1975, pp. 65–80.

60. Naaman, A. E. and Siriaksorn, A., Serviceability based design of partially prestressed beams. Part 1. Analytical formulation, *J. Prestressed Concr. Inst.*, Vol. 24, No. 2, March–April 1979, pp. 64–89.

61. Siriarksorn, A. and Naaman, A. E., Serviceability based design of partially prestressed beams. Part 2: Computerized design and evaluation of major parameters, *J. Prestressed Concr. Inst.*, Vol. 24, No. 3, May–June 1979, pp. 40–60.

17 Composite construction in steel and concrete

Gajanan M Sabnis

Howard University, Washington, DC,
USA

Contents

Notation

a	See Sections 4, 5 and 6 lists
A	Numerical factor depending on the properties of the composite beam and on the design requirements, usually taken as 2.7
A_b	Area of bottom flange of plate girder
A_B	Area of rolled steel beam
A_c	See Sections 2 and 6 lists
A_c'	Transformed effective area of cracked slab $= y_{cc}b/kn$
A_{Ds}	Steel area required to resist dead loads acting on the steel section alone
A_{Dc}	Steel area required to resist dead loads acting on the composite section
A_{LL}	Steel area required to resist live loads
A_p	Area of steel cover plate
A_s	See Sections 2, 3, 4, 5 and 6 lists
A_s'	Area of compressive reinforcement
A_{sr}	Total area of longitudinal reinforcing bars
A_t	Area of top flange of plate girder
A_w	See Sections 2 and 6 lists
b	See Sections 2, 4 and 6 lists
b'	Width of top flange of steel beam
b_E	See Sections 3 and 4 lists
b_s	Width of steel flange

C_R, C_L M_R/wl^2 and M_L/wl^2 respectively
C_S Width of one deck cell
d See Sections 2, 4 and 5 lists
d'_d Depth of the deck cell
d_s Stud diameter
d_w Depth of web of plate girder
d'_2 Moment arm between forces in concrete and steel section
d''_2 Moment arm between forces in steel sections in compression and tension
D See Sections 5 and 6 lists
e Eccentricity of thrust on a cross section
e_c Distance from top surface of steel beam or girder to center of gravity of effective concrete slab
E Modulus of elasticity
E_c Modulus of elasticity of concrete
E_{ct} Tangent modulus of elasticity of concrete
E_s Modulus of elasticity of steel
E_{st} Tangent modulus of elasticity of steel
EI Flexural stiffness quantity
EI_{tan} Tangent to load vs flexural curvature diagram
f_a Allowable stress
f_b See Sections 2 and 4 lists
f_c See Sections 2 and 6 lists
f'_c Compressive strength of concrete 6×12 in. (150×300 mm) cylinder
f'_{cu} Index compressive strength of standard concrete cubes
f_s Stress on steel
f_t Stress at outermost top steel fiber
f_w Allowable stress in weld
F_b Allowable stress in bending in steel section
F_y Yield strength of steel section
F_{yr} Yield strength of reinforcing steel bars
h See Section 2 and 6 lists
h_s Stud height
I See Section 2, 5 and 6 lists
I_b Moment of inertia of bottom flange of plate girder
I_B Moment of inertia of rolled steel beam
I_c Moment of inertia of composite section
I_g Moment of inertia of gross cross section
I_p Moment of inertia of cover plate
I_s Moment of inertia of steel section
I_t Moment of inertia of top flange of plate girder

I_w Moment of inertia of web of plate girder
I_x Moment of inertia about the x axis of a cross section
I_y Moment of inertia about the y axis of a cross section
k See Sections 2 and 6 lists
kl Effective length of a column, taken as the distance between points of inflection in the deflected shape of the column at buckling
kl_c Characteristic length of a column above which elastic buckling should be expected and below which inelastic buckling occurs
K_1, K_2 Constants to obtain allowable stresses in tension without and with shores, respectively
l Length of a member
L Span length
M Moment
M_D Service load moment caused by loads *before* concrete has reached 75% of its required strength
M_0 Moment capacity of a cross section in the absence of thrust
M_L See Sections 4 and 5 lists
M_R Right support moment due to applied loads
M_u Ultimate moment
n modular ratio E_s/E_c
N_1 Number of shear connectors required between points of maximum moment and zero moment
p Spacing of shear connectors
P Thrust
P_c Elastic buckling thrust capacity of a cross section
P_{cr} Thrust at which a column buckles
P_1 Live-load thrust
P_0 Thrust capacity of a cross section in the absence of moment (squash load)
P_u Ultimate thrust
q Allowable shear in connector as per AISC Table 1-11.4
Q First moment of inertia of section
S See Sections 2 and 5 lists
S_b Section modulus of bottom flange
S_{bc} Section modulus of outermost bottom steel fiber for composite section
S_{bs} Section modulus of outermost bottom steel fiber for steel section
S_{cc} Section modulus of outermost concrete fiber for composite section
S_s See Sections 2 and 4 lists

S_t	Section modulus of top flange	α	Ratio between moments at opposite ends of a beam column
S_{tr}	Section modulus for composite section		
S_{tc}	Section modulus of outermost top steel fiber for composite section	β	Coefficient relating the compressive strength of concrete in members to the compressive strength of control cylinders or cubes
S_{te}	Section modulus of outermost top fiber for steel section		
S'_{bc}	Section modulus for bottom flange of composite section without cover plate in beam with cover plate cut off some distance from support	β_d	A coefficient for concrete creep effects used by ACI Building Code; taken as the ratio between design dead-load moment and total design load moment
S'_{be}	Section modulus for bottom flange of rolled beam alone in beam with cover plate cut off some distance from support	β_1	Beta factor for different strengths of concrete in strength calculations $\beta = 0.85$ for $f'_c \leqslant 4000$ psi and varies with straight line with smallest value of 0.65 for $f'_c \geqslant 8000$ psi
t	See Sections 2 and 6 lists		
t_b	Thickness of bottom steel flange of plate girder	β_2	Beta factor; similar to β_1 equals 0.425 for $f'_c \leqslant 4000$ psi and reduces by 0.025 for every ksi of additional concrete strength
t_c	See Sections 5 and 6 lists		
l_p	Thickness of bottom steel cover plate on rolled beam		
t_s	Slab thickness	δ	Moment magnification factor
t_r	Thickness of top steel flange of plate girder	Δ_{LT}	Long term deflection due to sustained load
t_w	Thickness of web of plate girder	Δ'_{ST}	Short term deflection due to sustained load
V_h	See Sections 3 and 4 lists		
V'_h	Actual capacity of shear connectors used; less than V_h	Δ_X	Deflection due to load at location X
V_r	Range of shear force (i.e. between maximum and minimum shears)	ε_u	Strain at maximum stress in concrete (=0.003)
w	Uniformly distributed load on beam	ϕ	Member capacity reduction factor
		ρ	Ratio of steel to total concrete
w_c	Density of concrete		
W_r	Average width of rib	*Section 2 list*	
y_{be}	Distance from neutral axis of composite section to outermost bottom steel fiber	A_c	Transformed effective area of concrete slab $= bt/kn$
		A_s	Total area of steel section
y_{bs}	Distance from neutral axis of steel section to outermost bottom steel fber	A_w	Web area of steel section; also, weld area per inch of weld
		b	effective width of concrete slab
y_{cc}	See Sections 2 and 4 lists	d	depth of rolled steel beam or plate girder
y_{cs}	See Sections 2 and 4 lists		
y_{sb}	Distance between centroid of steel deck to extreme tension fiber	f_b	Stress at outermost bottom steel fiber
y_{tc}	Distance from neutral axis of composite section to outer most top steel fiber	f_0	Stress at outermost top concrete fiber
		h	maximum thickness of channel flange
y_{ts}	Distance from neutral axis of steel section to outermost top steel fiber	I	Moment of inertia
		k	Numerical factor depending on the type of loading; for temporary loads $k = k_{LL} = 1$, and for sustained loads $k = k_{Dc} = 3$
$\bar{y}_c$	Shift in neutral axis from addition of concrete slab		
$\bar{y}_s$	Shift in neutral axis from addition to steel cover plate		
Z_r	Allowable range of horizontal shear	S	Horizontal shear at junction of slab and beam
Z_s	Plastic section modulus for steel shape or tube	S_e	Section modulus of outermost bottom and top steel fiber for symmetrical steel section

S_{eff}	Effective section modulus as defined by AISC
t	Thickness of concrete slab or thickness of web of channel connector
y_{cc}	Distance from neutral axis of composite fiber to outermost top concrete fiber
y_{cs}	Distance from neutral axis of steel section to outermost top concrete fiber

Section 3 list

A_s	Area of steel beam
b_E	Effective width of a flange of composite member
V_h	Total horizontal shear to be resisted by connectors

Section 4 list

a	Depth of rectangular compression block
A_s	Area of steel beam
b	Width of concrete beam in strength calculations
b_E	Effective width of a flange of composite section
d	Depth of steel section
f_b	Actual stress in bending in steel section
M_L	Service load moment caused by loads *after* concrete has reached its 75% of required strength
S_s	Section modulus of steel section
V_h	Design horizontal shear for composite section
y_{cc}	Distance of extreme compression fiber to neutral axis
y_{cs}	Distance of extreme tension fiber to neutral axis.

Section 5 list

a, b	Fractions of span at section where deflection is to be calculated
A_s	Area of tensile reinforcement
b	Width of the transformed composite section
d	Effective depth of cross-section
D	Total depth of composite section.
I	Moment of inertia
L	Span
M_L	Left support moment due to applied leads
n	Ratio of moduli of elasticity of steel and concrete
S	Section modulus
t_c	Concrete depth above top of steel deck

Section 6 list

a	Depth of cross-section area on which rectangular stress block of concrete is assumed to act
A_c	Area of concrete in a cross section
A_g	Area of an entire cross section bounded by gross exterior boundaries
A_s	Area of steel in a cross section
A_w	Area of the web of a steel beam
b	Width of a cross section
D	Outside diameter of a cross section
f_c	Stress on concrete
f'_c	Index compressive strength of standard concrete cylinders
f_y	Yield strength of steel
h	Total thickness or depth of a cross section
I	Moment of inertia of a cross section
I_s	Moment of inertia of steel in a cross section
k	Coefficient of length (story height) of a column
t	Thickness of segment
t_c	Thickness of concrete cover

Summary

This chapter deals with the development of composite construction in steel and concrete which has taken place in the past four decades, including the latest (1981) design practice for such construction. Since composite construction is used in buildings as well as bridges, design aspects as applied to these applications are considered. The various components of a structure and their behavior are considered to make the treatment complete. The action between steel and concrete takes place in compression or tension, or in flexure, all three of which are fully considered. One must not disregard the fact that composite action between these two materials is possible due to the (almost) perfect bond between them and to the transfer of stresses from one to the other. Shear connectors are provided to ensure this; they form a very important aspect and are treated in detail. Finally, the recent development of using composite construction in slabs using light-gauge metal decks and some of the applications are presented. Detailed design examples are included.

1 Introduction

1.1 Types of composite construction

Composite structures are generally made up with the 'interaction of different structural elements and may be developed using either different or similar structural materials'. This is often referred to as 'composite construction', and includes steel–concrete beams, columns (whether fully encased, partially encased, or held together by means of suitable connection) and other structural components. The interaction of different structural components—including beams, frames, shear walls, columns, slabs, and panels of a structure made of similar materials—is sometimes referred to as composite. The most common construction in buildings and bridges considered in this chapter is the composite steel–concrete, wherein a steel beam and a reinforced concrete slab (cast *in situ* or precast) are so interconnected with shear connectors that they act together as a unit.

The steel beam may be fully encased in concrete, partially encased, or placed below the slab. In some cases, the beam may have a concrete haunch above it. If concrete encasement is monolithic and of at least a specified minimum thickness, its natural bond with the steel beam will provide some composite action and additional stiffness as well as strength. To insure composite action, shear connectors such as studs, steel bars, or rolled shapes can be welded to the top flange of the steel beam and embedded in the concrete slab.

The principle of composite action in the stressed-skin 'continuum'-type of structure is to consider combining the different elements of the structure into one complex but integrated continuum in order to resist loads and external environmental forces. Instead of assigning each structural member a single, isolated, and specific task, the continuum attempts to use composite action by uniting all the structural members by means of proper connections, thus creating a single structural member that acts as a multidirectional load-carrying element in the total structure, e.g. an orthotropic steel bridge plate deck in which the metal sheet is stressed in various ways. The concept can be used for a simple roof structure consisting of metal deck, purlins, steel truss, and wind bracing. Brandel [1] suggested reshaping the material of the truss chord into a metal sheet to fulfill simultaneously the function of deck, purlins, partial truss, and bracings. Buckling considerations will require the corrugation of such a metal sheet.

1.2 Historical development

The history of composite construction dates back to the 1920s and is intimately linked with that of reinforced concrete and reinforced brick masonry, which are other familiar forms of composite structure. Knowles [2] described the early development of composite construction in buildings as fireproof floors, jack arch construction, and the fireproofing of steel joists by embedding them in concrete. Scott [3] and Caughey and Scott [4] described the early tests on encased beams and steel beams–concrete slabs conducted in the United Kingdom in 1914. Gillespie et al. [5] reported a series of tests of I-beams encased in concrete, which were conducted under the direction of the Dominion Bridge Company in Canada.

The early research on composite construction for bridges in the United States was conducted at the University of Illinois, Urbana. The work was summarized by Siess [6]. Viest [7] described the historical development related to AASHO (now AASHTO) code specifications for bridges and German specifications using pre-stressed slabs on steel beams. Later research, done mainly at Lehigh University, has been reported by Fisher [8] and Slutter [9]. Johnson [10] described work in the UK with an extensive list of references. For further details on this and related composite construction, the reader is referred to Sabnis [11].

1.3 Composite versus noncomposite action

Composite construction became generally accepted by engineers for bridges during the 1950s and for buildings during the 1960s. Since then, research on continuous composite beams and connections has led to even greater economy in such design due to increased strength through continuity and stiffness resulting from composite action, and thus enhancing its use in practice. In composite construction, local buckling of the compression flange is prevented by the use of shear connectors between the steel beam and the concrete slab.

Research as reported by Furlong [12], Gardener and Jacobson [13], and Stevens [14] has shown that concrete-filled tubes and cased rolled shapes also possess higher shear strengths than do reinforced concrete columns of comparable size. Dobruszkes et al. [15] and Roderick [16] reported that the connections between columns and composite beams can be made very ductile.

Okamoto [17] has summarized Japanese earthquake research in structural connections. Experience in Japan has shown that encasing steel sections in reinforced concrete is particularly beneficial for earthquake-resistant design. More research studies are needed on the suitability of other types of composite construction for earthquake-resistant design.

The level of fire protection also influences the economic choice to be made among structural steel, reinforced concrete, and composite construction. In bridges and multistory garages with spans of 30–35 ft (9–10 m), where the vulnerability of steel to fire is not a problem, steel beams acting composite with concrete slabs are more economic. Encasing steel columns in concrete is economic since the casing provides a substantial increase in strength. On the other hand, the encasement of steel beams in concrete contributes little to the strength of the composite beam, and at the same time requires additional reinforcement to control cracks and to hold beams in place during fire. In such steel beams, therefore, lightweight coating materials are relied upon for fire protection.

In frame construction, composite beams, columns, composite construction, or their combination may be used. Composite columns help to reduce the effective slenderness of a steel column, thus increasing its buckling load. Furthermore, the concrete encasement also carries its share of the load according to the actual behavior of the composite columns under load. Sometimes, in an otherwise reinforced concrete frame building, in order to maintain a constant size of columns over the entire height, composite columns could be used in the lower part of the structure. A composite column can be constructed without the use of formwork by filling a steel tube with reinforced concrete. This is especially advantageous when fire protection for the steel is not a critical factor.

Many comparisons between composite and other types of construction have been made in the last 30 years. The benefits of composite construction may vary among structures, location, and relative costs of materials and labour in a particular country. Some comparisons are presented here, which indicate the economics of such construction.

Davies [18] compared steel beam sizes for a typical, simply supported span of 27 ft (8m) to those for an office floor beam carrying identical dead and live loads for unshored construction. There are savings brought about by the reduced structural floor depth and reduced story height which, in turn, means reduced building volume for heating, cooling and reduction in fireproof casing and savings in overall dead loads of the building, which reduces foundation costs.

Wilenko [19] showed that the relative weight of composite, shored construction is 73% and that of a composite, prestressed steel beam is 55% of that required for noncomposite construction in a building frame. He further claimed that the prestressing method adds little to the construction costs. Johnson et al. [20] reported design studies of composite frames for buildings designed both elastically and plastically according to British specifications as (see Table 1-1).

Iwamoto [21] reported on the savings in both weight and costs in the case of a continuous composite bridge construction in Japan, for which the design should

Table 1-1   Comparative weight and cost for a three-bay six-story building frame

Type of frame	Weight %	Height %
Elastic noncomposite	100	100
Plastic noncomposite	95	102
Elastic composite	86	91
Plastic composite	66	90

Table 1-2 Comparison of three-span continuous bridge for spans of 105 ft (32 m)

Type of beam	Weight (3)	Cost %
Noncomposite	100	100
Composite	87	92

also be carried out for eliminating tension in the concrete deck under negative bending. His findings are summarized in Table 1-2.

Studies were carried out by Siess [6] for simple bridge spans from 30 to 90 ft (9–27 m), with beam spacings of 5–7 ft (1.5–2.2 m). Table 1-3 presents these comparisons in terms of steel weight, which includes an allowance for shear connectors:

Table 1-3 Relative weights of simply supported bridge span

Type of beam	Relative weight %
Noncomposite rolled beam	100
Composite symmetric rolled beam	
Without flange plates	
Unshored	92
Shored	77
With flange plates on bottom flange	
Unshored	76
Shored	64
Composite, using unsymmetric rolled section, unshored	82
Composite, using double T-section, unshored	82
Composite, using welded section	
Unshored	69
Shored	40–60

It can be seen from the Table 1-3 that a significant weight saving is possible by shored composite construction in which the dead load of steel and concrete is supported by temporary shores until the concrete is cured. In contrast, in unshored construction, the dead load of steel beam and cast-in-place concrete is taken by the steel beam alone. However, in some cases, shoring may be difficult, especially in bridges or floors of buildings where finishing operations are hindered. Several other methods of achieving economic savings in composite construction involve some form of prestressing and are presented later.

The structural advantages of composite versus a non-composite construction may thus be summarized as follows:

(a) The depth of steel beam is reduced to support a given load.
(b) An increase in the capacity is obtained over that of a noncomposite beam, on a static ultimate load basis (fatigue effects may reduce this increase).
(c) For a given load, a reduction in dead loads and construction depth reduces in turn the story heights, foundation costs, paneling of exteriors, and heating, ventilating, and air-conditioning spaces, thus reducing the overall costs of buildings. In bridges, embankment costs would be reduced. The amount of reduction, however, will vary from case to case.

On the other hand, there are several disadvantages. The design methods are more complex for composite than for noncomposite construction. In addition, if recent research related to the effective width of composite beams, load–slip relations, cracking due to temperature differential, creep and shrinkage effects, and complex interaction effects from low loads to failure is to be incorporated in limit-states design, the increase in design time may become an inhibiting factor. Over the past 30 years, the very large amount of theoretical and experimental research, design applications, and construction work carried out has shown the efficiency and economy of composite construction.

1.4 Importance of shear transfer in composite action

Composite action between steel and concrete or between structural elements of the same material implies interaction between them and a transfer of shear at the connection. In reinforced concrete, this is taken care of by the natural bond between concrete and the deformed reinforcing bars. In the case of fully encased steel joists or beams, there is a large embedded area for a shear transfer. In the common type of composite beam, there is some shear transfer by bond and friction at the interface between steel beam and concrete slab. It can not be depended upon if there is a single overload or pulsating load that will destroy such bond and cause a separation of the slab from the beam. Hence, shear connectors are needed to give reliable composite action with two objectives:

(a) To transfer shear between the steel and the concrete, thus limiting the slip at the interface so that the slab–beam system acts as a unit to resist longitudinal bending (with one neutral axis for the composite section).
(b) To prevent an uplift between the steel beam and concrete slab, i.e. to prevent separation of the steel and concrete at right angles to the interface.

In the case of reinforced concrete monolithic beam–slab construction, the longitudinal shear is taken up by the concrete and steel stirrup in the web. However, in a steel–concrete composite beam, there is a distinct possibility of longitudinal shear failure in the concrete slab portion. The relatively wide concrete slab is required to receive shear force from the steel beam along a narrow interface. This may cause unacceptably high shear stresses on planes in concrete close to the beam. This is especially critical in haunched concrete slabs above the beam, since the haunch being narrower than the slab itself may be a source of weakness. Considerations of longitudinal shear in concrete place limits on haunch dimensions, as shown in Fig. 1-1. It is important to provide at least

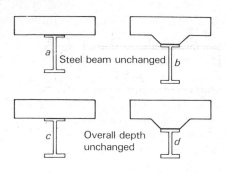

Overall depth of T-beam	$a < b$	$c = d$
Lever arm	$a < b$	$c < d$
Section of steel beam	$a = b$	$c > d$
Ultimate moment of resistance	$a < b$	$c = d$

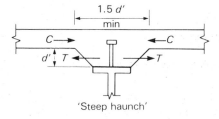

T = tension at roots of
    shear connectors
C = compression due to hogging
    $(-ve)$ moment in slab

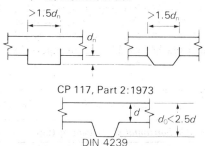

Figure 1-1   Haunch dimensions: longitudinal shear considerations (after Sabnis [11])

some transverse reinforcement at the lower face of the slab since it is in this area that stress concentrations occur at the shear connectors. Johnson [10] made a survey of a large number of load tests on composite beams up to and including failure and recommended some criteria for the design of steel for longitudinal shear.

Various types of shear connector have been used to resist longitudinal shear and uplift. They are rigid, flexible, bond-type, high-strength friction-grip bolts,

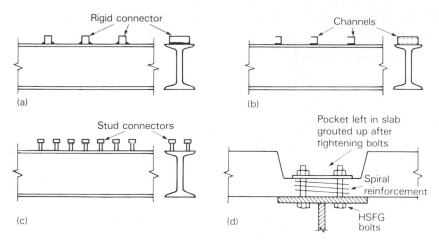

Figure 1-2   Types of shear connector
  (a) Rigid connector with steel bars
  (b) Flexible connector with channel
  (c) Flexible connector with studs
  (d) High-strength friction-grip (HSFG) bolt connector

and employ epoxy gluing between the two components, etc. Some of these are shown in Fig. 1-2. In a broad sense, connectors may be divided into two categories: rigid, and flexible. it must be pointed out that slip must occur before they are utilized; therefore, the terms are relative. The rigid type is the bar-like heavy connector. The flexible ones are the stud-and-channel type of connector. The rigid-bar or channel connector (Fig. 1-2a) is limited to shear transfer in one direction only, while the welded-stud connector (Fig. 1-2c) can resist and transfer shear in any direction perpendicular to the shank, making it the more useful connector. The welded-stud connector is particularly suitable for stringers in bridges.

Table 1-4   AISC design capacities for shear connectors (from AISC-1.11.4)[a]

| | Allowable horizontal shear load, q (kip) | | |
	$f'_c = 3000$ psi	$f'_c = 3500$ psi	$f'_c \geqslant 4000$ psi
$\frac{1}{2}$ in diameter, 2 in hooked or headed stud	5.1	5.5	5.9
$\frac{5}{8}$ in diameter, 2-$\frac{1}{2}$ in hooked or headed stud	8.0	8.6	9.2
$\frac{3}{4}$ in diameter, 3 in hooked or headed stud	11.5	12.5	13.3
$\frac{7}{8}$ in diameter, 3-$\frac{1}{2}$ in hooked or headed stud	15.6	16.8	18.0
3 in channel, 4.1 lb	4.3w[b]	4.7w	5.0w
4 in channel, 5.4 lb	4.6w	5.0w	5.3w
5 in channel, 6.7 lb	4.9w	5.3w	5.6w

[a] Applicable only to normal-weight concrete
[b] w is the length of channel in inches

Of the many types of shear connectors available, only channels, studs, and spirals are widely used in the United States. The AISC manual [22] lists allowable loads for both the channel and stud connectors (Table 1-4), and AASHTO [23] refers to the specifications on shear connectors as used in bridges. The ACI code [24] permits the use of closed loop spiral, which is quite useful in resisting the uplift.

1.5 Deflection characteristics

The deflection of a composite steel–concrete beam may be considered to be of three parts: (1) short-time dead load, (2) long-time creep and shrinkage, and (3) short-time live load. It is calculated using composite properties (transformed section) of steel and concrete and using a modular ratio, $n = E_s/E_c$. The modular ratio for deflection due to long-time creep and shrinkage (case 1), however, is taken to be higher than that for live load (case 2). In British specifications (CP117, [25]), the recommended value of n in case 1 and case 2 is 30 and 15, respectively, whereas in US practice it is recommended that a ratio of 3 (instead of 2, i.e. 30/15) be used to increase the value of n in case 1. (For example, see [9]). The higher modular ratio results in a smaller moment of inertia, and hence a larger calculated deflection, but only by a relatively small amount.

In computing deflections, it is necessary to take into account creep, shrinkage, thermal effects and the method of construction. When no temporary supports (props or shores) are used during casting and curing of the concrete slab, the dead loads are resisted by the steel beam alone. If the temporary supports are used during construction, then the dead loads are resisted by the composite section. If the dead loads are sustained, they cause creep, which is taken care of by increasing the conventional value of the modular ratio, n. This long-time effective modular ratio is taken as 2 and 3 times the short-time value for building and bridge design, respectively. Dead-load deflection may be one reason for excessive thickening of the slab in beams built without shores and excessive dishing of the slab beams with shores. Proper cambering must therefore be incorporated in design.

Proper design of composite floor systems is required to avoid the undesirable effects of vibrations. Steady-state vibrations are eliminated by insulating and damping the source. To alleviate transient vibrations, the following maximum depth-to-span ratios have been recommended by Viest [7]:

1 : 24 for ordinary building applications
1 : 20 for buildings where vibrations and shock are present
1 : 25 for bridges

Table 1-5 Reduction factors for connector capacities when using lightweight aggregate concrete (from AISC-1.11.4)

Air dry unit weight, (lb/ft^3)	90	95	100	105	110	115	120
Coefficient for $f'_c = 4.0$ ksi	0.73	0.76	0.78	0.81	0.83	0.86	0.88
Coefficient for $f'_c = 5.0$ ksi	0.82	0.85	0.87	0.91	0.93	0.96	0.99

In view of the damping characteristics of walls, partitions, and so on, depth is taken as the overall depth including the steel beam and the concrete slab. On the other hand, in open areas particularly susceptible to annoying vibrations, only the depth of the steel beam is considered in such limits.

1.6 Applications in buildings and bridges

From the discussion so far, it is clear that composite construction can have limitless possibilities and many applications. Although principles of composite construction do not vary in terms of application, the construction techniques and the applied loads influence the use of composite construction.

Some of the examples in buildings include Tears Towers, Chicago, One Shell Square, New Orleans, and Xerox Building Rochester, New York. Many more can be found in structures used as residential (apartment) buildings, office buildings, hotels, schools, multistory garages and combinations thereof.

Applications of bridges are found in many references including the *Handbook* [11] edited by the author. A few types are cited here to complete this section. These are

    (a) steel composite with concrete;
    (b) continuous steel-plate girder composite with concrete;
    (c) continuous welded plate girder composite with concrete;
    (d) Poured-in curved steel composite with concrete slab.

1.7 Applicable codes in composite construction

The design of structures using composite action between the two materials, namely steel and concrete, follow guidelines from the applicable codes. The major codes used in the United States are

    (a) AISC (structural steel in buildings) [22].
    (b) ACI (reinforced and prestressed concrete in buildings) [24].
    (c) AASHTO (highway bridges) [26].

Many other countries, Great Britain and West Germany in particular, also have their own code provisions for composite construction; however, specific reference herein will be made only to those just mentioned.

Although investigations of composite steel–concrete structures were carried out as early as the 1920s, the real impact on construction using such interaction did not occur until the mid-1940s, the main reason being the code provisions. The first formal code provisions for composite construction in the United States came about with the adoption of the AASHTO[1] specifications of 1944. Even with its limited provisions, this type of construction became more and more accepted in the highway bridges. At about the same time, i.e. near the end of World War 2, Germany developed its code of practice while achieving the most economic design using both steel and concrete for the construction of bridges and buildings destroyed in the war. It is interesting to note that the AASHTO code was much simpler in formulation and use than the German code; the complexity of the

1 Presently known as AASHTO (American Association of State Highway and Transportation Officials)

German code was a result of the inclusion of all possible methods to design the over-all structure as economically as possible due to the shortage of materials in the post-war years.

The inclusion of composite design for buildings occurred in the early 1950s when the AISC specifications were issued. Due to the earlier developments in highway bridges, composite construction in buildings was initially considered only for heavy industrial loads. However, starting with the 1961 editions of the AISC specifications, more provisions were made during different loading conditions. In the case of concrete-composite construction, the development took place following more and more use of precast and prestressed concretes. In combination with the cast-in-place concrete slab, they showed considerable advantages and economy, and the provisions were introduced in the ACI code for the first time in 1963 following recommendations of the ASCE–ACI Joint Committee 333 [31]. Later editions of the ACI codes [24] consisted of the more advantageous use of such construction with the development of light-weight concrete. It has become particularly important where long spans are required, whether in buildings or bridges, which cannot be achieved economically with non-composite structure.

AASHTO (1977) provides general instructions for designing composite structures for bridges between steel and concrete as follows: Sections 1.7.96 through 1.7.101 to steel concrete girders and 1.7.102 through 1.7.109 to box-girder composite structures.

AASHTO recommends the strength of a composite structure between steel and concrete, based on fatigue tests; however, it limits the ultimate strength value of the shear transfer by the lesser of the capacities of either steel or concrete based on the test results. The shear connectors are provided uniformly, based on the research by Slutter and Driscoll [27], which indicated no significant difference in the ultimate capacity of a beam under uniform loading and different spacing of these connectors. Later tests by Lew [28] also showed similar results. Lew pointed out that the uniform spacing of shear connectors showed a definitely lower strength in the case of transfer girders with very small shear-to-span ratio and recommended that the spacing may follow approximately the shear distribution in the span. AASHTO also deals with the design of multi-box as well as continuous girders, which have been used often in recent years. It also specifies the restriction of 6 ft (1.8 m) or 60% of spacing of the girders.

The 1980 edition of the AISC specifications [22] discusses the design of steel–concrete composite structure in buildings. In addition to giving the design details of this construction, it also gives useful design quantities related to section properties with standard thickness slabs of 4–5 in (100–125 mm) and suitable steel sections in the form of tables, which are helpful in the quick preliminary design of a composite section. Different values and conditions help the inexperienced engineer in making a reasonable guess and in reducing the number of design iterations before the final design of section. The type of construction—shored or unshored—should also be taken into account while using the aforementioned tables in AISC specifications.

The ACI code provisions [24] are similar to the AASHTO specifications discussed earlier. It allows the use of contact surfaces for shear transfer and specifies values of permissible shear stress in different surface conditions. The

ACI code, which is essentially the same as the AISC specifications, recommends the use of all loads acting on the beam regardless of the type of construction in computing total horizontal shear. The ACI code further provides the design information to calculate deflections and to maintain cracking control, similar to the rest of the reinforced concrete construction.

2 Properties of a composite section

2.1 Components of a section

In designing the steel–concrete composite section, it is generally necessary to determine the concrete slab thickness from other considerations and then to find out the steel section, either symmetric or unsymmetric, to make certain that stresses at no stage exceed the allowable limits. Thus, in this procedure one must: (a) obtain the thickness of the concrete slab as necessary from transverse load distribution between the steel beams; (b) determine effective width of this slab acting in a composite manner, as described in Sections 1.4 and 4.1, (c) determine the properties of this section; and (d) determine the stresses at various sections to ensure the safety. While the last step is covered in Section 4, this section deals with properties of the composite section.

It is simpler to assume the symmetric section so that standard tables can be used. However, when it is necessary to use a built-up section, it may be formed as welded girder or available I-section with plate at the bottom, thus forming an asymmetric section.

2.2 Detailed properties of a composite section

In order to compute stresses at a later stage, one needs the main flexural properties of a section, which are related to centroid, section modulus, moment of inertia of the section and so on. Twenty-five years ago, Viest, Fountain and Singleton [29] produced a reference on composite construction, which soon became out-of-print. The following material is presented from their work to enable the reader compute these properties.

A typical section of a plate girder, a standard section and a composite beam are shown in Fig. 2-1, with detailed dimensions. The following formulas can easily be written:

For a rolled section (Fig. 2-1a):

$$
\left.
\begin{aligned}
&K_s = A_p/A_s \\
&\bar{y}_s = \tfrac{1}{2}(d + t_p)K_s \\
&I_s = \tfrac{1}{2}(d + t_p)\bar{y}_s A_B + I_B \\
&\qquad \text{(neglecting small quantity } I_p) \\
&y_{ts} = d/2 + \bar{y}_s \\
&y_{bs} = d/2 + t_p - \bar{y}_s
\end{aligned}
\right\} \qquad (2\text{-}1)
$$

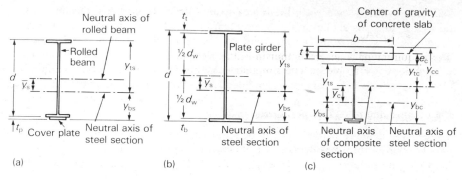

Figure 2-1   Details of the components of a composite beam (after Viest *et al.* [29])

(a) Rolled beam with cover plate
(b) Plate girder
(c) Composite beam

For plate girder (Fig. 2-1b)

$$\left.\begin{aligned}
\bar{y}_s &= \frac{\frac{1}{2}(d_w+t_b)A_b - \frac{1}{2}(d_w+t_t)A_t}{A_s} \\
I_s &= \tfrac{1}{4}(d_w+t_t)^2 A_t + \tfrac{1}{4}(d_w+t_b)^2 A_b + I_w - A_s\bar{y}_s^2 \\
&\quad \text{(neglecting small quantities } I_t \text{ and } I_b) \\
y_{ts} &= \tfrac{1}{2}d_w + t_t + \bar{y}_s \\
y_{bs} &= \tfrac{1}{2}d_w + t_b - \bar{y}_s
\end{aligned}\right\} \tag{2-2}$$

For composite section (Fig. 2-1c)

$$\left.\begin{aligned}
n &= E_s/E_c \\
k &= \text{numerical factor depending on the type} \\
&\quad \text{of loading, equal to 1 for temporary} \\
&\quad \text{loads, and 3 for sustained loads} \\
A_c &= bt/kn \\
k_c &= A_c/(A_c+A_s) \\
\bar{y}_c &= (y_{ts}+e_c)k_c \\
I_c &= (y_{ts}+e_c)\bar{y}_c A_s + I_s + A_c t^2/12 \\
y_{tc} &= y_{ts} - \bar{y}_c \\
y_{bc} &= y_{tc} + \bar{y}_c \\
y_{cc} &= y_{tc} + e_c + t/2
\end{aligned}\right\} \tag{2-3}$$

Equation 2-3 for composite properties can be used to determine stresses with sufficient accuracy even with the neutral axis in the slab, provided that

$$d/t \geqslant A_c/3A_s$$

This requirement is generally met in bridge design.

Equations 2-1 to 2-3 can be simplified suitably for a rolled section without a plate $(A_p = 0)$ and for symmetrical welded plate girders $(A_t = A_b$ and $t_t = t_b)$.

2.3 Approximate section properties

The equations developed earlier can be approximated and developed in the form of graphs. When the neutral axis of a composite section is located below the slab, the following properties can be developed from Eqn 2-3, when a *rolled beam* is used:

$$
\left.
\begin{aligned}
&\bar{y}_c/d = (\tfrac{1}{2} + e_c/d)k_c \\[2mm]
&\frac{I_s}{A_s d^2} = \left(\frac{1}{2} + \frac{e_c}{d}\right)^2 k_c + \frac{I_s}{A_s d^2} + \frac{1}{12}\frac{A_c}{A_s}\left(\frac{t}{d}\right)^2 \\[2mm]
&\frac{S_{bc}}{A_s d} = \frac{1}{\tfrac{1}{2} + (\bar{y}_c/d)} \cdot \frac{I_c}{A_s d^2} \\[2mm]
&\frac{S_{tc}}{A_s d} = \frac{1}{\tfrac{1}{2} - (\bar{y}_c/d)} \cdot \frac{I_c}{A_s d^2}
\end{aligned}
\right\} \tag{2-4}
$$

The last quantity in the second equation of 2-4 can be neglected, while $I_s/A_s d^2$ depends on the shape of the cross-section; for wideflange, its value is approximately 0.165, which gives

$$\frac{S_{bs}}{A_s d} = \frac{S_{ts}}{A_s d} = \frac{S_s}{A_s d} = 2\frac{I_s}{A_s d^2} \approx 0.33$$

and

$$\frac{I_c}{A_s d^2} \approx \left(\frac{1}{2} + \frac{e_c}{d}\right)^2 k_c + 0.165$$

The expressions for the section modulus result in

$$\frac{S_{bc}}{A_s d}, \frac{S_{tc}}{A_s d} \approx \frac{1}{\dfrac{1}{2} \pm \left(\dfrac{1}{2} + \dfrac{e_c}{d}\right)k_c}\left[\left(\frac{1}{2} + \frac{e_c}{d}\right)^2 k_c + 0.165\right] \tag{2-5}$$

where $k_c = A_c/(A_c + A_s)$.

The graphical solution of Eqn 2-5 for the bottom and top-section moduli are shown in Figs 2-2 and 2-3. Several values of e_c/d are shown for a practical range; for other values, interpolation is found to be satisfactory.

Similar approximations are possible when an asymmetric welded plate girder is used. Thus, if

$$d_w + t_b \approx d_w + t_t \approx d \approx d_w$$

the following approximate equations are developed for the bottom, S_{bc}, and top,

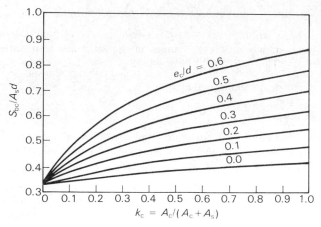

Figure 2-2   Rolled beams: bottom flange, Eqn 2-5 (after Viest *et al.* [29])

S_{tc}, flange-section moduli of the composite beam:

$$\left.\begin{array}{l}\dfrac{S_{bc}}{(A_w+2A_b)d_w}\approx\dfrac{\left(\frac{1}{2}k_b+\frac{e_c}{d_w}\right)^2\frac{k_c}{k_b}+\dfrac{S_{bs}}{2(A_w+2A_b)d_w}k_t}{\frac{1}{2}k_t+\left(\frac{1}{2}k_b+\frac{e_c}{d_w}\right)k_c} \\[4ex] \dfrac{S_{tc}}{(A_w+2A_t)d_w}\approx\dfrac{\left(\frac{1}{2}k_b+\frac{e_c}{d_w}\right)^2\frac{k_c}{k_t}+\dfrac{S_{ts}}{2(A_w+2A_t)d_w}k_b}{\frac{1}{2}k_b-\left(\frac{1}{2}k_b+\frac{e_c}{d_w}\right)k_c}\end{array}\right\} \qquad (2\text{-}6)$$

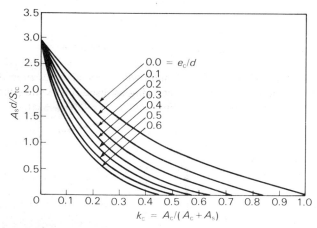

Figure 2-3   Rolled beams: top flange, Eqn 2-5 (after Viest *et al.* [29])

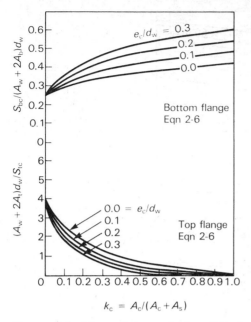

Figure 2-4 Plate girders: $A_t/A_w = 0.0$, $A_b/A_w = 0.5$ (after Viest *et al.* [29])

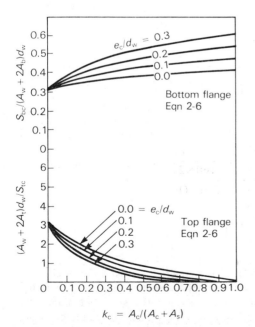

Figure 2-5 Plate girders: $A_t/A_w = 0.3$, $A_b/A_w = 0.5$ (after Viest *et al.* [29])

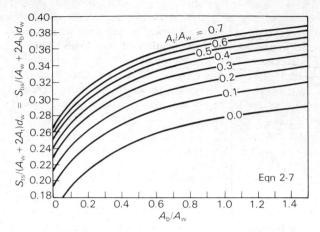

Figure 2-6 Plate girders, Eqn 2-7 (after Viest *et al.* [29])

where S_{bs} and S_{ts} for the steel section are given by

$$\frac{S_{bs}}{(A_w+2A_b)d_w}=\frac{S_{ts}}{(A_w+2A_t)d_w}\approx\frac{1}{2}\left(\frac{\frac{1}{6}+\frac{A_t}{A_w}}{1+2(A_t/A_w)}+\frac{\frac{1}{6}+\frac{A_b}{A_w}}{1+2(A_b/A_w)}\right) \tag{2-7}$$

and

$$\left.\begin{array}{l}k_b=\dfrac{1+2(A_b/A_w)}{1+(A_b/A_w)+(A_t/A_w)}\\[2.5ex]k_t=\dfrac{1+2(A_t/A_w)}{1+(A_b/A_w)+(A_t/A_w)}\\[2.5ex]k_c=\dfrac{A_c}{A_c+A_s}\end{array}\right\} \tag{2-8}$$

The graphical solutions of Eqns 2-6 and 2-8 for typical values of A_b/A_w and A_t/A_w are shown in Figs. 2-4 and 2-5 and of Eqn 2-7 in Fig. 2-6. With the availability of programmable calculators, it may be advantageous to program these equations for their easy use in the design office.

3 Design of shear connectors

3.1 Types of connector

The horizontal shear that develops between the concrete slab and the steel beam during loading must be resisted so that the composite section acts monolithically. Although the bond developed between the slab and the steel beam may be significantly high, it cannot be depended upon to provide the required interaction. Neither can the frictional force developed between the slab and the steel beam.

This natural bond between the concrete slab and the top surface of the steel beam, in the absence of a provision specially designed to preserve it, may be progressively destroyed due to the repeated application of live loads. This is particularly true for bridge structures where, in addition to the high impact values, there is also the possibility of a physical separation of the concrete slab from the steel beam due to uplift forces generated along the composite beam by certain dispositions of the live load. These uplift forces will also contribute to the rapid deterioration of the natural bond between the concrete and steel contact faces if

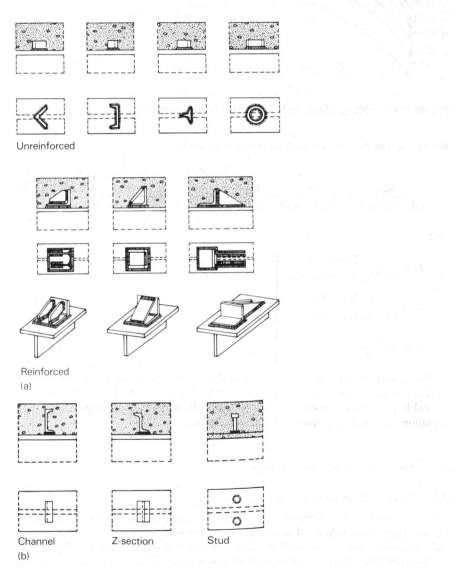

Unreinforced

Reinforced
(a)

Channel   Z-section   Stud
(b)

Figure 3-1   Various types of shear connector: (a) rigid and (b) flexible

devices designed to anchor down the concrete slab are not provided. In all such cases, mechanical shear connectors are provided to resist all the shear forces and to ignore the effect of natural bond altogether. This is a conservative assumption as the mechanical shear connectors will also help to preserve the natural bond to a great extent.

Mechanical shear connectors may be divided into three categories: rigid, flexible and bond. The rigid type of connectors (Fig. 3-1a) commonly used consists of short lengths of bars, angles or tees welded on to the top flange of the steel beam. This type of connector must, in general, also incorporate some device to anchor down the concrete slab against the uplift forces already mentioned. It consists of small $\frac{1}{4}$ in (6 mm) round bars threaded through holes provided in the T-connectors.

Among flexible type connectors (Fig. 3-1b), the most popular forms are the stud and the channel connectors. In these connectors, the enlarged head of the stud and the unattached flange of the channel provide their own holding-down device against uplift of the concrete slab.

3.2 Design of shear connectors

It may be noted that the connectors and the beam must resist the same ultimate load. However, under service loads, the beam resists dead and live loads, and when shores are used, the connectors resist essentially the live load. Using the working stress method, one might design the connectors only for live load; however, an increased factor of safety must be used, since the ultimate capacity would otherwise be inadequate.

AISC-1.11 [22] also uses an ultimate-strength concept but converts both the forces to be designed for and the connector capacities into the service-load range by dividing them by a suitable safety factor, with a nominal value of 2. Thus, for design under service loads,

$$V_h = \tfrac{1}{2}C_{max} = 0.85f'_c A_c/2 \qquad (3\text{-}1)$$

which is AISC formula 1.11-3, where $A_c = b_E t_S$, the effective area of concrete obtained by multiplying the effective width, b_E, and thickness of slab, t_S. Another form of this equation is the one using maximum tension resisted by steel A_s:

$$V_h = \tfrac{1}{2}T_{max} = \tfrac{1}{2}A_s F_y \qquad (3\text{-}2)$$

which is AISC formula 1.11-4. In Eqns 3-1 and 3-2, V_h is the horizontal shear to be resisted between the points of maximum positive moment and zero moment; the smaller of Eqn 3-1 and 3-2 is to be used.

AISC allowable values are given in Table 1-4 for hooked or headed studs and channels. Since the ultimate capacity expressions for hooked or headed studs are valid for $H/d_s \geqslant 4$, the values in Table 1-5 are also applicable to studs longer than the lengths indicated in the table.

When lightweight aggregate concrete is used, the connector values in Table 1-4 are to be multiplied by the coefficients given in Table 1-5.

The number, N_1, of connectors required is obtained by dividing the smaller value of V_h (Eqn 3-1 and 3-2) by allowable shear per connector:

$$N_1 = \frac{\text{smaller } V_h}{q} \tag{3-3}$$

where q is allowable load from Table 1-4.

AASHTO-1.7.100 uses the ultimate-strength concept directly (i.e. without dividing by a factor to convert the computation into a nominal service-load). However, the strength calculation is not used as the sole design procedure but rather as an additional check after determining the shear connectors required for the fatigue criterion. The fatigue requirement is an elastic procedure based on a slip limitation.

Adequately anchored, longitudinal reinforcing bar steel within the effective width of the concrete slab may be assumed to act compositely with the steel beam. The total horizontal shear to be resisted by the shear connectors between an interior support on a continuous beam and each adjacent point of contraflexure equals the maximum tensile force that can be developed in the reinforced concrete slab, i.e. neglecting the tensile capacity of the concrete,

$$T_{\text{slab}} = A_{sr} F_{yr}$$

where A_{sr} is the total area of longitudinal reinforcing bar steel at the interior support located within the effective flange width, and F_{yr} is the specified minimum yield strength of the longitudinal reinforcing steel.

In working stress design, the ultimate shear force, T_{slab}, developed between maximum negative moment and point of contraflexure is divided by 2 to bring it into the service-load range:

$$V_h = \tfrac{1}{2} T_{\text{slab}} = \tfrac{1}{2} A_{sr} F_{yr} \tag{3-4}$$

It is logically presumed that the tensile capacity of the reinforced concrete slab will be less than the tensile capacity of the steel beam, so that, for negative moment, only Eqn 3-4 is used.

In the elastic approach to shear connector design, the connectors are distributed according to the variation in horizontal shear between the slab and the steel beam. The connector capacities would be based on a limitation on slip. Formerly, design rules for both buildings and bridges used this approach with Eqns 3-1 and 3-2. When this elastic method was used where slip was limited, fatigue was not a controlling factor. The number of connectors required was conservatively large, indicating more than enough to develop the ultimate flexural strength of the composite member. If the shear connector requirements are reduced to the number required to develop the ultimate flexural strength, fatigue failure then may become the governing factor. Present bridge design considers both fatigue strength as well as ultimate strength.

The AASHTO requirements for fatigue are based largely on the work of Slutter and Fisher [30]. For fatigue, the 'range' of stress rather than the magnitude is the major variable influencing fatigue strength. Fatigue strength may be expressed as

$$\log N = A + BS_r \tag{3-5}$$

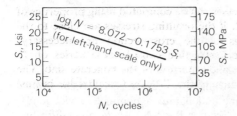

Figure 3-2 Fatigue strength of stud shear connectors (after Slutter and Fisher [30])

where S_r is the range of horizontal shear stress, N the number of cycles to failure, and A and B are empirical constants. The equation used for design is shown in Fig. 3-2.

Since the magnitude of shear force transmitted by individual connectors when service loads act on a structure agrees well with that predicted by elastic theory, the horizontal shear may be calculated as VQ/I. Fatigue is critical under repeated applications of service load; thus, it is reasonable to determine variation in shear stress using elastic theory. Using Eqn 3-1 for static load

$$\frac{VQ}{I} = \frac{\text{Allowable load, } q}{p} \tag{3-6}$$

For a cyclic load,

$$\frac{(V_{max} - V_{min})Q}{I} = \frac{\text{Allowable range, } Z_r}{p} \tag{3-7}$$

where p is the connector spacing. AASHTO-1.7.100(a) gives Eqn 3-7 as

$$S_r = V_r Q/I \leqslant Z_r \tag{3-8}$$

where

$$V_r = V_{max} - V_{min}$$

$$Z_r = d_s^2 \text{ for welded studs}$$

$$= 13\,000 \text{ for } 100\,000 \text{ cycles}$$

$$10\,600 \text{ for } 500\,000 \text{ cycles}$$

$$7850 \text{ for } 2\,000\,000 \text{ cycles.}$$

4 Flexural design of composite beams

4.1 Selection of beams for buildings and bridges

The selection of a cross-section may be made on the basis of dead and live loads (including impact) only. The dimensions of the slab are determined by the design for transverse load distribution and by the layout; only the steel section need be determined by trial.

With the steel section selected, the stresses can be computed using properties of the section determined as per Section 2. If the resulting stresses are close to the allowable values, no further revisions of the cross-section are necessary.

For design purposes, a composite-deck structure consists of a series of *T*-beams, each made up of one steel beam and a portion of the concrete slab. The width of slab assumed effective as the flange of the *T*-beam (except edge beams) must not exceed any of the following:

(a) one-fourth of the span of the beam;
(b) the distance center-to-center of the beams;
(c) twelve times the least thickness of the slab.

The flange of an edge beam is divided by the steel beam into two parts—inside and outside. The effective width of either part must not exceed one-twelfth of the span of the beam or six times the slab thickness. In addition, the effective width of the inside part must not exceed one-half the distance center-to-center of the beams, and the effective width of the outside part must not exceed the actual width.

With the provision of shear connectors between the steel beam and the concrete slab, one can assume a full composite section and these *T*-beams may be designed by the theory of the transformed section. It is customary to transform the effective cross-sectional area of the concrete slab into an equivalent steel area by dividing the effective slab area by the modular ratio *n*, and then the design calculations are carried out as for a monolithic section. The calculated stresses are kept lower than the allowable values for the respective materials.

Three types of load are considered in the design of composite construction similar to others:

(a) live loads;
(b) dead loads (including shoring);
(c) deformational loads (such as creep, shrinkage, expansion of concrete, and differential temperature changes).

Generally, the first two are adequate. The third is considered in some of the provisions by modifying the value of *n*.

The live loads on highway bridges are the truck loads. The distribution factors accounting for wheel loads are given in the AASHTO specifications for bridge design in the United States. For longer bridges, equivalent uniformly distributed lane loads with one concentrated load are often used in place of the wheel loads. Regardless of which type of loading is selected, the live load for bridges must include an allowance for impact.

In buildings, live loads are the occupancy loads and, generally, are considered to be uniformly distributed.

The live loads are always carried by the composite section. They are usually of short duration. Thus, when the live-load stresses are computed, the properties of the composite section should be evaluated with the effective concrete area transformed by dividing only by the modular ratio $n = E_s/E_c$. Long-term loading, such as live loads in warehouses and from storage tanks, involves the use of a multiplier for *n* to account for item (c) above.

4.2 Elastic approach to design

The actual stresses that occur under service load in a given composite member depend on the manner of construction. The slab formwork must be supported by either the steel beam acting alone or by temporary shoring which would also support the beam. When temporary shoring is used, service-load stresses will be lower than when such shoring is not used, since all loads will be supported by the composite section. If the system is built without temporary shoring, the steel beam alone must support itself and the slab without the benefit of composite action.

For economic construction, it is desirable to avoid the use of shoring wherever possible. It can be shown that, no matter which construction system is used, the ultimate-moment capacity is identical. It is a simple procedure, therefore, to design as if the entire load is to be carried compositely (i.e. assume shores are used) even when shores are not to be used. Strength is assured; however, it is necessary to insure that the stress in the steel beam does not approach too closely the yield stress under service-load conditions.

In order to resist loads compositely, the composite strength must be adequately developed. AISC-1.11.2.2 requires that 75% of the compressive strength, f_c', of the concrete must be developed before composite action may be assumed.

The 1980 AISC design procedure for flexure [22] may be summarized by the following four steps:

(1) *Select section as if shores are to be used*   The required composite-section modulus S_{tr} with reference to the tension

$$S_{tr}(\text{reqd}) = (M_D + M_L)/F_b \qquad (4\text{-}1)$$

where

M_D = the service-load moment caused by loads applied prior to the concrete
M_L = the service-load moment caused by loads applied after the concrete has reached 75% of its required strength
F_b = allowable service-load stress, $0.66F_y$ for positive moment regions (where sections are exempt from the 'compactness' requirements of AISC-1.5.1.4.1).

Lateral support is assumed to be adequately provided by the concrete slab and its shear connector attachments.

(2) *Check AISC formula 1.11-2*   When shores are actually not to be used, it must be assured that the service-load stress on the steel section is less than the yield stress. AISC-1.11.2.2 uses an indirect procedure for checking this. The section modulus of the composite section, S_{tr}, may not exceed (or be considered more effective than) the following:

$$S_{tr}\,(\text{effective}) \leq (1.35 + 0.35M_D/M_L)S_s \qquad (4\text{-}2)$$

In this formula, M_L refers to the moment caused by loads that are to be carried compositely, and M_D refers to the loads carried by the steel beam.

Equation 4-2 is developed based on the computation of service load stresses for construction with and without shores. Service-load tension stresses on the steel beam may be expressed in general as

$$f_b = \frac{M_D}{S_s} + \frac{M_L}{S_{tr}} \leqslant K_1 F_y \quad \text{without shores} \tag{a}$$

$$f_b = \frac{M_D + M_L}{S_{tr}} \leqslant K_2 F_y \quad \text{with shores} \tag{b}$$

where

S_s = section modulus of the steel beam referred to its bottom flange (tension flange)

S_{tr} = section modulus of composite section referred to its bottom flange (tension flange)

K_1, K_2 = constants to obtain the allowable stresses in tension without shores and with shores, respectively.

Divide Eqn (a) by Eqn (b), letting $kS_s = S_{tr}$:

$$\frac{K_1}{K_2} \geqslant \frac{M_D/S_s + M_L/KS_s}{(M_D + M_L)/S_{tr}} = \frac{KM_D + M_L}{M_D + M_L}$$

or

$$\frac{K_1}{K_2}(M_D + M_L) - M_L \geqslant KM_D$$

Divide by M_D,

$$K \leqslant \frac{K_1}{K_2}\left(1 - \frac{M_L}{M_D}\right) - \frac{M_L}{M_D}$$

Replacing K by S_{tr}/S_s gives the AISC formula in general terms:

$$S_{tr} \leqslant \left[\frac{K_1}{K_2} + \frac{M_L}{M_D}\left(\frac{K_1}{K_2} - 1\right)\right] S_s \tag{4-2}$$

The AISC value of $K_1/K_2 = 1.35$ is obtained if a compact section ($F_b = 0.66F$) is allowed to reach a service-load stress of $0.89F$ ($0.89/0.66 = 1.35$).

As is seen from Eqn 4-2 this limitation of stress is valid no matter what ratio of M_L to M_D is used.

(3) *Check the stress on the steel beam supporting the loads acting before concrete has hardened* The required section modulus of the steel beam is given by

$$S_s \text{ (reqd)} = M_D/F_b \tag{4-3}$$

where F_b may be $0.66F_y$, or some lower value if adequate lateral support is not provided. It is to be noted that Eqn 4-3 frequently controls the compression fiber (top in positive moment zone), particularly if a steel cover plate is used on the bottom.

(4) *Partial composite action* When fewer connectors are used than necessary to develop full composite action, an effective section modulus may be obtained by

linear interpolation. AISC-1.11.2.2 allows

$$S_{eff} = S_s + \sqrt{\frac{V'_h}{V_h}} (S_{tr} - S_s) \tag{4-4}$$

where V_h is the design horizontal shear for full composite action, V'_h is the actual capacity of connectors used (less than V), and S_s and S_{tr} are as defined for Eqns (a) and (b). For this case, S_{eff} is used to design calculations in place of that computed from beam dimensions, and is the quantity that may not exceed the value given by Eqn 4-2.

4.3   Ultimate strength design concept

The ultimate strength of a composite section depends on the yield strength and section properties of the steel beam, the concrete slab strength, and the interaction capacity of the shear connectors joining the slab to the beam. The provisions of AISC-1.11 are nearly all based on ultimate-strength behavior, even though all relationships are adjusted to be in the service-load range. These ultimate-strength concepts were applied to design practice as recommended by the ASCE–ACI Joint Committee on Composite Construction [31] and further modified as a result of research by Slutter and Driscoll [27]. The ultimate strength in terms of an ultimate-moment capacity gives a clearer understanding of composite behavior as well as providing a more accurate measure of the true factor of safety. The true factor of safety is the ratio of the ultimate-moment capacity to the actually applied moment. In both cases, whether the slab is termed 'adequate' or 'inadequate' compared to the tensile yield capacity of the beam, the connection between the slab and beam is considered adequate in the following development. Complete shear transfer at the steel–concrete interface is assumed.

In determining the ultimate-moment capacity, the concrete is assumed to take only compressive stress. Although concrete is able to sustain a limited amount of tensile stress, the tensile stress at the strains occurring during the development of the ultimate-moment capacity is negligible.

The procedure for determining the ultimate-moment capacity depends on whether the neutral axis occurs within the concrete slab or within the steel beam. If the neutral axis occurs within the slab, the slab is said to be adequate, i.e. the slab is capable of resisting the total compressive force. If the neutral axis falls within the steel beam, the slab is considered inadequate, i.e. the slab is able to resist only a portion of the compressive force, the remainder being taken by the steel beam. Figure 4-1 shows the stress distribution for these two cases.

Case 1: Slab adequate   Referring to Fig. 4-1a and assuming the Whitney rectangular stress block uniform stress of $0.85f'_c$ acting over a depth a, the ultimate compressive force, C, is

$$C = 0.85f'_c ab_E \tag{4-5}$$

The ultimate tensile force, T, is the yield strength of the beam times its area:

$$T = A_s F_y \tag{4-6}$$

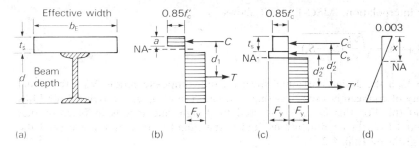

Figure 4-1   **Stress distributions at ultimate-moment capacity**

Equating the two we get

$$a = \frac{A_s f_y}{0.85 f'_c b_E} \qquad (4\text{-}7)$$

According to the ACI accepted rectangular stress block approach, the neutral axis distance, x, equals a/β_1, where $\beta_1 = 0.85$ for $f'_c = 4000$ psi (27.5 MPa). The ultimate-moment capacity, M_u, becomes

$$M_u = C d_1 \quad \text{or} \quad M_u = T d_1 \qquad (4\text{-}8)$$

Since the slab is assumed adequate, it is capable of developing a compressive force equal to the full yield capacity of the steel beam. Expressing the ultimate moment in terms of the steel force gives

$$M_u = A_s F_y (\tfrac{1}{2} d + t_s - \tfrac{1}{2} a) \qquad (4\text{-}9)$$

The usual procedure is to determine the depth of the stress block, a, by Eqn 4.7 and, if a is less than the slab thickness, t_s, to determine the ultimate-moment capacity by Eqn 4–9.

Case 2: Slab inadequate   If the depth, a, of the stress block as determined in Eqn 4-7 exceeds the slab thickness, the stress distribution will be as shown in Fig. 4-1b. The ultimate compressive force, C_c, in the slab is

$$C_c = 0.85 f'_c b_E t_s \qquad (4\text{-}10)$$

The compressive force in the steel beam resulting from the portion of the beam above the neutral axis is shown in Fig. 4-1b as C_s. The ultimate tensile force, T', which is now less than $A_s F_y$, must equal the sum of the compressive forces:

$$T' = C_c + C_s \qquad (4\text{-}11)$$

Also,

$$T' = A_s F_y - C_s \qquad (4\text{-}12)$$

Equating Eqns 4-11 and 4-12, C_s becomes

$$C_s = \frac{A_s F_y - C_c}{2}$$

or

$$C_s = \frac{A_s F_y - 0.85 f'_c b_E t_s}{2} \qquad (4\text{-}13)$$

Considering the compressive forces C_c and C_s, the ultimate-moment capacity M_u for case 2 is

$$M_u = C_c d'_2 + C_s d''_2 \qquad (4\text{-}14)$$

The moment arms d' and d'' are shown in Fig. 4-1b.

Whenever the case 2 situation occurs, the steel beam is assumed to accommodate plastic strain in both tension and compression at ultimate strength. Certainly, it is implied that such a steel section satisfy the requirements of 'compact' sections: that is, it should have proportions that insure its ability to develop its plastic moment capacity. Little research has been performed on case 2 situations because they rarely occur in practice.

4.4   Use of light-gauge metal decks in slab design

Light-gauge steel–concrete composite floor systems are in common use. Generally, such systems consist of a concrete slab on some type of cold-formed corrugated and/or ribbed decking. When composite construction is discussed, it is the use of metal deck that is commonly considered. There are, however, other uses for the deck:

(1) form the wet concrete
(2) working platform during construction
(3) diaphragm for the transfer of lateral loads
(4) part of a composite beam system.

This section has a twofold objective. First, a general overview of the background, research, and resulting design procedures for deck-concrete composite systems are presented. Second, examples are given to illustrate how composite systems can be designed. It is presumed that the reader is experienced in the basic rudiments of structural analysis and design.

The advantages of combining the structural properties of cold-formed light-gauge steel deck and concrete for use in floor systems for buildings were recognized many years ago. The most significant advantage was the reduced cost of the structure and foundations. Furthermore, the use of the deck, both as a platform for construction operations and as a form for the concrete, replaced the expensive conventional forming systems used previously. The deck, even in the early years, was recognized for its potential in channeling electrical and communications wiring through cellular construction. The steel ceiling formed by the underside of the deck facilitates the attachment of hanger supports for piping, duckwork, and suspended ceilings. Also, since the metal deck acts as the form for the wet concrete and generally does not require shoring, the time of construction for a structure may be reduced since each floor is independent and one need not wait for concrete to gain strength in order to support superimposed shoring leads. In addition, time is not required to remove shoring. The use of steel deck is accompanied by a nominal amount of temperature and shrinkage reinforcement, and the deck itself serves as the positive reinforcement once the concrete has hardened. (In building construction, top reinforcement is generally provided only when continuity is required.) The continuity of the deck is accompanied by the obvious aesthetic benefit of a crack-free bottom surface. Notwithstanding all of these advantages, the earliest use of light-gauge decks was in noncomposite applications. In order to present an unbiased treatment of the subject, it is reasonable to discuss the disadvantages of steel-deck composite construction. Such disadvantages seem centered on the characteristics of the deck. To insure good bond, the deck requires cleaning prior to placing concrete. Removal of oil and sawdust can be a problem. Furthermore, oil and water create a slippery and potentially dangerous working surface, tending to nullify one of the principal advantages. Extensive (and expensive) fire testing is required to obtain the fire rating called for by many national and local codes. Finally, the advantage of the availibility of the deck to act on a working platform can become a disadvantage if the deck is damaged by the temporary storage of heavy, concentrated loads, such as brick pallets.

Composite construction utilizing the light-gauge deck began with the Cofar deck (Granco Steel Products Co.). Rapid development of other deck systems followed. Much of the early work done with deck systems in terms of analysis, research, testing, and development of design procedures was on a proprietary basis, with each deck manufacturer working essentially independently.

The exact configuration of the various decks is proprietary with the various manufacturers. Widths of decks extend up to 30 in (75 cm), and depths generally vary from $1\frac{1}{2}$ to 3 in (38–76 mm). The heavy-gauge 0.07 in (2 mm) decks weigh up to 8 psf (380 kg/m^2).

It is important to note the various ways in which the different deck units achieve composite action. There are basically two ways in which such composite behavior is accomplished. First, many of the decks have ridges, corrugations, or other local 'deformations' that act as shear-transfer devices. These 'shear connectors' act in a mechanical manner to 'lock' the slab to the deck. The second class of decks have no deformations and are classified as 'smooth' decks, and composite action is accomplished only through the chemical bond that is achieved.

One of the most recent applications of composite construction using light-gauge steel deck is to use the deck as the form onto which the slab is cast, and to use the resulting combination to act compositely with beams, girders, and open-web joists. Significant research activities of Fisher [8] and Robinson [32] contributed heavily to current knowledge of the behavior of these members. Essentially, the composite slab-deck acts with the beam or joist to create T-beam action, stiffening the floor for a given design or permitting a lighter beam section where stiffness is not a consideration. Shown in Fig. 4-2 is a composite steel beam. Shear connectors are required to develop composite action between beam (or joist) and slab.

As in the design of any structural system, the design equations used for steel-deck composite systems are based on the various failure modes that may exist for the system. As stated previously, two primary failure modes exist for composite metal deck systems: (a) shear-bond failure, and (b) flexural failure.

The shear-bond failure is by far the most prevalent type of failure. It is characterized by the formation of diagonal cracks in the concrete slab in conjunction with bond slip between the metal deck and the concrete. As soon as slippage occurs, a significant drop in the load-carrying capacity results.

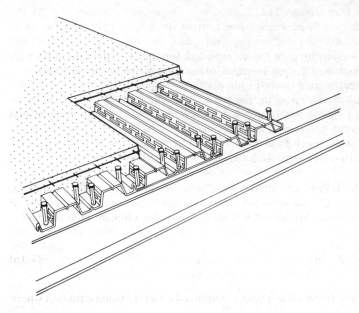

Figure 4-2 Details of combined slab, deck, and beam composite system with shear connectors

Flexural failures are identical to those of ordinary reinforced concrete flexural members. These failures can be either of the under-reinforced (steel yielding prior to concrete crushing) or over-reinforced type (concrete crushing prior to steel yielding).

Traditionally, design engineers have had to rely on experimental test results in order to determine the ultimate shear and the flexural capacities of steel-deck composite systems. The design procedure has typically been one of picking appropriate concrete thicknesses and steel-deck type and thickness from load tables provided by the various metal-deck manufacturers. Researchers have not attempted to derive generalized design equations. However, tentative design recommendations are currently being established by the American Iron and Steel Institute, based on extensive experimental investigations at several major universities. The results of the experimental investigations will be summarized here since presently they are the only existing design criteria available to the design engineer.

Two ultimate moment equations exist, one for over-reinforced beams and one for under-reinforced beams. The ultimate moment seldom controls the design or is the mode of failure in metal-deck systems. Based on strain compatibility and internal equilibrium, the balanced steel ratio, ρ_b, is defined as:

$$\rho_b = \frac{0.85\beta_1 f'_c}{F_y} \frac{87\,000(D-d_d)}{(87\,000+F_y)d} \tag{4-15}$$

where

$\beta_1 = 0.85$ for concrete with $f'_c \leqslant 4000$ psi and is reduced at the rate of 0.05 for each 1000 psi strength increase above 4000 psi (β_1 minimum = 0.65)

f'_c = compressive concrete strength, psi

F_y = steel-deck yield strength, psi

d = effective slab depth (distance from the extreme concrete compression fiber to the centroid of the steel), in.

d_d = depth of the steel deck, in

D = nominal out-to-out slab depth, in

The reader who is familiar with design of reinforced concrete will recognize ρ_b as the same balanced reinforcement ratio used for that system. For values of $\rho = (A_s/bd)$ less than ρ_b (under-reinforced sections), the ultimate moment is

$$M_u = \phi \frac{A_s F_y}{b} (d - \tfrac{1}{2}a) \qquad (4\text{-}16)$$

This equation is the conventional under-reinforced concrete beam equation where

$\phi = 0.90$

A_s = the net area of steel deck/unit width

$a = A_s/F_y/(0.85f'_c b)$, the depth of the equivalent rectangular stress block of width b.

In US usage, Eq. 4-16 may be written as

$$M_u = \phi \frac{A_s F_y}{12} (d - \tfrac{1}{2}a) \qquad (4\text{-}17)$$

in which M_u is ft lb/ft, A_s is the net area of steel deck per foot width, and $a = (A_s F_y/0.85f'_c) \times 12$, the depth of equivalent rectangular stress block of width $b = 12$ in.

It is assumed in Eqn 4-17 that the only reinforcing steel present is the steel deck itself. In addition, the normal assumption is made that the concrete reaches a strain of 0.003 in/in (mm/mm) at the outermost fiber. Caution should be used in applying this equation to a steel deck with a depth greater than 3 in (75 mm) because the steel deck may not have sufficient ductility to reach full yield at all fibers or it may have simply fracture prior to the concrete reaching a strain, ε_u, of 0.003 in/in. (mm/mm).

For over-reinforced systems, which are extremely rare, the ultimate moment (ft lb/ft) from Eckberg and Porter [33] is

$$M_u = \phi \frac{0.85\beta_1 f'_c bd^2 k_u}{12} (1 - \beta_2 k_u) \qquad (4\text{-}18)$$

where

$\phi = 0.75$ is recommended

$k_u = \sqrt{\rho\lambda + (\rho\lambda/2)^2} - \rho\lambda/2$
$\quad = E_s\varepsilon_u/(0.85\beta_1 f_c')$

$b =$ width of compression block

$E_s = 29\ 500\ 000$ psi

$\beta_2 = 0.425$ for $f_c' \leqslant 4000$ psi and is reduced at the rate of 0.025 for each 1000 psi strength increase over 4000 psi; f_c', d, and β_1 as defined previously.

More general equations for M_u may be written with suitable modifications in Eqn 4-18.

5   Deflection of a composite structure

5.1   Importance of deflection

Deflection in any structure is important since it affects its serviceability and the comfort of the occupants in buildings and the rideability of vehicles on bridges. In composite structure, deflection will usually result in 1/3 to 1/2 its value, when compared to that of the non-composite section of the same size. Furthermore, excessive deflection might cause partitions to separate and walls to crack. Both the short and long-term deflections are important since one of the components, i.e. the concrete, shrinks and creeps, and causes additional deflection.

The live load is generally carried by the composite section and therefore deflection also is determined by the properties of the composite structure. On the other hand, deflection due to the dead load depends on the method of construction. Thus, if the beam is shored, the dead load is *partially* carried by the beam; if the beam is unshored, it is *fully* carried by the beam and dead load deflections must also be checked.

5.2   Methods of deflection calculation

Methods of deflection calculation are based on the elastic approach, particularly for short-term deflection. Then, some approximation may be made for long-term effects. The following equations may be used for a typical case of composite structure:

5.2.1   *Beam with end moments plus uniform load*

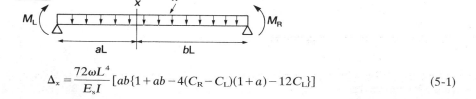

$$\Delta_x = \frac{72\omega L^4}{E_s I} [ab\{1 + ab - 4(C_R - C_L)(1 + a) - 12C_L\}] \tag{5-1}$$

in which a, b and ω (in kip/ft) are defined in the figure, and

E_s = modulus of elasticity of steel, ksi

I = moment of inertia of section, in^4

L = span, ft

$C_R = M_R / wL^2$

$C_L = M_L / wL^2$

For a simply supported beam, the above expression simplifies

$$\Delta_x = \frac{72\omega L^4}{E_s I} ab(1 + ab)$$

and, finally, the deflection at mid-span is given by

$$\Delta_{max} = \frac{45\omega L^4}{2E_s I}$$

$$= \frac{ML^2}{160 S_y} \tag{5-2}$$

for $E_s = 30\,000$ ksi, $M = \omega l^2/8$ and $I = S_y$ which is the AISC specification form for this equation.

To use the above equation in a two-component structure, these components (steel and concrete) are generally converted into one using the modular ratio, n, generally that for steel. Account must be taken of creep and shrinkage in the concrete. This inelastic (non-linear) behavior may be approximated by modifying the modular ratio, n, as was discussed earlier. AISC gives no indication. ASCE–ACI [31] and CP 117, Part 2: 1973 [25] recommend using half the value of the concrete modulus, E_c, when computing sustained load creep deflection. The 1980 edition of AASHTO [26] recommends $E_c/3$. Such arbitrary procedures give a good estimate of creep to approximately ±30%. Knowles [2] has mentioned values $(1 + \phi_N)$ and $(1.1 + \phi_N)$ for n, in which ϕ_N is the final creep coefficient.

The effect of variations in the value of the modular ratio on the section properties of a composite beam can be illustrated graphically, as shown in Fig. 5-1. It can be seen there is a little change in the bottom flange modulus of the

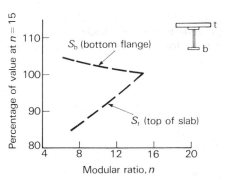

Figure 5-1   Change in section modulus, S, for changes in modular ratio, n (after Knowles [2])

section over a wide range of values. As the bottom flange stresses are generally critical in design, this variation in creep stresses is generally not critical in statically determinate beams.

As was mentioned earlier, light-gauge metal decks have been used for the formwork of slabs. Their deflection both before and after the concrete hardens should be checked for possible failures of the metal deck due to the 'ponding effect' if these deflections are excessive. A ponding failure occurs when, as the concrete is placed, the deck excessively deflects under the dead load. The deflected deck creates an obvious 'dip' in the surface, which requires more concrete to make a 'levelled' surface. The process continues until the deck fails. Deflection information is generally obtained from the manufacturer of the deck. However, the engineer should be able to calculate the deflection using the AISI procedure [34, 35] for the moment of inertia calculations and the ACI recommended procedure [24] for the deflection calculations. A typical composite section is shown in Fig. 5-2.

Deflections due to superimposed live and dead loads should also be checked when longer spans are used. The criterion suggested by several investigators is to base the composite deflections on the average moment of inertia of the uncracked and cracked sections, i.e. $\frac{1}{2}(I_u + I_c)$. The cracked moment of inertia may be written as

$$I_c = \frac{b}{3}(Y_{cc})^3 + nA_s(Y_{cs})^2 + nI_{sf} \qquad (5\text{-}3)$$

where

$$Y_{cc} = d\{[2\rho n + (\rho n)^2]^{1/2} - \rho n\}$$
$$\rho = A_s/bd$$
$$n = E_s/E_c$$

If $Y_{cc} > t_c$, then use $Y_{cc} = t_c$.

The uncracked moment of inertia is

$$I_u = \frac{bt_c^3}{12} + bt_c(Y_{cc} - 0.5t_c)^2 + nI_{sf} + nA_s Y_{cs}^2 + \frac{W_r b d_d}{C_s}\frac{d_d^2}{12} + (D - Y_{cc} - 0.5d_d)^2 \quad (5\text{-}4)$$

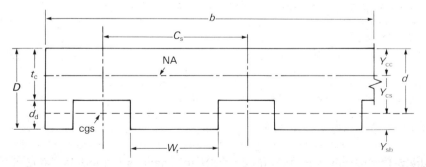

Figure 5-2 Transformed composite section

where

$$Y_{cc} = \frac{0.5bD^2 + nA_s d - (C_s - W_r)\dfrac{bd_d}{C_s}(D - 0.5d_d)}{bD + nA_s - \dfrac{b}{C_s}d_d(C_s - W_r)}$$

and C is the cell spacing (in), W the average rib width (in) and $d = D - Y_{sb}$

In many cases, deflections need not be checked since experience and span-to-depth ratios will give a good indication of whether a potential deflection problem exists. A span-to-depth ratio of 22 is recommended for simply supported spans. A quick check of deflection can be obtained by using the gross moment of inertia for the composite section. Using the equation

$$I_{gross} = \frac{1}{12}bh^3$$

where

$$h = D - Y_{sb}$$

will normally result in a conservative approximation for the short-term deflection of a simple span system. Long-term deflection considerations can be accounted for by the equation

$$\Delta_{LT} = \Delta'_{ST}[2 - 1.2(A'_s/A_s)] \tag{5-5}$$

where Δ'_{ST} is the short-term deflection due to sustained load, and A'_s is the area of compressive reinforcement, in^2/ft of width. The term within the brackets is a suggested multiplier used to account for the long-term effects and should be equal to or greater than 0.6. In general, assuming Δ_{LT} equal to twice the short-term deflection is conservative.

5.3 Specifications for deflections

Only a brief mention is made in AISC (Buildings) and AASHTO (Bridge) specifications. AISC (1978) recommends the following:

Limiting the depth/span ratio may also prevent many deflection problems. The AISC Commentary suggests a ratio of $F_y/800$ for fully stressed beams. This yields depth/span ratios as follows:

1/22 for $F_y = 36$ ksi

1/16 for $F_y = 50$ ksi

These ratios are offered as simple guidelines; however, the intent of the Specification is that a rational calculation of deflections should be made. Such calculations often reveal that smaller depth/span ratios are satisfactory. The depth used in the above ratios is the distance from the top of concrete to the bottom of the steel sections.

If it is desired to minimize the transient vibration due to pedestrian traffic when composite beams support large open floor areas free of partitions or other damping sources, the depth/span ratio of the steel beam should be not less than 1/20 for any grade of steel.

AASHTO Specifications (1977) state the following for deflection of composite members:

'The provisions of Article 1.7.6 in regard to deflections from live load plus impact also shall be applicable to composite girders.

When the girders are not provided with falsework or other effective intermediate support during the placing of the concrete slab, the deflection due to the weight of the slab and other permanent dead loads added before the concrete has attained 75 percent of its required 28-day strength shall be computed on the basis of noncomposite action.

1.7.6 *Deflection*

The term 'deflection' as used herein shall be the deflection computed in accordance with the assumption made for loading when computing the stress in the member.

Members having simple or continuous spans preferably should be designed so that the deflection due to service live load plus impact shall not exceed 1/800 of the span, except on bridges in urban areas used in part by pedestrians whereon the ratio preferably shall be 1/1000.

The deflection of cantilever arms due to service live load plus impact preferably should be limited to 1/300 of the cantilever arm except for the case including pedestrian use, where the ratio preferably should be 1/375.

When spans have cross-bracing or diaphragms sufficient in depth or strength to insure lateral distribution of loads, the deflection may be computed for the standard H or HS loading (M or MS) considering all beams or stringers as acting together and having equal deflection.

The moment of inertia of the gross cross-sectional area shall be used for computing the deflections of beams and girders. When the beam or girder is a part of a composite member, the service live load may be considered as acting upon the composite section.

The gross area of each truss member shall be used in computing deflections of trusses. If perforated plates are used, the effective area shall be the net volume divided by the length from center to center of perforations.

The foregoing requirements as they relate to beam or girder bridges may be exceeded at the discretion of the designer.' [From: *Standard Specifications for Highway Bridges*, 12th edition, Washington, D.C.: The American Association of State Highway and Transportation Officials, copyright 1977. Used with permission.

6 Steel–concrete composite columns

6.1 Why a composite column?

A composite column is a compression member that is constructed with load-bearing concrete plus steel in any form different from reinforcing bars. Seven examples of cross-sections of composite columns are shown in Fig. 6-1. Concrete-filled pipe and tubing represent a familiar form of a composite column. Quite often, the drilling 'mud' or mortar of caissons can serve to help carry load-bearing, as suggested by Fig. 6-1c. Although load-bearing concrete was used as

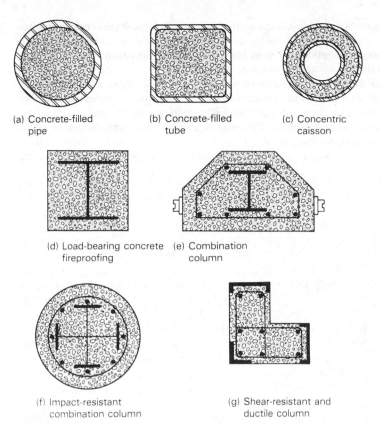

(a) Concrete-filled
pipe

(b) Concrete-filled
tube

(c) Concentric
caisson

(d) Load-bearing concrete
fireproofing

(e) Combination
column

(f) Impact-resistant
combination column

(g) Shear-resistant and
ductile column

Figure 6-1   Composite column cross-sections

fireproofing in the past, sprayed-on fireproofing has become more common since the 1930s. As the unit cost of steel increases, the use of load-bearing fireproofing may become necessary once again. The use of structural shapes to protect exposed corners of concrete columns in dock and traffic areas creates another form of composite column, as shown in Fig. 6-1e. Another form in Fig. 6-1g represents a seismic-resistant optimization of shear strength from structural shapes and ductility from a spirally-reinforced core to stabilize post-yield modes of local buckling of the shapes.

The first 'reinforcements' that were used for concrete columns were structural shapes. Shapes tied together with batten plates and rivets formed excellent reinforcing members. Obviously, the use of wired bars required less fabrication expense, and such reinforcement was recognized as more economic after steel producers developed bars with yield strengths considerably higher than those of rolled shapes (SSRC guide [36]).

Nevertheless, there remain numerous applications for which some forms of composite column provide excellent solutions to structural problems. When contrasted with plain steel columns, the composite column has superior load retention at increasing temperatures, considerably more resistance to local buckling, and greater stiffness and abrasion resistance. When compared with reinforced concrete, the composite column possesses much better strength and ductility in shear and generally in flexure also. Concrete-filled pipe or tube requires no column formwork. In earthquake-prone regions, the composite column can provide excellent ductility and load retention even after extensive concrete damage, which can be repaired if the overall structure can survive.

Reinforcing a concrete compression member with structural shapes or tubes does permit the designer the use of a greater percentage of steel in cross-sections than the maximum (8% permitted in conventionally reinforced concrete columns (ACI 318-77). The composite column can support more force per unit area than another reinforced concrete column of the same dimensions. The structural shape core of a concrete-encased column may support 80% of the thrust on a composite column, but the stiffening effect of the concrete encasement does permit the useful stress level on the core to be set at a level higher than that permitted on the bare steel shape and also makes it more effective against local and overall buckling.

Composite columns are a rather natural and logical structural form for compression members in fire-resistant construction. Encasing a structural shape in load-bearing concrete accomplishes the double benefit of providing the steel core with insulation from heat while supporting enough load to permit a reduction in the required size of the core. Composite column-load retention at high temperatures is least changed when lightweight aggregate is used because the kiln-produced aggregate remains stable at elevated temperatures (Malhotra and Stevens [37]). Concrete-filled steel tubes can continue to support service loads at high temperatures only if the concrete filling is strong enough to support the imposed loading without help from the softened tube. The concrete filling can conduct some heat away from the steel tubing exposed to fire, but the delay time prior to the steel softening is rather insignificant.

6.2 Behavior of a composite column

The behavior of steel–concrete compression members can be deduced on the basis of the following characteristics for steel and for concrete materials:

(a) Steel is eight to ten times as stiff as concrete until the steel is strained beyond its yield point, generally near 0.12% to 0.18% for commercial grades of structural steel.

(b) Steel tends to buckle locally after it yields in compression.

(c) Concrete cannot resist much tension strain without fracturing at strains greater than 0.1%.

(d) Concrete will sustain loads that create compression strains higher than 0.16%, but in the absence of some lateral confining pressure, concrete will display no stiffness when compressed to strains above 0.2%. In the absence of a strain gradient (adjacent fibers are strained less than surface fibers) or lateral confining pressure, concrete strained beyond 0.2% will spall and 'flow' laterally, typically with an explosive failure. In the presence of strain gradients, spalling and explosive failure can be delayed until surface strains from 0.2% to 0.5% occur, the maximum increasing with the gradient.

(e) Concrete strained less than 0.1% exhibits a Poisson ratio only one-third to one-half that of steel at the same strain. Concrete strained more than 0.16% exhibits a Poisson ratio greater than that of unyielded steel.

6.3 Design of an axially-loaded column

Truly concentric loads can exist only instantaneously for composite columns. The heterogeneous nature of concrete invariably permits one set of particles to respond to load more stiffly than others, and concentricity is destroyed. Under virtually concentric loads, commercial grades of structural steel reach their yield strength before concrete has exhausted its compressive strength. Concrete inside steel tubing prohibits post-yield wall buckling of steel by pushing outward against the tubing. Concrete encasement around structural shapes will prohibit local buckling of yielded steel. Consequently, strains on composite columns can be increased without loss of load after the steel yields, but longitudinal stiffness to resist more concentric load is virtually exhausted after the steel yields. Concrete confined laterally (as inside steel tubing) at strains above 0.3% eventually forces a post-yield outward buckling mode to occur in the tube wall.

Under the action of a *concentric compression* force, a composite column should shorten in length, and logically all components of the cross-section should undergo shortening by the same amount. With this assumption, the behavior of the column under increasing load could be derived from the simple superposition of forces that are caused by stresses associated with increases in overall strain. Typical stress–strain curves for steel and for concrete are shown in Fig. 6-2, along with the forces associated with the stresses at each strain for the assumed concrete and steel composite cross-section. When the forces on the concrete are added to those on the steel, the total behavior can be derived. The composite-section behavior contains a relatively linear portion from the origin to the strain value at which the proportional limit of steel is reached. After the steel begins to yield, the curve by its reduced slope reflects the loss of steel stiffness in the region from (1) to (2). After the concrete reaches its maximum strength and microcracking around aggregates progresses, there is a loss of total force and a measurable decrease in stiffness until the concrete fractures or spalls off entirely. Thereafter, only the yield strength of the steel core is available (at zero stiffness) to resist further compression strains.

Early investigators were able to observe the initial linear behavior as suggested to the left of (1) in Fig. 6-2. There was considerably less reporting of load–deformation behavior beyond the yield region (to the right of (1)) of composite columns. The nature of the residual stress in steel shapes and the tendency of concrete to creep when strains exceed 0.1% impaired the reproducibility of the few load–deformation functions that had been reported. Even the early investigators realized that the peak value of force, shown as P_0 in Fig. 6-2b, could be evaluated reasonably well simply by adding together the separate capacities of concrete and steel. The maximum thrust, or the squash load, on composite columns, P_0, can be estimated by adding the product of the steel area, A_s, and steel yield strength, f_y, to the product of concrete area, A_c, and concrete strength, f_c'. A coefficient β has been used to modify laboratory or material control values of concrete compression strength, f_c'. If 10 cm cubes are used for the concrete

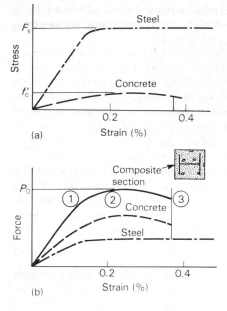

Figure 6-2   Axially concentric column load
(a) Stress–strain
(b) Force–strain

control, the values of β vary somewhat with f'_{cu}, the measured strength of each cube, but the quantity $\beta = 0.75$ is used frequently. If 6 in (15 cm) diameter cylinders, 12 in (30 cm) long are used for measuring f'_c, the coefficient $\beta = 0.85$ provides a reliable lower-bound value for concrete capacity. The maximum compression strength is expressed as

$$P_0 = A_s f_y + 0.85 f'_c A_c \tag{6-1}$$

Some tests of short lengths of concrete-filled pipe sections have revealed the possible development of loads higher than P_0 if the steel pipe, after yielding longitudinally, can continue to provide effective lateral confining pressures to the contained concrete. However, for columns more than a few diameters in height, the loss of longitudinal stiffness in the steel wall leads to subsequent buckling instability, and thrusts greater than P_0 cannot be realized.

The *buckling stability* of composite columns is analyzed at any strain level if the basic stress–strain behavior of the materials is known. A composite tangent-modulus form of strength–slenderness behavior represents a lower limit to column strength, P_c. A buckling load, P_c, can be expressed (Knowles and Park [38]) as

$$P_c = \pi^2 EI_{tan}/(kl)^2 \tag{6-2}$$

The quantity kl represents the effective length of the compression member (distance between the points of inflection in the buckled shape of the column), and the quantity EI_{tan} represents effective tangent-modulus stiffness for the cross-section. The effective tangent-modulus stiffness can be established by tests, but more commonly it must be estimated from the characteristic stress–strain properties of steel and concrete.

For purposes of design, the stress–strain functions for steel and for concrete, once expressed in analytic form, can be differentiated to obtain the slope of each as a function of strain. These slopes represent the tangent moduli for each material. Then, for any strain level, the products of tangent-modulus stiffnesses, E_{st} and E_{ct}, and the flexural shape factors, I_s and $\frac{1}{2}I_c$, can be combined to give an effective stiffness quantity EI_{tan}:

$$EI_{tan} = E_{st}I_s + \tfrac{1}{2}E_{ct}I_c \tag{6-3}$$

Only half the moment of inertia of concrete is suggested here because it should be assumed that something less than the full concrete cross-section can remain uncracked in pure flexure. At the same strain level for which EI_{tan} is evaluated, the corresponding thrust can be evaluated simply by adding together the products of the material stresses, f_s and f_c, and the material areas, A_s and A_c:

$$P_{cr} = f_sA_s + f_cA_c \tag{6-4}$$

After the values of EI_{tan} and P_{cr} have been evaluated, the effective length, kl, can be determined:

$$kl = \pi\sqrt{\frac{EI_{tan}}{P_{cr}}} \tag{6-5}$$

The designer must evaluate Eqn 6-5 for enough different strain levels to define an adequate strength–slenderness design curve. The calculations and graph for such a curve are given in Fig. 6-3 in order to illustrate the type and sequence of calculation necessary. The exponential equation that was used to represent A36 steel tubing for Fig. 6-3 will give misleadingly high values for f_s at very high strains, and an upper limit value of P_0 as expressed by Eqn 6-5 should be observed.

Note that the thrust–slenderness curve of Fig. 6-3 has an S-type form that is typical for any compression member. The graph starts with a thrust of P_0 at zero slenderness, and the thrust capacity decreases as the slenderness increases. The rate of decrease begins to lessen for ultimate thrusts less than $0.5P_0$.

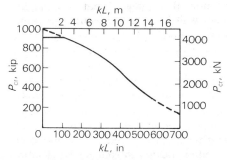

Figure 6-3 A load–slenderness curve for a concrete-filled pipe

Two familiar analytic functions, the parabola and the hyperbola, have been used to approximate column-strength curves. The fundamental parameters for such approximations involve only the effective long-column stiffness, EI, and the squash load, P_0. If P_{cr} is less than $0.5P_0$, long-column behavior is assumed, and EI_{tan} is taken simply as the low-stress initial value of flexural stiffness:

$$EI = E_s I_s + \tfrac{1}{2} E_c I_c \tag{6-6}$$

The ACI 318–77 [24] recommends that concrete stiffness be computed from the cylinder strength, f'_c, in psi and the concrete density, w_c, in pcf. Thus,

$$\left. \begin{array}{l} E_c = w_c^{1.5} 33\sqrt{f'_c} \quad \text{psi} \\[2mm] E_c = w_c^{1.5} 900\sqrt{f'_c} \quad \text{MPa } (w_c \text{ in kg/m}^3) \end{array} \right\} \tag{6-7}$$

Then, the effective length, kl_c, at which long-column action commences can be computed from Eqn 6-5 as

$$kl = \pi \sqrt{\frac{EI}{0.5P_0}} \tag{6-8}$$

The thrust capacity for P_{cr} values greater than $0.5P_0$ is made to fit the intermediate column portion by the relationship

$$P_{cr} = P_0 \left[1 - \frac{1}{2} \left(\frac{kl}{kl_c} \right)^2 \right] kl < kl_c \tag{6-9a}$$

$$P_{cr} = \pi^2 \frac{EI}{(kl)^2} \, kl > kl_c \tag{6-9b}$$

Generalized strength curves useful for design can be constructed from Eqns 6-2 and 6-9 if Eqn 6-6 is accepted for the effective flexural stiffness, and if safety factors are incorporated into the design curves. The strength–slenderness curves for $f'_c = 4$ ksi (28 MPa), concrete-filled, A36 (248 MPa) steel, round tubing are derived and displayed in Fig. 6-4. The dashed line of the graph in Fig. 6-4 was obtained from the theoretically more precise curve of Fig. 6-3. Correspondence between the 'precise' and the general curves indicates that the simplifications of Eqns 6-6 through 6-9 introduce negligible error into the procedure for developing generalized strength curves. For Fig. 6-3, the curves are parabolas with a horizontal tangent at P_0. The parabolas can be constructed to pass through the coordinate point kl/D and $P_0/2D$. To the right of the kl_c/D values, the curves are hyperbolas, and convenient points on each hyperbola were determined in order to construct the graph. At a slenderness ratio of $1.5kl_c/D$, the ordinates are $P_0/4.5A_g$, and, at $2kl_c/D$, they are $P_0/8A_g$.

Similar sets of generalized thrust–slenderness graphs can be constructed for any combination of steel yield strength, concrete strength, and cross-section shape. The graphs are applicable only for estimating thrust capacities for concentrically-loaded composite columns. A column can be considered to be concentrically loaded only if it is a part of a structure that is laterally braced against horizontal forces and only if the connections that introduce thrusts to columns are also not capable of introducing significant bending to the columns.

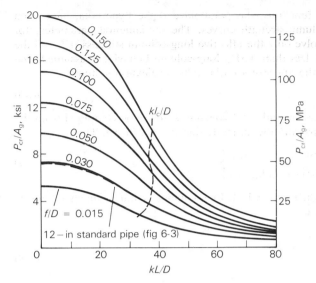

Figure 6-4 Strength–slenderness curves for A36 round tubes filled with concrete $f'_c = 4000$ psi (27.5 MPa)

6.4 Design of an eccentrically-loaded column

In the presence of eccentric compression loads, there are always some fibers subjected to less strain than that imposed on edge fibers. Much of the concrete and steel in eccentrically-loaded composite columns will separate from one another due to the differences in their Poisson's ratio. Concrete-filled steel tubes fail under eccentric compression much the same as they do in concentric compression except that only one face or side of the eccentric compression specimen experiences outward buckling of the tube wall when the concrete spalls at the same location. In all cases, the surface strains will reach at least 0.3% before the concrete spalls against the steel encasement.

Concrete-encased structural shapes can be said to 'fail' when the surface concrete spalls. In the presence of only a slightly eccentric thrust, it is possible that the surface concrete will fail at strains lower than the 0.3% generally expected because of the Poisson ratio separation between steel and concrete at lower strains. Lateral binding reinforcement, less effective than a column spiral, cannot restrain such failures of surface concrete. As the eccentricity of thrust increases, the probability of attaining surface strains above 0.3% is high enough that strength estimates based on such surface strain are accurate. Concrete-encased structural shapes may require special treatment of the thin segment of concrete adjacent to the flat face of a flange or box steel core, as shown in Fig. 6-5. If the width-to-thickness ratio of the edge concrete, b_f/t_c, is greater than 4, the edge concrete could become unstable and spall shortly after the surface strains reach 0.2%. Until laboratory studies reveal better techniques, a mechanical attachment such as one or more longitudinal reinforcing steel bars welded to the flat face, as suggested in Fig. 6-5, is encouraged.

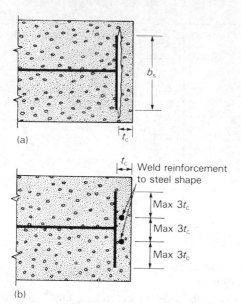

(a)

(b)

Weld reinforcement
to steel shape

Max $3t_c$

Max $3t_c$

Max $3t_c$

Figure 6-5   Concrete adhesion to steel surfaces
(a) Weak adhesion of concrete cover
(b) Mechanical aid to adhesion

The strength of composite cross-sections under the action of thrust P_u and bending moment M_u has been described in the form of interaction diagrams. Combinations of thrust capacity and the corresponding moment capacity are determined analytically on the basis of an assumed plane of strain, a limiting strain for concrete, and thereafter an integration of stresses that correspond with stress–strain characteristics for each material. In the absence of moment, the maximum thrust is, of course, the squash load, P_0. In the absence of thrust, the 'pure' moment capacity, M_0, is something larger than the plastic moment capacity for the steel alone in the cross-section. Except for the case of bending about the weak axis of a concrete-encased W, H, or I steel shape, an acceptable lower bound to pure bending capacity can be taken as the plastic flexural strength of steel alone. For weak-axis bending of encased steel shapes, an acceptable lower-bound value can be taken as the plastic bending strength of the steel flanges plus the flexural strength of a cross-section reinforced only by the web of the steel shape [39]. Values of P_u in the presence of M_u can be estimated from linear, parabolic, or elliptical functions that include the estimated points P_0 and M_0. A convenient and reliable analytic function takes the form

$$\left(\frac{P_u}{P_0}\right)^2 + \left(\frac{M_u}{M_0}\right) = 1 \tag{6-10}$$

A more precise though not demonstrably more accurate estimate of P_u and M_u values can follow the logic of reinforced concrete theory [42]. That logic involves the use of a limiting concrete strain, usually 0.3%, as the limit of concrete

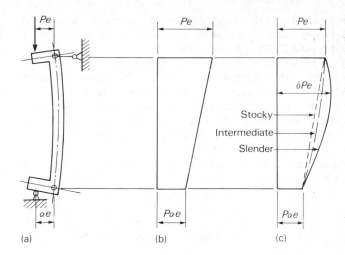

Figure 6-6 Moments caused by eccentric force
(a) Eccentrically loaded column
(b) Primary moments
(c) Total moments

strength. A selection of a neutral axis plus the designation of an extreme fiber at 0.3% strain defines a failure plane of strain for the selected neutral axis. Stresses on all steel elements and on concrete elements can be computed on the basis of characteristic stress–strain properties such as a trapezoidal function for steel and a parabola or equivalent rectangular stress block for compressed concrete. Furlong [39, 41, 42] has given detailed graphs to indicate the use of this approach.

When thrust is combined with flexural forces on relatively slender members, the thrust tends to increase the lateral displacement of the flexed or bent member, as illustrated in Fig. 6-6. The column in Fig. 6-6a supports a thrust P that is applied with an eccentricity e at one end and αe at the other end. Consequently, the moment at one end will be Pe, and at the other end $P\alpha e$, as illustrated in Fig. 6-6b. If the column is slender, the moment diagrams indicated in Fig. 6-6c can be developed. The long dashed line in this figure represents the moment diagram of an intermediate slenderness for which the maximum moment, Pe, remains at one end. The solid line is the moment diagram for a slender column, and the maximum moment, δPe, exists at some point between the ends of the column.

The amount of extra moment created by the lateral displacement of a slender compression member depends on the ratio, α, between the end moments, and also on the member slenderness as reflected by the column buckling capacity, P_c. A moment magnification factor, δ, can be estimated as

$$\delta = \frac{0.6 + 0.4\alpha}{1 - P/P_c} \geqslant 1 \tag{6-11}$$

where

$$P_c = \pi^2 EI/(kl)^2$$

The sense of α can be negative if the moments at each end create compression on opposite faces of the member. A minimum value $\alpha = -0.5$ is permitted by most building codes (e.g. ACI 318-77, AISC 1980).

The strength effect of slenderness can be estimated simply by checking against the short-column strength the combination of thrust P_u and slenderness-magnified moment M_u. If the values of δ as computed from Eqn 6-11 are less than 1, the calculations indicate simply that the end moment remains greater than any other moment between the ends of the member, as illustrated by the long dashed line of Fig. 6-6c. The column must support the thrust plus end moment, obviously, so factors less than 1 cannot be used.

Thrust–moment interaction graphs are perhaps the most precise description of strength against which the required thrust P_u and magnified moments δM_u can be compared. Most composite-column interaction graphs can fit adequately some linear analytic functions, and useful equations that relate thrust capacity and flexural capacity even for slender composite columns can be expressed with magnified moment values that are used for interaction ratios in the form of Eqn 6-10 to yield

$$\left(\frac{P_u}{P_{cr}}\right)^2 + \frac{\delta M_u}{M_0} = 1 \qquad (6\text{-}12)$$

In this equation, the magnification factor, δ, would be computed with the ultimate thrust P_u as the value of P in Eqn 6-11. The values of P_{cr} should be computed from Eqn 6-9, and the values of M_0 should be taken as the plastic bending capacity for steel alone except for the special case of weak-axis bending of encased shapes.

Slenderness effects for columns must be analyzed in terms of moment magnification factors, as suggested by Eqn 6-11. The ACI building code [24] requires the moment magnifier to be computed for the ultimate (augmented service load) thrust condition of loading, and values of EI must be evaluated for the 'softest' possible circumstance of column behavior in framed structures. It is permissible to use the EI value from either of the following two equations:

$$EI_1 = \frac{0.4E_c I_g}{1 + \beta_d} \qquad (6\text{-}13\text{a})$$

$$EI_2 = \frac{0.2E_c I_g + E_s I_s}{1 + \beta_d} \qquad (6\text{-}13\text{b})$$

with

E_c = modulus of elasticity for concrete (Eqn 6-7)
E_s = modulus of elasticity for steel (29 000 ksi or 200 000 MPa)
I_g = moment of inertia of the gross cross-section
β_d = creep factor equal to the ratio between design dead-load moment and total design moment.

Cross-section capacity as defined by interaction curves must be diminished by capacity reduction factors, ϕ, and the reduced strengths must exceed ultimate thrusts and magnified ultimate moments.

Lower limits to the thickness, t, of steel tubing for concrete-filled composite columns are set as

$$t > h\sqrt{f_y/8E_s} \quad \text{for circular sections of diameter } h \qquad (6\text{-}14a)$$

or

$$t > b\sqrt{f_y/3E_s} \quad \text{for each face of width } b \qquad (6\text{-}14b)$$

The analysis of capacity for composite columns is a vital but not a complete part of the design process. Safe design also requires that some allowances be made for probable discrepancies between real and anticipated material properties, member dimensions, and indeterminate frame forces. Finally, even the lower-bound estimates of capacity must be adequate to accommodate loads that exceed the largest probable service loads.

The procedures to design composite columns are discussed above. For actual examples, the reader is referred to other publications, e.g., Sabnis [11].

7 Typical examples

Two typical examples are presented to conclude this chapter. These examples cover detailed design procedures for flexure, shear and deflections using both the methods discussed earlier and the specifications.

Example 1

Design a typical floor beam of a building given: span length = 20 ft, beam spacing = $7\frac{1}{2}$ ft, slab thickness = 5 in.

Solution

(a) Bending moments

(1) Load applied before concrete hardens:

construction load

6 in slab (including finishes) = 0.75 kip/ft

beam weight (assumed) = 0.028 kip/ft

total = 0.778 kip/ft

$$M_D = \frac{wl^2}{8} = \frac{0.778 \times (20)^2}{8} = 39 \text{ ft kip}$$

(2) Load applied after concrete hardens:

live load = 0.06 kip/ft^2

$$M_L = \frac{wl^2}{8} = \frac{0.06 \times 20^2}{8} = 22.5 \text{ ft kip}$$

$$\frac{M_L}{M_D} = \frac{22.5}{39} = 0.58$$

(3) Maximum moment:

$$M_{max} = M_D + M_L = 39 + 22.5 = 61.5 \text{ ft kip}$$

(4) Maximum shear:

$$V_{max} = [(0.778) + (0.06)](\tfrac{20}{2}) = 10 \text{ kip.}$$

(b) Check effective width of concrete slab

$b = \frac{1}{4}l = \frac{1}{4} \times 20 \times 12 = 60$ in (governs)

$b = S = 7.5 \times 12 = 90$ in   [AISC-1.11.1]

$b = 16t + b_f = 80 + b_f$

(c) Required section modulus for M_{D+L}:

$$S_{tr} = \frac{12 \times 61.5}{24} = 30.75 \text{ in}^3$$ [AISC-1.5.1.4.1]

For M_D:

$$S_s = \frac{12 \times 39}{24} = 19.5 \text{ in}^3$$

(d) Select section and determine properties
  (1) Enter properties table for 5 in slab with $S_{tr} = 30.75$ in^3:

  Use W10×21

  (2) From properties table, interpolate for section property:

  $S_{tr} = 41.4$ $S_t = 164$ $S_{tr}/S_t = 0.25$

  $S_s = 21.5$ $y_b = 11.89$ $y_{bs} = 4.95$

  $W_s = 21$

(e) Check stresses concrete unshored:

  $S_{tr}/S_t = 0.25 < 0.51$ [AISC-1.11.2.2]

(f) Steel

	Furnished	Required		
total load	41·4 > 30·75		f_b	OK
dead load	21·5 > 19·5		f_b	OK
web shear	34·2 > 10		f_b	OK.

Therefore, f_b is acceptable.

(g) Check deflection

  (1) $\Delta_{DL} = \dfrac{M_D L^2}{160 S_s y_{bs}} = \dfrac{39 \times (20)^2}{160 \times 21 \cdot 5 \times 4 \cdot 95}$

  $= 0 \cdot 916 < 1 \cdot 5$ in   OK

  (2) $\Delta_{LL} = \dfrac{M_L L^2}{160 S_{tr} y_b} = \dfrac{22 \cdot 5 \times (20)^2}{160 \times 41 \cdot 4 \times 11 \cdot 89}$

  $= 0 \cdot 114 < L/360 = 0 \cdot 67$ OK

(h) Check formula

  $S_{tr} = (1 \cdot 35 + 0 \cdot 35 M_L/M_D) S_s$ [AISG1.11.2.2]

  $= (1 \cdot 35 + 0 \cdot 35 \times 0 \cdot 58) 21 \cdot 5$

  $= 33 \cdot 39 < 41 \cdot 4$

Use shoring
Shear connectors (for full composite action)

Use $\frac{3}{4}$ in $\phi \times 3$ in studs:
maximum stud diameter (Unless located directly over the web)

$$= 2.5t_f = 2.5 \times 0.34$$

$$\qquad\qquad = 0.85 > 0.75 \quad \text{OK} \qquad\qquad\qquad\qquad\qquad \text{[AISC-1.11.4]}$$

$$N_s = 21 \times 0.639 = 13.48 \,\text{(governs)}$$

$$N_c = 5 \times 69 \times 0.111 = 38$$

Use 28 $\frac{3}{4}$ in dia $\times$ 3 in studs equally spaced each side of the point of maximum moment four.

Example 2[1]

Determine the deflection of the $1\frac{1}{2}$ in (38 mm) composite metal deck shown below to support the applied loads. The deck consists of 20 gauge (0.91 mm thick) material. The manufacturer of the deck has published k and m values for this deck as 0.113 and 17 300, respectively. A 4 in (10 cm) normal-weight concrete slab is assumed.

Check also the deck for a 20 psf (960 N/m^2) uniform construction load and a 150 lb (660 N) concentrated construction load.
Deck properties

$$S^+ = 0.26 \text{ in}^3 \,(4260 \text{ mm}^3)$$

$$S^- = 0.28 \text{ in}^3 \,(4588 \text{ mm}^3)$$

$$I = 0.22 \text{ in}^4 \,(91\,570 \text{ mm}^2)$$

$$C_s = 6 \text{ in} \,(153 \text{ mm})$$

$$W_r = 1.95 \text{ in} \,(50 \text{ mm})$$

$$Y_{sb} = 1.0 \text{ in} \,(25 \text{ mm})$$

$$f'_c = 3,000 \text{ psi} \,(21 \text{ MPa})$$

$$A_s = 0.665 \text{ in} \,(17 \text{ mm})$$

$$d = 3.00 \text{ in} \,(75 \text{ mm})$$

$$F_y = 33\,000 \text{ psi} \,(231 \text{ MPa})$$

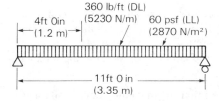

---

1 Taken from Sabnis [11]

Solution
Two types of deflection must be calculated: (a) deflection of the deck, and (b) deflection of the composite system.

(a) *Deflection of deck*   The maximum deck deflection will occur when construction loads are placed on only one side of the shore.

(1) Deflection due to the wet concrete dead load, assumed 38 psf (1820 N/m^2), using the simple beam-deflection equation, is

$$\Delta_{CL} = 0.0092 WL^4/EI$$
$$= \frac{0.0092(38)(5.5)^4}{29\,500\,000 \times 0.22} \times 1728$$
$$= 0.085 \text{ in } (2.1 \text{ mm})$$

(2) Deflection due to 20 psf (960 N/m^2) construction load is

$$\Delta_{CL} = 0.85(20/38) = 0.045 \text{ in } (1.1 \text{ mm})$$

(3) Deflection due to 150 lb (660 N) concentrated construction load is

$$\Delta_{CL} = 0.015 PL^3/EI$$
$$= \frac{0.015(150)(5.5)^3}{29\,500\,000 \times 0.22}(1728)$$
$$= 0.10 \text{ in } (2.5 \text{ mm})$$

(4) Compute the approximate ponding load of extra concrete due to the deflection occurring from the 150 lb (660 N) concentrated load. Assuming a parabolic deflection curve, the extra volume of concrete could be

$$\text{volume} = \tfrac{2}{3}\Delta_{CL} Lb$$
$$= \frac{2}{3}\frac{0.10}{12}(5.5)(1)$$
$$= 0.031 \text{ ft}^3 \ (877 \text{ cm}^3)$$
$$\text{weight} = 150(0.031)$$
$$= 4.65 \text{ lb } (20.4 \text{ N})$$

or approximately 1 psf extra load. Since this is a small load when compared to the dead load of 38 psf (1820 N/m^2) for this example, it will be neglected in further calculations. It must be emphasized, however, that, in many cases, especially for single-span unshored conditions, the ponding load can be significant.

From the preceding calculation, it can be seen that the maximum deflection due to the wet concrete would be less than 0.10 in (2.5 mm), which is considerably less than $L/180$ (0.37 in, 9 mm) or $\tfrac{3}{4}$ in (20 mm).

(b) *Deflection of composite system*

(1) Determine I_c

$$E_c = 57\,000\sqrt{f'_c} = 57\,000\sqrt{3000}$$

$$= 3\,122\,000 \text{ psi } (21\,500 \text{ MPa})$$

$$E_s = 29\,500\,000 \text{ psi } (203\,400 \text{ MPa})$$

$$n = 9.45, \text{ therefore } use\ n = 9$$

$$Y_{cc} = d\{2\rho n + (\rho n)^2]^{1/2} - \rho n\}$$

$$= 3.00\,\{[2(0.018)(9) + (0.018 \times 9)^2]^{1/2} - (0.018)(9)\}$$

$$= 1.29 \text{ in } (33 \text{ mm})$$

$$I_c = \frac{b}{3}(y_{cc})^3 + nA_s(y_{cs})^2 + nI_{sf}$$

$$Y_{cs} = d - y_{cc} = 3.00 - 1.25 = 1.71 \text{ in}$$

$$I_c = \tfrac{12}{3}(1.29)^3 + 9(0.665)(1.71)^2$$

$$+ 9(0.22) = 28.07 \text{ in}^4 \ (1168 \text{ cm}^4)$$

(2) Determine I_u

$$y_{cc} = \frac{0.5bd^2 + nA_{sd} - nA_{sd} - (C_s - W_r)\dfrac{bd_d}{C_s}(D - 0.5d_d)}{bD + nA_s - \dfrac{b}{C_s}d_d(C_s - W_r)}$$

$$y_{cc} = \frac{0.5(12)(4)^2 + 9(0.665)(3.00) - (6 - 1.95)\dfrac{(12)(1.5)}{6}[4 - (0.5 \times 1.51)]}{12(4) + 9(0.665) - \dfrac{12}{6}(1.5)(6 - 1.95)}$$

$$= 1.78 \text{ in } (45 \text{ mm})$$

$$I_u = \frac{bt_c3}{12} + bt_c(y_{cc} - 0.5t_c)^2 + nI_{sf} + nA_s y_{cs}^2 + \frac{W_{rb}d_d}{C_s}\left(\frac{d_d}{12}\right) + (D - y_{cc} - 0.5d_d)^2$$

$$= \frac{12(2.5)^3}{12} + 12(2.5)[1.78 - (0.5 \times 2.5)]^2 + 9(0.22) + 9(0.665)(1.22)^2$$

$$+ \frac{1.95(12)(1.5)}{6}\left\{\frac{1.5}{12} + [4 - 1.78 - (0.5 \times 1.5)]^2\right\}$$

$$= 48.31 \text{ in}^4 \ (2010 \text{ cm}^4)$$

Moment of inertia for deflection

$$I = \tfrac{1}{2}(I_c + I_u)$$

$$= \tfrac{1}{2}(28.07 + 48.31)$$

$$= 38.19 \text{ in}^4 \ (1590 \text{ cm}^4)$$

Check using I_{gross}

$$I_g = \tfrac{1}{12}bh^3 = \tfrac{1}{12}(12)(4-1.0)^3$$
$$= 27 \text{ in}^4 (1123 \text{ cm}^4)$$

(3) Short-term deflection*

$$\Delta'_{\text{ST}} = \frac{5WL^4}{384EI} + \frac{\rho_a}{3EIL} + \frac{PL^3}{48EI}$$

$$= \left\{ \frac{5(60)(11)^4}{384} + \frac{360(4)^2(7)^2}{311)} + \frac{365.75(11)^3}{48} \right\} \frac{1728}{3.122 \times 10^6 \times 38.18}$$

$$\Delta'_{\text{ST}} = 0.17 + 0.12 + 0.15$$
$$= 0.44 \text{ in } (11 \text{ mm})$$

Since, in the last expression, 0.15 in. (3.8 mm) is the deflection that would occur immediately after shore removal, it would normally not have to be considered in the short-term deflection criteria.

(4) Long-term deflection

$$\Delta_{\text{LT}} = \Delta'_{\text{ST}}[2 - 1.2(A'_s/A_s)]$$

Assuming no compression steel (mesh is negligible)

$$\Delta_{\text{LT}} = \Delta'_{\text{ST}}(2)$$

Assuming that one-half of the live load is sustained,

$$\Delta_{\text{LT}} = 2[0.17 + 0.12]$$
$$= 0.58 \text{ in } (14.7 \text{ mm})$$
$$\Delta_{\text{T}} = 0.44 + 0.58 = 1.02 \text{ in } (25.9 \text{ mm})$$

Deflections are excessive for the span. This would be expected since the span-to-depth ratio greatly exceeds the recommended value of 22 for a simple span.

Acknowledgements

This chapter is an adaptation and condensation of chapters from *Handbook of composite construction engineering*, edited by the present author and similar to the chapter from *Handbook of Structural Steel Design* both published by Van Nostrand Reinhold, New York, whose permission to reproduce copyright material is gratefully acknowledged.

Equations 2-1, 2-2, 2-3, 2-5, 2-6 and 2-10 and Figs 2-1, 2-2, 2-3, 5-1 , 5-2 and 5-9 are quoted from *Composite construction in steel and concrete* by I. M. Viest, R. S. Fountain and R. C. Singleton. Copyright © 1958 McGraw-Hill Book Company Inc. Used with the permission of McGraw-Hill Book Company.

References

1. Brandel, H., New concepts in structural design, Engel, H. (Ed.), *Structural systems*, Prager, New York, 1967, pp. 1–14.
2. Knowles, P. R., *Composite steel and concrete construction*, Halsted Press, New York, 1973, 200 pp.

3. Scott, W. B., The strength of steel joists embedded in concrete, *The Structural Engineer, (London)*, Vol. 26, 1925, pp. 201–219.

4. Caughey, R. A. and Scott, W. B., A practical method for the design of I-beams haunched in concrete, *Struct. Eng.*, Aug. 1929, p. 275.

5. Gillespie, P., Mackay, H. M. and Leluau, C., Report on the strength of I-beams haunched in concrete, *Eng. J. (Montreal)*, 1923, p. 265.

6. Siess, C. P. Composite construction for I-beam bridges, *Trans. Am. Soc. Civ. Eng.*, Vol. 114, 1949, pp. 1023–1045.

7. Viest, I. M., Review of research on composite steel concrete beams, *Proc. Am. Soc. Civ. Eng.*, Vol. 86, No. ST 6, June 1960, pp. 1–20.

8. Fisher, J. W., Design of composite beams with formed metal deck, *Eng. J.*, Vol. 96, No. 3, July 1970, pp. 88–96.

9. Slutter, R. G., Composite steel–concrete members, Tall, L. (Ed.), *Structural steel design*, Ronald Press, New York, 1974, Ch. 13.

10. Johnson, R. P., Research on steel–concrete composite beams, *J. Struct. Div. Am. Soc. Civ. Eng.*, Vol. 96, No. ST3, March 1970, pp. 445–460.

11. Sabnis, G. M. (Ed.), *Handbook of composite construction engineering*, Van Nostrand Reinhold, New York, 1979, 380 pp.

12. Furlong, R. W., Design of steel-encased concrete beam–columns, *J. Struct. Div. Am. Soc. Civ. Eng.*, Vol. 94, ST1, pp. 267–282.

13. Gardener, N. J. and Jacobson, E. R., Structural behavior of concrete filled steel tubes, *J. Am. Concr. Inst., Proc.*, Vol. 64, No. 7, July 1967, pp. 404–413, and Suppl. No. 2, Title 64–38.

14. Stevens, R. F., Encased stanchions, *Struct. Eng.*, Vol. 43, No. 2, Feb. 1965, pp. 59–66.

15. Dobruszkes, A., Janss, J. and Massonnet, C., *Experimental researches on steel concrete frame connections. Part 1: Frame connections comprising concrete columns and a composite beam. Part 2: Frame connections completely encased in concrete*, International Association for Bridge and Structural Engineering, Publication 29–11, 1969, pp. 67–100.

16. Roderick, J. W., Further studies of composite steel and concrete structures, Preliminary Report, *Proc. 9th congr. Int. Ass. Bridge Struct. Eng.*, Amsterdam, May 1972, pp. 165–172.

17. Okamoto, S., *Earthquake resistant design of concrete structures*, University of Tokyo Press and Halsted Press, New York, 1973.

18. Davies, C., *Steel–concrete composite beams for buildings*, Halsted Press, New York, 1975, 125 pp.

19. Wilenko, L. K., The application to large buildings of structural steelwork prestressed during erection, *Proc. conf. steel in architecture*, British Constructional Steelwork Association, London, 1969.

20. Johnson, R. P., Finlinson, J. C. H. and Heyman, J., A plastic composite design, *Proc. Inst. Civ. Eng.*, Vol. 32, Sep.–Dec. 1965, pp. 198–209.

21. Iwamoto, K., On the continuous composite girder, *Bull. Highw. Res. Bd.*, No. 339, 1962, p. 81.

22. American Institute of Steel Construction (AISC), *Specifications for design, fabrication and erection of structural steel for buildings*, New York, 1978.

23. American Association of State Highway and Transportation Officials (AASHTO), *Interim specifications* 1980.

24. ACI Committee 318, *Building code requirements for reinforced concrete*, ACI Standard 318–77, American Concrete Institute, Detroit, 1977, 103 pp.
25. British Standards Institution, CP117, *Composite construction in structural steel and concrete*. Part 1: 1965, *Simply-supported beams*. Part 2: 1973, *Beams for bridges*, London, 24 pp. and 32 pp. respectively.
26. American Association of State Highway and Transportation Officials (AASHTO), *Standard specifications for highway bridges*, 11th Edn, 1973, 12th Edn, 1977, 13th Edn, 1980, 449 pp.
27. Slutter, R. G., and Driscoll, G. C., Flexural strength of steel–concrete composite beams, *J. Struct. Div. Am. Soc. Civ. Eng.*, Vol. 91, No. ST2, April 1965, pp. 71–99.
28. Lew, H. S., *Effect of shear connector spacing on the ultimate strength of concrete-on-steel composite beams*, National Bureau of Standards Report No. 10246, Washington DC, 1970, 82 pp.
29. Viest, I. M., Fountain, R. S. and Singleton, R. C., *Composite construction in steel and concrete*, McGraw-Hill, New York, 1958 176 pp.
30. Slutter, R. G. and Fisher, J. W., Fatigue of shear connectors, *Highw. Res. Rec.*, No. 147, 1966, pp. 65–88.
31. ASCE–ACI Joint Committee on Composite Construction, Tentative recommendations for the design and construction of composite beams and girders for buildings, *J. Struct. Div. Am. Soc. Civ. Eng.*, Vol. 86, No. ST12, Dec. 1960, pp. 73–92.
32. Robinson, H, Composite beam incorporating cellular steel decking, *J. Struct. Div. Am. Soc. Civ. Eng.*, Vol. 95, No. ST3, March 1969, pp. 355–380.
33. Porter, M. L. and Eckberg, C. E. Jr, Design recommendations for steel deck floor slabs, *J. Am. Soc. Civ. Eng.*, Vol. 102, No. ST11, Nov. 1976, pp. 2121–2136.
34. American Iron and Steel Institute (AISI), *Specification for the design of cold-formed structural members*, 1968, pp. 1-5 to 1-30.
35. American Iron and Steel Institute (AISI), *Illustrative examples based on the 1968 edition of the specification for the design of cold-formed structural members*, 1968, pp. 4–5 to 4-35.
36. Structural Stability Research Council (SSRC), *Guide for the design of metal compression members*, 3rd Edn, Urbana, Illinois, 1977.
37. Malhotra, H. L., and Stevens, R. F., Fire resistance of encased steel stanchions, *Proc. Inst. Civ. Engrs*, (*London*), Vol. 27, Jan. 1964, pp. 77–98.
38. Knowles, R. B. and Park, R., Strength of concrete filled steel tubular columns, *J. Am. Soc. Civ. Engrs*, Vol. 95, Dec. 1969, pp. 2565–2568.
39. Furlong, R. W., AISC design logic makes sense for composite columns, too, *J. Am. Inst. Steel Constr.*, Vol. 13, 1976.
40. Ferguson, P. M., *Reinforced concrete fundamentals*, 3rd Edn, Wiley, New York, 1979, 724 pp.
41. Furlong, R. W., Concrete encased steel columns – design tables, *J. Struct. Div. Am. Soc. Civ. Eng.*, Vol. 100, No. ST9, Sept. 1974, pp. 1865–1882.
42. Furlong, R. W., Steel concrete composite columns, Sabnis, G. M. (Ed.), *Handbook of composite construction engineering*, Van Nostrand Reinhold, New York, 1979, Ch. 6, pp. 211–219.

18 Precast concrete: its production, transport and erection

J G Richardson

Cement and Concrete Association, Slough, England

Contents

Summary

The possibilities offered by precast concrete are examined. Two main types of precast elements are identified: structural elements and structural/visual elements. The factors governing the selection of precast concrete are discussed with reference to quality, accuracy and the required production facilities. Connections, fastenings and fixings are reviewed. Manufacturing processes, and the methods of handling, transporting and erecting precast elements are discussed. The emphasis throughout is on the practical aspects of the economic provision of sound, durable elements and their safe, speedy erection.

1 Introduction

Precasting removes many of the traditional restraints inherent with *in situ* work; for example, formwork erection, steel fixing, concreting and service installation need no longer be carried out on site. It also allows the uncoupling of construction activities, thereby removing the risk of a delay in any of the critical activities completely hindering progress. Element manufacture and the validation of the design of major components can begin early during the design of the overall structure. Standards, once established, can be monitored during production and modifications made if necessary before the elements are used in the structure.

Precast elements may be cast in the factory or on site, depending on the availability of materials, skilled labour and a suitable location. Economics or the shortage of local skills and facilities often dictate the factory production of precast elements and their subsequent transportation to site over thousands of kilometres.

2   Basic concepts

2.1   Considerations determining the use of precast concrete

Two types of element are considered in this chapter: (a) structural and (b) structural/visual.

2.1.1   *Structural elements*

These are used purely to meet structural requirements; examples include frame elements and large panels. Structural elements may be manufactured by a proprietary supplier or a contractor in the course of major works.

Proprietary elements are generally manufactured to national standards (e.g. CP 110) or in accordance with the recommendations of producer groups. Bespoke or special units are designed by specialist engineers.

In selecting elements and suppliers, the engineer must familiarize himself with the range of elements available, consulting published information on matters of span, depth, thermal and acoustic properties and the like. Care must be taken to ensure that the comparison takes account of quality, delivery and performance to similar standards.

Although all proprietary elements are guaranteed or warranted, the engineer must insist on the production of certificates covering materials quality and tests on fresh concrete and finished products. Certificates of calibration are also necessary for test and stressing equipment. Care over such details is essential to ensure that the elements supplied are in accordance with the quality standards established by the contract for the supply of goods.

In the case of bespoke or special elements, the engineer is responsible for all aspects of design, testing and performance, the establishment of design criteria, calculation, specification and the control of quality. Whereas proprietary elements are generally the product of standard moulds or casting machines, bespoke elements usually require the production and use of special, often expensive, moulds.

Discussion early in the course of design with those responsible for erection and for the supply of connections will yield valuable information for incorporation into the design. The early construction of a mock-up or trial assembly at the place of manufacture helps to identify problems of fit, connection appearance and so on. The designer must make a careful study of the availability of elements similar to those which he requires with a view to achieving economy by adoption, or adaptation, of standard elements.

2.1.2 *Structural/visual elements*

In the case of structural/visual elements, because of the special architectural detail, it is unlikely that any available elements will completely fill the full design requirements, bearing in mind local special details of sound and thermal insulation, service facilities and the like. The decision to precast in the form of structural/visual elements is often dictated by the need for quality, accuracy and high standards of finish, or for some special sculptured effect. Many features and textures are difficult if not impossible to produce and therefore call for precasting. For example, the *in situ* production of slender sections, sloping faces and intricate work would seriously slow down the work on a building structure.

The integration of the structural element and the facing element into one unit provides economy in erection. The in-process installation of ducts, conduits and fittings for services reduces the overall construction time while allowing the early enclosure of the structure, which provides sheltered working conditions for the finishing trades.

Structural/visual elements can yield economies by eliminating the need for falsework, minimizing the requirements for access scaffolding, and reducing the amount of handling and site work in finishing the concrete surfaces. The trend with structural/visual concrete has been towards the complete exposure of the unit in the form of an attractive structural frame.

Where there are particularly stringent requirements of form combined with the display of exotic aggregate, the structural/visual element may be cast with an integral or 'throughmix' of the exotic aggregate. This concrete must be subject to careful control to ensure that the required strength and durability are achieved.

2.2 Establishing standards of quality and accuracy

It is impossible to achieve complete uniformity in the appearance and physical properties of a series of mass-produced elements. Variations in temperature, concrete mix characteristics, workability, degree of compaction, mould configuration, surface texture and many other factors, can easily vitiate the quality of the finished product. The variability may be reduced. A greater degree of quality control and reference to established samples of finish can improve consistency. The best that can be achieved is some degree of consistency, some shade of the required colour, and some degree of accuracy—statistically speaking, some part of the normal distribution.

The designer of a structure introduces variability in handing, local changes of profile and similar detail. These changes may lead to inaccuracy, and where mould assembly and concreting technique are changed to meet some design modification, then surface finish and texture may also vary.

The following procedures ensure the establishment of design criteria and standards appropriate to the construction:

(a) The designer should study the performance of similar structures, consulting with specialist and colleagues with first-hand knowledge of the design problems and their solution.

(b) All relevant codes of practice, building regulations and digests, together with reports of research and testing, should be assembled and examined in detail.

(c) Outline details should be prepared, and calculation, simulation by computer and model testing carried out to examine the performance in terms of exceptional loading—winds, exposed conditions, etc.

(d) Proposals should be discussed with possible contractors or suppliers or trade associations to ensure that the chosen design is feasible in terms of casting, handling and erection.

(e) A search should be conducted for existing solutions to the problems encountered, with a view to adopting standard or near-standard elements to achieve economy.

(f) Element configuration must be determined with regard to casting, handling and erection techniques, bearing in mind the transport and erection aspects.

(g) The concrete technology should be examined, including the availability of materials, cement, aggregates and water and the provision of concrete capable of withstanding pollutants, acids, salts and other aggressive agents.

(h) The potential contractor and the supplier of the precast elements must be carefully selected on the basis of particular expertise, prior experience and the ability to provide technical support in detail or in the manufacturing technique. While expertise can be built up in the course of a project, the designer will be wise to capitalize on any existing experience and proven performance.

Discussion in the early stages of design can result in simplification and the establishment of detail which, as well as meeting the design criteria, is suited to the proposed methods of production and construction.

At the earliest possible stage, outline drawings should be made to enable the production of large-scale samples. It is advisable to include a sum in the bill of quantities to cover the manufacture of samples by several potential suppliers. The samples enable the validation of detail, the establishment of accuracy and fit and allow an assessment of the capabilities of suppliers to meet the quality requirements.

The assembly of the mock-up on a prepared foundation will allow upwards of 20 points of detail to be checked [1] and fed back into the design office where final details are to be prepared. The mock up should incorporate joints and connections typical of those which are critical to the performance of the eventual structure. Now the designer can produce the final drawings for the work. To a large extent, details of the design will have been validated although an opportunity exists for final revision and modification before the main production is set in hand. At this stage, some 50 activities into the programme of 250 or more activities leading to the first erection of elements on site, there is sufficient information to allow the preparation of meaningful detail.

2.2.1   Surface finish

The recommendations in the British code CP 110, Part 1:1972 [2] are extremely sound, based as they are to a large extent on CP 116, Part 2 1969 [3]. The recommendations are carefully framed and can be used as the basis of specification for the more standard structural elements. Section 6:10:6 of CP 110, relating to finishes, is realistic although the clauses on 'other types of finish' are vague and should be supplemented by specific detail of concrete mix proportion, method and degree of exposure.

Useful guidance on plain concrete is given by Schjodt [4], including quantitative methods for assessing surface finish. Blake [5] provides considerable information useful to the specifying authority and the producer alike, including recommendations for the avoidance of blemishes. He also describes the influence on surface appearance of mould texture and absorbency of mould surface. Gage [6] underlines the influence of form face and concreting method on the appearance of concrete, and also presents a variety of 'direct' and 'indirect' techniques.

It is essential that the specification governing finishes is related to actual attainable quality, and in this respect it is helpful to indicate where concrete elements of comparable acceptable standard may be inspected.

The *Formwork report* [7], while mainly concerned with formwork for *in situ* structures, is informative on the quality of concrete surface finish, making recommendations for both the specifying authority and the contractor.

2.2.2   *Accuracy of structural precast concrete*
Certain recommendations are made in CP 110 regarding permissible deviations in the accuracy of precast concrete for structural purposes. These recommendations, based upon those in CP 116, have proved to be attainable in practice and have resulted in successful construction.

The designer must bear in mind that concrete is essentially a mobile material, subject to shrinkage, creep and thermal movement. Consequently, units found to be quite acceptable immediately on removal from the mould can distort or be distorted to the stage where, at the time of erection into the structure, the changes in shape negate features of the design and may, indeed, cause damage to other parts of the construction.

Specifications for heavy and special precast work have been less well documented. In these circumstances, it is reasonable to specify as for *in situ* concrete. BS 5606: 1978, *Accuracy in building* [8] gives valuable assistance, based upon a statistical approach, in identifying and specifying deviations. The code identifies the considerations relating to the aesthetic, structural, economic and practical aspects of the achievement of accuracy.

An interesting approach to specification for accuracy is to be found in the Institution of Structural Engineers' manual *Structural joints in precast concrete* [9]. This recommends that joint design should cater for inaccuracy of manufacture and erection.

2.2.3   *Accuracy of structural/visual concrete*
When specifying and detailing this class of work, it is essential to consider the modular concept of design; the permissible deviations must be such that no unit can encroach upon the space allocated to another unit. CP 297 : 1972 [10] sets down recommendations for permissible deviations.

Structural/visual elements must be manufactured in accordance with these recommendations if the weather joints are to function as intended. Cladding recommendations also include the manufacture of special end or closure units where joints must be of 'reasonable evenness'. Where accuracy for

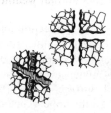

Rugged treatment of corners
results in ragged appearance of
joints – some architects like this

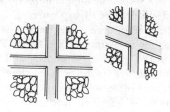

The use of chamfers and
margins 'tidies up' joints
and makes for a better production

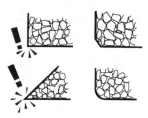

The introduction of chamfers
and rounds avoids ragged edges
and damage and assists in
concrete placement

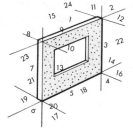

Accuracy of panel or slab
production can affect the
performance of any of 24
panels or elements – modular
concepts are critical

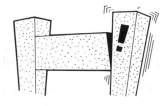

Movements are considerable – joints
and connections must cater for
inherent and induced innaccuracy,
thermal movement, creep and so on

Panels which are restrained
will bow and durability will
be adversely affected

Figure 2-1   Points of detail

structural/visual concrete is concerned, it is appropriate to follow the recommendations of CP 110, stating additional special requirements to ensure that weather joints work as intended.

Joints between structural/visual elements are often local to the connection and not continuous, as at the junction between cladding panels. Frequently, the joints in structural/visual concrete work are covered by some subsequent masking or by the casting of stitching concrete. Ideally, joints between structural/visual elements should be recessed or masked by features in such a way that they are in shadow. Limits on deviation, half the value of those set out in CP 110, ensure that visual/structural elements meet the functional requirements of construction.

The PCI manual [11] recommends specific position tolerances for items such as glazing gaskets and electrical outlets, which are cast integral with an element. Attention is drawn to the need for the designer of precast concrete to allow for inaccuracy in *in situ* concrete construction and other structural materials comprising the completed structure.

2.3 Specification for precast concrete

As well as accuracy and surface finish, the following general points must be observed when preparing specifications for precast elements:

(a) The specification must enable the contractor or supplier to understand the intentions of the engineer.

(b) The codes of practice with which the work must comply must be quoted, together with the codes governing materials, components and items such as test methods.

(c) Where required, the specifier should stipulate that shop drawings of moulds, lifting equipment, plant and factory layout and similar items should be made available for inspection by the engineer.

(d) Where shop or production drawings for individual elements are required to be produced by the contractor or supplier, these as well as the submission of bar-bending schedules and schedules of fixings and lifting devices must be covered by the specification.

(e) The concrete mix for the various types of unit must be specified (types of cement and aggregate with reference to sulphate and chloride content). Where specific national standards are appropriate, these should be noted. Certificates of compliance with the nominated standards should be required to be submitted to the engineer prior to the despatch of materials from the manufacturer's works.

(f) It is advisable that the conditions of storage of the aggregates, cement and similar materials are clearly set down, as these have a considerable effect on concrete quality—particularly in extremes of climate.

(g) The specification should include carefully prepared reference to the control of the usage of admixtures where these are permitted.

(h) The specification for the concrete must be appropriate to the type and location of the construction. It is important that suitable control tests are described and that the permissible water content is stated. Cube or cylinder tests, as appropriate, should be specified, together with clearly described criteria for compliance.

(i) Where large site or off-site production is involved, it will be necessary to describe the laboratory facilities and equipment essential to the establishment of the required degree of control.

(j) The specification must clearly indicate the standards of placing technique for the concrete in the elements and state the criteria by which the degree of compaction is assessed.

(k) The specification must include clauses governing curing arrangements, temperature and humidity control, setting limits to the period of curing.

(l) The stripping times for moulds, forms and temporary supports must be stated and standards established for the quality of concrete surface. Where the scale of on-site production is such that concrete may be supplied from a ready-mixed supplier, the specification must stipulate appropriate means for batching, mixing, delivering and handling the material, and for setting limits for accuracy. Where on-site batching and mixing are to be employed, the type of plant and requirements regarding minimum mixing and handling times must be stated.

(m) It is important that standards of workmanship and limits of accuracy for all preparatory work, the provision of foundations pockets, dowels, bolts, welding plates and similar connections to *in situ* work, are stated with reference to the appropriate national and/or trade standards. The techniques to be employed at construction joints and connections must be described with regard to bolting, welding, concreting and grouting, and the appropriate standards identified. Special techniques of connection incorporating stressing and force or torque measurement must be described in detail, together with relevant compliance criteria. The sequence of handling, the method of protection, marking and inspection prior to location in the structure, must be described with particular reference to the action to be taken in respect of damaged units.

(n) Where it is essential to the stability and function of the structure, the sequence of erection must be described and specific instructions incorporated for the proper handling of the elements.

(o) Special requirements must be stated regarding the casting techniques to be employed, such as the use of contact methods in manufacture and construction. In those instances where the results are dependent upon the methods employed, a method specification should be issued.

(p) The specification must cover stacking and handling techniques. Methods of protection—particularly of visual concrete—must be described.

(q) The specification must govern all aspects of the function of the finished structure, appearance, performance and accuracy as discussed in Section 2.2, establishing the standards of quality and accuracy.

(r) The specification for special components or materials should include instructions regarding relevant test procedures.

2.4 Factors governing the place of manufacture

The design of a structure, the scale of the elements incorporated in it, or any special geometric feature may require that elements be constructed on site. In major civil engineering works, the contractor undertakes site precasting in order to achieve the economies and advantages of precasting without losing the initiative in the construction process by becoming dependent upon an outside supply.

Special conditions may dictate the location of the casting facility. For example, where it is specified that the supply be carried by rail or water, the casting yard may well be at a railhead or dock. The availability of raw materials may dictate the establishment of a casting yard in a quarry or, again, at a railhead—a frequent occurrence in emerging countries. Where access to the site is difficult or limited, the casting of large elements may be carried out at, or near, the eventual place of use.

Often in major works, all available local labour is employed in the basic construction, so that decentralization of casting may serve to open up new sources of supply of labour.

2.5 Economic considerations, quantity, scale and programme time

Precasting provides a means for quantity production. Small components are normally unreinforced, pressed concrete, produced by machines which are sometimes automated to the stage where a factory can be run by one or two men. The men are mainly concerned with adjustment, breakdown and exceptions to normal manufacture. In the case of larger, reinforced products, the automation is limited by the steel reinforcement. Therefore, the accent is on the mechanization of individual parts of the process—cage fabrication, casting and curing. Flow production becomes batch production. Very large components enter into the category of jobbing production, similar to the process of *in situ* construction.

Generally, quantities are not critical and it may prove quite economic to precast as few as five elements in response to a special requirement, such as a finish that is unattainable by *in situ* techniques.

Large quantities imply massive daily production and the need for in-process inspection, testing and, where necessary, rejection as part of the quality control process. Production on this scale involves steel or concrete purpose-made moulds with the minimum of alteration between types, allowing greater accuracy to be achieved than where moulds have to be altered between casts. Conversely, small quantities may result in a multiplicity of mould components to allow the casting of varieties of the main element. Large numbers of elements of similar section offer opportunities for the use of extrusion, long-line beds or heavy steel and concrete moulds with their inherent stability.

Programme time frequently governs the selection of precast concrete. The preparation of a casting facility (where it is not possible to use proprietary supply or the facilities of an existing works) can be carried on while ground works proceed and foundations are prepared. Manufacture is carried out in time that is not critical to the overall duration of the contract. Uncoupling of manufacture and erection ensures that economic casting sequences can be developed with a minimum amount of production time being used up in mould changes.

3 Detail for manufacture

The designer must be concerned at all times with structural integrity. The further requirement of stability in the partially-built structure is also important. Handling and transportation introduce further considerations regarding the strength and ability of the unit to resist mechanical damage and handling stresses.

Modular coordination, the accuracy of individual elements, the ease and speed with which they can be assembled into the space allocated in the structure, the economy in materials which can be achieved through skilled design and the reduction of requirements for site skills, all affect the designer's choice of detail.

It is essential that the designer should have a clear understanding of the salient points of manufacture. He must understand not only the possibilities but also the limitations imposed by the production method.

3.1 Techniques available

3.1.1 *Wet-cast precast production*
The process of repetitive production of structural precast concrete is based upon the use of concrete which develops sufficient strength to allow demoulding and possibly handling at between 4 and 15 h.

For casting purposes, units are divided into groups or families and provision made in the design of the mould for its re-use to cast varying numbers of units. A timber-framed plyfaced mould may allow the casting of 30 or more units without major refurbishment; a light-steel mould may be used 100 times or more; and a heavy-steel mould over 1000 times. Glass-reinforced plastics and glass-reinforced concrete allow the production of highly geometric forms, but the number of uses may be limited to 30 or thereabouts.

Combinations of these materials can be used and, of course, dense concrete can be used as a backing to GRP, GRC or as a mould material in its own right. Concrete moulds can be readily formed, are stable and massive, and can be used several hundred times provided their corners are suitably rounded or strengthened and lead and draw are provided on the principal faces of the element being cast.

The degree of alteration which a mould must undergo in casting a group of elements is one of the factors deciding mould quantities. This must be balanced against the delivery programme in the light of available resources, such as skills, operatives available and concrete supply.

Preformed, preferably jig-assembled cages of steel reinforcement are assembled into the moulds, fixings and connections installed and the jig located. After the moulds have been checked, the concrete can be placed and compacted. Curing by enclosure, the use of membranes, steam or applied heat follows. When the concrete in the element achieves a strength between 5 and 10 MPa (700–1400 psi) demoulding and, in most instances, handling can take place.

Considerable care must be exercised over the selection of faces to be cast from the mould, as this choice governs the ease or otherwise of achieving compaction, surface finish and so on.

3.1.2 *Partial precasting*
Where an exceptional feature has to be incorporated into an otherwise standard or typical unit, partial precasting can yield economies. The technique, which must result from discussions between designer and manufacturer, is to precast exceptional features, corbel or end blocks. The precast feature incoporates projecting

bars or loops which provide the connection between the element proper and the feature.

3.1.3 *Tilting tables*

Steel tilting tables provide support to elements to resist the stresses imposed during early handling. Slender elements, sandwich elements incorporating thin veneers of stone, brick, marble slip and so on, and composites of normal and lightweight concrete and panels with a GRC skin are simply produced on tilting tables.

Concrete panels can be used to provide excellent tilting facilities. The merit of concrete is that, in emerging countries and low-budget situations, simple tilting tables may be cast at the workplace using local materials, with the possibility of breeding families of moulds [12].

Window openings, duct formers, feature formers, and so on, must be securely fastened to the table, sufficient lead and draw being allowed for the unit to clear from the mould during the stripping operation.

Vibratory compaction is generally achieved via external vibrators mounted under the table—heavy formers for edges or openings tend to dampen the vibration with the possibility of substandard compaction.

Although the newly cast panel receives adequate support during tilting, the designer should consider the forces imposed on the panel during stripping, particularly where the cast-in connecting bolts are used both for connection and lifting duties. Lifting loops, studs or anchors should be arranged within the simple reinforcing cage to resist rotational forces. Correct embedment depth and edge distances are essential.

3.1.4 *Tilting frames*

Quite apart from tilting tables which are integral parts of the mould arrangement, considerable use is made of tilting frames, both in works and on site. Elements of exceptional section or exceptional size require support during the handling process from the casting attitude to the loading or transport attitude. Elements are loaded into the frames, which may be fabricated from steel or concrete. Attachment is made as required to prevent slip or overturning during the rotation of the element, and the frame is then rotated.

3.1.5 *Battery casting*

Battery casting, using moulds of varying sophistication, is an economic way of producing substantial quantities of wall members, reinforced concrete floor units, balconies and units of similar plane configuration. Batteries can produce plane wall elements, angle wall elements, floor slabs cast on edge and balconies cast on edge. Any number of elements between two and 100 may be cast in each operation of the battery.

The cell dividers are generally constructed of steel or concrete. The casting machines have hydraulic closure and mechanical telfer beams, also specially-produced steel plate dividers which may be oil-filled or contain pipes for accelerated curing.

The thickness of the elements produced in a cell is governed by the width of the end formers. The stop ends are arranged to allow adjustment of the element

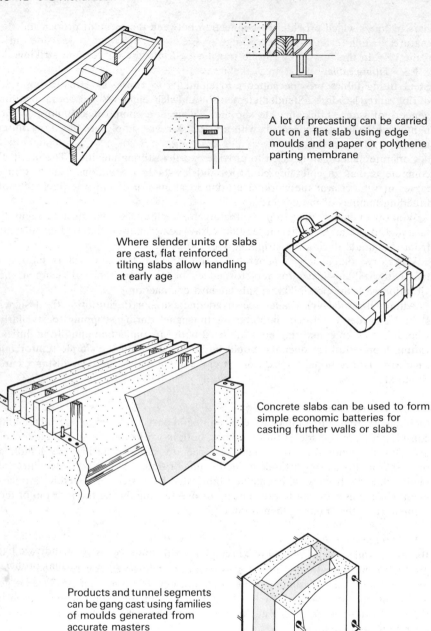

A lot of precasting can be carried out on a flat slab using edge moulds and a paper or polythene parting membrane

Where slender units or slabs are cast, flat reinforced tilting slabs allow handling at early age

Concrete slabs can be used to form simple economic batteries for casting further walls or slabs

Products and tunnel segments can be gang cast using families of moulds generated from accurate masters

Figure 3-1   Concrete in mould construction

length. Stools or pallets seated on a bed plate allow adjustment of the element height. Opening formers, service fittings and so on are attached to one side only of the cell. In this way, they can be transfer-cast into the concrete at striking.

Where low-cost construction is the target, battery casting offers considerable economies. A concrete battery can be generated from an accurately cast slab; the quality of the slab surface and the trowelling skills available determine not only the quality of the cell dividers but also the quality of the finish of any elements cast therefrom. The elements are cast, simply reinforced and highly finished; bearing plates are incorporated to allow installation into the gantry which provides the support upon which the cell members will be moved.

The gantry (again, concrete may be used in its construction) must offer a stable upper support with a suitable bearing rail for each end of the cell divider. The dividers are moved along the gantry by winch, jack or bar and wedged into position against dowel bars in the gantry beam and footing. Compaction is achieved in each cell by the use of an internal poker vibrator, although particular care is required to overcome the deadening effect of the dense concrete cell dividers.

Jigs will be used to support projecting dowel bars, cast-in billet connections or bolting angles, and these fitments must incorporate some means of bolting or clamping to resist displacement due to vibration. Lifting/continuity bolts or rods must be accurately located. Portal formers will provide the means for levelling and tying panels during erection. Cast-in sockets or ferrules facilitate the installation of push–pull props during erection.

3.1.6 Gang casting

Where a considerable number of elements of similar profile or section are required, such as piles, transmission poles, beams, lintels and the like, the gang-casting technique offers considerable economics as regards moulds and casting costs. The technique generally, but not essentially, employs concrete moulds cast from master units. In prestressed work, the gangs of moulds are linear, say four-gang units laid in six or eight rows wide by up to 200 m (656 ft) long. Abutments are formed at each end of the gang and wire storage reels, swifts or dispensers established to provide a continuous supply of wire.

The moulds, formed from concrete or steel, must have a lead or draw to facilitate the removal of the element. Where the gangs are constructed from individual steel-plate soffits and dividing plates, the plates are easily withdrawn by crane from between the elements, which may be of square or rectangular profile. The gang-casting technique can be developed into a combined gang and stack-casting arrangement by the partial withdrawal of divider plates, further units being cast against a suitable mould coating or parting agent.

The casting operation can be performed by discharge from agitation trucks or large dumper trucks—dispensing with the need to tie craneage, or hoisting capacity, to this part of the production activities. Compaction will best be achieved using internal vibrators. The gang mould can be serviced by gantry, tower or derrick crane, which will move steel cages into, and completed units out of the casting area.

3.1.7   *Segmental construction*

Although linear beam elements of I-section, box section, T and inverted T-section continue to provide economic means of construction for bridge decks, podia, and even large-scale parking houses, auditoria and so on, there has been a trend towards segmental construction.

Segments or sections taken transversely across the bridge deck have been precast, individually or using countercasting techniques. Elements have been produced both on site and in the works, and elements up to 120 tonnes weight have been transported by road to the construction location.

In the case of individually-produced segments, the joint between members will be of semi-dry or normal concrete compacted into a space of 25 to 300 mm (1–12 in) by mechanical means. The joint between countercast segments is made using synthetic resin applied by a gloved hand just prior to closure. Occasionally, some need arises to use a packing to correct minor problems of alignment; in this case, glass-fibre mat is embedded into the joint material.

The process of countercasting, now widely adopted for segmental construction, was used for many years in Europe to achieve a distribution of forces between linear-type beams which were subsequently stressed together using transverse tendons.

3.1.8   *Tilt-up construction*

The tilt-up techniques of precasting are gaining popularity—particularly in the USA and Australia [13]. This process is carried out by the preparation, at the construction site, of a series of forms, using the ground slab of the structure as a base. Elements are cast onto the base using a suitable parting agent. After they have been cast, the elements are rotated using cranes—and often quite intricate lifting equipment—into their final position as walls forming the enclosure. Casting is not limited to single-story structures and the finishes can comprise featured or exposed aggregate surfaces. The system is speedy (dispensing with much of the factory overhead) and can be carried out in virtually any location. Of course, where conditions are bad, some shelter must be provided and special attention paid to the curing process.

As well as the simple tilt-up technique using a ground slab as the soffit form, stack-casting techniques are sometimes employed. In stack casting, each unit provides the base mould for a succeeding element. The units are formed using side members which are climbed upon, cast-in fixings or tubes. Stacks are cast so that the first-required units are at the top of stack to reduce double handling.

The contractor retains the initiative and can take advantage of the economies of precasting with a minimum of expenditure on moulds and casting equipment. In many instances, the use of ready-mixed concrete provides further economy.

3.1.9   *The mass production of small elements*

One of the factors determining the use of precast concrete is the number of similar elements required. The designer can affect this factor by the manner in which he divides the structure into its precast component parts. Considerable economy can be achieved in this way in the design of tunnel and shaft or chimney segments, where Dolos, Stabit and sea-defence elements are to be produced in quantity, and where special blocks (even as large as 100 tonnes) are required in quantity.

Ideally, the manufacturing process, the supply of concrete, and the handling of the elements become parts of an integrated production system and the work becomes self-regulating with in-process checks incorporated to ensure the maintenance of quality.

The shape and size of a particular product will determine to a considerable extent the means of production. For example, where tunnel segments and similar units are required in large numbers, special steel or concrete moulds in gang form can be mounted on bogeys or rails which pass through cleaning, oiling and casting stations, into a curing chamber, in which the elements can be stored at an elevated temperature in conditions of humidity to ensure the development of the strength required for demoulding and handling.

Small units may be pressed in derivatives of the presses used for the manufacture of precast kerbs.

3.1.10   *Prestressed concrete*

The now vast range of elements produced by prestressing includes virtually every known form of construction, from ships, platforms, bridges, pressure vessels, storage tanks at one end of the scale, to the simple floor slab and lintel at the other [14]. The principles of construction are similar, but the details of tendons, anchorages and component configuration vary. The processes of manufacture involve clearly defined stages: mould assembly, steel reinforcement installation, and wire or tendon placement. Concreting and curing precedes or follows the stressing operations, depending on whether pretensioning or post-tensioning are employed. Frequently in the case of large elements, both pretensioning and post-tensioning are used, the latter generally being employed in the assembly process.

The tendons in post-tensioned concrete are arranged in the form of a catenary of special profile such that they combat the tensile and shear forces. Similar arrangements can be achieved in pretensioned work by the deflection of strand around strong points in the mould or on the casting bed [15]. As an alternative, debonding or sleeving of the tendon allows the force to be concentrated at various points within the concrete as required, avoiding the introduction of unwanted tensile forces for example, in the top of a beam.

Tendons may consist of wire, strand or bar, which may be used individually or in groups, but there is usually one feature which commends a particular type of tendon for a specific application.

3.1.10.1   *Ducts, anchorages and bearing plates*   The calculated profile of the duct must be indicated to the producer or constructor in terms of stations and offsets set out against the beam profile. The profiles must be maintained with specified limits of accuracy—CP 110 recommends ±5 mm (0.2 in). This can be achieved by the insertion of chairs or grillages securely attached to the reinforcing cage. Where tendons are inserted during casting, it is essential that the tendons are eased as the concrete hardens, to avoid the traps caused by cement paste which may infiltrate joints in the ducts. It is most important that ducts are located correctly in line with the anchor blocks or fittings.

When detailing the stressing points, the dimensions should be such that the nose of the stressing jack can bear against the plate and provide an axial load on the tendon system. The angles of the blocks vary at each tendon location both in plan and elevation. The lead and draw specified by the equipment supplier will allow the withdrawal of the formers without the concrete spalling in this critical area of the beam, and ensure that the recess will accommodate the nose of the jack.

3.1.10.2   *Camber or hog*   The camber in a prestressed element can be predicted. Camber will vary from one element to another due to differences in the tendon and duct location, concrete mix and compaction, curing technique, creep and relaxation. Generally, problems can be overcome by the grading of elements in terms of the amount of camber measured against a wire stressed from one end of an element to the other. The elements are placed in the structure in order of degree of camber, thus reducing the steps between units and grading out the differences from one side of a bridge span to the other or from one edge of a soffit to the other.

3.1.11   *The production of massive structural elements*
Where extremes of temperature or site conditions make conventional *in situ* construction difficult or impossible, precasting becomes imperative. For example, the larger parts of the foundation works for offshore concrete oil-production platforms have been successfully produced by precasting in a proprietary supplier's works, transporting by road, rail and sea and subsequently installing at the site.

Movement by flotation or partial support from ships or barges offers unlimited possibilities for the precasting of immense structural elements. One can visualize a situation where a structure or substantial part of it is precast on a suitable parting layer for subsequent jacking and removal to its eventual location.

Caissons (see Chapter 28) are exceptional pieces of precast concrete and means have been devised whereby precast caissons, cast in vertical lifts, are sunk by mining into location at the perimeter of, for example, a dry dock. The careful location and sinking of precast caissons reduces the amount of expensive *in situ* connection work required to complete the perimeter.

3.1.12   *Double-tee and single-tee and linear element production*
The wet-cast production of double-tee and special sections can be carried out extremely economically using either normal reinforced concrete or prestressed concrete. The units are ideal for precasting in that all faces may be designed with lead and draw without detriment to their appearance.

One-piece moulds reduce the errors likely to arise from constant assembly, stripping and reassembly. Standards of accuracy can be more readily maintained and, in many instances, the casting operation can be simplified. Accurate connections are best obtained by cast-in plates and angles connected by welding. Lifting studs, loops and anchors must be carefully installed into the reinforcing cage to resist the substantial forces which are required to part the large area of element and mould face.

Double-tee elements are used for both flooring and walling and can be produced incorporating a layer of thermal insulation between the beam and the slab component.

3.1.13 *Precast concrete permanent formwork*

In the construction of exceptional elements, such as columns of geometric profile or heavily-featured bridge abutments, precast permanent forms provide an economic solution to formwork problems. The use of such forms (which provide the permanent visual face of the concrete) permit the achievement of quality and shapes which would be difficult, if not impossible, to obtain using *in situ* techniques.

Thin elements, prestressed planking and the like have been used for slab supports between main beams, edge finishes to floor slabs, and so on, where technique, with repetitive casting from precast moulds, overcomes the problems of providing expensive geometric forms for *in situ* casting.

The use of thin prestressed planking as permanent forms introduces the need for supports at centres determined by the imposed load of fresh concrete—a need not always apparent to the site constructor. The designer must consider the pressures which develop during the filling process and design the precast forms accordingly, specifying, where necessary, the need for external support in the form of steel or timber yokes or strongbacks during the *in situ* operations. It is likely that, on examination of the pressures, arrangements utilizing cast-in sockets and tie bars will provide adequate support.

3.1.14 *Folded plates and parabolic shells*

These elements (see Chapter 32) have been used widely in Europe and to a more limited extent in the UK and USA. In certain parts of Italy and India, too, shell construction is much in evidence due to the enthusiasm generated by specialist engineers [16].

Shells are cast on steel, timber or concrete forms and in many instances the reinforcement is provided by prestressing tendons. Concrete is placed by machine and the thin-shell section is compacted by vibrators built into a vibratory screed guided by rails, which generate the outer shell form according to the engineer's design. Stripping is aided by the prestressing process and extremely large areas of roof enclosure are handled at one time with the aid of substantial articulated spreader bars.

3.1.15 *Box unit manufacture*

In this method of construction, extremely sophisticated mould arrangements are used to facilitate the casting of whole rooms or apartments, complete with services and surface finishes. Elements up to 90 tonnes have been erected into multistory buildings. One system in use in the UK utilizes fabrications of battery-cast walls and floors to form floored, three-sided enclosures, which are completely fitted and finished in the works for erection at site for housing, community centres, hotels, clinics, etc.

In practice, all connections for services can be made externally, the elements only being opened when occupation of the building commences.

3.1.16 *Jacking techniques*

Although essentially operations carried out on site and thus falling into the category 'in situ concrete', the processes of bridge jacking and lift slabbing employ precasting techniques.

Lift slabbing is an extension of the stack-casting process. Elements are cast one upon the other for subsequent separation and jacking into position up columns which slot into pockets in the slabs.

In bridge jacking, the segments are cast adjacent to an abutment at bridge level, and a launching nose is attached to bridge the gap between the abutment and the pier. As each successive element is cast into the back of the bridge, it is jacked forward using jacking bars pitched from the abutment, or by jacking from strongbacks behind the casting site. The consequent reduction in temporary works and falsework brings considerable economic advantage. Also, access can be maintained below the works at all times, which is essential where rail cuttings, for example, are being bridged.

Pipe and culvert jacking has provided an economic means of constructing pipelines, culverts, subways and pedestrian ways with minimal disturbance at ground level. Here, shafts are excavated to subway level, elements are lowered into the shaft and jacked from bearings provided at the shaft side.

3.1.17 *Stair units and steps*
The change of orientation of a precast element within its mould facilitates production and ensures sound compaction. In the production of stair flights and individual treads the following possibilities exist, each offering some advantage in meeting a particular requirement of the specification:

(a) Edge casting allows treads risers, soffits and one edge to be cast with ex-mould finish.

(b) Inverted casting allows the introduction of thin granolithic skins and the transfer of carborundum for non-slip purposes.

(c) where site precasting is being carried out, one static mould can be used, placed adjacent to the batching plant or concrete delivery point and stair elements may be cast as a non-critical job. The availability of precast stair flights which can be rapidly erected as the construction proceeds speeds up access to the working levels of a construction.

3.1.18 *Accelerated curing of precast elements*
Accelerated curing techniques are frequently employed in precast production in order to reduce establishment costs and to provide more economic use of expensive moulds and equipment.

The simplest means of accelerating production is to use hot water or steam in the mixing process, insulate skips and moulds, and thus take advantage of the improved rate of hydration of the cement and the gain in its strength. Further methods involve the use of heated casting beds and moulds, and heated cloches and chambers of high humidity.

Maturity expressed in terms of °C/h provides a guide to the strength achieved by concrete under a given curing regime. Various formulas have been published [17] which can be used to calculate maturity. The formula which most closely equates the maturity of the concrete to the strength achieved can be used in designing a curing cycle for a particular concrete. CP110, Part 1: 1972 [2] recommends a method for the calculation of maturity, the rate of increase of temperature (to a maximum of 80°C) and the rate of cooling.

It would appear that the maturity of concrete may prove to be an appropriate criterion for the stripping and handling times for precast elements provided sufficient work is done to ensure that the relationship between strength and maturity is clearly established [18].

3.2   Detail for structural precast concrete

A number of key points have been identified in the course of the production and erection of a variety of precast structures, including bridges, parking houses, industrialized buildings, precast frame structures and large-panel constructions [19]. This section covers the key points in respect of precast structural elements generally. Further points related particularly to structural/visual elements are dealt with in Section 3.3. Prestressed production is covered in Section 3.4. It will be evident that there is considerable interaction between the various items. For example, reinforcement cover and duct location will be factors to be considered in the case of both simply reinforced structural concrete and prestressed concrete.

Concrete cover to steel is of paramount importance, particularly where the precast element is to be used in exposed conditions. Table 19, Clause 3.11.2 in CP 110, Part 1: 1972 gives guidance on what is appropriate. Even in protected conditions, condensation or the ingress of salts may cause corrosion. Elements in situations where this is likely should be considered as subject to moderate exposure.

Loss of cover will result in the formation of expansive corrosion products capable of developing pressures within the concrete, which will spall away corners or produce areas of reduced cover. The depth of cover required is a function of the concrete quality and while relatively small depths of cover may be used with high-grade concrete, the designer must ensure that the cover is sufficient to resist the action of fire and should reinforce the concrete cover to main bars by the inclusion of a steel mesh.

The location of steel must be indicated clearly. The precaster uses concrete or plastic spacers between the steel reinforcement and the mould faces. A spacer should itself not include any metal prone to corrosion.

It is advisable to use the preferred shapes specified in national standards. BS 4416: 1969 [20] provides excellent guidance and indicates the method of calculating the lengths required for particular configurations of bar. Special attention should also be given to the specification of end cover to reinforcement and the avoidance of steel tying wire and clips encroaching on the concrete cover.

The profile accuracy should be specified in clauses which relate permissible deviation to the size and complexity of a unit. The recommendations of CP 110 regarding the reduction of permissible deviations to achieve improved performance in the structure are realistic, halving of the permissible deviations being acceptable in most contracts. The cost of precast elements will, however, be increased as tolerances are reduced and the designer should only specify extremely close tolerances where some particular point of connection or structural performance depends on the accuracy of the element.

The designer must be aware that the positions of fixings and fastenings depend to a great extent on sensible detail and on the care exercised by the producer (through the use of jigs and fixtures) to prevent their displacement during the casting process [2].

The stresses arising from the early handling of precast elements can be predicted and, where necessary, additional steel introduced to prevent excessive cracking. Large slender members can be expected to crack and permissible crack widths can be established and limited. However, the designer should be aware that, due to suction, friction and the trapping of concrete within returns and features, certain unpredictable stresses may be introduced into an element. In the case of units of exceptional slenderness or section, the designer should be prepared to inspect the first ones produced with a view to redesign to control cracking. The designer should specify the position of the lifting connections, indicating these clearly on detail drawings. They must be detailed with regard to the location of the centre of gravity of the concrete element, allowing for the insertion of gasket, shims and so on required to complete the structural connection or weatherproofing, while the element is suspended on the crane.

Lifting attachments cast into slabs must be concyclic. To avoid local high stresses, the attachments detailed must be suited to the direction of application of the lifting force, sufficiently well embedded into the element to avoid disruption during the early lifting process when the concrete is relatively 'green'.

The designer concerned with special structural units should consider the implications of the profile of a unit in terms of reinforcement location, concrete placement, and compaction, the installation of connections and lifting arrangements. Discussion with the supplier or, prior to contract, with a trade federation, will give guidance on the casting techniques most likely to be adopted.

It is generally easier to manufacture units which involve some reduction in profile or section which can be achieved by blockouts rather than corbels, ribs or features which require substantial alterations to the mould. Side profiles as cast are the simplest to change; soffit members are difficult, although corbels or steel can be allowed to project from the upper surface of the concrete.

Early in the preparation of final details, drawings of the profiles of all the units in one phase of a contract should be provided to the supplier, together with notes of projected repetition. Standardized and modular design will assist in achieving economy at all stages.

Where a particular profile is to be achieved from the connection of a number of adjacent precast units, it is sensible to specify that the elements be cast within a common mould. The units may be arranged for later dry packing or cast concrete connection and there is thus the possibility of some interchange of units.

At a connection, space must allow the introduction of concrete into and around the steel reinforcement *and compaction* to ensure structural integrity.

Discussions with a concrete specialist will result in the adoption of a concrete of appropriate mix proportions and aggregate size. The technique of placement should be described and regulated to ensure that air and workability fines are not trapped in corbels and features. Reduction of the aggregate size assists in placement, as does the introduction of superplasticizers or similar admixtures as flow promotors or water reducers. Air entrainment in the concrete also helps with placement and compaction and improves durability by combatting freeze–thaw effects. Where large elements are concerned, the specification should stipulate the rate and method of concreting. Retarders may be introduced into the concrete to control the rate of stiffening and ensure complete amalgamation of the constituent batches of concrete.

Where the element is to be cored for lightness, cavity formers of cardboard or metal foil are introduced into the unit. The displacement of formers can result in unpredictable high stresses in thin-wall members. The collapse of formers causes overweight units—a hazard in handling operations. Upward movement due to the buoyancy of the core may produce a substandard unit with unplanned thickening of its flanges, together with adverse effects on the cover to the steel and the load-carrying capacity of the unit.

During the casting operation, it is essential to avoid displacement of the fixings due to vibratory compaction.

3.3   Detail for structural/visual precast concrete

Many of the precast elements used in building and civil engineering combine the structural and visual attributes of concrete. A wide range of finishes can be achieved on the concrete by direct means, from the mould, or by indirect means whereby further work is carried out on the cast concrete. Surface finishes which are difficult or impossible to produce *in situ* can be produced by precasting.

CP 297 [10] describes the factors to be considered in the choice of finishes. Guidance is given on weathering, atmospheric pollution, frost action, efflorescence, lime bloom and corrosion staining. Reference is made to the incompatibility of certain materials and the problems of water draining, for example, from concrete onto glazing and etching the surface of the glass.

The nature of the finish selected for the visual component may dictate which surfaces are cast against a mould face. Brick slip, tile and mosaic can be simply transferred from moulded faces although post-application can also be employed. Design criteria have long been established for faced elements of this nature both as regards the bonding of the facing material and the counteraction of bow set up by differential movement, shrinkage, thermal movement and creep [21].

Where brick slip, stone or marble are used as a facing to cladding elements, there must be a positive key. This can be achieved by using slips with dovetail or similar format, or by the insertion of non-ferrous dowels into each attachment. The use of Nylon mesh or strapping is a means of ensuring mechanical fixity of the veneers, as is stainless steel wire passed through the individual elements. In faced and exposed aggregate elements, the detail at arrises should ideally incorporate chamfers or bull-noses to avoid local damage in handling and tooling operations.

Margins ensure that a formal straight line is maintained when the joint is viewed obliquely and that tooling and brushing can be neatly terminated in a straight line without loss of arris. In general terms, the larger the element, the more substantial must features be.

Features and geometric form cast from the mould reduce the need for manual skills. Whenever profiled elements are used or where a part of the concrete is removed by cutting, tooling or retardation of the surface for effect, then particular attention is required to ensure the maintenance of the correct cover to the reinforcement, 4 mm (0.16 in) of extra cover being allowed for light tooling and greater cover for exposure by retarder or heavy tooling.

While extremely slender sections can be precast, the designer must consider the stresses imposed by early handling, rotation and final erection of such units into the structure. As the visual effect is of extreme importance, great care must be exercised on the control of consistency of appearance. Units should be inspected at works and graded to avoid the placement adjacent to one another of elements of markedly different exposure colour or shade. Ideally, complete floors of elements should be viewed and marked for location in the structure.

A specification which excludes repair may be appropriate in exceptionally high-quality work. Where repairs are permitted, there should be a carefully prepared specification for, respectively, preparatory, structural and visual remedial work. Ideally, such a specification should be related to the appearance and performance of similar remedial work executed at a similar standard some ten years earlier, where some assessment of the performance of the repairs is possible. Where the repair is subject to stress, some suitable loading test is advisable.

Structural/visual concrete may be cast using two distinct mixes: an aggregate-rich one for the facing or surfaces, and normal grey structural concrete for the core backing of the element. The facing mix must be as strong and durable as the structural mix. The casting must be carried out in a way which ensures the production of a monolithic concrete element, and avoids dry joints at the interface between the two mixes. Complicated or heavily-featured elements can be produced more economically using a 'through' mix of special aggregate concrete—again designed to provide the required strength and durability.

For slender sections or where particularly exposed or aggressive conditions are to be encountered, galvanized reinforcement or composite steels incorporating a stainless steel coating should be specified.

Where a connection is to be cast between precast elements or between precast and *in situ* concrete, it is highly desirable that the connecting steel should be incorporated into one of the two elements to be joined. The projecting steel or exposed caged steel ensures a positive connection. Adjacent concrete may be ribbed-texture or aggregate-exposed to impove the bond between the component parts.

Weathering of the surface causes long-term changes in the appearance of concrete surfaces and the structure as a whole [22]. The strategic location of features and the orientation and scale of detail, such as striations, can assist in the control of the direction and speed of water movement across and down a unit.

3.4 Detail for prestressed concrete

This section covers general points of detail and specification.

Where proprietary standard elements are being used, the manufacturer's quotation should be checked for details of strength characteristics, accuracy, surface finish and similar details.

In selecting a section, the designer is advised to check standard sections obtainable from manufacturers. In most countries, groups of manufacturers have co-operated to offer preferred sections and have set up their casting facilities accordingly. Often, small alterations to a mould section can be made to incorporate a designer's special requirements and yet achieve economy in production.

In detailing prestressed elements, the designer must indicate the position of tendons in such a manner that the dimensions can be readily transferred to the mould or casting machine. Ducts for tendons are located by stop ends, by attachment to anchorage equipment and by grillages attached to the steel reinforcement. Where long-line production is used, it is frequently desirable to deflect

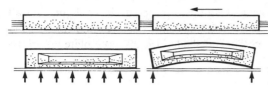

At transfer, units in long-line casting move towards the fixed abutment bearing plates and projections must be free to slide. Transfer of force into prestressed or post-tensioned concrete changes load pattern on base or falsework

Status of partially-tensioned elements must be indicated

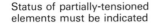

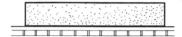

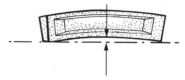

The manufacturer will grade units and deliver in sequence to avoid differential camber causing upset in soffits

Stacking battens should be at or near ends of most prestressed units, and directly above each other

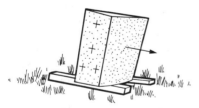

Bad ground or bad stacking can result in sideways deflection and creep

Most prestressed plank and flooring require propping until topping is complete and achieves specified strength. Avoid 'dumping' of concrete

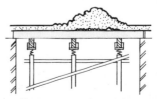

Figure 3-2   Prestressed concrete production and handling

tendons to form a catenary arrangement. In this case, the designer should consult with the manufacturer regarding the preferred system for strand or wire deflection. Some manufacturers prefer to use debonding techniques rather than strand deflection.

It is advisable that complicated sections of the structural element should be set out full size.

Any projections or details at the soffit of beams must be accommodated in pockets such that they are free to move on the application of force to the individual post-tensioned unit or line of pretensioned units.

The designer must ensure that the mould materials and construction will produce the required surface finish. Care must also be taken to avoid restraint of the concrete during the stressing operation. Some methods of manufacture, such as extrusion, preclude the inclusion of projecting bars so that where connection is required to adjacent units, welding plates or pockets for concrete stitches must be detailed.

The characteristics of the wire, strand or bars should be specified with reference to the appropriate standards. The initial stress in each strand, wire or bar and the corresponding extension must be stated. The manufacturer's certificate for each consignment of strand, including load–extension graphs, must be requested, together with tensioning records, including extension, gauge reading and pull-in at anchorages. Certificates of calibration of stressing equipment must also be requested.

In the post-tensioning process, the designer is advised to study carefully the recommendations made by the stressing equipment supplier as regards the anchorage detail, the continuity of line of duct, the angle with the end plate and the pocket size to accommodate the nose of the jack. Adequate safe access to the ends of units must be available to enable work to be carried out correctly.

It is essential that the concrete strength should be checked prior to the introduction of the prestressing force. For this purpose, the specifications must cover the testing of specimens made at the completion of casting operations.

In pretensioning, it is advisable to specify that all tendons are detensioned simultaneously. In post-tensioning, the sequence of stressing of individual tendons recommended by the manufacturer must be observed to avoid trapping against duct walls. The designer must issue instructions regarding the sequence of stressing complete tendons, stating where partial prestressing is allowable. It is advisable to stipulate that no tendon be trimmed before the stressing records have been checked to ascertain that the correct force has been introduced into each tendon.

Angle of skew and particularly 'hand' of skew must be clearly indicated to avoid confusion during forming and casting operations.

Ducts incorporated for subsequent main stressing operations, or for transverse or vertical stressing operations, should be proved by passing a bar or bars of similar configuration to that finally incorporated. Joints between ducts can be achieved prior to concrete placement using threaded metal or heat-shrink plastic connectors. The grouting of ducts must be covered by the specification. The FIP recommendations [23] must be observed to avoid the formation of air or water pockets, which can form the site of subsequent corrosive attack on the tendons. Precautions are required against the freezing of grout.

Cambers in prestressed units should be measured at the transfer of force and then weekly until erection. This check enables exceptions to be observed.

Units must be clearly marked with a reference number and the information necessary to ensure their correct location in the structure. The status of units which are partially tensioned must be indicated. Units of similar section cast with different forces or a different number of tendons must be clearly marked.

Where a prestressed element is dependent for its function upon some subsequently placed concrete or tendon, then the location of the supporting members during handling and erection, and up to the stage of completion, must be clearly indicated on the drawings and incorporated in the method statement.

All parties to a contract have a responsibility for safety and the recommendations of the FIP [24] provide useful guidance on these matters in respect of prestressed concrete.

4   Connections, fastenings and fixings

4.1   The concept of instant erection

The designer must so arrange the details that connections can be made rapidly and that 'wet' connections are avoided or made at a subsequent stage and in such a manner that the crane or handling equipment can be freed for other duties (consistent with safe working and the maintenance of structural stability).

The connections must be as simple to make as possible consistent with safety and stability. It is essential that sound fixings are incorporated both in *in situ* frame and supporting members and in the precast elements comprising a structure, to enable braces to be attached in the simplest manner. Welding must be undertaken by qualified welders who, as well as having recent experience of a similar work, must have prepared test pieces to a standard which satisfies the acknowledged testing authority. Bolted connections have the advantage that local adjustment can be made, although such adjustments must be subject to approval by the site engineer. The introduction at connections of packs of shims of steel, plastic or other materials must be carefully controlled as the efficiency of the joint can be impaired.

Throughout the erection, checks must be maintained for the stability of the structure and its ability to resist the loads imposed by winds (in exposed conditions), the construction and the plant and equipment.

The erection sequence must be so arranged that wind walls, shear walls, sway braces, etc., are installed as the erection proceeds. The precast erection must be carefully phased into the overall construction program. This is especially important where many operations depend upon the available craneage. Where the operation is extremely demanding in crane time, it will be advisable to introduce some independent means of handling the precast concrete. This may involve the installation of an erection gantry or traveller capable of being worked under and within the main site lifting equipment.

The most direct method is to arrange the deliveries so that erection takes place from the vehicle. However, a balance must be achieved between economic vehicle usage, crane utilization and the continuing supply of other materials into the work place [25]. In congested conditions and where erection is proceeding on several phases of a contract at one time, site stocks assist in uncoupling delivery and erection processes. The provision of stacking areas and buffer stocks on site is also worth while where considerable quantities of ancillary jointing materials, services, etc, are to be installed at the time of erection. Where panels can be placed for later lining, levelling and jointing, then high rates of placement and a rapid turn round of vehicles can be achieved.

4.2   Inbuilt disciplines: line, level and plumb

Connections can be designed such that elements are placed into an allocated space quickly and suitably braced to allow further operations to proceed. Final adjustment to line level and plumb is made a secondary operation. This technique is most applicable in, for example, industrialized building, where the construction comprises a multiplicity of relatively small units, say between 3 and 6 tonnes, which allows construction to proceed using cranes of medium capacity, usually mobile or on rail tracks.

An alternative technique favoured in the rapid construction of multistorey blocks, offices, schools and structures of a special nature (such as grandstands and sports stadia) is the use of column and beam or beam and slab combined in a single unit. Most structures can be sensibly designed in the form of elements combining what have previously been considered as separate components in the structure. Stairs and stair walls, stringer beams and stairs, balconies and floor slabs are but a few of the possibilities. It is also possible to frame together numbers of elements performing the same function, i.e. gangs of columns, gangs of frame elements. It is possible to make such groups of units 'self levelling' or at least 'self spacing' [26].

In the USA and in Europe, considerable use has been made of combined wall and beam elements of several stories height. Koncz [25] reported such work in the late 1960s and further developments have been reported by Morris [27].

4.3   Cast-in connections and fixings

A frequent cause of trouble in erection is dimensional discrepancy between *in situ* concrete and precast elements. Problems arise from misplaced foundation bolts, wrongly located pockets for columns, innaccuracy in the core structure and the wrong location of cast-in connections.

Fixing and connection design may be approached in two ways:

(a) First, by the introduction of disciplines into the *in situ* casting and the precasting by the use of jigs and templates to ensure success in obtaining extremely accurate (and expensive) fits.

(b) Alternatively, sufficient allowance must be made to accommodate discrepancies arising from the inherent and induced inaccuracy. The choice depends of course, upon the type of structure, the function of the element and considerations such as the economics of the contract, the visual requirements and so on.

4.3.1   *Connections between structural elements*
Connections must be designed to transmit the forces through the structure in the most direct way. The work in forming the connections must be broken down into

simple stages which can be carried out successfully under the prevalent conditions of weather and exposure.

Connections must provide stable fixings which allow instant removal of the crane for its use in other work. Ideally, a connection should provide this instant fixing, allow for adjustment *and* form a permanent joint between the units.

4.3.2 *Cast connections*

The simplest connection is that formed by cast concrete, generally between foundation and column base. In this type of connection, as with all other stitched cast connections, the concrete used must be carefully specified and controlled as regards its strength, durability and so on, to avoid the use of substandard material.

The spaces for stitching should be such that the concrete can be properly compacted in joints of 25–50 mm (1–2 in) width by grouting or by packing earth-dry concrete into place using mechanical caulking tools. Where a structural concrete mix is used in stitching, a 300 mm (1 ft) wide joint allows the introduction of a poker vibrator to achieve compaction.

In joints made by dry packing, any levelling bolts or bolts providing tie between members should be adjusted until the forces in the elements are transmitted uniformly over the hardened pack of the joint and not as point loads at the bolt positions. This will avoid local overstressing of cill and head members in system construction.

A formal inspection must be made of all steel projecting from elements which is to be incorporated into stitching concrete, to ensure that it is in the design location and is capable of providing the required tie force.

4.3.3 *Bolted connections and connections employing screw threads*

Connections that utilize threaded bars or bolts embedded in the concrete provide instant fixity. The bolts or threaded bars can be contained within the section of the elements to be joined and later protected by cast concrete.

A range of bolted or sleeved connections are used, including counter-threaded bars and sleeves which allow plumbing and levelling to be speedily carried out in column and beam connections. Torque-measuring nuts, washers or wrenches should be used to achieve the correct fixing. Site-swaged connections can be simply made. Hydraulically-operated cropping and swaging tools make these connections quickly and efficiently.

Column-to-foundation connections use cast-in bolts which pass through plates attached to the column. Where the column base can be concealed by flooring concrete or haunching, the plates can project around the column.

Bolted connections are used in many agricultural and commercial frame buildings and similar units, where halved or scarfed joints are employed. Felt or plastics washers should be incorporated in the joints and at washer positions to avoid point contact and cracking of the concrete.

Plates and angles are required in many situations where bolted connections are employed. Reference should be made to Section 4.3.7 for guidance on the protection of these important items.

4.3.4 *Dowelled connections*

In dowelled connections, bars projecting from the top of a column unit pass into tubes cast in the base of the next column unit. Gaskets are installed as the upper column is placed, the column being braced while further erection proceeds. In a subsequent operation, the bars are bonded by the injection (through small tubes) of a colloidal grout, sometimes by a resin mortar formulation. This joint reproduces the normal *in situ* construction technique and is rather expensive in labour and materials.

Where a beam is located onto a corbel, it is good practice to use dowels which project from the corbel into pockets in the beam. This arrangement limits the movement of the beam such that the necessary bearing area is maintained. This technique is frequently employed at the bearing connection between cladding panels and corbels, although in this instance sleeved dowels are used to avoid restraining the panels, which would otherwise bow or crack.

4.3.5 *Welded connections*
Welds are made between plates or angles embedded into concrete elements. Where the plates are cast flush into the concrete, it is often necessary to insert bridge pieces of bar. Careful checking of the welding procedure is essential and the specification must include clauses covering the conditions under which welding is allowed, the type of welds and welding equipment which may be employed, instructions, tack welding, preheating and such detail as weld profile, smoothness and the correction of faults. Limits must be set for the quality of test pieces with regard to inclusion, porosity and so on.

The detail for welded connections must be based upon the best practical experience. Where welding plates are used, lands or margins must be left and chamfers provided into which the required weld profile can be built.

4.3.6 *Connections made by post-tensioning*
The use of high-tensile steel bars and strands provides a quick and economic means for connecting and attaching elements, brackets, blocks and so on.

The initial force in the bar or tendon can be designed to exceed that resulting from loading, and the joint can be made directly on location using standard equipment. Pretensioned units bearing onto corbels on reinforced beams can be connected by subsequent post-tensioning. H-frame members can be connected vertically using welded plates and laterally by tendons running through a number of units. This type of connection is a logical extension of the prestressing process and is the essential ingredient of segmental and counter-casting techniques. Elements may be suspended from yokes built into columns, using threaded bar, the anchor nuts being taken to the required torque mechanically. Alternatively, the bars are stressed and nuts run into place to transmit the force into the joint.

4.3.7 *Coatings and the protection of steel fixings*
While fixings made from non-ferrous metals provide the greatest assurance against corrosion, it is still permissible in some countries to use carbon steel or malleable iron fixings. In these cases, coating with zinc in one of the following forms is recommended [28]: hot-dip galvanizing, zinc plating (electroplating), and sherardizing.

Hot-dip galvanizing provides good protection with a heavy but uneven coating. The largest of fixings can be coated by this method but the treatment leaves small holes, and threads require reworking with the danger of leaving metal exposed to corrosion. Components with holes of 25 mm (1 in) diameter or equivalent and with a minimum of 3 mm (0.1 in) clearance over the final hole size may be satisfactorily hot dipped and the process is generally used only for clips, washers and so on.

Zinc plating provides a thin, uniform and ductile coating on carbon steel or malleable iron. Small holes and threads can be satisfactorily coated. There is a maximum size restriction.

Sherardizing—the diffusion of hot (about 300°C) zinc dust onto the component surface—gives a thin, uniform, protective coating. Components with small holes can be successfully coated although this process is not recommended for threaded fixings. Most normal-sized fixings can be coated.

Where zinc-coated fixings are not accessible for maintenance, the use of non-ferrous fixings is advisable. The coating is liable to get chipped and scratched and rusted holes or threads are a weakness. The designer is warned of 'strain age embrittlement' under which some carbon steels become brittle and are liable to fracture if they are cold-worked prior to hot dipping.

5 Handling and transport

5.1 Stripping and storage

5.1.1 *Stripping*
Structural and visual elements are generally stripped (removed from their moulds) at time varying between $1\frac{1}{2}$ and 15 h after casting. The early removal of side members and features allows both quick re-use of a mould and ease of surface preparation. The cost of complex, and thus expensive, formers can be offset by using them with a number of moulds of the same family.

Normally-reinforced elements can usually be handled when the concrete strength reaches 5–10 MPa (700–1400 psi), as indicated by test specimens stored with the elements in the casting place. Prestressed elements can be moved from the casting bed only when the concrete has achieved adequate compressive and bond strength. In works with accelerated curing facilities, this condition may be achieved between 4 and 15 h.

The important aspect of stripping is that it must be achieved without damage to either the concrete element or the expensive mould on or in which it has been cast. Where slender units are to be removed from moulds, pockets should be provided for the insertion of jacks or slings at the correct lifting positions. Tilting tables and battery moulds both provide for the handling of "green" concrete units without the introduction of bending or handling stresses.

In the production of cladding elements, it is common practice to rotate the mould from the casting position into such a position that the mouldwork can be removed with the cladding panel suitably supported to avoid excessive stress.

5.1.2 *The stackyard and storage*
It is essential that the newly-cast elements are stacked with care, correctly supported on clean non-staining battens, bearers or spacers. The supports must be placed with regard to the design of the elements, whether precast or prestressed.

Care must be taken to provide suitable support for cantilever members and to ensure that all bearers between like members are located one above the other, to avoid damage to units during loading.

Where concrete elements are finished smooth top and bottom, they may, after the upper surface has been swept clear, be rested one upon the other. This is essential where layers of insulation material may be crushed by the point loading resulting from the introduction of battens. Large panels which are to be stored upright must be stacked either in vertical racks or against suitably designed A-frame supports. If panels are left in torsion for any period of time, the creep in the concrete will cause residual wind, which will render erection difficult and may make jointing impossible.

Creep must be considered where any long, slender member is stacked. The slightest amount of out-of-plumb stacking will result in sideways distortion, sufficient in extreme cases to prevent, for example, a bridge beam being placed immediately adjacent to another in the structure. The effects of creep and differential curing due to orientation in the stackyard can be considerable and are, of course, time dependent.

The stackyard area must be prepared to avoid local settlement with resultant damage to elements and the toppling of stacks.

Straddle lifts, mobile derricks and mobile tower cranes are suitable means of stacking and handling linear elements. Space below or adjacent to the plant can be fully utilized for stacking space. In the case of derrick cranes, the lifting capacity remains constant over a large swept area and elements can be arranged such that half their mass is outside the swept area, thus increasing available storage capacity.

When setting up stacks incorporating roadways for mobile cranes, it must be remembered that the area swept by the crane kentledge is greater than that of the tracks or chassis and appropriate allowance made in the width of the roadways.

5.2 Handling precast concrete

The connection between the lifting equipment and the concrete unit is most critical. The forces imposed on the equipment at the time of striking, while the concrete is relatively 'green', may well exceed the most stringent service loading conditions. The capacity of lifting equipment cast into concrete must therefore be factored down according to concrete strength at the time of use.

A variety of lifting connections exists, and includes bar, strand, plate, threaded connectors and studs. Many are proprietary fittings [1] which can be selected from a catalogue.

The simplest and perhaps cheapest way of connecting concrete elements and lifting equipment is a bolt or bar thrust through suitably located holes in the element. The lifting bar or bolt used must be a good fit in the hole and of such a diameter as to avoid bending under load. Large washers and pins or nuts should ensure that the connection to the crane is made close to the face of the unit and to avoid hooks or plates slipping from the bar during the lifting operation.

5.2.1 *Loops, hooks and plates*
It is normal and acceptable practice to use reinforcing bar, strand and, in the case of heavy elements, multiple assemblies of steel bar, strand and steel plate. The plate will be pierced to suit the hooks or lifting brothers chains, etc.

The permissible direction of application of force to the attachment must be indicated. Where considerable forces are encountered, the bearing and bond may be supplemented using embedded plates.

For massive precast elements such as bridge segments, it may be desirable to arrange to bolt shackle trunnion plates for substantial shackles to the element, using through-bolts or heavy embedded bolts.

The use of high-yield high-bonded steel in lifting equipment should be avoided in order to eliminate the possibility of failure due to reworking during handling.

5.2.2   Proprietary studs and cast-in connections

A wide range of ready-made components are available for connecting the lifting equipment and the precast element. Most are patented arrangements which offer some special facility for ease of embedment, speed of connection and so on. The capacity of these devices is established by tests and published by the manufacturers. Adequate safety factors must be applied and allowance made for the strength of the concrete in the element at the time of use. It must be clearly understood that not all such connecting equipment can be used for turning as well as lifting.

5.2.3   Eyebolts and threaded attachments

Care must be taken to ensure full engagement of the thread. Shrouded threads should be avoided except where the eyebolt attachment takes support from the tapered socket in which the threaded element is set. Sockets which are internally threaded for the insertion of lifting bolts, eyes, etc., must be plugged to avoid the ingress of grout, dirt or other extraneous material that will clog the thread. Where flexible strand-loop connections are used in conjunction with cast-in sockets, care should be taken either to ensure a direct tensile load or to use the special metal cup designed to protect the strand loop from damage by abrupt change of direction.

5.2.4   Location of lifting attachments

The fittings must be located so that neither they nor the concrete are subjected to excessive force or local disruption at the time of lifting. This entails ensuring that the connections are positioned according to whether the units are precast or prestressed.

With a heavily profiled element, it may be necessary to provide brackets for bolting to the element to relate the attachment to the centre of gravity, thereby facilitating turning or handling and eventual erection.

Lifting attachment to normal reinforced concrete elements is usually made at the one-fifth points from each end of an element. With prestressed elements, attachment is made at or near the ends of an element. In the case of a cantilever element, the connections should be made at the bearing positions.

5.2.5   Spreader beams and lifting rigs

Where long and slender elements are to be handled and, indeed, in all situations where lifting force is applied to thin panels, a spreader beam must be used. The beam, which allows the position of hooks or shackles to be adjusted relative to the line of the cranehook, ensures that the lifting force is applied axially to bolts and similar lifting connections. The suspension points can be adjusted to ensure that the element being handled is suspended at the required inclination. It is also possible in this way to allow for units of differing sections to be handled by one crane yet remain horizontal. The safe working load must be indicated on the beam and proof testing should be carried out at regular intervals. Various combinations of lifting beams are used for handling slabs and thin shells, secondary beams being incorporated in such a way as to avoid the introduction of torque into the element being lifted.

Where brother chains are used between a crane hook and a lifting connection, the chains should be as long as the clearance between the unit and the crane tackle will allow. Absolute care must be taken to ensure that the operatives understand the importance of proper connections.

5.3   Transport arrangements

5.3.1   *Road transport*
Many precasters run their own fleets of specialized vehicles; low loaders, trailers and tractors, semi-trailers and bogeys. They are generally fitted with special supports, cribs and A-frames to ensure the correct loading of elements to avoid unintended stresses.

Where prestressed elements—particularly deep I and inverted tees—are concerned, care must be taken to avoid torsion in the load. Also, special care must be taken to avoid the situation where, because of the inclination of the beam, the position of the centre of gravity of its mass and that of centroid of the prestressing force are such that tensile forces develop capable of damaging the beam.

The transportation of exceptional units requires properly engineered frames to support the units on the vehicle, trailer or turntable. Chains and turnbuckles are used to secure the load. All such equipment must be checked at frequent intervals en route. Trade practice generally dictates the way in which elements are loaded onto a vehicle; if they are required to be delivered to site a particular way up, this should be specified.

Tractors and trailers are usually the most convenient means of handling large quantities of elements. On a large site, a tractor may be required to shunt trailers into the off-loading position.

There are situations where site access dictates the use of a particular type of vehicle, rigid or articulated. This must be made clear to contractors or hauliers.

5.3.2   *Rail transport*
Where the precaster or the site is sufficiently near to a railhead, rail becomes an economic means of conveying concrete elements over distances. Where railway works are concerned, rail transport is often specified. The economics must be carefully examined, however, in the light of the increased amount of handling introduced between works and rail or rail and site. Attention is required to routing; the sequence of delivery to a rail bridge or railway working can be upset by the routing when it is possible for a train to be reversed end-for-end.

Where 'possession' times are limited, suitable arrangements must be made for a switching locomotive to handle the wagons at the site. Rail-mounted cranes are excellent in terms of lifting capacity, and the drivers are usually exceptionally highly skilled, but the manoeuvrability is limited. Care in loading is essential to avoid damaging the elements in the course of the switching operations. The restrictions of loading gauge often cause problems where the large panels are concerned. The almost universal use of containers makes the transport of smaller elements by rail an extremely economic operation.

5.3.3 *Transport by water*

In the cases of jetties, bridges and similar works over, and adjacent to water, the use of water-borne transport is frequently specified. In the construction of flood barriers and similar schemes, water is the most obvious and economic means of transporting concrete.

Remarkably, concrete is often exported from one country to another. Sleepers, piles, building elements and the like have all been exported at some time; often by barge and ship after transfer from road or rail. However, dock labour rarely understands the special requirements of slinging, handling and loading precast elements and careful supervision of these points is essential to eliminate damage. Ideally, elements should be closely packed to avoid movement. This is appropriate where most precast elements are concerned, although provision must be made for bolting or connecting slender elements one to another to avoid sideways vibration, movement and bow.

5.4 Receipt at the construction site

Arrangements for delivery and storage include the provision of suitable access to the site, vehicle standing space and unit stacking space.

Once the delivery route has been established and a survey made to ensure clearance below bridges, power cables and so on, then vehicle access to the site must be organized. The load-bearing capacity of roads and hard standings must be checked in the light of the high-point loadings, particularly from crane outriggers.

Clearance under scaffold, entrance ways through temporary works supporting structures and the like must be clearly established and markers provided to limit vehicle and bogey travel such that load overhang on bends does not interfere with temporary works. In cities and congested sites, care must be taken over the vehicle standing areas. In the case of work adjacent to rail or public roads, permission for 'possession' must be obtained from the police or the appropriate authority.

A policy decision will have established whether unit stock is to be held at site or whether units are to be erected direct from the vehicles. In the event of stock being held on site, it is essential to make careful arrangement regarding stacking and stockholding documentation.

Where part-castings are to be assembled into complete elements on site, appropriate space must be planned with room for manoeuvre and crane movement. In this instance, the assembly area should be separate from the delivery and stocking areas and must be treated as a workplace in terms of safety regulations.

Where elements are to be loaded onto temporary works, it is essential to obtain the permission of the temporary works co-ordinator. At that stage, a careful check of the location of bearings and platforms must be made to ensure that they comply with the engineer's intention.

The use of precast elements as permanent forms requires that care be taken to ascertain the direction of the resulting loads on the falsework. This check on falsework must be ongoing throughout the construction process.

6 Erection of precast concrete

The speed and efficiency of the erection process are dependent upon skilled design and careful manufacture. The performance of the completed structure is a function of the careful assessment of loads and the capability of the precast elements to sustain those loads. The designer, in his consideration of material selection, method of construction and means of connection, must at all times bear in mind the erection process and the need for simple direct connections and fixing which can be made under the conditions prevailing at the construction site.

6.1 The method statement

The most successful precast operations are those where the safe, speedy erection of elements into a finished structure has resulted in some specific gain in time, quality of construction and/or economy. This is generally the case where erection has been carried out according to a carefully prepared method statement.

Once the nature of the elements has been established and some assessment made of the weights to be handled, it becomes possible to visualize the most appropriate method of construction.

The precast operation must be considered in total. The place and methods of manufacture, the means of transport and, finally, the erection techniques, while capable of being uncoupled in terms of output, are closely interlinked.

It is essential that, at an early stage (ideally in the pre-contract stages), the designer, manufacturer and erector should discuss the projected method and determine what sequence of operations is to be adopted in erection, how the work is to be phased and so on. At this stage, site layouts, preliminary drawings and the profiles of the structural elements with an indication of their weights, should be made available. Information on transport requirements, restrictions on access, critical dates and time relative to the phases of construction and similar information must be provided.

The method statement should result from a co-operative effort on the part of those who will be concerned in the erection process and should include

> Results of mock-up and trial assembly
> Established standards of accuracy and finish
> Location of sample
> Program for manufacture
> Minimum stockholding
> Method of transportation
> Special vehicle requirements
> Delivery rate and sequence
> Site stockholding
> Site layout
> > stock
> > stores
> > offices
> > welfare facilities

Handling equipment
    cranes
    travellers
    hoists
    winches
    spreader bars, chains, etc.
Location of cranes and handling equipment
    placement and temporary support
    setting out
    lining and levelling
Means of connection
Means of weatherproofing
Site organization
    number of teams
    supervision
    administration
    safety and welfare
Responsibility for checking handover and acceptance

This method statement must make clear the designer's intentions regarding the joint and connection procedures particularly where extended or temporary bearings are to be used in the construction process.

6.2   Selection of plant and equipment

The selection of plant and equipment is governed by the size and weight of the indivisible components in the structure, and by the conditions in which the plant must operate (mobility, state of ground, access and similar restraints). Two other factors in the choice of plant are its availability in the contractors' yards and the contractors' experience in the use of particular types of equipment. For example, civil engineering contractors have, for many years, used scotch derrick cranes for a wide variety of services, including the handling operations in the site manufacture of concrete element.

Changes in the design of plant, particularly the use of hydraulics in lifting equipment, have made the erection of cranes a speedy process. In the past, a heavy crane with a lattice jib may have taken half a day to erect and involved the use of a smaller crane to lift the jib. Today, an hydraulic crane can be in action almost immediately on arrival at site. This speed of erection and, in all but the largest cranes, ability to travel with a load provide the flexibility essential to the erection of special precast structures.

In building construction, where the elements do not normally exceed 3–6 tonnes in weight, erection is usually carried out in the course of the construction process, making use of the site tower crane. Exceptional lifts, such as at the perimeter of the site or where large units exceed the capacity of the site cranes, are likely to be made by a mobile crane hired for that duty. It must be remembered that a tower crane's lifting capacity is considerably reduced as the reach is increased.

In civil engineering works and in site casting yards, scotch derricks, either static or rail-mounted, provide reliable heavy lifting facilities over the swept radius. The economy in gantry cranes lies in the fact that only the bridge and crab need be in steel; the gantry columns and beams supporting the rails can be of concrete and thus may be cast on site.

Where exceptional loads are involved and where the handling is such an integral part of the construction process that it limits the speed of erection, special travellers or erector beams are used. Their use requires special care in the detailing, plant locations and of the connections between the precast elements and the lifting equipment. Precautions must be taken to ensure that the kentledge on travellers is sufficient at all times to avoid their overturning under load, and that they are secured to the track and provided with physical stops. Connections made into concrete must be designed with regard to the strength of the concrete into which embedment is made.

Specialists in plant supply and hire should be consulted where some special problem arises. Such matters as the clearance under the hook of the crane and the stand-off distance required to allow clearance between the structure and the underside of the jib are equally as important as the lifting capacity at various radii.

6.3 Erection considerations in design

The process of erection depends on instant connection to allow the release of handling equipment for further duty. During the erection process, structural and structural/visual concrete elements are temporarily fixed prior to lining and levelling, and thus are vulnerable to impact loading, wind forces and so on.

The method of erection covered by the method statement must be strictly adhered to. In this way, the design provision for wind bracing, the transmission of horizontal shears in floors and the installation of the bracing essential to the maintenance of stability are assured.

The designer, in conjunction with those responsible, must incorporate fixings for bracing members and, where necessary, jigs and templates into the precast elements. Connections should ensure the stability of elements without the need for external support. Where supports are required, positive jig-cast components are the only sure way of achieving safety in the site operations. Bearings should be so arranged that anchorage is provided for the bearing element from the time of location, using dowels or welded or bolted connections. Such bearings are extended in the fixing and stitching phase by subsequent placement of the structural concrete.

Joint design must be considered in the light of the erection operations and the locations in which they must be completed. Bearing joints, for example, are best arranged such that the bearings throughout a complete story or lift of the construction can be levelled using the controlled insertion of approved shim material and the insertion of a gasket to avoid grout loss in stitching; speedy placement of the elements can then follow. The installation of site steel and transverse steel and, where appropriate, the early placement of joint concrete or topping complete that lift of the construction. It may be necessary in tall structures to specify the later removal of shims to avoid the development of point-loading situations.

In detailing the length of floor elements, a clearance must be allowed in order to avoid impact damage during site placement. Assembly drawings must clearly indicate the number and position of all joint steel added at site.

Where narrow plate column inserts are used to provide a bearing for beams, they must be checked to see that no distortion or buckling occurs under load.

Particular attention must be paid to the detailing of reinforcement at the ends of beams adjacent to bearings. Close tolerances should be imposed on all stirrups, hairpins and tail bars on bearing plates. The drawings must always indicate the position of such steel and the critical dimensions. End bearings, plate connections and the like may, during the course of erection, be subject to impact loads approaching the service loads for which the element is designed; therefore, they must be sufficiently robust. Bolts and projecting plate connections are also vulnerable and the erection sequence must be studied to ensure, where possible, that plates are flush with the concrete. In some systems, plates and brackets are bolted or welded to flush plates after erection.

Where recessed or socketed beam ends are used, there is a likelihood that inaccuracies in manufacture and erection may prevent the development of uniform bearing stresses. Steel-to-steel bearings avoid the risk of local spalling or crushing.

The detailing of corbels should follow the recommendations in *Structural joints in precast concrete* [9], especially those concerned with the support of the prestressed beam, in which there is a likelihood of increased horizontal forces developing due to the creep-shortening of the beam.

The process of making joints with grout or dry-packed concrete can be simplified by the provision of nibs or downstands on the upper member of an horizontal joint or at each side of a vertical plain or T-joint. This provides, in conjunction with a gasket, a seal for the grout or an abutment against which the dry pack may be driven to achieve the specified compaction.

Joints must be sufficiently wide to allow the proper compaction of wet-cast mix into the space provided. The concrete and grouting operations will be simpler to carry out if the upper face in an horizontal joint is inclined to allow the escape of entrapped air and water from the jointing concrete. If there is doubt that the concrete in the vertical joints between column members can be placed and compacted to completely fill the joint, it is sound practice to partially concrete the joint leaving a small gap, and then to pack the gap with dry mortar at a later stage.

Even with the most rigid inspection of products at the casting place, it is likely that elements will be delivered to site with some projecting bar or socket missing or misplaced. In these cases, there must be a clearly defined procedure which controls the work carried out in inserting drilled-in sockets or attaching steel members. All such work must be subject to inspection by a competent person prior to installation. The specification should include reference to suitable test procedures, as in the case of pull-out tests for stud fixings.

The load testing of elements in the casting yard or on the construction site is generally a simple matter, there being adequate supplies of concrete elements to provide test loads.

The designer must specify the age at which elements can be erected into the structure. In making this decision, he should consider the relatively high strengths of concrete used in most precast work. Generally, concrete elements may be erected within four days of manufacture and put under load once the jointing materials have achieved their specified 28-days characteristic strength, usually seven days from casting.

Where, during the erection process, some support is taken from slab edges or edge beams, a calculation must be made to establish whether extra reinforcement or additional external support is needed and the appropriate detail prepared.

Where adhesives or special resin preparations are specified for use in joints and connections, reference must be made to the manufacturer's recommendations regarding shelf-life, mixing procedure, pot life, conditions of application and curing.

The weatherproofing system must be so designed that its various parts can be satisfactorily installed without delay to the erection process. Gaskets, for example, must be provided in incremental width to cater for variability; also, provision must be made to ensure the continuity of membranes or insulation materials across joints. The simpler the weatherproofing, the better. Excellent results have been obtained by lap joints where sealants are put into shear rather than tension.

The designer should prepare a manual describing the method of connection, waterproof jointing and so on for incorporation into the method statement. One essential feature of the manual is a list of the methods and tools which must be excluded because they are likely to cause shock, impact or point loading on precast elements.

Tolerances on erection must be clearly stated in terms of plumb, deviation from line and joint size. The accuracy specified must be related to the service for which the structure is designed. Little has been published regarding acceptable deviation, although CP 110 [2] recommendations have been found to be realistic and provide a useful guide. It must be remembered that deviations are cumulative in their effect and that the location of oversize elements adjacent to undersize units compounds the problems of erection, connection and weather jointing.

References

1. Richardson, J. G., *Precast concrete production*, Cement and Concrete Association, London, 1973, 232 pp.
2. British Standards Institution, CP 110, Part 1: 1972 (as amended 1980), *The structural use of concrete*, London, 1980, 156 pp.
3. British Standards Institution, CP 116, Part 2: 1969 (as amended 1977), *The structural use of precast concrete*, London, 1969, 156 pp.
4. Schjodt, R., *Tolerances on blemishes of concrete*, CIB Report. No. 24, Rotterdam, 1975, 8 pp.
5. Blake, L. S., *Recommendations for the production of high quality concrete surfaces*, Cement and Concrete Association, London, 1981, 38 pp.
6. Gage, M. T., *Guide to exposed concrete finishes*, Architectural Press, London, 1974, 160 pp.
7. Institution of Structural Engineers and Concrete Society, *Formwork report*, Concrete Society Technical Report No 13, Cement and Concrete Association, London, 1977, 76 pp.

8. British Standards Institution, BS 5606: 1978, *Code of practice for accuracy in building*, London, 1978, 60 pp.
9. Institution of Structural Engineers, *Structural joints in precast concrete*, London, 1978, 56 pp.
10. British Standards Institution, CP 297: 1972, *Precast concrete cladding (non-loadbearing)*, London, 1972, 48 pp.
11. Prestressed Concrete Institute, *PCI manual for structural design of architectural precast concrete*, Chicago, 1977, 440 pp.
12. Richardson, J. G., *Practical formwork and mould construction*, Elsevier, London, 1976, 200 pp.
13. Cement and Concrete Association of Australia, *Tilt-up technical manual* Sydney, 1980, 23 pp.
14. Prestressed Concrete Institute, *PCI design handbook: precast, prestressed concrete*, (D. P. Jenny, ed.) 2nd Edn, Chicago, 1978, 400 pp.
15. Allen, A. H., *An introduction to prestressed concrete*, Cement and Concrete Association, London, 1981, 66 pp.
16. Ramaswamy, G. S., *Modern prestressed concrete design*, Pitman, London, 1976, 175 pp.
17. Neville, A. M., *Properties of concrete*, 3rd Edn, Pitman, London, 1981, 779 pp.
18. Kirkbride, T. E., Review of accelerated curing techniques, *J. Precast Concr.*, Vol. 2, No. 2, Feb. 1971, pp. 93–106.
19. Hartland, R. A., *Design of precast concrete*, Surrey University Press, London, 1975, 148 pp.
20. British Standards Institution, BS 4466: 1981, *Specification for bending dimensions and scheduling of reinforcement of concrete*, London, 1981, 16 pp.
21. Alexander, S. J. and Lawson, R. M., *Design for movement in buildings*, Construction Industry Research and Information Association, London, Technical Note 107, 1981, 54 pp.
22. Monks, W., *Visual concrete design and production*, Cement and Concrete Association, London, 1981, 28 pp.
23. FIP–RILEM, *Guides to good practice: grouts and grouting recommendations*, Cement and Concrete Association, London, 1975, pp. 15–22.
24. Concrete Society, *Safety precautions for prestressing operations*, Cement and Concrete Association, London, 1968, 8 pp.
25. Koncz, T., *Manual of precast concrete construction*, Vol. 1-3, Bauverlag, Wiesbaden and Berlin, 1967.
26. Associazione Italiana Tecnica Economica de Cemento (AITEC), *Industrializzazione dell' edilizia e prefabbricazione*, Rome, 1977, 418 pp.
27. Morris, A. E. J., *Precast concrete in architecture*, George Godwin, London, 1978, 571 pp.
28. Harrison, P., *Fixings for buildings—a design manual*, Harris and Edgar, London, 1977, 102 pp.

19 Prestressed concrete and partially prestressed concrete

R F Warner

University of Adelaide, Australia

Contents

Notation

The notation in this chapter generally conforms to ISO Standard 3898, although it has been necessary to introduce various special purpose terms.

A_c	area of concrete in a cross-section	V_p	vertical component of P at a section
A_p	area of prestressing steel		
A_s	area of reinforcing steel	w	distributed load
C	compressive force in concrete	W	concentrated load
d_p	depth of prestressing steel	α_x	sum of angular changes in direction of cable over length x
d_s	depth of reinforcing steel		
d_n	depth of neutral axis	β	wobble friction coefficient per metre of tendon length
d_z	depth of compressive force C		
D	overall depth of section	γ	rectangular stress block coefficient
E_c	elastic modulus for concrete	δ_d	instantaneous deflection increment due to dead load, or, where appropriate, self weight
E_p	elastic modulus for prestressing steel		
E_s	elastic modulus for reinforcing steel	δ_l	instantaneous deflection increment due to live load
e	eccentricity of prestressing force from section centroid	δ_p	instantaneous deflection increment due to prestress
f_c'	compressive strength of concrete	δ_{pdt}	total long term deflection due to dead load and prestress
f_{ps}	stress in prestressing steel when M_u acts	δ_{pdl}	instantaneous deflection due to combined effect of prestress plus dead and live load
f_{pu}	ultimate strength of prestressing steel		
f_{py}	yield stress (0.2% offset) of prestressing steel	ε_o	compressive concrete strain in top fibre
f_r	modulus of rupture of concrete, taken as $0.6\sqrt{f_c'}$ MPa	ε_{ce}	compressive strain due to prestress, in the concrete at the level of the prestressing steel
f_{sy}	yield stress of reinforcing steel		
f_t	tensile strength of concrete, taken as $0.33\sqrt{f_c'}$ MPa	ε_{cp}	tensile strain in the concrete at the level of the prestressing steel, when a moment M acts
f_{vy}	yield stress of stirrup steel		
I_c	second moment of area of full concrete section	ε_{pe}	strain in the prestressing steel before external load (dead or live) acts on the member
I_{cr}	second moment of area of cracked, reinforced section		
l	internal lever arm	ε_{py}	yield strain (0.2% offset) for the prestressing steel
L	span		
M	moment	ε_{sy}	yield strain for the reinforcing steel
M_{cr}	moment at flexural cracking	ε_u	value of ε_o when M_u acts
M_u	moment capacity of section	θ	inclination of prestressing cable to the horizontal
M_{dec}	decompression moment		
P	prestressing force	μ	curvature friction coefficient
S	shear force carried by the concrete in an uncracked section (equal and opposite to V_p)	ρ_p	A_p/bd_p = proportion of prestressing steel
		ρ_s	A_s/bd_s = proportion of reinforcing steel
T_p	tensile force in prestressing steel; horizontal component of P at a section		
		σ_b	compressive concrete stress in bottom fibre
T_s	tensile force in reinforcing steel	σ_{bd}	stress σ_b due to dead load
V	total shear force carried by a section	σ_{bp}	stress σ_b due to prestress
		$\sigma_{bI}, \sigma_{bII}$	upper and lower limits on compressive stress due to prestress in bottom fibre of concrete
V_{cm}	shear force at flexure-shear cracking		
V_{cw}	shear at web-shear cracking	σ_{gp}	concrete compressive stress at

	section centroid due to prestress	$\sigma_{oI}, \sigma_{oII}$	upper and lower limits on tensile
σ_o	concrete compressive stress at top		stress due to prestress in top fibre
	fibre		of concrete
σ_{op}	stress σ_o due to prestress		

Summary

Basic concepts and terminology relating to prestressing are dealt with in brief introductory sections. The advantages and disadvantages of prestressing as a design option are discussed, and the terms full prestressing and partial prestressing are explained. The behaviour of prestressed concrete flexural members, both at working load and at failure, is described. Methods of analysis are dealt with for uncracked and cracked members at working load and for the strength of members in flexure and shear. Inelastic deflections of cracked members are considered in some detail. Design procedures for fully prestressed and partially prestressed concrete members are treated separately, because of differing design requirements. Topics of special relevance to prestressed concrete which are treated include losses, anchorage, continuous construction, secondary moments and transformation of the cable profile. Indicative design data on the mechanical properties of prestressing steels are summarized. Because of variations in design criteria in the different national codes of practice, attention is focused throughout the chapter on basic concepts, rather than on design detail.

1 Introduction

The idea of introducing pre-compression into a concrete member, in order to delay or prevent cracking at service load and thereby improve structural performance, is almost as old as the idea of using steel reinforcement in the concrete to resist the internal tensile forces. However, the first practical applications of prestressed concrete were not made until the 1930s, when Eugene Freyssinet used very highly stressed, high-strength steel tendons to accommodate the large losses in prestress which result from creep and shrinkage of the concrete. Freyssinet's design approach was to provide sufficient compressive prestress in the concrete to eliminate cracking under full service load. This fitted well into the permissible stress design philosophy of the time, and became the basis of design for the many prestressed concrete structures which were built in the decades immediately following the Second World War.

More recently, as the attention of structural engineers has focused on the design objectives of strength and serviceability and on the philosophy of limit-states design, the principle of preventing cracking in all prestressed construction, while allowing cracking in any reinforced member, has been recognized to be inconsistent and often uneconomic. In the last decade, particularly in Europe, a considerable amount of research and development work has been carried out to prove the structural adequacy and economic viability of partially prestressed concrete [1, 2, 3]. In most English-speaking countries, code provisions [4, 5, 6] now allow for the design of both fully prestressed concrete and partially prestressed concrete structures.

In this chapter, the term partial prestressing refers to members which contain some prestress, but which are subject to cracking under service load. Such members usually contain significant quantities of reinforcing steel to provide adequate moment capacity and are, in effect, a prestressed reinforced form of construction. In the British concrete code, CP 110, a formal, three-tiered classification of prestressed concrete is adopted as follows:

Class 1: fully prestressed members in which flexural tensile stress does not occur at full design load;
Class 2: prestressed members in which flexural tensile stresses may occur, but are not large enough to cause cracking at full design load;
Class 3: partially prestressed members in which cracks of restricted width are allowed to occur at design load.

The choice of the level of prestress, as well as the form of construction, is an important design decision which should be made on the basis of desired structural performance and overall economy.

1.1   Prestressing as a design option

Prestressing is a design option which has associated with it a number of advantages, and some disadvantages. The designer must understand these advantages and disadvantages if he is to make effective use of the prestressing option. Some of the main design uses of prestressing are as follows.

Improved service load behaviour
In a flexural member, prestress can be used to delay cracking and to reduce the widths of cracks at higher load levels. Prestress can also be used to reduce or even eliminate deflections under service load. The service load behaviour of a concrete member can thus be controlled effectively by the introduction of an appropriate prestressing force. Prestressing can therefore be used to produce more slender members and to increase the maximum feasible spans for various types of construction.

Improved recovery after overload
Prestress causes cracks to close effectively as load is removed from a concrete member. Elastic recovery is thus much improved by prestress.

Improved strength in shear
The presence of a longitudinal precompression delays the formation of inclined cracks and hence improves the shear capacity of concrete flexural members. Shear capacity can also be increased very significantly by introducing vertical prestress into the member.

Improved resistance to fatigue
The fatigue resistance of a concrete member can be greatly improved by the introduction of longitudinal prestress. The effect of the prestress is to reduce the amplitude of the stress cycles in the tensile steel and hence increase the fatigue life of the member, as limited by steel fatigue. In practice, fatigue failure is almost always initiated in structural concrete by fatigue in the steel rather than in the concrete.

Efficient use of high-strength steel
Steel with a very high yield strength can rarely be used effectively in normal reinforced concrete construction because serviceability problems, such as excessive cracking and large deflections, occur long before the strength of the steel can be utilized. By introducing a high level of prestress, performance at service load can be controlled and a desirable pattern of behaviour over the full load range can be achieved. Prestressed concrete thus becomes a means for the efficient use of high-strength steel.

Although prestressing can be used to improve structural performance in various ways, it involves precise construction operations which in turn require the services of a skilled labour force. It also necessitates the use of special equipment and high-quality materials with corresponding attention to quality assurance. For this reason, prestressing will not always be viable economically. Like all design options, its use should be based on a careful analysis of cost effectiveness.

1.2 Design criteria

In the design of any concrete structure, whether prestressed or not, it is necessary to ensure that performance is satisfactory under service conditions and that there is adequate strength to resist unexpected overloads. Furthermore, the structure should have the additional properties of ductility and durability, as well as the overriding non-structural requirement of economy.

In the case of partially prestressed concrete, the structural requirements can be met by design criteria and calculation procedures which deal directly with service load behaviour and ultimate strength.

In the design of fully prestressed concrete, the situation is somewhat different, because crack prevention, and hence stress control, becomes a design requirement in addition to the requirements of strength and serviceability. It is therefore common design practice to choose the section and design details to satisfy permissible stress limits and then to check the strength and deflection requirements and any other special conditions.

Significant variations occur in the way the design criteria for prestressed concrete are formulated in the various national codes of practice. Attention will therefore be focused here on basic concepts. In specific instances, the requirements of the ACI Code [5] will be followed with regard to strength design. Thus, for flexural strength the design requirement will be expressed as

$$\phi(M_u) > M_{DU} \qquad (1\text{-}1)$$

where M_u is the flexural strength of the section, ϕ the capacity reduction factor,

0.9 for flexure, and M_{DU} the factored moment, representing the load effect at the design ultimate condition. For the loading case of dead plus live load,

$$M_{DU} = 1.4M_d + 1.7M_l \tag{1-2}$$

2 Concepts and terminology

Precompression can be introduced into a concrete structural member in a variety of ways, for example, by jacking against external rigid abutments, by using expansive cements, or by electrothermal stressing methods [7]. However, the most common method is by mechanically stressing high-strength steel tendons and anchoring them permanently against the concrete.

2.1 Pretensioning

Pretensioning is a method of mechanical prestressing whereby the concrete is cast around prestressing tendons which have been previously tensioned against external fixed abutments. When the tendons are released from their external restraints, they contract elastically and force the hardened concrete, which is bonded firmly to the steel, into compression.

Pretensioning lends itself to the factory production of large numbers of identical members on one long line of tendons in a prestressing bed. To maximize the use of the prestressing bed, the transfer of stress to the concrete must occur at the earliest possible age. Steam curing is frequently used to achieve a 24 h construction cycle.

2.2 Post-tensioning

In the post-tensioning process, small longitudinal ducts are cast in the concrete member and prestressing cables, placed in the ducts, are tensioned after the concrete has hardened. In short members, the cables may be initially anchored at one end and then jacked from the other end. In longer members, and especially in members with curved ducts, the cables may be jacked from both ends simultaneously in order to minimize friction losses.

In post-tensioned work, the on-site jacking operations are simplified by grouping the individual tendons into compact cables of large load capacity. Various proprietary systems have been developed for the stressing and anchorage of post-tensioned cables. At the detailed stage of design, information on end plates, anchorages, ducts and end block reinforcement will usually be obtained from the appropriate specialist company.

Post-tensioning is used particularly in on-site operations in the construction of large structural components, such as distribution beams, flat slab-floor systems, bridge girders, and in segmental bridge construction.

2.3 Cable eccentricity

In the design of a prestressed concrete flexural member, it is advantageous to locate the prestressing tendon eccentrically with respect to the centroid of the section of the member, so that the concrete stresses due to prestress vary from a

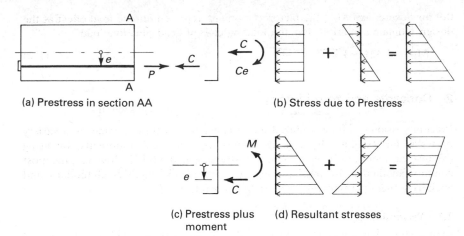

(a) Prestress in section AA   (b) Stress due to Prestress

(c) Prestress plus   (d) Resultant stresses
moment

Figure 2-1   Stresses in an uncracked section

maximum compression on the face closer to the tendon to a minimum compression (possibly tension) on the other face.

Such a distribution of prestress is effective in counteracting the tensile stresses which are induced in the lower fibres of the section by an applied moment, as in Fig. 2-1.

2.4 Stresses in an uncracked section

In Fig. 2-1a a tensile force P in the eccentric tendon induces an equal and opposite compressive force C at eccentricity e in the concrete at a typical section AA. The concrete stresses due to the force C and the equivalent moment Ce are added to give the resultant compressive stresses due to prestress (Fig. 2-1b):

$$\sigma = (C/A_c) + (Ce/I_c)y \qquad (2\text{-}1)$$

where A_c is the area of the concrete section, I_c the second moment of area of the concrete section, and y is the distance from the centroid to the point where the stress is calculated. It is positive when in the same direction as the eccentricity e.

A positive moment M in the section induces compressive stress above the centroid and tensile stress below:

$$\sigma = -(M/I_c)y \qquad (2\text{-}2)$$

2.5 Cable profile

It is usually desirable to vary the eccentricity of prestress along a flexural member. For example, in the end regions of a simply supported member where the applied moment is small, the eccentricity should also be small, while in the mid-span regions, where the applied moment is large, the eccentricity has to be large, with the cable close to the face of induced tensile stress. The term cable profile refers to the line followed along a member by the centre of gravity of the tensile cable force.

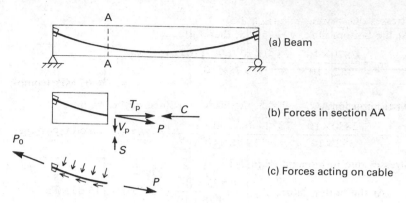

(a) Beam

(b) Forces in section AA

(c) Forces acting on cable

Figure 2-2 Forces due to prestress

In post-tensioned members, a curved profile can be achieved by careful location of the cable ducts prior to concreting. In the case of pretensioned tendons, sharp changes in direction can be effected by draping the tendons around cross bars which are held externally in place.

Figure 2-2 shows a post-tensioned beam with a curved cable profile. The tensile cable force P at section AA is inclined with horizontal and vertical components T_p and V_p. The forces acting on the cable are shown in Fig. 2-2c. The concrete stresses in section AA consist of longitudinal compressive stresses with a resultant C, which equilibrates T_p, and shear stresses with a resultant vertical force S, which equilibrates V_p. Except possibly near the ends of the beam, the slope of the tendon is quite small, and the forces C and T_p are usually taken to be equal to the tendon force P. Equation 2-1 can be written more conveniently (but inaccurately with regard to signs) as

$$\sigma_p = (P/A_c) + (Pe/I_c)y \qquad (2\text{-}3)$$

Example 1: Stresses in an uncracked section
The cross-section shown in Fig. 2-3a has an effective prestressing force of 2870 kN at depth 500 mm below the top fibre. The section is subjected to a dead-load moment of 575 kN m. The stresses at the section will be determined.

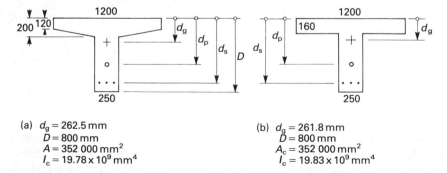

(a) $d_g = 262.5$ mm
   $D = 800$ mm
   $A = 352\ 000$ mm^2
   $I_c = 19.78 \times 10^9$ mm^4

(b) $d_g = 261.8$ mm
   $D = 800$ mm
   $A_c = 352\ 000$ mm^2
   $I_c = 19.83 \times 10^9$ mm^4

Figure 2-3 Beam cross-sections for numerical examples.

(a) Stresses due to prestress (Eqn 2-3):
At the bottom fibre, $y = +537.5$, therefore

$$\sigma_{bp} = \frac{2870 \times 10^3}{352 \times 10^3} + \frac{2870 \times 10^3 (500 - 262.5)}{19.78 \times 10^9} (+537.5)$$

$$= +26.67 \text{ MPa (comp)}$$

At the top fibre, $y = -262.5$, therefore

$$\sigma_{0p} = \frac{2870 \times 10^3}{352 \times 10^3} + \frac{2870 \times 10^3 (500 - 262.5)}{19.78 \times 10^9} (-262.5) = -0.90 \text{ MPa (tens)}$$

(b) Stresses due to moment (Eqn 2-2):

$$\text{At the bottom fibre, } \sigma_{bd} = -\frac{575 \times 10^6}{19.78 \times 10^9} (+537.5) = -15.63 \text{ MPa}$$

$$\text{At the top fibre, } \sigma_{0d} = -\frac{575 \times 10^6}{19.78 \times 10^9} (-262.5) = +7.63 \text{ MPa}$$

(c) Combined stresses:

$$\sigma_b = +26.67 - 15.63 = +11.0 \text{ MPa (comp)}$$

$$\sigma_0 = -0.90 + 7.63 = +6.7 \text{ MPa (comp)}$$

2.6 Load balancing

If friction losses are neglected, the prestressing force P is constant along a beam and the diagram of the bending moment caused by prestress, Pe, has the same shape as the cable profile. By choosing a cable profile which has the same shape as the bending moment diagram induced by an applied load, the designer can 'balance' the applied load by means of the prestress.

In Fig. 2-4, the cable eccentricity e varies parabolically from zero at the ends to a maximum value of e_0 at mid-span. This causes an upward camber. A uniformly distributed load w_b (Fig. 2-4b) induces a parabolically varying moment with a maximum mid-span value M_0:

$$M_0 = \tfrac{1}{8} w_b l^2 \tag{2-4}$$

If we set $Pe_0 = M_0$, the moment induced by the prestress is equal and opposite to the moment due to the load, not only at mid span but in all sections. The concrete is in a state of uniform compression and the beam has zero deflection. The load w_b is 'balanced' by the prestress. The parabolic tendon with internal tension P and end forces T supports the distributed load as indicated in Fig. 2-4d. The end forces T are in turn equilibrated by the compressive end forces C in the concrete.

The concept of load balancing is useful in the design of both determinate and indeterminate members [8] and will be considered again in later sections of this chapter.

Example 2: Load balancing
A beam with cross-section as in Fig. 2-3a is simply supported on a 13.5 m span. The cable details are to be determined which balance the self-weight of the beam plus a load of 25 kN/m distributed uniformly over the span.

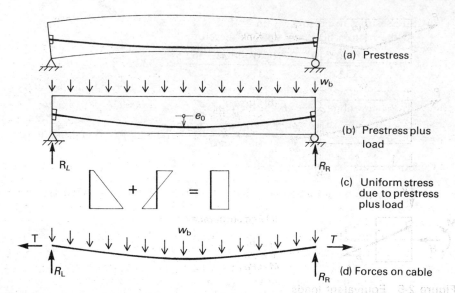

Figure 2-4   Load balancing

The load to be balanced is

$$w_b = 25 + 24 \times 0.352 = 33.5 \text{ kN/m}$$

The moment due to this load varies parabolically from zero at the ends to a maximum at mid-span of

$$M_0 = \tfrac{1}{8} w_b l^2 = 33.5 \times 13.5^2/8 = 763 \text{ kNm}$$

At the support e is zero. Allowing for cover, etc., the maximum eccentricity at mid-span is about 437 mm. Equating Pe_0 to M_0,

$$P \times 0.437 = 763 \text{ kN m} \qquad \text{therefore } P = 1745 \text{ kN}$$

To determine the cable details, the total losses are estimated as 25%. The required force is 1745/0.75, i.e. 2330 kN. From the design data in Table 12-2, a cable with 15 12.5 mm strands stressed initially to 85% of the breaking load supplies this force.

2.7 Equivalent load concept

In the load balancing concept, the cable profile is chosen so that the forces which are exerted on the concrete by the prestressing cable are equal and opposite to the externally applied loads. The forces acting on the concrete at the points of contact with the cable can be conveniently thought of as equivalent loads. Where a kink occurs in the cable profile (Fig. 2-5a), an equivalent point load W acts which is proportional to the prestressing force P and the angle of kink $\Delta\theta$;

$$W = P \, \Delta\theta \tag{2-5}$$

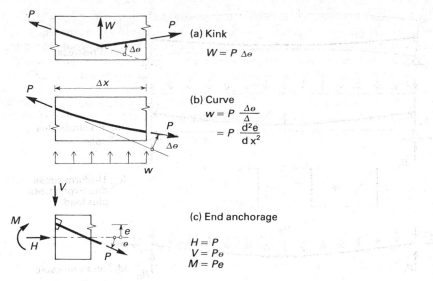

Figure 2-5   Equivalent loads

If the cable undergoes a smooth change in direction in the region between two sections (Fig. 2-5b), a distributed force acts:

$$w = P\frac{\Delta\theta}{\Delta x} = P\frac{d^2e}{dx^2} \tag{2-6}$$

where Δx is the length of the region, and $\Delta\theta$ the change in angle over length Δx. Equation 2-6 applies accurately when the cable follows a circular or parabolic shape but is a reasonable approximation for smoothly curved cables.

At the end of a member (Fig. 2-5c), an anchored cable with inclination θ and end eccentricity e imposes a horizontal force H, a vertical force V and a moment M. If θ is small, $H = P$, $V = P\theta$, and $M = Pe$. If the end cannot deflect, the vertical force V acts directly on the support and does not affect the moments or stresses in the beam.

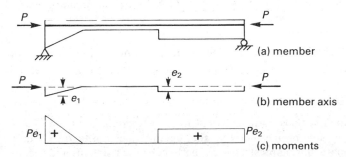

Figure 2-6   Equivalent loads: effect of location of longitudinal axis

The stresses induced in a member by a prestressing cable can thus be determined by analysing the concrete member when subjected to the equivalent load system. If the longitudinal axis of the member is not straight, additional moments are induced by changes in the axis position relative to the cable. As an example, Fig. 2-6 shows the additional moments produced in a haunch and at a sudden change in member depth.

2.8   Hyperstatic reactions and secondary moments

An eccentric prestressing force produces axial shortening, flexural deformation and lateral deflection. A simply supported member is free to deform under the effect of the prestress and usually deflects upwards, i.e. it cambers as in Fig. 2-4a. However, a statically indeterminate member cannot deform freely, because of the support constraints, and reactive forces may develop during the prestressing operation. In a two-span continuous member with a cable placed in the lower fibres (Fig. 2-7), the moment Pe tends to cause a deflection upwards and away from the interior support. If deflection is prevented at the support, a downward reaction R is induced, together with equilibrating reactions at the other supports. Such reactions depend on the shape of the cable profile, the magnitude of the prestressing force and on the bending stiffness of the member. These reactions are sometimes called hyperstatic reactions and the corresponding internal moments (M_2 in Fig. 2-7d) are called secondary moments. Methods for calculating hyperstatic reactions and secondary moments will be considered in Section 11, in connection with the design of indeterminate members.

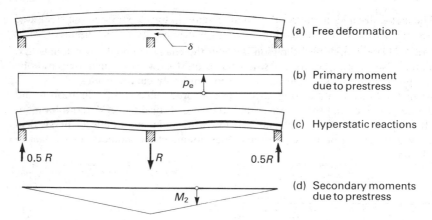

(a) Free deformation

(b) Primary moment due to prestress

(c) Hyperstatic reactions

(d) Secondary moments due to prestress

Figure 2-7   Hyperstatic reactions and secondary moments (indeterminate member)

2.9   Concordant cable

Although hyperstatic reactions normally occur in an indeterminate member, it is possible to choose a cable profile such that the member does not tend to deflect away from the supports during prestressing. The hyperstatic reactions and secondary moments are then zero, and such a cable is said to be concordant. In Fig.

2-7, a straight cable at the centroid of the member would be a simple example of a concordant cable.

3 Behaviour and analysis of beam sections

3.1 Uncracked sections

In an uncracked section, the stress due to prestress varies linearly with distance y from the centroid. To find the point G at which the stress is zero (Fig. 3-1), we set $\sigma = 0$ in Eqn 2-3:

$$y_0 = -\frac{I_c}{A_c}\frac{1}{e} \tag{3-1}$$

The negative sign means that G is on the side opposite to the cable. As the prestressing force P decreases gradually with time due to deferred losses, the stress line rotates about G (Fig. 3-1).

When the stresses due to an applied moment M are superimposed (Fig. 3-1d), the resultant concrete stresses are equivalent to a compressive force C, which is equal in magnitude to the prestressing force in the cable, but located at distance l above the cable. The internal forces in the concrete and the cable thus comprise a couple (Fig. 3-1f) which resists the external moment:

$$M = Cl \tag{3-2}$$

The above discussion ignores the small increase in tensile strain in the steel, and hence the increase in the tendon force P, which occurs as the moment M is applied. This effect is not usually important.

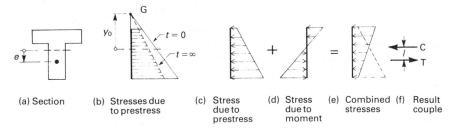

| (a) Section | (b) Stresses due to prestress | (c) Stress due to prestress | (d) Stress due to moment | (e) Combined stresses | (f) Result couple |

Figure 3-1   Stresses due to prestress and applied moment

3.2 Decompression moment and cracking moment

The compressive stress σ_{bp} in the bottom fibre due to prestress is obtained from Eqn 2-3 as

$$\sigma_{bp} = (P/A_c) + (Pe/I_c)y_b \tag{3-3}$$

where y_b is the distance from the centroid to the bottom fibre. The decompression moment M_{dec} is the external moment which induces zero stress in the bottom fibre of the concrete:

$$M_{dec} = +\sigma_{bp}(I_c/y_b) \tag{3-4}$$

In order to induce cracking in the section, the moment must be increased above M_{dec} by an amount sufficient to allow for the tensile strength of the concrete:

$$M_{cr} = +(\sigma_{bp} + f_r)I_c/y_b \tag{3-5}$$

A value commonly used for the modulus of rupture f_r is $0.6\sqrt{f'_c}$, the dimensions being MPa. In segmental construction, a value of zero may be appropriate for f_r.

Example 3: Decompression moment and cracking moment
The moments M_{dec} and M_{cr} will be determined for the section shown in Fig. 2-3a when a prestressing force of 2870 kN acts at depth 500 mm and $f'_c = 40$ MPa.
With $\sigma_{bp} = +26.67$ MPa (Example 1), we obtain from Eqn 3-4

$$M_{dec} = +26.67 \frac{19.78 \times 10^9}{537.5} \text{ N mm} = 981.5 \text{ kN m}$$

Taking $f_r = 0.6\sqrt{40} = 3.79$ MPa,

$$M_{cr} = (26.67 + 3.79) \frac{19.78 \times 10^9}{537.5} \doteqdot 1121 \text{ kN m}$$

3.3   Cracked section behaviour and analysis

As the applied moment rises above M_{cr}, fine hair cracks appear in the tensile face and gradually extend towards the middle of the beam. This behaviour is in contrast with the behaviour of a reinforced section without prestress, where there is a sudden upward extension of the crack when M_{cr} is exceeded. In an elastic analysis of a cracked reinforced section, the neutral axis position is independent of the applied moment; in a cracked prestressed section the neutral axis position is a function of M. As in the case of reinforced sections, the intact tensile concrete regions between adjacent cracks can contribute significantly to the flexural stiffness of a member, although this tension stiffening effect tends to die out as the moment rises into the high overload range.

In Fig. 3-2 the T-section is subjected to a moment M which is in excess of M_{cr}. Just prior to the application of the moment, the tensile strain in the prestressing steel is ε_{pe} and the compressive concrete strain at the tendon level is ε_{ce}. These strains can be evaluated from the prestressing force acting at this stage:

$$\varepsilon_{pe} = P/A_p E_p \tag{3-6}$$

$$\varepsilon_{ce} = \frac{P}{E_c} \left(\frac{e^2}{I_c} + \frac{1}{A_c} \right) \tag{3-7}$$

Upon application of the moment, the concrete compressive strain in the lower fibres reduces to zero and then becomes tensile. With M acting, the tensile strain in the reinforcing steel is ε_s and the strain in the concrete at the level of the tendon has changed from a compression of ε_{ce} to a tension of ε_{cp}. The strain

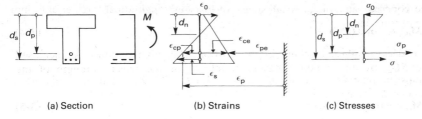

(a) Section	(b) Strains	(c) Stresses

Figure 3-2 Cracked section analysis

distribution in the section is assumed to be linear and hence defined by the neutral axis depth d_n and top fibre strain ε_0. For the reinforcing steel at depth d_s, the strain is

$$\varepsilon_s = \varepsilon_0(d_s - d_n)/d_n \tag{3-8}$$

The tensile strain in the concrete at the level of the prestressing steel is

$$\varepsilon_{cp} = \varepsilon_0(d_p - d_n)/d_n \tag{3-9}$$

where d_p is the depth of the prestressing steel.

The prestressing cable extends by the amount $(\varepsilon_{ce} + \varepsilon_{cp})$ during the application of M (Fig. 3-2), so that the total tensile strain in the tendon is

$$\varepsilon_p = \varepsilon_{pe} + \varepsilon_{ce} + \varepsilon_0(d_p - d_n)/d_n \tag{3-10}$$

Provided the materials remain elastic, the tensile forces in the reinforcing steel and prestressing steel are, respectively,

$$T_s = A_s E_s \varepsilon_s \tag{3-11}$$

$$T_p = A_p E_p \varepsilon_p \tag{3-12}$$

In the concrete compressive zone, the resultant compressive force is

$$C = 0.5\varepsilon_0 E_c b d_n \tag{3-13}$$

Equation 3-13 is valid provided the neutral axis is in the flange, $d_n < t$. The force C then acts at depth

$$d_z = \tfrac{1}{3}d_n \tag{3-14}$$

If the neutral axis is in the web of the T-section, the force C has to be reduced by an amount C_n:

$$C_n = 0.5\varepsilon_0 E_c (b - b_w)(d_n - t)^2/d_n \tag{3-15}$$

which can be regarded as a negative force acting at depth

$$d_{zn} = t + \tfrac{1}{3}(d_n - t) \tag{3-16}$$

The above equations allow the internal forces and hence M to be calculated from the strain distribution in the section, i.e. from ε_0 and d_n. Usually in a cracked section analysis M is known and the strain distribution has to be found by trial and error. It is convenient to begin with a trial value of d_n and find the corresponding value of ε_0. At all stages of loading, longitudinal force equilibrium must be satisfied:

$$C - C_n = T_s + T_p \qquad (3\text{-}17)$$

By substituting expressions already derived for the forces, an equation relating ε_o to d_n can be derived [9]:

$$\varepsilon_o = \frac{A_p E_p (\varepsilon_{ce} + \varepsilon_{pe})}{\left\{ \frac{1}{2} E_c \left[b_w d_n + (b - b_w) t \left(1 + \frac{d_n - t}{d_n} \right) \right] - \left[A_s E_s \frac{d - d_n}{d_n} + A_p E_p \frac{d_t - d_n}{d_n} \right] \right\}} \qquad (3\text{-}18)$$

This equation allows ε_0 to be determined for any chosen value of d_n in the range $D > d_n > t$. If the section is rectangular, b_w must be replaced by b. If the neutral axis is in the flange ($d_n < t$), Eqn 3-18 can be used provided the term in the denominator containing $(b - b_w)$ is set to zero and b_w is set equal to b. It should however be noted that the present analysis assumes linear elastic behaviour and that non-linear behaviour in either the concrete or the steel may well occur as the neutral axis rises into the flange.

With ε_0 determined for the trial value of d_n, the internal forces can be evaluated and the moment corresponding to d_n can be obtained:

$$M = T_p d_p + T_s d_s + C_n d_{zn} - C d_z \qquad (3\text{-}19)$$

From several trial values of d_n, it is possible to bracket the required moment and complete the calculation by interpolation or simple search. The above equations lend themselves to programming for a minicomputer or programmable calculator and the analysis can be carried out quickly and efficiently.

Example 4: Cracked section analysis
The beam section in Fig. 2-3a is idealized to that shown in Fig. 2-3b, and a cracked section analysis is carried out for M rising above M_{cr} into the overload range.

The steel reinforcement consists of six bars of 24 mm diameter ($A_s = 2700$ mm²) at depth 700 mm. The cable contains 19 strands of 12.5 mm diameter ($A_p = 1853$ mm²) stressed to 1133 MPa.

The initial tensile strain in the tendon and the compressive strain in the concrete at the level of the tendon are obtained from Eqns 3-6 and 3-7:

$$\varepsilon_{pe} = 0.0059; \qquad \varepsilon_{ce} = 0.00032$$

The calculations for the cracked section are conveniently carried out in tabular form. Decreasing values of d_n are chosen, beginning with D. For each d_n, ε_0 is calculated (Eqn 3-18) and hence σ_0. The strains in the steels (Eqns 3-8, 9 and 10), the steel and tendon stresses and finally the moment (Eqn 3-19) are then calculated.

d_n mm	ε_0 $\mu\varepsilon$	σ_0 MPa	σ_p MPa	σ_s MPa	M kN m
800	290	9.1	1175	−7.2	736
700	310	9.8	1179	0.0	790
600	340	10.8	1185	11.4	850
500	380	12.1	1196	30.6	925
400	450	14.2	1218	67.4	1035
300	590	18.7	1274	157.7	1261
250	760	24.0	1345	273.1	1533
200	1250	39.4	1562	622.7	2335

The calculations cease to be meaningful as d_n approaches 200 mm, since the stress in the reinforcement is in excess of the yield stress (410 MPa) and the concrete compressive stress is approaching $f'_c = 40$ MPa.

4 Deflections

In a prestressed concrete flexural member, the deflection consists of short-term increments due to dead load, live load and prestress, and a long-term increment due to creep under combined prestress and dead load. The short-term deflection increment due to prestress, δ_p, can be calculated from the equivalent loads which the cable exerts on the concrete, as discussed in Section 2. The beam usually cambers upward as a result of prestress, i.e. δ_p is negative.

The long-term deflection can be estimated by multiplying the short-term deflection due to combined prestress and dead (sustained) load by the creep term $(1 + \phi(t))$:

$$\delta_{pdt} = (\delta_d - \delta_p)(1 + \phi(t)) \tag{4-1}$$

where δ_d is the instantaneous deflection due to the dead load alone, and $\phi(t)$ the creep function, defined as the creep strain divided by the elastic strain, both strains being obtained from a constant stress creep test.

The total deflection of the member is then

$$\text{total } \delta = (\delta_d - \delta_p)(1 + \phi(t)) + \delta_l \tag{4-2}$$

where δ_l is the short-term deflection due to live load.

In an uncracked member, the short-term deflection increments due to dead load, live load and prestress can be determined independently using the un-cracked section bending stiffness $E_c I_c$.

The sections of maximum moment in a partially prestressed member will be cracked under full design load. As the load increases into the post-cracking range, deflections develop in a non-linear manner and the components cannot therefore be calculated independently and added. Branson [10] has shown that the initial deflection of a cracked prestressed beam can be calculated with reasonable

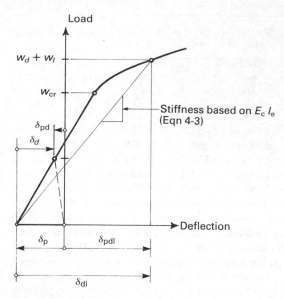

Figure 4-1   Short-term deflections

accuracy by means of the effective bending stiffness as used for cracked reinforced concrete sections:

$$E_cI_e = E_c\{I_c(M_{cr}/M)^3 + I_{cr}[1-(M_{cr}/M)^3]\} \tag{4-3}$$

where

M_{cr} = the cracking moment for the partially prestressed sections;
I_c = second moment of area for the gross section;
I_{cr} = second moment of area of the section regarded as a reinforced section (i.e. including the prestressing steel but ignoring the prestress).

As indicated in Fig. 4-1, the effective bending stiffness is used to calculate δ_{dl}, the increment in short-term deflection due to self-weight and external load in a cracked member. To obtain the total short-term deflection at full load, δ_{dl} must be added to the initial (negative) deflection due to prestress:

$$\delta_{pdl} = \delta_{dl} - \delta_p \tag{4-4}$$

The short-term increment in deflection due to dead load, δ_d, is calculated with E_cI_c if the beam is uncracked under the effect of dead load, and with E_cI_e otherwise. Note that I_e varies with M, which must be chosen to correspond to the load acting. The increment in deflection due to live load is

$$\delta_l = \delta_{dl} - \delta_d \tag{4-5}$$

The total deflection is obtained by adding δ_l to the long-term deflection due to dead load and prestress, as indicated by Eqn 4-2.

If the beam is designed so that the dead (sustained) load is balanced by the prestress, then $(\delta_d - \delta_p)$ is zero, and the long-term deflection is zero. The total deflection then is simply δ_l.

Example 5: Deflection of an uncracked beam
A beam with concrete cross-section as in Fig. 2-3a is simply supported on a 9.5 m
span. The tendon eccentricity varies parabolically from zero at the supports to
$e_0 = 390$ mm at mid-span. The tendon force is 1745 kN, provided by 17 12.5 mm
strands with $A_p = 1658$ mm^2. The elastic modulus for the concrete is $E_c =$
31 600 MPa; the long-term creep value is 2.0. The total dead load, including
self-weight, is 40 kN/m. The total deflection for a live load of 40 kN/m is
required. For the mid-span section, we take $f_r = 0.6\sqrt{f_c'} = 3.8$ MPa.

The bottom fibre stress due to prestress is obtained from Eqn 2-3: $\sigma_{bp} =$
23.5 MPa. From Eqn 3-5, the cracking moment is $M_{cr} = 1005$ kN m. The maximum moment under dead and live load is

$$M = \tfrac{1}{8}(40 + 40)9.5^2 = 903 \text{ kN m}$$

and so the beam is uncracked.

The equivalent load concept is used to obtain δ_p. The upward load equivalent
to the parabolic cable is:

$$w = -8\frac{Pe_0}{l^2} = -8\frac{1745 \times 10 \times 0.39}{9.5^2} = -60 \text{ kN/m}$$

For a simply supported beam under distributed load w the deflection is

$$\delta = \frac{5}{384}\frac{wl^4}{EI}$$

hence,

$$\delta_p = \frac{5}{384}\frac{(-60)9.5^4}{31.6 \times 19.78 \times 10^3} = -10.2 \text{ mm}$$

The short-term increment due to dead load is

$$\delta_d = \frac{5}{384}\frac{(+40)9.5^4}{31.6 \times 19.78 \times 10^3} = +6.8 \text{ mm}$$

The long-term deflection due to dead load and prestress (Eqn 4-1) is

$$\delta_{pdt} = (6.8 - 10.2)(1 + 2) = -10.2 \text{ mm}$$

The short-term deflection increment due to live load is $\delta_l = 6.8$ mm and the total
deflection is

$$\text{total } \delta = -10.2 + 6.8 = -3.4 \text{ mm (camber)}$$

Example 6: Deflection of a cracked member
The beam in Example 5 is subjected to a total dead load of 80 kN/m and a live
load of 40 kN/m. The beam section contains 2700 mm^2 of reinforcing steel at a
depth of 700 mm in addition to the prestressing cable. The deflections are to be
determined.

From Example 5, $\delta_p = -10.2$ mm (ignoring the presence of the reinforcing
steel).

Under dead load, the maximum moment is 903 kN m (see Example 5), and the beam is not cracked. From Example 5,

$$\delta_d = 2 \times 6.8 = 13.6 \text{ mm}$$

The long-term deflection is

$$\delta_{pdt} = (13.6 - 10.2)(1 + 2) = +10.2 \text{ mm}$$

Under full dead plus live load, the maximum moment is

$$M = \tfrac{1}{8}(80 + 40) \times 9.5^2 = 1353 \text{ kN m}$$

which is well in excess of the cracking moment. To calculate I_e for the cracked section, the simplified T-section in Fig. 2-3b is considered, with an equivalent quantity of reinforcing steel:

$$A_s' = 2700 + \frac{196}{200} \times 1658 = 4325 \text{ mm}^2$$

at depth

$$d_s' = \frac{1}{4325} \left(1658 \times \frac{196}{200} \times 652.5 + 2700 \times 700 \right) = 682 \text{ mm}$$

Using elastic reinforced concrete theory with $n = 6.3$ and $\rho = 0.0053$, we find that the elastic neutral axis position is at depth 155 mm, which is in the flange. For the cracked section, $I_{cr} = 9.06 \times 10^9 \text{ mm}^4$. From Eqn 4-3, we have

$$I_e = 19.78 \left(\frac{1005}{1353} \right)^3 + 9.06 \left(1 - \left(\frac{1005}{1353} \right)^3 \right) = 13.4 \times 10^9 \text{ mm}^4$$

hence

$$\delta_{dl} = \frac{5}{384}(120) \frac{9.5^4}{31.6 \times 13.4 \times 10^3} = 30 \text{ mm}$$

The total deflection (Eqn 4-5) is then

$$\text{total } \delta = 10.2 + (30 - 13.6) = 26.6 \text{ mm}$$

Example 7: Deflection with balanced dead load
The deflection of the beam will be calculated for the case of dead load and live load equal to 60 and 40 kN/m, respectively.

In this instance, the dead load is balanced by the prestress. The short-term and long-term deflections due to dead load and prestress are both zero.

With dead and live load acting,

$$M = \tfrac{1}{8} \times 100 \times 9.5^2 = 1128 \text{ kN m}$$

and the section is cracked. With $M_{cr}/M = 1005/1128 = 0.89$, we obtain

$$I_e = 19.78(0.89)^3 + 9.06(1 - 0.89^3) = 16.6 \times 10^9 \text{ mm}^4$$

Hence,

$$\delta_{dl} = \frac{5}{384} \cdot \frac{100 \times 9.5^4}{31.6 \times 16.6 \times 10^3} = 20.2 \text{ mm}$$

and with $\delta_d = 10.2$ mm, we obtain

total $\delta = 20.2 - 10.2 = 10$ mm

5   Flexural strength

Although prestressing steels do not have a distinct yield point like structural grade steel, a proof stress at the 0.2% offset can be clearly distinguished. Most flexural members with bonded prestressing steel are under-reinforced, in the sense that the tensile reinforcing steel and the prestressing steel in the section both reach the yield or proof stress when the moment capacity M_u of the section acts. In such cases, conditions at ultimate moment are similar to those in a reinforced concrete beam, the internal forces being as shown in Fig. 5-1.

The rectangular stress block concept can be used to evaluate the compressive force C, and the depth d_z at which C acts, in terms of the neutral axis depth d_n. When M_u acts, the neutral axis is usually in the flange of a T or I beam and so we have

$$C = 0.85 \gamma f_c' b d_n \tag{5-1}$$

$$d_z = 0.5 \gamma d_n \tag{5-2}$$

The following SI values for the stress block factor γ correspond to those used in ACI 318–77:

For $f_c' \leqslant 28$ MPa: $\gamma = 0.85$ $\tag{5-3a}$

For $f_c' > 28$ MPa: $\gamma = 0.85 - 0.00725(f_c' - 28)$ $\tag{5-3b}$

By equating forces in the section, we can evaluate d_n:

$$d_n = \frac{1}{0.85 \gamma f_c' b} (A_p f_{py} + A_s f_{sy}) \tag{5-4}$$

From Fig. 5-1, the moment capacity is

$$M_u = T_p(d_p - d_z) + T_s(d_s - d_z) \tag{5-5}$$

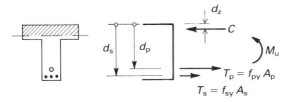

Figure 5-1   Flexural strength analysis

Substituting values, we have

$$M_u = A_p f_{py}(d_p - 0.5\gamma d_n) + A_s f_{sy}(d_s - 0.5\gamma d_n) \tag{5-5a}$$

A check is needed to ensure that both steels are at yield. For this purpose, the strain distribution is assumed to be planar and hence defined by the ultimate strain in the extreme compressive fibre, ε_u, and the depth of the neutral axis, d_n. As for reinforced concrete members, a constant value for ε_u is usually assumed:

$$\varepsilon_u = 0.003 \tag{5-6}$$

For the reinforcing steel to be at yield, ε_s must be at least as large as ε_{sy} and hence

$$d_n < d_s \frac{\varepsilon_u}{\varepsilon_u + \varepsilon_{sy}} \tag{5-7}$$

For the prestressing steel to be at yield,

$$d_n < d_p \frac{\varepsilon_u}{\varepsilon_u + (\varepsilon_{py} - \varepsilon_{ce} - \varepsilon_{pe})} \tag{5-8}$$

where ε_{py} is the yield or proof strain for the prestressing steel. Values for ε_{pe} and ε_{ce} can be obtained from Eqn 3-6 and 3-7. Expressions 5-7 and 5-8 must both be satisfied if Eqn 5-5 is to be used to calculate M_u.

In some design codes, including ACI-318, a semi-empirical approach is taken to the calculation of moment capacity. The stress in the prestressing steel when M_u acts, f_{ps}, is evaluated not as f_{py}, but by an empirical expression.
For members with bonded prestressing steel, the ACI expression is

$$f_{ps} = f_{pu}\left(1 - 0.5\frac{A_p f_{pu} + A_s f_{sy}}{bd_p f'_c}\right) \tag{5-9a}$$

For members with unbonded prestressing steel,

$$f_{ps} = f_{se} + 70 + \frac{f'_c}{100\rho_p} \tag{5-9b}$$

$$< f_{py}, \text{ and } f_{se} + 410 \tag{5-9c}$$

In the case of a non-rectangular section or a section with steel and cables distributed at various levels, an analysis for ultimate moment should be based on strain compatibility in the section [9].

Example 8: Moment capacity
The moment-carrying capacity of the section of Example 4 (Fig. 2-3b) is calculated. The following values apply:

$f_{py} = 1700 \text{ MPa};$ $\varepsilon_{py} = 0.0087$

$f_{sy} = 410 \text{ MPa};$ $\varepsilon_{sy} = 0.0021$

$f'_c = 40 \text{ MPa};$ $\gamma = 0.76$

From Eqn 5-4,

$$d_n = \frac{1}{0.85 \times 0.76 \times 1200}(1853 \times 1700 + 2700 \times 410) = 137.3 \text{ mm}$$

A check is made that the steels are at yield (Eqns 5-7 and 5-8):

$$d_s \frac{\varepsilon_u}{\varepsilon_u + \varepsilon_{sy}} = 415 \text{ mm}; \qquad d_p \frac{\varepsilon_u}{\varepsilon_u + (\varepsilon_{py} - \varepsilon_{ce} - \varepsilon_{pe})} = 274 \text{ mm}$$

As both values exceed d_n, the assumption of yield is correct. From Eqn 5-5a,

$$M_u = 1700 \times 1853(500 - 0.5 \times 0.76 \times 137.3) + 410$$
$$\times 2700(700 - 0.5 \times 0.76 \times 137.3)$$
$$= 2128 \text{ kN m}$$

Example 9: Moment capacity using semi-empirical expression for f_{ps}
The calculation of Example 8 is repeated using Eqn 5-9a to determine the tendon stress when M_u acts. For the tendon, $f_{pu} = 1890$ MPa.

$$f_{ps} = 1890\left(1 - \frac{1851 \times 1890 + 2700 \times 410}{1200 \times 500 \times 40}\right) = 1709 \text{ MPa}$$

Hence,
$$T_p = 1709 \times 1853 = 3167 \text{ kN}$$

Also
$$T_s = 410 \times 2700 = 1107 \text{ kN}$$

From Eqn 5-4,

$$d_n = \frac{1709 \times 1853 + 410 \times 2700}{0.85 \times 0.76 \times 40 \times 1200} = 138 \text{ mm}$$

From Eqn 5-2
$$d_z = 0.5 \times 0.76 \times 138 = 53 \text{ mm} = 0.053 \text{ m}$$

From Eqn 5-5
$$M_u = 3167(0.50 - 0.053) + 1107(0.70 - 0.053) = 2134 \text{ kN m}$$

6   Shear

Shear force has no significant effect on the behaviour of a prestressed concrete member unless or until inclined cracks form. If stirrup reinforcement is provided in the regions subject to inclined cracking, potential load paths exist for the transfer of shear force through the member by truss-like action. In the absence of proper shear reinforcement, such mechanisms cannot be relied on, and for design purposes the load capacity of the member is assumed to be restricted to the inclined cracking load.

An inclined crack can develop either as an extension of a vertical flexural crack which has previously formed at the tensile face (flexure–shear cracking) or, in a low-moment region, as a new inclined crack which forms in the web before any flexural cracks have appeared in the extreme fibre (diagonal tension or web-shear cracking).

6.1   Inclined cracking load

6.1.1   *Flexure–shear cracking* V_{cm}

The shear force V_{cm} which causes flexure–shear cracks to form in a beam can be obtained by first calculating the shear at flexural cracking, and then estimating the increment in shear which is needed to cause the crack to incline. If the shear at inclined cracking is to be determined for some section BB (Fig. 6-1), then strictly speaking M_{cr} should be calculated for an adjacent section AA at a short distance h away, in the direction of decreasing moment. The shear acting at BB when M_{cr} acts at AA is

$$V_0 = M_{cr}/k \qquad (6\text{-}1)$$

The factor k has the dimension of length. In most practical cases, h can be taken approximately as zero, so that k becomes the moment-to-shear ratio, M/V, for section BB. The ratio M/V will, of course, depend on the load configuration. In earlier codes such as ACI 318–71, a distance of $d/2$ was used for h, which gave a more complicated expression for k with little improvement in accuracy.

The additional shear force required to transform the vertical flexural crack at AA into the inclined crack at BB must be determined empirically. The expression used in ACI 318–77 and in AS 1481 [6] is

$$\Delta V = 0.05\sqrt{f_c'}\, b_w d_p \qquad (6\text{-}2)$$

For flexure–shear cracking at BB, the resultant expression becomes

$$V_{cm} = \frac{M_{cr}}{(M/V)} + 0.05\sqrt{f_c'}\, b_w d_p \qquad (6\text{-}3)$$

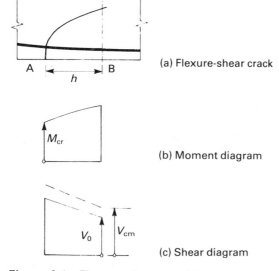

(a) Flexure-shear crack

(b) Moment diagram

(c) Shear diagram

Figure 6-1   Flexure–shear cracking

In Eqn 6-3, the ratio M/V is that applying when the maximum factored loads act, while M_{cr} is the total moment at flexural cracking, as given by Eqn 3-5.

In ACI 318–77, dead-load and live-load effects are separated and Eqn 6-3 is rewritten as

$$V_{ci} = \frac{\Delta M_{cr}}{(M_{max}/V_i)} + 0.05\sqrt{f_c'}\, b_w d_p + V_d \qquad (6\text{-}4)$$

where

V_{ci} = shear force required to cause flexure–shear cracking and has the same meaning as V_{cm};

V_d = shear due to unfactored dead load;

ΔM_{cr} = additional live load moment required to cause flexural cracking;

M_{max} = factored live load moment at the section;

V_i = factored live load shear acting with M_{max}.

In members with low levels of prestress, Eqn 6-1 and hence Eqn 6-4 become overly conservative. An empirically determined lower limit is accordingly allowed by ACI 318–77:

$$\text{minimum } V_{ci} = 0.14\sqrt{f_c'}\, b_w d_p \qquad (6\text{-}6)$$

The effect on V_{cm} or V_{ci} of the vertical component of the prestressing force, i.e. V_p in Fig. 2-2, is usually ignored.

6.1.2 Web-shear cracking V_{cw}

Web-shear cracks form in previously uncracked regions. Principal tensile stress calculations can therefore be used to estimate the shear acting when this type of crack appears. In Fig. 6-2a, the longitudinal and shear stress distributions are shown for a section to be checked for diagonal cracking.

The direct stress σ, due to prestress and applied moment M, varies with distance y from the centroid:

$$\sigma = (P/A_c) + (Pe/I_c)y - (M/I_c)y \qquad (6\text{-}7)$$

The shear stress τ also varies with y:

$$\tau = VQ/I_c b_w \qquad (6\text{-}8)$$

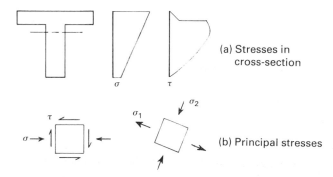

(a) Stresses in cross-section

(b) Principal stresses

Figure 6-2 Web-shear cracking

where Q is the first moment about the centroidal axis of the area which lies above (or below) the level y at which τ is being calculated, and b_w is the width of the section at level y.

From a Mohr's circle analysis, the principal tensile stress is obtained:

$$\sigma_1 = \sqrt{(\tfrac{1}{2}\sigma)^2 + \tau^2} - \tfrac{1}{2}\sigma \tag{6-9}$$

in which σ is compressive and σ_1 is tensile (Fig. 6-2b).

The principal tension σ_1 varies with y and may in fact reach a maximum value almost anywhere in the web. The situation is further complicated by the possibility of a varying M/V ratio at different load levels. To avoid lengthy trial calculations, it is usual to calculate V_{cw} as the shear which produces a principal tension of f_t, the tensile strength of the concrete, at the centroidal axis of the section, or at the web–flange junction if the centroid is in the flange. An appropriate value for f_t is $0.33\sqrt{f_c'}$, which is somewhat lower than the modulus of rupture.

The vertical component of the prestressing force, V_p, is often a maximum in regions prone to web shear cracking and is usually allowed for in determining V_{cw}.

In ACI 318–77, a simplified alternative to the principal stress calculation is allowed. V_{cw} may be obtained as follows:

$$V_{cw} = (0.29\sqrt{f_c'} + 0.3\sigma_{gp})b_w d_p + V_p \tag{6-10}$$

where σ_{gp} is the compressive prestress in the concrete at the centroid, or at the web-flange junction when the centroid is in the flange.

6.1.3 Inclined cracking Load V_c

The shear force V_c which causes inclined cracking in a small region of a prestressed member is taken to be the smaller of V_{cm} (or V_{ci}) and V_{cw}.

A simplified, conservative expression for the shear at inclined cracking is also given in ACI 318–77, which can be used in lieu of the above equations:

$$V_c = [0.05\sqrt{f_c'} + 4.8(V/M)d_p]b_w d_p \tag{6-10}$$

with the following upper and lower limits:

$$\text{maximum } V_c = 0.42\sqrt{f_c'}\, b_w d_p \tag{6-10a}$$

$$\text{minimum } V_c = 0.17\sqrt{f_c'}\, b_w d_p. \tag{6-10b}$$

The ratio V/M in Eqn 6-10 refers to the total factored design ultimate actions.

6.2 Design of stirrups

All regions in which inclined cracking occurs under design ultimate conditions require sufficient transverse steel to prevent premature failure in shear. In the calculation of shear reinforcement requirements, it is usually assumed that inclined cracks form at approximately 45°, and that the excess shear force, i.e. the shear force at design ultimate load minus the shear at inclined cracking, is carried by the shear reinforcement which is at yield.

In Fig. 6-3, the cross-sectional area of all legs of a vertical stirrup is A_v. If the

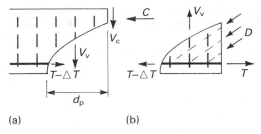

(a)   (b)

Figure 6-3   Design of stirrups

horizontal projection of the Inclined crack is taken equal to the depth d_p of the prestressing steel, the number of stirrups cutting the crack is d_p/s, where s is the horizontal stirrup spacing. With f_{vy} representing the yield stress in the stirrups, the transverse force carried by the stirrups is

$$V_v = A_v f_{vy}(d_p/s) \qquad (6\text{-}11)$$

Introducing a capacity reduction factor, $\phi = 0.85$, into the formal strength design requirement:

$$\phi(V_v + V_c) > V_{DU} \qquad (6\text{-}12)$$

we obtain the following design equation:

$$\frac{A_v}{s} > \frac{1}{\phi} \frac{V_{DU} - \phi V_c}{f_{vy}d_p} \qquad (6\text{-}13)$$

where V_{DU} is the factored design ultimate shear.

6.3   Details of shear design

6.3.1   *Conditions in end regions*
In the end regions of a post-tensioned member, it is usual for the tendon to be anchored in the vicinity of the section centroid. In such circumstances, it is desirable to include some longitudinal reinforcing steel in the lower fibres of the beam in order to allow truss action to develop and hence to allow the stirrups to become effective in resisting shear, as indicated in Fig. 6-4.

With reinforcing steel so provided, it is reasonable to replace d_p by d_s in Eqn 6-13. Even without such steel, ACI 318–77 allows d_p to be taken as 0.8 of the full depth D for use in Eqn 6-10.

In sections close to the end of a pretensioned member and within the transfer length, the prestress σ_{gp} is not at its full value. The web-shear cracking load

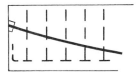

Figure 6-4   Shear reinforcement in end region

should in such cases be determined using the appropriate reduced value for the compressive prestress in the concrete.

6.3.2 *Prevention of web crushing*
The junction of a stirrup with the longitudinal steel acts very much like a joint in a truss (Fig. 6-3b), and an inclined compressive force is induced in the concrete which lies between adjacent inclined cracks. If an excess amount of shear reinforcement is used, the inclined compressive struts may crush. In ACI 318–77 a limit is therefore placed on the shear which may be carried by the stirrup reinforcement:

$$\text{maximum } V_v = 0.66\sqrt{f'_c}\, b_w d_p \qquad\qquad (6\text{-}14)$$

6.3.2 *Details of stirrups*
Various restrictions on spacing and area of stirrups are needed to ensure proper stirrup action. In ACI 318–77, the maximum spacing of stirrups is two-thirds of the full section depth. The minimum spacing is as for reinforced concrete members. The minimum quantity of shear reinforcement is as for reinforced members:

$$\text{minimum } A_v = 0.34(b_w s/f_{vy}) \qquad\qquad (6\text{-}15)$$

However, if the prestressing force is at least 40% of the tensile strength of the tensile reinforcement, the minimum area of shear reinforcement may, according to ACI 318–77, be taken as

$$\text{minimum } A_v = \frac{A_p f_{pu}}{80 f_{vy}} \frac{s}{d_p} \sqrt{\frac{d_p}{b_w}} \qquad\qquad (6\text{-}16)$$

Example 10: Design for shear
A beam with cross-section as shown in Fig. 2-3b is simply supported on a 9.5 m span. The tendon eccentricity varies parabolically from zero at the ends to a maximum of $e_0 = 390$ mm at mid-span. The tendon force is 1745 kN, provided by 17 12.5 mm strands with $A_p = 1658$ mm^2. The design dead and live loads are 60 kN/m and 65 kN/m.

The shear reinforcement requirements will be determined at the critical section close to the supports, and a check on inclined cracking will then be made throughout the span. The cross-section properties are given in Fig. 2-3b. Also $Q = 36.21 \times 10^6$ mm^3.

(a) *Critical section* The critical section is taken to be at 350 mm from the support. Should stirrups be required in the end regions, longitudinal steel will be placed at a depth of 700 mm.

(b) *Check for web-shear cracking at critical section* At the critical section, the cable eccentricity is $e = 55$ mm. The cable slope is $de/dx = 0.150$ rad. The compressive prestress at the centroid of the concrete section is $1745/352 = 4.96$ MPa. Evaluating f_t as $0.33\sqrt{40} = 2.09$ MPa and rearranging Eqn 6-9 to solve for τ,

$$\tau^2 = (\sigma_1 + \tfrac{1}{2}\sigma)^2 - (\tfrac{1}{2}\sigma)^2$$

we obtain $\tau = 3.84$ MPa. From Eqn 6-8:

$$V_{cw} = \frac{3.84 \times 19.83 \times 10^9 \times 250}{36.21 \times 10^6} = 526 \text{ kN}$$

The shear carried by the prestressing cable is

$$V_p = P(de/dx) = 1745 \times 0.150 = 262 \text{ kN}$$

The total web-shear cracking load in this region of low moment is also the inclined cracking load:

$$V_c = V_{cw} = 526 + 262 = 788 \text{ kN}$$

Hence,

$$\phi(V_c) = 0.85 \times 788 = 670 \text{ kN}$$

(c) *Design ultimate shear force*   For the design ultimate condition,

$$w_{DU} = 1.4 \times 60 + 1.7 \times 65 = 195 \text{ kN/m}$$

$$V_{DU} = 195 \times 0.5(9.5 - 0.7) = 858 \text{ kN}$$

(d) *Stirrup reinforcement requirement*   Stirrup reinforcement is required to carry the excess shear:

$$V_v = 858 - 670 = 188 \text{ kN}$$

Using U stirrups with 12 mm diameter bars ($A_v = 220$ mm^2) and a yield stress of 410 MPa,

$$s < \frac{0.85 \times 220 \times 410 \times 700}{188000} = 285 \text{ mm}$$

(e) *Stirrup spacing limitations*   In this case, V_v is less than the limit of $0.33\sqrt{f'_c}\, b_w d_p$ and so the maximum spacing is 0.67 times D, which is greater than 285 mm.

(f) *Inclined cracking shear at other sections*   It is convenient to tabulate the calculations for various sections in the beam.

x m	e mm	$\dfrac{de}{dx}$ rads	V_p kN	M_{cr} kNm	V_0 kN	V_{cm} kN	V_{cw} kN	V'_c kN	V_c kN	V_{DU} kN	V_v kN
0	390	0	0	973	0.0	55.3	551.8	188	155	0	0
1	373	0.035	61.1	943	87.5	142.8	612.9	188	155	195	63
2	321	0.069	120.4	851	183.4	238.7	672.2	188	239	390	187
3	234	0.104	181.5	700	309.7	365.0	733.3	315	365	585	275
4	113	0.138	240.8	490	598	653.3	792.6	465	654	780	224
4.4	55	0.15	261.8	387	1064.0	1119.3	788.0	465	788	858	188
4.75	0	0.164	286.2	291			838.0	465	838	926	214

In the table, x is distance measured from mid-span. Values for V_c are generally the smaller of V_{cm} and V_{cw}; however, at $x = 0$ and 1 the minimum value of 155 kN is obtained from Eqn 6-10b. Values of the cracking shear, calculated from Eqn 6-10 and indicated here by V'_c, are also given for purposes of comparison.

The table shows that stirrups are required throughout the span. At $x = 3$, a more severe design condition exists than at the critical section, the required stirrup spacing being

$$s = \frac{0.85 \times 220 \times 410 \times 700}{275} = 195 \text{ mm}$$

7 Anchorage

At the ends of a prestressed concrete member, permanent anchorage of the tendons to the concrete must be ensured. In post-tensioned construction, anchorage may be provided by external bearing plates or by special internal anchorages which are cast in the concrete. In pretensioned members, a direct transfer of stress between concrete and steel is achieved by bond.

In the anchorage region a complex state of stress exists whereby the concentrated compressive forces which are applied to the concrete gradually fan out and develop into the distributed pattern of prestress which exists in the interior regions of the beam. Large transverse tensile stresses develop in the anchorage zone and require careful attention, as they can lead to cracking, splitting and bursting of the concrete.

In post-tensioned members, where the tendons tend to be grouped into large cables, both vertical and horizontal transverse reinforcement will usually be provided in the anchorage zone to control cracking, should it occur. In thin web members, the end region may have to be thickened to form an end block.

Conditions in the end of a pretensioned member are somewhat different, because stress transfer occurs by bond. Nevertheless, the end region of a pretensioned member also requires careful consideration by the designer because of possible splitting due to transverse tension.

7.1 Anchorage of post-tensioned cables

Various methods of analysis of anchorage zones are reviewed in [11] by Abeles, who points out that good construction practice, coupled with the use of correctly located transverse reinforcement, is far more important than careful theoretical calculation of stresses and reinforcement quantities. In particular, Abeles suggests that:

(a) transverse steel should be provided as close to the end face as practicable, in the regions between anchorages;
(b) transverse steel should be provided on the axis of each anchorage, but at some distance away from the end face; and
(c) consolidation of the concrete is of utmost importance, and congestion of the steel must therefore be avoided.

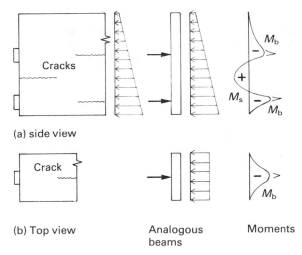

(a) side view

(b) Top view   Analogous   Moments
                      beams

Figure 7-1   End region, post-tensioned beam (spalling and bursting moments)

The need for transverse steel in these designated locations can be appreciated from the beam analogy of Fig. 7-1a, which applies to an end block with two separated anchorages. The beam, representing the end block, is subjected to concentrated forces at the anchorages. These are equilibrated by a distributed force which, in effect, is the prestress distribution across an interior section. Positive moment occurs between the anchorages, with negative moments over the anchorages.

The positive moment induces tensile stress on the outer face, which explains the tendency for horizontal, longitudinal cracking in the end face and the need for vertical reinforcement between anchorages. This moment is often referred to as a spalling moment.

The negative moments, or bursting moments, explain the development of tensile stress and cracks on the axis of the anchorages at some distance from the end face and the need for vertical reinforcement in this region.

The beam analogy also illustrates the need for horizontal transverse steel to control vertical cracks behind the anchorages which may occur due to bursting moments in a horizontal analogous beam (Fig. 7-1b).

Although the beam analogy cannot provide an accurate estimate of the bursting and spalling tensile stresses which occur locally, it can be used to estimate the amount of transverse steel required. The bursting and spalling moments are obtained from a static analysis of the analogous beam and the required tensile force is then calculated as the moment divided by the appropriate lever arm. In the case of a single anchorage, an internal level arm of about $D/2$ can be used. If n anchorages are more or less evenly distributed from the section, the lever arm can be taken as $D/2n$, so that

$$T = 2nM/D \qquad (7\text{-}1)$$

where M is the spalling or the bursting moment, as the case may be.

While the beam analogy can be used to determine where tensile steel is required, and in what quantities, in specific design applications the reinforcing details for the end regions will often be specified by the prestressing company which supplies the anchorage system. Such details are obtained from tests and previous experience, rather than from theoretical calculation.

If bearing plates form part of the anchorage, the possibility of concrete crushing behind the plate must be considered. For service load conditions, a limit on bearing stress which is frequently used is

$$f_b = 0.6 f'_c \sqrt{A'_b/A_b} \tag{7-2}$$

where A_b is the bearing area of the plate, and A'_b the maximum area of the end surface of concrete which has the same shape as, and is concentric with, A_b.

7.2 Anchorage of pretensioned tendons

In the end region of a pretensioned member, the tendon stress develops from zero to its full value over a short length L_t which is called the transfer length. The transfer length is affected by the surface properties and size of the tendon, as well as by its location in the section and the properties of the concrete.

For plain and indented wires, the transfer length is often assumed to be about 100 wire diameters, and 70 diameters for crimped wire. For seven-wire strand, L_t tends to be in the order of 50 diameters. These values may increase substantially in adverse circumstances, e.g. in tendons close to the top surface of the concrete, as cast.

If the tendons in a pretensioned beam are concentrated in a small region of the cross-section, bursting moments may produce horizontal or even vertical splitting in the end region. Static moments in the analogous beam (Fig. 7-1) can be used to check the need for transverse steel. The lever arm required for evaluating the tensile force T depends on the transfer length L_t.

8 Losses

It is convenient to consider prestress losses in the categories of immediate and deferred. Immediate losses occur during the transfer of stress to the concrete. The main cause of immediate loss in pretensioned members is elastic compression of the concrete during transfer. In post-tensioned members, immediate losses are also caused by friction between tendon and duct during the stressing operation, and by slip which occurs in the end grips when the tendon is anchored. Deferred losses develop gradually with time after the transfer of prestress to the concrete. They are caused by creep and shrinkage of the concrete and by stress relaxation of the prestressing steel.

The total loss in prestressing force is usually in the order of 20–25% for a post-tensioned member, with deferred losses accounting for the major part, e.g. about 15–20%. In pretensioned members, the concrete is prestressed at an earlier age and the total losses are somewhat higher, in the order of 25–30%, with deferred losses accounting for about 20–25%. While such approximate figures can be used in preliminary design calculations, more precise values for detailed calculations have to be obtained from an analysis of individual losses.

8.1   Elastic loss

In the manufacture of a pretensioned member, an initial tensile force P_0 in the tendons reduces to P_i as the concrete shortens elastically during the stress transfer operation. The decrease in stress in the tendons, $\Delta\sigma_p$, and the initial prestress in the concrete at the level of the tendon, σ_{ci}, are related as follows:

$$\Delta\sigma_p = \sigma_{ci}(E_p/E_c) \tag{8-1}$$

From Eqn 2-3,

$$\sigma_{ci} = P_i\left(\frac{1}{A_c} + \frac{e^2}{I_c}\right) \tag{8-2}$$

where e is the eccentricity of the tendon relative to the elastic centroid of the uncracked concrete.

With

$$\Delta\sigma_p = (P_0 - P_i)/A_p \tag{8-3}$$

the force P_0 required in the tendons prior to transfer can be calculated from the design prestressing force P_i:

$$P_0 = P_i\left[1 + \left(\frac{1}{A_c} + \frac{e^2}{I_c}\right)\frac{E_p}{E_c}A_p\right] \tag{8-4}$$

During post-tensioning operations, the cable is prestressed directly against the concrete, which compresses as the prestressing force increases. If there is only one cable, the elastic concrete shortening is automatically allowed for when the force in the tendon is brought to the required value. However, when several cables are stressed in sequence, the tensioning of the current cable causes elastic losses in all previously stressed cables.

Thus, when the kth cable at eccentricity e_k is stressed with force P_k, the reduction in force in cable j at eccentricity e_j is:

$$\Delta P_{kj} = P_k\left(\frac{1}{A_c} + \frac{e_k e_j}{I_c}\right)\frac{E_p}{E_c}A_p \tag{8-5}$$

An approximate estimate of overall elastic loss in a post-tensioned member can be obtained by assuming that the eccentricities are all about equal, and that a large number of cables are to be stressed. The average loss of tendon stress then becomes

$$\Delta\sigma_p = 0.5\sigma_{ci}(E_p/E_c) \tag{8-6}$$

8.2   Duct friction

During post-tensioning operations, friction between tendon and duct produces a fall-off in prestressing force with distance x from the jacking end. The friction effect is caused both by local irregularities in the duct, i.e. duct 'wobble', and by the angle changes in the duct which occur due to the cable profile. The fall-off in steel stress with distance x from the beam end is approximately exponential and can be represented algebraically as follows:

$$\sigma_{px} = \sigma_{pj}\exp[-\mu(\alpha_x + \beta x)] \tag{8-7}$$

where

σ_{pj} = stress at the jacking end;
μ = curvature friction coefficient;
α_x = sum of angular changes over distance x, in radians;
β = wobble friction coefficient per metre of tendon length.

The empirical parameters μ and β reflect the frictional effects between cable and duct. Performance records obtained during the stressing operations and from special stressing tests are the most reliable sources for values of μ and β. Approximate indicative values, taken from the Australian Prestressed Concrete Code [6], are given in Section 12.

Provided the exponential term is not too large, e.g. smaller than 0.3, then Eqn 8-7 may be simplified:

$$\sigma_{px} = \sigma_{pj}(1 - \mu\alpha_x - \mu\beta x) \tag{8-8}$$

If the cable is curved in three dimensions, then α_x should be taken as the sum of the absolute values of all angle changes.

8.3   Anchorage slip

During post-tensioning operations, some loss occurs as the prestressing force is transferred from the jack to the end anchorage because of slip of the cable and embedment of the anchorage. The amount of movement varies from almost zero, in systems which use threaded bars and nuts, to in excess of 5 mm in extreme cases. Anchorage losses are difficult to predict theoretically, and are best evaluated from information provided by the manufacturers of the various anchorage systems.

8.4   Stress relaxation

The stress relaxation that occurs in prestressing steel over an extended period of time can reduce the prestress significantly. Stress relaxation is affected by stress level in the steel, time and type of steel.

Experimental values of stress relaxation are frequently obtained from standard 1000 h relaxation tests, the result being expressed as a percentage loss of stress, R_{1000}. Thus, $R_{1000}^{0.7}$ refers to the percentage loss of stress at 1000 h, for an initial stress level of 0.7 of the steel strength.

For estimating total relaxation losses, it is usually assumed that the ultimate relaxation, R_u, is twice the 1000 h value. The actual relaxation loss will depend not only on the initial prestress in the steel, but also on the change in prestress due to long-term losses caused by concrete creep and shrinkage as well as steel relaxation. An approximate estimate is

$$R_d = R_u[1 - (2\Delta\sigma_p/\sigma_{pi})] \tag{8-9}$$

where $\Delta\sigma_p$ is the total reduction in stress.

Although high values of relaxation can be expected in stress-relieved steels, special low-relaxation steels are available with much improved properties. Typical design values are summarized in Table 12-3.

8.5 Creep loss

Creep strains for conditions of constant concrete stress in the working range can be calculated from the creep function $\phi(t)$ and initial elastic stress σ_{ci}. The long term creep strain for constant stress conditions is

$$\varepsilon_c^* = \phi^*(\sigma_{ci}/E_c) \qquad (8\text{-}10)$$

The long-term value of the creep function, ϕ^*, ranges from 0.5 up to about 3.0, and depends on climate, age at loading, concrete properties, etc. Design data need to come from local tests, although indicative values can be obtained from the CEB–FIP model code [12].

Creep losses are usually calculated on the simplifying assumption that the concrete stress level does not change significantly. Hence, by equating the reduction in strain in the prestressing steel to ε_c^*, we obtain

$$\Delta\sigma_p = \phi^* \frac{\sigma_{ci}}{E_c} E_p \qquad (8\text{-}11)$$

8.6 Shrinkage loss

Shrinkage strain occurs in concrete more or less independently of stress level. The long-term shrinkage strain, ε_{sh}^*, can be determined from local data, national code regulations or the CEB–FIP code. The corresponding reduction in tendon stress is

$$\Delta\sigma_p = \varepsilon_{sh}^* E_p \qquad (8\text{-}12)$$

8.7 Total losses

The total losses are calculated by summing individual components. Possible interactive effects among the losses are usually only considered in the evaluation of the relaxation losses.

9 Design of fully prestressed members

Fully prestressed members are designed to remain uncracked under full design load. Limiting stress conditions form the basis of the design calculations, but subsequent checks must be made to ensure that adequate strength is achieved and that deflections are not excessive.

Permissible stress limits given in ACI 318 for the design of prestressed concrete members are summarized in Table 9-1. The tensile stress limit of $0.5\sqrt{f_c'}$ is slightly less than the modulus of rupture of concrete and provides for the design of Class 2 structures as described in Section 1. The value $1.0\sqrt{f_c'}$ is well in excess of the tensile strength of concrete. This limit is used in the design of partially prestressed members and will be dealt with in Section 10. The limit of $0.25\sqrt{f_c'}$ at transfer can be exceeded if bonded reinforcement, calculated to carry the full tensile force in the uncracked section, is provided (ACI Code Commentary [13]).

Table 9-1   Permissible concrete stresses in flexural members

1. *Extreme compressive fibre stress*
   1.1 *at transfer, f_{ci}* $0.60f_c'$
   1.2 at service load, after losses, f_c $0.45f_c'$
2. *Extreme tensile fibre stress*
   2.1 at transfer, f_{ti}
        generally $0.25\sqrt{f_c'}$
        at ends of simply supported members $0.50\sqrt{f_c'}$
   2.2 at service load, after losses, f_t
        in precompressed zones $0.50\sqrt{f_c'}$
        in precompressed zones, but where calculations show
        that immediate and long-term deflections are not
        excessive (other than in two-way slabs) $1.00\sqrt{f_c'}$

9.1   Choice of section

In fully prestressed beams, the size of cross-section is governed principally by the variation in moment, M_v, which occurs after transfer. If the dead load is made up mainly of self-weight, which is applied at transfer, then M_v consists principally of the live-load moment. However, if the self-weight is small and the dead load is applied subsequent to transfer, then M_v will be the sum of the dead and live-load moments.

In the top fibres of a critical section in bending, the stress variation will be M_v/Z_0, which should be limited to $(f_c + f_{ti})$, where f_c and f_{ti} are the permissible stresses in compression and tension. A minimum top fibre section modulus thus applies:

$$\text{minimum } Z_0 = M/(f_c + f_{ti}) \tag{9-1}$$

This limit does not allow for the effect of deferred losses. Introducing a loss coefficient η, defined as the ratio of the prestressing force after losses to its initial values:

$$\eta = P_e/P_i \tag{9-2}$$

we can obtain the following approximate design expression [8]:

$$\text{minimum } Z_0 = 1.2M_v/(f_c + \eta f_{ti}) \tag{9-3}$$

A similar requirement applies to the bottom fibre:

$$\text{minimum } Z_b = 1.2M/(f_t + \eta f_{ci}) \tag{9-4}$$

In the design of a fully prestressed concrete flexural member, a convenient procedure is to choose an appropriate cross-section using Eqns 9-3 and 9-4, determine the cable force and eccentricity required to satisfy the permissible stress limits at the critical cross-sections, and then check strength and deflection requirements.

In the choice of prestress details to satisfy stress limits, the loading conditions

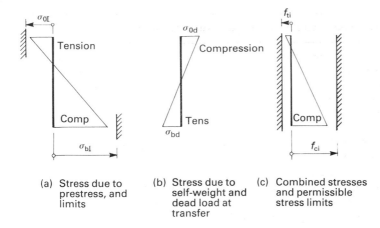

(a) Stress due to prestress, and limits

(b) Stress due to self-weight and dead load at transfer

(c) Combined stresses and permissible stress limits

Figure 9-1 Limits on prestress, just after transfer

which produce critical states of stress are usually as follows:

(a) just after transfer, when the initial prestress acts in combination with self-weight and possibly some dead load;
(b) after deferred losses occur and the member is subjected to full live load, dead load and self-weight load.

The state of stress in a typical cross-section for condition (a) is shown in Fig. 9-1. To ensure that the tensile stress limit f_{ti} is not exceeded, the tensile stress in the top fibre due to prestress alone must not exceed the limit σ_{0I}:

$$\sigma_{0I} = \sigma_{0d} + f_{ti} \tag{9-5}$$

where σ_{0d} is the compressive stress in the top fibre due to self-weight and dead load.

To ensure that the compressive stress limit f_{ci} is not exceeded, the compressive stress in the bottom fibre due to prestress (Fig. 9-1a) must not exceed the limit σ_{bI}:

$$\sigma_{bI} = \sigma_{bd} + f_{ci} \tag{9-6}$$

where σ_{bd} is the bottom fibre tensile stress due to self-weight and dead load.

Condition (b), with full live load and dead load acting, is considered in Fig. 9-2. A lower limit σ_{0II} must be placed on the top fibre tensile stress due to prestress in order to ensure that the allowable compressive stress f_c is not exceeded:

$$\sigma_{0II} = \sigma_{0dl} - f_c \tag{9-7}$$

where σ_{0dl} is the top fibre compressive stress due to full dead load and live load.

A lower limit σ_{bII} on the compressive prestress in the bottom fibre is also needed to ensure that under full load the tensile stress limit f_t is not exceeded:

$$\sigma_{bII} = \sigma_{bdl} - f_t \tag{9-8}$$

where σ_{bdl} is the tensile stress in the bottom fibre due to full dead and live load.

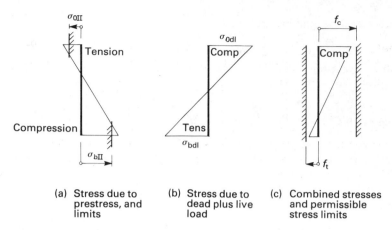

(a) Stress due to prestress, and limits

(b) Stress due to dead plus live load

(c) Combined stresses and permissible stress limits

Figure 9-2   Limits on prestress, at full design load

As indicated in Fig. 9-3, the four stress limits define a region within which the prestress line must be chosen in order to satisfy the allowable stress requirements. At transfer, the tensile limit σ_{0I} and the compressive limit σ_{bI} must not be exceeded; at full design load, after losses, the limits σ_{0II} (tension in the top fibre) and σ_{bII} (compression in the bottom fibre) are minima which must be reached.

The effect of the deferred losses is to rotate the prestress line about point G. For preliminary design calculations, approximate values for deferred losses can be taken, e.g. 17 and 22% for post-tensioned and pretensioned members, respectively.

With the initial and final prestressing lines chosen, appropriate values of prestress and eccentricity can be calculated. From the prestress at the centroid of

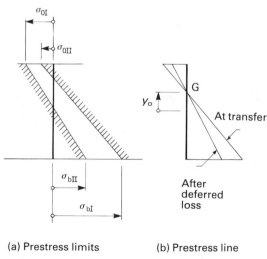

(a) Prestress limits

(b) Prestress line

Figure 9-3

the section, we evaluate the prestressing force:

$$P_e = A_c \sigma_{gp} \qquad (9\text{-}9)$$

The eccentricity of prestress e is calculated from y_0, the distance from the centroid to the point of zero stress:

$$e = \frac{1}{A_c} \frac{1}{(-y_0)} \qquad (9\text{-}10)$$

Values obtained for P_e and e must be checked to ensure that the section can accommodate the prestressing cable at the required eccentricity. If detailing becomes difficult, a larger section will be chosen.

9.2 Cable profile

With the value of the prestressing force determined from conditions at the critical section for bending, it is then necessary to choose an appropriate cable profile, i.e. values of e at cross-sections along the length of the member. This can usually be carried out by constructing stress diagrams as in Fig. 9-3 for several sections at regular spacing along the beam, and in each case determining an appropriate e value for the given P value. In the case of post-tensioned members, account should be taken of variations in P along the beam because of friction losses. The stress limits can be used to construct an admissible region within which the cable must lie. In choosing the cable profile for a post-tensioned member, it should be remembered that friction losses will be minimized if angle change is minimized.

9.3 Completion of design

To complete the design, losses have to be calculated and compared with the initially assumed values. If necessary, adjustments to the design calculations for section size and cable profile will be made. Checks must be made on strength in bending and shear, and longitudinal and transverse steel provided as required. Short-term and long-term deflections need to be calculated, and the end block and anchorage requirements checked.

10 Design of partially prestressed members

10.1 Basis of design

The design of a partially prestressed concrete member can be based directly on the design requirements of serviceability under working-load conditions and adequate strength under ultimate conditions. Little is to be gained from an allowable stress approach. The dimensions of critical cross-sections can be chosen from flexural strength requirements, while the magnitude of the prestressing force and the location of the prestressing cable can be chosen to control deflections, or, in special cases, to control cracking. The concept of load balancing, which has already been discussed, provides a simple and convenient means for choosing prestressing details to ensure zero deflection under any chosen load condition.

10.2   Cross-section dimensions

The overall dimensions of the section can be estimated from the requirement that M_u must be large enough to satisfy Eqn 1-1. Taking an average value d_{av} for the steel depths d_s and d_p, and estimating the internal lever arm as $0.85d_{av}$, we obtain the following design expression from Eqns 5-5 and 1-1:

$$bd_{av}^2 \geq \frac{M_{DU}}{\phi 0.85(\rho_s f_{sy} + \rho_p f_{py})} \tag{10-1}$$

where $\rho_s = A_s/bd_s$ and $\rho_p = A_p/bd_p$ are the steel proportions. With steel indexes defined as follows:

$$q_s = \rho_s(f_{sy}/f_c'); \qquad q_p = \rho_p(f_{py}/f_c');$$

Eqn 10-1 can be rewritten as

$$bd_{av}^2 \geq \frac{M_{DU}}{\phi 0.85 f_c'(q_s + q_p)} \tag{10-2}$$

From trial values for ρ_s and ρ_p, b and d_{av} can be estimated, and hence the section dimensions.

The proportion of reinforcing steel is likely to be small when the live load is a small proportion of the total load, and large when the live load is large. In the absence of other information, the steel quantities can be chosen such that $(q_s + q_p)$ is between 0.15 and 0.3.

The design ultimate moment M_{DU} should include an estimate for self-weight. When the section has been chosen, the self-weight can be checked and adjustments made as necessary.

The web thickness b_w must be adequate to accommodate the cables, the reinforcing steel and some stirrup reinforcement. Also, the nominal shear stress in the web,

$$v_u = V_{DU}/b_w d_s \tag{10-3}$$

should not be excessive. If v_u is less than the maximum nominal shear for a reinforced concrete member with stirrup reinforcement,

$$\text{maximum } v_u = \phi(0.83\sqrt{f_c'}) \tag{10-4}$$

the web thickness should be adequate to accommodate the shearing action.

It is usually advisable to keep the neutral axis at ultimate moment within the flange. This is accomplished by observing the following limit on flange thickness:

$$\text{minimum } t = \frac{(\rho_s f_{sy} + \rho_p f_{py})bd_{av}}{0.85\gamma f_c' b} \tag{10-5}$$

which is obtained from the rectangular stress block concept.

10.3   Details of prestressing cable

If deflection control is the prime serviceability consideration, an appropriate load condition can be chosen for which zero deflection is desirable. The prestressing force and cable profile can then be determined by applying the concept of load balancing. The condition of zero deflection may be appropriate for full dead load, or possibly for dead load plus a proportion of the live load.

If crack control is the prime serviceability consideration, the level of prestress can be chosen to control the decompression load, i.e. the load which is just sufficient to produce a state of zero stress in the 'tensile' face of the critical section of the member.

For example, if it is decided that a member should be crack-free under sustained dead load, then it would be appropriate to equate the design dead load to the decompression load. The required prestressing force, after all losses have taken place, is then

$$P = \frac{M_{dec}}{e_0 + Z_b/A_c}$$ (10-6)

where

e_0 = available eccentricity in the section;
Z_b = section modulus for the bottom fibre;
A_c = concrete area;
M_{dec} = maximum moment caused by the decompression load.

Although the concrete has some tensile strength, the first application of the full live load will usually cause cracking, at least in the region surrounding the section of maximum moment, so that these same cracks will close when the live load is removed, but will start to open again as soon as any live load is re-applied.

Provided care has been taken to place the reinforcing steel close to the concrete faces in the regions prone to cracking, the cracks which form should remain fine hair cracks even under full design load. For this reason, Walther [2] has suggested that the decompression load can be chosen to be significantly less than the full dead load, so that the tensile strength of the concrete is approached when the full dead load acts. This results in a structure in which the cracks remain permanently open in the regions of maximum dead load moment. Nevertheless, such cracks should not be as wide as those in a comparable design in reinforced concrete.

10.4 Reinforcing steel requirements

When the cable force has been determined, the area of prestressing steel, A_p, can be determined.

A check must be made of the reinforcing steel area required in the critical sections to carry the full design ultimate moment. Although an estimate of the required area A_s can be made by means of Eqn 5-5 with approximate values for lever arms l_p and l_s, the final check on the adequacy of the flexural strength should be made by means of Eqn 5-5a.

10.5 Crack control

A check should be made for excessive cracking at full service load. A direct calculation of crack width can be made but in practice an indirect stress check will usually be simpler to perform.

The hypothetical concrete tensile stress check, used by the British Code CP 110 and ACI 318–77, is based on the calculation of concrete stresses for an uncracked section. Limits placed on the calculated tensile stress give an indirect control on crack width. The permissible value of $1.0\sqrt{f'_c}$ listed in Table 9-1 is the ACI limit; a range of values, of this order, is given in CP 110. The limits in CP 110 depend on the acceptable crack width, the prestressing details, the concrete strength, and the section depth. The hypothetical concrete stress method is simple to use because it avoids the cracked section analysis but it has been criticized for its lack of rationality [2]. If this method is adopted, the preliminary choice of section size can be made with Eqns 9-3 and 9-4 and the design can be carried through as for a fully prestressed member.

An alternative approach to crack control, used in Europe and in the Australian Code, AS 1481, is to calculate stresses in the steel in the cracked section. A limit of 150 MPa is placed on the load-induced stress increment in the reinforcing steel and in the prestressing tendons, as a means of restricting crack widths. The cracked section analysis described in Section 3, although more complicated than the uncracked section analysis, lends itself to calculator programming and is quickly carried out.

10.6   Check for other serviceability requirements

If the cable profile and prestressing force have been chosen by load balancing to counteract the sustained load or a portion of the sustained load, then it is necessary to check the short-term deflection under full design load. The member is in a partially cracked state and an equivalent flexural stiffness should be used for the deflection calculation, as already discussed in Section 4.

If the prestressing details have been chosen from decompression moment calculations based on considerations of crack control, separate checks should be made of the short-term and long-term deflections under full sustained load and also of the short-term deflection under full design dead and live load.

10.7   Other design requirements

To complete the design, checks should be made of losses, end block reinforcement, and shear reinforcement. Procedures for carrying out these calculations have already been discussed.

Example 11: Design of a partially prestressed girder
A girder with an inverted T-section (Fig. 10-1a) is required to support a precast decking system. Design loads are 22 kN/m dead, plus self-weight, and 25 kN/m live. The girder is supported on a 10 m span with a 5 m cantilever at one end. The preliminary design of a partially prestressed member is carried out.

(a)   *Cross-section dimensions*   To support the precast decking, 125 mm flange outstands are required; a 200 mm web is used to allow good concrete compaction; concrete strength will be 40 MPa.

A self-weight estimate of 5 kN/m is made. The design ultimate moment at critical section C, Fig. 10-1, is

$$M_{DU} = \tfrac{1}{2} \times 5^2 [1.4(22+5) + 1.7 \times 25] = 1003 \text{ kN m}$$

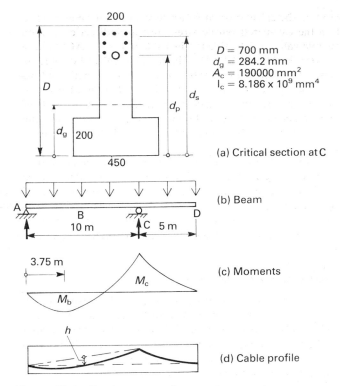

$D = 700$ mm
$d_g = 284.2$ mm
$A_c = 190000$ mm^2
$I_c = 8.186 \times 10^9$ mm^4

(a) Critical section at C

(b) Beam

(c) Moments

(d) Cable profile

Figure 10-1 Design example

With $b = 450$ mm and taking $(q_s + q_p) = 0.2$, we have (Eqn 10-2)

$$d_{av} = \sqrt{\frac{1003 \times 10^6}{450 \times 0.9(0.85 \times 40 \times 0.2)}} = 603 \text{ mm}$$

Allowing for cover, stirrups, duct, etc., we take $D = 700$ mm. For this section, the self-weight load is $0.19 \times 24 = 4.56$ kN/m.

Checking for shear at section C,

$$v_{DU} = \frac{5[1.4(22 + 4.56) + 1.7 \times 25]}{200 \times 603} = 3.3 \text{ MPa}$$

From Eqn 10-4,

maximum $v_u = 0.85(0.83\sqrt{40}) = 4.4$ MPa

and so the section should be adequate for shear.

Checking the flange thickness

minimum $t = \dfrac{0.2}{0.85 \times 0.76}\, 600 = 186$ mm

and $t = 200$ mm appears to be adequate.

(b) *Details of prestressing cable* The prestress will be used to balance self-weight plus full dead load. Zero short-term and long-term deflections under sustained load will thus be achieved.

In the cantilevered region, a parabolic cable will be used with zero slope and zero eccentricity at the free end. The load to be balanced is $w_b = 4.56 + 22 = 26.56$ kN/m. Taking the cable eccentricity at section C as e_0, we have

$$w_b = P(2e_0/l^2)$$

At section C, the available eccentricity is $e_o = 300$ mm. Substituting, we obtain $P = 1107$ kN after losses. Estimating the deferred losses at 17%, the initial prestresses is 1334 kN. From Table 12-2, a cable with 10 12.5 mm diameter strands provides the required force. Hence, $A_p = 1012$ mm^2; $\sigma_{pe} = 1094$ MPa; $\varepsilon_{pe} = 0.0057$; $T_{py} = 1650$ kN.

In the main span, a parabolic cable is used. The required sag at mid-span (Fig. 10-1d) is

$$h = \frac{wl^2}{8P} = \frac{26.56 \times 10^2}{8 \times 1107} = 0.3 \text{ m}$$

(c) *Reinforcing steel requirements at section C* For design ultimate conditions at C,

$$M_{DU} = 0.5 \times 5^2[1.4(22 + 4.56) + 1.7 \times 25] = 996 \text{ kN m}$$

With $T_{py} = 1650$ kN, $d_p = 584$ mm, and estimating $l_p = (584 - 100)$ and $l_s = (630 - 100)$, we obtain from Eqn 5-5

$$A_s = \frac{1}{530 \times 410}\left(\frac{996 \times 10^6}{0.9} - 1650 \times 10^3 \times 484\right) = 1400 \text{ mm}^2$$

Seven 16 mm bars are tried. A check for M_u shows that the section is satisfactory. A similar calculation for point B in the mid span region is also required.

(d) *Condition of prestress plus self-weight* If the construction sequence is such that the member initially carries only self-weight plus prestress, this condition is likely to be critical and must be checked. Prior to deferred losses the stresses in section C are (Eqns 2-2 and 2-3)

$$\sigma_b = \frac{1334}{190} - \frac{1334 \times 300}{8.186 \times 10^6}284.2 + \frac{57 \times 10^6}{8.186 \times 10^9}284.2 = -4.89 \text{ MPa (tens)}$$

$$\sigma_0 = \frac{1334}{190} + \frac{1334 \times 300}{8.186 \times 10^9}415.8 - \frac{57 \times 10^6}{8.186 \times 10^9}415.8 = 24.41 \text{ MPa (comp)}$$

The tensile stress is well above the tensile strength and cracks will appear under this load condition, if it occurs. Light reinforcing in the flange is required to control cracking until the dead load is applied. On the basis of the uncracked section analysis, the tensile force is 128 kN. Taking a working stress of 200 MPa, the required steel area is 128000/200 i.e. 640 mm^2, which satisfies ACI 318–77 requirements [13]. Four 16 mm bars are adequate.

The compressive stress is marginally in excess of the ACI limit of $0.6f'_c$, but an analysis taking account of the steel in the section gives a value within the ACI limit. However, the compressive stress limit is not in itself meaningful: at best it is a warning of possible excessive deformations, especially if application of the dead load is delayed.

(e) *Conditions at full service load* At full service load, the moment at section C is

$$M - 0.5 \times 5^2(22 + 4.56 + 25) - 644.5 \text{ kN m}$$

With $f_r = 0.6\sqrt{40} = 3.8 \text{ MPa}$, the cracking moment is

$$M_{cr} = (27.36 + 3.8)\frac{8.186 \times 10^9}{415.8} = 613 \text{ kN m}$$

The section is just cracked at full load. The deflection δ_1 is therefore calculated from the effective stiffness $E_c I_e$. The calculations follow those in Examples 6 and 7 and are not included here.

A cracked section analysis shows that the increment in steel stress is well within the limit of 150 MPa (details are not included).

(f) *Completion of design* In the preliminary calculations, a cable profile has been assumed with a sharp kink at C. To reduce friction losses, the kink will be replaced by a continuous curve over a short length around the support. Loss calculations can then be made for the real cable profile. Checks on stirrup requirements and end-block reinforcement are required to complete the design.

11 Continuous construction

Continuous construction is advantageous because moments and deflections at service load are reduced, while the load-carrying capacity and structural integrity are both improved, with a corresponding increase in structural safety. Continuous prestressed concrete construction is used extensively in floor systems in large-span buildings, in bridge girders, frames and various special applications.

As indicated in Fig. 2-7, hyperstatic reactions and secondary moments develop as prestress is applied to an indeterminate member. Secondary moments affect the stresses at service load but, provided the structure is ductile, have little effect on overload behaviour and load capacity. For design purposes, it is necessary to evaluate secondary moments so that their influence on deflections and cracking at service load can be investigated.

11.1 Pressure line

In the cross-section of a statically determinate member, prestress induces a compressive force C in the concrete at the level of the cable. In an indeterminate member, the effect of a secondary moment M_2 in the section is to displace the force C away from the cable by an amount l_2 where

$$l_2 = M_2/C \tag{11-1}$$

The term pressure line refers to a line running along the member and indicating the position of the C force in each section, taking account of the secondary moments.

11.2 Calculation of secondary moments

Hyperstatic reactions and secondary moments occur because of the deformations induced in the member by the prestress, as indicated in Fig. 2-7.

The equivalent load concept, introduced in Section 2, provides a convenient means for calculating secondary moments. Wherever the prestressing cable undergoes a change in direction, it exerts a force on the concrete member. Details are summarized in Fig. 2-5. By considering the continuous concrete member subjected to the equivalent force system induced by the cable, any desired method of structural analysis can be used to calculate the moments and reactions. It should be noted that this analysis gives the total moment due to prestress at each section:

$$M_p(x) = Pe(x) + M_2(x) \tag{11-2}$$

11.3 Linear transformations of cable profiles

Equivalent loads act only where a change of direction occurs in the cable profile. Such changes in direction, whether sudden or gradual, affect the secondary moments.

Certain changes can, however, be made to the shape of a cable profile without in any way affecting the total moments in the girder. Such changes, called linear transformations, cause no change in the shape of the cable (i.e. the angle changes in the cable remain unchanged) except at interior supports. In Fig. 11-1b, the cable profile has been obtained from that in Fig. 11-1a by introducing a kink at the interior support, but without otherwise changing the local cable curvatures. Apart from a concentrated force which acts directly on each support and therefore does not influence moments, the derived cable profile produces the same effects as the original profile, and the total moment $M_p(x)$ is therefore unaltered.

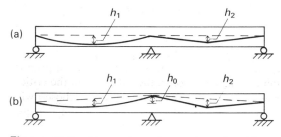

Figure 11-1 Linear transformation of cable

For a cable transformation to be linear, it is not permissible for a change in eccentricity to be made at an exterior support.

11.4 Collapse and ultimate strength

Prestressed concrete flexural members which have bonded tendons and which are not over-reinforced display ductile overload behaviour. The load capacity can thus be estimated by collapse-load analysis on the assumption of plastic hinge formation. Such methods are not commonly used for design calculations, which are rather based on the normal ultimate strength approach whereby elastic methods of analysis are used to evaluate the design ultimate moments and shears in critical sections. Ultimate strength theory is then used to check that the strength of local sections and regions is sufficient to withstand the design ultimate conditions.

11.5 Design procedure for continuous members

In the design of a continuous member, the step of prime importance is the choice of an appropriate cable profile. Many of the detailed calculations follow the same pattern as for determinate members.

In the case of partially prestressed continuous construction, the cable details can be chosen by use of the load balancing concept.

In the case of a fully prestressed member, allowable stress limits can be applied to the upper and lower extreme fibres at various sections along the member to determine the limits within which the cable profile should fall. Such stress calculations should include the stresses due to secondary moments. However, the load-balancing approach can also be used to choose a cable profile which balances an appropriate part of the total working design load.

The choice of the load to be balanced, i.e. the load at which zero deflections occur, may be a matter of trial and error in the design of both partially prestressed and fully prestressed continuous members.

An appropriate sequence for the design of a continuous member is as follows:

(1) Using design dead loads and live loads and an estimate of the self-weight load, determine by elastic analysis with gross cross-section properties the design dead-load and design live-load moments at critical sections.

(2) Choose dimensions for the critical cross-sections of the member. For a fully prestressed member, the cross-section is chosen by means of Eqns 9-3 and 9-4, with a preliminary estimate of cable friction and losses. For a partially prestressed member, the ultimate moment equation, 10-2, can be used.

(3) An appropriate cable profile is chosen. Initially, an idealized cable profile can be used to balance either the full sustained dead load, or a proportion of this load. To minimize the cable force, the cable sag in the critical span should be as large as possible. The real cable profile is chosen with kinks replaced, as appropriate, by curved portions. Details of the prestressing cable must be checked to ensure that it fits in the section.

(4) Friction losses are calculated, and details of the cable are adjusted as necessary.

(5) For a fully prestressed design, extreme fibre stresses have to be calculated at critical sections at full design load and also at transfer, and adjustments are made, as necessary, to the cable details or to the section properties to ensure that the permissible stresses are not exceeded.

(6) The moment capacity is checked at critical sections, and additional reinforcing steel is introduced, as needed.

(7) Shear capacity under design ultimate conditions is checked at critical sections, and stirrup reinforcement is introduced as required.

(8) Deflections at critical load stages are checked. In the case of fully prestressed construction, deflections are calculated from gross concrete sections. Partially prestressed members at full design load may be cracked, in which case the effective stiffness of Eqn 4-3 will be used to determine short-term deflections. Effective section stiffnesses for negative and positive moment regions may have to be averaged to allow mid-span deflections to be calculated.

(9) In the case of partially prestressed members, some check of crack widths at full design load may be necessary, either by means of fictitious tensile concrete stresses or by stress increments in the steel in cracked sections.

12   Appendix: design data

12.1   Mechanical properties of prestressing steel

Three types of prestressing steel find common use in prestressed concrete construction: high-strength wire, multi-wire strand and high-strength alloy steel bar. Some variation occurs from country to country in mechanical properties and in the grades of prestressing steel which are manufactured. The information presented in Tables 12-1 and 12-2 should therefore be regarded as indicative only;

Table 12-1   Typical properties of prestressing steel (selected tendon types)

Type of steel		Nominal diameter mm	Area mm²	E_p MPa×10⁻³	Yield stress (0.2% offset) MPa	Minimum breaking stress MPa	Remarks
Wire	(a)	5	19.6	200	1360	1600	Stress relieved
		7	38.5	200	1360	1600	
	(b)	5	19.6	193	1300	1700	Hard drawn
		7	38.5	193	1300	1700	
7-wire strand	(a)	7.9	38	193	1650	1850	Regular grade
		12.5	94	193	1580	1750	
		15.2	140	193	1500	1650	
	(b)	7.9	40	193	1650	1850	Super grade
		12.5	101	193	1630	1800	
		15.2	143	193	1580	1750	
	(c)	13.0	120	193	1700	1900	Compact grade
		15.2	165	193	1650	1800	
		18.0	223	193	1550	1700	
19-wire strand		18	210	193	1500	1750	Stress relieved
		25	423	193	1320	1560	
		32	660	193	1250	1500	
High-tensile alloy steel bar		23	415	172	933	1000	Super grade
		26	530	172	933	1000	
		29	660	172	933	1000	
		32	804	172	933	1000	
		35	962	172	933	1000	
		38	1140	172	933	1000	

Table 12-2 Typical properties of cables made up from seven-wire strands (12.5 mm diameter seven-wire super-grade strands)

No. of strands	A_p mm^2	Outer diameter of duct mm	Yield force kN	Minimum breaking force (kN)	Percentage of minimum breaking force (kN)			Minimum radius for cable curvature m
					65	75	85	
1	101	46	165	184	120	138	156	3.0
2	202	48	330	368	239	276	313	3.0
3	304	46	495	552	359	414	469	3.5
4	405	46	660	736	478	552	626	3.5
5	506	46	825	920	598	690	782	3.5
6	607	60	990	1100	718	828	938	3.5
7	708	60	1155	1290	837	966	1090	5.0
8	810	76	1320	1470	957	1100	1250	5.0
9	911	76	1485	1660	1080	1240	1410	5.0
10	1012	76	1650	1840	1200	1380	1560	5.0
12	1214	76	1980	2210	1440	1660	1880	5.0
19	1923	86	3135	3500	2270	2620	2970	5.0
27	2732	102	4455	4970	3230	3730	4220	8.0
37	3744	118	6105	6810	4430	5110	5790	8.0
42	4250	124	6930	7730	5020	5800	6570	8.0
55	5566	141	9075	10100	6580	7590	8600	8.0

Table 12-3 Typical values[a] of 1000 h stress relaxation, R_{1000}

Type of prestressing steel	Minimum initial stress	
	$0.70f_{pu}$	$0.80f_{pu}$
Stress-relieved wire		
normal relaxation	6.5	8.5
low relaxation	2.0	3.0
7-wire strand: regular and super		
normal relaxation	7.0	12.0
low relaxation	2.5	3.5
19-wire strand		
as-stranded	9.0	14.0
stress-relieved—normal relaxation	7.0	12.0
—low relaxation	2.5	3.5
Alloy steel bars	4.0	—

[a] Values taken from [6].

Table 12-4 Typical values[a] of friction curvature coefficient μ

Details of duct	μ
(a) Bright and zinc-coated metal sheathing	0.20^b
(b) Lead-coated metal sheathing	0.15^b
(c) Greased and wrapped coating	0.20
(d) External tendons over machined cast-steel saddles	0.15
(e) Unlined, preformed hole	0.50

[a] Values are taken from [6].
[b] If the tendons are rusted significantly, values of μ should be increased by 10% for case (b) and 20% for case (a).

Table 12-5 Typical values[a] of friction wobble term β

Details of duct	β
(a) Sheaths up to 50 mm diameter containing tendons other than bars	0.024–0.016
(b) Sheaths up to 50 mm diameter containing bars	0.016–0.008
(c) Sheaths of diameter between 50 and 90 mm	0.016–0.012
(d) Sheaths with internal diameter above 90 mm	0.008
(e) Bars in tape (any diameter)	0.008
(f) Unlined ducts formed by tubes	0.024
(g) Unlined ducts formed by bars	0.008

[a] These values are taken from [6]. They should be increased slightly if the tendon is not in the duct at the time of concreting.

final design calculations should be based on design data obtained from local manufacturers or distributors. Design data for materials and prestressing systems available in the United Kingdom are collected in [11]; similar data for the North American market are contained in [8].

Information on relaxation of prestressing steel for the calculation of losses is given in Table 12-3.

Typical values for friction loss coefficients for post-tensioning cables are contained in Tables 12-4 and 12-5.

12.2 Spacing of prestressing tendons

12.2.1 Pretensioned members

To ensure adequate anchorage development at the ends of pretensioned members, a minimum spacing of three times the diameter of the strand or four times the diameter of an individual wire is usually required. Tendons should also be separated sufficiently to allow the concrete to be placed and vibrated properly, a minimum spacing of one-and-one-third times the maximum aggregate size being commonly specified.

12.2.2 Post-tensioned members

In post-tensioned construction, the minimum spacing of cables and edge distances at the beam ends depend on the type of anchorage used and the dimensions of the jacking equipment. Data on proprietary systems should be obtained from the distributor. Spacing in end regions must permit adequate concrete compaction and should not be less than one-and-one-third times the maximum aggregate size.

In inner regions where the cables are curved, the spacing between cable ducts should be sufficient to prevent break-through of an outer cable into an inner duct. If this is properly allowed for, it is possible to bundle the cables to an extent consistent with the achievement of good concrete compaction.

12.3 Cover

The minimum concrete cover required to protect the tendons and reinforcing steel depends on the exposure conditions and fire resistance requirements, which are

Table 12-6   Cover requirements

| Type of structure | Minimum cover (mm) | | | |
	Condition 1	Condition 2	Condition 3	Condition 4
(a) *Buildings*				
(i) Post-tensioned section—				
(A) Ducts	25	40	40	50
(B) Tendons	25	40	40	50
(ii) Pre-tensioned sections made under factory conditions—				
Tendons	20	25	40	50
(b) *Bridges and Wharves*				
(i) All members—				
(A) Ducts	50	50	65	65
(B) Pre-tensioned tendons	40	40	50	50
(C) Reinforcement	40	40	50	50
(ii) Webs and thin slabs—				
(A) Ducts	40	40	50	50
(B) Pre-tensioned tendons	25	25	40	40
(C) Reinforcement	25.	25	40	40

Exposure conditions are as follows:

Condition 1: Concrete protected from weather and ground water;
Condition 2: Concrete cast in forms, but exposed to weather or ground water
Condition 3: Concrete deposited directly in contact with ground;
Condition 4: Concrete exposed to corrosive ground water, sea water, sea spray, liquids or vapours.

Note The above cover values, from [6], are slightly smaller than those in ACI 318, which vary from 75 mm for concrete in contact with the ground, to 20 mm for protected concrete.

usually stipulated in local codes of practice. The cover to be provided also depends on the manner of construction and the type of member. For example, in factory-produced pretensioned members, the usual limits for precast elements may apply, a typical value for minimum cover then being 12.5 mm for tendons of diameter not greater than 10 mm. Indicative values for minimum cover from [6] are contained in Table 12-6.

12.4   Maximum curvature in post-tensioning cables

The sharper the curvature in a prestressing cable, the higher are the bearing stresses, the larger will be the friction losses and the greater is the tendency for the cable to break through into an inner duct. In Table 12-2, the suggested minimum values for the radius of curvature of cables vary with the cable force.

References

1. Thürlimann, B., A case for partial prestressing, *Proc. symp. struct. concr.*, University of Toronto, May 1971. Reprinted as Report. No. 41, Institut für Baustatik, ETH, Zürich.
2. Walther, R. and Bhall, N. S., *Teilweise Vorspannung*, Deutscher Ausschuss für Stahlbeton, Heft 223, 1973.
3. Brøndum-Nielsen, T., Partial prestressing, *Proc. symp. prestressed concrete in short to medium span bridges*, FIP-Concrete Institute of Australia, Sydney Aug. 1976.
4. British Standards Institution, CP 110, Part 1: 1972 (as amended 1980), *The structural use of concrete*, London, 1980, 156 pp. + amend.
5. ACI Committee 318, *Building code requirements for reinforced concrete*, ACI Standard 318–77, American Concrete Institute, Detroit, 1977.
6. Standards Association of Australia, AS 1481, *Prestressed concrete code*, Sydney, 1974.
7. Ramaswamy, G. S., *Modern prestressed concrete*, Pitman, London, 1976, 175 pp.
8. Lin, T. Y. and Burns, N. H., *Design of prestressed concrete structures*, 3rd Edn, Wiley, New York, 1981, 646 pp.
9. Warner, R. F. and Faulkes, K. A., *Prestressed concrete*, Pitman, Melbourne, 1979, 336 pp.
10. Branson, D. E., Design procedures for computing deflections, *J. Am. Concr. Inst., Proc.*, Vol. 65, No. 9, Sept, 1968, pp. 730–742.
11. Abeles, P. W., Bardhan-Roy, B. K. and Turner, F. H., *Prestressed concrete designer's handbook*, 3rd Edn, Eyre & Spottiswoode Printers, Leatherhead, 1981, 556 pp.
12. CEB–FIP Committee, *Model code for concrete structures*, Comité Euro-International du Béton, English Translation, 1977.
13. ACI Committee 318, *Commentary on building code requirements for reinforced concrete*, American Concrete Institute, Detroit, 1977.

20 Plastic methods of analysis and design

M W Braestrup   M P Nielsen

Technical University of Denmark
Lyngby, Denmark

Contents

Notation

A	area
a	length of shear span
a, b, c	constants of catenary
b	width of beam (web)
C	compression stringer force
c	cohesion of concrete
D	dissipation; support diameter
d	distance from beam or slab face to reinforcement
d_0	diameter of loaded area (punch)
d_1	opening diameter of punching failure surface
e	lateral (elastic) displacement
f_c	uniaxial concrete compressive (cylinder) strength
f_c^*	effective concrete compressive strength
f_t^*	effective concrete tensile strength
f_y	yield stress of reinforcing steel
h	depth of beam or slab; thickness of wall
h_0	depth of conical punching failure surface
k, l, m	strength parameters of concrete
k_1, k_2	stress block factors for reinforced concrete section
M	bending moment in beam or slab
M_t	torque in beam
m	nondimensional bending moment
N	axial force; membrane force
N_0	axial force corresponding to zero extension
n	non-dimensional axial force
P	load
p	load per unit area; perimeter
Q_i	generalized stress
q_i	generalized strain rate
r	ratio of reinforcement; radius
S	stiffness
s	stirrup spacing

T	tension stringer force	κ	curvature rate
T_y	yield force of reinforcement	ν	effectiveness factor for compressive concrete strength
t	wall thickness of hollow section		
V	shear force	σ	normal stress
v	relative displacement rate	τ	shear stress
W_0	central deflection	Φ	reinforcement degree
W_I	rate of internal work	φ	angle of friction; direction of yield line in wall or slab; flexibility parameter
W_i	initial deflection		
W_l	rate of internal work per unit length or area		
w	deflection rate		
z	internal moment lever arm		

Subscripts

α	inclination of displacement rate; load distribution factor; constant	1, 2, 3	principal directions of stress and strain rate
		a	point A (support)
α_i	direction of reinforcing bars in wall or slab	b	point B (mid-span)
		c	concrete.
β	inclination of yield line	F	failure or yield
γ	shear strain rate; stirrup inclination	i	direction α_i
		n	normal component
Δ	middle surface extension; width of deforming zone	nt	tangential (shear) component
		s	steel
δ	rate of extension	u	ultimate
ε	strain rate	x, y	normal components in directions x and y
η	non-dimensional neutral axis distance		
		xy	tangential (shear) component in x, y-system
Θ	angle of rotation.		
θ	rotation rate; inclination of concrete compression		

Superscript

$'$	top reinforcement; derivative

Summary

The chapter explores the potential of using the theory of plasticity as a mathematical tool for the calculation of reinforced concrete members in the ultimate limit state. The relevance of regarding structural concrete as a plastic material is discussed, and rigid, perfectly plastic models for concrete and reinforcement are introduced. For concrete, the modified Coulomb failure criterion is adopted as yield condition, with the associated flow rule. The tensile concrete strength is neglected in most practical applications, and the compressive strength is represented by a reduced effective strength.

Based upon the above assumptions, yield conditions for beams, walls, slabs, and shells are derived, and methods of analysis and design are developed or reviewed for structural elements, viz:

– Slabs including membrane action and punching.
– Walls, including shear in joints.
– Beams subjected to bending, shear, and torsion.

The theoretical predictions are compared with experimental evidence, and remarkably good agreement is found, in spite of the primitive nature of the constitutive description.

1 Introduction

The study of the conditions at collapse is essential if we want to make predictions concerning the safety of structures. In doing so, structural engineers have always—explicitly or implicitly—made use of two fundamental principles of Nature. The first states that if there is a manner in which a structure can possibly collapse under a given load, it will do so. Thus, if the engineer is able to find just one collapse mode, he is convinced that the structure is unsafe. The other principle states that if there is a manner in which a structure can possibly carry a given load, Nature will do its best to find it. Thus, if the engineer can find just one way of transferring the loads from their points of application to some supports where they cease to be his problem, then he is fairly confident that the structure is indeed safe. In the theory of plasticity, these two principles are substantiated and refined, and they are formalized as the fundamental theorems of limit analysis or limit design.

The theory of plasticity is a branch of the strength of materials that can be traced back at least to Galileo. In our century, the theorems of limit analysis were formulated by Gvozdev [1] but his work was not widely known in the west till much later. The commonly used version was developed after the Second World War by Prager, Drucker, Greenberg, and Hodge [2–5], Hill [6], and Koiter [7]. Modern accounts of the theory of plasticity in the principal languages may be found in the monographs by Martin (English, [8]), Kachanov (Russian, [9]), Massonet and Save (French, [10]), and Reckling (German, [11]).

Whereas Gvozdev [1] formulated the theory with explicit reference to structural concrete, the school of Prager and Hill [2–7] was mostly concerned with metallic bodies. Concrete was regarded as a brittle material, generally unfit for plastic analysis, an exception being formed by cases where the strength is mainly governed by flexural reinforcement. A prominent example is the yield line theory for slabs, developed by Johansen [12] before the limit analysis theorems were formulated. The connection with the theory of plasticity was not firmly established till the 1960s, mainly through the work of Nielsen [13].

The implications of applying limit analysis to reinforced concrete structures were discussed by Drucker [14]. Chen and Drucker [15] considered a problem of plain concrete, using the modified Coulomb criterion with a non-zero tension cut-off (Section 2.2). The same constitutive model has since been applied by a research group at the Technical University of Denmark to treat a number of cases, mainly shear in plain and reinforced concrete, Nielsen et al. [16] Braestrup et al. [17]. Similar research has been carried out at various other institutions, notably by Müller [18] and Marti [19] at the Swiss Institute of Technology. In May 1979, a colloquium on plasticity in reinforced concrete was organized in Copenhagen, sponsored by the International Association for Bridge and Structural Engineering. Most of the results obtained so far are collected in the conference reports [20, 21].

A necessary condition for the validity of the limit analysis theorems is that the internal forces can be redistributed within the structure during loading to collapse. Thus, a certain ductility of the materials is essential. It is an open question whether concrete can be said to satisfy this requirement, but we regard an abstract discussion of this matter as rather futile. Instead, we propose that the theory be judged on its merits, and it would seem that plastic analysis does yield some surprisingly accurate and useful predictions on the strength properties of structural concrete.

In the first section below, we briefly summarize the theory of plasticity, and introduce its application to reinforced concrete. We proceed to consider yield conditions for beams, walls, slabs and shells. The remaining sections treat the analysis and design of structural members, i.e. slabs (bending, membrane action, punching), walls (in-plane forces, shear in joints), and beams (bending, shear, torsion). The review includes the classical methods, as well as results developed during the last decade. The predictions of the analyses are compared with experimental evidence.

2   Basic assumptions

2.1   Theory of rigid, perfectly plastic bodies

The theory of plasticity is concerned with the behaviour of structures at collapse, i.e. at the onset of large irreversible deformations. If elastic deformations and workhardening effects are neglected, the structure is called *rigid, perfectly plastic*. The mechanical state of the body is described by a set of stress measures Q_i (generalized stresses) and a set of strain measures q_i (generalized strain rates), such that the rate of internal work (dissipation) at a virtual deformation is given by

$$D = Q_i q_i \qquad\qquad (2\text{-}1)$$

The allowable stress states are defined by a *yield condition* of the form

$$F(Q_i) \le 0 \qquad\qquad (2\text{-}2)$$

In the space of generalized stresses, the equation $F(Q_i) = 0$ describes a surface which is called the *yield surface*, or *yield locus*, and is assumed to be convex.

Stress states outside the yield surface $(F(Q_i) > 0)$ cannot be sustained by the structure. For stress states inside the yield surface $(F(Q_i) < 0)$, the structure remains rigid. If the state of stress corresponds to a point on the yield surface, then deformations are possible, and they are governed by a flow rule. The *associated flow rule* or *normality condition* states that the yield function $F(Q_i)$ is a potential for the generalized strain rates, that is

$$q_i \sim \partial F / \partial Q_i$$

This means that if the strain rate state q_i is represented by a vector in generalized stress space, then this vector is an outwards directed normal to the yield surface. At a singular point (edge or corner), the strain rate vector is indeterminate, but confined by the normals to the adjacent parts of the yield surface. Note that the rate of internal work, as given by Eqn 2-1 is always uniquely determined by the strain rate vector.

Under the assumption of the idealized material behaviour represented by the existence of a convex yield surface and the validity of the associated flow rule, it is possible to prove the three limit analysis theorems, stated below.

The loading of the structure is supposed to be known apart from a scalar *load factor*, called the load. The yield load is the value of the load at collapse.

A *failure mechanism* is a field of generalized strain rates that are compatible and satisfy the kinematical boundary and continuity conditions.

A *statically admissible stress distribution* is a field of generalized stresses that are in equilibrium with the applied loads and satisfy the statical boundary and continuity conditions.

The upper bound theorem
A load for which a failure mechanism can be found satisfying the flow rule is greater than or equal to the yield load. An *upper bound* solution is derived by equating the rate of external work done by the loading with the rate of internal work dissipated in the structure, according to the flow rule. Using the equation for the yield surface, a corresponding field of generalized stresses can be found, but it need not constitute a statically admissible stress distribution.

The lower bound theorem
A load for which a statically admissible stress distribution can be found satisfying the yield condition is less than or equal to the yield load. A *lower bound* solution is derived by inserting the stress distribution into the statical equations, and satisfying the yield condition. By means of the flow rule, a corresponding field of generalized strain rates can be found, but it need not constitute a failure mechanism.

The uniqueness theorem
The lowest upper bound and the highest lower bound coincide, and constitute the *complete solution* for the yield load. The generalized strain rate and stress fields of the complete solution correspond to a failure mechanism and a statically admissible stress distribution.

As will be noted, the theory of perfect plasticity only involves the rates—or increments—of plastic strains, and does not predict the magnitude of the total deformations. However, when we regard the structure as rigid-plastic and only consider the instant of collapse, then the plastic deformations are the first and only to occur, and it becomes immaterial whether they are regarded as increments or not. Consequently, the use of superposed dots has been avoided, and although the term 'rate' is employed, the distinction from conventional 'small strains' is merely academic. An exception is formed by the analysis of *membrane actions* in slabs (Section 4.3), where we are interested in a history of plastic flow and not only incipient collapse.

2.2 Description of concrete

As will be apparent from Chapters 10 and 11, concrete is a rather complex material, and a complete constitutive description involves a large number of material parameters. It would therefore seem highly questionable to consider concrete as a plastic material—let alone as rigid, perfectly plastic. Nevertheless,

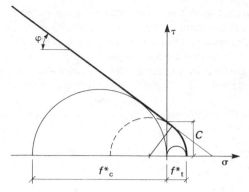

Figure 2-1 Modified Coulomb failure criterion

we base plastic analysis of concrete on the assumption:

> Concrete is regarded as a rigid, perfectly plastic material with the modified Coulomb failure criterion as yield condition and with the associated flow rule.

The modified Coulomb criterion is shown in Fig. 2-1 as the *failure envelope* of the Mohr's circles for sections subjected to the normal stress σ and the shear stress τ. The envelope is defined by the angle φ of internal friction, the tensile strength f_t^*, and the compressive strength f_c^*, which is related to the cohesion c by the formula

$$f_c^* = 2c\sqrt{k} \qquad \text{where} \quad k = \frac{1+\sin \varphi}{1-\sin \varphi} \qquad (2\text{-}3)$$

When the failure criterion of Fig. 2-1 is adopted as the yield condition, the yield surface can be determined in the space of principal stresses $(\sigma_1, \sigma_2, \sigma_3)$. For the cases of plane strain and plane stress, the yield loci are plotted in Fig. 2-2. They are found from the yield surface by projecting on, and intersection with, the plane $\sigma_3 = 0$. Figure 2-2 also illustrates the associated flow rule, the generalized strain rates being the principal strain rates $(\varepsilon_1, \varepsilon_2, \varepsilon_3)$.

Note that the yield surface is open towards the negative hydrostatic axis, hence, theoretically, arbitrarily high hydrostatic compression can be sustained. For plane strain, the slope of the yield locus is determined by the parameter k (Eqn 2-3). If the tensile strength is neglected ($f_t^* = 0$), the curve for plane stress reduces to the so-called *square yield locus*, Fig. 2-3.

The modified Coulomb criterion of Fig. 2-1 is obtained by combining the criteria of maximum shear stress and maximum normal stress, as first suggested by Dorn [22] in the case of cast iron. For concrete, the combination of Coulomb *sliding failure* and Rankine *separation failure* was introduced by Cowan [23] and Paul [24]. With a zero tension cut-off, the criterion was adopted for soil by Drucker and Prager [25] as a yield condition with associated flow rule. Chen and Drucker [15] introduced a finite tensile strength, and used the description for plastic analysis of concentrated loads on concrete blocks.

The constitutive model for concrete presented above is very crude in the sense

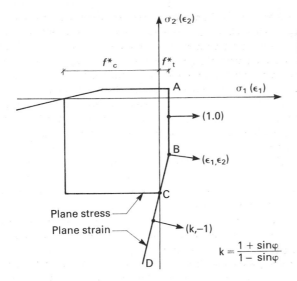

Figure 2-2   Yield loci for concrete in plane stress and plane strain

that it attempts to describe the strength properties and the deformations at failure by means of only three material parameters, namely, f_c^*, f_t^*, and φ. It would be purposeless to pretend that such a primitive description is particularly realistic. On the other hand, the model may be used as a mathematical tool for the treatment of strength problems in structural concrete, and the results may be compared with experimental evidence. In doing so, some care must be taken in the evaluation of the strength parameters.

It is characteristic for the plastic model that it assumes that the material is able to undergo arbitrarily large deformations at constant stress level. In reality, the ductility of concrete in compression is quite limited, and the stress–strain curve even has a falling branch. Consequently, the redistribution of stresses, which is a condition for the theorems of limit analysis, can only take place at the expense of losing strength. This is taken into account by assuming f_c^*, called the effective concrete strength, to be a certain fraction of the uniaxial compressive strength, f_c, estimated by conventional tests (cylinders, prisms, cubes, etc.). The ratio $\nu = f_c^*/f_c$ is called the *effectiveness factor*, and since it is primarily a measure of concrete

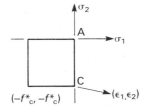

Figure 2-3   Square yield locus for concrete in plane stress

ductility, we would expect it to decrease with increasing strength level. Experimental investigations have shown that this trend may be described by assuming the effective strength, f_c^*, to be proportional to $\sqrt{f_c}$.

A theoretical estimate of the effective strength was obtained by Exner [81] by requiring the strain energy (i.e. the area under the stress–strain curve) corresponding to a particular limiting compressive strain to be identical for the actual and the idealized material. Based upon analytical expressions for experimentally determined stress–strain curves, the effectiveness factor is then found as a function of the peak stress, f_c, which is very similar to the square-root dependence mentioned above. Other investigations, however, indicate that the effective concrete strength depends not only on the corresponding compressive strain, but also on the magnitude of any co-existing shear strain (Collins and Mitchell [75]). As explained above, the effectiveness factor is primarily a measure of concrete ductility, but in addition it must also incorporate all the effects which are not explicitly accounted for in the theory, e.g. initial state of stress, stiffness of the materials, size effects, etc. Therefore, the effectiveness factor for a given type of structure will have to be evaluated by comparing the predictions of plastic analysis with test results.

Concrete in tension exhibits a behaviour which is almost brittle, hence the effective tensile strength, f_t^*, is very small. If $f_t^* = 0$ is assumed, the lack of ductility in tension becomes immaterial, and the theorems may be applied with confidence. Problems arise only if the strain rates change from tensile to compressive, but that is not relevant for simple yield-point analysis. Consequently, the tensile concrete strength is prudently neglected for all practical purposes, which means that reinforcement must be provided if tensile stresses cannot be avoided. It is a common belief that a tensile strength is necessary to explain the resistance of concrete against certain actions, notably shear. As will appear from the remainder of this chapter, this is not necessarily so.

The angle of friction appears to be fairly independent of the concrete quality, and ample experimental evidence suggests that we may assume $\varphi = 37°$, corresponding to $\tan \varphi = 0.75$ and $k = 4$.

2.3 Description of reinforcement

For the steel reinforcement, we introduce the assumption:

> The reinforcing bars are regarded as rigid, perfectly plastic, and able to resist forces in the axial direction only.

Thus, any dowel action of the bars is neglected. So, usually, are contributions from compressed reinforcement, because they are small in comparison with that of the surrounding concrete. The one-dimensional yield locus for steel subjected to the axial stress σ_s is illustrated in Fig. 2–4.

The tensile yield stress of the steel is termed f_y. For reinforcement without a

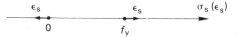

Figure 2-4 Yield locus for reinforcing steel

definite yield point, f_y is defined in a suitable manner, e.g. as the 0.2% offset strength.

The reinforcement is assumed to be either concentrated in lines (*stringers*) or continuously distributed over the section (*smeared*). In the latter case, the bars are assumed to be parallel and sufficiently closely spaced. The strength of a stringer is characterized by the yield force $T_y = A_s f_y$, where A_s is the cross-sectional steel area. The strength of *smeared reinforcement* is characterized by the equivalent yield stress, rf_y, where r is the reinforcement ratio, defined as

$$r = A_s/A_c \tag{2-4}$$

Here, A_c is the area of the concrete section perpendicular to the bars of area A_s. The actions of reinforcement in different directions are assumed to be independent.

Generally, problems with bond and anchorage are neglected. Thus, perfect bond between concrete and reinforcement is assumed in upper bound solutions. In lower bound solutions, any stress transfer is possible, including complete slip.

2.4 Yield lines and cracks

In the derivation of upper bound solutions, it is very convenient to use failure mechanisms where the deformations are localized in certain failure surfaces, separating two rigid parts of the body. The angle between the relative displacement rate, v, and the surface is α, where $-\pi/2 \leq \alpha \leq \pi/2$ (Fig. 2-5b). The intersection of the failure surface with the normal plane containing the displacement vector is a kinematic discontinuity called a yield line. The discontinuity is an idealization of a narrow zone of depth Δ with high strain rates, assumed homogeneous (Fig. 2-5a). Inspection of the deformation shows that in the normal plane the principal strain rates are

$$\varepsilon_1 = \frac{v}{2\Delta}(1+\sin\alpha) \quad \text{and} \quad \varepsilon_2 = -\frac{v}{2\Delta}(1-\sin\alpha) \tag{2-5}$$

The principal directions of strain rate coincide with the principal directions of stress, and are indicated on Fig. 2-5. The first principal axis bisects the angle between the deformation vector and the yield line normal.

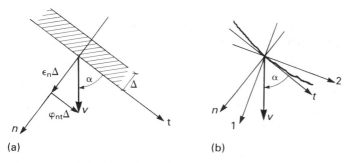

(a)   (b)

Figure 2-5   Yield line in plain concrete
  (a) Narrow zone with high straining
  (b) Kinematical discontinuity

Cracks in structural concrete are damages which may occur for a number of reasons. Under load, cracks tend to form perpendicular to the direction of the first principal stress. Thus, the yield line will only coincide with the crack direction provided the displacement rate is perpendicular to the yield line, which may then be termed a collapse crack, Müller [18, 26]. During a loading history leading to collapse, the principal axes (and the cracks) are likely to change directions, and at failure the latest formed cracks will generally be at an angle to the yield line. This means that shear stresses are transferred, presumably by aggregate interlock in old cracks and by crushing zones between cracks.

The transfer of shear in yield lines is expressed by the rate of work dissipated.

In the cases of plane stress ($\sigma_3 = 0$) or plane strain ($\varepsilon_3 = 0$), the rate of internal work per unit area of the discontinuity is

$$W_l = (\varepsilon_1 \sigma_1 + \varepsilon_2 \sigma_2)\Delta \qquad (2\text{-}6)$$

Inserting Eqns 2-5, we note that the dissipation is independent of the assumed depth Δ of the deforming zone.

The principal stresses (σ_1, σ_2) which are able to produce the principal strain rates ($\varepsilon_1, \varepsilon_2$) are determined by the flow rule and the yield condition, Fig. 2-2. Substituting into Eqn 2-6, we find

$$W_l = \tfrac{1}{2}vf_c^*(l - m \sin \alpha) \qquad (2\text{-}7)$$

for $\varphi \leq \alpha \leq \pi/2$ (plane stress or strain), and

$$W_l = \tfrac{1}{2}vf_c^*(1 - \sin \alpha) \qquad (2\text{-}8)$$

for $-\pi/2 \leq \alpha \leq \varphi$ (plane stress). Here, the parameters l and m are defined as

$$l = 1 - (k-1)f_t^*/f_c^* \qquad m = 1 - (k+1)f_t^*/f_c^* \qquad (2\text{-}9)$$

Note that Eqns 2-7 and 2-8 are identical for $\alpha = \varphi$. The derivation of the equations is explained in further detail in [16]. More general formulas for the dissipation in a Coulomb material are given by Jensen [27].

Equation 2-8 is valid for plane stress only, because the flow rule and the yield condition exclude yield lines with $\alpha < \varphi$ in the case of plane strain. To describe such deformations, it would be necessary to introduce a more sophisticated constitutive model, e.g., by assuming a curved failure envelope and/or a non-associated flow rule. For $\alpha = \pi/2$, Eqn 2-7 reduces to $W_l = vf_t^*$. Hence the resistance of concrete to cracking is negligible, but it picks up as soon as a tangential deformation is introduced. For pure shearing ($\alpha = 0$) in plane stress, Eqn 2-8 corresponds to a shear stress $\tau = \tfrac{1}{2}f_c^*$. Thus, to minimize the rate of internal work, concrete at failure will tend to crack (dilate) as much as possible. The efficiency of steel bars as reinforcement is to a large extent due to the restraint they offer against the dilation of the concrete.

Suppose a reinforcement stringer intersects a yield line at the angle β, where $0 \leq \beta \leq \pi$ and $\beta = 0$ corresponds to the same direction as $\alpha = 0$ (Fig. 2-6). The rate of strain ε_s in the stringer is then

$$\varepsilon_s = \frac{v}{\Delta} \sin \beta \cos (\beta - \alpha) \qquad (2\text{-}10)$$

The rate of internal work is determined by the flow rule and the yield condition,

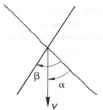

Figure 2-6   Yield line with reinforcing bar

Fig. 2-4:

$$\beta - \alpha \le \pi/2: \quad W_I = v T_y \cos(\beta - \alpha)$$
$$\beta - \alpha \ge \pi/2: \quad W_I = 0 \tag{2-11}$$

If the yield line is crossed by a band of smeared reinforcement, the contribution to the rate of internal work per unit area of the discontinuity is

$$\beta - \alpha \le \pi/2: \quad W_l = v r f_y \cos(\beta - \alpha) \sin \beta$$
$$\beta - \alpha \ge \pi/2: \quad W_l = 0 \tag{2-12}$$

2.5   Flexural yield lines

For a slab, the term yield line was coined by Ingerslev [28] to designate a curve of constant bending moment. Johansen [29] and Gvozdev [30] gave the concept a kinematic meaning as a line along which the deflection rate, w, is not twice differentiable. Such discontinuities may be termed *flexural yield lines* to distinguish them from the yield lines introduced in the preceding sections. The rotation rate, θ_n, in the yield line is defined as

$$\theta_n = \left(\frac{\partial w}{\partial n}\right)_- - \left(\frac{\partial w}{\partial n}\right)_+ \tag{2-13}$$

where n is the direction of the yield line normal. Thus, θ_n is the jump in the slope of the deflection rate, w, when the yield line is crossed with decreasing n. The yield line is termed positive or negative according to the sign of θ_n.

The relative rotation in a yield line is taking place about a neutral axis which is not necessarily at mid-depth. Thus, the rotation rate, θ_n, is accompanied by an extension rate, δ_n, of the middle surface. In a slab, δ_n is determined by the condition that the membrane force be zero, but when membrane action is included, δ_n becomes independent of θ_n. By further allowing a shearing discontinuity, δ_{nt}, of the middle surface, we arrive at a generalized flexural yield line, introduced for shells by Janas [31] and studied by Janas and Sawczuk [32].

3   Yield conditions for structural members

3.1   General assumptions

In the sections below, we shall consider the yield surfaces of beams, walls, slabs and shells, regarded as one and two-dimensional bodies defined by their middle surfaces. The yield conditions are formulated in terms of the appropriate stress resultants, which are regarded as generalized stresses. The corresponding generalized strain rates are assumed to completely define the deformations of the body, i.e., the Bernoulli and the Love–Kirchhoff assumptions are adopted. At each level parallel with the middle surface, the body is assumed to be in a state of plane stress.

The responses of the materials are described by the constitutive models introduced in the preceding sections. The tensile strength of the concrete is neglected, i.e., the square yield locus for plane stress (Fig. 2-3) is adopted. For the reinforcement, the yield locus of Fig. 2-4 is adopted, i.e., compressed reinforcement is neglected. The yield force of a reinforcing bar is $T_y = A_s f_y$, and all steel is regarded as having the same yield stress, f_y. In cases of varying reinforcement qualities, the nominal steel areas are assumed to be adjusted correspondingly.

3.2   Beams

If shear and torsion are absent or neglected, the static response of a plane beam is described by an axial force, N, and a bending moment, M. The corresponding generalized strain rates are the centre-line strain rate, ε, and curvature rate, κ, such that the dissipation per unit length of the beam is

$$D = N\varepsilon + M\kappa$$

At a kinematic discontinuity (*yield hinge*), the rate of internal work is given as

$$W_I = N\delta + M\theta \tag{3-1}$$

where δ and θ are the rates of extension and rotation in the hinge (Fig. 3-1).

The flexure is termed positive or negative according to the sign of θ. The bottom of the beam is identified as the face which is in tension for positive bending.

In the derivation below, the wording is related to a hinge, but the analysis is valid for distributed deformations as well.

For a given sign of the flexure, the deformation is defined by the parameter $\eta = \delta/h\theta$, which measures the (oriented) distance from the centre-line to the neutral axis. The two-dimensional yield surface (yield locus) of the beam defines the interaction between axial force and bending moment at failure. It consists of two parts, one for positive and one for negative bending, which can be determined by imposing a deformation and calculating the corresponding values of N and M. The stress distribution corresponding to a given deformation (Fig. 3-1) is determined by the constitutive models for concrete and reinforcement.

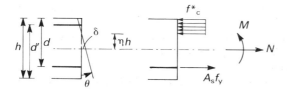

Figure 3-1   Yield hinge in beam or slab

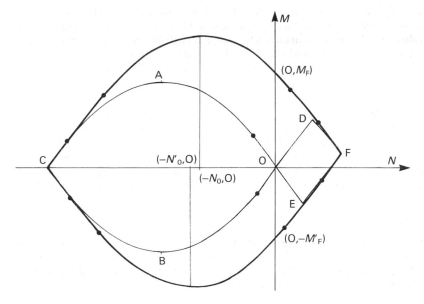

Figure 3-2   Yield locus for doubly reinforced section

Consider a rectangular beam of width b and depth h with a bottom steel area A_s at the distance d from the top face, and a top steel area A'_s at the distance d' from the bottom face (Fig. 3-1). The yield locus may conveniently be determined as a sum of contributions from the concrete and from the reinforcement (Fig. 3-2). The concrete contribution is the lenticular figure consisting of the two parabolas OAC and OBC, defined by

$$A: \quad N = -\tfrac{1}{2}bhf_c^* \qquad B: \quad N = -\tfrac{1}{2}bhf_c^* \qquad C: \quad N = bhf_c^*$$
$$M = \tfrac{1}{8}bh^2f_c^* \qquad\qquad M = -\tfrac{1}{2}bh^2f_c^* \qquad\qquad M = 0$$

The contribution from the reinforcement is the parallellogram ODFE, defined by

$$D: \quad N = A_sf_y \qquad\qquad E: \quad N = A'_sf_y$$
$$M = (d - \tfrac{1}{2}h)A_sf_y \qquad M = -(d' - \tfrac{1}{2}h)A'_sf_y$$

Note that in the absence of top reinforcement, the parallellogram degenerates to the line OD.

The resulting yield locus is found by adding the two contributions, i.e. as the frontier of the domain covered by OACB when the point O moves around ODEF (or vice versa). The yield locus is shown in Fig. 3–2.

The analytical expressions for the yield locus are quite complicated, due to the presence of the linear parts, which correspond to deformations with the neutral axis at the level of the reinforcement.

Of particular interest are the moments $M = M_F$ and $M = -M'_F$, corresponding to zero axial force (pure bending), and the axial forces $N = -N_0$ and $N = -N'_0$, corresponding to zero extension rate ($\eta = 0$) (Fig. 3-2). Introducing the *degrees* of reinforcement,

$$\Phi = \frac{A_s f_y}{bh f_c^*} \quad \text{and} \quad \Phi' = \frac{A_s' f_y}{bh f_c^*} \tag{3-2}$$

we find

$$\left. \begin{array}{l} M_F = \Phi\left(\dfrac{d}{h} - \tfrac{1}{2}\Phi\right)bh^2 f_c^* \\[2ex] M_F' = \Phi'\left(\dfrac{d'}{h} - \tfrac{1}{2}\Phi'\right)bh^2 f_c^* \\[2ex] N_0 = \tfrac{1}{2}bh f_c^* - A_s f_y \\[1ex] N_0' = \tfrac{1}{2}bh f_c^* - A_s' f_y \end{array} \right\} \tag{3-3}$$

The formulas for the yield moments M_F and M_F' assume that only the bottom and top reinforcement are active at, respectively, positive and negative pure bending.
An approximate yield condition is defined by

$$\theta > 0: \quad M - M_F + \frac{N}{bf_c^*}(N_0 + \tfrac{1}{2}N) \leq 0 \tag{3-4}$$

$$\theta < 0: \quad -M - M_F' + \frac{N}{bf_c^*}(N_0' + \tfrac{1}{2}N) \leq 0 \tag{3-5}$$

The yield locus consists of two parabolas passing through the points corresponding to pure bending and zero extension (Fig. 3-3). The approximate yield locus may conveniently be used for lightly reinforced sections (e.g., slabs), or in cases where the axial forces are known to be moderate.

The determination of the yield locus for a beam with bending moment and axial force is such a simple exercise that it is hard to credit any author in particular. The parabolic interaction curve was derived by Wood [33], the linear parts being included by Janas [34]. The approximate yield condition, Eqns 3–4 and 3–5, was suggested by Braestrup [35, 36].

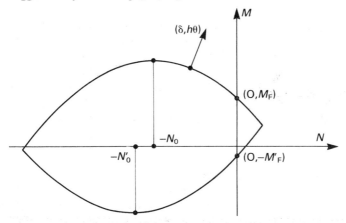

Figure 3-3   Approximate yield locus for section

3.3   Walls

The state of stress in a plane structure subjected to in-plane forces only may be described by the membrane forces N_x, N_y, N_{xy}, measured per unit length in a Cartesian coordinate system. The corresponding generalized strain rates are the membrane strain rates ε_x, ε_y, $2\gamma_{xy}$, the dissipation per unit area of the wall being

$$D = N_x\varepsilon_x + N_y\varepsilon_y + 2N_{xy}\gamma_{xy} \tag{3-6}$$

The rate of internal work per unit length of a yield line with the extension rate δ_n is

$$W_l = N_n\delta_n \tag{3-7}$$

where N_n is the normal component of the membrane force. Consider a wall of thickness h, and orthogonally reinforced in the x and y directions, the reinforcement ratios being r_x and r_y, respectively, defined by Eqn 2-4. As in the preceding section, the yield surface is composed of contributions from the concrete and from the reinforcement (Fig. 3-4).

In a coordinate system where the shear force is zero and the principal membrane forces are N_1 and N_2, the concrete is subjected to principal stresses $\sigma_1 = N_1/h$ and $\sigma_2 = N_2/h$, and the response is given by the square yield locus (Fig. 2-3). Thus, in the plane $N_{xy} = 0$, the contribution from the concrete is described by the square OABC, with side length $OA = hf_c^*$. Considering all possible directions of the principal axes, the total concrete contribution is found as the biconical yield surface OABCDE, shown in Fig. 3-4. The ordinates of points D and E are $N_{xy} = \frac{1}{2}hf_c^*$ and $N_{xy} = -\frac{1}{2}hf_c^*$, respectively. The contribution from the reinforcement is the rectangle OFGH, with side lengths $OF = hr_xf_y = N_{Fx}$ and $OH = hr_yf_y = N_{Fy}$. The surface reduces to a flat figure because the shear strength of the reinforcement alone is zero in the chosen coordinate system. The yield surface of the reinforced wall is obtained by adding the two contributions, i.e., as the frontier of the domain covered by the surface OABCDE as the point O moves around the rectangle OFGH (or vice versa). The resulting surface is symmetric about the plane $N_{xy} = 0$, and the part corresponding to $N_{xy} > 0$ is

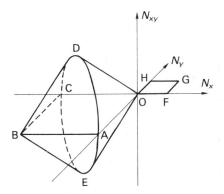

Figure 3-4   Yield surfaces for concrete and steel of orthogonally reinforced wall

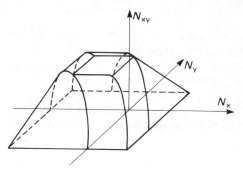

Figure 3-5   Yield surface for orthogonally reinforced wall

shown in Fig. 3-5. The surface consists of two cones, four cyclinders, and a flat face.

When the reinforcement strengths $r_x f_y$ and $r_y f_y$ are small compared with the concrete strength, it is feasible to use the approximate yield condition:

$$-(N_{Fx} - N_x)(N_{Fy} - N_y) + N_{xy}^2 \leq 0 \tag{3-8}$$

$$-(hf_c^* + N_x)(hf_c^* + N_y) + N_{xy}^2 \leq 0 \tag{3-9}$$

The corresponding yield surface is represented by the two cones in Fig. 3-5.

If the wall is reinforced in a number of directions at the angles α_i to the x-axis, and with the reinforcement ratios r_i, then the yield force N_F in a section defined by the angle φ (Fig. 3-6) is found to be

$$N_F(\varphi) = \sum_i N_i \cos^2 (\varphi - \alpha_i) \tag{3-10}$$

where $N_i = h r_i f_y$.

Defining the reinforcement parameters,

$$N_{Fx} = \sum_i N_i \cos^2 \alpha_i \qquad N_{Fy} = \sum_i N_i \sin^2 \alpha_i \qquad N_{Fxy} = \sum_i N_i \cos \alpha_i \sin \alpha_i$$

we can replace the yield condition 3-8 by the condition

$$-(N_{Fx} - N_x)(N_{Fy} - N_y) - (N_{Fxy} - N_{xy})^2 \leq 0 \tag{3-11}$$

The yield surface corresponding to the conditions 3-9 and 3-11 can be derived by

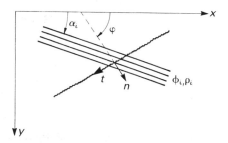

Figure 3-6   Reinforcement band and yield line in wall or slab

introducing the condition

$$-hf_c^* \le N_n \le N_F \tag{3-12}$$

and proceeding in the same way as that used for slabs in Section 3.4.

The yield surface of Fig. 3-5 was derived by Nielsen [37] in the case of isotropic reinforcement ($r_x = r_y$). Later, Nielsen [38] established the yield conditions 3-8 and 3-9 for orthotropic reinforcement. The case of arbitrary reinforcement was reduced to the orthotropic case by a suitable coordinate transformation. The yield condition 3-11, valid for any orientation of the coordinate system, was derived by Braestrup [35].

3.4 Slabs

A slab is a plane structure subjected to transverse loads. The state of stress in a slab element may be described by the moments M_x, M_y, M_{xy}, measured per unit length in a Cartesian coordinate system (Fig. 3-7). Note that M_{xy} is a component of the symmetric moment tensor, i.e., positive if the twisting moment vector on a section perpendicular to the x-axis is directed into the section. The corresponding generalized strain rates are the rates of curvature κ_x, κ_y, and $2\kappa_{xy}$, the dissipation per unit area of the slab being

$$D = M_x\kappa_x + M_y\kappa_y + 2M_{xy}\kappa_{xy} \tag{3-13}$$

The rate of internal work per unit length of a flexural yield line is

$$W_l = M_n\theta_n \tag{3-14},$$

where θ_n is the jump in slope, defined by Eqn 2-13, and M_n is the bending moment in the section.

A yield condition for the slab is found by assuming that the strength of a section depends upon the bending moment only. Thus, the yield condition for a section reduces to

$$-M_F' \le M_n \le M_F \tag{3-15}$$

Following Kemp [39], this may be called the *normal moment criterion*, and is visualized in Fig. 3-8. For an arbitrary section, the yield moments are functions of the angle φ (Fig. 3-6). If the slab is reinforced in the directions α_i, with the reinforcement degree Φ_i (bottom) or Φ_i' (top), then these functions, called the *polar diagrams*, are assumed to be

$$M_F(\varphi) = \sum_i M_i \cos^2(\varphi - \alpha_i) \tag{3-16}$$

$$M_F'(\varphi) = \sum_i M_i' \cos^2(\varphi - \alpha_i) \tag{3-17},$$

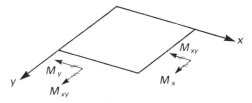

Figure 3-7 Definition of bending and twisting moments

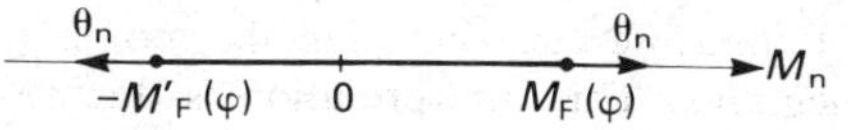

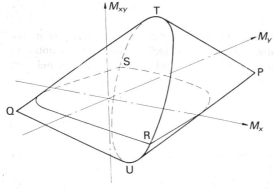

Figure 3-8   Normal moment criterion

where the band moments M_i and M'_i are identified with the yield moments M_F and M'_F given by Eqns 3-3. These formulas for the yield moment of a slab section were proposed by Johansen [12], see also Jones and Wood [40]. They are based upon the assumption that the differences in internal moment lever arm between the bands are neglible.

Equation 3-16 may be written

$$M_F(\varphi) = M_{Fx} \cos^2 \varphi + M_{Fy} \sin^2 \varphi + 2M_{Fxy} \cos \varphi \sin \varphi \qquad (3\text{-}18),$$

where

$$M_{Fx} = \sum_i M_i \cos^2 \alpha_i \qquad M_{Fy} = \sum_i M_i \sin^2 \alpha_i \qquad M_{Fxy} = \sum_i M_i \cos \alpha_i \sin \alpha_i$$

According to the transformation formulas, we have

$$M_n = M_x \cos^2 \varphi + M_y \sin^2 \varphi + 2M_{xy} \cos \varphi \sin \varphi \qquad (3\text{-}19)$$

Thus, the yield locus $M_n = M_F(\varphi)$ represents a plane in M_x, M_y, M_{xy}-space, defined by the parameter φ. The yield surface is found as the envelope of all possible planes as φ varies (see Braestrup [35, 41]) A similar argument applies to the negative moment equation $M_n = -M'_F(\varphi)$. The result is the biconical yield surface shown in Fig. 3-9. The corresponding yield conditions are

$$\theta_n > 0: \quad -(M_x - M_{Fx})(M_y - M_{Fy}) + (M_{xy} - M_{Fxy})^2 \leq 0 \qquad (3\text{-}20)$$

$$\theta_n < 0: \quad -(M_x + M'_{Fx})(M_y + M'_{Fy}) + (M_{xy} + M'_{Fxy})^2 \leq 0 \qquad (3\text{-}21)$$

valid for positive and negative bending, respectively. In the latter case, the primed strength parameters are defined analogously to their unprimed counterparts.

Equation 3-18 shows that we can interpret the parameters M_{Fx}, M_{Fy}, and M_{Fxy} as components of a symmetric resistance tensor $\overline{\overline{M}}_F$. Then, the yield condition

Figure 3-9   Yield surface for arbitrarily reinforced slab

3-20 simply states that the determinant of the difference between the moment tensor and the resistance tensor is non-negative. This interpretation is due to Gvozdev [42].

It is always possible to choose the coordinate system in such a way that either $M_{Fxy} = 0$ or $M'_{Fxy} = 0$. If the corresponding coordinate directions coincide, the biconical yield surface of Fig. 3-9 will have both vertices P and Q in the plane $M_{xy} = 0$, and was derived by Nielsen [13]. A special case is orthogonal reinforcement with the coordinate axes along the reinforcement directions. If, in particular, $M_{Fx} = M_{Fy}$ and $M'_{Fx} = M'_{Fy}$ (isotropic reinforcement), the yield surface may be represented as a square yield locus in the plane of principal moments (see Prager [5]).

3.5 Shells

A shell is a curved, two-dimensional body in which the state of stress is described by membrane forces N_x, N_y, N_{xy} and moments M_x, M_y, M_{xy}, measured per unit length in a Cartesian coordinate system. The corresponding generalized strain rates are the rates of membrane strain ε_x, ε_y, $2\gamma_{xy}$ and of curvature κ_x, κ_y, $2\kappa_{xy}$, the dissipation per unit area of the shell being

$$D = N_x\varepsilon_x + N_y\varepsilon_y + 2N_{xy}\gamma_{xy} + M_x\kappa_x + M_y\kappa_y + 2M_{xy}\kappa_{xy} \qquad (3\text{-}22)$$

When formulating the constitutive equations for a shell, it is customary to neglect the curvature, i.e., we consider a plane element subjected to the six stress resultants defined above. In a generalized flexural yield line (see Section 2.5), the rate of internal work per unit length is

$$W_l = N_n\delta_n + N_{nt}\delta_{nt} + M_n\theta_n \qquad (3\text{-}23)$$

If tangential deformations are not allowed, the rate of internal work reduces to

$$W_l = N_n\delta_n + M_n\theta_n \qquad (3\text{-}24)$$

The yield surface for a shell element may be found by the same procedure as that applied to slabs in the preceding section. In a yield line with extension rate δ_n and rotation rate θ_n only, the relationship between N_n and M_n is determined as for the beam with axial force (Section 3.2), e.g., as the approximate yield conditions 3-4 and 3-5. The yield moment of a section can then be expressed as a function $M_F(\varphi, \eta)$ of the direction φ and the neutral axis distance $\eta = \delta_n/h\theta_n$. For each set (φ, η), the normal moment criterion $M_n = M_F$ represents a hyperplane in the six-dimensional space of generalized stresses. The yield surface is the frontier of the region confined by these hyperplanes, as φ and η vary. Unfortunately, an analytical expression cannot be found, since the envelope of the hyperplanes is not convex, unless the relationship between M_n and N_n is linear. This was shown by Braestrup [35], who suggested an approximate analytical yield condition based upon a linearized M_n, N_n yield locus. Rajendran and Morley [43] developed a numerical procedure for the determination of points on the exact yield surface.

If tangential deformations are introduced in the yield lines, M_F will be a function of three parameters. A parametric representation of the corresponding yield condition was determined by Janas and Sawczuk [32]. By introducing appropriate approximations, Cookson [44] derived analytical expressions by determining the envelope of the corresponding hyperplanes. It has not yet been demonstrated, however, that the resulting yield surface is in fact convex.

4   Slabs

4.1   Yield line theory

The yield line theory for slabs is the earliest and most successful application of plasticity to structural concrete. More than half a century ago, Ingerslev [28] proposed an approach based upon the assumption of constant bending moments along certain lines. Johansen [29] developed the method into a theory in which the yield lines also had a geometrical significance as lines of relative rotation of slab parts. Also Gvozdev [30] introduced yield lines as kinematic discontinuities facilitating upper bound analysis. However, his paper was virtually unknown outside Russia, and the standard reference work on yield line theory remains the thesis by Johansen [12], which was eventually translated into English. Johansen [45] also established a collection of formulas covering most practical cases.

Whereas Gvozdev developed his concept of yield lines in connection with the theory of plasticity, Johansen regarded his contribution as an independent theory, in disagreement with Prager [5], who showed that yield line theory is an upper bound method. The constitutive model applied by Johansen [12] describes the strength of the slab by its yield moments, i.e., by the normal moment criterion, 3-15. The polar diagrams, Eqns 3-16 and 3-17, were derived by assuming the yield line to be formed by small 'steps' perpendicular to the reinforcing bars, an approach which is prone to misinterpretation. For isotropic slabs, the *'stepped'* *criterion* may be represented by the square yield locus (Section 3.4) which is generally known as Johansen's yield locus, although, ironically, Johansen never acknowledged that yield line theory had anything to do with limit analysis. In the 1960s, this question was the subject of some debate, and it was claimed, Jones and Wood [40], Wood [46], that the two theories could not be unified. Now, it is generally accepted that yield line theory is firmly based upon the theory of plasticity as a simple and ingenious method of computing upper bound solutions for slabs (Braestrup [41], Martin [8]). This connection has been explored in the work of Nielsen [13, 47]. The simplicity of yield line theory stems from the fact that it successfully describes the strength properties of a slab by means of the yield moments only, or in general by the components of the resistance tensors (Section 3.4). The calculation of the individual band moments M_i and M_i' may differ from country to country, depending upon the adopted *stress block factors* (Section 6.1), which are governed by the national codes of practice. The determination of a yield line theory solution may be divided into the following steps:

(a) Assume a yield line pattern corresponding to a failure mechanism defined by the vertical deflection rate at a particular point.

(b) Calculate the rate of internal work dissipated in the yield lines.

(c) Calculate the rate of external work done on the mechanism by the applied loading, and equate with the rate of internal work to determine an upper bound for the yield load.

(d) Modify the yield line pattern to find the lowest upper bound.

The actual calculations are described in many textbooks, e.g., Regan and Yu [48] Kong and Evans [49], and will not be considered in detail here.

Step (d) may be carried out analytically by minimization with respect to some geometrical parameters describing the mechanism, or by trial and error. A few attempts are sufficient, since it turns out that the upper bound is not very sensitive to the values of the geometrical parameters. On the other hand, it is possible that a completely different yield line pattern may give a much lower solution. Since there is no systematic way of finding the most dangerous mechanism, the application of yield line theory requires some skill and intuition.

One of the most controversial aspects of yield line theory is the concept of *nodal forces*, introduced by Johansen [12]. Instead of equating the rates of internal and external work, we may find the solution by considering the equilibrium of the individual slab parts. In order to arrive at the same result, it is necessary to introduce some self-equilibrated forces at the intersections between a yield line and other yield lines or a free boundary. Whereas the latter forces are connected with the Kirchhoff boundary conditions, the former simply ensure equilibrium of the slab parts for the mechanism of the lowest upper bound solution corresponding to the considered yield line pattern, as shown by Møllman [50]. For isotropic slabs, the rules for the determination of nodal forces are quite simple, but they become rather complicated for orthotropic slabs (Nielsen [47]). It is important to keep in mind that a nodal force formula corresponds to a degree of freedom for the yield line pattern. If the point of intersection is fixed—by a point load, a corner, or by symmetry—then the nodal forces are unknown. This was implied by Johansen, but not very well explained in his work, with considerable confusion as a result.

4.2 Lower bound methods

The principal disadvantage of yield line theory is the fact that it provides an upper bound solution only, and unless the prediction is backed by experimental evidence, we cannot be sure that it does not significantly overestimate the load-carrying capacity of the slab. It would therefore be preferable to provide a complementing lower bound analysis.

The determination of a lower bound solution may proceed as follows:

(a) Assume a moment distribution satisfying the statical conditions. Since the restrictions imposed by these are rather weak, we might as well consider a family of moment fields, defined by some statical parameters.

(b) Determine the unknown load factor from the statical equations. The load will be a function of the statical parameters.

(c) Establish restrictions on the static parameters by inserting the moment field into the yield condition, and requiring it not to be violated at any point.

(d) Optimize the lower bound by maximizing the load with respect to the static parameters, subject to the restrictions imposed by the yield condition.

If the lower bound thus obtained coincides with a yield line theory solution, then the *complete plastic solution* for the problem has been found. A summary of complete solutions for isotropic slabs has been given by Nielsen [47]. They can be extended to a class of orthotropic slabs by an affinity theorem first developed by Johansen [12] and extended to cover lower bound solutions by Nielsen [13].

A particularly interesting exact solution is the one requiring the minimum amount of reinforcement in a slab of given thickness, subjected to a given loading. The *optimum design* of reinforced concrete slabs was pioneered by Morley [82], who determined the solution for simply supported, rectangular slabs. Significant progress has been made since then, and the reader is referred to the exhaustive treatment given by Rozvany and Hill [83].

It is worthy of note that while only the yield moments are necessary in the yield line theory, a lower bound solution requires the complete yield condition of the slab. For this purpose, the analytical expressions given by Eqns 3-20 and 3-21 are most convenient.

In Cartesian coordinates, the equilibrium equation for the moments is

$$\frac{\partial^2 M_x}{\partial x^2} + 2\frac{\partial^2 M_{xy}}{\partial x\, \partial y} + \frac{\partial^2 M_y}{\partial y^2} + p(x, y) = 0 \tag{4-1}$$

The sign of the M_{xy}-term corresponds to the definition of the twisting moment of Fig. 3-7. In addition to Eqn 4-1, the moments must satisfy the Kirchhoff boundary conditions and the corresponding jump conditions. The latter are necessary in case of line loads, and enable discontinuous moment fields to be considered.

It is obvious from the procedure outlined above, that the determination of a lower bound solution is not as simple as upper bound yield line analysis. Primarily, this is due to the fact that it is not easy to specify a clever moment field which satisfies the static conditions, especially in cases of slabs with odd shapes and awkward boundary conditions. For rectangular slabs with different support conditions, a class of lower bound solutions based upon quadratic moment fields has been developed by Nielsen and Bach [51].

For the reasons mentioned above, the lower bound approach is not very suitable for the analysis of a given slab, but it may be used to design the reinforcement. A particularly simple application is the *strip method*, developed by Hillerborg [52] (see also Armer [53]). The basic idea is that the load is assumed to be carried by bending only, i.e. only twist-free moment fields are considered. Since actual slabs normally do carry quite a lot of load by twisting, a rather conservative solution would be expected. However, this is compensated through savings in reinforcement by providing only the amount necessary to resist the bending moments. These are determined by splitting the load into parts which are carried by individual systems of strips, designed as beams. Since the static equations are satisfied, a lower bound is obtained, and the solution even appears to be exact, as shown by Wood and Armer [54], Fernando and Kemp [55]. This means that there is no way a higher load could be carried with that particular reinforcement, but, of course, this does not guarantee that the solution is economic.

The number and directions of the strips are arbitrary, but normally it is feasible to consider strips in two orthogonal directions x and y. Then the equation of equilibrium, Eqn 4-1, may be replaced by the two strip equations:

$$\frac{d^2 M_x}{dx^2} + \alpha p(x, y) = 0$$

$$\frac{d^2 M_y}{dy^2} + (1 - \alpha)p(x, y) = 0 \tag{4-2}$$

The load distribution factor $\alpha = \alpha(x, y)$ is arbitrary, and is not even confined to the range $0 \le \alpha \le 1$. A rational basis for the choice of α may be to require that the deflection of the two intersecting strips be identical at any point. This so-called strip-deflection method was proposed by Fernando and Kemp [56].

The strip solution may not require any reinforcement in certain regions and directions. Then no advantage is taken of the minimum reinforcement specified by most codes of practice. This drawback is amended in the *bimoment method* developed by Gurley [57]. The approach explores the fact that the assumption of zero twisting moment enables a finite region of the slab to deflect into a hyperbolic parabola without any internal work being dissipated. This additional degree of freedom corresponds to an equilibrium equation involving moment of moments, i.e. bimoments. Introduction of the fourth equilibrium equation reduces the static indeterminacy of the slab, thus considerably facilitating design.

A particular application of the lower bound theorem is the use of the elastic moment field. In some cases, an elastic analysis will have to be performed anyway, e.g., to check the deflections under working load. Since the elastic solution satisfies the static conditions, we obtain a lower bound by providing reinforcement in such a way that the yield condition is satisfied everywhere. It may not be a very close lower bound, and the reinforcement pattern may be quite awkward because of the moment peaks often associated with elastic solutions. It should also be stressed that the design is still plastic, relying on the lower bound theorem. In the elastic stage, the curvatures are proportional to the moments, hence if differences in lever arms are neglected, yielding will always start at the moment peaks, irrespective of the amount of reinforcement. By the choice of a moment distribution which satisfies the equations of elasticity, the need for redistribution of moments may be reduced, but not eliminated (Section 6.1).

4.3 Membrane action

In pure bending of a reinforced concrete section with small steel proportions, the neutral axis at failure is close to the compressed face. Thus, flexural failure of a slab is accompanied by extensions of the middle surface (Sections 2.5 and 3.2). Normally, these deformations are not taken into account, i.e., the slab is regarded as a two-dimensional body with moments and curvatures only (Section 3.4). The strength of the slab is described by the yield moments, which are assumed to be independent of the deflections, the same being the case with the load-carrying capacity.

The reality is different, as indicated in Fig. 4-1, in which a typical load-parameter, p, is plotted against a typical deflection parameter, W_0. The flexural yield load is p_F. Curve (a) shows the load–deflection relationship for a slab which is restrained at the boundaries, which means that the middle surface extensions associated with pure flexural failure are prevented. As a result, the load-carrying capacity increases to several times the yield load, when compressive membrane forces are built up in the slab. This compressive membrane action is also known as the dome effect.

At the peak load, the neutral axes are close to mid-depth, and a stability failure occurs as the slab snaps through, the capacity dropping dramatically under increasing deflections. At the minimum load, the membrane forces change sign and the load-carrying capacity may eventually pick up again, due to the occur-

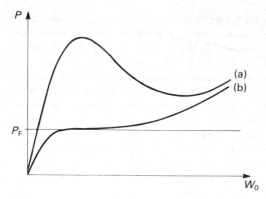

Figure 4-1 Membrane action in slabs
(a) Laterally restrained slab
(b) Laterally unrestrained slab

rence of tensile membrane forces. This behaviour is sometimes called the catenary effect.

Curve (b) in Fig. 4-1 shows the membrane effect for an unrestrained slab, where the implications are far less significant. The increase in load-carrying capacity is due to the formation of a central tensile membrane, carried by an outer compression ring.

It is characteristic that while tensile membrane action is associated with significant deformations, the dome effect occurs for deflections of the order of half the slab depth.

The dramatic effects of compressive membrane action have been known since the early days of reinforced concrete slab analysis, and the subject has received significant attention during the last 25 years. A review of the literature is given by Braestrup [36].

The high load-carrying capacity associated with the dome effect is contingent upon the compression supplied by the supports. If reinforcement is provided to achieve these restraining forces, such steel would normally be better employed as ordinary reinforcement in the slab. However, in many cases, slabs are laterally restrained by surrounding structures whose strength and stiffness derive from other considerations, and significant reductions in slab reinforcement may be achieved on account of the beneficial effect of compressive membrane action.

In order to predict the peak load, it is necessary to calculate the load–deflection relationship for a restrained slab. A significant step may be taken by rigid-plastic analysis, applying the following procedure:

(a) Assume a failure mechanism, and establish relationships between the distances ηh to the neutral axes in the yield lines.
(b) Determine the stress resultants in the yield lines as functions of the neutral axis distances by means of the yield condition and the flow rule.
(c) Eliminate the unknowns by the equations of equilibrium for the slab parts.

A particularly simple and illustrative example is furnished by a one-way spanning slab strip of length $2a$, depth h, and subjected to the uniformly distributed load p.

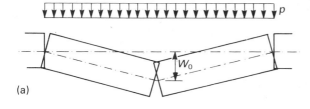

(a)

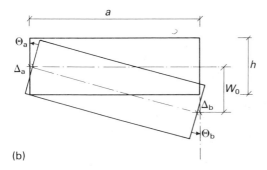

(b)

Figure 4-2 Failure of slab strip or circular slab
(a) Failure mechanism
(b) Deformations of slab half

A failure mechanism is indicated in Fig. 4-2a, Fig. 4-2b showing one half of the slab. At the central deflection W_0, the rotation and middle surface extension at the support are $\Theta_a = -W_0/a$ and Δ_a. At mid-span, the corresponding (half) values are $\Theta_b = W_0/a$ and Δ_b. The geometrical relation

$$(a - \Delta_a - \Delta_b)^2 + W_0^2 = a^2$$

reduces to

$$\Delta_a + \Delta_b = W_0^2/2a \qquad (4\text{-}3)$$

upon neglect of higher order terms. Differentiating with respect to some monotonically increasing function of time (e.g., the central deflection W_0), we find a relationship between the rate of deflection w_0 and the rates of extension δ_a and δ_b at a point in time characterized by the deflection W_0

$$\delta_a + \delta_b = W_0 w_0/a$$

Division by the rate of rotation, $\theta_b = -\theta_a = w_0/a$, yields

$$-\eta_a + \eta_b = W_0/h \qquad (4\text{-}4)$$

where $-\eta_a h$ and $\eta_b h$ are the distances to the neutral axes (Fig. 3-1). Equation 4-4 is the relationship referred to under (a) above. By adoption of the yield conditions 3-4 and 3-5, step (b) yields

$$\begin{aligned} n_a &= -\eta_a - n_0' & m_a &= -m_F' - \tfrac{1}{2}(n_0'^2 - \eta_a^2) \\ n_b &= \eta_b - n_0 & m_b &= m_F + \tfrac{1}{2}(n_0^2 - \eta_b^2) \end{aligned} \qquad (4\text{-}5)$$

Here, we have introduced the non-dimensional stress resultants

$$n = N/hf_c^* \qquad m = M/h^2 f_c^*$$

and analogous expressions for the strength parameters n_0, n_0', m_F and m_F' (Section 3.2). Equilibrium of the slab half requires

$$n_a = n_b \qquad \tfrac{1}{2}pa^2 = (-m_a + m_b + n_b W_0/h)h^2 f_c^* \tag{4-6}$$

Equations 4-4, 5 and 6 yield the solution:

$$n = \tfrac{1}{2}(n_0 + n_0' - W_0/h) \tag{4-7}$$

$$\tfrac{1}{2}pa^2 = (m_F + m_F' + n^2)h^2 f_c^* \tag{4.8}$$

where $n = n_a = n_b$. The equations show that the membrane force decreases linearly with the central deflection, whereas the load varies parabolically. Some numerical examples are plotted in Fig. 4-3. The upper curve corresponds to a fully clamped strip with equal top and bottom reinforcement, the middle curve to a restrained strip with bottom reinforcement only, and the lower curve to an unreinforced, restrained slab. The minimum load, corresponding to vanishing membrane force, is reached for a central deflection $W_0 = (n_0 + n_0')h$, i.e. close to the slab depth. The minimum capacity is equal to the flexural yield load for the unrestrained slab, which is not the case for two-way slabs (see below).

The analysis of the strip is due to Janas [34]. The derivation employs an upper bound technique, but the solution is exact as long as the membrane force is compressive, since a statically admissible stress distribution can be found (Braestrup [36]) For a circular slab, the solution (see below) was given by Morley [58].

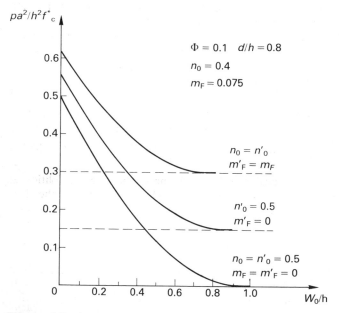

Figure 4-3   Load-deflection curves for rigidly restrained slap strips

Many authors have analysed the dome effect by the *deformation theory of plasticity*, i.e., by considering the yield condition as a plastic potential for the total deformations Θ and Δ. Consequently, the kinematic relationship 4-4 is replaced by the geometric relationship 4-3, and the central deflection corresponding to a given membrane force is overestimated by a factor 2. For certain deflections, the predicted load-carrying capacity is less than the unrestrained yield load. This is due to the fact that the concrete is assumed to exert full compressive stress as long as the total strains are compressive, even though they are decreasing in magnitude. The relationship between the deformation-theory and the flow-theory approaches was discussed by Braestrup [36].

The analysis described above may give a reasonable prediction of the decline of membrane forces and load-carrying capacity at moderate deflections, but it cannot furnish much information on the peak load. In rigid-plastic theory, the maximum load is obtained at zero deflection and falls off rapidly (Fig. 4-3), hence the slightest departure from perfect rigidity is bound to have a considerable impact. The traditional way of taking elastic effects into account is to assume an empirical value (e.g., $W_0 = \frac{1}{2}h$) for the deflection at the peak load, which is then determined as the corresponding point on the rigid-plastic curve.

For a slab strip (Fig. 4-2), the elasticity may be described by a lateral displacement $e = -N_a/S$, where S is a stiffness. The kinematic relationship 4-4 then becomes

$$-\eta_a + \eta_b + 2h\varphi \frac{\mathrm{d}n_a}{\mathrm{d}W_0} = \frac{W_0}{h} \tag{4-9}$$

Here, $\varphi = af_c^*/2hS$ is a flexibility parameter. The resulting first-order differential equation for the membrane force was solved by Janas [59]. For a circular slab, the solution was given by Morley (Braestrup and Morley [60]).

Figure 4-2 may also be taken as representing a radial section in a circular slab. If the in-plane compression and the lateral support flexibility are lumped together and described by a boundary stiffness S, then Eqn 4-8 represents the kinematic relationship. The procedure outlined above then yields the solution

$$\frac{1}{6}\frac{pa^2}{h^2 f_c^*} = m_F + m_F' + \frac{1}{4}(n_0 + n_0')^2 - \frac{1}{4}(n_0 + n_0')\frac{W_0}{h}$$

$$+ \frac{5}{48}\left(\frac{W_0}{h}\right)^2 - [\alpha \exp(-W_0/\phi h) - \frac{1}{4}\varphi]^2 \tag{4-10}$$

Here, α is a constant which is evaluated by specifying the boundary condition at the start of the deformation. To take account of the elastic curvature of the slab, we may require the membrane force n_a to be zero for an initial deflection $W_0 = W_i$, whence

$$\alpha = \frac{1}{2}(n_0 + n_0' + \frac{1}{2}\varphi - \frac{1}{2}W_i/h)\exp(W_i/\varphi h) \tag{4-11}$$

In Fig. 4-4, Eqn 4-10 is plotted for $\varphi = 0.468$ and for $\varphi = 0$, i.e. a rigid-plastic slab. In the former case, the load–deflection relationship is assumed to be linear for $W_0 \leq W_i = 0.03h$. The load is related to the unrestrained yield load p_F, where

$$p_F = 6(m_F + m_F')h^2 f_c^* $$

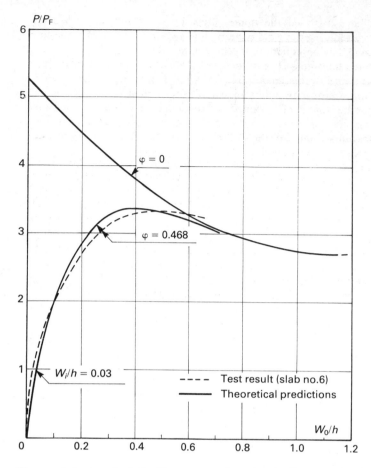

Figure 4-4   Load–deflection curves for elastically restrained circular slabs

For comparison, Fig. 4-4 also shows an experimental load-deflection curve for a slab with $\varphi = 0.468$, obtained in Cambridge, UK—see Braestrup and Morley [60], where the analysis is also discussed in further detail. It appears that the agreement is acceptable, and that the peak load is predicted reasonably well by this approach.

4.4   Punching shear

Punching shear failure may occur in flat slabs supported on slender columns or subjected to concentrated loads. The failure is characterized by the punching out of a concrete body limited by a failure surface running through the slab from the column or loaded area (Fig. 4-5). The main reinforcement is not very effective in preventing this type of failure, which consequently may reduce the ultimate load below the flexural capacity of the slab.

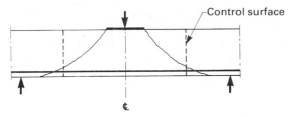

Figure 4-5 Punching shear failure of slab

The failure of slabs subjected to concentrated loading is very dependent upon the support conditions, especially the degree of restraint against lateral movements. Indeed, proper punching failure seems mainly to occur in cases where the load-carrying capacity has been enhanced by dome action (Section 4.3). This question was discussed by Braestrup [61], who also reviewed various approaches to the punching problem, including that of most building codes, which is based upon the nominal shear stress on a control surface around the loaded area (Fig. 4-5).

A plastic solution may be obtained by upper bound analysis, based upon the constitutive description introduced in Section 2. Consider a slab of depth h, supported along a circular perimeter of diameter D and centrally loaded by a load, P, applied on a circular punch (or column) of diameter d_0 (Fig. 4-6a). The

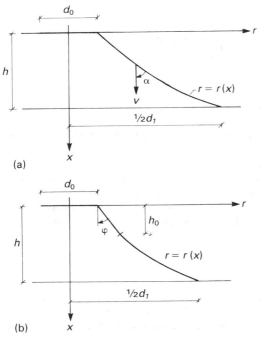

Figure 4-6 Punching failure
(a) Failure surface generatrix
(b) Optimal generatrix

deformations are assumed to be localized in a rotationally symmetric failure surface with the generatrix $r = r(x)$. The shear is constrained by the surrounding structure in such a way that the relative deformation rate is perpendicular to the slab. Due to the axial symmetry, each generatrix may be regarded as a yield line in a state of plane strain.

An upper bound solution is derived by equating the rates of external and internal work:

$$Pv = \int W_l \, dA$$

The dissipation W_l is given by Eqn 2-7, and the integral is taken over the entire failure surface. Substituting

$$dA = 2\pi r \frac{dx}{\cos \alpha} \quad \text{and} \quad \tan \alpha = \frac{dr}{dx} = r'$$

we find the upper bound

$$P = \pi f_c^* \int_0^h (l\sqrt{1+(r')^2} - mr')r \, dr \tag{4-12}$$

The material constants l and m are defined by Eqns 2-9. The lowest upper bound solution is found by variational calculus as the function $r = r(x)$ which minimizes the functional of Eqn 4-12 subject to the condition

$$\alpha \geq \varphi \quad \text{or} \quad r' \geq \tan \varphi \tag{4-13}$$

The solution may be written

$$0 \leq x \leq h_0: \quad r = \tfrac{1}{2}d_0 + x \tan \varphi \tag{4-14}$$

$$h_0 \leq x \leq h: \quad r = a \cosh\left[(x - h_0)/c\right] + b \sinh\left[(x - h_0)/c\right] \tag{4-15}$$

where

$$c = \sqrt{a^2 - b^2} \tag{4-16}$$

The optimal generatrix is sketched in Fig. 4-6b. It consists of a catenary curve, Eqn 4-15, joined at the level $x = h_0$ by a straight line, Eqn 4-14. The failure surface intersects the slab in a circle with the diameter d_1, which is determined so as to minimize the upper bound. The constants a, b, c and h_0 are determined by Eqn 4-16 and the boundary conditions

$$a = \tfrac{1}{2}d_0 + h_0 \tan \varphi \tag{4-17}$$

$$b/c = \tan \varphi \tag{4-18}$$

$$\tfrac{1}{2}d_1 = a \cosh\left[(h - h_0)/c\right] + b \sinh\left[(h - h_0)/c\right] \tag{4-19}$$

For certain values of d_0, d_1, and h, the optimal failure surface has no conical part, i.e., $h_0 = 0$. In this case, Eqn 4-18 reduces to the inequality $b/c \geq \tan \varphi$, and Eqns 4-16, 4-17 and 4-19 determine the constants a, b and c.

The lowest upper bound for the load is

$$P = \tfrac{1}{2}\pi f_c^*[h_0(d_0 + h_0 \tan \varphi)\frac{1-\sin \varphi}{\cos \varphi} + lc(h - h_0)$$

$$+ l(\tfrac{1}{2}d_1\sqrt{(\tfrac{1}{2}d_1)^2 - c^2} - ab) - m((\tfrac{1}{2}d_1)^2 - a^2)] \tag{4-20}$$

As a non-dimensional load parameter, we may take the quantity

$$\frac{\tau}{f_c} = \frac{P}{\pi(d_0 + 2h)f_c} \tag{4-21}$$

Thus, τ is the average shear stress on a control cylinder at a distance h from the loaded area.

The above plastic analysis was given by Braestrup *et al.* [62], who also described an iterative procedure which determines the upper bound solution for given material properties f_c^*, f_t^*, and φ, and for given geometrical quantities h, d_0, and D (see also Nielsen *et al.* [16]).

The results are not very dependent upon the angle of friction, and the value $\varphi = 37°$ is assumed (Section 2.2). On the other hand, the solution, particularly the opening diameter d_1, is very sensitive to the assumed ratio f_t^*/f_c^*. If the tensile strength is neglected, then the upper bound decreases with increasing value of d_1, which means that the optimal failure surface will extend as far as possible. When the support diameter D increases towards infinity, the ultimate load approaches zero asymptotically, corresponding to simple splitting of the concrete. Thus, in order to get realistic results, it is necessary to assign a finite tensile strength to the concrete. Then, the upper bound will be minimum for a finite value of d_1, i.e. a non-zero tensile strength has the effect of contracting the failure surface around the loaded area. A realistic tensile strength level, of the order of $f_t^* = f_c^*/10$, would correspond to a failure surface which was almost conical, and such failures are indeed observed by dynamic loading. At static tests, however, the failure surface is typically very flat, as seen in Fig. 4-7a.

In order to predict a failure surface like Fig. 4-7a, the assumed effective tensile strength should be only 0.25% of the effective compressive strength. Assuming $f_t^* = f_c^*/400$, the theoretical generatrix corresponding to the same geometry d_0/h is shown in Fig. 4-7b. With this extremely low effective tensile strength, the predicted ultimate loads are found to agree reasonably well with test results (Braestrup *et al.* [62]).

Since slabs are always supported at a reasonable distance from the loaded area, it is most prudent to neglect the tensile concrete strength and assume that the failure surface extends all the way to the support. Figure 4-8 shows the results of some tests on square slabs carried out by Taylor and Hayes [63]. The load parameter is plotted against the relative punch diameter and the results are compared with the theoretical curve for zero tensile strength and a support diameter equal to the slab span. The agreement is far from perfect, but on the other hand the predictions are not wildly off the mark.

The tests plotted in Fig. 4-8 were carried out on restrained slabs and are exceptional in the sense that most of the results exceed the theoretical prediction. For the majority of tests reported in the literature, the ultimate load is overestimated by plastic analysis, and to obtain qualitative agreement it is necessary to

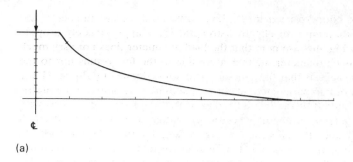

(a)

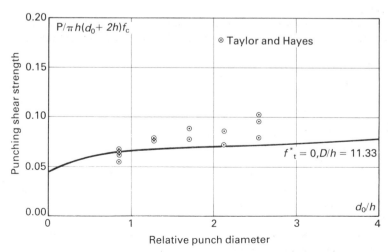

(b)

Figure 4-7 Pull-out failure
  (a) Predicted generatrix ($f_t^* = f_c^*/400$)
  (b) Experimental failure piece

Relative punch diameter

Figure 4-8 Punching test results versus theoretical prediction

introduce a reduced effective strength, f_c^*. Investigations of a great number of test results shows that the empirical effective strength, f_c^*, is approximately proportional to $\sqrt{f_c}$. From Fig. 4-8, we note that the load parameter does not vary much with the relative punch diameter, neither according to the test results nor to the plastic theory. This suggests that the average shear stress defined by Eqn 4-21 is a suitable choice as a design parameter. The principal difference between the plastic approach and the method of the building codes is that the former compares with the compressive concrete strength, whereas the latter uses a tensile strength measure. However, since the effective concrete strength varies with the concrete quality in much the same way as the tensile strength does, then this difference may to some extent be regarded as a matter of taste. In both cases, it is the ductility of the concrete which is the decisive factor. Comparison with published test results shows that the scatter of the two methods is of the same order of magnitude (Braestrup [61]).

5 Walls

5.1 Lower bound design

The term 'wall' is used to designate a flat, two-dimensional structural element, subjected primarily to in-plane forces. In contrast to the case of slabs, it is generally not easy to specify a clever failure mechanism for such a structure. Consequently, the plastic analysis of walls is normally carried out by lower bound methods. The same procedure may be used as outlined for slabs in Section 4.2, the term 'moment' being replaced by 'membrane force'.

As the yield condition, it is most convenient to use the expressions 3-11 and 3-9, which are valid for any arrangement of the reinforcing bars. However, if attention is restricted to orthogonal reinforcement, it is possible to derive explicit design formulas for the amount of steel necessary to resist a given field of membrane forces.

Consider a wall element subjected to the membrane forces $N_x = h\sigma_x$, $N_y = h\sigma_y$, and $N_{xy} = h\tau_{xy}$ (Fig. 5-1). The element is reinforced in the x and y-directions, the

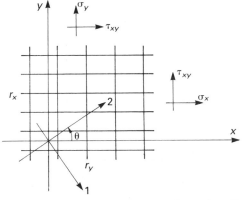

Figure 5-1 Orthogonally reinforced wall element

reinforcement ratios being r_x and r_y, respectively. The principal stresses in the concrete are σ_1 and σ_2, the second principal direction being at the angle θ to the x-axis, where $-\pi/2 < \theta \le \pi/2$. Denoting the stress in the reinforcement by σ_s, we find by equilibrium

$$\sigma_x = \sigma_1 \sin^2 \theta + \sigma_2 \cos^2 \theta + r_x \sigma_s \tag{5-1}$$

$$\sigma_y = \sigma_1 \cos^2 \theta + \sigma_2 \sin^2 \theta + r_y \sigma_s \tag{5-2}$$

$$\tau_{xy} = (\sigma_2 - \sigma_1) \cos \theta \sin \theta \tag{5-3}$$

We assume a state of plane stress in the element and neglect the tensile concrete strength. Thus, the constitutive assumptions of Section 2 require

$$-f_c^* \le \sigma_2 \le \sigma_1 \le 0 \quad \text{and} \quad 0 \le \sigma_s \le f_y$$

The case $\tau_{xy} = 0$ is trivial. We assume $\theta = \pi/2$ for $\sigma_x > \sigma_y$, $\sigma_1 = \sigma_2$ for $\sigma_x = \sigma_y$, $\theta = 0$ for $\sigma_x < \sigma_y$, and design the reinforcement in the x and y-directions by Eqns 5-1 and 5-2, respectively. Obviously, reinforcement is only required if the corresponding stress is tensile, in which case the concrete stress is zero.

For $\tau_{xy} \ne 0$, we initially assume a uniaxial state of stress in the concrete, i.e. $\sigma_1 = 0$. By Eqn 5-3, assuming $\theta \ne 0$, $\pi/2$,

$$\sigma_2 = \tau_{xy}(\tan \theta + \cot \theta) \tag{5-4}$$

Substituting, we find

$$r_x \sigma_s = \sigma_x - \tau_{xy} \cot \theta \tag{5-5}$$

$$r_y \sigma_s = \sigma_y - \tau_{xy} \tan \theta \tag{5-6}$$

Whence

$$(r_x + r_y)\sigma_s = \sigma_x + \sigma_y - \tau_{xy}(\tan \theta + \cot \theta) \tag{5-7}$$

From Eqn 5-3 it appears that θ and τ_{xy} are of the opposite sign, hence the last term in Eqn 5-7 is positive. The amount of reinforcement is represented by the quantity $r_x + r_y$, and minimum reinforcement is obtained for $\sigma_s = f_y$ and $\theta = -\pi/4$ when $\tau_{xy} > 0$, and $\theta = \pi/4$ when $\tau_{xy} < 0$. The concrete stress is given by Eqn 5-4.

However, since $r_x \sigma_s \ge 0$, Eqn 5-5 requires that $\sigma_x \ge -|\tau_{xy}|$. If this is not the case, θ is given by Eqn 5-5 with $r_x = 0$, and r_y is determined by Eqn 5-6. Similarly, for $\sigma_y < -|\tau_{xy}|$. Thus, we are led to the three following cases:

Case 1

$$
\begin{aligned}
&\sigma_x \ge -|\tau_{xy}| && \sigma_y \ge -|\tau_{xy}| \\
&r_x f_y = \sigma_x + |\tau_{xy}| && r_y f_y = \sigma_y + |\tau_{xy}| \\
&\theta = \pm \pi/4 && \sigma_2 = -2|\tau_{xy}|
\end{aligned}
\tag{5-8}
$$

Case 2

$$
\begin{aligned}
&\sigma_x < -|\tau_{xy}| && \sigma_x \sigma_y \le \tau_{xy}^2 \\
&r_x = 0 && r_y f_y = \sigma_y - \tau_{xy}^2/\sigma_x \\
&\tan \theta = \tau_{xy}/\sigma_x && \sigma_2 = \sigma_x + \tau_{xy}^2/\sigma_x
\end{aligned}
\tag{5-9}
$$

Case 3

$$\sigma_y < -|\tau_{xy}| \qquad \sigma_x\sigma_y \leq \tau_{xy}^2$$

$$r_x f_y = \sigma_x - \tau_{xy}^2/\sigma_y \qquad r_y = 0 \qquad\qquad\qquad (5\text{-}10)$$

$$\tan\theta = \sigma_y/\tau_{xy} \qquad \sigma_2 = \sigma_y + \tau_{xy}^2/\sigma_y$$

The requirement $\sigma_x\sigma_y \leq \tau_{xy}^2$ in Cases 2 and 3 ensures that $r_y \geq 0$ and $r_x \geq 0$. When $\sigma_x\sigma_y > \tau_{xy}^2$ (σ_x and σ_y being negative), then no reinforcement is necessary, and we assume a biaxial state of stress in the concrete. The principal stresses σ_1, σ_2 and the angle θ are determined by Eqns 5-1, 5-2 and 5-3 with $r_x = r_y = 0$.

The above design formulas were derived by Nielsen [37, 38, 64] and they correspond to stress regimes on the conical part of the yield surface (Fig. 3-5) described by the yield condition 3-8. These formulas correspond to the optimal concrete compression inclination, i.e. the value of θ giving the minimum reinforcement. For a given choice of θ, the reinforcement is designed by Eqns 5-5 and 5-6. If the reinforcement in one direction is known, these equations also determine the other reinforcement and the angle θ. In all cases, the compressive concrete stress, $-\sigma_2$, is found from Eqn 5-4, and it is a condition that $-\sigma_2 \leq f_c^*$. In order to establish a membrane force distribution which satisfies the static conditions, it is often convenient to regard the wall as separated into finite, triangular regions in which the stress field is homogeneous. Some of the static parameters may then be geometric quantities defining the boundaries between the homogeneously stressed regions. The conditions of continuity over the boundaries may be expressed analytically or graphically by means of Mohr's circles of stress. In this manner, a variety of lower bound solutions were developed by Nielsen [38]. In some cases, the solutions are backed up by corresponding upper bounds.

5.2 Free and constrained shear

In the analysis and design of walls, the special case of pure shear is important. Consider an orthogonally reinforced wall element subjected to the stresses $\sigma_x = \sigma_y = 0$, $\tau_{xy} = -\tau$ (Fig. 5-1). As found in Section 5.1 above, the optimal solution is obtained with $r_x = r_y$, corresponding to $\theta = \pi/4$. However, for given reinforcement ratios r_x and r_y, the shear strength is determined by Eqns 5-5 and 5-6:

$$\tau = f_y\sqrt{r_x r_y} \quad \text{for} \quad \tan\theta = \sqrt{r_y/r_x} \qquad\qquad (5\text{-}11)$$

Both bands of reinforcement are yielding and the concrete is splitting, thus the failure mechanism consists of collapse cracks (Section 2.4) parallel to the direction of concrete compression, i.e. at the angle θ. This situation is termed free shear, and the failure is visualized in Fig. 5-2a.

The compressive concrete stress is given by Eqn 5-4:

$$-\sigma_2 = f_y\sqrt{r_x r_y}\left(\sqrt{\frac{r_y}{r_x}} + \sqrt{\frac{r_x}{r_y}}\right) = (r_x + r_y)f_y$$

and the solution 5-11 is valid for $-\sigma_2 \leq f_c^*$, i.e., for $(r_x + r_y)f_y \leq f_c^*$. When this condition is violated, the element is over-reinforced, i.e. the inclined compressive

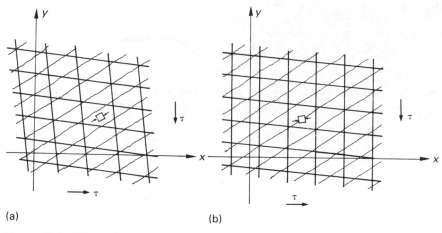

(a)   (b)

Figure 5-2   Shear failure of wall
 (a) Free shear
 (b) Constrained (over-reinforced) shear

stress reaches the concrete strength before the stronger reinforcement yields. The failure mechanism now consists of the yielding of the weaker reinforcement and sliding failure in the concrete, and the concrete compression is no longer parallel to the yield lines (Fig. 5-2b). The shear is constrained by the unyielding reinforcement. A situation of constrained shear is considered in Fig. 5-3 in the case of a finite tensile concrete strength. The element is assumed to be over-reinforced in the x-direction, i.e., $r_x \geq r_y = r$ and $r_x f_y \geq f_c^* - r f_y$, and only the active reinforcement is shown. The figure sketches a failure mechanism consisting of a single yield line at the angle β to the x-axis. This mechanism is most convenient for upper bound calculations, but the yield lines might as well be distributed, as illustrated in Fig. 5-2b. The displacement rate, v, is constrained to be perpendicular to the unyielding reinforcement in the x-direction. The corresponding stress distribution is also indicated. The direction of the concrete compression is defined by the angle θ,

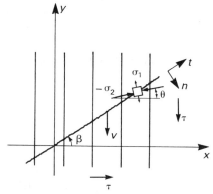

Figure 5-3   Constrained shear in wall

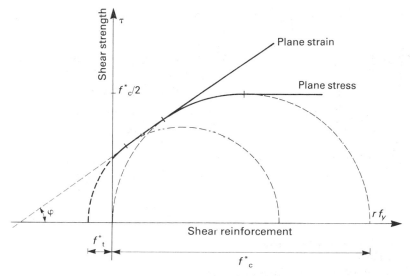

Figure 5-4 Constrained shear strength of wall

where $\theta = \beta/2$ (Fig. 2-5). The principal stresses, σ_1, σ_2, are given by the stress regime on the yield locus (Fig. 2-2), corresponding to the value of the angle $\alpha = \pi/2 - \beta$, which again depends upon the strength of the reinforcement.

The coinciding upper and lower bound solution is shown in Fig. 5-4, the ultimate shear stress τ being plotted as a function of the reinforcement strength rf_y. There are two curves, corresponding to plane strain and plane stress, respectively; the analytical expressions are given below.

Plane strain

$$\tau = \tfrac{1}{2}\sqrt{l^2 f_c^{*2} - (mf_c^* - 2rf_y)^2}$$

(5-12)

for $0 \leq rf_y \leq \tfrac{1}{2}(m - l \sin \varphi)f_c^*$, and

$$\tau = \frac{1 - \sin \varphi}{2 \cos \varphi} f_c^* + rf_y \tan \varphi$$

(5-13)

for $\tfrac{1}{2}(m - l \sin \varphi)f_c^* \leq rf_y$. The parameters l and m are defined by Eqns 2-9.

Plane stress

Equations 5-12 and 5-13 are valid for $0 \leq rf_y \leq \tfrac{1}{2}(1 - \sin \varphi)f_c^*$.

$$\tau = \sqrt{rf_y(f_c^* - rf_y)}$$

(5-14)

for $\tfrac{1}{2}(1 - \sin \varphi)f_c^* \leq rf_y \leq \tfrac{1}{2}f_c^*$, and

$$\tau = \tfrac{1}{2}f_c^*$$

(5-15)

for $\tfrac{1}{2}f_c^* \leq rf_y$.

Note that the curve for plane strain conforms with the modified Coulomb criterion (Fig. 2-1), Equation 5-15 for plane stress corresponds to the case where the element is over-reinforced in the y-direction, too, i.e., the failure mechanism consists in the crushing of the concrete without the yielding of any reinforcement.

Equations 5-14 and 5-15 correspond to the cylindrical and flat part, respectively, of the yield surface of Fig. 3-5. They were given by Braestrup [65], who considered constrained shear in the case of plane stress and vanishing tensile strength only. The general equations were derived by Jensen [27, 66] to describe the strength of joints (see Section 5.3) and applied to shear in walls by Braestrup et al. [17].

The value of the effective concrete strength, f_c^*, to be inserted into the formulas above depends upon the particular structure under investigation and will have to be assessed by comparison with experimental evidence. It is worth noting that tests have shown that the strength of concrete compressed in one direction decreases in the presence of reinforcement in a transverse direction. The effect increases with tension in the transverse bars, and reductions to approximately 80% of the cylinder strength have been reported.

5.3 Shear in joints

When two parts of a concrete body are cast against each other, the interface constitutes a weak section of the structure. Figure 5-5 shows such a construction joint with perpendicular reinforcement and subjected to the shear stress τ. The shear failure in a joint is constrained in the sense that the location of the yield line is fixed, whereas the direction α of the relative displacement rate, v, is variable. This situation is the opposite of the one considered in Section 5.2 (Fig. 5-3), but the equations are exactly the same (see Braestrup et al. [17]). Thus the upper bound solution for the shear strength is given by Eqns 5-12 to 5-15, plotted in Fig. 5-4, provided that the parameter r is interpreted as the ratio of transverse reinforcement. Since the failure is confined to the very interface, it is reasonable to assume a state of plane strain. However, to describe the weakness introduced by the joint, the strength parameters will be different from those of the surrounding structure. Unless the joint is especially smooth, the angle of friction, φ, is unchanged, but the effective compressive strength, f_c^*, describing the cohesion, and the effective tensile strength, f_t^*, will be reduced.

In the plane strain situation, the shear strength increases with increasing reinforcement (Fig. 5-4), but for a joint in a slender body, an upper limit is imposed by the strength of the surrounding monolithic concrete in plane stress.

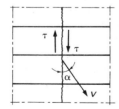

Figure 5-5 Shear in transversely reinforced joint

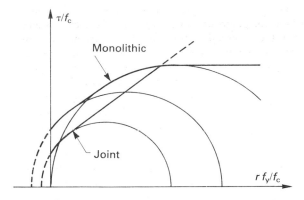

Figure 5-6 Shear strength of reinforced section (schematic)

The resulting relationship between joint shear strength and degree of transverse reinforcement is sketched in Fig. 5-6.

The curves in Fig. 5-6 represent the modified Coulomb failure criterion (Fig. 2-1). The abscissa is the normal compressive stress in the concrete balancing the tension in the reinforcement activated by the failure (Fig. 5-5). Thus, if the joint is not subjected to pure shear, then the normal component, σ, of the stress on the joint will have to be subtracted from the reinforcement term rf_y.

The shear strength of joints is usually calculated by the *shear-friction theory*, which predicts a straight line in the diagram of Fig. 5-6. The plastic approach gives a better agreement with tests on push-off specimens, (see Jensen [27, 66], Nielsen *et al.* [16], Braestrup *et al.* [17]). In particular, the analysis explains the existence of an upper limit for the shear strength, and the fact that the apparent angle of friction is steeper for small amounts of transverse reinforcement. The formulas may be extended to cases of reinforcement in several directions, not perpendicular to the joint, or to keyed joints between precast elements (Jensen [27]). Equations 5-12 and 5-13 apply, provided that the shear stress, τ, and the reinforcement ratio, r, are referred to the key area. Strength parameters for the mortar are inserted, the tensile strength being neglected in practical applications.

6 Beams

6.1 Bending

The use of plasticity has long been standard for the design of a reinforced concrete section to resist a given bending moment. A linear strain distribution is assumed (Fig. 6-1a) and the corresponding stresses are determined. Generally, the tensile concrete stresses are neglected, and the stress distribution in the compression zone is taken to be conform with the uniaxial stress–strain curve for concrete (Fig. 6-1c). The average stress and the level of the resultant are defined by the *stress block factors*, k_1 and k_2.

Many codes of practice allow the use of a rectangular stress block (Fig. 6-1d) with specified values of k_1 and k_2. Then the analysis reduces to the rigid-plastic

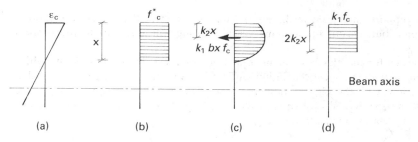

Figure 6-1   Stress blocks in section subjected to bending

case (Fig. 6-1b) with $k_1 = f_c^* / f_c$, which shows that, whereas the term 'effective concrete strength' may be new, the concept is not. The necessary reinforcement is then found by solving Eqn 3-3a with respect to Φ.

With a known strain in the reinforcement, the factor k_2 determines the maximum concrete compressive strain, ε_c. The limited strain capacity of concrete is traditionally taken into account by not allowing ε_c to exceed a specified value ε_{cu} (usually, $\varepsilon_{cu} = 3.5\%$). For $\varepsilon_c = \varepsilon_{cu}$, the section is termed over-reinforced (or balanced), which means that the reinforcing steel is not (or only just) yielding at ultimate. In the plastic approach, the lack of concrete ductility is accounted for by the effectiveness factor, ν, and by expressing ν as a function of the maximum concrete strain, ε_c, as explained in Section 2.2, the plastic analysis with a rectangular stress block can be pursued far into the 'over-reinforced' range.

If an axial force is introduced, we get an interaction diagram, as shown in Fig. 3-2. The curves corresponding to a compressive force may be modified to account for the limited concrete strain capacity (over-reinforced case). The interaction diagram can also be amended to take account of compression reinforcement.

The analysis of a given beam or frame structure may be carried out by standard plastic procedures, e.g., the yield hinge method, developed by Baker [67]. This is an upper bound approach, but normally the analysis is carried on until the exact solution is established, i.e., a corresponding statically admissible moment distribution is found. In principle, the procedures are the same as for steel frames, and the analysis may be formulated as a problem of linear programming.

A redistribution of moments is necessary to reach the stress state of the plastic solution, therefore adequate ductility (rotational capacity) must be present in the yielding sections. Some building codes contain rules limiting the difference between the plastic and the elastic moment distribution. This reflects the view that if the structure is designed only for the elastic moments, then no plastic rotations occur. This is generally not the case, as will appear from a simple example.

Consider a beam, clamped at both ends, and subjected to a uniformly distributed loading. According to elasticity theory, and assuming uniform stiffness, the negative moment at the support is twice as great as the maximum span moment. Hence, a so-called 'elastic' design would provide double as much reinforcement at the support as at mid-span (neglecting differences in internal moment arms). However, as long as the beam behaves elastically, the curvature is proportional to the moment. Hence, the support moment can be delivered by the same amount of reinforcement, since the strain will be double. If more reinforcement is provided at the support, the strain in the elastic stage will be less than twice the strain at mid-span, but, except for extreme cases, yielding will still start at the support sections.

A design for the elastic moments thus leads to a distribution at ultimate which is identical with the 'elastic' solution, but not before some yielding has occurred at the sections with the highest moments and stiffnesses. The example is due to the late Professor K. W. Johansen, who often used it as an argument against favouring 'elastic' over plastic design.

6.2 Shear

The design of beams against shear is much less developed than flexural design. The rules of most building codes are based upon the classical *truss analogy*, the inclination of the web concrete compression (strut inclination) being fixed at 45°. This approach does not take account of the ability of the beam to redistribute the stresses in such a way that a higher shear load can be carried with a flatter strut inclination.

Figure 6-2 shows the truss model of a shear span with a constant shear force, V, and an arbitrary strut inclination, θ. The longitudinal reinforcement and the compression zone are idealized as stringers in the distance h. Plane stress is assumed and the tensile concrete strength is neglected, i.e., the square yield locus, Fig. 2-3, is adopted.

Denoting the stirrup inclination by γ, the plastic solution for the shear strength, V_F, may be written

$$rf_y \leq (rf_y)_1: \quad V_F = bh\sqrt{rf_y \sin^2 \gamma (f_c^* - rf_y \sin^2 \gamma)}$$
$$+ bhrf_y \cos \gamma \sin \gamma \tag{6-1}$$

$$rf_y \geq (rf_y)_1: \quad V_F = \tfrac{1}{2}bhf_c^* \cot \frac{\gamma}{2} \tag{6.2}$$

Here,

$$(rf_y)_1 = \frac{1+\cos \gamma}{2 \sin^2 \gamma} f_c^* \quad \text{and} \quad r = \frac{A_s}{sb \sin \gamma} \tag{6.3}$$

where r is the shear reinforcement ratio (see Eqn 2-4), A_s the cross-sectional stirrup steel area, b the web width, and s the stirrup spacing measured along the beam axis (Fig. 6-2).

Equations 6-1 and 6-2 are termed the *web crushing criterion*, because shear failure is assumed to occur when and only when the strut inclination becomes so

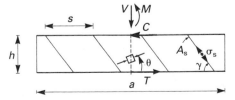

Figure 6-2 Truss model for beam subjected to shear

flat that the inclined compressive concrete strength reaches the effective strength. For $rf_y > (rf_y)_1$, the beam is over-reinforced in shear, i.e. the stirrups are not yielding at failure, which takes place by web crushing only. The strut inclination $\theta = \theta_F$ at failure is given by

$$rf_y \le (rf_y)_1: \quad \cot \theta_F = \sqrt{\frac{f_c^* - rf_y \sin^2 \gamma}{rf_y \sin^2 \gamma}} \tag{6-4}$$

$$rf_y \ge (rf_y)_1: \quad \cot \theta_F = \tan \gamma/2 \tag{6-5}$$

The corresponding failure mechanism consists of a yield line with the inclination β_F. Shear in beams is generally constrained by the longitudinal reinforcement, which normally does not yield at shear failure (see Braestrup *et al.* [68]). Thus, the relative deformation rate is perpendicular to the beam axis, whence $\beta_F = 2\theta_F$ (Fig. 2-5).

The upper bound analysis follows the same lines as those used for punching in Section 4.4. For a mechanism with a single yield line with inclination β and relative displacement rate v, the work equation yields

$$Vv = rf_y \sin(\beta + \gamma) \frac{bh}{\sin \beta} v \sin \gamma + \tfrac{1}{2} vf_c^* (1 - \cos \beta) \frac{bh}{\sin \beta}.$$

The second term is the contribution from the concrete, given by Eqn 2-8 (or Eqn 2-7 with $l = m = 1$), while the first term is the contribution from the stirrups, given by Eqn 2-12. The main reinforcement does not contribute to the rate of internal work, because the relative displacement rate is perpendicular to the beam axis (Eqn 2-11).

The work equation determines the upper bound:

$$V = bhrf_y \sin^2 \gamma (\cot \beta + \cot \gamma) + \tfrac{1}{2} bhf_c^* (\sqrt{1 + \cot^2 \beta} - \cot \beta)$$

The lowest upper bound is found by minimizing with respect to the kinematic parameter β. A minimum is found for $dV/d(\cot \beta) = 0$, the result being Eqn 6-1 for $\beta = 2\theta_F$, where θ_F is given by Eqn 6-4. The upper limit $rf_y = (rf_y)_1$ for the shear reinforcement is imposed by the condition $\beta < \pi - \gamma$, which expresses the fact that the failure mechanism must involve elongation of the stirrups. For $rf_y > (rf_y)_1$, the lowest upper bound is obtained with $\beta = \pi - \gamma$, which yields Eqn 6-2.

A lower bound solution is found by considering vertical equilibrium along a section inclined at the strut angle, θ:

$$V = bh\sigma_s (\cot \theta + \cot \gamma) \sin^2 \gamma$$

Vertical equilibrium in a horizontal section requires

$$\sigma_c \sin^2 \theta = r\sigma_s \sin^2 \gamma,$$

where σ_c is the inclined compressive concrete principal stress. Elimination of the strut inclination, θ, yields the lower bound:

$$V = bh\sqrt{r\sigma_s \sin^2 \gamma (\sigma_c - r\sigma_s \sin^2 \gamma)} + bhr\sigma_s \cos \gamma \sin \gamma$$

The highest lower bound is found by maximizing with respect to the static

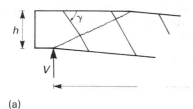

(a)

(b)

Figure 6-3   Shear failure of beam with stirrups
(a) Assumed failure mechanism
(b) Experimental failure

parameters σ_c and $r\sigma_s$. Equation 6-1 is found as a maximum for $\sigma_c = f_c^*$ and $r\sigma_s = rf_y$, as long as $rf_y \leq (rf_y)_1$. For $rf_y > (rf_y)_1$, the highest lower bound is found for $r\sigma_s = (rf_y)_1$, and the result is Eqn 6-2.

Rather than being concentrated in a single yield line, the deformations may be distributed over a parallellogram-shaped zone, as indicated in Fig. 6-3a. An experimental example of such a failure is shown in Fig. 6-3b.

In the case of vertical stirrups ($\gamma = \pi/2$), Eqns 6-1 and 6-2 reduce to:

$$rf_y \leq \tfrac{1}{2}f_c^*: \quad V_F = bh\sqrt{rf_y(f_c^* - rf_y)} \tag{6-6}$$

$$rf_y \geq \tfrac{1}{2}f_c^*: \quad V_F = \tfrac{1}{2}bhf_c^* \tag{6-7}$$

This is identical with the case of constrained shear in walls, (see Eqns 5-14 and 5-15).

Equations 6-6 and 6-7 were derived by Nielsen [69] as a lower bound solution, and Nielsen and Braestrup [70] showed that Eqns 6-1 and 6-2 represent coinciding upper and lower bounds.

For moderate strengths of the shear reinforcement, the effect of stirrup inclination is little, but for strong shear reinforcement Eqns 6-1 and 6-2 predict a substantial increase in shear strength for a suitable inclination of the stirrups. This effect is in accordance with test results.

In a plot of shear strength versus stirrup reinforcement, Eqns 6-6 and 6-7 are represented by a quarter circle and the horizontal tangent, (Fig. 5-4). In Fig. 6-4, the analysis is compared with reported test results (see Nielsen et al. [71]). The points are distributed rather well around the web crushing criterion with an effective concrete strength $f_c^* = \nu f_c = 0.74f_c$, f_c being the cylinder strength. The scatter is mainly due to differences in concrete quality and in testing procedures.

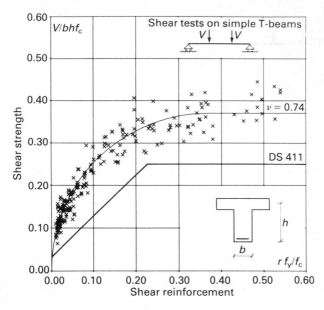

Figure 6-4   Results of 198 shear tests versus theoretical and building code predictions

If a single test series is picked out, the agreement is usually much better. The figure also shows the strength predictions of a typical building code (without the introduction of safety factors), which is seen to be quite conservative.

A prestressing of the beam merely introduces a different stress distribution at zero load and does not influence the strength prediction of the plastic solution. Some experimental investigations do, however, indicate a beneficial effect of prestressing, which can be taken into account by increasing the value of ν. The effectiveness factor also shows a moderate decline with increasing concrete strength level.

The shear strength of lightweight concrete beams also shows good agreement with the web crushing criterion, but the value of the effectiveness factor, ν, is typically 15–20% smaller than that for normal beams of the same strength.

As the stirrup reinforcement, rf_y, decreases, so does the optimal strut inclination, $\theta = \theta_F$, given by Eqn 6-4, and at a certain point the corresponding yield line with the inclination $\beta_F = 2\theta_F$ can no longer be contained in the beam. Thus, for deep beams, and beams with little or no stirrup reinforcement, the shear failure mechanism degenerates to a single yield line running from the load to the support. In this case, the failure may involve yielding of the longitudinal reinforcement (Fig. 6-5). For vertical stirrups, the corresponding upper bound solution is

$$\Phi \le \tfrac{1}{2}: \quad V_F = \tfrac{1}{2}bhf_c^*[\sqrt{(a/h)^2 + 4\Phi(1-\Phi)} - a/h] + barf_y \tag{6-8}$$

$$\Phi \ge \tfrac{1}{2}: \quad V_F = \tfrac{1}{2}bhf_c^*[\sqrt{(a/h)^2 + 1} - a/h] + barf_y \tag{6-9}$$

Here, Φ is the longitudinal reinforcement degree, defined by Eqn 3-2a, and a is

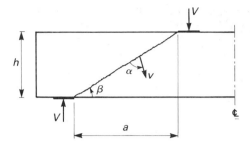

Figure 6-5 Shear failure mechanism with yielding of longitudinal steel (little or no stirrup reinforcement)

the length of the shear span, measured as the clearance between the load and support platens. When the stirrups are inclined, the solution is slightly more complicated (Braestrup [72]).

Equation 6-9 corresponds to the situation where the longitudinal reinforcement is not yielding, and the solution may be shown to be exact (Jensen [73]). In the case of no stirrups ($rf_y = 0$), also Eqn 6-8 is exact, a corresponding stress distribution being given by Nielsen et $al.$ [16]. This solution has been compared with a test series, and excellent agreement found, (Nielsen and Braestrup [74]). The effectiveness factor is of the order of $\nu = 0.60$, and rather strongly dependent upon the concrete quality. As in the case of punching shear (Section 4.4), the relationship between effective strength and cylinder strength may be described by a square root dependence.

In the presence of significant stirrup reinforcement, the tension stringer will generally be sufficiently strong to remain below yield, since a weaker longitudinal reinforcement would normally lead to flexural failure (see also below).

If the shear span is subjected to a uniformly distributed load, p, per unit area of a horizontal web section, then the shear strength may be defined as the ultimate shear force at the support, i.e., $V_F = bpa$. When the longitudinal reinforcement is not yielding, the upper bound solution is given by Eqns 6-1 and 6-2 if the quantity $rf_y \sin^2 \gamma$ is replaced by $rf_y \sin^2 \gamma + p$. An explicit expression for the shear strength is found by solving with respect to p. In the case of vertical stirrups, this result was given by Nielsen and Braestrup [70], who also derived a coinciding lower bound solution. The stress distribution has piecewise constant compressive concrete stresses and stirrup steel stresses, and a constant strut inclination, θ_F, which is found from Eqn 6-4 by the same substitution as above. The corresponding failure mechanism consists of a single yield line starting from the support with the inclination $\beta_F = 2\theta_F$.

A complete set of upper bound solutions for concentrated and distributed loading on beams with vertical, inclined or no stirrups, and with and without yielding of the tension stringer, has been given by Braestrup [72]. For a fixed strength, rf_y, of the shear reinforcement, the optimal (i.e., flattest possible) strut inclination is steeper by distributed than by point loading, and normally flatter the more inclined the stirrups are.

The analysis shows that the minimum tensile stringer strength to prevent yielding at shear failure is of the order of $\Phi = 1/2$, somewhat smaller for inclined stirrups. For weaker tension reinforcement, a slight reduction in shear strength is predicted, but, as mentioned above, such beams would normally fail in flexure. However, the mechanism of Fig. 6-5 highlights the importance of adequate anchorage of the main reinforcement, because a bond failure would reduce the tensile stringer force available in the shearing mechanism without affecting the flexural strength. It cannot be excluded that some of the failures reported in the literature as due to shear may have been governed by imperfect anchorage.

The design of stirrup reinforcement in beams subjected to shear may conveniently be based upon the truss model (Fig. 6-2). Assuming yielding of the stirrup steel, the equations of equilibrium give (Nielsen and Braestrup [71])

$$V = bhrf_y(\cot\theta + \cot\gamma)\sin^2\gamma \qquad (6\text{-}10)$$

$$M = h[T - \tfrac{1}{2}V(\cot\theta - \cot\gamma)] \qquad (6\text{-}11)$$

Equation 6-11 shows that the longitudinal reinforcement must be designed for a tensile stringer force, T, which exceeds the simple moment term M/h (see below). For $\theta = 45°$, Eqn 6-10 reduces to the traditional truss equation, but a more rational design is obtained by allowing the strut inclination, θ, to be chosen arbitrarily, between certain specified limits. This approach is called the *diagonal compression field* method—see Collins and Mitchell [75], who also proposed a range for θ based upon the stress–strain relationships of concrete and reinforcement. A similar method was advocated by Grob and Thürlimann [76].

By this design approach, the web crushing criterion reduces to an upper limit for the applied shear load, which for a chosen strut inclination, θ, amounts to

$$V \le bhf_c^*(\cot\theta + \cot\gamma)\sin^2\theta \qquad (6\text{-}12)$$

or equivalently to a restriction on the strut inclination, namely, $\theta \ge \theta_F$, where θ_F is given by Eqns 6-4 and 6-5.

The substitution of $rf_y\sin^2\gamma + p$ for $rf_y\sin^2\gamma$ in the case of uniformly distributed loading is equivalent to the replacement of the ultimate shear load $V_F = bpa$ by the load $V_F - bph\cot\theta$. This means that the shear reinforcement can be designed by Eqn 6-10, where V is the shear force at the distance $h\cot\theta$ from the support. This is also apparent from the equilibrium of a triangular body defined by a section emanating from the support, and inclined at the angle θ (Fig. 6-6).

The above argument is valid for any distribution of loading. This suggests the following general design procedure for beams with varying shear force:

(a) Choose a strut inclination, and start at the support.
(b) Calculate the shear force, V, at the distance $h\cot\theta$, and verify that it satisfies the condition 6-12, imposed by the strength of the web concrete.
(c) Design the shear reinforcement by Eqn 6-10.

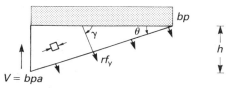

Figure 6-6 End section of beam with distributed loading

(d) Repeat the procedure from (b) until the shear force changes sign, or the shear can be resisted by any specified minimum stirrup reinforcement.
(e) Design the longitudinal reinforcement by Eqn 6-11.

In this way, a constant shear reinforcement is designed in intervals of length $h \cot \theta$. It is a condition that the loads be applied at the top of the beam. If this is not the case, additional stirrups capable of transferring the load to the top face must be provided. In addition to limits for the allowable strut inclination, θ, the values of the shear depth, h, and the effective concrete strength, f_c^*, will have to be specified, e.g. by adequate code rules. The design method is explained in further detail by Nielsen et al. [16] and Collins and Mitchell [75].

In the vicinity of a point of maximum moment, the concrete struts change direction, and Eqn 6-11 is not valid. It appears that if the load is applied at the top of the beam, then the moment does not exceed the maximum value corresponding to pure bending. Thus, the longitudinal reinforcement may be designed for a moment curve which is displaced the distance $\frac{1}{2}h$ ($\cot \theta - \cot \gamma$) towards the support (the so-called *shift rule*). Special attention must be given to the anchorages (see above), since a tensile stringer force is present even at a simple support with $M = 0$.

6.3 Torsion

Beams in torsion are traditionally analysed by means of a space truss model with 45° concrete struts, in analogy with the truss model for shear. If a variable strut inclination is assumed, then the ultimate torsional moment, M_{tu}, of a rectangular section with four corner bars at the distances b_0 and h_0 is

$$M_{tu} = 2b_0 h_0 \sqrt{\frac{4T_y}{p} \frac{A_s f_y}{s}} \qquad (6\text{-}13)$$

Here, T_y is the yield force of a corner bar, $p = 2b_0 + 2h_0$ is the perimeter, $A_s f_y$ is the yield force of a stirrup, and s is the stirrup spacing. In the case of different top and bottom reinforcement, T_y is the lesser yield force.

The concept of variable strut angle was introduced by Lampert [77], who also extended the analysis to combined bending and torsion. In the context of plasticity, Eqn 6-13 is a lower bound solution, but the truss model does not lead to a kinematically admissible failure mechanism (Müller [26]).

A kinematic approach to torsion and bending is the so-called *skew bending* theory, developed by Lessig (Gvozdev et al. [78]). Failure is assumed to take place in a spiralling crack, with rotation about an oblique neutral axis, parallel to one of the beam faces. The relationship between torsional and flexural moment at ultimate is determined by the equation of moment about the neutral axis, and the optimal hinge inclination is found by minimizing the applied moment. The method was extended and simplified considerably by Collins et al. [79].

In the case of pure torsion, the ultimate torque predicted is analogous with Eqn 6-13, but the depth h_0 is the internal moment lever arm, and the width b_0 and the perimeter p correspond to the centre-line of the stirrups, and not to the corner bars.

Instead of considering the moment equation in the skew crack (equilibrium method), the ultimate torque might be determined by equating the rates of external and internal work and minimizing with respect to the crack inclination (dissipation method). Apparently, this leads to slightly different results, as found by Elfgren *et al.* [80], who amended the discrepancy by introducing self-equilibrating 'nodal forces'. However, the reason for the difference is that, by rotation about an oblique neutral axis, the relative displacement rate is not perpendicular to the warped failure surface. Therefore, the failure mechanism will lead to stresses in the concrete (Section 2.4), which are not taken into account either in the equilibrium method or in the dissipation method. Consequently, the skew bending theory does not constitute a valid upper bound solution.

A correct failure mechanism for beams in torsion was determined by Müller [26]. It consists of two collapse cracks, joined by an inclined hinge. The upper bound solution, corresponding to the mechanism, is identical with the lower bound of Eqn 6-13, corresponding to the truss model. The analysis may be extended to include bending and shear, [21].

A slightly different approach to torsional design is due to Nielsen [38]. The torque, M_{tu}, is assumed to be delivered by a shear stress τ, uniformly distributed over a hollow section with the centre-line dimensions $b_0 \times h_0$ and thickness t:

$$M_{tu} = 2b_0 h_0 t \tau \tag{6-14}$$

Assuming *free shear* (Section 5.2), the ultimate shear stress, τ, is given by Eqn. 5-11a, where the ratios of longitudinal and transverse reinforcement are inserted, that is

$$\tau = f_y \sqrt{r_l r_s} \quad \text{with} \quad r_l f_y = 4T_y/pt \quad \text{and} \quad r_s f_y = A_s f_y/st$$

When these expressions are substituted, Eqn 6-14 reduces to Eqn 6-13. The strut inclination, θ, which is identical with the inclination of the collapse cracks, is given by Eqn 5-11b. Each face of the beam is treated as a shear wall, which is designed independently, and the method is easily extended to include bending and shear (see Nielsen [64]).

The analysis is based upon the yielding of all the reinforcement without the crushing of the concrete, and the compressive concrete stress depends upon the wall thickness, t. This represents no problem if the beam has a box section with the reinforcement near the centre-line. For a solid section, however, it is necessary to introduce a fictitious wall thickness in order to check if the beam is over-reinforced. Most building codes have empirical rules for the definition of the 'equivalent hollow section'.

A rational design procedure was recently proposed by Collins and Mitchell [75]. It is based upon the diagonal compression field approach, and limits for the strut inclination are established. The idea is that the twisting of a prismatic beam results in a curving of the plane faces. The compressive concrete strain is a function of the steel strains and the directions of the diagonal compression. Hence, it is possible to decide whether the beam is over-reinforced by assigning a value to the ultimate compressive concrete strain, in much the same way as is done for beams in flexure (Section 6.1).

In the rigid-plastic approach, the over-reinforced case is characterized by the fact that the shear flow is constrained by the unyielding reinforcement in the longitudinal or circumferential direction, or both. Müller's failure mechanism will have to be modified, the collapse cracks being replaced by kinematic discontinuities involving shearing.

7 Conclusion

We have reviewed some plastic methods of analysis and design of reinforced concrete members. The basic idea is to use the theory of plasticity as a mathematical tool, with rigid, perfectly plastic material models for the reinforcement and the concrete. While the rigid-plastic description of steel reinforcing bars is well established, the corresponding idealization of concrete is more controversial. The present approach is based upon the adoption of the modified Coulomb failure criterion as a yield condition, with the associated flow rule. The resulting constitutive model for concrete contains three parameters: the compressive strength, the tensile strength, and the angle of friction. In most practical applications, however, the tensile strength is neglected, and a standard value for the angle of friction is adopted. Consequently, a particular material is characterized only by its effective concrete strength, which is an estimate of the uniaxial compressive strength available at collapse of the member in question.

When the tensile concrete strength is neglected, the yield locus for concrete in plane stress reduces to the well-known square yield locus, which is the classical basis for analysis of under reinforced plane elements, and of beams and slabs in flexure. Thus the modified Coulomb condition is a generalization of the square yield locus, which makes the model applicable to arbitrary strength problems in concrete.

In spite of the primitive nature of the constitutive description, the plastic analysis of structural concrete yields predictions which show remarkably good agreement with experimental evidence. Thus, the method would seem to be a powerful engineering tool, and a simple and rational basis for design rules.

In addition to the analysis and design of structural members, the approach can also be used to treat special problems in structural concrete, such as pull-out of inserts, concentrated loads on blocks, the anchorage and bond of reinforcement. A more exhaustive treatment of the theory, its background and applications, are given in a monograph, Nielsen [84], and a thesis, Braestrup [85], both under preparation.

References

1. Gvozdev, A. A., The determination of the value of the collapse load for statically indeterminate systems undergoing plastic deformation, *Int. J. Mech. Sci.* Vol. 1, 1960, pp. 322–333. [English translation of Gvozdev's article in *Svornik trudov konferentsii po plasticheskim deformatsiyam*, Academy of Sciences, Moscow/Leningrad, 1938, pp. 19–30.]
2. Hodge, P. G. and Prager, W., A variational principle for plastic materials with strain-hardening. *J. Math. Phys.* Vol. 27, No. 1, 1948, pp. 1–10.
3. Drucker, D. C., Some implications of work hardening and ideal plasticity, *Q. Appl. Math.* Vol. 7, 1950, pp. 411–418.

4. Drucker, D. C., Prager, W. and Greenberg, H. J., Extended limit design theorems for continuous media, *Q. Appl. Math.*, Vol. 9, 1952, pp. 381–389.

5. Prager, W., The general theory of limit design, *Proc. 8th int. cong. theor. appl. mech.*, Vol. II, 1955, pp. 65–72.

6. Hill, R., *The mathematical theory of plasticity*, Clarendon Press, Oxford, 1950, 356 pp.

7. Koiter, W. T., Stress–strain relations, uniqueness and variational theorems for elastic-plastic materials with singular yield surfaces, *Q. Appl. Math.*, Vol. 11, No. 3, 1953, pp. 350–354.

8. Martin, J. B., *Plasticity: fundamentals and general results*, MIT Press, Cambridge, Mass., 1975, 931 pp.

9. Kachanov, L. M., *Osnovy teorii plastichnosti*, Izdatelstvo 'Nauka', Moscow 1969. (English edition: *Fundamentals of the theory of plasticity*, Mir Publishers, Moscow, 1974, 445 pp.)

10. Massonet, C. and Save, M., *Calcul plastique des constructions. Vol. 2: Structures spatiales*, Centre Belgo-Luxembourgeois d'Information de l'Acier, Bruxelles 1963, 473 pp. (English edition: *Plastic analysis and design of plates, shells and disks*, North-Holland, Amsterdam, 1972, 478 pp.)

11. Reckling, K.-A., *Plastizitätstheorie und ihre Anwendung auf Festigkeitsprobleme*. Springer-Verlag, Berlin, 1967, 361 pp.

12. Johansen, K. W., *Brudlinieteorier*. Gjellerup, Copenhagen, 1943, 189 pp. (English edition: Yield-Line Theory. London, Cement and Concrete Association, London, 1962.)

13. Nielsen, M. P., Limit analysis of reinforced concrete slabs, *Acta Polytech. Scand., Civ. Eng. Bldg Constr. Ser.*, No. 26, 1964, 167 pp.

14. Drucker, D. C., On structural concrete and the theorems of limit analysis, *Mem. Int. Ass. Bridge Struct. Eng.* Vol. 21, 1961, pp. 49–59.

15. Chen, W.-F. and Drucker, D. C., Bearing capacity of concrete blocks or rock, *Am. Soc. Civ. Eng.*, Vol. 95, No. EM4, Aug. 1969, pp. 955–978.

16. Nielsen, M. P., Braestrup, M. W., Jensen, B. C. and Bach, F., *Concrete plasticity. beam shear—punching shear—shear in joints*, Danish Society for Structural Science and Engineering, Copenhagen, Special Publication, 1978, 129 pp.

17. Braestrup, M. W., Nielsen, M. P. and Bach, F., Plastic analysis of shear in concrete, *Z. Angew. Math. Mech.*, Vol. 58, No. 6, June 1978, pp. 3–14.

18. Müller, P., *Plastische Berechnung von Stahlbetonscheiben und -Balken* Eidgenössische Technische Hochschule, Institut für Baustatik und Konstruktion, Zürich, Bericht Nr 83 (ETH Diss. 6083), July 1978, 160 pp.

19. Marti, P., *Zur plastischen Berechnung von Stahlbeton*, Eidgenössiche Technische Hochschule, Institut für Baustatik und Konstruktion, Zürich, Bericht Nr 104, Oct. 1980, 176 pp.

20. International Association for Bridge and Structural Engineering, *Plasticity in reinforced concrete. Introductory report*, Reports of the Working Commissions, Vol. 28, Oct. 1978, 172 pp.

21. International Association for Bridge and Structural Engineering, *Plasticity in reinforced concrete. Final report*. Reports of the Working Commissions, Vol. 29, Aug. 1979, 360 pp.

22. Dorn, J. E., The effect of stress state on the fracture strength of metals,

Fracturing of metals, American Society of Metals, Cleveland, Ohio, 1948, pp. 32–50.

23. Cowan, H. J., The strength of plain, reinforced and prestressed concrete under the action of combined stresses, with particular reference to the combined bending and torsion of rectangular sections, *Mag. Concr. Res.* Vol. 5, No. 14, Dec. 1953, pp. 75–86.

24. Paul, B., A modification of the Coulomb—Mohr theory of fracture, *J. Appl. Mech.*, Vol. 28, No. 6, June 1961, pp. 259–268.

25. Drucker, D. C. and Prager, W., Soil mechanics and plastic analysis or limit design, *Q. Appl. math.*, Vol. 10, 1952, pp. 157–165.

26. Müller, P., Failure mechanisms for reinforced concrete beams in torsion and bending, *Mem. Int. Ass. Bridge Struct. Eng.* Vol. 36 II, 1976, pp. 147–163.

27. Jensen, B. C., *Some applications of plastic analysis to plain and reinforced concrete*, Technical University of Denmark, Institute of Building Design, Copenhagen, Report No. 123, 1977, 119 pp.

28. Ingerslev, A., Om en elementaer beregningsmetode af krydsarmerede plader [On a simple analysis of two-way slabs], *Ingeniøren*, Vol. 30, No. 69, 1921, pp. 507–515. (See also: The strength of rectangular slabs, *Struct. Eng.*, Vol. 1, No. 1, 1923, pp. 3–14.)

29. Johansen, K. W., Beregning af krydsarmerede jernbetonpladers brudmoment, *Bygningsstat. Medd.*, Vol. 3, No. 1, 1931, pp. 1–18. (German version: Bruchmomente der Kreuzweise bewehrten Platten, *Mem. Int. Ass. Bridge Struct. Eng.*, Vol. 1, 1932, pp. 277–296.)

30. Gvozdev, A. A., Obosnovanie § 33 norm proektirovaniya zhelobetonnykh konstruktsii [Comments to par. 33 of the design standard for reinforced concrete structures], *Stroit. Prom.*, Vol. 17, No. 3, 1939, pp. 51–58.

31. Janas, M., Limit analysis of non-symmetric plastic shells by a generalized yield line method, Olszak, W. and Sawczuk, A. (Eds), *Non-classical shell problems* North-Holland, Amsterdam, 1964, pp. 997–1010.

32. Janas, M. and Sawczuk, A., Influence of position of lateral restraints on carrying capacity of plates, *Bull. Inf. Com. Eur. Béton*, No. 58, Oct. 1966, pp. 164–189.

33. Wood, R. H., *Plastic and elastic design of slabs and plates*, Thames & Hudson, London, 1961, 344 pp.

34. Janas, M., Large plastic deformations of reinforced concrete slabs, *Int. J. Solids & Struct.*, Vol. 4, No. 1, Jan. 1968, pp. 61–74.

35. Braestrup, M. W., *Yield lines in discs, plates, and shells*, Technical University of Denmark, Structural Research Laboratory, Copenhagen, Report R 14, 1970, 54 pp.

36. Braestrup, M. W., Dome effect in RC slabs: Rigid-plastic analysis, *J. Struct. Div. Am. Soc. Civ. Eng.*, Vol. 106, No. ST6, June 1980, pp. 1237–1253.

37. Nielsen, M. P., Yield conditions for reinforced concrete shells in the membrane state, Olszak, W. and Sawczuk, A. (Eds), *Non-classical shell problems*, North-Holland, Amsterdam, 1964, pp. 1030–1040.

38. Nielsen, M. P., On the strength of reinforced concrete discs, *Acta Polytech. Scand., Civ. Eng. Bldg Constr. Ser.*, No. 70, 1971, 261 pp.

39. Kemp, K. O., The yield criterion for orthotropically reinforced concrete slabs, *Int. J. Mech. Sci.*, Vol. 7, 1965, pp. 737–746.

40. Jones, L. L. and Wood, R. H., *Yield-line analysis of slabs*, Thames & Hudson/Chatto & Windus, London, 1967, 405 pp.
41. Braestrup, M. W., Yield-line theory and limit analysis of plates and slabs, *Mag. Concr. Res.*, Vol. 22, No. 71, June 1970, pp. 99–106.
42. Gvozdev, A. A., Sur le calcul par la méthode des lignes de rupture des dalles en béton armé pour disposition quelconque des armatures, *Bull. Inf. Com. Eur. Béton*, No. 56, 1966, pp. 152–155.
43. Rajendran., S. and Morley, C. T., A general yield criterion for reinforced concrete slab elements, *Mag. Concr. Res.*, Vol. 26, No. 89, Dec. 1974, pp. 212–220. Discussion, Vol. 27, No. 93, Dec. 1975, pp. 245–246.
44. Cookson, P. J., Generalized yield lines in reinforced concrete slabs, *J. Struct. Mech.*, Vol. 7, No. 1, 1979, pp. 65–82. (See also: *IABSE* (21), Aug. 1979, pp. 43–50.)
45. Johansen, K. W., *Yield-line formulae for slabs*, Cement and Concrete Association, London, 1972, 106 pp.
46. Wood, R. H., A partial failure of limit analysis for slabs, and the consequences for future research, *Mag. Concr. Res.*, Vol. 21, No. 67, June 1969, pp. 79–80. Discussion, Vol. 22, No. 71, June 1970, pp. 112–113.
47. Nielsen, M. P., The theory of plasticity for reinforced concrete slabs, *IABSE* (20), Oct. 1978, pp. 93–114.
48. Regan, P. E. and Yu, C. W., *Limit state design of structural concrete*, Wiley, New York, 1973, 325 pp.
49. Kong, F. K. and Evans, R. H., *Reinforced and prestressed concrete*, 2 edn, Nelson, London, 1980, 412 pp.
50. Møllmann, H., On the nodal forces of the yield line theory, *Bygningsstat. Medd.*, Vol. 36, No. 1, 1965, pp. 1–24.
51. Nielsen, M. P. and Bach, F., A class of lower bound solutions for rectangular slabs. *Bygningsstat. Medd.*, Vol. 50, No. 3, Sept. 1979, pp. 43–58.
52. Hillerborg, A., Jämnviktteori för armerade betongplattor [Equilibrium theory for reinforced concrete slabs]. *Betong*, Vol. 41, No. 4, 1956, pp. 171–181. (See also: A plastic theory for the design of reinforced concrete slabs, *6th Cong. Int. Ass. Bridge Struct. Eng.*, Prelim. Publ. 1960, pp. 177–186)
53. Armer, G. S. T., The strip method: a new approach to the design of slabs, *Concrete*, Vol. 2, No. 9, Sep. 1968, pp. 358–363.
54. Wood, R. H. and Armer, G. S. T., The theory of the strip method for design of slabs, *Proc. Inst. Civ. Eng.*, Vol. 41, 1968, pp. 285–311.
55. Fernando, J. S. and Kemp, K. O., The strip method of slab design: unique or lower-bound solutions? *Mag. Concr. Res.*, Vol. 27, No. 90, March 1975, pp. 23–29.
56. Fernando, J. S. and Kemp, K. O., A generalized strip deflection method of reinforced concrete slab design, *Proc. Inst. Civ. Eng.*, Part 2, Vol. 65, Mar. 1978, pp. 103–174.
57. Gurley, C. R., Bimoment equilibrium of finite segments of Hillerborg plate. *Mag. Concr. Res.*, Vol. 31, No. 108, Sept. 1979, pp. 142–150. Discussion, Vol. 32, No. 112, Sept. 1980, pp. 176–185. (See also: *IABSE* (21), Aug. 1979, pp. 153–157.)
58. Morley, C. T., Yield-line theory for reinforced concrete slabs at moderately large deflexions, *Mag. Concr. Res.*, Vol. 19, No. 61, Dec. 1967, pp. 211–222.

59. Janas, M., Arching action in elastic-plastic plates, *J. Struct. Mech.*, Vol. 1, No. 3, 1973, pp. 277–293.

60. Braestrup, M. W. & Morley, C. T., Dome effect in RC slabs: elastic-plastic analysis, *J. Struct. Div. Am. Soc. Civ. Eng.*, Vol. 106, No. ST6, June 1980, pp. 1255–1262.

61. Braestrup, M. W., Punching shear in concrete slabs, *IABSE* (20), Oct. 1978, pp. 115–136.

62. Braestrup, M. W., Nielsen, M. P., Jensen, B. C. and Bach, F., *Axisymmetric punching of plain and reinforced concrete*, Technical University of Denmark, Structural Research Laboratory, Copenhagen, Report R 75, 1976, 33 pp.

63. Taylor, R. and Hayes, B., Some tests on the effect of edge restraint on punching shear in reinforced concrete slabs, *Mag. Concr. Res.*, Vol. 17, No. 50, March 1965, pp. 39–44.

64. Nielsen, M. P., Some examples of lower-bound design of reinforcement in plane stress problems, *IABSE* (21), Aug. 1979, pp. 317–324.

65. Braestrup, M. W., Plastic analysis of shear in reinforced concrete. *Mag. Concr. Res.*, Vol. 26, No. 89, Dec. 1974, pp. 221–228. Discussion, Vol. 27, No. 93, Dec. 1975, pp. 247–248.

66. Jensen, B. C., Lines of discontinuity for displacements in the theory of plasticity of plain and reinforced concrete, *Mag. Concr. Res.*, Vol. 27, No. 92, Sept. 1975, pp. 143–150.

67. Baker, A. L. L., *The ultimate-load theory applied to the design of reinforced and prestressed concrete frames*, Concrete Publications., London, 1956, 91 pp.

68. Braestrup, M. W., Nielsen, M. P., Bach, F. and Jensen, B. C., *Shear tests on reinforced concrete T-beams. Series T*, Technical University of Denmark, Structural Research Laboratory, Copenhagen, Report R 72, 1976, 114 pp.

69. Nielsen, M. P., Om forskydningsarmering i jernbetonbjaelker [On shear reinforcement in reinforced concrete beams], *Bygningsstat. Medd.*, Vol. 38, No. 2, 1967, pp. 33–58. Discussion, Vol. 40, No. 1, 1969, pp. 60–63.

70. Nielsen, M. P. and Braestrup, M. W., Plastic shear strength of reinforced concrete beams, *Bygningsstat. Medd.*, Vol. 46, No. 3, Sept. 1975, pp. 61–99.

71. Nielsen, M. P., Braestrup, M. W. and Bach, F., Rational analysis of shear in reinforced concrete beams, *IABSE, Periodica*, 2/1978, *Proc.* P-15/1978, May 1978, 16 pp.

72. Braestrup, M. W., Shear capacity of reinforced concrete beams, *Arch. Inzyn. Lad.*, Vol. 26, No. 2, 1980, pp. 295–317. (See also: *Effect of main steel strength on shear capacity of reinforced concrete beams with stirrups*, Technical University of Denmark, Structural Research Laboratory, Copenhagen, Report R 110, 1979, 49 pp.)

73. Jensen, J. F., Plastic solutions for reinforced concrete beams in shear, *IABSE* (21), Aug. 1979, pp. 71–78.

74. Nielsen, M. P. and Braestrup, M. W., Shear strength of prestressed concrete beams without web reinforcement, *Mag. Concr. Res.*, Vol. 30, No. 104, Sept. 1978, pp. 119–127.

75. Collins, M. P. and Mitchell, D., Shear and torsion design of prestressed and non-prestressed concrete beams, *J. Prestressed Concr. Inst.*, Vol. 25, No. 5, Sept.–Oct. 1980, pp. 32–100.

76. Grob, J. and Thürlimann, B., Ultimate strength and design of reinforced

concrete beams under bending and shear, *Mem. Int. Ass. Bridge Struct. Eng.*, Vol. 36 II, 1976, pp. 105–120.

77. Lampert, P., *Bruchwiderstand von Stahlbetonbalken under Torsion und Biegung*, Eidgenössische Technische Hochschule, Institut für Baustatik und Konstruktion, Zürich, Bericht Nr. 26 (Diss. ETH 4445), Jan. 1970, 189 pp.

78. Gvozdev, A. A., Lessig, N. N. and Rulle, L. K., Research on reinforced concrete beams under combined bending and torsion in the Soviet Union, *Torsion in structural concrete*, American Concrete Institute, SP-18, 1968, pp. 307–336.

79. Collins, M. P., Walsh, P. F., Archer, F. E. and Hall, A. S., Ultimate strength of reinforced concrete beams subjected to combined torsion and bending, *Torsion in structural concrete*, American Concrete Institute, SP-18, 1968, pp. 379–402.

80. Elfgren, L., Karlsson, I. and Losberg, A., Nodal forces in the analysis of the ultimate torsional moment for rectangular beams, *Mag. Concr. Res.*, Vol. 26, No. 86, March 1974, pp. 21–28. Discussion, Vol. 27, No. 90, March 1975, pp. 42–45.

81. Exner, H., On the effectiveness factor in plastic analysis of concrete, *IABSE* (21), Aug. 1979, pp. 35–42.

82. Morley, C. T., The minimum reinforcement of concrete slabs, *Int. J. Mech. Sci.*, Vol. 8, 1966, pp. 305–319.

83. Rozvany, G. I. N. and Hill, R. D., The theory of optimal load transmission by flexure, *Adv. Appl. Mech.*, Vol. 16, 1976, pp. 183–308.

84. Nielsen, M. P., *Limit analysis and concrete plasticity*, Prentice-Hall, Engleword Cliffs, NJ (in press).

85. Braestrup, M. W., *Classical plasticity applied to plain and reinforced concrete*, Technical University of Denmark, Copenhagen, Department of Structural Engineering (in preparation).

21 Computer applications 1: use of large computers

R J Allwood   P J Robins

University of Loughborough,
England

Contents

Notation

A	cross-sectional area
a, b	overall dimensions of rectangular element
D	depth of member
d	displacement
E	Young's modulus
F	force
f	displacement within an element
G	shear modulus
G'	reduced shear modulus
H	horizontal force
I	second moment of area
K, k	stiffness
L, l	length of member
M	moment
M^*	moment of resistance
N	shape function
n	integer
P, p	force
U	element internal strain energy
u	displacement in x-direction
V	vertical force
v	displacement in y-direction
W	potential energy of nodal forces
w	displacement in z-direction
x, y	coordinates
α	constant
γ	shear strain
Δ	displacement
ε	linear strain
θ	rotation
ν	Poisson's ratio
ϕ	total potential energy
σ	stress
τ	shear stress

Summary

After outlining a minimum computer configuration, the problems of obtaining good engineering programs are discussed. Applications to continuous beam analysis and design and to frame analysis are described, summarizing both theory and typical programs. Examples are included to illustrate the range of potential applications. The finite element method of stress analysis is outlined, the use of a typical program described and advice given on the type of element to use. A flat-slab problem is taken as an example followed by a discussion of the difficulties of applying the method to the analysis of reinforced concrete structures. Particular attention is drawn to non-linear models of concrete behaviour, methods of allowing cracks to develop and the importance of allowing for bond slip between concrete and steel. In a brief discussion of reinforced concrete design and detailing, examples are given of simple detailing aids and of comprehensive design and detailing packages. There is a note on the rapidly developing scene of computer-aided draughting.

1   Introduction

What is a 'large' computer and what uses can a structural engineer make of it? We consider a large computer to be simply one with computing capacity which can only be fully exploited by hiring or purchasing other peoples' programs rather than by developing one's own programs from scratch. A minimum-sized computer justifying this label is available now (1983) for £20 000 ($35 000) and we describe in this chapter some of the major areas of application of such a computer (or, of course, larger computers). Because we consider the software to be more important than the hardware, this introductory section deals only briefly with the specification of the minimum-size computer but at length with the problems of locating, selecting and purchasing programs.

1.1   The minimum hardware

A minimum size of computer for the applications to be discussed would have the following features:

(a) Memory size. This is not easy to define without using computing jargon, but the following are roughly equivalent: enough memory to store 24 000 decimal numbers or 128 K bytes of addressable store. More memory would increase the problem-solving capacity and decrease the computing time.

(b) Backing store. Some device is necessary to act as an extension to the memory and to store copies of programs and data in a manner ready for instant use. The currently preferred devices are disks with magnetic coatings and two combinations of disks are suggested to provide a minimum of

10 million characters:

(1) 5M characters on a fixed disk+5M characters on an exchangeable disk, or

(2) 10M characters on a fixed disk+2M characters on exchangeable floppy disks.

(c) Printer. This should have the capacity to print 500 lines/min with up to 120 characters/line or it may prove to be a bottleneck.

(d) FORTRAN. Since most technical programs are written in the language FORTRAN, the computer must be able to run programs in this language, preferably the standard ANSI version.

(e) Graphics. For many engineering applications, a graphical output either on a screen or on paper is vital. The computer must therefore be able to be connected to a graphic display unit or a plotter.

(f) Terminals. It is easier to prepare data whilst sitting at a keyboard/display with the computer acting as prompter/checker rather than sitting at a desk filling in forms. However, the penalty of asking a computer to handle the transfer of data to and from several terminals while also solving problems is not trivial. The minimum specification set out here would be de-graded and such multi-user service would need more memory.

1.2 Software

Computer manufacturers no longer supply programs for the solution of technical programs. They only supply the utility programs needed to operate the computer, to run application programs and to store data or programs on the disks. Application programs to solve engineering problems may be purchased or leased from various sources to be run on one's own computer, or may be used only at a computer service bureau offering a problem-solving service on its own computer. The criteria for selecting which programs to use in either case overlap and so we discuss these principally in the context of the purchase of programs.

Finding application programs of a professional and reliable quality is not easy and structural engineers faced with this task are strongly advised to contact, in the first instance, one of the national organizations already established in several countries for the coordination of engineering programs. A number of these are set out in Table 1-1

Selection of suitable programs from those located is also not easy, especially as there may well appear to be many contenders. Hutton and Roston [1] list 993 programs for building applications available in 1979 in the United Kingdom under 12 subheadings ranging from management through structures to services.

The applications discussed in the other sections of this chapter will indicate the technical features which a potential user of a particular type of program should expect to find. Other, non-technical, factors which may be of equal importance for trouble-free use of the program concern the supplier, the product and its maintenance and the ease with which engineers can prepare the necessary data. Some of these are set out below as a checklist.

One factor, which is not however listed, is often the first concern of a newcomer to the software business scene. Who is responsible for the accuracy of the results

Table 1-1   Organizations coordinating information on computer programs for structural engineers

Country	Name and address of coordinating organization
UK	Construction Industry Computing Association Guildhall Place, Cambridge (formerly Design Office Consortium)

USA	CEPA
	  358 Hungerford Drive, Rockville, MD 20850
	APEC
	  Miami Valley Tower, Suite 2100, Fourth and Ludlow Streets, Dayton, OH 45402

Australia	ACADS 576 St Kilda Road, Melbourne
South Africa	Construction Industry Computer Information Centre, National Building Research Institute, PO Box 395, Pretoria 0001
Netherlands	CIAD Boerderij Noorderbosch, Voorweg 228, PO Box 74, 2700 Ab Zoetermeer

produced by program ABC when run on computer XYZ? The answer, which is usually disappointing, is that no one accepts responsibility for computer results. This is the consequence of the difficulties of guarding against wrongful use of a program outside its intended range, of faulty operation of the computer by its human operators, of failures in the computer or its various devices or of unauthorized alterations to the program. All professional computer bureaux or software suppliers will make their position clear by disclaimers typical of which is the following from the manual on GTSTRUDL [2], one of the most widely used structural analysis programs.

> The GTSTRUDL computer program is proprietary to, and a trade secret of, the Georgia Institute of Technology, Atlanta, Georgia.
> *Disclaimer*
> Neither Georgia Tech Research Institute nor Georgia Institute of Technology make any warranty expressed or implied as to the documentation, function or performance of the program described herein and the user of the program is expected to make the final evaluation as to the usefulness of the program in his own environment.

Such disclaimers should be kept in perspective since an engineer must accept responsibility for his work regardless of what services he may have employed to help him, e.g. drawing or RC detailing.

A number of non-technical factors which should be considered and some questions which could be asked before purchasing/leasing a program are set out below.

1.2.1 *Documentation*
If the author/supplier has confidence in the value of his program, he will have invested resources in preparing a comprehensible user manual and some operating instructions. Avoid programs with one sheet of hand-written instructions.

1.2.2 *Technical back-up by supplier*
Is the supplier an engineer and/or a provider of engineering services and will there be a continuous service to answer questions about the use of the program? How many staff are engaged in this activity? You will need this service particularly in the early stages of learning to use a program.

1.2.3 *Program language and media*
Is the program written in a common (preferably standard) language such as FORTRAN and will it be supplied as a listing or on a magnetic disk or tape? The former will create untold problems from errors generated during the transcribing operation. How big is the program and what size of memory will it need? How will the program be maintained and updated as improvements are made? Answers should be readily available to these questions.

1.2.4 *Portability experience*
Has the program been run on more than one type of computer? This seemingly mild requirement greatly increases the chances of eliminating errors and can be a quite stringent test of the program. If the program has not already been run on the type of computer you are proposing to use, have this demonstrated.

1.2.5 *Purchase/lease*
The software market-place is new but maturing rapidly. You may be able to negotiate a low price for a program but well-written programs are expensive investments and the supplier will want a fair return for his product and the service.

How you pay for the program has a bearing on your future relationship with the supplier. A single sum has the obvious financial advantage of simplicity but allows the supplier to disengage from future support. Leasing on the other hand establishes a continuing contact between user and supplier and virtually forces the latter to support the former to his satisfaction.

An alternative charging method of adding a royalty to the computer charge is only applicable to programs run at a computer service bureau.

1.3 Desirable features of application programs

Finally, in this introduction, we list some desirable facilities which any technical program should provide for a user. See also Table 1–2.

1.3.1 *Easy data preparation and good error checking*
The speed and economy of using a computer can be easily nullified by faulty runs due to data errors. These may be corrected quickly by a user working at a terminal but even the prospective efficiency of this mode of working is tempered by a program which finds one fault at a time. Worse still, of course, is the program

Table 1-2 Desirable features of engineering programs

General

Documentation	At least a user manual
Technical back-up	Provided by engineers
Program language and supply medium	FORTRAN or other standard language supplied on magnetic tape or disk
Portability experience	Run on at least two computers

User features

Easy data preparation	Free format: tabular layout
Good error checking	Continuation after first error found
Controlling calculation sequence	Commands
Saving data or results	SAVE/RESTART
Selective modification of data	Easy modification of design parameters
Selective printing of results	Reduce filing problem

with no error checking, which prints out wrong answers neatly and efficiently. In judging a program, look at the user manual for:

Logical layout and sequence of data	—	Tabular format is popular
Non-restrictive ways of writing data items	—	'Free' format
Interactive input	—	With immediate checking and prompting
Checking of data	—	With continuation after first error found

1.3.2 *Means of controlling the calculations performed*

Most application programs offer choices in their calculation sequences. Just how such options are offered varies, but an increasingly popular solution is by use of commands which the user includes in the input data. The commands are phrased in appropriate engineering terms and greatly help comprehension of the facilities. This technique was used in 1964 in the COGO program [3] and is the basis of the GENESYS [4] and ICES [5] libraries of civil engineering programs.

1.3.3 *Facilities for saving copies of input data or results*

Modern computers are provided with disks upon which data or results can be stored for as long as the user wishes. This feature can be useful in two ways. The input data so stored can be easily altered or edited to reflect engineering decisions and resubmitted for processing. The saved results can form the input to a further program for more processing. When using a substantial application program which performs its calculations in stages, it may be possible to save the intermediate results obtained after some initial stages are completed and 're-start' from that point at a later time. This can be helpful where the initial stages amount to reading and checking the input data and the later stages are the expensive computation components. In some analysis programs, it is possible to submit new 'load cases' for processing from an intermediate stage provided that stage has been saved, thus cutting computer costs.

1.3.4 *Selective modification of input data*
Whether or not the data is saved, as described above, an application program should have facilities to change selected items of data and allow re-processing. Engineers engaged in design need to be able to vary any parameter(s) of a problem. Programs should make this easy to do.

1.3.5 *Selective printing of input data/results*
It is too easy to be swamped by printed output from a computer. The size of paper used makes it difficult to file away with normal correspondence. Some facility to select the output required is very helpful, although it is probably a useful discipline always to insist on a print-out of the input data or at least an identification of the job.

2   Continuous beams: analysis and design

2.1   Introduction

The analysis and design of continuous reinforced concrete beams is a process frequently carried out in the design office. The distinction between analysis and design is, sensibly, often not clearly drawn in the mind of an engineer whose responsibility covers both these operations. Analysis entails the prediction of the behaviour of the beam under the specified loading conditions, while design encompasses the selection of the geometric and material configuration of the beam. Different methods of analysis should produce solutions which coincide but, on the other hand, different engineers will usually arrive at different 'correct' solutions to the same design problem. The rapid rise in the development of general-purpose structural analysis programs as opposed to the slower emergence of structural design programs is undoubtedly due in part to this distinction.

Any program which encompasses design must necessarily work to a set of design rules. In the United Kingdom, these rules, for reinforced concrete beams, will normally be based on compliance with CP 110 [6], although the programs may well differ in the way they interpret the code, and so in the results they produce. The Construction Industry Computing Association (formerly the Design Office Consortium) has recently published a report [7] evaluating seven computer programs for the analysis and design of continuous reinforced concrete beams in accordance with CP 110 and concludes that, despite adherence to the code, very different results can be obtained from these programs. This is to be expected since the code does not consist of a definitive set of design rules, but allows the designer a wide margin in which to apply his judgement and take decisions. So, in writing a continuous beam program, it is the programmer who exercises his engineering judgement and builds in decisions to his program, and consequently the results reflect his ideas and not necessarily those of the engineer who uses his program. This perhaps partially accounts for the wide variety and apparent duplication of programs in the field of continuous beam design; engineers may feel that other people's programs do not sufficiently reflect their own ideas of design judgement, and therefore they are tempted to write their own programs.

Before an engineer decides to develop his own program for the analysis and design of continuous beams, he should investigate those already available to see if any meets his needs and, just as important, agrees with his design philosophy. In this section, we outline the general types of program available and the range of facilities which a designer might reasonably expect a continuous beam program to contain. Of course, when selecting a program, other non-technical factors discussed in Sections 1.2 and 1.3, such as documentation and back-up, must also be borne in mind. The section concludes with a typical continuous beam problem solved using commercially available program CP 110–BEAMS/1, a subsystem of Genesys. It is not our intention to single out this program above others but merely to use it as an example of a program which undertakes the analysis and design of continuous reinforced concrete beams.

2.2 Types of continuous beam program

There are many available programs which perform differing ranges of tasks within the general context of continuous beam analysis and design. At one end of the spectrum are the programs which do little more than automate the structural analysis procedure, and at the other end are those which go right through the analysis and design tasks, producing at the end steel-fixing schedules. In addition, there are those which carry out only the detailing of the concrete. Continuous beam programs vary not only in the scope of the operations they perform but also in the extent to which they remove the design decisions from the engineers who use the programs. The author of a design program must decide at what level he should pitch his program, between the extremes of manual design and an all-embracing program which produces a complete design from a minimum of data.

If an engineer wishes to use a continuous beam design program merely as an aid to calculation, leaving him free to take his own design decisions, then an interactive program is more likely to give him the flexibility he requires. GLADYS, from Ove Arup and Partners, is an example of a reinforced concrete continuous beam program based on this philosophy. Some may consider this approach too time consuming and may favour developing programs with their own decisions and principles built in. This explains the development of continuous beam programs from within large firms. CONBEAM, from Scott Wilson Kirkpatrick and Partners, is an example of a program developed within a large organization and flexible enough to be used by engineers both from within the organization and from outside. Yet another alternative is to use a purpose built program written for general use; GENESYS BEAMS/1 is an example from this category.

In selecting a particular program, the potential user must first choose the type of program which suits best his own ideals, whether it be an interactive one to give him great flexibility in design, or a non-interactive one with design decisions built into the program, giving a quicker solution to the problem.

2.2.1 *Interactive programs*
In using a program from this category, the designer can expect a high degree of flexibility in application since it will be he and not the program that will take the main decisions. Interactive programs are available for use on both large computers and smaller desktop ones. One distinction worth mentioning is that the speed of response is likely to be better on a large computer—a slow response can be very irritating when working interactively. Furthermore, the smaller memory size of the desktop computer limits the extent of warning and diagnostic facilities that a program may contain, and such facilities are of great importance in an interactive program.

2.2.2 *Non-interactive programs*
For these programs, the designer needs to prepare all the data that the program requires for the analysis and design of a beam before the run is started. In non-interactive programs, the engineer does not have the freedom to converse with the program during its execution but he will usually have the opportunity of controlling the design to some degree by specifying preferred properties such as bar sizes and arrangements. Because the main design decisions are built into these programs, the user must satisfy himself that the decisions taken by the program agree broadly with his own design philosophy.

Non-interactive programs will be run either from a terminal in a time-sharing system or in batch mode. The time taken by the user to prepare and run a non-interactive program will generally be less than the time he would have to devote to an interactive one for the solution of the same problem.

2.3 Methods of analysis

The problem of analysing continuous beams lies in evaluating the most critical conditions for shear forces and bending moments in the continuous structure caused by combinations of loadings in the different spans. Beams may be continuous through several supports and monolithic with the supporting columns so that the loading on any span affects, and is affected by, the adjacent spans and columns. For a continuous beam which forms part of a frame, the loading on any one span will affect every beam and column in the frame, but it is usually considered accurate enough for the designer to consider the effects of adjacent spans and columns only (Fig. 2-1). If the supporting columns are flexible in relation to the beams they carry then it may be acceptable to regard such beams as continuous over the supports and free from rotational restraint: this is more likely to be valid for internal columns.

If a structure is statically determinate to the nth degree, then it is necessary to

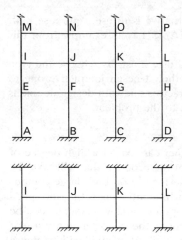

Figure 2-1 Idealization of beam IJKL

determine a set of n additional forces or displacements before the structure can be completely analysed; that is, before all the forces and displacements can be specified throughout the structure. This principle leads to the two basic approaches to the analysis of indeterminate structures: the flexibility (or force) method in which the unknown forces are first determined; and the stiffness (or displacement) method in which the unknown displacements are first evaluated, the corresponding forces being subsequently computed. In writing computer programs for the analysis of large or complicated frames, it is generally accepted that the stiffness method is the more convenient approach to adopt as the equations for solution are simpler to derive, involving general relationships between the member end forces and displacements. Moreover, it is usually possible to set up the equations without considering the exact degree of determinacy involved, whereas the flexibility method usually requires the precise determination of the degree of statical determinacy and a careful choice of the release system to be employed. The application of the stiffness method to frames is discussed more fully in Section 3.

Continuous beams can be analysed using any general method applicable to more general problems of frames, but because of the specialized nature of the continuous beam problem it is likely to prove more economic in effort, and therefore in program time, to use alternative methods more specifically suited to the particular problem of the continuous beam. As an illustration of this point, consider the five span continuous beam shown in Fig. 2-2a. If the generalized stiffness approach is used, then it will be necessary to set up equations and solve for the six unknown rotations at the supports. Alternatively, if the flexibility method is used, then not only can the number of unknowns (and therefore equations) be reduced, but also by careful choice of release system the computations in solving the equations can be made simpler. Four releases are necessary to render the structure determinate. Two alternative release systems are shown in Figs 2-2b and 2-2d: in the first, the internal support reactions are removed; in the second, moment releases are applied to the internal supports. Typical moment

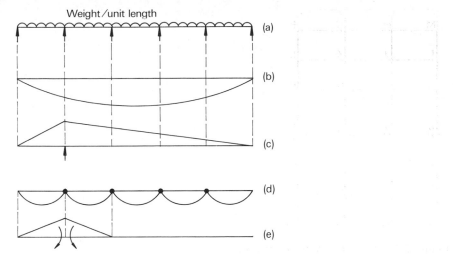

Figure 2-2   Continuous beam analysis

diagrams for the release forces for the two cases are shown in Figs 2-2c and 2-2e, respectively. Note that, in the second release system, the effect of each release is localized while in the first system the effect is felt over the entire structure. As well as making the computations simpler, the second system leads to a banded set of equations with leading diagonal terms appreciably larger than the off-diagonal terms, leading to a quicker and possibly more accurate solution.

Two special forms of the stiffness method which have found wide usage by engineers in the past for continuous beam analysis are worthy of mention here, since familiarity with these techniques may tempt engineers to adopt them for their programs; these are the slope deflection and moment distribution methods. The slope deflection method has enjoyed wide use for many decades; the technique involves the solution of a set of linear equations which are set up by considering moment equilibrium and slope compatibility at each support or node, and its relationship to the stiffness method is often overlooked. The moment distribution method of analysis is based on an iterative solution of the equilibrium equations at the joints, in which a trial-and-error procedure is used to iterate the solution until all the equations are satisfied by the final values. Perhaps because of its versatility and wide usage for hand computations, the authors of some of the commerically available continuous beam programs have chosen to use the method of moment distribution for their analysis (for example, GENESYS BEAM/1).

2.4   CP 110 in design

The range of design operations carried out by continuous beam programs is wide and a review of typical design features offered by purpose-written programs is given in Section 2.5. What is clear is that after the elastic analysis has been carried out, any design calculations must be in accordance with a particular set of rules and decisions. In the UK, these will normally be based on the recommendations of CP 110, and we shall concentrate our attention on design in accordance with

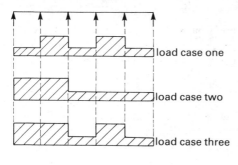

1.4 × dead load + 1.6 × live load

1.0 × dead load

Figure 2-3   Loading patterns

this code. Programs written in other countries for use by their own engineers will probably follow the design code of their particular country of origin.

For continuous beam design, the essential differences between CP 110 and the previous codes lie in four main areas: the concept of limit states (ultimate and serviceability); the introduction of independent safety factors for materials and loading; the specification of combinations of imposed loads to be considered when forming bending moment and shear force envelopes; and the explicit introduction of plasticity in reinforced concrete leading to a more detailed treatment of redistribution of moments.

Whatever the method of solution chosen for the elastic analysis, the beam must be loaded with chosen patterns of dead and live loads to produce envelopes of shear force and bending moment. Three typical patterns of loads are shown in Fig. 2-3, where the first two load combinations are the ones recommended by CP 110.

For the limit state of structural collapse, the ideal is that the loads multiplied by the load factors should cause plastic hinges to form in the beam, producing a mechanism leading to collapse. While it may be theoretically possible, for a given set of bending moments, to select an arrangement of reinforcement that would cause all hinges to form simultaneously, this is unlikely to be feasible when there are several different load patterns to be considered, therefore CP 110 allows the designer to redistribute the bending moments instead. This may be done for each load pattern independently and the reinforcement is then selected to withstand the largest adjusted bending moments so calculated. These adjustments may be made to span moments as well as support moments, subject to a maximum limit of 30%. The effect of redistribution may be viewed as a lowering or raising of the bending moment curve, Fig. 2-4.

The designer must take account of the effect of redistribution when selecting the main reinforcement. Where, say, a 30% reduction has been made to a bending moment value, this implies the formation of a hinge and so the designer must select the reinforcement so as to ensure that the neutral-axis depth factor,

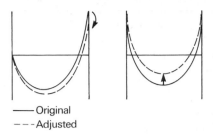

————Original
— — — Adjusted

Figure 2-4   Redistribution of moments

x/d, is sufficiently low for the necessary hinge rotation to occur (see CP 110, Clause 3.2.2.3). Any effects on column moments of redistribution must also be considered; this may mean redesigning the column.

These considerations and those concerning the serviceability limit states of cracking and deflection must be taken into account when selecting the beam section and reinforcement. Standard checks for local and anchorage bond must also be made to ensure compliance with the appropriate Clauses 3.11.6.1 and 3.11.6.2 of CP 110.

2.5   Packages: potential facilities

In this section, we discuss the range of options or facilities which a user can expect to find incorporated in a computer program written specifically for the analysis and design of continuous reinforced concrete beams. We do not suggest that every designer will require all these options or that every continuous beam program should include them; it is merely our intention that discussion of possible options will help the designer to select from the available range of programs one which most closely matches his needs and ideals. Even if in selecting the most suitable program compromises are still necessary, they may well be worth making in order to save the considerable time and expense involved in the development and documentation of a new program. Table 2-1 summarizes the main features discussed in this section.

2.5.1   *Idealization of the structure*
All structural analysis requires idealization of the real structure. Where a continuous beam forms part of a larger frame, an acceptable assumption is that all the columns framing into the beam are fixed at their remote ends (Fig. 2-1). Programs may allow a variety of intermediate and end-support conditions: ends of the beam may be fixed, free, simply supported or framed into columns; and at intermediate supports the beam may frame into columns, be supported on a single column or be simply supported (Fig. 2-5). Programs may contain the option of defining independently upper and lower columns at supports and may allow for various column profiles.

Most programs will offer a choice of beam sections and will allow different spans to have different profiles.

Table 2-1 Summary of features of continuous beam packages

Structural idealization	Supports (column, fixed, free . . .)
	Beam section (rectangular, T, I, . . .)
	Number of spans (upper limit?)
Loading	Choice of loading shapes
	Self-weight allowance
	Combinations of load cases considered
	Wind loading
	Redistribution of moments
Section design	Material properties
	Selection of reinforcement (bar sizes and layout)
	Serviceability limit states (deflection, cracking)
	Detailing
Input data and results	Shear force and bending moment envelopes
	Tension, compression, shear reinforcements
	Checks on bond, serviceability of limit states
Error messages	Scope of messages and warnings

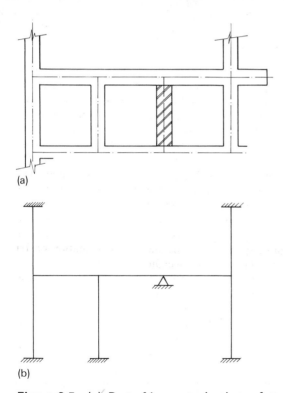

(a)

(b)

Figure 2-5   (a) Part of beam and column framework
(b) Equivalent line diagram for analysis

2.5.2 Load systems and analysis

A variety of elemental loads and load shape distribution can be expected, and superposition of different types within a span enables complicated load distributions to be modelled. Some programs automatically compute an allowance for the dead weight of the beam and add this to the other loads specified by the user. Analysis of the continuous beam loaded with predefined patterns of the specified dead and live loads is carried out to produce envelopes of shear force and bending moment for use in design. Some programs follow the recommendations of CP 110 (Clause 3.2.2.1), loading alternate and adjacent spans; others add patterns; and some allow the user to define his own patterns of loading.

It is rare to find a package which allows for side sway under vertical loading, though some may offer facilities for wind analysis which effectively takes side sway into account. Settlement of supports and axial deformation of columns are usually not permitted.

The redistribution of elastic bending moments in accordance with CP 110 was discussed in Section 2.4. Programs offer different limits on the value of maximum percentage reduction; some apply a single percentage, specified by the user, to all reductions; others will allow different percentages to be specified for each span or at each support. Redistribution at column supports can affect the column moments and so alter their design. Continuous beam programs must therefore either redistribute the moments on each side of a support by equal amounts leaving the column moments unchanged, or redistribute the beam moments by different amounts and inform the user of any change in column moments.

2.5.3 Section design

Available programs offer the designer varying degrees of flexibility in defining the material properties of the section. The concrete and reinforcement strengths are normally specified by the user, though there may sometimes be limits on this choice. Some programs allow the user to specify an ageing factor for the concrete so the ultimate local and anchorage bond stresses may be specified separately.

When designing in accordance with CP 110, programs usually assume that plane sections remain plane, that the concrete has no tensile strength, and that the stress-strain curve for compression is parabolic. Most programs determine and print the required arrangements of tension, compression and shear reinforcements. Checks for local and anchorage bond stresses should also be made. Some programs go beyond the design of specific sections and give information on bar layouts, curtailments and even bar bending and weight schedules.

The serviceability limit states of cracking and deflection are dealt with to varying degrees of thoroughness. Some programs check only compliance with allowable span-to-depth ratio and ignore cracking, while others may actually calculate and output deflections and crack widths, taking the latter into account when selecting bar sizes.

2.5.4 Input data and results

The requirements of data preparation and input are very different for the two categories of continuous beam programs, interactive and non-interactive. Non-interactive programs require all the design data to be prepared and input before the program run is commenced. With an interactive program, the user converses with the machine during the execution of the program and is required to input data and take decisions throughout the run.

Most programs reproduce the input data in the output for checking and reference. Results of the structural analysis usually include values of the shear force and bending moment envelopes, elastic and redistributed, at chosen intervals along each span (typically span/10). Information on support reactions, points of contraflexure, and shear force and bending moment values for the independent load cases can also be useful, though not all programs offer these facilities. Results from the structural design include required areas of tension, compression and shear reinforcements and arrangement of bars. Information on local bond may either be in the form of bond stress associated with the selection of the main reinforcement at particular cross-sections, required bonding perimeters at a cross-section, or sometimes simply a warning that the allowable bond stress is exceeded.

When studying the output, an engineer will usually need to cross refer to information spread over many pages. Clearly, the more concise the presentation of results, the easier his task is made. Output on standard 136-column line-printer paper can be inconvenient to use and file because of its size, so some programs produce output that can be cut down to A4 size sheets.

2.5.5 *Error messages and diagnostics*

There are basically two different types of error: those which are due to coding errors in the data and those which arise during the processing of the commands. Programs differ greatly in the range and style of the messages and warnings they issue to the user when an error occurs. From the user's standpoint, it is desirable for a program to have a wide range of explicit error messages and warnings.

2.6 Example use of a continuous beam program

We conclude this section with a simple continuous beam design example solved using CP 110–BEAMS/1, a subsystem of Genesys [8]. The four-span continuous beam used for this example is illustrated in Fig. 2-6. Though quite a straightforward example, the structure has been chosen to include a combination of support conditions.

The data for the, CP110–BEAMS/1 analysis and design of this continuous beam is shown in Fig. 2-7 and explanatory comments appear opposite in Table 2-2. The data falls into two distinct parts: the main part consists of a set of tables describing the members, the section properties, the supports and the loads; the second part contains a series of commands for execution by the program. These commands set parameters, such as steel and concrete properties, and cause the beam to be assembled from its components prior to being analysed and designed.

Some of the printed results from the analysis and design are shown in Figs 2-8, 2-9 and 2-10. After printing details of all the input data and values assumed, the program prints the results of the structural analysis. These include the ultimate bending moment envelope before redistribution and the ultimate shear force and bending moment envelopes after redistribution (Fig. 2-8). The results of the structural analysis are followed by a series of tables giving design information for

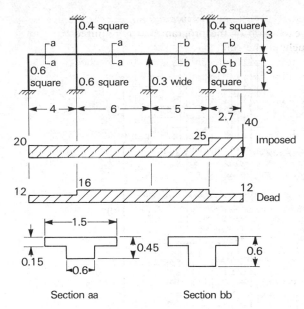

Figure 2-6 **Continuous beam example**

each span of the continuous beam. This information includes: design moments and shears; moment of resistance of the concrete section without compression reinforcement; shear resistance of the concrete only; and required areas of main and shear reinforcements (Fig. 2-9). These tables of design output are followed by a number of warnings and some supplementary design information (Fig. 2-10) which may be useful if the designer is to detail the beam himself.

Genesys CP 110–BEAMS/1 has detailing facilities and if detailing has been requested by the engineer, additional tables giving information on fixing, bending and weight schedules are printed. These features are described in Section 5.3.

3 Frame analysis

3.1 Statement of problem

Whereas the analysis of continuous beams by computer is a straightforward extension of manual methods, the analysis of frames has involved the development of new methods resulting in a spectacular step forward in structural analysis techniques. Even if we restrict the definition of a frame to two- or three-dimensional structures made from straight, uniform members connected together by rigid or pinned joints, then many frames previously intractable have become standard cases for modern frame analysis programs. We should also add that the frame geometry may be entirely irregular. Extensions to allow curved members, variable cross-sections, joints of specified stiffness, thermal expansion, etc., are seemingly limitless. The basic analysis assumes linear behaviour of members but material or geometric non-linearity (buckling and/or large deflections) can be

```
 c| Statement  |7  10   15   20   25   30   FORTRAN STATEMENT  35  40  45
 1|2      5|6|7 Number

*GENESYS
#START  'CP110-BEAMS/1'                                          ⎫
JOB EXAMPLE FOR HANDBOOK                                         ⎬ (a)
                                                                ⎭
*TABLES
'BEAMS'
BEAMREF    SPANS      SUPPREF    SECNO                           ⎫
'1/B1/3'   4,6        1,2,3      1                               ⎬ (b)
'3/B1/5'   5,2.7      3,2        2                               ⎭
'SECTIONS'
SECNO      SHAPE   DIMENSIONS               COVERS               ⎫
1          'T'     450,600,150,1500         35,40,40,40          ⎬ (c)
2          'T'     600,600,150,1500         =                    ⎭
'SUPPORTS'
SUPPREF    POSN   SHAPE    DIMENSIONS     HEIGHT                 ⎫
1          'L'    'R'      600,600        3                      ⎪
2          'U'    =        400,400        =                      ⎬ (d)
2          'L'    =        600,600        =                      ⎪
3          'L'    'S'      300            =                      ⎭
'LOADS'
BEAMREF    NATURE   TYPE   DISTANCES     LOAD                    ⎫
'1/B1/2'   'D'      'U'    0,4           20                      ⎪
=          'I'      =      =             12                      ⎪
'2/B1/3'   'D'      =      0,6           20                      ⎪
=          'I'      =      =             16                      ⎬ (e)
'3/B1/4'   'D'      =      0,5           20                      ⎪
=          'I'      =      =             16                      ⎪
'4/B1/5'   'D'      =      0,2.7         25                      ⎪
=          =        'P'    2.7           40                      ⎪
=          'I'      'U'    0,2.7         12                      ⎭

*MASTER
SET STEEL PROPERTIES FOR MAIN AS DESIGNATION ...                ⎫
    'HY' TYPE 1 AND SUB AS DESIGNATION 'R'                       ⎪
SET CONCRETE PROPERTIES AS GRADE 30 DENSITY 2500                ⎪
SET MAXIMUM BAR SIZE OF 20                                      ⎬ (f)
SET MAXIMUM REDISTRIBUTION PERCENTAGE 20                        ⎪
APPLY LOADS 'LOADS'                                             ⎪
DESIGN AND DETAIL                                               ⎪
*FINISH                                                         ⎪
*EXIT                                                           ⎭
```

Figure 2-7 Continuous beam data

handled at increased computer costs. It is in this area of analysis, particularly of three-dimensional frames, that the large computer can show its advantages over small computers.

The costs of elastic analyses are remarkably low and for any unusual structure, an engineer can confidently consider using one of the many standard frame analysis programs available. Some will be mentioned later, an example will be given and some basic features one can expect to find in a program will be listed.

Paradoxically, care should be taken before using an analysis program on regular office block structures. There may be justification if a shear wall or some unusual geometry is involved. But the multiple loading cases generated by the design requirements of adjacent and alternate-span loads on the beams coupled with the simple behaviour of beam-and-column structures make the use of general frame

Table 2-2   Notes on continuous beam data

Section	Comments
(a)	The command GENESYS initiates the job and the command START selects the particular subsystem required. The chosen title (Example for handbook) will be printed at the top of each page of results.
(b)	This table defines the spans of the beams and references the supports and cross-sections.
(c)	Each line of this table specifies a section by a reference number used to cross refer to the beams table. The section shape is defined by a letter, followed by the overall dimensions and the minimum covers to the reinforcement.
(d)	The supports table contains details of all the types of support contained in the beam, each type being defined under a reference number. Entries under POSN indicate the position of the column relative to the beam. The shape of the support is defined (R for rectangular, S for simple support, etc.) and the dimensions of the section of the supporting columns and their length are specified.
(e)	Characteristic loads are presented under the two categories of dead D and imposed I. The type of load, where it acts, and its value are all specified. Several load tables may be given. The loads specified all act simultaneously but separately from those in other tables.
(f)	The commands shown are typical for this type of problem, and others are available. The analysis and design are carried out and the results are printed.

analysis programs uneconomic. The continuous beam programs of Section 2 are of more value.

Most engineers accept that a linear analysis of a reinforced concrete frame is adequate despite the non-linear behaviour of a cracked section beam and the general adoption of design codes based upon limit state conditions.

In summary, one can expect a frame analysis program to handle frames and loadings of the following nature:

Frames	Two or three-dimensions, i.e. plane frames, grillages, space frames, with up to hundreds of joints
Geometry	Completely general
Members	Connecting two joints, straight, uniform section, elastic material properties
Joints	Rigid or pinned
Supports	Encastré, pinned, rollers
Loads	Point or distributed loads (not usually self-weight)
	Settlement
	Member expansion/contraction

```
'ULTIMATE BENDING MOMENT ENVELOPE (EXCLUDING WIND) AFTER REDIST.'
   BEAMREF    P0    P1    P2    P3    P4    P5    P6    P7    P8    P9   P10

  '1/B1/2'   -40   -10    10    13    12     7    -2   -15   -31   -52   -89
  '1/B1/2'    -8     3    19    35    43    43    33    16   -10   -45   -76

  '2/B1/3'  -150   -60    -3    18    30    33    28    13   -10   -50  -135
  '2/B1/3'   -70   -32    25    80   114   126   117    66    34   -40   -83

  '3/B1/4'  -135  -106   -84   -68   -60   -58   -62   -73   -91  -157  -239
  '3/B1/4'  -122   -63    16    37    47    53    43    18   -23   -79  -147

  '4/B1/5'  -383  -323  -269  -219  -174  -133   -97   -66   -39   -17     0
  '4/B1/5'     0     0     0     0     0     0     0     0     0     0     0

'ULTIMATE SHEAR FORCE ENVELOPE (EXCLUDING WIND) AFTER REDIST.'
   BEAMREF    P0    P1    P2    P3    P4    P5    P6    P7    P8    P9   P10

  '1/B1/2'    95    73    52    30     9   -12   -22   -32   -42   -51   -61
  '1/B1/2'    37    27    17     7    -2   -12   -34   -55   -76   -98  -119

  '2/B1/3'   182   146   110    74    38     3   -12   -27   -41   -56   -71
  '2/B1/3'    76    61    46    32    17     3   -33   -69  -105  -141  -177

  '3/B1/4'   154   122    91    60    28    -3   -17   -30   -43   -57   -70
  '3/B1/4'    43    30    16     3   -10   -24   -55   -86  -118  -149  -181

  '4/B1/5'   227   210   193   176   159   142   125   107    90    73    56
  '4/B1/5'     0     0     0     0     0     0     0     0     0     0     0
```

Figure 2-8   Results of structural analysis

The program will analyse the frame under a number of loading cases simultaneously and will print any or all of the following results for each case or for combinations of cases:

Deflections — Vertical and lateral deflections at each joint, rotations at each joint

Bending moments
Shear forces
Axial forces } Usually at the ends of members for each load case or combination

Reactions — At all supports

Stresses — Some programs will calculate stresses but extra input data is needed

3.2 Method of analysis

Virtually all frame analysis programs use the stiffness (i.e. displacement) method, unknown before computers. A generalized version of slope deflection, it is a simple repetitive operation very suitable for computers, quite unsuitable for manual methods. The method is not affected by the indeterminacy or otherwise of the structure but leads to a set of simultaneous equations, one for each displacement. Since there may be hundreds of displacements, yielding as many equations, it will be obvious why the method is unsuitable for manual solution. Unfortunately, this makes it less well known and therefore less well taught. An outline is

```
DESIGN OF BEAM 'B1'

'DESIGN MOMENTS AND SHEARS'
    BEAMREF  FORCE         END1         SPANMAX          END2

    '1/B1/2'  'MOM'    -19.8,    0.4    44.0,   0.0   -58.8,    0.0
    '1/B1/2'  'SHE'     78.5                           103.0

    '2/B1/3'  'MOM'    -98.5            126.3         -135.1
    '2/B1/3'  'SHE'    163.9                           176.8

    '3/B1/4'  'MOM'   -135.1,   0.0    53.0,  -57.5  -188.3,    0.0
    '3/B1/4'  'SHE'    153.8                           161.8

    '4/B1/5'  'MOM'   -317.3            0.0             0.0
    '4/B1/5'  'SHE'    208.3                            0.0


'AREAS OF MAIN REINFORCEMENT'
    BEAMREF  POSITION      END1         SPANMAX         END2
                          MM**2          MM**2         MM**2

    '1/B1/2'   'TOP'      142.5          0.0           430.0
    '1/B1/2'  'BOTTOM'      2.9         320.7            0.0

    '2/B1/3'   'TOP'      733.4          0.0          1023.6
    '2/B1/3'  'BOTTOM'      0.0         933.0            0.0

    '3/B1/4'   'TOP'      719.0         328.2         1015.0
    '3/B1/4'  'BOTTOM'      0.0         277.6            0.0

    '4/B1/5'   'TOP'     1767.3          0.0             0.0
    '4/B1/5'  'BOTTOM'      0.0          0.0             0.0


'REQUIRED AREAS OF SHEAR REINFORCEMENT PER METRE'
    BEAMREF        END1        END2
                  MM**2       MM**2

    '1/B1/2'        0.0        239.4
    '2/B1/3'      817.8        696.5
    '3/B1/4'      336.9        269.0
    '4/B1/5'      211.5          0.0
```

Figure 2-9 Required areas of reinforcement

```
WARNINGS FOR BEAM 'B1'

*WARNING B2560    SPAN/DEPTH RATIO OF SPAN '4/B1/5' IS  4.97

*WARNING B2780    CHECK CLEAR DISTANCE BETWEEN BARS IN TENSION
                  IN SPAN '1/B1/2'

*WARNING B2780    CHECK CLEAR DISTANCE BETWEEN BARS IN TENSION
                  IN SPAN '2/B1/3'

*WARNING B2780    CHECK CLEAR DISTANCE BETWEEN BARS IN TENSION
                  IN SPAN '3/B1/4'




'REQUIRED PERIMETERS FOR TENSILE REINFORCEMENT AT SUPPORTS'
    BEAMREF       END1        END2
                   MM          MM

    '1/B1/2'      71.3        93.6
    '2/B1/3'     149.0       160.7
    '3/B1/4'     101.2       106.4
    '4/B1/5'     137.0         0.0
```

Figure 2-10 Warnings and supplementary information

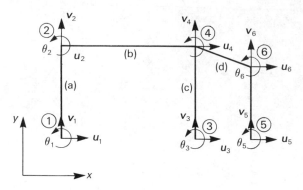

Figure 3-1 Plane frame analysis

given below but readers interested in studying the method may refer to [9, 10 or 11].

The objective of the stiffness method of analysis is to calculate first the displacements of the joints of the structure and from these to calculate the bending moments, forces and reactions. The displacements are always lateral and vertical displacements parallel to a set of coordinate axes and rotations about one or more axes. For a plane frame analysis, there are three displacements of a joint, as illustrated in Fig. 3-1, (For a space frame there are six displacements to be calculated at each joint.) Note that displacements are expected even at joints constrained by supports or at joints which an engineer would consider immovable in some directions, e.g. vertically at joint 2 of Fig. 3-1. In both instances, the computed displacements will be very small, reflecting the high stiffness of a foundation or of a member loaded axially. However, it is the consistent inclusion of all displacements at all joints which leads to the simple repetitive analysis process so suitable for computers.

On the frame in Fig. 3-1, the displacements are drawn in their positive directions matching the positive directions of the corresponding axes. This convention is consistent but, since it also applies to the applied loads, it yields the inconvenient result that *positive* vertical loads or displacements act *upwards* not downwards as every engineer would expect. This inconvenience is compensated for by a consistency which is invaluable when dealing with three-dimensional structures.

There are three stages in the process of calculating the joint displacements:

(1) Calculate, for each member separately, relationships giving forces and moments at each end in terms of the displacements at each end.
(2) Add together relationships (1) for all members and similar relationships for supports.
(3) Treat the results of (2) as a set of simultaneous equations to be solved for each set of applied loads.

A brief description of these steps is given below but many textbooks on the stiffness method are now available, e.g. Livesley [11], Martin [10] and McGuire and Gallagher [9].

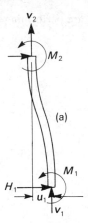

Figure 3-2   Member (a)

In Stage 1, each member is analysed individually and relationships calculated which give the forces and moments acting *on* the member in terms of the displacements of the ends of the member. Engineers are usually concerned with the reverse relationships (often called flexibility equations) which give displacements in terms of forces. The advantage of using stiffness relationships will be seen in Stage 2 when we assemble the equations for the complete frame.

The simplest approach to calculating stiffness equations is illustrated in Fig. 3-2, where we consider member (a) alone, isolated from the frame and subjected to one displacement only, u_1. We can easily calculate the forces and moments which must act *on* member (a) to cause that displacement pattern and if we do this in turn for all displacements, i.e. v_1, θ_1, u_2, v_2, θ_2, we can use superposition to add the results together. (It may be expected that u_1 and v_1 will be zero since joint 1 is a support but, as will be seen, there will be very small displacements there.)

We can find the forces that cause u_1 by several methods, such as moment area or Macaulay. The result is

$$H_1 = \frac{12EI}{l^3} u_1$$

$$M_1 = -\frac{6EI}{l^2} u_1$$

$$H_2 = -\frac{12EI}{l^3} u_1 \tag{3-1}$$

$$M_2 = -\frac{6EI}{l^2} u_1$$

where E is Young's modulus, I the second moment of area and l the length of the member.

This type of analysis can be repeated for each displacement in turn. Thus, the vertical displacement, v_1, will cause only compression of the member yielding expressions for the forces V_1 and V_2. The rotation θ_1 will cause the member to bend and moment area or a similar method is needed for the analysis.

The effects for all displacements can be added together since we are performing a linear elastic analysis and the result is

$$H_1 = \frac{12EI}{l^3} u_1 - \frac{6EI}{l^2} \theta_1 - \frac{12EI}{l^3} u_2 - \frac{6EI}{l^2} \theta_2$$

$$V_1 = \frac{EA}{l} v_1 - \frac{EA}{l} v_2$$

$$M_1 = \frac{-6EI}{l^2} u_1 + \frac{4EI}{l} \theta_1 + \frac{6EI}{l^2} u_2 + \frac{2EI}{l} \theta_2$$

$$H_2 = \frac{-12EI}{l^3} u_1 + \frac{6EI}{l^2} \theta_1 + \frac{12EI}{l^3} u_2 + \frac{6EI}{l^2} \theta_2$$ (3-2)

$$V_2 = \frac{-EA}{l} v_1 + \frac{EA}{l} v_2$$

$$M_2 = \frac{-6EI}{l^2} u_1 + \frac{2EI}{l} \theta_1 + \frac{6EI}{l^2} u_2 + \frac{4EI}{l} \theta_2$$

where in addition to previously defined variables, A is the section area of the member.

At this stage, the displacements u_1, v_1, etc., are unknown and we must look at these relationships as a set of six simultaneous equations with the displacements as the variables.

Before proceeding, we can simplify the representation of the following steps by rewriting the above equations after separating the coefficients and the displacements. Thus,

Displacements

	u_1	v_1	θ_1	u_2	v_2	θ_2

Coefficients

		u_1	v_1	θ_1	u_2	v_2	θ_2	
H_1		$+12EI/l^3$	$+0$	$-6EI/l^2$	$-12EI/l^3$	$+0$	$-6EI/l^2$	
V_1		$+0$	$+EA/l$	$+0$	$+0$	$-EA/l$	$+0$	
M_1	$=$	$-6EI/l^2$	$+0$	$+4EI/l$	$+6EI/l^2$	$+0$	$+2EI/l$	(3-3)
H_2		$-12EI/l^3$	$+0$	$+6EI/l^2$	$+12EI/l^3$	$+0$	$+6EI/l^2$	
V_2		$+0$	$-EA/l$	$+0$	$+0$	EA/l	$+0$	
M_2		$-6EI/l^2$	$+0$	$+2EI/l$	$+6EI/l^2$	$+0$	$+4EI/l$	

In Stage 2, the stiffness equations calculated for each member are added together, taking into account which forces and which displacements are referred to for each member. It is this stage which is made so simple by the choice of stiffness as the basic method. By way of illustration, consider two springs with stiffness relationships

$$p_1 = k_1 d_1 \qquad p_2 = k_2 d_2$$

If we join the springs together so that $d_1 = d_2$, then the new stiffness relationship is

$$(p_1 + p_2) = (k_1 + k_2)d_1$$

and we can note that $(p_1 + p_2)$ must be equal to whatever external forces is applied to the joined springs.

We illustrate the application of this principle to our frame by considering the stiffness relationships for members (a) and (b) and specifically the relationships which give the horizontal force, H_2, acting *on* each of those members at joint 2. Assuming the following structural properties, we can calculate the stiffness relationships directly from Eqn 3-3.

Thus, for member (a)

$$E = 15\,\text{GPa} \quad l = 2.0\,\text{m} \quad I = 16 \times 10^8\,\text{mm}^4$$

and for member (b)

$$E = 15\,\text{GPa} \quad l = 4.0\,\text{m} \quad A = 200 \times 10^3\,\text{mm}^2$$

yielding the following equations in units of kN and mm,

Displacements

u_1	v_1	θ_1	u_2	v_2	θ_2	u_4	v_4	θ_4

$$\begin{aligned} H_2^{(a)} &= \\ H_2^{(b)} &= \end{aligned} \quad \begin{array}{ccccccccc} -36 & 0 & 36\,000 & 36 & 0 & 36\,000 & & & \\ & & & 750 & 0 & 0 & -750 & 0 & 0 \end{array}$$
(3-4)

Adding these together yields

Displacements

u_1	v_1	θ_1	u_2	v_2	θ_2	u_4	v_4	θ_4

$$Fx_2 = \quad \begin{array}{ccccccccc} -36 & 0 & 36\,000 & 786 & 0 & 36\,000 & -750 & 0 & 0 \end{array}$$
(3-5)

where

$$Fx_2 = H_2^{(a)} + H_2^{(b)}$$

which by equilibrium must equal the horizontal force applied to joint 2.

It is a simple matter to write a computer program to perform this operation and to set up the three equilibrium equations for each joint. The numerical values of the coefficients are calculated from the structural properties of each member and the coefficients are then set out and added together in a table of N rows and columns, where N equals $3 \times$ number of joints of the frame. The result for our six-joint frame is the set of 18 simultaneous equations illustrated in Fig. 3-3, where, apart from the equation we have given in detail, the coefficients for each

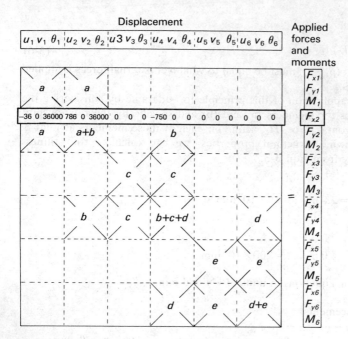

Figure 3-3 Assembly of stiffness equations

member are represented only by the letter assigned to the member. Blanks represent zero coefficients.

To complete Stage 2, the effect of the supports on the frame must also be added into the overall equations. A simple procedure is to consider each type of support as a set of very stiff springs representing the restraining action of the earth. Thus, the pinned support shown in Fig. 3-4 acting at joint 1 can be represented by

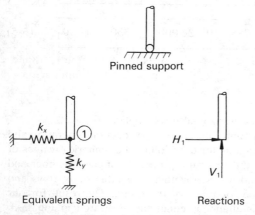

Figure 3-4 Method of dealing with supports

equivalent springs which have the following stiffness equations:

$$H_1 = K_x u_1 \qquad V_1 = K_y v_1 \tag{3-6}$$

The coefficients K_x and K_y may be chosen to represent the actual stiffness of the foundation if known, or set to an arbitrary high value such as 10^{20}. Other types of support can be similarly represented. Thus, a horizontal sliding support would have only a vertical spring while an encastré support would have an extra rotational spring. The support stiffness equations are added into the equilibrium equations of Fig. 3-3 just as the member equations were.

In Stage 3, the equations are solved for any set of applied forces yielding numerical values of the joint displacements, u_1, v_1, θ_1, etc. These can be substituted in the stiffness relationships (Eqn 3-3) for each member and multiplied out to give the forces and moments acting on each member. Similarly, the very small displacements which will be computed at each support can be substituted in Eqn 3-6 to give the reactions.

The power of the stiffness method springs from the ability of computers to assemble and solve hundreds or even thousands of equations very quickly and economically. The result is a new structural analysis tool of great value.

3.3 Facilities of typical frame analysis programs and potential errors

From the simple frame used in the next section, it will be clear that it is not necessary for a frame to be large to justify analysing it by computer. Further applications emphasizing this will be briefly described later but, in considering potential applications, an engineer should know of the types of structure which can be analysed by typical frame analysis programs and of the associated facilities likely to be provided. These are set out as a check list in Table 3-1 and are amplified in the following notes, which also warn of some traps for the user.

Two-dimensional frames or grids will present no problem to an engineer using an analysis program, although the latter will need torsion constants, which may be obtained from Roark [13]. Three-dimensional frames need care in specifying the direction of the major and minor bending axes of sections, as explained, for example, in Section 5.3 of [28]. Users' manuals should be consulted, particularly for the case of vertical members.

The type of connection between members may be either rigid or pinned, but again care should be taken with a fully pinned joint in an otherwise rigid-jointed frame. The computer may try to compute the angular deflection of the joint and fail. It is safe and satisfactory to leave one member at such a joint unpinned.

Facilities for applying loads usually cover most situations, but a common trap already referred to in the preceding example is to forget that *downwards* acting loads are *negative* if the y-axis points upwards. Self-weight facilities apply only to the structural members as defined for analysis purposes and further dead weight from floor finishes, walls, etc, must be specified separately.

All programs should print-out the input data to ensure a full record of the job is on file for future reference. A line drawing of the frame, plotted by the computer from the input data, provides an excellent check of the geometric elements of that data. This could be a perspective view of a 3-D frame. Similar plots of the deflected shape or of bending moments can be disappointingly confusing.

Table 3-1   Commonly available facilities of frame analysis programs

Types of frame	2-D rigid jointed, e.g. plane frames
	2-D pin jointed, e.g. trusses
	2-D grids, e.g. bridge decks, floors
	3-D rigid jointed, e.g. space frames
	3-D pin jointed, e.g. trusses, geodesic structures
Structural members	Straight with uniform section
	As above with rigid end extensions.
	Straight with varying sections[a]
	Connections—rigid or pinned
Supports	Rigid, pinned
	Roller along coordinate axes
	Rollers along inclined surface[a]
	Elastic, e.g. Neoprene cushion
Loads	Point force, uniformly distributed load, partial udl., varying distributed load in coordinate directions or local to member direction
	Point couple
	Settlement of supports
	Expansion/contraction of members
	Self-weight[a]
	Multiple load cases analysed simultaneously
	Factored combinations of load cases
General	Print-out of input data
	Plotting of frame for checking purposes
	Selective print-out of deflections, forces, moments and reactions
	Plotting of deflected shape and of moments[a]
	Print-out of forces and moments at intermediate points along members
	Automatic renumbering of joint numbers to minimize computing cost[a]

[a] Facility provided sometimes

3.4 Example use of a frame analysis program

There are many frame analysis programs available and in choosing one to illustrate how a structural engineer would use such a program we are not indicating that this particular program is, in our view, the best. STRUDL is a general-purpose structural analysis program very widely distributed throughout the world and which was developed from the MIT STRESS program [12]. We have used the GTSTRUDL version issued by Georgia Institute of Technology [2].

The simple frame used for the example is illustrated in Fig. 3-5 and we have deliberately used the program in a very simple mode to illustrate the analysis of this frame under one loading case. There are many further facilities in the program for tackling more sophisticated problems with, for example, tapering

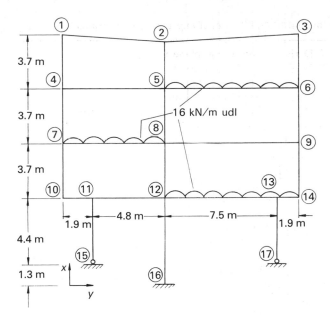

Figure 3-5   Plane frame example

beams, large joints, thermal expansion or for making dynamic analyses on three-dimensional frames.

Any frame analysis program requires the engineer to prepare data describing the geometry of the frame, the properties of the members and the loads to be applied. This data commonly takes the form of tables of (a) coordinates of the joints, (b) members and their properties, (c) information about supports, and (d) loading.

To enable these tables to be written out, there are a few preliminary tasks which are best done with the aid of an accurate sketch. First, number all joints from one upwards and write them on the sketch. The numbering for a large frame should follow a rational order to simplify the subsequent taking off of data. To minimize computing costs, the difference between joint numbers at the ends of any member should be kept to a minimum. Extra joints may be inserted at points where sudden changes in properties occur.

Secondly, choose a suitable origin for a set of xy-axes. It may be anywhere and often a joint at the lower left-hand side is convenient. Also choose a suitable unit of length and calculate the coordinates of all joints.

Finally, decide on the number of load cases to be analysed simultaneously (we have used one only) and the loads which will make up each case. Code of practice requirements to find the worst case of adjacent or alternate spans loaded will not be determined automatically by frame analysis programs. You must specify each load case and inspect the results to find the worst case.

3.4.1   *Example*

The small frame shown in Fig. 3-5 is a good example of the sort of problem which is difficult to analyse by conventional manual methods but which can so easily be

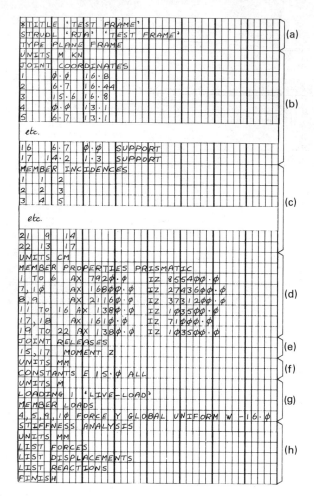

Figure 3-6 Data for STRUDL

solved using a frame analysis program. Although basically a simple two-bay frame, architectural requirements create analytical complications.

The data for the GTSTRUDL analysis for this frame is shown in Fig. 3-6 and explanatory comments appear in Table 3-2. Only one load case has been submitted.

For the engineer who has not used computers before, it may help to draw attention to the mixture of numeric data and English language statements that make up the data for this problem. The former make up the bulk of the actual data set out as tables, each preceded by an explanatory heading. The latter are commands to instruct the program about the task it is performing. Most modern programs follow this pattern, since the commands allow an engineer a great deal of flexibility in his use of the program. A good example of this is the command

Table 3-2   Notes on the frame data

Section	Comments
(a)	The command STRUDL initiates the job and the engineer's name and job title on that line will appear in the results. The command TYPE selects the type of structure such as truss, grid, plane, or space frame.
(b)	Units of metres (M) for all lengths, areas, etc., and of kilonewtons (KN) for all forces are selected and apply until changed. The joint coordinates are set out using the origin shown in Fig. 3-5. Note that any joint which is restrained must be noted as a SUPPORT (e.g. joints 15, 16 and 17)
(c)	In this table, the first number is a member reference number, the second and third numbers are the joints at the ends of that member.
(d)	The units of length only are changed to centimetres (CM) for convenience then the table gives member properties. The members are referred to by the reference numbers allocated in the previous table and it will be clear that giving consecutive numbers to members with the same properties helps to reduce the properties table.
(e)	Joints noted as SUPPORTS in the coordinate table are completely restrained unless selectively 'released' in a JOINT RELEASE table. This example makes joints 15 and 17 into pinned supports.
(f)	Changing length units to millimetres allows Young's modulus (E) to be specified as 15 GPa.
(g)	One load case only is specified very compactly as a uniformly distributed load of 16 kN/m acting vertically downwards. Note that because the positive y-axis points upwards, the *downward* acting load must be *negative*. A popular error.
(h)	The analysis is performed and selected results printed out in the units of millimetres and kilometres.

UNITS which is used several times in the data to change the units in which lengths or properties are given.

Some of the printed results from the analysis are shown in Fig. 3-7. These give the forces and bending moments acting on each member in two lines of results, one for each end. Care must be taken in interpreting the directions of these forces and moments since the sign convention used is strictly mathematical and not that customarily used by structural engineers. To make this clear, consider the bending moment diagram for member 5 which spans between joints 7 and 8. The printed bending moments at those joints are +20 707 and −39 786 kN mm respectively, but, because the sign convention is that a positive moment acts in an anticlockwise direction, both bending moments are 'negative', in the usual engineering convention, i.e. will develop compressive stresses on the underside of the member. Figure 3-8 shows both the sign convention and the resulting bending moment diagram obtained by super-imposing the computed end moments and the free bending moment diagram. This conversion of bending moment results is usually

```
TEST FRAME                                        PAGE     4


***************************
*RESULTS OF LATEST ANALYSES*
***************************


    PROBLEM - RJA        TITLE - TEST FRAME

    ACTIVE UNITS  MM    KN    RAD  DEGF  SEC


--------------------------------------------------------------------
---      LOADING - 1                 LIVE-LOAD
--------------------------------------------------------------------


    MEMBER  FORCES


MEMBER    JOINT    /------------------- FORCE ---------------------/
                       AXIAL            SHEAR Y          BENDING Z
1         1          -1.1518632        1.2740012        163.4953384
1         2           1.1918632       -1.2740012       8384.6258384
2         2           8.0759682        -.6548776       1697.8550960
2         3          -8.0759682         .6548776      -7531.0319129
3         4           4.9659205       -9.1407798      -2638.6239943
3         5          -4.9659205        9.1407798     -58604.6003377
4         5           -.9696106       76.6804848      92562.6190992
4         6            .9696106       65.7195152     -43786.3046642
5         7          -2.1463478       50.7524357      20707.6259222
5         8           2.1463478       56.4475643     -39786.3065707
6         8          -8.8142009        3.2406859      29760.9118238
6         9           8.8142009       -3.2406859       -918.8072779
7         10         -1.6977815      -42.9477703      -1878.0552739
7         11          1.6977815       42.9477703     -79722.7083037
8         11         -2.0997127       13.4462210      77954.2109426
8         12          2.0997127      -13.4462210     -13412.3503523
9         12          1.7408661       47.5743927      11493.8869205
9         13         -1.7408661       72.0251073    -101682.1915447
10        13          1.6879742       85.2067704     101449.4670614
10        14         -1.6879742      -62.8067704       2160.0114728
11        1           1.3361143        1.1217912       -163.4953384
11        4          -1.3361143       -1.1217912       4314.1228670
12        4          -7.8046654       -3.8441292      -1675.4988727
12        7           7.8046654        3.8441292     -12547.7792704
13        7          42.9477703       -1.6977815      -8159.8466517
```

Figure 3-7   Some results from the STRUDL analysis

Positive directions of forces and moments

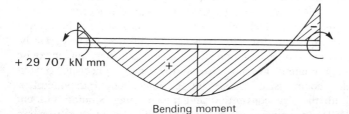

+ 29 707 kN mm

Bending moment

Figure 3-8   Results for member 7–8

necessary after using a frame analysis program, but some programs make the conversion automatically. The users' manual should be checked for this. Displacement and reaction results are quite simple to interpret.

3.4.2 *Costs*
It is not possible to publish formulas or guides for the cost of any analysis but an engineer considering using a computer bureau for such a job should obtain an estimate beforehand. It may be of interest that this sample frame cost £10 (US $20) to analyse. In general, the cost increases at least linearly with the number of joints or members, but only marginally with the number of load cases. It is therefore worth while submitting a generous set of load cases in one computer run when searching for worst possible load combinations.

3.5 Applications
In this section, we illustrate some of the ways in which a frame analysis program can be used to tackle some unusual problems.

3.5.1 *Variable section beams*
A few programs allow for the inclusion in a frame of members with varying cross-sections, but most do not. When this facility is absent, good results can be obtained by dividing such a member into several prismatic segments each with properties of the corresponding average cross-section.

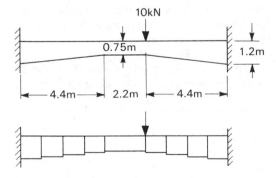

Figure 3-9 A haunched beam and its equivalent frame

The haunched beam in Fig. 3-9 was analysed after subdividing each haunch into two, four and eight segments. The reaction and moment at the right-hand support are compared in Table 3-3 with the exact values.

Table 3-3   Haunched beam results

No. of segments	2	4	8	Exact
Reaction, kN	6.67	6.71	6.72	6.72
Moment, kN mm	−19 574	−20 079	−20 210	−20 256

As can be seen, subdividing each haunch into four segments gives results within 1% of the exact value. A warning should be sounded however that a frame with very short and stiff segments combined with longer flexible members could yield inaccurate solutions. A general rule is to avoid the values of EI/l^3 varying by a ratio greater than 10^6 where E is Young's modulus, I the second moment of area, and l the length of the member.

3.5.2 Curved members
In a similar fashion, curved members can be represented by several straight segments and, again, good results can be obtained with few subdivisions. The pinned arch of Fig. 3-10 was analysed under three loading cases of a partial

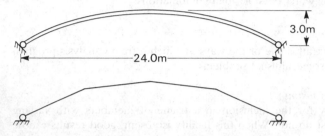

Figure 3-10 An arch and its equivalent frame

uniformly distributed load of 20 kN a central point load and a quarter point load both of 10 kN. The arch was analysed twice after being subdivided into four and eight straight segments. The horizontal and vertical reactions at the left-hand support are compared with the exact values in Table 3-4.

A beam curved in plan can be similarly analysed as a grid of straight, prismatic segments. A rigorous treatment of curved members is given in Section 3.6 of [11].

Table 3-4   Pinned arch results

No. of segments		4	8	Exact
Partial udl	V	12.59	12.62	12.59
	H	29.66	28.47	28.41
Central load	V	5.00	5.00	5.00
	H	16.32	15.62	15.58
Qtr point load	V	7.50	7.50	7.50
	H	11.65	11.18	11.15

V = vertical reaction (kN) at left-hand support
H = horizontal reaction (kN) at left-hand support

3.5.3 Prestressed frames
A particularly simple method of allowing for the effect of curved prestressing tendons is to use the equivalent load method [14].

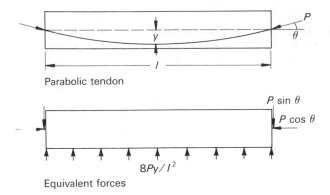

Parabolic tendon

Equivalent forces

Figure 3-11   A prestressing tendon and its equivalent force system

Apart from the point loads which represent the axial compression of a prestres-sing force, any change in angle of a tendon generates a transverse force acting normally to the tendon. Thus, the parabolically curved tendon of the beam in Fig. 3-11 produces a uniformly distributed reaction on the beam of $8Py/l^2$ and end forces as shown. Combining these equivalent loads with the dead and live loads is a simple matter using a frame analysis program. If some prediction can be made of the loss of tendon force due to friction, then this effect can also be represented by further axial forces along the line of the tendon equivalent to the incremental loss of prestressing force.

Prestressed frames can also be analysed by the same method. The portal frame in Fig. 3-12 has a parabolic girder tendon and straight column tendons. These generate the equivalent loads shown in the lower diagram and which may be taken as a load case to be combined with all applied loads. Both Figs 3-11 and 3-12 are based on diagrams in [14].

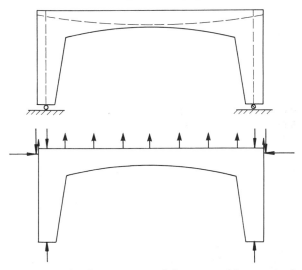

Figure 3-12   A prestressed frame and its equivalent forces

3.5.4 *Shear walls with openings*

MacLeod [15] has shown that shear walls with openings can be represented by an 'equivalent' frame using special structural members which have infinitely rigid end segments to simulate cross-wall sections. Thus, in Fig. 3-13, based on MacLeod's paper, the wall is represented by a frame in which the columns are normal members with properties corresponding to the outside sections of the wall. The horizontal beams have central flexible segments with properties corresponding to the cross-wall sections and rigid end segments to allow for the 'thick joint' effect of the connection with the column sections. Many frame analysis programs have the ability to include this type of member, which is useful whenever a large joint is encountered.

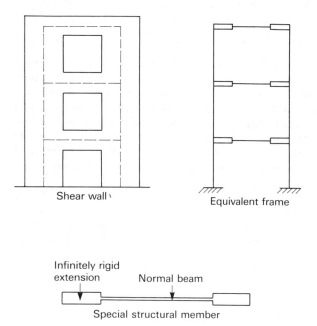

Shear wall

Equivalent frame

Infinitely rigid extension   Normal beam

Special structural member

Figure 3-13 A shear wall and its equivalent frame

4 Finite element method of stress analysis

4.1 Introduction

The finite element method is now widely accepted as a method of analysis in engineering practice. It is used in many engineering fields [17, 18], which include fluid mechanics and heat conduction as well as structural analysis. However, the finite element method originated in the structural field and there is little doubt that its commonest application still lies there [19].

During its development [20, 21] within the aircraft industry, it was initially regarded merely as a generalization of the well-established stiffness method of structural analysis. The significant step, introduced for the study of airplane wings, was the representation of the skin of the wing as an assemblage of thin plate elements. This was an important advance as it required the formulation of a stiffness relationship for elements of a continuum (the skin), and most of the subsequent development of the method (Clough [22] appears to have been the first to use the term 'finite element') has been directed towards its use for continuum problems. The basic finite element method is particularly suited to the analysis of elastic continua such as shear walls, slabs or shells. Many general purpose finite element programs are available which will provide solutions to such problems.

Analytical techniques aimed at accurately predicting the stresses and deformations in reinforced concrete members or structures are complicated due to many factors. Among these are the non-linear load–deformation behaviour of concrete, the progressive cracking of concrete under increasing loads, the bonding interaction between the steel and the concrete, and the time-dependent effects of creep and shrinkage. Because of these complexities, early analytical studies of reinforced concrete were based either on simplifying assumptions, such as linear elastic material behaviour, or on an empirical approach, using the results from large numbers of experiments. The use of the finite element method as a technique for the evaluation of the stress, strain, and the deflection and cracking of reinforced concrete structures has recently attracted increasing interest [23–27]. While the ability accurately to model the behaviour of reinforced concrete structures is of obvious practical value for design offices, it must be emphasized that the application of the finite element method in this area is still being researched and is outside the scope of the generally available finite element packages.

In this section, we commence with a brief description of the finite element method, followed by a discussion of the facilities offered by commercially available programs. We then go on to discuss the growing interest that has been shown in recent years in the application of the method to the analysis of reinforced concrete structures, particularly with respect to the influence of non-linear behaviour and cracking.

4.2 An outline of the finite element method

Stated simply, the finite element method is an approximation in which an actual continuum is replaced by an assembly of discrete elements, referred to as finite elements, connected together to form a structure, which may then be analysed by the standard stiffness method (already described in Section 3.2).

The finite element method may be thought of as an extension of matrix methods for skeletal structures to the analysis of continuum structures (Fig. 4-1). In the finite element method, the continuum is idealized as a structure consisting of a number of individual elements connected only at a limited number of points, known as nodes. The method assumes that if each element's load–deformation characteristics can be defined, then by assembling the elements the characteristics of the complete structure are reproduced. While the assembly of discrete elements can be analysed exactly, it must be remembered that an approximation, and therefore an error, is introduced into the solution of the original problem as soon as the continuum is replaced by these discrete elements. Clearly, successful

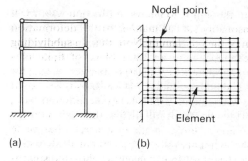

Figure 4-1 (a) Skeletal structure
           (b) Continuum structure

application of the method depends upon careful idealization of the continuum and correct formulation of the element load–deformation characteristics.

The analysis of a continuum by the finite element method may be thought of as a three-stage process [28]:

(1) Structural idealization. The continuum is idealized by using imaginary lines (or surfaces) to divide it into a number, often a large number, of two (or three) dimensional elements. These elements (Fig. 4-1b) are assumed to be interconnected at a number of nodal points, usually situated on their boundaries. The displacements of these nodal points become the basic unknowns of the problem, similar to the conventional structural stiffness method.

(2) Formulation of the element stiffness matrix. A function is chosen to define uniquely the state of displacement within each element in terms of its nodal displacements. This is done by using a displacement function to specify the pattern in which the element is to deform. On the basis of this displacement function, the element stiffness matrix, which relates the element nodal forces to the element nodal displacements, is derived using the principle of virtual work or the principle of minimum potential energy.

(3) Solution by standard matrix stiffness procedure. The overall structural stiffness matrix is assembled from the individual stiffnesses of the discrete elements and the nodal displacements are solved for the loading case under consideration. From these nodal displacements, the internal element displacements at any point, and hence the internal strains and stresses, may be determined.

Once the second stage of the process has been completed the solution procedure follows the standard stiffness method for discrete systems, described earlier in the section dealing with frame analysis. The choice of displacement function is perhaps the most crucial part of the finite element method, since it determines the performance of the element in the analysis; a good displacement function will lead to an element of high accuracy with converging characteristics, and, conversely, a wrongly chosen function can lead to results which converge towards an incorrect solution for successively finer mesh divisions of the continuum.

In selecting a displacement function, the aim is to choose a function which will allow the element to deform in a reasonably similar manner to the deformation developed in the corresponding region of the continuum. Note that, in subdividing a continuum—a concrete slab, for example—into an assemblage of finite elements, we are not just replacing the continuum with a large number of smaller pieces of concrete connected at their nodes only. Indeed, if we did so, we would find the behaviour of the assemblage of elements would be quite different from that of the original slab. In fact, what we do is to divide the slab into an assemblage of discrete elements which are no longer made of concrete, but made of such an imaginary material that the elements have displacement fields which closely correspond to those in the real structure. It is from these chosen displacement functions that the element stiffness relationships are developed.

The displacement function is commonly expressed either as a polynomial with undetermined coefficients which are subsequently evaluated in terms of the nodal displacements, or directly in terms of series of shape functions. Physically, the shape function associated with a nodal displacement gives us the displacement field over an element when that particular nodal displacement is given a unit value and all others are given zero values.

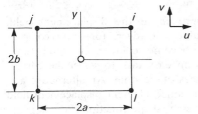

Figure 4-2   Rectangular element

Consider the four-noded rectangular element, *ijkl*, shown in Fig. 4-2. Suppose the displacement at any point (x, y) has two components, u and v, in the x- and y-directions, respectively. These displacements, u and v, uniquely define the internal displacements of the element. The simplest approach to the analytical description of this element's deformation behaviour is by means of a polynomial series:

$$u = \alpha_1 + \alpha_2 x + \alpha_3 y + \alpha_4 xy$$
$$v = \alpha_5 + \alpha_6 x + \alpha_7 y + \alpha_8 xy \tag{4-1}$$

or

$$\begin{bmatrix} u \\ v \end{bmatrix} = \begin{bmatrix} 1 & x & y & xy & 0 & 0 & 0 & 0 \\ 0 & 0 & 0 & 0 & 1 & x & y & x \end{bmatrix} \begin{bmatrix} \alpha_1 \\ \alpha_2 \\ \alpha_3 \\ \alpha_4 \\ \alpha_5 \\ \alpha_6 \\ \alpha_7 \\ \alpha_8 \end{bmatrix} \tag{4-2}$$

where α_1 to α_8 are constant coefficients whose values depend upon the nodal displacements. Note that the number of the unknown coefficients (eight) corresponds to the total nodal degrees of freedom. The evaluation of the coefficients in terms of the nodal displacements is accomplished by applying the polynomial expressions of Eqn 4-1 at each of the nodal points. For example, at node i, the nodal displacements in the x- and y-directions are given by

$$u_i = \alpha_1 + a\alpha_2 + b\alpha_3 + ab\alpha_4,$$
$$v_i = \alpha_5 + a\alpha_6 + b\alpha_7 + ab\alpha_8$$

Three further pairs of equations are obtained by applying Eqn 4-1 at the remaining nodes and these eight equations can be solved to give the unknown coefficients, α, in terms of nodal displacements (u_i, v_i, u_j, v_j, etc.). Substitution back into Eqn 4-1 gives expressions for the displacements u, v at any point (x, y) within the element in terms of the nodal displacements. For example, it will be found that the expression for the u-component is

$$u = \frac{1}{4ab}[(a+x)(b+y)u_i + (a-x)(b+y)u_j + (a-x)(b-y)u_k + (a+x)(b-y)u_j]$$

$$(4\text{-}4)$$

For simple elements, the polynomial approach to the displacement field, outlined above, proves useful and manageable. However, for more complex elements, the mathematical manipulations become cumbersome and time-consuming, and furthermore it is not always an easy task to select the correct polynomial terms for a given number of degrees of freedom. An alternative and often preferable approach is to express the assumed element displacement field directly in terms of the nodal displacements using 'shape functions'.

Suppose we define an element's displacement field by

$$\Delta = N_1(x, y)\Delta_1 + N_2(x, y)\Delta_2 + \cdots + N_n(x, y)\Delta_n + \cdots \qquad (4\text{-}5)$$

where Δ_n are the nodal displacements and $N_n(x, y)$ are the corresponding shape functions. The aim is to represent our displacement as a combination of simple functions. In selecting each of the shape functions, we choose a function which takes on the value of unity when evaluated at the coordinates of the node with which it is associated and is zero at any of the remaining nodes. With this as a basic requirement, the reason for calling $N_n(x, y)$ a shape function can now be seen: it represents the shape of the displacement field, Δ, when plotted over the surface of the element for $\Delta_n = 1$ and all other nodal displacements are held to zero. For a rectangular element, the displacement field corresponding to unit displacement at node j is shown in Fig. 4-3; this, of course, represents the shape function associated with node j.

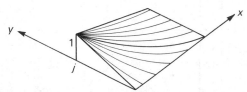

Figure 4-3   A shape function

Consider again the rectangular element of Fig. 4-2, but suppose a suitable polynomial function of Eqn 4-1 is not known. In shape function form, we may now define our displacement field as

$$u = N_1 u_i + N_2 u_j + N_3 u_k + N_4 u_l$$
$$v = N_5 v_i + N_6 v_j + N_7 v_k + N_8 v_l$$

or

$$\begin{bmatrix} u \\ v \end{bmatrix} = \begin{bmatrix} N_1 & N_2 & N_3 & N_4 & 0 & 0 & 0 & 0 \\ 0 & 0 & 0 & 0 & N_5 & N_6 & N_7 & N_8 \end{bmatrix} \begin{bmatrix} u_i \\ u_j \\ u_k \\ u_l \\ v_i \\ v_j \\ v_k \\ v_l \end{bmatrix} \tag{4-6}$$

where N_1 to N_8 are the shape functions for this eight degree-of-freedom element. Comparison of Eqn 4-6 with Eqn 4-4 suggests that suitable shape functions are

$$N_1 = \frac{(a+x)(b+y)}{4ab} \qquad N_2 = \frac{(a-x)(b+y)}{4ab}$$

$$N_3 = \frac{(a-x)(b-y)}{4ab} \qquad N_4 = \frac{(a+x)(b-y)}{4ab} \tag{4-7}$$

Each of the shape functions of Eqn 4-7 has a value of unity at one node and zero at all other nodes. The point to be made is that these equations represent an application of Lagrange's interpolation to a two-dimensional problem. In general, the polynomial form of the displacement function of Eqn 4-1 may be omitted; instead, the assumed displacement field may be written directly in terms of the nodal displacements using shape functions, as in Eqn 4-6. The necessary shape functions may be devised by intuition, trial, and familiarity with similar but simpler elements.

In general, a good displacement function (or a set of functions) should satisfy the following conditions:

(a) The displacement function, and its first derivatives, should be continuous within the element.
(b) The displacement function should not have any preferred directions. That is, under any set of loads having a fixed orientation to the element, the element's response should be independent of how it and its loads are orientated in global coordinates.
(c) The displacement function must allow the element to undergo rigid body movement without internal strain.
(d) The displacement function should allow for states of constant strain within the element.
(e) The displacement function should satisfy the compatibility of displacements along the boundaries with adjacent elements.

Once the displacement field has been specified, we may then derive the strains and stresses within the element and formulate the stiffness matrix. The shape function representing the element's displacement field, Eqn 4-6, may be written

$$[f] = [N][\Delta] \tag{4-8}$$

With the displacements known at any point within the element, the strains at any point may be determined. For the case of plane stress,

$$[\varepsilon] = \begin{bmatrix} \varepsilon_x \\ \varepsilon_y \\ \gamma_{xy} \end{bmatrix} = \begin{bmatrix} \dfrac{\partial u}{\partial x} \\[2mm] \dfrac{\partial v}{\partial y} \\[2mm] \dfrac{\partial u}{\partial y} + \dfrac{\partial v}{\delta x} \end{bmatrix}$$

or

$$[\varepsilon] = [B][\Delta] \tag{4-9}$$

where the matrix $[B]$ is obtained by the appropriate differentials of the displacement functions $[N]$ in Eqn 4-8.

Assuming a general elastic behaviour, the relationship between stress and strain will be of the form

$$[\sigma] = [D][\varepsilon]$$
$$= [D][B][\Delta] \tag{4-10}$$

where $[D]$ is an elasticity matrix containing the appropriate material properties. For the particular case of plane stress,

$$[D] = \frac{E}{1-v^2} \begin{bmatrix} 1 & v & 0 \\ v & 1 & 0 \\ 0 & 0 & \left(\dfrac{1-v}{2}\right) \end{bmatrix} \tag{4-11}$$

With the strains and stresses derived from the assumed displacement function, the stiffness of an element can now be formulated using the principle of minimum potential energy. Suppose the element is subjected to concentrated forces $[F]$ applied at the nodes. If these loads cause nodal displacements $[\Delta]$, then the potential energy of the forces in the deformed configuration is defined as

$$W = -[F]^T[\Delta] \tag{4-12}$$

The potential energy within the element, U, is equal to the internal strain energy, so

$$U = \frac{1}{2} \int_{vol} [\varepsilon]^T[\sigma]\, d(vol) \tag{4-13}$$

$$= \frac{1}{2}[\Delta]^T \left\{ \int_{vol} [B]^T[D][B]\, d(vol) \right\}[\Delta]$$

from Eqns 4-9 and 4-10.

The total potential energy of the element is

$$\phi = U + W$$

or

$$\phi = \tfrac{1}{2}[\Delta]^{\mathrm{T}} \left\{ \int_{\mathrm{vol}} [B]^{\mathrm{T}}[D][B]\,\mathrm{d}(\mathrm{vol}) \right\}[\Delta] - [F]^{\mathrm{T}}[\Delta] \qquad (4\text{-}14)$$

Now, the principle of minimum potential energy tells us that, for a prescribed set of displacements and strains, the equilibrium conditions are satisfied when the total potential energy is at its minimum. Differentiating Eqn 4-14 with respect to each of the nodal displacements, Δ, in turn and setting the result equal to zero, we obtain a set of simultaneous equations:

$$\begin{bmatrix} \dfrac{\partial \phi}{\partial \Delta_1} \\[2mm] \dfrac{\partial \phi}{\partial \Delta_2} \\[2mm] \vdots \\[2mm] \dfrac{\partial \phi}{\partial \Delta_n} \end{bmatrix} = 0 \qquad (4\text{-}15)$$

in which n is the total number of degrees of freedom of the element. Differentiating Eqn 4-14 as indicated in Eqn 4-15, we obtain

$$\left(\int_{\mathrm{vol}} [B]^{\mathrm{T}}[D][B]\,\mathrm{d}(\mathrm{vol}) \right)[\Delta] - [F] = 0$$

or

$$[F] = [k][\Delta] \qquad (4\text{-}16)$$

where $[k]$ is the stiffness matrix for the element and is defined by

$$[k] = \int_{\mathrm{vol}} [B]^{\mathrm{T}}[D][B]\,\mathrm{d}(\mathrm{vol}) \qquad (4\text{-}17)$$

This stiffness relationship may also be derived using the principle of virtual work.

So, we see that the stiffness matrix for an element, Eqn 4-17, may be obtained by appropriate manipulation of the element strain matrix $[B]$, derived from the shape function matrix $[N]$, and the element stress matrix $[D]$.

The complete structural stiffness matrix for the continuum is assembled from all the individual element stiffness matrices in the same way as described for discrete structures (Section 3.2).

4.3 Finite element programs: facilities and potential errors

There are said to be approximately 500 finite element programs available currently [29] and an engineer wishing to try the method should have no difficulty finding one at a computer bureau or through the sources listed in the introduction to this chapter. Widely available programs include ADINA, ANSYS, ASAS, ASKA, BERSAFE, MARC, NASTRAN, PAFEC 75 and SAP IV. A brief guide to the scope of these programs has been given by Robinson [30], from whom the addresses of the suppliers shown in Table 4-1 are taken.

Table 4-1 Some suppliers of finite element programs

Package	Suppliers
ADINA	Professor K. J. Bathe Acoustics and Vibration Laboratory Mechanical Engineering Department Massachusetts Institute of Technology Cambridge Massachusetts 02139, USA
ANSYS	Swanson Analysis Systems Inc. 870 Pine Drive Elizabeth Pennsylvania 15037, USA
ASAS	Atkins Research and Development Ltd Woodcote Grove Ashley Road Epsom Surrey KT18 5BW, UK
ASKA	IKO Software Service GmbH 7 Stuttgart 80 Vaihinger Strasse 49 West Germany
BERSAFE	Central Electricity Generating Board Research and Development Department Berkley Nuclear Laboratories Berkley Gloucestershire GL13 9PB, UK
MARC	MARC Analysis Research Corporation 314 Court House Plaza 260 Sheridan Avenue Los Angeles California 94306, USA
NASTRAN	MacNeal-Schwendler Corporation 7442 North Figueroa Street Los Angeles California 90041, USA
PAFEC 75	PAFEC Limited PAFEC House 40 Broadgate Beeston, Nottingham NG9 2FW, UK
SAP IV	Earthquake Engineering Research Centre College of Engineering University of California Berkeley, California, USA

The basic steps to be taken when solving a problem with any of these programs are similar to those described for a frame analysis job in Section 3.4. The engineer must prepare a sketch showing his chosen mesh of elements, the nodes must be numbered and nodal coordinates calculated. However, the step which precedes these and transcends all others in importance is that which appears most difficult to the new user of this technique. This is the selection of the appropriate type(s) of finite element to solve the problem. Whereas a frame is assembled from members which have two joints and well understood structural properties, a slab or solid body to be analysed by finite elements is represented by an assemblage of strangely shaped and unfamiliar components with up to 20 or more nodes and unfamiliar parameters. Thus, we start a discussion of finite element packages by listing the most common elements and making simple recommendations as to their use and value.

Choosing a finite element package should, however, not be based only on the availability of a wide range of elements. Because of the large number of nodes and elements likely to be necessary for an accurate result, the data needed can be voluminous. It has to be said, too, that some programs have less than convenient data preparation rules, e.g. the order in which the nodes of an element are specified. A 'first-time' user is strongly recommended to try a small problem under no time pressure to allow for some learning experience.

This task of preparing data for large problems has led to extra programs (pre-processors) being devised to generate the necessary data from a more compact form, particularly for common problem shapes. These may show the generated mesh on a graphic display screen or plot it on paper.

To a lesser extent, examination of the results of a finite element analysis can also be a major task and some post-processor programs have been devised to assist in this area. Many of these are designed to plot graphs of deflections, stresses, or bending moments along chosen sections or to plot crosses indicating directions and values of principal stresses or moments.

An engineer choosing a finite element package should consider what pre- and post-processor programs are offered in addition to the basic analysis program but no further remarks will be made about them here except to draw attention to the later example which uses a simple mesh generator.

4.3.1 Choice of element type

Part of the power of the finite element method is its application to seemingly very different types of structural problem such as shells, plates or solids. This stems from the mathematical development of the stiffness equations appropriate to each type of structure and it provides a first classification of element types. Table 4-2 lists the basic types of structure with some examples. Finite elements have been developed for all the types listed therein but potential users should note that the complexity (and therefore the cost) of an analysis will increase as one descends through that list.

A second classification of elements is by shape and number of nodes but this also involves the choice of mathematical technique used to set up the stiffness equations. To some extent, the range of shapes offered in a program will reflect the historical, research and practical interests of the programming team. To guide new users through this maze, it is suggested that the elements used be based on

Table 4-2 Structural types

Type of structural problem	Definition and example
Plane stress	2-D problems with zero *stress* normal to plane Example: stresses around holes in a concrete beam
Plane strain	2-D problems with zero *strain* normal to plane Example: stresses in a gravity dam
Plate bending or slabs	2-D problems with bending out of plane, sometimes extended to include in-plane stresses Example; flat slabs, bridge decks
Axisymmetric solids	3-D problems with both structure and loading formed by cylindrical roation about an axis. This allows of a 2-D treatment where an element represents a solid ring of material Example: cylindrical vessels
Shells	3-D problems with bending combined with in-plane stresses and curved in one or two directions Example: shell roof
Solids	General 3-D problems Example: stresses around openings in thick walled vessels

the assumed displacement approach as described above in Section 4.2. Other methods have been developed but have not generally shown a consistent superiority. This method includes the useful isoparametric approach which allows multi-noded, curved sided elements where appropriate.

In Table 4-3, we show for each type of structural problem, a few types of element which new users could consider using. These are well tested and reliable results will be obtained with an adequate mesh.

4.3.2 Mesh data

Dividing a structural problem up into a mesh of elements is not an easy task and no general rules exist. Much experience has been gained and advice should be sought from users where possible. Commercial pressures to cut computing costs tend towards coarser meshes but this can be unwise. The finite element method leads to an approximate solution in which the error should reduce as the number of nodes is increased. Convergence to an answer is assured as the mesh is made finer but no formula for estimating errors exists. Indeed, with many element types it is not possible to say whether computed stresses or displacements will converge from above or below.

One simple practical recommendation is to draw the most rectangular mesh possible. This simplifies the calculation of the coordinates and often facilitates the interpretation of results by allowing graphs of stresses or displacements to be more easily drawn. Further considerations are those of representing supports which are of finite size, e.g. large columns supporting a flat slab, and of allowing for patterns of distributed and point loads. This is discussed in a later paragraph.

Table 4-3   Suggested element types: a node is represented by a black bullet; u, v, w are deflections in x, y, z; and θ_x, θ_y, θ_z are rotations about x, y, z

Type of Strutural Problem	Types of element		Displacements at each node
Plane Stress & Plane Strain	curved sides if necessary	membranes and bar	u,v
Plate bending or slabs		plate bending and grillage beam	w, θx, θy
Axi-symmetric solids	curved sides if necessary	solid torus	u, v
Shells	Singly curved shells or folded plates treat as a facetted assemblage of plane triangles or rectangles triangles — — — — — — — — — — — — — — Doubly curved shells leave to experts!	membrane and plate bending	u, v, w θx, θy, θz
Solids	curved sides if necessary	solid bricks and prisms	u, v, w
Notation	• = node	u, v, w deflections in x, y, z θx, θy, θz rotations about x, y, z	

An important step in reducing computing costs is that of careful numbering of the nodes after the mesh has been designed. The rule is to number nodes so that the *difference* in node numbers for any element is minimized. Some programs have automatic renumbering to achieve this. If this facility is not available, it is usually best to number nodes in parallel lines across the shortest width of the problem.

4.3.3   Element data

After preparing coordinate data for all nodes, the elements must be identified in turn by listing their node numbers. Two possible error traps exist in this tedious task. First, for two-dimensional elements, the nodes will have to be listed in either clockwise or counter-clockwise order according to a given specification. Three-dimensional elements also have a required numbering direction. Secondly, the sequence in which the corner and mid-side nodes of multi-noded elements are to be listed is equally important and often not very obvious. Users should check on both points in their manuals.

Supports are usually specified in global directions and few programs allow supports in any other direction. Care should be taken when representing the constraint of an infinitely extending material, such as the soil below a foundation. It is surprising how far a mesh has to be taken before it can be safely constrained without generating adverse effects on the stresses. New elements which represent such 'infinite' regions are being developed.

Facilities for specifying loads are not as generous as for frame analysis programs. Point loads in global directions can only be applied to nodes. Surface loads can only be applied to the whole of an element never to a part, but may be specified in global or local axes, in and normal to the plane of two-dimensional elements. Body forces may be conveniently used to allow for dead load of the specified structural elements. Settlement at supported nodes and expansion/contraction due to shrinkage or temperature changes are usually allowed for as further load facilities.

The design of the basic mesh should allow for the proper modelling of the loads bearing the above considerations in mind.

4.3.4 *Results*

A substantial problem for the user of any finite element program is to be selective in choosing which results should be printed. It is imperative that a choice be available, else the quantity of paper becomes unmanageable. Possible results for each load case or combination of load cases are: nodal displacements, support reactions, stresses or moments calculated for each element, and stresses or moments averaged at each node.

The last is simply the average of the stresses calculated in all the elements that meet at a node and this provides a useful 'smoothing' of the results at the nodes.

A check list of commonly available facilities of finite element programs is shown in Table 4-4.

4.3.5 *Example use of a finite element program*

Rather than analyse a plane-stress problem, which was convenient for the description of the method in Section 4.2, we have chosen a flat-slab problem as an application likely to be of more interest in design offices. Figure 4-4 shows part of a flat-slab building with columns, a beam and a load bearing wall supporting the floor. A conventional analysis, in which strips of the floor are considered as continuous beams, is made difficult by the irregular spacing of the supports.

To analyse this problem, we have used the GENESYS SLAB-BRIDGE program [32] to illustrate a simple pre-processor program which generates all nodal coordinate and element data. However, rather than reproduce the data, we discuss below some of the considerations preceding the actual data preparation and illustrate some parts following the pattern of Section 4.3 when discussing finite element programs generally.

Before preparing data for a problem, a user should prepare a sketch such as that of Fig. 4-4, showing the structural features and the chosen mesh. In choosing the mesh illustrated, several factors were considered. The void was deliberately included in the mesh to produce a full rectangle which can be generated easily.

Table 4-4   Commonly available facilities of finite element programs

Types of structure	See Table 4-2
Mesh data	Pre-processors to generate mesh[a]
	Simplified data to generate regular meshes
	Graphical plot or display of mesh
	Automatic renumbering of nodes[a]
Element data	Range of element types—see Table 4-3
Supports	Global directions usual
	Other directions uncommon[a]
Loads	Concentrated loads at nodes only
	Distributed loads over *whole* of an element
	Body forces (useful for dead load)
	Settlement of supports
	Expansion/contraction[a]
General	Print-out of input data
	Selective print of results
	Post-processor to plot selected results on specified sections
	Natural frequencies and modes of free vibrations
	Response to time-varying force[a]

[a] Facility provided sometimes

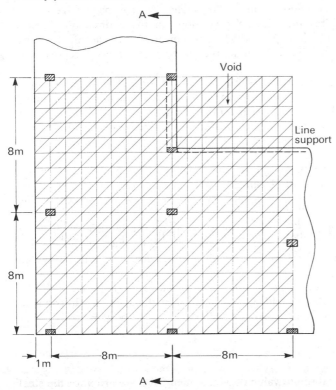

Figure 4-4   Flat slab example

Within the void region, the slab is given a thickness of 1/1000th of the true thickness. The mesh was also designed to match the position of the supports and to allow six to eight elements within the spans. The data to generate the nodal coordinates and element node information was three lines only, as below:

DIRECTION	NUMBER	STEP, M
'L'	17	1
'W'	16	1

This specifies a mesh of 17 steps of 1 m in length (L) by 16 steps by 1 m in width (W).

The triangular, plate-bending element used in this analysis is that of Bazeley *et al.* [33], and this assumes three unknown displacements or degrees of freedom at each node, often referred to as w, θ_x and θ_y, where w is the displacement normal to the x, y plane, θ_x the rotation about the x-axis, and θ_y the rotation about the y-axis. Supports must be defined in terms of restraining some or all of these displacements at specified nodes of the mesh. The columns of this problem were deemed to act as supports restraining w at the appropriate nodes, and their resistance to bending was included as a restraint to θ_x and θ_y. The wall was represented by a line of point supports.

The problem as analysed is only part of the complete structure and continuity with the adjacent parts must be represented along the top right-hand edges. This requires 'supports' of an unusual kind. Thus, along the right-hand edge of the mesh, continuity requires that there be no slope in the x-direction. This is represented in the data by a support which constrains the θ_y displacement, i.e. no rotation about the y-axis. Similarly, along the top edge continuity is represented by constraining θ_x.

Distributed loads can be applied, when using the SLAB-BRIDGE program, to rectangular areas enclosing many elements. Other programs achieve the same by asking for a list of elements to which a distributed load is to be applied. For convenience, the load of 20 kPa for the analysis was applied over the whole mesh and then cancelled over the region of the void by applying there a further load of −20 kPa.

The results which can be obtained from such an analysis include bending moments per metre width, displacements at nodes and reactions at supports. We illustrate in Fig. 4-5 the bending moment diagram along section AA of Fig. 4-4.

As well as calculating the bending moments per metre width, M_x and M_y, along the coordinate axes, a finite element analysis will yield values of the twisting moment, M_{xy}. This may be unfamiliar to engineers and thus attention is drawn to a means of providing reinforcement to resist M_x, M_y and M_{xy}.

Wood [34] has derived some simple rules to calculate M_x^* and M_y^*, the moments of resistance that should be provided for in the x and y-directions to resist all moments in the most economic fashion. Thus, for bottom steel

$$M_x^* = M_x + |M_{xy}|$$
$$M_y^* = M_y + |M_{xy}|$$

where $|M_{xy}|$ means the absolute value of M_{xy}. Similar rules are given for top steel

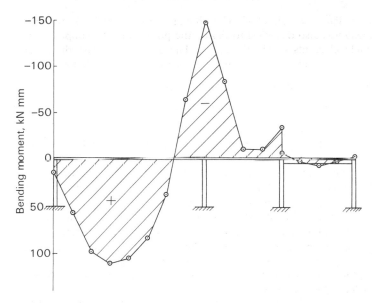

Figure 4-5 Bending moments along section AA

and also for the situations near corners, where negative values of M_x^* or M_y^* might arise. The rules are also extended to cover the case of reinforcement laid at other than right-angles.

Further discussion on this method and a useful set of slab examples compared with model experiments will be found in [31].

4.4 Application of the method to reinforced concrete structures

The finite element method, as outlined above, is based upon a linear elastic material. Reinforced concrete is not such a material and much research effort has been spent in developing the method to cope with the non-linear properties of concrete, with the effect of cracks and with the slip which occurs between steel bars and concrete.

These phenomena will be amplified below and some examples given in Sections 4.4.1 and 4.4.2, but because no generally available finite element package offers facilities for the proper representation of these factors, some remarks are also made about the use of simple linear analysis.

The behaviour of concrete under complex stress combinations is remarkably little understood, particularly in terms of relationships between stress and strain but also for stress failure criteria. Many researchers have turned to the experimental studies of Kupfer *et al.* [35] for basic data. Thus, Wanchoo and May [36] assumed linear elasticity but derived crushing and cracking criteria from Kupfer's data, while Phillips and Zienkiewicz [27] used non-linear stress–strain relationships in terms of a bulk modulus and a shear modulus and also biaxial failure envelopes derived from the same source of data.

An alternative approach is to assume a simple uniaxial stress–strain relationship and apply it to the principal stresses computed at each point in the concrete.

Krishnamoorthy and Panneerselvam [37] used the parabola-rectangle relationship of CEB–FIP. Robins and Kong [38] used the following relationship proposed by Desayi and Krishnan [39]:

$$\sigma = \frac{E\varepsilon}{1+(\varepsilon/\varepsilon_0)^2} \qquad (4\text{-}18)$$

where σ is the stress at any strain ε, ε_0 the strain at maximum stress σ_0, and E the initial tangent modulus $(=2\sigma_0/\varepsilon_0)$. As stresses develop, so these relationships are used to determine new, orthotropic moduli of elasticity in the principal stress directions.

The abrupt change in the properties of concrete as tension cracks form is commonly represented by assuming zero stiffness across the cracks and full stiffness along cracks. These elastic properties are then transformed to the global directions of x and y. The shear stiffness is reduced by some researchers by an arbitrary factor, arguing that aggregate interlocking preserves only some shear stiffness. General agreement is found between experiments and theory for crack patterns and load–deflection curves [27, 37, 38]. A particular application is described in Section 4.4.2.

The behaviour of steel reinforcement is well understood in its linear elasticity and defined yield criterion but not so well understood is the bond between steel and concrete. Upper limits for bond stress are confidently used by all designers but little is known about the relative slip between steel and concrete at lower bond stresses. Yet a proper representation of this element of elasticity is needed for correct modelling of reinforced concrete. Many researchers overlook this and assume infinite bond to exist. A better approach is to assume springs between bars and concrete but little data is available to provide the appropriate stiffnesses of those springs. The influence of bond springs on reinforcement stresses is shown in Section 4.4.1.

4.4.1 Bond

The bond between concrete and reinforcement is considered by designers only when checking that calculated bond stresses do not exceed permitted values. No attention need be paid to the relative slip between steel and concrete which must occur to develop the bond stress. Indeed, very little information is available about the bond stress/bond slip relationships needed to make such calculations and this is a major difficulty when allowing for bond behaviour in a finite element analysis. Experimental measurement of the phenomenon is not easy. Thus, simple pull-out tests put the concrete in compression whereas in practice the concrete is most likely to be in tension and cracked. Figure 4-6 shows three stress–slip relationships quoted in [40–42] and the variation between these is immediately obvious.

Finite element analyses which allow for bond use simple springs to represent the chosen stress–slip relationship. In Fig. 4-7, two springs are shown connecting a node on a reinforcing bar with two nodes in the concrete. The lengths of the two springs can be reduced to zero since only the resulting stiffness is needed for the analysis. Thus, the two concrete nodes become one with the same coordinates as the bar node. The stiffness of the horizontal spring is derived from the slope of the bond stress–slip curve assuming an average bond stress acting over the surface of the bar half-way to the adjacent bar nodes. Of course, this stiffness must be

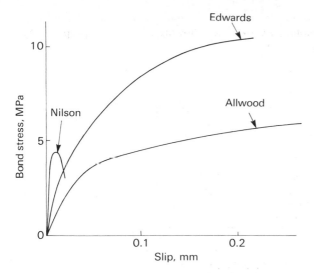

Figure 4-6   Three relationships between bond stress and bond slip

modified to follow the non-linear stress-slip relationship, usually by applying loads in increments. A large stiffness is given to the vertical spring to ensure that concrete and steel deflect together vertically.

An example showing the importance of representing bond correctly is illustrated in Fig. 4-8, taken from Allwood [40]. This shows an intersection of a beam and column in a multistory frame and the study was directed at determining the distribution of direct stress in the beam bars as they passed through the column. The beams on either side were loaded by a uniformly distributed load. The conventional view of this problem is that the column acts as a distributed reaction which 'rounds-off' the otherwise peaky form of the bending moment diagram in the region of the column. A finite element analysis of the problem was made with bond springs which represented the lowest of the three curves in Fig. 4-6. The computed steel stresses within the region of the column are shown, together with

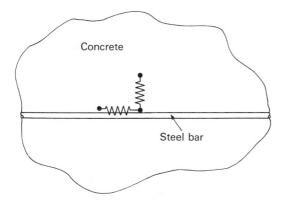

Figure 4-7   Springs simulating bond and dowel effect

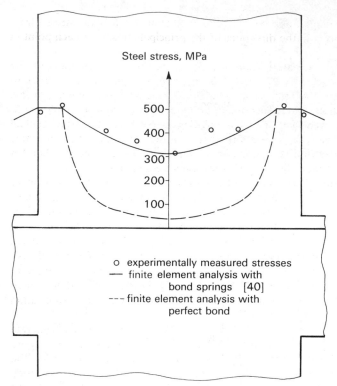

Figure 4-8   Steel stresses in a beam–column intersection

experimental measurements and also computed stresses assuming perfect, infinitely stiff bond. The importance of allowing for bond will be seen immediately as well as the good agreement obtained with the experimental results. In both finite element analyses, advantage was taken of the experimental results in allowing zero bond over the first 30 mm of the column, as shown from the measured stresses. This is probably due to tension cracks in the column at the root of the beam. An explanation for the rapid fall-off of steel stress within the column is found by considering each beam to be loaded separately. The fall-off of steel stress is then seen to be simply the natural decay of stress in an anchored bar due to the bond transferring the steel force to the concrete. The sum of the effect of both beams produces the stress distribution shown in Fig. 4-8.

4.4.2 Concrete non-linearity and cracking

One of the major factors that complicates the analysis of a reinforced concrete structure is the non-linear stress–strain characteristics of the concrete. Despite the widespread use of structural concrete, knowledge regarding concrete behaviour under all combinations of stress conditions is still incomplete. In particular, the majority of experimental investigations for biaxial and triaxial conditions have concentrated on strength characteristics to the exclusion of stress–strain relationships. Many researchers have made reference to the experimental studies of Kupfer et al. [35] for the biaxial stress–strain relationship and failure criterion of plain concrete. The simplest approach [37, 38] is to assume a uniaxial stress–strain relationship and apply it in the directions of the principal stresses at each point in the concrete.

To include the effect of material non-linearity in the finite element method, the most common approach [26, 37, 43, 44] is to apply the load to the structure incrementally, varying the stiffness at each load increment to account for the non-linearity. During each load increment, an iterative procedure is employed to take account of the non-linearity by calculating at each iteration the residual unbalanced nodal loads, and re-apply them to the structure. This iterative procedure is repeated until the unbalanced nodal loads are smaller than an acceptable value, and then the next load increment is applied.

Probably the single most important factor contributing to the non-linear behaviour of a reinforced concrete structure is the formation and propagation of tensile cracks, coupled with associated stress redistributions. Phillips and Zien-kiewicz [27] have pointed out that there are many situations where the behaviour up to useful limits of a structure can be predicted adequately by a study of crack propagation alone; the crushing of the concrete and the yielding of the reinforce-ment can often be regarded as secondary roles which occur in the later stages to precipitate final collapse. Robins and Kong [38, 45] successfully modelled deep beam behaviour with a finite element model which predicted crack propagation on a limiting tensile strain criterion.

The application of the finite element analysis to the study of reinforced concrete beams was first reported by Ngo and Scordelis [46]. Linear elastic analyses were performed on beams with predefined crack patterns. Nilson [41] extended the method to consider cracking by nodal separation. By physically splitting a node and allowing a separation to develop along a boundary between elements, cracking and crack propagation is, of course, restricted. Furthermore, nodal splitting results in computational inefficiency because of the continual change in the topology of the problem.

An alternative approach, adopted by Robins and Kong [38, 45], is to take the cracking and non-linear behaviour into account by altering the material property matrix used to describe the concrete during the development of the stiffness matrix. In the analysis, at each load increment, the principal stresses and strains are computed for each element. These are then compared with the chosen cracking criterion and if this is met or exceeded, a crack is assumed to develop through the element. Once cracked the element stiffness matrix is modified; the element is assumed to be unable to sustain any normal stress perpendicular to the direction of the crack, and its stiffness in this direction is reduced to zero. The effect of this simple approach to cracking is to produce cracked regions rather than discrete cracks, though, as Cedolin and Poli [44] have indicated, this consequence may be corrected by allowing a more gradual release of stress in the direction normal to the crack.

For the approach to cracking outlined above, the elasticity relationship (see Eqn 4-10) for a cracked element becomes

$$
\begin{bmatrix} \sigma_1 \\ \sigma_2 \\ \tau_{12} \end{bmatrix} = \begin{bmatrix} E & 0 & 0 \\ 0 & 0 & 0 \\ 0 & 0 & G' \end{bmatrix} \begin{bmatrix} \varepsilon_1 \\ \varepsilon_2 \\ \gamma_{12} \end{bmatrix}
\tag{4-19}
$$

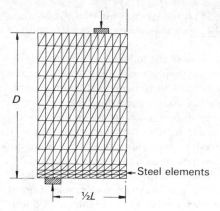

Figure 4-9 Finite element deep-beam model: $L/D = 1.0$, 288 elements, 195 nodes

where 1 and 2 are the directions parallel and perpendicular to the direction of the crack, and G' is a reduced value of the shear modulus, introduced to represent aggregate interlock. Some investigators [47] use a constant value for G', while others [44] choose to decrease G' with increase in crack width.

Phillips and Zienkiewicz [27] have used isoparametric elements under plane and axisymmetric conditions for the non-linear analysis of concrete structures, incorporating special features to simulate reinforcement, liners and prestressing cables. Although the isoparametric element is a more sophisticated element and fewer elements are required, the representation of the steel as bands within the element means perfect bond between the steel and the concrete is assumed.

Figure 4-9 shows a finite element model of a deep beam used by Robins and Kong [38]. They used plane-stress triangular elements and allowed cracks to pass through an element by modifying the element stiffness matrix in the manner

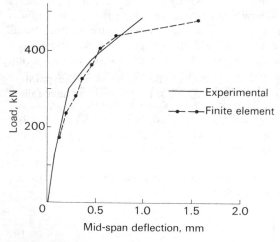

Figure 4-10 Load–deflection curves

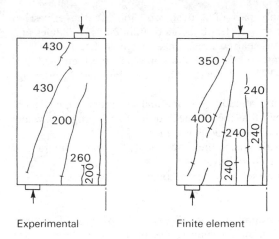

Experimental   Finite element

Figure 4-11   Crack patterns, showing cracking load in /kN and extent of crack

described above. The cracking criterion they used was one suggested by Rosenthal and Glucklich [48] for concrete under a biaxial stress system of tension and compression. The mid-span deflection and crack pattern of the finite element model are compared with the experimental behaviour in Figs 4-10 and 4-11.

That the finite element method can be used to model reinforced concrete structures has been ably demonstrated by many authors. However, it must be remembered that, ultimately, success in applying the method is dependent upon the availability of quantitative information on the load–deformation behaviour of concrete, on the bond stress–bond slip relationship and on the failure of concrete under multiaxial stress systems.

5   Design, detailing and graphics

5.1   Introduction

Many engineers believe that computers will only be used extensively in design offices when they are capable of tackling routine design work to the point of producing quality drawings and production schedules. Hardware costs have fallen, so that it is indeed possible now to hire for less than the cost of an engineer the appropriate equipment. However, experiences during the period 1970–1980 suggest that a great deal of work has still to be done before computer-aided-design (CAD) becomes commonplace. In the UK alone during that time, four major programs were launched for the automatic design and detailing of rein-forced concrete components, particularly beams. Only one survived the decade. A major change in design code was one factor that created difficulties but there are others:

(a) If the computer can only do part of a design, how can the work of the machine and designer be integrated to the satisfaction of the latter?

(b) Small programs doing only part of a designers' job are the easiest to introduce to an office but the greatest benefits of full details, schedules and drawings require the whole-hearted acceptance of a complete system.

(c) Drawings are the immediate 'product' of a design office and their quality is part of the firms image. Will this be tarnished by characterless drawings done by computer?

(d) Should design and detail programs use manual or more sophisticated methods? Experience shows that the engineer-user of such programs wants to know precisely what design decisions are in-built.

(e) How can programs allow designers their own 'style'? Clearly, interactive computing allows this as a possibility.

(f) How can one write programs which can accommodate changes in design codes?

The first question is a major stumbling-block. Ideally, the office CAD system should do the routine work and allow the man to use his skills on the irregular and difficult tasks. But no job splits into such categories. The routine section has a few difficult corners and man must interfere or edit the machine's work. This is difficult to coordinate and not satisfactory unless interactive computing with graphic displays is available to the designer. When design programs with interaction (and which are easy to operate) arrive, then we may see a greater acceptance of CAD. Perhaps the greatest stumbling-block to such progress is the exaggerated esteem which practising engineers have for computers and their mysteries and the low esteem which such engineers have of their own knowledge and skill. Real progress will be made when the balance is restored.

Until then, it is worth while to identify some of the existing developments to illustrate CAD programs at levels from the simplest aid to the full draughting system.

5.2 A simple design aid

A tedious task in any reinforced concrete design is the preparation of schedules specifying the bending of the bars and their weights. A popular yet very simple CAD program which helps a designer prepare these schedules is BARSHED [49, 50].

In its basic mode, the program reads data specifying a group of bars of given shape (specified by the BS 4466 shape codes) and given dimensions, and prints out standard bar-bending schedules with fixing information. The user quickly learns to take advantage of the many shorthand features such as that which allows groups of bars of different dimensions to be derived from one line of data. Thus, a set of floor bars at regular spacing only requires data for the spacing and the total reach and the appropriate number of bars are automatically calculated. Where such bars are required to vary in length, the lengths of the first and last bars are given and the program calculates the length of all intermediate bars. Other shorthand features allow stirrup and link sizes to be calculated from section profiles, lap lengths to be automatically added and fixing notes to be generated from a few codes.

5.3 Design and detailing programs

To extend the analysis of a basic structural component, such as a beam, to yield design information is an obvious step. All the continuous beam programs referred to in Section 2 have a capability to calculate the reinforcement needed and to print proposed bar sizes and patterns. Similarly, for other components such as columns, column bases and retaining walls, programs will be found in the survey by Hutton and Rostron [1].

The development of an analysis program to include reinforcement details requires the consideration of a number of factors, mostly not connected with computing directly. Those listed below are from the present authors' experience but these must be considered against the general considerations pertaining to CAD set out above.

(a) What irregularities, if any, are to be handled by the program? For example, changes in beam widths at column junctions, ribs on beams, holes in slabs, eccentricities at junctions, etc.

(b) What form of editing is to be allowed, e.g. by computer or by razor blade and rubber? If the latter, how are the bar-bending and weight schedules to be edited as well?

(c) What methods are to be used to obtain optimum designs? Engineers will want to know and understand these methods and will compare them with their own. An example is the redistribution of moments in a continuous beam program whereby full advantage is taken of the code of practice provisions. A computer can undertake a much more elaborate redistribution than a man can.

(d) What rules should be followed in arranging the patterns of bars? This is a very difficult problem with practical construction factors posing geometrical constraints. In particular, the intersection of separate components such as beams and columns creates a new dimension to this problem, which will be discussed in Section 5.4.

(e) How are the results to be presented to the designer and to the bar fixers? Computers can draw diagrams similar to those produced by a draughtsman but is that the best way to communicate this information?

As has been indicated in the opening remarks to this section, it is clear that no universally accepted solutions have yet appeared to these problems. Four approaches will therefore be described, showing some of the ideas employed to tackle these problems particularly in relation to the detailing of continuous beams.

The LUCID [51] project was addressed primarily at problems (a), (d) and (e) above. It was a collaborative study involving over 100 British firms and Loughborough University of Technology to develop standard details for reinforced concrete elements, in particular beams, columns and slabs. The objective was to develop a system suitable for manual or computer detailing and flexible enough to replace at least 25% of manual drawings. The developed system separates the drawing of a component into section outlines and parts of the reinforcement. Thus, a beam drawing is built up from four part details showing bottom span bars, shear links and reinforcement at the left-hand and right-hand support. Many versions of each part detail are pre-printed on transparent overlays and when put together can be copied to produce a partly finished drawing. Section outlines and bar details need to be added by hand. The total set of 300 overlays allows many combinations of bar arrangements.

The method of detailing follows the recommendations of the Concrete Society [52] that continuity at beam–column intersections be provided by loose top and bottom support steel thus solving problem (d) albeit at the cost of extra steel. The presentation of results—problem (e)—was, of course, by the standard drawings, which were not to scale. Doubts about their acceptability on site and as an image of the quality of a design office have led to continued development of LUCID towards computer-drawn, scaled drawings based on the same details.

This is also the approach of the BARD [53] suite of programs, which also use details based on the Concrete Society's recommendation. There are programs in the suite for beams, columns and slabs and they may be used interactively or in batch mode. The philosophy of the system is to start with the design information that would normally be passed to a detailer and to produce scaled drawings of high quality in ink on tracing paper. The hardware required for these programs is therefore a computer with a good plotter and conventional interactive terminals, not an unusual or expensive specification. An interesting decision taken during development was to *plot* both the drawing of an element i.e. a beam span, and its associated bar schedule on one A3 (297×420 mm) sheet of tracing paper. The relative slowness and therefore cost of the plotter when used as a printer are justified by the convenience of having both drawing and schedule on one sheet.

The programs refer to a computer file of 'standards' currently based on the recommendations of CP 110 and BS 4466, but which can be amended to conform with other codes, thus tackling a long-standing problem of design-oriented programs. The interactive data input allows default values to be offered to the user to reduce data entry and the command driven nature of the programs allow users to have their own 'style' of detailing. Manual alterations to the drawings may be anticipated by inputting information about extra bars, e.g. for ribs on beams, which will be included in the bar schedule and may be added later to the drawing.

This system provides a solution to most of the problems outlined above at the cost of manual intervention between the analysis–design stage and the detailing stage. The quality of the final product can be judged from Fig. 5-1. Best [53] reports that with the 'incorporation of a small number of hand amendments to the original BARD plots' approximately 75% of the detailing work above ground of several office blocks has been completed by computer.

In Section 2, we discussed programs for the analysis of continuous beams and illustrated the use of one of them, BEAMS/1 [8] from the Genesys library. Many of these programs calculate suitable bar arrangements necessary to provide suitable moments of resistance but few tackle the problems of laps and curtailment involved in working out final details. BEAMS/1 produces such details and contrasts with BARD by doing this without interactive dialogues with the detailer. The program is run in batch mode and arranges bars according to input data, specifying preferred options such as bar sizes and spacing, link sizes and types, and concrete properties, if the built-in values of these factors are not acceptable. This allows a detailer to retain substantial control of the result. The patterns of

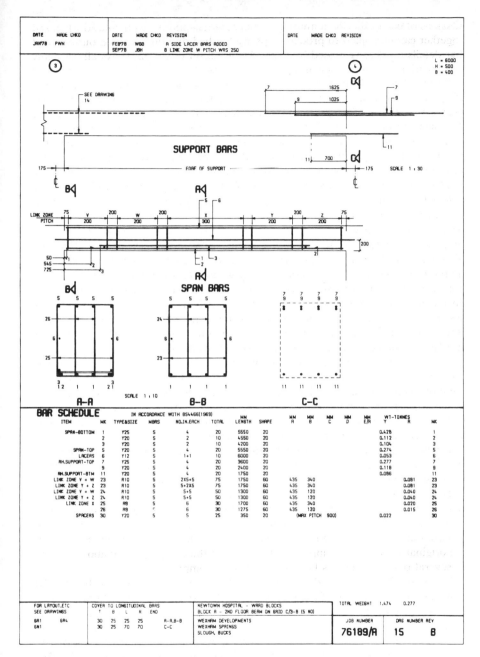

Figure 5-1   Reinforcement detail from BARD

reinforcement used by the program follow a flexible layout represented to users on six standard detail drawings, one of which is shown in Fig. 5-2. This illustrates diagrammatically the maximum reinforcement that *may* be detailed in the end spans of a continuous beam. When the program is used on a particular beam, the bars *actually chosen* for each span are printed out in a bar-fixing schedule identified by bar marks which include the location codes used on the appropriate standard drawing. Thus, bar mark 3B1 means mark 3 location B1, which in turn is identified as *bottom*, layer *one*. By careful choice of suitable mnemonics, the location codes become sufficiently meaningful to be used directly from the bar-fixing schedule.

This approach to the problem of communicating results to designers and bar fixers is a radical step from conventional scaled drawings but has been successfully applied in practice. The program is a good example of completely automatic detailing and produces, very quickly and economically, details as described above and bar bending and weight schedules for continuous beams. Editing both bar-fixing and bar-bending schedules is not difficult for small irregularities but must be done by hand.

Our fourth example of detailing represents the extension of the BEAMS/1 approach to encompass the analysis, design and detailing of complete buildings. In principle, the greatest benefits of computing should spring from situations where the maximum use is made of the input data; building design provides such a situation. Craddock [54] has described the CP 110 suite of programs for building design, also from the Genesys library, which includes the BEAMS/1 program already described. The focal point of this suite is a program which reads the basic data of a building and establishes a database of the positions of grid lines, levels of floors, locations and dimensions of components, loads, material properties, etc. From this base of information, various design and detail programs can function. The basic merit of this approach is that the designer is then free to design/detail sections of the building in any appropriate sequence, making alterations to component sizes and redesigning entirely as he wishes. However, an equally important benefit comes from setting up the database first. The program can make many checks on the input data to ensure that the building 'fits' together and indeed can produce quality general arrangement drawings at this stage. Figure 5-3 shows a general arrangement drawing of a simple building produced by the program.

Having established and verified the input data, the program can be used automatically to distribute floor loads to the beams and then to the columns, providing the basic loading information needed to analyse any component. The designer then selects a single component or group of components and analyses, designs and details them by the appropriate program. Beams, columns, solid one-way slabs, flat slabs and waffle slabs can all be handled. For the slabs, conventional detail drawings are produced by computer; for beams and columns, standard detail drawings as described for BEAMS/1 are used.

Because the database contains the data for all components, the program for detailing beams can take notice of the position of column bars and vice versa. This problem of avoiding bar clashing at component intersections is extremely challenging but important and, of course, not tackled directly by the programs which detail single components.

The successful use of such complete building design programs probably needs a whole-hearted commitment to its use. Changes will be necessary in office practice

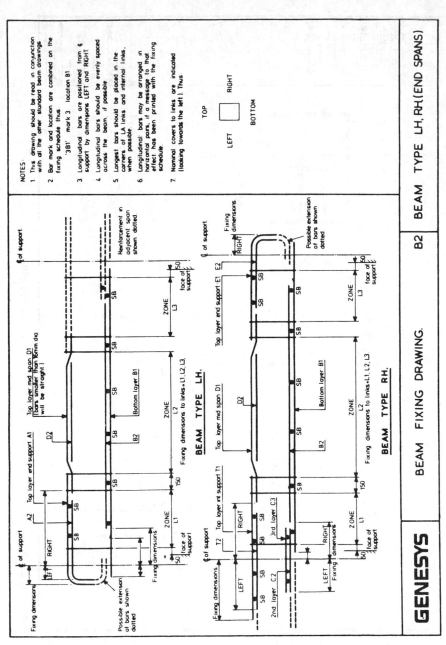

Figure 5-2 A Genesys standard detail drawing

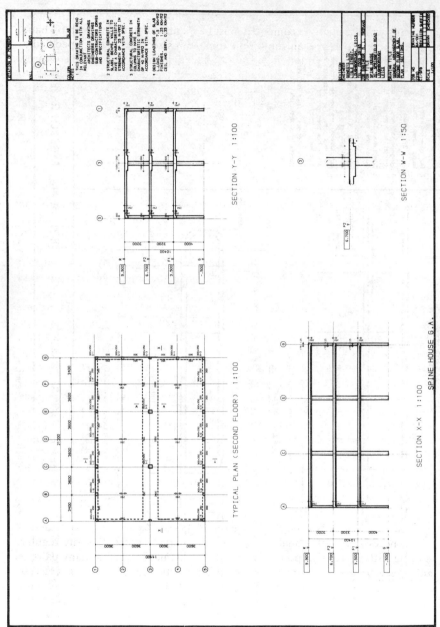

Figure 5-3 A computer-produced general arrangement drawing

to allow time at the right stage for data preparation and checking. Irregular components which cannot be computed must be identified and dealt with by hand. But, since these programs can run on the minimum size of computer specified at the beginning of this chapter, it is quite possible to have the system entirely in-house thus ensuring quick turn-round of computing and easier integration with manual tasks. For straightforward buildings, complete details have been produced in remarkably short times, which alone may be the needed justification in the market place of the eighties.

5.4 Draughting systems

A rapidly developing area of computer application is the production of general drawings with the aid of a computer. The cost trends of computing equipment and the increasingly scarce resource of skilled draughtsmanship have created an opportunity to exploit the combination of man and machine to produce a new design aid. The equipment needed will, of course, include a high-quality plotter, a graphics display screen and, probably, a digitizer—all connected to a dedicated computer likely to be a little more powerful than the minimum configuration spelled-out at the start of this chapter.

The software to drive this equipment, to allow a man to draw, edit and redraw easily, is not simple. A survey by Hamilton and Scoins [55] of six systems illustrates widely different approaches to providing an operator with a general-purpose tool. Hamilton [56] in another survey notes that the six systems were robust and reliable but offered 'many and varied forms of the user interface'.

Broadly, all systems recognize that a drawing is composed of basic elements such as lines, symbols and characters but also that groups of lines can be identified as occurring commonly. How that commonness is taken advantage of and how the basic elements are provided are the sources of variety. One feature of most systems is that the user controls the computer through a menu of commands, either displayed on a screen or drawn permanently on a digitizer. Picking out a command by cursor or pen causes a drawing task to be performed without the user needing to turn to a keyboard. The individual commands will range from 'draw a line from present pen position to a new point' to 'display a predefined object at a given point, rotated through a given angle'.

The facilities to draw are usually based on the concept of a grid of construction lines, not necessarily orthogonal, set at suitable intervals, not necessarily regular. A drawing is then created upon the grid without the need to input many pairs of coordinates. Commands to draw lines to the intersection of grid lines 'nearest' to a cursor or pen position thus eliminate errors due to human inaccuracies in lining up cursor or pen. The grid can be made to disappear before a plotted drawing is produced.

The facilities to take advantage of common 'objects' vary from simple groups of lines and symbols to complex hierarchies of objects defined in terms of other objects, even allowing for dimensions of such objects to be set by input data. Commands allow a copy of each such object to be positioned at a specified point in a drawing and to be enlarged, rotated or reflected. Drawings with a substantial amount of repetition will clearly be simpler to produce with these facilities.

One of the systems reviewed above is GIPSYS, which has been summarized by Williams and Parsons [57]. This system is specifically directed to producing general arrangement plans and reinforcement details for floor slabs and as well as having drawing commands the system can also calculate and print-out bar-bending and cutting schedules. This would seem a natural extension to any drawing system but probably leads to specialized applications. However, early experience by architects of draughting systems suggests that using a computer to produce drawings only is rapidly followed by the wish to perform some analyses on the stored geometric information. Selection of a draughting system should therefore be influenced by that capability.

Recently attention has been focused on simple draughting systems suitable for mounting on a desk-top micro-computer. These provide less comprehensive facilities but are easier to learn about and may prove useful 'starters'. Two such systems are described in the proceedings of CAD82 [58] which like its predecessors is a useful overview of the current state of the art.

The drawing office of the future may well contain a CAD draughting system. The managers of the office will need to consider how it should be introduced and integrated with manual practice. Particular problems will be the policy of whether the system should produce complete drawings or allow hand-completion of difficult details and how manual editing of drawings by erasure can be paralleled by appropriate editing of the electronic equivalent. Above all, the operators will need substantial training.

Acknowledgements

The authors wish to acknowledge the help and computer time given by Genesys Ltd and SIA Ltd, and the permission of the Director of GTICES Systems Laboratory, Georgia Institute of Technology, Atlanta, Georgia to quote the disclaimer from the *GTSTRUDL User's Manual*.

References

1. Hutton, G. and Rostron, M., *Computer programs for the building industry*. Architectural Press, London 1979, 312 pp.
2. Emkin, L. Z. *et al.*, GTSTRUDL user's manual, Georgia Institute of Technology, Jan. 1979, 240 pp.
3. Roos, D. and Miller, C. L., *COGO*-90 *engineering user's manual*, CESL, Massachusetts Institute of Technology, Oct. 1965, 69 pp.
4. Alcock, D. G. and Shearing, B. H., GENESYS—an attempt to rationalise the use of computers in structural engineering, *Struct. Eng.*, Vol. 48, No. 14, April 1970, pp. 143–152.
5. Roos, D., *ICES system design*, Massachusetts Institute of Technology, 1960, 312 pp.
6. British Standards Institution, CP 110: 1972 (as amended 1980), The structural use of concrete, London, 1972, 136 pp.
7. Construction Industry Computing Association (Design Office Consortium),

Computer programs for continuous beams—CP 110, Evaluation Report No. 2, Cambridge, 1978, 64 pp.

8. Genesys Ltd, *User manual for CP* 110—*Beams*/1, Loughborough, 1977, 105 pp.
9. McGuire, W. and Gallagher, R. M., *Matrix structural analysis*, Wiley, New York, 1979, 460 pp.
10. Martin, H. C., *Introduction to matrix methods of structural analysis*, McGraw-Hill, New York, 1966, 331 pp.
11. Livesley, R. K., *Matrix methods of structural analysis*, 2nd Edn, Pergamon Press, Oxford, 1975, 277 pp.
12. Fenves, S. *et al.*, *STRESS: a user's manual*, Massachusetts Institute of, Technology, Sept. 1964, 55 pp.
13. Roark, R. J., *Formulas for stress and strain*, 5th Edn McGraw-Hill, New York, 1975, 624 pp.
14. Nilson, A. H., *Design of prestressed concrete*, Wiley, New York, 1978, 526 pp.
15. Macleod, J. A., Lateral stiffness of shear walls with openings, *Proc. conf. tall bldgs.*, Southampton, 1966, pp. 223–244.
16. Macleod J. A. and Hosney, H. M., Frame analysis of shear wall cores, *J. Struct. Div. Am. Soc. Civ. Eng.*, Vol. 103, No. ST 10, Oct. 1977, pp. 2037–2047.
17. Zienkiewicz, O. C., *The finite element method*, 3rd Edn, McGraw-Hill, London, 1977, 787 pp.
18. Spooner, J. B., Finite element analysis: development toward engineering practicality, *Chart. Mech. Eng.*, Vol. 55, No. 8, Aug. 1976, pp. 96–101.
19. Clough, R. W., Areas of application of the finite element method, *Comput. & Struct.*, Vol. 4, No. 1, 1974, pp. 17–40.
20. Argyris, J. H. and Kelsey, S., *Energy theorems and structural analysis*, Butterworth, London, 1960, 85 pp.
21. Turner, M. J., Clough, R. W., Martin, H. C. and Topp, L. J., Stiffness and deflection analysis of complex structures, *J. Aero. Sci.*, Vol. 23, No. 9, 1956, pp. 805–823.
22. Clough, R. W., The finite element in plane stress analysis, *Proc. 2nd ASCE conf. electronic comput.*, Pittsburgh, Sept. 1960, pp. 345–378.
23. *Finite element methods in engineering, Proc. Int. Conf. Finite Element Methods*, Unisearch Ltd, Kensington, Australia, 1979, 802 pp.
24. Krishnamoorthy, C. S. and Panneerselvam, A., Finite element program for non-linear analysis of reinforced concrete framed structures, *Comput. & Struct.*, Vol. 9, No. 5, Nov. 1978, pp. 451–461.
25. Dunder, V. and Ridlon, S., Practical applications of finite element method, *J. Struct. Div. Am. Soc. Civ. Eng.*, Vol. 104, No. ST1, Jan. 1978, pp. 9–21.
26. Schnobrich, W. C., Behaviour of reinforced concrete structures predicted by the finite element method, *Comput. & Struct.*, Vol. 7, No. 3, June 1977, pp. 365–376.
27. Phillips, D. V. and Zienkiewicz, O. C., Finite element non-linear analysis of concrete structures, *Proc. Inst. Civ. Eng.*, Vol. 64, Part 2, March 1976, pp. 59–88.
28. Coates, R. C., Coutie, M. G. and Kong, F. K., *Structural analysis*, 2nd Edn, Nelson, Sunbury-on-Thames, 1980, 579 pp.

29. Fredricksson, B. and Mackerle, J., *Structural mechanics, finite element computer programs, surveys and availability*, Linköping Institute of Technology, Sweden, 1978.

30. Robinson, K., *Survey of finite element software*, Computing Division, Rutherford Laboratory, Oxford, 1979, 14 pp.

31. Ministry of Transport, *Report on slab bridge analysis by computer*, Vols I and II, HM Stationery Office, London, 1968, 236 pp.

32. Genesys Ltd, *SLAB-BRIDGE/1—user manual*, Loughborough, 1976, 104 pp.

33. Bazeley, G. P., Cheung, Y. K., Irons, B. M. and Zienkiewicz, O. C., Triangular elements in bending—conforming and non-conforming solutions, *Proc. conf. matrix methods struct. mech.*, Air Force Institute of Technology, Wright Patterson Air Force Base, Ohio, Oct. 1965, pp. 547–576.

34. Wood, R. H., The reinforcement of slabs according to predicted fields of moments, *Concrete*, Vol. 2, No. 2, Feb. 1968, pp. 69–76.

35. Kupfer, H., Hilsdorf, H. and Rüsch, H., Behaviour of concrete under biaxial stress, *J. Am. Concr. Inst., Proc.*, Vol. 66, No. 8, Aug. 1969, pp. 656–666.

36. Wanchoo, M. K. and May, G. W., Cracking analysis of reinforced concrete plates, *J. Struct. Div. Am. Soc. Mech. Eng.*, Vol. 101, No. ST1, Jan. 1975, pp. 201–215.

37. Krishnamoorthy, C. S. and Panneerselvam, A., A finite element model for non-linear analysis of r.c. framed structures, *Struct. Eng.*, Vol. 55, No. 8, Aug. 1977, pp. 331–338.

38. Robins, P. J. and Kong, F. K., Modified finite element method applied to r.c. deep beams, *Civ. Eng. Publ. Wks Rev.*, Vol. 68, No. 808, Nov. 1973, pp. 963–966.

39. Desayi, P. and Krishnan, S., Equation for the stress–strain curve of concrete, *J. Am. Concr. Inst., Proc.*, Vol. 61, No. 3, March 1964, pp. 345–350.

40. Allwood, R. J., Reinforcement stresses in a reinforced concrete beam–column connection, *Mag. Concr. Res.*, Vol. 32, No. 112, Sept. 1980, pp. 143–146.

41. Nilson, A. H., Nonlinear analysis of reinforced concrete by finite element method, *J. Am. Concr. Inst., Proc.*, Vol. 65, No. 9, Sept. 1968, pp. 757–766.

42. Edwards, A. D. and Yannopoulos, P. J., Local bond–stress–slip relationships under repeated loading, *Mag. Concr. Res.*, Vol. 30, No. 103, June 1978, pp. 62–72.

43. Nam, C. and Salmon, C. G., Finite element analysis of concrete beams, *J. Struct. Div. Am. Soc. Civ. Eng.*, Vol. 100, No. ST12, Dec. 1974, pp. 2419–2432.

44. Cedolin, L. and Poli, S. D., Finite element studies of shear critical r.c. beams, *J. Eng. Mech. Div. Am. Soc. Civ. Eng.*, Vol. 103, No. EM3, June 1977, pp. 395–410.

45. Robins, P. J., *R. C. deep beams studied experimentally and by the finite element method*, PhD Thesis, University of Nottingham, 1971, 258 pp.

46. Ngo, D. and Scordelis, A. C., Finite element analysis of reinforced concrete beams, *J. Am. Concr. Inst., Proc.*, Vol. 64, No. 3, March 1967, pp. 152–163.

47. Lin, C. and Scordelis, A. C., Nonlinear analysis of r.c. shells of general form, *J. Struct. Div. Am. Soc. Civ. Eng.*, Vol. 101, No. ST3, Mar. 1975, pp. 523–538.

48. Rosenthal, I. and Glucklich, J., Strength of plain concrete under biaxial stress, *J. Am. Concr. Inst., Proc.*, Vol. 67, Nov. 1970, pp. 903–914.
49. Anchor, R. D., Scheduling and automation, *Consulting Eng.*, Vol. 35, Feb. 1971, pp. 97–100.
50. Computer Consortium, *BARSHED—user's manual*, London, 60 pp.
51. Jones, L. L., LUCID—an aid to structural detailing, *Struct. Eng.*, Vol. 53, No. 1, Jan. 1975, pp. 13–22.
52. Concrete Society, *Standard reinforced concrete details*, London, 1973, 28 pp.
53. Best, B. C. and Hollington, M. R., An interactive computer system for reinforcement drawings, *Struct. Eng.*, Vol. 58A, No. 11, Nov. 1980, pp. 351–4.
54. Craddock, A., GENESYS as applied to detailed design of reinforced concrete structures, *Struct. Eng.*, Vol. 56A, No. 10, Oct. 1978, pp. 277–282.
55. Hamilton, I. and Scoins, D., *Computer draughting in construction*, Construction Industry Computing Association (Design Office Consortium) Cambridge, England, 1980, 152 pp.
56. Hamilton, I., A general review of computer drawing techniques for structural engineering, *Struct. Eng.*, Vol. 58A, No. 11, Nov. 1980, pp. 345–346.
57. Williams, G. T. and Parsons, T. J., A computer aided draughting system for reinforced concrete, *Struct. Eng.*, Vol. 58A, No. 11, Nov. 1980, pp. 346–350.
58. *CAD82, Fifth int. conf. and exhibition on computers in design engineering*, Butterworths, Guildford, 1982 709 pp.

22 Computer applications 2: small computers

J W Bull

University of Newcastle upon Tyne, England

Contents

Notation

The notation used in this chapter follows CP 110.

A_{sv}	cross sectional area of the two legs of a link
b	width of section
d	effective depth of tension reinforcement
DL	characteristic dead load (Appendix I)
F_{BL}	load bond stress
G_k	characteristic dead load
h	overall depth of action in the plane of bending
$KN-M/M$	kilo newton-meter per meter width (Appendix I)
$KN/M\uparrow2$	kilo newton per square meter
$LCASE$	load case (Appendix II)
LL	characteristic live-load (Appendix I)
LX	length in the x direction (Appendix I)

LY	length in the y direction (Appendix II)	V	shear force due to ultimate load
N	ultimate axial load at section considered	V_c^*	ultimate shear resistance of concrete
$N/MM \uparrow 2$	newtons per square millimeter	W_k	characteristic wind load
S_v	spacing of links along a member		

Summary

The use of small computers for the analysis and design of structural concrete is discussed. A broad outline is given of the decision-making process before and after the introduction of a small computer into a structural concrete design team. A section is devoted to explaining the three types of small computer hardware, namely, minicomputers, microcomputers, and programmable calculators. As hardware and software are sold separately, software availability and programming languages are also discussed. A multistory building is analysed using all three types of small computer and samples of input and output are given. A short section is included to indicate areas of future hardware and software development.

1 Introduction

1.1 The ground rules

All too often, structural engineers find they have bought a small computer which does not do everything they had expected. This is a common problem because a skilful computer salesman can sell a package 'solution' to problems he does not really understand and the engineer may not properly understand his own needs, or the effect that a small computer will have on his ways of working.

So, the ground rules are

(a) understand the problem;
(b) establish the objectives of the solution.

Once the engineer has established the optimum solution, he can then assess the impact of achieving the solution using a computer, against the other available alternatives.

1.2 Benefits of a small computer

The three obvious advantages of a small computer are

(a) It has great arithmetic speed and almost total lack of error in executing a wide range of programs. The speed gives the structural engineer the opportunity to calculate rather than speculate. For example, he can test a hypothetical design and modify it in the light of his results until the optimum solution is formed.
(b) It can be placed in the work location (e.g., the design office).

(c) It allows many tasks to be excuted in real time (i.e. computer output occurring virtually simultaneously with the input). Consequently, a large number of jobs which previously had been done by hand, or in batch mode, can be economically handled on a small computer.

The small computer's ability to hold and rapidly scan large amounts of information can save time in bar scheduling and in producing bar-bending lists and cutting schedules. The small computer can also produce bills of quantities very quickly and be used to compare tenders. During the construction stage, design alterations can be checked in hours rather than days, thus reducing construction delays.

Drawings and schedules can also be updated by editing the data input rather than by constant alteration of the original drawings. In this way, all design changes can be recorded and are available for discussion during the payments phases.

The small computer's limitations are basically imposed by today's technology in electronics and software. There is on the one hand the need to keep capital costs low and on the other the need to increase the capacity to interchange information between the CPU, the input/output devices and other peripherals. This means there is little redundancy available if a machine fault occurs, and unsuspected errors will invalidate results.

1.3 Basic concepts

1.3.1 *Definition of a small computer*
What is a small computer? It has—like a large, mainframe computer—a central processor unit (CPU), Fig. 1-1. The small computer is a general-purpose, electronic digital calculator with a stored program and the capability to manipulate symbols as well as numbers. Connected to the CPU is some increment of memory. This memory element is necessary for the temporary storage of data and program instructions. The third element of the computer comprises the input and output devices. These provide the means of communicating with, and obtaining data from, the CPU and its memory. These three basic elements or building blocks comprise the computer *hardware*. Finally, there must also be the *software* to instruct the computer what to do.

The term 'small' refers to such features as physical size, processing capability, minimum available configuration, price and software support. The common physical characteristics of a small computer are

(a) Small physical size ranging up to 100 kg.
(b) Low power consumption and single-phase supply, 240 V or 115 V, 50 or 60 Hz.
(c) Environmental conditions requiring little more than reasonable temperature and humidity.
(d) Main memory ranging up to 128k words.

This concept is further developed in Sections 2.2 and 2.3.

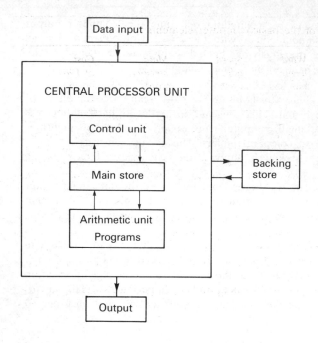

Data input	Cards, tapes, teletype or a visual display unit (VDU)
Central processor unit	The control unit controls operations in the memory, the arithmetic unit and all peripherals according to the program instructions
Output	Line printer VDU paper tape or a plotter

Figure 1-1 Diagrammatic representation of a small computer system

Given this definition, is a hand-held calculator a computer? The calculator has a CPU to carry out instructions, a memory for temporary data storage, (a programmable calculator also has instruction control storage for read-in memory), and a keyboard input. The display output completes the hardware. The software is buried inside the calculator in the read-in memory and the predefined instruction set of the CPU.

Therefore, a pocket calculator qualifies as a computer.

1.3.2 *The various classes of small computer*

The distinctions between the various classes of small computer are as blurred as the distinction between a small computer and a large computer. To understand their differences, one must examine some of the major elements of a computer, as shown in Table 1-1.

(a) CPU word length. The longer the word length, the more sophisticated are the instructions which can be decoded and executed.

Table 1-1   Comparison of the basic computer elements

Type of computer	Word length bits	Speed add cycle μs	Main memory words	Cost at 1981 prices £
Small computers				
Programmable calculator	2 to 4	2000 to 20	32 to 1k	50 to 500
Micro computer	4 to 16	30 to 3	512 to 8k	500 to 10 000
Mini computer	12 to 32	4 to 0.4	4k to 2M	5000 to 75 000
Large computers	32 to 64+	1 to 0.1 μs	500k to 10M	50 000 to 10M+

(b) Execution speed. Microcomputers are principally designed with low cost in mind and tend to be slower than minicomputers.

(c) Memory. The amount of memory with any computer depends upon the type of computer, e.g. mini or micro. The larger the computer, the larger the memory it can handle.

(d) Cost. The cost depends not only upon the type of computer chosen, but also upon the level at which the system begins. For example, whether it has VDUs, printers, etc.

1.3.3   Software

A computer can do nothing until instructed to solve a specific problem. Each step in the problem-solving process must be exactly defined in such a way that the computer can read, understand and execute the instructions in a logical and controlled way. These instructions are called *programs* and the term applied in general to programs is software.

A program is written in one of a number of languages, which for the structural concrete engineer will probably be BASIC or FORTRAN. The program is submitted to the computer via an input device. The computer translates the program into machine language, follows the program instructions and then outputs the results.

The problem of output validation is a very real one for the structural engineer. He must be familiar with how a program works and the assumptions made in that program, for those assumptions may affect the program's usefulness.

1.4   Efficient computer use

The efficient use of a computer depends upon two factors: the organization of the computer facility, and the efficient utilization of the computer itself.

To organize the computer facility, one needs a computer manager who should be a structural engineer trained in computer techniques.

One of the many problems the computer manager will encounter could be the reluctance of some engineers to use the computer. He must therefore ensure that the computer is part of the office scene and promote its use and advantages to his structural engineering colleagues. The manager should offer to do some of the design work on one of his colleague's job, as rarely will the offer be refused, especially if the work is dull and repetitive as, for example, bar scheduling. Once a structural engineer has been shown that certain tasks can be done more efficiently by a computer, he will very often change to that method. The computer staff must comply with any job request quickly, otherwise the structural engineer will revert to his manual methods. The hardware and software documentation must be readily available, easy to understand and continually updated. Each amendment sheet must be inserted as it is received.

Once the structural engineer is converted to the small computer, he must learn to work it efficiently. He must familiarize himself with operating systems, the methods of input and how to ensure the output is in a form which he can understand and use. Frequent use of the small computer is the best way to achieve this.

It will be found that a few months will be needed to ensure that the small computer works smoothly and for all the structural engineers in the office to become familiar with its powers and limitations. In that time, most of the structural engineers will begin to accept the small computer and be making use of it on many of their own jobs.

2 Hardware

2.1 Basic features

2.1.1 *Central processor unit*
The central processor unit (CPU) controls and performs the processing tasks required by the structural engineer as given by the program instructions. The program is carried in the internal working store (or core store). When the structural engineer commands the computer to start processing, the program instructions are transferred individually and sequentially from the internal store to the control unit and then acted upon. The actual processing is performed by the arithmetic unit (see Fig. 1-1).

2.1.2 *Basic input and output*
The input of data and programs is accomplished mainly by the use of a visual display unit (VDU). The VDU has the advantage that each line of input data can be seen and checked as it is fed in. Other forms of data and program input are punched cards, punched paper tape and magnetic cassette. The programs and data after being fed through the input devices are passed on to the CPU.

The results may be output to a line printer which presents them in tabular form. Alternatively, they may be plotted by a graph plotter. One of the advantages of a VDU is that output can be shown either in tabular or graphic form and then transferred to hard copy when required. The advantage of hard-copy printers is that all input and output can be recorded and checked for accuracy, but as the printer has many more mechanical parts than a VDU, it will be slower.

Visual representation of data can seldom completely replace numerical output, because most diagrams produced by graphical systems are composed of averaged values extracted from raw data and then used as input. The exception is where the structural engineer is dealing in absolute values, such as the cross-sectional areas of reinforcing bars.

2.1.3 *Storage*
The most expensive and significant factor in evaluating the cost of a computer is its internal storage capacity and access time to read in and read out data.

Computers are generally referred to as having a main storage capacity of, say, 32K, where K here does not mean exactly 1000, but represents $2^{10} = 1024$ eight *bit bytes*. The larger the K value of the internal working store, the more sophisticated the machine configuration and the larger the problems it can handle.

An internal working storage is expensive and as a program may not need all the input data at one time, an external backing store will often be incorporated. This backing store is used for storing input data to be used later in a program, for parts of a program not in use, and for long-term storage of files. The most common type of storage is the magnetic disk.

2.1.4 *Graphics devices*
Graphical output can be broken down into two areas: first, the use of existing line printers with specialist software which will produce the more usual graphs (these can also be output to a VDU); secondly, specialist flat bed and drum plotters and software to produce engineering drawings. The main problem with the production of engineering drawings by computer is twofold: (i) the software is often cumbersome and requires long running time and (ii) the data input must be complete and unambiguous. This usually indicates a large database and hence a computer that can only be used for producing a particular type of drawing. Also, the draughting system may be too expensive to use unless there is a large throughput of work.

2.2 The minicomputer

The application of large mainframe computers has been discussed in Chapter 21. The next size of computer down from the mainframe is the minicomputer. It is very difficult to define exactly where to draw the dividing line between a mainframe computer and a minicomputer. This is due mainly to the increasing sophistication and power of the minicomputer. One way of defining a minicomputer is to state that it normally requires no extra air conditioning other than that found in a normal office, that it will operate off a normal single-phase power supply and that it will be no larger than the average office desk. It will be able to support a full range of peripherals and to be programmed in a variety of high-level languages.

The minicomputer is not meant to rival the mainframe computer but, in many ways, to mimic it. It has the same internal structure as the mainframe but is much simplified, slower in operation and has reduced facilities. For example, the minicomputer has a smaller internal memory, which reduces its speed of processing and limits the number of peripheral devices it can operate. When buying a minicomputer, one initially gets the basic central computing unit and then the peripheral devices (memory disks, printer, input readers, etc.) to make the minicomputer work. Also the software is usually bought separately.

The main limitation of the minicomputer is its comparatively small memory capacity, which means that the software must be written to circumvent this disadvantage. A small memory capacity also means a longer processing time. It is necessary, therefore, to choose a minicomputer which is *known* to be capable of satisfactorily processing structural concrete programs.

These are two ways of overcoming the problem of limited memory capacity. One is to use a minicomputer that has an especially large internal memory, rather like the mainframe computer, but this is costly. The other is to use a minicomputer with a 'virtual storage' facility. This is usually accomplished by using a peripheral disk memory as an extension of the machine's internal memory.

The most important peripheral device necessary to the minicomputer is the disk memory storage system. This may be a cartridge disk, which looks like a long-playing record and is permanently contained in a plastic case, or it may be a floppy disk, which is about the size of a 'single' record (178 mm diameter) and kept permanently in a cardboard sleeve. Floppy disks are very cheap, easy to use and store. It is advisable to make a copy of all disks used each day at the end of the day in case one or more of them 'crashes' during use. That way, only one day's work will be lost.

Communicating the input to the minicomputer was previously done through a teleprinter type terminal, which was cheap and easy to use. Recently, the teleprinter has been supplanted by the visual display unit (VDU). The VDU is cheaper and more reliable than a teleprinter type terminal. The VDU displays about 25 lines on its screen but as each new line is printed, the top line disappears and is not recorded, as it is on a teleprinter type terminal. Input devices may also be used for output but as they are slow (or do not produce hard copy, as in the case of the VDU), a higher speed printer is normally used.

Apart from structural analysis, the other use for a minicomputer is graphical output. It must be remembered that engineering drawings form a significant part of the engineer's work load. These drawings can be produced interactively as the structural engineer gives the commands or non-interactively by submitting the data in batch mode. The usual procedure is to purchase a minicomputer specifically for graphics input and output. For output, this means a full-sized flat-bed or drum plotter that can draw on AO-sized paper. Flat-bed plotters work by moving a stylus over a static sheet of paper (any paper can be used) to give accurate drawings. On a drum plotter, the pen and paper both move, but the directions of movement are a little more limited than those in a flat-bed plotter. Both types of plotter output not only the actual drawings, but also the titles, dimensions and notes usually associated with detailing.

As graphical minicomputers are designed specifically for graphical output, they must be used intensively to become cost-effective. This may cause work-scheduling problems, as the graphical minicomputer can only work on one drawing at a time. However, two advantages are that a drawing can be physically produced in about 30 min after the data has been submitted and that the technician using the machine can also be doing another job while periodically checking the graphical output.

The specific problem of a multi-story building is analysed in Section 5 using (a) a programmable calculator (described more fully in Section 2.4), (b) a microcomputer (described more fully in Section 2.3) and (c) the Olivetti P6066 minicomputer, which is now described.

2.2.1   Olivetti P6066
This computer has ready-to-use software, uses the programming language BASIC, and is built in modular form for further expansion. The system comprises

an arithmetic/logic unit, a central random access memory (RAM), a dual-drive floppy disk for rapid access to programs, data and the operating system, and an extended keyboard for data, programs and system commands input. It also has a console and alphanumeric 'strip' display (for user guidance and for establishing the operating modes) and a printer, which can act as a plotter for high-resolution graphics.

The system is programmable in BASIC, the simplest high-level language to use, and it also provides for calling a range of subroutines. During program development, each instruction is checked and shown on the display and an editor analyses and allows corrections and the merging of programs and data. The BASIC is comparable with ECMA Standard BASIC and offers many extensions, including string manipulation, matrix operations, random access file handling, output formats selected by a program and graphic functions.

The assembler language allows direct access to the CPU, giving the possibility of using the system's potential to the utmost in those parts of a BASIC program where it can make a significant contribution, for example, in certain specialized calculations, and some input/output operations. The modular operating system, which is disk-based, controls all handling of libraries, commands, utility programs, languages, editor, subroutines and programs. In fact, as the system is modular and expandable, any of its configurations can be adapted to both the existing and future needs of the structural concrete engineer.

2.3 The microcomputer

The next size down from the minicomputer is the microcomputer, which, in its basic form, is about the size of an office typewriter and is operated from a normal single-phase power supply. Data input is usually via a keyboard or from a prerecorded cassette tape. Output is onto a screen or to some form of printer or graphical display device. Programming is usually via a high-level language, such as BASIC. The problem-solving capability ranges from simple calculations to the solution of relatively complex analysis problems calling for immediate answers. Standard library programs are available from the hardware vendor for the more common structural concrete problems.

The microcomputer is slower than the minicomputer, but it is very useful where the cost of a minicomputer cannot be justified. If the microcomputer is used interactively, delays of over 10 s may occur and this may be annoying to the user in some circumstances. A way to overcome this difficulty is to run the microcomputer in batch mode, where the program is read in, followed by the data in a block and the microcomputer is then left to run the program and produce the results on a printer.

A standard microcomputer configuration might consist of a 48K bytes user memory with typewriter keyboard, cassette input and TV monitor output. To extend the storage capacity and facilitate the retrieval of programs, two floppy disks which hold 100K bytes of storage could be used with the facility of adding more disk drives. High-resolution graphics can be used, and the addition of a colour card makes a very good facility for displaying results in coloured graphical form. Another add-on feature may allow the use of FORTRAN 77 or PASCAL, as well as BASIC. The microcomputer can also easily cope with printers, paper-tape readers and plotters.

The microcomputer configuration described in Section 5 comprised the CBM 8032 microcomputer and the CBM 8050 disk drive. The CBM 8032 is a combined CPU and VDU. It has a memory capacity of between 32K and 96K, depending upon the amount of add-on memory. The machine uses BASIC as its main programming language and this is stored in a read only memory (ROM) for immediate use on 'powering up' the microcomputer.

The CBM 8050 is a twin-disk drive unit using 135 mm diameter mini-floppy diskettes storing 550K characters per drive. It is an intelligent peripheral and contains microprocessors and memory and needs only the minimum of instruction from the CBM 8032 CPU.

The CBM 8032 was connected to an 80 character/line silent printer.

2.4   The programmable calculator

The simplest electronic calculator is the battery/mains-operated pocket calculator. It has fixed, built-in programs that enable it to carry out the four basic arithmetic functions of addition, subtraction, multiplication and division. Some of these 'simple' pocket calculators have additional operations such as squares, square roots, trigonometrical functions and logarithmic functions.

The next stage in sophistication is to provide a memory block, which allows the user to load a selected program to control a sequence of calculations. Although these programmable calculators are slow in operation when compared with the minicomputer and microcomputer and have a small memory, they are very convenient and can perform many of the simple and repetitive calculations and reference to tables that often plague the structural concrete engineer.

In the United Kingdom, the programmable calculator is, unfortunately, still widely regarded as a purely 'personal' calculator to be purchased by the structural engineer who wishes to use one. In mainland Europe and the USA, many companies realize the potential of programmable calculators and purchase them as standard office equipment for their structural engineers. For example, the Hewlett Packard, HP 67/97 calculator with program can automatically analyse an 11-span continuous-beam system and calculate the support moments, while the HP 41C will undertake the calculations required to investigate a helical staircase and plot the resulting moment and shear force diagrams, using a method originally written for an IBM 1620 computer in FORTRAN.

The use of two programmable calculators are described in Section 5. These are the Hewlett Packard HP 97 and HP 41C. The HP 97 is a desk-top version of the HP 67, measures 230×200 mm and is equipped with a thermal printer. The programs are recorded on magnetic cards 71×11 mm, each of which can hold a program of up to 224 steps. The same magnetic cards are also used to input data and store information. Longer programs than 224 steps can be used, provided the calculator has been instructed to expect another card containing the next stage of the program.

The HP 41C, although resembling the HP 67 in size (150×80 mm), has much greater potential and many more facilities. The HP 41C can be linked to a detachable card reader and a separate printer, which itself contains sufficient pre-programming to permit quite complex graphical plotting on the 160 mm wide thermal paper. By adding up to four plug-in memory modules, up to 2000 program lines can be stored. Also, a program may be halted while running, and fresh data inserted via the card reader, to replace part or all of the information already stored.

2.5 Evaluating and choosing the hardware

The successful use of small computers by the structural concrete engineer can only occur if he has the correct computer to run the programs he requires, backed up by adequate support.

The first stage is for him to establish precisely what he wishes to use his small computer for, remembering that the machine will be used solely by him for structural engineering purposes. As the hardware may involve a large financial commitment, it needs to be evaluated by a structural engineer familiar with the problems to be solved and a hardware expert who knows what hardware is available. In many cases, a structural engineer with knowledge in computing can combine the two fields of expertise.

The evaluation procedure will usually follow a set pattern. After making an assessment of the need, potential suppliers of hardware are initially identified from directories [1] and information services [2]. The potential suppliers are then contacted to discuss the capabilities of their equipment. Following discussions, one company will be then invited to make formal proposals.

These points are further discussed below.

2.5.1 *Assessment of need*

Is the intention to computerize the analysis and design process completely or only partially? If the structural engineer requires brief or straightforward calculations, for example, the bending moments and areas of steel reinforcement for a simply supported beam, then a programmable calculator is adequate. If he has to analyse a multistory building purely for bending moments and shear forces and then to break down the remaining part of the design into small modules, as is usual when using manual methods, then a microcomputer is the best option. If, however, he frequently has to perform long and complex calculations for major design works and use interactive programs for structural concrete design, then a minicomputer is the best option. If he also needs computer graphics for the production of standard drawings, then specialist hardware must be purchased and consideration given to whether the company can also use the computer for management information and other design purposes when the drawing load is not fully utilizing the available capacity.

With respect to input, for simple calculations the numbers can be punched in via the keyboard on a programmable calculator and the program kept on a magnetic plate. If programs are long, they may be recorded on disks and either a microcomputer or minicomputer can be used. If the structural engineer wishes to retain large amounts of data, for example, bar-bending schedules and drawing details, a large amount of storage will be required and a minicomputer is the best option.

With respect to output, if the structural engineer requires engineering drawings, a drum printer or flat-bed printer is required. If the output is to be in the form of headed tables, a hard copy printer is required. If, however, limited amounts of output will be required and can be hand written onto calculation sheets then a microcomputer with VDU is sufficient.

Ease of access to hardware is very important and should be considered in relation to the amount of computation to be done:

(a) When processing small calculations, a programmable calculator, being immediately accessible, is ideal.

(b) When analysing a multistory frame, for example, it may take many hours to sort the data required for the program and so a few hours delay waiting for the use of a microcomputer is quite acceptable. Usually, the structural engineer will want to use the microcomputer interactively and have the machine to himself when running the program. In this case, a few hours delay will be amply rewarded by the use of the *interactive facilities* which allow alterations to data and design using fully the structural engineer's experience.

(c) For complex and lengthy analysis and design, a 24 h waiting time will be acceptable and in this case a minicomputer will be used where job streaming is important to obtain maximum mini-computer useage.

2.5.2   *Hardware suppliers*

Potential suppliers have now to be contacted and a range of costs discussed (Table 2.1). Not only is there the initial purchase price to be considered, but various rental charges, leasing, maintenance and running costs. Also, there may be additional costs for site alterations and hardware installation.

The supplier's capabilities must also be considered. Where is the hardware manufactured and where is the manufacturers' warehouse? Programmable calculators are usually available off the shelf or in a few days. The delivery of a microcomputer may take up to several weeks, because it may only be available from the manufacturer's regional office. A few months may elapse before the installation of a minicomputer, which may only be available from the national headquarters of its manufacturer.

It is especially important that there is a service contract for the chosen minicomputer or microcomputer, and that service is available from a local organization. This will ensure that a faulty device can be temporarily replaced while being repaired. The speed of repair will be increased if many similar systems are being maintained by the same repair organization, as familiarity with hardware increases maintenance efficiency.

Table 2.1   Hardware and software costs

Type of small computer	Typical costs	Costs of hardware and software in this chapter	
		Hardware	Software
Programmable calculator	£50–£500	£500	£50
Microcomputer	£500–£10 000	£1900	£500
Minicomputer	£5000–£75 000	£6500	£1250
Graphics plotter	£3000–£30 000	Not used in this chapter	

It is a fact that the more a structural engineer uses hardware, the more uses he will find for it, and very quickly a computer system will need extending. Often, such extensions will finally result in the complete or partial replacement of a system. Hence, a point always worth pressing with a supplier is the trade-in value of the present installation.

Having looked at a number of alternatives, the structural engineer must now discuss details with one hardware supplier. The structural engineer now has a clear idea of the type of machine he requires (programmable personal calculator, micro- or minicomputer), the work for which it is required, and the peripherals needed to operate the system. He must negotiate a satisfactory system.

The structural engineer, when evaluating and choosing the hardware option, should discuss his problems and possible solutions with other structural engineers who already have small computer systems. Much valuable experience and information can be gleaned in this way.

2.5.3 *Specific details worth checking*

Although the vast majority of vendors and users are honest, it is usual for both parties to have some form of contract to codify the obligations and contingencies. For example, the supplier's obligations should include delivery terms, performance undertakings and spares availability. The user's may include payment details and certain specified tests. Contingencies are matters both parties hope will never arise but usually include terms covering non-delivery or non-payment within a specified period.

When obtaining sales literature, ask for the vendor's standard form of agreement. Remember this agreement can be altered by both parties before signing and do ensure that any letters of intent are all subject to contract.

It is unlikely that a maintenance contract or manufacturer's guarantee will cover the cost of repairing damage from external sources or negligence. In the case of a hirer, the contract will probably make the user responsible for certain loss and damage. Therefore, the structural engineer must decide whether he wants insurance cover and whether this is to be limited to material damage to, say, fire, theft, electrical fault, etc., or is to include consequential loss, such as the loss of data and financial loss resulting from a breakdown.

Most computer insurers are willing to arrange a policy to meet a particular need, and the structural engineer must not hesitate to be specific in his insurance requirements.

3 Software

3.1 Introduction

Small-computer manufacturers provide less software support than large-computer manufacturers for the following reasons:

(a) As the small-computer market is a new market, software development has lagged behind hardware development.

(b) The low cost of small computers does not encourage small-computer manufacturers to undertake expensive software development.

(c) Small computer multiplicity and versatility make standard applications software impossible to produce.

(d) Small-computer vendors have found that software is not essential for hardware sales.

(e) Software programming effort is very much greater for a small computer than for a large computer as efficient computer use is more important for a small computer.

Software for the small computer is available in three different ways. First, one can buy a small computer with ready-to-use software as a package. Secondly, the computer and the programs may be bought separately from the computer vendor. Thirdly, one can buy a small computer and then go to any one of the many software suppliers and obtain programs as separate items. Each way has its own advantages and disadvantages, which are discussed later.

3.2 Program languages

Program languages are categorized into three distinct levels: machine languages, assembly languages and high-level languages.

3.2.1 *Machine languages*

The instructions which a computer's control unit decodes and executes are a binary representation of the order code of the machine. Programming in machine language is an exacting task and requires a full and precise knowledge of a small computer's operation codes and painstaking accounting of the contents of every location in the store area of the computer at every stage of its operation. To escape this form of programming, the job of checking and allocating addresses is passed to the computer via the use of assembly languages. Machine languages are not used by structural engineers.

3.2.2 *Assembly languages*

Assembly languages are at a higher level than machine language. To escape the exacting requirements of machine language, the small computer itself is used to translate between the high-level languages and machine code. This translation is performed through an assembly language (by an assembler), and the language is generally specific to a make or model of small computer.

3.2.3 *High-level languages*

High-level languages are based on a set of grammatical rules that allow the programmer to express his instructions in line-by-line statements in simple English-language phrases and arithmetic notation. The high-level language is the one a structural engineer will most usually use.

The high-level source program is translated into a machine-language object program by processing it under the control of a computer program, called the compiler. The compiler translates each statement into an equivalent set of machine language instructions, incorporating any standard subroutine required and also linking the parts of the program.

There is a growing number of high-level programming languages available. Each language is used as a means of describing programs and for thinking about them. The language will also influence the way the structural engineer thinks about a problem. A number of high-level languages are discussed below.

3.2.3.1 *Commercial languages* In commercial languages, the emphasis is on file handling and the manipulation of textual terms. There is little numerical capacity other than decimal arithmetic.

The major language in this field is COBOL, which is the standard choice for most commercial data processing. However, for small business machines, RPG/2 is widely gaining ground.

3.2.3.2 *General-purpose languages* PL/1, ALGOL 68 and PASCAL are languages whose area of application is not restricted to any one class of work.

3.2.3.3 *Real-time languages* These are designed to increase the run time efficiency of compiled programs, to facilitate the controlled sharing of data areas and for handling program interrupts. The two main languages in this area are CORAL 66 and RTL/2.

3.2.3.4 *Conversational languages* They are designed for remote terminal on-line use, where the user is at an on-line terminal throughout the processing of his program. The two main languages are APL and BASIC. APL is popular as a means of implementing financial and managerial problems as it can operate on arrays of numbers and text characters. BASIC is popular with structural engineers as it is easy to implement on small computers and is suitable for structural engineering problems.

3.2.3.5 *Scientific languages* These provide powerful mathematical notations, algebraic functions and array or matrix manipulation. The two most widely used scientific languages are ALGOL, which tends to be used in Europe, and FORTRAN.

FORTRAN is the most widely used language for scientific, technical and mathematical program writing. It incorporates a wide range of mathematical functions and has the facility to build up complex expressions.

In general, BASIC is used for the smaller numerical problems, while FORTRAN is used for very much larger ones.

3.2.4 *Languages for small computers*

The early exploitation of small computers was entirely in assembly languages and to a large extent this is still the case. However, the increasing software development for small computers has meant that BASIC is universally available on small computers and the use of FORTRAN is very rapidly increasing.

3.3   Program limitations

There are six distinct areas in which an engineer must consider the limitations of his program. These are

(1) choice of small computer;
(2) assumption made in developing the program;
(3) program input;
(4) processing of the program;
(5) program output;
(6) hardware/software compatibility.

3.3.1   *Choice of small computer*

In theory, any program written in a high-level language can be easily transferred to another computer. However, in all computers a compiler is required to translate the program from the high-level language into the machine instructions that drive the computer. These instructions vary from machine to machine and very often programs are not transferable.

These variations in machine instructions arise because of the way in which a computer communicates with its peripheral devices, e.g. data stores, printer, VDU, etc., and this is machine dependent.

Each make of computer has advantages and disadvantages compared to its competitors, and users will want to exploit fully the advantages of a specific machine. It is exactly these advantages which make the program non-transferable.

For these reasons, a program written for one machine will not, in general, run on another machine without conversion. It is paradoxical that the more sophisticated and efficient a program is, the more conversion it will need. Consequently, if a program is to be converted to another machine, this should be done by the original writer.

3.3.2   *Assumptions made in developing the program*

The assumptions made in developing a program will materially affect the results obtained. Parameter sets may be closely or loosely defined, depending upon the accuracy required from the results. For example, specifying approximate concrete beam sizes will give results that indicate the stresses to be expected in the concrete. But once a beam size has been decided upon, the amount of reinforcement or prestress can be accurately assessed. Also, the fact that a program conforms to CP 110 [3] is no guarantee that all the possible alternatives have been considered. For example, CP 110 divides prestressed concrete members into three classes and the structural engineer must check that, if he is designing a Class 1 member, Class 1 criteria are being used in that program.

The choice of structural idealization requires a high level of structural engineering judgement of which pure computer programmers have no experience. For a simple framework, the structural engineer can idealize each member as a line and each joint as a node. For a continuous slab, a grillage model may be used, where a number of physical components are idealized as one member. Complex slabs can only be analysed on a large computer. Another assumption is that stress is proportional to strain and that displacements are small. Thus, large changes in the geometry of a structure will be neglected when stresses and displacements are calculated.

A further point to be noted is that as no two engineers will idealize a structure in the same way; the structural engineer must expect to have a variety of answers that are 'correct'. For this reason, structural engineering experience is needed to pick out the unexpected answer.

3.3.3 *Program input*
To produce an accurate answer to a design problem, a large amount of data input is required, using as much known data as possible. An examination of the kind of database a program uses will indicate the program's limitations. No structural engineer wants to spend more time studying the documentation and the rigid data input requirements than he devotes to solving the problem. He wants to input data in an easy and convenient form and in such a way that he can easily check the input data.

3.3.4 *Processing the program*
When a program is written by a non-engineer, it will be done with the efficiency of computer processing in mind. The input and output will also be adjusted to suit the computer being used. This is not the essential requirement of the structural engineer. He requires engineering convenience. The input must be easily identified and checked, and the output must be in a readily understood form, which may, however, be comparatively inefficient in the use of computer print-out resources. For example, most structural concrete calculations are done in the UK on white A4 sheets of paper with a readily understood format of CP 110 references, while the usual computer print-out is on larger two coloured paper with no CP 110 reference.

3.3.5 *Program output*
Program output must list the data input for easy checking by the structural engineer, together with any error messages. For example, has a number been given for each input value or has one number been omitted? Are the results in the form required and how accurate are they? Many simple programs will give notional results while a large complex program may give all-embracing answers. Check the program output against a problem with a known 'hand-method' answer. If the answer is within accepted tolerances, the program can be used more widely.

3.3.6 *Hardware/software compatibility*
Some programs developed for mainframe computers can be made available for mini- or microcomputers. But the main problem that the structural engineer will meet concerns the minimum small-computer configuration and internal memory needed to support the program. The structural engineer may find that a program he requires cannot be run on his hardware unless he either expands his existing configuration or learns how to alter the program. In short, ensure before buying a program that it will run on your existing hardware.

3.4 Program writing
The structural engineer may write a program himself, he may commission somebody else to write it for him, or he may obtain a ready-made program.

However, he must remember that, although the cost of hardware has rapidly fallen over the past few years, the cost of producing software has risen (in absolute terms) due to increased salaries and service costs.

It may seem obvious for each user to write his own program because the exact requirements can be specified and the program written for particular hardware, but considerations of cost and lead time often militate against this course. For example, a program with 1000 instructions may take two-man months to perfect and thus a better approach might be purchase a software package. This will be discussed in Section 3.5.

The writing of a program can be broken down into a three systematic phases. First a systems analysis (feasibility study) is performed. For example, can the multistory problem used in Section 5 be completely analysed by one program? Obviously not, but a number of programs can be used to break down the work into phases. Secondly, an analysis of the existing solution must be made to reveal any shortcomings. This part usually indicates that the existing non-computer method does not allow the structural engineer to quickly alter his arrangements of beams and columns. The new program will have to incorporate this facility. Thirdly, and finally, a detailed program specification will be produced.

Once the systems analysis phase is complete, the detailed program specification must be developed into a source program. This is accomplished by writing the required program statements in an appropriate language. For the minicomputer and microcomputer, the language will be BASIC or FORTRAN. The program itself is divided into segments or sub-routines that deal with specific parts such as data input, data checking, processing and output. The program is then fed into the computer with a set of test data via a keyboard and the program compiled into its object form. At this stage, the structural engineer will edit and correct his program until it runs correctly and gives the expected answer.

However, not all combinations of data will produce valid results and for this reason engineering judgement must always be applied when using results of computer programs.

The final part of program writing is documentation. A record must be available on how to use the program and interpret the results, as well as on the design theory used. The documentation of software may take a considerable effort to produce, but is nevertheless essential as it will warn the user what the program's validity limits are and also assist in program improvement.

3.5   Choosing software

A computer is only as good as the software it uses, and it can be difficult to obtain useful and reliable software. The problem is due to the very rapidly increasing capabilities of the hardware which has outrun software development. For example, even software written only five years ago is likely to be of limited capability and crude in the way it uses the small computer. As time passes, the same structural engineering problems are being tackled again and again with increasingly up-to-date software. The structural engineer can be deceived as to the usefulness of a program by means of carefully arranged documentation. How, then, can an engineer ascertain whether a packaged program is suitable?

The first step is to find out what programs are available and this can be done by studying some of the many guides to computer usage that are available in the construction industry [1, 4, 5, 6]. These guides usually give descriptions of the programs, the machines they can be used on and the marketing organization. Another, and probably more up-to-date way of finding programs, is to check the advertisements of engineering periodicals and to contact at least one of the independent bodies dealing with computing in structural engineering, for example the Construction Industry Computing Association.

Having obtained a list of suitable programs, one must check to see if they are usable. Some programs may not be available in the country of use, some may be obsolete and others that are suitable may have insufficient back-up available. A further problem that frequently arises with the use of small computers is that many programs are written for specific machines and this again limits program choice.

The next stage is to contact the vendors to obtain details of the program. Normally, one will receive a few pages which are intended to make a good impression on the potential buyer and to give a general idea of the type of problem that the program can deal with. The next stage is to obtain a user's manual. The manual will describe the actions of the program and indicate its scope and validity. It will also detail the assumptions made in the program and perhaps indicate which assumptions can be altered. For example, the type of stress–strain curve used for the concrete and whether this can be modified.

It must be emphasized that the correlation between the quality of program and the quality of documentation will vary. The only way to check whether a program is suitable is to see it used on the vendor's problems and then for the structural engineer to try it on his own problems.

A vendor's demonstration of a program will bring out the program's good points and gloss over the bad. The data input will already have been prepared and the cost in computer time will have been minimized as the problem will be a standard test. One way to overcome this form of presentation is to ask the vendor to run one of your own typical problems to which the solution is known. When the program has been run, check thoroughly its input data. Is it easy to input or is it tedious, repetitive and obscure? Is the input easy to understand? Remember, if the input is difficult, the program will not be used.

Check the run time of the program, both with the vendor's problems and your own submitted problem. You can expect your problem to require more computer time than the vendor's but the difference should not be excessive. Compare it against your usual manual-methods times.

The output must be well presented and short. Pages should be numbered and have titles and columns. Do not accept a program on the vendor's promise that it will be updated. You are buying a program now and you want the best that is immediately available. Also, a program that has been used for some time will have other users on whose experience you can draw.

The support and maintenance given to a program must also be established. No matter how well a program is written, there is a high chance that some form of after-sales support will be required. An error in a program can go undetected for many years before a particular set of data triggers the fault and produces erroneous results. For example, the present author has used a well-known (but now corrected) bridge deck analysis program that under one specific deck load produced support reactions that were very much higher than the applied loading.

This was not obvious from the output but was only noticed due to a fortuitous random 'reaction check'.

Even after all this checking, it still may not be possible for the program to be error-free. Find out how long a program has been available, on how many jobs it has been used, and what sort of results it has given. If very few jobs have been run, beware. Also find out how large the program is and the amount of effort involved in its writing. If it has only 1000 lines, it may not be very versatile and the input and output may be difficult to understand. A program of 3000 lines is usually well thought out, as the programming effort increases very rapidly with the number of lines. Check which language a program is written in. Both BASIC and FORTRAN are widely used for the small computer.

A problem that arises in seeking specific details of a program is copyright, or more precisely, lack of copyright [7]. In most countries, including the UK, copyright laws do not apply to computer programs and hence the vendor will normally insist on protecting himself by keeping program details secret, and will not supply listing before purchase. This can be a problem because the engineer has to ensure that the program is what he needs.

When you have decided to buy a particular program, ensure that there is a maintenance contract that will (a) guarantee the sorting out any program problems that may arise and (b) give you the right to any program modifications subsequently produced. Check what training courses and manuals are included, because a program is useless without this back-up.

3.6   Software acquisition and availability

Software is available from computer manufacturers, independent software companies and computer users [1, 4, 6].

Independent software companies can offer significantly better programs than hardware companies since they are only interested in selling software, but there are risks. For example, can a software supplier provide after-sales service and will the company still be in business when its support is required? Does the company have a structural engineering background and thus understand engineering requirements? Is it a general software-producing company or does it specialize in programs for small computers? (The latter is more likely to understand the nature of small-computer restrictions and sell its product at a more realistic price.) When acquiring software, the structural engineer must start with simple programs and then build them up as confidence and problems complexity allows.

There are basically three ways that a structural engineer can acquire programs:

(1) develop his own programs;
(2) use a software development company;
(3) obtain an existing software system.

3.6.1   *Own development*
The structural engineer has the advantage here that he can specify and control the engineering assumptions and limitations of the program. He has control of costs and can modify the program as required. Also, he has no problems of documentation fees, copyright or restricted use. But, as it is unusual for a structural engineer to stay with a company for more than a few years, any programs he

develops will decline in use after he has left due to differences in styles of programming and lack of documentation. Two further points to be considered are (a) in-house program development takes longer than contract development and (b) the engineer developing the program becomes unavailable for other tasks.

3.6.2 *Software development companies*
From the structural engineer's point of view, this is the easiest way of obtaining software. An agreement is reached between the structural engineer and vendor over cost, time scale and program requirement. Care must be taken in choosing a supplier that is familiar with the requirements of the relevant codes of practice.

3.6.3 *Existing software systems*
There are many sources of package programs, but as a first step the structural engineer should consult an up-to-date reference book on computers [1, 4, 5] or consult an organization which has a special interest in structural concrete [2]. Again, one must be sure that the program documentation is adequate for the structural engineer to fully understand the capabilities and restrictions of the programs. For example, can the program deal with a multi-span beam and accept redistribution of moment as per Clause 3.2.2.3 of CP 110? The structural engineer should be able to ascertain the validity of the program for his work. He should be able to find out the engineering theory behind the program, e.g. the ACI code [12] or CP 110 [3], and the way in which the program works to obtain the answer.

Most organizations levy a charge for the supply and use of their programs, and various payment schemes are in operation. For example, a program may be rented or leased, in which case restrictions are usually put on who may use the program and on the making of alterations to it. Consequently, the user must have an effective maintenance contract to back up the program. A further method, which is very similar to renting or leasing, but with some legal differences, is to use a program under licence. Usually, the user pays a royalty each time the program is used.

One further point must also be considered. No program can ever guarantee to give the right answer every time. There will almost always be a set of circumstances that will upset a program and make it give a ridiculous answer. For this reason, it remains the engineer's responsibility to satisfy himself that the answers he obtains are within acceptable limits.

If the structural engineer obtains an answer that appears to be wrong, he must run the program again. If the answer he obtains is still wrong, he must start checking his data input and program limitations.

3.7 Criteria for successful software
It is not sufficient, once having purchased a program, to hand it over to any structural engineer, assuming he will know the program and immediately use it efficiently. The program must be marketed in any design office in the same way as all new products.

For the small computer, it is usual for the person who looks after the hardware also to keep records and documentation of new programs. In fact, one engineer with an enlightened interest in computer aided design should always be made responsible for the software application. When a new program has been acquired, there must be an initial period of training and support by the supplier. The supplier must give demonstrations to the users of the program and help the person responsible for looking after the software to acquire a deeper understanding of it. It is also necessary for the supplier to be available immediately to sort out any problem in the first few weeks following the program's introduction.

Having a person within the design office who is familiar with the software helps enormously with the introduction of a program. He can not only sell the program to the other structural engineers and indicate in which areas it can be useful to them, but also immediately sort out any problems and indicate more efficient modes of program usage.

Once a program has been introduced, it must be used quickly and constantly, not only to gain a fast and adequate financial return on the outlay but also to surmount the psychological barrier of something new.

3.8 Description of software used in Section 5

In Section 5, the problem of a multistory building is solved using (a) a minicomputer, (b) a microcomputer and (c) a programmable calculator. The three software packages used are described below.

3.8.1 *Minicomputer*

The software used for the minicomputer was DECIDE [9], which is a series of interconnected programs for structural reinforced concrete design to CP 110 [3] and written for the Olivetti P6066 personal minicomputer of 32K core size and over. The program covers the design process from analysis to bar curtailment and provides the designer with maximum flexibility in use. The designer can proceed in almost the same manner as he would by hand, using the computer interactively to speed up the routine and tedious calculation work.

The programs can be used either independently or for operating on data generated by a previous program. The system is so designed that, from almost any stage in the design, the engineer can go back to any previous stage and re-start the design. The results from a frame analysis carried out by programs that are not part of DECIDE can be used as input for later parts of the program. This allows the program to become a useful tool for checking designs.

The sequence of the program system is shown in Fig. 3-1. The main series of programs, from SUBFRAME or CONTINUOUS BEAM to SHEAR is for the analysis and design of beams in reinforced concrete framed structures.

The program for slab design, SLAB, is complete in itself in calculating the moments and steel areas for two-way spanning slabs. The two programs for column design are SLENDER COLUMN and COLUMN SECTION.

A further facility is that the data can be recorded to disk at any stage of the design process for later use.

The program can be described as follows:

(a) CONTINUOUS BEAM This program analyses continuous beams up to seven spans with options for cantilevers at each end. Upper and lower columns may also be considered in the analysis and redistribution of moments may be

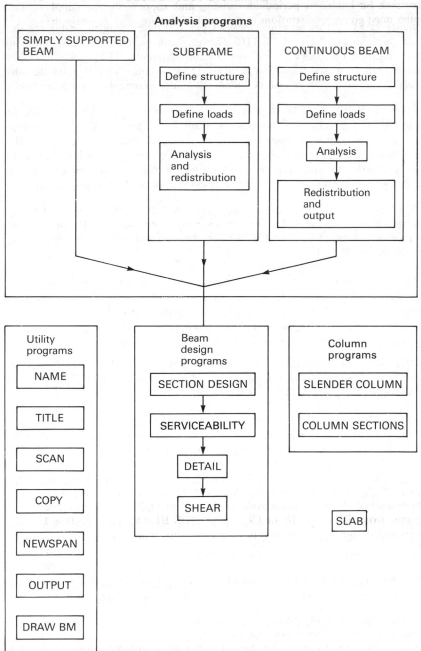

Figure 3-1   DECIDE P6066 system plan

carried out. The bending moment and shear-force envelopes can be stored on a floppy disk for later detailed design.

(b) SUBFRAME This program analyses plane frames using the subframe described in Clause 3.2.2.1 of CP 110. The program generates bending-moment and shear-force envelopes as well as column moments. The redistribution of moments can be carried out interactively. The program can also handle any combination of uniformly distributed load, triangular, trapezoidal and point loads.

(c) SIMPLY SUPPORTED BEAM This program has two uses: first, as an alternative to the subframe analysis program for analysing simply supported beams and secondly as a means of interfacing other analysis programs to the system. Used in the second mode, it enables a simply supported bending moment to be corrected to any specified end moments and the results put into a bending-moment envelope. Any number of load cases can be handled in this way.

(d) SECTION DESIGN This program calculates the steel areas required to resist flexure in rectangular, T or L-sections, paying due regard to the neutral axis provisions of Clause 3.2.2.3 of CP 110 and the minimum reinforcement percentage provisions of 3.11.4.

(e) SERVICEABILITY This program checks deflections by means of span/effective depth ratios. If the member being checked slightly exceeds the permissible ratio, an interactive routine is used to modify the steel areas. Maximum bar spacings are calculated which will ensure satisfactory crack control. For this, the equations in Appendix A of CP 110 are used.

(f) DETAIL This program is used interactively to curtail the reinforcement according to a bending-moment envelope and in compliance with the provisions of Clauses 3.11.7.1 and 3.11.6.2 of CP 110.

(g) SHEAR This program calculates the shear reinforcement requirement and local bond stresses at any specified section of a beam or may be used automatically to check the shear and local bond at tenth points.

(h) SLENDER COLUMN This program calculates additional moments for slender column design according to Clauses 3.5.7.1 and 3.5.7.2 of CP 110.

(i) COLUMN SECTION DESIGN This program calculates the steel areas required to resist unaxial or biaxial bending combined with vertical load on a symmetrically reinforced rectangular column.

(j) SLAB This program designs two-way spanning continuous slabs of the type dealt with in Clause 3.4.3.2 of CP 110. The calculations include

(1) design ultimate moments;
(2) loads from slab onto supporting beams;
(3) steel areas, making allowance for minimum steel area provisions;
(4) checking span/effective depth ratio;
(5) maximum bar spacings to satisfy Clause 3.11.8.2.

It also includes an option to specify the reinforcement over one support and the ultimate moments at other sections are adjusted to allow for this.

(k) OUTPUT   This program prints the contents of any data file requested by the user. The program is of value in reminding a user where he left off in the design of a structure and produces a copy of the completed calculations.

(l) NEWSPAN   This program changes the file currently being used by the design programs for the input or output of data to a new file, specified by the user. This program is used when a number of data files has been set up by CONTINUOUS BEAM, each holding data of one span. The user may then select data files through NEWSPAN in order to design the spans using the string of design programs on a span by span basis.

(m) COPY   This program copies the contents of one data file onto another data file. This is of value, as it avoids the necessity of recreating the data for each trial design.

(n) TITLE   This program prints on request a predefined title block.

(o) NAME   This program allows the contents of the title block to be defined.

(p) SCAN   This program prints out the current status of the files in the system.

(q) DRAW BM   This program draws the bending moments and shear-force envelopes currently held in any specified file.

3.8.2 *Microcomputers*

The software used for the multistory frame analysis was STRIDER 1, which is part of a series of engineering software products developed by Claremont Controls Ltd [10] as part of their constructional software suite.

STRIDER 1 uses a developed form of the stiffness method of analysis [11] to solve the problem of rigid-jointed frame structures with in-plane loading. STRIDER 2 and 3 are for the analysis of two-dimensional pin-jointed frames with in-plane and two-dimensional rigid-jointed frames with orthogonal loading, respectively. Data input is straightforward, using a specially provided editing facility for the creation and modification of the data files, which contain program instructions in an easily understood format.

Each program can handle in excess of 50 joints (nodes) and 100 loaded members. Multiple load cases are catered for with combinations and factors as well as batch mode operation, when several jobs may be processed consecutively. This allows several structures to be analysed in a single run, with full combinations of live, dead and wind loading being included in each run.

For each data file, the program will plot the frame on the VDU screen in order that the structural engineer can check the structure coordinates. The program then proceeds to form the stiffness matrix and solves the equations for each load case. The output may be sent to a printer for hard copy or to the VDU screen.

The program description is as follows:

(a) STRIDER 1 consists of two main operating environments. The first is for the preparation of the structure data files and takes the form of an editing program. The second performs the analysis working from the previously constructed data files.

(b) The data files represent the structure in detail. For each data file, it is necessary to give the global nature of the problem, the nodal locations, the member positions together with the type of loading and member properties, and the nodal loading, settlement and restraint.

(c) The program will output the global vertical, horizontal and rotational displacements at each node, together with the axial force, bending moment and shear at each end of every member.

3.8.3   Programmable calculator

This is the area in which a vast amount of growth is possible. The current UK situation is that suitable proprietary software to CP 110 [9] is not readily available. This is because the two principal makes of machine, Hewlett Packard and Texas Instruments, use their own software which emanates from the USA and produces designs in accordance with the ACI code [12].

As far as Europe is concerned, Hewlett Packard SA (Geneva) has published a book [13] of structural engineering programs for the Hewlett Packard HP 41 which, when implemented with the Quad-memory (2K) will automatically carry out a slope deflection analysis on a frame having up to 56 joints and five stories. Many more programs are available, and in the case of Hewlett Packard a Users' Program Library is also available [14]. However, for Section 5, the programs developed by both Steedman [15] and Hewlett Packard SA were used.

The beam analysis and moment diagrams given in Section 5 were produced using the aforementioned Swiss program on an HP 41C in about 8 min. The shear calculations and plots have been included from Hewlett Packard's Structural Engineering Pac, which is a 4K ROM module that plugs into the calculator. The other program used is loaded from magnetic cards.

An HP 97 was also used to calculate the moments and shears on beams. The software used consisted of three programs. The first program calculated the fixed-end moments on a beam. The second program calculated the final support moments in a continuous beam system of up to seven spans. The third program calculated the bending moment and shear forces across a span. By using programs 2 and 3, the final moments and shears throughout each individual span were obtained.

4   Using the small computer

4.1   Nature of structural-concrete applications

The main phases to be considered are

(1) preliminary design;
(2) final design;
(3) detailing;
(4) detailed drawings.

4.1.1 *Preliminary design*

Having decided the outline shape of the structure and to use structural concrete, the structural engineer is faced with the task of obtaining a general outline of the locations and sizes of the beams and columns he wishes to use. He draws a diagram of the frame, the layout plan, together with his assessment of the sizes of the main elements and their loading. This information is data input for the relevant program and the resulting forces, moments, stresses and deflections are output. These results are compared with the allowable stresses and deflections and any element overstressed must be altered. This procedure is followed until all stresses and deflections are within the allowable limits.

4.1.2 *Final design*

At this stage, an analysis is made to ensure that the structure complies with the relevant code of practice, that the correct concrete properties are used, and that the physical factors such as headroom and structural morphology are considered.

The structure is again submitted to the computer and the effect of different load conditions upon the interaction and rigidities of the various structural members can be checked in more detail. The member sizes are confirmed or redetermined as are the amounts of reinforcement.

At this stage the small computer has given a first complete picture of the project and elaborated all the salient features.

4.1.3 *Detailing*

At this stage, the sizes of the concrete elements and the areas of reinforcement have been decided. By using a small computer, the structural optimum of concrete strength, bar diameters and arrangement can be found. Also, a limited number of alternative bar diameters and arrangements can be kept to assist in bar scheduling and ordering.

From the bar sizes, shapes and locations come the bar-bending schedules and cutting lists. From the bar sizes and shapes, the small computer can output bar-bending schedules and cutting lists, floor by floor or bay to bay, depending upon the construction phases.

4.1.4 *Detailed drawings*

The small computer can be used to draw and detail general arrangement drawing. It can also be used to detail beams, slabs and columns. The advantages are that the drawing is readily understandable as it looks like a normal engineering drawing and that alterations can be quickly made.

4.2 Organizing the computer facility

Once a small computer has been installed, the engineer must establish effective operating procedures and the support and maintenance of hardware and software. For this, dedicated management is essential and the number of staff required depends upon the size of the small-computer installation.

Basically, the organizing of the computer facility must encompass the control of operation, job scheduling, hardware operation, user education and support, and systems updating.

To ensure that these areas are covered, various job areas must be identified:

(a) The overall responsibility for the computer facility must be given to one person, usually called a computer manager.

(b) The next area of responsibility is to ensure the hardware and software meet the needs of the engineer. This area is usually covered by a systems analyst.

(c) The programming effort has to be coordinated and a programmer fluent in BASIC and FORTRAN is usually necessary.

(d) To ensure efficient usage and job scheduling, an operations manager is required.

(e) The work of job recording, documentation updates and office organization requires general clerical staff.

The actual number of staff required will depend upon the size of installation. The programmable calculator requires one person for a few hours each week to complete all the requirements described above. However, for a minicomputer with a full range of hardware, and servicing a design office of 40 engineers a staff of four or five may be required.

It must also be remembered that any computer facility once installed will lead to additional applications not originally contemplated. For this reason, the computer facility must be planned to cope with increased work loads.

4.3 Defining the work load

When considering the purchase of a small computer, the structural engineer must understand his requirements and establish the objectives of his solution. He must decide whether he wishes to deal with straightforward calculations for small design projects (such as beam analysis), complex calculations arising from a frame analysis, or even the production of standard drawings.

On the other hand, the structural engineer could look at the benefits he wishes to gain by using a small computer. He may want to

(a) Investigate a number of alternative designs and optimize his solution.
(b) Make small design alterations and be able to check the effect of the alterations very quickly.
(c) Reduce design times so that the customer gets prompt service.
(d) Reduce repetitive tasks and make more time available for other jobs.

By defining the work load in terms of requirements and objectives, the whole design process will become well documented and a design procedure formalized. From this, each area of possible computer activity can be identified and the computing requirements defined.

4.4 Modes of use

With the programmable calculator, the mode of use is immediate. The relevant program is put into the machine, the data keyed in, the program run and the output immediately displayed. However, with the microcomputer and the minicomputer, there is a number of alternative modes of use, such as batch processing, time sharing, interactive mode and conversational mode.

The most usual mode of use is batch processing. Data is submitted in a block and the program run. The time taken between submitting the program and the return of the results depends upon how many jobs the computer is expected to run. The turn round time may be a few minutes on a microcomputer if the structural engineer has immediate access. If, however, a number of programs are waiting to be run on a minicomputer and each one has a special requirement for output or peripherals, the turn round time may be a number of hours.

In many cases, the structural engineer will want to take decisions which depend upon the results of earlier processing; then he will want contact with the machine as the program is running. This contact is usually achieved through a terminal. A small computer used in this way may be dealing with a number of users, each through his own terminal. This is called a multi-user or time-sharing system.

Very often when parts of a structure are being designed, a number of alternative arrangements need to be tested. For this the interactive mode is used. Usually, the program is run with its initial data and then the data altered and the program re-run. This is a very quick method for checking and altering an initial design prior to running the complete analysis, and for refining a final design where some last-minute alteration is required.

The conversational mode is usually used to edit data before a program is run in batch mode and for checking data and program files.

4.5 The need for application support

The structural engineer needs application support to help him understand how to use the small computer and its software. This support depends upon the active participation of people. As many structural engineers will need to use the machine, one person should be responsible for technical support. This support will be required in a number of areas. Assistance will be required purely in running the programs and in maintaining the existing computer system. Once a structural engineer has used a program, he will invariably wish to extend its use and some form of support for program development is needed. As solving a structural concrete problem is a modular process, a structural engineer will wish to know which programs are the best for his use and how to arrange them for convenience of running. For this, systems and program documentation is necessary and good documentation can save much time and frustration.

Lastly, training in hardware and software usage is vital. Following the introduction of either a new computer system or computer, user training is necessary and this must be made easy and readily available.

5 Analysis and design

5.1 Introduction

This section is devoted to the small-computer solution of a multi-story building (Fig. 5-1). The problem chosen has already been used as an illustrative example by Kong and Evans [16]. In this way, the interested reader can compare the non-computer method of Kong and Evans with the small-computer method given

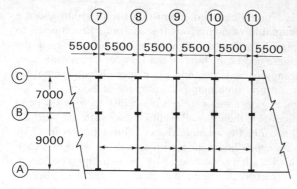

(a) Typical floor plan

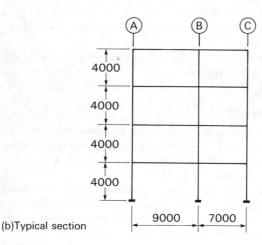

(b) Typical section

Design information

Exposure condition:	internal	Mild
	external	Moderate
Fire resistance		1 hour
Dead loads:	partitions and finishes	1.5 kN/m²
	external cladding	5 kN/m²
Imposed loads:	roof	1.5 kN/m²
	floors	3 kN/m²
Allowable soil bearing pressure		200 kN/m²
Characteristic strengths		
Concrete:	f_{cu} (columns)	40 MPa
	(other members)	30 MPa
Reinforcement:	f_y (main bars)	410 MPa
	f_{yv} (links)	250 MPa

Member sizes

All columns	380 × 380
Main beams	550 × 350
Edge beams	375 × 325

Figure 5-1   Floor plan and cross-section of multistory building

here. The problem was solved in three ways: by minicomputer, by microcomputer solution, and by programmable calculator.

5.2 Minicomputer solution

An Olivetti P6066 was used together with the DECIDE system of interconnecting programs (see Section 3.8).

The analysis and design strategy considered the structural elements in the following order: floor, main floor beams and then columns.

5.2.1 SLAB

SLAB produces design calculations for rectangular restrained two-way spanning slabs as defined in CP 110, Clause 3.4.3.2. The calculations include

(1) preliminary shear check to Tables 5 and 14 of CP 110;
(2) the calculation of design ultimate moments;
(3) the calculation of loads from slab onto supporting beams;
(4) the calculation of steel areas;
(5) the checking deflections by span/effective depth ratios
(6) the calculation of maximum bar spacings to satisfy the provisions of Clause 3.11.8.2.

When the program is used in the most straightforward manner, the resulting moments will be almost identical to those obtained from Table 13 in CP 110. There are, however, refinements built into the program which can lead to economies in detailing. First, it is possible to specify the areas of steel across any side and use these to provide specified support moments. This allows the support moments to be tailored to a convenient bar layout.

Secondly, where the program finds that a steel area less than the minimum area is required at a particular section, the moment corresponding to the minimum area is substituted back as a specified moment. This can lead to reductions in the subsequently calculated steel areas.

The slab analysed is 7×5.5 m and has to be designed to carry a characteristic live load of 3 kN/m^2 and a dead load, including partitions and finishes, of 5.77 kN/m^2. Sides 1, 2 and 4 are continuous and side 3, because it represents an outside edge, is discontinuous. The overall slab depth is 180 mm with effective steel depths of

> *Shorter span direction*
> Top steel 155 mm (155 mm)
> Bottom steel 155 mm (143 mm)
> *Longer span direction*
> Top steel 143 mm (143 mm)
> Bottom steel 143 mm (155 mm)

A minimum area of edge steel of $745 \text{ mm}^2/\text{m}$ width was also specified.
The computer output is shown on pages 22-41 to 22-42.

5.2.2 CONTINUOUS BEAM

This program produces bending-moment and shear-force envelopes for continuous beams of up to seven spans plus two cantilevers. The columns above and below the beam being analysed may be included in the analysis. If they are included, they are assumed fixed at their remote ends. Redistribution to reduce the support moments may be carried out.

The input requirements are a title, the number of spans, the length and second moment of area of each member (optionally, a subroutine may be used to calculate the second moments of area of sections), the dead and imposed load on each span and the partial safety factors required for the dead and imposed loads. To reduce the amount of data which has to be keyed in, facilities exist for the automatic repetition of member data or loading data where the data supplied for one member can be applied to subsequent members. If redistribution is to be carried out, the modified moments for the relevant supports will be required as input.

All data input is recorded on the print-out together with a summary of the elastic support and mid-span moments from each critical load case and the shear and bending moment envelopes for each span (including cantilevers). The percentages of redistribution carried out at each support will also be printed.

The total program consists of the following four separate sub-programs; each of the first three will automatically call up the next program in the sequence when it has reached the end.

CONB   This is used to input the data which defines the frame of beams and attached columns.

CONB2   This is used to input the characteristic loads for each span of the beam.

CONB3   This takes in the partial safety factors for the dead and imposed loads, carries out elastic analyses for the various load patterns, which are generated automatically, and prints out the support and column moments for the critical load patterns.

CONB4   This carries out the redistribution and, if required, prints out the bending-moment and shear-force envelopes for each span (including cantilevers).

The load cases used in the elastic analyses are:

(a) Alternate spans carrying the maximum dead and imposed loads and the remainder carrying only the dead load.

(b) Pairs of adjacent spans carrying the maximum dead and imposed loads and the remainder carrying only the dead load.

The total number of load cases considered will be equal to the number of spans plus cantilevers plus one. The summary will give, for each member, the left-hand and right-hand support moments and the mid-span moments from each of the following load cases:

Alternate A   Odd numbered spans loaded with maximum load.
              Even numbered spans loaded with minimum load.
Alternate B   Odd numbered spans loaded with minimum load.
              Even numbered spans loaded with maximum load.

Adjacent A   Span in question and the one on its left carrying maximum load.

Adjacent B   Span in question and the one on its right carrying maximum load.

The moments in the columns (if any) will also be printed for the load cases where the load is a maximum on either side of the column and on both sides of the column.

When using redistribution, the program will reduce the support moments on each side of the joint to the specified value for any load cases which give support moments greater than the specified value. Exceptions to this procedure are

(a) if a reduction greater than 30% of the maximum moment at a support is asked for, only a 30% reduction will be carried out;
(b) the support moments of cantilevers are not reduced.

The final results of the analysis will be printed out for each span, including cantilevers. The output for each span will consist of

(a) The percentages of the redistribution carried out at the supports. A value of zero is taken for mid-span since the program cannot reduce this moment.
(b) The bending-moment and shear-force envelopes for the member, defined at one-tenth intervals along the span.

As all data generated by CONB, CONB2 and CONB3 are stored on disk, modified analyses can be obtained without the necessity for a complete re-run.

Error checks are built into the loading routines, which check that the load parameters specified are acceptable. For example, they will check that the distance given from the support for a point load does not place it beyond the end of the span. If any of the parameters are found to be unacceptable, the program will return and ask for all the parameters for that particular load to be repeated.

The continuous beam analysis uses the data given in Fig. 5-1, where the columns above and below the beam are 380×380 mm, 4 m long with beam 4 having a span of 9 m and beam 7 a span of 7 m. The beams are considered as T-beams. The characteristic live load is 16.5 kN/m and the characteristic dead load is 34.8 kN/m. The input and output are given on pages 22–43 to 22–47.

5.2.3   *SECTION DESIGN*

This program uses the information sent to disk from CONTINUOUS BEAM to calculate the reinforcement required to resist the moments in the beam sections. The required steel areas are assessed using CP 110's simplified rectangular stress block [17]. Both the tension and the compression steel are each limited to a maximum of 0.04 *bd*. The program will handle hogging or sagging moments on rectangular, T or L-sections with the flange either at the top or at the bottom. The program can also be used completely independently of other programs in the DECIDE system.

The bar-sizing routine is interactive and allows the structural engineer to choose bar numbers and sizes. When the program is being used as part of the DECIDE system and it is intended to use the detailing and shear programs, the use of the sizing routine is essential as the routine stores the bar-size information in the manner required for use by the later program.

The input and results shown in pages 20–48 to 20–51 use the output obtained from CONTINUOUS BEAM. Pages 22-48 to 22-49 give the result for the 9 m span (span 4) and pages 22-49 to 22-51 give the results for the 7 m span (span 7).

5.2.4 SERVICEABILITY

This program checks the code provisions for cracking and deflection control. Deflections are checked using the span/effective depth ratio provisions of Clauses 3.3.8.1 and 3.3.8.2 of CP 110. The method adopted in the program for dealing with cracking is an extension of that used to produce Table 24 in CP 110. The information given consists of a maximum bar spacing for each of the critical sections of the beam. This value may differ from that given in Table 24, but will always be conservative compared with the crack-width restrictions stated in Appendix A of the Code.

If, when the span/effective depth ratio is checked, it is found that the actual ratio slightly exceeds the permissible ratio, the reinforcement areas can be adjusted interactively to produce a satisfactory solution.

In some cases, no serviceability check is required and the program may be omitted. The detail program may then be used immediately after the section design program.

The span/effective depth ratio is as given in Table 8 of CP 110 but the connection required for flanged beam ratios required by Clause 3.3.8.2 of the Code is incorporated in the program and no user action is required. The program will calculate the multipliers required to take account of the effects of tension and compression steel and, if appropriate, the flanged beam correction. These factors will all be printed out together with the calculated permissible ratio and the actual span/effective depth ratio. If the permissible ratio is greater than the actual value, the program will then pass on to the calculation of the maximum permissible bar spacings.

If the actual ratio exceeds the permissible ratio, one of a number of possible courses of action will follow.

If the actual and permissible ratios differ so radically that the steel areas cannot be adjusted sufficiently to produce a permissible ratio, the program will state this.

If the actual ratio only slightly exceeds the permissible ratio, the following will be printed:

(i) TRY INCREASING A'S TO (value)

The values of steel area printed out give the minimum extra area of total reinforcement required to satisfy the span/effective depth ratio provisions. Alternative steel arrangements will be possible. After printing out the required new areas of steel, the program will jump to a bar-sizing routine, which is operated in an identical manner to that in the section design program.

After the successful completion of the span/effective depth ratio check, the program will jump into the crack-checking section. The bottom cover will be requested and the program will compute and print out the maximum bar spacing which will ensure that the crack-width limits are not exceeded at the three critical sections.

The results shown on page 22-51 use the results for the 7 m span.

5.2.5 *DETAIL*

This program carries out bar curtailment in accordance with Table 22 and Clause 3.11.7.1, (1) and (3) of CP 110. It is used in the flow of the main system when bar curtailment is required for a bending-moment diagram or envelope already generated.

The details of the beam, including the bending-moment envelope, will be on disk from a previous program sequence. The only additional input is the sizes and numbers of the bars to be curtailed at the four sections considered.

The program calculates the four curtailment points in the order A, B, C, D, as shown on page 22-53.

The section is assumed to be under-reinforced and the tension steel is assumed to have yielded. This assumption is conservative as, in the SECTION DESIGN program, the neutral axis depth is always kept as $\frac{1}{2}d$ or less.

The rib breadth is used in the lever arm calculation regardless of whether the section is a rectangle, T or L. This is a conservative assumption.

The machine will detect cases when it is impossible to curtail the amount of steel required if the moment in the beam is at all locations too high. The program assumes, for its full operation, that the moment diagram is hogging at the supports and sagging at mid-span. The program will detail simply supported beams if the two end cases are by-passed when giving data. Similarly, the support steel of a beam without a mid-span sagging moment may be detailed if the two mid-span cases are by-passed.

The example given on pages 22-52 to 22-53 uses the details for the 7 m span already held on disk.

5.2.6 *SHEAR*

This program is used as an aid to the design of a beam in shear. The design is in accordance with Clause 3.3.6.1 of CP 110, and is intended to be used in the flow of the main programs.

As a large number of options is open to the engineer when he treats shear, this program must be viewed as a design aid rather than a complete design unit. The program is in two parts, the first of which is automatic and considers a list of predetermined critical sections. The second part considers any sections that the designer specifies.

The first part of the program requires no data beyond that from previous section, apart from the characteristic strength of the stirrup steel.

The output gives the distance from the left-hand support to the section considered, the shear stress at that section, the nominal stress from Table 5 of CP 110 appropriate to the steel percentage at the section, the value of A_{sv}/s_v for the section considered and the local bond stress at the section.

The method of shear design in CP 110 is complicated because the allowable shear stress depends on the tension steel percentage. This gives problems when designing sections away from supports, as at such places it is not always clear whether the top or bottom steel is in tension. Strictly speaking, the shear force diagrams appropriate to the various loading conditions should all be considered. However, this program uses only the shear force diagrams based on the worst of the maximum unredistributed and redistributed shears.

With reference to CP 110, Clause 3.11.4.3, the machine will also recognize a section where the shear stress is less than half the nominal stress allowed for that section and will print an identifying mark by the relevant line.

The shear output for the 7 m span is given on page 22-54.

5.2.7 SUBFRAME

This program analyses framed structures supporting vertical loads using the simplified subframe defined in CP 110, Clause 3.2.2.1. The bending moments resulting from analysis of the subframe may be redistributed at the ultimate limit state subject to the provision of CP 110, Clause 3.2.2.3.

Basically, the program consists of two parts. The first part deals with the definition of the structure, while the second calculates the influence of loads on the structure defined by the first part. In the first part, each member in the subframe must be specified by its length and the second moment of area of its cross-section (or relative second moment of area since it is only the relative stiffnesses of the members which are of importance). To assist in this procedure, the program contains a subroutine for the calculation of section properties. The second part calculates and prints out, for each load case, the support moments for the central beam (member 4) together with the moments in the four columns at the beam–column joints. The support moments may be redistributed interactively and the resulting bending moments and shears are then superimposed upon the results from previous load cases to produce shear-force and bending-moment envelopes. After the final load case, these envelopes are both printed out and remain stored for use by subsequent programs.

This part of the program was not used as CONTINUOUS BEAM had already been used.

5.2.8 SLENDER COLUMN

This program calculates the additional moments for slender-column design according to CP 110, Clauses 3.5.7.1 to 3.5.7.3. The data required for the column would, for the most part, already have been calculated in the rest of the program flow when SUBFRAME was used. The moments and axial loads produced by SLENDER COLUMN or CONTINUOUS BEAM are used as input for COLUMN SECTION DESIGN.

In Clause 3.2.2.1 of the CP 110, it is stated that the subframes may be used to give column moments provided that the centre span is larger than (or equal to) the side spans.

The basic loading that should be considered to produce the column moments for imposed loads is with the most unfavourable span adjacent to either end of the column loaded with live loads (CP 110, Clause 3.5.2). These moments may conveniently be selected from the subframe analysis already carried out. The designer, therefore, has only to select the worst case to produce the basic loading for a braced column.

For braced columns, it is necessary to consider the combinations of live load and wind load as specified in CP 110.

Column loads may be tabulated floor by floor by either taking the worst beam shears from the subframe analysis and adding in the column weight, or by calculating loads from the tributary areas of the floors, assuming all beams and slabs to be simply supported.

Details of the column section, the effective length, and the particular case of loading are all input together with the values of the column load and primary moments.

The program computes the additional moments according to Clauses 3.5.7.1 and 3.7.7.3. The largest primary moment, the additional moment and the reduced primary moment for braced columns are all output.

This program has not been used as SUBFRAME was not used.

5.2.9 COLUMN SECTION

This program is used to calculate the areas of steel required to resist axial load combined with uniaxial or biaxial bending of symmetrically reinforced rectangular sections. Simplifications have been made to the most rigorous procedures laid down in CP 110.

However, the resulting steel areas rarely differ significantly from the design charts. An approximate method has been developed for biaxial bending [9] and is used in this program. The difference between these results and CP 110 is unlikely to exceed 5%.

If slender columns are being designed, CP 110, Clause 3.5.7.4, permits the reduction of the additional moment by a factor K, which is calculated and printed out for each section designed and can be used to make iterative adjustments to the moments. The program automatically checks the code requirements regarding minimum eccentricity and steel ratios.

The example given on pages 22-55 to 22-56 uses the axial loads and moments from Kong and Evans [18] which are reproduced in Table 5.1.

Table 5.1   Total internal column axial loads and moments

Column location	Total axial load kN	Design bending moment kN m
Foundation to 1st floor	2581	107
1st floor to 2nd floor	1982	107
2nd floor to 3rd floor	1330	107
3rd floor to roof	629	109

5.3 Microcomputer solution

The microcomputer used was a Commodore CBM 8032 with CBM 8050 disk-drives and an 80-column printer. The program was STRIDER 1, which analyses rigidly-jointed plane frames under in-plane loading. This program does not analyse structures to CP 110, but allows a typical portion (Fig. 5.1) of the multistory building to be loaded and analysed as a unit to give the structural engineer a more realistic idea of the moments, forces and deflections he can expect in his proposed structure. Usually, a structure would be analysed as separate parts, such as columns and beams, and column–beam interaction would not be taken into account.

5.3.1 *STRIDER* 1 *input*
The input and output data are shown on pages 22-57 to 22-64.

(a) Line 1020 tells the program how many members are in Fig. 5.1, how many nodes there are, and how many load cases for each member have to be considered.
(b) Lines 1100 to 1210 give the cross-sectional area, second moment of area and value of Young's modulus for all of the columns.
(c) Lines 1230 to 1540 give details of the beam members and all their combinations of dead and imposed loads.
(d) Lines 1560 to 1600 detail the restraints at the foundation nodes.
(e) Line 1660 indicates that that the loads also act in combination.

5.3.2 *STRIDER* 1 *output*
The output on pages 22-59 to 22-60 gives a global vertical, horizontal and rotational displacement at every node for each load case.

On pages 22-61 to 22-64, member axial, shear and bending forces are given for each end of every member for each load case.

Local axes are defined as having the positive x-axis along the member from the low node to the high node with the positive y-axis perpendicular and anti-clockwise from this. Rotation is anti-clockwise positive.

5.4. Programmable calculator solution

5.4.1 *Programs*
Using either an HP 41C or an HP 97 programmable calculator, the 9 m and 7 m floor beams were analysed as a two-span continuous beam using either the Hewlett Packard programs or the programs of Steedman [15].

The examples given on page 22-65 are for the two-span continuous beam. In case 1, there is 75.1 kN/m on the 9 m span and 34.8 kN/m on the 7 m span; in case 2 the loads are interchanged. A plot of the bending moment diagrams is also given.

On page 22-66, case 3 has both spans loaded with 75.1 kN/m and a bending moment plot is also given. In Case 4, the shear forces on the 9 m span are printed out for intervals of span/8 for a loading of 75.1 kN/m run on both the 9 m and 7 m spans.

Pages 22-67 and 22-68 show the input and output for three programs. Program 1 on page 22-67 calculates the fixed-end moments for both the 9 m and 7 m spans for both 34.8 kN/m and 75.1 kN/m loads. Program 2 on page 22-67 calculates the final support moments using the fixed-end moments obtained from program 1 for the two-span continuous beam.

Program 3 on page 22-68 calculates the final bending moments and shearing forces at one-eighth points across each span for both dead load and total load.

Program 3 uses the support moments of program 2 to obtain the final moments and shears throughout each individual span.

6 The future

6.1 Hardware development

The principal improvements resulting from the current rapid advances in hardware technology are reduced hardware cost, greater processing power and better long-term storage facilities. Performance and cost improvements in processors and main storage derive from developments in semiconductor technology. It is appropriate to note that in the last 20 years the number of logical elements that can be placed on a single chip (i.e. its integration level) has increased by a factor of 10 every five years.

Such advances in integration level permit two distinct paths to be followed by manufacturers: the production of more powerful large computers and the production of more powerful small computers. In fact, small computers are reliving the history of the development of large computers, but at a vastly accelerated rate.

Developments in backing storage (or memory) are less easy to predict. Cost/performance improvements in magnetic-disk storage are continuing, but there is a growing gap between the performance of such storage and that of semiconductor storage. Some new technologies such as bubble memories will plug this gap and provide very fast backing stores of modest size.

Small-scale backing stores are in a particular state of flux—tape cassettes are already being superseded by floppy disks, which may in turn be ousted by bubble memories and/or small versions of conventional, magnetic disks. One other development with less predictable impact is the video disk, which will permit extremely cheap storage and fast access to huge amounts of data, but does not allow data, once written, to be subsequently modified.

6.2 Software development

There have been extensive software developments in recent years, and a huge growth of application packages. Developments range from improved high-level languages (of which PASCAL is the most widely used) to very sophisticated suites of display-oriented interactive tools for assisting all phases of structural design. However, such new tools may be difficult to combine with the desire of existing users to preserve the value of their past software investment and to use a particular language learnt many years ago. The most successful examples of software development tools are often in design offices using new updated hardware.

Despite such developments, software costs are continuing to rise. This is hardly surprising, since the main function of software is to fit general-purpose mass-manufactured hardware to some specialized application, such as structural design. In contrast to hardware, software development costs cannot be amortized over many subsequent identical production runs as each structural design is unique.

6.3 Conclusion

The pace of development of the small computer will rapidly increase and will have a profound effect upon the structural engineer in the following areas:

(a) Very much more enhanced hardware will become available, extending the size of problem the structural engineer can handle.

(b) Software capabilities will increase, allowing the structural engineer to attempt to solve a wider variety of problems.

(c) For effective use of hardware and software, the structural engineer will have to consider more carefully and codify his methods of working.

(d) The structural engineer will be able to get away from the detailed calculations and time-consuming chores. However, he will have to have a more finely tuned feeling for a structure and how it should behave, because a computer is only as good as the person operating it.

Acknowledgement

The author wishes to acknowledge the assistance given by British Olivetti Ltd, for their assistance in making available their P6066 minicomputer and the DECIDE software; Keystroke Computing Ltd for assisting in the running of their STRIDER 1 software; Jacys Computing Services for developing part of and running the programmable calculator software.

References

1. *Computer users' year book*, CUYB, Bournemouth 1981, 1200 pp.
2. Computer Applications and Method Group, Institution of Civil Engineers, London.
3. British Standards Institution, CP 110: 1972 (as amended 1980), *The structural use of concrete*, London, 1972, 156 pp.
4. *International directory of software 1980–81*, CUYB, Bournemouth, 1981, 1105 pp.
5. Hutton, G. and Rostron, M. (ed) International Dictionary of Computer Programs for the Construction Industry, Architectural Press, London, 1979, 240 pp.
6. Harrison, H. B., *Structural analysis and design: some minicomputer applications*, Parts I and II, Pergamon Press, Oxford, 1980, each 217 pp.
7. Perry, L., Computer programs—art or science? *Pract. Comput.*, Vol. 4, No. 4, April 1981, pp. 104–107.
8. British Standards Institution, CP 114: 1957, *The structural use of reinforced concrete in buildings*, London, 1957, 96 pp.
9. Taylor, H. P. J. and Beeby, A. W., *DECIDE P6060 design of reinforced concrete to CP 110*, British Olivetti Ltd, London, 1978, 171 pp.
10. *Constructional software user manual*, Claremont Controls Ltd, Newcastle upon Tyne, 1981, 86 pp.
11. Coates, R. C., Coutie, M. G. and Kong, F. K., *Structural analysis*, 2nd Edn, Van Nostrand Reinhold, Wokingham, 1982, 579 pp.
12. ACI Committee 318, *Building code requirements for reinforced concrete*, ACI Standard 318–77, American Concrete Institute, Detroit, 1977, 104 pp.
13. Hewlett Packard SA, *Libro di applicazioni ingegneria strutturali*, Volumes I and II, Hewlett Packard SA, Geneva, 1981, 98 pp and 118 pp.

14. *Hewlett Packard user's program library*, Hewlett Packard, SA, Geneva, 1981.
15. Steedman, J. C., JACYS Computing Services, Steyning, 1981.
16. Kong, F. K. and Evans, R. H., *Reinforced and prestressed concrete*, 2nd Edn, Van Nostrand Reinhold, Wokingham, 1982, 412 pp.
17. Beeby, A. W. and Taylor, H. P. J., The use of simplified methods in CP 110. Is rigour necessary?, *Struct. Eng.*, Vol. 56A, No. 8., Aug. 1978, pp. 209–215.

Appendix 1   Minicomputer data for Section 5

SLABS SUPPORTED ON FOUR SIDES

Handbook of Structural Conrete — SLAB

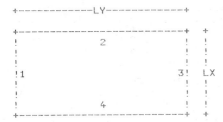

```
DIMENSIONS                          MATERIAL PROPERTIES


SHORTER SPAN      5.5 M      CONCRETE STRENGTH     30 N/MM^2
LONGER SPAN       7 M        STEEL STRENGTH        410 N/MM^2
SLAB THICKNESS    180 MM
EFFECTIVE DEPTHS
      TOP STEEL, SIDES 1 AND 3  143
      TOP STEEL, SIDES 2 AND 4  155
      BOTTOM STEEL, LONG SPAN   143
      BOTTOM STEEL, SHORT SPAN  155

                  EDGE CONDITIONS

      SIDE 1 CONTINUOUS   745 MM^2/M, TOP STEEL
      SIDE 2 CONTINUOUS   745 MM^2/M, TOP STEEL
      SIDE 3 DISCONTINUOUS
      SIDE 4 CONTINUOUS   745 MM^2/M, TOP STEEL

      CHARACTERISTIC LOADS ON SLAB

            DEAD        5.77 KN/M^2
            IMPOSED     3 KN/M^2

      LOADS ON SUPPORTING BEAMS

PEAK LOAD IS SAME FOR ALL BEAMS AND OCCURS 2.75 M FROM END
      MAX PEAK LOAD (DL+LL)  35.41
      MIN PEAK LOAD (DL   )  15.86

MAXIMUM POSSIBLE SHEAR STRESS ON SOLID SLAB   =   .24 N/MM^2
MINIMUM POSSIBLE PERMISSIBLE SHEAR STRESS     =   .43 N/MM^2
NO FURTHER CHECK FOR SHEAR REQUIRED
```

MINIMUM STEEL AREAS MM^2

MIN STEEL AREAS IN MIDDLE STRIPS						MIN STEEL AREAS IN EDGE STRIPS					
SIDES				BOTTOM		SIDES				BOTTOM	
1	2	3	4	SHORT	LONG	1	2	3	4	SHORT	LONG
540	540	540	540	232	216	540	540	540	540	216	216

DESIGN OF CRITICAL SECTIONS

SECTION	MOMENT KN-M/M	STEEL AREA MM^2/M	MAX BAR SPACING MM
TOP SIDE 1	35.06	745	429
TOP SIDE 2	38.24	745	465
TOP SIDE 3	0	540	429
TOP SIDE 4	38.24	745	465
BOTTOM SHORT SPAN	12.56	232	465
BOTTOM LONG SPAN	10.77	216	429

DEFLECTION CHECK

BASIC RATIO	26
MULTIPLIER	2
PERMISSIBLE RATIO	52
ACTUAL RATIO	35.48

DEFLECTIONS OK

```
RUN CONB
**** FORMALLY CORRECT PROGRAM ****
TITLE    (UP TO 52 CHARACTERS)?
MULTI-STOREY BUILDING
           MULTI-STOREY BUILDING


                        CONTINUOUS BEAM ANALYSIS



NO OF SPANS?
2
```

```
MEMBER 1 LGTH    (0=NONE)?
0
MEMBER 2 LGTH    (0=NONE)?
4
INERTIA (0 IF UNKNOWN)?
0
DEPTH?
380
FLANGE DEPTH (0 FOR RECT SEC)?
0
BREADTH?
380
MEMBER  2, AREA=  144400 MM^2, N AXIS DEPTH=  190 MM, I=  17.38
REST OF UPPER COLS SAME (Y/N)?
Y
MEMBER 3 LGTH    (0=NONE)?
4
INERTIA (0 IF UNKNOWN)?
0
DEPTH?
380
FLANGE DEPTH (0 FOR RECT SEC)?
0
BREADTH?
380
MEMBER  3, AREA=  144400 MM^2, N AXIS DEPTH=  190 MM, I=  17.38
REST OF LOWER COLS SAME (Y/N)?
Y
MEMBER 4 LGTH ?
9
```

```
INERTIA (0 IF UNKNOWN)?
0
DEPTH?
550
FLANGE DEPTH (0 FOR RECT SEC)?
180
RIB BREADTH?
350
O/A BREADTH OR 0 FOR T, 1 FOR L?
0
MEMBER  4, AREA=  419300 MM^2, N AXIS DEPTH=  174 MM, I=  90.29
REST OF SPANS SAME (Y/N)?
N
MEMBER 7 LGTH ?
7
INERTIA (0 IF UNKNOWN)?
0
DEPTH?
550
FLANGE DEPTH (0 FOR RECT SEC)?
180
RIB BREADTH?
350
O/A BREADTH OR 0 FOR T, 1 FOR L?
0
MEMBER  7, AREA=  368900 MM^2, N AXIS DEPTH=  186 MM, I=  84.79
MEMBER 10 LGTH  (0=NONE)?
0
```

MEMBER NO	LENGTH M	MOMENT OF INERTIA
1	MEMBER ABSENT	
2	4.00	17.38
3	4.00	17.38
4	9.00	90.29
5	4.00	17.38
6	4.00	17.38
7	7.00	84.79
8	4.00	17.38
9	4.00	17.38
10	MEMBER ABSENT	

```
ALTER MEMBER NO (0=NONE)?
0
**** FORMALLY CORRECT PROGRAM ****
```

LOADING
———————

MEMBER 4
——————————

IMPOSED LOADS

```
MEMBER 4 LIVE LOAD TYPE?
1
UDL  W(KN/M)?
16.5
                    UNIFORMLY DISTRIBUTED LOAD OF 16.5 KN/M
REPEAT ON FOLLOWING SPANS (Y/N)?
Y
                    ***REPEATED ON FOLLOWING SPANS***
MEMBER 4  MORE LIVE  LOADS (Y/N)?
N

      DEAD LOADS
MEMBER 4 DEAD LOAD TYPE?
1
UDL  W(KN/M)?
34.8
                    UNIFORMLY DISTRIBUTED LOAD OF 34.8 KN/M
REPEAT ON FOLLOWING SPANS (Y/N)?
Y
                    ***REPEATED ON FOLLOWING SPANS***
MEMBER 4  MORE DEAD  LOADS (Y/N)?
N


            MEMBER 7
            -----------

      IMPOSED LOADS
MEMBER 7 LIVE LOAD TYPE?
1
UDL  W(KN/M)?
16.5
                    UNIFORMLY DISTRIBUTED LOAD OF 16.5 KN/M
MEMBER 7  MORE LIVE  LOADS (Y/N)?
N

      DEAD LOADS
MEMBER 7 DEAD LOAD TYPE?
1
UDL  W(KN/M)?
34.8
                    UNIFORMLY DISTRIBUTED LOAD OF 34.8 KN/M
MEMBER 7  MORE DEAD  LOADS (Y/N)?
N


CHANGE LOAD ON MEMBER NO (0=NONE)?
0
**** FORMALLY CORRECT PROGRAM ****
LIVE LOAD FACTOR?
1.6
DEAD LOAD FACTOR?
1.4


LIVE LOAD FACTOR= 1.6
DEAD LOAD FACTOR= 1.4


            SUMMARY OF ELASTIC MOMENTS
```

		ALTERNATE A	ALTERNATE B	ADJACENT A	ADJACENT B
SPAN 4	L.H.	-67.52	-258.50	0.00	-222.81
	MID	92.23	353.11	0.00	304.35
	R.H.	-452.72	-556.45	0.00	-689.67
SPAN 7	L.H.	-607.60	-469.97	-736.41	0.00
	MID	465.75	144.56	417.09	0.00
	R.H.	-301.33	-93.52	-269.85	0.00

COLUMN MOMENTS

MEMBER	2	33.76	129.25	0.00
MEMBER	3	-33.76	-129.25	0.00
MEMBER	5	77.44	-43.24	23.37
MEMBER	6	-77.44	43.24	-23.37
MEMBER	8	-150.66	-46.76	0.00
MEMBER	9	150.66	46.76	0.00

```
**** FORMALLY CORRECT PROGRAM ****
RESULTS TO DISK (Y/N)?
Y
FILE NO FOR FIRST SPAN?
1
REDISTRIBUTION (Y/N)?
N
MEMBER  4
```

MEMBER 4
————

BENDING MOMENT AND SHEAR FORCE ENVELOPES

FRACTION OF SPAN	DISTANCE FROM L.H SUPPORT	HOGGING MOMENT KNM	SAGGING MOMENT KNM	SHEAR FORCE KN
0.0	0.0	-258.5	0.0	304.9
0.1	0.9	-14.5	20.8	237.3
0.2	1.8	0.0	170.6	169.7
0.3	2.7	0.0	291.0	102.1
0.4	3.6	0.0	352.5	34.5
0.5	4.5	0.0	353.1	51.9
0.6	5.4	0.0	292.9	119.5
0.7	6.3	-41.2	171.8	187.1
0.8	7.2	-150.2	0.0	254.7
0.9	8.1	-369.2	0.0	322.3
1.0	9.0	-689.7	0.0	389.9

```
RESULTS IN FILE  1
```

MEMBER 7

MEMBER 7

BENDING MOMENT AND SHEAR FORCE ENVELOPES

FRACTION OF SPAN	DISTANCE FROM L.H SUPPORT	HOGGING MOMENT KNM	SAGGING MOMENT KNM	SHEAR FORCE KN
0.0	0.0	-736.4	0.0	592.5
0.1	0.7	-358.5	0.0	487.3
0.2	1.4	-121.8	42.6	382.2
0.3	2.1	0.0	257.3	277.0
0.4	2.8	0.0	398.3	171.8
0.5	3.5	0.0	465.8	66.7
0.6	4.2	0.0	459.6	61.4
0.7	4.9	0.0	379.8	166.6
0.8	5.6	0.0	226.4	271.8
0.9	6.3	-0.7	22.3	376.9
1.0	7.0	-301.3	0.0	482.1

RESULTS IN FILE 2

SUPPORT REACTIONS AND MOMENT TRANSFER

			ALTERNATE A	ALTERNATE B	ADJACENT
JOINT 1 - 4	MOMENT		67.5	258.5	0.0
	REACTION		113.8	304.9	0.0
JOINT 4 - 7	MOMENT		154.9	86.5	46.7
	REACTION		769.0	668.5	982.4
JOINT 7 - 10	MOMENT		301.3	93.5	0.0
	REACTION		482.1	189.8	0.0

```
RUN SECDES
**** FORMALLY CORRECT PROGRAM ****
IS BME ON DISK (Y/N)?
Y
FILE NUMBER ?
1
PROGRAM WILL USE FILE 1
OVERALL DEPTH (MM)?
550
EFFECTIVE DEPTH (MM)?
499
DEPTH TO TOP STEEL (MM)?
48
FLANGE DEPTH (0 FOR RECT SEC)?
180
BOTTOM BREADTH (MM)?
350
TOP BREADTH OR 0 FOR T,1 FOR L?
0
CONCRETE STRENGTH (N/MM^2)?
30
STEEL STRENGTH (N/MM^2)?
410
```

                        SECTION DESIGN

         SECTION DIMENSIONS   MATERIAL PROPERTIES
         ------------------   -------------------

```
OVERALL DEPTH           550 MM      CONCRETE STRENGTH    30 N/MM^2
EFFECTIVE DEPTH         499 MM      STEEL STRENGTH      410 N/MM^2
DEPTH TO TOP STEEL       48 MM
FLANGE OR SLAB DEPTH    180 MM
EFFECTIVE TOP BREADTH  1610 MM
BREADTH AT BOTTOM       350 MM
```

	MOMENT KN-M	REDISTRIBUTION %	BOTTOM STEEL MM^2	TOP STEEL MM^2
L.H. SUPPORT	-258.5	0	0	1651
SPAN	353.11	0	2054	0
R.H. SUPPORT	-689	0	2099	4698

                    L.H. SUPPORT SECTION

```
SECTION              1

    TOP AREA = 1651 BAR SIZE?
25
 4 NO 25 = 1963 CF 1651 NO?
4
             TOP    4 NO 25 MM

             AREA SUPPLIED =   1963 MM^2

        BOTTOM          NIL
```

```
                    SPAN   SECTION
                    --------------------------------
SECTION          2

               TOP              NIL

BOTTOM AREA = 2054 BAR SIZE?
32
  3 NO 32 = 2412 CF 2054 NO?
3
          BOTTOM     3 NO 32 MM

               AREA SUPPLIED =   2412 MM^2


               R.H. SUPPORT  SECTION
               --------------------------------
SECTION          3

    TOP AREA = 4698 BAR SIZE?
25
  10 NO 25 = 4908 CF 4698 NO?
4
               TOP     4 NO 25 MM

    TOP AREA = 2735 BAR SIZE?
32
  4 NO 32 = 3216 CF 2735 NO?
4
               TOP     4 NO 32 MM

               AREA SUPPLIED =   5179 MM^2

BOTTOM AREA = 2099 BAR SIZE?
32
  3 NO 32 = 2412 CF 2099 NO?
3
          BOTTOM     3 NO 32 MM

               AREA SUPPLIED =   2412 MM^2

CHANGES? IF SO, SEC NO (0=NONE)          ?
0
DATA TO DISK (Y/N)?
Y
DATA SENT TO FILE 1
END OF SECTION DESIGN
RUN SECDES
**** FORMALLY CORRECT PROGRAM ****
IS BME ON DISK (Y/N)?
Y
FILE NUMBER ?
2
PROGRAM WILL USE FILE 2
OVERALL DEPTH (MM)?
550
EFFECTIVE DEPTH (MM)?
499
DEPTH TO TOP STEEL (MM)?
48
FLANGE DEPTH (0 FOR RECT SEC)?
180
```

```
BOTTOM BREADTH (MM)?
350
TOP BREADTH OR 0 FOR T,1 FOR L?
0
CONCRETE STRENGTH (N/MM^2)?
30
STEEL STRENGTH (N/MM^2)?
410
```

SECTION DESIGN

SECTION DIMENSIONS		MATERIAL PROPERTIES	
OVERALL DEPTH	550 MM	CONCRETE STRENGTH	30 N/MM^2
EFFECTIVE DEPTH	499 MM	STEEL STRENGTH	410 N/MM^2
DEPTH TO TOP STEEL	48 MM		
FLANGE OR SLAB DEPTH	180 MM		
EFFECTIVE TOP BREADTH	1330 MM		
BREADTH AT BOTTOM	350 MM		

	MOMENT KN-M	REDISTRIBUTION %	BOTTOM STEEL MM^2	TOP STEEL MM^2
L.H. SUPPORT	-736.41	0	2450	4989
SPAN	465.75	0	2771	0
R.H. SUPPORT	-301	0	0	1981

L.H. SUPPORT SECTION

```
SECTION             1

    TOP AREA = 4989 BAR SIZE?
32
 7 NO 32 = 5629 CF 4989 NO?
4
              TOP    4 NO 32 MM

    TOP AREA = 1773 BAR SIZE?
25
 4 NO 25 = 1963 CF 1773 NO?
4
              TOP    4 NO 25 MM

              AREA SUPPLIED =  5179 MM^2

BOTTOM AREA = 2450 BAR SIZE?
32
 4 NO 32 = 3216 CF 2450 NO?
4
           BOTTOM   4 NO 32 MM

              AREA SUPPLIED =  3216 MM^2
```

SPAN SECTION

```
SECTION            2
```

```
                TOP              NIL
BOTTOM AREA = 2771 BAR SIZE?
32
 4 NO 32 = 3216 CF 2771 NO?
4
            BOTTOM    4 NO 32 MM

                AREA SUPPLIED =   3216 MM^2


            R.H. SUPPORT  SECTION
                 ---------------------------
SECTION          3

   TOP AREA = 1981 BAR SIZE?
25
 5 NO 25 = 2454 CF 1981 NO?
5
            TOP    5 NO 25 MM

            AREA SUPPLIED =   2454 MM^2

        BOTTOM            NIL
CHANGES? IF SO, SEC NO (0=NONE)          ?
0
DATA TO DISK (Y/N)?
Y
DATA SENT TO FILE 2
END OF SECTION DESIGN
RUN SERVIC
**** FORMALLY CORRECT PROGRAM ****
PROGRAM WILL USE FILE 2
                DEFLECTION CHECK


BASIC RATIO (SEE MANUAL)?
26
            BASIC SPAN/EFFECTIVE DEPTH RATIO    26
            TENSION STEEL MULTIPLIER           1.45
    .       COMPRESSION STEEL MULTIPLIER       1.00
            MULTIPLIER FOR FLANGED BEAM        0.80

            HENCE PERMISSIBLE RATIO =         30.12
            ACTUAL RATIO =                    14.03


        PERMISSIBLE RATIO > ACTUAL RATIO
        THEREFORE DEFLECTIONS OK
        ----------------------------------
```

```
        MAXIMUM BAR SPACINGS TO ENSURE CRACK WIDTHS LESS THAN 0.3 MM


COVER (MM)?
35
        COVER = 35 MM

        LEFT-HAND SUPPORT... 207 MM



        MID-SPAN............ 526 MM
        RIGHT-HAND SUPPORT.. 359 MM


END OF SERVICEABILITY
RUN DETAIL
**** FORMALLY CORRECT PROGRAM ****


                        DETAIL
```

```
**** FORMALLY CORRECT PROGRAM ****
PROGRAM WILL USE FILE 2
BAR TYPE (P,D OR D2)?
D


                    DEFORMED BARS
        ANCHORAGE BOND STRESS =  2.2 N/MM^2



                TENSION STEEL DETAILS
        -----------------------------------

        LEFT HAND SUPPORT - A    4 NO 32
                            +    4 NO 25

        MID SPAN - CD            4 NO 32

        RIGHT HAND SUPPORT - B   5 NO 25
```

CURTAILED BARS

```
LHS (A) SIZE?
32
HOW MANY CURTAILED?
4
LHS (A) SIZE?
25
HOW MANY CURTAILED?
4
MID SPAN LHS (C) SIZE?
32
HOW MANY CURTAILED?
4
MID SPAN RH (D) SIZE?
32
HOW MANY CURTAILED?
4
RHS (B) SIZE?
25
HOW MANY CURTAILED?
5
```

```
        LEFT HAND SUPPORT  - A      4 NO 32
                             +      4 NO 25

        MID SPAN L - C              4 NO 32

        MID SPAN R - D              4 NO 32

        RIGHT HAND SUPPORT - B      5 NO 25
```

CALCULATED CURTAILMENT POINTS

```
        DIMENSION A        2.602

        DIMENSION B        5.098

        DIMENSION C         .201

        DIMENSION D        7.499
```

```
CHANGE CURTAILMENT ( Y OR N )?
N
DATA SENT TO FILE 2
LAST CURL'T TAKEN OVER TO SHEAR
END OF DETAIL
```

```
RUN SHEAR
**** FORMALLY CORRECT PROGRAM ****
PROGRAM WILL USE FILE 2
SHEAR STEEL STRENGTH  N/MM^2?
250
                              SHEAR




         CHARACTERISTIC STRENGTH OF SHEAR STEEL =  250 N/MM^2
                PERMISSIBLE LOCAL BOND STRESS =  2.8 N/MM^2


         DISTANCE        VC              V            ASV/SV        FBL
         FROM LHS M     N/MM^2        N/MM^2          MM          N/MM^2
AUTOMATIC ( Y OR N )?
Y
           0.00          0.95          3.37          3.90         1.64

           0.70          0.95          2.77          2.94         1.35

           1.40          0.95          2.18          1.98         1.06

           2.10          0.87          1.59          1.15         1.38

           2.80          0.87          0.98          0.69         0.85

           3.50          0.87          0.38          0.69         0.33      +

           4.20          0.87          0.35          0.69         0.30      +

           4.90          0.87          0.95          0.69         0.83

           5.60          0.87          1.56          1.11         1.35

           6.30          0.78          2.15          2.20         1.91

           7.00          0.78          2.74          3.16         2.44

ANY MORE SECTIONS ( Y OR N )?
N


         £ = INADEQUATE SHEAR CAPACITY
         * = INADEQUATE LOCAL BOND CAPACITY
         + = SHEAR STRESS LESS THAN 1/2 PERMISSIBLE


END OF SHEAR
```

```
RUN COLSEC
**** FORMALLY CORRECT PROGRAM ****
                SYMMETRICAL RECTANGULAR COLUMN SECTIONS


TITLE    (UP TO 52 CHARACTERS)?

TITLE·   (UP TO 52 CHARACTERS)??
TEST
            TEST
```

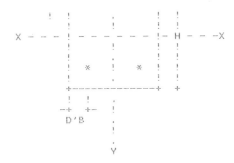

```
OVERALL DEPTH, H (MM)?
380
BREADTH, B (MM)?
380
D'H (MM)?
50
D'B (MM)?
50
STEEL STRENGTH (N/MM^2)?
410
CONCRETE STRENGTH (N/MM^2)?
40
        SECTION DIMENSIONS                      MATERIAL PROPERTIES
        ------------------                      -------------------

OVERALL DEPTH               380 MM      CONCRETE STRENGTH      40.0 N/MM^2
BREADTH                     380 MM      STEEL STRENGTH         410  N/MM^2
DEPTHS TO STEEL   D'H    50 MM
                  D'B    50 MM
```

AXIAL LOAD KN	MOMENT ABOUT X AXIS KN-M	MOMENT ABOUT Y AXIS KN-M	TOTAL STEEL AREA MM^2	K

```
AXIAL LOAD (KN)?
2581
MOMENT ABOUT X AXIS (KN-M)?
107
MOMENT ABOUT Y AXIS (KN-M)?
0
     2581.00      107.00       0.00        2380       0.34

MORE LOAD CASES (Y/N)?
Y
AXIAL LOAD (KN)?
1982
MOMENT ABOUT X AXIS (KN-M)?
107
MOMENT ABOUT Y AXIS (KN-M)?
0
     1982.00      107.00       0.00        1444       0.56

MORE LOAD CASES (Y/N)?
Y




AXIAL LOAD (KN)?
1330
MOMENT ABOUT X AXIS (KN-M)?
107
MOMENT ABOUT Y AXIS (KN-M)?
0
     1330.00      107.00       0.00        1444       0.91

MORE LOAD CASES (Y/N)?
Y
AXIAL LOAD (KN)?
629
MOMENT ABOUT X AXIS (KN-M)?
109
MOMENT ABOUT Y AXIS (KN-M)?
0
      629.00      109.00       0.00        1444       1.00

MORE LOAD CASES (Y/N)?
N
DATA TO DISK (Y/N)?
N


END OF COLUMN SECTION
```

Appendix 2 Microcomputer data for Section 5

```
-----------------------------------------------------------------------
                             STRIDER  1
-----------------------------------------------------------------------

YOUR REFERENCE
DATE RUN
DATA FILE
JOB NUMBER


-----------------------------------------------------------------------
                             DATA
-----------------------------------------------------------------------

1000 ;
1000 'TEST FRAME'
1010 ;
1020 *MEMBER 20 *NODE 15 *LOADCASE 3
1030 ;
1040 *CO-ORDS 1 3        0     0 7000     0 16000      0
1050 *CO-ORDS 4 6        0  4000 7000  4000 16000   4000
1060 *CO-ORDS 7 9        0  8000 7000  8000 16000   8000
1070 *CO-ORDS 10 12      0 12000 7000 12000 16000  12000
1080 *CO-ORDS 13 15      0 16000 7000 16000 16000  16000
1090 ;
1100 *MEMBER 1 4 *AREA 144400 *I-VALUE 1.7376E9 *E-VALUE 31000
1110 *MEMBER 4 7
1120 *MEMBER 7 10
1130 *MEMBER 10 13
1140 *MEMBER 2 5
1150 *MEMBER 5 8
1160 *MEMBER 8 11
1170 *MEMBER 11 14
1180 *MEMBER 3 6
1190 *MEMBER 6 9
1200 *MEMBER 9 12
1210 *MEMBER 12 15
1220 ;
1230 *MEMBER 4 5 *I-VALUE 4.8526E9 *AREA 192500 *E-VALUE 28000
1240 *LOADCASE 1 *Y-DIR *UDL -34.8
1250 *LOADCASE 2 *Y-DIR *UDL -75.1
1260 *LOADCASE 3 *Y-DIR *UDL -75.1
1270 *MEMBER 5 6
1280 *LOADCASE 1 *Y-DIR *UDL -75.1
1290 *LOADCASE 2 *Y-DIR *UDL -34.8
1300 *LOADCASE 3 *Y-DIR *UDL -75.1
1310 *MEMBER 7 8
1320 *LOADCASE 1 *Y-DIR *UDL -34.8
1330 *LOADCASE 2 *Y-DIR *UDL -75.1
1340 *LOADCASE 3 *Y-DIR *UDL -75.1
1350 *MEMBER 8 9
1360 *LOADCASE 1 *Y-DIR *UDL -75.1
1370 *LOADCASE 2 *Y-DIR *UDL -34.8
1380 *LOADCASE 3 *Y-DIR *UDL -75.1
1390 *MEMBER 10 11
1400 *LOADCASE 1 *Y-DIR *UDL -34.8
1410 *LOADCASE 2 *Y-DIR *UDL -75.1
1420 *LOADCASE 3 *Y-DIR *UDL -75.1
1430 *MEMBER 11 12
1440 *LOADCASE 1 *Y-DIR *UDL -75.1
1450 *LOADCASE 2 *Y-DIR *UDL -34.8
1460 *LOADCASE 3 *Y-DIR *UDL -75.1
1470 *MEMBER 13 14
1480 *LOADCASE 1 *Y-DIR *UDL -30.6
1490 *LOADCASE 2 *Y-DIR *UDL -56.04
1500 *LOADCASE 3 *Y-DIR *UDL -56.04


------------------------------------------------------- PAGE  --  1
```

```
1510 *MEMBER 14 15
1520 *LOADCASE 1 *Y-DIR *UDL -56.04
1530 *LOADCASE 2 *Y-DIR *UDL -30.6
1540 *LOADCASE 3 *Y-DIR *UDL -56.04
1550 ;
1560 *RESTRAINT
1570 *LOADCASE 1
1580 *NODE 1 *RESTRAINT *Y-DIR *X-DIR
1590 *NODE 2 *RESTRAINT *Y-DIR *X-DIR
1600 *NODE 3 *RESTRAINT *Y-DIR *X-DIR
1610 ;
1620 ;
1630 *OUTPUT
1640 *XTRALOADCASES 1
1650 ;
1660 *COMBINATION *LOADCASE 4
1670 ;
1680 *NUMBER-IN-COMBO 2        1 1.6 2 1.4
1690 ;
1700 *FINISH
```

                        NODAL   DEFLECTIONS

NODE	LCASE	GLOBAL X	GLOBAL Y	GLOBAL Z	LCASE	NODE
1	1	0	0	9.83879979E-04	1	1
	2	0	0	8.74361693E-04	2	
	3	0	0	1.27018933E-03	3	
	4	0	0	2.79831434E-03	4	
2	1	0	0	1.71478202E-03	1	2
	2	0	0	-2.05053393E-04	2	
	3	0	0	1.03259713E-03	3	
	4	0	0	2.45657648E-03	4	
3	1	0	0	-9.52224746E-04	1	3
	2	0	0	-6.95767704E-04	2	
	3	0	0	-1.12521804E-03	3	
	4	0	0	-2.49763438E-03	4	
4	1	-2.38827319	-.356181272	-1.76555067E-04	1	4
	2	-.0221974778	-.825831854	-1.73207528E-03	2	
	3	-1.65002825	-.7991399	-1.30285749E-03	3	
	4	-3.85231357	-1.72605463	-2.7073935E-03	4	
5	1	-2.36218274	-1.67089917	-1.65792699E-03	1	5
	2	.0352901609	-1.51885287	3.83639165E-04	2	
	3	-1.59301184	-2.15475125	-8.70435392E-04	3	
	4	-3.73008616	-4.79983269	-2.11558835E-03	4	
6	1	-2.2353015	-1.07993404	3.58092562E-03	1	6
	2	.0927450394	-.500798172	1.32197663E-03	2	
	3	-1.46723931	-1.06848386	3.35086557E-03	3	
	4	-3.44663935	-2.42901191	7.58024828E-03	4	
7	1	-4.27304069	-.625213558	-2.62429865E-04	1	7
	2	.0770021475	-1.43389074	-1.37774074E-03	2	
	3	-2.87378127	-1.38995538	-1.11638899E-03	3	
	4	-6.7290621	-3.00778873	-2.34872482E-03	4	
8	1	-4.27471007	-2.89317523	-1.27733374E-03	1	8
	2	.0696527072	-2.63248752	2.83360351E-04	2	
	3	-2.87963594	-3.72595389	-6.76129532E-04	3	
	4	-6.74202233	-8.3145629	-1.64702949E-03	4	
9	1	-4.29390646	-1.87399781	2.72127186E-03	1	9
	2	.0615640143	-.87496732	1.06964927E-03	2	
	3	-2.89743922	-1.85512083	2.58523565E-03	3	
	4	-6.78406072	-4.22335075	5.85154395E-03	4	
10	1	-5.81016481	-.798666262	-2.86741043E-04	1	10
	2	.108132651	-1.81825213	-1.41788323E-03	2	
	3	-3.90748768	-1.76262117	-1.17470348E-03	3	
	4	-9.14487798	-3.823419	-2.44382219E-03	4	
11	1	-5.80187237	-3.68042304	-1.3451004E-03	1	11
	2	.11708781	-3.35147752	3.3676344E-04	2	
	3	-3.89835832	-4.73013941	-6.9972258E-04	3	
	4	-9.11907286	-10.5807454	-1.68069182E-03	4	
12	1	-5.7794434	-2.37702704	2.72524628E-03	1	12
	2	.136449807	-1.11785841	1.02912519E-03	2	

	3	-3.87605867	-2.35320468	116	2.597234E-03	3	
	4	-9.05607971	-5.36824504		5.80116931E-03	4	
13	1	-7.33618156	-.875301094		-5.84088253E-04	1	13
	2	.10144424	-1.97729749		-1.70019805E-03	2	
	3	-4.91740545	-1.91567249		-1.47629C6E-03	3	
	4	-11.5958686	-4.16869824		-3.31481847E-03	4	
14	1	-7.37904244	-4.03502262		-1.60121234E-03	1	14
	2	.0195591155	-3.67827708		1.23153845E-04	2	
	3	-5.00019636	-5.16988834		-9.47717117E-04	3	
	4	-11.7790851	-11.6056241		-2.38952436E-03	4	
15	1	-7.56012165	-2.58788		3.53382922E-03	1	15
	2	-.0718542039	-1.22863575		1.78529936E-03	2	
	3	-5.18078907	-2.56161966		3.40944352E-03	3	
	4	-12.1967905	-5.86069805		8.15354586E-03	4	

```
-----------------------------------------------------------------------
                            MEMBER   FORCES
-----------------------------------------------------------------------
```

MBR	LC	NODE	LOCAL X	LOCAL Y	LOCAL Z	NODE	LC	MBR
1	1	1	398602.462	-7813.44127	-.0333251953	1	1	1
		4	-398602.462	7813.44127	-31253765	4		
	2	1	924188.428	-17549.6614	-.0172071457	1	2	
		4	-924188.428	17549.6614	-70198645.7	4		
	3	1	894317.462	-17324.8391	-.439697266	1	3	
		4	-894317.462	17324.8391	-69299355.8	4		
	4	1	1931627.74	-37071.032	-.0638427734	1	4	
		4	-1931627.74	37071.032	-148284128	4		
2	1	4	301074.031	-27903.0948	-54649765.3	4	1	2
		7	-301074.031	27903.0948	-56962614.1	7		
	2	4	680478.7	-61815.1433	-128401897	4	2	
		7	-680478.7	61815.1433	-118858676	7		
	3	4	661181.604	-61227.472	-124966003	4	3	
		7	-661181.604	61227.472	-119943885	7		
	4	4	1434388.63	-131186.152	-267202281	4	4	
		7	-1434388.63	131186.152	-257542329	7		
3	1	7	194110.921	-26617.6788	-52907973.6	7	1	3
		10	-194110.921	26617.6788	-53562741.7	10		
	2	7	430138.832	-56156.0741	-111771573	7	2	
		10	-430138.832	56156.0741	-112852723	10		
	3	7	417050.287	-56719.3795	-112653473	7	3	
		10	-417050.287	56719.3795	-114224045	10		
	4	7	912771.838	-121206.79	-241132961	7	4	
		10	-912771.838	121206.79	-243694199	10		
4	1	10	85762.0399	-33002.8674	-62001538.4	10	1	4
		13	-85762.0399	33002.8674	-70009931.3	13		
	2	10	177987.662	-63051.5463	-122301328	10	2	
		13	-177987.662	63051.5463	-129904857	13		
	3	10	171279.731	-63748.9875	-123436682	10	3	
		13	-171279.731	63748.9875	-131559268	13		
	4	10	386401.992	-141076.753	-270424321	10	4	
		13	-386401.992	141076.753	-293882690	13		
5	1	2	1869903.26	-22709.1244	-.137634277	2	1	5
		5	-1869903.26	22709.1244	-90836497.4	5		
	2	2	1699748.25	3963.78473	-7.11250305E-03	2	2	
		5	-1699748.25	-3963.78473	15855138.9	5		
	3	2	2411382.12	-12813.4987	-.331481934	2	3	
		5	-2411382.12	12813.4987	-51253994.7	5		

```
------------------------------------------------------------- PAGE  --  5
```

	4	2	5371492.76	-30785.3004	-.234863281	2	4	
		5	-5371492.76	30785.3004	-123141201	5		
6	1	5	1367849.14	-78607.2362	-162339693	5	1	6
		8	-1367849.14	78607.2362	-152089251	8		
	2	5	1246268.54	13820.1783	28990751.2	5	2	
		8	-1246268.54	-13820.1783	26289961.9	8		
	3	5	1758332.88	-44234.6389	-91085878.3	5	3	
		8	-1758332.88	44234.6389	-85852677.4	8		
	4	5	3933334.57	-106423.328	-219156458	5	4	
		8	-3933334.57	106423.328	-206536856	8		
7	1	8	881009.025	-68396.1545	-135879736	8	1	7
		11	-881009.025	68396.1545	-137704882	11		
	2	8	804621.71	13005.3376	25291527.8	8	2	
		11	-804621.71	-13005.3376	26729822.5	11		
	3	8	1123784.01	-38080.5546	-75843395.8	8	3	
		11	-1123784.01	38080.5546	-76478822.7	11		
	4	8	2536084.84	-91226.3745	-181999439	8	4	
		11	-2536084.84	91226.3745	-182906060	11		
8	1	11	396832.389	-75443.4411	-147437976	11	1	8
		14	-396832.389	75443.4411	-154335788	14		
	2	11	365721.387	8305.12488	19486802	11	2	
		14	-365721.387	-8305.12488	13733697.5	14		
	3	11	492123.025	-44405.9667	-85472339.8	11	3	
		14	-492123.025	44405.9667	-92151527.1	14		
	4	11	1146941.77	-109082.331	-208619239	11	4	
		14	-1146941.77	109082.331	-227710084	14		
9	1	3	1208554.18	30522.6081	.0756835938	3	1	9
		6	-1208554.18	-30522.6081	122090432	6		
	2	3	560443.234	13585.8762	.0216064453	3	2	
		6	-560443.234	-13585.8762	54343504.6	6		
	3	3	1195740.29	30138.3663	.217834473	3	3	
		6	-1195740.29	-30138.3663	120553465	6		
	4	3	2718307.22	67856.3996	.169677734	3	4	
		6	-2718307.22	-67856.3996	271425598	6		
10	1	6	888636.765	106510.37	224597181	6	1	10
		9	-888636.765	-106510.37	201444298	9		
	2	6	418732.693	47994.9644	99387870	6	2	
		9	-418732.693	-47994.9644	92591987.7	9		
	3	6	880325.432	105462.137	221234553	6	3	
		9	-880325.432	-105462.137	200613996	9		
	4	6	2008044.6	237609.542	498498508	6	4	
		9	-2008044.6	-237609.542	451939660	9		
11	1	9	562940.011	95013.8618	189974203	9	1	11
		12	-562940.011	-95013.8618	190081245	12		

	2	9	271819.419	43150.7361	86847185.8	9	2	
		12	-271819.419	-43150.7361	85755758.8	12		
	3	9	557405.637	94799.9533	189438332	9	3	
		12	-557405.637	-94799.9533	189761481	12		
	4	9	1281251.21	212433.209	425544784	9	4	
		12	-1281251.21	-212433.209	424188054	12		
12	1	12	235965.547	108446.327	206003953	12	1	12
		15	-235965.547	-108446.327	227781356	15		
	2	12	123970.922	54746.4213	99309898.8	12	2	
		15	-123970.922	-54746.4213	119675787	15		
	3	12	233237.205	108154.967	205372396	12	3	
		15	-233237.205	-108154.967	227247473	15		
	4	12	551104.164	250159.114	468640184	12	4	
		15	-551104.164	-250159.114	531996271	15		
13	1	4	-20089.6455	97528.4311	85903530.3	4	1	13
		5	20089.6455	146071.569	-255804513	5		
	2	4	-44265.4818	243709.735	198600544	4	2	
		5	44265.4818	281990.265	-332582395	5		
	3	4	-43902.6357	233135.861	194265359	4	3	
		5	43902.6357	292564.139	-402264335	5		
	4	4	-94115.107	497239.119	415486410	4	4	
		5	94115.107	628500.881	-874902574	5		
14	1	5	-75987.7654	355982.565	508980703	5	1	14
		6	75987.7654	319917.435	-346687614	6		
	2	5	-34409.0883	171489.459	287736505	5	2	
		6	34409.0883	141710.541	-153731375	6		
	3	5	-75323.7709	360485.132	544604208	5	3	
		6	75323.7709	315414.868	-341788019	6		
	4	5	-169753.149	809657.347	1.21720023E+09	5	4	
		6	169753.149	710262.653	-769924108	6		
15	1	7	1285.42179	106963.114	109870588	7	1	15
		8	-1285.42179	136636.886	-213728793	8		
	2	7	5659.06908	250339.873	230630250	7	2	
		8	-5659.06908	275360.128	-318201143	8		
	3	7	4508.09443	244131.325	232597358	7	3	
		8	-4508.09443	281568.675	-363628084	8		
	4	7	9979.37208	521616.803	498675290	7	4	
		8	-9979.37208	604123.197	-787447668	8		
16	1	8	11496.5042	350203.253	501697779	8	1	16
		9	-11496.5042	325696.747	-391418501	9		
	2	8	4844.22827	166286.72	266619653	8	2	
		9	-4844.22827	146913.28	-179439174	9		
	3	8	10662.1875	352980.203	525324157	8	3	
		9	-10662.1875	322919.797	-390052328	9		

	4	8	25176.3249	793126.613	1.17598396E+09	8	4	
		9	−25176.3249	726793.387	−877484445	9		
17	1	10	−6385.17618	108348.886	115564280	10	1	17
		11	6385.17618	135251.114	−209722077	11		
	2	10	−6895.4724	252151.175	235154052	10	2	
		11	6895.4724	273548.825	−310045827	11		
	3	10	−7029.60817	245770.561	237660727	10	3	
		11	7029.60817	279929.439	−357216800	11		
	4	10	−19869.9448	526369.863	514118522	10	4	
		11	19869.9448	599370.137	−769619480	11		
18	1	11	−13432.4614	348925.526	494864935	11	1	18
		12	13432.4614	326974.474	−396085199	12		
	2	11	−11595.6849	165351.505	263829202	11	2	
		12	11595.6849	147848.495	−185065658	12		
	3	11	−13355.0121	351731.565	519167963	11	3	
		12	13355.0121	324168.435	−395133877	12		
	4	11	−37725.8978	789772.949	1.16114478E+09	11	4	
		12	37725.8978	730147.051	−892828239	12		
19	1	13	33002.878	85762.0417	70009931.3	13	1	19
		14	−33002.878	128437.958	−219375640	14		
	2	13	63051.5459	177987.671	129904858	13	2	
		14	−63051.5459	214292.329	−256971163	14		
	3	13	63749.0002	171279.74	131559269	13	3	
		14	−63749.0002	221000.26	−305581086	14		
	4	13	141076.769	386402.005	293882691	13	4	
		14	−141076.769	505509.995	−710760652	14		
20	1	14	108446.326	268394.452	373711427	14	1	20
		15	−108446.326	235965.548	−227781356	15		
	2	14	54746.4213	151429.076	243237466	14	2	
		15	−54746.4213	123970.925	−119675786	15		
	3	14	108154.969	271122.793	397732613	14	3	
		15	−108154.969	233237.207	−227247472	15		
	4	14	250159.111	641431.83	938470736	14	4	
		15	−250159.111	551104.17	−531996271	15		

Appendix 3 Programmable calculator data for Section 5

JACYS COMPUTING SERVICES

HP 41C OUTPUT

CASE 1			CASE 2		
BEAM OF 2 TO 5 SPANS			BEAM OF 2 TO 5 SPANS		
(A, B)		SPANS = 2.00			SPANS = 2.00
9 m span		L1 = 9.00	9 m span (A, B)		L1 = 9.00
load on 9 m span		W1 = 75.10	load on 9 m span		W1 = 34.80
7 m span (B, C)		L2 = 7.00	7 m span (B, C)		L2 = 7.00
load on 7 m span		W2 = 34.80	load on 7 m span		W2 = 75.10
moment at A		M1 = 0.00	moment at A		M1 = 0.00
moment at C		M3 = 0.00	moment at C		M3 = 0.00
final	A	M1 = 0.00	final	A	M1 = 0.00
support	B	M2 = −520.75	support	B	M2 = −399.22
moments	C	M3 = 0.00	moments	C	M3 = 0.00
	AB	M12 = 522.30		AB	M12 = 181.01
span	BC	M23 = 32.39	span	BC	M23 = 282.03

PLOT OF MOMENT	PLOT OF MOMENT
X (UNITS = 1)	X (UNITS = 1)
Y (UNITS = 1)	Y (UNITS = 1)

```
          −521          522              −399          282
                  0                              0
        ------------+----------          ------------+----------
  0.0          x                    0.0          x
  0.9            x                  0.9             x
  1.8              x                1.8            x
  2.7                x              2.7             x
  3.6                 x             3.6             x
  4.5                 x             4.5             x
  5.4               x              5.4            x
  6.3             x                6.3           :x
  7.2          :x                  7.2       x   :
  8.1     x    :                   8.1     x    :
  9.0x         :                   9.0x         :
  9.7  x       :                   9.7     x    :
 10.4     x                       10.4           x:
 11.1       x  :                  11.1         :   x
 11.8         x:                  11.8         :    x
 12.5          x:                 12.5         :      x
 13.2          x                  13.2         :      x
 13.9          :x                 13.9         :      x
 14.6          :x                 14.6         :    x
 15.3          :x                 15.3         :  x
 16.0          x                  16.0       x
```

JACYS COMPUTING SERVICES
HP 41C

<table>
<tr><td colspan="2">CASE 3</td><td colspan="2">CASE 4</td></tr>
<tr><td colspan="2">BEAM OF 2 TO 5 SPANS</td><td colspan="2">SHEAR</td></tr>
</table>

	SPANS = 2.00	Distance	X = 0.00
9 m span (A, B)	L1 = 9.00	shear	Y = 280.06
load on 9 m span	W1 = 75.10		
7 m span (B, C)	L2 = 7.00	$\dfrac{l_{AB}}{8}$	X = 1.13
load on 7 m span	W2 = 75.10		Y = 195.58
	M1 = 0.00		
	M3 = 0.00	$\dfrac{l_{AB}}{4}$	X = 2.25
			Y = 111.09

final   A   M1 = 0.00
support   B   M2 = −628.74 $\dfrac{3l_{AB}}{8}$ X = 3.38
moments   C   M3 = 0.00   Y = 26.60

      AB   M12 = 478.51 $\dfrac{l_{AB}}{2}$ X = 4.50
span   BC   M23 = 199.33   Y = −57.89

                                     $\dfrac{5l_{AB}}{8}$ X = 5.63
PLOT OF MOMENT   Y = −142.37

        X (UNITS = 1) $\dfrac{3l_{AB}}{4}$ X = 6.75
        Y (UNITS = 1)   Y = −226.86

       −629   479 $\dfrac{7l_{AB}}{8}$ X = 7.88
            0   Y = −311.35

```
 0.0        ẋ       -----    l_AB    X = 9.00
 0.9          x                     Y = −395.84
 1.8             x
 2.7               x
 3.6               x
 4.5               x
 5.4             x
 6.3          x
 7.2      x¦                 PLOT OF BEAM
 8.1   x
 9.0x                        X (UNITS = 1)
 9.7  x                      Y (UNITS = 1)
10.4   x  ¦                 −396          280
11.1     x¦
11.8     ¦x                    0
12.5      ¦   x            ------------r----------
13.2      ¦     x
13.9      ¦     x        0.00        ¦        x
14.6      ¦   x         1.13         ¦       x
15.3      ¦ x          2.25        ¦   x
16.0      ẋ           3.38        ¦x
                       4.50     x¦
                       5.63   x  ¦
                       6.75  x   ¦
                       7.88  ẋ   ¦
                       9.00x     ¦
```

JACYS COMPUTING SERVICES

HP 97 OUTPUT

PROGRAM 1

TO CALCULATE FIXED END MOMENTS

Span AB	9.0000	GSBA
Dead Load on AB	34.80000	GSBB
		GSBe
		R/S
FEM_{AB}	−234.9000	* * *
		R/S
FEM_{BA}	234.90000	* * *
Span AB	9.0000	GSBA
Total Load on AB	75.10000	GSBB
		GSBe
		R/S
FEM_{AB}	−506.9250	* * *
		R/S
FEM_{BA}	506.9250	* * *
Span BC	7.0000	GSBA
Dead Load on BC	34.8000	GSBB
		GSBe
		R/S
FEM_{BC}	−142.1000	* * *
		R/S
FEM_{CD}	142.10000	* * *
Span BC	7.0000	GSBA
Total Load on BC	75.1000	GSBB
		GsBe
		R/S
FEM_{BC}	−306.6583	* * *
		R/S
FEM_{CB}	306.6583	* * *

PROGRAM 2

FINAL SUPPORT MOMENTS

			GSBA
		9.00	1/X
	K_{AB}		R/S
	K_{BC}	7.00	1/X
	K_{BC}		R/S
	Fixity A	0.00	GSBB
	Fixity C	0.00	R/S
Case 1	FEM_{AB}	506.925	GSBC
	FEM_{BA}	506.925	GSBD
	FEM_{BC}	142.10	GSBC
	FEM_{CB}	142.10	GSBD
			GSBE
	M_A	0.00	* * *
	M_B	−520.97	* * *
	M_C	0.00	* * *
Case 2	Restart		GSBe
	FEM_{AB}	234.90	GSBC
	FEM_{BA}	234.90	GSBD
	FEM_{BC}	306.6583	GSBC
	FEM_{CB}	306.6583	GSBD
			GSBE
	M_A	0.00	* * *
	M_B	−399.44	* * *
	M_C	0.00	* * *
Case 3	Restart		GSBe
	FEM_{AB}	506.925	GSBC
	FEM_{BA}	506.925	GSBD
	FEM_{BC}	306.6583	GSBC
	FEM_{CB}	306.6583	GSBD
			GSBE
	M_A	0.00	* * *
	M_B	−628.96	* * *
	M_C	0.00	* * *

PROGRAM 3

JACYS COMPUTING SERVICES   FINAL BENDING MOMENTS AND SHEARS

Total load on AB

L_{AB}	9.00	GSBA
load	75.10	GSBe
M_A	0.00	ENT ↑
M_B	520.97	GSBE
At A Shear	280.06	***
moment	0.00	***
At $\frac{1}{8}l$	195.58	***
	267.55	***
At $\frac{1}{4}l$	111.09	***
	440.05	***
At $\frac{3}{8}l$	26.60	***
	517.50	***
At $\frac{1}{2}l$	−57.89	***
	499.90	***
At $\frac{5}{8}l$	−142.37	***
	387.26	***
At $\frac{3}{4}l$	−226.86	***
	179.56	***
At $\frac{7}{8}l$	−311.35	***
	−123.18	***
At B Shear	−395.84	***
moment	−520.97	***

Total load on BC

l_{BC}	7.00	GSBA
load	75.10	GSBB
M_B	520.97	ENT ↑
M_C	0.00	GSBE
At B Shear	337.27	***
moment	−520.97	***
$\frac{l}{8}$	271.56	***
	−254.60	***
$\frac{l}{4}$	205.85	
	−45.74	
$\frac{3l}{8}$	140.14	***
	105.63	***
$\frac{l}{2}$	74.42	***
	199.50	***
$\frac{5l}{8}$	8.71	***
	235.87	***
$\frac{3l}{4}$	−57.00	***
	214.75	***
$\frac{7l}{8}$	−122.71	***
	136.12	***
At B	−188.43	***
	0.00	***

Dead load on AB

l_{BC}	9.00	GSBA
load	34.80	GSBB
M_A	0.00	ENT ↑
M_B	520.97	GSBE
At A Shear	87.71	***
moment	0.00	***
$\frac{l}{8}$	59.56	***
	89.03	***
$\frac{l}{4}$	20.41	***
	134.02	***
$\frac{3l}{8}$	−18.74	***
	134.96	***
$\frac{l}{2}$	−57.89	***
	91.87	***
$\frac{5l}{8}$	−97.04	***
	4.72	***
$\frac{3l}{4}$	−136.19	***
	−126.47	***
$\frac{7l}{8}$	−175.34	***
	−301.70	***
At B Shear	−214.49	***
moment	−520.97	***

Dead load on BC

l_{BC}	7.00	GSBA
load	34.80	GSBB
M_B	520.97	ENT ↑
M_C	0.00	GSBE
At B Shear	196.22	***
moment	−520.97	***
$\frac{l}{8}$	165.77	***
	−362.68	***
$\frac{l}{4}$	135.32	***
	−230.87	***
$\frac{3l}{8}$	104.87	***
	−125.78	***
$\frac{l}{2}$	74.42	***
	−47.34	***
$\frac{5l}{8}$	43.97	***
	4.46	***
$\frac{3l}{4}$	13.52	***
	29.62	***
$\frac{7l}{8}$	−16.93	***
	28.13	***
At C Shear	−47.38	***
moment	0.00	***

23 Models for structural concrete design

Richard N White

Cornell University, Ithaca, NY, USA

Contents

Notation

a/c	aggregate/cement ratio in mix	r	number of dimensions needed to express a physical problem
D	thermal diffusivity	S_i	scale factor for quantity $i = i_p/i_m$
E	elastic modulus	S_E'	modulus scaling factor for reinforcement
E_c	elastic modulus of concrete		
F	dimension of force	s/g	sand/gravel ratio in mix
f_c'	uniaxial compressive strength of concrete, measured on cylinders	T	dimension of time
		t	time; model dimension
h	model dimension	w/c	water/cement ratio in mix
L	dimension of length		
l	model dimension	α	coefficient of thermal expansion
m	model (when used as subscript); number of dimensionless ratios	δ	displacement
		ε	strain
M	dimension of mass	ρ	density
n	number of parameters in model study	θ	dimension of temperature temperature
p	prototype (when used as subscript)	σ	stress
Q	concentrated load	ν	Poisson's ratio
q	distributed surface load intensity	ω	frequency

Summary

The physical models approach for determining the behavior of concrete structures is presented. This approach is a complement to analytical modeling to be used for unusually complex and monumental structures, both in design and in research. An introduction to modeling terminology and classification is followed by a definition of the many distinct steps involved in planning, executing, and interpreting a successful model study. Similitude requirements are presented for elastic, strength, dynamic, and thermal models. A brief discussion of quantities to be measured and the appropriate instrumentation is given. Major sections are devoted to elastic models and to strength models, with coverage of model materials and fabrication techniques, loading techniques (both static and dynamic), how to choose the correct geometrical scaling factor, examples of modeling applications to a variety of structural problems, and reliability of results.

1   Introduction

The physical models approach for determining the behavior of a concrete structure is an important complement to analytical modeling. It is particularly useful in resolving difficult questions in the design of complex and monumental structures, and as a research tool to help build up the necessary base of physical evidence for developing and substantiating improved analytical models.

The basic concepts of physical modeling of concrete structures are presented in this chapter. An introduction to models terminology and classification is followed by a definition of the many distinct steps involved in planning, executing, and interpreting a successful model study. Similitude requirements, which form the basis for model design, loading, and interpretation of results, are presented in Section 2. A very brief discussion of experimental techniques is given in Section 3. The next two sections, 4 and 5, are on elastic models and strength models; they form the heart of the chapter. The chapter contains a number of illustrative examples as well as a brief look into the future of physical modeling of reinforced and prestressed concrete structures.

A *structural model* is defined by ACI Committee 444 (*Models of concrete structures*) as 'any physical representation of a structure or a portion of a structure. Most commonly, the model will be constructed at a reduced scale.' This rather broad definition encompasses a variety of types of model, which may be conveniently classified by type of desired response (*elastic* or *post-elastic* up to failure; the latter is commonly called a *strength* model), and by type of loading (*static, dynamic,* or *thermal*). The loading history is dictated by the circumstances of each individual modeling study, but the decision to utilize the elastic or strength model must be made by the responsible engineer. An *elastic* model is designed and built for predicting elastic response only and is useful for substantiating elastic analyses and for portraying the working load behavior of reinforced and prestressed concrete structural forms that have relatively little cracking. The *strength* model is intended to enable the prediction of post-cracking behavior, with reasonable accuracy, as the loads increase all the way to failure. The latter application places severe demands on the model, particularly with regard to materials used in building the model. Concrete and reinforcing must be reproduced at a reduced scale and special fabrication techniques are needed to produce models of adequate dimensional tolerances.

Other classes of models are often seen in the literature, including:

(a) *indirect*—an experimental approach for getting influence lines and surfaces, which will not be treated here;

(b) *shape*—a wind-tunnel model of a structure on which wind pressures may be measured;

(c) *aeroelastic*—a model used to determine wind-induced stresses and deformations in a structure;

(d) *photomechanical*—in which either the *photoelastic effect* or an *interference pattern* may be interpreted to give stresses, strains, or displacements.

The use of reduced-scale structures for research studies is so common that it hardly needs mention. Design models, on the other hand, are not nearly so common because most engineers are accustomed to being able to 'get their answers' analytically. The use of models in design practice is usually confined to those situations where analytical results for the idealized mathematical model of the structure are considered to be inadequate. The type and degree of analytical inadequacies vary considerably, of course, and may arise from sources such as the following:

(a) Basic doubts are met in attempting to apply existing analytical techniques to new and complex structural forms.

(b) Existing methods are not sufficiently well developed and documented to treat behavior at load stages approaching failure of the structure.

(c) Simplification of the three-dimensional structure to a two-dimensional structure is questionable.

(d) The consequences of potential failure are so great that a confirmation of analytical predictions is considered essential.

Examples of each of these categories will be given later in the chapter.

Reference [1] has a listing of types of structure that are candidates for possible structural model studies during the design phase. It includes shell roof forms of complex configuration and boundary conditions, nuclear reactor vessels and other reinforced and prestressed concrete pressure vessels, dams, undersea (offshore) structures, complex bridge configurations such as the multi-cell prestressed box-girder bridge, structural slabs with unusual boundary or loading conditions, or with irregular geometry produced by cut-outs, ordinary framed structures subjected to complex loads and load histories such as blast, earthquake, and wind, and finally, new building structural systems involving the interaction of many components.

Figure 1-1 shows a model of a reinforced concrete cooling tower in the laboratory of the Cement and Concrete Association, England. This model was built from model concrete and small-scale reinforcement and was used to study the behaviour of the thin shell structure under simulated wind action. The loading system will be described later in this chapter.

Figure 1-1 Strength model of a reinforced concrete cooling tower (By courtesy of Cement and Concrete Association)

The modeling process involves careful planning to execute the sequential steps involved in each study. The necessity for careful planning is further accentuated by the unfortunate fact that many design model studies are initiated too late, that is, after all other efforts to obtain the desired results have been exhausted.

The planning process begins by defining the scope of the modeling study; this should include decisions on what is needed and what is not needed, and on the required level of accuracy. Costs and time requirements, which are perhaps the two main deterrents to wider use of models in design, both increase rather sharply as the model becomes more sophisticated and acceptable error levels are tightened. For example, strength models are more expensive than elastic models and usually must be built at substantially larger scales, which in turn increases the costs of the loading system.

The next step in planning is a similitude analysis to set requirements for geometry, materials of construction, loading, and interpretation of results. A final decision on geometric scale and on the selection of model materials is made at the end of this phase. This is followed by planning of the model fabrication process, which should also include specification of strength tests on model materials, and then design of the necessary loading equipment. Appropriate instrumentation devices and recording equipment must be specified and ordered if not already available. Finally, a schedule for fabrication, testing, and interpreting the results should be drawn up, including identification of critical path elements.

One final comment on terminology is appropriate in this introductory section of the chapter—symbols, strength tests (such as the cylinder strength of concrete), units, and other terminology will conform to that used by the American Concrete Institute in its codes and publications.

The reference list at the end of the chapter is necessarily quite short. Many additional references are cited in [2–4, 16].

2 Similitude requirements

Similitude requirements for structural models, which may be derived from basic concepts of dimensional analysis, have been treated by many authors [2–5]. The discussion given here is directed at how to (a) quickly establish similitude requirements, and (b) apply the requirements to typical structural modeling problems.

Similitude in modeling is needed in designing the model (*geometry* and *materials*), in establishing model *loading* intensities and *time* histories, and in *predicting* the results of the prototype (full-size) structure from the measured model response. Each of these categories of similitude relations will be defined in this chapter.

Similitude (or scaling) *relations* are derived by establishing a complete set of pertinent dimensionless ratios from the various physical parameters involved in the particular study at hand, and then forcing these dimensionless ratios to be equal in model and prototype. Scaling laws come directly from the forced equalities. For a problem that has n parameters expressible in r different basic fundamental measures (dimensions such as length, force, time, and temperature), it can be shown that there will be $m = n - r$ dimensionless ratios and hence $(n - r)$ scaling relationships to be enforced in the model study. Static, mechanically loaded structural problems are expressible in terms of force F and length L, or $r = 2$; for dynamic mechanical systems, the necessary dimensions are length L, time T, and either force F or mass M, and $r = 3$ for this category of structures. Transient thermal modeling requires the use of temperature units θ in addition to length L, time T, and either F or M; hence $r = 4$. These three cases with $r = 2, 3$, or 4 cover all possible applications discussed in this chapter.

The n pertinent physical parameters must be prescribed by the engineer for each individual study. If a significant parameter is overlooked, the modeling will in general be incorrect. If unnecessary parameters are included, the resulting similitude relations will be overly restrictive. Thus, there is no substitute for careful study of each structural problem prior to moving ahead with a physical modeling effort.

2.1 Static elastic models

Models intended to predict prototype response in the elastic range as produced by non-thermal mechanical loads are the simplest of all structural models. As an example, consider a static elastic model of a prototype reinforced concrete shell carrying its own self-weight plus uniformly distributed surface loading and some concentrated loads, as shown in Fig. 2-1. The pertinent physical parameters are shell thickness t, shell plan dimensions l and w, shell rise h, concrete modulus E_c, concrete Poisson's ratio ν, concrete density ρ, distributed load intensity q, and concentrated load intensity Q_i for each of the i loads.

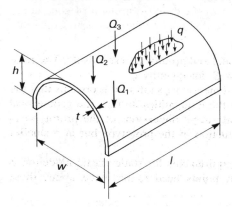

Figure 2-1 Shell model geometry

Quantities to be determined in the model study include stress σ, strain ε, and displacement δ.

Note that since only elastic stresses and displacements can be predicted by this type of model, the material strengths of the concrete and reinforcing steel are irrelevant. Also, since an elastic approach is customarily based on the assumption of homogeneous isotropic materials, the amount and distribution of reinforcement is not considered to be relevant. The only possible influence of the reinforcing steel is in how it might affect the value of E_c for the composite reinforced concrete material in the shell. Cracking of the concrete and subsequent reduction of stiffness cannot be modeled with this type of model, which will be discussed in more detail in Section 4.

The parameters may be expressed in the dimensions of force, F, and length, L, as follows:

Quantity	t	l	w	h	E_c	ν	ρ	q	Q	σ	ε	δ
Force, F	0	0	0	0	1	–	1	1	1	1	–	0
Length, L	1	1	1	1	–2	–	–3	–2	0	–2	–	1

The numbers in the table represent the exponents on the units of length L and force F; that is modulus E_c is in units such as pounds per square inch, or F/L^2, or FL^{-2}. Two quantities are dimensionless—Poisson's ratio ν and strain ε.

Since $n = 12$ and $r = 2$, there will be $(12-2) = 10$ dimensionless ratios and 10 scaling factors to be enforced. A proper (but not unique) set of 10 dimensionless ratios is given by:

$$\frac{t}{l} \quad \frac{w}{l} \quad \frac{h}{l} \quad \nu \quad \frac{\rho l}{E} \quad \frac{q}{E} \quad \frac{Q}{E l^2} \quad \frac{\sigma}{E} \quad \varepsilon \quad \frac{\delta}{l}$$

The first scaling factor is determined by setting the ratios

$$(t/l)_{\text{model}} = (t/l)_{\text{prototype}}$$

and solving for

$$t_m = t_p(l_m/l_p)$$

where the subscripts m and p denote model and prototype, respectively. The ratio (i_p/i_m) will be defined as the scale factor S_i for quantity i.

By inspection of the second and third dimensionless ratios, it is obvious that the same scale factor S_l must be applied to these quantities. In fact, the geometrical scale factor S_l must be applied uniformly to all dimensions of the prototype to produce a complete geometrical reproduction of the prototype but at a smaller scale.

A more complete set of similitude requirements for static elastic modeling is given in Table 2-1. Several important points need to be made about these modeling laws:

(a) The strain scaling factor S_ε is always unity, that is, model and prototype strains must be equal in a true, undistorted model.

(b) The only requirements on model material for an elastic model is that it be linear elastic within the stress history it sees, and that $S_\nu = 1$ (equal values of Poisson's ratio in model and prototype). The latter requirement is often relaxed in applications where Poisson's ratio has only a minor influence on results.

(c) All stress-related and loading similitude requirements contain the scale factor S_E. By using a model material with low modulus E, S_E becomes large, and model load intensities are sharply reduced.

(d) Self-weight dead load stresses cannot be scaled if the same material is used in model and prototype. Since the density scale factor $S_\rho = S_E/S_l = \rho_p/\rho_m$, then the required density of model material is

$$\rho_m = \rho_p(S_l/S_E)$$

If $S_E = 1$, then $\rho_m = S_l\rho_p$ and similitude on density (and on self-weight stresses) is violated unless $S_l = 1$, which is not a very practical solution since it means a full-scale model. This difficulty in self-weight similitude is true for all structural modeling applications. It can be met in two different ways: (i) select a model material with values of E and ρ such that the relation $\rho_m = \rho_p(S_l/S_E)$ is satisfied, or (ii) apply additional loads to the model to augment the model self-weight to the proper level of total load. The latter approach is feasible for linear and surface structures but may be impossible to implement in a three-dimensional solid structure such as a dam.

Table 2-1 Similitude requirements for static elastic models

Parameter	Dimensions	Scale factor
Geometry		
Linear dimension, *l*	L	S_l
Area, *A*	L^2	S_l^2
Moment of inertia, *I*	L^4	S_l^4
Section modulus, *S*	L^3	S_l^3
Linear displacement, δ	L	S_l
Angular displacement, β	—	1
Materials and related parameters		
Stress, σ	FL^{-2}	S_E
Modulus of elasticity, *E*	FL^{-2}	S_E
Poisson's ratio, ν	—	1
Density, ρ	FL^{-3}	S_E/S_l
Strain, ε	—	1
Loading		
Concentrated force, *Q*, and shear force, *V*	F	$S_E S_l^2$
Pressure or uniformly distributed load, *q*	FL^{-2}	S_E
Line load, *w*	FL^{-1}	$S_E S_l$
Moment, *M*, or torque, *T*	FL	$S_E S_l^3$

Distorted models, in which one or more important similitude requirements are not met, may be useful in some instances. A general treatment of this topic is beyond the scope of this handbook. One example is given in Section 2.2 and there is additional discussion in [1–5].

2.2 Static strength models

The strength model for reinforced concrete structures is intended to portray behavior through the failure stage. Hence, the models must be built from model concrete and small-scale reinforcement with properties similar to prototype materials. In fact, the failure criteria for model and prototype concretes should be identical, but this is usually relaxed to the simpler requirements of similar stress–strain curves for model and prototype concrete in both uniaxial tension and compression.

It is theoretically possible to satisfy all similitude requirements for reinforced concrete strength models by using a model concrete that fails at the same strain level as the prototype concrete but at a different stress level, i.e. $S_\varepsilon = 1$ and $S_\sigma = S_E$. This would require a model reinforcement with the same modulus scale factor[1] $S_E' = S_E$, which means that steel reinforcement could not be used in the model since S_E' automatically equals 1 for steel reinforcement. It is not practical to use other metals and the use of stress scaling factors other than unity will lead to distortions. Hence, only two categories of similitude are useful in reinforced

1 S_E' refers to model reinforcement and S_E to model concrete

concrete strength models:

(a) $S_\sigma = S_\varepsilon = 1$, or model and prototype materials (both concrete and reinforcing steel) have identical strengths, failure strains, and moduli. The corresponding similitude requirements are given in Table 2-2 for this *practical true model*.

(b) The concrete strain scale factor $S_\varepsilon \neq 1$, or strains at failure are distorted. This is permissible in structural models when response is not influenced significantly by the absolute magnitude of strains, but in any geometrically non-linear behavior (such as in beam-columns, shell instability, etc.) a strain distortion will introduce errors that are usually not acceptable. If the distorted similitude approach is used, the model steel reinforcement strain at yield must satisfy the requirement $S'_\varepsilon = S_\varepsilon$; since $S'_E = 1$ for the reinforcement, this strain scaling requirement means that the reinforcement stress scale factor, S'_σ, must also be equal to $S'_\varepsilon = S_\varepsilon$. The model concrete strength can differ from the prototype strength by the independent scale factor, S_σ. Stress–strain curves for this case of distorted modeling are given in Fig. 2-2 and corresponding similitude requirements are summarized in Table 2-2. The correct total force in model reinforcing is provided by using a distorted area of model steel.

Table 2-2   Similitude requirements for reinforced concrete strength models

Parameter	Dimensions	Practical true model scale factor	Distorted model scale factor
Geometry Identical to Table 2-1 with the following exceptions:			
Area of reinforcing steel, A_r	L^2		$S_\sigma S_l^2 / S_\varepsilon$
Linear displacement, δ	L		$S_\varepsilon S_l$
Angular displacement β	—		S_ε
Materials and related parameters			
Concrete strength, f'_c and f'_t	FL^{-2}	1	S_σ
Concrete strain, ε_c	—	1	S_ε
Modulus of concrete, E_c	FL^{-2}	1	S_σ / S_ε
Poisson's ratio of concrete, ν_c	—	1	1
Concrete density, ρ_c	FL^{-3}	$1/S_l$	S_σ / S_l
Reinforcing steel strength, f_y	FL^{-2}	1	S_ε
Reinforcing steel strain, ε_y	—	1	S_ε
Modulus of reinforcing steel, E_s	FL^{-2}	1	1
Bond stress, u	FL^{-2}	1	[a]
Loading Identical to Table 2-1 with $S_E = 1$ for the practical true model			

[a] Function of distorted model reinforcing selection

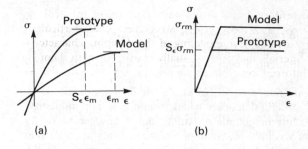

Figure 2-2 Distorted strength models: (a) concrete and (b) reinforcement

In both categories of similitude, the model concrete tensile strength properties must meet the same similitude requirements as the compressive strength if tension-sensitive phenomena, such as cracking and diagonal tension strength, are to be modeled. Since deformations of the model structure are tied rather strongly to degree of cracking, the tensile similitude requirements normally are crucial.

Finally, the bond strength between concrete and model reinforcement must be properly modeled. This rather difficult task is discussed in Section 5.3.

2.3 Dynamic models

The dynamic response of a typical structure is dependent upon inertial forces, resisting or restoring forces, and damping forces, which in turn are functions of the local mass and acceleration, the stiffness of the structure, and the properties of the materials of construction and the way in which the structural elements are connected together. In addition, in some types of structure the self-weight effects are important and the acceleration of gravity must be included along with the mass density of the material.

Several classes of dynamic modeling for concrete structures are in use, including the vibration of elastic structures under mechanical forces, wave action, wind-induced loadings, response to seismic loads, and the effects of blast loadings. The similitude requirements for each of these loadings are different and a full treatment of each is beyond the scope of this chapter.

Similitude requirements for elastic vibrations are similar to those given in Table 2-1 with two additional parameters—time t and frequency ω. For true models, including all gravity effects, the time scale factor $S_t = \sqrt{S_l}$, and the frequency scale factor $S_\omega = 1/\sqrt{S_l}$. Thus, a 1/100 scale model ($S_l = 100$) will have a natural frequency ten times higher than the corresponding prototype structure.

Similitude requirements for wind-tunnel modeling leads to conflicting require-ments and practical aeroelastic modeling requires that either Reynold's number or Froude's number (related to gravity stresses) be neglected. The corresponding similitude requirements are given in [3]. Reynold's number is usually not impor-tant for sharp-edged structures but should be modeled for rounded structural surfaces such as chimneys. Froude's number is important where an appreciable part of the total structural stiffness is derived from gravity forces, such as in suspension and cable-stayed bridges, but it is normally not critical for ordinary buildings, free-standing towers, and most bridge structures. One particularly important consideration here is to achieve modeling of the damping characteristics of structural concrete. Materials that are essentially elastic but with damping characteristics similar to reinforced concrete or to prestressed concrete will be discussed in Section 4.4.

Seismic effects modeling is complicated by the fact that ordinary seismic design implies some inelastic behavior of the structure; hence the strictly elastic model is not fully realistic. True elastic models that satisfy full similitude of inertial, restoring, damping, and gravitational forces must meet the similitude requirements of Table 2-1, the time and frequency scaling laws given above, and the damping characteristics of the model material must be identical to those of the prototype. Simultaneous satisfaction of the scaling of material density and damping is usually impossible, and the models engineer is forced either to use artificial mass simulation or to neglect gravity force effects. In the latter case, identical materials are used in model and prototype and all materials-related scale factors (second category in Table 2-1) become unity. The loading scale factors are as in Table 2-1 with S_E set equal to unity, and the time-dependent scale factors on seismic acceleration a, velocity v, time t, and frequency ω are:

$$S_a = 1/S_l \quad S_v = 1 \quad S_t = S_l \quad S_\omega = 1/S_l$$

With the same materials being used in model and prototype, this type of model can be extended beyond the elastic range to determine failure behavior.

2.4   Thermal modeling

Thermal effects modeling in concrete structures has been done for a variety of problems, including arch dams subjected to variable atmospheric and water temperatures, nuclear reactor vessels under operating and accident conditions, and reinforced concrete structures subjected to fire effects. Similitude requirements for elastic structural response are identical to those in Table 2-1 with the addition of scaling factors for four additional parameters: temperature θ, time t, coefficient of linear expansion α, and thermal diffusivity D (equal to the thermal conductivity divided by the product of the specific heat capacity and the specific weight of the material). These new scaling factors are easily shown to be

$$S_\theta = 1/S_\alpha \quad S_t = S_l^2/S_D \quad S_\alpha = S_\alpha \quad S_D = S_D$$

It is seen that the addition of quantities that need two new dimensions (time and temperature) leads to the specification of two additional independent scaling factors. Here, the two thermal property quantities of coefficient of expansion and thermal diffusivity have been selected as appropriate for prescribed scaling.

The time scale factor comes from the dimensionless ratio commonly called Fourier's number (tD/l^2); it leads to very short times and permits the modeling of long-time thermal effects, such as the annual temperature variation in a dam, in a matter of hours when S_l is large.

The time-dependent nature of the loading system must follow the time similitude requirements, of course, and this fact may seriously complicate the modeling study.

3 Experimental stress analysis

The instrumentation phase of a model study often consumes a substantial portion of the total project budget (in both time and expenses). Some appreciation of measuring devices and their auxiliary equipment will be of great help in planning any modeling project. Three key questions must always be answered: what to measure, where to measure, and how to measure. The temptation to 'over-specify' the number of strain gages and other devices at this point in the planning process is very real, but it can lead to unnecessary delays and extra costs in terms of devices, testing personnel, reduction of data, and interpretation time.

The key to successful modeling in a design context is to measure no more and no less than what is needed to achieve a safe and economic design. On the other hand, research models usually require more elaborate instrumentation in order to capture all relevant data.

Coverage of experimental stress analysis will be limited here to a very brief listing of quantities to be measured and the associated devices and supporting instrumentation. More complete coverage is given in [3, 6–8].

3.1 Quantities to be measured and the associated instrumentation

The phrase 'experimental stress analysis' is somewhat of a misnomer in that stresses are hardly ever measured experimentally; instead, stresses are calculated from measured strains and known values of material properties. In addition to strain, other primary quantities to be measured in static modeling include linear and angular displacements, forces and reactions, intensity of distributed pressure loads, and cracks in strength models. Techniques satisfactory for short-term tests (a few days or less) may be unsuitable for long-term experiments where creep behavior must be determined—the new element of instrument stability becomes crucial.

Dynamic modeling requires the measurement of accelerations and velocities as well as strains, displacements, and loads, all as a function of time. Dynamic sensing and recording devices must be highly sophisticated to maintain an acceptable accuracy while measuring quantities over very short time periods. Finally, thermal modeling requires the sensing of temperatures, usually as a function of time.

A tabulation of experimental quantities and associated instrumentation follows:

Strain Strains must be measured on the surface of concrete and reinforcing steel and inside the concrete. The electrical resistance strain gage is by far the most versatile surface gage, particularly for measuring strain over short gage lengths. Three electrical resistance strain gages cemented to the surface of a concrete nuclear pressure vessel model are shown in Fig. 3-1. In addition to showing the installation of typical gages, this figure also points out the fact that, in the general case, three independent strains, measured in different directions, are needed to determine the magnitudes and directions of strains (and stresses) in a structure. Mechanical gages that measure the distance between two metal plugs embedded in the surface of the model structure should not be overlooked when a longer gage length is satisfactory. Embedded gages include Carlson strain meters,

Figure 3-1   Electrical resistance strain gages mounted on concrete (By courtesy of Oak Ridge National Laboratory)

vibrating wire gages, and various custom devices using electrical resistance strain gages. All these devices (except the mechanical gage) require substantial supporting electrical circuitry. Modern data acquisition systems controlled by microcomputers enable rapid strain sensing and reduction of data from many dozens of gages on a single model.

Displacement   The simplest displacement device is the mechanical dial gage, but it must be read and recorded manually. Electrical devices, such as linear variable differential transformers (LVDTs), direct current differential transformers (DCDTs), and linear potentiometers, offer automatic sensing and recording but at a substantially higher initial cost. Angular displacements are sensed with mechanical inclinometers made with level bubbles and micrometers, and with rotational LVDTs and DCDTs.

Forces, reactions, and loads   The standard method of determining concentrated forces and reactions is to use load cells instrumented with electrical resistance strain gages and calibrated to convert electrical output into force magnitude. Vacuum loads (see Section 4.3) are best sensed with a simple water manometer.

Cracking   Cracking in reinforced concrete models, including the length and width of cracks, may be determined visually, acoustically, or electrically. Visual methods are enhanced by applying whitewash (lime and water) or a brittle lacquer to the model surface before testing; photoelastic coatings may also be utilized.

The acoustic emission technique is an important new method that electrically senses the noise produced by material cracking. The other electrical approaches use either electrical resistance strain gages or a special crack detection coating that changes its electrical properties upon cracking.

Discussion of the instrumentation for dynamic and thermal models is beyond the scope of this book, and the interested reader is referred to [3, 4, 6–9].

4 Elastic models

Elastic models were defined in Section 1 and their similitude requirements were given in Section 2. In this section, the various roles of the elastic model in contemporary structural engineering for reinforced concrete will be treated and illustrated, and such topics as materials, fabrication, loading techniques, and reliability of results will be covered.

4.1 Applications to design problems

The role of the elastic model in design is changing constantly as computer-based analytical modeling capabilities are gradually extended to cover more and more structural analysis problems. In recent years, the elastic model has been used quite widely to help define the behavior of a broad array of reinforced and prestressed concrete structures, including stress distribution in complex bridges, shell roofs, and unusual shaped slabs, stresses in prestressed concrete pressure vessels, thermal and dynamic response of arch dams, response of large structures (buildings, towers, and bridges) to wind, seismic, and wave action, and stability of folded plates, shells, and dams. Another important use has been in investigating portions of structures to get a better definition of load sharing between parts of a three-dimensional configuration. Elastic models also have seen considerable application in the development of improved structural theories where detailed experimental evidence was needed to either implement or check a new mathematical model.

The elastic model cannot predict post-cracking behavior of concrete structures. Hence, it is best applied to structures that have relatively little cracking, such as structures subjected to low-level loadings, prestressed concrete structures, and shells.

4.2 Materials and fabrication

The simple similitude requirements for elastic model materials are given in Section 2. Nearly all metals have a linear elastic response over an appreciable stress range and hence are candidates for elastic model making, but they suffer several disadvantages, including high E values (which means high loading intensities on the model because prototype loads are always multiplied by the ratio E_m/E_p) and difficulty in forming and fabrication. Aluminum has had some applications in model studies of shells and folded plates and a few steel models have been used in research studies for shell structures.

Certain types of plastic are the most popular materials for elastic models in spite of the fact that their stiffness under load is time-dependent. Two popular model plastics are *methyl methacrylates* (Perspex, Plexiglas, Lucite) and *polyvinylchlorides* (usually referred to as PVCs). These *thermoplastic* materials have an instantaneous elastic modulus of about 400 000 psi (2.8 GPa), which gives a very attractive stress scaling factor S_E of about 8 for typical concrete prototypes. They are readily available in a variety of thicknesses, easily worked and joined with commercially available cements, and are moderate in cost. Flat sheets of the plastic may be softened by heating and then formed against a mold of complex curvature by the vacuum-forming process.

There are also several useful *thermosetting* plastics that can be cast into molds in much the same way that concrete is placed. Epoxy resins, phenolics, amino resins, and polyesters are all in this category of plastics. Epoxies and polyesters are particularly versatile in that various types of filler may be blended into the plastic to adjust physical properties, including Poisson's ratio. Thermosetting plastics are preferred in applications where a carefully controlled variable thickness model is needed. The costs of the complete mold (all surfaces of the structure) must be balanced against the costs of machining solid plastic to variable thickness, as would have to be done with the thermoplastics since they cannot be cast.

Most plastics have a very high ratio of strength to modulus, which means that in typical modeling applications only a small fraction of the linear portion of the stress–strain curve is used. It also means that multiple repeated tests on the same model may be possible even if the model has large deformations, such as in the elastic instability of shell models.

Most plastics do have two major disadvantages: their Poisson's ratio is usually about twice that of concrete and their modulus of elasticity is time-dependent. The simplest way to combat the latter problem is to establish the E-value of the model material as a function of time of loading and then measure all strains in the model as a function of time as well. This does involve using many different values of E when converting strains into stresses, but the problem is minimized with modern computerized data acquisition and reduction systems. The low thermal conductivity and high coefficient of expansion of plastics may also cause problems in using electrical resistance strain gages because the current in the gage heats the plastic and causes local thermally-induced strains that are superimposed on the desired load-induced strains. This problem is best overcome by using strain-gage circuitry that has very short power pulses to minimize the amount of current that the gage sees.

Physical properties of plastics should be obtained from the manufacturer's literature and from uniaxial and bending tests on specimens cut from the same plastic as used in each individual model. Thicknesses and properties vary from sheet to sheet in the thermoplastics, and the properties of thermosetting plastics depend upon the degree of polymerization achieved. Reference [3] has a complete chapter on plastics properties, measurement of plastics properties, and use of plastics in structural models.

4.3 Static loading techniques

The very light load intensities needed on models made from plastic may necessitate special loading systems. For example, if $S_E = 8$ for a Perspex model of a reinforced concrete shell roof, then a live load of 40 psf (2 kN/m^2) on the prototype shell

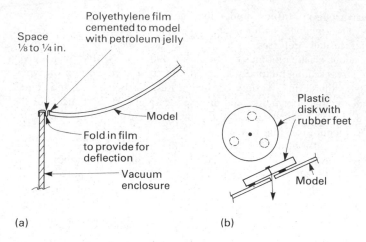

(a)   (b)

Figure 4-1   Loading techniques
(a) Vacuum loading system
(b) Load pad for shell models

is only $40/8 = 5$ psf (0.24 kN/m^2) on the model shell, and a square model 3 ft on a side would have a total design live load of only 45 lb (200 N).

The simplest method of loading a surface structure is with a vacuum acting on the underside of the model, as illustrated in Fig. 4-1a. An enclosure must be built for the vacuum source. It can be sealed off at the junction with the model by a thin sheet of polyethylene film attached to the enclosure and the model with petroleum jelly. A vacuum cleaner and an adjustable port in the side of the enclosure to control the load is all that is needed in this approach. The sole disadvantage of the vacuum load is that the load acts normal to the model surface and thus may not be satisfactory for gravity loads on sloping roofs.

Discrete loads on a surface structure model may be applied by suspending dead weights on strings that pass through the model and are tied off against small load pads. A typical three-footed load pad used on plastic shell models is shown in Fig. 4-1b. A 'catching system' located immediately under the loads should always be employed to catch the weights if the model fails. The disadvantage of this type of load for modeling live loads is that it cannot be quickly added or removed from the model. It is ideal for providing the augmented dead load needed to satisfy self-weight similitude requirements (as discussed in Section 2.1).

Dead weights can also be applied to the surface of a model but care must be exercised to avoid bridging between the individual weights. In addition, safety is a problem because the weights will fall if the model should collapse.

A great deal of ingenuity, common sense, and concern for safety must go into designing any loading system. Model studies on structural stability are particularly sensitive to loading magnitude and direction, and the load system must be designed and built such that it does not offer any artificial restraint to the model, thereby raising its capacity to an unconservative level.

4.4 Wind effects models

Small scale models are now used extensively in the design of high-rise buildings, tall communication towers, cooling towers, major bridges, large shells, and other unusual structures. Wind action has produced many undesirable effects, both structural and environmental, in a number of major projects in recent years, and adequate analytical modeling of wind effects remains an elusive goal.

Very-small-scale rigid models (on the order of 1/2000) can be built and tested to determine wind pressures and flows for an entire section of a large city or urban area. Larger rigid-shape models of individual structures can give more detailed data on local pressures on the structure and on the effect of a proposed new structure on adjacent structures and spaces. Still larger-scale aeroelastic models are used to model the actual dynamic response of a structure and its interaction with the flow of air (such as vortex shedding and aerodynamic oscillations). These instrumented models, which can be built to model both stiffness and damping as well as geometry, provide the designer with magnitudes and directions of wind-induced moments, shears, and displacements. They can also give detailed information needed for the proper design of glass and cladding and help to establish the probability of glass damage in postulated severe wind storms.

Special wind tunnels are needed to simulate properly the atmospheric boundary layer with its high gradient of wind velocities near the earth's surface. These wind tunnels are called *boundary-layer* or *environmental* wind tunnels. The velocity profiles for two different types of prototype terrain are modeled in the University of Western Ontario's boundary-layer wind tunnel, shown in Fig. 4-2, which also shows two different scale models of reinforced concrete cooling towers. They achieve the necessary boundary-layer effect either by having a long tunnel with the boundary layer developing naturally as the air is forced over a rough floor, or by having a shorter tunnel with passive devices, such as grids or spires, to produce a thick boundary layer before the moving air flows over any of the simulated rough terrain. In either case, the model loading and data-sensing equipment

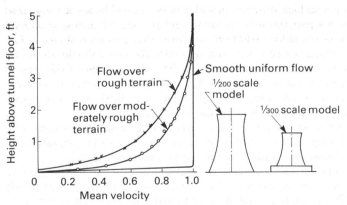

Figure 4-2   **Mean velocity profiles in boundary-layer wind tunnel tests on cooling towers (University of Western Ontario)**

needed for wind effects studies is highly specialized and this type of modeling is normally done only in regional laboratories that have both the equipment and specially trained personnel. Wind-tunnel modeling is both an art and a science, with the art portion developing only by extensive experience in the laboratory. The same argument can be made for other types of modeling, but it is even more pertinent for any type of dynamic modeling study.

Some comments are in order on materials for wind-tunnel modeling. The rigid models are usually made from blocks of softwood because only the overall shape of the structure is of interest. Aeroelastic models must be made from materials that satisfy the similitude requirements outlined earlier. A very important parameter to consider is the damping inherent in structural concrete. Damping coefficients vary with the degree of cracking and the model designer may try to reflect the expected level of cracking in a particular structure when selecting the model material. Filling epoxies are one of the most versatile model materials because they can be blended to have different stiffness and damping properties and can be cast into the necessary shapes for any structure. Investigators at the University of Western Ontario's boundary-layer tunnel have used an epoxy called Devcon for wind effects modeling of a number of reinforced and prestressed concrete structures. The unit weight of Devcon is about 140 lb/ft^3 (2243 kg/m^3) and its elastic constants are $E = 700\,000$ psi (4.8 GPa) and $\nu = 0.23$. It has a critical damping ratio of about 0.8%, which is similar to that for uncracked normal-weight structural concrete.

4.5   Other dynamic response models

In addition to the aeroelastic models described in the preceding section, elastic response models may be used to study the influence of vibratory equipment (such as compressors, generators, and turbines) on supporting concrete structures and foundations. Wave action on offshore concrete structures is a very complex problem that occasionally must be approached experimentally for the determination of both the wave-induced loads and the resulting structural response. And pseudo-elastic response of important structures to seismic action may require experimental study on reduced scale models.

As in aeroelastic models, these other dynamic response models require sophisticated and expensive equipment for loading the model and for sensing and recording the dynamic response data. The very short time periods typical in any reduced scale model study, where the prototype times are always reduced by $\sqrt{S_l}$ as a minimum, places more severe demands on the instrumentation than corresponding full-scale dynamic tests on prototypes.

The harmonic forcing function of oscillating equipment may be introduced into the model with small-scale shakers consisting of two eccentric masses driven by an electric motor. Seismic-type loads are best modeled by attaching the model to a shake table that can be driven to a prescribed acceleration or displacement time history, or by exciting the model with an electromagnetic oscillator that can introduce suitable random motions into the model.

4.6   Choice of scale

The geometrical scale of a model is crucial from several respects:

(a) If the model is too large, it may require high loads and expensive loading equipment plus a large space in the laboratory.

(b) If it is too small, the fabrication becomes overly difficult and normal instrumentation techniques may be inadequate.

(c) The scale factor may be prescribed in advance by limited availability of model material dimensions (such as the thickness of plastic sheets), or by the minimum feasible thickness of a particular type of elastic model.

While the optimum geometrical scale factor is unique for each model study as well as being a function of the experience and practices of the particular laboratory doing the study, the following guidelines should be useful in preliminary planning for model studies:

Type of structure	Scale factor
Shell roofs	1/200 to 1/50
Slab structures	1/25
Bridges	1/30 to 1/20
Wind effects	1/300 to 1/50
Concrete pressure vessels	1/100 to 1/50
Local details	1/10 to 1/4
Arch dams	1/400

With the exception of the wind effects item, the suggested scale factors are for static loadings. Dynamic models of complete structures may be at somewhat smaller scales.

4.7   Examples of elastic modeling

Because of space limitations, only several examples, all related to design applications, are given here. In each case, the laboratory doing the model study is given in parentheses at the end of the paragraph.

A half-section of a slice through the Westway elevated road model is shown in Fig. 4-3. This model, cut from a single sheet of Perspex, was post-tensioned and the resulting stresses from the stressing operation were determined from strain gages on the model. The boundary conditions on the right side of the section are designed to achieve the desired symmetry condition. This model study was followed by a strength model of the entire bridge. (Cement and Concrete Association, England)

Figure 4-4 shows a 1/40 scale epoxy resin model of St Mary's Cathedral in San Francisco. This structure consists of eight hyperbolic paraboloid shells joined together to have a cruciform shape at the top and a vertical space for stained glass on each of the four sides. As with the first example cited above, the elastic model study was supplemented by a strength model (in this case at a scale of 1/15) to determine the ultimate load capacity of the structure. (ISMES, Bergamo, Italy)

An elastic model for determining the seismic response of a high-rise reinforced concrete shear wall building is shown in Fig. 4-5. This model was built from epoxy resin at a scale of 1/40. Prior to testing the model with a simulated seismic input at its base, it was tested for natural vibrations characteristics, where it was found that the torsional stiffness had a strong effect on the overall dynamic response of

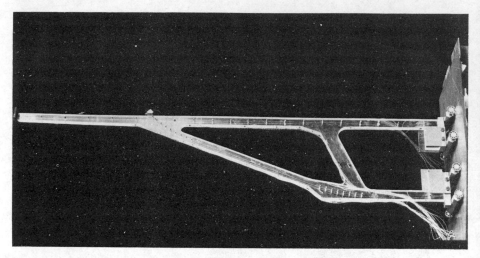

Figure 4-3   Elastic model of section of Westway elevated roadway, London, England (By courtesy of Cement and Concrete Association)

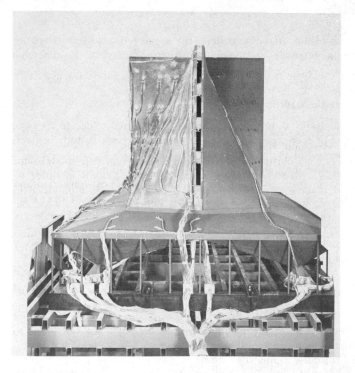

Figure 4-4   Elastic model of St Mary's Cathedral, San Francisco (By courtesy of ISMES)

Figure 4-5   Elastic dynamic model of Parque Central building, Caracas (By courtesy of ISMES)

the building. The prototype structure is the Parque Central building in Caracas, Venezuela. (ISMES, Bergamo, Italy)

Another design model is illustrated in Fig. 4-6. This 1/400 scale-model dam was made from a plaster–diatomite mix with a very low modulus in order to satisfy the dead-load similitude requirements. This material is essentially elastic in its response to loads up to close to failure and hence is well-suited for the determination of elastic stresses. In addition, some indication of the failure capacity of the structure can also be determined from this model. For thermal effects studies, the entire model can be covered with a waterproof membrane and upstream and downstream cavities can be constructed to house heaters and liquid. (LNEC, Lisbon, Portugal)

A good example of an elastic wind-tunnel model is the 1/450 scale model of the 1815 ft high CN Communications Tower in Toronto, Canada. The 1715 ft high prestressed concrete shaft of the tower was modeled aeroelastically using Devcon as the model material, and the 100 ft steel antenna on top of the tower was simulated with a metal spline. This model study provided the designers with values of wind-induced bending moments and shears and showed how important the dynamic components of the response were. (University of Western Ontario Boundary Layer Wind Tunnel, London, Canada)

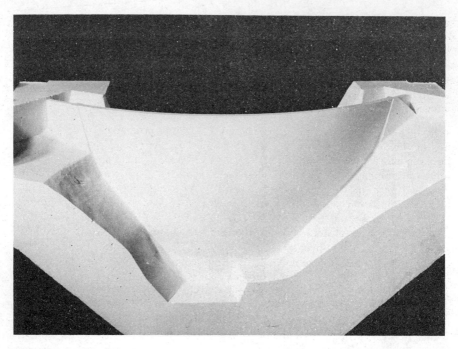

Figure 4-6   Model of doubly curved arch dam (By courtesy of LNEC)

4.8   Reliability of results

Assuming that the basic similitude has been done correctly and that the corresponding planning and model design is correct, then the accuracy of an elastic model depends upon the following factors.

Model material properties   Most plastics have different elastic modulus values in different directions as well as having some sensitivity of properties to heat-forming processes. In addition, the modulus of plastics tends to change quite sharply with temperature (and even with humidity in some cases) and an experiment must be run at a constant temperature with all data reduction being done with the corresponding value of the modulus. The inability to satisfy Poisson's ratio similitude with most plastics also introduces errors of various magnitude into model results for concrete structures.

Fabrication accuracy   Thermoplastics often have surprisingly large variations in thickness and the models engineer is well-advised to ascertain the tolerances on available plastics before deciding on a particular type. Experience has shown that PVC sheets are generally more uniform in thickness than methyl methacrylates. Vacuum forming a sheet to a non-developable surface introduces non-uniform changes in thickness. Models made from thermosetting plastics will have some variation in thickness because it is impossible to build perfectly accurate molds.

Loading techniques   Errors in loading arise from a number of sources: wrong magnitude, wrong direction, inaccuracies in attempting to achieve self-weight similitude, the effects of discretizing a uniformly distributed load, unintentional loads arising from improperly restrained boundaries, inability to follow the scaled load time history in a dynamic test or in a transient thermal experiment, etc. It is probably fair to say that this part of the modeling process has more uncertainties than any other part. Extreme care must be exercised in every phase of loading and nothing can be taken for granted, even the most trusted calibrations.

Instrumentation and measurement There are so many potential error sources in measurement that some engineers will certainly claim this category of problems to be even more overwhelming than those experienced in loading. They include inherent limitations in mechanical devices such as dial gages, randomness in the electrical properties of strain gages, random and systematic errors in electrical devices, improper calibrations, wrong voltages applied to circuits, improper bonding of strain gages to the model, insufficiently stiff supporting system for displacement transducers, spurious thermal effects under electrical resistance strain gages attached to plastics (as discussed in Section 4.2), reading or recording data incorrectly, and difficulties in getting true strain readings on very thin plastic models because the strain gage provides local stiffening. The latter effect is essentially invisible to the novice experimentalist and can only be determined by separate tensile specimen calibration tests on identical thicknesses of plastic instrumented with the same gages as used on the model.

In spite of these many potential error sources, with extreme care most of them can be eliminated and the elastic model can give results with very small errors (on the order of a few per cent from the 'true' result). It may be unreasonably expensive in terms of both time and money to achieve this level of accuracy, however, and some notion of acceptable accuracy limits should always be established before the model study is started. Considering how poorly the real service loads (and probable overloads) are known for nearly all structures, it is suggested that errors up to about 10 to 15% are fully acceptable in most design situations. This level of accuracy is readily achievable with elastic models without resorting to unusually expensive and time-consuming techniques.

In any discussion of reliability, it must be appreciated that any mathematical model will have certain inaccuracies; in many cases, they are quite large but go unnoticed because of the false security that the many digits in the calculated answer tends to convey. The real power of the physical model is its ability to portray the behavior of the full structure with actual boundary conditions. A properly done model study will eliminate two major sources of error that are inherent in most analytical solutions: idealization of a three-dimensional structure into a series of two-dimensional structures, and assumed boundary conditions with regard to fixity, hinging action, etc.

5 Strength models

This section of the chapter contains information on strength models for reinforced and prestressed concrete structures, following the same order of topics as in Section 4 on elastic models. More attention is given to materials and to fabrication because obtaining proper materials and casting them into acceptable model geometries are often the two most difficult parts of the strength-modeling process.

5.1 Applications to design problems

The same classes of structures listed in Section 4.1 are also candidates for strength modeling with the possible exception of wind effects models and certain types of seismic model where failure conditions are not being investigated.

The strength model is invaluable in assessing the true factor of safety against collapse; in fact, this is one of the prime uses for the strength model in both design and research. It also plays an important role in helping to establish post-cracking structural behavior (displacements, cracking, degradation of stiffness in dynamic response, etc.) at service-load levels and beyond.

Strength models used in design situations in recent years have included a wide variety of bridge forms, buildings of unusual complexity, large shell structures, cooling towers, prestressed concrete pressure vessels, protective structures, and dams. Reference [16] contains considerable information on strength modeling.

5.2 Model concretes

The similitude requirements for model concrete properties are given in Section 2.2, where it is pointed out that the general requirements are usually relaxed to satisfying similitude for uniaxial stress–strain characteristics in both compression and tension. Compressive strength and stiffness are both of prime importance, of course, and tensile behavior is also crucial in that it controls cracking and has a strong influence on deformations and on bond strength.

Model concrete is sometimes called *microconcrete* to help distinguish it from mortar. Two types of model concrete are used: Portland cement-based mixes and gypsum-based mixes. Both contain sand and small aggregate. The first category has the advantage of using materials that are essentially identical to prototype materials, while the second has the dual advantage of reaching design strength very rapidly (on the order of one day) and being easier to place because it has a high ratio of gypsum to aggregate. Some investigators have also reported that it is easier to model the tensile strength of typical prototype concrete with gypsum mortars.

Model concrete properties are always a function of local materials and it is essential to do a series of trial batches to arrive at the desired properties. Reference [3] has extensive information on proportioning mixes. Typical proportions for various strength small-scale Portland cement model concretes, with good workability characteristics, are tabulated below.

f'_c psi	Aggregate/ cement	Water/ cement
2500	4.0	0.83
3000	3.75	0.72
4000	3.25	0.60
5000	2.75	0.55
6000	2.5	0.50
7000	2.25	0.40

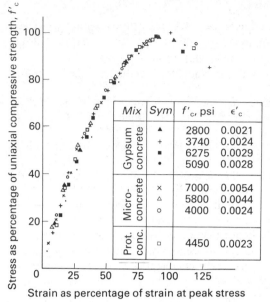

Strain as percentage of strain at peak stress

Figure 5-1   Stress–strain curves for model concrete

A blended mixture of crushed quartz sand with the proportions of 20% No. 10 (0.0787 in)[1] 20% No. 16 (0.0469 in), 35% No. 25 (0.0278 in), 25% No. 35 (0.0197 in), and 10% No. 70 (0.0083 in) has been used to produce a model concrete with $f'_c = 3000$ psi (21 MPa) for 1/10 scale models. In this mix, the aggregate/cement (a/c) ratio was 3.25 and the water/cement (w/c) ratio was 0.80.

Very high-strength cement-based mix proportions are also reported in [3]. Type III high-early strength Portland cement was used. A mix with $f'_c = 10\,000$ psi (69 MPa) at 28 days, as measured on 2 in. by 4 in (50 by 102 mm) cylinders, was obtained by using $w/c = 0.50$, $a/c = 3.8$, and a sand/gravel (s/g) ratio of 1.0. The very high-quality, washed and kiln-dried river sand passed an ASTM No. 3 (0.265 in) mesh and was retained on an ASTM No. 100 (0.0059 in) mesh, and the river gravel had a maximum size of $\frac{1}{4}$ in. These aggregates, which were intended to be used as filter sands, were of exceptional quality. A 12 500 psi (86 MPa) mix with acceptable workability had $w/c = 0.40$, $a/c = 2.5$, and $s/g = 1.0$.

Gypsum-based mixes made with Ultracal 60 (product of US Gypsum Company) are highly workable and cure to design strength in 24 h or less, after which time they must be surface-sealed to prevent additional drying and excessive brittleness. A 3000 psi (21 MPa) mix may be made from 1 part Ultracal, 1 part sand, and 0.31 parts water.

It is possible to produce model concretes with uniaxial compressive stress–strain curves that have the same geometrical shape as prototype concretes (Fig. 5-1). Extensive experience in proportioning and testing model concretes has led to the

1 ASTM sieve sizes. The equivalent British standard sieve sizes are respectively Nos. 8, 14, 22, 30 and 72. The metric apertures are respectively 2, 1.18, 0.71, 0.5 and 0.212 mm.

following guidelines:

(a) Use high-quality aggregate that is not simply a direct linear scaling of the prototype aggregate gradation; instead, reduce the amount of fines substantially to reduce the required cement and water content.

(b) Use natural, rounded aggregates, if available, to reduce the tensile strength of the model concrete.

(c) Use a maximum aggregate size no greater than about one-fifth the minimum dimension of the model, and less than about 80% of the minimum reinforcement spacing.

(d) Conduct strength tests on reduced-scale cylinders with dimensions on the order of 1 in diameter by 2 in long, or on 2 in by 4 in cylinders.

(e) Compute modulus E_c from strains measured with electrical resistance strain gages bonded to cylindrical compression specimens.

(f) Measure tensile strength by the split cylinder test procedure, using the same size cylinders as used for compressive tests, and with loading strips made of $\frac{1}{16}$ or $\frac{1}{8}$ in thick pressed fiberboard (Masonite) or similar material.

5.3   Model reinforcing

The use of small-scale steel reinforcement is the only practical way to model prototype reinforcing bars and prestressing tendons. Properties to be considered include: (a) yield and ultimate strength in tension, and yield strength in compression, (b) shape of stress–strain curve, strain-hardening behavior, and ductility, and (c) bonding characteristics. For true models with no distortion, the stress and strain scaling factors must be unity and the model and prototype reinforcement must have identical properties. The modeling of strain-hardening characteristics is important when severe reversing load effects are to be studied, but it is not critical for models loaded monotonically to failure. Bond-strength modeling is one of the more difficult problems facing the models engineer; generally, deformed model reinforcement is necessary although some investigators claim good success with rusted wire reinforcement.

Deformed 6 mm (and larger) prototype reinforcement should be used for large-scale models. For smaller models, a variety of types of deformed wire have been utilized at diameters ranging from about 0.20 in (5 mm) down to 0.03 in (0.75 mm), as illustrated in Fig. 5-2. Some commercially available welded-wire fabrics are made of deformed wires (at diameters of about 0.10 in and above) with either rectangular or elliptical patterns embossed in the wire, as shown on the left-hand side of Fig. 5-2. A $\frac{1}{6}$ scale reproduction of the $\frac{1}{2}$ in diameter GK60 deformed bar used in the UK is described in [11], where it is shown that excellent similitude was achieved for both local bar geometry and strength properties. Some investigators have used threaded steel rods for model reinforcement.

The cold working done by the common deforming methods leads to major changes in the stress–strain characteristics of the wire, and a heat treatment is normally necessary. Thus, the obtaining of a suitable model reinforcement involves some trial-and-error work on the necessary heat treatment. Full annealing at about 870°C followed by slow cooling through the critical range tends to produce very low yield strengths for the mild steel typically used in making wire.

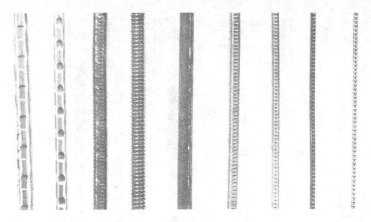

Figure 5-2   Model reinforcement. Left to right: two commercially deformed bars, one laboratory-deformed bar, one threaded bar, one plain bar, arÍd four laboratory-deformed bars

Process annealing, which is done at 480 to 650°C, has been found to be quite successful for producing yield strengths on the order of 50 000 psi (345 MPa) [3].

The modeling of bond is seriously complicated by the fact that the bond phenomena at the prototype level is not fully understood. Since bond depends on both surface adhesion and mechanical action of the deformations bearing against the concrete, a precise geometric scaling of deformations does not guarantee bond-strength similitude. One useful approach to judge the adequacy of model reinforcement bond properties is to conduct pull-out tests on small specimens with wires embedded to different L/d ratios. If the average bond stress at pull-out compares well with similar prototype specimen bond failure stresses, the model reinforcement is judged to be adequate. Another useful comparative test is to embed single model bars in thin sections of model concrete and subject them to tension. As shown in Fig. 5-3, the resulting primary crack spacings vary widely. These tests [2, paper SP24-6] reveal that: (a) plain wires are clearly inadequate, (b) threaded bars have excessive bond strength, (c) rusted wires have highly variable bond properties, (d) commercially available deformed wires tend to be slightly low in bond strength, and (e) deformed wires with surface textures similar to those on the right-hand side of Fig. 5-2 are adequate.

Stirrups, ties, and the very small wires used in shell models need not be deformed.

Small-scale prestressing tendons may be chosen from a variety of sources, including; (a) individual strands of prestressing wire that are used in making up twisted-strand cables for prototype prestressing, (b) piano wire, (c) stainless-steel twisted-strand cable normally used for aircraft and power boat controls, and (d) bicycle spoke material, complete with threaded ends for ease in post-tensioning. The latter two types normally have strength properties somewhat below that of conventional prestressing steels. Friction grips at relatively small sizes are available for anchoring post-tensioned elements. Pretensioning elements may have severe bond problems unless a small twisted-strand cable is used. Some success

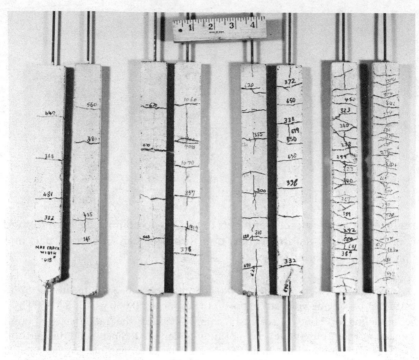

Figure 5-3 Cracking patterns in tensile specimens. Left to right: smooth, deformed, deformed, and threaded rod

has been achieved with individual strands roughening by cutting a light thread on their surface. Any type of prestressing force loss at the anchorages presents a rather severe problem in that the tendons are extremely short and have very little length over which to distribute the loss.

5.4 Fabrication techniques

In addition to following the usual rules of careful workmanship, there are several key points to be made on the fabrication of concrete models. Forms are best made from plastic (such as Plexiglas) because close tolerances can be maintained and release from the model surface is easy. Very small models should be cast on a form that is attached to a table vibrator. Careful curing is crucial because most models have a high ratio of surface area to volume and hence can dry out quickly if not properly sealed or moistened.

Because of the tight tolerances needed at small scales, it is worthwhile to produce ties and stirrups by bending them around precise metal jigs made from steel pins set into a steel plate. Fabrication of model beam and column reinforcement cages is best done by tying the intersections of longitudinal and transverse steel. Welding, brazing, soldering, and the use of quick-setting epoxies are all possible alternatives to tying, but the welding process in particular must be very carefully controlled to prevent undesirable effects on the small wires.

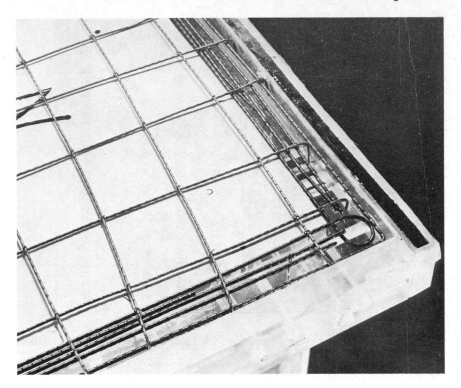

Figure 5-4 Reinforcing and plastic form for hyperbolic paraboloid shell model

Edge beam and shell reinforcement for a $\frac{1}{4}$ in thick hyperbolic paraboloid shell model is shown in Fig. 5-4. The edge beam forms were made of Plexiglas and the shell surface from PVC vacuum-formed to the correct shape against a plaster mold. One of the key problems in casting models is keeping the reinforcement in the correct location. In a shell, this may be done by using scaled spacers under the intersections of the reinforcement mesh and then tying the entire mesh down against the form with fine wires passing through the form.

5.5 Loading techniques

Live-load intensities on surface-type concrete strength models, such as shells, will be similar to prototype loads inasmuch as S_E will be approximately unity. Hence, a shell model with 20 ft² (1.85 m²) surface area will need a total live load of 2000 lb (9 kN) to model a 100 psf (4.8 kN/m²) load (say 2.5 times a normal design live load of 40 psf). Additional live load must be applied to compensate for lack of self-weight similitude; this would be about 48 psf (2.3 kN/m²) for a 1/20 scale model of a 4 in thick prototype shell. This total load of about 3000 lb (13 kN) is best applied by vacuum if the shell is flat, as described in Section 4.3, or with a mechanical whiffle tree loaded with a single hydraulic ram. The latter loading

Figure 5-5   Whiffle tree loading system for folded plate model (By courtesy of University of Texas)

system is shown on a folded plate structure in Fig. 5-5. It can be made as strong as necessary by using steel plates and bolts for the whiffle tree. It has the disadvantage of being very difficult to take back to a zero load state—the suspended weight of the system is always present.

Concentrated loads are normally applied with hydraulic rams. Seismic-type loadings require special equipment similar to that discussed briefly in Section 4.5, but the required force levels will be substantially higher than those needed for most elastic models.

Concrete pressure vessel models must be pressurized internally. If the prototype loading is a gas pressure, then it would naturally be desirable to use a gas pressurization in the model, but the safety issues involved may prevent this and force the substitution of a liquid as the pressurizing agent. Extreme caution must be exercised in pressure tests because of the possibility of explosive failure modes.

Chapter 7 of Sabnis *et al.* [3] is devoted to loading techniques for models and contains many other examples of successful model loading methods.

5.6   Choice of scale

Optimum scale factors for strength models vary among various laboratories, based on their loading capabilities and their experience in dealing with small reinforcement and microconcrete. General guidelines include the following:

(a) Minimum model beam and column dimensions should be on the order of 1 in; larger dimensions are preferred.

(b) Minimum shell thickness should be about $\frac{1}{4}$ in. Reinforced shells as thin as 0.12 in have been built but thickness control and positioning of the reinforcement is exceedingly difficult.

With these minimums in mind, suggested ranges of scale for typical modeling are tabulated below.

Type of structure	Scale factors
Shell roofs	1/30 to 1/10
Slab structures	1/10 to 1/4
Bridges	1/20 to 1/4
Concrete pressure vessels	1/20 to 1/4
Local details	1/8 to 1/3
Arch dams	1/75

As in elastic modeling, these ranges of scale factors are primarily for static loadings. Dynamic tests of complete structures may have to be done at even smaller scales than the small end of the ranges listed.

5.7 Examples of strength models

A number of representative examples of strength models of concrete structures, drawn from both research and design situations, are presented here. The research model results help to give a perspective of the level of accuracy achievable with strength models. Additional examples are given [2–4, 16].

5.7.1 Reinforced concrete frames

The behavior of 1/10 scale model frames under combined gravity load and statically equivalent seismic loading is reported in [12]. Prior to testing two three-story, two-bay model frames, considerable attention was given to producing model materials with properties that closely met the similitude requirements. A series of reversing load tests were done on a portion of the frame located between assumed points of inflection at the mid-lengths of frame members (see sketch of this exterior beam-column joint on Fig 5-6).

Nine cycles of reversing load at ductility factors up to 5 were applied at the end of the cantilever to simulate earlier tests done at the Portland Cement Association Laboratories in Skokie, Illinois. The moment-rotation characteristics of the hinging region in the beam is compared for one model and its full-scale prototype in Fig. 5-6. Major cracks in the prototype joints, as well as overall load–deflection response, were also modeled very successfully. (Cornell University Structural Models Laboratory, Ithaca, NY)

Models of this size, and slightly larger, have been used successfully in dynamic model studies at a number of different laboratories. Quite often in research applications, the small-scale structure is not a model of any particular prototype but instead is simply regarded as a small structure. However, for the results to be applicable for extrapolation to the behavior of full-scale structures, the model materials must still satisfy the general similitude requirements given earlier.

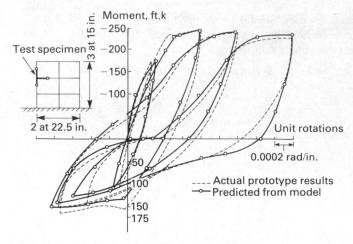

Figure 5-6 Comparison of predicted and actual moment–rotation response of prototype beam–column joint

5.7.2 *Prestressed concrete pressure vessels*

The medium-scale and small-scale strength model played a major role in the engineering and construction of a large number of gas-cooled reactors built in Europe. These heavy-walled structures were usually post-tensioned to achieve a state of triaxial compression under normal working conditions. Behavior at overloads was predicted primarily from the results of literally hundreds of model tests at scales ranging from 1/24 to 1/3. Load histories included pressure and thermal effects and creep behavior was studied in a number of models. A typical model test is shown in Fig. 5-7 for a 1/20 scale model of a relatively thin prestressed vessel. (ISMES, Bergamo, Italy)

The same modeling techniques are applicable to other types of prestressed concrete pressure vessel, and to offshore structures that may see accidental pressure loadings. Such model studies are invaluable in building up an understanding of how concrete pressure vessels fail and what their true reserve capacity is. Shear failures of the end closures of cylindrical pressure vessels are particularly difficult to predict analytically, and considerable investigation of this problem has been done on both complete vessels and on 'head region' models that are clamped against steel base plates with the post-tensioning elements and pressurized to test only the end-closed failure capacity.

5.7.3 *Highway bridges*

A 1/10 scale strength model study of half of the proposed three-span (440–750–440 ft) continuous curved box-girder bridge over the Potomac River is shown in Fig. 5-8. The prototype design featured post-tensioned segmental construction with a maximum bridge depth of 60 ft over the two interior piers. Structural complexities included curved side spans, very slender webs, curved fascia elements, prestressing in all three coordinate directions, and a long central span. The model study was undertaken to determine bridge performance under overload

Figure 5-7 Prestressed concrete pressure vessel model (By courtesy of ISMES)

conditions and the ultimate load capacity. An elastic cross-section model similar to that shown in Fig. 4-3 was also used to help understand the transverse load distribution characteristics of the bridge. The model construction, testing, and results are given as a case study in [3]. (Portland Cement Association Laboratory, Skokie, Illinois)

A 1/30 scale dynamic model study of a long curved reinforced concrete bridge is described in [14]. This model (shown in Fig. 5-9) was designed to represent the behavioral characteristics of a prototype bridge that collapsed under seismic loads in California. The joints between bridge segments, shown at the top of the figure, were key elements in the prototype failure. Horizontal and torsional motions of the bridge segments cause large impacting forces at the joints and potential

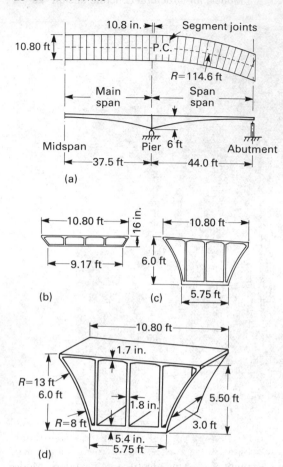

Figure 5-8   Three Sisters Bridge model
(a) Plan and elevation
(b) Near mid-span
(c) Near pier
(d) Dimensions near pier

destruction of the seat where one span bears on the next. The model study was intended to help produce new design criteria for this type of joint.

Lead weights were attached to the bridge spans to satisfy dynamic similitude requirements. The entire model structure was loaded dynamically on a 20 ft square earthquake simulator (shake table). (Earthquake Engineering Research Laboratory, University of California, Berkeley)

5.7.4 *Shells and buildings*

A 1/10 scale model of a compound hyperbolic paraboloid roof structure was used as one of four models in the design of the TWA Maintenance Hangar Facility in Kansas City. The prototype shells were square in plan and were supported at two

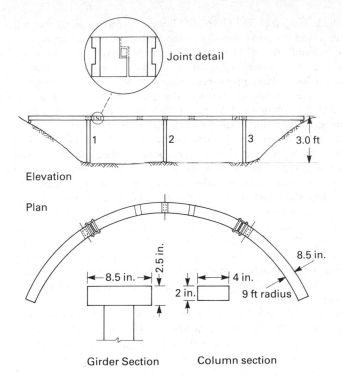

Joint detail

Elevation

1   2   3   3.0 ft

Plan

2.5 in.

8.5 in.

8.5 in.

8.5 in.   4 in.

2 in.   9 ft radius

Girder Section   Column section

Figure 5-9 Dynamic strength model

points over a 320 ft span. The strength model (Fig. 5-10) had a thickness of only 0.30 in over much of its surface, which was cast against a wooden form. Vibration of the stiff model concrete mix was accomplished using electric hand sanders. Model reinforcing was used on a one-for-one basis (all prototype bars were reproduced) as established by the preliminary design calculations.

The model was used to determine the load–deflection behavior under a series of increasing load intensities and also to determine the failure mode and ultimate load capacity of the shell. The design was modified using the results of the model study to produce a prototype with a higher factor of safety but at little added cost.

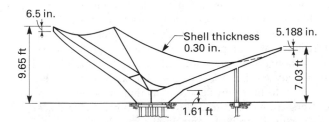

6.5 in.

9.65 ft

Shell thickness
0.30 in.

5.188 in.

7.03 ft

1.61 ft

Figure 5-10 Elevation view of TWA hanger model

Three other models, as described in [3] and [15], were also used in the overall design of the facility: a 1/300 scale solid-shape model of the entire complex of four shells was used to study the wind flow and wind pressures; a 1/100 scale shape model of a single shell was also tested in the wind tunnel to get more detailed information on local wind pressures and velocities; and a 1/50 scale elastic model was used to determine elastic behavior under many loading conditions, including the heavy maintenance equipment that was to be suspended from the shell. (Wiss, Janney, Elstner, and Associates, Northbrook, Illinois).

The last model example was introduced earlier in Section 1. Shown in Fig. 1-1, this model was used to study the response of hyperbolic cooling towers to wind forces. The diagonally inclined wires visible on the right of the model were used to induce additional gravity-type forces in order to meet the self-weight similitude requirements. The loading shown on the left of the model consists of radial forces against the model surface (static simulation of wind pressures) applied with articulated loading devices that reacted in tension against a stiff reaction core located inside the model. The model was constructed in a patchwork approach that minimized overall shrinkage effects. Both the construction and testing techniques used on this model illustrate well the fact that considerable ingenuity and common sense are vital in successful model studies. (Cement and Concrete Association Laboratory, England)

A number of important buildings have been modeled with strength models in recent years, including the Pirelli skyscraper in Milan, where the model concrete was made with pumice in order to help meet the self-weight similitude, and the Roman Catholic cathedral in Liverpool (England), which consists of a series of large sloping concrete frames located radially around the central spire portion of the church.

5.8   Reliability of results

Much of the discussion in Section 4.8 on the reliability of elastic models is also pertinent here, particularly the sections on loading techniques and instrumentation and measurement difficulties. There are several new factors involved with strength models, however, including dimensional tolerances (concrete dimensions and location of reinforcement), model concrete strength (particularly in tension), size effects, and bond strength. Chapter 9 of Sabnis *et al.* [3] treats these and other reliability topics in detail, and the conclusions from that reference will be summarized here.

Dimensional accuracy is strictly a function of workmanship. Errors on the order of ± 2 to 5% are typical for the smallest model beams and columns (about 1 in overall thickness), while slab and shell thicknesses may vary by as much as $\pm 15\%$ for nominal $\frac{1}{4}$ in thickness and ± 5 to 10% for nominal $\frac{3}{4}$ in thickness. Errors in positioning model reinforcement are comparable to the ranges quoted above.

Model concretes tend to show higher strength with decreasing specimen size; hence, control tests should always be done on small cylinders or cubes. In addition, model concretes tend to have tensile strengths that are higher than the scaled prototype tensile strengths, which can lead to errors in predicting cracking, deflections, shear failures, torsional strength, and bond strength.

Flexural strength can be modeled at 1/10 scale with essentially negligible error.

There is lack of agreement on the modeling accuracy for shear strengths, but with careful choice of model concrete, model predictions should be within 15 to 20% of prototype values, but usually on the higher side.

Strength models tend to have fewer visible cracks than do prototype concrete members, but load–deflection behavior can still be modeled to within ±10 to 20% accuracy. The achievable precision of large-displacement, reversing-load modeling has been shown for a 1/10 scale frame member in Fig. 5-6. Tests on two identical prototypes would normally not show any better agreement than this. The reader is cautioned that considerable attention must be given to model concrete tensile strength, to post-yield characteristics of the model reinforcement, and to bond strength simulation if accuracies of this type are to be reached.

In summary, small-scale strength models built and tested with care will predict strengths within 10% of prototype values if the failure is controlled by reinforcement yield, and within 20% for failures related to concrete tensile strength. Load–deflection modeling should fall within this range (10 to 20%). This level of accuracy is more than adequate for design purposes; the upper levels of possible error may be a problem in research.

6   Summary and the role of the model in the future

The physical model approach is an important complement to analytical solutions in reinforced and prestressed concrete design; it also enables researchers to study the true behavior of concrete structures loaded to ultimate capacity. Experimental structural analysis is both an art and a science—a successful model study involves many distinct steps, and careful planning and patient execution of details are crucial. Each model study is a small engineering project in itself and it must have a proper budget for both time and expenses.

The increasing sophistication of computer analysis means that the range of applications of physical models will continue to change, particularly in design, where only the most complex problems will be studied experimentally. But the need for improved understanding of true structural response will become more and more important as designs are pared down to minimum levels in our never-ending effort to improve structural efficiency. Also, rational design methods for severe load environments will become more essential. Engineers and architects will evolve more daring structural forms made of higher strength materials. These developments cannot proceed by mathematical modeling alone, and the physical model will always have a limited but very important role in the engineering of structural concrete.

References

1. ACI Committee 444, *Models of concrete structures*, American Concrete Institute, Detroit, March 1972, 180 pp.
2. White, R. N. (Ed.), *Models for concrete structures*, American Concrete Institute, SP-24, 1970, 495 pp.

3. Sabnis, G. M., Harris, H. G., White, R. N. and Mirza, S. M., *Structural modeling and experimental techniques*, Prentice-Hall, Englewood Cliffs, NJ, 1982, 500 pp.
4. Fumagalli, E., *Statical and geomechanical models*, Springer-Verlag, New York, 1973, 182 pp.
5. Murphy, G., *Similitude in engineering*, Ronald Press, New York, 1950, 305 pp.
6. Dally, J. W. and Riley, W. F., *Experimental stress analysis*, McGraw-Hill, New York, 1978.
7. Dove, R. C. and Adams, P. H., *Experimental stress analysis and motion measurements*, Prentice-Hall, Englewood Cliffs, NJ, 1964, 515 pp.
8. Hendry, A. W., *Elements of experimental stress analysis*, Pergamon Press, 1977, 193 pp.
9. Hart, G. (Ed.), Dynamic response of structures: experimentation, observation, prediction and control, *Proc. speciality conf.*, American Society of Civil Engineers, Jan. 1981, 952 pp.
10. Sabnis, G. M. and White, R. N., A gypsum mortar for small-scale models, *J. Am. Concr. Inst., Proc.*, Vol. 64, No. 11, Nov. 1967, pp. 767–774.
11. White, I. G. and Clark, L. A., Bond similitude in reinforced microconcrete models, *Proc. seminar on reinforced and prestressed microconcrete models*, Institution of Structural Engineers/Building Research Establishment, Garston, May 1978.
12. Chowdhury, A. H. and White, R. N., Materials and modeling techniques for reinforced concrete frames, *J. Am. Concr. Inst., Proc.*, Vol. 74, No. 11, Nov. 1977, pp. 546–551.
13. Chowdhury, A. H. and White, R. N., Multistory reinforced concrete frame models under simulated seismic loads, *Reinforced concrete structures subjected to wind and earthquake forces*, American Concrete Institute, SP-63, 1980, pp. 275–300.
14. Williams, D. and Godden, W. G., *Experimental model studies on the seismic response of high curved overcrossings*, University of California, Berkeley, Report No. EERC 76-18, June 1976.
15. Guedelhoefer, O. C., Moreno, A. and Janney, J. R., Structural models of hangars for large aircraft, *Proc. Symp. ACI Canadian Chapter on models in structural design*, Montreal, 1972, pp. 71–109.
16. Garas, F. K. and Armer, G. S. T., *Reinforced and prestressed microconcrete models*, Construction Press, Lancaster, 1980, 387 pp.

24 Environmental design

Henry J Cowan
University of Sydney, Australia

T Jumikis
Woolacott, Hale, Corlett, and Jumikis, Consulting Engineers, Sydney, Australia.

Contents

Notation

c_p	Specific heat per unit mass
d	Thickness
h	Boundary layer heat transfer coefficient
H	Height
k	Thermal conductivity
L	Length
Q	Quantity of heat
T	Temperature
U	Thermal transmittance
VSA	Vertical shadow angle
α	Thermal diffusivity
λ	Latitude
ρ	Density

Summary

The structural and environmental systems of a building interacted closely before the nineteenth century. Although loadbearing walls are now rarely used, it is still possible to use the structural concrete to improve the environmental performance of a building, by taking advantage of the high thermal inertia of concrete and its good sound insulating properties.

The thermal properties of concrete are explained. Sunshading, sunlight penetration, thermal inertia, thermal insulation, and reflectivity of concrete are then described. The theory of variable heat transfer is briefly discussed, and applied to passive solar design.

The linear movement of concrete due to temperature and moisture is then considered, and suitable control joints are illustrated.

The final section deals with the insulating qualities of concrete against airborne sound.

1 Cost of the structural system and its interaction with the environmental system prior to the 19th century

In the great buildings of the Middle Ages and the Renaissance, the walls were generally load-bearing, and therefore part of the structure.

Buildings had only primitive heating, and no artificial cooling or ventilation. In summer, solid masonry buildings were often pleasantly cool on a hot afternoon because of the thermal inertia of the thick masonry walls and roofs: the structure performed a secondary function by acting as a thermal store. In most parts of the world, it is only necessary to reduce the temperature by a few degrees below outside shade temperature to achieve pleasant conditions in summer, so that the thermal inertia was quite effective.

Winter heating presented greater problems because a much larger temperature change might be required for thermal comfort. The thermal inertia of the walls was equally helpful in reverse, but not sufficient in cool climates. The great height of many monumental buildings of the past created additional problems. Even today, some medieval cathedrals are unheated, and considered unheatable at a reasonable cost. People dressed warmly, and on cold days took along a hand- or a foot-warmer. In a 13th century manuscript, preserved in the Bibliothéque Nationale in Paris, the Gothic masterbuilder Villard de Honnecourt described a hand-warmer in the form of a brass apple made from two fitting halves:

> Inside this brass apple there must be six brass rings, each with two pivots. The two pivots must be alternated in such a way that the brazier remains upright, for each ring bears the pivots of the others. If you follow the instruction and the drawings, the coals will never drop out, no matter which way the brazier is turned. A bishop may freely use this device at High Mass; his hands will not get cold as long as the fire lasts. [1, p. 66].

Residential buildings were heated, but the cost of the heat source was not high: it might be an open fire-place or a stove. Only in Ancient Rome for about five centuries and in north-east Asia from late-medieval times was hot air conveyed in ducts to heat the building [2, pp. 91–92; 3, p. 239].

The structure also played its part in sunshading and in natural ventilation. Colonnades which provided summer shading sometimes contained buttresses to absorb horizontal reactions from the roof. Openings in the load-bearing walls were often arranged with great skill in the traditional architecture of the countries of the Middle East and the Mediterranean Sea to give the best natural ventilation possible, as surviving buildings testify. Ventilation ducts were not used before the 19th century.

Windows for natural lighting were generally openings in load-bearing walls. Artificial lighting prior to the 19th century was mainly for the wealthy, because tallow, vegetable oil and fish oil were all edible and too precious to burn in large quantities.

During the 19th and 20th centuries there was a rapid increase in the standard of living of the majority of the population in most countries of North America, Europe and Australasia, which resulted in much higher standards of heating and artificial lighting, and the introduction of artificial ventilation, air conditioning and passenger elevators. Load-bearing walls were, for most long-span and multistory buildings, replaced by non-load-bearing walls. Structural systems became more efficient and used less structural material. As a result, the cost of the structure no longer dominates the cost of the building as it did prior to the 19th century, except for long-span buildings. For multistory buildings, the structure accounts for 15 to 40% of the total cost. The building services frequently cost twice as much as the structure.

The coordination of the structural and the environmental systems can therefore produce significant economies. It is generally not worthwhile to optimize the cost of the structural system if this increases the cost of the environmental system (Section 11).

2 Using structural concrete to improve the environmental performance of buildings

The use of structural concrete for integral sunshades can produce worthwhile savings. Since sunshades are a conspicuous feature of the facade of the building, and thus have an important bearing on its appearance, a realistic assessment of the economics of sunshading should allow for its positive or negative effect on the aesthetics of the facade.

The thermal insulation and thermal inertia of structural concrete can also contribute to improving the interior environment of a building, or alternatively to reducing the cost of heating or air conditioning if the same standard of thermal comfort is maintained.

Structural concrete has good insulation against airborne sound, but is, by itself, of little value against impact sound. Thus, the cost of slightly thicker concrete walls, roofs or floors can, in some circumstances, be offset against the cost of providing an adequate standard of insulation in noisy surroundings (Section 10).

The use of structural concrete in reducing the cost of lighting is only indirect, but it can be significant. If good sun shading permits the admission of daylight without glare, it may reduce the need for keeping venetian or other blinds drawn during daylight hours, and thus decrease the demand for artificial lighting. This reduces the cost of energy, and in addition it reduces the heat load created by the artificial lights, and thus the cost of air conditioning in summer.

3 Thermal properties of concrete

The properties relevant to the use of structural concrete for the improved thermal performance of a building are its thermal conductivity, which measures its value as an insulating material, and its heat capacity per unit volume, which measures its thermal inertia. The insulating properties of concrete are poor: better than those of metals and a little better than those of most building stones, much inferior to those of the lightweight insulating materials. However, a great improvement can be made by adding a layer of an insulating material (Section 5).

Table 3-1 Thermal conductivity, k, of selected materials (From [10] pp. 213–5; [12] pp. 9, 32; and [15] pp. IV, 23–24)

Material	Thermal conductivity, k W/mK
Copper	386
Steel	43
Granite	2.8
Limestone	1.5
Concrete	1.2
Brick	0.8
Sand	0.4
Gypsum plaster	0.10
Water	0.6
Timber (pine)	0.14
Paper	0.13
Glass or mineral wool	0.039
Urea formaldehyde foam	0.038
Expanded polystyrene	0.034

Concrete has a relatively high heat capacity per unit volume (or thermal inertia), slightly better than that of the rock stores favoured for passive solar energy (see Section 8). It is the product of the density, ρ, in kg/m^3 and of the specific heat per unit mass, c_p, of a material in J/kgK. Thus, the heat capacity per unit volume, ρc_p, is measured in MJ/m^{3}K.

Table 3-2 Density, specific heat per unit mass, and heat capacity per unit volume of selected materials (From [10] pp. 213–214; [12] p. 9; [14] pp. 136–138)

Material	Density, ρ kg/m³	Specific heat per unit mass, c_p J/kg K	Heat capacity per unit volume, ρc_p MJ/m³ K
Copper	8900	390	3.48
Steel	7800	470	3.67
Granite	2650	900	2.39
Limestone	2200	860	1.89
Concrete	2100	900	1.89
Brick	1800	840	1.51
Gravel (30–50 mm)	1800	920	1.66
Sand	1500	800	1.20
Gypsum plaster	1300	840	1.09
Water	1000	4180	4.18
Timber (pine)	650	2300	1.50
Fibre insulating board	380	1500	0.57
Glass or mineral wool	80	700	0.14

4 Sunshading, sunlight penetration and thermal inertia

There are a few regions remote from the equator where overheating of interiors in summer never occurs. A larger circumequatorial region has minimum temperatures that are at or above the level of comfort. However, the great majority of buildings employing structural concrete are in regions where the sunshading is desirable in summer and sunlight penetration is desirable in winter.

It is possible to design sunshades which keep the sun out when it is not wanted, and allow it to enter when additional heat is useful; however, the resulting sunshades may be of complex shape, and thus the theoretically perfect solution is not necessarily optimal.

Sunshades may be of non-structural concrete or some other material, and these are outside the scope of this book. Sunshading may also be accomplished by balconies or other projections of structural floors (Fig. 4-1) or by load-bearing walls of appropriate shape (Fig. 4-2).

Sunlight penetrates unshaded windows, because glass is mainly transparent to the radiation of the solar spectrum. However, glass is mainly opaque to the longer-wavelength radiation re-radiated by the walls and floors in the absence of ventilation, so that the heat is retained in the room (see also Section 7). This is undesirable in hot weather, and desirable in cold weather. The window thus acts as a solar collector of relatively low efficiency, but also of negligible cost.

The thermal inertia of the concrete floor can be utilized if a hard, heat-absorbing floor surface, such as marble paving, tiles, or terrazzo, is used. The structural

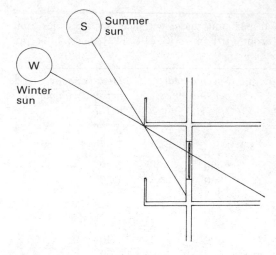

Figure 4-1 Use of balcony slabs of structural concrete for sunshading. If the interior floor is faced with a hard, heat-absorbing surface, such as marble paving, tiles or terrazzo, the structural concrete acts as a heat store which releases the heat during the night. The thermal inertia of the concrete floor is greatly reduced by a carpet which acts as an insulator; however, the value of the heat stored may not be sufficient to warrant the use of a hard floor surface if the client is likely to prefer a carpet

concrete then acts as a heat store, and the heat collected by it during the day is released during the night, thus increasing the efficiency of the solar collector. Carpet is an insulator, and it greatly reduces the usefulness of the floor as a heat store.

Whether the floor is given a hard finish or a carpet is largely a matter of custom. In many hotels, residential and commercial buildings in Italy, marble, tiles, or terrazzo are regarded as very desirable floor finishes. In Australia, in a similar climate, and in a similar type of occupancy, a carpet is usually preferred. It is doubtful whether the saving in fuel resulting from the use of a hard surface would compensate people who prefer a carpet; but the choice is at least worthy of consideration.

It is a simple matter to calculate the required projection of a balcony or floor slab on the facade of a building facing south in the northern hemisphere in the subtropics or the temperate zone (north in the southern hemisphere), assuming that it is desired for the sunlight penetration to commence at noon on one equinox, and terminate at noon on the other (Fig. 4-3). If λ is the latitude of the place, then at noon on the equinox on a wall facing precisely south (or north in the southern hemisphere) the vertical shadow angle (VSA) is

$$VSA = 90° - \lambda \qquad (4\text{-}1)$$

If the distance from the soffit of the projecting floor slab or balcony to the window sill is H, then the required projection of the floor slab or balcony beyond the wall

Figure 4-2 Load-bearing reinforced concrete wall with integral sunshades: Greater Pacific House, Sydney, Australia, a 20-story building with load-bearing precast concrete walls with integral sunshades. The building faces approximately north–south, and the same wall units are used on all four facades, so that the shading is a compromise. The precast units measure 2.4×3.2 m, and weight 12 700 kg [5]

is

$$L = H \tan (90° - \text{VSA}) = H \tan \lambda \qquad (4\text{-}2)$$

Although the highest average daily temperature occurs in most places some time after the summer solstice, the above equations are sufficiently close for a sunshade which wholly excludes sunlight for about two months in the hottest part of summer, fully admits it for about two months in the coldest part of winter, and partially admits it at other times of the year.

Assuming that the distance H from the soffit of the balcony or projecting floor slab is 1.5 m, the projection L required (Fig. 4-4) at the latitude of London, England (51°) for a wall facing exactly south is $L = 1.85$ m, and the projection L required for a wall facing precisely north at the latitude of Sydney, Australia (34°) is 1.00 m. However, if the wall faces east or west of south (north), the required

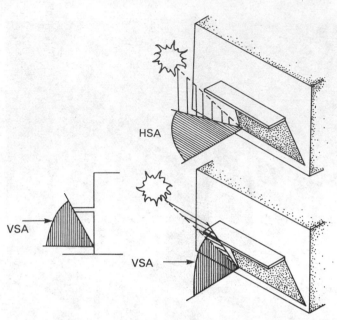

Figure 4-3 Vertical (VSA) and horizontal (HSA) shadow angles. At the equinox (21 March and 22 September) the sun's path crosses the equator, its declination is 0°, and therefore its altitude at noon is $90° - \lambda$, where λ is the latitude of the place. When the facade faces south (north in the southern hemisphere), the vertical shadow angle is equal to the altitude

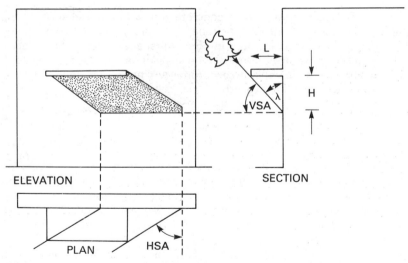

Figure 4-4 If sunlight penetration is to commence and cease at noon on the equinoxes for a facade facing exactly south (north in the southern hemisphere), the projection L required for height H from soffit to sill is $L = H \tan \lambda$

DEC. 22   12 noon

Figure 4-5   A heliodon or sun-machine consists of a table for the model of the building or room, and a lamp representing the sun. There are two principal types. In one, the table is stationary and the lamp moves; this type can be made direct-reading, that is, the machine can be set directly to the latitude, time of day, and time of year. For the second type (illustrated), the model is rotated about two axes, and the lamp is stationary. It is necessary to calculate the altitude and azimuth of the sun, but the result is more accurate because the lamp can be placed at a greater distance so that its light more nearly resembles the parallel rays of the sun

projection L increases rapidly, and use of a projecting floor slab or balcony soon becomes impracticable.

The design of sunshades which do not face directly south or north has been discussed in many books and papers, for example [6, 7]. It can be done by model analysis, using a heliodon (Fig. 4-5). This is a simple and quick method if a suitable model of the building exists, and if a heliodon is readily available. It can be done by one of several graphic methods, or by superposing a transparent sunpath diagram on a drawing of the building [6, 7]. The quickest method is to use a computer program which produces a graphic output [7].

Frequently, the precise solution requires a sunshading device which is expensive to make or unattractive in appearance, and a simplified version of the ideal solution has to be used. This is particularly so if the sunshade is part of the concrete structure (Fig. 4-2).

5 Thermal insulation

For the purpose of determining the heat transfer through a wall or floor, it is convenient to divide it into the layers of the different materials of which it is composed, and the boundary layers on each side of the wall.

If Q is the quantity of heat passing through a unit area of wall or floor (in W/m^2) and the temperatures on the two faces are T_0 and T_i, then the thermal transmittance

$$U = Q/(T_0 - T_i) \text{ W/m}^2 \text{ K} \tag{5-1}$$

Examining the individual layers of the wall (Fig. 5-1) when the temperature on either side of the wall is steady,

$$Q = h_0(T_0 - T_1) \tag{5-2}$$

$$= \frac{k_1}{d_1}(T_1 - T_2) \tag{5-3}$$

$$= \frac{k_2}{d_2}(T_2 - T_3) \tag{5-4}$$

$$= \frac{k_3}{d_3}(T_3 - T_4) \tag{5-5}$$

$$= h_i(T_4 - T_i) \tag{5-6}$$

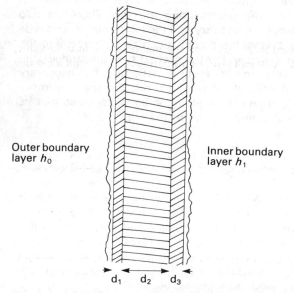

Outer boundary layer h_0

Inner boundary layer h_1

d_1 d_2 d_3

Figure 5-1 Wall or roof consisting of three layers of different materials, whose thickness is d_1, d_2, and d_3 respectively

where h_0 and h_i are the boundary layer heat transfer coefficients (in W/m^2K) outside and inside the wall, or above and below the roof; k_1, k_2, and k_3 are the thermal conductivities of the materials in layers 1, 2, and 3 (in W/mK), and d_1, d_2, and d_3 are the thicknesses of layers 1, 2 and 3 (in metres).

Combining these equations

$$\frac{1}{U} = \frac{1}{h_0} + \frac{d_1}{k_1} + \frac{d_2}{k_2} + \frac{d_3}{k_3} + \frac{1}{h_i} \qquad (5\text{-}7)$$

The thermal conductivities of common building materials are available in tables in various handbooks and textbooks, for example [8] Ch. 22; [9] pp. 155–8; [10] pp. 213–4. Some are reproduced in Table 3-1.

The boundary-layer heat-transfer coefficient depends on the temperature difference between the surface concerned and the air in contact with it, and the velocity of the air, which in turn determines whether the air flow is laminar or turbulent. The theory is complex, but values of h for various inclinations of surface, various surface emissivities, and for various wind velocities are listed in the *ASHRAE handbook of fundamentals*, Table 1, pp. 22.11–12, [8].

For non-reflective surfaces, the value of h ranges from 6 to 9 W/m^2K in still air. It is much larger in a high wind; for example, for a wind velocity of 10 m/s, the value of h for a non-reflective wall surface is approximately 50 W/m^2K. However, for these high values, the effect of the term $1/h$ becomes insignificant in relation to the terms d/K, so that it has little effect on the value of U in Eqn 5–7.

Taking as an example a structural concrete wall 150 mm thick, with a k-value of 1.2, and the heat transfer coefficients outside and inside $h_0 = 20\ W/m^2K$ and $h_i = 8\ W/m^2K$, we obtain from Eqn 5-7

$$\frac{1}{U} = \frac{1}{20} + \frac{0.15}{1.2} + \frac{1}{8} = 0.30$$

and $U = 3.3\ W/m^2K$.

If the wall thickness increased to 200 mm, the thermal transmittance (U-value) is reduced only to 2.9 W/m^2K. A 50 mm layer of insulating material with a k-value 0.04 is much more effective:

$$\frac{1}{U} = \frac{1}{20} + \frac{0.05}{0.04} + \frac{0.15}{1.2} + \frac{1}{8} = 1.55$$

and $U = 0.65\ W/m^2K$.

An increase in insulation may or may not be desirable, as will be shown in Section 6. It depends on the climate of the place, and the occupancy of the building.

6 Variable heat transfer

In Section 5, it was assumed that the temperature differential between the two sides of a wall or floor remains constant. When the temperature varies, it becomes necessary not merely to consider the thermal insulation of the structure, but also its thermal inertia. As noted previously, the thermal insulation of concrete is poor, but its thermal inertia is good.

The external temperature generally reaches its lowest point at dawn, and its highest in the early afternoon (2 to 4 pm). The interior temperature is kept constant in buildings whose heating or air conditioning is automatically controlled, but fluctuates in most other buildings. Buildings such as schools and offices that are used only for part of the time are often allowed to cool during night. Residential units whose occupants all go to work may cool during the day in winter.

When a building is first heated in cold weather after a cooling period, a considerable part of the heat is needed to heat the concrete structure because of its relatively high thermal inertia. When heating is reduced or discontinued, the heat stored by the structural concrete is released gradually. Thus the thermal inertia reduces the fluctuations in the internal temperature, but it also slows the response of a cold room on heating.

The flow of heat under variable conditions is controlled by two material properties: its thermal conductivity, k (Table 3-1), and its thermal inertia, or heat capacity per unit volume, ρc_p (Table 3-2). The ratio of these two quantities

$$\alpha = k/\rho c_p \tag{6-1}$$

is called the diffusivity of the material. The larger the diffusivity, the more rapidly is a temperature change propagated through the material.

The product of the conductivity and the heat capacity, $k\rho c_p$, does not have a special name, but it is also of interest.

A high thermal inertia is desirable to smooth out the response to the daily variations of temperature. A low thermal conductivity is also desirable to reduce the overall heat loss (or heat gain in summer). Thus, the value of $\alpha = k/\rho c_p$ should be as low as possible.

Table 6-1   Values of $k/\rho c_p$ and $k\rho c_p$ for selected materials (From Tables 3-1 and 3-2)

Material	Thermal conductivity, k	Heat capacity per unit volume ρc_p	Diffusivity, $k/\rho c_p$	$k\rho c_p$
	W/mK	MJ/m³ K	mm²/s	N²/mm²s K
Copper	386	3.48	110	1340
Steel	43	3.67	12	158
Granite	2.80	2.39	1.17	6.7
Limestone	1.50	1.89	0.79	2.83
Concrete	1.20	1.89	0.63	2.86
Brick	0.80	1.51	0.52	1.21
Sand	0.40	1.20	0.33	0.48
Gypsum plaster	0.10	1.09	0.092	0.109
Water	0.60	4.18	0.144	2.51
Timber (pine)	0.14	1.50	0.093	0.210
Fibre insulating board	0.048	0.57	0.084	0.027
Glass or mineral wool	0.039	0.14	0.279	0.005

If the heating is only intermittent, because a room or a building is not in use for part of the time, as discussed earlier in this section, the time taken to reach an acceptable temperature may be of greater importance than a low value α. The diffusivity, α, has the dimensions mm^2/s; given walls or roofs of equal thickness, the lowest diffusivity gives the shortest time of response. However, in lightweight construction with a low thermal inertia, the necessary insulation can be achieved with less thickness, so that the response is faster; a reinforced concrete frame with lightweight cladding panels is therefore preferable to an insulated reinforced concrete load-bearing wall. The solution for homogeneous and simple composite walls has been derived by Billington [11, pp. 52–83], who showed (p. 55) that the product $k\rho c_p$ should be as low as possible when rapid heating is required, that is, the wall should consist mainly of low-density insulating material. Billington also calculated (p. 68) the velocity of propagation of the temperature wave during the daily 24 h cycle of temperature through walls consisting of various homogeneous materials. His values are given in Table 6-2.

Table 6-2 Propagation of temperature wave during a 24 h period

Material	Velocity mm/h
Concrete	35
Brick	30
Timber	17
Fibre insulating board	15

The same considerations apply when the exterior temperature is above the desired interior temperature, and the building is cooled by air conditioning.

There are, however, two important differences between the heating and the cooling of buildings. Buildings are heated by the expenditure of energy, even in primitive societies, although the indoor temperatures are often lower than in industrially developed countries. Cooling by the expenditure of energy is still the privilege of the upper class in under-developed countries, and it is likely to remain so for some time. It may become less common in industrial countries as the price of energy increases and greater use is made of the fabric of the building in environmental design ('passive' solar energy, see Section 8). Furthermore, the difference between the desired indoor temperature and the maximum exterior temperature is much less than for the minimum. In a temperate climate, it is less than 10°C for a hot day, but more than 20°C for a cold night. In an extreme climate, it is about 15°C for a hot day, and 50°C for a cold night.

We must distinguish between hot-arid climates, which have low humidities and high shade temperatures, sometimes above the temperature of the human body, and hot-humid climates, in which the shade temperature is always less than the temperature of the human body, but the humidity can be very high.

The traditional method of construction in hot-arid countries employs thick walls, and sometimes thick roofs of stone, brick, adobe, or mud. These have a high thermal inertia, and in this respect an equal thickness of concrete performs as well. The interior temperature during the hottest time of the day is reduced by the structure, although a higher night-time temperature must be accepted. This is a disadvantage only in very hot climates. In that case, a good, if somewhat more expensive solution, is the use of concrete living quarters, and a bedroom built from light panels with a low thermal inertia.

Reference has already been made to use of sunshades, and to the utilization of the thermal inertia of structural concrete floors by sunlight penetration in winter (Section 4).

In hot-humid climates, the difference between the maximum daytime and the minimum night-time temperature is much smaller, and a structure with a high thermal inertia does not improve the thermal environment during the hottest time of the day. Indeed, since the relatively high night temperature and very high humidity often make sleep difficult, a high thermal inertia is undesirable.

When air conditioning is not used, thermal comfort in hot-humid climates depends mainly on air movement. This may be produced by designing a building to catch a prevailing breeze, or by fans, widely used in countries such as China and India, where air conditioning is a luxury.

To sum up, concrete walls and roofs with a high thermal inertia are useful for buildings which are kept at a constant indoor temperature by heating or air conditioning. They are also useful in hot-arid climates for buildings which are not air conditioned. A high thermal inertia is undesirable when heating is only intermittent, and also in buildings in hot-humid climates which are not air conditioned, so that the concrete structure should be restricted to a frame.

7   Reflectivity of concrete surfaces

The solar radiation received by the surface of a building is partly absorbed and partly reflected. The sun is very hot, and the radiation emitted by it has wavelengths of less than 3 μm. This heats the surfaces on which it falls, but as they remain much cooler than the surface of the sun, the radiation emitted by them has a much longer wavelength (Section 4). For the same wavelength, the absorptivity and emissivity of a surface is the same, but the absorptivity and emissivity are different for different wavelengths. Thus the absorptivity of solar radiation and the emissivity of the longer-wavelength radiation are not the same, and materials vary greatly in this respect.

Thus concrete, black paint and white paint all have the same emissivity, that is, they are cooled equally at night by radiation to the sky. However, the proportion of the solar radiation which they absorb during the day is quite different. A surface painted black becomes much hotter on a sunny day than a surface painted white; concrete is intermediate between the two. Evidently, the reflectivity of solar radiation by concrete can be improved by using exposed white aggregate or white paint. However, it depends on the climate whether a high reflectivity is desirable. The absorption of solar radiation reduces the demand for heating, if heating is required.

Table 7-1 Reflectivity and absorptivity of solar radiation, and emissivity of long-wavelength radiation (From [6] p. 98; [13] p. 108)

Material	Reflectivity of solar radiation	Absorptivity of solar radiation	Emissivity of long-wavelength radiation
Bright galvanized steel	0.75	0.25	0.25
Bright aluminium foil	0.95	0.05	0.05
Aluminium paint	0.50	0.50	0.50
White paint	0.70	0.30	0.90
Green paint	0.30	0.70	0.90
Black paint	0.10	0.90	0.90
White marble	0.55	0.45	0.95
Grey concrete	0.35	0.65	0.90
Brick	0.40	0.60	0.90

8 'Passive' solar energy

In the fast-growing literature on solar energy, its applications are divided into 'active' and 'passive' uses, depending on whether a solar collector is used to convert solar energy into hot water, hot air, steam or electricity, or whether the building itself is used as a solar collector. The active applications of solar energy are outside the scope of this book, because the use of this energy is the same for buildings of any structural material. However, passive solar energy could be integrated into the design of structural concrete because of the high thermal inertia of concrete.

As far as the authors are aware, no building has been designed so far (1982) in which *structural* concrete has been utilized for the passive collection of solar energy. Indeed, the number of buildings designed for this purpose, including all methods of construction, remains less than 1000, and for many the extra cost of utilizing passive solar energy will not be compensated by fuel savings for many years.

Passive solar buildings have been described and their design discussed in many papers and several books (for example [16] pp. 65–107; [17]).

However, as the cost of energy increases relative to other commodities, passive solar buildings may become more common. Passive solar design employs the same principle as the sunshaded window illustrated in Fig. 4-1, which admits the sun in winter and excludes it in summer; however, the heat or coolness is stored for later use. This requires a large volume of material with high thermal inertia. The simpler designs employ a wall of adobe, concrete blocks, or site-cast concrete behind a large area of glass (Fig. 8-1). Evidently, if reinforced concrete is used, this thermal storage wall, which has appreciable load-bearing capacity, can be used to support the building in the same way that the service core containing the

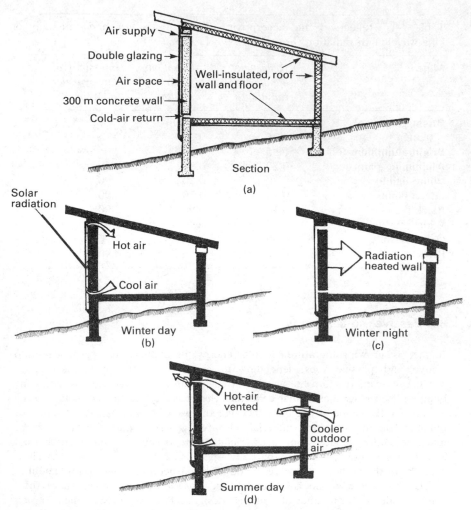

Figure 8-1 House at Odeillo, Southern France, designed by Dr Felix Trombe, built in 1967. The house has a large, double-glazed window on the south face. However, there is no view from this window, because immediately behind it, with only a narrow passage for air circulation, there is a 300 mm concrete wall, painted black. This is now commonly called a Trombe wall (a). Heating and cooling is accomplished by natural ventilation (b, c, and d).

During the summer, the overhang of the roof shades the south wall completely, and there are vents opening outward which draw cool air from the north side of the house.

In winter, the south wall stores heat during the day which is radiated at night. The total time of transmission of heat through the wall is about twelve hours. The thick concrete wall provides sufficient insulation on the south face. The vents through the wall are closed at night, and the glazing reduces heat loss by radiation from the wall (From [17] p. 636)

lifts is used to resist a large part of the weight and wind load in a multistory building.

More complex passive solar designs employ a thermal store in contact with the ground into which heat or coolness is pumped from the building when it is in excess of requirements, and from which it is withdrawn by the pump when needed. These designs require a mechanical ventilation system. The thermal store most frequently used consists of a pile of rocks or gravel. Table 3-2 shows that the heat capacity per unit volume of concrete is about the same as that of rocks or gravel, and in some buildings the reinforced concrete foundation or lowest floor could be utilized as a thermal store, provided that due allowance is made for the reduction in the strength of the concrete by the pipes which would be needed to circulate the air through it.

9 Linear movements and control joints

Expansion and contraction joints are located in concrete structures for the purpose of controlling the accumulation of linear movements in the structure. The average effects of concrete drying shrinkage, as well as the average temperature and moisture change in the structure, cause structural elements to expand or contract in their own plane. Differential concrete drying shrinkage, and gradients of temperature and moisture through the depth of structural elements, cause the elements to flex by bending and warping. Neither expansion nor contraction joints in a structure control flexural movements, which will be discussed in the next section.

The great majority of concrete structures are constructed using Portland cement concrete without expanding additives. Such structures suffer considerable drying shrinkage, normally in excess of 0.5 mm/m. If there is no restraint to movement, stresses do not develop. In practice, the restraints to movement are considerable. In reinforced concrete structures, steel reinforcing provides internal restraint to concrete drying shrinkage, and partially controls shrinkage movements. The external restraints to linear movement are the structural supporting elements which ultimately bear upon the ground. The centroid of external restraints is the point to which the movement of all parts of the structure under consideration is orientated.

Weather-protected structural building elements experience significant concrete drying shrinkage movements. By comparison, the linear movements of such elements as a result of temperature and moisture changes are generally insignificant.

Joints designed for weather-protected structural elements to control their accumulated shrinkage movement are therefore acting in one direction only. The gap between the two concrete surfaces across the joint is subject to widening with time. The joint is appropriately termed a contraction joint. If joint sealant is needed, it must be capable of bonding effectively to both concrete surfaces, and of stretching without rupture with the passage of time.

When structural building elements are exposed to weather, the linear movement due to temperature and moisture changes can be as great or greater than that due to the concrete drying shrinkage. The gap between two concrete surfaces

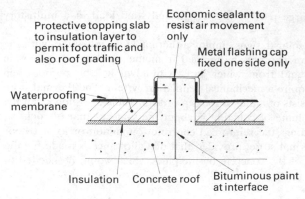

Figure 9-1   Joint for roof slab. The prime objectives are watertightness and ease of maintenance

across such joints is subject to closing as well as opening. It is appropriate to call them expansion joints. If joint sealant is required, it must be capable of bonding effectively to both concrete surfaces, of stretching and compressing, with complete recovery, and preferably have unidirectional strain response, i.e. have a zero Poisson's ratio.

Joint design details vary. It is important to define the objectives and priorities clearly for each and every case. The requirements may demand satisfaction of some or all of the following:

(a) shear transfer capability;
(b) watertightness;
(c) edge-wear resistance;
(d) ease of maintenance;
(e) airtightness.

Figure 9-1 illustrates a joint design detail, with watertightness and ease of maintenance as the prime objectives. Airtightness is also desirable, but not critical. Figure 9-2 illustrates an entirely different joint. Shear transfer capability

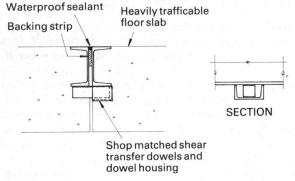

Figure 9-2   Joint for floor. The prime objectives are shear transfer capability and edge-wear resistance. The secondary objective is watertightness

and edge wear resistance are the dominant requirements. Watertightness is desirable, but not critical.

The design details of joints for the purpose of separating different construction materials are a particularly important matter in domestic buildings. Such buildings are often constructed using reinforced concrete floor and roof slabs which obtain their support from load-bearing masonry walls. False ceilings are rarely used, and the joint between the finished wall surfaces and the finished underside of concrete slabs is exposed to view. If the selected joint detail performs poorly, and extensive cracking between the two construction materials results, complaints from occupants must be expected. On the other hand, indiscriminate use of expensive joint details, guaranteed to separate effectively the different construction materials in a visually acceptable way, causes appreciable construction cost increases. The selection of joint details should therefore be carried out with care, economy of use, and proper understanding of the functions it must perform.

Load-bearing masonry walls can either be constructed of clay or silica-lime bricks, or concrete blocks. Clay brick walls expand with time, whilst the other two materials cause walls to shrink. Reinforced concrete floor slabs experience significant drying shrinkage immediately after their construction upon the load-bearing walls. Linear differential movement between walls and slabs must then be expected. In the case of reinforced concrete roof slabs, additional movements occur in consequence of temperature differences and moisture changes. If concrete roof slabs are used, great care is needed. Even the addition of a thick layer of insulating material, covered by a waterproofing membrane, does not stop the roof slab from seasonal temperature movements. In most cases, the insulation layer is not thick enough to stop noticeable daily temperature movements. Even with small restraint on movement, roof slabs tend to show cracks at wall-to-ceiling joints, which spoil their appearance.

One way of minimizing cumulative linear movement effects is to break wall continuity. This can often be achieved easily by extending the door frames to the underside of concrete slabs. Story-height door frames are structurally highly desirable in cases where the walls rest on structures with large spans. Breaks in the continuity of the walls help to hide the ill-effects of creep deflection of the long-span reinforced concrete supporting structure. Furthermore, the lateral strength of long walls should be checked to provide adequate lateral stiffness.

Two types of joint detail between concrete slabs and masonry load-bearing walls are illustrated. Figure 9-3 shows an inexpensive joint. In dormitory-type buildings, where wall finishes in natural, untreated materials are acceptable, it is possible to articulate the junction between masonry and concrete. Any fretting as a result of linear differential movement between the two materials collects dust on top of the wall in the recess spaces and is hidden from view. Care exercised at this early construction stage pays handsome dividends. The exposed masonry walls are protected from dirt and spillage during the construction and painting of ceilings. Upon removal of the plastic sheet covering, the masonry walls show neat and crisp lines.

Figure 9-4 illustrates an entirely different type of joint. Where walls and ceilings are rendered, set and painted, it is best to use cornices. It is also best to fix cornices to walls and leave a small gap between them and the ceiling finish.

In both cases, consideration must be given to the type of separation strip best

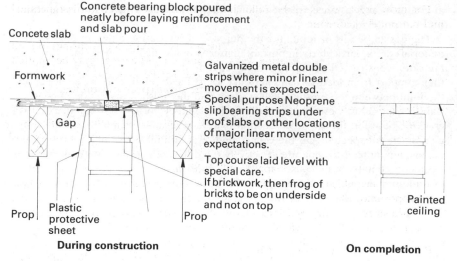

Figure 9-3 Separation of slab from wall

used between the walls and slabs. Factors which influence separation strip selection decisions are:

(a) extent of weather protection;
(b) distance from centroid of slab lateral restraint;
(c) intensity of gravity loads;
(d) likely wind exposure during slab construction.

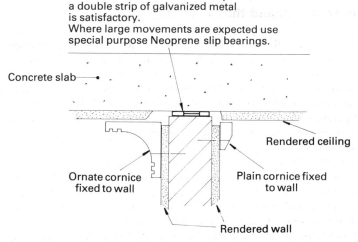

Figure 9-4 Junction of slab and wall

For most practical cases, the following separation strip selection is functional and reasonably economic:

Roof slabs	Use special purpose Neoprene slip-bearings on all bearing walls. Neoprene slip-bearings may be omitted locally on walls close to the centroid of slab lateral restraint, and plastic sheet cover substituted.
Floor slabs which support one-story construction	Use double layer of galvanized metal strips or like separation material on all bearing walls. Double-layer separation strips may be omitted locally on walls close to the centroid of slab lateral restraint, and plastic sheet cover substituted.
Floor slabs which support two or more story construction	No separation strips are necessary, unless strong winds are likely to occur during slab construction and slab curing. Plastic sheet cover over all bearing walls is then an appropriate specification.

Plastic sheet covering of load-bearing walls is often specified, even in cases where small linear differential movements are expected, due to the expected performance of concrete slabs immediately following their casting. In dry and windy climates, any failure to properly wet-cure the slabs will cause them to warp. Many cases of slab corner lifting have been observed. The bond between the concrete and the top masonry course is sometimes greater than the bond between the lower courses of masonry, which causes lifting of the top course or courses and cracking of the lower mortar bed joints. In a humid climate, or in construction projects where wet-curing of the concrete can be achieved with confidence, no separation of load-bearing walls and slabs need be made, provided the differential linear movements of the structural parts under consideration in their final service conditions are expected to be small.

10 Flexural movements and their control

This section is concerned with the performance of reinforced and prestressed concrete slabs in consequence of their exposure to natural or man-made climatic conditions.

Structural concrete slabs experience differential rates of drying and varying conditions of exposure to heating and cooling throughout their useful life. Gradients of temperature and moisture through slab thickness are induced. Even under conditions of relative external stability, concrete drying rates vary when changes in slab thickness occur.

Unrestrained concrete expands with temperature rise or increase in moisture content. Similarly, it shrinks when cooled or dried. Concrete slabs are cast either on the ground or on forms. In both cases, the underside of slabs is protected from wind and sun, and prevented from an excessive rate of drying. Slab top surfaces do not have such protection. In practice, the top surface of slabs dries out at a faster rate than desirable. Generally, wet-curing is inconvenient. When it is adopted, the period of its maintenance is too short. Curing agents rarely, if ever, match the benefits of wet-curing. Furthermore, rapid surface drying results when windy conditions coincide with hot weather. Consequently, most slabs in their early life experience a gradient of moisture conditions through their thickness which causes top surfaces to shrink at a faster rate than their undersides. Unrestrained slabs must be expected to curve, and corners of isolated slabs will lift.

The performance of slabs under service conditions depends not only upon the level of stress induced by superimposed loads, but also upon their surface exposure. In weather-protected situations, and outside the special conditions listed hereafter, the influence of variations in internal temperature and humidity upon slab performance is minimal and may be safely ignored. The structural designer should be on guard when he is called upon to design exposed concrete roof slabs, or within the structure being designed it is intended to house the following functions:

(a) cool rooms; ·
(b) steam and fog rooms,
(c) drying rooms;
(d) activities which generate sustained heat, such as boiler rooms.

The conditions are further aggravated in cases where cyclic heating, cooling, wetting or drying is anticipated.

Suspended slabs subjected to anticipated severe differential surface exposure conditions need to be investigated for three different situations. These are

(a) Gradients of temperature and moisture which create slab bending moments of like sign to those induced by gravity loads.
(b) Conditions opposite to (a) above.
(c) Crack propagation as a consequence of cyclic stress level changes or reversals.

For case (a), the bending moment increases may be sufficient to warrant increase in reinforcement content. The disposition of reinforcement does not change.

For case (b), the bending moments induced by gravity loads are reduced. In the case of reinforced concrete slabs, the reduction must be severe enough to cause reversal of moment before a need arises to change the reinforcement pattern. However, should significant reversal occur, the normally unreinforced parts of slabs need to be reinforced, and full strength laps of reinforcement used. It is of particular importance to recognize that prestressed slabs suffer severely under these conditions. Normally, draped cables are used. When the self-weight bending moments are matched by prestress uplift bending moments, the reserve for moment reversal from secondary effects is lost. It follows that slabs which may be subjected to bending moment reversals in consequence of differential surface exposure conditions cannot be prestressed in the normal manner. Unless the slabs are heavily reinforced, the draping of cables must be severely limited.

Stress-level reversals in reinforced concrete slabs tend to induce numerous cracks which propagate. The durability of slabs is severely and adversely affected. The slabs also become more flexible, often sagging to a surprising degree.

A number of known conditions of failure in identical situations are illustrated in

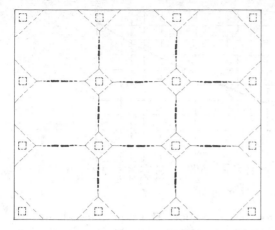

Figure 10-1   Crack pattern induced in roof slab. When the (hogging) moments induced by temperature and moisture gradients exceed the tensile strength of the concrete, the underside of a prestressed concrete roof slab has a crack pattern as shown in this figure. This situation arises particularly when load balancing is used, and the full load of the concrete slab is balanced by the drape of the cables

Fig. 10-1. Roof slabs are sometimes used to accommodate car parking. Since vehicular traffic cannot be permitted upon built-up waterproofing membranes, it is appropriate to use exposed prestressed concrete slabs. Waterproof conditions must result as well as satisfactory conditions of surface wear. If the allowance for temperature and moisture gradient effects is inadequate, the slabs will crack on the underside along the lines indicated in Fig. 10-1.

The structural designer can and must control the ill-effects of flexural movements induced by temperature and moisture gradients. Recognition of the shape of unrestrained curvature of slabs is the starting point. Unrestrained movements create end slopes in slabs, which become discontinuous in multispan slab structures. The forcing of slab continuities creates bending moments. These are superposed on gravity load moments, wind moments and earthquake moments. When due allowance for the secondary moments is made, the structure will perform in a satisfactory manner.

11   Insulation against airborne sound

Insulation against airborne sound, that is, the noise produced by people talking or by television sets, depends almost entirely on the mass of the insulating material. Concrete, being comparatively heavy, compares favourably with lighter methods of construction.

The scale used for measuring sound pressure is the decibel (dB), which roughly fits the human perception of the loudness of sound. It is not an absolute measure

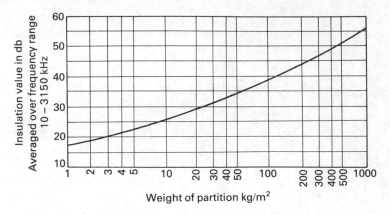

Weight of partition kg/m²

Figure 11-1   The 'mass law' for sound insulation. The experimental relation between sound attenuation and mass is an almost straight line. The decibel scale is logarithmic (see Eqn 11-1). An increase in the sound insulation value by 6 dB means that the sound has been reduced by half. In the range 200–1000 kg/m², a doubling of mass produces a sound attenuation of approximately 5 dB (Reproduced from [4] p. 140)

but a relative one. The difference in loudness between two sound pressures p_1 and p_2, expressed in decibels, is

$$20 \log_{10} (p_1/p_2) \tag{11-1}$$

Thus 6 dB means that one sound is antilog $(6/20) = 1.995$, or approximately twice as loud as the other.

The sound insulation of a wall or partition against airborne sound is almost proportional to its mass; this is known as the mass law. In the region of the common thicknesses of structural concrete walls and floors (200–1000 kg/m²), a doubling in thickness increases the sound insulation by approximately 5 dB (Fig. 11-1), that is, the proportionality is not exact.

Structural concrete is therefore an excellent material for the attentuation of airborne sound. It may be economic to increase the thickness of concrete floors where airborne sound is a problem, and to determine the thickness of reinforced concrete floors on this basis rather than by the more customary criteria of strength and deflection. The structural system can then be adjusted to the greater depth of the concrete floor slabs by increasing some spans, or by using less reinforcement.

Structural concrete is not as useful for insulation against noise from motor cars and aircraft. An increase in the thickness of load-bearing concrete walls is not as effective because the noise enters largely through windows. This is so, to a lesser extent, even when windows are sealed.

Structural concrete provides little insulation against impact noise, for example, the noise produced by people walking about on the floor above. The best method of reducing impact noise is to stop the noise from getting into the structure, for example, by means of a carpet, or by the use of a floating floor, that is, a concrete or timber raft laid on a resilient material on top of the structural concrete floor.

It should be noted that the requirements for impact sound insulation are almost entirely opposite to those for utilizing the thermal inertia of the structural concrete floor, discussed in Section 4.

12 Optimization of environmental design

Evidently, the optimal designs for the various environmental requirements can be in conflict with one another, and each can be in conflict with the optimal structural design. In particular, a thicker concrete structure can provide improved thermal and acoustic performance, but it is not always immediately obvious whether this is worthwhile. The decision involves a judgement on the part of the designer. This can be greatly aided by assigning values to each of the requirements, and plotting their interaction by a suitable computer program [18] that draws an area containing all the solutions providing acceptable results, for example, for the structural design and the thermal performance. One boundary of this area provides the optimal values for both criteria, although none may be optimal for each criterion considered separately.

References

1. Bowie, T. (Ed.), *The sketchbook of Villard de Honnecourt*, Indiana University Press, Bloomington, 1968, 144 pp.
2. Cowan, H. J., *The master builders*, Wiley, New York, 1977, 299 pp.
3. Cowan, H. J., *Science and building*, Wiley, New York, 1978, 374 pp.
4. Parkin, P. H., Humphreys, H. R. and Cowell, J. R., *Acoustics, noise and buildings*, 4th Edn, Faber, London, 1979, 297 pp.
5. *Greater Pacific House*, Constrl Rev., Vol. 46, Nos 8 and 9, Aug–Sept. 1973, pp. 20–21.
6. Harkness, E. L. and Mehta, M. L., *Solar radiation control in buildings*, Applied Science Publishers, London, 1978, 271 pp.
7. Saleh, A. M., The design of sunshading devices, Cowan, H. J. (Ed.), *Solar energy applications in the design of buildings*, Applied Science Publishers, London, 1980, pp. 33–94.
8. *ASHRAE handbook of fundamentals*, American Society of Heating, Refrigerating and Air-Conditioning Engineers, New York, 1977, 37 chapters.
9. McGuiness, W. J. and Stein, B., *Mechanical and electrical equipment for buildings*, 5th Edn, Wiley, New York, 1971, 1011 pp.
10. O'Callaghan, P. W., *Building for energy conservation*, Pergamon Press, Oxford, 1978, 232 pp.
11. Billington, N. S., *Thermal properties of buildings*, Cleaver–Hume Press, London, 1952, 208 pp.
12. Faber, O., Kell, J. R. and Martin, P. L., *Heating and air-conditioning of buildings*, Architectural Press, London, 1971, 562 pp.
13. Givoni, B., *Man, climate, and architecture*, 2nd Edn, Applied Science Publishers, London, 1976, 483 pp.

14. Jennings, B. H., *Environmental engineering*, International Textbook Company, Scranton, 1970, 765 pp.
15. Porges, J., *Handbook of heating, ventilating and air conditioning*, 7th Edn, Newnes–Butterworth, London, 1976, 15 sections.
16. Szokolay, S. V., *Solar energy and building*, 2nd Edn, Architectural Press, London, 1977, 174 pp.
17. Lee, Kaiman, *Encyclopedia of energy-efficient building design: 391 practical case studies*, Environmental Design and Research Center, Boston, Mass., 1977, Two volumes, 1023 pp.
18. Radford, A. D. and Gero, J. S., Tradeoff diagrams for the integrated design of the physical environment of buildings, Cowan, H. J. (Ed.), *Solar energy applications in the design of buildings*, Applied Science Publishers, London, 1980, pp. 197–223.

Part IV Construction

25 Maintenance, repair and demolition of concrete structures

Jack R Janney

Consultant to Wiss, Janney, Elstner and Associates, Northbrook, Illinois and Kellogg Corporation, Denver, Colorado, USA

Contents

Summary

The various causes of distress to concrete structures are discussed. Emphasis on the importance of analyzing these causes in order to develop effective remedial measures is made. Suggested methods of effecting repairs to columns, beams and slabs are shown. Concrete mix type and use of polymers in conjunction with Portland cement are treated as alternative methods of repairing concrete structures, the selection depending upon the exposure the repaired members will experience in service.

A number of the tools available for the non-destructive assessment of structural conditions are described. Tests such as full scale load tests, structural model tests and laboratory mock-ups which may be helpful to properly assess need for repair are also described and discussed.

Methods of demolishing concrete structures are described with emphasis on the safety and damage mitigation measures which should be considered.

1   Introduction

Concrete structures are designed and built to last. Many do indeed last a great length of time, possibly until they become obsolete for continued use. Many others, however, become in need of restoration or repair, often before useful life has been expended and much sooner than hoped. Early requirement for remedial attention can come about due to improper design or construction or because faulty materials were employed. Need for such attention, however, is usually the result of deteriorating man-made or natural mechanisms or from wear and tear arising out of use.

This chapter deals with the causes of deficiency as well as methods of maintenance and repair. Without thorough understanding of cause, effective maintenance and repair cannot be achieved with dependability.

Structural maintenance and repair demand sound understanding of structural behavior. A good understanding of the significance of defects to the structural integrity and future service is imperative. Most cost-effective restoration methods cannot be developed otherwise. In fact, perceptive engineering and analysis of conditions can be even more important to effective restoration than to the design of new construction. Ill-conceived maintenance and repair *can* be counterproductive and accelerate deterioration, further shortening the life of a structure.

Generally, disturbance of structural components as they exist before restoration should be minimized. The greater the amount of material removed, the greater the risk of adverse secondary effects. Repairs to concrete structures should only be accomplished if the safety or serviceability of a structure require that they be done. Cosmetic repair is excluded from this admonition and is not within the scope of this chapter.

2 Deterioration

2.1 Environmental conditions

Most concrete structures combine two materials, concrete and reinforcement. Either or both can be adversely affected by the environment to which they are exposed. This environment may be natural or produced by man in connection with construction or use of a structure. Natural environmental conditions which can be detrimental to concrete or reinforcement are water, especially sea water, freezing or freeze-thaw cycles, ocean atmosphere and chemicals incompatible with concrete or reinforcement existing naturally in soils or water.

Many man-produced environmental exposures cause deterioration of both concrete and reinforcement. Detailed discussion of all such exposures is not within the scope of this treatise. The most common which are responsible for most severe damage, however, are various kinds of acids, sulfate compounds and deicing salts.

2.2 Chemical attack

The presence of deteriorating chemicals should be identified and established as the principal cause of deterioration before effective repair methods can be developed. Laboratory chemical analysis and petrographic examination of concrete samples removed from the structure may be required to establish what needs to be done (Section 9.5). In this case, samples should be removed from affected areas as well as from undamaged areas, for comparative purposes. The analysis, as a general rule, should include at least:

(1) Concentration of suspected chemical versus depth of concrete.
(2) Detection of adverse reaction products to determine character of damage and likelihood of further damage.

Chemicals which penetrate concrete either through cracks or by permeation may not necessarily be detrimental to the concrete but rather only attack reinforcement. Such is the case with most deicing salts applied to flat surfaces to remove ice and snow. Deicing salts can be applied directly to surfaces or transported by automobiles which pick up snow containing salts and deposit it on parking surfaces. Parking structures, pavements, bridge decks and other reinforced concrete structures exposed to deicing salts are especially vulnerable to reinforcement corrosion.

Structures exposed to environmental conditions such as tanks used to store liquid chemicals or plants utilizing chemicals which are aggressive to concrete or steel as part of a manufacturing process, require special attention and special repair techniques.

Many concrete structures along sea coasts suffer deterioration due to reinforcement corrosion when protection against exposure to salt laden humid atmosphere is inadequate. Structures partially immersed in sea water are vulnerable to severe damage especially in freezing climates.

2.3 Freeze-thaw damage

Freeze-thaw action is responsible for considerable damage to concrete, Fig. 2-1. If concrete is susceptible to such damage, the deterioration is greatly accelerated when deicing salts are applied. Concrete which is properly air entrained, contains aggregates that are not sensitive to freezing action and is otherwise of good quality, will not normally suffer freeze-thaw damage. The procedures for making good freeze-thaw resistant concrete are treated in detail in Chapter 27.

Fresh concrete that is allowed to freeze before adequate curing will suffer damage the extent of which will depend upon its age at freezing, the depth of freezing and the exposure following freezing. Instances have been recorded of very fresh concrete being frozen for a great length of time. After thawing with sufficient continuing warmth, it cures in much the same way that it would have if it had not frozen.

Figure 2-1   Freeze-thaw damage to concrete surface

2.4 Water

Of all the elements which contribute to deterioration of concrete and reinforcement, water is common to most deleterious reactions. Water is important to the development of the strengths of concrete both in mixing and curing. After curing, however, if concrete could be kept completely dry, most damage which is experienced by both concrete and steel would not occur. This understanding is important to those planning or designing maintenance or repair programs. Many repairs have been accomplished which were not lasting because the fundamental problem, the presence of water, was not, or could not be, eliminated or minimized. The adverse effects of the presence of water or moist atmosphere is usually the most important consideration of the effectiveness of remedial work.

3   Use considerations

The use to which a reinforced concrete structure is exposed or a change in use can be cause for the need of repair. This exposure may be from gravity loads, (static or dynamic) hydrostatic loads, earth pressure or from natural lateral forces such as wind or earthquakes. For our purposes we will consider fire damage a use exposure.

3.1   Changed loading requirements

Buildings or bridges which have been designed for one use are often considered for a different use in which the loading requirements will be increased. This requires, in most cases, an engineering analysis to determine what must be done to increase strength so that the structure can sustain heavier loads.

3.2   Fatigue

A great amount of attention has been given to the matter of fatigue of structural materials. Structural steel does indeed suffer distress quite frequently due to stress cycles. Fatigue is seldom a serious problem, however, with reinforced concrete. This subject is treated in detail in Chapter 16.

3.3   Design versus actual loads

Most buildings and bridges seldom, if ever, are loaded to the extent assumed by designers. Certain types of structures such as water tanks, silos or warehouses, on the other hand, are indeed loaded as assumed in design. The incidence of failure is, therefore, much greater percentage-wise for such structures. Special engineering attention must be given to fully loaded structures when repair is necessary. Tanks, bins and silos containing materials which can attack reinforcement are especially challenging with respect to developing repair methods when such attack is experienced.

3.4   Natural forces

Natural forces have historically been a common source of collapse or severe damage to structures. The risk of such damage is dependent upon the location of the structure as well as upon design and construction. The natural forces are produced by earthquakes, high winds, snow and rain. There have been few instances of failure or serious damage to reinforced concrete cast-in-place structures resulting from snow, rain or wind. More incidences of failure or severe damage from these natural causes have occurred to structures built of precast concrete. This is because such construction usually lacks structural redundancy

Figure 3-1   Ponding failure of precast roof

which exists in cast-in-place construction (Fig. 3-1). On the other hand, many concrete structures have suffered severe damage as a result of earthquakes. In most instances, the damage is concentrated in vertical members such as columns, walls or at connections (Fig. 3-2).

3.5   Fire damage

Severe fire causes damage to both concrete and reinforcement. Time and temperature are both important variables which combine together to result in deterioration of concrete. Concrete in which the Portland cement is well hydrated and which has a low moisture content can withstand a hot fire for a greater length of time than new concrete with a high moisture content. The destruction of concrete which takes place when free water within turns to steam can be very rapid. If this phenomenon takes place, the damage is evidenced by spalling and will be essentially restricted to that which is visible (Fig. 3-3). On the other hand dry concrete may suffer damage that is not visible in the form of spalling. Experienced investigators can estimate the temperature to which concrete has been exposed by an analysis of color. A petrographic examination is recommended if the depth of damage to concrete is important to one's investigation.

Reinforcement can be damaged by exposure to high temperature. This damage

Figure 3-2   Earthquake damage to reinforced concrete column *Courtesy*:
B. Bresler

Figure 3-3   Fire damage to reinforced concrete

can be aggravated if the reinforcement is 'quenched' with water while in a state of elevated temperature during fire fighting operations. Intermediate or higher strength conventional reinforcement will suffer damage if it is exposed to temperature of about 600°–700°F (315–370°C) or higher. High strength tendons used to prestress concrete are especially vulnerable to damage when exposed to elevated temperatures. In general, temperatures in excess of 500°F (260°C) will damage prestressing steel.

Several techniques can be employed to measure the level of tension in prestressing tendons. If that level is found to be approximately what it should be, one is safe in assuming that serious damage has not been experienced. The extent of damage to reinforcement is best established by removing samples and submitting them to laboratory tests to determine physical properties.

Reinforced concrete structures suffering severe fire damage will exhibit large deflections.

3.6 Floors

Slabs in factories or manufacturing facilities often receive abusive loads, usually from vehicles used to transport heavy equipment or materials as part of a manufacturing process. Most slabs on-grade, even though designed by code, are relatively light unless the designer has addressed the possibility of excessive loads and designed the slab accordingly. As a result, vehicular traffic can break slabs up sometimes rather badly especially if the slab was not cast with adequate attention paid to expansion and contraction joints. Concrete shrinks, and if provisions are not made to accommodate this shrinkage, cracking will occur. Heavy loads continually passing over cracked concrete can cause it to break up, often necessitating repair.

4 Design deficiencies

4.1 Significance of admitted errors

Errors or deficiencies in design, which are brought to light after a structure is completed or during construction, frequently give rise to the need for developing strengthening or repair procedures. Design deficiencies, however, may or may not actually reduce the strength or factor of safety of a structure below acceptable levels. Careful analysis should be undertaken to establish if apparent design error has, indeed, resulted in unsatisfactory strength.

4.2 Selection of design concrete strength

Many designers still specify low strength concrete just because a code permits it without sufficient regard to durability. The resulting savings might be questioned in view of the future maintenance costs that will become necessary if exposure conditions are adverse. The durability and resistance to deterioration of a structure in most cases is much greater when higher strength concretes are used.

4.3 Concrete cover

Minimum code requirements for concrete cover over reinforcement are often insufficient under conditions of very high humidity or other adverse environmental exposure. In the experience of the writer, corrosion of reinforcement has almost always been the result of inadequate cover or low strength concrete or both when the exposure is moisture only and not complicated by the presence of aggressive chemicals (Fig. 4-1).

4.4 Thermal forces

Probably the most common deficiency in the design of reinforced concrete structures is failure to give enough thought to the forces produced by changes in

Figure 4-1   Reinforcement corrosion due to inadequate cover and low strength concrete

temperature. The cracking and distress that result from this oversight are, in most cases, difficult and costly to remedy (Fig. 4-2).

4.5   Reinforcement details

Usually the most serious problems attributable to design deficiencies have to do with improper attention paid to reinforcement details. This is especially true with respect to reinforcement designed for regions of post-tensioning anchorages, bearings in precast construction, discontinuities in members to accommodate mechanical systems and unusual requirements for load distribution.

4.6   Other common design deficiencies

Other design errors which frequently surface include improper account for torsional forces, especially with precast concrete, consideration of creep deflection in slab construction and provisions to accommodate shrinkage, thermal movement and sub-grade compaction for slabs on-grade. Unacceptable cracking will occur in post-tensioned slabs not intended to crack when improper account is taken in design of the interaction between slab and walls and columns which support the

Figure 4-2   Distress at bearing due to thermal shortening

slab. A very common design error which is encountered in precast concrete construction and causes distress at the supports of precast members is to provide that both ends be rigidly attached to nonflexible supporting members. Ponding failures and deflection problems have occurred as a result of improper assessment of concrete creep and shrinkage losses in prestressed members.

5   Construction deficiencies

5.1   Reinforcement

Most deficiencies arising from faulty construction are: (1) misplacement of reinforcement, (2) installation of improper size or grade of reinforcement, or (3) providing insufficient lap length at splices. Any of these construction errors, if significant, can seriously reduce strength and factor of safety.

5.2   Concrete

Construction errors having to do with concrete placement include: (1) excessive water, (2) inadequate curing, (3) improper consolidation, (4) addition of incorrect amounts of admixtures to mix, (5) frozen concrete, (6) fly ash mistakenly used for Portland cement and (7) excessive mixing time between batching and placing.

Generally, construction errors resulting in low strength concrete or improperly consolidated concrete are not as serious to the strength of a structure as errors in reinforcement placement or amount. There are exceptions, of course, some of which are: (1) unacceptably reduced punching shear capacity in slabs, (2) unsatisfactory bearing strength, (3) excessive deflections due to low strength concrete, unsatisfactory durability and (5) appearance.

5.3 Construction loads

Cracking of floor or roof members or even more serious distress including failure can be experienced if a constructor permits excessive construction materials to be temporarily stored on these members. This can be especially critical when the concrete strength is low either due to inadequate early curing or if insufficient time has elapsed since casting and the time loads are imposed upon the structure.

5.4 Erection of precast members

Improper erection procedures for precast construction can result in distress or collapse. Some frequently encountered errors include: (1) improper account given to erection sequence, (2) careless setting of members resulting in inadequate bearing, (3) improper measures taken to provide temporary stability to members being erected, (4) improper sequence of making structural connections and (5) improper stacking of precast members being stored prior to erection.

6 Materials deficiencies

6.1 Reinforcement

Conventional reinforcement is seldom found to be deficient with respect to material properties unless it has been damaged by fire, fatigue stresses, overstressing from excessive loads or has been welded improperly. In fact, the physical properties are usually found to exceed those specified.

6.2 Portland cement

Portland cement may occasionally be used which has properties adversely affecting quality of construction. Some of these are: (1) flash set, (2) false set, (3) color variation, (4) strength properties, (5) air content from air entrained cement, and (6) inadequate sulfate resistance.

6.3 Aggregates

Aggregates are more likely to be the ingredient in concrete that is deficient. Some of the properties of aggregates which can cause problems with concrete quality resulting in need for repair are: (1) alkali reactivity, (2) high shrinkage characteristics, (3) low strength, (4) low resistance to freeze-thaw cycles, (5) tendency to cause pop-outs in freezing climates, (6) low abrasion resistance and (7) poor gradation or shape.

6.4 Admixtures

Admixtures sometimes lose their effectiveness due to damage from shelf life or storage exposure problems. Infrequently they are found to have been improperly manufactured. Some admixtures contain chloride ions and as such must be regarded as potentially corrosive to reinforcement.

7 Assessment of conditions

7.1 Cause

Before repair of concrete structures is considered, it is imperative that the cause of the distress be firmly established. Without doing this, repairs may be completely ineffective and, in some cases, can be even counterproductive. The two most common causes for the need of repair are corrosion of reinforcement and deficient concrete. The latter may be the result of deterioration damage from external or internal sources or may just be a result of poor concreting procedures or deficient materials used during original construction.

7.2 Corrosion of reinforcement

The pressures built up as corrosion products develop on reinforcement are so great that a significant amount of corrosion must crack the surrounding concrete. Corrosion products occupy approximately eight times the volume of the original steel. Thus, if cracking is not present, one can conclude that corrosion, if present at all, is very insignificant.

It is also important to realize that the cause of the corrosion may be no longer present. If corrosion is inactive, and if the extent is not serious enough to result in structural deficiency, significant repairs may not be necessary. In this case one may only consider sealing cracks.

Active corrosion can often be established by measuring the voltage which accompanies corrosion with a copper copper-sulfate half-cell (Section 9.4). It is difficult to judge the extent of corrosion from appearance. Very light corrosion is obvious, of course; deformations on the reinforcing bars will still be present. Corrosion serious enough to completely erode the steel will also be evident by eye. Loss of area between these extremes is difficult to perceive. Bars can appear severely damaged when, in fact, the reduction in area may be minimal. Such situations require sand blasting or other methods of cleaning the corrosion in order to measure the loss of cross-sectional area in critical locations.

It is important to determine what chemicals present with water have promoted corrosion of reinforcement. These chemicals will most likely be chlorides. Other oxidizing agents may be present depending upon exposure from use or nearby equipment. The presence of these chemicals can be established from chemical analysis of samples removed from the concrete. The depth of penetration and the quantity of the oxidizing chemicals should be established. Petrographic examination and chemical analysis of concrete samples are very helpful for these determinations and other evaluations (Section 9.5). One usually is interested in assessing concrete strength, so the samples that are removed for chemical and petrographic analysis can sometimes be also employed for strength determination.

7.3 Post-tensioning

Contractors often do not correlate jacking forces with tendon elongation to establish that the post-tensioning forces required are actually delivered. Unexpected cracks and unusual deflections could be evidence that such has occurred.

7.4 Faulty concrete

Symptoms which lead one to suspect that concrete may have become frozen are excessive deflection, excessive cracking and dusting. Concrete that cracks excessively during very early age may indicate the presence of excessive water reducing retarder in mix. Too much water or inadequate curing results in low-strength concrete which is also manifested by excessive deflections and/or cracking. Excessive air entrainment also produces low-strength concrete.

Improper placing of concrete is usually evident immediately upon removal of forms. Honeycombing or rock pockets result from segregation and improper vibration. A thorough engineering analysis may reveal that honeycombing or rock pockets, unless the condition is very extensive, do not adversely affect structural integrity. In such cases, cosmetic repair may only be required.

Cavitation from water or waterborne debris is visibly evident but will require diver inspection if under water. Acid attack, freeze-thaw damage and abrasion damage will also be visibly evident.

7.5 Structural damage

Structural damage from foundation settlement, improper placement of or insufficient reinforcement, fire, accidental external forces (gravity or dynamic) or any other cause may not be easy to assess. Some of the methods described in Section 9, as well as perceptive engineering analysis may be required.

8 Repair techniques

8.1 Crack repair

Cracks in concrete normally come about as a result of any or a combination of: (1) drying shrinkage, (2) thermal forces, (3) flexural, and (4) shear or torsional stresses.

Cracking can also result from the development of internal forces such as those produced when unsound aggregates react with cement or when corrosion products build-up on reinforcement.

Cracks which result from external causes can be effectively repaired by epoxy injection. Other polymerized resins such as polyesters, polyurethanes and plastic calking materials and cement grout are also employed for repairing cracks. This work requires experience and proper equipment. Cracks as fine as 0.0005 inch (0.0127 mm) have been effectively repaired at 40°F (4.4°C) by epoxy injection. Selection of the epoxy materials is very important. Manufacturers or applicators who are experienced in this method of repair should be consulted before the material is specified. In general, the following properties of epoxy should be requested:

(1) Low viscosity.
(2) Adequate pot life; 30 minutes or more unless materials are mixed in line.
(3) Ability to bond in presence of moisture.
(4) Good bonding strength, good tensile strength and high heat deflection temperature (at least 30°F (17°C) above maximum service temperature).

In order to be effective, in most cases, cracks should be nearly completely filled. This is best accomplished by using pressure. Pressures of 50–400 psi (0.35–2.76 MPa) are usually required for satisfactory repair. Finer cracks and/or deeper cracks require the higher pressures.

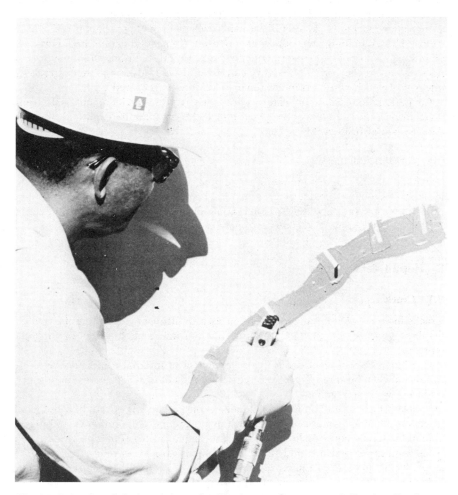

Figure 8-1 Crack being injected with epoxy *Courtesy*: Adhesive Engineering Company

Figure 8-2 Core showing extent of epoxy injection *Courtesy*: Adhesive Engineering Company

The normal procedures for epoxy injection are as follows:

(1) Seal cracks on both faces if possible so that epoxy will not flow out when it is injected under pressure. This sealing may be accomplished with epoxy or tape—the choice depending on the pressures which are to be employed.
(2) Entry ports also used for flow detectors are installed along the length of the crack which is to be injected (Fig. 8-1). These ports vary in configuration and are either epoxied in place or fastened to the concrete with drilled inserts. The type and spacing will depend on the depth of crack, width of crack, viscosity properties of the epoxy and application pressure.
(3) Epoxy is pumped into an entry port until it flows from the adjacent port at which time the pump is moved to that port and the operation is repeated. Constant pressure should be held at the last port without flow of epoxy for several minutes to make sure that the material is not escaping and that the crack is full.

Similar procedures, each designed to fit the materials employed for crack filling, must be established for other resins or grouts. Persons experienced and skilled in accomplishing injection or grouting repairs should be employed whenever possible. They are best qualified to select methods of sealing cracks, entry port type, spacing, method of attachment to concrete surfaces, pump type, required pressures and means of handling the epoxy materials with respect to temperature control, batch size and timing. Ineffectual filling can be a complete waste of time and money.

Early in the repair process exploratory coring should be done to check visually the degree of crack filling that is being accomplished (Fig. 8-2). This is especially important if repair is to be extensive. Non-destructive means such as pulse velocity or radiographs can sometimes be employed.

Before a decision is made to repair cracks, an analysis should be made of the need. If the need is established, an analysis should be made of the probability of successful repair.

8.2 Honeycomb or cavitation

Sometimes inadequate consolidation of concrete during placement results in 'honeycombing' or desegregation. A careful study should be made to identify the need for repair. If appearance is not of concern, the need should only exist if the structural integrity or durability of the structure requires that poorly consolidated or segregated concrete be removed and replaced. The act of removing concrete in preparation for patching materials can cause more damage than would result if the affected region were left alone. There are many concrete structures in which large honeycombed or segregated regions in non-critical locations have remained unattended for a great number of years without adverse effect on the life or performance of the structure.

When a determination has been made that repairs to large voids in concrete must be made, the repair technique should be designed to fit the work that the patch will be expected to do as part of the structure being repaired. For example, if the distressed region is located in a girder at a point where high shear stress will result when the member is loaded, the replacement concrete must be effectively tied to the parent concrete with reinforcement and every effort must be expended to assure that the patching material bonds well to the concrete against which it is placed. On the other hand, if the portion to be replaced will not be stressed very much in compression, tension or shear, such as will usually be the case in a foundation wall not receiving soil or other lateral loads, only provisions to assure good bond between the undamaged part of the structure and the patch need be made.

Usually honeycombing exists principally where it is visible and is most common

Figure 8-3   Honeycomb concrete and repair techniques

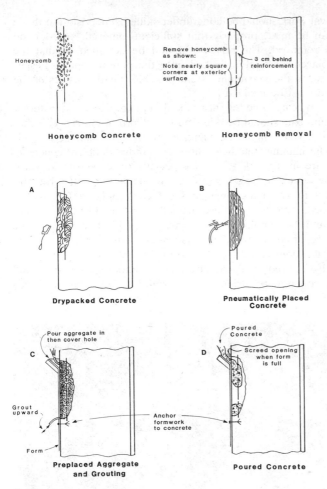

Figure 8-3 (*contd.*)

against vertical surfaces. Honeycombing or serious segregation is seldom experienced in horizontal pours. Rarely does serious honeycombing exist which is hidden from view behind the visible surfaces of concrete. If such a condition is suspected, however, the presence of voids may most easily be detected by using pulse velocity or pulse echo instruments (Section 9.2). If access problems prevent the ready use of these instruments, coring may be necessary. Coring is very positive but is expensive if extensive coverage is required and core holes require filling in most cases. The techniques to repair honeycombed or segregated concrete which have proven most successful are as follows.

(1) Remove unsound concrete with the smallest power tool which will do the job in a reasonable time. Excessively large hammers can damage sound concrete, and thus the repair can be counterproductive to the intent of the

work. The removal work should be done under skilled supervision so that a determination can be made properly that sufficient material, and not too much, has been removed. The removal should be designed so that the subsequently replaced material is 'locked' into the parent concrete, especially at the edges, to the maximum extent possible. Feather edges of the replaced concrete must be avoided.

If reinforcing is to be exposed for encasement in the replacement material, the concrete behind bars should be removed to a depth of at least $1\frac{1}{2}$ inches (38 mm), (Fig. 8-3). Care should be exercised to prevent damage to reinforcement with the hammer bits being used to remove deficient concrete.

If large areas are in need of removal especially on expansive, vertical surfaces, the use of explosives has proven to be a fast, effective and economical method for removing damaged or defective concrete (Fig. 8-4). If this method is used, experts in the use of explosives should be consulted. Instruments should be employed to monitor the blast vibrations the structure being repaired as well as adjacent structures receive. Damage from these vibrations can thus be controlled.

(2) Reinforcement, if required, is placed next. Dowel bars may be embedded into the parent concrete by drilling holes slightly larger than the bars

Figure 8-4 Face of wall from which deteriorated concrete has been removed with explosives *Courtesy*: U.S. Army Corps of Engineers

to be placed and embedded with epoxy or grout materials. The depth of holes should be established by laboratory tests employing the grouting or polymer materials to be used for embedding the bars or should be a minimum of 16 bar diameters if testing is not practicable or data are not available from which embedment lengths can be determined. Dowel bars can also be fastened to parent concrete with drilled inserts proven by test data to have the capability of developing the strength of the bar. Core tests may be necessary to establish the strength of the undamaged concrete, a quantity necessary to the determination of the capacity of drilled inserts.

The size and spacing of the dowel reinforcement should be established so that this steel can accommodate the entire shear force which will develop across the interface of the patched and parent concrete.

Other reinforcement which may be used in conjunction with the dowel reinforcement should be tied to the dowel steel. Welding, unless very carefully engineered and supervised, is not recommended.

(3) The chipped surface of the parent concrete must be cleaned before applying the patch concrete. Sand blasting is an effective way of cleaning and removing small loose pieces of concrete from the parent concrete and from the reinforcement. After cleaning and just prior to placing the patch concrete, the chipped surface may be moistened. If the patching material can be placed with a hydraulic head or if it can be vibrated, such treatment is all that is necessary.

If the patch concrete is very dry, such as dry-pack or if it is to be more or less plastered into place, epoxy-bonding agents are recommended. The epoxy-bonding agent should be of low viscosity, be able to harden in the presence of mortar and have a pot life sufficiently long to prevent premature hardening before the patch concrete is in place. The biggest danger in using bonding agents is losing control of time and placing patch concrete over hardened materials. The results are obvious. Other polymers are sold as bonding agents and some are effective. Care should be exercised in selection and proof of performance should be furnished before these materials are employed as bonding agents.

(4) In designing the mixture to be used to replace removed concrete, the relative shrinkage of the parent concrete and patching materials must be considered. If the member being repaired is relatively new and has not experienced very much of the drying shrinkage which it will eventually undergo, the patching concrete mix should be designed to approximate that of the parent concrete as much as possible. The same materials and consistency should be employed if possible. By this procedure, differential shrinkage between the base concrete and patching concrete may be minimized.

On the other hand, if the concrete being repaired is of such an age that all or most of the drying shrinkage has taken place by the time the repair is accomplished, patching materials should be selected that will shrink as little as possible. Shrinkage is minimized by proper selection of materials. The aggregate should be selected from sources that have proven to produce concrete with the lowest shrinkage available. The greatest strength and lowest shrinkage will result with patching concrete having low unit water content. Superplasticizing admixtures are very beneficial to patching materials for which minimum shrinkage and high strength are desired. Remember that concretes with high cement content shrink more than those with lower cement content.

8.3 Dry pack

Small surface voids are often best repaired with dry pack mortar. A typical mix design for such mortar is:

One part Portland cement to three parts mortar or fine concrete sand. Water/cement ratio should be from 0.33 to 0.4.

Slump measurements are meaningless for dry pack material. Just enough water should be added to the mix so that a ball can be formed by hand with some effort with the surface of the ball having little or no sheen from moisture after forming. Dry packing requires considerable tamping to assure good compaction and contact between the patching material and parent concrete.

8.4 Pneumatically applied mortar

Placement of pneumatically applied concrete is a versatile building or concrete repairing technique. It may be used with or without formwork for horizontal, vertical and overhead surfaces in any desired shape or thickness; for curved or folded roof sections; for coatings over brick masonry, concrete steel or rock; for strengthening of concrete columns, slabs and concrete and masonry walls; for repair of reservoir linings, dams, tunnels, wharves, piers, pipe, etc.

The material can be built up to any thickness and may be placed with or without fibers. Each layer should be built up by several passes until a thickness of from 1 inch to 3 inches (25–76 mm) is reached. Initial set should be allowed before additional layers are applied. Material can be placed over 1500 feet horizontally (458 m) and over 320 feet (97 m) vertically from the gun for small ornamental repairs with a $\frac{1}{2}$ inch (12.7 mm) nozzle, or high volume $4\frac{1}{2}$ cubic yards (3.4 m^3) per hour with appropriate equipment.

Proper preparation of surfaces must be accomplished. Remove all deteriorated concrete or masonry and clean by sandblasting. The surface to receive the material must be well wetted just before application.

The material should be applied by a nozzleman with proven experience and under the direction of an experienced foreman. A good practice is to have the nozzleman shoot preconstruction test panels in boxes about 18 inches by 18 inches by 3 inches (45 cm × 45 cm × 7.5 cm) for compression tests.

Portland cement-to-sand ratios of 1:3 to 1:4$\frac{1}{2}$ should be used. Coarse aggregate up to 3/4 inch (19 mm) may be used. Sand should be ASTM C33 Gradation II. Finer sand (ASTM C33 Gradation I) may be used if necessary and if properly tested.

A natural gun finish is recommended; however, a float, screed or trowel finish can be accomplished if done by skilled workmen.

Underwater repair of honeycombed concrete or concrete damaged by cavitation has successfully been accomplished by using prepacked methods. By this method the region to receive replacement concrete is formed and filled with gap graded aggregates. The water filling the spaces between the aggregate particles is replaced by pumping Portland cement–sand mortar or neat cement into the aggregate mass. The water within the space to be filled must be static, as significant flow will wash the mortar from the space. This work requires underwater helmeted divers at the point of deposit. Skill and experience of workmen is imperative to the successful execution of this type of repair. Prepacked methods have also been successfully employed for repairs out of water.

8.5 Polymer mortars

The US Army Corps of Engineers has done considerable work toward the development of techniques to repair cavitation resulting from abrasion. Of interest is the concept of applying special polymer mortar coatings to repaired areas to serve as armor against abrasion. Properly applied and cured, these protective coats can prove very effective. The best results are obtained with typical mixes within the following ranges of proportions:

Abrasion-resistant sand—such as quartzite, silica, etc.—3 to 5 parts sand to one part epoxy which sets up in presence of moisture and has good pot life. After application, these mortars must be maintained at a temperature of 60°F (15°C) or better for at least 24 hours.

Polymer-impregnated and polymer-Portland cement concretes also provided good resistance to abrasion in tests conducted by the Corps of Engineers; though not as good as polymer-bound mortars, polyurethane and epoxy being the best.

8.6 Flat surface repairs and overlays

Pavement, bridge decks, floor slabs and other flat work are repaired by replacing deteriorated concrete with or without additional overlay. The procedures described in Section 8.2 for removing deteriorated concrete and preparing surfaces of parent slabs to receive new concrete apply to flat work. Because the workmen removing concrete are handling power tools downward, it is easier than when they are working overhead or on vertical surfaces. Care must be exercised, therefore, to prevent removal of too much concrete. It is the opinion of the author that excessive removal can do more harm than good. Only concrete that is clearly bad should be removed except that which is necessary to expose reinforcing bars with at least $1\frac{1}{2}$ inch (38 mm) clear distance under each bar to be encased with replacement concrete.

Remembering that the causes of deterioration most commonly are a result of freeze-thaw damage, corrosion of reinforcement, chemical attack on the concrete surface or excessive cracking of slabs on-grade, the repair must be designed to minimize further deterioration after repair.

If the cause of damage is freeze-thaw action, the replacement unmodified concrete should be properly air entrained and of high strength and low permeability. Corrosion of reinforcement will have probably resulted because the original construction either did not provide enough cover, and/or the concrete quality was poor or the surface was exposed severely to deicing salts or similar chemicals. The repair concrete should be placed to correct cover deficiencies and should be of high strength. An overlay is usually required to provide increased cover.

If environmental exposure is severe, the repair is best done by employing a combination of fill concrete of good quality and an overlay designed to act as armor or an impermeable membrane.

It is important to realize that overlay or replacement concrete will reflect the cracks that exist in the base slab unless special preventive provisions are made. The following procedures have been successfully employed to prevent this occurrence on slabs on-grade after deteriorated concrete, if required, has been removed:

(1) Provide contraction joints by sawcutting the base slab in a grid small enough to assure that thermal volume changes will be accommodated by these joints. Approximately twenty feet by twenty feet (6 m × 6 m) is suggested as a starting point.

(2) Epoxy inject existing cracks when the temperature is at or near the lowest that the slab is normally expected to experience. In the case of exterior slabs or pavement, the temperature should be at the lowest point which will still permit the epoxy to polymerize within a reasonable time. Slabs on-grade should be injected using the in line mixing method in order to employ the minimum pot life possible to minimize leakage into the sub-grade materials.

(3) If the depth of removed concrete is more than two inches, fill concrete of high quality is placed with or without bonding agents to about the original grade.

(4) If a special thin layer of overlay concrete or epoxy mortar is to be employed, it should be placed immediately after the fill concrete is in place.

(5) Joints must be provided in the fill concrete or overlay material directly over those which were previously cut in the parent slab.

(6) The finish may then be accomplished as desired.

Overlays which change the floor elevation are only practical when the floor level can be raised without creating unacceptable functional problems with respect to usage of the floor, and if a majority of the slab area is being repaired. Obviously, such overlays are not practical if only isolated sections of the slab under consideration are in need of repair.

Slab or pavement overlays can be designed with a wide range of materials. Some are listed below in ascending order of cost:

(1) Portland cement concrete mixes should be designed to provide good strength, result in minimum drying shrinkage and be properly air entrained if exposed to cycles of freezing and thawing.

(2) Portland cement concrete containing superplasticizers can be used to produce overlays of very high strength and low permeability. Using these plasticizers will permit the placement of concrete with water/cement ratios as low as 0.33. Care must be taken in designing the concrete mix, especially with respect to the addition of the superplasticizing agents. They should be added to the mix just prior to placing and in such a way as to assure that the admixture is very thoroughly disbursed throughout the mix in order to minimize slump loss problems. More control tests, i.e. slump, strength tests and air content measurements should be made than for normal concrete

placement. In fact, these tests are recommended for each batch. Furthermore, small test patches should be cast in advance of the work so that the mix proportions, mixing procedures, placing and finishing technique can be refined to produce the best results.

(3) Latex modified Portland cement concrete properly designed and placed results in very strong, impermeable overlays, which are resistive to aggressive attack of many chemicals and freeze-thaw damage. Mixing and placing can be difficult and consultation and coordination with the manufacturer of the latex materials is strongly recommended. Test sections should be made in advance of the repair work so that all procedures, mixing, transporting, placing, finishing and curing can be carefully established. If the justification for this more expensive method of repair is to provide resistance to aggressive environment, heavy wear or protection against intrusion of water, laboratory tests to assess performance under exposure to the appropriate adverse conditions are often very cost-effective, especially if the repair to be accomplished is extensive. It should be noted that latex modified concretes cannot be air entrained. This concrete must be finished rapidly as it has a short working life and also curing with burlap, polyethelene sheeting or similar ptotection must be in place immediately after finishing. Placing at night is common practice to minimize plastic shrinkage caused by sun and wind.

(4) Polymer impregnated concretes can provide a very tough surface from about $\frac{1}{4}$ to $1\frac{1}{2}$ inches (6-38 mm) in thickness. This technique of producing high strength, chemical resistant and impermeable concrete is in a state of development at the time of this writing. Many monomers are available and have been or are being tested for impregnation into concrete. Consultation with suppliers or consultants experienced in this work is recommended before attempting to employ this method of providing essentially an armor coat to the surface of replacement concrete or overlay.

In general, the procedure is as follows:

(1) After the replacement or overlay concrete has cured, the surface must be thoroughly dried. One way of accomplishing this is to use heat lamps to bring the temperature of the surface to between 250°F and 350°F (120°C–175°C) for several hours.

(2) Allow concrete to cool gradually to from 75° to 100°F (24°C–38°C) preferably within 25°F (14°C) of ambient temperature.

(3) Flood area with monomer in amounts specified by manufacturer but generally from 0.05 to 0.15 US gallons per square foot. The monomer is allowed to soak into the concrete for several hours. It must be covered with a plastic membrane to prevent evaporation and to control fumes.

(4) After soaking, heat is reapplied to the surface for sufficient time to bring about polymerization. The time of heat curing and curing temperature should be established by the manufacturer.

Concretes or mortars bound with epoxy resins have proven to provide a very hard tough wearing surface. This is the most expensive of the methods discussed here. The most common procedure for overlay is to design a mixture consisting of hard sand such as silica sand in proportions by weight ranging from three to six parts sand to one part resin. Graded, aggregate concretes bound by epoxy resins are used in special cases for filling larger voids. The manufacturer's recommendations should be followed with respect to mixing, handling and curing. The surface over which the epoxy mortar is to be applied should be clean and dry. Curing time and temperature should be controlled and will vary between epoxy materials and ratio of resin to hardener. It is very important to understand that the coefficient of expansion of epoxies is much greater than that of Portland cement concrete. If epoxy mortars are used as protective coatings or overlays, they are best employed within an enclosure, such as in a building, or at locations where ambient temperature variations will be less than 30°F.

As with all concretes, repair and overlay Portland cement concretes and mortars should receive adequate curing after placement. Wet curing at temperatures greater than 60°F (15°C) for a minimum of three days is imperative to successful repair. Longer curing times up to seven days are recommended when conditions permit. If timing is critical and the cost justified, curing times may be reduced by using high early strength cements or elevated moist curing temperatures up to 120°F (49°C).

Special care is required to assure that repair or overlay concretes bond well to the parent concrete to which they are applied. Many bonding agents, including epoxies, polyurethanes, polyesters, and acrylics are on the market for this purpose. Many of them do a good job under favorable and carefully controlled conditions. However, improper application of bonding agents with respect to time between application and placing plastic concrete against the bonding agent can have exactly the opposite effect of preventing bond. For this reason many feel that, unless absolute control is assured, the greatest success is likely by proper preparation of the parent concrete and placing well consolidated repair or overlay concrete without interface material of any kind. Good results can be obtained by the following:

(1) Thoroughly clean and roughen surface of parent concrete to receive repair concrete. Sandblasting is an effective way to accomplish both. Mechanical tools are often used to roughen smooth surfaces.
(2) If the concrete or mortar to be placed is low slump, the prepared parent concrete surface should be moistened just prior to placement. Do not permit enough water on flat surfaces so that standing water exists when the repair or overlay concrete is placed. Excessive water should be removed with a broom or similar device.
(3) Repair or overlay concrete is then placed and consolidated against the interface surface by vibration or other means.

Some suggest coating the interface surface of the parent concrete with neat cement grout just prior to placement. The writer does not recommend this procedure because cement grout can dry very quickly and once again timing becomes critical.

Regardless of the technique used for overlaying, the new surface should provide that water be drained in such a way that puddles will not occur.

8.7   Column or pier repair

Reinforced concrete columns or piers may need strengthening generally for one of the following reasons:

(1) Concrete deterioration or low strength original concrete.
(2) Corrosion of reinforcement or inadequate amount included in design or placed during construction (Fig. 8-5).
(3) Load increases over those originally provided for in original design due to unanticipated change in use.

Several methods of repairing or strengthening vertical load carrying members have been successfully accomplished. One is described here for guidance to those faced with the task of designing such repair.

Columns and piers derive their load carrying capacity from the interaction of concrete and reinforcing. Repair or strengthening methods must, therefore, assure that added concrete and/or reinforcement act with the existing materials.

Figure 8-5   Column with insufficient reinforcement and inadequate cover

Figure 8-6 illustrates a column which has suffered corrosion of the vertical and tie reinforcement. Assume this corrosion came about because the original concrete strength was low and the concrete cover was inadequate to protect the steel from a very humid atmospheric exposure. First, the strength of the core concrete must be determined by coring or a combination of coring and pulse velocity measurements. Next, determine the loss of steel area that has resulted from the corrosion. In order to make this determination, it is necessary to remove the

Figure 8-6   Method of repairing columns

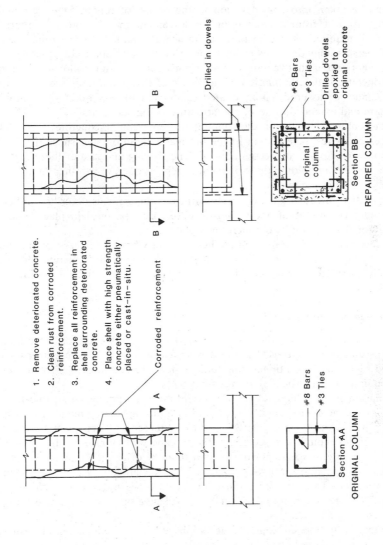

Figure 8-6 (contd.)

corrosion products from the reinforcement. This cleaning is required before new concrete is placed around the steel anyway. After the strength of the concrete, the amount of steel and the yield point have been established, it is possible to determine the amount of added reinforcement and concrete needed to bring the column up to required strength. The fact that the original concrete and the reinforcement is stressed under dead load and the new concrete and steel will not be must be taken into consideration in the design of repair. Steel required to replace that lost from corrosion or reinforcement in addition to the original amount, if required, is tied in place as shown. All added vertical reinforcement should be surrounded by ties spaced and sized to meet applicable code requirements ignoring ties that remain in the original column after removal of deteriorated concrete. If the elimination of the cause of corrosion, moisture or oxidizing agent cannot be assured, epoxy coated reinforcement may be used to minimize corrosion and extend the life of the structure being repaired.

If the column is long or the amount of replacement concrete represents a considerable percentage of that in the original column or pier, shear ties such as are shown in Fig. 8-6 are recommended. The spacing and size of these drilled in dowels to assist in holding the original and repair concrete together should be an engineering determination.

Preparation of the original concrete surface is the same as described in Section 8.2. Under conditions of good control, epoxy may be used in place of drilled dowels for bonding the two concretes together.

Replacement concrete may either be cast and vibrated in place or pneumatically placed. As stated previously, the use of pneumatic concrete should be carefully considered and applied by persons skilled and experienced in the use of this material.

8.8   Beam and girder repair

The procedures for repairing beams and girders are similar to those described for columns and piers. A typical method is shown in Fig. 8-7. The stirrups shown around replaced or additional longitudinal steel are recommended regardless of the requirements for shear reinforcement.

Replacement or added longitudinal reinforcement greater than one-half the length of the member should be anchored into the supporting member or original beam concrete as shown in Fig. 8-7. Concrete placed in forms placed around the original girder should be placed from the top and well vibrated. If the final width of the beam or girder is somewhat greater than the original, it may be expeditious to place the concrete through holes cored in the slab supported by the member being repaired. The core holes must be spaced to assure complete filling of the forms. The holes serve not only as means of placing the concrete but also as vent holes allowing air to escape.

Pneumatically applied concrete accomplished by skilled persons is usually a more practical method of replacing or adding concrete to beams and girders, Fig. 8-7.

8.9   Elevated structural slabs

The procedures for replacing deteriorated top surface concrete are discussed in Sections 8.1 and 8.6. Top slab, beam or girder reinforcement may be found in

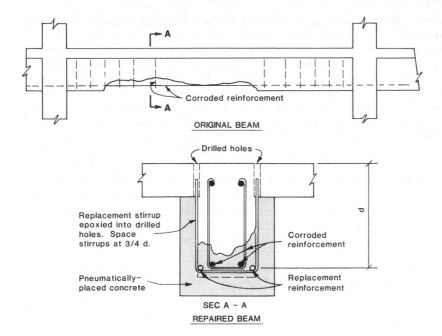

Figure 8-7   Beam repair

need of replacement or require the addition of supplemental reinforcement because of loss of steel due to corrosion. New bars should lap existing bars an amount equal to at least 1.5 times the requirement established by code for the construction being repaired, Fig. 8-8. Considerations for replacement or overlay concrete for the top surface of elevated structural slabs will be found in Section 8.6.

Replacement or supplemental reinforcement placed near the bottom surface of elevated structural slabs should be firmly tied in place and anchored to the slab.

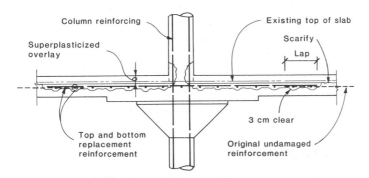

Figure 8-8   Structural slab repair

Pneumatically placed concrete usually offers the best solution for concrete replacement on the lower surfaces of slabs.

The most important evaluation which must be made concerning flat slab or flat plate construction is punching shear at columns. The author is of the opinion that punching shear stresses should be limited to about one half the amount allowed in new design at ultimate load for structures which have suffered deterioration. If punching shear stresses exceed this amount, a shear head or auxiliary drop panel may be built to reduce shear stresses to acceptable levels. An example of this solution is illustrated in Fig. 8–9.

The added drop panel solution must be carefully designed to assure good interaction between the new concrete and the bottom surface of the existing slab. Because of settlement of concrete while it is plastic and drying shrinkage after it has set, poor bond may result. Addition of vertical shear reinforcement as shown in Fig. 8-10 in sufficient quantity to develop fully the horizontal shear is strongly recommended.

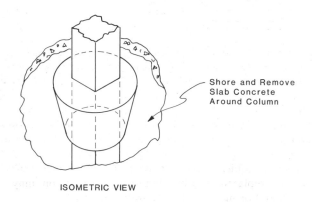

ISOMETRIC VIEW

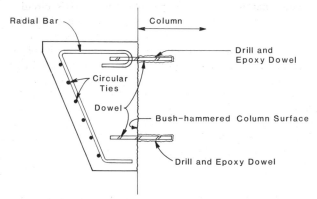

DETAILS OF SHEAR–FRICTION COLLAR

Figure 8-9 Example of shear friction collar

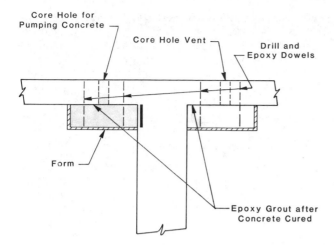

Figure 8-10 Added drop panel

8.10 Silos and tanks

Silos and tanks require special attention when repairs or maintenance are being considered. This is because these structures are usually loaded as designed in contrast to bridges or buildings which seldom, if ever, receive design live loads. Furthermore, concrete silos and tanks are often exposed to a severe environment even if they are only holding water. Materials stored in concrete tanks or silos, other than water, can attack the concrete and/or reinforcement much more aggressively. Rapid deterioration is frequently experienced when such aggressive materials are present.

Tank or silo repairs usually require a dual approach. First, deficient concrete and reinforcement must be replaced and, second, added protection may be desirable on the inside if liquid or moist materials are being stored.

If the structure is isolated, that is not connected to others, and is circular, banding with straps or prestressing tendons is an effective method of reinforcing (Fig. 8-11). Of course, this is done after the deteriorated concrete has been removed, corroded reinforcement cleaned and replacement concrete is in place. Removal of concrete, cleaning reinforcement and replacing concrete should be done as described in Section 8.2.

The inside surface of tanks or silos which have suffered deterioration from the aggressive action of liquid materials stored within may require, after deteriorated concrete and reinforcement have been replaced, a lining of membrane material bonded to the repaired inner surface. The membrane must be resistant to the chemicals stored in the containing structure. It must be applied in such a way that cracks do not develop or joints do not open permitting stored liquid to penetrate and attack the concrete or steel.

If the material being stored is abrasive as well as chemically aggressive, the inside surface might be coated with epoxy mortar after all cracks in the existing concrete have been epoxy injected. The difference in thermal coefficient of

Figure 8-11 Tank under repair by circular post-tensioning *Courtesy*: Schupach Suarez Engineers, Inc.

expansion between Portland cement concrete and epoxy-bound concrete must be taken into account when this method is considered.

High strength pneumatically applied concrete (6000 psi or greater) has been successfully employed to repair both the inside and outside surface of tanks and silos if the exposure is not overly severe.

8.11 Precast concrete

Most problems which develop and create need for repair with the use of precast concrete have to do with bearing. This difficulty is about equally divided between distress in the support member (Fig. 8-12) or in the supported member (Fig. 8-13). There are several causes for this kind of distress which include:

(1) Inadequate bearing area.
(2) Insufficient or improperly placed reinforcement in either the supporting or supported member.
(3) Both ends of the supported member rigidly attached to the support by welding or other methods.

Figure 8-12   Ledge support failure

The most frequently employed methods of repairing distressed bearing are illustrated in Fig. 8-14. Each circumstance will dictate if auxiliary bearing need be provided to carry the entire load or only part of it. Regardless of the method employed, the fact that pretensioned prestressed concrete members have little bending capacity in the pretension force transfer zone (about four feet at each end) must be considered. If the distress has come about because both ends of members are firmly attached at the supports, it will be necessary that one end be released before repairs are made.

8.12   Concrete poles

The most common source of deterioration of concrete light standards distribution poles or transmission towers is freeze-thaw damage at the ground line, anchor bolt corrosion or foundation distress. Poles knocked down by wind or earthquake are so badly damaged that replacement is the only solution. Foundation or anchor bolt problems are best solved by replacement. Poles suffering freeze-thaw damage at the ground line can be repaired as illustrated in Fig. 8-15 with good quality air entrained concrete.

Figure 8-13   Distress at bearing

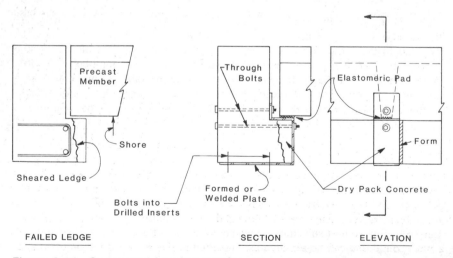

FAILED LEDGE   SECTION   ELEVATION

Figure 8-14   Suggested bearing repair

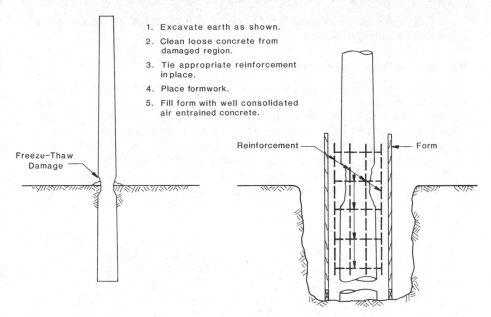

1. Excavate earth as shown.
2. Clean loose concrete from damaged region.
3. Tie appropriate reinforcement in place.
4. Place formwork.
5. Fill form with well consolidated air entrained concrete.

Freeze–Thaw Damage

Reinforcement

Form

Figure 8-15   Suggested repair of concrete pole at ground line

9 Tools

Many tools and instruments are available which can be invaluable as aids in assessing conditions of a structure prior to designing repair or restoration methods. Most of these should be used by persons skilled in application and interpretation of results. Some of the more commonly employed are discussed below.

9.1 Impact hammer

The impact or Swiss hammer is a device that delivers a controlled impact against the surface of concrete and measures the rebound. Properly used and calibrated with a few cores removed and tested from the concrete under study, it can provide useful data with respect to the strength of concrete. The following procedure is recommended for most effective results:

(1) Cover the area to be evaluated by taking 24 readings at each location on a grid covering the area under consideration. Omit the two highest readings and the two lowest readings. Try to avoid taking readings over air bubbles, small voids or exposed aggregate.

(2) Obtain an average of the remaining 20 readings.

(3) Remove a core from the location of the highest average, one from the location of the lowest average and one from the location of a mid-average set.

(4) Develop a curve relating the compressive strength of the cores to the impact hammer readings.

9.2 Pulse velocity measurement

The pulse velocity method employs an instrument which measures the velocity of sound through concrete. It is based on the principle that the velocity of sound through a material is related to the density of the material; denser materials producing higher velocities. Concrete strength also varies with the density of concrete. Thus, sound travels faster through concrete with high strength than it does through concrete of low strength. A signal is sent through concrete of known thickness from one face, and the time required for the signal to be received at the other face is used to calculate the velocity of the sound through the material.

The proposed procedure for using the pulse velocity measuring technique is similar to that described for the impact hammer:

(1) Cover the area under consideration by taking readings over a grid.
(2) Remove a core from the location of the highest reading; one from the location of the lowest reading and one from the location of a mid-range reading.
(3) Plot a curve relating the compressive strength of the cores to the pulse velocity.

Other techniques such as Windsor Probe have been used to estimate concrete strength with variable success.

9.3 Steel locations

Instruments can be obtained which use a magnetic field to locate reinforcement in concrete. If the cover is known, the size can be estimated or if the size is known, the cover can be estimated. The limitations of this instrument are:

(1) If the cover is greater than about four inches, the results are questionable.
(2) If several layers of reinforcement at various depths exist, the meter is only effective for the outer layer.
(3) If the quantity of steel is great or if the bars are bundled, the meter will detect the presence of steel and cannot be used to identify concrete cover or bar size.
(4) These instruments detect the presence of magnetic metal. Form ties, steel conduit and other embedded material are often mistakenly identified as reinforcement.

It is advisable to expose just enough of the reinforcement with a chipping hammer to verify the data obtained with the magnetic detector.

9.4 Corrosion detection

Corrosion of reinforcement is usually most easily detected by the presence of cracks at the level of bars caused by the expansion of the corrosion products. These cracks will show rust stains if the corrosion has advanced to a significant degree.

The major exception to the obvious presence of cracking resulting from corrosion of reinforcement is in the case of top steel in slabs. Corrosion in this case will cause delamination of the concrete at the level of the bars and is often not readily evident.

A simple method of detecting delamination in flat slabs is to drag a heavy chain over the area. A distinct difference in sound will be perceived as the chain goes over an area of delamination near the top surface. A map such as shown in Fig. 9-1 can be developed from this process in preparation for the development of repair specifications. Some coring is recommended to verify chain drag results.

If the delamination is deep within the slab, pulse velocity instruments can be used to detect the presence of this cracking if both top and bottom surfaces of the slab are accessible to the transducers employed with this instrument. If a delamination crack exists between the transmitting and receiving transducer, no signal will be received. The measurements can be taken over a grid and a map similar to that shown in Fig. 9-1 can be developed.

If only one surface of the slab is accessible, a technique known as pulse echo can sometimes be used to detect the presence of delamination at locations remote from the exposed surface. This instrument sends a sound signal from one surface and measures the time taken for the sound wave to reflect back from the remote surface which may be a delamination crack or the original remote surface. If the slab is of uniform thickness, apparent difference in thickness detected by this technique can indicate the presence of delamination. This technique should only be employed by persons experienced and skilled in reading and interpreting the instrument. Verification by a modest amount of coring is recommended.

Conditions which existed to promote corrosion of reinforcement may have changed at the time of inspection and active corrosion may not be taking place. It is often desirable to make a determination if the system producing corrosion is

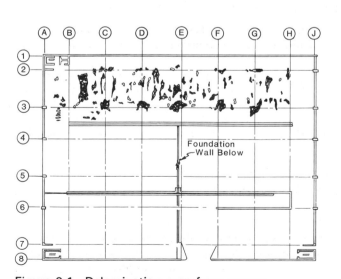

Figure 9-1 Delamination map from survey

active, or detect locations where such activity is taking place. Corrosion of reinforcement in concrete is brought about by the activity of a galvanic cell. Thus, when corrosion is active, the voltage being generated by the galvanic cell can be measured.

This is usually done with a copper copper-sulfate half-cell. The procedures for the use of this technique are described in ASTM C 876.

9.5 Laboratory physical, petrographic and chemical analyses

In the course of evaluating conditions which are responsible for deterioration of concrete, it is often very helpful to remove samples of deteriorated concrete as well as sound concrete and submit them to a competent laboratory for petrographic or chemical analysis or both.

Some of the important information that can be obtained from these tests is:

(1) Estimate of cement content.
(2) Air content.
(3) Evidence of flash set.
(4) Air void system.
(5) Estimate of water/cement ratio.
(6) Aggregate reactivity.
(7) Freeze damage.
(8) Aggregate gradation and type.
(9) Concrete strength and modulus of elasticity.
(10) Concentration of chloride or other chemicals deleterious to concrete or reinforcement.
(11) Distribution of these chemicals with respect to depth of concrete.
(12) Presence of and amount of admixtures.
(13) Character of consolidation during placement.
(14) Internal cracking and depth of surface cracks.
(15) Presence and character of cold joints.

9.6 Structural models

Sometimes the time and expense of building a scaled structural model of the structure and of simulating repairs is justified. Such a model can be load tested, exposed to accelerated deteriorating conditions, subjected to thermal cycles and other simulated force systems to evaluate performance.

9.7 Full scale load tests

When the condition of original structure due to low strength concrete, missing or misplaced reinforcement, deterioration of concrete or deterioration of reinforcement is in doubt, load tests are often conducted. Load tests may be deemed desirable after repairs to establish that the repairs are effective. These tests may be conducted in accordance with the provisions of the ACI building code or similar documents. Most codes provide for the application of test loads of prescribed magnitude and require only measurement of a minimum number of deflections.

The writer is of the opinion that strains should be measured on reinforcement at critical locations. Adequate strain data can be used to estimate the loads the structure is capable of sustaining above those applied during the load test. An exception is capacity in punching shear.

Test loads can be applied with dead weights such as sand, concrete blocks, metal blocks or similar materials. It is important that the structure be shored when using dead load because if failure occurs, the load cannot be removed in time to prevent collapse. Of course, the shoring should have enough space between the bottom of the loaded structure and top of shoring to permit deflections to take place. This method is usually the least expensive of those available.

If multiple pattern loading is desired, the load test system of applying dead load becomes more expensive as it is costly to move the loading from location to location.

Hydraulic rams can be used to apply test loads. While this system is usually more expensive than the dead load system, it has several very distinct advantages:

(1) Loads can be applied rapidly assuring fast strain and deflection response. The data, therefore, will be more reliable than those obtained with a long period of time between recording, and temperature effects will also be minimized.

Figure 9-2 Load test being conducted with hydraulic rams

(2) The loads can easily be removed so that recovery after each load increment can be observed.

(3) Pattern loads can easily be designed with a properly devised system of hydraulic valving.

(4) If distress occurs in the structure when test loads are applied, the load can be released rapidly, and thus collapse can usually be prevented.

Other tests in the laboratory that may be helpful are structural tests reproducing the moments, shears or torsions which the member to be repaired is expected to receive in service. Data regarding ultimate strength may also be obtained in the laboratory from such tests.

The most challenging problem in designing a loading system employing hydraulic rams is providing the system against which the rams must react. Figure 9-2 shows a load test being conducted on a stadium with the forces delivered by the rams reacting against dead load positioned on-grade below the structure. Soil anchors may be used for reaction devices.

Atmospheric pressure can also be used to apply uniform loading. This method usually proves to be the most expensive. The space from which air is to be withdrawn must be sealed with plastic sheets as illustrated in Fig. 9-3. All surfaces surrounding this space will be loaded with this system and those which are not intended to be tested must be braced or shored to resist this unwanted loading. A large capacity pump is required for air withdrawal. If circumstances justify the cost, this is a marvelous way of applying uniform loads up to about 1000 pounds

Figure 9-3   Load test being conducted with atmospheric pressure

per square foot ($48 \, kN/m^2$). The load can be rapidly applied with good control within about one pound per square foot. Incremental loading and unloading to obtain recovery data is easy and the load can be released rapidly if distress develops in the structure during the test.

9.8 Laboratory mock-ups

Repair methods employing materials or techniques not proven for the conditions under consideration can be tested with respect to methodology in a laboratory or in the field. It may be desirable to subject the test specimens to accelerated environmental exposure in a laboratory to evaluate the effectiveness of the repair method. Freeze-thaw cycles, heat-cool cycles, wet-dry cycles, exposure to high concentration of salt or other chemicals are some of the tests that can be made relatively inexpensively.

10 Demolition

Occasionally repair of concrete structures is not possible and demolition becomes necessary. This may be the case for any of the following reasons:

(1) Deterioration has progressed to the state that rehabilitation is not economically feasible.
(2) The structure has become obsolete functionally and cannot be converted to other use.
(3) The land on which the structure stands is valuable and must be made available for more cost-effective construction.
(4) The structure has been badly damaged by fire, earthquake or other disastrous event.

Demolition of large or complicated structures presents a greater engineering challenge in many respects than design. This is because of the dangers attendant to demolition. These dangers vary, the degree increasing as the density and size of structures near to the one to be demolished increases.

10.1 Preconstruction survey

If other structures are near the one to be removed, preconstruction surveys should be conducted before demolition work is started. This procedure is necessary to offset allegations of damage to adjacent structures which may not be warranted.

A preconstruction survey consists of a careful inspection, noting existing conditions such as cracks in plaster, concrete, broken panes of glass and other deficiencies representative of those which might result from excessive ground motion, air blast or flying debris produced by the demolition. Sketches, notes and photographs should be used to document pre-demolition conditions.

10.2 Disassembly

The logical way to demolish an existing structure is to disassemble opposite to the way it was originally constructed. In this way one would start at the top and work downward. This is not normally a practical solution, however, for monolithically cast-in place construction. The arduous task of breaking concrete where members are joined together in order to expose reinforcing for cutting is very time-consuming and expensive. Tools which are available to break the concrete and cut steel are:

(1) Jack hammers.   (4) Diamond saws.
(2) Hydraulically actuated wedges.   (5) Rock drills.
(3) Thermic cutting equipment.   (6) Water jets.

10.3   Wrecking balls

The most common way of demolishing reinforced concrete structures in congested urban areas is with the use of wrecking balls. This method involves the use of a steel ball weighing several tons suspended from a crane. The ball is raised and allowed to fall freely against horizontal surfaces or is swung as a pendulum against vertical surfaces. Concrete is broken away from reinforcement and considerable cutting of steel with oxyacetylene torches is often required.

Wrecking with these heavy balls must be carefully planned to avoid uncontrolled collapse which may damage adjacent structures or result in personal harm to workmen. This method can produce considerable noise and dust. Some control of the dust is sometimes accomplished by sprinkling the work area with water. Vibrations which are transmitted through the earth sometimes result in complaints from neighbors, but it is very unlikely that the vibrations will be intensive enough to damage adjacent construction unless it is actually connected to the structure being demolished.

10.4   Explosives

The most rapid and least expensive method of demolishing reinforced concrete structures is with the use of explosives. This method is used frequently to remove bridges, silos and other structures which are in rural locations and remote from other structures or facilities which could be damaged. It is more infrequently employed in crowded areas, but properly designed and executed can be a very cost-effective means of rapid demolition. It is absolutely imperative that demolition with explosives be planned and accomplished by persons skilled and experienced in this technique. The writer was once asked to investigate the accidental collapse of an eight-story structure which was to be demolished by persons having no experience in the use of explosives for demolition work. In an effort to make the task easier, the person in charge decided to weaken the steel columns at ground level by cutting part way through them with torches prior to setting the charges. The building collapsed as shown in Fig. 10-1 while this cutting was being done and before the explosives were even on site. Several workmen were killed. Fortunately, the building collapsed laterally into an adjacent vacant lot rather than into the street in front or against a building which was also adjacent.

Demolition by explosion requires not only careful analysis of the response of the structure frame to the detonation of charges, but provisions to protect against damage to underground and overhead utilities.

Figure 10-1   Structure demolished with improper planning and technique

10.5   Vibration monitoring

Regardless of the method used, if structures are located in the near vicinity, the owners or occupants are likely to claim damage from the demolition activity; also actual damage could occur. Monitoring vibration in the adjacent structures with seismic instruments or accelerometers is strongly recommended during the course of demolition work. Monitoring noise levels may be advisable under certain conditions also. The location of these instruments and interpretation of the data should be the responsibility of experienced individuals who are qualified to testify in the event of litigation arising out of alleged damage claims. Constant monitoring of these instruments can provide guidance and assist the demolition contractor in regulating his activity to maintain vibration and noise levels within acceptable limits. The uppermost consideration in all demolition work is, of course, personal safety.

Acknowledgement

The author wishes to acknowledge valuable assistance provided by Mr R E Philleo of the US Army Corps of Engineers, Mrs Donald Pfeifer and William Perenchio of Wiss, Janney, Elstner and Associates as well as Jan Hodous of Kellogg Corporation in preparing this chapter.

References

1. American Concrete Institute, ACI 318-77: Building code requirements for reinforced concrete, Detroit, Michigan 1977, 103 pp.
2. Perkins, P. H., *Concrete structures: repair, waterproofing and protection*, Applied Science Publishers, London, 1976, 302 pp.
3. Wiss, J. F. Construction vibrations: state-of-the-art, *J. Geotech. Eng. Div., Am. Soc. Civ. Eng.*, Vol. 107, No. 6T2, Feb. 1982, pp. 167–181.
4. American Society for Testing Materials, Section C876-80, *Standard test method for halfcell potentials of reinforcing steel in concrete*, 1981, pp. 554–556.
5. Johnson, W. and Ghosh, S. K. Demolition and dismantling mainly of building structures: an overview for use by lecturers, *Int. J. Mech. Eng. Educ.*, Vol. 8, No. 3, Mar. 1980, pp. 111–126.
6. Minimizing detrimental construction vibrations, *American Society of Civil Engineers National Convention and Exposition*, Portland, Oregon, 14–18 Apr., 1980, 176 pp.
7. Pinjarkar, S. G., Osborn, A. E. N., Koob, M. J. and Pfeifer, D. W., Rehabilitation of parking decks with superplasticized concrete overlay, *Concr. Int.*, Vol. 2, No. 3, Mar. 1980, pp. 56–61.
8. Pfeifer, D. W., Steel corrosion damage on vertical concrete surfaces, Part 1: Causes of corrosion damage and useful evaluation tests, *Concrete Construction*, Vol. 26, No. 2, Feb. 1981, pp. 91–93.
9. Pfeifer, D. W., Steel corrosion damage on vertical concrete surfaces, Part 2: Repair and restoration, *Concrete Construction*, Vol. 26, No. 2, Feb. 1981, pp. 97–101.
10. Lasalle, M. *Rehabilitation, repairs and strengthening of concrete structures 1953–1979, Limited Bibliography No. 237*, Construction Technology Laboratories, Skokie, Illinois 1979, 73 pp.
11. Peterson, C. A. and Pinjarkar, S. G., *Performance, load test, and repair of a severely deteriorated flat slab parking structure—a case study, Preprint 81-137*, American Society of Civil Engineers, New York City, 1981, 14 pp.
12. US Army Corps of Engineers, Concrete removal using explosives. Concrete Structures Repair and Rehabilitation. C-80-2, Aug. 1980, 8 pp.
13. *Cracks in concrete: causes, prevention, repair*, (A collection of articles from *Concr. Constr.*) Concrete Construction Publications Inc., Elmhurst, Illinois, March 1973, 44 pp.
14. McDonald, J. E., *Maintenance and preservation of concrete structures, Report 2: Repair of erosion-damaged structures*, US Army Engineer Waterways Experiment Station, Vicksburg, Mississippi, 1980, 306 pp.
15. *Concrete repair techniques, methods and materials*, (A collection of articles from *Concr. Constr.*) Concrete Construction Publications, Inc., Elmhurst, Illinois, Feb. 1973, 56 pp.
16. A guide to repair of concrete, *Concr. Constr.* Vol. 22, No. 3, March 1977, pp. 123–178.

26 Concrete construction in hot climates with particular reference to the Middle East

D J Pollock

E A Kay

Sir William Halcrow and Partners, London, England

Halcrow International Partnership, Dubai, UAE

Contents

Summary

The manufacture of good-quality concrete in any location requires care and skill. In hot regions, and particularly those like the Middle East which are very dry, there is a need for extra skill and care. This chapter concentrates upon those aspects of concrete construction which are recognized as being essential if durable concrete construction is to be achieved in this region. The use of some local and imported materials is considered as is the hostility of the conditions in which the concrete is made and which it is required to withstand in service. The performance of some Middle Eastern concretes is reviewed by means of a subjective classification of deterioration and the underlying factors are discussed. Simple methods of investigating commonly found defective structures are reported. Case histories from the Middle East are given and the aspect of reinforcement corrosion is highlighted. A repair philosophy is touched upon as is the potential development for more durable concretes in the future.

1 Introduction

1.1 The hot arid environment

This chapter deals with practical aspects of the production of durable, high-quality structural concrete in a hot arid environment. Amongst the topics discussed are the problems of locating and producing suitable aggregates in many parts of the Middle East, lessons to be learned from the performance of existing structures, designs, specifications and site procedures and the protection and repair of structures. The chapter is based on experiences gained in the Middle East but the concepts are applicable in other regions with similar climatic conditions.

Hot arid environments occur at various locations right around the world, mostly between latitudes 35°S and 35°N, and are typified by conditions found in many parts of the Middle East (Fig. 1-1). Temperatures are extremely high in the summer—often exceeding 40°C. Precipitation is low, being measured in tens of millimetres, is often confined to a few days of the year and is greatly exceeded by evaporation. Humidities inland tend to be low but in coastal locations they can be extremely high and variable.

The climatic regime is accompanied in many places by highly saline ground conditions and these two factors combine to produce an environment which is not conducive to the production of good quality concrete and which is extremely hostile to concrete structures in service.

1.2 Climate

Meteorological data for Dubai on the western coast of the Oman peninsula for 1981 are summarized in Fig. 1-2. The points which are of particular note in relation to concrete production are

(a) For any concrete mix, the high temperatures lead to lower slump for a given

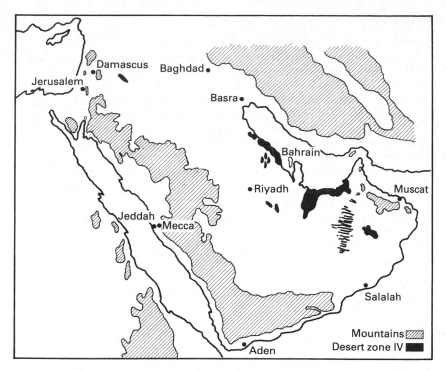

Figure 1-1   The **Middle East showing main areas of mountains and** desert zone IV

water content and to rapid loss of workability. Cold joints and honeycombing due to poor compaction are therefore more common than in temperate climes. There is, therefore, a strong temptation to add more water to the mix to improve workability. The high concrete temperatures also mean that hydration proceeds more rapidly with a consequently greater temperature rise and risk of cracking as the concrete cools from the hydration peak temperature. Though early strengths are usually higher (as is the case with autoclaved concrete), the long-term strength and durability are reduced; this will not usually be noticed from the results of standard quality control methods which employ tests on specimens stored and cured under controlled temperature conditions.

Reinforcement cages which are left in the direct rays of the sun soon become too hot to touch with bare hands. Concrete which comes into contact with steel at such a temperature sets very quickly and may result in a porous, poorly compacted region surrounding the bars.

(b) The high daily and annual temperature ranges (night frosts may occur in inland areas in winter) can impose large thermal stresses on structures.

(c) Low humidities in association with drying winds and high temperatures result in rapid loss of moisture from the exposed surfaces of fresh concrete pours. This loss of moisture often results in plastic shrinkage cracking.

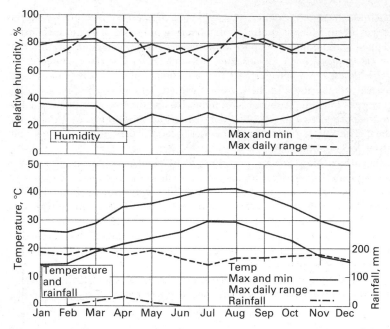

Figure 1-2 Summary of meteorological data for Dubai for 1981

Water curing is made difficult for similar reasons and on thin members may result in insufficient water being available for complete hydration. When water curing ceases, concrete dries out very rapidly and extensive shrinkage cracking may result. Potable water is often at a premium and the temptation to use brackish or sea water for curing reinforced concrete must be resisted as it leads to high concentrations of salts in the concrete close to the embedded steel.

(d) Relative humidity may change from 100% to less than 10% within a space of 6 hours. The associated moisture content and volume changes in concrete may be sufficient to cause cracking.

(e) At times of high relative humidity, condensation on structures often contains dissolved salts. With repeated cycles of condensation and evaporation concentrations can build up in the structures to an extent where they are instrumental in causing the corrosion of embedded steel.

1.3 Geology

The geological history of an area, its present day relief and its climate, all contribute to the properties of the available natural materials and also contribute to the environment to which constructed projects are exposed. In some locations, for example near the coast, all of these factors combine to produce particularly difficult conditions.

Hot deserts are typically comprised of intermontane basins which can be anything from a few metres to hundreds of kilometres wide. Some are centrally draining and do not have outlets to the sea.

With much idealization, Fookes and Knill [1] developed a conceptual model of a typical desert basin which is useful for engineering purposes. The model consists of four defined sedimentation deposition zones each characterized by different weathering or transporting agencies and consequently by surface materials of different sizes and other properties. Erosion takes place in the mountains due to physical weathering and the erosion products spread out in fans from the mountain slopes into the centrally draining basins or towards a coast. The deposition zones tend to run in bands parallel to the mountains and grade into one another. The surface sediments tend to become progressively finer with distance from the mountains.

In many locations, the bedrock is composed of chalks and young limestones which were deposited under marine conditions. These often contain horizons of evaporite rocks which release various salts into the groundwater. Subsequently, the salts become deposited at the desert surface by evaporation. In general, the water table is nearest the surface in the centre of the desert basin (or coast where the centre of the basin is occupied by the sea). This factor has particular significance for the production of chemically clean concrete aggregates and for the durability of concrete structures.

Figure 1-3 is a diagrammatic representation of the Fookes and Knill model. A discussion of the relevant factors in each zone is given below.

Zone I: the mountains

Physical erosion, which is still active, has caused broken rock slopes to build up against the mountains and coarse stream-flow debris fills the floors of the valleys. The main transporting agencies are therefore gravity and water. Individual grain sizes vary from large boulders to medium gravel. The material tends to be fresh and fairly angular.

Physical properties of aggregates from this zone will obviously depend to a great extent on the parent rock but it is usually possible to obtain good material from either hard rock quarries or borrow pits in the valleys.

Zone II: the apron fan

Gravel is carried out of the valleys by intermittent sheet and stream flows during storms. It is formed into interfingering fans which eventually extend right around the mountains. These apron fans can extend outwards for several kilometres. Surface deposits are cobble to sand grading, of sub-rounded shape and tend to become finer with distance from the mountains.

It can take a significant period even in geological terms for material to cross the apron fan and during this time the material is open to the natural processes of weathering. These processes can develop a mantle of weaker material with high porosity around each individual cobble or pebble and also adjacent to bedding and fracture planes.

Aggregates from this source are processed by screening alone or by crushing and screening. The former process results in a product which may be unsound and absorptive because it does not get rid of weathered material. Crushing produces large amounts of dust from the weathered layers and this dust is difficult to remove by subsequent screening alone.

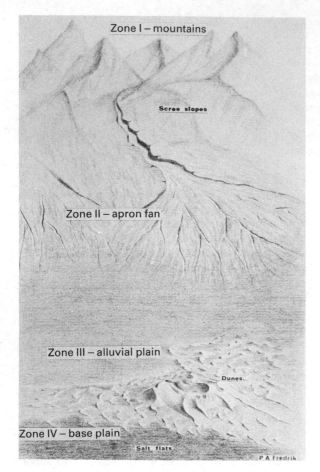

Figure 1-3 Diagrammatic presentation of the Fookes–Knill model [1]

Zone III: the alluvial plain

This is usually the most extensive zone and consists of material washed from the apron fan principally by sheet flow but sometimes by stream flow in shallow channels. In many places, water flow reaches the plain only very occasionally at times of heavy cloud bursts.

Transport by wind also occurs and sand dunes are common.

Wide variations in grading occur locally but the material is mainly from pebble to silt size.

Few opportunities exist for winning concrete aggregates in this zone. It may be possible to obtain naturally occurring sand-size material in some of the fluvial deposits but the wind blown material is usually too fine and single-sized for use in concrete.

Significant salt concentrations may occur where contaminated material has been blown over from zone IV.

Zone IV: the base plain
This occupies the central region of the centrally draining basin and margins the coastal area where the central region is occupied by the sea. Wind is the main agency of transport and sands and silts dominate the area. One of the principal features is the presence of the water table near the ground surface. The water table usually represents the base level for wind erosion as wind cannot carry sands and silts when they are damp.

The only major source of concrete aggregates is sand from the beach and near-beach dunes.

In areas where the water table is close to ground level, capillary moisture movement to the surface occurs readily and continual evaporation takes place. Salts dissolved in the groundwater precipitate out at the surface and a thick salt crust can build up.

Capillary rise often reaches ground level and moisture can then climb up vertical concrete elements and stockpiles of aggregates which become contaminated by the dissolved salts.

In zone IV, the ground conditions are therefore extremely aggressive to concrete.

Historically, many of the centres of population have developed in coastal locations because of the difficulties of overland transport and the pattern of modern construction has followed this trend.

1.4 The Middle East

The conditions described above are common to any of the world's arid areas. The construction boom in the Middle East faced some additional problems not the least of which were associated with the pace of the development itself. There was little time to experiment with the best ways of processing the naturally occurring materials or to prove that the materials themselves were entirely suitable.

Congestion in the ports meant that it was not always possible to obtain supplies of fresh cement and reinforcement.

In order to implement the large number of projects, it was necessary to import expertise in the form of consulting engineers and contracting companies. Many of these were from Europe, USA, Japan or other countries with mainly temperate climates and they brought with them designs, specifications and construction techniques which were not always appropriate to the more exacting conditions found in the Middle East.

There was little indigenous labour and workmen also had to be imported (mainly from the Indian sub-continent). The vast majority were unskilled and had to learn modern construction techniques 'on the job'.

The topics of appropriate design and construction practices and site investigation and routine testing for aggregates are covered in later sections of this chapter.

2 Materials

2.1 General

Some of the problems of obtaining suitable materials for reinforced concrete were discussed briefly in the first section of this chapter. They are covered in more detail in this section where practical steps are described for ensuring that the materials, particularly aggregates, are of the necessary quality to produce concrete that will be durable in the face of the formidable environment.

In order to assure the quality of the raw materials for even a medium-sized project involving reinforced concrete, a considerable testing program is required. Many construction projects are remote from the nearest commercial testing house and it follows that an adequate laboratory must be provided on site.

2.2 Water

By definition, fresh/potable water is in short supply in hot arid regions. However, in recent years, much more water has become available because of the construction of desalination plants.

Town main supplies are usually acceptable for concrete but in many cases they rely on well fields and may contain relatively high concentrations of dissolved salts.

Where sufficient quantities of domestic piped water are not available, supplies may have to be drawn from natural sources. Such water often carries suspended matter and dissolved salts and requires careful checking. The quality may vary through a pumping cycle. Sources which are initially acceptable can deteriorate and a routine of weekly testing for chloride and sulphate content should be established to detect any seasonal or other fluctuations. This need not always be by classical chemical methods as testing kits, e.g. Qantabs, are now available which give rapid results to sufficient accuracy. In general, for reinforced work the chloride and sulphate contents should be kept below 360 mg/l (as chloride ion) and 600 mg/l (as SO_3). The allowable concentrations of these contaminants in the mix are also subject to overall limits. It should also be remembered that any alkali metal salts in the water may contribute to alkali–aggregate reaction if reactive aggregates are used.

Another useful test for a proposed water supply is that given in BS 3148: 1980 [2]. In this test, the initial setting times and compressive strengths of mixes made with both the proposed water and distilled water are compared. The initial setting times should not differ by more than 30 min and the average compressive strength of cubes made with the proposed water should be not less than 80% of the average strength of cubes made with distilled water. AASHTO T26 [3] also gives guidelines on the quality of water suitable for use in concrete.

Water for curing should be to similar standards as the mixing water as evaporation may cause potentially hazardous concentrations of salts to build up in the concrete close to the reinforcement.

2.3 Reinforcement

Reinforcement is imported into the Middle East from all parts of the world and since most of it is transported by sea some may travel as deck cargo. Certification may be to an unfamiliar standard or non-existent. In many cases, there is no possibility of alternative supplies and the site is remote from the nearest testing house.

When a consignment of steel is received, the first step is close visual scrutiny to check

(a) Whether the bundle tags relate to those mentioned in the certification.
(b) For deformed bars, whether the ribs follow a regular defined pattern at constant spacing (particularly for cold worked bars).
(c) The extent of corrosion—check particularly for glistening deposits and pitting which can indicate active corrosion cells—probably due to chlorides.

If there are sufficient doubts about the origin and quality of the steel, it is worth while setting up a rudimentary tensile testing frame on site using jacks and steel sections. It may be necessary to cast the ends of the test specimens in concrete to transmit the forces. Even if it is not possible to develop the full tensile strength of the larger diameters, it should be possible to test to working load. The normal site bar-bending machine can be employed for improvised bend and re-bend tests. As a guide, the bars should be bent round formers of approximately half the diameter of those to be used for the works. The bends should be examined for cracking and other signs of distress.

Once the overall acceptability of the reinforcement has been established, the prime consideration becomes that of ensuring that its surface is corrosion-free at the time when it is surrounded by concrete. Methods of storage and site practices to ensure this are discussed in a later section of this chapter.

2.4 Cement

Cement is also imported from many parts of the world and, because of its nature, is of much more variable quality. Strengths vary considerably as may also the water required to produce a mix with a given slump.

Many of the imported cements contain fillers of pozzolanas. Though this may have certain advantages in terms of increased resistance to sulphate attack and possibly also alkali reactivity, such cements lead to slower strength gain and also require prolonged and thorough water curing or a friable, dusty surface can result. In the prevailing conditions, it is most unlikely that adequate curing can be achieved. Pozzolanic cements can be detected by high insoluble-residue test results.

Cement imported by sea is usually several months old by the time it arrives at its destination. Cement in bulk can be discharged direct into silos and handled fairly rapidly but cement in bags is often unloaded by labour-intensive methods which result in much damage. The bags may be stored under conditions which are far from ideal, until sold. Bags need to be at least five-ply to be able to resist the elements. The cement may be partially air-set and in some cases cement has been sieved and re-bagged.

It is strongly advisable to carry out check tests on imported cements before purchase. Strength test, loss on ignition, insoluble residue and water requirement for standard consistency probably give the best indication of the quality of the product.

Many good cements are now produced by countries in the Middle East. New sources should be monitored using the tests mentioned above, until production has settled down into a regular pattern.

Care needs to be exercised in the choice of cement to be used for concrete in a particular situation. In the extremely severe ground conditions which exist in zone IV, some protection against sulphate attack is obviously necessary. If a sulphate-resisting cement type, i.e. with low tricalcium aluminate (C3A) content, is used, the total chloride content of the mix has to be kept to a very low value. It may not be possible to attain such low limits with the locally available aggregates. In many cases, the choice of a moderately sulphate resisting cement (approximating to ASTM C 150 Type II [4]) with C3A content in the range 4–8% is the best compromise and may give better protection from reinforcement corrosion induced by penetration of the concrete by external chlorides.

2.5 Aggregates

The geological factors which control the properties of aggregates which are available in hot arid regions were outlined in the first section of this chapter and are discussed at length in Fookes and Higginbottom [5]. The main problems associated with aggregate production and use are

(a) Aggregates which contain high percentages of deleterious salts. Sulphates, if present in sufficient quantities, can attack the cement products in concrete while even relatively low chloride contents can lead to corrosion of embedded reinforcement.
(b) Dusty aggregates which result in greater water demand to produce a concrete of given workability. The increase in water/cement ratio has a direct effect on durability. Variations in dust contents can cause unexpected large changes in mix properties and affect the performance of admixtures. Dust contents can be reduced by simple techniques on site. As the aggregates are usually dry, air blowers acting on conveyor belts are very effective.
(c) Weathered and unsound aggregates which break down under physical attack from salts in the ground or sea water.
(d) Aggregates of poor shape or grading, e.g. some beach sands composed of shell fragments and aggregates crushed by inappropriate processes. These tend to make harsh concrete mixes or mixes which have a tendency to bleed.
(e) Aggregates with high absorptions or high shrinkage values.

These factors need to be taken into account when investigating new aggregate sources and also when planning regular routine test programs.

Before a quarrying or borrowing operation is undertaken, investigations are required to determine the suitably of the source material and to check that there is a sufficient quantity for economic exploitation.

2.5.1 Investigations of sources
Guidelines for the evaluation of potential aggregate sources have been given by Fookes and Collis [6] from which Table 2-1 has been adapted.

Rock quarry For a rock quarry, the first step is an assessment by a qualified geologist who will survey the site and study existing geological information. The

Table 2-1 Guidelines for evaluation of aggregate sources. (Adapted from Fookes and Collis [6])

Phase	Brief description of work
1 Desk study	Obtain any geological information, study air photos, plan investigation, provide the investigating team with clear terms of reference
2 Walk-over survey	Field examination, surface sampling of potential deposits, sample local aggregate workings and processed aggregates
3 Examination of existing concrete	Study the performance of any concrete structures in similar environments to the proposed structure
4 Laboratory testing	Investigation and tests on samples from phases 2 and 3 to determine the physical, chemical and mineralogical properties of the aggregates and to assess their performance within concrete
5 Review	Consideration of evidence obtained to date leading to choice of borrow area
6 Trial pitting and testing	Dig trial pits in selected area to determine variability of physical and chemical properties and to confirm total quantities. Select processing plant
7 Production trials	Winning and processing trials. Sampling and testing of product for final approval

survey should reveal any areas and zones of weathering, intrusion or faulting and the possible associated localized regions of unsuitable material. Following the geologists' report, a pattern of boreholes should be drilled, avoiding any major suspect regions. The boreholes should extend to below the proposed working level and cover an area which will provide adequate rock for the required purpose. Cores should be recovered for the full depth of the borehole and subjected to petrographic examination (ASTM C 295, [7]). Tests should be carried out on samples (crushed where necessary) to determine acid-soluble sulphate and chloride contents, shrinkage characteristics, potential alkali reactivity, water absorption, specific gravity, crushing value and sulphate soundness loss. A suggested list of required test values is given in Table 2-2.

Borrow areas in gravel The fan gravels show up clearly on air photographs and a study of such photographs is of great assistance in locating suitable deposits.

When proving a borrow pit in the mountain valleys or apron fans, trial pits should be dug by machine to approximately 1 m below the proposed depth of working. The trial pits should be disposed over the full area likely to be exploited. Representative samples (20 kg, 45 lb) should be taken at approximately 1 m (3 ft)

Table 2-2 Required test values for coarse aggregates

Test	Method	Value	Remarks
Grading	BS 812, Part 1	BS 882, Table 1	
	ASTM C 136	ASTM C 33, Table 2	
Sulphate soundness	ASTM C 88	Loss not exceeding 10%	5 cycles magnesium sulphate
Specific gravity	BS 812, Part 2/ASTM C 127		Value depends on type
Water absorption	BS 812, Part 2/ASTM C 127	Less than 2.0%	
Clay, silt and dust	BS 812, Part 1	Not exceeding 1.0%	Decantation or wet sieve
	ASTM C 117		method
Organic impurities	ASTM C 40		
Sulphate content	BS 1377, Test 9	Less than 0.4% by wt ⎫	Subject to overall
Chloride content	BS 1881, Part 6	Less than 0.05% by wt ⎬	limits on total mix
Aggregate crushing value	BS 812, Part 3	Less than 20% ⎭	
10% fines value	BS 812, Part 3	Not less than 5 tons	
Aggregate impact value	BS 812, Part 3	Not exceeding 45%	
Elongation index	BS 812, Part 1	Not exceeding 35%	
Flakiness index	BS 812, Part 1	Not exceeding 20%	
Potential alkali reactivity	ASTM C 227	Expansion not exceeding 0.05% at 3 months 0.10% at 6 months	Very slow test—3 months for reliable results
Los Angeles abrasion	ASTM C 131 or C 535	Not exceeding 50%	
Aggregate shrinkage	BRE Digest No. 35		
Petrographic examination	ASTM C 295	Not exceeding 0.05%	

vertical centres and also where there are obvious changes in the nature or colour of the material. The samples should be subjected to the same range of tests given for the rock quarry samples.

The trial pits should be logged, even if only a simple engineering classification is used.

It is advisable to make an early assessment of yield as dust contents can be as high as 30–40%.

Borrow areas for beach sand   The winning of beach sand requires strict control because of the possibility of high chloride and sulphate levels. It is probably true to say that the use of contaminated beach sand has been one of the single greatest contributors to the well-publicized rapid deterioration of concrete structures in the Middle East. Nonetheless, it is possible to obtain material with suitably low chloride and sulphate contents and in some areas there may be few alternative sources because of the paradoxical scarcity of any other suitable concreting sand.

The best sources are usually the dunes immediately behind the beach but even these need careful investigation and control.

When proving a new source, trial pits should be excavated at relatively close centres and the level of water table and capillary rise zone noted. For the first few exploratory pits, samples for chemical analysis (chloride and sulphate) should be taken at 200 mm (8 in) centres vertically and at obvious changes in the bedding patterns or moisture content e.g. capillary zone. Larger samples should be taken at 1.5 m (5 ft) centres vertically for determination of grading, water absorption, specific gravity and other physical properties. Later sampling may be reduced in frequency to a level appropriate to the variability in the initial results.

Existing sources   If an existing quarry or borrow pit controlled by others is to be used, the source should be visited to obtain an impression of the overall degree of control that is being exercised and particularly keeping in mind many of the points raised above. The stockpiles should be sampled and subjected to similar tests. The sampling and testing should be repeated on a daily basis for a week to gain an impression of variability and control.

In summary, the important factors for aggregate supplies are

(a) Fresh, sound aggregates characterized by good crushing values (ACV less than 20% by BS 812 [12]), absorptions less than 2% and sulphate soundness losses less than 10% when tested by ASTM C 88 [8]. In general, the finer fractions will contain higher percentages of weathered material and crushed products are to be preferred to uncrushed.

(b) Low chloride and sulphate contents—less than 0.4% by weight for sulphates for all aggregates, less than 0.05% (NaCl) by weight for chlorides for coarse aggregate and less than 0.5% for chloride for sand. Beach sands require particularly careful checking. Washed sands are available in some areas. Washing removes chlorides if properly executed but probably will have little effect on the less soluble sulphates. The sulphate content of coarse aggregates can be reduced by proper scalping and screening, as the sulphates are friable.

(c) When faced by an unfamiliar source of aggregate, possibly in a new area, the first and most instructive step is to examine the oldest existing structures in which similar aggregates were used. The structures should be examined for the typical signs of unsoundness, corrosion of reinforcement or possible alkali reactivity which have been described elsewhere in this chapter.

2.6 Admixtures

In the interests of durability in the harsh environment, water/cement ratios of concrete mixes have to be kept low. The high temperatures and dusty aggregates lead to lower initial slumps for a given water/cement ratio and rapid loss of workability. Effective plasticizers, superplasticizers and workability retention aids are very useful admixtures under these conditions as they help to ameliorate some of these problems. Air-entraining admixtures are also used to increase durability by reducing the number of interconnecting pores. (See also Chapter 9.)

Some problems have been noted in the use of admixtures with some of the aggregates available in the Middle East.

Superplasticizers based on melamine formaldehyde have been known to be associated with excessive bleeding when used to produce flowing concrete. They have been used successfully to produce concrete of normal workability at a low water/cement ratio.

Retarders based on hydroxylated carboxylic acid also seem to induce excessive bleeding with some aggregates and those based on lignosulphonic acids are to be preferred.

Since admixtures based on lignosulphonic acid also entrain a little air (2–3%) they can be considered the best choice for most applications in the Middle East.

3 Performance of concrete in the Middle East

3.1 Introduction

The case histories quoted in Section 6 indicate some of the most usual deficiencies in concrete structures and the apparently early age at which they become visible. This is true compared with concrete in more temperate climes; the startling rate of deterioration of many concrete structures may, however, be used as a tool in understanding the processes. The rates of deterioration of some 89 structures were considered [9] and the underlying causes considered. The case histories are from seven countries in the Middle East, UAE, Oman, Libya, Saudi Arabia, Jordan, Bahrain and Egypt. Although it could be argued that mass concrete might not be structural, case histories for mass concrete are included here.

The rates of deterioration of the various structures are probably more rapid than average because the points plotted represent structures which were deemed to require investigation. Since the results only represent a subsample, the average performance of structures should be better than shown. The wide scatter on Fig. 3-1 gives some insight into the probable variations in performance and the complexity of causes.

Particular care should be taken in using the plots in Fig. 3-2.

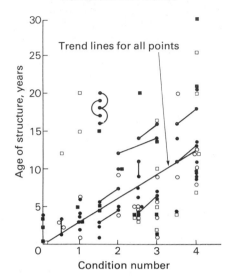

■ Mass marine
● Reinforced marine
□ Mass non-marine
○ Reinforced non-marine

Trend lines for all points

Age of structure, years

Condition number

Definitions of condition number					
	0	1	2	3	4
Reinforced concrete	No defects	Rust staining and minor cracking	Onset of corrosion cracking	Moderately severe cracking some spalling	Severe cracking and spalling Terminal condition
Mass concrete	No defects	Minor surface weathering or pop-outs or cracks	Moderate surface weathering Isolated cracks	Moderate to severe weathering or inter-connected cracks	Severe loss of concrete surface Disruptive cracking Terminal condition

Figure 3-1 Deterioration versus age of various types of structure

3.2 Rates of deterioration

To allow data to be collected and collated, it is necessary to define the degree of deterioration. These arbitrary definitions are shown in Fig. 3-1 and are related to a condition number from 0 (no defects) to 4 (a terminal condition).

Obviously, sweeping generalizations are necessary when the functions of structures vary so much. What may be serviceable in a floor slab of a domestic dwelling may not be in a bridge deck, especially when it is known that the axle weights of road vehicles are uncontrolled.

The factors affecting the rates of deterioration of the structures all vary in degree if not in type and the opportunity to investigate all diligently would allow a better appreciation of deterioration rates. A breakdown of the case histories into a number of simple exposure categories is, however, quite instructive. The types considered are marine and non-marine mass and reinforced concrete. Figure 3-2 indicates the rate of deterioration for these various categories.

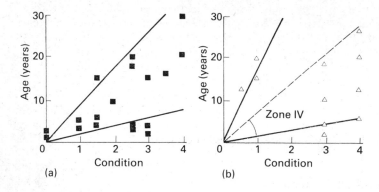

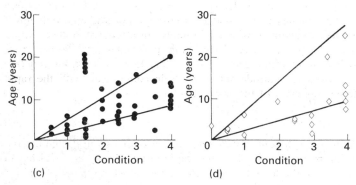

Figure 3-2 Deterioration of structures under various conditions
(a) Mass marine: a broader scatter, i.e. showing several variable influences
(b) Mass non-marine: very wide scatter, i.e. showing a highly variable set of influences; quicker rates of deterioration occur in desert zone IV
(c) Reinforced marine: a fairly narrow scatter, i.e. showing the dominating influence of one factor affecting all similarly (secondary chloride ingress)
(d) Reinforced non-marine: a fairly tight scatter but not as tight as (c), i.e. showing bigger variations in influence due to primary chlorides and deficiencies in the mix

The results for reinforced marine structures are generally more tightly grouped than for the other structures. The rusting of the reinforcement caused by sea water penetrating the concrete is a dominant factor which partially masks bad workmanship variations, salty aggregates in the original mix and the other features which could cause difficulties. Severe disruption can be caused at as early an age as six years; bad workmanship and specification, of course, play a big part in the rapid process. *In situ* concrete construction is particularly difficult to achieve without building in some undesirable chemistry from the sea.

By contrast, the scatter of results for non-marine reinforced concrete probably indicates that one factor does not dominate. The causes of deterioration here may well be chloride in the mix or indeed deficiencies in mix design. The scatter would be much greater and the rate of deterioration smaller if more of the case histories lay beyond zone IV; in zone IV salts are drawn to the ground surface.

In zone IV, it is also more difficult to make concrete structures without incorporating salty contaminants on aggregates, on any reinforcing and on construction joints. Thus, not only is it probable that reinforced concrete is of a lesser quality, but also the hostility of some of the service conditions is greater than in zones I, II and III, where salts are much less in evidence.

4 Design and specification

4.1 Introduction

There is not usually much scope for major changes in basic design for a structure of a particular type and function. However, design and specification require particular attention in the aggressive conditions found in hot arid environments if durable structures are to result. It is mainly a matter of attention to detail rather than of wholesale changes in basic design concepts; unless, of course, mass concrete is used instead of reinforced concrete. Very low quality mass concrete has apparently survived for considerable periods in comparison with the rapid decline of reinforced concrete of the same quality.

4.2 Design

The particular factors which contribute to changes of emphasis in design are the prevalence of reinforcement corrosion, the rapid drying shrinkage which can cause excessive cracking, the large thermal movements which must be catered for and the aggressive ground conditions.

4.2.1 *Corrosion of reinforcement*

As discussed elsewhere, corrosion is often associated with high chloride levels in concrete, either because they were included with the mix ingredients or because they have subsequently penetrated from the environment. The most obvious way of avoiding this problem is to use mass (plain) concrete in preference to reinforced concrete wherever possible. This does not always result in large increases in the dimensions of members. As an example, the use of small, high-strength, interlocking paving blocks is an apt, economic and elegant alternative to traditional reinforced concrete slabs.

Since it has been found that chlorides can penetrate concrete in severe exposure locations to considerable depths, the cover to reinforcement needs to be greater than the values shown, for example, in CP 110: 1972 [10].

Suggested values to be used in various exposure conditions are shown in Table 4.1. The values given in the table are not related to concrete strength, as it has been found that chloride penetration is relatively insensitive to that variable. The cover values are related to water/cement ratios and the use of air entrainment.

Mix specification must be considered along with cover to steel. This is highlighted by advice in ACI 201.2R 1977 [11], which recommends, under certain exposure conditions, an extra 13 mm ($\frac{1}{2}$ in) of cover if the water/cement ratio is raised from 0.40 to 0.45. Cover of 75 mm is recommended by ACI for the harshest conditions while AASHTO (quoted in [11]) suggest 100 mm unless the concrete is made by precasting—in this instance the cover may be reduced to 75 mm.

It is sometimes argued that greater cover to reinforcement leads to wider surface cracks and that attempts to reduce corrosion by this method are therefore self-defeating as the cracks allow the aggressive elements to penetrate more easily. However, this is not entirely valid, as, although the increased cover results in wider cracks at the surface, the crack width at the level of the reinforcement is not altered; the conditions near the bars are in general improved by greater cover.

Construction joints in structures in particularly aggressive exposure conditions need to be sealed. The large shrinkage and thermal movements cause the construction joints to open and allow air, water and salts to attack the steel unless they have been adequately sealed.

Structures in the intertidal zone, or which are subject to intermittent wetting and drying by salty water, should be designed with positive falls. Otherwise, ponding and evaporation usually cause significant chloride levels to build up very rapidly.

4.2.2 Rapid shrinkage

In conditions of high temperatures and low humidity, concrete can dry out very quickly. The associated volume changes occurring before the full tensile strength has developed, can cause substantial cracking. Floors, walls and roofs are particularly susceptible.

There is a number of steps which can be taken to minimize or control the effects of shrinkage. Additional steel can be provided but this must have the necessary concrete cover. Other measures include restrictions on the maximum size of pour and the minimum time between casting adjacent bays. The shape of individual pours should also be carefully examined and abrupt reductions in cross-sectional area or other causes of locally high stress eliminated if possible. If such situations cannot be avoided, the cracking should be encouraged to take place at controlled locations by the use of weakened sections (dummy joints) which can be subsequently sealed if necessary.

Any cracks which do develop in aggressive zones should be sealed as quickly as possible to prevent the entry of water, salts, air or other corrosive elements. Epoxy resins are not always suitable for this sealing as subsequent shrinkage may cause new cracks. Early sealing with a pouring grade latex emulsion can be followed by a permanent seal.

Table 4-1 Recommendations for concrete mix details for various Middle East exposure conditions

Exposure condition	Type of structure	Minimum cement content (kg/m³) related to maximum aggregate size			Maximum water/cement ratio	Cement type	Minimum cover mm	External protection
		10 mm	20 mm	40 mm				
Superstructure concrete where the structure is not in desert zone IV	Mass	280	230	200	0.6	OPC (high C3A)	—	
	Reinforced	330	280	240	0.52		30	
Superstructure concrete well above capillary rise zone where the structure is in desert zone IV	Mass	280	230	200	0.6	OPC		
	Reinforced	370	320	280	0.52		40	
Permanently below sea or groundwater. Underground but above the influence of capillary rise zone	Mass	280	230	200	0.6	SRPC (low C3A)	—	Tanking with thick membrane if below groundwater
	Reinforced	400	350	310	0.5	ASTM Type II (medium C3A)	50	
Within groundwater capillary rise zone	Mass	350	300	260	0.5	SRPC	—	Tanking with thick membrane
	Reinforced	420[a]	370[a]	330[a]	0.42	ASTM Type II	75	
Intertidal and splash zone	Mass	400	350	310	0.47	SRPC	—	Surface coating required for reinforced concrete if minimum cover cannot be achieved
Subject to frequent wetting and drying with saline water	Reinforced	420[a]	370[a]	330[a]	0.42	ASTM Type II	100	

[a] With entrained air

4.2.3 Large thermal movements

The greater range of diurnal and annual thermal movements needs to be taken into account at the design stage. Expansion joints should be provided at closer centres than in temperate climates and may have to be designed to accommodate larger movements.

It may even be necessary to provide movement joints at positions where they would not be required under less exacting conditions. A particular example of this is at the junction between flat roof slabs and walls. Exposure to solar radiation causes the roofs to undergo considerable lateral thermal expansion and severe cracking at roof–wall junctions is common.

4.2.4 Aggressive ground conditions

Reinforced concrete foundations in desert zone IV are subject to attack from both sulphates and chlorides in significant concentrations. Sulphates can be countered by the use of sulphate resisting cement but this can be a retrograde step in the presence of chloride ions, particularly if they were incorporated in the mix ingredients. In these circumstances, it is often preferable to accept a slightly lower degree of sulphate resistance in the cement (ASTM Type II, for instance) in conjunction with external tanking. Many proprietary systems of tanking are now available. The systems supplied as rolls consisting of polyester or polyethylene sheets with a bitumen backing (sometimes fibre reinforced) have been found to be superior to paint-on systems.

In extremely aggressive conditions where the water table is high, there is a case for tanking foundations even when made with sulphate-resisting cement as it eliminates the associated problem of salt being carried up vertical members by capillary rise from the groundwater.

Precasting is also to be encouraged as it allows structural elements to be fabricated in a relatively controlled environment and can be organized to allow the concrete to achieve a degree of maturity before exposure to the elements.

As a rule, contamination of the surface of the concrete should be kept at bay for as long as possible. It is well known [13] that the water permeability of concrete decreases with age. For a cement paste with a water/cement ratio of 0.7 and taking the value of permeability of fresh concrete as a comparator, the data of Table 4-2 emerge.

There is a spectacular decrease in permeability; even a few days' extra protection in the early life of concrete makes a great difference to the ability of the concrete to keep out water. The air permeability or diffusivity is also very

Table 4-2

Age of concrete, days	5	6	8	13	24	Ultimate (calculated)
Ratio of permeability	2×10^{-4}	0.5×10^{-4}	2×10^{-5}	2.5×10^{-6}	0.5×10^{-6}	3×10^{-7}

Table 4-3

Age of cement pastes, days	4	7	14	24
Diffusivity ratio	0.13	0.09	0.04	0.03

important as rusting is enhanced by air and moisture. Nilsson [14] has shown that, for tests in cement paste, the diffusivity reduced rapidly with age. The results with diffusivity at 1 day counting as unity are as shown in Table 4-3.

Although not so dramatic as the water permeability, the diffusivity, which is a measure of the ability of air and moisture to move, also decreases with age.

4.3 Specification

There are many items in concrete specifications which apply on a world-wide basis. Reinforcement, cement, water and aggregates have to be to an acceptable minimum standard if durable structures are to be constructed in any environment. Strength tests on concrete have to be carried out in a standard way for quality control purposes. Shuttering and falsework have to be provided to give the required surface finish and structural members of the required size within given dimensional tolerances. However, in the particularly harsh environment found in areas with hot arid climates, there are certain aspects of concrete production where additional emphasis and guidance are necessary.

Many of these aspects are discussed elsewhere in this chapter. In Section 2, particular attention is drawn to the methods of obtaining and testing materials, and the site practices for ensuring that they remain free from contamination are described in Section 5. It is also important adequately to specify the measures required to limit concrete temperature at the time of placing and to ensure that it is adequately cured and protected in the early stages. These aspects are also covered in Section 5.

One subject that is not discussed elsewhere in this chapter is the specification of concrete mix proportions for durability. Table 4-1 gives tentative minimum cement contents and maximum water/cement ratios for various exposure conditions likely to be encountered in hot arid environments. The recommendations are based loosely on Table 48 of CP 110 and the authors' own experience in the Middle East [15]. When used in association with the cover to reinforcement and other recommendations given in this chapter, they should lead to concrete of adequate durability. Some codes require even lower water/cement ratios and, while this is very desirable, the combinations quoted here are often practically possible and satisfactory when taken along with other recommendations here.

To achieve adequate workability with the local materials and water/cement ratios given in Table 4-1, it is almost inevitable that a plasticizing admixture will be required. Taking into account the difficulties of retaining workability in hot conditions and the problems associated with bleeding and plastic cracking, it is essential that the plastic properties of the concrete are fully investigated at the trial mix stage. Cores are particularly useful for detecting plastic settlement voids under reinforcement, honeycombing due to difficulty in compaction, etc. and the specification should provide for the cutting of such cores from each mix in the early stages of the contract. On larger sites, trial pours with typical reinforcement details can be included in the trial mix procedure specifically for the purpose of investigating the plastic properties of the mix.

5 Site practice

All of the time spent on careful design and thorough specification and selection of materials is wasted if the procedures on site are not also geared to the harsh environment. It is important that the supervisory staff not only understand what is required but also why it is required and in addition appreciate the likely consequences of deviating from an apparently laborious specification.

Particular attention must be paid to the storage and handling of materials, to reducing the temperature of concrete at the time of placing and subsequently to curing and protection. Attention paid to looking after early age concrete in these climates is amply rewarded.

5.1   Reinforcement

Reinforcement often arrives on site in the Middle East in a corroded condition. This is sometimes due to a prolonged journey by sea from its country of manufacture, but rusting soon starts after its arrival in the area because of the aggressive environment. The problem is particularly serious if chloride-laden dust is allowed to settle on the bars and pitting takes place. Rapid changes in temperature and humidity can cause bars to be damp early in the day and hence capable of collecting salt-laden dust.

Good storage arrangements are needed on site to reduce the possibility of this happening. Reinforcement should be stored off the ground and adequately supported so that the ends of the bundles do not droop and touch the ground. The storage area should be covered but the practice of using plastic sheets in close contact with the bars is not recommended as it promotes condensation. An ideal arrangement might consist of a concrete hard-standing remote from dusty site roads and with vertical wooden racks for storing the bars and a light shading structure.

The usual argument in more temperate climes is that rust does not impair bar performance. This, of course, is in relation to the bond between the steel and concrete. The argument for cleaning rusty bars under Middle Eastern conditions relates to the likely presence of salts and corrosion cells. John [16] has reported that, in experiments with bars having a coating of rust, the rusty bars apparently corroded some eight to fifteen times faster than non-rusty steel when each was embedded in concrete.

Under Middle East conditions it is almost inevitable that some form of cleaning will be required. In order to obviate prolonged arguments on site as to what constitutes an acceptable standard of rust-free steel, it is best if some form of cleaning is specified in the documents for all reinforcement.

When bars are pitted, cleaning by hand-held wire brush is ineffective in most instances. Mechanical wire brushing is often little better, usually only polishing the surface of the rust and not removing scaly deposits or reaching the bottom of corrosion pits. Chemical methods using acids can be effective but may not be economic for smaller sites and cannot be used on cages once they are in position or on starter bars from previous pours. Grit or sandblasting has been found to be highly effective and is to be preferred under most circumstances.

It may be possible to clean the steel after bending and before fixing if only a short period of time is to elapse until concreting. Often this is not possible and the steel has to be given a thorough final clean when made up into cages just before concreting. Deflecting nozzles are required to adequately clean the underside of bars but the contact points between crossing bars need careful attention and bunching of bars makes cleaning impossible.

5.2 Cement

Because of the high relative humidity and damp ground conditions, storage of cement on site has to be carefully controlled. Cement in silos, so long as they are weatherproof, poses few problems. Cement in bags should be stored inside a light building with a concrete floor. The bags should be stacked on wooden pallets no more than seven bags high. Different types of cement should be separated as should different deliveries of cement so that the oldest cement is used first. If possible, purchase of cement should be planned and phased to ensure a short storage period.

If prolonged storage cannot be avoided, regular sampling and testing should be put in hand to ensure that the cement remains in good condition. A good indicator is the loss on ignition test (BS 4550, Part 2: 1970 [17]) with higher values indicating a deterioration in the condition of the cement. Acceptable values are quoted in BS 12: 1978 [18] as 3% for temperate climes and 4% for tropical climes.

5.3 Aggregates

A great deal of effort is often required to produce good aggregates in the difficult geological and environmental conditions in the Middle East; it is therefore worth making careful arrangements to ensure that they do not become contaminated while stored on site.

It is common to see aggregates stockpiled directly on the ground. In regions with a high water table, groundwater and dissolved salts can rise into stockpiles by capillary action. There is also the possibility that the salty, naturally occurring materials in the ground might be scooped up along with the aggregates. Stockpiles should therefore always be sited on a concrete apron slab with positive drainage away from the aggregates. The storage area should be remote from heavily used site roads or any operation which generates dust.

As explained in a previous section, much dust is produced during crushing operations on the local materials. If aggregates are to be processed on site, due account should be taken of the direction of the prevailing winds when siting the crusher and stockpiles. Dust tends to accumulate at the bottom of stockpiles and storage bins. They should be cleaned out at regular intervals, say once per month, so that the dust does not build up to unacceptable levels.

Deliveries of aggregate should be sampled and tested regularly paying particular attention to those tests which indicate contamination or unsoundness. If beach sand is being used it is necessary to test each delivered load for chloride content.

5.4 Concrete production

High concrete temperatures during mixing and placing lead to undesirable properties in both the plastic and hardened states. Some of these are listed in Table 5-1.

Rapid loss of workability and a tendency to plastic cracking are quickly evident. The effects of reduced strength and durability are less immediate and do not normally show up in routine quality control tests which are carried out on specimens stored under controlled temperature and humidity conditions.

A generally accepted figure for the upper limit on concrete temperature is 32°C. There is a number of practical steps which can be taken to assist in achieving this figure. Perhaps the most obvious is to confine concrete production to the cooler parts of the day. If night work is not feasible, concreting should only take place in the early morning and late evening.

Table 5-1   Undesirable effects of hot weather on concrete

Plastic state	Hardened state
More water required for a given workability	Reduction in durability because of increased water demand
Rapid loss of slump	Poorly compacted concrete, tendency to cold joints
Rapid evaporation of mixing water	Rapid evaporation of curing water
Difficulty in controlling air entrainment	Increased permeability
Greater tendency to plastic shrinkage	Reduction in durability because of cracking
Accelerated set	Reduced strength
	Reduced bond to reinforcing steel
	Greater dimensional changes
	Increased permeability and cracking lead to a greater risk of reinforcement corrosion
	Deleterious reactions proceed at a faster rate

5.4.1 *Materials*

Mixing water has the greatest effect per unit weight of any of the constituents on the temperature of the mix because of its much higher specific heat. Chilling plants have been found to be very effective in conjunction with shading and insulating of tanks and pipes. The addition of flake ice in place of part of the mix water is doubly effective because of the benefits derived from latent heat. Care must be taken that no ice crystals remain at the end of the mixing cycle.

Good results can also be obtained by reducing the temperature of the aggregates as they constitute such a large proportion of the mix. Reductions in temperature can be achieved by shading the stockpiles; dark coloured aggregates tend to be significantly hotter than light-coloured aggregates. Shade-netting has been used extensively for this purpose because of its light weight. Spraying stockpiles with water also reduces their temperature, but this has to be carefully controlled if excessive variations in the moisture content of the aggregates are to be avoided.

5.4.2 *Plant*

Wherever possible the aggregate hoppers, cement silos, water tanks, conveyor belts, weighing gear and batching plants should be shaded. If this is not possible, they should be painted white and kept clean. Mixer trucks, skips and pumps should also be painted white.

The work performed on the concrete while mixing and agitating generates heat. Mixing time should therefore be kept to the minimum consistent with adequate uniformity and quality.

5.4.3 *Placing*

Great advantage can be gained by shading the workplace to reduce the temperature of the forms and reinforcement. It also improves the working conditions and can result in a better standard of workmanship.

Formwork, reinforcement and subgrade can also be kept cool by spraying with water but should be free from standing or adhering water at the time of concreting.

Transit time from mixer to pour location and the time during placing operations should be kept to the absolute minimum possible.

The most important consideration is good organization and planning so that the operation can be completed at the fastest possible rate and allow protection and curing to be initiated as soon as possible. Delivery of concrete to the pour should be phased so that it can be placed promptly on arrival. This applies particularly to the first batch which should not be ordered until everything is completely ready. Manpower and plant should be sufficient to allow operations to continue uninterrupted and spares should be available in the event of mechanical breakdown.

Retempering, the addition of water to retrieve lost workability, must not be allowed as it will obviously affect strength and durability.

5.5 Curing and protection

It is impossible to overemphasize the importance of effective curing and protection of concrete produced in environments such as the Middle East.

In conditions of high temperature and hot drying winds, water can evaporate from the surface of fresh concrete at a considerable rate (3 kg/m^{2}h (0.6 lb/ft^2 h) under adverse conditions). Severe plastic cracking will result unless adequate measures are taken to prevent this loss. The ACI [19] recommended that 1 kg/m^{2}h (0.2 lb/ft^2 h) is an upper allowable limit for hot weather concrete.

As hydration proceeds, sufficient water must be present to permit the continued development of gel formations which can then fill up the pore spaces until virtually no interconnecting spaces remain. Only if a continuous supply of water is available will the concrete develop its full potential strength and impermeability and hence durability.

Under Middle East conditions, the use of proprietary curing membranes alone is not to be recommended for the production of high-class concrete unless there is insufficient fresh water available for wet curing. Most curing membranes cannot be applied effectively until all surface water has evaporated—at this stage plastic cracks have often already started to develop. Thorough water curing gives much better results in terms of strength, durability and reduction in shrinkage cracking. Automatic systems using sprinklers or fine sprays are to be preferred due to the poor quality of workmanship and low level of supervision experienced on many sites. However, a judgement must be made on what curing method is to be adopted based upon the probable quality of water curing. Wetting and drying cycles caused by poor water curing can produce much worse concrete than concrete sprayed with a membrane.

5.5.1 *Horizontal surfaces*
Immediately after it has been finished, the surface should be covered with clear plastic sheeting in contact with the concrete. If there is a delay between placing and finishing, cover the surface with plastic sheeting in the intervening period.

After initial set has taken place, remove the plastic sheet and cover the surface with wet hessian. Replace the plastic sheet over the hessian. Keep the hessian in a damp condition for at least 7 days by frequent application of water. At the end of the initial wet curing period, remove the plastic and hessian and apply a pigmented, resin-based curing compound by spray.

If possible, shading and wind breaks should be provided during the wet-curing period to control the temperature and to counter the effects of drying winds.

Where specific surface textures are required (e.g. brush-marked finish), the above procedure may not be appropriate. In this case, the best technique may be to apply a spray-on curing membrane at the earliest opportunity and start wet curing immediately initial set has taken place.

5.5.2 *Vertical surfaces*
As soon as is practicable, ease off the side shutters and allow the curing water from the top surface to trickle down the sides.

Immediately after stripping, wrap the member in damp hessian, over-wrap the hessian with plastic sheeting held intimately and firmly in place. Keep the hessian in a damp condition continuously for a period of at least 7 days. The plastic and hessian can then be removed and a proprietary pigmented resin based curing compound applied.

The purpose of the polythene sheet in the curing procedures is to reduce evaporation losses. The reduction in evaporation results in less water being consumed in the curing process and also reduces the possibility of the build-up of any salts which may be present in the curing water. Even water which is acceptable as a concrete ingredient in terms of chloride content could cause alarming increases in the chloride concentration in the surface concrete, if it were subject to continuous evaporation and replenishment.

In no circumstances should brackish or sea water be used for curing reinforced concrete.

In many areas there may not be sufficient fresh water available to permit adequate wet curing. If this is the case, a proprietary membrane curing system will have to be accepted as second choice. It is important that a product with as high a curing efficiency as possible is used.

An excellent guide to the effects due to hot weather and the methods of overcoming them is given in the American Concrete Institute's *Recommended practice for hot weather concreting* [19].

6 Investigations into defective concrete

6.1 Introduction

There are many reasons for investigating concrete. The most usual during construction is to check the quality of the concrete; this aspect is not dealt with here. After the construction period, the most usual reason for investigations is to determine the causes of deterioration and the most suitable measures of enhancing the serviceability of the structure.

The deficiencies most common in concrete made in the Middle East are shown in Table 6-1 [20]. It is important to be able to determine the cause of deficiencies so that appropriate remedial or maintenance measures can be selected. A simple way of collecting data from a concrete structure is to make a visual examination and record the observed features on a scale drawing or map. A list of terminology for defects is shown in Table 6-2 and is based partly on the *ACI manual of concrete practice*. Cracks are a most useful surface feature; for Middle Eastern concrete, cracks as small as 0.1 mm (0.004 in) should be noted. The vast variety of types is well categorized by Mercer [21].

A diagnosis of the condition and any underlying cause of deterioration can be made by a study of maps and by comparison of maps made at intervals. Other field procedures are listed in Table 6-3 [20]. They are all relatively easy techniques requiring relatively inexpensive and easily transported equipment. Although one examination of the nature described in Table 6-3 often uncovers deficiencies, it may require repeated examination at intervals to allow sound prediction of the rate of deterioration. This aspect is considered in Section 3.

6.2 Techniques for investigating concrete

6.2.1 *Equipment*
Table 6-3 lists the activities involved in mapping the external features of a concrete member or structure. The equipment required is listed in Table 6-4.

Table 6-1   Common defects in Middle East concrete and frequent contributory factors

Defect	*Major contributing factors*
Plastic settlement cracks	Mix design Materials—aggregates of poor shape and grading Workmanship
Plastic shrinkage cracks	Workmanship—poor curing Environment—drying winds
Cold joints	Workmanship—inadequate concrete supply Environment—foreshortened setting time at high temperatures
Hydration thermal shrinkage cracks	Design—location of joints and size and sequence of pours Workmanship—shutter type and striking time Climate—high mix temperatures
Structural distress	Design Workmanship—failure of concrete to gain specified strength
Controlling drying shrinkage cracks	Climate—shrinkage takes place rapidly Workmanship—over-wet concrete, poor curing
Uncontrolled drying shrinkage cracks	Design—failure to provide sufficient reinforcement, inappropriate pour layout Climate—shrinkage takes place rapidly Workmanship—over-wet concrete, poor curing Materials—high shrinkage aggregates
Thermal cracks	Climate—high daily and seasonal temperature range
Salt weathering	Mix design—low strength, porous concrete Environment Workmanship—poor curing
External sulphate attack	Environment Mix design—low strength, porous concrete, wrong cement type
Corrosion of reinforcement	Materials—salty aggregates Environment—contamination of stored materials, ingress of chlorides after construction Workmanship—salt water for curing
Internal sulphate attack	Materials—aggregates containing sulphates Environment—contamination of stored materials
Alkali reactivity	Materials—reactive aggregates, cement chemistry

Table 6-2   Terminology for defects. (Adapted from *ACI manual of concrete practice*)

Defect	Notes
Cracks	Can be classified by direction, e.g. longitudinal, transverse, vertical, diagonal, random and by width:
	fine less than 1 mm[a]
	medium 1–2 mm
	wide greater than 2 mm
	Or by pattern, e.g. map cracking (or pattern cracking or crazing), checking (shallow cracks at closely spaced irregular intervals), D-cracking (progressive formation of fine cracks often near edges), Isle-of-Man (pattern of short fine cracks radiating from a point, often precedes pop-outs)
	Or by cause, e.g. corrosion, plastic settlement, shrinkage, etc.
Disintegration	Break up into small fragments
Distortion	Abnormal deformation of an element
Efflorescence	Deposit of salts on surface usually from within
Encrustation	Hard crust or coating on the surface
Exudation	Discharge of liquid or gel from within through a crack
Pits	Small cavities on the surface
Popouts	Breaking away of small portion of surface due to internal pressure leaving a conical depression
	Can be classified by size
Erosion	Loss of surface or section due to abrasive action
Scaling	Local loss of surface. Can be classified by depth
Spall	Fragment detached from surface of concrete by external force or internal pressure. Can be classified by size
Hollow area	Section of surface which sounds hollow when struck
Dusting	Development of a powdery surface
Corrosion	Deterioration due to chemical or electrolytic attack
Lamination	Separated surface layer caused by shallow planar crack running almost parallel to the surface
Bleeding channel	Mainly vertical channels on surface or within mass caused by upward migration of water
Sand run	Sandy streak on formed surface caused by bleeding
Voids	Small holes on surface or within mass—initially filled with water
Laitance	Weak, mortary top surface common on overwet or overworked concrete
Honeycombing	Series of voids where mortar fails to fill the spaces between the coarse aggregate particles
Sand pocket	Area of sand within mass

[a] Cracks as small as 0.1 mm may need to be logged [20].

Table 6-3   Field procedures used in investigations

Activity	Details	Remarks
Phase A: external examination		
Surface mapping	Decide on a suitable reference grid and mark it on the concrete surface. Draw a dark coloured line alongside all cracks indicating the end with a short line at right angles and the date	The extent of visible cracking can vary under different lighting conditions
Covermeter survey	Locate the reinforcement using a proprietary covermeter. Mark the position and depth of bars if possible	This is not always possible if the bars are bunched or the member is heavily reinforced
Sounding survey	Tap the surface with a hammer on a closely spaced grid pattern. Mark the limits of any hollow sounding areas	
Photography	Take a colour photograph of the location	Include a scale, job description card and colour card
Drawing	Prepare an accurate scale drawing showing the features mentioned above	Indicate any areas of dampness, encrustations, colour variations, honeycombing, weathering, popouts, exudations, spalling, pitting, rust staining or efflorescence
Review	Review data and decide an exploratory and testing work in phase B	Crack patterns or other features may be recognized which will assist in determining appropriate exploratory work and location of sampling
Phase B: physical testing and sampling		
Exposure of reinforcement	Select typical or particular areas and break out to expose the reinforcement	Note the colour and condition of the reinforcement immediately. Take photographs, make sketches
Potential survey	Use a copper/copper sulphate half-cell and a high impedance voltmeter on a previously marked grid	The method and apparatus are described in ASTM C 876 [22]

Table 6-3   (*contd.*)

Activity	Details	Remarks
Dust sampling	Use an electric percussion drill to obtain samples of concrete dust	Test samples for moisture content, acid soluble chloride content or acid soluble sulphate content depending on circumstances. It is often necessary to obtain and test samples from different depths
Investigate peculiarities	Break out any features such as rust stains or hollowness	Test breakout for depth of carbonation using phenolphthalein
Coring	Cut cores in concrete. If it is structurally acceptable, cores through horizontal reinforcement can confirm the presence of plastic settlement voids	Check the indications of mix properties, segregation and effectiveness of compaction. Mark the cores clearly with their orientation and log them in detail

6.3   Crack types

Some of the cracks related to forms of deterioration common in the Middle East are shown in Table 6-5 [20]. These crack types occur on structures all over the world but, because of the combination of aggressive ground conditions, material and climatic factors, they tend to propagate and progress more rapidly in the Middle East. Two categories of crack have been identified and can be defined as progressive and non-progressive.

Non-progressive cracks are those which once formed do not continue to develop unless an agency other than the original cause is also present. Examples from Table 6-5 are plastic settlement and plastic shrinkage, crazing, thermal hydration shrinkage.

Progressive cracking, on the other hand, continues without any other cause than the reason for the initial cracking. Examples from Table 6-5 are structural distress (load variations), corrosion related cracking, alkali reactivity and diurnal temperature effects.

In Middle Eastern concrete, cracks which are non-progressive allow air, moisture and soluble salts to penetrate the concrete and a different mechanism then takes over. The reported difficulties with highway bridges in the USA where salt is used to de-ice the carriageways is a particular example. The age of the concrete when the cracks first appeared must be determined if at all possible; crack patterns alone may give an inadequate understanding of the reasons for the deterioration.

Table 6-4   Equipment required for field work

Activity	Equipment
Crack mapping	Indelible markers of two different colours (one for crack locations, one for reinforcing bar positions). Flexible 3 m scale (e.g. steel tape). A magnifying glass with a magnification power of ten and a graticule, or a transparent strip with lines of various thicknesses (to measure cracks). A powerful torch
Covermeter survey	A covermeter able to detect the location and the depth of bars. The measurement of depth requires the bar diameter to be known (the depth of cover should be checked by breaking out as well)
Sounding survey	A 1 kg (approx) hammer
Photography	A single lens reflex camera with flash. Shots vary from general views to close-ups of bars. A 35 mm lens is a good compromise
Exposure of reinforcement	Hand tools or jackhammer. For weak concrete and small cover, hand tools may be adequate Tough clean plastic bags with seals (for samples of concrete or steel)
Potential survey	Two copper/copper sulphate half-cells (one as a spare) and a high impedance voltmeter—see ANSI–ASTM, C 876 [22]. (Other cell types have been used for underwater work e.g. silver/silver nitrate.) For rapid recording of many results, automatic data loggers which store information on magnetic tape; the tape can be subsequently analysed by computer and potential maps made
Dust sampling	Hand-held percussion electric drill capable of making holes up to, say, 8 mm diameter. Plastic cups to collect dust from overhead holes
Strength testing	Rebound hammer or ultrasonic devices
Coring	Coring machine with various diameter core barrels

6.4   Examples of investigations

Surveys and inspections of structures are undertaken in most parts of the world. Details of methods often used are given in the *ACI manual* [23]. The methods used in the investigations described below were discussed earlier in this section.

The case histories below are a useful illustration of the difficulties caused by the salty conditions which occur widely in the Middle East.

There are many reasons for the deterioration of concrete and these can be

Table 6-5   Crack types and related features and patterns

Type of crack	Approx age at which cracks become visible	Features/patterns
Plastic settlement	Day 1	Occur over horizontal bars in plastic state. Settlement on either side of bar. Void under bar
Plastic shrinkage	Day 1	Over rebar. In mass concrete are somewhat random
Crazing	Between day 1 and 20	Tiny cracks (0.05 mm) with a pattern spacing probably less than 10 mm
Thermal hydration shrinkage	Between week 1 and month 50	Occur during hydration and are often impossible to recognize at the investigation stage unless they were noted during construction. May be related to pour sequence and relative cross-sections
Structural distress	Between month 1 and 200	At an angle to reinforcement (shear) although can be over rebar or at right angles if excessive bending strain produces tension
Restrained drying shrinkage	Between month 1 and 100. Max rate month 10 to 40	Large polygonal loops related to rebar spacing but tend to be displaced from bars
Unrestrained shrinkage	Between month 1 and 100. Max rate month 10 to 40	Near and over rebar and may be related to pre-existing plastic or thermal cracks mid pour
Corrosion related	Between month $1\frac{1}{2}$ and 200	Over bars or may form crack in plane of bars which does not reach surface
Alkali reactivity	Between month 3 and 300	Popouts followed by short cracks radiated from a central point (Isle-of-Man cracks) which develop with time to mapwork cracks, and often ooze a clear fluid
Diurnal temperature effects	Between month 7 and 100	Typically at slab to wall connections

related to a lack of appreciation of design details, improper use of construction materials, or poor construction practice. A proper liaison between designer, constructor and maintenance engineer generally does not exist in the developing countries of the Middle East; the investigator probably has more defects and less reference to authoritative data than would be the case in many parts of the developed world. An excellent introduction to the synthesis of design construction and maintenance is given by Johnson [24].

6.4.1 Case history 1

The site consisted of a number of structures of various types constructed in a coastal region of the United Arab Emirates. The geological classification was zone IV as defined in Section 1.3. The structure was about 18 months old when an investigation was initiated although cracks had been in evidence earlier. Examination of the construction records revealed that the mix proportions were as follows:

Free water/cement ratio	0.57 to 0.60
Cement content	335 kg/m^3 (SRPC BS 4027) [25]
Maximum aggregate size	20 mm (gabbro gravel from zone III) [1]
Sand	the finest allowed by BS 882 [26] (carbonate sands from the beach)

Immediately, two aspects were thought to be likely causes of defects: salts in the sand won from the beach and high and variable dust contents of gravel aggregates improperly screened. The water/cement ratio and cement content were inappropriate for the hostility of the local conditions. A maximum water/cement ratio of 0.42 and minimum cement content of 370 kg/m^3 (623 lb/yd^3) with air entrainment would have been appropriate for the parts of the structure near the ground or subject to frequent wetting and drying. However, the cracking at an early age could not be attributed to the inadequate concrete specification.

On an elevated thin cantilever walkway, plastic settlement cracks were diagnosed on the top surface when the cracks were seen to occur directly over reinforcing bar locations. Coring through the bars confirmed that the concrete had settled around the bars and a gap below the bar was apparent. These cracks were known to have been present within a few hours of concreting but they had not been sealed or any protective measures taken. Curing of the slabs was by curing membrane, which was probably not applied soon enough after finishing. Plastic settlement cracking of thin slabs is a feature of hot desert environments and can be avoided by good mix design and careful placing and curing as well as ensuring that there is a good top cover over the reinforcement; in thin slabs low cover (say less than 50 mm) can cause the bar to act as a crack inducer.

Although the cracks described above were initially of the non-progressive type, the structure was splashed regularly with saline water which penetrated the cracks and hence corrosion of the reinforcement was likely even if the cracks were sealed after the investigation. In such circumstances, sealing may help by reducing the availability of oxygen at the corroding reinforcement. The cracks were sealed and the deterioration of the slab continued. There have been attempts to remove chlorides from concrete by applying electrical potentials and although such methods are theoretically feasible little evidence has been published of their success in practice.

6.4.2 Case history 2
At another location on the same site, much cracking was observed in thin-walled circular tanks. The cracks were mapped along with reinforcement locations. In this case, the cracks were not related to the position of the reinforcement and the chloride content of the concrete, and the condition of the reinforcement at selected break-outs suggested that the cracks were not due to corrosion of the embedded steel. The eventual diagnosis was that the cracks were due to drying shrinkage controlled by the well distributed and adequate reinforcement, i.e. a case of controlled drying shrinkage.

Cracking due to changes in volume on hydration is present in all concrete structures to some degree. Provision of reinforcement and other design procedures to control cracking are covered by codes in various countries, for example [10], [27] and [28]. If inadequate or poorly distributed steel is provided, then uncontrolled drying shrinkage cracking occurs. Because of the extreme drying conditions, the rate at which cracking occurs tends to be greater than in damper climates. The thickness of the member also affects the rate of drying. It is often difficult to distinguish between thermal hydration cracks and drying shrinkage cracks as they tend to become apparent in the same age range (Table 6-5).

6.4.3 Case history 3
On the same site, a regular vertical crack pattern was observed on the thin walls of a low, large-diameter tank. It was possible that the cracks were related to the vertical reinforcement. However, this did not prove to be the case and unrestrained drying shrinkage cracking was diagnosed. The non-progressive nature of the cracks was confirmed by low chloride levels.

There were some areas within the structures described above where the concrete had chloride contents greater than those permitted by CP 110 [10]. By checking the distribution of the chloride ions with depth, it was possible to discover whether salts were introduced with the mix ingredients or whether they had penetrated after construction. The latter case gives a typical chloride 'profile' with a maximum concentration near the surfaces in contact with the salty ground or atmosphere.

The initial assessment of the deterioration due to high and variable dust contents in the gravel aggregates was borne out by the variations in water/cement ratio from the examination of site records and the presence in some areas of defects which were obviously due to high water contents.

Perhaps the greatest single cause of deterioration of Middle East concrete structures is excessive salt concentrations due to the use of beach sand. Traditional building practice with no embedded steel gave no indication of the difficulties likely to be encountered with reinforced concrete. High salt concentrations in reinforced concrete structures leads to corrosion-related cracking.

6.4.4 Case history 4
The example of corrosion-related cracking given here occurred during the construction of a multistory building. Cracks were noted in small columns only four

months after they were poured. Exposure of the reinforcement showed that active gel-producing rust was present in some locations. As it was possible that the crack patterns in other members were not well developed because of the extreme youth of the concrete, hundreds of samples were taken to determine the *in situ* chloride content. The environment was not particularly salty and the cracking did not occur in the areas near the ground where contamination from external sources was most likely.

Analysis of the results of chloride tests related to the dates of construction showed that the worst areas were poured mainly at the week-ends when supervision was at a minimum. The availability of clean sand was limited and it was concluded that a contaminated source local to the site had been used when stocks became low.

The decision whether to cut out individual members was made on the basis of their chloride content as determined by testing dust samples or the presence of cracks.

Structural distress due to underdesign or mishandling during construction happens world-wide. The deterioration rate after the initial damage is, however, markedly variable and dependent upon the conditions in which the member exists subsequently. Minor cracks are often self-healing in the right conditions. Identified cracks along reinforcement should usually be sealed unless they are smaller than say 0.1 mm (0.004 in) wide or the exposure condition is mild. The topic of repairs is dealt with in another section.

6.4.5 *Case history 5*

Much has been published about the hostility of the Middle Eastern climatic and other environmental features and its deleterious effect on the durability of reinforced concrete. Even for well-specified and well-made concrete, the rate at which deleterious salts invade the pores constitutes a serious hazard. This is especially significant in the most hostile zones. An example is shown in Fig. 6-1 from [15]. The structure is a culvert mostly below ground level. The mix design was as follows:

Free water cement ratio	0.45
Cement content	370 kg/m^3 (OPC to BS 12)
Maximum aggregate size	20 mm (gabbro gravel from zone III)
Sand	blend of washed beach sand and washed wadi fines
Admixture	4% entrained air

The base slabs and wall kickers were poured first and then the walls. Brackish water was used for curing and the culvert was used to collect the output from a dewatering operation for the construction of a foundation nearby. The water flooded the channel to a maximum depth of about 150 mm (6 in) and after one year rusting became apparent near the kicker. The theoretical cover to the steel was 50 mm. Investigation of the chloride content of the concrete at three elevations produced the curves in Fig. 6-1 [15]. If it is assumed that the curve at the highest elevation represents the effects of the brackish curing water, then the amount by which the other curves exceeds these values can be taken as the ingress from the outside of the concrete subsequent to curing. Although the specification

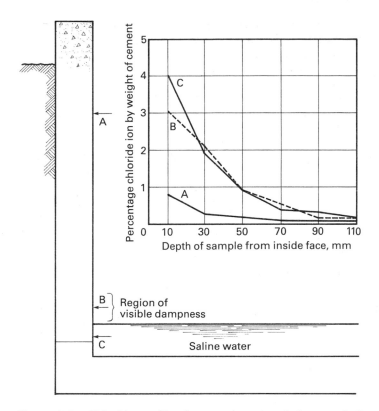

Figure 6-1   Chloride profiles from various levels in a partly flooded channel

of the concrete was not to an adequate standard, the apparent rate at which salts penetrated gives great cause for concern.

Carbonation of concrete, which proceeds from the surface inwards, also has a direct effect on the probability of reinforcement corrosion. Carbonation is the reaction of carbon dioxide with calcium hydroxide from the cement, which thus lowers the alkalinity of the concrete. The high alkalinity is responsible for the normally passive state of embedded steel. At elevated temperatures and with the greater radiation in the Middle East (see Fig. 6-2), carbonation probably occurs more quickly. Simple tests using phenolphthalein have indicated a few mm depth of carbonation after 5 years for unexposed locations and up to 30 mm in exposed areas after 10 years.

The reaction of certain siliceous aggregates with alkalis in the concrete, mainly from the cement, can lead to deterioration. This form of deterioration is well known in parts of the USA and Europe. The damage occurs because the reaction products are many times larger than the reactants and the expansive forces generated disrupt the concrete. Examples of this form of distress have been found in the Middle East. Pop-outs and cracks of the three-legged variety, which are classical symptoms of alkali–aggregate reaction, were discovered on a structure when it was about 18 years old. Further examination has shown the gel, which is a

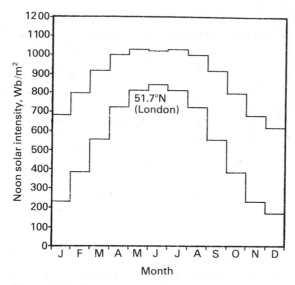

Figure 6-2 Comparison of theoretical noon solar intensities on horizontal surfaces at latitudes of London and Middle East (Webers/m²)

reaction product, to be present. The aggregates were silica rich and had been chosen because of their hardness. In the Middle East, the elevated temperature and alkaline environment may tend to enhance the rate at which the reaction takes place.

6.5 Concluding remarks

The case histories illuminate some of the commonly found durability defects. There are, of course, others and poorly designed details make a significant contribution which is discussed elsewhere in this chapter. The lessons gleaned from examining deteriorating structures have been used to produce designs and specifications which should lead to better quality construction. Many of the data have been analysed so that better predictions of the behaviour of concrete structures with time can be made.

7 Repairs

7.1 Principles

Repairs to concrete structures may become necessary at any stage in their lifetime and for a variety of reasons. The general topic of repairs to concrete is covered elsewhere in this handbook (Chapter 25). In this section, techniques and repair methods which are particularly relevant to the hot arid environments are discussed.

In such severe environments, most repairs become necessary because of deterioration of the concrete itself rather than because of damage inflicted by physical force. In these circumstances, it is always vital to determine the underlying reason for the deterioration as in many cases this information is fundamental to the choice of mode of repair. For instance, repair methods for concrete cracked by plastic settlement would not necessarily be appropriate to concrete cracked by reinforcement corrosion although superficially the crack patterns might look similar.

In the extremely hostile environment in desert zone IV [1], it is necessary to consider cold joints, areas of honeycombing, cracks, etc., as an effective reduction of cover to the reinforcement which must be reinstated.

7.2 Some repair methods

7.2.1 *Cracks*

In Section 6.3, a distinction has been made between progressive and non-progressive types of cracking. Under normal temperature climatic conditions, there is often no need to treat non-progressive cracks (unless they are unsightly) since they are, by definition, in a stable condition. In extremely aggressive conditions, it is essential to seal non-progressive cracks as they present a pathway by which the agents of corrosion can reach the reinforcement. The high diurnal and annual temperature ranges must be kept in mind when considering sealing methods and choosing materials. Considerable movements can take place across cracks. Filling the cracks with a rigid material (for example, by injection of epoxy resin) may only result in the formation of new cracks under the influence of subsequent thermal strains because the percentage elongation over relatively narrow cracks can be very severe.

Emulsions based on styrene butadeine or latex polymers are sometimes used to fill cracks as they are flexible but on wider cracks they may only coat the faces rather than totally filling the voids because of the reduction in volume associated with polymerization. Surface coating of polyurethane, acrylic or epoxy resin can be used for sealing narrow cracks but they may suffer damage from abrasion or become embrittled when exposed to the strong ultraviolet element in sunlight. Where cracks are moving because of continued shrinkage or temperature effects, most thin coatings quickly re-crack. Fibreglass mats laid into the coating can assist in overcoming this problem.

A method of treating isolated cracks which has been successfully employed is to cut a chase along the line of the crack and fill it with an elastomeric sealing compound.

7.2.2 *Surface defects*

Surface defects, such as honeycombing, blow holes, deep sand runs and cold joints, have to be made good in reinforced concrete where they constitute a reduction in effective cover to the steel. Epoxy mortars are often used for small repairs of this type with a good success rate. Special hot-weather grades have to be employed because of the much reduced pot-life at elevated temperatures.

7.2.3 *Reinforced concrete*

Corrosion of reinforcement is more often than not associated with high salt concentrations in the concrete, which are present because the mix ingredients were contaminated or because chlorides have penetrated the cover concrete, either through the mass or by way of cracks. The resulting corrosion products occupy several times the volume of the steel before corrosion and expansive pressures are generated which crack the concrete. There is little to be gained by attempting to seal these corrosion-related cracks. At best, it may stifle the process for a short time by blocking the direct access of salt, moisture and air. At worst, it may assist in spreading the corrosion to previously unaffected areas because of the essentially electrolytic nature of the reaction.

The only effective long-term remedy is to break out the contaminated concrete. This is not always as drastic a measure as it may appear at first sight. In most reinforced concrete members, the bars are concentrated in a zone close to the surface and it may only be necessary to remove a layer which extends 25–50 mm behind the reinforcement. Concrete should be broken out until there is at least 300 mm (1 ft) of uncorroded reinforcement in all directions and ideally the half cell potential of the area should be checked as described in Section 6. The edge of the repair area should be cut square with a disc or saw to ensure that feather edges are avoided.

Once the concrete has been removed, the full extent of the corrosion on the reinforcement can be appreciated. It is quite often surprising how little corrosion is required to generate extensive cracking. The reinforcement should be cleaned by sand or grit blasting using deflected nozzles to reach behind the bars. Mirrors are useful for inspecting the back faces. Any badly pitted bars should be replaced. Where this is not possible because the bars pass into the concrete at the edge of the repair area, cleaning has to be particularly thorough. It is often difficult to clean out all corrosion pits by grit blasting alone. Though the process seems successful at first, the pits begin to exude fresh rust products within a few hours. Alternate grit blasting and washing with liberal doses of fresh water have been found to overcome all but the most stubborn cases. A policy of waiting for 24 h to check on changes in the cleaned bar should be adopted where possible.

To ensure a durable repair, epoxy bonding agents are applied to the surface of the existing concrete before the new concrete is placed.

Mix design for the replacement concrete requires careful attention. The aggregates and proportions should be kept as close to the original as possible to avoid problems with differential thermal movements. The very large radiation and shade temperatures shown elsewhere in this chapter put repairs under stresses much greater than those in temperate climes. The water/cement ratio should be kept as low as possible (less than 0.4 if possible) to reduce shrinkage. The mix must be such as to produce an impermeable cover to the steel and therefore must be workable enough to compact.

For repairs to large areas of slabs or walls, the individual pours should be as near square as possible and restricted in size to a maximum of 2×2 m (6.5×6.5 ft); the size may need to be adjusted in relation to the thickness of the repair.

Thorough water curing is particularly important for thin concrete repairs.

8 The future

The preceding sections of this chapter have touched on some of the obvious lessons to be gained from working and studying the construction and performance of concrete made in hot climates and specifically the Middle East.

Research work regarding the making of durable concrete is certainly required in both the field and laboratory—mainly for concretes in the most aggressive zones. It may be that research work already undertaken has yet to filter through to practising field engineers. The aspect which is obvious from field studies is the rapid rate at which salts can penetrate concrete, particularly fresh concrete; ways to combat it are required. The value of drying shrinkage used in designs needs to be reviewed, perhaps by laboratory and field studies. The ideal concrete would be free from internal defects and hence be relatively impermeable and less likely to shrinkage.

Because the significance of permeability of concrete is now becoming obvious in the more hostile environments, many attempts to improve on this aspect have been made in the field. They fall under the headings of mix improvements, admixtures and coatings to steel and concrete. A reliable site test for permeability would allow concrete to be designed to achieve particular values just like strength is used at present.

Keeping the water/cement ratio significantly below 0.4 while ensuring a workable and plastic mix makes a considerable contribution in conjunction with adequate cement contents. At present, however, it is expected this will only improve the durability of the concrete marginally. Admixtures to achieve such low water/cement ratios are obviously required.

The use of specialized admixtures to reduce the risk of corrosion is discussed in Chapter 4; field use has yet to be fully evaluated but may be significant. Much work on coating of reinforcement may not be applicable in countries where a relatively low level of technology exists. However, an internal coating on the bar has obvious advantages over an external application.

Extensive field trials during the early 1980s may lead to durable and effective coatings. Even if the coatings fail after some time, they will have provided an ideal environment for the concrete to hydrate and fill pores as it ages. One obvious difficulty with coatings is that the physical properties are different from the substrate concrete; making a covering stay in place is difficult. One hazard is that moisture drawn through pin-holes in otherwise impermeable membranes may cause a build-up of deposits at the hole and subsequent forcing off of the membrane.

Concrete overlays with a very low water/cement ratio of say less than 0.32 have been investigated for bridge decks; their use for other structures is obviously limited. The blocking of the pores by modifying the concrete (e.g. latex or epoxy modified) are probably costly solutions; however, the cost of repairs, and consequential losses of income, may make such solutions acceptable in certain cases.

Because on many smaller sites the temperature of fresh concrete is unlikely to be controlled, research is necessary on the structure and properties of concrete made under these conditions.

Good site practice allied to appropriate design can give concrete a satisfactory life in most cases. There still remains a doubt, however, concerning reinforced concrete in the most extreme conditions, because of the rapid penetration of salts; ways of extracting these by electrical or chemical methods could prolong the life of some concretes. The techniques are still to be tested by application to a significant number of structures.

References

1. Fookes, P. G. and Knill, J. L., The application of engineering geology in the regional development of northern and central Iran, *Eng. Geol.*, Vol. 3, 1969, pp. 81–120.
2. British Standards Institution, BS 3148: 1980, *Tests for water for making concrete*, London, 1980, 9 pp.
3. American Association of State Highway and Transportation Officials, *Standard method of test for quality of water to be used in concretes*, AASHTO T26, Washington, DC, 1972, 5 pp.
4. American Society for Testing and Materials, ASTM C 150-78a, *Specifications for Portland cement*, Philadelphia, 1978, 7 pp.
5. Fookes, P. G. and Higginbottom, I. E., Some problems of construction aggregates in desert areas, with particular reference to the Arabian Peninsular, *Proc. Inst. Civ. Eng.*, Parts 1 and 2, Vol. 68, Feb. 1980, pp. 39–90.
6. Fookes, P. G. and Collis, L., Cracking and the Middle East, *Concrete*, Vol. 10, No. 2, Feb. 1975, pp. 14–19.
7. American Society for Testing and Materials, ASTM C 295-65, *Recommended practice for petrographic examination of aggregates for concrete*, Philadelphia, 1965, 19 pp.
8. American Society for Testing and Materials, ASTM C 88-73, *Method of test for soundness of aggregates by use of sodium sulphate or magnesium sulphate*, Philadelphia, 19 pp.
9*. Fookes, P. G., Pollock, D. J. and Kay, E. A., Concrete in the Middle East: rates of deterioration, *Concrete*, Vol. 15, No. 9, Sept. 1981, pp. 12–19.
10. British Standards Institution, CP 110, Part 1: 1972 (as amended 1980), *The structural use of concrete*, London, 1972, 156 pp.
11. ACI Committee 201, Guide to durable concrete, ACI 201-2R, *ACI manual of concrete practice*, American Concrete Institute, Detroit, 1980, pp. 201-1 to 201-37.
12. British Standards Institution, BS 812, *Methods for sampling and testing of mineral aggregates, sands and fillers*, (four parts) London 1975 and 1976, 68 pp.
13. Neville, A. M., *Properties of concrete*, 3rd Edn, Pitman, London, 1981, 779 pp.
14. Nilsson, L. O., *Hygroscopic moisture in concrete-drying, measurements and related material properties*, Report TV BM-1003, Lund Institute of Technology, 1980, 162 pp.
15*. Kay, E. A., Fookes, P. G. and Pollock, D. J., Deterioration related to chloride ingress, *Concrete*, Vol. 15, No. 11, Nov. 1981, pp. 22–28.
16. John, D. G., Novel electrochemical techniques for investigating steel/concrete

systems, *Proc. conf. The failure and repair of corroded reinforced concretes*, Oyez IBC, London, 1981, pp. 59–86.

17. British Standards Institution, BS 4550, Part 2: 1970, *Methods of testing cement. Chemical tests*, London, 1970, 48 pp.

18. British Standards Institution, BS 12: 1978, *Specification for ordinary and rapid-hardening Portland cement*, London, 1978, 4 pp.

19. ACI Committee 305, *Recommended practice for hot weather concreting*, American Concrete Institute, Detroit, 1977.

20*. Pollock, D. J., Kay, E. A. and Fookes, P. G., Crack mapping for investigation of Middle East concrete, *Concrete*, Vol. 15, No. 5, May 1981, pp. 12–18.

21. Mercer, L. B. in Troxell, E. G. and Davis, H. E., *Composition and properties of concrete*, McGraw-Hill, New York, 1968, 434 pp.

22. ANSI–ASTM, C 876-77. *Standard test method for half cell potentials of reinforcing steel in concrete*, American Society for Testing and Materials, Philadelphia, 1977, 7 pp.

23. American Concrete Institute, *ACI manual of concrete practice*, Part 1, Detroit, 1980, pp. 201–301.

24. Johnson, S. M., *Deterioration, maintenance and repair of structures*, McGraw-Hill, New York, 1965, 373 pp.

25. British Standards Institution, BS 4027: 1980, *Sulphate-resisting Portland cement*, London, 1980, 4 pp.

26. British Standards Institution, BS 882, 1201, Part 2: 1973, *Aggregates from natural sources for concrete (including granolithic)*, London, 1973, 16 pp.

27. British Standards Institution, BS 5337: 1976, *Code of practice for the structural use of concrete for retaining aqueous liquids*, London, 1976, 16 pp.

28. ACI Committee 224, Control of cracking in concrete structures, *J. Am. Concr. Inst., Proc.*, Vol. 69, No. 12, 1972, pp. 717–753.

* References [9], [15] and [20] have been combined as *Concrete in the Middle East*: Part 2, Viewpoint Publications, Eyre and Spottiswoode, London, 1982, 33 pp.

27 Concrete production, quality control, and evaluation in service

Robert E Philleo,

Chief, Structures Branch, Office of the Chief of Engineers, Washington, D.C. USA

Contents

Notation

C temperature, °C
f_{cr} required average strength
f_c' specified design strength
h overall thickness of member
l span of member supported at the ends or twice the span of cantilever members
M maturity, degree C–hours
t a characteristic of the normal distribution curve and the number of degrees of freedom
Δt duration of curing at temperature C, hours
$\bar{x}$ mean value of strength distribution
σ standard deviation

Summary

The quality of concrete in place is a function of the properties of its constituent materials, the accuracy of batching, the techniques used to transport, place, and consolidate it, and the adequacy of curing after placing. The quality can be controlled by rigorous inspection and by making appropriate tests of the fresh concrete. Tests most commonly made are for workability and air content. Recently tests which provide a rapid analysis of the ingredients have been introduced. The efficacy of quality control can be assessed by a statistical analysis of tests on the fresh concrete and on strength tests of the hardened concrete. Such analysis is facilitated by using either Shewart or Cusum control charts. For evaluating older structures load tests and core tests are cumbersome and expensive. Fortunately there are now available a number of economical non-destructive tests which can be conducted in place.

1 Introduction

Concrete used in a construction project will be no better than the concrete represented by the mixture proportions selected for the project. The goal of the concrete production facility is to produce on a continuing basis the concrete envisioned by the designer and to do so with a minimum of variability. Most specifications are now written in such a manner that there is an economic incentive to reduce variability to the lowest practicable level. Acceptance of the concrete is based on a 'characteristic strength', which is defined usually as the 5th or 10th percentile of the universe of concrete strengths being produced by the plant. Thus, the concrete must be proportioned for an average strength high enough to insure that 90 or 95% of all test results may be expected to equal or exceed the specified characteristic strength. In US practice, the characteristic strength is the design strength, f_c'. Such a requirement has made concrete plant operators at least dabblers in the mathematics of statistics. It is apparent that the required average strength is a function of the variability of the product and, therefore, of the degree of control. Poor control requires proportioning for a high average strength; a high average strength requires a high cement content. Money spent to improve control may be repaid in a saving of cement. Since cement is by far the most expensive ingredient of concrete, significant reductions can make a substantial sum of money available for plant improvement. As a result, sophisticated materials handling equipment and automatic controls, once thought to be unnecessary embellishments in concrete plants, are now commonplace. Each proposed improvement, however, should be analyzed for its contribution to reduction in variability in order to decide if it is economically justified.

2 Batching and mixing

2.1 Handling of aggregates

Uniformity of concrete production begins with the handling of aggregates. Of all the ingredients of concrete, only the aggregates present significant problems in handling. The problems are of two types: variation in moisture content and variation in grading. The first exists primarily in fine aggregates or occasionally in the smallest size of coarse aggregate. The simplest solution is to allow the material to remain in a free-draining storage pile until a stable moisture content is achieved. The time required and the stable level obtained depend on the grading and particle shape, but natural sands usually achieve a level not exceeding 6% within 24 h. Manufactured fine aggregate might take a bit longer and might retain as much as 8% moisture. If coarse aggregate becomes overly wet during processing and handling, the water is most easily removed by passing the aggregate over a dewatering screen before delivering it to the batch plant bins.

The principal weapon for maintaining uniformity of grading of coarse aggregate is the complete separation of size fractions all the way from initial processing until batching. Specifications for important work frequently require separating aggregate on the 19 mm ($\frac{3}{4}$ in) sieve and on all larger sieves with openings equal to 19 mm multiplied by a power of 2. An alternate form of the specification starts the sieves with the 12.5 mm ($\frac{1}{2}$ in) sieve. Thus, two common groupings of aggregate sizes are the following

  4.75 to 19 mm (No. 4 to $\frac{3}{4}$ in)
  19 to 37.5 mm ($\frac{3}{4}$ to $1\frac{1}{2}$ in)
  37.5 to 75 mm ($1\frac{1}{2}$ to 3 in)
  75 to 150 mm (3 to 6 in)

  4.75 to 12.5 mm (No. 4 to $\frac{1}{2}$ in)
  12.5 to 25 mm ($\frac{1}{2}$ to 1 in)
  25 to 50 mm (1 to 2 in)
  50 to 100 mm (2 to 4 in)

In each case, the larger sizes are used only for mass concrete such as that placed in dams. For such concrete, the specifications frequently require rescreening of the aggregate before it is placed in the batch plant bins. The rescreening devices are commonly placed on top of the batch plants so that each particle falls into the proper bin after having passed through the appropriate screen. This procedure is effective in removing objectionable undersize material from each aggregate fraction. Although rescreening is a powerful tool for maintaining acceptable grading, even for aggregate used in structural concrete, it is capable of serious abuse. The device is usually arranged so that the oversize from each screen falls into the bin for the next larger size and the undersize falls into the bin for the next smaller size, with all the undersize from the smallest screen wasted. With this arrangement, it is necessary to operate the device so that all sizes of coarse aggregate are fed onto the top screen simultaneously. If it is not possible to operate in this manner, the screening device must be arranged so that all the undersize or oversize from each size fraction is wasted or returned to the stockpile. Otherwise there will be an undesirable accumulation of poorly graded material in the bins adjacent to the bin for which material is being processed at any given time.

While it is not good practice, there are still plants in which only one size of coarse aggregate is batched. It is necessary, then, to minimize variation in grading over a particle size range from 4.75 to 25 mm (No. 4 to 1 in) or 4.75 to 37.5 mm (No. 4 to $1\frac{1}{2}$ in). Regardless of the range, aggregates should be handled between processing and batching in such a manner as to maintain the integrity of the

Incorrect methods of stockpiling aggregates cause segregation and breakage

(a)

Preferable
Crane or other means of placing material in pile in units not larger than a truck load which remain where placed and do not run down slope.

Objectionable
Methods which permit the aggregate to roll down the slope as it is added to the pile or permit hauling equipment to operate over the same level repeatedly.

Limited acceptability – generally objectionable

Pile built radially in horizontal layers by bulldozer working from materials as dropped from conveyor belt. A rock ladder may be needed in setup.

Bulldozer stacking progressive layers on slope not flatter than 3:1 unless materials strongly resist breakage. These methods are also objectionable

(b)

Correct
Chimney surrounding material falling from end of conveyor belt to prevent wind from separating fine and coarse materials. Openings provided as required to discharge materials at various elevations on the pile.
Wind Separation

Incorrect
Free fall of material from high end of stacking permitting wind to separate fine from coarse material.

Unfinished or fine aggregate storage (dry materials)

(c)

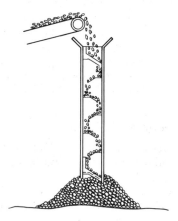

When stockpiling large sized aggregates from elevated conveyors. Breakage is minimized by use of a rock ladder.

Finished aggregate storage

Note: If excessive fines can not be avoided in coarse aggregate fractions by stockpiling methods used, finish screening prior to transfer to batch plant bins will be required.

Figure 2-1   Correct and incorrect methods of handling and storing aggregates (Courtesy of American Concrete Institute)

grading. The critical part of the operation is the establishment of and the removal of material from stockpiles. Stockpiles should be built up in horizontal or nearly horizontal layers. Bulldozers or other equipment should not be permitted to work directly on the piles because they will cause breakage of the material and will introduce dirt into the pile. Dirt is also minimized by establishing the piles on concrete surfaces so that there is no contamination with the underlying ground when the piles are low. The cardinal rule in transferring aggregate from one location to another is that it should fall vertically into place and not be permitted to roll down the edge of a stockpile. Furthermore, the distance through which a vertical fall takes place should be minimized in order to prevent separation by the wind and breakage upon impact. Where fall from a great height is necessary, the material should be contained within a circular drop chute or a rock ladder should be installed. Further details may be found in [1,2]. Figure 2-1, which illustrates good and bad handling procedures, is taken from [1].

The preferred method for transporting aggregate from storage piles to the batch plant bins is by a conveyor belt operating through a reclaiming tunnel under the storage piles. The material falls vertically through a gate beneath the center of each pile onto the belt. The withdrawal from the centerline of the pile should produce a grading representative of that in the pile. If rescreening is used and all sizes of aggregate are to be loaded onto the belt at the same time, the piles should be arranged so that the largest size is nearest the batch plant so that the large particles will not foul other gates. In small plants, aggregate may be transferred from storage piles to batch plant bins directly by a front-end loader. The procedure is deficient in that the material is always being taken from the edges of the pile where the coarser material tends to congregate, but if the operator works all accessible parts of the pile uniformly segregation may be minimized. For this loading procedure, a paved surface beneath the pile is absolutely essential.

2.2 Batching

The purpose of a batch plant is to deliver the ingredients of concrete to the mixer in the proper proportions and in a reasonably homogeneous mixture so that the work to be done by the mixer is minimized. Most concrete operations utilize central, or one-stop batch plants, wherein all the materials are delivered to the mixer at a single point. Within the ready-mixed concrete industry, there are some multi-stop plants in which a transit mixer is driven from batcher to batcher to receive fine aggregate, one or more sizes of coarse aggregate, cement, water, and admixtures. More than one of the ingredients may be added at a single stop. Concrete from a multi-stop plant usually requires more mixing than that from a single-stop plant.

Within the batch plant, aggregate handling techniques continue to be important. To the greatest extent possible, aggregate should move vertically through the plant. Horizontal movement is always a potential source of segregation. Figure 2-2 illustrates some principles of batch plant layout. Handling of materials other than aggregate is not particularly critical. Cement, pozzolan, and powdered admixtures should be stored in weathertight but ventiliated structures to prevent absorption of moisture. It is important that completely separate handling and storage facilities be provided for cement and pozzolan, when pozzolan is to be

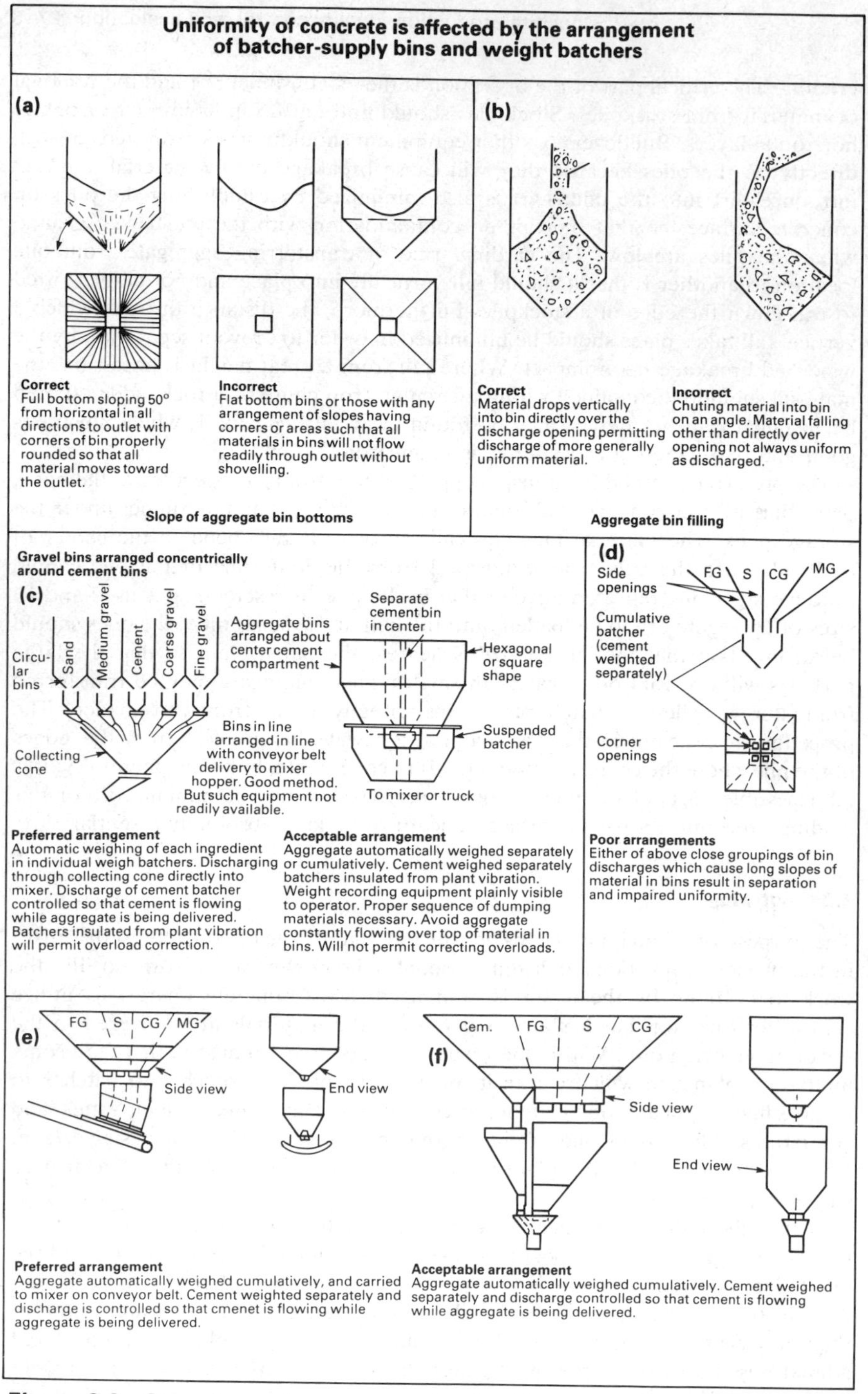

Uniformity of concrete is affected by the arrangement of batcher-supply bins and weight batchers

(a)

Correct
Full bottom sloping 50° from horizontal in all directions to outlet with corners of bin properly rounded so that all material moves toward the outlet.

Incorrect
Flat bottom bins or those with any arrangement of slopes having corners or areas such that all materials in bins will not flow readily through outlet without shovelling.

Slope of aggregate bin bottoms

(b)

Correct
Material drops vertically into bin directly over the discharge opening permitting discharge of more generally uniform material.

Incorrect
Chuting material into bin on an angle. Material falling other than directly over opening not always uniform as discharged.

Aggregate bin filling

(c)

Gravel bins arranged concentrically around cement bins

Circular bins — Sand, Medium gravel, Cement, Coarse gravel, Fine gravel

Collecting cone

Aggregate bins arranged about center cement compartment

Bins in line arranged in line with conveyor belt delivery to mixer hopper. Good method. But such equipment not readily available.

Separate cement bin in center

Hexagonal or square shape

Suspended batcher

To mixer or truck

Preferred arrangement
Automatic weighing of each ingredient in individual weigh batchers. Discharging through collecting cone directly into mixer. Discharge of cement batcher controlled so that cement is flowing while aggregate is being delivered. Batchers insulated from plant vibration will permit overload correction.

Acceptable arrangement
Aggregate automatically weighed separately or cumulatively. Cement weighed separately batchers insulated from plant vibration. Weight recording equipment plainly visible to operator. Proper sequence of dumping materials necessary. Avoid aggregate constantly flowing over top of material in bins. Will not permit correcting overloads.

(d)

Side openings — FG, S, CG, MG

Cumulative batcher (cement weighted separately)

Corner openings

Poor arrangements
Either of above close groupings of bin discharges which cause long slopes of material in bins result in separation and impaired uniformity.

(e)

FG, S, CG, MG

Side view   End view

Preferred arrangement
Aggregate automatically weighed cumulatively, and carried to mixer on conveyor belt. Cement weighted separately and discharge is controlled so that cmenet is flowing while aggregate is being delivered.

(f)

Cem., FG, S, CG

Side view

End view

Acceptable arrangement
Aggregate automatically weighed cumulatively. Cement weighed separately and discharge controlled so that cement is flowing while aggregate is being delivered.

Figure 2-2 Correct and incorrect methods of batching (Courtesy of American Concrete Institute)

used, in order to prevent contamination. Cement and pozzolan are usually stored in high silos at the batch plant from which they can be batched by gravity assisted as necessary by a screw conveyor, air slide, or rotary feeder. Storage of water is necessary only when temperature control of the concrete requires either heating or cooling of the water. Liquid admixtures should be stored in watertight drums or tanks protected from freezing. Most liquid admixtures, especially those mixed in powder form with water at the plant, require agitation in storage to prevent settlement of the ingredients.

Cement, pozzolans, aggregate, and powdered admixtures should be measured by weight. Water and liquid admixtures may be measured by weight or volume. The accuracy of batching is usually controlled by the specifications which apply to the work. There are three components of accuracy to be considered: inherent accuracy of the scales; accuracy of cut-off of the flow of material into a weighing hopper; and the effect on accuracy of individual components when cumulative weighing is permitted. Cumulative weighing, a common feature of one-stop plants, is a procedure whereby materials are batched sequentially into a common weighing hopper. While it is possible to control the accuracy of the amount of total material in the hopper after any weighing operation, the accuracy for any given component is a function of the batching errors both for the material already in the hopper and the batching of the material in question, and also a function of the ratio of its own weight to the amount of material already in the hopper before it is batched. The first material batched is the most accurately batched. Therefore, when specifications permit the weighing of cement cumulatively with aggregates, it should be required that the cement be weighed first. Then preferably each aggregate should be weighed in ascending order of weight. Water should not be batched cumulatively. All components are batched individually in high-production plants. This procedure permits accurate control of the weight of each material.

Another distinction among batching systems must be made. They may be manual, semi-automatic, or automatic. In a manual plant, an operator opens a gate at the bottom of each material storage hopper, permittting the material to flow into the weighing hopper, and shuts off the flow of material when the desired weight of material is in the hopper. A semi-automatic system is similar except that the flow is stopped automatically when a sensing device in the weighing system detects that the desired weight has been achieved. It is necessary to set the desired weights for each ingredient in a control unit before weighing begins. The system is adaptable to a situation in which repetitive batches require identical weights, but it is a time-consuming system when weights must be changed frequently. In an automatic system, a selector switch establishes required batch weights from any of several pre-selected mixture proportions, and the entire batching sequence is activated by a single switch. Usually, such systems contain interlocks which prevent batching when the weighing hoppers fail to empty from the preceding batching cycle and prevent discharging when batching of any ingredient is not within the prescribed tolerance. When a batch plant is established for a single project, it is permissible to use a fairly simple selector system with, perhaps, half a dozen sets of mixture proportions in the system. In a commercial ready-mixed concrete plant, however, where literally several hundred sets of mixture proportions might be required, a punched-card system provides unlimited flexibility.

Based on the above information, it is now possible to establish weighing and batching tolerances. In the USA, scales are generally required to meet the requirements of the National Bureau of Standards *Handbook 44* [3] or variations of them. By these requirements, a weighing error may never exceed 0.2% of scale capacity. Compliance is checked by placing test weights on the scales. Batching tolerances most widely stipulated in the USA are those of the Concrete Plant Manufactures Bureau [4]. They require the following:

(a) *Cumulative batchers*

Cement, pozzolans, and aggregates: ±1% of the required cumulative weight of material being weighed, or ±0.3% of scale capacity, whichever is greater.

Admixtures: ±3% of the required cumulative weight of material being weighed, or ±0.3% of scale capacity, whichever is greater.

(b) *Individual batchers*

Cement and pozzolans: ±1% of the required weight of material being weighed, or ±0.3% of scale capacity, whichever is greater.

Aggregates: ±2% of the required weight of material being weighed, or ±0.3% of scale capacity, which is greater.

Water: ±1% of the required weight of material being weighed, or ±0.3% of scale capacity, whichever is greater.

Admixtures: ±3% of the required weight of material being weighed, or ±0.3% of scale capacity.

On important work, a graphical or digital recorder may be required to produce a permanent record of each ingredient batched. A commonly used specification [5] permits a recorder tolerance of ±2% of batcher capacity.

The aspect of batch plant operation which is the leading cause of non-uniformity of concrete is inadequate compensation for moisture in aggregates. The water/cement ratio of concrete, the most important parameter controlling its properties, is based on added mixing water and on surface water carried by the aggregates, but not on water absorbed by the aggregates. Surface moisture must be accounted for by reducing the batched water by the amount of the surface moisture. If any fraction of the aggregate does not contain all the absorbed water it is capable of holding, the batched water must be increased to cover the deficiency. While handling the aggregates in such a manner so as to achieve a nearly uniform moisture content reduces the size of the problem, it is still necessary to keep track of the moisture content of each fraction of aggregate, with emphasis on fine aggregate and the smallest size of coarse aggregate, and to adjust batch weights accordingly. The problem is especially serious during rainy weather and in situations in which steam is added to aggregate bins during cold weather to prevent freezing of the concrete. The moisture content of samples removed from the bins may be determined from either of two ASTM test methods, C70–79 and C566 [7]. In addition, it is common to have an automatic sensing device near the bottom of the bin to give an instantaneous estimate of the moisture content of the material being batched. Most devices are based on electrical resistance, but measurement of dielectric properties and neutron absorption also can estimate moisture content. Even after calibration, these instruments usually give only a rough approximation of the actual moisture content, but they are useful in indicating when an abrupt change has taken place. In some highly sophisticated plants, batch weights are automatically adjusted for moisture content. A more common approach is a dial on the control panel on which the operator enters the moisture content as a result of which both water and aggregate weights are adjusted. If the plant contains no automatic adjustment features, the operator must compute the necessary adjustments. For example, if there is an increase of 2% in surface moisture in a fine aggregate for which the nominal batch weight is 1000 kg (2200 lb), the operator must increase the batch weight of fine aggregate by 20 kg (44 lb) and decrease the batch weight of water a like amount.

2.3 Mixers

Concrete may be central mixed, shrink mixed or truck mixed. If it is to be central mixed, a stationary mixer is included in or adjacent to the batch plant. If the plant is on-site, mixed concrete is delivered directly to the forms by one of the methods described in Section 3. If the plant is off-site, the mixed concrete is delivered to the construction project in an agitator truck or in a specially-shaped non-agitating dump truck. Shrink mixing consists of mixing in a central stationary mixer until the ingredients are intermingled. The partially mixed concrete is discharged into a ready-mix truck where mixing is completed. For truck-mixed concrete, there is no mixer at the batch plant. The ingredients are placed directly in the truck-mounted mixer, and all mixing takes place there.

Stationary mixers may be horizontal axis, vertical shaft or tilting. The form of the mixer is not important so long as the product complies with reasonable mixer performance standards. The ASTM standard C94–78a, *Specification for ready-mixed concrete* [7], requires that samples taken from the beginning and end of discharge shall yield results such that, for five of the required six tests, the differences in test results do not exceed the following values:

Weight per cubic metre (cubic foot) calculated to an air-free basis	16 kg/m^3 (1.0 lb/ft^3)
Air content, volume per cent concrete	1.0
Slump	
If average slump is 100 mm (4 in) or less	25 mm (1.0 in)
If average slump is 100–150 mm (4–6 in)	38 mm (1.5 in)
Coarse aggregate content, portion by weight of each sample retained on a 4.75 mm (No.4) sieve, per cent	6.0
Unit weight of air-free mortar, per cent	1.6
Average compressive strength, per cent	7.5

Shrink mixed concrete is required to be truck mixed until it can meet the above uniformity requirements. Truck-mixed concrete must be mixed 70 to 100 revolutions at the mixing speed stipulated by the manufacturer and then must meet the above criteria. If it does not, the mixer may be rejected, or, at the option of the purchaser may be used if a reduced load, more efficient charging sequence, or a longer mixing time will permit the requirements to be met.

3 Transportation of concrete

There are many ways to move concrete from the mixer to the forms. The choice usually depends on the quantity of concrete to be moved, the topography of the

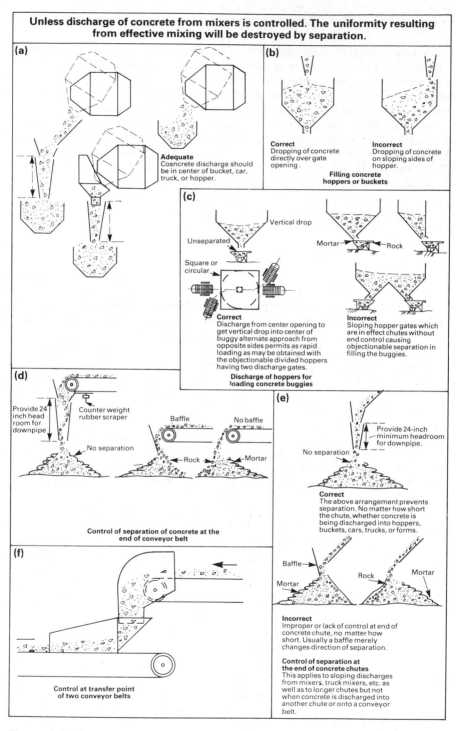

Figure 3-1 Correct and incorrect methods of handling concrete (Courtesy of American Concrete Institute)

construction site, the proximity of the mixer to the forms, and contractor preference. The objective is to preserve the concrete in a well-mixed condition and to protect it from adverse effects of weather. On small projects, a ready-mix truck can get close enough to all parts of the job to permit placing all the concrete from the truck chute. In fact, transit mixers are designed with a high discharge to facilitate this operation. But on almost all projects some transportation device must be used between the mixer and forms. It is important that the transporting device be capable of handling concrete of the workability required for placing. It should not be necessary to provide extra workability to provide for the requirements of transportation. As in handling aggregate, concrete should be dropped vertically when being passed from one vessel to another. Any horizontal component of motion tends to produce segregation. Any wet-batch hopper through which concrete passes should have a bottom discharge and should conform to the shape requirements for buckets given Section 3.1. General principles for handling concrete are illustrated in Fig. 3-1.

3.1 Buckets and cranes

A convenient method for delivering concrete in fairly low structures with unrestricted access to the forms is to drop concrete from the mixer into a concrete bucket and to transport the bucket directly to the form by crane. The bucket should have steep side slopes and a discharge gate with a clear opening equal to no less than one-third the maximum interior horizontal area, and the minimum dimension of the clear gate opening should be five times the nominal maximum aggregate size. For low slumps, the side slopes should be not less than 70° from the horizontal. For medium slumps, the angle may be as low as 58° [8]. The discharge gates should be controlled so that they may be opened or closed at any time during the discharge. Lay-down buckets are common in building construction and are satisfactory provided they meet the above requirements when they are hoisted to the vertical position. In addition to direct placement of concrete, buckets are frequently used to transport concrete to a hopper from which it is placed by some other method.

3.2 Buggies

When concrete has reached the level of the placement, a convenient way to distribute it is by buggies, which may be either hand-powered or motorized. Buggies are normally filled from a bottom-dump hopper which receives concrete from a crane or elevator. Smooth rigid runways should be provided for the buggies. Operating them on an uneven or bumpy surface will produce vertical segregation of the concrete. Even on a smooth surface, the maximum recommended delivery distance is 60 m (200 ft) for hand buggies and 300 m (1000 ft) for motorized buggies. This method of delivery is well adapted to construction in which there is a need for concrete over a large horizontal area, such as the floors, beams, and columns in a building.

3.3 Belt conveyors

The belt conveyor is a simple method for moving large quantities of concrete from mixer to form. It is also an excellent means for segregating the concrete if certain precautions are not taken. Problems associated with belts are

(a) Too little tension in belt with a consequent segregation as the concrete is jarred in passing over each idler.
(b) Too steep a slope so that coarse aggregate tends to separate from the concrete and roll down the slope.
(c) Too slow a belt speed with the result that the large exposed concrete surface on the belt is vulnerable to evaporation loss because of the long period of exposure.
(d) Segregation produced by concrete moving off the end of the belt with a large horizontal component of velocity.
(e) Loss of mortar as concrete is discharged from the belt.

Accordingly, a conveyor belt system is not satisfactory unless it includes high-speed belts with adjustable tension, with effective discharge hoppers at the end of each belt, and with an effective mortar scraper at the discharge end arranged so that all the mortar goes into the discharge hopper. Well-designed systems have belt speeds of about 150 m/min (500 ft/min) and easily adjusted belt tension. They should be arranged so that nowhere in the system is there a slope which produces segregation, and provisions should be available to cover the belts if they are to be used in rain or on long runs during daylight hours in an arid climate. There are self-contained truck-mounted belt systems with a high degree of flexibility. A more common approach is the use of several portable units in series. At the final transfer point, a unit may be used which is free to translate and rotate so that concrete may be deposited anywhere within the physical range of the belt. Belt systems are satisfactory for distances of up to 450 m (1500 ft).

3.4 Pumping

The past generation has seen a spectacular increase in the transportation of concrete by pumping. While pumping has been in use in heavy construction, primarily in tunnel lining, for 50 years, only in recent years has lightweight versatile equipment been available which has made it attractive for structural concrete. There are three generic types of pump: piston, squeeze, and pneumatic. They are illustrated schematically in Fig. 3-2. The piston pump is the most common and is capable of the greatest capacity. Synchronized inlet and outlet valves are arranged so that, on the backward stroke of the piston, the inlet valve opens and concrete is drawn into the cylinder from the hopper, while on the forward stroke the concrete is forced into the pipeline through the open outlet valve. Commonly, there are two pistons operating alternately to produce a more nearly steady flow of concrete. The squeeze pump forces the concrete through a flexible hose by means of hydraulically powered rollers that press against the hose. In the pneumatic pump, concrete and compressed air are fed alternately into the feeding hopper. The air forces the concrete through the line to the discharge box, which bleeds off the air and remixes the concrete as it is discharged. This type of pump is most likely to segregate the concrete.

Pipelines consist either of rigid pipe or heavy-duty flexible hose. Rigid pipe is preferred because it offers less resistance to the flow of concrete. However, in truck-mounted self-contained assemblies, flexible units joining rigid pipes are

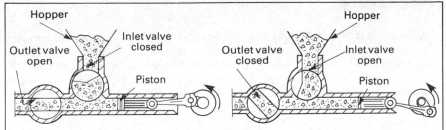

Inlet valve opens while outlet valve is closed and concrete is drawn into cylinder by gravity and piston suction. As piston moves forward inlet valve closes, outlet valve opens, and concrete is pushed into pump line.

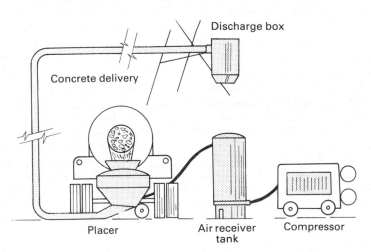

Compressor builds up air pressure in tank, which forces concrete in placer through the line.

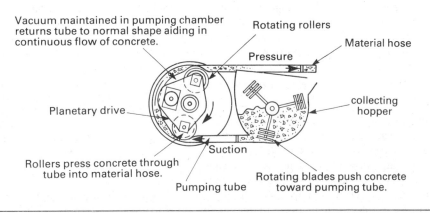

Figure 3-2   Schematic drawings of concrete pumps (Courtesy of American Concrete Institute)

effectively used to give a high degree of versatility to the system. Aluminum pipe should not be used. If the concrete abrades the surface, the abraded particles react with alkalis in the cement to produce hydrogen gas and might increase the air content of the concrete to undesirably high levels. Pipes are available in diameters from 75 to 200 mm (3–8 in) and generally in 3 m (10 ft) lengths. The diameter should preferably be at least three times the nominal maximum aggregate size.

It should not be necessary to change mixture proportions to accommodate a pump. However, pumpability is sensitive to fine aggregate grading. If a fine aggregate is deficient in fines or has a particularly poor particle shape, it might be well to blend it with a fine natural sand, especially when small-diameter lines are to be used. For small lines, the fine aggregate should have 15 to 30% passing the 0.3 mm (No. 50) sieve and 5 to 10% passing the 0.15 mm (No. 100) sieve. Another problem arises from pumping lightweight aggregate concrete. If the aggregate particles are not saturated when they enter the pump, the pump pressure will drive water into the particles with an apparent reduction in mixing water and loss of slump. In some cases, the pipe may become plugged. In the state of California, most lightweight aggregates are either vacuum saturated or thermally saturated prior to use. These procedures solve the pumpability problem, but they reduce resistance to freezing and thawing until the concrete has had several weeks to dry out. If concrete is to be exposed to freezing without an opportunity to dry first, these procedures should not be used. Most lightweight aggregates may be rendered pumpable by two or three days of sprinkling of storage piles.

Concrete moves through the lines as a cylindrical plug separated from the pipe wall by a thin lubricating film of mortar or grout. As pumping starts, excess mortar should be placed in the line. This can be accomplished by omitting or reducing the coarse aggregate in the first batch of concrete.

Pumping capacities and distances vary widely as a function of type of equipment and diameter of line. Many units are now available which, without modification, will pump concrete through a 100 mm (4 in) line at a rate of 60 m³/h (78 yd³/h) for a horizontal distance of 240 m (800 ft), or a vertical distance of 60 m (200 ft), or any combination of the two. Specially modified units have pumped concrete to a height of 300 m (1000 ft) above the pump.

4 Placing and consolidation

4.1 Lowering of concrete in forms

Getting concrete into the forms is a simple operation if adequate advance preparation has been made, but it is probably the most critical part of the entire construction operation so far as the integrity and appearance of the final structure are concerned. Honeycomb (rock pockets adjacent to a formed surface), sand streaking, unsightly cold joints, excessive air bubbles in a finished wall surface, and diagonal cracks emanating from the top corners of window or door openings are all the result of improper placing practices. Concrete should be lowered vertically into its final position in the forms. Horizontal movement, either by allowing concrete to fall in an uncontrolled manner or by movement of concrete over excessive distances in the forms by vibrators, runs the risk of segregation and its resultant honeycomb. Vibrators are for consolidating concrete, not for moving it. In wall forms, drop chutes, attached to hoppers at the top, should be used to lower concrete to the bottom of the forms. Such a procedure prevents the concrete from being segregated by the reinforcing steel as it travels down through the form. Thus, splashing of mortar both on the steel and the form is prevented. The drop chutes should be spaced no further that 5 m (17 ft) apart. If concrete must be placed under windows, and this spacing cannot be otherwise maintained, it should be placed through openings in the window form to insure complete filling under the window. The concrete surface should be raised about 0.5 m (20 in) at a time, and the level throughout the entire wall section being placed should be kept approximately horizontal. It is advisable to place construction joints at the level of the top of window and door openings. When this procedure cannot be followed, concrete should be placed up to the top of the openings and allowed to rest until initial settlement has taken place. The time depends on weather conditions, but usually 2 h is adequate. Placing should be resumed while it is still possible for a vibrator to penetrate the original concrete. Such a procedure prevents the diagonal cracks which might otherwise emanate upward from the corners of the openings. If slabs are placed by buggies, each buggy should deposit concrete toward previously placed concrete and not away from it.

4.2 Vibration

As concrete is deposited in the form, it usually contains some quantity of entrapped air. If it is allowed to harden in this condition, it will be unsightly, weak, and poorly bonded to the reinforcement. The usual procedure for removing the air is the insertion of internal vibrators. In walls, vibrators should be inserted vertically through the concrete just placed and a few centimeters into the previously placed layer. The vibrator should be kept in place until air bubbles cease escaping from the surface. The distance between insertions should be about $1\frac{1}{2}$ times the radius of the area visibly affected by vibration. In thin slabs, the vibrator should be sloped as necessary to embed it completely in the concrete. The physical characteristics of the vibrator, such as head diameter, frequency, and amplitude of vibration, depend on the characteristics of the concrete to be vibrated. Table 4-1, taken from [9], contains the detailed information needed for vibrator selection. That reference also contains much supplementary information on consolidation of concrete. Where a superplasticizer is used to produce 'flowing concrete', very little vibration may be required. However, it should not be assumed that concrete will completely fill beneath openings without some vibration.

4.3 Exposed surfaces

The quality of exposed surfaces is a function of the form material and consolidation technique. Forms should be of quality material, should create the desired architectural effect, and should be mortar-tight. Usually, forms are coated with a material which prevents bonding between the form and the concrete surface and which extends the number of uses of a given form panel. Although tight forms are necessary for a good appearance, they make it more difficult to eliminate the

Table 4-1 Range of characteristics, performance, and applications of internal vibrators (Courtesy of American Concrete Institute)

Column	1	2	3	4	5	6	7	8	9
				Suggested values of			Approximate values of		
	Group	Diameter of head in (cm)	Recommended frequency, min^{-1} (Hz)	Eccentric moment, in lb (cm kg)	Average amplitude in (cm)	Centrifugal force lb (kgf)	Radius of action, in (cm)	Rate of concrete placement yd^3/h (m^3/h)	Application
	1	$\frac{3}{4}$–$1\frac{1}{2}$ (2–4)	10 000–15 000 (170–250)	0.03–0.10 (0.035–0.12)	0.015–0.03 (0.04–0.08)	100–400 (45–180)	3–6 (8–15)	1–5 (0.8–4)	Plastic and flowing concrete in very thin members and confined places. May be used to supplement larger vibrators, especially in prestressed work where cables and ducts cause congestion in forms. Also used for fabricating laboratory test specimens.
	2	$1\frac{1}{4}$–$2\frac{1}{2}$ (3–6)	9000–13 500 (150–225)	0.08–0.25 (0.09–0.29)	0.02–0.04 (0.05–0.10)	300–900 (140–400)	5–10 (13–25)	3–10 (2.3–8)	Plastic concrete in thin walls, columns, beams, precast piles, thin slabs, and along construction joints. May be used to supplement larger vibrators in confined areas.
	3	2–$3\frac{1}{2}$ (5–9)	8000–12 000 (130–200)	0.20–0.70 (0.23–0.81)	0.025–0.05 (0.06–0.13)	700–2000 (320–900)	7–14 (18–36)	6–20 (4.6–15)	Stiff plastic concrete less than 3 in (8 cm) slump in general construction such as walls, columns, beams, prestressed piles, and heavy slabs. Auxiliary vibration adjacent to forms of mass concrete and pavements may be gang mounted to provide full width internal vibration of pavement slabs.

4	3–6 (8–15)	7000–10 500 (120–180)	0.70–2.5 (0.81–2.9)	0.03–0.06 (0.08–0.15)	1500–4000 (680–1800)	12–20 (30–51)	15–40 (11–31)	Mass and structural concrete of 0–2 in (5 cm) slump deposited in quantities up to 4 yd³ (3 m³) in relatively open forms of heavy construction (powerhouses, heavy bridge piers and foundations). Also auxiliary vibration in dam construction near forms and around embedded items and reinforcing steel.
5	5–7 (13–18)	5500–8500 (90–140)	2.25–3.50 (2.6–4.0)	0.04–0.08 (0.10–0.20)	2500–6000 (1100–2700)	16–24 (40–61)	25–50 (19–38)	Mass concrete in gravity dams, large piers, massive walls, etc. Two or more vibrators will be required to operate simultaneously to melt down and consolidate quantities of concrete of 4 yd³ (3 m³) or more deposited at one time in the form.

Notes

Column 3 While vibrator is operating in concrete.

Column 4 Computed by formula in Fig. A.2 in Appendix A.†

Column 5 Computed or measured as described in Section 15.3.2. This is peak amplitude (half the peak-to-peak value), operating in air.

Column 6 Computed by formula in Fig. A.2 in Appendix A, using frequency of vibrator while operating in concrete.

Column 7 Distance over which concrete is fully consolidated.

Column 8 Assumes insertion spacing is $1\frac{1}{2}$ times the radius of action, and that vibrator operates two-thirds of time concrete is being placed.

Columns 7, 8 These ranges reflect not only the capability of the vibrator but also differences in workability of the mix, degree of deaeration desired, and other conditions experienced in construction.

† The references in these notes are to the original source [9].

surface voids, frequently called 'bug holes', from the surface. Form coatings of high viscosity also tend to hold voids at the surface. To eliminate bug holes, the distance between vibrator insertions should be reduced, and the time of each insertion should be increased. There should be a row of insertions as close to the form as possible without touching it. If the vibrator touches a wood form panel, it may mar the surface and disfigure the concrete. Where more than usual problems in eliminating voids exist, hand spading between the form and concrete can be helpful.

4.4 Special placing methods

4.4.1 *Slip forming*
Vertical slip forming has not been commonly used in the past for buildings. Its use has largely been confined to silos, grain elevators and chimneys. It has now, however, become fairly common to slip form the cores of high-rise buildings. The core normally contains elevators, stairways, and shafts for electrical and mechanical systems. Occasionally, all the walls of buildings are slip formed. The system consists of inner and outer forms which are raised by jacks as placing progresses. The concrete emerging from the bottom of this extrusion process must be strong enough and sufficiently consolidated to support the structure above. Forms for building construction are usually 1.2–1.5 m (4–5 ft) high with the outside about 15 cm (6 in) higher than the inside to prevent concrete from falling from the form. Jacks, or yokes attached to jacks, are spaced at about 2 m (6.6 ft) centers. Jacks may be hydraulic, pneumatic, electric, or screw. Concrete having a slump of about 7 cm (2.8 in) is placed in layers 15–20 cm (6–8 in) thick. It should be vibrated, but the vibrator should barely penetrate previously placed concrete. Rate of slipping depends on the rate of setting of the concrete, which in turn is a function of weather conditions, but a rate usually attainable is 30 cm/h (12 in/h). Where concrete is to be exposed, the surface, which is still amenable to a rubber-float finish as it emerges from the bottom of the form, may be finished from a platform suspended from the form. Or the surface may be left until slipping is completed, when a sack-rubbed or cement plaster finish may be applied. A complication in all building construction is the necessity to incorporate openings in the walls. Framed openings may be inserted as slipping progresses, or an oversized blockout may be left in the wall during slipping, after which the actual openings may be accurately positioned and concreted or grouted into place. A complete discussion of the subject is given in [10].

4.4.2 *Underwater placing*
Concrete may be safely placed under water if precautions are taken to prevent the the intrusion of water into the concrete. Covered buckets are available for the purpose, but by far the most positive method for maintaining the integrity of the concrete is the tremie method. Pumping may also be used if positive means are assured to keep the delivery end of the pipe embedded in concrete as the pipe is raised. A tremie consists of a pipe, 25–30 cm (10–12 in) in diameter, with a hopper at the top. The entire assembly is supported by a cable so that it can be raised as the concrete level rises. In large placements, tremies should be spaced not more than 10 m (33 ft) apart. The bottom of the tremie should be buried in

1.5 m (5 ft) of concrete at all times. Concrete is started by placing a tight-fitting plug or 'go-devil' at the top of the pipe and forcing it down to the bottom and out into the water by fresh concrete dumped into the hopper. The concrete should be very workable, with a slump of 15 cm (6 in) and a relatively high proportion of fine aggregate.

5 Curing and stripping

5.1 Temperature control

Cement will hydrate and cause concrete to gain strength at any temperature above −10°C (14°F). As a first approximation, it may be assumed that during the early life of concrete there is a unique relationship between strength and maturity where maturity is defined as the integrated product of time and temperature above −10°C (14°F). Thus,

$$M = \sum (C + 10)\Delta t$$

where M is the maturity in °C h, C is the temperature in °C, and Δt is the duration (in hours) of curing at temperature C.

The temperature at which concrete is maintained depends on the age at which particular strength levels are required. It should be kept above 0°C (32°F) to prevent damage by the freezing of water, and for most structural projects progress will be unduly delayed if the temperature does not exceed 10°C (50°F). Temperature control is achieved by insulation, which conserves the heat of hydration, or by enclosing the structure and applying heat. Large temperature gradients should be avoided. This problem is particularly acute when forms are stripped or protection is removed. The removal should be accomplished in such a manner that the temperature gradient at the surface does not exceed 2.8°C/cm (12.5°F/in). The gradient may be determined in typical sections by measuring the temperature at the surface and at a location 5 cm (2 in) inside the surface. During hot weather, this requirement is most easily satisfied if the placing temperature of the concrete is maintained as low as practical.

5.2 Criteria for form stripping

Forms of structural elements such as walls, columns, and sides of beams, may be removed as soon as the stripping operation will not mar the concrete. For normally-cured concrete, stripping at one day is common. Steam-cured concrete may be stripped in a few hours. Forms of the soffits of arches, beams, girders, and floor slabs must be left in place until sufficient strength has been developed to prevent failure or excess deflection. The designer of the structure should stipulate the strength required for stripping. Normally, a strength sufficient to carry the dead weight with a factor of safety of 2 is adequate. There are several ways to determine when this strength has been obtained. Specifications frequently require strength-test specimens to be cured under conditions identical to those of the prototype. The objection to this method is that it is a practical impossibility to achieve identical conditions. An improved approach is to use one or more of the non-destructive tests, discussed in Section 7.2, suitably calibrated to standard test

Table 5-1   Duration of protection for percentage of design strength required

Percentage of design strength required	At 10°C (50°F) days			At 21°C (70°F) days		
	Ordinary Portland cement	Modified low-heat cement	High-early strength cement	Ordinary Portland cement	Modified low-heat cement	High-early strength cement
50	6	9	3	4	6	3
65	11	14	5	8	10	4
85	21	28	16	16	18	12
95	29	35	26	23	24	20

specimens. If the maturity function of a given concrete has been obtained and plotted, it is possible to estimate the strength in the structure by maintaining an internal temperature history. There are, in fact, maturity meters available which provide a read-out directly in degree-hours. Many specifications implicitly use the maturity concept by requiring a given number of days at a given temperature. Table 5-1, taken from [11] provides information for such a specification.

5.3   Types of curing

When the relative humidity falls below 90%, the hydration of cement ceases or proceeds at a very low rate. The objective of curing is to maintain the internal relative humidity of concrete well above 90%. This may be done by providing a source of water at the surface or by preventing the evaporation of water from the concrete. If no positive curing measures are applied and the ambient relative humidity is below 90%, hydration at the surface will cease and the zone of ceased hydration will slowly penetrate into the mass. Since this penetration is slow, the concrete in the inerior of a thick section may well develop sufficient strength to safely carry the design load, but the concrete at the surface will be non-durable and non-resistant to abrasion.

5.3.1   Moist-curing

The most effective means of curing is that in which water is continuously added to the surface. Since concrete tends to dessicate itself internally as the cement hydrates, moist-curing replaces the needed water to keep the relative humidity near 100%. Thus, the concrete in the structure is in substantially the same state as the strength-test specimens used to judge the adequacy of the concrete. Moist curing is accomplished by ponding, covering horizontal surfaces with damp earth, sand, straw or sawdust, sprinkling, or covering with saturated burlap. A disadvantage of moist-curing is that it requires frequent attention by workmen and inspectors to insure that intermittent drying is not taking place. Another disadvantage where architectural appearance is important is the appearance of iron staining from reinforcing bars which are partially exposed during construction.

5.3.2   Membrane curing

There are two general methods for preventing the evaporation of water from concrete without adding water. They are the enveloping of the concrete in waterproof paper or plastic and the forming of a water-retaining membrane on the surface of the concrete by spraying the surface with a liquid membrane-forming compound. These have the advantage of requiring very little inspection after they are placed, but they permit the relative humidity within the concrete to drop and, thus, are not as efficient as moist-curing. When sheet material is used, it is important to insure that joints between adjacent sheets are well secured so that construction traffic or wind will not expose portions of the concrete. The sheets should be inspected periodically to determine whether any sheets have become torn. Since the liquid membrane-forming compounds were developed originally for use in arid regions where water is available only at great expense, they commonly contain a white pigment which, by virtue of its heat reflecting ability, prevents excessive temperature rise as well as evaporation. The pigment also is a very convenient inspection tool in that it is easy to determine by sight what areas have been covered and the uniformity of application. Pigmented compounds, however, are not acceptable where architectural appearance is important. For such uses, unpigmented compounds are available. They can be specified to contain fugitive dyes which color the concrete as they are applied but disappear in a few days. This feature facilitates inspection of the application. Usually, membranes produce some discoloration of the concrete surface. On important architectural work, they should be tested on a sample panel before being approved. If a satisfactory material cannot be found, the concrete should be moist-cured or sealed in sheets of plastic or waterproof paper. Curing compound should be applied by power sprayer to formed surfaces immediately after forms are removed and to unformed surfaces as soon as the free water has disappeared. The normal application rate is 5 m²/l (200 ft²/US gal).

5.3.3 Steam-curing

Atmospheric steam-curing is a technique to increase dramatically the strength development during the first day. It is a direct application of the maturity concept and combines elevated temperatures with favorable curing conditions. It is primarily used in the precast industry to provide a quick reuse of casting beds and forms, but it is also possible to enclose portions of cast-in-place structures and to introduce steam into the inclosure. The high early strength is usually achieved at the expense of later-age strength.

For most cements the optimum temperatures for steam-curing is in the range 65–80°C (150–175°F). Strength is usually improved if the application of steam is delayed for a period of 4–6 h after casting. The rate of temperature rise has been found to be critical. If no presteaming period is used, the rate of temperature rise should not exceed 11°C/h (20°F/h). If there is a presteaming period, the rate may be as high as 33°C/h (60°F/h). After steaming, the rate of temperature drop should not exceed 33°C/h (60°F/h). The length of time the concrete should be held at the maximum temperature depends on the degree of maturity desired. Usually, it is possible to work out a complete cycle which does not exceed 24 h.

It is also possible to steam cure at temperatures higher than 100°C (212°F). Since this procedure requires an autoclave to withstand pressures well above atmospheric, its use is confined to masonry blocks and small precast units. Hydration products formed in this environment are somewhat more crystalline than the colloidal products formed at atmospheric pressure. Thus, the autoclaved product is considerably more volume-stable than its non-autoclaved counterpart, although it is not as strong. Autoclaved products are also more sulphate resistant. Steam-curing is dealt with more extensively in [12].

6 Quality control

There is some confusion as to the precise meaning of terms such as 'quality control' and 'quality assurance'. In the most general sense, quality assurance includes everything done to assure the owner of a structure that the concrete in the structure complies with the plans and specifications. Quality control is that portion of the system whereby the constructor controls his or her operations so that the concrete may be expected to be satisfactory. The remainder of the system consists of inspections and acceptance tests performed by someone responsible to the owner to determine whether the concrete is in fact acceptable. Hence, tests on the fresh concrete may be made both by contractor quality control personnel and by the owner's inspectors. Strength tests should normally be made only by the owner's representative. While they are excellent for acceptance purposes, the information is obtained too late to be of use in quality control. Regardless of how responsibilities are distributed, quality is controlled and maintained partly by inspection of construction operations and partly by selecting pertinent tests, performing them on random samples in accordance with a coherent sampling plan, and analyzing the results to determine whether the concrete in the structure is significantly different from that contemplated in the design. It is the latter component of the system with which this section is concerned.

6.1 Tests for evaluating fresh concrete

6.1.1 *Workability*
The property of concrete of greatest interest to those handling it in construction is its workability. Hence, the most popular test for fresh concrete, and on many projects the only test, is a workability test. The specifications normally provide a limiting value for workability. The value should be set to accommodate the most difficult placing conditions anticipated. In US practice, the slump test, as described in ASTM C143-78 [7], is universally used as a field measure of workability. In British practice, both the slump test and the compacting factor test, as described in BS 1881, Part 2 [13] are used. The British slump test is similar but not identical to the US test. Some specifications limit the variability of slump. The specifications in [5], for example, stipulate that the slump shall not depart more than 38 mm ($1\frac{1}{2}$ in) from the target value. The control of concrete by slump is rendered difficult when a superplasticizer is used to produce 'flowing' concrete. For such concrete control, workability is effected primarily by control of the batch proportions.

6.1.2 *Air content*
Entrained air is required in concrete which will be exposed to freezing in a saturated condition and in mass concrete to impart workability. It is used primarily in highways and runways rather than in vertical structures, but it is not

uncommon in buildings in the northern part of the USA. Specifications, where it is used, usually require about 6% air (by volume). It is measured in the field by (a) comparing the unit weight of concrete with the theoretical unit weight of air-free concrete (ASTM C138-81 and BS 1881, Part 2); (b) manipulating the air out of the concrete by rolling it in water and noting the decrease in volume (ASTM C173-78); and (c) applying pressure to the concrete, noting its decrease in volume, and determining through suitable calibration the original air content by an application of Boyle's law (ASTM C231-78 and BS 1881, Part 2). Except for lightweight concrete the third method is by far the most widely used. Lightweight concrete must be tested by one of the other methods. The specifications in reference [5] restrict variation in air content to ±1.5% from the stipulated value.

6.1.3 *Rapid analysis of ingredients*

A long-time dream of concrete technologists has been a test which would quantitatively analyze the ingredients of concrete within a few minutes after sampling. Of particular importance is the determination of the quantities of cement and water so that an assessment of water/cement ratio might be made. Many tests have been proposed. There are two which enjoy significant use at present.

The first in the Rapid Analysis Machine (RAM) developed by the Cement and Concrete Association in Great Britain [14]. A concrete sample of between 5 and 8 kg (11 and 18 lb) is placed in an elutriation column where rising water separates the ingredients and carries off the fine material in a slurry. Precisely one-tenth of the slurry is automatically sampled and directed through a vibrating 150 μm (No. 100) sieve into a conical vessel where it is stirred and dosed with a flocculating agent. The supernatant liquid is removed, leaving a fixed volume of solid and liquid whose weight is proportional to the weight of solid it contains. Assuming that all the solid is cement, the quantity of cement expressed as a fraction of weight of concrete can easily be obtained. When aggregates contain a significant amount of material passing the 150 μm (No 100) sieve, a correction can be made in the apparent cement content. A test can be completed in 10 min. A recent development provides a capability for determining water content by a procedure similar to that described below.

A procedure for determining both cement and water contents was proposed by Kelly and Vail in 1968 [15] and refined for field use at the Construction Engineering Research Laboratory in Illinois [16]. Water content is determined by adding 500 ml of 0.5 N sodium chloride solution to 1 kg (2.2 lb) sample of concrete and 500 ml of distilled water to another. The samples are agitated and 50 ml of liquid pipetted from each. The sodium chloride concentration of each liquid sample is determined by titration. The difference in concentration is a measure of the original water content. Cement is determined by breaking down a 1 kg (2.2 lb) sample of concrete in a known amount of water in an agitating washing machine and withdrawing a 125 ml sample of cement–water slurry. The calcium content of the slurry is determined by titration or flame photometer. If the calcium content of the cement is known, the results may be related to the cement content of the concrete. The method loses accuracy when the aggregate contains calcium. A calibration for the aggregate effect must be included. Both water and cement determinations can be made in 15 min.

6.2 Tests for evaluating hardened concrete

Hardened structural concrete is evaluated almost exclusively by testing for compressive strength. Although the modulus of elasticity and creep coefficient may be important design parameters in some structures and extensive testing to determine these properties may be carried out as part of the design process, such tests are rarely required as a part of the quality control or acceptance regime. Likewise, although most structures contain many flexural members, design is based on compressive strength; therefore, flexural strength tests are not normally required. In US practice, compressive strength is normally determined by breaking a 150×300 mm (6×12 in) cylinder in accordance with ASTM C39-72 (1979) [7]. British practice calls for a 150 mm (6 in) cube tested at right angles to the position as cast and in accordance with BS 1881, Part 3 [13]. The normal test age for determining compliance with design strength is 28 days, although in some areas where pozzolan is included in high-strength concrete, the test age is 56 days.

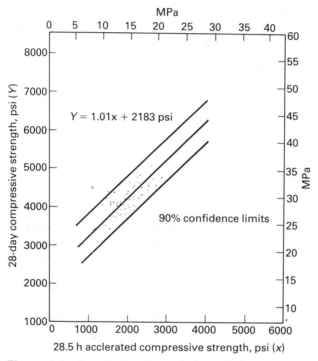

Figure 6-1 Correlation between accelerated and 28-day strength: number of tests = 76, correlation coefficient = 0.85, standard error of estimate S(yx) = 364 psi (2.5 MPa) (Courtesy of American Concrete Institute)

Cement	CSA Type 10 (ASTM Type I)
Coarse aggregate	Quartzite, limestone, sandstone
Fine aggregate	Cruched sand and natural sand
Admixtures	Air-entraining agent and water reducing, set retarding admixture

The most significant development in recent years has been the development of accelerated strength tests whereby the quality of the concrete may be assessed in one or two days. Three such tests have been standardized in ASTM C684-74 (1979) [7]. In the first, cylinders are covered with a rigid plate and immediately after casting are placed in a water bath at a temperature of 35°C (95°F). At $23\frac{1}{2}$ h, the cylinders are removed, demolded, and capped. They are tested at 24 h. In the second method, the cylinders are cured normally for 23 h. Then they are immersed in boiling water for $3\frac{1}{2}$ h. After at least an hour's cooling, they are capped and subsequently tested at $28\frac{1}{2}$ h. In the third method, the strength of a cylinder is accelerated by its own heat of hydration. Immediately after casting it is placed in an insulated container of stipulated properties and allow to remain until an age of 48 h. It is then removed, capped, and tested at 49 h. All the tests give strengths about half of the 28-day strength as may be expected from maturity considerations. Since usually the 28-day strength is the legally required strength, the principal use of the accelerated strength to date has been as a predictor of 28-day strength. For this purpose, a correlation such as the one shown in Fig. 6-1, taken from [17], has been developed for each concrete of interest. But since accelerated strength has been shown to be as good a predictor of late-age strength as the 28-day strength is [18], and therefore as good an indicator of concrete quality, there is a rationale for accepting the accelerated test result itself as a design and acceptance parameter without reference to the 28-day strength. Such acceptance already exists in Norway [19] and is being seriously considered in the UK and Canada.

6.3   Assessment of variability

6.3.1   *Statistical approaches*
All populations of test results have some inherent variability. An understanding of whether a desired degree of quality is being obtained requires an ability to recognize whether the difference between the average of a set of test results and a stipulated value is a real difference or a chance difference which is an artifact of the variability. The mathematics of statistics is designed to deal with this matter. The most frequent applications of statistics in concrete technology are the formulation of acceptance criteria for strength tests and the computation of a required overdesign to meet those criteria. The codes in most countries are based on such procedures. A good general discussion of the statistical approach is contained in ACI 214-77 *Recommended practice for evaluation of strength test results of concrete* [20]. The strength used in design is a 'characteristic' strength selected so that a fixed percentage of the population of test results may be expected to exceed that value. Most national codes require 95% of the results to exceed the design strength. In North America, the requirement is 90%. In either case, the required overdesign may be easily calculated. It is necessary only to know the standard deviation of the population of concrete strengths of the source from which the concrete will be provided. There has been some argument over the past two decades as to whether concrete strengths from a given source are best characterized by standard deviation or coefficient of variation, which is the standard deviation expressed as a percentage of the mean of the population. The latest edition of ACI 214 recommends that standard deviation be used. Evidence

demonstrates that, in concrete plants producing concrete at different strength levels, standard deviation is a nearly constant parameter. Accordingly, the required average strength may be calculated as follows:

$$f_{cr} = f_c' + t\sigma$$

where:

f_{cr} = required average strength;
f_c' = specified design strength;
t = 1.65, when 5% low strengths are permitted;
   = 1.28, when 10% low strength are permitted;
σ = standard deviation.

This equation is sufficient to define the distribution contemplated in the codes. However, as samples are taken from the distribution, half of them will fail to meet the requirement for mean value. Thus, an additional overdesign is customary in order to meet acceptance criteria with a low probability of rejection.

Statistical procedures are also used for the quality control of aggregate and the properties of fresh concrete. The best known approaches are the control charts discussed in the following sections.

6.3.2 Shewart control charts

Simple control charts, called Shewart charts, are easily constructed to determine whether either the value of a property under consideration or its variability is under control. A property is considered under control when values depart from the mean no more than might be expected by chance. Three horizontal lines are drawn on the chart for the property itself, representing the target value for the property and an upper and lower control limit. The control limits are drawn so that the probability of a value's falling above the upper limit or below the lower is a low value, such as 0.05. The same three lines can be drawn on the chart for standard deviation, but since a standard deviation below the lower control limit is seldom a matter of concern, frequently only the upper control limit is drawn. A lower control limit is useful, however, in indicating when the standard deviation has decreased significantly and needs to be recalculated. It is customary to plot averages of groups of tests rather than individual tests. Erratic individual results might cause an operator to overreact, but when the average of several tests indicates lack of control, it is likely that a real problem is present. Also, it is necessary to base the chart for standard deviation on groups of tests since it is impossible to calculate a standard deviation from a single test result. Normally, standard deviation is estimated from the range of the values under consideration. If the number of values does not exceed 5, the standard deviation calculated as a multiple of the range is quite accurate. Hence, control charts are usually expressed in terms of ranges rather than standard deviations. The parameters in Table 6-1 may be used to calculate control limits. Figure 6-2 is an example of a control chart for the fineness modulus (FM) of fine aggregate. Since values above and below the mean are both matters of concern, control limits have been selected so that the probability of being either above the upper control limit or below the lower control limit is 0.025; that is there is a 0.95 probability of staying between the limits. A single point outside the control limits is a warning that something

Table 6-1   Factors for establishing limits on control charts

| Sample size | Control limits for property being measured | | |
	Upper control limit for range a	One-sided test, probability = 0.95 of not exceeding on limit b	Two-sided test, probability = 0.95 of staying between both limits c
2	2.21	1.16	1.39
3	2.93	0.95	1.13
4	3.32	0.82	0.98
5	3.58	0.73	0.88

Upper control limit for range $= a\sigma$

One sided chart:

Control limit $= \bar{x} + b\sigma$ or

$\bar{x} - b\sigma$

Two sided chart:

Upper control limit $= \bar{x} + c\sigma$

Lower control limit $= \bar{x} - c\sigma$

where σ is the standard deviation and $\bar{x}$ the mean value

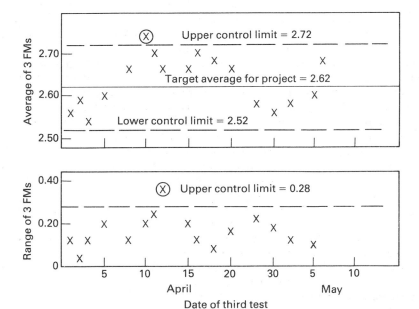

Figure 6-2   Control chart for fineness modulus of fine aggregate

may be adversely affecting the process. Two consecutive points outside the limits provide near certainty that the process is out of control and needs correction. While Shewart control charts have served their purpose very well, they require extreme values in order to be operative and do not indicate when small changes in the mean or standard deviation have occurred. If values are running consistently above or consistently below the target value, it is possible to determine, by appropriate statistical calculations, whether a significant change has occurred in either the mean or standard deviation; but that information cannot be conclusively deduced from the charts. Sometimes, there are important economic implications in being aware of a change as soon as possible. The graphical technique described in the following section accomplishes this purpose.

6.3.3 *Cusum method*

Cumulative sum (cusum) charts are widely used in the UK to analyze strength test results. They are usually based on early-age test results, such as 7 days, in order to provide a control functon as well as an analysis function. They have also been applied to accelerated test results [21]. They may be applied both to test results and to standard deviations. As test results become available, they are tabulated. From each result, the target value is subtracted and the difference tabulated. Starting with the first value, the cumulative sum of the differences is compiled as each test value is added to the table. For examining standard deviation, the range of each pair of consecutive test results is calculated and the target range subtracted from it. The cumulative sum is plotted on the graph rather than the test result itself. If the plotted points produce an essentially horizontal display, there is no significant departure from the target value. If the points form a pattern which consistently rises or falls, there may be a significant departure from the target. To determine whether the departure is significant, a 'V-mask' is placed on the chart, as shown in Fig. 6-3, with the mid-point of the vertical line resting on the most recently plotted point. If all the points fall between the diverging lines of the mask, no significant change has occurred. If the line connecting the points goes outside the action line, a significant change can be presumed to have occurred and to have occurred at about the date where the line-crossing occurred. By maintaining charts for both strength and range, the

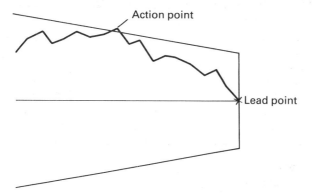

Figure 6-3 Cusum mask

operator can determine whether the change has taken place in the standard deviation or in the population mean. A change in the strength plot without a corresponding change in the range plot suggests that there has been a change in the mean value. An advantage of the system is that precise variability data are not required to start the process. The system is self-correcting as test data are collected. With some loss of accuracy, most concrete plants using the system plot all their production on a single graph by correcting all strength values to theoretical values for a particular cement content through the use of correlation factors compiled within the plant. The V-masks may be constructed as follows.

For the strength mask, the height of the vertical line is 16.2σ (8.1σ on each side of the center line) and the slope of the action line is $\sigma/6$. The horizontal unit used is the distance between consecutive plotted points.

For the range mask, the height of the vertical line is 17.2σ (8.6σ on each side of the center line) and the slope of the action line is $\sigma/10$.

A complete discussion of the method may be found in [22].

7 Evaluation of structures

While concrete delivered to structures is normally accepted on the basis of strength tests and concrete in structures is accepted on the basis of compliance with specifications for curing, protection, and form stripping, sometimes the concrete in structures under construction, and frequently the concrete in older structures in service, must be evaluated in place. The need arises during construction when strength data are missing, strength results are deficient, or the concrete has been mishandled after placing. The evaluation may be by visual inspection, by non-destructive tests, by removing and testing cores from the structure, or by load tests.

7.1 Condition surveys

A detailed check list for the visual inspection of a structure is given in the *Guide for making a condition survey of concrete in service* prepared by ACI Committee 201 [23]. The applicable portions of the check list depend on the type of structure and the nature of the problem. For structural concrete, the following excerpts from that guide are likely to be pertinent:

Overall alignment of structure
   Settlement
   Deflection
   Expansion
   Contraction
Portion showing distress (beams, columns, walls, floors, etc.)
Surface condition of concrete
   Spalls
   Popouts
   Air pockets
   Sand streaks
   Honeycomb

Soft areas
Cracks
    Locations and frequency
    Type and size
    Leaching, stalactites
Extent of corrosion or chemical attack
    Stains
Exposed steel
Drainage
    Flashing
    Weepholes
    Slopes to drains

7.2 Non-destructive tests

There is a number of tests which are non-destructive, or nearly so, and which can be applied to concrete *in situ*. They are used to determine safe stripping time and to provide information useful in the evaluation of questionable concrete. Unfortunately, these tests do not measure compressive strength. But since the test results change as the cement hydrates, as does compressive strength, it is reasonable on a given concrete to expect a correlation between non-destructive test results and strength test results. These tests are used with greatest confidence when a complete suite of correlation tests has been conducted in the laboratory before an attempt is made to interpret field results. When it is necessary to examine questionable concrete in the field in the absence of correlation data, tests may be conducted both on the questionable concrete and on concrete which has been found acceptable. A comparison of results provides some bases for evaluation. Certainly, if the questionable concrete produces as good results as the acceptable concrete, there is little reason to doubt the quality of the questionable concrete. One note of caution is necessary. The test results are all to some degree sensitive to the moisture content of the concrete. The correlation test should include the range of moisture levels likely to be encountered in the field, and field tests in the absence of correlation should be restricted to tests on concrete of similar moisture contents. A complete discussion of tests currently in use in given in [24]. The most common tests are summarized below.

7.2.1 *Pulse velocity*

The test which samples the concrete most comprehensively is the pulse velocity test, which measures the velocity of a mechanical pulse between two hand-held piezoelectric or magneto-strictive transducers placed on the surface of the concrete. An electronic timing circuit measures the time of transit of the pulse. An independent measure is made of the distance between the transducers, and the quotient of the two figures yields velocity. The method is limited only by the range of the instrument and the accessibility to the surface by transducer holders. Usually in structural concrete, neither is serious. While heavy duty instruments are available with a range of 15 m (50 ft) of concrete, the portable units best adapted to structural work have a range of about 2 m (6.6 ft). Best results are obtained when the transducers are on opposite sides of a member, but they may be placed on perpendicular surfaces or even on the same surface. In the latter case, little of the concrete is sampled. Since a pulse will not travel across a crack but may travel around a crack if the distortion of its path is not too great, the technique may be used both to locate and to explore the extent of cracks. It is excellent for assessing the uniformity of concrete. Velocity levels depend somewhat on the nature of aggregate, but most mature concrete in good condition has a velocity in excess of 4200 m/s (14 000 ft/s). In reinforced concrete, the results are not significantly affected by the reinforcing steel unless most of the path between the transducers is occupied by a single bar. The test has been standardized in ASTM C597-71 [7] and BS 4408, Part 5 [25].

7.2.2 Rebound hammer
There are several devices which measure the coefficient of restitution between steel and concrete. Probably the most popular of all non-destructive tests is the Schmidt rebound hammer, a spring-loaded device which drives a steel hammer against a concrete surface with a standard amount of energy and measures the distance of the hammer rebound. The method has been standarized in ASTM C805-79 [7] and BS 4408, Part 4 [25]. They require ten readings in each test area. The rebound is somewhat sensitive to the inertia of the concrete in that thin slabs and massive blocks of equal strength concrete might give slightly different readings, but the method can provide a good assessment of the near-surface layer of concrete.

7.2.3 Penetration resistance
The Windsor probe, which has been standardized in ASTM C803-79 [7], fires a steel projectile into the concrete by a powder charge of carefully controlled energy release. The penetration of the projectile is measured. In practice, the length protruding from the concrete is measured and subtracted from the total length in order to determine penetration. The method receives its greatest use in determining stripping time.

7.2.4 Pull-out test
The full-out test, standardized in ASTM C900-78T [7], actually measures strength. In its normal form, it requires embedment of a metal insert in concrete as the concrete is placed. The insert consists of a cylindrical head and a shaft which extends to the surface. A center-pull jack applies a pull-out force through a rod threaded into the shaft. The reaction to the load is supplied by a concentric metal ring bearing on the concrete surface with an inside diameter larger than that of the insert cylinder. A concrete failure is forced approximately on the surface of the frustum of a cone formed by the insert cylinder, the bearing ring and the failure surface. A recent development has made it possible to install an insert after the concrete has hardened.

7.2.5 Break-off tests
For the break-off method, a tubular disposable form is inserted in the concrete. When the form is removed, an intact core of concrete, still attached at its base, remains. A radial force is applied at the surface to produce flexural failure at the base of the core. The method has not been standardized but it is described in [26].

7.3   Core test

Usually, when a dispute as to the adequacy of strength in a structure arises, cores are drilled and tested. The core has an advantage over non-destructive tests in that it provides a direct measure of compressive strength, the parameter which forms the basis of strength specifications. It has the disadvantage that it seldom is economically feasible to get the quantity of data that is routinely available from non-destructive tests. Furthermore, it is not always possible to obtain cores at random location because of the possibility of damage to the structure. Thus, decisions made on the basis of core test results are likely to be of questionable statistical validity. The strength criterion to be applied to cores is not obvious since the environment of the cored concrete throughout its life is likely to have been somewhat different from that of a standard test cylinder or cube moist-cured at standard temperature. The procedures mandated by ACI Committee 318 *Building code requirements for reinforced concrete* [27] require that three cores be drilled for each strength test which falls more than 3.4 MPa (500 psi) below the design strength. The concrete is accepted if the average of the three is at least 85% of the design strength provided that no single core has a strength less than 75% of the design strength.

7.4   Load tests

When all tests of the concrete itself leave the adequacy of the structures in doubt, the structure may be accepted by the building authority if it passes a load test. Detailed instruction for such a test is given in [27]. When a portion of a structure is to be tested, the portion of the dead load not already acting shall be placed on the structure 48 h before the live load is added. The structure shall be loaded to 85% of the total design dead plus live load in four equal increments applied in such a manner as to avoid shock to the structure and arching of the loading materials. After the test load has been in position for 24 h, initial deflection readings shall be taken, the test load shall immediately be removed and final deflection readings taken 24 h after removal. A flexural member is considered satisfactory if there is no visible evidence of failure and if either of the following conditions is satisfied:

(a) The measured maximum deflection is less than $l^2/20\,000h$.
(b) The measured maximum deflection exceeds $l^2/20\,000h$ but the deflection recovery within 24 h after removal of the test load is at least 75% of the maximum deflection for non-prestressed concrete or 80% for prestressed concrete.

Here, l is the span of the members supported at the ends or twice the span of cantilever members, and h is the overall thickness of the member.

A non-prestressed member failing to show 75% recovery may be retested not earlier than 72 h after removal of the first test load and shall be considered satisfactory if it shows no visible evidence of failure and recovers at least 80% of the maximum deflection in the second test. Prestressed concrete construction shall not be retested.

References

1. ACI Committee 304, Recommended practice for measuring, mixing, transporting and placing concrete (ACI 304-73), *American Concrete Institute manual of concrete practice*, Part 2, 1980, pp. 304–1 to 304–40.
2. Waddell, J. J., *Concrete construction handbook*, 2nd Edn, McGraw-Hill, New York, 1974, Ch. 19.
3. National Bureau of Standards, *Specifications, tolerance, and other technical requirements for commercial weighing and measuring devices*, NBS Handbook 44, 1980, 205 pp.
4. Concrete Plant Manufacturers Bureau, *Concrete plant standards of the CPMB*, 6th Edn., Silver Spring, Maryland, 1977.
5. US Department of the Army, *Civil works construction guide specification: concrete*, CW-03305, Corps of Engineers, Office of the Chief of Engineers, Washington, DC, 1978, 101 pp.
6. Strehlow, R. W, *Concrete plant production*, Concrete Plant Manufacturers Bureau, Silver Spring, Maryland, 1973, 112 pp.
7. American Society for Testing and Materials, *Concrete and mineral aggregates: manual of concrete testing*, Annual Book of ASTM Standards, Part 14, 1981.
8. US Department of the Army, *Civil works construction guide specification: cast-in-place structural concrete*, CW 03301 Corps of Engineers, Office of the Chief of Engineers, Washington, DC, 1978, 65 pp.
9. ACI Committee 309, Recommended practice for consolidation of concrete (ACI 309-72), *American Concrete Institute manual of concrete practice*, Part 2, 1982 pp. 309–1 to 309–40.
10. Waddell, J. J., *Concrete construction handbook*, 2nd Edn., McGraw-Hill, New York, 1974, Ch. 34.
11. ACI Committee 306, Cold weather concreting, *American Concrete Institute manual of concrete practice*, Part 2, 1982, pp. 306R-1 to 306R-22.
12. Mindess, S. and Young, J. F., *Concrete*, Prentice-Hall, Englewood Cliffs, New Jersey, 1981, Ch. 11.
13. British Standards Institution, BS 1881. 1970, *Methods of testing concrete*, London 1970.
14. Forester, J. A., Black, D. F. and Lees, T. P., *An apparatus for the rapid analysis of fresh concrete to determine its cement content*, Cement and Concrete Association Technical Report 42.490, 1974.
15. Kelly, R. T. and Vail, J. W., Rapid analysis of fresh concrete, *Concrete*, Vol. 2, No. 4, April 1968, pp. 140–145.
16. Howdyshell, P. A., *Evaluation of a chemical technique to determine water and cement content of fresh concrete*, Construction Engineering Research Laboratory, Champaign, Ill., Technical Manuscript M119, 1975.
17. American Concrete Institute, *Accelerated strength testing*, SP-56, 1978, 319 pp.
18. Philleo, R. E., Lunatics, liars, and liability, *J. Am. Concr. Inst., Proc.*, Vol. 73, No. 4, Apr. 1976, pp. 181–183.
19. *Norwegian Standard for Concrete*, NS 3474.
20. ACI Committee 214, Recommended practice for evaluation of strength test

results of concrete (ACI 214-77), *American Concrete Institute manual of concrete practice*, Part 2, 1982, pp. 214–1 to 214–14.

21. Grant, N. T. and Warren, P. A., *A cusum-controlled accelerated curing system for concrete strength forecasting*, RMC Technical Centre, Egham, Surrey, Technical Report No. 79, 1977.

22. British Ready Mixed Concrete Association, *Code for ready mixed concrete*, Shepperton, 1975.

23. ACI Committee 201, Guide for making a condition survey of concrete in service, *American Concrete Institute manual of concrete practice*, Part 1, 1982, pp. 201.1R-1 to 201.R-14.

24. Malhotra, V. M., *Testing hardened concrete: non-destructive methods*, American Concrete Institute, Monograph No. 9, 1976, 204 pp.

25. British Standards Institution, BS 4408. 1971, *Recommendations for non-destructive methods of test for concrete*, London, 1971.

26. Johansen, R. I., In-situ strength evaluation of concrete: the breakoff method, *Concr. Int.*, Vol. 1, No. 9, Sept. 1979, pp. 45–51.

27. ACI Committee 318, Building code requirements for reinforced concrete (ACI 318-77), *American Concrete Institute manual of concrete practice*, Part 3, 1982, pp. 318–1 to 318–103.

Part V Structures

28 Caisson foundations

A J Mitchell

Edmund Nuttall, Scottish Division, Broxburn, West Lothian, Scotland, UK

Contents

Summary

The origins, development, design and construction of different types of caisson foundation are discussed. A case study is presented of the New Redheugh Bridge—a 680 m (2231 ft) prestressed concrete bridge constructed in 1982 across the River Tyne at Newcastle upon Tyne. The foundation for the Redheugh Bridge is discussed with reference to the reasons for selecting pneumatic caissons; the design and construction of the shoe; site preparations, towage and floating-in procedures; the compressed-air work; the sinking procedures and results; bottoming out and ground treatment. This is followed by a review of the design and construction of the foundations of several other recently built bridges, illustrating the development of techniques and the problems encountered. The chapter concludes with a discussion of caissons in tunnelling, pumping stations, docks, harbours and offshore structures.

1   Origins and development

Where heavy localized loads have to be carried to great depths, through water or unsuitable strata, the caisson foundation offers many advantages. The technique has its origins in the ancient bridge-building methods used in India, Egypt and Burma. In the dry season, when the rivers were low, hollow, timber, brick and masonry piers, or wells, would be built on the river bed. The sand and gravel beneath the piers would then be excavated by hand, by swimmers. The pier would gradually sink as the process continued. The tools employed were crude adaptations of farm implements with which founding depths down to 6 m (20 ft) were achieved. The obvious limitation was the ability of these 'skin divers' to hold their breath.

For some time, the use of caissons was limited to bridge foundations. Great advances were made between the end of the 19th and the early part of the 20th centuries. Structural steelwork was used to form the framework and outer skin of caissons, which were much larger than their predecessors. They were also more robust. Excavation was undertaken by mechanical grabs suspended from steam derricks, or by sand pumps. These innovations quickly pushed the maximum possible founding depth down to 60 m (200 ft) and beyond. In 1915, the piers of the Hardinge Bridge over the Lower Ganges were sunk by these means to a depth of 48 m below low water. Similar depths were achieved at the Willingdon Bridge, Calcutta, in 1931 and the Lower Zambezi Bridge in 1935. But, a year later, these depths were all surpassed when the foundations of the San Francisco–Oakland Bay Bridge were sunk to a depth of 69 m (226 ft) below low-water level. At the time of writing, caissons are being sunk to a depth of 105 m (344 ft) below the Jumana River in Bangladesh.

These great depths could not be achieved using the traditional brickwork or masonry steining. It was too light. Concrete became accepted as the ideal material to give the required weight and stiffness within the steel skin plates. More recently, the use of reinforced concrete has eliminated the need for an extensive internal structural steel framework.

The open grabbing technique of caisson sinking sometimes came up against obstructions or thin hard strata which were difficult to negotiate. This led to the use of compressed air in a development of the diving bell principle, whereby a working chamber was established at the point of excavation. The hard material could then be removed by hand, by men working in a 'dry' environment. The earliest use of the pneumatic caisson seems to have been at a pit shaft in Chalonnes by the French engineer, Triger (1839). By 1860 it was widely used in the construction of bridge foundations, notably by Cubitt and Wright at Rochester in 1851.

At the beginning of the 20th century, a completely separate line of caisson development began. The original idea of building jetties and breakwaters from precast reinforced concrete boxes is credited to Professor Kraus of Delft. His plans for the harbour at Valparaiso, Chile, in 1903 were, due to an earthquake, not executed. However, his designs were adapted and used, two years later, at Talcahuano by the Dutch contractors HBG. These box caissons are designed to be sunk onto a levelled and consolidated strip of sea bed. The technique has subsequently advanced to the stage where 'production line' methods may be employed.

2 Caisson design

The design of a caisson must not only be suited to the loads it is to carry in use, it must also provide for a suitable method of sinking and be able to withstand the loads that will be imposed during the sinking process. It follows that a very thorough site investigation programme, with in situ and laboratory testing is the first requirement.

2.1 Caisson types

2.1.1 *Open caissons*
The earliest caissons were of this type. Rectangular or circular in cross-section, with a vertical axis, these may have one single dredging well or, in the case of very large structures, be divided into many separate wells. Being open both at the top and bottom, they are suited to excavation by grab or suction dredging equipment and therefore used where the subsoil consists of soft silts, clays and granular material. *Monoliths* are open caissons having very thick walls, designed to resist the overturning moments arising from horizontal loading. When conditions are favourable, the open caisson offers a very economic solution. Figure 2-1 illustrates the use of a traditional open caisson well foundation, for a modern prestressed concrete bridge in Pakistan.

2.1.2 *Pneumatic caissons*
Although it is possible for divers to be used when open caissons encounter obstructions, or dense strata, these conditions are far more effectively dealt with by hand excavation in compressed air. A working chamber is provided at the

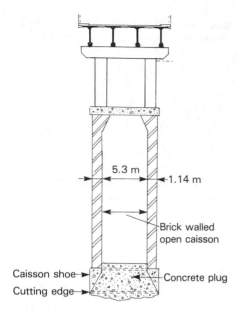

Figure 2-1 Cross-section through well foundation, Jhelum Bridge (after Lee [13])

bottom of the caisson by an *air deck*, which is usually a permanent feature. The chamber is provided with the necessary services from the surface and access is via a vertical shaft which is entered at the top by an air lock. Excavated soil and construction materials also pass through the air lock. By balancing the air pressure against the hydrostatic head, the pneumatic caisson can be sunk with greater control than by open grabbing. The method is, however, limited by the physiological effects on the compressed air workers, to a maximum working pressure of 3.5 bar (51 psi).

2.1.3 *Floating caisson*

The use of compressed air to improve the control of caissons, excavated by open grabbing or dredging, was developed in the USA. The caisson is constructed with a symmetrical arrangement of identical dredging wells. Interchangeable steel air domes are provided which will fit any combination of wells. As excavation proceeds in open wells, others are kept under different pressures to control the effective weight and trim of the caisson. In this way, the characteristic tilting and plunging during the early stages of sinking may be limited.

2.1.4 *Box caissons*

The name caisson is derived from the French 'caisse' meaning a box or casing [1]. The earliest box caissons were made of wood, and inside these masonry piers were built. The modern box caisson is of reinforced concrete construction, closed at the bottom. It is located by controlled sinking onto a prepared bed, which will consist of a granular carpet overlaying a firm stratum, or bearing piles where the stratum is too soft to provide support. The effects of subsequent erosion must be investigated and scour mats may be necessary to preserve the integrity of the foundation.

2.2 Overall Geometry

The caisson plan area is determined by the loadings to be carried and the assumed bearing capacity of the founding stratum. *Skin friction* may be taken into account in certain circumstances. However, Tomlinson [2] gives the following conditions, in which the contribution of skin friction to carrying capacity should be ignored:

  (a) if the depth of the caisson below founding level is less than its least width;
  (b) if the ground above founding level is liable to be scoured away;
  (c) if compressible backfilling is placed between the foundation structure and the walls of the excavation;
  (d) if, in sinking a caisson, the soil is undercut behind the cutting edge, thus forming a gap around the walls which becomes filled with loose or softened material;
  (e) if the pier or caisson is surrounded by soft clay or fill material;
  (f) if the soil is liable to sink away from the foundation due to drying action.

A pneumatic caisson allows close inspection of conditions at the founding level and bearing capacity may be more accurately assessed. In the case of an open caisson, samples recovered by divers or inspection by underwater television will not be so reliable and the necessarily more cautious assumptions will lead to the use of a larger caisson area.

Where excessive skin friction is likely to be a problem during sinking, a circular cross-section offers the minimum surface area. However, it is more difficult to control the tilt of a cylindrical caisson and in stiffer soil conditions, arching or wedging across a chord, inaccessible to a grab, may delay the sinking of an open caisson.

Open caissons should be provided with a minimum of three *dredging wells* so that, by varying excavation rates, the tilt can be controlled. The minimum size of each well will depend upon the dredging method to be used. If this is to be grabbing, then clearance should be provided for a 2 or $3\,m^3$ ($4\,yd^3$) clamshell in the fully open position. Wall thickness will be influenced by the overall weight required to penetrate the anticipated strata. However, it must be noted that in dense sand or stiff clay it is important that the grab should dig as close to the edge of the excavation as possible. In this case, additional weight may have to be provided by kentledge.

The *working chamber* of a pneumatic caisson should be 3 m in height. The access shaft should be centrally placed and one shaft should be provided for every $100\,m^2$ ($1076\,ft^2$) of excavation base area [3]. If the working chamber and air deck can be cast in one concrete pour, with no construction joints, this will add to the strength and air tightness of the most highly stressed area of the caisson.

2.3 Detailed design

The exterior walls of the caisson should be truly perpendicular to the base and it is important that a smooth, true finish is achieved. Bulges and corrugations will add considerably to the skin friction to be overcome during sinking. For the same reason, the exterior walls should be 'set in' up to 50 mm (2 in) at the top of the working chamber or *shoe*. This provides a nominal annular space into which a lubricant such as *bentonite* slurry may be injected. Between the cutting edge and the 'set in', the exterior wall may be subjected to locally concentrated loads as the caisson pushes aside boulders or other obstructions. It should therefore be sheathed in 12 mm ($\frac{1}{2}$ in) thick mild steel plate from the base to the 'set in'.

At the base of the wall is the *cutting edge*. Here, the stress concentrations are greater when, for example, the caisson is 'hung up' on a boulder. The design must allow for a large proportion of the overall weight to be supported on a localized area. Mild steel skin plate, 20 mm thick, is normally used, with closely placed stiffeners and tangs to anchor the plating securely to the concrete. Provision should be made for the reinforcement to pass through the stiffners so that the construction is truly composite. For a pneumatic caisson, it is extremely useful to have a mild steel skirt projecting some 150 to 200 mm (6–8 in) below the shoe. This allows the *hydrostatic head* to be balanced below the normal level of excavation and reduces loss of air under the cutting edge. However, this skirt is vulnerable and must be made very stiff. Should it begin to peel away from the shoe, it may cause great damage.

The inner wall of the working chamber will be sloped or canted to give added stiffness and provide resistance against excessive penetration. The angle may be chosen to suit the type of ground anticipated. Stiff clays and dense sands will demand a fine angle of attack up to 30° to the vertical. If the ground is very soft, it may be necessary to prevent the shoe from sinking in too far, and an angle nearer 45° to the vertical will be chosen. The inner canted wall should also be protected from local damage by mild steel plate, to at least the maximum anticipated penetration. In open caissons, the full height of the working chamber should be protected, as damage by the grab must also be prevented. The final design of the shoe must be suited to the predominant ground conditions through which the caisson must pass, but must be able to cope with all the ground types anticipated.

Large caissons will have internal cross walls in order to give the required rigidity. These are stopped at or above the top of the working chamber and at this point a V-shaped cutting edge may be provided.

The development of rapid shuttering systems and advances in pumping techniques have made reinforced concrete more economic in most situations than structural steelwork for caisson construction. However, where a shallow draft is necessary for floating the shoe into position or initial excavation passes through a soft river mud, a system of horizontal trusses in a steel skin may still be the best way to provide a light structure with the stiffness required, reinforced concrete being used to complete the structure *in situ*. Concrete mixes must be designed for maximum density and durability. Cement-rich mixes are to be avoided for heavy sections unless the heat of hydration (which may cause cracking) can be dissipated safely.

Vertical reinforcement must carry the tensile loads, which occur when the top of the caisson is held by skin friction while the cutting edge is undermined and loaded with kentledge. Adequate shear reinforcement must be provided to resist the unequal horizontal loadings caused by tilting and tilt correction procedures.

2.4 Design responsibilities

Whilst the consulting or designing engineer will wish to ensure that the caisson fulfils the requirements of the overall structure, there is good reason to leave as much detail as possible to the contractor. The latter should be permitted to use his own experience to determine the working methods to be adopted, and these should be allowed to influence the design of the cutting edge and caisson shoe. The contractor may also wish to detail the concrete lifts and reinforcement schedules to suit his planned rate of working. Greater efficiency is bound to result from this approach, but the contractor will need to be given adequate information concerning the overall design requirements.

3 Construction

The remarks in this section refer to open and pneumatic caissons. Floating and box caissons are discussed later in the chapter.

3.1 Alternative sinking methods

A most enlightening debate took place in 1950 at the Institution of Civil Engineers (England) on the relative merits of methods of sinking bridge foundations [4]. At that time it was quite common to provide open caissons with facilities for changing to compressed air, should the need arise. However, advances in other foundation techniques, such as large bored piles, and improvements in site investigation techniques have tended to limit the use of the pneumatic caisson to very specific circumstances. In favourable ground conditions, the open grabbing technique, requiring simple equipment and little manpower will always be more economic. Under these circumstances, only the danger of *subsidence* to nearby structures, due to an uncontrolled inrush of ground, would justify the more expensive alternative. However, where troublesome obstructions are known to exist, or the need to inspect, sample or treat the founding stratum is paramount, then the pneumatic method will provide the required conditions.

3.2 First-stage construction

The lower portion, incorporating the shoe, of the caisson will, in the case of a structure ashore, be built *in situ* over the spot where it is to be sunk. Where the caisson is to be located in water, several options exist.

If the site is close inshore and the water is not too deep, a temporary *sand island* may be constructed on which the caisson may be set up (Fig. 3-1). This will take the form of a cofferdam, filled with granular material and large enough to accommodate the shoe, with working space for the necessary equipment. In order to avoid excessive excavation, the surface of the granular fill may be kept below the outside water level and maintained in a dry condition by a well-point dewatering system. When the shoe has been completed, the cofferdam is flooded and sinking commences. This method requires ground conditions suitable for an unbraced cofferdam and sufficient space to construct it, clear of navigational limits.

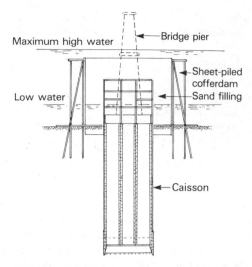

Figure 3-1 Cross-section through the Baton Rouge Bridge foundation illustrating the sand-island technique (after Blaine [11])

An advantage offered by the lighter, structural steelwork, shoe is that it may be assembled over the intended position on a simple trestle structure, supported by bearing piles. When complete, it is lowered by jacks or winches onto a prepared level bed of tremie concrete. Where the site consists of a tidal mudflat, it may even be possible to assemble the prefabricated shoe, between tides, on a concrete raft. Precautions will need to be taken to ensure that it does not float out of position on the ensuing high tide.

When the above options are not available and particularly when the site is offshore, the caisson shoe must be constructed in a dry dock or similar facility. A careful study must be made of the tow route to the site, with particular attention being paid to soundings, currents, tidal ranges and liaison with the authorities which control shipping.

3.3 Location and control

Once sinking commences, three associated problems will be encountered: tilt, twist and lateral shift. They are more pronounced and problematic during the early stages, for the following reasons:

(a) The caisson is not fully embedded in the ground.
(b) The caisson is usually passing through soft layers offering little lateral resistance.
(c) Tidal fluctuations make it very difficult to control the effective weight of the caisson.

It is usually worth while carrying out some predredging to remove light flocculent material and establish a level uniform bearing area before the shoe is positioned. An estimate of the bearing resistance must be made to determine the necessity for any further bed preparation. This may consist of a granular blanket, unreinforced tremie concrete slab or timber bearing piles which will be cut off as the caisson descends. The dredging must cover a sufficiently wide area to prevent differential lateral earth pressures developing.

The weight of the caisson, in the early stages, should be kept to the minimum required to maintain a steady rate of sinking. Before commencing, an estimate of skin friction and weight required at each level is made, based on all available subsoil information. Weight is added to the caisson to correspond with this pre-estimate and the results closely monitored.

Sometimes, the likely direction of shift can be anticipated, e.g. towards the centre of a deeply veed river channel. Where space permits, this can be resisted by a temporary gravity structure such as a sand-filled sheet pile box. Alternatively, the caisson can be restrained in a sheet-piled enclosure or held between four spud-legged towers.

Once excavation commences, the rate of sinking and the attitude of the caisson must be continuously and accurately monitored. Prompt action to correct tilt is essential and it should be borne in mind that, with the cutting edge buried in the early stages, the conditions giving rise to problems will not be seen before the tilt occurs. Indeed, they may be outside the caisson altogether. Careful variation of excavation levels across the base, combined with good control of *effective weight*, is the best method of correction. More drastic measures, involving blocking off part of the cutting edge, have been known to work but are more likely to give rise to a lateral shift which can never be recovered.

As work progresses, soil samples should be taken and compared with available borehole logs, in order to predict the conditions below the cutting edge. Strenuous efforts should be made to bring the caisson plumb as it passes from soft ground into a stiffer, or more dense, stratum.

Although difficult to predict, in most cases it will be possible to 'bottom out' at within 300 mm (12 in) of the desired plan position. It is essential that a tolerance of this order be allowed for in the design.

3.4 Skin friction

The increase of caisson weight, by raising the walls, should be planned such that the effective weight is always in excess of the frictional resistance. Otherwise, the working cycle will have to be disrupted by the tedious business of adding kentledge to maintain progress and further disrupted when the kentledge is removed to allow the walls to be extended once more.

The effect of skin friction must not be underestimated. Records of caissons already sunk in the same area may give clues, but it is vital to discover all the circumstances. The accurate measurement of skin friction is very difficult; there is nearly always some weight taken on the cutting edge. The presence of obstructions bearing on the outside of the caisson and the irregularities of the exterior surface will add considerable resistance, and the scouring action of escaping air will affect the interface between soil and caisson.

Nevertheless, the figures in Table 3-1 given by Terzaghi and Peck [5], based on the loads required to start caissons which have become stuck, offer some guidance. Some very high values of skin friction were recorded by Yang [6] in a detailed study of the Verrazano Narrows Bridge foundations. He states that, in

Table 3-1   Values of skin friction during the sinking of caissons (after Terzaghi and Peck [5])

Type of soil	Skin friction, kN/m²
Silt and soft clay	7–29
Very stiff clay	48–192
Loose sand	12–34
Dense sand	34–67
Dense gravel	48–96

medium dense sands and gravel, values between 84.75 and 95.4 kN/m² (12–14 psi) were experienced. However, after considering results from a wide range of observed values, Tomlinson [2] concludes that the range to be expected is from 9.6 to 29 kN/m² (1.4–4 psi).

In Section 2.3 it was noted that the caisson walls should be 'set in' up to 50 mm at the top of the shoe. If this gap can be kept filled with a low-friction material, then the effects of skin friction, above the shoe, can be greatly reduced. Bentonite slurry has been commonly used for this purpose. Being thixotropic, it has some resistance to *wash out* or dissipation.

Closely spaced injection nozzles should be connected to a header tube cast into the caisson walls. The slurry is injected at regular intervals in volumes which ensure that the annular space is kept full. Friction reductions of up to 40% have been claimed for this procedure. However, the advantage can be quickly lost if escaping air scours away the slurry or if the caisson passes through an open gravelly layer through which the bentonite can escape.

3.5   Excavation methods

3.5.1   *Open caissons*
The usual method of excavation in cohesive soils is by large clamshell grabs operated by a derrick or crawler crane. Granular material can sometimes be more economically removed by sand pump or airlift. In either case, the excavation of material close to the cutting edge can be facilitated by the use of jetting nozzles incorporated in the caisson walls. Differential operation of jetting nozzles is also a useful way of controlling tilt in granular materials.

Sand pumps and airlifts produce large volumes of water which must be handled.

3.5.2   *Pneumatic caissons*
In cohesive soils, excavation is essentially by hand, assisted by small pneumatic clay spades. (The method has not changed in 100 years.) The spoil is loaded into bottom opening skips of 0.9 m³ (1.2 yd³) capacity.

Granular materials may be amenable to more rapid removal by a form of air lift which is boosted by the over-pressure of the working chamber. The use of high-pressure water jets with sand pumps is a recent successful development.

3.5.3 *'Hanging' caissons*

When the excavation level progresses below the bottom of the shoe, and the cutting edge is exposed without downward movement of the caisson, it is said to be 'hanging'. Further excavation runs the risk of a sudden uncontrolled drop. This may be accompanied by an inrush of material from outside the caisson, and create excessive tilt. Furthermore, if the descent is abruptly halted by one part of the cutting edge coming to rest on a boulder, the caisson shoe may be split by the impact.

In these circumstances, a pneumatic caisson may usually be freed by *blowing down*. That is, by a rapid reduction of working chamber air pressure, which has the effect of suddenly increasing the effective weight of the structure. This must be done in a controlled way, and precautions are necessary to safeguard the health of those working in the compressed air.

Where there is no danger to adjacent structures, open caissons in granular soils are sometimes freed by substantial reduction of the internal water level. This creates *boiling* and a corresponding reduction in the bearing pressure of the ground around the caisson shoe. Large inflows of soil usually occur.

If neither of the above remedies are available or effective, one must resort to the application of more kentledge; a time consuming and tedious business.

4 Case study: pneumatic caissons for the New Redheugh Bridge, Newcastle upon Tyne

In 1980, Edmund Nuttall Ltd, in a joint venture with HBM, commenced the construction of two pneumatic caissons which support the main 160 m (525 ft) span of this prestressed concrete bridge. The client, Tyne & Wear County Council, urgently required this new structure to carry both an important traffic route and major water and gas supplies between Newcastle and Gateshead.

4.1 Foundation design

The bridge is part of a high-level structure crossing the River Tyne with a navigational clearance of 26 m (85.3 ft) above MHWS. An average pressure of 1000 kN/m^2 (145 psi) was to be carried by the rock below each caisson. The design was based on information gained from six boreholes sunk at the site of each pier and records of caissons sunk for two previous bridges immediately adjacent.

Below the recent soft river deposits, the substrata consisted of three main types: boulder clay, sands and gravels, and coal measures. The boulder clay in its natural condition is extremely hard, although the top layer could be expected to be softened by the proximity of the river. It was known to contain some large boulders. The sands and gravels were described as compact to dense, and the coal measures consisted of alternating layers of mudstone, siltstone, sandstone and coal. Some of the shallow coal seams adjacent to the river were known to have been worked and a doubt existed about the presence of voids in the coal measures below the foundations of the bridge. Piezometric head measurements showed the sand and gravel layers to be in hydraulic continuity with the river water. Water pressures equivalent to tidal head levels could be expected.

The need to carry the heavy foundation loads to rock was self evident. Since bearing piles might be hung up on boulders or fail to penetrate the dense granular layers, caisson foundations were chosen by the consulting engineers, Mott, Hay & Anderson. There were several sound reasons for their choice of pneumatic caissons: '

- (a) It was essential to prove that the upper coal measure was in a sound undisturbed condition.
- (b) If voids were located, they could be grouted with more certainty from a working chamber at rockhead.
- (c) It was essential to prevent any inrush of material from outside the caisson, which might have affected the stability of the adjacent bridge.
- (d) If the caisson came down on a boulder, the obstruction would be more easily removed.
- (e) None of the strata through which the caisson was to pass, would be easily excavated by grab.

The King Edward VII Bridge (1902), 300 m (984 ft) downstream, had been constructed on three caissons of plan area 308 m^2 (3315 ft^2) [7]. All were successfully accomplished and yielded valuable information on excavation rates (360 m^3/wk), foundation loadings (750 kN/m^2, 109 psi), skin friction (27–37 kN/m^2, 4–5.4 psi), and highlighted possible problems with hydrogen sulphide gas.

Most recently, the Tyne Bridge (1929) employed steel caissons 25.6 × 8.5 m (84 × 28 ft) in plan [8]. These were generally successful apart from one serious incident where a horizontal split developed between the working chamber and the concrete pier above. The working chamber descended 400 mm (15 in) leaving the upper portion held by skin friction. This gap developed to a maximum of 0.5 m (1.6 ft) and required expensive remedial action. It is noteworthy that corrugated galvanized steel sheeting was used as shuttering on this structure. The resulting surface texture is likely to have contributed to the friction problem.

4.2   Construction of the caisson shoe

Each main pier was to be founded on a reinforced concrete caisson 11 m (36 ft) in diameter. The centres of the caissons were to be located 9.25 m (30 ft) from the edge of the navigation limits and the bed level varied from 1 to 3 m below low-water level across the caisson site. Working space onshore was extremely limited.

Due to ground conditions, unfavourable for piling, a sand island would have required a substantial double skinned or cellular cofferdam. There was neither the room nor the time for this approach.

Construction of the shoe on a piled platform above its intended location was a viable option. The completed section could then be lowered by jacks onto the river bed to commence sinking. The main disadvantage was that the assembly of the shoe could not start until the piled platform was completed. The contractors were looking for a method which would save some valuable program time.

A dry dock was located 12 km (7$\frac{1}{2}$ miles) downstream of the bridge site where the two caisson shoes could be built under the ideal conditions. This allowed site preparations, piled access jetties and preparatory dredging to be undertaken concurrently during the first 12 weeks of the project.

The consulting engineers specified the overall dimensions of the caissons and gave typical reinforcement details, but the detailing of the reinforcement and the design of the caisson shoe was left to the contractors. The caissons were to be 15 m (49 ft) high, terminating at the nominal dredged-river-bed level, in a 3 m thick capping slab. The walls were only 1 m thick. The contractors did not alter the shape of the shoe, which had a 3 m high working chamber with the inner walls inclined at the very fine angle of 25° to the vertical, terminating in a stiffened steel cutting edge. The cutting edge, designed by the contractors, was chosen to suit the predominantly stiff material through which the caisson was to pass. It would be necessary to reduce the bearing area to a minimum at times, to overcome skin friction, and the 125 mm (5 in) downstand was required to allow dry excavation at cutting-edge level. Care would have to be taken to prevent damage when entering the rock. The contractors' design included a reinforced concrete air deck, into which the bottom section of access trunking was cast.

The cutting edge, weighing 6 tonnes, was prefabricated, in three sections (Fig. 4-1). High-quality birch plywood was used to make the accurate curved shuttering which was required to give a regular smooth surface. The shoe was constructed to a diameter of 11.1 m (36.4 ft) to provide a 500 mm 'set-in' immediately above the working chamber. Above this change in diameter, 16 bentonite injection nozzles were located, each protected from the ingress of debris by a soft rubber diaphragm, which acted as a simple one-way valve. All the necessary operational services were cast into the air deck, each with a duplicate access point for emergency use.

The section of caisson which could be built in the dock (Fig. 4-2) was limited to a height of 4.7 m (15.4 ft) by the available clearance over the sill at high water spring tide. This corresponded to a draught of 6.65 m (21.8 ft), which was approaching the maximum considered advisable for the up-river tow.

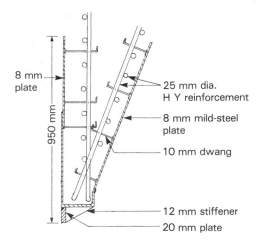

8 mm plate

950 mm

25 mm dia. H Y reinforcement

8 mm mild-steel plate

10 mm dwang

12 mm stiffener

20 mm plate

Figure 4-1 Steel cutting edge for the Redheugh caisson

Figure 4-2 Concreting the caisson shoe in dry dock

The concrete was designed to give a 28-day strength of 30 MPa (4350 psi) and sulphate-resisting Portland cement was used with a sea-dredged gravel; 32 tonnes of high-yield reinforcement, mainly 25 mm diameter, were incorporated. Two concrete pumps with hydraulic booms were used to ensure that no cold joints occurred as the concrete was poured in one operation.

While the concrete was curing, preparations were made for towing. In order to

Figure 4-3 The caisson shoe approaches the site of the new bridge

give the required freeboard, stiffened steel shells—*tubbings*—were bolted around the top of the concrete. These 2 m high rings were made up in 12 segments of 10 mm mild-steel plate with stiffeners at 500 mm (20 in) vertical centres. Two rings were required at this stage. The total weight of the caisson, ready for towing, was 593 tonnes (Fig. 4-3).

4.3 Site preparation and location of the shoe

While work progressed in the dry dock, the caisson site was prepared. Two temporary access jetties were constructed and a sheet-piled wall was installed to safeguard the stability of the existing river bank. The soft river deposits were dredged to produce a level bench on the firm boulder clay at −5.5 OD. On top of this, a 0.5 m thick granular blanket was spread and levelled. Calculations revealed that the bearing capacity of the ground would be sufficient to support the caisson shoe while the walls were raised to their full height.

Prior to the tow up river, a thorough study of soundings and obstructions was carried out, including depths alongside any quays which might be needed for emergency stops. The safe towing speed in restricted waters was unknown and deep water was severely limited in the upper reaches, at the end of the 12 km journey. A detailed method statement was needed to cover all eventualities.

In order to keep the draught of the shoe to a minimum, the working chamber was kept under a controlled air pressure and six 5 tonne *flotation bags* were secured around the outside. A compressor was carried onboard to replace any losses.

Two tugs of 1450 bhp were used: one towing and one steering from astern. The 17 tonne bollard pull of the lead tug was never fully employed as the safe towing speed was less than 3 knots. At greater speeds, the caisson began to yaw excessively in the narrow river channel. The slow towing speed and obstructions in the river resulted in the 12 km journey being undertaken in two high tides.

Two 5 tonne winches were used to pull the caisson into position against three, accurately positioned, 600 mm (2 ft) diameter piles. Once secured against the *guidance piles*, the air was slowly exhausted from the working chamber and the shoe settled gently on the river bed. At this stage, the caisson shoe sat plumb within 30 mm of its true position (Fig. 4-4).

After the first two tides, the caisson settled and adopted a tilt of 1.2 m (4 ft) across its 11 m (36 ft) diameter. This was subsequently discovered to be due to a large boulder, just below the dredged level, on which the cutting edge was lodged. Ballast was applied to stabilize the situation and divers using water jets of 20 000 kN/m^2 (2900 psi) nozzle pressure, assisted by grabbing, exposed the obstruction, which was broken up by hydraulic bursting equipment.

Once re-established in its plumb position, the top 10 m of the caisson were constructed in 2 m high lifts. The tubbing units were used as outside shutters and raised as the work progressed. It is important to note that the contractors' assessment of the ground conditions were crucial to the success of this operation. It would have been impossible to make a good job of the 10 m high walls if the caisson had plunged downwards or tilted out of plumb during this process.

The wall construction was terminated below cap slab level. At this point,

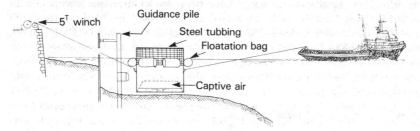

Figure 4-4   Manoeuvring the caisson into position

reinforcement couplers were cast in so that, if a deeper founding level proved necessary, the caisson walls could be extended.

4.4 Arrangements for working in compressed air

The establishment and operation of the plant required for compressed-air working are very costly. For this project, each caisson required four large compressors. Two electric 28 m³/min (1000 ft³/min) compressors supplied the air for pressurizing the working chamber, whilst a third diesel-powered machine of similar capacity stood by, in case of power cuts. The air from these three, which was to be breathed by the miners, passed through a scrubber and after-cooler before entering the low-pressure air-supply system. The fourth compressor, powered by electricity, supplied air at a pressure of 1035 kN/m² (150 psi) and a rate of 14 m³/min (500 ft³/min) for the air tools required. The equipment was suitable for operation up to the maximum anticipated working pressure of 240 kN/m² (35 psi).

Access to the working chamber was by a *Gowring blister lock* and air trunking, as described by Wilson and Sulley [3]. Although the equipment has not changed in the intervening period, the controls required for work in compressed air are now far more strict. The 8.5 m (28 ft) long decompression chamber was set up alongside a medical air lock on the quayside. The compressed air workers, on leaving the working chamber, had to adopt a *decant procedure*. This involved rapid decompression in the blister lock, a smart walk to the decompression chamber and recompression to the maximum working pressure of their shift. This had to be achieved in 5 min. Thereafter, controlled decompression would take place.

Associated with the coal measure, was the possible occurrence of highly inflammable methane gas. The oxygen-enriched atmosphere of a compressed air chamber intensified the risk of fire and/or explosion. Four sensor heads were installed in each caisson. The first two were set to operate a light signal as soon as the gas was detected and the second pair activated an audible signal when the level of methane was such that the chamber should be evacuated (25% of the lower explosive methane/air mixture). All electrical equipment in the chamber was flameproof, no inflammable materials were used and the no-smoking rule was strictly applied.

The history of the use of compressed air in civil engineering is marred by the frequent occurrence of *decompression sickness*. The permanent disability of workers and many fatalities have resulted from a relative ignorance of the physiological effects. However, significant advances have been made since the Illinois and St Louis Bridge was built in 1870 at a cost of 14 lives. The Medical Research Council Decompression Sickness Panel, under the chairmanship of Professor D. N. Walder is based at Newcastle University. The contractor took advantage of the locally available expert advice in their planning and the training of their own supervisory staff. In return, the Panel were able to carry out on site studies into the physiological effects on the workers involved. The third edition of the Panel's report *Medical code of practice for work in compressed air* [9] was in its draft form at the time. By agreement with the Health and Safety Executive, this code was applied wherever possible, and in particular, the *Blackpool decompression tables* were used. Dr J. D. King, a member of the Panel, has commented [10]: 'Perhaps the most significant addition to the regulations contained in the code is a call for regular examinations of major joints for the detection of *aseptic necrosis* of bone.' All workers and supervisory staff were X-rayed and thoroughly examined by a physician before starting work in compressed air, and re-examined at monthly intervals during the working period.

Continued medical fitness was made a condition of employment and cooperation with the physicians and medical research workers was good. However, the initial screening of potential employees eliminated many applicants and considerably prolonged the recruiting period.

Just as important as the medical back-up was the meticulous on-site organization, rigorous record keeping and a determination to enforce a disciplined approach to the regulations.

Before work started, the MRC doctors anticipated an incidence of decompression sickness between 1 and 1.5% of all exposures. The actual results for the two caissons were as follows:

Number of exposures to compressed air (maximum pressure 204 kN/m²)	1965
Number of type I bends	11
Number of type II bends	0
Number of cases of aseptic necrosis reported	0
Incidence of decompression sickness	0.56%

No cases of permanent ill health or disability have been reported.

4.5 Sinking the caisson

After use as exterior shuttering, the steel tubbing units were re-erected on top of the completed concrete walls. In this position, they were to provide a watertight cofferdam in which the caisson cap could be constructed. However, for the moment, they projected 15 m (49 ft) above high water level and access was by a scaffold tower alongside (Figs 4-5 and 4-6). This was inconvenient but the full-height caisson and tubbing extension would not require extending during sinking, and so a possible source of delay was eliminated.

On entering the caisson, the miners found only 1.5 m headroom and the first few shifts were spent in establishing more working space. Then, fixed lighting, telephone and bell-signal communications were set up.

The spoil was hand mucked into the 0.9 m³ (1.2 yd³) bottom-opening skip and, on a given signal, hoisted out by a 40 tonne crane fitted with a non-twist rope. The air pressures required in the chamber, to balance the water head, were calculated in advance, taking account of tidal variation. The results were plotted on a chart which was followed by the compressor attendant. The caisson tilt was measured by a water level externally and a pendulum device inside the working chamber.

During the early stages of sinking, the rise and fall of the tide had a great influence on the effective weight of the caisson. Rapid downward movements of 400 mm (16 in) occurred around low tide, even when the cutting edge was well buried.

With so little of the structure embedded, it was difficult to control the tilt. The differential level across the 11 m diameter shoe was of the order of 450 mm (4%)

Figure 4-5 The south caisson during sinking

at this stage. Efforts to correct this usually resulted in a tilt in the opposite
direction, so that the caisson oscillated or 'wobbled' during its early descent.

As the whole of the shoe became embedded, these movements were 'damped'
by the passive soil resistance. After penetrating 4 m, the tilt of the caisson was
kept to less than 1%. The friction also increased and kentledge was added by
pumping water into the void above the air deck, the level being adjusted to give
the required weight as the tide rose and fell. The adjustment of water kentledge is
faster than individually handled billets of iron or concrete blocks.

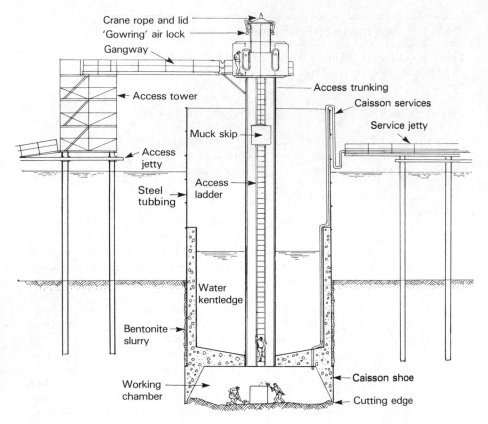

Figure 4-6 Cross-section through the caisson during sinking

When granular materials were encountered, excavation was accelerated by the use of a modified 150 mm air lift known as a 'snorer'. This removed a sand–water solution extremely rapidly—(using a large volume of air in the process). The snorer was also used for routine dewatering duties.

Bentonite injection was commenced as soon as the nozzles were below the river bed. It was not possible to quantify the friction reduction but a significant improvement was noticed on the second caisson, when injection was commenced somewhat lower than intended. The rubber diaphragms prevented choking of the nozzles and injection was continued until sinking was complete.

The assessments of skin friction made on the north caisson, assuming the bentonite reduced skin friction by 20% above the point of injection, are given in Table 4-1.

Strenuous efforts were made to level the caissons as they entered the top surface of the rock and little change of tilt occurred thereafter. Excavation, now with light pneumatic tools, was continued up to, but not beyond the cutting edge. This created a 'tight fit' into the rock which was advantageous for stability but increased sinking resistance.

Table 4-1 Skin friction on north caisson

Level of cutting edge below river bed level	Skin friction	Ground conditions around shoe
5.5 m	33 kN/m^2	Hard boulder clay and dense sand
10.0 m	38 kN/m^2	Dense sandy gravel and cobbles
12.0 m	36 kN/m^2	Top in gravel, bottom in mudstone

Maximum kentledge was provided by the tubbing full of water and yet a stage was reached when no movement occurred, even at low tide. The caisson was 'hanging', supported by skin friction and the cushion of air in the working chamber. The solution was to reduce the chamber pressure very quickly, a method known as *blowing down* the caisson. By lifting the snorer clear of the water, chamber pressure was reduced by between 40 and 100 kN/m^2 (6–14.5 psi). Excavation below the cutting edge was carefully controlled to allow the caisson to sink 300 mm or so on each blow down. Blowing down was, in this case, usually carried out in rock. Very great care must be exercised when the technique is applied in 'soft' ground. In order to prevent an inrush of ground and water, possibly accompanied by excessive tilting, small pressure reductions should be tried and excavation should not advance more than 500 mm (20 in) lower than the cutting edge. If this is unsuccessful, more kentledge should be applied. The safety and health (vis-a-vis exposure to differing pressures) of the miners must be carefully considered when blowing down.

Table 4-2 Sinking rates

Material excavated	Average advance in metres per 8 h shift	
	North caisson	south caisson
Boulder clay	0.20	—
Sand	0.33	0.38
Gravel and boulders	0.24	0.21
Coal	—	0.33
Mudstone and siltstone	0.18	0.16

4.6 Bottoming out

The rock was highly weathered at the surface. After a 1 m thick coal seam, where hydrogen sulphide gas was encountered, the rock quality improved and appeared quite satisfactory at the founding level. At this point, the rock consisted of fine horizontal laminations of siltstone and sandstone.

As the founding level was approached, the working pressure was allowed to fall below the theoretical point of balance with no significant inflows of water. It appeared that the care taken to eliminate overbreak had resulted in a partial seal into the rock. Instead of the theoretical 248 kN/m² (36 psi) maximum pressure, the maximum reached was 193 kN/m² (28 psi). This increased the maximum working time possible per shift and reduced decompression times accordingly.

After the formation had been trimmed, the caisson was gently blown down to its final position, 26 mm below its planned reduced level and within 75 mm (3 in) of its intended plan position. A concrete haunch was packed around the cutting edge, into which 50 mm grout tubes were set at 2 m intervals. Through the 150 mm thick blinding concrete, 50 mm pressure relief pipes were set at 1 m centres.

The north caisson was so well sealed that it was possible to drop the working pressure below the minimum level at which medical supervision and decompression is required (93 kN/m², 13.5 psi). Exploratory drilling then took place, the intention being to check the condition of two coal seams, thought to lie 10 and 20 m (33 and 66 ft) respectively below the caisson formation.

A pattern of 75 mm diameter open holes was arranged on a 5 m grid to thoroughly check the coal seams for old workings. Holes were inclined so as to cover a circle of 25 m (82 ft) diameter at the level of the first seam. Drilling was carried out by carriage-mounted hydraulic drill driven by a 10 hp power pack. In addition to these 20 m long holes, one hole 40 m long was drilled vertically near the centre of the caisson. All these probes revealed sound rock, free from workings, but finely laminated in horizontal planes. The holes were grouted up but there was no 'take up' by the bedrock. Tests on cores recovered revealed a uniaxial compressive strength in excess of 40 MPa (5800 psi).

The annular space around the caisson was grouted from the pipes set below the cutting edge and through the bentonite injection points. When grouting was complete, the working chamber was returned to atmospheric pressure and the caisson plugged with 37.5 MPa (5440 psi) concrete. The concrete was pumped into the chamber and distributed through a 360° swivel joint. The first section of access tubbing was filled to provide a pressure head and vibrators inserted through all the spare access holes in the air deck. The good workability and rapid placing by pump eliminated cold joints and voids. Very little back grouting was required.

5   Bridge foundations

5.1   The problem of scour

The Mississippi River, USA, has imparted many lessons to the aspiring bridge engineer, none more chastening than those recorded by Blaine concerning the construction of the Baton Rouge Bridge [11]. The terms of the contract specified that the two mid-channel piers were to be constructed at the same time. The contractor chose to use the convenient sand island method and the result of these two decisions was a reduction in river width at the bridge site from 730 m (at low water) to 635 m (798 to 694 yd). Current velocities up to 12.9 km/h (8 mile/h) were known to occur, prior to the width reduction. At pier 4, 15 m (49 ft) of scour occurred undercutting the sheet piling, and the sand island disappeared in 3 min. A further three months was required to restore the caisson to its plumb position and re-establish the sand island.

Engineers continue to underestimate the effects of scour, particularly in the use of the sand island method of caisson construction. The south main pier of the Humber Bridge was delayed three months whilst several thousand tonnes of chalk boulders were dumped into scour cavities, which threatened the stability of sheet piling, forming the sand island around the twin caissons [12].

The severest cases of scour occur in rivers subject to very large peak flows, caused by monsoons or snow melts. A conservative calculation of the maximum potential scour depth is essential and stability calculations will include the maximum scour condition. Foundations for the Sutlej River Bridge [13] were carried to a depth of 35 m (115 ft) to ensure adequate penetration below the predicted 23 m (75 ft) maximum scour level.

Scour protection systems, based on synthetic fabrics, have been developed for use in connection with offshore oil production installations. These should be considered for the protection of bridge foundations subject to scour.

5.2 Deep-water locations

Although normally avoided, it is sometimes necessary to locate bridge foundations in deep water. A caisson is a natural choice in this situation, but the problem of keeping it in place, during the early stages of sinking, demands a special solution when the sand island cofferdam is no longer feasible. The caisson will have to be serviced by floating plant and the use of a multiple-cable mooring system will seriously restrict access, require continuous adjustment and be susceptible to damage by shipping. A better solution, giving excellent tilt control and positional accuracy, is the construction of temporary mooring dolphins using the 'piled jacket' technique. Tubular steel frames are prefabricated ashore. Vertical tubes are made large enough to accommodate bearing piles which are driven through them to fix the dolphin in position. Diagonal bracing must be capable of resisting caisson docking forces and the bearing piles may require grouting into the jacket after driving.

This technique was employed for the Construction of the Mackinac Bridge [14], where caissons of 35 m diameter were located in up to 42 m of water using four dolphins connected by horizontal trusses. The steelwork for each dolphin weighed 160 tonnes and the horizontal trusses a total of 80 tonnes.

5.3 Open caisson sinking

The problems associated with open caissons are not usually to do with excavation. The method is only selected where the material to be removed is amenable to grabbing or dredging. Sometimes, a calculated risk is taken with regard to a thin layer of clay, but even this can be broken up with the powerful jetting equipment now available.

The usual problem is a very simple one; the caisson does not have enough mass. There is, of course, a clash of interests here between the designer, who wants to retain maximum bearing capacity for his superstructure, and the contractor, who wants to get his caisson down with the least possible difficulty. Internal cross walls

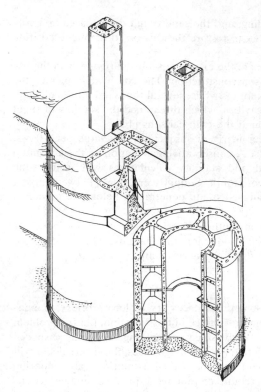

Figure 5-1 Humber Bridge: Barton Tower caissons

provide the best means of adding weight to a caisson. They serve to divide up the dredging well, so helping with tilt control, and stiffen the structure at the same time. If they are stopped off 2 or 3 m above the cutting edge, they do not add to the bearing resistance, as would a thickening of the outer walls. A good example of this approach is the design of the twin caissons for the Barton Pier of the Humber Bridge (Fig. 5-1).

However, the 24 m diameter × 32 m deep (79 × 105 ft) Barton caissons were not sunk without incident. After starting off well, with sinking rates up to 0.5 m per 12 h shift, they both became lodged in an isolated band of stiff Kimmeridge clay [15]. A total additional load of 5500 and 6500 tonnes respectively was required to get the two caissons moving again. This was provided, with great difficulty, by a permanent extension to the caisson walls, involving redesign of the pier, and 2900 tonnes of kentledge. The foundations are reported to have been delayed by three months as a result. The lesson is clear: open caissons should be designed with provision for adding substantial additional weight, either in permanent or temporary form.

5.4 Floating caissons

The use of captive air, to float caissons into position, is a commonly used and straightforward operation. The chief consideration is to ensure that the metacentric height is sufficient for the tow conditions.

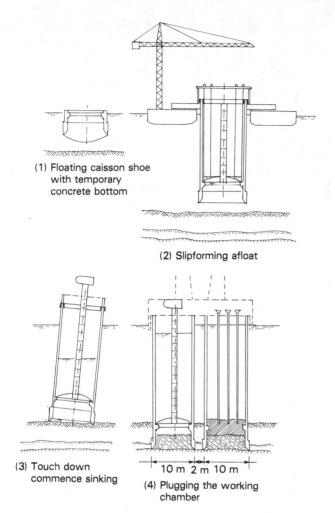

(1) Floating caisson shoe
with temporary
concrete bottom

(2) Slipforming afloat

(3) Touch down
commence sinking

10 m   2 m   10 m

(4) Plugging the working
chamber

Figure 5-2   Fray Bentos International Bridge caisson sinking sequence

The value of this technique is nicely illustrated by its use in the construction of the Fray Bentos Bridge (Fig. 5-2), across the Rio Uruguay [16]. The main piers are each founded on a group of four 10 m (33 ft) diameter caissons which stand in approximately 20 m of water, having been sunk through 5.5 m of overburden by the pneumatic method. The contractor elected to slipform the caissons, using floating plant, immediately above the foundation position. To avoid the need for special docks or slipways, he cast a 120 mm thick hemispherical concrete shell onto the underside of the steel cutting edge, which enabled it to float in shallow water. Once on location in deeper water, the air deck was fitted, the temporary shell broken out, and the slipforming progressed in the conventional manner.

A similar approach was used on a much bigger scale for the largest pier, supporting the main 375 m (1230 ft) cable-stayed span, of the Mississippi Bridge at Luling, 19 km (12 mile) from New Orleans [17]. This 61 m long×25 m

(200×82 ft) wide caisson was to penetrate 30 m (98 ft) into the river bed by open dredging through 40 wells. Due to a steel shortage (1974), the caisson shell could only be fabricated 6 m deep. In order to complete the structure in concrete, afloat, the contractor fitted 5 m diameter domed airtight caps over the dredging wells. Careful control of the air pressure beneath the domes allowed the trim and buoyancy to be maintained. The air domes were broken out when the caisson touched the river bed.

A much more sophisticated use of compressed air, to control the trim of a caisson, whilst sinking it by the open dredging method, was used on the Tagus River Bridge [18]. This technique was developed by the Tudor Engineering Company for the substructures of the San Francisco–Oakland Bay Bridge and used by them in the design of the Tagus main piers (Fig. 5-3).

The largest pier was designed to found on bedrock 79 m (259 ft) below the water surface, after sinking through 33 m (108 ft) of soft mud and 18 m (59 ft) of sand, by open grabbing. The great depth and uncertain bearing capacity of the mud layer made it imperative that the attitude of the caisson be controlled,

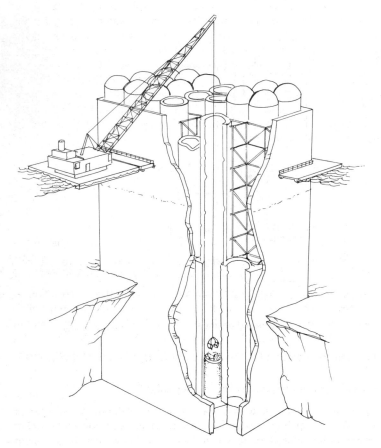

Figure 5-3 Sinking the main Tagus caisson (after Riggs [18])

throughout its full-height construction and through most of its sinking phase. The problem was complex because three factors were changing at the same time. Concrete was being added to the structure, the immersed volume was changing and the base was being influenced by the mud layer. A detailed method statement was produced which controlled the pouring of the concrete, the movement of the air domes and the dredging between the 28 wells. Air pressures were around 125 kN/m^2 (18 psi) in the sealed wells.

The second main pier, though slightly smaller, required even closer control. It had a warped cutting edge designed to fit a sloping rockhead, the maximum difference in level between opposite corners being 7 m (23 ft). This irregular shape had a natural list which was controlled by pressurizing the appropriate dredging wells.

5.5 Pneumatic caissons on a piled foundation

The new Lillebaelt Bridge [19] is a suspension bridge of 600 m span, whose main piers consist of caissons supported on precast concrete piles. Two hundred and six 380×480 mm (15×19 in) piles, approximately 31.5 m (103 ft) long, were driven into the channel bed some 19 m (62 ft) below mean water level. The 56 m long×22 m wide (184×72 ft) caissons were built in a construction dock on the site and floated into position over the piled foundations. The caissons (Fig. 5-4) incorporated water ballasting compartments which enabled the sinking to be accurately controlled. Operations carried out in the working chamber consisted of trimming piles, removing soft material and placing the high-quality concrete plug. Working pressures were around 200 kN/m^2 (29 psi). The caisson performed the function of a high-quality pile cap in a location where a very large cofferdam would otherwise have been required.

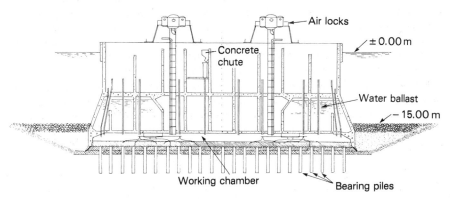

Figure 5-4   Foundation caisson for the Lillebaelt Bridge (after Meldner [19])

5.6 Pneumatic caissons: underpinning the cutting edge

This example concerns yet more hard-won experience in bridging the Mississippi River, this time for a pipeline suspension bridge (Fig. 5-5) located 130 km (81 mile) below St Louis, and constructed in 1955 [20]. The 19 m long×8.5 m wide (62×28 ft) caisson on the Missouri side was intended to reach bedrock 44 m

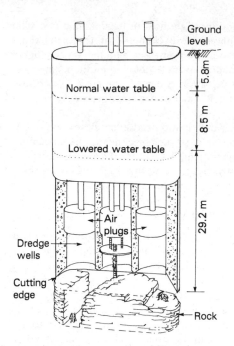

Figure 5-5   Mississippi Pipeline Bridge (after Newell [20])

(144 ft) below groundwater level. As this would involve air pressures well over the permitted maximum of 350 kN/m² (51 psi), deep wells were installed to lower the groundwater by 15 m (49 ft). The caisson was sunk initially by open grabbing, converting to pneumatic for the last 7.5 m because a very uneven rock surface was known to exist. The deep wells were damaged by subsidence during the open grabbing phase, and could only draw down the groundwater by 9 m.

After sinking to within 1.6 m (5 ft) of its intended founding level, the cutting edge was seriously damaged by contact with a ridge of rock. A painstaking *underpinning* job (Fig. 5-6) was then undertaken between the cutting edge and the steeply sloping bedrock. The vertical sand face was exposed in small areas of less than 0.1 m² (1 ft²) and quickly protected with 'puddled' clay to prevent loss of

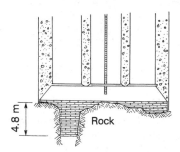

Figure 5-6   Detail of the underpinning on the Mississippi Pipeline Bridge

air. The wall was then built up with cast *in situ* reinforced concrete blocks
0.6×0.3×1.5 m long (2×1×5 ft), starting from the areas where the cutting edge
could be supported on solid rock. The maximum depth of underpinning achieved
was 5 m below the cutting edge, whereupon the maximum allowable air pressure
had been reached. At this point, the caisson was bottomed out with a reinforced
concrete slab.

6 Onshore structures

One of the most interesting recent developments in the use of concrete caissons,
has been in the field of underground railways. Two examples follow.

6.1 The Amsterdam Metro

Most of the buildings in Amsterdam that are more than 60 years old are founded
on the upper of two sand layers. Many are founded on timber piles which must
remain immersed in the groundwater to avoid rot. The water table is within 1 m
of the surface.

In order to avoid unacceptable settlements, the extension to the Metro constructed in 1978, from the Central Station to Amstel station [21], had to be
founded in the lower sand layer, some 12 m (39 ft) below the surface. Moreover,
it had to pass between many of these fine old buildings with minimum risk of
damage to their foundations. No form of groundwater lowering could achieve this
and so the pneumatic-caisson solution was chosen (Fig. 6-1).

Sections of the running tunnel varied in length from 32.35 to 39.35 m (106–
129 ft) in length and were 10.4 m high by 10.3 m wide (34.1×33.8 ft). Special

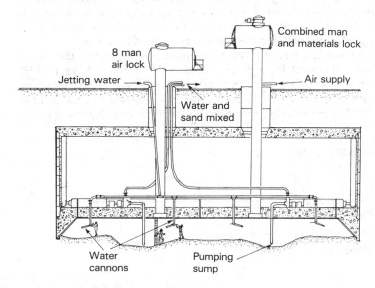

Figure 6-1 Amsterdam Metro caisson

station units were 35.67 m long, 17.48 m wide and 13.70 m high (117×57.3×45 ft).

All sections were cast above ground and detailed attention was given to concrete mixes and techniques in order to achieve a waterproof structure [22]. No external membrane was used. Four concrete mixes were devised for use in different parts of the caisson, all intended for placement by pump. Heat of hydration was kept low by limiting the cement content to 275 kg/m^3 (464 lb/yd^3) and concrete density improved by the use of an extremely fine glacial sand. Phosphate retarders were used to delay the setting of the base slab by up to 24 h and the walls, where retardation was undesirable, were cooled by internal water circulation (800 to 1000 1/h). By these methods, concrete cracking was largely avoided.

The working chamber, only 2.1 m (7 ft) in height, was formed by laying permanent concrete shutter slabs against a preformed sand mould. On top of this, the tunnel section was constructed, with temporary concrete end walls for the sinking phase. Access to the working chamber was by two air locks but the inside of the tunnel section remained at atmospheric pressure. The distance between adjacent caissons before sinking was 650 mm (26 in).

High pressure-water cannons, delivering 750 m^3/h (2000 US gal/h) at a pressure of 850 kN/m^2 (123 psi), were fitted to the working chamber roof. These were able to break up all the soils encountered. The resulting slurry was pumped away to a stilling basin on the surface. The jetting and sand pump motors were able to operate to maximum efficiency in 'free' air, inside the tunnel. Timber piles and old foundations were frequently encountered and these were removed through the air locks. Allowed tolerances were 175 mm (7 in) longitudinal, 200 mm (8 in) transverse and 100 mm (4 in) vertical, but the final results were normally within 30 mm (1.2 in) of the true position.

After plugging the working chamber, the connections between adjacent sections were made. This required local ground freezing, using liquid nitrogen.

6.2 West Berlin subway extension

The line of an extension to the underground railway, constructed in 1968, crossed an infilled glacial valley [23]. The fill material was generally soft and partly organic, quite unsuitable for foundation purposes. The natural water table was 6 m (20 ft) below the surface but the dense sand on which the structure was to be supported was at depths of up to 30 m. The convential cut-and-cover technique was ruled out because temporary lowering of the groundwater would have threatened the stability of the adjacent buildings.

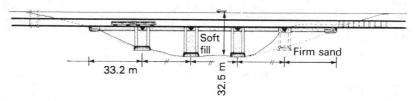

Figure 6-2 West Berlin underground railway

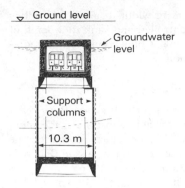

Figure 6-3 Section showing the tunnel on the caisson

Many alternatives were proposed by contractors and a method employing two types of caisson was selected. Four foundation caissons were to be sunk, by pneumatic methods, down to the dense sand layer and between these would span the concrete tunnel sections. The 170 m (558 ft) wide valley was to be crossed by what amounted to a buried five-span bridge (Fig. 6-2).

The pier caissons (Fig. 6-3) consisted of four reinforced concrete 1.5 m (5 ft) square piles connected by thin walls to give a plan area 7.1×10.3 m (23.3×33.8 ft). They were constructed to full height before sinking began.

Spoil removal was by water jetting at 500 m³/h (1300 US gal/h) and a pressure of 600 kN/m², (87 psi) and bentonite slurry was used to lubricate the outer surface. The caissons sank easily under their own weight. A maximum air pressure of 260 kN/m² (38 psi) was reached in the deepest caisson 32.5 m (107 ft) below ground level.

Ground settlements did occur up to 8 m (26.2 ft) from the caissons with 10 cm (4 in) being recorded at a distance of 4 m. The following causes were given:

(a) rocking of the caisson (change of tilt) in the early stages;
(b) pressure changes due to bentonite injection;
(c) compaction of loose soil under the cutting edge;
(d) friction between soil and caisson;
(e) inflows of soil during blow down.

Tunnel sections 33.2 m (109 ft) long were also constructed on the surface and sunk as caissons (Fig. 6-4). They consisted of an outer and inner shell with a waterproofing layer between. The tunnel cross-section was given additional

Figure 6-4 Cross-section of the support caisson

clearances of 50 mm (2 in) in height and 200 mm in width to allow for sinking inaccuracies. These caissons were more easily controlled and sank to a depth of 15 m (49 ft) without producing significant settlements. Connections between adjacent tunnel sections were made in open excavations.

6.3   Huntly power project: cooling-water pumphouse

This 1000 MW thermal power station is situated beside the Waikato River, New Zealand [25]. A high water table and highly permeable substrata made more conventional foundation techniques unattractive. The designers recognized the savings in time and money made possible by the use of an open caisson (Fig. 6-5) and, after a successful small scale trial, construction began in 1976.

The reinforced concrete caisson walls were built to a height of 13.5 m (44.3 ft) on the surface and the weight of the 64 m long × 24 m wide (210 × 79 ft) structure, ready for sinking, was 16 000 tonnes. It was divided into four cells 19.1 × 11.8 m (62.7 × 38.7 ft) in plan, with a smaller central cell 19.1 × 8.2 m (62.7 × 27 ft).

Excavation of the highly permeable pumiceous sands was by submersible electric pumps assisted by jetting nozzles set into the caisson walls. This combination was very effective but gave rise to the washing in of a considerable amount of material from outside the caisson. In order to limit this action, a sheet pile wall was constructed to form a cut-off around the caisson and secure firm standing for the 60 tonne crawler cranes and other plant which serviced the construction.

Vertical steel bearing piles were driven to depths between 36 and 47 m (118 and 154 ft) at 2 m intervals around the perimeter, inside the sheet piled cut-off. These served both to restrain the structure against sliding and to give added vertical support, when the founding level was reached.

Jetting nozzles were set in two rings at 1 m centres. The upper ring was normal to the sloping inside face of the cutting edge and the lower ring, set into the base of the walls, sloping inwards and downwards at 45°. The jets were coupled to a ring main broken into quadrants to allow selective use for control of caisson tilt. This worked very well for most of the operation.

Difficulties were encountered with a layer of firm silt 400 mm (16 in) thick. Localized undermining of this layer by the pumps resulted in sudden failures of the working face accompanied by inflows beneath the caisson walls. A long jetting probe in combination with a powerful 200 mm diameter airlift was finally successful in breaking up this layer.

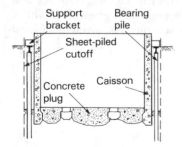

Figure 6-5   Huntly power project: the cooling-water caisson

The submersible pumps proved very successful, excavating up to 40 m /h (52 yd³/h) and coping with material up to 70 mm (2¾ in) in size. However, the water they removed had to be constantly replaced to maintain a level not more than 2 m below the outside groundwater level. This prevented boiling and maintained the stability of the sheet piled cut off.

Particular attention was paid to the planning of the reinforced concrete plug to ensure sound construction.

6.4 Complex structures

The onshore caissons described above are all of regular shape and simple construction. Where irregular shapes such as office foundations are sunk as caissons, some internal stiffening of beams and columns will be required. An example is the Nikkatsu International Building in Tokyo [25], where extensive instrumentation was used to check tilt and wall loadings.

7 Ports and harbours

7.1 Development and application of the box caisson

Advances in reinforced concrete technology at the beginning of the 20th century made the construction of a thin-walled box possible. A design developed by Dutch engineers Kraus and Van Hemert, for caissons, constructed and launched on their sides, was used in 1904 for the development of a military harbour at Talcahuano, Chile [26]. These units were 10.35 m high, 6.5 m wide and 10.0 m long (34×21.3×33 ft), with a large buttressed bottom plate to give stability when covered with backfill. The average wall thickness was 0.20 m (8 in) and each box weighed approximately 215 tonnes (Fig. 7-1).

The large wall areas were cast as slabs and the completed boxes could be

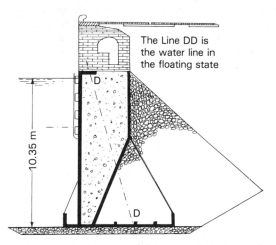

The Line DD is the water line in the floating state

Figure 7-1 The Kraus-designed floating caisson

launched into relatively shallow water (5 m or less). Very simple shore establishments were needed. Once they had been towed into location, the caissons were tilted by water ballasting, the tapered lower section ensuring verticality. Permanent ballasting, to resist overturning and sliding, was provided by dredged sand.

With the introduction of larger vessels, depths demanded at the quayside increased and bigger caissons were required. To obtain the required stiffness, the boxes were divided into cells and intermediate longitudinal diaphragms were required. It was no longer economic to build these designs as tilting caissons.

The development of the Port of Rotterdam in the years 1920 to 1930 saw great advances. To begin with, the boxes were designed with economy of materials in mind. They had sloping walls of tapering cross-section. However, towards the end of this period, very large contracts were awarded (one company alone, HBM, had 4228 m (4624 yd) of quay wall under construction simultaneously in 1930) and alternative labour-saving designs were offered, using vertical walls of uniform cross-section.

The widespread use of steel sheet-piling ended what some engineers regarded as the 'golden years' of caisson construction. Nevertheless, the technique has clear advantages in certain circumstances. Rapid development of large harbours in Dubai and Dammam, Saudi Arabia, have seen the establishment of 'production line' techniques and a rate of progress inconceivable by other techniques. the construction of a yacht marina at Brighton, situated on an exposed part of the south coast of England, was achieved by using box caissons placed in limited periods of calm weather. The extension to one of the world's largest coal-handling ports at Bay Point, Australia, incorporated caissons constructed at a remote harbour and installed with minimum disturbance to ships using the existing port facilities.

7.2 Construction methods

The methods employed for the construction of box caissons depend upon the size of the overall project. Small schemes do not justify the construction of specialized facilities and use is normally made of existing dry docks, floating docks and quays.

One of the first purpose-made construction jetties was built by HBM in Rotterdam in 1928 (Fig. 7-2). It consisted of two overhead lattice girder bridges, carrying a materials conveyor, travelling cranes and winches, all servicing six parallel building berths and two materials unloading berths. The base sections were constructed in two reinforced concrete floating docks.

The establishment of an efficient onshore slipforming yard was the common starting point for the major developments at Dubai, Dammam and Brighton. Casting bays were set up on each side of a central rail spine. Base sections were cast on transportation bogies, then the slipform gantries were moved into position. The gantries provide all the services required: shutter support, materials hoisting, personnel hoisting, etc. After curing, the caissons are moved sideways onto the main rail line ready for placing.

At Dubai [27], 162 caissons of between 3000 and 3500 tonnes each were cast in 95 weeks using ten casting positions and two gantries. Concreting at the rate of 25 m³/h (33 yd³/h), the slipforming of the 18 m (59 ft) high boxes was completed in 48 h. A similar arrangement was used at Dammam [28]. The 199 caissons 20 m

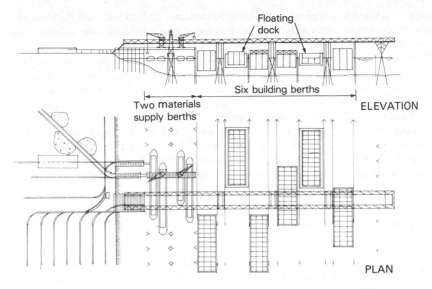

Figure 7-2   The HBM-designed caisson building jetty of 1928

long by 14 m wide by 16 m (65.6×46×52.5 ft) high were cast in 15 bays serviced by three gantries, which produced 20 units per month.

The more complicated design of the Brighton caissons [29], incorporating horizontal slabs, and the contractor's decision to slipform only on dayshift, resulted in a slower rate of production. Each caisson (Fig. 7-3) took a total of eight weeks to complete, the 12 casting bays being serviced by one gantry.

Concrete of characteristic strength 31.5 MPa (4572 psi) with a cement content of 140 kg/m³ (692 lb/yd³) was used at Brighton for the slipforming. In warm

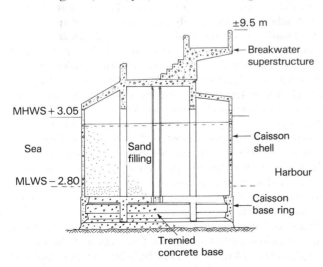

Figure 7-3   Brighton Marina caisson

summer conditions, a lignosulphate-based plasticizer was added. The high workability and early strength required for the base ring and deck was achieved by
increasing the cement content to 440 kg/m^3 (742 lb/yd^3).

A completely different approach was taken for the new loading facility at Hay
Point, Queensland [30]. The caissons were assembled at a dry dock in Mackay
Harbour, 29 km (18 mile) to the north. All the caisson walls and bulkheads were
precast horizontally on the dock floor. Floors and roofs were cast *in situ* and the
completed assembly post-tensioned together. All handling and loading equipment
was installed on the caissons prior to being towed to the site. The longest caissons
were 46.2×41.9×8.5 m (151×137.5×28 ft) with 12.2 m (40 ft) square columns
18.3 m (60 ft) high at each corner which carried the 4000 tonne/h shiploader. All
were positioned in an advanced state of construction and the first ship was able to
dock 47 days after the last caisson was placed.

7.3 Preparation of the sea bed

This is a critical operation in the box-caisson construction technique. It is strongly
influenced by the unknowns of weather and substrata conditions and the most
serious delays usually occur as a result of some misjudgement or omission in the
preparation of the sea bed for caisson placement.

In 1916, three caissons forming part of the new Ijmuiden quay at Surabaya,
[26] disappeared into the harbour mouth during the backfilling operations.
Through heroic efforts of the resourceful engineers, the boxes were recovered,
largely intact, and restored to their intended locations. But the nine months it
took to retrieve them could not be regained. Subsequent borings revealed that the
sand filling below the caissons was contaminated by a 1 m thick layer of soft mud.

Confidence in the use of box caissons was further shaken when, during the
reconstruction of the war-damaged Port of Rotterdam, some caissons moved
forward 0.5 m during backfilling. The cause was again traced to contamination of
the sand foundation by silt. After this, even greater attention was paid to bed
preparation and the caissons were fitted with a steel skirt, which projected down into
the sand to give greater resistance to sliding.

The method adopted for clearing unsuitable material from the sea bed depends
upon the dredging technique most suitable for the site conditions. If large-scale
operations, well ahead of caisson placement, are anticipated, a study of sea-bed
movement is necessary to establish the amount of deposition which may occur
between the two operations. In every case, final clearance by airlift and checking
by divers is an absolute necessity.

At Dammam, a 2 m deep trench was dredged along the caisson line. This was
backfilled with 750 mm (30 in) of coarse aggregate and 250 mm of sand, carefully
placed by bottom dumping hopper barges, and levelled by a template operated
from a pontoon. The final levelling was controlled by divers. This gave a caisson
founding level 1 m below the new harbour bed.

The caissons at Brighton were held in position approximately 0.5 m above the
cleared chalk bed. Adjustable side shutters were slid down into position and sand
bags used to seal any gaps. A tremie concrete plug was then poured. The mix for
this concrete initially contained 410 kg/m^3 (692 lb/yd^3) of OPC cement, but the
heat of hydration gave rise to vertical cracking in the stiffening ring, at the caisson

base. The mix design was amended to use 350 kg/m³ (590 lb/yd³) of sulphate-resisting cement for the lower portion of the plug, where early bearing strength was required, with 70% of the cement being replaced by ground blast-furnace slag in the mix for the upper portion.

7.4 Transportation and sinking

For the convenience of constructing the caissons in ideal onshore conditions, the problem of getting these very large structures into the water must be overcome. The enormous size of the projects at Dubai and Dammam (Fig. 7-4) justified the installation by the contractors of a 5000 tonne shiplift (*synchrolift*). This consisted of a platform, supported between two piled jetties, which could be raised or lowered by two banks of seven synchronized winches. The caissons were transferred to the lift by means of the rail line running through the casting yard. As the platform was lowered into the water, the caisson cells were ballasted with water. They finally floated clear with a draught of 11 m (36 ft).

When floating caissons into position, it is usual to install a demountable assembly incorporating all the necessary mooring winches, deballasting pumps, power packs, water-level monitoring and control gear. Submersible pumps are installed in each of the caisson cells so that water levels can be adjusted to give precise control of trim. If necessary, the caisson can be refloated and the operation recommenced. A positional tolerance of ±50 mm can be achieved.

Caissons are interlocked by vertical nibs of concrete which protrude at each end, the space between, usually about 0.5 m, being filled with grouted Nylon bags. At Dammam, the quay-wall caissons were divided by a central longitudinal diaphragm. The cells on the seaward side were filled with seawater whilst those at the rear were ballasted with dredged sand to give additional resistance to horizontal loads.

The Brighton caissons were positioned by a method which placed less reliance on calm sea conditions (Fig. 7-5). Anticipating a downtime due to bad weather of 56%, the contractors opted for a caisson-placing crane and transfer car. They were able to obtain equipment previously used at Hanstholm, Denmark.

Based on two 53 m (174 ft) long box girders which supported a portal crane, the assembly ran on a 6 m (19.7 ft) gauge track. Working from the last completed caisson, the crane subframe was keyed into the deck and jacked clear of its supporting bogies. In this position, the main box girders were cantilevered

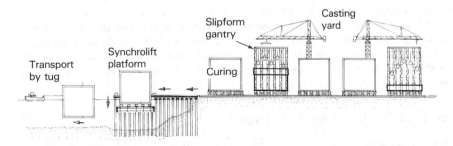

Figure 7-4 Caisson production sequence at Dammam

Figure 7-5   Caisson-placing crane and transfer car used on the Brighton Marina

forward over the new caisson location. The next caisson to be placed was brought along the crane tracks on a transfer car and correctly orientated. The portal crane moved back, lifted the caisson and travelled forward. Twenty-four hours of calm seas (less than 1 m wave-height) were required for the caisson placing and plugging sequence. When the forecast was satisfactory, the portal crane lowered the caisson into position, its leading edge restrained by a robust nose guide whilst the trailing edge was keyed into the caisson behind.

This method is obviously limited as to the size and weight of caisson that can be handled by the crane, but it has the great merit of minimizing the period during which the unballasted, partially complete caisson is at risk to storm damage.

Scour protection to the foundation of box caissons is a factor which must be examined. If it is deemed necessary, it should be carried out as soon as possible after the caisson has been placed.

8   Caissons in offshore construction

8.1   Introduction

Recent developments in the design and construction of reinforced and prestressed concrete gravity structures for hydrocarbon exploitation are described in Chapter

34 of this book. It is sufficient here to note that these are a development of the humble box caisson, brought about by the pressing economic need to recover the resources which lie beneath the sea bed.

Wherever such a need arises, requiring engineering on a grand scale to be executed in a limited time span, technology often takes a leap forward. While many of the problems are purely logistical, the assembly of skilled teams of engineers and the large sums of money involved frequently result in development of new techniques.

Such a scheme is currently under construction in the Netherlands. The Eastern Scheldt storm surge barrier is part of the Delta Plan, conceived as a response to the disastrous floods which devasted the south-west part of the country in 1953, in which 1800 people were drowned.

8.2 The Eastern Scheldt storm surge barrier

Sixty-six caissons play a vital part in this structure, by supporting the piers from which the storm gates are suspended. The barrier is in three parts with a total length of 2835 m (3100 yd) and is designed to hold back surge tides corresponding to a high water level which has a 2.5×10^{-4} probability of being exceeded in any one year [31].

The piers support a water-retaining structure resisting a differential head of up to 4.2 m (13.8 ft), and the road deck which runs across the top of the barrier. Pier design was originally based on the use of pneumatic caissons but it was discovered that the firm pleistocene sand lay up to 35 m (115 ft) below sea-bed level. The loosely packed holocene sand above could not be relied upon to resist the overturning moments on the gate structure, which increased with depth. The risk to men in a working chamber 35 m below sea level, should a sudden storm arise, was unacceptable.

8.3 Pier design and construction

The final pier design was based on a tapered box caisson giving the optimum base area with minimum weight and resistance to tidal flow. Above the box is a vertical shaft in which the gate guides are incorporated. The base dimensions are 50 m long by 25 m (164×82 ft) wide while the height varies from 35 to 45 m (115–148 ft) depending upon the location. The structural concrete members were designed in accordance with the semi-probabilistic method with an envisaged service life of 200 years. The external form of all the piers has been kept as close as possible to a standard shape for ease of construction and in order to accommodate the placing vessel.

The piers are under construction in a 1 km square dock formed in the artificial Schaar island. The dock is divided into three sections so that the piers will be released in complete batches, each section being flooded only once, by dredging through the enclosing dam. The construction period for each pier is 240 working days, the structure being divided into seven pours.

Tests were carried out to establish the relationships between initial concrete temperature and cement types with rate of gain of strength and the build-up of heat of hydration [32]. Designers were concerned about the effects of *thermal cracking* during the construction stage, its effect on the durability of the structure, and the corrosion of the reinforcing steel. A superplasticizer additive was used to reduce the water/cement ratio to 0.42 in the design mix, which contained 350 kg/m³ (590 lb/yd³) of cement and gave a 28-day cube strength of 55 MPa (7982 psi). The base slab, which is up to 2 m thick, required insulation for 7 days to prevent unacceptable tensile stresses developing. Other parts of the structure were cooled by circulating water.

8.4 Development of special equipment

One of the most important aspects of this project has been the development of specialized equipment, to undertake precise engineering functions in the hostile offshore conditions.

After preliminary dredging, the loose sand below the dam has been compacted by the barge Mytilus using a bank of five vibrating needles. Working in water depths up to 35 m (115 ft), layers of sand up to 15 m (49 ft) thick have been successfully treated. This barge is 88.25 × 32.9 m (289.5 × 108 ft) and is equipped with swell compensators.

Immediately following ground treatment, a tough protection mat is laid by the barge Cardium (cockle), shown in Fig. 8-1. This is equipped with a pair of sophisticated levelling beams, which smooth the sand surface whilst removing loose material by a combination of fine water jets and suction dredging—known as 'dust panning'. The barge then lays a prefabricated *filter mat* 200 m long by 42 m wide (656 × 138 ft) along the axis of the pier. Using finer nozzles on the other side of its levelling beam, it cleans off any sand which has been deposited on top of the filter mat and lays a protection mat over the area where the pier will be located. The mats consist of three layers of carefully graded sand and gravel, held in place between wire reinforced, plastic fabric sheets. They are made up in a dockside factory and wound onto floating rollers which are towed out to Cardium.

Perhaps the most innovative piece of equipment, whose successful operation will prove crucial to the success of the whole operation, is the pier transportation and placing ship Ostrea (oyster). This vessel (Fig. 8-2) has a U-shaped hull 87.25 m (287 ft) long with a beam of 47 m (151 ft). It is intended that she will

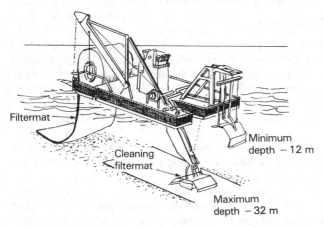

Figure 8-1 Cardium matlaying vessel

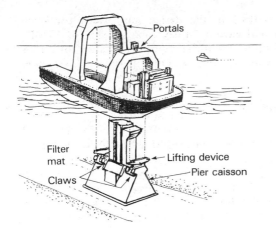

Figure 8-2 Ostrea pier-lifting ship

manoeuvre inside the flooded construction basin, without the aid of tugs, over the pier to be lifted. To achieve this, she is equipped with a propeller and jet tube at each corner and a propulsive power of 8000 hp. Four hydraulic lifting claws, suspended from the two portal frames, will be fastened to projections on the pier base. The piers will have submerged weights up to 9320 tonnes, when lifted clear of the dock bottom, and a draught of 11 m. The tug-assisted tow to the pier sites will be up to 23 km (14.4 mile) in length and tide and sea conditions will be critical. At her destination, Ostrea will link up with an accurately positioned mooring barge.

The actual pier-placing procedure will be undertaken in good weather when tidal currents are slack. Several alternative procedures have been planned and it remains to be seen how critical the sea conditions will be. Certainly the intended positional tolerances of 300 mm (12 in) on the x and y-axis with an allowed rotation about the vertical z-axis of 7 mm/m (0.25 in/yd) are better than those which have been achieved in most offshore situations to date.

8.5 The harnessing of tidal power

The search for inexhaustible, or at least renewable, supplies of power has received greater attention in recent years, since those countries with oil reserves realized the true value of their wasting assets and began to charge a realistic price.

Tidal power has long been recognized as a source which could be developed and the technology now exists to harness it. This has already been done at La Rance, in Brittany, France, and currently attention is focused on the River Severn in south-west England.

According to Stokes and Street [33], the cofferdam technique used at La Rance accounted for 30% of the project's total cost. They proposed the use of caissons for the Severn tidal barrage on the grounds of hydraulic engineering, improved construction time and economy.

There can be no doubt that, when this imaginative scheme is eventually undertaken, the experience gained and the techniques developed at the Eastern Scheldt will be of immense value.

References

1. Carson, A. B., *Foundation construction*, McGraw-Hill, New York, 1965, 424 pp.
2. Tomlinson, M. J., *Foundation design and construction*, 4th Edn, Pitman, London, 1980, 793 pp.
3. Wilson, W. S. and Sulley, F. W., *Compressed air caisson foundations*, Institution of Civil Engineers, Works Construction Division Paper No. 13, 1949, 30 pp.
4. Wilson, W. S. and Smith, H. S., *The relative methods of sinking bridge foundations*, Institution of Civil Engineers, Works Construction Division Meeting, Introductory Notes and Discussion, Dec. 1950, 32 pp.
5. Terzaghi, K. and Peck, R. B. *Soil Mechanics in engineering practice*, 2nd Edn, Wiley, New York, 1967, 563 pp.
6. Yang, N. C., Conditions of large caissions during construction, *Highway Res. Rec.* No. 74, pp. 68–82.
7. Davis, F. W. and Kirkpatrick, C. R. S., The King Edward VII Bridge, Newcastle upon Tyne, *Min. proc. Inst. Civ. Eng.*, Vol. 174, 1907–1908, Paper No. 3742, pp. 158–221.
8. Anderson, D., Tyne Bridge, Newcastle. Institution of Civil Engineers. Minutes of proceedings volume 230 1929/30 Paper 4771. pp. 167–202.
9. Walder, D. N. *et. al.*, *Medical code of practice for work in compressed air*, 3rd Edn, Construction Industry Research and Information Association, Report 44, 1982, 52 pp.
10. King J. D., Coping with compressed air sickness down the hole, *Tunnels and Tunnelling*, Jan./Feb., 1980, pp. 64–69.
11. Blaine, E. S., Practical lessons in caisson sinking from the Baton Rouge Bridge, *Eng. News Rec.* 6 Feb. 1947, pp. 213–215.
12. Hayward, D., Humber Bridge, *New Civ. Eng.*, 10 April 1975, pp. 22–24.
13. Lee, D. H., Two prestressed concrete bridges in Pakistan, *Civil Engineering*, Oct. 1969, pp. 61–62.
14. Boynton, R. M., Mackinac Bridge foundations constructed at record speed by unusual methods, *Civil Engineering*, May 1956, pp. 301–306.
15. Hayward, D., Stubborn caisson sinks under 6500^T, *New Civ. Eng.* 17 July 1975, pp. 16–17.
16. International Bridge Fray Bentos—Puerto Unzué, Uruguay/Argentina *Hochtief* Nachr. 49, No. 3/4, 1976, p. 27.
17. Massive caissons sunk for record bridge, *Eng. News Rec.*, 11 Aug 1977, pp. 20–23.
18. Riggs, L. W., Tagus River Bridge, *Civil Engineering*, Feb. 1966, pp. 41–45.
19. Meldner, V., Pneumatic caisson work for the bridge accross Lillebaelt, *Baumasch. Bautech.*, July, 1971, pp. 289–295.
20. Newell, J. N., Pneumatic caisson pier for world's longest pipeline suspension bridge, *Civil Engineering*, May 1956, pp. 51–55.
21. Pause, H. and Hillesheim, F. W., Underground Railway, Amsterdam, *Tech. Ber. P. H. Holzmann*, No. 4, 1975, p. 50.
22. An underground railroad is built on site above ground, *Concr. Constr.*, Sept. 1974, pp. 453–456.

23. Scheibner, D. V., The caisson technique used for underground bridging of a mud-containing ice age trough. *Strasse-Brucke-Tunnel*, Vol. 23, No. 1, 1971, pp. 23–30.

24. Brown, A. S. and Cox, D. D., Design and sinking of a large concrete ciasson at the Huntly Power Project, *New Zealand Eng.*, 15 Feb. 1980, pp. 28–32.

25. Mason, A. C., Open caisson method used to erect Tokyo office building, *Civil Engineering*, Nov. 1952, pp. 46–49.

26. Mesman, T. J. *et. al.*, *Caisson construction*, Hollandsche Beton Groep. nv. (HBG) Rijswijk, 1977, 72 pp.

27. Dubai dry dock slides to success, *Constr. Ind. Int.*, Dec. 1977, pp. 24–29.

28. Brown, K., Caissons at the Dammam site, *Constr. News*, 25 May 1978, pp. 28–31.

29. Llewellyn, T. J. and Murray, W. T., Harbour works at Brighton marina: construction, *Proc. Inst. Civ. Eng.*, Vol. 66, Part 1, May 1979, pp. 209–226.

30. Expansion of coal loading facility at Hay Point, *Port and Harbour Authority*, No. 57, Sept. 1976, pp. 161–162.

31. Kooman, D. *et al.*, Design of the Eastern Scheldt storm surge barrier, *Cement*, No. 12, 1979, pp. 12–15.

32. Berg Pvd, Concrete technology aspects of the Eastern Scheldt storm surge barrier, *Cement*, No. 12, 1979, pp. 64–67.

33. Stokes, C. J. and Street, R. D. J., Turbine caissons for the Severn barrage, *BHRH Fluid Eng*, Sept. 1981, pp. 167–176.

29 Bins and silos

John E Sadler Mostafa H Mahmoud
F Thomas Johnston & Abdul Q Ghowrwal

R S Fling & Partners Inc., Columbus,
Ohio, USA

Contents

Notation

B outlet size required to prevent material arching or binding

BB ratio of vertical shear stress to vertical pressure of flow channel

C factor to be used in determining pressures in the converging part of the flow channel

D outlet size required to prevent ratholing

D_f diameter of flow channel

D_s silo diameter

f_c unconfined compressive strength of bulk solid

ff flow factor of hopper or flow channel

FS factor of safety to insure self-cleaning

h distance from the apex of the converging section of the flow channel to the point under consideration within that section

h_0 distance from the apex of the converging section of the flow channel to the transition between the cylindrical and the converging sections

h_1 distance from top of hopper to point under consideration

K ratio of the horizontal pressure to the vertical pressure in the silo under static conditions

K_0 initial ratio of the horizontal pressure to the vertical pressure in the silo under static conditions

l coefficient; $l = 0$ for plane flow; $l = 1$ for conical flow

m percentage increase in coefficient of lateral pressure due to vertical pressure

n percentage increase in unit weight due to vertical pressure

p horizontal static pressure in silo

p_f horizontal pressure in flow channel

p_s horizontal static pressure in the material

q vertical static pressure in silo

R hydraulic radius of silo

s angle of repose of bulk solid

T hoop tension in wall

u coefficient of wall friction ($\tan \phi'$)

V vertical pressure at any point within the hopper or the converging part of the flow channel

V_f total vertical material friction load on the silo wall

V_0 vertical pressure at the top of the hopper or the converging part of the flow channel

W pressure normal to the hopper surface or the boundary of the converging part of the flow channel

W_m total weight of material in silo

z distance from top of material at wall to point under consideration

α half-apex angle of the convergent part of the flow channel

β angle between major principal plane and boundary of the convergent part of the flow channel

γ unit weight of bulk solid

γ_0 initial unit weight of bulk solid

δ effective angle of internal friction

θ hopper slope angle measured with the horizontal

$\bar{\theta}$ slope of converging part of flow channel measured with the vertical

Φ piping factor

ϕ angle of internal friction of the bulk solid

ϕ' angle of material–wall friction

Summary

The design of silos should always be based on a thorough understanding of the flow characteristics of the bulk solid and the flow pattern which may develop in the silo. Both the flow characteristics of the solid and the configuration of the silo bottom play a major role in determining the flow pattern in the silo.

Under static conditions, the pressures in a silo are usually computed using Janssen's formula. During flow, the pressures within the flow channel may be either higher or lower than the pressure in the static material around the flow channel. In the case of concentric discharge, the silo wall is usually subjected to hoop tension only. In case of eccentric discharge, the wall is usually subjected to both hoop tension and horizontal bending.

The silo wall thickness should be estimated considering the vertical material friction load. The hoop reinforcement is usually computed considering the hoop tension and horizontal bending in the wall. The silo bottom should be designed to support the vertical pressure of the solid. The foundations should be analyzed considering the cyclic nature of silo loading and unloading and, in some cases, the interaction between the structure and the supporting strata.

Silo construction techniques, construction supervision and inspection, and lessons learned from observations of silos in service are essential for a successful silo installation.

1 Introduction

The first large silos were constructed over 120 years ago for storing grain. Today, silos are increasing rapidly in height, diameter, storage capacity, and the variety of stored materials. Designers and builders are employing higher strength steels and concrete, prestressing concepts, and a great variety of withdrawal systems having high throughput.

Before 1860, designers assumed that granular material behaved as a quasi-liquid and exerted pressures similar to hydrostatic pressure. These assumptions caused overdesign for static horizontal loads and bottom loads, but failed to account for the vertical material friction load on the walls. Early in the 1880s, Roberts [1] discovered that the pressure on the bottom of a grain silo did not increase after the material depth reached twice the silo width. The friction between the granular material and the wall transferred weight to the wall. In 1885, Janssen [2] derived formulas for granular material pressures on the silo walls and bottom. The first period of growth in silo design, starting with Roberts [1] in 1882 and ending with Ketchum [3] in 1909, was characterized—for that time—by experimental precision and great clarity of formulation. In the middle period, Reimbert [4] introduced the 'antidynamic tube', and modified Janssen's formula based on a hyperbolic function. Reimbert's formula is used in many codes around the world. Few other contributions were made until 1965, as investigators primarily tried to relate flow pressures to Janssen's static pressure.

Since 1965, many excellent studies have appeared in the literature. The experimentation reflects the precision of modern technology, and the formulations reflect the clarity of the early investigators.

Several investigators questioned and investigated many of Janssen's basic assumptions of a constant ratio of vertical to horizontal pressure, and constant density. Cowin and Sundaram [5] derived a modified Janssen formula reflecting linear variations in both variables.

One of the most important silo developments in the last two decades is the increased understanding of granular flow. Jenike [6] defined mass and funnel flow and derived differential equations for mass flow. Johanson [7] used the method of characteristics to determine the stresses in converging flow channels. Walker [8], followed by Walters [9], developed the most practical and simplest approaches for calculations of mass-flow pressures. Clague and Wright [10], and Bransby et al. [11] have experimentally measured the pressures created by mass flow. Johanson [12] and Williams [13] developed formulas for computing discharge rates from mass-flow conical hoppers.

Presently, investigators are modifying existing theories on mass flow to predict the pressures which develop during funnel flow. The authors of this chapter are currently using such pressures to design silos for funnel and eccentric flow, employing finite element methods.

2   Material flow considerations

A silo installation should always be considered as a part of a total system which may consist of preparation, handling, loading, unloading, and transportation. The system requirements, the operation constraints, and the material flow characteristics should form the basis of selecting the optimum number and size of silos and the most desirable silo flow pattern. In fact, the flow characteristics should also be considered in all system components coming in contact with the material, including chutes, hoppers, transfer points, rail cars, etc. Therefore, a thorough understanding of the flow characteristics of the solid and the flow patterns which may develop in a silo is essential for developing economic and reliable silo installations.

The hopper bottom configuration required to maintain the desired flow pattern in the silo should be established considering the flow characteristics of the material and the methods of filling and withdrawal. The design loads and the structural design of the silo should be based on the characteristics of the material and the flow pattern.

2.1   Flow patterns and definitions

Mass flow is a flow pattern which occurs when the entire silo contents are set in motion whenever any solid is withdrawn from the outlet. This pattern results in a first-in, first-out flow of solid in the silo. The flow rate is uniform and almost independent of the height of solid in the silo. The entire volume of solid is live with no stagnant or dead zones. Mass flow silo bottoms should be sufficiently steep and smooth to ensure that the entire silo contents flow whenever the outlet is activated.

Funnel flow is a flow pattern which occurs when the solid flow is confined to a channel that forms within static (stagnant) material. This pattern results in a first-in, last-out flow of solid in the silo. The flow rate is usually erratic except for free-flowing solids. Funnel-flow silos can have significantly dead storage because the solid in the static zone may pack. With properly sloped bottoms and adequate outlet sizes, dead storage can be eliminated, provided the silo is emptied periodically.

Expanded flow is a flow pattern which combines the features of mass flow and funnel flow. Flow usually starts within channels above the outlets, similar to funnel flow, while the rest of the solid is static. As the silo empties, the static solid sloughs into the flow channel and the silo eventually empties. An expanded-flow silo has a hopper consisting of two sections. The lower section is a mass-flow hopper selected to prevent arching and ratholing of the solid. The upper section is a self-cleaning hopper which has sufficient slope to force the static solid to slide towards the mass-flow section as the silo empties. This flow pattern usually results in the most economic silo storage provided the silo is emptied periodically to avoid solid degradation. The mass-flow section minimizes the outlet size and ensures a uniform flow rate. The self-cleaning hopper maximizes the silo storage capacity.

Free gravity flow is the feature that all the solid in a silo can be withdrawn without the use of flow aid devices.

A *free-flowing* solid is a material which develops very little or no cohesion after being subjected to consolidation pressures in a silo.

A *non-flowing* solid is a material which develops high cohesion after being subjected to consolidation pressures in a silo. Some non-flowing solids are not suitable for free gravity flow, or for silo storage.

Live capacity is the maximum volume of solid which can be withdrawn from the silo.

Dead storage is the volume of solid which remains in the silo upon emptying. Dead storage is the difference between the total volume of solid in a full silo and its live capacity. In general, dead storage in silos should be avoided.

A *flow channel* is a zone within the silo that extends upward from an outlet through which the solid flows.

Static or *stagnant* material is the zone of solid around the flow channel. With improperly sloped bottoms, part or all of the static material can be dead storage, resulting in a drastic loss of the live capacity.

Ratholing (piping) occurs when a stable vertical hole forms within the solid above the silo outlet. Ratholing will obstruct the flow of solid and will result in a drastic loss of the live capacity of the silo. Ratholing may also subject the silo structure to severe loading conditions. Ratholing can be avoided by a properly-selected hopper configuration.

Arching (bridging) occurs when the solid forms a stable dome or arch across the silo outlet. Arching will obstruct the flow of the solid out of the silo. Arching can be avoided by adequately sized outlets.

A *flow aid device* is a device used for breaking arching or ratholing near a silo outlet, or for assisting in reclaiming dead storage.

A *feeder* is a device for controlling the rate of flow of the solid from a silo outlet. Types of feeder include vibratory, reciprocating, belt, screw, apron, table, star, and rotary plow feeders. Selecting a feeder is usually based on experience, solid flow characteristics, and the required flow rate. In some cases, an adjustable mass-flow gate will provide excellent control of flow rate.

2.2   Selecting a bottom configuration

The silo (hopper) bottom configuration should be selected by taking into consideration the flow characteristics of the solid, the size (diameter) of the silo, the solid withdrawal system, and the desired flow pattern in the silo.

The number of outlets generally depends on the size (diaméter) of the silo and the solid withdrawal system. Small-diameter silos may have one or two outlets. Large-diameter silos may have as many as seven outlets or more. High withdrawal rates may be obtained from one or two large-size outlets or from several smaller outlets. For example, direct flood-feeding of coal into rail cars at the rate of 13 000 ton/h from 70 ft (21 m) diameter silos in western US coal fields has been achieved using only one outlet. On the other hand, withdrawal of coal at the rate of 3000 to 4000 ton/h from seven outlets in 70 ft (21 m) diameter silos has been typical for other barge and train loading systems.

Regardless of the number of outlets selected or the desirable solid flow pattern, the size of the hopper outlets should be large enough to prevent arching or ratholing and to maintain the desired flow rate. The shape of the outlet will depend on the configuration of the hopper and on the type of feeder. Circular, square, and rectangular outlets are commonly used.

If a mass-flow pattern is necessary, steep and smooth hoppers are required. The hopper slope should be selected by taking into consideration the effective angle of internal friction of the solid, and the angle of friction between the solid and the hopper surface.

For free-flowing solids, a funnel-flow pattern may be all that is needed. In this situation, a self-cleaning hopper slope should be selected to help eliminate dead storage. The hopper slope is usually based on the angle of friction between the solid and the hopper liner.

For moderately cohesive solids, an expanded-flow pattern may be adequate. The lower part of the hopper should have a steep, smooth surface meeting mass-flow requirements, while the upper part of the hopper should have a self-cleaning slope. The size of the transition between the lower and the upper parts of the hopper should be larger than the stable rathole diameter.

2.3   Flow rate

Hopper outlets should be sized not only to prevent flow obstruction, but also to provide the desired solid withdrawal rate. The outlet size should be the larger of the size required to prevent obstruction and the size required to maintain the required withdrawal rate.

2.4   Flow aid devices

Some bulk solids gain significant strength and cohesion after storage for an extended period of time. To initiate free gravity flow under this condition may require very large hopper outlets and steep hopper slopes. As an alternative to large outlets and steep slopes, a flow aid device may be used for breaking arches and stable ratholes which may form after extended storage. The hopper configuration could then be based on the strength and cohesion of the solid under continuous-flow conditions only.

Flow aid devices include vibrators, live (vibrating) bin bottoms, low and high-pressure air blasters, and aerated bin bottoms. In general, flow aid devices are usually effective in initiating the flow of solids which gain cohesion after extended time in storage, but not for maintaining the continuous flow of solid in a silo. The use of vibration can, in some cases, result in the packing of the solid, which hinders flow.

2.5 Feeder compatibility

To maintain the desirable flow pattern, silo outlets should be fully effective and live. This condition can only be achieved if the feeder inlet has the same size as the hopper outlet and the feeder draws the solid uniformly through the outlet.

Any constriction in the transition chute from the hopper outlet to the feeder inlet will have a detrimental effect on the silo flowability and should be avoided.

2.6 Segregation

Bulk solids tend to segregate during filling according to their particle size. The fine particles usually form a column which follows the solid trajectory, while the coarse particles roll towards the silo wall. In general, the fine particles will be more cohesive and less free flowing than the coarse particles. The difference in the flowability of the fine and the coarse fractions sometimes results in flow patterns significantly different from those which may develop without segregation. In some cases, preferential flow channels form within the coarse, free-flowing zone, subjecting the silo to eccentric discharge loads.

Segregation can be minimized by filling the silo using a rotary distributor chute to move the solid trajectory away from the center. However, in order to avoid excessive localized wall abrasion, the solid trajectory should never impact the wall. Remixing of the segregated solid may be achieved by using a mass-flow pattern, or by simultaneous discharge from multiple outlets.

The accumulation of fines above a silo outlet, as a result of segregation, should be considered when establishing outlet size and hopper slopes. The fines will usually be more cohesive than the coarse fraction, requiring larger-size outlets and steeper hopper slopes.

2.7 Self-ignition

Some materials, including low-grade and high-sulfur coals, may heat up and self-ignite while in storage. Whether it is an open-storage pile or a silo, the oxidation of coal takes place due to the presence of air and moisture. The oxidation process generates heat and the rate of heat generation is a function of the rate of supply of oxygen. If the rate of heat generation exceeds the rate of heat dissipation, heat will build up and the temperature within certain parts of the storage pile or silo may reach the ignition temperature of the coal.

In silos where wet fines accumulate near an outlet for an extended time, the oxidation process accelerates due to the presence of moisture and the infiltration of air through the outlet. The silo, as in the case of open-storage piles, may promote a chimney effect and the flow of air supplies more oxygen to the coal. If the heat generation rate exceeds the heat dissipation rate, self-ignition may occur.

To minimize the potential for self-ignition, expanded-flow silos should be totally emptied periodically, and permanent dead storage should be avoided.

2.8   Abrasion

Abrasion occurs in silos as the solid slides along a wall or a hopper surface. The severity of abrasion depends on the velocity and abrasiveness of the solid, the pressure it exerts on the surface, and on the hardness of the surface.

In mass-flow silos, the solid continuously slides along the hopper surface during flow, resulting in a higher level of abrasion as compared to funnel-flow or expanded-flow silos. In most cases, the highest level of abrasion has been observed to occur near the outlet of mass-flow hoppers where the velocity of flow is highest.

In funnel-flow or expanded-flow silos, where flow channels do not form at the wall, abrasion is usually limited to the hopper outlet area. However, in cases where flow channels form at the wall, significant localized wall abrasion has been observed. Significant wall abrasion has also been observed where the solid trajectory during filling impacts the silo wall.

2.9   Hopper liners

Materials typically used as hopper liners include stainless steel, mild steel, and abrasion-resistant steel. In some situations, a smooth troweled concrete surface would be adequate. In some applications, ultra-high molecular weight (UHMW) polyethylene may be appropriate. The selection of a hopper liner should be based on its angle of friction with the solid, the presence of a corrosive environment, and on abrasion considerations.

The angle of friction between the solid and the liner should be established by tests. These tests should consider the effects of surface abrasion on the friction by testing unabraded and abraded liner samples. In some cases, friction/abrasion tests have been performed on liners to simulate performance over a 15 to 20 year service life.

Mass-flow hoppers usually require a stainless steel 304[1] liner having a 2B surface finish. Self-cleaning hoppers may be lined with stainless steel, abrasion-resistant steel, or mild steel. Even though the cost per square foot of stainless steel liner is higher than that of other liners, it usually results in flatter slopes and may provide overall cost savings in hopper and silo quantities.

2.10   Operations considerations

To minimize degradation and potential dead storage, funnel-flow and expanded-flow silos should be emptied periodically. The frequency of emptying such silos should be based on the effect of extended time-in-storage on the degradation and the flow characteristics of the solid.

---

1 American Iron and Steel Institute Specifications

To minimize wall abrasion in multiple-outlet silos, where flow channels can form at the wall, withdrawal of the solid from outlets near the wall should be limited to cleaning purposes when the silo is mostly empty. Restricting the withdrawal of solid from off-center outlets to times when the silo is mostly empty will also minimize the effects of eccentric discharge loading on the wall.

3 Material properties

Selecting material design parameters for flow and pressure calculations is one of the most critical steps in silo design. Whenever possible, the selection should be based on a combination of experience with the material and laboratory testing. The laboratory testing should, whenever possible, be conducted to estimate the necessary solid flow and strength properties. Usually, these properties will vary with the gradation of the material, the surface moisture content, the ambient temperature, and the time in storage.

The fine fraction of a solid is usually more cohesive and less free flowing than the coarse fraction. Therefore, the flow testing should be conducted on samples representing the fine fraction of the material in order to obtain hopper slopes and outlet sizes which can handle the potential accumulation of fines in the hopper zone. However, the design loads and the structural analysis should be based on testing conducted on representative samples consisting of both the fine and the coarse fractions, since higher lateral pressures and flatter flow channels will usually develop due to the presence of the coarse fraction.

The material design parameters should be selected by taking into consideration expected changes in the physical and chemical properties of the solid, including its gradation, moisture content, chemical composition (ash content), etc.

Silo installations at mine sites or material preparation plants may be designed by taking into consideration the properties of the various seams and product preparation, gradation, and moisture. Silos at transfer and shipping terminals and at end-user facilities, may have to be designed for a wider variation in material properties to account for variability of materials.

In general, the more the variability of material properties, the more expensive the silo installation will be. Both the owner of the facility and the designer should jointly define a range of material properties which would result in a satisfactory performance of the silo at an acceptable cost.

3.1 Flow properties of bulk solids

The flow properties of a bulk solid necessary for selecting silo hopper slopes and outlet sizes are

(a) the bulk density, γ

(b) the effective angle of internal friction, δ

(c) the angle of internal friction, ϕ

(d) the angle of material–wall friction, ϕ'

(e) the unconfined compressive strength, f_c

Usually these flow properties are a function of the consolidating pressure encountered in the hopper. For a given bulk solid, these properties will likely vary with changes in moisture, temperature, gradation, and time in storage.

The effective angle of internal friction, δ, is a measure of the shear strength and cohesion of the material. For free-flowing materials, this angle will be equal to the angle of internal friction, ϕ. For cohesive solids, δ will be greater than ϕ.

The unconfined compressive strength, f_c, of the solid is also another measure of its cohesion. It determines the ability of a solid to arch across an outlet. Free-flowing bulk solids have very low unconfined compressive strength, while cohesive solids have a high strength.

The angle of material-wall friction, ϕ', is a measure of the sliding resistance between a solid and a hopper liner. The wall friction angle will usually vary with the normal pressure exerted by the solid on the liner. These properties are usually determined by conducting laboratory flowability tests, utilizing direct shear testing equipment as given by Jenike [14].

3.2   Determination of hopper slopes

Mass-flow hopper slopes can be computed using charts given by Jenike [14]. These charts give the relationship between the angle of material–wall friction, ϕ', and the slope of the hopper wall, θ, for various values of the effective angle of internal friction, δ.

Self-cleaning hopper slopes should be computed by considering the equilibrium of a wedge of the static material on the hopper. If the resistance to sliding is primarily derived from friction along the hopper surface, the hopper slope, θ, may be computed by the following formula:

$$\theta = \arctan\,(FS \tan \phi')$$

where FS is a factor of safety to insure self-cleaning. However, if the resistance to sliding is derived from both friction along the hopper surface and hoop compression within the bulk solid, as in the case of conical or pyramidal hoppers, steeper slopes will be required.

3.3   Determination of outlet dimensions

The outlet dimension, B, required to prevent arching may be computed by the following formula [14]:

$$B > \frac{f_c}{\gamma}(1+l)$$

For slot openings, B is the width of the opening and $l = 0$. For circular or square openings, B is the diameter or the side and $l = 1$. Once the hopper slope is selected, the critical flow factor of the hopper can be determined from Jenike [14]. The value of f_c can be determined using the critical flow factor and the flow function of the solid, as shown in Fig. 3-1.

The outlet dimension, D, required to prevent ratholing in funnel-flow or

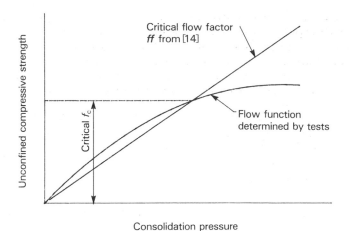

Figure 3-1 The critical value of f_c is determined by the intersection of the critical flow factor line and the flow function

self-cleaning hoppers may be computed by the following [14]:

$$D > 4 \frac{f_c}{\gamma} \Phi$$

where Φ is given in Fig. 3-2 as a function of the angle of internal friction, ϕ. The value of f_c can be computed using the critical flow factors for ratholing given in Fig. 3-3 in a manner similar to Fig. 3-1. Details of outlet dimension computation procedures are given in Jenike [14].

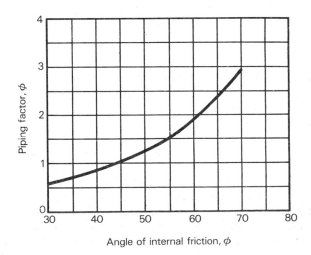

Figure 3-2 Piping factor

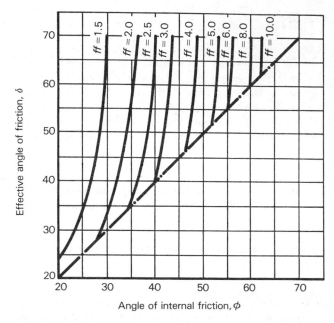

Figure 3-3   Piping flow factors

3.4   Design and bidding information

Performance specifications for silo installations built on a turnkey basis should cover all material flow considerations. The owner should perform all the necessary material testing, select the desirable flow pattern, and establish the silo and hopper configuration before finalizing the bid package. The performance specifications should include the following:

Silo and hopper configuration
Hopper slopes and outlet sizes
Hopper liner material, surface finish, and liner thickness
Feeder type, size, and capacity
Needed flow aid devices
Material design parameters and summary of test data
Design flow pattern
Operational constraints (e.g. eccentric discharge)
Slope and liner materials for chutework

Without the detailed information outlined above, the turnkey performance specifications may not adequately cover material flow considerations in a manner which will ensure responsive bids.

Silo installations developed on a design-bid basis will usually be completely designed before bidding. However, the design drawings and specifications should still include all the information included in the turnkey specifications in the form of general design notes.

4 Static (filling) pressures

Once the flow characteristics of the solid are established and the desirable hopper configuration is selected, the next step is to estimate the pressures exerted by the solid under static conditions before withdrawal of material from the outlet. Several methods for computing static pressures are available. In the USA, Janssen's formula [2] is used by most designers. In other parts of the world, Reimbert's formula [4] is used. Recently, finite element techniques have been used for estimating pressures exerted by compressible materials in silos [15].

4.1 Janssen's formula

The static pressure exerted by a solid in a silo may be computed using classical Janssen's formula. The vertical pressure, q, and horizontal pressure, p, are computed as follows;

$$q = \frac{\gamma R}{uK} [1 - \exp(-uKz/R) + \tfrac{2}{3}\gamma R \tan(s) \exp(-uKz/R)]$$

$$p = Kq$$

where γ is the unit weight of the bulk solid, R is the hydraulic radius of the silo, u the coefficient of wall friction, K the ratio of the horizontal pressure to the vertical, s the angle of repose of the solid, and z the distance from the top of the material at the wall to the point under consideration.

The classical Janssen formula was originally derived assuming that γ, K, and u are constants along the silo height. This assumption is satisfactory for relatively incompressible materials. For compressible materials, Cowin [14] derived a version of Janssen's formula where γ and K are assumed to be linear functions of the vertical pressure, q, having the form

$$K = K_0(1 + mq)$$

$$\gamma = \gamma_0(1 + nq)$$

This form appears consistent with test data performed on a variety of compressible materials.

Whenever practical, Janssen's material parameters δ, K, and u should be selected based on tests conducted on representative samples at a pressure range similar to the range expected in the silo. The variation of γ and K with vertical pressure can be determined by direct measurement of the vertical and lateral pressures in a specially instrumented, laterally confined, compression test on a sample of the solid. The coefficient of wall friction, u, can be determined from direct shear tests. Testing should be performed on samples having moisture and particle size within the range expected in the silo.

4.2 Finite element techniques

In some situations, it may be necessary to obtain more detailed information on the pressure distribution in silos. Examples include materials having significant time-dependent properties and silos where the interaction between the materials and the silo wall or floor may have an effect on the pressure distribution. In such cases, finite element techniques provide a useful tool.

The finite element model should simulate the filling of the silo by using incremented construction techniques (building the model in layers). The analysis should consider material non-linearity. In the case of very compressible materials, the analysis should consider geometric non-linearity (large deformations) as well as material non-linearity. Consideration should be given to adequate modeling of sliding friction at the silo wall. The material parameters for the finite element model can be obtained from tests similar to those conducted to obtain the parameters of Janssen's formula or from triaxial tests. Details of finite element modeling of static silo pressures are given by Mahmoud and Bishara [15].

4.3   Hopper pressures

Static pressures within the silo hopper zone may be computed using a combination of Janssen's formula and Walker's theory [8]. The vertical static pressure, V_0, at the top of the hopper is computed using Janssen's formula. The vertical static pressure, V, at any point within the hopper may be estimated by

$$V = V_0 + \gamma h_1$$

where γ is the unit weight of the material and h_1 is the distance from the top of the hopper to the point under consideration.

The pressure normal to the hopper wall can be estimated using Walker's initial filling pressure [8] within the converging section:

$$W = V \frac{\sin 2\alpha \cos \phi'}{\sin (\phi' + 2\alpha) + \sin \phi'}$$

where α is the hopper half-apex angle and ϕ' is the angle of material–wall friction. Most solids have a degree of compressibility which usually results in hopper pressures less than those given using Walker's initial filling formula. This formula applies when $\sin (\phi' + 2\alpha) \sin \delta > \sin \phi'$ [20], otherwise

$$W = V \frac{1 - \cos 2\alpha \sin \delta}{1 + \sin \delta}$$

4.4   Pressures on gates and feeders

Static pressures on gates and feeders may be estimated using the procedures outlined above for static pressures in hoppers. However, the compressibility of most solids coupled with the flexibility of gates or feeders reduces the static pressure to a level very close to two to four times the flow pressure. A detailed discussion of feeder loads is given by Reisner and Rothe [16].

5   Flow pressure

The technical literature refers to pressures in silos during flow as dynamic pressure, overpressure, or flow pressure. Both the terms dynamic pressure and overpressure can be misleading as the pressure in a silo during flow can be either higher or lower than the static pressure. Therefore, the term 'flow' pressure appears to be more accurate in describing the state of pressures in silos during flow than 'dynamic' pressure.

In general, the pressure within the flow channel can be computed by one of several methods available in the literature, e.g. Jenike [6], Walker [8], and Walters [9]. Walker's formulation [8] in conjunction with Janssen's formula appears to result in pressures close to available experimental data [10]. The most critical task in computing flow pressures is to define the configuration of the flow channel.

5.1   Flow channel configuration

The first step in estimating flow pressures is to decide on the geometry of the flow channel(s). In single-outlet silos, one flow channel is to be considered. In multiple-outlet silos, several flow channels may have to be considered.

The silo hopper configuration, material segregation, and the flow pattern that develops in the silo usually form the basis for selecting the flow channel. In general, the flow channel will initiate at the outlet of a funnel-flow hopper, or at the top of a mass-flow hopper and will expand upward at a relatively steep angle which depends on the effective angle of internal friction, δ, of the solid, as shown in Fig. 5-1 given by Giunta [17].

The design must recognize that segregation in silos, where coarse free-flowing particles slide towards the wall, coupled with the proximity of an outlet to a silo wall, will influence the configuration of the flow channel. In this case, asymmetric flow channels, which tend to follow the path of least resistance to flow, develop in the silo. These are usually called preferential flow channels. The silo designer should be aware that solid properties change from time to time, resulting in corresponding changes in flow channel configurations. These potential changes should be considered in the design.

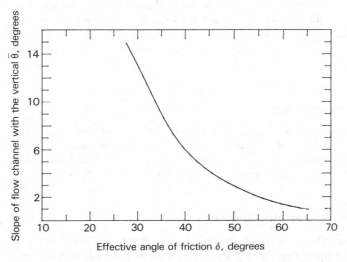

Figure 5-1   Slope of flow channel versus effective angle of friction

5.2   Flow pressure calculations

Once the configuration of the flow channel is established, the vertical and lateral pressure in the flow channel during flow is computed. In the cylindrical part of the flow channel, the pressures may be estimated using Janssen's formula for a 'pseudo' silo having a diameter equal to that of the flow channel. If the flow channel forms at the silo wall, a weighted average of the coefficient of material-wall friction should be used in the computation. This weighted average should be based on the angle of friction between the solid and the wall and the angle of internal friction of the solid.

In the convergent part of the flow channel, the pressure may be estimated using the formulas developed by Walker [8] and given below:

$$V = \frac{\gamma h}{C-1}\left[1-\left(\frac{h_0}{h}\right)^{C-1}\right] + V_0\left(\frac{h}{h_0}\right)^{C}$$

$$W = \left[\frac{1+\sin\delta\,\cos 2\beta}{1-\sin\delta\,\cos 2(\alpha+\beta)}\right]V$$

$$C = 2BB/\tan\alpha \quad \text{(conical or pyramidal)}$$

$$C = BB/\tan\alpha \quad \text{(long wedge)}$$

$$BB = \frac{\sin\delta\,\sin 2(\alpha+\beta)}{1-\sin\delta\,\cos 2(\alpha+\beta)}$$

$$\beta = \tfrac{1}{2}[\phi' + \arcsin(\sin\phi/\sin\delta)]$$

where V is the vertical pressure, W the pressure normal to the boundary of the flow channel, α the half-apex angle of the convergent part of the flow channel, ϕ' the angle of material-wall, δ the effective angle of internal friction, γ the solid bulk density, V_0 the vertical surcharge pressure acting at the top of the convergent section of the flow channel, h the distance from the apex of the flow channel to the point under consideration, and h_0 is the distance from the apex of the flow channel to the transition between the cylindrical and the convergent sections of the channel. The designer should consider the condition at the initiation of flow where the vertical surcharge pressure V_0 equals the vertical static pressure in the silo.

5.3   Pressures on feeders and gates

Pressures on feeders and gates after the initiation of flow may be computed using Walker's equations [8]. It should be recognized that once flow pressures develop within the flow channel they remain, even if the flow stops. A detailed discussion of pressures on feeders and gates is given by Reisner and Rothe [16].

6   Design loads: concentric discharge

The design wall loads should account not only for the static conditions, but also for conditions during flow. The classical design philosophy is based on increasing the static pressures by a factor to account for conditions during flow. These factors, referred to in ACI 313-77 [18] as overpressure factors, are rather empirical. Recent design trends are based on a more rational approach which considers the flow channel configuration and the interaction between the static and flowing solid.

6.1 Empirical overpressure coefficients

In current silo design codes, design forces are usually based on static pressures modified by empirical coefficients to account for overpressures during emptying. The ACI 313-77 overpressure factors are an example of such coefficients. These empirical coefficients are usually determined either by trial and error on full size silos, or from tests on model silos. The coefficients are further modified empirically by designers in an attempt to cover adverse flow or pressure conditions they have encountered in their practice from time to time.

6.2 Flow pattern considerations

As an alternative to empirical overpressure coefficients, a more rational approach to estimating design forces is based on flow considerations. The designer establishes possible flow channel configurations and static flow pressures as outlined in Sections 4 and 5. If the flow channel expands upward and reaches the silo wall, as in the case of mass-flow silos, or funnel-flow silos with a large height-to-diameter ratio, a zone of overpressure will develop just below the transition between the cylindrical and convergent parts of the channel. By considering a range for the angle of the flow channel, design pressures may be computed for different locations along the wall height. If the silo configuration and outlet location are such that the flow channel never expands to the wall, then the overpressure at the transition will be dampened by the static material around the channel. In this case, the design hoop tension can be computed by statics as follows:

$$T = \tfrac{1}{2} p_f D_f + p_s (D_s - D_f)$$

where p_f is the horizontal pressure in the flow channel, D_f the diameter of the flow channel, p_s the static pressure, and D_s the silo diameter.

6.3 Fibrous materials

Silos for the storage of non-free-flowing fibrous materials, like silages, rely on mechanical devices to unload the material from a central outlet. The action of these devices sometimes can create a loosened mass of material, or a cavity at the bottom of the silo. The static material usually arches across the silo diameter and spans the cavity. The design pressures and forces in the zone affected by the unloader operation can be estimated by considering the equilibrium of the static material. More details regarding design forces for these silos can be found in the International Silo Association's design standards [19].

7 Design loads: eccentric discharge

The design wall loads in the case of eccentrically located hopper outlets can be established following the same philosophy as given in Section 6. However, in addition to hoop tension, the engineer must estimate wall bending. Recent design trends consider the interaction between the static and the flowing solid rather than the direct use of empirical overpressure factors.

7.1   Empirical overpressure coefficients

Empirical overpressure coefficients for eccentric discharge are much less reliable than in the case of concentric discharge. For example, ACI 313-77 [18] suggests a 25% increase in design pressures for silos with outlets located at the wall. The overpressure coefficients do not account for the horizontal wall bending associated with eccentric discharge.

7.2   Flow pattern considerations

A more reliable approach to estimating wall forces due to eccentric discharge is based on silo and flow pattern configurations. The designer establishes the configuration of the flow channel and the static and flowing pressures as outlined in Sections 4 and 5. Several locations along the wall height need to be considered. To account for possible variations in flow channel configurations, the converging boundary of the channel should be varied at least 4° above and 4° below the computed slope based on testing.

Placing an outlet in or near the wall, and segregation during filling can lead to eccentric discharge. Eccentric discharge can be symmetric with respect to the center of the silo, if the outlets are symmetrically located. Segregation will usually result in the placing of coarse free-flowing solid adjacent to the wall, and fine less-flowing solid in the center of the silo, resulting in preferential flow channels forming at the silo wall. In general, a vertical flow channel of nearly constant cross-section will form above the outlet. The presence of coarse material at the wall may result in significant lowering of the flow channel of more than 4° or may introduce preferential flow channels. The pressure in the static material may increase above that computed by Janssen's formula, as a large portion of the weight of the solid in the flow channel is transmitted to the static material by shear along the boundary of the channel. This increase in pressure should also be considered in the design.

The wall should be designed for the hoop tension and the horizontal bending associated with eccentric discharge by considering the most critical flow channel configuration. Design hoop tension and horizontal bending moment may be estimated using standard methods of structural analysis. A more realistic approach is to use finite element techniques to incorporate the effects of material–wall interaction. This approach usually results in bending moments significantly lower than those predicted by other methods of analysis which do not account for the interaction. The designer should consider the conditions at the initiation of flow as well as during flow.

8   Wall design

Silo walls are subject to many combinations of vertical and horizontal loads. Silo walls must support vertical loads from material friction, gravity dead loads of the wall and roof, and resultant vertical loads from earthquake, wind, and conveying equipment. Silo walls are also subject to horizontal pressures from static and flowing material, and horizontal and vertical bending moments produced by variations in pressure between the static and flowing material. Vertical loads generally dictate the wall thickness in the larger silos. However, circumferential bending from eccentric discharge may control the minimum wall thickness in both large and small silos.

8.1 Wall thickness

The preliminary wall thickness should be selected based on vertical load. Janssen's formula is used for the determination of vertical material friction load. Whenever possible, a range of material parameters should be established from tests performed on representative samples of the bulk solid. If not, a range of values for these parameters should be established based on past tests or experience of the designer. As an alternative to using a range of parameters, the vertical material friction load may be estimated using the following formula [18]:

$$V_f = W - 0.8q\pi D_s^2/4$$

where V_f is the total vertical material friction load on the silo wall, W the total weight of material, q the vertical static pressure at the bottom of the silo, and D_s the silo diameter. The factor 0.8 is used to calculate a conservative value of the material friction load in the absence of reliable estimates of material parameters.

The maximum vertical wall load usually occurs at the base of the silo where material friction load is a maximum, and the net perimeter of the wall is a minimum. Other locations may be more critical depending on opening sizes and locations in the silo wall. If the openings in the silo wall are 180° apart, the total vertical load can be assumed to be evenly distributed around a net perimeter of the silo wall. If the openings are not symmetrically located, the load should be distributed so that the effective centroid of the wall segment loads coincides with the center of the stored material.

The vertical wall loads produced by earthquake or wind should be calculated and combined with the wall dead load, the roof dead and live load, the equipment load, and the material friction load using the appropriate load factors. The material friction load should be considered as live load.

In the preliminary calculations for the wall thickness, the concrete stress under the design wall load should not exceed $0.55f$, where f is the permissible design vertical stress for walls in which buckling is not the controlling design criterion. The silo wall thickness should also be checked for elastic stability. To account for creep in wall stability calculations, a 50% reduction in the modulus of elasticity of concrete should be used. The factor of safety against buckling should be at least three.

In addition to vertical load considerations, the silo wall thickness should provide adequate shear strength to resist forces from earthquake, wind, equipment, and soil–structure interaction. The reduction in the shear strength of the wall due to circumferential hoop tension should be considered, and if necessary, additional circumferential and vertical reinforcing should be added.

Eccentric flow channels can produce significant circumferential bending moments in the silo wall because of the variations in pressure between the static and flowing material. Stable vertical flow channels adjacent to a wall may require that the wall be increased in thickness to reduce the amount of horizontal reinforcing required for bending. We suggest maintaining a minimum thickness of 10 in (254 mm) with reinforcing in both faces if eccentric flow is anticipated.

8.2   Wall reinforcement

The minimum ratio of vertical reinforcement to gross concrete area shall not be less than 0.002. The vertical reinforcing should also be sufficient to resist any vertical bending, uplift, tension, and the effects of a temperature differential through the wall.

The minimum ratio of horizontal reinforcement to gross concrete area shall not be less than 0.0025. If possible, horizontal reinforcement should be placed adjacent to both the interior and exterior faces of the silo wall and have a minimum cover of 2 in (50 mm). The interior cover should be increased to a minimum of 3 in (75 mm) if the stored material is moist and contains any substances capable of corroding the reinforcement.

The designer should check the vertical crack widths under both static and flow conditions if the stored material is hygroscopic, or subject to deterioration when exposed to moisture. The designer should consider crack width and its effect on concrete durability if the stored material contains sufficient moisture to penetrate and saturate the silo wall. Crack width requirements may limit the tensile stress in the hoop reinforcement to a value less than that based on strength requirements only.

Unless the hopper is designed to independently resist material pressures, the wall adjacent to the hopper should be designed for a minimum pressure equal to the resultant horizontal component of the initial pressures normal to the hopper. In mass flow silos, the wall should be checked for the horizontal component of the dynamic mass-flow pressures normal to the hoppers.

In silos with multiple outlets, the designer should consider the possibility of eccentric flow due to use of only one outlet. Very large circumferential bending moments can be created by eccentric flow channels having large cross-sections. Again, the silo wall should not be reinforced for less than 1.35 times Janssen's static horizontal pressure. In some cases, it may be more economic to place different amounts of reinforcing in the inner and outer faces of the wall to meet the varying requirements of bending moments.

8.3   Thermal stresses

The storage of materials with temperatures significantly greater or less than ambient can cause horizontal and vertical bending moments in silo walls. Additional horizontal and vertical reinforcement may be required in the colder face of such walls. The additional reinforcement required for thermal stresses can be determined using ACI 313-77, Section 4.5.4 [18], or by other rational, or analytical methods. In situations where the thermal loading is predominantly long term, a reduced modulus of elasticity of concrete may be used in computing the thermal moments because of the creep of concrete.

8.4 Post-tensioned reinforcement

The design of post-tensioned silos subjected to eccentric discharge should be based on the combined effects of circumferential hoop tension and bending moment. Usually, a combination of prestressed and non-prestressed reinforcement will be the most economic. The use of empirical overpressure coefficients as a basis for designing such silos can be unconservative because there may not be enough prestressing in the wall to adequately resist the extreme fiber tensile stress due to combined tension and bending.

9 Bottom design

Various types of silo bottoms are used. Each type is largely a function of the type and configuration of the hopper system. Figures 9-1 through 9-4 show the types most commonly used. The primary function of the silo bottom is to support the load from the stored solid, the weight of the hopper, and any suspended equipment.

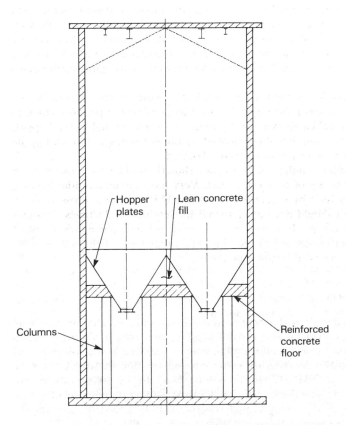

Figure 9-1 Elevated floor on columns

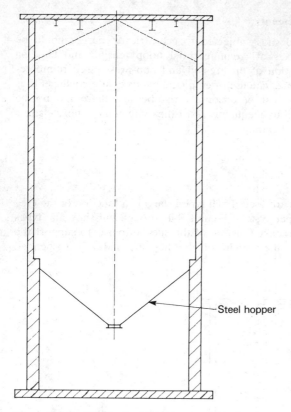

Figure 9-2   Conical steel hopper attached to concrete ring girder integral with the silo wall

9.1   Design loads

The load from the stored solid may be calculated using Janssen's formula with proper adjustments to account for the collapse of arched materials, material impact, earthquake, and reduced friction between the material and the wall. The collapse of arched materials, material impact during filling, and the effect of earthquake loads are usually included by using empirical overpressure coefficients similar to those given in ACI 313-77 [18]. Reduced friction between the solid and the wall is usually included by using a reduced lateral pressure ratio and a reduced coefficient of wall friction in Janssen's formula.

9.2   Elevated floor on columns

To maintain a uniform silo wall thickness to simplify wall construction, it is sometimes desirable to support the silo bottom on columns independent of the wall. The column spacing is usually established considering the following requirements:

(a) clearance for mechanical equipment for the reclaim system;

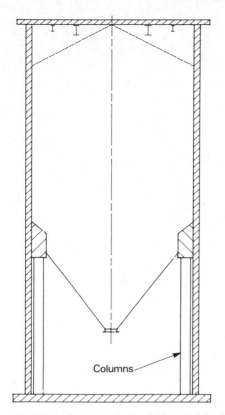

Figure 9-3 Conical steel hopper attached to a concrete ring girder independently supported on columns

(b) forming system and design spans for permanent forms supporting the bottom slab;

(c) clearance for the slipform work deck;

(d) design loads.

The bottom usually consists of a structural slab and a lean concrete hopper fill. The structural slab may be poured in one or two lifts, depending on the design of the form system. If adequate shoring is provided to support the weight of the wet concrete, the slab may be poured in one lift. If a permanent steel beam and deck form system is provided, the slab is usually poured in two lifts. The permanent form system is designed to support the first lift. The thickness and reinforcing of the first lift should be adequate for supporting the weight of the two lifts. The construction joint between the two lifts should provide adequate composite action to ensure that the structural slab will support the total load on the silo bottom.

Silo-bottom structural slabs may be analyzed as two-way floor systems. However, the presence of discharge outlets would require certain portions of the floors to be designed as beams. In determining the required structural slab thickness, the existence of the hopper openings in the vicinity of the punching perimeter of a

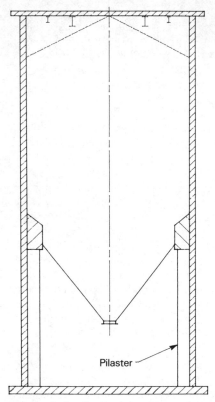

Figure 9-4   Conical steel hopper attached to a concrete ring girder independently supported on pilasters

column must be considered. If structural steel beams and decking are used as a permanent form, the shear capacity of the steel beams may be included in the analysis, provided these beams are continuous over the columns. In cases where, due to column and outlet locations, a portion of the slab must act as a ring girder around a large outlet, torsional shear stresses at the outlet should be investigated. If the slab is constructed in two lifts, the horizontal construction joint should be checked for adequate composite action.

As an alternative, silo bottoms may be analyzed using finite element techniques. A finite-element plate-bending analysis of the structural slab would be helpful in case there is an unusual outlet and column layout. A finite element three-dimensional solid analysis would be helpful in case the designer needs to consider the stiffeners and strength of the lean concrete fill or if the structural slab is too deep to permit the use of a plate-bending model. In either case, the elastic properties of concrete should be carefully selected to reflect changes in concrete strength, long term creep effects, and reduction in the shear rigidity due to cracking. The finite element model should be set to adequately simulate the flexibility of the columns supporting the slab.

9.3 Hopper suspended from ring girder

Small-diameter single-outlet silo bottoms may consist of a structural steel conical hopper suspended from a reinforced concrete or a structural steel ring girder. This bottom support system is usually economic in silos not exceeding 5000 tons capacity and 50 ft (15 m) diameter. This silo bottom support system may consist of any of the following arrangements:

(a) Conical hopper attached to a concrete ring girder integral with the silo wall. The wall below the girder may have to be thicker than the wall above the girder. This support system may be used for either slipformed or jumpformed silos. Figure 9-2 shows this type of arrangement.

(b) Conical hopper attached to a ring girder independently supported by columns. A constant wall thickness is used for the entire silo height. This support system may be used for either jumpformed or slipformed silos. Figure 9-3 shows this type of arrangement.

(c) Conical hopper attached to a ring girder supported on pilasters which are poured monolithic with the silo wall. This support system can only be used in slipformed silos. Figure 9-4 shows this type of arrangement.

(d) Conical hopper attached to concrete ring girder which is supported by the silo wall through concrete shear lugs poured monolithic with the concrete ring girder. This support system may be used for slipformed or jumpformed silos.

Ring girders which are not monolithic with silo walls and are supported on either columns, pilasters, or lug pockets should be investigated for bending, shear, and torsion in addition to axial compression. The buckling of the ring should also be investigated.

9.4 Bin bottoms on tunnel walls and inert fill

The bottoms of multiple-outlet silos may be constructed from lean concrete fill supported on a system of tunnel walls and compacted inert fill. Earlier versions of this bottom support system consisted of tunnel walls and compacted inert fill; the bottom slopes were formed by the inert fill taking its natural angle of repose. In general, the use of sloping inert-fill hoppers should be limited to silos for the storage of free-flowing solids where segregation and degradation can be tolerated.

The tunnel layout is usually established by taking into consideration equipment layout and clearances. A joint should be provided in the lean concrete fill to accommodate differential settlement between the tunnel walls and the inert fill. This joint becomes extremely critical in the case where the silo and tunnel walls are supported on deep foundations resulting in large differential settlement between the walls and the fill.

The tunnels should be designed to support the vertical material load, the dead load of the hopper, the friction load between the walls and the inert fill, and the lateral pressure of the fill. In some situations, it may be necessary to design the tunnel for material arching on the tunnel roof.

10   Foundation design considerations

Unlike most structures, silos are usually subjected to the full design live load every time they are loaded. Furthermore, loading and unloading result in cyclic application of the live load. The foundation design should be based on a geotechnical investigation which considers the following:

(a) The effect of silo diameter on the extent of stress influence in the ground. The borings should be sufficiently deep to provide adequate information on the strength and compressibility of the supporting strata.

(b) The effect of the compressibility of the supporting strata on the initial, cyclic and total settlement.

(c) The effect of ground loading on other structures in the area, including tunnels and pits.

(d) Negative skin friction on caissons and piles.

(e) The lateral and rotational flexibility of the soil–foundation system for the purpose of earthquake analysis.

10.1   Shallow foundations

Shallow foundations consisting of mats or spread footings may be used provided the supporting strata have adequate strength and low compressibility. Mat foundations may be analyzed as a plate on an elastic foundation. A continuous mat under a cluster of closely spaced silos may have to be analyzed as a plate on an elastic half-space to adequately model the effects of differential settlement.

10.2   Deep foundations

Deep foundations consisting of drilled piers, or piles, may be required for supporting large-capacity silos located on weak soil strata. The decision to use shallow or deep foundations should be based on the subsurface conditions and on the cost of their foundation system. Analysis techniques of silo clusters on deep foundations are essentially the same as for the case of mat foundations.

10.3   Soil–structure interaction

Monolithically constructed silo clusters are inherently stiff. The foundation, walls, and roof of the cluster act as a composite deep beam. In cases where the stiffness of the cluster is significantly higher than the stiffness of the supporting strata, the distribution of foundation bearing pressure will approach the case of an infinitely rigid structure on an elastic half-space. This distribution results in minimizing the differential settlement within the cluster but will introduce internal forces in the silos which must be considered in the design. This behaviour will occur whether the silos are supported on shallow or deep foundations. The magnitude of the internal forces induced in the structure will depend on the relative stiffness of the soil–structure system; the stiffer the supporting strata, the lower is the magnitude of the induced internal forces. The analysis of silo clusters under these conditions should be performed using three-dimensional finite element techniques. The load–deformation characteristics of the foundations should be estimated, whenever possible, from *in situ* tests or pile load tests.

11 Roof design

Silo roofs should be designed to support a variety of loads which may include some or all of the following:

(a) roof dead and live loads and equipment loads;
(b) wind and earthquake forces from attached superstructures;
(c) in-plane forces from soil structure interaction in the case of monolithic silo clusters;
(d) thermal loads from hot stored material;
(e) positive or negative internal pressure.

Roofs may be tied down to silo walls to restrain potential wall movements due to eccentric discharge and to help to transfer all the roof load to the wall. Dead-load deflections of the roof beams should be accounted for when providing drainage for the roof slabs. Roof beams should also be checked for induced vibrations.

Beam bearings must be designed for maximum reactions and should be placed as concentric to the silo wall as possible. For hot stored material in silos, the beam bearings should not be fastened to the beams. The bearing surfaces must be as smooth as possible to reduce the build-up of large horizontal frictional forces. Restraint to vertical thermal movements at beam bearings should be avoided.

12 Slipform wall construction

Slipform concrete construction is the continuous placement of concrete wall through the use of a 4 ft (1.2 m) high form lifted in small, but continuous, increments while concrete and reinforcing is placed in the top of the open form.

The slipform assembly consists of a vertical form on each face of the wall, working deck, yoke, jackrods, and jacks. The forms are fabricated and erected to conform to the configuration of the walls, and are held in position by the yokes which are attached to the form wales. The working deck is supported from the upper wale and the assembly is raised by the jacks as they climb up the jackrods and lifts against the yoke. The working deck provides a platform for the workmen placing the concrete and reinforcing in the forms. The movement of all the jacks is controlled from a single console located on the working deck. Figure 12-1 shows a typical section through a slipform assembly.

In the USA, slipform construction is usually economic for silo diameters larger than 50–60 ft (15–18 m). Slipform silos should have a fairly regular cross-sectional plan with walls of uniform thickness. To accommodate variations in vertical wall load, while maintaining a uniform wall thickness, higher strength concrete may be used in the lower part of the wall.

The mix design of slipform concrete must accommodate changes in weather conditions encountered during construction. For relatively thin walls, the concrete slump should be in the range of 3 in (75 mm) to 5 in (125 mm). The maximum size of aggregate should be $\frac{3}{4}$ in. Since the rate of set of the concrete determines the rate of form movement, the concrete mix may have to include additives that can accelerate or retard the concrete set. This will allow a uniform rate of form

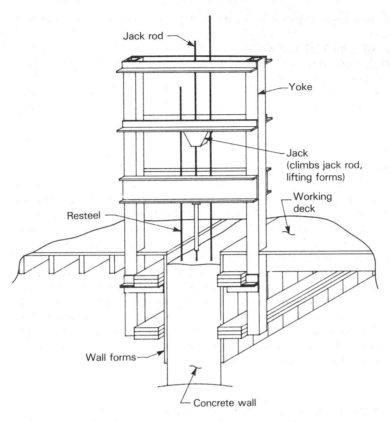

Figure 12-1 Section through a slipform assembly

movement to be maintained under varying conditions of ambient temperatures and humidity. However, the use of admixtures containing chlorides must be avoided to prevent possible corrosion of the reinforcement.

13 Jumpform wall construction

Unlike slipform, jumpform wall construction is not continuous and is interrupted between each lift. Each lift may consist of either 2 ft (0.60 m) or 4 ft (1.20 m) of depth. However, the placing of the concrete in each lift must be continuous and the formation of cold joints within the depth of the lift should be avoided. In jumpform construction, construction joints will occur between lifts. For a large variety of stored materials, these joints need to be cleaned and properly prepared so water does not leak through the joints. In some situations, a water stop may be required.

In jumpform construction, the lower-level forms are stripped and reused for the next lift. However, the lowest level of forms must not be stripped until the concrete in contact with the forms directly above the level being stripped has attained sufficient strength to safely support the loads imposed by the form and scaffold system.

Jumpform construction, in general, is usually economic for diameters not exceeding 40 ft. However, with the advent of the newer jumpform techniques, up to 72 ft diameter silos can be constructed. The prolonged construction period and the formation of construction joints between lifts would still appear to make the jumpform construction of large diameter silos somewhat less of a viable option.

14 Hopper construction

Hopper surfaces must be protected during construction from concrete spillage and marring. All welds on the exposed surfaces of the hopper plates which may interfere with flow must be ground smooth. Any protrusions in the exposed hopper surfaces must also be removed. These precautions are necessary so as to avoid obstructions in the direction of material flow.

The thickness of hopper plates should be selected with allowance for the effects of abrasion and corrosion. In situations where both abrasion and corrosion will occur, their combined effect must be considered. Whenever lean concrete fill is used, the hopper plates usually serve as forms. The plates should be designed to prevent excessive bulge due to the pressure of wet concrete. The lean concrete fill is usually placed in lifts to minimize the pressure on the plates and to provide adequate anchorage in prior lifts.

15 Construction supervision and inspection

The silo designer and construction supervisors should meet prior to construction to discuss silos having complicated patterns of reinforcement or concrete placement. If special patterns of wall reinforcement have been specified for eccentric withdrawal, the designer should emphasize the necessity for accurate location and placement within the wall. The builder should review the drawings prior to the meeting and prepare questions relating to the clarity and intent of the design drawings. The builder should also establish any special requirements for the control of the concrete setting time and any other special concrete requirements.

The supervisors and foremen of each shift should understand the drawings thoroughly and the reasons for any special reinforcement placement. Supervisory personnel should also thoroughly understand the reasons for controlling water, slump, vibration, and the effects that concrete 'lifts' and 'fallouts' have on the integrity of silo walls.

Before starting slipform operations, supervisors should prepare a schedule of the elevations and locations of blockouts, openings, changes in reinforcing steel, and location of inserts on the inside and outside of silo walls. Each slipform operation should have marked rods or poles at one or more locations to indicate the elevation of the slipform deck at any time. The silo centerlines and any other points of special significance should be marked on the circumference of the form.

The constructor should constantly monitor the orientation of the silo centerlines and the plumbness of the slipform operation.

Silo owners should insist on constant inspection during silo construction. Contracts should specify construction inspection adequate to ensure high-quality materials and construction methods. Inspectors should monitor the critical parameters controlling concrete quality, such as cement content, mixing water control, slump, mixing time, waiting time, methods of placement, and adequacy of vibration. Inspectors on the slipform deck should constantly monitor the length and stagger of the reinforcing laps. In general, checking the spacing between horizontal hoops is not adequate to ensure that sufficient quantities of reinforcement are placed in the wall. The quantity of reinforcement on the ground and the work deck should be counted at specific intervals, based on the position of the slipform deck. Keeping track of the total quantity of steel, and the position of the deck, is an accurate method for determining the amount of reinforcement actually placed in the silo wall. Bar counts should be taken at about 5 ft (1.5 m) intervals to determine the reinforcement being placed.

Contractors should not be allowed to 'float' or place reinforcement directly on the wet concrete since this makes it very difficult to control the spacing of the reinforcement and the length of the splices. The circumferential reinforcement should be tied lightly to the jackrods above the concrete and below the structural members supporting the form.

16 Serviceability considerations

The major cause of silo problems are construction-related errors, such as missing reinforcement, inadequate reinforcement splice lengths, aligned splices, low-strength porous concrete, and lifts and fallouts which are left unrepaired and covered up by cement finishers. However, design-related problems do occur. A summary of problems affecting serviceability follows.

16.1 Interstices

In some situations, the interstice between monolithic silo clusters is utilized for storage. Designers, sometimes, analyze the interstitial walls as fixed or pinned circular arches with no shear or bending moments. A conventional structural analysis of the interstitial walls will reveal the presence of shear forces and bending moments in addition to axial forces. The failure of interstitial walls due to bending and shear have been observed to occur in several silo installations after five years in service.

16.2 Dynamic overpressure factors

The use of empirical 'dynamic overpressure' factors for design does not ensure that the silo will have an adequate service life. The walls of silos having eccentric flow can be subjected to large horizontal bending moments. If the designer places all of the reinforcing near the outside face of the wall, the bending moment at the center of the flow channel (tension at inside face of wall) can cause the horizontal reinforcing to reach yield each time the silo is emptied. Any combination of a lack of reinforcement, inadequate splice length and incorrect splice alignment can cause deterioration of bond at the splice points. Deterioration of the bond at the splice points has been observed in a large coal silo after 15 years in service. The wall split vertically at the level of the circular steel, separating the wall into two cylindrical segments. Such a failure is progressive in nature. In this case, the wall separated for a height of 70 ft (21 m), with splitting nearly 20 ft (6 m) wide in one location. No corrosion of any reinforcement was observed. A similar condition could develop if the designer places all of the reinforcing near the inside face of the wall.

16.3   Eccentric discharge

In silos subjected to eccentric discharge, the placement of horizontal reinforcement near the outside face of the wall results in relatively wide cracks at the inside face of the wall within the flow-channel zone. Field experience indicates that the concrete in the cracked zone could be subjected to severe abrasion and deterioration. It appears that repeated flexing of the wall enhances the abrasion and deterioration. In one such case, 3 in (75 mm) of wall abrasion resulted over a period of 15 years. Of course, such abrasion in a thin-walled silo could reach the reinforcing and significantly reduce its effectiveness.

16.4   Corrosion

Recent investigations have revealed severe corrosion problems in several slipform concrete silos storing coal. The silos had been in service for eight to twelve years. Corrosion of the horizontal and vertical reinforcing caused vertical, cylindrical lamination of the wall through the horizontal reinforcement. The expansion forces created during the formation of iron oxide produces large tensile splitting forces in the wall. Once the wall splits, the splices of the circular reinforcing cease to function. Corrosion of a large enough area can lead to total silo failure.

16.5   Effect of chlorides

The corrosion of the reinforcement is usually related to chlorides. In one case, excess calcium chloride was used as an accelerator during construction. However, in several cases the chlorides came from the stored material, which was very wet coal. The chloride content at the inside face of the concrete was seven to fifteen times higher than the chloride content at the outside face. The chloride solution penetrated to the outside face reinforcement through vertical and horizontal cracks at the inside face. The vertical cracks on the inside were caused by eccentric flow adjacent to the wall. The horizontal cracks were 'cold' joints produced by the construction methods used.

16.6   Stored material characteristics

Designers should determine the chemical and physical characteristics of the stored material and design the silo in accordance with those characteristics. The designer should consider withdrawal methods, circumferential bending, abrasion, and applicable crack width when designing silos which store a substance which will corrode steel in the presence of moisture and oxygen. The walls should not be permitted to crack from bending and tension if post-tensioned reinforcement is used. Concrete additives which contain chloride or any other substance which destroys protective oxide films should not be used in silo structures.

16.7   Eccentric flow and differential settlement

The designer should also consider the effects of eccentric flow and differential settlement on roof-beam bearings and wall keys. Wall movements at the roof level from eccentric withdrawal and differential settlement have been observed to fail beam bearings where relative movement between the roof and wall was permitted. The designer should eliminate movement at the beam-bearing points by anchoring the wall to the roof slab or by designing the beam bearings to accommodate the forces created by movement.

16.8   Inherent stiffness

The inherent stiffness of silo clusters can lead to large vertical foundation loads at the extremities of the cluster. Piling, foundations, silos on tunnel walls, and pilasters at extremities can fail under these large loads. The designer should analytically model such monolithic silo clusters and design for these loads. As an alternative, the designer may elect to separate the silos. In this case, the associated conveyor trusses and adjacent structures should be designed to accommodate the horizontal and vertical movements due to differential settlement and tilting of the separated silos.

References

1. Roberts, I., Pressure of stored grain, *Engineering, Lond.*, Vol. 34, 27 Oct. 1882, p. 399.
2. Janssen, H. A., Versuche über Getreidedruck in Silozellen, *Z. Ver. dt. Ing.*, Vol. 39, 31 Aug. 1895, pp. 1045–1049.
3. Ketchum, M. S., *The design of walls, bins, and grain elevators*, McGraw-Hill, New York, 1909.
4. Reimbert, M. and Reimbert, A., *Silos—traité théorique et pratique*, Editions Eyrolles, Paris, 1956.
5. Cowin, S. C. and Sundaram, V., The effect of material compressibility on static bin pressures, *Powder Technol.*, Vol. 25, 1980, pp. 225–227.
6. Jenike, A. W. and Johanson, J. R., Bin loads, *J. Struct. Div. Am. Soc. Civ. Eng.*, Vol. 94, No. ST4, April 1968, pp. 1011–1041.
7. Johanson, J. R., Stress and velocity fields in the gravity flow of bulk solids, *J. Appl. Mech.*, Series E, Vol. 86, Sept. 1964, pp. 499–506.
8. Walker, D. M., An approximate theory for pressures and arching in hoppers, *Chem. Eng. Sci.*, Vol. 21, 1966, pp. 975–997.
9. Walters, J. K., A theoretical analysis of stresses in axially-symmetric hoppers and bunkers, *Chem. Eng. Sci.*, Vol. 28, 1973, pp. 779–789.
10. Clague, K. and Wright, H., *Pressures in bunkers*, American Society of Mechanical Engineers, Material Handling Division, Paper No. 73-MH-4, 1973.

11. Bransby, P. L., Blair-Fish, P. M. and James, R. G., An investigation of the flow of granular materials, *Powder Technol.*, Vol. 8, 1973, pp. 197–206.

12. Johanson, J. R., Methods of calculating rate of discharge from hoppers and bins, *Trans. Am. Inst. Min. Metall. Eng.*, Vol. 232, March 1965, pp. 69–80.

13. Williams, J. C., *The rate of discharge of coarse granular materials from conical mass flow hoppers*, School of Powder Technology, University of Bradford, 1974.

14. Jenike, A. W., Storage and flow of solids, *Bull. Utah Eng. Exp. Stn*, No. 123, 1964.

15. Mahmoud, M. H. and Bishara, A. G., Using finite elements to analyze silo pressure, *Agric. Eng.*, Vol. 57, No. 6, June 1976, pp. 12–15.

16. Reisner, W. and Rothe, M., *Bins and bunkers for handling bulk materials— practical design and techniques*, Trans Tech Publications, 1976, 280 pp.

17. Giunta, J. S., *Flow patterns of granular materials in flat-bottom bins*, American Society of Mechanical Engineers, Materials Handling Division, Paper No. 68 MH-1, 1968.

18. ACI Committee 313, *Recommended practice for design and construction of concrete bins, silos, and bunkers for storing granular materials*, ACI Standard 313-77, American Concrete Institute, Detroit, 1977, 40 pp.

19. International Silos Association, *Recommended practice for the design and construction of bottom unloading monolithic concrete farm silos*, West Des Moines, Iowa, 1981.

20. Walker, D. M. and Blanchard, M. H., Pressures in experimental coal hoppers, *Chem. Eng. Sci.*, Vol. 22, 1967, pp. 1713–1745.

30 Reinforced concrete chimneys

A B Cassidy　　　　　M Hartstein

Pullman Power Products Corporation,
New York, USA

Contents

Notation

A_c	area of concrete
A_s	area of steel
B	fraction of critical damping; half the opening angle
B_1	factor defined in ACI 318–77, Section 10.2.7
C	numerical coefficient for base shear; temperature difference between sunny side and shaded side of chimney
C_b	coefficient of thermal conductivity of the chimney uninsulated lining or insulation around steel liner
C_c	coefficient of thermal conductivity of the concrete of the chimney shell
C_L	lateral lift coefficient
C_s	coefficient of thermal conductivity of insulation filling the air space between lining and shell
c'	ratio of inside face circumferential reinforcing steel area to the outside circumferential reinforcing steel area.
D	outside diameter of chimney shell at top; dead load
D_b	mean diameter of uninsulated lining or insulation around steel liner
D_{bi}	inside diameter of uninsulated lining or insulation around steel liner
D_c	mean diameter of concrete chimney shell; critical chimney diameter
D_{ci}	inside diameter of concrete chimney shell
D_{co}	outside diameter of concrete chimney shell
D_1	outside diameter of chimney shell at base
D_s	mean diameter of space between lining and shell

D_x — mean diameter of concrete chimney shell at a level designated as x

d_m — mean diameter of concrete chimney shell

d_o — outside diameter of concrete chimney shell

E_c — modulus of elasticity of concrete

E_s — modulus of elasticity of reinforcement

e — eccentricity

F_h — lateral force applied to a level designated as h

F_L — lateral force per unit height

f'_c — concrete cylinder compressive strength

f_{cw} — concrete stress at outside diameter of shell due to wind plus dead load

f'_{cw} — concrete stress at mean diameter of shell due to wind plus dead load

f''_{CTV} — maximum vertical stress in the concrete occurring at the inside of the chimney shell due to temperature

f_y — reinforcement yield strength

H — height of chimney above base; total height of chimney

h — height above base to a level designated as h

h_x — height above base to a level designated as x

J — numerical coefficient for base moment

J_x — numerical coefficient for moment at a level designated as x

K — number of openings entirely in compression zone

K_e — E_s/f_y

K_r — coefficient of heat transfer by radiation between outside surface of lining and inside surface of concrete chimney shell

K_1 — coefficient of heat transmission from gas to inner surface of chimney lining when chimney is lined, or to inner surface of chimney shell when chimney is unlined

K_2 — coefficient of heat transmission from outside surface of chimney shell to surrounding air

K_s — coefficient of heat transfer between outside surface of lining and inside surface of shell for chimneys with ventilated air spaces

k — ratio of distance between inside of chimney shell and neutral surface resulting from vertical temperature, to the total shell thickness, t

k' — ratio of distance between inside of chimney shell and neutral surface resulting from circumferential temperature, to the total shell thickness, t

k_{comb} — ratio of distance between the inner surface of the chimney shell and the neutral surface resulting from combined wind, dead loads and temperature, to the total shell thickness, t

L — live load

M — moment at the base

M_T — moment at the top

M_x — moment at any level x

m — mass per unit height of chimney; pf_y/f'_c

n — ratio of the modulus of elasticity of the reinforcement to the modulus of elasticity of the concrete; natural frequency of chimney

P — mass density of air (0.00238 lb s^2/ft^4)

p — ratio of the total area of vertical reinforcement to the total area of concrete of chimney shell at section under consideration

p' — ratio of the cross-sectional area of the circumferential outside face reinforcing steel per unit of height to the cross-sectional area of the chimney shell per unit of height.

q — ratio of actual to modal values in vortex shedding calculations

R — thermal resistance, h ft^2 °F/Btu; mean radius of chimney shell

r — mean radius of chimney shell at section under consideration

r_q — ratio of heat transmission through chimney shell to heat transmission through lining for chimneys with ventilated air spaces

S — Strouhal number, a constant

T — fundamental period of vibration of the chimney; maximum specified design temperature of gas inside chimney; taper of chimney

T_o — minimum temperature of outside air surrounding chimney

T_x — temperature differential across shell

t — thickness of chimney shell at section under consideration

t_b — thickness of uninsulated lining or insulation around steel liner

t_m — mean thickness of chimney shell at

	section under consideration
t_s	thickness of air space or insulation filling the space between lining and shell
t_x	thickness of chimney shell at a level designated as x
U	use factor which varies from 1.3 to 2.0
V	total shear at the base; volume of section
V_c	critical wind velocity
W	weight of chimney above section under consideration; total weight of chimney without lining for earthquake design; wind load
W_1	total weight of chimney including

	corbel supported lining
w_h	that portion of W or W_1 which is assigned to the level designated as h
z'	ratio of distance between the inner surface of the chimney shell and circumferential outside face reinforcing steel to the total shell thickness, t
α	one-half the central angle subtended by the neutral axis as a chord on the circle of radius r
$\sum w_h h$	summation of the products of all $w_h h$ for the chimney
ϕ	mode shape; capacity reduction factor

Summary

This chapter covers the design and construction of reinforced concrete chimneys to comply with the requirements of ACI 307–79 and ACI 318–77. The advantages and disadvantages of various types of chimney liner are discussed. Recommended loadings and design procedures in accordance with accepted codes are demonstrated by sample calculations. Loads include static wind and earthquake combined with thermal effects; dynamic wind loads due to vortex shedding; dynamic earthquake either by time history or in accordance with a specified response spectrum. Formulas are presented for ultimate strength design of chimneys with provisions for one or two openings in a section; and with recommended load factors and capacity reduction factor. Also covered are the effects of the sun; construction methods including jumpforms and slipforms; and maintenance and repair of chimneys.

1 Introduction

Chimneys for small plants have been built using radial brick or steel. However, for larger installations, reinforced concrete has become the material of choice for the exterior shell since it can be designed and erected most economically to resist the loads imposed upon it. In recent years, chimneys have grown taller to meet governmental restrictions on air pollution. Often more than one flue is incorporated in one reinforced concrete shell, resulting in very large diameters. This chapter will therefore discuss the problems of reinforced concrete chimneys only.

1.1   Types of liner

In most cases, the flue gases are too hot, too abrasive or too corrosive to be allowed to impinge directly on the concrete chimney shell. Therefore, liners must be installed to resist these effects. In the past, many chimneys were built with brick liners supported about every 9 m (30 ft) by concrete projections from the inside face of the concrete shell, called corbels. These functioned well as long as the chimneys operated under natural draft. Under these conditions, there was always a slight vacuum inside the liner. Any leakage through the bricks or their mortar joints would bring fresh air into the inside of the liner, doing little harm. However, when larger boilers were fitted with induced draft fans to increase their efficiency, a positive pressure was created inside the liner and the gas leakage progressed from the inside out. Corrosive acids in the flue gas then condensed and attacked both the brick and the concrete shell.

To remedy this condition, freestanding acid-proof brick liners were built with acid-proof mortars. Lead pans protected by acid-proof brick were installed in the floor inside the liner and in the annular space between the brick and the concrete shell to collect the acid, and drains were installed to remove the acid.

In many cases, the annular space was pressurized to slightly above the gas pressure to prevent gas leakage through the brick. This was a viable solution for chimneys up to about 180–215 m (600–700 ft) tall. Above this height, the cost of the heavy brick walls required to support self-weight became prohibitive. In addition, if the chimney had to withstand earthquake forces greater than those for Zone 2, it became almost impossible to design the brick for these forces.

The next step was the use of insulated steel liners for the taller chimneys. These had the advantage of being gas-tight and could be designed for the most severe earthquake conditions. At the same time, they were more economic than brick liners for the taller chimneys. The steel liners enjoyed great success until stricter air pollution agencies dictated the use of desulfurization units (scrubbers) in the gas stream before it entered the chimney. The desulfurization process removed up to 90% of the sulfur compounds in the gas stream. But during the process, which consists of using a chemical spray, the temperature of the gases was decreased to below the dew point of sulfuric acid. The remaining acid in the gas stream then condensed on the steel liners, and in a few weeks corroded large holes in the plate. Various protective coatings were tried, but none was found to be successful in an economically feasible range.

Glass-fiber reinforced polyester (FRP) liners then came into use. They are corrosion resistant and strong, but lose strength rapidly at high temperatures. They can be used where the gas temperatures do not exceed about 177°C (350°F).

1.2   Loading conditions and methods of design

After the general arrangement of the chimney has been determined, including the choice and configuration of the liner, the loading conditions and design procedures must be determined. Generally, engineering offices specializing in chimney design and construction use computer programs to perform most of the design calculations. However, before an engineer can use a computer program intelligently, he should understand the calculation procedures which are performed. Therefore, detailed hand calculations will be presented in this chapter for a typical chimney.

1.3 Other aspects

Construction methods are described in Section 4, and the maintenance and repair of chimneys are touched on in Section 5.

The notation and units will, in general, follow the usage of the current ACI Building Code.

2 Loading conditions

2.1 General

Chimneys must resist environmental conditions such as wind or earthquake loads. Maps are available which, based on past history, give the recommended design wind velocity and earthquake intensity for most locations. For the USA, such maps are included in *American national standard building code requirements for minimum design loads in buildings and other structures* and in the *Uniform building code*. The *National building code of Canada* provides similar information for Canadian locations.

2.2 Static wind loads

2.2.1 *Formulas specified by ACI 307*
Design wind pressures versus height zones are specified by ACI 307-79 [1] based on map areas found in the *Uniform building code* for US locations. For example, if a location falls in a map area designated as 40 psf, the following wind pressures on the chimney projected area shall be used:

Height zone, ft	Wind pressure, psf
0–100	26
100–500	36
500–1200	42
1200–Over	48

Using these wind pressures, the wind moments at various heights may be calculated. The dead loads at these heights are also obtained. Then, using the formulas given in ACI 307-79, the stresses in the concrete and reinforcing steel may be calculated.

2.2.2 *Sample designs*
Typical calculations for a 550 ft (167.64 m) reinforced concrete chimney with an independent brick liner are shown in Examples 2-1 to 2-8.

Example 2-1 shows the column and brick liner configuration as well as the design conditions.

Example 2-2 shows the derivation of the formulas for calculating the volume and the center of gravity of a column section using Simpson's rule. It can be shown that, for the typical situation where the wall thicknesses are small compared to the diameters, the volumes and centers of gravity calculated using Simpson's rule give accurate results. (See Examples 2-3 to 2-6)

Example 2-7 shows a systematic arrangement for calculating the column dead loads applying the formulas from Example 2-2. The centers of gravity are included since they are required for the earthquake calculations.

Example 2-8 shows a convenient layout of the calculations for the shears and moments caused by wind acting on the column. The wind pressures are taken on the projected area of the column. For any section, this area is a trapezoid which can be considered as broken up into two triangles by a diagonal line. This is the basis for the calculations shown.

2.2.2.1 *Dead load plus wind: no openings* Example 2-9 demonstrates a typical calculation for the stresses at a section which has no openings. At 310 ft (94.5 m) down, the wind moment is greater than the earthquake moment. Therefore, the calculation is made using the wind moment. Note that the steel ratio, p, is determined by a trial-and-error procedure. For any chosen value of p together with the fixed value of e/r, a value of α may be chosen from the ACI 307-79 chart, figure 5.1. This leads to a determination of the maximum concrete and steel stresses. If the stresses exceed the allowable values, the value of p is increased. If the stresses are too low, p is decreased. This process is continued until satisfactory values are obtained. The calculations shown are the final results of this process.

2.2.2.2 *Dead load plus wind: at openings* Example 2-10 shows similar calculations for the case where two openings exist 180° apart. The calculations would be identical if only one opening existed. The reason is that, in accordance with ACI 307-79, the vertical steel cut by the opening must be replaced at the sides of the opening. Therefore, if the load is assumed to act towards one opening, the other opening will be in the tension zone. The concrete in the tension zone is ignored and, since the reinforcing steel cut by the opening in the tension zone has been replaced, the opening in the tension zone may also be ignored.

Example 2-1: column and brick liner configuration
The design data are

(a) Wind pressure map area according to *Uniform Building Code* is 40 psf
(b) Earthquake zone 1, use factor of 2
(c) Concrete $f_c' = 4000$ psi
(d) Reinforcing steel ASTM A615, Grade 60
(e) Design gas temperature is 285°F
(f) Minimum ambient temperature is −2°F

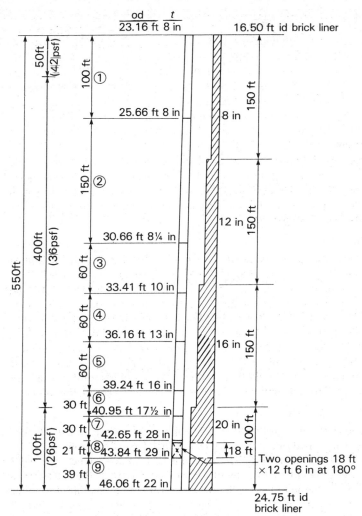

Example 2-2: *volume and center of gravity of a column section using Simpson's rule*

Mean diameter $d_M = \frac{1}{2}(D+d)$, and mean thickness $t_M = \frac{1}{2}(T+t)$.

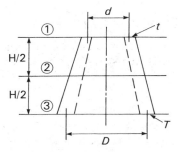

item	A/π	Simpson's multiplier	V/π	h	Vh/π
(1) Top area	dt	1	dt	H	dtH
(2) Mean area	$d_m t_m$	4	$4d_m t_m$	$H/2$	$2d_m t_m H$
(3) Bottom area	DT	1	DT	0	0
			$dt + DT + 4d_m t_m$		$H(dt + 2d_m t_m)$

Volume is given by

$$\tfrac{1}{6}\pi H(dt + DT + 4d_m t_m)$$

Center of gravity above level (3) is given by

$$\frac{\tfrac{1}{6}\pi \times H(dt + 2d_m t_m)}{\tfrac{1}{6}\pi H(dt + DT + 4d_m t_m)}$$

$$= \frac{(dt + 2d_m t_m)H}{(dt + DT + 4d_m t_m)}$$

Example 2-3: volume of a column section (exact solution)
Assume wall thicknesses are small compared to diameters.

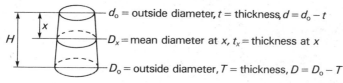

d_o = outside diameter, t = thickness, $d = d_o - t$

D_x = mean diameter at x, t_x = thickness at x

D_o = outside diameter, T = thickness, $D = D_o - T$

From the diagram, it follows that

$$D_x = d + \frac{x}{H}(D - d) = \frac{dH + (D - d)x}{H}$$

$$t_x = t + \frac{x}{H}(T - x) = \frac{tH + (T - t)x}{H}$$

Therefore, the volume, V, is

$$V = \int_0^H \pi D_x t_x \, dx = \frac{\pi}{H^2} \int_0^H [dH + (D - d)x][tH + (T - t)x] \, dx$$

$$V = \frac{\pi}{H^2} \int_0^H [dtH^2 + dH(T - t)x + tH(D - d)x + (D - d)(T - t)x^2] \, dx$$

$$V = \frac{\pi}{H^2} [dtH^3 + \tfrac{1}{2}dH(T - t)H^2 + \tfrac{1}{2}tH(D - d)H^2 + \tfrac{1}{3}(D - d)(T - t)H^3]$$

$$V = \frac{\pi}{H^2} [dtH^3 + \tfrac{1}{2}dTH^3 - \tfrac{1}{2}dtH^3 + \tfrac{1}{2}tDH^3 - \tfrac{1}{2}tdH^3$$

$$+ \tfrac{1}{3}DTH^3 - \tfrac{1}{3}DtH^3 - \tfrac{1}{3}dTH^3 + \tfrac{1}{3}dtH^3]$$

$$V = \frac{\pi}{H^2} [\tfrac{1}{3}dtH^3 + \tfrac{1}{6}dTH^3 + \tfrac{1}{6}DtH^3 + \tfrac{1}{3}DTH^3]$$

$$V = \tfrac{1}{6}\pi H[2(dt + DT) + dT + Dt]$$

Example 2-4:   moment of a column section about its top (exact solution)
With reference to the diagram in Example 2-3, the moment about the top is given by

$$M_T = \int_0^H \pi D_x t_x x \, dx = \frac{\pi}{H^2} \int_0^H [dH+(D-d)x][tH+(T-t)x]x \, dx$$

$$= \frac{\pi}{H^2} \int_0^H [dtH^2 x + dH(T-t)x^2 + tH(D-d)x^2 + (D-d)(T-t)x^3] \, dx$$

$$= \frac{\pi}{H^2} \left[\frac{dtH^4}{2} + dH(T-t)\frac{H^3}{3} + tH(D-d)\frac{H^3}{3} + (D-d)(T-t)\frac{H^4}{4} \right]$$

$$= \pi H^2 \left[\frac{dt}{2} + \frac{dT}{3} - \frac{dt}{3} + \frac{tD}{3} - \frac{td}{3} + \frac{DT}{4} - \frac{Dt}{4} - \frac{dT}{4} + \frac{dt}{4} \right]$$

$$= \tfrac{1}{12}\pi H^2 [dt + dT + Dt + 3DT]$$

Example 2-5:   center of gravity of a column section from bottom of section (exact solution)
From the results in Examples 2-3 and 2-4,

$$CG_B = H - \frac{\tfrac{1}{12}\pi H^2 (dt + dT + Dt + 3DT)}{\tfrac{1}{6}\pi H[2(dt + DT) + dT + Dt]}$$

$$= H - \frac{H}{2} \frac{(dt + dT + Dt + 3DT)}{[2(dt + DT) + dT + Dt]}$$

$$= H \frac{(4dt + 4DT + 2dT + 2Dt - dt - dT - Dt - 3DT)}{2[2(dt + DT) + dT + Dt]}$$

$$= \frac{(3dt + DT + Dt + dT)}{[2(dt + DT) + dT + Dt]} \times \frac{H}{2}$$

Example 2-6:   validity of Simpson's rule
Here, we show that Simpson's rule gives the same results as the exact solution for the volume and center of gravity of a column section, for the case where the walls are thin compared to the diameters.

$$d_m = \tfrac{1}{2}(D+d) \quad t_m = \tfrac{1}{2}(T+t)$$
$$d_m t_m = \tfrac{1}{4}(D+d)(T+t) = \tfrac{1}{4}(DT + Dt + dT + dt)$$

The volume is given by

$$V = \tfrac{1}{6}\pi H(dt + DT + 4d_m t_m)$$
$$V = \tfrac{1}{6}\pi H(dt + DT + DT + Dt + dT + dt).$$
$$= \tfrac{1}{6}\pi H[2(dt + DT) + dT + Dt]$$

which checks with Example 2-3.

The center of gravity is given by

$$\frac{(dt + 2d_m t_m)H}{(dt + DT + 4d_m t_m)}$$

$$= \frac{[dt + \frac{1}{2}(DT + Dt + dT + dt)]H}{(dt + DT + DT + Dt + dT + dt)}$$

$$= \frac{(3dt + DT + Dt + dT)}{[2(dt + DT) + dT + Dt]} \times \frac{H}{2}$$

which checks with Example 2-5.

Example 2-7: dead-load calculations
With reference to the figure in Example 2-1:

$dt + DT$ $+ 4d_m t_m$		$dt + 2d_m t_m$	Volume ft^3	Section dead load tons	Cumu- lative dead load tons
22.49×0.67	$= 15.07$	15.07			
24.99×0.67	$= 16.74$		CG $= \dfrac{46.88}{95.43} \times 100 = 49.13$		
$23.74 \times 0.67 \times 4$	$= 63.62 \times \frac{1}{2}$	$= 31.81$			
	$\overline{95.43}$	$\overline{46.88}$			
		$\pi \times 95.43 \times \dfrac{100}{6} = 4997$		375	375

①————————————————————————————————

24.99×0.67	$= 16.74$	16.74			
29.97×0.69	$= 20.68$		CG $= \dfrac{54.11}{112.17} \times 150 = 72.36$		
$27.48 \times 0.68 \times 4$	$= 74.75 \times \frac{1}{2}$	$= 37.37$			
	$\overline{112.17}$	$\overline{54.11}$			
		$\pi \times 112.17 \times \dfrac{150}{6} = 8810$		661	1036

②————————————————————————————————

29.97×0.69	$= 20.68$	20.68			
32.58×0.83	$= 27.04$		CG $= \dfrac{68.23}{142.81} \times 60 = 28.67$		
$31.28 \times 0.76 \times 4$	$= 95.09 \times \frac{1}{2}$	$= 47.55$			
	$\overline{142.81}$	$\overline{68.23}$			
		$\pi \times 142.81 \times \dfrac{60}{6} = 4486$		336	1372

③————————————————————————————————

$dt + DT$ $+ 4d_m t_m$	$dt + 2d_m t_m$	Volume ft^3	Section dead load tons	Cumulative dead load tons

32.58×0.83	$= 27.04$	27.04		
35.08×1.08	$= 37.89$	$CG = \dfrac{91.66}{194.16} \times 60 = 28.33$		
$33.83 \times 0.955 \times 4$	$= \dfrac{129.23 \times \frac{1}{2}}{194.16}$	$= \dfrac{64.62}{91.66}$		

$$\pi \times 194.16 \times 16 \times \frac{60}{6} = 6100 \qquad 457 \qquad 1829$$

④

35.08×1.08	$= 37.89$	37.89		
37.91×1.33	$= 50.42$	$CG = \dfrac{125.86}{264.24} \times 60 = 28.58$		
$36.50 \times 1.205 \times 4$	$= \dfrac{175.93 \times \frac{1}{2}}{264.24}$	$= \dfrac{87.97}{125.86}$		

$$\pi \times 264.24 \times \frac{60}{6} = 8301 \qquad 623 \qquad 2452$$

⑤

37.91×1.33	$= 50.42$	50.42		
39.49×1.46	$= 57.66$	$CG = \dfrac{158.39}{324.03} \times 30 = 14.66$		
$38.70 \times 1.395 \times 4$	$= \dfrac{215.95 \times \frac{1}{2}}{324.03}$	$= \dfrac{107.97}{158.39}$		

$$\pi \times 324.03 \times \frac{30}{6} = 5090 \qquad 382 \qquad 2834$$

⑥

39.49×1.46	$= 57.66$	57.66		
40.32×2.33	$= 93.95$	$CG = \dfrac{208.92}{454.13} \times 30 = 13.80$		
$39.91 \times 1.895 \times 4$	$= \dfrac{302.52 \times \frac{1}{2}}{454.13}$	$= \dfrac{151.26}{208.92}$		

$$\pi \times 154.13 \times \frac{30}{6} = 7133 \qquad 535 \qquad 3369$$

⑦

40.32×2.33	93.95	93.95		
41.42×2.42	100.24	$CG = \dfrac{288.08}{582.46} \times 21 = 10.39$		
$40.87 \times 2.375 \times 4$	$\dfrac{388.27 \times \frac{1}{2}}{582.46}$	$= \dfrac{194.13}{288.08}$		

$$\pi \times 582.46 \times \frac{21}{6} = \quad 6404 \times 10.39 = \qquad 66\,538$$

$$-18 \times 12.5 \times 2.38 \times 2 \quad = \underline{1071 \times} \quad 9.00 = - \underline{9\,639}$$

$$ 5333 \qquad\qquad 56\,899$$

$$CG(\text{corrected for opgs}) = \frac{56\,899}{5333} = 10.67$$

$$\phantom{CG(\text{corrected for opgs}) = } 5333 \qquad 400 \qquad 3769$$

⑧

$dt + DT$ $+ 4d_m t_m$		$dt + 2d_m t_m$	Volume ft^3	Section dead load tons	Cumu- lative dead load tons
41.42×2.42	$= 100.24$	100.24			
44.23×1.83	$= 80.94$		$CG = \dfrac{283.27}{547.24} \times 39 = 20.19$		
$40.87 \times 2.125 \times 4$	$= 366.06 \times \frac{1}{2}$	$= 183.03$			
	$\overline{547.24}$	$\overline{283.27}$			
		$\pi \times 547.24 \times \dfrac{39}{6} = 11\,175$		838	4607

⑨

Total volume of concrete $= 61\,425$ ft^3 (2275 yd^3).

Example 2-8: *wind loads*
With reference to the figure in Example 2-1

	Shears	*Moments*
$23.16 \times 50/2 \times 42/2000 =$	$12.16 \times 50 \times \frac{2}{3} =$	405
$24.41 \times 50/2 \times 42/2000 =$	$12.82 \times 50 \times \frac{1}{3} =$	214
	$\overline{24.98 \text{ tons}}$	$\overline{619 \text{ t ft 50 ft-down}}$

	$24.98 \times 50 \quad =$	$1\,249$
$24.41 \times 50/2 \times 36/2000 =$	$10.98 \times 50 \times \frac{2}{3} =$	366
$25.66 \times 50/2 \times 36/2000 =$	$11.55 \times 50 \times \frac{1}{3} =$	193
	$\overline{47.51 \text{ tons}}$	$\overline{2\,427 \text{ t ft 100 ft down}}$

	$47.51 \times 150 \quad =$	$7\,127$
$25.66 \times 150/2 \times 36/2000 =$	$34.64 \times 150 \times \frac{2}{3} =$	$3\,464$
$30.66 \times 150/2 \times 36/2000 =$	$41.39 \times 150 \times \frac{1}{3} =$	$2\,070$
	$\overline{123.54 \text{ tons}}$	$\overline{15\,088 \text{ t ft 250 ft down}}$

	$123.54 \times 60 \quad =$	$7\,412$
$30.66 \times 60/2 \times 36/2000 =$	$16.56 \times 60 \times \frac{2}{3} =$	662
$33.41 \times 60/2 \times 36/2000 =$	$18.04 \times 60 \times \frac{1}{3} =$	361
	$\overline{158.14 \text{ tons}}$	$\overline{23\,523 \text{ t ft 310 ft down}}$

	$158.14 \times 60 \quad =$	$9\,488$
$33.41 \times 60/2 \times 36/2000 =$	$18.04 \times 60 \times \frac{2}{3} =$	722
$36.16 \times 60/2 \times 36/2000 =$	$19.53 \times 60 \times \frac{1}{3} =$	391
	$\overline{195.71 \text{ tons}}$	$\overline{34\,124 \text{ t ft 370 ft down}}$

	$195.71 \times 60 \quad =$	$11\,743$
$36.16 \times 60/2 \times 36/2000 =$	$19.53 \times 60 \times \frac{2}{3} =$	781
$39.24 \times 60/2 \times 36/2000 =$	$21.19 \times 60 \times \frac{1}{3} =$	424
	$\overline{236.43 \text{ tons}}$	$\overline{47\,072 \text{ t ft 430 ft down}}$

	Shears		*Moments*
	236.43×20	$=$	$4\,729$
$39.24 \times 20/2 \times 36/2000 =$	$7.06 \times 20 \times \frac{2}{3} =$		94
$40.38 \times 20/2 \times 36/2000 =$	$7.27 \times 20 \times \frac{1}{3} =$		48
	250.76 tons		$51\,943$ t ft 450 ft down

	250.76×10	$=$	$2\,508$
$40.38 \times 10/2 \times 26/2000 =$	$2.62 \times 10 \times \frac{2}{3} =$		17
$40.95 \times 10/2 \times 26/2000 =$	$2.66 \times 10 \times \frac{1}{3} =$		9
	256.04 tons		$54\,477$ t ft 460 ft down

	256.04×30	$=$	$7\,681$
$40.95 \times 30/2 \times 26/2000 =$	$7.98 \times 30 \times \frac{2}{3} =$		160
$42.65 \times 30/2 \times 26/2000 =$	$8.32 \times 30 \times \frac{1}{3} =$		83
	272.34 tons		$62\,401$ t ft 490 ft down

	272.34×21	$=$	$5\,719$
$42.65 \times 21/2 \times 26/2000 =$	$5.82 \times 21 \times \frac{2}{3} =$		81
$43.84 \times 21/2 \times 26/2000 =$	$5.98 \times 21 \times \frac{1}{3} =$		42
	284.14 tons		$68\,243$ t ft 511 ft down

	284.14×39	$=$	$11\,081$
$43.84 \times 39/2 \times 26/2000 =$	$11.11 \times 39 \times \frac{2}{3} =$		289
$46.06 \times 39/2 \times 26/2000 =$	$11.68 \times 39 \times \frac{1}{3} =$		152
	306.93 tons		$79\,765$ t ft 550 ft down

Example 2-9: wind stress—no openings
Consider section 3, at an elevation of 240 ft, i.e. 310 ft down:

$$D = 33.41 \qquad M = 23\,523 \text{ t ft}$$

$$t = 0.83 \qquad W = 1372 \text{ tons}$$

$$2R = 32.58$$

$$R = 16.29 \qquad \frac{e}{R} = \frac{23\,523}{1372 \times 16.29} = 1.05$$

From the ACI 307-79 chart (figure 5.1), for $n = 8$, $\beta = 0$, $\alpha = 71°$ (1.2392 rad) $p = 0.005$. Therefore,

$$np = 0.040$$
$$1 - p = 0.995$$

$$A = 1 - \cos \alpha \qquad\qquad = 0.6744$$
$$B = \sin \alpha - \alpha \cos \alpha \qquad = 0.5421$$
$$C = \pi \cos \alpha \qquad\qquad = 1.0228$$
$$D = (1 + \cos \alpha)/(1 - \cos \alpha) = 1.9655$$

$$f'_{cw} = \frac{WA}{2Rt[(1-p)B - npc]} = \frac{1372 \times 0.6744}{32.58 \times 0.83[0.995 \times 0.5421 - 0.040 \times 1.0228]}$$

$$f'_{cw} = 68.6 \text{ t/ft}^2 \times \frac{2000}{144} = 953 \text{ psi}$$

$$f_{cw} = f'_{cw}\left[1 + \frac{t}{2RA}\right] = 953\left[1 + \frac{0.83}{32.58 \times 0.6744}\right] = 989 \text{ psi} < 1000$$

which is all right.

$$f_{sw} = nf'_{cw}D = 8 \times 953 \times 1.9655 = 14\,985 \text{ psi} < 15\,000$$

which is also all right.

The area of concrete is

$$A_c = 32.58 \times 0.83 \times \pi \times 144 = 12\,233 \text{ in}^2$$

The area of steel required is

$$A_s = 0.005 \times 12\,233 = 61.2 \text{ in}^2$$

The area of steel supplied is

outside: 116 – No. 6 at 11 in ±, 51.0

inside: 51 – No. 4 at 24 in ±, $\dfrac{10.2}{61.2 \text{ in}^2}$

Example 2-10:   wind stresses—with openings
Consider section 8, at an elevation of 39 ft, i.e. 511 ft down. There are two $12\frac{1}{2}$ ft openings 180° apart:

$$D = 43.84 \qquad M = 68\,243 \text{ t ft}$$

$$t = 2.42 \qquad W = 3\,769 \text{ tons}$$

$$2R = 41.42$$

$$R = 20.71 \qquad \frac{e}{R} = \frac{68\,243}{3769 \times 20.71} = 0.874$$

$$\sin \beta = 6.25/20.71 = 0.3018$$

$$\beta = 17.57°(0.3067 \text{ rad})$$

$$\cos \beta = 0.9534$$

From the ACI 307-79 charts (figures 5.2 and 5.3), for $n = 8$, $\beta = 18°$, $\alpha = 77°(1.3439 \text{ rad})$ with $p = 0.003$. Therefore,

$$np = 0.024$$

$$1 - p = 0.997$$

$$1 - p + np = 1.021$$

$$A = \cos \beta - \cos \alpha \qquad = 0.7284$$

$$B = \sin \alpha - \alpha \cos \alpha \qquad = 0.6720$$

$$C = \pi \cos \alpha \qquad = 0.7069$$

$$D = (1 + \cos \alpha)/A \qquad = 1.6818$$

$$E = \sin \beta - \beta \cos \alpha \qquad = 0.2328$$

$$F = \cos \beta (\cos \beta - \cos \alpha) = 0.6945$$

$$f'_{cw} = \frac{WA}{2Rt[(1-p)B - (1-p+np)E - npC]}$$

$$f'_{cw} = \frac{3769 \times 0.7284}{41.42 \times 2.42[0.997 \times 0.6720 - 1.021 \times 0.2328 - 0.024 \times 0.7069]}$$

$$f'_{cw} = 65.9 \text{ t/ft}^2 \times \frac{2000}{144} = 915 \text{ psi}$$

$$f_{cw} = f'_{cw}\left(1 + \frac{t}{2RF}\right) = 915\left(1 + \frac{2.42}{41.42 \times 0.6945}\right) = 992 \text{ psi} < 1000$$

which is all right

$$f_{sw} = nf'_{cw} \times D = 8 \times 915 \times 1.6818 = 12\,311 \text{ psi} < 15\,000$$

which is also all right.

The area of concrete is

$$A_c = (41.42\pi - 25.4)2.42 \times 144 = 36\,494 \text{ in}^2$$

The area of steel,

$$A_s = 0.003 \times 36\,494 = 109.5 \text{ in}^2$$

2.3 Static earthquake design

Various formulas have been developed to determine static loads which, when applied to a structure, will simulate the shears and moments induced in the structure by the dynamic vibrations of actual earthquakes. They generally give a method of determining the total shear force and formulas for distributing the shears vertically. The total shear force is usually a function of the total weight of the structure. It also depends on the earthquake zone. These are zones of approximately equal probability of seismic intensity based on a statistical analysis of past history. The natural frequency of vibration also affects the magnitude of the total shear. In addition, a use factor is usually included. This factor depends on how important it is to keep the installation in operation after an earthquake. For example, the Commentary to ACI 307-79 states: 'It was believed that chimneys designed for $U = 2.0$ would be relatively damage-free and service-able after a strong shock. The lower limit of 1.3 was intended to produce safe chimneys although their serviceability may be impaired.'

In distributing the shear vertically, most codes find it necessary to account for the 'whipping' action which causes large moments in the upper portion of the chimney. This is done by applying a portion of the total shear as a concentrated load at the top of the chimney.

Various codes follow the principles indicated above, but differ in their details. They are the *Uniform building code*, American National Standards Institute,

ANSI A58.1, *National building code of Canada,* and American Concrete Institute, ACI 307-79. The first two include seismic risk zones for the USA, the third seismic risk zones for Canada.

2.3.1   Sample design for static earthquake

Example 2-11 shows the formulas and sample calculations of the static earthquake loading for our sample problem. These are based on the procedures recommended by ACI 307-79.

Example 2-11:   *earthquake loads on column*
The calculations given below are based on the procedures recommended in ACI 307-79.

   With reference to the figure in Example 2-1:

$$E = 57\,000\sqrt{f_c'} = 57\,000\sqrt{4000} = 3.605 \times 10^6 \text{ psi (see ACI 318-77)}$$

$$T = \frac{1.8H^2}{(3D_1 - D)\sqrt{E}} = \frac{1.8(550)^2}{(3 \times 46.06 - 23.16)\sqrt{3.605 \times 10^6}} = 2.49 \text{ s}$$

$$c = 0.1/\sqrt{T} = 0.0738$$

$$J = 0.6/\sqrt[3]{T} = 0.443; \text{ use } 0.45 \text{ minimum value}$$

$$J_x = J + (1-J)(h_x/H)^3 = 0.45 + 0.55(h_x/550)^3$$

$$V = ZUCW = 0.3 \times 2 \times 0.0738 \times 4607 = 204.0 \text{ tons}$$

Note that $z = 0.3$ for zone 1, and $u = 2.0$, upper limit of use factor.

$$F_{top} = 0.15\,V = 30.6 \text{ tons} \qquad 0.85\,V = 173.4 \text{ tons}$$

The following table gives vertical distribution of earthquake shears

Section	W_h tons	h ft	$W_h \times h$ t ft	$F_h = \dfrac{w_h h}{\sum w_h h} \times 0.85\,V$ tons
Top				30.6
1	375	499.13	187 174	39.2
2	661	372.36	246 130	51.6
3	336	268.67	90 273	18.9
4	457	208.33	95 207	19.9
5	623	148.58	92 565	19.4
6	382	104.66	39 980	8.4
7	535	73.80	39 483	8.3
8	400	49.67	19 868	4.2
9	838	20.19	16 919	3.5
			827 599	204.0

The moments induced in each section are

$$
\begin{array}{llll}
30.6 \times 100 & = & 3\,060 & \\
39.2 \times 49.13 & = & \underline{1\,926} & \\
& & 4\,986 \times [0.45 + 0.55(450/550)^3] = 3746\ \text{t ft } 100\ \text{ft down} \\
69.8 \times 150 & = & 10\,470 & \\
51.6 \times 72.36 & = & \underline{3\,734} & \\
& & 19\,190 \times [0.45 + 0.55(300/550)^3] = 10\,348\ \text{t ft } 250\ \text{ft down} \\
121.4 \times 60 & = & 7\,284 & \\
18.9 \times 28.67 & = & \underline{542} & \\
& & 27\,016 \times [0.45 + 0.55(240/550)^3] = 13\,392\ \text{t ft } 310\ \text{ft down} \\
140.3 \times 60 & = & 8\,418 & \\
19.9 \times 28.33 & = & \underline{564} & \\
& & 35\,998 \times [0.45 + 0.55(180/550)^3] = 16\,893\ \text{t ft } 370\ \text{ft down} \\
160.2 \times 60 & = & 9\,612 & \\
19.4 \times 28.58 & = & \underline{554} & \\
& & 46\,164 \times [0.45 + 0.55(120/550)^3] = 21\,038\ \text{f ft } 430\ \text{ft down} \\
179.6 \times 30 & = & 5\,388 & \\
8.4 \times 14.66 & = & \underline{123} & \\
& & 51\,675 \times [0.45 + 0.55(90/550)^3] = 23\,378\ \text{t ft } 460\ \text{ft down} \\
188.0 \times 30 & = & 5\,640 & \\
8.3 \times 13.80 & = & \underline{115} & \\
& & 57\,430 \times [0.45 + 0.55(60/550)^3] = 25\,885\ \text{t ft } 490\ \text{ft down} \\
196.3 \times 21 & = & 4\,122 & \\
4.2 \times 10.39 & = & \underline{44} & \\
& & 61\,596 \times [0.45 + 0.55(39/550)^3] = 27\,730\ \text{t ft } 511\ \text{ft down} \\
200.5 \times 39 & = & 7\,820 & \\
3.5 \times 20.19 & = & \underline{71} & \\
\underline{204.0}\ \text{tons} & & 69\,487 \times 0.45 & = 31\,269\ \text{t ft } 550\ \text{ft down} \\
\text{shear} & & &
\end{array}
$$

2.4 Temperature stresses

2.4.1 *Formulas for temperature stresses*
In order to calculate the temperature stresses in the concrete column, it is first necessary to solve the heat transfer problem of heat flow from the hot gases through the lining, the air space and through the concrete column to the ambient air. It is general practice to use the approximate procedure given by ACI 307-79 rather than to perform a complete heat balance study for each particular chimney. This procedure will give the temperature differential across the chimney column with sufficient accuracy.

ACI 307-79 also provides the formulas for calculating the temperature stresses in both the concrete and reinforcing steel, both alone and combined with other forces. These formulas are intentionally very conservative. The stresses vary almost directly with the concrete modulus of elasticity. The instantaneous value for this modulus is normally used, although it is well known that the thermal gradient is applied as a long-time load and that, due to shrinkage and creep, the modulus of elasticity of concrete for long-time loads may be reduced by a factor of two to three times. The temperature stresses would also be reduced by corresponding factors.

2.4.2 Sample design

The sample calculations included herein, however, follow exactly the formulas given in ACI 307-79. Example 2-12 shows the calculation for T_x for a lined chimney with a ventilated air space between the lining and shell in accordance with ACI 307-79, equation (22). However, the calculations have been arranged to show the basis for the formula. That is, that the total thermal resistance from the hot gas to the ambient air is equal to the sum of the component thermal resistances, and that the temperature drop across any of the component resistances is proportional to the magnitude of that resistance. Thus, we can obtain a complete temperature profile as well as the temperature drop across the concrete column, T_x.

Example 2-13 shows typical calculations for vertical temperature stresses alone and also combined with wind loads.

Similarly, Example 2-14 demonstrates the procedure for calculating circumferential temperature stresses.

Example 2-12:   *temperature differential across shell*
Consider the thermal effects at section 3, at an elevation of 240 ft, i.e. 310 ft down.

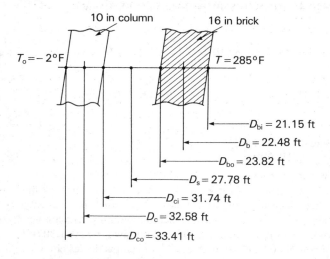

The gas velocity is

$$\frac{834\,000 \text{ ft}^3/\text{min} \times \dfrac{\text{min}}{60 \text{ s}}}{\frac{1}{4}\pi(21.15)^2 \text{ ft}^2} = 40 \text{ ft/s}$$

From ACI 307-79 (figure 5.15),

$K_{1c} = 4.8$

$K_{1r} = 2.0$

$K_1 = \overline{6.8} \text{ Btu h}^{-1} \text{ ft}^{-2} \,^\circ\text{F}^{-1}$

and

$$K_s = 285/150 = 1.90 \text{ Btu h}^{-1} \text{ ft}^{-2} \text{ °F}^{-1}$$

Thermal resistances, R h ft² °F Btu⁻¹		ΔT^a °F	Temp profile °F
		285	gas
$\dfrac{1}{r_q K_1} = \dfrac{1}{0.5 \times 6.8}$	$= 0.29$	13	
		272	int lining
$\dfrac{t_b D_{bi}}{r_q C_b D_b} = \dfrac{16 \times 21.15}{0.5 \times 6 \times 22.48}$	$= 5.02$	229	
		43	ext lining
$\dfrac{D_{bi}}{K_s D_s} = \dfrac{21.15}{1.90 \times 27.78}$	$= 0.40$	18	
		25	int column
$\dfrac{t D_{bi}}{C_c D_c} = \dfrac{10 \times 21.15}{12 \times 32.58}$	$= 0.54$	25	
		0	ext column
$\dfrac{D_{bi}}{K_2 D_{co}} = \dfrac{21.15}{12 \times 33.41}$	$= 0.05$	2	
	$\overline{}$	-2	ambient
	6.30		

$$^a \Delta T = \frac{R}{\sum R}(T - T_0) = R\frac{287}{6.30}$$

Temperature drop across column, $T_x = 25\text{°F}$.

Example 2-13: vertical temperature stresses
Consider again section 3, 310 ft down:

$$p = 51.0/12\,233 = 0.004\,17$$

$$c = 10.2/51.0 = 0.200$$

$$z = (10 - 2.875)/10 = 0.7125$$

$$k = -pn(c+1) + \sqrt{[pn(c+1)]^2 + 2pn[z + c(1-z)]}$$

$$= -0.004\,17 \times 8(1.2) + \sqrt{(0.0400)^2 + 2(0.004\,17)8[0.7125 + 0.2(1 - 0.7125)]}$$

$$= 0.1902$$

The vertical temperature stresses alone are

$$f''_{CTV} = 0.000\,0065\,kT_x E_c = 0.000\,0065 \times 0.1902 \times 25 \times 3.605 \times 10^6$$

$$= 111 \text{ psi}$$

$$f_{STV} = 0.000\,0065(z-k)T_x E_s = 0.000\,0065 \times 0.5223 \times 25 \times 29 \times 10^6$$

$$= 2461 \text{ psi}$$

The combined effect of temperature stress and wind load is calculated as follows:

$$K_{comb} = -pn(c+1) + \left\{[pn(c+1)]^2 + 2pn[z+c(1-z)]\right.$$

$$\left. + 2K[1+pn(c+1)]\frac{f'_{cw}}{f''_{CTV}}\right\}^{\frac{1}{2}}$$

$$= -0.0400 + \left\{0.0016 + 0.051\,37 + 2\times(0.1902)\times1.04\times\frac{953}{111}\right\}^{\frac{1}{2}}$$

$$= 1.8173 > unity$$

$$f''_{cw-comb} = f'_{cw} + \frac{f''_{CTV}}{K}\left\{\frac{2pn[z+c(1-z)]+1}{2[1+pn(c+1)]}\right\}$$

$$= 953 + \frac{111}{0.1902}\left\{\frac{0.051\,37+1}{2\times1.04}\right\}$$

$$= 1248 \text{ psi}$$

$$f_{sw-comb} = \frac{f_{STV}}{(Z-K)}\left[z+pn(c+1) - \left\{[pn(c+1)]^2 + 2pn[z+c(1-z)]\right.\right.$$

$$\left.\left. - 2pn(z-k)(c+1)\frac{f_{sw}}{f_{STV}}\right\}^{\frac{1}{2}}\right]$$

Check expression raised to power $\frac{1}{2}$:

$$0.0016 + 0.051\,37 - 2(0.004\,17)8(0.5223)(1.200)\frac{14\,985}{2461} = -0.2016$$

For negative value, use equations 26A, 28A and 46A in ACI 307-79. Thus,

$$K = \frac{c(1-z)+z}{(c+1)} = \frac{0.2(0.2875)+0.7125}{1.200} = 0.6417$$

$$f_{STV} = L(z-k)T_xE_s = 0.000\,0065 \times 0.0708 \times 25 \times 9 \times 10^6$$

$$= 334 \text{ psi}$$

Therefore,

$$f_{sw-comb} = 14\,985 + 334 = 15\,319 \text{ psi}$$

Example 2-14:   circumferential temperature stresses
Take again section 3, 310 ft down.
   The horizontal reinforcement is as follows:

   outside area: $A_{so} = 0.24$ in^2, No 4 at 10 in

   inside area: $A_{si} = 0.20$ in^2, No. 4 at 12 in

Therefore,

$$p' = 0.24/10 \times 12 = 0.002 \qquad np' = 8 \times 0.002 = 0.016$$

$$c' = 0.20/0.24 = 0.833 \qquad z' = (10 - 2.25)/10 = 0.775$$

$$k' = -p'n(c' + 1) + \sqrt{[p'n(c' + 1)]^2 + 2p'n[z' + c'(1 - z')]}$$

$$= -0.016(1.833) + \sqrt{(0.029\ 33)^2 + 2 \times 0.016[0.775 + 0.833(0.225)]}$$

$$= 0.1486$$

$$f''_{CTC} = Lk'T_xE_c = 0.000\ 0065 \times 0.1486 \times 25 \times 3.605 \times 10^6$$

$$= 87\ \text{psi}$$

$$f_{STC} = L(z' - k')T_xE_S = 0.000\ 0065 \times 0.6264 \times 25 \times 29 \times 10^6$$

$$= 2952\ \text{psi}$$

2.5 Circumferential stresses due to radial wind pressure distribution around a chimney

Although the overall effect of wind loads on the chimney column may be simulated by applying a pressure acting on the projected area of the column, there are also other effects. It has been found that the wind pressure varies around the circumference of the chimney cross-section. There is a positive pressure on the windward side and suction on the leeward side as well as on the sides perpendicular to the wind. This uneven wind distribution causes circumferential bending moments. These moments cause tensile stresses on both the inside and outside faces of the concrete column. (These are often referred to as ovalling stresses.) Prior to ACI 307-79, checking for this phenomenon was not required by any code. Many cases of vertical cracking probably resulted from this omission. Because of this effect, the present code requires calculations for this problem and also requires inside as well as outside steel for the full height of the chimney.

2.5.1 Sample calculations

Example 2-15 shows the ACI 307-79 procedure applied at a typical column section.

Note that the neutral axis for ovalling with compression inside is in the same location as that for circumferential temperature stresses. Therefore, the stresses for these two conditions may be added to obtain the combined stresses.

Example 2-15: *ovalling stresses*

Section 3, 310 ft down is again taken as the example.

(a) Compression at inside, tension at outside. For this case, p', c', z', k' are the same as for circumferential temperature stresses (see Examples 2-14).

$$f''_{cwc} = \frac{wr^2}{106t^2} \left[\frac{k'}{(k')^3 + 3np'[(z'+k'-1)^2c'+(z'-k')^2]} \right]$$

$$= \frac{36(16.29)^2}{106(0.833)^2}$$

$$\times \left[\frac{0.1486}{(0.1486)^3 + 3 \times 0.016[(0.775 + 0.1486 - 1)^2 \times 0.833 + (0.775 - 0.1486)^2]} \right]$$

$$= 864 \text{ psi}$$

$$f_{swc} = nf''_{cwc}[z'/k'-1] = 8 \times 864 \left[\frac{0.775}{0.1486} - 1 \right] = 29\,136 \text{ psi}$$

(b) Compression at outside, tension at inside.

$$K'_o = -p'n(c'+1) + \sqrt{[p'n(c'+1)]^2 + 2p'n[c'z'+1-z']}$$

$$= -0.016(1.833) + \sqrt{(0.029\,33)^2 + 2 \times 0.016[(0.833 \times 0.775) + 1 - 0.775]}$$

$$(-0.029\,33)$$

$$K'_o = 0.1401$$

$$f_{cwc} = \frac{wr^2}{92t^2} \left[\frac{k'_o}{(k'_o)^3 + 3np'[(z'+k'_o-1)^2 + c'(z'-k'_o)^2]} \right]$$

$$= \frac{36(16.29)^2}{92(0.833)^2}$$

$$\times \left[\frac{0.1401}{(0.1401)^3 + 3 \times 0.016[(0.775 + 0.1401 - 1)^2 + 0.833(0.775 - 0.1401)^2]} \right]$$

$$= 1091 \text{ psi}$$

$$f''_{swc} = nf_{cwc}[z'/k'_o - 1] = 8 \times 1091 \left[\frac{0.775}{0.1401} - 1 \right] = 39\,553 \text{ psi}$$

(c) Ovalling: compression inside combined with temperature stresses

concrete stress inside $= 864 + 87 = 951$ psi

steel stress outside $= 29\,136 + 2952 = 32\,088$ psi

2.6 Dynamic wind loads (vortex shedding)

2.6.1 *Discussion of the phenomenon*

When a steady wind blows across a cantilevered cylinder, vortices are shed alternately from one side, then the other, causing fluctuating forces to act in the direction perpendicular to the wind. A resonant condition occurs when the frequency of the vortex shedding is the same as the natural frequency of vibration of the chimney.

2.6.2 *Rumman and Maugh formulation modified according to the National Building Code of Canada* (1977)

Rumman and Maugh [3] determined the critical wind velocity at resonance from the well-known relationship

$$V_c = nD_c/S$$

where

V_c = critical wind velocity (ft/s)
n = natural frequency of chimney (Hz)
D_c = critical chimney diameter (ft)
S = Strouhal number

Then, assuming that the fluctuating lateral wind force varied sinusoidally, that is,

$$F_L = \tfrac{1}{2}C_L P V_c^2 D_c \sin(2\pi nt)$$

where

F_L = lateral force per unit height
P = mass density of air (0.00238 lb s²/ft⁴)
C_L = lateral lift coefficient

the maximum displacements, shears and moments may be obtained by multiplying their respective modal values by a multiplier, q:

$$q = 1.507 \times 10^{-5}\, \frac{C_L D_c^2 \int_0^H D\phi\, dh}{BS^2 \int_0^H m\phi^2\, dh}$$

where

B = fraction of critical damping
m = mass per unit height of chimney (lb s²/ft²)
ϕ = mode shape (ft)
H = total height of chimney (ft)

Rumman and Maugh, lacking experimental data, made the following conservative assumptions:

(a) $C_L = 0.66$, the same as the drag coefficient.
(b) Although the value of V_c varies with the chimney outside diameter, it is calculated using the value of D_c at about two-thirds the chimney height and assumed constant over the entire height of the chimney.
(c) $S = 0.2$.
(d) $B = 0.05$.

With these assumptions, the dynamic wind loads frequently controlled the design of chimneys. However, the 1977 *National building code of Canada* incorporated results obtained from wind-tunnel tests and measurements on full-scale structures as follows:

(a) $C_L = 0.2$ for circular cylinders.
(b) $S = 0.25$ for $Re > 2 \times 10^5$, where Reynolds' number, $Re = (V_c D/16) \times 10^5$. Most chimneys fall into this range.

(c) For tapered chimneys, when the critical diameter, D_c, is assumed at a particular elevation, the height over which the corresponding resonant wind force acts is determined by the height of chimney over which the diameter only changes by $\pm5\%$ from the value D_c. Therefore, for tapered chimneys, there is a range of critical wind velocities corresponding to the variation in diameter each acting over a different portion of the chimney height. By trial and error, the critical diameter, D_c, which results in the largest value of q may be determined. For any value of D_c, the vertical height over which the resonant force acts is from height H_1 to height H_2. The upper integral in the formula for q becomes

$$\int_{H_1}^{H_2} D\phi \, dh$$

(d) For reinforced concrete structures, $B = 0.02$.

Using this approach and the latter values for C_L, S and B, it was found that the vortex shedding effect on chimneys was less by a factor of about 5 to 7 times and this phenomenon usually did not control the chimney design.

2.6.3 Sample calculation

Examples 2-16 to 2-19 are the results obtained for our sample problem from a computer program using the Rumman and Maugh formulation modified in accordance with the recommendations of the 1977 *National building code of Canada*. Example 2-16 shows the first-mode frequency and the corresponding modal deflections, moments and shears. Example 2-17 indicates that the maximum value for q (0.1087 rounded to 0.11 ft) was obtained by letting the resonant wind force act on the vertical distance from the top of the chimney where the diameter, 23.16 ft, is 95% of the effective diameter, 24.38 ft, down a vertical distance of 82.62 ft, where the diameter, 25.60 ft, is 105% of the effective diameter. The effective velocity was obtained from:

$$V_c = nD_c/S = \frac{2.7073}{2\pi} \times \frac{24.38}{0.25} = 42.01 \text{ ft/s}$$

The final moments and shears were obtained by multiplying the modal values in Example 2-16 by 0.1078, the maximum value of q. As shown in Examples 2-17 and 2-18, the minimum permissible steel ratio, 0.0025, is required throughout. That is, vortex shedding did not control the design.

2.6.4 Random vibration theory

The latest work on vortex shedding by Davenport, Vickery and others [4–8] has shown that experimental data can be predicted more closely by assuming the lateral wind forces act randomly rather than periodically. The phenomenon is then formulated using statistical methods. It is expected that this approach will soon be incorporated into building and chimney codes.

Example 2-16:   *chimney dynamics*—1
The computer output reproduced below gives the first-mode frequency and the corresponding deflections, moments and shears for the sample problem

	FIRST MODE		
DISTANCE	FREQ. = 2.7073 RAD/SEC		
FROM BASE	SHAPE	MOMENTS	SHEARS
(FT.)	(FT.)	(FT-T)	(TONS)
550.00	1.0000	0	0
450.00	0.6702	3699	70
300.00	0.2577	19856	135
240.00	0.1491	28545	152
180.00	0.0749	38025	163
120.00	0.0294	48053	170
90.00	0.0158	53191	172
60.00	0.0070	59376	173
57.00	0.0063	58896	173
39.00	0.0030	67024	173
0.0	0.0	68815	174

Example 2-17:   *chimney dynamics*—2
The computer output reproduced below gives the maximum moments and shears for the sample problem

	EFF. DIA. = 24.38 FT.	EFF. VEL. = 42.01 FT/SEC.
	MAX. DFFL. = 0.11 FT.	
DISTANCE		
FROM BASE	MOMENTS	SHEARS
(FT.)	(FT-T)	(TONS)
550.00	0	0
450.00	402	7
300.00	2159	14
240.00	3103	16
180.00	4134	17
120.00	5224	18
90.00	5783	18
60.00	6347	18
57.00	6403	18
39.00	6744	18
0.0	7432	18

2.7 Dynamic earthquake loads

2.7.1 *Time history*
Measurements of the acceleration versus time taken during actual severe earthquakes have been reduced to digital form. These accelerograms together with the modal properties of a chimney permit a numerical solution of the differential

Example 2-18: chimney dynamics—3

The computer output reproduced below gives the results of the intermediate stress calculations for the sample problem

DISTANCE FROM BASE (FT.)	NO. OF OPGS	VERTICAL STEEL RATIO	TOT VERT LOAD (TONS)	LOAD FACTOR (HIGH) = 1.5000				LOAD FACTOR (LOW) = 1.0000			
				DESIGN MOMENTS (FT-T)	NEUTRAL AXIS (FT.)	CONCRETE STRESS (PSI)	STEEL STRESS (PSI)	DESIGN MOMENTS (FT-T)	NEUTRAL AXIS (FT.)	CONCRETE STRESS (PSI)	STEEL STRESS (PSI)
550.00	0	0.00250	0	0	0.0	0	0	0	0.0	0	0
450.00	0	0.00250	373	603	60.75	123	0	402	84.88	114	0
300.00	0	0.00250	1031	3238	50.73	310	0	2159	68.61	279	0
240.00	0	0.00250	1368	4655	55.25	312	0	3103	74.74	281	0
180.00	0	0.00250	1827	6201	62.83	292	0	4134	85.48	264	0
120.00	0	0.00250	2451	7837	75.12	284	0	5224	103.20	259	0
90.00	0	0.00250	2833	8675	83.40	282	0	5783	115.23	259	0
60.00	0	0.00250	3368	9521	92.03	201	0	6347	127.96	186	0
57.00	2	0.00250	3475	9605	58.12	199	0	6403	66.28	162	0
39.00	2	0.00250	3768	10116	61.34	196	0	6744	69.95	160	0
0.0	0	0.00250	4603	11223	122.38	303	0	7482	172.52	284	0

Example 2-19:   *chimney dynamics*—4

The computer output below gives the steel and concrete stresses (psi) for the sample problem

LOAD FACTORS   HIGH =   1.50   LOW =   1.00
STEEL ALLOW. STRESSES (PSI) IIGH = 60 000.0   LOW = 30 000.0
CONC. ALLOW. STRESSES (PSI) HIGH = 3 200.0   LOW = 1 600.0

DISTANCE FROM BASE (FT.)	NO. OF ORGS	VERTICAL STEEL RATIO	STEEL STRESSES		CONCRETE STRESSES	
			HIGH LOAD FACTOR	LOW LOAD FACTOR	HIGH LOAD FACTOR	LOW LOAD FACTOR
550.00	0	0.00250	0	0	0	0
450.00	0	0.00250	0	0	123	114
300.00	0	0.00250	0	0	310	279
240.00	0	0.00250	0	0	312	281
180.00	0	0.00250	0	0	292	264
120.00	0	0.00250	0	0	284	259
90.00	0	0.00250	0	0	282	259
60.00	0	0.00250	0	0	201	186
57.00	2	0.00250	0	0	199	162
39.00	2	0.00250	0	0	196	160
0.0	0	0.00250	0	0	303	284

NOTE—FOR SECTIONS CONTAINING OPENINGS, THE STRESSES ARE BASED ON NET SECTION.

equation of motion. In effect, one can simulate subjecting a chimney to an actual earthquake and find the maximum moments and shears induced in the chimney. By using several accelerograms from different earthquakes, any unusual peaks and valleys in the response from any one earthquake are smoothed out, resulting in a more general representation of probable ground motion.

2.7.2   *Response spectrum*

Another approach is to use the accelerograms obtained in a particular area to generate curves giving the maximum response versus frequency for various damping ratios. The response is given as the spectral (or maximum) displacement, velocity or acceleration. For a particular chimney, the modal properties are first obtained. Then, for each mode and a chosen damping ratio, the spectral displacements may be found from the response spectrum. This, together with the modal properties and the mass distribution in the chimney, permits the calculation at each section of the maximum moment and shear for each mode. The total response at each section may then be calculated as the square root of the sum of the squares of the individual modal responses.

2.8   Sun effects

2.8.1   *Deflection due to sun effects*

When the sun causes a temperature differential between the sunny side and the shaded side of a chimney, the chimney will deflect towards the shaded side. If we assume that the temperature drop across any section is proportional to the distance along the diameter, the maximum deflection will be caused with no longitudinal stresses occurring in the shell. Example 2-20 demonstrates the calculations for the deflection of our sample chimney due to sun effects. Deflections due to the sun are usually less than those caused by wind or earthquake. However, it is important to be aware of this possibility during construction. Plumbing operations should be scheduled during early mornings or late afternoons to avoid error.

2.8.2 Circumferential stresses due to sun effects

Circumferential stresses in the shell are also produced by sun effects. However, their magnitude has been found to be small enough to be negligible. This is demonstrated in Example 2-21.

Example 2-20:   sun effects—top deflection

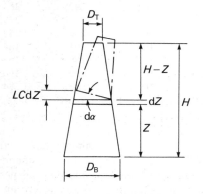

Consider a tapered chimney for which

D_T = top outside diameter

D_B = bottom outside diameter

H = total height

T = taper = $(D_B - D_T)/H$

C = temperature difference between sunny side and shaded side

S = horizontal deflection at top

Then

$$d\alpha = \frac{LC\,dZ}{D} \qquad D = D_B - TZ$$

$$S = \int_0^H d\alpha\,(H - Z) = \int_0^H \frac{LC(H - Z)\,dZ}{D_B - TZ}$$

or

$$S = LC \int_0^H \frac{(H - Z)\,dZ}{(D_B - TZ)}$$

From a table of integrals

$$\int_0^H \frac{(a+bz)}{(a'+b'z)}\,dz = \frac{bz}{b'} + \frac{ab'-a'b}{(b')^2} L_N(a'+b'z)$$

Therefore,

$$a = H \quad b = -1 \quad a' = D_B \quad b' = -T$$

$$S = LC\left[\frac{-Z}{-T} + \frac{H(-T) - D_B(-1)}{(-T)^2} L_N(D_B - TZ)\right]_0^H$$

$$S = LC\left[\frac{Z}{T} + \frac{D_B - HT}{T^2} L_N(D_B - TZ)\right]_0^H$$

$$S = \frac{LC}{T}\left[H + \frac{D_T}{T} L_N(D_T/D_B)\right]$$

For the sample problem

$$L = 5.65 \times 10^{-6} \text{ per }°F \quad C = 25°F \quad D_B = 46.06 \text{ ft}$$

$$D_T = 23.16 \text{ ft} \quad T = (46.06 - 23.16)/550 = 0.041\,64$$

$$S = \frac{5.65 \times 10^{-6} \times 25}{0.041\,64}\left[550 + \frac{23.16}{0.041\,64} L_N\left(\frac{23.16}{46.06}\right)\right] = 0.57 \text{ ft (7 in)}$$

Example 2-21: sun effects—circumferential moments
Consider the case of a diametric temperature distribution. Assume a section is cut at A and that free expansion due to temperature causes point A to move to A'.

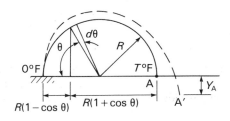

$$T_\theta = \frac{R(1 - \cos\theta)}{2R} T = \frac{T}{2}(1 - \cos\theta)$$

$$Y_A = \int_0^\pi LT_\theta R\,d\theta \cos\theta = \tfrac{1}{2}LRT\int_0^\pi (1 - \cos\theta)\cos\theta\,d\theta$$

$$Y_A = \tfrac{1}{2}LRT\int_0^\pi (\cos\theta - \cos^2\theta)\,d\theta$$

$$Y_A = \tfrac{1}{2}LRT[\sin\theta - \tfrac{1}{2}\theta - \tfrac{1}{4}\sin 2\theta]_0^\pi = \tfrac{1}{2}LRT(-\tfrac{1}{2}\pi)$$

$$Y_A = -\tfrac{1}{4}\pi LRT$$

Let

d_{YY} = vertical deflection at A due to a unit vertical load at A

d_{RY} = rotation at A due to a unit vertical load at A

d_{RR} = rotation at A due to a unit moment at A

then

$$d_{YY} = \int_0^\pi \frac{[R(1+\cos\theta)]^2}{EI} R\, d\theta = \frac{R^3}{EI} \int_0^\pi (1 + 2\cos\theta + \cos^2\theta)\, d\theta$$

$$d_{YY} = \frac{R^3}{EI} \left[\theta + 2\sin\theta + \frac{\theta}{2} + \frac{\sin 2\theta}{4}\right]_0^\pi = \frac{R^3}{EI}(\pi + \pi/2) = \frac{3}{2}\frac{\pi R^3}{EI}$$

$$d_{RY} = \int_0^\pi \frac{R(1+\cos\theta)}{EI} R\, d\theta = \frac{R^2}{EI}[\theta + \sin\theta]_0^\pi = \frac{\pi R^2}{EI} = d_{YR}$$

$$d_{RR} = \int_0^\pi \frac{R\, d\theta}{EI} = \frac{R}{EI}[\theta]_0^\pi = \frac{\pi R}{EI}$$

Let V_A be the vertical force at A, and M_A the moment at A, required to make $Y_A = 0$ and the rotation at A, $R_A = 0$. Then,

$$Y_A = -\frac{\pi}{4} LRT + d_{YY}V_A - d_{YR}M_A = 0$$

$$-\frac{\pi}{4} LRT + \frac{3}{2}\frac{\pi R^3}{EI} V_A - \frac{\pi R^2}{EI} M_A = 0 \qquad (E1)$$

$$R_A = d_{RY}V_A - d_{RR}M_A = 0$$

$$\frac{\pi R^2}{EI} V_A - \frac{\pi R}{EI} M_A = 0 \qquad (E2)$$

$$(E1) - R\times(E2) = -\frac{\pi}{4} LRT + \frac{1}{2}\frac{\pi R^3}{EI} V_A = 0$$

Therefore,

$$V_A = \frac{1}{2}\frac{LEIT}{R^2}$$

From (E2), $M_A = RV_A$, therefore,

$$M_A = LEIT/2R$$

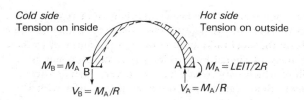

Cold side
Tension on inside

Hot side
Tension on outside

$M_B = M_A$

B

$V_B = M_A/R$

A

$M_A = LEIT/2R$

$V_A = M_A/R$

Sum of moments at B is

$$V_A \times 2R - M_A + M_B = 0$$
$$2RV_A - RV_A + M_B = 0$$
$$M_B = -RV_A$$
$$M_B = -M_A$$

For sample problem at an elevation of 240 ft:

$$od = 33.41 \text{ ft}$$
$$t = -0.83 \text{ ft}$$
$$2R = \overline{32.58} \text{ ft}$$
$$R = 16.29 \text{ ft}$$

For $f'_c = 4000$ psi,

$$E = 57\,000\sqrt{4000} = 3.6 \times 10^6 \text{ psi}$$
$$I = \tfrac{1}{12} \times 12(10)^3 = 1000 \text{ in}^4/\text{ft}$$

For $T = 25°F$,

$$M_{\text{circ}} = \frac{LEIT}{2R} = \frac{5.65 \times 10^{-6} \times 3.6 \times 10^6 \times 1000 \times 25}{2 \times 16.29 \times 12}$$
$$= 1301 \text{ lb in}$$

For uncracked section,

$$S = \tfrac{1}{6} \times 12(10)^2 = 200 \text{ in}^3$$

concrete stress $= \pm 1301/200 = 6.50$ psi, which is *negligible*.

3 Ultimate strength design

3.1 Definition

Ultimate strength design of reinforced concrete structures is a procedure in which members are proportioned to just reach failure (or ultimate strength) after their expected service loads are increased by specified load factors, and the capacities of the members are reduced by capacity reduction factors.

3.2 Advantage

Vertical stresses in a chimney shell increase at a faster rate than an increase in the overturning moment due to the shift in the neutral axis. The rate of increase in stress is variable. It depends on the steel ratio and the ratio of the moment to the vertical load. Therefore, chimneys designed by working stress methods may have factors of safety against failure which vary from very low to very high at various sections of the same chimney. Since failure will occur at the section with the lowest factor of safety, the excess material at sections with higher factors of safety is wasted. And since the true factors of safety are not known, the lowest factor may be dangerously low.

In ultimate strength design, every section may be proportioned for the same factor of safety against overloads, the specified load factor. Similarly, each section may be provided with the same factor of safety against possible understrength of the concrete or reinforcing steel, the capacity reduction factor.

3.3 Formulas

The formulas for ultimate strength design presented below are based on Cannon's design charts [9]. Modifications have been made to the original formulation to

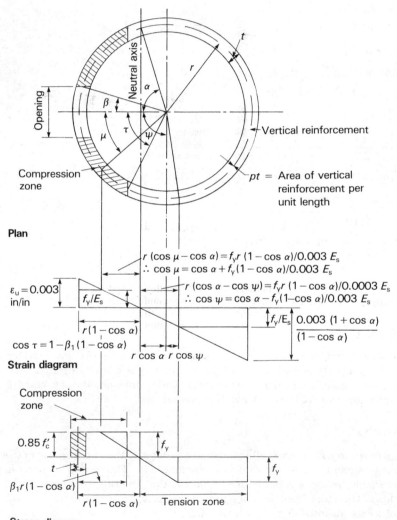

Figure 3-1   Strain and stress diagrams

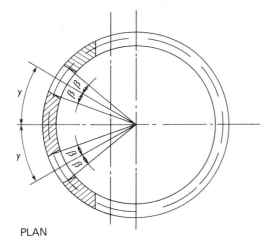

PLAN

Figure 3-2 Two openings in compression zone. (The dimensions not shown are the same as in Fig. 3-1)

include:

(a) Compression reinforcement.
(b) Two openings completely or partly in the compression zone.
(c) Values of B_1 less than 0.85 when f'_c exceeds 4000 psi according to ACI 318-77.

Assumptions conform to the requirements of ACI 318-77 [10] considering an equivalent rectangular concrete stress block in the compression zone. No allowance has been made for the area of concrete replaced by steel in the compression zone.

The nomenclature for non-standard items is shown on Figs 3-1 to 3-3 or is described below.

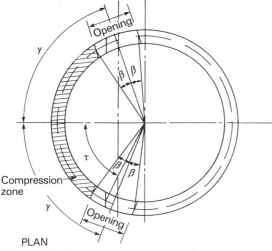

PLAN

Figure 3-3 Two openings partly in compression zone. (The dimensions not shown are the same as in Fig. 3-1)

3.3.1   *General equations*

The following equations apply:

$$\frac{W}{rtf'_c} = K_1 = 1.70\text{``}A\text{''} + 0.006K_e m\text{``}Q\text{''} + 2m\text{``}V\text{''} \tag{3-1}$$

where

$W = $ applied vertical load

$r = $ average radius of section

$t = $ thickness of section

$\text{``}A\text{''} = \tau - KB$

$\text{``}Q\text{''} = \dfrac{\sin \psi - \sin \mu - (\psi - \mu) \cos \alpha}{1 - \cos \alpha}$

$\text{``}V\text{''} = \mu + \psi - \pi$

$\cos \tau = 1 - B_1(1 - \cos \alpha)$

$\cos \psi = \cos \alpha - \left(\dfrac{1 - \cos \alpha}{0.003}\right) f_y / E_s$

$\cos \mu = \cos \alpha + \left(\dfrac{1 - \cos \alpha}{0.003}\right) f_y / E_s$

and where B is half the opening angle, B_1 a factor defined in ACI 318-77, Section 10.2.7, $K_e = E_s/f_y$, K is the number of openings entirely in compression zone, and $m = pf_y/f'_c$.

$$M_u / Wr\phi = K_3 = \cos \alpha + K_2 / K_1 \tag{3-2}$$

$$K_2 = 1.70\text{``}R\text{''} + 0.003K_e m\text{``}H\text{''} + 2m\text{``}K\text{''} \tag{3-3}$$

where

$M_u = $ ultimate moment capacity of section

$\text{``}H\text{''} = [(\psi - \mu)(1 + 2 \cos^2 \alpha) + \tfrac{1}{2}(4 \sin 2\alpha + \sin 2\psi - \sin 2\mu)$

$\qquad - 4 \cos \alpha (\sin \alpha + \sin \psi - \sin \mu)]/(1 - \cos \alpha)$

$\text{``}K\text{''} = \sin \psi + \sin \mu + (\pi - \psi - \mu) \cos \alpha$

$\text{``}R\text{''} = \sin \tau - (\tau - KB) \cos \alpha - \tfrac{1}{2}K[\sin (\gamma + B) - \sin (\gamma - B)]$

$\gamma = $ half angle between center-lines of two openings

and

for no openings $\qquad\qquad\qquad\qquad\quad K = \gamma = B = 0$

for one opening in compression zone $\quad K = 1, \gamma = 0$

for two openings in compression zone $\quad K = 2$

3.3.2   *Two openings partly in compression zone*

This condition exists when

$$\gamma + B > \tau \quad \text{and} \quad \gamma - B < \tau$$

For this case, let $\delta = \gamma - B$. Then in Eqn 3-1, $\text{``}A\text{''} = \delta$; and in Eqn 3-3,

$$\text{``}R\text{''} = \sin \delta - \delta \cos \alpha$$

3.3.3 *Openings in the tension zone*

Openings in the tension zone may be ignored since the tensile strength of the concrete is neglected and the bars cut by the openings are replaced at the sides of the openings.

3.3.4 *Calculation procedure*

At any section of a chimney, the following items are given: W, r, t, f'_c, f_y and p. If openings exist at the section B, K and γ may also be given. By trial and error, a value of α may be found which satisfies Eqn 3-1. K_2 and K_3 may then be evaluated from Eqns 3-2 and 3-3. The ultimate capacity of the section is then determined as $M_u = K_3 Wr\phi$.

3.3.5 *Limitation*

The angle, α, should never be permitted to be less than 23.4°. This requirement is necessary to keep the maximum steel strain within the range of the minimum 7% elongation permitted by ASTM A615, as shown below.

The maximum steel strain is given by

$$0.003\left(\frac{1+\cos\alpha}{1-\cos\alpha}\right) = 0.07$$

Therefore, $\cos\alpha = 0.91781$ and $\alpha = 23.4°$.

3.4 Comparison with working stress designs

In order to insure a uniform factor of safety at all sections and to maintain the safety inherent in working stress designs, the following combinations of load factor, U, and capacity reduction factor, ϕ, are proposed:

(a) The required strength, U, to resist dead load, D, live load L, and wind load, W, shall be the greatest required strength U of either

$$U = 1.4D + 1.7L + 1.7W \quad \text{or} \quad U = 0.9D + 1.7W$$

(b) For earthquake loads or forces, E, the load combinations of (a) shall apply except that $0.83E$ shall be substituted for W.

(c) The design strength of a section shall be taken as the nominal strength multiplied by a capacity reduction factor, ϕ, equal to 0.70.

4 Construction

For chimneys in the 'medium size' range (up to 700 ft, 213 m), the jumpform method of construction is normally used. For taller chimneys, or for installations requiring short schedules, the continuous slipform process is preferred.

4.1 Jumpform method

Jumpforming is the traditional—and most economic—way to construct a chimney. A construction derrick is first erected on the completed foundation. After the first three $7\frac{1}{2}$ft (2.3 m) sections of concrete have been poured, the derrick is suspended from steel stirrups embedded in the concrete column. The derrick, in turn, supports the other work components: the work deck and scaffolds on which the building activity takes place, and the concrete forms and hoisting cable which carry men and materials up to the work level.

For each $7\frac{1}{2}$ ft section of concrete, the derrick is raised and then the outside concrete forms are raised to their new position. Next, reinforcing steel bars are placed and tied into a strong skeletal structure. Then, the inner forms are raised and secured. Wall thickness and diameter dimensions are re-checked and, finally, the concrete is poured and allowed to set.

The next day, after the concrete hardens, the same sequence is repeated for the next section. Each new lift reduces the diameter and wall thickness to produce the configuration required by final design. This procedure is followed, section upon section, until the final height is achieved.

4.2 Slipform method

Slipforming is a continuous process, with concrete being placed 24 h a day.

The same basic concrete construction principles apply to both jumpforming and slipforming except that the operation of the concrete forms is much more complex in slipforming. In its simplest terms, the slipforms, working decks, scaffolds and central hoisting tower climb as a unit—by means of hydraulic jacks—up steel jack rods embedded in the concrete wall.

A typical slipform consists of two 4 ft (1.2 m) high concentric steel rings which are the inside and outside concrete forms. The forms are made up of telescoping curved steel plates braced and reinforced by adjustable steel yokes and spindles. While steel reinforcing bar is being placed and concrete poured on a continuous basis, the entire construction system—forms, work decks, scaffolds and tower—climbs up the jack rods a few millimeters at a time, controlled from a central hydraulic console.

The spindles are re-adjusted every 10 in (254 mm) of height to predetermined dimensions. Computers calculate the exact adjustments months before field construction begins. By the time construction is complete, workers will make hundreds of thousands of adjustments to the slipform system.

5 Maintenance and repair

Since chimney performance is vital to the operation of a plant, it is important to keep the chimney in good operating condition. A common mistake is to take the chimney for granted until a malfunction causes a costly shutdown for repairs. To maintain efficient and uninterrupted operation, a complete inspection by experienced personnel should be scheduled at least every two years. This permits early detection of problems, when corrections can be made during operation or with only minor interruption of service.

References

1. ACI Committee 307, *Specification for the design and construction of reinforced concrete chimneys*, ACI 307-79, American Concrete Institute, Detroit, 1979, pp. 47.
2. Rumman, W. S., Earthquake forces in reinforced concrete chimneys, *J. Struct. Div. Am. Soc. Civ. Eng.*, Vol. 93, No. ST6, Dec. 1967, pp. 55–70.
3. Maugh, L. C. and Rumman, W. S., Dynamic design of reinforced concrete chimneys, *J. Am. Concr. Inst., Proc.*, Vol. 64, No. 9, Sept. 1967, pp. 558–567.
4. Davenport, A. G., The application of statistical concepts to the wind loading of structures, *Proc. Inst. Civ. Eng.*, Vol. 19, August 1961, pp. 449–472.
5. Davenport, A. G., The response of slender line-like structures to gusty winds, *Proc. Inst. Civ. Eng.*, Vol. 23, November 1962, pp. 389–408.
6. Davenport, A. G., Note on the distribution of the largest value of a random function with application to gust loading, *Proc. Inst. Civ. Eng.*, Vol. 28, June 1964, pp. 187–196.
7. Vickery, B. J. and Clark, A. W., Lift or across-wind response of tapered stacks, *J. Struct. Div. Am. Soc. Civ. Eng.*, Vol. 98, No. ST1, pp. 1–20.
8. Vickery, B. J., A model for the prediction of the response of chimneys to vortex shedding, *Proc. 3rd Int. Chimney Design Symp.*, Munich, 1978, pp. 157–162.
9. Cannon, R. W., *Ultimate strength design chart of reinforced concrete chimneys*, Civil Engineering Design Research Report, Tennessee Valley Authority (undated).
10. ACI Committee 318, *Building code requirements for reinforced concrete*, ACI 318-77, American Concrete Institute, Detroit, 1977, 103 pp.
11. Pinfold, G. M., *Reinforced concrete chimneys and towers*, Viewpoint Publications, London, 1975, 233 pp.
12. Reynolds, C. E. and Steedman, J. C., *Reinforced concrete designer's handbook*, 9th Edn, Viewpoint Publications, London, 1981, 505 pp.

31 Piles and piled foundations

M J TOMLINSON

Chartered Civil Engineer, Isleworth, England

Contents

Notation

A_b	base area of pile
A_s	embedded shaft area of pile
B	width of pile
c_b, c_u	undrained shear strength of clay
E	Young's modulus

E_d	deformation modulus	Q_p	ultimate bearing capacity of pile
F_1, F_2, F_3	partial load factors		
e	height of point of application of load above ground	Q_s	ultimate skin friction capacity of pile
f_y	tensile yield strength of steel reinforcement	q	bearing pressure
		q_s	ultimate unit skin friction on pile
H	horizontal load on pile		
I	moment of inertia	W_p	weight of pile
K_s	coefficient of lateral earth pressure	z_f	depth below ground level to point of fixity
L	length of pile	$\alpha, \alpha_{11},$	
$N_{c, q}$	bearing capacity factors	α_{12}, α_{13}	adhesion factors
p_d	effective overburden pressure	δ	angle of wall friction
Q_a	allowable pile load	ϕ	angle of shearing resistance
Q_b	ultimate base resistance of pile	ρ	settlement

Summary

Types of driven and bored and cast-in-place piles are described. Driven types of pile include jointed and unjointed precast and prestressed concrete sections and driven and cast-in-place piles. Information is given on methods of pile installation by driving with various types of hammer, and by drilling for cast-in-place types using rotary auger rigs. The techniques for placing concrete in the various types of cast-in-place piles are described.

Design methods for the pile section, pile cap and ground beams are given together with methods for evaluating the bearing capacity of piles and pile groups in various types of soil and rocks. The importance of the pile installation technique on the bearing capacity of the pile is stressed. The durability of concrete piles in aggressive environmental and ground conditions is discussed and information is given on concrete mix design and other protective methods for these conditions. The chapter concludes with information on control of pile installation, load testing and non-destructive testing.

1 Introduction

Piles are slender structural members designed to transmit foundation loading through weak compressible soils or fill materials on to stiff or dense soil strata or on to rock, thereby preventing excessive settlement of the superstructure. Piles are also used in river and marine works to carry the loading from wharf structures and jetties and also to absorb the impact forces from berthing ships and waves.

Reinforced concrete is extensively used for piling because the materials are normally locally available and relatively cheap, the concrete can be moulded into a variety of structural shapes to suit the loading conditions, and good-quality well-compacted concrete is durable in sea water and in aggressive ground conditions.

2 Types of pile

2.1 Classification

Piles are divided into two principal classes. These are displacement piles and non-displacement piles. The first type comprises those piles which cause bodily displacement of the ground into which they are driven. They include precast concrete piles, which may be of solid or hollow section and can be formed and driven in a single length, or in short sections which are joined together as the pile is driven into the ground. Displacement piles also include those types in which a steel tube is driven to the required penetration depth and is filled with concrete in stages as the tube is withdrawn. Alternatively, the tube can be left in the ground and filled with concrete. Other types of displacement pile are steel *H*-sections, steel tubes driven with open ends, steel box sections and timber baulks.

Non-displacement piles are those types in which a hole is drilled to the required depth; a reinforcing cage is lowered into the hole, and then concrete is placed to complete the pile.

The preformed type of displacement pile has the advantage that the material of the pile can be inspected before it is driven into the ground, thus ensuring compliance with the specified requirements for density and strength. Pile installation is not affected by the presence of groundwater, and the pile shafts can be extended through water, which is particularly advantageous in marine structures. The ability to produce well-compacted high-quality precast concrete makes the displacement pile suitable for installation in aggressive soil conditions and in very soft and peaty clays where 'squeezing' of the clay can cause distortion or 'necking' in the unset concrete of the non-displacement type of pile.

The driven and cast-in-place type of displacement pile has the advantage that the length can be easily adjusted to suit the varying level of the bearing stratum, whereas the precast pile must be cut down or extended if the required length from pile cap to toe level differs from the length as precast. The installation of driven and cast-in-place piles is unaffected by groundwater because the withdrawable tube is driven with a closed end which excludes the groundwater from contact with the shaft until the concrete is placed. This type of pile has considerable advantages, namely, that an enlarged base can be formed, the reinforcement in the pile shaft is not determined by considerations of handling and driving stresses, and noise and vibration can be reduced in some types by using an internal drop hammer instead of driving on to the head of the pile.

The bored and cast-in-place type of non-displacement pile has the advantages that the length can readily be varied to suit the varying level of the bearing stratum; soil or rock removed from the pile borings can be inspected, tested if necessary, and compared with the borehole records; *in-situ* loading tests or penetration tests can be made in the pile boreholes; very large diameters (up to 6 m) and long lengths can be formed in favourable ground; and drilling tools can be used to break up boulders and other obstructions allowing these piles to be installed in conditions where it would be impossible to drive displacement piles.

The bored pile also has the particular advantages that the amount of reinforcing steel is not determined by handling or driving stresses, the installation operations do not cause ground heave or appreciable noise or vibration, and installation is possible in conditions of low headroom.

It is sometimes possible to combine the advantages and eliminate the disadvantages of one or the other of the displacement or non-displacement pile types by designing a composite form of pile. This could, for example, consist of a preformed element installed in a drilled hole, or a part driven and part bored pile.

2.2 Precast concrete piles

Precast concrete piles have their principal use in marine and river structures, that is, in situations where the cast-in-place pile is impractical or uneconomic. The reinforcement in the shaft of precast concrete piles which is necessary to withstand the stresses caused by lifting the pile from the casting bed and to withstand the driving stresses can be used to economic advantage in marine structures where the piles are subjected to lateral loads from wind and wave action and the impact of berthing ships.

Precast concrete piles can be economic for land foundations, compared with cast-in-place piles, where the savings in cost provided by the rapid installation of large numbers of piles in easy driving conditions can outweigh the costs of the extra reinforcement required for driving and handling, and for setting up a precasting yard on site, or the cost of transporting the piles from an off-site factory.

The piles can be designed and manufactured in ordinary reinforced concrete, or in the form of pretensioned or post-tensioned prestressed concrete elements. The ordinary reinforced concrete pile is likely to be preferred for a project requiring a fairly small number of piles where the cost of setting up a prestressing production line on site is not justifiable and where the site is too far from an established factory for economic transport of the elements to a site assembly yard.

Precast concrete piles in ordinary reinforced concrete are usually square or hexagonal and of solid cross-section for short to moderate pile lengths, but weight can be saved by designing long piles as hexagonal, octagonal or circular sections with a hollow interior.

If nominal mixes are adopted, a concrete with a minimum 28-day cube strength of 25 MPa (3600 psi) is suitable for hard to very hard driving and for all marine construction. A cube strength of 20 MPa (2900 psi) is suitable for normal or easy driving conditions. Prestressed concrete piles are usually made with concrete having a minimum 28-day works cube strength of 41 MPa (5950 psi). Working stresses in concrete and steel are usually laid down in codes of practice and are generally of the order of 0.25 to 0.3 times the unconfined compression strength of the concrete and 0.4 to 0.5 times the yield strength of the reinforcing steel (see Section 4.1).

For piles carrying axial compression loading only, the proportion of main reinforcing steel is determined by the bending stresses caused by lifting the pile from the casting bed. Design data for various lifting conditions are tabulated in Section 4.2.

Generally, code of practice limitations on working stresses in steel and concrete and the minimum requirements for main and link steel reinforcement are prescribed with the principal object of avoiding unseen breakage of piles during driving. The design of concrete mixes to avoid deterioration in aggressive soil and groundwater conditions is discussed in Section 7.

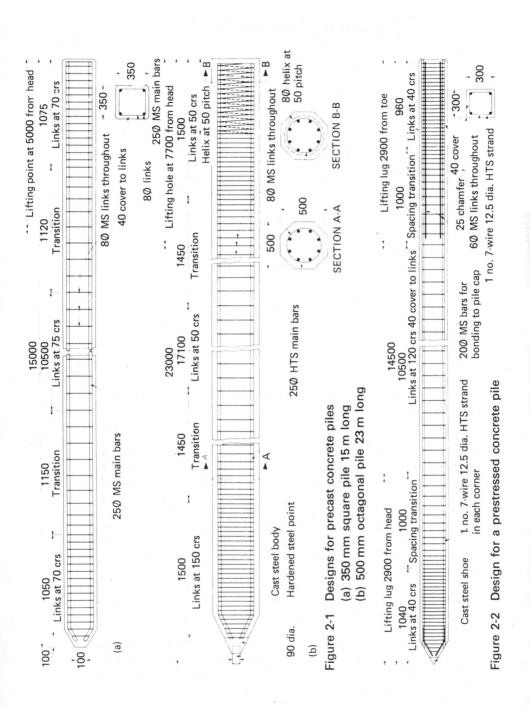

Figure 2-1 Designs for precast concrete piles
(a) 350 mm square pile 15 m long
(b) 500 mm octagonal pile 23 m long

Figure 2-2 Design for a prestressed concrete pile

Typical designs for precast and prestressed concrete piles are shown in Figs 2-1 and 2-2.

The problem of handling precast concrete piles in long lengths can be overcome by the use of jointed sections. These are usually factory-made units in high-quality concrete incorporating a joint of proprietary design. The Swedish Herkules pile is of hexagonal cross-section. Units of 220 and 305 mm across the flats are suitable for working loads of 700 and 1200 kN respectively. The Balken pile, also of Swedish origin, is a square section 235 and 275 mm in size for working loads of 640 and 980 kN respectively. The British designed West's Hardrive pile has a 285 mm square section for working loads up to 800 kN, and a 280 mm hollow circular section which can be loaded up to 300 kN.

The proprietary joints are designed to be made on site rapidly and without welding and to have a bending and tensile strength not less than that of the body of the pile.

When using jointed piles it is the normal practice to select a continuous length which will be suitable for the lower range of penetration depth expected from the borehole information, up to a maximum which can be easily handled and transported to the site. The maximum length of standard units varies from 10 to 13 m (33–43 ft). Various standard lengths can then be joined to the first length driven until the final penetration depth is reached. It is claimed that the Herkules pile can be driven to penetrations of up to 90 m (295 ft) below ground level.

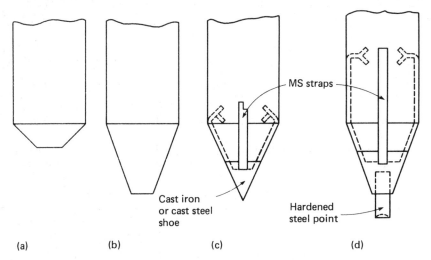

Figure 2-3   Shoes for precast (including prestressed) concrete piles

 (a) For driving through soft or loose soils to shallow penetration into dense granular soils or firm to stiff clays
 (b) Pointed end suitable for moderately deep penetration into medium-dense to dense sands and firm to stiff clays
 (c) Cast-iron or cast steel shoe for seating pile into weak rocks or breaking through cemented soil layers
 (d) 'Oslo' point for seating pile into hard rock

Metal shoes at the toes of precast concrete piles are needed only when it is necessary to split and drive through hard layers or other obstructions in the soil or to key the toe of the pile on to a sloping hard rock surface. Elsewhere, a square end or a blunt point without a metal shoe is satisfactory. Some typical designs for pile shoes are shown in Fig. 2-3.

2.3   Driven and cast-in-place piles

The driven and cast-in-place pile is a form of displacement pile which is installed by driving, to the desired penetration, a thick-wall steel tube with a closed end. A reinforcing cage is then placed in the tube followed by the placing of concrete as the tube is withdrawn in stages, each lift of semi-dry concrete being tamped by a heavy rammer. Alternatively, the tube can be filled completely with a highly workable concrete followed by extracting the tube with the aid of a vibrator attached to the head. The closed end of the pile can be formed by a plug of gravel or dry concrete, or by a precast concrete or flat steel plate shoe. Where the plug is formed from gravel or dry concrete, the tube is driven by a heavy internal drop hammer. On reaching the desired penetration depth, the tube is restrained from further sinking by suspension ropes, then the hammer is operated again to drive out the plug and to drive out additional charges of concrete to form an enlarged base, followed by the concreting of the pile shaft. The Franki pile is a well-known proprietary base-driven type of pile and stages in its installation are illustrated in Fig. 2-4.

Where a shoe is used to close the bottom of the tube, it is the usual practice to drive the pile by a hammer acting on the top of the tube. Proprietary piles of this type are the GKN, Vibro, Delta and Vibrex piles. Enlarged bases can be formed with the top-driven piles by partial withdrawal followed by re-driving of the tube to tamp out the concrete at the toe.

The enlarged base of the driven and cast-in-place pile is advantageous where the working load on the pile is carried principally in end-bearing on a stiff or compact soil formation.

The possibility of defective pile shafts being caused by 'necking' of the concrete in soft, squeezing clay soils was mentioned in Section 2.1. Necking in squeezing soils can be overcome by inserting a light-gauge spirally-welded or corrugated steel casing into the drive tube on completion of driving. This forms a permanent casing to the pile. It has the drawback that some skin friction may be lost if the void left by pulling the drive tube remains unfilled or only partly filled by loose soil. Also, great care is needed to avoid lifting the permanent casing when extracting the drive tube, particularly in long piles when the tube may deviate from the vertical.

The permanent casing can be driven with an internal collapsible mandrel which is extracted and the casing is then filled with concrete. The Raymond constant section and step-taper piles are of this type.

Where relatively short penetration depths and easy driving conditions are expected, the permanent casings can be driven without a mandrel before filling with concrete. The light-gauge spirally-welded BSP cased pile can be driven either from the top or on to a plug of dry concrete at the base. Care is needed in both methods to avoid bulging or splitting of the casing if the driving becomes hard

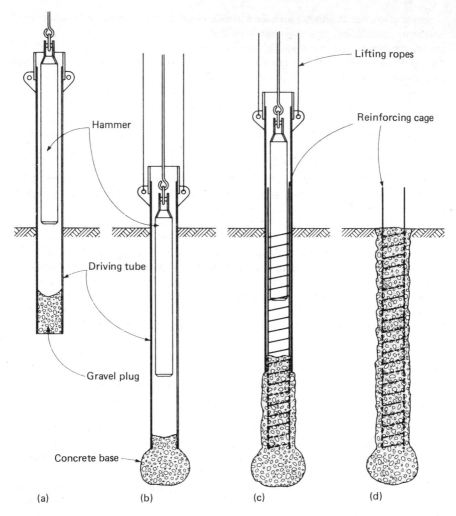

Figure 2-4   Stages in installing a Franki pile
(a) Driving pile tube
(b) Placing concrete in piling tube
(c) Compacting concrete in shaft
(d) Completed pile

(see Section 3.1). The West's shell pile consists of hollow cylindrical precast concrete segments threaded on to an internal mandrel. The pile shoe is driven by the mandrel and the segments by light blows of the hammer on the top segment. On reaching the required penetration depth, the mandrel is extracted, a reinforcing cage is inserted and the shaft filled with concrete.

The shell and step-taper piles have the advantage that the lengths of permanent casing can be readily adjusted to suit the level of the bearing stratum and for all permanently cased piles the interior of the casing can be inspected for soundness and verticality before the concrete is placed.

Maximum working loads on driven and cast-in-place piles are determined by the allowable compressive stress on the concrete in the shaft. Codes of practice usually limit this stress to 0.25 to 0.3 times the crushing strength of the concrete. Although reinforcing steel is not needed to withstand handling or driving stresses, it is the usual practice to provide a light cage of reinforcement to deal with the low bending stresses caused by minor eccentricity of load on the pile head or minor deviations of the pile from the vertical.

2.4 Bored and cast-in-place piles

In its simplest form, the bored and cast-in-place type of non-displacement pile consists of drilling an unlined hole to the required depth by mechanical auger, inserting a reinforcing cage, then placing concrete to fill the hole.

When rotary augers are used in cohesionless soils and very weak or shattered rocks, it is necessary to support the walls of the borehole by means of a bentonite slurry. Alternatively, mechanical equipment is available for drilling pile boreholes with ground support by lengths of steel casing with screwed or mechanically locked joints. Equipment for installing bored piles is described in Section 3.2.

There are risks of settlement of the ground with possible damage to existing structures when installing bored piles in groups. This is due to the cumulative effects of creep or, at worst, collapse of soils into the pile boreholes, and it is most liable to occur in loose water-bearing granular soils, particularly if a 'shell' or baler is used to remove the soil. Loss of ground can be eliminated or greatly reduced by drilling under a bentonite slurry using rotary methods.

Another useful method of drilling to avoid loss of ground is to use a continuous flight auger with a hollow stem. As the auger is being rotated down, the soil within the spiral flights acts as a support to the walls of the borehole. On reaching the required depth, a sand–cement grout or concrete is pumped down the hollow stem as the auger is slowly withdrawn. In the United Kingdom, the proprietary Dowsett Prepacked and the Cementation Company's Cemcore and Concore piling systems use the hollow-stem flight auger.

The disadvantage of loosening a granular soil and the consequent loss of bearing capacity at the toe of a bored pile are overcome, in the case of the Bauer pile, by a grouting technique. The soil is first grouted around the completed shaft of the pile for a distance of a few metres and the grout is allowed to harden. Grout is then injected under pressure into the space between a permanent circular steel plate forming the base of the pile and a rubber membrane bonded to the underside of this plate. The pressure from the grout injection forces down the membrane thus compressing the soil beneath the pile toe.

Another proprietary bored piling system is the Cementation Company's Prest-core pile (Fig. 2-5) in which precast concrete cylindrical units are assembled on a hollow pipe stem and lowered down a borehole fully lined with steel casing. Vertical steel reinforcing bars are then threaded through holes in the precast units, and this is followed by pumping cement grout down the pipe stem to fill the annulus around the precast units as the casing is withdrawn. This type of pile is useful in soft, squeezing soils and in chemically contaminated ground where dense, high-quality concrete is needed to prevent disintegration of the pile shaft.

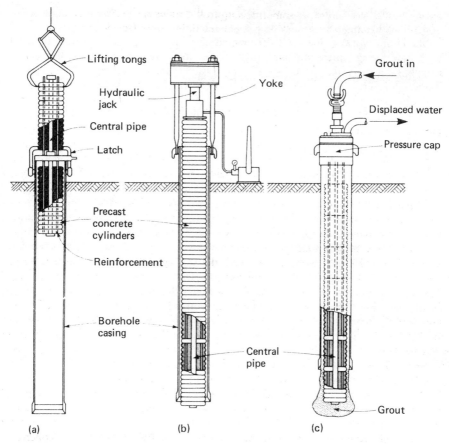

Figure 2-5   Stages in installing a Prestcore pile
        (a) Lowering precast concrete cylinders down borehole casing
        (b) Bedding down cylinders by jacking
        (c) Grouting pile shaft through pressure cap

Where large-diameter bored piles carry only axial compression loads, it is the usual practice to provide only a short reinforcing cage at the pile head to bond into the pile cap or column base. However, a full-length cage is desirable in small or medium-diameter piles (up to 600 mm, 2 ft) to accommodate bending stresses caused by deviation of the pile head from the correct position or deviation in verticality (see Section 3.4).

Care is necessary in forming the shafts of bored piles in soft, squeezing clays and peats. If necessary, light-gauge steel permanent casing can be used to avoid necking of the pile shaft as described for driven and cast-in-place piles.

Maximum working loads for axially loaded piles are determined from considerations of the allowable compressive stress on the concrete in the pile shaft. Codes of practice generally limit the allowable stress to 0.25 to 0.33 times the crushing strength of the concrete, or sometimes up to 0.4 times the crushing strength if the concrete is confined in a durable permanent lining tube (see Section 4.1).

3 Methods of installing concrete piles

3.1 Driving precast concrete and driven and cast-in-place piles

Precast concrete piles, including both jointed and unjointed types, are usually driven by means of piling frames, which consist of rigid steel leaders which support the pile while it is above the ground and help to keep the pile at the correct alignment, either in the vertical position or at the specified angle of rake. The leaders also guide the pile hammer and keep it in correct position concentric with the longitudinal axis of the pile. They can be adjusted to the vertical or to forward or backward raking positions by means of hydraulically operated stays.

For pile driving in marine works, the leaders and winch are mounted on a fixed or wheeled carriage which is carried by staging or on a platform. Where steam-operated piling hammers are used, the carriage also carries a steam boiler. The following types of hammer are used to drive precast concrete piles or the tubes of driven and cast-in-place piles: drop hammer, single-acting hammer (steam or air-operated), diesel hammer, and double-acting hammer.

The drop hammer is simple to use and maintain because it has no working parts. The mass of the hammer generally ranges between 1 and 5 tonne. It is operated by the winch mounted on the piling frame. It is quite widely used for driving precast concrete piles but strict control must be exercised over the driving operations in order to avoid an excessive height of drop which might cause breakage of the pile. The internal drop hammer is used for driving on to the plug of concrete at the bottom of the temporarily or permanently cased driven and cast-in-place pile.

The single-acting hammer is the most suitable type for driving precast concrete piles because the energy of blow is limited by the stroke of the hammer.

The diesel hammer is suitable for driving the heavy withdrawable tube of the driven and cast-in-place pile and for driving precast concrete piles in moderately easy driving conditions. Use of this hammer for driving precast concrete piles in ground consisting of alternating hard and soft layers may cause breakage of the pile.

The double-acting hammer is generally used for driving the withdrawable tube of the driven and cast-in-place pile in fairly easy to moderate ground conditions.

It is essential to cushion the hammer blow on the head of precast concrete piles in order to avoid spalling at the head or unseen breakage below ground level. The helmet which is placed on top of the pile has a recess in the upper part into which is placed the dolly. An elm block can be used for the dolly in easy driving conditions, but a hardwood block set end-on to the grain, or a resin block reinforced with laminations of cotton fibre must be used for hard driving. Packing in the form of thin hardwood strips, coiled rope, hessian, wallboard or asbestos fibre is placed between the underside of the dolly and the top of the pile to cushion further the blow (Fig. 3-1).

A recent development which is helpful in avoiding excessive impact energy on

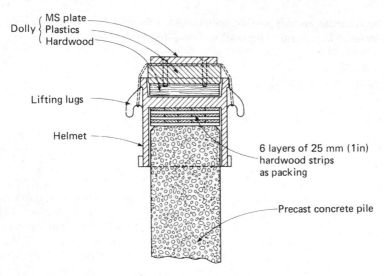

Dolly { MS plate
Plastics
Hardwood

Lifting lugs

Helmet

6 layers of 25 mm (1in)
hardwood strips
as packing

Precast concrete pile

Figure 3-1   Dolly and helmet for precast concrete pile

concrete piles is the pile driving analyser [1]. It consists of accelerometers and strain transducers, which are attached to the top of the pile. The energy transferred to the pile, the compression and tension forces in the pile and the acceleration and velocity of the impact forces are displayed on an oscilloscope and recorded on magnetic tape.

Difficulties may be experienced where piles are driven in large groups. In clays, the displaced soil may cause ground heave, which lifts the piles off their end bearing on a hard stratum, or lateral ground displacement may damage existing buried structures. A careful watch must be kept for ground heave, including taking levels on the tops of piles after driving. Any risen piles should be redriven after completing all piles in a particular group. Ground heave effects can be minimized by preboring for part of the shaft length but skin friction is lost over the prebored length.

In granular soils, the effect of driving piles in groups is to densify the soil, making it increasingly difficult to drive the piles to the required penetration. These effects can be minimized by commencing to drive at the centre of the group and working outwards in all directions.

The rigs used for installing driven and cast-in-place piles are similar to those used for precast concrete piling, except that they incorporate means of filling the drive tube with concrete as it is extracted. These may consist of vertical rails forming a guide for a concrete skip, or an auxiliary winch to raise the skip, which is guided by hand into the discharge position at the pile head.

Where the steel drive tube is left permanently in position, a simple timber or steel trestle can be used to guide the pile. The internal drop hammer is guided by the tube. It is essential to renew the concrete plug at the bottom of the tube at frequent intervals during driving. When this is not done, the concrete becomes dense and hard and no longer cushions the blow, with the result that the tube may split above or over the length of the plug.

Noise emission in pile driving can be reduced by shrouding the hammer with absorbent material, or preferably by surrounding pile and hammer by a sound-absorbent box with some reduction in driving efficiency. Driving piles by internal drop hammer is advantageous in noise control since airborne and surface vibration waves are greatly reduced.

3.2 Equipment for installing bored and cast-in-place piles

The mechanical rotary augers referred to in Section 2.4 consist of a base machine, which is usually a standard crawler crane, on which is mounted a rotary table driven by a hydraulic power pack. The square-section 'kelly' passes through the rotary table and terminates in the drilling tool, which may be a spiral plate auger for drilling in clay or a bucket with a hinged bottom flap for removing loose materials such as broken rock. The bucket may be provided with teeth for loosening weak rock. In strong rock formation, a heavy chisel is used to break up the rock, which is then removed by the bucket. Drilling to depths of up to 44 m (144 ft) is effected by telescoping the kelly in three sections, and with extension drill stems it is possible to drill to depths of up to 90 m (295 ft). Enlarged bases are formed by means of a belling bucket which consists of hinged arms which are forced out mechanically to ream out the sides of the previously drilled straight-sided hole to form a bell-shaped base. For shaft diameters of 1 m or larger, enlarged bases can be formed by hand excavation in stable cohesive soils.

Rotary augers are the most efficient means of drilling in cohesive soils or weak rocks which can stand unsupported while drilling to the full depths, or until a length of casing can be suspended in the hole. As mentioned in Section 2.4, support to the walls of the borehole in cohesionless soils or fractured rocks can be provided by means of a bentonite slurry. This necessitates the provision of equipment for preparing and continuously agitating the slurry in tanks, and pumps for circulating the slurry from the holding tanks to the pile borehole and back to cleaning tanks. Disposal of the contaminated and unused slurry also creates problems on urban sites.

Another method of drilling in cohesionless soils or loose rocks is to use winch-operated grabs with continuous support to the ground by steel casing which is allowed to sink down as the hole is advanced by grabbing. The equipment includes a casing oscillator, which imparts a semi-rotary motion to the casing (thereby preventing it from sticking to the ground) and which incorporates hydraulic arms to raise and lower the casing.

Small-diameter pile boreholes (up to 600 mm, 2 ft) can be drilled by cable percussion methods using clay cutters and balers operated by a tripod-mounted winch. This equipment is useful for working in confined spaces and conditions of low headroom.

3.3 Concreting cast-in-place piles

The mix proportions which are suitable for placing concrete in the shafts of driven and cast-in-place piles depend on whether the drive tube is filled completely with concrete before it is withdrawn, or whether an internal drop hammer is used to compact the concrete as the drive tube is raised in stages. In the former case, the concrete must be easily workable so that it will slump outwards to fill the void left by the tube as it is pulled out, and to prevent it from sticking and jamming causing the concrete and reinforcing cage to be lifted with the tube. A slump of 100 to 150 mm (4–6 in) is suitable. Such a mix is self-compacting and does not require ramming or vibration. Where the concrete is compacted by an internal rammer, a mix of almost earth-dry consistency must be used in order to prevent the rammer from becoming buried in the concrete.

The problem of squeezing soils was mentioned in Section 2.3. Necking of the shaft can be avoided by using a permanent light-gauge steel casing which is left in place as the drive tube is withdrawn, but this adds considerably to the cost of the pile and may cause difficulties with jamming between the casing and the drive tube as the latter is pulled out. It is often possible to avoid necking by a careful concreting technique with observations of the level of the concrete at the top of the pile while withdrawing the drive tube and by comparing the quantity of concrete used with the calculated volume of the pile shaft.

An easily workable self-compacting mix with a slump of 100 to 150 mm is suitable for filling permanently cased piles.

Before placing concrete in the shafts of bored and cast-in-place piles, a loosely fitting lining tube should be in position to prevent collapse of the weaker shallow soil or rock into the borehole. Loose debris should be cleaned from the bottom of the hole and any small accumulations of water should be baled out. Any lumps of soil adhering to the walls of the borehole or to the inside of the lining tube should be removed. The time interval between the final clean-up and the placing of the concrete should not exceed 6 h. If there is appreciable delay before placing the concrete, the borehole should be plumbed to ensure that no soil has fallen from the sides.

If groundwater seepage causes small amounts of water to accumulate in the bottom of the borehole, a dry mix should be used for the first charges of concrete. Thereafter, the easily workable self-compacting mix should be used as described above for the driven and cast-in-place pile.

An inflow of water not exceeding, say, a few centimetres in 5 min can be dealt with as described above by baling out the water and then forming a plug of dry concrete in the bottom of the borehole. However, a strong inflow of water can wash out the cement from the unset concrete. In such cases, it is preferable to allow the borehole to fill with water up to its normal rest level or preferably to top up the borehole to ground level to assist in stabilizing the base. The concrete is then placed through water using a tremie pipe which extends to the bottom of the borehole. The pipe is lifted in stages as the level of concrete rises always keeping the bottom of the pipe buried in the concrete.

Concrete for underwater placing through a tremie pipe should have a slump between 125 and 180 mm (5–7 in) and a minimum cement content of 400 kg/m^3 (674 lb/yd). Placing concrete in a deep and large-diameter pile may take a long time and a retarder should be used in the mix to prevent early setting and sticking of the mix in the tremie pipe and to prevent the jamming of the lining tubes as they are lifted.

Where concrete is placed in a borehole drilled under a bentonite slurry, the contaminated slurry must be pumped out and replaced by a clean light slurry immediately before placing the concrete through a tremie tube.

Bottom-opening skips are unsuitable for concreting the shafts of bored piles and their use should not be permitted.

Problems in installing bored piles have been described in a useful publication by the Construction Industry Research and Information Association [2].

3.4 Positional tolerances

Provision must be made in foundation design for the heads of piles to deviate from their intended position and for the axis of piles to deviate from the truly vertical position or specified rake. Driven piles tend to move out of position during installation because of obstructions in the ground or the tilting of the pile frame. Bored piles can deviate from the intended alignment as a result of the 'wandering' of the plate auger on the flexible drill stem. Misalignment affects the design of pile caps and ground beams (see Section 5) and may cause interference between adjacent piles close-spaced in groups. Excessive bending moments in piles may be caused if there are gross deviations from the vertical.

The UK code of practice, CP 2004 : 1972, *Foundations*, permits pile heads to be out of position by not more than 75 mm (3 in) at the level of the piling rig and the maximum deviation from the vertical or specified rake is 1 in 75. The American Concrete Institute recommends that the position of the pile head should be within 75 to 150 mm (3–6 in) for the normal usage beneath a structural slab. The axis may deviate by up to 10% of the pile length for completely embedded vertical piles or for all raking piles, provided that the pile axis is driven straight. For vertical piles extending above the ground surface, the maximum deviation is 2% of the pile length, except that 4% can be permitted if the resulting horizontal load can be taken by the pile cap structure. For bent piles, the allowable deviation is 2–4% of the pile length, depending on the soil conditions and the type of bend (e.g. sharp or gentle). Severely bent piles should be evaluated by soil mechanics calculations, or checked for acceptability by loading tests.

Where cast-in-place concrete is used, allowance must be made in the level of termination of the pile head for the removal of laitance or contaminated concrete which rises to the surface particularly where the concrete is placed under water. Fleming and Lane [3] recommended the following tolerances:

Concrete placed under water $+1.5$ to $+3$ m

Concrete cast in dry uncased holes $+7.5$ to $+300$ mm

Concrete encased in holes, the greater of

(a) $+75$ to $+300$ mm $+$ (cased length)/15

or

(b) $+75$ to $+300$ mm $+ \dfrac{\text{depth to casting level} - 900 \text{ mm}}{10}$

4 The structural design of concrete piles

4.1 Allowable stresses

In the UK, the working stresses in the concrete and steel in precast piles are limited by CP 2004 : 1972, *Foundations*, to the values given in Table 4-1.

The American Concrete Institute recommends an allowable compression stress

Table 4-1

Nominal mix using Portland cement	Allowable compression stress MPa	Allowable tensile stress in steel
1:1:2	7.86	138 MPa (20 000 psi) for mild steel bars not exceeding 40 mm in diameter or 0.55 × guaranteed yield or proof stress for high bond bars
1:1½:3	6.55	
1:2:4	5.24	

in the concrete of one-third of the unconfined compression strength, and an allowable tensile stress in the reinforcing steel of $0.5f_y$ or 207 MPa (30 000 psi). It is recommended that these stresses should be reduced by 10% for trestle piles, and piles supporting piers, docks, and other marine structures.

If nominal mixes are adopted, a 25-grade concrete with a minimum 28-day cube strength of 25 MPa (3600 psi) is suitable for hard to very hard driving and all marine construction. For normal or easy driving, a 20-grade concrete with a minimum 28-day cube strength of 20 MPa (2900 psi) is satisfactory.

For driven or bored and cast-in-place piles, CP 2004 limits the compression stress in the concrete to one-quarter of the cube compression strength with the proviso that the concrete should not be leaner than 1:2:4 nominal mix.

The American Concrete Institute recommendations are more comprehensive. The compression stress in the concrete is limited to one-third of the unconfined compression strength for uncased piles or 0.4 times this strength where the concrete is confined within a permanent steel shell not more than 432 mm (17 in) in diameter in 14 gauge (US) or greater steel either seamless or spirally welded with a tensile strength of not less than 206 MPa (30 000 psi). The shell must not be exposed to corrosion and should not be allowed to carry part of the working load. Corrugated steel shells are not regarded as load-bearing. The steel reinforcement should have a maximum allowable stress of $0.4f_y$ or not greater than 206 MPa. If a structural steel core is provided the allowable compression or tensile stress should not exceed $0.5f_y$ or a maximum of 172 MPa (25 000 psi).

4.2 Designing precast concrete piles for lifting and handling

As stated in Section 2.2, it is necessary to design precast concrete piles to withstand the bending stresses caused by lifting the piles from a horizontal position on the casting bed or stacking area and transporting them to the driving rig. The static bending moments induced by lifting and pitching piles at various points on their length are given in Table 4-2.

The design charts shown in Figs 4-2a to f have been prepared by the Cement and Concrete Association. These show the bending moments due to self-weight which are induced when square and octagonal piles of various cross-sectional dimensions are lifted from the pick-up positions shown in the diagrams. Also shown on the charts are horizontal lines representing the ultimate resistance

Table 4-2

Condition	Figure reference	Maximum static bending moment
Lifting by two points at $L/5$ from each end	4-1a	$WL/40$
Lifting by two points at $L/4$ from each end	4-1b	$WL/32$
Pitching by one point $3L/10$ from head	4-1c	$WL/22$
Pitching by one point $L/3$ from head	4-1d	$WL/18$
Pitching by one point $L/4$ from head	4-1e	$WL/18$
Pitching by one point $L/5$ from head	4-1f	$WL/14$
Pitching from head	4-1g	$WL/8$
Lifting from centre	4-1h	$WL/8$

moment of each pile section for main reinforcing bars of various sizes. These moments were calculated to conform to CP 110 for concrete having a characteristic strength of 25 MPa (3630 psi), mild steel reinforcement having a characteristic strength of 250 MPa (36280 psi) and with 40 mm (1.6 in) of concrete cover to the link steel. The possibility of the pile being rotated on its longitudinal axis during lifting was taken into account.

The required concrete compression strength is usually determined from considerations of stresses induced by driving (see Section 4.3), or of durability in an aggressive environment (Section 7).

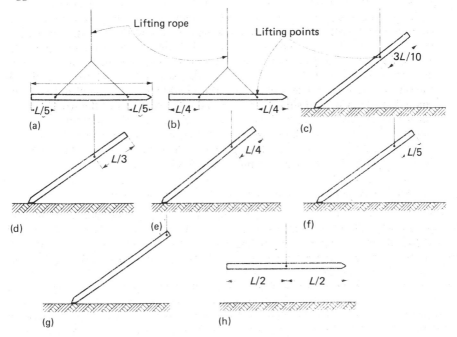

Figure 4-1 Methods of lifting concrete piles

The lengths and cross-sectional dimension of piles are determined from the resistance provided by the ground (see Section 6). Then, for the given length and cross-section, the pick-up point is selected having regard to the type of piling rig and available craneage. The economics of lifting from various positions should be studied. The bending moment due to the *ultimate* dead load corresponding to the selected pick-up point is then read off from the appropriate curve. Next, the ultimate dead load bending moment is compared with the ultimate moments of resistance for various sizes of reinforcement, as shown on the charts, and the appropriate reinforcement is selected. A partial safety factor of 1.4 on the bending moment due to dead load is suitable for normal designs and this value has been used in the derivation of the charts in Fig. 4-1 and Tables 4-3 and 4-4. These tables show the maximum lifting lengths for three pick-up conditions and also the link steel required to conform to CP2004.

After selecting the diameter of the reinforcing bars from Fig. 4-2, it is then necessary to check the serviceability limit state for the design lifting condition. The crack width is the only criterion which needs to be considered. A surface crack up to 0.3 mm wide is permitted by CP 110 for buried concrete, but for severe exposure conditions the crack widths near the main reinforcement should not generally be more than 0.004 times the cover to this reinforcement. Severe conditions include piles exposed to sea water.

Table 4-3   Maximum lengths of square section precast concrete piles for given reinforcement

Pile size mm	Main reinforcement mm	Maximum length in metres for pick up at Head and toe	0.33 × length from head	0.2 × length from head and toe	Transverse reinforcement Head and toe	Body of pile
300 × 300	4 × 20	9.0	13.5	20.0	6 mm at	6 mm at
	4 × 25	11.0	16.0	24.0	40 mm crs	130 mm crs
350 × 350	4 × 20	8.5	13.0	19.0	8 mm at	8 mm at
	4 × 25	10.5	16.0	23.5	70 mm crs	175 mm crs
	4 × 32	12.5	19.0	28.5		
400 × 400	4 × 25	10.0	15.0	22.5	8 mm at	8 mm at
	4 × 32	12.5	18.5	28.0	60 mm crs	200 mm crs
					or	
	4 × 40	15.0	22.0	33.0	10 mm at	
					100 mm crs	
450 × 450	4 × 25	10.0	14.5	22.0	8 mm at	8 mm at
	4 × 32	13.0	18.5	27.0	60 mm crs	180 mm crs
	4 × 40	15.0	22.0	32.0	or	or
					10 mm at	10 mm at
					90 mm crs	225 mm crs

Notes
Piles designed in accordance with CP 110 and CP 2004
Characteristic strength of mild steel reinforcement 250 MPa
Cover to link steel 40 mm.
Characteristic strength of concrete 25 MPa

Table 4-4   Maximum lengths of octagonal section precast concrete piles for given reinforcement

Pile size mm	Main reinforcement mm	Maximum length in metres for pick up at			Transverse reinforcement	
		Head and toe	0.33 × length from head	0.2 × length from head and toe	Head and toe	Body of pile
400	8 × 16	9.5	14.0	21.5	8 mm at	8 mm at
	8 × 20	11.0	16.0	25.5	60 mm crs or 10 mm at 90 mm crs	180 mm crs or 10 mm at 200 mm crs
500	8 × 16	9.0	13.0	19.5	8 mm at	8 mm at
	8 × 20	10.5	15.0	24.0	50 mm crs or 10 mm at 75 mm crs	150 mm crs or 10 mm at 240 mm crs
600	8 × 20	10.0	15.5	22.5	10 mm at	10 mm at
	8 × 25	12.0	18.5	27.5	65 mm crs or 12 mm at 100 mm crs	200 mm crs or ⎰ 12 mm at ⎱ 300 mm crs[a] ⎰ 12 mm at ⎱ 240 mm crs[b]

Notes
Piles designed in accordance with CP 110 and CP 2004
Characteristic strength of mild steel reinforcement 250 MPa
Cover to link steel 40 mm
Characteristic strength of concrete 25 MPa

[a] for 25 mm main bars
[b] for 20 mm main bars

4.3   Designing precast concrete piles to resist driving stresses

Working stresses under static loading should be selected to ensure that heavy driving is not required to achieve the depth of penetration necessary for the design bearing capacity. For example, if the design load has a factor say of 3 on the calculated ultimate bearing capacity, the pile will have to be driven by a dynamic force equivalent to three times the static working load at the final stages of penetration. If the working stresses are high under static loading conditions, the driving stresses may then become excessive. It is for this reason that codes of practice limit the allowable stresses under static loading to a small proportion of the crushing strength of the concrete as noted in Section 4.1.

Codes of practice also seek to limit the driving stresses by limiting the ratio of the weight of the pile to the weight of the hammer and by restricting the height of drop of the hammer. For an example, CP 2004 states that the heaviest practical hammer shall be adopted, and that the stroke of a single-acting hammer or drop hammer should not exceed 1.2 m (47 in) or preferably 1 m (39 in).

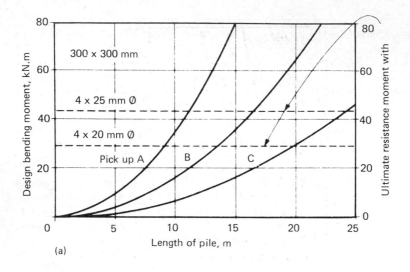

(a)

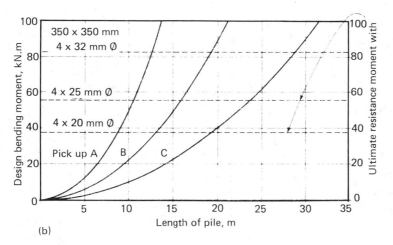

(b)

Figure 4-2   Required lifting points for reinforced concrete piles of various cross-sections

The amount of transverse reinforcement has an important influence on the ability of the pile to withstand hard driving without splitting or shattering. The proportions of transverse reinforcement shown in Tables 4-3 and 4-4 are those suitable for fairly easy driving conditions where the volume of link steel at the head and toe of the pile is 0.6% of the gross volume, and in the body of the pile 0.2% of this volume. For hard driving conditions, it may be preferable to increase these proportion to 1.0% and 0.4%, respectively.

Driving stresses can be calculated by means of the stress-wave theory [4], which divides the length of the pile into a number of elements, usually 1.5 to 3 m long. Each element is represented by a weight joined to the adjacent element by a

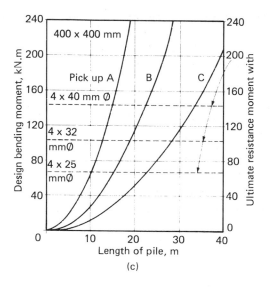

(c)

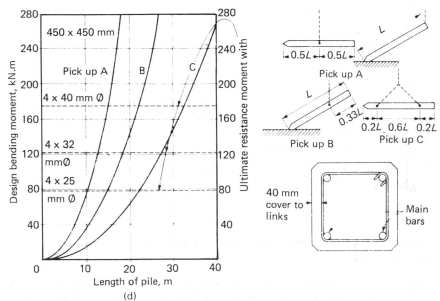

(d)

Figure 4-2 (continued)

spring. The hammer, helmet and packing are also represented by separate elements joined to each other and to the pile by springs. The soil beneath the pile toe is assumed to provide a resisting force to the downward motion of the pile when struck by the hammer. In the case of friction piles, the soil alongside the shaft acts as a damping force to the stress wave. For each blow of the hammer, and each element in the pile–hammer system, the displacement of the element,

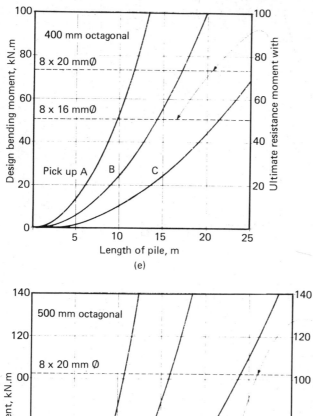

Figure 4-2   (continued)

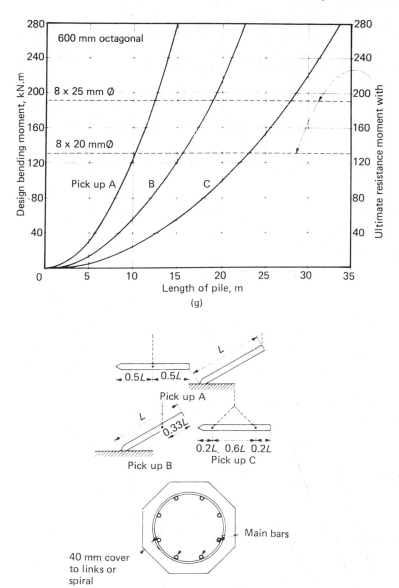

Figure 4-2   (continued)

the spring compression of the element, the force exerted by the spring, the accelerating force and the velocity of the element in a given time interval, are calculated. The output of the computations by this theory is the compressive or tensile force on the pile at any required point between the head and toe. Their accuracy depends on the correctness of the assumptions made for the energy of the hammer blow, the elastic modulus and the coefficient of restitution of the packing beneath the hammer, and the elastic compression of the soil. Because these elastic modulus values change continuously during the driving of a pile, the stress-wave method can never give exact values throughout all stages of driving.

4.4 The design of axially-loaded piles as columns

Axially-loaded piles which terminate at or below ground level at a pile cap or ground beam will not buckle provided that the permissible working stresses on the pile cross-section are not exceeded. Thus, such piles need not be considered to act as long columns for structural design purposes. However, column strengths must be considered where piles are extended above the soil line, as in jetties or bridge trestles.

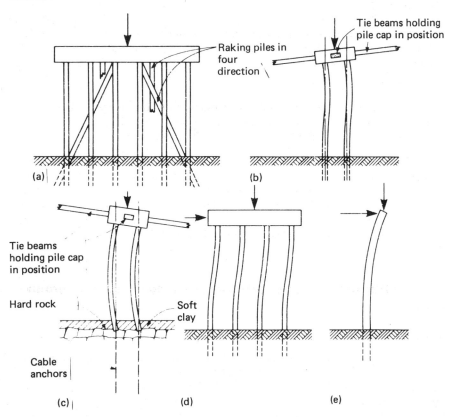

Figure 4-3 Conditions of restraint for vertical piles
(a) Restrained at top and bottom in position and direction
(b) Restrained at bottom in position and direction; restrained at top in position but not in direction
(c) Restrained at top and bottom in position but not in direction
(d) Restrained at bottom in position and direction; restrained at top in direction but not position
(e) Restrained at bottom in position and direction; unrestrained at top in position or direction

For small-diameter piles, it is often assumed that the point of virtual fixity of a pile occurs at a depth of 1.5 m in good ground and 3 m in poor ground such as soft clay or silt. Liquid mud is assumed to act as water. The column strength of the pile above this point of fixity is then calculated as for a short column and a reduction factor is applied to the calculated ultimate load to allow for the slenderness of the column. The slenderness is defined as the ratio of the effective length to the breadth or to the radius of gyration. Effective lengths are defined in CP 110 as

Properly restrained at both ends in position and direction $\Big\}$ $0.75L$

Properly restrained at both ends in position and imperfectly restrained in direction at one or both ends $\Big\}$ Between $0.75L$ and L depending on the efficiency of the directional restraint

Properly restrained at one end in position and direction and imperfectly restrained in position and direction at the other end $\Big\}$ Intermediate between L and $2L$ depending on the efficiency of the imperfect restraint

A pile embedded in the soil can be regarded as properly restrained in position and direction at the point of its virtual fixity in the soil (see Section 6.12). The restraint at the upper end depends on the design of the pile cap and the restraint provided to the cap by connections with adjacent pile caps or structures. Some typical cases of restraint of piles are shown in Figs 4-3a–e.

5 The structural design of pile caps, capping beams and ground beams

5.1 Pile caps

Pile caps have the function of transferring load from a member in the superstructure on to a group of piles to ensure that there is, as far as possible, equal load carried by each pile. The caps also accommodate deviations in pile position, and by rigidly connecting the heads of the piles by a deep and massive cap loads between the piles can be redistributed if one or more of the piles should fail by the yielding of the soil or faulty construction.

Isolated pile caps must incorporate at least three small-diameter piles. Caps for two piles must be interconnected by ground beams in a line transverse to the common axis of the pair, while caps for single small diameter piles must be interconnected by ground beams in two directions. Single large-diameter piles do not necessarily require caps. Instead, the tops of the piles may be cut down and the pile reinforcement bonded to the starter bars of the column reinforcement. Horizontal wind forces on the superstructure are transferred to the pile heads through the ground-floor or basement structure.

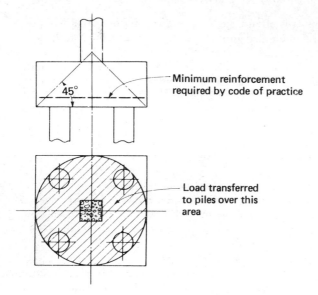

Figure 5-1 Load transfer from column to deep four-pile cap

Load is transferred to the heads of deep and relatively small caps only in compression as illustrated in Fig. 5-1. The bending and shearing forces are negligible, requiring only the minimum proportion of steel as required by codes of practice. The reinforcement should be placed in two directions at the bottom of the cap.

Where a large number of piles support a heavily-loaded column, the cap is designed as a solid slab carrying concentrated loads from the piles, as shown in Fig. 5-2. Where the slab is designed by simple beam theory, the bending moments are taken as the sum of the moments acting from the centre of the pile to the face

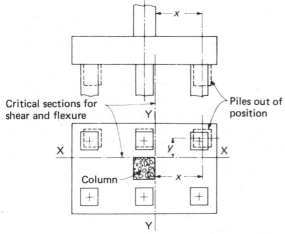

Figure 5-2 Calculation of bending moments and shearing forces on a rectangular pile cap

of the nearest column or column base. Allowance must be made for deviations in the position of the pile heads up to the specified maximum (see Section 3.4).

Alternatively, the slab forming the pile cap may be designed by the truss analogy. This method is appropriate for the smaller pile caps where the depth-to-width ratios of the deep members are outside the range of validity of beam theory. The basis of the truss analogy is the plastic behaviour of reinforced concrete structures at the ultimate limit state which induces paths of resistance within a structure. If a system of forces can be found within a structure which is in equilibrium with the applied load and nowhere exceeds the material strengths, then that load is a lower bound to the strength of the structure.

Suitable truss formations within a pile cap can be constructed by assuming that concrete struts run between the centre of the column down to points above the centres of the piles at the bottom reinforcement. It is recommended that no strut should have an inclination flatter than 30°. The tie forces are assumed to run between piles around the perimeter of the cap for up to four piles. The truss is analysed and reinforcement is placed in concentrated bands in the direction of the tie forces. About 80% of the steel is placed in this manner, the remaining 20% being placed in the centre of the cap to control cracking. However, the minimum spacing of bars in small caps may require the steel to be distributed uniformly across the width. Because tie forces are constant between the piles there must be no curtailment of reinforcement within the cap and full anchorage must be provided beyond the pile centre line. This usually requires the bars to be extended up the side faces of the cap as shown in Fig. 5-3. Examples of pile cap design by the truss analogy are given in [5].

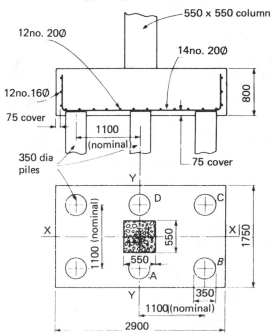

Figure 5-3 Arrangement of reinforcement in a six-pile cap carrying a 3000 kN column load

Punching shear must be considered, the critical perimeters being around a column and around an outer pile.

CP 2004 requires the cover to all reinforcement to be at least 75 mm for concrete cast against the ground, or 40 mm for concrete cast on to a blinding layer. In a marine environment, the cover should be at least 70 mm or preferably 80 mm.

5.2 Capping beams and ground beams

Pile capping beams have the function of distributing the load from walls or closely-spaced columns on to rows of piles. For heavy wall loading accompanied by transverse bending moments, the piles are placed in transverse rows surmounted by a wide capping beam. For lightly-loaded walls, the piles can be placed in a staggered pattern on each side of the wall, or in a single row. In the latter case, the capping beam must be restrained against rotation and sidesway by tying it to transverse capping beams located beneath cross walls in the superstructure.

It is the usual practice to design capping beams by conventional methods, but economies can sometimes be achieved by taking into account the composite action between the beam and the wall above [6].

Where buildings are sited in areas of swelling clays, provision should be made for the uplift pressure on the underside of the pile capping beams. Where the weight of the wall with or without the additional weight of a suspended ground floor is insufficient to counteract the uplift pressure, either a void should be left beneath the beam or the concrete should be cast upon a slab of compressible material such as cellular cardboard.

Ground beams are provided to act as ties or compression members between adjacent pile caps, thereby restraining them against sidesway or buckling of the piles caused by lateral or eccentric loading (see Section 4.4). Ground beams and capping beams may be subjected to lateral soil pressure if there is a tendency to sideways movement of the piles under vertical loading. Beams may also be subjected to bending in a vertical plane as a result of differential settlement between adjacent pile groups.

The passive resistance of the soil in contact with the sides of pile caps and ground beams should not normally be used as a means of restraining a structure against movement caused by lateral forces. In clays, the soil may shrink away from the substructure. The possibility of trenching alongside the substructure must also be considered. In any case, appreciable yielding of the soil must take place before its passive resistance can be mobilized. This movement may be sufficient to cause fracture of slender vertical piles.

6 The load capacity of piles and pile groups

6.1 General theory

The basis of all calculations to determine the bearing capacity of isolated piles from knowledge of the properties of the soil or rock, (the 'static' method of calculation) is that the total ultimate resistance of the pile to compression loading, Q_p, is the sum of the ultimate base resistance, Q_b, and the ultimate shaft resistance, Q_s, minus the weight of the pile, W_p. These components are each separately evaluated and used in the equation

$$Q_p = Q_b + Q_s - W_p \qquad (6\text{-}1)$$

In the above equation, W_p is generally neglected except in piles for marine structures where the pile may extend for a considerable distance above the soil line.

The effect of the pile installation method on the bearing capacity of the ground at the toe of the pile and on the skin friction mobilized on the pile shaft must be considered. Thus, driven and bored piles each require a different approach to evaluation of the two components. Having calculated the total ultimate pile resistance, Q_p, this is divided by an arbitrary load factor to obtain the allowable pile load, Q_a. The magnitude of the load factor is selected by experience to a value which will limit the settlement of the pile or pile group at the working load which can be tolerated by the superstructure. The load factor also takes into account the reliability of the information on the soil characteristics, the uncertainty of the design method, and whether or not any extensive load testing programme will be undertaken in advance of construction (see Section 6.8).

Where the design method has been reliably established and adequate information is available on the ground conditions, and provided that loading tests are made to confirm that settlements at the working load are tolerable, then a load factor of 2.5 has been shown by experience to be adequate. Using this load factor, the settlement of an isolated pile at the working load should not exceed 10 mm, for pile diameters up to about 600 mm (2 ft).

6.2 Driven precast concrete piles in cohesive soils

A reliable correlation has been established between the skin friction or adhesion which can be mobilized on the shaft of piles in cohesive soils and the undrained shear strength of these soils.

The total shaft skin friction, Q_s, is calculated from

$$Q_s = \alpha \bar{c}_u A_s \qquad (6\text{-}2)$$

where α is an adhesion factor, $\bar{c}_u$ the average undrained shear strength of the clay over the length of the pile shaft, and A_s is the area of shaft in contact with the soil.

The value of the adhesion factor depends on the length-to-diameter ratio of the pile, and the stratification of the soil. Where piles are driven through soft clays to obtain their bearing capacity in an underlying stiff clay, it has been found that a skin of soft clay is carried down to a limited depth into the stiff clay stratum. This reduces the skin friction available from the stiff clay but the deeper the penetration of the pile the less is the effect of this weak skin. This is shown by the set of curves relating the adhesion factor to the undrained shear strength in Fig. 6-1a. The converse occurs where the clay is overlain by a sand stratum which forms a skin of high frictional value over a limited depth of penetration. The design curves for this case are shown in Fig. 6-1b. Where a pile is driven into a stiff clay which extends from the surface downwards, the soil is displaced away from the sides of the pile forming a gap over a limited depth below the soil line.

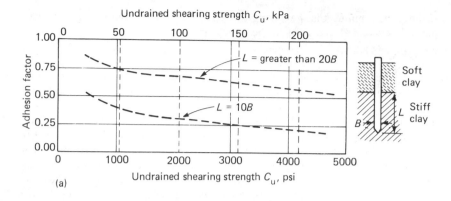

(a)

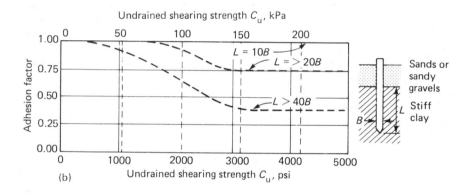

(b)

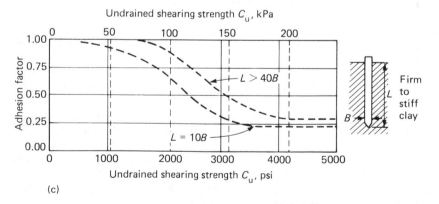

(c)

Figure 6-1   Design curves for adhesion factor for piles driven into clay soils
(a) Soft clays over stiff clays
(b) Sands or sandy gravels over stiff clays
(c) Firm to stiff clay without overlying strata

Again, the deeper the penetration the less is the significance of the gap and the corresponding adhesion factors are shown in Fig. 6-1c.

The base resistance, Q_b, is calculated from

$$Q_b = N_c c_b A_b \qquad\qquad (6\text{-}3)$$

where N_c is a bearing capacity factor which is usually taken as 9 provided that the pile has been driven to a penetration of at least five diameters into the bearing stratum, c_b is the undrained shear strength at the pile base, and A_b is the area of the base.

Equation 6-2 can be used to calculate the uplift resistance of tension piles but a larger load factor, say 3 or more, is required to obtain the allowable load because some tests have shown that the skin friction mobilized in uplift is less than that mobilized in compression, particularly where the uplift loads are maintained over a long period.

The effects of elapse of time between pile driving and the stage when piles are required to carry their working loads are of significance in soft to firm clays. An increase in carrying capacity of 200% or more may be achieved over a period between 1 and 30 days after driving in these soils. However, time effects do not appear to be of great significance in stiff to very stiff clays. The increase in carrying capacity is due principally to the dissipation of excess pore pressure in the soil caused by the displacement of the pile, that is to say the effective stress acting on the sides of the pile shaft is increased and therefore the effective skin friction is correspondingly increased. Research work [7] is being undertaken to develop methods of calculating the skin friction on piles in cohesive soils based on the principles of effective stress, but as yet these methods are not in general use.

6.3 Driven and cast-in-place piles in cohesive soils

Equations 6-2 and 6-3 can be used to calculate the skin friction and end-bearing resistance respectively of driven piles in cohesive soils. However, the down-drag effects which influence the adhesion factor, (α), for precast concrete piles are not present in the case of the driven and cast-in-place pile. This is because the dragged-down clay skin is formed on the drive tube. After the tube has been extracted, the concrete is cast against the soil which has been remoulded and compacted by the entry of the tube. The adhesion factor is thus affected to some extent by the remoulding which has taken place, but it is also affected by chemical and physical changes at the interface between the concrete and the soil as water is absorbed from the wet concrete by the soil, followed possibly by re-transfer of moisture from the soil to the hardening concrete.

In a comprehensive review of pile-bearing capacity in boulder clays made by the Construction Industry Research and Information Association [8], it was shown that the adhesion factor for driven and cast-in-place piles in these clays reduced progressively with increased stiffness of the clay, as shown in Table 6-1. In the absence of other information, the values in Table 6-1 can be used for other types of clay.

Table 6-1 Adhesion factors for driven and cast-in-place piles in boulder clay

Undrained cohesion kN/m^2	Adhesion factor, α
80	1.0
120	0.75
160	0.5
200	0.4

6.4 Bored and cast-in-place piles in cohesive soils

Equations 6.2 and 6.3 can again be used to calculate the skin friction and end-bearing resistance, respectively, of bored and cast-in-place piles in cohesive soils. The methods of forming the pile borehole (see Section 3.2) do not normally cause any appreciable remoulding of the soil around the pile shaft, but there is likely to be swelling and softening of the clay in the time interval between completing the borehole at any level and the placing of the concrete. Thereafter, the skin friction is influenced by chemical and physical changes at the pile–soil interface as described for the driven and cast-in-place pile.

The values in Table 6-2 for the adhesion factor were recommended by Skempton [9] for London Clay.

O'Neill and Reese [10] made separate allowances for the softening effects of drilling and of concreting in the equation

$$\alpha_{(average)} = (\alpha_{11}\alpha_{12}\alpha_{13})\alpha_2\phi \tag{6-4}$$

where

$\alpha_{11} = 0.65$ (for cylindrical shaft) for drilling effects
$\alpha_{12} = \alpha_{13} = 1 - (25/L) \times 0.1$ (where L is measured in feet) for concrete effects
$\alpha_2 = 1.0$ (a general adhesion factor)
$\phi = 0.6$ (when casing and bentonite are used) or 1.0 when concrete is placed in a dry hole

Weltman [8] showed that the adhesion factor for bored piles in boulder clay is reduced with increasing stiffness of the clay, as shown in Table 6-3.

Table 6-2 Adhesion factors for bored and cast-in-place piles in London Clay

Installation condition	Adhesion factor, α
Short piles in heavily fissured zone	0.3
Normal conditions	0.45 (but mobilized skin friction not to exceed 96 kN/m^2)
Shafts of piles with enlarged bases and elsewhere when drilling time is prolonged	0.3

Table 6-3 Adhesion factors for bored and
cast-in-place piles in boulder clay

Undrained cohesion kN/m^2	Adhesion factor, α
80	0.85
120	0.6
160	0.4
200	0.35

6.5 Preformed displacement piles in cohesionless soils

The unit skin friction on the shaft of a precast concrete pile or of a pile formed by
placing concrete in a steel tube driven to the required level and left in place, in a
cohesionless soil (e.g. a soil in the range from fine silty sands to coarse gravelly
sands) can be calculated from

$$\text{Ultimate unit skin friction} = q_s = K_s p_d \tan \delta \qquad (6\text{-}5)$$

where K_s is a coefficient of lateral earth pressure, p_d is the effective overburden
pressure at the depth under consideration, and δ is the angle of wall friction
between pile and soil. The values of K_s and δ are related to the relative density
and angle of shearing resistance, ϕ of the soil [11] as shown in Table 6-4.

The angle of shearing resistance of the soil is best determined by field tests. The
two types most suitable are the standard penetration test made in a borehole (test
19, BS 1377 : 1975, *Methods of test for soil for civil engineering purposes*), or the
static cone penetration test as described in BS 5930 : 1981, *Code of practice for site
investigations*. The correlations between the results of these tests and the angle of
shearing resistance and relative density of a sand are given in Table 6-5.

Equation 6-5 is not valid for pile penetration depths greater than 20 diameters.
For penetrations below this depth, the unit skin friction should be assumed to be
constant at the level calculated for a penetration depth of 20 diameters, but in no
case should the *average ultimate* unit skin friction calculated over the whole
penetration depth of the pile be higher than 107 kN/m^2 (2.2 kips/ft^2).

Equation 6-5 can be used to calculate the skin frictional resistance to uplift
loads on piles but the ultimate skin friction should be reduced by one-half to
allow for the different mechanism of failure where uplift loads are carried.

Table 6-4 Values of δ and K_s related to soil density

Material of pile shaft	δ	Values of K_s	
		Low relative density	High relative density
Concrete	$3/4\phi$	1.0	2.0
Steel	$20°$	0.5	1.0

Table 6-5   Relationship between angle of shearing resistance standard penetration test N value and static cone resistance of cohesionless soils

Standard penetration test N value blows/0.3 m	Static cone resistance kg/cm²	Angle of shearing resistance, φ, degrees	Relative density
5–10	0–10	28–30	Low
10–15	10–15	30–32	Medium
15–22	15–40	32–34	Medium
22–30	40–70	34–36	Medium
30–38	70–130	36–38	High
38–45	130–210	38–40	High
45–55	210–300	40–42	High
55–68	300–400	42–44	High

The base resistance of the preformed driven pile is calculated from

Ultimate base resistance $= Q_b = N_q p_d A_b$ (6-6)

where N_q is a bearing capacity factor and the other terms are as previously defined. The values of N_q are related to the angle of shearing resistance of the soil. The values of Berezantsev [12], shown in Fig. 6-2, are normally employed for pile calculations.

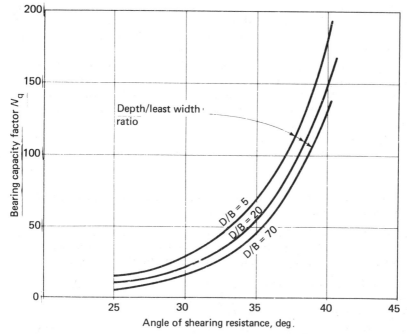

Figure 6-2   Bearing capacity factor, N_q

Equation 6-6 is again not valid for penetration depths greater than 20 diameters. For deeper penetrations, the ultimate base resistance should be assumed to be constant at the value calculated for the 20 diameters depth, but the ultimate unit base resistance *at any level* must not exceed a peak value of $10.7\,\text{MN/m}^2$.

6.6 Driven and cast-in-place piles in cohesionless soils

Equations 6-5 and 6-6 can be used to calculate the ultimate unit skin friction and base resistance of driven and cast-in-place piles. If extraction of the drive tube causes slumping and loosening of the soil around the shaft, the values of K_s and δ should be selected as corresponding to conditions of low to medium relative density. Where an enlarged base is provided, the base area, A_b, should be estimated from the volume of concrete used in forming the base. Piles driven into cohesionless soils derive a high proportion of their total resistance from the base component in Eqn 6-1. Therefore, the driven and cast-in-place pile with an enlarged base is a very economic type to use in loose to medium-dense soils. It is unlikely that any worthwhile base enlargement can be achieved by hammering out the concrete plug in a dense cohesionless soil. The peak values of average ultimate skin friction and base resistance of $107\,\text{kN/m}^2$ and $10.7\,\text{MN/m}^2$ respectively also apply to driven and cast-in-place piles.

6.7 Bored and cast-in-place piles in cohesionless soils

Equations 6-5 and 6-6 can be used to calculate the ultimate unit skin friction and base resistance, respectively, of bored and cast-in-place piles. Where cable percussion rigs are used for drilling the pile borehole (see Section 3.2), the action of the 'shell' or baler is to loosen the soil over the full depth of drilling and beneath the pile toe. Therefore, the values of K_s, δ, and N_q in the two equations should be those corresponding to low relative density ($\phi = 28$ to $30°$).

Where auger equipment is used for drilling, the values corresponding to the undisturbed soil can be used with a reduction factor of 0.7 applied to the calculated unit skin friction to allow for the effects of drilling with or without support to the soil by a bentonite slurry. No reduction factor need be applied to the calculated base resistance.

The peak values of average ultimate skin friction and base resistance of $107\,\text{kN/m}^2$ and $10.7\,\text{MN/m}^2$ respectively also apply to bored and cast-in-place piles.

6.8 Selection of load factor in relation to settlement at working load of piles in cohesive and cohesionless soils

It is necessary to divide the ultimate resistance of a pile calculated from Eqn 6-1 (or as determined by a loading test taken to failure) by a load factor to obtain the design working load on a pile. A load factor is required for the following reasons:

(a) To provide for natural variations in the strength and compressibility of the soil.
(b) To provide for uncertainties in the calculation methods adopted.
(c) To ensure that the working stresses on the materials forming the pile shaft are within safe limits.

(d) To ensure that the total settlement(s) of the single isolated pile or the group of piles are within tolerable limits.

(e) To ensure that differential settlements between adjacent piles or within groups of piles are within tolerable limits.

In the case of small diameter piles (up to 600 mm diameter or width), the results of a large number of loading tests have shown that the settlement of the single pile at the working load will not exceed about 10 mm if a load factor of 2.5 is adopted on the ultimate pile load calculated as described in the preceding sections of this chapter. This order of settlement is tolerable for most building and civil engineering structures provided that the settlement of the pile group is not excessive.

For large-diameter piles, the proportion of load carried by the pile base increases and the settlement increases with the increase in base diameter. The ultimate shaft resistance may be fully mobilized at a settlement of about 10 mm, the allowable load then being determined by the load factor on the base resistance. Thus, for large-diameter piles, it is the usual practice to adopt partial load factors for shaft and base resistance when the allowable load is determined by the lesser of the load as determined by

$$Q_a = Q_p/F_1 \qquad (6\text{-}7)$$

and

$$Q_a = \frac{Q_s}{F_2} + \frac{Q_b}{F_3} \qquad (6\text{-}8)$$

In the case of bored piles in clay, F_1 is frequently taken as 2, and the partial factors F_2 and F_3 as unity and 3, respectively. It is then necessary to calculate the settlement of the pile base from the expression for a circular or square pile:

$$\text{Settlement} = \rho = 0.6q/E_d \qquad (6\text{-}9)$$

where q is the bearing pressure corresponding to the calculated load carried by the pile base at the working load, B is the pile diameter and E_d a deformation modulus which is determined either from plate-bearing or pressuremeter tests made at the bottom of the pile borehole or from empirical correlations between this modulus and properties of the soil, such as the undrained shear strength of clays, and the standard penetration test N-value or static-cone resistance of cohesionless soils. For long piles, the elastic compression of the pile shaft must be added to the calculated base settlement to obtain the total settlement at the pile head.

6.9 Piles bearing on rock

Failure in the sense of undergoing large settlements with relatively small increase in load is unlikely to occur where piles are terminated in rock strata. Only in the case of weak, highly-weathered rocks is failure likely to take place in this way and it is possible to predict the ultimate load by calculation, assuming that weathering has reduced the rock to a soil-like consistency when the methods described in Sections 6.1 to 6.7 may be used. Alternatively, empirical correlations between base resistance and skin friction and simple physical properties of the rock, such as its unconfined compression strength or weathering grade, can be adopted.

Where piles are terminated on strong rocks, the allowable load is determined by the allowable working stress on the material of the pile shaft, or in the case of driven piles by the stresses induced when driving the pile to the desired level.

Where piles are driven into strong rocks, the rock is shattered around the shaft, and the skin friction is either neglected or calculated on the assumption that it behaves as a loose granular material. In the case of bored piles in weak rocks, it is good practice to assume that the whole of the load is carried in skin friction on the rock socket. This is because it is difficult to remove all soft and loose debris from the bottom of pile boreholes drilled by mechanical equipment. The maximum allowable base load (calculated from considerations of base settlement using Eqn 6-9) should be adopted only in the case of large-diameter piles where the base can be inspected and, where necessary, cleaned manually. The allowable skin friction on the rock socket depends on the roughness at the concrete–rock interface, this in turn depends on the strength and fracture frequency of the rock, and the occurrence of clayey weathered rock which might smooth out the asperities caused by drilling and in the worst case form a complete soft clay coating on the socket.

A correlation between rock-socket skin friction and the unconfined compression strength of piles in sandstone, mudstone and shale has been established by Williams and Pells [13]. Skin friction and base resistance values for other rock formations have been listed elsewhere [16].

The settlement of piles founded in rock is best determined by loading tests. Where all or a proportion of the load is permitted to be carried by the base, the settlement can be calculated from Eqn 6-9.

6.10 Piles in filled ground: negative skin friction

Where piles are installed in filled ground, consolidation of the fill under its own weight produces a down drag (negative skin friction) on the pile shaft which is an addition to the working load on the pile head. Similarly, if the pile is installed in a soft compressible natural clay formation and fill is placed on this formation, then downdrag will take place not only over the length of shaft within the fill but also over the length over which the soft clay is consolidating under the superimposed weight of the fill.

The magnitude of negative skin friction can be calculated by the methods described in Sections 6.2 to 6.7. The length of shaft over which downdrag takes place depends on the relative movement between the pile and the supporting stratum. Where the pile terminates on a hard incompressible stratum (Fig. 6-3a), the movement between the consolidating strata and the pile is a maximum and the negative skin friction is mobilized over most of the length embedded in these strata. However, where the pile toe yields under the combined working load and downdrag load the relative pile–soil movement may not be enough to mobilize the negative skin friction and the latter is calculated over a lesser shaft length, as shown in Fig. 6-3b.

It is possible to reduce the negative skin friction on the shafts of precast concrete piles by coating them with a soft bitumen. However, the cost of

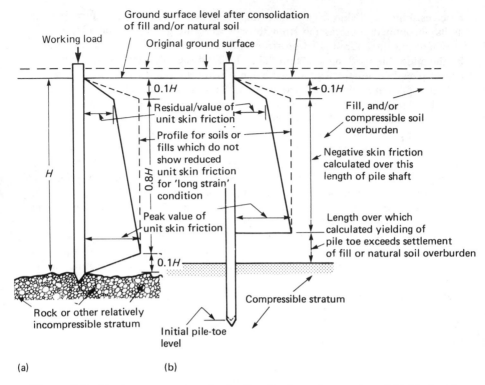

Figure 6-3 Design curves to obtain distribution of negative skin friction in pile shaft

   (a) For piles terminated on rock or relatively incompressible stratum
   (b) For piles terminated in compressible stratum

providing this 'slip coat' is usually greater than the cost of providing an additional length of shaft within the supporting stratum to carry the combined working load and downdrag load.

   If the negative skin friction has been calculated by conservative methods, it may be over-conservative to adopt a load factor of 2.5 or 3 on the combined working load and downdrag load. it is often sufficient to adopt these load factors on the working load only and then to check that there is still a reasonable load factor on the combined load.

6.11 The behaviour of pile groups

Only if piles are taken down to a bearing on a hard incompressible stratum is the settlement of a group of piles equal to that of a single pile under the same working load as each pile in the group. Where piles are terminated in compressible soils or rocks, there is interaction between the piles in the group and, in the case of large groups, the settlement can be many times greater than that of the

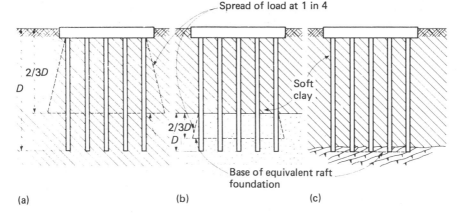

Figure 6-4   Load transfer to soil from pile group
        (a) Group of piles supported predominantly by skin friction
        (b) Group of piles driven through soft clay to combined skin
            friction and end bearing in stratum of dense granular soil
        (c) Group of piles supported in end bearing on hard rock
            stratum

single pile. This is because the ground beneath the group is stressed to a much greater depth than that beneath the single pile.

In order to calculate group settlement, it is the usual practice to assume that the group behaves as an equivalent raft foundation. The overall plan dimensions and the level of the base of the raft foundation depends on the manner in which the piles receive their support from the ground, as shown in Fig. 6-4. These overall dimensions having been determined, the settlement of the group is calculated by conventional soil mechanics methods [14], assuming the equivalent raft to form the base of a flexible or semi-rigid block foundation.

Consideration must be given to the effects of ground heave where piles are driven in groups in stiff cohesive soils or of loss of ground where bored piles are installed in groups in cohesionless soils or soft clays.

When calculating the settlement of pile groups in filled ground, the fill layer should be assumed to act as an overall load superimposed on the ground surface over the full extent of the filled area and not as an addition in downdrag to the working load on the piles.

6.12   Piles carrying horizontal or inclined loads

Foundation piles are frequently required to carry inclined loads which are the resultant of the dead load of the structure and horizontal loads from wind, water pressure or lateral earth pressure on the structure. Where the horizontal component of the load is small in relation to the vertical component, it can be safely carried by vertical piles. Thus, special provision for horizontal wind forces on piles supporting buildings is made only for very tall structures.

However, horizontal forces may be large in piled structures such as wharves and jetties or in piled foundations to retaining walls, travelling cranes, and chimneys.

These forces may be carried by a combination of vertical and inclined, or raking, piles. Raking piles are an efficient method of carrying inclined loads because a large proportion of the horizontal component is carried by the pile in an axial direction. A useful angle of rake is possible only with precast concrete piles which can be driven to an inclination of 1 horizontal to 2 or 3 vertical. Such flat angles are impossible with bored or driven and cast-in-place piles and in these cases it is necessary to calculate the resistance of the vertical pile to horizontal loading, which in the case of small-diameter piles is relatively low.

The ultimate lateral resistance of short rigid piles (length/diameter ratio less than 10 to 12) can be calculated from the passive resistance of the soil to yielding by rotation of the pile. The method of Brinch Hansen [15] is suitable for such piles. However, in the case of long piles, the passive resistance of the soil is very large and failure of the pile will take place in bending before the stage of passive soil failure occurs.

A simple method of calculating the ultimate horizontal load on long piles is to assume that bending takes place about a point of virtual fixity which is assumed to be at a depth of 1.5 m (5 ft) below the ground surface in a dense granular soil or stiff clay or 3 m in a soft clay or silt. The deflections of vertical piles under horizontal loading can be similarly calculated from Eqns 6-10 and 6-11:

$$\text{Deflection at head of free-headed pile} = \frac{H(e+z_f)^3}{3EI} \qquad (6\text{-}10)$$

$$\text{Deflection at head of fixed-headed pile} = \frac{H(e+z_f)^3}{12EI} \qquad (6\text{-}11)$$

where

H = horizontal load
e = height of point of application of load above ground surface
z_f = depth below ground surface to point of vertical fixity
E = elastic modulus of material forming the pile shaft
I = moment of inertia of the pile cross-section

The above simple methods should be used only for relatively low horizontal load components or where inclined loading is not critical to the design. Where appreciable inclined loads are carried, the ultimate resistance and deflections should be calculated by more sophisticated methods which are described elsewhere [16].

7 The durability of concrete piles

7.1 Investigations

The deterioration of concrete in piled foundations may occur as a result of the presence of aggressive substance in soils, in rocks, in the groundwater, and in sea or river water. Piles in river or marine structures are also exposed to aggressive atmospheric conditions, and to abrasion from shifting sand, shingle, ice or driftwood.

Precautionary measures against deterioration and damage to piles can be expensive and it is essential to make detailed investigations of the presence or otherwise of harmful substances in the ground and a careful assessment of the risks of aggressive action in order to determine the extent of any precautions. In particular, adequate information is required on the range of fluctuation of river or sea levels, or of the groundwater table.

In normal soil conditions, it is usually sufficient to limit chemical analyses of soil and groundwater samples to the determination of pH values, sulphate content and chloride content. Where the sulphate content exceeds 0.5% in soils, it is advisable to determine the water-soluble sulphate content, expressing this in grammes of SO_3 per litre of water extracted [17]. It is important that the selection of samples for sulphate content determination should be sufficiently numerous to provide an overall appreciation of the conditions which might give rise to sulphate attack on the concrete of the pile shaft. Some clays contain gypsum crystals and it is evident that soil specimens containing numerous crystals would show a very high concentration of sulphates. Yet gypsum crystals are usually scattered at random or concentrated only at specific levels within the soil mass. Thus, it would be wasteful to make recommendations for expensive precautions against attack by high concentrations of sulphates if these were representative only of a small proportion of the soil in contact with the pile. The best indication of the likelihood of sulphate attack is given by the sulphate content of the groundwater. Samples of the latter should be taken at various levels over the intended depth of the piles and they should not be diluted by drilling water. Whenever possible, the groundwater samples should be taken towards the end of the dry season.

The pH value of the soil or groundwater provides a crude indication of the risk of aggressive action by organic acids, dissolved carbon dioxide, or acidic industrial waste. The lower the pH value, the greater the concentration of acids. Exposure of the sample to the air may cause a rapid reduction of the pH value. Therefore, if low values are measured in the laboratory, a re-check should be made on site with immediate determination of pH value by a portable meter.

Where fill material is present, particularly household refuse or industrial wastes, full chemical analyses are required to identify potentially aggressive substances. It is important to obtain representative samples. Preferably, they should be obtained from trial pits in which the zones of chemically contaminated material can be identified by their appearance.

7.2 Precautions against sulphate attack on concrete piles

The naturally-occurring sulphates in soils are those of calcium, magnesium, sodium and potassium. The basic mechanism of the attack on concrete is a reaction with hydrated calcium aluminate in the cement paste to form calcium sulphoaluminate. This reaction is accompanied by an increase in molecular volume of the minerals, resulting in the expansion and finally the disintegration of the hardened concrete. Ammonium sulphate attacks concrete very severely. It does not occur naturally but may be present from the use of artificial fertilizers, or infill material containing waste from gasworks.

Attack is most severe when flowing groundwater is present. The flow brings fresh sulphates to continue and accelerate the chemical reaction. The risks of

Table 7-1 Classification of sulphates in soils affecting concrete in piled foundations and recommended precautions

	Concentration of sulphates expressed as SO_3			Types of cement and limiting mix proportions[a] for dense fully-compacted concrete and other protective measures
Class	In soil			
	Total SO_3 %	SO_3 in 2:1 aqueous extract g/l	In groundwater parts/100 000	For precast concrete piles and pile shells, precast concrete caps and ground beams
1	Less than 0.2	—	Less than 30	OPC: Min = 300 kg/m³ (500 lb/yd³) Max w/c = 0.55
2	0.2–0.5	—	30–120	(a) Above highest water level OPC: Min = 310 kg/m³ (525 lb/yd³) Max w/c = 0.50 (b) In contact with fluctuating water level OPC: Min = 350 kg/m³ (600 lb/yd³) or SRPC: Min = 310 kg/m³ (525/yd³) Max w/c = 0.50
3	0.5–1.0	1.9–3.1	120–250	(a) Above highest water level OPC: Min = 380 kg/m³ (650 lb/yd³) or SRPC: Min = 340 kg/m³ (575 lb/yd³) Max w/c = 0.50 (b) In contact with fluctuating water level SRPC: Min = 350 kg/m³ (600 lb/yd³) Max w/c = 0.50
4	1.0–2.0	3.1–5.6	250–500	(a) Above highest water level SRPC: Min = 380 kg/m³ (650 lb/yd³) Max w/c = 0.45 (b) In contact with fluctuating water level in lower range of SO_3 and favourable cations use SRPC: Min = 390 kg/m³ (650 lb/yd³) Max w/c = 0.45 For higher range of SO_3 and unfavourable cations use SSC: Min = 370 kg/m³ (625 lb/yd³) or provide permanent sheathing in metal or plastics
5	Over 2.0	Over 5.6	Over 500	(a) Above highest water level SRPC: Min = 390 kg/m³ (650 lb/yd³) Max w/c = 0.40 (b) In contact with fluctuating water level provide permanent sheathing in metal or plastics over SRPC concrete Min = 390 kg/m³ (650 lb/yd³) Max w/c = 0.40

OPC Ordinary Portland cement
SRPC Sulphate-resisting Portland cement
SSC Supersulphated cement
Min Minimum recommended cement content
[a] The minimum cement contents recommended above are suitable for
  (i) Precast concrete with low workability [12–25 mm ($\frac{1}{2}$–1 in) slump].
  (ii) Cast-in-situ concrete in pile shells, pile caps and ground beams with medium workability [50–75 mm (2–3 in) slump].
  (iii) Cast-in-situ piles with high workability [100 mm (4 in) slump].

Types of cement and limiting mix proportions[a] for dense fully-compacted concrete and other protective measures	Types of cement and limiting mix proportions[a] for dense fully-compacted concrete and other protective measures
For concrete placed in thin steel shells in dry conditions[b]; reinforced concrete in pile caps and ground beams[c]	*For concrete in driven-and-cast-in-situ and bored-and-cast-in-situ piles[d]*
(a) Above highest water level OPC: Min = 300 kg/m³ (500 lb/yd³) Max w/c = 0.55 (b) In contact with fluctuating water level OPC: Min = 310 kg/m³ (525 lb/yd³) Max w/c = 0.55	(a) Above highest water level OPC: Min = 330 kg/m³ (550 lb/yd³) Max w/c = 0.55 (b) In contact with fluctuating water level OPC: Min = 370 kg/m³ (625 lb/yd³) Max w/c = 0.55.
(a) Above highest water level OPC: Min = 330 kg/m³ (550 lb/yd³) Max w/c = 0.50 (b) In contact with fluctuating water level OPC: Min = 350 kg/m³ (600 lb/yd³) or SRPC: Min = 310 kg/m³ (525 lb/yd³) Max w/c = 0.50	(a) Above highest water level OPC: Min = 370 kg/m³ (625 lb/yd³) Max w/c = 0.50 (b) In contact with fluctuating water level OPC: Min = 380 kg/m³ (650 lb/yd³) or SRPC: Min = 340 kg/m³ (575 lb/yd³) Max w/c = 0.50
(a) Above highest water level OPC: Min = 400 kg/m³ (675 lb/yd³) or SRPC: Min = 350 kg/m³ (600 lb/yd³) Max w/c = 0.50 (b) In contact with fluctuating water level SRPC: Min = 390 kg/m³ (650 lb/yd³) Max w/c = 0.50	(a) Above highest water level OPC: Min = 400 kg/m³ (675 lb/yd³) or SRPC: Min = 350 kg/m³ (600 lb/yd³) Max w/c = 0.50 (b) In contact with fluctuating water level SRPC: Min = 390 kg/m³ (650 lb/yd³) Max w/c = 0.50[e]
(a) Above highest water level OPC: Min = 400 kg/m³ (675 lb/yd³) or SRPC: Min = 350 kg/m³ (600 lb/yd³) (b) Max w/c = 0.45. In contact with fluctuating water table as for precast concrete but external sheathing to consist of polyethylene, hot bitumen spray, bituminous paint, trowel-applied mastic asphalt or adhesive plastics sheet	(a) Above highest water level and soil free from seepage water SRPC: Min = 400 kg/m³ (675 lb/yd³) Max w/c = 0.45 (b) In contact with fluctuating water level in lower range of SO_3 and favourable cations use SRPC: Min = 390 kg/m³ (650 lb/yd³) Max w/c = 0.45 For higher range of SO_3 and unfavourable cations place concrete in durable metal or plastics sleeve left in place
(a) Above highest water level OPC: Min = 400 kg/m³ (675 lb/yd³) or SRPC: Min = 350 kg/m³ (600 lb/yd³) Max w/c = 0.40 (b) In contact with fluctuating water level SRPC: Min = 390 kg/m³ (650 lb/yd³) Max w/c = 0.40 with permanent external sheathing as above	(a) Above highest water level and soil free from seepage water SRPC: Min = 390 kg/m³ (650 lb/yd³) Max w/c ratio = 0.40 (b) Cast-in-situ piles are unsuitable for installation below the water table

If concrete of a lower workability than (iii) is required by the specialist piling contractor, the cement content may be reduced provided that the concrete can be compacted to a dense impermeable mass. In no case must the maximum water/cement ratio be exceeded.

[b] The precautions assume that the shells may be ruptured or lost by corrosion.

[c] The precautions assume fairly massive sections where the concrete can be vibrated.

[d] The precautions assume that the driving tube or borehole casing is extracted during or after placing the concrete.

[e] A higher cement content may be required if the concrete is placed under water by tremie pipe.

attack are low above the highest groundwater level, where the damage may be limited to an unsustained attack on the concrete surface.

Free sulphuric acid may be formed in natural soil or groundwater by the oxidation of pyrites in some peats, or in ironstone or alum shales. It may also be present in some industrial wastes. The effect of sulphuric acid on hardened concrete is similar to that of sulphate attack.

A dense well-compacted concrete provides the best protection against attack by sulphates on piles, pile caps and ground beams. For this reason, precast concrete piles are the most favourable type in aggressive ground conditions. However, this type of pile may not be suitable because of other environmental conditions. Hence, the concrete mixes for the cast-in-place types must be designed to provide the required degree of impermeability and resistance to aggressive action.

Although general recommendations have been made by the Building Research Establishment [17] in the UK for precautions against sulphate attack, these are not suitable in all cases for concrete piles, particularly for concrete cast-in-place where a highly-workable, self-compacting concrete is frequently required. The precautions set out in Table 7-1 are more suitable for piled foundations including the ground beams and pile caps.

7.3 Precautions against attack by organic acids on concrete piles

Naturally-occurring organic acids or carbon dioxide dissolved in groundwater may cause leaching of concrete exposed to groundwater flow. Conditions are severe in some European countries where reliance is generally placed on obtaining a dense impermeable concrete as a means of resisting attack rather than by using special

Table 7-2   Classification of aggressiveness of liquids of mainly natural composition according to DIN 4030

Line	Examination	Degree of aggressiveness		
		Slight	Severe	Very severe
1	pH value	6.5–5.5	5.5–4.5	below 4.5
2	Lime-dissolving carbonic acid (CO_2), mg/l determined by marble test according to Heyer	15–30	30–60	over 60
3	Ammonium (NH_4), mg/l	15–30	30–60	over 60
4	Magnesium (Mg), mg/l	100–300	300–1500	over 1500
5	Sulphate (SO_4), mg/l	200–600	600–3000	over 3000

Notes

The maximum aggressiveness is applicable to the evaluation of water even if it is only achieved by one of the values in lines 1 to 5. If two or more values lie in the upper quarter of a range (in the lower quarter for the pH) the aggressiveness is increased by one stage. This increase does not apply to sea water.

Greater aggression occurs at higher temperatures or if concrete is subject to abrasion by swift-flowing or agitated water. The aggressiveness declines in low temperatures or if water is present only in small volumes and is quiescent, e.g. in soils having low permeability ($K < 5–10$ m/s).

Table 7-3 Precautions in concrete mixes for degrees of aggressiveness as recommended in DIN 1045

Degree of aggressiveness to DIN 4030	*Precautions recommended in* DIN 1045
Slight	Permeability to DIN 1048 $e_{max} \leqslant 50$ mm Water/cement ratio $\leqslant 0.60$
Severe	Permeability to DIN 1048 $e_{max} \leqslant 30$ mm Water/cement ratio $\leqslant 0.50$
Very severe	Use protective coating on concrete
From 400 mg/l of SO_4 in water From 3000 mg/kg of SO_4 in soil	Use sulphate-resisting cement to DIN 1164

cement [18]. The precautions set out in Tables 7-2 and 7-3 are based on practice in West Germany and Holland.

7.4 Precautions against attack by sea water or chlorides in the ground water

Where reinforced concrete structures are permanently immersed in sea water, there will be little or no risk of disintegration of the concrete provided that a rich mix (not leaner than $1 : 1\frac{1}{2} : 3$) is used, and the concrete is well compacted to give a dense impermeable mass. The sulphate content of sea water expressed as SO_3 is about 230 parts per 100 000, which is in excess of the figure regarded as marginal between aggressive and non-aggressive where sulphates are present in the groundwater on land sites. However, sea water is not markedly aggressive to concrete because sodium chloride has an inhibiting or retarding effect on the expansion of the sulphates. Normal Portland cement can be used for marine structures, but the use of sulphate resisting cement is an added precaution against cracking of the concrete followed by corrosion of the reinforcement. The minimum cover to the steel should be 50 mm (2 in). Sea water should not be used for mixing reinforced concrete.

The deterioration of concrete in marine structures is most likely to occur in the 'splash zone', and in areas of marine growth, frost attack, and erosion by waves or moving shingle and sand at sea-bed level. Disintegration can be minimized by the use of rich, well-compacted dense concrete, as described in the previous paragraph.

7.5 Precautions against attack by industrial wastes

Severely aggressive conditions may be present in filled ground containing industrial wastes, and it may not be possible to design concrete mixes to withstand the aggressive action. However, the concrete in bored and cast-in-place piles can be protected by casting it inside a rigid PVC sleeve which is left in place as the guide tube or drive tube is extracted. Thick flexible plastics sheeting made up into cylindrical sleeves has also been used for this protective casing, but there are difficulties in holding the sheeting in place without tearing as the temporary tube is pulled out.

Pile caps and ground beams can be protected by heavy-duty polythene sheeting turned up against the sides of the excavation or formwork. Where necessary, the protection to the sides of these structures, when cast within formwork, can be provided by self-adhesive plastics sheeting.

8 Pile testing

8.1 Loading tests on piles

Loading tests on piles are required as a means of verifying the validity of the method used to calculate the ultimate carrying capacity of the pile. This verification may be needed where the soil-test information is meagre or difficult to interpret. It is particularly necessary in weak rocks where published information on skin friction values related to simple geological classifications is lacking (see Section 6.9).

Loading tests are also necessary to check that the settlement at the working load is within limits which are tolerable to the designer of the superstructure.

Two types of axial compression or tension test can be performed on piles. These are the constant rate of penetration (CRP) test and the maintained load (ML) test.

In the CRP compression test, the pile is jacked into the soil, the load being adjusted to give a constant rate of downward movement to the pile which is maintained until failure point is reached. Failure is defined either as the load at which the pile continues to move downward without further increase in load, or the load at which the penetration reaches a value equal to one-tenth of the diameter of the pile base. A penetration rate of 0.8 mm/min (0.03 in/min) is suitable for friction piles in clay, for which the total penetration to reach failure is likely to be less than 25 mm (1 in). Considerably larger movements, possibly up to one-fifth of the base diameter, may be needed to achieve ultimate load conditions on piles end-bearing in granular soils, and a jacking rate of 1.5 mm/min (0.06 in/min) in suitable.

The CRP test has the advantage that it is rapidly performed and is thus useful for preliminary testing when failure loads are unknown, and when the design is based on a load factor against ultimate failure. However, the method has the disadvantage that it does not give the insight into pile behaviour that can be obtained from cycles of loading and unloading in various stages up to the maximum test load. Much useful information is given on the yielding of the pile in skin friction and base resistance as a result of study of the load–settlement curves obtained when the load is cycled.

In the ML test, the load is applied in increments, generally of about one-fifth of the maximum test load, and the load is maintained at each increment until settlement of the pile head has ceased or has slowed down to a rate not exceeding 0.25 mm in one hour and still decreasing. The load is returned to zero at various stages. If the tentative working load is known at the preliminary testing stage, the test load can be returned to zero at this load and at other multiples, say 1.5, 2 and, of course, after reaching failure load or the maximum test load.

Table 8-1   Loading stages for maintained loading test
on pile to 1.5 times working load

Stage	Increments	
1	0.25	× working load
2	0.5	× working load
3	0.75	× working load
4	1.0	× working load
5	0.75	× working load
6	0.5	× working load
7	0.25	× working load
8	0	× working load
9	1.0	× working load (hold for 24 h)
10	1.25	× working load
11	1.5	× working load (hold for 6 h)
12	1.25	× working load
13	1.0	× working load
14	0.75	× working load
15	0.5	× working load
16	0.25	× working load
17	0	× working load

If at the preliminary pile stage it is specified that the loading should be taken to failure, then the ML procedure can be used up to that increment when the pile is seen to be yielding, when the load should be removed, after which the pile should be reloaded to failure using the CRP method.

Proof load-tests on working piles are best made by the ML method. Suitable increments for a proof load of 1.5 times the working load are given in Table 8-1.

It is usual to specify limiting total and residual (i.e. total settlement minus the elastic recovery on removal of load) settlements at the working load and at the proof load. The specified values are determined from considerations of the tolerable total and differential settlements of the superstructure, taking into account any group effects. Realistic figures of total settlement should be specified, particularly in the case of long piles when the elastic compression of the material of the pile shaft should be taken into account.

Using either the CRP or ML test, the compression load is applied to the pile head by jacking against a beam or pair of beams loaded with kentledge. Alternatively, the beams can be restrained by anchoring their ends to pairs of tension piles. The load on the pile should be measured by a pressure cell rather than by relying on the pressure gauge of the hydraulic jack. The settlement of the pile head is measured by three or four deflectometers attached to a reference beam supported well clear of the kentledge stack or tension piles. A loading testing rig using kentledge is illustrated in Fig. 8-1, and typical load–time–settlement curves in Fig. 8-2.

Uplift tests are performed on tension piles by jacking against beams taking their

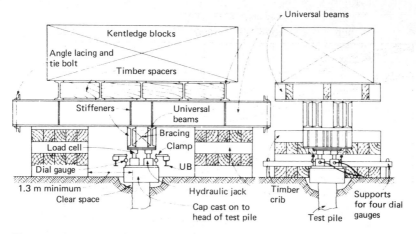

Figure 8-1   Testing rig for compression test on a pile using kentledge for reaction

reaction against the ground. The ML test is the preferable type with the loads returned to zero after each increment.

Lateral load tests can be performed simply by jacking apart a pair of piles. They should be extended above ground level so that the angle of deflection as well as the amount of deflection can be measured.

In British practice, the procedure for making load tests on piles is dealt with in an Institution of Civil Engineers' model specification [19]. Compression and tension test procedure is also specified in ASTM D 1143–74 and ANSI/ASTM D 3689–78, respectively.

8.2   Integrity tests on piles

Although loading tests may reveal gross defects in the workmanship of pile installation, there may be other defects such as cracking or honeycombing in the concrete of the shaft which would be detrimental to the long-term performance of the piles but would not necessarily be apparent from the settlement curves of a loading test. In any event, load testing as a means of checking workmanship is costly and time consuming.

Non-destructive integrity test methods are available. There are two types in general use. The first is the dynamic method, in which a single hammer blow or shock, or continuous vibration, is applied to the head of the pile and the response of the pile is measured by oscilloscope or vibrograph. The second method involves the use of ultrasonic or radiometric loggers lowered down tubes installed in the piles at the time of their construction.

Either method can be used for precast concrete piles (but not the jointed types), driven and cast-in-place, or bored and cast-in-place piles.

The various proprietary techniques available for integrity testing have been reviewed in detail by the Construction Industry Research and Information Association [20].

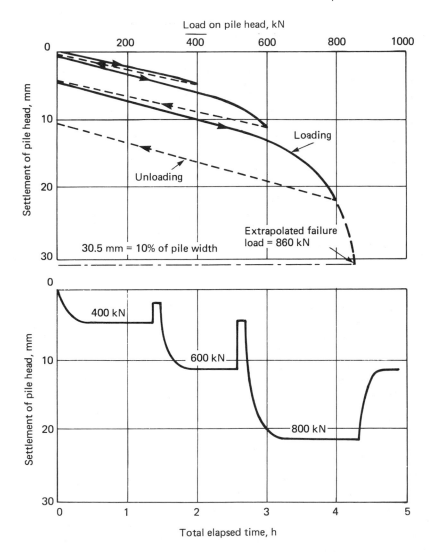

Figure 8-2 Load–settlement and time–settlement curves for maintained load compression test on a 305×305 mm pile

Acknowledgment

The illustrations and tables in this chapter were originally published in the author's book *Pile design and construction practice* [16], and are reproduced by kind permission of the publishers, Eyre and Spottiswoode. Reference should be made to this book for further information on pile installation techniques, design and testing.

References

1. Gravare, C-J. and Goble, F. R., Pile driving construction control by the Case method, *Ground Eng.*, Vol. 13, No. 2, March 1980, pp. 20–25.
2. Thorburn, S. and Thorburn, J. Q., *Review of problems association with the construction of cast-in-place concrete piles*, Construction Industry Research and Information Association, Report PG2, 1977, 42 pp.
3. Fleming, W. G. K. and Lane, P. F., Tolerance requirements and constructional problems in piling, *Proc. conf. on behaviour of piles*, Institution of Civil Engineers, London, 1977, pp. 175–8.
4. Smith, E. A. L., Pile driving analysis by the wave equation. *Proc. Am. Soc. Civ. Eng.*, Vol. 86, No. SM4, 1960, pp. 35–61.
5. Clarke, J. L., *Behaviour and design of pile caps with four piles*, Cement and Concrete Association Report 42.489, 1973, 19 pp.
6. Wood, R. H. and Simms, L. G., *A tentative design method for the composite action of heavily loaded brick panel walls supported on reinforced concrete beams*, Building Research Establishment Current Paper 26/29, July 1969, 9 pp.
7. Burland, J. B., Shaft friction of piles in clay — a simple fundamental approach, *Ground Eng.*, Vol. 6, No. 3, 1973, pp. 36–42.
8. Weltman, A. J., *Piling in boulder clay and other glacial tills*, Construction Industry Research and Information Association, Report PG5, 1978, 78 pp.
9. Skempton, A. W., Cast-in-situ bored piles in London clay, *Géotechnique*, Vol. 9, No. 4, 1959, pp. 153–73.
10. O'Neill, M. W. and Reese, L. C., Behaviour of bored piles in Beaumont clay, *Proc. Am. Soc. Civ. Eng.*, Vol. 98, No. SM2, 1972, pp. 195–213.
11. Broms, B., Methods of calculating the ultimate bearing capacity of piles, a summary, *Sols–Soils*, Vol. 5, No. 18–19, 1966, pp. 21–31.
12. Berezantsev, V. G. *et al.*, Load bearing capacity and deformation of piled foundations, *Proc. 5th int. conf. Soil Mech. and Found. Eng.*, Paris, Vol. 2, 1961, pp. 11–12.
13. Williams, A. F. and Pells, P. J. N., Side resistance rock sockets in sandstone, mudstone and shale, *Can. Geotech. J.*, Vol. 18, 1981, pp. 502–513.
14. Tomlinson, M. J., *Foundation design and construction*, 4th Edn., Pitman, 1980, 793 pp.
15. Hansen, J. Brinch, The ultimate resistance of rigid piles against transversal forces, *Bull. Danish Geotech. Inst.*, No. 12, 1961, pp. 5–9.
16. Tomlinson, M. J., *Pile design and construction practice*, Eyre and Spottiswoode, 1977, 425 pp.
17. *Concrete in sulphate bearing soils and ground waters*, Building Research Establishment Digest No. 250, 1981, 4 pp.
18. Eglinton, M. S., *Review of concrete behaviour in acidic soils and ground waters*, Construction Industry Research and Information Association, Technical Note 69, 1976, 52 pp.
19. Institution of Civil Engineers, *Piling model procedures and specification*, London, 1978, 161 pp.
20. Weltman, A. J., *Integrity testing of piles, a review*, Construction Industry Research and Information Association, Report PG4, 1977, 36 pp.

32 Shell roofs

C B Wilby

University of Bradford, England

Contents

Notation

There is not an internationally agreed nomenclature for the analysis and design of shell roofs. Nor is there a nationally agreed nomenclature suitable for this purpose. The nomenclature most commonly used and which is most suitable for shells is as given, for example, by Timoshenko [1]. This will be used in this chapter; it is listed below.

A	a constant (see Eqn. 10-2)	h	thickness of shell
a	radius of spherical dome in Section 8, and dimension on Fig. 9-1	k	a constant (see Eqn 9-1)
		l	span
B	a constant (see Eqn 10-2)	N_x	stress resultant at $x = x$ (see Fig. 6-1)
b	dimension on Fig. 9-1, and dimension on Fig. 10-1	N'_x	$N_x + (\partial N_x/\partial x).\, dx$, i.e. value of N_x
$C_1(\phi)$	a function of ϕ		at $x = x + dx$
$C_2(\phi)$	a function of ϕ	$N_{x\phi}$	stress resultant at $x = x$ and $\phi = \phi$
$C_3(\phi)$	a function of ϕ		(see Fig. 6-1)
$C_4(\phi)$	a function of ϕ	$N'_{x\phi}$	$N_{x\phi} + (\partial N_{x\phi}/\partial x)\, dx$, i.e. value of
c	value of z when $x = a$ and $y = b$		$N_{x\phi}$ at $x = x + dx$
E	Young's modulus of shell	N_{xy}	stress resultant (in-planar shear) at
g	$1 - H_0/H$		$x = x$ and $y = y$ (see Fig. 9-1)
H	dimension on Fig. 10-2	N_y	stress resultant (tension in direc-
H_0	dimension on Fig. 10-2		tion 0y) at $y = y$ (see Fig. 9-1)

N_θ stress resultant at $\theta = \theta$ (see Fig. 6-5)

N'_θ $N_\theta + (\partial N_\theta/\partial\theta)\,\mathrm{d}\theta$, i.e. value of N_θ at $\theta = \theta + \mathrm{d}\theta$

$N_{\theta\phi}$ stress resultant at $\theta = \theta$ and $\phi = \phi$ (see Fig. 6-5)

$N'_{\theta\phi}$ $N_{\theta\phi} + (\partial N_{\theta\phi}/\partial\theta)\,\mathrm{d}\theta$, i.e. value of $N_{\theta\phi}$ at $\phi = \phi$ and $\theta = \theta + \mathrm{d}\theta$

N_ϕ stress resultant at $\phi = \phi$ (see Figs 6-1, 6-5 and 6-8)

N'_ϕ $N_\phi + (\partial N_\phi/\partial\phi)\,\mathrm{d}\phi$, i.e. value of N_ϕ at $\phi = \phi + \mathrm{d}\phi$

$N_{\phi x}$ stress resultant at $\phi = \phi$ (see Fig. 6-1)

$N'_{\phi x}$ $N_{\phi x} + (\partial N_{\phi x}/\partial\phi)\,\mathrm{d}\phi$, i.e. value of $N_{\phi x}$ at $\phi = \phi + \mathrm{d}\phi$

$N_{\phi\theta}$ stress resultant at $\theta = \theta$ and $\phi = \phi$ (see Fig. 6-5)

$N'_{\phi\theta}$ $N_{\phi\theta} + (\partial N_{\phi\theta}/\partial\phi)\,\mathrm{d}\phi$, i.e. value of $N_{\phi\theta}$ at $\theta = \theta$ and $\phi = \phi + \mathrm{d}\phi$

p total load on shell per unit of curved surface area

q total load per unit area of curved surface

q_0 self-weight load per unit surface area

R total loading on a symmetrically loaded shell of revolution about a vertical axis, above the horizontal section defined by an angle ϕ, under consideration (see Fig. 6-8)

r radius of shell (see Fig. 6-1)

r_0 radius of horizontal circle (see Fig. 6-5)

r_1 radius of curvature of generating curve at point under consideration (see Fig. 6-5)

r_2 radius of principal curvature in plane perpendicular to other plane of principal curvature

u displacement in longitudinal direction x at $x = x$ (see Fig. 6-3)

v displacement in transverse tangential direction at $\phi = \phi$ (see Fig. 6-3)

w displacement in radial direction at $x = x$ and $\phi = \phi$ (see Fig. 6-3)

X component of total loading in plane of differential element (see Figs 6-1 and 6-5) per unit area of element

Y component of total loading in plane of differential element (see Figs 6-1 and 6-5) per unit area of element

Z component of total loading normal to plane of differential element (see Figs 6-1 and 6-5) per unit area of element

$X(x)$ a function of x, independent of y

$Y(y)$ a function of y, independent of x

x, y, z coordinates in three dimensions, at right-angles to one another

θ horizontal angle locating point under consideration (see Fig. 6-5), but for Eqns 9-5 and 9-6, θ is angle of slope of differential element with respect to axes Oy and Oz (see [1])

ϕ angle between point under consideration and vertical (see Figs 6-1, 6-5 and 6-8), but for Eqns 9-5 and 9-6 ϕ is angle of slope of differential element with respect to axes Ox and Oz (see [1])

ω angle between Ox and Oy axes (see Fig. 9-1)

Summary

Different types of reinforced concrete shell roofs are described and designated and the practical uses and economics of these shells are discussed. The methods of construction, design and analysis of shells are narrated. A first requirement is usually the membrane theory. A membrane analysis is given for shells of arbitrary and cylindrical shape and of revolution.

For practical designs, tables and graphs are reviewed. Digital and electronic analogue computer and finite element methods of analysis and design and the problem of instability are considered. Individual sections are included for domes, hypars and conoids.

The practical designer can ascertain whether or not to use a shell and decide upon which type. He can then find out which method of design to use and if any tables, graphs or programs exist to help him. If interested further he can study the analytical basis of his designs.

Many references are given which have been specially selected to be of the most help to the designer with the topics already described in this synopsis.

1 Introduction

In this section, a brief description of the most popular types of shells used, the reasons for their use (excluding economics, which are dealt with in another section), their designation and application will be given. A photograph will show an example of a certain type of shell used for a certain purpose. Other purposes for which this type of shell is often used will be mentioned. Because of the limitations upon space in this chapter, references will be given to photographs in other publications, for the benefit of the reader interested in a particular type of shell, perhaps with a usual or unusual application. These references will be spread, to some extent, internationally to help readers internationally. For example, for a particular type there may be three references quoted; one British, one American and one by a non anglo-american author. Sometimes, alternative references may be given for the same photograph or scheme as the reader will usually be able to locate one easier than one of the others.

1.1 Types of shells

Various types of shells in popular use will be described in this sub-section.

Consider Fig. 1-1; diagram (a) shows the cross-section of an *in-situ* reinforced concrete roof construction, the slab spanning between the beams shown, which are of 10 m span and simply supported. Diagram (b) shows how the latter mentioned slab can be folded so that it is stronger in that, normal to its plane, it carries only a component of the vertical roof loading and not its full amount. It can therefore be thinner. Although the slab is thinner, the overall depth of construction is much greater than that for the beam–slab construction of diagram (a) and it can therefore span further. This is a 'folded plate', 'hipped plate' or

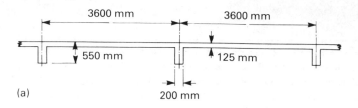

(a)

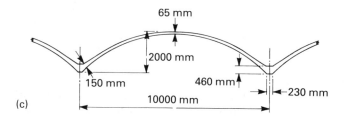

(b)

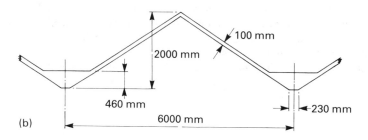

(c)

Figure 1-1   Cross-sections of _in-situ_ reinforced concrete roof constructions

'prismatic' construction, and it spans 20 m (simply supported). Diagram (c) now curves the slab of diagram (a) above to a cylindrical shape (the soffit is cylindrical). Because the slab is arching, it can now be thinner still, much of the roof loading being carried by compressions in the arch, reducing the bending moments. As for diagram (b), its substantial overall depth allows it to span further than the construction of diagram (a). This (c) is a 'cylindrical shell roof' or 'barrel vault roof' and it spans 20 m (simply supported).

In comparing the roofs of Fig. 1-1, it can be seen that the constructions (a), (b) and (c) are mainly 125, 100 and 65 mm thick, respectively. Thus the shell is of a lighter-weight construction capable of spanning further and using less concrete. With roofs, the self-weight of the concrete is the main proportion of the total loading; in cases (a), (b) and (c) the self-weight would be about 83, 80 and 70% of the total loading respectively for the concrete thicknesses of 125, 100 and 65 mm, respectively. Also concrete is a heavy structural material. Therefore, the amount

Figure 1-2   Blackburn College of Technology, England

of concrete is very important when it is required to use it to span larger distances.

The dimensions shown in Fig. 1a-c were assessed approximately using, respectively, Table 7.1, Art. 9.1.2 Proportioning of cylindrical shells and Art. 9.2.4 Design tables for folded plates in [2].

Figure 1-2 shows part of a scheme of cylindrical shell roofs designed by the author. This scheme at Blackburn College of Technology (England) includes many different spans and proportions of 'ordinary cylindrical shell roofs', called 'ordinary' to distinguish them from 'N-light shell roofs'. References [3] and [4] show photographs of an ordinary cylindrical shell over machinery for the manufacture of leathercloth in the UK. Each of these shells has a special feature in that it has steel sleeves through the 'valley beams' (see Fig. 5-1 for description) and bolts projecting from the soffit of the shell to enable light-weight pipes, electrical conduits, pieces of machinery, etc., to be supported from the roof in any location. In this type of factory, the optimum positions of many of these fitments, etc., are decided from time to time after occupation of the factory. The structural steel roof in competition with this shell-roof scheme naturally gave this facility, which was demanded by the mechanical engineers of the concern. Reference [5] shows a photograph of ordinary cylindrical shells used in an unusual way; essentially this roof at St Louis Airport, USA, is the conjunction only of four shells, making an interesting structure architecturally. Reference [5] also shows a scheme of ordinary cylindrical shell roofs, at Barcelona Airport, Spain, without 'end diaphragms' or 'end stiffeners' (see Fig. 5-1 for description), an uncommon feature, just with the valleys (see, Fig. 5-1 for description) tied together at the supports instead of diaphragms. References [6] and [7] show the same photograph of a scheme of prestressed concrete ordinary shell roofs in the UK.

Ordinary shell roofs of the type shown in Fig. 1-2 can span up to about 36 m simply supported, without the need for prestressing.

Overhead cranes have sometimes been supported from 'edge beams' (see Fig. 5-1 for definition) by making a boot lintel feature of the bottom of each edge beam. Otherwise (Denton Gas Works, UK, designed by the author for his former employer), the crane can be supported from corbels from the columns carrying an ordinary cylindrical shell at diaphragm supports and from intermediate columns, for this purpose, between the diaphragm supports and supporting the shell edges.

Figure 1-3 and [3] show a scheme of N-light (north-light) shell roofs designed by the author. This scheme, a printing works for Sharpe's at Bradford (England), was required to maintain an attractive and clean internal environment, the former to impress important visitors and the latter to protect the expensive machinery for high-quality printing which is very sensitive to dirt. This particular scheme had half the normal number of columns in the north-to-south direction. This necessitated top truss members which were kept low down so that normally they could not, for aesthetic reasons, be seen on the top of the shell roofs. In practice, N-light schemes, for reasons of aesthetics and natural lighting, limit substantially the normal span between valleys. That is why the scheme just described halved the number of supports in the north-to-south direction. In the longitudinal direction, the shells were continuous over three spans.

Very large spans have been constructed in reinforced concrete by using large

Figure 1-3 North-light scheme for a printing works in Bradford, England

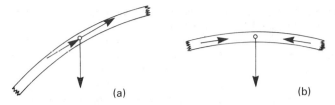

Figure 1-4   In-planar forces supporting loads

arches, the roof between comprising cylindrical shells for lightness in weight as opposed to slabs. (An arch is a very efficient reinforced concrete member for large spans as it tends to carry loading by axial compression rather than bending, often eliminating resultant tension. Concrete is, of course, very economic for carrying compression but not tension, as this has to be resisted by reinforcement, the concrete having cracked and thus become useless for resisting tension.) To keep the. total height of the construction down for aesthetic and economic reasons, the 'stiffener beams' (see Fig. 5-1 for description) are integral with the large arches. Thus, the arches have considerable depths and can span remarkably great spans for reinforced concrete roof construction. One such scheme, an aircraft hangar in S. Dakota (USA) comprises arches of 102 m span and at 7.5 m centres. Between these arches, there is a 127 mm thick shell. References [6] and [7] show the same photograph of a scheme at the Bank of England, in which large arches carry N-light shell roofs, the stiffener beams of the N-light shells being integral with the arches.

In the case of the cylindrical shell, in the curved direction each element of loading is supported by forces, as shown in Fig. 1-4a,b. These are called 'in-planar' forces, that is forces in the plane of the shell. A thin shell is strong at resisting these in planar forces but not strong in bending. The curvature is thus a great asset in providing in-planar resistance to supplement very substantially the weak bending resistance, to carry the loading.

Now, a doubly curved shell has this benefit of curvature in three dimensions. Hence, each element of loading is supported very effectively by 'in-planar' forces in three dimensions. This means that doubly-curved shells are more rigid, can be thinner and thus lighter in weight than cylindrical shells, already described.

Figure 1-5 shows a dome (a doubly-curved shell) over a chemical-waste disposal tank in the UK, designed to resist explosive gases under pressure. An external post-tensioning system, suitably protected against corrosion, was designed and the construction then supervised by the author for the consultants because a mistake had been made in providing inadequate reinforcement in the lower periphery, i.e. to resist 'ring tension', and the dome was spreading out minutely but sufficiently for cracks to develop in the shell of the dome, enabling explosive gas to leak. Reference [5] includes a photograph of a spherical roof covering a building of rectangular shape in plan to give an attractive auditorium in the USA.

Figure 1-6 shows a 'hyperbolic paraboloidal' or 'hypar' shell roof (a doubly-curved shell) over a market in Huddersfield. A different aspect of the same roof scheme is illustrated in [3]. Reference [3] also shows a photograph of a model of this inverted umbrella type of hypar shell tested under the supervision of the author [8].

Figure 1-5   Chemical-waste disposal tank with dome cover, England

Figure 1-6   Hypar shell roof over a market in Huddersfield, England

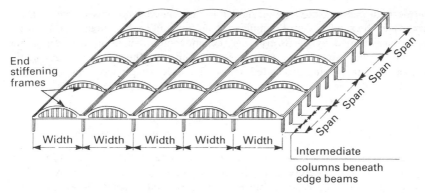

End
stiffening
frames

Width | Width | Width | Width | Width

Span Span Span Span Span

Intermediate
columns beneath
edge beams

Figure 1-7   Scheme of conoidal shell roofs (From: C.B. Wilby, *Concrete for structural engineers*, by permission of Newnes–Butterworth, London, 1977)

References [6] and [9] show photographs of a scheme including four hypars with their lower extremities tied together with a prestressed concrete tie and also a scheme of hypars, where each group of four looks like two folded plates; both schemes are in the UK. In [5] and [10] there are photographs of various interesting types of hypar used in the USA and India, respectively.

Figure 1-7 indicates a scheme of conoidal shell roofs (conoids). A photograph of a scheme of conoidal shell roofs over a motor-car factory in India can be found in [10], which also shows an electrical equipment factory in India whose roof scheme looks as though it is covered with conoidal shells but is actually composed of tilted, ordinary cylindrical shells.

1.2   Designation of shells

A classification of shell roofs was proposed by Bharucha [11] but is too long and comprehensive to include in this chapter. Shells may be classified from various viewpoints [11]: (a) shells of single or double curvature, (b) shells of translation or rotation, (c) shells that can or cannot be ruled, and (d) single form or combination shells. Of these, (a) is regarded as the most important [11] classification and Billington [12] states that the most general classification is by Gaussian curvature. Most engineers are familiar with curvature being the reciprocal of the radius of curvature for two-dimensional structures. For three-dimensional shells, the Gaussian curvature at any point is the product of the principal curvatures [13].

Shells of positive Gaussian curvature [11, 12], 'synclastic shells' are such that the surface curves away from a tangent plane at any point on the surface and lies completely on one side of this plane, e.g., spherical domes and elliptic paraboloids.

Shells of negative Gaussian curvature [11, 12], 'anticlastic shells', are formed by two families of curves, each curved in opposite directions; e.g. hypars, conoidal shells or conoids [13] and hyperbolas of revolution.

Shells of zero Gaussian curvature [11, 12], or singly-curved shells, lie between shells of positive and negative Gaussian curvature, e.g. cylinders and cones.

Classification (b) is fairly popular. Shells of rotation [11, 12] are formed by the rotation of a curve about any straight line. The curve and axis of rotation are normally in the same plane. Examples are domes and tanks. Shells of translation [11, 12] are formed by the translation (i.e. movement) of a curve in one plane such that its ends move along two curves which may be the same or different; the plane of the curve being translated, called the generating curve, always remains parallel to its original position. Examples are cylindrical shells, elliptic paraboloids and hypars.

1.3   Applications

Generally, a prime reason for using shells seems to be aesthetic. The photographs in Section 1.1 and its references indicate a range of aesthetic possibilities. With good architectural treatment, most interesting and attractive-looking roofs can be produced—a delightful change from the box type of buildings which are a bore to our present environment. Shells can easily be made to look attractive. The only type of shell scheme which the author regards as being ugly is that which gives the impression of giant oil drums laid on their sides on the landscape. This type of scheme involves ordinary cylindrical shells covered with black or very dark green roofing felt. The author prefers the avoidance of this type of colour scheme and the use of white, light green or even pink mineral-finished roofing felts. Light-coloured felts also resist transmission of radiant heat better than black or dark coloured felts.

Probably the second most important reason for using shells is that a good robust roof is produced, which can span comparably large spans to those possible with lighter constructions using strong materials, e.g. structural steelwork supporting light-weight roof sheeting. In the UK, a shell will often be 65 mm thick. It will contain four layers of reinforcement in parts; its minimum reinforcement may be two layers of fabric, one near each surface, and a layer of reinforcement between these fabrics. It will be covered with either 25 mm thick corkboard (or 50 mm thick vermiculite) and three-layer built-up roofing felt, the top layer being mineral finished. The soffit may be plastered, or not, and painted with emulsion or distemper. This quality of roof is thus superb compared to any roof sheeting fastened to roof purlins. An alternative reinforced concrete roof of large span not using shell construction would have to be constructed by using prestressed concrete beams, close together and supporting a one-way continuously spanning *in-situ* reinforced concrete (or hollow tile or precast concrete) slab. The overall depth of this construction is much less than that of a shell and its weight is greater, hence it is much less of a practical proposition for spanning large spans because of both strength and deflection limitations. The author once checked the design of a roof comprising *in-situ* post-tensioned rectangular beams cast between hollow concrete caissons. It surprisingly achieved a simply supported span of 17.7 m with an overall depth of only 381 mm but its deflection was considerable and because of this it would not have been acceptable in the UK but it was satisfactory in the country concerned.

Ordinary cylindrical shells have been used where large (for reinforced concrete) spans were required for factories, bus stations, etc. They have also been used where shorter spans would have been satisfactory, for offices, schools, colleges, etc.

North-light cylindrical shell roofs have more limited spans than ordinary cylindrical shell roofs because they are not as strong or rigid in the north-to-south direction and also because of the aesthetic layout required because of the north lighting in the east-to-west direction; for example, if the N-lights are long and deep, one either has undesirably deep glazing areas which become relatively expensive per unit area of glazing or obtains less lighting per unit area covered. They have been built over colleges, schools, waterworks, factories, etc.

Conoids give a good distribution of light over the area covered and have been used as alternatives to N-light cylindrical shells. They are often more attractive, use less materials and are thus lighter in weight but are more expensive for shuttering and steel fixing, and very slightly more expensive for concreting. They have been used, for example, over factories [10] and the testing laboratories at Delft.

Domes have been used greatly over circular tanks for sewage, water and chemical waste (Fig. 1-5). Although bad for acoustics, it was fashionable to use them over certain prestigious libraries constructed in the UK before the Second World War. Domes, not necessarily spherical, have been used over many prestigious buildings, such as exhibition halls, temples, mosques and synagogues.

Cylindrical shells tended to be favoured because the shutter was straight in one direction and conformed to the simplest curve—a circle—in the other direction.

Hypars found favour in that they were and had the advantages of doubly-curved shells (see Section 1.1) and yet could be formed with straight boards or planks. Most are formed with straight boundaries. In time, the fact that the straight boards needed to be tapered or tapered occasionally became rather a disenchantment. However, they are possibly the easiest of the doubly-curved shells to construct. They have been used for many aesthetically attractive structures over factories, garages, markets, churches, exhibition halls, etc.

2 Economics

There are many different types of construction which can be used for roofs. These may sometimes be compared only on first cost if this consideration is so desperately overriding. In this instance, the type of shell satisfactory in the UK would never be less expensive than competitive forms of construction. For the general roof area covered, Big 6 corrugated asbestos sheeting with a fibre-board lining for insulation, or equivalent, would be much cheaper than the 65 mm thick shell described in the second paragraph of Section 1.3. It would also be lighter in weight, helping the economy of the supporting structure.

First cost is not the only practical consideration and often not the most important. The quality of the roof matters with regard to its aesthetics and its resistance to watertightness, heat transmission, fire, small knocks and forces, for example, from maintenance men, craftsmen engaged in internal alterations, etc. The life of the roof and its maintenance cost are also important.

If one wanted a column layout of say 9 m by 18 m and to obtain comparable (even though not as good) qualities to the shell by using wood wool slabs supported by precast concrete purlins and frames and assuming there were say 20 of these bays (9 m by 18 m), then ordinary cylindrical shells would most probably be the cheaper solution of these two alternatives. In addition the shell would be safer for workmen to walk and work upon than the wood wool slabs. It is likely that the shell could be beaten on initial cost in the UK if structural steel work were used with steel purlins supporting say a patent metal sheeting. However, this latter construction would require periodic painting, its fire resistance would be relatively poor and its life would probably be less.

The author designed, estimated and constructed some barrel shells like the ones just mentioned, in the UK. The alternative steelwork solution was slightly less in cost even if one allowed for the cost of painting it for fifteen years. However painting could only be effected in a two week period each year, the works holiday. The people responsible for maintenance found the task of engaging a tremendous number of structural steel painters for this short period in the particular locality, at a time of full/over employment, exceedingly difficult, so the shells were used. All the windows had reinforced concrete frames. The shells were emulsion-painted after construction, and nineteen years later had not been redecorated; their condition was regarded as satisfactory, and during this period nothing was spent on maintenance. This is an accurate case study; a normal profit was made on the shell construction which was well done, and the steelwork price was as expected and accurately assessed.

In the early days of shell-roof popularity in the UK, there was—in the author's experience—a particularly wide variation in contractors' prices for constructing shells. All contractors thought that curved shuttering was a formidable and expensive proposition. At one extreme, some would price very high, either over-reacting to this problem of curved shuttering or being realistic about their firms' lack of ability to be efficient with a type of structure which none of their personnel had tackled before. However, at the opposite extreme, some would believe that this type of construction was a coming thing and that they should gain experience with it, even if they had to pay to some extent for this experience. Hence, they would price very optimistically and not worry too much if there was a risk of making very little profit.

The latter type of contractor would win the contract. Often, he would set about the construction in most unskilled, uneconomic ways and fight hard not to lose money or, at worst, to contain his losses. Then one would find that the next time this contractor quoted for a shell, he would be sky high because he was basing his costing on a most inefficient operation; his first attempt.

It is thus difficult to generalize about costing from case studies unless one knows a great amount of detail in each case, e.g. whether a profit or a loss was made.

Shells became easier to design mainly for the few who knew how to design them and had experience of this work. This knowledge did not spread easily because of the complexity of shell analysis compared to the mathematical limitations of the average designer or engineering graduate. Shells also became easier and cheaper to construct. The curved shuttering became less expensive because various firms specialized in this work.

In the last shell scheme which the author designed, estimated and constructed, the final cost per unit area of the sub-let curved shuttering was the same as the cost per unit area of the flat shuttering. Admittedly, the latter price was not the most competitive, but even so this demonstrates that the cost of curved-shuttering is not excessive. In estimating shells, the author tries to see how the cost will be distributed and where the major expenditure will be incurred. Generally, it is difficult to point to any particular item of extravagance. However, in the author's experience the cost of a valley beam (Fig. 5-1) for a scheme of cylindrical shells tends to be very high indeed. The valley beam plus local thickening of the shell, special reinforcement at this location, etc., can account for about one third of the cost of the whole shell plus the valley beam (excluding end stiffeners and columns). So if it is possible to manage without valley beams, it is better to do so.

The difficulty of generalizing about UK costs becomes worse when attempting to generalize internationally. It is complicated by the fact that there is a different family of less expensive shells which can be used in some countries. These are not in total construction as watertight (maintenance is effected when necessary) and as resistant to heat transmission, fire and localized forces and minor knocks from, for example, maintenance men, etc. These shells are often much thinner than the 63.5 mm (2.5 in) minimum thickness generally used in the UK. Practice in the USA is similar to the UK. In some countries, hypars have been constructed only 32 mm thick and without finishes such as roofing felt, thermal insulation or plaster. In some of these, it is very hot and seldom rains. When it does, the rainfall is often very powerful for a short duration. Then, the sun comes out and rapidly makes the concrete very hot and dry. If the storm exposes any leaks, they are grouted up on the top of the roof. A restaurant, for example, may be inexpensively but attractively decorated before the beginning of each tourist season by distempering or emulsion painting the concrete ceiling; the odd leak causes little difference to the general effect. The floor is tiled, so that any leakage can be quickly mopped up and no damage is done.

The UK and similarly cold and wet areas of the world pose a different problem. In the UK, building surfaces, such as concrete and brickwork, are rarely very hot and dry. They are usually damp and there is so much rain-fall that it takes a long (by UK standards) freak heat-wave to dry out the ground to more than about 150 mm depth. During this process, water vapour is evaporated continuously so that the humidity is high in the best hot weather as well as when it is raining or drizzling—which it does frequently day and night. If a leak occurred in a shell roof in the UK then, usually, expensive finishes under the roof would be damaged. In a restaurant, for example, fitted carpets and underfelts would be damaged. Not only would the architect, structural designer and contractor have to make the roof watertight for the next six years, but they would have to make good the damaged decorations and furniture.

So, in some countries, the standards reasonable for the climatic conditions, way of life, etc., allow shell roofs to be cheaper than in the UK. Also, in some of these countries, the ratio of the cost of structural steel-work to that of reinforced concrete is greater than that in the UK. Furthermore, the cost of labour relative to materials is lower than that in the UK, so doubly-curved shells, for example, which are very efficient at using the minimum reinforced concrete materials for any large span, can be more economic than in the UK.

For work in the USA, Ketchum [14] gives a (two-part) figure—prepared from a report made by his firm, Ketchum and Konkel—relating cost of structure to span, for spans between 9 and 30 m, for bays of 6 and 12 m width, respectively. The comparison is made between ordinary cylindrical shells, hypars and a structure with steel frame, purlins and deck (or covering or sheeting). The cylindrical shells are about 20 and 8% more expensive than the steelwork for the 6 and 12 m bay widths, respectively. However, Ketchum [14] reports that an insurance charge is required in the USA for the steel structure. This then makes the costs of the cylindrical shell and steelwork structures the same for the 6 m bay width; for the 12 m bay width, the cylindrical shells are about 8% less in cost than the steelwork roofs. Ketchum gives a table of quantities for various folded plates, barrel (ordinary cylindrical) shells, hyperbolic paraboloid inverted umbrella shells and hyperbolic paraboloid dome shells. With this table, a reader anywhere in the world can use the various costs relevant to where his structure is to be built and obtain an estimate of the total final cost of each of these types of construction. The roofs were designed for a firm in Denver, Colorado, 'for a uniform live load of 1.436 kN/m^2 (30 psf) so for the smaller live loading used in the south or on the west coast [of the USA] the reinforcing steel quantities may be too high and must be reduced' in the table. The hypars are square in plan and are only considered for spans between 9 and 18 m. Generally, they are about 8% cheaper than the steelwork structures which have a bay width of 12 m.

Ketchum makes remarks for the USA similar to those made earlier in this sub-section regarding the high overall quality of concrete shell roofs relative to alternative roofs in steelwork. He also makes observations for the USA similar to those made earlier in this sub-section on the great variations in prices quoted for shell schemes by contractors.

Whitney [15] agrees with Ketchum and the present author regarding the high quality of shells and cites the further advantages of 'superior wind and blast resistance and low amortization costs'.

Ketchum [14] considers forming (shuttering) costs and concludes that, in the USA,

> Unless forming systems can be devised that will permit a pour of reasonable size with five to six re-uses, then it is better to stick to a single-use form. . . . The trade secret in reducing the cost of single-use forms is to persuade the contractor that he should not write off all the cost of form material against this single job and to design the forms so that there is a minimum of plywood forming material so most of it may be salvaged. A large percentage of the shell structures designed by the writer's [Ketchum's] firm have been built with single-use forms often at the option of the contractor and in preference to movable forms.

This is opposite to the experience of the author both with firms of designers and designer-contractors in the UK. This is probably because timber is very expensive in the UK and also because the labour force and capital investment in shuttering, scaffolding and plant would be less in the UK, the construction time longer and more use and value obtained from the aforementioned capital investment. Based on his experience in the UK, the author recommends that good shutters be made and re-used as many times as possible. Hired specialist shuttering is often economic, even though it has often to be transported up to about 200 miles. This is because the same shutters are used many times. In the UK, the salvage value of timber shuttering is nil: transportation and cleaning of the timber are involved and joiners are very reluctant to make shutters with second-hand timber—it takes them much longer and the shutters thus cost more to fabricate.

However, Whitney [15] cites USA experience favouring as many uses as possible being obtained from the same shutters.

Ketchum [14] seems to be particularly impressed with the low cost of hypars in the USA. Considering a column grid 12×12 m and taking the cost of a hypar umbrella shell as 100 units per unit area covered, then the pro-rata cost for various structural systems would be: timber frame 91, steel frame with wood joists 78, steel frame with wood purlins 77, steel frame with steel purlins and deck 104, steel frame with open web joists 90, and prestressed concrete 101. This comparision favours hypars when the maintenance cost and relative qualities are taken into account.

When the author worked for a firm of designer-contractors he designed and estimated certain schemes for precasting shell roofs, and is convinced that, in the UK, any precast shell would be much more expensive than an *in-situ* shell. This is because of the costs involved in precast work which do not occur in *in-situ* work, namely, factory overheads, handling, stacking at factory, loading, transportation, unloading, stacking at site, and erection. Ketchum's paper [14] seems to indicate less difference in price between *in-situ* and precast construction for the USA. Generally, *in-situ* appears to be slightly cheaper than precast in the USA but Ketchum has the opinion that barrel shells of spans up to 15–18 m might be economic to precast.

Earlier in this sub-section, it was mentioned that not many designers, because of their mathematical limitations, became able to design shells. Ketchum comments similarly for the USA and goes on to say 'shells are not used in many situations where they could be the best solution'. The author's UK experience fully agrees with this statement.

Whitney [15, 16] gives examples of the costs of shells in the USA. Generally, these favour shells compared to alternative constructions for long spans, particularly those requiring good quality roofs. Whitney [15] concludes with a good piece of advice:

> The conception of an economical concrete shell structure must be based primarily on construction considerations. After it has been determined how it can be built most cheaply, the structural design follows logically. The cost lies not so much in the quantity of concrete and reinforcement steel needed, as in the cost of the process of getting them into place.

The author has experienced an economic use of shells where large concrete spans were required over a service reservoir. Large spans were used because of bad ground conditions, i.e. the fewer columns there were, each of which required an expensive foundation, the less was the total cost of the scheme.

In many underdeveloped countries, labour is less expensive relative to materials than in the more advanced countries. This point is made by Chatterjee [17] with regard to India. He gives details of factors affecting the costs of a considerable range of shells constructed in India and these should be of use to those designing and/or constructing shells in underdeveloped countries. The shells quoted range in thickness from 50 to 100 mm.

3 Design

Designing shells comprises assessing their dimensions and then analysing them elastically for internal forces and moments.

Recent British and US codes of practice have superseded the designing of sections by elastic theory only, but these and other codes do not deal with shells. It can be argued that there is insufficient research and practical experience to make the change from the long established practice of designing shell sections elastically, to resist forces and bending moments obtained by elastic analysis, to a limit state analysis like the present British and US codes of practice recommendations for reinforced concrete slabs, beams and frames.

The established practice of shell design [2, 3, 10, 12, 13, 18, 19] has been found to be satisfactory and reliable over a period of many years. Where cracks have occurred, this has led to local thickenings [4] being used. The elastic analysis endeavours to control cracks by limiting steel and concrete stresses. Deflections can be calculated from the elastic analysis and these should always be considered when, for example, walls and windows may be damaged by shell deflections. In the case of windows or fragile walls, etc., built up to the shells, sliding arrangements need to be devised so that shells can deflect freely without damaging them. The established practice has a good record with regard to safeguarding against failure. The author does not know of any failures where proper elastic design has been used; indeed, all the indications are that shells designed elastically are particularly robust and, even though we probably would not design them thinner than 64 mm in the UK, they often seem to be conservatively designed at this thickness with regard to ultimate strength.

If, for a shell of the dimensions assessed, the reinforcement is too heavy to fit into the sections, or if the concrete stresses or deflections are too great, then the dimensions of the shell have to be altered and the analysis repeated until a satisfactory solution is found.

The dimensions can be initially assessed from experience and/or looking at similar shell solutions reported in architectural and structural engineering publications, and/or using approximate designs such as the beam analogy [20], or the beam–arch approximation, for cylindrical shells.

For ordinary and N-light cylindrical shells, [2] and [19] give proportions for shells, in British Imperial (USA) and SI (metric) units, respectively, where the width-to-length ratio is 1 to 2. This ratio is regarded as an optimum, bearing in mind both economy and appearance. Reference [19] then gives an approximate but fairly accurate method of estimating the reinforcement in these shells. This method yields the kind of accuracy required by firms which design and supply reinforcement. Contractors and consultants do not usually require this accuracy, as the item in the bill of quantities will be subject to final measurement, whereas the designer–suppliers usually quote a lump sum price for the reinforcement, so [2] gives a simple-to-use group of graphs, for estimating the reinforcement. This is reproduced in Fig. 3-1. Curves SH and AH are for normal and N-light roofs respectively, where high-yield reinforcement is used apart from mild steel for L-bars [5], stirrups and column bars. Curves SM and AM are for normal and N-light shells respectively, where mild steel is used throughout except for shell fabrics and their spacer bars [5].

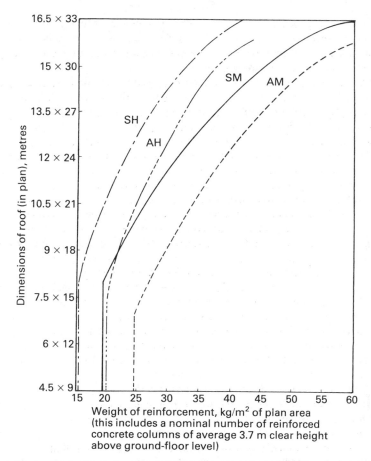

Figure 3-1 Graphs for estimating the reinforcement in cylindrical shells. Curves SH and AH are respectively for normal and north-light shells using high-yield reinforcement apart from mild steel for *L*-bars, stirrups and column bars. Curves SM and AM are respectively for normal and north-light shells using mild steel throughout except for shell fabrics and their spacer bars (From: C. B. Wilby, *Concrete for structural engineers,* by permission of Newnes–Butterworth, London, 1977)

Reference [14] includes a table giving quantities of steel and concrete for typical interior bays of shell roof structures: (a) barrel shells 15–30 m span, width 9 m and thickness 89 mm; (b) hyperbolic paraboloid inverted umbrella shells, 9–18 m square, and of thickness 57 mm; (c) hyperbolic paraboloid dome shells, 12–30 m square, and of thickness 57 mm.

4 Analysis

A common way of analysing a shell is to consider an infinitesimal element of it with all the forces (longitudinal and shear), moments and torsions in three dimensions which could conceivably occur on the element; in other words, a differential element. These forces and moments are equated to one another by general equations of static equilibrium, and are related to strains, stresses and displacements, e.g. see [12], or for cylindrical shells only [2].

When the differential equation finally obtained is applied to the case of a uniformly distributed loading, the result is very similar to that obtained from a much simpler statically determinate membrane analysis. This latter is an analysis which assumes the shell has negligible bending rigidity and that all the loading is carried by in-planar forces only, i.e. forces and shears in the plane of the shell. For example, Billington [12, pp. 171–172] shows for a cylindrical shell the negligible error involved in using the membrane theory. This is a mathematical explanation. A physical explanation would be to imagine, say, a cylindrical shell as analogous to a large number of folded plates [2]. Now, the loading is carried transversely between folds and the consequent reactions at each fold are resisted by the adjacent two plates acting as beams. If the folds are close together, analogously to a cylindrical shell, then the transverse bending moments are negligible for the negligible spans between folds and the loads are carried entirely by plate action, i.e., in-planar forces.

The loadings considered for shells are often approximated to uniformly distributed loadings. In the UK, it is common to consider the snow load of 0.75 kN/m² ^{2} (16 psf) of horizontal area as a load per unit of curved surface area, so that it can be dealt with in the same way as the self-weight loading. Wind loading is often suction and not as great as the self-weight of a reinforced concrete (not a laminated timber) shell so that it can normally be ignored in the UK; similarly for the USA. Refer to Section 2 for points regarding the use of thinner shells in other countries and the loadings used in various parts of the USA.

Thus, for practical purposes, shells are often just analysed for a uniformly distributed loading per unit of curved surface area, comprising self-weight and snow load, and to begin with this loading is assumed to be carried by the membrane analysis (i.e. by in-planar forces only). This analysis would only be adequate if, at the boundaries where a shell became discontinuous, the supports provided forces equal and opposite to the membrane forces in the shell at these locations and if the support displacements were the same as those determined at these locations by the membrane theory applied to the shell. Not many practical shells adhere adequately to these conditions. For example, ordinary and N-light cylindrical shell roofs with edge beams (see Fig. 5-1 for description) free to deflect vertically and horizontally do not, but a fairly impractical shell, a hemisphere, supported vertically only by, say, a tank wall all the way round, could be analysed by membrane theory only.

For the majority of shell roofs which cannot be analysed by membrane theory only, therefore, after the membrane theory has been used for determining the in-planar forces, the edge displacements, displacements in three dimensions, are determined according to this membrane theory. The support to the edges of the shell, be it an edge beam, ring beam, etc., has to resist the membrane forces at the edge of the shell in addition to its self-weight. In doing this, it will, almost without exception, realize different displacements to those of the shell edge. As the edge or ring beam, etc., and the shell do not part company, unknown equal and opposite forces must be considered as acting between them by the analysis. For the shell, these are called 'edge loads' or 'line loads'. They are line loads occurring at the shell edges or boundaries. Now, whereas we said previously that uniformly distributed loads over the whole shell surface could be resisted with negligible error by in-planar or membrane forces, these edge loads cannot be considered as being resisted by in-planar forces without significant error. They cause beinding moments, torsions, longitudinal and shear forces in three dimensions and the analysis is indeterminate for edge loads, not determinate as it was for membrane forces. To simplify the analysis, unit line or edge loads are considered one at a time and in each case the forces and moments at various locations in the shell and edge displacements are determined.

Also, unit line loads equal and opposite to the unit edge loads just mentioned are applied to the edge or ring beam, etc., and the displacements obtained at the junction with the shell in three dimensions. Bending moments, etc., for the design of the edge member when the unknown line forces are determined later can be determined at this stage.

Then, equations are constructed to obtain the unknown line loads to match up the displacements of the shell edge and edge or ring beam, etc., in three dimensions.

Once these equations are solved, the forces and moments due to the edge loads on the shell are, to obtain resultant forces and moments for our design, added to the membrane forces due to the uniformly distributed shell loading. The edge members are designed for these line loads as well as to carry the forces to resist the edge shell membrane forces and the self-weight of the edge members.

The designer without prior knowledge of shell analysis/design will appreciate that the above is arduous and complicated out of all relation to the simplified design of slabs and frames, e.g. [2]. Thus, if need be a shell has to be designed in a tightly limited time (i.e. not allowing time for considerable study), there must be recourse to specialists and/or design tables and/or computer programs (see later).

If a designer were to ignore the difference in displacements of a shell and its supporting edge member due to the shell carrying its total uniformly distributed loading by membrane forces, and the edge member carrying its self-weight and providing reactions to the membrane forces at the edges of the shell, he would obtain a structure where the equilibrium of the forces is all right but the ignoring of the displacements would require considerable redistribution of the forces, involving much cracking. Hopefully, the redistribution would not be so severe as to require complete disintegration of one portion to allow the necessary forces to be carried by another portion. It would be wrong, according to the author's experience of tests, to assume that such considerable redistribution can easily or normally take place. The author has experienced the sensitivity of various types of shells to different modes of failure; for example, a shell can fail in one direction without being anywhere near failure in another direction.

However, many doubly-curved shells, particularly the hypar, have been designed in this way, often without even assessing any displacements. The amount of cracking which occurs depends upon how capable the designer is at placing nominal reinforcement, extra to the membrane design requirements, in the places where the shell and/or edge member is most likely to crack. Then, there is the consideration of whether the cracking really matters if the structure is a piece of sculpture (hypars have been used in this way) or a roof over a building in a very hot country (see Section 2 regarding acceptable occasionally leaking roofs).

Although the analysis of a shell is commonly split into a membrane and edge (or line) load problem, the author found it necessary to keep the problem as one combined analysis when using the electronic analogue computer for analysing cylindrical shells, see [2, 4, 21, 22].

5 Construction

The contractor needs to know where he may make joints in the structure. There are two types of joint, namely, expansion and construction joints.

Expansion joints are made so that the structure is not too long in any horizontal direction as to be troubled by temperature movement. The coefficient of thermal expansion of both concrete and steel is about 0.000009 per °C. A reinforced concrete member of length, say, 20 m, in a UK location where the maximum annual range of temperature is 30°C, would experience a change in length of 5.4 mm, or nearly 3 mm at each end. If, for example, a cylindrical shell of this length were constructed between two existing buildings, then sheets of a special compressible material can be fastened to the existing walls in suitable locations and the shells cast against these sheets.

For a scheme of ordinary cylindrical shells, if the transverse end or stiffener beams (see Fig. 5-1, for descriptions) have flat soffits, then every, say, 36 m, a sheet of compressible board about 25 mm thick will be used to make a discontinuity in the end stiffener and this will be continued to split the valley or valley beam for about 0.1 of the span from each end. Where each end stiffener beam is jointed, the column will be doubled. These can be constructed by casting one column, fastening 25 mm thick compressible board to the relevant face and then casting the second column against this board. The crowns of the central parts of the very flexible shells just rise and fall in accordance with temperature variations. This was not possible for the rigidly straight soffited end stiffener beams. Had the end stiffeners been slender arches and the crowns of these had been able (i.e. without damage to walls) to rise and fall with temperature movement, then they may have continued for a distance of, say, 60 m without an expansion joint. This joint would then be taken completely through the structure (i.e. splitting the whole valley or valley beam).

Thus, where curvature allows the temperature movement of a shell and/or arch to be accommodated by rise and fall at its crown rather than by creating a tremendous horizontal force, then expansion joints can be placed further apart than would happen with straight work.

The distance between expansion joints for reinforced concrete buildings varies somewhat between different designers and can depend upon the layout of the building in plan. The author's experience has generally been to place expansion joints at about every 36 m for buildings of rectangular shape in plan. Thus, for cylindrical shells, expansion joints should be made every 36 m or less in the straight direction, and the same in the direction of the stiffener beams if they have flat soffits but, say, every 60 m for flexible construction as just described.

Similar ideas can be applied to doubly-curved shells. An inverted umbrella type of hypar has straight horizontal edges, therefore joints are required every 36 m or less, whereas a spherical dome is flexible, and so the joints can be more widely spaced. In these cases 'joints' are shell boundaries.

Construction joints serve two purposes. First, they enable the structure to be cast in practical amounts; secondly, they relieve shrinkage stresses [2]. Designers in the UK specify and detail expansion joints but commonly ignore construction joints or deal with them by using rather loose general statements. The designer usually cannot know where the contractor eventually appointed will consider the construction joints should be made for optimum efficiency. The contractor may often agree the position of construction joints with the designer or ask the designer where he should make them.

Although expansion joints for temperature movement automatically relieve shrinkage locally to some extent, they do not relieve shrinkage generally. But shrinkage [2, 19] is different from temperature movement in that the concrete moves but the steel does not, except for a small amount of compression strain induced in it, away from shrinkage cracks, by the concrete shrinking. Taking a shrinkage coefficient of 0.0005, from and with the reservations of [2], then for a reinforced concrete member of length 20 m, the concrete will want to move 10 mm, or 5 mm at each end (cf. a temperature movement of 3 mm cited at the beginning of this section) to avoid developing internal tensile stresses, and ideally the concrete should freely slide along the reinforcement. In practice, this cannot happen. The shrinkage coefficient of 0.0005 multiplied by a value of Young's modulus for concrete of, say, 28 000 MPa (4.06×10^6 psi) [2] would mean a tensile stress of 14 MPa (2030 psi) developing in the concrete if the reinforcement completely restricted its shrinkage. Now, the ultimate tensile stress of this concrete would be, say 2.8 MPa (406 psi) [2] so it would certainly crack. In practice, cracks occur periodically at right-angles to a reinforcement bar, the bond [2] stress between bar and concrete being nil at a crack (slip having occurred locally) and having a maximum value half-way between two cracks. Bond resistance depends upon various factors [2] and is greater for bars which mechanically bond to the concrete. (Such bars are now widely used in the UK.) Then there is the problem of differential shrinkage. When a thin shell dries out in a drying wind and sunshine, the surface concrete shrinks very much faster than the concrete in contact with, say, a steel shutter, where the moisture will be well retained and thus this concrete may not shrink at all. This action can cause deep surface cracking, some of which may at some later date, after stripping the soffit shutter, penetrate the complete thickness of the shell.

The author recalls studying various mathematical works about 1948 concerning a differential element and partial differential equations for trying to predict shrinkage stresses, etc. These analyses have never given practically useful results nor have they agreed adequately with experimental behaviour. This is because of the already mentioned practical difficulties. For practical design purposes, therefore, the elastic shell analysis has been used ignoring temperature and shrinkage effects. Temperature and shrinkage effects have been dealt with adequately enough to obtain satisfactory shells in practice by detailing expansion and construction joints, respectively. It should be remembered that shell construction led the analytical and experimental work which the more academic would have deemed adequate to justify construction. This is certainly not an unusual way of making progress, judging from history and present practice in many technologies. Analyses can, of course, be concocted for both temperature and shrinkage effects but have not been necessary to produce satisfactory shell structures.

With all reinforced concrete structures, it is to some extent debatable where to place construction joints to do the least harm to the structure. Ideally, there should be no joints at all, yet they can help in reducing shrinkage stresses to some extent. What are the principles? Of first importance, a construction joint should be made where the shear stresses are the least and then, of second importance, where the bending moment is a minimum. In a beam, there are horizontal as well as vertical shear stresses. For economy of construction, a T-beam, where the top flange is an *in-situ* slab, is often jointed just below the slab, where the horizontal shear stresses are a maximum, hence suitable stirrups have to be present to take this horizontal shear which the joint cannot be relied upon to resist. Similarly, joints have to be made in shells at undesirable locations and suitable dowel bars used to make up for the weakness of the joint.

Sometimes, a joint may be not straight but provided with, say, a tongue to help key and resist shear.

When a joint is at a section of high bending moment, there is no difficulty as regards bending strength. In flexural compression, the joint causes no weakness, the surfaces just push together and take compression. In tension, the reinforcement takes the tension but the crack size will be increased at the joint. So, joints can be made where there are bending moments but ideally, if the bending moments are less, the crack size will be more favourable.

Shells tend to crack at changes of shape, e.g. at the junctions between shells and edge beams. Hence, shells are usually thickened [2, 4] towards these junctions and construction joints made where the shells begin to thicken, well away from the junctions. Thus, an edge member is cast integral with the thickening, and starter bars protrude from the end of the cast to lap with the bars of the shell when it is cast and to act as dowel bars to strengthen the construction joint.

The shell itself will often be cast in portions to reduce shrinkage stresses and to enable construction. A thin shell can dry out very rapidly if a drying wind and/or strong sunshine occur just after casting. This results in a very high shrinkage in the liquid-cum-solid state and can cause very wide and deep cracks say 10 minutes after placing, i.e., well before the concrete is hard enough to cover with a polythene sheet. The author has experienced this kind of severe cracking in the UK, just because the sun came out whilst trowelling.

Construction joints are placed along the length of edge beams, valley beams, end arches, etc., in the same way as for normal reinforced concrete construction, as described earlier in this section.

Figure 5-1 shows a suitable system of construction joints (broken lines) for a scheme of barrel shells.

The shuttering of shells is the most important constructional problem. Some pertinent points have already been made in Section 2 regarding the re-use of formwork. However, for accurate work (say to ±3 mm), shell formwork needs to be strong and well made to withstand steel fixing and concreting operations without suffering distortion. This means that it is expensive and capable of lasting for many jobs (hired formwork is an extreme example of multiple re-use). Therefore, generally the objective is to obtain as many uses as possible out of each shutter. Again, as described in Section 2, the author has found the use of specialist shuttering sub-contractors economic.

One such system comprises T-shaped steelwork sections which, for a barrel shell, are bent to the required radius at the works and then sent to the site

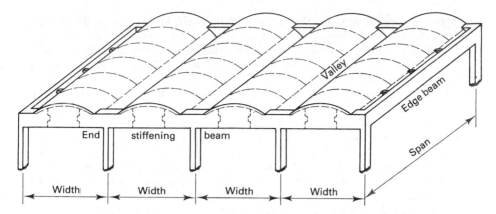

Figure 5-1 System of construction joints (broken lines) for a scheme of barrel shells (Adapted from: C. B. Wilby, *Design graphs for concrete shell roofs,* Applied Science Publishers Ltd, London, with permission)

together with the necessary steel plates. These plates are laid on the T-sections so that adjacent plates abut on the top flange of each T-section and at their unsupported sides. After casting, a small whisker of mortar between each plate pulls off the soffit; these locations can be improved by minimum rubbing with a carborundum stone. This soffit is good enough for emulsion or distemper painting and the slight pattern of the plates is aesthetically pleasing. Another system of hired shuttering uses plywood instead of steel plates.

With a scheme of barrel shells (Fig. 5-1), it is possible to cast the first shell and the first valley, terminating the concreting at the first longitudinal construction joint in the second shell just after the first valley. The framework of the shuttering needs to be in place for the second shell but it only requires a portion of the soffit sheeting. Then, when the concrete is sufficiently hard, the shuttering from the first shell can be moved to the third shell and the soffit sheeting removed from it to complete the shuttering of the second shell and to shutter a portion of the third shell. Then, the second cast completes the second shell and casts a portion of the third shell. Whilst the operation is continued, it is necessary always to keep the last cast valley propped. This is a way to maximize the use of the shuttering for multi-barrel shells. With this scheme, when the end stiffeners have horizontal soffits level with the valley soffits, the deep inner shutters to the sides of the end stiffeners can be lifted out of the large, long strip voids required for metal-framed lantern-lights at the crown. But if the lights are individual—see, for example, frontpiece photograph in [4] or Plate 3 in [3]—and say about 1.2 m square or circular, then this is not possible and the shutter has to be made in three or four pieces which will unfasten easily so that they can be man-handled beneath the cast shell into their next position.

For short wide shells continuous in the span direction, moving shutters are sometimes used. This was done over a turbine house in the UK, where movable steel shuttering used crane rails which were supported by reinforced concrete beams supported by reinforced concrete brackets from the columns. These crane rails were later to support an overhead crane. As the headroom was considerable, this arrangement saved an enormous amount of scaffolding. At each support, there was an arch, therefore the shell shutter had to be lowered below this, moved forward and then jacked up again.

Sometimes, shells continuous in the longitudinal direction are favoured with upstand diaphragms at the supports, so that a movable shutter can very easily slide forward with the minimum difficulty.

In the early days of shells in the UK, cylindrical shells were favoured because the shuttering was straight in one direction and to the simplest curve for setting out—namely a circle—in the other direction. They were also thought to be much simpler than doubly-curved shells to construct and design.

In Mexico, Felix Candella created many architecturally very attractive structures with hypars. He appeared to demonstrate that many of these were doubly-curved shells formed by straight planks supported on straight boundaries, had the strength advantage of doubly-curved shells over singly-curved cylindrical shells and were adequately designed by very simple analyses, much simpler than had been developed for cylindrical shells.

Although the shuttering comprises straight planks, these need to be tapered. Ripping down the whole length of each plank is very expensive. This cost can be reduced by tapering more severely only one in every group of four or five planks. However, after this kind of procedure had been going on for some years in the UK, the author came across one contractor who preferred to shutter hypars the curved way for economic reasons, i.e., obliterating the principal economic reason for using hypars in the UK.

If the valley of a cylindrical shell were cast strictly horizontally, for optical reasons it would appear to sag. There would be, of course, a slight deflection due to self-weight, but it would be exaggerated visually, Aesthetically this is undesirable, so it is best to calculate the total deflection under live and dead loading and to give the valley shutters slightly more camber than this quantity. Often a rule-of-thumb guide can be used, such as that given in Table 5-1, where OCS and NCS are abbreviations for ordinary and north-light cylindrical shells, respectively.

Casting concrete shells is more difficult and consequently less economic as the surfaces are steeper. In the UK, a $1:1.5:3$ mix by dry volumes of cement:sand:gravel (10 mm down, i.e. maximum size of particles is 10 mm) has been found to be satisfactory for shells. The concrete is built up from the lower parts of a shell, full thickness as one goes, until the crown is reached. The concrete is as dry as possible consistent with full compaction. Poker and sausage vibrators may be put in the concrete to help the placing around reinforcement and in difficult corners etc., but these will only be able to compact by vibration a very small amount of the concrete. So, essentially the concrete will have to compact itself easily under the weight of patting with a shovel and trowelling only. To compact easily, it must not be too dry. As the concrete is built up the slopes, wetness will rise to the top and it may be possible to mix the concrete drier because it will receive

Table 5-1

Span, m	15	18	24
OCS camber, mm	25	35	50
NCS camber, mm	35	50	75

wetness from the already laid concrete when worked into position. Should it unexpectedly rain during concreting, the concrete added toward completion may need to be of an exceedingly dry mix because, by the time the rain has mixed with it during placing and wetness has risen from the previous concrete, the freshly laid concrete is adequate in water content. The shell may contain two layers of fabric reinforcement, one near each face. To prevent the concrete flowing easily down the shell slopes, extra 'spacer bars' [19] are used which not only help the casting but also help the fixing of the fabrics, strengthen the end laps of the fabrics and act as bars to help distribute cracks due to shrinkage and temperature, i.e. encourage a large number of smaller cracks. For maximum bulk for cost, the author found that the square twisted bar was the most economic and had a mechanical bonding action. A guide to the choice of spacer bars [19] for cylindrical shells is as follows: for shells up to a span of 18 m, use 8 mm square twisted (or 10 mm diameter) bars at 380 mm centres; for shells of longer spans, use 10 mm square twisted (or 12 mm diameter) bars at 380 mm centres. Spacer bars are briefly mentioned in [2] and [3]. The edge and stiffener beams and columns are cast with the same type of specification as non-shell work.

Reinforcement fixing tends to be thought of as difficult because of the sloping curved shell surfaces. However, the cost of steel fixing is very small compared to the total cost of the contract and the difference in cost between fixing reinforcement on shell surfaces and to, say, normal slabs is not very great. Consequently, the overall cost effect is fairly insignificant in the UK with skilled fixers.

Thermal insulation on top of the shell is, in the UK, usually either 50 mm thick vermiculite concrete or 25 mm thick corkboard. A good practical tip with the former is to waterproof the surface with about a 3 mm thick layer of cement–sand. This is because, if it rains before the roofing felt is stuck to its surface, and this is not waterproofed, the vermiculite concrete will be saturated. Then, in hot weather, the water evaporates, causing the roofing felt to bubble and become unstuck. In any case, every precaution is taken to lay the roofing felt during a dry period—but this can be difficult in unpredictable climates, such as the UK's.

Corkboard does not have this trouble. It is stuck down with bitumastic by those also laying the roofing felt. The felting follows immediately. All this is effected rapidly when the concrete shell is dry.

The author has found that laying corkboard is to the benefit of the subcontractor, and that laying vermiculite rather than corkboard is more economic for the concrete contractor because it dispenses with the subcontractor's services. Even to a consultant specifying the respective duties of the contractors, the corkboard tends to be more expensive overall, but this should be checked as prices vary considerably from time to time.

6 Generalized membrane analysis

The reasons for desiring a membrane analysis are described in Section 4. It is usual—see for example [5, 23, 24, 25]—to consider three membrane analyses: (a) cylindrical shells, (b) shells of revolution (e.g. those used for spherical domes), and (c) shells of arbitrary shape (e.g. those used for hypars and conoids).

6.1 Cylindrical shells

The membrane analysis is only part of the total analysis for cylindrical shells (see Section 4) and can very rarely be used on its own. This analysis assumes that Poisson's ratio is zero.

Figure 6-1 shows a differential element of the surface of a cylindrical shell roof. Consider the longitudinal force per unit length of shell, or 'stress resultant', N_x at $x = x$; at $x = x + dx$, this increases to

$$N'_x = N_x + \frac{\partial N_x}{\partial x} dx$$

Similarly, $N_{\phi x}$ at $\phi = \phi$ increases to

$$N'_{\phi x} = N_{\phi x} + \frac{\partial N_{\phi x}}{\partial \phi} d\phi$$

at $\phi = \phi + d\phi$. Similarly, for the other stress resultants shown in Fig. 6-1.

The total loading on the element has components X, Y and Z, as shown in Fig. 6-1, all taken as forces per unit area. (X and Y are in, and Z normal to, the plane of the element.) Resolving in the x-direction,

$$\left(N_x + \frac{\partial N_x}{\partial x} dx\right) r \, d\phi - N_x r \, d\phi + \left(N_{\phi x} + \frac{\partial N_{\phi x}}{\partial \phi} d\phi\right) dx$$

$$-N_{\phi x} \, dx + Xr \, d\phi \, dx = 0 \quad (6\text{-}1)$$

where, as N_x is a force per unit length and the length of the side of the element upon which it acts is $r \, d\phi$, it gives a force on this side of the element of $N_x r \, d\phi$, and as $N_{\phi x}$ is a force per unit length and it acts on the side of the element of length dx, then it gives a force on this side of $N_{\phi x} \, dx$; and similarly for the other stress resultants; again X is force per unit area and the area of the element is $r \, d\phi \, dx$, so this gives a total force of $Xr \, d\phi \, dx$.

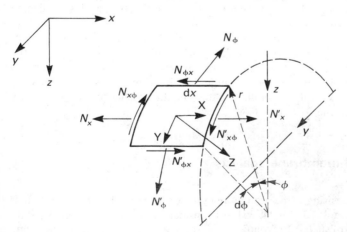

Figure 6-1 Differential element of the surface of a cylindrical shell roof

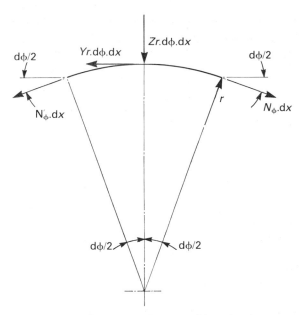

Figure 6-2 Forces to be considered when resolving perpendicularly to plane of element

Cancelling like terms and dividing throughout by $r\,d\phi\,dx$ gives

$$\frac{\partial N_x}{\partial x} + \frac{1}{r}\frac{\partial N_{\phi x}}{\partial \phi} + X = 0 \qquad (6\text{-}2)$$

Resolving in the y-direction gives a similar equation to Eqn 6-1 which, with cancelling and dividing throughout by $r\,d\phi\,dx$, gives

$$\frac{\partial N_{x\phi}}{\partial x} + \frac{1}{r}\frac{\partial N_\phi}{\partial \phi} + Y = 0 \qquad (6\text{-}3)$$

Figure 6-2 shows the forces to be considered when resolving perpendicularly to the plane of the element, namely,

$$\left(N_\phi + \frac{\partial N_\phi}{\partial \phi}d\phi\right) dx \sin\left(\tfrac{1}{2}d\phi\right) + N_\phi\, dx \sin\left(\tfrac{1}{2}d\phi\right) + Zr\, d\phi\, dx = 0$$

In the limit, as $d\phi \to 0$, $\sin\left(\tfrac{1}{2}d\phi\right) \to \tfrac{1}{2}d\phi$. Therefore

$$N_\phi\, dx\, d\phi + \frac{1}{2}\frac{\partial N_\phi}{\partial \phi}dx(d\phi)^2 + Zr\, d\phi\, dx = 0$$

Therefore

$$N_\phi + \frac{1}{2}\frac{\partial N_\phi}{\partial \phi}d\phi + Zr = 0$$

In the limit $\partial N_\phi/\partial\phi \to 0/0$, which is neither 0 nor ∞ but a normal quantity, and $\mathrm{d}\phi \to 0$, therefore

$$N_\phi + Zr = 0 \tag{6-4}$$

Then, taking moments about the normal to the centroid of the element,

$$(N_{x\phi}r\,\mathrm{d}\phi)\tfrac{1}{2}\,\mathrm{d}x + (N'_{x\phi}r\,\mathrm{d}\phi)\tfrac{1}{2}\,\mathrm{d}x = (N_{\phi x}\,\mathrm{d}x)\tfrac{1}{2}r\,\mathrm{d}\phi + (N'_{\phi x}\,\mathrm{d}x)\tfrac{1}{2}r\,\mathrm{d}\phi$$

Dividing throughout by $\tfrac{1}{2}r\,\mathrm{d}\phi\,\mathrm{d}x$ gives

$$N_{x\phi} + \left(N_{x\phi} + \frac{\partial N_{x\phi}}{\partial x}\,\mathrm{d}x\right) = N_{\phi x} + \left(N_{\phi x} + \frac{\partial N_{\phi x}}{\partial\phi}\,\mathrm{d}\phi\right)$$

In the limit, $\partial N_{x\phi}/\partial x$ and $\partial N_{\phi x}/\partial\phi \to 0/0$ and $\mathrm{d}x$ and $\mathrm{d}\phi \to 0$, therefore

$$N_{x\phi} = N_{\phi x} \tag{6-5}$$

which is similar to the well-known equality of complementary shear stresses for two dimensional work, e.g. beams. The four equations 6-2 to 6-5 can be solved for the four stress resultants N_x, N_ϕ, $N_{x\phi}$ and $N_{\phi x}$.

With regard to longitudinal strain, the element of length $\mathrm{d}x$ (Fig. 6-3) has a longitudinal displacement of u at $x = x$ and of $u + (\partial u/\partial x)\,\mathrm{d}x$ at $x = x + \mathrm{d}x$. Hence, the change in length is $(\partial u/\partial x)\,\mathrm{d}x$ and the strain $\partial u/\partial x$. If the shell thickness is h, the corresponding stress is N_x/h. Therefore, using Hooke's law and assuming Poisson's ratio is zero,

$$\partial u/\partial x = N_x/(Eh) \tag{6-6}$$

where E is Young's modulus for the concrete.

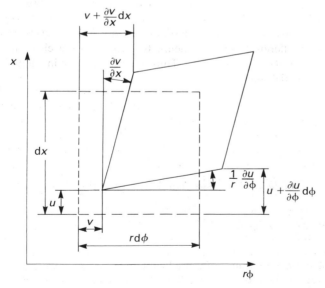

Figure 6-3 Longitudinal strain

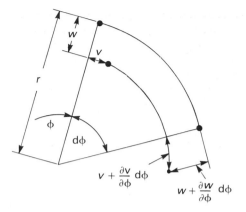

$$v + \frac{\partial v}{\partial \phi}\, d\phi$$

$$w + \frac{\partial w}{\partial \phi}\, d\phi$$

Figure 6-4 Tangential strain

Figure 6-3 shows the element distorted by shear stress. From this figure, the shear strain is

$$\frac{1}{r}\frac{\partial u}{\partial \phi} + \frac{\partial v}{\partial x}$$

As the elastic shear modulus is equal to $E/2$ and the shear stress is $N_{\phi x}/h$ or $N_{x\phi}/h$ then, applying Hooke's law and assuming Poisson's ratio is zero,

$$\frac{1}{r}\frac{\partial u}{\partial \phi} + \frac{\partial v}{\partial x} = 2\frac{N_{\phi x}}{Eh} \tag{6-7}$$

The tangential strain can be considered by reference to Fig. 6-4. The increase in length of the element due to the tangential displacements v and $v + (\partial v/\partial \phi)\, d\phi$ of its ends is $(\partial v/\partial \phi)\, d\phi$. Because of the radial displacements, w, of these ends (neglecting the very small difference between them), the length of the element decreases by an amount $r\, d\phi - (r - w)\, d\phi = w\, d\phi$. Thus, the total change in length of the element in the tangential direction is

$$(\partial v/\partial \phi)\, d\phi - w\, d\phi$$

Dividing this by the initial length $r\, d\phi$, the strain is

$$\frac{1}{r}\frac{\partial v}{\partial \phi} - \frac{w}{r}$$

The tangential stress is N_ϕ/h, hence, assuming Poisson's ratio is zero and applying Hooke's law,

$$\frac{1}{r}\frac{\partial v}{\partial \phi} - \frac{w}{r} = \frac{N_\phi}{Eh} \tag{6-8}$$

Now we have to explain the importance of using load harmonics (i.e. expressing loads and forces in terms of Fourier's series). It is difficult to explain to most engineers why this has to be done when a uniformly distributed load seems so simple to deal with. Indeed, beginners designing shells often just accept the use of load harmonics without really understanding why this procedure is necessary.

According to Evans and Wilby [19], under the heading Load harmonics: 'To solve the various mathematical equations employed in the design of shells it is necessary to express the loading on the shell in terms of a Fourier's series'.

And Chronowicz states [26]: 'The analysis of the deformations of the membrane is considerably simplified if the uniformly distributed loading is expressed by Fourier's series'.

Gould maintains [5]: 'Since we have a partial differential equation to solve, we try the standard method of separation of variables. Specifically, we apply the Fourier series technique, whereby all loadings and dependent variables are taken in the form of a Fourier's series'.

According to the ASCE's manual on the design of concrete shell roofs [18]:

'Although it is possible to obtain the internal forces produced by any surface loads, because of mathematical difficulties, the corrective line loads [see Section 4] applied along the longitudinal edges are expressible only as a Fourier's sine or cosine series. . . . To avoid confusion in the application of the line loads [see Section 4]. . . it is expedient to regard the surface load on the shell as the sum of partial loads (i.e. terms of a Fourier series). . . comparing the values obtained by the partial loads with those obtained by the uniform load, it is seen that good agreement is obtained for the longitudinal force N_x. The shear and transverse force components obtained by the sum of the first two sinusoidal loads are approximately 10% and 15% too small. In designing shells by partial loads, this discrepancy should be borne in mind and adjustments should be made to the final values'.

Tottenham [27] states:

'The basic differential equation for the edge load problem [see Section 4] is only soluble by considering the edge load in the form of a Fourier series. . . . Fortunately the effect of the first term is predominant and this is the only one considered here. . . . The loads we have applied have been the first term in the Fourier series for unity'.

According to Gibson and Cooper [28]:

'To be of direct operational use the uniformly distributed loading must be put in the form of a Fourier half range series . . . only the first term of the Fourier series will be utilised as this is dominant. . . . This (latter) tends to slightly overestimate the magnitude of the moments and consequently errs on the side of safety'.

Essentially, doubly integrating a cosine gives the same cosine. This does not happen with, for example, an algebraic polynomial; the double integral becomes more complex and unlike the original expression. This device of using Fourier's series has been used for solving many elastic analyses, e.g. folded plates, grillages, columns with lateral loading, soil foundation problems, problems in continuum mechanics, etc.

Continuing the quotation from [19]:

'In practice (in the UK) shells are often designed to carry their self weight and a snow load of 0.75 kN/m² (16 psf) of curved surface area. This is slightly more conservative than taking the snow as 0.75 kN/m² of plan area, but simplifies the design, as all the loading can then be considered as per unit of curved surface area. Thus, assuming the origin of x to be at the centre of the span l the loading can be expressed as:

$$p \frac{4}{\pi} [\cos (\pi x/l) - \tfrac{1}{3} \cos (3\pi x/l) + \tfrac{1}{5} \cos (5\pi x/l) \dots] \tag{6-9}$$

where p is the total load on the shell per unit of curved surface area.

Experience shows that generally it is satisfactory to consider only the first term of this series; hence the loading is considered as $(4/\pi)p \cos (\pi x/l)$ per unit area of the surface'.

For the differential element, the vertical loading on a unit of area is thus $(4/\pi)p \cos (\pi x/l)$. On the same unit of area, the component perpendicular to the element (Fig. 6-1) is

$$Z = (4/\pi)p \cos (\pi x/l) \cos \phi \tag{6-10}$$

and on the same unit of area the component tangential to the element (Fig. 6-1) is

$$Y = (4/\pi)p \cos (\pi x/l) \sin \phi \tag{6-11}$$

Substituting these values in Eqn 6-4 gives

$$N_\phi = -r(4/\pi)p \cos (\pi x/l) \cos \phi \tag{6-12}$$

Inserting this value of N_ϕ in Eqn 6-3 gives

$$(\partial N_{\phi x}/\partial x) + (4/\pi)p \cos (\pi x/l) \sin \phi + \frac{4}{\pi} p \cos (\pi x/l) \sin \phi = 0$$

Therefore,

$$(\partial N_{\phi x}/\partial x) = -(8/\pi)p \cos (\pi x/l) \sin \phi \tag{6-13}$$

Many designers [19, 27] use $N_{\phi x}$ in this form, i.e. as a stress, rather than as $N_{\phi x}$. However, integrating Eqn 6-13 gives

$$N_{\phi x} = -2 \frac{l}{\pi} \frac{4}{\pi} p \sin (\pi x/l) \sin \phi + C_1(\phi) \tag{6-14}$$

where $C_1(\phi)$ is a function of ϕ, but an arbitrary constant as far as x is concerned. Then, Eqn 6-2, at $X = 0$, gives

$$\frac{\partial N_x}{\partial x} - \frac{2}{r} \frac{l}{\pi} \frac{4}{\pi} p \sin (\pi x/l) \cos \phi + \frac{1}{r} \frac{\partial}{\partial \phi} C_1(\phi) = 0$$

Therefore,

$$N_x = -\frac{2}{r} \frac{l^2}{\pi^2} \frac{4}{\pi} p \cos \frac{\pi x}{l} \cos \phi - \frac{x}{r} \cdot \frac{\partial}{\partial \phi} C_1(\phi) + C_2(\phi) \tag{6-15}$$

where $C_2(\phi)$ is a function of ϕ, but an arbitrary constant as far as x is concerned.

In the case of a symmetric shell, when $\phi = 0$, i.e. at the crown, $N_{\phi x}$ must be zero as it is an asymmetric (or antisymmetric) force. Therefore, from Eqn 6-14,

$$C_1(\phi) = 0 \qquad (6\text{-}16)$$

In the case of a simply supported shell, the longitudinal forces, N_x, must be zero at the ends, i.e. where $x = \pm l/2$, therefore Eqn 6-15 gives

$$C_2(\phi) = 0 \qquad (6\text{-}17)$$

From Eqns 6-6 and 6-15, using also Eqns 6-16 and 6-17,

$$\frac{\partial u}{\partial x} = -\frac{2}{Ehr}\frac{l^2}{\pi^2}\frac{4}{\pi} p \cos(\pi x/l) \cos\phi$$

Therefore,

$$u = -\frac{2}{Ehr}\frac{l^3}{\pi^3}\frac{4}{\pi} p \sin(\pi x/l) \cos\phi + C_3(\phi) \qquad (6\text{-}18)$$

where $C_3(\phi)$ is an arbitrary constant as far as x is concerned. At the centre of the simply supported shell, when $x = 0$, u must be zero, therefore, from Eqn 6-18,

$$C_3(\phi) = 0 \qquad (6\text{-}19)$$

From Eqns 6-7, 6-18 and 6-14, using also Eqns 6-16 and 6-19,

$$\frac{1}{r^2}\frac{2}{Eh}\frac{4}{\pi} p \frac{l^3}{\pi^3} \sin(\pi x/l) \sin\phi + \frac{\partial v}{\partial x} = -\frac{4}{Eh}\frac{l}{\pi}\frac{4}{\pi} p \sin(\pi x/l) \sin\phi$$

Therefore,

$$\frac{\partial v}{\partial x} = -\frac{2}{Eh}\frac{l}{\pi}\frac{4}{\pi} p\left(2 + \frac{1}{r^2}\frac{l^2}{\pi^2}\right) \sin(\pi x/l) \sin\phi \qquad (6\text{-}20)$$

whence,

$$v = \frac{2}{Eh}\frac{l^2}{\pi^2}\frac{4}{\pi} p\left(2 + \frac{1}{r^2}\frac{l^2}{\pi^2}\right) \cos(\pi x/l) \sin\phi + C_4(\phi) \qquad (6\text{-}21)$$

where $C_4(\phi)$ is an arbitrary constant as far as x is concerned. At the end of the simply supported shell, when $x = l/2$, the end diaphragm, or stiffening beam or frame, restrains the shell, so that v is zero. Therefore, from Eqn 6-21,

$$C_4(\phi) = 0 \qquad (6\text{-}22)$$

From Eqns 6-8, 6-21 and 6-12

$$\frac{2}{Ehr}\frac{l^2}{\pi^2}\frac{4}{\pi} p\left(2 + \frac{1}{r^2}\frac{l^2}{\pi^2}\right) \cos(\pi x/l) \cos\phi - \frac{w}{r} = -\frac{r}{Eh}\frac{4}{\pi} p \cos(\pi x/l) \cos\phi$$

Therefore,

$$w = \frac{2}{Eh}\frac{l^2}{\pi^2}\frac{4}{\pi} p\left(2 + \frac{1}{r^2}\cdot\frac{l^2}{\pi^2}\right) \cos(\pi x/l) \cos\phi + \frac{r^2}{Eh}\frac{4}{\pi} p \cos(\pi x/l) \cos\phi$$

whence,

$$w = \frac{1}{Eh} \frac{4}{\pi} p \left[r^2 + 2 \frac{l^2}{\pi^2} \left(2 + \frac{1}{r^2} \frac{l^2}{\pi^2} \right) \right] \cos \phi \cos (\pi x/l) \tag{6-23}$$

Thus, using the first term of Eqn 6-9 for the loading, Eqns 6-12 to 6-15, 6-18, 6-21 and 6-23—using also Eqns 6-16, 6-17, 6-19 and 6-22 for the arbitrary constants (all zero)—give the values of N_ϕ, $\partial N_{\phi x}/\partial x$, $N_{\phi x}$, N_x, u, v and w, respectively. Allowing for the different notation and the opposite sign of $N_{\phi x}$ between the sign conventions, these equations are exactly the same as those given by Tottenham [27].

6.2 Shells of revolution

A surface of revolution or rotation (see also Section 1.2) is formed by rotating a plane curve, the 'generating curve' or 'meridian', about an axis (here considered vertical for convenience) in the plane of the curve, the 'meridian plane'. A differential element is shown in Fig. 6-5; its sides are off meridians and its top and bottom are off horizontal circles, located by angles θ (the 'meridian angle') and ϕ, respectively, ϕ being the angle between the normal and the axis of rotation. The planes containing the element's principal curvatures, of radii r_1 and r_2, are the meridian plane and the plane perpendicular to it, respectively. The radius of the horizontal circle shown is r_0. Therefore, the lengths of the sides of the element are $r_1\, d\phi$ and $r_0\, d\theta = r_2 \sin \phi\, d\theta$ and its surface area is $r_0 r_1\, d\phi\, d\theta$ or $r_1 r_2 \sin \phi\, d\phi\, d\theta$.

The main loading on, say, a reinforced concrete dome comprises its self-weight and snow load, which together can be considered as a uniform load per unit of surface area (Section 6.1). Because of the assumed symmetry of loading, there will be no shearing forces, as these are asymmetric forces, on the sides of the element.

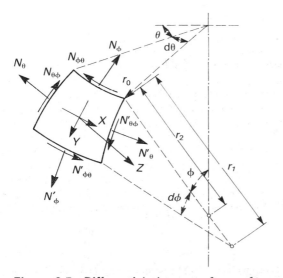

Figure 6-5 Differential element of a surface of revolution

As shown in Fig. 6-5, N_ϕ and N_θ are forces per unit length, and X and Y are external loads per unit area, all tangential to the element, whilst Z is the external loading per unit area normal to the element.

Resolving the forces for the element in the direction of the tangent to the meridian, let us first consider the contribution due to N_ϕ:

$$-N_\phi r_0\, d\theta + \left(N_\phi + \frac{\partial N_\phi}{\partial \phi}\, d\phi\right)\left(r_0 + \frac{\partial r_0}{\partial \phi}\, d\phi\right) d\theta$$

$$= N_\phi \frac{\partial r_0}{\partial \phi}\, d\phi\, d\theta + r_0 \frac{\partial N_\phi}{\partial \phi}\, d\phi\, d\theta + \frac{\partial r_0}{\partial \phi}\frac{\partial N_\phi}{\partial \phi} (d\phi)^2\, d\theta$$

Eventually, this will be divided by $d\phi\, d\theta$, then the quantities $\partial r_0/\partial \phi$, $\partial N_\phi/\partial \phi$ become in the limit zero/zero, that is normal quantities, e.g. rate of change of radius, rate of change of N_ϕ, but the last term of the above expression will contain $d\phi$, which in the limit is zero. Therefore, the above becomes

$$N_\phi \frac{\partial r_0}{\partial \phi}\, d\phi\, d\theta + r_0 \frac{\partial N_\phi}{\partial \phi}\, d\phi\, d\theta = \frac{\partial}{\partial \phi} (N_\phi r_0)\, d\phi\, d\theta$$

The contribution due to N_θ is a force $N_\theta r_1\, d\phi$ tangential to the horizontal circle of radius r_0. Resolving this and its incremental force N_θ' $(= N_\theta + (\partial N_\theta/\partial\theta)\, d\theta)$ at the other end of the element (Fig. 6-6) in a horizontal direction gives

$$N_\theta r_1\, d\phi\, \sin\left(\tfrac{1}{2}\, d\theta\right) + N_\theta' r_1\, d\phi\, \sin\left(\tfrac{1}{2}\, d\theta\right)$$

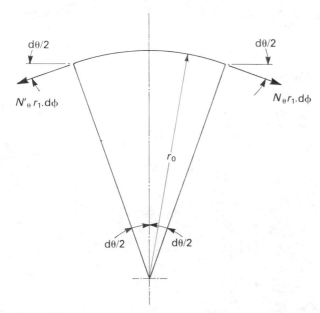

Figure 6-6 Tangential forces acting on element of surface of revolution

and as in the limit when $d\theta \to 0$, $\sin(\frac{1}{2} d\theta) \to \frac{1}{2} d\theta$ this becomes

$$N_\theta r_1 \, d\phi \frac{d\theta}{2} + \left(N_\theta + \frac{\partial N_\theta}{\partial \theta} d\theta\right) r_1 \, d\phi \frac{d\theta}{2} = N_\theta r_1 \, d\phi \, d\theta + \frac{r_1}{2} \frac{\partial N_\theta}{\partial \theta} d\phi (d\theta)^2$$

Eventually, this expression will be divided by $d\theta \, d\phi$, then the quantity $\partial N_\theta / \partial \theta$ becomes in the limit zero/zero, i.e. a normal quantity (the rate of change of N_θ with respect to θ), but the last term contains $d\theta$, which in the limit is zero. Therefore, the above expression becomes

$$N_\theta r_1 \, d\phi \, d\theta$$

Then, referring to Fig. 6-7, the component of this force in the direction of the tangent to the meridian is

$$(N_\theta r_1 \, d\phi \, d\theta) \cos \phi$$

The contribution due to $N_{\theta\phi}$ is

$$-N_{\theta\phi} r_1 \, d\phi + \left(N_{\theta\phi} + \frac{\partial N_{\theta\phi}}{\partial \theta} d\theta\right) r_1 \, d\phi = \frac{\partial N_{\theta\phi}}{\partial \theta} r_1 \, d\theta \, d\phi$$

Adding the above forces and the component of the external force in the direction of the tangent to the meridian gives

$$\frac{\partial}{\partial \phi} (N_\phi r_0) \, d\phi \, d\theta - N_\theta r_1 \cos \phi \, d\phi \, d\theta + \frac{\partial N_{\theta\phi}}{\partial \theta} r_1 \, d\theta \, d\phi + Y r_0 r_1 \, d\phi \, d\theta = 0$$

Therefore,

$$\frac{\partial}{\partial \phi} (N_\phi r_0) - N_\theta r_1 \cos \phi + \frac{\partial N_{\theta\phi}}{\partial \theta} r_1 + Y r_0 r_1 = 0 \tag{6-24}$$

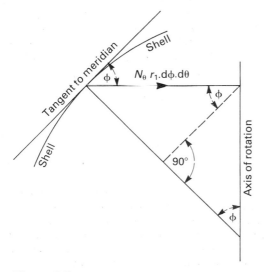

Figure 6-7

Resolving the forces for the element tangentially to the horizontal circle of radius r_0, we see the contribution due to N_θ is

$$-N_\theta r_1\, d\phi + \left(N_\theta + \frac{\partial N_\theta}{\partial \theta}\, d\theta\right) r_1\, d\phi = \frac{\partial N_\theta}{\partial \theta}\, r_1\, d\theta\, d\phi$$

The contribution due to $N_{\phi\theta}$—a force of $N_{\phi\theta} r_0\, d\theta$ in the tangential direction to the horizontal circle of radius r_0—is

$$-N_{\phi\theta} r_0\, d\theta + \left(N_{\phi\theta} + \frac{\partial N_{\phi\theta}}{\partial \phi}\, d\phi\right)\left(r_0 + \frac{\partial r_0}{\partial \phi}\, d\phi\right) d\theta$$

$$= N_{\phi\theta}\frac{\partial r_0}{\partial \phi}\, d\phi\, d\theta + r_0\frac{\partial N_{\phi\theta}}{\partial \phi}\, d\phi\, d\theta + \frac{\partial N_{\phi\theta}}{\partial \phi}\frac{\partial r_0}{\partial \phi}\, d\phi(d\phi)^2$$

Eventually, this expression will be divided by $d\theta\, d\phi$, then the last term contains $d\phi$, which in the limit is zero. Therefore, the expression becomes

$$\frac{\partial}{\partial \phi}\,(r_0 N_{\phi\theta})\, d\phi\, d\theta$$

There is a contribution due to $N_{\theta\phi}$ because the horizontal components of $N_{\theta\phi} r_1\, d\phi$ and $N'_{\theta\phi} r_1\, d\phi$, namely, $N_{\theta\phi} r_1\, d\phi \cos \phi$ and $N'_{\theta\phi} r_1\, d\phi \cos \phi$ (Fig. 6-7) contains a horizontal angle $d\theta$ (Fig. 6-5). Therefore, the contribution is

$$\left(N_{\theta\phi} + \frac{\partial N_{\theta\phi}}{\partial \theta}\, d\theta\right) r_1\, d\phi \cos \phi \sin d\theta$$

In the limit, when $d\theta \to 0$, $\sin d\theta \to d\theta$, and as the above expression will be eventually divided by $r_1\, d\phi$, the second term will then contain $d\theta$, which in the limit becomes zero. Therefore, the expression becomes

$$N_{\theta\phi} r_1\, d\phi\, d\theta \cos \phi$$

Adding the above forces and the component of the external force in the direction tangential to the horizontal circle of radius r_0 gives

$$\frac{\partial N_\theta}{\partial \theta}\, r_1\, d\theta\, d\phi + \frac{\partial}{\partial \phi}\, r_0 N_{\phi\theta}\, d\phi\, d\theta + N_{\theta\phi} r_1\, d\phi\, d\theta \cos \phi + X r_0 r_1\, d\phi\, d\theta = 0$$

Therefore,

$$\frac{\partial N_\theta}{\partial \theta}\, r_1 + \frac{\partial}{\partial \phi}\,(r_0 N_{\phi\theta}) + N_{\theta\phi} r_1 \cos \phi + X r_0 r_1 = 0 \qquad (6\text{-}25)$$

Resolving the forces for the element normal to its plane, let us first consider the contribution due to N_ϕ. A figure similar to Fig. 6-6 is relevant thus resolving normal to the element gives

$$(N_\phi r_0\, d\theta) \sin (\tfrac{1}{2} d\phi) + \left(N_\phi + \frac{\partial N_\phi}{\partial \phi}\, d\phi\right)\left(r_0 + \frac{\partial r_0}{\partial \phi}\, d\phi\right) d\theta \sin (\tfrac{1}{2} d\phi)$$

In the limit when $d\phi \to 0$, $\sin (\tfrac{1}{2} d\phi) \to \tfrac{1}{2} d\phi$, so this expression becomes

$$N_\phi r_0\, d\theta\, d\phi + \frac{N_\phi}{2}\frac{\partial r_0}{\partial \phi}\, d\theta(d\phi)^2 + r_0\frac{\partial N_\phi}{\partial \phi}\, d\theta(d\phi)^2$$

Eventually, this expression will be divided by $d\theta\,d\phi$, then the quantities $\partial r_0/\partial\phi$ and $\partial N_\phi/\partial\phi$ become in the limit zero/zero, i.e. normal quantities (rates of change of r_0 and N_ϕ respectively with respect to ϕ) but the last two terms will contain $d\phi$, which in the limit is zero. Therefore, the above expression becomes

$N_\phi r_0\,d\theta\,d\phi$

The contribution due to N_θ—a force of $N_\theta r_1\,d\phi$ tangential to the horizontal circle of radius r_0—gives, as previously, a resultant force of $N_\theta r_1\,d\phi\,d\theta$ in a horizontal direction. Referring to Fig. 6-7, the component normal to the generating curve is

$(N_\theta r_1\,d\phi\,d\theta)\sin\phi$

Adding the above forces and the component of the external force in the direction of the normal to the plane of the element gives

$N_\phi r_0\,d\theta\,d\phi + N_\theta r_1\,d\phi\,d\theta\sin\phi + Zr_0r_1\,d\phi\,d\theta = 0$

Therefore,

$$N_\phi r_0 + N_\theta\sin\phi + Zr_0r_1 = 0 \tag{6-26}$$

Taking moments about the normal to the centre of the element, or because of complementary shear stresses,

$$N_{\theta\phi} = N_{\phi\theta} \tag{6-27}$$

Thus, Eqns 6-24 to 6-27 inclusive enable N_ϕ, N_θ, $N_{\theta\phi}$ and $N_{\phi\theta}$ to be obtained. Allowing for very slightly different notation, these equations are the same as those derived by Flügge [23], who also applies them to various loading conditions. If the generating curve has a point of inflection, e.g. certain oriental domes, Flügge describes how ϕ presents difficulties and instead of it he uses the distance along the curve.

The author found that several works, excluding Flügge [23], used a diagram equivalent to Fig. 6-5, but showing r_1 terminating where it hits r_2, which, of course, is incorrect, and/or made certain mistakes in precisely describing directions of forces or their components. The authors concerned undoubtedly know what they are doing but such deficiencies in communication could, without this warning, waste a considerable amount of the beginner's time.

If the loading is symmetric, the above equations can be used putting $N_{\theta\phi} = N_{\phi\theta} = 0$ because these are asymmetric stress resultants and thus cannot exist for a symmetric shell with symmetric loading. However, a much simpler method is to consider the equilibrium of a portion of a shell above a horizontal section defined by an angle ϕ (Fig. 6-8). If the total loading on the portion is R, then resolving vertically gives

$$2\pi r_0 N_\phi\sin\phi + R = 0 \tag{6-28}$$

which yields N_ϕ. Then, N_θ can be obtained from Eqn 6-26.

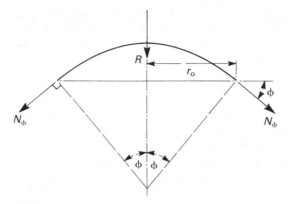

Figure 6-8

6.3   Shells of arbitrary shape

We must mention briefly a general membrane theory for shells of arbitrary shape. (In the preceding sections, the particular shapes treated allowed advantage to be taken of simpler theories.)

This was expounded by Flügge [23] and then applied to hypars, etc.

Two concepts are involved which have not been encountered so far in this chapter: firstly, the idea of taking and dealing with the components of all stress resultants in the horizontal plane; and secondly, the use of stress functions, which are a device to enable the solution of the partial differential equations.

7   Cylindrical shells

These can be designed and analysed as described in Sections 3 and 4. Now, with regard to the line load problem, this causes, as well as the in-planar (membrane) forces, bending moments with accompanying shears and torsions in the shell. Various theories have been evolved for this edge load problem. The one which considers all such forces, moments and torsions is due to Donnell [29] in the USA, Jenkins [30] in the UK and Vlassov [31] in the USSR. It is also described in [28]. A theory which considers only in-planar forces and transverse bending moments with accompanying radial shears is due to Finsterwalder [32] and is also described in [26]. Various assumptions were made by Schorer [33] to reduce the work involved in using this theory, because computers were not available to designers at the time. Tottenham [27] developed a method of using Schorer's theory, which tabulated the coefficients which needed to be obtained by using considerable arithmetical accuracy, and enabled the remainder of the calculations to be performed to slide-rule accuracy. Thus, a designer could use a slide rule and the tables for the whole of his design. Reference [27] gives the tables and a design for a symmetric practical shell. It also derives Schorer's equation and discusses the practical range of applicability of Schorer's theory. About and prior to 1954 in the UK, the theories used for designing practical shells were principally those given by [27, 30, 32, 33]. The author believes that to date more shells have been designed in the UK by these latter two methods than any other method. The author has shown [34] how to use Tottenham's method for designing N-light shell roofs.

Schorer's theory has been criticized for making its assumptions. It is not suitable for very short shells, where the longitudinal bending moments with accompanying radial shears are important, i.e. the shell spans to some extent like a one-way spanning slab between the end diaphragms, arches or frames. The author regards it as the simplest of the reasonably respectable elastic theories and suitable for a considerable range of practical designs, borne out by experience in the UK. Yet there are simpler methods than this, see [35].

The elastic analysis of practical cylindrical shells is explained in [19]. This also gives the Schorer equations adapted for unit edge loads and gives all the algebraic equations (not published completely elsewhere) in Table 19 for the coefficients in these equations for both symmetric and asymmetric shells. These could easily be programmed for solution with a desk computer. Reference [19] also deals with cylindrical shells with edge beams, valley beams and with numerous shells side by side.

7.1 Design tables and graphs

Without a computer program, the calculations are very arduous. The least arduous of the elastically respectable methods, by Tottenham [27], for an end (or external) shell and a valley, of a scheme of barrel shells, takes a minimum of 8 h and 4 h, respectively, of solid uninterrupted artithmetic before any checks are encountered. Usually, the checks show errors. Locating these is exceedingly time consuming and can easily take as long again. In the last scheme the author designed in this way, he had an assistant working independently with whom he cross-checked the arithmetic after every half hour or so. In this way, the above minimum times were slightly greater and with double labour, but the checks were correct and no searching for errors was involved. A minimum N-light [34] scheme takes about four times as long as the symmetric shell scheme just discussed, and the work is tedious rather than stimulating. All these times are for a skilled designer. A practising engineer having to learn about the method would need to allow much more time.

Hence the desirability of design tables.

The tables compiled by Schorer [33] are mainly not satisfactory in practice because of the large differences in the coefficients and the correct interpolation is non-linear. Tottenham's [27] tables overcome this difficulty. However, both of these give only partial assistance with the complete analysis. The tables in [18] are more complete in this respect, but there are significant differences between the coefficients and again the correct interpolation is non-linear. The author found that to try to use these for a particular practical shell design without interpolation was unsatisfactory and he realized that this would generally be the case.

The tables in [36] are published with adjacent German and English texts and common mathematical work. They are hard to follow and, because the notation is not collected in a single list, it is very time consuming trying to find out what various symbols mean. From the examples, considerable arithemetical work is obviously required. It is also obvious from the coefficients that large differences exist and one knows that correct interpolation is non-linear.

The tables in [37] give complete analyses. They deal with a limited range of symmetric shells. Unfortunately, they are for shells 76 mm (3 in) thick, whereas UK companies would generally use shells 64 mm (2.5 in) thick in most of the cases covered and designers in other countries would generally use shells thinner than 76 mm. However, an appendix gives some multiplying factors which may be used to convert the results to apply to shells 64 mm thick. In the author's opinion, some of the shells included have undesirable proportions in other practical respects.

Because of the shortcomings of past tables produced by authors who took on a tremendous task, including learning computing, the author and Dr Khwaja have published tables [3] which give complete solutions for a large range of shells of practical proportions for schemes of ordinary and north-light cylindrical shells. A complete example is given using the tables for a scheme of north-lights five side by side, each 30 ft (9.1 m) wide and all of span 60 ft (18.3 m). The example includes calculations of the necessary reinforcement and its practical disposition. Complete analyses are given for shells very near together so that, if a shell is required, say, 32 ft 4 in wide by 62 ft 8 in span, one consults the values of forces and moments in the following shells 32 ft wide by 62 ft span, 34 ft wide by 62 ft span, 32 ft wide by 64 ft span, and 34 ft wide by 64 ft span, and takes the highest values for designing the reinforcement. In this way, interpolation is not necessary and the solution is economically satisfactory. The shells have practical proportions known from experience to be satisfactory. One is limited to certain satisfactory geometries of the shells themselves but not to the areas they can cover, which can be of any rectangular shape or shapes.

The author [4] gives design graphs for complete designs of practical shell roofs. The range is much more limited than that of Wilby and Khwaja [3]. In practice, it has been found necessary to thicken shells towards their valleys and edge beams. Reference [4] gives graphs for shells where this has been both taken and not taken into account. The former should be used and give a better practical design than previous tables. Reference [4] also gives an example using the graphs for a single-bay barrel-shell, 8 m wide by 16 m span, which includes the design of edge beams and end stiffeners, and reinforcement is also calculated.

Reference [2] shows, with an example, how continuous shells can be designed using simply supported solutions.

7.2 Analogue computer

Wilby and Bellamy demonstrated [4, 21, 22] that the analogue computer is a very useful tool for helping to design ordinary and north-light shells. It is much faster (instantaneous results) than the digital computer, particularly with regard to integrations. One simply has an 'integrator' unit whose input terminal can be connected to a cathode-ray oscilloscope or an x–y plotter to record a relationship between input voltage and time, and whose output terminal can be connected to a second oscilloscope or plotter to record the integral of this relationship between voltage and time. In addition, or as an alternative to the plotter, one can couple up a digital voltmeter which can easily be read at any time.

Time is taken as analogous to the angle ϕ and the voltages at different parts of the analogue circuit are taken as analogous to the values of the forces and moments in a shell.

The computer has slide rule accuracy, which is good enough for the results. Being instantaneously fast, it has a useful semi-design ability. If a shell is analysed and a certain force or moment is found to be undesirably great, ones inclination might be to alter a certain dimension, but this might not do much to solve the problem. The analogue computer, however, can be adjusted so that one can find the most powerful influence on the problem, which might be a peculiar relationship between certain dimensions. This gives a design as opposed to just an analysis facility.

The analogue computer was much cheaper than a digital computer. It still is much cheaper than a main-frame computer but doubtfully much cheaper than modern desk digital computers.

It enabled [22] fewer assumptions to be made in the theory than the theories of [29–31], see Section 7.

7.3 Instability

Consider a thin, longitudinal strip of a shell, say 64 mm (2.5 in) thick and 18 m (60 ft) long; its length-to-thickness ratio is therefore 281 : 1. This would undoubtedly be unsuitable as a reinforced concrete column yet in some locations it would have to withstand compression. Its curvature saves it from buckling—the more curvature the better, from this point of view.

The shells of practical dimensions described in [2–4, 19, 21] are known from experience to be satisfactory with regard to buckling. These are shells at least 64 mm thick and of reasonable curvatures and overall depth-to-span ratios. It is only when thinner shells of greater span-to-overall depth ratios and with less curvatures are used that buckling becomes a problem. Early guidance was given by Lundgren [20]. Professor Haas, at Delft, has been interested and has pursued experimental work on this problem Reference [38] outlines the problem and cites [20, 39, 40]. Furthermore, buckling is given significant treatment by Flügge [23].

8 Domes

Spherical domes are the most common for functional purposes because of the simpler shuttering.

For a spherical dome of radius a and a total load per unit area of curved surface, q, it is easy to obtain [1] the following membrane forces from Eqns 6-28

$$N_\phi = -aq/(1 + \cos \phi) \tag{8.1}$$

$$N_\theta = aq\left(\frac{1}{1 + \cos \phi} - \cos \phi\right) \tag{8.2}$$

From these equations, N_ϕ will always be negative (i.e. compression) and increase as ϕ increases, whilst N_θ is negative for $\phi < 51°50'$ but positive (tension) for $\phi > 51°50'$. The latter condition is commonly the case in practice.

Because of the dome's double curvature, there is less worry in designing it by membrane theory only than there is in the case of the cylindrical shell, and many domes have been designed in the UK in this way [41]. That is, by providing edge supports designed strong enough to withstand the N_ϕ forces at the edge of the dome.

The membrane can be acceptable for a hemi-spherical dome supported vertically at its periphery. If the dome is less than a hemisphere and supported vertically, an edge ring force can be provided by post-tensioning so that the resultant of these vertical and horizontal forces is vectorially equal and opposite to N_ϕ.

In practice, the peripheries are often built in to a 'ring beam' which is supported vertically. If the design provides for the forces as already described, then the desire of the junction point between the ring beam and the dome to deflect differently for the two members will tend to cause cracking, unless the situation is as described in the last paragraph.

A more acceptable design is to consider unknown equal and opposite forces and moments at the junction point, and determine these (see Section 4) so that the above mentioned deflections agree. This is lucidly explained by Chatterjee [10] and a useful example is given. He does not derive the equations for the deflections, etc. These equations are derived in [1]. Gol'denweiser [42] discusses the limitations of the membrane theory.

Instability (Section 7.3) is not as great a problem as with cylindrical shells because of the double curvature. It does not seem to cause trouble for normal domes, i.e. ones which are not particularly shallow. Buckling is considered analytically by Flügge [23], and summarized by Chatterjee [10].

9 Hypars

Figure 9-1 shows a surface of a hypar which can be expressed [3] as

$$z = kxy \sin \omega \tag{9-1}$$

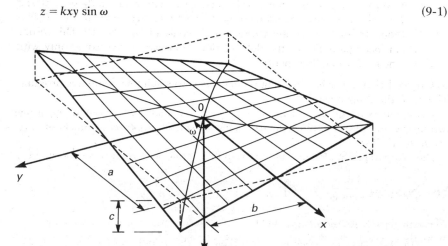

Figure 9-1 Surface of the hypar $z = kxy \sin \omega$ (From *Concrete shell roofs* by courtesy of the authors and Applied Science Publishers Ltd, London)

When $x = a$ and $y = b$, $z = c$ and if $\omega = 90°$, then

$$z = \frac{c}{ab} xy \tag{9-2}$$

Its intersections with the vertical planes $x =$ constant and $y =$ constant are respectively the straight lines $z =$ (const.) y and $z =$ (const.) x, the 'generators'. Using Eqn 9-2 in the general membrane equations [23] for shells of arbitrary shape (see Section 6.3) enables the membrane forces to be obtained. Timoshenko [1] considers a vertical load, such as snow load, of q per unit of horizontal area and this gives $N_x = N_y = 0$ and

$$N_{xy} = -\frac{q}{2k} \tag{9-3}$$

Then he considers a self-weight load of q_0 per unit surface area and this gives

$$N_{xy} = 0.5 q_0 \left(x^2 + y^2 + 1/k^2\right)^{0.5} \tag{9-4}$$

$$N_x = -\frac{q_0 y \cos\theta}{2 \sin\phi} \log \left[\frac{x + (x^2 + y^2 + 1/k^2)^{0.5}}{(y^2 + 1/k^2)^{0.5}}\right] \tag{9-5}$$

$$N_y = -\frac{q_0 x \cos\phi}{2 \cos\theta} \log \left[\frac{y + (x^2 + y^2 + 1/k^2)^{0.5}}{(x^2 + 1/k^2)^{0.5}}\right] \tag{9-6}$$

Billington [12] demonstrated that there is normally little error in assuming the self-weight, q_0, as an increase in the value of q so that the simpler solutions can be used. Essentially, this means that the shell mainly resists only the shear stress resultant of Eqn 9-3. Then, according to Pflüger [43]:

'Of course these shears must be received by appropriate stiffeners at the edges. The edge members are subject to certain strains while the shell is without strains in the x and y directions because of $N_x = N_y = 0$. This results in an incompatibility of the deformations that can be overcome only with the help of a bending theory'.

Reference [3] gives such a bending theory, and gives suitable computer program listings in the Appendices.

Many hypars have been designed, and some have cracked considerably, using membrane theory only. Worked examples indicating reinforcement details are given in [10, 38].

The last paragraph of Section 8 on instability applies to hypars as well as domes.

10 Conoids

According to Wilby and Naqvi [13]

'In the Cartesian system of coordinates, the middle surface of a shell is expressed by the equation

$$z = X(x)Y(y) \tag{10-1}$$

where the functions $X(x)$ and $Y(y)$ are independent of y and x respectively.

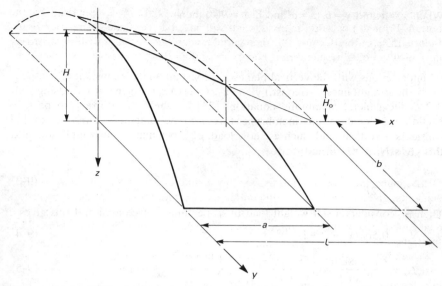

Figure 10-1   Half of a parabolic conoid which might be used as a roof of width *2b* and span *a* (From: C. B. Wilby and M. M. Naqui, *Reinforced concrete conoidal shell roofs,* by permission of Eyre and Spottiswoode Publications Ltd, Leatherhead)

If, however,

$$X(x) = Ax + B \tag{10-2}$$

where A and B are constants, then any section of the surface where y is constant is a straight line, and therefore a surface can be generated by straight lines. Such surfaces are called 'ruled surfaces'. A ruled surface is completely defined by the constants A and B, and the function $Y(y)$ is called the 'directrix' of the surface. A conoid belongs to this group of surfaces of which the most commonly used directrix is a parabola. Thus the equation of the parabolic conoid shown in Fig. 10-1 is

$$z = -H\left(1 - \frac{x}{l}\right)\left(1 - \frac{y^2}{b^2}\right) \tag{10-3}$$

where $g = 1 - H_0/H$'.

Figure 10-1 shows half of a conoid which might be used as a roof of width $2b$ and span a.

Reference [13] goes on to consider edge beams and give a large number of computer-generated tables which may be used for designing conoids with edge beams. Then it gives two practical examples using the tables, namely, for unsupported and propped edge beams, and drawings give reinforcement details. In [2], the addendum and corrigenda to these tables are published and an example is given using the tables of a scheme of numerous conoidal shells covering a large area.

The background to [13, 44, 45] is given in the PhD [46] supervised by the author. Prior to this work, many conoids will have been designed using membrane theory [47] only. Because of their double curvature, there is less worry in designing conoids by membrane theory only than there is the case of cylindrical shells. However, one is likely to experience more cracking than using the method of [13].

The last paragraph of Section 8 on instability applies to conoids as well as domes.

11 Method of finite elements

Of the many methods of analysis (e.g. finite difference, relaxation), that of finite elements, given modern powerful computers, is at present generally considered to be a method applicable to a wide range of problems (in structures, hydraulics, soil mechanics, etc.) of the continuum mechanics type. It is a method only suitable for solution with computers.

Programs are commercially available world-wide. Some can be adapted for a variety of applications; for example, a program developed by a UK firm of civil engineering consultants is also widely used by mechanical and aeronautical engineers, who would claim to be able to adapt it adequately to any shape of shell.

Writing and adapting such programs becomes the work of specialists who usually have no experience of reinforced concrete shell design, construction or research, but are very experienced at elastic analysis using the finite element method and computers. For the beginner to set about learning how to write a finite element computer program for a shell design would not be practical if he were designing a job wanted in a normal time, because the program writing for this kind of work is very time consuming. The cost of producing the program would normally only be retrieved if the program were marketed for several designers to use.

For information on the finite element method (assisted by computers) applied to shells, the reader is referred to [24, 48, 49].

12 Computer programs

In addition to the finite element packages just mentioned, other computer packages based on the methods previously outlined in this chapter may well be more available. For example, computer programs were used for cylindrical shells, hypars and conoids in [3, 13], the program listing being published for the hypar in [3].

13 Design practicalities

Reinforced concrete shells, with or without prestressing, should be designed by those expert and experienced in structural concrete design, detailing and construction. It is also an advantage if the designer also has appropriate research or testing experience of how structural concrete behaves.

One difficulty is that many of the books and papers on shells are written by experts, with or without computing expertise, in the elastic analysis of shells for mechanical, aeronautical or structural problems, who have little or no experience of concrete as outlined above.

Experienced designers know that, to design many of the structures with which they are presented, or which they wish to create, they have to make assumptions which would occasionally horrify some analysts with little or no practical experience. Sometimes they have to make sure that forces are adequately resisted without the proper matching up of deflections. They then have to rely on some plastic action, which, from experience, they know will occur, and they will also have to detail reinforcement by experience to ensure adequate distribution of the cracks which will be caused. Often, a design is required within a very limited time and may involve situations which a pure analyst would take years to solve and then doubt whether his solution were really good enough to use for design.

A similar philosophy has to apply to the practical design of structural concrete shells. Often, architectural ideas present problems which have not previously been solved or for which the existing solutions are not presented in a form suitable for use in the time available to the designer. For example, an industrialist may wish to utilize the whole area of a non-rectangular portion of his site, or want natural roof-lighting or special loads hung from the roof. Such demands will almost certainly pose special problems for the designer because it is unlikely that he can resort to ready-made solutions.

Laboratory and field tests made by the author and others show that shells are inherently strong and allow more liberties to be taken with strength than non-shell structures. Cracks can be troublesome, hence the use of special thickenings and nominal reinforcement, see for example [2, 3, 4, 13, 19].

References

1. Timoshenko, S. and Woinowsky-Krieger, S., *Theory of plates and shells*, 2nd Edn, McGraw-Hill, New York, 1959, 580 pp.
2. Wilby, C. B., *Concrete for structural engineers*, Newnes-Butterworth, London, 1977, 212 pp.
3. Wilby, C. B. and Khwaja, I., *Concrete shell roofs*, Applied Science Publishers, London, 1977, 327 pp.
4. Wilby, C. B., *Design graphs for concrete shell roofs*, Applied Science Publishers, London, 1980, 170 pp.
5. Gould, L. G., *Static analysis of shells*, Lexington Books, Massachusetts, 1977, 401 pp.
6. Wilby, C. B., *Post-tensioned prestressed concrete*, Applied Science Publishers, London, 1980, 265 pp.
7. Wilby, C. B., *Prestressed concrete beams*, Applied Science Publishers, London, 1969, 97 pp.
8. Khwaja, I., *Theoretical analysis and experimental behaviour of hyperbolic paraboloidal shells*, PhD Thesis, University of Bradford, England, 1968, 263 pp.
9. Wilby, C. B., *Elastic stability of post-tensioned prestressed concrete members*, Edward Arnold, London, 1964, 46 pp.

10. Chatterjee, B. K., *Theory and design of concrete shells*, Edward Arnold, London, 1971, 244 pp.
11. Bharucha, J. N., Proposed classification of shell roofs, *Ind. Concr. J.*, Vol. 33, No. 12, Dec. 1959, pp. 419–421.
12. Billington, D. P., *Thin shell concrete structures*, McGraw-Hill, New York, 1965, 332 pp.
13. Wilby, C. B. and Naqvi, M. M., *Reinforced concrete conoidal shell roofs*, Cement and Concrete Association, London, 1973, 87 pp.
14. Ketchum, M. S., Economic factors in shell roof construction, *Proc. world conf. shell struct.*, San Francisco, 1962, National Academy of Sciences/National Research Council, Washington, DC, Publication No. 1187, 1964, pp. 97–102.
15. Whitney, C. S., Economics, *Proc. conf. thin concr. shells*, Massachusetts Institute of Technology, Cambridge, Mass., 1954, pp. 22–24.
16. Whitney, C. S., Cost of long span concrete shell roofs, *J. Am. Concr. Inst.*, Vol. 21, No. 10, June 1950, pp. 765–776.
17. Chatterjee, B. K., Economics of shell roof construction in India, *Ind. Concr. J.*, Vol. 33, No. 12, Dec. 1959, pp. 456–460.
18. American Society of Civil Engineers, *Design of cylindrical concrete shell roofs*, Manuals of engineering practice, No. 31, New York, 1952, 177 pp.
19. Evans, R. H. and Wilby, C. B., *Concrete: plain, reinforced, prestressed and shell*, Edward Arnold, London, 1963, 252 pp.
20. Lundgren, H., *Cylindrical shells*, Danish Technical Press, Copenhagen, 1960, 360 pp.
21. Wilby, C. B. and Bellamy, N. W., *Elastic analysis of shells by electronic analogy*, Edward Arnold, London, 1961, 56 pp.
22. Wilby, C. B., A proposed "exact" theory for analysing shells, and its solution with an analogue computer, *Proc. Inst. Civ. Engrs*, Vol. 22, No. 3, July 1962, pp. 291–308.
23. Flügge, W., *Stresses in shells*, 2nd Edn, Springer-Verlag, New York, Heidelberg, Berlin, 1973, 525 pp.
24. Gibson, J. E., *Thin shells*, Pergamon Press, Oxford, 1980, 289 pp.
25. Paduart, A., *Shell roof analysis*, CR Books, London, 1966, 97 pp.
26. Chronowicz, A., *The design of shells*, 3rd Edn, Crosby Lockwood, London, 1968, 343 pp.
27. Tottenham, H., A simplified method of design for cylindrical shell roofs, *Struct. Eng.*, June 1954, pp. 161–180.
28. Gibson, J. E. and Cooper, D. W., *The design of cylindrical shell roofs*, Spon, London, 1954, 186 pp.
29. Donnell, L. H., *Stability of thin-walled tubes under torsion*, NACA Report 479, 1934, 24 pp.
30. Jenkins, R. S., *Theory and design of cylindrical shell structures*, Ove Arup, London, 1947, 75 pp.
31. Vlassov, V. Z., Some new problems on shells and thin structures, *Izv. Akad. Nauk SSSR*, No. 1, 1947, pp. 27–53. [English translation in NACA TM 1204].
32. Finsterwalder, U., Die Theorie der kreiszylindrischen Schalengewölbe System Zeiss-Dywidag, *J. Bridge Struct. Eng.* 1932, pp. 127–152.
33. Schorer, H., Line load action on thin cylindrical shells, *Proc. Am. Soc. Civ. Eng.*, Vol. 61, No. 3, Mar. 1935, pp. 281–316.

34. Wilby, C. B., A method of designing north-light shell roofs, *Ind. Concr. J.* Vol. 35, No. 1, Jan. 1961, pp. 6–10.

35. Paduart, A. and Dutron, R. (Eds), Simplified calculation methods of shell structures, *Proc. colloq. simplified calc. methods*, Brussels, 1961, North-Holland, Amsterdam; and Interscience, New York, 1962, 529 pp.

36. Rüdiger, D. and Urban, J., *Circular cylindrical shells*, Teubner Leipzig, 1955, 270 pp.

37. Gibson, J. E., *Computer analysis of cylindrical shells: design tables for cylindrical shell roofs calculated by automatic digital computer*, Spon, London, 1961, 259 pp.

38. Ramaswamy, G. S., *Design and construction of concrete shell roofs*, McGraw-Hill, New York, 1968, 641 pp.

39. Paduart, A., *Introduction au casul et à l'execution des voiles minces en béton armé*, Centre d'Information de l'Industrie Cimentière Belge, Eyrolles, Paris, 1961, 95 pp.

40. Bradshaw, R. R., Some aspects of concrete shell buckling, *J. Am. Concr. Inst., Proc.*, Vol. 60, No. 3, 1963, pp. 313–328.

41. Reynolds, C. E., *Reinforced concrete designers' handbook*, Cement and Concrete Association, London, 1971, 362 pp.

42. Gol'denweizer, A. L., *Theory of thin elastic shells*, GITTL, Moscow, 1953, p. 423.

43. Pflügger, A., *Elementary statics of shells*, 2nd Edn, English translation, Dodge Corporation, New York, 1961, 122 pp.

44. Wilby, C. B. and Naqvi, M. M., Structural analysis of conoidal shells, *Struct. Eng.*, Vol. 50, No. 5, May 1972, pp. 97–201.

45. Wilby, C. B. and Naqvi, M. M., A flexural analysis of conoidal shells for reinforced concrete roofs, *Ind. Concr. J.*, Vol. 49, No. 7, July 1975, pp. 200–205.

46. Naqvi, M. M., *Theoretical analysis and experimental behaviour of conoidal shells*, PhD Thesis, University of Bradford, England, 1969, 206 pp.

47. Soare, M., *Membrane theory of conoidal shells*, Library Translation No. 83, Cement and Concrete Association, London, 1959, 33 pp.

48. Kraus, H., *Thin elastic shells*, Wiley, New York, 1967, 476 pp.

49. Ashwell, D. G. and Gallagher, R. H. (Eds), *Finite elements for thin shells and curved members*, Wiley, London, 1976, 268 pp.

33 Nuclear reactors

M F Kaplan

University of Cape Town, South Africa

Contents

Abbreviations

ACI	American Concrete Institute
AGR	advanced gas-cooled reactor
ASME	American Society of Mechanical Engineers
ASTM	American Society for Testing and Materials
BWR	boiling water reactor
CCV	concrete containment vessel
CRV	concrete reactor vessel
DBA	design basis accident
DBE	design basis earthquake
DBT	design basis tornado
FIP	Fédération Internationale de la Précontrainte
GCFR	gas-cooled fast reactor
HCDA	hypothetical core disruptive accident
HTGR	high-temperature gas-cooled reactor
HTR	high-temperature reactor
IAEA	International Atomic Energy Agency
ISO	International Standards Organization
LMFBR	liquid-metal fast-breeder reactor
LOCA	loss-of-coolant accident
LWR	light water reactor
MHE	maximum hypothetical earthquake
NPP	nuclear power plant
OBE	operating basis earthquake
PCCV	prestressed concrete containment vessel
PCPV	prestressed concrete pressure vessel
PCRV	prestressed concrete reactor vessel
PWR	pressurized water reactor
RCCV	reinforced concrete containment vessel

SMIRT	Structural Mechanics in Reactor	THTR	thorium high-temperature
	Technology		reactor
SSE	safe shutdown earthquake		

Summary

A survey is made of the design, construction, testing and surveillance of prestressed concrete pressure/reactor vessels (PCPV/PCRVs) and concrete containment vessels for nuclear reactors. The development and features of PCPV/PCRVs are described with particular reference to the 'integral design' concept and multicavity vessels for gas-cooled reactors as well as concrete pressure vessels for lightwater and liquid-metal fast-breeder reactors. Reference is made to standards and codes, materials, design and analysis, fabrication and construction, inspection and tests during construction, penetrations and closures, insulation and cooling, protection against overpressure, and testing and surveillance of the completed structure. The development and features of concrete containment vessels are described and reference is made to standards and codes, materials, design and analysis, and testing and surveillance. In the section on design and analysis, mention is made of loads, load combinations, earthquakes, aircraft and missile impact, external explosions, design basis accidents, probabilistic analysis and design. Reference is also made to concrete structures for radiation shielding and the effects of elevated temperature and nuclear radiation on the properties of concrete.

1 Prestressed concrete pressure/reactor vessels

1.1 Development and features

1.1.1 Introduction
The first industrial nuclear power plant was opened in October 1956 at Calder Hall in England. It has four gas-cooled nuclear reactors each with a net electrical output of 45 MWe.[1] The fuel consists of natural uranium rods encased in a magnesium alloy called Magnox and this type of reactor is referred to as a Magnox reactor. The core of the reactor, through which the carbon dioxide gas coolant flows at a considerable pressure and temperature, is contained in a steel pressure vessel. A large concrete biological shield, consisting of walls 2.14 m thick, surrounds the steel pressure vessel to protect the operators and the public from the effects of radioactivity from the reactor.

Further Magnox nuclear power plants of increasing power output, requiring larger pressure vessels at higher pressures, were built in the United Kingdom. The reactor core was contained in a steel pressure vessel and steel gas ducts lead to

1 MWe is used throughout the chapter to designate the *net electrical power* which is generated, i.e. 1 MWe is 1 MW net electrical power.

and from heat exchangers, or boilers, which were outside the concrete biological shield. The construction of steel pressure vessels gave rise to problems because of concern about the uniformity in quality of the increasingly thick steel plates which were required and the difficulties in welding these thick plates. The plates for the steel pressure vessels for the Sizewell A 290 MWe reactors were, for example, 10.49 mm thick.

It became apparent that there was a practical limit to the one-piece thickness of the walls of steel pressure vessels. The massive concrete structure needed for biological shielding also appeared to be sufficient to support not only a prestressing system to withstand the internal pressure but also a considerable thermal load. Prestressed concrete pressure vessels (PCPVs), also referred to as prestressed concrete reactor vessels (PCRVs), promised to have economic advantages over steel vessels. There also appeared to be an advantage from the safety aspect, as in the event of overpressure or overheating, gradual failure, typical of prestressed concrete structures, was likely to occur instead of sudden rupture.

Research into PCPVs for nuclear reactors proceeded independently in the 1950s in Britain and France. The first nuclear power reactors with PCPVs came into operation at Marcoule in France in 1959. Only the core of the reactor was contained in the horizontal cylindrical pressure vessel, which was 33.65 m in length; the internal diameter was 14 m and the concrete wall had a uniform thickness of 3 m. The ends were concave domes. The output of each of the two reactors was 37 MWe and the normal working and designed ultimate pressures were 1.47 and 6.37 MPa (213 and 924 psi), respectively. A larger 480 MWe power producing concrete pressure vessel reactor, EDF3, was built at Chinon and came into operation in 1967. This pressure vessel also only contained the reactor core and did not include the heat exchangers. The normal working and designed ultimate pressures were 3.04 and 7.60 MPa (441 and 1100 psi), respectively. The pressure vessel was a short vertical cylinder with flat ends, the internal diameter being 19.00 m and the internal height 21.25 m. The minimum wall thickness was 5.04 m and the minimum top and bottom cap thicknesses were 6.91 and 5.00 m, respectively.

1.1.2 The integral design concept

In Britain, consideration was given to the substitution of the steel pressure vessel by one in concrete which would also only contain the reactor core. It was, however, concluded that this did not lead to significant advantages as regards feasibility and cost. The overall integrity of 'reactor only' vessels was also questioned as several major components of the main circuit lay outside the prestressed concrete vessel.

Gradually, the 'integral design' concept was evolved in which the reactor core, the heat exchangers, and the gas circulators are all contained within a single prestressed concrete vessel. The whole of the gas pressure circuit would be contained in a structure which could not fail suddenly, and this meant a high standard of safety because of warning of failure. The plant would also be compactly grouped, and the thickness of the concrete required for structural purposes was sufficient for the biological shielding against radiation. This meant that the whole concept was more economic.

Construction of the nuclear power plant at Oldbury, England, commenced in

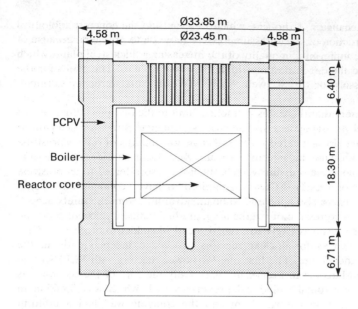

Figure 1-1 Oldbury PCPV for Magnox reactor

1962. It was the first commercial implementation of the 'integral design' concept and the first of the British nuclear power plants in which prestressed concrete pressure vessels were used. The plant is of 560 MWe capacity from two gas-cooled Magnox type reactors, and the reactor core, the heat exchangers and the coolant-gas circuit are contained in the concrete pressure vessels.

The Oldbury vessels are short vertical cylinders, 18.30 m high internally, and with an internal diameter of 23.45 m (Fig. 1-1). The concrete wall contains 22 layers of prestressing tendons arranged in a helical system at 45°. There are many large and small penetrations in the vessel, which has a steel liner internally to prevent leakage of the carbon dioxide gas coolant. The liner and the concrete behind it are protected by thermal insulation, and a cooling system consisting of a network of pipes is placed in the concrete immediately behind the liner. The normal working and design ultimate pressures were 2.41 and 7.85 MPa (350 and 1138 psi), respectively. This power plant was commissioned in 1968.

Construction of the Wylfa (Anglesey, Wales) nuclear power plant began in 1963 and was the last of the Magnox type to be built in Britain. The output of each of the two reactors is 590 MWe and, as at Oldbury, the reactor core, coolant-gas circuit and heat exchangers are housed in concrete pressure vessels. In contrast with the vessels at Oldbury, the cavity of the cylindrical pressure vessels at Wylfa is spherical, as it was considered that this shape was more economic. The internal diameter of the spherical cavity is 29.25 m and the minimum wall thickness 3.36 m with 16 vertical ribs to carry the external circumferential prestressing hoop tendons and anchorages. There are many large penetrations in the vessel and a steel liner, 19 mm thick, is anchored to the inside surface of the concrete vessel to prevent leakage of the gas coolant. The mean temperature of the concrete, 15 cm behind the liner, is limited to 35°C and the temperature cross-fall through the vessel wall is approximately 15°C. To maintain these temperatures, a cooling system is provided which consists of a series of 4 cm square cooling water pipes welded to the outside of the liner and metallic insulation attached to its inner surface. The working and designed ultimate pressures were almost the same as for the Oldbury vessels, and the plant came into commercial operation in 1971.

Construction of the Dungeness B nuclear power plant in England commenced in 1965. It has two reactors, giving a total net output of 1200 MWe and was the first commercial nuclear power plant based on the advanced gas cooled reactor (AGR) system which replaced the Magnox reactors in the United Kingdom. The fuel is enriched uranium dioxide and the canning material is stainless steel. The reactor core, heat exchangers and gas circulators are all contained in a concrete cylindrical pressure vessel. The working pressure is 3.30 MPa (479 psi) and the designed ultimate pressure 8.99 MPa (1300 psi). It is anticipated that this plant will come into operation during 1982.

The Hinkley Point B and the Hunterston B plants which followed in Britain were virtually identical in concept, design and construction. The output of each of the two reactors of the AGR type, at each plant, was 625 MWe. The pressure vessels are short vertical cylinders with an internal diameter of 18.90 m and internal height of 19.40 m (Fig. 1-2). The normal working and designed ultimate pressures are 4.03 and 10.60 MPa (584 and 1537 psi), respectively. The Hinkley Point station was commissioned in 1975 and the Hunterston station in 1976.

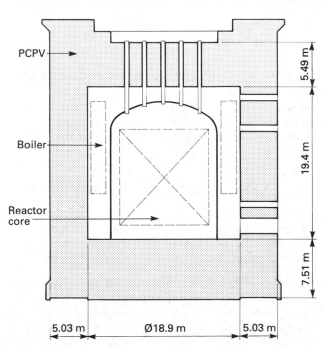

Figure 1-2 Hinkley Point B PCPV for AGR

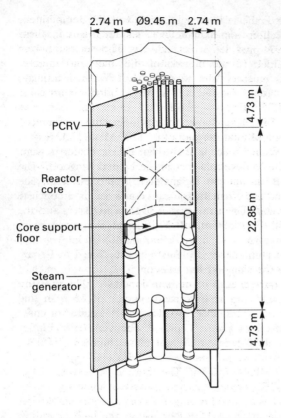

Figure 1-3   Fort St Vrain PCRV for HTGR (after Jones and Hedgecock [37])

Table 1-1   PCPV dimensions

Plant	Int dia m	Int ht m	Minimum thickness Wall m	Top cap m	Bot cap m
Oldbury	23.45	18.30	4.58	6.40	6.71
Wylfa	29.25[a]		3.36	3.66	3.36
Dungeness B	19.95	17.70	3.81	6.33	5.95
Hinkley Point B | Hunterston B }	18.90	19.40	5.03	5.49	7.51
St Laurent 1 and 2	19.00	36.30	4.75	5.70	6.00
Bugey 1	17.08	38.25	5.49	7.46	7.46
Fort St Vrain	9.45	22.85	2.74	4.73	4.73
Uentrop THTR	15.90	15.75	4.45	5.10	5.10

[a] Spherical PCPV.

Developments in France led from the 'reactor only' vessel for EDF3 at Chinon, to integral PCPV designs for St Laurent-des-Eaux 1 and 2 and for Bugey 1, which were commissioned in 1969, 1971 and 1972, respectively. The nuclear power station built at Vandellos in Spain (1972) was identical with St Laurent-des-Eaux 2. The outputs of the St Laurent reactors were 487 and 515 MWe, respectively, the normal operating pressures being 2.60 and 2.76 MPa (377 and 400 psi) and the designed ultimate pressure 7.35 MPa (1066 psi) in both cases. The reactor at Bugey 1, with a net output of 540 MWe, has a normal operating pressure of 4.48 MPa (650 psi) and a designed ultimate pressure of 11.87 MPa (1722 psi).

An interesting feature of the French reactors is that the heat exchangers were placed below the reactor core. This reduced the diameter of the vertical cylindrical vessel in relation to its height, the internal height being approximately twice the internal diameter, whereas in the British cylindrical vessels the diameter had generally been larger than the height. The arrangement of the reactor core above the heat exchangers means that the direction of the flow of the gas coolant through the reactor core is reversed, and there is no published information indicating whether or not the cost of dealing with this more than offsets any economies which may be due to the shape of the vessel.

The 330 MWe Fort St Vrain nuclear station, commissioned in 1973, was the first in the USA to use a prestressed concrete reactor vessel (PCRV). In this reactor, the heat exchangers or steam generators are also below the reactor core, which is supported by a concrete floor located at about the mid-height of the single cylindrical cavity. This is a high-temperature gas-cooled reactor (HTGR), the coolant being helium. The working pressure is 4.86 MPa (704 psi) and the designed ultimate pressure 12.11 MPa (1756 psi). The thermal protection of the PCRV from heating by the HTGR's primary coolant posed an important technological problem. This problem has existed for all gas-cooled reactors employing a PCRV, but is particularly severe in the HTGR because of the high operating temperatures. The PCRV cavity liner is of steel 19 mm thick (Fig. 1-3).

In West Germany, the first order for a PCPV was placed in 1972 for a 300 MWe thorium high-temperature reactor (THTR) at Uentrop and which used spherical fuel elements [1]. It followed the integral design concept and the vessel has an internal diameter of 15.9 m and internal height of 15.75 m.

The main dimensions of the PCPVs mentioned in this subsection are given in Table 1-1.

1.1.3 *Multicavity vessels*

In the integral design concept, the heat exchangers, which required regular statutory examination, were almost totally inaccessible and could only be removed with very great difficulty. It was therefore considered that it would be advantageous if the heat exchanger or boiler units could be removed from the PCRV during normal service for maintenance and inspection and, if necessary, for replacement. This could be done by providing vertical cavities or pods within the walls of the vessel into which the boilers could be placed and removed if required. The large closures for sealing the openings were expensive as they had to have high integrity. The vertical pods also complicated the shape of the vault, thus adding to the liner and insulation problems which were being experienced in vessels which had a single vault or cavity. The structural analysis of the vessel also became more complicated.

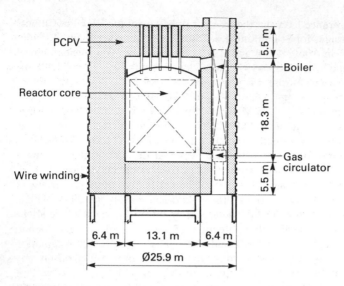

Figure 1-4   Hartlepool multicavity PCPV for AGR (after Jago *et al.* [2])

It was, however, decided to use 'podded' or multicavity PCPVs for the nuclear power plants to be built at Hartlepool and Heysham in England.

The Hartlepool and Heysham PCPVs [2] are cylindrical with a main pressure void 13.10 m in diameter and 18.29 m high. The outside diameter of the vessel is 25.91 m and the height 29.26 m. The minimum wall thickness is 6.40 m and the minimum top and bottom cap thickness is 5.49 m. The boilers are housed in eight symmetrically placed cavities 2.74 m in diameter, which extend through the full height of the vessel walls and are linked by gas ducts to the top and bottom of the pressure void. A gas circulator is mounted below each boiler and a novel feature of the design is the use of wire winding to provide the hoop stress (Fig. 1-4).

The Hartlepool and Heysham reactors are identical and each reactor (there are two at each power station) has an electrical output of 622 MWe. The normal operating pressure is 4.03 MPa (584 psi) and the designed ultimate pressure 11.09 MPa (1608 psi). It is anticipated that they will come into operation in 1982.

In West Germany, work has been in progress to develop a multicavity PCRV to provide the primary containment for an integrated system consisting of a high-temperature nuclear gas-cooled reactor, heat exchangers and either one or several helium gas turbines in direct circuit. The complete helium circuit, consisting of reactor, turbines, heat exchangers and gas ducts are contained in a PCRV [3]. The vertical cavities, which are not arranged symmetrically, are of different lengths and have diameters which vary from 2 to 5.8 m and each is subject to different pressure loadings. In addition, the bottom cap contains horizontal cavities for the turbo-set, 5.8 m in diameter, and for the generator, which is about 9 m in diameter. The vessel will have a 'warm' liner at a temperature between 150 and 200°C. The design and development of multicavity PCRVs for gas-turbine high-temperature gas-cooled reactors is also being carried out in the USA [4].

In the USA and Europe, multicavity PCRVs are being designed for gas-cooled high-temperature thermal reactors (HTGR) and for gas-cooled fast-breeder reactors (GCFR). A 300 MWe demonstration plant for a GCFR has been investigated [5] in which the PCRV is a cylinder 25.6 m in diameter and 24.5 m in height. The helium coolant is circulated at 9.0 MPa (1305 psi) outlet pressure downwards through the core and down through the steam generator. The support of the reactor core and its control are from above the core, while refuelling is carried out from below the core through penetrations in the PCRV.

The helium-coolant operating pressure in a GCFR is about 80 per cent higher than that in an HTGR but the reactor cavity for the GCFR is relatively smaller than that for an HTGR of the same power. In the multicavity vessels being developed for the Gas Breeder Reactor Association [6], the pods for the GCFR vessel are much larger relative to the central vault size than those in the HTGR vessel. For both the HTGR and GCFR designs a hoop wound and vertical tendon prestressing arrangement has been adopted.

In Austria, the concept of a PCPV with an elastic 'hot' liner and a temperature regulated concrete vessel wall is being developed, as it is felt that this would allow the possibility of inspecting and repairing the liner which has been prone to operational problems. The feasibility has been investigated of a multicavity PCPV, with an elastic hot liner, for an HTGR [7]. The normal operating and design temperatures of the liner were taken to be 300 and 350°C, respectively. To maintain the liner in the elastic compression range, the concrete in the wall of the vessel was designed to be at a temperature of 140°C.

Two types of PCPV, with a hot liner, have been investigated for a GCFR [8]: (a) a multicavity podded-boiler type, and (b) a satellite type consisting of a central reactor vessel surrounded by satellite vessels which are rigidly connected to the central vessel. The walls of both types consist of an elastic hot liner, a thermal barrier of insulating concrete and then prestressed concrete containing a temperature regulation system. A leak detection and evacuation system is located in the

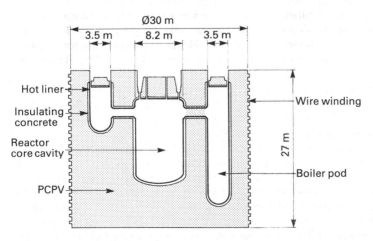

Figure 1-5 Multicavity PCPV with 'hot' liner for a GCFR (after Kumpf *et al.* [8])

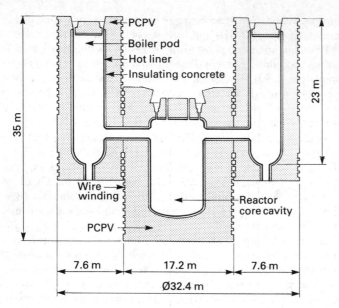

Figure 1-6 Satellite type PCPV with 'hot' liner for a GCFR (after Kumpf *et al.* [8])

thermal insulating concrete. The vessels have been designed for a liner temperature of 275°C and a gas pressure of 13.8 MPa (2000 psi). The designed liner temperature of 275°C decreases within the thermal barrier to 95°C and this level is controlled by means of the temperature regulation system in the prestressed concrete.

The multicavity vessel has an external diameter of 30 m and an overall height of 27 m. There are ten pods, 3.5 m in diameter, in the vessel wall and the central core cavity is 8.2 m in diameter. A large revolving concrete closure weighing about 3500 kN (390 US tonf) seals the central cavity (Fig. 1-5).

The satellite version consists of six separate satellite pods, 3.5 m in diameter, which are rigidly connected to the central vessel and which has a central cavity also 8.2 m in diameter (Fig. 1-6).

The investigations in Seibersdorf, Austria, have indicated that both of the above PCPV designs with a hot liner are feasible for gas-cooled fast reactors.

1.1.4 *Light water reactors*
Many years of research and development have been carried out in West Germany and in the Scandinavian countries with the cooperation of France, Italy and the United Kingdom, to design PCPVs for light water reactors (LWRs). The Scandianavian program has reached the stage where further structural work on the model and on theoretical aspects have not been considered by its sponsors to be necessary. A prototype vessel has however not been built [9]. In West Germany, a PCPV has been designed [10] with the object of using it instead of steel pressure vessels for pressurized water reactors (PWRs). The dimensions and shape of the vessel were based on a PWR of 1300 MWe. The investigations concerning the

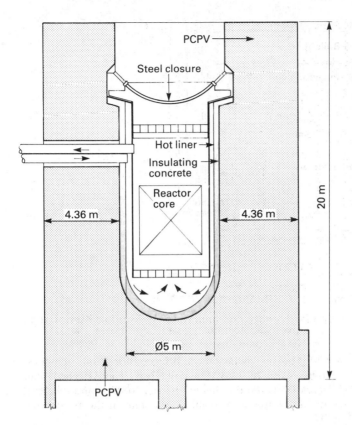

Figure 1-7 PCPV with 'hot' liner for a PWR (after Jungmann *et al.* [10])

reference design were carried out to show the feasibility of a PCPV with an operating pressure of 16.2 MPa (2350 psi), and with a hot liner at a temperature of 300°C. The concept of a hot liner is similar to that which is being developed in Austria for gas-cooled reactors (Fig. 1-7).

The development of PCPVs has not been without its problems, but they have given satisfactory performance over a long period of time and are considered to have advantages in regard to safety. The use of PCPVs for LWRs has, however, not made much progress.

1.1.5 *Liquid-metal fast-breeder reactors*

In the liquid-metal fast-breeder reactors (LMFBRs) which have been designed, the sodium coolant is contained in a stainless steel tank in which the reactor core, heat exchangers and pumps are suspended. The normal operating pressure, due to the pressure of the blanket of argon gas above the sodium, is very low, being about 0.035 MPa (5 psi). The steel tank, the reactor core, heat exchangers, pumps and pipework have been designed to be suspended from the top cap of a prestressed concrete structure. The reason for the use of a PCRV instead of a steel structure for this purpose is because, in the event of a hypothetical core

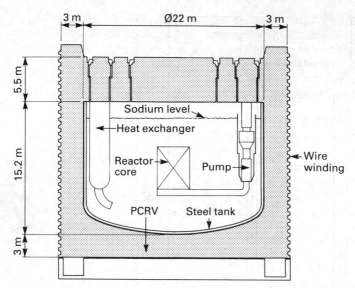

Figure 1-8   PCRV for an LMFBR (after Machertas *et al.* [11])

disruptive accident (HCDA), considerable energy can be released and a PCRV has the ability to absorb larger amounts of energy [11]. A PCRV is also considered to be more economic than a steel vessel for this purpose. In Britain, an LMFBR has been under development during the past 25 years and investigations have shown that a PCRV is viable for a 1320 MWe LMFBR [12] (Fig. 1-8).

1.2   Standards and codes

PCPVs for nuclear reactors are concrete structures which require particularly high integrity. The first standard specification, BS 4975, for PCPVs for nuclear reactors was published in Britain in July 1973 [13] and covers 'the design, construction, inspection and testing of single cavity and multicavity PCPVs for nuclear reactors including components which are necessary to maintain the structural integrity and leaktightness of the vessels'.

As mentioned in BS 4975, the safety of concrete pressure vessels depends on:

(a) a correct assessment of the loads which may be applied to the vessel;
(b) the proper design and construction of the vessel to withstand the loads;
(c) the adoption of proper operational procedures and adequate monitoring and inspection of the vessel during its service life.

The first requirement is not dealt with in BS 4975, as it is regarded as the responsibility of the purchaser and operator of the plant to specify the loads for which the PCPV is to be designed. In doing so, the purchaser must meet any statutory or other regulations. In Britain, nuclear installations must be licensed under the Nuclear Installations Act of 1965 and 1969 and the licensee will be the ultimate purchaser and operator of the plant which includes the PCPV.

The third requirement is also not covered by BS 4975, as the adoption of proper operational procedures, adequate monitoring and inspection of the vessel are regarded as being the responsibility of the operator of the plant. Brief reference to surveillance of the vessel is, however, made in the standard. The standard contains nine sections and nine appendices but does not include administrative arrangements for quality assurance and quality control, which are normally dealt with in a separate contractual arrangement between the purchaser and the manufacturer.

In the design section of BS 4975, there are many areas where only general guidance is given as the standard is based on the principle that the designer is an expert in the field of PCPV design and that it is intended to give him guidance. Extensive reference is, however, made to the British Standard Code of Practice, CP 110, for the structural use of concrete which is referred to as the Unified Code or UC [14].

Section III, Division 2, of the American Society of Mechanical Engineers (ASME) *Boiler and Pressure Vessel Code* [15] deals with concrete reactor vessels and containments. This part of the code was prepared under the sponsorship of the American Concrete Institute (ACI) and the ASME, and in this chapter it will be referred to as the ASME Code. The final draft was published in April 1973 and became mandatory in the USA on 1 July 1975. It is a comprehensive document which covers almost all aspects of concrete reactor vessels (CRVs) and concrete containment vessels (CCVs). The code contained three main subsections: CA—General Requirements; CB—Concrete Reactor Vessels; CC—Concrete Containments. Subsection CA was later deleted and merged with Subsection NA of Section III, Division I to form Section III, Subsection NCA.

Subsection NCA, which deals with the administrative control requirements for both Subsections CB and CC, is a unique feature of the ASME Code that differs from BS 4975 and documents pertaining to CRVs which have been prepared in France [16] and West Germany [17]. This subsection deals with administrative control requirements and the certification of quality assurance programs and which involves industry, technical societies and national and state governments before a 'Certificate of Authorization' is awarded. The philosophy which underlies the inclusion of Subsection NCA in the ASME Code is that the design and the allowable stresses in the materials depend on the quality control exercised during fabrication, construction and testing.

Subsection CB covers the requirements for concrete reactor vessels and establishes rules for materials, design, fabrication, construction, testing and examination, protection against overpressure, marking, stamping, and preparation of reports.

The International Atomic Energy Agency (IAEA) is developing international standards for nuclear power plants with particular reference to developing countries. The standards currently being prepared deal with licensing organization, siting, design, operation and quality assurance for nuclear power plants. The Fédération Internationale de la Précontrainte (FIP) has also appointed a commission on nuclear construction.

The International Standards Organization (ISO), through Technical Committee 85, Subcommittee 3, which deals with nuclear power reactor technology, is also developing nuclear standards which are of interest to the structural engineer, and Working Group 10 is dealing with concrete pressure vessels for nuclear reactors.

The view has, however, been expressed [18] that

> Standards developed by ISO tend to be a compromise between the various national industrial standards developed in the leading industrial nations. As such, they tend to be used in smaller countries which do not have a well developed standards activity of their own and are particularly useful in providing a common bid basis for suppliers who would otherwise use their own national standards as a bid basis. However, it is doubtful whether international standards will ever effectively replace national standards of the leading industrial countries as a basis for detailed design.

1.3 Materials

Both BS 4975 and the ASME Code cover the requirements for concrete and concrete materials, materials for reinforcing, materials for prestressing systems, materials for liners and welding materials.

In BS 4975, the requirements for materials for concrete and concrete mixes are for the most part covered by reference to the Unified Code [14], while in the ASME Code these are generally covered by reference to the relevant American Society for Testing Materials (ASTM) standard and/or American Concrete Institute recommendations. In general, requirements for materials for concrete for PCPVs do not differ greatly from the requirements for materials for other prestressed concrete structures while concrete mixes should be designed to meet the requirements of the structure.

Concrete in PCPVs is, however, subjected to elevated temperature and nuclear radiation. Prestressed concrete pressure vessels are also massive structures. For these reasons, the effects of temperature and nuclear radiation on concrete properties, such as uniaxial and multiaxial strength, elasticity, creep, shrinkage and expansion, and behaviour when subjected to dynamic loads, changes in moisture content and elevated cyclic temperature variations, need to be known. Further reference to the properties of concrete is made in Section 4 of this chapter.

In regard to the reinforcement, prestressing system and welding materials, BS 4975 refers to the relevant British Standards and makes a general statement to the effect that their properties shall be demonstrated by the manufacturer to be suitable for use in the construction of the PCPV. As far as the liner is concerned, BS 4975 states that 'the manufacturer shall give particulars of the materials he proposes to use for the liner, together with a statement of the properties required which are relevant to the design, fabrication and life of the liner taking account of the operational environment'.

The ASME Code, while referring to the relevant existing US standards concerning these materials, deals with the requirements for material for liners, and with welding, in much greater detail than BS 4975.

1.4 Design and analysis

1.4.1 General structural design requirements
BS 4975 requires that the suitability of the PCPV must be substantiated for all possible conditions of reactor operation including the effects of fault conditions.

The integrity of the structure must therefore be determined to ensure that adequate margins of strength over the design pressure are available. The design pressure must generally be 1.1 times the working pressure and the permissible stresses shall be such that the vessel will exhibit essentially elastic behaviour at pressures well in excess of the working pressure. The ultimate pressure with the liner intact is to be not less than 2.5 times the design pressure and the overload behaviour of the vessel must be such as to give advance warning of failure.

The ASME Code [15] states that the PCRV must be designed so that an essentially elastic response is ensured under normal service conditions. As the ultimate structural capacity of the PCRV is approached, the vessel response must be gradual, observable, and predictable and in a manner such that the minimum ultimate load capacity is clearly developed. The minimum ultimate structural capacity must be such as to be able to resist twice the loads associated with the maximum cavity pressure.

It is thus clear that the PCPV must be designed for essentially elastic response to pressures above working pressure under normal service conditions, and also to have an adequate factor of safety against failure.

One of the main features of a PCPV is that when in operation it is subjected to continuous elevated temperatures and nuclear radiation, and in both BS 4975 and the ASME Code, attention is drawn to the fact that in designing the PCPV specific consideration must be given to such matters as (a) creep and shrinkage, (b) steel relaxation, (c) effects of nuclear radiation on the properties of the liner, concrete and reinforcing and prestressing components, (d) thermal effects on the properties of the liner, concrete and reinforcing and prestressing steels, and (e) the effects of moisture and age on the properties of concrete.

1.4.2 *Loads*
Contrary to most other branches of structural engineering, the structural engineer does not determine the pressure and temperature loads to be adopted for PCPVs. These loads are established by the designer of the nuclear reactor and are controlled by means of very sophisticated instrumentation. As mentioned above, BS 4975 states that in Britain it is the responsibility of the purchaser to specify the loads for which the PCPV has to be designed.

There are many types of load and load combination which must be considered in the design of a PCPV. In the ASME Code, six different categories of load combinations are mentioned (Table 1-2).

BS 4975 does not detail the different loading combinations and simply states that the PCPV should be analysed for all relevant combinations of mechanical and thermal loads which can arise under normal operating conditions during the life of the structure. In Appendix A of the British Standard a simple qualitative description of the effects of a typical sequence of loadings on the vessel structure is given in order to indicate an approach to the stress analyses of the structure. The ultimate purchaser and operator of a nuclear power plant in Britain must satisfy the Nuclear Installations Inspectorate that the PCPV is safe by means of reports which include details concerning the loading conditions to which the PCPV may be subjected.

There has been much discussion in recent years concerning the 'gas in cracks' loading problem arising from leaks in the steel liner and which could result in the

Table 1-2 Load combination categories, ASME Code [15]

Load combination category	Loads which should be considered for inclusion
Construction	Loads resulting from PCRV fabrication and construction or preoperational testing
Normal	Loads resulting from system start-up, power range operation, shut down and servicing
Abnormal	Loads resulting from unplanned or unscheduled events due to such causes as operator error, equipment failure, and unusual off-site electrical problems
Severe environmental	Loads resulting from severe environmental effects which could occur infrequently at the site, e.g. the operating basis earthquake selected for the site, wind, and flood
Extreme environmental	Loads which recur at extremely long intervals, e.g. the safe shutdown earthquake, tornado and tsunami
Failure	Loads resulting from postulated component failures of extremely remote possibility, e.g. specified pressurized crack conditions, penetration closure failure, tendon failure, steam impingement on prestressing anchorage assemblies

reactor coolant gas penetrating into cracks or construction joints in the prestressed concrete. The area subjected to pressure is thus increased and this could reduce the ultimate load factor. The PCPV could thus fail without the pressure or temperature in the vessel cavity being in an overload state.

It is difficult to quantify and analyse the loadings due to gas in cracks because they depend on factors such as the permeability of the concrete and the flow characteristics of the gas through the vessel wall. Unless liner integrity against significant leakage can be guaranteed at a pressure of up to 1.5 times design pressure, BS 4975 requires a calculation assuming that the pressure diminishes linearly from the inside to the outside of the vessel unless other decay laws are agreed with the purchaser. Calculations for this hypothetical form of loading must generally indicate that the ultimate pressure is not less than 1.5 times the design pressure.

The ASME Code requires that specified pressurized crack conditions must be considered for inclusion in the failure category load combinations, in the analysis of which a limit is placed on the prestressing tendon and liner strains.

The French Decree 103 [16] also refers to 'gas in cracks' loading and specifies an assumed linear pressure gradient from the interior to the exterior of the PCPV.

High-temperature gas-cooled fast reactors operate at higher pressures and have relatively thicker walls. This poses a bigger problem in regard to gas in cracks loading and it has been suggested that the problem could be limited, if not completely eliminated, by providing a ventilation system in the concrete immediately behind the liner. Research in Scandinavia has indicated that for a PCPV for a boiling water reactor (BWR), which has a thick wall and a relatively small cavity working at high pressure, the forces in a 'vented' vessel may be reduced to a significant extent.

1.4.3 *Analysis*

BS 4975 requires that stress analyses should be carried out for at least four loading conditions and the method of analysis selected for each loading condition shall take appropriate account of the time and temperature dependent characteristics of the concrete. Particular attention must also be given to the calculation of deformations in the PCPV.

Permissible principal stresses in the concrete for each of the four loading conditions are specified and a distinction is drawn between 'field' stresses and 'local' stresses. Consideration is also given to multiaxial compressive stresses in the concrete. The permissible stresses are such that the PCPV will exhibit essentially elastic behaviour at pressures which are considerably greater than the design pressure and sufficient prestress must be provided so that there is a net compressive force across all sections of the vessel under service load conditions.

Adequate margins of strength over the design strength must be assessed by means of ultimate load analyses and the British Standard requires two separate ultimate load analyses to be carried out: (a) ultimate load with liner intact, and (b) ultimate load with leaking liner, allowing the gas coolant under pressure to penetrate any cracks in the concrete, thus increasing the pressure loading on the structure. In case (a), it has to be shown that the ultimate pressure is not less than 2.5 times the design pressure, and in case (b) not less than 1.5 times the design pressure. In case (b), it is assumed that the pressure diminishes linearly from the inside to the outside surface of the vessel and also that the prestressing system remains intact. Strongly dissenting views are held on the merits of retaining the ultimate load design approach [19].

The first PCPVs were built in the United Kingdom on the basis of a calculated ultimate load substantiated by overpressure tests on models. Design and analysis methods, however, came under increasing scrutiny and the analysis of stresses under service or normal operation conditions received much attention. Finite difference, finite element and dynamic relaxation methods of elastic analysis were applied to the calculation of stresses in these thick-walled structures under normal operating conditions at elevated temperatures. Attention then returned to overload and fault conditions and, consequently, to inelastic or non-linear analysis. BS 4975 contains very few firm requirements in regard to methods of analysis General guidance is however given in Appendix B of the standard on methods of elastic and inelastic stress analysis which may be adopted. The three main methods are the finite element, finite difference and dynamic relaxation methods. Although all of these methods have been shown to give sufficiently accurate results for PCPV design, it seems as if the finite element method is considered to be the most effective.

At the fifth Structural Mechanics in Reactor Technology (SMIRT) Conference in Berlin in 1979, no less than 90 papers were presented dealing with methods for structural analysis. In one paper [20], reference is made to the growing role of computer-based systems in the design and analysis of nuclear reactors and the result of an international survey of computer systems based on the finite element method is summarized. It is stated that the domination of this method results largely from its potential for, and proven success in, solving problems involving various types of physical systems subjected to arbitrary environmental conditions. Further improvements are anticipated which will enhance the analyst's ability to cope with increasingly complex geometries, materials, and operating conditions over extended time periods.

A development largely concurrent with that of the PCPV has been the introduction of limit state design in most areas of structural engineering. There are, however, many difficulties in expressing the design of PCPVs in limit state terms because of the lack of the data required, particularly in regard to establishing the probabilities of occurrence of the less likely faults, and therefore the associated loads and overloads. BS 4975 mentions that a limit state method of analysis could be a more logical design approach but states that 'as the methods of applying limit state design to PCPVs have not yet been established, they shall only be adopted with the permission of the purchaser'.

The ASME Code [15] specifies the analytical design procedures for each of the six load combination categories mentioned in Table 1-2, e.g. for construction and normal category load combinations, the analytical methods used 'shall account for the thick section geometric characteristics of the CRV, and the stress, strain, and deformation characteristics of the CRV liner and embedments' and 'the assumption of linear elastic stress–strain properties for concrete and steel is permissible'. For the extreme environmental and failure load categories, 'the analytical methods used for the investigation of the strength margins in the CRV shall be capable of predicting with reasonable accuracy the ultimate load behaviour of the type of CRV under investigation'.

The ASME Code also specifies four 'condition' categories of stress corresponding to various safety situations that describe four plant conditions of varying probability of occurrence [21]. The permissible 'average' and 'point' stresses for concrete and reinforcing steel are given for each condition. The stress limits for prestressing steel are specified and provision is also made to account for the effects of a multiaxial compressive stress field.

The strength, creep and shrinkage behaviour of concrete is affected by elevated temperatures and by time. The difficulties of calculating stresses in the concrete at different ages are considerable particularly when it is also realized that a PCPV is a thick-walled structure which is stressed not necessarily in two but also in three dimensions. These difficulties are aggravated by the lack of data on material properties under these conditions.

In this context, it is interesting to note what Hannah [9] has said in connection with PCPV design:

> It seems to me doubtful whether we can profitably pursue calculation refinement much farther. The power and sophistication of the analyst and his computer has surely outstripped the quality of data being used, particularly that on concrete material characteristics. It is idle to pretend that we have done, or can do, more than scratch the surface of these multi-dimensional concrete properties. When one adds that previous stress history

and elevated temperature experience will undoubtedly affect such properties and when further one recalls the infinite range of concrete materials and
mixes, this vast limitation on data availability is all too clear.

Not only more data on material properties is needed but improved constitutive
descriptions of the behaviour of materials is also required [22]. There are, for
example, many mathematical models for the behaviour of concrete. These models
have used both hypo-elastic and elasto-plastic descriptions of the material, which
are however based on varying and sometimes quite different assumptions. It must
also be appreciated that if mathematical models of materials are to be used in
practice, they must not only represent the actual behaviour of the material but
they must also be amenable to computational methods.

1.4.4 *Liners*

The internal concrete surface of a PCRV for gas-cooled reactors is generally
covered by a steel liner about 20 mm in thickness. The purpose of the liner is to
serve as an impermeable or gas-tight membrane to prevent leakage of the coolant
from the pressure vessel. If the coolant penetrates the liner it could give rise to
the 'gas in cracks' problem previously mentioned. It could also cause pressurization of steel–concrete interfaces, which could result in a back pressure on the
liner when the reactor vessel is depressurized. If the leakage is sufficiently great,
hot gas could also reach the prestressing ducts, causing possible damage to the
prestressing tendons.

The liner is anchored to the concrete by means of studs or ribs which are
welded to the outside of the liner and which are embedded in the concrete.
Cooling tubes through which water is circulated are also welded to the outside of
the liner to control the temperature of the concrete in the vessel walls. In the
PCRVs which have been built, the 'cold'-liner system has been adopted, the liner
being kept 'cold' by means of insulating materials fixed to the inside surface of the
liner to form a thermal barrier. The liner is also used to serve as internal
formwork for the concrete in the vessel wall.

The ASME Code specifies liner design analysis procedures and requires a stress
analysis and, if necessary, experimental methods to show that the liner together
with its anchorages, brackets and attachments will comply with the specified
design requirements when subjected to the specified loadings. In contrast with the
requirements for concrete containment vessels, the ASME Code does not prohibit
the use of the liner as a strength element in the structural analysis of a PCRV.
BS 4975 does not cover the design and analysis of liners in such detail as the
ASME Code, although it is interesting to note that particular attention is drawn to
the effects of nuclear radiation on the properties of the liner and its attachments.

One of the problems with existing insulated PCPV cold-liner systems is that
they are very difficult to inspect and repair. Reference has already been made to
the work in progress to develop a hot-liner system for PCPVs. The concept of a
PCPV with an elastic hot liner, together with arrangements to adjust wall
temperature, has also been investigated [23]. By balancing temperatures over the
whole vessel wall, consisting of the liner, the thermal barrier of insulating
concrete, and the structural concrete, the resulting liner stresses can be limited.

The improvement of the conventional liner cooling system into a temperature control system, distributed over the vessel wall, permits the adjustment of concrete temperatures by cooling and heating arrangements and presents the possibility of achieving an optimal temperature distribution and thus the minimization of thermal stresses in the vessel wall. This could mean improved safety of the PCPV.

1.4.5 *Models*

In the design of PCRVs, considerable use has been made of models [24]. In referring to methods of analysis, BS 4975 states that in order to establish confidence in an analytical technique it should, if necessary, also be checked against measurements from models or previously completed vessels. No mention is made of the scale or size of the model. The ASME Code requires structural modelling 'where analytical procedures to predict CRV ultimate strength and behaviour in the range approaching failure are not established, and where a model of a prototype with characteristics similar to those of the current design has not been constructed and tested in accordance with Article CB-3340'. It also requires that the model 'shall maintain similitude, including that of material, to the prototype design and be geometrically similar with respect to the principal dimensions of the prototype'. The scale ratios for models for different purposes are indicated. For example, when the model is to be used to indicate elastic response, or the ultimate failure mode, then the scale of the model should be 1 to 14 or larger, whilst if it is to be used to indicate long-term temperature response, then the scale should be 1 to 4 or larger. The French Decree [16] states that, in order to establish elastic response and the ultimate failure mode of each new vessel design, the scale should be at least 1 to 6. The West German requirements are that model tests need to be carried out only when analytical methods are unable to predict the ultimate load-bearing capacity of the PCRV [17].

Models used in the design of PCRVs have been divided into the following categories [24]: (a) small-scale epoxy resin models to ascertain elastic response, (b) microconcrete models to determine modes of failure, (c) large-scale conventional concrete models in which the maximum aggregate size is about 1 cm and (d) models of specific portions of the PCRV, such as the end caps.

When PCRVs were first being designed, model tests were of great importance. The development of analytical methods and the acquisition of experimental data has reduced the need for expensive model tests. Most designers now accept that, for elastic response, tests on models can be replaced by analytical techniques, although the same does not yet apply to ultimate failure behaviour or to PCRVs with complex configurations. It has also been pointed out [5] that the majority of the specified loading conditions can no longer be applied to a model and the mean and peak stresses, which are closely specified, cannot be measured with the accuracy required. For the design of complex components such as end caps which have many large and small penetrations, tests on models are, however, considered to be a very helpful design tool.

1.5 Fabrication and construction

1.5.1 *Introduction*

When considering the use of PCPVs for nuclear reactors, one of the advantages which was claimed over welded-steel pressure vessels was that their fabrication and construction did not require any abnormal new fabrication and construction techniques. To a large extent, this has been shown to be true and the requirements for the concrete, reinforcement, prestressing system and liner are mostly covered by reference to existing standards and codes.

1.5.2 Concrete

PCPVs are large structures with vertical walls and end caps of considerable thickness which necessitate the casting of concrete in large bays. Careful consideration has therefore to be given to the location of vertical and horizontal construction joints and the design of the concrete mix so as to minimize temperature stresses and thermal movements due to the heat of hydration. High-strength concrete is generally required, which means mixes rich in cement and the use of low-heat cement often becomes necessary.

The workability of the concrete is also an important consideration in order to ensure satisfactory compaction around the numerous prestressing tendon ducts, liner anchorages, cooling pipes, and steel reinforcing which, in some areas, is very closely spaced. In the PCPV for the Fort St Vrain nuclear power plant, it was decided to use preplaced aggregate concrete for the bottom head and the core support floor because of the highly congested and confined space for construction [25].

One of the main problem areas is, however, the construction of the thick concrete top cap of a PCPV in which are embedded a great number of standpipes for fuelling and control of the reactor [26]. In the top caps of the Hartlepool and Heysham PCPVs, there are 405 standpipes with a minimum diameter of 324 mm and set on a square pitch of 460 mm. At the Dungeness B station, there are 465 standpipes in the top cap within a circle of 10 m. The standpipes are 322 mm in diameter and 6.4 m in length, the minimum ligament distance between each standpipe being 70 mm [27]. To prevent misalignments which could affect the fuelling and control operations of the reactor, the positional tolerances are very tightly specified and this requires extraordinary accuracy in construction. The top cap must also be structurally sound with no cracks or voids in the concrete and this means that control over the casting of the concrete goes beyond what is usually needed for conventional concrete structures.

1.5.3 Prestressing system

1.5.3.1 *Introduction* The linear prestressing systems which have been used for PCPVs are the wire, strand, and bar type systems, in which the tendons respectively consist of a number of parallel wires, a number of strands each consisting of a number of wires, and a number of high-tensile strength bars grouped together. The sizes of tendons have become larger and ultimate loads of approximately 10 MN per tendon are now not unusual.

The development of internationally agreed acceptance tests for prestressing tendons and systems for use in nuclear construction has been mentioned, and it has been said that it is possible that this may emerge from the deliberations of the FIP Commission on Nuclear Construction [28].

1.5.3.2 *Corrosion protection* There has been much discussion concerning the corrosion protection of prestressing tendons for PCPVs. At present, protection of prestressing steel against corrosion is provided either by encapsulating the tendons in a grease or organic petrolatum-based material containing corrosion inhibitors, or in a cement grout. There is considerable controversy as to which method is to be preferred. BS 4975 states that 'tendons shall not be grouted unless there are special reasons for so doing which have been agreed with the purchaser' and in Appendix G it is mentioned that in the United Kingdom it is required 'that it should be possible to check the loads in prestressing tendons throughout the life of the vessel and to withdraw sample tendons for inspection'. Although tendons in ducts in PCPVs built in the UK have not been grouted, the British Standard recognizes that it is possible that changed circumstances may make it advantageous to grout and comments are made on the special factors which have to be considered if grouting is envisaged. Current practice in the USA is to use ungrouted prestressing tendons, mainly because nuclear licensing requirements are such that reactor vessels and containments must be designed to allow for periodic inspection of all important areas, and for appropriate surveillance of the structure.

Arguments in favour and against both grouted and ungrouted tendon systems have been advanced and a recent experimental investigation in the USA concluded that 'both Portland cement grout and organic petrolatum-based materials, when properly applied, provide positive exclusion of corrosive agents, even under severe conditions' [29].

In the majority of PCPVs and prestressed concrete containments which have been built, the prestressing tendons are protected against corrosion by means of grease mainly because, for a small increase in cost, friction in the ducts is minimized, surveillance is possible and dismantling and replacement can be done if necessary. The prevention of voids in cement grouting is also very difficult to achieve in practice. On the other hand, the view is held that cement grouting seems to offer the best protection to the steel, provided the work is done properly and is strictly supervised on the site. Some cable ducts should, however, be filled with grease so that periodical checks on tension in the tendons can be made [30].

During a statutory inspection of the PCPVs at the Wylfa nuclear power station in 1971, it was found that the exposed prestressing hoop tendons were extensively pitted, with pit depths up to 0.3 mm. Corrosion of a similar type had previously been found on the ducted tendons at Dungeness B and was ascribed to stray electrical currents. The pitting of the non-ducted hoop tendons at Wylfa is considered, in the main, to have involved the chloride ion [31].

The arguments for and against grouting of prestressing tendons for PCPVs and containments have not yet been resolved.

1.5.3.3 *Circumferential winding* The first PCRVs were prestressed by means of circumferential and vertical tendons housed in ducts in the concrete walls and by cross-head tendons. In the mid-1960s, because of the large quantities of steel being used for the circumferential prestressing of PCRVs, attention was given to the development of circumferential wire-winding techniques.

In 1966, Taylor Woodrow Construction Ltd started the design and development of a wire-winding system which has been used in the construction of the Hartlepool and Heysham pressure vessels. The high-tensile steel wire is wound under tension into channels which are preformed in the wall of the vessel [32].

BBRV developed a winding system and completed the construction of equipment for this purpose in 1971. The continuous strand, which forms each layer, is anchored at both ends [33]. The General Atomic Co. in the USA has also developed a strand-winding system for applying high circumferential prestress to PCPVs [34]. During construction, the outer cylindrical surface of the vessel is formed using precast concrete panels with steel lined channels. Layers of high-tensile steel strand are wrapped under tension into these channels, using a machine that is supported from an overhead trolley system and which travels round the circumference of the vessel. The entire strand winding process is handled by only three operators.

With the increasing size of PCRVs and the high densities of circumferential prestress which are required, there are many advantages to be gained by using a circumferential winding procedure for hoop prestressing. Circumferential winding on the outside surface of the vessel wall makes more space available in the wall for other purposes, such as the inclusion of heat exchangers. Circumferential prestress can also reduce the need for conventional tendons in the cross-heads which contain many penetrations and which are very congested. The cost of wire winding is also considered to be less, one reason being that expensive anchorage components are not required.

Wire or strand-wound circumferential prestressing systems also have disadvantages, but it seems as if the advantages are greater than the disadvantages, and that for PCRV designs which are amenable to winding it is advantageous to use this technique in combination with conventional tendons. Strip-wound prestressing systems, which could have even greater advantages, are also being developed.

1.6 Inspection and tests during construction

For purposes of safety and reliability, it is important that rigorous quality control is practised in the construction of concrete pressure vessels for nuclear reactors to ensure the quality of the materials and the standards of workmanship. The amount of time spent on, and the cost of, doing this is very much more than for conventional civil engineering structures. In the construction of the PCRV for the 300 MWe THTR being built in West Germany, the cost of the quality control of the materials per cubic metre of concrete placed was approximately ten times that for normal prestressed concrete structures [1].

The inspection and testing requirements for the concrete, reinforcement, prestressing system and liner are stated in both BS 4975 and the ASME Code [15]. For the concrete, reinforcement, and liner, BS 4975 refers to the appropriate existing British Standards. Because ungrouted tendons make the safety of the PCPV particularly dependent on the integrity of the prestressing system, BS 4975 states that design proving tests should be carried out on previously unproven systems. For both ducted and wound tendons, the tests should demonstrate that tensioning, retensioning and detensioning can be carried out not only safely and expeditiously, but also with sufficient precision. Tests to determine the efficiency of the anchorages, the ultimate strength of the tendons, and evidence that components transferring the anchorage load to the vessel concrete have adequate strength, should also be carried out.

In Article CB-5000 of the ASME Code, the quality control, examination and testing of concrete materials and concrete are generally covered by reference to the appropriate ASTM requirements. In the same Article, the requirements for the examination and testing of reinforcement, prestressing systems and liners are stated. As far as the prestressing is concerned, the examination of tendon fabrication, the placing and tensioning of tendons, installation of ducts and bearing plates, and the injection of the corrosion inhibiting materials are specified. In regard to liners, the required examination of welds is specified in some detail.

1.7 Penetrations and closures

The end slabs, particularly the top head slabs, of PCPVs can have a large number of penetrations, both large and small. In multicavity PCPVs, the top head slab usually contains a large number of small penetrations for refuelling and for control rods, as well as large penetrations for the heat exchangers, feed-water and steam pipes. The bottom head slab may also contain penetrations, e.g. for the gas circulators.

The design of heads with penetrations is very complex as the strength of such a head must be such that the overall strength and integrity of the PCPV is not reduced. The reduction of groups of small penetrations, and large penetrations, to axisymmetric geometries is not easy, and the calculation of local stress variations can be very difficult. This often necessitates the use of experimental methods and models to prove that a design is adequate in spite of advanced methods of analysis.

PCRVs also have closures, or plugs, which are usually designed so that they can be removed, e.g. the closures which are designed to permit the replacement of heat exchangers or boilers.

The heat exchangers in the Hartlepool and Heysham multicavity pressure vessels are housed in eight separate vertical penetrations 2.74 m in diameter, and which extend over the full height of the vessel walls. The penetrations are sealed at the top with removable closures so that the heat exchangers can be replaced, and also at the bottom, where there is access for the gas circulators. The top closure also has penetrations for nine major feed-water and steam pipes, and the original design consisted of a concave steel dome. Investigations into the strength of wire-wound, biaxially-stressed concrete disks led however to the adoption of prestressed concrete closures. The concrete is contained within a light steel membrane to provide a smooth moisture-tight surface on to which the prestressing wire is wound. A load factor, well in excess of three times the vessel design pressure, was demonstrated by model tests. The Hartlepool/Heysham prestressed concrete closures were the first to be used in nuclear pressure vessel construction [35]. Closures have, however, generally consisted of concave or convex steel domes or a steel plug made up of several steel plates.

1.8 Insulation and cooling

The coolant outlet temperatures in nuclear reactors are high and can vary from about 400°C for Magnox reactors to over 800°C for high-temperature reactors. Elevated temperatures affect the strength, creep and other properties of concrete, as well as the behaviour of the liner and the prestressing tendons. The general practice has been to reduce and maintain the maximum temperature of the liner and the concrete behind it to below 100°C. Some of the factors which affect the specification of the temperature levels in the liner and the adjacent concrete, and thus the temperature gradients in the concrete in the PCPV, are the thermal stresses and dimensional changes in the concrete, and the thermal strains in the liner.

Consideration must be given to thermal effects not only during normal operating conditions, but also during the initial heating up of the PCPV, and to the effects of reactor transient conditions. During initial heating up, temperature distributions could occur which may give rise to more severe stresses than under normal operating conditions. At Wylfa, Hartlepool and Heysham arrangements were made to obviate this [36]. Transient conditions in the reactor cause changes in temperatures in the PCPV, leading to changes in the thermally induced stresses in the vessel concrete and liner. Significant tensile stresses could occur in the concrete under these conditions, and arrangements should be made to alter the liner cooling-water temperature to reduce these stresses.

For Magnox reactors and AGRs, the temperature of the liner and the concrete immediately behind it has generally been limited to temperatures between 60 and 80°C. In the Fort St Vrain HTGR, the reinforced-concrete core support floor, which is encased in steel, is exposed to both core inlet and outlet gas temperatures, the latter of which could, at certain locations, be as high as 1085°C. In a potential accident, the transient peak temperature to which the core support floor could be exposed is about 1650°C. The general average and maximum concrete temperatures adjacent to the steel casing were set at 54 and 66°C respectively, while the maximum local concrete temperature adjacent to the steel case was specified as 93°C [37].

Large temperature differences between the reactor coolant and the inside face of the vessel must therefore be maintained. The procedure has generally been to fix an insulating blanket on the coolant side of the liner and to weld cooling-water pipes on the concrete side. The main types of insulating material which have been used are metallic foil and ceramic fibre, which are fixed to the liner by a system of cover plates and studs. The insulation and cooling system for a PCRV forms a high proportion of the total cost of the vessel. It has been stated [36] that for the Hartlepool and Heysham vessels the cost of the optimized temperature control system was about 30% of the total vessel cost.

Pressure fluctuations can cause vibration of the hot face components of the insulation system which can fail due to fatigue and severe wear. These components are very difficult to inspect and repair and reference has already been made to the concept of a PCPV with a hot liner and a temperature-regulated vessel wall which is being developed for HTGRs and GCFRs. For the HTGR, the liner was designed for a temperature of 350°C, decreasing within the layer of insulating concrete behind the liner to a temperature of 140°C [7]. For the GCFR, the liner was designed for a temperature of 275°C, decreasing within the thermal barrier to 95°C [8]. Reference has also been made to work in West Germany to develop a PCPV with a hot liner for PWRs [10]. Because of difficulties encountered in the development of the heat insulation for gas-cooled reactors, the principle of 'hot wall' insulation has also been investigated in France [38] with particular reference to PCRVs for BWRs. The research indicated that, with the liner at a temperature of 286°C, an acceptable maximum temperature in the first layer of the structural concrete could be obtained with an insulating 'hot concrete' thickness of about 20 cm.

1.9 Protection against overpressure

The ASME Code [15] requires that pressure-relief devices and associated pressure-sensing elements shall be provided in PCRVs to protect them, while in service, from the consequences of pressure and coincident temperature that exceed the conditions for which the vessel was designed in accordance with the design specification. The requirements for acceptable overpressure protection devices and their installation are stated in Article CB-7000 of the Code.

BS 4975 states that automatic pressure-relief devices shall be installed in the PCPV which comply with the requirements of Section 6 of BS 3915, which deals with steel pressure vessels for nuclear reactors. In deciding on the number and capacity of the pressure-relief devices, consideration must be given to there being adequate redundancy, bearing in mind the reliability of the individual relief devices. In the event of a pressure overload in service which exceeds 1.1 times the design pressure, careful consideration must be given to 'all relevant factors' before the vessel structure is deemed fit for further service.

1.10 Testing and surveillance of completed structure

1.10.1 *Vessel proof or structural integrity test*
To eliminate risks arising from errors in design and analysis, or during construction, tests must be carried out to establish the integrity and performance of the PCPV before it is put into service. Proof tests cannot indicate the behaviour of the vessel during all stages of service life, such as overloading due to incorrect operation. They can, however, indicate whether there is correlation between the calculated and measured deflections and strains, and this is of great value in demonstrating the adequacy of the design.

BS 4975 requires that the PCPV shall be proof tested at 1.15 times the design pressure unless otherwise agreed between the purchaser and manufacturer. The instrumentation should, as a minimum, include strain and deflection measurements and the criteria for an acceptable test must be agreed between the purchaser and manufacturer. BS 4975 also states that consideration should be given to the carrying out of leak and vacuum tests on the completed vessel.

The ASME Code [15] also requires that the test pressure must be at least 1.15 times the design pressure, and instrumentation requirements for the measurement of deflection, strain, tendon force, temperature and pressure are indicated. The PCPV shall be considered to satisfy the structural acceptance test if (a) the conventional reinforcement does not yield, (b) there are no signs of permanent damage, (c) the deflection recovery 24 h after depressurization is not less than 80%. and (d) the maximum deflections do not exceed the predicted values by more than 30%. Leak testing is referred to in connection with the examination of welds in the liners. The French Decree [16] requires that a proof test at 1.1 times the design pressure be carried out.

Descriptions and results of proof or structural integrity tests on PCPVs have been reported [39, 40]. The leak tightness of a PCPV is also a very important consideration because fission products can be released to the primary coolant and the description and results of leakage tests on the Fort St Vrain HTGR have, for example, also been reported [41].

1.10.2 *Surveillance*

The results of pressure proof tests have indicated that calculated design values for deflections and strains compare reasonably well with the observed values for normal service and slight overpressure conditions of the PCPV. Proof tests are, however, of comparatively short duration, and it is considered necessary to arrange for continued surveillance and inspection of the PCPV while in service to ensure structural safety even though this could place constraints on the operation of the vessel.

BS 4975 requires that PCPVs be monitored and inspected periodically during their service life. As far as the concrete structure is concerned, reference is made to the observation of deflections, concrete strains, moisture contents, cracking and concrete temperatures. Ungrouted tendon and/or tendon anchorage loads should also be measured. Attention is also drawn to the surveillance of the liner, insulation, cooling system, closures and pressure relief valves during the service life of the vessel.

Experience of in-service surveillance and monitoring of PCPVs has been described and the results compared with design expectations [42–45]. In general, it may be concluded that the surveillance of PCPVs for gas-cooled reactors has indicated that they have performed within their original design margins and that they are notable for their safety characteristics.

The large-scale instrumentation and regular monitoring of PCPVs is, however, expensive and requires much effort and the question has been asked whether it is really needed and/or satisfactory. Suggestions have also been made as to how better value may be obtained from the costs of instrumentation and the effort of monitoring [46].

2 Concrete containment vessels

2.1 Development and features

The main purposes of containment structures for nuclear reactors are (a) to prevent the inadvertent escape of radioactive materials into the environment during normal service operation and in the event of possible, but unlikely, accidents; (b) to protect the reactor system from damage due to external hazards, such as aircraft and missile impact, tornadoes and explosions; and (c) to provide biological shielding against nuclear radiation.

The first containment structures were of welded steel-plate construction, spherical in shape. They were followed by cylindrical steel structures with hemispherical top domes. These steel structures, which were designed to resist pressure and to provide leaktightness, were surrounded by reinforced concrete structures, about 1 m in thickness, to provide biological shielding.

The development of larger and more powerful reactors increased the amount of energy which could be released in the event of an accident, and the indications were that concrete containment structures could be more economic.

The first concrete containment vessels built in the USA were of reinforced concrete, with a steel liner to provide leaktightness. The first was constructed at Parr, South Carolina, in 1963. It was followed by the Connecticut Yankee containment structure, which had a volume of about 57 000 m³ (74 550 yd³), which is typical of the volume of present day containment vessels. These reinforced concrete containment vessels (RCCVs) were cylindrical in shape with a hemispherical dome and a flat base slab.

The design of prestressed concrete containment vessels (PCCVs) commenced in the USA in 1966. An example of the first generation of PCCVs for PWRs is the Palisades containment structure, completed in 1969, which was the first containment structure built in the USA in which the wall and dome were fully prestressed [47]. It is a 700 MWe plant, and the containment was designed for an internal pressure of 0.38 MPa (55 psi), based on the maximum energy released in the event of an accident. The internal diameter of the 1.07 m thick cylindrical wall structure was 35.4 m and it had six buttresses for the anchorages of the prestressing tendons. The torospherical dome was joined to the wall by a heavy ring girder. The inside height was 57.9 m, and the 6.4 mm thick mild-steel leak-resistant liner was anchored to the inside surface of the concrete. The design and basic shape of the second generation of PCCVs in the USA remained unchanged, except that they had three buttresses instead of six.

The third generation of PCCVs have a hemispherical dome which simplifies the layout of the prestressing tendons, and the absence of a ring girder of considerable mass at a great height is beneficial in high seismic areas. The internal diameter and height are approximately 43 and 63 m, respectively. The prestressing tendons have generally not been grouted.

Containments for BWRs usually include a pressure suppression system which allows the steam which may escape from the primary coolant circuit to be fed into a pool of water, where it condenses and causes pressure oscillations in the pool and vibrations in the surrounding walls. The containment system for all BWRs has a dry-well containing the reactor and the reactor cooling system, and a wet-well which contains the suppression pool.

The Mark I containment vessel for BWRs is of steel construction. The Mark II containment vessel is of concrete construction, and consists of a large cylindrical bottom portion, a smaller cylindrical top portion, with a torospherical head and a conical transition section between the bottom and top cylinders (Fig. 2-1). The Mark III containment for BWRs in the USA normally consists of a primary containment comprising a steel cylinder and dome anchored to the concrete base mat of a cylindrical concrete structure which surrounds the primary containment [48].

From the design point of view, both prestressed concrete containment vessels and reinforced concrete containment vessels have advantages and disadvantages. The opinion has been expressed that PCCVs have an advantage in resisting the pressure load from a loss-of-coolant accident (LOCA) while RCCVs have advantages over PCCVs in withstanding impact and other dynamic loads [49]. It has also been stated that, in the USA, PCCVs are in the majority mainly because the uncracked state of prestressed concrete is assumed to render it better able than reinforced concrete to resist membrane shear resulting from an earthquake, and thus obviating the need for any special seismic reinforcement [50].

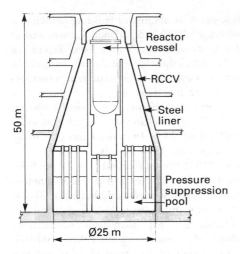

Figure 2-1 Mark II RCCV for a BWR (after Stevenson [50])

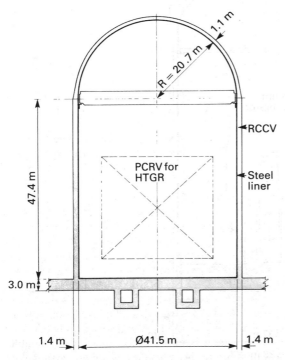

Figure 2-2 RCCV for an HTGR (after Stevenson [50])

HTGRs in the USA are at present enclosed in containment buildings which are, however, not designed to withstand pressure. They are secondary containment buildings which permit the treatment of any gas which has leaked from the reactor pressure vessel before being released to the atmosphere (Fig. 2-2). It is considered that, in future, HTGRs in the USA will be enclosed in containment structures designed to withstand pressure loads and to be leaktight [48].

In 1964, the first concrete containment vessel in France for a nuclear power reactor was built at Monts d'Arree [51]. It is a cylindrical PCCV, 46 m in diameter and 56 m in height, with a wall 0.6 m thick. Its inner leaktight lining, 500 μm thick, consists of several layers of tar plus Epikote mix.

In 1970, France ordered Westinghouse PWR 900 MWe units for Fessenheim and also for Tihange in association with Belgium. The Fessenheim containment consists of a single primary PCCV with an internal diameter and height of 37 and 51.3 m, respectively. The vessel has a 6 mm thick steel liner and a raft of reinforced concrete. Four similar units were ordered for Bugey after Fessenheim I and II, and this led to the concept of standard containments for PWR 900 MWe units, independent of site, except in the design of the base slab to suit site conditions. Twelve units were ordered in 1974 for Tricastin, Gravelines and Dampierre. The PCCVs were cylindrical with a steel liner and a reinforced

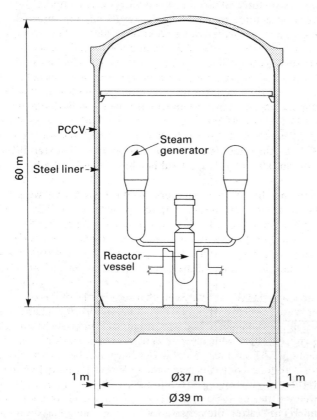

Figure 2-3   PCCV for a PWR (after Costaz and Picaut [30])

concrete base slab. (Fig. 2-3). The Tihange containment consists of a primary prestressed concrete structure lined on the inside with a 6 mm thick steel sheet, and a secondary structure of reinforced concrete. The function of the primary structure is to confine radioactive material which might be released from the reactor, and the function of the secondary structure is to collect any leakage through the primary structure and to protect it from external hazards.

It was intended that the construction of the first BWRs in France, at St Laurent, be commenced in 1975. A Mark III reinforced concrete containment vessel, with a 6 mm thick steel liner, was to be used after it had been decided that the cost of a double containment design, with a steel primary containment, was too high. It was considered that the small pressure in the event of an accident did not require the use of prestressed concrete and the RCCV was designed with an internal diameter of 36.6 m, internal height of 54.5 m and a wall 0.9 m thick [51].

The PWR containments under construction in France in 1975 were of prestressed concrete with a steel liner. The almost absolute leaktightness requirement of the liner makes proper control of its fabrication important and its erection difficult. It is therefore also costly. In 1970, investigations commenced in France to develop a containment system which would eliminate the steel liner, and a new containment system has been designed for 1350 MWe PWRs [30]. It consists of an inner prestressed concrete structure without liner and which is surrounded by an outer reinforced concrete structure. In the preliminary design, the internal diameter of the inner vessel was 42 m, with a wall thickness of 0.9 m. The outer structure is regarded as a second barrier against leakage, and has been designed to withstand external hazards such as wind loads and explosions. The thickness of the wall is about 0.5 m and it is separated from the inner structure by an annular space 2 m wide which leaves sufficient room for installing the prestressing cables. The design of this multi-barrier containment system without liner was licensed by the French authorities in 1974 (Fig. 2-4).

The Super-Phenix 1200 MWe LMFBR, being built at Creys-Malville on the river Rhone in France, has a cylindrical RCCV, with an external diameter of 66 m, wall thickness of between 0.9 and 1.0 m, a total height of 90 m, and a base slab 5.5 m thick.

Until 1975, the containments for nuclear reactors in West Germany were constructed of steel. The first German nuclear power plant with a PCCV, designated KRB II, is the Gundremmingen II plant, which has a BWR [52]. The KRB II containment is cylindrical in shape with internal and external diameters of 29 and 31 m, respectively, and an external height of 39.2 m. A steel liner, 8 mm in thickness, is anchored to the interior surface of the concrete, and the prestressing tendons are grouted.

The high-temperature reactor (HTR) at the Schmehausen II nuclear power plant was the first HTR in West Germany with a PCCV. It is cylindrical in shape, with a spherical dome, the internal diameter being 40 m and the internal height 56.7 m. The dome has a rise of 10 m, whilst the thickness of the wall and the dome is 1.6 m. It has a steel liner 6 mm thick [52]. Studies have also been carried out for the containment of a sodium-cooled fast reactor in West Germany [53].

In Sweden and Finland, there are a number of BWR nuclear power plants with electrical outputs of between 440 and 1050 MWe. The reactors at all of these plants have concrete containment vessels, the containment system being based on the pressure–suppression system principle. The vessels consist basically of a

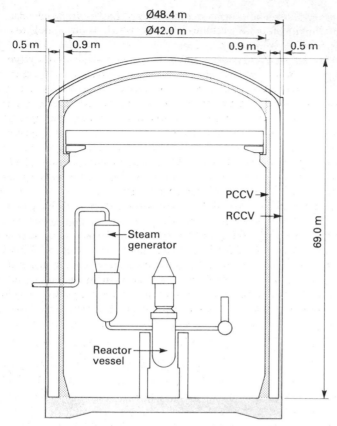

Figure 2-4 Multi-barrier concrete containment without steel liner for a 1350 MWe PWR (after Costaz and Picaut [30])

prestressed concrete cylindrical wall, with either prestressed or reinforced concrete base and roof structures. The primary leaktight barrier is a steel liner attached to the concrete [54].

In the UK, studies have been made for the design of containments for an HTGR which has been under development for many years [55]. The emphasis has been on an HTGR design which does not require a pressure containment to meet current safety requirements. However, designers in the UK are now required to take into account hazards arising from external causes such as aircraft crashes and industrial explosions, and it is likely that the allowable releases of radioactive gases will be further reduced. Various forms of containment, taking these factors into account, have been studied and a large circular pressure-tight containment with an internal diameter of about 82 m and a height of 107 m is favoured. Since the maximum internal design pressure is relatively small, the possibility of using a plastic sealing membrane instead of a steel liner has been considered. Although the containment is being developed specifically for the HTGR, it is felt that it may also be suitable for other types of reactor systems, such as the LMFBR.

The containment systems for the Canadian heavy-water (CANDU) nuclear power plants are either of the single-unit or multi-unit type [56]. In the single-unit type, a containment building and a vacuum building, which is at a negative pressure, are provided for each reactor.

In the multi-unit type, up to four reactor containment buildings, each of which contains a reactor and associated steam generators, are connected to the same vacuum building. When a LOCA occurs in a reactor building, the steam which is released is led to the vacuum building through a duct and the overpressure in the reactor building is reduced to below atmospheric pressure. The buildings are of reinforced or prestressed concrete with either plastic, such as vinyl or epoxy resin, or with steel liners.

Because of frequent severe earthquakes in Japan, containment vessels there have been built of steel. It has however been decided also to build concrete containment vessels.

The commercial generation of electric power from controlled nuclear fusion is still far from reality, but design studies for a demonstration fusion power reactor are in progress. In Italy, these studies aim at defining the main features of a tokamak-type fusion power reactor, including the structural stability of the outer concrete containment vessel. The first proposal is to enclose the reactor inside a cylindrical reinforced-concrete vacuum building which is 60 m in diameter and which has a hemispherical upper dome. The internal height of the proposed structure is 100 m and detailed calculations for a 2 m wall thickness have been carried out [57].

2.2 Standards and codes

Reference has been made in Section 1.2 of this chapter to the ASME *Boiler and Pressure Vessel Code*, Section III, Division 2, in which Subsection CB deals with prestressed concrete reactor vessels, and Subsection CC with prestressed or reinforced concrete containment structures.

Subsection CC, which deals specifically with concrete containments, was prepared because existing building codes and national standards in the USA were not directly applicable and did not deal adequately with such matters as loads due to earthquakes, aircraft crashes and the effects of elevated temperature. The United States is one of the few countries which has prepared a code specifically for concrete containment structures. In the United Kingdom, BS 4975 was prepared specifically for prestressed concrete pressure vessels. In France, Electricité de France has prepared standard structural design rules for containments [58], while in Japan a draft of a technical standard for concrete containments for nuclear power plants was published in July 1977 by the Agency of Natural Resources and Energy in the Ministry of International Trade and Industry. The fundamental principles of this draft standard are based on the ASME Code, Section III, Division 2, with modifications which are appropriate to circumstances in Japan.

In West Germany, it was appreciated that containment structures have functions and are subjected to abnormal loads which differ from those for pressure vessels, and that the design criteria were therefore not the same. The usual standards for reinforced and prestressed concrete structures were not adequate for containment structures. Guidelines for design criteria for the first two PCCVs in Germany for the Gundremmingen II (BWR) and the Schmehausen II (HTR) nuclear power plants were, however, developed and it was intended that they would serve as the basis for a standard for containments in Germany [52]. The Nuclear Safety Standards Commission (KTA) in Germany has also completed some standards, and prepared the drafts of others, relating to the design of nuclear power plants [59].

In Canada, a standard has been prepared for concrete containment structures for CANDU nuclear power plants [60].

Reference has also been made to the work of the International Standards Organization (ISO) and Working Group 3 of Subcommittee 3 of ISO Technical Committee 85 is dealing with concrete containment structures for nuclear power plants.

2.3 Materials

In the ASME Code, Article CC-2000, dealing with materials for concrete containment structures, is similar to Article CB-2000 dealing with materials for PCRVs. As far as concrete properties are concerned, those which affect the design of the containment must be stated in the construction specification. Properties such as strength, modulus of elasticity, thermal conductivity, creep and shrinkage are referred to, and the age and temperature at which these properties shall be determined must also be specified. Further reference to the properties of concrete is made in Section 4 of this chapter.

2.4 Design and analysis

2.4.1 Loads

In the design of the first concrete containment structures built in the USA in 1963, the loads that were considered were simply the design pressure load within the containment, and the thermal effects and loads generated by the accident on which the design was based, i.e. the design basis accident (DBA). In about 1967, earthquake loads and tornado loads, for sites east of the Rocky Mountains, were included [50]. Since then, the loads to be considered in the design and analysis of containment structures have increased.

The ASME Code refers to two categories of loads: service loads, and factored loads.

Service loads are defined as loads encountered during construction and in the normal operation of a nuclear power plant and include: (a) normal loads which occur during normal plant operation and shutdown, (b) construction loads, (c) test loads applied during structural integrity or leak rate tests, and (d) severe environmental loads which may be expected during the life of the plant. Severe environmental loads include earthquake loads (E_0) generated by the operating basis earthquake (OBE). The OBE for a site is that which produces the vibratory ground motion for which those features of the power plant, necessary for continued operation without undue risk to the health and safety of the public, are designed to remain functional. The maximum vibratory ground acceleration of the OBE is equal to at least one half of the safe shutdown earthquake (SSE)—see below.

Factored loads are defined as loads which are encountered infrequently and include: (a) severe environmental loads which could be imposed infrequently during the life of the plant, (b) abnormal loads which are developed during the design basis accident, (c) extreme environmental loads that are possible but highly improbable, such as earthquake loads (E_{ss}) generated by the SSE, and tornado loading (W_t). The SSE for a site is that which produces the vibratory ground motion for which those features of the power plant, necessary to shut down the reactor and maintain the plant in a safe condition without undue risk to the health and safety of the public, are designed to remain functional.

In the USA, environmental events such as tornadoes and earthquakes, as well as abnormal internal occurrences such as high-energy pipe breaks, are considered for inclusion in the design basis events. The inclusion of accidental external explosions, aircraft impact and turbine missile impact depends mainly on the likelihood of their occurrence [61].

In West Germany, consideration has been given to categorizing loads into service loads, abnormal loads, accident loads and extreme external loads [52]:

(a) *Service loads* are loads which occur during construction, testing and normal operation.

(b) *Abnormal loads* are loads caused by unplanned or unscheduled events such as internal disturbances of short duration. Their probability of occurrence is more than once during the service life of the plant, and include increased pressures and temperatures over large areas inside the containment.

(c) *Accident loads* are loads which have a very low probability of occurrence and include loads due to a maximum loss of coolant, flooding of the containment and missile-like impacts of fittings or piping.

(d) *Extreme external loads* are loads due to earthquakes, aircraft impact forces and explosions.

2.4.2 *Load combinations*

The ASME Code specifies the following load combinations and applicable load factors for which the containment shall be designed:

Service

(1) Test			$D + L + F + P_t + T_t$
(2) Construction			$D + L + F + T_o + W$
(3) Normal			$D + L + F + T_o + R_o + P_v$
(4) Severe environmental	(a)		$D + L + F + T_o + E_o + R_o + P_v$
	(b)		$D + L + F + T_o + W + R_o + P_v$

Factored

(5) Severe environmental	(a)		$D + 1.3L + F + T_o + 1.5E_o + R_o + P_v$
	(b)		$D + 1.3L + F + T_o + 1.5W + R_o + P_v$
(6) Extreme environmental	(a)		$D + L + F + T_o + E_{ss} + R_o + P_v$
	(b)		$D + L + F + T_o + W_t + R_o + P_v$
(7) Abnormal	(a)		$D + L + F + 1.5P_a + T_a + R_a$
	(b)		$D + L + F + P_a + T_a + 1.25R_a$
(8) Abnormal/Severe environmental	(a)		$D + L + F + 1.25P_a + T_a + 1.25E_o + R_a$
	(b)		$D + L + F + 1.25P_a + T_a + 1.25W + R_a$
	(c)		$D + L + F + T_o + E_o + H_a$
	(d)		$D + L + F + T_o + W + H_a$
(9) Abnormal/Extreme environmental			$D + L + F + P_a + T_a + E_{ss} + R_a + R_r$

where

$\quad$ D = dead loads

$\quad$ E_o = operating basis earthquake load

$\quad$ E_{ss} = safe shutdown earthquake load

$\quad$ F = prestressing force loads

$\quad$ H_a = load on the containment from internal flooding

$\quad$ L = live loads

$\quad$ P_a = accident/incident maximum pressure load

$\quad$ P_t = test pressure load

$\quad$ P_v = sub-atmospheric minimum pressure loads

$\quad$ R_a = piping loads due to increased temperature resulting from the design accident

$\quad$ R_o = piping loads during operating conditions

$\quad$ R_r = local effects on the containment due to the design basis accident

$\quad$ T_a = accident/incident temperature effects

$\quad$ T_o = operating temperature effects

$\quad$ T_t = test temperature effects

$\quad$ W = wind loads

$\quad$ W_t = tornado load

It will be seen that the ASME Code requires that the loads due to an operating basis earthquake (OBE), and the loads due to the design basis accident (DBA) be combined. The loads due to a safe shutdown earthquake (SSE) and the design basis accident are also combined. In France, the practice is not to combine loads due to an OBE and the DBA, the reason given being that the plant is designed to withstand the effects of an OBE. The effects of an SSE and the DBA are, however, combined because, although an SSE is a highly unlikely event, it is nevertheless possible, and it is considered that arrangements must therefore be made to maintain the integrity and the stability of the structure [30].

In regard to load combinations, all loads that could act concurrently have to be considered in West Germany. Although it is not normally required that extreme external loads be combined with abnormal or accident loads, the possibility of it occurring must be investigated in each case, and combinations of earthquake loads and abnormal loads which could arise due to an earthquake must be considered [52]. All buildings for nuclear power plants are considered from the aspect of the safety relevance of the systems installed in the buildings, and different buildings have to be designed to withstand the effects of different load combinations [59].

The draft code for concrete containments prepared in Japan in July 1977 by the Ministry of International Trade and Industry stipulates that the containments shall resist all load combinations which could occur during the life of the structure without causing excessive deformation and without affecting its leaktightness. This includes the combination of pressure due to internal accident and loads due to earthquakes.

In the design of containment structures for nuclear reactors, seismic loads have been combined with pressure due to some level of accident. Stevenson [18] has pointed out that although in Japan and Canada pressure and earthquake loads are also combined, the maximum effect of both accident and earthquake loads are not considered simultaneously and the opinion is expressed that the combination of the maximum effect of accident and earthquake loads is apparently not rational. Allowable stresses due to the SSE or equivalent earthquake load are limited in Japan, Canada and West Germany to more conservative values than have been used in the USA, and the probability of failure due to earthquakes has therefore been reduced. Stevenson has suggested that this should also be considered as an alternative to the current combined SSE and LOCA approach in the USA.

2.4.3 *Earthquakes*

Earthquake design for the first nuclear power plants (NPPs) was generally in accordance with conventional national building codes. In 1963, a design basis earthquake (DBE) was defined in Japan and the USA as the largest earthquake which had been experienced at the site during recorded history, and earthquake designs for NPPs were based on the DBE with an increase in the stresses normally allowed. At about the same time, the concept of a maximum hypothetical earthquake (MHE), which was the largest earthquake considered to be possible at the site, was developed. In the USA, the MHE was generally taken to be twice the DBE, while in Japan it was taken to be from 1.25 to 1.5 times the DBE.

In the USA in 1968, the MHE was renamed the safe shutdown earthquake (SSE) and the DBE was called the operational basis earthquake (OBE). The maximum vibratory ground acceleration of the OBE was to be equal to at least half of that for the SSE. For sites without well established faulting or seismic activity, the SSE was defined as the epicentral movement of the largest earthquake within a 200 mile radius of the site during recorded history. For sites with well established faulting or seismic activity, the SSE was the postulated maximum earthquake at a location on a 'capable fault' at its closest proximity to the site. A 'capable fault' was defined as a fault at which significant surface movement may have occurred within the previous 35 000 years.

Countries such as Canada and West Germany have followed the USA in this regard but have limited the allowable SSE stresses to about 80% of the yield stress [18].

A probabilistic methodology has been developed [62] which indicates that the current regulatory criteria in the USA for determining fault capability can result in extremely conservative requirements for seismic design spectra for NPP projects and the Lawrence Livermore Laboratory at the University of California is currently conducting a comprehensive seismic research program for the US Nuclear Regulatory Commission [63]. This work is being done to estimate the conservatism in the current US–NRC seismic safety requirements for NPPs, and the hazards due to seismically induced accidents are being determined using a probabilistically based systems model of a nuclear power plant.

The Nuclear Safety Standards Commission (KTA) in West Germany has initiated work on KTA 2201—*Design of NPPs against seismic events*, which deals with basic principles, characteristics of seismic events, design of structural components, design of electrical and mechanical components, seismic instrumentation, and measures subsequent to earthquakes [64].

In the design of NPPs for earthquake loads, consideration has been given to underground and floating NPPs and also to the problems of shock-isolated plants in which attempts are made to isolate structures from the effects of earthquakes [65].

2.4.4 *Aircraft and missile impact*

The US–NRC requires that NPPs be designed to resist accidental aircraft impact if the aircraft crash could result in a release of radioactivity exceeding the Code of Federal Regulations guidelines, with a probability greater than 10^{-7} per year. Only two NPPs in the USA have been designed to resist aircraft impact, one at Three Mile Island, which was in close proximity to the Harrisburg commercial airport, and the other at the Seabrook Nuclear plant [66]. Other NPPs in the USA have not been designed to resist aircraft impact because, by appropriate selection of the site, the probability of impact was reduced to below that specified. In many countries, especially in Western Europe, where NPPs are near airports and large cities, containment structures must normally be designed specifically to withstand accidental aircraft impact.

Countries in Europe have generally adopted the impact of a military fighter aircraft travelling at flight speed as the limiting design condition. Other countries have followed the USA criteria of the impact of a commercial aircraft, usually of the Boeing 720 or 707 class, at landing speed. The military aircraft design condition is more severe than the impact of a large passenger plane as military aircraft are much more compact and do not have a large fuselage which is able to absorb most of the kinetic energy by deformation. They also fly at high speeds and at low altitudes. The impact model recommended in West Germany is for the crash of a Phantom aircraft at 215 m/s.

The design and analysis of concrete structures to withstand the effects of aircraft impact involves a complex interaction of many parameters. In analysing the effects of aircraft impact on concrete structures, procedures have been used which are essentially the same as for seismic analysis. The behaviour of structures subjected to aircraft impact loads is, however, significantly different from that under seismic loads which are transmitted upwards through rock and soil into the structure. Aircraft impact loads are also uni-directional and of much shorter duration than seismic loads.

The application of dynamic seismic analysis techniques to the aircraft impact problem has been questioned, and investigations have been undertaken to attempt to resolve some of the basic underlying issues and to identify the most important and relevant topics for future investigations [67].

Structural elements of NPPs are often required to withstand not only the effects of aircraft impact but also of missiles which could be impelled by external sources, such as tornadoes, or by internal sources such as high-pressure systems and rotating machinery, e.g. turbine blades and valve stems. Much research has been done in this connection [68] and many papers on this subject have been presented at SMIRT conferences. In Britain, a coordinated program exists for acquiring data and developing calculation methods, and it covers a wide range of missile types. Several organizations in Britain participate and there are also collaborative research programmes with other European nuclear safety organisations [69].

2.4.5 *External explosions*

In the design of containment structures, it may be necessary to consider the effects of an accidental external explosion in close proximity to the NPP because of the induced overpressure wave or air blast, which is generally the most important of the phenomena associated with such explosions.

In the USA, the decision as to whether an accidental explosion should be included in the design basis for a NPP was originally based on a comparison of the peak-reflected pressure from an air blast, due to the explosion, with the maximum dynamic wind pressure developed by a tornado. If the proposed accidental explosion produced a peak-reflected pressure less than the postulated maximum dynamic wind pressure of the prescribed tornado, then the accidental explosion need not be considered in the design basis. It became evident, however, that several important factors and parameters were neglected in a comparison between dynamic tornado wind pressure and air-blast reflected pressure, and the comparison was abandoned by the US–NRC. An accidental explosion may now be excluded from the design basis of an NPP either if it produces a peak blast overpressure of 0.007 MPa (1 psi) or less, or alternatively, the probability of its being exceeded is less than about 10^{-7} per year. If air blast due to an external explosion is to be included in the design basis, the load combination required by the US–NRC is $C = D + L + T_o + R_o + B$, where B is the blast load effect. The effect of the combined load, C, must be less than the ultimate strength of the section under consideration. Investigations have indicated that the design of NPPs in the USA will only begin to be controlled by accidental explosions if they result in overpressures greater than 0.007 MPa [61].

In West Germany, the criteria for the design of nuclear structures to resist external explosions includes a loading requirement of 1 atm (0.10 MPa) pressure acting for at least 0.1 s. This is much more severe than the 0.2 atm (0.02 MPa) pressure, with a rise time of 0.007 MPa/s, which is the pressure loading typically associated with tornadoes in the USA [18].

2.4.6 *Design basis accidents*

Although most countries consider postulated, or potential, accident design criteria, which are typically divided into inside and outside containment accidents, they are not necessarily all included in the design and analysis of structures for NPPs. As has been pointed out [18], 'postulated accidents only become design basis accidents when the probability of occurrence or the consequence of such accidents are deemed to be of sufficient magnitude to be considered in design'. The effects of a turbine missile impact have, for example, generally not been included as a design basis accident because of this, and exclusion criteria have been suggested for use in eliminating postulated hazards from being included as a design basis [18].

2.4.7 *Analytical procedures*

The ASME Code [15] states that the methods of analysis to be used for containment structures shall be based on accepted engineering mechanics principles which are appropriate to the geometry of the containment. Consideration must also be given to the effects of the cracking of the concrete and, in the case of prestressed concrete containments, to the effects of creep. The liner shall also not be used as a strength element, and the strength of the containment is entirely dependent on the structural concrete. It is also stated that containments are normally thin shell structures and that the acceptable basis for predicting internal forces, displacements, and stability of thin shells, shall be elastic behaviour. Although shell analysis may be based on membrane theory, additional attention must be paid to the bending and shear forces at penetrations, to the intersection of the cylindrical shell with the base mat, to discontinuities, and also to the stresses and strains caused by variations in temperature. Reference is also made to the use of model tests.

In nuclear containment structures, the main penetrations usually consist of one penetration for equipment and one for personnel. These penetrations are generally circular cut-outs in the shell which are stiffened by a ring beam, and their design was at first based on isotropic shell analysis with membrane forces applied to the ring beam. This was followed by linear finite element analysis. Three-dimensional non-linear finite element analysis is now being used in the design of large penetrations [70]. The design and analysis of the junction of the base mat or raft, with the cylindrical wall, where very large bending moments and shears occur, is also a difficult part of the design of containment structures.

The development of finite element, finite difference, and other methods of analysis has, however, made it possible to analyse concrete containments in great detail.

2.4.8 *Probabilistic analysis and design*

The design, analysis and construction requirements for structures for nuclear power plants are generally considered to be more stringent than those for conventional structures. The loads which are considered in the design of nuclear structures, and the properties of structural materials when subjected to these loads, are also different to those for ordinary structures. It is recognized that these loads, and the structural properties of materials, are random and increasing attention has been given to probabilistic methods for the analysis and design of nuclear structures. According to Ravindra and Walser [72], the objectives of probabilistic design are '(a) to recognize and consistently treat the uncertainties inherent in the applied loads, structural resistances and the methods of analysis, and (b) to provide a balanced design considering the consequences and the acceptable risks for other systems in a nuclear plant'.

In the evaluation of loads, current practice varies from a completely probabilistic analysis to a totally deterministic approach. Some loads are derived on a probabilistic basis, e.g. the design basis tornado (DBT) has been defined as a tornado which exceeds a specified velocity with a probability of 10^{-7} per year. Whether nuclear structures should or should not be designed to withstand the effects of aircraft and missile impact and of blast loading is being based on probabilistic considerations. The effects of natural phenomena are, however, also being postulated on a deterministic basis, e.g. the SSE motion at a site is derived by postulating the historical maximum earthquake to occur in the region closest to the plant site. There is, nevertheless, a trend towards adopting a probabilistic approach to seismic risk analysis and to obtain an indication of the annual probability of exceeding various levels of peak ground earthquake accelerations at an NPP site. Similar risk analysis procedures are being developed for other kinds of environmental loading [72].

Investigations into the effects of combinations of loads on a probabilistic basis are also being carried out. For this purpose, as well as for a probabilistic evaluation of loads, data on the frequency of occurrence, duration and intensity of different loads is required.

Research is also in progress to develop probabilistic structural design criteria for nuclear structures as present values of load factors and allowable stresses have apparently been decided in an arbitrary manner [72]. The stringent quality control systems and analysis procedures for nuclear structures are generally not reflected in the design criteria. The properties of structural materials such as concrete and steel vary, and the probabilistic distribution of these properties, such as strength, is required so that codes based on probabilistic principles may be developed. The same applies to information concerning failure modes for nuclear structures.

Although much further work needs to be done in the development of probabilistic structural analysis and design, research in this field could have a significant influence on the analysis and design of nuclear structures in the future.

2.5 Testing and surveillance of completed structure

2.5.1 *Structural acceptance test*
A structural acceptance or integrity test is usually carried out before the reactor is fuelled for the first time. The purpose of the test is to demonstrate that the containment structure responds as predicted theoretically to the specified internal pressure loads and to provide evidence that the quality of construction and the materials is adequate. The structural response data in the structural acceptance test is used to evaluate the methods of structural analysis and the properties of the materials assumed in the analysis, not only in regard to response to pressure loads, but also to the response to other loads which cannot be reasonably simulated in tests, e.g. earthquake loads and loads due to temperature gradients. If there is good correlation between the measured and predicted response to internal pressure loads, then it provides confidence in regard to the prediction of structural response to other design loads [73]. The ASME Code [15] requires the containment structure to be pressurized by increasing the internal pressure from atmospheric pressure to at least 1.15 times the design pressure in five or more approximately equal pressure increments. Measurements or observations as required by the designer are made after each pressure increment and corresponding decrement during depressurization. Prototype containment structures must be instrumented for the measurement of strains as well as gross deformations; for non-prototype structures, strain measurements are usually not required. Radial and vertical deformations are measured at specified locations and crack patterns

and widths in selected areas are also mapped. The ASME Code specifies the requirements if the concrete containment structure is to be regarded as satisfying the structural acceptance test.

Structural acceptance and integrity tests on prestressed concrete containments have confirmed that both strain and dimensional change were predictable by the analytical methods used: those structures which have been tested have also performed as expected, thus indicating that the quality of the construction was satisfactory [73].

2.5.2 Leakage-rate test

One of the main functions of containment structures is to prevent the uncontrolled release of radioactive materials to the environment. Leaktightness tests are therefore required not only before the reactor is allowed to operate but also during the life of the nuclear power plant to make sure that the containment will continue to perform this function. The initial integrated leakage-rate test is usually carried out in conjunction with the structural acceptance or integrity test to minimize disruption of construction of the containment vessel and also because the test preparations are similar. To verify leaktightness, the structure is generally pressurized to the design pressure, which is usually that calculated to occur subsequent to a postulated LOCA. Containment leakage-rate testing requirements for water-cooled reactors are specified by the US–NRC in Table 10, *Code of Federal Regulations*, Part 50, Appendices A and J. American National Standard ANSI N45.4, *Leakage rate testing of containment structures for nuclear reactors*, also deals with this subject.

The maximum allowable leakage rate varies from containment to containment but it is generally between 0.1 and 0.5% by weight of the contained air per 24 h. A safety factor is usually incorporated by stipulating that the maximum allowable leakage rate during the test shall be not greater than 75 per cent of the specified maximum allowable leakage rate. These leakage rates are very small and special methods have been developed to ascertain the integrated leakage rate with a high level of confidence [74].

2.5.3 Surveillance

Unbonded post-tensioned systems are used in PCCVs and it is considered to be desirable to monitor the prestress force level and the strength and condition of post-tensioning systems periodically over the life of the structure. In-service surveillance of the post-tensioning systems in PCCVs has therefore been included in the operational requirements of nuclear power reactors. Surveillance establishes confidence in the performance of the post-tensioning system and provides the opportunity for timely remedial measures if found to be necessary.

Tendon surveillance is generally conducted at specified intervals after the structural acceptance test, and for surveillance purposes, each containment structure is provided with extra tendons in addition to the number required for design prestress levels.

A surveillance program has been suggested and surveillance of the post-tensioning systems in containment structures has indicated that the systems are in good condition, that the predicted prestress force levels are being maintained and that the tendons have not deteriorated [75].

3 Concrete structures for radiation shielding

The radioactivity of fission products in a nuclear reactor is very great and neutron and gamma radiation must be reduced or attenuated by means of a radiation, or biological shield, to a level which will not cause harm to personnel in the operating zones of the reactor. To attenuate gamma radiation, it is the mass of the shielding material which is most important. Attenuation of neutron radiation is accomplished chiefly by causing the neutrons to lose energy or to be slowed down in collisions with the nuclei of other atoms. The greatest energy loss occurs when the target nucleus has a mass which is nearly the same as that of the neutron; hydrogen nuclei have approximately the same mass as neutrons and are therefore the most efficient. The relatively high efficiency of hydrogen and oxygen in the slowing down process makes the presence of these elements very desirable.

A radiation shield sometimes acts both as a barrier to radiation and as a structural support, and concrete is considered to be an excellent shielding material. It has good structural properties and because of the presence of water, both free and chemically bound, it is a good neutron shield. The relatively high density of concrete, particularly if heavy aggregates are used, also makes it suitable for gamma-ray attenuation.

In designing the shield, stresses due to temperature assume great importance. The inner face of the concrete is exposed to heat transferred by convection, conduction and radiation from the reactor core. Heat is also generated within the concrete as it absorbs neutron and gamma radiation. The temperature gradient and the calculation of thermal stresses during operation therefore depend upon the heat escaping from the core of the reactor and the heat released within the shield due to neutron and gamma absorption.

The thickness of a concrete shield for a nuclear reactor is governed primarily by the nuclear radiation it is required to attenuate and not by structural loads. The thickness thus depends on the intensity and energy of the radiation source, the permissible level of radiation outside the shield and the radiation shielding properties of the concrete in the shield.

Temperature distribution across the shield is generally non-linear and thermal stresses due to temperature gradients, and the degree of restraint, must be determined. Secondary stresses due to shrinkage, creep and heat of hydration must also be considered.

Where it is necessary to reduce the thickness of radiation shield, generally because of space considerations, high-density concrete has been used. A wide variety of materials has been used as aggregate in heavy concrete, e.g. barytes, magnetite, limonite, ferrophosphorous and ilmenite. The density of concrete made with these aggregates varies between 3540 and 3940 kg/m^3 (221–246 lb/ft^3). Iron and steel shot have also been used as aggregates and produce concrete with a density between 5470 and 5960 kg/m^3 (341–372 lb/ft^3). Caution is, however, required in the use of ferrophosphorous aggregates [76].

As mentioned, high density is of primary importance for the attenuation of gamma radiation, but for neutron shielding, light atoms, such as hydrogen, are required in the concrete. In the design of high-density concrete mixes for radiation shields in nuclear reactors, a compromise must therefore be made to meet these conflicting requirements and the optimum composition of the concrete requires careful consideration.

Heavy aggregate concretes also give rise to problems in construction. Special falsework and shuttering is required and problems concerning segregation arise in the course of placing the concrete. It is important that voids in radiation shields be prevented and, to this end, use has been made in construction of the grout intrusion or preplaced aggregate method.

The ASME Code specifies that the mix proportions of heavyweight aggregate concrete shall be in accordance with ACI-211.1 *Recommended practice for selecting proportions for normal and heavyweight concrete*. The requirements for compressive strength, durability and other specific properties shall be as required by the construction specification and the aggregates shall conform to ASTM C-637, *Aggregates for radiation-shielding concrete*.

ACI Special Publication, SP-34, contains the papers presented at an international seminar on concrete for nuclear reactors, held in Berlin in 1970. It includes an annotated bibliography on concrete for radiation shielding which lists 365 selected references pertaining to the types and properties of concretes used in radiation shields [77]. Reference should also be made to the *Engineering compendium on radiation shielding* [78].

4 Properties of concrete

4.1 Effects of elevated temperature

4.1.1 *Compressive and tensile strength*

In concrete structures for nuclear reactors, the moisture content of the concrete varies. Moisture content has an important effect on the properties of concrete and tests are usually carried out on unsealed and sealed test specimens. In unsealed specimens, the moisture is allowed to migrate and evaporate while in sealed specimens this is prevented. In general, the compressive strength of both unsealed and sealed concrete specimens is reduced when subjected to elevated temperatures and the higher the temperature, the greater the reduction [79–81]. At 400°C, compressive strength can be reduced to 50% of the strength of unheated specimens. Sealed specimens usually undergo greater reduction in compressive strength than unsealed specimens.

Some investigators have observed increased strength of the order of 10 to 25% for unsealed specimens at temperatures up to 100°C [81, 82]. It has also been reported that the reduction in compressive strength, between 20 and 200°C, can be represented by two straight lines, one for unsealed and the other for sealed specimens. The results indicate a reduction in strength at 150°C of 20% for unsealed specimens, and 45% for sealed specimens [83]. Another paper [84] gives the results of compressive strength tests on concrete cured in water for 90 days prior to exposure at temperatures of 60, 105 and 180°C. Under sealed conditions, limestone aggregate concrete showed a loss of strength of 40% after 42 days exposure to a temperature of 180°C, whereas quartz aggregate concrete showed a

20% increase in strength. Unsealed limestone and quartz aggregate concretes increased in strength by approximately 10% after exposure to a temperature of 180°C for 42 days.

Large variations can occur in the results of compressive strength tests and many factors must be taken into account when evaluating the results of tests of concrete at elevated temperatures [85]. Depending on the aggregate, compressive strengths at 200°C can vary between an increase of 10% and a decrease of 15%; at 300°C, they can vary between no decrease and a decrease of 25%; and at 400°C, they can vary between a decrease of 10% and a decrease of 50% [86]. Cyclic temperature changes, which occur in PCRVs, are important and can lead to further decreases in compressive strength [79].

Decreases in tensile or flexural strength, due to the effects of elevated temperatures, are greater than for compressive strength and at 400°C decreases in flexural strength can vary between 55 and 70% depending on the aggregate [87]. It also appears that decreases in tensile strength of sealed specimens are greater than for unsealed specimens [88].

4.1.2 *Modulus of elasticity and Poisson's ratio*

Most of the test results reported indicate that the modulus of elasticity, E, of concrete decreases as temperature increases and that the decrease is greater for sealed than for unsealed specimens. It has been indicated that between 20 and 200°C the decrease in E is approximately linear. At 150°C, the decrease for unsealed specimens is about 20 per cent of the value at 20°C, and about 45 per cent for sealed specimens [83]. At 400°C, the decrease in E is generally between 45 and 70%, but reductions of up to 85 per cent have been reported. Reductions in E are generally greater than those for compressive and tensile strength, while temperature cycling causes further reductions. The E of unsealed specimens is considered to be related largely to the absence of evaporable water, which constitutes an incompressible phase. E is consequently reduced because strains are increased. The reduction in E of sealed specimens, i.e. even when evaporable water is retained, may be due to chemical changes in the hydrated cement which result in a weaker matrix.

Some tests have indicated a slight increase in E between 20 and 50°C and then a decrease at higher temperatures. This increase was generally observed for concrete at early ages and corresponds to an acceleration of the maturation of the concrete [89]. The sharp decrease in E which has been reported at approximately 80°C, particularly with calcareous aggregates, has been attributed to the larger dislocations which occur in the concrete due to the lower thermal expansion of the aggregate in comparison with that of the cement matrix. Another view is that the decrease is not necessarily induced by the aggregate but is due to a critical relative humidity within the concrete.

The results of the few tests reported on the effects of elevated temperatures on the Poisson's ratio, μ, of concrete are scattered and even contradictory. Some investigators have found a random variation of μ with temperature while others have found it to be independent of temperature. In the present state of knowledge, it seems as if it can be assumed that μ of concrete remains constant at elevated temperatures.

4.1.3 *Creep*

Creep in concrete increases at elevated temperatures and follows the same general pattern as creep at 20°C, i.e. it is approximately an exponential function of the time under load and, up to stress/strength ratios of 0.50, it is an approximately linear function of the applied stress. The creep of sealed concrete specimens increases as the temperature is increased. Sealed specimens show less creep than unsealed specimens.

Below 100°C, creep is very sensitive to the moisture condition of the concrete. Many investigators agree that for unsealed specimens the maximum creep rate occurs at temperatures between 50 and 80°C, and is about two to three times greater than the creep rate at 20°C. There is no agreement as to whether the creep rate increases further up to a temperature of 100°C, and some are of the opinion that the creep rate at temperatures of about 100°C decreases to a value which is approximately equal to the 20°C rate, after which the creep rate again starts to increase. Specimens which have been pre-dried at 105°C have shown little creep below 100°C, but creep then increased with increasing temperature [89]. This indicates that creep is very small if measured at temperatures below that at which the evaporable water has been removed. Above 100°C, chemical and physical effects play an increasingly important role and factors such as differential expansion between the aggregate and mortar matrix can lead to weakening and disruption of the concrete and thus an increase in creep. At 400°C, creep may be ten times greater than at 20°C.

4.1.4 *Thermal expansion and shrinkage*

It is not easy to evaluate the results of tests to determine the coefficient of thermal expansion, α, of concrete, because when unsealed concrete is heated the absolute expansion is reduced by shrinkage due to the loss of evaporable and possibly also the non-evaporable water. Also, if the temperature is high, or is applied too rapidly, microcracking may occur due to different coefficients of expansion of the cement paste and the aggregates. However, by using the results of thermal cycling tests, it is possible to deduce the approximate shrinkage and hence determine α. Investigations have indicated that α remains approximately constant up to 250°C, varies between 250 and 300°C and then again becomes approximately constant at higher temperatures [79, 80, 89].

The coefficient of thermal expansion for different concretes depends on many factors such as the aggregate type, mix proportions, temperature level, and whether the concrete is sealed or unsealed during heating. Values of α range from 5×10^{-6} to 15×10^{-6} per °C. The mineral composition and structure of the aggregate is a major factor in determining the coefficient of thermal expansion of concrete because the aggregate usually occupies between 60 and 80% of the total volume of hardened concrete. The effect of the coarse aggregate appears to be more pronounced than the fine aggregate fraction and concrete with quartzite aggregate has a higher α than concrete made with other aggregates such as limestone [90].

Because of their large dimensions, PCRVs may dry at a very slow rate. The coefficient of thermal expansion of concrete at different moisture contents is therefore relevant. Not much information is available but the indications are that α for sealed concrete specimens is less than that for unsealed specimens. For specimens pre-dried at 180°C, the coefficient increased slightly with increasing temperature, whereas sealed specimens showed a decrease in α at higher temperatures [90].

Elevated temperature considerably accelerates shrinkage in concrete and it has been shown [89] that for unsealed specimens, at a temperature of 100°C, all the shrinkage occurred within 20 days, which was ten times faster than at 20°C. It has also been shown that shrinkage is slowed down by an increase in the thickness of the concrete through which the moisture escapes [91].

4.2 Effects of nuclear radiation

Concrete in PCRVs and in biological shielding structures is subjected to significant levels of nuclear radiation. The types of radiation which may affect concrete are (a) thermal, or slow, neutrons and fast neutrons from the core of the reactor, and (b) gamma radiation which may be emitted as a consequence of other nuclear reactions, such as the capture of neutrons by steel within the reactor vessel. Exposure of concrete to nuclear radiation results in an increase in the temperature of the concrete and increases of up to 250°C have been reported. Elevated temperatures may in themselves affect the properties of concrete and this must be considered when assessing the results of investigations into the effects of radiation on the properties of concrete. A comprehensive review [92] of the effects of nuclear radiation on the properties of concrete refers to the difficulty of comparing the results of data from different tests because of differences in the materials used in making the concrete, the mix proportions, specimen size, and in the effects of temperature.

Several investigations have shown that aggregates increase in volume when exposed to nuclear radiation and that the volume change varies for different aggregates. It has also been concluded that the increase in volume of irradiated concrete is mainly due to the increase in volume of the aggregates. Nuclear radiation has been observed to decrease the compressive and tensile strength and the modulus of elasticity of aggregates, whereas this has not been proved to be the case for neat cement paste specimens. It seems that changes in the properties of concrete due to nuclear radiation are mainly the result of changes in the aggregates which may be attributed to their mineralogical composition [93].

Available information indicates that, for integrated neutron fluxes greater than 10^{19} neutrons/cm^2 (n/cm^2), the volume of concrete may increase and that the compressive strength, tensile strength and modulus of elasticity may be reduced. The creep of concrete does not appear to be greatly affected by low dosages of radiation but, at dosages which result in lower strength, the creep may be increased. The coefficients of thermal expansion and thermal conductivity of concrete are not affected significantly by nuclear radiation; the same applies to the shielding properties of concrete except that the increase in temperature could cause loss of moisture.

In biological shields, the integrated neutron radiation flux may exceed 5×10^{19} n/cm^2. In these structures, the shielding effectiveness of the concrete is generally of more importance than its load-carrying capacity. Some deterioration in the mechanical properties of concrete can therefore be tolerated particularly since the effectiveness of the concrete as a biological shield does not seem to be greatly affected by nuclear radiation except for moisture loss due to increased temperature. The radiation dose also decreases rapidly as the distance from the surface exposed to radiation increases, and the possibility of nuclear radiation effects on the mechanical properties of concrete in most of the concrete shield is consequently reduced considerably.

In a PCRV, the radiation dose after many years of service depends on the type of reactor and on its construction. It has been estimated [94] that after 30 years of service a PCRV may be exposed to an integrated thermal neutron flux of 6×10^{19} n/cm^2 and a fast neutron flux of 2 to 3×10^{18} n/cm^2.

Exposure to radiation in PCRVs may thus be high enough to cause some deterioration in the properties of concrete and care must therefore be exercised in the choice of concrete-making materials and in the design of such structures. Only the inner few metres of the thick walls and end slabs of a PCRV may, however, be subjected to significant levels of radiation during the life of the vessel.

In Appendix H of BS 4975, it is stated that the effects of nuclear radiation on concrete are considered to be insignificant for doses up to 0.5×10^{18} n/cm^2. At higher values, it must be shown that the overall integrity of the vessel and its components will not be unacceptably affected by the resulting changes in the concrete properties. The dose of 0.5×10^{18} n/cm^2 which has been set in Clause 3.2.1.11 of BS 4975 is at a level below which no significant changes in concrete properties are considered likely. In the ASME Code, the radiation limit for concrete is set at 10×10^{20} nvt. The term nvt (neutrons, velocity, time) is frequently used instead of n/cm^2 for neutron flux.

4.3 Multiaxial strength

Multiaxial stress distributions are induced in the thick concrete walls and end slabs of PCPVs. Under some combinations of stress, concrete can safely sustain loads at stresses which are greater than the uniaxial compressive strength. Other stress combinations may result in failure of the concrete at stresses which are less than the uniaxial compressive strength.

In BS 4975, the permissible principal compressive stress for various loading conditions is specified. If, however, any one of a set of three simultaneously induced principal compressive stresses exceeds the allowable principal compressive stress, it may be admissible if assessed in a manner which is indicated. The ASME Code also provides for adjustments in permissible stresses in a multiaxial compression field.

Existing knowledge on the behaviour of concrete under multiaxial stress states is incomplete and only covers particular load combinations. The results reported also do not always agree. Experimental data have been compared with some of the failure criteria which have been proposed [95], and the definition and location of the elastic limit is still open to discussion. The requirements of the ASME Code were found to be in close agreement with the curves defining the elastic limit, but it is not certain whether a larger region than that defined by the ASME Code could be accepted as allowable stress states.

For biaxial stress conditions, it has been indicated that compressive strength may be increased by approximately 25% [96]. In a triaxial state of stress, it is considered that the strength of concrete can be increased by much larger margins.

There has been much research into the strength of concrete under triaxial stress conditions and a number of failure criteria have been proposed. All of these criteria are applicable to a limited extent but a universally accepted failure surface is not yet available.

Although many mathematical models, which have been proposed to simulate stress–strain behaviour of concrete under multiaxial stress conditions, have yielded promising results when applied to simple problems, more work appears to be necessary before they can be useful in the analysis of practical problems [97].

Acknowledgements

The author and publishers gratefully acknowledge permission from the American Society of Mechanical Engineers to reproduce material from Section III, Division 2 of the *ASME Boiler and Pressure Vessel Code*, 1980. Material from BS 4975: 1973 is reproduced by permission of the British Standards Institution, 2 Park Street, London W1A 2BS from whom complete copies of the standard can be obtained.

References

1. Bremer, F., Design and construction of the PCPV for the 300 MWe THTR nuclear power station in West Germany, *Proc. int. conf. experience in the design, construction and operation of PCPVs and containments for nuclear reactors*, York, Sept. 1975. Institution of Mechanical Engineers, pp. 305–16. [*York Conference*].[1]
2. Jago, J. *et al.*, Important aspects of the construction of the Hartlepool and Heysham pressure vessels, *York Conference*, pp. 283–290.
3. Schöning, J. and Schwiers, H. G., The characteristics of the PCRV of the HHT demonstration plant, *Trans 5th int. conf. struct. mech. reactor technol.*, Berlin, 1979, North Holland, Paper H6/4 [*SMIRT-5*].[2]
4. Ople, F. S., Latest developments in prestressed concrete vessels for gas cooled reactors, *SMIRT-5*, Paper H3/1.
5. Macken, T. *et al.*, PCRV design and development for the GCFR, *York Conference*, pp. 489–497.
6. Morgan, P. L. T. and Bradbury, J. N., Multicavity PCPVs for HTR and GCFR systems, *York Conference*, pp. 545–553.
7. Schwiers, H. G. *et al.*, Feasibility of a multicavity PCPV with elastic hot liner for HTR, *SMIRT-5*, paper H6/3.
8. Kumpf, H. *et al.*, Advanced prestressed concrete pressure vessels for gas-cooled fast breeder reactors, *SMIRT-5*, Paper H6/2.
9. Hannah, I. W., Structural engineering of prestressed reactor pressure vessels, *Nucl. Eng. & Design*, Vol. 50, No. 3, Nov. 1978, pp. 443–462.
10. Jungmann, A. *et al.*, Reference design of a PCPV for pressurised water reactors, *York Conference*, pp. 509–519.

1 All subsequent references to these *Proceedings* are abbreviated to *York Conference*.
2 All subsequent references to these *Transactions* are abbreviated to *SMIRT-5*.

11. Marchertas, A. H. *et al.* Analysis and application of prestressed concrete reactor vessels for LMFBR containment, *Nucl. Eng. & Design*, Vol. 49, 1978, pp. 155–173.
12. Bradbury, J. N. and Morgan, P. L. T., Containment for LMFBR (CFR), *York Conference*, pp. 563–579.
13. British Standards Institution, BS 4975: 1973, *Specification for prestressed concrete pressure vessels for nuclear reactors*, London, 1973.
14. British Standards Institution, CP 110: 1972 (as amended 1980), *Structural use of concrete*, London, 1980, 156 pp. + amend.
15. American Society of Mechanical Engineers, *Code for concrete reactor vessels and containments*, Boiler and Pressure Vessel Code, Section III—Division 2, 1980.
16. Nuclear reactor vessels of prestressed concrete with metal reinforcements, *Journal officiel de la République Française*, Decree 103: 6119–6128, June 1970.
17. Deutsches Institut für Normung, *Reactor pressure vessels made of prestressed concrete. A study for the development of a standard*, West Berlin, June 1972.
18. Stevenson, J. D., Current summary of international extreme load design requirements for nuclear power plant facilities, *Nucl. Eng. & Design*, Vol. 60, No. 2, Sept. 1980, pp. 197–205.
19. Hannah, I. W., Prestressed concrete pressure vessels in the United Kingdom, *York Conference*, pp. 1–8.
20. Gloudeman, J. F., Integrated computer-based systems—survey and outlook, *SMIRT-5*, Paper M1/1.
21. Northup, T. E., The ACI–ASME Code for concrete reactor vessels and containments (USA), *York Conference*, pp. 125–131.
22. Bathe, K.-J. *et al.*, On constitutive modelling in finite element analysis, *SMIRT-5*, Paper M2/1.
23. Nesitka, A. *et al.*, Assessment and structural analysis of a PCPV with hot liner and adjustable wall temperature, *SMIRT-5*, Paper H6/5.
24. Goodpasture, D. W. *et al.*, Design and analysis of multicavity prestressed concrete reactor vessels, *Nucl. Eng. & Design*, Vol. 46, 1978, pp. 81–100.
25. Ople, F. S., Preplaced aggregate concrete application on Fort St Vrain PCRV construction, *York Conference*, pp. 317–326.
26. Hughes, A. N. *et al.*, Principles of design and construction for the top caps of prestressed concrete reactor pressure vessels, *York Conference*, pp. 327–333.
27. Dunlop, J. N. and Grigsby, R. E., The design and construction of the top cap of the Dungeness 'B' concrete pressure vessel, *York Conference*, pp. 335–344.
28. Long, J. E., International approval policies for nuclear tendons, *York Conference*, pp. 589–595.
29. Naus, D. J., Grouted and non-grouted tendons for prestressed concrete pressure vessels, *SMIRT-5*, Paper H3/6.
30. Costaz, J. L. and Picaut, J., Multi-barrier concrete containment without steel liner, *York Conference*, pp. 533–537.
31. Fountain, M. J. *et al.*, Corrosion protection of prestressing tendons, *York Conference*, pp. 237–243.
32. Brunton, J. D. and Thomas, W. D., Wire-winding of Hartlepool and Heysham reactor pressure vessels, *York Conference*, pp. 217–228.

33. Thorpe, W. and Speck, F. E., BBRV post-tensioning systems as applied to reactor containments and prestressed concrete pressure vessels, *York Conference*, pp. 201–210.

34. Matt, P., The General Atomic strand winding machine, *York Conference*, pp. 211–216.

35. Crowder, R. *et al.*, Design and construction of the prestressed concrete boiler closures for the Hartlepool and Heysham pressure vessels, *York Conference*, pp. 145–157.

36. Tate, L. A. *et al.*, The temperature control of prestressed concrete reactor pressure vessels, *York Conference*, pp. 83–92.

37. Jones, H. and Hedgecock, P. D., Thermal protection system for the concrete core support floor at Fort St Vrain, *York Conference*, pp. 111–118.

38. Merot, J. P. and Lacroix, R., Principle of 'hot wall' insulation for prestressed concrete reactor vessels, *York Conference*, pp. 521–531.

39. McD.Eadie, D. and Bell, D. J., Proof pressure tests of the PCPVs at Hinkley Point 'B' and Hunterston 'B', *York Conference*, pp. 425–434.

40. Ople, F. S. and Neylan, A. J., Construction, testing and initial operation of Fort St. Vrain PCRV, *York Conference*, pp. 291–304.

41. Neylan, A. J., Leaktightness in HTGRs—experience at Fort St Vrain, *York Conference*, pp. 179–184.

42. Irving, J. *et al.*, Experience of in-service surveillance and monitoring of prestressed concrete pressure vessels for nuclear reactors, *York Conference*, pp. 443–456.

43. Beaujoint, N. and Guery, A., Experience in operation and inspection of prestressed concrete pressure vessels belonging to Electricité de France, *York Conference*, pp. 457–464.

44. Hornby, I. W., The strain behaviour of a prestressed concrete reactor pressure vessel after 12 years operation, *SMIRT-5*, Paper H3/5.

45. Aird, H. M. *et al.*, Surveillance of prestressed concrete pressure vessels under commissioning and operational conditions at Hunterston 'B' power station, *SMIRT-5*, Paper H3/4.

46. Browne, R. D. and Bamforth, P. B., The value of instrumentation in the assessment of vessel performance during construction and service, *York Conference*, pp. 411–424.

47. Reuter, H. R. and Whitcraft, J. S., Development and optimisation of containment structure concepts, *York Conference*, pp. 33–39.

48. Walser, A., An overview of reactor containment structures, *Nucl. Eng. & Design*, Vol. 61, Nov. 1980, pp. 113–122.

49. Bartley, R. and Davies, I. L., Aircraft impact design for SGHWR containment, *York Conference*, pp. 55–62.

50. Stevenson, J. D., Overview of concrete containment design practice in the USA, *York Conference*, pp. 9–17.

51. Costaz, J. L. and Moreau, P., Review of French containment vessels, *York Conference*, pp. 19–24.

52. Schnellenbach, G., Proposed design criteria for containments in Germany, *York Conference*, pp. 581–587.

53. Langhans, J., Experience in the safety assessment of the SNR-300 containment, *SMIRT-5*, Paper J1/9.

54. Engelbrektson, A. and Bøye-Møller, K., Performance of Scandinavian BWR-containments during pressure tests, *SMIRT-5*, Paper J6/5.

55. George, B. V. and Roberts, A. C., Alternative containment concepts for the high temperature gas cooled reactor, *York Conference*, pp. 555–562.

56. Smith, E. C. and Kenthol, J., Leakage testing of Candu containments, *York Conference*, pp. 185–193.

57. Biggio, M. *et al.*, Structural engineering problems in FINTOR conceptual designs, *SMIRT-5*, Paper N1.1/5.

58. Electricité de France, *Cahier de spécifications techniques, étude des enceintes de confinement des tranches types PWR 900*, 1st Edn, 12 May 1975.

59. Kellerman, O., Trends and summaries of extreme load design criteria in Germany, *Nucl. Eng & Design*, Vol. 60, Sept. 1980, pp. 206–209.

60. Canadian Standards Association, Standard N287: *Concrete containment structures for CANDU nuclear power plants*.

61. O'Brien, J., US regulatory requirements for blast effects from accidental explosions, *Nucl. Eng. & Design*, Vol. 46, 1978, pp. 145–150.

62. Wheaton, R. *et al.*, Probabilistic evaluation of the SSE design spectrum for a nuclear power plant site: a case study, *SMIRT-5*, Paper K2/3.

63. Cummings, G. E. and Wells, J. E., Systems analysis methods used in the seismic safety margins research program, *SMIRT-5*, Paper K3/2.

64. Philip, G. and Bork, M., KTA 2201—seismic design standards in the Federal Republic of Germany, *SMIRT-5*, Paper K2/7.

65. Hadjian, A. H., Engineering of nuclear power facilities for earthquake loads, *Nucl. Eng. & Design*, Vol. 48, 1978, pp. 21–47.

66. Riera, J. D., A critical reappraisal of nuclear power plant safety against accidental aircraft impact, *SMIRT-5*, Paper J9/2.

67. Kamil, H. *et al.*, An overview of major aspects of the aircraft impact problem, *Nucl. Eng. & Design*, Vol. 46, 1978, pp. 109–121.

68. Kennedy, R. P., *A review of procedures for the analysis and design of concrete structures to resist missile impact effects*, Nuclear Sciences Group, Holmes and Narver, Anaheim, California, Sept. 1975.

69. Brown, M. L. *et al.*, Local failure of reinforced concrete under missile impact loading, *SMIRT-5*, Paper J8/7.

70. Sarne, Y. and Long, K. N., Design and analysis of penetrations in reinforced concrete nuclear structures, *York Conference*, pp. 69–73.

71. Sarne, Y., and Reeves, C. F., Design and analysis of reinforced concrete containment structures for the base shear force, *York Conference*, pp. 63–68.

72. Ravindra, M. K. and Walser, A., Probabilistic design of nuclear structures: a summary of state of the art and research needs, *Nucl. Eng & Design*, Vol. 50, Oct. 1978, pp. 115–122.

73. Hill, H. T. *et al.*, Structural integrity test of prestressed concrete containments, *York Conference*, pp. 465–471.

74. Brown, R. H. and Cranston, G. V., Integrated leakage rate testing of prestressed concrete containments, *York Conference*, pp. 195–200.

75. Rotz, J. V., In-service surveillance of unbonded post-tensioning systems in prestressed containment structures, *York Conference*, pp. 473–478.

76. Davis, H. S., Concrete for radiation shielding—in perspective, *Proc. int.*

seminar on concrete for nuclear reactors, American Concrete Institute, SP-34, 1972, pp. 3–28.

77. Polivka, M., Annotated bibliography on concrete for radiation shielding, *Proc. int. seminar on concrete for nuclear reactors*, American Concrete Institute, SP-34, 1972, pp. 1639–1722.

78. *Engineering compendium on radiation shielding*, Springer, Berlin, 1968.

79. Bertero, V. and Polivka, M., Influence of thermal exposure on mechanical characteristics of concrete, *Proc. int. seminar on concrete for nuclear reactors*, American Concrete Institute, SP-34, 1972, pp. 505–531.

80. Crispino, E., Studies on the technology of concrete under thermal conditions, *Proc. Int. seminar on concrete for nuclear reactors*, American Concrete Institute, SP-34, 1972, pp. 443–479.

81. Campbell-Allen, D., Low, E. W. E and Roper, H., An investigation on the effect of elevated temperatures on concrete for reactor vessels, *Nucl. Struct. Eng.*, Vol. 2, No. 4, Oct. 1965, pp. 382–388.

82. Seki, S. and Kwasumi, M., Creep of concrete at elevated temperatures, *Proc. int. seminar on concrete for nuclear reactors*, American Concrete Institute, SP-34, 1972, pp. 591–638.

83. Rodriguez, C. *et al.*, Model—including thermal creep effects—for the analysis of three-dimensional concrete structures: comparison with tests, *SMIRT-5*, Paper H4/8.

84. Kottas, R. *et al.*, Strength characteristics of concrete in the temperature range of 20°C to 200°C, *SMIRT-5*, Paper H1/2.

85. Kaplan, M. F. and Roux, F. J., Variations in the properties of concrete at elevated temperatures, *Trans. 6th int. conf. struct. mech. reactor technol.*, Paris, Aug. 1981, Paper H1/2.

86. Blundell, R. *et al.*, *The properties of concrete subjected to elevated temperatures*, Construction Industry Research and Information Association, Report No. 9, London, 1976, 78 pp.

87. Zoldners, N. G., *Effect of high temperatures on concretes incorporating different aggregates*, Canadian Mines Research Branch, Report R64, 1960, 48 pp.

88. Lankard, D. R. *et al.*, The effects of moisture content on the structural properties of Portland cement concrete exposed to temperatures up to 500°F, *Temperature and Concrete*, American Concrete Institute, SP-25, 1971, pp. 59–102.

89. Maréchal, J. C., Contribution à l'étude des propriétés thermiques mécaniques du béton en fonction de la température, *Ann. Inst. Tech. Bat. et Trav. Publ.*, No. 274, Oct. 1970.

90. Ziegeldorf, S. *et al.*, Thermal expansion of concrete for nuclear structures, *SMIRT-5*, Paper H2/1.

91. Ross, A. D. and Parkinson, J. D., Shrinkage in concrete pressure vessels, *Nucl. Eng. & Design*, Vol. 5, 1967, pp. 150–160.

92. Hilsdorf, H. K., Kropp, J. and Koch, H. J., The effects of nuclear radiation on the mechanical properties of concrete, *The effect of nuclear radiation on the mechanical properties of concrete*, American Concrete Institute, SP 55, 1978, pp. 223–251.

93. Seeberger, J. and Hilsdorf, H. K., Effect of nuclear radiation on mechanical properties of concrete, *SMIRT-5*, Paper H2/3.

94. Houben, J. A., De bestraling van mortelproefstukken, *Second conf. on prestressed concrete reactor pressure vessels and their thermal insulation*, Commission of the European Communities, Brussels, 1969, pp. 170–183.
95. Robutti, G. *et al.*, Failure strength and elastic limit for concrete: a comprehensive study, *SMIRT-5*, Paper H2/5.
96. Kupfer, H. *et al.*, Behaviour of concrete under biaxial stresses, *J. Am. Concr. Inst., Proc.*, Vol. 66, Aug. 1969, pp. 656–666.
97. Meyer, C. and Bathe, K.-J., Non-linear analysis of concrete structures in engineering practice, American Society of Civil Engineers Annual Convention, Florida, Oct. 1980.

34 Offshore concrete structures

S Fjeld

Det norske Veritas, Oslo, Norway

C T Morley

University of Cambridge, England

Contents

Notation

A	cross-sectional area
A_i	coefficient in non-linear wave theory
A_p	tip area of skirt
A_s	side area of skirt
a	acceleration of water particles
c	coefficient in Miner's hypothesis on fatigue; wave speed
c_d	drag coefficient
c_{de}	effective drag coefficient in linearized theory

c_m	added mass coefficient	p^*	peak value probability density
c_s	slamming coefficient	q_c	cone resistance
D	diameter	R	radius; penetration resistance; autocorrelation function
d	water depth; depth of penetration of skirts	Re	Reynolds number
E_d	drained Young's modulus for soil	r	response
E_u	undrained Young's modulus for soil	$\bar{r}_m$	most probable largest range for response r
E_t	tangent modulus	S	mean square spectral density (for sea elevation)
f	frequency (Hz); force per unit length	S_{rr}	mean square spectral density for response r
f_{sp}	stress in steel at proportional limit	St	Strouhal number
f_{tm}	mean tensile strength of concrete	s	standard deviation
G	shear modulus	s_u	undrained soil shear strength
G_0	shear modulus for soil at small strains	T	time period
g	acceleration due to gravity	T_z	mean apparent wave period
H	wave height (trough to crest)	t	thickness
H_c	wave height parameter	U	group velocity
H_s	significant wave height	v	velocity of water particles
h	wave height (from mean sea level)	w	ambient water pressure; radial displacement
H_h	hydrodynamic transfer function	w_0	imperfection magnitude
H_m	mechanical transfer function	X	mean sea level above datum
$\bar{H}_m$	most probable largest wave height	x	sea-surface elevation above datum
I	rotational inertia	x, y, z	rectangular coordinates
i	index number	z	depth
K	Keulegan–Carpenter number	α, β, γ	parameters in extreme wave height theory
k	wave number $2\pi/\lambda$		
k_f	friction coefficient for skirts	β	slenderness number
k_p	tip coefficient for skirts	γ_w	unit weight of water
k_s	reduction factor	γ_f	partial safety factor on loads
L	length	γ_m	partial safety factor on material strengths
M	modulus of uniaxial deformation		
M_ϕ	hoop moment per unit width	ε	spectral width
m	number of waves in buckling mode; modulus number for soil	η	magnification factor; surface displacement in wave theory
m_i	ith moment of spectral density curve	κ	curvature
N_i	number of cycles to failure for ith block	λ	wavelength
		ν	kinematic viscosity; Poisson's ratio
N	number of mean periods in a sea state record	ϕ	velocity potential
		ρ	mass density of water
N_ϕ	hoop force per unit width	ρ_a	mass density of air
n_i	number of cycles in ith block	ρ_{min}	minimum steel proportion
P	probability	σ'	effective stress
p	pressure; probability density	σ_{tc}	critical stress on tangent modulus theory
p_e	extreme value probability density		
p_{tc}	critical pressure on tangent modulus theory	ω	frequency (rad/s)

Summary

The history and current technology of offshore structures in reinforced and prestressed concrete are surveyed, with special emphasis on fixed gravity structures (whose weight holds them in place on the sea-bed) for hydrocarbon exploitation. Functional requirements for such structures are outlined, and the severe environmental conditions to which they may be subjected are discussed in some detail. There follow sections on the durability of concrete at sea, the construction and installation of large marine structures, and foundation design. The choice of structural configuration to meet the various requirements is discussed, and methods of static and dynamic analysis and design of structural elements are presented. The chapter concentrates mainly on problems and techniques which either are particularly prominent offshore or are not often encountered in land-based structures.

1 Introduction

This chapter summarizes the technology of the design and construction of concrete offshore structures for use in the North Sea to support the equipment and men needed for the exploitation of oil and gas discoveries. It concentrates mainly on special structural problems related to the marine environment and on methods and techniques which either are not commonly used for land-based structures or are not commonly emphasized in academic courses for intending civil engineers. In the main, the chapter deals with *fixed gravity structures* whose own weight holds them in place on the sea-bed, but some of the technology described may be applicable to other types of marine structure, such as ships, barges, harbour works and breakwaters, or the more novel articulated or submarine structures which are being developed for deep waters.

Forms of concrete have been used in marine civil engineering works from the earliest days. The Romans used pozzolanic cement mortars and concretes widely in harbour works and breakwaters, and John Smeaton in 1757 developed a lime–pozzolana cement mortar for joints in the masonry of the third (and first successful) Eddystone lighthouse. The rowing boat built by Lambot in 1848 is thought to have been the first reinforced-concrete structure. His mesh-covered iron skeleton plastered with dense cement mortar has been developed into the modern *ferrocement* which is widely used for yachts and small boats [1].

According to Morgan [1, 2] 69 large seagoing reinforced-concrete vessels were built in the United States during the first war and 488 during the second, the largest being the 132 m long Selma built in 1919. Presumably, the idea was to save steel, but the hurriedly executed designs in traditional form rather disappointingly used almost 60% of the steel required for an equivalent all-steel vessel, and gave rather low cargo capacities as a proportion of displacement (up to only 55% at best, for the Askelad constructed in Norway in 1917). This problem of high hull weight, due to concrete's relatively low ratio of strength to weight, has stimulated the development of lightweight aggregate concrete for floating structures [3].

However, the early ships, some of which remain in use as breakwaters to this day, certainly demonstrated that well-made concrete structures can be remarkably durable and watertight in a hostile marine environment. Expenditure on maintenance is correspondingly low, particularly so in large structures if the concrete is prestressed to eliminate cracking. Prestressed concrete barges were constructed in Germany in 1943 and in Hawaii in 1960.

Many concrete marine structures, completed and proposed, were described at the FIP symposium in Tblisi in 1972, including reinforced concrete floating dry docks of up to 8000 tons capacity in use in the USSR since 1929, and a 350 m long floating dry dock of 100 000 tons lifting capacity, constructed segmentally as a cellular caisson of prestressed lightweight concrete [4]. Floating breakwaters and airports have been designed in cellular prestressed concrete, and large rectangular prestressed-concrete boxes have been constructed to form seagoing barges carrying small factories or refineries, or liquid natural gas in steel containers. In future, the favourable properties of concrete at cryogenic temperatures [5] may be exploited in LNG carriers.

Applications of concrete in coastal and harbour engineering are too numerous to mention in detail, ranging as they do from quays formed in ordinary reinforced column-beam-and-slab construction on concrete piles, through cast *in situ* defence works such as sea-walls and groynes, to large precast dolos or hammerhead units placed as armour on the seaward faces of mounded rock-fill breakwaters.

This chapter mainly deals with concrete used *offshore*, where the sea-bed (or *mudline*) is below the level of the lowest astronomical tide, so that foundations are always submerged. Since the 1920s, platforms cast *in situ* on driven concrete piles have been used successfully for oil production from Lake Maracaibo, Venezuela, in up to 40 m of water. Notable early offshore concrete structures included the Nab tower placed in the Solent in 1920, and the naval forts, floated out and sunk on to station in 13 m of water in the Thames estuary in 1941 (see Section 2.1). Later in the war, numerous hollow concrete pontoons were floated across the Channel and sunk on the shore of Normandy to form Mulberry harbours. Some of these pontoons and forts are still on station with their structural concrete in good condition after 40 years.

Great impetus was given to the development of offshore structures by the discoveries in the late 1960s of oil and gas reservoirs beneath the North Sea. In many cases (see Section 2.3), concrete gravity structures were chosen in preference to the tubular steel 'jacket' structures more traditional in offshore hydrocarbon exploitation. A number of factors combined to give concrete the advantage: its cheapness and greater durability at sea; the presence of heavily overconsolidated soils near the mudline, making practicable large gravity structures supported on spread foundations; the development of slipforming and to a lesser extent of pumpable concrete simplifying the construction of large concrete structures; and the need for storing a large volume of oil between tanker visits, which could conveniently be done in the large hollow caisson needed for the spread foundation. Above all, a concrete gravity structure could be constructed inshore in shallow waters, floated out as a single unit, and made safe within a few days of being sunk on station—thus needing a much shorter *window* of calm weather than was required for a steel-jacket structure, which could not be regarded as safe against storms until a lengthy pile-driving operation had been completed (and whose deck structure then had to be erected).

Since the Ekofisk platform was placed in 70 m of water in 1973, 13 fixed concrete structures of various types have been positioned in the North Sea, culminating in the giant Ninian Central platform (of 600 000 tons displacement at float-out, which was placed in 136 m of water in May 1978), and Statfjord B, placed in 1981. Platforms in concrete have also been placed off the Brazilian coast in relatively shallow water. It seems likely that in waters over 300 m deep or so, rigid structures fixed to the sea-bed and supporting platforms above the water will not always prove economic, either in concrete or in steel, partly because of the sheer size necessitated by the large wave forces and moments attracted, and partly because of problems with structural dynamics. Novel structures are therefore being developed for deep waters, for example articulated *compliant structures*, which will deflect markedly in heavy seas and greatly reduce the wave forces, or completely submerged *subsea structures* placed on the sea-bed well below the surface zone where the major wave forces arise. Much of the proven technology of fixed structures which is outlined below will be of relevance to the design of concrete offshore structures of novel form.

The contractual arrangements for the construction of platforms for oil exploitation have differed markedly from those traditional in large-scale civil engineering works in the English-speaking world, where a contractor usually works under the supervision of an independent consulting design engineer advising a governmental client. Rather there have been all-in 'package deals' between the oil company and a single large company acting as both designer and contractor. Because of the great dangers, of loss of life and of pollution, attendant upon oil extraction, legislation (in the UK, the Offshore Installations (Construction and Survey) Regulations Act 1974) has tended to require independent *certification* and *quality assurance*, i.e. the review and supervision of design, calculation and construction. Usually, this has been by one of the ship classification societies, such as Lloyd's or Det norske Veritas, who issue their own rules [6] to supplement governmental guidance notes [7] and international codes such as the FIP recommendations [8].

This survey of the technology of offshore concrete structures begins in Section 3 with an outline of the functional requirements which a platform may have to meet. This is followed by a section on the marine environment and the forces due to waves, wind, etc., which a structure must be designed to withstand. Section 5 deals with the durability of concrete in sea-water and with the steps necessary in design and construction to ensure that concrete marine structures have adequate life. There then follow two sections on the construction and installation of offshore structures and on the design of spread foundations on the sea-bed.

The remainder of the chapter deals with the design of structures to meet these requirements and sustain these loadings. Section 8 outlines the steps which might be taken in deciding upon a suitable structural configuration; Section 9 deals with static and dynamic analysis of the complete structure or of substantial parts of it, and Section 10 with the detailed design of structural elements. But before plunging into the details of the technology, we begin in the following section with a survey of some of the more notable offshore concrete structures completed since 1940.

2 Notable offshore concrete structures

2.1 Wartime structures

In 1941–2, in an operation with remarkable similarities to later activities in North Sea oil exploitation, four naval forts were constructed at Gravesend and floated out complete to be sunk in up to 13 m of water in the Thames estuary [9]. Each fort incorporated a boat-shaped base pontoon, 51×27 m in plan and 4.3 m deep, with outer vertical walls 300 mm thick and a flat bottom 225 mm thick stiffened by ribs. After being constructed of cast *in situ* reinforced concrete in a dry dock, each pontoon was floated out, with a draught of 2.2 m, to a second tidal berth where the twin 18 m high reinforced concrete towers were added. These were of 7.3 m external diameter and 300 mm wall thickness, and incorporated seven decks, cast on precast permanent formwork, to provide accommodation for equipment, stores and the 100-strong crew. The steel main deck for the guns and military equipment was placed in 6 ton sections using derricks, and concreted into the tops of the towers.

After being fitted-out in a third inshore berth, each fort (now with 3.8 m draught) was towed to its station by three 1000 hp tugs, complete with crew and stores so that it could go into action immediately after being placed on

Figure 2-1 Completed naval fort being sunk in position *Courtesy* J A Posford

the sea-bed. The fort was tilted (Fig. 2–1) by admitting water to the forward compartment: eventually water spilled over the open sides of the pontoon, which plunged to the sea-bed bow first in about 20 s, the rate of descent being somewhat controlled by buoyancy chambers with small vents. A reinforced concrete cylindrical buffer softened the impact at the bow, and the stern settled on the sea-bed within a further 30 s. Apparently no attempt was made to prepare the sea bed for the forts, of which one was placed on clay, one on shingle and two on sand. The boat-shaped pontoons were aligned with the tidal currents to reduce scour.

Later in the war over 200 reinforced concrete Phoenix cellular caissons were constructed in Britain, floated across the Channel, and sunk in up to 16 m of water to form breakwaters round the Mulberry harbours on the Normandy beaches [10]. Each unit was rectangular, approximately 62 m long, 17 m wide and 18 m high, constructed of cast *in situ* reinforced concrete walls and slabs varying in thickness from 225 to 375 mm. Using model tests, the units were designed to be stable when floating with a draught of 6 m, and to withstand waves of 2.5 m height and 60 m wavelength. Some of the units were floated from dry dock part complete, so that building could continue afloat.

2.2 Lighthouses

Two major problems were encountered by the designers of the Kish Bank Lighthouse, a single circular concrete caisson placed in 21 m of water for the Commissioners of Irish Lights in 1963 [11]. With a design wave 13.7 m high, the diameter at the water-line on station needed to be kept small to minimize wave forces on the structure, but the water-line diameter during float-out needed to be large, to ensure stability. A tapering profile was adopted, 31.7 m diameter at the base, narrowing to 18.3 m diameter at the sea surface. The second difficulty was with the foundation; a level gravel bed on the sandbank had to be laboriously prepared by divers, and precautions against scour were required.

Foundation difficulties also occurred at the Royal Sovereign Lighthouse [12], which was placed in 18 m of water in 1968, designed to withstand a 15 m high wave with 11 s period. The sea-bed, strewn with boulders, had patches of soft clay between hard spots of oolitic limestone, and had to be cleared, levelled, and covered with a layer of gravel protected against scour. The water-line diameter problem was overcome by floating the structure on to station in two stable parts, the wide base and telescoped tower, and the concrete cabin on pontoons. The base, which was in ordinary 40 MPa reinforced concrete with a prestressed tower of 6.1 m diameter in 55 MPa concrete, was floated out, using compressed air for extra buoyancy, and sunk on to the prepared bed. The cabin was mated with the tower using the tidal range, and the tower then jacked out to its full extent, giving a final overall height of 49 m.

2.3 Ekofisk and Ninian Central

A great step forward was taken with the placing of the huge Ekofisk artificial island [13] in 70 m of water, 270 km from the Norwegian coast, in June 1973. It consists essentially (Fig. 2-2) of a cellular oil-storage caisson, approximately 52 m square in plan by 90 m high, surrounded by a perforated Jarlan wall breakwater, almost a circle of 92 m diameter in plan, connected to the same cellular sand-filled

Figure 2-2   Ekofisk: perforated Jarlan wall *Courtesy* Construction News

base slab. The perforations in the wall produce turbulence, which dissipates the energy of the waves and thus reduces the total wave force on the structure, to an extent determined by model tests.

The 6 m deep cellular base slab, designed to cope with differential settlement and high local foundation reactions, was built in a temporary dry dock near Stavanger and floated out to adjacent deep water. Here, the caisson proper, which had 500 mm thick arch walls and diaphragms and was prestressed horizontally and vertically, was constructed by slipforming all nine cells together, the structure sinking steadily to its final draught of 66 m (displacement 215 000 t). Floating stability was ensured by adding ballast in the base, and by temporarily blocking some of the perforations in the Jarlan wall, which was constructed in a mixture of precast and *in situ* work during the slipforming. Ordinary Portland cement concrete was used throughout, with air entrainment and a specified characteristic cylinder strength of 40 MPa.

The caisson, complete with decks and equipment, was towed 320 km to its station by five tugs of total power 50 000 hp (37 MW), and was sunk in a few

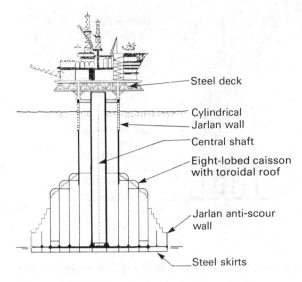

Steel deck

Cylindrical
Jarlan wall

Central shaft

Eight-lobed caisson
with toroidal roof

Jarlan anti-scour
wall

Steel skirts

Figure 2-3 Ninian Central platform [14]

hours by ballasting with water, penetrating about 0.5 m into the upper layer of silty sand in the unprepared sea-bed.

Jarlan walls are also incorporated in the Ninian Central platform [14], the largest concrete offshore structure built to date (600 000 tons displacement at tow-out), which was placed in 136 m of water east of the Shetland Islands in May 1978. The base slab, of 140 m diameter and 1.5 m thickness, was constructed in dry dock at Loch Kishorn in Western Scotland, together with the anti-scour Jarlan wall and the lower parts of the radial and lobate walls of the caisson (Fig. 2-3). After the base had been floated out into adjacent deep water, the draught gradually increased as construction continued by (independent) slipforming in concrete of 50 MPa specified characteristic cube strength, the toroidal roof to the caisson being formed of massive precast elements with *in situ* joints.

After completion of the slipforming, the construction of the Jarlan wall for the wave zone using 300 t precast elements, and the horizontal and vertical prestressing, the structure was water-ballasted down for the floating-over and fixing of the steel deck structure. All the deck equipment and accommodation modules were then added inshore (total payload 23 000 t) and the structure deballasted and towed out to station. Here, the structure was ballasted down on to the clay sea-bed, where it is held in place by its own weight and by the horizontal resistance provided by steel *skirts* which penetrate 4 m into the soil.

2.4 Condeep and Sea Tank

Although Jarlan perforated walls have been used in the first and the largest of the North Sea platforms, they are not incorporated in the dominant Condeep [49] and Sea Tank [15] designs (Fig. 2-4) in which wave forces are minimized by keeping member-diameters small in the wave zone. Since the first Condeep Beryl A was

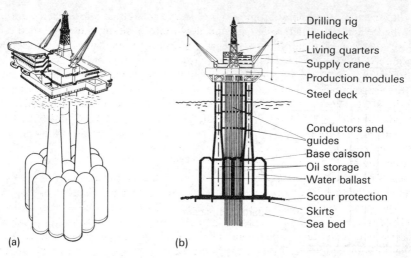

(a)   (b)

Figure 2-4   Concrete gravity structures with caisson: (a) Condeep, (b) Sea Tank

placed in 117 m of water in July 1975, eight other large concrete platforms of this type have been positioned in the North Sea. All these structures incorporate a large cellular foundation caisson in which oil can be stored, with a small number of circular shafts supporting above the waves a steel deck housing production equipment and accommodation. All were constructed in a broadly similar way: initial construction, in a temporary dry dock, of a base tray which is then floated out into sheltered deep water, construction of the caisson and columns by slipforming while afloat, deep submergence in sheltered water for deck mating, tow-out on to station in the North Sea, ballasting down to the sea-bed, where skirts penetrate the upper strata, and finally grouting beneath the base. In most cases, the oil and water within the caisson are kept at lower pressure than the external sea-water, to guard against leaks and to keep the caisson concrete in compression.

Of course, there are important differences between the Condeep and Sea Tank designs, and between individual structures of each type. Thus, the Condeep caisson is basically hexagonal in plan, consisting of a number of contiguous cylindrical cells (with substantial continuity between the cells where they touch), with water ballast and oil storage in the main cylinders but with the intermediate triangular *star cells* open to the sea. By contrast, the Sea Tank and Andoc designs have a roughly square caisson, divided into square cells by a number of plane internal diaphragms, but with the outer walls arched against the external pressure. The Condeep cells have domed roofs cast *in situ*, while the other structures have pyramidal, flat or domed roofs to the caisson, sometimes incorporating precast permanent formwork. The shafts of the Condeeps are of relatively large diameter, so that oil conductors and risers (and some utilities) can be accommodated within them. Sea Tanks have supporting columns of fairly small diameter, so that conductors have to be carried externally, supported by special guide frames.

All the structures incorporate some form of skirt to penetrate a few metres into the sea-bed when the structure is ballasted down into position, but some of the skirts are of stiffened steel, others of precast concrete. (A structure on firm sand will require only shallow skirts to prevent hydraulic pumping and scour, while a structure on soft clay overlying firmer soils will require deeper skirts.) In several cases, the first contact with the sea-bed was taken on a few dowels, protruding below the skirts and designed to eliminate horizontal displacement in the final stages. The Sea Tank structures have flat base slabs, whereas the Condeep cells are domed downwards against foundation pressures, but in all cases the space between the structure and the sea-bed was filled with cement grout to form a monolithic foundation.

It should be emphasized that the main problems in the construction of these giant offshore platforms are often not so much in design theory or in technology as in organization and logistics [15]. The main challenges are due to the sheer scale of the operation, and very careful planning is required to ensure the speedy and successful completion of operations, such as slipforming, which might have to continue without major interruption for many weeks, employing up to 500 men on structures with a wall length of perhaps 2 km. However, such matters are outside the scope of this chapter, the remainder of which deals with theoretical and technological aspects of design.

3 Functional requirements

Before commencing a design, it is necessary to be clear about the requirements which the offshore structure must satisfy if the equipment and operations it supports are to proceed correctly and safely. These might include requirements on the space to be provided for processing equipment, utilities or living quarters, the loads to be sustained by the structure during ordinary operation of the equipment, loads which the structure might have to withstand in abnormal situations such as equipment breakdown or accident, and safety requirements such as fire resistance. Loads arising from equipment operation should be classified into those which regularly occur and therefore are treated as ordinary live loads, and those occurring in exceptional circumstances, for instance, an explosion or a ship collision, for which lower load factors may be justified.

Naturally, requirements will differ according to circumstances. Here, we consider briefly as an example the main requirements for an oil production platform, which will typically need to accommodate equipment for drilling, processing and additional oil recovery, as well as utilities, storage and living quarters (see Fig. 2-4). Table 3-1 gives some approximate basic figures for three typical platforms.

3.1 Drilling

An oil platform may have from 15 to 50 wells, perhaps 3 m apart, each composed of an outer 0.8 m diameter conductor enclosing concentric pipes with cement–mortar–grouted annuli. The drilling area on the platform is normally fitted with a movable derrick. Several North Sea platforms have external conductors braced by steel guides, but often, for protection against waves and ship impact, the conductors are taken down inside a main shaft. Since the oil and gas may have

Table 3-1   Typical functional requirements

	Statfjord A	Murchison	Tor
Main data			
Gas/oil ratio	178	110	262
Wells	42	25	18
Transport	Loading buoy	Pipe	Pipe
Storage (million barrels)	1.2	0	0
Water cleaning, (m³/day)	56 000	5 000	900
Power supply (MW)	66	43	5.6
Oil processing			
Capacity (barrels/day)	300 000	165 000	107 000
Reservoir pressure (bar)	70	45	88
Gas compression			
Capacity (million Nm³/day)	8.8	1.3	—
Water injection			
Capacity (barrels/day)	340 000	150 000	—
Volume (m³)			
Deck	50 000	33 000	⎫ 17 500
Production equipment	80 000	100 000	⎬
Quarters	12 000	12 000	4 000
Dry weight (t)			
Utilities, pipes,			
electrical	23 500	6 500	2 900
Steel in deck	9 500	4 500	800
Steel in modules	8 150	8 700	1 800
Quarters, helideck	2 000	1 900	800

temperatures up to 100°C, the shaft must be designed for thermal stresses or the water inside the shaft must be cooled. For obvious safety reasons (blow-out, fires, etc.), the drilling area should be sited well away from living quarters.

3.2   Processing equipment

Depending on the characteristics of the reservoir, the hydrocarbons produced and the onward transportation equipment, the platform will house heavy equipment, such as separators, heat exchangers, scrubbers and compressors, for separating and treating oil, gas, condensate, water and impurities. The reliability of this equipment and the associated high-pressure piping, carrying large quantities of combustible material, must be given high priority, as well as the fire protection of the structure. Even without fire, a rupture of a high-pressure pipe (at perhaps 15 MPa, or 2000 psi) in a closed or water-filled compartment, might jeopardize a platform by causing substantial outward pressure on the walls (either the walls must be designed accordingly, or this situation avoided altogether).

Additional compressors and conductors may be required to inject gas or water into the reservoir so as to increase production.

3.3 Utilities

The equipment required for services and utilities, such as power supply, fire fighting, control rooms and offices, sewage, heating and ventilation, and cooling water, can be as large and heavy as the process equipment, e.g. a power supply might weigh 4000 t. To save deck weight and space, on some platforms a main shaft has been kept dry for utilities. This solution has recently been less popular, perhaps because of difficulties of access for maintenance and replacement of the equipment as well as the possible effects of accidental flooding of the utility shaft. In any case, shafts are already crowded by conductors, risers, ballast and bilge pipes and so on.

3.4 Oil storage

Most concrete platforms have been provided with a hollow caisson for oil storage, perhaps one million barrels (i.e. one to two weeks production) to allow continuous production from the reservoir over an emergency such as pipeline shutdown or weather restrictions on tanker loading. Significant thermal stresses can be caused by the contrast in temperature between the ambient sea-water (about 5°C) and the stored oil (35 to 40°C to ensure smooth flow).

3.5 Serviceability requirements

To allow safe functioning of equipment and men, limits will be imposed on the maximum amplitude of vibration, wave-induced motion or settlement of the structure. To date, these conditions have easily been met by fixed concrete gravity structures—but there may be problems with tethered or compliant structures or with fixed structures in very deep water.

3.6 Function-related accidents

It is important to make sure that if an accident or a breakdown occurs in the operation of a platform, it does not produce an effect disproportionate to the cause by overloading the structure of the platform. The possible effects of riser-rupture and of fire have already been mentioned; other possible accidents related to the intended function of the platform should also be considered [16], for example, gas explosions and the dropping of objects [17] from the deck on to the roof of the caisson (which it is now usual to protect by a layer of lightweight concrete).

The possible effects of ship collision can be important for the design of concrete offshore structures, especially the part near the water-line [18]. Even small impacts from supply boats moored to a platform in moderate seas can cause high local horizontal loads on shafts supporting the deck, and the provision of fenders is often considered. The possibility of a heavier collision with an out-of-control supply boat or a larger tanker should not be disregarded.

But the main feature of offshore structures is that, in addition to live loads arising from their function or from function-related accidents, they may also be subjected to very large environmental loads; and to these we now turn.

4   Environmental conditions and loads

A major difference between land-based and offshore structures lies in the hostile environment often encountered offshore—wind, waves, currents, even ice in some regions. Extensive records and sophisticated statistical methods are needed to predict for each structure the worst conditions to which it may be subjected (i.e. those conditions with a sufficiently small probability of occurrence during the design life). Horizontal forces on offshore structures due to wind and waves can be much larger, relative to the structure's weight, than is common on land, with important consequences for design, and the fluctuating nature of environmental loads can lead to dynamic problems and fatigue.

The major component of environmental loading is usually due to waves. It would be presumptuous to attempt a detailed survey here, and we limit ourselves to a basic introduction to wave statistics and wave loading, followed by some remarks on other environmental loads, leaving the interested reader to delve further into the references quoted.

4.1   Ocean wave statistics

4.1.1   *Basic statistical concepts*
Each record, such as that sketched in Fig. 4-1, of sea surface elevation above chart datum at a fixed position is, of course, unique. But over time periods of order 1 h and surface areas of order 100 km^2, the sea surface is usually *stationary* statistically, i.e. certain time-averaged quantities (e.g. the mean level, X, or the *mean square deviation*, s^2, from it) are approximately the same when computed from different (sufficiently long) records. A stationary sea-state is characterized by two statistical averages, the *mean apparent period*, T_z, between upcrossings of the mean sea level, and the *significant height*, H_s, the mean of the one-third greatest trough-to-crest heights H. T_z and H_s tend to correspond to the wave period and height reported by experienced visual observers [19].

Also obtainable from the records are the *probability*, $P(x_1)$, that the surface elevation $x(t)$ does not exceed some given value x_1 and the *probability density*,

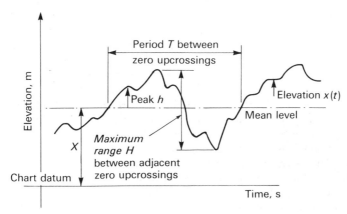

Figure 4-1   Typical record of sea-surface elevation

$p(x_1)$—usually [20] the normal **or** Gaussian distribution function—such that $p(x_1)\delta x_1$ is the proportion of time that $x(t)$ spends in a narrow band of width δx_1. Further analysis leads to the peak value distribution, $p^*(x_1)$, giving the probability that a maximum of $x(t)$ occurs within a narrow band at x_1.

The *mean square spectral density*, $S(\omega)$, of a stationary sea-state indicates how on average the variance s^2 of elevation is distributed across the range of frequencies ω (rad/s). For records extending over $-T/2 < t < +T/2$, the *autocorrelation function*, $R(\tau)$, is defined by

$$R(\tau) = \lim_{T \to \infty} \frac{1}{T} \int_{-T/2}^{+T/2} x(t)x(t+\tau)\,dt \qquad (4\text{-}1)$$

and $S(\omega)$ is obtained [21] as the *Fourier transform* of $R(\tau)$:

$$S(\omega) = \frac{1}{\pi} \int_{-\infty}^{+\infty} [R(\tau)\cos\omega\tau]\,d\tau; \qquad \omega > 0 \qquad (4\text{-}2)$$

Typical dimensionless plots of $S(\omega)$ are given in Fig. 4-2. Efficient computer programs are now available [22] for rapidly evaluating $S(\omega)$ via the Fourier transform of the record itself, given in discrete digital form. The spectral density plays a very important part in calculations (Section 9.2) of the response of a structure to random excitation by the sea.

Also of importance are the *moments*, m_n, of the area under the curve $S(\omega)$ about the axis $\omega = 0$, defined by

$$m_n = \int_0^\infty \omega^n S(\omega)\,d\omega; \qquad m_0 = \int_0^\infty S(\omega)\,d\omega = s^2 \approx \frac{H_s^2}{16} \qquad (4\text{-}3)$$

The mean apparent period, T_z, is related to the moments by

$$T_z = 2\pi\sqrt{m_0/m_2} \qquad (4\text{-}4)$$

and a measure of the *spectral width* is ε, given by

$$\varepsilon^2 = 1 - m_2^2/m_0 m_4 \qquad (4\text{-}5)$$

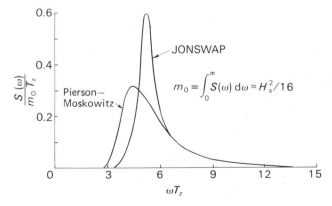

Figure 4-2 Dimensionless plots of standard sea spectra

A narrow spectrum, with all motion close to one frequency, has $\varepsilon \to 0$, while $\varepsilon \to 1$ for a wide spectrum with significant intensity at all frequencies.

4.1.2   Short-term wave statistics

Numerous data on wave records have been published since 1945, mainly for the North Atlantic area [23, 24, 25]. For design purposes in open sea, it has become usual to use the modified Pierson–Moskowitz spectrum [26]:

$$\frac{S(\omega)}{m_0 T_z} = \frac{2}{\pi^2} \left(\frac{\omega T_z}{2\pi}\right)^{-5} \exp\left[-\frac{1}{\pi}\left(\frac{\omega T_z}{2\pi}\right)^{-4}\right] \tag{4-6}$$

If high-frequency components above $\omega T_z = 15$ are excluded, the width parameter ε is 0.63. This spectrum (Fig. 4-2) is for *fully-developed* seas, where the wind has blown over a sufficient distance, or *fetch*, as to whip up the waves fully (and where low-frequency *swell* from disturbances elsewhere is negligible).

The spectrum resulting from the JONSWAP [25] investigation of fetch-limited seas has five parameters, only two (m_0 and T_z) clearly fetch-dependent [27]: the 'mean' dimensionless curve in Fig. 4-2 is independent of fetch and has $\varepsilon = 0.57$ (truncating at $\omega T = 15$). It is not clear how the JONSWAP spectrum is transformed into the Pierson–Moskowitz as the sea-state becomes fully developed, but the difference is perhaps only structurally important when the peak frequency is close to a resonant frequency of the structure.

In theory [20, 28], the crest heights, h, and the trough-to-crest ranges, H, follow the Rayleigh probability distribution only for $\varepsilon \to 0$, but experimental data suggest that this distribution (Fig. 4-3), with the density function

$$p(H_1) = (H_1/4m_0) \exp(-H_1^2/8m_0) \tag{4-7}$$

applies to wave heights H in all sea-states. The significant height, H_s, is then [20] very close to four times the standard deviation (s or $\sqrt{m_0}$) of the surface elevation.

The probability density, $p_e(H_1)$, of the maximum height, or *extreme value*, H_m in a long narrow-spectrum record was studied by Longuet-Higgins [20]. For large

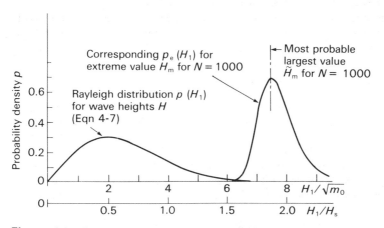

Figure 4-3   Rayleigh and extreme-value distributions

N, where N is the duration of the record as a multiple of T_z,

$$p_e(H_1) = \frac{H_1}{4m_0} u \exp(-u); \qquad u = N \exp(-H_1^2/8m_0) \qquad (4\text{-}8)$$

This result, usually accepted even for fairly large ε, is plotted in Fig. 4-3 for $N = 1000$. The *most probable largest* value of H, indicated by the maximum p_e, is given approximately for a record of large length N by

$$\tilde{H}_m = H_s(\tfrac{1}{2} \log_e N)^{\frac{1}{2}} \qquad (4\text{-}9)$$

Commonly [6] design is based on the most probable largest value of an environmental parameter over a period, or on the extreme value with a certain *return period* (average interval between exceedances), the chosen period being related to the anticipated life of the structure (or to the duration of a short construction or installation phase). Unfortunately the equations above apply only to sea-states which are statistically stationary, i.e. to *short-term* periods of a few hours.

4.1.3 Long-term statistics

In principle, it should be possible to calculate a long-term spectrum, $S_L(\omega)$, for the entire life of a structure by averaging the short-term spectra weighted according to the probability of the corresponding significant height H_s $(=4\sqrt{m_0})$ and T_z. Thus,

$$S_L(\omega) = \int_0^\infty \int_0^\infty S(\omega; H_{s1}, T_{z1}) p(H_{s1}, T_{z1}) \, dH_{s1} \, dT_{z1} \qquad (4\text{-}10)$$

Here, $p(H_{s1}, T_{z1})$ is the joint probability density function for H_s and T_z which are quite strongly correlated: this might be found from a *scatter diagram* of occurrences of H_s and T_z in certain ranges at the site, recorded or predicted [29] from wind speed and fetch. Quantities such as $\tilde{H}_m$, the most probable largest wave during the design life, would be calculated from $S_L(\omega)$.

Although other methods are available [30, 31], $\tilde{H}_m$ is commonly estimated by the following method [19], based on data suggesting that the long-term probability distribution of significant height H_s, for all T_z, is of the two-parameter Weibull type

$$P(H_s < H_{s1}) = 1 - \exp[-(H_{s1}/H_c)^\gamma] \qquad (4\text{-}11)$$

The parameters γ and H_c can be determined from the straight line obtained by plotting recorded probabilities P against H_{s1} on special Weibull probability paper (Fig. 4-4).

For given H_s, the individual wave heights, H, follow a Rayleigh distribution, and the long-term probability that an individual wave height $H < H_1$ can be obtained by combining this with Eqn 4-11, giving according to Nordenstrøm [19] approximately another two-parameter Weibull distribution (also plotted on Fig. 4-4):

$$P_L(H < H_1) = 1 - \exp[-(H_1/\beta H_c)^\alpha] \qquad (4\text{-}12)$$

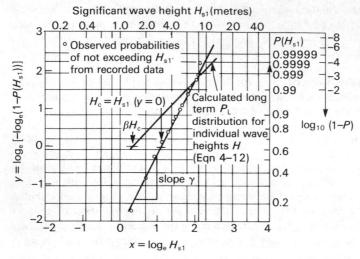

Figure 4-4   Long-term wave statistics on Weibull probability paper

where α and β are given approximately by

$$\alpha = 0.279 + 0.523\gamma - 0.06\gamma^2$$
$$\beta = 0.361 + 0.079\gamma - 0.007\gamma^2 \tag{4-13}$$

For a large number N of mean periods $\bar{T}_z$, the most probable largest value $\tilde{H}_m$ becomes

$$\tilde{H}_m = \beta H_c (2.3 \log_{10} N)^{1/\alpha} \tag{4-14}$$

Calculations of this type predict 100-year values of $\tilde{H}_m$, ranging from about 10 m in summer in the southern North Sea to around 30 m for all seasons off the central Norwegian coast.

A more recent trend [32] has been to recognize that high waves tend to occur in groups in storms, and to consider the long-term average frequency of storms combined with the probability of at least one wave in a storm exceeding some given height.

4.1.4.   *Directional spectra*

For a short-term stationary sea, the spectral density, $S(\omega)$, for the surface elevation, x, at one point might be considered as the sum of many *directional spectra*, $S(\omega, \theta)$, each corresponding to waves travelling at some angle θ to, say, the wind direction, where in the absence of better information we may take

$$S(\omega, \theta) = \frac{2}{\pi} S(\omega) \cos^2 \theta \qquad -\frac{\pi}{2} \leqslant \theta \leqslant +\frac{\pi}{2} \tag{4-15}$$

so that

$$S(\omega) = \int_0^{2\pi} S(\omega, \theta) \, d\theta \tag{4-16}$$

If calculations are carried out with directional spectra, the calculated response must be integrated over all θ. Since Eqn 4-15 concentrates most of the wave energy near $\theta = 0$, it is usually sufficient to simplify calculations by assuming the waves to be *long-crested*, so that $S(\omega)$ represents a series of water waves all travelling in the same direction $\theta = 0$.

There remains the question of wave direction relative to the structure. The most probable largest elevation-range, $\bar{H}_m$, during the design life, as calculated above, may correspond to a wave in any direction, and will be somewhat larger than the most probable largest wave coming from a restricted sector (perhaps the worst sector for structural response). This problem is complicated [32] and conservatively one may consider a long-crested wave of height $\bar{H}_m$ coming from the worst sector—the so-called *extreme design wave*.

4.2 Regular waves

Whichever approach is to be used for designing a structure in waves, via either an extreme design wave or spectral densities for surface elevation, one needs to know the velocity and acceleration of water particles in regular waves of given frequency on the open sea well away from the structure. Here, we begin with the linear theory of sinusoidal waves of small amplitude, Airy waves, and go on later to non-linear theories for higher waves, and to consideration of the effect of obstructions on the wave pattern.

4.2.1 Linear theory of gravity waves

In linear wave hydrodynamics, which is often adequate in practice, the sea water is assumed homogeneous and incompressible, with negligible viscosity and surface tension. The flow is then *irrotational* and the fluid velocities are the gradients of a *velocity potential*, ϕ, which satisfies Laplace's equation $\nabla^2 \phi = 0$, with zero vertical velocity $\partial\phi/\partial z$ at the sea-bed, and at the surface the pressure and fluid velocity properly related to the displacement η (Fig. 4-5). In the linearized theory [33], these two latter conditions are combined into

$$g\frac{\partial\phi}{\partial z} = \frac{\partial^2\phi}{\partial t^2} \quad \text{at } z = 0 \tag{4-17}$$

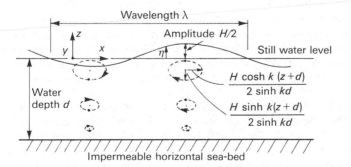

Figure 4-5 Long-crested wave on linear theory

One solution is

$$\phi = \frac{H\omega}{2} \cos{(kx - \omega t)} \frac{\cosh{k(z+d)}}{k \sinh{kd}} \quad ; \quad k = 2\pi/\lambda \tag{4-18}$$

which corresponds to a sinusoidal wave

$$\eta = \tfrac{1}{2}H \sin{(\omega t - 2\pi x/\lambda)} \tag{4-19}$$

long-crested in the y-direction (Fig. 4-5) of amplitude $H/2$ and wavelength λ, travelling in the positive x-direction.

The water particles at given position x move in phase round ellipses. At the surface, the horizontal velocity, v, and acceleration, a, are given by

$$v = \frac{\pi H g}{\lambda \omega} \sin{(\omega t - kx)}; \qquad a = \frac{\pi H g}{\lambda} \cos{(\omega t - kx)} \tag{4-20}$$

and both diminish with depth $-z$ as ϕ does. The excess pressure over still-water hydrostatic is

$$p = \frac{\gamma_w H}{2} \frac{\cosh{k(z+d)}}{\cosh{kd}} \sin{(\omega t - kx)} \tag{4-21}$$

where γ_w is the unit weight of water.

The waveform travels over the sea surface at the *wave speed* $c = \lambda \omega/2\pi$; and wavelength λ depends on frequency ω because of the important [33] *dispersion relation*

$$\omega^2 = \frac{2\pi g}{\lambda} \tanh{\left(\frac{2\pi d}{\lambda}\right)} \tag{4-22}$$

The average wave energy per unit area of the sea surface is $\gamma_w H^2/8$ and is transported by the wave at the *group velocity*

$$U = \frac{c}{2}\left[1 + \frac{2kd}{\sinh{2kd}}\right] \tag{4-23}$$

For a stationary random sea of long-crested waves, the average energy in a frequency band $d\omega$ is given by $\gamma_w S(\omega)\,d\omega$, where $S(\omega)$ is the mean square spectral density of surface elevation.

For waves on *deep water*, with depth d greater than about one third of the wavelength λ, the equations for linear theory take a particularly simple form; the water particles travel at constant speed in circles, with speed and radius diminishing exponentially with depth; the hyperbolic function in Eqn 4-22 tends to unity and the group velocity, U, tends to half the wave speed, c.

4.2.2 Non-linear wave theories

Linear theory is adequate for waves of small height, H, up to perhaps 5% of the wavelength, λ. Extreme design waves can be appreciably higher: for example, a deep-water wave of 30 m height and period 15 s (at the low end of the likely range) will have $\omega = 0.42$ rad/s and wavelength $\lambda = 350$ m from Eqn 4-22, giving $H/\lambda = 0.086$. In this case, non-linearities, arising mainly at the free surface, should be taken into account.

In the perturbation theory [34] due to Stokes, the waveform is regular but no longer purely sinusoidal and is analysed as a Fourier series, with accuracy improving as more terms are taken. The velocity potential is taken as

$$\phi = \sum_{n=1}^{m} HA_n \cosh [nk(z+d)] \cos (nkx - n\omega t) \qquad (4\text{-}24)$$

and the coefficients A_n for given depth, wave height and wavelength are calculated from the non-linear free-surface boundary conditions. Commonly, terms up to $m = 3$ or 5 are taken, giving the Stokes' third-order or fifth-order theories. The waveform is no longer symmetrical about the still-water level, the crests being more pronounced than the troughs; in deep water, there is a theoretical maximum height $H = 0.142\lambda$ at which the wave crest is a sharp $120°$ angle, just about to break. The water particles describe circular paths with a superimposed drift velocity, in the direction of wave travel, diminishing exponentially with depth.

In shallow water, with the parameter $H\lambda^2/d^3$ between 1 and 32, non-linear *cnoidal waves* are possible [33], so called because they take the form of the square of the Jacobian elliptic function, *cn*. Again the crests are more pronounced than the troughs, but do not become sharp-angled; for high values of $H\lambda^2/d^3$, the profile becomes a series of humps separated by long flat regions. Another theory sometimes used for high waves on deep water is Gerstner's trochoidal theory, but the resulting flow is not irrotational. Often, linear wave theory is used for high waves, despite its lack of precision, because of the computational advantages it brings.

4.2.3 *Modification of waves: diffraction*
As regular waves approach a coastline through gradually-shallowing water, the frequency tends to remain constant but the wavelength decreases according to Eqn 4-22. The wave speed also decreases, so that the wave crests tend to become aligned with the depth contours of the beach. The energy density increases, as does the height; coupled with the decreasing wavelength, this leads to a great increase in steepness, H/λ, until the wave breaks.

Of more importance here is the scattering or *diffraction* of small-amplitude waves (for which linear theory is appropriate) by an obstruction such as an offshore structure. Onto the potential ϕ_0 of an incident wave is superimposed a potential ϕ_s for the scattered wave, the condition of zero velocity normal to the surface of the rigid immovable structure imposing

$$\frac{\partial \phi_s}{\partial n} = -\frac{\partial \phi_0}{\partial n} \qquad (4\text{-}25)$$

on the submerged surface of the structure, where n is a local coordinate measured along the normal into the fluid. The potential ϕ_s must satisfy the sea-bed and free-surface conditions, and also a *radiation condition* [33, 35] ensuring that well away from the structure ϕ_s represents only scattered waves travelling outwards from the structure.

Solution of the diffraction problem is required to estimate the forces on a large structure due to waves (Section 4.3.2) and to study interaction between adjacent structures, since the wave climate at one structure may be radically affected by the presence of a neighbour. A numerical approach is usually adopted and suitable computer programs are widely available. One possibility is to adopt a method based on the Green's function, which gives the potential function, ϕ_s, throughout the sea for a radiating source of unit strength at a given point. The surface of the structure is imagined divided into a large number of small plane panels, each with an unknown source strength: the source strengths are determined by satisfying the normal velocity condition 4-25 at each panel, and all the required information about the scattered waves can then be obtained. Alternatively, a finite element method may be used, either in two dimensions [34] when the structures all consist of vertical prisms piercing the sea surface, or in three dimensions [35].

4.3   Wave forces on fixed structures

4.3.1   *Small structures*
For structures whose significant dimensions are small compared to the length of the waves (e.g. for a vertical cylinder of diameter D less than about 0.2λ), the energy radiated in scattered waves is small and the forces on the structure may be determined directly from the properties of the incident wave assumed undisturbed. The water pressure, p, (Eqn 4-21) might be important in some cases, and the net force on the structure can be estimated from the velocity and acceleration which the water would have at the centre of the structure, if the structure were not present. Thus, the horizontal force per unit length of a vertical cylinder is given by the well-known *Morison equation* [36].

$$f = \rho A (1 + c_m) a + \tfrac{1}{2} \rho D c_d v \, |v| \tag{4-26}$$

Here, ρ is the mass density of water, A the cross-sectional area of the cylinder, D the projected width of the cylinder normal to the wave direction, and v and a the horizontal velocity and acceleration of the water particles.

The first *inertia* term includes both the *Froude–Krylov* force, the resultant of the pressure distribution in the undisturbed wave needed to accelerate the water in the volume now occupied by the structure, and an *added mass* term with the coefficient c_m. This coefficient [6, 34] depends mainly on the shape of the structure and may be calculated by diffraction theory in simple cases (e.g. $c_m = 1.0$ for a circular shaft). Experiments [37] show that c_m depends also on Reynolds number $Re = v_{max} D / \nu$ and the Keulegan–Carpenter number $K = v_{max} T / D$ (T the wave period, ν the kinetic viscosity) taking the tabulated values [38] for low K and high Re.

The second *drag* term in Eqn 4-26 is of particular importance for small-diameter bodies such as risers or conductor tubes, and is related to the viscosity of the water and to boundary-layer separation. The coefficient c_d (around 1.0 for a circle) depends mainly on Reynolds number and shape, but is also affected by K and surface roughness.

Morison's equation may be used for estimating wave forces on the main shafts of offshore concrete structures, where drag is usually not important. Results for a typical shaft of diameter 10 m in a sinusoidal deep-water wave of height 20 m and period 15 s are given in Fig. 4-6.

A difficulty arises in that the equations of linear hydrodynamics for small-amplitude waves apply up to the still-water level $z = 0$, but a substantial zone may

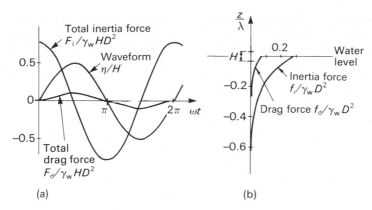

Figure 4-6 Wave forces on a vertical cylinder: cylinder diameter $D = 10$ m, wave height $H = 20$ m, period $T = 15$ s, $\lambda = 350$ m (deep water) (a) Total force on a vertical cylinder (b) Variation of force per unit length with depth

be covered and uncovered in a design wave of large amplitude. One could consider the wave profile at short intervals in time, and apply Morison's equation to those parts of the structure which are currently wetted, using either linear wave theory or (despite Lighthill's criticism [39]) non-linear Stokes' theory.

For inclined cylinders, the force normal to the axis may be estimated from Eqn 4-26 if the velocity and acceleration normal to the axis are inserted. A general three-dimensional body will have three mutually-perpendicular principal axes for fluid inertia, each with its own added mass, and Morison's equation may be applied along each principal axis.

4.3.2 Large structures
Inertia effects usually dominate the wave forces on structures whose dimensions are a significant proportion of the wavelength ($D > 0.2\lambda$ for a vertical circular cylinder). The scattered wave energy is now significant, and the effective added-mass coefficient in Morison's equation is less than when $D/\lambda \to 0$. The usual method of estimating wave forces is to use linear diffraction theory, to predict the scattered wave potential, ϕ_s, and then to integrate the transient hydrodynamic pressure, $-\rho \, \partial\phi/\partial t$, over the surface of the structure. Various computer programs are available [34, 35, 40]. The large caissons of gravity structures are usually treated by this method, and the scattered waves are then taken into account when using Morison's equation to calculate forces on shafts and risers.

4.3.3 Other wave effects
In appropriate circumstances, one must also take into account other forces due to waves. Thus, horizontal members near the still-water level are subjected to *slamming*, i.e. impulsive upward forces when the member is suddenly immersed. The force per unit length may be estimated [6] as $\frac{1}{2}\rho c_s D v^2$, where v is the water velocity normal to the member and c_s is a coefficient, usually at least 3. Members in this area are also subjected to impulsive *slapping* forces from breaking waves, and the high dynamic pressures may affect design.

Significant fluctuating *lift forces*, normal to the direction of wave advance, may occur on small-diameter members due to regular *vortex shedding* at a frequency, f, given by the Strouhal number, $St = fD/v_{max}$, which is a function of Reynolds number, Re. The object in design will be to ensure that the natural frequency of the member is well away from the vortex-shedding frequency so that resonant vibrations do not occur.

4.4   Other environmental conditions and loads

4.4.1   *Changes in still-water level: tides and surges*

Changes in still-water level relative to a fixed structure can occur for several reasons, and must be allowed for in calculating buoyancy forces, hydrostatic pressure on hollow members, and in deciding upon the platform elevation required to provide a specified *air gap* between the bottom of the deck structure and the crest of the highest likely wave. There will be an *astronomical tidal range* (found from Admiralty charts, about 3 m in the North Sea), storm surge (induced by low atmospheric pressure and wind) and geotechnical effects such as skirt penetration, foundation settlement and inclination, and sea-floor lowering due to oil or gas reservoir depletion.

4.4.2   *Current due to tide and wind*

Most-probable largest values of tidal current speed in 100 years can range up to 1.5 m/s in parts of the North Sea, at the ocean surface [6]. Wind-induced drift velocities at the surface are at 45° to the wind direction [34] and of order 2% of the mean hourly wind speed 10 m above the sea (giving 100-year wind-induced currents of order 1 m/s). Current velocities diminish more slowly with depth [6] than do water velocities due to waves.

The usual method of calculating forces due to current is to add the current velocity vectorially to the velocity due to waves, before using Morison's Eqn 4-26. Thus, currents induce drag, but no inertia forces, and are particularly important for members of small dimensions. However, because the velocity decreases so slowly with depth, tidal currents may give significant forces on large caissons even when the wave forces are inertia-dominated and calculated by diffraction theory.

Currents can also give rise to vortex shedding from slender members, can modify the amplitude and wavelength of surface waves [34], and can give rise to standing waves behind a fixed structure, which may have some effect on adjacent structures.

4.4.3   *Wind and wind forces*

On fixed offshore structures, the wind loads are typically less than 10% of the total environmental loads, but they may vitally affect stability during float-out. A designer in concrete will not normally be interested in the fine details of the wind loading on the deck structure, which may well be constructed in steel, so much as the total wind force and moment transmitted from the deck to the concrete supporting structure and the wind force on exposed concrete shafts.

All the techniques available for estimating wind forces on land-based structures may be used for offshore construction (except that high waves may modify the wind near the sea surface). In dynamic analysis, a short-term spectral density of wind speed might be used [41]. More commonly, design is based on the wind speed (averaged over some suitable time interval) with an adequate return period, determined by analysis of records (as in Section 4.1). Estimates are available [7] through meteorological offices, usually of the mean hourly wind speed 10 m above the sea, which might in the North Sea be of the order of 40 m/s for a 100-year return period. Empirical formulas [6] give the corresponding wind speed at different heights or averaged over shorter time intervals (in structural design, the averaging interval should be comparable with the natural period of the structure, perhaps a few seconds).

Wind forces have negligible inertia components, and a common approach is to use a *basic wind pressure*, $\frac{1}{2}\rho_a v^2$, applied with empirical drag coefficients, c_d, to the various exposed parts of the structural components [6, 34]. Due to turbulence, the wind speed will not be uniform in plan, and for structures with dimensions large compared with 'turbulence wavelength' the spatial variation of wind speed must be taken into account [42]. The direction of the wind relative to the structure is, of course, important, and often the best estimates of wind forces are obtained from careful model tests in wind tunnels. Regular vortex shedding can occur in steady winds.

4.4.4 *Other environmental conditions*

Apart from waves and wind, other environmental conditions and variables may have to be taken into account in appropriate cases. In Arctic water, forces due to pack ice may be of such potential importance that the structure must be specially designed so that the ice is broken up and the forces imposed on the structure reduced in consequence. Deck structures may become coated with ice, especially in the zone accessible to spray from the sea, with consequent increase in dead weight and in cross-sectional area and wind force.

The ambient temperature of the air and the water may affect the behaviour of structural materials, and temperature changes may cause significant strains (especially of the deck structure; the temperature of the sea-water fluctuates rather little, except near the surface).

The density of sea-water, and hence the buoyancy forces on the structure, may be affected by changes of salinity. Salinity, temperature, oxygen content and pH value also affect the likely amount of marine fouling—colonization by marine organisms, possible up to 100 m deep. In some cases, these organisms may grow up to 200 mm thick, and the consequent increase in wave force may be significant.

In seismically active areas, a structure will have to be designed to withstand a *design earthquake* and a careful dynamic analysis (Section 9.2) will doubtless be required.

5. Durability of reinforced concrete offshore

Prolonged use of concrete offshore has shown that it can be a remarkably durable material requiring little maintenance, but careful design and good workmanship are required to ensure adequate useful life. The greatest problems occur in the *splash zone*, where the structure is subject to repeated wetting and drying. (Usually this zone is defined by the range of spring tides and the wave with a six-month return period.) Above the splash zone is the *atmospheric zone*, where conditions are similar to those in land-based structures, except that the lower part may be subject to salt-water spray. Below the splash zone is the *submerged zone*. In this section, we briefly describe the main processes which can lead to degradation of concrete or corrosion of steel, and measures taken in practice to combat them.

5.1   Durability of plain concrete

The sulphates in sea-water can attack [43] the calcium hydroxide ($Ca(OH)_2$) or tricalcium aluminate hydrate (from C_3A in cement) in hardened cement paste, and magnesium ions may replace calcium [44], leading to disruption or softening of the concrete. In submerged concrete, the more soluble components, especially $Ca(OH)_2$, may leach from the matrix, reducing alkalinity. These processes may be combated to some extent by controlling the quality and composition of the *cement*—control of the C_3A content and cement replacement by slag or pozzolan may be advantageous [44].

However, the dominating requirement to ensure durability is to keep the permeability of the concrete low, by using sound dense and properly graded aggregate, rich mixes with a minimum cement content of at least $320 \, kg/m^3$ ($540 \, lb/yd^3$)—$400 \, kg/m^3$ in the splash zone [6]—a water/cement ratio not more than 0.45, and good concreting practice and workmanship.

Water-reducing *admixtures* (e.g. superplasticizers based on high polymer naphthalene condensates) will almost certainly be needed to obtain adequate workability (allowing proper placing and compaction) at low water/cement ratios. To avoid thermal stresses which may lead to cracking and corrosion, the curing temperature in large pours must be kept low (maximum 70°C) by careful mix design to keep the cement content low (in adiabatic conditions the temperature will rise perhaps 11°C for each $100 \, kg/m^3$ ($170 \, lb/yd^3$) of cement).

To produce sound dense concrete, attention must be paid to good workmanship, including proper handling to avoid segregation, full consolidation, and avoidance of cracking due to slipforming, plastic shrinkage, etc. The quality of construction joints is of particular importance; water stops should normally be fitted. At horizontal joints, a retarder may be sprayed on when concreting ceases, and laitance should be removed by water jetting or sand blasting, taking care not to cut too deeply into the hardened concrete. The quality of the concrete placed against the joint is important; it is good practice to use at least 250 mm (10 in) of a high-workability 'oversanded' mix to ensure good compaction. At vertical joints, loose or porous concrete should be removed and the joint shaped to provide a good mechanical key and easy access for the poker vibrator.

Concrete in the splash zone may be disrupted by alternate freezing and thawing. This may be combated by keeping permeability low and strength high, and by careful use of admixtures designed to entrain air in pores with the correct size and spacing.

5.2   Durability of reinforcement

5.2.1   *The mechanism of corrosion*
The corrosion of reinforcing or prestressing steel is an electrolytic process [45]. If a reinforcing bar is placed in an electrolyte, conditions may vary along the bar due

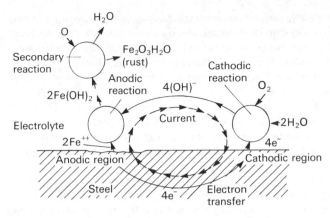

Figure 5-1 Idealized corrosion reaction [45]

to impurities, etc., so that parts of the bar surface become anodic (more negatively charged) and parts cathodic (more positive). Galvanic currents then flow and a chemical reaction proceeds (Fig. 5-1), producing rust: a *corrosion cell* has been set up. However, concrete normally provides a high degree of protection against corrosion of embedded reinforcement, because it provides an alkaline environment of high pH in which the corrosion reaction is inhibited and replaced by *passivation* (the formation of a coherent coating of ferric oxide, Fe_2O_3, protecting the bar from further reaction). If the pH value drops, perhaps due to carbonation of the concrete, the passive film is disrupted and corrosion will begin, proceeding at a rate which depends on the electrical resistivity of the concrete and on the rate at which oxygen is supplied (especially at the cathode). Depassivation will also occur when the concentration of chloride ions at the steel surface reaches a certain threshold value (dependent on the pH), so that chlorides should be eliminated from the mix.

The corrosion products occupy a substantially greater volume than that of the steel removed, and the rust therefore tends to burst the concrete surrounding the reinforcing bar, producing cracking along the bar, disruption of the cover, and eventually complete exposure of the bar to the sea (when corrosion will continue rapidly).

In the atmospheric zone, carbonation and/or ingress of chlorides first occurs at cracks, leading to depassivation at the crack of a small steel zone which becomes the anode of a corrosion cell. Portions of the bar still protected by sound alkaline concrete form the cathode. After a year or two, corrosion is found at cracks only, leading to the supposition that corrosion rates depend on crack width. However, eventually all the bar becomes depassivated and crack widths have little noticeable correlation with long-term corrosion [45]. The long-term corrosion rate is governed by the quality of the cover concrete at the cathode, its thickness and especially its permeability. (This argument about the insignificance of cracks applies to cracks normal to a reinforcing bar: cracks along a bar can occur [46], can lead to corrosion, and should be avoided.)

In the submerged zone, some little time is taken [47] for significant amounts of chloride from the sea to reach the steel, through concretes of the high quality usually specified for offshore structures. Thereafter, since the highly-saturated concrete has low electrical resistivity (in contrast, for example, to repair epoxy), the corrosion rate is controlled by the oxygen supply to the cathode. The amount of dissolved oxygen in sea-water is low, and the permeability of saturated concrete to O_2 is low—and so, therefore, is the corrosion rate. The significance of cracks is further reduced in submerged concrete since they fill after a few months with deposits of magnesium hydroxide and calcium carbonate. Similar deposits also seem to reduce the general permeability of concrete to sea-water [48]. All practical experience indicates that the rate of corrosion of steel embedded in completely submerged concrete of good quality is negligible (except possibly in tropical waters). In the splash zone, however, careful precautions must be taken against corrosion.

5.2.2   *Design against corrosion of reinforcement*

In the submerged zone, the main protective measure is to provide cover of low permeability (Section 5.1). In some codes [6], a limit on permeability is given, and cover thicknesses are often laid down. Crack widths need not be checked, but enough reinforcement should be provided to avoid yielding when the concrete cracks and so ensure proper crack distribution. Additionally in the splash zone, where intermittent wetting and drying and plentiful oxygen could cause rapid corrosion, and where the closing of cracks by soft deposits does not occur, crack widths should be limited in the traditional way (e.g. by limiting steel stress). If possible, reinforcement should be detailed to avoid cracking along bars. Protective coatings have not hitherto been found entirely satisfactory.

5.3   Protection of exposed steel

Usually, the reinforced concrete will have a higher electrochemical potential than nearby exposed steel, which therefore tends to act as an anode and corrode seriously unless protected.

Corrosion can be inhibited by *electrical insulation* of the exposed steel from the reinforcement, for example, by providing adequate clearance, a thick (3–5 mm) insulating coating between the exposed item and the concrete, or plastic sheathing round adjacent reinforcing bars. In practice, it is extremely difficult to provide insulation which is 100% effective.

Exposed steel items are therefore commonly afforded *cathodic protection* by *sacrificial anodes*, kept at a negative potential relative to the exposed steel so that they corrode first. Extra anodes should be provided to allow for current drainage to the reinforcement (up to 1 mA/m^2 initially, but falling drastically after a year or two in sea-water). The aim is not to protect cathodically the reinforcement itself, which is protected by the concrete cover, but the current drainage may assist areas of substandard cover, etc. Cathodic protection should be planned by a specialist, who will avoid under-protection on some parts and provide back-up anodes in case some are consumed too rapidly.

6   Construction and installation

The construction of recent tower-and-caisson platforms has begun with the building of a base tray in a graving dock. The tray is then floated out to deeper sheltered waters where construction continues afloat, followed by deep sub-mergence for mating with the deck structure. The platform is then towed out to its permanent location and ballasted down until the skirts penetrate the sea-bed. Installation is completed by underbase grouting and final *in situ* construction.

6.1 Construction inshore

During all the construction phases, attention must be paid to good concreting practice and workmanship, and especially to construction joints, grouting of ducts, and tolerances. At float-out of the base tray, an underbase clearance of 0.5 m is normally required. Draught may be reduced by an underbase air cushion—the platform should remain stable and seaworthy if air pressure is lost in one compartment.

The structure will gradually submerge as construction continues afloat. Careful calculations are necessary of stress changes due to the increasing loads and water pressures. Construction is often by simultaneous slipforming of all cells, an operation which demands the utmost skill in planning and detailing. Stiff formwork is necessary to maintain the required geometrical accuracy (e.g. to maintain strength against buckling of cell walls in compression). Continuous construction of complete reliability is necessary if cold joints are to be avoided (wheelbarrows have often been preferred to concrete pumps). When precast concrete is used as permanent formwork (e.g. for caisson roofs), high water pressures may come to act on the inner precast shell during submergence, because of the greater permeability of the *in situ* concrete; drainage should be provided.

Deep submergence for deck mating may be the most critical operation in the lifetime of the structure, with very high stress levels and little reserve buoyancy. In some cases, pressure differentials across the caisson walls have been reduced by introducing compressed air at up to 500 kPa (70 psi) into the caisson, either by large compressor or by preventing air egress during ballasting. The safety of the platform after an accidental loss of air pressure should be fully considered. The deck is floated over the base which is deballasted until the parts can be welded. The connection should be designed with adequate tolerances, taking into account elastic deformation of the two structures during load transfer, so as not to overstress the platform shafts.

During construction, the platform is normally anchored by catenary moorings, designed with a weak link so that overloading will not damage the structure.

6.2 Floating and towing

6.2.1 Stability requirements
In all floating conditions, a platform should have:

(a) A metacentric height (corrected for free surface and air cushion) of at least 1 m.
(b) A heel not exceeding 5° under the wind of appropriate return period (say 1 year) plus any towing or mooring loads.
(c) Righting energy (area under the righting-moment/rotation curve) up to capsize, downflooding or collapse at least 1.4 times the corresponding overturning energy under wind of return period 1 year.

(d) Ability to withstand accidental rapid load increases during transfer of heavy weights.

(e) Ability to remain afloat and stable, without progressive flooding, if one compartment is open to the sea due to damage, etc.

Appropriate inclining tests should be carried out before critical marine operations.

Concrete platforms have generally been able to withstand a static 10° heel, twice the maximum expected under tow. Floating on the caisson (which may not be possible after deck-mating) gives good stability and large metacentric height, but high accelerations at the shaft tops. Usually, the structure is ballasted down (passing in calm weather the point of minimum stability as the caisson top submerges) to float with the water-line through the columns. Heel under tow in a storm with a 10-year return period should not exceed 5°. The damage stability requirement (e) can be significant for concrete structures with plane internal bulkheads.

6.2.2 *Towing capacity*

Acceptable speed in calm weather is normally obtained by providing sufficient towing power to hold against a Beaufort 8–9 gale (5 m seas, 40 knot winds, 1 knot wind-induced current). To hold in more severe storms, the required effective power goes up by a factor of about 1.5 per Beaufort number. A tug's effective bollard pull will not exceed 75% of its static bollard pull, except in emergencies, and will be much less in severe weather. A large ocean-going tug will often develop more effective pull, when most needed, than a smaller, higher-powered vessel. In good European practice, tow-wires 600 to 1000 m long are used with breaking loads two to three times the static bollard pull (the higher safety factor for the smaller tugs). Where long catenaries cannot be used, shock loading may be greater and safety factors of 3 to 5 are needed.

Motion during a tow may be checked against *critical motion curves* showing allowable angles of roll or pitch at various frequencies. If necessary, the tugs should alter speed or course to reduce motion by altering the apparent wave period or relative direction.

6.2.3 *Sea-room and clearance*

Sea-room is needed to avoid collision or grounding, and will include a distance in which the platform may drift safely in a storm (perhaps of changing direction) under inadequate towing power. Required sea-room depends on effective towing force, tow-line strength, and water depth requirements. Under-keel clearance often needs careful study, with allowance for heel and other motions, tow-line pull, water density changes, surges, and surveying errors. If the draught is too great to allow the 6–16 m under-keel clearance typically required, air cushions under the base or temporary buoyancy tanks may be needed.

6.3 Installation

Installation of a concrete gravity platform [49] involves *positioning, submergence* to sea-bed, *penetration* of skirts and dowels, and *underbase grouting*. Proper

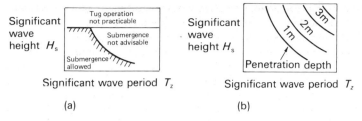

Significant wave period T_z

(a)

Significant wave period T_z

(b)

Figure 6-1 (a) Start criteria for submergence
(b) Acceptable sea-states during penetration

control requires instrumentation [50] to provide data on (a) the environment (waves, wind, current), (b) platform behaviour (position, draught, ballast levels, penetration), and (c) applied loads (soil pressure) and stresses set up (unless the design is sufficiently conservative).

During final approach to the target, the tug boats will be arranged in a star formation for optimum control of the platform. At least two independent navigation systems should be provided. A platform to be placed near other platforms may be positioned using distances (measured electromagnetically) from these platforms. Alternatively acoustic transponders placed on the sea-floor and surveyed-in previously may be used as reference marks.

Prediction of platform motion during submergence to the sea-bed is a difficult task. A Condeep structure was predicted to have a first mode of vibration with a fairly high period (between 30 and 60 s). Submergence should not begin in a sea-state of similar period unless wave heights are small (Fig. 6-1a; Fig. 6-1b indicates acceptable sea-states for various penetration depths). On grounding, skirts and dowels may be subjected to large forces, with dowels causing little reduction of the horizontal pressure on skirts.

Variation of penetration over the site, due perhaps to a sloping sea-bed or varying soil properties, calls for eccentric ballasting to control platform inclination. Limited capacity of internal cell walls may limit the ballasting moment obtainable (see, for example, Fig. 6-2, where the maximum moment occurs at only 50% of the total ballast weight): careful monitoring and control of ballasting is essential.

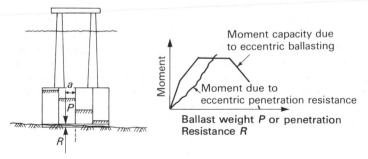

Figure 6-2 Available ballasting moment to cope with eccentric penetration resistance

Even when ballasting is complete, extensive voids will probably remain between the structure and the seabed, due to unevenness of seafloor and/or structure, a sloping sea-bed, incomplete penetration of skirts, etc. The trapped water is often replaced by a grout, lean in cement, matching the properties of the underlying soil [50], in order to improve resistance to horizontal sliding, to achieve an even distribution of soil reaction over the base area, to limit settlements under repeated wave loading, and/or to eliminate piping at the platform edges due to excessive pressures in the trapped water under high wave moments. Pipes are provided for introducing the grout into the underbase compartment (sometimes before ballasting is complete), in an operation which differs markedly from other grouting operations in civil engineering.

In situ construction after penetration and grouting should be kept to a minimum, since an important economic advantage of the concrete gravity structure is that it may be built in sheltered waters inshore and floated out complete in a short weather window.

6.4 Mechanical systems

The structure must, of course, be equipped with all the systems necessary to maintain complete control of the marine operations. A separate study, of normal operations and emergency situations, may well be required to determine the systems to be provided and specify them in detail. Consideration must normally be given to main and emergency power supply, electrical distribution, control systems for machinery and valves, instrumentation, bilge and ballast arrangements, and compressed air systems. Shortly before the start of a critical operation, all systems and equipment involved should be tested to demonstrate that they have adequate capacity and reliability.

7 Foundations

The main geotechnical problems for offshore gravity structures are (a) the performance and penetration of skirts, (b) stability of the foundations and sea-bed, (c) settlements and dynamic displacements, (d) the local contact forces against the base, and (e) the effects of repeated loading on soil properties, as well as scour and scour protection. Local geological conditions and the risk of seismic activity must, of course, be properly assessed as part of the design process. Soils encountered in the North Sea have usually been dense sands or very stiff over-consolidated clays, horizontally bedded, often with a soft layer at the mudline.

7.1 Skirts and penetration

A platform on soft clay overlying firm beds will require deep concrete or steel skirts (Fig. 7-1), taken down into sufficiently firm soil to give an adequate factor of safety against sliding on a horizontal surface through the skirt tips. The skirts must also be deep enough to prevent local failure under horizontal forces, with yield of the soil beside the skirts and flow of material round the skirt tips. A structure on firm sand will require only shallow skirts, to resist scour and to counteract

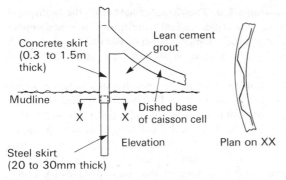

Figure 7-1 Typical caisson skirts

hydraulic effects. (Under large moments due to waves, high water pressures may occur under the platform near the edges, either in trapped water before grouting or in pore water afterwards; skirts reduce the risk of piping or internal erosion by decreasing the hydraulic gradient.)

In some cases, protruding dowels (hollow pipes of diameter 1 to 3 m and wall thickness 10 to 50 mm) are provided to limit horizontal motion of the platform before touchdown of the skirts. Bearing capacity formulae may normally be used to analyse and design dowels and skirts under horizontal loading, with due allowance for the extent of the failure zone.

The penetration resistance of skirts and dowels is important during the installation of the platform: an overestimate may be needed for the design of the ballasting arrangements. Calculations may be based on the results of cone penetration tests in the field, supported by appropriate laboratory tests. For steel skirts or dowels, the unit cone resistances, q_c, from a number of tests are used to establish an average $\bar{q}_c$ profile with depth z. The penetration resistance, R, at depth d is then given as the sum of point resistance and accumulated skin resistance:

$$R = k_p(d)A_p\bar{q}_c(d) + A_s \int_0^d k_f(z)\bar{q}_c(z)\,\mathrm{d}z \qquad (7\text{-}1)$$

where A_p and A_s are the total tip and side areas respectively. The empirical friction coefficient k_f, in North Sea experience, varies between 0.03 and 0.05 in very stiff silty clay and between 0.001 and 0.003 in dense sand. The tip coefficient k_p correspondingly varies between 0.4 and 0.6 and between 0.3 and 0.6. For the upper metre or so, the coefficient may be reduced by 50% to allow for trench formation at the skirts due to water outflow and lateral movement at touchdown. The internal skin resistance for a dowel will be limited by the tip resistance for a solid cylinder.

The penetration resistance of wider concrete skirts is preferably estimated by bearing capacity formulae [52] perhaps combined with conventional earth-pressure and friction theory [53].

7.2 Stability

The critical loading condition for the foundation of an offshore gravity structure is a combination of vertical and horizontal forces, the latter predominantly induced by waves (or occasionally by earthquake). The dynamic nature of the horizontal forces and the large dimensions of the structure make undrained analyses important for both clays and fine-grained sands. Deterioration of the soil shear strength under repeated loading must be considered as well as the susceptibility of near-surface soil to scour.

Large gravity structures depend on their weight to resist horizontal loads, which are often so large that special consideration must be given to sliding modes of failure (along the base or a horizontal surface through the skirt tips, or a combined mode with sliding along the base and local failure at the skirts, or sliding along a soft layer below the platform). For structures on more than one footing, safety against lifting from one foot must be investigated. With a single footing, bearing-capacity failure under the edge of the foundation will precede overturning. Calculations must verify that the platform has acceptable safety against the worst example of each of these types of failure mode, under the most critical loading conditions. Standard bearing-capacity formulae are not recommended for the final design of foundations with a high ratio of horizontal to vertical loading: here one should revert to effective or total stress analysis from first principles [54] or to the method of slices.

The risk of sea-bed failure should also be assessed, especially for structures on a sloping bed. Such failures are most likely in loose deposits of fine sand and coarse silt, or very soft clays, and may be triggered by sea-bed pressure variations due to waves.

Calculations must be based on extensive data from the actual site. Stability analyses require high-quality triaxial or simple shear test results on samples subjected to a prior stress history to bring them back to actual field conditions. Care must be taken during sampling, handling, etc., to minimize disturbance of the sample, which may cause differences between the laboratory and *in situ* shear strengths. In certain 'slickensided' soils or in fissured over-consolidated clays, the strength on a large scale *in situ* may be lower than the strength determined in the laboratory. Field investigations, especially cone penetration tests, are an invaluable supplement to laboratory investigations in assessing the *in situ* shear strength.

Since the wave loading is of short duration, shear strengths under undrained conditions are most relevant for clays. In saturated sand, the undrained strength depends largely on the initial voids ratio. A loose sand which contracts on shearing may loose its strength completely (liquefy). Corresponding strength increases in dense (dilating) sands, accompanied by negative pore pressures and often by large strains, are not commonly relied upon in practice.

7.3   Long-term deformation

Large total settlements, due to soil deformation and/or reservoir depletion, may reduce unacceptably the air gap between wave crests and platform deck. Differential settlements and tilt, perhaps caused by non-uniform soil conditions or loading, may be incompatible with proper operation of the platform and the process equipment. The rate of settlement affects foundation safety, which will usually improve as consolidation proceeds. Foundation displacement, horizontal or vertical, occurring after the installation of conductors and similar equipment will affect their design, and care must be taken to minimize the stresses in conductors due to downward skin friction.

Any recognized method of settlement analysis is considered adequate bearing in mind the uncertainties involved in assigning values to material properties. More refined methods, such as finite element methods, are at present not expected to improve the overall result. At present, no reliable method is available to predict the effect of repeated loads on settlement, and it seems best to rely on empirical evidence from similar structures.

Initial settlements in clay may be estimated using Poisson's ratio $\nu = 0.5$ and an undrained Young's modulus, E_u, between 250 and 1000 times the undrained shear strength, s_u, for normally consolidated clays (between 250 and $500s_u$ when over-consolidated). Final settlements in over-consolidated clays may be calculated [55] using a drained Young's modulus, E_d, between 100 and $150s_u$ and $\nu = 0.1$: in normally consolidated clays, a uniaxial deformation modulus, M, about ten times the effective vertical stress, σ', is recommended.

The stiffness of sand will depend largely on the relative density (or voids ratio), which may be estimated from cone penetration tests and driving resistance during sampling. For dense North Sea sands, a uniaxial modulus, $M = m\sqrt{\sigma'\sigma_a}$, is proposed, with $\sigma_a = 100$ kPa and the modulus number, m, between 400 and 1000 depending on the state of consolidation. Drained moduli, E_d, will be between three and five times the cone penetration resistance, q_c.

7.4 Reaction forces on the base structure

The unevenness of the sea-bed and the shape of the base structure may lead to the development of high local contact forces on the base of a gravity structure. With dense and stiff soil at the mudline, as sometimes encountered in the North Sea, these local forces may govern the design of the base structure. Any one of the following is adopted: (a) the base structure is designed for the largest contact forces compatible with the soil conditions and the sea-bed topography; (b) instrumentation is provided to monitor the contact forces during ballasting, which is interrupted for grouting when a preset design force is approached; or (c) wide and deep skirts are provided to ensure that the main base slab does not come into contact with the sea-bed during installation.

If the base structure is domed (Fig. 7-1), lateral soil displacement due to localized contact with the dome may cause horizontal loads on skirts. Boulders may have a similar effect, and awkward loading on the skirts may be caused by geometrical imperfections leading to non-axial penetration.

7.5 Effects of repeated loading

Repeated loading due to wave action affects the values of the soil parameters required for many of the analyses described earlier, affecting the stiffness and undrained shear strength after cyclic loading has ceased, as well as the risk of total failure due to the cyclic loading itself. Repeated loading effects are of equal concern in sand and clay—the soil conditions determining where the emphasis is for a particular project. Experience indicates that liquefaction of sand is not a serious problem for the dense sands found in the North Sea although some build-up of pore pressure has been recorded during storms.

The undrained shear strength of a clay subjected to cyclic loading may be less than for a non-cycled sample, the reduction depending on the applied cyclic strain and the number of cycles. For cyclic (or permanent) shear strains less than 3%, generated by less than 1000 cycles, the reduction may be up to 25%. This effect must be considered in all stability calculations where relevant.

Repeated loading may also cause a significant reduction of shear modulus, primarily with large shear strains. Where the cyclic shear strain is less than 0.1% the effect is normally insignificant. Previous repeated loading may affect the cyclic behaviour of soil in a later storm. It is usual to assume that any excess pore pressures are drained away between the storms.

8 Structural configuration

There should be no standardized configurations for offshore structures. A designer must treat each project as a prototype and consider different structural configurations in working towards an optimal solution. What is optimal will depend on the structure's purpose, the water depth, environmental and soil conditions, methods and sites available for construction, etc.

The structure must be designed not only to withstand the hostile environment but also to fulfil certain functional requirements, which will vary according to the

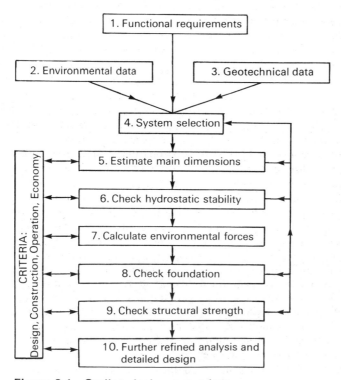

Figure 8-1   Outline design procedure

purpose of the platform (see Section 3). The choice of basic design concept is greatly influenced by the soil conditions at site and the water depths available while the structure is afloat. When considering economy, the costs of maintenance, operation and removal (if required) should not be underestimated in comparison with the costs of construction and installation.

The design procedure is outlined in Fig. 8-1. As many criteria must be satisfied, the designer must accept a trial-and-error procedure and change the estimated dimensions if checks reveal problems.

8.1 Fixed platforms of traditional type

Some existing fixed gravity platforms are illustrated in Figs 2-3 and 2-4.

8.1.1 Main dimensions

The deck should be elevated so as to leave an air gap of 1 to 3 m above the crest of the non-linear design wave, allowing for tides and surges (Section 4.4.1), increased crest elevation due to the caisson and to suction below the deck as the wave passes, settlement and subsidence, and tolerances on water depth and penetration. A concrete deck may be economic if adequate stability in the floating phases can be provided. The appropriate fixity of the deck-shaft connection should be carefully investigated.

Oil companies have not hitherto favoured designs with less than three shafts. For the upper shaft, the diameter is often governed by the space needed for equipment and utilities (though these will be difficult of access and vulnerable to flooding); the thickness is governed by wave-slamming pressures (up to 0.3 MPa, or 40 psi) and ship-collision criteria. Conductors from the wells are sometimes carried externally, but risers (pipes transporting oil or gas) are usually placed inside the shafts for protection against ship impact. A riser rupture may result in high pressures inside the shaft, so that risers might better be placed outside; a gas riser should never be guided through a water-filled compartment. The lower shaft dimensions are governed mainly by strength and stiffness requirements; to combat fatigue, no tension should be allowed at the tower–caisson connection.

An alternative would be a single large shaft so strong that even a large tanker collision could not jeopardize the structure. A perforated Jarlan wall (Section 2.3) surrounding a central shaft gives excellent protection against external impact.

Usually, the shaft(s) have projected up from a large caisson, which acts as a foundation and gives buoyancy and stability during floating phases. The caisson dimensions may be determined by floatability requirements (with a large deck weight at tow-out) or by the specified oil storage volume. The danger of leakage may be reduced by maintaining a low pressure inside the caisson (causing favourable compressive stress in the walls) or by surrounding the oil by a barrier of water-filled cells. At the wave position corresponding to maximum horizontal force on the shafts, the water pressure (Eqn 4-21) on the caisson roof will tend to reduce the net overturning moment on the foundation. This may be exploited by choosing a low caisson with a large base area. Existing platforms tend to have high caissons (up to 40% of water depth) perhaps because of the relative cheapness of slipformed walls. The foundation requirements will affect the caisson/footing area and the total platform weight (i.e. the necessary ballast) as well as the skirts (Section 7.2.1).

As an alternative to the single, large caisson, widely-separated hollow shafts have been proposed on separate footing slabs but connected into a single stiff body by rugged trusses or girders (of steel or concrete). A larger deck area could be supported, allowing wider safety zones between processes. Oil storage might be in the shafts, which would be dimensioned to ensure floatibility.

Dynamic response should be checked to evaluate the feasibility of a proposed structure. Excessive movements or excessive dynamic magnification of the quasi-static forces may rule out a design. Fatigue, especially of the deck structure, should be checked allowing for dynamic magnification for the numerous waves of moderate height in the frequency range 0.05–0.20 Hz. The natural frequency of the structure should be well clear of this range; for bottom-supported structures, the natural frequency will be higher, whereas compliant and floating structures have lower natural frequencies.

The dynamics of a platform without a caisson might cause problems in deep water. Inclined towers, which can easily be constructed by modern slipforming, have been proposed to increase stiffness as a counter-measure.

8.1.2 Secondary dimensions and details

Figure 8-2 indicates those parts of the structure whose design is governed by the various temporary conditions during construction (Section 6) which are more important for offshore structures than for land-based ones.

The size of the available dock might well affect the base slab area or the draft at float-out. If the bottom slab is cast in two steps to reduce float-out draft, water pressure at launching will govern the thickness of the first step. An air cushion will cause lateral forces on the skirts, and release of the air may cause severe stresses. The design of certain walls may be governed by the ballasting program after launching: a small swell can cause significant stresses in this phase.

The very large water pressures at deep immersion for deck mating are usually taken on vaulted external caisson walls of dimensions providing adequate resistance to implosion (considering second-order effects, creep, etc.). The hoop stresses on the cell walls will be reduced because the stiff roof and base of the caisson take much of the external compression: the cells walls must therefore take significant bending moments. The requirements for stability when floating with a flooded compartment may govern the design of internal caisson walls (Section 6.2.1).

Applied forces at installation will affect the design of dowels and skirts (Section 7.1), and of the base slab (against load contact forces). Underbase suction (up to 0.2 MPa to improve penetration), and the heat of hydration of underbase grout, can cause significant stresses.

Loads in the operating phase will govern the design of the shafts or towers and of the foundations. Skirts should be designed to take horizontal forces due to waves augmented by an allowance for the horizontal earth-pressures-at-rest being greater beneath the bottom slab than outside the structure. For caissons with stored oil, temperature effects will govern the prestressing of the inner walls and roof. The caisson design should be checked for accidental loss of intended internal underpressure.

For static strength, the only parts governed by the operating conditions are the deck, towers and skirts. All other elements are preloaded (and therefore tested)

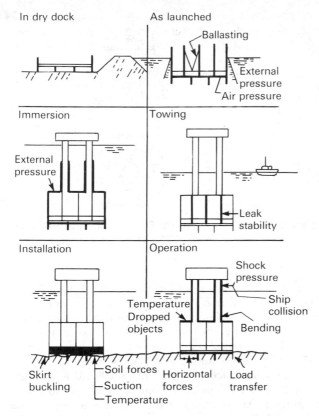

Figure 8-2 Temporary conditions governing design

during construction to higher stress levels than are experienced in operation. Thus, once completed, offshore concrete structures of this type have high strength reserves and negligible risk of failure.

8.2 Articulated and floating structures

To reduce horizontal wave or seismic forces, an *articulated tower* may be adopted (Fig. 8-3), tied to the sea-bed at a joint and kept upright by buoyancy. Acceptable motion characteristics are obtained by giving this 'inverted pendulum' a natural period of about 60 s, well above main wave periods. Often, a simple vertical concrete pipe is proposed, with partition walls in the upper part where ship collision and other damage is possible, to stand two-compartment flooding without loss of stability or overloading of the joint. The construction phases would, in principle, be similar to those for fixed structures.

The joint is a key element and many concepts have been put forward, e.g. universal cardan joints, ball joints, tensioned rods, and hemispherical shells covered by rubber pads, allowing for tower inclinations of up to 25°. In seismic areas, the joint may be severely affected by vertical accelerations.

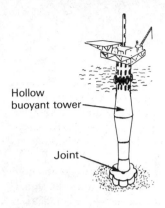

Hollow
buoyant tower

Joint

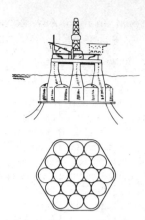

Figure 8-3 An articulated tower

Figure 8-4 A floating platform: CONPROD

Most floating concrete platforms proposed so far have been very similar in configuration to bottom-supported structures (Fig. 8-4), without footings but with catenary moorings. To damp the resulting heave motions, which might hamper oil production, the base may be fitted with a projecting horizontal slab. An anchored vertical spar (like an articulated tower) will have better motion characteristics and may be used as an oil storage and loading buoy.

Fixed or floating caissons may have steel storage tanks for liquid natural gas inside the cells but almost completely separated from the concrete structure. Thus, accidental leaks from the tanks do not affect the structure, damage to the structure does not affect the LNG tanks, and sea-water circulating inside the caisson maintains a constant external temperature. Liquefaction or degasification plant may be located on deck or in the shafts.

8.3 Subsea systems

For marginal fields, in deep waters (where inspection and maintenance of production systems may be difficult) or for early production, there is a tendency to propose the provision of 'dry' chambers on the sea-bed, with the internal pressure at one atmosphere. These chambers will have to withstand large external pressures, but relatively small forces due to waves and currents. Concrete would seem to be a feasible and economic material for such chambers, perhaps in a double-skin steel–concrete–steel sandwich [56].

One possibility is a torus-shaped concrete structure, where the production drilling is performed through the central hole from a semi-submersible rig. Thus, the wellheads are protected against dropped objects, trawls and anchors, etc. The oil would be piped from the wellheads into the ring-shaped chamber of the torus where valves, manifolds, pumps, etc., could be kept in a dry atmosphere, either permanently (perhaps for specially vulnerable equipment) or when necessary for inspection and maintenance.

9   Structural analysis: static and dynamic

9.1   Static analysis

Complex geometries, complicated and numerous loading cases on the structure, and complex foundation conditions make complete structural analysis of an offshore structure impossible without the use of numerical techniques such as the finite element or finite difference methods on large computers. Stress fields caused by local as well as global effects are considered, yielding detailed information for all essential parts of the structure.

In most gravity structures, there is much repetition of one or more fundamental components, e.g. the cell in a caisson. To save computer time, programs for global analysis should take advantage of this repetition by *substructuring* i.e. modelling the repeated component only once in detail. A typical model of the caisson and lower towers might have 8000 finite elements and 150 000 degrees of freedom, and consider up to 30 fundamental load cases.

A linear elastic analysis is usually made, based on the properties of the gross concrete section, so that further load cases can be treated by superposition. Cracking and non-linear behaviour are not allowed for, and ultimate strength design reduces to a formal check of cross-sections against a reasonable (but probably not actual) system of forces in equilibrium (in the spirit of the lower bound theorem of plasticity). Some peaks of stress predicted by linear-elastic theory may be smoothed out [6], preserving equilibrium. If these procedures are not thought adequate in local areas where non-linear behaviour is important (e.g. at a dome–cylinder intersection under hydrostatic load), analysis in further detail by special iterative programs will be needed, using results from the global analysis as input.

9.2   Dynamic analysis

In practice, dynamic analysis of an offshore structure is carried out by numerical methods on modern computers, though with much simplified discretization and far fewer degrees of freedom than in the main statical model. The real structure with its distributed mass, forces and stiffness is simulated numerically, often by a system of straight beam elements of appropriate stiffness connected at nodal points where the mass of the structure is assumed to be lumped (Fig. 9-1).

Interaction between the structure and the surrounding water must, of course, be modelled correctly (Section 9.2.3), as well as the dynamic properties of the foundation soil (Section 9.2.2). A fairly simple model will be sufficient to evaluate overall dynamic response, but the analysis must be adequate to allow evaluation of the dynamic stresses in all affected parts of the structure including the deck, external conductor bracing, etc.

Important modes of vibration may include twisting, and bending in a diagonal direction, as well as more straightforward modes. In such cases, a three-dimensional model of the structure will be needed, with rotational inertia of the deck structure properly represented.

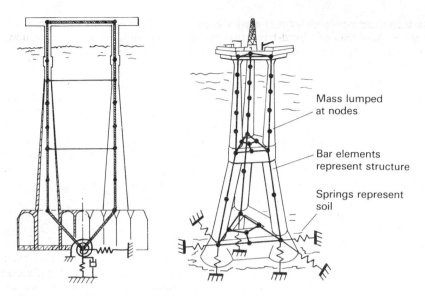

Figure 9-1   Dynamic models

9.2.1   *Structural properties*

It is normally best to base the stiffness on the gross concrete section and the short-term modulus of elasticity for concrete (not the dynamic modulus appropriate to high-frequency vibration). If the more flexible parts of the structure will be appreciably cracked at relevant load levels, the associated reduction in stiffness should be allowed for.

Internal structural damping (due either to hysteresis in the materials or to energy dissipation at joints) is usually introduced into a dynamic analysis by treating the motion as the superposition of vibrations of the entire structure in its (undamped) normal modes. Each mode is allocated a certain percentage of critical damping, the percentage being chosen to allow for stress level, types of joint, limit state under consideration, etc. Table 9-1 gives some recommendations, based mainly on practical experience at full scale. For complex fluid–soil–structure interactions, a rational distribution of dashpots or viscous damping terms across the structure is needed [57], modelling damping in the local material. Eatock

Table 9-1   Recommended coefficients for structural damping

Material	Proportion of critical damping, %			
	Ultimate limit state		Fatigue	
	Dry	Submerged	Dry	Submerged
Steel	1.0	1.0	1.0	1.0
Reinforced concrete	2.5	3.0	2.0	2.0
Prestressed concrete	1.5	2.0	1.0	1.0

Taylor [58] proposes regarding the motion as a sum of vibrations in the normal modes of the undamped fixed-base structure in air; damping coefficients can now be assigned with more confidence.

9.2.2 *Soil properties*

The soil properties required for a dynamic analysis are the cyclic shear modulus, G, the internal damping, and Poisson's ratio (0.5 for effectively undrained conditions). The small-strain shear modulus, G_0, depends mainly on the voids ratio and the mean effective stress but also on the over-consolidation ratio and the plasticity index. It should preferably be determined *in situ*, e.g. by measuring the shear-wave velocity, but may also be measured in the laboratory on undisturbed samples or predicted from well-established empirical relationships [6]. However, the shear modulus, G, depends strongly on amplitude of shear strain (Fig. 9-2). This strain dependency must be accounted for in the analysis, and should preferably be confirmed by laboratory tests, such as cyclic torsional shear tests, cyclic triaxial tests, or resonant column tests.

The foundation properties are often represented [59] by a few springs and dashpots attached to the lower nodes of the structure (Fig. 9-1); the foundation block is treated as rigid, and rotational degrees of freedom are included. Approximate values for the spring constants, etc., can be found [60] by considering a rigid massless disk of radius r_0, resting on a semi-infinite elastic half-space of averaged soil stiffness G, forced to vibrate at some frequency ω. Unfortunately, although the half-space is elastic, the derived parameters are frequency dependent, due to *radiation damping* (the continued radiation of energy away from the foundation). Some recommended lumped parameter values are listed in Table 9-2. Methods of removing frequency dependence by introducing an effective soil mass vibrating in phase with the foundation, and of treating non-circular foundations are given in [6], Appendix G. Radiation damping may be substantially reduced in layered soils [6].

In some approaches, e.g. the *boundary element method* [34], a number of nodes are considered along the interface between the soil and a flexible foundation, with the soil constants again coming from standard theory for a linear-elastic half-space. Alternatively, the foundation may be represented by a three-dimensional finite-element mesh modelling the real soil strata; proper representation of the soil boundaries well away from the structure (to allow for radiation damping) can be troublesome.

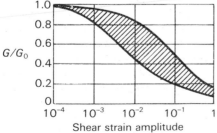

Figure 9-2 Bounds on the reduction of shear modulus with increasing cyclic strain amplitude [6]

Table 9-2   Recommended lumped parameters for a linear-elastic half-space excited harmonically through a rigid massless pad of radius R at its horizontal surface. The half-space is taken to have uniform shear modulus G, Poisson's ratio 0.5, and uniform mass density ρ. The half-space is represented by a mass–spring–dashpot system with the following values[a] (mainly taken from [60])

Type of motion	Effective inertia	Spring constant	Damper coefficient
Vertical translation	$1.44\rho R^3$	$8GR$	$6.8R\sqrt{\rho G}$
Horizontal translation	$0.16\rho R^3$	$5.33GR$	$3.0R^2\sqrt{\rho G}$
Rocking	$0.96\rho R^5$	$5.33GR^3$	$0.64R^4\sqrt{\rho G}$
Torsion	$0.24\rho R^5$	$5.33GR^3$	

[a] For excitation frequency $\omega \approx 1.0\sqrt{G/\rho R^2}$; for other frequencies see [60].

The internal (material, hysteretic) damping in soils increases markedly with cyclic shear strain amplitude, but the relationship [6] varies widely with soil type, effective shear stress, voids ratio, number of cycles, etc. This damping should preferably be assessed by laboratory tests on representative soil samples: it may be included in the analysis by using the lumped parameters [60] for a visco-elastic half-space.

9.2.3   Hydrodynamic effects
Hydrodynamic effects may be incorporated by modifying Morison's Eqn 4-26 for the force on a vertical cylinder in waves to allow for a significant horizontal velocity $\dot{x}$ of the structure. The force per unit length is now

$$f = \rho A a + c_m \rho A (a - \ddot{x}) + \tfrac{1}{2}\rho D c_d (v - \dot{x})\,|v - \dot{x}| \qquad (9\text{-}1)$$

where v and a are the horizontal fluid velocity and acceleration and the first term on the right is the Froude–Krylov force. Unfortunately, the last term on the right is non-linear in the velocities, and a solution can involve great mathematical difficulties, which there is space to mention only briefly. One approach is to solve the structural equations in the *time domain* by step-to-step integration of the response to an assumed train of waves, perhaps the 100-year design wave considered as a regular wave: unfortunately, it is not easy to apply the results to random wave climates. An alternative is to *linearize* the drag term in Eqn 9-1, so that if k represents the stiffness and m the mass per unit length, the equation of motion of an element becomes (simplifying the matrix form)

$$(m + c_m \rho A)\ddot{x} + \tfrac{1}{2}\rho D c_{de}\dot{x} + kx = (1 + c_m)\rho A a + \tfrac{1}{2}\rho D c_{de}v \qquad (9\text{-}2)$$

The term $c_m \rho A$ is the added mass of fluid effectively vibrating with the structure, and the terms on the right represent the applied fluid forces causing vibration. The equivalent drag coefficient, c_{de}, giving *hydrodynamic damping* must be considered as an 'average' value over the vibration period, and consequently depends on frequency as well as the period and dimensions of the body. Values for c_{de} might be estimated [61] by careful comparison of time-domain solutions of non-linear and linearized equations.

If the dimensions of a structural member are so large that Morison's equation cannot be used, it will be necessary to consider both waves scattered by the stationary body (Section 4.2.3) and further scattered waves due to the motion of the structure. Drag will probably be negligible, and thus linear equations will result [59] with a fluid mass to be added to that of the structure and a (often negligible) damping term corresponding to *hydrodynamic radiation damping* (analogous to radiation damping in the soil).

9.2.4 *Spectral analysis*

If the structural equations are linearized, one may bring to bear all the techniques of linear dynamic analysis in matrix form, suitable for computers. The actual motion may be treated by the *superposition* of uncoupled motions in the various *normal modes* of vibration of the structure at its natural frequencies (either undamped or, preferably, damped). (This procedure is somewhat questionable if springs or dashpots representing the soil are frequency-dependent.)

Of special importance for random excitation by waves is the technique of spectral analysis in the *frequency domain* by which the spectral density, $S(\omega)$, of sea-surface elevation (Section 4.1.1) is transformed into the spectral density of some important response of the structure (perhaps the stress at a section prone to fatigue). The principle is illustrated in Fig. 9-3. A train of regular waves of frequency ω is considered, and the ratio, H_h, of the hydrodynamic load on the structure to the wave amplitude is computed by appropriate theory. The frequency ω is then varied and the *hydrodynamic transfer function*, $H_h(\omega)$, is obtained. Since the forces on the platform towers reinforce or cancel, depending

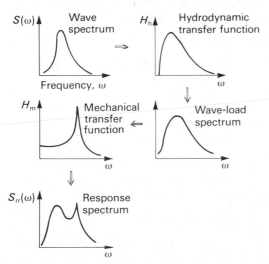

Figure 9-3 Principles of spectral analysis

on the tower spacing and the wavelength, which is a unique function of ω (Eqn 4-22), $H_h(\omega)$ for total load is a rather wavy function, generally decreasing with ω because rapid short waves attenuate markedly with depth and have low inertia coefficients (Section 3.3.2). The wave-load spectrum is found by combining $H_h(\omega)$ and $S(\omega)$.

Computing the response, r, to unit wave load in forced vibration at various frequencies gives the *mechanical transfer function*, $H_m(\omega)$, which combined with the wave-load spectrum gives the spectral density $S_{rr}(\omega)$, of the response. Alternatively, the response may be computed directly for waves of different frequencies, giving a single transfer function between the wave and response spectra.

The response spectral density is important since the area under the graph of $S_{rr}(\omega)$ against ω is the variance s^2 of the response (Eqn 3-3). Also, the most probable largest value of the response range in a chosen sea-state is given by

$$\tilde{r}_m = 4s\sqrt{\tfrac{1}{2}\log_e N} \qquad (9\text{-}3)$$

where N is the number of mean zero-up-crossing periods T_z in the sea-state (see Eqn 4-9).

Methods of solving non-linear equations in the frequency domain are slowly being developed [61].

9.3 Parametric studies and typical results

Because of the importance of dynamic effects, and the difficulty of assigning values to significant material properties such as the soil shear modulus and the damping, it is important to conduct *parametric studies*. Calculations are carried out for ranges of possible values of important parameters, to make sure that the worst possible cases have been covered.

A typical four-leg concrete platform had, as its first mode, bending in the diagonal direction with a period of 3.8 s. The next modes included vertical vibrations, twisting, and another diagonal bending mode. The response spectra at

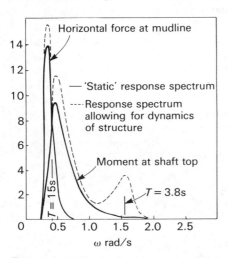

Figure 9-4 Typical response spectra

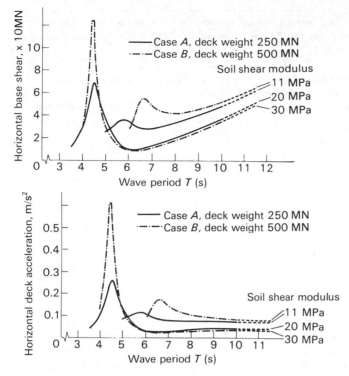

Figure 9-5 Parametric studies

mudline and shaft top to a Pierson–Moskowitz wave spectrum with $T_z = 15$ s are shown in Fig. 9-4 at an arbitrary scale. The proportional increase in area under the curve, from the full 'quasi-static' spectrum to the dotted line indicates the influence of dynamic effects, which clearly increases from the bottom of the structure to the top.

The differing response at the base and the deck is further illustrated in Fig. 9-5, which shows some results of a (deterministic) parametric study of horizontal shear at the base and horizontal deck acceleration in waves of steepness $H/\lambda = 1/20$. Whatever the deck weight, the response at resonance is by far the largest for the stiffer soil, although the natural period is lower. This is due to the reduced contribution of the soil to damping—an effect which would have been overlooked if a value for composite damping had been assigned as input. Deck accelerations do not much reduce with increasing wave period, and the unfavourable effect of soil stiffness is more marked for the deck than the base.

10 Design of structural elements

10.1 Aims and methods

Most offshore concrete structures must satisfy the requirements of a code of practice or similar document, perhaps the rules of a certification authority [6], or

Table 10-1 Specified partial safety factors, γ_f, for loads. The unbracketed numbers are taken from the FIP recommendations [8]; the bracketed numbers from the DnV rules [6]

Limit state	Permanent or dead load, P	Live or imposed load L	Deformation loads (prestress, temperature), D	Environmental loads, E Normal (1 month)	Extreme (100 years)
Serviceability (deflection, crack width)	1.0 [1.0]	1.0 [1.0]	1.0 [1.0]	1.0	[0.5]
Ultimate (a) normal environment	1.2 [1.3]	1.6 [1.3]	1.1 or 0.9 [1.0]	1.4	[0.7]
Ultimate (b) extreme environment	1.1 or 0.9 [1.0]	1.3 or 0.9 [1.0]	1.1 or 0.9 [1.0]		1.3 [1.3]

the latest recommendations of FIP, [8], each supplementing a national code or government guidance [7]. Most codes allow fully probabilistic design or traditional working-stress design, but the dominant modern approach is semi-probabilistic limit state design in which partial safety factors are applied to characteristic (5th percentile) loads and material strengths to ensure sufficiently low probability of occurrence of limit states, of collapse, excessive deflection, fatigue, etc.

In the FIP recommendations [8], the specified material factors, γ_m, are 1.0 for both concrete and steel when checking deflection, 1.5 and 1.15 when checking strength; the DnV rules [6] specify identical factors, with reduction allowed in special cases. The load factors, γ_f, recommended by FIP and DnV are given in Table 10-1 for the various types of load and the main limit states, for the structure in its final operating condition. The main difference between FIP and DnV is that in the DnV rules 'ordinary' environmental loads are obtained by applying a factor of 0.7 to the 100-year values, to avoid a second statistical analysis for a shorter return period.

Other limit states, of fatigue (FLS) and progressive collapse (PLS), are also treated in the DnV rules, as are short construction phases. Large accidental loads (due to earthquakes, ship collision, dropped objects) are usually treated separately, with $\gamma_f = 1.05$ and relaxed conditions on the imposed and environmental loads taken to act at the same time.

Usually, the force distribution within the structure is determined by linear elastic analysis (Section 9.1), even at the ultimate limit state, with stiffness based on the gross concrete section. The strength at cross-sections is then checked, using non-linear short-term stress–strain curves at the ultimate limit state (ULS). Plastic theories, such as yield-line theory, may be used [6] where appropriate, and material and geometrical non-linearities may be important for slender compression members (Section 10.3).

In general, the limit state design of offshore structures is rather similar to the design of land-based structures, though applied loadings are of a rather different type and some special problems assume greater prominence offshore.

10.2 Effects of ambient water pressure

External hydrostatic pressure can benefit submerged concrete structures by keeping them in compression, but the magnitude of the pressure can govern design for deep submergence, which may cause implosion of shells. In addition, the water pressure may come to act not upon the external surface, as one first assumes, but within the concrete.

Whenever a crack occurs, water pressure, w, will come to act on crack faces which were previously carrying tension. To prevent local overstressing of the reinforcement, concentrating deformation at the first crack, all tension zones must have a minimum proportion [6] of steel

$$\rho_{min} = (f_{tm} + w)/f_{sp} \tag{10-1}$$

where f_{tm} is the mean tensile strength of the concrete and f_{sp} the proportional limit of the steel. In design against shear, water pressure in cracks is thought likely [6] to counteract the favourable short-span effect for loads applied near a support.

The minor damage [62] which occurred to Statfjord A during deep submergence was apparently caused by a small unanticipated zone of high tensile stress in the corners of the caisson star cells (Fig. 10-1). The transverse reinforcement was too light to prevent water pressure from extending a crack, in two cases right across the plane AB. The fracture mechanics of the situation, with vertical shear stresses on the plane AB apparently accumulating at the head of the crack, evidently caused the crack to bend over in a vertical plane and eventually cause a leak into the caisson.

It is now customary, in detailed non-linear analyses of parts of offshore structures, to allow for water pressure in cracks, but the possible effects of water pressure in pores [63, 64] do not seem to be of practical significance.

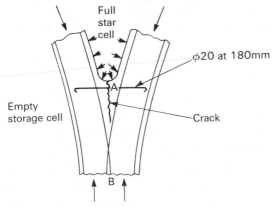

Figure 10-1   Plan view of cracking in a star cell

10.3 Instability and second-order effects

The high pressure-difference across the caisson walls at deep submergence is often
resisted by cylindrical cells or arched external walls, which may become unstable
under external pressure or be sensitive to second-order effects due to geometry
change under load (caused perhaps by material non-linearities such as creep).
Since concrete failing in compression loses stiffness markedly, the final failure
mode of any shell under external pressure will involve a gross change of shape,
leading to leaks and *implosion*. For thick-walled shells, the average hoop stress in
the wall at implosion will be close to the material strength, but in thin-walled
structures the average hoop stress at failure may be much lower.

10.3.1 *Test results and empirical approaches*

A report by Haynes [65], surveying his long series of tests on spheres and
cylinders up to 250 mm thick, suggests that circular cylinders with a ratio of
thickness, t, to external radius, R, greater than 0.12 may be considered as
thick-walled. Regardless of initial imperfections or length/radius ratio, the average hoop stress on an unreinforced wall at implosion will, in a short-term test,
reach the uniaxial (cylinder) strength of the concrete (perhaps more for short
cylinders with $L/R < 4$). If t/R exceeds 0.13, the hoop stress at failure will be less,
depending on t/R, L/R and the initial imperfections. Haynes gives semi-empirical
formulae and recommends an overall factor of safety (γ_f and γ_m combined, plus an
allowance for long-term loading) of 2.5 (3.0 for human occupation).

Few of Haynes' models were reinforced, and he recommends that any hoop
reinforcement should be disregarded in design against implosion, unless it is tied
laterally. Unfortunately, links across the thickness of the wall are difficult to
provide in practice, and might encourage leaks; it is common practice to count on
the strength of light hoop reinforcement in design against external pressure, even
when links are not provided.

A single 1/5 scale model of one cell from the Statfjord A Condeep, tested at
Trondheim in 1976, failed at rather more than design pressure, but the result is of
doubtful relevance to the cylinder-implosion problem because the failure seems to
have initiated in the dome or at the intersection between cylinder and capping
dome where shear forces were high. The performance of full-scale reinforced
cylinder and arch walls has been monitored in practice during deep submergence
and found to be satisfactory, though the true factor of safety of course remains
unknown. Shells for water over 200 m deep, needing heavy reinforcement
perhaps impractical in bar form, may incorporate inner and outer skins of thin
steel plate separated by concrete [56].

10.3.2 *Calculation methods*

Since test results are scarce, particularly for shells with complex shape or loading,
for the design of the shell itself and the assessment of second-order effects
transmitted to supports, a fully accurate calculation of response is desirable,
allowing for initial imperfections and material non-linearities, and considering the
equilibrium of the shell in its deformed position. Finite element programs are
available for this purpose [66] but tend to be rather expensive for routine use in
design.

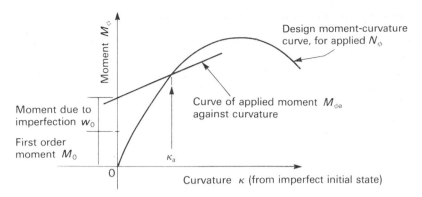

Figure 10-2 Principle of adjusted equilibrium method [67]

One powerful simplified approach [67], based on the idea that a shell is acceptable if for a certain acceptable state of deformation the design resistances exceed the design loading effects, is illustrated for cylinders in Fig. 10-2. A curve of resisting hoop moment, M_ϕ, against curvature, κ, is compared with a second line relating applied moment, $M_{\phi a}$, to curvature, plotted from

$$M_{\phi a} = M_0 + k_s N_\phi (w_0 + w) \qquad (10\text{-}2)$$

where M_0 is the first-order moment for a perfect linear-elastic shell, w the radial displacement and k_s a reduction factor. Initial out-of-roundness amplitude w_0 (related to specified construction tolerances, perhaps radius/200) is assumed, distributed round the shell in a sine curve with m waves (the buckling mode used [6] to estimate k_s). If, for all m, the two curves on Fig. 10-2 intersect at an acceptable deformation, the design is acceptable.

Alternatively, the *tangent modulus* method is sometimes used. A *slenderness number*, β, is calculated from a linear eigenvalue analysis of the perfect shell as the reciprocal of the hoop strain at buckling. The critical hoop stress for a perfect shell in non-linear material is then

$$\sigma_{tc} = E_t / \beta \qquad (10\text{-}3)$$

where E_t is the tangent modulus for the reinforced wall (at the most stressed section). Iteration gives the pressure, p_{tc}, at which Eqn 10-3 is satisfied, and finally the first-order moments and imperfections in the real shell are magnified by $\eta = 1/(1 - p/p_{tc})$, p being the actual pressure, to produce 'applied moments' for use in design.

Whichever calculation method is used, duration of load and possible creep effects must be considered.

10.4 Temperature effects

Temperature-change effects have produced some of the most difficult detailing problems encountered in offshore concrete structures, since the temperature strains often cannot occur freely. An isolated reinforced concrete cylinder, at first cooled by sea-water and then heated internally by oil at up to 50°C, will crack axially and circumferentially on the external surface, where minimum reinforcement (Eqn 10-1) is needed to ensure crack distribution. If the inner surface is cooled again, it may crack due to restraint by the now warmer outer parts—so that steel is also needed near the inner surface.

Despite Table 10-1, imposed thermal deformations may often be ignored at ULS (or PLS) on the grounds that the mode of failure is known to be ductile. By contrast, thermal effects may be very important at the serviceability limit state. For a single sustained temperature change, thermal stresses are usually calculated for the final steady-state temperature distribution using linear-elastic theory and assuming the structure is uncracked. However, the stress magnitude will be roughly proportional to the structural stiffness which may be much reduced if the structure is cracked (due to the thermal stress itself or to some other loading condition). In appropriate cases, a semi-empirical reduction factor [6] may be applied to the thermal stress, depending for each cross-section on reinforcement percentage, maximum stress resultants due to thermal and non-thermal loads, etc. Alternatively, thermal stress may be calculated using no-tension theory for concrete, and then increased empirically to allow for tension stiffening.

Further problems can occur if the temperature is cycled, as when a caisson cylinder is regularly filled with hot oil followed by cool sea-water. Even with substantial hoop prestress, the combined effects of temperature-cycling and temperature-dependent creep can cause tensile hoop stress near the inside surface [68]. Substantial local stresses occur at the horizontal interface between cold water and hot oil—and temperature-dependent creep may cause unexpected redistribution of axial prestress. The significance of these effects in practice is as yet unclear (further research on temperature effects is needed), but at least the minimum prescribed reinforcement (Eqn 10-1) must be provided in zones with substantial changes of temperature.

10.5 Reinforcement design

The detailing of reinforcement for given stress resultants (including prestress as a 'deformation' load) may, in principle, be as for load-based structures—despite a tendency to use higher strength materials offshore. Two problems are particularly prominent offshore, the first relating to reinforcement design for thin plate or shell elements subjected to up to six stress resultants (per unit width) causing stresses in the element plane, including membrane shear N_{xy} and twisting moment M_{xy} (Fig. 10-3). For the ULS, an appropriate plastic theory neglecting concrete tension is embodied in Nielsen's equations [69] and their extensions. Thus, for membrane forces (at least one principal force tensile) orthogonal reinforcement symmetrical about mid-depth should be provided to take the forces:

$$F_x = N_x + |N_{xy}| \cot \theta$$
$$F_y = N_y + |N_{xy}| \tan \theta \qquad\qquad (10\text{-}4)$$

where θ may be chosen at will (45° for minimum reinforcement of the element, considered in isolation). Special formulas apply if either F_x of F_y turns out negative, or if skew steel is required. For combinations of moment and force, the element may be regarded as a sandwich [70] with Nielsen's equations applying to equivalent 'membrane forces' in the outer layers. At the serviceability limit state,

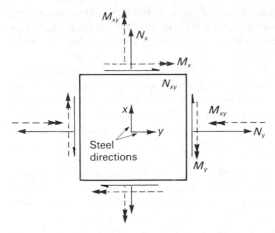

Figure 10-3 Slab element subjected to combined stress resultants per unit width

where the aim of design is somewhat uncertain (Section 5), an approach [71] based on neglecting tension and limiting crack widths may be appropriate. Programs are gradually becoming available for *post-processing* finite element analyses to produce required steel percentages.

Apart from difficulties with the stress peaks predicted by finite element analysis, which have to be smoothed over a reasonable zone for use in practical design, problems have also occurred in design for transverse shear at ULS in the presence of large moments and axial forces, a situation not dealt with very thoroughly in the standard codes. The method proposed by DnV [6] is to regard the shear resistance (never greater than one quarter of the compressive strength on the full cross-section) as being the sum of three components, two due to the concrete and longitudinal steel and to the shear reinforcement, which are estimated in the standard fashion, and a third component, V_{pr}, due to the axial force (including prestress) given by

$$V_{pr} = \frac{M_0}{M_{fd}} V_{fd} \qquad (10\text{-}5)$$

where M_0 is the applied moment (about the principal axis of the uncracked section) which combined with the axial force would give zero concrete stress at the extreme fibre. V_{fd} and M_{fd} are design applied shear force and moment at the section considered.

10.6 Fatigue

The FIP recommendations [8] imply that fatigue is unlikely to be critical in offshore structures made of concrete—a possible advantage over welded steel. Fatigue failure of plain concrete, even after several million cycles, is not thought likely [72] if the compressive stress range is less than 40% of the compressive strength for zero minimum stress (permitted stress range declining linearly to zero at a minimum stress of 75% of strength). These conditions are likely to be satisfied if the design requirements at other limit states (ULS, SLS) have been met. Neither is fatigue of high-yield reinforcement thought likely if the stress range (superposed on a minimum stress up to 40% of yield) is less than one third of the yield strength.

Unfortunately, offshore structures can be subject to greater fluctuating components of load than are common on land, so that previous experience may not be relevant, and some recent tests [73] on submerged specimens under hydrostatic pressure, with a fluctuating bending moment opening and closing flexural cracks, gave markedly low fatigue lives. Apparently, this is due to the 'pumping' of water in and out of the cracks, washing out cement and leading to the deterioration of bond.

These results have led Det norske Veritas to demand that important submerged members exposed to fatigue, such as concrete towers, be prestressed to eliminate membrane tension in any load condition, to eliminate tensile stress everywhere under ordinary environmental loads, and to limit reinforcement stresses and compressive stresses in zones subject to cracking. The modern approach is to say that if certain simple conditions (e.g. on the stress range in concrete in compression or in steel, or on the maximum and minimum stress resultants as a proportion of the ultimate strength at a section) are satisfied, then no further check on fatigue life is necessary. If these simple conditions are not satisfied, a more refined calculation of fatigue life must be made.

For concrete, a general formula may be used [6] for the number, N, of cycles to failure under a loading effect varying steadily between S_{min} and S_{max}, where S may be compressive stress or transverse shear force, namely,

$$\log_{10} N = C\left[1 - \frac{S_{max}}{S_{ref}}\right]/\left[1 - \frac{S_{min}}{S_{ref}}\right] \tag{10-6}$$

Here, S_{ref} is a reference value, perhaps the design compressive strength (adjusted for stress gradient) or the design shear resistance, and C takes different values for different situations (less for submerged members). Recommendations are given [6] for reinforcement and for prestressing tendons.

Design against fatigue offshore involves considering the statistics of the stress cycles caused by wind and waves during the design life (perhaps 20 years). For each sea-state, the spectral density for stress (Section 9.2.4) gives the mean up-crossing period T_z (Eqn 4-4) and hence the number of cycles, and the associated stress range may often be assumed to follow a Rayleigh distribution (Eqn 4-7). By considering different sea-states, the actual randomly-varying stress may then be represented by a small number, say $m = 10$, of blocks of constant-amplitude stress cycles, each with associated S_{max} and S_{min} and number, n_i, of cycles during the design life of the structure. Using Miner's hypothesis, a design is considered satisfactory if

$$\sum_{i=1}^{m} \frac{n_i}{N_i} \leq c \tag{10-7}$$

where N_i is the number of cycles (Eqn 10-6) to cause failure under the S_{max} and S_{min} associated with the ith block of cycles. There is constant debate about the appropriate value of c: for concrete, $c = 0.4$ is often used, corresponding to aiming for a fatigue life of two to two and a half times the design life.

10.7 Other design details

For lack of space, it has not been possible to discuss more than a few of the design problems which can arise in offshore concrete structures. Further problems can, for example, occur when devices of greater or lesser importance, ranging from steel skirts and dowels, through mooring and towing attachments, to fixings for conductors and risers, have to be attached to the concrete structure proper. Here, the basic design principle must be that any failure of the attachment does not cause damage to the main structure, so that the adjacent parts of the concrete structure must be designed to take appreciably greater forces and/or moments than the attached device (at its maximum likely strength) can transmit. Any measures taken to protect the attachment from corrosion must be designed so as not to increase corrosion in the main structure (Section 5-3).

Offshore structures tend to be constructed of higher strength materials and with greater sophistication than are common on land, and the cost of repairs is likely to be high. Workmanship must therefore be of the highest quality and a designer must detail his structure to assist the production of high-class work. Numerous codes give guidance to this end, and the need for good workmanship must be borne in mind right from the start in planning construction offshore.

11 Conclusion

A short state-of-the-art survey of this sort attempting to cover offshore concrete structures will inevitably suffer from at least two major defects (and may of course contain many more). The first defect stems quite simply from the very wide range of technical problems which may be encountered in the design and construction of offshore structures, and the rapid developments which are occurring in a number of areas of relevant technology. The papers prepared for the annual Offshore Technology Conference at Houston typically fill several large volumes, and interesting research results are continually appearing in the proceedings of other conferences, in journals and in books. Not all offshore technology is directly relevant to design in concrete, but much of it is, and it is not easy to keep abreast of all the interesting developments. We are conscious that this short survey has covered many important matters in less depth than they deserve, and omitted some altogether.

Furthermore, the 'art' of offshore concrete construction is by no means static. Although the line of descent from the naval forts of the 1940s, through the lighthouses, to the large North Sea structures is fairly clear, great strides forward were taken very rapidly in the 1970s under the pressure of the need to develop North Sea resources economically; novel and impressive structures were proposed, detailed and constructed, and the associated technical problems were identified and solved with remarkable speed. We have attempted in this chapter to outline the design and technology of large fixed concrete gravity structures of the type developed during the 1970s. But there has recently been a pause in the deployment of these structures, and it seems very likely that offshore concrete design will take a new turn. Concrete structures have been proposed for entirely submarine applications, for tethered buoyant platforms and towers, for wave-energy devices of various sorts, for Arctic sites, among pack-ice, and for other uses offshore. Each of those designs will encounter its own problems, to which we hope the technology surveyed in this chapter may be of some relevance.

References

1. Morgan, R. G., *Concrete floating and submerged structures*, Concrete Society, London, 1973, 56 pp.
2. Morgan, R. G., Concrete ships, *Proc. FIP symp. concr. sea struct.*, Tblisi, 1972, pp. 114–119.
3. FIP Fédération Internationale de la Précontrainte, *Lightweight aggregate concrete for marine structures*, State-of-art report, 1978, 25 pp.
4. Levi, F. *et al.*, Prestressed concrete floating drydock with a lifting capacity of 100,000 t, *Proc. FIP symp. concr. sea struct.*, Tblisi, 1972, pp. 31–35.
5. Turner, F. H., *Concrete and cryogenics*, Viewpoint Publications, London, 1979, 100 pp.
6. Det norske Veritas, *Rules for the design, construction and inspection of offshore structures*, Oslo, 1977, 67 pp.+supplements.
7. Department of Energy, Offshore installations: guidance on design and construction, 2nd Edn, HM Stationery Office, London, 1977, 116 pp.
8. Fédération Internationale de la Précontrainte, *Recommendations for the design and construction of concrete sea structures*, 3rd Edn, Cement and Concrete Association, London, 1979, 59 pp.
9. Posford, J. A., The construction of Britain's sea forts, *Symposium on the civil engineer in war*, Vol. 3, Institution of Civil Engineers, London, 1948, pp. 132–163.
10. Wood, C. R. J., 'Phoenix', *Symposium on the civil engineer in war*, Vol. 2, Institution of Civil Engineers, 1948, pp. 336–368.
11. Hansen, F. J., North Sea structures—a new breed?, *Struct. Eng.*, Vol. 51, No. 1, Jan., 1973, pp. 17–26.
12. Antonakis, C. J., A problem of designing and building for a structure at sea, *Proc. Inst. Civ. Eng.*, Part 1, Vol. 52, 1972, pp. 95–126.
13. Marion, H. and Mahfouz, C., Design and construction of the Ekofisk artificial island, *Proc. Inst. Civ. Eng.*, Part 1, Vol. 56, 1974, pp. 497–511.
14. Draisey, D. *et al.*, The Ninian Field concrete gravity platform, *8th FIP int. congr.*, London, May 1978, 12 pp.
15. Derrington, J. A., Construction of McAlpine/Sea Tank gravity platforms at Ardyne Point, Argyll, *Proc. conf. design constr. offshore struct.*, Institution of Civil Engineers, London, 1977, pp. 121–130.
16. Dowrick, D. J., Modes of failure of concrete platforms, *Concrete in the oceans*, Technical Report No. 2, Cement and Concrete Association, London, 1979, 35 pp.
17. Jensen, J. J. Impact of falling loads on submerged concrete structures, *Proc. int. symp. offshore Struct.*, Vol. 1, Paper 11, Rio de Janeiro, Oct. 1979.
18. Davies, I. L., A method for the determination of the reaction forces and structural damage arising in ship collisions, *J. Petrol. Technol.*, Vol. 33, No. 10, Oct. 1981, pp. 2006–2014.

19. Nordenstrøm, N., *Methods for predicting long term distributions of wave loads*, Det norske Veritas, Report No. 71-2-s, Part I, 1972, 59 pp. plus appendices.
20. Longuet-Higgins, M. S., On the statistical distribution of the heights of sea waves, *J. Marine Res.*, Vol. 11, No. 3, 1952, pp. 245–266.
21. Crandall, S. H. and Mark, W. D., *Random vibration in mechanical systems*, Academic Press, New York, 1963, 166 pp.
22. Newland, D. E., *An introduction to random vibrations and spectral analysis*, Longman, London, 1975, 285 pp.
23. National Academy of Sciences, *Ocean wave spectra*, Prentice-Hall, Englewood Cliffs, NJ, 1963, 357 pp.
24. Hogben, N. and Lumb, F. E., *Ocean wave statistics*, HM Stationery Office, London, 1967, 263 pp.
25. Hasselmann, K. *et al.*, Measurement of wind-wave growth and swell decay during the Joint North Sea Wave Project, *Dt. Hydrogr. Z.*, A8, 12, 1973, pp. 1–95.
26. Pierson, W. J. and Moskowitz, L., A proposed spectral form for fully-developed wind seas based on the similarity theory of S. A. Kitaigorodskii, *J. Geophys. Res.*, Vol. 69, No. 24, 1964, pp. 5181–90.
27. Ewing, J., Some results from the Joint North Sea Wave Project, *Symp dynam. marine vehicles and structures in waves*, Institution of Mechanical Engineers, London, 1975, pp. 41–46.
28. Cartwright, D. E. and Longuet-Higgins, M. S., The statistical distribution of the maxima of a random function, *Proc. R. Soc.*, Vol. 237A, No. 1209, 1956, pp. 212–232.
29. Darbyshire, M. and Draper, L., Forecasting wind-generated sea waves, *Engineering*, Vol. 195, 5 Apr. 1963, pp. 482–484.
30. Ochi, M. K., On prediction of extreme values, *J. Ship Res.*, Vol. 17, No. 1, 1973, pp. 29–37.
31. Mansour, A. E. and Faulkner, D., On applying the statistical approach to extreme sea loads and ship hull strength, *Trans. R. Inst. Nav. Archit.*, Vol. 115, 1973, pp. 277–314.
32. Battjes, J. A., Encounter probability of extreme structural response values based on multi-parameter descriptions of the physical environment, *Proc. 2nd int. conf. behav. off-shore struct.*, London, 1979, Vol. 3, pp. 609–616.
33. Lighthill, M. J., *Waves in fluids*, Cambridge University Press, 1978, 504 pp.
34. Brebbia, C. and Walker, S., *Dynamic analysis of offshore structures*, Newnes-Butterworths, London, 1979, 323 pp.
35. Zienkiewicz, O. C., Bettess, P. and Kelly, D. W., The finite element method for determining fluid loadings. *Numerical methods in offshore engineering*, Wiley, New York, 1978, pp. 141–183.
36. Morison, J. R., O'Brien, M. P., Johnson, J. W., and Schaat, S. A., The force exerted by surface waves on piles, *Petrol. Trans. Am. Inst. Metall. Eng.*, Vol. 189, TP2846, 1950, pp. 149–154.
37. Sarpkaya, T., Vortex shedding and resistance in harmonic flow about smooth and rough circular cylinders, *Proc. 1st int. conf. behav. off-shore struct.*, Trondheim, 1976, Vol. 1, pp. 220–235.
38. Hogben, N., Wave loads on structures, *Proc. 1st int. conf. behav. off-shore struct.*, Trondheim, 1976, Vol. 1, pp. 187–219.

39. Lighthill, M. J., Waves and hydrodynamic loading, *Proc. 2nd int. conf. behav. off-shore struct.*, London, 1979, Vol. 1, pp. 1–40.

40. Garrison, C. J., Hydrodynamic loading of large offshore structures, *Numerical methods in offshore engineering*, Wiley, New York, 1978, pp. 87–140.

41. Davenport, A. G., The application of statistical concepts to the wind loading of structures. *Proc. Inst. Civ. Eng.*, Vol. 19, 1961, pp. 449–472.

42. Davenport, A. G., Gust loading factors, *J. Struct. Div. Am. Soc. Civ. Eng.*, Vol. 93, No. ST3, 1967, pp. 11–34.

43. Browne, R. D. and Baker, A. F., Performance of concrete in a marine environment, *Developments in concrete technology*, Vol. 1, Applied Science Publishers, London, 1979, pp. 111–149.

44. Regourd, M., Physico-chemical studies of cement pastes, mortars and concretes exposed to sea water, *Performance of concrete in marine environment*, American Concrete Institute, SP-65, 1980, pp. 63–82.

45. Beeby, A. W., Cracking and corrosion, *Concrete in the Oceans*, Technical Report No. 1, Cement and Concrete Association, London, 1978, 77 pp.

46. Regan, P. E. and Hamadi, Y. D., Axial tensile tests of reinforced concrete. *Concrete in the oceans*, Technical Report No. 4, Cement and Concrete Association, London, 1981, pp. 35–82.

47. Browne, R. D., Mechanisms of corrosion of steel in concrete, *Performance of concrete in marine environment*, American Concrete Institute, SP-65, 1980, pp. 169–204.

48. Haynes, H. H., Permeability of concrete in sea water, *Performance of concrete in marine environment*, American Concrete Institute, SP-65, 1980, pp. 21–38.

49. Eide, O. *et al.*, Installation of the Shell/Esso Brent B Condeep production platform, *Proc. 8th offshore technol. conf.*, Houston, 1976, Vol. 1, pp. 101–114.

50. DiBiagio, E., Myrvoll, F. and Borg Hansen, S., How successful have performance monitoring programs been for gravity base structures? *Proc. 2nd int. conf. off-shore struct.*, London, 1979, Vol. 3, pp. 709–727.

51. Det norske Veritas, *Underbase grouting of a gravity-type structure*, Oslo, Technical Note E7/1, 1977, 8 pp.

52. Caquot, A. and Kerisel, J., Sur la terme de surface dans le calcul des fondations en milieu pulvérulent, *Proc. 3rd int. conf. soil mech.*, Zurich, Vol. 1, 1953, pp. 336–337.

53. Foss, I. and Warming, J., Three gravity platform foundations, *Proc. 2nd int. conf. behav. off-shore struct.*, London, 1979, Vol. 2, pp. 239–256.

54. Janbu, H., Design analysis for gravity platform foundations, *Proc. 2nd int. conf. behav. off-shore struct.*, London, 1979, Vol. 1, pp. 407–426.

55. Kjekstad, O. and Lunne, T., Soil parameters used for design of gravity platforms in the North Sea, *Proc. 2nd int. conf. behav. off-shore struct.*, London, 1979, Vol. 1, pp. 175–192.

56. Montague, P. and Goode, C. D., Some aspects of double-skin composite construction for sub-sea pressure chambers, *Proc. 2nd int. conf. behav. off-shore struct.*, London, 1979, Vol. 2, pp. 415–424.

57. Vugts, J. H. and Hayes, D. J., Dynamic analysis of fixed offshore structures: a review of some basic aspects of the problem, *Eng. Struct.*, Vol. 1, No. 3, 1979, pp. 114–120.

58. Eatock Taylor, R., A linear analysis of interaction problems in offshore platforms, *Proc. 2nd int. conf. behav. off-shore struct.*, London, 1979, Vol. 2, pp. 61–86.

59. Sigbjørnsson, R., Bell, K. and Holand, I., Dynamic response of framed and gravity structures to waves, *Numerical methods in offshore engineering*, Wiley, New York, 1978, pp. 245–280.

60. Veletsos, A. S. and Verbic, B., Basic response functions for elastic foundations. *J. Eng. Mech. Div. Am. Soc. Civ. Eng.*, Vol. 100, No. EM2, 1974, pp. 189–202; see also *Int. J. Earthquake Eng.*, Vol. 2, 1973, pp. 87–102.

61. Fish, P. and Rainey, R. C. T., The importance of structural motion in the calculation of wave loads on an offshore structure, *Proc. 2nd int. conf. behav. off-shore struct.*, London, 1979, Vol. 2, pp. 43–60.

62. Naesje, K. *et al.*, Structural damage in an offshore concrete platform, the cause and the cure, *FIP Notes*, No. 76, 1978, pp. 8–17.

63. Morley, C. T., Theory of pore pressure in reinforced concrete cylinders, *FIP Notes*, No. 79, 1979, pp. 7–15.

64. Haynes, H. H. and Highberg, R. S., *Long-term deep-ocean test of concrete spherical structures—results after 6 years*, Civil Engineering Laboratory, Port Hueneme, California, Technical Report R869, 1979, 51 pp.

65. Haynes, H. H., *Design for implosion of concrete cylinder structures under hydrostatic loading* Civil Engineering Laboratory, Port Hueneme, California, Technical Report R874, 1979, 81 pp.

66. Holand, I., Ultimate capacity of concrete shells, *Proc. 1st int. conf. behav. off-shore struct.*, Trondheim, 1976, Vol. 1, pp. 744–764.

67. Olsen, O., Implosion analysis of concrete cylinders under hydrostatic pressure, *J. Am. Concr. Inst. Proc.*, Vol. 75, No. 3, Mar., 1978, pp. 82–85.

68. England, G. L., Andrews, K. R. F., Moharram, A. and Macleod, J. S., The influence of creep and temperature on the working stresses in concrete oil-storage structures, *Proc. 2nd int. conf. behav. off-shore struct.*, London, 1979, Vol. 1, pp. 281–298.

69. Nielsen, M. P., Limit analysis of reinforced concrete slabs, *Acta Polytech. Scand.*, Civil Engineering and Construction Series No. 26, 1964, 167 pp.

70. Morley, C. T. and Gulvanessian, H., Optimum reinforcement design of concrete slab elements, *Proc. Inst. Civ. Eng.*, Part 2, Vol. 63, 1977, pp. 441–454.

71. Hallingstad, O. K., Olsen, T. O., and Støve, O. J., Practical design of reinforcement in plates and shells, *Proc. 2nd int. conf. behav. off-shore struct.*, London, 1979, Vol. 1, pp. 267–280.

72. ACI Committee 215, Considerations for design of concrete structures subjected to fatigue loading, *J. Am. Concr. Inst., Proc.*, Vol. 71, No. 3, Mar., 1974, pp. 97–121.

73. Waagaard, K., Fatigue of offshore concrete structures: design and experimental investigations, *Proc. 9th offshore technol. conf.* Houston, 1977, Paper 3009.

35 Water-retaining structures

B P Hughes

University of Birmingham, England

Contents

Notation

A_c	area of concrete
A_s	total area of steel
A_{st}	area of steel in tension
a_{cr}	distance of point (at crack) to surface of nearest longitudinal bar
b	breadth
b_t	breadth of section at level of tension reinforcement
c	concrete cover
d	effective depth
d_n	depth of neutral axis from elastic analysis (as for ε_1)
E_c	concrete modulus
E_s	modulus of elasticity for steel
F	force
f_b	mean bond strength
f_{ct}	concrete stress in tension
$*f_{ct}$	design concrete early strength in tension
f_{cu}	characteristic concrete cube strength
f_{dcb}	design stress for concrete in bending
f_{dst}	design stress for steel in tension
f_s	steel stress
f_{sc}	steel stress in compression
f_y	characteristic yield strength of reinforcement
h	overall depth of section
l	length
M_d	design (service) moment of resistance
n_w	number of welded intersections
P_c	Euler critical load
s	spacing
s_{max}	maximum crack spacing

T_1	temperature fall in concrete from hydration peak
T_2	seasonal temperature fall from initial ambient
w	crack width
w_{max}	maximum crack width
y	coordinate
α_c	effective coefficient of thermal expansion of concrete
α_e	modular ratio for elastic design (E_s/E_c)
γ_f	partial safety factor for load
γ_m	partial safety factor for material strength
δ_n	neutral axis depth factor for elastic design
δ_z	lever arm factor for elastic design
ε_1	strain at the level considered
ε_{cs}	concrete drying shrinkage strain
ε_m	average strain
ε_{sc}	steel strain in compression
ε_{te}	equivalent thermal strain
ε_{ult}	ultimate tensile strain in concrete
ζ	coefficient
ξ	coefficient
ρ	steel ratio $(=A_{st}/bd)$
ρ_c	total steel ratio based on overall depth (A_s/bh)
ρ_{crit}	critical steel ratio (based on overall depth)
ρ_e	equal-strength steel ratio (ESR) for elastic design
$\sum u_s$	sum of the effective perimeters of the tension reinforcement
ϕ	bar diameter
ψ	coefficient

Summary

The design of water-retaining structures for serviceability and collapse is considered and the special requirements for water-tightness and durability suitably emphasized. Cracking due to external loads and to early thermal and shrinkage effects are shown to be major design criteria. The former generally determines the primary design of the concrete section, rather than the ultimate strength requirements, although both strength at collapse and deflections can sometimes influence the design. Early thermal and shrinkage cracking is controlled by alternative options using various combinations of distribution reinforcement and movement joints, as described in the chapter. Workmanship, materials and testing of water retaining structures are also discussed.

1 General design considerations

1.1 Introduction

The purpose of design is the attainment of acceptable probabilities that the structure being designed will not become unfit for its intended use [1, 2]. Water-retaining structures as considered in this chapter include reinforced and prestressed concrete service reservoirs and tanks used for the storage of water and other aqueous liquids. The principles are, of course, often applicable to other structures but special considerations outside the scope of the present text may also be appropriate. For example, in reinforced and prestressed concrete aqueducts and other structures for the conveyance of liquids, special considerations should be given to dynamic forces and to the detailing of joints to resist scour, etc. (see, for example, [3]) which are not discussed here.

The essential function of any water-retaining structure is to support the various possible combinations of loading and to maintain a suitably watertight condition throughout its design life. The design should therefore ensure, in particular, that overturning, buckling of slender elements and material failure will be highly improbable, that leakage will be minimal and that durability will be adequate.

1.2 Stability and flotation

The resistance to sliding and overturning at all locations, and under all loading conditions, must be adequate. For small tanks without joints, the overturning forces on any wall are readily balanced by those on the opposite wall. In large tanks and reservoirs, the use of cantilever walls depends on adequate passive resistance of the ground under the wall to both sliding and overturning. If the soil is inundated with groundwater, the ground resistance can be inadequate and the wall needs to be tied to the opposite wall, either by spanning horizontally to the side walls, or directly by a system of beams or ties. The designer should choose the particular conditions of loading which cause the greatest bending moment, shear force or direct force, as appropriate, on the element or at the section under consideration.

In the design of external walls, full allowance must be made for any active soil pressure and surcharge from vehicles acting on the outside when the tank is empty, and only an allowance for the minimum permanent active soil pressure on the outside vertical face when the tank is full. (In theory, sufficient passive soil resistance could often be mobilized by the limited movement which occurs in the relatively stiff wall to augment the active pressure and completely counteract the water load. In practice, however, it would be very unwise to design a wall, which is very stiff relative to the soil, to have little flexural strength and to rely on the passive soil resistance being both adequate and suitably distributed to resist the water load directly at all locations.)

A structure which is built below ground level will tend to become a sump into which any ground or surface water will drain. As indicated above, the backfill around the walls cannot be consolidated as well as the natural ground so that water will tend to accumulate, especially in cohesive soils. The structure must clearly resist flotation whenever it is empty and the water table is temporarily or permanently above the floor of the structure.

Flotation may be prevented by increasing the weight of the floor, providing nibs on the external walls, or a combination of both. Where nibs are used to mobilize the weight of the external backfill, allowance must be made for the reduced effective weight of the submerged backfill and the floor must span between the walls. The effective weight of the structure should exceed the uplift due to the groundwater by a factor of not less than 1.1. A form of drainage should be provided around the external walls wherever groundwater can effectively drain away. Where contamination of the retained liquid can be accepted, the provision of pressure-relief valves can be considered. They are, however, susceptible to jamming and mechanical failure and in consequence the load factor of safety, even when they fail, should still marginally exceed unity.

1.3 Settlement

Serious differential settlement, subsidence and movement can occur in foundations subject to mining, ground movement, geological faults and subsoils of varying compressibility. The consequences of the movements must be allowed for wherever sites with good and reasonably uniform ground conditions cannot be selected. Division of the structure into smaller units and the provision of additional joints help both to reduce and to accommodate the anticipated differential movement between units. Notice that flexible joints should be provided at all connections for outlet pipes and fittings where differential movement can occur. Superimposed prestressed tendons can reduce the anticipated tension due to settlement in a reinforced concrete design. Where subsidence is likely to be severe and progressive the maximum flexibility for the anticipated movements should be incorporated in the design. For example, periodical correction of the levels may more easily be made to a water tower if it is provided with only three points of support.

Where very poor ground conditions occur to a shallow depth only, it may be economic to over-excavate and then backfill with selected material well compacted in layers. Otherwise, piles may be necessary.

1.4 Degrees of exposure and cover

BS 5337 [2] considers three classes of exposure as follows.

Class A   Exposed to a moist or corrosive atmosphere or subject to alternate wetting and drying.
Class B   Exposed to continuous or almost continuous contact with liquid.
Class C   Not exposed as severely as for either Class A or B.

Notice that the soffit of the roof structure over a reservoir or tank is subjected to repeated condensation and evaporation, whereas the legs of a water tower and the superstructure over an open tank are not exposed to contact with the liquid.

The water level in a reservoir can fluctuate continuously and the upper levels of the inside face of a wall can frequently be exposed to Class A conditions. Lower levels, however, will rarely dry out and Class B is more appropriate. Since degrees of exposure are linked to durability, which in turn is linked to minimum cement content (see later), the designation of two different degrees of exposure can imply two different grades of concrete for a single member, which is usually impracticable. In this case, it is suggested that the entire wall uses the same grade of concrete but is designed essentially for Class B, particular attention being paid to the sequence of construction and the detailing of the reinforcement at the high levels. The longitudinal restraint for early thermal effects can reduce significantly with height wherever one end of a bay is always free at the time of casting, since restraint is then provided, essentially, by the hardened foundation. Similarly, if bar diameters are not reduced and spacings are not increased when detailing the reinforcement towards the top of the wall, (where Class A is, strictly, applicable) then in practice the combined effects of reduced restraint and maintaining the same reinforcement in the top of the wall can provide reduced crack widths which should be appropriate for Class A exposure. Internal columns supporting a roof will rarely be subjected to tensile stresses under service conditions and concrete for Class B can normally be used.

Class C exposure may be considered for the outside face of a retaining wall where it is above the water table, the wall thickness exceeds 225 mm and the appearance is not important. For good appearance, Class A may be necessary and designed for accordingly—including a higher grade of concrete—even though Class B is adequate for the internal face. If the roof slab thickness does not exceed 225 mm, then the external as well as the internal face should be Class A: this can often be desirable anyway if severe exposure to freeze–thaw conditions can occur. If the thickness exceeds 225 mm, then Class B or Class C can be considered for the external face of the roof slab.

The minimum cover of concrete to all steel, including all links, spacers, sheathings, etc, should be at least 40 mm for Classes A and B. In especially corrosive situations with aggressive liquids, where abrasion or erosion is anticipated, or where greater tolerances may be necessary, the cover should be increased.

1.5 Durability

Careful consideration should be given to possible deterioration of the concrete, embedded steel, jointing materials and linings due to stored liquids or aggressive soils and groundwater. Clause 6.3.3 and Table 4.9 of CP 110, Part 1 : 1972 [1] recommend that where sulphates are present, up to 0.5% in the groundwater or 2% SO_3 in the soil, then sulphate-resisting Portland cement can be used and that for higher limits a suitable protective coating should also be used. In some cases, special cements or even impermeable linings may also be necessary when storing certain processed liquids and sewage. Jointing materials in these cases must resist biological attack.

Moorland waters often contain dissolved carbon dioxide, organic acids and salts from the catchment area which may attack the concrete. Where necessary, the minimum cement content of the mix and the cover can be increased. The recommendations given in BS 5337 [2] concerning concrete quality, as defined by concrete grade, minimum cement content and achieving good compaction, are intended to ensure a low permeability concrete meeting the durability requirements for the three degrees of exposure.

1.6 Impermeability

Permeable concrete not only permits leakage and corrosion of embedded steel but also offers poor resistance to leaching, chemical attack and frost damage. Provided that good compaction and curing is achieved throughout, grade 25 concrete with at least the minimum cement content specified in BS 5337 ensures a degree of impermeability which is generally adequate. Impermeability can be improved where necessary by reducing the water/cement ratio (i.e. increasing the grade) or by introducing a pozzolanic admixture such as fly ash. The latter increases the resistance to chemical attack and corrosion of embedded steel but not the resistance to freezing and thawing. Permeability tests of the concrete are not normally carried out because, provided that the impermeability is adequate, leakage in practice is essentially due to cracks and honeycombed patches in the concrete and to incorrect location, design or poor construction of joints.

1.7 Crack control

The cracks in the concrete are a major source of leakage, leaching of lime and consequent efflorescence, and marring of the appearance of the structure. Furthermore, where cracks follow the line of the reinforcement, resistance to corrosion of the steel is reduced. Thus, wherever transverse reinforcement is on the outside of the reinforcement which is normal to the cracks, a stress concentration is provided which encourages the formation of the crack along the transverse reinforcement rather than merely parallel to it. (A crack which crosses a bar at right angles causes much less reduction in resistance to corrosion of the steel.) Controlling the widths of cracks to acceptable levels is therefore a necessary prerequisite for controlling leakage, corrosion of embedded steel and disfigurement of the concrete surface. BS 5337 therefore specifies the maximum design crack width according to the degree of exposure: 0.1 mm for Class A, 0.2 mm for Class B, and 0.3 mm for Class C.

1.8 Design selection

Basic design decisions concerning reservoir shape, column spacings and wall type depend upon the site layout and ground conditions and the operating requirements such as capacity, top and bottom water levels, division walls, valve chamber location and access.

The plan shape providing minimum wall length for a given capacity and height is a square. However, if a division wall is required then the best plan shape becomes a rectangle with a ratio of 3 to 2. If movement joints are to be provided these are influenced by the following considerations.

(a) Ground slabs are conveniently cast and compacted by a screeder/compactor beam in continuous strips up to about 6.5 m wide.

(b) Joint spacing in walls should not exceed 7.5 m for Option 2 (see Section 4) or about 5 m for Option 3 (see Section 3).

(c) Roof slabs should preferably be provided with four layers of reinforcement so that, even in the form of two mats of fabric reinforcement, the minimum practical thickness is about 175 mm if 40 mm minimum cover is to be provided on both faces. Alternatively, if a roof thickness of 225 mm (or, strictly, 230 mm) is used, then Class C exposure can sometimes be assumed for the upper face. Thicknesses of 175 and 230 mm in continuous two-way spanning slabs can conveniently span 5 m and nearly 7 m, respectively, for typical coverings of 150 mm soil, 150 mm gravel (drainage), insulation ($5\ kN/m^2$, $104\ lbf/ft^2$) and a live loading of $1.5\ kN/m^2$ ($31\ lbf/ft^2$).

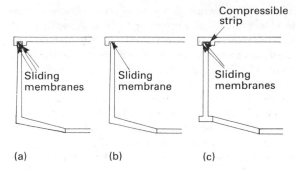

(a)   (b)   (c)

Figure 1-1   Three types of reservoir wall:
(a) Propped cantilever with pinned-joint to roof
(b) Propped cantilever with sliding joint to roof
(c) Simply supported wall

Since movement joints in walls and ground slabs should line up with each other and with the columns, suitable joint spacings are 6.5 and 5 m with roof slab thicknesses of 225 and 175 mm, respectively. Some typical types of reservoir wall are shown in Fig. 1-1 as follows:

(a) Propped cantilever with pinned joint to roof
(b) Propped cantilever with sliding joint to roof
(c) Simply supported wall

It may appear at first that (a) must be the most economic since the roof can act as an additional tie. (Notice that even if there is a movement joint in the roof slab parallel to the wall, it will be sufficiently remote from the wall itself for the roof slab to act as a very deep beam which readily resists the in-plane forces and provide the top of the wall with very effective support.) However, if the wall is backfilled then (a) need not be the most economic since even when the roof slab is protected with a covering of soil or with light reflective chippings, seasonal temperature variations of ±15°C can readily occur. If the length between walls (or expansion joints) is typically 50 m, the movement at each end can be approximately ±3 to ±5 mm, i.e. total movements of 6 to 10 mm can occur, which are sufficient to mobilize passive resistance in any backfill which may be placed behind the wall. The stiffness of the roof slab in compression is high, hence the wall can be subjected to relatively high flexural stresses by the passive resistance of the backfill. The high stiffness of the roof slab can be effectively reduced by introducing a flexible bearing strip between the top of the wall and the roof slab, as shown in Fig. 1-1c. However, if a sliding joint is used, as in Fig. 1-1b, the maximum force which can be exerted in either direction at the top of the wall is restricted to the frictional force between wall and roof slab. Furthermore, if shear keys are used on centre lines of the reservoir to ensure that the thermal movements in the roof develop from these positions along the lengths of the walls, progressive creep of the roof slab relative to the walls cannot occur and the external lug of the roof slab should never bear against the outer face of the wall for a correctly designed clearance. For maximum economy, the frictional force should be large enough to provide a substantial propping force inwards to resist the water load and yet induce relatively little passive resistance in the backfill when it acts outwards as the roof slab expands and slides over the top of the wall. Notice also that a sliding joint with a significant but limited propping force limits secondary moments induced at the base of the cantilever wall, which is why the simply supported wall shown in Fig. 1-1c should be more economic than the propped cantilever with pinned joint to roof shown in Fig. 1-1a. Notice that the shear and sliding forces at the base of the simply supported wall are less than the corresponding forces for either of the cantilever walls. Shallow tanks should be avoided and sloping the floor around the perimeter reduces the wall height and the overall cost, for a given capacity. For example, for reservoir capacities of only about 5000 m^3 (177 000 ft^3) and with sloping floors around the perimeter, a wall height of about 4 m and depth of 5 m or more can be very suitable. For reservoir capacities of over 10 000 m^3, costs tend to decrease with increasing depth up to 7 m or more. If internal division walls are not required for compartmentation then circular reservoirs, in either reinforced or prestressed concrete, can be the most economical [19].

1.9 General design details

Some general comments on particular design details should be noted as follows. Although sloping the floor inwards from the perimeter walls increases the reservoir capacity for a given wall height, the slope should preferably be limited to about 15° to avoid the need for temporary support during casting. Again, the general falls to reservoirs to facilitate cleaning of the floor or drainage of the roof should preferably be made the same (about 1 in 150) so that wall heights can be kept constant.

Where floor slabs are founded on rock, the columns can be founded on the floor slab. Otherwise, the column bases must be isolated from the floor to avoid introducing additional shear stresses and bending moments in the latter. Surface type water bars can conveniently be placed across the underside of the floor slab and the column base if both floor and base are founded at the same level. Where floor slabs need to resist sliding the slab and oversite concrete should not be separated by a membrane.

Where the internal angle formed by two slabs meeting at a corner tends only to close, either of the details shown in Fig. 1–2a, b is satisfactory. However, in water-retaining structures, the internal angle, in general, also tends to open under the water load and the reinforcement should be detailed as indicated in Fig. 1–2c [4]. U-bars are used in vertical planes for the inside faces and L-bars are used in horizontal planes for the outside faces.

1.10 Bending moments and shear forces in walls and floors

The triangular pressure distribution of the liquid on the walls is resisted by a combination of horizontal and vertical bending moments. In small deep tanks, the horizontal bending moments are more significant, whereas in the central lengths of large reservoirs the horizontal bending moments can become insignificant or

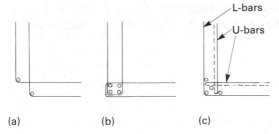

Figure 1-2 Reinforcement details at corners

even zero if vertical movement joints are introduced and the water pressure is resisted entirely by the vertical cantilever action of the wall.

The determination of the horizontal and vertical bending moments for rectangular panels with different edge conditions subjected to triangularly distributed loads are most readily determined by means of coefficients and reference can be made elsewhere, such as to the handbook for BS 5337 [5], and to tables which were originally produced by the Portland Cement Association [6]. Similarly, for circular tanks, reference can be made to tables of coefficients indicating the elastic distribution of ring tension and vertical bending moments throughout the wall height for different edge conditions and for different tank geometries [6].

In the case of suspended tank floors which are continuous with the walls, it should be noted that the maximum sagging moment in the floor is not necessarily given when the tank is full. For example, if the wall height subjected to a triangularly distributed load is y, the cantilever moment at the base is proportional to $w_g y^3/6$. The free mid-span sagging moment for a span L is proportional to $w_g L^2/8$. If the span in the other direction is large, then the maximum sagging moment is given simply by

$$w_g y L^2/8 - w_g y^3/6$$

which is seen to be a maximum when $y = L/2$ and not necessarily when y is a maximum.

1.11 Thermal stresses

Thermal stresses due to dissimilar temperatures occurring in different parts of the structure can occur in the mature concrete for a variety of reasons. For example, solar gain can cause the external faces of the roof or the walls to be at higher temperatures than either the corresponding inner faces or the adjacent parts of the structure. Conversely, if warm water enters the tank when the external air is cold, a temperature gradient can be produced in the reverse direction. If very warm liquids are likely, then the maximum temperature gradient should be assessed, making suitable allowance for the surface resistivity at any concrete–air interface (see, for example, [7]).

Consider, as an example, a 300 mm thick wall with an inside face temperature, T_0, of 10°C subjected to a rapid rise in temperature due to solar gain to give a value of T_s at the outside face of 40°C. If the short-term concrete modulus, E_c, is

28 GPa (4.1×10^6 psi) and the coefficient of thermal expansion of the concrete, α, is 12×10^{-6} per °C then the differential thermal strain of the outer face relative to the inner face is

$$\alpha(T_s - T_0) = 360 \times 10^{-6}$$

If fully restrained, the extreme fibre stresses, as given by elastic theory for an uncracked section, would be

$$\pm\tfrac{1}{2}\alpha(T_s - T_0)/E_c = \pm 5.0 \text{ MPa (730 psi)}$$

This theoretical tensile stress would be excessive for the concrete and cracks would occur at the inner face.

If there is no restraint the curvature, $1/r$, of the wall would be given by

$$1/r = (1 + 360 \times 10^{-6})/(r + 0.3)$$

or the radius of curvature, r, equals 833 m.

If an initially straight length, l, of wall becomes an arc of a circle of radius r, the ends will move inwards relative to the mid-length of the wall by a distance of $l^2/8r$. Hence, for a 1 m length of the above wall, the value becomes

$$l^2/8r = 1/(8 \times 833) = 0.15 \text{ mm}$$

In the case of a continuous cantilever wall, the top of the wall can deflect freely and no vertical thermal stresses should arise. In the horizontal direction, vertical cracks due to early thermal effects (see Section 3) can be present already, typically at about 1 m spacing, and can therefore readily permit lateral displacements of the order of ± 0.075 mm at, and midway between, the cracks, respectively. Thus, this relatively small amount of bowing in the wall can in practice relieve the high thermal stresses which can theoretically occur in very stiff and rigid structures.

Near the base of the wall, where the lateral displacements can induce greater resistance, the wall in general will be protected from solar gain by backfill. If not, then the wall expansion will be restricted by the relatively massive base to give high compressive stresses in the outer face, with perhaps some additional tensile cracking in the inner face. These additional vertical cracks will tend to be fine and of temporary duration and associated horizontal cracks in the inner face may also occur. However, provided that the wall has been well-designed for both early thermal cracking in the immature concrete and for structural load cracking in the mature concrete, any temporary increase in crack widths due to solar effects can generally be ignored. Similarly, where the thermal expansion due to the entry of warm liquids does not exceed that necessary to close the early thermal contraction cracks, the thermal stresses are likely to be small. However, where hot liquids can enter, the resulting thermal movements and stresses should be carefully considered and, if necessary, designed for accordingly.

1.12 Summary of the causes of cracking

Since the control of cracking in water-retaining structures is of especial importance, the various causes of cracking can be usefully summarized as follows.

(a) *Plastic shrinkage and settlement in very immature concrete.* Moisture loss from the surface of the fresh concrete can give rise to very high contraction strains and extensive cracking and crazing of the surface. Rapid protection from sun and wind immediately after placing and compaction of horizontal surfaces in particular is essential. Vertical movement of water within the fresh concrete after it has been placed, known as 'water-gain' [8], can give rise to vertical settlement of the fresh concrete and quite wide settlement cracks. Such cracking can often be eliminated by re-vibration of the concrete an hour or so after placing, while it is still workable and capable of responding to vibration.

(b) *Early thermal contraction in the immature concrete.* Ways of both reducing and controlling the early thermal contraction due to the heat of hydration of the cement, with both reinforcement and movement joints, are considered in detail later in this chapter.

(c) *Long-term drying shrinkage and thermal movements in the mature concrete.* These additional contraction strains can usually be considered in relation to the cracking patterns and movements already established in the immature concrete.

(d) *Flexural, direct tension or shear stresses in the mature concrete.* Cracking due to flexure and direct tension should be considered as a major criterion for the design of the concrete sections. Furthermore, the structural design must ensure that excessive cracking, due to either shear stresses or abnormal loading due to settlement, does not occur.

(e) *Volume changes due to physical or chemical attack.* Volume changes in the mature concrete or the embedded steel can cause progressive cracking and deterioration of the structure. Precautions should be taken to ensure that loss of serviceability caused by corrosion of steel reinforcement due to inadequate or porous concrete cover, freeze–thaw cycles and sulphate attack of the concrete, etc., does not occur.

Vigilance in both design and construction is essential if cracking due to any of the above causes is not to render a water-retaining structure unserviceable.

2 Design requirements and analysis

2.1 Introduction

The single feature which perhaps characterizes water-retaining structures is the serviceability requirement for controlled cracking. The ultimate limit states and the other serviceability limit states must still be satisfied, but the fineness of the crack widths which are normally permitted for water-retaining structures effectively makes this—the crack width limit state—the primary design consideration. Thus, although the usual approach for most structures is to design for collapse and then ensure compliance with various other limit states, in the case of water-retaining structures the requirements for collapse can conveniently be considered indirectly in terms of serviceability conditions, as indicated below, and then checked after the main design parameters have been determined from the serviceability requirements.

2.2 Ultimate limit states

The usual requirements for the ultimate limit states, such as resistance to overturning, buckling and material failure, etc., must be satisfied as for all structures. However, since the limit state of collapse, i.e. material failure under static loading in the structure as designed, is generally less critical than the crack width serviceability limit state, it can be conveniently considered indirectly in terms of service stresses. Thus, conventional elastic design can be used with suitable design service stresses to ensure compliance with the limit state of collapse. This approach necessarily means that the design for collapse will generally be conservative; but since the crack width requirement is usually the crucial consideration, the economics of the resultant design should be largely unaffected.

Let the required design stresses under service loads be f_{dcb}, the design concrete stress in bending, and f_{dst}, the design steel stress in tension.

These design service stresses are used in conjunction with conventional elastic theory and a triangular stress block in the concrete. They therefore need to take into account not only the partial safety factors appropriate for collapse but also the collapse design using a rectangular or similar stress block with a probable reduction in lever arm factor. Assuming a typical lever arm factor for elastic design of 0.86 and a minimum value for ultimate load design of 0.75, a suitable allowance is seen to be about $0.86/0.75 = 1.15$ (see below for expressions for f_{dcb} and f_{dst}).

The partial safety factor for load, γ_f, is taken as 1.6 for live loads and water loads and 1.4 for dead loads. Although it can be argued that the water load can be determined as accurately as a dead load for normal situations, any accident involving silting up, etc., can rapidly increase the effective density of the water and hence BS 5337 recommends that γ_f for water be taken as 1.6. If 1.6 is also taken for dead loads where these may occur in a water-retaining structure, a single value of γ_f applies for all types of load. The partial safety factor for material strength, γ_m, is 1.5 for concrete and 1.15 for steel, hence suitable design service stresses would appear to be as follows.

$$f_{dcb} = f_{cu}/(\gamma_f \times \gamma_m \times 1.15) = f_{cu}/2.76$$

$$f_{dst} = f_{yt}/(\gamma_f \times \gamma_m \times 1.15) = f_{yt}/2.12$$

where $1.15 = 0.86/0.75$ is the ratio of the elastic lever-arm factor to ultimate-load lever arm factor, as explained above.

Traditional elastic design can now be used to determine the ESR (equal-strength steel ratio) values for singly reinforced concrete beams. Using results given elsewhere [8], the neutral axis depth factor, δ_n, equal-strength steel ratio, ρ_e, and design (service) moment of resistance, M_d, are given by

$$\delta_n = \alpha_e/(\alpha_e + f_{dst}/f_{dcb}) \tag{2-1}$$

$$\rho_e = A_{st}/bd = \tfrac{1}{2}\delta_n f_{dcb}/f_{dst} \tag{2-2}$$

$$M_d/bd^2 = \tfrac{1}{2}f_{dcb}\delta_n\delta_z = f_{dst}\rho_e\delta_z \tag{2-3}$$

where α_e is the modular ratio E_s/E_c and δ_z the lever-arm factor. Furthermore, if the long-term value of E_c is assumed to be given (in MPa units) by $4550(f_{cu})^{\frac{1}{3}}$ [1]

Table 2-1 Typical parameters for ESR beams with design concrete stress of $f_{cu}/2.76$ and design steel stress of $f_{yt}/2.12 = 200$ MPa (29 000 psi)

f_{cu}	α_e	δ_n	δ_z	ρ_e	M_d/bd^2
25	15.0	0.40	0.87	0.91	1.58
30	14.1	0.43	0.86	1.17	2.01

and $E_s = 200$ GPa (29×10^7 psi), then $\alpha_e = 15.0$ and 14.1 for $f_{cu} = 25$ and 30 MPa, respectively. Table 2-1 gives values of δ_n, δ_z, ρ_e and M_d/bd^2 for ESR sections using steel $f_{yt} = 425$ MPa (62 000 psi) and concrete $f_{cu} = 25$ and 30 MPa.

Elastic design using the above design stresses can be conveniently used for the initial design of a section and then modified as necessary to satisfy the crack width limit state, as indicated later. Any subsequent check for collapse should almost invariably indicate that the ultimate limit state is satisfactory for static loading.

2.3 Deflection

The limits for deflection should normally be those for non-liquid-retaining structures since only in exceptional circumstances would deflection be critical with respect to freeboard, redistribution of load or drainage. CP 110 [1] gives recommendations for calculating deflections but such calculations need not normally be required if the recommended span/effective-depth ratios are not exceeded. For example, for rectangular reinforced cantilevers not exceeding 10 m span:

$$\text{span/effective-depth ratio} \not> \xi\psi\zeta(7)$$

where 7 is the appropriate basic ratio of effective span to effective depth and ξ, ψ and ζ are multipliers for amounts of tension reinforcement, compression reinforcement and concrete or beam characteristics, respectively. For design purposes, it is usually convenient to carry out a check for deflection after the section has been designed for the crack width limit state. Modification of the reinforcement will seldom be required but if necessary an estimated value of a suitable span/effective-depth ratio can be included as an initial design requirement [8].

2.4 Crack width requirements

The restriction of crack widths to acceptable values has already been emphasized as the major criterion for the design of the structural sections. However, it should also be appreciated that the requirement to control crack widths applies at all stages in the life of the structure, i.e. when the concrete is immature and when it is mature and subjected to normal working loads. Design for controlled cracking in the immature concrete is considered in Sections 3 and 4. Although compliance with the crack width criteria is usually more difficult when considering contractions due to early thermal and other effects, any required modifications to the design involve adjustments to the reinforcement, inclusion of movement joints and similar details rather than changes in section sizes. Thus, only cracking in the mature concrete due to the applied loading need be considered here in order to formulate the initial design.

2.5   Flexural cracks in mature concrete

The formula given in BS 5337 indicates a crack width which is likely to be exceeded only by about 5% of the cracks, unlike the similar formula in CP 110 for which almost 20% excessives are to be expected. It is applicable to steel strains up to $0.8f_y/E_s$ and concrete stresses up to $0.45f_{cu}$ and can be expressed as

$$w_{max} = 4.5a_{cr}\varepsilon_m/K_1 \tag{2-4}$$

where

$K_1 = 1 + 2.5(a_{cr} - c)/(h - d_n)$
a_{cr} = distance from point considered to the surface of the nearest longitudinal bar
c = cover to the tension steel
h = overall depth of member
d_n = depth of neutral axis from elastic analysis as for ε_1
$\varepsilon_m = \varepsilon_1 - K_2$
ε_1 = average strain at level of point considered calculated from elastic analysis ignoring the stiffening effect of the concrete in the tension zone (i.e. ignoring factor K_2)
$K_2 = (0.7 \times 10^{-3})b_t hy/A_s(h - d_n)f_s$
b_t = width of the section at centroid of tension steel
y = distance from neutral axis to level of point considered
A_s = area of tension reinforcement
f_s = service stress in the reinforcement (in MPa units)

Notice that an elastic analysis is required so that the neutral axis depth has been denoted by d_n to emphasize that it will generally not equal the neutral axis depth, x, as given by collapse design. Notice also that elastic analysis provides the values of ε_1 and f_s which are essential prerequisities for calculating w_{max}. Thus, elastic design as outlined earlier can conveniently be incorporated in a design procedure to calculate w_{max}. BS 5337 recommends that the long-term concrete modulus, taken as half the instantaneous value (e.g. $\frac{1}{2} \times 9100(f_{cu})^{\frac{1}{3}}$) should be used and that if the shrinkage strain exceeds 0.06%, ε_m should be increased by $0.5\varepsilon_{cs}$.

2.6   Direct tension cracks in mature concrete

Bate *et al.* [9] have indicated that a more suitable formula for determining crack widths in mature members subjected to tension is given by

$$w = K_3 a_{cr}\varepsilon_m \tag{2-5}$$

where

$K_3 = 1.6 + 1.4(c/\phi)(A/B)^{\frac{1}{2}}$
A, B = longer and shorter sides respectively of a prism of concrete which may be assumed to surround a particular bar (see Fig. 2-1)

The formula as given referred to CP 110, so that for 5% excessives it will underestimate the value of w.

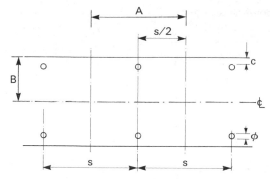

Figure 2-1 Concrete 'prisms' for direct tension crack calculations

2.7 Reinforced concrete

The serviceability requirements for cracking in BS 5337 have been seen to depend on the exposure conditions and are essentially as follows for both flexural and direct tension:

Class A exposure $w \leqslant 0.1$ mm
Class B exposure $w \leqslant 0.2$ mm
Class C exposure $w \leqslant 0.3$ mm

For Class A and B exposures, the above can be considered as the crucial design requirement provided that a reasonable steel ratio is used. Elastic design ensures a fairly reasonable and consistent choice of steel ratio, hence low steel ratio sections with low ultimate strengths, as sometimes indicated in Appendix 1 of [5], for example, can be avoided. Even for Class C exposure any calculations for collapse in flexure should prove satisfactory for sections designed as indicated earlier. The possibility of excessive shear, however, must still be checked.

Notice that, when steel is being curtailed according to the crack width requirement, the steel ratio will tend to decrease rapidly, even for a wall which tapers. In this case, a check for collapse in flexure at each point of curtailment becomes much more necessary, as indicated in the design example given later. Conversely, however, further checks for shear at the curtailed sections are unlikely to be critical in slabs with no shear reinforcement at the most heavily loaded section.

The normal anchorage (average) bond stresses given in CP 110 apply generally for BS 5337 also, except for horizontal bars subjected to direct tension. In this case, because of water-gain effects and a lack of transverse shear the bond stresses should not exceed 0.7 times the values given in Table 22 of CP 110. These are reproduced in Table 2-2 for convenience.

Redistribution of moments is not permitted for water-retaining structures—unlike CP 110. Cover to *all* steel and reinforcement should not be less than 40 mm. For particularly aggressive liquids or where erosion or abrasion are important, this cover should be increased.

The minimum areas of reinforcement to be provided should satisfy the requirements for crack control in the immature concrete and will therefore exceed the minimum given in CP 110. (See Sections 3 and 4)

Table 2-2 Ultimate anchorage bond stresses as in CP 110. For horizontal bars in direct tension reduce by 30%.

Bar type	Concrete grade	Tension	Compression
Plain	25	1.4	1.7
	30	1.5	1.9
Deformed type 1	25	1.9	2.4
	30	2.2	2.7
Deformed type 2	25	2.5^a	3.1^a
	30	2.9^a	3.5^a

[a] Equal to 1.3×type 1 values

2.8 Prestressed concrete

The serviceability requirements for prestressed concrete for the different exposure conditions to BS 5337 are given in Table 2-3, together with the corresponding class of member given in CP 110. The design of prestressed concrete members for Class A or Class B exposure can therefore generally follow that for Class 1 or Class 2 members to Section 4 of CP 110, Part 1:1972, since the serviceability requirements concerning the limit states of decompression, and crack initiation, are equally applicable and generally control the design. The ultimate strengths in flexure, shear and torsion should of course be checked in all cases.

Table 2-3 Serviceability requirements for prestressed concrete

Exposure class (to BS 5337)	Prestressed concrete	
	Class (to CP 110)	Tensile stress or crack width
A	1	No tensile stress
B	2	No visible cracking
C	3	Crack width $\not> 0.2$ mm

Guidance concerning the special requirements for the design of cylindrical concrete tanks as given in BS 5337 can be summarized as follows:

(a) Jacking force in circumferential tendons

$\not> 70\%$ of characteristic strength.

(b) Concrete compressive stress in circumferential direction when tank is full, after allowance for all losses of prestress

$\not< 1.0$ MPa (145 psi)

(c) Concrete tensile stress at any point when tank is empty

$\not>$ 1.0 MPa generally or

$\not>$ 0 (e.g. 1 MPa compression) if tank is left empty for prolonged periods, or frequently filled and emptied.

(d) Principal compressive stress in the concrete

$\not>$ $0.33 f_{cu}$.

Where the foot of the wall is nominally free to slide, the maximum stress in the circumferential prestressing steel occurs when the reservoir is full and the foot is taken as frictionless. However, the bending moment in the vertical direction should be assessed assuming a restraint (i.e. a frictional force in the nominally free-to-slide foot) equal to one-half of that provided by a pinned foot. Where the foot is restrained, the maximum circumferential steel stress should still be assessed as above but the bending moment in the vertical direction should be determined for a pinned, or fixed, foot as appropriate.

During the circumferential stressing operation, a flexural tensile stress can be induced in either face which can be taken as numerically equal to 0.3 times the ring compressive stress.

The vertical tension stresses, due to either the restraint of the wall foot or the partially wound condition for the ring stressing, can be designed for using either vertical prestressing or vertical reinforcement (i.e. reinforced concrete).

Prestressing wire may be wound around the outside of the walls and subsequently protected with pneumatic mortar, or placed inside the walls and grouted where increased protection from corrosion is desired (e.g. in very corrosive environments and by the sea).

Bending moments, tensions and shears can again be determined using the tables originally produced by the Portland Cement Association [6] and reproduced in part, with amendments to the notation, in [5].

2.9 Detailed design of sections

The manual calculation of w_{max} using the code formula is necessarily tedious and the use either of standard tables such as those given in [5] or of a computer is to be recommended. If the latter is used, then an interactive form of solution, in which the designer is presented with results and required to take conscious decisions assisted by this information, can be preferable to a fully automatic process. The interactive approach has the advantage that not only does the designer remain more in control but also relatively inexpensive microcomputers can usually be used. The author has therefore taken this a stage further and written a program for a programmable pocket calculator attached to a small printer (a Texas Instruments TI59 with a PC100 printer). This calculator is of humble capacity and cost compared with the sophisticated computer or even of many microcomputers. Programming for this pocket calculator has the advantage of requiring only simple algebraic logic, rather than a special high-level language, and the disadvantage that extended programs often have to be made very efficient in terms of length and storage and, preferably, also in terms of operation (the time for each operation is measured in ms rather than μs) if they are to fit into the capacity available and produce solutions within a convenient time. Since the capacities of all sizes of machines are continually increasing, it is likely that future trends will be more and more towards small inexpensive personal microcomputers. It therefore seems appropriate to outline briefly the above program for the TI59 using only 40 registers and 720 'lines'.

(a) *Preliminary wall design* The required steel and concrete grades, maximum crack width, maximum and minimum bar spacings, maximum bar size and concrete cover are entered and the program uses ESR design as described earlier to list all combinations of standard bar sizes (maximum of two sizes), for each depth of section, which satisfy the specified limits of bar spacing and size. The listing commences with crack widths which may exceed the maximum permitted value (for information and additional guidance) and progresses to increasing overall thickness of section until an uncracked section (i.e. $w < 0$) is indicated.

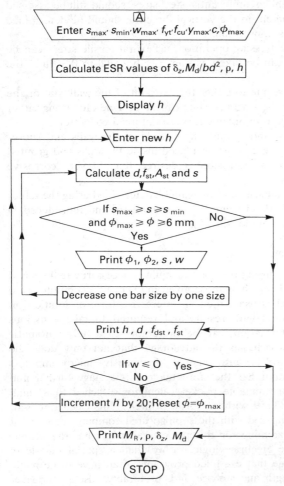

Figure 2-2 Flow chart for Preliminary Wall Design Program A

(b) *Wall detail design* With the aid of the tabulated data output from the preliminary design a particular bar spacing, s, and bar sizes, ϕ_1 and ϕ_2, are selected and re-input to determine the required overall thickness, h, which is necessary, with the selected values of s, ϕ_1 and ϕ_2, to ensure compliance with the specified crack width and the design stresses. This program does not assume ESR design so that any reasonable choice of s and ϕ (or s, ϕ_1 and ϕ_2) can be made using the preliminary data purely for guidance.

Using the output value of h, the final value can be selected (e.g. rounded to nearest 10 or 20 mm) and re-input, together with s, ϕ_1 and ϕ_2. The delay, i.e. the order of stopping off ϕ_1 and ϕ_2, is also entered at this stage. The output then gives the height at which ϕ_1 and ϕ_2 can be reduced by one bar size at a time for the given spacing s. The wall may be of constant h or taper uniformly from h at the base to h_{min} at the top. The programs are illustrated in simplified form in Figs 2-2 and 2-3 and by Example 2-1 below.

Example 2-1

The cantilever wall for a reservoir is to retain a maximum height of 3.88 m of water and is to be designed for Class A exposure (0.1 mm maximum design crack width) with 32 mm maximum bar diameter and 40 mm minimum concrete cover,

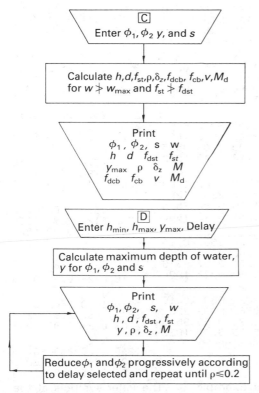

Figure 2-3   Flow charts for Wall Design/Detail Programs C and D

using grade 425 reinforcement and grade 25 concrete. Assume required bar spacing of between 225 and 300 mm, and the use of a programmable pocket calculator. The wall is to taper uniformly with height to $h_{min} = 250$ mm at the top of the wall and all the heights at which bar sizes can be reduced should be determined for the convenience of detailing the wall.

Solution Using the approach described in the text, a suitable input for the Preliminary Wall Design Program A would be

$$s_{max} = 300 \text{ mm} \quad s_{min} = 225 \text{ mm} \quad w_{max} = 0.1 \text{ mm} \quad f_{yt} = 425 \text{ MPa}$$

$$f_{cu} = 25 \text{ MPa} \quad y = 3.88 \text{ m} \quad c = 52 \text{ mm} \quad \phi_{max} = 32 \text{ mm}$$

The displayed value of 318 for h was adjusted up to the nearest 20 mm for convenience and re-input as 320 with the output as shown in Fig. 2-4. The first block of five results shows all combinations of ϕ_1, ϕ_2 and s which satisfy the specified limits of ϕ and s for a section thickness, h, of 320 mm which is printed on the last line for the block. Similarly, for the second block of three results for

WALL DESIGN		A			WALL DESIGN		BC		WALL DETAIL		D
GF1	GF2	RS	RW		GF1	GF2	RS	RW	GF1	GF2	RS	RW
RH	RD	RFDS	RFST		RH	RD	RFDS	RFST	RH	RD	RFDS	RFST
MR	GR	GDZ	M		RY	GR	GDZ	M	RY	GR	GDZ	M
					RFDC	RFCB	RV	MD				

GF1	GF2	RS	RW
20.0	32.0	287.	0.33
16.0	32.0	257.	0.32
12.0	32.0	234.	0.31
10.0	32.0	225.	0.30
25.0	25.0	251.	0.31
320.	258.	229.	215.
20.0	32.0	266.	0.23
16.0	32.0	239.	0.22
25.0	25.0	233.	0.22
340.	278.	229.	185.
25.0	32.0	288.	0.15
20.0	32.0	248.	0.14
360.	295.	229.	164.
25.0	32.0	270.	0.07
20.0	32.0	232.	0.06
380.	315.	229.	144.
25.0	32.0	254.	0.00
400.	335.	229.	127.
32.0	32.0	299.	-0.06
420.	352.	229.	115.
1.52	0.76	0.87	95.5

GF1	GF2	RS	RW
25.0	25.0	250.	0.09
389.	324.	229.	169.
3.88	0.60	0.88	95.5
9.05	5.81	0.36	130.
32.0	32.0	300.	0.09
362.	294.	229.	140.
3.88	0.90	0.86	95.5
9.05	6.16	0.40	151.
32.0	32.0	250.	0.09
344.	276.	229.	126.
3.88	1.16	0.84	95.5
9.05	6.51	0.42	160.
32.0	20.0	250.	0.10
376.	310.	229.	157.
3.88	0.71	0.87	95.5
9.05	5.97	0.37	138.
20.0	20.0	250.	0.10
438.	376.	229.	222.
3.88	0.33	0.90	95.5
9.05	5.35	0.31	102.

GF1	GF2	RS	RW
25.0	25.0	250.	0.09
390.	325.	229.	169.
3.88	0.60	0.88	95.7
20.0	25.0	250.	0.10
384.	321.	229.	185.
3.73	0.50	0.89	85.4
16.0	25.0	250.	0.10
380.	318.	229.	197.
3.62	0.43	0.89	77.9
20.0	16.0	250.	0.10
373.	312.	229.	223.
3.41	0.32	0.90	65.3
16.0	16.0	250.	0.02
363.	303.	229.	229.
3.15	0.26	0.91	51.3
12.0	16.0	250.	-0.09
354.	295.	229.	229.
2.88	0.21	0.92	39.2
12.0	12.0	250.	-0.30
342.	284.	229.	229.
2.56	0.15	0.93	27.5

Figure 2-4 Calculator output for Example 2-1. The letter symbols at the head of each column are explained in the text

$h = 340$ mm and so on, until w as calculated becomes less than zero (i.e. section is uncracked). The values of M_R, ρ, δ_z and M appropriate to all Program A results are then printed on the final line of the last block for Program A. Capital roman letters only are available on the small PC100 printer associated with the TI59 pocket calculator, hence the notation used for representing the standard symbols in terms of this single alphabet follows that suggested elsewhere [8], where, essentially, a prefix R indicates lower-case roman letters and a prefix G indicates lower-case greek letters of similar sound or shape. Thus, the headings shown represent the following:

ϕ_1(mm)	ϕ_2(mm)	s(mm)	w(mm)
h(mm)	d(mm)	$f_{ds}(=f_{dst}\text{MPa})$	f_{st}(MPa)
$M_R(=M_d/bd^2 \text{ MPa})$	ρ(%)	$\delta_z(=z/d)$	M(kN m)

The output from Program A indicates that 25 mm bars at 250 mm spacing in a 380 mm deep section should give a crack width of about 0.1 mm. Hence, values of $\phi_1 = 25$ mm, $y = 3.88$ m, $\phi_2 = 25$ mm and $s = 250$ mm were input into Wall Design Program B to obtain the output shown by the first block only in the second column of Fig. 2-4 and indicating that $h = 389$ mm. The values of $h_{max} = 390$ mm, $h_{min} = 250$ mm, $y_{max} = 3.88$ m and delay $= 3$ were then input into Wall Detail Program D with the output given by all the blocks in the final column in Fig. 2-4. Thus, 25 mm bars at 250 mm are only just adequate at $y = 3.88$ m, and alternate 25 mm bars cannot be replaced with 20 mm or 16 mm bars below $y = 3.73$ m or 3.62 m, respectively. The remaining 25 mm bars in conjunction with 16 mm bars cannot be replaced with 20, 16 or 12 mm bars below $y = 3.41$, 3.15 and 2.88 m, and so on. If a delay of 4 had been selected, then the output would have indicated the heights at which alternate 25 mm bars could be replaced by 20, 16 and 12 mm bars, respectively, followed by the heights at which alternate 25 mm bars could be reduced successively in conjunction with alternate 12 mm bars, and so on.

In this case, it would be convenient to curtail alternate 25 mm bars at $y = 3.62$ m and $y = 2.88$ m lapping onto 16 and 12 mm bars, respectively. The overlap at design stress is 47ϕ for the smaller bar for Type 2 deformed bars and 61ϕ for Type 1. If 12 mm bars at 250 mm are used in the opposite face for the full height, then the vertical steel would be adequate for favourable early thermal cracking conditions, as discussed in Sections 3 and 4; otherwise, the steel in the opposite face should be at least alternate 16 mm and 12 mm bars at 250 mm spacing.

The maximum shear stress at collapse is given in the first block of the second column in Fig. 2-4 as 0.36 MPa, which compares with a critical shear stress v_c from CP 110 for $f_{cu} = 25$ MPa and $\rho = 0.60\%$ of 0.53 MPa, hence shear is not critical in this case. Similarly, for deflection, the CP 110 multipliers are $\xi = 1.62$, $\psi = 1.64$ ($\rho_c \nless 0.15\%$) and $\zeta = 1.0$, hence

$$3.88/0.325 = 11.9 \nless \xi\psi\zeta(7) = 11.8$$

which it is, so there is no need to calculate the deflection. The moment of resistance, M_d, given by $M_u/1.6 = 130$ kN m, is also satisfactory since it exceeds the working load moment of 95.5 kN m.

The second column in Fig. 2-4 gives a number of alternatives for the design of the section at the base of the wall, ranging from 32 mm at 250 mm with $h = 344$ mm to 20 mm at 250 mm with $h = 438$ mm. The approximately ESR design which was selected is seen to give a very reasonable compromise in the amount of steel and size of section. If preferred, however, a smaller section with a much higher steel ratio can be considered on the basis that the high steel ratio can be rapidly reduced with height.

Notice that it is only when the steel ratio is made extremely high that the concrete stress could become critical, and only when the steel ratio is low that the steel stress can become more critical than the crack width requirement. For the wide range of sections with moderate steel ratios, the crack width requirement governs.

3   Design for continuity

3.1   Introduction

All concrete structures are subject to thermal, creep and drying shrinkage effects but these can be of little consequence where reinforcement is already provided for structural loads due to gravity and other causes. However, where the external loading is small and only nominal reinforcement is considered necessary, dimensional changes in restrained members can become the major cause of cracking. If long, continuous walls contain insufficient distribution reinforcement, they can often be rendered unserviceable by wide cracks. The reinforcement should therefore ensure that not only is the cracking fully controlled, but also that the widths of the cracks which occur do not exceed the design widths for the particular conditions of exposure. This section considers, in detail, the use of distribution reinforcement to control cracking due to early thermal effects in long, continuous concrete members.

3.2   Early thermal movement

All concrete expands after placing due to the heat of hydration of the cement. The initial expansion to peak temperature is soon followed by the early thermal contraction as the concrete cools to the temperature of its surroundings. Thus, early thermal movement takes place within a very few days of placing the concrete and any attempts to control the cracking must consider this essential fact, since the mechanisms controlling the behaviour in the immature concrete are very different from those considered for the mature concrete. Since the crack pattern due to the early thermal contraction is formed first, allowances can be made for the subsequent shrinkage and seasonal temperature variations, as indicated later.

3.3   Early thermal cracking behaviour

Early thermal movement in a restrained concrete member produces tensile stresses which can readily exceed the tensile strength of the concrete and initiate a crack which relieves the restraint. In the case of plain concrete, a single crack can continue to develop and relieve the tension so that no further cracks are introduced elsewhere.

Uncontrolled cracking, resulting in the formation of a few isolated but very wide cracks, is clearly undesirable. Controlled cracking, on the other hand, can be achieved by introducing a modest amount of reinforcement in the form of small-diameter bars or fabric whose function is to distribute the cracking and produce many fine cracks, rather than a few wide cracks. Compared with the flexural tension cracking in the mature concrete considered earlier, there are two major differences. First, there is now no strain gradient across the section. Second, the ratio of bond strength to tensile strength for the immature concrete (when early thermal cracking occurs) is much lower than that for mature concrete, so that very extensive bond creep can be assumed.

3.4 Fully restrained contraction cracks

As the heat of hydration is dissipated, the first crack is formed and local slipping starts between the concrete and steel. The extent of the slipping and the probable spacing between cracks depends upon the steel ratio and the properties of the concrete and reinforcement present. If the ends are fully restrained, the steel cannot be in compression until the concrete cracks, since the overall strain is zero. When the concrete fractures, the steel at the crack strains locally in tension, resulting necessarily in compression strains elsewhere in the steel. The uncracked lengths of concrete try to contract, as shown in Fig. 3-1.

Consider a contraction strain in the concrete, ε_c, due to early thermal effects, which exceeds the ultimate tensile strain, ε_{ult}, which can be resisted by the immature concrete. For a fully restrained member, the concrete must crack and the number of cracks will depend on the bond characteristics between the concrete and the reinforcement—ranging from a very large number of fine hair

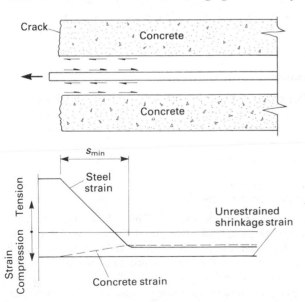

Figure 3-1 Probable strain distribution in concrete and steel adjacent to a crack

cracks for near-perfect 'no-slip' bond to more widely spaced and wider cracks where extensive bond slipping (or 'bond creep') occurs along the bars.

3.5 Critical steel ratio

The equilibrium of the portion of the member adjacent to the crack (e.g. between sections 1 and 2 in Fig. 3-2) is given by

$$A_s f_{st} = A_c f_{ct} - A_s f_{sc}$$

or

$$f_{ct} = \rho_c (f_{st} + f_{sc})$$

The critical steel ratio, ρ_{crit}, is given when the tension stresses in the concrete and steel reach their maximum values simultaneously. Hence, when $f_{ct} = {}_*f_{ct}$ and $f_{st} = f_{yt}$

$$\rho_{crit} = {}_*f_{ct}/(f_{yt} + f_{sc}) \tag{3-1}$$

$_*f_{ct}$ and f_{yt} can be readily estimated whereas f_{sc} cannot. Neglecting f_{sc} (and so introducing a small margin of safety) gives

$$\rho_{crit} = {}_*f_{ct}/f_{yt} \tag{3-2}$$

Notice that ρ_{crit} is inversely proportional to the grade of steel and hence there is a strong incentive to use higher grades for distribution reinforcement since the cost of high-yield steel of grade 425 and above is typically only 5 or 10% higher, per tonne, than grade 250. Occasionally, the much greater yield strain to failure of the lower grade steel may be desirable but, in general, the economic advantages of the higher grade steel can be considered paramount.

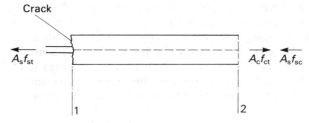

Figure 3-2 Forces acting adjacent to a crack

3.6 Controlled thermal and shrinkage cracking

The heat of hydration for all walls up to about 500 mm thick and with early removal of formwork, is largely dissipated after about three days. This is the critical period which, for a fully restrained wall, must be fully controlled by the steel alone. Hence, $_*f_{ct}$ may be taken as the tensile strength of the concrete at 3 days only, which is approximately $0.12(f_{cu})^{0.7}$.

The upper lift of a wall is usually restrained along its base by the lift below. At later ages, therefore, the steel crossing the early thermal cracks can be assisted from being overstressed by local base restraints, which can reverse direction adjacent to any cracks which tend to widen significantly in the mature concrete due to seasonal or daily thermal movements. The shear force along the base is therefore an additional term which could be included in Eqn 3-1 as follows:

$$\rho_{min} = (_*f_{ct} - R/A_c)/(f_{yt} + f_{sc}) \tag{3-3}$$

where ρ_{min} is the minimum steel ratio required to give controlled cracking at later ages and R the force due to local restraint at later ages. Even if R is negligible (e.g. as in a suspended floor slab), f_{sc}, although small, cannot be zero remote from the first crack (for overall length compatibility of steel reinforcement). Again, during the contraction prior to the first (early thermal) crack, tensile strains in the immature concrete would be variable along the length of the bay with the concrete having been weakened in potential, subsequent crack positions already. It may therefore be assumed that, for practical purposes, if Eqn 3-2 is satisfied controlled cracking occurs, not only in the immature concrete but also subsequently in the mature concrete for seasonal thermal movements and long-term drying shrinkage as well.

Consider a wall or floor slab cast and then exposed to the UK climate while construction proceeds elsewhere. The concrete, after a hot summer's day, could set at 30 to 40°C and subsequently fall to 0°C during the following winter. The anticipated thermal strain for such cases could approach 300 microstrain. The autogenous shrinkage of the concrete would be of the order of 75 microstrain only. Results indicate that the apparent shrinkage is small for normal structures exposed to normal climatic conditions and drying from one face only, even for one or two years [10, 11]. This is probably due to a combination of only partial drying out and to tensile creep greatly reducing the resultant shrinkage strain.

Where structures are protected from the weather, long-term drying shrinkage can become much more significant. For example, a continuous ground slab cast within a building experiences continuous drying out and, conversely, the immature slab is protected from significant thermal variations of solar gain by day followed by radiation losses at night. Initial cracking may be extensively delayed and occur in the mature concrete. However, even in this case, the total contraction strain is not so dissimilar and Eqn 3-5 (see later) should give a high (and therefore safe) estimate, since s_{max} tends to decrease with maturity of concrete, and ρ_{crit} from Eqn 3-2 may still be adequate for the reasons indicated above.

3.7 Low steel ratios

If $\rho_c < \rho_{crit}$, the strength of the concrete exceeds the strength of the steel and the latter continues to yield in tension. It follows that the cracks tend to be widely spaced and unevenly distributed along the wall. A few very wide cracks occurring at quite random positions are clearly very undesirable.

3.8 Moderate steel ratios

If $\rho_c > \rho_{crit}$, the strength of the steel exceeds the strength of the concrete so that cracks occur at intermediate sections and the maximum stress in the steel is less than the yield stress. The subsequent cracks in the concrete form when the bond force between steel and concrete equals the tensile strength of the concrete. That is,

$$f_b s \sum u_s \geq *f_{ct} A_c$$

where

$\sum u_s$ = total perimeter of steel reinforcement
u_s = perimeter of one bar ($= \pi \Phi$ for round bar)
f_b = mean bond strength between concrete and reinforcement adjacent to a crack

Now, for round bars of diameter ϕ,

$$\sum u_s = \pi \phi \times A_s / (\tfrac{1}{4} \pi \phi^2) = 4 A_s / \phi$$

Hence, substituting for $\sum u_s$ in the above, putting $A_s = \rho_c A_c$ and rearranging gives

$$s \geq *f_{ct} \phi / (f_b 4 \rho_c)$$

The minimum crack spacing, s, is given by the equality sign and the maximum crack spacing is twice this value (since a further crack can form once the development length on both sides of the potential crack position equals the minimum value of s), hence

$$\frac{*f_{ct} \phi}{f_b 2 \rho_c} \geq s \geq \frac{*f_{ct} \phi}{f_b 4 \rho_c} \tag{3-4}$$

where

$*f_{ct}/f_b \approx 1$ for plain round bars at early ages
$\approx \tfrac{4}{5}$ for Type 1 deformed bars at early ages
$\approx \tfrac{2}{3}$ for Type 2 deformed bars at early ages

3.9 High steel ratios

If high steel ratios are necessary for structural reasons, then $\rho_c \gg \rho_{crit}$ and s becomes very small in accordance with Eqn 3-4, which is valid for gradually applied stresses in immature concrete. If tensile loads are applied quickly to the mature concrete, little 'bond creep' can occur and an even larger number of cracks tend to form, resulting in further intermediate cracks between the existing early thermal cracks.

Example 3-1
The horizontal reinforcement in a reinforced concrete wall, 240 mm thick, consists of 10-mm diameter high-yield ribbed bars at 150 mm centres near each face. Calculate the probable crack spacings for the wall if it is fully restrained at each end. Assume the following values: $*f_{ct} = 1.5$ MPa, $f_b = 2.25$ MPa, $f_{yt} = 425$ MPa

Solution

$$\rho_c = 2 \times \tfrac{1}{4} \pi \times 10^2 / (240 \times 150) = 0.436\%$$

$$\rho_{crit} = 1.5/425 = 0.353\%$$

Hence, cracking is controlled by the steel since $\rho_c > \rho_{crit}$. Substituting in Eqn 3-4 and simplifying gives

$$0.764 > s > 0.382$$

The probable minimum and maximum crack spacings are

380 mm (minimum) and 760 mm (maximum)   [Answer]

3.10 Cracks at construction joints

If concrete is cast between two existing bays, the new bay undergoes a rise in temperature (due to heat of hydration), causing compression and therefore considerable creep in the immature concrete. When the temperature subsequently falls, the concrete has had time to harden so that its rate of creep is less and tension stresses are induced. The bond strength between new and mature concrete is generally less than the tensile strength of the concrete, hence the first crack is most likely to occur at the construction joint. Such cracks therefore tend to be wide and fully developed. They can become excessive if $\rho_c < \rho_{crit}$.

3.11 Intermediate cracks

The overall thermal contraction of a bay can be satisfied by relatively large cracks at one (or possibly both) construction joints at the end of the bay if either yielding or extensive bond slipping can occur. However, provided that $\rho_c > \rho_{crit}$, these first cracks will be restricted in width to the fully developed value and the surplus contraction strain transferred to adjacent intermediate cracks. The possible development of intermediate cracks before the first crack is fully developed is unlikely. That is, intermediate cracks do not usually appear (if at all) until after the formation of a fully developed crack at the construction joint. Thus, the worst cracks are usually observed either at the construction joints or at points of stress concentration (e.g. large changes in section).

3.12 Fully developed crack width

The crack spacing formula, Eqn 3-4, assumes that the actual contraction strain due to thermal and shrinkage movement is sufficiently large to produce a crack at all the positions which are theoretically possible. In practice, the actual contraction strain, as indicated earlier, should usually be very much less, so that many of the cracks predicted by the theory do not in fact occur and only a few cracks are likely to become fully developed. Nevertheless, it is the width of a fully developed crack which is of interest since it is the cracks of maximum width which cause most concern.

The total contraction is given by the sum of the thermal strain and the shrinkage strain. The tensile stress in the concrete builds up from zero at a cracked section to a maximum at a section which is a distance of s_{min} (or more) from the crack. Hence, if the steel ratio exceeds ρ_{crit} and the total contraction strain exceeds the ultimate tension strain, the mean tension strain in uncracked concrete of length s_{min} adjacent to the crack is $\frac{1}{2}\varepsilon_{ult}$ (see Fig. 3-1). This same reasoning applies to lengths of s_{min} on each side of the crack. Hence, the maximum crack width is given by development lengths of s_{min} ($=\frac{1}{2}s_{max}$) on each side of the crack. Thus, the mean tension strain over the length s_{max}, which is bisected by the crack, is $\frac{1}{2}\varepsilon_{ult}$ and the maximum final crack width is given by

$$w_{max} = s_{max}[(T_1 + T_2)\alpha_c + \varepsilon_{cs} - \tfrac{1}{2}\varepsilon_{ult}] \qquad (3\text{-}5)$$

where

T_1 = fall in temperature from the hydration peak
T_2 = seasonal fall in temperature from the initial ambient
α_c = effective coefficient of thermal expansion of the concrete
ε_{cs} = drying shrinkage strain

The effective coefficient, α_c, for the immature concrete is greatly reduced by creep effects and can be taken as about half the value for mature concrete. The seasonal fall, T_2, occurs in mature concrete and should therefore be associated with a higher α_c. However, since the bond strength for mature concrete is also much higher, T_2 should also be associated with a much lower value of s_{max}. The net result is that $s_{max}\,\alpha_c$ is approximately the same for either T_1 or T_2. Also, since ε_{cs} for external members is usually small and approximately equal to $\frac{1}{2}\varepsilon_{ult}$, Eqn 3-5 can be further simplified to give

$$w_{max} = s_{max}(T_1 + T_2)\alpha_c \qquad (3\text{-}6)$$

Example 3-2
Determine the probable maximum crack width for the wall described in Example 3-1 if the coefficients of expansion of the mature concrete and the steel are given by $\alpha_c = 12$ and $\alpha_s = 10$ microstrain per °C. Assume that the concrete temperature falls by 30°C and that creep effects effectively halve the thermal contraction of the immature concrete.

Solution The probable maximum crack spacing is 0.764 m, as before, and the ends are assumed to be fully restrained. The contraction strain of the concrete is given by

$$\tfrac{1}{2} \times 12 \times 30 = 180 \text{ microstrain}$$

Hence, since $\rho_c > \rho_{crit}$, s_{max} is known and Eqn 3-6 gives

$$w_{max} = 0.764 \times 180 \times 10^{-6} = 0.14 \text{ mm} \qquad \text{[Answer]}$$

3.13 Distribution reinforcement for Class B exposure

It has been shown that, if the width of a fully developed crack is not to be excessive, Eqn 3-6 must be satisfied as well as $\rho_c \not< \rho_{crit}$.

Table 3-1 shows the required steel ratios from Eqn 3-6 for different types of reinforcement in continuous slabs for a contraction strain of 180 microstrain and a crack width of 0.2 mm (i.e. Class B exposure). Notice that even a substantial amount of large-diameter structural reinforcement may not necessarily ensure that early thermal cracking is adequately distributed, since the required steel ratio increases linearly with bar diameter as expected from Eqn 3-4. For example, consider a retaining wall with counterforts which has 1% of 32 mm diameter plain round horizontal bars as structural reinforcement. Table 3-1 shows that the maximum crack width would be nearly 0.3 mm (i.e. 1.44×0.2 mm) and therefore

Table 3-1   Distribution steel required for 180 microstrain and crack widths not exceeding 0.2 mm

Bar size (mm)	6	8	10	12	25	32
Plain round	0.27%[a]	0.36%[a]	0.45%[a]	0.52%	1.13%	1.44%
Deformed (Type 1)	0.22%[a]	0.29%	0.36%	0.44%	0.90%	1.15%
Deformed (Type 2)	0.18%[a]	0.24%[b]	0.30%[c]	0.36%	0.75%	0.96%

[a] Too low even for grade 25
[b] Too low even for grade 25 if $f_y = 425$ MPa
[c] Too low for grade 30 if $f_y = 425$ MPa

too wide. For small-diameter bars, ρ_{crit} can become more critical than the crack width requirement.

3.14   Fabric reinforcement

If welded fabric (or mesh) reinforcement is used with transverse wires of at least equal diameter welded to the wires in tension, then it can be assumed that not less than 20% of the maximum force in the bar can be transferred to the immature concrete at each welded intersection within the bond development length of s_{min}.

Thus, if n_w = number of welded intersections within s_{min}, the force which must be transmitted by normal bond is reduced by $(1-0.2n_w)$, i.e. from Eqn 3-4,

$$s_{min} = (1-0.2n_w) \frac{*f_{ct}\phi}{f_b 4\rho_c} \qquad (3\text{-}7)$$

However, it should be noted that the values of s_{min} obtained must be compatible with the assumed value of n_w and that the most critical position of a crack is at a welded intersection, since the particular intersection then no longer assists the bond transfer, as shown in Fig. 3-3.

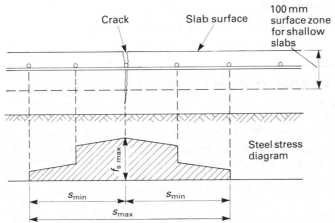

Figure 3-3 Probable stress distribution in welded-fabric reinforcement adjacent to a crack

Example 3-3

The reinforced concrete cantilever walls to a reservoir are typically 300 mm wide and fully restrained longitudinally, apart from partial contraction joints at 7.5 m centres. The continuous horizontal reinforcement in the walls consists of 12 mm high-yield ($f_{yt} = 425$ MPa) medium-bond ($f_{ct}/f_b = 0.8$) bars at 200 mm centres near each face. Difficulties are experienced on the site with excessive early thermal cracking and it is estimated that the peak hydration temperature rise is 45°C, the effective coefficient of expansion/contraction (allowing for all other effects) is 5 microstrain per °C, and the tensile strength of the immature concrete is 1.4 MPa.

There is the opportunity of obtaining 'Fabrik' purpose-made wall sheets providing the same reinforcement but with welded intersections with at least equal-diameter 12 mm cross wires (bars) at 200 mm centres. Determine the predicted crack widths

    (a) for existing 12 mm bars;
    (b) for similar 12 mm bars but with welded intersections at 200 mm centres which can each take 20% of the maximum force in the bar;
    (c) as for (a) but where the partial contraction joints were mistakenly omitted and the seasonal temperature variation is estimated at 20°C.

Solution Early thermal contraction strain is $45 \times 5 \times 10^{-6} = 225 \times 10^{-6}$. Seasonal variation is accommodated by partial contraction joints (see Section 4),

$$\rho_c = 2 \times \tfrac{1}{4}\pi(12)^2/(300 \times 200) = 0.377\%$$
$$\rho_{crit} = 1.4/425 = 0.329\% < \rho_c; \text{ satisfactory}$$

(a) Equating bond strength to concrete strength,

$$2 \times \pi(12) \times f_b \times s_{min} = 300 \times 200 f_{ct}$$

or $s_{min} = 637$ mm and $s_{max} = 1273$ mm hence

$$w_{max} = 1273 \times 225 \times 10^{-6} \text{ mm} = 0.29 \text{ mm} \qquad \text{[Answer]}$$

(b) As above, but allowing for welded intersections gives

$$s_{min} = 0.637(1 - 0.2n_w) \text{ m}$$

if $n_w = 1$,

$$200 \leqslant s \leqslant 400 \quad \text{and} \quad s_{min} = 453 \text{ mm}$$

if $n_w = 2$,

$$400 \leqslant s \leqslant 600 \quad \text{and} \quad s_{min} = 340 \text{ mm}$$

Neither of these s_{min} values can be valid and hence $s_{min} = 400$ mm, since the second welded intersection is not fully utilized. Hence,

$$w_{max} = 0.800 \times 225 \times 10^{-6} = 0.18 \text{ mm} \qquad \text{[Answer]}$$

(c) Adding a seasonal contraction strain of $20 \times 5 \times 10^{-6} = 100 \times 10^{-6}$ due to the omission of contraction joints, then

$$w_{max} = 1.273 \times 325 \times 10^{-6} = 0.41 \text{ mm} \qquad \text{[Answer]}$$

3.15   Thick sections

The appearance of the cracking in a thick wall in direct tension will be similar to a thin wall. That is, cracks will occur at the construction joints at either end and, possibly, one, two or more cracks within the bay depending upon the bay length, etc. The development of cracking in thick walls, however, is progressive from the surface and this fact can be utilized to effect economies in reinforcement. Figure 3-4 shows a thick base slab where the conditions for the top face can be similar to either face of a wall. Although the temperature gradients at the surface depend upon the air temperature and surface insulation, if any, BS 5337 states that a 'surface zone' thickness of 250 mm can be considered in all cases and reinforced accordingly. Thus, a wall (with faces on both sides exposed to significant thermal gradients) whose thickness exceeds 500 mm need therefore only contain an amount of distribution reinforcement given by ρ_{crit} based on a thickness of 500 mm. This assumes a reinforcement depth of 125 mm or less so that, as the surface cools, the surface zone behaves as a continuous, axially-reinforced, thin (250 mm thick) wall. Hence, once the cracks have developed in the surface, they can extend into the core, as the latter cools, without the need for additional distribution reinforcement [13]. The temperature rise (and hence the subsequent contraction) which should be considered for a thick section restrained at the ends should be that for the core.

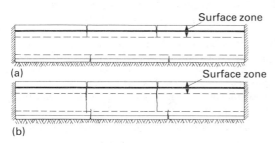

Figure 3-4   Controlled cracking in a thick base slab (a) initially and (b) finally

Example 3-4
The continuous walls to a very deep tank are 700 mm wide near the base and contain horizontal 16 mm high-yield ($f_{yt} = 425$ MPa) ribbed (high-bond, $f_{ct}/f_b = \frac{2}{3}$) bars at 100 mm centres near each face. Assuming an early temperature fall of 40°C and a seasonal temperature fall of 20°C, an effective coefficient of thermal expansion of the concrete of 6 microstrain per °C, an early tensile strength of the concrete of 1.3 MPa and a 'surface zone' thickness of 250 mm, determine the predicted maximum crack width.

Solution For each surface zone,

$$\rho_c = \tfrac{1}{4}\pi(16^2)/(250 \times 100) = 0.804\%$$

$$\rho_{crit} = 1.3/425 = 0.306\% < \rho_c; \text{ satisfactory.}$$

From Eqn 3-4,

$$s_{max} = \tfrac{2}{3} \times 16 \times 10^{-3}/(2 \times 0.804\%) = 0.663 \text{ m}$$

Therefore,

$$w_{max} = (40 + 20) \times 6 \times 0.663 \times 10^{-6} = 0.24 \text{ mm} \qquad \text{[Answer]}$$

Notice that, despite the high steel ratio (0.57% on overall thickness) and use of high-bond bars, the predicted crack width still exceeds 0.2 mm because of the large temperature changes. Reductions in crack width can be obtained by introducing movement joints, increasing the reinforcement especially with fabric, and reducing the temperature fall.

3.16 Early-age temperature rise data

The rate of hydration of cement increases with temperature. Thus, the heat of hydration is liberated more quickly, and the subsequent temperature rise is greater, in summer than in winter. The lowering of the temperature of the freshly placed concrete, to below that of the surrounding mature concrete, can prevent the early onset of cracking by reducing the actual temperature rise and by offsetting some of the rise against the lower initial temperature relative to the surroundings.

The early thermal temperature rise values given in BS 5337 for T_1 based on site data are as follows for 340 kg/m³ OPC concrete in 400 mm thick sections.

Winter concreting (10°C ambient, timber shutters removed after 15 h): 20°C
Summer concreting (20°C ambient, steel shutters, some solar gain):   30°C

The actual temperature of the concrete at placing can be assumed to be about 5°C above ambient. Where other or less onerous placing conditions can be assured, then the data in Table 3-2 based on computer predictions [12] can be used.

For thick sections (1000 mm and above) with timber formwork left in position,

Table 3-2   Predicted temperature rises, T_1, for different conditions

Mean ambient temp. °C	Cement content kg/m³	T_1 values for OPC concretes (°C)					
		Steel formwork		18 mm plywood removed after 15h		Ground slabs (no solar gain)	
		Section width		Section width		Section width	
		300 mm	500 mm	300 mm	500 mm	300 mm	500 mm
10	290	8	14	14	19	11	18
	340	9	17	17	23	13	22
	400	11	22	21	30	16	28
15	290	10	17	18	24	13	21
	340	12	21	23	30	16	26
	400	15	27	29	39	21	34

Fitzgibbon's approximate guide [13] of 12°C temperature rise per 100 kg of OPC usually gives T_1 values to within ±5°C. A temperature differential between surface and core of over 20°C can induce additional cracking in both the surface zone and the core of a thick section. As a general guide, therefore, the early removal of timber shutters and the use of steel shutters are desirable for all sections with end restraint and undesirable for free-standing thick sections. Remove end restraints whenever possible but especially in the case of thick sections.

4 Design for movement

4.1 Introduction

Thermal and shrinkage cracking in continuous members, fully restrained at their ends, has been considered in Section 3. However, there are other forms of restraint and degrees of restraint: the degree of restraint can be reduced by introducing movement joints, and the actual form of the external restraint has a fundamental influence on the cracking behaviour. Movement joints include expansion joints as well as complete and partial contraction joints. Thick sections, which illustrate one different form of restraint to continuous thin members with full end restraint, have already been discussed. Other forms of restraint include warping at the ends of walls with base restraint, ground slabs with base restraint and vertical end restraint.

4.2 Design options

The design options which are available for controlling early thermal and other contractions can be summarized as in Table 4-1. Option 1 assumes that all movement is controlled by the reinforcement and accommodated by the formation of closely spaced cracks, as indicated in Section 3 and further considered later. Option 2 assumes that only the early thermal movement is accommodated by crack formation and that once the reinforced concrete member has matured it can expand and contract as a unit between adjacent movement joints. Option 3 assumes that the primary control of all movement—early or otherwise—is by means of movement joints and not, primarily, by means of distribution reinforcement.

4.3 Unrestrained contraction in reinforced concrete

Consider a mature reinforced concrete member with perfect bond between concrete and steel, as shown in Fig. 4-1. If the drying shrinkage strain for the plain concrete is ε_{cs}, then the inclusion of reinforcement produces a tension strain, ε_{ct}, in the concrete and a compression strain, ε_{sc}, in the steel such that

$$\varepsilon_{cs} = \varepsilon_{ct} + \varepsilon_{sc} = (f_{ct}/E_c) + (f_{sc}/E_s) \tag{4-1}$$

Equating internal forces and assuming no external restraint,

$$A_c f_{ct} = \rho_c A_c f_{sc} \tag{4-2}$$

Table 4-1 Design options for control of early thermal contraction and restrained shrinkage

Option	Type of construction and method of control	Movement joint spacing	Minimum steel ratio %	Comments
1	Continuous Design for full restraint	No joints but expansion joints at wide spacings may be desirable in walls and slabs which are not protected from solar gain	ρ_{crit}	Use small-size bars at close spacing to avoid high steel ratios well in excess of ρ_{crit}
2	Semi-continuous Design for partial restraint	(a) Complete joints $\not> 15$ m (b) Alternate partial and complete joints $\not> 11\frac{1}{4}$ m (c) Partial joints $\not> 7\frac{1}{2}$ m	ρ_{crit}	Use small-size bars but less steel than in Option 1
3	Close movement joints Design for freedom of movement	(a) Complete joints $\not> 4.8$ m $+ w/\varepsilon$ (b) Alternate partial and complete joints $\not> \frac{1}{2}s_{max} + 2.4$ m $+ w/\varepsilon$ (c) Partial joints $\not> s_{max} + w/\varepsilon$ where w = maximum crack width $\varepsilon = \varepsilon_{cs} + \varepsilon_{te} - 100 \times 10^{-6}$	$\frac{2}{3}\rho_{crit}$	Avoid small-size bars which restrict the joint spacing for Options 3b and 3c

Notes

(i) In all cases, the actual steel ratio to be provided must not be less than that required for structural loading.

(ii) ρ_{crit} and s_{max} are as given by Eqns 3-2 and 3-4 and ρ refers to surface zones only for thick sections and ground slabs.

(iii) In wall slabs over 200 mm thick, reinforcement to $\frac{1}{3}\rho_{crit}$ should be placed near each face.

(iv) Steel ratios for Options 1 and 2 will generally exceed ρ_{crit} in order to satisfy Eqn 3-5

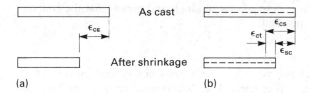

Figure 4-1 Strains before and after shrinkage:
(a) Plain concrete
(b) Reinforced concrete

Substituting for f_{ct} in Eqn 4-1 and putting $\alpha_e = E_s/E_c$ gives

$$f_{sc} = \varepsilon_{cs} E_s/(1 + \alpha_e \rho_c) \tag{4-3}$$

4.4 Differential thermal movement

If there is no external restraint, it is only the differential movement between concrete and steel which produces stress. However, since the coefficient of thermal expansion for the steel is very approximately equal to that for the concrete, any differential movement should be fairly small and cracking is highly unlikely. The combined shrinkage and differential thermal strain is given by

$$\varepsilon_{cs} + (\alpha_c - \alpha_s)T$$

where T is the fall in temperature, and this expression replaces ε_{cs} in Eqn 4-3 if differential thermal strains are significant. Contrast this situation with the cracking which occurs due to thermal contraction of the concrete in restrained members.

Example 4-1
Calculate the theoretical steel and concrete stresses in a reinforced concrete column if $E_s = 200\,\text{GPa}$, $\alpha_e = 15$, $\rho_c = 1\%$ and $\varepsilon_{cs} = 250$ microstrain. Assume (a) no restraint and (b) full restraint.

Solution
(a) From Eqn 4-3

$$f_{sc} = 250 \times 200 \times 10^3/(1 + 0.15) = 43.5\,\text{MPa} \qquad \text{[Answer]}$$

From Eqn 4-2

$$f_{ct} = 0.01 \times 43.5 \times 10^3 = 435\,\text{kPa} \qquad \text{[Answer]}$$

Notice that these stresses are relatively small even for an extremely high shrinkage strain. Although ε_{cs} and α_e can vary within wide limits, the stresses in unrestrained members, in practice, can be readily sustained by both the concrete and the steel.

(b) $f_{sc} = 0$, and \qquad\qquad\qquad\qquad\qquad [Answer]

$$f_{ct} = (200/15) \times 250 \times 10^3 = 3.33\,\text{MPa} \qquad \text{[Answer]}$$

The calculated stress for the concrete is far too high and means that the concrete must crack in at least one position.

4.5 Types of joint

There are essentially two basic types of joint: construction and movement. A construction joint (or day joint) is introduced as a convenience in construction, although the engineer should specify on his drawings where they are to be situated. No provision is made for subsequent relative movement and indeed measures should be taken to achieve continuity with good bond between new and old concrete (see Fig. 4-2 and Section 5).

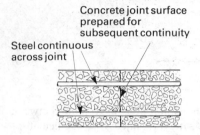

Figure 4-2   Construction joint

A movement joint is a specially formed joint intended to accommodate relative movement between adjoining parts of a structure, special provision being made for maintaining the watertightness of the joint. Movement joints include the following types.

   (a) *Expansion joint.* A movement joint which has complete discontinuity in both reinforcement and concrete and which can accommodate expansion in excess of the initial contraction is known as an expansion joint. Normally, such a joint accommodates an initial gap between the adjoining concrete sections of the joint, which then opens or closes to take up the subsequent expansion or contraction movement (Fig. 4-3).

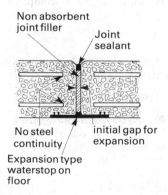

Figure 4-3   Expansion joint

(b) *Contraction joint.* This is a movement joint intended to permit contraction of the concrete with a deliberate partial or complete discontinuity but with no initial gap between the concrete on each side of the joint (Fig. 4-4). A gap can develop due to contraction of the concrete which can accommodate some subsequent limited expansion. A distinction must be made between a complete contraction joint, in which the concrete and the reinforcement are both completely interrupted, and a partial contraction joint in which only the concrete is interrupted while the reinforcement is continued (completely or partially) through the joint.

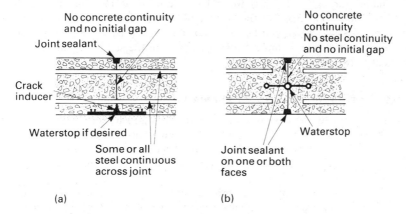

Figure 4-4 Contraction joints: (a) partial and (b) complete

(c) *Induced contraction joint.* A contraction joint where the concrete is continuous at the time of casting and the reinforcement is either partially or completely discontinuous. It is provided with a reduced concrete cross-section at which a crack should be induced—this then provides a discontinuity in the concrete and the joint functions as a contraction joint as indicated in (b).

(d) *Sliding joint.* This is a movement joint in which special provision is made to allow relative movement in the plane of the joint. There is complete discontinuity in both the concrete and the steel reinforcement. A typical example is the joint between the wall and the floor in some cylindrical tanks.

(e) *Sliding layer.* This is a special type of sliding joint which permits movement over a considerable area, such as between a floor and a sub-floor or blinding layer.

4.6 Positioning of construction joints

Where movement joints are included in the design, they can provide a natural choice for a break in construction since the joint must be formed anyway and special preparation of the concrete stop-end is neither necessary nor desirable if the construction joint actually coincides with a movement joint.

Conversely, however, there are situations where it is not convenient to form a discontinuity in the concrete at a movement joint by means of a stop-end. For example, where closely spaced joints occur as for Option 3, it may be convenient to form a stop-end at each second or third joint, or even less frequently still according to the lengths of slab which can conveniently be cast in a day.

An alternative, therefore, is to use an *induced* contraction joint whereby the concrete cross-section is reduced by surface recesses or other means and the steel is also made either partially or completely discontinuous. The weakened section should therefore crack and so produce a concrete discontinuity in the desired position for the movement joint. A 50% partial joint (i.e. 50% only of reinforcement continuous through the joint) for a close movement joint layout would therefore have only $\frac{1}{3}\rho_{crit}$ crossing the joint and the spacing for a 50% partial joint/50% partial joint system can be assumed to be given by Option 3b i.e. an alternate 100% partial joint/complete joint system with the same $\frac{2}{3}\rho_{crit}$ horizontal reinforcement between joints.

4.7   Design for full restraint (BS 5337 Option 1)

This option is essentially the design for continuity discussed in Section 3, with or without occasional movement joints. If all movement joints (including expansion joints) can be omitted entirely, then other savings become possible. Ground slabs should be laid directly on the oversite concrete and the reinforcement need only satisfy Eqns 3-2 and 3-5.

However, if all movement joints are omitted entirely, then it should be verified that the continuous slab will not be subjected to excessive solar gain or other thermal stresses such that buckling and crushing of the concrete can occur. An estimate of the possibility of buckling can be based on the Euler critical load, P_{crit} $(=\pi^2 EI/l^2)$ for a pin-ended strut as follows.

If $f_{cu}/(\gamma_m \gamma_f) = P_{crit}/A$ and short-term $E_c = 9100(f_{cu})^{\frac{1}{3}}$ (E_c and f_{cu} in MPa units), the expansion strain causing crushing failure is given by

$$\varepsilon_{cu} = f_{cu}/(\gamma_m \gamma_f E_c) = f_{cu}^{\frac{2}{3}}/(9100\gamma_m \gamma_f)$$

If $f_{cu} = 25 \times 10^6$ and $\gamma_m \gamma_f = 1.5 \times 1.4$, then $\varepsilon_{cu} = 447 \times 10^{-6}$. If this very high strain were achieved then the corresponding (minimum) length of a pin-ended slab which would heave upwards would be given by the above as $l = 125h(f_{cu})^{-\frac{1}{3}}$, i.e. if $h = 150$ mm and $f_{cu} = 25$ MPa, then $l = 6.4$ m. However, this strain would be very improbable in the UK, since the expansion is relative to the temperature when the concrete hardens (not the minimum) and any slight drying shrinkage is also deductable. Even in hot climates, expansion joints could be omitted in the floor and walls of a covered reservoir, for example, if only the roof is exposed to the extremes of solar energy. Notice that for walls any tendency to buckle should produce bowing which reduces the stresses and any tendency to buckle further.

Care should be taken to ensure that expansion joints, if provided, function correctly: for example, ground slabs with expansion joints should be laid on a separating layer of polythene or similar material. Movement joints in walls should generally coincide with corresponding joints in floors, otherwise sliding joints become necessary.

Where joints in ground slabs are widely spaced, the strength of the reinforcement *acting alone* at a cracked section should exceed the *lesser* of (a) the frictional drag of the slab length nearest to the movement joint [14] and (b) the tensile strength of the *mature* concrete (i.e. $\rho \nless 2\rho_{crit}$ approximately)

Thus, sections remote from a joint (e.g. at $\frac{1}{2} \times 75$ m for 75 m spacing) need satisfy (b) only, while sections nearer a joint need satisfy (a) only (and $\rho \nless \rho_{crit}$ etc.).

For all continuous ground slabs, the construction joints at the sides of each continuous strip can conveniently be reinforced by placing a narrow strip of mesh symmetrically along the side forms and at the appropriate depth (e.g. using overlay side-form strips), so as to provide an overlap of not less than three meshes with the fabric in the long strips on either side.

4.8   Design for partial restraint (BS 5337 Option 2)

If movement joints are included in a long wall slab, then the restraint adjacent to each joint will be virtually zero for complete movement joints and partly reduced even for a partial movement joint.

If complete movement joints are provided at 15 m spacing, or partial movement joints at 7.5 m spacing, then all movements in the mature slab due to seasonal or daily variations can be assumed to be accommodated at the movement joints. Thus, only the contraction strain due to the fall from the hydration peak temperature, T_1 need be considered when calculating the widths of cracks in the slab between the movement joints.

Thus, for concrete with a coefficient of thermal expansion of 12 microstrain per °C, the contraction strains in the UK are typically as follows.

Summer concreting: $\frac{1}{2} \times 12 \times 30 \times 10^{-6} = 180 \times 10^{-6}$
Winter concreting: $\frac{1}{2} \times 12 \times 20 \times 10^{-6} = 120 \times 10^{-6}$

In general, it will be necessary to assume summer concreting, since the time of construction cannot usually be predicted at the design stage.

If few or no movement joints are provided, then the full seasonal movement must be allowed for and continuous construction with full restraint assumed, as indicated earlier. Since the ratio of bond strength to concrete tensile strength is much higher for mature concrete, the crack spacing is reduced. However, the creep strains are also much less for mature concrete, as indicated earlier, hence the increase in crack width for a seasonal temperature fall, T_2, for continuous construction can be taken simply as $\alpha T_2(s_{max})$, i.e. using the same effective α and s_{max} as for immature concrete.

4.9   Movement capacity at partial contraction joints

It is assumed in BS 5337 that if partial contraction joints are provided at not more than 7.5 m centres, the long-term movements in the mature concrete due to seasonal variations can be accommodated at the joints. For a 20°C temperature fall, the contraction in a mature reinforced concrete member amounts to approximately $7.5 \times 20 \times 12 \times 10^{-6} = 1.8$ mm

If the distribution steel at the joint is stressed to its 0.2% proof stress, then the bond length on each side of the joint which would need to become fully debonded and remain 'live' in order to accommodate the seasonal contraction and expansion is given by

$\frac{1}{2} \times 1.8 \times 10^{-3}/(2 \times 10^{-3}) = 450$ mm

Some elastic extension undoubtedly occurs within the central length of the reinforced concrete member which could offset the fact that some bond transfer must occur within the 'live' zone. A high stress in round bars, adjacent to the joint, can cause local debonding and the high stress can then progress as far as is necessary along the bar to accommodate the movement. In the case of ribbed bars, however, complete debonding is strongly resisted since fracturing of the concrete locally around the ribs becomes necessary. Very high steel stresses are therefore likely to be induced at and near partial contraction joints where ribbed bars are used.

4.10 Design for no cracking between movement joints

The probable maximum spacing of cracks, s_{max}, can be calculated and the probable minimum spacing of cracks, s_{min}, is simply $\frac{1}{2}s_{max}$. Hence, if joints are provided at a spacing which does not exceed s_{max}, as shown in Fig. 4-5, the possibility of a crack occurring midway between two joints is eliminated. With such a close spacing of joints, the cracking need not be entirely controlled by the reinforcement and hence the steel ratio in this case can be reduced to $\frac{2}{3}\rho_{crit}$. By taking advantage of the reduced steel ratio of $\frac{2}{3}\rho_{crit}$, and using round bars of as large a diameter as possible (consistent with not exceeding about 300 mm bar spacing) the crack spacing, and hence the joint spacing, can be made as large as possible.

Table 4-2 shows the maximum spacing of 100% partial contraction joints (i.e. 100% reinforcement passing through the joint for typical cases, including those for no cracking between joints. Notice how very close these spacings are, especially for a narrow wall with a double layer of reinforcement (i.e. a 250 mm wall). Notice also that the widest spacings can be given by a hot-rolled plain round bar of 425 MPa (62 000 psi) characteristic strength to BS 4449. These bars are called 'special' in Table 4-2 since they are not normally produced.

Site evidence indicates that the effect of a free joint is generally given, for a wide range of wall thicknesses and vertical steel ratios, and horizontal steel ratios not less than $\frac{2}{3}\rho_{crit}$, by assuming that the end 2.4 m adjacent to the complete contraction joint (or a free standing end) is free to contract inwards. Thus, for a succession of complete contraction joints the joint spacing can be up to 4.8 m and no cracks should occur.

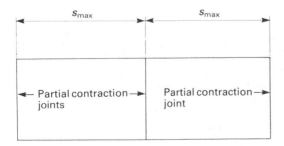

Figure 4-5 Partial-joint spacings which eliminate random intermediate cracks entirely

Table 4-2 Maximum spacing of partial contraction joints with 100% continuity of $\frac{2}{3}\rho_{crit}$ distribution bars at about 300 mm centres and grade 30 concrete for zero, 0.1 and 0.2 mm crack widths,

Reinforcement				Partial joint spacing		
Wall thickness	Layers	Type	Size	No cracks	0.1 mm cracks	0.2 mm cracks
200 mm	Single	Round MS	16 mm	2.30 m	2.86 m	3.41 m
		Special	12 mm	2.94 m	3.50 m	4.05 m
		Type 1	12 mm	2.35 m	2.91 m	3.46 m
250 mm	Double	Round MS	12 mm	1.73 m	2.29 m	2.84 m
		Special	8/10 mm	2.23 m	2.79 m	3.34 m
		Type 1	8/10 mm	1.78 m	2.34 m	2.89 m
500 mm	Double	Round MS	16/20 mm	2.62 m	3.04 m*	3.45 m*
		Special	12/16 mm	3.50 m	3.92 m*	4.33 m*
		Type 1	12/16 mm	2.80 m	3.22 m*	3.63 m*
Over 500 mm	Double	Round MS	20 mm	1.92 m	2.34 m*	2.75 m*
		Special	16 mm	2.61 m	3.03 m*	3.44 m*
		Type 1	16 mm	2.09 m	2.51 m*	2.92 m*

180 microstrain contraction assumed except where shown * (240 microstrain). Yield strengths taken as 250 MPa (36 000 psi) for MS or 425 MPa (62000 psi) for Special and Type 1, respectively

4.11 Cracking between close movement joints

It has been shown above that a length of wall of s_{min} adjacent to a partial contraction joint, and of 2.4 m (approximately) adjacent to a complete contraction joint, can be assumed to contract away from the joints. Assuming that all the movement for a length, as given for no cracking, is still accommodated at the adjacent movement joints, any increase in joint spacing introduces a fully restrained central length which gives rise to an intermediate crack between the joints. The width of the crack (assuming that it is inadequately controlled by the reinforcement) is directly proportional to the restrained central length, hence, for a crack width of up to 0.1 mm and 180 microstrain contraction, the joint spacings for no cracks in Table 4-2 can be increased by 0.1/0.180 = 0.556 m. The corresponding increase for 240 microstrain is 0.417.

Similar increases in length can also be applied to the spacing of 4.8 m between complete movement joints, as shown in Table 4-3. Notice that the crack widths

Table 4-3 Suggested maximum spacings of successive complete contraction joints for given crack widths and 180 microstrain contraction

Wall thickness	No cracks	0.1 mm crack	0.2 mm crack
<400 mm	4.8 m	5.3 m	5.9 m
⩾500 m	4.8 m	5.2 m	5.6 m

increase very rapidly with increase in joint spacing, so that the positioning and spacing of movement joints is very critical.

Example 4-2
Determine the maximum distance between 100% partial contraction joints for the 400 mm thick retaining walls of a water retaining structure for which $\rho \not< \frac{2}{3}\rho_{crit}$ and the bar spacing must not exceed 300 mm if

    (a) no cracks are permitted;
    (b) maximum crack width is 0.1 mm;
    (c) as for (b) but for complete contraction joints.
    (d) Also determine the maximum lift height for an infill bay for which the vertical steel near the ends need not be increased if maximum crack width is still to be 0.1 mm. Assume high yield bars, $f_{yt} = 425$ MPa and grade 25 concrete, for which $_*f_{ct} = 1.15$ MPa and $_*f_{ct}/f_b = 0.8$, and an effective total contraction of 180 microstrain.

Solution
(a) $\frac{2}{3}\rho_{crit} = \frac{2}{3} \times 1.15/425 = 0.180\%$
    Hence, minimum bar area is

$$0.2 \times 0.3 \times 0.0018 = 108 \text{ mm}^2$$

    Use 12 mm at 300 EF, giving 0.188%. Therefore,

$$s_{max} = 0.8 \times 12/(2 \times 1.88) = 2.553 \text{ m}$$

    and required partial joint spacing also equals 2.55 m   [Answer]
(b) Contraction of each end length of $\frac{1}{2}s_{max}$ is assumed to be taken by the joints, hence additional restrained central length for an 0.1 mm crack is given by

$$(0.1/180) \times 10^{-3+6} = 0.556 \text{ m}$$

    Therefore, required joint spacing is $2.55 + 0.56 = 3.11$ m   [Answer]
(c) For complete contraction joints, 2.4 m at each end can move inwards. Hence, joint spacing for 0.1 mm crack is

$$2 \times 2.4 + 0.556 = 5.36 \text{ m} \qquad [\text{Answer}]$$

(d) Ignoring gravity effects, only a height of 2.4 m of freshly placed concrete will slide down the joint. (See vertical tension later.) Hence, if the lift height exceeds $2.4 + 0.556 = 2.96$ m, cracks at a horizontal construction joint could exceed 0.1 mm.   [Answer]

4.12   Design for freedom of movement (BS 5337 Option 3)

The above considerations form the basis of the Option 3 approach for movement joints which are sufficiently closely spaced to control the early thermal cracking in the concrete. In such circumstances, only a very nominal amount of distribution reinforcement is necessary to hold the concrete together in thin walls and Eqn 3-2 need no longer apply. Where any reinforcement passes through a partial contraction joint, the freedom of movement is restricted by that reinforcement and the length of wall which can contract away from the joint is again assumed to be given by s_{min} from Eqn 3-4. Where Eqn 3-2 is not satisfied, the reinforcement may yield at the joint but if the base restraint is mobilized over the length s_{min} until the total force transferred to the concrete over the length of s_{min} is sufficient to crack the concrete, a crack can occur at not less than s_{min} from the joint. However, if s_{min} exceeds 2.4 m, then clearly an end length exceeding 2.4 m is inappropriate since the base restraint is then more critical. Hence, the end length which can be assumed to contract away from the joint, it is suggested, should be the lesser of

2.4 m and s_{min}

Three possibilities for Option 3 are included in Table 4-1: (a) complete movement joints at each end of the bay; (b) a complete movement joint at one end and a 100% partial movement joint at the other; (c) 100% partial movement joints at each end.

4.13   Cracking due to vertical restraints

The forms of restraint considered so far have been essentially end restraint due to direct tension in thin and thick walls and the reduction or elimination of horizontal end restraint by means of vertical movement joints. However, there are also vertical effects associated with both infill bays and free-standing bays, as illustrated in Fig. 4-6, which should be considered and designed for accordingly.

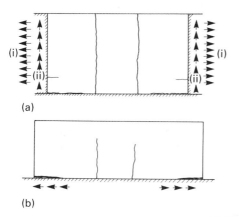

(a)

(b)

Figure 4-6   (a) Tension cracking in a wall due to (i) end restraint and (ii) vertical restraint
(b) Cracking in a free-standing bay due to base restraint and warping

Figure 4-6a shows that for the vertical reinforcement the contraction strain at the end of an infill bay will be governed by the restraint at the interface in much the same way (if gravity is ignored!) as the horizontal base restraint adjacent to either a complete contraction joint or a free-standing end. The vertical reinforcement near the ends for an infill bay with a lift height exceeding 2.4 m should be sufficient to provide acceptable crack widths assuming full vertical restraint at all depths exceeding (as a safe estimate) 2.4 m. Thus, in cases where the construction sequence is entirely unknown, vertical reinforcement should equal ρ_{crit} and provide for half the contraction strain (see below) to a depth of 2.4 m and for the full contraction strain at greater depths. Any increase in the vertical height value of 2.4 m would need further site evidence.

Figure 4-6b shows the form of cracking which can occur in a bay with freedom of movement at both ends. The shear forces exerted by the hardened base induce extensive creep movements in the immature concrete which give rise to two effects. First, the movements gravitate towards localized planes of weakness; hence a relatively few wide cracks occur rather than a large number of fine cracks. (Fine cracks only would form if the immature concrete were attached to the mature base in such a way that little or no bond creep were possible.) Secondly, the base restraint reduces from a maximum at the base to a negligible value at some height, l_w, above the base since the upper region of the wall can be virtually unrestrained.

For a simplified analysis, assume that the restraint reduces linearly with height l_w such that contractions in the length of the wall increase with height. Further, if a vertical section, distance l_w from one end of the wall, remains vertical and cleavage along the base only is assumed, then the wall tends to warp as shown in Fig. 4-7. If Δl_w is the contraction in the wall at height l_w, the angle θ which the end of the wall makes with the vertical is given by $\Delta l_w / l_w$. Assuming small angles, the vertical movement at the end of bay is $l_w \theta / 2 = \frac{1}{2} \Delta l_w$, hence the vertical contraction strain is only *half* the horizontal contraction strain.

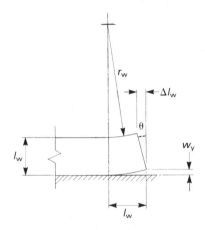

Figure 4-7 Warping movement at end of a bay with base restraint only

The horizontal component of the contraction is suitably controlled if horizontal reinforcement is provided as for walls with end restraint. The vertical component is controlled if the vertical reinforcement also satisfies the requirements for ρ_{crit}. The height l_w in Fig. 4-7 is given by the lesser of s_{max} for the vertical reinforcement and, provided that the steel is suitably anchored at the top of the wall (e.g. inner and outer face bars are looped together at the top), by the actual wall height, h_w. The warping cracks can occur either as horizontal or as inclined cracks.

Example 4-3

A counterfort retaining wall, 350 mm thick and 8.5 m high, contains the equivalent of 25 mm diameter horizontal bars at 225 mm spacing (one face). The vertical reinforcement consists of 16 mm bars at 300 mm in each face. What are the probable maximum widths of horizontal and vertical cracks if the thermal contraction is approximately 300 microstrain? Assume mild steel reinforcement (f_{yt} = 250 MPa) and grade 30 concrete with an early tensile strength of 1.3 MPa. Suggest suitable modifications if crack widths should not exceed 0.3 mm and the wall is cast in long free-standing bay lengths.

Solution The actual steel ratios are 0.623% (horizontal) and 0.382% (vertical).

$$\rho_{crit} = *f_{ct}/f_{yt} = 1.3/250 = 0.52\%$$

(High because mild steel and high-quality concrete are proposed). Hence, the vertical steel is inadequate and uncontrolled (i.e. wide) horizontal or inclined cracks could occur near the free ends of the bays. However, if high-bond, high-yield steel is used instead of mild steel for the vertical bars, ρ_{crit} is then less than 0.32% and the vertical reinforcement can now control the cracking.

For horizontal mild steel bars,

$$s_{max} = 25 \times 100/(2 \times 0.623) = 2.0 \text{ m}$$

and for vertical high yield high bond bars,

$$s_{max} = \tfrac{2}{3} \times 16 \times 100/(2 \times 0.382) = 1.4 \text{ m}$$

Hence, w_{max} = 0.60 mm for vertical cracks (i.e. horizontal bars), and w_{max} = 0.21 mm for horizontal cracks (i.e. vertical bars).

The vertical cracks are excessive. If high-bond horizontal bars of smaller diameter are considered, then w_{max} for 16 mm high-bond bars with the same horizontal steel ratio of 0.623% becomes less than 0.26 mm for the vertical cracks and would be satisfactory.   [Answer]

4.14 Small tanks

If Option 3 is applied to a small tank, it is necessary to consider how each corner can be pulled towards the centre when determining the largest size of tank which can be constructed without introducing any movement joints.

Let F be the force which can be exerted by the immature concrete wall section without cracking and let a be the length over which a straight length can be drawn (i.e. a = 2.4 m). If two lengths of wall meet at right angles at a corner, the component of each force exerted along the bisecting diagonal at 45° is simply $\Delta F/\sqrt{2}$, i.e. the resultant for both components is $2\Delta F/\sqrt{2} = \sqrt{2}\Delta F$. This force can draw in and distort the corner locally to enable the wall lengths, a, to be drawn towards the mid-lengths of each side.

Thus, by reference to Fig. 4-8, it can be seen that if the sides of a square tank do not exceed $2a$ in length, the corners can effectively be drawn towards the centre of the tank and any horizontal reinforcement can be designed to Option 3 wherever the structural loading requirements are not more critical.

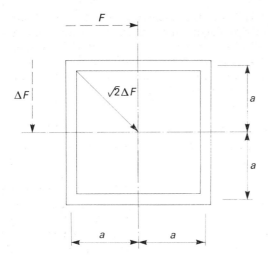

Figure 4-8 Contraction movement for small tank

4.15 Floor slabs

Reference has already been made to the use of a sliding layer, such as polythene sheeting, between a ground slab and the sub-base, and of a Neoprene or rubber strip, or similar, between the roof and the top of the wall. The jointing material must not be squeezed out excessively under the loading and it should provide for the necessary sliding while maintaining a reasonably constant coefficient of friction.

All three options to BS 5337 can be applied to floor slabs as well as to walls, as indicated earlier. However, since a sliding joint between wall and either wall kicker or ground slab is preferably avoided, any movement joints in wall and floor should coincide and the wall can then be made monolithic with the ground slab. Hence, whichever design option is used for the walls is preferably also applied to the ground slab and vice versa. Thus, if complete movement joints are required for the walls (i.e. expansion joints for Option 1 or any complete movement joints for Option 2), they should coincide with similar joints in the ground slab and a sliding layer under the ground slab is then essential.

If expansion joints are unnecessary and either Option 1 or Option 2 with partial contraction joints are used for the walls and ground slab, the sliding layer can be omitted and the ground slab can be cast directly upon the oversite concrete. Where the direct tension to be resisted by the mature ground slab is small, partial contraction joints with only 50% reinforcement continuous through the joints can be used—otherwise 100% partial contraction joints will be necessary in the ground slab. If desired, 50% partial contraction joints may still be used in the walls but their spacing will be governed by the joint spacing for the ground slab. The bottom unexposed surface of floor slabs in contact with, and restrained by, the ground is subjected usually to only mildly changing temperature conditions. Only the top exposed surface is critical and since: (a) the reinforcement can be placed near to the surface in both directions (a wall often has relatively heavy reinforcement in one direction) and (b) the moderating effect of the strain gradient in the lower layers can be relied upon, the thickness of this surface zone can be less than that for a wall slab. Similarly, for the mature slab, daily fluctuations in temperature have little effect at depths greater than about 250 mm from the top surface so that, here again, the largest fluctuations and temperature gradients are restricted to a thin surface zone. Hence the recommendations of BS 5337 for ground slabs.

Table 4-4 Grade 425 reinforcement areas for grade 30˙ concrete ground slabs on a sliding layer with complete movement joints up to 15 m apart

Slab thickness mm	Surface zones		Reinforcement mm²/m	
	Top	Bottom	Top	Bottom
200 and under	100	nil	306	—
300	150	nil	459	—
400	200	50	612	153
500 and over	250	100	765	306

The recommendations of BS 5337 for the assumed surface-zone thicknesses for slab thicknesses, h, between 200 and 500 mm can be taken as: $\frac{1}{2}h$ (top) and $\frac{1}{2}h - 150$ (bottom). Table 4-4 shows the appropriate surface zones, together with the minimum reinforcement areas, for some typical thicknesses. Notice that fabric reinforcement can be placed centrally in a surface zone which is only 100 mm thick and still provide a minimum cover of 40 mm. Table 4-5 shows some suitable fabrics satisfying BS 5337 for some typical slab thicknesses.

Table 4-5 Grade 485 reinforcement areas and suggested round-wire fabrics for grades 25 and 30 concrete ground slabs on a sliding layer with complete movement joints up to 15 m apart and a design contraction strain of 180 microstrain

Slab thickness mm	ρ_{crit} area mm²/m		Suggested wire fabrics			
			Class B		Class A	
	Grade 25	Grade 30	Grade 25	Grade 30	Grade 25	Grade 30
200 and under	237	268	A252	X283[a]	X283[a]	X283[a]
250	296	335	X335[c]	X335[c]	X385[b]	X385[b]
300	356	402	X393	X402[d]	X385[b]	X503[e]

[a] 6 mm at 100 mm spacing both ways
[b] 7 mm at 100 mm spacing both ways
[c] 8 mm at 150 mm spacing both ways
[d] 8 mm at 125 mm spacing both ways
[e] 8 mm at 100 mm spacing both ways

Road slab recommendations contrast with Table 4-5 as follows.

The longitudinal reinforcement recommended for long lengths of road slab exceeds that indicated in Table 4-5. Where long lengths between movement joints occur, then the increased steel ratios typical for road construction practice [14] can be used to overcome the frictional drag exerted on the mature slab. However, the reinforcement on this basis can become very substantial for very long lengths and provision of controlled cracking in the mature concrete (e.g. $\rho \simeq 2\rho_{crit}$) can become a more economic alternative as indicated earlier. Furthermore, if movement joints are omitted entirely and sliding is prevented throughout, any increase for the longitudinal reinforcement, either to overcome frictional drag or to control cracking in the mature slab, is unnecessary but the contraction strain assumed for both directions must of course include a term for the seasonal contraction in Eqn 3-6, as for all Option 1 continuous construction.

The virtually trivial amounts of transverse reinforcement commonly used in road construction practice fall far short of the amounts required to control early thermal cracking in reservoir floors, advocated in Table 4-4. Furthermore, unreinforced road slabs of 250 and 300 mm thickness would appear to indicate that no reinforcement (other than a substantial amount of dowel bars) is required for ground slabs on a sliding layer. Too close a comparison with road construction practice, however, can be misleading for two reasons. First, air-entrained concrete of fairly low workability is invariably used for road construction and is therefore more cohesive than other fresh concretes. Second, the site control is generally such that any plastic shrinkage for the low workability concrete is negligible due to the rapidity with which curing membranes and protective covers can be applied. In a reservoir floor, air-entrained concrete need not be used and even if it is, it is not uncommon for some plastic shrinkage to occur for other than the very best quality control. If the air-entrained concrete is pumped, the workability needs to be much higher than that necessary for normal road construction, and under these circumstances it is extremely difficult to prevent plastic shrinkage occurring for summer placing in exposed positions. One example of a 150 mm thick grade 30 non-air-entrained concrete reservoir ground slab, with minimum reinforcement to road construction standards and a sliding layer with partial contraction joints 4.3 m apart, was found to contain cracks up to 0.7 mm wide—due, no doubt, to the superposition of plastic shrinkage on the normal early thermal contraction. With minimum reinforcement to BS 5337, the maximum crack widths would have been little more than 0.2 mm despite the deficiencies in quality control. This was confirmed by the lack of serious cracks transverse to the longitudinal (main reinforcement) direction for all slabs, whether laid in either long or short 4.4 m bays, since the longitudinal reinforcement was at least equal to the minimum suggested in BS 5337.

4.16 Practical precautions

Experience has indicated that large variations in the cracking behaviour of essentially similar members can readily occur on site. Members designed to earlier codes have commonly contained appreciably less steel than the critical, with movement joints at spacings which could often offer little relief for the early thermal cracking and variations in local restraints became paramount. If concrete was placed during cold weather, or the movement joint spacings were sufficiently close, there would be little tendency to crack. However, as the warmer weather arrived, much consternation could arise on site as the cracking became more and more severe and traditional site reactions—such as increasing the cement content and the concrete strength—merely exacerbated the situation. Notice also that while elaborate procedures for curing the concrete for as long as possible can only delay the eventual drying out (or partial drying out) of the concrete, relatively simple procedures can be adopted to obtain an actual reduction in the probable thermal strain. The emphasis in curing therefore should be on reducing the temperature of the freshly placed concrete, rather than on wet-curing the concrete for extended periods after it has reached its peak temperature due to the heat of hydration.

When placing the concrete in summer, it is desirable to reduce the temperature of the concrete as much as possible, and preferably down to 20°C if ordinary Portland cement is being used. Rapid-hardening Portland cement should be avoided. Recommended precautions therefore include the following.

(a) Shield the concrete materials and the concrete from direct sunlight wherever possible by erecting canvas canopies, with open sides, over the aggregate piles etc. The point of concrete placing should also be shielded if possible. If necessary, also consider adding ice to the mixing water or using evaporative cooling of the aggregates. Evaporative cooling of the water by recirculation in shower-tray cascades can also be very effective.

(b) Fully effective water-curing of the concrete should be undertaken for about 30 h (i.e. till the end of the day after placing). This minimizes both shrinkage and compressive creep strain. The heat of hydration must be dissipated quickly if the maximum temperature rise (which can occur after 14 h or even earlier) is to be reduced. Thus, steel shutters are preferable to timber for all sections with end restraint due to their superior thermal conduction. Once the shutters have been removed, it is important to avoid thermal shock at the surface of the concrete, i.e. intermittent torrents of ice-cold water (rather than continuous spraying of cool water) must be avoided. Thick sections without end restraint are best insulated to minimize temperature gradients.

(c) Adopt 'successive bay construction' rather than 'alternate bay construction' for equal length bays, since the former enables one end to contract before the next bay is cast against it. The infill bays for alternate bay construction receive no relief in restraint at all and hence, if used, should be very short.

5 Materials, workmanship and testing

The specifications for materials, workmanship and testing for water-retaining structures generally require compliance not only with good practice for building structures but also with certain other requirements, especially in relation to durability and impermeability, which are more restrictive. The special requirements are discussed and particular reference is made to jointing materials, testing for watertightness and pneumatically applied mortar.

5.1 Concrete materials

Ordinary, sulphate-resisting and low-heat Portland cements, Portland blast-furnace cement and supersulphated cement can all be used in water-retaining structures. Rapid-hardening Portland cement is not necessarily precluded for use in cold weather but the occasions when it can offer a special advantage are likely to be rare. In general, the lower the rate of heat evolution the better, since this directly reduces the early thermal cracking.

Where sulphate-resisting cement is required, the oversite concrete as well as the structural concrete should be not less than grade 25. Less dense 'blinding' concrete would deteriorate much more rapidly and soon permit exposure of the structural concrete itself to sulphate attack. Otherwise, the oversite concrete in non-sulphate bearing soils and non-aggressive ground water can be grade 15.

Most good-quality normal aggregate of 40 or 20 mm maximum size, reasonable shape and grading can be used; lightweight and heavy aggregates are excluded. Thus, natural and crushed rock aggregates and crushed blast-furnace slag can be used. Aggregate absorption should generally be less than 3%. Marine aggregates are excluded for prestressed concrete and total sodium chloride concentration for aggregates should not exceed 0.32% by weight of the cement in the concrete mix. Hollow shells or shells of unsuitable shape in quantities which could adversely affect the quality and especially the impermeability of the concrete should be excluded. The shell content, expressed as calcium carbonate, should not in any case exceed 2, 5, 10, 15, or 30% by weight of dry aggregate for aggregates of 40, 20, 14, 10 mm and fine nominal sizes, respectively.

Calcium chloride or admixtures containing chlorides are excluded because of the risk of increasing the corrosion rate of the embedded reinforcement. However, admixtures which contain small amounts can be considered acceptable if the total chloride content remains within the sodium chloride limit for marine aggregate.

5.2 Concrete properties and composition

The concrete used for reinforced concrete will normally be either grade 25 or 30. The lowest grades for prestressed concrete are 30 for post-tensioning and 40 for pretensioning, which are normal for concrete structures generally.

The minimum cement contents for Class A and Class B exposure correspond to those for 'severe' and 'moderate' exposures for concrete structures generally and are given for convenience in Table 5-1.

Table 5-1   Minimum cement content in kg/m^3 to ensure durability

Class of exposure	Reinforced concrete Nominal maximum size of aggregate		Prestressed concrete Normal maximum size of aggregate	
	40 mm	20 mm	40 mm	20 mm
A	320	360	320	360
B	260	290	300	300

Impermeability can be drastically affected by an inadequate cement content. Durability of the concrete, is not noticeably affected by the cement content. Both impermeability and durability are, however, affected by the water/cement ratio, hence specified minimum cement contents provide a useful indirect control for durability as well as a direct control for impermeability.

Air entrainment dramatically improves durability and, since it also effectively increases the very fine fraction of the mix (entrained air bubbles are only of the order of 0.1 mm in diameter), there would seem to be a very strong case for allowing reduced cement contents when effective air entrainment is used. For example, concretes of 40 and 20 mm maximum aggregate size with average air contents of 4 and 5%, respectively, and minimum cement contents just satisfying those for Class B exposure in Table 5-1 could be deemed to be more than satisfactory for Class A exposure. This could help to overcome the problem which occurs for the concrete at the top of a Class B wall which does not satisfy Class A. It has already been pointed out that, if no adjustment is made in the design of the reinforcement for reduced restraint at the top of a wall, reduced crack widths can be expected in practice at the top of the wall where Class A exposure is appropriate. Furthermore, because of the increased cohesion provided by air entrainment, it is found in practice that early thermal crack widths tend to be finer than those for normal concretes. Thus, for both durability and impermeability a wall with air entrained concrete which satisfies the requirements for Class B exposure near the base of the wall can reasonably be accepted for Class A exposure conditions at the top of the wall.

High cement contents exceeding 400 and 550 kg/m^3 for reinforced and prestressed concrete, respectively, should be avoided because of increased early thermal cracking and, to a lesser extent, long-term shrinkage.

5.3 Concrete placing, finishing and curing

General considerations for normal concretes as discussed in Chapters 4 and 27, and elsewhere, apply also to concrete for water-retaining structures. The workability of the fresh concrete must be related to the equipment and methods of handling and compaction to be used. It is essential that the concrete is placed without segregation, is fully compacted, completely fills out all formwork and surrounds all waterstops, reinforcement, tendons and ducts, etc. A target concrete workability of a Vebe time of 4 s (or a Modified Vebe time of 10 s) is generally appropriate for water-retaining structures where immersion vibrators are used in moderately congested reinforced sections and concrete may also be pumped [8]. The corresponding slump will, of course, vary greatly with different materials, mix proportions and additives (e.g. air entrainment) but the appropriate target slump for normal site control can readily be determined from a batch after the mix with a Vebe time of 4 s has been designed.

The water-retaining faces will generally require a Type D finish, i.e. a high quality concrete with Type B finish which is then improved by carefully removing all fins and other projections, thoroughly washing down and then filling the most noticeable surface blemishes, due to entrapped air and water migration, with a cement and fine aggregate paste matching the colour of the concrete. The release agent should not permanently stain or discolour the concrete surface. A Type B finish requires the use of properly designed forms of closely-jointed wrought boards, steel or other suitable material imprinting the joint pattern and slight grain on the concrete surface. Small blemishes may be expected, but the surface should be free of large blemishes, voids or honeycombing.

The ties used to secure and align the formwork should generally not pass completely through the structure to a water-retaining face unless effective precautions can be taken to ensure watertightness after their removal. The cover to the ends of any embedded ties should be not less than at least 40 mm, as required for any other embedded steel, using an effective sealant in the gap which provides adequate protection against corrosion.

A not unusual feature of water-retaining structures is the occurrence of steeply sloping surfaces, exceeding 15° to the horizontal. Such slopes require extra care to achieve full compaction of the concrete. Slopes in excess of 35° are best achieved with a top shutter. For intermediate slopes, techniques for providing extra support for the concrete range from the inclusion near the top surface of fabric, or even expanded metal, reinforcement; manual compaction of a suitably designed mix by beating the surface with bass brooms; to a climbing top shutter used in conjunction with immersion (*not* clamp-on) vibrators. Concreting should, of course, start at the bottom of the slope and proceed upwards in all cases.

Reference has already been made in Section 1 to the importance of controlling plastic shrinkage and settlement cracking. Rapid evaporation of moisture from the surface of the plastic concrete must be prevented so that the rate of migration of moisture to the surface at least equals the rate of evaporation from the surface. The former depends on a complex combination of factors which offer little scope for direct control on site. The latter, however, depends not only on the wind velocity, air temperature and relative humidity but also on two factors which can be directly controlled: concrete temperature and degree of exposure of the fresh surface to wind and sun. In hot climates, it is possible for plastic shrinkage cracking to occur on vertical surfaces protected by the shuttering and hence even thick protecting membranes must themselves be protected from the direct rays of the sun. However, in more temperate climates, immediate protection with plastic sheeting is usually adequate, with generous overlaps and the sheeting held well down around the perimeter as soon as relatively small horizontal areas have been surface finished and completed.

Plastic shrinkage cracking and crazing occur as fairly fine cracks, often fairly straight and parallel in one direction at 50 to 100 mm spacing, with few random cracks in other directions, and varying in length from 50 to 500 mm, or even more if cracks are extended by other causes. Cracks are usually shallow but in severe cases and when combined with early thermal effects, cracks can extend through the depth of a slab. The occurrence of hot sunny weather or strong drying winds can often suddenly produce this type of cracking on sites which have previously been trouble free. Severe cases should be treated as soon as possible with a Portland cement grout well brushed into the cracks and the treated surface protected with plastic sheets for 2–4 days. Air entrainment tends to reduce plastic shrinkage cracking even more than early thermal cracking.

Plastic settlement cracking has also been referred to earlier and is due, primarily, to the vertical migration of water in the plastic concrete. The downward settlement of the plastic concrete can then be resisted either by the surface of the formwork or by horizontal reinforcing bars. The former tends to provide a crack which is widest at the surface and rarely penetrates more than 20–30 mm. It occurs typically at a change in section of a column, e.g. at the base of either a mushroom head or an intersecting beam. The latter produces more serious cracks which are usually widest at the reinforcement and less obvious at the surface although they must always penetrate to the reinforcement which resists the movement. Links in the upper third of a column are a common cause of this type of cracking. Relatively rigid reinforcing bars in the top face of slabs or beams can cause longitudinal cracks along the lines of the bars which in this case are widest at the top horizontal surface but again always penetrate to the steel. Re-vibration of the concrete after much water migration has taken place, but while it is still workable, is an effective means of eliminating plastic settlement. Redesign of the mix by reducing the water content or increasing the aggregate fines can also help. For floor slabs, rotary power floated secondary compaction can eliminate both plastic shrinkage and settlement cracking and give a smooth surface finish. Where very severe plastic settlement cracks have occurred, then repair, either by epoxide resin penetration or by cutting out a V shaped notch and filling with a high-quality mortar, may be necessary. In less severe cases, rubbing a neat cement grout into the cracks and carefully cleaning off the adjacent surfaces where colour of the surface is important may be adequate. Where crack widths do not exceed the values appropriate to the class of exposure, it may be assumed that autogenous healing will eventually occur for all cracks subsequently subjected to moist conditions.

All concrete should be cured for not less than 4 days after casting because of the special requirements for watertightness in liquid-retaining structures. Only curing membranes with a reasonably high water-retention efficiency should be used to prevent early drying out of the concrete. Early protection to walls may be provided by the shuttering but, on removal, the concrete should be covered either with polythene which is well held in place against the concrete surface around the edges, or with hessian which is constantly maintained in a wet condition for not less than 2 days after casting and is left in place and saturated at least twice daily for at least a further 2 days. The importance of completely covering horizontal surfaces of fresh concrete, immediately each bay is finished, with polythene or a similar material has already been emphasized. Protection of large horizontal surfaces from the direct rays of the sun should be provided by efficient tentage, for at least the day of casting and preferably for the following day as well, to minimize early thermal cracking, as indicated earlier. Vertical surfaces can sometimes be protected from the sun's rays by resting the shutters against the top of the concrete after they have been removed (or merely eased away) and the polythene has been placed in position; otherwise some other form of screen is necessary. BS5337 suggests that, in some cases, extended protection from wind and sun may be necessary beyond the minimum of 4 days.

5.4 Shotcrete

Shotcrete, or gunite, i.e. pneumatically applied mortar, can be used as a structural concrete around normal reinforcement, as an outer layer or external cover to prestressing steel in cylindrical tanks, as a rendering, or for repairs to pockets or areas of defective concrete which have been cut out. Gunite is a mixture of cement and sand which is shot through a nozzle under pressure in the form of a spray onto the surface to be covered and the force of the impact compacts the material. The mortar can be deposited in a continuous flow which builds up the thickness into a compact and dense mass.

Guniting is a special operation which requires specialist equipment. When used instead of normal reinforced concrete construction, the excavation should be kept free of water and carried out as carefully as possible so that no formwork need be used, except for ring beams at the top and any intermediate beams which may be necessary to support the wall panels. Heavy hessian fixed to timber supports, at about 1 m centres, can provide a satisfactory dividing layer between the structural shell and the sides of the excavation. The hessian is given a preliminary spray with gunite to stiffen it and provide a suitable surface to which the main gunite can be applied after the reinforcement has been placed in position. The gunite for the floor can be applied on a 150–200 mm layer of large shingle or hardcore (usually reject aggregate) to the required thickness, which may vary typically from a minimum of 100 mm in the central area of the floor to a maximum of 250 mm adjacent to the walls. The angle between wall and floor is usually formed to a large radius since this is structurally preferable. A single-layer fabric reinforcement is usually used in conjunction with stouter vertical bars in the walls to hold the fabric in position. The cover to this fabric reinforcement can often be reduced to as little as 25 mm. This contrasts with the situation where gunite is used as an outer layer or external cover to prestressing tendons or larger-diameter bars, for which BS 5337 requires the usual minimum of 40 mm cover, i.e. where gunite is used in conjunction with reinforced or prestressed concrete construction it should be regarded as providing no greater protection than the concrete itself. It is usual to gun wall panels to their full height and thickness, together with an adjoining area of floor, in one operation. The length of the panel is largely governed by the capacity of the equipment and the output of the operator.

Construction joints can be formed either by feather edging throughout the thickness, or by guniting up to a stop-end-batten placed above the reinforcement to produce a plain butt joint down to the reinforcement and a tapered joint below.

Gunite construction is usually lighter than its corresponding normal reinforced concrete construction and hence problems with flotation and the unavoidable use of pressure relief valves become more likely.

The use of fabric reinforcement in conjunction with a reduced cover of as little as 25 mm in gunite construction means that the 'surface zone', when considering early thermal contraction, is little more than 50 mm and hence good early thermal crack control is provided. Furthermore, the more rapid heat dissipation from the thinner sections of the gunite generally more than offsets the higher cement content. Movement joints should therefore never be necessary and Option 1, i.e. continuous construction, should always be the most appropriate for gunite.

The usual method is to force premixed mortar from the digestor (i.e. upper chamber of the cement gun) along the hose by compressed air to the nozzle where, in the dry-mix process, water is added before discharge and controlled by the operator. The sand grading should generally conform with Zones 1 and 2 of Table 2 of BS 882 (see Chapter 4). The volumes of sand in the mix to 50 kg cement should lie between the limits of 0.07 and 0.11 m^3. The external cover to prestressing cables is usually built up in two layers, with the final coat using an amended mix as follows—50 kg cement: 4.5 kg hydrated lime: 0.1 m³ sand finer than 2.36 mm nominal aperture size sieve.

The velocity of the mortar leaving the nozzle should be uniform and the water content adjusted so as to achieve minimum rebound of the sand, which is usually about 10–15% for floors and 15–30% for walls.

Curing is even more important for gunite than for normal concrete. BS 5337 requires that it is protected from strong sunshine and drying wind for at least 7 days. The gunite should be thoroughly wetted as soon as it has hardened just sufficiently to avoid damage and well covered with polythene held down at the edges to keep the gunite continuously wet for 7 days or more. Further information on gunite is given in [15].

5.5 Reinforcement

The requirements for reinforcement and prestressing tendons are essentially the same as for other concrete structures. There is a greater tendency to use smaller diameter bars and fabric because of the finer crack widths which are required.

5.6 Jointing materials

Jointing materials may be classified as

(a) Joint fillers
(b) Waterstops
(c) Joint sealants
(d) Joint stress inducers

All jointing materials should accommodate repeated movements without becoming too soft at any time during the summer, or too brittle in winter, whether the tank or reservoir is full or empty. The material must not be dislodged or permanently distorted or extruded and must be capable of resisting the fluid pressure at all times. Thus, it must be insoluble in the retained liquid and remain effectively unaffected by exposure to light or by evaporation of any solvent or plasticizer incorporated during manufacture. Depending on their application, jointing materials may need to be non-toxic and taintless (e.g. for potable water) or resistant to chemical or biological attack (e.g. sewage disposal works).

Convenience in construction should also be considered: ease of handling, placing and fixing of strip materials, formation of the intersection of joints between walls and floor, and thorough compaction of the concrete all around the joint.

(a) Joint fillers
Joint fillers are compressible sheets or strips used in expansion joints to provide the initial gap and act as a base for the sealant. The joint filler is cut to shape (or length) and attached to the first placed concrete face, ready to receive further concrete which will form the other face of the expansion joint. The joint filler, with the surface sealing groove, provides complete separation between the joint faces and should be thick enough to ensure that the required initial gap is achieved after the concrete for the second stage has been placed.

Non-absorbent, non-rotting cork-based joint fillers are preferred to fibrous joint fillers, which can absorb moisture which prevents good continuous adhesion with the joint sealant, when the latter is subsequently applied. Since the proportional expansion or contraction of the filler is inversely proportional to the initial width, the performance of a given material improves with increasing initial joint width. Ideally, the filler should either adhere permanently to the faces of the joint, or have sufficient elastic precompression to follow any retraction of the concrete faces. This would prevent the joint sealer which is subsequently applied in the surface recess from being forced, under water pressure, to penetrate between filler and concrete when movement occurs. A waterstop should generally be provided, as shown in Fig. 4–3, to prevent the filler from being either dislodged or penetrated by the sealant. Specifications and testing methods for joint fillers are given in ASTM specifications D545, D994, D1751 and D1752 [16].

(b)   Waterstops

Waterstops are essentially preformed strips of durable impermeable material which span the joint and are embedded in the concrete on either side. They should allow for movement of the joint in service but should remain watertight in all conditions. They are available in many forms, but the most usual are moulded or extruded rubber or PVC for the internal dumb-bell type for fixing vertically, as in Fig. 4-4b, and PVC for the surface type used as the base of floor slabs, as in Fig. 4-4a. The choice of material depends largely on the required movement and pressure head: rubber can accommodate most movement while good quality PVC is perhaps the most durable. Internal PVC waterstops are suitable only for small extensions and small heads up to 10 or 15 m of water.

The width of a waterstop should be sufficient to ensure that the water under pressure does not force a path around it. In practice, it is found that the distance of an embedded type waterstop from the nearest exposed concrete face should equal at least half the width of the waterstop. Hence, a 225 mm wide internal waterstop requires a section not less than 250 mm thick. It should be possible to fully compact the concrete all round the waterstop when it is fixed in position. The edges (or side bulbs) of an internal type waterstop should be supported by tying them to the adjacent reinforcement to prevent them vibrating and creating a void in the concrete during compaction. Flexible type waterstops should never be used in a flat horizontal position because of the difficulty of achieving full compaction of the concrete under the waterstop. If the preferred external PVC type waterstop cannot be used, then a rigid, metal-strip waterstop of plane or corrugated section can be used in this position.

A correctly formed construction joint does not require a waterstop. However, where special or adverse conditions exist metal strip waterstops are well suited to joints with no relative movement. In a wall kicker, for example, lengths up to about 3 m of 150 × 3 mm mild steel water bar can be vibrated into the top of the kicker on completion of the cast.

The waterstop should also be capable of being effectively butt-jointed on site by vulcanizing or mechanical welding techniques. Ts and crosses and especially any jointing of waterstops of dissimilar materials are best produced as factory-made specials which can then be simply butt-jointed on site.

(c) *Joint seals*

Joint seals are placed on the outside of joints to prevent the ingress of foreign material, such as grit, which could prevent free movement of the joint. The seal on the water face also provides the first impermeable barrier to the penetration of the liquid. There are effectively three main types: *in-situ* joint sealants, preformed joint sealants and preformed surface-type water bars, which also act as joint seals. Both types of joint sealant are used in grooves on the outside of the joint. Both types of preformed joint seal are usually based on Neoprene rubber which is very durable under a wide range of aggressive conditions. A channel-shaped Neoprene surface-type water bar can span the joint and be fixed with an adhesive primer into grooves formed (or cut) parallel to the sides of the joint and need not be cast-in as for conventional PVC water bars.

A wide range of materials can be used as *in-situ* joint sealants, which include mastics and thermoplastic and thermosetting compounds. Hot-applied thermoplastics are commonly used for horizontal joints, where they can be poured and trowelled or gunned into position. They include rubber-bitumen compounds and conform to BS 2499: 1973, *Hot applied joint sealants for concrete pavements*. A cold-applied material such as rubber-asphalt is covered by BS 5212: 1975, *Cold poured joint sealants for concrete pavements*. Surface grooves for contraction joints should be slightly tapered to be wider at the surface (20–25 mm), and of similar depth, and should be thoroughly cleaned out to remove any loose material by brushing or grit-blasting followed by blowing out with compressed air. In the case of expansion joints, it is important to ensure that solid particles do not become embedded in the joint filler at the bottom of the groove and so restrict subsequent movement. The concrete at the sides of the groove should then be dried by heat (e.g. gas torch) before applying the primer, which is usually required to assist the bond, followed by the sealant itself.

For vertical wall joints not only should the compound adhere tenaciously to the concrete throughout any movement but also the requirement not to slump or extrude excessively from the joint is much more onerous. Thermosetting compounds such as polysulphides, silicone rubbers, polyurethanes and epoxies have non-slumping properties and increased extensibility, which are much more suitable for all vertical movement joints as well as all expansion joints. Polysulphides are covered by BS 4254: 1967 (as amended 1975), *Two-part polysulphide-based sealants for the building industry*, and BS 5215: 1975, *One-part gun-grade polysulphide-based sealants*. ASTM specifications for joint sealants include D1190, D1191, D3405-8 and D1850-1 [16].

(d) *Joint stress inducers*

Where a contraction joint is obtained from an induced joint, it is essential that the concrete cracks to form the concrete discontinuity in the desired position. Where stop-ends are formed at contraction joints, it can be assumed that adhesion of new concrete is sufficiently weak for cracking to occur at the interface, especially if the joint interface is coated with a bitumen rubber latex. In the case of an induced joint, a weakness must be created if the crack is to occur where required. For a complete contraction joint, the reinforcement is discontinued and hence surface grooves at both faces, or a surface type waterstop with crack inducer, as

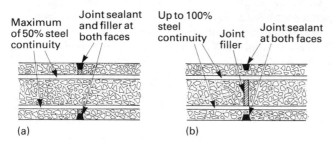

Figure 5-1   Induced partial contraction joints for (a) 50% and (b) 100% steel continuity

shown in Fig. 4-4a, should create sufficient stress concentrations to initiate the crack as required. For a partial contraction joint greater precautions are necessary. If no more than 50% of the reinforcement is continuous at the joint, then crack inducers which merely penetrate the cover to the steel on both faces should be sufficient, i.e. narrow widths of joint filler fixed at the base of the surface groove and securely tied to the reinforcement as shown in Fig. 5-1a. Where 100% of the reinforcement is continuous at the joint, over 50% of the concrete should be made discontinuous and, in this case, a wider joint filler strip can be positioned in the centre of the section and securely tied to the reinforcement at each side, as shown in Fig. 5-1b. A waterstop is usually unnecessary for a 100% partial contraction joint but, if desired, a surface type can be used provided that a minimum of continuous concrete is maintained, as in Fig. 5-1b. As an alternative to a waterstop, either internally or on the water face of a wall, inflatable tubes can be inserted in the centre of the wall (Fig. 5-2) to provide a duct which acts as a crack inducer. The tube can be deflated for easy withdrawal after casting and, when the early thermal contraction and perhaps some seasonal contraction has taken place, the duct can be filled with an expanding grout. Any cracking which subsequently occurs around half the perimeter of the duct should be fairly fine and ensure that, in conjunction with the joint sealant, a water tight joint is achieved.

5.7 Inspection and testing

Where required, structural inspection and testing for liquid-retaining structures can follow the same procedures as for other structures. Testing for watertightness, however, should normally be carried out for all water-retaining structures on

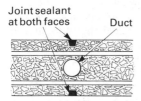

Figure 5-2   Induced partial contraction joint using inflatable duct former

completion and after cleaning, by the simple expedient of filling with water and measuring the fall in level over a period of time.

If the structure is above ground, any initial leakage through walls will be readily apparent. If the structure is below ground, backfilling should preferably be delayed until testing is complete. If not, then measures should be taken to reduce the external water table either to its probable lowest level in service or to below the floor slab of the structure.

The structure should be filled with water at an approved rate and left standing for a minimum period of three days to allow for absorption. The water level should then be topped up and recorded and allowed to stand for a further period of seven days, during which the drop in level, after allowing for evaporation and rainfall, should not exceed 1/1000 of the average water depth of the full tank. If the fall exceeds this limit, the period of the test may be extended for a further seven days and if the fall is then not excessive the structure can be accepted as satisfactory. However, any remaining visible leakage on the outside faces of the structure should also be stopped and any making good of cracks in the wall should preferably be carried out from the inside face. Where there are no obvious serious defects and the visible leakage is small, a period of saturation with a minimum differential pressure between the two faces, to allow autogeneous healing to proceed, may be all that is necessary to eradicate visible damp patches. Cracks 0.2 mm wide will generally require either time for self-healing or some simple treatment, such as painting, before the remote face is likely to remain apparently dry.

The main difficulty with the watertightness test is the allowance for evaporation for the 7 days, which can quite easily vary from 2 to 20 mm or more, depending on temperature, humidity and wind speed. The loss due to evaporation, or the gain due to rainfall, can be estimated by mooring a floating liquid-tight container on the surface of the reservoir over the 7-day test period and measuring the loss, or gain, from within the container.

Roofs should be tested on completion by covering with water, preferably to a minimum depth of 25 mm for a period of 24 h. Where this is impracticable, the roof should be wetted by continuous hosing for at least 6 h. A satisfactory result is given if no leaks or damp patches are visible at the end of either test.

References

1. British Standards Institution, CP 110, Part 1: 1972 (as amended 1980), *The structural use of concrete*, London, 1980, 156 pp. + amend.
2. British Standards Institution, BS 5337: 1976 (as amended 1982), *The structural use of concrete for retaining aqueous liquids*, London, 1976, 16 pp. + amend.
3. Bureau of Reclamation, *Design of small dams*, 2nd Edn, US Dept Interior, Washington, DC, 1974, 816 pp.
4. Concrete Society, *Standard reinforced concrete details*, Report of Working Party, 1973, 28 pp.
5. Anchor, R. D., Hill, A. W. and Hughes, B. P., *Handbook on BS 5337; 1967*, Viewpoint Publications, London, 1979, 60 pp.

6. Portland Cement Association, *Circular concrete tanks without prestressing*, Skokie, Illinois, 1941, 32 pp.
7. Creasy, L. R., *Prestressed concrete cylindrical tanks*, Wiley, New York, and Contractors Record, London, 1961, 216 pp.
8. Hughes, B. P., *Limit state theory for reinforced concrete design*, 3rd Edn, Pitman, London, 1980, 697 pp.
9. Bate, S. C. C. *et al.*, *Handbook on the unified code for structural concrete (CP 110: 1972)*, Cement and Concrete Association, London, 1972, 153 pp.
10. Evans, E. P. and Hughes, B. P., Shrinkage and thermal cracking in a reinforced concrete retaining wall, *Proc. Inst. Civ. Eng.*, Vol. 39, Jan. 1968, pp. 111–25.
11. Hughes, B. P. and Miller, M. M., Thermal cracking and movement in reinforced concrete walls, *Proc. Inst. Civ. Eng.*, Vol. 45, Feb. 1970, Supplement 3, Paper 7254S, pp. 65–86; Discus. Supplement 19, pp. 443–451.
12. Harrison, T. A., *Early-age temperature rises in concrete sections with reference to BS 5337: 1976*, Cement and Concrete Association Interim Technical Note, 5 Nov. 1978, 15 pp.
13. Concrete Society, *Large pours for RC structures*, Birmingham, Sept. 1973, 91 pp.
14. Road Research Laboratory, *A guide to structural design of pavements for new roads*, Road Note 29, 3rd Edn, HM Stationery Office, London, 1970, 30 pp.
15. ACI Committee 506–66, *Recommended practice for shotcreting*, American Concrete Institute, Detroit, 1972 (reaffirmed 1978) 26 pp.
16. ASTM Committee D-4, *Concrete and mineral aggregates: (including manual of aggregate and concrete testing)*, Annual Book of ASTM Standards, Part 14, American Society for Testing and Materials, Philadelphia, 1981, 868 pp.
17. Green, J. K. and Perkins, P. H., *Concrete liquid-retaining structures*, Applied Science Publishers, London, 1980, 355 pp.
18. Anchor, R. D., *Design of liquid-retaining concrete structures*, Surrey University Press, Blackie Publishing Group, Glasgow, 1981, 160 pp.
19. Walmsley, J., Service reservoir design, with particular reference to economics and the 'simply supported wall' concept, *The Structural Engineer*, Vol. 59A, No. 9, Sep. 1981, pp. 285–291.

36 Bridges

A C Liebenberg

Liebenberg & Stander, Cape Town, South Africa

Contents

Notation

A_{ef}	effective torsion area
A_{eq}	equivalent shear area
A_s	area of reinforcing steel
A_{sh}	shear area
a, b	diagonal and off-diagonal position indicators of the independent bending stiffness terms of the member matrix respectively
B	expected present value of the overall benefits derived from the existence of the structure (positive utility)
C	constant damping matrix
C_i	initial cost
C_m	capitalized normal maintenance costs
C_p	capitalized prime costs $= (C_i + C_m)$
C_t	total capitalized costs (negative utility)
d_{ef}	diameter of largest circle within the boundaries of A_{ef}
E	Young's modulus or modulus of elasticity
E_c	initial tangent modulus of concrete
E_d	expectation of damages
E_{dyn}	dynamic modulus of elasticity
E_{long}	long-term modulus of elasticity
E_S	Young's modulus of steel
E_{sec}	secant modulus of concrete
E_t	time-dependent modulus of elasticity
F	force on a member; safety factor
$\boldsymbol{F}$	internal member stress resultant vector
F_r	probability distribution function of the structural capacity, R
F_x	equivalent static earthquake load
f	foundation factors used in conjunction with earthquake loading
$\boldsymbol{f}$	member fixed-end vector
f_s	probability density function of the load effect, S
G	shear modulus
h_{ef}	equivalent wall thickness used to define torsion stiffness
I	importance factor of structures used in conjunction with earthquake loading; moment of inertia or second moment of area of a section (sectional inertia)
I_d	effective sectional inertia for an applied deflection
I_{eq}	equivalent sectional inertia
I_r	effective sectional inertia for an applied rotation
J	torsional moment of inertia
$\boldsymbol{K}$	stiffness matrix
k	member stiffness
$\boldsymbol{k}$	member stiffness matrix
k_d	deflection stiffness in bending
k_{gh}	peak horizontal ground acceleration
k_r	rotational stiffness in bending
k_t	torsional stiffness
L	length of a member
L_w	wind-loaded length
M	bending moment
$\boldsymbol{M}$	constant mass matrix
m	mass
n	number of degress of freedom in a structure
n_0	fundamental natural frequency
$\boldsymbol{p}$	stress resultant
p_f	probability of failure
p_m	probability of exceeding a serviceability limit state
p_n	probability of reaching an ultimate limit state
p_s	probability of survival
R	redistribution factor
$\boldsymbol{R}$	external nodal point load vector
R_D	resistance of a section for design purposes
S_D	force on a section for design purposes
S_a	the seismic response factor for a structure read from the response spectrum
S_m	capitalized cost of damage or loss due to non-compliance with serviceability criteria
s	stirrup spacing
T	natural period of vibration
$\boldsymbol{T}$	transformation matrix
t	time
U_n	capitalized cost of reaching an ultimate limit state
$\boldsymbol{U}$	time dependent displacement vector
$\dot{\boldsymbol{U}}$	time-dependent velocity vector
$\ddot{\boldsymbol{U}}$	time-dependent acceleration vector
u_{ef}	polygonal perimeter length of section in torsion
Δ	an increment
γ_i	modal participation factor used for earthquake loading
δ	deflection of a member or support
η	relaxation coefficient of concrete
θ	rotation of a joint
λ	fixity factor used to determine member stiffness

μ	ductility factor of the structure	Ψ_{ij}	normalized mode component of
ϕ or ϕ_t	coefficient of creep of concrete to		the ith mode in the direction of
	be used for long-term load		the jth degree of freedom
	effects	ω	radial frequency

Summary

This chapter describes the historical development of bridges leading to modern structural configurations, systems and materials of construction. The design criteria that relate to the functional requirements of concrete bridges, including the nature of the site and those actions and effects that influence the design, are enumerated and the importance of aesthetical and environmental considerations explained. These and other factors determine the choice of configuration, materials and methods of construction and their evaluation form part of the design process to obtain optimum results. It is shown that modern design philosophy based on probabilistic and statistical concepts has had a fundamental influence on the drafting of new codes of practice. An outline is given of modern methods of analysis and design including the application of computers with reference also to other publications. The various configurations and components of concrete bridges that are currently used, as well as modern methods of construction and erection, are described. These include slab, beam and box-girder superstructures, arches and cable-stayed bridges. Methods of estimating and optimization of costs are examined and the importance of contractual procedures and maintenance emphasized.

1 Introduction

1.1 General

Although this chapter deals mainly with concrete bridges, reference is made to bridges constructed of other materials whenever this is necessary for a correct perspective of bridge technology as a whole. Limited space does not permit great depth of analysis or wide coverage of the subject matter, so that it has been necessary to supplement the contents by reference to publications by others. Reference is also made to other chapters of this handbook to avoid unnecessary duplication. The general approach to design is based on ISO 2394, *General principles for the verification of the safety of structures* [1] as expounded in Volumes I and II of the *International system of unified standard codes of practice for structures* [2] published by the Comité Euro-International du Béton in 1978. As it is a model code that has not yet found general acceptance, reference is also made to specific codes that are being applied in practice.

The units used in this chapter are those of the SI (International System of units), but where it is useful, imperial dimensions are also given.

1.2 A summary of the historical development of the art and science of bridge building

1.2.1 *Ancient history*

Bridges have fascinated mankind from the earliest recorded times. Apart from their civil and military importance, they are considered by many as symbolical of civil engineering achievement. Although the origins are uncertain, the evolution of the various types of bridge structures probably covers a period exceeding 5000 years. One can speculate with fair certainty that the bridge types that are described in the next section were foreshadowed in the earliest constructions to span rivers and gorges. These included the primitive use of timber logs or stone lintels as beams or slabs, stone clapper bridges, boulders in crude masonry arches and ropes made of creepers, vines or woven natural fibres in small suspension bridges. The subsequent development by empirical methods can clearly be related to the advancement of successive civilizations and their knowledge of materials of construction. Various authors have covered the major periods in the history of bridges in some detail [3–5] and made reference to source material dating back more than four millenia. The history of the development of bridges makes most interesting reading and an in-depth study of the above references is essential for any prospective bridge designer. Only a few salient developments are given here.

The earliest bridge on record was that built on the Nile by Menes, the first king of the Egyptians, about 2650 BC. No details of this bridge are known but a remarkable bridge, with a timber deck on stone piers as described by Diodorus Siculus, was built over the Euphrates in Babylon 4000 years ago. Primitive suspension bridges, where the traveller slid along a single cable made of strands of bamboo rope twisted together, were made in India. The first true suspension bridge consisted of three cables with two on each side to act as handrails to enable the passenger to walk on the third which was tied below. The oldest extant chain bridge is thought to be that over the river Tchin-tchin in China. India is noted for its early use of iron suspension chain bridges. Another form of primitive bridge in timber was a girder type on floating supports, which was the forerunner of the pontoon bridges used in modern warfare.

The development of the brick or masonry arch can be traced back to the Chaldeans and Assyrians but was apparently developed independently in the Western world. Until the time of the Roman conquest of Persia, it appears that these arches mostly took the form of corbelled arches with a pointed profile built of brick or stone in horizontal courses. These have been found widely distributed over various parts of the ancient world including China, the Middle East and Mexico. The first arches with voussoirs were most probably built in Egypt, where tombs dating from 1800 BC were discovered with roofs of elliptical profile. The Persians built arches with ogival or pointed as well as elliptical profiles. The Etruscans, the immediate predecessors of the Romans, developed the semicircular arch built with voussoirs. There is a difference of opinion amongst historians as to whether the Romans absorbed this knowledge during the Roman Conquest or learned about the arch directly from Eastern people. The Greeks had developed an elementary theory of statics and Archimedes (287–212 BC) understood the basic conditions of equilibrium. There is, however, no evidence from ruins that they built true stone arches. Even the Romans apparently never had a full

Figure 1-1   Pont du Gard—Agrippa's masterpiece, 19 BC. (Sketch based on photograph in [6])

understanding of how arches acted to resist the forces generated by the applied loads. However, they have gone down in history as the greatest builders of stone-masonry arches, almost without exception of circular profile and of comparatively small span and heavy proportions. From experience, observation and deduction within the limits of their understanding, they perfected the art of construction to the extent that many examples of their arches, such as the masterpiece of Agrippa, the Pont du Gard, 19 BC (Fig. 1-1), are to this day visible evidence of their creative engineering ability.

The Greeks and Romans used timber for less permanent structures. Timber trestle bridges were developed for military purposes. They built the first timber bridge across the Thames in London.

Perhaps the most important discovery of the Roman builders was that of natural cement, pozzuolana, a volcanic sand found at Pozzuoli near Naples, that forms a hydraulic cement when mixed with ordinary lime and was the forerunner of modern concrete.

Bridge builders in those ancient times were practical men who unfortunately did not readily avail themselves of the ideas of philosophers, who were interested in the early developments of applied science. This lack of communication delayed the proper understanding by bridge builders of the structures that they were developing by empirical methods, for many centuries.

1.2.2   *The Middle Ages*

A great deal of the knowledge of structural engineering accumulated by the Romans was lost during the Middle Ages, so that little progress was made and hardly any structural innovation took place. The art did, however, spread more widely throughout Europe. Chinese timber arch bridges built in the tenth century are still in use. How far the technical as distinct from empirical knowledge of bridge designers had developed by the end of this period is not known.

1.2.3   *The Renaissance and the Age of Reason*

The revival in the arts and sciences under the leadership of men such as Leonardo da Vinci (1452–1519) that occurred during the Renaissance had a fundamental influence on engineering and especially on bridge design. It heralded the birth of what can be considered to be the modern bridge in that it brought about a profound marriage between the conceptual understanding of structures based on theoretical considerations and the practice of bridge construction. This elevated bridge building to an applied science. Practical builders could no longer ignore the ideas of theoretical designers. Several of the great names associated with bridge building in the ensuing periods were to be men who combined both abilities.

The most notable bridges which were the earliest manifestation of this creative upsurge in the 16th century were the Santa Trinità Bridge in Florence, the picturesque Rialto Bridge spanning the Grand Canal in Venice and the Pont Notre Dame and Pont Neuf in Paris.

Using the ideas of men such as Leonardo da Vinci [6] and the results of their experimental work, it was now possible, for example, to determine the correct lines of thrust and the resultant forces acting in arches. The notion of the principle of virtual displacements made it possible to analyse various systems of pulleys and levers such as are used in hoisting devices. Da Vinci also studied the strength of materials such as iron wires by carrying out breaking tests by ingenious methods, and carried out tests on beams and columns. Although he understood the method of moments, he did not adequately solve the problem of the internal resistance to bending of a structural member. It is probable that he invented the timber truss.

Galileo (in about 1600 AD) made great strides by establishing various relationships between beam dimensions and their resistance to loads and came to the important conclusion [6].

> You can plainly see the impossibility of increasing the size of structures to vast dimensions either in art or in nature; likewise the impossibility of building ships, palaces, or temples of enormous size in such way that their parts will hold together. This can only be done by employing a material which is harder and stronger than usual, or by enlarging the size or shape of the members. (From S. P. Timoshenko, *History of Strength of Materials*, © 1953, published by McGraw Hill Book Company, New York and reproduced with permission).

He realized the advantages of hollow sections. He understood the concept of stress but did not solve the problem of the bending resistance of beams.

Academies of science originated in Italy in 1560 and the 17th century saw similar developments in various other countries, e.g. England, France and Germany.

The knowledge of what became known as the *strength of materials* [6] advanced significantly in the 17th century with important contributions by Robert Hooke (1635–1703), whose major discovery was the law named after him; Mariotte (1620–1684) for his studies in bending theory; Jacob (1654–1705) and John Bernoulli (1667–1748), the two famous brothers who produced theories for the deflection of beams (based on the differential and integral calculus developed separately by Newton (1642–1727) and Leibnitz (1646–1716)), and also formulated the principle of virtual displacements; Daniel Bernoulli (1700–1782), the son of John Bernoulli, for his investigations into the shapes of curves which a slender elastic bar will take up under various loading conditions and for deriving the differential equations governing lateral vibrations of prismatic bars; and Leonard Euler (1707–1783) who extended Daniel Bernoulli's work and furthermore derived the famous formula for the critical buckling load of compression members. He also explored the theory of deflection and vibration of membranes.

During the 18th century, the scientific results of the preceding 100 years found practical applications and scientific methods were gradually introduced in various fields of engineering. There was a rapid spread of knowledge through the publication of various books and the calculus became a useful tool to many practising engineers.

In 1713, Parent (1666–1716) published two very important memoirs in which he for the first time clarified the importance of the position of the neutral axis and correctly solved the internal stress distribution in a beam.

The first engineering school in the world was formed by the Corps des Ingénieurs des Ponts et Chaussées in Paris in 1716 and at that time the first books on structural engineering were published. The influence of these movements could soon be noticed in the evolution of new designs such as Perronet's design of the Neuilly Bridge over the Seine, which had piers with a thickness of less than one-ninth the span of the arches.

The longest span truss bridge ever built in timber was the Wittengen Bridge over the River Limmat in Baden, Germany, in 1758. It was destroyed by the French army in 1800 [7].

C. A. Coulomb (1736–1806) played a very important role in this period, covering a wide field. His main contributions to structural engineering were his experimental and theoretical work on the bending of beams and torsion of rods, the stability of retaining walls and the theory of arches. At this time, experimental testing to obtain the mechanical properties of structural materials became well established.

1.2.4 *The Industrial Revolution*
The later decades of the 18th century saw the first application of the products of the Industrial Revolution to structural engineering in England. The first cast-iron bridge to be built in the world, the famous Coalbrookdale Bridge, a semicircular arch with a span of 30.5 m (100 ft) over the River Severn, built by Abraham Darby and probably based on designs by Thomas Pritchard and John Wilkinson, was completed in 1779 [8]. The first iron bridge in America was built 60 years later.

In the early part of the 19th century, very significant advances were made in the theory of the mechanics of materials by Navier (1785–1836), but it took several decades before engineers began to understand them satisfactorily and to use them in practical applications. This work heralded a new period in engineering and was probably the beginning of modern structural analysis. Navier was the first to evolve a general method of analysing statically indeterminate problems. His work was followed by the major contributions of other famous mathematicians, the scientists and engineers whose works have been well documented [6] and form the basis of modern structural engineering.

1.2.5 *Early iron and steel bridges*
The 19th century saw the rise of the great iron and steel bridges, in which Great Britain took the lead, and included a large variety of configurations. Initially, the

proposals were mainly in cast iron and some very bold designs were proposed by Watt, Telford, the Stephensons, Brunel and others. A cast iron arch of 183 m (600ft) span proposed by Telford, was never built due to various objections and it was many years before an arch of comparable span was built in steel. After doing tests on the tensile strength of malleable iron in 1814, Telford developed a design for a suspension bridge with a main span 305 m (1000 ft) long and two side spans of 152 m (500 ft). He retained this idea for the design of the now famous 177 m (580 ft) span suspension bridge over the Menai Straits, completed in 1826. Several major suspension bridges were built in this period but, unlike Menai (re-chained and strengthened in 1938–39), few survive today.

The earlier suspension bridges were designed to carry only horse-drawn roadway traffic and some of these were inadequately designed to resist wind forces and the dynamic effects of traffic and were consequently subjected to disquieting movements. Several failures occurred, including the Broughton chain bridge at Manchester in 1831, due to the oscillations set up by a troop of soldiers marching in step to a drum and fife. Various other bridges had failed in windstorms in England, Germany, France and America. A notable case was that of the Brighton chain pier in 1833. The first suspension bridge which carried a railway was built in 1835 by Sir Samuel Brown over the River Tees, with a span of 85.6 m (281 ft). It was actually designed for road traffic and suffered excessive sagging under the heavier weight of the trains. It consequently had a very brief life due to fatigue failure. Various lessons were learnt from these failures and the need for more adequate stiffening girders was realized. An example of an early solution was Robert Stephenson's original design for the Britannia Bridge across the Menai Straits, with four continuous spans of 70, 140, 140, 70 m (230, 460, 460, 230 ft) respectively. It was to have consisted of two wrought-iron tubes through which the trains were to travel and which were to be supported by cables from above. On the basis of model tests done by Fairbairn and theoretical analyses, the tubes were subsequently found to be sufficiently strong and stiff on their own and the suspension cables were omitted. This was the beginning of plate girder construction. The bridge was opened to traffic in 1850. It was damaged by fire in 1970 and reconstructed with steel arches replacing the tubes [7]. I. K. Brunel [9] designed the Royal Albert Bridge at Saltash, completed in 1859, in which he combined the principles of arch and suspension bridge to form the two 139 m (455 ft) main spans. For the arch he used a massive wrought iron tube. He also designed the Clifton suspension bridge with a span of 214 m (702 ft), completed in 1864 after his death. The 122 m (400 ft) span Albert Bridge over the Thames (1873) combined a catenary suspension system with stays.

The advance of railways had given bridge building a major impetus, so much so that more than 25 000 bridges were built in Great Britain in 70 years. The heavier and more concentrated railway live loading, with its hammer blow and other dynamic effects, required more rigid structures and arches and cantilever trusses replaced suspension bridges in the medium to long-span range. In the USA, a large number of masonry arches and patented trusses, mainly of composite wood and wrought iron [10] were built for the rapidly expanding railroad system.

In the second half of the 19th century, steel superseded iron in the superstructures of bridges although it took time before it was generally accepted with confidence. It introduced the era of large steel bridges. One of the famous early bridges to be built in steel was the St Louis Bridge over the Mississippi River, designed by J. B. Eads and completed in 1873 [11]. This was the first big bridge of steel arches to be erected by the modern cantilever method. It consisted of three arches with a central span of 158 m (520 ft).

In 1879 the tragic collapse of the Tay Bridge occurred. This badly shook the confidence of engineers and led to a greater appreciation of the magnitude of wind forces. The disaster was caused by lack of aerostatic stability, whereas the early suspension bridge failures were blown down because (as is now realized) of insufficient aerodynamic rigidity. There was insufficient knowledge of the nature and magnitude of wind forces and the response of structures. The history of failures demonstrates the courage, perhaps occasional foolhardiness and almost inexplicable lack of knowledge of previous failures, of engineers who have ventured beyond their field of experience and understanding.

Significant advances had however been made in construction materials and methods of design and the understanding of structures developed rapidly due to the prior establishment of the theory of elasticity on a sound basis by men such as Navier, Cauchy (1789–1857), Poisson (1781–1840), Lamé (1795–1870), Clapeyron (1799–1864) and Saint-Venant (1797–1886), to mention only a few. These men started a process which has led to the modern methods of analysis of structures of great complexity (see Section 4.1). The history of this period is particularly interesting and has been well researched by others [6]. Experimental work had also become a well-established discipline due to the pioneer work of William Fairbairn (1789–1874), Eaton Hodgkinson, Weisbach and others. The progress that was made with major contributions by Clerk-Maxwell (1831–1879) and Karl Culman (1821–1881) was such that Sir Benjamin Baker could by 1880 design his masterpiece, the Forth Bridge, by calculation alone. He had to resort to experiments, however, to determine the magnitude of the wind forces.

A large range of interesting steel bridges was built in Europe in this and the ensuing period, including movable bridges (bascules, swing spans, and the vertical lift types). With the development of the New World and the British Empire, the construction of roads and railways required a large number of bridges. In the USA and Canada, Australia and New Zealand, many of the early bridges were built in timber. In Southern Africa, the early bridges were mainly of the through type with steel trusses fabricated in Great Britain and transported in sections by sea and ox waggon to the sites. Several of these bridges built towards the end of the 19th century still survive although they are rapidly being replaced by modern bridges. Several notable steel bridges were built in South America, Africa and India.

The Forth Bridge, which when completed in 1887 introduced the era of long-span cantilever truss bridges, was designed by Sir Benjamin Baker in partnership with Sir John Fowler. It consisted of two main spans of 521 m (1710 ft) made up of two 207 m (680 ft) cantilever arms and a 125 m (350 ft) suspended span, all of truss configuration with the main compressive members being tubular steel struts.

This was followed in 1909 by the Queensboro Bridge over the East River in New York City, with a main span of 360 m (1182 ft), and in 1918 by the Quebec Railway Bridge over the St Lawrence River in Canada, with a main span of 549 m (1800 ft), which at that time was the longest span in the world. The tragic collapse

[3] of the first design in 1907 and the collapse of one suspended span of the revised design during construction in 1916 were further examples in the long line of collapses which were to be repeated several times more in spite of the lessons learnt. Another notable steel cantilever bridge, opened in 1943, is the Howrah Bridge over the Hooghly River in Calcutta, which has a single main span of 457 m (1500 ft), with two short anchor arms [4].

Following on the success of the St Louis arch bridge in 1873, a large number of large steel arches have been built. Most famous are the 152 m (500 ft) span Victoria Falls spandrel-braced arch over the Zambezi River, designed by Sir Ralph Freeman, Snr (1880–1950), under the direction of G. A. Hobson; the 298 m (977 ft) span Hell Gate Bridge over the East River in New York, designed by Gustav Lindenthal and O. H. Ammann and completed in 1916; the 504 m (1652 ft) span Bayonne Bridge across the Kill van Kull at New York, designed by Ammann and completed in 1931; the Sydney Harbour Bridge completed in 1932 with a span of 503 m (1650 ft), which had been conceived by Dr J. J. C. Bradfield after studying the Hell Gate Bridge and was designed for the contractor by Sir Ralph Freeman (Snr), who also designed the 329 m (1080 ft) span Birchenough arch bridge over the Sabi River in Zimbabwe, which was completed in 1935. This latter bridge has very slender and beautiful proportions. It was probably the first instance of the use of wind-tunnel tests on scale models to establish the wind forces on a structure. In the same period, Freeman designed the 320 m (1050 ft) span Otto Beit suspension bridge over the Zambezi at Chirundu (completed in 1939). This is another example of a slender structure built at minimal cost and in less than two years in a relatively undeveloped region. It was also the first suspension bridge of over 1000 ft span to be designed and built outside North America, and its parallel-wire cables were formed by an unique method.

1.2.6 *Prewar long-span steel suspension bridges*
In the USA, the art of bridge building had made significant advances under the leadership of several able designers and builders, amongst whom John A. Roebling (1806–1869) can be considered the inventor of long-span suspension bridges. He designed the famous Brooklyn Bridge (1883), which, at 486 m (1595 ft), was the first long-span bridge to be built of steel and in which he used inclined stays as well as the vertical hangers to support the superstructure. Due to his untimely death, it was completed by his son Washington in 1883. The success of this bridge was largely responsible for a renewed confidence in suspension bridges and it became the forerunner of many of the great suspension bridges in America, such as the George Washington Bridge with a main span of 1067 m (3500 ft) over the Hudson River, designed by O. H. Ammann and completed in 1931. The famous Golden Gate Bridge with a main span of 1280 m (4200 ft), was completed in 1937. The Verrazano Narrows Bridge, with a main span only 18 m (60 ft) greater, was completed 28 years later.

1.2.7 *The birth of reinforced concrete*
It was towards the end of the 19th century that reinforced concrete[1] was for the first time significantly applied to bridge construction. Although crude attempts at

---

1 See also Chapter 1, Section 2; Chapter 2, Section 2 and Chapter 17, Section 1.2.

reinforcing various forms of concrete are evident in the ruins of ancient Rome, these ideas lay dormant for 18 centuries until Smeaton's experiments in 1750 [12]. Louis Joseph Vicat (1786–1861) carried out research on cement mortars prior to 1818 and Joseph Aspdin (1779–1855) patented Portland cement in 1824 [3]. J. L. Lambot built a rowing boat of concrete reinforced with a rectangular mesh of iron rods in 1848, and Francois Coignat constructed a concrete building in 1852 by casing an iron skeleton framework in concrete [12]. The real innovator, however, was Thaddeus Hyat (1816–1901), who carried out experiments in the USA in the 1850s with flat wrought-iron bars on edge which he bonded to the concrete with transverse rods inserted through holes therein [3]. He published a book in 1877 on his experimental work. He recognized the importance of the near-equality of the coefficients of thermal expansion of concrete and iron and assumed a modular ratio of 20 [12].

The concept of a composite structural material, combining the high crushing strength of concrete with the high tensile strength of iron, was, however, largely developed in France around the turn of the century by François Hennebique, Edmund Coignet, N. de Tedesco and Armand Considère. The real credit for laying the foundation of the reinforced concrete industry in about 1867 seems to belong to Joseph Monier. W. B. Wilkinson had patented a reinforced concrete floor system in England in 1854, but his ideas were not widely adopted [3, 12].

An early example of a reinforced concrete beam bridge is Homersfield Bridge in Suffolk, with a 15 m (50 ft) span, completed in 1870 [7]. It was Wayss of Germany who made a first attempt at a design theory. Together with Freitag, he designed and constructed some 320 arched reinforced concrete bridges between the years 1887 and 1891 with spans of up to 40 m. M. Koenen was probably the first to publish, in 1886, an analysis of the behaviour of a reinforced concrete beam. A period of rapid development of various reinforced concrete systems followed in Europe and the USA. Hennibique designed and built the first notable reinforced concrete arch, the Pont de Chatellerault in 1898 [4] with spans of 40, 50 and 40 m and a rise-to-span ratio of approximately one-tenth. The depth of the arch at the crown is only about one-hundredth of the span. His Risorgimento Bridge in Rome (1913), with a span of 100 m (305 ft), is considered to be a masterpiece for that period.

As with other structures, reinforced concrete, however, also experienced several failures in the pioneering days.

1.2.8 *Pre-war reinforced and prestressed concrete*

Whereas steel bridges dominated in the long-span range, reinforced concrete gradually made progress in the short-span ranges, so that by 1900 Hennebique alone had been responsible for the design and construction of at least 100 concrete bridges. Melan introduced the system in which self-supporting reinforcement, mainly in the form of rolled-steel sections, acted as temporary support to the shuttering and wet concrete. In Germany, the firm of Wayss & Freitag published the first authoritative textbook *Der Eisenbeton*, written by Prof. E. Mörsch, which dealt with fundamental aspects of reinforced concrete design. By 1897, such agreement of fundamentals had been reached that it was possible for Charles Robert to give the students of the Ecole des Ponts et Chaussées a course of instruction on the design of reinforced concrete [12]. Standard specifications

Figure 1-2   Maillart's Salgina Bridge in Switzerland, completed in 1930. Its span is 90 m. (Sketch based on photograph in [5]).

Figure 1-3   Maillart's Schwandbach Bridge in Switzerland, completed in 1933. Its span is 37.5 m. (Sketch based on photograph in [42])

for reinforced concrete were introduced by the Swiss Institute of Engineers and Architects in 1903, to be followed by various other European countries.

Robert Maillart (1872–1940) of Switzerland made a considerable impact with the bridges he designed (Figs 1–2 and 1–3) and, with François Hennebique of France, played a leading role in establishing reinforced concrete in bridge building. Although largely an intuitive designer who did only a minimum of calculations, Maillart's mastery of both the technical and aesthetic aspects of design have left an indelible mark on the engineering and architectural professions [3, 4, 13–15].

The early decades of this century saw the general acceptance of reinforced concrete and many short and medium-span bridges were being built in this composite of materials [16].

The autobahn developed in pre-war Germany, and the autostrada in Italy, were good examples of the benefits of reinforced concrete applied to freeway bridge structures. Various reinforced concrete arches, varying in span from 145 m (475 ft) to 192 m (630 ft), were built in Europe before 1943 when the famous Sandö arch, designed by S. Höggbom, was completed over the Ångerman River in Sweden, with a span of 264 m (866 ft), a thickness-to-span ratio of 1 : 100 and a rise-to-span ratio of 0.151. It was successfully completed after the initial centering, which consisted of a timber-framed tied arch, had failed and was replaced by a timber trestle supported on piles. This bridge was a very significant achievement at that time in terms of its span, low rise, slender dimensions and the high strength of concrete. It was only exceeded in span 20 years later by the Arrábida arch in Portugal.

1.2.9 The birth of prestressed concrete

Although the idea of prestressing masonry or concrete probably dates far back, its development only became feasible when high-tensile steels became available.[1] Eugene Freyssinet (1879–1962) can be considered to be the father of prestressing [17, 18] in that he played a leading role in inventing and developing one of the first effective systems which was applied to a wide range of concrete structures. He first used the method in 1907 in the construction of a tie joining the abutments of a test arch of 50 m span, but it took many years before prestressing was accepted by the profession. In fact, it was only in 1934 when he saved the new Ocean Terminal at Le Havre from collapse due to settlement, by post-tensioning applications, that the authorities became convinced of its uses. Being a practically minded man, he was closely involved in the construction of some of his bridges, whereby he gained a clear understanding of relevant problems such as that of deferred elastic strains and the loss of stress incurred due to various factors. His bridges, especially those over the Marne, were as epoch-making as those of Maillart, whose intuitive and practical approach he shared. His bridge at Orly (Fig. 1–4), with a main span of 53.3 m (175 ft), completed in 1959, is a masterpiece in composition of piers and deck.

Major contributions to the early theory and practice of prestressing were also made by Gustav Magnel in Belgium, by Ulrich Finsterwalder in Germany (Fig. 1–5), and by E. Hoyer in Germany, who developed the long-line pretensioning method for producing factory-made prestressed concrete.

1 See also Chapter 19, Section 1

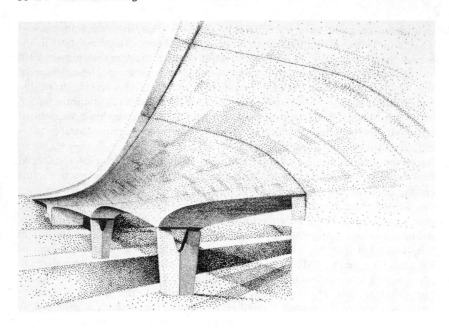

Figure 1-4 Freyssinet's bridge at Orly, completed in 1959. Its main span is
53.3 m. (Sketch based on photograph in [3])

1.2.10 *The post-war period and modern times*

After the Second World War, as the economies of countries were revived, a
resurgence in the building of a great diversity of bridges occurred in Europe to
replace the large number destroyed. The development of the modern motorways
generated a great demand in many countries for grade separation structures and
bridges over gorges and rivers. In addition to the large number of short and

Figure 1-5 Gänstor-Brücke at Ulm, completed in 1950. (Sketch based on
photograph in [5])

medium-span bridges in reinforced and prestressed concrete, many new major structures have arisen. Eleven major steel cantilever truss bridges [7, 19] in the span range of 305 m (1000 ft) to 510 m (1673 ft) have been built, with the longest span in Japan and the others in the USA. Nine steel arches in the span range of 305 m (1000 ft) to 518 m (1700 ft) have been built, all in the USA, except one each in England, Canada and Czechoslovakia [20]. The world's only bridge of any size built of aluminium alloy is the fixed arch of 88 m (290 ft) span crossing the Saguenay River at Arvida, in Canada.

The British designers Freeman, Fox and Partners, under the leadership of Sir Gilbert Roberts (1899–1979), evolved an extremely light but aerodynamically stable steel box-girder deck in conjunction with a triangulated suspender cable system in the process of developing designs for very long span suspension bridges in the 900–1500 m (2953–4921 ft) range. This form was first applied to the Severn (1966) and then to the Bosporus in Turkey (1973) [21], culminating in the Humber Bridge (1981) [22], which to date has the longest span in the world—1410 m (4626 ft). Other notable suspension bridges have been built since the war in the USA, Portugal, Venezuela, Japan, Canada, France, Denmark and West Germany. The maximum span to date with a concrete deck is the Save River Bridge in Mozambique, with a span of 210 m (689 ft) [7].

Steel I-girder and especially steel box-girder bridges proved to be very economic and elegant solutions at many sites. Many of these structures in the span range of 80 m (252 ft) to 300 m (984 ft) were built, mainly in the European countries with well-established steel industries, where West Germany has taken the lead. Some imposing structures were built over the Rhine, including the bridges at Koblenz, 235 m (771 ft) span; Wiesbaden–Schierstein, 205 m (673 ft) span; the Köln–Deutz, 185 m (607 ft) span, as well as several at other sites in Germany. The largest span girder bridge to date, however, is the Niteroi in Rio de Janeiro, Brazil, with a main span of 300 m (984 ft). More than two dozen major steel box-girder bridges (excluding cable-stayed) have been built in various countries since the war. The improvement in the quality of structural steels generally and the development of low-alloy steels with higher yield stresses increased the competitiveness of these steel structures. In Britain, a superior high yield stress steel (BS 968 : 1962) was produced and in the USA the weldable T1 steel had an even higher yield stress of 40 ton/in^2 (618 MPa). Welding both in the shop and at site replaced riveting, but the use of high-strength friction-grip bolts was developed for making connections in the field where welding facilities were not available or convenient.

Composite steel I- or box-girders with cast *in situ* or precast reinforced concrete deck slabs, proved to be very competitive in many cases.[1]

This period also saw the revival of the cable-stayed girder bridge [23], which in concept dates back to 17th century Venice but is generally credited to Löscher (1784) in the form of a completely timber bridge. Redpath and Brown in England and the Frenchman Poyet, early in the 19th century designed bridges with steel-wire cable- and steel bar-stays respectively. The first concrete structure to utilize cable-stays was the Tempul aqueduct with a main span of 60 m (198 ft), crossing the Guadalete River in Spain, designed by Prof. Torroja and completed

---

1 See also Chapter 17, Section 1.3

in 1925 [24]. However, the first modern cable-stayed bridge with a steel deck, designed by F. Dischinger, a German engineer, was built at Strömsund in Sweden in 1955 with a main span of 183 m (599 ft) and a fan-type cable configuration supported on twin-column bents. In spite of initial scepticism, more and more cable-stayed bridges have been and are being built at sites suitable for this configuration.

The Strömsund was followed by the Theodor Heuss (North)Bridge over the Rhine at Düsseldorf in 1958, also with a steel box-girder deck and a main span of 260 m (853 ft), with side spans of 108 m (354 ft) and a fan-type cable configuration. The Severins Bridge at Cologne (1959) has a main span of 302 m (990 ft). The Lake Maracaibo Bridge with spans of 235 m (771 ft), designed by Prof R. Morandi and constructed in 1962, (Fig 1-21), is generally considered to be the first modern concrete cable-stayed bridge [24]. These bridges were followed by a large number with both steel and concrete deck constructions. The longest span to date with a steel deck is that of the Geislingen Bridge near Nuremberg in West Germany, with a main span of 651 m (2137 ft) and a fan-type cable configuration. The longest span to date with a concrete deck is that of the 1300 m (3650 ft) Brotonne Bridge (1977) over the Seine between Rouen and Le Havre (Fig. 1-6), with a main span of 320 m (1050 ft) [25], followed by Prof. Morandi's 282 m (925 ft) span Wadi-Kuff (1971) at Beida, Libya, in which single sets of cables were used (Fig. 1-7).

Figure 1-6   Brotonne Bridge over the Seine, completed in 1977. Its main span is 320 m. (*Courtesy* Freyssinet International)

Figure 1-7 Prof. Morandi's cable-stayed bridge at Wadi-Kuff, Libya. Completed in 1971, it has a main span of 282 m. (*Courtesy* Prof. Ing. Riccardo Morandi)

A competition design submitted for the Danish Great Belt Bridge had a clear span of 345 m (1131 ft) [23], but was not awarded the first prize. A very novel proposal is that of the Ruck-A-Chucky bridge in California [26], which is an horizontal arc-shaped bridge in plan that will be suspended by cables anchored to the steep banks of the canyon (Fig. 1-8). The cable-stays fan out two-dimensionally so as to provide the necessary support for the curved concrete deck. The proposed span is 384 m (1260 ft) measured on the straight. The deck has a radius in plan of 457 m (1500 ft).

The most significant development in the post-war years has been the growth in the application of reinforced and prestressed concrete[1] to bridge construction. A large number of prestressed concrete girder bridges have been constructed with various cross-sections including precast and composite construction as well as in-situ cast box-girders.

The free-cantilever system of prestressed concrete bridge construction was first applied in 1950 to a bridge across the Lahn River at Boldwinstein in Germany. Two years later this method was used to construct the 114 m (375 ft) main span bridge over the River Rhine at Worms [27] (Fig. 1-25b). Since then, more than 100 bridges have been constructed, including prestressed concrete box-girder bridges

1 See also Chapter 19, Section 1

Figure 1-8   Ruck-A-Chucky Bridge in California. (Sketch provided by Prof. T. Y. Lin)

Figure 1-9 Medway Bridge, England, with a main span 152 m. (*Courtesy* Freeman, Fox and Partners)

such as the Medway Bridge in England with a main span of 152 m (500 ft) (Fig. 1-9) and the bridge spanning the Rhine at Bendorf near Koblenz (Fig. 1-25a), with a central span of 208 m (682 ft), which was the longest of its type in the world at that time (1963). Other notable structures of this type are the Urato Bridge at Shikoku in Japan (1972) with a main span of 230 m (754 ft), the Hamana-Chasi Bridge in Japan, 240 m (787 ft), and the Koror–Babelthuap balanced cantilever bridge, 241 m (790 ft), in the Palau Islands in the Pacific, completed in 1977. The first major prestressed concrete bridge in the USA, built in 1950, was the 49 m (160 ft) centre span Walnut Lane Bridge in Philadelphia, designed by G. Magnel, the Belgian pioneer of prestressed concrete. This resulted in prestressed concrete being well accepted in the States in spite of initial scepticism [28] and many major bridges have since been constructed by American engineers.

The decks of many cable-stayed girder bridges are of reinforced and prestressed concrete construction. A significant feature of the construction of these bridges has been the diversity of construction methods that have been developed to increase their competitiveness. The various types of concrete girder and cable-stayed bridge are described, with reference to modern examples and methods of construction, in Section 5. The methods of analysis used in design are described in Section 4.

Figure 1-10   Bloukrans Bridge, South Africa, which is scheduled for completion in 1983. The span of the arch is 272 m.

Since the Sandö Bridge, five reinforced concrete arches of larger span have been built. The world's longest-span concrete arch, the mainland to Island-of-Krk Bridge near Zagreb, Yugoslavia [30], with a span of 390 m (1280 ft), was built by a cantilever method using temporary composite trusses anchored to the banks. The trusses were progressively constructed from opposite ends and incorporated the arch ribs and spandrel columns in conjunction with temporary upper chords and diagonals consisting of steel members and post-tensioned steel tendons. The Bloukrans Bridge in South Africa [29], with a span of 272 m (892 ft) (Fig. 1-10), has been built by the suspended cantilever method using temporary suspension and tie-back (anchor) cables supported on temporary towers and anchored into the rock banks. This method of construction, which was first applied on a smaller scale on two arch bridges in Wales and on a larger scale on the 198 m (650 ft) Van Stadens gorge arch bridge in South Africa, is described in greater detail in Section 5. It has made the long-span arch competitive again after the more costly methods of construction with centering used on the 270 m (886 ft) span Arrábida arch at Porto, Portugal, in 1963, on the 290 m (951 ft) span Amizade arch over the Paraná-joki, Brasilia–Paraguay, in 1964, and on the 305 m (1000 ft) span Gladesville arch in Australia, in 1964 [7] (Fig. 1-11). The construction of the arch on the route between Caracas and La Guaira, in Venezuela, on articulated formwork

Figure 1-11   Gladesville Arch, Australia, completed in 1964. the span of the arch is 304.8 m. (*Courtesy* Maunsell and Partners)

supported by cables from towers on the abutments, can be considered a compromise between the two methods. In the late 1960s, prominent engineers had declared the supremacy of the cable-stayed girder bridges in the span range of 150–400 m, but the elimination of the need for expensive centering for arches has changed the picture considerably.

The longest span prestressed concrete truss to date is the Mangfall Bridge, in West Germany, with a main span of 108 m (354 ft), completed in 1960.

The long history of failures extending into modern times and including long-span bridges such as the 853 m (2800 ft) span Tacoma Narrows suspension bridge in the USA, which failed so dramatically in 1940 due to aerodynamic instability [23], and various steel box-girder cable-stayed and other bridges which collapsed during construction in more recent times [31], has taught the bridge-building fraternity expensive but invaluable lessons and led to greater efforts in the form of both experimental and theoretical work to understand the forces that bridges are subjected to and their response thereto [32–35]. In time, all this experience has culminated in a new and sounder philosophical approach to design and construction practice, as discussed in Section 3.

Throughout its history, the availability of specific materials of construction has played a major role in determining the course of development of structural engineering. This is true today and the tendency for structural engineers to specialize in the use of one or other of the specific materials, e.g. structural steel rather than reinforced and/or prestressed concrete, or vice versa, has historical grounds in the development of virtually separate industries with different technologies. This specialization has been further reinforced by the sponsorship provided by the manufacturing or constructional concerns, of societies, institutes and publications that propagate the use of their particular material products and provide a very good information service as well as donations and bursaries for the education and training of engineers and technicians. These activities are praiseworthy in spite of being motivated by vested interests, as knowledge and competition are generally increased thereby. Furthermore, the zeal of a specialist designer using specific material products only, may result in innovations that would otherwise never have been conceived. There are many examples thereof and many famous engineers have become associated with particular materials of construction.

Consequently, there are today these separate engineering disciplines in modern bridge design and construction, often with very little cross-communication. Although historically structural steel had a big lead, the applications of concrete have steadily increased. In bridge design it can today provide better solutions in a majority of cases over a very wide range of circumstances, with the exception of very large spans exceeding at present approximately 400 m. In the general interest however, it remains necessary that the solution of the problems posed by any project be given the widest possible consideration and that alternatives be considered, as discussed in Section 5.9.

1.3 Modern concrete bridges: a summary of the different configurations and structural aystems

It is traditional to subdivide a bridge into (a) the superstructure, (b) the substructure, and (c) the foundations, as there is usually a clear division between these parts in slab, beam, girder, frame and truss bridges. This does not, however, always apply to arch-type and suspension or cable-stayed girder bridges.

For the purpose of classifying bridges, it is therefore more convenient to describe the configurations of the *primary structural systems* (Fig. 1-12) which are constituted of the main load-carrying members that support the *deck structural system* by spanning between the substructures or foundations. For the purposes of this chapter, the following definitions shall apply.

(a) The *bridge deck structural system* is that part of the superstructure that directly supports and may be integral with the members that form the deck surface. It may be furnished with some of the following:

   (1) balustrades (parapets) and crash barriers;
   (2) highway surfacing, pedestrianways (footways) or raised sidewalks, traffic islands or railway tracks on ties and ballast;
   (3) expansion joints;
   (4) drainage systems;

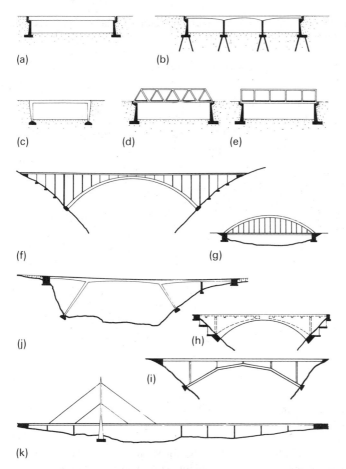

Figure 1-12   Some examples of primary structural systems
Group 1   (a, b) Slabs, grids, beams and girders, and (c) portal frame
Group 2   (d) Trusses and (e) Vierendeel girder
Group 3   (f) Open spandrel arch, (g) tied arch, (h) solid spandrel arch, (i)
          funicular arch and (j) strut frame
Group 4   (k) Cable-stayed girder bridge

    (5) service ducts or brackets carrying cables or pipes;

    (6) fixtures, brackets or recesses for supporting lighting poles or signboards.

    Items (1) to (6) are described in Section 5.8.

  (b) The *substructure* comprises those portions of the piers, columns, or abutments and their capping beams and bearings that act as direct support to the deck structural system.

  (c) The *foundations* support substructures or, in the case of structures such as arches and cable-stayed bridges, provide direct support to the primary structural systems and may consist of concrete footings, spread foundations or rafts bearing directly on soil or rock, or capping slabs supported on caissons or piles.

Foundations and substructures are described in greater detail in Section 5.4.

1.3.1 *Configurations of primary structural systems*

Although there are countless variations in bridge types, they can broadly be classified into four groups if defined in terms of the configurations of the primary structural systems that mainly resist gravitational loads:

(1) Slabs, coffered slabs, grids, beams, girders, cantilevers and frames in which loading induces mainly bending, torsional and shear forces; (Figs 1-12a-c).

(2) Trusses, including cantilevered trusses[1] and frames in which linear elements are joined together to form systems of interacting members in which either compressive or tensile extensional forces predominate (Figs 1-12d,e).

(3) Arches and related types in which the configuration and shape of the main load-resisting members are such that compressive forces predominate (Figs 1-12f–j).

(4) Cable-suspended structures, i.e. suspension bridges and cable-stayed bridges in which tensile cables, supported over towers, form the main supporting system acting in conjunction with the deck structural system (Fig. 1-12k), and related types such as prestressed concrete ribbon bridges (Fig. 1-22).

Combinations of the above systems have been tried, but usually with unsatisfactory aesthetic results (see Section 1.4). This statement only refers to the primary structural role of these systems resisting mainly gravitational loads, as in most structures these systems also have 'secondary' roles in resisting, for example, wind forces such as the bow-girder action of an arch acting in conjunction with bending in the horizontal plane of the deck structural system. There are, however, exceptions such as the Vierendeel girder which combines system (1) and an incomplete system (2) to produce what is visually a very satisfying simplification. A good example is the 200 m (656 ft) span Rio Colorado Bridge in Costa Rica [26], completed in 1972 (Fig. 1-13), which is a precast, post-tensioned, cable-suspended concrete bridge which combines systems (4), (3) and (1). The two inclined piers function as the towers for the suspended cables and, because of their inclination, they also provide beneficial arch and cantilever effects. Unlike a conventional suspension bridge, the roadway deck is situated above rather than below the cables.

The factors affecting the selection of the most suitable configuration for the primary structural system, the proportions of the structural members, the deck system and the reinforcing and/or prestressing systems, are referred to in Sections 2.5 and 5.

---

1 Cantilever truss bridges are often considered as a separate group because of the significant era of longspan cantilever truss bridges. Trusses are sometimes classified with beams.

Figure 1-13 Rio Colorado Bridge, Costa Rica. (*Courtesy* Prof. T. Y. Lin)

1.3.2 *Bridge deck structural systems*

In small and medium-span structures, the deck which carries the highway or railway is usually the primary structural system in the form of various types of solid, voided or cofferred slab (Figs 1-14a–e), or monolithic slab and beam constructions (Fig. 1–14f). Decks may be continuous or simply supported over more than one span and may be cantilevered with drop-in sections in one or more alternate spans (Fig. 1-14g). The spans may have various degrees of skew depending on the crossing.

Simply supported (non-continuous) deck structures extending over several spans require movement joints which may provide a poor riding surface unless costly joint systems are provided. In the case of slab and beam construction, this may be overcome by reducing the number of joints by making the deck slabs of groups of spans continuous over the supports (Fig. 1-14h). Provision must then be made for longitudinal expansion and contraction by using either flexible columns or sliding bearings. With this technique, prefabricated beams can be used where centering is not feasible, e.g. over traffic, water, etc., and where achieving continuity of the beams is difficult.

Drop-in spans provide a useful solution where expansion and contraction joints are required in a long continuous structure, especially if the movements cannot be

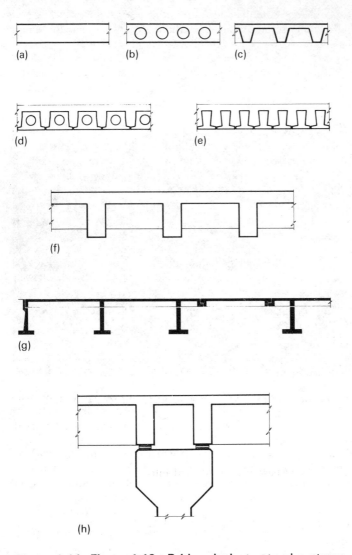

Figure 1-14 Figure 1-12 Bridge deck structural systems

    (a)   Solid slab
    (b)   Voided slab
    (c)   Coffered slab
    (d, e) Contiguous voided or solid beams with reinforced con-
           crete infilling
    (f)   Monolithic slab and beam construction
    (g)   Continuous deck with drop-in section
    (h)   Detail of continuous slab with beam discontinuity

accommodated at the piers or abutments. In continuous multispan bridges, longer spans are normally possible for the same construction depth as that for simply supported, non-continuous spans, due to the distribution of the bending moments. Haunched slabs or beams usually result in savings in materials, but may be undesirable for constructional, aesthetic or other reasons. It is usual to limit the end spans to approximately 85% of the internal span length in order to approximately equalize bending moments. This type of structure normally has a higher factor of safety against collapse than non-continuous decks if designed in accordance with existing codes of practice, due to additional redundancies with the possibility of the redistribution of bending moment from the over-stressed section to adjacent sections with reserve capacity. Continuous-deck systems are normally sensitive to differential support settlement and these effects must be properly evaluated.

For larger spans, the deck may be a secondary part of the structure such as in arch bridges, but is often integrated with the primary structural systems as in various girder or truss type structures and suspension or cable-stayed bridges. The lateral stiffness and strength of a continuous deck can play a major role in resisting wind, earthquake and other horizontal actions by spanning effectively as a horizontal girder between abutments and other intermediate substructures.

Bridge-deck structural systems in reinforced and prestressed concrete that are currently in use can be broadly subdivided into:

(a) the members that form the surface of the deck and which may or may not be part of the primary system;
(b) primary structural systems which may or may not be integral with the deck surface members.

Deck surface members and primary systems may be classified into the following groups which may be of cast *in situ*, precast or composite construction with ordinary reinforcement, partially or fully prestressed (pretensioned or post-tensioned).

1.3.2.1 *Type (a) systems* These can be divided into:

(1) *Slabs*

    (i) Solid slabs, acting approximately as isotropic plates (Fig. 1-14a).
    (ii) Voided slabs, acting approximately as anisotropic or orthotropic plates (Fig. 1-14b).
    (iii) Cofferred slabs, acting approximately as anisotropic or orthotropic plates (Fig. 1-14c).
    (iv) Precast beams of various cross-sectional shapes of constant or varying depth, contiguously joined together for the full depth by *in situ* reinforced concrete and transverse prestressing to effectively act as a voided or solid slab with the upper surface forming the deck (Figs 1-14d,e).

(2) *Beams*

    (i) Longitudinal stringer beams with webs spaced apart integral with the deck slab which acts as an upper flange with or without crossbeams or

diaphragms (Fig. 1-14f). The beams may be precast to form a composite construction with *in situ* cast diaphragms and deck slab.
(ii) Longitudinal and transverse beams forming a grid system integral with the deck slab.

1.3.2.2   *Type (b) systems* These can be divided into:
(1) Inverted longitudinal beams, trusses or girders with various arrangements and sizes of components at the outer edges of or between carriageways, in conjunction with one of the above deck systems, thereby forming a through or semi-through bridge which is fully or partially integral with the deck (Fig. 1-15a).
(2) A single central longitudinal spine beam or T-beam, truss or girder of various arrangements and sizes of components supporting one of the deck systems described in Section 1.3.2.1. The deck spans transversely but may also act as a composite or monolithic upper flange to the primary system (Fig. 1-15b).
(3) Two or more longitudinal beams (single or multi-webbed), trusses or girders, with various arrangements and sizes of components spaced apart, with or without crossbeams, diaphragms or bracing, supporting one of the deck systems described in Section 1.3.2.1, thereby forming a deck bridge fully or partially monolithic (Fig. 1-15c). For large spans, orthotropic prestressed concrete girder bridges with ribbed thin webs and decks may be competitive.
(4) A single longitudinal box-beam, or several box-beams with or without cantilevered top flanges, consisting of one of the following:

   (i) a single unicellular box (double webbed) (Fig. 1-15d);
  (ii) a single multicellular box (Fig. 1-15e);
 (iii) twin or multiple unicellular boxes with or without crossbeams or diaphragms (Figs. 1-15f);
 (iv) twin or multiple multicellular boxes with or without crossbeams or diaphragms (Fig. 1-15g).

(5) Solid or voided slab deck supported directly on piers with or without haunches or drop-heads, e.g. Elztalbrücke built in 1965 (Fig. 1-15h).
(6) Frames, which in this context will be defined as structures with more than one member, which members may be in one or more planes and which rely for stability on the flexural or torsional rigidity at all or some of the connections between members. Other connections may have various degrees of freedom. This definition does not include the case of strut-frames or funicular arches, which are classified under arch-type structures. There is a large variety of configurations of frames that are used either as secondary members or as parts of the primary structural system, such as portal frames (single or multiple), Vierendeel girders, trestle piers, spill-through abutments and towers for suspension or cable-stayed girder bridges. Frames are generally suitable for short-span structures such as flyovers on freeways or bridges over small rivers, but some medium-span bridges such as the prestressed concrete portal frames designed by Freyssinet over the Marne have been competitive (see Section 1.2 and Fig. 1-5).

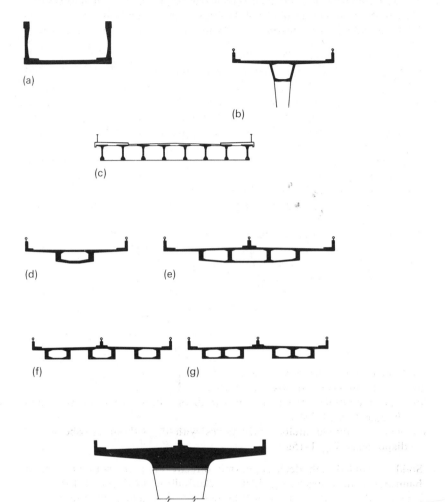

Figure 1-15   Primary structural deck systems
(a) Through girder bridge
(b) Spine beam
(c) Longitudinal beams
(d) Single unicellular box
(e) Single triple-cellular box
(f) Triple unicellular boxes
(g) Twin double-cellular boxes
(h) Solid deck

Relatively flexible frames or articulated designs of statically determinate structures on bearings allowing rotation are suitable for sites where large differential settlements are anticipated.

1.3.3 *Substructures directly supporting deck structures*

At the ends of bridges, the deck structures are supported directly on abutments which can be of various types (Figs 1-16a-h):

(a) Mass concrete gravity abutment.

(b) Closed end abutments with:

   (1) solid or voided walls acting as cantilevers or struts or diaphragms, or forming part of a cellular construction with wingwalls or side walls or part of a rigid portal frame;

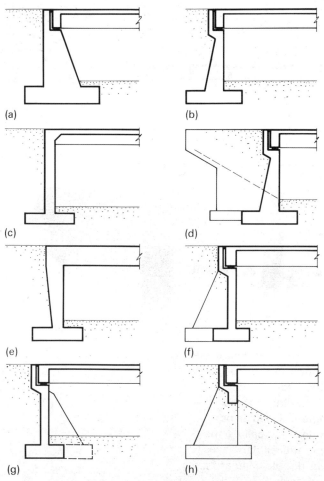

(a)

(b)

(c)

(d)

(e)

(f)

(g)

(h)

Figure 1-16 Abutment types: (a) mass gravity, (b) cantilever, (c) diaphragm, (d) wingwalls, (e) portal frame, (f) counterforted, (g) buttressed, (h) spill-through

(2) counterforted walls;
(3) buttressed walls;
(4) combinations of (2) and (3).

(c) Open-end or spill-through abutments with breast (trestle) beams supported on columns.

Abutments may be of the seat-type supporting the superstructure on bearings or be monolithic with the superstructure.

In multispan bridges, the intermediate piers can be of the following types:

(a) Solid or voided walls with or without capping beams.
(b) Single solid or voided columns with or without caps or of the mushroom type.
(c) Trestle piers or bents with or without capping beams supported on solid or hollow columns which can be of a great variety of shapes, depending on whether the bridge spans a river or gorge, or overpasses a roadway or railway, and on aesthetic considerations.
(d) Specially shaped columns of which there are a great variety, e.g. V-shaped, forked, etc. (Fig. 1-17).

As with abutments, piers may be of the seat-type with bearings, or monolithic with the superstructure. Substructures are supported on various types of foundation, depending on the nature of the subsoil, such as spread footings, rafts, caissons or piles, as referred to in Section 5.4.

Figure 1-17 Wynhol Viaduct, Clevedon Hills, England. (*Courtesy* Freeman, Fox and Partners)

1.3.4   *Arches and arch-type structures*
With suitable foundation conditions, the arch has since early times (see Section
1.2) proved to be a competitive structural system in bridge engineering. With the
correct profiling, symmetrically distributed loads such as dead loads induce mainly
compressive forces, making concrete, with its high compressive strength/cost
characteristics, an ideal construction material. Unsymmetrical loading due to
traffic, as well as temperature effects and shrinkage and creep shortening effects
(Section 5.6), however, induce bending moments in the arch for which reinforcement must be provided. Due to the high compressive stress component, the
amount of reinforcement required for the serviceability condition of the bridge is
relatively low compared with other structural systems which rely mainly on flexure
as a system of support.

In order to function as an arch, the foundations must be restrained from moving
apart. For this reason, large arches require stable banks to provide this support.
Smaller arch structures have, however, been constructed on relatively poor
foundations using ties between the foundations to provide horizontal support. In
arch bridges, the span-to-rise ratio is an important factor which controls the
structural behaviour and economics of the structure and which must be carefully
considered for each site, taking into account the cost and proportioning of other
elements of the bridge.

There are a large variety of arch-type structures of which the following are
representative:

1.3.4.1   *True arches*   These may be classified as

(a) open spandrel arches (Fig. 1-12f);
(b) solid spandrel arches (Fig. 1-12h);
(c) tied arches (bow-string) which are usually through bridges (Fig. 1-12g);
(d) funicular arches (Fig. 1-12i).

Arch cross-sections may be of solid construction, but larger sections are
invariably uni- or multicellular. Depending on the width-to-span ratio of the
structure, arches may be single ribbed, double or multi-ribbed, with or without
crossbeams and diaphragms.

Small arches have been built to circular, elliptical or parabolic profiles, but in
modern times it is more usual to express the profile in a suitable mathematical
form (usually a power series) that corresponds as closely as possible with the dead
load thrust line. In the case of small arches, aesthetic requirements may dictate a
deviation from the ideal profile. Arches may be encastred with no hinges, or may
have two or three hinges, depending usually on the nature of the founding
conditions.

Open spandrel arches are commonly used over rivers and deep gorges in the
span range of the order of 100 m (328 ft) up to 300 m (984 ft). In the 1960s, many
bridge designers favoured the newly developed cable-stayed structures in this
span range because of the cost of temporary arch centering. The suspended
cantilever method developed in the late 1960s, which is referred to in Section 1.2
and described in Section 5.6, has once more made arches competitive for specific
sites such as deep gorges and sea straits. The new mainland to Isle-of-Krk bridge
near Zagreb in Yugoslavia has a span of 390 m (1280 ft). Solid spandrel arches

Figure 1-18   Strut-frame over freeway

are rarely built today, with the exception of small-span structures with a specific aesthetic requirement.

Tied arches are suitable as through bridges over rivers in the span range of the order of 100 to 200 m, where a maximum clearance for yachts or ships is required without raising the road or railway level unduly, or where relatively poor founding conditions do not provide adequate thrust resistance. Funicular arches may minimize the material costs, but are not necessarily the most economic and rarely provide a satisfactory aesthetic solution. Multiple arches have the advantage that dead-load thrusts are balanced, except at the end abutments.

1.3.4.2   *Related arch-type structures*   A fifth type is the strut-frame (inclined leg frame) (Fig. 1-12). In the case of these structures, the thrust line deviates considerably from the profile of the linear structural members and bending forces are consequently a major consideration. Strut-frame bridges are often built as overpass structures over double carriageway highways where a central column is not desirable (Fig. 1-18) and have been constructed over deep gorges with spans approaching 200 m measured between the springing points of the struts. The prestressed concrete strut-frame bridge over the Gouritz River in South Africa has a clear span of 171 m (561 ft) between springing points (Fig. 1-19).

Figure 1-19   Strut-frame over Gouritz River, South Africa

1.3.5   *Cable-suspended concrete structures*

1.3.5.1   *Suspension bridges* Modern suspension bridges with draped cables and vertical or triangulated suspender hangers, are usually economical in the span range exceeding 300 m and are more competitive with the lighter steel deck constructions. The maximum span to date with a concrete deck is the Save River Bridge in Mozambique, with a span of 210 m (689 ft) [7]. The towers, substructures and minor approach spans are usually more economical in concrete. Concrete towers were used for the world's longest steel deck suspension bridge with a span of 1396 m over the Humber river in England [22]. A proposal for a 3000 m span bridge has been considered to cross the Messina Straits between Sicily and Italy.

1.3.5.2   *Cable-stayed bridges* Modern cable-stayed bridges [23] are at present considered to be the most interesting development in bridge design. These structural systems are statically highly indeterminate and have become practically feasible by the development of sophisticated methods of analysis, coupled with the advent of improved electronic computers during the last three decades. The latest designs have extremely slender members and are competitive over the span range of the order of 100 to 700 m with concrete decks, and even greater spans in excess of 1000 m with steel decks. The towers, which are usually single or twin columns (pylons) or frames of various shapes including A-frames, are in most cases competitive in concrete. Examples are given in Sections 1.2 and 5.7. The cable arrangements are classified in terms of:

(a) the elevational arrangement, e.g. single, radiating, harp, fan and star, or combinations thereof and may be asymmetrical about the pier depending on the relative span lengths (Figs 1-20a-f);
(b) the transverse arrangement, e.g. single plane—vertical (central or eccentric); double plane—vertical; or double plane—sloping;
(c) the number of cables, e.g. single, double, triple, multiple or combined.

Although earlier designs were based on fewer support cables encased in concrete and spaced at greater distances apart (Fig. 1-21), modern trends are for more cables at closer spacing which allow a considerable saving in deck material. The cable configuration is dependent on the deck structural system (and vice versa) and the type of loading. Cables arranged in a single plane will generally require a torsionally stiff deck, whereas for cables in two parallel or sloping planes a solid slab may be adequate depending on the width of deck (see Section 5.7).

1.3.5.3   *Combinations of suspension and cable-stayed bridges* Examples of such combinations were discussed in Section 1.2 and have in more recent times been suggested by Dischinger [23] and also by the designers of the proposed Messina Straits Bridge.

1.3.5.4   *Stressed ribbon bridges* These consist of prestressed concrete slabs or ribs with a sagging profile anchored into the abutments and subsoil or rock. In the case of the Genf–Lignon Bridge [7] it has been successfully used as a pedestrian and pipeline bridge with a span of 136 m (446 ft) over the Rhône River in France.

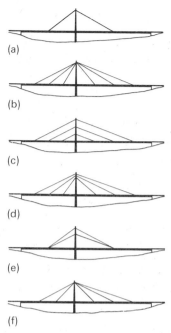

(a)

(b)

(c)

(d)

(e)

(f)

Figure 1-20   Cable-stay configurations: (a) single, (b) radiating, (c) harp, (d) fan, (e) star and (f) asymmetrical radiating

Figure 1-21   Maracaibo Bridge, Venezuela. (*Courtesy* Prof. Ing. Riccardo Morandi)

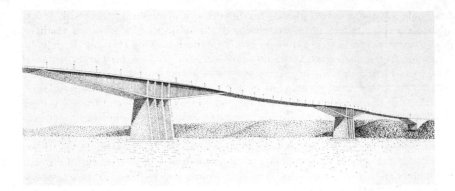

Figure 1-22 Stressed-ribbon bridge. Proposal for the Bosporus crossing by Dr. Ing. U. Finsterwalder

The longest span, however, is that of the bridge carrying a belt conveyor at the cement works of Holderbank–Wildegg in Switzerland, with a span of 216.4 m (710 ft) and a 14.75 m (48.4 ft) sag. It consists of steel cables with precast concrete deck elements clipped to them [7].

A proposal for a 2 km (1.25 mile) long stressed ribbon bridge has been made by Dr Finsterwalder, to cross Lake Geneva. This bridge is a combination of suspended concave spans with a radius of about 2500 m (8200 ft) and supported convex parts near the piers, which have varying radii of 3000 m (9800 ft) or more. A limited prestressing of the stressed ribbon is accomplished so that the dead load does not produce tensile stresses [27]. The central and end spans are almost 460 m (1500 ft) long and alternate with 200 m (650 ft) spans. The stressed ribbon is a reinforced concrete slab about 254 mm (10 in) thick, with about 25% of reinforcing steel. He also proposed a stressed ribbon design for the Bosphorus Bridge (Fig. 1-22).

1.4 The aesthetic evaluation of bridges

During the earlier periods in the development of bridge structures, especially masonry arches, the methods of construction were closely related to those for buildings. This probably accounts for the high standard in aesthetic design that was generally achieved. Michelangelo is recorded to have said: 'A bridge ought to be built as though it were intended to be a cathedral, with the same care and the same materials.' The significance of this statement is apparent if one bears in mind that bridge structures are visually prominent and often dominate the surrounding landscape.

It is quite apparent that the divorce of bridge engineering from architecture during the Industrial Revolution had a negative influence on bridge design. Some of the iron and steel bridges designed during the latter part of the 19th century indicated an almost total lack of sensibility to aesthetic considerations. Although there have been many outstanding designs by engineers during that and subsequent periods, this cannot unfortunately be said of all bridge structures. Although

few engineers today would dispute the importance of aesthetics, they still generally have a singular lack of understanding of the subject and tend to see aesthetic
design as a simple extension of engineering design. This fact can largely be
attributed to the almost total separation of disciplines in most tertiary educational
systems, which results in engineers very rarely receiving formal education or
training in art or the appreciation of aesthetic values. This deficiency is compounded by the need for specialization and the consequential demands on the
time of practising engineers in order to achieve excellence in the purely technical
aspects of engineering.

Aesthetics, being a subject belonging to philosophy and the arts, differs
essentially from the disciplines that constitute modern engineering. It follows that
an understanding of aesthetics does not come naturally to most engineers.
Fortunately, there are exceptions who, because of their domestic and scholastic
backgrounds, through the influence of parents, friends or teachers, acquire an
appreciation of art at an early age. Generally, however, engineers develop a
predominantly logical approach to design without the intuitive sensibility and
judgement that is essential for the appreciation and meaningful evaluation of the
aesthetic aspects of their work. They have consequently over the years applied
whatever innate abilities they may have had with greatly varying degrees of
success.

The subject of aesthetics has, since the time of the earliest philosophers such as
Plato, Aristotle and Plotinus, been the cause of much controversy and even today
there is no universally accepted theory of aesthetics. The development of art and
architecture has gone through many phases which also included various attempts
at the formulation of principles and rules. These included the well-known rule of
the golden ratio during the Graecian period, which was an age of formalism, and
various subsequent attempts to produce geometrical formulations for beautiful
forms or shapes. The 18th century philosopher David Hume related beauty to
utility. However, Emmanuel Kant was probably the first to assess aesthetics to be
equal to and independent of reason and ethics. Most artists and architects today
appear to agree that there are no rules by which one can create or measure the
quality of art or architecture. Even if the relevant values were absolute in the
Neo-Platonic sense, they would remain elusive to analysis in terms of our highly
complex processes of perception. According to Herbert Read [36] in discussing
the meaning of art:

> Many theories have been invented to explain the workings of the mind in
> such a situation, but most of them err, in my opinion, by overlooking the
> instantaneity of the event. I do not believe that a person of real sensibility
> ever stands before a picture and, after a long process of analysis, pro-
> nounces himself pleased.

Yet Roger Scruton [37] concludes:

> Like all decorative arts, architecture derives its nature not from some
> activity of representation or dramatic gesture, but from an everyday preoc-
> cupation with getting things right, a preoccupation that has little to do with
> the artistic intentions of romantic theory. My thesis has been that the
> aesthetic sense is an indispensable part of this preoccupation, and that the

resulting 'aesthetics of everyday life' is as susceptible of objective employment as any other branch of practical reason.

Perhaps the most realistic 'working' definition for acceptable norms of aesthetic design is that which expresses them in terms of the 'average opinion' of a section of society that is constituted of leading professionals, practitioners, critics and other interested persons. Such 'common wisdom' which may be considered to represent 'good taste', is, however, not easily expressed in precise terms and will tend to vary in time.

Bridges in this context can be classified under architecture and any discourse on the aesthetic aspects of the latter would to a great extent apply to bridges which form a sub-class that are, however, less complex than most architectural creations. Bridge designers should therefore, to the extent that they are able, acquaint themselves with current trends in architectural thought relating to bridges. The most important attribute required of aesthetic appreciation and design is that of imagination of a rather special kind which is probably inherited, but can be substantially developed if nurtured by study, observation and much practice.

The type of mental activity that constitutes aesthetic judgement is probably not entirely unlike that which is necessary for innovative (creative) engineering design, which also requires an imaginative ability and is achieved largely by 'lateral' (non-logical) thinking [39]. 'Vertical' logic, which is suitable for applying or extending rules, is however, essential for testing the validity of creative ideas. Innovative engineering design and aesthetic judgement are nevertheless different in essence as the latter is largely based on ideas selected from a wide range of impressions which cannot readily be quantified. An intuitive ability in the one field therefore does not necessarily imply an equivalent competence in the other.

However, according to Scruton [38] (*The Aesthetics of Architecture* © 1979 by Princeton University Press):

> Architectural tastes in so far as they are tastes in the aesthetic sense, inevitably make way for deliberation and comparison. And here deliberation does not mean the cultivation of a vast and varied experience, like that of the over-travelled connoisseur. It denotes not the fevered acquisition of experience, but rather the reflective attention to what one has. A man might know only a few significant buildings—as did the builders of many of our great cathedrals—and yet be possessed of everything necessary to the development of taste. It suffices only that he should reflect on the nature of those choices that are available to him and on the experience that he might obtain.

There have always been gifted engineers with a good understanding of the subject, as was illustrated by a series of lectures on *The aesthetic aspect of civil engineering design*, published by the Institution of Civil Engineers in 1944. In one of these, Sir Charles Inglis gave a very enlightening analysis [40] of the aesthetic qualities of bridges spanning a period of 2000 years, from the Pont du Gard in France (Fig. 1-1) to the Golden Gate Bridge in San Francisco. In his lecture, Oscar Faber [41] explained that beauty could not be derived from some simple shortcuts, but depended upon a large number of qualities of which the following are a few: harmony, composition, character, interest, the expression of function, the expression of construction, rhythm, colour and texture of materials.

The famous Maillart (Section 1.2 and Figs 1-2 and 1-3) consulted no other designer, but practised three principles: to work within the constraints of the relatively new materials of reinforced concrete, to apply his original insight into deck-stiffened behaviour and to achieve minimum cost. There was no imposition of aesthetic rules in his designs, but there was a strong desire for aesthetic results [42].

During the last decade or two, engineers have shown a renewed interest in aesthetics, coupled with an awareness of the need to relate their structures to the environment. This is reflected in numerous recent lectures and papers on the subject [42–47]. The general tendency is to formulate rules for aesthetic design.

It is not the intention to decry the work of those that have in the past and recently produced such design rules, as there is no doubt that they serve a useful purpose, especially for the novice. It must be remembered, however, that these rules or laws have been deduced from past results and do not necessarily have a fundamental basis. They only work to the extent that they define some visual properties of bridges which are aesthetically satisfactory and have withstood the tests of time, not unlike classical art. Generalizations should consequently be treated with great care. Every design can best be considered to be unique and even where such rules are applied, an imaginative adjustment will invariably result in some improvement.

Most of the authors [43, 48–50] who have suggested guidelines or rules are generally in agreement with the most important aspects. These are discussed below. It is not always easy to differentiate between the more generally accepted principles of composition and what may be mere working rules.

1.4.1 *Unity of form and harmony*
This is a prerequisite of all artistic or architectural works and even more so in the case of bridge structures. It is best achieved in modern bridges by simplicity of design and pureness of structure with continuous straight or smoothly curved lines and consisting of few and simple elements that are in harmony amongst themselves and with the environment. It requires the organization or arrangement of the members of a bridge in such manner that it excites within the observer a sense of wholeness generated by some central or dominating idea in the composition. Duality, such as that created by the voidal spaces in a two-span bridge (Fig. 1-23a), has a disturbing effect on the observer in that he has difficulty in resolving a central focal point. Even a powerful motif created by increasing the visual mass of the central pier does not entirely solve the problem.

By going to the other extreme of expressing the deck and abutments more prominently than the central pier, as shown in Fig. 1-23b, the voidal duality may, however, be partially broken. Three-span bridges can be improved by making the central span longer than the approach spans (Fig. 1-23c). Four spans need careful treatment, as a form of duality can in some cases become apparent. The relations of pier dimensions to deck spans and depths and the treatment of the abutments, all become important, as shown in Figs. 1-23d–i. In multispan bridges, equal spans usually present few problems as any span becomes a minor component of the whole and the rhythmic repetition can be very satisfying. There is, however, the danger of such bridges creating a sense of boredom and some feature providing relief in the form of contrast may improve the design (Fig. 1-21).

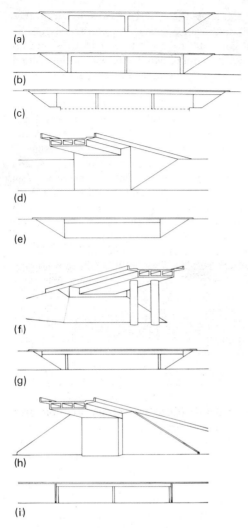

Figure 1-23   (a) Duality, (b) reduced duality and (c) effect of span ratios

The treatment of the abutments is very important. If they are too large in proportion to the other elements, they may mar the blending of the bridge with the banks of a gorge or the embankments of a roadway (Figs 1-23d,e). Side spans with smaller spill-through abutments are usually an improvement in roadway underpass bridges (Figs 1-23f,g and Fig. 1-24). The shaping of the abutment face also needs careful consideration in relation to the configuration of the underside of the deck (Fig. 1-23h).

In addition to the above, the general problem of relating bridge structures in an harmonious way to the environment, needs careful consideration of the location and siting of the bridge as well as that of the approaching roads so as to avoid any unnecessary spoiling (Figs 1-25a,b).

Figure 1-24 Motorway bridge in England. (*Courtesy* Freeman, Fox and Partners)

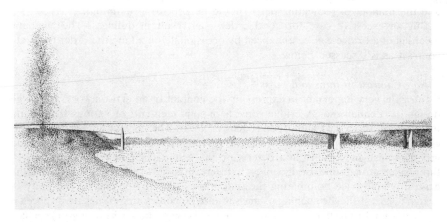

Figure 1-25a Bridge over the Rhine at Bendorf. (Sketch based on photograph in [5])

Figure 1-25b   Bridge over the Rhine at Worms. (Sketch based on photograph by Dyckerhoff and Widmann)

1.4.2 *Good order*

Good order is an aspect of unity of form and has been explained by Leonhardt [43] as

> . . . order of systems and order of direction of lines and edges. For bridges, order of systems means to choose a beam, an arch, a suspension bridge or a frame, but never to mix systems. Disorder cannot lead to beauty.

Whereas unity refers to wholeness, good order can be considered as one of the means towards attaining such wholeness. In practical terms, it means shaping, proportioning and aligning elements of the simplest forms possible in such manner that lines and edges are parallel or diverge, curve and intersect in a pleasing manner so that good order is manifest and that any appearance of haphazardness is avoided. A smooth flow of lines should connect different parts of the structure and the number of projecting parts should be kept to a minimum.

But an excessive care for good order may result in dullness. The aesthetic meaning of a bridge can be enhanced by occasionally breaking the order to create contrast as is discussed below.

1.4.3 *Contrast in form and mass*

Contrast is very important in expressing the content of an artistic work. In bridge design, it is equally so. A bridge consists of solid elements of various shapes and forms, contrasting also with open spaces or voids. The correct application of this concept of voidal or 'negative' spaces as a real part of the visual design, can add great interest and beauty and determine the essential character of a bridge. A three-span strut-frame bridge over a freeway is a very good example of the central voidal space dominating the real masses, thereby enhancing the unity of design (Fig. 1-18). The shaping of masses and the contrasting of negative spaces is thus very important because the discerning observer who looks at the bridge as a whole, will read these shapes as integral features which may very effectively break the monotony of repetition and be in sympathy with the landscape. The changing shadows during the day and with seasons can furthermore have an interesting effect on the relations of these shapes. Engineers with a narrow-minded analytical approach based purely on their structural knowledge will perhaps not see these forms so readily, but once their attention has been drawn to it, a new interest is usually stimulated.

The application of textured surfaces or colouring has been applied with varying success.

1.4.4 *Good proportion*

As described in Section 1.2, the proportions of the members of a structure have since the earliest of times been considered of importance and there was much debate on the matter by the philosophers of ancient Greece. Good proportions depend on certain effects of variation and contrast. However, the subject remains a controversial one due to the various attempts to produce geometric rules for good proportions, as discussed previously in this subsection. Whether we are psychologically conditioned to these rules or not, the fact cannot be disputed that some of these rules such as the 'golden ratio' do give pleasure to many people. The history of bridges, however, clearly demonstrates that concepts of good proportion can change with time and are largely influenced by improved materials of construction. In practice, good proportions for a structure and its members can best be established by trial and error, using judgement in comparing various proposals.

The proportioning of details requires great care, but it is the total effect achieved by combining members of various proportions that is most important. The ability to achieve a composition that is satisfactory comes only by practice and studying the works of others. A knowledge of art in general and its history of development is helpful to enable one to relate modern bridge design with older works that have been successful.

1.4.5 *Appearance of strength and stability*

The users of bridges require to feel safe and consequently it must not only in fact be strong and stable, but it becomes an aesthetic requirement that its visual form must generate a sense of security by appearing to be so.

By comparison with stone arches, the earlier steel structures were considered ugly, together with many products of the Industrial Revolution, and it took some time for the general public to adapt to the new materials and forms of construction. A lack of understanding of materials strong in tension initially lead to confusion and a sense of insecurity. This reaction was repeated in a somewhat different way with the introduction of prestressed concrete with invisible reinforcement or tendons which provided no obvious evidence of how a brittle material like concrete could resist bending forces. The whole situation changes, however, if the effect of reinforcement or prestressing in concrete is understood or is accepted by the uninformed after the reliability becomes evident. From the aesthetic point of view, a very definite preference for slender structures has developed. This has probably also been nurtured by a sense of achievement in the structural sense. Members should, however, not be unduly slender in all cases, even if modern materials and methods make it feasible, as it may aesthetically 'weaken' the structure.

1.4.6 *The statical form of the structure should be clearly expressed*

Provided care is taken to avoid the extremes sometimes understood under functionalism, the compliance with the requirements of Section 1.4.5 will normally dictate an articulated whole of interdependent structural parts. The structural requirements of strength and stiffness imply the equilibrium of forces and compatibility of deformations, thereby establishing a unifying law which contributes to the requirements of Section 1.4.1. The statical form can be emphasized to advantage by 'dissociating' certain parts that have different modes of structural function in resisting gravitational forces, even though they may interact effectively in resisting other actions such as, for example, wind and earthquake forces. Figure 1-10, which illustrates an arch in which the deck has been clearly separated from the arch, is an example where this approach has been adopted with good results. The Salgina Gorge arch designed by Maillart is a masterpiece in which an exactly opposite approach has been used to express the interaction between arch and deck in a smaller structure and in which the unified structural form is logical and aesthetically satisfying (Fig. 1-2).

Functional 'honesty' or 'sincerity', related to the properties of materials and the structural configuration, does not necessarily lead to beautiful structures, as discussed under Section 1.4.5.

1.4.7 *Ornamentation*

Ornamentation is seldom used in modern designs and is usually not required. It is, however, unwise to dogmatize that a bridge is only perfect if nothing can be omitted. Stark functionalism can be overdone. There are many good examples where some ornamentation has enhanced the beauty of a bridge and its sensible application must not be confused with misguided attempts to disguise the true function of structures. On the other hand, there are cases where it may be an improvement to disguise unsightly components, such as cable anchorages, bearings or unsightly discontinuities in the structure (Fig. 1-26).

1.4.8 *The need to obtain advice*

The experience and insight that constitute a fully developed aesthetic sensibility is not readily attained. It is therefore imperative to consult an architect, artist or sculptor qualified and experienced in the design of solid forms, and preferably with some understanding of bridge structures and experience in the aesthetic design of bridges. As with most problems in engineering, the young enthusiast in bridge design may feel he also has all the necessary understanding to appreciate aesthetics. He would nevertheless be advised at least to work on the design with

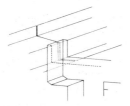

Figure 1-26 An example of an unsightly arrangement effectively disguised

engineers more experienced in aesthetic design. In this connection, it must however be emphasized that the approach of designing a structure in accordance with engineering requirements and then attempting to improve the appearance by minor alterations or by ornamentation, is to be decried in the strongest of terms. A beautiful design can only be achieved if the aesthetic design is developed as an essential part of the total concept. There must therefore be a full involvement of the architect, or whoever is consulted, from the very beginning when the basic form is being conceived.

The above does not imply that individual engineers cannot develop a fully mature ability in aesthetic design. There are ample examples of engineering works which compare well with the best works that architects have produced.

1.5 Factors that determine the suitability and competitiveness of bridges

The suitability of a bridge is often evaluated in terms of the purely functional requirements or utility (see Section 3.2), such as its reliability in the sense of performance, safety and durability, with total capitalized cost—including an estimate of probable maintenance costs for its serviceable life—being a prime consideration. As the functional requirements of a bridge can be fairly well defined in terms of specifications and drawings, the above mentioned procedures are reasonably objective and have in the past been accepted as a satisfactory method.

Since the 1960s, people have become more aware of the need to preserve what is referred to as the 'quality of life'; a term which is not easily defined, but amongst other things relates to the attainment of certain social and aesthetic standards and freedoms for mankind, while preserving as much of the beauty of the natural environment and its resources as is feasible and keeping it free of pollution. Likewise, engineers have come to recognize the importance and value of these considerations that extend beyond those directly related to engineering technology as practised in the past. Although there have always been exceptional designers who gave careful consideration to these matters, it must be admitted that this awareness has largely been generated by a strong reaction from various groups of nature lovers, ecologists, architects and others who can in this regard collectively be classed as environmentalists, even although some of them have at times gone beyond reasonable bounds. It is nevertheless a fact that functional and economic considerations tend to predominate because of the prime needs of our society and it remains for us to do the best with whatever means are available to satisfy the broader requirements of aesthetics as discussed in Section 1.4, as well as complying with reasonable demands related to environmental preservation. Unfortunately, many of these considerations cannot be quantified accurately because of their subjective nature.

Various procedures have, however, been developed for doing so-called 'impact studies' to assess the effects of a project on the environment and the inhabitants of the affected area. Various authorities require Impact Statements which are usually considered by interdisciplinary committees prior to approval of the project. A discussion of the subject is beyond the scope of this chapter and reference should be made to published work [51–58]. It has in many cases been found that with proper preplanning and recognition of the abovementioned factors, the overall design can be vastly improved with minimal extra expenditure.

It is consequently highly desirable that competitiveness should not be assessed entirely in terms of financial cost and that the concept of economy should be broadened to include these matters. The various procedures that are used for obtaining tenders or bids for proposed bridges, or for negotiating contracts, are given in Section 5.10. These usually only make provision for obtaining a product that satisfies specified functional requirements, as defined above, which serve as the basis for comparative assessments of alternatives. In many cases, the other subjective factors discussed herein may tacitly play an important role.

There are further factors that may influence the choice of materials of construction that do not relate directly to the actual cost as contained in a tender, but that are considered beneficial to local industry by the responsible authorities. This is quite justifiable if economic considerations are seen in a broader context.

It is consequently necessary that all those factors that will be taken into consideration should be clearly defined and that methods and standards of assessment should be clarified. Where consultants are appointed, their brief should very clearly state the broader criteria of design. Consultants on their part should where necessary take the initiative in these matters.

2 Design criteria for bridges

2.1 General

In order to carry out a brief for the design of a bridge, it is necessary that all design criteria be clearly defined or established. The extent of the designer's duties and responsibilities will depend on the information provided by the client body. This can vary from a bare minimum to elaborate codes of practice and design manuals as well as standard documents of general conditions of contract, general specification clauses, tender documents and procedures related to contract administration and supervision. As the standards applied by various authorities in different countries vary greatly, general principles only will be covered in this section, with reference to specific documents where it is necessary to illustrate certain points.

2.2 Class of bridge

The classification of bridges can conveniently be done in terms of their functional purpose, the configuration and structural systems, the materials and methods of construction used and the nature of the site. Under the heading of Functional Purpose would be indicated: the type of traffic, for example, class of highway, railway, airport runway, cycle or pedestrian traffic and the applicable geometric standards of the roadway; the type of pipeline or conveyor; the nature of the space that it must span, for example, an underpass bridge spanning a number of railway lines and service roads, or a bridge spanning a deep gorge or a wide river in flat terrain. Under Configuration, a description of the primary structural system that resists gravitational loads is usually adequate, but it may be necessary to elaborate the detail of secondary structural systems (see Section 1.3). Under Materials of Construction, the main classifications for concrete bridges would be under ordinary reinforced concrete, prestressed concrete or composite. The type of reinforcement, the type, method and degree of prestressing would be indicated as well as whether all the concrete is cast *in situ* or, if not, which elements are precast. If composite, the nature of the other material would be described.

The method of construction (see Section 5.3) is often important in bridge classification as well as the features of the site that would affect the design and construction method (see Section 2.3).

2.3 The nature of the site, topographical, environmental and foundation conditions

The broader implications of the impact of a bridge structure and the road system of which it is part, on the environment, have been referred to in Section 1.5. From a purely engineering point of view, one cannot overstress the importance of a proper inspection, survey and investigation of the site topography and subsoil conditions and of the probable hydraulic behaviour of a canal, river or estuary (where applicable). The malfunction or failure of many bridges in the past has been attributed to inadequate site investigations.

The practical procedures that are usually adopted for the selection of the most suitable site are described in Section 5.2. The actions that may be imposed on the projected structure will depend on the class of bridge and the site location, which will determine the probable nature and magnitude of actions due to natural causes. These are described in Section 2.4.

The soil or rock on which the structure is founded, or which is retained by parts of it, should be considered as extensions of the structure in order to ensure that soil–structure interaction is taken into account.

The properties of the soil should be established to a degree of accuracy that is sufficient in terms of the design assumptions made in order to achieve the required reliability. Upper and lower limits of estimated settlement, swelling or heaving, should be used to establish extreme effects on the structure.

The proper investigation of the subsoil conditions is thus essential. The nature and extent of such investigations will depend on the nature of the subsoil and the type of structure and its foundations and should preferably be planned in stages in conjunction with geotechnical experts in order to optimize the expenditure in terms of risk, as discussed in general terms in Section 3.2. The details of the necessary procedures, exploratory fieldwork and laboratory testing are beyond the scope of this chapter and reference should be make to suitable texts on the subject. The results of preliminary investigations may result in a revision of the planned investigations, or even a reconsideration of the type of foundations and even of the type of structure. Where short or long-term settlements are predicted, it is essential that the structure be designed to accommodate these or that the type of foundation or its founding depth be determined so as to limit the movements to within the permissible range for the particular structure envisaged. The possibility of scour action around piers or abutments in structures spanning flowing water should be investigated and the design adapted accordingly. The banks of gorges or embankments may require protection, for which various techniques are available, or stabilization by retaining structures or anchoring techniques.

2.4 Actions and effects on bridges

2.4.1 *Definitions of actions* [2]

An action is an assembly of concentrated or distributed forces (*direct actions*); or imposed or constrained deformations (*indirect actions*); applied to a structure due to a single cause. An action is considered to be a single action if it is stochastically independent, in time and space, of any other assembly of forces, or imposed or constrained deformations, acting on the structure. Actions can be qualitatively classified according to their variation in time, or space, or according to their dynamic nature.

2.4.2 *Classification of actions* [59]

It is convenient to subdivide direct actions into

(A) *Principal actions*, which are gravitational forces;
(B) *Supplementary actions*, which are usually applied separately in combination with associated principal actions on the basis of risk considerations.

Principal actions include

(a) *Permanent and long-term actions*, such as

(1) dead loads;
(2) superimposed dead loads;
(3) earth pressure due to retained fill;
(4) water pressure of excluded or retained water.

(b) *Transient and variable actions*, such as primary live loads due to

(1) vehicle traffic loading;
(2) railway loading;
(3) footway loading;
(4) cycle loading.

(c) *Short-term actions* such as erection loads. Dynamic or impact effects are usually included or allowed for by equivalent static impact factors.

Supplementary actions include

(a) *Transient secondary forces* due to primary live loads of highway traffic such as

(1) centrifugal forces;
(2) longitudinal braking and traction forces;
(3) forces due to accidental skidding;
(4) impact due to vehicle collision with bridge balustrades or parapets;
(5) impact due to vehicle collision with bridge supports;

and due to primary live loads of *railway traffic*, such as

(6) lurching effects which result from the temporary transfer of part of the live loading from one rail to the other;
(7) nosing which allows for lateral loads applied by trains to the tracks;

(8) centrifugal load effects on curved tracks;
(9) longitudinal loads due to traction and application of brakes;
(10) derailment loads.

(b) *Transient forces* due to natural causes, such as

(1) wind action;
(2) flood action;
(3) earthquake action (seismic forces).

Indirect actions include imposed deformations or restraint actions of long-term effects due to

(a) creep and shrinkage of concrete;
(b) parasitic prestress and prestrain;
(c) differential settlement;

and of short term effects due to

(d) temperature range;
(e) temperature gradient.

Reduced loads or the relieving effects of actions are taken into account (where applicable) with reduced partial load factors (see Section 3.4).

The responsible government authorities usually specify the actions (loads) that are to be applied in the design of bridges in their area of jurisdiction and require compliance with specific codes of practice covering the structural design. Only a few general comments shall be made herein.

Unfortunately, there is as yet little uniformity between the codes used in various countries. This is especially true of the design actions and various comparative analyses [60–62] of, for example, highway traffic loading as specified in countries such as the United Kingdom, France, Switzerland, the Federal Republic of Germany, Italy, the Netherlands, Belgium, the United States of America, Canada, Australia, South Africa and Japan, which show an astonishingly large variation in the type of loading models used and the magnitude of axle and total loadings. The order of these differences far exceeds real traffic differences. Various attempts made by members of the International Association for Bridge and Structural Engineering (IABSE) have not been successful in obtaining greater uniformity [63]. It would therefore appear that some countries are less conservative in terms of normal traffic, i.e. those vehicles that comply with the relevant regulations but then have to impose more severe limitations on abnormal loading.

2.4.3 *Highway traffic loading*

Traffic loading can conveniently be subdivided into three main classes [64]: normal traffic, abnormal vehicles and superloads. Most codes make provision for normal traffic and some form of abnormal (or military) loading. In some countries [59, 65–68], provision is also being made for superload vehicles which consist of slow-moving, multi-wheeled trailers with controlled hydraulic suspension and steering (either drawn by haulers or self-propelled) with payload capacities of up to 400 tonnes and a gross mass of over 600 tonnes. These vehicles use special routes and bypassing sub-strength bridges may be costly. Economic analyses indicate that the upgrading of new and even existing bridges, where feasible, can be justified on selected routes.

The loading models used for normal traffic vary considerably. Some authorities use groupings of wheel or axle loads, representing one or more actual vehicles, plus a uniformly distributed loading. Others apply formula loadings that reduce in intensity for increased loaded lengths, in conjunction with one or more knife-edge loads or wheel loads which are applied according to specified rules in order to approximately simulate the effects of actual traffic. These loadings either include impact effects, or otherwise require multiplication by an equivalent static impact factor. Various studies [69–73] have been made of the results of statistical information about the distribution of vehicles along highways. Insufficient statistical information is, however, at present available about traffic in general. It is not a simple random phenomenon, but is conditioned by the characteristics of a particular route, the type of traffic and such effects as the tendency for a queue to form behind a heavy vehicle which tends to accumulate other heavy vehicles which cannot overtake as readily as lighter vehicles. Furthermore, traffic behaviour is subjected to human direction and manipulation which may result in heavy closely-spaced vehicles in convoy. It can readily be shown that, except for the very small loaded lengths, the worst loading condition occurs under jam-packed (bumper to bumper) conditions caused by a traffic blockage and that the dispersion of traffic at speed, caused by the increased vehicular interspacing, more than off-sets the effects of impact.

Accidental impact loading is usually specified in the form of equivalent static loads to be applied at specified levels against balustrades and piers. This is a useful way to ensure a minimum degree of robustness, but it does not give a true indication of the actual effect of impact due to vehicles at speed. The effect is dependent on the impulsive response of the balustrades or piers and the load–time characteristics of the colliding vehicle. A correct dynamic analysis is highly complex so that present designs are largely based on tests, but progress is being made with theoretical studies of concrete structures under impact and impulsive loading.

2.4.4 *Wind action*

Wind action and its effects on a bridge depend on its geographical location, the local topography, the height of the relevant bridge member above ground and the horizontal and cross-sectional dimensions of the element under consideration. The maximum pressures or suctions are due to gusts that cause local and transient fluctuations about the mean values. The natural frequency of the bridge or section of bridge under consideration can also in large flexible structures have an influence on the wind action. The accurate determination of these effects is extremely complex. Reference should be made to suitable texts on the structure and behaviour of wind, its effects on and interaction with various types of structural elements and the aero-elastic response of structures thereto [32–35, 74, 75].

Wind actions are variously specified by equivalent static, or semi-dynamic methods. In these methods, the dynamic pressure head acting on the relevant part of the structure is either determined in terms of the maximum probable mean hourly wind speed, or the maximum probable wind gust speed based on short period gusts of 3 to 10 s, in a specified period (usually the expected lifespan of the bridge variously taken as up to 120 years). The effects on relevant members of the structure are determined by using various coefficients and factors which allow for the degree of exposure (depending on the surface roughness of the terrain, the degree of shielding and the effective height of the relevant member); local funnelling effects in the case of bridges in valleys or over gorges or where the bridge is sited to the lee of a range of hills or on an escarpment causing local acceleration of the wind; gusting effects dependent on the size and horizontal length or the height of the member to which the wind loading is being applied and a drag coefficient which depends on the shape of the relevant part of the structure and the effects of frictional drag thereon.

The procedures for doing more accurate dynamic simulations have, in recent years, been greatly improved through the use of computers. In the case of very long slender wind-sensitive structures or members, reference should be made to suitable texts that describe methods of doing dynamic analyses based on the statistical approach in determining wind forces and the aero-elastic structural response, as well as the possibility of aerodynamic instability as a result of wind-excited oscillations. It may in such cases also be necessary to do wind tunnel tests, but the natural frequency of vibration of concrete bridges is usually of a sufficiently high order to require only one of the simplified methods specified in various codes. This is not necessarily true of the cables of cable-stayed concrete bridges which have been known to develop excessive wind-excited oscillations. The degree of sophistication of the analysis required for concrete bridges and concrete components can usually be related to the probable maximum mean hourly wind speed appropriate to a return period equal to the expected lifespan of the bridge, the fundamental natural frequencies and the wind loaded lengths of critical members.

The following rule can be used as a very rough guide only:

> A concrete structure or any concrete member of the structure is unlikely to be susceptible to the dynamic effects of wind action if $n_0 > 4v/L_w$ or $n_0 > 0.5$ Hz, where n_0 is the fundamental natural frequency and L_w the wind-loaded length respectively of the structure or member being considered and v is the mean hourly wind speed.

2.4.5 Earthquake action

Similar approaches are usually adopted for earthquake design and reference should be made to relevant codes and texts [76–79]. See Section 4.6 and Chapter 15.

2.4.6 Restraint actions

The effects considered above all result from body forces or externally applied actions. Restraint actions on the contrary are due to induced dimensional changes or deformations which generate reactions in constrained structural members. The stress effects caused by these induced deformations depend on the stiffness of the relevant structural members and the rigidity of the restraints. On account of the non-linear behaviour of concrete as the ultimate limit state is approached and the resultant reduction in the effective stiffness of the relevant members, the restraint effects constitute a larger portion of the stress effects for a serviceability limit state condition than for an ultimate limit state condition. As most codes recommend that analysis should in general be performed for service actions imposed on structures consisting of members whose properties are evaluated on the basis of elastic material behaviour, an over-estimation of the effects of restraint action may be made in the case of certain ultimate combinations.

2.5 Factors affecting the choice of configuration, materials and methods of construction

Almost every major bridge that is designed seems to be unique. Except for a degree of standardization that has at times been achieved on certain projects for bridges in the short-span range, for example, freeway structures, bridge designers tend to produce results that reflect their personal preferences. Progress in knowledge, improved quality of materials and advanced methods of analysis and construction do naturally lead to a greater scope in possible alternatives, but personal ambition to innovate is probably also a strong motivating factor. This in itself is not necessarily a bad thing provided it does not lead to innovation for the sake of innovation and does not disregard the economic implications. It would appear that almost every useful type of configuration that is conceivable has been utilized to date, but almost endless variations or combinations thereof are certainly still available to be exploited.

Experience has shown that certain configurations (see Section 1.3) are most suitable under specific circumstances that can be related to the functional requirements, the site conditions, the magnitude of span(s) required and the materials and the methods of construction used. These factors are all interrelated so that there are usually several alternative solutions that may be competitive. *The objective of the designer is therefore to find the combination that gives an optimum solution in terms of the functional requirements and cost, as well as the broader issues referred to in Section 1.5 and based on the design philosophy described in Section 3 or such other as is prescribed.*

The procedure to be adopted in order to achieve this result is typical of most design procedures (see Fig. 3-1) which require the development of several conceptual designs incorporating all feasible alternatives and by the evaluation and comparison of these, reducing the number of possible solutions in stages by the process of elimination (see also Section 5.9). Past experience may enable a designer to commence with a greatly reduced number of alternatives. It is not easy to describe the conceptual process which, unless it amounts to merely adapting existing designs, requires innovation, the nature of which is briefly discussed in Section 1.4. As there are so many alternatives and details that are relevant to this discussion, it will only be feasible to discuss the more important factors in broad outline and to give general guidelines.

The interrelationships referred to above are important in the selection of the exact location and it may be necessary to develop conceptual designs at alternative sites. The practical procedures for site investigation and selection are described in Section 5.2. The practical details and construction methods referred to below are described in Section 5.

The profile and size of the space that must be spanned, together with the nature of the founding conditions, will usually narrow down the selection of the config-uration considerably. In the span range beyond approximately 350 m, the dead load is a governing factor so that only steel cable-suspended bridges have been built to date although concrete towers and substructures were used for the Humber suspension bridge with the longest span in the world of 1410 m (4626 ft). Cable-stayed bridges with steel decks are at present competitive in the range of 100–500 m (330–1650 ft).

In the range of 100–350 m, cable-stayed bridges with concrete decks are able to overcome the dead load disadvantage where the material and construction costs are lower. New developments will in all probability increase the range considera-bly (see Section 5.7). Suspension bridges with draped cables and suspension hangers, apart from being the most competitive solution in the immediate past for spans in excess of 500 m, have the added advantage that centering is not necessary for the deck construction which makes them eminently suitable for large spans over wide rivers or estuaries where the deck sections can be floated and hoisted into position to assemble the deck in a predetermined sequence. Cable-stayed bridges have a similar advantage and are also very suitable for construction over deep gorges, the deck normally being constructed by the suspended cantilever method with the cable-stays being applied after completion of the related segment of deck.

Concrete arches over gorges appear to be competitive up to spans approaching 400 m if the suspended cantilever method of construction is used, as described in Section 5.6.

In the medium-span range, concrete girder bridges with a wide range of cross-sections are suitable for construction by the cantilever method (with or without temporary suspension) in which travelling falsework carriages are used. A wide range of specialized travelling steel girders that span the piers are available to support falsework for *in situ* casting of concrete as well as support for precast sections placed by cranes or hoisted into position. The systems are advanced ahead of the concrete superstructure and eliminate support by centering except perhaps at midspans, say of larger spans. The use of these and other systems, described in greater detail in Section 5.3, in preference to centering, will depend largely on the height of the superstructure above ground, the total length of the bridge, which determines the number of repetitive uses, and the availability or cost of the specialized girders.

The risks involved in constructing any particular bridge may play a significant part in the assessment of the cost of construction and should thus be minimized.

The factors that determine the choice of the materials of construction are closely related to most of those factors already discussed in Section 1.5 and in this section. Apart from the structural properties, special reference must be made to durability (see also Chapter 4) in the particular circumstances of exposure such as the possibility of weathering, i.e. corrosion of steel, deterioration, discoloration, cracking and spalling of concrete, as well as fatigue under repetitive loading, any of which factors may not only have a significant effect on maintenance costs, but may disfigure or shorten the useful life of the bridge. The surface texture of the material and the variety of forms or shapes that can be constructed are very significant from an aesthetic point of view. The suitability of concrete in compari-son to other materials is discussed in Section 2.6.

In all these considerations, the location and accessibility of the site, the availability of materials, the scale of operations and the time factor are very important. The contractor's capital resources and his general turnover of work of a similar nature will determine the scale of economic operations, namely, how much he can invest in specialized temporary works, machinery and equipment in order to expedite the completion of the permanent structure. It is thus necessary to know what the capabilities and resources of available contractors are as this may have a crucial influence on the configuration of the bridge and the methods of construction. It is accordingly essential to take as many of these factors as possible into account when doing the conceptual design.

Various authors of papers and textbooks have prepared graphs comparing various types of bridge over the total practical span ranges. With the abscissa being span lengths, the ordinate values are either the cost in a specific currency at a specific date or the weight of steel for steel structures, or the volume of concrete for concrete structures, with percentages of reinforcement sometimes indicated. The cost may be for the superstructure only or may also include the substructure and the foundations. These graphs however have limited usefulness in giving a very approximate indication of relative costs. Practical experience has shown that if the optimum solution is to be found, every project requires a separate analysis of all feasible alternatives except that previous experience under the particular circumstances can reduce the initial work drastically. The factors that influence costs as described previously are many and varied and change with time. They may also differ substantially between countries or even districts because of local industry and vested interests.

Figure 3-1 is a greatly simplified abstraction of the design process, but demonstrates the method quite clearly. A large number of additional cycles of analysis and feedback are required to optimize the details.

2.6 The suitability of concrete as a bridge material

The properties and many uses and advantages of concrete in its various forms are described in several other chapters. It is especially suitable as a bridge material as has been proven by the ever-increasing use thereof in most parts of the world relative to other structural materials. The reasons therefore are mainly functional and economic, but the aesthetic advantages are considerable. It can be readily moulded into almost any form that is desired and, if constructed with care, has a texture which blends well with the natural environment. The basic requirements to satisfy aesthetic standards, as described in Section 1.4, can, in general, more readily be satisfied in concrete. This applies to almost all configurations in the short and medium-span range and for arches up to 400 m span. However, even concrete cable-stayed decks can be very pleasing to the eye as they blend in well with the concrete towers and substructures while contrasting effectively with steel cables. On the other hand, the Humber suspension bridge is a good example where a steel-decked suspension bridge has been very satisfactorily combined with concrete substructures and towers.

Although it is a brittle material, the properties can be transformed from that of ordinary reinforced concrete, through partial prestressing to full prestressing, to have the required structural properties at the lowest cost. Prestressing is usually applied only in one or two directions corresponding approximately with principal tensile stresses with ordinary reinforcement taking transverse, shear, torsional, secondary and bursting stresses. In the case of structural components involving high tensile stresses in several or varying directions, prestressing has, however, been applied in multiple directions.

Concrete as a material can thus be moulded into virtually any shape that is required, can be coloured by using suitable admixtures and given various surface textures for aesthetic effects. By the use of suitable reinforcement, it can be given triaxial structural properties to suit the specific requirements (see Chapter 10). It can furthermore be cast *in situ*, precast in a factory or yard and transported to the site and placed on the works in a size and form to suit the circumstances and the contractor's resources. It can be combined with structural steel to form a composite construction (see Chapter 17) in which form it very suitably forms the deck of a steel girder bridge.

In regions far from steel producing industrial areas and where concrete aggregates and cement are readily available, it has the advantage of a big saving in transportation costs compared to structural steel.

Although concrete may suffer various forms of deterioration as described in Chapter 4, the record for well-designed and constructed concrete bridges is good and could with greater care to detail, be even better. Especially in areas with very severe climates and aggressive environments, great care is required to ensure a high quality of concrete with adequate cover to the steel reinforcement. It is essential to avoid recesses that might accumulate soil and moisture which could more readily lead to corrosion of the reinforcement if the concrete has areas of lower density or large cracks in the vicinity. Although they cannot always be avoided, good design and detailing can minimize the number and size of cracks. Shrinkage and temperature effects are major causes of cracks and full account should be taken of their effects with the provision of movement joints where necessary.

2.7 Properties of reinforced/prestressed concrete as a material of construction

The structural properties of concrete as seen by the designer are described in Chapters 3, 4, 10, 11, 12, 16, 19 and 20. Reference should be made by the designer to the codes of practice that are applicable to the project he is involved in for the basic assumptions and definitions that prescribe the material properties and design methods. Where outdated codes are still in use, it is recommended that safety aspects be checked in terms of the latest proposals contained in the CEB–FIP model code for concrete structures as contained in Volume II of [2], pp. 53–78 (including the notes), or such other practical code that is based on the principles contained therein (see Section 3.4 and Chapter 40).

3 Design philosophy

3.1 General

The very notion of structural design has, even before it was properly understood, always implied that safety was of prime concern. The concept of safety has developed from intuitive understanding, to the modern philosophy of structural reliability [80] which represents a significant advance in putting structural engineering on a more rational basis. Progress has not, however, always been rapid and misconceptions are common even today. The lack of proper understanding of the functional behaviour of structures due to imposed loads or influences, does not, however, in general, appear to have greatly impeded engineers in the development of new and great structures to meet the ever-increasing demands of society. The remarkable enterprise and courage of engineers are demonstrated by the history of the development of bridges as related in Section 1.2. They had to rely almost entirely on empirical knowledge of structural design prior to the second half of the 19th century. These early designers did not, however, always succeed at the first attempt and many failures were experienced.

The history of the development of suspension bridges and the systematic study of wind forces, initiated by the Tay Bridge disaster in 1879, is well documented. In spite of this work, the spectacular failure of the Tacoma Narrows Bridge occurred because of lack of understanding of the nature of the torsional oscillations induced by wind action. The extensive research which was initiated as a result of this failure, accomplished the dual purpose of determining the mechanism of the wind action which caused the failure and ensuring the aerodynamic stability of the proposed design for the new bridge. These failures illustrate that man has always been prepared to undertake the design and construction of large engineering structures beyond his experience and full understanding. On the average, these structures had a reserve of strength resulting from the so-called factor of ignorance that was usually applied.

The development of modern high-strength steels which do not have a corresponding increase in fatigue properties and which are subject to the added risk of brittle failure, and the tendency to use very high-strength concretes, coupled with the greatly enhanced ability of the modern engineer to incorporate every part of the potential structural capacity in the strength analysis—thereby eliminating previously unaccounted for reserves—have made a sound understanding of structural behaviour and the nature of the applied forces imperative.

In spite of the accomplishment of major structural engineering feats in the past, the concepts of safety and risk have not until recently been very well understood. From an initial intuitive understanding based mainly on experience gained from studying failures, a misguided notion of the safety of structures was unfortunately fostered by the comprehensive deterministic design procedures that were developed since the theory of structural analysis was placed on a sound footing by men such as Navier. It was generally believed that the factors of safety which were applied to breaking or yield stresses of materials in order to determine the safe working loads, were an almost absolute measure of safety dependent only on the accuracy of linear structural theory and the material parameters. Furthermore, most major failures resulted from gross or conceptual errors so that the shortcomings of these deterministic procedures were not self-evident.

In spite of the development of wide-ranging research and testing techniques, the working stress method was until recently accepted as a satisfactory design method. Many practising engineers failed to realize that it did not result in consistent levels of safety.

The introduction of ultimate load checks based on an approximate theoretical simulation of the behaviour of structures at failure, although still deterministic, greatly improved the situation, but was not generally applied although the application of a degree of redistribution of bending moments in the working stress methods became common practice. However, there had been an awareness of these inconsistencies by men such as Prof. A. Pugsley [81] who suggested the application of ultimate load methods with multiple load factors for various types of loading.

3.2 Structural reliability and utility

Further attempts to rationalize the situation on theoretical lines led to the classical theory of structural reliability, which can be defined as the 'measured chance' that a structure will support the loads to which it is subjected. Theory of structural reliability and the concepts of 'risk of failure' and structural 'utility', as developed by A. M. Freudenthal [82] and others, attempt to take into account the many uncertainties in the assumed applied forces, the properties of the materials and the construction techniques being used, by quantifying the relevant values in terms of statistical concepts and predicting the expected shortest operational life of an engineering structure by the application of probability theory. In classical reliability theory, the probability of survival or reliability is $p_s = 1 - p_f$, where p_f (the probability of failure) $= p(R < S) = \int_0^\infty F_r(x) f_s(x) \, dx$, where F_r and f_s are respectively the probability distribution and density functions of R and S, the structural capacity and load effect respectively.

On the basis of this idealized theory, greater clarity of the fundamental principles of structural engineering has been achieved, but although this development has in a qualitative sense been a major breakthrough, it was not of immediate practical significance because of the difficulties in accurately quantifying probability levels due to lack of knowledge of the distributions of the relevant probability density functions.

The extended reliability concept by Ang [83] is an improvement on the above formulation as it eliminates the sensitivity of the semi-probabilistic methods to the shapes of the tails of the probabilistic distributions by defining failure as the probability that $R < NS$, where N is a 'judgement factor', and is necessarily greater than 1.0 to take account of unknown uncertainties.

The principles of reliability analysis, related more directly to safety, have been well documented and further research is an on-going activity. The last two decades have seen the growth of a very substantial literature on the theory of structural reliability based on the statistical mean values and coefficients of variation which are more readily determinable than the extreme values of the above mentioned probability density functions which are necessary for the determination of the probabilities of failure in classical theory.

The first-order second-moment reliability analysis proposed by Cornell [84, 85], and further developed by Lind and others, is generally considered to be an improvement on the above proposals. Cornell's format provides a method based on distribution-free assessments and within this simplified framework, approximately consistent reliabilities can be obtained. Since the early work of Cornell, there has been considerable progress in the development of these methods known as Level 2 methods. Level 2 is a probabilistic design process used principally in assessing appropriate values for the partial safety factors in the Level 1 method; thus it is intended primarily as a tool for code-drafting committees. The Level 1 method is the basis of various modern codes using the limit state design method. It is not possible to deal adequately with reliability analysis in this chapter. For an excellent summary the reader is referred to Esteva [86] and for fuller details, to the CIRIA Report 63 of July 1977 [87].

Ultimately, the aim is to achieve optimal structural reliability by maximizing effectiveness expressed in terms of an objective function (utility function linear with money).

3.2.1 *Structural utility function*

$$\text{Structural utility} = B - C_t$$
$$= B - C_p - E_d$$
$$= B - (C_i + C_m) - \sum (S_m \times p_m) - \sum (u_n \times p_n)$$

where

$\quad B$ = expected present value of the overall benefits derived from the existence of the structure (positive utility)

$\quad C_t$ = total capitalized costs (negative utility)

$\quad = (C_p + E_d)$ = loss function

$\quad C_p$ = capitalized prime costs $= (C_i + C_m)$

$\quad C_i$ = initial cost

$\quad C_m$ = capitalized normal maintenance costs

$\quad E_d$ = expectation of damages

$\quad S_m$ = capitalized cost of damage or loss due to non-compliance with serviceability criteria

$\quad p_m$ = probability of exceeding a serviceability limit state

$\sum p_m S_m$ = risk of exceeding a serviceability limit state

$\quad U_n$ = capitalized cost of reaching an ultimate limit state

$\quad p_n$ = probability of reaching an ultimate limit state

$\sum p_n U_n$ = risk of reaching an ultimate limit state

The generalized reliability can be defined as

$$R = \frac{C_p}{C_t} = \frac{C_t - E_d}{C_t} = 1 - \frac{E_d}{C_t}$$

The various values that make up an important part of the terms denoted by B, S_m and U_n are, however, at present irreducible to a form that can be accurately quantified in practice. These concern human life and various subjective values that fall broadly in the domain of the aesthetic. Whether the 'economic probability' basis of the above comprehensive objective function can in practice be developed into a form that will be quantitatively meaningful is debatable in the light of the subjective content of some of the judgments that are an essential part of the process of assessing the relevant factors. In this connection, Bayesian analysis [84, 88] promises to become a useful tool for the rationalization of the decision-making processes that depend on both subjective and objective information. There are also legal and ethical implications arising out of design procedures based on those concepts that need careful consideration and the fact that a structure is at risk during its planned lifetime has yet to be accepted by many controlling authorities and clients and even by the design profession as a whole. Despite these seemingly insurmountable problems, this comprehensive approach is nevertheless a sound and necessary formulation and it is important to go through the exercise of doing such evaluations even if only on a qualitative basis.

A useful result can be derived from a simplified version of the utility function applicable to a single member–single load case as follows. The probability of ultimate failure can be expressed as a decreasing function of capitalized prime cost or $p_n(C_p)$. It can be shown that the cost of failure, U_n, is usually only weakly dependent on C_p as the consequential losses can be far more serious and, in the range of high reliability, only minor variations in C_p are required to ensure significant increases in reliability. U_n can thus be considered to be constant. According to Lind and Davenport [89], the probability of failure may be written as

$$p_n = \exp\left[-f(C_p)\right] = \exp\left[-f(C_p(F))\right] = \exp\left[-g(F)\right]$$

where $g(F)$ is a function that depends on the distribution of the safety factor F.

This function, if developed in terms of several mathematical distributions that arise out of various assumptions regarding the distribution of strength and resistance (normal, lognormal, extreme or Weibull type distributions), clearly illustrates the sensitivity of p_n to assumed distributions. More important is the fact that $\ln(p_n)$ is nearly linear as a function of the safety factor so that $p_n \cong b \exp\left[-C_p/c\right]$, where b is a constant and c is the attenuation cost, i.e. the cost of reducing the probability of failure by a factor of $e(\cong 2.7)$.

We can therefore express the total cost as

$$C_t = C_p + U_n p_n \text{ (for single member–single load case)}$$
$$= C_p + u_n b \exp\left(-C_p/c\right)$$

The derivative of the total cost is

$$dC_t/dC_p = 1 - (U_n b/c) \exp\left(-C_P/c\right)$$
$$= 1 - p_n U_n/c$$

Equating this to zero gives the useful result that the total cost is minimum when the expected cost of failure, $p_n U_n$ (i.e. risk), equals the attenuation cost, c, i.e. $p_{no} U_n = c$, where the subscript o refers to the optimum.

For the further development of the theory applicable to multiple failure mechanisms etc., the relevant references [89] should be consulted. Theoretically, the same principles could be applied in the formulation of contractual clauses covering construction tolerances on dimensions, strength and other material properties by penalizing the contractor for defective work on the basis of a loss function as an alternative to necessarily requiring replacement or strengthening of the defective portions [90].

3.2.2 The uniqueness of bridges

The reliability analysis applied to electronic systems and small mechanical systems is a sub-discipline of mathematical statistics in that the multiple replication testing of component parts is possible. In bridges, the problem is different. Although it may be possible to proof-load component members or parts of bridge structures, the characteristics of the whole system can only very approximately be established by model testing. In the design of large structures, system reliability over and above member reliability is thus very important as it is an elementary principle that larger systems are less reliable than smaller systems. In the context of bridges, larger systems would imply more and longer members without any substantial increase in redundancy. The development of structural reliability analysis for the design of bridge structures is thus very important if significant progress is to be made in the design of large bridges.

3.2.3 The concept of risk $(\sum p_m S_m + \sum p_n U_n)$

For the purposes of design, the optimal level of risk could ideally be established by maximizing the relevant utility function or, in a less comprehensive form, minimizing the loss function, C_t. As mentioned above, there are, however, practical problems that have as yet not been resolved. The idealized procedures of classical reliability theory applied to structures are not amenable to accurate solutions due largely to the lack of statistical knowledge about the extremes (tails) of the probability density functions of the applied forces and the material properties and physical processes that determine the behaviour of structures. There are, however, various considerations that give positive guidance to the order or risk that is reasonable in terms of present socio-economic standards.

In the first place, it is an accepted principle of code theory [89] that new codes should be calibrated on the basis of existing standards proven by past experience so as to effectively obtain the same reliability and to adjust values only as additional knowledge and information are gained or as the socio-economic demands change.

Secondly, it has been demonstrated very convincingly by Ligtenberg [91] that an upper bound to practical reliability is in fact not determined by what are usually considered to be random quantities (i.e. dimensions, material properties, loads, etc.), but by other risks outside of this 'normal' range of fluctuations, such as those due to gross structural errors, fires, explosions, collisions, severe corrosion and various types of abnormal or catastrophic excitation. An analysis done by him of damage or failure due to these 'abnormal' events in Holland, indicates probabilities of occurrence of the order of 10^{-2} to 10^{-3} within the useful life of a building structure, as against the low figures of 10^{-4} to 10^{-6} usually assumed for structural collapse due to the random events referred to above. In the light of these studies, it would appear that the latter values are rather conservative.

It is however clear that the profession and the general public will not at this point in time be prepared to accept a code which has as its basis a procedure whereby the risk of collapse of permanent structures such as buildings and bridges is knowingly increased relative to present standards which require that such risk shall be virtually non-existent. On the other hand, it can be shown that the cost of most structures is not very sensitive to variations of the probability of failure on the basis of classical reliability theory and that the cost paid for being slightly conservative is not very great. This statement must, however, not be misinterpreted, for injudicious over-design can be very wasteful without necessarily improving safety.

Our immediate conclusion is therefore that, although reliability is of prime concern, the ultimate refinement of the design of a specific structure is not principally related to reliability, but rather to other factors such as its configuration and the materials and the methods of construction, all of which have a more significant influence on the cost. In many structures, only certain members are critical in the sense that failure of a member could cause total collapse of the whole or a large portion of a structure. Failure of the balance of the members may at worst cause only local damage. It therefore stands to reason that those critical members, or combinations of members, should have lower probabilities of failure than the less critical in order to maximize the utility function.

Similar arguments apply to larger members which theoretically have an increased probability of failure and to large structures such as tall buildings and bridges consisting of a large number of members with many possible modes of failure. Redundancy, on the other hand, has the effect of reducing such risks. The importance of system reliability as against member reliability is thus clear and it may be necessary in specific cases, where a risk exceeds an acceptable limit, to modify the design so as to comply with the fail–safe principle.

Where, however, certain critical members such as, for instance, the piers of a bridge have a very low level probability of over-loading due to traffic loads but may fail with catastrophic results due to the impact of a heavy vehicle or a ship in the case of bridges spanning navigable rivers or estuaries or shipping lanes, even a probability of failure based on design for normal loads as low as 10^{-6} may be meaningless unless special precautions in the form of protective structures or islands are taken to protect the piers against collision. Piers should therefore be designed more conservatively because of the increased implications of failure and it is clear that in order to achieve more consistent levels of safety, cognizance must be taken of abnormal events. Whereas it is reasonable to reduce safety levels for elements that are less critical, it is essential to establish system reliability on a proper assessment of risk.

These principles are equally important in the design of temporary works for large bridges where the assessment of risk is very important in optimizing constructional costs.

3.2.4   *The risk of gross errors*

Gross errors as distinct from random and systematic errors, will, however, remain the most serious threat to the safety of structures. The underlying causes can be roughly classified as due to the following:

(1) A conceptual misunderstanding of one or more aspects of structural function on the part of the designer or analyst.
(2) The use of incorrect assumptions as the basis for the design.
(3) Gross computational errors.
(4) A breakdown in communication (specifications, drawings, instructions).
(5) Undetected flaws in materials and serious omissions or errors in the execution of the work (workmanship).

All these errors, except those in Category 1, can be eliminated by systematic control using checking procedures. Category 1 errors, however, may require a much higher level of activity in that their elimination presupposes the recognition

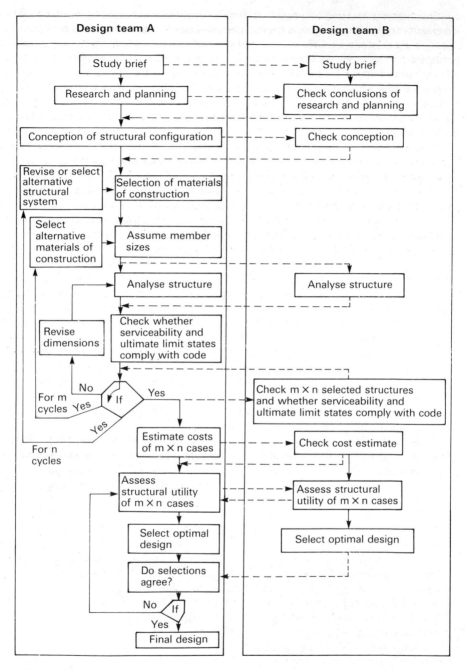

Figure 3-1 Simplified flow chart for limit state structural design with independent checking

and complete understanding of all the critical problems and an ability to analyse and solve them in terms of the specific requirements. Major structures such as bridges should consequently only be entrusted to experienced engineers of proven ability.

The other categories can be eliminated by developing systematic checking and feedback procedures which are incorporated into all stages from research and planning through conception and design to the detailing and construction stages. The normal design processes can be represented by the simplified flow chart shown in Fig. 3-1.

The addition of independent checking procedures, shown in broken lines, if executed in a systematic routine manner, would greatly reduce these risks. Unfortunately, these procedures can be costly.

If the cost of design were to be considered as another term of the utility function, it would be clear that such cost should not necessarily be proportional to the cost of the structure. In fact, increased expenditure on design reduces structural cost and risk. In the case of major structures, the latter factor may be a prime consideration.

3.2.5 *Integrated reliability analysis of bridges*

Whereas in conventional design the various aspects of analysis are separated by specifying loads and other influences without regard to the behavioural characteristics of the structure and vice versa, reliability analysis recognizes the interrelationships that demand that an integrated analysis be carried out.

This applies particularly to dynamic effects and a sophisticated analysis which integrates the action of the applied forces with the dynamic response becomes essential if meaningful predictions of the probability of failure are to be made. Analysis at this level involves rather complex problems in the theory of stochastic processes and probabilistic structural dynamics. Considerable progress in this field has been achieved by Davenport and others, who have developed the spectral analysis method for the effects of wind forces on building structures [92]. The ultimate objective has been to abandon the safety-factor approach of conventional design methods based on low occurrence intervals, which does not guarantee a consistent level of reliability, and rather to base the analysis on the response characteristics associated with a much longer recurrence interval. This load level would then be associated with a safety factor which is much closer to unity [93].

In theory, it is necessary to develop probabilistic models of the various possible combinations of load actions and other influences on the structure and to establish all possible modes of failure. In bridges, these modes of failure can be very complex, consisting of brittle, ductile or mixed types of behaviour of modes which may vary from extensions of a 'weakest link' model to multi-load combinations, to models that may suffer collapse due to progressive conditional failures [94].

It is, however, usually possible to identify the more significant modes and to use a simplified model of the structure. Nevertheless, even such simplified models are far more complex than the fundamental reliability case where a load system defined by one single parameter acts on a structure with only one potential failure mode. Although the more refined models required to represent bridges can be reduced to what are conceptually fairly simple systems, their analysis remains highly complex. The main object of research in this field at present is to establish typical relationships between system and member reliability for the different levels of damage ranging from cracking and deflections to ultimate collapse, and to choose safety factors consistent with knowledge of determinate structural phenomena as well as information on random properties of loads and resistances. In the meantime, a significant advance has been made by the so-called 'engineering' approach, which is based on probabilistic concepts but which aims to maintain the simplicity of existing design codes.

3.3   Codified design

The object of codes of practice should be to maximize utility from the viewpoint of society but, for the reasons given above, the present state of knowledge is such that interim codes will have to be expressed in less advanced terms in such a manner as to facilitate their evolution towards a more ideal form.

The CEB model code [2] and the practical codes such as the British Standards Institution's CP 110: 1972, *The structural use of concrete*,[1] and BS 5400: Parts 1–10[2] [95], are Level 1 semi-probabilistic codes based on the limit state design method. These codes represent an advance in that the uncertainty of design data is recognized and expressed in statistical terms and the use of different partial factors to check the structure for the serviceability and ultimate limit states under various conditions of loading and other influences, reduces the inconsistencies that are apparent in the working stress methods of previous codes. There are nevertheless shortcomings. The characteristic or nominal values for the material strengths and design forces, which are arbitrarily defined in terms of normal distributions, do not necessarily ensure a consistent level of reliability, especially in the case of certain members subjected to stress reversals. The partial factors were introduced largely to differentiate between influences that exhibit greatly varying margins of safety and which could not be covered adequately as well as consistently by a single global factor. This is, however, a great improvement in that satisfactory levels of safety are ensured.

3.4   Modern codes of practice for bridges

Codes of practice prepared by various authorities in different countries and intended for general use in the design and construction of concrete structures, while mainly orientated towards building structures, are discussed and compared in Chapter 40. Some authorities responsible for the construction of bridges, or institutes promoting the use of concrete, have accordingly produced specialized codes such as the *Standard specifications for highway bridges* (12th Edition, 1977) adopted by the American Association of State Highway and Transportation Officials and the *ACI manual of concrete practice: bridge analysis and design ACI 343R-77* in the USA. In the United Kingdom, the BSI has published BS 5400, *Steel, concrete and composite bridges*, Parts 1 to 8 and 10 (1978–1982). In the case of the aforementioned AASHTO and ACI codes, reinforced concrete members are designed either with reference to service loads and allowable stresses or, alternatively, with reference to load factors and nominal strengths multiplied by

1   Amended to November 1980.
2   All parts except Part 9 have been published between 1978 and 1982.

strength (or capacity) reduction factors. The capacity reduction factors depend on the statistical variance of the type of resistance being considered.

The codes that are currently applicable in most countries are still based on the working stress or factor of safety method with a deterministic philosophy. Many authorities are, however, in the process of reviewing the situation in the light of proposals such as that of the CEB model code, (see also Chapter 40). The UK has been the forerunner in this respect by producing BS 5400. It consists of ten parts:

(1) General statement;
(2) Specification for loads;
(3) Code of practice for design of steel bridges;
(4) Code of practice for design of concrete bridges;
(5) Code of practice for design of composite bridges;
(6) Specification for materials and workmanship, steel;
(7) Specification for materials and workmanship, concrete, reinforcement and prestressing tendons;
(8) Recommendations for materials and workmanship, concrete, reinforcement and prestressing tendons;
(9) Code of practice for bearings;
(10) Code of practice for fatigue.

As a code it is based on the same philosophy and in detail is similar to CP 110, which is a great advance on previous codes. These codes were initially not readily accepted by the practising fraternity in general because of the increased complexity and have not immediately resulted in more economic structures because of the calibration procedures used for the partial factors, as explained in Section 3.2. There are differences of opinion of the consistency of certain design procedures, such as for determining the shear and torsional resistance, which are largely empirical and give very different results from the more recent procedures proposed by Thürlimann [96, 97].

The CEB model code allows the use of both methods as alternatives, apparently reflecting a difference of opinion amongst the responsible committee members. The so-called 'accurate' approach is based on a simplified theoretical model of beams which are considered to consist of discrete but contiguous compressive members acting in the approximate direction of the principal compressive stresses to form a highly redundant shear or torsional truss system in conjunction with tensile reinforcement or prestressing. Compatibility equations for the simplified systems are satisfied.

4 Methods of analysis of bridge structures

4.1 General

A brief account is given in Section 1.2 of the historical development of the capacity of engineers to analyse bridge structures from the early crude efforts based largely on intuition, to the more systematic empirical procedures which were developed rather spasmodically during specific periods of civilization by the accumulation of rules based on experience and tests. The early engineers were practical men who failed to benefit fully from such primitive theories as were developed by their contemporaries amongst the philosophers. However, the mathematicians and physicists of the 17th century and onwards, played a major role in producing theoretical formulations that eventually made it possible, by the end of the 19th century, for bridges to be designed and analysed almost entirely on the basis of theoretical methods using parameters for material properties and applied forces that had been predetermined by experiment. Initially, many of these theories had been grossly over-simplified abstractions which at best gave only very approximate solutions. The range of practical configurations was accordingly limited and it was common practice to make structures statically determinate by the insertion of hinged joints or other discontinuities in order to simplify the analysis.

Following on the initial developments referred to above, the theory of elasticity, as applied to structures, developed rapidly and solutions were found for many of the relatively complex problems presented by structures with redundant members. The most significant development in the linear structural analysis of statically indeterminate structures has been the influence coefficient method based on the work of Clerk Maxwell, Alberto Castigliano (1847–1884) and especially H. Müller-Breslau in Germany (1886) and Camillo Guidi [6] in Italy. The extreme complexity of many types of structure has, however, provided serious barriers to progress because of the intractability of the mathematical equations of analysis and engineers were forced to rely on greatly oversimplified theories in order to obtain workable solutions. It is an historical fact that engineers have seldom balked at any problem that barred progress for lack of precise theories of analysis, but they have unfortunately also paid a heavy price in failures. Venturing into bigger and unknown terrain by extrapolations is extremely risky without a complete understanding of the relevant problems. Excluding cases of gross negligence, almost all recent major collapses fell into this class of error. Many of these were due to effects which in terms of simplified theories were considered to be of secondary importance but actually resulted in serious overstressing or instability.

The advent of numerical and especially iterative methods such as the moment distribution methods of Hardy Cross and the relaxation methods of Southwell, introduced a new era to analysis. Whereas the period preceding these innovations was characterized by a great variety of special methods, each of which provided the best solution for a particular problem, it was now possible to adopt a more general method. Although the advent of the electronic computer in the 1950s gave this approach tremendous momentum, the more significant development resulting therefrom was the return to a generalized mathematical formulation of equations, which had now become feasible for extremely complex problems. The matrix algebra which had been developed almost a century earlier by Arthur Cayley as an exercise in pure mathematics, presented an ideal means of formulating problems for processing by computer and, together with the very compact notation of modern mathematics and the invariance principle as used in tensor analysis, has been a major advance in obtaining general methods that are applicable to a much greater range of problems. Whereas initially most methods of analysis were formulated in terms of flexibility matrices, the advent of large-capacity computers has made the more powerful stiffness approach feasible.

The natural result of these developments was the emergence of integrated engineering computer systems [151] which provide a very comprehensive facility for the solution of interrelated problems. In spite of the inherent advantages of these programs, a considerable number of special programs are being written for repetitive problems because of the increased economy or because of the specific nature of the problem. Whereas the aforementioned systems, which operate on large-capacity computers, are used to solve complex problems, the bulk of engineering design problems encountered in bridges can be effectively processed on the minicomputers or self-contained programmable calculators (see Chapters 21 and 22).

Whereas the mathematical models which served as abstract representations of structural behaviour in the past were in many ways oversimplified because of the limited means of analysis available, closer approximations to the prototype are now feasible with the application of methods such as the finite element and finite strip methods of which there are several versions and which are developing very rapidly in scope and refinement to something which, in conjunction with large capacity computers, is potentially the ultimate tool for analysing structural behaviour in complex three-dimensional structural members (see Chapter 38). The limit of accuracy of this technique is at present only dependent on available computer capacity and the economics of analysis. Likewise the application of interactive graphics and computer-aided design (CAD) is developing into the ultimate design tool by eliminating many of the analytical stages of design and detailing as at present constituted.

It is generally accepted practice to use linear elastic methods in the analysis of bridge structures. These results can readily be applied to the serviceability condition of limit state analysis, but require appropriate factoring for the load effects of the various load combinations for the ultimate limit state. For critical cases, it may however be necessary to do non-linear analyses, e.g. in cases where buckling or large deflections may generate additional stresses or cause a member to approach instability. Where the assumption of partial or total plasticity at cross-sections where maximum stress conditions occur, implies a more general non-linear behaviour of the structure, it may require further investigation.

4.2 Mathematical simulation of material properties

The basic assumptions which determine the mathematical properties of structural concrete and steel are usually defined in codes of practice issued by the responsible authority. These properties relate *inter alia* to the stress–strain relations, the effects of cracking of concrete and the modes of failure and the yield and ultimate strengths of steel and concrete respectively. The assumptions and properties recommended by the CEB–FIP are described in the *Model code for concrete structures: structural design and detailing* [2]. Further information can be obtained in the relevant chapters of this book. For further information relevant to concrete bridges, reference should be made to [2, 59, 68, 95, 98, 99].

For linear elastic analysis of structures subjected to static or quasi-static loading of short duration, it is accepted practice to use the elastic concrete properties of the monolithic concrete to determine the member stiffnesses for both the serviceability and ultimate limit states. The effective Young's modulus of concrete used for analysis is taken to be the secant modulus which is approximately

$$E_{sec} = 0.9E_c \tag{4-1}$$

where E_c is the initial tangent modulus of concrete.

The torsional stiffness, k_t, of bridge deck members can play an important role in the analysis of bridge structures and must be given more consideration than is normally required for beam members in buildings. Due to the presence of microcracks, the torsional stiffness is substantially reduced relative to the elastic stiffness of the section unlike the flexural stiffness which may only reduce by a few per cent when the tensile zone of the concrete starts to develop cracks. This reduction in stiffness can manifest itself already at the serviceability level of loading and is fully developed at an ultimate limit state level of loading. Using the formulation for torsional stiffness

$$k_t = GJ/L \tag{4-2}$$

where G is the shear modulus, J the torsional moment of inertia of the section in the uncracked state, and L the length of the member.

The CEB–FIP model code [2] recommends the following values for G in the absence of more accurate methods:

$$G_I = 0.30E_{sec}/(1+1.0\phi) \tag{4-3}$$

$$G_{II_m} = 0.10E_{sec}/(1+0.3\phi) \tag{4-4}$$

$$G_{II_t} = 0.05E_{sec}/(1+0.3\phi) \tag{4-5}$$

where I refers to the uncracked state of the section, II_m to the state when flexural cracks have formed, and II_t to the state when torsional and shear cracks have developed; E_{sec} is the secant modulus defined in Eqn 4-1 and ϕ the coefficient of creep of concrete to be used for long-term loading.

When the level of torsional and longitudinal reinforcement is known, a more accurate expression can be used for state II:

$$G_{II}J = \frac{E_s A_{ef}^2}{\dfrac{u_{ef}s}{2A_s} + 1.5\dfrac{E_s}{E_{sec}}\dfrac{u_{ef}}{h_{ef}}(1+\phi)} \tag{4-6}$$

where E_s is Young's modulus of the reinforcing steel, A_s the area of a stirrup or a fraction of that area which balances the torsional moment, and s the stirrup spacing, and referring to Fig. 4-1 the equivalent hollow section properties are

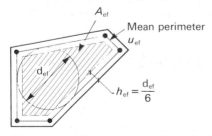

Figure 4-1   Equivalent hollow section

defined as:

A_{ef} = area enclosed by the polygonal perimeter length, u_{ef}, which is formed by joining the centres of the longitudinal reinforcement which are enclosed by the torsional stirrup (A_s)

$h_{ef} = d_{ef}/6$ = equivalent thin walled thickness

d_{ef} = the diameter of the largest circle which can be inscribed within the boundaries of A_{ef}.

For linear elastic analysis of structures subjected to dynamic loading, the dynamic modulus should be used. This approximates to

$$E_{dyn} = 1.25 E_{sec} \qquad (4\text{-}7)$$

When the loading is of a long-term nature or when stress dependent material non-linearity affects the member stiffness, then the non-linear properties of concrete are required. References [2, 68, 95] deal with this matter in detail. The equivalent long-term modulus generally takes the form

$$E_{long} = E_t = E_c/(1 + \phi_t) \qquad (4\text{-}8)$$

where ϕ_t is the creep factor of concrete.

Generally, the non-linear material properties have to be assessed for each member at each load and/or time step of the non-linear analysis in order to update the member stiffness coefficients. The relevant analytical procedures are dealt with in Sections 4.3 and 4.5. In the case where non-linear material properties are used for analysis, it is important that the unfactored stress–strain relationship be used, i.e. the characteristic or nominal values.

In some types of concrete structure, the contractor prefers to use a concrete mix which gives a 28-day strength considerably in excess of the specified characteristic strength in order to achieve the high early strengths he requires to save construction cycle time (incrementally launched decks, *in situ* segmental construction, columns cast by sliding formwork, etc). Cognizance has to be taken of these practices in order to assess realistic material properties for analytical purposes, especially when the displacements of the structure are required. The design engineer has to be aware of the fact that displacements are proportional to the absolute values of the structural stiffness while the stress resultants (in an indeterminate structure) are dependent on the relative structural stiffness of the respective structural elements.

In finite element analysis, additional concrete properties are required [100, 101]. Especially in the cases of triaxial stress problems and anisotropy, specialist literature should be consulted for the required properties and their mathematical simulation.

4.3 Mathematical modelling of the structural components and discretization of bridge structures

In the mathematical modelling of structural components, it is current practice to obtain discretized analytical elements that deform the same way when subjected to loading as the real structural component. This implies that the model element must have the same stiffness properties as the structural member.

When it comes to the modelling of the whole structure, the discretization of the structural members into an adequate number of elements has to be decided on. The following factors influence this decision:

(a) Type of structural configuration, i.e. arch structure, continuous box-girder on monolithic supports, etc.

(b) Type of loading the structure will be subjected to, i.e. dead and live loads acting in only one plane, or earthquake loading in any direction.

(c) The nature of output required, i.e. longitudinal beam action only or beam action and transverse slab bending or stress concentrations due to indirect supports, etc.

(d) Number of output points required. Most programs supply stress resultants only at the ends of members, therefore finer subdivision may be necessary for particular load types (i.e. prestress).

(e) In finite element analysis, it must be determined how many elements are required for a sufficient solution. Unless the particular element types have been used before, it is essential to test a similar mesh configuration with known results before using them in the bridge model.

(f) More elements usually require more nodes, which implies that the analytical model has more degrees of freedom. This has a direct influence on the cost of the analysis and must accordingly be optimized relative to the degree of accuracy thereof.

In order to illustrate the choice of discretization and modelling options, the structure shown in Fig. 4-2a has been simulated by several valid models as shown in Figs 4-2b-e. The advantages and shortcomings of each of these models can easily be verified in terms of the factors listed above.

In the case where many different types of action are imposed on a bridge structure, as is often the case, it may seem advantageous to use several different models of the structural configuration best suited for the purpose of analysing the effects of a particular type of action. The limit state design method, however, requires an investigation of many different combinations of the respective stress resultants. Instead thereof, it is often advisable rather to use an all-encompassing model and to subject it to all the different types of loading. The stress resultants can then be readily combined by the computer program as required for design purposes.

In some cases of modelling, it may be necessary to use hypothetical elements which ensure the correct global stiffness relationship but display meaningless local stress resultants. A typical example of such a model element is encountered in grillage analysis of box-type bridge decks [68, 111]. Here, the transverse stiffness of the combined top and bottom slabs of the box section has to be modelled by a single member. The required inertia for rotational stiffness and deflection (shear) stiffness, however, are quite different. Even so, the correct stiffness can, however, be achieved by using an equivalent inertia and appropriate shear area in the modelled element. It is often necessary in the case of such a 'coupled' element model, to resort to a small equivalent member analysis to ensure correct displacement (or stiffness) characteristics. This is shown below by way of example.

Figure 4-3a shows a typical single-cell box cross-section. Figures 4-3b,c show the skeletal structure required to evaluate the stiffnesses for transverse 'plane-frame'

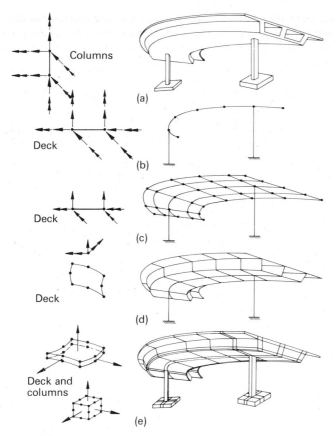

Figure 4-2 Choice of structure discretization and modelling

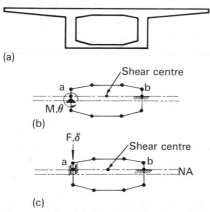

Figure 4-3 (a) Typical single-cell box cross-section
(b) Evaluation of rotation stiffness
(c) Evaluation of deflection stiffness

deformations. The structure in Fig. 4-3b is *fully* fixed at both supports except for the rotational release at support 'a' at which a moment M is applied. The resultant rotational stiffness is found from the expression

$$k_r = 4EI_r/L = M/\theta \tag{4-9}$$

where E is Young's modulus, I_r the effective sectional inertia for an applied rotation, L the length of member between effective support points, and θ the rotation of support 'a' due to an applied moment M. Hence I_r is evaluated.

The structure in Fig. 4-3c is *fully* fixed at both supports except for the deflection release at support 'a' at which a force F is applied as shown in Fig. 4-3c. The resultant deflection stiffness is found from the expression

$$k_d = 12EI_d/L^3 = F/\delta \tag{4-10}$$

where I_d is the effective sectional inertia for an applied deflection, and δ the deflection of support 'a' at right angles to the member axis due to a force F applied in the same direction. Hence I_d is evaluated.

Since most frame analysis programs do not allow for rotational *and* displacement inertia, but allow for shear area, an equivalent set of inertia and shear area can be evaluated as follows.

Allowance for shear area in the stiffness matrix generally takes the form

$$k_{a_i} = 4\frac{EI}{L}\lambda_i \tag{4-11}$$

$$k_{a_j} = 4\frac{EI}{L}\lambda_j \tag{4-12}$$

$$k_{b_{ij}} = 2\frac{EI}{L}\lambda_{ij} \tag{4-13}$$

where i and j refer to the supports at ends i and j of the member respectively and a and b are the diagonal and off-diagonal position indicators of the independent bending stiffness terms of the member matrix.

For prismatic members

$$\lambda_i = \lambda_j = \frac{B+1}{4B+1}$$

$$\lambda_{ij} = \frac{1-2B}{4B+1}$$

where $B = 9I/A_{sh}L^2$ (and $B \to 0$ as $A_{sh} \to 0$, not ∞!), in which A_{sh} is the shear area

For equivalent properties

$$B = I_r/I_d - 1$$

and hence λ_i and λ_{ij} can be determined.

In programs where the coefficient λ cannot be input directly, an equivalent inertia and shear area can be evaluated, that is

$$I_{eq} = 4I_r - I_d$$

and

$$A_{eq} = 9 I_{eq}/BL^2$$

Specialized bridge structures such as arch bridges, strut-frames and cable-stayed bridges require more detailed treatment in modelling and discretization for both the complete structure and the various construction stages. Some of these requirements will be discussed in Sections 4.4 to 4.6 (inclusive).

Few design codes give guidance in modelling [2, 59] and it is therefore advisable to refer to the modern literature for guidance [99, 102–110].

4.4 The simulation of actions on bridge structures

The actions to which bridges may be subjected are described and classified in Section 2.4. Chapters 11, 15, 19, 20 and 23 should also be consulted and the references [32, 60, 61, 68, 111–115] may prove useful. For the purpose of analysis, the following re-classification is useful:

(a) *Long-term actions*. These consist of dead load, superimposed dead load, earth pressure, water pressure, long-term restraint actions due to creep and shrinkage effects, secondary (parasitic) prestress and prestrain effects and differential settlement.

(b) *Short-term quasi-static actions*. These consist of all those short-term and variable loads which have negligible dynamic effects on a bridge structure. The following types of action normally fall into this category: erection loads, primary loads and secondary forces due to railway, highway and cycle traffic and pedestrian loads, flood action, short-term restraint actions due to temperature range and temperature gradient.

(c) *Short-term dynamic actions*. Here we consider only railway loading, earthquake action and wind action even though they can in many cases be simulated by equivalent static loads, obviating rigorous dynamic analysis.

In most codes, very little guidance is given on the simulation and effective application of actions to the bridge structures. Methods of simulation are dealt with in detail in the references given in the following subsections. The following guidelines are intended to cover cases of load application where most problems appear to arise.

4.4.1 *Load simulation adapted to the structural configuration*

Most bridges are analysed by plane-frame, plane-grid or space-frame methods with finite element analyses reserved for special structures and the local analysis of complicated stress flow in a structural element where the plane section assumption does not hold. Most load simulation problems arise from the discretized nature of the skeletal structure that is used to simulate the bridge. A typical example is a twin-beam superstructure without transverse beams (see Fig. 4-4). A plane grid simulation of this deck requires the connecting slab to be represented by a series of transversely spanning beam elements. It is normal to use from five to eleven such beam elements per span. The imposed gravitational loading acting on the superstructure (DL and LL), if placed on the transverse members, will result in correct stress resultants for these members but incorrect

Figure 4-4 Twin-beam superstructure without crossbeams

sectional forces in the adjacent longitudinal beams. In fact, the results are only correct at the centre of each longitudinal beam element. This may be unacceptable in the case where such output is automatically coupled to design programs. A remedy to this dilemma would be to subject the longitudinal members to the continuously distributed load consisting of the equivalent end forces derived from the loads applied to the slab members.

However, in such a case, the stress resultants in the equivalent slab members are incomplete, since they only consist of the distributed load effects. The solution is to apply both forms of loading to the structure, each in a separate loadcase and always using the output from those members which have been loaded. The support reactions and sectional forces outside the loaded area are identical in both loadcases—thus resulting in a useful check for the correct evaluation of the equivalent loading.

In the case of genuine grid-type superstructures, it is advisable to resort to automatic load generator programs to generate the equivalent member loading due to traffic actions. The programs are usually coupled to an influence surface (or equivalent influence line) generator program for specified critical sections in the structure. In this case, the structure can be loaded in such a way as to automatically result in maximum and associated load effects at the predetermined sections for design purposes.

4.4.2 *Equivalent prestress loading*
Prestress loading is covered by Chapters 19 and 21 and references [2, 59, 68, 95, 111, 114, 116, 117]. In modern design codes based on limit state design, the primary prestress effects and the secondary (or parasitic) prestress effects have to be factored differently in the design formulation $S_D \leqq R_D$. This is required since the primary prestress effects form part of the resistance R_D of the section while the parasitic prestress effects form part of the action effects S_D (see [2]), Vol. II, p. 76). It is therefore required to obtain the parasitic prestress effects separately from the primary effects in the analysis. In the normal prestress load simulation method [68, 111, 116, 117], the equivalent prestress load is evaluated by the process shown in Figs 4-5a,b. The equivalent loading shown in Fig. 4-5b will result in stress resultants due to combined primary and parasitic effects. If,

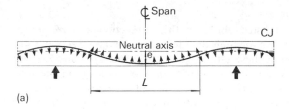

(a)

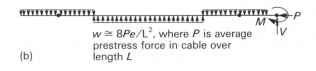

$w \cong 8Pe/L^2$, where P is average
prestress force in cable over
(b) length L

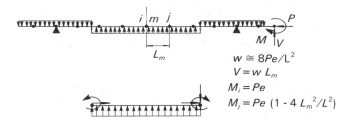

$w \cong 8Pe/L^2$
$V = w\,L_m$
$M_i = Pe$
$M_j = Pe\,(1 - 4\,L_m^2/L^2)$

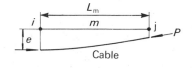

(c)

Figure 4-5 (a) Typical prestress profile showing forces exerted by prestress cable on the beam
(b) Equivalent prestress loading for primary + parasitic output
(c) Equivalent prestress loading for parasitic output

however, the primary prestress forces at *each member* end are added to the
equivalent loading as shown in Fig. 4-5c, the output will consist only of restraint
(or parasitic) effects. A check for the correctness of the input is that the equivalent
load of *each member* must be self-equilibrating while in Fig. 4-5b the sum of the
equivalent loading on all members must be in equilibrium.

Another reason for keeping the primary and parasitic prestress effects separate,
is the difference in their variation with time observed in stage-by-stage analysis of
continuous structures. The primary prestress is only affected by the time-
dependent losses while the parasitic prestress effects are affected by both time-
dependent losses and the variation in relative structural stiffness with time. This
phenomenon will be covered in more detail in Section 4.6.

4.4.3 *Restraint action simulation*

The restraint effects have been listed above in this subsection and reference should be made to applicable codes for their detailed definition (see Section 3.4). For a more detailed description of these effects, refer to Chapters 11 and 21 and [2, 59, 68, 95, 111]. They all develop due to the inability of statically indeterminate structures to deform freely. The equivalent member loads required to produce these parasitic effects in the structural analysis are evaluated on the basis that they would produce the free deformation in each member when subjected to its equivalent load in isolation from the rest of the structure.

The equivalent fixed-end forces for any restraint action are therefore evaluated by reversing the free member end deformations and summing the forces developed at the fixed member ends. These equivalent forces are therefore dependent on the member stiffness, which in turn may be dependent on the member stresses and time. Realistic parasitic effects can therefore only be obtained if the applicable Young's modulus and the effective section properties are used. It is therefore imperative that the member stiffnesses used for the evaluation of the equivalent restraint loading be the same as those used for the subsequent analysis! If the results of the analyses are to be used for a serviceability limit state, it is usually sufficiently accurate to use member properties based on the nominal dimensions of the cross-sections and on the elastic short or long-term moduli. However, if the results of the analysis are to be used in an ultimate limit state load combination, cognizance has to be taken of the effect of reduced member stiffness due to plasticity and cracking of concrete.

Methods of non-linear analysis for bridge structures are referred to in Section 4.5 and the application of non-linear analysis in Section 4.6.

4.4.4 *Simulation of wind action*

Details of the evaluation of wind action on structures can be found in [2, 32, 59, 79, 95, 115]. The evaluation of wind action on a bridge structure is extremely complex and several methods have been devised to simplify the load evaluation. The wind pressure consists of a steady-state pressure on which gust and eddy effects are superimposed. Besides the geographic, topographic and geometric influences on wind action, the natural frequency of the structure also plays a role on the fluctuating portion of the pressure. In practice, however, only wind-sensitive and/or large bridges need to be investigated for interaction with wind. Whether the pressure intensity is derived from empirical tables or by semi-deterministic wind–structure interaction, the result is a distributed member loading applied to the structure without any further simulation problems. Consideration must, however, be given to the eccentricity of the line of action relative to the centroidal member axis due to such effects as:

(a) shear centre and centroidal axis not coinciding;
(b) wind load on non-structural members such as balustrades;
(c) wind load on traffic.

The analytical procedures required for the evaluation of wind effects on bridge structures are referred to in Section 4.5 and the application of wind effects to bridge structures in Section 4.6.

4.4.5 *Simulation of earthquake action*

Details of the assessment of earthquake action on structures can be found in Chapters 15 and 21 and in [59, 76, 78, 79, 114]. The structural response of a bridge structure depends on the seismic characteristics of a particular earthquake and on the stiffness characteristics (natural frequencies) of the structure. Equivalent static methods of evaluating nominal earthquake forces have been specified by certain applicable codes in areas where earthquake action is likely. In cases where structures are either exposed to higher intensity earthquakes or where structures are more vulnerable to earthquake action, semi-dynamic simulation of these effects has been prescribed in the form of elastic and inelastic design spectra.

Multi-degree-of-freedom structures are best analysed by computer programs which cater for the above methods. In the case of very vulnerable structures, rigorous dynamic analysis is required using characteristic earthquake accelerograms. Specialized dynamic analysis computer programs are available which can perform elastic (linear) and non-linear analyses. The output consists of the structural response (displacements and stress resultants) for a series of time steps during the excitation phase. This can be very voluminous and has to be assessed with caution. Only the maximum effects are, however, required for design purposes. The analytical procedures and practical applications required are referred to in Sections 4.5 and 4.6 respectively.

4.5 Modern analytical methods applicable to bridges

A summary with brief descriptions of computerized methods used in modern bridge analysis is given here under the following subheadings:

(1) Linear static stiffness analysis of skeletal structures;
(2) Finte difference analysis;
(3) Finite element analysis;
(4) Finite strip analysis;
(5) Dynamic stiffness analysis;
(6) Non-linear stiffness analysis.

Some of these analytical methods have been covered in greater detail in Chapters 21 and 38.

4.5.1 *Linear static stiffness analysis of skeletal structures*

A structure can be discretized as a set of skeletal (one-dimensional) elements or members if it is considered that the second and third member dimensions play a negligible role in the solution (stress flow) of the structure when subjected to a particular set of loads. It must be recognized that the stiffness matrix of a one-dimensional (or beam) element is just a special case of a general finite element stiffness formulation. At the same time, the linear static portion of the matrix formulation is only a subset of the general non-linear dynamic finite equilibrium equation [118]:

$$\boldsymbol{M}^{t+\Delta t}\ddot{\boldsymbol{U}}^{(i)} + \boldsymbol{C}^{t+\Delta t}\dot{\boldsymbol{U}}^{(i)} + {}^{t}\boldsymbol{K}\,\Delta\boldsymbol{U}^{(i)} = {}^{t+\Delta t}\boldsymbol{R} - {}^{t+\Delta t}\boldsymbol{F}^{(i-1)} \tag{4-14}$$

where $i = 1, 2, 3 \ldots$ are iteration indices.

At time $t + \Delta t$,

$$^{t+\Delta t}\boldsymbol{U}^{(i)} = {}^{t+\Delta t}\boldsymbol{U}^{(i-1)} + \Delta\boldsymbol{U}^{(t)} \tag{4-15}$$

and

$$M = \text{constant mass matrix}$$
$$C = \text{constant damping matrix}$$
$${}^{t}K = \text{tangent stiffness matrix at time } t$$
$${}^{t+\Delta t}R = \text{external nodal point load vector due to the body forces, surface}$$
$$\text{loads and concentrated loads at time } t + \Delta t$$
$${}^{t+\Delta t}F^{(i-1)} = \text{nodal point force vector equivalent to the element stresses that}$$
$$\text{correspond to the displacements } {}^{t+\Delta t}U^{(i-1)}$$
$${}^{t+\Delta t}U^{(i)} = \text{vector of nodal point displacements at the end of iteration } (i),$$
$$\text{and time } t + \Delta t$$

The derivatives with respect to time are denoted by superior dots and the superscript (i) indicates an iteration index.

The linear static stiffness portion of this equation can be extracted as

$$KU = R$$

where K is the linear stiffness matrix, U the displacement vector, and R the load vector due to joint loads and equivalent member loads (fixed end forces).

The basic steps involved in a linear static stiffness analysis are as follows.

(I) *Data preparation* This is done by an engineer and/or a generation program:

(1) Line diagram. Decide on the type of skeletal modelling.
(2) Numbering of joints and members and deciding on support conditions.
(3) Member properties are evaluated and member end connectivities defined.
(4) Loads are evaluated and applied to joints and members.

(II) *Calculations* Ten stages are involved:

(1) Degree-of-freedom table is generated.
(2) Member location table is generated.
(3) Member direction cosines are calculated.
(4) Member *transformed* stiffness matrices are calculated k_{XYZ}.
(5) Main structure stiffness matrix K_{XYZ} is assembled using (1), (2), (4), above.
(6) Fixed end reactions are evaluated for all loaded members f_{XYZ}.
(7) Column matrix of joint loads R_{XYZ} is assembled.
(8) Deformation matrix U_{XYZ} is solved for using $KU = R$.
(9) Appropriate joint deformations are back-substituted into the transformed member stiffness equilibrium equation:

$$p_{XYZ} = k_{XYZ}U_{XYZ} + f_{XYZ}$$

to yield the global axis stress resultants p_{XYZ}.
(10) Stress resultants in local (or member) axis system are evaluated from $p_{xyz} = Tp_{XYZ}$, where T is the transformation matrix.

XYZ here refers to the global or structure axis system, xyz to the local or member axis system, and the bold italic type denotes a matrix or tensor.

The above basic steps are required in all static stiffness solution techniques, varying only in the degree of complexity of the particular discretization and solution scheme.

4.5.2 *Finite difference analysis*

Finite difference mesh solution procedures have been applied successfully to various structural analysis problems. The traditional difference approach with its boundary formulation problems was found to be very limited in its range of applicability. However, bridge decks which could be simulated by orthotropic plate theory have been successfully and economically solved by this approach. Some finite strip formulations employ finite differences in the transverse direction [123]. In some modern applications of finite difference formulations, the final product (solution procedure) is virtually indistinguishable from procedures based on finite element formulations [135]. Certain of these techniques are also used in dynamic analysis dealt with in Section 4.5.5. Methods based on the finite difference technique are covered in [135,137].

4.5.3 *Finite element analysis*

The finite element method of stress analysis has been dealt with in Chapter 21, Section 3 and in Chapter 38. Further information can be found in [105–109, 118–139, 152]. This is a form of structural analysis which is developing at a fast rate, resulting in rapid improvement of elements, solution techniques and analytical procedure. It is therefore advisable to refer to the latest developments in relevant literature, such as *International Journal for Numerical Methods in Engineering* and *Computers and Structures* published by John Wiley & Sons of New York and Pergamon Press, Oxford, respectively.

The general finite element equilibrium equation 4-14 can form the basis of most spatial and temporal finite element discretization formulations.

4.5.4 *Finite strip analysis*

The finite strip method of analysis has been dealt with in Chapter 21, Section 4 and in Chapter 38. Further information can be obtained from [108, 109, 119, 123, 125, 127, 137]. The finite strip method of analysis lends itself to the analysis of bridge structures due to the prismatic shape and elongated form of the superstructure. Relative to the finite element method of analysis, this method is less general in application but more economic on computer time. Many different formulations have been used in the form of infinite series over the length of the structure and displacement functions transverse to the superstructure. As for the finite element methods, the finite strip formulations have also been derived from basic variational principles of mechanics. One interesting aspect of the finite strip method is that it lends itself to the substructuring of matrices with further reduction in computative effort. This method has been successfully applied to straight, skew and curved plate and folded-plate structures.

4.5.5 *Dynamic stiffness analysis*

Chapter 15 deals with dynamic analysis with particular reference to the action of earthquakes. For further information on dynamic analysis methods refer to [118, 129, 132, 139].

The finite element equilibrium equation 4-14 can form the basis for this method of analysis. In the absence of non-linearity, it can be rewritten as

$$M^t\ddot{U} + C^t\dot{U} + K^tU = {}^tR \tag{4-16}$$

where

M = constant mass matrix
C = constant damping matrix
K = linear stiffness matrix
${}^tR$ = time dependent load vector
${}^tU$ = time dependent displacement vector
${}^t\dot{U}$ = time dependent velocity vector
${}^t\ddot{U}$ = time dependent acceleration vector

There are a number of time integration schemes employing mode superposition which give satisfactory solutions of a discretized system of equations defining a structure based on Eqn 4-16. These can be found in [118, 129, 132, 139].

In the case of bridges, dynamic stiffness methods are mainly used for earthquake analysis. Earthquake actions are usually applied in the form of base excitation expressed as an acceleration time function. Analysis for earthquake action has been dealt with in greater detail in Chapter 15. The application of earthquakes to bridge structures is briefly summarized in Section 4.6.

4.5.6   Non-linear stiffness analysis

Non-linear stiffness analysis can be separated into time-dependent and static formulations. Furthermore, one must distinguish between material and geometric non-linearity although both phenomena are often dealt with in the same analysis since the deformations due to material plasticity often give rise to large displacements which require updating of the structure geometry in order to take account of the 'P–Δ' effects. In the case of time-independent stiffness analysis, Eqn 4-14 can be rewritten as

$$K \Delta U^{(i)} = R - F^{(i-1)} \quad i = 1, 2, 3 \ldots \tag{4-17}$$

where

$\Delta U^{(i)} = U^{(i)} - U^{(i-1)}$
K = tangent stiffness matrix
R = external nodal point load vector
$F^{(i-1)}$ = nodal point force vector equivalent to the element stresses that correspond to the displacements $U^{(i-1)}$
$U^{(i)}$ = vector of nodal point displacements at the end of iteration (i)

Equilibrium iteration ($i = 1, 2, 3 \ldots$) is often necessary in this kind of analysis in order to ensure convergence of load incrementation techniques.

Time-dependent non-linear stiffness analysis can be represented in the form of the equilibrium equation 4-14. The load vector R can be time-dependent and a solution at specified time intervals is required in order to determine maximum structural response during the interval of time-dependent loading.

A form of time-dependence in bridge structures is encountered in stage-by-stage analysis involving time-dependent material properties. Since inertial effects do not play a role in this type of analysis, no dynamic formulation is required. Therefore, this type of analysis is usually performed in a static mode with stepwise varying properties. A typical application of this type of analysis is dealt with in Section 4.6.

The basic and advanced development of non-linear stiffness analysis theory and methods of analysis can be obtained from specialist publications [128–134].

4.6 Computer applications to bridge analysis

The methods of analysis described in other chapters apply equally to bridge structures. The differences which do occur are not in the analytical theory, but in the application thereof. This will be illustrated by considering a few examples of action effects on bridge structures as found in the evaluation of earthquake effects, wind effects, the long-term effects of stage construction and the assessment of non-linear effects in bridge structures.

4.6.1 *The application of earthquake action to bridge structures*
Chapter 15 describes the nature of earthquake action on structures in general and the methods applied to evaluate the structural response thereto. In Section 4.4, the simulation of these effects on bridge structures is summarized and in Section 4.5 the basis of modern dynamic analysis is given. References [59, 76, 78, 79, 114] deal with particular aspects that are relevant to bridges. A method of evaluation that is suitable for application to concrete bridges is summarized below.

4.6.1.1 *Outline of method* The method is based on the average response spectrum of structures which was derived from the response of single-degree-of-freedom structures when subjected to a range of recorded earthquakes. The elastic average response spectrum bounds are shown in Fig. 4-6 normalized to a ground acceleration of 1.0g. For single-degree-of-freedom structures, the design spectrum is entered with the natural period of the structure. The maximum acceleration, S_a, is then read off at the applicable damping ratio. Typical damping ratios are listed in Table 4-1. This method can be used as a manual method on small structures or as a computerized method on larger structures. In a multi-degree-of-freedom system that has independent or uncoupled modes of deformation, each mode responds to the excitation as an independent single-degree-of-freedom system. Generally, only the response in the fundamental natural frequency is required in the orthogonal orientations of the bridge structure. The equivalent static load to which the bridge is subjected to can then be calculated from

$$F_x = k_{gh} I f S_a m_x \tag{4-18}$$

where

k_{gh} = peak horizontal ground acceleration (see Table 4-2)
I = importance factor of the structure which generally lies in the range 1.0 to 1.3
f = foundation factors which are given in Table 4-3
S_a = the seismic response factor for the structure read from the response spectrum in Fig. 4-6
m_x = the mass of the structure or part of the structure considered

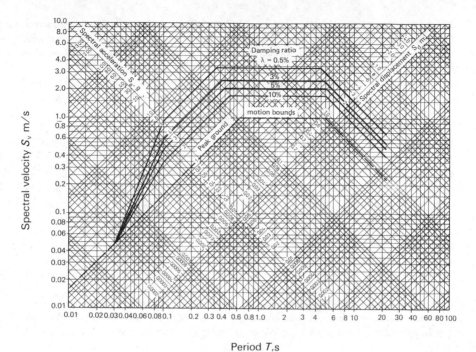

Period T,s

Figure 4-6   The elastic average response spectrum bounds, normalized to a ground acceleration of 1.0g

Table 4-1

Stress level	Type and condition of structure	Damping ratio
Working stress no more than about half yield	(a) Welded steel, prestressed concrete, well reinforced concrete (only slight cracking)	0.02
	(b) Reinforced concrete with considerable cracking	0.03–0.05
	(c) Bolted and/or riveted steel	0.05–0.07
At or just below yield point	(a) Welded steel, prestressed concrete under full prestress (slight cracking)	0.05
	(b) Prestressed concrete in cracked state	0.07
	(c) Reinforced concrete	0.07–0.10
	(d) Bolted and/or riveted steel	0.10–0.15

Table 4-2

Modified Mercalli Intensity at epicentre MM	Maximum ground acceleration, k_{gh}, at epicentre
ii–iii	0.003g
iv–v	0.01g
vi	0.03g
vii–viii	0.1g
ix	0.3g
x–xi	1.0g

The dynamic earthquake loading is in this way reduced to an equivalent static loading.

4.6.1.2 *Adjustment for elastic-plastic behaviour* For a structure designed to undergo elastic-plastic deformation under earthquake action, the average elastic response spectrum may be modified as follows in order to obtain the inelastic design spectrum.

The elastic spectrum, for any given damping ratio, is modified along the displacement bound region by multiplying the bound line by a factor $1/\mu$, and along the acceleration bound region by multiplying the bound line by a factor $1/\sqrt{2\mu-1}$

A typical conversion is shown in Fig. 4-7. Note that the period at the corner points is unchanged. The ductility factor μ is defined as the total elastic-plastic deformation of the structure divided by its total elastic deformation at yield.

Table 4-4 gives values of μ for typical structural configurations.

Table 4-3

Type and depth of soil	f
Rock, dense and very dense coarse-grained soils, very stiff and hard fine-grained soils; compact coarse-grained soils and firm and stiff fine-grained soils from 0 to 15 m deep	1.0
Compact coarse-grained soils, firm and stiff fine-grained soils with a depth greater than 15 m; very loose and loose coarse-grained soils and very soft and soft fine-grained soils from 0 to 15 m deep	1.3
Very loose and loose coarse-grained soils and very soft fine-grained soils with depth greater than 15 m	1.5

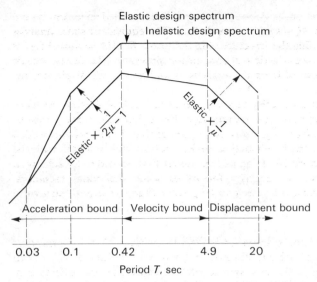

Figure 4-7   The transformation from elastic to elastic-plastic design spectrum. Note that the corner period values are unchanged

Table 4-4

Case	Type or arrangement of resisting elements	Structural ductility factor, μ
1	Structural steel unbraced bridges, piers and superstructures adequately designed to resist the total lateral forces by bending of the members in accordance with their relative rigidities considering the interaction of the various parts	4
2	Structures with braced flexural piers of structural steel, or substructures consisting of slender reinforced concrete columns and superstructures in braced structural steel, or in reinforced or prestressed concrete, adequately designed to resist the total lateral force in accordance with their relative rigidities considering the interaction of the various parts	3
3	Structures with piers or abutments of shear wall proportions and superstructures in monolithic reinforced or prestressed concrete adequately designed to resist the total lateral force in accordance with their relative rigidities considering the interaction of the various parts	2

The maximum acceleration, S_a, is then read off at the modified spectrum bound and used in the equation $F_x = k_{gh} If S_a m_x$ as before. The equivalent static analysis can be performed by loading the structure with the forces F_x. The sectional forces obtained are the true elastic-plastic response of the structure. The elastic-plastic displacements can be obtained from the analysis by multiplying the displacements by the ductility factor μ.

The maximum structural ductility factor μ given in Table 4-4 must be used, except where tests and calculations demonstrate that higher values are justified.

For structural ductility factors μ equal to or greater than 5, this recommended procedure should not be used, but a time series analysis with realistic material behaviour should be performed. It should be noted that, in order to achieve a given structural ductility, the ductility factors of some individual members, particularly those of girders, may need to be greater than the overall structural ductility.

4.6.1.3 *Application to multi-degree-of-freedom structures* For symmetrical structures, the influence of seismic disturbances is to be considered along both principal axes of symmetry. For non-symmetrical structures, seismic effects may be considered along two arbitrarily chosen orthogonal axes. For both symmetrical and non-symmetrical structures, the seismic effects along the two orthogonal axes may be considered independently of one another. Where it is deemed necessary to consider the influence of vertical seismic motions, the average vertical design spectrum may be taken as two-thirds the average horizontal seismic response spectrum (Fig. 4-6). It should be noted that the relevant resonance frequencies of the structure and its components in the vertical direction are generally quite different from those in the horizontal direction. The response caused by vertical excitation shall be combined with the response caused by horizontal excitation as specified below.

The structure is first analysed to determine its natural modes of vibration and the corresponding natural frequencies (or periods). This generally involves the solution of an eigenvalue formulation of the structure of the form

$$\left((m) - \left[\frac{1}{\omega^2} \right] (K) \right) \{x\} = 0$$

where

      (m) = diagonal matrix of lumped masses representative of the dead and superimposed dead load imposed on the structure
      (K) = stiffness matrix of the structure
      $\{x\}$ = general displacement vector of all degrees of freedom of the structure
      $[1/\omega^2]$ = the inverse square vector of the natural radial frequencies ω

Computation of structural response Each mode is first assumed to behave independently in the earthquake response. The structural response acceleration S_{ai} (in units of gravitational acceleration) can thus be evaluated independently for each mode i with period T_i in the way shown in Sections 4.6.1.1 and 4.6.1.2. Each mode will participate in the total combined response by an amount $\gamma_i \Psi_{ij} S_{ai}$. The 'modal participation factor' γ_i is evaluated for the ith mode as follows:

$$\gamma_i = \frac{\sum \bar{m}_j \Psi_{ij}}{\sum_{mj} (\Psi_{ij})^2}$$

where

> $\sum \equiv \sum_{j=1}^{n}$
> m_j = lumped mass free to move in the direction of the degree-of-freedom j
> $\bar{m}_j = m_j$ when the direction of the ground motion is parallel to the direction of the degree-of-freedom j. Otherwise,
> $\bar{m}_j = 0$
> Ψ_{ij} = normalized mode component of the ith mode in the direction of the jth degree-of-freedom
> n = number of degrees-of-freedom of the structure

The generalized set of equivalent static forces acting on the structure can thus be written as:

$$F_{ij} = k_{gh} If \gamma_i \Psi_{ij} S_{ai} m_j.$$

4.6.1.4 *Combination of modes* By the process described in Section 4.6.1.2, the maximum response of each mode of the multi-degree-of-freedom structure can be found. The maximum values of combined effects are obtained by taking the square root of the sum of the squares of the effects from each mode. The first three natural modes of vibration of the structure are normally sufficient to obtain maximum combined effects.

4.6.2 *The application of wind action on bridge structures*
The nature of wind action and its simulation is described in Chapter 23 and in Sections 4.4 and 4.5 of this chapter, and can be studied in more detail in [32, 59, 74, 75, 79, 92, 95, 115]. The effect of wind on bridge structures differs from that on tall buildings, for which most of the field research has been done. The methods of analysis based on this work are consequently not directly applicable to bridges. Most concrete bridge structures are, however, not very sensitive to the gustiness of wind because of the stiffness (short natural periods of vibration) or size (low cross-correlation of gusts) of bridge members (see Section 2.4). Realistic equivalent static wind loads can therefore be evaluated with sufficient accuracy for most types of bridge structure. The application of wind loads to bridge structures has been dealt with in great detail in the codes referred to in [59, 95]. Some large concrete bridge structures, as well as slender ones, are susceptible to wind gusting effects and should be analysed taking account of their natural frequencies and corresponding gust factors as described in [32] or [79]. The natural frequency or period of the bridge structure or of its slender component members requires evaluation. This is best done by an eigenvalue solution performed on a computer. The structural discretization and mass distribution of members is the same as that required for dynamic earthquake analysis, as illustrated in Fig. 4-8.

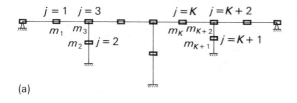

(a)

(b)

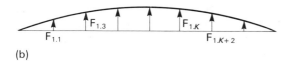

(c)

Figure 4-8 (a) Idealized bridge structure for transverse excitation
(b) Forces due to first transverse mode $i = 1$
(c) Forces due to second transverse mode $i = 2$

4.6.3 Stage-by-stage construction analysis

A summation of the load effects at all construction stages of a bridge evaluated by elastic analysis does not produce the correct final forces or displacements due to the tendency of 'green' concrete to undergo creep deformation during construction. The exact solution of the redistribution of stresses due to creep effects requires considerable computational effort. There are various inherent inaccuracies in the assumptions on which the creep effects are based so that the methods applied should be of a comparable order of accuracy in order to reduce the work to a realistic level. The two effects to be evaluated are: (a) redistribution of sectional forces, and (b) member (or nodal) deflections. While the effects under (a) are only dependent on the relative stiffness of members (and changes thereof with time), those under (b) are dependent on the absolute or total stiffness of the structure over a period of time. It is therefore often necessary, for slender bridge structures, to use more accurate methods of analysis to obtain member deflections than is required to evaluate changes in member forces.

The phenomenon of creep is defined by the relevant codes such as [2] and [95] and the methods of evaluating creep effects have been widely dealt with in technical publications [153–156, 163]. The following is a summary of the effects.

When a structure is constructed in 'one go', strains and deformations due to creep (and shrinkage) are proportional to the corresponding elastic effects. No tendency therefore exists for the redistribution of forces since no restraints are built up. This does not, however, apply to a structure built in stages. Here, two considerations are important:

(a) The sum of the elastic effects during stage construction is not equal to the effects from a 'one go' (or built-in-one) structure. The differences between the two states are reduced by creep, i.e. the structural forces tend to redistribute from the 'sum of stages' to the 'one go' state.

(b) Since at any particular stage of construction concretes with different creep characteristics occur in different parts of the structure due to the differences in age and time under load, the creep deformations are not directly proportional to their respective elastic deformations. This results in creep redistribution. The younger concrete undergoing larger creep deformation transfers (or sheds) internal forces (stresses) to older concrete undergoing less creep deformation. If the creep which takes place during the construction period is small relative to the creep which takes place after completion, then a simplified approach can be applied:

$$M_F^i = \sum M_s^i - (\sum M_s^i - M_{OG}^i)R \tag{4-19}$$

where

M_F^i = final moment at point i

$\sum M_s^i$ = sum of all stage moments at point i evaluated by elastic analysis

M_{OG}^i = the moment at point i due to a 'one go' analysis using elastic (short term) or long-term member properties

R = redistribution factor

$= \phi/(1 + \eta\phi)$

and

ϕ = average total creep factor of the concrete members

η = relaxation coefficient which can be taken as 0.85

This operation can easily be programmed by using appropriate scale factors on the computer output of the respective load cases. If, however, the creep effects during the construction period cannot be neglected, then the following procedure should be followed.

Using the notation ${}^{t_r}M_{mi}^n$ to describe moment at the position m due to load case i at time t_r on the structural system n:

Load case $i = 1$

Construction stage $n = 1$

$${}^{t_1}M_{m_1}^1 = {}^{t_0}M_{m_1}^1 \tag{4-20}$$

where t_0 is the time of loading of load case $i = t_1$, and t_1 is the time at start of first creep interval t_1 to t_2.

Construction stage $n = 2$

Start:

$${}^{t_1}M_{m_1}^2 = {}^{t_1}M_{m_1}^1 \tag{4-21}$$

End:

$${}^{t_2}M_{m_1}^2 = {}^{t_1}M_{m_1}^1 - ({}^{t_1}M_{m_1}^1 - M_{m_1}^{2*})R_{1,2} \tag{4-22}$$

where

$$R_{1,2} = \frac{\Delta\phi}{1 + \eta\Delta\phi}$$

$\Delta\phi$ = change of ϕ factors in the time interval $t_2 - t_1$

$$= \phi_{t_2} - \phi_{t_1}$$

$$\eta = 0.85$$

* denotes an analysis performed with a different effective Young's modulus $^{t_2}E = E/(1 + \phi_{t_2})$ for each member

At this stage, $\phi_t = \phi_{t_2}$.

Construction stage $n = 3$
Start:

$$^{t_2}M^3_{m_1} = {}^{t_2}M^2_{m_1} \tag{4-23}$$

End:

$$^{t_3}M^3_{m_1} = {}^{t_2}M^2_{m_1} - ({}^{t_2}M^2_{m_1} - M^{3*}_{m_1})R_{2,3} \tag{4-24}$$

$$R_{2,3} = \frac{\Delta\phi}{1 + \eta\Delta\phi}$$

where

$$\Delta\phi = \phi_{t_3} - \phi_{t_2}$$

$$\eta = 0.85$$

$^{t_3}E = E/(1 + \phi_{t_3})$ for each member.

In general, one can write the expression for load case i at time t_r and construction stage n:

$$^{t}M^n_{m_1} = {}^{t_{r-1}}M^{n-1}_{mi} - ({}^{t_{r-1}}M^{n-1}_{mi} - M^{n*}_{mi})R_{n-1,n} \cdots \tag{4-25}$$

where

$$R_{n-1,n} = \frac{\Delta\phi}{1 + \eta\Delta\phi}$$

$$\Delta\phi = \phi_{t_r} - \phi_{t_{r-1}}$$

$$\eta = 0,85$$

$^{t}E = E/(1 + \phi_{t_r})$ for each member.

This operation can also be computerized using an appropriate summation algorithm to combine the respective output files of the computer analyses.

Care must be taken if the displacements are required from such an analysis. Since this is a stepwise non-linear analysis, the principle of superposition does not hold for displacements. They are best evaluated separately from first principles using the appropriate sectional forces and long term sectional properties.

4.6.4 *Non-linear effects in bridge structures*
The improvements in methods of analysis and design of structures have increased the accuracy of assessment of load effects and strength capacities. This has brought about a tendency towards the design of very slender bridges, with the result that certain effects like non-linear structural behaviour that could previously be neglected, may now be as important as some of the primary load effects. In concrete bridge structures, these non-linear effects are usually negligible at service load level. Since it is accepted practice to perform linear elastic analysis at *service* load level and then combine the results with appropriate load factors to obtain ultimate load combination effects, this can result in faulty *ultimate* load capacity estimates. The assessment of the non-linear effects has been subject to much investigation and development. The state of the art has now reached a level where most of the required effects can be assessed with sufficient accuracy for the purposes of the design engineer. Certain design codes provide the designer with information in order to assess non-linear effects [2, 95, 98]. More information can be obtained from technical publications [100, 101, 140–150].

Depending on the accuracy required, it may be necessary to ensure that the selected non-linear method of solution allows for some or all of the following effects:

(a) creep in concrete including the effects of mixed short and long-term loads;
(b) shrinkage in concrete;
(c) temperature effects at any level of stress in the concrete and steel;
(d) presence of cracks in concrete;
(e) partial foundation fixity;
(f) effects of initially built-in forces (e.g. prestress);
(g) biaxial bending;
(h) accurate simulation of cyclic load effects in concrete.

In addition, the analysis may have to allow for combined (interactive) material and geometric non-linearity. The specification of members should allow for arbitrary shapes including unsymmetric box-type cross-sections with arbitrary arrangements of reinforcement. Incremental loading should be allowed for with selected output options at any stage of loading up to failure or instability.

The stress–strain relation and its effect on the simulation of member stiffness under variable loading play an important part in (a), (b), (c), (d), (f), (g) and (h) above and the selected program should be tested for all the above effects before using it for actual bridge analysis.

In bridge structures, it is often possible to isolate a non-linear portion of the structure and analyse it separately in more detail. The procedure to be followed can take the following form:

(a) Analyse the bridge as a whole for a particular loading configuration (i.e. dead load, prestress, live load, temperature, shrinkage, etc.) using a linear analysis.
(b) Isolate the portion of the structure which is likely to behave non-linearly (e.g. a column assembly).
(c) Subdivide the isolated portion into sufficient elements so that it can reproduce the non-linear effects.

(d) Subject the isolated portion to incremental loading using a non-linear program.

Note: It is important to impose the correct forces on the boundary planes of the isolated portion of the structure at each loading increment. The boundary planes must therefore be chosen at positions where the stress resultants are small and vary linearly in proportion to the loading. The results of such an analysis can be sufficient for design purposes if care is taken in the choice of the boundaries of isolation.

The CEB–FIP code [2] describes a related procedure called model-column method in Section 14.4.3.2 of the code.

4.7 Model testing

Model testing of structures in general is described in Chapter 23. The methods and procedures covered therein are equally applicable to bridge structures, but a few aspects that are peculiar to bridge structures require further mention. Whereas building structures are usually clad with materials and finishes that can have a significant effect on the structural behaviour, this rarely applies to bridges. Furthermore, because of its elongated form, it can usually be simulated more readily. The type of model and its detailing depends on the problem being investigated. Parts only of the structure need be modelled in some cases.

In addition to using the materials and procedures described in Chapter 23, various tests [157–161, 168] have in recent years been done on large reinforced and prestressed concrete models of bridges up to a scale of 1 in 2.82. Such tests can accordingly give a very accurate simulation of the whole process of loading, from construction, at various stages, up to completion and thereafter.

Wind-tunnel tests may be essential in the case of certain long-span cable-stayed bridges. A full model is usually scaled down to 1:200 or higher and sectional models in the range 1:30 to 1:50.

5 Design and construction practice for concrete bridges

5.1 General

Modern design and construction practice for concrete bridges is the result of the historical process described in Section 1.2 and it is clear that the developments, particularly in the last three decades, have very greatly increased the understanding of structures and the actions they are subjected to. This improved knowledge, together with the powerful methods of analysis that are now available, have had a marked influence on modern practice, which is described in general terms in this section. Other chapters in this book should be consulted for more detailed information about the properties of materials, design and analysis, construction, reinforcing steel and prestressing details, as well as other publications [68, 111, 162–168]. The more realistic definition and control of material properties using statistical concepts has greatly improved the quality of construction and consequently the scope and reliability of bridges. Innovations in methods of construction have in turn influenced the competitiveness of specific configurations. The conceptual designer must consequently take cognizance of all feasible methods of construction which in the case of long-span bridges may radically influence the design.

5.2   Selection and investigation of bridge sites

Usually, the approximate site at which a bridge is to be built is determined by considerations related to the overall planning of the highway, railway or pipeline route and the bridge engineer has the responsibility of finding the best location within the area that the flexibility of the route planning allows. There are, however, many examples where a particularly suitable bridge site was the starting point for the location of the highway, railway or pipeline approaches. This usually applies in the case of river or gorge crossings where suitable bridge sites may exist only at a limited number of places. In the evaluation, the bridge cannot be isolated but must be considered as part of the total system being designed.

In recent years, an awareness of aesthetic considerations and the need for the conservation of the environment and objects of historical value have influenced the location and construction of roads and freeways (see Sections 1.4 and 1.5). The optimum solution in terms of functional considerations, with due regard to environmental impact studies, may indicate sites which are not necessarily the most suitable from the pragmatic engineering point of view. It nevertheless remains the responsibility of the engineer to obtain sufficient information to determine the most suitable position for the crossing within the constraints imposed on him and to conceive the best structure in terms of the site conditions, so as to achieve the requirements described in Section 1.5.

In order to evaluate a probable bridge site, many aspects have to be considered, of which the most important are summarized below.

Before proceeding with the investigation, it is advisable to obtain all available information on aspects which will affect the eventual design, including design criteria (see Section 2). Use should be made of the following site data (where applicable):

(a) route location maps;
(b) topographical maps, including any available aerial and terrestrial surveys and river cross-sections;
(c) local geological or soil maps and any relevant geological and geotechnical information available from neighbouring works;
(d) aerial and terrestrial photographs;
(e) weather and flood records or oceanographic and tidal information where applicable;
(f) location of existing bridges over the relevant river and performance data during floods;
(g) information on ice, debris and channel stability;
(h) expropriation plans;
(i) services location plans.

5.2.1   *Site investigation*
The extent and detail of the investigation of a bridge site depends largely on the class of bridge (see Section 2.2) and the nature of the sites that are available. The

investigation may comprise various stages, starting with a study of topographical maps, surveys, photographs and other sources of information and by a process of elimination, selecting the most promising sites to be investigated in further detail. Such studies will require knowledge and experience in the interpretation of geological features from aerial or terrestrial photographs and the advice of specialists may have to be obtained.

Where a project may be situated in a foreign country, the engineer should familiarize himself with local customs and the facilities available and investigate the laws and regulations pertaining to the procurement, importation and use of plant, materials and labour as these matters may eventually influence the selection of the most suitable structure for a particular site.

A detailed inspection and sufficient information of each of the selected sites is required before a realistic comparison can be made.

5.2.2 Investigation and visual inspection of the site

At this stage, the engineer should have some idea of the various alternative structures that could be considered. Useful information can only be obtained by a visual inspection if the engineer is properly prepared to observe the relevant features at the site.

A detailed checklist could be drawn up and should cover the following items:

(a) Topography: nature of the approaches to the site, the space to be spanned, suitable abutment and foundation positions, accessibility.

(b) Vegetation.

(c) River and flood characteristics (recorded flood levels, flood marks, scour damage, direction of flow, tendency of river to meander, etc.).

(d) Oceanographic and tidal information (if relevant).

(e) Surface geology: rock and soil types: discontinuities in rock strata, i.e. global and local stability of the rock and soil masses; faulting; folding; bedding; joints; weathering; water table and drainage.

(f) Any seismic records or information that may be available.

(g) Obstructions such as structures, pipes, cables or overhead wires that may interfere with operations.

(h) Degree of exposure to weather: wind, rain, snow, icing, and the degree of corrosiveness of the atmosphere.

(i) Available services: electricity, communication services, water supply, highways and railways, access to site and suitable locations for offices, construction plant, stores, yards and camps.

(j) Availability of materials of construction.

The investigation and interpretation of the surface geology will require a thorough knowledge of engineering geology and normally requires the services of a specialist.

At completion of this preliminary investigation, the engineer might be in a position to select the most suitable site which should then be investigated in more detail.

The detailed investigation should be carefully planned to avoid duplication of work or inadequate information on any particular aspect. This may comprise a detailed topographical survey, the compiling of data for hydrologic and hydraulic analyses, sub-surface foundation investigations and the investigation of material sources for sand, stone and water.

5.2.3 *Topographical survey of site*
For bridge design, accurate contour plans showing all surface detail of the site, the approaches and surrounding areas, are indispensable. The survey of the site should be tied in with the trigonometric survey coordinate system of the responsible authority.

On difficult terrain, specialized survey techniques comprising aerial and/or terrestrial photography backed up with conventional trigonometric survey methods, may be required. In the case of bridge structures over water, special survey and depth sounding methods may have to be employed, depending on the extent of the areas to be surveyed, the depth of water and the tidal and current flow patterns. Over deep water, initial seismic surveying methods, subsequently substantiated by direct depth sounding measurements, may be useful.

5.2.4 *Hydrologic and hydraulic analyses*
The hydrologic and hydraulic analyses consist of using various approved methods of establishing maximum probable floods at the site within specific return periods in order to size the waterway under the proposed bridge to pass a design flood of a magnitude and frequency consistent with the risk that is acceptable for the type or class of highway or railway. The procedures are usually prescribed by the authorities.

5.2.5 *Sub-surface soil investigation*

5.2.5.1 *On land*   Where alternative sites have to be investigated and compared, the sub-surface investigations may initially be of an exploratory nature only in order to establish which site has the most favourable foundation conditions, i.e. depth and consistency of suitable load-bearing strata. On sites where the number of alternatives is limited to one or two structures, the initial investigation may be planned to cover the actual foundation areas. A detailed investigation would usually only be proceeded with once a structure has been chosen and the actual foundation positions established.

During the preliminary investigation, it should be attempted to retrieve continuous core samples of the substrata which will enable the engineer to form a reliable conception of the various soil and rock formations and to identify areas of weakness and potentially unstable zones. Once the general structure of the sub-surface formations have been determined, the engineer will be in a better position to establish suitable foundation positions and to proceed with the planning of the bridge.

The detailed sub-surface investigation required for a particular site will depend largely on the conditions at site and the type of structure which is being considered. In all cases, the potential foundation problems should, where possible, be identified at the earliest feasible stage and the detailed investigation for the particular structure planned accordingly.

The optimum amount which should be expended on the investigation of a site will depend on the nature and variability of the soil, the effect on the structure of probable differential settlements and the risk and cost implications of possible foundation failure (see Section 3.2). The detailed foundation investigation may comprise some of the following procedures:

(a) *Shallow investigations* in hand or mechanically dug holes or trenches. Tests include

  (1) Plate loading or horizontal jacking tests to determine the load-bearing capacity and the *E*-modulus of soil or rock;
  (2) Shear vane—for shear strength of soil;
  (3) Density tests in soil—standard test or dynamic cone method;
  (4) Recovery of disturbed and undisturbed samples for laboratory analysis.

(b) *Deep investigations in soils*, which include

  (1) Wash boring with 'down the hole' tests; recovery of disturbed samples, standard penetration tests, pressuremeter tests, shear vane and screw-plate tests.
  (2) Cone penetration and sleeve resistance tests, Dutch cone penetrometer, equivalent point and shaft perimeter resistances of piles, structural properties including shear strength of soils and approximate density of soils.
  (3) Core boring; as (1) above, but with the recovery of undisturbed samples for laboratory analysis; or by large-diameter auger-drill for down-the-hole inspection, testing or sampling.

(c) *Deep investigations in rock* by rotary core drilling recovering representative rock samples for visual inspection and laboratory analysis.
(d) *Laboratory testing*, which includes

  (1) Grading analysis, Atterberg indicator tests;
  (2) Density, compactibility, consolidometer, direct shear or triaxial compression tests;
  (3) Tests on rock samples: point load, unconfined compression and aggregate suitability tests where required.

5.2.5.2 *Over water* Sub-surface investigations conducted over water may comprise seismic survey methods to establish the overall geology of the site below the waterline. Seismic surveys should be substantiated by core drilling conducted from suitable floating drilling platforms.

It is advisable to use specialists in the field of geology and geotechnical engineering for advice and guidance on all aspects of site investigation where the engineer lacks sufficient competence.

5.3 Methods of construction and erection of bridge superstructures

As indicated in Section 2.5, the available construction techniques may have a significant influence on the selection of the most suitable class of bridge for a particular site. It is therefore important that the designer should at an early stage be aware of the various construction techniques that are economically viable, as well as their advantages and disadvantages.

There are three main methods of construction for concrete

(a) *in situ* casting in formwork in position on the works;
(b) precasting off the works and subsequent transportation and erection;
(c) composite, being a combination of parts as in (a) and (b);

and four main forms of erection:

(a) on centering, i.e. stationary falsework supported directly at ground level or in the form of fixed girders or arches, or travelling falsework supported on the substructure or, when necessary, also on intermediate towers;
(b) by cantilevering from previous sections or substructures with or without suspended cable support;
(c) horizontal incremental jacking;
(d) by vertical hoisting, lifting or jacking.

5.3.1 *Stationary falsework*

Probably the oldest form of construction of concrete structures is by the *in situ* placing of concrete on temporary stationary falsework. Despite the fact that labour and economic conditions vary from country to country, its use is probably still the most widespread means of construction for short to medium-length bridges (up to about 300 m, 1000 ft) of moderate height (about 10 m, 33 ft). While closely-spaced timber centering was used in the past (some of the timber falsework for arches—now obsolete—were in fact masterpieces of engineering in their own right [5]), most stationary falsework today consists of standardized or proprietary steel sections used in towers, frames, girders, etc. Catalogues of various systems showing sizes of struts, frames, girders, couplings and accessories and safe load tables, are readily available in most countries. Points requiring close attention are: adequate founding of towers and frames on prepared beds or concrete footings where necessary and adequate bracing against lateral loads and buckling. The probable settlement of formwork during the placement of concrete should be allowed for and the necessary precamber built in to allow for long-term deformations. Special precautions such as differential jacking in gradual stages may be necessary to avoid cracking of the semi-hardened concrete. Mechanical jacks such as screw jacks, wedge type jacks or sandpots are suitable. Simple hydraulic jacks that cannot be locked are generally not suitable for supporting long-term loading due to possible leakage.

Some advantages of a stationary falsework system are

(a) It can be erected and dismantled by semi-skilled labour under supervision.
(b) It is readily adaptable to intricate and variable geometry of the superstructure.
(c) Sections are standardized and suitable for multiple re-use.
(d) Sections are of limited size and readily transportable.

Some disadvantages of the system are

(a) The system is cumbersome and slow, therefore, to achieve a reasonable rate of construction, staging may have to be erected for one or two spans ahead of the span under construction.

Figure 5-1 Mainbrücke Bettingen, Frankfurt. (*Courtesy* Polensky and Zöllner)

(b) It requires relatively firm founding conditions which may render it unsuitable for certain sites. It is not very suitable for river crossings where the additional risk of floods may exist.

(c) In built-up areas the need to maintain traffic may be a problem unless a special underpass section is provided. This is more costly and includes the risk of vehicular collision.

Examples of stationary formwork are shown in Figs 5-1 and 5-2. This form of staging is most commonly used for *in situ* concreting, but can also be used for precast segments [169].

5.3.2 *Travelling falsework*

Some of the disadvantages of stationary falsework can be overcome by the use of travelling formwork. The simplest form is longitudinally travelling formwork. For limited lengths of construction, the staging consists of scaffold towers with beams or girders supporting the formwork. On stripping, the formwork is lowered, the soffit shutters in the line of the piers swing aside and the whole assembly is pulled forward by winching to the next span. For this method, firm ground conditions and a reasonably constant height above ground level are essential. If the spans are simply supported, the staging extends for one span. If a continuous superstructure is constructed in stages, the construction joint is usually provided at approximately the point of contraflexure. When the formwork–falsework assembly is transported to the next span, it is rigidly connected to the cantilever of the previous span (Fig. 5-3).

Figure 5-2   Bridge over the Vaal River, South Africa

If the ground conditions are poor, the ground level is variable or the bridge is high above the ground, movable casting girders which are supported off the permanent sub- or superstructure can be a viable alternative solution. Casting girders can be classified into three types:

(a) the girder is above the final superstructure;
(b) the girder is below the final superstructure;
(c) the girder extends above and below the final superstructure.

The principal advantage of the 'girder above' solution is that it can be used for spans where the height of construction above ground level may be small. It is also suitable for the launching of precast elements. A major disadvantage is that the formwork normally requires temporary suspension rods passing through the superstructure.

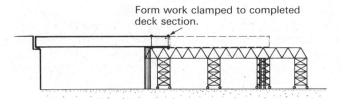

Figure 5-3   Travelling formwork on rollers

The 'girder below' solution has the advantages that no suspension rods pene-trate the superstructure and the deck area can be kept free of obstructions. Major disadvantages are that the height of construction above ground level must exceed that of the formwork and that the piers normally require adaptation to support and permit launching of the girder [170]. If the spans are large, use can be made of auxiliary temporary towers to limit the girder height, or the girder can be stiffened by means of prestraining with a king post and adjustable inclined ties, as in Fig. 5-4.

The girder shown in Fig. 5-4 was used to construct a continuous bridge in 40 m (130 ft) stages. The main casting girder consisted of four steel trusses. The central two trusses were provided with lighter structural steel launching noses and tail ends. The total length of girder from nose to toe was 92.73 m (304 ft) and its mass 190 tonnes (210 US tons). In a typical span, the side girders were supported on structural steel brackets stressed with prestressing rods to the outside of the pier stub columns. The construction technique was as follows.

(1) After the curing of the deck, the two main beams were post-tensioned longitudinally.

Figure 5-4 Draaiberg Bridge near Cape Town, South Africa. (*Courtesy* R. Meuwese)

(2) The girders were lowered by wedge jacks onto rollers at the piers. This also released the shutters from the deck soffit.

(3) The king post and ties were uncoupled and fixed to the girder underside.

(4) The beam soffit shutters were swung aside to clear the pier stub columns and the girder launched forward by means of hydraulic jacks. Cross-bracing had to be removed and replaced as it passed the piers.

(5) The girder was set up on the wedge jacks, king post and ties reinstated, shutters realigned and precamber set ready for the construction of the next stage.

Another variation of the 'girder under' operates on the slide-rule principle (Fig. 5-5). The system consists of three girders, a central launching and casting girder and two outer casting girders. The central girder is supported on the pier between stub columns and extends over three piers. The outer girders extend for the span of the stage under construction. At the rear end they are supported by means of a cross-girder on the completed section of the bridge. At the front end they are supported by means of a cross-girder on the central launching girder and on temporary brackets stressed to the front pier. During launching the outer girders are supported on the central girder at the front and the completed deck at the rear. The temporary pier brackets are transported forward with the outer girders during launching.

Figure 5-5 Vinxtbachtalbrücke, Cologne. (*Courtesy* Polensky and Zöllner)

There are many other variations which are beyond the scope of this handbook. For further details refer to [170–173].

Launching casting girders have proved suitable in the 30–60 m (100–200 ft) span range. Some advantages of the system are

(a) Superstructure construction can proceed irrespective of the ground conditions and any obstructions below superstructure or girder level.

(b) Superstructure construction can proceed at a rapid rate, as fast as one stage per week depending on size and resources, although two weeks per stage or longer is more common.

(c) With suitable design of cover to the girder, the superstructure can be constructed under site factory conditions.

Some disadvantages of the system are

(a) Launching casting girders require a fairly high capital investment, the cost of which must often be amortized on one or only a limited number of structures. Typical masses of steel per ton of concrete are 300–600 kg (660–1320 lb) [170]. The material content of the girder shown in Fig. 5-4 relative to the structure, was 250 kg steel/1000 kg concrete (552 lb/2208 lb).

(b) Launching girders are not readily adaptable to changes in superstructure cross-section and sharp radii of curvature.

(c) Launching casting girders, for reasons of economy, are usually highly stressed and flexible relative to the bridge superstructure. Therefore, to ensure a satisfactory final profile of the superstructure, the deflection of the girder during casting should be accurately assessed so that the correct precamber can be applied. Temperature effects, if significant, should be taken into account. The casting procedure and sequence should be chosen in such a manner that the hardening concrete does not suffer damage due to excessive imposed strains. Concrete set retarders might be advisable. Figure 5-6 shows the casting sequence employed to overcome this problem on a particular structure where the achievable casting rate was slow (about 80 m³/day, 105 yd³/day).

Note: It is important that the section of new concrete adjacent to the completed structure be cast when most of the deformation of the girder has taken place. Re-vibration of the concrete in this area at completion of the cast may be advisable.

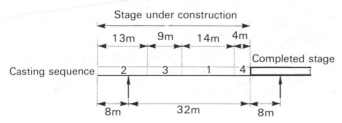

Figure 5-6 Casting sequence employed to avoid excessive imposed strains in hardening concrete

5.3.3   *Free cantilever construction*
Free cantilever construction with cast *in situ* concrete was originally developed in West Germany in 1950 for long-span river crossings, while precast segmental construction was pioneered in France [174] (see Section 1.2).

5.3.4   *In situ free cantilever construction*
In its basic form, *in situ* free cantilever construction proceeds as shown in Fig. 5-7. After construction of the deck section over the pier, construction proceeds by means of cantilevering sections, usually symmetrically about the pier to control the unbalanced moment acting on the pier. The length of segment cast depends on economic factors determined by the mass of concrete cast in one cycle and the cost of the carriage. Segments of 3–5 m are common. After the hardening of the concrete, the segment is post-tensioned to the completed portion to carry the cantilever moment. As construction proceeds, the continuity cables are progressively inserted and stressed across various segments. Before the last concrete that closes the central gap is cast, the two cantilevers are locked together by various methods to avoid damage to the hardening concrete due to movements induced

Figure 5-7 Medway Bridge, England. (*Courtesy* Freeman, Fox and Partners)

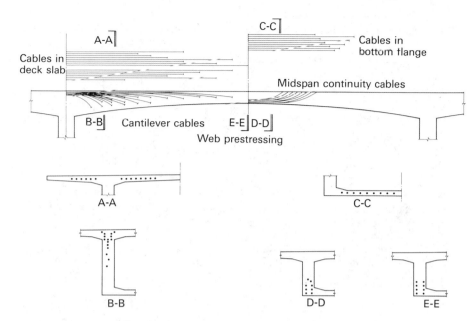

Figure 5-8 Prestressing details

by temperature changes or external actions. As the concrete in the gap hardens, the sagging moment cables are inserted and stressed in stages (Fig. 5-8). In this way, construction proceeds from span to span and the structure is made continuous. A major advantage of *in situ* cantilever construction is the continuity of reinforcement across the construction joints. The method of construction has been successfully used for spans of up to 240 m (787 ft)—for example, the Hamana–Chasi Bridge in Japan.

It is important for stability that the deck be fixed to the piers during construction. If the pier height is suitable, it can be made monolithic with the superstructure. Alternatively, the deck can be stabilized with temporary supports or the bearings at the top or bottom of the piers can be fixed during construction. The latter for instance was adopted for the Mosel Bridge at Schweich [175].

A variation of the free cantilever construction method is that of using launching girders which can serve as supports during construction while also being used for the transport of men, materials and the carriage. The method was first used during the construction of the Siegtal Bridge [176].

To limit the cantilever moments, the superstructure can be constructed in stages, e.g. the box section is constructed by the free cantilever method and the deck cantilever slabs constructed in subsequent operations (Eschachtal and Kochertal Bridges) [177]. In certain instances, the use of temporary auxiliary towers offers a practical solution (Medway Bridge in England, Fig. 5-7). Auxiliary pendulum towers and temporary tie-back cables have been used for the cantilever construction of continuous bridges. The method is eminently suited for the construction of arch bridges and strut-frames over deep gorges (see Section 5.6 and Figs 5-9 and 5-10).

Figure 5-9   Bloukrans Bridge, South Africa

5.3.5   *Precast segmental free cantilever construction*
The segmental construction proceeds similarly to the *in situ* free cantilever construction with the following major differences. The superstructure is divided into segments which are precast in suitable sizes as dictated by transport and handling facilities. These segments are then transported to site and erected, commencing with the section over the pier and proceeding by cantilever construction from both sides of the piers. An important feature is the treatment of the joints between the elements. There are two basic methods:

(a) The joint is cast *in situ* of adequate size to permit sufficient overlap for the effective continuity of longitudinal reinforcement. This method was employed for the construction of the Oosterschelde Bridge in the Netherlands [5].

(b) The more common form of joint treatment is by 'glueing' the segments together and forming a very thin joint. No ordinary unstressed reinforcement can accordingly cross the joint. To achieve the close tolerances required, match-casting is done by casting the element under construction against the preceding one. The elements are then erected in sequence. Prior to fitting, the contact areas are coated with a bonding agent, generally an epoxy, sometimes a cementitious product. This bonding agent is not usually

Figure 5-10   Gouritz River Bridge, South Africa

taken into account in the structural design as contributing to the ultimate strength of the section. However, in the case of a segmental bridge constructed on staging over the Europa Canal near Nuremberg, the bonding agent was considered as contributing and tests carried out to assess its effectiveness [178]. The functions of the bonding agent are: (1) to lubricate the matching segment faces so as to facilitate alignment; (2) to take up any minor misalignment which, despite match-casting, occurs mainly due to differential shrinkage, creep and temperature; and (3) to ensure that the joints are solid and watertight.

The faces of the joints were previously provided with shear keys which served to align the segments and transfer shear stresses and diagonal compressive stresses. Shear keys can give rise to excessive stress concentrations. The modern method of profiling the joints is by means of sawtoothing and providing shallow keys for alignment purposes in the webs [174, 179] (See details of the Sallingsund Bridge [180]). The fact that the unstressed reinforcement is discontinuous over the construction joints requires full prestressing for moments due to loads, temperature and secondary moments. For the former reason, the theoretical ultimate moment of resistance is reduced [181]. On the other hand, the construction method is speedy and this serves to offset the additional cost of prestressing. In both the *in situ* and precast cantilever construction methods, the deflections due to loading, creep, differential shrinkage and temperature, have to be carefully assessed so that both the structural and geometric requirements are complied with [182].

5.3.6 *Incremental launching*
The incremental launching method, which was developed by Leonhardt and Baur [183], consists of constructing the bridge superstructure in stages by casting successive segments at one end of the bridge. The segments are usually 15–30 m (50–100 ft) long. After the segment under construction has gained sufficient strength, it is prestressed for the launching stage and the completed sections of the bridge launched one stage forward and the procedure repeated. During launching, the section undergoes complete stress reversals as it progresses from a cantilever to the first support and thereafter over the following spans to its final position. For launching from one end, the bridge profile should be constant, preferably straight but horizontal and vertical curvature of constant radius can be accommodated within limits. The deck slab should have a constant superelevation without any transition areas. Minor variation in deck width, especially near the bridge ends, can be accommodated.

Because of the stress reversals during launching, a box section with nearly symmetrical section properties is the most efficient. Other sections, such as a double tee [184], have been used. To cope with the launching stresses, the section is usually post-tensioned axially with cables in both the top and bottom flanges. To limit the cantilever stresses, the front end is usually fitted with a lightweight structural steel launching nose. If the spans are excessively long or variable, temporary auxiliary towers may be necessary. The towers should be designed so that their stiffness is compatible with the launching requirements. Use can be made of jacks to control the support reactions and thus the stresses in the superstructure. The additional permanent state cables are usually threaded in or fixed subsequently and post-tensioned after the bridge has reached its final position. Some of the cables required during launching could be recovered, but it is rarely economic to do so. As the final set of cables is therefore, in effect, required to cater for the range of stresses in the completed structure, this method requires more prestressing. Under favourable conditions, this is more than compensated for by saving in centering and formwork and by the speed of erection. The capital investment in equipment is not very large, with the result that the method has proved to be very competitive.

Figure 5-11 shows a diagrammatic elevation of a bridge under construction by

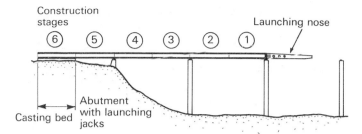

Figure 5-11 Incremental launching

the incremental launching system. The first section is poured on the casting bed behind the abutment on the fill. Provision for accurately controlling the soffit levels is essential. To ensure a smooth launching operation, the soffit and edge of the superstructure must be to the correct profile without kinks and discontinuities. The launching nose is stressed to the first section after it has been axially stressed in preparation for launching. One method of launching is by means of hydraulic jacks fixed to the abutment, which pull the bridge segments forward by means of stressing rods which are temporarily anchored to the section just completed. An alternative method [185] involves using a proprietary system of jacks in which the superstructure at the abutment is lifted and then jacked forward, lowered, the launching jacks retracted and the operation repeated to move the superstructure forward in small increments. The piers, abutment and supports of the section under construction are provided with temporary sliding bearings made up of steel or concrete with a stainless steel surface and side guiding plates to keep the bridge on line. Steel reinforced Neoprene bearing plates, usually of 13 mm ($\frac{1}{2}$ in) nominal thickness and coated with Teflon on one side to facilitate sliding over the stainless steel, are placed on each bearing. As the plates move forward, new ones are inserted immediately behind. When the front ones come free, they are re-used at the back. The coefficient of friction is approximately 2%, but the launching equipment and structure should be designed for at least 4% friction [183]. If the bridge is on a downward slope, the launch is usually in the downward direction. Slopes greater than about 2% require braking devices. The piers must be designed so that the temporary bearings can be replaced by permanent ones. Modern practice is to adapt the permanent bearings to serve as temporary sliding bearings [184]. The usual cycle per stage is one week.

5.3.7 Hoistings, lifting or jacking
The hoisting, lifting or vertical jacking of precast beam units is commonly used. Depending on the equipment available and economic considerations, use may be made of cranes or vertical jacks in conjunction with towers, girders or slings. Cross beams and decks may then be cast *in situ* or precast and similarly placed in position. Entire spans have been precast and jacked into position in this way, e.g. the Seven Mile Bridge in the Florida Keys (Fig. 5-12).

A variation of the jacking technique is to construct the bridge, such as a subway crossing, alongside the embankment under which the subway is to pass and then to progressively jack the completed structure into the embankment excavation.

Figure 5–12 Seven Mile Bridge in the Florida Keys, June 1981. (*Courtesy* Figg and Muller Engineers Inc.)

The end of the structure which is jacked into the embankment is normally provided with cutting edges not unlike tunnelling [186].

5.4 Bridge foundations and substructures

The functions of the bridge substructure are to support the superstructure at the required levels and safely carry the dead and live loads from the superstructure through the foundations to the subsoil and to provide stability against all actions

Figure 5-13   Clichy Bridge, France. (*Courtesy* Freyssinet International)

on the structure. Substructures are either designed to restrain or to accommodate horizontal and rotational movements of the superstructure. Where soil investigations predict settlements or heave of an appreciable order, the effects thereof are to be taken account of in the design of the total structure. In some bridges, monolithic construction of the superstructure and substructures is feasible. Substructures and foundations are important and substantial parts of bridges, both functionally and costwise. The aesthetic treatment of substructures in relation to the whole structure is very important (Fig. 5-13).

5.4.1   *Abutments*
Abutments or bank seats support the ends of bridges. The various types of abutment commonly used are listed and illustrated in Section 1.3. They are usually constructed by conventional means of *in situ* concrete. The superstructure may be integral with the abutment, as in portal frames or diaphragm walls, or may be pinned or supported on bearings with various degrees of freedom. In the case of small structures, the superstructure can be fixed to the abutments by means of dowels which permit small relative rotations, but restrain large horizontal displacements (Fig. 5-14). An abutment must be adequately designed to resist all combinations of forces acting on it, including the earth pressure of the fill it retains. Simplified procedures are prescribed in codes by most authorities. These may under certain circumstances underestimate the resultant forces due to earth

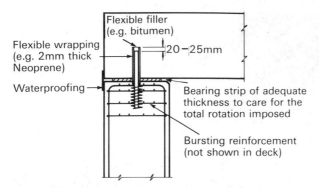

Figure 5-14 Dowel detail for small structure

pressures, which may exceed that of the 'earth pressure at rest'. This applies particularly to backfill between counterforts. The fill is compacted in layers, each layer imposing increased pressures on the wall which will exceed active pressures if the walls do not yield. The fill is furthermore compacted by traffic and in the case where horizontal movement between super- and substructure is restrained, the expansion of the superstructure may induce further increases in horizontal earth pressure. Care is therefore required in the design of wall elements to ensure that the basic design assumptions are realized or otherwise to allow for probable increased forces.

When the horizontal movement is too large to permit the superstructure to be pinned to both abutments, the deck is supported on bearings that allow freedom of movement and rotation. In the case of medium length bridges, it is preferable to fix the deck at one abutment and to provide for freedom of movement at the other. This saves one expansion joint (see Section 5.8). Beyond a certain length of deck, longitudinal movement may have to be allowed for at both abutments and fixity provided at one or more of the intermediate supports.

No decks should depend entirely on friction for horizontal stability. Unless sufficient horizontal support is provided, it is imperative in areas where floods or earthquakes may occur or in the case of sloping decks, to allow for 'stops' beyond the limits of movements required for temperature and other effects.

5.4.2 Wingwalls

The wingwalls of the abutments are extensions that serve to contain the fill from spilling forward or sideways at the ends of the abutment wall. In closed abutments, they can be either parallel to the roadway or continuously in line with the abutment wall, or skewed between these planes (Fig. 5-15). In spill-through abutments, wingwalls (if used) usually cantilever from the breast beam and limit the amount of spill that takes place.

The walls of closed abutments act as deep beams vertically and horizontally, and as slabs in resisting forces normal to their planes. Wingwalls primarily act as cantilever slabs supported at one or two edges depending on their size and type. Design tables for assessing the moments are available in various publications, e.g. Bareš [187] and Stein [188]. For reasons already stated, the horizontal earth

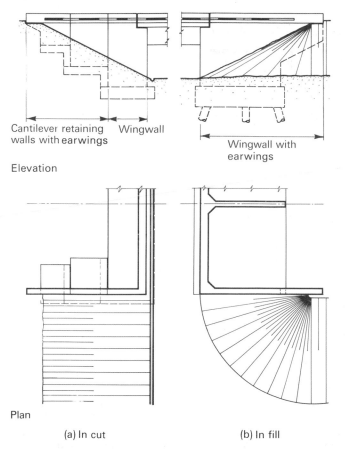

Elevation

Cantilever retaining   Wingwall
walls with earwings

Wingwall with
earwings

Plan

(a) In cut   (b) In fill

Figure 5-15   Closed abutments

pressures should not be underestimated. Leonhardt [181] recommends between 1.5 and 3 times active pressure. If the wingwall moments or deflections become excessive, splitting the wingwalls into separate walls might be advisable (Fig. 5-15a).

5.4.3   Approach slabs

The approach slabs for bridges are constructed under the roadway of highway bridges to avoid a sudden step between the end of the bridge and the fill due to settlement of the latter. The approach slabs are supported on the abutment at one end and rest on the fill towards the other end. The length of approach slab depends on the expected settlement of the fill and the acceptable change of gradient in the road profile. Figure 5-16 shows two types of approach slab.

5.4.4   Drainage

The fill behind closed abutments and walls should be drained to prevent build-up of water pressure on the walls and pumping and failure of the fill. In addition, the

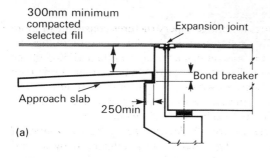

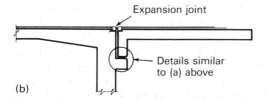

Figure 5-16   Two types of approach slab
(a) Buried in fill
(b) On fill

surface water from the deck might have to be led away via the piers or abutments. Figure 5-17 shows a detail for draining the fill. Surface water from the deck, namely, stormwater run-off plus seepage water through expansion joints, should be led via drainage pipes to discharge outlets. Externally mounted pipes are usually aesthetically unacceptable. Pipes cast into the concrete should be sufficiently rigid and strong to prevent damage and blocking during casting. They should be detailed to allow cleaning by rodding, i.e. have no sharp bends and be sufficiently large to prevent clogging up. A diameter of 100 mm (4 in) is probably a minimum practical size. In areas subject to freezing temperatures, the pipes might have to be sleeved to prevent the concrete from bursting due to the freezing of blocked water.

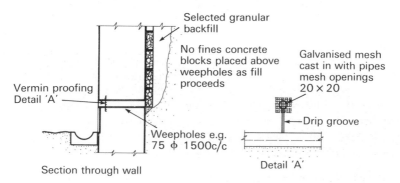

Figure 5-17   Abutment detail for draining fill

5.4.5 *Piers*

Piers form the intermediate supports for bridge superstructures consisting of more than one span. They are usually constructed in *in situ* concrete, either by conventional means or, for tall piers, by continuous or intermittent sliding of the formwork. Precasting is rarely used. The superstructure may be monolithic with the pier or may be supported on bearings with varying degrees of freedom. The same variations may apply at the connection with the foundation and altogether there are many possible combinations to suit the requirements.

The number and spacing of piers are usually determined by the economics of the total structure. The choice of pier type and shape is influenced by several factors. Aesthetic considerations are very important but other considerations may dictate the basic shape. This applies particularly if the bridge crosses flowing water. The shape and orientation of the piers affect the flow characteristics through the river crossings [189]. Wall-type piers with shaped noses are preferred for their hydraulic efficiency. In rivers where the likelihood of floating debris exists, closely spaced piers are normally not desirable. Figure 5-19 shows typical river piers, A to D, offering decreasing resistance to flow. It has however, been observed that sharp edges on piers as in shape D, are more prone to trapping debris, such as branches, than well-rounded ends as in type B. This increases the resistance to flow. Wall-type piers are usually required to be not too slender for aesthetic reasons. Type C, due to shadow effects, can from an angle appear to be as thin as the nose ends. Piers near navigable channels are normally massive to resist impact from shipping. Additional barrier islands are essential in the case of large bridges spanning major shipping routes (Fig. 5-18).

Wall-type piers are not very suitable for road crossings as they give rise to a visual tunnel effect. Isolated columns are preferable. They should be limited in number in the transverse direction for improved appearance and sight distance.

Figure 5-18 Dauphin Island Bridge. (*Courtesy* Figg and Muller Engineers Inc.)

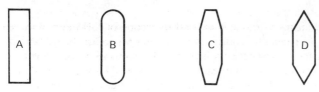

Typical sections of piers in rivers

Figure 5-19   Typical sections of piers in rivers

However, impact on slender columns can cause catastrophic damage. Impact effects on columns are extremely complex and research on the subject is an ongoing process [190]. Reinforcing the compression zone against destruction has proved successful. Large-diameter columns on the other hand look very clumsy. Figure 5-20 shows various pier shapes and cross-sections. See also Figs 1-17 and 5-13.

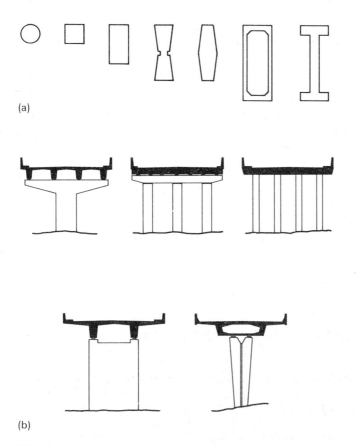

(a)

(b)

Figure 5-20   Typical (a) pier and (b) bridge cross-sections

5.4.6 *Foundations*

An in-depth discussion of foundations is beyond the scope of this handbook and the reader is referred to specialized literature covering the geotechnical aspects (soil and rock mechanics). Foundations can be classified into two types: 'shallow' (up to say 6 m below ground level) and deep. Examples of the former are spread footings, strip footings and raft foundations. They are founded directly on the substrata. Loose subsoil can be consolidated artificially by mechanical means such as vibroflotation and fissured rock by grouting, rock anchors, etc.

Deep foundations may be in the form of piles or caissons. They transmit the forces down to load-bearing strata (end bearing) or utilize skin friction or a combination of both. In river crossings, they must be adequately founded well below the probable scour depth. Small-diameter piles in soil that may be subject to scour should be checked for buckling. See also Chapter 31 on piles.

It is advisable to use specialists in the field of geology and geotechnical engineering on all sites that may present problems.

5.5 Slab, beam and box-girder superstructures in bridges

5.5.1 *General*

The choice of superstructure is affected by several considerations. Listed here are some of the more important ones: (a) topography of site; (b) geometry of superstructure, i.e. straight, curved, skew, varying or constant width, etc; (c) required permanent open space under deck; (d) length of spans; (e) founding conditions; (f) type of live load and its magnitude: road, railway, seismic; (g) available depth of construction; (h) available construction methods; (i) restrictions during construction, such as maintenance of traffic.

Increasing the length of span increases the cost of the superstructure but for multi-span structures there is a reduction in the number of piers and foundations. Generally, for a given set of circumstances, it is found that there is a range of economic span lengths. Figure 5-21 shows the relative cost of a road-over-river

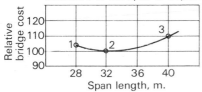

(1) Four beam, total length 588m, $l/h = 14$
(2) Eight beam, total length 576m, $l/h = 20$
(3) Two beam, total length 584m, $l/h = 15$
where l/h is the span-to-depth ratio

Figure 5-21   Relative cost of road-over-river bridge with deep piled foundations
1 Four-beam, total length 588 m, $1/h = 14$
2 Eight-beam, total length 576 m, $1/h = 20$
3 Two-beam, total length 584 m, $1/h = 15$
where $1/h$ is the span-to-depth ratio

bridge with deep-piled foundations in relation to span length. The curve is flat over a reasonable span range. The designer can thus usually optimize both cost and aesthetic requirements. Other considerations are the degree of continuity of spans, i.e. simply supported, partially continuous, or fully continuous. Helminger [191] has tabulated the absolute and relative material content for 188 bridge superstructures built in Bavaria in the early to mid-1970s.

The various deck configurations that are commonly used are listed in Section 1.3. The design and detailing of slabs and beams in general are dealt with in Chapters 12, 13, 16, 19 and 20.

5.5.2 *Slabs*

5.5.2.1 *Solid flat slabs* The solid concrete flat slab is commonly used for short, simply supported spans and up to medium lengths for continuous spans. It is especially suitable for skew crossings or shapes that are variable in plan. Spans up to approximately 15–17 m (49–56 ft) are commonly constructed in reinforced concrete. Larger spans are usually prestressed. Leonhardt [181] considers the practical limit for solid slabs to be approximately 20 m (66 ft) for simply supported spans, 30 m (98 ft) for continuous spans and 36 m (118 ft) for continuous spans with haunches. Larger spans have, however, been built [191].

Solid flat slabs can be regarded as isotropic, i.e. having the same physical properties, such as stiffness, in all directions. The effect on the stiffness of varying reinforcement in different directions is usually negligible under service loads. A slab is considered to be anisotropic when these properties differ substantially in various directions. If the properties differ in two specific orthogonal directions, the slab is orthotropic.

A slab loaded perpendicular to its plane can be subject to bending and twisting moments in all directions. These moments and twisting moments have to be resolved into principal moments and their directions determined for evaluation of reinforcement. To do this, the moments about two perpendicular axes and the twisting moment at any specific point is required, i.e. M_x, M_y, M_{xy}. In rectangular one-way spanning slabs, the principal moments due to uniformly distributed loads are approximately parallel and perpendicular to the edges of the slab at midspan and over supports in continuous slabs. In skew slabs and slabs of odd shape, this is not necessarily the case. Under non-uniform superimposed loads, the direction depends on the distribution and location of the loads.

Slabs are usually analysed by the following methods (see Section 4):

(a) Influence surfaces or tables such as those published in [192–199].
(b) Grid analysis. The layout of the grid and correct simulation of member properties is important [111].
(c) Finite strip method in certain cases only.
(d) Finite element and finite difference methods.
(e) Model analysis (see Chapter 23).

Methods (c), (d) and (e) can be relatively expensive and should only be used when the complexity or importance of the structure warrants their use.

In skew slabs, the end support reactions differ with the degree of skewness and stiffness and spacing of bearings. Stiff linear support conditions give rise to high

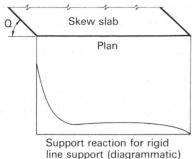

Support reaction for rigid
line support (diagrammatic)

Figure 5-22 Support reaction for rigid line support (diagrammatic)

support reactions at the obtuse corner with high hogging moments (Fig. 5-22). The reactions at the acute corners can be negative. Skew slabs often require shear [181] reinforcement. These effects can be reduced by the use of 'soft' bearings and/or by varying the spacing of bearings [181, 198].

In continuous skew slabs with large width-to-span ratios, the magnitudes of the moments do not greatly exceed those in equivalent continuous rectangular slabs. In narrow continuous skew slabs, the effect of the support arrangements on the moments and support reactions is not much more pronounced than for simply supported skew slabs [181]. Where circumstances permit, the designer should investigate the alternative possibility of replacing skew slab decks with rectangular decks as the cost of the resulting increased span lengths may be more than offset by the savings due to the simplicity of deck reinforcement and smaller abutments.

Rectangular slabs can conveniently be reinforced in the directions of the principal moments. In skew slabs, this requires fanning of the reinforcement to approximately coincide with the varying directions of the principal moments. This is not always practical. It might be advantageous to place the reinforcement parallel to the free edges and support lines or to use a skew or orthogonal mesh [181, 198], but this would result in an increased amount of reinforcement. Where the directions of reinforcement differ significantly from the directions of the principal moments, these have to be transformed to determine the amount of reinforcement [200, 198].

The effect of prestressing slabs can be simulated by loading the slab with the vertical components of the radial forces due to the curvature of the prestressing cables. The effects of axial force and, where applicable, the eccentricities, are then considered separately.

5.5.2.2 *Hollow slabs* When the required depth and resulting dead load become excessive, slabs are usually made hollow. Commonly used void formers made of polystyrene, hollow reinforced cardboard, compressed hardboard, sheet metal, timber, inflated rubber tubes, etc, are cast into the deck to form voids of circular or rectangular cross-section or as required (Fig. 5-23).

Longitudinally, the stiffness and strength is not materially affected. The transverse stiffness is, however, reduced. The exact reduction depends on the shape and size of the voids, the width of the ribs and the thickness of the top and

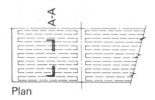

Plan

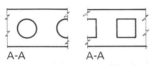

A-A   A-A

Alternative void shapes

Figure 5-23   Voided slab

bottom flanges. According to Hambly [111], if the width and depth of the voids is less than 60% of the overall width and structural depth respectively, the slab still behaves like a plate. Hollow slabs are usually provided with crossbeams at least at the supports and at midspan for effective transverse distribution of moments and shears. If the voids become large, the slab acts like a cellular deck, i.e. deformations due to 'shear' become important and the deck no longer behaves like a simple plate.

5.5.2.3   *Coffered slabs*   These act like a grid and are suitable for decks where the moments are primarily sagging (Fig. 5-24).

5.5.2.4   *Precast beams*   Where it is inconvenient or impossible to erect staging such as over existing roads or railways, precast beams can be laid contiguously next to each other. Plate action is usually achieved by casting *in situ* concrete strips between the beams with or without a thin *in situ* top slab. The slabs are transversely reinforced or post-tensioned. Ingenious methods of achieving 'dry

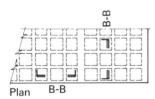

Plan   B-B

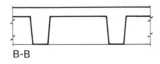

B-B

Figure 5-24   Coffered slab

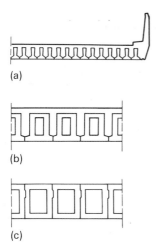

(a)

(b)

(c)

Figure 5-25   Precast beam slab decks
(a) T-beams with *in situ* concrete topping
(b) 'Tophat' beams with *in situ* topping
(c) Contiguous beams

joints' have been developed such as the Dywidag 'contact method' [201]. Figures 5-25a-c show various cross-sections of precast slab construction.

5.5.3 *Beams*

A traditional form of superstructure consists of precast I- or T-beams, with precast or *in situ* crossbeams and deck slabs. In this form, the primary structural

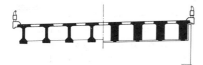

Figure 5-26

elements act as a grillage, the deck slab spanning between beams and acting compositely with the beams as top flanges. Crossbeams are usually provided at the supports and at the one-third points of the span. Figures 5-26 and 5-27 show two versions. Unless post-tensioning is applied where precast elements are joined together, the gap provided for *in situ* concrete must be sufficient to provide adequate bond for the overlapping reinforcement.

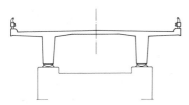

Figure 5-27

Precast post-tensioned I-beams have a useful span range of 20–35 m (66–115 ft) whereas precast post-tensioned T-beams are suitable for span ranges up to 45 m (148 ft). The main limitations are transport and handling problems.

With the demand for speed of construction and the widespread increase in labour costs, simplicity of construction has become ever more important and crossbeams have become less popular. Various deck profiles suitable for *in situ* casting on travelling formwork have become popular. These include single spine or twin beam systems, usually without crossbeams. See also Section 5.6 and Fig. 5-28. In the case of twin-beams, crossbeams should, however, be provided at the end supports both for torsional fixity of the beams and to keep the slab moments and deflections at the ends within acceptable limits. In the span and over intermediate supports in continuous structures with twin- or multi-beams, the slab can be designed to provide transverse stability by interaction in bending with the main beams. Torsional resistance of the deck system in these cases is resisted by a combination of torsion in the beams and a warping action of the beam and slab system. Figure 5-28 shows a typical cross-section.

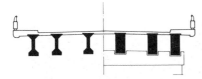

Figure 5-28

The design of beams is covered in Chapters 12, 13, 16, 19 and 20 and in most national codes (see Section 3) and will not be repeated here. See also [181].

Upstand beams are used in through bridges where maximum clearance is required. Cross-sectional shapes commonly used are rectangular or I-beams. They are not aesthetically very pleasing and are only really suitable for narrow structures. Figure 5-29 shows a pedestrian bridge across a railway line where the authorities required solid balustrades at least 1.450 m ($4\frac{3}{4}$ ft) high, which were used as beams.

Figure 5-29   Pedestrian bridge with upstand beams

5.5.4   *Box-girders*

Box-girders are extremely versatile and suitable for many applications. Their principal advantages are:

(a) The relatively large torsional stiffness of the closed section.
(b) Good sectional properties for bending resistance. The difference between the upper and lower section moduli at any section is not as great as for a T-beam slab deck. This results in less creep deformation in prestressed structures, and the lower flange provides a larger compression zone at the

supports and more space to accommodate the tensile reinforcement at midspan.

(c) Numerous arrangements of box sections are feasible to suit any deck width. The slab thickness and corresponding web spacing can be varied to suit (see Section 1.3).

(d) Its bending and torsional stiffness makes it suitable for bridges curved in plan.

For the above reasons, box-girders are eminently suitable for long spans. The main span in an alternative proposal for the Great Belt Bridge is 325 m (1066 ft) long [5].

The box section can be constructed as a single or multicellular box, although the tendency for large bridges is to use a single cellular box with large cantilever deck slabs which can be stiffened with ribs or struts, for example the Eschachtal Bridge [177] (see also Fig. 5-30).

In addition to simple beam action, single-box sections and twin or multiple box sections, interconnected by slab decks, are subject to distortion and warping. The

Figure 5-30 Hammersmith Flyover, London. (*Courtesy* Maunsell and Partners)

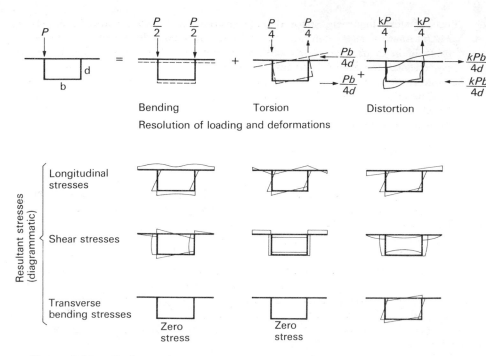

Resolution of loading and deformations

Figure 5-31a   Deformations and stress patterns (diagrammatic)

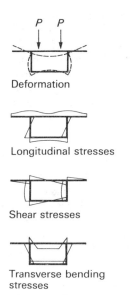

Deformation

Longitudinal stresses

Shear stresses

Transverse bending stresses

Figure 5-31b   Deformations and stress patterns (diagrammatic)

effects of shear lag, distortion and warping are shown diagrammatically in Figs 5-31a,b. As with any beam and slab system, shear lag effects may have to be considered in the flanges of box-girders, including the cantilever deck slabs. In regions where temperature gradients between the inside and outside of boxes may be large (>20°C, 36°F), ventilation should be provided. Box-girder reinforcement should be adequately detailed to ensure the integrity of the webs and slabs.

For the analysis of box girders, see Section 4 and [111, 202]; for reinforcement and prestressing, see [181].

5.6 Arch bridges and related types

5.6.1 *General*
Various configurations of arch type bridges are summarized in Section 1.3. For reasons of space, it is not possible to deal with all these types in any detail. This section will therefore deal mainly with the modern form of concrete arch bridge, namely, the open arch with spandrel columns. The strut-frame type is described in [203].

Before the advent of the computer, arch bridges were designed using classical theory based on the principles of elasticity and the theory of small displacements. In order to reduce the considerable effort in computation that was necessary to solve the equations which govern the elastic behaviour of an arch, various assumptions and simplifications to the structure and the theoretical modelling thereof were usually made [204].

With the development of modern computing methods and a clearer understanding of the reliability of structures, it has become practicable to simulate the total structure and the various actions and effects thereon much more realistically, with the result that it is now possible to design concrete arch bridges to closer tolerances with greater reliability. This, together with the efficient use of materials with improved strength properties, has led to slender structures, making it possible to eliminate hinges and to improve the economics of arch bridges in general. In concrete, other bridge configurations were, however, preferred in the 1950s and early 1960s because of the high cost of centering required for arch construction. Some engineers predicted that concrete arches would be superseded by other more economical structures such as cable-stayed girder bridges, even at sites ideally suited to arch construction. However, the development of the suspended and truss cantilever methods once more made concrete arches competitive up to spans approaching 400 m (1300 ft); see Section 1.2. These methods are described later in this subsection.

5.6.2 *Structural configuration of concrete arch bridges*
Modern medium to large-span concrete arch bridges appear to have optimum characteristics in the form of open spandrel arches composed of the following members (see Figs 1-10, 1-11 and 1-14):

(a) A continuous prestressed or reinforced concrete deck supported by the abutments and at regular intervals by spandrel and approach columns, which are usually also dimensioned to provide lateral support to the deck.

(b) A reinforced concrete arch which, on large arch bridges, is usually designed as a hollow box-section with full fixity at the springings. Solid arches may,

however, be more economic on shorter spans. Although hinges are normally avoided, they could be beneficial under specific conditions for shorter
spans.

(c) The proportioning and arrangement of the above members are interdependent because of their interaction in resisting various actions.

The following factors require careful consideration.

5.6.3 *The influence of the site on the dimensional and geometric requirements*

On a major bridge structure, especially arch bridges spanning gorges, the exact
horizontal and vertical alignments of the bridge should not only be located in
terms of road geometric requirements. The total cost is determined by the
dimensions of the structure, as well as by the nature of the approaches and the
foundation conditions. It is therefore essential that a detailed foundation investigation be conducted before the structural configuration is finalized. The level of the
roadway is an important factor that can influence the configuration. It will, for
instance, affect the span-to-rise ratio of the arch, the height of columns, the
position of abutments and therefore also the length of the deck. All these factors
influence the overall economics of the structure, which should be balanced against
the cost of cutting or fill for the approach roads. See Section 5.2.

The foundations of the arch are parts of the primary support system and both
should preferably be at the same elevation. The positions of the foundations of
the approach columns must be carefully considered and investigated, especially on
steep banks, to ensure stability.

Where use is made of the suspended cantilever method for construction of the
arch, suitable anchorage conditions for the tie-back system must be available. This
requires sound rock formations.

The width of the deck is an important structural feature and is determined by
the number of carriageways, traffic lanes, shoulders and sidewalks. It affects the
relative transverse arrangement of deck members and columns, and therefore the
cross-section of the arch, all of which contribute to the lateral stability and
flexural stiffness of the structure and affect the economics of the bridge (Fig.
5-32). Future widening of the deck may need special consideration in the location

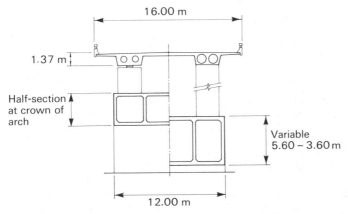

Figure 5-32 Section through arch and deck (Bloukrans Bridge, South Africa)

of the bridge and in the design and arrangement of the supporting elements. On dual carriageway structures, consideration must be given as to whether the bridge should be separated into two parallel structures or combined in one structure. On long-span arch bridges there are normally structural and economic advantages in a single wider structure with greater transverse stiffness than two separate structures.

The shape of the arch should preferably be symmetrical with respect to the crown of the arch in order to avoid permanent displacement or sway in the longitudinal direction of the bridge. A difference in the number of spans on the embankments and hence the length of the deck on both sides of the arch due to an unsymmetrically shaped valley or river bed is normally not a problem provided that the deck is symmetrically supported relative to the crown of the arch and provided that the effects of longitudinal displacements are adequately provided for in the bearings and columns.

With an unequal number of spans on the embankments, however, transverse bending of the deck due to wind, earthquake and eccentric live load will not be symmetrical about the crown of the arch. An uneven distribution of loading along the structure must be considered and carefully analysed, especially the displacement and rotational effects on the columns and bearings (See Fig. 5-35).

5.6.4 Design aspects

5.6.4.1 *Foundations* The foundations of an arch bridge should be able to resist the thrust from the arch without excessive settlement in any direction. It may be feasible to improve the load-bearing characteristics of the rock by the injection of cement in the highly fissured zones. Consideration must be given to the construction stages when large eccentricities of the thrust may result in high edge pressures. On large arch bridges, it is most important to establish the upper limit of probable settlement of the foundation. This can be estimated by means of two

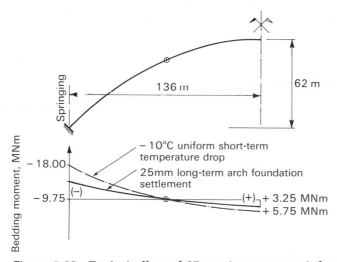

Figure 5-33 Typical effect of 25 mm long-term arch foundation settlement compared with unit temperature effect (Bloukrans Bridge, South Africa)

or three-dimensional displacement analyses in which the properties of the rock mass, joints and other discontinuities are taken into account [205–207]. The amount of foundation settlement that can be accommodated by the arch will depend on the shape, the sectional properties and flexural stiffness of the arch and the quality of concrete and amount of reinforcement (Fig. 5-33).

5.6.4.2 *Columns* Structural considerations dictate that the columns should be as light and slender as possible, especially on the arch where the concentrated dead-load reactions of the columns result in long-term local peak bending moments in the arch (Fig. 5-34).

Due to the relatively large longitudinal movements of the continuous deck when subjected to temperature changes, it is advantageous if the columns are made flexible in the longitudinal direction of the bridge, thereby reducing bending moments in the columns and in the case of the taller columns, making hinges or sliding bearing unnecessary. Slender columns are also beneficial for providing the necessary flexibility in order to reduce the effects of earthquakes (see Section 4.6). If the deck is subjected to ground accelerations, large inertial forces will result from its considerable mass. Rigid connections restraining movement between the deck and the foundations should therefore be avoided. All connections between the deck and the columns and other supports, should be designed to accommodate the maximum relative displacements that may develop during seismic disturbances.

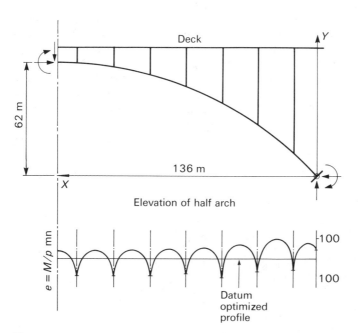

Figure 5-34 Eccentricity M_{DL}/P_{DL} in arch due to dead-load effects at $t = \infty$ (Bloukrans Bridge, South Africa)

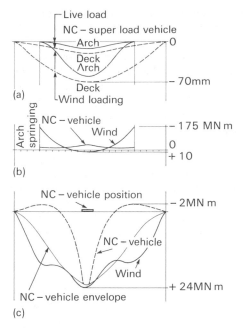

Figure 5-35   Transverse behaviour of the Bloukrans Bridge, South Africa
(a) Transverse displacements
(b) Transverse bending moment in arch
(c) Transverse bending moment in deck

Care should however be taken to ensure that the structure as a whole has adequate flexural stiffness and strength in the longitudinal direction to withstand the effects of longitudinal wind action, braking and acceleration forces due to traffic. The required stiffness of the columns in the transverse direction of the bridge will depend on the relative stiffnesses of the deck and arch. Under the action of transverse wind loads, earthquake effects and eccentric traffic loading, the deck acts as an horizontal girder spanning between the abutments and interacting through the columns with the arch, which behaves like a bow-girder (Fig. 5-35). In determining the minimum column dimensions, stability against buckling must be considered. This may require non-linear analyses of the individual members or the structure as a whole, as described in Section 4.5. The spacing of columns is an extremely important factor which must be carefully investigated. The normal principles governing the design and economics of girder bridges generally apply to the approach spans, but additional factors have to be considered for the spans over the arch. The optimum spacing of the spandrel columns will not only depend on the economics of the deck system and that of the columns, but also on the influence of the spacing of the columns on the cost of the arch. The final spacing selected should, however, also satisfy the basic structural requirements for continuous deck systems. It is aesthetically preferable to have a column on each side of the crown and not *on* the crown of the arch. For constructional and economic reasons, it is advantageous to position columns on the arch foundation—preferably behind the springing. This generally improves the stability of the arch foundation and eliminates two sets of column foundations.

5.6.4.3 *The arch* The economy of an arch is related to the profile which, if correctly designed, will ensure that the selfweight of the structure induces mainly compressive forces. This requirement also ensures that the deformation of the arch due to creep of the concrete is limited to mainly axial shortening of the arch rib which does not give rise to large secondary forces in the rest of the structure. This is achieved by making the arch shape correspond as closely as possible with the line of thrust due to the dead loads, thereby minimizing the eccentricity of the thrust on the arch.

From an aesthetic point of view and for constructional reasons, it is preferable to have a smooth-arch profile. The above requirement can therefore only be partially satisfied by selecting a smooth profile which results in minimum deviations between the centroid of the arch and the line of thrust. On large arches, the most suitable profile is normally a power function which is fitted to equalize the positive and negative areas of the bending moment diagram along the arch rib, thereby minimizing permanent deflection of the arch rib (Fig. 5-34).

Other long-term effects are shrinkage of the arch rib and foundation settlement of the arch springings. Provision can be made to counteract the long-term stress effects due to creep and shrinkage by constructing a predetermined profile which would initially set up selfweight bending moments approximately equal but of opposite sign to the moments due to creep and shrinkage. Provision in the arch profile to reduce live-load moments is not recommended. Due allowance should also be made to allow for deformations that would result from the construction procedure in order to achieve the required profile in the long term.

On slender concrete arches the non-linear effects due to the deformations of the arch under load (P–Δ effects) as well as the non-linear strain effects on concrete when subjected to high compressive stresses, should be investigated. The resistance against buckling under heavy eccentric loads should be calculated by means of a non-linear analysis up to failure, considering both geometric and material non-linearity of the whole structure.

The actual dimensioning of the arch cross-section and reinforcement is strongly influenced by the construction method.

5.6.4.4 *The superstructure* The deck structure is usually designed as a continuous member supported by columns, with expansion joints at each end. Concrete decks on arch bridges may be in prestressed or reinforced concrete, and may be constructed by various construction methods as applicable to slab, beam or girder bridges.

The degree of flexibility and hence deformation of the arch which supports the deck, will increase the moment range at any particular section due to the various combinations of loads. This may result in a system where reinforced concrete is more economic than a fully prestressed section. This will, however, depend on the stiffness of the arch, the relative stiffness of the deck, the column spacing and the type of superimposed live loads. The spans on the banks are less highly stressed than the spans over the arch except over the first column off the arch, which is supported directly on a bank, where high moments can develop due to the deformation of the arch. The transverse forces on the deck and the resultant behaviour has already been described.

The design of the deck is influenced by the construction method as is normally the case with all continuous deck systems constructed in stages. The redistribution of moments in the structure due to the creep effects on the concrete in the main structural elements, namely, the deck, the columns and the arch, must therefore be considered.

5.6.5 *Analytical procedures*
Section 4 describes the relevant analytical methods and gives references for further details. In long-span arch bridges the following systems have to be analysed.

5.6.5.1 *Analysis of the construction stages* The construction stages of an arch bridge may be divided into the following main stages:

(a) Construction of the foundations, approach span columns, abutments and approach deck spans.
(b) Construction of the arch.
(c) Construction of the spandrel columns and deck spans over the arch.

Depending on the construction techniques adopted for the arch and the deck, these stages may be executed in sequence or may be combined [208, 209]. During the construction stages, the structure undergoes various structural changes during which bracing and staying may be required until the permanent structural system is completed and the bridge becomes self-supporting. It may be necessary to analyse a large number of these structural systems at the construction stages for the load effects specified by the applicable codes of practice. Temperature effects on suspension cables are important.

5.6.5.2 *Analysis of the completed structure* An analysis is made of the completed structure subjected to:

(a) Long-term dead load effects resulting from the construction stages. This analysis should include long-term effects developing from creep, prestress, shrinkage and support settlement and should be an extension of the stage-by-stage analysis of the bridge covering all the construction stages.
(b) Traffic loading as specified by the relevant design codes. For this analysis, the completed structure is normally simulated by a space frame with short-term elastic stiffness properties.
(c) On slender members, the effect of non-linear material properties as well as geometric non-linearity of the loaded structure should be analysed for the post elastic stress range up to failure to investigate instability due to buckling when subjected to unsymmetric loading (see Section 4.5). This analysis may comprise the whole structure as a two-dimensional frame or individual members with slender proportions (see Section 4.5).
(d) Temperature effects including axial and gradient temperature effects for the serviceability limit state of cracking. (Restraint forces due to temperature effects need not be considered for the ultimate limit state.)

(e) Wind loading as prescribed by the relevant design codes for the particular area. On longer slender arch-type bridges, dynamic or semi-dynamic analysis may be necessary using a spaceframe simulation

(f) Earthquake or seismic effects as prescribed by the relevant design codes for the particular area.

5.6.6   *Planning and design of the construction stages of arch bridges*
The construction of the arch by means of the various cantilever methods requires careful preplanning with special bracing and staying to be provided for the safe erection of the arch.

It is not in the scope of this book to deal with the details of the various alternative construction methods and the reader is referred to specialized literature dealing with specific projects [30, 208, 209, 210, 212].

The suspended cantilever and cantilever truss construction methods employed on some large arch bridges have been mentioned in Sections 1.2 and 1.3. A brief résumé of these methods will be given here.

5.6.6.1 *The suspended cantilever method* [203, 210–212] The arch is constructed in stages from both embankments using travelling formwork carriages on which each new segment is constructed. The travelling formwork carriages are supported from the previously completed segments and are moved forward for the next construction stages once the concrete has reached the required strength. As the cantilevered portions are extended, they are supported by suitably placed suspension cables which are stressed to predetermined forces in order to provide the required support to maintain the correct arch profile and bending moments at each particular stage (Fig. 5-36). The suspension cables are supported by temporary towers which are tied back and anchored into the rock. A construction cycle of approximately 7 days per segment can be achieved.

Temperature effects on the cables are monitored and allowance is made in the forces to which the cables are stressed to allow for any temperature deviations above or below a chosen datum temperature.

The main construction phases can be summarized as follows:

(1) The suspended cantilever construction stages for each segment of the arch.
(2) The profile adjustment prior to closure of the arch.
(3) Closure of the arch.
(4) Removal of the arch carriages.
(5) Removal of the suspension cables.

Each of these stages has to be preplanned to follow a detailed procedure and analysed accordingly so as to obtain the correct profile of the arch at completion, allowing for subsequent creep and shrinkage.

It is normally attempted to utilize the steel reinforcement that is required for the permanent service condition of the arch. It may, however, be necessary to increase the arch reinforcement in certain sections to increase the bending capacity for the construction stages. Although these moments can be accurately controlled by the regular adjustment of the cables, the effort and cost required for the additional adjustment of cable forces must be evaluated against the cost of extra reinforcement. The closure procedure involves the temporary locking of the

5.7.3 *Span properties of cable-stayed bridges*

Cable-stayed bridges are suitable for most sites but as the superstructures are supported from above by means of inclined cables they can be constructed without centering and are especially ideal for river crossings, deep gorges, and over wide busy traffic lanes and railway lines. Cable-stayed bridges can be broadly classified into the following types:

    (a) two-span symmetric or asymmetric (Fig. 5-38a,b);
    (b) three-span (Fig. 5-38c and Fig. 5-39);
    (c) multispan (Fig. 5-38d).

In the two-span asymmetric bridges, the average length of the main span normally constitutes ±70% of the total deck length. In the three-span arrangement, the centre span on average constitutes approximately 55% of the total deck length. In multi-span cable-stayed bridges, all spans are normally equal except the end spans.

The drop-in spans joining consecutive multi-span stayed structures may comprise up to 20% of the length of the total span between pylons for designs with single or few stay cables. In bridges with multi-cable-stays, the drop-in spans may reduce to approximately 8% of the total span between pylons.

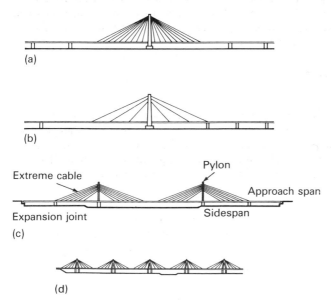

(a)

(b)

Extreme cable

Pylon

Approach span

Expansion joint

Sidespan

(c)

(d)

Figure 5-38 Arrangement of spans
    (a) Two-span symmetric with radiating cables (Ludwigshaven, West Germany)
    (b) Two-span asymmetric with fan-type cables. (Cologne, West Germany)
    (c) Three-span with fan-type configuration
    (d) Multispan with radiating cables (Ganga, India)

Figure 5-39   Pasco–Kennewick Intercity Bridge, USA. (*Courtesy* Leonhardt, Andrä and Partner)

In structures with viaduct approaches, it is normally advantageous to anchor back stays at the approach piers; this normally stiffens the main span when subjected to unbalanced live loads, allowing a shallower section. With multi-stay designs, this is not possible with all cables but the same effect is achieved by anchoring the extreme cable which has an increased area and load capacity, above a rigid support (pier or abutment), Fig. 5-38a.

5.7.4   *Geometry and properties of the cable-stays*

The number of stay cables, their size and their geometry may vary considerably for different bridges depending on the spans, the deck structural system and design loads. The arrangement of stays controls the design of the deck and pylons to a large degree, as mentioned in Section 1.3. Stay cables may consist of parallel wire, parallel strands or ropes, single strands or ropes, locked coil strands or solid bars [23].

In modern cable-stayed bridges, cables are normally constructed from standard strand or parallel wire. The wires are manufactured by the cold drawn process, of specially heat-treated steel rods using successive dies to improve the internal structure of the steel and thereby the tensile strength properties. The ultimate strength of the wire may vary between 1400 and 1850 MPa $(2–2.68 \times 10^6$ psi) depending on the properties of the rope, strand or wire.

Wires and cables are normally zinc-coated for protection to conform with a specific grade. Painting, rust-preventive compounds or plastic jacketing are used as added protection prior to and during erection. The mechanical properties of the finished product must be carefully controlled. Steel ropes and strands normally require pre-stretching to ensure true elastic properties throughout the stress range. Normally, pre-stretching up to 50–55% of the ultimate strength of the ropes or strands is required to produce consistent elastic properties, which are essential in the design of the structure. Values for Young's modulus of elasticity (E-modulus) should be determined from long lengths of cable which will yield more reliable values than shorter test specimens. The values should be determined for a stress range between 10 and 90% of the pre-stretching loads. The E-value varies with the type of cable, depending on whether it is strand, rope or parallel wires. The E-value is normally based on the gross metallic area of the cable including the zinc protection coatings and may vary from as low as 165 500 MPa (24×10^6 psi) for rope cables to 190 000 MPa (28×10^6 psi) for parallel wire cables. The allowable working stress for parallel wire cables is normally limited to approximately 50–55% of the yield strength or 0.2% proof stress.

End anchorages and saddles on the pylons must be designed to develop 100% of the strength of the cables which they must anchor or support. Cables may be stopped off and anchored in the pylon or may be taken continuously through over saddles with adequate radius to avoid excessive bending stresses in the cables. Continuous cables are clamped to the saddles after tensioning to avoid sliding over the saddles.

The stay cables are part of the primary support system in a cable-stayed bridge and must therefore be protected against any type of corrosion. The most vulnerable sections are at the anchorages.

While earlier cable-stayed bridges had cables that were encased in concrete (see Figs 1-7 and 1-21), various systems have since been developed for protection after installation, such as:

(a) multiple-layered plastic coating impregnated with glass-reinforced acrylic resins;
(b) elastomeric coating of liquid Neoprene, multiple coats of uncured Neoprene sheets and top coats of Hypolon paint
(c) polyethylene or polypropylene ducts filled with corrosion protective epoxy enriched cement grout [213, 214], Fig. 5-40a.
(d) cement-grouted steel ducts [215].

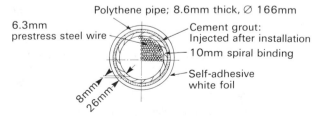

Figure 5-40a **Construction of parallel wire stay-cable: Pasco–Kennewick Bridge (Leonhardt and Andrä)**

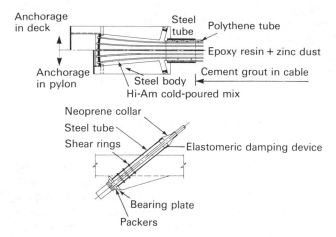

Figure 5-40b Cable anchorage in girder: Pasco–Kennewick Bridge (Leonhardt and Andrä)
   (1) Long section through head
   (2) Anchorage in girder

During the life of the bridge, the stay cables are subjected to a considerable number of stress fluctuations and must be protected against fatigue. The effects of vibrations due to wind must also be carefully considered and rectified if of significance. Apart from complying with fatigue resistant design criteria for the cables, special anchorages have been developed to limit stress concentration in the anchor zone of the wires, thereby considerably improving the fatigue resistant properties of stay cables [213, 216], Fig. 5-40b.

5.7.5   *Erection procedures*
Cable-stayed bridges may be erected by three basic methods: construction of centering, the push-out (incremental launching) method and the cantilever method [23].

   The construction of centering is generally used where the clearance between the superstructure and the underlying terrain is low enough to make temporary staging economically viable, where suitable founding conditions exist, and where it will not interfere with land or water traffic or be exposed to dangerous floods. The incremental launching method has been used in some special cases where the cantilever method could not be applied due to inadequate clearances. With this technique, which is described in Section 5.3, the pylon with supporting stay cables, which form part of the primary structural system, whether these are temporary or permanent, have to be launched together with the deck. Launching may require intermediate temporary supports to reduce the effective span during construction. The cantilever method is probably most often used with cable-stayed bridges, especially where the positions of temporary supports below the deck are limited or where traffic and other obstacles have to be crossed. The technique may comprise free cantilevering or a combination of free and sus-pended cantilever methods (see Section 5.3). The permanent stay cables are

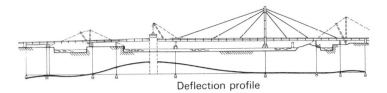

Deflection profile

Figure 5-41 Free and suspended cantilever construction (*Stahlbau*, Vol. 26, No. 4, April 1957)

utilized, sometimes in conjunction with additional temporary cables, to stay the cantilevered structure as it is constructed in stages using travelling formwork carriages or erection carriages in the case of precast segmental construction. Temporary piers may be used as intermediate supports where this is economically advantageous (Fig. 5-41). Designs with multiple stays have the advantage that they can be economically erected by means of the cantilever method.

During the erection stages, it is usually necessary to take the non-linearity of the cables due to sagging into account. Approximate formulas to allow for equivalent linear moduli of elasticity have been developed [23].

Concrete decks can be of *in situ* or precast construction and additional temporary prestressing may be required during the erection process. The various alternative possibilities are numerous depending on the local conditions and the design of the bridge—the reader is therefore referred to specialized literature.

5.7.6 *Design considerations*

5.7.6.1 *Deck cross-sections* The design of the deck cross-section is largely controlled by the spans, the width of the roadway and the longitudinal and transverse arrangement of the stay-cables (see Section 1.3).

Alternative systems are illustrated in Fig 5-42.

Designs with cables in a single plane require torsionally stiff girders, as illustrated in Fig. 5-42a. Systems with cables in two planes, which support the edges of the deck, can have relatively flexible and slender cross-sections (Fig. 5-42b). For wide decks, the slab may require intermediate transverse stiffening by means of transverse girders placed at regular intervals (Fig. 5-42c).

In cable-stayed bridges where the stay-cables are anchored in the deck side-spans, known as a self-anchored system, the stay-cables cause compression in the main girder which, in concrete girders, is utilized as permanent prestressing of the deck. In this arrangement, expansion joints are normally provided at abutments (Fig. 5-38c). Additional prestressing may, however, be required near the end supports where the total effect of the horizontal components of the cable forces is less and which may cause local tension in the deck.

It is necessary to consider the forces in the superstructure during all stages of construction, making due allowance for the effects due to shrinkage and creep. The final stress condition under long-term loads, can be controlled by pre-calculated adjustment of the suspension cables during installation. In concrete cable-stayed bridges, the secondary forces which develop due to creep are caused by two effects: one develops due to the relaxation of bending moments which are

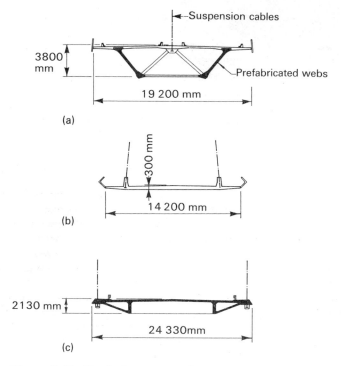

Figure 5-42   Typical cross-sections
(a) Deck with precast web: Brotonne Bridge, France
(b) Proposed cross-section for a narrow deck: design project
(c) Deck in precast concrete: Pasco–Kennewick Bridge, USA

induced in the system and the other by creep shortening of the concrete members due to axial forces. Shrinkage will have similar effects in the axial direction as the creep shortening effects. The redistribution of bending moments can be controlled to be similar to that which develops in continuous systems on rigid supports, bearing in mind that an elastically supported system with moments which are equal to the 'one-go' moments of a rigidly supported continuous girder, will not be subjected to redistribution of bending moments due to creep. This implies that the moments in the girder of a concrete cable-stayed bridge will not change if the vertical components of the stay-cable forces are so arranged that they are identical to the 'one-go' reactions of an equivalent rigidly supported continuous girder—if the axial shortening effects are not considered [213].

Due to the creep and shrinkage shortening of the concrete members, bending moments develop in the girder and pylons similar to the effects of a gradual support settlement in a continuous girder. It is possible, by means of adjustment of the stay-cables, to induce bending moments in the girder of which the effects will be opposite and approximately equal in magnitude to those caused by the axial shortening due to creep [213].

5.7.6.2 *Non-linear behaviour of cable-stays* In cable-stayed bridges there may be cases when some of the stay-cables behave non-linearly due to the catenary effect of a cable under low tension. This effect is due to the change in the rate of increase of the tension force as the sagging profile straightens out or vice versa. As the behaviour of the structure is dependent on the relative stiffnesses of the various structural elements, i.e. girder, pylon and cables, it will be affected by the effective stiffnesses of the cables, which depend on the initial pretension.

Normally, the structure starts behaving linearly at sag ratios of less than 1 : 80. During construction when dead load is applied in stages, the cable forces build up from an initial low pretension. During these stages, the cables therefore behave non-linearly. It is desirable that when all dead loads have been applied, the 'initial sag ratio' should be well below the above value.

In large span concrete cable-stayed bridges, the dead loads constitute the largest portion of the total load resulting in relatively high pretension forces under dead load. Hence, the catenary effect in the cables when live loads are applied to the structure can, in many cases, be ignored, i.e. the structure behaves linearly during the application of live load [23].

Depending on the structural configuration, the slenderness of individual members and the concrete stress levels, it may be necessary to perform non-linear analyses to investigate the stability of all members under loads approaching the required ultimate values.

5.7.6.3 *The behaviour of stayed bridges under wind loading* The results of dynamic analyses of various cable-stayed structures, especially designs with multi-stays in two planes, have shown that cable-stayed bridges are aerodynamically far more stable than suspension bridges. This can be attributed to 'system damping' effects which come into play as soon as the amplitude of the dynamically induced displacements become large enough to be structurally significant. These 'damping systems' can be attributed to the following phenomena:

(a) The non-linear spring stiffness of the cables which reduces with increased upward deflection of the deck due to a reduction in cable tension causing increased cable catenary. (Each cable also behaves differently depending on its geometric properties.)
(b) The interaction between the various individual systems, namely, the cables, pylon(s) and deck girder which all vibrate at different natural frequencies, resulting in interference of oscillations which prevents resonant conditions with large amplitudes of deflections developing.

Multi-cable-stayed concrete bridges (Figs 5-39 and 5-43) are complex, sophisticated structures which demand experience and technical expertise both on the designer's part, as well as the contractor's, As a bridge type, it offers many advantages and the development of long-span multi-stayed decks with closely spaced cables in two lontitudinal planes is very promising because of the requirement of only a relatively light and flexible deck and the inherent stability because of the system damping effects due to the dynamic interaction of its members.

Figure 5-43 Donaubrücke, Metten (*Courtesy* Ingenieurbüro Grassl)

5.8 Bridge bearings, movement joints, balustrades, surfacing, other furniture and special details

Although bridge furniture is functionally essential and aesthetically important, it is often given belated consideration in the design of bridges. While furniture such as balustrades, bearings, movement joints, signposting, lighting and drainage can constitute a relatively small proportion of the total cost of a bridge, careful consideration of details is required at the early stages of the design. The addition of some of these items at a later stage can result in substantial redesign or disfigurement of the bridge.

5.8.1 *Bearings*[1]

The function of a bearing is to transmit loads or forces in specific directions from one member of the structure to another, providing restraints against specific directional or rotational movements while allowing freedom for other movements to take place. Bridge bearings typically may be required to transmit vertical loads from the superstructure to a substructure with freedom of any required combination of movements in the horizontal plane and rotations about components of any of three orthogonal axes.

1 Part 9 of reference [95], which has been published in draft form for public comment, covers the subject matter in detail.

Movements of the superstructure relative to the substructures can be classified under (a) short-term, (b) long-term and permanent movements, and (c) variable or transient movements. Axial expansion or contraction and rotations due to temperature range and temperature gradient, as well as erection loads, belong to category (a). Superimposed dead loads, earth pressure, differential settlement of foundations, creep, relaxation of prestressing and shrinkage effects, belong to category (b). Movements or rotations due to applied actions of traffic loading, wind forces of earthquake excitation belong to category (c).

The degree of articulation of the structure and freedom of movement at joints by means of bearings requires careful consideration in order to minimize stresses due to the aforementioned causes without unnecessarily eliminating useful redundancies or increasing the risk of instability.

The quantitative assessment of movements and forces has been dealt with in Section 4. It must be stressed that they should not be underestimated. The cost of enlarging the top plate of a sliding bearing initially, or of providing greater fixity, is relatively small.

5.8.2 Types of bearing
Bridge bearings can be classified as follows:

(a) elastomeric bearing pads (unreinforced or laminated), with or without sliding plates;
(b) pot bearings;
(c) spherical, cylindrical, pin or leaf knuckle bearings;
(d) concrete hinges;
(e) steel (linear or point) rocker bearings;
(f) single or multiple (cylindrical or non-cylindrical) steel roller bearings;
(g) multiple steel ball bearings;
(h) guide bearings.

Only some of the types under (a) to (d) will be described here, being those most commonly used in concrete bridges.

5.8.2.1 Elastomeric bearings pads
An elastomer is a substance which can be stretched at room temperature to at least twice its original length and, if released, returns to its approximate original length quickly. For bridge bearings, the term elastomer applies to natural and synthetic rubbers. Bridge bearings are required to have good durability and ageing resistance, such as Neoprene which has good resistance to sunlight and ozone. They are usually painted with a synthetic rubber paint or encased entirely in 3 mm Neoprene. Elastomeric bearings may be available ex-stock in standard sizes with given load and displacement characteristics. More usually, they are individually designed and purpose-made, often from standard sheets (See Fig. 5-44a).

Rubber, when confined, is almost incompressible, i.e. for practical purposes, its volume does not alter under load, therefore for it to deflect under load it must be able to bulge laterally. High deflection under compression is generally not desirable in bridges. This is overcome by laminating thin rubber sheets between steel plates. Typical thicknesses of rubber are between 5 and 15 mm (0.2–0.6 in), while the steel plates are usually between 1.5 and 4 mm (0.06–0.16 in) thick.

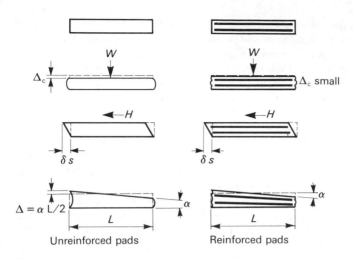

Figure 5-44a   Elastomeric bearing pads
W = vertical or normal load; H = horizontal or transverse load; Δ_c = compression; δ_s = shear deformation; α = angle of rotation in radians.

Thinner plates can be used between the intermediate layers than for the outer plates. These plates restrain the rubber and are under considerable tension ensuring that the lateral strain in the rubber is largely restricted to a reduced bulge at the edges. As can be seen from Fig. 5-44a, the laminations increase the vertical stiffness and the resistance to rotation. They do not, however, affect the shear stiffness. These bearings should not be subject to transverse (usually vertical) tension and the compression Δ_c due to load should be greater than the maximum theoretical tensile strain due to rotation $\alpha L/2$. Bearings subject to transverse tension are prone to delaminate. Fixed bearings should have an adequate factor of safety to resist horizontal forces. If the friction between the bearing and the sub- and superstructure is insufficient, this can be remedied by using dowels, as shown in Fig. 5-14. As bearings should be replaceable, it is recommended that these dowels do not pass through but are rather positioned alongside them, so that the required rotation is not restricted. Elastomeric bearings should preferably be placed normal to the maximum load and well bedded. If the inclination of the resultant load can vary considerably, the resultant components of shear must be taken into account. Design recommendations for rubber bearings are given in [95, 217, 218]. An innovation of the pure elastomeric bearing is to combine this with a sliding plate, as shown in Fig. 5-44b.

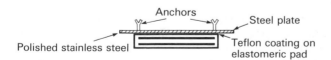

Figure 5-44b   Elastomeric sliding bearing

5.8.2.2 *Pot bearings* A pot bearing ingeniously utilizes the fact that rubber is incompressible if confined. In principle, it consists of a steel pot containing a layer of natural rubber or Neoprene which supports a piston that is shaped so that it can rotate about the horizontal axes. The load is applied via the piston through the rubber and pot to the substructure. Effective sealing of the pot at the perimeter of the piston face is essential as the rubber, under high pressures approaching 30 MPa (4350 psi), acts like a viscous fluid. Due to its low shear modulus, it offers little resistance to rotation of the piston, thus approximately centering the vertical load on the piston face. Translation, if required in one or more directions, is catered for by Teflon and stainless steel sliding plates. Figure 5-45 shows a locating bearing permitting 5 mm (0.21 in) movement in any direction before acting as a bearing fixed against sliding. Figure 5-46 shows a unidirectional pot bearing.

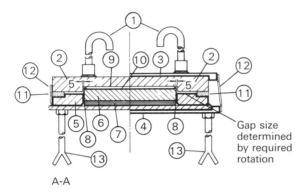

A-A

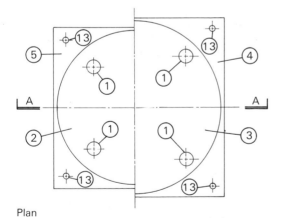

Plan

Figure 5-45 Locating pot bearing (based on Nova–Kreutz bearing): (1) upper fixing bolts; (2) top plate; (3) top adaptor plate; (4) bottom adaptor plate; (5) pot; (6) piston plate; (7) elastomer pad; (8) slip seal; (9) stainless steel disc; (10) Teflon disc; (11) pot seal; (12) failing strap; (13) lower fixing bolts with lock and locating nuts

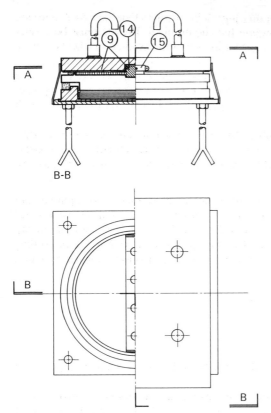

Part sectional plan A-A

Figure 5-46 Unidirectional pot bearing (based on Nova–Kreutz bearing). Component parts as for Fig. 5-45 except the following: (9) stainless steel sliding plates; (14) keyway with Teflon sliding surface; (15) keyway wiper seal. Also, there is no top adaptor plate

Before selecting or specifying the bearing, the bridge designer should define the following parameters:

(a) Type of bearing, i.e. fixed, unidirectional or multidirectional.
(b) Maximum and minimum vertical loads.
(c) Maximum and minimum horizontal loads and their directions, if applicable. In standard bearings, the horizontal load is usually limited to 10% of the maximum vertical load. Should it be required to resist a greater horizontal force, it is often more economic to design a special bearing rather than select a larger standard bearing.
(d) Range of rotation about any axis. This must be realistically determined. Specifying too small a value may impair the function of the bearing. Too large a value increases the cost unnecessarily. The thickness of the rubber is determined by the acceptable moment due to rotation. The thicker the rubber, the smaller the moment [181]. Standard bearings usually allow up to 0.010 rad, but the designs with up to 0.020 rad are common.

(e) Movement range and preset if required. Presets less than 20 mm (0.8 in) are usually allowed for by enlarging the sliding plate. Presets must be clearly specified and marked on the bearing and the installation checked by a competent person on site.
(f) Permissible concrete stresses. Many standard bearings are designed for a peak concrete stress of 20 MPa (2900 psi). Reducing this stress can significantly increase the cost of the bearings as the plates are required to be both larger and thicker.
(g) The requirements for adaptor plates. These are normally an optional extra. Adaptor plates facilitate the removal of the bearing subsequent to installation, with a minimum of jacking of the superstructure. They also reduce the concrete stresses.

Criteria such as Teflon stress, elastomer stress, elastomer shore number, are generally determined by the designer of the bearing. Common limiting values are: 30–40 MPa (4350–5800 psi); 25–30 MPa (3600–4350 psi); 55–60 Shore hardness. The resistance to sliding (coefficient of friction μ) between Teflon and the sliding plate (usually polished stainless steel) decreases with increased normal pressure. Commonly accepted values are:

p MPa	μ
10	0.055
20	0.040
30	0.030
40	0.025

(See codes and manufacturers' catalogues.) In addition, the sliding surfaces are coated with a silicon grease.

Bearings rely on uniform support conditions for satisfactory performance. Beddings of epoxies or semi-dry cement mortar rammed well home have proved satisfactory. Grouting between concrete and bearing is not recommended due to the risk of air pockets and voids being formed.

5.8.2.3 *Spherical bearings* Spherical bearings are suitable for large rotations. Permissible rotations up to 0.045 rad are standard. The resistance to rotation is higher in spherical bearings than in pot bearings. Figure 5-47 shows a Glacier

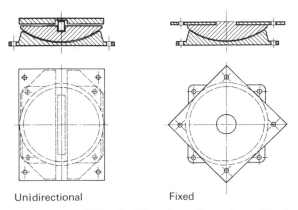

Unidirectional   Fixed

Figure 5-47 Spherical bearings (based on Glacier bearing)

fixed and unidirectional bearing. General comments of Section 5.8.2.2 apply to this section as well.

5.8.2.4 *Concrete hinges* In principle, concrete hinges are simple. It is, however, essential that they are properly detailed and that the workmanship is of a high standard. In a concrete hinge, the section of the member is reduced to form a throat (see Fig. 5-48). In the rectangular hinge, the concrete in the throat is under biaxial stress, whereas in the circular hinge it is under triaxial stress, thus permitting high axial stress in the concrete without impairing the safety of the hinge. The British Ministry of Transport allows working stresses up to twice the cube stress. The small dimensions of the throat permit rotation to take place with little restraint. It is not necessary to provide vertical reinforcement across the hinge except in exceptional circumstances. If reinforcement is provided through the hinge, it should consist of small-diameter bars located in the line of bending in the centre of the throat. These increase the resistance to rotation. Small rotations are accommodated by increased concrete strain. Large rotations cause the concrete to crack. Tests have shown that this does not impair the functioning of the

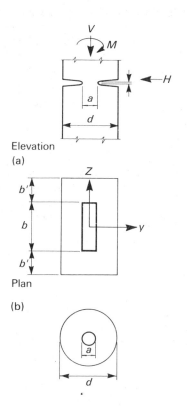

Elevation
(a)

Plan

(b)

Figure 5-48   Concrete hinges
(a) Rectangular hinge: rotation about z-axis only
(b) Circular hinge: rotation about any horizontal axis

Figure 5-49 Concrete rocker bearing

hinge, even with reversals of rotation causing the concrete to crack on either side of the hinge. The corners of the throat should be well rounded. A parabolic shape is considered ideal to prevent the concrete from spalling off. For this reason, it is also advisable to constrict the throat in the transverse direction in the rectangular hinge. The resistance to transverse shear is limited and is a function of the axial compression. It can be enhanced by the use of dowels. It is essential that the throat is under residual compression under all load cases, if necessary by means of prestressing across the axis of rotation, if possible outside the throat area. The member on each side of the throat is subject to high splitting forces which must be adequately catered for by suitable reinforcement, similarly to a prestressed concrete endblock. Detailed design procedures for concrete hinges are given in [200, 219]. Concrete design procedures can also be used as rocker bearings, thus allowing translation as well as rotation (See Fig. 5-49). As the restricted throat area and congestion of reinforcement make concreting difficult, concrete hinges are often precast.

5.8.2.5 *Other bearings*   The bearings as listed under (e) to (h) in Section 5.8.2 are more often used in steel bridges, but may be applicable to large concrete bridges. Certain bearings, such as the pot and spherical bearings, can be adapted to alter their function during their lifetime, e.g. a temporarily fixed bearing can be altered to be a permanently uni- or multidirectional bearing or vice versa (useful for stage construction). The reader is referred to manufacturers' literature on the installation and use of these bearings. Another type of bearing is that used for incremental launching described in Section 5.3.

5.8.3 *Movement joints*
The main functional requirements of joints are as follows:

(a) To bridge the variable gap due to expansion or contraction in the carriageway in such a manner that under all circumstances no objectionable discontinuities occur in the riding surface of the carriageway. The riding quality should not cause inconvenience to any class of road user for which the structure is designed. Where applicable, this includes pedestrians, cyclists and animals.
(b) To withstand the loads due to traffic, vibration, impact, shrinkage, creep, settlement, etc., without causing excessive stresses in the joint or elsewhere in the structure.
(c) Not to constitute a traffic hazard, e.g. skidding.
(d) Not to give rise to excessive noise or vibration.

(e) Either to be sealed against the ingress of water and debris, or to have provision for drainage and cleaning.

(f) To be easy to inspect and maintain.

(g) To be economic in their total cost, including initial capital outlay and maintenance.

It is probably safe to say that no other single component of a bridge has caused as many service problems as expansion joints. Low-cost joints have often proved to be expensive in maintenance and replacement costs, quite apart from the inconvenience and disruption of traffic.

5.8.3.1 *Suitability of joint types for movement ranges* There are two types of gap: covered and open. In the context used here, a covered gap is one which is structurally bridged such that loads are carried across the gap. It does not relate to the water resistance of the joint. An open gap is one which is not structurally bridged, e.g. a gap filled with compressed bitumen-impregnated high-density polyurethane foam might be watertight, but the foam cannot transmit loads, therefore the gap is an open gap.

It has been stated that for joint movements [219, 220]:

(1) of less than 10 mm (0.4 in) the surfacing can be taken across the joint but a bond breaker should be applied to limit the crack widths in the surfacing; of less than 5 mm, no special precautions need be taken;

(2) the maximum open gap acceptable for vehicles is 65 mm ($2\frac{1}{2}$ in) covered gap joints should be provided for pedestrian, animal and bicycle traffic;

(3) open gap joints should have a minimum gap of 6 mm ($\frac{1}{4}$ in).

Small movements, i.e. 10–40 mm (0.4–1.6 in), are economically catered for by means of flexible sealers such as polysulphide, or inserts such as pre-compressed impregnated foam or pre-formed Neoprene strips. Joints for larger movements are usually specially designed. A range of proprietary types are available. These are usually made of steel in the form of combs with interlocking teeth, articulated sliding plates that move in under an overlapping end plate, or various versions of an assembly of transversely positioned rolled steel or extruded aluminium channels, I-sections or plates with compressible hollow or folding rubber infill sections in between, but these will not be described herein. The technical literature provided by the suppliers should be consulted.

The two-part polysulphide sealant is suitable for movements in compression and tension of 25% of the sealant width as installed. The permissible shear movement is higher, namely ±50% of the joint width as installed (see Fig. 5-50a).

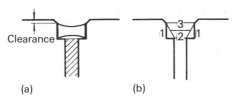

(a)   (b)

Figure 5-50 Polysulphide joint
(a) Sealant under tension
(b) Placing sequence for wide sealant

In horizontal joints, the limiting vertical movement is therefore usually the maximum acceptable step between the two sides of the joint. Where the movement is predominantly horizontal, i.e. across the joint, the width to depth ratio should be 2 : 1 (minimum depth 10 mm). The maximum width of a joint which can be installed in one application is about 25 mm (1 in). Greater widths can be sealed by applying the compound in stages, allowing each application to set before applying the next (Fig. 5-50b).

The surface preparation of a joint prior to installation of the compound is vital. Surfaces must be sound, clean and dry. The back-up material in the joint is important; it regulates the depth of sealant and serves as a bond breaker allowing free movement of the sealant. Surfaces must be primed and the joint installed strictly in accordance with the manufacturer's recommendations. An important point is to take cognizance of the suction action of tyres travelling at speed across the joint. The joint must be adequately recessed so as not to be whipped out by traffic. A clearance of 15–20 mm (0.6–0.8 in) is recommended. If the top of the joint is too deep below the surface, it will fail as a result of debris. Polysulphide itself is not trafficable. Its use in carriageways has been limited. It has, however, been extensively and successfully used in sealing kerbs and balustrades.

The movement range of bitumen-impregnated high-density polyurethane foam depends on the functional requirements of the joint. It has no tensile resistance and for satisfactory function depends on remaining under adequate compression. Table 5-1 shows the claimed characteristics with respect to compression. The recommended vertical shear movement is ±10% of the seal depth. In road carriageway crossings, the minimum recommended depth of seal is 75 mm (3 in). The material should not be used where there is the likelihood of dirt or debris being forced into it. Petroleum products dissolve the bituminous impregnation. Ultraviolet light degrades the upper surface. The depth of degradation is generally small so that the joint function is not impaired. Taking cognizance of the maximum acceptable 'open gap' joint width of 65 mm ($2\frac{1}{2}$ in) and desirable water resistance quite apart from cost, the total movement range is 30 mm (1.2 in). The top of the joint should be recessed 20 mm (0.8 in) below the roadway surface.

Compression seals are available in sizes to cater for movements in excess of 80 mm. However, with a maximum open gap of 65 mm, the movement range is approximately 30 mm. The seals are dependent on a minimum compression for satisfactory performance, the minimum compression normally being 10% and the maximum 55% of the uncompressed width. The depth-to-width ratio of the uncompressed seal should be at least 1 : 1. As above, depth of installation and

Table 5-1

Compression, %	Effect
50	Dust seal
$33\frac{1}{3}$	Draught seal
25	Water resistant
20	Watertight
15	Hydrostatic and vacuum seal
10	Liable to compression set

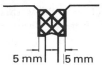

Figure 5-51   Neoprene compression seal

surface preparation are critical. Also the correct type of seal should be chosen. For bridge joints, the edges of the seal should be positively seated (see Fig. 5-51). The seals cater for a shear movement of up to an angular rotation of 10%.

5.8.3.2 *Nosings*   The aforementioned joints are generally protected by nosings. These are usually of concrete, epoxy or polyester mortar or steel. These nosings receive a severe pounding under traffic and should be robust enough to withstand these forces [181, 219, 220]. Their tops should be level with the top of the wearing course to minimize impact loading. To ensure smoothness of ride, it is common practice to first lay the wearing course then cut out and construct the nosing. For concrete and epoxy/polyester nosings, the surface preparation is vital to ensure adequate bonding. Transverse joints should be supplied at sufficient intervals—900 mm (3 ft) for epoxy nosings—to cope with differential shrinkage and temperature. Most epoxies are moisture sensitive. Polyesters are exothermic. The manufacturers' recommendations should be strictly adhered to. Correctly used, epoxy is an excellent material, yet failure rates of epoxy nosings have been extremely high mainly due to faulty workmanship.

5.8.4   *Balustrades, handrails and other details*

5.8.4.1 *Function*   The main function of balustrades or handrails is to prevent traffic and pedestrians respectively from crashing and falling over the edge. Another function of balustrades is to minimize the consequences of an accident. Vehicles should not overturn or be thrown into the face of oncoming traffic. Opinions as to the correct solution differ. Figures 5-52a,b show two concrete balustrades. In the former type, the flexible approach guardrails extend for the length of the bridge. The latter is shaped so that when a vehicle strikes the concrete balustrade at an oblique angle, the wheel travels up the lower slope and

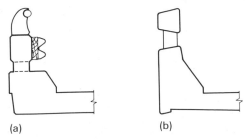

(a)   (b)

Figure 5-52   Concrete balustrade sections
(a) Approach guard rails continue over bridge
(b) Approach guard rails end at bridge

is then deflected back and downwards. The maintenance costs of this type of balustrade are low and damage to the vehicle and its occupants usually less than for flexible types that function on the principle of absorbing energy by yield deformation of the components. Leonhardt [181] has proposed cables strung to spring-loaded posts with shear notches at the bottom so that during an impact the posts will shear and not be flattened, the cables thereby restraining the vehicle. All types of balustrade should be designed for the specified factored impact loads, but should fail prior to the load-carrying component of the deck to which they are attached.

Balustrades are visually a dominating part of the bridge. They should therefore not only be functional, but aesthetically pleasing. For long-span structures such as suspension bridges, the wind effects on the balustrades are significant so that the projected area and shape should be optimized.

5.8.4.2 *Kerbs* The kerb configuration is usually specified by the road authority. In cases where the deflection of the deck may vary during construction, it is recommended that kerbs be constructed only after the deck structure is sufficiently advanced.

5.8.4.3 *Wind deflectors* On high bridges exposed to strong winds, balustrades may be fitted with wind deflectors.

5.8.4.4 *Noise abaters* The noise level generated by traffic on main roads may reach intolerable proportions. Large sections of freeways in heavily built-up areas in Japan, for instance, have been provided with noise deflectors. These might alleviate the noise problem for riparian owners, but they are aesthetically displeasing and create a tunnel effect for the motorist. It still appears that, from an aesthetic and noise pollution point of view, elevated motorways in heavily built-up areas are unacceptable to many. Sunken freeways may offer a better solution but are costly. This is an aspect which requires the attention of town planners.

5.8.5 *Surfacing*
The most common surfacing for bridge decks is bituminous premix 40–75 mm (1.6–3 in) thick. In countries subject to frost, especially when de-icing salts are used, effective waterproofing to the deck is essential [181, 221]. In areas not subject to frost, the common procedure is to clean the concrete surface by washing and brushing, to allow the surface to dry and then to uniformly cover it with a spray-grade bituminous emulsion. After the tack coat has dried, the premix is applied.

5.8.6 *Drainage*
Drainage details should be considered at an early design stage. The frequency of outlets is dependent on the intensity and duration of rainfall and the slope of the surface and slope and size of drainage pipes. To avoid standing water, a minimum deck slope of 2% is recommended. This is not possible at changes in superelevation. These areas therefore require special consideration. For bridges in open countryside, the water can be drained directly through scupper outlets in the deck.

The inlets thereto should be detailed so that they do not present a hazard to animals or pedestrians. In built-up areas, water will generally have to be led to stormwater systems. Pipes cast into concrete should be 150 mm (6 in) diameter unless they are readily accessible for cleaning, which is unusual. Exposed pipework should be avoided as this can spoil the appearance of a bridge. If drainage pipes cannot be accommodated in the concrete, they should be placed in recesses under the deck. All bends should be provided with rodding eyes. In areas subject to frost, drainage pipes should be designed so that if they freeze up, they do not cause structural damage. Premix itself is not waterproof. The water in the premix should also be drained to avoid pumping and failure of the premix.

5.8.7 *Services*
The requirements of signposting, street lighting and the need to carry services such as electric cables, telephone cables, water pipes, etc. across the bridge, should be established early on and adequate provision for their support and accommodation made. It is often technically and aesthetically extremely difficult to make provision for these services when the design is well advanced or construction has commenced.

5.9 Estimation and optimization of costs by value engineering

In previous sections, various values related to the design of bridges have been discussed including the importance of reliability and utility together with aesthetic excellence and the importance of environmental aspects. The total cost, which by implication is a very important part of the utility function (see Section 3.2), needs full consideration from the very first stages of the conceptual design.

At the initial stages, cost valuations are done by estimating. The assumptions made and method used will depend on the stage of development of the design. Accurate estimating is difficult unless it is based on considerable experience of similar work. Cost graphs can be prepared for total structures or parts of structures in terms of specific variables and concomitant parameters. Interpolation can give useful results if the cost graphs are based on experience covering a wider range of structures than the one under consideration. Extrapolation is often necessary but needs great care. Unknown aspects of new projects should best be analysed in fair detail although this may be very difficult if complex operations are involved. It is reasonably easy to estimate material costs, but plant and manhour content as well as overheads are often difficult to quantify without previous experience. Even with previous experience, the application of cost extrapolation to larger or more complex works can be very inaccurate.

It is consequently necessary to do regular updating of estimates as the design proceeds and the details become better defined. When the design is fully developed and a more complete schedule of quantities can be prepared, it remains to determine more reliable cost rates. Estimates should always be done at present costs and the projected additional costs due to escalation indicated separately to avoid misunderstanding as to the probable costs at time of tendering. If the contract documents make provision for contract price adjustment formulas to allow for escalation due to inflation or the opposite, then the employer should be advised on the probable end cost as well.

During the total period of design and construction, it should be the objective to optimize costs in terms of utility as defined in Section 3.2. This is naturally part of the design process as depicted in Fig. 3-1 and requires knowledge, experience, proficiency and the ability to innovate.

The concept of value engineering [222] or value analysis began during the Second World War when shortages of materials and labour necessitated changes in method, materials and traditional designs. Many of these changes resulted in superior performance at a lower cost.

Value engineering or analysis as defined by Miles [223] is 'an organized, creative approach which has for its purpose the effective identification of unnecessary costs, i.e. costs which provide neither quality nor use nor life nor appearance nor customer features'.

Although the techniques inherent in this process have in the past largely been applied to multi-disciplinary works, it equally applies to civil engineering. It is applicable to the total process from conceptual design to construction of bridges.

The considerable literature on the subject should be studied by designers and constructors.

5.10 Contractual procedures

Contractual procedures differ from country to country and according to the requirements of a particular project. The following are definitions of the types of contract based in British practice.

(a) A *fixed price contract* is one wherein a price is agreed and fixed before construction starts. It may or may not be a firm price contract. Usually, it has priced bills of quantities for basis.

(b) A *firm price contract* is a fixed price contract which does not allow for its prices to be adjusted for fluctuations in normal market prices of labour and materials.

(c) A *lump sum contract* is a fixed price contract wherein the contractor agrees to be responsible for executing the whole of the contract work for a stated total sum. This type of contract would normally be chosen only when few variations are expected. Usually, it is based upon drawings and specification.

(d) A *measurement contract* is a fixed price contract wherein prices are fixed in advance for units of work to be measured later. These are recorded in a schedule of prices. The total work to be done is not usually decided in full detail before the contract so the total price cannot be ascertained until the work is measured, unless a provisional bill of approximate quantities is produced.

(e) A *cost-reimbursement contract* is one wherein the actual prime costs of labour, material, sub-contracts, use of plant, etc. will be paid for at net cost to the contractor plus a fee. The fee is intended to pay the contractor for his management costs, overheads and profit. The fee may be a sum wholly or partly fixed in advance or may be a percentage of the prime cost. The contract should define what is to be deemed a prime cost and what is to be deemed as being paid for in the fee.

(f) A *target price contract* is one where the contractor undertakes not to exceed an agreed ceiling figure. The financial basis of the contract is somewhat similar to a cost-reimbursement contract with a sharing (between client and contractor) of any savings achieved below the target figure.

(g) A *negotiated contract* is one which is negotiated between the client or his agents and a selected contractor. The price basis may be any of the foregoing.

(h) A *managed contract* is a form of negotiated contract wherein the contractor plays a more positive role in the time/cost control of the contract in the client's interest.

(i) A *packaged contract* is similar to a managed contract but the contractor undertakes all responsibility for the design and eventual performance of the bridge.

The measurement contract with a provisional bill of approximate quantities is commonly used by government authorities. Although many authorities have a bridge design capability, it is common practice to appoint consultants for the design of selected bridges. The consultants' major duties are normally the following:

(a) To prepare a preliminary design(s), cost estimates and reports including, where applicable, feasibility and environmental impact studies based on sufficient investigations and surveys.

(b) On acceptance of (a), to proceed with detailed design and preparation of contract documents. These usually consist of conditions of contract, such as those prepared by the ICE [224] or by FIDIC [225], specifications, plans and schedules of quantities.

(c) To call for tenders either by open tender (advertised), or selective tendering where each prospective tenderer is invited. Alternative designs may or may not be considered. This is stated in the invitation to tender.

(d) To evaluate tenders, which may include some or all of the following: to evaluate the rates, to assess the implications of qualifications of tender, to report on tenderers' technical competence and resources, their ability to comply with the completion date and their financial standing. At times of high inflation and high interest rates, cash flow studies in terms of the tenderers' construction programs are usually required to determine the effective tender prices.

(e) To advise on and arrange the award of the tender and the signing of the agreement between the employer and the contractor.

(f) To supervise construction. Usually, resident engineers are appointed for full-time supervision at the site in order to ensure the quality of materials and workmanship. They should be fully briefed of their duties. Regular site meetings with representatives of the client and the contractor, and technical meetings with representatives of the contractor and subcontractors, should be held and minutes kept thereof. Full records of all observations, recordings and measurements are necessary. A daily logbook should be kept.

(g) To check the measurement of quantities and control the valuation of claims and the preparation of payment certificates [226].

There are variations to the above, for instance, tenders may be called after stage (a), without a detailed design, in which case the design parameters and requirements, any restrictions, e.g. openings for shipping, preference for large spans, etc., are clearly set out and tenderers are invited to submit tenders on the preliminary design proposal(s), or on alternative designs. In this way, a particular contractor's specialized knowledge or construction equipment may be of advantage. This procedure requires adapted contract documents depending also on whether the employer commissions a consultant or not. The allocation of responsibility for design, the preparation and approval of detailed drawings and the checking, quality control and quality assurance of the construction and the procedures for measurement and payment, are matters which require clarification therein.

An extreme version of contract, namely, the 'packaged contract', places full responsibility for design and construction, including supervision and decisions on compliance with standards of acceptance, on the contractor himself, who is then fully responsible for the eventual performance of the bridge.

There are strong differences of opinion about the merits of these methods, largely because of the subjective element in decisions relating to quality and standards and the natural tendency for the contractor in a competitive situation to minimize costs within the terms of the contract. Theoretically, the latter procedure may result in an increase of risk as defined in Section 3.2.

It is beyond the scope of this handbook to enter into a detailed analysis of the pros and cons of the various national and international contractual procedures. The reader is referred to [227, 228] for comprehensive studies in this respect.

5.11 The monitoring of bridge behaviour

Under this heading we do not understand the routine type of control involving inspection for defects as referred to in Section 5.12, but rather the monitoring of bridge behaviour to gain information about its performance when subjected to various actions for comparison with the results of theoretical analyses and laboratory experiments. Such studies are particularly useful for the rating of bridges for special permit loadings in terms of a specified expected future life. Whereas concrete bridges do not, under normal loading, often exhibit fatigue failure, the regular overloading at and beyond permitted loads exceeding service loads, can cause cracking and deterioration.

Such studies can best be done by selecting particular bridges for study that are for one or other reason prone to fatigue or malfunction. This usually involves the instrumentation of the bridge in order to record deflections and strains at selected positions and to record traffic loadings by methods such as weigh-in-motion systems with portable strain transducers. Failing such sophisticated methods, it is extremely difficult to obtain reliable information about load spectra. Visually observed or photographic records of traffic can be used as basic data to generate random traffic by computer [73]. In this manner, probable traffic load spectra can be determined and used in conjunction with observations in the field of bridge behaviour to predict its expected useful future life.

The regular recording of bridge deflection profiles to study effects that are time related, such as creep and shrinkage, is useful in checking the validity of the theories that were applied in the design or for the derivation of improved formulations. The reduced scale of most laboratory experiments involving creep and shrinkage effects makes such monitoring in the field especially necessary because of the scale effects with respect to volume/surface ratios.

In the case of long-span slender concrete cable-stayed bridges, the monitoring of the effects of wind action would serve a useful purpose in confirming theoretical procedures. The same would apply to earthquake effects in regions of high risk.

5.12   The field inspection, maintenance, repair, rehabilitation and demolition of concrete bridges

In spite of the importance of bridges, their maintenance is often grossly neglected with costly results which could have been avoided by timeous action. It is therefore essential that the responsible authorities should have programs and procedures to ensure adequate maintenance. Such programs do, however, require additional personnel and can be relatively costly, but if optimized can usually be justified.

The first essential is a regular inspection routine and many authorities have bridge inspectors' training manuals [229] to provide guidelines for the training of bridge inspectors, as well as handbooks and checklists for reference in the field. Current bridge inventory information and full records of inspections, preferably on standard forms, should be kept. A properly conceived bridge-inspection reporting system is essential. Where defects are observed, full reports should be prepared on the basis of detailed notes and sketches made in the field, as well as photographs taken. These reports, together with all available historical information, should serve as the basis for further studies, computations, statistical evaluations and diagnostic investigations. Where any doubts exist, it would be essential for such routine inspections to be followed up by qualified engineers to establish or confirm the true nature of the problem. It may furthermore be advisable to consult the engineer responsible for the original design. A repair procedure, if necessary, can then be prepared.

Safety precautions are of the utmost importance and long-span and high bridges should, where feasible, have built-in provision to facilitate inspection and maintenance work, e.g. box-section arches, and girder bridges should have access to all interior compartments, with adequate means of climbing where steep inclines or great heights are involved. Permanent ladders and walkways are desirable, but ropes or chains attached to hooks may in some cases be adequate. In the case of confined spaces, pre-entry air tests for oxygen content or the presence of other toxic fumes or gases may be necessary with approved devices [229] and mechanical ventilation applied continuously during occupancy. Inspection of high parts of bridges may be hazardous and adequate scaffolding, ladders, platforms, or a truck-mounted bucket on a hydraulically operated boom should be used. Life belts should be worn when working at heights above 6 m (20 ft) or above traffic.

The maintenance, repair and demolition of concrete bridges is dealt with in Chapter 25.

Research is currently being done in various countries on methods of rating and evaluating the remaining life of bridges. The statistical information gathered during routine maintenance inspections will play an important role in the updating of such assessments. It is, of course, no precise science as many aspects cannot be quantified precisely. The greatest benefit that can possibly be derived from such studies is that appropriate measures of rehabilitation can timeously be taken to prolong the useful life of the bridge. The rehabilitation of bridges severely damaged, e.g. by ship collision, can be a major operation requiring specialized knowledge and experience. There have been several examples in recent years, such as the Great Belt Bridge disaster in Denmark, reported on in the IABSE *Proceedings* P-31/8/, which deals with the investigation of ship collision problems. Another type of damage that is comparable thereto is that due to seismic action as experienced in many parts of the world in recent years.

Acknowledgements

The author wishes to acknowledge the assistance received from the following partners in preparing the first drafts of the subsections listed: Mr V. Trümpelmann (5.2, 5.6 and 5.7); Mr R. D. Kratz (4.2 to 4.6) and Mr W. Göhring (5.3 to 5.5, 5.8 and 5.10); and from Mr S. Roux for researching the references, Messrs M. Minnaar and M. Quarmby for preparing the line illustrations and his son Louis Liebenberg for the pen sketches in Section 1, and all those who supplied photographs, as acknowledged.

References

1. International Organization for Standardization, ISO 2394, *General principles for the verification of the safety of structures*, Geneva, 1973, 5 pp.
2. Comité Euro-International du Béton (CEB), *International system of unified standard codes of practice for structures*, Vol. I: *Common unified rules for different types of construction and material*, 123 pp., and Vol. II: *CEB–FIP model code for concrete structures*, Bulletin d'information, N.124/125-E, 1978, 348 pp.
3. Hopkins, H. J., *A span of bridges*, David and Charles, Newton Abbot, 1970, 288 pp.
4. Smith, H. Shirley, *The world's great bridges*, Rev. Edn, Phoenix House, London, 1964, 250 pp.
5. Wittfoht, H., *Triumph der Spannweiten*, Beton Verlag, Düsseldorf, 1972, 314 pp.
6. Timoshenko, S. P., *History of strength of materials*, McGraw-Hill, New York, 1953, 452 pp.
7. Stephens, J. H., *The Guinness book of structures*, Guinness Superlatives, London, 1981, pp. 11–60.
8. Pevsner, N., *Pioneers of modern design*, Penguin, London, 1968, pp. 126, 183.
9. Pugsley, Sir Alfred (Ed.), *The works of Isambard Kingdom Brunel*, Cambridge University Press, Cambridge, 1980, pp. 163–182.
10. Jackson, D. C., Railroads, truss bridges, and the rise of the civil engineer, *Civ. Eng. ASCE*, Oct. 1977, pp. 97–101.
11. Williams, J. W., Jr, James B. Eads and his St Louis bridge, *Civ. Eng. ASCE*, Oct. 1977, pp. 102–106.

12. Hamilton, S. B., *A note on the history of reinforced concrete in buildings*, National Building Studies Special Report, No. 24, DSIR, HM Stationery Office, London, 1956, 30 pp.

13. Robertson, H., *Modern architectural design*, Architectural Press, London, 1952, pp. 29, 38, 127.

14. Siegel, C., *Structure and form in modern architecture*, Crosby Lockwood, London, 1962, pp. 160–167.

15. Richards, J. M., *An introduction to modern architecture*, Pelican, London, 1965, pp. 134–135.

16. Menn, C., Fifty years of bridge building in reinforced concrete, *Proc. Int. Ass. Bridge Struct. Eng. symp. on bridges*, Zürich, 1979, pp. 29–40.

17. Freyssinet, E., Prestressed concrete: principles and applications, *J. Inst. Civ. Eng.*, Vol. 33, Feb. 1950, pp. 331–380.

18. Freyssinet, E., *The birth of prestressing*, Cement and Concrete Association, London, 1954, Library Translation No. 59.

19. Virola, J., The world's greatest cantilever bridges. *Acier-Stahl-Steel*, No. 4, April 1969, pp. 164–170.

20. Virola, J., The world's greatest steel arch bridges, *Int. Civ. Eng.*, Vol. II, No. 5, Nov. 1971, pp. 209–228.

21. Wex, B. P., Case study—the Humber Bridge, *Proc. Int. Ass. Bridge Struct. Eng. symp. on bridges*, Zürich, 1979, pp. 109–120.

22. Wex, B., Stephens, J., Wood, B. and Glen, H., Humber Bridge, *New Civ. Eng.*, Supplement, May, 1981, 28 pp.

23. Podolny, W., Jr and Scalzi, J. B., *Construction and design of cable-stayed bridges*, Wiley, New York, 1976, 506 pp.

24. Podolny, W., Jr, The evolution of concrete cable-stayed bridges, *Concr. Int.*, Vol. 3, No. 8, Aug. 1981, p. 35, pp. 43–47.

25. De Ing, G. A. (Presenter), Il Ponte di Brotonne sulla Senna tra Rouen e Le Havre, *L'Industria Italiana del Cemento*, No. 9, 1979, pp. 499–518.

26. Scordelis, A. C. and Lin, T. Y., Cable-suspended concrete structures. *Proc. Conf. Int. Ass. Shell and Spatial Struct.*, Alma Ata, 13–16 Sept. 1977, 23 pp.

27. Finsterwalder, U., Prestressed concrete bridge construction, *J. Am. Concr. Inst., Proc.*, Vol. 62, No. 9, Sept. 1965, pp. 1037–1046.

28. Billington, D. P., Historical perspective on prestressed concrete, *J. Prestressed Concr. Inst.* Sept–Oct. 1976, pp. 48–71.

29. Liebenberg, A. C., Highway bridge over the Bloukrans Gorge—the longest arch bridge in Africa, *Concrete/Beton*, CSSA, No. 18, 1980, pp. 13–16.

30. Candrlic, V., Cantilever construction of large concrete arch bridges in Yugoslavia, *Concr. Int.*, Vol. 3, No. 8, Aug. 1981, pp. 43–47.

31. Report of the Royal Commission into the failure of West Gate Bridge, Government Printer, Melbourne, Australia, 1971, 143 pp.

32. The modern design of wind-sensitive structures, *Proc. seminar of Institution of Civil Engineers, London, 18 June 1970*, Construction Industry Research and Information Association, London, 1971, 139 pp.

33. Hay, J. S., *An introduction to some current theories on the aerodynamic behaviour of bridges*, Transport and Road Research Laboratory, Dept. of Environment, Dept. of Transport, London, Supplementary Report 542, 1980, 50 pp.

34. Simiu, E. and Scanlan, R. H., *Wind effects on structures: an introduction to wind engineering*, Wiley, New York, 1978, Ch. 6.
35. Biétry, J., Sacré, C. and Simui, E., Mean wind profiles and change of terrain roughness, *J. Struct. Div. Am. Soc. Civ. Eng.*, Vol. 104, St 10, pp. 1585-1593.
36. Read, H., *The meaning of art*, Penguin/Faber & Faber, London, 1949, p. 29.
37 and 38, Scruton, R., *The aesthetics of architecture*, Methuen, London, and Princeton University Press, Princeton, N.J., 1969, 259 pp.
39. De Bono, E., *The mechanism of mind*, Penguin, Harmondsworth, 1969, pp. 236–245.
40. Inglis, Sir Charles, *The aesthetic aspect of civil engineering design*, (Third Lecture), Institution of Civil Engineers, London, 1945, pp. 43–49.
41. Faber, O., *ibid.*, (First Lecture), pp. 1–23.
42. Billington, D. P., Bridge aesthetics: 1925–1933, *Final report 11th congr. Int. Ass. Bridge Struct. Eng.*, Vienna, Aug.–Sept. 1980, pp. 47–52.
43. Leonhardt, F., Aesthetics of bridge design, *J. Prestressed Concr. Inst.*, Vol. 13, No. 1, Feb. 1968, 19 pp. (Reprint).
44. Harley-Haddow, T., Structures in architecture, *Struct. Eng.*, Vol. 59A, No. 6, June 1981, pp. 193–201.
45. Slater, R. E., Bridge aesthetics, *Final report 11th congr. Int. Ass. Bridge Struct. Eng.*, Vienna, Aug.–Sept. 1980, pp. 115–120.
46. Murray, J., Visual aspects of motorway bridges, *Proc. Inst. Civ. Eng.*, Vol. 70, Part 1, Nov. 1981, pp. 755–778.
47. Elliott, A. L., Aesthetics of highway bridges, *Civ. Eng. ASCE*, June 1968, pp. 64–66.
48. Tarnos, T. P., Relativity and optimization of aesthetic rules for structures, *Final report 11th congr. Int. Ass. Bridge Struct. Eng.*, Vienna, Aug.–Sept. 1980, p. 59.
49. Schmid, W. A., Proportionen in der Natur und im Menschenwerk—wir messen, sehen und hören, *ibid.*
50. Yasuji, Tahara and Yoshio, Nakamura, On the manual for aesthetic design of bridges, *ibid.*, pp. 101–108.
51. Pearson, J. R., Impact statements—present and potential, *Engineering Issues*, Oct. 1973, pp. 449–455.
52. Bella, D. A. and Williamson, K. J., Interdisciplinary research and environmental management, *Engineering Issues*, July 1978, pp. 193–202.
53. Committee on Impact Analysis of the Water Resources, Planning and Management Division, ASCE, The civil engineer's responsibility in impact analysis, *J. Water Resour. Plann. Mngmnt Div. Am. Soc. Civ. Eng.*, Vol. 104, No. WR1, Nov. 1978, pp. 253–263.
54. Jones, L. R., Environmental impact reports, problems and opportunities, *Civ. Eng. ASCE*, May 1975, pp. 71–73.
55. Wanielista, M. P., McLellon, W. M. and Smith, L. L., Environmental impact statements and civil engineering, *Engineering Issues*, July 1975, pp. 321–327.
56. Ad hoc Committee on Community and Environmental Problems Associated with Urban Highway Proposals, *Report*, National Association of Australian State Road Authorities, 1974, 49 pp.

57. Urban Transportation Division, ASCE *Proc. environ. impact spec. conf.*, Chicago, 21–23 May 1973, American Society of Civil Engineers, New York, 393 pp.

58. Weck, T. L., Environmental impact of transportation projects, *Environmental impacts of international civil engineering projects and practices*, American Society of Civil Engineers National Convention, San Francisco, 17–21 Oct., 1977, pp. 1–28.

59. National Institute for Transport and Road Research, *Technical methods for highways: code of practice for the design of highway bridges and culverts in South Africa*, TMH 7—Parts 1 and 2, Pretoria, 1981, 100 pp.

60. Seni, A., Comparison of live loads used in highway bridge design in North America with those used in Western Europe. *Symposium on concrete bridge design*, American Concrete Institute, SP-26, Vol. 1, 1969, pp. 1–34.

61. Rajagopalan, K. S., Comparison of loads around the world for the design of highway bridges, *Symposium on concrete bridge design*, American Concrete Institute, SP-26, Vol. 1, pp. 35–48.

62. Liebenberg, A. C. *Report to the National Transport Commission of the Republic of South Africa on proposals for a uniform specification of live loading due to traffic on national road bridges*, Liebenberg & Stander, Cape Town, March 1974, pp. 9–217.

63. International Association for Bridge and Structural Engineering, British Group, *Bridge loading colloquium*, Cambridge, England, April 1975. (Unpublished, but available from IABSE.)

64. Liebenberg, A. C., The simulation of normal traffic loading on highway bridges, *Bridge loading colloquium*, Cambridge, England, April 1975, International Association for Bridge and Structural Engineering, British Group. (Unpublished but available from IABSE.)

65. Basler, K., Belastungsvorschriften für Brücken der Schwertransportenstrassen, *Schweiz. Bauztng*, Vol. 85, No. 20, 18 May 1967.

66. VSS–Arbeitsgruppe 13, *Versorgungsnetz. Bericht an die kantonalen Baudirektoren zu technischen Lösungen*, 6 Dez. 1967, Berne, Switzerland.

67. SURCH, 71:1, Service d'Etudes Techniques des Routes et Autoroutes (SETRA), *Application des nouvelles charges réglementaires des ponts routes—notice générale*, DOA B, June 1972, 60 pp.

68. Pennels, E.; *Concrete bridge designer's manual*, Cement and Concrete Association, London, 1978, 164 pp.

69. Ivy, R. J. *et al.*, Live loading for longspan highway bridges, *Am. Soc. Civ. Eng. Trans.* Paper No. 2708, Separate No. 198, June 1953, pp. 981–1003.

70. Mitchell, S. and Borrmann, G. F., Vehicle loads and highway bridge design, *J. Struct. Div. Am. Soc. Civ. Eng.*, Vol. 83, No. ST4, July 1957, pp. 1–21.

71. Lynch, R. L., *Research of traffic loads on bridges*, Kentucky Dept of Highways, Division of Research, Lexington, Feb. 1968.

72. Rey, E., Adjunkt, Eidg. Amt für Strassen- und Flussbau, Bern: Strasse und Verkehr 7/70. Die Verkehrslasten der Strassenbrücken mit Normalbelastung, *Strasse und Verkehr*, No. 7, 6 July 1970.

73. Buckland, P. G. *et al.*, Traffic loading of longspan bridges, (TRR 655). *Bridge engineering*, Vol. 2, 1978, pp. 146–154.

74. Davenport, A. G., Gust loading factors, *J. Struct. Div. Am. Soc. Civ. Eng.*, Vol. 93, No. ST3, June 1967. pp. 11–34.

75. Davenport, A. G., The response of slender, line-like structures to a gusty wind, *Proc. Inst. Civ. Eng.* Vol. 23, Nov. 1962, pp. 389–408.

76. Blume, J. A., Newmark, N. M. and Corning, L. H., *Design of multistorey reinforced concrete buildings for earthquake motions*, Portland Cement Association, Skokie, 1961, 318 pp.

77. Newmark, N. M., Blume, J. A. and Kapur, K. K., Seismic design spectra for nuclear power plants, *J. Power Div. Am. Soc. Civ. Eng.*, Vol. 99, No. PO2, Nov. 1973, pp. 287–303.

78. Newmark, N. M. and Rosenblueth, E., *Fundamentals of earthquake engineering*, Prentice-Hall, Englewood Cliffs, NJ, 1971, 640 pp.

79. Associate Committee on the National Building Code, *National building code of Canada*, NRCC 17303 and Supplement NRCC 17724, National Research Council of Canada, Ottawa, 1980, 547 and 293 pp. respectively.

80. Lovelace, A. M., Keynote address, *Proc. int. conf. struct. safety and reliability*, Smithsonian Institute, Washington, DC, April 1969, pp. 3–4.

81. Pugsley, A. G., *The safety of structures*, Arnold, London, 1966, 156 pp.

82. Freudenthal, A. M., Safety, reliability and structural design, *J. Struct. Div. Am. Soc. Civ. Eng.*, Vol. 87, No. ST3, March 1961, pp. 1–16.

83. Ang, A. H. S., Probability considerations in design and formulation of safety factors, *Proc. symp. concepts of safety of structures and methods of design*, International Association for Bridge and Structural Engineering, London, 1969, pp. 13–23.

84. Cornell, C. A., Bayesian statistical decision theory and reliability-based design, *Proc. int. conf. struct. safety and reliability*, Smithsonian Institute, Washington, DC, April 1969, pp. 47–68.

85. Cornell, C. A., Structural safety specifications based on second-moment reliability analysis, *Proc. symp. concepts of safety of structures and methods of design*, International Association for Bridge and Structural Engineering, London, 1969, pp. 235–245.

86. Esteva, L., Summary Report, Technical Committee No. 10: Structural Safety and Probabilistic Methods, *Proc. int. conf. planning and design tall bldgs*, Vol. 1b, Lehigh University, Bethlehem, Pennsylvania, Aug. 1972, pp. 1043–1066.

87. Construction Industry Research and Information Association, *Rationalization of safety and serviceability factors in structural codes*, CIRIA Report 63, London, July 1977, 226 pp.

88. Turkstra, C. J., *Applications of Bayesian decision theory: structural reliability and codified design*, SM Study No. 3, University of Waterloo, Solid Mechanics Division, Ontario, 1970, Ch. 4, pp. 49–71.

89. Lind, N. C. and Davenport, A. G., Towards practical application of structural reliability theory: probabilistic design of reinforced concrete buildings, American Concrete Institute, SP-31, Detroit, Michigan, 1972, pp. 63–110.

90. Rosenblueth, E., Technical Committee No. 10, Report No. 2, Code specifications of safety and serviceability, *Proc. int. conf. planning and design tall bldgs*, Vol. 1b, Lehigh University, Bethlehem, Pennsylvania, Aug. 1972, pp. 931–959.

91. Ligtenberg, F. K., Structural safety and catastrophic events, *Final report symp. concepts for safety of structures and methods of design*, International Association for Bridge and Structural Engineering, London, 1969, pp. 25–33.

92. Davenport, A. G., The treatment of wind loading on tall buildings, *Proc. symp. tall bldgs*, Dept Civil Engineering, University of Southampton, April 1966, pp. 3–44.

93. Davenport, A. G., Structural safety and reliability under wind action, *Proc. int. conf. struct. safety and reliability*, Smithsonian Institute, Washington, DC, April 1969, pp. 131–145.

94. Moses, F. and Tichy, M., Safety analysis for tall buildings, *Proc. int. conf. planning and design tall bldgs*, Lehigh University, Bethlehem, Pennsylvania, Aug. 1972, Technical Committee No. 10, State-of-the-Art Report No. 5, pp. 993–1005.

95. British Standards Institution, BS 5400, Parts 1–10: 1978 to 1981, *Steel, concrete and composite bridges*, London.

96. Lampert, P. and Thürlimann, B., Ultimate strength and design of reinforced concrete beams in torsion and bending. *IABSE Publications*, Vol. 31-I, 1971, 107–131.

97. Grob, J. and Thürlimann, B., Ultimate strength and design of reinforced concrete beams under bending and shear. *IABSE Publications*, Vol. 36-II, 1976, pp. 105–120.

98. American Concrete Institute, *ACI manual of concrete practice*, Detroit, 1981, Part 4, Ch 3.

99. Parrot, L. J., *Simplified methods of predicting the deformation of structural concrete*, Cement and Concrete Association, London, Oct. 1979, Development Report 3, 11 pp.

100. Gerstle, K. H., *et al.*, Behaviour of concrete under multiaxial stress states, *J. Eng. Mech. Div. Am. Soc. Civ. Eng*, Vol. 106, No. EM6, Dec. 1980, pp. 1383–1403.

101. Bathe, K. J. and Ramaswamy, S., On three-dimensional non-linear analysis of concrete structures, *Nucl. Eng. and Design*, Vol. 52, 1979, pp. 385–409.

102. Hobbs, D. W., *Shrinkage-induced curvature of reinforced concrete members*, Cement and Concrete Association, London, Nov. 1979, Development Report 4, 19 pp.

103. Meredith, D. and Witmer, E. A., Non-linear theory of general thin-walled beams, *Comput. Struct.*, Vol. 13, June 1981, pp. 3–9.

104. Dittmann, G. and Bondre, K. G., Die neue Rheinbrücke Düsseldorf-Flehe/Neuss-Uedesheim: statische Berechnung des Gesamtsystems, *Bauingenieur*, Vol. 54, 1979, pp. 59–66.

105. Douglas, M. R., Parekh, C. J. and Zienkiewicz, O. C., Finite element programs for slab bridge design, *Symposium on concrete bridge design*, Vol. I, American Concrete Institute, SP-26, 1971, pp. 117–141.

106. Davies, J. D., Parekh, C. J. and Zienkiewicz, O. C., Analysis of slabs with edge beams, *ibid.*, pp. 142–165.

107. Dravid, P. S. and Mehta, C. L., Bridge grids with integral slabs—a new analytical approach, *ibid.*, pp. 166–181.

108. Cheung, Yau-Kai, Orthotropic right bridges by the finite strip method, *ibid.*, pp. 182–205.

109. Cheung, Yau-Kai, Analysis of box girder bridges by the finite strip method, *ibid.*, pp. 357-378.

110. Kollbrunner, C. F. and Basler, K., *Torsion*, Springer Verlag, New York, 1966, 280 pp. [In German].

111. Hambly, E. C., *Bridge deck behaviour*, Chapman and Hall, London, 1976, pp. 187–203.

112. Niyogi, P. K. *et al.*, Statistical evaluation of load factors for concrete bridge design, *Symposium on concrete bridge design*, Vol. I, American Concrete Institute, SP-26, 1971, pp. 75–111.

113. Nayak, G. C. and Davies, J. D., Influence characteristics for slab bridges, *ibid.*, pp. 75–111.

114. FIP, *Recommendations for the design of aseismic prestressed concrete structures*, Cement and Concrete Association, London, Nov. 1977, 28 pp.

115. Sigbjornsson, R. and Hjorth-Hansen, E., Along-wind response of suspension bridges with special reference to stiffening by horizontal cables, *Eng. Struct.*, Vol. 3, Jan. 1981, pp. 27–37.

116. Pfaffinger, D. D., Prestressing load cases in linear and non-linear finite element analysis, *Proc. ADINA conf.*, Cambridge, Mass., Aug. 1979, pp. 363–382.

117. Sawko, F. and Wilcock, B. K., Automatic design of pre-stressed concrete bridges by electronic computer, *Symposium on concrete bridge design*, Vol. II, American Concrete Institute, SP-26, 1971, pp. 690–713.

118. Bathe, K. J. and Gracewski, S., On non-linear dynamic analysis using substructuring and mode superposition, *Comput. & Struct.*, Vol. 13, Dec. 1981, pp. 699–707.

119. Cusens, A. R., Box and cellular girder bridges: a state-of-the-art report, *Symposium on concrete bridge design*, Vol. I, American Concrete Institute, SP-26, 1971, pp. 272–319.

120. Aneja, I. and Roll, F., Experimental and analytical investigation of a horizontally curved box-beam highway bridge model, *ibid.*, pp. 379–410.

121. Pama, R. P. and Cusens, A. R., A load distribution method of analysing statically indeterminate concrete bridge decks, *ibid.*, pp. 599–633.

122. Salse, E. A. B., Analysis and design of prestressed concrete circular bow girders for bridge structures, *Symposium on concrete bridge design*, Vol. II, American Concrete Institute, SP-26, 1971, pp. 714–740.

123. Du Preez, R. J., Review of basic theory and methods of analysis of continuous structures, *Symp. bridge deck analysis*, South African Institution of Civil Engineers, Kyalami Ranch, South Africa., May 1973, Paper No. 2.

124. McMillan, C. M., Analysis of slab bridges by finite elements, *ibid.*, Paper No. 4.

125. Du Preez, R. J., Analysis of slab bridges by the finite strip method, *ibid.*, Paper No. 3.

126. Houlding, S. W., Analysis of box section bridges, *ibid.*, Paper No. 13.

127. Du Preez, R. J., Analysis of box girders by folded plate and strip methods, *ibid.*, Paper No. 14.

128. Freese, C. E. and Tracey, D. M., Adaptive load incrementation in elastic–plastic finite element analysis, *Comput. & Struct*, Vol. 13, June 1981, pp. 45–54.

129. Bathe, K. J., Ram, E. and Wilson, E. L., Finite element formulations for large deformation dynamic analysis, *Int. J. Numer. Methods Eng.* Vol. 9, 1975, pp. 353–386.

130. Meyer, C. and Bathe, K. J., Non-linear analysis of concrete structures in engineering practice, *American Society of Civil Engineers Annual Convention and Exposition*, Hollywood-by-the-Sea, Florida, Oct. 1980.

131. Bathe, K. J. and Ramaswamy, S., On three-dimensional non-linear analysis of concrete structures, *Nucl. Eng. & Design*, Vol. 52, 1979, pp. 385–409.

132. Bathe, K. J., *Finite element procedures in engineering analysis*, Prentice Hall, Englewood Cliffs, NJ, 1982, 735 pp.

133. Ferritto, J. M., An evaluation of the ADINA concrete model, *Proc. ADINA conf.*, Cambridge, Mass., Aug. 1979, pp. 650–680.

134. Hannus, M., Modelling reinforced concrete structures with ADINA, *ibid.*, pp. 681–691.

135. Strang, G. and Fix, G. J., *Analysis of the finite element method*, Prentice-Hall, Englewood Cliffs, NJ, 1973, 320 pp.

136. Collatz, L., *The numerical treatment of differential equations*, Springer Verlag, New York, 1966, 535 pp.

137. Cusens, A. R. and Pama, R. P., *Bridge deck analysis*, Wiley, New York, 1975, 278 pp.

138. Zienkiewicz, O. C., *The finite element method*, 3rd Edn, McGraw-Hill, Maidenhead, 1977, 787 pp.

139. Przemieniecki, J. S., *Theory of matrix structural analysis*, McGraw-Hill, New York, 1968, 480 pp.

140. Salse, E. A. B., Analysis and design of prestressed concrete circular bow girders for bridge structures, *Symposium on concrete bridge design*, Vol. II, American Concrete Institute, SP-26, 1971, pp. 714–740.

141. Causevic, M. S., Computing intrinsic values of flexural vibrations of cable-stayed bridges, *Final report 11th congr. Int. Ass. Bridge Struct. Eng.*, Vienna, 1980, pp. 603–608.

142. Daiguji, H., Okui, D. and Yamada, Y., Interactive and atuomated design of cable-stayed bridges, *ibid.*, pp. 615–620.

143. Arkin, A., Bunch-Nielsen, T. and Petersen, S. E., Computer design of complex building structures, *ibid.*, pp. 621–626.

144. Bien, J., Kmita, J. and Machelski, C., Eine effektive Methode zur Spannungsanalyse, *ibid.*, pp. 637–646.

145. Fanelli, M. and Giuseppetti, G., Mathematical analysis of structures: usefulness and risks, *ibid.*, pp. 637–646.

146. Geradin, M., Idelsohn, S. and Hogge, M., Computational strategies for the solution of large non-linear problems via quasi-Newton methods, *Comput. & Struct.*, Vol. 13, June 1981, pp. 73–82.

147. Ito, Y. M., England, R. H. and Nelson, R. B., Computational methods for soil/structure interaction problems, *ibid.*, pp. 157–162.

148. Cheng, F. Y., Inelastic analysis of three-dimensional mixed steel and reinforced concrete seismic building systems, *ibid.*, pp. 189–196.

149. Hsu, C-T. T., Mirza, M. S. and Sea, C. S. S., Non-linear analysis of reinforced concrete frames, *ibid.*, pp. 223–227.

150. Noor, A. K., Survey of computer programs for non-linear structural and solid mechanics problems, *ibid.*, pp. 425–465.

151. Cope, R. J. and Bungey, J. H., Examples of computer usage in bridge design, *J. Struct. Div. Am. Soc. Civ. Eng.*, Vol. 101, No. ST4, April 1975, pp. 779–793.

152. Avramidis, I. E. and Pohlmann, G., Berechnung von ebenem Bogensystemen, *Bautechnik*, Vol. 6, 1980, pp. 193–195.

153. Mitschke, H., Beitrag zur Schnittkraftumlagerung am abschnittsweise hergestellten Durchlaufträger aus Spannbeton, *Bauingenieur*, Vol. 42, 1967.

154. Trost, H. and Wolff, H-J., Zur wirklichkeitsnahen Ermittlung der Beanspruchungen in abschnittsweise hergestellten Spannbetontragwerken, *Bauingenieur*, Vol. 5, 1970.

155. Trost, H., Mainz, B. and Wolff, H-J., Zur Berechnung von Spannbetontragwerken im Gebrauchszustand unter Berücksichtigung des zeitabhängigen Betonverhaltens, *Beton- und Stahlbetonbau*, Vol. 9, 1971, pp. 220–225 and Vol. 10, 1971, pp. 241–244.

156. Schade, D., Alterungsbeiwerte für das Kriechen von Beton nach den Spannbetonrichtlinien, *Beton- und Stahlbetonbau*, Vol. 5, 1977.

157. Chung, H. W. and Gardner, N. J., Model analysis of a curved prestressed concrete cellular bridge, *Symposium on concrete bridge design*, Vol. I, American Concrete Institute, SP-26, 1971, pp. 320–356.

158. Aneja, I. and Roll, F., Experimental and analytical investigation of a horizontally curved box-beam highway bridge model, *ibid.*, pp. 379–410.

159. Scordelis, A. C., Bouwkamp, J. G. and Wasti, S. T., Structural response of concrete box girder bridge, *J. Struct. Div. Am. Soc. Civ. Eng.*, Vol. 99, ST10, Oct. 1973, pp. 2031–2048.

160. Scordelis, A. C. and Larsen, P. K., Structural response of curved reinforced concrete box girder bridge, *J. Struct. Div. Am. Soc. Civ. Eng.*, Vol. 103, No. ST8, Aug. 1977, pp. 1507–1524.

161. Bouwkamp, J. G., Scordelis, A. C. and Wasti, S. T., Failure study of a skew box girder bridge model, *Final Report 11th congr. Int. Ass. Bridge Struct. Eng.*, Vienna, 1980, pp. 855–860.

162. Lin, T. Y. and Burns, N. H., *Design of prestressed concrete structures*, 3rd Edn, Wiley, New York; 1981, 752 pp.

163. Neville, A. M., *Creep of concrete: plain, reinforced and prestressed*, North Holland, Amsterdam, 1970, 622 pp.

164. Rowe, R. E., *Concrete bridge design*, Wiley, New York, 1962, 336 pp.

165. Leonhardt, F., *Vorlesungen über Massivbau*, Vol. 1–6, Springer Verlag, Berlin, 1979.

166. Leonhardt, F., *Prestressed concrete: design and construction*, 2nd Edn, Wilhelm Ernst & Sohn, Berlin, 1964, 677 pp.

167. Lee, D. J. and Richmond, C. J., Bridges, Blake, L. S. (Ed.), *Engineer's reference book*, 3rd Edn, 1975, Newnes–Butterworth, London, pp. 18-2 to 18-71.

168. American Concrete Institute, *Symposium on concrete bridge design*, SP-26, two volumes, Detroit, 1971.

169. Büchting, F., Brücke über den Europakanal aus fertigteilen mit Verbindungsfugen aus Epoxidharz, *Bauingenieur*, Vol. 51, 1976, pp. 136–143.

170. Schlub, P., Formwork launching girders, *IABSE Periodica*, 4, 1981; *IABSE Surveys*, 5–18, 1981, pp. 45–68.

171. Wittfoht, W. *et al.*, Bridges and formwork; launching girders, *IABSE*

Periodica, 2, 1981; *IABSE Structures*, C-17, 1981 pp. 29–43.

172. Landgraf, Th., Lehrgerüstarbeiten für den Mittelteil der Hochstrasse Südumgehung Oldenburg—Bauwerk OL—53. *Bauingenieur*, Vol. 53, 1978, pp. 349–354.

173. Wittfoht, H., VII Internationaler Spannbetonkongress, New York, 1974. 'Brücken—Herstellungserfahren (Konstruktion) und bermerkenswerte Bauwerke'—Bericht, *Beton- und Stahlbetonbau*, Vol. 70, 1975, pp. 16–24.

174. Muller, J., Ten years of experience in precast segmental construction, *J. Prestressed Concr. Inst.*, Vol. 20, No. 1, Jan.–Feb. 1975, pp. 28–61.

175. Schambeck, H., Brücken aus Spannbeton: Wirklichkeiten, Möglichkeiten, *Bauingenieur*, Vol. 51, 1976, pp. 285–298.

176. Wittfoht, H., Die Siegtalbrücke Eiserfeld im Zuge der Autobahn Dortmund–Giessen, *Beton- und Stahlbetonbau*, Vol. 65, 1970, pp. 1–10.

177. Sommerer, R., Wösener, K. und Nehse, H., Die Eschachtalbrücke im Zuge der BAB Stuttgart-Westlicher Bodensee, *Bauingenieur*, Vol. 53, 1978, pp. 117–126.

178. Buchting, F. und Moosbrugger, P. Brücke über den Europakanal aus fertigteilen mit Verbindungsfugen aus Epoxidharz, *Bauingenieur*, Vol. 51, 1976, pp. 137–143.

179. Neuere Entwicklungen im Spannbetonbau mit Kunstharz verklebten Fertigteilen, *Bauingenieur*, Vol. 52, 1977, pp. 456–458.

180. Nelsen Pedersen, H. *et al.*, Vom Bau der Sallingsund Brücke, *Beton- und Stahlbetonbau*, Vol. 72, 1977, pp. 77–85.

181. Leonhardt, F., *Vorlesungen über Massivbau*, Vol. 6: *Grundlagen des Massivbrückenbaus*, Springer Verlag, Berlin, 1979.

182. Barker, J., Construction techniques for segmental concrete bridges, *J. Prestressed Concr. Inst.*, Vol. 25, No. 4, July–Aug. 1980, pp. 66–86.

183. Leonhardt, F. and Baur, W., Erfahrungen mit dem Taktschiebe-verfahren im Brücken und Hochbau, *Beton- und Stahlbetonbau*, Vol. 66, 1971, pp. 161–167.

184. Bonatz, P., Raquet, V. und Seifried, G., Brücken Trichtinbachtal und Ramersdorf—zwei neuartige Anwendungen des Taktschiebe-verfahrens, *Beton- und Stahlbetonbau*, Vol. 74, 1979, pp. 29–35.

185. Leonhardt, F. und Baur, W., *ibid.* (Ref. 183).

186. Stumpp, A., Herstellung einer Unterführung unter Bundesbahngleisen nach dem hydraulischen Vorpressverfahren. *Beton- und Stahlbetonbau*, Vol. 70, 1975, pp. 113–115.

187. Bareš, R., *Tables for the analysis of plates, slabs and diaphragms based on the elastic theory*, Bauverlag, Wiesbaden, 1968.

188. Stein, D., Brückenflügel als an zwei benachbarten Rändern eingespannte Platte, *Beton- und Stahlbetonbau*, Vol. 3, 1966, pp. 62–68.

189. US Department of Transportation, *Hydraulics of bridge waterways*, Federal Highway Administration, Washington, DC, 1970, pp. 1–111.

190. RILEM, CEB, IABSE and IASS, *Proceedings of the inter-association symposium on concrete structures under impact and impulsive loading*, Bundesanstalt für Materialprüfing, Berlin (West), 2–4 June 1982.

191. Helminger, E., Der absolute und spezifische Materialaufwand für Brückentragwerke aus Spannbeton im Zuge neuzeitlicher Strassen, *Bautechnik*,

Vol. 1, 1978, pp. 5–15; Vol. 3, 1978, pp. 83–91; Vol. 4, 1978, pp. 133–143.

192. Bareš, R., *Tables for the analysis of plates, slabs and diaphragms based on the elastic theory*, Bauverlag, Wiesbaden, 1968.

193. Rüsch, H. and Hergenroder, A., *Influence surfaces for moments in skew slabs*, Cement and Concrete Association, London, 1961.

194. Krug, S. and Stein, P., *Influence surfaces of orthogonal anisotropic plates*, Springer Verlag, Berlin, 1961, 193 pp.

195. Pucher, A., *Influence surfaces of elastic plates*, Springer Verlag, Vienna and New York, 1964, 93 charts.

196. Molkenthin, A., *Influence surfaces of two span continuous plates with free longitudinal edges*, Springer Verlag, Berlin, 1971, 165 pp.

197. Homberg, H. and Trenks, K., *Drehsteife Kreuzwerke: Formeln und Zahlentafeln zur berechnung von Trägerrosten Orthotropen Platten und Zellwerken Systeme mit veränderlicher Trägerhöhe*, Springer, Berlin, 1962, 318 pp.

198. Rüsch, H., *Berechnungstafeln für schiefwinklige Fahrbahnplatten von Strassenbrücken*, Wilhelm Ernst & Sohn, Berlin, 1967, 89 pp.

199. Homberg, H. and Ropers, W., *Fahrbahnplatten mit veränderlicher Dicke*, Springer Verlag, Berlin, 1965, 104 pp.

200. Leonhardt, F. and Mönnig, E., *Vorlesungen über Massivbau*, Vol. 2: *Sonderfälle der Bemessung im Stahlbetonbau*, 2nd Edn, Springer Verlag, Berlin and New York, 1975, pp. 91–98.

201. Seidel, O., Die Dywidag-Spannbetonkontaktbauweise, *Beton- und Stahlbetonbau*, Vol. 68, 1973, pp. 257–264.

202. Maisel, B. I. and Roll, F., *Methods of analysis and design of concrete box beams with side cantilevers*, Cement and Concrete Association, London, 1974, Technical Report 42.494, 176 pp.

203. Liebenberg, A. C., Trümpelmann, V. and Kratz, R. D., The design and analysis of the new highway bridge over the Gouritz River Gorge in South Africa, Separate publication by CSSA distributed at *8th int. congr. Fédération Internationale de la Précontrainte*, London, 1978.

204. Taylor, F. W., Thompson S. E. and Smulski, E., *Concrete: plain and reinforced*, Vols. 1 and 2, Wiley, New York, 1948, 969 pp and 688 pp.

205. Deere, D. U. and Hendron, A. J., Design of surface and near surface construction in rock, *Proc. 8th symp. rock mech.*, Minneapolis, Sept. 1966.

206. Crouch, S. L., *Analysis of stresses and displacements around underground excavations: an application of the displacement discontinuity method*, University of Minnesota, Dept Civil and Mineral Engineering, Nov. 1976.

207. Poulos, H. G. and Davis, E. H., *Elastic solutions for soil and rock mechanics*, Wiley, New York, 1973.

208. Candrlic, V., Zur weiteren Entwicklung des Freivorbaus von weitgespannten Massivbogenbrücken, *Beton- und Stahlbetonbau*, Vol. 76, Jan. 1981, pp. 1–6.

209. Miyazaki, Y., Design and construction of the Hokawazu Bridge, *Prelim. Rept. 10th congr. Int. Ass. Bridge Struct. Eng.*, Tokyo, Sept. 1976, pp. 21–26.

210. Liebenberg, A. C., The longspan across the Van Staden's Gorge, *Civ. Eng. Contract.*, (*South Africa*), Oct. 1967, pp. 18–27, 45.

211. Liebenberg, A. C., Trümpelmann, V. and Kratz, R. D., The largest concrete arch in Africa, Separate publication by CSSA distributed at *9th int. congr. Fédération Internationale de la Précontrainte*, Stockholm, 1982.

212. Wossner, K. *et al.*, Die Talbrücke Rottweil–Neckarbing in Zuge der A 81 BAB Stuttgart—Westlicher Bodensee, *Beton- und Stahlbetonbau*, Oct. 1979, pp. 117–126.

213. Leonhardt, F., Zellner, W. und Svensson, H., Die Spannbeton-Schrägkabelbrücke über den Columbia zwischen Pasco und Kennewick im Staat Washington, USA, *Beton- und Stahlbetonbau*, 2, 3, 1980, pp. 29, 64.

214. Leonhardt, F., The Columbia River Bridge at Pasco-Kennewick, Washington, USA, *Proc. 8th int. congr. Fédération Internationale de la Précontrainte*, London, 1978, Part 2, pp. 143–153.

215. Mathivat, J., The Brotonne Bridge, *ibid.*, pp. 164–172.

216. Walter, R. and Luigi, R., Cable-stayed bridges, *Proc. symp. modern trends in bridge design and construction*, Concrete Society of Southern Africa, Johannesburg, July, 1981.

217. Department of the Environment, *Provisional rules for the use of rubber bearings in highway bridges*, HM Stationery Office, London, 1971, pp. 3–18.

218. International Union of Railways, *Code for the use of rubber bearings for rail bridges*, 3rd Edn, Paris, 1969, UIC Code 772 R.

219. Lee, D. J., *The theory and practice of expansion joints for bridges*, Cement and Concrete Association, London, 1971, 65 pp.

220. Leonhardt, F. and Mönnig, E., *Vorlesungen über Massivbau*, Vol. 2: *Sonderfälle der Bemessung im Stahlbetonbau*, 2nd Edn, Springer Verlag, Berlin and New York, 1975, pp. 91–98.

221. MacDonald, M. D., *Waterproofing concrete bridge decks: Materials and methods*, Transport and Road Research Laboratory, Crowthorne, Berkshire, 1974, TRRL Report 6.36, 14 pp.

222. Barrie, D. S. and Mulch, G. L., The professional CM team discovers value engineering, *J. Construct. Div. Am. Soc. Civ. Eng.*, Vol. 103, No. CO3, Sept. 1977, pp. 423–435.

223. Miles, L. D., *Techniques of value analysis and engineering*, 2nd Edn, McGraw-Hill, New York, 1972, 320 pp.

224. Institution of Civil Engineers, *Conditions of contract and forms of tender, agreement and bond for use in connection with works of civil engineering construction*, 5th Edn, London, 1979, 51 pp.

225. Fédération Internationale des Ingénieurs-Conseils, Conditions of contract (international) for works of civil engineering construction with forms of tender and agreement, The Hague, 1963, 1976 printing, 31 pp.

226. Reynolds, C. J., *Measurement of civil engineering work*, Granada, London, 1980, 132 pp.

227. Duncan Wallace, I. N., *Building and civil engineering standard forms: a commentary on the four principal RIBA forms of contract, the FASS form of subcontract, the RIBA fixed fee contract and the ICE contract*, Sweet & Maxwell, London, 1969, 473 pp.

228. Duncan Wallace, I. N., *The international civil engineering contract: a commentary on the FIDIC international standard form of civil engineering and building contract*, Sweet & Maxwell London, 1974, 197 pp.

229. US Department of Transportation, *Bridge inspector's training manual*, Federal Highway Administration, Bureau of Public Roads, Washington, DC, 1970, approx. 350 pp.

37 Tall buildings I

A Coull
University of Glasgow, Scotland

B Stafford Smith
McGill University, Canada

Contents

Notation

The major symbols used in this chapter are listed below. The suffixes used with certain of these symbols are defined locally in the text.

A	cross-sectional area of member	L	clear opening between walls
A_f	area of foundation	M	bending moment
a	clear opening between walls	m	relative stiffness parameter
c	distance from centroidal axis	N	axial force
E	modulus of elasticity	P	applied load
F	design function for wall-frame structure	Q	shear force
		q	shear flow in connecting medium
G	shear modulus	S	shear force
H	total height	s	relative stiffness parameter
h	story height	t	thickness of slab or member
I	second moment of area	v	Poisson's ratio
I_f	second moment of area of foundation	W	depth of wall
		w	lateral load intensity
K	design function for coupled shear walls	X	depth of building
		x	height
K_v	vertical foundation stiffness	Y	bay width
K_θ	rotational foundation stiffness	Y_c	effective width of slab
k	modulus of sub-grade reaction	y	lateral deflection
l	horizontal distance	Z	flange wall width

z	relative height (x/H)	β, γ, μ	geometrical parameters
α	relative stiffness parameter	β_f	foundation geometrical parameter
α_f	relative foundation stiffness parameter	θ	rotation
		σ	stress

Summary

This chapter, the first of two devoted to tall concrete building structures, is concerned primarily with approximate methods of analysis, leaving the second to cover the more versatile and powerful matrix-computer techniques. As the first, it sets the scene by presenting a historical review of the development of efficient and economic structural forms for such structures, ranging from the earliest skyscrapers to modern modular-tube systems.

A brief discussion is given initially of the more important factors affecting the design and analysis of high-rise structures and the selection of an appropriate mathematical model to include the dominant structural actions. Approximate methods of analysis, based generally on continuum techniques, are then presented for the primary structural elements (frames and walls) which comprise such structures, and for two-dimensional combinations of elements involving either pin-connected or bending-resistant connections. Consideration is also given to core structures in which warping effects are significant, and to framed-tube systems. Finally, the action of floor slabs coupling shear walls is considered, and design curves presented to show the influence of wall shape and spacing on the effective slab bending stiffness.

1 Introduction

What constitutes tallness in a building is obviously a subjective quality, and to a large extent it will reflect the particular environment concerned. To a resident of New York, the answer to such a question would probably be quite different from that given by an inhabitant of a small country town, whose prominent five or ten-story building would be submerged amid the skyscrapers of Manhattan. The concept of tallness will also reflect the particular attitude or discipline of the professional questioned. A planner might relate the quality to its dominance of the skyline or the traffic it generates, whereas a fire prevention specialist would probably relate it to the height of his tallest fire-fighting appliance.

From the structural engineer's point of view, a tall building may be defined as one in which lateral forces due to wind or earthquake play an important or dominant role in the structural design. Low-rise buildings are generally designed to resist gravitational loads, and the influence of wind forces only checked subsequently, since most building design codes allow some overstress due to the transient nature of the wind. However, the structure of a high-rise building must be designed *ab initio* to resist both horizontal as well as vertical forces, and an optimum system sought to minimize the influence of the former.

2 Scope of chapter

The fundamental aspects of materials technology, design, and construction, which apply equally well to tall buildings as to other structures, have been dealt with in Parts I–IV of this *Handbook*. Consequently, this chapter restricts itself to considerations of the subject which are peculiar to high-rise buildings. In particular, a description is given of the various structural forms which have been developed specifically to give economic solutions for tall concrete buildings, and guidance is given on the approximate methods which have been devised for their analysis.

3 Development of structural forms for tall buildings

The first steps towards the modern multistory building appear to have been taken in the Bronze Age, with the appearance of houses with second stories, this feature apparently arising simultaneously with the emergence of proper cities. Even today, there appears to be an intrinsic relationship between the tall building and the city. Multistory buildings were considered a characteristic of ancient Rome, and four and five-story wooden tenement buildings were common. Those built after the great fire of Nero used the new burnt brick and concrete materials in the form of arch and barrel vault structures, which replaced the earlier post and lintel construction.

Throughout the following centuries, the two basic materials used in building construction were timber and masonry, although the former lacked the strength required for buildings of more than about 16 m (52 ft) in height, and always presented a fire hazard. The latter had the advantages of high compressive strength and fire resistance, but suffered from its high weight, which tended to overload the lower supports. The limits of this form of construction became apparent in 1891 in the 16-story Monadnock Building in Chicago, which required the lower walls to be over 2 m (6.5 ft) thick, and was the last tall building in the city for which load-bearing masonry walls were employed.

The socio-economic problems which followed the industrialization of the 19th century, allied to the insatiable demand for space in the US cities, gave a big impetus to tall building construction. However, the growth could not have been sustained without two major technical innovations during the middle of that century, namely, the development of new higher strength and structurally more efficient materials, wrought iron and subsequently steel, and the introduction of the elevator to facilitate vertical transportation.

The new materials allowed the development of lightweight framed or skeletal structures, permitting greater heights and more and larger openings in the building. The forerunner of the steel frame which appeared in Chicago around 1890 may well have been a seven-story iron-framed Manchester cotton mill, built in 1801, in which the contemporary I-beam shape appears to have been used for the first time. The Crystal Palace, built for the London International Exhibition of 1851, used a completely autonomous iron frame, with columns of cast iron and beams of cast or wrought iron. One of the notable features of this design was the large-scale approach towards mass-production techniques to facilitate fabrication and erection. The first wrought-iron frame structure to appear in the United States was the lighthouse at Black Harbor, Long Island, built in 1843. Subsequently, a number of buildings appeared employing an interior skeleton of cast iron columns and wrought iron beams, with load-bearing masonary façade walls.

Although the first elevator appeared in 1851, in a New York hotel, its potential in high-rise buildings was apparently not realized until its incorporation in the Equitable Life Insurance Company Building in New York in 1870. For the first time, this made the upper stories as attractive a renting proposition as the lower ones, and in so doing made the taller-than-average structure financially viable.

The first example of a high-rise building totally supported by a metal framework was Jenny's 11-story Home Insurance Building in Chicago in 1883, in which the masonry façade walls were only self-supporting. The first tall building to appear with an all-steel frame was the second Rand-McNally Building, designed by Burnham and Root, and erected in Chicago in 1889. It had nine stories, and stood some 36 m (118 ft) in height. The same architects were responsible for the 20-story Masonic Temple in Chicago in 1891, in which diagonal bracings were introduced into the facade-frames to form vertical trusses, the forerunner of modern shear wall construction. Their introduction was necessary because at that height it was appreciated that wind forces were an important design consideration.

Improved steel design methods and construction techniques allowed steel-framed structures to grow steadily upwards, although progress slowed down during the period of the First World War. In 1909, the 50-story Metropolitan Tower Building was constructed in New York, followed in 1913 by the 60-story, 241 m (792 ft) high, Woolworth Building. This golden age of American skyscraper construction culminated in 1931 in its crowning glory, the Empire State Building. Its 102 stories rose to a height of 381 m (1250 ft) which has now increased to 449 m (1472 ft) with the addition of a TV aerial. The building used 57 000 t (US) of structural steel, nearly 53 500 m^3 (70 000 yd^3) of concrete, and was designed and built in the record time of 17 months.

Although reinforced concrete construction began to be adopted seriously around the turn of the century, it does not appear to have been used properly for multistory buildings until after the end of the First World War. The inherent advantages of the composite material were not at that time fully appreciated, and the early systems were developed purely as imitations of steel structures. An early landmark was the 16-story Ingalls Building in Cincinnatti, Ohio, (1903), which was not superseded until 1915 when the 19-story Medical Arts Building in Dallas was hailed as the world's tallest reinforced concrete building. Thereafter, progress was slow and intermittent, and when the Empire State Building was completed, the Exchange Building in Seattle had attained a height of only 23 stories.

The economic depression of the 1930s put an end to the great skyscraper era, and it was not until some years after the end of the Second World War that the construction of high-rise buildings recommenced, bringing with it new structural and architectural solutions. However, modern developments have produced new structural layouts, improved material qualities, and better design and construction techniques rather than significant increases in height.

Design philosophies altered during the period of recession and war. The earlier tall buildings were characterized by having heavy structural elements and being very stiff due to the high in-plane rigidities of the interior partitions and facade cladding with low areas of fenestration. However, modern office blocks tend to be characterized by light demountable partitions to allow planning flexibility of occupancy, exterior glass curtain walls, and lighter sections as a result of high-strength concrete and steel materials, whilst non-load-bearing infills have given way to load-bearing walls which simultaneously divide and enclose space. As a result, much of the hidden reserve of the earlier buildings has disappeared, and the basic structure must now provide both the required strength and stiffness against vertical and lateral loads. Consequently, the last three decades have seen major changes in structural framing systems for tall buildings.

The building frame was traditionally designed to resist the gravitational loads which are always present and form the reason for its very existence. These loads derive from the self-weight of the vertical and horizontal structural components, including the cladding, and the superimposed floor loadings. These will give rise to necessary minimum cross-sectional areas, based on allowable stress levels, for the vertical column and wall elements, in the design. Since the story functions and loadings tend to be sensibly regular, a uniform floor design will result. The axial stress in, and thus the required cross-sectional area of, the vertical elements, goes up in proportion to the weight carried, and so the curve of minimum required volume of material (or some related efficiency parameter such as structure weight or cost) will increase roughly linearly with height, as shown by the lower curve in Fig. 3-1.

If the chosen structural system were such that the gravity load system could simultaneously satisfy the lateral load capacity and stiffness requirements, no penalty would be exacted for the wind force design. If the degree of overstress caused by lateral forces were within the additional 30% or so normally allowed in structural design codes, no additional structure weight would be required for the influence of building height.

Khan [1], however, has suggested that the relationship for traditionally designed frames will take the form of the upper curve in Fig. 3-1. The difference between the two curves will then indicate the extra cost, or penalty, which must

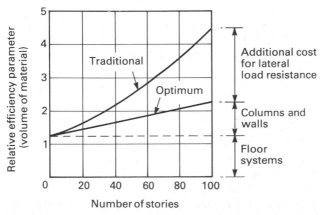

Figure 3-1 **Influence of building height on structural efficiency**

be paid for the additional height, and shows graphically the possible influence of wind forces on the structural design.

In the past three decades, therefore, designers have sought to evolve structural systems which will reduce as far as possible the difference between the two curves, while simultaneously fulfilling the primary building function. A suitable arrangement of the vertical column and wall elements, allied to the horizontal floor system, is required which will provide an economic method of resisting lateral forces and minimizing the additional height premium.

Although the provision of load paths for gravitational forces is limited, there is considerable scope for organizing the structural system to resist lateral forces as efficiently as possible. This may be achieved by the judicious disposition of the vertical elements and their interconnection by horizontal structural components in order to resist moments by axial forces rather than bending moments in these vertical elements.

In general, different structural systems have evolved for residential and office buildings, which reflect their differing functional requirements. However, some notable buildings have been constructed in which the two categories have been mixed, in a deliberate attempt to revitalize moribund city centre areas.

The basic functional requirement of a residential building is the provision of discrete dwelling units for groups of individuals. These have common requirements of living, sleeping, cooking and toilet areas, which must be separated by partitions which offer fire and acoustic insulation between dwellings.

Framed structures may be usefully employed for residential buildings, since the presence of permanent partitions allows the column layout to correspond to the architectural plan. However, these depend on the rigidity of the joints for their resistance to lateral forces, and tend to become uneconomic at heights above 20–25 stories, depending on the overall dimensions, when wind forces begin to control the design, and it becomes increasingly difficult to meet stiffness requirements. Since their introduction in the late 1940s, shear walls, acting either independently or in the form of core assemblies, have been used extensively as additional stiffening elements for traditional frame structures. Typical planforms for tall buildings, whose load-resisting structure consists of interacting shear walls and frames are shown in Fig. 3-2.

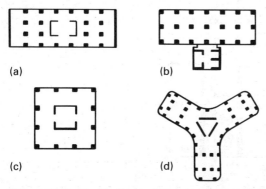

(a)   (b)

(c)   (d)

Figure 3-2 Typical planforms of tall buildings incorporating frames and shear walls

In order to provide adequate fire and acoustic insulation between dwellings, infill panels of brickwork or blockwork are introduced into the frames. Although techniques exist (see, for example, [2]) for assessing the influence of these infill panels on the strength and stiffness of the frame, they are generally assumed to be non-load-bearing, in view of the designer's fear that they may be either removed or perforated for a change of function at some future date, as well as the difficulty of achieving a tight fit between an infill panel and the surrounding frame. Consequently, later trends were to utilize the walls which are required for space division in a structural context, and omit the relatively heavy infills which could not be employed in a load resisting capacity. This has led to the development of the shear wall building, in which structural walls are used to divide and enclose space, while simultaneously resisting both vertical and horizontal loads. These walls are generally of precast large panel or reinforced concrete *in-situ* construction, but concrete blockwork and brickwork have also been employed, allied to precast floor slab construction. Since the functional plan requires a large number of division walls between dwellings, it is frequently found that the minimum thickness required for fire and acoustic insulation will be adequate for structural requirements also.

Functional requirements for this form of building have given rise to the slab block of cross-wall construction, in which horizontal movement of occupants is achieved by long corridors running along the length of the building, with apartments positioned on either side, or to point blocks in which apartments are grouped around the area of vertical transportation, lifts and stairwells. Typical planforms for the two systems are shown diagrammatically in Fig. 3-3. In each case, the basic structure consists of orthogonal systems of shear walls, connected by floor slabs and perhaps lintel beams spanning across door, window or corridor openings, to form a stiff structure. Structural cores, which consist of assemblies of walls connected along their vertical edges to form open or partially closed box sections enclosing lift shafts and stair wells act as additional strong points in such buildings, and can play a major role in resisting lateral forces.

In a design, the shear walls must be sufficiently stiff to meet the imposed deflection criterion, and, in addition, should be so arranged that the tensile stresses caused by wind forces are less than the compressive stresses produced by the weight of the building. How a careful arrangement of walls can improve structural efficiency can be illustrated simply by considering the structure shown in

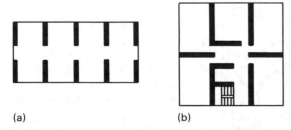

(a)　　　　　　　　　　(b)

Figure 3-3 Typical planforms for slab and point blocks of shear wall construction

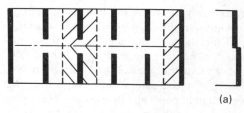

(a)

Figure 3-4   Cross-wall structure

Fig. 3-4, which consists of a series of cross-walls and two flank walls running across the width of the building. As a reasonable approximation, each wall will carry the vertical loads associated with the surrounding tributary area shown hatched in the figure, so that the compressive stresses in the cross walls will be roughly twice those in the flank walls, if they are of the same thickness. However, if all walls deflect equally under the action of the wind forces, as a result of the high in-plane rigidity of the floor slabs, the bending moment and associated stresses in each wall will be proportional respectively to its moment of inertia and section modulus. Consequently, the maximum tensile stresses in the flank walls will be roughly four times those in the cross-walls. The flank walls may then be subjected to unacceptable tensile stresses. A more efficient structure could be achieved by splitting each flank wall into two units, perhaps by forming an architectural feature by having them out of alignment, as shown in Fig. 3-4a. The flank walls would then be subjected to roughly the same wind moments as the cross-walls, and the tensile wind stresses reduced by a factor of more than four. A further beneficial redistribution of forces would occur if the cross-walls were connected together by lintel beams, to form systems of coupled shear walls, which are considerably stiffer and would then attract a greater proportion of the wind moment. A suitable choice of connecting beam stiffness would lead to an optimum arrangement for resisting the combined system of lateral and vertical forces.

Shear wall structures are well suited for resisting seismic loadings, and have performed well in recent disasters. They tend to beome economical as soon as lateral forces affect the design and proportioning of flat plate or framed systems. However, they do possess the disadvantage of an inherent lack of flexibility for future modifications, while discontinuities are frequently required at the critical ground level area to provide a different architectural function on the ground floor, and special detailing becomes necessary.

Since the first use of shear wall construction in conjunction with flat plate floor slabs, it has become the major form of construction for apartment buildings and other residential blocks of any height.

A relatively recent innovation which is particularly suitable for residential blocks is the staggered wall-beam system [3], depicted in Fig. 3-5. The structure consists of a series of parallel bents, each comprising columns with perforated story-height walls between them, in alternate bays. Each wall panel acts in conjunction with, and supports, the slab above and below to form a composite I-beam. By this device, large clear areas are created on each floor, yet the floor slabs span only half the distance between adjacent wall beams, from the bottom of one to the top of the next. The wind shears are transmitted through the floor slabs from the wall

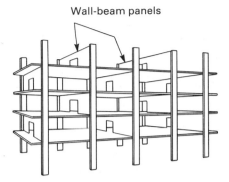

Wall-beam panels

Figure 3-5   Staggered wall-beam structure

beams on one story to those on the next. Similar systems are possible with staggered trusses rather than staggered walls.

The essential functional requirement of an office building is the provision of areas unobstructed as far as possible by walls or columns to allow each occupant to design the partitioning and space enclosure most suitable for his particular business organization. The partition layout will generally alter when tenants change, and this necessitates flexibility in the distribution of the various services to any particular floor. As a result, services tend to be carried vertically within one or more service cores, and a distribution network run beneath the structural floor slab to the entire floor area.

By judicious planning of the column layout to maximize the open floor areas, shear wall-frame interactive structures may also be employed for office blocks, although the presence of the columns may make it difficult to achieve the desired planning flexibility.

Possibly the simplest method of creating open floor areas is to use a central concrete shear core, which carries all essential services and which is designed to resist all lateral forces (Fig. 3-6a). The floor system spans betwen the central core and the exterior façade columns, and a large unobstructed floor area is created between the two vertical components. The exterior columns can be designed to be effectively pin-connected at each floor level, so that they transmit vertical forces only, in conjunction with the interior core. These exterior columns are frequently precast to form a sculptured façade [4]. Another possibility is to provide a core at each end, especially if the building is slender. However, in order to support the floor slabs in the interior, it is then necessary to provide a spine beam running

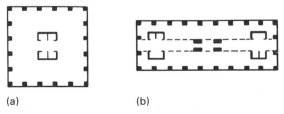

(a)   (b)

Figure 3-6   Typical planforms of office blocks with interior service cores

between the cores, which will require additional supporting interior columns. A representative example planform is shown in Fig. 3-6b. If the floor spans are long, it may become economic to introduce additional columns in the interior to reduce the span of the slabs.

In some situations, a different architectural arrangement is desired at ground level, which precludes the columns being taken right down to ground level. In that case, heavy cantilevers are required to collect the column loads from the levels above and transmit them to the central core.

An alternative approch is to introduce a roof truss in either prestressed concrete or steel construction, at the top of the core. The floor slabs may then be supported between the core and a system of steel hangers suspended from the roof truss. The system has the architectural advantage of lightness of façade, and can simplify construction on a congested city site. The core may be slipformed, and the floor slabs cast on site and simply hoisted into position [4]. However, there is the inherent structural disadvantage that the core is subjected to the entire weight of the building, compressive forces are high at roof level, and settlement may pose problems. Intermediate level trusses will assist in carrying the external tie forces and reduce the extensions of the hangers.

A further increase in lateral stiffness can be achieved if the central core or shear wall system is tied to the exterior columns by deep (usually story height) flexural members or trusses at the top and possibly at other intermediate levels (Fig. 3-7). The effect of these connections is to create an overall framed system, which mobilizes the axial stiffness of the exterior columns to resist wind forces. The objective is to cause the structure to act more as a vertical cantilever beam, and so resist the wind by axial forces in, rather than by bending of, the columns. The larger lever arm involved ensures that large moments of resistance may be produced by relatively low column forces.

The first reinforced concrete building to utilize this concept was the 51-story Place Victoria Building in Montreal (1964), in which an X-shaped core is linked at four levels by story-high girders to the massive corner columns [4].

In both systems, the stories which have been sacrificed to make space for the outrigger trusses may be utilized to house mechanical plant for the building.

As buildings become taller, the use of a core on its own to resist lateral forces will lead to unusually large cores, occupying too large a ratio of a given floor area, and leading to uneconomic solutions. The efficiency can be increased substantially if the outer façade is replaced by a rigidly-jointed framework, which can be used to resist lateral as well as vertical forces. The outer shell then acts effectively as a

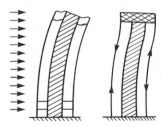

Figure 3-7   Stiffening effect of top girder connecting core to exterior columns

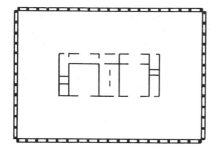

Figure 3-8 Planform of tube-in-tube structure

closed box-like structure, whose faces are formed of rigidly-jointed frame panels, or as a highly perforated tube, whose cross-sectional shape is maintained by the floor slabs acting as horizontal diaphragms.

A combination of the framed-tube concept with the shear wall–frame interaction concept yields the structural form termed the tube-in-tube system, in which an exterior closely spaced column system is constrained by the floor slabs to act in collaboration with a very stiff shear core enclosing the central service area. The first design application of this form of shear wall-frame interactive behaviour appears to have been in the 38-story Brunswick Building in Chicago, completed in 1962. The cross-sectional structural floor plan is shown in Fig. 3-8. The exterior frame has perimeter columns spaced at 2.84 m (9.3 ft) all round the building, and the interior shear core, consisting of an assembly of plane walls, acts effectively as an interior tube core [1]. A clear unobstructed span of 11.6 m (38 ft) is thereby created between the interior and exterior tubes, the floors consisting of either ribbed or waffle slabs. In this case, the lateral forces are resisted by both the interior core and outer framed tube, in proportion to their stiffnesses. The large lever arm involved between opposite normal faces of the exterior tube gives rise to an efficient moment-resisting structure, akin to an ordinary tubular structural component.

This system was also used successfully in 1970 in the 52-story One Shell Plaza Building in Houston, Texas, which at 218 m (715 ft) is the world's tallest lightweight concrete building [5].

While the system is very useful in the creation of flexible spaces in office buildings, it is less suitable for very tall apartment buildings. An alternative solution using the framed-tube concept was devised first [1] for the 43-story De Witt Chestnut Apartment Building in Chicago in 1965. In this case, the exterior columns were closely spaced at 1.68 m (5.5 ft) centres and, when rigidly connected to 600 mm (2 ft) deep spandrel beams, gave rise to a relatively stiff exterior perforated tube which was designed to resist all wind forces. A system of interior columns at approximately 6 m (20 ft) spacing was provided to support the flat plate type of floor construction. The closely spaced exterior columns in this form of construction allow simpler methods of fixing the window glazing directly to the columns themselves.

From the point of view of construction economy, it is claimed that the framed tube compares favourably with shear wall construction for medium-rise buildings,

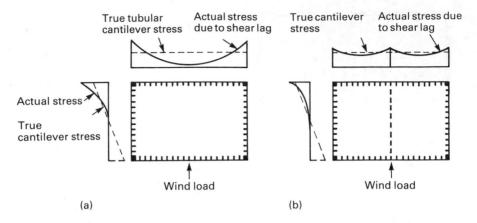

Figure 3-9   Reduction of shear lag effect in modular tube structure

but possesses definite advantages for taller buildings up to about 60 stories, although it has been used in the 110-story twin towers of the World Trade Centre, in New York, constructed in steel [5].

The closely-spaced columns in a framed tube may pose problems in gaining access to the building at ground level, and some structural rearrangement may be necessary in that region. Several columns may be run into one at regular intervals, as in the World Trade Centre, or a deep girder may be provided at first-floor level to transfer column forces to more widely spaced first-floor columns [1].

The pure framed tube has the disadvantage that under bending action, a considerable degree of shear lag occurs in the faces normal to the wind, as a result of the flexibility of the spandrel beams. This has the effect of increasing the stresses in the corner columns, and of reducing those in the inner columns of the normal panels (Fig. 3-9a), and results in a loss of efficiency in the desired pure tubular action of the structure. Warping of the floor slabs, and consequently deformations of interior partitions and secondary structure will occur, which may become of importance in design.

One technique which has been employed to help overcome this problem is to add substantial diagonal bracing members in the planes of the exterior frames. The exterior columns may then be more widely spaced, and the diagonals, aligned at some 45° to the vertical, serve to tie together the exterior columns and spandrel beams to form facade trusses. Consequently, a very rigid cantilever tube is produced. External diagonals have been used in different configurations in the 100-story John Hancock Building in Chicago [3], and the 41-story Boston Company Building in Boston. The diagonals, however, pose their own special problems in the design of the curtain wall system. Although the technique has been used only in steel construction so far, there appears to be no intrinsic reason why it should not be a feasible solution for tall concrete structures.

For very tall buildings, the shear lag effect may be greatly reduced by adding additional interior web panels across the entire width of the building in each direction to form a modular tube or bundled-tube system. The additional stiffening of the structure produced by the interior webs increases the local stress levels at the exterior frame junction and thereby reduces substantially the non-uniformity of column forces caused by shear lag (Fig. 3-9b). The structure may be regarded as a set of modular tubes which are interconnected with common panels to form a perforated multi-cell tube, in which the frames in the wind direction resist the wind shears. The system is such that modules can be omitted at different heights to reduce the cross-section and still maintain the structural integrity. Any torsion arising from the resulting unsymmetry is readily resisted by the closed-sectional form of the modules.

The best known example of this form of construction is the 109-story, 442 m (1450 ft) high, Sears Tower in Chicago, the world's tallest building. Completed in 1974, the basic cross-sectional shape consists of nine 22.86 m (75 ft) square modular tubes, for an overall floor area 68.58 m (225 ft) square, which continues up to the 50th floor. Step backs, produced by a termination of one or more of the modular tubes, then occur at floors 50, 66 and 90, creating a variety of floor configurations [5].

An alternative possibility, yielding the same general form of structural behaviour, is to use shear walls to form the interior webs of the framed tube and create an alternative form of multi-cellular construction. This approach has been adopted for the 74-story, 262 m (860 ft) high Water Tower Place Building, Chicago (1975), the world's tallest concrete building. The 64-story tower which rises from a 12-story base is a slender tube of cross-sectional dimensions 67×29 m (220×95 ft) which is bisected by an internal transverse perforated shear wall to form a two-cell structure. The building is a multi-purpose one and encompasses an hotel and apartments in addition to office space.

General guidelines have been presented by Khan [5] to indicate the range of heights over which the various structural systems are most economically feasible.

4 Design and analysis considerations

Innovations leading to more efficient and economic buildings are only possible through a thorough understanding of the relationship betweeen the functional, structural and service requirements, the nature and behaviour of alternative structural forms, and the problems of construction. In this chapter, we are concerned solely with structural aspects, for which the primary design considerations which are special to tall buildings are as follows:

(a) Lateral load effects—the need to provide adequate strength and stability against wind or earthquake forces, and sufficient stiffness to meet lateral drift criteria and avoid unacceptable dynamic building motions.
(b) Time-dependent behaviour of columns and walls, due particularly to creep and shrinkage.
(c) Differential column movement due to stress variations and to temperature effects on fully or partially exposed exterior columns.
(d) Soil-structure interaction.

In the selection of an efficient and economic structural solution, the most important factors which must be considered, in addition to those listed above, are building function, fire protection and service requirements, and cost and method of construction.

No unique solution generally exists, but, as discussed in Section 3, all systems will encompass different combinations of vertical and horizontal components:

(a) Vertical elements, consisting of columns (either singly or connected by beams to form rigidly-jointed frameworks) and walls (either singly as vertical cantilevers, of various cross-sectional shapes, or connected by beams to form coupled wall assemblies, or jointed along their vertical edges to form open or partially closed, by lintel beams, box-shaped cores).

(b) Horizontal elements, consisting of beams and slabs, and possibly shallow trusses, to form floor and roof structures, and horizontal trusses to form outrigger systems.

These can be found in many different combinations and configurations, depending on the function and layout of the building. However, the structural design will involve an analysis of different combinations of the various components and assemblies, and the remainder of this chapter is largely devoted to a description of the simplified techniques which have been devised to deal specifically with the most common structural components of tall buildings, when subjected to lateral forces.

These methods are generally based on the replacement of the real two-dimensional structural assembly by an equivalent one-dimensional substitute system, which may be solved to give the variation of the primary actions in the major height direction. The modelling is achieved by a smearing technique which replaces the horizontal structural elements by a uniform substitute system.

Once the distributions of forces and deformations are established, the design process is no different from that required for other concrete structures.

Unless the axial stiffnesses of the vertical members are highly non-uniform, it is fairly straightforward to assess approximately the actions due to gravitational forces, and attention is directed particularly to the effects of lateral forces. It is beyond the scope of this chapter to consider the detailed nature and action of wind and earthquake forces (see [6]), and lateral forces are treated only as equivalent quasi-static loads.

In spite of the obvious limitations of applying elastic theory to concrete, the magnitude of the analytical problem involved is such that the distribution of load between the elements in a tall concrete building is invariably determined by an elastic analysis, regardless of the eventual method of design of the members. Although ultimate load methods of analysis may eventually become common-place, their development is as yet incomplete.

An important first step in the analysis of a tall building is the selection of an appropriate idealized model, to include all the significant load-resisting elements and their dominant modes of behaviour. It is usually assumed that the slabs are rigid in their own plane, so that each floor is subjected to a rigid-body movement in plan. Consequently, the vertical elements at any floor level undergo the same horizontal and rotational components of displacement in the horizontal plane.

In order to simplify the analytical problem as far as possible, it is useful to decide in which of the following categories the structure lies:

(a) Symmetric overall plan with parallel identical assemblies of walls, columns, etc. (as in Fig. 3-3a) subjected to a symmetric load system. Since all plane assemblies behave similarly, the analysis of one only is sufficient.

(b) Symmetric overall plan comprising non-identical plane assemblies (as in Fig. 3-2a), subjected to symmetric loading. In this case, the structure may again be analysed as a plane system by assembling the frames in series with connecting rigid pin-ended links at each floor level. The links simulate the behaviour of the floor slabs in constraining the assemblies to deform identically, and allow the resulting redistribution of load. If any assemblies are oblique to the axis of symmetry, only the appropriate component of stiffness must be used. The analysis may be performed using the techniques described in succeeding sections.

(c) Symmetric plan as in (a) and (b), but subjected to eccentric loading. The load may then be replaced by a concentric lateral load and a twisting moment, whose effects can be considered separately and superimposed. The former can be treated as in (a) or (b), whilst the torsional moment can be considered as an equivalent system of pairs of concentrated forces at each floor level applied at convenient corresponding frames on opposite sides of the axis of symmetry [7]. By transforming the stiffnesses and displacements of the other frames into the same locations, the entire structure can be assembled in the same plane, and a plane analysis performed as described in (b).

(d) Non-symmetric plan (as in Fig. 3-3b). In this case, the structure will generally undergo simultaneous bending and torsional displacements, and a three-dimensional analysis is required to determine the load distribution among the elements.

Once the overall form of action has been established, it is necessary to consider the dominant modes of behaviour of the various components and assemblies with a view to selecting the most appropriate representation and method of analysis. In the following sections, the continuum approach is used to analyse the different standard structural forms, whilst the more powerful and flexible matrix techniques are described in Chapter 38.

Fortunately, the functional requirements for tall buildings tend to lead to symmetric structures, which eases the problems of analysis. However, the same approach may be used, with caution, on a representative system to give an approximate assessment of the structural behaviour of structures which are roughly symmetric.

5 Structural frames

The most fundamental component of a tall building is the rigid frame, which achieves its lateral stiffness from the rigidity of the joints between columns and beams or slabs.

The arrangement of the stiffness method of analysis in matrix form allows a concise and systematic approach to the analysis of rigidly jointed frames by digital computation. The method provides the most flexible and powerful technique for the analysis of frames and associated structures and programs are commercially available which demand little more of the engineer than the specification of the structural geometry, stiffness, and loading. The size of problem which may be tackled is limited only by the capacity of the available computer.

However, simplified methods suitable for hand calculation are still valuable for preliminary design analyses, being particularly useful for furnishing qualitative assessments of different systems. In addition, if the frame is combined with other stiffer structural elements, such as shear walls or cores, an approximate representation of its structural behaviour may be sufficiently accurate for a realistic estimation of the load distribution amongst the various components of a complex three-dimensional assembly.

Approximate analyses of frames may be performed most conveniently by effectively reducing the degree of static indeterminacy by suitable moment releases. By recognizing the dominant mode of racking deformation of a laterally loaded structure, it is possible to make realistic predictions of the resulting points of contraflexure in both beams and columns. The degree of indeterminacy is reduced by the number of points of inflexion, or zero bending moment, assumed.

The modes of deformation of laterally loaded portal frames with very rigid and very flexible beams are shown in Fig. 5-1a and 5-1b, respectively, along with the associated bending moment diagrams. In the first case, if column axial deformations are neglected, the beam deflects horizontally as a rigid body and, for compatibility, the columns bend in double curvature with a point of inflection at mid-height. In the second case, the beam acts effectively as a pin-ended strut, although at its ends it must bend to match the rotations at the tops of the columns. The moments at the column bases are now roughly twice as large as those in the first case and the points of inflection are near the tops of the columns.

However, portals and multistory frames normally have girders which are considerably stiffer than the columns, on account of the primary vertical forces which they must resist, and the point of contraflexure will usually be nearer the

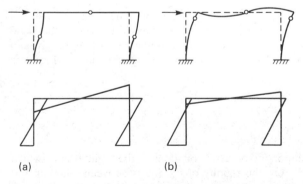

(a)   (b)

Figure 5-1   Mode of deformation and associated bending moment diagram for laterally loaded portal frame with (a) stiff girder and (b) flexible girder

mid-height position. In addition, in a multistory frame, the points of inflection will tend to move downwards as rotations occur at the bases.

As a result, the two main techniques which have been devised for the approximate analysis of multistory frames, the Cantilever and Portal methods, make use of the same assumption that points of contraflexure occur at the mid-height positions of all columns and at the mid-span positions of all beams. Apart from single-bay frames, additional assumptions are required to reduce the problem to a statically determinate system, and it is here where the two techniques differ substantially.

In the Cantilever method, the underlying philosophy is to recognize the bending action of the entire frame as the dominant one, as opposed to the panel shearing action. Consequently, the additional assumption is made that, as the columns act effectively as the fibres in a beam, the axial stress in each column is proportional to its distance from the centroidal axis, or centre of gravity, of the column areas as a whole. This enables the axial force, tensile or compressive, in each column to be determined from the known wind moment at each mid-story level. The centroid and second moment of area, I, of the concentrated column areas are first calculated and individual column stresses evaluated from the simple beam formula Mc/I, in the which M and c are respectively the applied wind moment and the distance of the column in question from the centroid of the group. Once the column axial forces are known, the other stress resultants may readily be determined for the released structure with known positions of zero moment.

In the Portal method, on the other hand, it is considered that the behaviour of the structure is dominated by shearing actions across the panels, in a similar manner to a low-rise frame. It is then assumed that each bay acts as a simple portal frame, and the total shear force at each story level is distributed between the bays in proportion to their spans. With equal bays, this assumption results in the exterior columns carrying half the shear of the interior ones, with no vertical axial force in the interior columns. The combined assumptions of points of inflection and shear distribution between columns results in a statically determinate system, allowing all forces to be determined from conditions of equilibrium at each level.

Once the distribution of forces throughout the frame has been determined, the lateral deflection at any level may be obtained readily by the principle of virtual work. If, for example, bending deformations only are considered, the horizontal deflection, y, at any specified level is given by

$$y = \int \frac{M m_v}{EI} \, ds$$

where M is the calculated moment at any position on the structure for which EI is the local flexural rigidity, m_v is any virtual moment diagram in equilibrium with a unit load applied at the specified level in the direction of the required deflection, and the integral is evaluated over the entire structure. It is generally assumed that axial deformations of the beams may be ignored, so that the horizontal deflections are equal at any particular level.

If the frame is rigidly built in at the base, the simplest virtual moment diagram is obtained by assuming that the horizontal unit load is carried down one exterior column, giving a linear moment diagram for that column. The integration need then be taken over that column only. For standard codified loadings, the bending moment diagram consists of straight lines or simple curves, and the integration can be carried out rapidly by means of tables of product integrals given in many texts on structural analysis.

Of the two methods, the Portal technique is generally preferred as being simpler to apply and more in accord with true frame behaviour. The method neglects the influence of axial deformations in the columns, which may be significant in tall structures, particularly near the base. However, the actual deformations associated with the calculated system of moments and axial forces, from either method, may always be determined by the principle of virtual work in order to assess their relative importance.

If the structure is regular in form, and the member stiffnesses form an orderly pattern, both methods tend to produce reasonable results, with the merit that they are both equilibrium solutions. However, the assumptions inherent in the Portal method result in errors in the vicinity of the base and top, and at setbacks or where significant changes in member stiffness occur. In the lower levels, the points of contraflexure tend to occur at higher levels in the columns due to the increasing influence of the rigid base, and some cognisance of this fact may be included in the analysis, if desired, if only to examine its relative influence on the force distribution near the base.

It is generally assumed that the Portal method yields reasonable results for frames up to 25 stories in height, and the Cantilever method for tall buildings of 25 to 35 stories and moderate height-to-width ratio.

Recently, a more accurate analysis of laterally loaded multistory frames, in which axial deformations of the columns are included, has been presented by Chan *et al.* [8]. In addition to the usual assumptions of points of contraflexure in the beams and columns, the alternative additional assumption made is that the axial deformations of the columns have a hyperbolic sine variation across the width of the building. An energy analysis yields a linear governing differential equation of the second order, which can be integrated to give closed-form solutions for any particular lateral load distribution. Design curves have been presented to enable rapid assessments to be made of column axial forces and shears. However, the approach does not appear to have been employed to deal with the interaction between frames and shear walls or cores.

5.1   Replacement of tall frame by an equivalent shear cantilever

Due to the racking action which occurs over each story height, plane frames in tall buildings behave essentially as shear components, since, as shown in Fig. 8-1, the overall mode of deformation is more akin to a shearing than a bending action. The tall frame component may then be replaced for analysis by an equivalent 'shear cantilever', with effective shearing rigidity, GA, such that, when subjected to lateral forces, the substitute beam deflects only in shear and has no bending stiffness (Fig. 5-2b). The overall shear stiffness will depend on the individual member stiffnesses, the frame configuration, and the rigidity of the joints.

By making use of the earlier assumptions regarding the positions of the points of contraflexure in both beams and columns, the effective shearing rigidity, GA, may

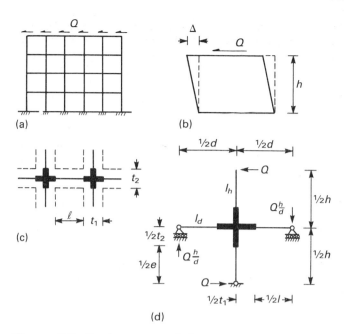

Figure 5-2 Replacement of frame panel segment by equivalent shear cantilever element

be determined. The rigidity of the shear beam must be chosen so that, under the action of the same system of shearing forces, the horizontal deflection of both frame panel and beam should be the same.

To achieve an accurate model, the finite size of a joint must be considered if the beams are of relatively short span, although this effect is more important in the case of framed-tube structures with closely spaced columns and deep spandrel beams than in the normal building frame (Section 11). The size of the joint may be incorporated by assuming that short stiff arms exist at each node, of span equal to the width of the column, and height equal to the depth of the beam, as shown in Fig. 5-2c for a single-story segment of a frame panel. The forces on a typical interior frame segment, bounded by the assumed points of contraflexure and subjected to a horizontal shear force Q, are shown in Fig. 5-2d, in which the appropriate boundary conditions are also indicated.

The horizontal displacement, Δ, of the segment of the equivalent shear cantilever, of story height h, subjected to the same shear force Q is given by

$$\Delta = (Q/GA)h$$

The horizontal deflection of the frame segment may be calculated, and on equating the two displacements, the effective shearing rigidity, GA, established.

When including the bending and shearing deformations of both beams and columns and the shearing deformations of the finite joint, the effective shearing rigidity, GA, becomes

$$GA = 1/K \qquad (5\text{-}1)$$

where

$$K = \frac{e^3}{12EhI_h}\left[1 + \frac{I_h}{I_d}\frac{h^2}{d^2}\frac{l^3}{e^3}\right] + \frac{1}{G'}\left[\frac{hl}{d^2 A_{r_b}} + \frac{e}{hA_{r_c}}\right] + \frac{h}{G'A_jt_2}$$

in which, referring to Fig. 5-2,

$I_d, I_h =$ second moments of area of beams and columns;
$d =$ bay width;
$l =$ clear distance between columns;
$e =$ clear distance between spandrel beams;
$A_{r_b}, A_{r_c} =$ effective cross-sectional areas for shear of beams and columns;
$A_j =$ cross-sectional area of joint parallel to shearing force Q;
$t_2 =$ height of stiff arm at joint;
$G' =$ shear modulus of joint material.

If the deflection due to the shearing deformations of the joint is assumed negligible, the last term in the expression for K is omitted, whilst if shearing deformations of the members are neglected, the second term may then be omitted also.

If the relative size of the joint is small and the rigid arms are omitted, the value of K reduces to that frequently quoted in the literature for normal building frames, namely,

$$K = \frac{h^2}{12EI_h}\left[1 + \frac{I_h}{I_d}\frac{d}{h}\right]$$

Similar expressions may readily be developed for structures in which the beam lengths, or their stiffnesses, are not equal [9].

Corresponding expressions for an exterior column may readily be deduced, by the simple expedience of omitting the contribution of one of the beams and its associated stiff arm at the joint.

Equation 5-1 gives the contribution of a single column to the shearing rigidity, GA, of the shear cantilever and the total rigidity of the frame is the sum of the individual GA terms for all columns in a typical story.

The governing equation for the equivalent shear cantilever subjected to a laterally distributed load of intensity w then becomes

$$-GA\frac{d^2y}{dx^2} = w(x) \tag{5-2}$$

where y is the lateral deflection at height x, which may be obtained by double integration. For example, if the cantilever is subjected to a uniformly distributed loading of intensity w, integration of Eqn 5-2 yields

$$y = \frac{wH^2}{GA}\left[\left(\frac{x}{H}\right) - \frac{1}{2}\left(\frac{x}{H}\right)^2\right]$$

However, the main advantage of the analogy lies in its use in considering the interaction between frames and other structural components such as walls, coupled shear walls and cores in the complete building.

6 Shear walls

The other basic component of a tall building is the structural or shear wall, which occurs in a variety of cross-sectional shapes. Although the term shear wall is widely used, it is something of a misnomer as far as multistory buildings are concerned. Over the height of a single story, a wall panel may appear to be very deep, but when viewed in the context of the complete building, it will appear as a slender cantilever beam. When subjected to lateral forces, a shear wall will be dominated by its flexural behaviour and shearing effects will generally be insignificant.

If an individual wall has a height-to-width ratio of more than about 5, ordinary engineers' beam theory will give an acceptable estimate of the stresses and deflections due to horizontal forces. However, if the ratio is lower, or if large variations in width occur, or if the wall is perforated by openings for doors, windows or corridors, the fundamental assumption of beam theory that plane sections remain plane is no longer valid and a more sophisticated analysis is necessary to obtain an accurate estimate of the stresses and deflections, particularly near structural discontinuities. The only feasible techniques for the analysis of irregular structures are the frame analogy and finite element methods, which are discussed in the next chapter.

In the majority of practical situations, wall elements are constrained to act together through the action of the floor slabs or by lintel beams at each floor level. If the bending stiffness of the connecting members, or their wall connections, is low, so that they behave effectively as pin-ended links, as in Fig. 6-1a, the total lateral moment and shear force at any level will be shared between the walls in proportion to their flexural rigidities, EI, provided these bear a constant ratio to each other throughout the height of the building.

For a planar system of n linked walls, subjected to any lateral load intensity w, the intensity of the applied loading w_r, bending moment M_r and shear force S_r on any particular wall r will be given by

$$w_r = k_r w \qquad M_r = k_r M \qquad S_r = k_r S \qquad (6\text{-}1)$$

where k_r is the stiffness ratio, $E_r I_r / \sum_{i=1}^{n} (E_i I_i)$, for wall r, and M and S are the applied moment and shear force at that level. In most cases, the elastic modulus, E, will be the same for all elements and may be omitted from the formulas.

If the walls are geometrically dissimilar, a similar assumption, although frequently used in practice, may lead to gross errors and a more accurate analysis is advisable.

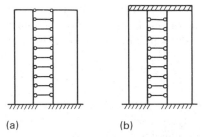

(a)   (b)

Figure 6-1 Use of top beam to stiffen system of linked shear walls

With independent walls, for which beam theory applies, each wall element is subjected to linearly distributed bending stresses across the section, and resultant tensile and compressive forces on opposite sides of the neutral axis. For a rectangular cross-section of depth d, the lever arm for these forces is $2d/3$, so that there is a limited resistance to wind moments and the maximum tensile stress in the concrete will be as large as possible. The efficiency of the structure can be increased greatly if axial forces as well as bending moments can be induced into the walls by some device. The lever arm for axial forces in a pair of walls will be equal to the distance between their centroidal axes, and a relatively small axial stress will give rise to a disproportionately larger moment of resistance and hence reduce the maximum tensile stress in the concrete.

This situation arises with walls connected by lintel beams at each story level to form a system of coupled shear walls which, depending on the stiffness of the connecting members, will behave more as a composite unit, with a performance nearer to that of a beam of depth equal to the overall depth of the pair of walls (Section 7).

Another simpler possibility is to introduce a single stiffening beam, which may be incorporated in the roof structure. If the beam is rigidly connected to the top of each wall, shears will be induced when it is constrained to bend with the walls and these will in turn introduce axial forces, tensile in one wall and compressive in the other (Fig. 6-1b).

The example given in [10] illustrates that even a relatively modest roof beam can be very effective in reducing both lateral deflections and maximum tensile stresses in a system of linked shear walls. For example, for two equal walls, 4 m (13 ft) deep and 0.25 m (10 in) thick, spaced a distance 2 m (6.6 ft) apart and of height 40 m (131 ft), a top beam of the same thickness as the walls and of only 0.5 m (20 in) depth is sufficient to reduce the top deflection by 42% and the maximum tensile stress by 16%.

A residential building whose structure consists of a parallel set of uniform walls may be represented by an equivalent two-dimensional linked wall system and the load and moment in each wall are again given by Eqn 6-1. Since each load is proportional to the flexural rigidity of the wall, it follows that the resultant of these loads must pass through the centroid of the flexural rigidities of the walls.

If the line of action of the applied load, P, does not pass through the centroid, the structure will rotate as well as translate. The load may then be replaced by an equal load, P, passing through the centroid and a twisting moment Pe, where e is the distance between the line of action of the applied load and the centroid.

The load in wall r will then be equal to

$$P_r = P_{r_1} + P_{r_2}$$

where P_{r_1} is the load caused by P acting through the centroid and P_{r_2} is the load produced by the applied torque. Considerations of equilibrium and compatibility show that P_{r_2} is given by

$$P_{r_2} = Pe \, \frac{c_r E_r I_r}{\sum\limits_{i=1}^{n} (E_i I_i c_i^2)}$$

where c_i is the distance of wall i from the centroid.

If the building contains an unsymmetric group of randomly oriented walls, it is necessary to carry out a corresponding calculation in which both the rotation and translations of the cross-section in the two coordinate directions are considered along with the corresponding three conditions of equilibrium.

7 Coupled shear wall structures

Walls containing openings present a more complex design problem. Openings normally occur in vertical rows to accommodate doorways, windows, and corridors in the essentially regular layout of a tall residential building. The connection between the wall sections is provided by either connecting beams which form part of the wall or floor slabs, or a combination of both. Such assemblies are normally termed coupled shear walls.

The presence of the moment-resistant connections greatly increases the stiffness and efficiency of the wall assembly, in a similar manner to that of a stiffening top beam on a system of linked walls discussed earlier. When the walls deflect, shears and moments are induced in the connecting beams or slabs which are constrained to bend with them if they are rigidly connected together, and these in turn induce axial forces in the walls, tensile in one and compressive in the other. The wind moment is then resisted both by bending moments and axial forces in the walls, with a consequent increase in structural effiiciency. The effect of the finite width of a shear wall subjected to bending is to impose a substantial vertical displacement as well as a rotation to the end of each connecting beam. This action causes a much greater stiffness to be reflected from the connecting member to the system than in a column supported structure and emphasizes the importance of including the influence of the wall width in any analysis.

The analysis of such structures is most economically carried out by the use of either a frame analogy approach or a simpler substitute continuous system. In the former, the walls and beams are represented by line members of corresponding stiffness along their centroidal axes, the influence of the finite width of the walls being incorporated by a stiff arm connecting the end of the beam to the centroidal axis of the wall. By formulating the stiffness matrix for displacements of the column nodes of the 'wide-column frame', a solution may be achieved using standard matrix procedures (see Chapter 38).

The second approach is more suitable for hand computations and is most appropriate for structures in which the walls are of uniform section throughout the height and are connected by regularly spaced uniform beams, although a limited number of discontinuities can be incorporated. Its simplicity, its usefulness in the production of design curves and charts, and the fact that when allied to the services of a small computer it can be developed to give solutions to relatively complex structural assemblies, have contributed to its wide use in the design of tall buildings.

In order to illustrate the technique, consider the typical plane structure of Fig. 7-1, which consists of two slender unequal shear walls connected at each floor level by a regular set of lintel beams. For simplicity, the loading is restricted to a uniformly distributed lateral load of intensity w per unit height. However, the

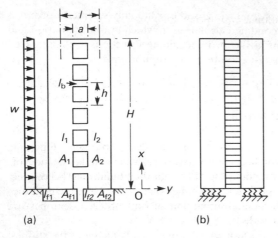

(a)   (b)

Figure 7-1 Real and substitute coupled shear wall structure on elastic foundations

formulation is expressed in terms of a general applied bending moment, M, which might be due to any form of lateral load.

In order to obtain a solution by the continuum approach, the following basic assumptions are made:

(a) Plane sections of both walls and beams which are plane before bending remain plane after bending.

(b) Due to the high in-plane stiffness of the beams and floor slabs, both walls are assumed to deflect equally in the horizontal plane. As a result, the end slopes of the beams, which are assumed connected rigidly to the walls, will be equal and, as may be seen by an examination of the slope–deflection equations, they will then deflect with a point of contraflexure at the mid-span position. This assumption has been shown to yield accurate results, unless one wall is relatively slender, with a stiffness approaching that of the connecting beams.

(c) The connecting beams are of equal stiffness and are uniformly spaced throughout the height. It is then assumed that the discrete set of beams may be replaced by a continuous connecting medium of equivalent stiffness (Fig. 7-1b). If the story height is h, a set of beams of flexural rigidity EI_b may be replaced by a substitute set of 'laminas' of flexural rigidity EI_b/h per unit height, by effectively smearing the stiffness of each beam over half a story height above and below its centroidal axis.

(d) If the connecting beams are 'cut' along the vertical axis through the point of contraflexure, the only internal stress resultants at these positions will be a set of shear forces and axial forces in the beams. The further assumption is then made that the set of discrete shear forces and axial forces in the beams may be replaced in the substitute medium by a continuous shear distribution, or shear flow, of intensity q per unit height and by an axial force distribution of intensity n per unit height.

The overall condition of rotational equilibrium is

Applied moment $= M = \frac{1}{2}w(H-x)^2$

$$= M_1 + M_2 + Nl \tag{7-1}$$

where M_1 and M_2 are the bending moments in the two walls, l is the distance between the centroidal axes and N is the axial force in each wall, given by

$$N = \int_s^H q\,dx \qquad \left(\frac{dN}{dx} = -q\right) \tag{7-2}$$

On using the compatibility condition for vertical displacement at the line of contraflexure of the connecting medium, together with the equilibrium conditions for the walls, it may be shown [11] that the behaviour of the structure may be expressed as a linear differential equation of the second order in one of the chosen stress resultants, for example, the axial force N, as

$$\frac{d^2N}{dx^2} - \bar{\alpha}^2 N = -\bar{\gamma}^2 M \tag{7-3}$$

in which

$$\bar{\alpha}^2 = \bar{\gamma}^2 \left\{\frac{AI}{A_1 A_2 l} + l\right\}$$

$$\bar{\gamma}^2 = \frac{12 I_b l}{a^3 h I}$$

where a is the clear opening between walls, I_1 and I_2 are the second moments of area of the walls, $I = I_1 + I_2$, A_1 and A_2 are the cross-sectional areas of the walls, $A = A_1 + A_2$, and I_b is the second moment of area of the connecting beams.

The influence of shearing deformations in relatively deep connecting beams may always be included [12] by using an effective second moment of area I_c, rather than the true value I_b.

Alternatively, the behaviour may be expressed in terms of the lateral deflection, y, as a fourth-order governing equation:

$$\frac{d^4y}{dx^4} - \bar{\alpha}^2 \frac{d^2y}{dx^2} = \frac{1}{EI}\left(\frac{d^2M}{dx^2} - \bar{\beta}^2 M\right) \tag{7-4}$$

in which

$$\bar{\beta}^2 = \frac{AI}{A_1 A_2 l}\bar{\gamma}^2$$

In each case, the equation may readily be integrated to give a closed-form solution in terms of constants of integration, which are determined from the known boundary conditions at the base and top of the structure. Once the solution is obtained in terms of N or y, the other forces and displacements follow from the internal equilibrium and force–displacement relationships.

The general solution is given in Table 7-1 for a structure which is free at the top and supported at the base on either individual elastic foundations or a portal

Table 7-1 Formulas for coupled shear walls subjected to uniformly distributed load

Quantity	Value	
$\dfrac{EI}{wH^4}y$	$\dfrac{\beta^2}{24\alpha^2}z^4 - \dfrac{\beta^2}{6\alpha^2}z^3 + \dfrac{1}{2\alpha^2}\left(\dfrac{\beta^2}{\alpha^2}+\dfrac{\beta^2}{2}-1\right)z^2 + C_1 + C_2 z + C_3\sinh\alpha z + C_4\cosh\alpha z$	
$\dfrac{M_p}{wH^2}$	$\dfrac{\beta^2}{2\alpha^2}(1-z)^2 - \dfrac{1}{\alpha^2}\left(1-\dfrac{\beta^2}{\alpha^2}\right) + \alpha^2(C_3\sinh\alpha z + C_4\cosh\alpha z)$	
$\dfrac{Nl}{wH^2}$	$\left(1-\dfrac{\beta^2}{\alpha^2}\right)\left\{\tfrac{1}{2}(1-z^2)+\dfrac{1}{\alpha^2}\right\} - \alpha^2(C_3\sinh\alpha z + C_4\cosh\alpha z)$	
	Elastic foundation	Portal or column base
C_1	$C_3\tanh\alpha - \dfrac{\alpha^2-\beta^2}{\alpha^6}\operatorname{sech}\alpha$	$\dfrac{r_H r_L}{\alpha^2} - \dfrac{\alpha^2-\beta^2}{\alpha^6}\operatorname{sech}\alpha + \left(\tanh\alpha + \dfrac{r_H r_L\alpha^3}{\alpha^2-\beta^2}\right)C_3$
C_2	$\dfrac{1}{\alpha_f}\left(1-\dfrac{\beta^2}{\alpha^2}+\dfrac{m\beta_f}{2}\right) - \left(1-\dfrac{\alpha^2}{\alpha_f}\right)\alpha C_3$	$\dfrac{r_L}{\alpha^2} + \left(\dfrac{\alpha^2 r_L}{\alpha^2-\beta^2}-1\right)\alpha C_3$
C_3	$\dfrac{s\dfrac{\alpha_f}{\alpha^2}\left\{\left(1-\dfrac{\beta^2}{\alpha^2}\right)(\operatorname{sech}\alpha - 1)+\dfrac{\beta^2}{2}\right\}-\left(1-\dfrac{\beta^2}{\alpha^2}\right)-\dfrac{m\beta_f}{2}}{\alpha^2(s\alpha_f\tanh\alpha+\alpha)}$	$\dfrac{\dfrac{1}{\alpha^2}\left\{\left(1-\dfrac{\beta^2}{\alpha^2}\right)(\operatorname{sech}\alpha-1)+\dfrac{\beta^2}{2}\right\}-\left(1-\dfrac{\beta^2}{\alpha^2}\right)r_1 - \tfrac{1}{2}(1-r_L)+r_H r_L}{\alpha^2(r_1\alpha+\tanh\alpha)}$
C_4	$-C_1$	$\dfrac{\alpha^2-\beta^2}{\alpha^6}\operatorname{sech}\alpha - C_3\tanh\alpha$

frame with the different possible configurations shown in Fig. 7-2. The necessary structural parameters r_I, r_L, and r_H, used in the general solution for the latter are defined in the figure.

The table gives, in non-dimensional form, the horizontal deflection y, the total bending moment M_p (equal to $M_1 + M_2$) in the walls and the axial force, N, in each wall at any level.

Since both walls deflect equally, the bending moments are proportional to their

	r_I	r_L	r_H
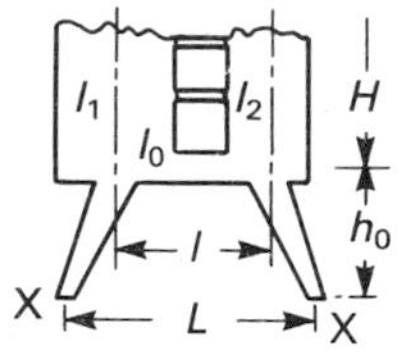	$\dfrac{l_0 h}{l_b H}$	$\dfrac{l}{L}$	$\dfrac{h_0}{H}$
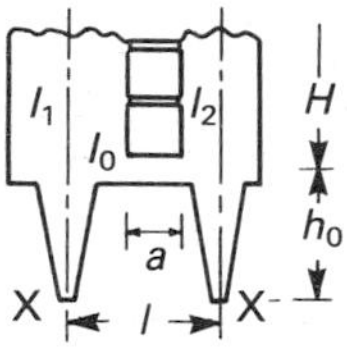	$\dfrac{l_0 h}{l_b H}$	1	$\dfrac{h_0}{H}$
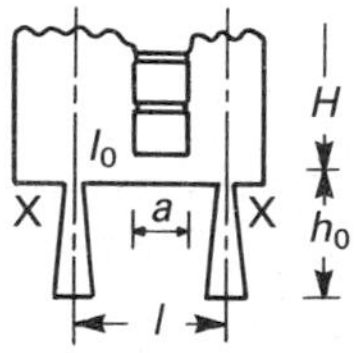	$\dfrac{l_0 h}{l_b H}$	1	0
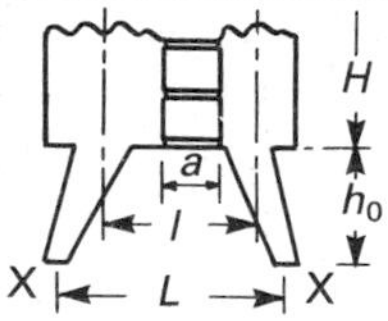	0	$\dfrac{l}{L}$	$\dfrac{h_0}{H}$
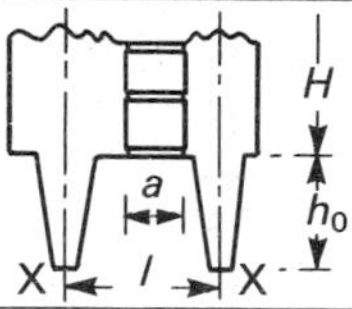	0	1	$\dfrac{h_0}{H}$
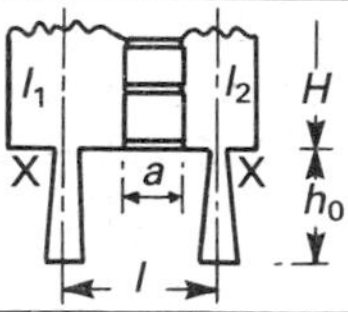	0	1	0

Figure 7-2   Base support parameters for coupled shear walls

flexural rigidities, so that

$$M_1 = \frac{I_1}{I} M_p \qquad M_2 = \frac{I_2}{I} M_p$$

The solution applies equally well to shear walls with two symmetric bands of openings [11], since in that case the shear flow, q, will be the same in each connecting medium and there will be no axial force in the central wall.

In the case of elastic foundations, it is assumed that the walls are supported on rectangular bases, whose cross-sectional areas are A_{f_1} and A_{f_2} and second moments of areas I_{f_1} and I_{f_2}, resting on a soil whose modulus of sub-grade reaction is k.

For simplicity, the solution has been presented in terms of a non-dimensional height coordinate and non-dimensional structural parameters defined as

$$z = \frac{x}{H} \qquad s = \frac{EI}{kI_fH} \qquad m = \frac{E}{kH}$$

$$\alpha^2 = \bar{\alpha}^2 H^2 \qquad \beta^2 = \bar{\beta}^2 H^2 \qquad \gamma^2 = \bar{\gamma}^2 H^3$$

$$\alpha_f = \frac{12I_b}{ha^3} \left\{ \frac{l^2}{I} + \left(\frac{1}{A_{f_1}} + \frac{1}{A_{f_2}} \right) \frac{I_f}{I} \right\} H^2$$

$$\beta_f = \frac{12I_b}{ha^3} \left\{ \frac{1}{A_{f_1}} + \frac{1}{A_{f_2}} \right\} H^2$$

$$I_f = I_{f_1} + I_{f_2}$$

Where the shear walls are discontinued at first-floor level for architectural reasons and supported on portal frames or columns, it is necessary to make further assumptions about their mode of behaviour in order to derive suitable base-boundary conditions. In the present case, points of contraflexure have been assumed to occur at the positions marked X in Fig. 7-2. The overall analysis gives the applied forces and moments on the portal frame. Since the support system is effectively statically determinate, the internal force distribution follows directly.

Once the axial force function, N, is known, the shear flow in the continuous medium follows by differentiation, as in Eqn 7-2.

The forces in the discrete structure may then be deduced from those in the corresponding substitute system. For example, the shear force, Q_k, in any particular connecting beam at level x_k may be obtained by integrating the shear flow, q, over half a story height above and below the level concerned, that is

$$Q_k = \int_{x_k - \frac{1}{2}h}^{x_k + \frac{1}{2}h} q \, dx \qquad (7\text{-}5)$$

or from the corresponding values of the axial force N, namely,

$$Q_k = N(x_k - \tfrac{1}{2}h) - N(x_k + \tfrac{1}{2}h)$$

In this way, a complete picture of the force distribution in the real structure may be achieved.

It may be observed from the solutions that the deflections and forces in the structure depend upon a very limited set of parameters, the height $z(= x/H)$, the relative stiffness parameters α, β, γ, which are related, and the parameters defining the flexibility of the support system.

When the shear forces in the walls are evaluated, it is generally found that a specific value exists at the top of each, indicating that a concentrated horizontal interactive force must exist in the connecting medium at that position. This force vanishes in the particular case when the two walls are of equal size.

Although, for simplicity, the solution for elastic foundations has been expressed in terms of the sub-grade modulus k, this is not essential, provided the vertical and rotational stiffnesses K_v and K_θ respectively can be estimated for any particular form of foundation system. The same formulas would then apply with kA_{f_1} replaced by K_{v_1}, and kI_{f_1} by K_{θ_1}, etc.

Corresponding general solutions have been given in [11] for two other standard load cases, a triangularly distributed load with maximum intensity at the top and a point load at any height. The former may be used in conjunction with a uniformly distributed load to simulate any trapezoidal wind pressure distribution, or in conjunction with a point load at the top to give a quasi-static simulation of horizontal earthquake inertia forces on the building. The latter will yield a set of flexibility influence coefficients for the system for use in the analysis of more complex three-dimensional structures [12], or, alternatively, a set of point loads at different levels may be employed to simulate any complex distributed loading. General solutions are also available for a lateral loading expressed as a polynomial series in the height coordinate [13].

A particular but very common practical situation occurs when the walls are built into a common rigid foundation, when the deflection and slope at the base become zero. The solutions for forces and deflections may then be obtained by putting k equal to infinity and hence s, m and β_f equal to zero in Table 7-1.

In that case, the expressions for the forces q, N and M_p, in non-dimensional form, are found to be functions only of the non-dimensional height, z ($= x/H$), and the relative stiffness parameter, α, for any particular load form. The deflection y depends in addition on the parameter β, but, if attention is directed at the important design values of the maximum lateral deflection y_H at the top, the solution is again a function of two parameters only.

This feature enables design curves or charts to be produced, enabling the significant forces and deflection to be evaluated rapidly for any configuration and particular load form.

Physically, the form of stress distribution across the system of walls will depend on the stiffness of the connecting beams. If they are very flexible, the structure will behave as a system of linked walls, with individual linear bending stress distributions across each wall, whereas, if they are very stiff, the structure will behave as a single composite dowelled beam, with a linear bending stress distribution across the entire section. These limiting conditions are defined respectively by low (less than about 1) and high (greater than about 12) values of the stiffness parameter α.

Simple sets of design curves may then be produced which will indicate the general behaviour of the structure within these wide limits, and will allow the stresses and deflections to be evaluated rapidly.

For the wall system of Fig. 7-3a, the stress distribution at any section under the actions of the bending moments M_1 and M_2, and the axial forces N, will be as

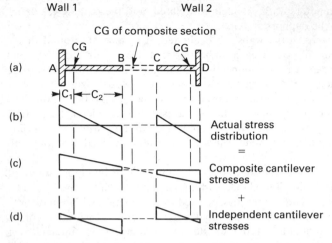

Figure 7-3 Superposition of stress distributions due to composite and individual cantilever actions to give true stress distribution in walls

shown in Fig. 7–3b. Taking tensile stresses as positive, the maximum extreme fibre stresses in wall 1 will be,

$$\sigma_A = \frac{M_1 c_1}{I_1} + \frac{N}{A_1} = (M - Nl)\frac{c_1}{I} + \frac{N}{A_1}$$

$$\sigma_B = \frac{-M_1 c_2}{I_1} + \frac{N}{A_1} = -(M - Nl)\frac{c_2}{I} + \frac{N}{A_1}$$

(7-6)

Similar expressions hold for stresses σ_C and σ_D in wall 2.

The stress distribution, which consists in reality of a superposition of a uniform axial stress on a linear bending stress distribution in each wall, may be considered to be derived from an alternative superposition of two pure bending stress distributions relating to the two limiting cases discussed above, namely, (a) a bending stress obtained on the assumption that the wall system acts as a single composite cantilever, the neutral axis being situated at the centroid of the two wall elements, as shown in Fig. 7-3c; and (b) two linear bending stress distributions obtained on the assumption that the walls act independently, with the neutral axes at the centroidal axes of the individual walls, as shown in Fig. 7-3d.

Suppose then that K_1 is the percentage of the load carried by independent cantilever action, and that K_2 is the percentage $(100 - K_1)$ carried by composite cantilever action. Consider then the two component stress distributions:

(a) *Composite cantilever action* (Fig. 7-3c) The total bending moment at any section is equal to $MK_2/100$, where M is the applied wind moment, and the extreme fibre stresses in wall 1 will be given by

$$\sigma_A = \frac{M}{I'}\left(\frac{A_2 l}{A} + c_1\right)\frac{K_2}{100}$$

$$\sigma_B = \frac{M}{I'}\left(\frac{A_2 l}{A} - c_2\right)\frac{K_2}{100}$$

(7-7)

where I' is the second moment of area of the composite cantilever, given by

$$I' = I_1 + I_2 + \frac{A_1 A_2}{A} l^2$$

(b) *Individual cantilever action* (Fig. 7-3d) Since the moment in each wall is proportional to its flexural rigidity, the bending moments in walls 1 and 2 are given by,

$$M_1 = \frac{I_1}{I} \frac{K_1}{100} M \qquad M_2 = \frac{I_2}{I} \frac{K_1}{100} M$$

and the extreme fibre stresses in wall 1 become

$$\sigma_A = \frac{M_1 c_1}{I_1} = \frac{c_1}{I} \frac{K_1}{100} M$$

$$\sigma_B = -\frac{M_1 c_2}{I_1} = -\frac{c_2}{I} \frac{K_1}{100} M$$

(7-8)

Similar expressions for each hold for wall 2.

On equating the corresponding stresses at the extreme fibre positions from Eqns 7-6, 7-7 and 7-8, and using the general solution for N in the first equation, expressions are readily derived for the proportional functions K_1 and K_2. These are tabulated for three standard load cases in [14, 15].

The proportions of composite and individual cantilever action required to produce the true stress distribution at any level are functions only of the stiffness parameter, α, and the height ratio, z. The form of these functions for the case of a uniformly distributed loading are shown in Fig. 7-4, for a range of values of the parameter α covering the range of practical situations.

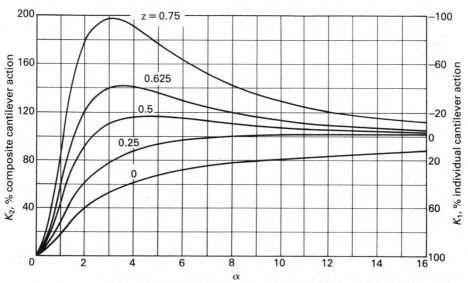

Figure 7-4 Variation of wall bending moment factors

Corresponding curves for a triangularly distributed loading and a concentrated point load at the top are given in [15]. It is found that the former are very similar in form to those of Fig. 7-4, but the latter rise continuously to become asymptotic to a value of K_1 of zero at high values of α.

Differentation of the axial force function, N, gives the shear flow, q, in the substitute medium, which may be expressed in terms of a third function, K_3, for a uniformly distributed loading as

$$q = w\frac{H}{l}\frac{1}{\mu}K_3$$

in which

$$\mu = 1 + \frac{A}{A_1 A_2}\frac{I}{l^2} = \frac{H}{l}\frac{\alpha^2}{\gamma^2} = 1 + \frac{H}{l}\frac{\beta^2}{\gamma^2}$$

The variations of the function K_3 with the height ratio z, for different values of the stiffness parameter α, are shown in Fig. 7-5. The curve of K_3(max) has been

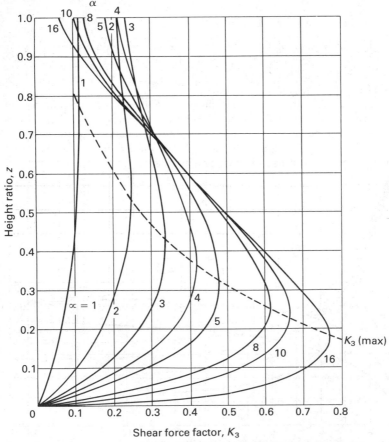

Figure 7-5 Variation of connecting beam shear force factor

superimposed as a broken line to indicate the maximum value of the shear flow in any particular situation, and will serve to indicate closely the position of the most highly stressed connecting beam.

Corresponding curves have been given in [15] for the other two standard load cases. The curves for a triangularly distributed load are of the same form, but the maximum value of the shear flow always occurs at the top in the presence of a horizontal load at the top.

The shear force, Q_k, in any particular beam at level z_k is then given by the area underneath the curve of q between half story height levels above and below the beam position (see Eqn 7-5).

The maximum value Q_{max} can be assessed rapidly from the curve of $K_3(max)$, and will be a little less than $q_{max}h$; the greater the number of stories, the smaller will be the difference. The maximum bending moment in any connecting beam will then be equal to $\frac{1}{2}Q_{max}a$.

In a similar way, the maximum deflection at the top of the uniformly loaded structure may be expressed conveniently in the form

$$y_{max} = \frac{1}{8}\frac{wH^4}{EI}K_4$$

where K_4 is a function of the two parameters α and μ, and indicates directly the stiffening influence of the connecting beams since the term $wH^4/8EI$ is the maximum deflection at the top of a pair of linked walls. The variation of the function K_4 with the stiffness parameter α, for various representative values of the parameter μ, is shown in Fig. 7-6. The form of curves for the other two standard load cases, given in [15], are almost identical.

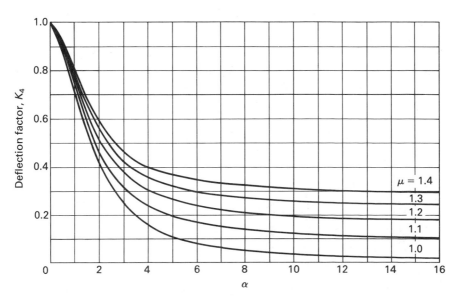

Figure 7-6 Variation of maximum deflection factor

The physical interpretation of the magnitudes of the parameters α and μ is discussed in [15].

Similar curves may readily be produced for other load forms [13].

Although the continuum technique lends itself most simply to the analysis of uniform regular structures, it can accommodate a limited number of discontinuities by matching piecewise uniform solutions using such techniques as the transfer matrix or matrix progression methods, including the analysis of elastoplastic behaviour [16]. It is also readily applicable to the dynamic analysis of shear wall structures [17], and the analysis of complex three-dimensional interconnected wall assemblies [18].

8 Interaction between walls and frames

When walls, either acting independently or assembled together to form cores, are used to brace a framed structure, it may be acceptable to neglect the lateral stiffness of the frame and assume that the horizontal load is carried entirely by the walls. This initial approach is adopted frequently when one or more structural cores act as strong points in the building. The load carried by a core, and hence the resulting lateral deflection, will be overestimated; an examination of the forces on the frame associated with such displacements will then indicate if the initial assumption was justified.

If the frame portion is relatively stiff, it is necessary to determine the distribution of the horizontal forces between the wall and frame elements. If allowed to deform independently under the action of similar horizontal loads, the cantilevered wall and frame adopt different configurations, as shown in Fig. 8-1. Consequently, when the two components are constrained to deflect together by the floor slabs, they will have to be 'pushed apart' at the top, and 'pulled together' near the base. A redistribution of horizontal load must then take place throughout the height, with heavy interactions near the top and bottom (Fig. 8-1c). The frame will thus resist proportionately more load near the top, whilst the wall will tend to pick up a larger proportion in the lower regions of the building.

An approximate continuum-based method of analysis may again be derived to

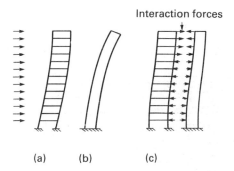

Interaction forces

(a) (b)   (c)

Figure 8-1 Interaction between frame and shear wall constrained to act together by floor system

yield simple formulas and design curves for both preliminary and final design calculations [9].

The composite wall–frame structure is assumed to consist of a combination of a vertical flexural beam and a shear cantilever. The equations governing the behaviour of the two components are

$$\text{in flexure:} \quad EI\frac{d^4y_b}{dx^4} = w_b(x)$$

$$\text{in shear:} \quad -GA\frac{d^2y_s}{dx^2} = w_s(x)$$

(8-1)

in which the subscripts 'b' and 's' refer to the flexural and shear cantilevers respectively, w is the distributed lateral load intensity, and EI and GA are the flexural and shear rigidities of the two beams.

The two components are assumed linked by floor slabs so that they have identical horizontal deflections throughout the height. It is then assumed that the discrete set of links may be replaced by an equivalent pin-ended continuous connecting medium, so that only horizontal forces are transmitted between the two. Consequently, there will exist a distributed horizontal interactive force of intensity $n(x)$ per unit height, together with a top concentrated interaction force of magnitude Q to maintain horizontal compatibility. As with the system of coupled shear walls considered earlier, the presence of the latter may be deduced subsequent to the overall analysis.

On using the overall condition of horizontal equilibrium, in conjunction with Eqns 8-1, a governing fourth-order differential equation is obtained, of the general form

$$\frac{d^4y}{dx^4} - \bar{\alpha}^2\frac{d^2y}{dx^2} = \frac{w}{EI}$$

(8-2)

in which w $(= w_b + w_s)$ is the intensity of the applied distributed load, and $\bar{\alpha}^2$ is the relative stiffness ratio, given by $\bar{\alpha}^2 = GA/EI$.

Equation 8-2 may readily be integrated to give a closed-form solution for any applied loading. The four constants of integration which arise in the solution may be determined from the known boundary conditions at the base and the top of the structure.

In the particular case of a structure which is free at the top and rigidly built in at the base, and subjected to a uniformly distributed load of intensity w, the complete solution becomes, in non-dimensional form,

$$y = \frac{1}{8}\frac{wH^4}{EI}\frac{8}{\alpha^4}\left\{\frac{\alpha\sinh\alpha + 1}{\cosh\alpha}(\cosh\alpha z - 1)\right.$$
$$\left. - \alpha\sinh\alpha z + \alpha^2[z - \tfrac{1}{2}z^2]\right\}$$
$$= \frac{1}{8}\frac{wH^4}{EI}F_1(z, \alpha)$$

(8-3)

The solution is expressed in this form to emphasize the dependence of the horizontal deflection on two non-dimensional parameters only, the height ratio $z(= x/H)$ and the relative stiffness $\alpha(= \bar{\alpha}H)$.

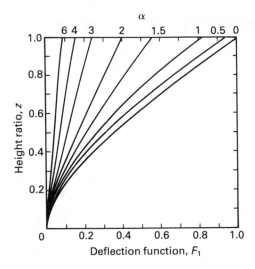

Figure 8-2 Variation of deflection function

Since the term $\frac{1}{8}wH^4/EI$ represents the maximum lateral deflection at the top of a uniformly loaded cantilever of flexural rigidity EI, the function F_1 is a measure of the stiffening influence of the frame component. If F_1 is large, of the order of unity, the stiffness of the building is derived almost entirely from the wall component, whereas if F_1 is small, the frame component provides a major part of the lateral stiffness of the building.

The solution is in a convenient form for the preparation of design curves, and Fig. 8-2 shows the variation of the design function, F_1, with height, for different values of the relative stiffness parameter, α. The curves show graphically the influence of the frame stiffness on the horizontal deflection.

Once the deflection is known, the other force components follow from the force–displacement relationships. The values of the bending moment and shear force on the flexural cantilever, expressed in design form in terms of the maximum applied bending moment and shear force at the base of the structure, then become

$$M_b = \tfrac{1}{2}wH^2 \cdot F_2(z, \alpha)$$
$$S_b = wH \cdot F_3(z, \alpha)$$

in which the moment and shear functions are

$$F_2 = \frac{2}{\alpha^2} \left\{ \frac{\alpha \sinh \alpha + 1}{\cosh \alpha} \cosh \alpha z - \alpha \sinh \alpha z - 1 \right\}$$

$$F_3 = \frac{1}{\alpha} \left\{ \alpha \cosh \alpha z - \frac{\alpha \sinh \alpha + 1}{\cosh \alpha} \sinh \alpha z \right\}$$

These functions again indicate graphically the influence of the frame component on the force distribution in the wall element, and are shown in Figs 8-3 and 8-4.

The corresponding bending moment and shear force on the frame will be given

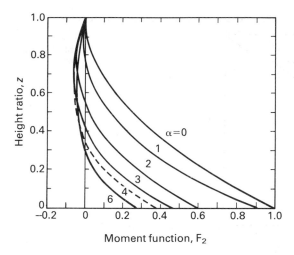

Figure 8-3 Variation of wall bending moment function

by

$$M_s = \tfrac{1}{2}wH^2(1-z)^2 - M_b$$
$$S_s = wH(1-z) - S_b$$

Solutions are given in [9] for the other two standard load cases of a triangularly distributed loading and a concentrated point load at the top, which may be used to produce corresponding sets of design curves. Solutions are also given for the interactive forces in the connecting links.

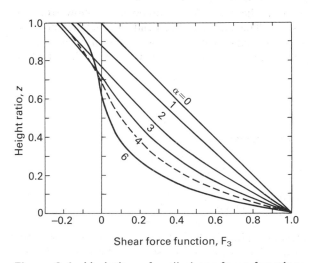

Figure 8-4 Variation of wall shear force function

9 Interaction between walls, coupled shear walls and frames

For residential buildings, whose structures consist of three-dimensional regular symmetric assemblies of coupled shear walls acting in conjunction with relatively stiff service cores, a similar solution to that obtained for wall–frame structures may readily be derived.

The two components will be constrained to act together by the set of floor slabs, which may be treated as an effective pin-ended connecting medium. By using the continuum approach to obtain the force–displacement relationships for the coupled shear walls, in conjunction with the overall conditions of equilibrium and horizontal displacement compatibility, it is found that the structural behaviour may again be expressed in the form of a governing fourth-order differential equation in the common horizontal deflection, or a second-order equation in one of the force components [19].

Closed-form solutions may be obtained for standard load cases, and design curves produced for the rapid assessment of forces and deflections. The curves are similar in form to those obtained for wall–frame structures.

If the symmetric structure is subjected to unsymmetric loading producing bending and torsion, the two may be considered independently. A solution for the latter is readily achieved using an equivalent planar structure by the reduction technique outlined in [7].

A similar approach yields simple closed-form solutions for structures consisting of coupled shear walls and frames [20], or a general combination of coupled shear walls, frames and cantilevered walls.

10 Core structures

A vertical core may serve alone in supporting the structure, as in a suspended building, or may act in conjunction with other walls or frames. If twisting is insignificant, the core can be analysed as a shear wall by the continuum approach. However, if from eccentric loading or structural asymmetry, the building is subjected to a torque, the core may be expected to play a useful role in resisting the twist. The commonly used E, H or U-shaped open section arrangement of the core walls, and the low thickness/width and width/height ratios of the walls cause the core to behave as a thin-walled beam, which will warp as well as bend. Since the base of the core is invariably restrained by a stiff foundation, axial and shear warping stresses will be induced. These may be of such magnitude as to merit consideration in the design.

The torsional stiffness of the core, and the magnitude of the warping stresses, will be modified by the bending actions of the lintel beams and floor slabs, which span the opening at each floor level and partially close the cross-section. With the presumption of a fixed position of a shear centre for the cross-section, a solution may be achieved using Vlasov's theory of thin-walled beams [21], or using the folded-plate approach [22], allied to the use of the continuous medium technique to define the restraining action of the lintel beams. However, as for the analysis of coupled shear walls, the method is limited to uniform walls with regular beam systems, with a limited number of discontinuities, subjected to standard loading patterns.

11   Framed-tube structures

As discussed earlier, a framed-tube is essentially a perforated box comprising four orthogonal frame panels of closely spaced columns connected by spandrel beams around the perimeter at each floor level. The outer tube is designed to resist all lateral forces.

In a preliminary approximate analysis, it is sometimes assumed that the side frames parallel to the wind carry all lateral loads, so that a plane frame analysis can be used. However, the frames normal to the wind direction are constrained by the floors to deform as flanges in the same mode as the side frames, and can play a significant part in resisting wind loads. In this case, these normal frames are subjected mainly to axial forces and the side frames are subjected to shearing actions. The primary action is complicated by the flexibility of the spandrel beams allowing a shear lag which increases the stresses in the corner columns, and reduces those in the inner columns of the normal frame. The major interactions between the two types of frame are the vertical shear forces at the corners. By recognizing the dominant modes of action of the panels, a more accurate assessment of the structural behaviour may again be achieved using a plane frame analysis. The side and normal frames may be considered to lie in the same plane, and are connected in series by fictitious linking members whose stiffnesses are appropriately chosen to allow only vertical forces to be transmitted between frames. This may be performed most conveniently by incorporating a hinge release in the fictitious members, or alternatively by assuming high shear stiffness and low (zero) bending stiffness values in the stiffness matrices for these shear transfer members. If the framed tube is symmetric in both coordinate directions, conditions of symmetry require only one-quarter of the structure to be analysed.

If a symmetric framed-tube structure is subjected to bending and twisting forces, the two may be considered separately. In the case of pure torsion, the stiff floors ensure that the cross-section shape is maintained at each level, so that the applied torque is resisted primarily by the shearing resistance of each plane frame around the periphery. The main interactive forces between the frame panels are again the vertical shear forces at the corners. Since the rotation of each frame relative to the centre of the building is the same, the shearing deformation in the plane of each frame is equal to the product of the rotation and the distance from the centre. Knowing the shear stiffness of each frame, and using overall conditions of compatibility and equilibrium, the horizontal forces on each frame and the vertical interaction forces at the corners can be established.

A simple method for the preliminary analysis is possible if each framework panel of columns and spandrel beams is replaced by an equivalent uniform orthotropic plate to form a substitute closed-tube structure. The properties of the orthotropic plate must be chosen so that the two elastic moduli in the horizontal and vertical directions represent the smeared axial stiffnesses of the beams and columns, respectively, and the shear modulus represents the shear stiffness of the framework. A suitable shearing rigidity, *GA*, which models the racking behaviour of the frame panel may be derived using the technique discussed in Section 5.

By making further simplifying assumptions regarding the stress distributions in both side and normal panels, the behaviour of the structure may be defined by a linear second-order governing differential equation [24]. Closed-form solutions are then possible, leading to design curves for standard applied load distributions causing both bending and torsion of the structure.

Preliminary calculations may also be performed using the influence curves presented by Khan and Amin [23]. The curves were based on a series of numerical studies performed using the equivalent plane frame technique described above.

11.1 Tube-in-tube structures

If the floors are effectively pin-connected to both outer tube and core, horizontal forces only will be transmitted under the action of wind forces. The approximate solution may be employed to derive flexibility influence coefficients for the outer tube, and hence, by writing the conditions of equilibrium and horizontal deflection compatibility at a number of specific levels, an assessment may be made of the distribution of forces between the two components.

12 Bending stiffness of floor slabs connecting shear walls

The continuum method of analysis has been developed for shear walls connected by lintel beams. The same results may be used for shear walls coupled by floor slabs provided that the equivalent width of the slab, which acts effectively as a wide coupling beam, or its corresponding bending stiffness, can be assessed. Unlike a beam, however, the coupling stresses are not uniform across the width of the slab, and, in order to achieve a safe design, the magnitude and distribution of the bending stresses developed in the slab must be determined. It is also necessary to estimate accurately the interactive shearing forces at the wall–slab junction, particularly near the inner edges of the walls, where high stress concentrations occur.

The resistance of a floor slab against the displacements imposed by the deforming shear walls is a measure of its coupling stiffness, which can be defined in terms of the displacements at its ends and the forces producing them. Thus, referring to Fig. 12-1a,b, the stiffness of the slab may be defined either as a rotational stiffness M/θ or as a translational stiffness N/δ, since the two are related.

Due to the non-uniform bending across the slab, the force–displacement relationship must be evaluated from a two-dimensional plate bending analysis. In view of the awkward boundary conditions involved, the most convenient technique for such an analysis is the finite element method. A system of rigid body displacements along the slab-wall junction of the form shown in Fig. 12-1a, or 12-1b, may be imposed, and the resulting nodal forces and moments determined, giving the overall moment–rotation or load–displacement relationship. This may

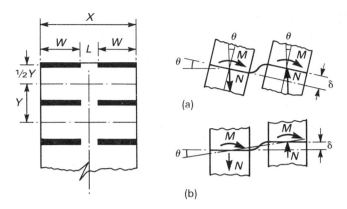

Figure 12-1 Structural actions of coupled shear wall–slab system subjected to lateral bending or differential vertical movement

be carried out for any wall cross-sectional shape, using an appropriate finite element mesh to give the desired level of accuracy.

For convenience, the rotational and translational slab stiffnesses are expressed in the form of non-dimensional stiffness factors K and K_δ, defined as

$$K = \frac{M}{\theta} \frac{1}{D} \quad \text{and} \quad K_\delta = \frac{NL^2}{\delta\, D}$$

where D is the flexural rigidity of the slab, and L is the clear opening between walls.

In order to relate to the analysis of walls coupled by beams, it is advantageous to assume that a strip of slab acts effectively as a beam in connecting the walls, enabling the slab stiffness to be defined simply in terms of the geometric and material characteristics of the equivalent beam. By equating the rotational and translational stiffnesses of the slab to those of the equivalent stiff-ended beam, assuming the beam remains rigid over the depth of the wall, the effective width, Y_e, may be expressed as,

$$\frac{M}{\theta} = \frac{6EI}{L^3}(L + W)^2$$

and

$$\frac{N}{\delta} = \frac{12EI}{L^3}$$

where W is the width of the wall, and $I\ (= \frac{1}{12}Y_e t^3)$ is the second moment of area of the beam, of width Y_e and depth t.

The effective width may then be expressed in terms of the stiffness factors K and K_δ as

$$\frac{Y_e}{Y} = \frac{K}{6(1-v^2)} \frac{L}{Y} \left(\frac{L}{L+W}\right)^2$$

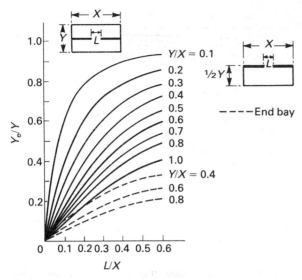

Figure 12-2   Influence of bay width and opening size on effective slab width for plane walls

or

$$\frac{Y_e}{Y} = \frac{K_\delta}{12(1-v^2)} \frac{L}{Y}$$

where Y is the bay width or longitudinal wall spacing (Fig. 12-1) and v is Poisson's ratio for the slab material, usually taken as 0.15 for concrete.

Based on this approach, Fig. 12-2 gives effective slab widths for a system of plane coupled walls, for a typical interior bay in a building with the slab continuous in each longitudinal direction. These results may be used directly, using interpolation, for design calculations. Numerical studies [25] have shown that the effects of variations in the absolute wall lengths in a pair of coupled walls may be disregarded if the results are presented in the non-dimensional form shown. In addition, the effect of dissimilar wall lengths in a pair of coupled walls can also be neglected provided that the ratio of the length of the shorter wall to that of the opening is greater than about 0.5. Corresponding values for an end bay, in which one edge is continuous and the other assumed free, are indicated by broken lines on Fig. 12-2. Since the effective width is then found to lie between 44 and 47% of that for the corresponding doubly-continuous slab, a convenient rule-of-thumb approach would be to use a constant value of 45%.

Corresponding curves of effective slab width for walls with flanges formed from interior corridor walls are shown in Fig. 12-3. These design curves are a comprehensive set derived from a series of individual curves used to examine the relative influence of the ratios L/X, Z/Y, and Y/X, in which X and Z are the overall depth of the building and the flange width, respectively. Since the regions of the slab behind the flanges are very lightly stressed, and play little part in the structural actions, the absolute wall length has a negligibly small influence on the

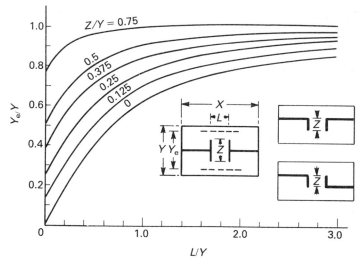

Figure 12-3 Influence of flange width and opening size on effective slab width for coupled flanged wall structures

effective width. The values given by the curves may be regarded as accurate provided the wall-slab proportions are such that $[L/X+(Y-Z)/X] \leqslant 1$, which is true for most practical situations. In unusual cases, where the ratio is greater than unity, and L/X is greater than 0.4, the effective widths tend to be overestimated, although the error is less than 10%.

Corresponding generalized design curves are presented in Fig. 12-4 for the case of a plane wall coupled to a T-shaped wall. The curves may be considered

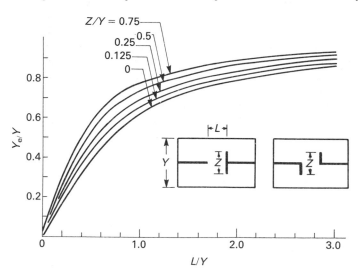

Figure 12-4 Influence of flange width and opening size on effective slab width for coupled planar-flanged wall structures

accurate for walls of normal proportions such that $(L/X + Y/X) \leqslant 1$. When the ratio exceeds unity, and L/X is greater than 0.4, the values again tend to be overestimated, although the errors are less than 10%. It may be observed that the omission of one flange from a wall configuration results in a disproportionately large reduction in the effective width when the ratio L/Y is small.

Studies [25] have shown that any external flanges may be neglected since they contribute a negligible increase in the slab stiffness.

Coupled L-shaped walls may occur in different configurations. For configurations in which the two flanges are directly opposite, regardless of the alignment of the two webs, the effective width may be obtained from the design curves for T-shaped walls (Fig. 12-3) using the same flange width. In the third possible configuration, in which the webs are coupled in line, but the flanges are offset, or disposed skew-symmetrically, numerical studies [25] have shown that the effective slab width for the coupled L-shaped walls is given sufficiently accurately by the curves for the planar-flanged wall configurations (Fig. 12-4) provided that the flange width used is the sum of the two flange widths for the L-shaped walls. Although the flanges are not directly cross-coupled, they still have a considerable influence on the slab stiffness as a result of their restraining action.

The design curves have been established for walls of zero (line) thickness, which is generally sufficiently accurate for most practical configurations. In certain circumstances, however, particularly when the opening is small, the influence of the finite thickness of the wall may become significant [25]. The data have been derived on the assumption of a rigid joint between floor slab and shear wall. In practice, cracking under the influence of the high local stresses may have a considerable influence on the slab stiffness [26].

In a similar manner, flat plate structures subjected to lateral loads may be represented by rigidly jointed plane frameworks provided the floor slabs may be replaced by connecting beams of equivalent stiffness. Design data for the effective coupling stiffness of floor slabs connecting rectangular columns have been presented in [27].

13 Creep, shrinkage and temperature effects

Neglect of the differential movements which occur in tall buildings due to creep, shrinkage, and temperature can lead to distress in non-load-bearing partitions as well as overstressing of the horizontal elements. The effects of the time-dependent volume changes and temperature changes are cumulative, and increase with the height and length of the building. The creep of the vertical members is influenced particularly by the amount of reinforcing steel, as well as by the size, while the influence of relative thermal movement in exposed columns is most pronounced in regions of high diurnal and annual climatic changes. A detailed consideration of such problems is beyond the scope of this short chapter, and the reader is referred to the comprehensive report in [28], and to other chapters in the *Handbook*, e.g., Chapters 4, 11, 38 and 40.

14 Conclusions

Developments in the analysis and design of tall buildings have been a major growth area in the past two decades, and the published literature in the field is now very extensive. In a short chapter, it has not proved possible to do other than discuss briefly the developments in structural form which have taken place, and to give guidance on the approximate techniques which have been devised for their analysis.

A comprehensive discussion of all aspects of the design of high-rise concrete buildings has been given in [28].

References

1. Khan, F. R., Current trends in concrete high-rise buildings, *Tall buildings*, Pergamon Press, Oxford, 1967, pp. 571–590.
2. Stafford Smith, B. and Riddington, J. R., Design of masonry infilled frames for bracing structures, *Struct. Eng.*, Vol. 56B, No. 1, 1978, pp. 1–7.
3. Fintel, M., Staggered transverse wall beams for multi-story concrete buildings, *J. Am. Concr. Inst.*, *Proc.*, Vol. 65, No. 5, May 1968, pp. 366–378.
4. Frischmann, W. W. and Prabhu, S. S., Planning concepts using shear walls, *Tall buildings*, Pergamon Press, Oxford, 1967, pp. 49–79.
5. Khan, F. R., Structural design of tall buildings, *Proc. nat. conf. tall bldgs*, New Delhi, India, 1973, pp. 106–128.
6. American Society of Civil Engineers, *Tall building criteria and loading*, Monograph on planning and design of tall buildings, Vol. CL, New York, 1980, pp. 888.
7. Coull, A. and Stafford Smith, B., Torsion analysis of symmetric building structures, *J. Struct. Div. Am. Soc. Civ. Eng.*, Vol. 99, No. ST1, Jan 1973, pp. 229–233.
8. Chan, P. C. K., Heidebrecht, A. C. and Tso, W. K., Approximate analysis of multi-storey multibay frames, *J. Struct. Div. Am. Soc. Civ. Eng.*, Vol. 101, No. ST5, May 1975, pp. 1021–1035.
9. Heidebrecht, A. C. and Stafford Smith, B., Approximate analysis of tall wall-frame structures, *J. Struct. Div. Am. Soc. Civ. Eng.*, Vol. 99, No. ST2, Feb. 1973, pp. 199–221.
10. Coull, A., Stiffening of linked shear walls, *Proc. Inst. Civ. Eng.*, Vol. 53, Part 2, Dec. 1972, pp. 579–583.
11. Coull, A. and Mukherjee, P. R., Coupled shear walls with general support conditions, *Proc. conf. tall bldgs*, Kuala Lumpur, 1974, pp. 4.24–4.31.
12. Coull, A. and Irwin, A. W., Analysis of load distribution in multi-story shear wall structures, *Struct. Eng.*, Vol. 48, No. 8, Aug 1970, pp. 301–306.
13. Coull, A. and Adams, N. W., Analysis of the load distribution in multi-story shear wall structures, *Response of multistory concrete structures to lateral forces*, American Concrete Institute, SP-36, 1973, pp. 187–216.
14. Coull, A. and Choudhury J. R., Stresses and deflections in coupled shear walls, *J. Am. Concr. Inst.*, *Proc.*, Vol. 64, No. 2, Feb. 1967, pp. 65–72.
15. Coull, A. and Choudhury, J. R., Analysis of coupled shear walls, *J. Am. Concr. Inst.*, Vol. 64, No. 9, Sept. 1967, pp. 587–593.
16. Gluck, J., Elasto-plastic analysis of coupled shear walls, *J. Struct. Div. Am. Soc. Civ. Eng.*, Vol. 99, No. ST8, Aug. 1973, pp. 1743–1760.
17. Tso, W. K. and Chan, H. B., Dynamic analysis of plane coupled shear walls, *J. Eng. Mech. Div.*, *Am. Soc. Civ. Eng.*, Vol. 97, No. EM1, Feb. 1971, pp. 33–48.

18. Rosman, R., Analysis of spatial concrete shear wall systems, *Proc. Inst. Civ. Eng.*, 1970 Supplement, Paper 7266S, pp. 131–152.

19. Coull, A., Interactions between coupled shear walls and cantilevered cores in three-dimensional regular symmetrical cross-wall structures, *Proc. Inst. Civ. Eng.*, Vol. 55, Part 2, Dec. 1973, pp. 827–840.

20. Arvidsson, K., Interaction between coupled shear walls and frames, *Proc. Inst. Civ. Eng.*, Vol. 67, Part 2, Sept. 1979, pp. 589–596.

21. Rosman, R., Torsion of perforated concrete shafts, *J. Struct. Div. Am. Soc. Civ. Eng.*, Vol. 95, No. ST5, May 1969, pp. 991–1010.

22. Kanchi, M. B. and Dar, G. Q., Torsion of rectangular shear cores of tall buildings, *Bull. Ind. Soc. Earthquake Tech.*, Vol. 10, 1973, pp. 100–112.

23. Khan, F. R. and Amin, N. R., Analysis and design of framed-tube structures for tall concrete buildings, *Response of multistory concrete structures to lateral forces*, American Concrete Institute, SP-36, 1973, pp. 39–49.

24. Coull, A. and Bose, B., Simplified analysis of framed-tube structures, *J. Struct., Div. Am. Soc. Civ. Eng.*, Vol. 101, No. ST11, Nov. 1975, pp. 2223–2240.

25. Wong, Y. C., *Interaction between floor slabs and shear walls in tall buildings*, PhD Thesis, University of Strathclyde, 1978.

26. Schwaighofer, J. and Collins, M. P., Experimental study of the behaviour of reinforced concrete coupling slabs, *J. Am. Concr. Inst., Proc.*, Vol. 74, No. 3, Mar. 1977, pp. 123–127.

27. Wong, Y. C. and Coull, A., Effective slab stiffness in flat plate structures, *Proc. Inst. Civ. Eng.*, Vol. 69, Part 2, Sept. 1980, pp. 721–735.

28. American Society of Civil Engineers, *Structural design of tall concrete and masonry buildings*, Monograph on planning and design of tall buildings, Vol. CB, New York, 1978, 938 pp.

38 Tall buildings 2

Y K Cheung

University of Hong Kong, Hong Kong

Contents

Notation

A	cross-sectional area of member	J	torsional rigidity
E	modulus of elasticity	L	clear opening between walls
G	shear modulus	u,v,w	displacement along x, y, z-axes
h	story height	v	Poisson's ratio
I	second moment of area	$\theta_x,\theta_y,\theta_z$	rotation around x, y, z-axes
		ρt	mass per unit volume
		ω	natural frequency

Summary

In this chapter, the various computer methods (including the frame method, the finite element method and the finite strip method) used for the analysis of tall buildings are discussed in detail. Numerical examples for different types of structures (including shear walls, shear cores, plane frames and frame-shear wall structures) are presented and wherever possible, comparison of the accuracy of results due to different methods is made. Special structural features such as local deformation at beam-wall junction, rigidity of floor slabs, difference in behaviour between bending mode of shear wall and shear mode of frame have all been incorporated into the analysis.

While most of the contents are connected with static analysis, some efforts have been made to introduce readers to the dynamic analysis of tall buildings. However, due to space limitation, only free vibration problems have been dealt with.

1 Method of analysis

In the previous chapter, the characteristics of tall building structures have been discussed and various approximate methods presented. In this chapter, the powerful computer methods of analysis will be introduced. Such methods will be able to deal with complex problems involving coupled shear walls, frame-shear walls, shear wall on portal frame or on elastic foundation, etc.

The frame method [1, 2] was originally developed for the analysis of skeletal frame structures, using the standard frame member stiffness matrix. However, with the incorporation of a transformation procedure to incorporate the effect of the rigid ends which account for the finite dimension of the much stiffer shear walls, a modified frame program can also be used for the analysis of both frame and coupled shear wall systems. The frame method is simple and effective for frame structures, although the problem usually becomes very large for spatial structures and the solution is very often intractable or uneconomic.

The finite element method [3], being the most powerful tool of analysis available, can be applied to the analysis and design of any type of tall building structure with different combinations of loadings and for various support conditions. The standard procedure for the method is to divide the frames, shear walls, spandrel beams, and floor slabs into a large number of small elements, using different elements for the different structural components.

Usually, more than one lower-order elements [4] are used to represent each shear wall between two floor levels, and the number of nodes and elements required to model a structure is therefore very large, and consequently the resulting computer analysis could be very costly. It is for this reason that the standard finite element method is still not popular among tall-building design engineers.

Recently, higher-order elements [5] have been used to model shear walls. Since only one to two elements are required for the whole height of the tall wall structure, the method obviously has great potential and is capable of further development.

The finite strip method [6] has been successfully applied to the static, dynamic and stability [7, 8, 9] analyses of tall building structures and has proved to be a versatile and economic tool. In this method, a shear wall is modelled as an assemblage of strip elements and a frame-shear wall structure as an assemblage of strip and line elements, with each element always stretching from the base to the top of the building. Using such an approach, the number of degrees of freedom for a building becomes, to a large extent, independent of the height.

1.1 Frame method: static

1.1.1 *Structural idealization and matrix analysis procedure*

Consider a plane coupled shear wall as shown in Fig. 1-1. The structure can be idealized as a frame in which the walls and beams are represented by the line elements along their centroidal axis. A wall element between two consecutive floors is usually represented by a frame element (Fig. 1-2), the stiffness matrix of

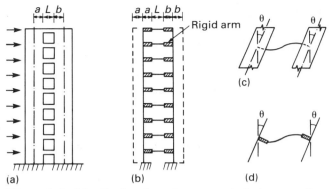

Figure 1-1   Idealization for plane coupled shear wall
(a) Plane coupled shear wall
(b) Frame idealization
(c) Actual structure
(d) Analogous structure

which can be found in a number of texts [10] and is given below.

$$[S_w] = \begin{bmatrix} u_1 & v_1 & \theta_1 & u_2 & v_2 & \theta_2 \\ \dfrac{EA}{h} & & & & & \\ 0 & \dfrac{12EI}{(1+\alpha)h^3} & & SYM & & \\ 0 & \dfrac{6EI}{(1+\alpha)h^2} & \dfrac{(4+\alpha)EI}{(1+\alpha)h} & & & \\ -\dfrac{EA}{h} & 0 & 0 & \dfrac{EA}{h} & & \\ 0 & \dfrac{-12EI}{(1+\alpha)h^3} & \dfrac{-6EI}{(1+\alpha)h^2} & 0 & \dfrac{12EI}{(1+\alpha)h^3} & \\ 0 & \dfrac{6EI}{(1+\alpha)h^2} & \dfrac{(2-\alpha)EI}{(1+\alpha)h} & 0 & \dfrac{-6EI}{(1+\alpha)h^2} & \dfrac{(4+\alpha)EI}{(1+\alpha)h} \end{bmatrix}$$

(1-1)

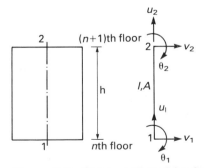

Figure 1-2   Shear wall and its frame member model

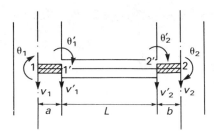

Figure 1-3   Coordinate systems and nodal parameters for a connecting beam

The term $\alpha = 12EIk/h^2GA$ deals with shear deformation effects and, by putting it equal to zero, Eqn 1-1 becomes identical with the standard stiffness matrix. k is the shape factor, and is equal to 1.2 and 1.11 for a rectangular and a circular section respectively.

On the other hand, the beams are usually assumed to be axially rigid and therefore only four degrees of freedom are associated with each beam element (Fig. 1-3). In addition, the presence of the rigid arm due to the finite dimension of the wall implies that the node of a beam does not coincide with the node of a connecting wall element, and therefore a coordinate transformation has to be carried out. Referring to the nodal displacement parameters of the beam element and wall element as $\{D'\}$ and $\{D\}$ respectively, and given that the beam stiffness matrix is

$$[S'_b] = \begin{bmatrix} \dfrac{12}{L^3} & & SYM & \\[2mm] \dfrac{6}{L^2} & \dfrac{(4+\alpha)}{L} & & \\[2mm] -\dfrac{12}{L^3} & -\dfrac{6}{L^2} & \dfrac{12}{L^3} & \\[2mm] \dfrac{6}{L^2} & \dfrac{(2-\alpha)}{L} & -\dfrac{6}{L^2} & \dfrac{(4+\alpha)}{L} \end{bmatrix} \dfrac{EI}{(1+\alpha)} \qquad (1\text{-}2)$$

it is now possible to write

$$\{D'\} = [H]\{D\} \qquad (1\text{-}3a)$$

where

$$[H] = \begin{bmatrix} 1 & a & 0 & 0 \\ 0 & 1 & 0 & 0 \\ 0 & 0 & 1 & -b \\ 0 & 0 & 0 & 1 \end{bmatrix} \qquad (1\text{-}3b)$$

It follows that the stiffness matrix of a beam with rigid arms can be given by

$$[S_b]=[H]^T[S_b'][H]$$

$$
\begin{matrix}
v_1 & \theta_1 & v_2 & \theta_2
\end{matrix}
$$

$$
= \begin{bmatrix}
\dfrac{12}{L^3} & & & & \\[2mm]
\dfrac{6}{L^2}+\dfrac{12a}{L^3} & \dfrac{4+\alpha}{L}+\dfrac{12a}{L^2}+\dfrac{12a^2}{L^3} & & SYM & \\[2mm]
-\dfrac{12}{L^3} & \dfrac{-6}{L^2}-\dfrac{12a}{L^3} & \dfrac{12}{L^3} & & \\[2mm]
\dfrac{6}{L^2}+\dfrac{12b}{L^3} & \dfrac{2-\alpha}{L}+\dfrac{6(a+b)}{L^2}+\dfrac{12ab}{L^3} & \dfrac{-6}{L^2}-\dfrac{12b}{L^3} & \dfrac{4+\alpha}{L}+\dfrac{12b}{L^2}+\dfrac{12b^2}{L^3}
\end{bmatrix} \dfrac{EI}{(1+\alpha)}
$$

$$(1\text{-}4)$$

The beam and wall stiffness matrices are assembled in the standard manner and solved for the displacements, which in turn can be used to yield the nodal forces of all the members.

1.1.2 *Effect of local deformation at beam–wall junction*

In the frame method, the beams are assumed to be connected perpendicular to the axis of each wall. However, it has been demonstrated by Michael [11] that a small local rotation exists at the connecting point due to the significant difference between the depth of the beam and wall members, and that the effective stiffness of a beam is thereby reduced. Such flexibility effects can be approximately accounted for by assuming the beam span to be $(L+\frac{1}{2}d_1+\frac{1}{2}d_2)$, with d_1 and d_2 being the beam depths at the two ends. At the same time, the length of the rigid arm should be correspondingly reduced to keep the total distance between the centre lines of the two columns constant. It should be noted that an assumption in the analysis was that the half wall width was at least three and a half times the beam depth, and therefore, for smaller wall depths, a different correction factor should be used.

Using triangular finite elements, Bhatt [12] computed the top deflections of a cantilever beam of any depth/span ratio under a point end load. In this way, the bending and shear deformations, as well as the local rotation for a finite shear wall width, have all been taken into account. By comparing the computed deflection and that due to the original cantilever deflection formula (bending only), a factor β can be determined such that the effective span of any spandrel beam can be taken as βl. Note once more that shear deformation has already been taken into account in the computation of β.

Table 1-1 shows the values of β for different span/beam depth and wall width/beam depth ratios. All the results were computed from a fine mesh of 8-node isoparametric elements.

1.1.3 *Three dimensional analysis*

For smaller cores which can be treated as a thin walled frame member (Fig. 1-4) and by neglecting warping effects, the stiffness matrix [13] presented below should be used.

Table 1-1 Values of β for computing equivalent beam span, βl

		Half of wall width, $\frac{1}{2}b$						
		3.50	3.00	2.50	2.00	1.50	1.00	0.50
Cantilever span, $\frac{l}{2}$ (Half of beam clear span)	4.00	1.00	1.10	1.10	1.10	1.10	1.09	1.07
	3.50	1.12	1.12	1.12	1.12	1.11	1.11	1.09
	3.00	1.14	1.14	1.14	1.14	1.14	1.13	1.11
	2.50	1.17	1.17	1.17	1.17	1.17	1.16	1.13
	2.00	1.22	1.22	1.22	1.22	1.22	1.20	1.17
	1.50	1.32	1.32	1.32	1.31	1.30	1.29	1.24
	1.00	1.52	1.52	1.51	1.51	1.49	1.47	1.41
	0.90	1.59	1.59	1.58	1.58	1.56	1.53	1.46
	0.80	1.68	1.68	1.68	1.67	1.65	1.62	1.54
	0.70	1.80	1.80	1.79	1.79	1.76	1.72	1.64
	0.60	1.97	1.96	1.96	1.95	1.92	1.87	1.77
	0.50	2.20	2.20	2.19	2.18	2.14	2.09	1.97

When using Eqn 1-5, the transformations due to the following factors might have to be carried out:

(a) The directions of the principal axes of the section may be different from those of the global axes.
(b) The shear centre (for the lateral forces) and the centre of gravity (for the axial force) may be situated at different points.

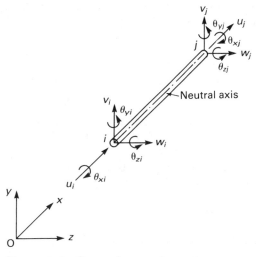

Figure 1-4 Space frame element

$$
[S]=
\begin{array}{cccccccccccc}
u_i & v_i & w_i & \theta_{xi} & \theta_{yi} & \theta_{zi} & u_j & v_j & w_j & \theta_{xj} & \theta_{yj} & \theta_{zj}
\end{array}
$$

$$
[S]=
\begin{bmatrix}
\dfrac{EA}{l} \\[2ex]
0 & \dfrac{12EI_z}{l^3(1+\alpha_y)} \\[2ex]
0 & 0 & \dfrac{12EI_y}{l^3(1+\alpha_z)} \\[2ex]
0 & 0 & 0 & \dfrac{GJ}{l} \\[2ex]
0 & 0 & \dfrac{-6EI_y}{l^2(1+\alpha_z)} & 0 & \dfrac{(4+\alpha_z)EI_y}{l(1+\alpha_z)} \\[2ex]
0 & \dfrac{6EI_z}{l^2(1+\alpha_y)} & 0 & 0 & 0 & \dfrac{(4+\alpha_y)EI_z}{l(1+\alpha_y)} \\[2ex]
\dfrac{-EA}{l} & 0 & 0 & 0 & 0 & 0 & \dfrac{AE}{l} \\[2ex]
0 & \dfrac{-12EI_z}{l^3(1+\alpha_y)} & 0 & 0 & 0 & \dfrac{-6EI_z}{l^2(1+\alpha_y)} & 0 & \dfrac{12EI_z}{l^3(1+\alpha_y)} \\[2ex]
0 & 0 & \dfrac{-12EI_y}{l^3(1+\alpha_z)} & 0 & \dfrac{6EI_y}{l^2(1+\alpha_z)} & 0 & 0 & 0 & \dfrac{12EI_y}{l^3(1+\alpha_z)} \\[2ex]
0 & 0 & 0 & \dfrac{-GJ}{l} & 0 & 0 & 0 & 0 & 0 & \dfrac{GJ}{l} \\[2ex]
0 & 0 & \dfrac{-6EI_y}{l^2(1+\alpha_z)} & 0 & \dfrac{(2-\alpha_z)EI_y}{l(1+\alpha_z)} & 0 & 0 & 0 & \dfrac{6EI_y}{l^2(1+\alpha_z)} & 0 & \dfrac{(4+\alpha_z)EI_y}{l(1+\alpha_z)} \\[2ex]
0 & \dfrac{6EI_z}{l^2(1+\alpha_y)} & 0 & 0 & 0 & \dfrac{(2-\alpha_y)EI_z}{l(1+\alpha_y)} & 0 & \dfrac{-6EI_z}{l^2(1+\alpha_y)} & 0 & 0 & 0 & \dfrac{(4+\alpha_y)EI_z}{l(1+\alpha_y)}
\end{bmatrix}
$$

SYM

(1-5)

In addition, the standard transformation from local coordinate system to global coordinate system must be carried out.

1.2 Finite element method: static

1.2.1 *Finite element procedure*

The finite element method can be considered as a generalized displacement method for two and three-dimensional continuum problems; indeed, the frame member discussed previously can be easily incorporated into the analysis as a one-dimensional member. The finite element procedure for the analysis of tall building structures can be summarized as follows:

(a) The frame is divided into one-dimensional finite elements in which the joints are treated as nodes and the standard frame stiffness matrix is used for each member.

(b) The shear wall (or the floor slab if bending of the floor is to be taken into account) is divided into a number of rectangular elements by fictitious lines.

(c) The elements are assumed to be interconnected at a discrete number of points situated on the element boundaries. The nodal displacement parameters usually include displacements and their slopes for the plane stress elements, and displacements, slopes, and curvatures for the bending elements.

(d) A stiffness matrix is derived from virtual work or other variational principles after displacement functions have been chosen to approximate the actual displacement fields inside the element.

(e) All the stiffness matrices are assembled and the equations are solved in the usual manner.

1.2.2 *Structural idealization*

The shear wall is divided into a number of elements such that there will be one to two elements per story height. For general core structures, the flat shell element, which is simply made up of a plane stress element and a bending element, is used in the analysis, although it is obvious that the bending component will be omitted for the case of plane shear walls. In general, the lower order bilinear plane stress element without rotational degrees of freedom should not be used because of its excess stiffness and also because of the difficulty in coupling up with the spandrel beam elements. For higher order elements, however, the use of the rotational degrees of freedom is no longer essential. Since each element will now represent the shear wall occupying many stories, the spandrel beams are either modelled by an equivalent continuum or alternatively their stiffness matrices are transformed to the higher-order element coordinate system through compatibility conditions.

1.2.3 *Lower-order plane stress elements*

1.2.3.1 *MacLeod's rectangular element* [4] In MacLeod's rectangular element (Fig. 1-5a), θ_{zi} is assumed equal to $-\partial u/\partial y$ or $\partial v/\partial x$ at alternate nodes. The displacement, u, is assumed linear in x and quadratic in y, whereas v is assumed quadratic in x and linear in y.

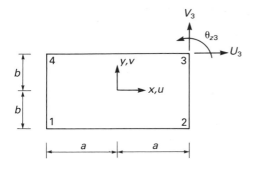

(a) MacLeod's element   (b) Abu-Ghazaleh's element

$$\theta_{z1} = \theta_{z3} = \frac{-\partial u}{\partial y}$$
$$\theta_{zi} = \frac{1}{2}\frac{\partial v}{\partial x} - \frac{\partial u}{\partial y}_i$$

$$\theta_{z2} = \theta_{z4} = \frac{\partial v}{\partial x}$$

or   (c) Sisodiya and Cheung's element

$$\theta_{z1} = \theta_{z3} = \frac{\partial v}{\partial x}$$
$$\theta_{zi} = -\frac{\partial u}{\partial y}_i$$

$$\theta_{z2} = \theta_{z4} = \frac{-\partial u}{\partial y}$$

Figure 1-5   Lower-order rectangular elements

1.2.3.2   *Abu-Ghazaleh's rectangular element* [14]   In this case (Fig. 1-5b)

$$\theta_{zi} = \frac{1}{2}\left(\frac{\partial v}{\partial x} - \frac{\partial u}{\partial y}\right)_i$$

The displacement u is linear in x and cubic in y. However, the shape functions do not satisfy the constant strain criterion [15].

1.2.3.3   *Rectangular element by Sisodiya and Cheung* [16]   This element (Fig. 1-5c) has been referred to as beam-type element and is therefore more suitable for the analysis of shear walls when compared with the previously mentioned elements. The nodal parameters are u_i, v_i and $\theta_{zi} = -(\partial u/\partial y)_i$, and the displacement u is linear in the x-direction and cubic in the y-direction, while v is linear in both directions. Thus, u would correspond to the horizontal deflections of the shear wall while v would yield a linear distribution of vertical stresses on its horizontal sections. A detailed description of this element will be given in the following sections.

(a) Displacement functions. The displacement field is given by

$$\begin{Bmatrix} u \\ v \end{Bmatrix} = \sum_{i=1}^{4} \begin{bmatrix} f_2 & 0 & f_3 \\ 0 & f_1 & 0 \end{bmatrix}_i \begin{Bmatrix} u_i \\ v_i \\ -\left(\frac{\partial u}{\partial y}\right)_i \end{Bmatrix}$$

$$= \sum_{i=1}^{4} [N]_i \{\delta\}_i \qquad\qquad\qquad (1\text{-}6)$$

where

$$f_{1i} = \tfrac{1}{4}(1+\xi\xi_i)(1+\eta\eta_i)$$
$$f_{2i} = \tfrac{1}{2}(2+\eta\eta_i-\eta^2)f_{1i}$$
$$f_{3i} = \tfrac{1}{4}\eta_i(2b)(1-\eta^2)f_{1i}$$

and

$$\xi = x/a \qquad \eta = y/b$$

ξ_i, η_i are simply the values of ξ and η at node i, and they always take up the value of ±1.

(b) Strain–displacement relationship

$$\{\varepsilon\} = \left\{ \begin{matrix} \varepsilon_x \\ \varepsilon_y \\ \gamma_{xy} \end{matrix} \right\} = \sum_{i=1}^{4} [B]_i\{\delta\}_i \tag{1-7}$$

$$[B]_i = \begin{bmatrix} \dfrac{\partial f_2}{\partial x} & 0 & \dfrac{\partial f_3}{\partial x} \\[2mm] 0 & \dfrac{\partial f_1}{\partial y} & 0 \\[2mm] \dfrac{\partial f_2}{\partial y} & \dfrac{\partial f_1}{\partial x} & \dfrac{\partial f_3}{\partial y} \end{bmatrix}_i \tag{1-8}$$

(c) Stress–strain relationship

$$\{\sigma\} = \left\{ \begin{matrix} \sigma_x \\ \sigma_y \\ \tau_{xy} \end{matrix} \right\} = \sum_{i=1}^{4} [D][B]_i\{\delta\}_i \tag{1-9}$$

$$[D] = \frac{Et}{1-\nu^2} \begin{bmatrix} 1 & \nu & 0 \\ \nu & 1 & 0 \\ 0 & 0 & \dfrac{1-\nu}{2} \end{bmatrix} \tag{1-10}$$

(d) Stiffness matrix

$$[S^p]_{ij} = \int_{-a}^{+a}\int_{-b}^{+b} [B]_i^T[D][B]_j \, dx \, dy \tag{1-11}$$

$$= \frac{Eab}{1-\nu^2} \begin{bmatrix} (S_{11})_{ij} & (S_{12})_{ij} & (S_{13})_{ij} \\ (S_{21})_{ij} & (S_{22})_{ij} & (S_{23})_{ij} \\ (S_{31})_{ij} & (S_{32})_{ij} & (S_{33})_{ij} \end{bmatrix}$$

where

$$(S_{11})_{ij} = \frac{\xi_{ij}}{a^2}\left((1+\eta_{ij})\frac{13}{70} + (1-\eta_{ij})\frac{9}{140}\right) + \frac{1-\nu}{2}\frac{\eta_{ij}}{b^2}\left(\frac{1+\xi_{ij}}{5} + \frac{1-\xi_{ij}}{10}\right)$$

$$(S_{21})_{ij} = \nu\frac{\xi_i\eta_i}{4ab} + \frac{1-\nu}{2}\frac{\xi_i\eta_j}{4ab}$$

$$(S_{31})_{ij} = \frac{\xi_{ij}\eta_i b}{a^2}\left((1+\eta_{ij})\frac{11}{210} + (1-\eta_{ij})\frac{13}{420}\right) + \frac{1-\nu}{2}\frac{\eta_j}{b}\left(\frac{1+\xi_{ij}}{30} + \frac{1-\xi_{ij}}{60}\right)$$

$$(S_{22})_{ij} = \frac{\eta_{ij}}{b^2}\left(\frac{1+\xi_{ij}}{6} + \frac{1-\xi_{ij}}{12}\right) + \frac{1-\nu}{2}\frac{\xi_{ij}}{a^2}\left(\frac{1+\eta_{ij}}{6} + \frac{1-\eta_{ij}}{12}\right)$$

$$(S_{32})_{ij} = \nu\frac{\xi_i\eta_{ij}}{12a} - \frac{(1-\nu)}{2}\frac{\xi_j\eta_{ij}}{12a}$$

$$(S_{33})_{ij} = \xi_{ij}\eta_{ij}\frac{b^2}{a^2}\left((1+\eta_{ij})\frac{2}{105} + \frac{1-\eta_{ij}}{70}\right)$$

$$+ \frac{1-\nu}{2}\left(\frac{1+\xi_{ij}}{3} + \frac{1-\xi_{ij}}{6}\right)\left((1+\eta_{ij})\frac{2}{15} - \frac{1-\eta_{ij}}{30}\right)$$

$$(S_{12})_{ij} = (S_{21})_{ji}$$

$$(S_{13})_{ij} = (S_{31})_{ji}$$

$$(S_{23})_{ij} = (S_{32})_{ji}$$

and

$$\xi_{ij} = \xi_i\xi_j \qquad \eta_{ij} = \eta_i\eta_j$$

1.2.4 *Lower-order bending rectangular element*

The rectangular bending element shown in Fig. 1-6 with three degrees of freedom (one deflection and two rotations) at each node is one of the simplest bending elements. It has been used extensively in plate analysis and the results are in general quite satisfactory [17].

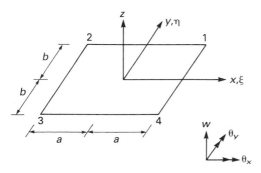

Figure 1-6   Rectangular bending element

Explicit form of $[D][B]$ in Eqn 1.14a

$$\frac{K}{4ab}\begin{bmatrix}
6p^{-1}+6vp & 8va & -8b & -6vp & 4va & 0 & -6p^{-1} & 0 & 0 & 0 & 0 & 0 \\
6p+6vp^{-1} & 8a & -8vb & -6p & 4a & 0 & -6vp^{-1} & 0 & 0 & 0 & 0 & 0 \\
-(1-v) & -2(1-v)b & 2(1-v)a & (1-v) & 0 & -2(1-v)a & (1-v) & 0 & -(1-v) & 0 & 0 & 0 \\
-6vp & -4va & 0 & 6p^{-1}+6vp & -8va & -8b & 0 & 8va & 8b & -6p^{-1} & 4va & -4b \\
-6p & -4a & 0 & 6p+6vp^{-1} & -8a & -8vb & 0 & 8a & 8vb & -6vp^{-1} & 4a & -4vb \\
-(1-v) & 0 & 2(1-v)a & (1-v) & -2(1-v)b & -2(1-v)a & 0 & 2(1-v)b & 2(1-v)a & -(1-v) & 0 & 0 \\
-6p^{-1} & 0 & 4b & 0 & 0 & 4b & 6p^{-1}+6vp & 8va & 8b & -6vp & -8va & 8b \\
-6vp^{-1} & 0 & 4vb & 0 & 0 & 4vb & 6p+6vp^{-1} & 8a & 8vb & -6p & -8a & 8vb \\
-(1-v) & 0 & 0 & 0 & -2(1-v)b & 0 & (1-v) & 2(1-v)b & 2(1-v)a & -(1-v) & 2(1-v)b & -2(1-v)a
\end{bmatrix}$$

where $p = a/b$

38–12

$$[S^*]=$$

Lower‑triangular (symmetric) matrix, read by columns (C1 … C12); entry (i,j) shown for $i \ge j$:

	C1	C2	C3	C4	C5	C6	C7	C8	C9	C10	C11	C12
r1	$60p^{-2}+60p^2-12\nu+42$											
r2	$30p^2+12\nu+3$	$20p^2-4\nu+4$										
r3	$-(30p^{-2}+12\nu+3)$	-15ν	$20p^{-2}-4\nu+4$									
r4	$30p^2-3\nu+3$	$-30p^2+3\nu-3$	$-15p^{-2}+12\nu+3$	$60p^{-2}+60p^2-12\nu+42$								
r5	$-15p^{-2}+12\nu+3$	$10p^2+\nu-1$	0	$-(30p^2+12\nu+3)$	$20p^2-4\nu+4$							
r6	$-60p^{-2}+30p^2+12\nu-42$	0	$10p^{-2}+4\nu-4$	$-(30p^{-2}+12\nu+3)$	-15ν	$20p^{-2}-4\nu+4$						
r7	$15p^2-12\nu-3$	$15p^2-12\nu-3$	$30p^{-2}-3\nu+3$	$-30p^2+3\nu-3$	$15p^2+3\nu-3$	$30p^{-2}-3\nu+3$	$60p^{-2}+60p^2-12\nu+42$					
r8	$-30p^2+3\nu-3$	$10p^2+4\nu-4$	0	$15p^{-2}+3\nu-3$	$10p^2+4\nu-4$	0	$30p^2+12\nu+3$	$20p^2-4\nu+4$				
r9	$-30p^{-2}-30p^2-12\nu+42$	0	$10p^{-2}+\nu-1$	$-30p^{-2}-30p^2-12\nu+42$	0	$10p^{-2}+\nu-1$	$-(30p^{-2}+12\nu+3)$	-15ν	$20p^{-2}-4\nu+4$			
r10	$15p^2+3\nu-3$	$-15p^2-3\nu+3$	$15p^{-2}+3\nu-3$	$-15p^2-3\nu+3$	$-15p^2-3\nu+3$	$15p^{-2}+3\nu-3$	$30p^2-3\nu+3$	$-30p^2+3\nu-3$	$-15p^{-2}+12\nu+3$	$60p^{-2}+60p^2-12\nu+42$		
r11	$-15p^{-2}-3\nu+3$	$5p^2-\nu+1$	0	0	$5p^2-\nu+1$	0	$-15p^{-2}+12\nu+3$	$10p^2+\nu-1$	0	$-(30p^2+12\nu+3)$	$20p^2-4\nu+4$	
r12		0	$5p^{-2}-\nu+1$	$-60p^{-2}+30p^2+12\nu-42$	0	$5p^{-2}-\nu+1$	$-60p^{-2}+30p^2+12\nu-42$	0	$10p^{-2}+4\nu-4$	$-(30p^{-2}+12\nu+3)$	-15ν	$20p^{-2}-4\nu+4$

Eqn 1-15c

(a) Displacement functions

$$w = \sum_{i=1}^{4} [N]_i \{\delta\}_i \qquad\qquad (1\text{-}12a)$$

in which

$$[N]_i = \tfrac{1}{8}[(\xi\xi_i + 1)(\eta\eta_i + 1)(2 + \xi\xi_i + \eta\eta_i - \xi^2 - \eta^2),$$
$$a\xi_i(\xi\xi_i + 1)^2(\xi\xi_i - 1)(\eta\eta_i + 1),$$
$$b_{\eta i}(\xi\xi_i + 1)(\eta\eta_i + 1)^2(\eta\eta_i - 1)] \qquad (1\text{-}12b)$$

and

$$\{\delta\}_i = \left\{ \begin{array}{c} w_i \\ \left(\dfrac{\partial w}{\partial y}\right)_i \\ -\left(\dfrac{\partial w}{\partial x}\right)_i \end{array} \right\}$$

(b) Strain–displacement relationship

$$\{\varepsilon\} = \left\{ \begin{array}{c} -\dfrac{\partial^2 w}{\partial x^2} \\[1mm] -\dfrac{\partial^2 w}{\partial y^2} \\[1mm] 2\dfrac{\partial^2 w}{\partial x\,\partial y} \end{array} \right\} = \sum_{i=1}^{4} [B]_i \{\delta\}_i \qquad (1\text{-}13)$$

(c) Stress–strain relationship

$$\{\sigma\} = \left\{ \begin{array}{c} M_x \\ M_y \\ M_{xy} \end{array} \right\} = [D]\{\varepsilon\} = \sum_{i=1}^{4} [D][B]_i \{\delta\}_i \qquad (1\text{-}14a)$$

where

$$[D] = K \begin{bmatrix} 1 & \nu & 0 \\ \nu & 1 & 0 \\ 0 & 0 & \dfrac{1-\nu}{2} \end{bmatrix}$$

$$K = \frac{Et^3}{12(1-\nu^2)} \qquad\qquad (1\text{-}14b)$$

The explicit form of $[D][B]$ in Eqn 1-14a is given on page 38–12.

(d) Stiffness matrix

$$[S^b] = \frac{K}{60ab}[T]^T[S^*][T] \tag{1-15a}$$

$$[T] = \begin{bmatrix} [T_s] & & & SYM \\ [0] & [T_s] & & \\ [0] & [0] & [T_s] & \\ [0] & [0] & [0] & [T_s] \end{bmatrix}$$

$$[T_s] = \begin{bmatrix} 1 & 0 & 0 \\ 0 & 2b & 0 \\ 0 & 0 & 2a \end{bmatrix} \tag{1-15b}$$

$$[S^*] = [\text{see page 38–13}] \tag{1-15c}$$

1.2.5 Higher-order plane stress elements [5]

Several higher order elements together with their displacement functions are given in Table 1-2. Such elements are ideally suited for the modelling of shear walls and it will be shown later in examples that only one or two elements are required to take up the whole height of the building, thereby achieving a very economic solution when compared with the lower-order elements, and at the same time managing to cope with abrupt wall thickness changes effectively.

(a) Displacement functions. For an element with m nodes, the displacement function for u or v can be written as

$$f = a_1 + a_2 x + a_3 y + a_4 x^2 + a_5 xy + a_6 y^2 + \dots$$
$$= [P]\{A\} \tag{1-16}$$

By carrying out a nodal parameter–polynomial constant transformation, Eqn 1-16 can be rewritten as

$$\begin{Bmatrix} u \\ v \end{Bmatrix} = \begin{bmatrix} [P][C_N]^{-1} & [0] \\ [0] & [P][C_N]^{-1} \end{bmatrix} \begin{Bmatrix} \{u_N\} \\ \{v_N\} \end{Bmatrix} = \begin{bmatrix} [N] & [0] \\ [0] & [N] \end{bmatrix} \begin{Bmatrix} \{u_N\} \\ \{v_N\} \end{Bmatrix} \tag{1-17}$$

where

$$[C_N] = \begin{vmatrix} 1 & x_1 & y_1 & x_1^2 & x_1 y_1 & y_1^2 \cdots \\ 1 & x_2 & y_2 & x_2^2 & x_2 y_2 & y_2^2 \cdots \\ \vdots & \vdots & \vdots & \vdots & \vdots & \vdots \\ 1 & x_m & y_m & x_m^2 & x_m y_m & y_m^2 \cdots \end{vmatrix} \tag{1-18}$$

The shape function $[N]$ is given in Table 1-2.

(b) Strain–displacement relationship

$$\{\varepsilon\} = \begin{Bmatrix} \varepsilon_x \\ \varepsilon_y \\ \gamma_{xy} \end{Bmatrix} = \begin{bmatrix} \dfrac{\partial[P]}{\partial x} & [0] \\ [0] & \dfrac{\partial[P]}{\partial y} \\ \dfrac{\partial[P]}{\partial y} & \dfrac{\partial[P]}{\partial x} \end{bmatrix} \begin{bmatrix} [C_N]^{-1} & [0] \\ [0] & [C_N]^{-1} \end{bmatrix} \begin{Bmatrix} \{u_N\} \\ \{v_N\} \end{Bmatrix} \tag{1-19}$$

Table 1-2   Higher-order elements and displacement functions

Element type	Shape function	Polynomial function
	$\frac{1}{2}(9\eta^3 - 9\eta^2 + 2\eta)$ $-\frac{9}{2}(3\eta^3 - 4\eta^2 + \eta)$ $\frac{9}{2}(3\eta^3 - 5\eta^2 + 2\eta)$ $-\frac{1}{2}(9\eta^2 - 18\eta^3 + 11\eta - 2)$	1 $x \quad y$ $xy \quad y^2$ $xy^2 \quad y^3$ xy^3
	$\frac{1}{3}(32\eta^4 - 48\eta^3 + 22\eta^2 - 3\eta)$ $-\frac{16}{3}(8\eta^4 - 14\eta^3 + 7\eta^2 - \eta)$ $4(16\eta^4 - 32\eta^3 + 19\eta^2 - 3\eta)$ $-\frac{16}{3}(8\eta^4 - 18\eta^3 + 13\eta^2 - 3\eta)$ $\frac{1}{3}(32\eta^4 - 80\eta^3 + 70\eta^2 - 25\eta + 3)$	1 $x \quad y$ $xy \quad y^2$ $xy^2 \quad y^3$ $xy^3 \quad y^4$ xy^4
	$\frac{1}{24}(625\eta^5 - 1250\eta^4 + 875\eta^3 - 250\eta^2 + 24\eta)$ $-\frac{25}{24}(125\eta^5 - 275\eta^4 + 205\eta^3 - 61\eta^2 + 6\eta)$ $\frac{25}{12}(125\eta^5 - 300\eta^4 + 245\eta^3 - 78\eta^2 + 8\eta)$ $-\frac{25}{12}(125\eta^5 - 325\eta^4 + 295\eta^3 - 107\eta^2 + 12\eta)$ $\frac{25}{24}(125\eta^5 - 350\eta^4 + 355\eta^3 - 154\eta^2 + 24\eta)$ $-\frac{1}{24}(625\eta^5 - 1875\eta^4 + 2125\eta^3 - 1125\eta^2 + 275\eta - 24)$	1 $x \quad y$ $xy \quad y^2$ $xy^2 \quad y^3$ $xy^3 \quad y^4$ $xy^4 \quad y^5$ xy^5

Left-side node shape functions to be multiplied by $(1-\xi)$
Right-side node shape functions to be multiplied by (ξ)
$\xi = x/a, \quad \eta = y/b$

(c) Stress–strain relationship

$$\{\sigma\} = \left\{ \begin{array}{c} \sigma_x \\ \sigma_y \\ \tau_{xy} \end{array} \right\} = [D]\{\varepsilon\}$$

$$[D] = t \begin{bmatrix} \dfrac{E_x}{1 - \nu_x \nu_y} & \dfrac{\nu_x E_y}{1 - \nu_x \nu_y} & 0 \\[2ex] \dfrac{\nu_x E_y}{1 - \nu_x \nu_y} & \dfrac{E_y}{1 - \nu_x \nu_y} & 0 \\[2ex] 0 & 0 & G \end{bmatrix}$$

(1-20)

with $[D]$ being defined by a modified form of Eqn 1-10 to take care of orthotropic properties.

(d) Stiffness matrix

$$[S^p] = \begin{bmatrix} [C_N]^{-1^T} & [0] \\ [0] & [C_N]^{-1^T} \end{bmatrix} \begin{bmatrix} I_{11} & I_{12} \\ I_{12}^T & I_{22} \end{bmatrix} \begin{bmatrix} [C_N]^{-1} & [0] \\ [0] & [C_N]^{-1} \end{bmatrix} \tag{1-21a}$$

where

$$\left. \begin{array}{l} I_{11} = \dfrac{E_x t}{1 - \nu_x \nu_y} \displaystyle\int\!\!\int_A \dfrac{\partial[P]^T}{\partial x} \dfrac{\partial[P]}{\partial x} \, dx \, dy + Gt \displaystyle\int\!\!\int_A \dfrac{\partial[P]^T}{\partial y} \dfrac{\partial[P]}{\partial y} \, dx \, dy \\[4mm] I_{12} = \dfrac{\nu_x E_y t}{1 - \nu_x \nu_y} \displaystyle\int\!\!\int_A \dfrac{\partial[P]^T}{\partial x} \dfrac{\partial[P]}{\partial y} \, dx \, dy + Gt \displaystyle\int\!\!\int_A \dfrac{\partial[P]^T}{\partial y} \dfrac{\partial[P]}{\partial x} \, dx \, dy \end{array} \right\} \tag{1-21b}$$

and

$$I_{22} = \dfrac{E_y t}{1 - \nu_x \nu_y} \displaystyle\int\!\!\int_A \dfrac{\partial[P]^T}{\partial y} \dfrac{\partial[P]}{\partial y} \, dx \, dy + Gt \displaystyle\int\!\!\int_A \dfrac{\partial[P]^T}{\partial x} \dfrac{\partial[P]}{\partial x} \, dx \, dy$$

The above integrals can be computed automatically in the subroutines by using the following formula:

$$\int\!\!\int x^m y^n \, dx \, dy = \frac{1}{m+1} \frac{1}{n+1} x^{(m+1)} y^{(n+1)} \tag{1-21c}$$

Hence, the whole family of elements given in Table 1-2 can be readily derived and automatically formed by the same computer subroutine.

Alternatively, the formulation can start from the shape function $[N]$, in which case the stiffness matrix will be

$$[S^p] = \begin{bmatrix} I_{11}^* & I_{12}^* \\ I_{12}^* & I_{22}^* \end{bmatrix} \tag{1-22}$$

in which I_{ij}^* can still be given by Eqn 1-21b, but with $[P]$ changed to $[N]$ in all the expressions. Note that the amount of integration work for I_{ij}^* is significantly higher than that for I_{ij}.

1.2.6 *Higher-order bending element*
The elements given in Table 1-2 can also be used as bending elements, although the nodal parameters will have to be changed to w_i, θ_{xi} and θ_{yi} instead of u_i and v_i [18]. The formulation of this set of higher-order elements is based on Mindlin's theory [19], in which the Kirchhoff's normality concept is abandoned, and the effect of shear deformation is included by assuming independent functions for the normal deflection w and the two rotations θ_x and θ_y. This procedure is valid for thin and moderately thick plates and has the great advantage that the bending and plane stress elements will be completely compatible with each other at non-coplanar junctions.

(a) Displacement functions

$$w = \sum_{i=1}^{m} N_i W_i$$

$$\theta_x = \sum_{i=1}^{m} N_i \theta_{xi} \qquad\qquad (1\text{-}23)$$

$$\theta_y = \sum_{i=1}^{m} N_i \theta_{yi}$$

in which m is the number of nodes in each element, and the shape functions N_i is given in Table 1-2. θ_x and θ_y are the rotations around the x and y axes respectively.

(b) Strain–displacement relationship

$$\{\varepsilon\} = \left\{ \begin{array}{c} -\dfrac{\partial \theta_y}{\partial x} \\[2mm] -\dfrac{\partial \theta_x}{\partial y} \\[2mm] -\left(\dfrac{\partial \theta_y}{\partial y} + \dfrac{\partial \theta_x}{\partial x} \right) \\[2mm] \dfrac{\partial w}{\partial x} - \theta_y \\[2mm] \dfrac{\partial w}{\partial y} - \theta_x \end{array} \right\} = \sum_{i=1}^{m} \begin{bmatrix} 0 & 0 & \dfrac{-\partial N_i}{\partial x} \\[2mm] 0 & \dfrac{-\partial N_i}{\partial y} & 0 \\[2mm] 0 & \dfrac{-\partial N_i}{\partial x} & \dfrac{-\partial N_i}{\partial y} \\[2mm] \dfrac{\partial N_i}{\partial x} & 0 & -N_i \\[2mm] \dfrac{\partial N_i}{\partial y} & -N_i & 0 \end{bmatrix} \left\{ \begin{array}{c} w_i \\ \theta_{xi} \\ \theta_{yi} \end{array} \right\}$$

$$= \sum_{i=1}^{m} [B]_i \{\delta\}_i \qquad\qquad (1\text{-}24)$$

(c) Stress–strain relationship

$$\{\sigma\} = \left\{ \begin{array}{c} M_x \\ M_y \\ M_{xy} \\ S_x \\ S_y \end{array} \right\} = \begin{bmatrix} D_x & D_1 & 0 & 0 & 0 \\ D_1 & D_y & 0 & 0 & 0 \\ 0 & 0 & D_{xy} & 0 & 0 \\ 0 & 0 & 0 & G'_x & 0 \\ 0 & 0 & 0 & 0 & G'_y \end{bmatrix} \{\varepsilon\}$$

$$= \sum_{i=1}^{m} [D][B]_i \{\delta\}_i \qquad\qquad (1\text{-}25)$$

where $G'_x = \kappa^2 G_x t$, $G'_y = \kappa^2 G_y t$, and S_x, S_y are the transverse shears, κ^2 can be taken approximately as $\pi^2/12$.

(d) Stiffness matrix

$$[S^b]_{ij} = \int_0^a \int_0^b [B]_i^T [D][B]_j \, dx \, dy$$

$$= \int_0^a \int_0^b
\begin{bmatrix}
G_x'\left(\dfrac{\partial N_i}{\partial x}\dfrac{\partial N_j}{\partial x} + G_y'\dfrac{\partial N_i}{\partial y}\dfrac{\partial N_j}{\partial y}\right) & -G_y'\dfrac{\partial N_i}{\partial y}N_j & -G_x'\dfrac{\partial N_i}{\partial x}N_j \\[3em]
-G_y'N_i\dfrac{\partial N_j}{\partial y} & D_y\dfrac{\partial N_i}{\partial y}\dfrac{\partial N_j}{\partial y} + D_{xy}\dfrac{\partial N_i}{\partial x}\dfrac{\partial N_j}{\partial x} + G_y'N_iN_j & D_1\dfrac{\partial N_i}{\partial y}\dfrac{\partial N_j}{\partial x} + D_{xy}\dfrac{\partial N_i}{\partial x}\dfrac{\partial N_j}{\partial y} \\[3em]
-G_x'N_i\dfrac{\partial N_j}{\partial x} & D_1\dfrac{\partial N_i}{\partial x}\dfrac{\partial N_j}{\partial y} + D_{xy}\dfrac{\partial N_i}{\partial y}\dfrac{\partial N_j}{\partial x} & D_x\dfrac{\partial N_i}{\partial x}\dfrac{\partial N_j}{\partial x} + D_{xy}\dfrac{\partial N_i}{\partial y}\dfrac{\partial N_j}{\partial y} + G_x'N_iN_j
\end{bmatrix}
dx \, dy \qquad (1\text{-}26)$$

1.2.7 *Flat shell element*
For the analysis of non-planar shear walls, flat shell elements have to be used. Such flat shell elements can be obtained by simply combining a plane stress element with a bending element since there is no coupling between the bending and membrane actions. Thus, the stiffness matrix of a flat shell element can be in general written as

$$[S]_{ij} = \begin{bmatrix} [S^p]_{ij} & [0] \\ [0] & [S^b]_{ij} \end{bmatrix} \qquad (1\text{-}27)$$

In some cases, the coefficients may have to be rearranged because of the ordering of the nodal displacement parameters.

1.3 Finite strip method: static

1.3.1 *Introduction*
In finite strip analysis, the solid shear walls in the structure are divided into a number of strip elements while the connecting frames and spandrel beams between the walls are treated either as an orthotropic medium with equivalent elastic properties, or simply as discrete beams spanning between the strip and line elements, the latter representing the long columns having the same height as the frame. A typical frame shear wall assembly and the associated structural elements are shown in Fig. 1-7.

In the continuum approach, the connecting frame is idealized as an equivalent continuum, and the points of contraflexure of both the beams and columns are

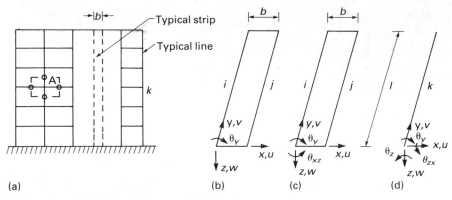

(a)   (b)   (c)   (d)

Figure 1-7   (a) Typical frame shear wall assembly
(b) Lower-order strip element
(c) Higher-order strip element
(d) Finite line element

assumed to be located at their respective mid-spans. Based on this, the equivalent properties of a continuum can be computed from the properties of the relevant beams and columns (Fig. 1-8).

$$E_{xc} = \frac{b_1 + b_2}{c_1 + c_2} \frac{E/t}{(b_1/A_{b_1} + b_2/A_{b_2})}$$

$$E_{yc} = \frac{c_1 + c_2}{b_1 + b_2} \frac{E/t}{(c_1/A_{c_1} + c_2/A_{c_2})}$$

$$G_c = \frac{3E/t}{\dfrac{c_1 + c_2}{b_1 + b_2}\left(\dfrac{b_1^3}{I_{b_1}} + \dfrac{b_2^3}{I_{b_2}}\right) + \dfrac{b_1 + b_2}{c_1 + c_2}\left(\dfrac{c_1^3}{I_{c_1}} + \dfrac{c_2^3}{I_{c_2}}\right)}$$ (1-28)

$$\rho_c = \frac{(b_1 A_{b_1} + b_2 A_{b_2} + c_1 A_{c_1} + c_2 A_{c_2})\rho/t}{(b_1 + b_2)(c_1 + c_2)}$$

$$\nu_{xc} = \nu_{yc} = 0$$

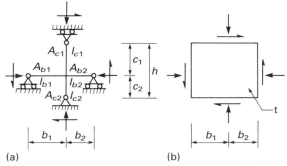

(a)   (b)

Figure 1-8   Equivalent wall modelling
(a) Actual joint, Section A in Fig. 1-7a
(b) Equivalent wall segment

where E_{xc}, E_{yc} and G_c are the Young's modulus and shear modulus of the continuum, and ρ_c the equivalent mass density, and t the assumed thickness.

For the case of a panel of spandrel beams, such as those used in connecting two shear walls, the equivalent continuum properties are

$$E_{xc} = EA_b/th$$
$$E_{yc} = 0 \tag{1-29a}$$
$$G_c = 12EI_b/thl^2$$

where h is the height of the particular story, and l is the modified span of the beam (see Section 1.1.3).

Shear effects can be introduced by using a reduced second moment of area of the beam, i.e.

$$I_{red} = \frac{I}{(1+\alpha)} \tag{1-29b}$$

where α has been defined already in Section 1.1.1.

In the alternative (discrete beam) approach, the long column in the structure is idealized by a line element which may be thought of as a strip element with only one nodal line while the standard stiffness and load matrices (see Eqn 1-2) of each connecting beam is individually computed and transformed to the adjacent strip or line elements through compatibility conditions at the two ends of the beam. Such a transformation process will be discussed in detail in Section 1.3.5. The formulation of the line element (long column) stiffness matrix does not follow the standard finite element procedure and will also be presented in detail in the same sub-section.

1.3.2 Finite strip procedure

The general finite strip method can be found in detail in a text by Cheung [6]. Briefly, this method is essentially a special form of the finite element method. However, unlike the standard finite element method, which uses exclusively polynomials as the displacement functions, the finite strip method calls for the use of polynomials in some directions (for example, horizontal directions for strip element applied to tall building analysis) and continuously differentiable smooth series in other directions (for example, the vertical direction in the case of strip elements for tall buildings), with the stipulation that such series function should satisfy *a priori* the boundary conditions at the ends of the strip. Thus, the general form of the displacement function is given as the product of polynomials and series. For the shear wall strip elements, the displacement function is given as

$$f = \sum_{m=1}^{r} \psi_m(x) Y_m \tag{1-30}$$

where Y_m can be of one of the forms listed below

(a) Fixed lower end

$$Y_m^f = \sin(\mu_m y) - \sinh(\mu_m y) - \alpha_m[\cos(\mu_m y) - \cosh(\mu_m y)]$$

and

$$\left. \begin{array}{l} \alpha_m = (\sin\mu_m + \sinh\mu_m)/(\cos\mu_m + \cosh\mu_m) \\ \mu_m = 1.875, 4.694, \ldots, (2m-1)\pi/2 \end{array} \right\} \tag{1-31}$$

(b) Elastically-supported lower base (elastic foundation characterized by translational and rotational springs, portal frames, etc.)

$$\left.\begin{array}{l} Y_1^e = 1 \\ Y_2^e = y \\ Y_m^e = Y_{(m-2)}^f \qquad m = 3, 4, \ldots \end{array}\right\} \tag{1-32}$$

The general displacement functions for rectangular strips bounded by nodal lines i and j, as shown in Fig. 1-7b, c, can be expressed as follows.

1.3.2.1   *Lower-order strip*   This is shown in Fig. 1-7b.

In-plane case

$$\left.\begin{array}{l} u = \displaystyle\sum_{m=1}^{r} [(1-\xi)u_{im} + \xi u_{jm}]Y_m \\ v = \displaystyle\sum_{m=1}^{r} [(1-\xi)v_{im} + \xi v_{jm}]Y'_m \end{array}\right\} \tag{1-33a}$$

or

$$\left\{\begin{array}{l} u \\ v \end{array}\right\} = [N^p]\{\delta\} = \sum_{m=1}^{r} [N^p]_m \{\delta\}_m \tag{1-33b}$$

Bending case

$$w = \sum_{m=1}^{r} [(1-3\xi^2+2\xi^3)w_{im} + \xi b(1-2\xi+\xi^2)\theta_{yim}$$
$$+ (3\xi^2-2\xi^3)w_{jm} + \xi b(\xi^2-\xi)\theta_{yjm}]Y_m \tag{1-33c}$$

or

$$w = [N^b]\{\delta\} = \sum_{m=1}^{r} [N^b]_m \{\delta^b\}_m \tag{1-33d}$$

where

$$\xi = \frac{x}{b} \qquad Y'_m = \frac{dY_m}{dy}$$

1.3.2.2   *Higher-order strip*   For this strip element (Fig. 1-7c), u and w are the same as those given by Eqns 1-33a and 1-33c. However, due to the introduction of the in-plane rotation variable, the displacement, v, takes up a form similar to Eqn 1-33c:

$$v = \sum_{m=1}^{r} [(1-3\xi^2+2\xi^3)v_{im} + \xi b(1-2\xi+\xi^2)\theta_{zim}$$
$$+ (3\xi^2-2\xi^3)v_{jm} + \xi b(\xi^2-\xi)\theta_{zjm}]Y'_m \tag{1-34}$$

For both the in-plane and bending cases, the displacement field for a strip can be written as

$$\{f\} = [N]\{\delta\} = \sum_{m=1}^{r} [N]_m \{\delta\}_m \tag{1-35}$$

Once the displacement functions are known, the stiffness and load matrices are formulated in accordance with the standard finite element procedure. The stiffness matrix is given by

$$[S] = \begin{bmatrix} [S]_{11} & [S]_{12} & & [S]_{1r} \\ [S]_{21} & [S]_{22} & & [S]_{2r} \\ \vdots & \vdots & & \vdots \\ [S]_{r1} & [S]_{r2} & \cdots & [S]_{rr} \end{bmatrix} \tag{1-36a}$$

with

$$[S_{ij}]_{mn} = \iint [B_i]_m^T [D] [B_j]_n \, dx \, dy \tag{1-36b}$$

where i, j refer to the nodal numbers and m, n refer to the term numbers.
The load matrix is given by

$$\{F\} = \iint [N]^T \{q\} \, dx \, dy$$

$$= \iint \begin{bmatrix} [N]_1^T \\ [N]_2^T \\ \vdots \\ [N]_r^T \end{bmatrix} \{q\} \, dx \, dy \tag{1-36c}$$

with

$$\{F\}_m = \iint [N]_m^T \{q\} \, dx \, dy \tag{1-36d}$$

where $\{q\}$ represents the distributed force acting on the strip element.
The various matrices have been formulated explicitly and are listed in the following paragraphs. The integrals I_a, I_b, etc., are given separately in the Appendix, p. 38-50.

1.3.2.3 *Lower-order strip element* (Refer to Fig. 1–7b)

In-plane case

$$[S^p]_{mn} = \frac{1}{6b} \begin{bmatrix} 6I_a + 2b^2 I_b & -3b(I_b + I_d) & -6I_a + b^2 I_b & 3b(I_b - I_d) \\ -3b(I_b + I_c) & 6I_b + 2b^2 I_e & 3b(-I_b + I_c) & -6I_b + b^2 I_a \\ -6I_a + b^2 I_b & 3b(-I_b + I_d) & 6I_a + 2b^2 I_b & 3b(I_b + I_d) \\ 3b(I_b - I_c) & -6I_b + b^2 I_e & 3b(I_b + I_c) & 6I_b + 2b^2 I_e \end{bmatrix} \tag{1-37a}$$

Bending case

$$[S^b]_m =$$

$$\frac{1}{420b^3}
\begin{bmatrix}
\begin{array}{c}(5040I_1-504I_2\\-504I_3+156I_4\\+2016I_5)\end{array} &
\begin{array}{c}(2520I_1-462I_2\\-42I_3+22I_4\\+168I_5)b\end{array} &
\begin{array}{c}(-5040I_1+504I_2\\+504I_3+54I_4\\-2016I_5\end{array} &
\begin{array}{c}(2520I_1-42I_2\\-42I_3-13I_4\\+168I_5)b\end{array} \\
\hline
\begin{array}{c}(2520I_1-42I_2\\-462I_3+22I_4\\+168I_5)b\end{array} &
\begin{array}{c}(1680I_1-56I_2\\-56I_3+4I_4\\+224I_5)b^2\end{array} &
\begin{array}{c}(-2520I_1+42I_2\\+42I_3+13I_4\\-168I_5)b\end{array} &
\begin{array}{c}(840I_1+14I_2\\+14I_3-3I_4\\-56I_5)b^2\end{array} \\
\hline
\begin{array}{c}(-5040I_1+504I_2\\+504I_3+54I_4\\-2016I_5)\end{array} &
\begin{array}{c}(-2520I_1+42I_2\\+42I_3+13I_4\\-168I_5)b\end{array} &
\begin{array}{c}(5040I_1-504I_2\\-504I_3+156I_4\\+2016I_5)\end{array} &
\begin{array}{c}(-2520I_1+462I_2\\+42I_3-22I_4\\-168I_5)b\end{array} \\
\hline
\begin{array}{c}(2520I_1-42I_2\\-42I_3-13I_4\\+168I_5)b\end{array} &
\begin{array}{c}(840I_1+14I_2\\+14I_3-3I_4\\-56I_5)b^2\end{array} &
\begin{array}{c}(-2520I_1+42I_2\\+462I_3-22I_4\\-168I_5)b\end{array} &
\begin{array}{c}(1680I_1-56I_2\\-56I_3+4I_4\\+224I_5)b^2\end{array}
\end{bmatrix}$$

(1-37b)

Consistent force vectors $\{F^p\}_m$ and $\{F^b\}_m$

Types of forces	In-plane case $\{F^p\}_m$	Bending case $\{F^b\}_m$
Body and surface forces	$\dfrac{b}{2}\begin{Bmatrix} I_{\bar{x}} \\ I_{\bar{y}} \\ I_{\bar{x}} \\ I_{\bar{y}} \end{Bmatrix}$	$\dfrac{b}{12}I_{zy}\begin{Bmatrix} 6 \\ b \\ 6 \\ -b \end{Bmatrix}$
Line forces	$\begin{Bmatrix} I_{xi} \\ I_{yi} \\ I_{xj} \\ I_{yj} \end{Bmatrix}$	$\begin{Bmatrix} I_{zi} \\ I_{myi} \\ I_{zj} \\ I_{myj} \end{Bmatrix}$
Concentrated forces	$\begin{Bmatrix} f_{xi} \\ f_{yi} \\ f_{xj} \\ f_{yj} \end{Bmatrix}$	$\begin{Bmatrix} f_{zi} \\ f_{myi} \\ f_{zj} \\ f_{myj} \end{Bmatrix}$

(1-37c)

1.3.2.4 *Higher-order strip element*   Refer to Fig. 1-7c. Only the in-plane case is given since the bending case can be obtained from Eqn 1-37b.

$$[S^p]^{mn} = [T]^T[S^p]^*_{mn}[T]$$ (1-38a)

where

$$[T] =
\begin{bmatrix}
-1/b & 0 & 0 & 1/b & 0 & 0 \\
\hline
1 & 0 & 0 & 0 & 0 & 0 \\
\hline
0 & 1 & 0 & 0 & 0 & 0 \\
\hline
0 & 0 & 1 & 0 & 0 & 0 \\
\hline
0 & -3/b^2 & -2/b & 0 & 3/b^2 & -1/b \\
\hline
0 & 2/b^3 & 1/b^2 & 0 & -2/b^3 & 1/b^2
\end{bmatrix}$$

(1-38b)

$$[S^p]^*_{mn} =$$

$$\begin{bmatrix} bI_a + b^3I_b/3 & b^2I_b/2 & bI_d & b^2(I_b+I_d)/2 & b^3(2I_b+I_d)/3 & b^4(3I_b+I_d)/4 \\ b^2I_b/2 & bI_b & 0 & bI_b & b^2I_b & b^3I_b \\ bI_c & 0 & bI_e & b^2I_e/2 & b^3I_e/3 & b^4I_e/4 \\ b^2(I_b+I_c)/2 & bI_b & b^2I_e/2 & bI_b+b^3I_e/3 & b^2I_b+b^4I_e/4 & b^3I_b+b^5I_e/5 \\ b^3(2I_b+I_c)/3 & b^2I_b & b^3I_e/3 & b^2I_b+b^4I_e/4 & 4b^3I_b/3+b^5I_e/5 & 3b^4I_b/2+b^6I_e/6 \\ b^4(3I_b+I_c)/4 & b^3I_b & b^4I_e/4 & b^3I_b+b^5I_e/5 & 3b^4I_b/2+b^6I_e/6 & 9b^5I_b/5+b^7I_e/7 \end{bmatrix}$$

$$(1\text{-}38c)$$

Consistent force vector $\{F^p\}_m$

Types of forces	In-plane case $\{F^p\}_m$
Body and surface forces	$\dfrac{b}{12}\begin{Bmatrix} 6I_{\bar{x}} \\ 6I_{\bar{y}} \\ bI_{\bar{y}} \\ 6I_{\bar{x}} \\ 6I_{\bar{y}} \\ -bI_{\bar{y}} \end{Bmatrix}$
Line forces	$\begin{Bmatrix} I_{xi} \\ I_{yi} \\ I_{mzi} \\ I_{xj} \\ I_{yj} \\ I_{mzj} \end{Bmatrix}$
Concentrated forces	$\begin{Bmatrix} f_{xi} \\ f_{yi} \\ f_{mzi} \\ f_{xj} \\ f_{yj} \\ f_{mzj} \end{Bmatrix}$

$$(1\text{-}38d)$$

With the plane stress and bending stiffness known, it is a simple matter to combine the two of them into a flat shell strip stiffness matrix as given in Eqn 1-27.

1.3.3 Rigid section constraint

The role of the floor slab has been discussed in the previous chapter and it is noted that, due to the high in-plane rigidity of the floor, all structural components connected to the slab are forced to move as a rigid section in the horizontal plane, i.e. the horizontal displacement parameters, u_{km}, w_{km} and θ_{ykm}, at all nodal lines are no longer independent, but can be set to some related values, and advantage

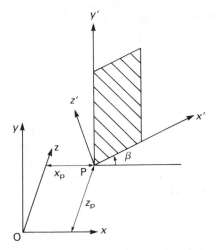

Figure 1-9 Rigid section transformation

should be taken of this fact to reduce substantially the number of degrees of freedom associated with the structure being analysed. Referring to Fig. 1-9, if the global displacements and rotations of a chosen reference point (say the origin O) on the cross-section for the mth term of the series are denoted by u_{om}, w_{om} and θ_{yom}, then the local displacements and rotations of any other point P with coordinates x_p and z_p may be expressed in terms of the displacements at the reference point as

$$
\begin{Bmatrix} u'_{pm} \\ w'_{pm} \\ \theta'_{ypm} \end{Bmatrix} = \begin{bmatrix} \cos\beta & \sin\beta & x_p\sin\beta - z_p\cos\beta \\ -\sin\beta & \cos\beta & x_p\cos\beta + z_p\sin\beta \\ 0 & 0 & 1 \end{bmatrix} \begin{Bmatrix} u_{om} \\ w_{om} \\ \theta_{yom} \end{Bmatrix}
\tag{1-39}
$$

In the above equation, the primed system refers to the local coordinate system, while the non-primed system refers to the global coordinate system.

For a strip with sides i and j, a transformation matrix can be constructed from Eqn 1-39:

$$
\begin{Bmatrix} u'_{im} \\ v'_{im} \\ w'_{im} \\ \theta'_{yim} \\ u'_{jm} \\ v'_{jm} \\ w'_{jm} \\ \theta'_{yjm} \end{Bmatrix} =
\left[
\begin{array}{ccc|cc}
\cos\beta & \sin\beta & x_i\sin\beta - z_i\cos\beta & 0 & 0 \\
0 & 0 & 0 & 1 & 0 \\
-\sin\beta & \cos\beta & x_i\cos\beta + z_i\sin\beta & 0 & 0 \\
0 & 0 & 1 & 0 & 0 \\
\hline
\cos\beta & \sin\beta & x_j\sin\beta - z_j\cos\beta & 0 & 0 \\
0 & 0 & 0 & 0 & 1 \\
-\sin\beta & \cos\beta & x_j\cos\beta + z_j\sin\beta & 0 & 0 \\
0 & 0 & 1 & 0 & 0
\end{array}
\right]
\begin{Bmatrix} u_{om} \\ w_{om} \\ \theta_{yom} \\ v_{im} \\ v_{jm} \end{Bmatrix} = [T_R] \begin{Bmatrix} u_{om} \\ w_{om} \\ \theta_{yom} \\ v_{im} \\ v_{jm} \end{Bmatrix}
\tag{1-40}
$$

The stiffness and load matrices of a flat shell strip can now be transformed to correspondingly smaller matrices:

$$[S]_{mn} = [T_R]^T [S']_{mn} [T_R] \tag{1-41a}$$

$$\{F\}_m = [T_R]^T \{F'\}_m \tag{1-41b}$$

Note that the stiffness matrix can be partitioned into four submatrices such that

$$[S]_{mn} = \begin{bmatrix} [A] & [B] \\ [B]^T & [C] \end{bmatrix} \tag{1-41c}$$

and that the submatrix $[A]$ for every strip is connected with the same displacement parameters u_{om}, w_{om} and θ_{yom} and will therefore be assembled into the same location in the overall structural stiffness matrices.

1.3.4 Rigid section strip element

Equation 1-41a represents the formation of a 5×5 matrix through the triple product of 5×8, 8×8 and 8×5 matrices, and is therefore computationally inefficient. Significant economies can be made if the rigid section constraint is applied directly to the displacement functions, so that a 5×5 stiffness matrix of a rigid section strip element can be formed directly. For a rigid section strip element, it is obvious that the following relations would hold true in the local coordinate system:

$$u_{im} = u_{jm}$$

$$w_{im} = w_{jm} - b\theta_{yim} \tag{1-42}$$

$$\theta_{yim} = \theta_{yjm}$$

It readily follows that Eqn 1-33 can now be simplified to the following form:

$$\left. \begin{aligned} u &= \sum_{m=1}^{r} u_{im} Y_m \\ v &= \sum_{m=1}^{r} [(1-\xi)v_{im} + \xi v_{jm}] Y'_m \end{aligned} \right\} \tag{1-43}$$

$$w = \sum_{m=1}^{r} [w_{im} + x\theta_{yim}] Y_m \tag{1-44}$$

and it also follows that the stiffness matrices will be of the order $(3r \times 3r)$ and $(2r \times 2r)$, instead of the two $(4r \times 4r)$ given previously, thus resulting in significant savings in the formulation efforts.

The transformation matrix $[T]$ is now of the order 5×5 and is given by

$$\begin{Bmatrix} u'_{im} \\ v'_{im} \\ v'_{jm} \\ w'_{im} \\ \theta'_{yim} \end{Bmatrix} = \begin{bmatrix} \cos\beta & \sin\beta & x_i\sin\beta - z_i\cos\beta & 0 & 0 \\ 0 & 0 & 0 & 1 & 0 \\ 0 & 0 & 0 & 0 & 1 \\ -\sin\beta & \cos\beta & x_i\cos\beta + z_i\sin\beta & 0 & 0 \\ 0 & 0 & 1 & 0 & 0 \end{bmatrix} \begin{Bmatrix} u_{om} \\ w_{om} \\ \theta_{yom} \\ v_{im} \\ v_{jm} \end{Bmatrix} \tag{1-45a}$$

or

$$\{\delta'\}_m = [T]\{\delta\}_m \qquad (1\text{-}45\text{b})$$

Once again, the stiffness and load matrices in global coordinates are given by

$$[S]_{mn} = [T]^\mathrm{T}[S']_{mn}[T]$$
$$\{F\}_m = [T]^\mathrm{T}\{F'\}_m$$

The various matrices for the lower order strip element (Fig. 1-7b) have been formed explicitly and are given below:

In-plane case

$$[S^p]_{mn} = \frac{1}{6b}\begin{bmatrix} 6b^2I_b & -6bI_b & 6bI_b \\ -6bI_b & 2b^2I_e+6I_b & b^2I_e-6I_b \\ 6bI_b & b^2I_e-6I_b & 2b^2I_e+6I_b \end{bmatrix} \qquad (1\text{-}46\text{a})$$

Bending case

$$[S^b]_{mn} = \frac{b}{6}\begin{bmatrix} 6I_4 & 3bI_4 \\ 3bI_4 & 2b^2I_4+24I_5 \end{bmatrix} \qquad (1\text{-}46\text{b})$$

Consistent force vectors $\{F^p\}_m$ and $\{F^b\}_m$

Types of forces	In-plane case $\{F^p\}_m$	Bending case $\{F^b\}_m$
Body and surface forces	$\dfrac{b}{2}\begin{Bmatrix} 2I_{\bar{x}} \\ I_{\bar{y}} \\ I_{\bar{y}} \end{Bmatrix}$	$\dfrac{b}{2}I_{\bar{z}}\begin{Bmatrix} 2 \\ b \end{Bmatrix}$
Line forces	$\begin{Bmatrix} I_{xy} \\ I_{yi} \\ I_{yj} \end{Bmatrix}$	$\begin{Bmatrix} I_{zy} \\ I_{my} \end{Bmatrix}$
Concentrated forces	$\begin{Bmatrix} f_x \\ f_{yi} \\ f_{yj} \end{Bmatrix}$	$\begin{Bmatrix} f_z \\ f_{my} \end{Bmatrix}$

$$(1\text{-}46\text{c})$$

1.3.5 Alternative approach in the treatment of beams and columns

The continuum approach presented at the beginning of this section is fairly accurate in giving equivalent stiffness properties. However, from the solved displacement parameters it is difficult to obtain the internal forces of the beam and column members directly. The alternative approach to be discussed in this sub-section not only overcomes the above-mentioned deficiency but also will yield more accurate results since no assumption has been made for the position of the contraflexure points.

1.3.5.1   *Stiffness matrix of a long column (finite line)*   The displacement fields for a higher-order line element (Fig. 1-7d) at a nodal line, k, may be expressed as

$$u = \sum_{m=1}^{r} u_{km} Y_m$$

$$v = \sum_{m=1}^{r} v_{km} Y'_m$$

$$w = \sum_{m=1}^{r} w_{km} Y_m$$

$$\theta_x = \sum_{m=1}^{r} \theta_{xkm} Y'_m \tag{1-47}$$

$$\theta_y = \sum_{m=1}^{r} \theta_{ykm} Y_m$$

$$\theta_z = \sum_{m=1}^{r} \theta_{zkm} Y'_m$$

However, an exceedingly large number of terms will have to be used if all the contraflexure points near the centre point of each story are to be reproduced, and therefore a different approach must be resorted to. In this alternative approach, only a few terms of the series are used, and the structural properties of the line element are formulated not in accordance with the standard finite element procedure, but through the transformation of the properties of each individual column element (Eqn 1-5) occupying one story height. For the adopted coordinate systems of the column element and the line element as shown in Fig. 1-10a, the relationship of the displacements between the two systems for node i of the column element at height y_i can be easily verified to be

$$u_{bi} = \sum_{m=1}^{r} v_{km} Y'_m(y_i)$$

$$v_{bi} = \sum_{m=1}^{r} u_{km} Y_m(y_i)$$

$$w_{bi} = \sum_{m=1}^{r} w_{km} Y_m(y_i)$$

$$\theta_{xbi} = \sum_{m=1}^{r} \theta_{ykm} Y_m(y_i) \tag{1-48a}$$

$$\theta_{ybi} = \sum_{m=1}^{r} \theta_{xkm} Y'_m(y_i)$$

$$\theta_{zbi} = \sum_{m=1}^{r} \theta_{zkm} Y'_m(y_i)$$

or

$$\{\delta_{bi}\} = \sum_{m=1}^{r} [T_i]_m \{\delta_k\}_m \tag{1-48b}$$

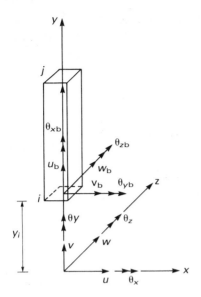

Figure 1-10a Coordinate systems for column element and line element

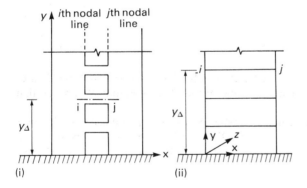

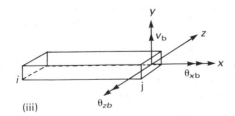

Figure 1-10b Transformation of connecting beams
        (i) Coupled shear wall
        (ii) Beams and line elements
        (iii) Typical beam element and global coordinate system

where

$$\{\delta_k\}_{mn} = [u_{km} \quad w_{km} \quad \theta_{ykm} \mid v_{km} \quad \theta_{xkm} \quad \theta_{zkm}]^T \tag{1-48c}$$

and

$$[T_i]_m = \begin{bmatrix} u_{km} & w_{km} & \theta_{ykm} & v_{km} & \theta_{xkm} & \theta_{zkm} \\ 0 & 0 & 0 & Y'_m(y_i) & 0 & 0 \\ Y_m(y_i) & 0 & 0 & 0 & 0 & 0 \\ 0 & Y_m(y_i) & 0 & 0 & 0 & 0 \\ \hline 0 & 0 & Y_m(y_i) & 0 & 0 & 0 \\ 0 & 0 & 0 & 0 & Y'_m(y_i) & 0 \\ 0 & 0 & 0 & 0 & 0 & Y'_m(y_i) \end{bmatrix} \tag{1-48d}$$

For each column, the transformation becomes

$$\begin{Bmatrix} \{\delta_{bi}\} \\ \{\delta_{bj}\} \end{Bmatrix} = \sum_{m=1}^{r} \begin{bmatrix} [T_i]_m \\ [T_j]_m \end{bmatrix} \{\delta_k\}_m$$

$$= \sum_{m=1}^{r} [T]_m \{\delta_k\}_m \tag{1-49a}$$

and once again the transformed stiffness and load matrices take up the form

$$[S]_{mn} = [T]_m^T [S_b][T]_n$$
$$\{F\}_m = [T]_m^T \{F_b\} \tag{1-49b}$$

where $[S_b]$ and $\{F_b\}$ are the standard column stiffness matrix and load matrix, respectively.

Finally, the stiffness and load matrices of a line element are established after having summed up the contributions from all the storey columns making up that particular line element.

As before, all submatrices connected with the first three displacement parameters u_{km}, w_{km} and θ_{ykm} (compare with Eqn 1-41c) will be assembled into a common area of the overall structural stiffness matrix when rigid section constraint is imposed.

1.3.5.2 *Transformation matrices for connecting beams* It has been already stated that in the alternative approach (discrete beam model) the stiffness properties of each connecting beam are lumped onto those of the connecting strip nodal lines or long columns (line elements).

Consider a typical beam element at a height y_Δ from the base of a shear wall or frame, as shown in Fig. 1-10b. The compatibility conditions at end i of the beam (taking into account the rigid section constraint) require that

(a) when connected to a strip element (Fig. 1-10b(i))

$$v_{bi} = \sum_{m=1}^{r} v_{im} Y'_m(y_\Delta)$$

$$\theta_{zbi} = -\sum_{m=1}^{r} u_{im} Y'_m(y_\Delta) \tag{1-50a}$$

or

$$\left\{ \begin{matrix} v_{bi} \\ \theta_{zbi} \end{matrix} \right\} = \sum_{m=1}^{r} \begin{bmatrix} 0 & Y'_m(y_\Delta) \\ -Y'_m(y_\Delta) & 0 \end{bmatrix} \left\{ \begin{matrix} u_{im} \\ v_{im} \end{matrix} \right\} \qquad (1\text{-}50b)$$

(b) when connected to a line element (Fig. 1-10b(ii))

$$v_{bi} = \sum_{m=1}^{r} v_{im} Y'_m(y_\Delta)$$

$$\theta_{xbi} = \sum_{m=1}^{r} \theta_{xim} Y'_m(y_\Delta) \qquad (1\text{-}51a)$$

$$\theta_{zbi} = -\sum_{m=1}^{r} \theta_{zim} Y'_m(y_\Delta)$$

or

$$\left\{ \begin{matrix} v_{bi} \\ \theta_{xbi} \\ \theta_{zbi} \end{matrix} \right\} = \sum_{m=1}^{r} \begin{bmatrix} Y'_m(y_\Delta) & 0 & 0 \\ 0 & Y'_m(y_\Delta) & 0 \\ 0 & 0 & -Y'_m(y_\Delta) \end{bmatrix} \left\{ \begin{matrix} v_{im} \\ \theta_{xim} \\ \theta_{zim} \end{matrix} \right\} \qquad (1\text{-}51b)$$

Both Eqn 1-50 and Eqn 1-51 can be written as

$$\{\delta_{bi}\} = \sum_{m=1}^{r} [T]_m \{\delta_i\}_m$$

For a beam element with ends 1 and 2 connected to nodal lines i and k respectively, the transformation becomes

$$\left\{ \begin{matrix} \{\delta_{b1}\} \\ \{\delta_{b2}\} \end{matrix} \right\} = \sum_{m=1}^{r} \begin{bmatrix} [T]_m & [0] \\ [0] & [T]_m \end{bmatrix} \left\{ \begin{matrix} \{\delta_i\}_m \\ \{\delta_k\}_m \end{matrix} \right\} \qquad (1\text{-}52a)$$

or

$$\{\delta_b\} = \sum_{m=1}^{r} [T_b]_m \left\{ \begin{matrix} \{\delta_i\}_m \\ \{\delta_k\}_m \end{matrix} \right\} \qquad (1\text{-}52b)$$

and therefore again one obtains

$$[S]_{mn} = [T_b]_m^T [S_b][T_b]_n \qquad (1\text{-}52c)$$

$$\{F\}_m = [T_b]_m^T \{F_b\} \qquad (1\text{-}52d)$$

where $[S_b]$ and $\{F_b\}$ are the stiffness and load matrices of a beam element with the relevant displacement parameters given in Eqn 1-50 and Eqn 1-51.

The alternative approach (discrete beam model) presented above can be applied to the case of a shear wall on a portal frame. In this case, the stiffness matrices of all the frame members forming the portal frame are transformed onto the various connecting nodal lines (Eqn 1-50), with y_Δ taken as zero and the displacement function series Y_m given by Eqn 1-32. Note that the height of the shear wall will now be calculated from the neutral axis of the horizontal frame member to the top end of the structure.

1.3.6 *Overall structural stiffness matrix formulated according to rigid section constraint criterion*

The overall structural stiffness matrix for a tall building is shown in Fig. 1-11. Due to the adoption of the rigid section constraint, all displacement parameters are now coupled to the common displacement parameters, u_{om}, w_{om} and θ_{yom}, through off-diagonal coefficients. It is quite obvious that the ordinary band matrix solver is no longer suitable here and special techniques should be resorted to. A simple way of reducing the size of the problem is to eliminate the v displacement parameters of all the internal nodal lines in each individual core section by static condensation. In this way, for an individual core section formed by several strip elements, there will only be the three common displacement parameters, u_{om}, w_{om} and θ_{yom}, plus the v_m displacements of two remaining nodal lines. Using five terms of the series ($r = 5$), the matrix size for an individual core section will only be 25×25.

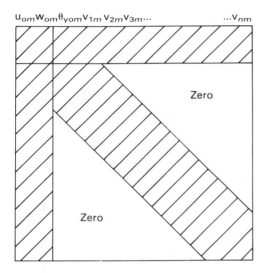

Figure 1-11   Overall structural matrix

1.4   Finite strip method: free vibration

1.4.1 *Mathematical formulation*

In the matrix theory of free vibrations [6], the end product of the mathematical formulation is invariably a set of eigenvalue equations of the form

$$([S] - \omega^2[M])\{\delta\} = 0 \tag{1-53}$$

where $[S]$ is the stiffness matrix of the structure, $[M]$ the mass matrix of the structure, ω the natural frequencies of the structure, and $\{\delta\}$ a vector giving the corresponding mode shapes.

Various schemes have been developed for solving eigenvalue equations and both standard and advanced techniques can be found in a number of texts [20, 21].

The formulation of the consistent mass matrix is based on the standard finite element procedure. Given a general displacement function in the form of Eqn 1-35, the consistent mass matrix of a strip can be written as

$$[M] = \int\int \rho t [N]^T [N] \, dx \, dx$$

$$= \int\int \rho t \begin{bmatrix} [N]_1^T[N]_1 & [N]_1^T[N]_2 & [N]_1^T[N]_3 & \cdots & [N]_1^T[N]_r \\ [N]_2^T[N]_1 & [N]_2^T[N]_2 & [N]_2^T[N]_3 & \cdots & [N]_2^T[N]_r \\ [N]_3^T[N]_1 & [N]_3^T[N]_2 & [N]_3^T[N]_3 & \cdots & [N]_3^T[N]_r \\ \vdots & \vdots & \vdots & & \vdots \\ [N]_r^T[N]_1 & [N]_r^T[N]_2 & [N]_r^T[N]_3 & \cdots & [N]_r^T[N]_r \end{bmatrix} dx \, dy$$

(1-54a)

and the basic unit submatrix is

$$[M]_{mn} = \int\int \rho t [N]_m^T [N]_n \, dx \, dy$$ (1-54b)

1.4.2 *Consistent mass matrix for a flat shell strip*
For a given strip the mass matrices corresponding to in-plane action $[M^p]$ and to bending action $[M^b]$ are

$$[M^p] = \int\int \rho t [N^p]^T [N^p] \, dx \, dy$$

$$[M^b] = \int\int \rho t [N^b]^T [N^b] \, dx \, dy$$ (1-55a)

where $[N^p]$ and $[N^b]$ have been defined in Eqns 1-33b, d.

The mass matrix for a flat shell strip is a simple combination of the two matrices, and takes up the form of

$$[M] = \begin{bmatrix} [M^p] & [0] \\ [0] & [M^b] \end{bmatrix}$$ (1-55b)

The mass matrices for rigid section and non-rigid section strips are listed below.

1.4.2.1 *Lower-order strip element*   Refer to Fig. 1-7b

(a) Nonrigid section
  (i) In-plane case

$$[M^p]_{mn} = \frac{b}{6} \begin{bmatrix} 2I_{ma} & & & & \text{SYM} \\ 0 & 2I_{mb} & & & \\ I_{ma} & 0 & 2I_{ma} & & \\ 0 & I_{mb} & 0 & 2I_{mb} \end{bmatrix}$$ (1-56a)

(ii) Bending case

$$[M^b]_{mn} = \frac{bI_{ma}}{420} \begin{bmatrix} 156 & \vdots & \text{SYM} & & \\ \hline 22b & \vdots & 4b^2 & \vdots & \\ \hline 54 & \vdots & 13b & \vdots & 156 & \vdots \\ \hline -13b & \vdots & -3b^2 & \vdots & -22b & \vdots & 4b^2 \end{bmatrix}$$ (1-56b)

(b) Rigid section

(i) In-plane case

$$[M^p]_{mn} = \frac{b}{6} \begin{bmatrix} 6I_{ma} & \vdots & \text{SYM} & & \\ \hline 0 & \vdots & 2I_{mb} & \vdots & \\ \hline 0 & \vdots & I_{mb} & \vdots & 2I_{mb} \end{bmatrix}$$ (1-56c)

(ii) Bending case

$$[M^b]_{mn} = \frac{bI_{ma}}{6} \begin{bmatrix} 6 & \vdots & 3b \\ \hline 3b & \vdots & 2b^2 \end{bmatrix}$$ (1-56d)

2 Plane shear walls

2.1 Static analysis

For this class of problems, all of the methods presented in this chapter will yield accurate results.

2.1. Coupled shear walls

Two examples will be given here. In the first example, a plane shear wall with one row of openings (Fig. 2-1a) was analysed by the frame method and by conventional finite element method [4]. The mesh divisions for the two methods are shown

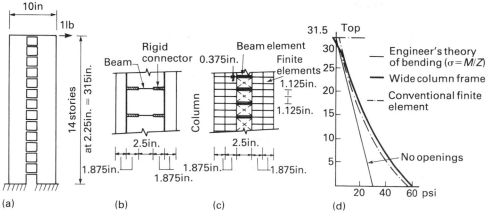

Figure 2-1   Edge stress for shear walls (after MacLeod [4])
(a) Wall
(b) Wide column frame analogy for wall (84 DOF)
(c) Finite element analogy for wall (504 DOF)
(d) Edge stress

in Figs 2-1b and 2-1c respectively, while the vertical stress at the outer edge of the wall is plotted in Fig. 2-1d. Note that in this case, in which the stiffness of each spandrel beam is relatively weak compared with that of the wall, the engineer's bending theory gives stresses which are non-conservative and therefore should not be used.

In the second example, a non-symmetrical shear wall (Fig. 2-2a) of uniform section with several rows of openings is analysed by lower-order finite strips [8] (Fig. 2-2b) and by higher-order finite elements [5] (Fig. 2-2c). The results are presented in Fig. 2-2d and are compared with the theoretical and experimental results from Coull and Subedi [22]. They are all in good agreement.

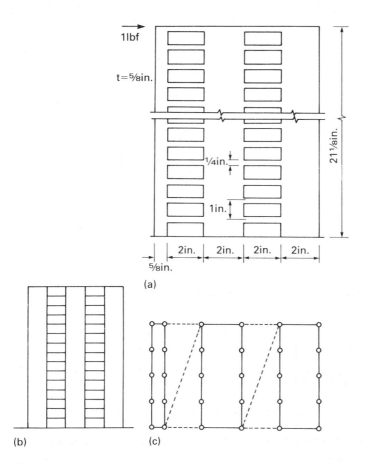

(a)

(b)   (c)

Figure 2-2   Non-symmetrical shear wall with two bands of openings and idealized models

        (a) Non-symmetrical shear wall ($E = 4.6 \times 10^5$ psi)

        (b) Finite strips (5 terms, 60 DOF)

        (c) Higher-order elements (60 DOF)

        (d) Deflections and stresses (psi)

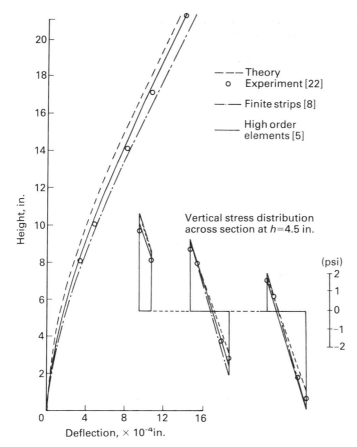

Figure 2-2(d) Deflections and stresses (psi)

2.1.2 *Shear wall on portal frame*

A shear wall on portal frame [9] acted upon by a concentrated load at the top (Fig. 2-3a) was analysed by using higher-order strip elements (Fig. 2-3b) to model the wall, and frame members to model the columns and beams. The structure was also analysed by using beam-type rectangular elements (Fig. 1-5c) and the mesh is shown in Fig. 2-3c.

For the present finite strip analysis, Y_m is given by Eqn 1-32 and the frame member stiffness matrices are transformed and coupled onto the finite strip stiffness matrices by virtue of Eqn 1-52c. Rigid section constraint has not been assumed in the analysis.

The lateral deflection profile obtained from finite strip and finite element analysis agree well with experimental results (Fig. 2-3d). The stress variations at the extreme fibres of the right column and in the wall at 4 in from ground level are shown in Fig. 2-3e. The three sets of results are also in good agreement.

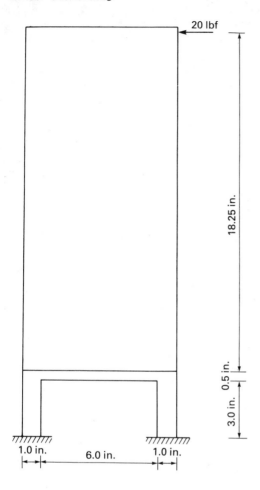

20 lbf

18.25 in.

0.5 in.

3.0 in.

1.0 in.

6.0 in.

1.0 in.

Figure 2-3a   Solid shear wall on portal frame: $E = 4.6 \times 10^5$ psi, $\nu = 0.377$, wall thickness = 0.5 in., frame thickness = 1.5 in.

2.2   Dynamic analysis

A free vibration analysis [9] of a coupled plane shear wall (Fig. 2-4) is presented here. Both the lower-order finite strip and finite element analyses (using beam-type plane stress elements) are carried out and the results compared. Rigid section constraint has been adopted and from Table 2-1 it can be seen that all three sets of results agree very well with each other. The mesh used here is really finer than necessary and sufficiently good results can be obtained with one strip representing each wall.

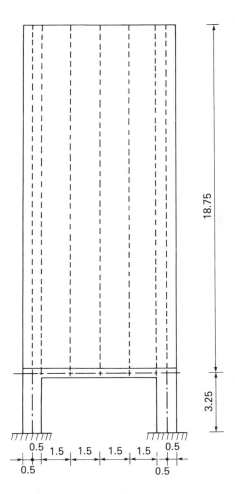

Figure 2-3b Finite strip mesh (all dimensions are in in.)

3 Non-planar shear walls (shear cores)

A three-dimensional coupled shear wall with an open section is subjected to a concentrated load at the top (Fig. 3-1a). The structure was tested and studied by Tso and Biswas [23] using the continuum approach in conjunction with Vlasov's thin-walled beam theory. In this section, this structure was analysed by finite strips and by higher order finite elements.

3.1 Lower-order finite strips

The discrete beam model (Fig. 3-1b) is used to represent the spandrel beams connecting the two angle sections. Rigid section and non-rigid section analyses

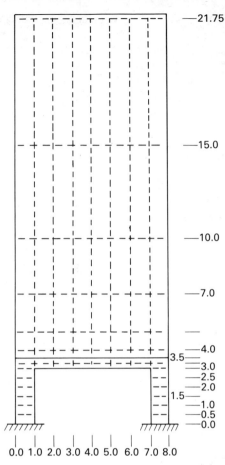

Figure 2-3c   Finite element mesh (all dimensions are in in.)

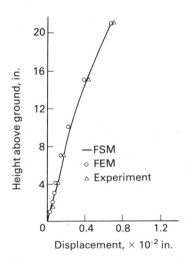

Figure 2-3d   Deflection profile for solid shear wall on portal frame

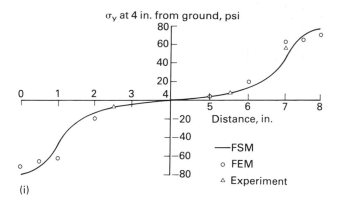

(i)

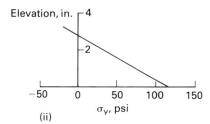

(ii)

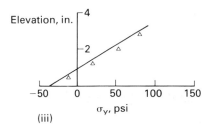

(iii)

Figure 2-3e   Stress variation for solid shear wall on portal frame
(i) In wall at 4 in. from ground
(ii) Outer edge of right column
(iii) Inner edge of right column

have been carried out [8] and it can be seen that the difference between the two sets of results is quite small.

3.2   Higher-order finite elements

In this analysis [24], the spandrel beams are represented by a continuous medium with equivalent properties, and five elements defined by six vertical nodal lines each with five nodes are employed to approximate the core wall structure (Fig. 3.1c). Rigid section constraint is adopted and, by neglecting the out-of-plane

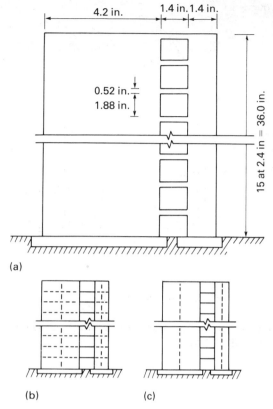

(a)

(b)   (c)

Figure 2-4   (a) Coupled plane shear wall structure with unequal width:
$E = 60.5 \times 10^4$ psi, $\nu = 0.38$, $\rho = 1.12 \times 10^{-4}$ lbf s^2 in^{-4}
(b) Finite element mesh
(c) Finite strip mesh

Table 2-1   Natural frequencies of a coupled shear wall

Type of analysis	Spandrel beams	No. of terms	Angular frequencies				DOF
			ω_1	ω_2	ω_3	ω_4	
Lower-order finite strips	Discrete beam model	3	374.5	1789.7	3154.1	4207.7	21
		4	373.8	1788.7	3151.3	4132.7	28
		5	373.8	1778.7	3150.6	4131.2	35
	Continuum model	3	375.7	1790.7	3131.4	4197.0	21
		4	375.0	1789.6	3130.1	4128.9	28
		5	374.9	1780.8	3129.9	4128.0	35
Beam-type finite elements			373.6	1756.5	3165.2	4097.6	130

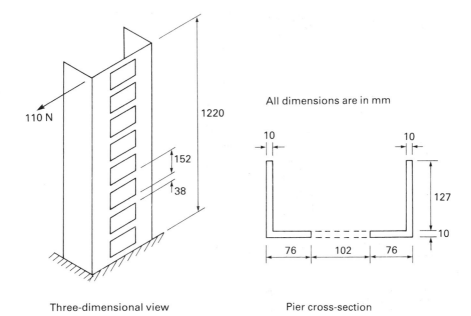

All dimensions are in mm

Three-dimensional view   Pier cross-section

Figure 3-1a Non-planar shear walls (all dimensions are in mm). Material data: $E = 2760$ MPa, $\nu = 0.148$, $G = 1020$ MPa

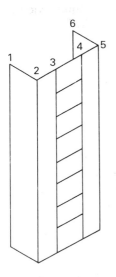

Figure 3-1b Finite strip mesh with discrete beams

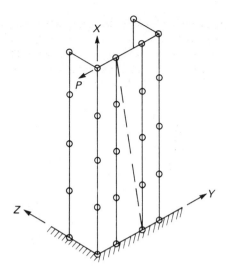

Figure 3-1c   Higher-order finite element mesh

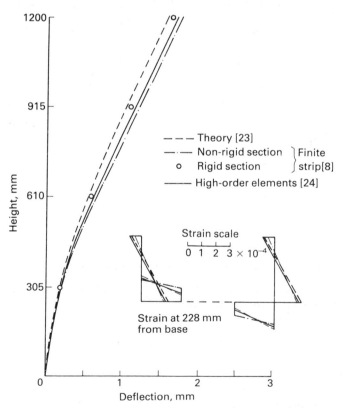

Figure 3-1d   Deflections and strains of the non-planar coupled shear wall

bending action of the walls, the total number of degrees of freedom for the problem amounts to only 45.

The transformation matrix in connection with a node i and a reference node O on the same horizontal plane is similar to Eqn 1-39, although since bending action has been neglected, there is no need to retain the out-of-plane displacements for the local coordinate system. Thus,

$$\begin{Bmatrix} u_i' \\ v_i' \end{Bmatrix} = \begin{bmatrix} \cos \beta & \sin \beta & x_i \sin \beta - z_i \cos \beta & 0 \\ 0 & 0 & 0 & 1 \end{bmatrix} \begin{Bmatrix} u_o \\ w_o \\ \theta_o \\ v_i \end{Bmatrix}$$

$$= [T]_i \begin{Bmatrix} u_o \\ w_o \\ \theta_o \\ v_i \end{Bmatrix} \tag{3-1}$$

and

$$[S]_{ij} = [T]_i^T [S']_{ij} [T]_j \tag{3-2}$$

The transformed submatrices are then assembled into their respective locations in the overall stiffness matrix.

The torsional resistance of the structure is included by the incorporation of the following simple stiffness matrix

$$\begin{matrix} \theta_1 & \theta_2 & \theta_3 & \theta_4 & \theta_5 \\ \begin{bmatrix} C & -C & & & \\ -C & 2C & -C & & \\ & -C & 2C & -C & \\ & & -C & 2C & -C \\ & & & -C & C \end{bmatrix} \end{matrix} \tag{3-3}$$

where C is the torsional rigidity and is given by

$$C = GJ/l$$

with

$$J = \sum J_{walls}$$
$l =$ distance between two consecutive nodes in the vertical direction.

The deflections and stresses are presented in Fig. 3-1d, and the agreement between the different sets of results is excellent.

4 Plane frames

4.1 Static analysis

A ten-story plane-frame structure subjected to the loading system shown in Fig. 4-1a is analysed by the frame method and by the finite strip method. In the latter

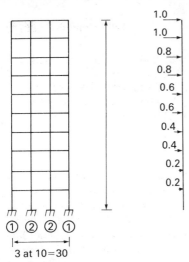

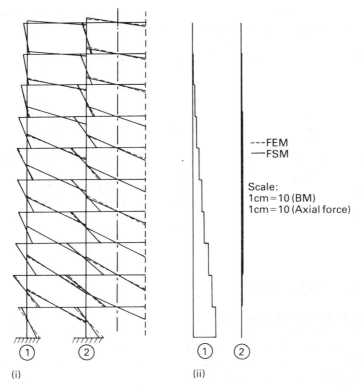

Figure 4-1a Three-bay plane frame structure: $E = 1$, $\rho = 1$, beam and column section $= 1 \times 1$

Figure 4-1b Stress resultants in members of a three bay plane frame structure
   (i) BMD plotted on compression side
   (ii) Axial forces in columns 1 and 2 (Linear dimensions are not to scale)

38–46

analysis, the finite line (long column) elements are used in conjunction with the discrete beam model. The bending moments of all members and the axial force of the columns are shown in Fig. 4-1b. It can be seen that there is perfect agreement between the two sets of results everywhere else except for the moments of the ground floor columns, where there are some discrepancies.

4.2   Vibration analysis

The free vibration of the plane frame shown in Fig. 4-1a is considered here. Both the continuum model using lower-order strips and line elements with discrete beam model are used in idealizing the structure, and rigid section constraint is imposed. The frequencies computed are tabulated in Table 4-1 and they agree well with those obtained from matrix frame analysis. From the point of computa-tional economy, the continuum model is more attractive in vibration analysis since less computation effort and expenditure are required in producing results which are generally acceptable for design purpose.

Table 4-1   Natural frequencies of plane frame structure

Type of model	No. of terms	DOF	Angular frequencies $\times 10^{-3}$ rad/s		
			ω_1	ω_2	ω_3
	3	18	0.742	2.264	4.360
Continuum	4	24	0.729	2.264	3.907
	5	30	0.727	2.218	3.906
Line element	3	27	0.761	2.324	4.360
with discrete	4	36	0.747	2.323	4.042
beam	5	45	0.745	2.283	4.037
Frame analysis		120	0.739	2.263	3.956

5   Frame shear wall structure

5.1   Response of structural components

Typical frame shear wall structures are shown in Fig. 5-1a, and it can be seen that the frame can be hinged or rigidly jointed to the wall. It is well-known that a frame deflects predominantly in a shear mode (Fig. 5-1b) while a cantilever wall will deflect predominantly in a bending mode (Fig. 5-1c), and interacting forces between the two structural components (Fig. 5-1d) will develop as a result of the imposition of equal deflections at each story level. The interacting forces tend to pull back the cantilever toward the top end, and will therefore help in reducing the maximum deflection of the structure.

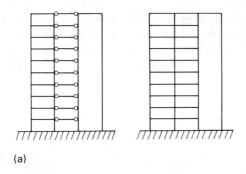

(a)

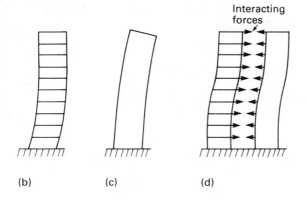

Interacting
forces

(b)   (c)   (d)

Figure 5-1   (a) Frame shear wall structures
             (b) Frame shear mode
             (c) Cantilever bending mode
             (d) Combined frame shear wall mode and interacting forces

5.2   Example

Figure 5-2a shows a 20 story frame shear wall structure [25]. The structure consists of a solid wall 9 in. thick and 35 ft in width, coupled to two columns 24×30 in. in cross-section by 20 slabs 9 in. thick with an equivalent width of 130 in. over a span of 35 ft. The elastic and shear moduli of the material are 3×10^6 and 1.2×10^6 psi, respectively. The deflection and the stress resultants in various members from both the frame method [25] and the finite strip method using non-rigid lower-order strip and line elements with the discrete beam model are plotted against the height of the structures, as shown in Fig. 5-2b. All results are in very good agreement.

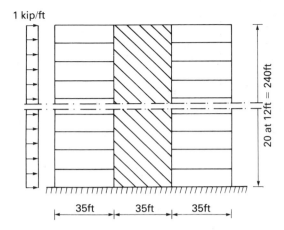

Figure 5-2a Coupled frame shear wall structure

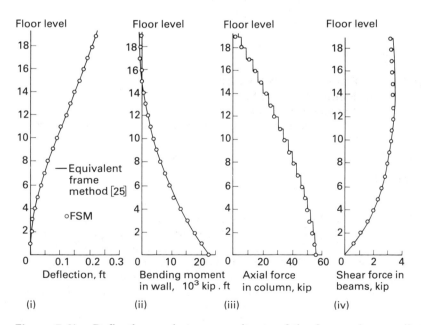

Figure 5-2b Deflection and stress resultants of the frame shear wall structure

Appendix

Integrals for Eqns 1-37, 1-38, 1-46, 1-56

$$I_a = \sum E_{1r} t_r [Y_m Y_n]_r$$

$$I_b = \sum G_r t_r [Y'_m Y'_n]_r$$

$$I_c = \sum E_{3r} t_r [Y''_m Y_n]_r$$

$$I_d = \sum E_{3r} t_r [Y_m Y''_n]_r$$

$$I_e = \sum E_{2r} t_r [Y''_m Y''_n]_r$$

$$I_1 = \sum D_{xr} [Y_m Y_n]_r$$

$$I_2 = \sum D_{1r} [Y''_m Y_n]_r$$

$$I_3 = \sum D_{1r} [Y_m Y''_n]_r$$

$$I_4 = \sum D_{yr} [Y''_m Y''_n]_r$$

$$I_5 = \sum D_{xyr} [Y'_m Y'_n]_r$$

$$I_{ma} = \sum (\rho t)_r [Y_m Y_n]_r$$

$$I_{mb} = \sum (\rho t)_r [Y'_m Y'_n]_r$$

$$I_{\bar{x}} = \sum t_r \bar{X}_r [Y_m]_r + \sum X_{sr} [Y_m]_r$$

$$I_{\bar{y}} = \sum t_r \bar{Y}_r [Y'_m]_r + \sum Y_{sr} [Y'_m]_r$$

$$I_{\bar{z}} = \sum t_r \bar{Z}_r [Y_m]_r + \sum Z_{sr} [Y_m]_r$$

$$I_{xy} = \sum X_{1r} [Y_m]_r$$

$$I_{zy} = \sum Z_{1r} [Y_m]_r$$

$$I_{xi} = \sum X_{1r} C_{ui}(x) [Y_m]_r$$

$$I_{yi} = \sum Y_{1r} C_{vi}(x) [Y'_m]_r$$

$$I_{zi} = \sum Z_{1r} C_{wi}(x) [Y_m]_r$$

$$I_{my} = x \sum Z_{1r} [Y_m]_r$$

$$I_{myi} = \sum Z_{1r} C_{\theta yi}(x) [Y_m]_r$$

$$I_{mzi} = \sum Y_{1r} C_{\theta zi}(x) [Y'_m]_r$$

$$I_{xj} = \sum X_{1r} C_{uj}(x) [Y_m]_r$$

$$I_{yj} = \sum Y_{1r} C_{vj}(x) [Y'_m]_r$$

$$I_{zj} = \sum Z_{1r} C_{wj}(x) [Y_m]_r$$

$$I_{myj} = \sum Z_{1r} C_{\theta yj}(x) [Y_m]_r$$

$$I_{mzj} = \sum Y_{1r} C_{\theta zj}(x) [Y'_m]_r$$

$$f_x = \sum_{\alpha=1}^{A} X_{c\alpha} Y_m(y_\alpha)$$

$$f_z = \sum_{\alpha=1}^{A} Z_{c\alpha} Y_m(y_\alpha)$$

$$f_{xi} = \sum_{\alpha=1}^{A} X_{c\alpha} C_{ui}(x) Y_m(y_\alpha)$$

$$f_{yi} = \sum_{\alpha=1}^{A} Y_{c\alpha} C_{vi}(x) Y'_m(y_\alpha)$$

$$f_{zi} = \sum_{\alpha=1}^{A} Z_{c\alpha} C_{wi}(x) Y_m(y_\alpha)$$

$$f_{my} = x \sum_{\alpha=1}^{A} Z_{c\alpha} Y_m(y_\alpha)$$

$$f_{myi} = \sum_{\alpha=1}^{A} Z_{c\alpha} C_{\theta yi}(x) Y_m(y_\alpha)$$

$$f_{mzi} = \sum_{\alpha=1}^{A} Y_{c\alpha} C_{\theta zi}(x) Y'_m(y_\alpha)$$

$$f_{xj} = \sum_{\alpha=1}^{A} X_{c\alpha} C_{uj}(x) Y_m(y_\alpha)$$

$$f_{yj} = \sum_{\alpha=1}^{A} Y_{c\alpha} C_{vj}(x) Y'_m(y_\alpha)$$

$$f_{zj} = \sum_{\alpha=1}^{A} Z_{c\alpha} C_{wj}(x) Y_m(y_\alpha)$$

$$f_{myj} = \sum_{\alpha=1}^{A} Z_{c\alpha} C_{\theta yj}(x) Y_m(y_\alpha)$$

$$f_{mzj} = \sum_{\alpha=1}^{A} Y_{c\alpha} C_{\theta zj}(x) Y'_m(y_\alpha)$$

where

$$\sum = \sum_{r=1} = \text{summation of the values of all } R \text{ sections from the bottom to the top of the strip or line elements}$$

$$[\ ]_r = \int_{y_r}^{y_{r+1}} (\ )\,dy \quad \text{of section } r$$

$$(\ )' = d(\ )/dy$$

$C_{ui}(x)$ = component of shape function in x-coordinate associated with displacement u at nodal line i

X_{sr} = surface pressure in x-direction

Y_{lr} = line load in y-direction

$Z_{c\alpha}$ = concentrated load in z-direction

A = total number of concentrated forces acting on a strip or line elements

References

1. MacLeod, I. A., Lateral stiffness of shear walls with openings, *Tall buildings*, Pergamon Press, Oxford, 1967, pp. 223–244.
2. Jenkins, W. M. and Harrison, T., Analysis of tall buildings with shear walls under bending and torsion, *Tall buildings*, Pergamon Press, Oxford, 1967, pp. 413–449.
3. Zienkiewicz, O. C., *The finite element method*, 3rd Edn, McGraw-Hill, New York, 1977, 787pp.
4. MacLeod, I. A., New rectangular finite element for shear wall analysis, *J. Struct. Div. Am. Soc. Civ. Eng.*, Vol. 95, No. ST3, 1969, pp. 399–409.
5. Chan, H. C. and Cheung, Y. K., Analysis of shear walls using higher order elements. *Bldg and Environ.*, Vol. 14, No. 3, 1979, pp. 217–224.
6. Cheung, Y. K., *Finite strip method in structural analysis*, Pergamon Press, Oxford, 1976, 232 pp.
7. Cheung, Y. K. and Kasemset, C., Approximate frequency analysis of shear wall frame structures, *Int. J. Earthquake Eng. Struct. Dynamics*, Vol. 6, 1978, pp. 221–229.
8. Cheung, Y. K. and Swaddiwudhipong, S., Analysis of frame shear wall structures using finite strip elements, *Proc. Inst. Civ. Eng.*, Part 2, Vol. 65, No. 8116, Sept. 1978, pp. 517–535.
9. Swaddiwudhipong, S., *A unified approximate analysis of tall buildings using generalized finite strip method*, PhD Thesis, University of Hong Kong, 1979, 264 pp.
10. Ghali, A. and Neville, A. M., *Structural analysis*, Chapman and Hall, London, 1978, 779 pp.
11. Michael, D., The effect of local wall deformations on the elastic interaction of cross walls coupled by beams, *Tall buildings*, Pergamon Press, Oxford, 1967, pp. 253–270.
12. Bhatt, P., Effect of beam–shear wall junction deformations on the flexibility of the connecting beams, *Bldg Sci.*, Vol. 8, No. 2, 1973, pp. 149–151.

13. Przemieniecki, J. S., *Theory of matrix structural analysis*, McGraw-Hill, New York, 1968, 468 pp.
14. Abu-Ghazaleh, B. N., *Analysis of plate-type prismatic structures*, PhD Thesis, University of California, Berkeley, 1965.
15. Sisodiya, R. G., *Finite element analysis of bridges*, PhD Thesis, University of Calgary, 1971, 162 pp.
16. Sisodiya, R. G. and Cheung, Y. K., A higher order in-plane parallelogram element and its application to skewed girder bridges, Rockey, K. C., *et al.* (Eds), *Developments in bridge design and construction*, Crosby Lockwood, London, 1971, pp. 304–317.
17. Zienkiewicz, O. C. and Cheung, Y. K., The finite element method for analysis of elastic isotropic and orthotropic slabs, *Proc. Inst. Civ. Eng.*, Vol. 28, 1964, pp. 471–488.
18. Cheung, Y. K. and Chan, H. C., A family of rectangular bending elements, *Computers and Struct.*, Vol. 10, 1979, pp. 613–619.
19. Mindlin, R. D., Influence of rotary inertia and shear on flexural motions of isotropic, elastic plates, *J. Appl. Mech.*, Vol. 18, No. 1, 1951, pp. 31–38.
20. Jennings, A., *Matrix computation for engineers and scientists*, Wiley, New York, 1977, 330 pp.
21. Bathe, K. J. and Wilson, E. L., *Numerical methods in finite element analysis*, Prentice-Hall, Englewood Cliffs, NJ, 1976, 528 pp.
22. Coull, A. and Subedi, N. K., Coupled shear wall with two and three bands of openings, *Bldg Sci.*, Vol. 7, No. 2, 1972, pp. 81–86.
23. Tso, W. K. and Biswas, J. K., General analysis of non-planar coupled shear walls, *J. Struct. Div. Am. Soc. Civ. Eng.*, Vol. 99, No. ST3, 1973, pp. 365–380.
24. Chan, H. C. and Cheung, Y. K., Lateral and torsional analysis of spatial wall systems using higher order elements, *Struct. Eng.*, Vol. 58B, No. 3, Sept. 1980, pp. 67–70.
25. Chan, P. C. K. and Heidebreckt, A. C., Stiffening of shear walls, *Can. J. Civ. Eng.*, Vol. 1, No. 1, 1974, pp. 85–96.

Part VI Practical Considerations

39 The effect of structural design on safety, economy, workmanship, completion time and performance of the finished structure

G M J Williams

Scott Wilson Kirkpatrick & Partners, Basingstoke, England

Contents

Summary

This chapter draws together those general matters that the designer needs to take into account when specifying concrete construction. In particular it describes his role in ensuring that a structure meets the requirements of all interested parties, especially the owner, the architect and the engineers responsible for the building services.

Since variations in materials, labour available, working practices and on-site conditions can all influence the detailed development of a construction project it is the designer's function to see that the design concept is right to begin with and that a proper balance is maintained when compromises become necessary between performance, cost, and time.

The chapter discusses especially structural safety on site, stability in construction, structural performance, and fixings into concrete, always with an eye to what is most economical in time and cost.

1 Introduction

The designer of a structure should always appreciate that its owner will finally judge it from three fundamental standpoints: first, its performance throughout its lifetime judged according to his needs; secondly, its cost, both initially and later, judged by the costs that he was led to expect when he committed himself to build it; and finally the time that was required to achieve it, judged again by the construction time that he was advised at the outset. Thus, one of the designer's basic responsibilities must be to ensure that his client fully appreciates his intentions in each of these areas and, furthermore, that the client appreciates the various risks and uncertainties that could frustrate the original intentions. This will include such risk factors as variations in cost during construction, the extent to which difficulties might arise in developing the design in detail, and unexpectedly adverse weather conditions.

Most design concepts will be a compromise between the needs to fulfil each of these basic requirements—for instance, between first cost and maintenance costs, between performance requirements and cost, between construction cost and construction time, and so on. Again, the designer should ensure that this client understands and accepts the nature of each of these compromises. Many of the differences that arise after a building's completion between its owner and its designer arise from misunderstandings about these compromises.

The designer will also appreciate that skilled workmanship is required before his design can be realized, and the quality of the available workmanship varies greatly from one locality to another. Therefore, the designer must adapt his design to suit the resources and the degree of skill that are available where it is to be built, and at all stages of the design he should try to avoid practical construction difficulties. No structure can be erected to comply absolutely with every word of the contract specification or with every detail on the drawings, and although it might be possible legally to insist on the demolition and rebuilding of offending work, it is often impracticable to do so. The designer must then exercise discretion whether to accept the work as it is, imperfect as it may be, or after repairs short of complete replacement. For instance, if the strength of the concrete is in question, there will be many parts of the structure where the full specified strength will not be required. However, only the designer is in a position to decide when substandard work can be accepted and it is difficult for him to delegate this function. Therefore, designers should always be involved in supervising the construction of their work, and the practice of separating the design function from the supervision of construction is to be deplored.

Every aspect of this handbook will bear to some extent on the subject of this chapter, and no attempt is made to repeat what is contained elsewhere. Rather, the chapter attempts to draw attention to those items of a general nature that should be taken into account by a designer, although they might not be explicitly referred to in textbooks on design or in design codes of practice.

2 Design concepts

Probably the most important phase of the designer's work is settling the design concept at the outset, because the choice of the wrong concept will not be readily recovered at later stages. Generally, the concept of a structure will depend on other than purely structural considerations, because few structures are erected for their own sake, but to support something else. In most cases, the structure will contribute a relatively small part of the total cost of the entire project, so that there may have to be departures from the cheapest structure to provide greater savings in other elements of the entire building. Thus, and particularly at the conceptual stage, the designer should liaise closely with the other parties involved in the design, including the owner, the architect and the engineers responsible for the building services, to ensure that the structure will meet their requirements.

Again, construction work is usually very dependent on local labour and materials, which are not readily transported over long distances in the required quantities. Thus, from the outset, the designer should ensure that his design is compatible with the locally available resources—that it does not make too great demands on the labour and plant available, and that suitable materials, particularly aggregates, are to be had. Aggregates are usually natural material and in consequence are subject to great variations. The unconsidered use of unfamiliar aggregates has given rise to various problems in the past, including excessive aggregate shrinkage, chemical reactions between aggregate and cement, the introduction of various salts, frost susceptibility and problems of insufficient strength and unsuitable particle shape.

While progress and the introduction of new methods and processes should not be unnecessarily restricted, designers should be cautious about departing from long-established practice in a locality. Any defects that might be inherent in new forms of concrete construction often take several years to become manifest—sometimes a decade—and by that time a new method offering a marked advantage could have come into wide use. The consequent remedial works can then be widespread and expensive. It is a wise rule only to adopt novel and relatively untried concepts when there is a clearly identified and substantial advantage to be had through their use that will offset the risk of unforeseen difficulties arising—and even so the designer should ensure that his clients fully appreciate the relative risks and advantages of the unorthodox approach.

3 The structure

3.1 The substructure

One of the first considerations of the designer, and certainly among the most far-reaching in its effect on the project, is the choice of foundation. This will depend on a proper site investigation, a full treatment of which is beyond the scope of this handbook, but from the construction point of view shallow foundations are the easiest to form. The depth of excavation required for any particular foundation must be determined from the nature of the soil or rock providing the foundation and from the loading of the proposed building. To keep the excavation as shallow as possible, it may be possible to spread the load over the larger area at a shallower depth. If pad footings go below a depth of 1.5 m, the excavation will have to be shored or the side battered. When the size and spacing of bases have been determined, the designer must consider the construction process involved. If, in the construction of the bases, shoring is required or over-dig is needed to form a batter to the sides, an alternative strip or raft design might be cheaper and quicker. The contractor might use larger machinery to excavate a strip or raft with a saving in construction time. Each situation must be considered on its merits and the strip or raft-type foundation becomes more attractive if bases are in close proximity. Where deeper excavations are required to achieve the necessary bearing capacity, the possibility of incorporating a basement as part of the substructure may be considered. The decision to construct a basement is important, as the floor area there is often the most expensive to provide in the building.

On open sites, it may be possible to batter the surrounding ground away from the foundations and this has the advantage that there are no temporary retaining loads to cater for and dewatering may be achieved by well-point or sump drainage outside the boundary of the building.

The retaining walls around basements can be built from sheet piles, bored piles or can be concrete diaphragm walls. The bored piles and diaphragm walls obviously will be a permanent part of the substructure, whereas sheet piles can sometimes be withdrawn after forming the back shutter for basement retaining walls. This method is simple if the cantilever action of the retaining wall is relied on to support the excavation until the floors are constructed, but as excavations become deeper the cantilever moments produced and the depth of passive pressure required will mean additional support must be provided to the wall. Support of retaining walls in deep excavations can be achieved by

(a) temporary shoring of the piles from inside the site;
(b) supporting sheet piles by using ground anchors outside the site boundaries;
(c) propping of the sheet piles using the substructure as a support.

Temporary shoring is placed into position as the excavation proceeds. The two basic methods available are

(a) the provision of raking supports from ground left undisturbed in the centre of the excavation;
(b) the use of framing across the complete width of the excavation.

Both the variations of temporary shoring have the same disadvantages in that the members used for shoring impose a severe restraint on movement within the excavation. This is particularly so where a framing method is used and these frames become increasingly expensive and cumbersome as the width of the excavation increases. If props are used from a central bund, the permanent retaining wall built around the boundary of the site must support the surrounding ground so the temporary shores can be removed and the central ground excavated. This will mean the retaining wall will be more substantial than would normally be the case. An alternative method to internal shoring is the use of ground anchors.

Ground anchors have a great advantage in that they achieve their support by anchoring into rock or ground outside the excavation. The substructure is left unobstructed and there can well be a considerable saving on construction time due to ease of operation. The actual ground anchors can be installed quickly and can be used with retaining walls made of sheet or bored piles and diaphragm walls. A disadvantage with ground anchors is when they are used in more developed areas they may well extend under adjoining developments and great care must be given to the possible effect of the anchors on these structures or an adjacent road.

The third method of supporting deep excavation is by using the permanent substructure to carry the propping loads. The basis of this method is that horizontal propping loads are carried by the floors of the basements. Vertical elements are formed first from ground level. The floors are then progressively built down from ground level and the earth is excavated from underneath the completed floor. Using this method, the shoring forces provided by the temporary frames in previous methods are designed into the permanent works and are carried by the plate action of the floors. A disadvantage of this method is that excavation is carried out under completed floors, which restricts the removal of excavated material but, nevertheless, when a deep excavation and a number of basement levels are required, it is often cheaper to use the permanent basement floors as struts during construction.

A variation of this method which has proved to afford speedier construction is the 'up and down' construction technique (Fig. 3-1). The main steps in this construction technique are illustrated below. The central load-bearing elements of the building are constructed as the first operation. When this has been done, the superstructure construction can proceed simultaneously with the substructure construction, greatly reducing the overall construction time. Where buildings have central cores as the main load-bearing elements, great rates of progress can be achieved by slipforming the core and 'hanging' floors from a truss at the top of the core, leaving lower levels completely free of superstructure construction.

The treatment of walls surrounding a basement must be considered in relation to the client's requirements and the intended use of the area. If it is to be used for

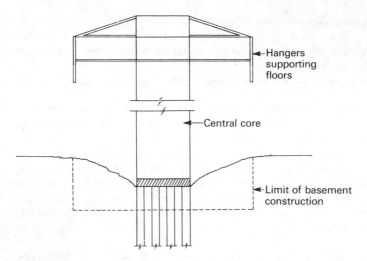

Figure 3-1 Up and down construction

car parking or if the ground is always dry, it may be adequate to rely on the structural concrete alone to provide waterproofing, usually with water stops in the construction joints. For other uses, it may be necessary to upgrade the water-proofing requirements by providing either tanking or forming a drained cavity inside the structural retaining wall.

To be reliably watertight in itself, a basement wall should be built of dense impermeable concrete. To this end, the concrete mix should have a high cement content and as low a water/cement ratio as is practical for thorough compaction. Concrete water-retaining structure codes can be adopted to eliminate as much cracking as possible. If normal stresses for structural concrete are assumed, the recommended minimum areas of reinforcement may have to be increased to ensure crack widths are kept to acceptable levels. It is recommended that the distance between construction joints does not exceed 5 m (16 ft) in walls and 10 m (32 ft) in slabs to minimize cracking, due to initial restrained thermal shrinkage.

In theory, basement walls in wet ground could be built to meet the require-ments for water-retaining concrete structures, but in practice it is difficult to ensure that they are entirely watertight when the complexities of most basement layouts are encountered. It is more usual to use the higher stresses allowed for normal structural concrete, often with generous secondary reinforcement to resist local cracking, and closely-spaced construction joints. Water-resistant membranes of asphalt or bitumen can be applied as tanking to the outside of the basement. The membrane is usually applied externally to the wall and care is required to prevent it being ruptured. To achieve this, the membrane can be sandwiched between 50 mm (2 in) thick blinding layers under floors or be protected by blockwork walls in the case of vertical members. This type of basement protection is usually more expensive than using drained-cavity construction and the introduc-tion of another trade to the program may involve a longer construction period. It is also difficult to exactly locate any damage in the membrane when the work is finished, as water can travel through joint surfaces and other weaknesses in the wall and appear on the inside of the basement some distance from the defect in the membrane.

The alternative common method of damp-proofing a basement is by providing an inner blockwork or brickwork wall inside the concrete wall. Any water coming through the wall is collected between the walls. By providing ventilation to the cavity, any dampness is taken away and does not affect the usable basement area. When this type of construction is used, it is important to remedy any defects in the concrete wall before the inner skin is built. This will be particularly important in the case of diaphragm walls.

Many considerations enter into the choice of a piled foundation, including the choice between the various types of pile available. Reference should be made to specialized textbooks on foundations and piling (see also Chapter 31).

3.2 Floor design

The superstructure of the building provides the main working areas. Each building will have its own function and will have to satisfy a number of requirements dictated by its intended use. When these have been established, the structural engineer must decide, from the options open to him, the framing and floor type for the building. Generally, there will not be an unrestricted choice of structural form open to the structural engineer—it will be secondary to architectural requirements, and sometimes to the building services.

As an initial consideration, from the construction viewpoint the more times an operation may be repeated the better. Repetition of operation has the advantage that for, *in situ* work, the same formwork can be used repeatedly. Timber formwork can be used ten or more times before the need for replacement arises and if, for instance, liftshaft and floor soffits maintain the same configuration, the maximum economy can be realized from the formwork. To this purpose, it may be necessary to oversize certain members to maintain the overall pattern and hence achieve an overall economy. Where operations are repeated to a considerable extent, the possibility of precasting members should be considered, as this may well save on overall construction time (see Chapter 18).

Another advantage of repeating operations is that the labour force becomes progressively familiar with the sequence of work. One example of a cycle of operations is one floor to another. Provided the sequence of operations and the member sizes (of, say, liftshaft, staircase and floor arrangements) remain the same, any difficulties experienced on the first cycle of operation should be eliminated from subsequent operations. This should lead to a progressive reduction of the turn round time as the construction proceeds.

The same principle of repetition of operation can also be applied to other aspects of the work. Detailing reinforcement, for instance, is one example where the designer can assist with the ease of site operation. If reinforcement can be repeated from floor to floor, the overall number of bar types, sizes and length can be greatly reduced. This will help considerably with the storage of reinforcement before fixing and the ease with which required bars can be located. If beam reinforcement is not only repetitive, but the bars are detailed so that cages can be made up away from the construction site and dropped into position, it is often even more economical. In floor areas where bars are detailed at the same centres and in beams where links are positioned at the same centres, considerable savings can be achieved in fixing times. To achieve a repetition in detailing, the designer may well consider that it is beneficial for the overall economy of the building to design for the worst case of floor loading and make other floors the same, provided the difference between them is small. The actual difference in reinforcement costs may indeed be minimal in such situations, *but* the construction benefits considerably from the repetition of operation.

When the requirements of a building have been established, the overall structural design concept will have to be formulated; and when the type of floor to be used is being considered in relation to the structural system, it may prove that the most economic structural layout is not the most economic overall layout when mechanical and architectural requirements are considered. Generally, it is an advantage to have the minimum floor-to-floor height, thus reducing the height of internal and external partitions. If the structural configuration is such that services have to add to the structural depth sooner than be incorporated within it, the overall economic structure may not yet have been achieved.

Another consideration in the early design is the type of soffit the architect may require. If finishes can be applied directly to the soffit of the floors, the need for a false ceiling may be eliminated. Not only can an architect make good use of a flat ceiling but some coffered and troughed ceilings have been shown to advantage.

The most common form of construction for floor slabs is the beam and slab method. The building is formed into an overall frame system where slabs span one or two ways onto drop beams which frame into vertical elements. This method has become so common, as it is the most adaptable with the most economic slab and beam depth. The overall frame produced is also a useful means of catering for structural loads and transferring them to the foundations. In situations where it is not possible to build a degree of repetition into a structure, formwork has to be remade to form the required member sizes. The disadvantage of the system is that it is not often possible to incorporate large openings in beams to allow for the distribution of mechanical services and the resulting soffit may not be aesthetically pleasing and may require a false ceiling. The alternative method is to produce a flat slab.

A flat slab overcomes the previously mentioned disadvantage of the drop beam and slab construction in that it provides a surface to accept ceiling finishes and the omission of drop means that a free choice of service routes is left open to the mechanical engineer. The overall structural system, however, is not as stiff as beam and slab. The use of beams with their greater depth means that there will be less deflection than flat slab construction and, in many cases, the deflection of the slab will limit the spans. Where heavy floor loading provisions are required, columns may have to be placed at close centres and the benefit of flat slabs is not realized under these conditions. Where loads are lighter, as in the case of apartment blocks, flat-slab construction can be a considerable advantage.

The shear force in flat slabs is usually high around the supporting columns and to cater for this force, column heads and drops may be required. The advantage of having a completely smooth soffit is lost in this case, but the space is still available between drops for mechanical and electrical services. The increase in the allowable shear force can also mean that longer spans may be achieved and the soffit produced is usually quite acceptable for car parks or for industrial uses.

The overall construction depth of both beam and slab construction increases considerably when the spacing of columns is increased and, as more of the carrying capacity of the structure is taken up by its own weight, methods must be considered to lighten the slab construction.

The most common way of doing this is to provide a ribbed slab spanning either one way or two ways. Using beam strips to span between columns, the overall construction depth of the ribbed slab and supporting beam can often be made the same. Again, when a regular pattern is produced, the soffit may be acceptable from an aesthetic consideration but large-sized ductwork and pipework will generally have to be carried underneath the soffit.

The process of forming the ribbed profile out of timber formwork is quite a complicated process and quite time consuming. Plastic moulds are available on the market to form either the trough in a one-way spanning slab or the waffle effect in a two-way spanning slab. During the design process, consideration can be given to the size and shape of moulds in deciding on the spacing of ribs. The advantage to the construction of using moulds is quite considerable. A flat formwork soffit is required to receive the moulds or, in some cases, only bearers need be provided. The supporting system required can be based on a flat soffit and does not have to follow the complicated profile of the ribs, as the moulds are capable of supporting the dead load of the concrete. Where the construction depth is fairly shallow and the need exists to reduce weight, hollow clay pots may be used as a void former. The weight of the pot means that the weight saving of forming a waffle or trough floor is not completely realized but the smooth soffit produced means that ceiling finishes can be directly applied.

Where the designer has managed to maintain a flat soffit to a slab, or in the case of ribbed floors has used moulds, the contractor has the option of using 'table-forms' or 'flyingforms' (Fig. 3-2). The method consists of making up a form-work and falsework soffit support system which can be used from floor to floor.

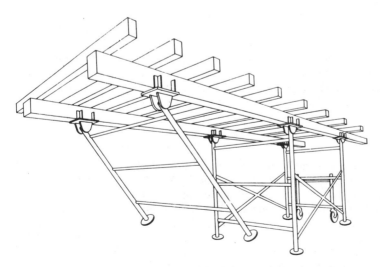

Figure 3-2 Tableform or flyingform floor-support system

When a floor has been cast and the formwork has been released, sections of formwork are then withdrawn by crane through the outside of the building and placed onto the complete slab ready for the slab above. Where the use of tableform is a possibility, details of the structure need to be considered to allow the formwork to be removed as easily as possible. Downstand edge beams are usually not possible with this formwork system and have to be replaced with upstand beams in the external wall of the floor above. As rapid turnrounds can be achieved using this system, there is an obvious advantage in being able to strike the formwork as soon as possible and this should be considered in the choice of concrete mix. The most economic application of tableforms is in multistory structures where a degree of repetition occurs but the system may also be chosen for low-rise buildings provided there is again a degree of repetition. Although the main application is to flat soffits, tableform may be used with beam and slab construction, provided no transverse beams are used in the area where this type of formwork is used.

An alternative *in situ* floor construction which has proved particularly suitable for tall buildings is the use of a profiled steel deck acting compositely with a concrete topping. The steel deck provides the formwork for the concrete, which may be a normal or lightweight mix. The floor slab produced in this manner is relatively lightweight, resulting in lower foundation and column loads. As only minor support is required to the deck while the concrete is placed in position, rapid turnrounds from floor to floor can be achieved. A disadvantage lies in the fact that fire protection will be needed beneath the profiled steel deck.

One specialist method of construction is the lift-slab technique. There must be an early decision in the design process in favour of this system, as quite a number of special details are involved. After the foundations have been finished, concrete or steel columns are erected. Floor slabs are then cast around the columns at ground level using the ground surface as the initial soffit. A separating membrane is used between each slab and when the required number of slabs has been cast they are raised into position by means of jacks attached to the tops of columns. A special connection detail is required between columns and floor slabs, which has to be cast in both elements. The advantage of the system is that slabs are cast at ground level and as previous slabs or the ground have been used to provide the soffit, formwork is largely eliminated from the construction process. The floor slabs produced can incorporate waffles, but basically to achieve the stacking of floors the soffit must be flat.

3.3 Precasting

When choosing the structural form of a building, consideration will be given to precasting at least some of the structural elements. Buildings can consist of almost all precast elements jointed together or only some of the elements can be precast.

For a building to consist of almost entirely precast units, the structure must be divided into suitable-sized elements that can be cast, carried to site if necessary, erected and adjusted in line and level. The continuity of the structure must be considered at an early stage, because the connections between the various units are usually the most critical element in the design. The decision to use precasting techniques should be made early and designing repetition into a building at an early stage is much more effective than trying to modify sections to obtain repetition when the overall concept of the structure has been formulated (see Chapter 18).

The final structural configuration will have to be divided into manageable elements which, when connected together, can resist any expected vertical and lateral design loading. Special and expensive joint details usually result when precast frames are required to be moment resisting and it is for this reason that the designer must decide to what extent the building is precast. Almost total precasting will result in an expensive building, but the construction time will be short. If *in situ* elements are used in the building, such as *in situ* shear, lift or staircase walls to resist lateral forces, building costs will be reduced, but the construction time will increase. The resulting economy using precast elements usually comes by way of reducing the construction time and this must be balanced against the increase in building costs in deciding the optimum level of precasting for any particular structure.

Construction time is usually reduced by reducing the site work—precast units are usually made away from the work and are transported to site and erected as required. The operation of erection and jointing can be undertaken by a smaller work force than the equivalent *in situ* concrete construction, and work can often proceed in weather conditions unsuited to *in situ* work. The units used in the construction will have already reached their design strength and units can be loaded as soon as they have been erected and fixed in position.

Against this saving in construction time, there are factors to consider that may increase the cost of precast construction over *in situ*, including

(a) Higher strength concrete is often used than that which is required in *in situ* design, to facilitate early striking of units.

(b) As frames of precast units usually have reduced continuity, the members are often larger than those in continuous *in situ* construction.

(c) The joints can be expensive, if they are properly designed to resist corrosion and fire, and, in the case of precast cladding units, to be waterproof and windproof.

The optimum arrangement for many buildings is a combination of precast and *in situ* elements.

Not only engineering concepts figure in the decision whether to precast or not—architectural requirements will also merit consideration. Situations arise where, to obtain the highest quality of architectural finish, exposed parts of a structure are best precast. In units made under factory conditions, the highest standards of surface finish can be obtained consistently, and any blemished units can be rejected, whereas with *in situ* work most blemishes have to be accepted after repairs, which are never fully satisfactory. Fair-faced precast units can then either be used alone as non-structural cladding panels, or they can be used as permanent formwork for *in situ* reinforced concrete, thereby providing full continuity.

The floor system in a building has possibilities where precasting can often offer a viable alternative to *in situ* work. Many proprietary precast floor-slab units are available and adaptable to most floor spans. Usually, the segmented nature of flooring units means that some form of topping may be required. This may also be desirable in achieving monolithic action between floor slabs and the surrounding structure. Generally, precast floor units should be considered where floor panels are of repetitive rectangular shapes with few large holes, story heights are above average and fast construction is required with the minimum site labour.

Another aspect of building construction that may warrant precast units is the production of unusually shaped members. The time involved in construction and trying working conditions may mean precasting is the simpler solution. A staircase, for instance, has an awkward shape and has to be formed and poured in confined working conditions. Bearing this in mind, the possibility of precasting the stairflights will often prove attractive to a contractor.

When the decision has been in favour of precast units, the designer must consider the best method of forming joints. Where the joints are to provide a moment connection, special structural consideration must be given to the method of joining, as has already been stated. The types of joint will not be considered in this chapter, but joints involving welding or bolting may prove to be of advantage regarding length of construction times, as the alternative of using grout-filled joints means that the grout needs time to achieve its design strength. Where wall units and cladding panels are used, the fixing mechanism must incorporate practical construction considerations, particularly when the units serve non-load-bearing purposes.

The strength of the joints should be calculated for all possible forces on them, including those from accidental loadings such as vehicle impacts and domestic gas explosions. In general, it should be possible for any one joint to fail in a precast structure without a catastrophic collapse of the entire structure.

3.4 Slipforming

Slipforming is a construction method that was originally used in silos and chimney construction, but has been increasingly applied to the construction of building cores. Its popularity stems from the overall economy it can offer for high-rise buildings and a considerable saving in construction time.

Provided the design of the building core lends itself to this method of construction, then, for walls of 30 m (98 ft) or over, slipforming can be considered to be in contention with more traditional construction for the most economic method of working. Where a development includes a number of similar high-rise blocks, the economic height of 30 m for slipforming may be reduced.

The basis of the slipforming technique is to produce a continuous casting system for the core walls. Formwork is assembled at ground level to suit the proposed arrangement of staircase or lift-shaft walls. After the pattern has been established, the complete arrangement of formwork is raised, at a rate of about 300 mm/h, by a series of jacks. To climb up, the jacks use either cast-in rods or recoverable rods, positioned in the core walls, and as the jacks rise, so does the attached formwork. The process of concreting and steel fixing goes on around the clock and, apart from unforeseen difficulties, there is no reason why the complete height of the walls cannot be achieved as a continuous operation.

Using slipforming, cores of high-rise buildings can be completed in a matter of weeks. The actual operation of casting the concrete is usually a fraction of this, with the organization and assembly of equipment taking the majority of time in the construction period. If this technique of constructing the core is coupled with up and down construction methods, there can be a great saving in the length of construction time. The speed at which the core can be constructed means that the placing of flooring systems from the top down can proceed earlier. After the initial assembly of the formwork necessary for slipforming, the work to the core walls and flooring systems does not greatly interfere with basement excavation and construction.

When slipforming is undertaken, it will mean quite a large initial outlay on the part of the contractor. Large quantities of plant will be required to provide both the slipforming formwork mechanism and the services required during the operation. Generators, lighting systems, hoists, craneage, maintenance crews, labour force and facilities for the work have to be available on a 24 h basis.

Slipforming systems and expertise are provided by a specialist subcontractor with the main contractor providing the labour force. A great deal of prior planning will be needed to acquaint the labour force with this special type of construction to ensure trouble-free operations.

Supply of materials will also be required on a 24 h basis. If site conditions mean that there is no available space for concrete to be site-mixed, then suitable arrangements will have to be made with a ready-mixed supplier. Arrangements will also have to be made for protecting the placed concrete in cases of adverse weather conditions.

The other major problem to be tackled is the one of alignment. As the slipforming operation has become more popular, this problem has largely been solved. With the use of laser levels and optical plumbs, high levels of accuracy can be achieved, quite often higher than in traditional methods.

The time required to complete a building's core and the reduction of the length of the construction program are the main advantages slipforming can offer a contractor. The process means that no joints will be produced unless a halt occurs and there will be a saving on formwork materials. The high quality of the concrete surface produced by the use of metal shutters and the absence of joints may well prove beneficial from the architectural viewpoint and reduce the amount of required finishes. Although an expensive form of shift working is required during the 24 h slipforming operation, an overall saving in the labour force compared with traditional construction can be realized.

In most cases of specialized construction methods, the designer has his part to play. The core structure of a building formed by slipforming offers an inherently strong core that has been constructed monolithically. In forming the layout of walls for the core, the designer will have to maintain the same arrangement throughout the core height to allow slipforming. Tapering walls can be formed by modifications to the slipforming formwork, but the cost and difficulty in operation make this form of wall more expensive.

An important part of the design procedure must lie in the connection of floors to the core. Floor connection can be accommodated by casting in starter bars or by limiting floor connection to the core to selected points. Where starter bars are cast-in, they have to be arranged so that they do not interfere with the slipforming operation. This is usually achieved by casting in bars parallel to the walls and subsequently pulling them out before the floor is cast. The designer must satisfy himself that the floor–core connection is adequate to transfer all lateral and vertical load from the rest of the structure to the core. Where a large number of voids is formed in the core, for door, window or floor connections, the overall stability of the core in its temporary state must be considered. It may even be necessary to introduce extra walls or temporary propping to give the required stability in the temporary state.

The design of a suitable concrete mix is another early consideration. The design of the concrete mix will not be considered in this chapter, but the designer will have to ensure that the concrete strength, work stability, slump and initial setting time is suitable for the casting operation. A test of the concrete mix will probably be required to ensure the smooth running of the operations.

3.5 Prestressing

In buildings, prestressed concrete is most easily applied to precast construction. The introduction of prestressing techniques to *in situ* work generally involves a disruption of site working and introduces another aspect of construction in the work program. Extra equipment and a specialized labour force are required; whereas, if the prestressed element is in the form of a precast unit, normal building procedure can be followed. However, in simple multistory buildings with flat plate floors, prestressed *in situ* floor construction is frequently economic. Specialist sources should be referred to for the design of such construction.

Using prestressed precast units means that all the advantages of precast sections are realized with the advantage of reducing the length of the construction program. The additional advantage of prestressing is that, when considering floor and beam units, shallower construction depth can be produced compared with reinforced concrete. Prestressing achieves this by reducing the amount of deflection. A prestressed cable will lose some of its tensile force during the life of a building, but this is much less than the deflection producing creep that occurs in reinforced concrete. The resulting deflection will, therefore, be much less in prestressed than in reinforced concrete and for the same deflections to be produced, a thinner horizontal prestressed section will be required. The overall effect of using prestressed flooring systems can well result in minimizing the floor-to-floor height, thus producing an economy in the quantity of vertical elements and helping to reduce foundation loads. Possibly, the lightest flooring system that can be obtained utilizes voided prestressed precast decking made from lightweight aggregate. As greater span can be achieved using prestressed units, the need for a secondary beam system may be eliminated with 'off the shelf' prestressed units providing the flooring areas between primary beams.

Where concrete members are in tension, prestressing can be used to maintain an uncracked section. This can find a particular application when suspended construction is envisaged. Where the floor systems are hung from concrete trusses, supported by a core, high-tensile forces are produced in some members. Prestressing these members is one method of dealing with these forces and maintaining an uncracked section. The actual hangers holding the floor can also consist of prestressed concrete members.

Another application of prestressing is to provide stability into precast frames. By prestressing the precast vertical and horizontal units together, a degree of stability is produced that will enable the frames to carry lateral loads.

4 Workmanship considerations

4.1 Presentation of information

The clear presentation of information to the builder is essential if the construction process is to incorporate all the design requirements. The method of presenting this information should be standardized wherever possible to eliminate any misunderstanding. Under the British system, general arrangement drawings, reinforcement details, with complete bar bending information, and specifications are provided to the builder by the designer as the basis for construction. The US system differs in the fact that the design is presented by the designer to the contractor, who then usually produces the reinforcement-placing drawings and bending schedules.

At the tender stage, a contractor should be given as much information as possible. Any special requirements of the design should be made known to the contractor so that he can make allowance in his tender sum. Special requirements may include the provision of construction joints in specified positions to suit architectural requirements, areas of floor that need to be propped through more than one story height, any temporary construction loads designed into the structure, and other similar relevant information.

General-arrangement drawings are usually on a metric scale of 1:100 and should show all the dimensions required for formwork. Any major holes through the structure should be shown and dimensioned on the general arrangement. Minor holes and chases not particularly affecting reinforcement may be shown on the reinforcement-placing drawings, provided the builder is warned that they may be there and they are not confused with other information.

Reinforcement-placing drawings are shown at 1:50 scale or at a larger scale, if necessary, with larger-sized sections and all dimensions relating to the fixing of bars should be shown. Again, wherever possible, holes should be indicated and any bars detailed on other drawings, but pertinent to the area in question, should be indicated with a suitable reference. Clear mention should be made on the drawing of the relevant bending schedules to be used.

Care should be taken to ensure that all the reinforcement-placing drawings and bending schedules in a project are consistent in style and in the presentation of information, so that site labour will not be confused by occasional drawings that have a different presentation. In Britain, the detailing procedures published jointly by the Institution of Structural Engineers and the Concrete Society are strongly recommended.

4.2 Elimination of congested reinforcement

The codes of practice give the maximum areas of reinforcement that should not be exceeded for any particular concrete member. It is left to the designer's decision as to whether the arrangement and spacing of reinforcement bars enables successful compaction of the concrete in the member. Where laps occur, the reinforcement areas will be double that in other parts of the member in question and the staggering of laps may be necessary to reduce the congestion.

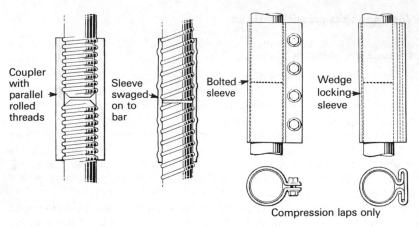

Coupler with parallel rolled threads

Sleeve swaged on to bar

Bolted sleeve

Wedge locking sleeve

Compression laps only

Figure 4-1   Patent splices

The use of mechanical splice connectors, as in Fig. 4-1, may be considered for heavily-stressed members as the staggering of laps will be achieved over a much shorter distance with less congestion. The coupler approximately doubles the diameter of the bars being spliced but the lapping of the bars is achieved over a much shorter length than by conventional means. A typical area of heavy congestion of reinforcement is where primary and secondary beams cross a column. In this instance, only essential beam reinforcement should cross over the column position and any lapping of beam bars should be located in the spans. The designer should ensure that adequate space is left between beam top bars or through the side of beams at the slab–beam connection to allow for placing and compacting concrete.

4.3 Detailing of laps

Generally, laps should be placed at areas of low stress. Under normal situations, beam reinforcement is lapped at approximately third-span positions and column laps are above floor level. The special requirements for buildings designed to resist earthquake loadings will not be considered in this chapter.

The setting-out dimensions for the reinforcement are shown on the relevant drawings and the main problem is providing adequate transfer of setting-out marks to the formwork, to enable the bars to be positioned accurately. If columns under floors are out of plumb, or starter bars have been incorrectly positioned, the cover to the reinforcement in the upper columns will be less than that required. The starter bars can be bent back to their correct position or new bars grouted into pre-drilled holes in the correct position. The bending of bars to the correct position must be closely observed and the radius of any bends made to correct the errors should not be less than that required under the normal design rules. The lack of cover resulting from the repositioning of bars must be considered in relation to the reduced fire resistance of the section, and the increased liability to corrosion.

4.4 Construction joints

Construction joints are potentially a source of weakness in concrete construction and should be located in areas of low shear stress. Traditionally, slab and beam construction joints are positioned in the middle third of the span and those for walls and columns above floor level. It is often desirable, from the construction point of view, to have a slab or beam joint at the face of a wall or column, thus allowing the vertical member to be constructed ahead of the horizontal member. If this practice is to be followed, adequate pockets or other means should be provided in the wall to ensure a bearing for the horizontal member and allowance made to enable concrete to flow into the pocket to make a good joint.

For vertical members, a kicker is often positioned above a floor surface, around which the profile of the column can be formed. To ensure an adequate amount of concrete is available for reasonable compaction, the depth of a kicker should be over 75 mm (3 in) in depth. The concrete in the kicker should be of the specified strength for concrete generally and should be properly cured.

To prepare the joint for the adjacent concrete pour, the aggregate will have to be exposed if one pour does not follow on the previous. This may be done by wire brushing and washing newly-placed concrete or by point tooling or scabbling the surface of concrete that has been allowed to harden. Any loose material should be removed from the joint and a clean surface made to receive the new concrete. The use of joggle or keyed joints for horizontal members is not usually considered necessary.

Vertical concrete formworks should be assembled so the shutters are as tight as possible against the kickers. To stop any grout leaks, it may be necessary to tape the bottom of the formwork. The surface to receive the concrete should be prepared in the manner previously described and should be free of any surplus water. Any water standing in the bottom of the formwork will either be mixed in with the newly placed concrete hence producing a weaker mix locally or, if it is not mixed in with the concrete, will run out after the striking of the shutter, leaving a void.

In the case of water-retaining structures, water bars are often prescribed to minimize water flowing through the joint. Dumbbell water bars cast inside the concrete member may not be completely effective in eliminating this problem, as there is often difficulty compacting concrete around this type of water bar. This trouble does not arise with external water bars. Joints in water bars must be vulcanized or cemented according to the maker's instructions.

4.5 Size of concrete pours

The size of concrete pours is often limited by specification and is generally a principle agreed between designer and builder. The main problem associated with large pours is ascertaining whether the size of the pour is a practical proposition. A great deal of preparatory work and organization of materials and labour is required to guarantee a smooth construction operation. The increasing use of ready-mixed concrete has provided an adequate source of concrete for large pours and modern pumping techniques mean a rapid rate of placing the material. The formwork and reinforcement placing must be completed prior to the pour and shift labour organized for the concreting operation. The advantage of a large pour to a contractor is the reduction in the length of construction joints. The provision of formwork for stop ends is often a difficult operation in areas of high reinforcement. This formwork has to be struck and the joint prepared for the next pour. All these operations present a finite time element to the contractor and the omission of construction joints can result in the reduction of construction time. Generally, however, individual pours in ordinary buildings should not exceed some 10 m (33 ft) in any dimension unless special steps are taken to control or accommodate shrinkage, although individual pours up to 30 m (98 ft) are done with special precautions.

To avoid surface cracking, the temperature gradient between the centre of the hydrating concrete and the surface should not exceed 20°C. The heat of hydration at the centre of a newly-placed member is directly proportional to the minimum cement content and the consideration of lower-strength concrete mixes for thick members may well prove beneficial. The temperature gradient from the centre of a concrete member to its surface may be reduced by providing an insulating layer to the surface. This will have the effect of increasing surface temperature and hence reducing the temperature difference between this area and the centre of the member. Where limited concrete surface cracking is acceptable, the size and position of the cracks can be controlled by the positioning of reinforcement. Chemical retarders may be used in the concrete to prevent hardening of already placed concrete before adjacent concrete can be poured and compacted. In all large pours, adequate provision should be made for the implementation of a suitable curing method to areas of concrete as soon as pouring and compacting have been completed.

5 Temporary stability during construction

Provided the recommendations of codes of practice and national standards are followed, the safety and stability of completed structures are generally satisfactory. There have been cases where failures of completed buildings have occurred, but these tend to be far fewer in number than failures occurring during the construction phase. The failure of completed buildings usually occurs over a considerable time due to the nature of interconnecting structural elements, whereas failures at the construction phase can occur without prior warning and can result in a considerable loss of life.

The stability of a structure must be considered at each step in the construction process. It is imperative that the designer has a method of construction in mind and that the basis of this method is clearly understood by the builder. If, as is often the case, a contractor proposes an alternative method of construction, the design philosophy will have to be completely reviewed to ensure the stability of the structure is adequate at each stage of construction.

5.1 Site investigation

The first step in collecting information is the site investigation. The decisions made at this stage are obviously of great importance to the structure. Often, the ground-bearing pressures realized during construction are the highest that particular parts of the building will encounter. This can be particularly true when cranes and hoists are positioned within the perimeter of a building and the design of the foundations must allow for these temporary loads.

Where excavations are involved, the determination of the active pressure of the ground is an important factor. Temporary shorings are usually weaker than the completed structure and the adequate strength of these items of temporary work should be of particular concern to the contractor. The supporting of temporary shoring from permanent parts of the new structure is a subject that must be referred to the design team. A mound of existing ground, within the excavation, may be used to supply the desired support and the adequacy of the ground to provide a safe support should be thoroughly investigated by specialist engineers. The contractor at all stages of the contract should clearly detail his proposals in respect of temporary work and they should be forwarded to the engineer for his consideration and approval. It should be remembered that if shoring has to be left in excavations for a prolonged period, the loads in it will tend to increase.

5.2 Demolition of existing buildings

The first stage in the construction phase may well involve the demolition of an existing building. The order of demolition of structural members should not be undertaken without due regard to the interaction of such members and the structural concept of the building as a whole. The demolition of a building in a highly built-up area may well affect surrounding buildings; settlements and movements over the years may well mean one building relies to some extent on an adjacent building for its support.

When demolition is finished, the site must be left in a stable condition. It is quite common for basement retaining walls to be propped by ground floor slabs and the removal of the ground-floor slab may mean the stability of surrounding roads and buildings are adversely affected. Temporary shoring will often be required until the new structure can provide the necessary stability.

5.3 Safety and stability during construction

The understanding of the method of construction by the contractor and designer is vital to the safety and stability of a building during the construction phase. Quite often, a structure will experience the heaviest imposed loads during the construction phase when concrete is at its youngest and weakest. The standard method of supporting formwork for *in situ* floor systems is by means of props onto lower completed floors. Where live loads are lighter than dead loads, the design imposed load will be exceeded for the floor providing the props. To alleviate this problem, the imposed load can be increased to the value of the general floor dead load.

When the tableform (flyingform) method of shuttering is used, rapid turnround times from floor to floor can be achieved, but the speed with which the floor-to-floor cycle can be completed may well mean that concrete has not reached its design strength when it is required to support the next floor. Bearing this in mind, it may be necessary to provide props from two completed floors. Another method that may be considered to overcome the problem, is using a higher grade of concrete than would normally be required, to produce an adequate early strength. The monitoring of the early concrete strength is important in these situations and site storage and testing of cubes can be undertaken to determine when an adequate strength has been reached. Under conditions where a longer floor-to-floor cycle time results from slower construction methods, the concrete will have had time to attain a higher strength. Not enough time may have elapsed for the full design strength to be realized but the designer may find the stresses produced, in a floor-supporting formwork for the floor above, acceptable.

Where precast flooring systems are used, the stability of the floor needs to be checked at each stage of operation. The bearing provided for the precast units must be sufficient to allow for any constructional tolerances and any shortening of the members due to deflection in the temporary state. When precast floor systems are used with *in situ* topping, the strength of the unit should be checked to ensure it has enough strength to withstand the load of the topping plus a nominal live load from the building operations determined by construction conditions. The loading of completed sections of floor with construction machinery and materials before the whole span has been completed may well result in overloading and failure and should be avoided.

Overloading on all floor systems at the construction stage can well cause excessive stress and failure of concrete that has not reached full design stress. Overloading usually results from the stacking of building material, the erection of scaffolding or the positioning of heavy plant on newly-completed areas and cases have occurred where heavy, precast concrete elements or steel girders have been laid on floors while adjustments are made. This obviously is a practice that should be avoided. To prevent this form of problem, a diligent site supervision must be maintained and the programming of deliveries should be so as to minimize the requirement for site storage.

Mention has already been made of the necessity to confirm the temporary stability of vertical elements when they are formed by the slipforming technique. The strength of the completed vertical elements should be checked for the loading from the construction equipment, without the stabilizing influence of the floors. The walls will often be weaker, as recesses will have been left for floor and beam connections.

Columns are constructed ahead of floors and they must be designed to be stable in a temporary cantilever state before the higher floor tie is constructed. This is particularly true when loads can be applied to the tops of columns by floor elements before the column is tied in. One example of this is in precast or composite construction methods, where the top of the column can supply a seating for slab or beam units. The case may also arise where certain ties to the columns are not constructed until a later date in the construction program. This may well mean, for a period, the column has an effective length between supports exceeding that assumed in the design of the completed building. The designer will have to check the temporary stresses to ensure that they are acceptable in such circumstances.

For precast columns, the strength of the column will depend on the strength of the joints used. Where the joint is formed using *in situ* concrete or by grouting, the concrete or grout must be allowed to reach adequate strength before loads are applied to the column.

When the lift slab technique is used, the stability of the columns must be carefully considered. Jacks on top of the columns lift the slabs and these lifting loads are a loading case to be considered in the column design.

Vertical elements in some cases will require temporary guys for stability and calculations for these and any fixing details should be supplied to the engineer for his approval. The same procedure should also be followed for any falsework required, whether it be required for vertical or horizontal elements.

Safety features for the work force can be designed into the structure in a number of cases. Where tableform is used, the falsework system is generally extended outside the building to provide a safety handrail to the floor under construction above the falsework. The loading from external scaffolding ties to the structure can be allowed for in the design and if the possibility exists that the contractor requires cradles or safety nets, then these loads, too, can be provided for. It is also possible to cast in fixings for safety harnesses should they be required for particular aspects of the work. Unfortunately, the work force is often casual about safety requirements and the contractor may have to bring pressure to bear to ensure safety equipment is used.

5.4 Site supervision

A high degree of site supervision has to be maintained throughout the contract to ensure that the desired strength of materials and the standards of workmanship are attained. The collecting of material samples for testing is a process that must be undertaken over the whole of the construction phase, as the use of lower strength materials than allowed for in the design can be a source of structural failure. The other aspect of supervision comes into ensuring that the materials used are placed in a correct manner and are protected as prescribed by the specification. Allowable tolerances for the completed works are established at the commencement of the design work and if, for any reason, these tolerances are exceeded, the philosophy of the design will have to be reappraised. Any eccentricities produced as a result of exceeded tolerances will result in larger moments being produced and the effect of these must be investigated.

5.5 Reversal of loading pattern

Another problem that can result in the construction phase is the reversal of loading pattern on some structural elements. Some examples of this are as follows:

(a) Lifting points in precast members can produce stresses not considered in the final design.

(b) Cantilever slab and beams should be supported until adequate sections of supporting slab and beams have been constructed to resist the moment. Some columns may have tension forces in them produced by cantilever moment before the main structural load is applied.

(c) The assumed fixity of beams and slabs to vertical elements may not be realized with only the vertical element constructed underneath the slab or beam.

The failure of plant and equipment has often resulted in loss of life during the construction phase. The safety aspect of plant is a specialist field and will not be considered in this chapter.

5.6 Stability of adjacent buildings

A major consideration in highly developed areas is maintaining the stability of adjacent buildings, both during the construction phase and also in the long term. In cases where a new building will generate heavier foundation loads or the new basement level is lower than surrounding foundations, measures will be required to ensure the stability of the adjacent structure. The transfer of existing loads to lower levels can be achieved by concrete underpinning the existing foundations, using soil stabilizing techniques on the soil under the existing foundation or providing a piling system to carry the existing foundation loads to a lower level. Where a high water level exists, it may also be necessary to provide some form of grout curtain to stop water weakening the soil under the existing foundation as it tries to percolate into the new excavation. Concrete underpinning is probably the most common method employed to stabilize existing adjacent buildings. The underpinning operation consists of removing sections of soil from under existing foundations and using concrete fill in its place to carry the load to lower levels. Dry pack or pressure grouting is undertaken on top of the new concrete infill and below the existing foundation to minimize any settlement. Dry packing is preferable to grouting, as better control can be exercised during the operation and smaller shrinkages are produced.

A detailed investigation of the nature of the soil supporting the existing foundations is required to ensure the load-carrying capacity of the soil is adequate to carry the load. Where the new foundations are a great deal deeper than the adjacent existing foundation, the concrete underpinning must be designed to carry both the vertical foundation load and also the lateral load due to soil pressure.

6 Performance considerations

6.1 Cracking and concrete durability

The cracking of concrete can result from a number of causes. The main ones are structural distress, shrinkage and temperature effects.

Structural distress has the most important implications as regards the safety of a building. As higher material design stresses are more readily accepted in modern design, particularly in limit state design, a certain amount of cracking must be expected in the tensile zone of structural members. By limiting the spacing of reinforcement and estimating crack widths, the location and extent of cracking can be predicted. It is when the extent and position of structural cracking exceed these limits that the assumption of the design must be brought into question. Excess structural cracking can be caused by settlements, overloading or the failure of concrete to reach its design strength.

Shrinkage cracking is probably one of the most common forms of cracking in building. Concrete shrinks as a result of the drying out process and can go on for a number of years. The effect of shrinkage is resisted locally by reinforcement in the concrete and when the tensile stress acceptable in the concrete is exceeded cracking will result. The cracking produced when concrete members shrink away from areas of fixity can be much more severe. When the overall structural concept is considered, it may be appropriate to provide movement joints between areas of fixity to allow shrinkage to take place or allow for a section of the connecting member to be constructed at a much later date when a degree of shrinkage under free-end conditions has taken place. The rate of shrinkage is determined by the relative humidity of the surroundings. A drying wind will cause a high rate of initial surface shrinkage and the immature concrete will not have attained adequate tensile stress to accommodate the shrinkage stress and cracking will result. The provision of a cover of polythene, damp hessian or a sprayed membrane at construction stage are probably the most common methods of protecting the concrete from such conditions.

Cracking due to thermal effect falls into two categories: (a) cracking due to the heat of hydration; (b) cracking due to climatic effects. While concrete is curing in the first few days after it has been placed, high internal temperatures are produced. A temperature gradient will be established from the hottest parts in the centre of the concrete member to the air temperature on the outside faces. Where there is a large temperature difference, as will result in cold conditions, the differing expansions will produce cracking. This can be prevented by ensuring that the temperature gradient does not exceed a level that causes excessive stress in the concrete. This is achieved by providing insulation to the concrete immediately after it has been poured.

Seasonal and diurnal temperature effects cause concrete members to expand and contract. The inside of a building is usually maintained by heating and air conditioning at approximately the same temperature throughout the year and it is the members subjected to external influences that can suffer distress. External cladding must be carefully designed to allow for temperature movements and the other main area for careful consideration is the roof. The roof of a structure is subject to the full effects of seasonal and diurnal temperature changes, and expansion and contraction of the concrete may cause cracking of both the concrete and the roof's finishes. The watertightness of the roof could be affected as a result of these cracks.

The durability of concrete can be directly related to its permeability. The higher the permeability, the more susceptible the concrete is to deterioration. Permeability can be reduced by having a high cement content and a water/cement ratio as low as possible, but adequate to allow good compaction. Many substances attack concrete to some extent, but the more harmful external attacks come from salt weathering or sulphates. Both forms of attack result in the deterioration of the concrete surface and the loss of cover to reinforcement.

Where water and oxygen can reach the reinforcement in concrete, as a result of a high concrete permeability, inadequate cover or by means of cracks, the reinforcement will eventually corrode. The expansion involved in this process will result in cracking and rust staining followed by the spalling off of concrete in more serious conditions. The corrosion of reinforcement is a serious problem, causing widespread damage to concrete and, where corrosion is expected to be a problem, it may be necessary to consider galvanizing or bitumen painting the reinforcement or even using stainless steel reinforcement.

Internal deterioration of concrete has resulted from internal sulphate attack or by alkali reactivity. Both these effects can result in total disintegration of the concrete over a number of years and can be avoided by careful selection of aggregates.

6.2 Settlement

It is quite common that the structural configuration of a building will mean that certain vertical load-carrying elements will support more load than others. At foundation level, the loading pattern will be transferred to the ground, either directly or by piles. As the ground is compressible, different loading will produce different settlements in the foundation. The problem will be further accentuated by a change in soil characteristics over the site area. Modern developments in foundation construction often mean poorer areas of ground can be developed and this may be at the cost of greater degrees of building settlement. The problem of differential settlement may be overcome by

(a) Founding on relatively incompressible ground or founding at a lower level with the use of piles. In this context, it must be born in mind that, where individual shallow foundations are used for economic reasons, an assessment must be made of the possible differential settlements.
(b) Providing a raft type of foundation.
(c) Providing movement joints between areas of a structure that have major differences in foundation loadings.

Where a certain amount of differential settlement is expected, the effect of support displacement to horizontal members should be carefully considered. Most structural analysis methods rely on the fact that supports remain rigid and a detailed analysis based on this criterion may well be partially invalidated by settlements. Where unexpectedly large settlements occur, structural cracking and a certain amount of distress to cladding may result.

Buildings erected in areas of mining subsidences are often subjected to large differential movement and special design and placing of movement joints are required in these instances.

6.3 Staining and wearing of exposed surfaces

The basis of weathering design must stem from the fact that rain will run down the surface of a building, taking with it any impurities that have accumulated on horizontal surfaces. In areas that receive a high quantity of rainfall, these impurities will be completely removed by the rain, but in areas that receive less rain, the evaporating water will leave the impurities behind as staining.

When smooth-faced *in situ* concrete is used to provide the external facade, the variation in compaction and material from one place to another will produce different rates of absorbency and, as a result, will create uneven weathering patterns. It is for this reason, and the possibility of cracking, that this form of surface has fallen into disfavour. Exposed aggregate surfaces have been used quite extensively over the years and have advantages over plain concrete. Water running over the exposed aggregate surface is distributed over a wide area so a certain amount of streaking is eliminated. The rough nature of the surface also means that the particles of aggregate act as drips and have the effect of throwing the water away from the surface. Reeded, striated and other forms of pattern moulding have also been used successfully and again serve to distribute the rain over a large area and along routes the designer intends.

Concrete, over the years, has provided a durable surface for a wide variety of floor use. The decision to use concrete as a wearing floor surface, without additional topping, is one that must be based on the intended use of the area. Where concrete provides the wearing surface, there is a direct relationship between the abrasion resistance and the strength of the concrete. Good-quality concrete can prove resistant to mild chemical attack and impact loads, but special treatment may be necessary where severe chemical attack is expected. The ability of the concrete to provide a durable wearing surface is enhanced at the construction stage by sealing any surface irregularity by power floating or trowelling, curing the concrete to prevent surface cracks due to initial drying shrinkage, and by the application of one of a range of chemical surface hardeners. Care should be taken to avoid excess cement in floor surfaces that will create dust throughout the life of the building.

Another method of treatment of the concrete surface to achieve a finish suitable for industrial use is to remove the weaker surface layer of the newly poured concrete to expose the denser concrete using a grinder. For severe abrasive conditions and high impact load, special hard stone aggregates, or metallic aggregates, or steel wire fibres may be introduced to the surface concrete. When a separate mix is used in this way, its thickness may be reduced to 20 mm (0.8 in) if it is laid before the structural slab beneath has set so that there is a complete bond. Otherwise, the surface layer should either be sufficiently thick to resist cracking within itself (75–100 mm, 3–4 in) or it should be bonded to the concrete below with a bonding agent. Construction joints in the finish should coincide with those in the slab below.

6.4 Deflection under load

Structural deflection is the natural reaction of a member being loaded and is not in itself of great consequence. Partitions, finishes and cladding can all be damaged by excess deflection and it is for this reason that the deflection pattern must be assessed. In buildings where expensive cladding and partitions have been chosen, the need for restricting deflection will be more important than in buildings where cheaper finishes have been used. For certain situations, it may be necessary to decide on a certain figure for deflection, sooner than the acceptance of allowable deflection on a span-to-depth ratio. The deflection is limited by increasing the size of members and limiting stress, but this will generally result in a heavier and more expensive building.

The determination of deflection is a complicated process. Deflection is directly proportional to stress and the prediction of the realistic superimposed load and its pattern is often difficult. An assessment will also have to be made for the way in which creep will affect the stress-carrying capacity of the concrete. As has been mentioned previously, a structure is often subjected to high stresses during the construction stage before the concrete has developed its full design strength and excess loading at this time will affect the future creep pattern and add to long-term deflection. The time allowed before an element is to be loaded or the time a floor must remain propped must be given careful consideration at design stage.

Where stiff partitions are built in the direction of the floor span, the central deflection will result in the partition having to span over a central section. The ability of the partition to do this will be affected by door or window openings and the inclusion of movement joints in the length of the wall may be necessary to avoid damage. A movement joint may also be required along the head of the partition to prevent the deflecting floor above loading the partition. When dismantlable types of partition are used, with frequent vertical joints, high structural deflections may be allowable.

The addition of vertical load to columns will result in the shortening of the member and will affect any form of cladding attached to it. Adequate allowance should be allowed for this type of vertical movement in the horizontal joint between units.

Cantilever beam and slab are often required for architectural reasons to support the external facade. As the deflection pattern of a cantilever is the most difficult to predict and will be affected by the loading pattern of adjacent spans, adequate provision should be made for deflection in external cladding elements.

6.5 Noise and vibration

Traditional heavyweight partitions and external wall construction have served to regulate the noise levels in buildings. Modern developments have meant that

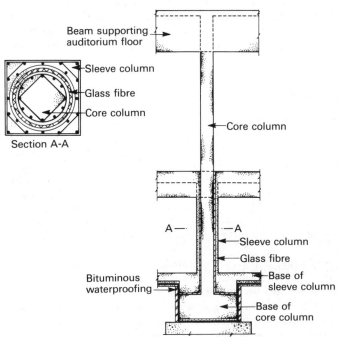

Figure 6-1 Details used for the Royal Festival Hall, London

lightweight thermal insulating materials are used for external walls and this means a higher noise level penetration from outside the building. The choice, on economic grounds, of lightweight partitions and the thinner structural sections, which can be used as the permissible concrete and reinforcement stresses rise, results in a higher noise level transference between various parts of the building. Some of this noise transference can be reduced by careful selection of room finishes such as carpets and false ceilings but where expected noise levels are high it may be necessary to isolate parts of a building with denser structural members and partition walls.

As most modern buildings are air conditioned, the plant room is an area which has grown in size and one which can transmit both vibration and noise to other parts of the building. By surrounding such areas with heavier partitions and providing antivibration mountings to machinery, a substantial improvement can be made to suppress noise transference. Where a high level of soundproofing and vibration isolation is required, it may be necessary to build a room within a room. This technique involves building an inner and outer layer of slabs, walls and ceilings and isolating the inner construction with vibration-absorbent membranes.

Vibrations can be transmitted to a building by underground railways, road traffic and the like. When a building is to house delicate machinery, the prospect of isolating the whole building from external influences may be considered. This can be achieved by designing columns as pin ended and separating foundations and columns by Neoprene bearings (see Fig. 6-1).

7 Fixings into concrete

A wide variety of methods of attaching to concrete is available to the design engineer. After calculating the possible loading and appropriate safety factors required, a choice of fixing can be made.

7.1 *In situ* concrete fixings

When *in situ* concrete fixings have been selected, they are attached to the inside of the formwork and concrete placed around them. They must be securely attached to the formwork to resist the accidental contact of a concrete vibrator. If the fixing is found to be in an area of honeycombing, when the formwork is removed the load-carrying capacity of the fixing must be brought into question. The accurate location in line and level is difficult using *in situ* fixings in walls, as the fixing is often attached to the formwork before the formwork is placed in position. Where *in situ* fixings are used for ceiling fixings, setting out can be done accurately and reinforcement arranged accordingly.

7.2 Fixings inserted after construction

Post-drilled fixing can be of the mechanical type or may be of the bonded type. A common problem that exists for post-drilled fixing is that reinforcement bars may be at the same location as the required hole. For this reason, as much tolerance as possible should be allowed in the location of post-drilled fixings. Bonded fixings are produced with epoxy or polyester resin adhesives. The hole drilled for the fixing must be thoroughly cleared of dust and, as the adhesives have a limited shelf life, the last recommended date for use should be checked. The placing of either a bonded or mechanical anchor near an edge or adjacent to another should be inspected to ensure safe loads are not exceeded or failure planes do not overlap.

The possibility of corrosion for both *in situ* and post-drilled fixings should be investigated. Even if fixings are galvanized, any chipping of the protective coat can result in corrosion and staining of the concrete, particularly when fixings are in exposed conditions. Stainless steel fixings or one of the other types of corrosion resistant fixing may be required, although care should be exercised to ensure galvanic action is not set up between different metals in damp conditions.

References

1. British Standards Institution, CP 2004: 1972 (as amended 1975) *Foundations*, London, 1972, 160 pp.
2. Institution of Structural Engineers, *Design and construction of deep basements*, 1975, 64 pp.
3. Hanna, T. H., *Design and construction of ground anchors*, 2nd Edn, Construction Industry Research and Information Association, London, 1980, 70 pp.
4. Institution of Civil Engineers, *Diaphragm walls and anchorages*, proceedings of conference London, Sept. 1974, Telford, London, 1975, 233 pp.
5. Institution of Civil Engineers, *A review of diaphragm walls*, record of seminar, 9–10 Sept. 1976, London, Telford, London, 148 pp.
6. British Standards Institution, CP 102: 1973 (as amended 1978), *Protection of buildings against water from the ground*, London, 1973, 28 pp.
7. Coffin, F. G., Beckmann, P. and Pearce, T., *Guide to design of waterproof basements*, Construction Industry Research and Information Association, London, 1978, 38 pp.
8. ACI Committee 515, *A guide to the use of waterproofing, dampproofing, protective, and decorative barrier systems for concrete*, ACI 515. R-79, American Concrete Institute, Detroit, 1979, 41 pp.
9. Prestressed Concrete Institute, *PCI manual for structural design of architectural precast concrete*, Chicago, 1977.
10. Batterham R. G., *Slipform concrete*, Construction Press, Lancaster, 1980, 112 pp.
11. Institution of Structural Engineers and Concrete Society, *Standard method of detailing reinforced concrete*, London, 1973, pp. 28.
12. ACI Committee 315, *Details and detailing of concrete reinforcement*, ACI 315-80, 1980, 50 pp.; *Manual of engineering and placing drawings for reinforced concrete structures* (Part 2 of *ACI detailing manual*), ACI 315. R-80, 1980, 56 foldouts and commentary, American Concrete Institute, Detroit.
13. Birt, J. C., *Curing concrete—an appraisal of attitudes, practices and knowledge*, 2nd Edn Construction Industry Research and Information Association, London, 1981, 33 pp.
14. American Concrete Institute, *Durability of concrete*, SP-47, 1975, 385 pp.

15. ACI Committee 201, *Guide to durable concrete*, ACI 201. 2R-77, American Concrete Institute, Detroit, 1977, 37 pp.
16. American Concrete Institute, *Corrosion of metals in concrete*, SP-49, 1975, 137 pp.
17. Deacon, R. C., *Concrete ground floors: their design, construction and finish*, 2nd Edn, Cement and Concrete Association, London, 1974, 30 pp.
18. Alexander, S. J. and Lawson, R. M., *Design for movement in buildings*, Construction Industry Research and Information Association, London, Technical Note 107, 1981, 54 pp.
19. Paterson, W. S., *Selection and use of fixings in concrete and masonry*, Construction Industry Research and Information Association, London, Guide 4, 1977, 30 pp.

40 Specification for concrete and reinforcement: comparison of national code provisions

J Rygol

Sir William Halcrow and Partners, London, England

Contents

Notation

d_b	nominal diameter of reinforcement bar
E	modulus of elasticity
E_c	modulus of elasticity of concrete
E_{cm}	mean value of modulus of elasticity of concrete
E_{cq}	dynamic modulus of elasticity of concrete
$E_c(t_0)$	modulus of elasticity of concrete at age t_0
E_{c28}	modulus of elasticity of concrete at age 28 days
E_s	modulus of elasticity of steel reinforcement
f	stress; strength
f_c	concrete compressive strength
f_{ck}	characteristic compressive strength of concrete
f_{ckj}	characteristic compressive strength of concrete at age j days
f_{cm}	mean compressive strength of concrete
f_{ct}	tensile strength of concrete
f_{ctm}	mean tensile strength of concrete
f_{cu}	characteristic cube strength of concrete

f_k	characteristic strength
f_m	mean strength
f_r	modulus of rupture of concrete
f_y	yield strength or characteristic strength of reinforcement
f_{ykc}	characteristic strength of reinforcement in compression
f_{ykt}	characteristic strength of reinforcement in tension
$f_{0.2}$	0.2% proof stress
f'_c	specified cylinder compressive strength of concrete
f'_{ct}	cylinder splitting tensile strength of concrete
W_c	unit weight of concrete
ε	strain
ε_c	concrete strain
ε_{cc}	concrete creep strain
$\varepsilon_{cc}(t, t_0)$	concrete creep strain from age of loading t_0 to age t
ε_{co}	concrete strain corresponding to maximum compressive stress σ_{co}
ε_{cs}	concrete shrinkage
$\varepsilon_{cs}(t, t_0)$	concrete shrinkage from age t_0 to age t
ε_{cu}	ultimate compressive strain of concrete
ε_s	steel strain
ν	Poisson's ratio of concrete

σ	standard deviation; stress	ϕ	reinforcement-bar diameter; creep function
σ_c	concrete compressive stress		
σ_{co}	maximum compressive stress in concrete	φ	creep coefficient (=creep strain/elastic strain)
σ_{ct}	concrete tensile stress	$\varphi(t, t_0)$	creep coefficient corresponding to interval between age of loading t_0 and age t
σ_s	reinforcement stress		

Summary

This chapter presents a summary of the specification for concrete and reinforcement as laid down by the CEB–FIP Model Code (1978), the British code CP 110:1972 (amended 1980) and the ACI Building Code 318–77. The recommendations of the three codes regarding concrete grading, strength, stress–strain relation, modulus of elasticity, Poisson's ratio, creep, shrinkage and thermal expansion are set out in parallel for easy comparison, as are the corresponding recommendations regarding reinforcement, including ductility and weldability.

1   Codes of practice and related standards

The comparison in this chapter is confined to three major codes: (a) CEB–FIP Model Code [1]; (b) British code CP 110 [2, 3]; (c) ACI Building Code [4, 5].

The CEB–FIP Model Code cites ISO 3893 [6] and 3898 [7]; the British code cites [8–39]; the ACI Building Code cites [40–62].

2   Specification for concrete

2.1   CEB–FIP Model Code

2.1.1   *Concrete grade*

Designs should be based on a *grade* of concrete which corresponds to a specific value of the characteristic strength. The grade can be selected from Table 2-1, taken from ISO 3893 [6].

Concrete of Grade C12/15 shall not be used for reinforced concrete work unless that use is suitably justified. Grades higher than C50/55 relate to special concretes and the rules given in the CEB–FIP Model Code are not applicable to these.

Unless otherwise stated, the characteristic compressive strength f_{ck} refers to cylinder tests only.

Table 2-1 Concrete grades according to characteristic compressive strength: CEB–FIP

Concrete grade	Compressive strength at 28 days	
	Cylinders $\phi 150 \times 300$ mm MPa	Cubes 150×150 mm MPa
C12/15	12.0	15.0
C16/20	16.0	20.0
C20/25	20.0	25.0
C25/30	25.0	30.0
C30/35	30.0	35.0
C35/40	35.0	40.0
C40/45	40.0	45.0
C45/50	45.0	50.0
C50/55	50.0	55.0

2.1.2 Characteristic strength

The CEB–FIP Model Code is based on the compressive strength measured on cylinders of 150 mm (5.9 in) diameter and 300 mm (11.8 in) height at the age of 28 days stored in water at $20 \pm 2°C$ ($68 \pm 4°F$), in accordance with ISO 4012.

The *characteristic strength*, f_{ck}, is defined as that strength below which 5% of all possible strength measurements for the specified concrete may be expected to fall.

In some calculations—for example, for the purpose of estimating the modulus of longitudinal deformation—the mean compressive strength, f_{cm}, is obtained from the specified characteristic strength, f_{ck}, by means of the following formula:

$$f_{cm} = f_{ck} + 8 \quad \text{MPa} \tag{2-1}$$

2.1.3 Tensile strength

In the CEB–FIP code the term *tensile strength*, unless otherwise specified, refers to the strength in axial tension. An estimate of the mean tensile strength, f_{ctm}, can be obtained from Table 2-2.

The f_{ctm} values in Table 2-2 are given approximately by the formula

$$f_{ctm} = 0.3[f_{ck}]^{\frac{2}{3}} \tag{2-2}$$

where f_{ctm} and the characteristic compressive strength, f_{ck}, are in MPa units.

Table 2-2 Relation between mean tensile strength, f_{ctm}, and characteristic compressive strength, f_{ck} (MPa): CEB–FIP

f_{ck}	12	16	20	25	30	35	40	45	50
f_{ctm}	1.6	1.9	2.2	2.5	2.8	3.1	3.4	3.7	4.0

The mean flexural tensile strength, f_{cmm}, may be estimated from the formula:

$$\frac{f_{cmm}}{f_{ctm}} = 0.6 + \frac{0.4}{4\sqrt{h}} \not< 1 \qquad (2\text{-}3)$$

where h is the height (in metres) of the flexural member under consideration.

2.1.4 Stress–strain relation

The stress–strain relation is of the form shown in Fig. 2-1.

Depending on the nature of the constituents and the rate of strain, the abscissa of the peak lies between −0.002 and −0.0025, the ultimate strain, ε_{cu}, varies between −0.0035 and −0.007, and the ultimate stress between $0.75f_c$ and $0.25f_c$, where f_c is the strength in compression.

The shape of the stress–strain diagram cannot be defined precisely. The designer may use a particular shape which complies with particular design conditions if he can justify its use or, alternatively, he may use an appropriate idealized diagram.

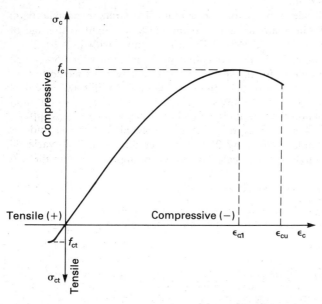

Figure 2-1   Schematic stress–strain diagram for concrete: CEB–FIP

$f_c(f_{ct})$ strength in compression (tension)

ε_c concrete strain

ε_{cu} ultimate concrete strain

$\sigma_c(\sigma_{ct})$ concrete compressive (tensile) stress

2.1.5 *Modulus of longitudinal deformation*

An estimate of the mean value of the *secant modulus*, E_{cm}, can be obtained from Table 2-3. The range of variation may extend from $0.7E_{cm}$ to $1.3E_{cm}$. The E_{cm} values given in Table 2-3 correspond to cases where the compressive stress σ_c is between about $0.4f_{ck}$ and $0.5f_{ck}$, where f_{ck} is the characteristic compressive strength. For $\sigma_c \approx 0.1f_{ck}$, the E_{cm} values may be increased by 10%. The relation between the f_{ck} and E_{cm} values in Table 2-3 is given by the formula

$$E_{cm}(GPa) = 9.5[f_{ck}(MPa) + 8]^{\frac{1}{3}} \tag{2-4}$$

As a rule, since f_{ck} is given for the age of 28 days, the E_{cm} values given in Table 2-3 are the 28-day values.

Table 2-3   Relation between characteristic compressive strength, f_{ck} and secant modulus, E_{cm}

f_{ck}, MPa	12	16	20	25	30	35	40	45	50
E_{cm}, GPa	26	27.5	29	30.5	32	33.5	35	36	37

2.1.6 *Poisson's ratio*

A value between 0 and 0.2 can be adopted for Poisson's ratio, ν. Poisson's ratio for elastic strains is to be taken equal to 0.2; it can be assumed to be 0 when cracking is allowed for the concrete in tension.

2.1.7 *Creep and shrinkage*

In cases where great accuracy is not required, the mean values given in Table 2-4 may be accepted as representative values of the final *creep coefficient* and final *shrinkage* for a concrete subjected to a stress σ_c not exceeding $0.4f_{ckj}$, where f_{ckj} is the characteristic compressive strength at age j (days) when the loading is applied. The values in Table 2-4 also apply to concrete subjected to tension, and are representative values liable to vary by 20% in either direction. This variation should be introduced into the design calculations where appropriate, on the safe side.

For $\sigma_c < 0.4f_{ckj}$, it may be assumed that

(1) Creep deformation varies linearly with the applied stress σ_c.
(2) Creep deformations due to stress increments are additive.

Assumption (1) leads to

$$\varepsilon_{cc}(t, t_0) = \frac{\sigma_{co}}{E_{c28}} \varphi(t, t_0) \tag{2.5}$$

where

$t_0 =$ age at which the concrete is loaded
$\varepsilon_{cc}(t, t_0) =$ creep deformation at a time $t > t_0$
$\sigma_{co} =$ constant stress applied at time t_0
$E_{c28} =$ modulus of elasticity (longitudinal deformation) of concrete at the age of 28 days
$\varphi(t, t_0) =$ creep coefficient.

Table 2-4 Final values of the creep coefficient, $\varphi(t_\infty, t_0)$, and the shrinkage, $\varepsilon_{cs}(t_\infty, t_0)$: CEB–FIP

Notional size[a]: $2A_c/u$	Humid atmospheric conditions outside $(RH \approx 75\%)$		Dry atmospheric conditions inside $(RH \approx 55\%)$	
	Small $\leqq 200$mm	Large $\geqq 600$ mm	Small $\leqq 200$ mm	Large $\geqq 600$ mm
Creep: $\varphi(t_\infty, t_0)$ Age of commencement of loading:				
fresh ($t_0 = 3$–7 days)	2.7	2.1	3.8	2.9
medium ($t_0 = 7$–60 days)	2.2	1.9	3.0	2.5
mature ($t_0 > 60$ days)	1.4	1.7	1.7	2.0
Shrinkage: $\varepsilon_{cs}(t_\infty, t_0)$ Age of concrete at the instant from which the shrinkage effect is being considered:				
fresh ($t_0 = 3$–7 days)	0.26×10^{-3}	0.21×10^{-3}	0.43×10^{-3}	0.31×10^{-3}
medium ($t_0 = 7$–60 days)	0.23×10^{-3}	0.21×10^{-3}	0.32×10^{-3}	0.30×10^{-3}
mature ($t_0 > 60$ days)	0.16×10^{-3}	0.20×10^{-3}	0.19×10^{-3}	0.28×10^{-3}

[a] A_c denotes the area of cross-section and u the perimeter in contact with the atmosphere.

Considerable deviations from assumption (1) are observed where stress variations are accompanied by a reduction of the strain (e.g. in the case of reduced loading). The total strain at time t under constant stress is then equal to the initial strain at time t_0 plus the creep strain:

$$\varepsilon_{total}(t, t_0) = \sigma_0 \left[\frac{1}{E_c(t_0)} + \frac{\varphi(t, t_0)}{E_{c28}} \right] \tag{2-6}$$

where $E_c(t_0)$ is the value of the modulus of longitudinal deformation at age t_0. The term

$$\phi(t, t_0) = \frac{1}{E_c(t_0)} + \frac{\varphi(t, t_0)}{E_{c28}}$$

is called the *creep function*.

Appendix (e) of the CEB–FIP code gives a more accurate method of creep assessment.

2.1.8 *Coefficient of thermal expansion*
The coefficient of thermal expansion of concrete may be taken as 10×10^{-6} per °C $(5.6 \times 10^{-6}$ per °F).

2.1.9   *Density*
In the calculations the unit weight of plain concrete may be taken as $24 \, \text{kN/m}^3$ ($150 \, \text{lb/ft}^3$). For reinforced concrete containing a normal percentage of reinforcement, the unit weight may be taken as $25 \, \text{kN/m}^3$ ($156 \, \text{lb/ft}^3$).

2.2   British code CP 110

2.2.1   *Concrete grade*
Designs should be based on a *grade* of concrete [64] which corresponds to a specific value of the characteristic cube strength, f_{cu} (see Table 2-5).

The grade of concrete required depends partly on the particular use and the characteristic strength needed to provide the structure with adequate ultimate strength and partly on the exposure conditions and the cover provided to the reinforcement.

The characteristic strength [64] is that determined from 150 mm test cubes at an age of 28 days after mixing. Procedures for making and testing the cubes are given in BS 1881 [13–18].

Table 2-5   Grades of concrete: CP 110

Grade	20	25	30	40	50
Characteristic strength, f_{cu} N/mm^2	20.0	25.0	30.0	40.0	50.0

2.2.2   *Characteristic strength*
CP 110 defines the characteristic strength [64] of concrete as that 28-day cube strength below which not more than 5% of the test results may be expected to fall.

In current British practice the characteristic strength, f_k, is the mean strength, f_m, less 1.64 times the standard deviation σ:

$$f_k = f_m - 1.64\sigma \tag{2-7}$$

Designs may be based on the characteristic strength or, if appropriate, the strength given in Table 2-6 for the age of loading. It should be noted that Table 2-6 is relevant only to concretes made with ordinary Portland cements.

2.2.3   *Tensile strength*
CP 110 gives no information regarding the tensile strength [64] of concrete.

A method for measuring the *indirect tensile strength*, sometimes referred to as the *splitting tensile strength* [64], is described in BS 1881 [13–18]. The test consists essentially in loading a standard concrete cylinder (300×150 mm diameter) across a diameter until failure occurs, by splitting across a vertical plane. The splitting tensile strength, f_{ct}, may be taken as [64]

$$f_{ct} = \frac{2F}{\pi dl} \quad \text{MPa} \tag{2-8}$$

Table 2-6   Strength of concrete: CP 110

Grade	Characteristic concrete cube strength, f_{cu} N/mm²	Cube strength at an age of				
		7 days MPa	2 months MPa	3 months MPa	6 months MPa	1 year MPa
20	20.0	13.5	22.0	23.0	24.0	25.0
25	25.0	16.5	27.5	29.0	30.0	31.0
30	30.0	20.0	33.0	35.0	36.0	37.0
40	40.0	28.0	44.0	45.5	47.5	50.0
50	50.0	36.0	54.0	55.5	57.5	60.0

where F is the maximum applied force (N), d the cylinder diameter (mm) and l the cylinder length (mm). The determination of the *flexural strength* of concrete is described in BS 1881 [13–18]. The test consists essentially of testing a plain concrete beam under symmetrical two-point loading applied at one-third-span points. The flexural strength is calculated as M/Z, where M is the bending moment at the section where rupture occurs and Z is the elastic section modulus of the beam. The flexural strength so calculated is often referred to as the *modulus of rupture* [64], and is only a hypothetical stress based on the assumption of linear-elastic behaviour up to the instant of rupture. The modulus of rupture overestimates the true flexural strength of the concrete, and is for use as a comparative measure for practical purposes only [64]. The modulus of rupture is usually about $1\frac{1}{2}$ times the splitting tensile strength.

2.2.4   Stress–strain relationship

CP 110 gives no information about the 'actual' stress–strain relationship for the concrete.

It is generally recognized that the stress–strain relationship is substantially linear in the initial stages of loading, but thereafter the gradient of stress to strain decreases with increasing load until the maximum stress is reached. In the region of the maximum, there is a phase in which the strain increases considerably while the stress remains almost constant. After that there appears a 'falling branch', with the stress decreasing as the strain increases, and spalling occurs on the surface, until finally the specimen ruptures completely.

The typical stress–strain curves (Fig. 2-2) can be simplified [64] to shapes comprising an initial rising parabolic branch (maximum stress σ_{co} at strain ε_{co}) and a subsequent flat plateau terminating at maximum limiting strain ε_{cu}.

2.2.5   Modulus of elasticity

The modulus of elasticity [64] is primarily dependent on the crushing strength of the concrete. For concrete made with natural aggregate and having a density of 2300 kg/m³ (144 lb/ft³) or more, the static or dynamic modulus of elasticity, relevant to the serviceability limit states, may be taken from Table 2-7 for concretes of various compressive strengths. If a more accurate value is required

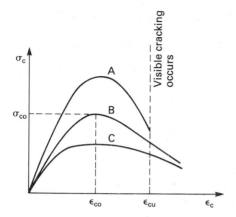

Figure 2-2   Typical stress–strain curves for concrete in compression: CP 110

A   High-strength concrete
B   Medium-strength concrete
C   Low-strength concrete

for particular materials and a particular mix, tests should be made in accordance with BS 1881 [13–18].

The secant modulus [64] as determined from a laboratory test depends on the rate of application of the load. In testing, BS 1881 specifies that the stress should be applied at a rate of 15 N/mm^2 (15 MPa, 2200 psi) per minute and the secant modulus shall be that corresponding to a stress equal to one third the cube strength plus 1 N/mm^2 (1 MPa, 145 psi). These specifications may appear rather arbitrary, but they are chosen to yield a value which is of practical interest in structural design [64]. In Table 2-3 the term *static modulus* refers to the secant modulus determined in accordance with BS 1881, as described above. The term *dynamic modulus* refers to the modulus of elasticity determined by an electrodynamic method described in BS 1881, in which the first natural mode frequency of the longitudinal vibration of a standard beam is measured. The

Table 2-7   Modulus of elasticity: CP 110

Compressive cube strength, f_{cu} N/mm^2	Static modulus, E_c Mean value kN/mm^2	Typical range kN/mm^2	Dynamic modulus, E_{cq} Mean value kN/mm^2	Typical range kN/mm^2
20	25	21–29	35	31–39
25	26	22–30	36	32–40
30	28	23–33	38	33–43
40	31	26–36	40	35–45
50	34	28–40	42	36–48
60	36	30–42	44	38–50

Note: 1 N/mm^2 = 1 MPa; 1 kN/mm^2 = 1 GPa.

longitudinal vibration subjects the test beam to very small stresses only; hence the dynamic modulus is often taken as being roughly equal to the initial tangent modulus and is therefore higher than the secant modulus [64]. The dynamic modulus method of test is more convenient to carry out than the static method and CP 110 gives the following approximate formula for calculating the static (secant) modulus, E_c, from the dynamic modulus, E_{cq}:

$$E_c(\text{kN/mm}^2) = 1.25 E_{cq}(\text{kN/mm}^2) - 19 \qquad (2\text{-}9)$$

Such an estimated value will generally be correct within $\pm 4 \text{ kN/mm}^2$ (± 4 GPa, ± 580 ksi).

For sustained loading conditions, appropriate allowance for shrinkage and creep should be made.

2.2.6 *Poisson's ratio*
For the serviceability limit states, Poisson's ratio may be taken as 0.2. Poisson's ratio usually lies within the range of 0.1 to 0.2 and is slightly lower for high strength concretes. In British practice, for design calculations a value of 0.15 is usually assumed [64].

2.2.7 *Creep and shrinkage*
CP 110 gives no information on creep [64] and drying shrinkage [64] and refers to specialist literature.

2.2.8 *Coefficient of thermal expansion*
CP 110 gives no information on the coefficient of thermal expansion. In British practice a value of 10×10^{-6} per °C (5.6×10^{-6} per °F) is commonly used.

2.2.9 *Density*
CP 110 gives no information on the density [64] of reinforced concrete. In British practice the unit mass of reinforced concrete is usually taken as 2400 kg/m^3 (150 lb/ft^3).

2.3 ACI 318 building code

2.3.1 *Concrete grade*
The design should be based on a *grade* of concrete which corresponds to a specific value of *compressive strength* of concrete, referred to as compressive cylinder strength, f'_c.

As specified by ASTM C31 [49], tests for f'_c shall be made on cylindrical specimens of height equal to twice the diameter. The standard specimen shall be the 6×12 in (152×305 mm) cylinder when the nominal maximum size of the coarse aggregates does not exceed 2 in (51 mm). When the nominal maximum size of the coarse aggregates exceeds 2 in, the diameter of the cylinder shall be at least three times the nominal maximum size of the coarse aggregate in the concrete. Unless required by the project specifications, cylinders smaller than 6×12 in shall not be made in the field.

During the first 24 h after moulding, all test specimens shall be stored under conditions that maintain the temperature immediately adjacent to the specimens in the range of 60–80°F (16–27°C) and prevent loss of moisture from the specimens.

For checking the adequacy of the laboratory mix proportions for strength, or as the basis for acceptance or for quality control, the test specimens shall be removed from the moulds at the end of 20 ± 4 h and stored in a moist condition at $73.4 \pm 3°F$ ($23 \pm 1.7°C$) until the moment of test.

For determining when forms may be removed or when a structure may be put in service, the test specimens shall be stored as near to the point of use of the structure as possible, and shall receive, in so far as practicable, the same protection from the elements on all surfaces as is given to the portions of the structure which they represent. To meet these conditions, specimens made for the purpose of determining when a structure may be put in service shall be removed from the moulds at the same time as the removal of framework.

Similar curing procedures shall be adopted for beams to be tested for flexural strength.

As specified by ASTM C39 [50], the test specimens for compressive strength shall be subjected to load continuously and without shock. In testing machines of the screw type, the moving head shall travel at a rate of approximately 0.05 in (1.3 mm) per minute when the machine is running idle. In hydraulically operated machines, the load shall be applied at a constant rate within the range 20–50 psi (0.14–0.34 MPa) per second.

Unless otherwise specified, f'_c shall be based on 28-day tests.

In present US practices, 28-day cylinder strengths in the range of $f'_c = 2500$–6000 psi (17.2–41.4 MPa) are usually specified for reinforced concrete structures, values between 3000 and 4000 psi (20.7 and 27.6 MPa) being the most common.

2.3.2 Characteristic strength

The ACI 318 building code [4] does not use the term characteristic compressive strength of concrete.

The requirements for the proportioning of concrete mixtures and the criteria for acceptance are based on the philosophy that the code is intended primarily to protect the safety of the public.

The design is based on the *specified compressive strength* defined as the compressive cylinder strength of concrete, denoted by f'_c as explained further below.

(a) *Proportioning on the basis of field experience*   The concrete mix selected must yield an average strength appreciably higher than the design strength, f'_c. When a concrete production facility has a record based on at least 30 consecutive strength tests that represent similar materials and conditions to those expected, the required average compressive strength used as a basis for selecting concrete proportions shall exceed the required f'_c at designated test age by at least the value given in Table 2-8.

Table 2-8   Average strength requirements: ACI 318

| Standard deviation | | Required minimum |
psi	MPa	average strength
300 or less	2.1 or less	f'_c + 400 psi (2.8 MPa)
300–400	2.1–2.8	f'_c + 550 psi (3.8 MPa)
400–500	2.8–3.5	f'_c + 700 psi (4.8 MPa)
500–600	3.5–4.1	f'_c + 900 psi (6.2 MPa)
>600	>4.1	f'_c + 1200 psi (8.3 MPa)
Unknown	Unknown	f'_c + 1200 psi (8.3 MPa)

(b)   *Proportioning by laboratory trial batches*

(1) When laboratory trial batches are used as the basis for selecting concrete proportions, strength tests shall be made in accordance with ASTM C39 [50] and ASTM C192 [56].
(2) When laboratory trial batches are made, the air content shall be within ±0.5% and the slump within ±0.75 in (19 mm) of the respective maxima permitted by the specifications.

(c)   *Tests of laboratory-cured specimens*

(1) Samples for strength tests shall be taken in accordance with ASTM C172 [55].
(2) Cylinders for strength tests shall be moulded and laboratory-cured in accordance with ASTM C31 [49] and tested in accordance with ASTM C39 [50].
(3) The strength level of concrete is considered satisfactory as long as averages of any three consecutive strength tests remain above the specified f'_c and no individual strength test falls below the specified f'_c by more than 500 psi (3.5 MPa).
(4) If either of the requirements of (3) above are not met, steps shall be taken immediately to increase the average of subsequent strength test results.

(d)   *Tests of field-cured specimens*

(1) Field-cured cylinders shall be cured under field conditions in accordance with ASTM C31 [49].
(2) Field-cured test cylinders shall be moulded at the same time and from the same samples as laboratory-cured cylinders.
(3) Procedures for protecting and curing concrete shall be improved when the strength of field-cured cylinders at the test age designated for measuring f'_c is less than 85% of that of companion laboratory-cured cylinders. When laboratory-cured cylinder strengths are appreciably higher than f'_c, field-cured cylinder strengths need not exceed f'_c by more than 500 psi (3.5 MPa) even though the 85% criterion is not met.

(e) *Core tests* Concrete in an area represented by cylinder core tests shall be considered structurally adequate if the average of three cores is equal to at least 85% of f'_c and if no single core is less than 75% of f'_c. To check testing accuracy, locations represented by erratic core strengths may be re-tested.

(f) *Watertight concrete* For concrete made with normal-weight aggregate, the water/cement ratio shall not exceed 0.50 by weight for exposure to fresh water and 0.45 by weight to sea water.

2.3.3 Tensile strength

The tensile strength of concrete is important in several respects. Thus the shear and torsion resistance of reinforced concrete beams appears to depend primarily on the tensile strength of the concrete. Also, the conditions under which cracks form and propagate on the tension side of reinforced concrete flexural members depend strongly on the tensile strength.

The tension properties of concrete can be measured in terms of the modulus of rupture (beam tests) or the so-called split-cylinder test.

2.3.3.1 Modulus of rupture

The procedure for determining the modulus of rupture, f_r, using simple beams with third-point loading is described in ASTM C78 [52], and with centre-point loading in ASTM C293 [58]. In each case, the specimen shall have a test span within 2% of being three times its depth as tested.

For third-point loading, if the fracture initiates in the tension surface within the middle third of the span length, the modulus of rupture is calculated from the formula

$$f_r = \frac{Pl}{bd^2} \tag{2-10}$$

where f_r is the modulus of rupture (psi or MPa); P is the maximum applied load indicated by the testing machine (lbf or N); l is the span length (in or mm); b is the average width of specimen (in or mm); and d is the average depth of specimen (in or mm).

If the fracture occurs in the tension surface outside of the middle third of the span length by not more than 5% of the span length, the modulus of rupture is calculated from the formula

$$f_r = \frac{3Pa}{bd^2} \tag{2-11}$$

where a is the average distance between line of fracture and the nearest support measured on the tension surface of the beam (in or mm).

If the fracture occurs in the tension surface outside of the middle third of the span length by more than 5% of the span length, the results of the test should be discarded.

For centre-point loading the modulus of rupture is calculated from the formula

$$f_r = \frac{3Pl}{2bd^2} \tag{2-12}$$

The weight of the beam is not included in the above calculations.

2.3.3.2 *Splitting tensile strength* The procedure for determining the *splitting tensile strength* of cylindrical concrete specimens is described in ASTM C496 [60].

The test specimens shall conform in size, moulding and curing requirements set forth in either ASTM C31 [49] (field specimens) or ASTM C192 [56] (laboratory specimens). Drilled cores shall conform to the size and curing requirements set forth in ASTM C42 [51].

The load should be applied continuously and without shock at a constant rate of 100–200 psi (0.69–1.38 MPa) per minute tensile stress until failure of the specimen.

The splitting tensile strength of the specimen is calculated from the formula

$$f'_{ct} = \frac{2P}{\pi dl} \tag{2-13}$$

where f'_{ct} is the splitting tensile strength (psi or MPa); P is the maximum applied load indicated by the testing machine (lbf or N); l is the length (in or mm); and d is the diameter (in or mm).

The modulus of rupture is larger than the strength of concrete in uniform axial tension. It is thus a measure of, but not identical with, the real axial tensile strength. The results of the split-cylinder tests likewise are not identical with (but are believed to be a good measure of) the true axial tensile strength. The results of all types of tensile test show considerably more scatter than those of compression members. Tensile strength, whichever way determined, does not correlate well with the compressive strength, f'_c. It appears that a reasonable estimate for the split-cylinder strength, f'_{ct}, is given by 6 to 7 times $\sqrt{f'_c}$ (f'_c being expressed in psi), that is 0.50 to 0.58 times $\sqrt{f'_c}$ (f'_c being expressed in MPa) for normal-weight concretes.

The true tensile strength, f'_t, for normal-weight concretes appears to be of the order of 0.5 to 0.7 times f'_{ct} and the flexural tensile strength, f_r (modulus of rupture), from 1.25 to 1.75 times f'_{ct}. The smaller of the foregoing factors apply to higher-strength concretes, and the larger to lower-strength concretes.

2.3.4 *Stress–strain relationship*

A compressive stress–strain relationship for concrete may be obtained by appropriate strain measurements in cylinder tests (ASTM C39 [50]) or on the compression side in beams (ASTM C78 [52] and ASTM C293 [58]).

All stress–strain curves have somewhat similar character. They consist of an initial, relatively straight elastic portion in which stress and strain are closely proportional, then begin to curve to the horizontal, reaching the maximum stress σ_{co} (i.e. the compressive strength) at a strain of approximately $\varepsilon_{co} = 0.002$, and finally show a descending branch, as shown in Fig. 2-2. Concretes of lower strength are less brittle, i.e. fracture at a larger maximum strain ε_{cu}, than high-strength concretes.

The shape of the stress–strain curve for various concretes, or the same concrete under various conditions of loading, varies considerably. The descending branch of the curve, probably indicative of internal disintegration of the material, is much more pronounced at fast than at slow rates of loading. Also, the peaks of the curves, i.e. the maximum stress is reached, are somewhat smaller at slower rates of strain.

2.3.5 *Modulus of elasticity*
The static modulus of elasticity, E_c, for concrete may be taken as

$$E_c = 33 W_c^{1.5} \sqrt{f_c'} \quad \text{psi} \tag{2-14}$$

for unit weights of concrete W_c between 90 and 155 lb per cu.ft. For normal weight concrete, E_c may be taken as

$$E_c = 57\,000 \sqrt{f_c'} \quad \text{psi} \tag{2-15}$$

In SI units, i.e. E_c and f_c' in MPa and W_c in kg/m^3, Eqn 2-14 becomes

$$E_c = 0.043 W_c^{1.5} \sqrt{f_c'} \tag{2-16}$$

and for normal-weight concrete

$$E_c = 4730 \sqrt{f_c'} \tag{2-17}$$

A procedure for determining the modulus of elasticity is described in ASTM C469 [59] (static modulus of elasticity) and ASTM C215 [57] (dynamic modulus of elasticity).

2.3.6 *Poisson's ratio*
A procedure for determining the Poisson's ratio is described in ASTM C469 [59] (static Poisson's ratio) and ASTM C215 [57] (dynamic Poisson's ratio).

In US practice, it is commonly accepted that the Poisson's ratio, for concrete at stresses lower than $0.7 f_c'$, falls within the limits of 0.15 and 0.20, with 0.17 a representative value.

2.3.7 *Creep and shrinkage*
ACI 318 [4] states that estimations of creep and shrinkage shall be based on a realistic assessment of such effects [64] occurring in service, but gives no information about how to assess them. In the commentary [5] on ACI 318, it is explained that the term 'realistic assessment' is used to indicate that the most probable values rather than the upper bound values of the variables should be used.

The determination of the creep of moulded concrete cylinders subjected to sustained longitudinal compressive load is specified in ASTM C512 [61].

Creep deformations for a given concrete are practically proportional to the magnitude of the applied stress provided the applied stress does not exceed about one-half of the cylinder strength, f_c'; at any given stress, high-strength concretes show less creep than lower-strength concretes. With elapsing time, creep proceeds at a decreasing rate and practically ceases after 2 to 5 years at a final value which, depending on concrete strength and other factors, attains 1.5 to 3 times the magnitude of the instantaneous strain ε_{inst}. Creep also depends on the average ambient relative humidity, being about twice as large for 50% as for 100% humidity.

2.3.8 *Coefficient of thermal expansion*
ACI 318 [4] gives no information as to the value of the thermal coefficient of concrete, α_t.

The coefficient of expansion varies somewhat, depending on the type of aggregate and richness of the mix.

In US practice, it is usually assumed that the coefficient of expansion varies within the range from 4×10^{-6} to 6×10^{-6} per °F (7.2×10^{-6} to 10.8×10^{-6} per °C). A value of 5.5×10^{-6} per °F (9.9×10^{-6} per °C) is generally accepted as satisfactory for calculating stresses and deformations caused by temperature changes.

2.3.9   *Density*
ACI 318 [4] gives no information as to the value of density of plain concrete or reinforced concrete.

A procedure for determining the density of concrete is specified in ASTM C138 [53].

The unit weight of so-called stone concrete, i.e. concrete with natural stone aggregate, varies from about 140 to 152 lb/ft^3 (2240–2440 kg/m^3) and can generally be assumed as 145 lb/ft^3 (2320 kg/m^3).

3   Specification for reinforcement

3.1   CEB–FIP Model Code

3.1.1   *Reinforcement grade*
The steels covered by the CEB–FIP code [1] can be sub-divided according to

(a) Method of production:
   (1) hot-rolled steel (natural steel);
   (2) cold-worked steel (either by torsion and/or tension, or by cold drawing and/or rolling);
   (3) special steel (e.g. hardened and tempered steel).
(b) Surface properties:
   (1) plain smooth bars or wires (including welded mesh);
   (2) high-bond bars or wires (including welded mesh).
(c) Weldability:
   (1) not capable of being welded;
   (2) weldable under some conditions;
   (3) weldable steel.

The calculations should be based on the *nominal section*, determined from the *nominal diameter*. The preferred nominal diameters for bars is 5, 6, 8, 10, 12, 16, 20, 25, 32, 40 and 50 mm. Preferred diameters for wires used in welded mesh lie in the range 4 to 12 mm in steps of 0.5 mm.

Euronorm 80 defines three graduations corresponding to the following grades: S 220, S 400, S 500, where the numbers denote the minimum limit of elasticity in MPa.

3.1.2   *Characteristic strength*
The characteristic strength, f_{yk}, is defined as the 5% fractile of the yield stress, f_y, or 0.2% proof stress, (denoted by $f_{0.2}$).

If the steel supplier guarantees a minimum value for f_y or $f_{0.2}$, that value may be taken as the characteristic strength.

The tensile strength obtained from tests on a bar should also show that

$$f_{st} \geq 1.1 f_{yk} \tag{3-1}$$

$$f_{st} \geq 1.05 f_{y,obs} \tag{3-2}$$

where $f_{y,obs}$ denotes the limit of elasticity determined during the tensile test.

In principle, the design should be based on a grade of steel corresponding to a specified characteristic strength, f_{yk}.

For steels totally or partially cold-worked by means of axial tension, it may be that

$$|f_{ykc}| < f_{ykt} \tag{3-3}$$

where f_{ykc} is the characteristic strength in compression and f_{ykt} that in tension. The value of f_{ykc} to be used in the design calculations should then be stipulated in the approval documents.

3.1.3 Stress–strain relationship

As a simplification, the actual stress–strain relations can be replaced by bilinear or trilinear diagrams chosen so that the approximations are on the safe side.

In the absence of more accurate information, the bilinear diagram in Fig. 3-1 can be used for mild steel or cold-worked steel (by drawing or rolling).

For steel cold-worked by axial torsion and/or tension, an idealized diagram can be used, containing a linear section with a slope E_s and curved sections defined by reference to the limit of elasticity in tension f_{ykt}, and in compression f_{ykc} (Fig. 3-2, where σ_s, f_{ykt} and f_{ykc} are given in MPa, $f_{ykt} > 0$, $f_{ykc} < 0$ for the signs).

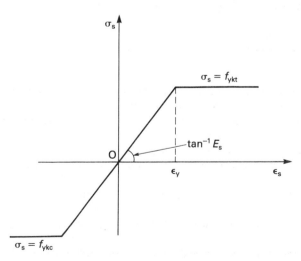

Figure 3-1 Simplified stress–strain diagram for reinforcement: CEB–FIP

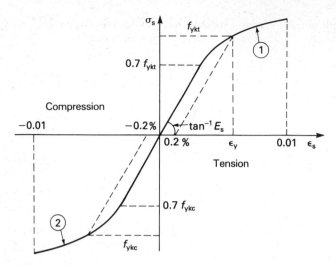

Figure 3-2   Idealized stress–strain diagram for reinforcement: CEB–FIP

(1) $\varepsilon_s = \dfrac{\sigma_s}{E_s} + 0.823 \left[\dfrac{\sigma_s}{f_{ykt}} - 0.7 \right]^5$

(2) $\varepsilon_s = \dfrac{\sigma_s}{E_s} - 0.823 \left[\dfrac{\sigma_s}{f_{ykc}} - 0.7 \right]^5$

3.1.4   *Modulus of longitudinal deformation*

For all reinforcement for reinforced concrete, the modulus of elasticity (longitudinal deformation), E_s, can be assumed to be equal to 200 kN/mm² (200 GPa, 29 000 ksi).

3.1.5   *Coefficient of thermal expansion*

The coefficient of thermal expansion is assumed to be equal to 10×10^{-6} per °C (5.6×10^{-6} per °F).

3.1.6   *Ductility*

It should be shown that the ductility of the steel is adequate for its utilization in the event of redistribution of stresses. This requirement can be regarded as satisfied if the unit elongation of the steel at failure, measured over a length of 10ϕ or 5ϕ, is not less than $\varepsilon_{10} = 8\%$ or $\varepsilon_5 = 12\%$ respectively (ϕ is the diameter of the bar).

3.1.7   *Weldability*

The weldability of reinforcement steel depends mainly on the method of manufacture (hot-rolling or cold-drawing), on its chemical composition, and on its diameter. From the point of view of weldability, the steels are classified as follows:

Class a (not weldable): steel that cannot be welded with acceptable results.

Class b (weldable in certain conditions): steel which can be welded with acceptable results, by using special methods or normal methods accompanied by special safety precautions.

Table 3-1 Classification of hot-rolled weldable types of steel: CEB–FIP

Class	Welding by continuous weld	Tack welding
b	$C \leqslant 0.35\%$ $C_{rep} \leqslant 0.60\%$	$6 \leqslant \phi \leqslant 16: C \leqslant 0.28\%$ $16 < 25: C \leqslant 0.26\%$ $\phi > 25: C \leqslant 0.24\%$ $C_{rep} \leqslant 0.54\%$
c	$C \leqslant 0.24\%$ $C_{rep} \leqslant 0.52\%$	$6 \leqslant \phi \leqslant 16: C \leqslant 0.22\%$ $16 < \phi \leqslant 25: C \leqslant 0.20\%$ $\phi > 25: C \leqslant 0.18\%$ $C_{rep} \leqslant 0.48\%$

C_{rep} denotes the carbon equivalent value defined by

$$C_{rep} = C + \frac{Mn}{6} + \frac{Cr + Mo + V}{5} + \frac{Ni + Cu}{15}$$

where C, Mn, Cr, Mo, V, Ni, Cu are the percentage contents by weight of carbon, manganese, chromium, molybdenum, vanadium, nickel and copper, respectively.

Class c (weldable): steel which can be welded with the usual methods with acceptable results.

The usual welding methods are: (a) electric flash welding; (b) electric resistance welding; (c) electric arc welding (by means of coated electrodes or under a protective gas cover); (d) high-pressure gas welding.

All reinforcement which is not exclusively intended for fixing is regarded as load carrying.

Strength and ductility are considered satisfactory if in a tensile test the welded joint can withstand a force giving at least 4% strain at the maximum stress outside the area influenced by welding, and in any case not less than 1.1 times the force corresponding to the limit of elasticity for the bars being joined.

Weldable hot-rolled steels can be classified according to their chemical composition, the type of welding and their diameter (see Table 3-1). To avoid possible confusion, it is preferable to stock only one type of steel on a site.

3.2 British code CP 110

3.2.1 Reinforcement grade
The steels are classified according to
    (a) Method of production:
        (1) hot-rolled steel;
        (2) cold-worked steel;
        (3) hard-drawn steel wire.

(b) Surface properties:
   (1) plain smooth bars or wires (including welded mesh);
   (2) deformed bars (bond classification).
      Type 1. A plain square twisted bar, each with a pitch of twist not greater than 14 times the nominal size of the bar.
      Type 2. A bar with transverse ribs with a substantially uniform spacing not greater than 0.8ϕ (and continuous helical ribs where present), having a mean area of ribs (per unit length) above the core of the bar not less than 0.15ϕ mm^2/mm, where ϕ is the nominal bar size.
(c) Weldability.

Table 3-2   Strength of reinforcement: CP 110

Grade	Nominal size mm	Specified character-istic strength, f_y N/mm^2
Hot-rolled steel, grade 250 (BS 4449)	All sizes	250
Hot-rolled steel, grade 460/425 (BS 4449)	Up to and including 16 Over 16	460 425
Cold-worked steel, grade 460/425 (BS 4461)	Up to and including 16 Over 16	460 425
Hard-drawn steel wire and fabric (BS 4482 and BS 4483)	Up to and including 12	485

The grades of reinforcements and their specified characteristic strengths, f_y, are given in Table 3-2. The range of preferred nominal sizes of bars is given in Table 3-3, the nominal size being defined as the diameter of a circle with an area equal to the effective cross-sectional area of the bar.

The nominal density of reinforcing steel can be taken as 7850 kg/m^3 (490 lb/ft^3), i.e. 0.00785 kg/mm^2 per metre run.

Table 3-3   Preferred sizes: CP 110

Nominal size, (mm)	6	8	10	12	16	20	25	32	40	50

3.2.2   *Characteristic strength*
The specified characteristic strength, f_y, of hot-rolled steel bars is the value of the yield stress below which shall fall not more than 5% of the test results of the material supplied (see BS 4449) [27].

Table 3-4   Tensile properties: CP 110

Grade	Nominal size of bar mm	Specified characteristic strength N/mm²	Minimum elongation on gauge length, L_0 %
250	All sizes	250	22
460	6 up to and including 16	460	12
425	Over 16	425	14

The specified characteristic strength, $f_y = f_{0.2}$, of cold-worked steel bars complying with the requirements of BS 4461 [28] is based on either (a) not more than 5% of the 0.2% proof-stress test results falling lower than the values in Table 3-3 or, where applicable (b) there being no test results below the 0.2% proof stress; the proof stress is here defined as the tensile strength which shall be between 5% and 10% greater than the actual yield stress measured in the tensile test. In this latter case, the actual yield stress shall be not less than $B(2.1 - A)$ MPa, where A is the ratio of the measured tensile strength to the actual yield stress and $B = 460$ for bars of nominal size 6 to 16 mm or 425 for bars of nominal size over 16 mm.

Compliance with the requirements of BS 4449 [27] and BS 4661 [28] is satisfied if

(a) not more than two test results of yield stress out of the last 40 consecutive results are less than the specified characteristic strength;
(b) no test results are less than 0.93 of the specified characteristic strength.

In calculating the yield stress of reinforcing bars, the area as determined from the actual mass of the bars is used.

The minimum elongation on gauge length $L_0 = 5.65\sqrt{S_0}$, where L_0 is the gauge length of the test piece and S_0 is the original cross-sectional area of the test piece, is given in Table 3-4.

The relationship between the characteristic strength of the reinforcement, f_y, and the mean yield strength, f_m, is

$$f_y = f_m - 1.64\sigma \qquad (3\text{-}4)$$

where σ is the standard deviation.

Design may be based on the characteristic strength or a lower value if necessary to reduce deflection or control cracking.

3.2.3   Stress–strain relationship

CP 110 gives no information about the 'actual' stress–strain relationship for the steel reinforcement.

Typical 'actual' stress–strain curves are shown in Fig. 3-3; these may be considered applicable for both tension and compression.

Both mild steel bars and hot-rolled high-yield bars have definite yield points.

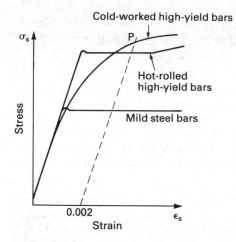

Cold-worked high-yield bars

Figure 3-3   Typical stress–strain curves for reinforcement: CP 110

Cold-worked high-yield bars do not have a definite yield point, and the characteristic strength refers to the 0.2% proof stress (point P in Fig. 3-3).

3.2.4   Modulus of elasticity

For all reinforcement for reinforced concrete, the modulus of elasticity, E_s, is usually taken as 200 kN/mm^2 (200 GPa, 29 000 ksi) in design.

3.2.5   Coefficient of thermal expansion

CP 110 gives no information on the thermal coefficient of expansion. In British practice, a value of 10×10^{-6} per °C (5.6×10^{-6} per °F) is commonly used.

3.2.6   Ductility

CP 110 gives no direct information on ductility requirements for steel reinforcement, but refers to *bend tests* described below.

The bend-test requirements shall be based upon bends through 180° of full-size bars (all sizes) around pins (formers) with diameters of twice the nominal size of the bar for Grade 250 and three times the nominal size of the bar for Grade 460/425.

The bend test shall be carried out on test pieces having a temperature between 5 and 20°C (41–68°F) and in such a way as to produce a continuous and uniform bending deformation (curvature) at every section of the bend.

3.2.7   Weldability

The weldability of reinforcement steel depends mainly on the method of manufacture (hot-rolling or cold-drawing), on its chemical composition, and on its diameter.

Welding on site should be avoided if possible, but where suitable safeguards and techniques are employed and provided that the types of steel (including 'weldable' and 'readily weldable' reinforcement as defined in BS 4449 [27] and

BS 4461 [28]) have the required welding properties, it may be undertaken. Generally, however, all welding should be carried out under controlled conditions in a factory or workshop.

Welding may be used for

(a) Fixing in position, for example, by welding between crossing or lapping reinforcement or between bars and other steel members. Metal-arc welding or electric resistance welding may be used on suitable steel.
(b) Structural welds involving transfer of load between reinforcement or between bars and other steel members. Butt welds may be carried out by flash butt welding or metal-arc welding or electric resistance welding may be used.

Structural welds should not occur at bends in reinforcement. Unless otherwise agreed by the engineer, joints in parallel bars of the principal tensile reinforcement should be staggered in the longitudinal direction. For joints to be considered as staggered, the distance between them must not be less than the end anchorage length for the bar. The length of run deposited in a single pass of a welded lapped joint should not normally exceed five times the size of the bar. If a longer length of weld is required, it should be divided into sections and the space between runs made not less than five times the size of the bar.

Since cold-worked steels tend to lose strength after heating, it is desirable to keep all welding remote from these sections of reinforcement subjected to maximum stresses in service. Welding should also be avoided where reinforced concrete members are to be subjected to large numbers of repetitions of substantial loads. The fatigue strength of beams in which the links have been welded to the main bars can be reduced by as much as 50%.

The chemical composition of the steel based on cast analysis shall meet the requirements given in Table 3-5.

Hot-rolled steels supplied in accordance with BS 4449 shall be capable of being

Table 3-5   Chemical composition of steel: CP 110

	Hot-rolled steel (BS 4449)		Cold worked steel (BS 4461)
	Grade 250 % max	Grade 460/425 % max	Grade 460/425 % max
Carbon	0.250	0.400	0.250
Sulphur	0.060	0.050	0.060
Phosphorus	0.060	0.050	0.060

$$\text{Carbon equivalent value} = C + \frac{Mn}{6} + \frac{Cr + V + Mo}{5} + \frac{Cu + Ni}{15} \qquad (3\text{-}5)$$

where C, Mn, Cr, V, Mo, Cu, Ni, are the percentage contents by weight of carbon, manganese, chronium, vanadium, molybdenum, copper and nickel, respectively.

welded on site provided that the requirements of BS 5135 [39] and the manufacturer's recommendations are followed.

(a) All steel of Grade 250 shall be considered readily weldable.
(b) Steels of Grade 460/425 shall be considered weldable only if the cast analysis gives a carbon equivalent not greater than 0.51% when derived from Eqn 3-5 in Table 3-5. Particular note of the requirements of BS 5135 shall be taken in the exceptional cases where the carbon equivalent value derived from Eqn 3-5 in Table 3-5 exceeds 0.42%.

Cold-worked steels supplied in accordance with BS 4461 are readily weldable provided that the requirements of BS 5135 and the manufacturer's recommendations are observed. The carbon equivalent shall be not greater than 0.42% when derived from the Eqn 3-5.

3.3 ACI 318 Building Code

3.3.1 *Reinforcement grade*
Reinforcing steels are classified according to the method of production, surface properties, and weldability.

Reinforcing steels shall conform to one of the ASTM specifications for concrete reinforcement.

Reinforcement shall be deformed reinforcement, except that plain reinforcement may be used for spirals.

3.3.1.1 *Reinforcing bars* The *grade* designation of a reinforcing steel bar is stated by a number which corresponds to its minimum yield strength expressed in ksi. The standard grade designations are 40, 50, 60, and 75.

Reinforcing bars are available in a large range of diameters, from about $\frac{1}{4}$ to about $1\frac{3}{8}$ in (6–35 mm) for ordinary applications, and in two heavy bars of about $1\frac{3}{4}$ and $2\frac{1}{4}$ in (45 and 57 mm). These bars, with the exceptions of the $\frac{1}{4}$ in size, are furnished with surface deformations for the purpose of increasing the bond strength between steel and concrete. Different bar producers use different patterns.

Bar sizes are designated by numbers, 2 to 11 being commonly used and 14 and 18 representing the two special large size bars previously mentioned. Designation by number, instead of by diameter, has been introduced because the surface deformations make it impossible to define a single easily measured value of the diameter. The numbers are chosen so that the value in the unit place in the designation equals the number of $\frac{1}{8}$ in (3.2 mm) in the diameter. Thus, a 7 bar has a nominal diameter of $\frac{7}{8}$ in, and similarly for the other sizes.

In order for bars of various grades and sizes to be easily distinguished, which is necessary to avoid accidental use of lower-strength or smaller-sized bars than called for in the design, all deformed bars are furnished with rolled-in markings. These identify the producing mill (usually an initial), the bar size number (3 to 18), the type of steel (*N* for billet, *A* for axle, a rail sign for rail steel, and an additional marking for identifying the higher-strength steels). These markings consist either of one or two longitudinal lines for 60 and 75 ksi (414 and 517 MPa) yield steels, respectively, or the numbers 60 and 75.

3.3.1.2 *Steel wires and welded wire fabrics* Welded wire fabric (WWF) is a prefabricated reinforcement consisting of parallel series of cold-drawn wires welded together in square or rectangular grids. Each wire intersection is electrically resistance-welded by a continuous automatic welder. Smooth wires, deformed wires, or a combination of both, may be used. Welded deformed wire fabric uses wire deformations plus the welded intersections for bond and anchorage.

Both wire for welded wire fabric and wire for reinforcement are specified by cross-sectional area. Gauge numbers were used exclusively for many years. The industry changed to a letter–number combination. The letters W and D are used in combination with a number. The letter W designates a smooth wire and the letter D denotes a deformed wire. The number following the letter gives the cross-sectional area in hundreds of square inches (1 square inch = 645 mm²). For instance, wire designation $W4$ would indicate a smooth wire with a cross-sectional area of 0.04 in² (26 mm²); a $D10$ wire would indicate a deformed wire with a cross-sectional area of 0.10 in² (65 mm²).

The nominal cross-sectional area of a deformed wire is determined from the weight (mass) per foot (metre) of wire rather than the diameter.

Spacings and sizes of wires in welded wire fabric are identified by 'style' designations. A typical designation (with the meaning of the symbols explained) is

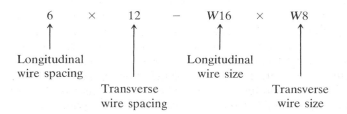

A welded deformed wire fabric style would be noted in the same number by substituting D-number wire sizes for the W-number wire sizes shown. The 'style' gives spacings and sizes of wires only and does not show any other information such as width and length of sheet or roll. The terms 'longitudinal' and 'transverse' are related to the manufacturing process and have no reference to the position of the wires in a concrete structure.

In the manufacture of welded wire fabric, longitudinal wires move continuously through the automatic welding machine. Transverse wires are individually welded at right angle as the fabric advances through the welders.

3.3.2 *Specifications and strength requirements*
The specifications for reinforcement published by the American Society for Testing and Materials (ASTM) are generally accepted for construction in the United States.

Reinforcing bars ACI 318 requires reinforcing bars to conform to one of the following ASTM specifications, with a few important exceptions noted below:

(a) ASTM A615: *Specification for deformed and plain billet-steel bars for concrete reinforcement,* [45].
(b) ASTM A616: *Specification for rail-steel deformed and plain bars for concrete reinforcement,* [46].
(c) ASTM A617: *Specification for axle-steel deformed and plain bars for concrete reinforcement,* [47].
(d) ASTM A706: *Specification for low-alloy steel deformed bars for concrete reinforcement* [48].

Exceptions to these specifications required by Sections 3.5.3.2 and 3.5.3.3 of ACI 318–77 are as follows [4]:

(1) For ASTM A615, A616 and A617, yield strength, f_y, shall correspond to that determined by tests on full size bars.
(2) For ASTM A615, A616 and A617, bend test requirements shall be based upon bends of full-size bars around pins with diameters specified in Table 3-6. If Nos 14 and 18 bars as used in the structure are required to have bends exceeding 90°, specimens shall be bend tested 180° with other criteria the same as for 90°.
(3) Deformed reinforcing bars with a specified yield strength, f_y, exceeding 60 000 psi (414 MPa) may be used, provided f_y shall be the stress corresponding to a strain of 0.0035 and the bars otherwise conform to the listed ASTM specifications and the exceptions as noted.

Both items (1) and (2) above will be automatically provided for if the supplementary requirements of ASTM A615 are included by the specifier. Bars meeting these requirements may be identified by the letter S rolled into the surface of the bar in place of the N used for standard ASTM A615 steel.

Bar and mats for concrete reinforcement are required to conform to ASTM A184: *Specification for fabricated deformed steel bar mats for concrete reinforcement* [41].

Table 3-6   Bend test requirements: ACI 318

Bar designation	Pin diameter for bend test[a] d_b = nominal diameter of specimen	
	Grade 40	Grade 60
3, 4, 5	$3\frac{1}{2}d_b$	$3\frac{1}{2}d_b$
6, 7, 8	$5d_b$	$5d_b$
9, 10, 11	$5d_b$	$7d_b$
14, 18 (90°)	—	$9d_b$

[a] Test bends 180° unless noted otherwise

3.3.3 *Stress–strain relation*
ACI 318–77 [4] does not directly give information on the stress–strain relations of reinforcement, though some information can be obtained through the ASTM standards listed in Section 3.3.2 (see also Figs 3-1 and 3-2).

3.3.4 *Modulus of elasticity*
The modulus of elasticity, E_s, for nonprestressed reinforcement may be taken as 29×10^6 psi (200 GPa).

3.3.5 *Coefficient of thermal expansion*
ACI 318–77 [4] does not directly give specific information on the coefficient of thermal expansion. In US practice, a value of 6.5×10^{-6} per °F (11.7×10^{-6} per °C) is commonly used.

3.3.6 *Ductility*
ACI 318–77 [4] gives no direction information on ductility, but refers to bend tests as specified in Section 3.3.2 above.

3.3.7 *Weldability*
Welding of crossing bars is not permitted for assembly of reinforcement unless authorized by the engineer.

Reinforcement to be welded shall be indicated on the drawings and welding procedure to be used shall be specified. ASTM steel specifications, except for ASTM A706 [48], shall be supplemented to require a report of material properties necessary to conform to the welding procedures specified in *Reinforcing steel welding code* (AWS D12.1) [63] of the American Welding Society.

Acknowledgements

The author and publishers gratefully acknowledge permission of the American Concrete Institute to refer to the provisions of the ACI 318-77 *Building code requirements for reinforced concrete,* and of the Comité Euro-International du Béton to quote from the *CEB-FIP Model Code for concrete structures,* 3rd edition, 1978. Portions of various ASTM specifications listed in the References are copyright, American Society for Testing and Materials, 1916 Race Street, Philadelphia, PA 19103, and are adapted with permission. Extracts from British and International Standards are reproduced by permission of the British Standards Institution, 2 Park Street, London W1A 2BS from whom complete copies can be obtained.

References

1. *CEB–FIP model code for concrete structures,* 3rd Edn, 1978. (English edition from Cement and Concrete Association, London).
2. British Standards Institution, CP 110, Part 1: 1972 (as amended 1980) *The structural use of concrete,* London, 1980, 154pp.
3. Bate, S. C. C. *et al., Handbook on the unified code for structural concrete (CP 110: 1972),* Cement and Concrete Association, London, 1972, 153pp.

4. ACI Committee 318; *Building code requirements for reinforced concrete*, ACI 318–77, American Concrete Institute, Detroit, 1977, 103pp.
5. *Commentary on 'Building code requirements for reinforced concrete (ACI 318–77)'*, American Concrete Institute, Detroit, 1977, 132pp.
6. International Organization for Standardization, ISO 3893, *Concrete classification by compressive strength*, Geneva, 1977, 1p.
7. International Organization for Standardization, ISO 3898, *Basis for design of structures—notations and general symbols*. Geneva, 1976, 4pp.
8. British Standards Institution, CP 3, Chapter V, Part 1: 1967, *Dead and imposed loads*, London, 1967, 20pp.
9. British Standards Institution, CP 3, Chapter V, Part 2: 1972, *Wind loads*, London, 1972, 52pp.
10. British Standards Institution, BS 12: 1978, *Specification for ordinary and rapid-hardening Portland cement*, London, 1978, 4pp.
11. British Standards Institution, BS 146, Part 2: 1973, *Portland-blast furnace cement*, London, 1973, 8pp.
12. British Standards Institution, BS 1370: 1979, *Specification for low heat Portland cement*, London, 1979, 4pp.
13. British Standards Institution, BS 1881, Part 1: 1970, *Methods of sampling fresh concrete*, London, 1970, 16pp.
14. British Standards Institution, BS 1881, Part 2: 1970, *Methods of testing fresh concrete*, London, 1970, 36pp.
15. British Standards Institution, BS 1881, Part 3: 1970, *Methods of making and curing test specimens*, London, 1970, 28pp.
16. British Standards Institution, BS 1881, Part 4: 1970, *Methods of testing concrete for strength*, London, 1970, 28pp.
17. British Standards Institution, BS 1881, Part 5: 1970, *Methods of testing hardened concrete for other than strength*, London, 1970, 40pp.
18. British Standards Institution, BS 1881, Part 6: 1971, *Analysis of hardened concrete*, London, 1971, 36pp.
19. British Standards Institution, BS 4027: 1980, *Specification for sulphate-resisting Portland cement*, London, 1980, 4pp.
20. British Standards Institution, BS 4246, Part 2: 1974, *Low heat Portland-blast furnace cement*, London, 1974, 12pp.
21. British Standards Institution, BS 4248: 1974, *Supersulphated cement*, London, 1974, 24pp.
22. British Standards Institution, BS 4408, Part 1: 1969, *Electromagnetic cover measuring devices*, London, 1969, 12pp.
23. British Standards Institution, BS 4408, Part 2: 1969, *Strain gauges for concrete investigations*, London, 1969, 16pp.
24. British Standards Institution, BS 4408, Part 3: 1970, *Gamma radiography of concrete*, London, 1970, 4pp.
25. British Standards Institution, BS 4408, Part 4: 1971, *Surface hardness methods*, London, 1971, 12pp.
26. British Standards Institution, BS 4408, Part 5: 1974, *Measurement of the velocity of ultrasonic pulses in concrete*, London, 1974, 20pp.
27. British Standards Institution, BS 4449: 1978, *Specification for hot-rolled steel bars for the reinforcement of concrete*, London, 1978, 12pp.
28. British Standards Institution, BS 4461: 1978, *Specification for cold-worked*

steel bars for the reinforcement of concrete, London, 1978, 8pp.

29. British Standards Institution, BS 4466: 1981, *Specification for bending dimensions and scheduling of bars for the reinforcement of concrete*, London, 1981, 16pp.
30. British Standards Institution, BS 4482: 1969, *Hard-drawn mild steel wire for the reinforcement of concrete*, London, 1969, 12pp.
31. British Standards Institution, BS 4483: 1969, *Steel fabric for the reinforcement of concrete*, London, 1969, 12pp.
32. British Standards Institution, BS 4550, Part 0: 1978. *Methods of testing cement—general introduction*, London, 1978, 4pp.
33. British Standards Institution, BS 4550, Part 1: 1978, *Sampling*, London, 1978, 6pp.
34. British Standards Institution, BS 4550, Part 2: 1970, *Chemical tests*, London, 1970, 48pp.
35. British Standards Institution, BS 4550, Part 3: 1978, *Physical tests*, London, 1978, 38pp.
36. British Standards Institution, BS 4550, Part 4: 1978, *Standard coarse aggregate for concrete cubes*, London, 1978, 2pp.
37. British Standards Institution, BS 4550, Part 5: 1978, *Standard sand for concrete cubes*. London, 1978, 6pp.
38. British Standards Institution, BS 4550, Part 6: 1978, *Standard sand for mortar cubes*, London, 1978, 2pp.
39. British Standards Institution, BS 5135: 1974, *Metal-arc welding of carbon and carbon manganese steels*, London, 1974, 48pp.
40. American Society for Testing and Materials, ASTM A82–79, *Standard specification for cold-drawn steel wire for concrete reinforcement*, Philadelphia, 1979, 4pp.
41. American Society for Testing and Materials, ASTM A184–79, *Standard specification for fabricated deformed steel bar mats for concrete reinforcement*, Philadelphia, 1979, 3pp.
42. American Society for Testing and Materials, ASTM A185–79, *Standard specification for welded steel wire fabric for concrete reinforcement*, Philadelphia, 1979, 6pp.
43. American Society for Testing and Materials, ASTM A496–78, *Standard specification for deformed steel wire for concrete reinforcement*, Philadelphia, 1978, 5pp.
44. American Society for Testing and Materials, ASTM A497–79, *Standard specification for welded deformed steel wire fabric for concrete reinforcement*, Philadelphia, 1979, 6pp.
45. American Society for Testing and Materials, ASTM A615–80, *Standard specification for deformed and plain billet-steel bars for concrete reinforcement*, Philadelphia, 1980, 6pp.
46. American Society for Testing and Materials, ASTM A616–79, *Standard specification for rail-steel deformed and plain bars for concrete reinforcement*, Philadelphia, 1979, 6pp.
47. American Society for Testing and Materials, ASTM A617–79, *Standard specification for axle-steel deformed and plain bars for concrete reinforcement*, Philadelphia, 1979, 6pp.
48. American Society for Testing and Materials, ASTM A706–81, *Standard*

specification for low-alloy steel deformed bars for concrete reinforcement, Philadelphia, 1981, 6pp.

49. American Society for Testing and Materials, ASTM C31–80, *Standard method of making and curing concrete test specimens in the field,* Philadelphia, 1980, 6pp.

50. American Society for Testing and Materials, ASTM C39–80, *Standard method of test for compressive strength of cylindrical concrete specimens,* Philadelphia, 1980, 4pp.

51. American Society for Testing and Materials, ASTM C42–77, *Standard method of obtaining and testing drilled cores and sawed beams of concrete.* Philadelphia, 1977, 5pp.

52. American Society for Testing and Materials, ASTM C78-75, *Standard method of test for flexural strength of concrete (using simple beam with third-point loading),* Philadelphia, 1975, 3pp.

53. Ameican Society for Testing and Mateials, ASTM C138-77, *Standard method of test for unit weight, yield, and air content (gravimetric) of concrete,* Philadelphia, 1977, 4pp.

54. American Society for Testing and Materials, ASTM C150–81, *Standard specification for Portland cement,* Philadelphia, 1981, 7pp.

55. American Society for Testing and Materials, ASTM, C172–77, *Standard method of sampling fresh concrete,* Philadelphia, 1977, 3pp.

56. American Society for Testing and Materials, ASTM C192–77, *Standard method of making and curing concrete test specimens in the laboratory,* Philadelphia, 1977, 3pp.

57. American Society for Testing and Materials, ASTM C215-60, *Standard method of test for fundamental transverse, longitudinal, and torsional frequencies of concrete specimens,* Philadelphia, 1976, 5pp.

58. American Society for Testing and Materials, ASTM C293-79, *Standard method of test for flexural strength of concrete (using simple beam with centre-point loading),* Philadelphia, 1979, 3pp.

59. American Society for Testing and Materials, ASTM C469-65, *Standard method of test for static modulus of elasticity and Poisson's ratio of concrete in compression,* Philadelphia, 1975, 5pp.

60. American Society for Testing and Materials, ASTM C496–71, *Standard method of test for splitting tensile strength of cylindrical concrete specimens,* Philadelphia, 1971, 6pp.

61. American Society for Testing and Materials, ASTM C512-76, *Standard method of test for creep of concrete in compression,* Philadelphia, 1976, 5pp.

62. American Society for Testing and Materials, ASTM C595–81a, *Standard specification for blended hydraulic cements,* Philadelphia, 1979, 11pp.

63. American Welding Society, AWS D12.1–75, *Reinforcing steel welding code,* Miami, 1975.

64. Kong, F. K. and Evans, R. H., *Reinforced and prestressed concrete,* 2nd Edn, Nelson, Walton-on-Thames/Van Nostrand Reinhold, Wokingham, 1980, 412pp.

41 Structural design: practical guidance and information sources

J C Steedman

JACYS Computing Services, Steyning, England

Contents

Notation

A_s area of tension reinforcement
A'_s area of compression reinforcement
A_{sc} total area of reinforcement in column section
b breadth of section
D density of soil
d effective depth of section
d' depth to compression reinforcement
F earth pressure
f_c limiting stress in concrete for elastic design
f_{cu} characteristic concrete cube strength
f_y characteristic strength of reinforcement
h height of retaining wall; overall depth of section
h_{max} larger overall dimension of section
h_{min} smaller overall dimension of section
J torsional coefficient
k proportion of compression reinforcement
l_x length of shorter side of rectangular slab
l_y length of longer side of rectangular slab
M bending moment
M_x bending moment on shorter span of two-way slab
M_y bending moment on longer span of two-way slab
N axial load
n number

S	surcharge	α	angle of respose
T	torsional moment due to ultimate	α_e	modular ratio
	load	β	angle of inclination of retained
v	torsional shear stress		material
w	distributed load per unit area	ϕ	angle between back of retaining
x	depth to neutral axis		wall and horizontal

Summary

The purpose is to provide information on practical concrete design and to direct the reader to some further sources of important information.

Design aids such as tables, charts, nomograms and calculator or micro-computer programs play a considerable and valuable role in routine office practice. Each type of aid has particular advantages and disadvantages and these are discussed. Some of the more common problems encountered when designing frames, beams, slabs and columns, and various specialized types of structure, are reviewed. The analysis of sections to resist bending, shearing, torsion and to meet serviceability requirements is also dealt with. Particular attention is paid throughout to the relationship between the theoretical assumptions that are made in order to undertake an analysis, and the probable behaviour of the actual structure or member.

1   Introduction

In this final chapter, the emphasis is somewhat different to that of the previous contributions, since it is principally concerned with the carrying-out in the simplest, quickest and most convenient way, concomitant with the requirements of rigour and accuracy, of the design processes and calculations that form the bulk of a concrete designer's day-to-day duties. The notation employed conforms, where applicable, to CP 110 [1], the principal UK design code.

Practical structural design is a complex procedure requiring great skill and understanding on the part of the engineer. This is particularly true where a composite material such as reinforced concrete is used. After calculating the moments and forces acting on a idealized (and often linear, two-dimensional) structure, the designer must use his judgement, expertise and experience to apply the results to an actual construction that may differ considerably from the idealized model. As an example, Table 1-1 lists some of the deviations between theory and practice that may occur in the case of a framed building.

In view of the complexity and unpredictability of the actual behaviour of concrete structures, the task of the designer is not so much to predict exactly the forces that act on a particular structure but rather to so design the structure that it will adequately resist the forces that may occur.

This is *practical* design. For example, while it is often useful and sometimes essential to analyse and design a multistory framed structure as a single entity, it is frequently quite satisfactory to design the beams at each floor level as an

Table 1-1

Theoretical assumption	Behaviour in practice
Single, homogeneous, isotropic material	Composite action of concrete (itself a composite of aggregate–sand–cement) with steel bars spaced at discrete intervals. Cracking may occur due to shrinkage, settlement, failure to determine service moments and forces accurately, rendering theoretical assumptions of behaviour invalid
In the case of a small structure, virtually no joints provided	Many daywork (i.e. construction) joints
Concrete uniform throughout	Variations of strength (and therefore of elastic modulus) occur among individual batches in addition to differing mixes perhaps being specified for different parts of the structure
Analysis based on linear structure	All members have appreciable dimensions, considerably restricting free joint rotation, etc
Calculated moment of inertia normally based on gross dimensions of concrete section	Actual reinforcement provided to meet design requirements influences moment of inertia of sections
Structure unloaded until construction is complete	Floor-by-floor construction ensures that progressive joint rotations and axial deformations of columns occur due to progressively increasing dead loads
Uniform value for Poisson's ratio (in slabs, for example)	If cracks occur due to the reasons mentioned above, Poisson's ratio becomes zero. Elsewhere, incorrect choice of value (which depends on various factors (principally the age of the concrete and the type of aggregate used) may, in extreme cases, result in the adoption of moments of the wrong sign
Non-interaction of partitions, cladding, etc	Unless very carefully detailed, their interaction influences structural behaviour
Two-dimensional behaviour normally only considered	Three-dimensional behaviour occurs, thereby restricting deformations, and hence moments, due to longitudinal rigidity
Axial and shear deformation may safely be ignored	Due to creep from sustained loading, such deformation may be considerable, expecially in tall structures
Settlement of the foundations does not occur	Unless founded directly upon rock, all foundations settle to some extent. Settlement may take place as dead loads, resulting from the construction of successive floors, are applied, thus rotating the joints in the lower floors of the building before construction is complete
For slabs that are freely supported or are continuous over rigid unyielding supports	The supporting beams (with which the slabs are monolithic) deflect and rotate due to loading

individual system, basing the analysis on the assumption that the beams, together with the columns directly above and below (assumed fixed at their far ends) act as individual systems. Often, still smaller sub-systems may be valid, as described later. The same situation applies to many other structures or parts of a structure.

2   Design aids

Practical design office procedures often involve the replacement of more rigorous and more exact methods by procedures that, while both simpler and quicker, are sufficiently accurate for practical design purposes, provided that the specified limitations to their use are strictly observed. If this is done, many tedious design calculations can be eliminated by the use of design aids in the form of tables, charts, etc. Each type of aid has both advantages and disadvantages, and these merit some discussion, since they are not at all well understood.

Ideally, each designer should produce as many of his own design aids as possible, since, by so doing, he will gain a special understanding and familiarity with the particular subject and the interrelationships between the variables involved that it is impossible to obtain in any other way. Nevertheless, however desirable this aim, it is not possible practically. The optimum version of a chart (or program) may be the product of many previous attempts, discarded for various reasons, and to produce a comprehensive series of charts (e.g. those given in [6] for limit state design to BS 5337) may take several months of work.

The use of graphs and tables has relatively recently been supplemented by calculator routines and the like. As well as offering the possibility of completely automatic structural design (see Chapters 21 and 22), microcomputers and programmable calculators can be used interactively as a form of design aid, as described below. However, they may never replace conventional aids completely as, for example, it will probably always be simpler and quicker to select a suitable bar arrangement visually from a table than to use a computer program.

2.1   Tables

Tabulated values have the particular advantage that, provided the known values correspond exactly to those included in the table, the likelihood of errors occurring when reading the table is remote (far less than when graphs are used). The principle disadvantage is that, if the values do not correspond, interpolation must be used. Linear interpolation is not difficult if only one or two variables are involved, although the chance of errors occurring is considerably increased. However, the representation of four or more variables requires the use of one or more sets of tables, and interpolation between values in different tables greatly complicates the task. Furthermore, if the tabulated values do not, in fact, show a linear relationship, linear interpolation may introduce considerable inaccuracy.

If a programmable calculator or a computer is available, a simple routine to handle interpolation between several sets of values can be prepared quite easily, which will reduce the possibility of errors being introduced from the actual calculations, although it is still easy to enter the values obtained from the tables incorrectly or in the wrong sequence.

2.2 Graphs

Cartesian graphs have the particular advantage over other forms of design aid that they display pictorially the relationships involved between two variables and thus enable them to be appreciated 'at a glance'. A single relationship between two variables produces a single curve on the graph. A third variable necessitates a family of curves, while the introduction of a fourth normally introduces the need for a set of graphs and thus the advantages of pictorial display are somewhat lost. Interpolation along each curve is extremely simple and the possibility of error remote. When a set of curves is involved, interpolation is more complex but still simpler than in the case of tables. If interpolation is necessary between more than three variables, and a set of graphs is needed, this becomes difficult and, because of the need to read each value required from a different graph, is actually more error-prone than when tables are used.

For example, when preparing the design charts that form Parts 2 and 3 of CP 110 [1], Beeby and Taylor of the Cement and Concrete Association undertook research to investigate the likelihood of errors arising when using various forms of graph. This research, which is reported in [78], indicates a rather greater probability than one might expect, and the authors give details of the layout, method of captioning scales, etc., recommended to reduce the possibility of such errors to a minimum.

One problem with graphs is that, unless considerable thought and care are exercised during their preparation, the space actually used to represent the variables may only be a small proportion of the total space available. There is nothing intrinsically wrong in this, except that it clearly does not make the best possible use of the space provided. A typical example of such a graph is that shown in Fig. 2-1, which is taken from Part 2 of CP 110 [1]. Although it illustrates

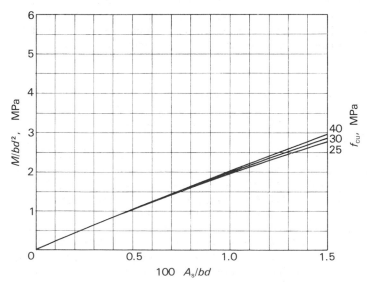

Figure 2-1 An example of the inefficient use of space in plotting graphs. Compare with Figure 2-2

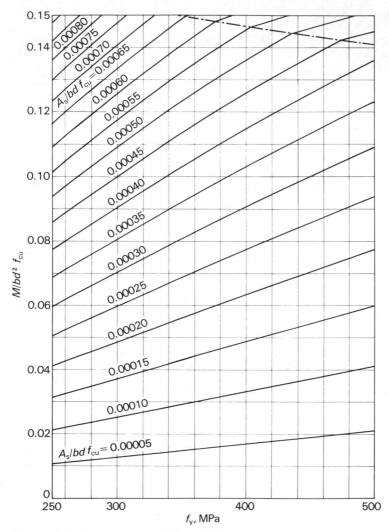

Figure 2-2   An example of the efficient use of space. Compare with Figure 2-1

the relationship between the percentage of steel and the resistance moment extremely well, less than 2% of the available space is actually being used. By contrast, the graph shown in Fig. 2-2, which is taken from [81] and represents a very similar relationship between variables, utilizes the entire available space to represent the relationships concerned.

Problems may be encountered when producing design graphs, where curves cross or lie on top of each other, or where the relationship is such that, with linear scales, an unacceptable graph results. Such difficulties may be overcome by using different scales involving power, logarithmic or reciprocal relationships. Graph paper with one or both axes graduated in this way may be obtained commercially

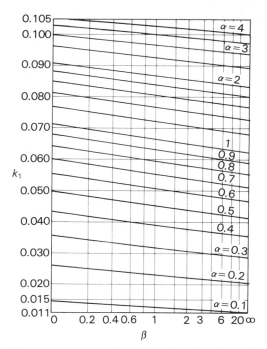

Figure 2-3 Use of purpose-designed graph paper to produce near straight-line plots

but it is often better to draw up a suitable grid oneself. A short automatic calculation sequence on a programmable calculator simplifies the plotting of such graphs, converting the actual values to be plotted into cartesian co-ordinates. A typical example of what can be achieved in this way is shown in Fig. 2-3, which is reproduced from Table 44 in [6]. Here, the scales along both axes have been adjusted to transform quite complex curves into near 'straight-line' relationships, simplifying the reading of the resulting graphs and interpolation between the 'curves' provided.

2.3 Nomograms

Nomograms or alignment charts seldom achieve the understanding or use by engineers that they deserve, yet basically they offer many advantages over conventional graphs. The simplest nomogram relates two variables by means of a single scale (Fahrenheit and Centigrade divisions along opposite sides of a thermometer scale are an excellent example) and corresponds to a cartesian graph representing a single-line relationship. The ease in reading, and the reduction in the chance of misreading are obvious. To relate three variables (i.e. a graph displaying a family of curves), three separate scales are required, which are then related by laying a straight-edge across the nomogram so that it intersects two of the scales at known values, the unknown term being read off where the third scale is intersected.

By the use of grids, it is sometimes possible to relate four or more variables on a single nomogram, thus replacing a whole set of graphs. Alternatively, additional variables can be easily catered for by employing 'pivot lines', though these add considerably to the complexity of the final nomogram and their use may thus be counterproductive.

The disadvantages of nomograms are that they are difficult to construct and that it is not always possible to relate three variables practically on a single diagram. A further criticism, that they are subject to error if the material on which they are printed changes dimensionally due to humidity variations, etc., is true but is seldom of practical importance.

2.4 Programmable calculators and microcomputers

In order to minimize the number of aids and yet to accommodate the widest possible range of actual values, and to cater for the use of different types of unit, it is usual to prepare design aids in non-dimensional terms as far as possible. This tends to introduce errors through converting actual values into their non-dimensional equivalents in order to use the design aid, and back again.

The advent of the programmable calculator avoids this hazard since the programs accept true loads, moments, etc., and respond by displaying actual steel areas, and so on. Most programs are valid for any consistent set of units, and a single program will cater for any number of variables, thus replacing a whole series of design charts. Suitable programmable calculators for concrete design are described in more detail in [6]. With the simplest types, the program must be keyed in each time it is used—a time-consuming and error-prone operation. With more-sophisticated machines, the programs can be stored on, and read in from, magnetic cards or tape.

A problem with calculators capable of displaying numbers only is that no relatively direct prompting information can be displayed by the machine. Also, unless the user is extremely familiar with the particular program in use, the operating instructions provided (especially regarding data entry) must be scrupulously followed. More recent machines can display text, and so messages can be included to prompt for the correct data. However, such facilities are limited in a calculator by constraints on the memory available. Even with the smallest microcomputers, such restraints are relatively minimal and it is possible, not only to prompt the entry of data, but to query values that appear to be incorrect due to the use of the wrong terms.

There can be little doubt that, in the future, the programmable calculator together with the microcomputer will form the concrete designer's chief design aid, largely replacing the books of graphs and tables previously needed. Graphical representations will always have a place since they enable relationships to be appreciated at a glance in a form that no other type of representation can achieve, but such representations can be easily displayed by using the computer's graphical facilities.

Details of a number of structural analysis and concrete design programs for microcomputers and sophisticated programmable calculators are given in [50, 51]. Further information can be obtained from the present author.

The remainder of this chapter briefly reviews certain aspects of reinforced concrete design, and directs the reader to some of the key sources containing more detailed information and/or useful design aids. The space devoted to each subject by no means relates to its importance: most major topics have their own chapters elsewhere in this Handbook and are, at most, only briefly mentioned here. The bibliography provided is also not intended to be comprehensive since such an attempt would fill the entire chapter. The criteria for inclusion are that the writer knows, works from, and finds particularly useful the reference quoted, and/or that it contains valuable aids for day-to-day design that are difficult or impossible to obtain elsewhere.

3 Frames

Theoretically, unless special stiffening walls are provided (as discussed below) an ordinary framed building is a three-dimensional space frame where the frame stiffness is provided by the resistance of the joints to rotation. Such an analysis is possible using computer methods, although a large machine is needed for all but the smallest structures, in order to cater for the large number of unknowns. Where buildings are longer than their width, the usual practice is to assume that the longitudinal strength is only nominal, and the structure is designed as a series of two-dimensional frames with the main beams running transversely. The floor and roof loads are transferred to the columns via these beams, and, in addition, the frames provide the lateral resistance to wind forces.

A complete analysis of such a frame may be undertaken using, for example, the matrix stiffness method or a method such as influence coefficients. The classical method of obtaining a solution using such procedures is to invert the corresponding matrix. This matrix may be large and thus take up appreciable amounts of computer storage but, since much of it consists of zero values, methods have been devised of 'banding' the significant data and thus rearranging it to occupy significantly less space (see, for example, [8]). With these large matrices, great precision is needed during the inversion process to prevent rounding-off errors, introduced during the repetitive calculations, from growing and swamping the results.

This need for considerable amounts of storage combined with great precision is the prime factor preventing very small computers from being employed for frame analysis using matrix inversion. However, instead of inverting the stiffness matrix, the corresponding simultaneous equations can also be solved by using iterative procedures such as the Gauss–Seidel method. With this method a five-story, 35-joint frame can be analysed entirely automatically on a programmable calculator (Quad-memory Hewlett Packard HP41C). The principal drawback is that such an analysis (involving about 15 iterations) takes nearly 80 min. (In comparision, similar calculations on an Apple microcomputer require about 7 min.)

If the structure is much higher than about six stories, simple analytical methods, in which axial and shearing deformations are assumed negligible and may safely be ignored, become increasingly incorrect and axial deformations, at least, must be taken into consideration. The minimum requirement for such analyses is then a minicomputer rather than a smaller machine, and the time needed to investigate even a single loading condition becomes not inconsiderable. To find the critical moments resulting from the worst combinations of imposed and wind loads is more lengthy. The load distribution required may be fairly obvious for a simple 'checkerboard' frame, but less so for a more irregular structure.

The need for the frame to resist lateral forces is often avoided by locating substantial solid walls at the ends or at intermediate points along the building. These 'shear walls' provide all the lateral stability necessary, the wind loads being transferred to them by the action of the floors as large beams lying on their sides. If the otherwise solid walls are pierced with fairly substantial openings, their design becomes a specialized matter. They can be analysed by general finite element methods [60], but specialist procedures have also been developed: see [61, 62], both of which include sets of design charts.

Towers can be designed in a somewhat similar manner, the overall strength being provided by a central core that cantilevers vertically from the foundations. The floors themselves cantilever outwards from this core and only light cladding is needed at the faces of the building. The design of such tall buildings is dealt with in detail in Chapters 37 and 38.

As mentioned in Section 1, the engineer must judge for himself to what extent the analysis of the idealized continuous monolithic frame discussed above resembles the behaviour of the actual building. In practice, the deviations from the idealization probably seldom make it worth while attempting an 'exact' analysis and many codes permit certain simplifications to be made. For example, if the columns are not slender, CP 110 permits the effects of sway to be neglected even if the frame provides all resistance to lateral loads. The analysis is then undertaken in two stages. In the first, the effects of lateral load only are considered, assuming the points of contraflexure to occur at mid-story heights (this assumption is actually only true for a frame if side-sway does not occur). The results of such an analysis are then considered in conjunction with those due to vertical loads only, determined floor by floor (as described later), and the most severe combinations determined.

Except in structures that are tall relative to their width, the effects of wind on the moments in the beams are relatively insignificant. If this assumption is valid, the series of beams at a particular floor level together with the columns immediately above and below the level considered (assumed fully fixed at their far ends) may be analysed as an individual system. This simplified method of analysis is recommended, for example, by CP 110 when a frame is subject to vertical loads only, and forms part of the procedure described above, suggested for frames supporting both vertical and lateral loads, where the frame provides lateral stability.

Because of the number of variables involved, graphical aids are generally not very useful for such investigations. Analysis using moment distribution methods (either the well-known Hardy Cross method or so-called 'precise' variants—see [6]) is simple and rapid. Automatic routines for programmable calculators for systems of up to 50 bays, and many microcomputer programs for systems of any size are available (see [51]).

If the lateral stability of the frame is assured by some other means, further simplification is possible. For example, CP 110 permits each floor system to be considered as a series of separate subframes, each subframe comprising a beam flanked by the interconnected members. The latter are assumed to be fully fixed at their far ends and, in addition, the stiffnesses of the outer beams are taken as only one-half of their true values.

As before, such a system may be analysed using moment-distribution methods, but alternatively the corresponding slope–deflection equations can be rearranged to produce expressions giving the support moments corresponding to critical combinations of dead and imposed load directly. Such expressions are tabulated in [6] and design graphs based on them are included in [7]; corresponding calculator programs are also available [51].

Included in [52] is a valuable discussion concerning the effects of the various idealizations of the actual frame described above, and many sound and useful comments on the differences between theory and actual behaviour are given in [53].

4 Beams

If the frame incorporates walls or bracing to withstand all lateral forces, the beams and columns could be well considered entirely separately. For example, in such a case CP 110 provides simple (but uneconomic) bending-moment and shearing-force coefficients for beams, that apply if the dead load does not exceed the imposed load and the system consists of not less than three spans differing by not more than 15% of the length of the greatest and supporting 'substantially uniform' loads.

As an alternative to employing these coefficients, a more rigorous analysis may be made, considering the beam as continuous over, and capable of free rotation about, the supports. For systems consisting of two, three, or more equal spans and supporting uniform or symmetrically-arranged concentrated loads, coefficients giving the maximum and minimum mid-span and support moments due to dead and imposed loading are included in various textbooks (see, for example, [6, 7]). Such moments can also be obtained for systems comprising any number of unequal spans and subjected to any arrangement of loading by a large number of different methods; some of these are described in [6]. Of these, the most popular is moment distribution. While this method may be used manually, calculator programs for the automatic analysis of systems of many spans are available. For example, using the program provided with the Hewlett Packard structural analysis ROM cartridge with a Quad-memory HP41C calculator, the support moments for a system of up to 67 spans may be found automatically. This program uses an algorithm based on the iteration of the slope–deflection formulas, rather than conventional moment distribution.

Such methods are usual only where members are prismatic. Non-prismatic members considerably complicate the analysis, even if a basically simple method, such as moment distribution, is used. Luckily, non-prismatic members seldom occur in building frames, because the use of splays increases the complexity of the formwork required. Tables giving the modified rotational and stiffness factors for non-prismatic members are available [53, 64], but computer methods are preferable here since the program can both evaluate the factors required and undertake the analysis automatically, thus minimizing any chance of error when entering data.

If a continuous structure is subjected to systems of substantial moving loads, the use of influence lines becomes essential. For structures comprising two to five equal, or symmetrically unequal, spans, sets of influence lines are readily available—see, for example, [6]. Again, programs for the production of these and also less-standard cases using microcomputers and programmable calculators have been prepared.

5   Slabs

Slabs spanning in one direction are identical to single-span or continuous beams and may be considered as such. When designing two-way slabs, either an analysis under service loading assuming that the slab behaves as an elastic plate, or behaviour at collapse is considered on the basis of limit state principles. In either case, the slab is normally assumed to be freely supported on, or continuous over (but separate from), knife-edge supports that do not deflect or rotate. In practice, of course, the slabs are usually supported on (and are often constructed monolithically with) beams that deflect substantially and also rotate. Further assumptions made in plate analysis (and sometimes otherwise also) are that a slab is a two-dimensional structure made from a homogeneous, isotropic material having a constant elastic modulus and Poisson's ratio; furthermore, membrane action is normally ignored. However, any cracking will modify the effects of Poisson's ratio (the assessment of a suitable value for which is difficult, anyway) and the curtailment and wide spacing of bars will nullify the assumptions of isotropy and homogeneity. Designers must bear in mind these deviations from the calculated idealized behaviour and try to allow for them in practical design. It appears in practice that the effects of three-dimensional behaviour, membrane action, and other indeterminate influences normally increase the strength of slabs beyond that determined by calculation.

5.1   Elastic analysis

The 'exact' theory of bending in plates spanning in two directions was first developed by Navier and Lagrange in the early 19th century, the work for freely-supported plates being completed by Pigeaud and others. The exact theoretical expressions so derived are complex (especially where the loading considered is not extremely simple) and so close arithmetical approximations are employed in practice. Many such expressions, together with typical values calculated assuming a Poisson's ratio of 0.2 or 0.3, are given in [11]. Bares [10] has gathered a large number of results of elastic analyses for slabs of various shapes due to numerous loadings and with a range of values of Poisson's ratio.

For a single freely-supported slab, where the corners are not prevented from lifting or twisting, Grashof and Rankine developed the following formulas, by equating the deflections at the intersection of central strips spanning in each direction;

$$M_x = \frac{wl_x^2}{8}\left[\frac{(l_y/l_x)^4}{1+(l_y/l_x)^4}\right] \qquad M_y = \frac{wl_x^2}{8}\left[\frac{(l_y/l_x)^2}{1+(l_y/l_x)^4}\right]$$

where w is the applied load on a rectangular slab having sides of length l_x and l_y. This expression, which gives greater bending moments than those that actually occur, has been used to derive the appropriate approximate coefficients provided in [1].

Much important work on the behaviour of slabs was undertaken in the 1920s in the USA by Westergaard. The results of this research formed the basis for the slab design requirements in the ACI code until 1971, and Westergaard's theoretical analyses, modified by subsequent test results, were used to produce the design coefficients in the UK codes up to and including the last version of CP 114—see [48]. An alternative method for the approximate elastic design of two-way slabs was developed by Marcus in Germany. The principles are similar to those of Grashof and Rankine already mentioned, but simple corrections are included to cater for corner restraint and torsional resistance. It is claimed that the resulting moments vary by only 1 or 2% from those given by complex analyses using elastic plate theory. Brief details are given in [6] and Marcus's method is described in greater detail in [48], where its extension to systems of continuous panels, proposed by Loser, is also explained.

The methods so far described deal, in general, with uniform loads covering the entire panel. For concentrated and partial loads, the complexity of elastic analysis increases considerably. The best-known set of design data for partial loading on single rectangular slabs is that produced by Pigeaud. These curves are reproduced in [6], where details of their extension to systems of continuous slabs are also given.

The exact analysis of a slab of arbitrary shape and support conditions and subjected to generalized loading is extremely involved but such a problem is well suited to automatic calculation procedures. As the use of computers has increased, numerical techniques such as finite differences and finite elements have been used increasingly to obtain solutions to such problems. As yet, the work undertaken is still mainly towards the production of sets of solutions to standard complex problems and there is little sign so far of the development, for everyday office use, of a generalized method for the elastic analysis of any arbitrary slab with any support conditions subjected to any loading. (Current finite element techniques [60], although theoretically suitable, still require too much specialized knowledge and computer power to fill this need.) This situation will undoubtedly change very soon. One interesting possibility lies in developments of the design method proposed by Fernando and Kemp [23] discussed below.

5.2 Collapse theory

The two best-known methods for analysing slabs plastically are Johansen's yield-line method and Hillerborg's 'strip method'. Due to the high degree of indeterminacy, it is normally impossible to calculate the precise ultimate resistance of a slab by collapse theory. Instead, separate 'upper-bound' and 'lower-bound' solutions may be determined. To achieve the former, a collapse mechanism is first assumed. When the slab deforms in accordance with this basis, the energy absorbed in developing the ultimate moments along the lines where yielding occurs is equated to the work done by the action of the applied load to produce such deformations. According to the upper-bound theorem [54], the load that is obtained from such an analysis is the maximum that the slab will support before collapse occurs. The problem with such an analysis is that no investigation is made regarding the conditions between the assumed yield lines. Since the moments set up in such areas may exceed the ultimate resistance of the slab, such an analysis does not tell the designer what minimum load the slabs can support. This is a fundamental limitation of upper-bound solutions, which are the type that result from yield-line analysis, although with experience it is normally not difficult to postulate the most likely type or types of collapse mechanism [54].

Lower-bound [54] solutions, on the other hand, give values of loads which the slab can certainly support. With such methods, a distribution of moments is postulated which ensures that the resistance of the slab is not exceeded and that equilibrium is satisfied at all points.

Collapse methods usually assume that the supports at the slab edges are rigid and unyielding. In practice, however, these supports are often beams having finite strength, and it is possible for collapse mechanisms to form in which the yield lines pass through the beams concerned, so that they form part of the collapse mechanism. This possibility must be taken into account when using collapse methods to analyse continuous beam-and-slab systems.

With yield-line analysis, the designer first postulates an appropriate collapse mechanism for the slab being considered, assuming that any yield lines needed are straight and can only change direction where they intersect another yield line, that the yield line separating two elements of a slab passes through the intersection of their rotational axes, and that all the reinforcement intercepted by a yield-line yields at that point. Any variable dimensions defining the yield-line pattern are then adjusted until the resulting pattern produces the maximum ultimate resistance for a given loading.

The two principal methods used to establish the controlling resistance are known as the 'virtual work' (or 'work') method and the 'equilibrium' method. The latter, which is not a true equilibrium method but merely a variant of the work method, has the advantage that the resulting equations contain sufficient information to eliminate all the unknown variables and thus lead to a direct solution. However, it is generally more limited in scope and not widely used; for details, see [15].

In the virtual work method, various means are used to solve the basic work equation. For relatively simple shapes and loadings, the equation may be expressed algebraically, and then differentiated to establish the critical values of the unknowns. Solutions to many standard problems obtained in this way are tabulated in [14]. However, since, as already indicated, collapse analysis is especially useful when a corresponding elastic analysis is virtually impossible (e.g., for irregularly shaped slabs), standard solutions are seldom available for such circumstances. Nevertheless, the validity of the standard formulas has been somewhat extended by the development of various affinity theorems. These permit slabs that are continuous over their supports to be analysed as if they were freely supported at the edges, skewed slabs to be transformed to their rectangular equivalents, and so on. Further superposition theorems enable the effects of complex combined loads (which would otherwise require the adoption of a unique yield-line pattern) to be represented by the sum of simpler individual loading cases (and hence simpler yield-line patterns). These theorems are described and illustrated in [6, 13–15, 17–22].

Alternatively, trial values are substituted for the unknown variables and the resulting resistances plotted, the maximum resistance then being read from the graph so constructed. Another possibility is to tabulate the terms involved in the work equation using actual numerical values, and then employ trial and adjust-ment, based on 'feedback' from previous calculation cycles. This method is explained in detail and illustrated in [17], and more briefly in [6]. Trial-and-adjustment processes are well suited to automatic computer analysis. A relatively simple example of a program to analyse rectangular slabs with corner levers (see below), which utilizes iterative procedures, is included in [50].

Tests and elastic analyses of slabs show that the negative moments along the edges reduce to zero near corners and increase rapidly away from such points. With slabs that have fixed or continuous edges, negative yield lines thus do not extend into the support intersections but instead form across the corners to produce, in conjunction with additional pairs of positive yield lines, additional triangular slab elements known as corner levers. A similar mechanism occurs if the slab is freely supported, causing the corner to lift from the support. If these mechanisms are substituted for the original yield lines extending to the support intersections, the overall strength of the slab is decreased. The formation of corner levers is stated to be more likely and more significant where acute-angled corners, heavy concentrated loads and fixed, continuous or unsupported edges occur and top steel is omitted from corners. (The influence of these individual factors is additive, heavy loads being the most influential.) Formulas to establish the dimensions, and thus the effects, of corner levers are derived in [15]; see also [6].

Although Johansen's yield-line method enables the maximum resistance corres-ponding to a specific failure pattern to be determined, it does not indicate whether the pattern considered is the critical one, nor does it apportion the loads carried by the slab to their respective supporting beams. For this, a lower-bound solution is required. Such a solution is provided by the strip method devised by Hillerborg and described in [16], which achieves a good economic arrangement of reinforce-ment. Hillerborg's original theory, which is now known as the simple strip method, assumes that, at failure, no load is carried by the torsional resistance of the slab, all support being provided by flexural bending in one of the two principal directions. The theory leads to simple solutions giving full information concerning the distribution of moments over the entire slab to resist a unique collapse load with the reinforcement being disposed economically in bands.

Unfortunately, however, simple strip theory cannot be used when concentrated loads and/or point supports occur, and difficulties also arise when investigating slabs with free edges. To extend the scope of his original method, Hillerborg developed his advanced strip method, which employs complex moment fields to overcome such difficulties, although the simplicity and directness of his original concept have been somewhat clouded as a consequence.

An interesting feature of both collapse theories discussed above is that, unlike elastic analyses, since cracking is assumed to have occurred, they take account of the actual arrangement of steel within the slab. However, they both suffer from the disadvantage that, since they are based solely on conditions at failure, they permit unwary designers to assume distributions of moments that may differ drastically from those that actually occur due to service loading, and the resulting designs may thus be susceptible to early cracking. A development intended to eliminate this difficulty, as well as to overcome the limitations that arise from simple strip theory, is the strip deflection method proposed by Fernando and

Kemp [23]. With this method, the load distribution in each principal direction is neither selected arbitrarily by the designer (as in the case of the Hillerborg method) nor by choosing the proportions of steel provided in each direction and that at mid-span compared with that over each support (as in the yield-line method), but is calculated to ensure the compatibility of deflections in mutually orthogonal strips. The method leads to the solution of sets of simultaneous equations (a typical number being eight) and thus requires the use of a small computer or similar device.

Some codes of practice provide ultimate bending-moment coefficients for designing freely-supported and continuous two-way slabs that have corners restrained from lifting and reinforced to resist torsion. For example, the coefficients in CP 110 are based on yield-line analyses but have been modified to cater for the non-uniform spacing of the reinforcement recommended in the code.

Such coefficients are intended for use where slabs are supported on substantially rigid beams. Where beams are formed within the slab depth (i.e. so-called flat-slab construction), CP 110 provides different design coefficients. This type of floor may be thickened around the column heads (which may themselves be splayed) by forming drop panels, or a uniform slab thickness may be maintained throughout.

Alternative design methods are available in addition to the method utilizing coefficients mentioned above. For example CP 110 permits a less empirical procedure for designs that do not conform to the rather restrictive rules which must be observed when employing the empirical coefficients. The moments and shearing forces are then calculated by assuming that the structure is formed of transverse and longitudinal frames. This design method is described and illustrated with examples in [7], but since it usually leads to the need for more reinforcement, the empirical method should be adopted where all the necessary requirements can be met.

The current American Concrete Institute requirements, which are based on an elastic analysis, do not differentiate between systems with or without clearly defined supporting beams and the same general design principles, which are based on elastic-analysis principles, apply to all types of two-way slab. For details, see [2, 24].

6 Columns

The bending moments in columns may be determined from an analysis of the entire structural frame or a simplified subframe as described earlier. Alternatively, if a column is axially loaded or supports a symmetric beam arrangement, the design may be based on empirical expressions that include nominal allowances for tolerances, as described below. Though the design recommendations in different national codes are essentially based on the same broad principles, to assist the reader a specific code is used as the main reference in the explanation that follows; for various reasons the present author has chosen CP 110 [1]. When the loading is not symmetric, CP 110 permits bending in the column to be ascertained by undertaking a simple distribution of moments at each individual joint, assuming that the members concerned are fully fixed at their further ends, that the stiffnesses of the beams are taken as one-half of their true values, and that the imposed load is so arranged as to cause the maximum moment in the column.

To design column sections it is necessary to know whether the column is braced (in which case the lateral stability of the entire structure in the direction considered is provided by walls or some other bracing to resist all lateral forces in that direction) or unbraced, and also the effective height. The latter, which normally ranges from 0.75 to twice the actual height, depends on the efficiency of directional restraint at the ends, and may either be read from a table provided in CP 110 or calculated from formulas. For a more detailed explanation, see [26].

Although, until relatively recently, sections subjected to bending with or without axial loads were always designed according to elastic (i.e. modular ratio) theory, it was realized long ago that this method was inapplicable for axially-loaded columns. Consequently, the strength of such columns was determined by summing the resistances provided by the concrete and steel separately. In the British code CP 114, the predecessor to CP 110, implied 'load factors' (i.e. partial safety factors for materials) were included in the design process by specifying lower permissible stresses in the materials when in direct compression. With CP 110, these partial safety factors are embodied in the design formulas by including separate numerical factors that also incorporate an allowance for constructional tolerances and some moment unbalance. These expressions thus contain an anomaly in that it is not possible to investigate the influence of different partial safety factors (as may be done when considering the effects of abnormal loads, according to CP 110).

CP 110 also does not make clear whether, when carrying out the calculations, allowance should be made for the area of concrete displaced by the reinforcement in compression. Although this was normally done in the UK (but seldom in the USA or on the continent of Europe) with modular-ratio design, such a refinement is considered unnecessary when designing sections for bending with or without axial load according to limit state principles, and there seems a good case for extending this assumption to the design of axially-loaded sections also.

The simplicity of the formulas and the large number of variables involved make design tables or charts relatively impracticable, although nomograms may be useful for checking purposes [7]. Simple routines may be prepared for programmable calculators [51] that either determine the steel required in a given column to support a specified load or evaluate the load carried by a given section containing given reinforcement.

The foregoing expressions only apply to short braced axially-loaded columns. If the column is unbraced, it must be designed for a minimum moment equal to 5% of that obtained by multiplying the axial load by the minimum cross-sectional dimension.

If the ratio of effective height to thickness about either major axis exceeds 12 (or 10, if lightweight-aggregate concrete is being used), the column is classified as 'slender' rather than 'short'. Before CP 110, the maximum permissible load that could be supported by such a column was determined by multiplying the value for the corresponding short column by a factor relating to the slenderness ratio. In CP 110, this procedure is replaced by the 'additional-moment concept', which is based on CEB recommendations; a detailed description of the background to these is given in [73]. Unlike its predecessor, this method applies to members subjected to bending about one or both axes, as well as to axially-loaded columns.

Briefly, the method consists of determining an additional moment corresponding to the column slenderness and adding this to any initial applied moment. The section is then designed to resist the resulting combined moment and load using the normal sets of design charts such as those given in Parts 2 and 3 of CP 110. Now having found the amount of steel required, the value of the additional moment can normally be reduced somewhat by multiplying it by a further factor which is related to the properties of the section just designed. The inclusion of this factor reflects the fact that, as the loading on the column more nearly approaches that of axial load only, the likelihood of column buckling decreases. The revised additional moment is now added to the initial moment and the design process repeated. This cyclic procedure may be continued until adjustments become insignificant.

This 'cycling' is tedious to undertake manually even if suitable graphical aids [6] are employed. (Of course, if absolute economy is not important, it need not be done since each cycle should decrease the amount of steel required.) However, it is easy to undertake using a computer and is also possible with the most sophisticated programmable calculators [51].

By resolving the action of two moments acting mutually at right-angles to each other into a single equivalent moment, a rectangular section subjected to biaxial bending may, of course, be regarded as an irregular section subjected to uniaxial bending. Although such a problem can theoretically be solved directly using design aids, a large number of relatively complex charts would be needed to cope with the many variables involved. Pannell [80] devised such a series for designing biaxial columns using the load factor method, the direct predecessor of limit state design in the UK.

CP 110 avoids the need for such complexity by envisaging a three-dimensional interaction diagram derived from the load–moment interaction diagrams for uniaxial bending and thrust provided in CP 110. Insufficient space is available here to discuss the derivation of the various expressions employed in the code method, but this is clearly explained in [72], and further useful information is included in [63]. Suitable design procedures to meet the code requirements are described in [6, 7]. As with slender columns, to determine the section requiring the minimum reinforcement, a reasonable amount of rather tedious trial and adjustment is needed even if design charts, such as those provided in [6], are available. This adjustment process can be carried out very rapidly with a computer, and suitable programs are available.

If a slender column is subjected to biaxial bending, the above procedures must be combined. In other words, the initial moment (if any) about each principal axis must be modified to cater for both the slenderness and the appropriate section modifier relating to that direction, and the code requirements regarding biaxial bending must also be observed simultaneously. Using manual methods, the resulting calculations are very time-consuming indeed if attempts are made to find the minimum steel area possible. Even with computer aid, it is difficult to define a design procedure that ensures strict compliance to all conditions in all circumstances, and even if some approximation is permitted, in critical situations several hundred cycles of the basic iteration loop, occupying several minutes on a microcomputer, may be needed to achieve a solution.

To avoid the complexity of the present CP 110 design method, Beeby [72] has proposed a much simpler method for the direct design of biaxially-loaded columns, and it seems very likely that this procedure will be included in the revised version of CP 110 in preparation at the time of writing. The method requires the determination of an equivalent uniaxial moment, and results in the design process described in [6]. Its brevity makes it suitable for inclusion in calculator programs [51] and a comparison between designs produced by the rigorous CP 110 method and Beeby's simplification shows [72] that, except with low axial loads and percentages of steel, the latter is always slightly conservative but, in normal circumstances, the amount of steel calculated by the two methods seldom differs by more than 5%.

Another method of treating biaxial bending, which, while more complex than Beeby's proposal, is considerably simpler than the rigorous CP 110 method, is described in [63].

6.1 Walls

The distinction between walls and columns is sometimes unclear, and the design process usually basically similar. CP 110 defines a wall as a section where the greater cross-sectional dimension is more than four times the lesser. The criteria for 'short' and 'slender' walls correspond to those for columns, the relevant design methods to be used in such cases also being similar. CP 110 specifies differing minimum amounts of reinforcement, however: at least 1% must be provided in a column (although a lower percentage is permissible if, for architectural reasons, the section has to be larger than would otherwise be necessary structurally), but this is reduced to 0.4% for a wall. If less than this amount is provided, the wall is defined as unreinforced, and CP 110 also incorporates design information for short and slender, braced and unbraced walls of 'plain' concrete; further details are included in [6, 7]. When designing in accordance with CP 110, note that, if the area of vertical steel provided is between 0.4 and 1%, although the section is considered as reinforced for structural purposes, it counts as plain concrete from the viewpoint of fire resistance.

7 Roof structures

If the roof is flat, as is common in multistory building construction, it is similar to, and may be designed in the same way as, a floor slab. Indeed, if the possibility exists that additional floors may be added to the building at some future time, the roof should be considered as another floor, for design purposes. Asphalt, or a proprietary finish, will almost certainly be required to ensure that rainwater is kept out, and some thermal insulation may well be needed to restrict condensation forming on the underside. A drainage slope must be incorporated into the roof structure itself or provided by adding a sloping finishing screed.

Inclined planar roofs are uncommon in reinforced concrete buildings. If such a roof is required, it is usually constructed of lightweight materials, but precast purlins may be incorporated.

So-called 'non-planar' roofs may, in fact, comprise planar slabs in the form of prismatic or hipped-plate construction. Otherwise, they consist of singly- or doubly-curved shells. The design of such shell structures is considered in detail in Chapter 32 and only brief notes follow here.

Singly-curved surfaces such as, for example, cylindrical (barrel vault) or segmental shells, are termed 'developable' surfaces, since they can theoretically be 'opened up' into flat plates without shrinking or stretching. As such, they are understandably less stiff than doubly-curved or their equivalent prismatic surfaces, but the formwork is correspondingly simpler to construct. If a doubly-curved surface curves in the same general direction throughout, as in the case of a dome, the surface is termed 'synclastic'. Conversely, 'anticlastic' structures, which are more rigid and less liable to buckling than their counterparts, have saddle-shaped surfaces. The hyperbolic paraboloid is the best-known example of the latter. Here, the doubly-curved shell is generated by two intersecting sets of straight lines, thus facilitating formwork construction—see [30, 34–36].

In the UK, shell-roof construction was principally in vogue in the 1950s and early 1960s: its popularity dwindled with the rise in labour costs relative to that of materials (although other factors are also involved). Since this period preceded the availability of inexpensive computing power, a plethora of books and papers appeared during this time describing complex methods for analysing specific types of the specialized structure. More recently, general methods of investigating three-dimensional forms of any arbitrary shape by means of a computer have been developed, and the need for particular specialized solutions is less necessary. Further information is given in [29–31].

The most common type of shell roof is the cylindrical vault. Here, the thin curved slab is assumed to offer no resistance to bending and not to deform under applied uniform loads. Except near stiffening edges and ends, it is thus only subject to membrane forces, namely, longitudinal and tangential direct forces and a shearing force. Cylindrical shells may be divided arbitrarily into 'short' shells (where the ratio of length to radius is less than about 0.5), 'long' shells (where this ratio exceeds 2.5) and 'intermediate'. For short shells, beam action is relatively insignificant compared to membrane action, and, for preliminary design purposes, it is sufficient to consider the latter only. For long shells, the converse is true.

For intermediate shells, no simple short-cut method for preliminary analysis is available, although attempts have been made to reduce the mathematics involved (see, for example, [32]). However, Bennett [33] has developed an empirical method of designing long and intermediate shells, based on the analysis of 250 actual roofs. The method involves the use of simple formulas or graphs embodying empirical constants.

8   Culverts and subways

For limited drainage purposes, culverts can be formed from circular precast concrete pipe sections. The units should be bedded on concrete and, where passing beneath a road or railway, surrounded by a sufficient thickness of reinforcement. Enough longitudinal reinforcement should be provided to resist any bending due to settlement and unequal vertical loading. Techniques are available for driving extensive lengths of concrete pipe culvert from one side of an embankment to another by using jacks, thus causing the minimum interference to traffic flow above.

Larger culverts can be either circular or square and may be formed from precast units or cast *in-situ*. The loading to which the culvert is subjected depends on the depth beneath the road, etc., at which it is located. If wheels or other concentrated loads bear directly on the top, similar loads to those that act on bridge decks may be considered. Most bridge codes specify a severe ('abnormal') load or isolated concentrated load as an alternative to the basic loading train, and it is this load that should be taken into account. If the culvert is located at a substantial depth below road level, a suitable reduction may be made for the dispersal of load with depth. Current UK requirements regarding all aspects of rail and road loading are included in Part 2 of BS 5400.

In addition to vertical imposed loads on the roof, and the corresponding reactions beneath the base, rectangular culverts and subways are subject to lateral pressures due to the retained soil and, in the case of a culvert, the contained water. Account must be taken of all critical combinations of these loads and the members designed to resist the appropriate combinations of moment and axial force. If cast *in-situ* construction is used, the possibility of lateral loads acting before the roof is formed must be considered.

Further details regarding the design of culverts and subways, together with formulas for the moments at the corners of rectangular sections due to various types of loading, are included in [6].

9 Arches

Although principally used in bridge construction, arches are sometimes employed to form roofs. Of the three main types, the three-hinged arch can easily be analysed by simple statics. Both two-hinged and fixed arches are statically indeterminate, but the latter implies that the abutments are incapable of deflection and rotation as well as resisting thrust. Such onerous conditions are more likely to be applicable to bridge construction.

The design of fixed arches involves trial and adjustment, since the actual shape and dimensions of the arch structure affect the analysis. However, various stratagems have been devised (some of which are described briefly in [6]) to minimize the calculation: an excellent example is [73], which embodies 33 design tables that, it has been estimated, save 70% of the calculation time required by conventional methods.

In addition to the basic types mentioned, more complex and more general arch shapes are possible, for which computer solution is expedient.

10 Stairs

Stairs have unique structural properties in a building. They often form a principal feature at the entrance and, as such, are clearly visible to all who enter. Since, unlike the floors they serve, the edges of the stair slab are frequently revealed, it is often worth making considerable efforts to produce as slender and elegant a structure as possible, even at the cost of using extra reinforcement since, in relation to the structure as a whole, the amount of materials involved in their construction is insignificant.

Recent developments in the use of computers mean that it is now a fairly routine matter to analyse a stair structure of any shape using a generalized space-frame program. However, when using familiar stair configurations (helical, free-standing, dog-leg, etc.) the work involved in setting up the necessary expressions to use such a program may be such that it is simpler and quicker to employ formulas derived using strain-energy principles for the specific arrangement concerned. Details of some of the formulas are given in [6, 82–84], and the former also includes design graphs for helical and saw-tooth stairs. Computer and calculator programs have been produced for helical, dog-leg and saw-tooth stairs [51].

These analyses determine the forces acting along the stair structure. It is then necessary to reinforce each section concerned to resist the three moments and three forces that occur. In theory, a complex analysis is required but normally in practice the majority of these values are insignificant and the resistance required is provided by the inherent strength of the concrete. Thus, account need only be taken of the principal (vertical) bending moment and shearing force and, possibly, the torsional moment also. Such assumptions may not, however, be true where the stair section consists of a cast *in-situ* spine beam supporting treads of precast concrete or some other material, and in such cases special methods may be needed to design such concrete sections.

When designing stairs it must be remembered that the three-dimensional nature of the structure formed from intersecting shallow–wide elements provides rigidity and induces stresses that are not predicted by the idealized line model on which the analysis is based. The triangular tread elements provide additional rigidity. Special care must be taken when detailing the reinforcement to prevent cracks forming at the concentrations of stress which are produced at re-entrant angles, etc.

11   Roads, pavements and runways

The design of concrete roads is based as much on experience as on theory since the effects of expansion and contraction due to moisture and thermal changes, of weather, friction with the sub-base, fatigue, and the need to span weak areas are difficult to assess. The provision of joints controls some of the stresses that arise due to these causes, but since stresses due to traffic are greatest near the edge or corner of a slab, from this point of view it is desirable to minimize the number of effective edges and corners.

The design of major roads and motorways in the UK should be in accordance with the recommendations of Road Note 29 [55]. This document provides recommendations concerning slab thickness, amount of reinforcement required and joint spacing, depending on the amount of commercial vehicles using the road. For less-important concrete roads, provided the foundation is firm and stable, experience has shown that reinforcement is not essential but its use is normally considered a worth-while precaution. The usual amount of mild steel provided is between 3 and 5 kg/m^2 in a single layer about 60 mm beneath the top surface but heavily trafficked roads may need a similar amount of steel near the slab bottom as well.

Tests undertaken by the American Association of State Highway Officials (the organization responsible for highway design in the USA) show that road damage is approximately proportional to the fourth power of the applied axle load, and recent UK research [59] suggests that this relationship may be to the fifth or even sixth power. Thus, private cars do not contribute significantly to damage and UK design is based on the number of commercial vehicles carried. The procedure for determining the number of equivalent standard axles is described in [56], and this supersedes the simpler method of evaluation given in [55]. Further brief details are given in [6]. More detailed guidance on UK road design is given in [57].

The design of aircraft runways is dealt with in detail in [58].

12 Chimneys

Since chimneys are dealt with in detail in an earlier chapter, they are discussed only briefly here. In the USA *Specification for the design and construction of reinforced concrete chimneys* (ACI 307–79) is available [4], which includes charts for designing annular sections on modular-ratio principles with values of 8 and 10 for E_s/E_c, together with design formulas and additional charts. No similar 'official' document is available in the UK but Pinfold [41] treats the subject comprehensively, providing charts and data for both modular-ratio ($E_s/E_c = 12$) and ultimate limit state design. The bases for the latter charts are slightly simplified versions of the stress–strain relationships for concrete and steel specified in CP 110.

Modular-ratio design charts for $E_s/E_c = 15$ are included in [42, 81, 6]; the theoretical bases adopted to produce the various sets of charts differ in the assumptions made. The last reference also provides conversion scales to permit any other value of E_s/E_c to be employed. Calculator and microcomputer programs are available [51] to design annular sections in accordance with the formulas derived in [42, 81].

In addition to the design of the overall structure to resist wind and seismic forces combined with self-weight, temperature gradients in the shaft give rise to additional stresses that must be taken into account during design. All the references previously cited discuss such matters in varying detail, and it forms a major topic in the slightly old, but still valuable, reference [42]. To make allowance for such stresses during preliminary design, it is prudent to adopt lower limiting material stresses or strengths when designing for longitudinal loads and moments. There is also some indication [41] that the strength of concrete diminishes when it is subjected to high temperatures for long periods, and the adoption of a lower limiting concrete strength than might otherwise be necessary should reflect this.

13 Containers

13.1 Reservoirs and tanks

Water-retaining structures are dealt with in another chapter; they are thus only briefly mentioned here. Tanks are usually either circular or rectangular in plan. Calculation of the forces and moments in the wall of a circular tank according to elastic theory is considered in some detail in [73], where details are also given of an analytical method for determining the moments devised by Reissner. Coefficients for the latter are also included in [6]. Tables to calculate the ring tension and bending moment in a tank wall are included in [74].

For rectangular tanks, the results of a large number of elastic analyses are collected in [10], and design curves based on what the present author considers are the most reliable of these are included in the current edition of [6]. This subject forms a considerable part of [11], where many tables are provided. The appropriate table in [74] is taken (unacknowledged) from earlier editions of [6]. It should be used with caution since the original sources of all the values included are unclear although some, due to Buchi, were previously recorded in [75]. Coefficients for the design of tank walls are also provided in [76].

In the UK, the design of walls for liquid-containing tanks must meet the requirements of BS 5337. Some information regarding the requirements of the two design methods sanctioned by this document is included in Section 20.1, and the design procedure is discussed in detail in [6, 77]. Direct design using the limit state method is impractical unless suitable microcomputer or calculator trial-and-adjustment procedures are to hand. Furthermore, because of the many variables involved, the usual form of non-dimensional graph or table is less helpful than usual, and recourse to a comprehensive set of aids (each only applicable to a single bar size, crack width and cover thickness) is needed. Such aids, in graphical and tabular form (the theoretical bases differ slightly), are included in [6] and [74] respectively, and in a different and less directly useful form in [77]. Reference [6] also includes formulas and design graphs for the alternative (i.e. modular-ratio) design method also permitted by BS 5337.

13.2   Bunkers and silos

These are considered in detail elsewhere in this *Handbook*. In shallow structures, the pressure increases linearly with depth, as behind a retaining wall. However, in a structure that is deep compared with its width, the frictional resistance between the wall and the contained material is activated by the slight settlement of the latter, and the forces on the wall modified accordingly. The best-known theory accounting for such behaviour is that developed by Janssen.

Unloading from the bottom (as is normally done) a silo or bunker containing a free-flowing material modifies the stable internal force structure and leads to one of two entirely different types of behaviour (mass flow or core flow) occurring. Both often induce greater local pressures than occur under stable conditions or during loading: this phenomenon is discussed in detail in [6].

The American Concrete Institute has produced ACI 313-77, *Reinforced concrete practice for the design and construction of concrete bins, silos and bunkers for storing granular materials* [4]. A great deal of useful information is also summarized in Section 16 of [24]. This document also describes the design of the trapezoidal slabs for hopper bottoms; an alternative method of designing such slabs is described in [6].

14   Deep beams

As the ratio of depth to span increases, the stress distribution within a beam differs from that normally assumed, and is seriously influenced by the particular arrangement of applied loads and supports. For this reason, special design procedures are necessary when the clear span-to-depth ratio becomes less than 2 to 3 for simply-supported spans and 2.5 to 4 for continuous beams.

The design of deep beams is not discussed specifically in CP 110; however, the ACI and the Portland Cement Association have both developed design methods, while the CEB recommendations [3] include proposals based on extensive research by Leonhardt and Walther. All these methods are summarized in [65], where Kong, Robins and Sharp develop their own design method. This is based largely on the results of many tests and, unlike many proposals, also applies to deep beams having web openings. This work also provided a basis for the excellent and comprehensive guide for designing deep beams issued by CIRIA [66]. The Swedish Concrete Committee has also put forward design recommendations and details are included in [67].

15 Earth-retaining structures

The classical earth-pressure theory for cohesionless granular materials was originally proposed by Coulomb in 1773. In 1856, Rankine suggested a modified version which, by neglecting cohesion and friction between the retained material and the wall, greatly simplified the calculation involved. Rankine's method is still that most often employed in general calculations for earth pressures, although numerous other expressions have been proposed. Many years' research has led Reimbert [9] to propose the following comprehensive expression:

$$\text{Force } F = \frac{Dh^2}{2}\left(\frac{\pi - 2\alpha}{\pi + 2\alpha}\right)^2 \left(1 \pm \frac{2\beta}{\pi}\right)\left(1 + \frac{S}{Dh}\right)\left(\frac{\phi - \alpha}{\frac{1}{2}\pi - \alpha}\right)$$

where D is the bulk density, h the vertical height of wall, α the angle of repose, β the angle of inclination of the retained material (positive when above horizontal), S the surcharge per unit area, and ϕ the angle of the back of the wall from the horizontal. To calculate the minimum and maximum passive resistances, the above values of F must be multiplied by $[(\pi + 2\alpha)/(\pi - 2\alpha)]^n$ and $[(\pi + 2\alpha)/(\pi - 2\alpha)]^{n+1}$ respectively, where n is 1 when considering rotational resistance and 2 for resistance to sliding.

All the foregoing methods assume a plane rupture surface. Theoretically, such an assumption does not satisfy statical equilibrium, for which some other surface must be considered. The most popular is a so-called 'log sandwich', comprising an upper plane wedge combined with a logarithmic spiral zone. Rosenfarb and Chen have developed extremely complex expressions to determine active and passive pressure based on this assumption; details are given in [37]. These expressions involve two unknown variables and automatic computer or calculator iteration is needed to find the critical values and hence the maximum active or passive thrust.

Research indicates that the Rankine formula overestimates the true pressure, and hence its use errs on the side of safety, thus introducing a margin for error in allowing for other indeterminate factors such as pressure from groundwater. In view of its simplicity, for practical design purposes it is probably not worth while employing more complex procedures unless, for some special reason, it is important to attempt to ascertain the true pressure that may occur.

If the slope of the retained material is not uniform or if concentrated loads act behind the wall, a method due to Culmann or the 'trial-wedge' procedure can be used. The pressure is calculated using a trial-and-adjustment process, either semi-graphically or automatically by means of a computer or calculator; details are given in [5, 37].

The pressure resulting from cohesive soil is less predictable than that due to granular materials. Because of cohesion, the pressure should theoretically be lower, but saturation and/or the swelling of clays may induce much greater values. The most commonly adopted expression is that due to Bell, details of which are given in [5, 6]. If the backfill is irregular in shape, a trial-wedge solution is possible [37].

The simplest type of retaining wall is a vertical cantilever, the base either projecting back beneath the filling or forward in front of the wall, or some intermediate combination may be adopted. The first arrangement is usually the most economic. Preliminary calculations are often necessary to determine convenient dimensions before embarking on a more detailed design; these may be made by ignoring, for the time being, the thickness of the base and stem, and investigating the stability of the idealized line structure. A design procedure using graphs for this is described in [6]. Alternatively, calculator programs are available that enable a number of trial solutions to be investigated in as many minutes. The large number of variables involved limit the use of graphs or tables when designing retaining walls. Programs for the complete design of walls using a calculator or microcomputer are readily available and frequently incorporate the automatic design of the required concrete sections. Such comprehensiveness may be a mixed blessing, however.

The formation of splays, where the wall meets the base, reduces the stress concentration at this critical point as well as increasing the effective depth of the steel, although such an arrangement may give rise to constructional difficulties.

Beyond a certain height, simple cantilevering becomes uneconomic and it is cheaper to provide counterforts at suitable intervals. The slab between each counterfort then spans horizontally. Account should also be taken of the fixity adjoining the base, the slab thus being continuous over two opposite sides, and virtually fully fixed and free, respectively, along the others, while subjected to a triangularly distributed load. This slab may be designed using elastic methods, or yield-line or strip analysis may be employed. For even larger structures, it is worth-while introducing horizontal beams running between the counterforts, with the slab spanning vertically or in both directions. The beam spacing can be graduated to keep the maximum moment in each beam similar, and also equalize the moments in the slabs.

Counterforts are designed as vertical T-section cantilevers with the tension steel located in the sloping face. In the unlikely event of shearing reinforcement being needed, U-bars with varying amounts of overlap can be used to make up the horizontal links required. Such steel should be provided to counter the tendency for the flange of the T-beam to be torn from the web. Careful detailing is required here and throughout the counterforts to restrict tension cracks occurring, since the ingress of water from the ground may lead ,to corrosion of the bars.

With isolated walls, the possibility of the retained bank or cutting slipping beneath the base of the retaining wall and carrying the entire wall away must always be carefully considered.

Sheet-pile walls provide an alternative to a reinforced concrete retaining wall where a satisfactory bearing stratum does not occur at a practicable depth. In their simplest form, the piles act as simple vertical cantilevers, their heads often being linked by a cast *in-situ* capping beam. To reduce the moments in higher walls, this beam can be tied back to anchor blocks that are set well beyond the retained material, by means of steel bars or cables. Due to the flexibility of sheet piling compared to a rigid retaining wall, some redistribution of the bending moments may be considered, according to Stroyer; details are given in [6]. A method of designing cantilevered, anchored, fixed and strutted sheet-pile walls, devised by Rowe, together with a comprehensive analysis of UK, Danish and German methods, is described in a valuable CIRIA committee report [47].

Basement walls may be designed as cantilevers propped at the top by the ground floor of the building. Three critical loading conditions may have to be considered: when the wall is complete and is supporting backfill but before the supporting floor is constructed; when the ground floor prop is complete; and when the wall also supports the loads from the building above. Details and examples are given in [7]. A major hazard in basement construction is the ingress of water and if the upward pressure is high it may even be necessary to investigate the possibility of the basement floating before the load from the superstructure is applied. Unless other means are adopted to seal the basement, consideration should be given to designing it to meet the requirements of a water-containing structure (i.e. BS 5337 or its equivalent).

CIRIA has published a useful guide [85] for the design of waterproof basements.

16 Foundations

The particular type of foundation most suitable for a given situation depends on site conditions, the nature of the ground, and the location of a suitable bearing stratum, and it may be worth while comparing the possible effectiveness and cost of widely different types. It is important to ensure, not only that excessive settlement or failure does not occur, but also that, as far as possible, any long-term settlements are similar throughout the structure. This is especially important where, for example, one part of a building is much higher than the rest and the supporting ground is liable to long-term settlement.

The simplest foundations are isolated bases. For very small bases, or if the bearing capacity of the ground is high (e.g. rock), a simple rectangular block of plain concrete will suffice, the thickness being such that the load can be transferred from column to ground through the block at an angle of dispersion of not less than 45° to the horizontal.

For bases larger in area than 1 m^2, a pyramidal base with splayed top faces becomes economic. If the depth-to-width ratio is such that a dispersion angle of at least 45° can be maintained, no bending moments need be considered and only nominal steel is needed. Otherwise, reinforcement must be provided to resist the upward moment caused by the bearing resistance of the soil. In the UK, CP 110 states that the maximum moment that should be considered is that due to all the forces to one side of the section. If the base supports a concrete column, the critical section occurs at the column face; if a steel section is supported, the moment should be determined about the centre-line of the base. Previous UK codes stipulated that the critical plane for shearing was located at a distance from the column face equal to the effective depth, but in CP 110 a distance of 1.5 times the effective depth is specified. In addition, sufficient resistance to punching shear must be provided, as required where concentrated loads act on slabs.

When designing all types of foundation, but especially separate bases, where the shearing force changes rapidly, particular attention must be paid to the need to satisfy local-bond requirements. This often leads to the need to provide smaller-diameter bars at closer spacings than would otherwise be preferred. Anchorage-bond lengths also require extra consideration, not only regarding the main reinforcement that provides resistance to bending but also for any column starter bars. Satisfying the latter requirement may necessitate a thicker base being needed than would otherwise be so.

If, due to site constraints, a base cannot be positioned centrally under a column, the resulting distribution of ground pressure beneath a rectangular base will not i e uniform. The same situation will arise where a base is subject to moments and/or horizontal shearing forces as well as vertical loads if these additional forces are not constant (in which case the centroid of the base can be displaced to counteract the resulting eccentricity). Extreme variations of pressure should be avoided as far as possible since they may lead to differential settlement occurring, and the possibility of providing a base that is trapezoidal, rather than rectangular, in plan may be well worth investigating.

Where, to support trestle or gantry towers, the bases occur in pairs with the moments and shearing forces acting in the same sense on both bases simultaneously, it may be practicable to design them as concentrically loaded, linking them together by a relatively flexible tie beam to obviate the effects of eccentricity. A similar expedient, which can be adopted where a base of sufficient size cannot be positioned centrally under a load due to site restrictions, is to devise a so-called 'balanced' foundation where the otherwise eccentrically-loaded outer base is tied back to a concentrically-loaded inner base.

If the base sizes required for adjoining columns are such that independent bases would join or overlap each other, combined foundations can be provided. Depending on the loadings of the various columns, such foundations may be concentrically and eccentrically loaded at different times, and the designer must examine the effects of each combination of loading. For such investigations, automatic computer and programmable calculator procedures have been developed that are very helpful [38].

If the vertical loads are closely spaced in one direction only, it is usual and convenient to provide a strip base similar to a wall footing, the moments and shears being calculated and the base designed as an inverted continuous beam. This simple design method assumes that the base is sufficiently rigid to distribute the pressure beneath so that it varies uniformly throughout its length. However, for moments to occur, strains, and hence deflections, must be set up and these will lead to some redistribution of the ground pressure occurring. As a result, pressures will increase beneath the loads and decrease elsewhere, and such a situation will lead to the resulting moments being reduced. The reductions may be such that, midway between the loaded points, the direction of the actual moments may differ from that calculated and such effects must be borne in mind when providing reinforcement. If the resistance of the ground is insufficient to withstand the extra pressure, increased settlement occurs beneath the load concentrations and a more uniform pressure distribution is re-established. Such behaviour is discussed in detail in [7].

Where columns are closely spaced both ways, the loads are high, the allowable bearing pressure is low, or a combination of these circumstances occurs so that little space would occur between independent bases, a single foundation raft or mat may be provided. The simplest design of such a raft is basically similar to an inverted reinforced concrete floor. The column spacing and loads determine the shearing forces and moments and thus the thickness. If this does not exceed about 300 mm, a simple solid slab is normally most convenient, although a thickened edge may be necessary to prevent the weathering of the ground beneath. For greater thicknesses, beam-and-slab construction becomes more economic and, if the total thickness required exceeds about 1 m (or a level top surface is essential), cellular construction comprising top and bottom slabs separated by intermediate ribs can be adopted.

As in the case of strip footings, in order to achieve the moments determined by simple statics, the corresponding deflections would be sufficiently large to violate the assumption of uniform pressure and probably also sufficient to cause failure of the supported structure. Various procedures have thus been proposed to design composite foundations more rationally, taking into account the elastic deformation of the soil. One such procedure is the soil-line method described in [63], which makes use of the modulus of subgrade reaction of the soil. Although mathematically only approximate, the results obtained are sufficiently accurate for practical purposes.

Different types of reinforced concrete foundation are discussed in detail in specialist books such as [37, 38] and a comprehensive introduction to the subject, with useful data and formulas, is included in [6].

16.1 Foundations for machinery

In designing foundations to support machinery, it is important to minimize the effects of vibration. The bearing pressure adopted should be considerably less than the maximum that would otherwise be possible for that type of ground, especially if it wholly or largely consists of clay. The transmission of vibration to the adjacent structure, either directly or via the ground, is normally prohibited and, in addition to isolating the machine base completely from the surrounding structure, this may necessitate placing an insulating lining between the base and the ground or, in severe situations, supporting the base on springs.

Part 1 of CP 2012, *Foundations for reciprocating machines*, includes a step by step account of the design of a reinforced concrete base for a reciprocating machine. Comprehensive information concerning the dynamics of machine foundations may be found in [40] and specialized information concerning the design of reinforced concrete bases to support vibrating equipment on raft, piled and massive foundations is given in [39], which includes 21 design charts.

The design of all structures in contact with the ground is a sphere where practical experience plays a particularly important role, and references such as [67–70], all of which are written by experts who have spent many years in this field, contain a wealth of valuable advice.

17   Piles and pile caps

Two principal types of concrete pile can be identified, precast and cast *in-situ,* although combinations of both forms (i.e. a cylindrical shell formed of precast units through which concrete is forced to form the core and enlarged base) are also available.

Precast piles more than 30 m in length are not unknown, although lengths of up to 20 m are more usual. Such piles achieve support by direct bearing on a lower stratum of greater strength, or by frictional resistance, or by a combination of both means. The safe load that may be supported depends on that which the pile can carry as a column (with some partial lateral restraint throughout) and also that which causes excessive settlement to occur. The principal stresses that must be resisted are normally those caused by lifting and driving. In the past, when all piles were driven using a simple falling ram or drop hammer, many empirical formulas were devised to calculate the safe bearing capacity of a pile and the stresses due to driving set up therein, the best-known of these being the Hiley formula. Although such 'dynamic' formulas are still useful for predicting the stresses that occur during driving, and thus for designing precast piles, modern pile-driving equipment has rendered them inapplicable as regards the calculation of ultimate bearing capacity and the prediction of settlement under working loads. For such purposes, they have been superseded by so-called 'static' formulas founded on soil mechanics theory. To date, no such simple basic design method has achieved general acceptance, although various empirical procedures have been suggested, details of which are given in [43]. Also given in [43] is detailed information on different proprietary forms of cast *in-situ* pile (which normally depend for much of their support on the formation of an enlarged foot). More limited information is included in [6].

To support loads greater than can be carried by an individual pile, a group is required and if, under all circumstances, the centres of gravity of load and pile group do not coincide, differential forces will be set up in the individual piles. In addition, the bearing capacity of a group is less than the sum of the capacities of the individual piles, due to the overlapping of the zones of stress surrounding each pile. These, and similar specialized matters, are discussed in specialist works such as [43]. The calculations required are readily programmed for computer solution— see [37, 38], for example.

Pile caps are designed to resist the moments and shears set up when transferring the vertical and horizontal loads to the individual piles forming a group. Two design methods are common. Either the cap can be analysed as a short deep beam (Section 14) transferring load primarily by bending, or alternatively it may be considered as an imaginary space frame with inclined lines of compressive force extending from the underside of the load to the pile heads, the latter being linked together by reinforcement acting as the tension members of the frame. The size of the column being supported is usually ignored but Yan [44] has developed design expressions that take account of this. Whittle and Beattie [45] have proposed a series of standardized pile caps.

Recent research into the behaviour and design of pile caps is summarized briefly in [6], which also includes data to design caps using the space-frame theory. Design matters requiring particular consideration include the treatment of both direct and 'punching' shear, and the need to provide adequate anchorage bond for the main tension steel.

18 Seismic resistance

The design of structures to resist the effects of earthquakes is discussed in detail in Chapter 15. Opinions generally differ as to whether the disruptive effects of tremors are best resisted by designing structures as flexible, rigid or semi-rigid. The effect of an earth tremor is basically equivalent to the action of a horizontal thrust additional to the normal loading (but without wind). Codes for earthquake-resistant construction exist in several countries including the USA [2], and generally the requirements incorporated in recent documents are more complex than their predecessors.

The satisfactory behaviour of structures that were designed to withstand arbitrary seismic forces and actually subjected to severe earthquakes has been attributed to the following principal causes. Yielding occurred at critical sections, thus increasing the period of vibration and enabling greater amounts of energy to be dissipated. So-called non-structural partitions assisted in resisting the shock and absorbed energy as they cracked. Also, due to the yielding of the foundations, the response of the structure was less than had been anticipated. It is generally considered uneconomic to design a structure to withstand a major earthquake elastically and current design procedures therefore assume that a structure will possess sufficient strength and ductility to withstand such tremors by responding inelastically. The likelihood of this being achieved is greater if the interconnections between members are designed specifically to ensure that their ductility is sufficient.

A detailed discussion of the philosophy of earthquake-resistant design of concrete structures is included in [24].

19 Fire resistance

Fire resistance is dealt with comprehensively in Chapter 14; it is therefore only briefly discussed here. The fire resistance of an element or structure is usually specified as the length of time it remains serviceable when subjected to fire. This resistance is controlled principally by the types of steel and concrete used, the thickness of concrete covering the reinforcement, and the size and shape of the members.

Section 10 of CP 110 [1] specifies requirements for the minimum dimensions and cover thicknesses for various types of element of reinforced and prestressed concrete. Similar but differing information was included in its predecessor, CP 114 [47]. The UK *Building Regulations* (1976) also include tables specifying minimum dimensions and cover thicknesses to achieve prescribed periods of fire resistance, these requirements generally being similar to, or less than, their equivalents in CP 110. The by-laws governing construction within the Greater London area incorporate another set of limiting values that again show minor variations. Yet another, and possibly the most up-to-date, set is that presented in the important joint report on fire resistance produced by the Institution of Structural Engineers and the Concrete Society [49].

The actual period of fire resistance needed depends, according to the *Building Regulations*, on the size of the structure and the use to which it is put.

Neither the corresponding ACI code [2] nor the CEB recommendations [3] consider fire resistance. In the USA, fire resistance ratings are specified in various model building documents and the requirements controlling the design of sections to achieve such resistance are embodied in the relevant American Society for Testing and Materials (ASTM) documents.

20 Analysis of sections

20.1 Bending

When analysing reinforced concrete sections, the single basic assumption that is almost always adopted (and certainly always is in practical design) is that sections that are plane before bending remain so after bending. In other words, the strain at any given section is proportional to the distance of that section from the neutral axis. It is also assumed that no slip occurs between the reinforcement and the surrounding concrete. When undertaking calculations to determine the strength of a given section, it is also generally assumed that the tensile strength of the concrete is zero.

In order to analyse a section, it is now necessary to make some assumption regarding the relationship between stress and strain in the materials. When investigating sections subjected to service loading, the usual assumption is that this relationship is linear (i.e. that both materials behave purely elastically). This is the basis of the well-known modular-ratio or elastic-strain method.

If a section is being analysed to determine its resistance to cracking, it is usual to take some account of the strength of the concrete that is in tension, and in fact this value is usually a principal criterion in designing the section. For deflection calculations, too, allowance needs to be made for the additional stiffness resulting from the tensile strength of the concrete.

When assessing the resistance of a concrete section immediately prior to failure, research has shown that the assumption of a linear relationship between stress and strain is no longer valid. The stress–strain relationship for steel resembles that shown in Fig. 20-1a. For practical design purposes, this may be represented by a simplified bilinear (Fig. 20-1b) or trilinear (Fig. 20-1c) relationship. According to the requirements of CP 110, where a trilinear relationship is specified (Fig. 20-1d), the limiting strength permitted in steel in compression is normally less than that in steel subjected to tension. A bilinear relationship is proposed in the CEB recommendations [3].

The distribution of compressive stress in concrete as failure is approached has been found experimentally to resemble that shown in Fig. 20-2a. For practical design purposes, various simplifications to this shape have been suggested. One concept, due to Whitney, is to ignore the actual relationship between stress and strain and to replace the actual concrete 'stress block' by a rectangle having a similar area and with a centroid at approximately the same depth (Fig. 20-2b).

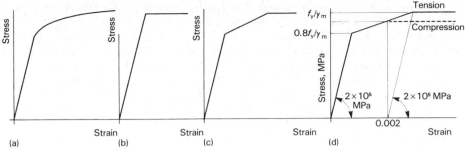

Figure 20-1   (a) **Stress-strain relationship for steel**
            (b) **Bilinear representation of (a)**
            (c) **Trilinear representation of (a)**
            (d) **Trilinear representation according to CP 110, where f_y is the characteristic strength of steel and γ_m the partial safety factor for steel**

This is the basis of assumptions in both the US and UK design codes, and also in the recommendations of the CEB. Figure 20-2c shows the rectangular stress block assumed in CP 110.

In addition, CP 110 permits the alternative assumption of a stress block that consists of the combination of a rectangle and a parabola (Fig. 20-2d). This implies that the stress–strain relationship for the concrete is similar (i.e. the initial relationship between stress and strain is parabolic but, beyond a certain limiting

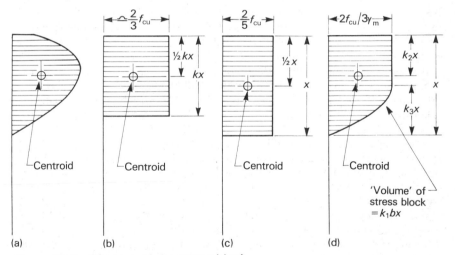

Figure 20-2   Concept of the stress block
            (a) **The distribution of compressive stress in concrete as failure is approached**
            (b) **Rectangular stress block to replace (a)**
            (c) **Stress block assumed in CP 110**
            (d) **Combined parabola and rectangle, also permitted in CP 110 and CEB**

strain, no increase in stress occurs). The CEB also permits this type of stress–strain relationship to be assumed, although the method of actually defining the relationship differs slightly from that in CP 110.

If a section subjected to bending only is analysed assuming that the tensile strength of the concrete is neglected, it is normally possible to formulate expressions for the direct design of rectangular concrete sections. For limiting material strengths and applied moment per unit breadth, a minimum depth can be determined at which the tension steel must be located. If a lesser value is adopted, the compressive resistance provided by the concrete is inadequate and reinforcement in compression is needed. A greater 'effective depth' may be employed. In such a case, the amount of tension steel necessary is reduced and the maximum stress in the concrete falls below its limiting value.

Since direct design using formulas is possible, graphical design aids are not particularly helpful in this situation. Indeed, since it is necessary to reduce the variables to non-dimensional units in order to take account of the number involved, direct design is preferable. The appropriate formulas are easily programmed for analysis using calculator or computer and, since iterative methods are not needed, solution is virtually instantaneous. If the program is carefully designed, it is possible to juggle with the variables involved, changing individual values only each time, and produce a dozen interim solutions in almost as many seconds, leading to an optimum design.

In the case of a flanged section, solution is not so simple, unless the neutral axis (i.e. the level at which the strain is zero) falls within the flange. (In such a case, since the concrete in tension is ignored, the shape of the section below the neutral axis is immaterial.) With modular-ratio design, if the neutral axis falls within the web, it is usual to ignore the small area of concrete in compression between the neutral axis and the underside of the flange. The effect of this assumption is negligible in a normal T-beam, but may be less so where the section considered is part of a hollow-tile or coffered floor.

If a non-linear stress–strain relationship is adopted for the concrete, direct design is not possible unless the true relationship is ignored and a uniform rectangular stress block assumed. (The design procedure required in such a case using CP 110 is illustrated in Table 106 of [6]). Due to the large number of variables involved, the preparation of sets of design charts is impracticable. Design by means of iterative methods (as described below) using a programmable calculator or computer is quite simple and efficient, however, and such methods can also be used for modular-ratio analysis to take account of the compression in the web of the section.

When a section is subjected to axial force as well as bending, two simultaneous equations of equilibrium must be satisfied, and direct design is now impossible. With linear stress–strain relationships for steel and concrete, families of design curves may be prepared for each shape of section considered. By expressing the axial load and bending moment in terms of the section dimensions and the limiting steel or concrete stress, the charts may be made non-dimensional. The usual format is to plot load terms and moment terms along the x and y-axes, relating these to series of curves representing the percentage (or proportion) of steel and the neutral-axis position (i.e. the relationship between the maximum steel and concrete stresses in terms of the modular-ratio adopted). The remaining variables to be considered are the actual modular-ratio and the position of the steel relative to the overall section depth. Many sets of such curves have been produced—see, for example, [6, 41, 42, 81].

If the stress–strain relationships for the materials are non-linear, as in ultimate limit state or strength design, the problem of producing suitable design charts is complicated. With CP 110, for example, since the shape of the stress block depends on the strength of the concrete, separate sets of charts are theoretically needed for each individual concrete strength considered; such sets are provided in Parts 2 and 3 of CP 110. However, [6] shows that if non-dimensional charts are prepared using a concrete strength of 30 MPa, the charts are valid for practical design purposes for concrete strengths of from 20 to 50 MPa. For example, if $N/bhf_{cu} = 0.5$, $M/bh^2f_{cu} = 0.25$ and $d/h = 0.9$, the proportions of reinforcement required (in terms of f_{cu}/f_y) with f_{cu} equal to 20, 30 and 50 MPa, respectively, are 43 010, 43 208 and 43 546. These represent differences of −0.54% and +0.78% from the 30 MPa value, which are well within the limits of accuracy to which graphs can be plotted.

Such an analysis is ideal for automatic solution using a programmable calculator or computer. The procedure is to combine the two equilibrium equations into a single expression. The most suitable way to do this is normally to rewrite the load expression in terms of the area of steel required, and then to substitute this into the moment expression (or vice versa).

As a relatively simple example, consider the modular-ratio analysis of a rectangular section. The equilibrium equations are

$$N = \frac{bxf_c}{2} + \frac{A'_s(x - d')f_c(\alpha_e - 1)}{x} - \frac{A_s(h - x - d')f_c\alpha_e}{x}$$

and

$$M = \frac{bxf_c}{2}\left(\frac{h}{2} - \frac{x}{3}\right) + \frac{A'_s(x - d')f_c(\alpha_e - 1)(\frac{1}{2}h - d')}{x} + \frac{A_s(h - x - d')f_c\alpha_e(\frac{1}{2}h - d')}{x}$$

where the notation is as shown in Fig. 20-3.

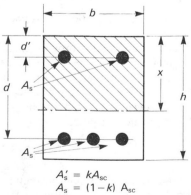

$$A'_s = kA_{sc}$$
$$A_s = (1 - k)\,A_{sc}$$

Figure 20-3

The latter expression may be rewritten as

$$A_{sc} = \frac{M - \frac{1}{2}bxf_c(\frac{1}{2}h - \frac{1}{3}x)}{(\frac{1}{2}h - d')xf_c[(\alpha_e - 1)k(x - d') + \alpha_e(1 - k)(h - x - d')]}$$

if $A'_s = kA_{sc}$ and $A_s = (1 - k)A_{sc}$, and substituted for A_{sc} in the former to give

$$\text{unbalance} = (N - \tfrac{1}{2}bxf_c)(\tfrac{1}{2}h - d') \; [k(x - d')(\alpha_e - 1) + (1 - k)(h - x - d')\alpha_e]$$
$$- [M - \tfrac{1}{2}bxf_c(\tfrac{1}{2}h - \tfrac{1}{3}x)][k(x - d')(\alpha_e - 1) - (1 - k)(h - x - d')\alpha_e].$$

This expression will be satisfied only if the correct value of x is inserted. This value is obtained by commencing the calculations with an initial arbitrary value of x of, say, $h/2$ and determining the value of the right-hand side of the equation. If this is lower than the correct amount, the value of x is incremented, if greater it is decremented. This procedure is repeated automatically a number of times, the adjustment to x being reduced each time the difference between the true and the calculated value changes sign. Eventually, a situation will be reached when either the change in x or the difference between the calculated and the 'target' value is negligible. At this point, the iterative procedure is terminated and the required area of steel calculated using the expression derived above.

This procedure has been described in reasonable detail since it forms the basis of almost all section design using computer techniques. Once the equilibrium equations have been correctly formulated, the production of the program is relatively simple since the same basic program can be employed for any shape of section and method of analysis. If non-linear relationships between stress and strain are employed, the use of subsidiary equations is necessary to relate the strain at a given point to the corresponding stress in the steel or concrete.

Certain codes of practice, such as those controlling the design of liquid-containing structures, may be based on slightly different basic requirements to those considered above. In such cases, the concrete is assumed to provide some strength in tension also, but the maximum tensile strength in the concrete is restricted to a low value to ensure that any cracking is minimized. Such requirements form part of the 'alternative' design method sanctioned by the relevant UK code BS 5337. The resulting design equations can still be developed and combined in a similar manner to that described above.

When undertaking such a design, it is necessary to analyse a section of assumed size according to both the strength and cracking requirements, and to calculate the reinforcement needed to satisfy each condition, eventually providing the greater of the two amounts. For most given combinations of values, a unique solution exists, where the same section dimensions and reinforcement satisfy both sets of limiting design requirements simultaneously (i.e. the so-called 'balanced' situation). This condition can be determined by using computer iteration, although the resulting section may be unsuitable on practical grounds.

20.2 Shearing and torsion

The calculation of shearing forces is usually straightforward. Below a limiting value (which in CP 110 depends on the amount of longitudinal steel present at the section being considered), the concrete is assumed capable of resisting the shearing stress without aid, and only nominal shearing reinforcement is needed. With greater shearing stresses, assistance in the form of inclined or vertical steel becomes necessary to account for the difference between the applied shearing force and the resistance provided by the concrete alone. (This requirement is less onerous than that of earlier codes which, if the limit was exceeded, required the entire shearing resistance to be provided by the reinforcement.)

The shearing reinforcement is usually provided in the form of vertical or inclined links (sometimes termed stirrups or binders). If these extend round any longitudinal compression bars and thus effectively restrain them from buckling, they serve a dual purpose. The resistance provided by such links may be supplemented by bending up some of the main longitudinal tension reinforcement, where the bars are no longer required to provide bending resistance, and in unusual circumstances it may be apposite to incorporate extra inclined bars specially. However, CP 110 restricts the assistance that may be contributed by inclined bars to not more than one-half of that required from all the shearing reinforcement.

Shearing reinforcement is normally designed on the assumption that the member behaves, for such purposes, as a lattice girder or pin-jointed truss, where the links or inclined bars forming the shearing reinforcement act as the tension members while compression is resisted by the inclined compressive force in the concrete. The representation of the actual behaviour of a member after cracking has commenced is rather poorly represented by this so-called 'truss analogy', but tests indicate that the resulting designs are conservative.

In order to obtain a better understanding of the action of shearing forces on reinforced concrete, a great deal of research has been undertaken recently and is still continuing. Various theories have been proposed to explain the behaviour after cracking has commenced and to provide a more rational basis for designing shearing reinforcement but none has yet found general acceptance to replace the relatively empirical process already described. Perhaps the most promising is the 'truss block' method which is discussed in some detail in [63]. A useful and extensive, if now slightly outdated, general review of the various theories is given in [27], but perhaps the most interesting and useful discussion is provided in [25].

The possibility of failure occurring due to the concrete crushing is prevented by imposing an upper limit ($0.75\sqrt{f_{cu}}$ in CP 110) to the shearing stress that can be imposed on a section irrespective of the amount of shearing reinforcement provided.

Unlike shear, where the converse tends to be true, the principal problem with torsion is concerned with analysis rather than design. Although torsional stresses occur in many members, the loads inducing these stresses are supported by bending moments and axial and shearing forces, and not by torsional resistance itself. CP 110 states that, if the torsional resistance or stiffness of a member is ignored when analysing a structure, it is normally unnecessary to design members for torsion since adequate resistance is provided by the nominal shearing reinforcement required. In the case of a beam-and-slab floor, for example, almost all the torsion induced in the supporting beams due to unsymmetric slab loading is relieved by the beams twisting and redistributing the moments in the slabs.

Past experience has shown that, in general, torsion due to such structural indeterminacy or continuity may safely be ignored since the lack of sufficient torsional resistance does not endanger the structure; this is not true where torsion results from other causes. However, Allen [26] has pointed out some exceptions to this general rule, and framed structures should be carefully examined to see if such behaviour occurs.

Calculations to determine torsional moments must be made about the shear centre (termed the flexural centre in US literature) of the section employed. Expressions to determine the position of the shear centre are given in Table 14 of [12], which contains much other useful information concerning torsional analysis, including formulas for beams curved in plan. When undertaking torsional analysis, CP 110 recommends that the shear modulus of the section should be taken as 40% of the modulus of elasticity, and also that the torsional constant should be one-half of the St Venant value calculated for a plain concrete section. For a rectangular section, the St Venant value, J, is given by

$$J = \left\{ \frac{1}{3} - 0.21 \frac{h_{max}}{h_{min}} \left[1 - \frac{1}{12} \left(\frac{h_{max}}{h_{min}} \right)^4 \right] \right\} h_{max}^3 h_{min}$$

Flanged sections are investigated by dividing them into their component rectangles in such a way that the sum of the values of $h_{min}^3 h_{max}$ for the components is as great as possible. This situation normally occurs when the widest rectangle comprising the section is as long as possible, but in cases of doubt all possible arrangements should be examined.

CP 110 recommends that the torsional stresses, v_t, are calculated on the assumption that a plastic distribution of stress occurs across the section, using the expression

$$v_t = 2T/[H_{min}^2(h_{max} - \tfrac{1}{3}h_{min})]$$

where T is the ultimate torsional moment.

For flanged sections, the component rectangles previously determined are each considered to withstand a proportion of the total torsional force equal to their contribution of $h_{min}^3 h_{max}$ to the sum of these values. If the resulting stress exceeds the limiting value specified in CP 110, torsional reinforcement is required. This consists of a combination of rectangular closed links, and longitudinal bars evenly distributed near the perimeter of the section, with at least one bar at each corner. It is often convenient to combine the area of longitudinal steel necessary with that required to resist bending and to provide a bar arrangement to meet both needs. Similarly, the torsional links can be combined with those needed to resist shear and a mutually satisfying system adopted. Examples of the design of torsional reinforcement are given in [6, 7, 54, 26], the last providing a valuable background to the CP 110 requirements. For more comprehensive information concerning torsion in structural concrete, see [28, 54].

20.3 Bond

To meet the assumptions implicit in concrete design, that the composite section behaves as though it were a single material and that full design stress is attained in the reinforcement, it is imperative to prevent significant slipping occurring between concrete and steel. Two aspects of the associated bond between these materials are especially important. First, it is essential to ensure that an adequate length of bar is provided from the point at which a given stress occurs. Such anchorage bond lengths depend on the type of bar surface as well as various other factors. In addition to plain bars, CP 110 recognizes two types of deformed bar. Bars having transverse ribs meeting specified limits, or other bars shown by test to display the necessary properties, are classified as Type 2 and 30% higher bond stresses are permitted. Appropriate bond lengths can be conveniently read from tables (see [6, 7], for example), although it should be remembered that because of the large number of variables involved, these lengths usually only correspond to the *maximum* force in the bar. If the actual stress is lower, the length indicated may be reduced pro rata. Simple calculator routines can easily cope with all the variables concerned and offer a rapid and convenient way to calculate anchorage bond lengths.

To reduce the length otherwise required for bars in tension, it is possible to achieve some form of mechanical anchorage by forming a hook or right-angled bob at the bar end. This practice is perhaps less common now than hitherto since it is less easy to form such anchorages than was true when plain mild steel was the only type of reinforcement available.

If a bend must be formed in a bar carrying a substantial tensile force, the radius must be such that the concrete within the bend does not crush. Suitable expressions to calculate the minimum radius possible are provided in codes such as CP 110.

The other important condition concerning bond occurs where the tensile force in a bar changes very rapidly over a short distance. Since the resulting local bond stress is determined by dividing the shearing force at a section by the product of the perimeters of the tension bars and their depth, such a critical situation is more likely to arise where the shear is large but the steel area needed to resist bending is unexceptional. This is more likely in foundation design, at the supports of freely-supported beams, and in similar situations. CP 110 specifies limiting local bond stresses and, as before, higher values relate to deformed bars with enhanced bonding properties. Also, as before, the number of variables concerned makes 'hard' design aids, such as tables or graphs, of little use here, and local bond investigations are most quickly and efficiently made using calculator routines.

20.4 Serviceability requirements

The principal criteria that must be considered involve limiting cracking, deflection and vibration. In the UK, in limit state design to CP 110, cracking need not be considered specifically if certain requirements regarding maximum bar spacing are satisfied. If these requirements are not met, as an alternative, Appendix A3 of CP 110 describes a method of calculating the 'design surface crack width', which must not exceed a value of 0.3 mm to meet the requirements of the code. These requirements are discussed in more detail in [7, 26, 54]. A design procedure that uses graphs is presented in [7].

The limit state requirements regarding cracking in the UK code for water-containing structures, BS 5337, employ basically the same design method as that described in CP 110. However, for surfaces in proximity to liquid, the maximum permitted crack widths are smaller, 0.1 mm and 0.2 mm for Class A (a moist or corrosive atmosphere, or where alternate wetting and drying occurs) and Class B (continuously or almost continuously in contact with liquid) respectively, and the design expressions are themselves modified to reduce the probability of an individual crack width exceeding the limiting value from 1 in 5 to 1 in 20.

Excessive deflection is also undesirable. CP 110 limits long-term deflections (including the effects of all time-dependent variables such as creep and shrinkage) to span/250 and restricts the deflection that occurs after the completion of partitions, etc., to span/350. Simplified rules are provided that meet these requirements by determining a limiting span-to-effective-depth ratio that must not be exceeded. This ratio is found by modifying a basic value corresponding to the support conditions and span length by factors determined by the amount of, and stress in, the reinforcement. With flat slabs, flanged sections and/or lightweight concrete, further multipliers are employed.

As an alternative to this simplified method, CP 110 describes a rigorous procedure for calculating deflections, which is described in detail in [7, 26], the former including 22 charts to facilitate the calculations necessary. The complex procedure is tabulated for clarity in [6]. It involves first undertaking an elastic analysis with service loads to obtain the moments and forces. Curvatures are then calculated on the dual assumptions of cracked and uncracked sections and the greater value (usually the former) is adopted. The long-term curvature is now obtained by summing the instantaneous curvature due to total load, the long-term curvature due to permanent load and any curvature due to shrinkage, and subtracting the instantaneous curvature due to permanent load. Finally, the recommended method of calculating the deflection is to plot a diagram of the curvature along the member and integrate it twice, although, alternatively, it can be evaluated directly using an appropriate deflection coefficient.

CP 110 states that there are four principal factors which are difficult to allow for but which considerably influence the accuracy of the results. These are the actual support restraint, the precise loading (particularly the proportion which is long-term), whether or not the member is cracked, and the effects of finishes and partitions. Although it is claimed that, if such points are correctly considered, both short- and long-term deflections are predicted reasonably accurately by this method and although much computational drudgery can be minimized by using automatic techniques, it is clear that attempting to predict deflections by such methods is seldom a worthwhile undertaking.

References

1. British Standards Institution, CP 110: 1972 (as amended 1980) *The structural use of concrete*, Part 1: Design, materials and workmanship, 154 pp.; Part 2: Design charts for single reinforced beams, doubly reinforced beams and rectangular columns; Part 3: Design charts for circular columns and pre-stressed beams, London, 1972.
2. ACI Committee 318–77, *Building code requirements for reinforced concrete*, American Concrete Institute, Detroit, 1977, 236 pp., including *Commentary*.
3. CEB–FIP, *International recommendations*, 3rd Edn, Comité Euro-International du Béton, Brussels, 1978, 123 + 348 pp.

4. American Concrete Institute, *ACI manual of concrete practice*, Part 2: Notation and nomenclature, structural design, structural specifications and structural analysis, 1982, 508 pp.; Part 4: Bridges, substructures, sanitary and other special structures, 1982, 784 pp.

5. Institution of Structural Engineers, *Earth retaining structures: civil engineering code of practice* 2, London, 1951, 224 pp.

6. Reynolds, C. E. and Steedman, J. C., *Reinforced concrete designer's handbook*, 9th Edn, Viewpoint Publications, London, 1981, 505 pp.

7. Reynolds, C. E. and Steedman, J. C., *Examples of the design of reinforced concrete building to CP 110 and allied codes*, Viewpoint Publications, London, 1978, 322 pp.

8. Coates, R. C., Coutie, M. G. and Kong, F. K., *Structural analysis*, 2nd Edn, Nelson, Sunbury-on-Thames, 1980, 579 pp.

9. Reimbert, M. L. and Reimbert, A. M., *Retaining walls, anchorages and sheet piling*, Volume 1, Trans Tech Publications, Clausthal, 1974, 284 pp.

10. Bares, R., *Tables for the analysis of plates, slabs and diaphragms and beams based on elastic theory*, 2nd Edn, Bauverlag, Wiesbaden, 1971, 626 pp.

11. Timoshenko, S. P. and Woinowsky-Krieger, S., *Theory of plates and shells*, 2nd Edn, McGraw-Hill, New York, 1959, 580 pp.

12. Roark, R. J. and Young, W. C., *Formulas for stress and strain*, 5th Edn., McGraw-Hill, New York, 1965, 624 pp.

13. Johansen, K. W., *Yield-line theory*, Cement and Concrete Association, London, 1962, 181 pp.

14. Johansen, K. W., *Yield-line formulae for slabs*, Cement and Concrete Association, London, 1972, 106 pp.

15. Jones, L. L. and Wood, R. H., *Yield-line analysis of slabs*, Thames and Hudson/Chatto and Windus, 1967, 405 pp.

16. Hillerborg, A., *Strip method of design*, Viewpoint Publications, London, 1975, 225 pp.

17. Pannell, F. N., Basic application of virtual-work methods in slab design, *Concr. Constr. Eng.*, Vol. 61, No. 6, June 1966, pp. 209–216.

18. Pannell, F. N., Economical distribution of reinforcement in rectangular slabs, *Concr. Constr. Eng.*, Vol. 61, No. 7, July 1966, pp. 229–233.

19. Pannell, F. N., Edge conditions in flat plates, *Concr. Constr. Eng.*, Vol., 61, No. 8, Aug. 1966, pp. 290–294.

20. Pannell, F. N., General principles of superposition in the design of rigid-plastic plates, *Concr. Constr. Eng.*, Vol. 61, No. 9, Sept. 1966, pp. 323–326.

21. Pannell, F. N., Design of rectangular plates with banded orthotropic reinforcement, *Concr. Constr. Eng.*, Vol. 61, No. 10, Oct. 1966, pp. 371–376.

22. Pannell, F. N., Non-rectangular slabs with orthotropic reinforcement, *Concr. Constr. Eng.*, Vol. 61, No. 10, Nov. 1966, pp. 383–390.

23. Fernando, J. S. and Kemp, K. O., A generalized strip deflexion method of reinforced concrete slab design, *Proc. Inst. Civ. Eng.*, Vol. 65, Part 2, March 1978, pp. 163–174.

24. Fintel, M. (Ed.), *Handbook of concrete engineering*, Van Nostrand Reinhold, New York, 1974, 802 pp.

25. Regan, P. W. and Yu, C. W., *Limit state design of structural concrete*, Chatto and Windus, London, 1973, 325 pp.

26. Allen, A. H., *Reinforced concrete design to CP* 110—*simply explained*, Cement and Concrete Association, London, 1974, 227 pp.
27. Institution of Structural Engineers. *The shear strength of reinforced concrete beams*, London, 1969, 170 pp.
28. American Concrete Institute, *Torsion of structural concrete*, SP-18, 1968, 505 pp.
29. Gibson, J. E., *The design of shell roofs*, 3rd Edn, Spon, London, 1968, 300 pp.
30. Gibson, J. E., *Thin shells: computing and theory*, Pergamon Press, Oxford, 1980, 289 pp.
31. Chronowicz, A., *The design of shells*, Crosby Lockwood, London, 1959, 202 pp.
32. Tottenham, H. A., A simplified method of design for cylindrical shell roofs, *Struct. Eng.*, Vol. 32, No. 6, June 1954, pp. 161–180.
33. Bennett, J. D., *Some recent developments in the design of concrete shell roofs*, Reinforced Concrete Association, London, 1958, 24 pp.
34. Bennett, J. D., *The structural possibilities of hyperbolic paraboloids*, Reinforced Concrete Association, London, 1961, 25 pp.
35. Faber, C., *Candela: the shell builder*, Architectural Press, London, 1963, 240 pp.
36. Portland Cement Association, *Elementary analysis of hyperbolic paraboloid shells*, PCA Structural and Railways Bureau, ST85, 1960, 20 pp.
37. Bowles, J. E., *Foundation analysis and design*, 2nd Edn, McGraw-Hill, New York, 1977, 750 pp.
38. Bowles, J. E., *Analytical and computer methods in foundation engineering*, McGraw-Hill, New York, 1974, 519 pp.
39. Irish, K. and Walker, W. P., *Foundations for reciprocating machines*, Cement and Concrete Association, London, 1969, 103 pp.
40. Barkan, D. D., *Dynamics of bases and foundations*, McGraw-Hill, New York, 1962, 434 pp.
41. Pinfold, G. M., *Reinforced concrete chimneys and towers*, Viewpoint Publications, London, 1975, 233 pp.
42. Taylor, C. P. and Turner, L., *Reinforced concrete chimneys*, 2nd Edn, Concrete Publications, 1960, 82 pp.
43. Tomlinson, M. J., *Pile design and construction practice*, Cement and Concrete Association, London, 1977, 413 pp.
44. Yan, H. T., Bloom base allowance in the design of pile caps, *Civ. Eng. Publ. Wks. Rev.*, Vol. 49, No. 575, May 1954, pp. 493–495; Vol. 49, No. 576, June 1954, pp. 622–623.
45. Whittle, R. T. and Beattie, D., Standard pile caps, *Concrete*, Vol. 6, No. 1, Jan. 1972, pp. 34–36; Vol. 6, No. 2, Feb. 1972, pp. 29–31.
46. Fairhurst, Sir W. A., *Arch design simplified*, Concrete Publications, London, Reprinted 1954, 61 pp.
47. Construction Industry Research and Information Association, *A comparison of quay wall design methods*, CIRIA Technical Note 54, London, 1974, 125 pp.
48. Scott, W. L., Glanville, Sir W., and Thomas, F. G., *Explanatory handbook on the British Standard Code of Practice for reinforced concrete CP 114: 1957*, 2nd Edn, Cement and Concrete Association, London, 1965, 172 pp.

49. Joint Committee of Institution of Structural Engineers and The Concrete Society, *Designing and detailing of concrete structures for fire resistance*, London, 1978, 59 pp.
50. Clark, D., *Computer aided structural design*, Wiley, Chichester, 1978, 225 pp.
51. Jacys Computing Services, *Catalogue of computer software for structural engineering*, Steyning, 1982, 16 pp.
52. Beeby, A. W., *The analysis of beams in plane frames according to CP 110*, Cement and Concrete Association Development Report No. 1, London, 1978, 34 pp.
53. Cross, H. and Morgan, N. D., *Continuous frames of reinforced concrete*, Wiley, New York, 1932, 343 pp.
54. Kong, F. K. and Evans, R. H., *Reinforced and prestressed concrete*, 2nd Edn, Nelson, Sunbury-on-Thames, 1980, 412 pp.
55. Department of Transport, *A guide to the structural design of pavements for new roads*, Road Note 29, 3rd Edn, HM Stationery Office, London, 1970.
56. Department of Transport, *Road pavement design*, Technical Memorandum H6/78, London, 1978, 4 pp. + appendices and amendments.
57. Department of Transport/Cement and Concrete Association, *A guide to concrete road construction*, 3rd Ed., HM Stationery Office, 1978, 82 pp.
58. Sargious, M., *Pavements and surfacings for highways and airports*, Applied Science Publishers, London, 1975, 619 pp.
59. Department of Transport, *Commercial traffic: its estimated damaging effect 1945–2005*, TRRL Report LR 910, Crowthorne, 1979.
60. Zienkiewicz, O. C., *The finite element method*, 3rd Edn, McGraw-Hill, London, 1977, 787 pp.
61. Magnus, D., *Pierced shear walls*, Concrete Publications, London, 1968, 28 pp.
62. Coull, A. and Choudhury, J. R., Stresses and deflections in coupled shear walls, *J. Am. Concr. Inst., Proc.*, Vol. 64, Feb. 1967, pp. 65–72; Analysis of coupled shear walls, *J. Am. Concr. Inst., Proc.*, Vol. 64, Sept. 1967, pp. 587–593.
63. Baker, A. L. L., *Raft foundations*, 3rd Edn, Concrete Publications, London, 1957, 148 pp.
64. Portland Cement Association, *Handbook of frame constants*, Chicago, undated, 32 pp.
65. Kong, F. K., Robins, P. J. and Sharp, G. R., The design of reinforced concrete deep beams in current practice, *Struct. Eng.*, Vol. 53, No. 4, April 1975, pp. 173–180.
66. Ove Arup & Partners, *The design of deep beams in reinforced concrete*, CIRIA Guide 2, London, 1977, 131 pp.
67. Manning, G. P., *Design and construction of foundations*, Concrete Publications, London, 1961, 240 pp.
68. Szechy, C., *Foundation failures*, Concrete Publications, London, 1961, 140 pp.
69. Wentworth-Sheilds, F. E., Gray, W. S. and Evans, H. W., *Reinforced concrete piling and piled structures*, 2nd Edn, Concrete Publications, London, 1960, 149 pp.
70. Lee, D. H., *An introduction to deep foundations and sheet piling*, Concrete Publications, London, 1961, 260 pp.
71. Cranston, W. B., *Analysis and design of reinforced concrete columns*, Cement

and Concrete Association Research Report No. 20, London, 1972, 28 pp.

72. Beeby, A. W., *The design of sections for flexure and axial load according to CP 110*, Cement and Concrete Association Development Report No. 2, London, 1978, 31 pp.

73. Manning, G. P., *Reinforced concrete reservoirs and tanks*, Cement and Concrete Association, London, Reprinted 1972, 384 pp.

74. Anchor, R. D., Hill, A. W. and Hughes, B. P., *Handbook on BS 5337: 1976 (The structural use of concrete for retaining aqueous liquids)*, Viewpoint Publications, London, 1979, 60 pp.

75. Reynolds, C. E., *Basic reinforced concrete design*, Volumes 1 and 2, Concrete Publications, London, 1962, 488 pp.

76. Portland Cement Association, *Circular concrete tanks without prestressing*, Information Sheet IS 072.01D, Skokie, 1942.

77. Threlfall, A. J., *Design charts for water retaining structures*, Viewpoint Publications, London, 1978, 66 pp.

78. Beeby, A. W. and Taylor, H. P. J., *Development and production of design charts using a digital plotter*, Cement and Concrete Association Paper PP/64, London, 1970, 11 pp.

79. Steedman, J. C., *Charts for limit-state design to CP 110: uniform rectangular concrete stress-block*, Cement and Concrete Association, London, 1974, 20 pp.

80. Pannell, F. N., *Design charts for members subjected to biaxial bending and thrust*, Concrete Publications, London, 1966, 51 pp.

81. Manning, G. P., *Reinforced concrete design*, 3rd Edn, Longmans, London, 1966, 505 pp.

82. Cusens, A. R. and Kuang, Jing-Gwo, A simplified method of analysing free-standing stairs, *Concr. Constr. Eng.*, Vol. 60, No. 5, May 1965, pp. 167–172, 194.

83. Cusens, A. R., Analysis of slabless stairs, *Concr. Constr. Eng.*, Vol. 61, No. 10, Oct. 1966, pp. 359–364.

84. Santathadaporn, Sakda, and Cusens, A. R., Charts for the design of helical stairs, *Concr. Constr. Eng.*, Vol. 61, No. 2, Feb. 1966, pp. 46–54.

85. Construction Industry Research and Information Association, *Guide to the design of waterproof basements*, CIRIA Guide 5, London, 1978, 38 pp.

Index

1